T0350537

THE UNIVERSITY OF ARIZONA SPACE SCIENCE SERIES
RICHARD P. BINZEL, GENERAL EDITOR

Europa
Robert T. Pappalardo, William B. McKinnon,
and Krishan K. Khurana, editors, 2009, 727 pages

The Solar System Beyond Neptune
M. Antonietta Barucci, Hermann Boehnhardt, Dale P. Cruikshank,
and Alessandro Morbidelli, editors, 2008, 592 pages

Protostars and Planets V
Bo Reipurth, David Jewitt, and Klaus Keil, editors, 2007, 951 pages

Meteorites and the Early Solar System II
D. S. Lauretta and H. Y. McSween, editors, 2006, 943 pages

Comets II
M. C. Festou, H. U. Keller,
and H. A. Weaver, editors, 2004, 745 pages

Asteroids III
William F. Bottke Jr., Alberto Cellino, Paolo Paolicchi,
and Richard P. Binzel, editors, 2002, 785 pages

TOM GEHRELS, GENERAL EDITOR

Origin of the Earth and Moon
R. M. Canup and K. Righter, editors, 2000, 555 pages

Protostars and Planets IV
Vincent Mannings, Alan P. Boss,
and Sara S. Russell, editors, 2000, 1422 pages

Pluto and Charon
S. Alan Stern and David J. Tholen, editors, 1997, 728 pages

**Venus II—Geology, Geophysics, Atmosphere,
and Solar Wind Environment**
S. W. Bougher, D. M. Hunten,
and R. J. Phillips, editors, 1997, 1376 pages

Cosmic Winds and the Heliosphere
J. R. Jokipii, C. P. Sonett,
and M. S. Giampapa, editors, 1997, 1013 pages

Neptune and Triton
Dale P. Cruikshank, editor, 1995, 1249 pages

Hazards Due to Comets and Asteroids
Tom Gehrels, editor, 1994, 1300 pages

Resources of Near-Earth Space
John S. Lewis, Mildred S. Matthews,
and Mary L. Guerrieri, editors, 1993, 977 pages

Protostars and Planets III
Eugene H. Levy and Jonathan I. Lunine, editors, 1993, 1596 pages

Mars
Hugh H. Kieffer, Bruce M. Jakosky, Conway W. Snyder,
and Mildred S. Matthews, editors, 1992, 1498 pages

Solar Interior and Atmosphere
A. N. Cox, W. C. Livingston,
and M. S. Matthews, editors, 1991, 1416 pages

The Sun in Time
C. P. Sonett, M. S. Giampapa,
and M. S. Matthews, editors, 1991, 990 pages

Uranus
Jay T. Bergstralh, Ellis D. Miner,
and Mildred S. Matthews, editors, 1991, 1076 pages

Asteroids II
Richard P. Binzel, Tom Gehrels,
and Mildred S. Matthews, editors, 1989, 1258 pages

Origin and Evolution of Planetary and Satellite Atmospheres
S. K. Atreya, J. B. Pollack,
and Mildred S. Matthews, editors, 1989, 1269 pages

Mercury
Faith Vilas, Clark R. Chapman,
and Mildred S. Matthews, editors, 1988, 794 pages

Meteorites and the Early Solar System
John F. Kerridge and Mildred S. Matthews, editors, 1988, 1269 pages

The Galaxy and the Solar System
Roman Smoluchowski, John N. Bahcall,
and Mildred S. Matthews, editors, 1986, 483 pages

Satellites
Joseph A. Burns and Mildred S. Matthews, editors, 1986, 1021 pages

Protostars and Planets II
David C. Black and Mildred S. Matthews, editors, 1985, 1293 pages

Planetary Rings
Richard Greenberg and André Brahic, editors, 1984, 784 pages

Saturn
Tom Gehrels and Mildred S. Matthews, editors, 1984, 968 pages

Venus
D. M. Hunten, L. Colin, T. M. Donahue,
and V. I. Moroz, editors, 1983, 1143 pages

Satellites of Jupiter
David Morrison, editor, 1982, 972 pages

Comets
Laurel L. Wilkening, editor, 1982, 766 pages

Asteroids
Tom Gehrels, editor, 1979, 1181 pages

Protostars and Planets
Tom Gehrels, editor, 1978, 756 pages

Planetary Satellites
Joseph A. Burns, editor, 1977, 598 pages

Jupiter
Tom Gehrels, editor, 1976, 1254 pages

Planets, Stars and Nebulae, Studied with Photopolarimetry
Tom Gehrels, editor, 1974, 1133 pages

Europa

Europa

Edited by

Robert T. Pappalardo
William B. McKinnon
Krishan K. Khurana

With the assistance of

Renée Dotson

With 85 collaborating authors

THE UNIVERSITY OF ARIZONA PRESS
Tucson

in collaboration with

LUNAR AND PLANETARY INSTITUTE
Houston

About the front cover:

Europa. This landscape view aims to characterize some features typical of the satellite. The viewer is in a large fractured valley, lined by cliffs on both sides. The floor is fresher ice, and has been broken by new fracturing (foreground). Periods of compression have thrown up ice boulders along the fracture (based loosely on shattered Arctic sea pack ice witnessed by the artist in Barrow, Alaska). Jupiter, along with its volcanic moon, Io, are prominent in the sky. Angular height of picture is about 45°. *Painting by William K. Hartmann, Planetary Science Institute, Tucson, Arizona.*

About the back cover:

This mosaic of the Conamara Chaos region on Jupiter's moon Europa clearly indicates relatively recent resurfacing. Irregularly shaped blocks of water ice were formed by the breakup and movement of the existing icy crust. The icy blocks were shifted, rotated, and tipped within a mobile material that was either liquid water, warm mobile ice, or some combination thereof. The presence of young fractures cutting through this region indicates that the surface has since frozen again into solid, brittle ice. The mosaic shows high-resolution image data inset into a lower-resolution context of this tumultuous region. The background image was obtained during Galileo's sixth orbit of Jupiter in February 1997. Five very-high-resolution images obtained during the spacecraft's twelfth orbit in December 1997 provide an even closer look. The scene has been colorized using lower-resolution images also obtained during Galileo's twelfth orbit. *Mosaic constructed by Ryan Sicilia, Seattle, Washington.*

The University of Arizona Press
in collaboration with the Lunar and Planetary Institute

♾ This book is printed on acid-free, archival-quality paper.
Manufactured in the United States of America

Library of Congress Cataloging-in-Publication Data

Europa / edited by Robert T. Pappalardo, William B. McKinnon, Krishan Khurana ; with the assistance of Renee Dotson with 85 collaborating authors.
p. cm. — (The University of Arizona space science series)
"In collaboration with Lunar and Planetary Institute, Houston."
Includes bibliographical references and index.
ISBN 978-0-8165-2844-8 (hardcover : alk. paper)
1. Europa (Satellite) 2. Europa (Satellite)—Geology. I. Pappalardo, Robert T. II. McKinnon, William B. III. Khurana, K.
QB404.E97 2009
523.9'85—dc22

2008053778

Contents

PART III: INTERIOR, ICY SHELL, AND OCEAN

PART IV: EXTERNAL ENVIRONMENT

PART V: ASTROBIOLOGY AND PERSPECTIVES

List of Contributing Authors

Acknowledgment of Reviewers

The editors gratefully acknowledge the following individuals for their time and effort in reviewing chapters for this volume:

Nadine Barlow
Richard P. Binzel
Lorenzo Bruzzone
Robert Carlson
Tim Cassidy
Julie Castillo-Rogez
Roger Clark
John Cooper
Andrew Dombard
Janusz Eluszkiewicz
Paul Feldman
Olivier Grasset
Robert Grimm
Steven Hauck
Paul Helfenstein
Karl Hibbitts
Wing Ip
Konstantin Kabin
Ozgur Karatekin
Don Korycansky
Jere Lipps
Craig Manning
Isamu Matsuyama
Tom McCord
Alfred McEwen
Christopher McKay
H. Jay Melosh
Michael Meltzer
Scott Murchie
Fritz Neubauer
David O'Brien
Greg Ojakangas
Chris Paranicas
Bill Paterson
G. Wesley Patterson
Stan Peale
Cynthia Phillips
Carl Pilcher
James Roberts
David Sandwell
Richard Selesnick
David Senske
Adam Showman
Slava Solomatov
Nicole Spaun
David Stevenson
Richard Thomson
Gabriel Tobie
Rob Tyler
Stuart Weidenschilling
Dale Winebrenner
Kevin Zahnle
Michael Zolensky

Foreword

In 1977 two very disparate exploration paths began to gradually converge, almost unnoticed by the science community at the time. As the Alvin submersible descended slowly into the dark depths of the ocean above the East Pacific Rise, two Voyager spacecraft were ascending atop pillars of smoke and flame from the Kennedy launch complex on the Florida peninsula, bound for the outermost regions of the solar system. Both of these exploration platforms were destined to make discoveries that would profoundly alter our views of where and under what conditions life can survive and, indeed, possibly originate.

Over the subsequent three decades, Alvin and its scientific siblings have charted the hitherto unimagined richness of seafloor hydrothermal vent systems, teeming with organisms that thrive in the absence of sunlight at temperatures previously thought to be above the limits for life. At the same time, outer planet spacecraft, first the Voyager mission, followed by Galileo and Cassini, were discovering that the icy moons of the outer solar system are far from geologically dead, inactive bodies, but rather complex worlds where radioactive heating and tidal forces may have resulted in global oceans hidden beneath their icy surfaces, redefining the concept of a "habitable zone" for planetary bodies. This parallelism of terrestrial ocean exploration and planetary discoveries was first enunciated by University of Washington oceanographer John Delaney (the chief scientist for numerous Alvin expeditions), when he invited me and other planetary scientists to Seattle to discuss evidence for an ocean on Europa. Delaney has continued to use this as an example of the value of cross-disciplinary research in geosciences.

The chapters in this book review the current state of knowledge about Europa, where Galileo data now point to a subsurface ocean some 100 kilometers deep, above a rocky ocean floor that may resemble the terrestrial seafloor environments first glimpsed by Alvin investigators over 30 years ago. Europa is the smallest of the four moons of Jupiter discovered by Galileo in 1610, approximately the size of Earth's Moon. For most of the 400 years since its discovery it was remarkable only for its apparently higher reflectivity and for its regular orbital "dance" about Jupiter, with its period locked by gravity in strict ratios to the length of the days of its larger siblings, Io and Ganymede. With the Voyager flybys in 1979, the Galilean satellites went overnight from points of light in astronomers' telescopes to *terra cognita*, each world exhibiting its own unique characteristics. Although volcanic Io stole the "Voyager satellite show" with its over-the-top pyrotechnics, Europa emerged as "most enigmatic." This moon's smooth, fractured surface and paucity of large impact craters suggested an active geologic history and lent credibility to theoretical models of subsurface liquid water oceans for large icy satellites.

Voyager discoveries laid the foundation for subsequent exploration, and Europa became one of the highest-priority targets for the Galileo mission, which arrived at Jupiter in 1995. The first high-resolution views of Europa's surface showed vistas of ice rafts and fractured surfaces strikingly reminiscent of terrestrial ice packs. After seven years of exploration, Galileo's findings included gravity data indicating the satellite has a layer of water and/or ice about 100 kilometers thick, spectroscopic data suggesting both salts and radiation produced chemicals on the surface, and magnetic signatures interpreted as a subsurface global layer of conducting saltwater. The data pointed to a global ocean with a volume twice that of all Earth's oceans.

Each new stage of exploration has brought with it tremendous increases in our knowledge of the icy satellites and Europa. In 1967, my professor, Bruce Murray, assigned me the task of reporting everything that was known about the Galilean satellites to our class in planetary science. After a little (very little) time in the library, my report was about 15 minutes long. Following the Voyager Jupiter encounters, a 962-page book, *Satellites of Jupiter* (edited by D. Morrison, University of Arizona Press, 1982) was required just to review the Voyager results, with a 1021-page follow-on just four years later to include further research on Jupiter's satellites and Saturn and Uranus results: *Satellites* (edited by J. A. Burns and M. S. Matthews, University of Arizona Press, 1986). Since the Galileo mission, there have been hundreds of scientific papers and scores of reviews and book contributions related to the icy satellites.

This book concentrates on current research related to Europa. In it you will find a compilation of what is known about this intriguing world. The contributions here also raise many questions and point to areas of controversy and debate. There are many things we do not know with certainty and we have many questions still. How old is the surface that we see today? What is the composition of the non-ice material on the surface and does some of it originate in the ocean? How often does material from the subsurface ocean reach the surface? How is the surface reshaped and modified, and are these processes gradual or short-lived events? How deep is the ocean? How thick is the icy shell and does it vary in thickness? Are the physical and chemical conditions in Europa's ice, seafloor, and ocean environment suitable for life of some kind to exist or originate? These questions and many others are discussed here and form the foundation for setting the objectives for the next stage of exploration. NASA and the European Space Agency (ESA) are currently studying a joint exploration of the Jupiter system and Europa to addresses these objectives, including a Europa orbiter. Sometime in the next two decades, a new generation of scientists will began unraveling Europa's secrets, answering old questions . . and, of course, raising new ones.

Torrence V. Johnson
NASA Jet Propulsion Laboratory/California Institute of Technology
Pasadena, California
November 2008

Preface

Four hundred years ago, Galileo pointed his spyglass toward the planet Jupiter, and the universe has been different ever since. Over several historic nights in January 1610, Galileo Galilei observed and recorded the relative locations of the points of light that he termed the Medicean stars. He laid out his observations for the public in his book *The Starry Messenger,* demonstrating support through these and his other celestial observations for the Sun-centric hypothesis. Decidedly nonintuitive, the heliocentric hypothesis had been suggested in antiquity by Aristarchus, but it was the deep and detailed proposal by Copernicus in 1543 — notably published upon his deathbed, with a preface disavowing its radicalness by claiming it a thought experiment — that would illuminate Renaissance Europe. Simon Marius would ultimately be awarded status as co-discoverer of the four jovian satellites, for evidence that he observed them around the same time as Galileo, and it is Marius' classical names for the moons that we use today. But it was Galileo who published and promoted their discovery, and who explained the observations in the greater scientific context. For these reasons, the Galilean satellites acknowledge the man who changed forever the face of the heavens and our place within.

The twin Voyager fly-throughs of the Jupiter system in 1979 hinted at Europa's distinctiveness. They revealed a lined surface and few definite craters, with signs of lateral surface motions. Calculations suggested tidal heating might plausibly be able to maintain an interior ocean greater than that of the Earth's beneath a thin ice shell. Then, beginning in 1996, the *Galileo* spacecraft changed our perception of the solar system's habitable zones. It detected magnetic induction and imaged distinctive surface features, pointing to a global subsurface ocean beneath Europa's icy shell *today*. Magnetic induction was also found at Callisto, and inferred through Ganymede's intrinsic magnetic field. But *Galileo* was not designed for divining oceans, and innumerable questions about Europa and its probable ocean remain, as detailed within the chapters of this book.

As the pages of this book were being finalized and proofed, NASA announced its intention to return to Europa with an orbiting spacecraft, fulfilling a key recommendation of the 2003 Planetary Science Decadal Survey and the dreams of the many planetary scientists who have urged such a mission since *Galileo*'s discoveries first hit home. The Jupiter Europa Orbiter, as the craft is currently termed, would first explore the jovian system from planetary orbit, then would enter Europa orbit to scrutinize the satellite from its deep core to its plasma-sputtered tenuous atmosphere using a modern suite of remote-sensing and *in situ* instrumentation. Contingent on the outcome of its Cosmic Vision Programme process, the European Space Agency may expand the exploration effort by sending a spacecraft to simultaneously explore Ganymede and the jovian system, better placing Europa in its planetary context. The Japan Aerospace Exploration Agency and the Russian Federal Space Agency have also expressed interest in joining the international exploration flotilla.

Europa is the 36th book in the University of Arizona Press Space Science Series, the first to concentrate on a single moon. As editors of this volume, we thank General Editor Richard Binzel for pushing this book forward at every turn, and for believing in our ability to deliver an outstanding result. We are more than grateful to Renée Dotson for her tireless efforts in the compilation and production of the book. We thank each of the chapter authors for their time, effort, and dedication. We thank the chapter reviewers for their invaluable role in making this book a reality. Admiration and appreciation go to William Hartmann for his original painting that graces the book's cover. We are grateful to Curt Niebur of NASA Headquarters for his support, moral and financial, the latter reducing the cost of color production and thus purchase price. We thank Stephen Mackwell and the Lunar and Planetary Institute for entrusting us with this project. We offer our deepest gratitude to Mollie Doherty, Kate McKinnon, and Amber Williamson for their support and perpetual patience through this process. Portions of this work performed by R.T.P. were carried out at the Jet Propulsion Laboratory, California Institute of Technology, under a contract with the National Aeronautics and Space Administration.

After more than a decade of study of the Galileo data, Europa is coming to be revealed as a diverse world that varies both in appearance and substance from place to place, and perhaps over time as well. Geological processes may allow exchange of material between its interior ocean and its radiation-altered oxidant-rich surface, either directly (e.g., by melting) or indirectly (e.g., by convection). Unlike other large icy satellites, Europa's total H_2O layer is probably thin enough that its vast ocean directly contacts its rocky mantle, rather than being transformed by pressure into the high-density ice polymorphs that floor the oceans of Ganymede, Callisto, and Titan. Tidal kneading of Europa's ice — and possibly its mantle — is the engine that could maintain the ocean through heating, crack the brittle ice carapace, trigger ice convection and plate-like motions, and shuttle material between surface and ocean. Chemical energy could feed an internal ocean from both above (photochemistry or radiation-produced oxidants) and below (hydrothermal systems), making Europa arguably the best place in our solar system to search for extant life beyond Earth. The Europa of today might or might not be similar to the Europa of 100 million years ago, if its level of activity is cyclical, being tied to Io's orbital evolution. More than any known world, Europa's interior and crustal geophysics, its surface geology, its ocean chemistry, the external environment, and its planetary brethren are intricately and inherently linked.

The authors of this volume aim to take a step back, collating what we think we know, what we don't know, and what we'd like to know about Europa. We intend that this book will serve to inspire professional researchers and engineers, planetary science students, and the interested public alike as we collectively plan the next stage in the exploration of Europa and the jovian system. When a spacecraft ultimately enters Europa orbit, this volume will transition from a reference compendium to a historical document, as knowledge about the satellite expands to new realms. Until then, readers are invited to delve into these pages, the multitude of sources cited within, and the icy depths of this ocean world.

Robert T. Pappalardo, William B. McKinnon,
and Krishan K. Khurana
August 2009

Part I:

History, Origin, and Dynamics

The Exploration History of Europa

Claudia Alexander and Robert Carlson
NASA Jet Propulsion Laboratory/California Institute of Technology

Guy Consolmagno
Specola Vaticana

Ronald Greeley
Arizona State University

David Morrison
NASA Ames Research Center

Almost 400 years after Galileo's momentous discovery of Europa (January 1610) with a one-inch magnifier, the New Horizons spacecraft took images of Europa (February 2007) with a high-quality CCD imager. In the intervening years, theoretical work coupled with improvements in technology have led scientists from one paradigm to another. In Galileo's era, the movements of Europa, logged by hand, were part of a paradigm shift regarding Earth as the center of the solar system. Another paradigm shift was the importance of sputtering, a process that may be critical to Europa as an object of biological interest, and that was treated like science fiction when the physics were first proposed in the 1970s. In the late twentieth century the *Galileo* spacecraft collected evidence that suggested Europa to have a subsurface ocean. The evolution of thought coupled with the improvements in technology are presented in this chapter.

1. INTRODUCTION

Europa, the smallest of the Galilean satellites, was the least-well-observed target during the two Voyager flybys of the 1970s. Post-Voyager, this small ice-covered world catapulted into prominence as a place in the solar system with extensive water oceans that may be habitable. The beginning of this Europa revival can be traced to the *Nature* paper by *Squyres et al.* (1983) suggesting that the linear fracture-like markings and absence of craters revealed by Voyager were evidence for liquid water and active resurfacing. Consequently, Europa emerged as a prime target for exploration by the *Galileo* mission of the 1990s, and the primary focus of the first of *Galileo*'s two extended missions. Today Europa is a candidate for a dedicated orbiter mission, with eventual landings on its surface. Many astrobiologists consider Europa, among leading candidates, to be a more interesting potential abode for life even than Mars, since the discovery of life in its oceans would constitute compelling evidence for an independent origin of life, a second genesis within our solar system, whereas depending upon the context, martian life might have significant ties to "seeding" from Earth.

Until the advent of the "space age" circa 1950, the only information we had about Europa was its mass, diameter, surface brightness in various colors, and orbital commensurability with the other large satellites of Jupiter. Nonetheless, these were sufficient to allow astronomers to arrive, eventually, at a global picture of Europa not far removed from the perspective afforded by contemporary spacecraft. This chapter traces the history of Europa studies over four centuries, from Galileo (the man) to *Galileo* (the mission).

2. FROM DISCOVERY TO THE BIRTH OF MODERN PLANETARY SCIENCE

From its discovery in 1610 through World War II, Europa languished in the backwaters of astronomy. While discovery of Jupiter's satellite system played a major part in establishing the Copernican cosmology and founding the new science of planetary dynamics, when reasonable data on size and mass were available, the astronomers of subsequent centuries did not bother to speculate about the physical or chemical properties of the satellites of Jupiter. Even when the tools of astrophysics became available in the early twentieth century, planetary scientists seem to have been among the last to embrace new techniques.

2.1. Discovery

Europa was one of the three jovian satellites discovered in the opening weeks of the era of telescopic astronomy. In January 1610, Galileo Galilei used the "*occiale*" to observe the heavens — a telescopic invention that was de-

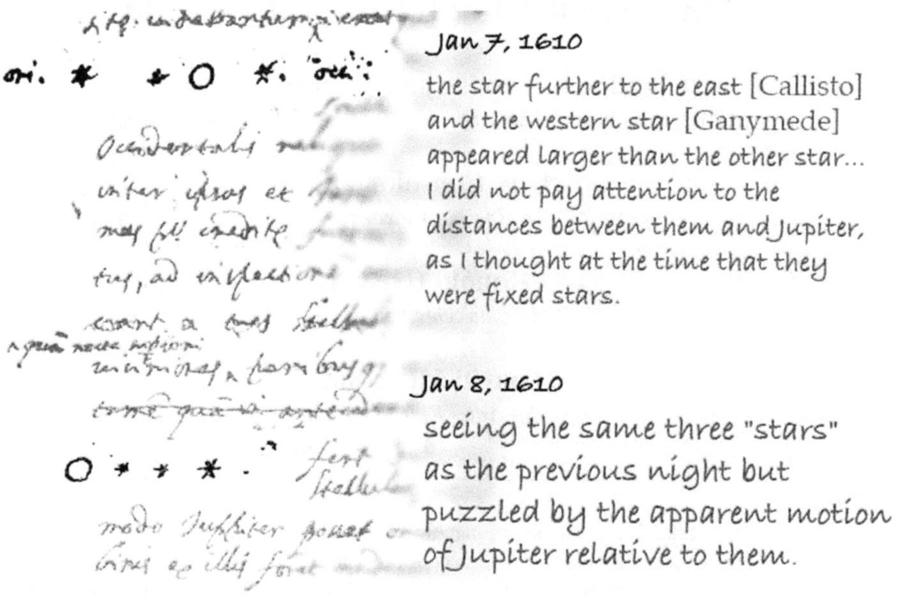

Jan 7, 1610
the star further to the east [Callisto] and the western star [Ganymede] appeared larger than the other star... I did not pay attention to the distances between them and Jupiter, as I thought at the time that they were fixed stars.

Jan 8, 1610
seeing the same three "stars" as the previous night but puzzled by the apparent motion of Jupiter relative to them.

Fig. 1. This figure shows a sample of notes from Galileo's observing run in 1610, the first image of Europa, as drawn in Galileo's handwritten draft of *Sidereus Nuncius. Top (Jan. 7):* The "star" immediately to the left of Jupiter is in fact both Io and Europa, which were in conjunction just after sunset in Padua the night of January 7, 1610, when this figure was drawn. (The other "stars" are Callisto, to the left, and Ganymede, to the right.) "Ori" and "Occ" represent east and west. *Bottom (Jan. 8):* Europa is visible as the middle of the three "stars"; Io is closer to Jupiter, while Ganymede is indicated by the larger star to the right. (Callisto is near greatest elongation that night, so is far to the east.) English translation of Galileo's notes from G. Consolmagno. The facsimile illustrated here is taken from Volume III of *Le Opere di Galileo Galilei*, published by G. Barbèra in Florence in 1892.

signed to serve maritime and land enterprises. His tool was primitive by today's standards — about one inch of aperture, obtaining a magnification of about 20 times that of the naked eye. Nevertheless, over the nights of January 7–15, 1610, Galileo conducted observations that would contribute to changing the paradigm about the solar system forever.

Figure 1 shows a representation of the notes Galileo took as he made these historic observations. Galileo noted that "three little stars, small but very bright, were near the planet." On his second night of observing, Galileo noted that the three little stars were "all west of Jupiter, and nearer together than on the previous night." Subsequent observations confirmed that the satellites (three little stars) went around Jupiter in the prograde fashion. This was the first example of one heavenly body orbiting another, thereby providing an argument for the controversial heliocentric theory of Copernicus. The first findings from Galileo's study of the Jupiter system were published just 60 days after the observations were made, in March 1610 in the short monograph entitled *Siderius Nuncius.* Bavarian astronomer Simon Marius also made observations of the moons in this timeframe, without publishing his results. Although Galileo proposed calling them the Medician satellites, in honor of his Medici patrons, over time, the term "Galilean satellites" has come to refer to the four major moons of Jupiter discovered by Galileo during that fortnight, including Europa. Simon Marius in 1614 suggested, based upon myths of the four lovers of Jupiter, the names of the four satellites as we know them today. A few decades later (in the 1670s), they were used by Danish astronomer Ole Roemer for early deductions of the speed of light.

Present in Galileo's observations is the first hint of the resonances that would play an important role later. The term "resonance" refers to the orbital commensurability of Io, Europa, and Ganymede, respectively, of 4:2:1. In the time it takes Ganymede to orbit Jupiter once, Europa has gone around twice, and Io four times. The interplay of their mutual gravitational effects, resulting in distortion of their orbit trajectories, was demonstrated by *Laplace* (1805) and is commonly called the "Laplace resonance." The resonance pumps, or accelerates, the body in orbit, such that the eccentricities of the satellite orbits cannot erode to a circular state. This circumstance introduces tidal deformations into the physical structure of these small bodies. For more discussion of how the orbital eccentricity of a body leads to tidal deformation, the reader is referred to section 5.8, the tail end of section 3.6, and the chapter by Sotin et al. For Io and Europa, an intimate connection exists between their orbital commensurabilities and their internal evolution that does not exist with the other satellites. As noted by *Greenberg* (2005), "everything interesting about Europa follows from the fact that the eccentricity [of its orbit] is not zero."

2.2. Mass

The masses of Europa and the other Galilean satellites, as a fraction of Jupiter's mass, were first deduced by *Laplace* (1805) in his treatise *Celestial Mechanics*, using an elaborate theory of the mutual perturbations of those moons. His published result for Io was low by a factor of 4, but the mass of Europa, 2.3×10^{-5} of the mass of Jupiter, was within 10% of the modern value. (He overestimated the mass of Ganymede by 13%, and underestimated Callisto by 25%.) At the end of that chapter, he pointed out that the masses of Io and Europa were "almost equal [to that of the Earth's Moon]."

Laplace's calculations depended on careful choice of eclipse timing data, which he himself noted were problematical at that time. These data were improved with observations of the moons by Baron Marie-Charles Damoiseau and others during the first half of the nineteenth century (see *Sampson,* 1921). Nonetheless, reference papers and popular textbooks of the period continued to use Laplace's mass values.

Mass determinations including more modern values were reviewed by *Brouwer and Clemence* (1961), *Kovalevsky* (1970), and *Duncombe et al.* (1974). The prespacecraft average given by *Morrison and Cruikshank* (1974) is 480 ± 10 in units of 10^{20} kg.

2.3. Diameter

An accurate diameter of Europa was far more difficult to obtain than its mass in the prespacecraft era. Through the years, techniques for determining this diameter have included timing satellite eclipses and occultations; the filar micrometer (a measuring microscope using movable threads to measure the size of Europa's disk as seen in a telescope's eyepiece); a double-image micrometer used by Dollfus at Pic du Midi; a diskmeter used by Kuiper at the Mt. Wilson and Palomar telescopes (including the Hale 5 m); and optical interferometry employed by Michelson, Hamy, Danjon and others (*Dollfus,* 1975).

Secchi (1859) provided an early result, listing the diameter of Europa of 3330 km, 6% too large. He also listed masses of the satellites in terms of an Earth mass, based on the *Laplace* (1805) work. Combined with the Europa mass estimate, a calculation can produce a density of 2.5 g/cm^{-3} (20% low). Secchi himself never calculated this density nor discussed the implications of these numbers.

Astronomy books popular in their day, by *Herschel* (1861) and *Newcomb* (1878), presented similar diameters and masses for the Galilean moons, but they neither calculated densities nor attempted interpretation of composition from mass values. Consideration of Europa as a place, a planet-sized body with a geological history, was apparently not grounds of contention for nineteenth-century astronomers.

2.4. Brightness, Albedo, and Shape

In the field of photometry, otherwise known as the quantitative measurement of light, the most rudimentary observation is the integrated brightness as a function of time. This integrated brightness is, for a rotating body, a periodic function, where the amplitude — the change in the intensity of collected light over a rotation period — can be used to infer not only the rotation period of a body but its shape. In the nineteenth century, the perceived variability of the brightness and shape of these objects, as reported by a number of reputable observers, constituted the primary controversy related to the nature of the Galilean moons. *Secchi* (1859), for example, said of Ganymede that "one can see various spots and a notable flattening, from which I conclude that it spins with a very short period. It seems at some times to be quite flattened, and at other times round . . ." (translation by G.J.C.). Secchi suggested that this could be explained if it precessed very rapidly, wobbling like a child's top. Likewise, *Newcomb* (1878) reported that "the light of these satellites varies to an extent which it is difficult to account for, except by supposing very violent changes constantly going on their surfaces."

Pickering et al. (1879) produced the first accurate relative photometry of these moons. Their photometer used a series of prisms to view simultaneously the images of the satellite (or star) and a calibration object, whose brightness could be dimmed to match the objects' brightness by crossed Nichols or a variable aperture. Their conclusion was straightforward: "It has been thought by many astronomers that the light of the satellites of Jupiter was variable. This view is not sustained by the present measurements . . ." The inherent difficulty of visually estimating the brightness of these moons so close to the overwhelming light of Jupiter is the reason suggested by Pickering et al. for the misleading reports of their contemporary nineteenth-century observers.

These authors also produced a table of relative albedos; the actual values depended on which set of diameters they used, but a typical set of values gave the ratio of the albedos relative to Io for Europa, Ganymede, and Callisto as 1.21, 0.625, and 0.359. They did not notice or comment on the curious fact that the more dense moons, Io and Europa, are significantly brighter than the less-dense moons of Ganymede and Callisto.

Although this work disproved the reported brightness changes, the idea that the moons were highly elliptical was still widely accepted. Thirty years later, *Pickering* (1908) reported on further observations, including measurements of the moons' diameters made with a filar micrometer. He found a significant difference between the polar and equatorial axes of all the satellites, and he discussed at length the question of "elliptical shape" of the moons reported by others before him. Searching the literature, he noted that from 1850 to 1895 Io was observed to be "elliptical" by 13 observers, and Callisto by 10 observers. (By contrast, Europa was reported to be "elliptical" in shape only by two observers, the fewest of the four satellites.) From today's perspective, we suspect that apparent differences between polar and equatorial diameter were due to a brighter polar cap on Ganymede and a darker cap on Io, whereas the disk of Europa is more uniform.

Pioneering photoelectric photometry was carried out by *Stebbins* (1927) and *Stebbins and Jacobsen* (1928); these data were subsequently converted to the UBV photometric system and augmented by *Harris* (1961). The Stebbins observations of 1927 demonstrated that all four Galilean satellites were in synchronous rotation, keeping the same face toward Jupiter, and therefore tidally locked. The largest lightcurve amplitude dichotomy was found for Europa, particularly in the blue and ultraviolet, suggesting an asymmetry between leading and trailing hemispheres. *Burns* (1968) was the first to suggest that asymmetric magnetospheric fluxes and radiation effects could lead to the hemispherical differences in color observed by Harris. Much more extensive photometric studies were carried out by in the late 1960s and early 1970s by *Johnson* (1971), *Johnson and McCord* (1971), *Morrison et al.* (1974), and later by others (see section 3.2. and chapter by Carlson et al.).

2.5. Density and Bulk Composition

One of the most striking aspects of the Galilean satellites to modern planetary scientists is the dichotomy between the two inner satellites, with their lunar-like densities, and the outer two, which have densities too low for a rocky composition. Today we recognize that density is a measure of bulk composition, and thus we see this dichotomy as a fundamental challenge to theories of the origin and evolution of these objects, e.g., that all four satellites evolved from the same gaseous cloud of volatiles (the composition of which includes water vapor and carbon dioxide, volatiles that form ice at low temperatures; see section 3.6).

One of the stranger bits of history comes from *Chambers* (1861, p. 61), who was perhaps the first to calculate densities for the satellites. Unfortunately, his work was flawed by a mathematical error; although he presented, reasonably, Europa's diameter as 2191 miles (3500 km, 12% large) and used Laplace's mass, his tabulated densities are for some reason consistently low by a factor of 11.5; thus every moon is given a density significantly below the density of water. For Europa, Chambers reported a density of 0.171 g cm^{-3}. He did not comment on the significance of these remarkably low densities.

Pickering (1908) was the first to consider density and albedo as clues to composition. He used a Europa diameter of 3540 km and calculated a density of 1.91 g cm^{-3}. Pickering also commented: "It thus appears that . . . the three inner satellites are therefore composed of some light colored material . . . It will be noted that the densities given . . . are extremely small for solid bodies . . . [S]uch figures almost necessarily imply that these bodies are either enveloped in clouds supported in suitable atmospheres, or that they are composed of dense swarms of meteorites. Their density and brightness are what we should expect if they were composed of loose heaps of white sand . . . Aside from the impossibility that such small light bodies should possess dense cloud-laden atmospheres, their varying albedos preclude the former supposition." Although Pickering's suggestion that Europa was a swarm of white sand strikes us as unlikely, he does get credit for being the first person to assemble the essential data and attempt to interpret it in terms of the possible composition and structure of these moons, concluding that they are low-density bodies with bright surfaces.

The first published suggestion that these moons could be made of ice came in an almost off-handed way by *Jeffreys* (1923). In a paper arguing that the gas giant planets were colder than previously suggested, and made substantially of ices, he comments that "the necessity of supposing that the outer planets are composed of something lighter than terrestrial rocks receives additional support when we consider their satellites . . . The densities of [Io and Europa] are comparable with those of terrestrial rocks, but those for [Ganymede and Callisto] are too small, and are again comparable with the density of ice."

This suggestion of an icy composition for these moons was influential. In a widely used astronomy textbook, *Russell et al.* (1945) wrote that of the Galilean satellites, Io and Europa "are therefore not far from the same size as our moon [with densities of] 2.7, 2.9 [g cm^{-3}]. It is therefore probable that [they] are masses of rock, like our own satellite. Jeffreys suggests that the third and fourth may be composed partly of ice or solid carbon dioxide."

3. FROM THE BIRTH OF MODERN PLANETARY SCIENCE THROUGH THE PIONEER MISSIONS

The situation at mid-century was summarized by Cecilia Payne-Gapschkin in her 1956 college textbook: "The density of Io and Europa is not very different from that of our moon, and they are probably rocky bodies. Callisto is very likely a chunk of ice. Perhaps Ganymede is partly rock, partly ice. All four satellites . . . can hardly have atmospheres; surface gravity and velocity of escape are too low" (*Payne-Gapschkin,* 1956). No one had yet noted the contradiction between lunar-type density and ice-like albedos of Io and Europa — the key to recognizing their unique properties, although the explanations were entirely different for these two small worlds.

In the 1950s Gerard P. Kuiper (University of Chicago and McDonald Observatory) and Fred L. Whipple (Harvard College and Smithsonian Astrophysical Observatory) initiated the renaissance of planetary astronomy and its transformation into the much broader modern field of planetary science. They began the trend toward thinking of members of the solar system as individual worlds that could be studied by the quantitative tools of astrophysics. With the creation of NASA in 1958, the potential for a radical jump in our knowledge of the solar system occurred, and both Kuiper and Whipple were among the first to propose space science missions to the Moon and planets.

A summary of what was known about solar system objects shortly past mid-century was published in the University of Chicago volume *Planets and Satellites* (*Kuiper and Middlehurst,* 1961). Fifteen years later, a comprehensive review of prespacecraft knowledge of all the satellites (52 then known) was published in *Space Science Reviews* by *Morrison and Cruikshank* (1974). Along with our own recollections of this period, much of the information in this section is derived from this 1974 discussion. We have also used the popular history on the Pioneer missions: *Pioneer Odyssey* (*Fimmel et al.,* 1977).

3.1. Early Ideas on Composition

In 1963, Whipple wrote: "It is thought-provoking . . . that the two inner Galilean satellites should be so much like the Moon while the two outer ones should contain such a large proportion of lighter materials. These facts must undoubtedly constitute an important clue to the formation, not only of the satellites, but of Jupiter itself. All four of the Galilean satellites differ in one respect from the Moon: they are all much better reflectors of light . . . Their surfaces, therefore, must be quite different in character from that of the Moon, at least partially covered with some H_2O ice and also other frozen gases."

Discussions of surface composition date to Kuiper's suggestions that low-density objects condensed at low temperatures from the volatiles in the solar nebula, predominantly H_2O. While Kuiper was exploring the Galilean moons from an astronomical perspective, Harold C. Urey was fitting them into his groundbreaking picture of the solar system as a chemical system. In his 1952 book *The Planets*, he speculated that the outer solar system was filled with water ice. Noting that Ganymede and Callisto had densities similar to water ice, he wrote "[T]he interiors of objects of similar mass [to the Moon] regardless of their original temperatures must have risen above the melting point of ice in their interiors, and hence the water of the Jovian moons must all be at or near their surfaces. In fact, water flows instead of terrestrial lava flows may occur from time to time" (*Urey,* 1952, pp. 161–162).

Cameron (1973) proposed from more detailed modeling that the composition in the solar nebula changed from higher-density, rocky material near the Sun to ice-rich material beyond Jupiter. Urey's student John Lewis provided quantitative models of the chemical equilibrium condensation of such a nebula in a series of paper in the early 1970s (*Lewis,* 1971a,b, 1972, 1973), a subject we will return to when we discuss thermal evolution in sections 3.6 and 5.8. All these solar nebula chemical models were consistent with a composition of the satellites of Jupiter of roughly half rock and half ice — a description that matched Callisto and Ganymede but not Io and Europa.

The effort to actually determine which ices were present began with low-resolution infrared spectral observations. In one of the first applications of new lead-sulfide infrared detectors to planetary astronomy, *Kuiper* (1957) published an abstract describing the spectra of the Galilean satellites. He noted that the spectra of JII and JIII, Europa and Ganymede, were reduced in relative intensity beyond 1.5 μm compared to JI and JIV. His interpretation was that "H_2O snow" covered the surfaces of Europa and Ganymede, but not Io or Callisto. He also noted that the presence of H_2O was consistent with Europa's high visual albedo, while Ganymede's darker surface could be contaminated by silicate dust. *Moroz* (1965) was the first to publish spectra of the Galilean satellites, from which he found that Europa and Ganymede's spectra resembled those of the martian polar cap and the rings of Saturn. He therefore assumed that the surfaces of Europa and Ganymede were largely covered by H_2O ice.

3.2. Infrared Spectra and Thermal Properties

Beginning in the mid-1960s, NASA began to support the construction and operation of a new generation of optical telescopes to be used for planetary studies. These included several instruments on Mauna Kea, Hawaii, which was being recognized as one of the best sites in the world to pursue the new field of infrared astronomy. In the early 1970s groundbased studies using these and other telescopes yielded direct measurements of the temperatures of the Galilean satellites, infrared spectra that were diagnostic of surface composition, and improved multicolor photometric lightcurves.

High-precision multicolor spectrophotometry in the visible and near infrared (from 0.3 to 1.1 μm) were obtained for the Galilean satellites by *Johnson* (1971), *Johnson and McCord* (1971), and *Wamstecker* (1972). Longer-wavelength infrared detectors and interferometric spectrometers made possible the extension of these reflectivity curves to a wavelength of 4 μm by *Pilcher et al.* (1972) and *Fink et al.* (1973). From an analysis of the depths of the water ice absorption bands, *Pilcher et al.* (1972) concluded that between 50% and 100% of Europa's surface was comprised of H_2O ice. *Pilcher et al.* (1972) further concluded that both the albedos and the infrared band depths of Europa, Ganymede, and Callisto could be understood in terms of variations in only one parameter, the fraction of exposed non-ice surface material. The variations with rotation in visible and UV reflectance, however, and in particular the strong minimum in Europa's brightness aligned along its trailing side (*Morrison et al.,* 1974), suggested some exogenic effects, perhaps involving interaction with the jovian magnetosphere. For a comprehensive review of these data and their interpretations, see *Johnson and Pilcher* (1977).

Thermal radiation from the Galilean satellites was first measured in the 8–14-μm atmospheric window by *Murray* (1975) and *Low* (1965). They found the lowest brightness temperature for Europa (120 K), consistent with its very high albedo. *Morrison and Cruikshank* (1973) extended the infrared data to 20 μm and also detected a temperature variation with orbital phase in which the leading side of Europa was warmer, consistent with its lower visual albedo.

Observations of the thermal response of the surface to the relatively rapid changes in isolation during satellite eclipses can be used to determine the thermophysical properties of the satellites (*Morrison et al.,* 1972; *Morrison and Cruikshank,* 1973). Europa was the least-well observed by this technique, but its thermal inertia was similar to that of Ganymede (about 2–3 g K s^{-1}), indicating a low-conductivity, porous surface material. These eclipse observations also gave the first hints of Io's volcanic activity, although they were not correctly interpreted at the time.

3.3. Magnetospheres

Along with photometric observations of the jovian system, in the mid-twentieth century groundwork was laid for understanding of the plasma environment that influences the Galilean satellites. Hannes Alfvén, the Swedish Nobel laureate for his work in plasma physics, first spoke in 1937 about the pervasiveness of electric currents in the plasma of interplanetary space: "Space is filled with a network of currents that transfer energy and momentum over very large distances." In 1939, Alfvén proposed a theory of magnetic storms to explain aurorae and other aspects of plasma dynamic phenomena in near-Earth space, including the existence of hydromagnetic waves (*Alfvén,* 1942), that would play so prominent a role decades later in interpreting the magnetic signal at Europa.

For the first forays into the Jupiter system, concerns about the radiation environment were paramount. In a groundbreaking discovery, *Burke and Franklin* (1955) had measured decametric radio signal bursts from Jupiter at 22.2 MHz. Soon an incredible spectrum of radio signals was being recorded from Jupiter, covering 24 octaves. Commensurate with the discovery of Earth's Van Allen belts, Jupiter was suddenly recognized as not just a gas giant surrounded by the vacuum of space, but a planet with a substantial magnetic field containing extensive energetic plasma, capable of sustaining radio emission, emanating from its own Van Allen belts. The radio signals suggested radiation similar to that emitted by pulsars — namely synchrotron radiation where electrons are trapped in so strong a magnetic field that they move at relativistic speeds and emit powerful radio signals (*Drake and Hvatum,* 1959; *Roberts and Stanley,* 1959). The signals suggested that the inner magnetosphere also contained cyclotron radiation, a type of radiation emitted by electrons that are not moving at relativistic speeds, but with high energy nonetheless. Another fundamental discovery came in 1964 when some of the jovian emission was attributed to a complicated electrodynamic coupling between Jupiter and its moon Io (*Biggs,* 1964). A discussion of the discoveries in the jovian magnetosphere from this period and their relation to the interpretation of phenomena related to the electrodynamics of Io (precursors to understanding the same phenomena near Europa) may be found in *Physics of the Jovian Magnetosphere* (*Dessler,* 1983). A growing realization of the magnitude and efficacy of the jovian magnetosphere provided much of the motivation for the Pioneer missions to the outer solar system.

3.4. The Pioneer Missions

Following the successful flight of Mariner 2 to Venus in 1962, NASA planners began to consider ambitious missions to the outer solar system. The key to accessing these distant planets was gravity assists based on close flybys of Jupiter. The positions of the outer planets in the early 1980s would offer an exceptional opportunity for a "grand tour" with flights to Jupiter, Saturn, Uranus, Neptune, and even Pluto. Such a fortuitous alignment would not repeat for more than a century. However, the grand tour could not be achieved without flying spacecraft deep into the jovian magnetosphere in order to achieve the desired gravitational "slingshot" effect. Thus an argument was made for a precursor mission, designed to investigate two potentially lethal problems: concentrations of dust in the asteroid belt, and damage to electronics from high-energy particles in the jovian magnetosphere.

This precursor, called Pioneer and assigned to NASA Ames Research Center, received Congressional approval in 1969. Its primary objectives were to (1) explore the interplanetary medium beyond the orbit of Mars; (2) investigate the nature of the asteroid belt, assessing possible hazards to missions to the outer planets; and (3) explore the environment of Jupiter, including its inner magnetosphere. The choice of a spinning spacecraft carrying two magnetometers, a plasma analyzer (for the solar wind), a charged-particle detector, and an ion detector reflected this emphasis. The Pioneers also carried three rudimentary remote-sensing instruments: an imaging photopolarimeter, ultraviolet photometer, and infrared radiometer, some of which struggled to perform in the hard radiation environment.

Pioneer 10 was launched on February 27, 1972, followed by Pioneer 11 a year later. These were the first human artifacts to leave Earth with sufficient energy to escape the solar system entirely. Both spacecraft transited the asteroid belt without incident. The Pioneer 10 Jupiter encounter, which lasted several weeks, took place in December 1973. Pioneer 10 obtained the first spacecraft view of Europa, constructed from the photopolarimeter data, shown in Fig. 2. Although it was very low resolution, on the order of 200 km $pixel^{-1}$, the image did show a heterogeneous surface. Based on these results, Pioneer 11 was targeted for an even closer pass by Jupiter a year later, 40,000 km above the jovian cloud tops.

As shown in Fig. 2, the photometry obtained on the satellites was crude and rudimentary. Careful tracking of the

Fig. 2. The first spacecraft image of Europa, obtained in December 1973 by Pioneer 10.

spacecraft did provide improved masses for the satellites, yielding a refined density for Io of 3.53 g cm^{-3} and for Europa of 2.99 g cm^{-3}, suggesting that these two inner satellites really did not have the same compositions and structure.

3.5. Formation of the Galilean Satellites

The post-Pioneer perspective on the Galilean satellites was presented in the first of what would be four comprehensive multiauthor books published by the University of Arizona Press. *Planetary Satellites*, edited by Joseph Burns with 33 collaborating authors and 598 pages (*Burns,* 1977), was based on a conference held at Cornell University in 1974; the last section of *Jupiter,* edited by Tom Gehrels (*Gehrels,* 1976), was based on a conference held at the University of Arizona in 1975. These books documented in detail the subjects lightly summarized in this section. A useful popular history is *Voyage to Jupiter* (*Morrison and Sanz,* 1980).

In some ways the jovian system can be considered a miniature solar system. The regularity of the Galilean satellite orbits suggests that they were formed within a circumjovian nebula analogous to the solar nebula that gave rise to the planetary system, including a decrease in density with increasing distance from the primary. Thus the most fundamental properties of the satellites are closely linked to the formation of Jupiter itself.

Theory suggested that Jupiter's formation was characterized by three main stages (*Pollack and Fanale,* 1982): (1) early, slow contraction, in quasihydrodynamical equilibrium; (2) rapid hydrodynamical collapse, triggered by the dissociation of hydrogen; and later (3) slow contraction leading to gradual cooling (e.g., *Bodenheimer,* 1974). The circumjovian nebula probably developed during stage 2 when the collapsing proto-Jupiter shrank inside the current orbits of the Galilean satellites. *Cameron and Pollack* (1976) proposed that accretionary processes occurring in the disk led to formation of the satellites, by processes analogous to those that took place in the solar nebula.

Constraints on the temperature conditions in the circumjovian nebula were provided by the trend in bulk composition of the satellites, as first modeled by J. Pollack of NASA Ames and his colleagues (*Pollack and Reynolds,* 1974; *Cameron and Pollack,* 1976; *Pollack et al.,* 1976). The solar nebula, cooling with distance from the Sun, was modeled to have a zone cool enough that volatiles such as water vapor would finally condense. The transition from the zone where rocky material predominated, to that where volatiles would readily condense, was designated the "snow line." By analogy, the density dichotomy among the satellites suggested that there was a snow line in the jovian nebula between Ganymede and Europa. Presumably the high luminosity of the proto-Jupiter inhibited the condensation of H_2O inside the orbit of Ganymede.

In this emerging view, Europa's history bore some similarities to that of Earth, with its bulk composition dictated by the moderately high temperatures in the part of the nebula where it formed, followed by the acquisition of a surface veneer of H_2O ice from impacting comets or other volatile-rich planetesimals. Observations, however, gave no hint of the thickness of Europa's veneer of ice.

3.6. Interiors and Thermal Evolution

Thermal modeling had been initially driven by theoretical concepts, not new data. John Lewis (now at the Massachusetts Institute of Technology) continued Urey's line of research on the cosmochemical nature of the extended solar system with a series of papers (*Lewis,* 1969, 1972) that outlined the various types of ice that should be in stable equilibrium with a cold solar nebula of cosmic abundances. *Lewis* (1971a,b) presented a simple heat-balance calculation showing that if the radiogenic heat from the rocky fraction of an ice-rock moon in cosmic proportions were in steady state with the outflow of heat via conduction, then temperatures would rise well above the melting point of ice in bodies as small as 1000 km radius. He further recognized the role of high-pressure ice forms in the evolution of these bodies, the importance (and difficulty) of modeling convective heat transport, and the likely alteration of the surfaces due to evaporation of the ices, the infall of meteoritic material, and the formation of minor species from "photolytically labile material."

Lewis in turn directed a student who developed a more detailed model (*Consolmagno,* 1975) for the thermal evolution of icy moons. This model took into account the decaying levels of radionuclide abundances, phase changes between different forms of ice (as well as the melting of that ice), and the heat released during the formation of a rocky core once the ice melted. Convection was assumed for the molten regions of the body but, it was argued, was less likely to be important for the ice (or, at least, difficult to model) given the uncertain and non-Newtonian properties of dirty water ice with multiple high-pressure phases. Assuming Europa was made of 90% rocky material with the density of an ordinary chondrite and 10% H_2O, the model (*Consolmagno,* 1975) predicted that on Europa "a thin crust of ice covers a convecting region of water, which is cooling off the upper layers of the silicate core."

In an appendix, Consolmagno speculated on the importance of the chemical evolution of the rocky core with the liquid water mantle. He closed this argument by writing, "Given the temperatures of the interiors, and especially of the silicate layers through which liquid will be percolating, the possibility exists of simple organic chemistry taking place involving either methane from the ice or carbon in the silicate phase. However, we stop short of postulating life forms in these mantles." Similar arguments on the melting of the icy Galilean moons (although without the speculation about life) were presented in *Consolmagno and Lewis* (1976) and *Fanale et al.* (1977), who agreed that Europa would likely differentiate within the first 500 m.y., with a liquid water layer (ocean) underlying a water ice crust, both together forming a layer on the order of 100 km depth over the rocky core.

As all these authors acknowledged, thermal modeling required more knowledge than was actually available at the time for the initial composition, radioactive heat sources, temperature of formation, and the thermal and physical behavior of unconstrained mixtures of ices and rocky material, and furthermore required speculation about the total amount of anhydrous vs. hydrous silicates in the initial infall of material to form the satellite. Uncertainty as to how much accretional energy was retained has always made it difficult to assess how effective this heat source was at early differentiation. Lack of data never prevented theoreticians from speculating, however!

It was Io that first drew public attention to the thermal evolution of the Galilean satellites. In a paper dramatically published in *Science* a few days before the Voyager 1 encounter, Stan Peale of the University of California, Santa Barbara and Pat Cassen and Ray Reynolds of NASA Ames Research Center calculated the expected heating of Io from tidal stresses (maintained by the Laplace resonance) and predicted currently active volcanism (*Peale et al.,* 1979). Their prediction was almost immediately confirmed by Voyager images of Io, first of volcanic flows on the surface and then by spectacular plumes of gas rising from ongoing eruptions identified a few days after the flyby. We will return to the analysis of the interior and implications of tidal heating for Europa in section 5.8.

4. THE VOYAGER ERA

The Voyager "grand tour" followed closely upon Pioneers 10 and 11. The years around the Voyager missions saw a revolutionary change in our understanding of the Galilean satellites. In many respects, these small satellites were the stars of the two Voyager Jupiter encounters, emerging as full-fledged worlds with unique geological histories. Europa was seen as a rocky world with a young surface layer of plastic ice and enigmatic surface features that hinted at a dynamic thermal history. And after the dramatic discovery of active volcanism on Io, these two satellites would never again be considered as "twins."

4.1. The Voyager Missions

The back-to-back Voyager missions represented a scaled down but still highly capable implementation of the grand tour concept. Built and operated by the Jet Propulsion Laboratory (JPL) of the California Institute of Technology, the Voyagers were large three-axis-stabilized spacecraft with a fully articulated scan platform, permitting detailed study of the planet and satellites. Voyager 1 was launched in September 1977 on a path to Jupiter and Saturn and Titan; Voyager 2 undertook the 13-year trip to Jupiter, Saturn, Uranus, and Neptune. Both eventually followed the Pioneers in leaving the solar system entirely.

Voyager imaging data determined the diameters (and hence densities) of the Galilean satellites to yield what are essentially the modern values (for Europa: diameter 3100 km; density 3.0 g cm^{-3}), and their surface features were extensively mapped. Europa was the least well imaged of the Galilean satellites, an unavoidable result of its orbital position during the two flybys in March and July 1979. The best data were obtained by Voyager 2 on July 8, 1979, at a closest range of 206,000 km, when a fraction of one hemisphere was mapped at a resolution of about 2 km $pixel^{-1}$. Its bright (high-albedo) surface was found to be remarkably smooth and almost free of craters. The surface consisted primarily of uniformly bright terrain crossed by long linear markings and a few very low ridges (elevation up to a few hundred meters). Voyager scientists characterized it as looking like "a white billiard ball faintly crossed by lines applied with a pen" or looking "cracked like a broken eggshell" (*Morrison and Sanz,* 1980).

The absence of craters or surface relief (Fig. 3) suggested to Voyager scientists either recent resurfacing or a "soft" icy surface, while the long "cracks" seemed to imply a brittle crust subject to tectonic stresses. While the ice crust was estimated as roughly 100 km thick, there was little consideration to the possibility that much of this material might actually be liquid today. However, the suggestion was made that there might be episodic heating of Europa, perhaps analogous to the tidal energy source currently so active on Io. Perhaps the most enigmatic features, visible only near the terminator on the best images, were cycloid ridge systems with wavelength on the order of 100 km and total length of more than 1000 km.

The Voyager encounters were an unforgettable experience for the participants, who included nearly 100 members of the press, about half of whom encamped at JPL for a month in both March and July 1979, as well as the then Governor of California, Jerry Brown. In those days NASA initiated an open communications policy with daily press

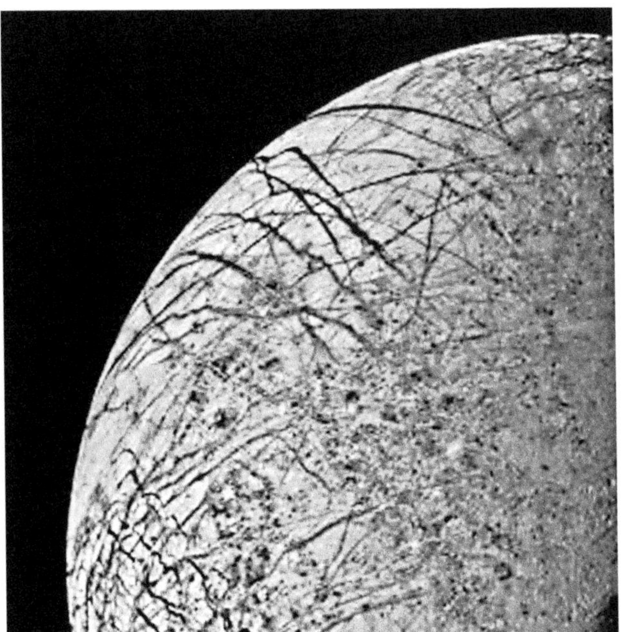

Fig. 3. Among the first close-up images of the linear crack-like features of Europa's surface obtained from Voyager 2, July 9, 1979, at a range of 246,000 kilometers.

conferences, informal science seminars, and frequent personal exchanges involving mission staff and scientists. Some members of the press, such as Henry Cooper (who was then writing for *The New Yorker*), Kelley Beatty of *Sky and Telescope*, and Jonathan Eberhart of *Science News*, were given some access to the science work areas where the data were being analyzed and discussed. The spacecraft raw images (coming in at a rate of one every 90 seconds) were also displayed "live" on the JPL internal TV net, so that everyone present could participate in the shared excitement of discovery. Popular books written in this timeframe include *The New Solar System* (*Beatty et al.,* 1988).

The circumstances in which the imaging resolution and other data rapidly improved as the spacecraft approached its targets allowed a unique compression of the scientific method, carried out on a public stage. Data such as satellite images that were received one day were interpreted by each science team to a level appropriate for the press conference the following day. But everyone knew that by the time these results were presented, together perhaps with ideas about their interpretation and significance, new and better data were already in the pipeline. In the final days before each encounter, the imaging resolution doubled from one day to the next. Within a week, hypotheses could be formed, tested, and rejected or modified several times. This was a heady experience never equaled in any other space science mission.

Carl Sagan had famously written that only one generation was privileged to witness the transition from planets (and satellites) as mere points of light in the night sky to real worlds, each with its own unique history. For the Galilean satellites, that transition took place within just the five months that encompassed the Voyager encounters.

4.2. Non-Ice Material and Processes

In the years immediately preceding Voyager's Jupiter flybys, groundbased observers discovered the neutral torus and plasma torus associated with Io, the first being discovered by R. A. Brown's observation of Io's sodium cloud and the second by Kupo et al.'s 1976 discovery of ionized sulfur in Jupiter's magnetosphere (see review by *Thomas et al.,* 2004). Voyager *in situ* and ultraviolet measurements characterized the density and spatial extent of this iogenic plasma and demonstrated the intimate coupling between the satellites and the jovian magnetosphere. These observations prompted *Eviatar et al.* (1981) to suggest that ion implantation was occurring on Europa's trailing side; that Io was painting that hemisphere with sulfur. Concurrent with the groundbased torus observations, the International Ultraviolet Explorer (IUE) was being used to characterize the ultraviolet reflection spectra of Jupiter's satellites. *Lane et al.* (1981) examined trailing-leading side ratio spectra, finding a broad absorption feature centered at 280 nm that they identified with SO_2 and suggested was formed by ion implantation.

Voyager images enabled mapping of Europa's surface, in optical pass bands from the ultraviolet to orange regions, and then study of the distribution of chromophores. Sulfur had earlier been suggested (cf. *Johnson and McCord,* 1971; *Wamsteker,* 1972), and analyses by *Johnson et al.* (1983), *McEwen* (1986), and *Nelson et al.* (1986) found distributions consistent with implantation. The sulfur chemistry was not completely understood, and the association of dark material with geological features was (and remains) unexplained. Possibilities include enrichment by geological processes or emplacement of material from below, or both (see chapter by Carlson et al.).

In addition to observational work, two separate communities began laboratory work that would bear fruit for planetary exploration in the decades to come. Experimental work showed that ion bombardment of minerals and salts darkened surfaces and caused spectral shifts (*Hapke,* 2001; *Nash and Fanale,* 1977). Ion bombardment also ejects molecules from the surface by a process known as sputtering. *Brown et al.* (1978) and *Lanzerotti et al.* (1978) showed that ion bombardment of low-temperature ices breaks chemical bonds, creating new species, and efficiently sputters the parent and daughter molecules. The production efficiency for producing O_2 was found to be strongly temperature dependent, but at temperatures relevant to the surface of Europa. H_2O, H_2, and O_2 proved to be important components of the ejecta. On the basis of this work, *Johnson et al.* (1982) predicted the presence of an O_2 sputter-produced atmosphere on Europa (see section 4.4).

4.3. First Thoughts on Habitability

Consolmagno (1975), *Consolmagno and Lewis* (1976, 1977, 1978), *Reynolds et al.* (1983), and others, perhaps most famously Arthur C. Clarke (inspired by conversations with some of these investigators) in his 1982 novel *2010*, opened questions about the habitability of Europa. Two scenarios presented themselves — life on, and life under, the surface. Surface life, subjected to the radiation environment and near-vacuum conditions, seemed highly unlikely. When Consolmagno first informally suggested life in a subcrustal ocean, he was reminded by Carl Sagan (personal communication) that life on Earth depended on sunlight for its energy, a source not available in the dark oceans.

An important change in the understanding of the energetics of life occurred when terrestrial life forms were discovered associated with black smokers deep in Earth's oceans (*Corliss et al.,* 1979; *Jannasch and Wirsen,* 1979; *Corliss et al.,* 1981; *Sullivan and Palmisano,* 1981). The discovery dramatically demonstrated that chemotropic life *could* thrive on a geochemical source of reduced compounds, although it was later pointed out that such processes took place in the presence of an ocean saturated in molecular oxygen produced by photosynthesis. For Europa, comparable oxygen- and oxidant-producing reactions may be induced by radiolysis at the surface that, with subduction, could form a radiation-driven ecosystem (*Chyba,* 2000; *Chyba and Hand,* 2001; chapter by Hand et al.).

In those early days, speculation about the habitability of a proposed europan ocean included assumptions such as inferred abundances of nutrients by comparison with the Murchison meteorite, assuming Europa to be of carbonaceous chondritic origin (*Oro et al.,* 1992; *Reynolds et al.,* 1983). The metabolic pathways of chemotropic microorganisms were only moderately well understood. Radiation and insolation were inferred based upon (incomplete) sampling and characterization of the radiation environment near the L-shells crossed by Europa (*Eviatar et al.,* 1981). Quantitative discussion of biosignatures would have to wait until higher-resolution data on the geochemistry of Europa could be obtained.

4.4. Post-Voyager Perspectives on Thermal History, the Presence of an Atmosphere, and Europan Electrodynamics

The prediction by *Peale et al.* (1979) that tidal heating forces would result in a volcanic Io, and the subsequent discovery by Voyager of the geysers Pele and Loki in full eruption, electrified the planetary science community. Europa's youthful surface and global tectonic activity were hypothesized to originate, like those of its neighbor Io, from tidal interactions with its innermost and outermost orbital neighbors. *Reynolds and Cassen* (1979) argued that the ice layer would be unstable to convection and that thermal convection would freeze a liquid layer in a time short compared to the lifetime of the satellite (C. Alexander at the time was the student who ran many of Reynolds' convection models). But the calculations were critically sensitive to material parameters and assumed initial conditions, because the heat production and subsolidus heat removal processes were of comparable magnitude. For example, the *Squyres et al.* (1983) solution (next discussion) was based upon an assumption that hydrated silicates do not lose appreciable strength until dehydration occurs and liquid water is released — an assumption that impacts the calculated ability of the body to flex and respond to tidal forces.

Squyres et al. (1983) introduced the paradigm of a relatively thick ice crust, potentially overlying a liquid water layer that remained to be proved. The paradigm included the notion that tidal heating could maintain a liquid ocean layer, but heating rates would be slow enough to maintain a frozen crust tens of kilometers thick, and that convection would be the dominant mechanism for heat transport from the ocean to the surface. Alternatively, *Helfenstein and Parmentier* (1985) suggested that nonsynchronous rotation (an ocean or "soft ice" layer that effectively decoupled the surface from the interior) might provide a fracturing mechanism to explain the geologic features revealed by Voyager.

Early speculation about a tenuous atmosphere on Europa, generated by sublimation, just as it is on a comet, was effectively damped by the measurements of the low surface temperature of Europa. Even when suggestions were made concerning migrating ice on Ganymede, Europa seemed too cold for such processes to take place. Alternatively, *Johnson et al.* (1982) predicted a bound O_2 atmosphere for Europa, with a column density of ~2–3 × 10^{19} m^{-2}, generated by sputtered water ice.

After the discovery of electrodynamic signals associated with Jupiter's interaction with Io (*Bigg,* 1964), *Neubauer* (1980), in a seminal paper, studied the motion of a large conducting body through space, and determined that energy would be radiated away in the form of Alfvén waves. *Reynolds et al.* (1983) were the first to discuss an electrical Europa, although they assumed charge separation due to a $\mathbf{J} \times \mathbf{B}$ force (where **J** is the current and **B** the magnetic field) that would generate an electric field and associated currents closing through Jupiter's ionosphere. The *Galileo* mission would exploit the interactions between Europa and the magnetosphere to provide the most compelling evidence for liquid water in its interior, looking for the evidence for the presence of a conducting layer in these Alfvén waves and other electrodynamic phenomena first discussed at this time.

In the years after Voyager, the final two of the four comprehensive volumes (see section 3.5), published by the University of Arizona Press and featuring the Galilean satellites, were released. *Satellites of Jupiter*, edited by David Morrison with 47 collaborating authors and 972 pages (*Morrison,* 1982), summarized the state of knowledge after the flybys. A still more mature perspective was reflected in *Satellites,* edited by Joseph Burns and Mildred Shapley Matthews, with 45 collaborating authors and 1021 pages (*Burns and Matthews,* 1986).

5. THE *GALILEO* SAGA

For planetary scientists, the decade after Voyager was characterized by budget threats (in 1981 the Office of Management and Budget argued for terminating all planetary exploration and redirecting JPL into defense work) and the loss of the shuttle Challenger in January 1986, an event that halted all deep space launches, including that of the Hubble Space Telescope (HST), and forced a major redesign of the *Galileo* mission to Jupiter. After the rapid-fire successes of the Mariner and Pioneer missions, crowned by the triumphs of Viking (to Mars) and Voyager in the late 1970s, this was a severe disappointment. Many young scientists who had entered planetary studies thinking that there would be at least one new mission per year faced a new and unpleasant reality. However, when the next generation of planetary missions such as *Galileo* and Magellan did fly, they produced data of unprecedented quality and quantity. Reviews of the *Galileo* mission as a whole can be found in *Johnson et al.* (1992), *Barbieri et al.* (1997), *Harland* (2000), *Fischer* (2001), *Cruikshank and Nelson* (2007), and *Meltzer* (2007). Technical details of the science investigations are comprehensively described in the book *The Galileo Mission* (*Russell,* 1992).

5.1. The *Galileo* Mission

The mission that would come to be known as *Galileo* was under discussion long before Voyager was launched. At NASA Ames Research Center, a science working group studied a Jupiter Probe mission that combined a spinning spacecraft (with heritage from Pioneers 10 and 11) with an entry probe. This team focused on investigating jovian atmospheric composition and the magnetosphere. A parallel study at JPL investigated the potential of a three-axis-stabilized spacecraft like Voyager that could orbit Jupiter and undertake detailed study of the jovian satellites and the dynamics of the planet's atmosphere. Both teams argued for a long-lived mission that could make many orbits of Jupiter, mapping out the magnetosphere and providing multiple close flybys of the satellites. These two concepts were merged in a study led by James van Allen that recommended to NASA a multidisciplinary mission called Jupiter Orbiter Probe (JOP). While there was obvious value in pursuing so many objectives with a single mission, these choices also planted seeds of conflict that would emerge when difficult choices had to be made in planning trajectories and prioritizing investigations. Given *Galileo*'s limitation on computing power and onboard memory (which used tape recorders to store data), this turned out to be a very challenging mission from an operational perspective.

The 16 instruments carried by the *Galileo* orbiter included a magnetometer mounted on a boom to minimize interference with the spacecraft; a plasma instrument for detecting charged particles (PLS); a Plasma wave detector (PWS) to study waves generated by the particles; an energetic particle detector (EPD) for measurements of plasma at the very-high-energy end of the spectrum; and a detector of cosmic and jovian dust. It also carried the Heavy Ion Counter (HIC), an engineering experiment added to assess the potentially hazardous charged particle environments surrounding the spacecraft. The remote sensing instruments included the camera system [SSI, the first charge-coupled device (CCD) on a deep-space project], designed to obtain images of Jupiter's satellites at resolutions from 20 to 1000 times better than Voyager's best; the Near Infrared Mapping Spectrometer (NIMS) to make multispectral images (image cubes) for chemical analysis, the first imaging spectrometer ever flown; an Ultraviolet Spectrometer (UVS) to study gases; and a Photo-Polarimeter Radiometer (PPR) to measure radiant and reflected energy. The payload also included an extreme ultraviolet detector (EUV) associated with the UV spectrometer on the scan platform.

The JOP mission was ready to be proposed to Congress as a "new start" even before the September 1977 Voyager 1 launch. In March 1977 funds were approved following an unprecedented roll-call vote in the Senate that dealt with this specific mission alone. The official start of the project, now named *Galileo*, was in October 1977, with a planned launch in January 1982 using the space shuttle.

The launch of *Galileo* was repeatedly delayed, primarily by slips in the space shuttle schedule and changing configuration of the associated upper stage launch vehicles that were required to accelerate the spacecraft from low-Earth orbit and on to Jupiter. Following the Challenger accident in 1986 and subsequent decisions not to use the high-energy Centaur upper stage with the shuttle, the mission was reconfigured for launch in October 1989. Because the available solid-fuel IUA upper stage was less powerful than a Centaur, the new trajectory included flybys of Earth (twice) and Venus. Concerns about heating close to the Sun led the mission planners to delay deployment of the main 4.8-m high-gain antenna until after the Venus flyby. It then failed to open fully, eliminating all high-gain transmissions from the spacecraft. As a consequence, less than 1% of the originally planned data could be returned from the mission. Heroic changes in software and adoption of sophisticated data compression schemes bought back about a factor of 10, but throughout its lifetime *Galileo* was "data starved," requiring very careful planning to ensure maximum science return (*O'Neil,* 1997).

5.2. Satellite Tours

Galileo was in the Jupiter system from December 1995 through September 2003, almost a full Jupiter year. The "tour" for the *Galileo* orbiter was designed to enable detailed observations of the satellites and Jupiter. Jupiter orbit insertion (JOI) was achieved with a close flyby of Io, with subsequent orbits anchored on satellites with the dual purpose of studying each satellite and using the satellite gravity to modify the orbit for the next loop. Each orbit was designated by the flyby satellite. Thus, orbit "G1" was the first orbit in the mission and Ganymede was the target for the close flyby; "C3" was the third orbit with a flyby of Callisto; etc. Substantial data could also be collected for the other satellites from a greater distance; for example, the first images of Europa were taken on orbit G1. Even though these far-encounter views were only equivalent to the Voyager resolution (typically 1–2 km pixel^{-1}), they were use-

TABLE 1. Europa data takes during *Galileo*'s prime and extended missions.

Orbit	Mission	Date	Altitude (km)	Hemisphere of Closest Approach	Local Time
G1	Prime	June 27, 1996	156,000	Between trailing and antijovian	
G2	Prime	Sept. 6, 1996	673,000	Trailing (Upstream)	
C3	Prime	Nov. 4, 1996	41,000	Antijovian	
E4	Prime	Dec. 19, 1996	692	Leading (Wake region)	16.7 (night sector)
E6	Prime	Feb. 20, 1997	586 km	Trailing (Upstream region)	12.9 (night)
G7	Prime	April 5, 1997	24,600	Between antijovian and leading	
C9	Prime	June 25, 1997	1,200,000	Between leading and subjovian	
C10	Prime	Sept. 17, 1997	621,000	Trailing (Upstream)	
E11	Prime	Nov. 6, 1997	2043	Leading (Downstream)	11.0 (day)
E12	GEM	Dec. 16, 1997	201	Trailing (Upstream)	14.7 (day)
E13	GEM	Feb. 10, 1998	3562	*(Gravity only)*	
E14	GEM	March 29, 1998	1644	Trailing (Upstream)	14.4 (day)
E15	GEM	May 31, 1998	2515	Leading (Downstream)	10.1 (day)
E16	GEM	July 21, 1998	1834	(No data obtained because of spacecraft safing)	
E17	GEM	Sept. 26, 1998	3582	Leading (Downstream)	9.9 (day)
E18	GEM	Nov. 22, 1998	2271	(No data obtained because of spacecraft safing)	
E19	GEM	Feb. 1, 1999	1439	Trailing (Upstream)	9.8 (night)
I25	GEM	Nov. 26, 1999	8860	Subjovian	
E26	GMM	Jan. 3, 2000	351	Trailing (Upstream)	2.9 (night)
G28	GMM	May 20, 2000	593,321	Antijovian	
I33	GMM	Jan. 17, 2002	1,002,152	Subjovian	

Excellent images were taken on encounters where Europa was not the prime target of G1, G2, C3, G7, C9, C10, I25, G28, and I33, some with resolutions as good as 420 m pixel^{-1}, and often in color. GEM = *Galileo* Europa Mission; GMM = *Galileo* Millennium Mission. Local time is relative to Jupiter; neither remote sensing nor fields and particles (F&P) data was collected on E1, E16, and E19 for anomaly and telemetry limitation reasons. Data extracted from *Kurth et al.* (2001).

ful because of the high quality of the *Galileo* CCD camera and the opportunity to view the targets at different phase angles and rotational phases.

For the first two years, the *Galileo* orbiter made close flybys only of Callisto and Ganymede, using these satellites to shrink and circularize the orbit. The strategy for imaging Europa was to acquire enough information during this time to be able to plan specific close passes as the spacecraft orbit permitted. The team implementing this imaging strategy was a target planning and sequence design team led by Ron Greeley and associates at Arizona State University, during the prime mission, and co-organized with James Head, at Brown University, for the extended mission. All such remote sensing sequences were severely impacted, of course, by the failure of the high-gain antennae to deploy. This was particularly problematic for imaging and other high-data-rate instruments. Each picture was a precious commodity, resulting in considerable debate within the team and the *Galileo* Project Science Group over the allocation of data downlink resources.

For Europa, medium-resolution (few hundred meters per pixel) images provided regional views, and high-resolution (tens of meters per pixel) images provided samples of key terrains and features, set within the context of the regional view. For morphology, low-Sun-angle illumination was desirable to enhance terrain characteristics, while color and photometry objectives were best met with high-Sun views.

In all, the *Galileo* spacecraft encountered Europa 12 times (see Table 1). The first three encounters took place during the prime mission, and the bulk of the remaining encounters occurred during the *Galileo* Europa Mission (GEM). The final encounter was part of the *Galileo* Millennium Mission (GMM), an extension to allow the Jupiter system to be studied by two spacecraft at once: *Galileo* and Cassini, passing by in the year 2000 on its way to Saturn.

Ironically, just as the *Galileo* spacecraft was arriving at Jupiter, policy makers were gathering to decide how best to destroy the spacecraft to protect Europa. Even before the 1989 launch, the *Galileo* project plan was revised to include the following: "In addition, the Project will supply data obtained bearing on the biological interest of the Jovian satellites to the [NASA] Planetary Protection Officer in a timely manner. This information will be provided by letter before the end of mission and while the spacecraft is con-

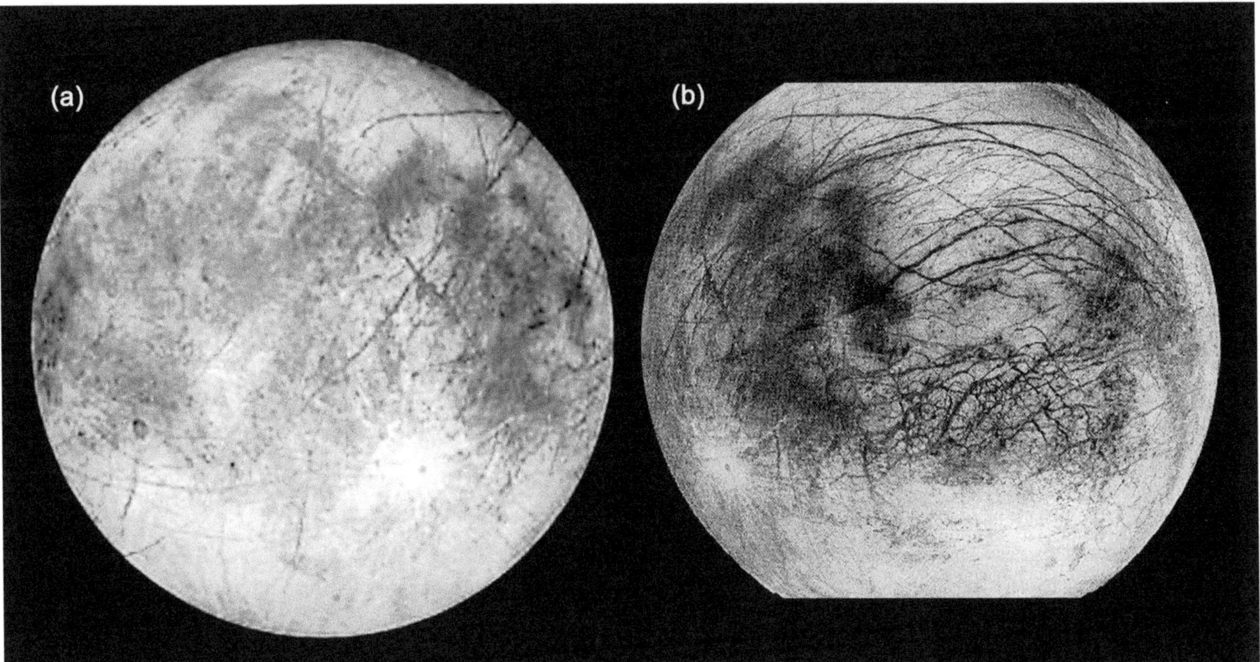

Fig. 4. **(a)** The trailing hemisphere of Europa — so designated because the hemisphere facing the camera happens to be the trailing side of the moon's orbital motion around Jupiter and also subtends the stream of magnetospheric plasma, that flows faster than the orbital motion of Europa. The impact crater Pwyll is the central point on the lower right of the image from which the white rays are splayed. The longitude of the center of this hemisphere is approximately 290°W. **(b)** This image shows Pwyll rotated to the west (left) about 70° from the previous image such that the trailing hemisphere is oriented to the left and the leading hemisphere to the right; the Conamara Chaos region is now shown on the left; the Cadmus and Minos reqions are in the upper right; the "wedges" region is just below the equator and west of the antijovian (center) point. The longitude of the center of this hemisphere is 220°W.

trollable. If the Planetary Protection Officer finds that a satellite should be protected further than Category II requirements call for, the Project will negotiate options that will preclude an impact of that satellite by the Orbiter." In April 1999, after the final GEM Europa encounter, a National Research Council board was convened to make recommendations on how to best prevent forward contamination of Europa (*National Research Council,* 2000), and the project began considering the logistics of the final disposition of the spacecraft.

5.3. Geology Unique to Europa

The hypothesis of a europan ocean was considerably strengthened in the extended GEM. Figure 4 shows the hemisphere(s) in which most of the features relevant to the efforts to deduce the presence of an ocean are to be found.

With the first close flyby of Europa on E6, images were obtained with resolution up to 21 m/pixel. These represented a huge increment in detail, but because of the constrained data budget they could not always be placed in their low-resolution context. This problem of high-resolution "postage stamp" images without full context was a challenge throughout the mission. These high-resolution images showed the "triple-band ridges" (Fig. 5) and "chaotic terrains" (Fig. 6) in detail, both of which indicated the high probability of sub-ice activity at the time that the features formed. Chaotic terrain was particularly intriguing, because it seemed to indicate places where the crust has been extensively disrupted by internal processes. Reconstructions of the geometry of the chaos terrain showed that the various pieces could be fit back together like a jigsaw puzzle.

Even before the high-resolution images were obtained, the concept of a young, active sub-ice ocean was beginning to emerge, as reported by Greeley at the conference "The Three Galileos: The Man, The Spacecraft, and the Telescope" (*Barbieri et al.,* 1997). Held January 1997 in Padova, Italy, the sessions convened in the same university venue as Galileo's lectures, and the conference culminated with a visit to Pope John Paul II in Rome, arranged by Guy Consolmagno and the Director of the Vatican Observatory, George Coyne.

By the end of 1997, the *Galileo* science team was in consensus that Europa is a fascinating object that merited more extensive exploration, leading to the GEM extension of operations through repeated flybys of this moon. Summaries of the geology of Europa as it was understood at the time can be found in *Carr et al.* (1998) and *Greeley et al.* (1998). A series of papers from the imaging team, initiated in 1997 and published in 1998, addressed detailed aspects of the "notion of the ocean" within Europa, including a paper by *Sullivan et al.* (1998) on "wedges"; an analysis by *Pappalardo et al.* (1998) of solid-state convection; and a paper by *Geissler et al.* (1998) on evidence for non-

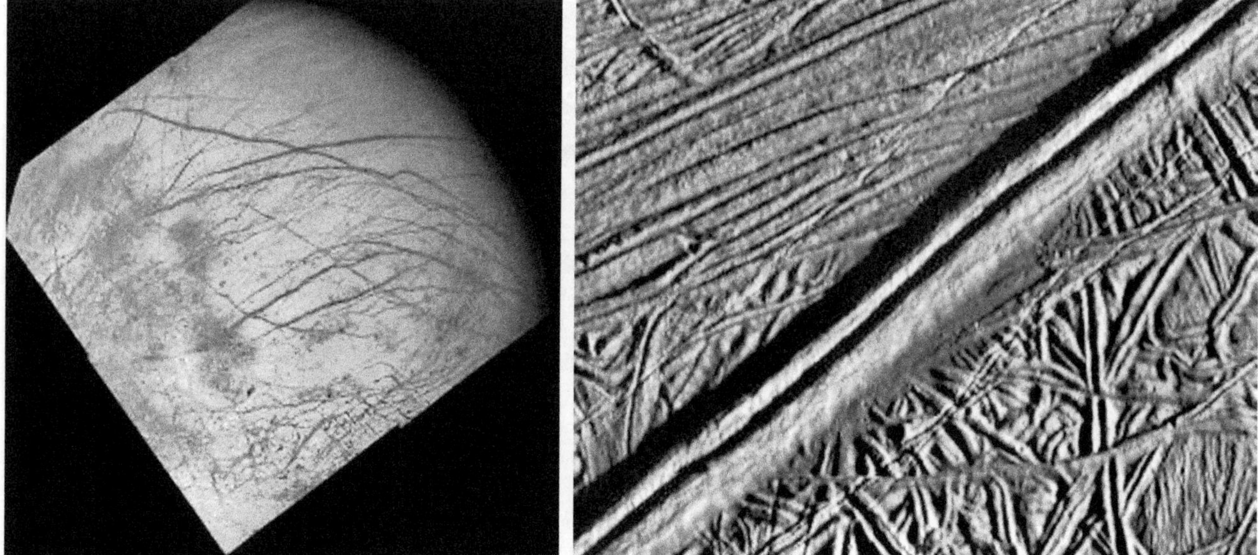

Fig. 5. Ridges and triple bands. Original image taken during the G1 encounter, subsequent high-resolution images taken on E6.

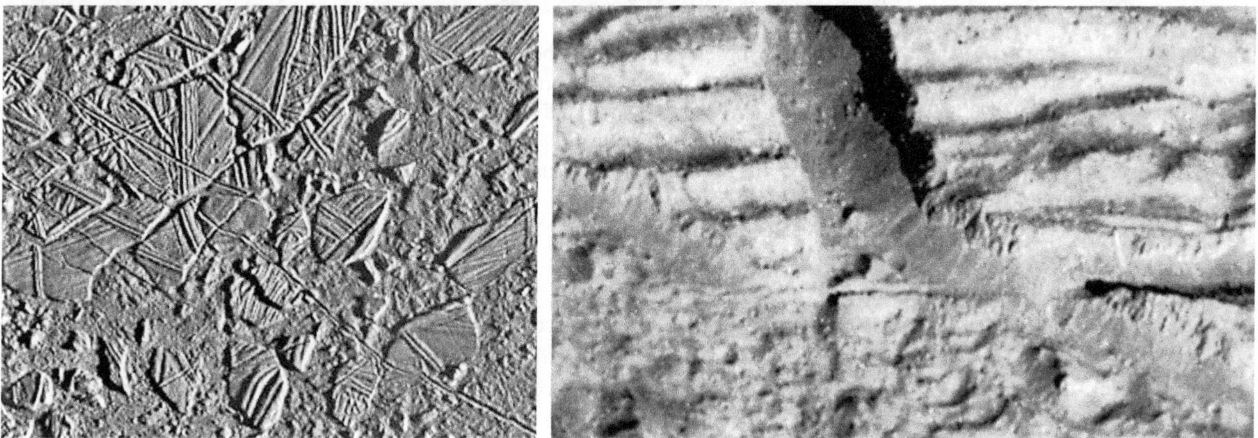

Fig. 6. The Conamara Chaos region. The lower-resolution image was taken on E6, the high-resolution image on E12.

synchronous rotation. This was followed by a paper by *Pappalardo et al.* (1999) on the evidence for an europan ocean.

The mounting evidence for nonsynchronous rotation was based in part on images that show strike-slip faults (horizontal movement along fractures in the crust). More than 100 such faults were identified by Greenberg's team [see *Greenberg et al.* (2002) for a complete list of citations], who noted that left-lateral offsets occur preferentially in the northern hemisphere, while right-lateral offsets dominate in the southern hemisphere. These features correlated well with the stress field — a map of the extensional and compressional forces that drive surface deformation. The crack pattern on the surface was found to be consistent with the induced stress field found by models of nonsynchronous rotation and the presence of a sub-ice ocean. In addition, cycloidal ridges, first seen by Voyager, was evidence that diurnal tides were important drivers for surface deformation (*Hoppa et al.,* 1999; *Greenberg et al.,* 2002).

However, the evidence of an ocean based on surface features simply referred to the time when the features formed, and did not necessarily suggest that an ocean exists today. This consideration required an assessment of the age of the europan surface, coupled with data from the *Galileo* magnetometer.

5.4. Age of the Europan Surface

Very few impact craters are seen on Europa's surface. The frequency of impact craters superposed on planetary surfaces provides an estimate of surface ages, based on models of the flux of impacting objects. The impacting population for the jovian satellites is presumably very different from that of the inner solar system, which is dominated by asteroidal sources. The impact flux in the inner solar system has been calibrated using crater counts on the lunar maria, for which chronological dates are known (e.g., *Hartman,* 1981). It is presumed that the primary impacting population in the jovian system is heliocentric (not planetocentric) and consists of comets, i.e., of debris from the Kuiper belt and Oort cloud. To date a relatively young surface like that of Europa, one needs to know the flux of small (kilometer-scale) comets. Observations from Earth indicate that very few kilometer-scale comets penetrate the inner solar system, so estimating the flux at Europa involves a chain of assumptions. One must also take into account the considerable focusing of incoming projectiles by the jovian gravity, and this in turn depends on the dynamics of this population (*Moore et al.,* 1998).

Flux uncertainties produced initial estimates for the age of Europa's surface ranging from more than 2 b.y. to less than 10 m.y. With refinements in models for the impacting flux, most planetologists now regard the surface to be 30–70 m.y. old (*Zahnle et al.,* 2003), a geological blink of the eye in comparison to most planetary surfaces (*Carr,* 1999). Consequently, it seems unlikely that an ocean sufficiently extensive to account for the surface features would have frozen in such a short period of time geologically. The *Galileo* imaging team [see *Pappalardo et al.* (1999) and *Greeley et al.* (2004) for citations related to this growing consensus] therefore concluded that an ocean is probably present today; this conclusion was further supported by the magnetometer data discussed below. Questions remain as to the thickness of the ice shell over the ocean, which is one of the key issues to be addressed by a future mission (see section 6).

5.5. Inference of an Induced Magnetic Field

The most powerful arguments for a current liquid water ocean on Europa were derived from measurements of interactions with the jovian magnetosphere. *Kargel and Consolmagno* (1996) suggested that evidence of an ocean rich in dissolved salts and thus electrically conducting might be visible in the magnetic field around Europa. Saltwater, although a poor conductor compared to metals like copper wire, is much more conductive than some natural nonmetallic solids expected at Europa, such as water ice, and salty hydrated silicate-ice polyphase aggregates. On a planetary body, when found physically near the surface and not deeply buried, currents engendered in an idealized salty brine, with the conductivity of terrestrial seawater, should be capable of producing a magnetic signature detectable above the surface (*Zimmer et al.,* 2000), if that liquid is located physically near the surface and not buried too deeply, with a signature that is slightly different in character than that produced by a magnetic core of iron and nickel, like that of Earth.

A dipolar magnetic perturbation was observed in the magnetic field data from *Galileo*'s initial pass by Europa on E4 (*Kivelson et al.,* 1997, 2000). The magnetic field depression expected in the presence of the ambient jovian magnetic field of 450 nT was 100 nT. The actual measured depression was about 50 nT. With additional passes, those sampled during GEM, it became clear that the magnetic moment of Europa changed sign in phase with the changing sign of the ambient jovian magnetic field's radial component. Such a change would be expected in the presence of Alfvén wings, the plasma structure first predicted by Alfvén in the 1930s for a conducting object moving through a magnetic field. As modeled by *Neubauer* (1998), Alfvén wings would create perturbations in the radial direction. The observed deflection provided compelling evidence of the presence of a global-scale conducting shell located within 100 km of the surface [see *Kivelson et al.* (2004) and the chapter by Khurana et al. for an extended discussion].

However, although the symmetry of the Alfvén wings, with offsets in the radial direction toward and away from Jupiter, appear generally consistent with asymmetries for an induced magnetic response (*Neubauer,* 1998) due to the presence of a highly conductive material (liquid, salty water) beneath the crust, it has been shown (*Saur et al.,* 1998) that the signature can also be reproduced solely using electric currents in the thin exosphere. In 1998, the signature of the footprint of an Alfvén/Birkeland-style current was discovered in the aurora of Jupiter (*Clarke et al.,* 2002), demonstrating that there is an electrodynamic response of Europa in the jovian magnetosphere that mimics that of Io, where Io is capable of producing a palpable auroral signature at Jupiter with atmospheric currents alone (*Clarke et al.,* 2004). It should also be noted that a salty ocean with electrical conductivity strong enough to support current requires that the mole fraction of the active brine component be sufficiently high and it is not clear, with current work, that the required high fractions exist in nature (*McCarthy et al.,* 2006; *Grimm and Stillman,* 2008). For more on the salinity of a potential europan ocean, see section 6.

5.6. Europa's Sputtered Atmosphere and Exosphere

The interaction of magnetospheric plasma with the surface results in the creation of an atmosphere. Observations of an atmosphere at both Ganymede and Europa were confirmed independently, just prior to *Galileo* orbit insertion by groundbased observers using the Goddard High-Resolution Spectrograph of the Hubble Space Telescope. Molecular oxygen was observed in the atmospheres of Europa and Ganymede (*Hall et al.,* 1995, 1998) and separately in the near-surface of Ganymede (*Spencer et al.,* 1995) and later Europa (*Spencer and Calvin,* 2002). The 1356- and 1304-Å emission lines suggested that the emission resulted from the electron impact dissociation of O_2, a process that leaves one oxygen atom in an excited state (see *McGrath et al.,* 2004; chapter by McGrath et al.).

With *Galileo*, the principal sputtering agents were shown to be S and O at energies of hundreds of keV (*Paranicas et al.,* 2001, 2002), as predicted pre-*Galileo* by, e.g., *Johnson* (1990). *Ip* (1996), in modeling the process, included a secondary source of sputtering, as if a 20% fraction of exospheric ions were recycled to Europa's surface. His resulting column density was higher than those of the Johnson group, as well as values derived from Hubble observations, but in subsequent models Johnson chose to include the "back-sputtering" element as well. Table 2 shows a comparison of estimates of the column density of Europa's neutral atmosphere, and current derived from the modeled electron distribution and calculated Pedersen and Hall conductivities.

Although most of the material in the atmosphere remains bound to Europa, a portion is ejected with sufficient energy to escape, forming a neutral cloud in Europa's orbital path.

TABLE 2. Evolution of derived atmospheric column densities for Europa.

Source	Column density (m^{-2})	Alfvén Current (A)
Johnson (1990)	5×10^{17}	
Hall et al. (1995)	1.5×10^{19}	
Ip (1996)	1×10^{22}	5×10^5
Hall et al. (1998)	$\sim 2–14 \times 10^{18}$	
Saur et al. (1998)	5×10^{18}	7×10^5

First observations of the sodium component of this cloud were made by *Brown and Hill* (1996). There were other hints of the presence of the torii prior to the Cassini flyby. In the GMM portion of *Galileo*'s extended mission, in January 2000, Cassini made critical observations of the torii of Europa using the imaging portion of the magnetospheric imaging instrument, capable of imaging energetic neutral atoms (ENAs) — those atoms produced in a collision between a cold neutral and an energetic particle. Measurements of ENAs enabled the derivation of the presence of a neutral gas cloud estimated at 9×10^{33} molecules, a volume that rivals that produced by Io, with an implied composition of H, H_2, O, and O_2 (*Mauk et al.*, 2003). A complementary measurement by the ultraviolet instrument reported an almost entirely hydrogen composition, oxygen a very minor component, with a diameter of just 0.3 jovian radii hydrogen (*Hansen et al.*, 2005).

Studies of the sodium/potassium content of the neutral cloud (*Brown,* 2001) suggested that sodium had a local source, possibly from a salty subsurface ocean, or an extrajovian, micrometeorite source, rather than the implantation of iogenic material (*Leblanc et al.*, 2002), for which the plasma interaction with Europa is not well known (see chapter by Carlson et al.). In other ways the Europa torus exhibits morphological differences from that of its neighbor, Io, with the trailing cloud being brighter and more extensive than the leading cloud. Observations of the asymmetries of the sodium cloud (*Burger and Johnson,* 2004) showed that the sodium and oxygen lifetimes were very similar despite different destruction mechanisms, proving sodium to be an excellent tracer of other species lost from Europa's atmosphere.

Exospheric phenomena related to the interaction of Europa's extended atmosphere with the jovian magnetosphere were summarized by *Kurth et al.* (2001). A region of electron cyclotron harmonics spans the entire leading (upstream) side of the moon, at a distance of approximately 2 Europa radii. Closer to the moon is a region characterized by the presence of whistler-mode hiss or chorus. The existence of most of these plasma-wave phenomena suggest underlying sources of free energy in the plasma distribution function near the moon.

5.7. Surface Materials

While previous groundbased measurements demonstrated the presence of water ice on Europa, observations by *Galileo*'s spacebased NIMS instrument provided access to wavelength regions inaccessible to groundbased telescopes. An example is the water band at 3.1 μm that is so strong that H_2O ice behaves like a metal and produces a specular reflection feature. The shape of this reflectance peak near 3.1 ìm is diagnostic of the lattice order in the top micrometers of the surface that can vary from crystalline to amorphous. The presence of amorphous ice on Europa's surface was found by *Hansen and McCord* (2004). Amorphous ice can be crystallized by heating, while the disruption caused by particle radiation can amorphize crystalline ice. Model comparisons show that the surface ice is predominantly amorphous on Europa, although below a depth of <1 mm, the ice is predominantly crystalline. The occurrence of crystalline and amorphous ice on all three satellites, coupled with the falloff of radiation flux with distance in the region, suggests competing processes of disruption and crystallization for the surface ice. The surface ice of Callisto seems to be dominated by thermal crystallization, and Ganymede exhibits patterns of both kinds of ice; as the icy moon most embedded in the radiation belts, the high degree of amorphous ice at the surface of Europa seems to suggest the surface of Europa is dominated by radiative disruption.

Carbon dioxide, previously found on Ganymede and Callisto (*Carlson et al.,* 1996; *McCord et al.,* 1998a), was discovered on Europa's leading side by (*Smythe et al.,* 1998). A recent reanalysis of the NIMS data shows the presence of CO_2 on the trailing side (*Hansen and McCord,* 2007). The presence of CO_2 in the Galilean satellites was surprising since this molecule is very volatile and cannot exist as an ice at Galilean satellite temperatures. During the *Galileo* era, another unexpected and volatile compound, molecular oxygen, was found in the icy Galilean satellites' surfaces (*Spencer and Calvin,* 2002) as well as in atmospheres on Ganymede and Europa (see section 5.6). Molecular oxygen is certainly produced by radiation, but the origin of the CO_2 is unknown. Compositional data from the Linear Etalon Imaging Spectral Array (LEISA) instrument, taken in 2007 when New Horizons made its Jupiter gravity assist, provided near-complete spatial coverage in the spectral range 1.25–2.5 μm. The data filled large gaps in the coverage by *Galileo*, including the Jupiter-facing hemisphere. Spectra of water bands, obtained by LEISA, and distorted by a non-ice component were found to be distributed in a fairly symmetrical way about the apex of the trailing hemisphere (270°W, 0°N), supporting the contention of an exogenic origin for at least some of this material (*Grundy et al.,* 2007).

Radiation effects on Jupiter's satellites, first noted by *Burns* (1968) and later by several others, have long been invoked to explain many phenomena, but the magnitude of such effects were unknown. Unlike photolysis, where chemical bonds are broken under the influence of photon flux, in radiolysis, chemical bonds are broken under the influence of energetic particles. Voyager results provided hints of the importance of radiolysis for Europa's surface but quantitative descriptions were not developed. One of the earliest volumes related to this process, a post-Voyager summary of the field, is given by *Johnson* (1990).

Predictions of radiolytic products for icy satellite observations were given by *Johnson and Quickenden* (1997) and included molecular oxygen and hydrogen peroxide (H_2O_2) as expected radiolytic products. Early in the *Galileo* mission, infrared spectra of Europa were obtained close to the satellite, but long-wave measurements were severely compromised by radiation-induced noise. By obtaining Europa spectra from great distances, at nearly the orbit of Ganymede, an infrared feature of H_2O_2 was identified and corroborated by ultraviolet spectra (*Carlson et al.,* 1999a). Since hydrogen peroxide is rapidly dissociated by near-UV solar radiation, a large production rate was implied, leading to the conclusion that radiolysis is a major factor in determining Europa's surface chemistry (see chapter by Carlson et al.). The fluxes of ionizing radiation at the Galilean satellites were compiled by *Cooper et al.* (2001) and are updated in the chapter by Paranicas et al.

Gardening by micrometeoroid impact competes with radiolysis and sputtering, and buries the radiolytic products quite rapidly, forming a fluffy regolith (*Buratti,* 1995) with large effective areas that can react with and trap O_2 and other atmospheric species. Since the regolith depth is large compared to the penetration depth of most of the impacting particles, the lower layers are shielded and the regolith can store and protect oxidants for eventual delivery to Europa's ocean by crustal processes (*Prockter and Pappalardo,* 2000).

Radiolysis, while forming interesting chemical products, unfortunately also hides the chemical history and ultimate source of material. The origin of hydrated species, observed by NIMS and earlier by *Pollack et al.* (1978) and *Clark and McCord* (1980), is an important case. One suggestion for these hydrated species is hydrated salts, upwelled and emplaced on the surface from the ocean below (*McCord et al.,* 1998b). An alternative suggestion is hydrated sulfuric acid (*Carlson et al.,* 1999b), which is the stable end-product of the radiolysis of sulfurous material in H_2O ice, but here the source of sulfur is hidden — it could be endogenic salts, exogenic sulfur ions, or any other form of sulfur. The hydrate is associated with disrupted surface areas in a manner that suggests an endogenic source for the sulfur (*McCord et al.,* 1999), although surface heating processes (diapers and shear heating) can produce lag deposits that may explain the geological associations without invoking an endogenic source (*Fagents et al.,* 2000; *Fagents,* 2003). The nature of Europa's hydrated species is an important but as-yet-unresolved question. Europa's sodium and potassium extended atmosphere may be key to resolving the debate, but Europa's plasma interactions need better understanding to accurately determine the iogenic Na and K fluxes onto Europa.

5.8. Interior and Tidal Heating

Radiogenic heating introduces conduction and convection, processes that deform the interior of a body. The overall shape of the body (the ice shell) will also deform (over longer timescales) as a result of rotation about its axis, the distribution of mass within, and tides. The degree to which any of these processes dominates depends upon certain rheological properties of ice such as viscosity, melting point, degree of impurity, dislocation creep, etc., collectively known as the lagging response of Europa's material. In the case of tides, the minimum kinetic state would be that of a circular orbit, all spins aligned, and rotation of the moon synchronized with its orbital motion. The pace at which this final state is achieved depends critically on the lagging response of the material.

The orbital commensurability, first noticed by Galileo, in which Io orbits Jupiter four times in the time it takes Ganymede to go around once, contributes to a very important phenomenon: Conjunction (the configuration where Io, for example, would over take Europa, and all three line up with Jupiter) always occurs in accord with the major axes of the (elliptical) orbits, but the conjunction of Io with Europa always occurs on the opposite side of Jupiter from that of Europa with Ganymede. Because of this resonance, the moons have a forced component of orbital eccentricity; the orbits are not merely elliptical, but in fact the repeated gravitational pull in the same geometry results in a small but nonzero tug on the body that not only prevents the orbit from circularizing, but prevents the material of the body from dissipating heat in such a way that the minimum kinetic energy configuration can be achieved. Thus the resonances maintain the orbital eccentricities of the Galilean satellites, leading to an enhanced component of tidal dissipation, particularly for the innermost moons.

The thermal evolution of Europa, especially as related to tidal heating, was expected to directly influence the characteristics of the ice shell and the putative ocean, which would in turn affect the amount of heating produced by those tides. A leading model by *Yoder* (1979) and *Yoder and Peale* (1981) suggested that the orbital interactions among Io, Europa, and Ganymede evolved through time. For example, all three satellites might have been temporarily captured into low-order Laplace-like resonances, and then evolved into their present state with increasing rates of heating through time. Alternatively, *Greenberg* (1982, 1987) argued that the Laplace resonance is primordial and that heating rates were even higher in Europa's early history than they are today. Post-*Galileo* models, such as those of *Canup and Ward* (2002) and *Peale and Lee* (2002), provide insight into both possibilities, but the issues of timing and rates remain largely unconstrained. However, refinements in values for density and moment of inertia (*Anderson et al.,* 1998) have enabled updated models of the interior structure and differentiation (*McKinnon,* 1996; *Schubert et al.,* 2004; chapter by Schubert et al.), as well as the inferred thermal evolution. Although a great many uncertainties remain, these results generally support the notion that a partial melting of the rocky interior has occurred, which could drive basaltic volcanism at the rock-ocean interface, even today.

Unfortunately, only the very latest geological record on Europa is preserved on the surface, and there are no surface clues to the properties of the subsurface beyond about 40 to 90 m.y. ago. It is possible that the tidal heating rates vary through time, even for the modern epoch. Geological mapping of the visible part of Europa's history (*Greeley et al.,* 2000, 2004; *Thomas et al.,* 2007) is based on the identification of specific features and placement of their formation in a relative time sequence (see chapter by Doggett et al.). Initial results suggest that the earliest events involved formation of extensive ridge systems, driven by global-scale tectonic deformation. Formation of chaos terrain, low domes, and smaller features came later and is thought to reflect local upwelling of warmer ice and/or water from the ocean, but not the global-scale processes indicated by the earlier extensive ridge formation. Regardless of the debate within the community on the details of specific feature formation and timing, there is general agreement that most of the surface features represent processes within and below the ice shell, and that the ice shell is likely to have been relatively thin at the time of the formation of the features. The nature of the ice represented by the young surface features today are expected to be representative of conditions over a much longer period of time, although changes in ice thickness on a dramatic scale as expected for freezing or melting on long timescales have been modeled (*Mitri and Showman,* 2005).

Many of the issues regarding interior structure and evolution, and the possible links to the surface, require new geophysical measurements and observations of the surface. For example, measuring the amplitudes of the diurnal tides and the gravity field would constrain models of the ocean depth and thickness of the ice shell. If the tidal amplitude is large (i.e., 30 m), then a thin shell (less than tens of kilometers) is present today, but if the amplitude is ~1 m, then the water is probably completely frozen (*Moore and Schubert,* 2000). Presently we have less than only about 10% of the surface of Europa imaged in sufficient resolution to assess the existence of ancient surface features; exploration of the solar system has shown that many bodies have terrain dichotomies, such as the ancient cratered highlands dominating the lunar farside vs. the younger mare regions on the nearside. Until Europa is mapped globally under uniform conditions of illumination and resolution, we will not know the full extent of the "visible" history. These geophysical, geological, and other measurements will require a spacecraft in orbit around Europa.

6. ASTROBIOLOGY AND EUROPA

Emerging evidence for Europa's global and perhaps habitable ocean is often cited to justify new programs to search for life beyond Earth (*Morrison,* 2001, 2006). The potential astrobiological signature in the *Galileo* infrared data is contained in components of the europan disrupted terrain — the dark lineae seen in Fig. 4 that may be oceanic in origin. These regions are found to contain highly hydrated molecules (*McCord et al.,* 1998a; *Carlson et al.,* 2005), producing distorted and asymmetric H_2O-related features in the spectra at 1.5 and 2.0 μm. Quantitative analysis of signatures for various (abiotic) laboratory mixtures, including, e.g., blends of bloedite, hexahydrite, epsomite, and mirabilite (hydrated magnesium, sulfur, and sodium salts) and hydrated sulfuric acid, only partially reproduce the observed spectra (*Dalton,* 2007; chapter by Carlson et al.). The stability of hydrated salt minerals epsomite, mirabilite, and natron under Europa environmental conditions was studied by *McCord et al.* (2001). The darker lineae contain a reddish-brown pigmentation of unknown composition but related to the hydrated species. In addition, the dark terrain exhibits a possible narrow absorption feature at 2.05 μm that is potentially due to the presence of an amide (*Dalton et al.,* 2003). Post-*Galileo* laboratory work is being done to test for infrared biosignatures with many organic and astrobiological materials that might constitute the content of those deposits.

Meteoritic and cometary delivery of carbon to Europa over geologic time are predicted to eventually result in both inorganic (i.e., sodium carbonates, CO_2, CO) and organic (i.e., amino acids, formaldehyde) carbon compound production through a combination of endogenic and exogenic processes (*Zolotov and Shock,* 2004). These processes could have led to a number of different potential ocean compositions. Models of the ocean capture a wide range of salinities with compositions ranging from magnesium dominant, to sodium dominant with a hydrogen component, to an ocean of moderately alkaline sulfate-carbonate composition overlying a moderately oxidized magnetite-bearing silicate mantle (*Kargel et al.,* 2000; *Zolotov and Shock,* 2001; *McKinnon and Zolensky,* 2003; *Hand and Chyba,* 2007). Thus, instead of an NaCl ocean resembling the composition of the terrestrial ocean, the europan ocean could be a $MgSO_4$ ocean, Na_2SO_4, or a sulfurous and sulfuric acid ocean. The spectroscopic evidence for the anion (sulfate) is present (*McCord et al.,* 1998b); however, the nature of the cation is disputed. *McCord et al.* (2002) prefer the presence of both salts and the sulfuric acid where sodium sulfate would be part of the material from the ocean along with magnesium sulfate, Na is lost easily due to radiation (where it forms the sodium exosphere), and H substitutes. This reaction does not happen easily for Mg-sulfate. *Orlando et al.* (2005) match Europa hydrate spectra with a model that includes sodium, and propose a rather specific composition for the ocean: sodium- and magnesium-bearing sulfate salts mixed with sulfuric acid. Alternatively (see section 5.7 and chapter by Carlson et al.), the cation could be hydrogen, with the hydrate being hydrated sulfuric acid.

Also at issue is the manner in which the oxidant is delivered to the ocean. Chemotropic life requires chemical disequilibria, i.e., renewable sources of both reduced and oxidized substrates. If the icy crust is more than a few tens of meters thick, photosynthesis is unlikely to be the dominant metabolic pathway. For respiration and/or fermentation to be viable, oxidants produced radiolytically at the surface must be delivered to the subsurface ecosystem through planetary dynamics of some sort: diapirism, convection, or other circulation (*Chyba,* 2000; *Chyba and Hand,* 2001; chapter by Zolotov and Kargel).

Although some of the composition models mentioned above predict the ocean water of Europa could be considered fresh, with such a wide range of predicted salinities, current knowledge about the capacity for survival for biological systems in a high salinity environment becomes important. Terrestrial halophilic microorganisms have been shown to be capable of surviving at NaCl saturation (*Oren,* 2002; *Lin et al.,* 2006; *d'Hondt et al.,* 2004). Among these microbes are samples from each domain of life: Archaea, Bacteria, and Eucarya. Metabolic pathways for these microbes include oxygenic and anoxygenic photosynthesis (*Dunaliella salina, Halorhodospira halphila*), aerobic respiration (*Halobacterium salinarum*), and fermentation (*Halobacterium salinarum*). Terrestrial methanogens are capable of surviving in solutions near NaCl saturation if methanol or methylated amines are available (*Oren,* 2001). Microbes such as *Halobacterium sodomense* are known to survive in solutions of high Mg^{2+} concentration (*Oren,* 1994). Thus on Europa, although the habitability of the ocean may be dependent upon the potentially high salinity requirements of a current-bearing ocean, and although questions as to the origin of such life remain unanswered, conditions of extreme salinity would not seem to preclude life as we know it (*Hand and Chyba,* 2007).

7. THE FUTURE

Future astrobiological exploration of Europa could proceed in stages: (1) An orbiter could measure the thickness of the ice shell and identify interesting locations where the crust is thin, where intrashell water is close to the surface, or where there are indications of recent geological activity. (2) A small surface lander could access directly the postulated organic material in the surface ice, and perhaps distinguish between exogenic and endogenic sources. (3) Adding drilling capability with a larger lander might permit study of compounds below the level where magnetospheric processes rapidly modify the chemistry. (4) Perhaps someday we can design something that can melt its way through the ice shell and directly explore the global ocean, ultimately even searching for living communities around hydrothermal vents at the base of the ocean.

Europa is perhaps the best candidate within the solar system for evidence of a second genesis, life that formed independently of that on Earth. The inner planets exchange material in the form of crater ejecta, as we know from the presence of Moon rocks and Mars rocks in our meteorite collections. Therefore, when we look for life on Mars, we might find a distant cousin that shares a common ancestor with terrestrial life. Finding such related life, with an independent 4-G.y. history of evolution on Mars, while exciting and important, would not answer some basic questions relating to fundamental vs. contingent nature of the origin and properties of life. Life on Europa, in contrast, would represent an independent origin as well as evolutionary history.

Acknowledgments. We wish to acknowledge and thank R. T. Reynolds for helpful discussions. As Reynolds says, to be among the first to explore a planetary body is an experience that no one else can have. We wish to thank T. McCord, C. Pilcher, and M. Meltzer for thorough and insightful reviews. The research in this chapter was carried out at the Jet Propulsion Laboratory, California Institute of Technology, under a contract with the National Aeronautics and Space Administration.

REFERENCES

Alfven H. (1942) Existence of electromagnetic-hydrodynamic waves. *Nature, 150,* 405.

Anderson J. D., Schubert G., Jacobson A., Lau E. L., Moore W. B., and Sjogen W. L. (1998) Europa's differentiated internal structure: Inferences from four Galileo encounters. *Science, 281,* 2019–2022.

Barbieri C., Rahe J., Johnson T., and Sohus A., eds. (1997) *The Three Galileos: The Man, The Spacecraft, The Telescope.* Kluwer, Boston.

Beatty K., Petersen C. C., and Chaikin A., eds. (1988) *The New Solar System.* Cambridge Univ., Cambridge.

Biggs E. K. (1964) Influence of the satellite Io on Jupiter's decametric emission. *Nature, 203,* 1008–1010.

Bodenheimer P. (1974) Calculations of the effects of angular momentum on the early evolution of Jupiter. *Icarus, 23,* 319–325.

Brouwer D. and Clemence G. M. (1961) Orbits and masses of planets and satellites. In *Planets and Satellites* (G. Kuiper and B. Middlehurst, eds.), pp. 31–94. Academic, New York.

Brown M. E. (2001) Potassium in Europa's atmosphere. *Icarus, 151,* 190–195.

Brown M. E. and Hill R. E. (1996) Discovery of an extended sodium atmosphere around Europa. *Nature, 380,* 229–231.

Brown W. H., Lanzerotti L. J., Poate J. M., and Augustyniak W. M. (1978) Sputtering of ice by MeV light ions. *Nature, 300,* 423–425.

Burger M. H. and Johnson R. E. (2004) Europa's neutral cloud: Morphology and comparisons to Io. *Icarus, 171,* 557–560.

Burke B. F. and Franklin K. L. (1955) Observations of a variable radio source associated with the planet Jupiter. *J. Geophys. Res., 60,* 213–217.

Burns J. A. (1968) Jupiter's decametric radio emission and the radiation belts of its Galilean satellites. *Science, 159,* 971–972.

Burns J. A., ed. (1977) *Planetary Satellites.* Univ. of Arizona, Tucson.

Burns J. A. and Matthews M. S., eds. (1986) *Satellites.* Univ. of Arizona, Tucson.

Buratti B. (1995) Photometry and surface structure of the icy Galilean satellites. *J. Geophys. Res., 100(E9),* 19061–19066.

Cameron A. G. W. (1973) Formation of the outer solar system. *Space Sci. Rev., 14,* 383–391.

Cameron A. G. W. and Pollack J. (1976) On the origin of the solar system and of Jupiter its satellites. In *Jupiter* (T. Gehrels, ed.), pp. 64–84. Univ. of Arizona, Tucson.

Canup R. M. and Ward W. R. (2002) Formation of the Galilean satellites: Conditions of accretion. *Astron. J., 124,* 3404–3423.

Carlson R. W., Smythe W., Baines K., Barbinis E., Becker K., Burns R., Calcutt S., Calvin W., Clark R., Danielson G., Davies A., Drossart P., Encrenaz T., Fanale F., Granahan J., Hansen G., Herrera P., Hibbitts C., Hui J., Irwin P., Johnson T., Kamp L., Kieffer H., Leader F., Lellouch E., Lopes-Gautier R., Matson D., McCord T., Mehlman R., Ocampo A., Orton G., Roos-Serote M., Segura M., Shirley J., Soderblom L., Stevenson A., Taylor F., Torson J., Weir A., and Weissman P. (1996) Near-infrared spectroscopy and spectral mapping of Jupiter and the Galilean satellites: Results from Galileo's initial orbit. *Science, 274,* 385–388.

Carlson R. W., Anderson M. S., Johnson R. E., Smythe W. D., Hendrix A. R., Barth C. A., Soderblom L. A., Hansen G. B., McCord T. B., Dalton J. D., Clark R. N., Shirley J. H., Ocampo A. C., and Matson D. L. (1999a) Hydrogen peroxide on the surface of Europa. *Science, 283,* 2062–2064.

Carlson R. W., Johnson R. E., and Anderson M. S. (1999b) Sulfuric acid on Europa and the radiolytic sulfur cycle. *Science, 286,* 97–99.

Carlson R. W., Anderson M. S., Mehlman R., and Johnson R. E. (2005) Distribution of hydrate on Europa: Further evidence for sulfuric acid hydrate. *Icarus, 177,* 461–471.

Carr M. H. (1999) Mars: Surface and interior. In *Encyclopedia of the Solar System* (P. R. Weissman et al., eds.), Academic, San Diego.

Carr M. H., Belton M. J. S., Chapman C. R., Davies M. E., Geissler P., Greenberg R., McEwen A. S., Tufts B. R., Greeley R., Sullivan R., Head J. W., Pappalardo R. T., Klaasen K. P., Johnson T. V., Kaufman J., Senske D., Moore J., Neukum G., Schubert G., Burns J. A., Thomas P., and Veverka J. (1998) Evidence for a subsurface ocean on Europa. *Nature, 391,* 363–365.

Chambers G. F. (1861) *A Handbook of Descriptive and Practical Astronomy.* John Murray, London. 514 pp.

Chyba C. F. (2000) Energy for microbial life on Europa. *Nature, 403,* 381–382.

Chyba C. F. and Hand K. P. (2001) Life without photosynthesis. *Science, 292,* 2026–2027.

Clark R. N. and McCord T. B. (1980) The Galilean satellites: New near-infrared spectral reflectance measurements (0.65–2.5 μm) and a 0.325–5 μm summary. *Icarus, 41,* 323–329.

Clarke A. C. (1982) *2010: Odyssey Two.* Ballantine, New York.

Clarke J. T., Ajello J., Ballester G., Jaffel L. B., Connerney J. E. P., Gerard J.-C., Gladstone G. R., Grodent D., Pryor W., Trauger J., and Waite J. H. Jr. (2002) Ultraviolet auroral emissions from the magnetic footprints of Io, Ganymede and Europa on Jupiter. *Nature, 415,* 997–1000.

Clarke J. T., Grodent D., Cowley W. H., Bunce E. J., Zarka P, Connerney J. E. P., and Satoh S. (2004) Jupiter's aurora. In *Jupiter — The Planets, Satellites, and Magnetosphere* (F. Bagenal et al., eds.), pp. 329–362. Cambridge Univ., Cambridge.

Consolmagno G. J. (1975) *Thermal History Models of Icy Satellites.* M.S. thesis, Massachusetts Institute of Technology, Cambridge. 202 pp.

Consolmagno G. J. and Lewis J. S. (1976) Structural and thermal models of icy Galilean satellites. In *Jupiter* (T. A. Gehrels, ed.), pp. 1035–1051. Univ. of Arizona, Tucson.

Consolmagno G. J. and Lewis J. S. (1977) Preliminary thermal history models of icy satellites. In *Planetary Satellites* (J. A. Burns, ed.), pp. 492–500. Univ. of Arizona, Tucson.

Consolmagno G. J. and Lewis J. S. (1978) The evolution of icy satellite interiors and surfaces. *Icarus, 34,* 280–293.

Cooper J. F., Johnson R. E., Mauk B. H., Garrett H. B., and Gehrels N. (2001) Energetic ion and electron irradiation of the icy Galilean satellites. *Icarus, 149,* 133–159.

Corliss J. B., Dymond J., Gordon L. I., Edmond J. M., von Herzen R. P., Ballard R. D., Green K., Williams D., Bainbridge A., Crane K., and van Andel T. H. (1979) Submarine thermal springs on the Galapagos Rift. *Science, 203,* 1073–1083.

Corliss J. B., Baross J. A., and Hoffman S. E. (1981) An hypothesis concerning the relationship between submarine hot springs and the origin of life on Earth. *Oceanol. Acta (SP),* 59–60.

Cruikshank D. P. and Nelson R. M. (2007) A history of the exploration of Io. In *Io After Galileo: A New View of Jupiter's Volcanic Moon* (R. Lopes and J. Spencer, eds.), Springer-Praxis, New York.

Dalton J. B. (2007) Modeling Europa's surface composition with cryogenic sulfate hydrates. In *Ices, Oceans, and Fire: Satellites of the Outer Solar System,* Abstract #6097. Lunar and Planetary Institute, Houston.

Dalton J. B., Mogul R., Kagawa H. K., Chan S. L., and Jamison C. S. (2003) Methodologies and techniques for detecting extraterrestrial (microbial) life: Near-infrared detection of potential evidence of microscopic organisms on Europa. *Astrobiology, 3,* 505–529.

Dessler A., ed. (1983) *Physics of the Jovian Magnetosphere.* Cambridge Univ., New York.

D'Hondt S., Jørgensen B. B., Miller D. J., Batzke A., Blake R., et al. (2004) Distributions of microbial activities in deep sub seafloor sediments. *Science, 306,* 2216–2221.

Dollfus A. (1975) Physical studies of asteroids by polarization of the light. In *Physical Studies of Minor Planets* (T. Gehrels, ed.), pp. 95–116. NASA SP-267, NASA Scientific and Technical Information Office, Washington, DC.

Drake F. D. and Hvatum H. (1959) Non-thermal microwave radiation from Jupiter. *Astron. J., 64,* 329–330.

Duncombe R. L., Klepczynski W. J., and Seidelmann P. K. (1974) The masses of the planets, satellites, and asteroids. *Fund. Cos. Phys., 1,* 119–165.

Eviatar A., Siscoe G. L., Johnson T. V., and Matson D. L. (1981) Effects of Io ejecta on Europa. *Icarus, 47,* 75–83.

Fagents S. A. (2003) Considerations for effusive cryovolcanism on Europa: The post-Galileo perspective. *J. Geophys. Res., 108(E12),* 13-1, DOI: 10.1029/2003JE002128.

Fagents S. A, Greeley R., Sullivan R. J., Pappalardo R. T., and Prockter L. M. (2000) Cryomagmatic mechanisms for the formation of Rhadamanthys linea, triple band margins, and other low-albedo features on Europa. *Icarus, 144,* 54–88.

Fanale F. P., Johnson T. V., and Matson D. L. (1977) Io's surface and the histories of the Galilean satellites. In *Planetary Satellites* (J. A. Burns, ed.), pp. 379–405. Univ. of Arizona, Tucson.

Fimmel R. O., Swindell W., and Burgess E. (1977) *Pioneer Odyssey.* NASA SP-396, NASA Scientific and Technical Information Office, Washington, DC.

Fink U., Dekkers N. H., and Larson H. P. (1973) Infrared spectra of the Galilean satellites of Jupiter. *Astrophys. J. Lett., 179,* L155–L159.

Fischer D. (2001) *Mission Jupiter: The Spectacular Journey of the Galileo Spacecraft.* Copernicus Books, Springer-Verlag.

Galilei G. (1610) *Siderius Nuncius.* Translated by E. S. Carlos (1929) in *A Source Book in Astronomy* (H. Shapley and Howarth, eds.), McGraw-Hill, New York.

Gehrels T., ed. (1976) *Jupiter.* Univ. of Arizona, Tucson.

Geissler P. E., Greenberg R., Hoppa G., Helfenstein P., McEwen A., Pappalardo R. T., Tufts R., Ockert-Bell M., Sullivan R., Greeley R., Belton M. J. S., Denk T., Clark B., Burns J., Veverka J., and the Galileo Imaging Team (1998) Evidence for non-synchronous rotation of Europa. *Nature, 391,* 368–370.

Greeley R., Sullivan R., Klemaszewski J., Homan K., Head J. W. III, Pappalardo R. T., Veverka J., Clark B. E., Johnson T. V., Klaasen K. P., Belton M., Moore J., Asphaug E., Carr M. H., Neukum G., Denk T., Chapman C. R., Pilcher C. B., Geissler P. E., Greenberg R., and Tufts R. (1998) Europa: Initial Galileo geological observations. *Icarus, 135,* 4–24.

Greeley R., Figueredo P. H., Williams D. A., Chuang F. C., Klemaszewski J. E., Kadel S. D., Prockter L. M., Pappalardo R. T., Head J. W. III, Collins G. C., Spaun N. A., Sullivan R. J.,

Moore J. M., Senske D. A., Tufts B. R., Johnson T. V., Belton M. J. S., and Tanaka K. L. (2000) Geologic mapping of Europa. *J. Geophys. Res., 105,* 22559–22578.

Greeley R., Chyba C. F., Head J. W. III, McCord T. B., McKinnon W. B., Pappalardo R. T., and Figueredo P. (2004) Geology of Europa. In *Jupiter — The Planets, Satellites, and Magnetosphere* (F. Bagenal et al., eds.), pp. 329–362. Cambridge Univ., Cambridge.

Greenberg R. (1982) Orbital evolution of the Galilean satellites. In *Satellites of Jupiter* (D. Morrison, ed.), pp. 65–92. Univ. of Arizona, Tucson.

Greenberg R. (1987) Galilean satellites: Evolutionary paths in deep resonance. *Icarus, 70,* 334–347.

Greenberg R. (2005) *Europa The Ocean Moon: Search for an Alien Biosphere.* Springer-Praxis Books in Geosciences, Chichester.

Greenberg R., Geissler P., Hoppa G., and Tufts B. R. (2002) Tidal-tectonic processes and their implications for the character of Europa's icy crust. *Rev. Geophys., 40,* 1.1 to 1.33.

Grimm R. E. and Stillman D. E. (2008) Electrical properties of saline ices, salt hydrates, and ice-silicate mixtures: Applications to solar system exploration. In *The Science of Solar System Ices (ScSSI): A Cross-Disciplinary Workshop,* Abstract #9062. Lunar and Planetary Institute, Houston.

Grundy W. M., Buratti B. J., Cheng A. F., Emery J. P., Lunsford A., McKinnon W. B., Moore J. M., Newman S. F., Olkin C. B., Reuter D. C., Schenk P. M., Spencer J. R., Stern S. A., Throop H. B., Weaver H. A., and the New Horizons Team (2007) New Horizons mapping of Europa and Ganymede. *Science, 318,* 234, DOI: 10.1126/science.1147623.

Hall D. T., Strobel D. F., Feldman P. F., McGrath M. A., and Weaver H. A. (1995) Detection of an oxygen airglow of Europa and Ganymede. *Nature, 393,* 677.

Hall D. T., Feldman P. F., McGrath M. A., and Strobel D. F. (1998) Detection of an oxygen airglow of Europa and Ganymede. *Astrophys. J., 499,* 475.

Hand K. P. and Chyba C. P. (2007) Empirical constraints on the salinity of the europan ocean and implications for a thin ice shell. *Icarus, 189,* 434–438.

Hansen C. J., Shemansky D. E., and Hendrix A. R. (2005) Cassini UVIS observations of Europa's oxygen atmosphere and torus. *Icarus, 176,* 305–315.

Hansen G. and McCord T. B. (2004) Amorphous and crystalline ice on the Galilean satellites: A balance between thermal and radiolytic processes. *J. Geophys. Res., 109(1),* E01012, DOI: 10.1029/2003JE002149.

Hansen G. and McCord T. B. (2007) Widespread CO_2, and other non-ice compounds on the anti-jovian and trailing sides of Europa from Galileo/NIMS observations. *Geophys. Res. Lett., 35,* L01202, DOI: 10.1029/2007GL031748.

Harland D. M. (2000) *Jupiter Odyssey: The Story of NASA's Galileo Mission.* Springer Praxis Books in Geosciences, Chichester.

Hartmann W. K., Strom R. G., Weidenschilling S. J., Blasius K. R., Woronow A., Dence M. R., Grieve R. A., Diaz J., Chapman C. R., Shoemaker E. M., and Jones K. L. (1981) Chronology of planetary volcanism by comparative studies of planetary cratering. In *Basaltic Volcanism on the Terrestrial Planets,* pp. 1049–1127. Pergamon, New York.

Harris D. L. (1961) Photometry and colorimetry of planets and satellites. In *Planets and Satellites* (G. P. Schubert and B. M. Middlehurst, eds.), Univ. of Chicago, Chicago.

Hapke B. (2001) Space weathering from Mercury to the asteroid belt. *J. Geophys. Res., 106,* 10039–10073.

Helfenstein P. and Parmentier E. M. (1985) Patterns of fracture and tidal stresses due to non-synchronous rotation: Implications for fracturing on Europa. *Icarus, 53,* 415–430.

Herschel J. (1861) *Outlines of Astronomy,* 4th edition, p. 331. Longman, Brown, Green, and Longmans, London.

Hoppa G. V., Tufts B. R., Greenberg R., and Geissler P. E. (1999) Formation of cycloidal features on Europa. *Science, 285,* 1899–1902.

Ip W-H. (1996) Europa's oxygen exosphere and its magnetospheric interaction. *Icarus, 120,* 317–325.

Jannasch H. W. and Wirsen C. O. (1979) Chemosynthetic primary production at East Pacific sea floor spreading centers. *Bioscience, 29,* 592–598.

Jeffreys H. (1923) The constitution of the four outer planets. *Mon. Not. R. Astron. Soc., 83,* 350–354.

Johnson R. E. (1990) *Energetic Charged-Particle Interactions with Atmospheres and Surfaces.* Springer-Verlag, Berlin.

Johnson R. E. and Quickenden T. I. (1997) Photolysis and radiolysis of water ice on outer solar system bodies. *J. Geophys. Res., 102,* 10985–10996.

Johnson R. E., Lanzerotti L. J., and Brown W. L. (1982) Planetary applications of ion induced erosion of condensed-gas frosts. *Nucl. Instr. Meth., 198,* 147–157.

Johnson T. V. (1971) Spectral geometric albedo of the Galilean satellites 0.30 to 2.5 mm. *Astrophys. J., 169,* 589–594.

Johnson T. V. and McCord T. B. (1971) Galilean satellites: Narrowband photometry 0.30 to 1.10 mm. *Icarus, 14,* 94–111.

Johnson T. V. and Pilcher C. B. (1977) Satellite spectroscopy and surface compositions. In *Planetary Satellites* (J. A. Burns, ed.), pp. 379–405. Univ. of Arizona, Tucson.

Johnson T. V., Soderblom L. A., Mosher J. A., Danielson G. E., Cook A. F., and Kupperman P. (1983) Global multispectral mosaics of the icy Galilean satellites. *J. Geophys. Res., 88,* 5789–5805.

Johnson T. V., Yeates C. M., and Young R. (1992) Galileo mission overview. *Space Sci. Rev., 60(1–4),* 3–21.

Kargel J. S. and Consolmagno G. J. (1996) Magnetic fields and the detectability of brine oceans in Jupiter's icy satellites. In *Lunar and Planetary Science XXVII,* pp. 643–644. Lunar and Planetary Institute, Houston.

Kargel J. S., Kaye J. Z., Head J. W. I., Marion G. M., Sassen R., Ballesteros O. P., Grant S. A., and Hogenboom D. L. (2000) Europa's crust and ocean: Origin, composition, and the prospects for life. *Icarus, 148,* 226–265.

Kivelson M. G., Khurana K. K., Joy S., Russell C. T., Southwood D. J., Walker R. J., and Polanskey C. (1997) Europa's magnetic field signature: Report from Galileo's pass on 19 December 1996. *Science, 289,* 1340–1343.

Kivelson M. G., Khurana K. K., Russell C. T., Volwerk M., Walker R. J., and Zimmer C. (2000) Galileo magnetometer measurements: A stronger case for a subsurface ocean at Europa. *Science, 289,* 1340–1343.

Kivelson M. G., Bagenal F., Kurth W. S., Neubauer F. M., Paranicas C., and Saur J. (2004) Magnetospheric interactions with satellites. In *Jupiter — The Planets, Satellites, and Magnetosphere* (F. Bagenal et al., eds.), pp. 329–362. Cambridge Univ., Cambridge.

Kovalevsky J. (1970) Détermination des masses des planètes et satellites. In *Surfaces and Interiors of Planets and Satellites* (A. Dollfus, ed.), pp. 1–44. Academic, New York.

Kuiper G. P. (1957) Infrared observations of planets and satellites. *Astron. J., 62,* 245.

Kuiper G. P. and Middlehurst B. M., eds. (1961) *Planets and Sat-*

ellites. Univ. of Chicago, Chicago.

Kupo I., Mekler Y., and Eviatar A. (1976) Detection of ionized sulphur in the jovian magnetosphere. *Astrophys. J. Lett., 205,* L51–L54.

Kurth W. S., Gurnett D. A., Persoon A. M., Roux A., Bolton S. J., and Alexander C. J. (2001) The plasma wave environment of Europa. *Planet. Space Sci., 49,* 345–363.

Lane A. L., Nelson R. M., and Matson D. L (1981) Evidence for sulfur implantation in Europa's UV absorption band. *Nature, 292,* 38–39.

Lanzerotti L. J., Brown R. H., Poate J. M., and Augustyniak W. M. (1978) On the contributions of water products from Galilean satellites to the jovian magnetosphere. *Geophys. Res. Lett., 5,* 155–158.

Laplace (1805) *Mechanique Celeste 4,* Courcier, Paris. Translated by N. Bowditch (1966), New York.

Leblanc F., Johnson R. E., and Brown M. E. (2002) Europa's sodium atmosphere: An ocean source? *Icarus, 159,* 132–144, DOI: 10.1006/icar.2002.6934.

Lewis J. S. (1969) The clouds of Jupiter and the NH_3-H_2O and NH_3-H_2S systems. *Icarus, 10,* 365–378.

Lewis J. S. (1971a) Satellites of the outer planets: Their physical and chemical nature. *Icarus, 15,* 175–185.

Lewis J. S. (1971b) Satellites of the outer planets: Thermal models. *Science, 172,* 1127.

Lewis J. S. (1972) Low temperature condensation from the solar nebula. *Icarus, 16,* 241–252.

Lewis J. S. (1973) Chemistry of the outer solar system. *Space Sci. Rev., 14,* 401–411.

Lin L.-H., Wang P-L., Rumble D., Lippmann-Pipke J., Boice E., Pratt L. M., Sherwood Lollar B., Brodie E. L., Hazen T. C., Andersen G. L., DeSantis T. Z., Moser D. P., Kershaw D., and Onstott T. C. (2006) Long-term sustainability of a high-energy, low-diversity crustal biome. *Science, 314,* 479–482.

Low F. J. (1965) Planetary radiation at infrared and millimeter wavelengths. *Bull. Lowell Obs., 6,* 184–187.

Mauk B. H., Mitchell D. G., Krimigis S. M., Roelof E. C., and Paranicas C. P. (2003) Energetic neutral atoms from a trans-Europa gas torus at Jupiter. *Nature, 421,* 920–922.

McCarthy C., Cooper R. F., Kirby S. H., and Durham W. B. (2006) Ice/hydrate eutectics: The implications of microstructure and rheology on a multi-phase europan crust. In Lunar and Planetary Science XXXVII, Abstract #2467. Lunar and Planetary Institute, Houston.

McCord T. B., Hansen G. B., Clark R. N., Martin P. D., Hibbits C. A., Fanale F. P., Granahan J. C., Segura M., Matson D. L., Johnson T. V., Carlson R. W., Smythe W. D., Danielson G. E., and the NIMS Team (1998a) Non-water-ice constituents in the surface material of the icy Galilean satellites from the Galileo near infrared mapping spectrometer investigation. *J. Geophys. Res., 103,* 8603–8626.

McCord T. B., Hansen G. B., Fanale F. P., Carlson R. W., Matson D. L., Johnson T. V., Smythe W. D., Crowley J. K., Martin P. D., Ocampo A., Hibbitts C. A., and Granahan J. C. (1998b) Salts on Europa's surface detected by Galileo's near infrared mapping spectrometer. *Science, 280,* 1242–1245.

McCord T. B., Hansen G. B., Matson D. L., Johnson T. V., Crowley J. K., Fanale F. P., Carlson R. W., Smythe W. D., Martin P. D., Hibbitts C. A., Granahan J. C., Ocampo A., and the NIMS Team (1999) Hydrated salt minerals on Europa's surface from the Galileo NIMS investigation. *J. Geophys Res., 104,* 11827–11851.

McCord T. B., Orlando T. M., Teeter G., Hansen G. B., Sieger M. T., Petrik N. G. and Van Keulen L. (2001) Thermal and radiation stability of the hydrated salt minerals epsomite, mirabilite and natron under Europa environmental conditions. *J. Geophys. Res., 106,* 3311–3319.

McCord T. B., Hansen G. B., Matson D. L., Johnson T. V., Crowley J. K., Fanale F. P., Carlson R. W., Smythe W. D., Martin P. D., Hibbitts C. A., Granahan J. C., Ocampo A., and the NIMS Team (2002) Brines exposed to Europa surface conditions. *J. Geophys Res., 107(E1),* 5004, DOI: 10.1029/2000JE001453.

McEwen A. S. (1986) Exogenic and endogenic albedo and color patterns on Europa. *J. Geophys. Res., 91,* 8077–8097.

McGrath M. A., Lellouch E., Strobel D. F., Feldman P. D., and Johnson R. E. (2004) Satellite atmospheres. In *Jupiter — The Planets, Satellites, and Magnetosphere* (F. Bagenal et al., eds.), pp. 329–362. Cambridge Univ., Cambridge.

McKinnon W. B. (1996) Evolution of the interior of Europa, and the plausibility of active, present-day hydrothermal systems. In *Europa Ocean Conference Abstracts,* pp. 50–51. San Juan Institute, San Juan Capistrano, California.

McKinnon W. B. and Zolensky M. E. (2003) Sulfate content of Europa's ocean and shell: Evolutionary considerations and some geological and astrobiological implications. *Astrobiology, 3,* 879–897.

Meltzer M. (2007) *Mission to Jupiter: A History of the Galileo Project.* NASA SP 2007-4231, NASA Scientific and Technical Information Office, Washington, DC.

Mitri G. and Showman A. P. (2005) Convective-conductive transitions and sensitivity of a convecting ice shell to perturbations in heat flux and tidal-heating rate: Implications for Europa. *Icarus, 177,* 447–460.

Moore J. M., Asphaug E., Sullivan R. J., Klemaszewski J. E., Bender K. C., et al. (1998) Large impact features on Europa: Results of the Galileo nominal mission. *Icarus, 135,* 127–145.

Moore M. H. and Hudson R. L. (2000) IR detection of H_2O_2 at 80 K in ion-irradiated laboratory ices relevant to Europa. *Icarus, 145,* 282–288.

Moore W. B. and Schubert G. (2000) The tidal response of Europa. *Icarus, 147,* 317–310, DOI: 10.1006/icar.2000.6460.

Moroz V. I. (1965) Infrared spectrophotometry of satellites: The moon and the Galilean satellites of Jupiter. *Astron. Zh., 42,* 1287, translated in *Soviet Astron. AJ, 9,* 999–1006.

Morrison D. (1982) The satellites of Jupiter and Saturn. *Annu. Rev. Astron. Astrophys., 20,* 469–495.

Morrison D. (2001) The NASA astrobiology program. *Astrobiology, 1,* 3–13.

Morrison D. (2006) Review of four astrobiology texts. *Astrobiology, 6,* 87–91.

Morrison D. and Cruikshank D. P. (1973) Thermal properties of the Galilean satellites. *Icarus, 18,* 224–236.

Morrison D. and Cruikshank D. P. (1974) Physical properties of the natural satellites. *Space Sci. Rev., 15,* 641–739.

Morrison D. and Sanz J. (1980) *Voyage to Jupiter.* NASA SP-439, NASA Scientific and Technical Information Office, Washington, DC.

Morrison D., Morrison N. D., and Lazarewicz A. R. (1974) Four color photometry of the Galilean satellites. *Icarus, 23,* 399–416.

Morrison D., Cruikshank D. P., and Murphy R. E. (1972) Temperatures of Titan and the Galilean satellites at 20 microns. *Astrophys. J. Lett., 173,* L142–L146.

Murray J. B. (1975) New observations of the surface markings on Jupiter's satellites. *Icarus, 25,* 397–404.

Nash D. B and Fanale F. P. (1977) Io: Surface composition model based on reflectance spectra of sulfur/salt mixtures and proton irradiation experiments. *Icarus, 31,* 40–80.

National Research Council Space Studies Board (2000) *Preventing the Forward Contamination of Europa.* National Academy, Washington, DC.

Nelson M. L., McCord T. B., Clark R. N., Johnson T. V., Matson D. L., Mosher J. A., and Soderblom L. A. (1986) Europa: Characterization and interpretation of global spectral surface units. *Icarus, 65,* 129–151.

Neubauer F. M. (1980) Nonlinear standing Alfven waves current system at Io: Theory. *J. Geophys Res., 85,* 1171–1178.

Neubauer F. M. (1998) The sub-Alfvénic interaction of the Galilean satellites with the jovian magnetosphere. *J. Geophys. Res., 103(E9),* 19843–19866.

Newcomb S. (1878) *Popular Astronomy.* MacMilland and Co., London, 337 pp.

O'Neil W. J. (1997) Project Galileo — The Jupiter mission. In *The Three Galileos: The Man, The Spacecraft, The Telescope* (C. Barbieri et al., eds.), Kluwer, Boston.

Oren A. (1994) The ecology of the extremely halophilic archaea. *FEMS Microbiol. Rev., 13,* 415–440.

Oren A. (2001) The bioenergetic basis for the decrease in metabolic diversity at increasing salt concentrations: Implications for the functioning of Salt Lake ecosystems. *Hydrobiologia, 466,* 61–72.

Oren A. (2002) Diversity of halophilic microorganisms: Environments, phylogeny, physiology, and applications. *J. Ind. Microbiol. Biotechnol., 28,* 56–63.

Orlando T. M., McCord T. B., and Grieves G. A. (2005) The chemical nature of Europa surface material and the relation to a sub-surface ocean. *Icarus, 177,* 528–533, DOI: 10.1016/j.icarus.2005.05.009.

Oro J., Squyres S. W., Reynolds R. T., and Mills T. M. (1992) Europa: Prospects for an ocean and exobiological implications. In *Exobiology in Solar System Exploration* (G. Carle et al., eds.), pp. 103–126 (see NTRS document N93-18545 06-51).

Pappalardo R. T., Head J. W., Greeley R., Sullivan R. W., Pilcher C., Schubert G., Moore W., Carr M. H., Moore J. M., Belton M. J. S., and Goldsby D. L. (1998) Geological evidence for solid-state convection in Europa's ice shell. *Nature, 391,* 365–368.

Pappalardo R. T., Belton M. J. S., Breneman H. H., Carr M. H., Chapman C. R., Collins G. C., Denk T., Fagents S., Geissler P. E., Giese B., Greeley R., Greenberg R., Head J. W., Helfenstein P., Hoppa G., Kadel S. D., Klaasen K. P., Klemasewski J. E., Magee K., McEwen A. S., Moore J. M., Moore W. B., Neukum G., Phillips C.B., Prockter L. M., Schubert G., Senske D. A., Sullivan R. J., Tufts B. R., Turtle E. P., Wagner R., and Williams K. K. (1999) Does Europa have a subsurface ocean? Evaluation of the geological evidence. *J. Geophys. Res., 104,* 24015–24055.

Paranicas C. R., Carlson R. W., and Johnson R. E. (2001) Electron bombardment of Europa. *Geophys. Res. Lett., 28,* 673–676.

Paranicas C. R., Ratliff J. M., Mauk B. H., Cohen C., and Johnson R. E. (2002) The ion environment near Europa and its role in surface energetics. *Geophys. Res. Lett., 29,* 14127.

Payne-Gaposchkin C. (1956) *Introduction to Astronomy.* Prentice-Hall, Englewood Cliffs. 299 pp.

Peale S. J. and Lee M. H. (2002) A primordial origin of the Laplace relation among the Galilean satellites. *Science, 298,* 593–597.

Peale S. J., Cassen P., and Reynolds R. T. (1979) Melting of Io by tidal dissipation. *Science, 43,* 65–72.

Pickering E. C. (1908) Brighter satellites of Jupiter and Saturn. Part I, Chapter VI. *Ann. Harvard College Observ., 61,* 72–85.

Pickering E. C., Searle A., and Upton W. (1879) Photometric observations. Part II. Chapter VIII. Satellites of Jupiter. *Ann. Harvard College Observ., 11,* 239–246.

Pilcher C. B., Ridgway S. T., and McCord T. B. (1972) Galilean satellites: Identification of water frost. *Science, 178,* 1087–1089.

Pollack J. B. and Fanale R. W. (1982) Origin and evolution of the Jupiter satellite system. In *Satellites of Jupiter* (D. Morrison, ed.), pp. 872–910. Univ. of Arizona, Tucson.

Pollack J. B. and Reynolds R. W. (1974) Implications of Jupiter's early contraction history for the composition of the Galilean satellites. *Icarus, 21,* 248–253.

Pollack J. B., Grossman A. S., Moore T., and Graboske H. C. (1976) The formation of Saturn's satellites and rings as influenced by Saturn's contraction history. *Icarus, 29,* 35–48.

Pollack J. B., Witteborn F. C., Erickson E. F., Strecker D. W., Baldwin B. J., and Bunch T. E. (1978) Near-infrared spectra of the Galilean satellites: Observations and compositional implications. *Icarus, 36,* 271–303.

Prockter L. M., Pappalardo R. T., and Head J. W. III (2000) Strike-slip duplexing on Jupiter's icy moon Europa. *J. Geophys. Res., 105,* 9483–9488.

Reynolds R. T. and Cassen P. (1979) On the internal structure of the major satellites of the outer planets. *Geophys. Res. Lett., 6,* 121–124.

Reynolds R. T., Squyres S., Colburn D., and McKay C. (1983) On the habitability of Europa. *Icarus, 56,* 246–254.

Roberts J. A. and Stanley G. J. (1959) Radio emission from Jupiter at a wavelength of 31 centimeters. *Publ. Astron. Soc. Pacific, 71,* 485–496.

Russell C. T., ed. (1992) The Galileo mission. *Space Sci. Rev., 60.*

Russell H. N., Dugan R. S., and Stewart J. Q. (1945) *Astronomy, Vol. I,* 2nd edition. Ginn and Co., Boston. 470 pp.

Sampson R. A. (1921) Theory of the four great satellites of Jupiter. *Mem. R. Astron. Soc. LXIII,* 1–270.

Saur J. D., Strobel D. F., and Neubauer F. M. (1998) Interaction of the jovian magnetosphere with Europa: Constraints on the neutral atmosphere. *J. Geophys. Res., 103,* 19947.

Schubert G., Anderson J. D., Spohn T., and McKinnon W. B. (2004) Interior composition, structure and dynamics of the Galilean satellites. In *Jupiter — The Planets, Satellites, and Magnetosphere* (F. Bagenal et al., eds.), pp. 329–362. Cambridge Univ., Cambridge.

Secchi P. A. (1859) *Quadro Fisico del Sistema Solare.* Topgraphia Delle Belle Arti, Rome. 190 pp.

Smythe W. D., Carlson R. W., Ocampo A., Matson D., Johnson T. V., McCord T. B., Hansen G. E., Soderblom L. A., and Clark R. N. (1998) Absorption bands in the spectrum of Europa detected by the Galileo NIMS instrument. In *Lunar and Planetary Science XXIX,* Abstract #1532. Lunar and Planetary Institute, Houston (CD-ROM).

Spencer J. R. and Calvin W. M. (2002) Condensed O_2 on Europa and Callisto. *Astron. J., 124,* 3400–3403.

Spencer J. R., Calvin W. M., and Person M. J. (1995) Charge-coupled-device spectra of the Galilean satellites: Molecular-oxygen on Ganymede. *J. Geophys. Res., 100,* 19049–19056.

Squyres S. W., Reynolds R. T., Cassen P., and Peale S. J. (1983) Liquid water and active resurfacing of Europa. *Nature, 301,* 225–226.

Stebbins J. (1927) The light variation of the satellites of Jupiter and their applications to measures of the solar constant. *Lick Obs. Bull., 13,* 1–11.

Stebbins J. and Jacobsen T. S. (1928) Further photometric measures of Jupiter's satellites on Uranus, with tests for the solar constant. *Lick Obs. Bull., 13,* 180–195.

Sulliivan C. W. and Palmisano A. C. (1981) Sea-ice microbial communities in McMurdo Sound. *Antarct. J. U.S., 16(5),* 126–127.

Sullivan R., Greeley R., Homan K., Klemaszewski J., Belton M. J. S., Carr M. H., Chapman C. R., Tufts R., Head J. W., Pappalardo R. T., Moore J., and Thomas P. (1998) Episodic plate separation and fracture infill on the surface of Europa. *Nature, 391,* 371–373.

Thomas N., Bagenal F., Hill T. W., and Wilson J. K. (2004) The Io neutral clouds and plasma Torus. In *Jupiter — The Planets, Satellites, and Magnetosphere* (F. Bagenal et al., eds.), pp. 561–592. Cambridge Univ., Cambridge.

Thomas T. C., Greeley R., Figueredo P., and Tanaka K. (2007) Global geology and stratigraphy of Europa. *EOS Trans. AGU, 88,* Fall Meeting Supplement, Abstract P24-12651.

Urey H. C. (1952) *The Planets: Their Origin and Development.* Yale Univ., New Haven. 245 pp.

Wamsteker W. (1972) Narrow-based photometry of the Galilean satellites. *Comm. Lunar Planetary Lab., 9,* 171–177.

Yoder C. F. (1979) How tidal heating in Io drives the Galilean orbital resonance locks. *Nature, 79,* 767–770.

Yoder C. F. and Peale S. J. (1981) The tides of Io. *Icarus, 47,* 1–35.

Zahnle K., Schenk P., Levinson H., and Dones L. (2003) Cratering rates in the outer solar system. *Icarus, 163,* 263–289.

Zimmer C., Khurana K. K., and Kivelson M. G. (2000) Subsurface oceans on Europa and Callisto: Constraints from Galileo magnetometer observations. *Icarus, 147,* 329–347.

Zolotov M. I. and Shock E. L. (2001) Composition and stability of salts on the surface of Europa and their oceanic origin. *J. Geophys. Res., 106,* 32815–32827.

Zolotov M. I. and Shock E. L. (2004) A model for low-temperature biogeochemistry of sulfur, carbon, and iron on Europa. *J. Geophys. Res., 109,* E06003, DOI: 10.1029/2003JE002194.

Formation of Jupiter and Conditions for Accretion of the Galilean Satellites

P. R. Estrada and I. Mosqueira
SETI Institute

J. J. Lissauer, G. D'Angelo, and D. P. Cruikshank
NASA Ames Research Center

We present an overview of the formation of Jupiter and its associated circumplanetary disk. Jupiter forms via a combination of planetesimal accretion and gravitational accumulation of gas from the surrounding solar nebula. The formation of the circumjovian gaseous disk, or subnebula, straddles the transitional stage between runaway gas accretion and Jupiter's eventual isolation from the circumsolar disk. This isolation, which effectively signals the termination of Jupiter's accretion, takes place as Jupiter opens a deep gas gap in the solar nebula, or the solar nebula gas dissipates. The gap-opening stage is relevant to subnebula formation because the radial extent of the circumjovian disk is determined by the specific angular momentum of gas that enters Jupiter's gravitational sphere of influence. Prior to opening a well-formed, deep gap in the circumsolar disk, Jupiter accretes low specific angular momentum gas from its vicinity, resulting in the formation of a rotationally supported compact disk whose size is comparable to the radial extent of the Galilean satellites. This process may allocate similar amounts of angular momentum to the planet and the disk, leading to the formation of an *ab initio* massive disk compared to the mass of the satellites. As Jupiter approaches its final mass and the gas gap deepens, a more extended, less massive disk forms because the gas inflow, which must come from increasingly farther away from the planet's semimajor axis, has high specific angular momentum. Thus, the size of the circumplanetary gas disk upon inflow is dependent on whether or not a gap is present. We describe the conditions for accretion of the Galilean satellites, including the timescales for their formation and mechanisms for their survival, all within the context of key constraints for satellite formation models. The environment in which the regular satellites form is tied to the timescale for circumplanetary disk dispersal, which depends on the nature and persistence of turbulence. In the case that subnebula turbulence decays as gas inflow wanes, we present a novel mechanism for satellite survival involving gap-opening by the largest satellites. On the other hand, assuming that sustained turbulence drives subnebula evolution on a short timescale compared to the satellite formation timescale, we review a model that emphasizes collisional processes to explain satellite observations. We briefly discuss the mechanisms by which solids may be delivered to the circumplanetary disk. At the tail end of Jupiter's accretion, most of the mass in solids resides in planetesimals >1 km in size; however, planetesimals in Jupiter's feeding zone undergo a period of intense collisional grinding, placing a significant amount of mass in fragments <1 km. Inelastic or gravitational collisions within Jupiter's gravitational sphere of influence allow for the mass contained in these planetesimal fragments to be delivered to the circumplanetary disk either through direct collisional/gravitational capture, or via ablation through the circumjovian gas disk. We expect that planetesimal delivery mechanisms likely provide the bulk of material for satellite accretion.

1. INTRODUCTION

The prograde, low-inclination orbits and close spacing of the regular satellites of the four giant planets indicate that they formed within a circumplanetary disk around their parent planet. The fact that all of the giant planets of our solar system harbor families of regular satellites suggests that satellite formation may be a natural, if not inevitable, consequence of giant planet formation. Thus to understand the origins of the jovian satellites, we must understand the formation of Jupiter and its circumplanetary disk.

Observations provide some basic constraints on Jupiter's origin. Jupiter must have formed prior to the dissipation of the solar nebula because H and He (the planet's primary constituents) do not condense at any location in the protoplanetary disk; these light elements were therefore accumulated in gaseous form. Protoplanetary disks are observed to have lost essentially all their gases in $<10^7$ yr (*Meyer et al.,* 2007), e.g., by photoevaporation (*Shu et al.,* 1993; *Dullemond et al.,* 2007), implying that Jupiter must have completed its formation by then. A fundamental question of giant planet formation is whether Jupiter formed as a result of a gravitational instability that occurred in the solar nebula gas, or through accretion of a solid core that eventually became large enough to accumulate a gaseous envelope directly from the nebula. The evidence supports the latter formation scenario in which the early stage of planetary growth consists of the accumulation of planetesimals, in a manner analogous to the consensus treatment of terrestrial planets (*Lissauer and Stevenson,* 2007; cf. *Durisen et al.,* 2007). The heavy element core grows sufficiently large (~5–15 $M_\oplus$, where $M_\oplus = 5.976 \times 10^{27}$ g is an Earth mass) that it is able to accrete gas from the surrounding nebula (*Hubickyj et al.,* 2005). The planet's final mass is determined by either the opening of a gas gap in the protoplanetary disk, or the eventual dissipation of the solar nebula. Jupiter's atmosphere is enhanced in condensible material by a factor of ~3–4 times solar; Jupiter as a whole is enhanced in heavy elements by a factor of ~3–8 compared to the Sun. These observations suggest that the jovian subnebula could also have been enhanced in solids with respect to solar composition mixtures, although the way in which the disk was enhanced may differ from that of the planet (section 4.2).

The enhancement of solids in Jupiter's gaseous envelope, the composition of its core, and the source material that went into making the jovian satellite system likely reflect the composition of planetesimals that formed within its vicinity as well as farther from the Sun in the circumsolar disk. The composition of outer solar system bodies such as comets contain grains with a wide range of volatility (*Brownlee et al.,* 2006; *Zolensky et al.,* 2006). Dust samples from Comet 81P/Wild 2 brought to Earth by the Stardust space mission show a great diversity of high- and moderate-temperature minerals, in addition to organic material (e.g., *Brownlee et al.,* 2006; *Sandford et al.,* 2006). The implication from these observations is that there was a significant amount of radial mixing within the early solar nebula leading to the transport of high-temperature minerals from the innermost regions of the solar nebula to beyond the orbit of Neptune. These high-temperature minerals could have been transported through ballistic transport away from the midplane (*Shu et al.,* 2001), or turbulent transport in the midplane (e.g., *Scott and Krot,* 2005; *Cuzzi and Weidenschilling,* 2006; *Natta et al.,* 2007). Based on the "X-wind" model, Shu and colleagues predicted that there would be transport of CAIs (calcium-aluminum-rich inclusions) from near the Sun to the outer edge of the solar system where Comet Wild 2 formed. Once transported, these minerals were incorporated into the first generation of planetesimals that went into making Jupiter and the outer planets. The presence of supersolar abundances of more volatile species such as argon (which requires very low temperatures to condense) in Jupiter's atmosphere likely implies that, in addition to outward transport of high-temperature material, there was also inward transport of highly volatile material within solid bodies (e.g., *Cuzzi and Zahnle,* 2004) (see section 2.4). Similar migration might also have characterized circumplanetary nebulae at some stage in their evolution.

The formation of giant planet satellite systems has been treated as a poorly understood extension of giant planet formation (see, however, early work by *Coradini et al.,* 1981, 1982). Thanks in great part to the success of the Voyager and Galileo missions, the jovian satellite system has been studied extensively over the last three decades; the emphasis has been on the Galilean satellites Io, Europa, Ganymede, and Callisto. The cosmochemical and dynamical properties of the jovian satellites may provide important clues about the late and/or postaccretional stages of Jupiter's formation (*Pollack and Reynolds,* 1974; *Mosqueira and Estrada,* 2003a,b; *Estrada and Mosqueira,* 2006).

The abundances and the chemical forms of constituents observed in the jovian satellite system are a diagnostic of their source material (i.e., either the source material condensed in the solar nebula or it was reprocessed/condensed in the subnebula). However, interpretation of their cosmochemical properties is complicated because objects the size of the Galilean satellites could have been altered through extensive resurfacing due to differentiation, energetic impacts, and other forms of postaccretional processing. Callisto and Ganymede have mean densities of 1.83 and 1.94 g cm^{-3}, respectively; each may consist of ~50% rock and ~50% water-ice by mass, with Ganymede being slightly more rock rich (*Sohl et al.,* 2002). In contrast, the solids component of solar composition material may be closer to ~60% rock, ~40% water-ice by mass, although the ice/rock ratio in the solar nebula is uncertain since it depends on the carbon speciation (*Wong et al.,* 2008). This variation from solar composition may indicate that Callisto and Ganymede were enhanced overall in water-ice at the time of their accretion. However, the total amount of water in Europa (3.01 g cm^{-3}) is only ~10% by mass (*Sohl et al.,* 2002). If the initial composition was like that of its outer neighbors, its current state requires additional mechanisms to explain its water loss. On the other hand, Io has a mean density of 3.53 g cm^{-3} consistent with rock. Thus, a key feature of the Galilean satellite system is the strong monotonic variation in ice fraction with distance from Jupiter.

Yet, despite this remarkable radial trend in water-ice fraction, the largest of the tiny moons orbiting interior to Io, Amalthea, has a density so low that even models for its composition that include high choices for its porosity require that water-ice be a major constituent (*Anderson et al.,* 2005). The lack of midsized moons, which could provide additional clues to the relative abundances of rocky and icy material, complicates our attempts to understand the cosmochemical history of the jovian system. Satellites in the jovian system are found to contain evidence of hydrocarbons through their spectral signatures in the low-albedo surface materials (e.g., *Hibbitts et al.,* 2000, 2003) and possibly CN-bearing materials (*McCord et al.,* 1997). Whether these materials are intrinsic or extrinsic (e.g., a coating from impactors) in origin is not known. Contrast this with the saturnian system, where the average density of the regular, icy satellites (~1.3 g cm^{-3}) may allow us to infer a nonsolar composition or the presence of a significant amount of low-density hydrocarbons (e.g., *Cruikshank et al.,* 2007, 2008).

A caveat is that the saturnian midsized satellites may have been subject to disruptive collisions (see section 4.5.1).

Is there accessible information on Ganymede and Callisto that can distinguish between material processed in the jovian subnebula and material processed in the solar nebula? Both bodies show surface deposits of water ice (*Calvin et al.,* 1995), and CO_2 is found in small quantities on both these Galilean satellites (*Hibbitts et al.,* 2001). While the surface water ice is probably native material, the CO_2 is largely associated with the non-ice regions, and the CO_2 spectral band is shifted in wavelength from its nominal position measured in the laboratory, indicating that it is probably complexed with other molecules (*Chaban et al.,* 2006). The CO_2 is likely to have been synthesized locally by the irradiation of water and a source of carbon, perhaps from micrometeorites. If the chemistry in the subnebula occurred at pressures in the range 10^{-3}–10^{-4} bar as the gas cooled, CO gas and water would have condensed early on (e.g., *Fegley,* 1993). The condensation of water may trap CO (*Stevenson and Lunine,* 1988) if the pressure is high, thus altering the C/O ratio in the remaining gas and affecting formation and growth of mineral grains. Unfortunately, molecules such as CH_4, NH_3, and native CO_2 and CO, which would give important clues to the conditions in which the Galilean satellites formed and would help distinguish between gas-poor and gas-rich models, for example, are not detected on the surfaces of any of these bodies (although we note that CO and CH_4 are too volatile in general to be stable on the surfaces of these bodies).

Although one can think of Jupiter and its satellites as a "miniature solar system," there are several significant differences between accretion in a circumplanetary disk and the solar nebula. In the circumplanetary disk, thermodynamical properties, length scales, and dynamical times are quite different from their circumsolar analog. Most of these differences can be attributed to the fact that giant planet satellite systems are compact. For instance, the giant planets dominate the angular momentum budget of the solar system; however, the analogy does not extend to the jovian system where the majority of the system angular momentum resides in Jupiter's spin. Moreover, dynamical times, which can influence a number of aspects of the accretion process, are orders of magnitude faster in circumplanetary disks than in the circumsolar disk (accretion of planets in the habitable zones of M dwarf stars are an intermediate case) (*Lissauer,* 2007). As an example, Europa's orbital period is $\sim 10^{-3}$ times that of Jupiter.

Over the last decade, there have been a number of global models of the jovian satellite system designed to fit the observational constraints provided by the Voyager and Galileo missions (*Mosqueira and Estrada,* 2000, 2003a,b; *Estrada,* 2002; *Canup and Ward,* 2002; *Mousis and Gautier,* 2004; *Alibert et al.,* 2005a; *Estrada and Mosqueira,* 2006; and others). Key constraints for models of the Galilean satellite system are as follows:

1. The mass and angular momentum contained within the Galilean satellites serve as firm constraints on the system. Aside from explaining the overall values, one must also address why there is so much mass and angular momentum in Callisto, yet no regular satellites outside its orbit.

2. The increase in ice fraction with distance from the planet for the Galilean satellites (a.k.a., compositional gradient) is often attributed to a temperature gradient imposed by the proto-Jupiter. However, it is debatable whether the compositions of Io and Europa, specifically, are primordial or altered by mechanisms (e.g., tidal heating, hypervelocity impacts) that would preferentially strip away volatiles. Moreover, Amalthea's low density strongly suggests that water-ice is a major constituent of this inner, small satellite.

3. The moment of inertia of Callisto suggests a partially differentiated interior provided hydrostatic equilibrium is assumed (*Anderson et al.,* 1998, 2001). This would imply that Callisto formed slowly ($\gtrsim 10^5$ yr) (*Stevenson et al.,* 1986). However, it may be possible that a nonhydrostatic component in Callisto's core could be large enough to mask complete differentiation (*McKinnon,* 1997).

4. The nearly constant mass ratio of the largest satellites of Jupiter and Saturn (and possibly Uranus) to their parent planet, $\mu \sim 10^{-4}$, suggests that there may be a truncation mechanism that affects satellite mass (*Mosqueira and Estrada,* 2003b).

This chapter is organized as follows. In section 2, we present a detailed summary of the currently favored model for the formation of Jupiter. In section 3, we discuss the formation of the circumjovian disk. In section 4, we address the possible environments that arise for satellite formation during the late stages of Jupiter's accretion. We touch on the processes involved in the accretion of Europa and the other Galilean satellites, the methods of solids mass delivery, timescales for formation, and discuss the key constraints in more detail. In section 5, we present a summary.

2. FORMATION OF JUPITER

Jupiter's accretion is thought to occur in a protoplanetary disk with roughly the same elemental composition as the Sun; i.e., primarily H and He with ~1–2% of heavier elements. A lower bound on the mass of this solar disk of ~0.01–0.02 $M_\odot$ (where $M_\odot = 1.989 \times 10^{33}$ g is the mass of the Sun) has been derived from taking the condensed elemental fractions of the planets of the solar system and reconstituting them with enough volatiles to make the composition solar. This is historically referred to as the "minimum mass solar nebula" (MMSN) (*Weidenschilling,* 1977b; *Hayashi,* 1981). The radial distribution of solids is found by smearing the augmented mass of the planet over a region halfway to each neighboring planet. This approach yields a trend in surface density (by design, both in solids and gas) with semimajor axis, r, of $\Sigma \propto r^{-3/2}$ between Venus and Neptune (*Weidenschilling,* 1977b). The radial temperature of the photosphere of the solar nebula in these earlier models has been commonly taken to be that derived from assuming that the disk is optically thin (to its own emission), and dust grains are in radiative balance with the solar luminosity ($T \propto r^{-1/2}$) (*Pollack et al.,* 1977; *Hayashi,* 1981). This temperature dependence is one that might be expected from a scenario in which most of the dust has accumulated into larger bodies.

In this picture, it is implicitely assumed that the planets form near their present locations. This means that planetary formation is characterized by "local growth" — i.e., the planets accreted from the reservoir of gas and solids that condensed within their vicinity. In order to explain the abundances above solar of heavy volatile elements in Jupiter's atmosphere, the distribution of solids at the location of Jupiter is often assumed to be enhanced by a factor of several above the MMSN. However, the jovian atmosphere contains enhancements above solar in highly volatile noble gases such as argon (see section 2.4), so that even with a modification in the surface density, a local growth model likely does not do a good enough job of explaining the composition of Jupiter's atmosphere because it would require local temperatures much lower than what is predicted by an optically thin nebula model.

Efforts to match the spectral energy distributions of circumstellar disks around T Tauri stars have spawned a number of alternative disk models that lead to different thermal disk structures (e.g., *Kenyon and Hartmann,* 1987; *Chiang and Goldreich,* 1997; *Garaud and Lin,* 2007) that are considerably lower than what was assumed previously. Even with these lower-temperature models, though, the implication is that Jupiter would need to have accreted material from much farther out in the circumsolar disk. If planetesimals migrated over such large distances, it suggests that the $r^{-3/2}$ trend in surface density derived for the MMSN model is not a particularly useful constraint. These uncertainties in the most fundamental of disk properties underlie some of the reasons why the formation of giant planets remains one of the more scrutinized problems in planetary cosmogony today (see, e.g., recent reviews by *Wuchterl et al.,* 2000; *Hubbard et al.,* 2002; *Lissauer and Stevenson,* 2007).

Of the two formation models for giant planets that have received the most attention, the preponderance of evidence supports the formation of Jupiter via core nucleated accretion, which relies on a combination of planetesimal accretion and gravitational accumulation of gas. The alternative, the so-called *gas instability model,* would have Jupiter forming directly from the contraction of gaseous clump produced through a gravitational instability in the protoplanetary disk. In the gas instability model, the formation of Jupiter is somewhat akin to star formation. Although numerical calculations have produced ~1 M_J (where $M_J = 1.898 \times 10^{30}$ g is a Jupiter mass) clumps given sufficiently unstable disks (e.g., *Boss,* 2000; *Mayer et al.,* 2002), these clumps do not form unless the disk is highly atypical (very massive, and/or very hot) (*Rafikov,* 2005). Furthermore, unless there are processes that keep the disk unstable, weak gravitational instabilities lead to stabilization of the protoplanetary disk via the excitation of spiral density waves. These waves carry away angular momentum that spread the disk, lowering its surface density. Given the lack of observational support, along with theoretical arguments against the formation of Jupiter via fragmentation (*Bodenheimer et al.,* 2000a; *Cai et al.,* 2006), the gas instability model for Jupiter will not be considered in this chapter. For more in-depth discussions of the gas instability model, see *Wuchterl et al.* (2000), *Lissauer and Stevenson* (2007), and *Durisen et al.* (2007). In the core nucleated accretion model (*Pollack et al.,* 1996; *Bodenheimer et al.,* 2000b; *Hubickyj et al.,* 2005; *Alibert et al.,* 2005b; *Lissauer et al.,* 2009), Jupiter's formation and evolution is thought to occur in the following sequence: (1) Dust particles in the solar nebula form planetesimals that accrete, resulting in a solid core surrounded by a low-mass gaseous envelope. Initially, runaway accretion of solids occurs, and the accretion rate of gas is very slow. As the solid material in the planet's feeding zone is depleted, the rate of solids accretion tapers off. The "feeding zone" is defined by the separation distance between a massive planet and a massless body on circular orbits such that they never experience a close encounter (*Lissauer,* 1995). The gas accretion rate steadily increases and eventually exceeds the accretion rate of solids. (2) Proto-Jupiter continues to grow as the gas accretes at a relatively constant rate. The mass of the solid core also increases, but at a slower rate. Eventually, the core and envelope masses become equal. (3) At this point, the rate of gas accretion increases in runaway fashion, and proto-Jupiter grows at a rapidly accelerating rate. The first three parts of the evolutionary sequence are referred to as the nebular stage, because the outer boundary of the protoplanetary envelope is in contact with the solar nebula, and the density and temperature at this interface are those of the nebula. (4) The gas accretion rate reaches a limiting value defined by the rate at which the nebula can transport gas to the vicinity of the planet (*Lissauer et al.,* 2009). After this point, the equilibrium region of proto-Jupiter contracts, and gas accretes hydrodynamically into this equilibrium region. This part of the evolution is considered to be the transition stage. (5) Accretion is stopped by either the opening of a gap in the gas disk as a consequence of the tidal effect of Jupiter, accumulation of all nearby gas, or by dissipation of the nebula. Once accretion stops, the planet enters the isolation stage. Jupiter then contracts and cools to the present state at constant mass. In the following subsections, we present a more detailed summary of this formation sequence.

2.1. From Dust to Planetesimals

Sufficiently far away from the Sun where temperatures are low enough to allow for the condensation of solid material, grain growth begins with sticking of submicrometer-sized dust, composed of surviving interstellar grains and condensates that formed within the protoplanetary disk. These small grains are dynamically coupled to the nebula gas, and collide at low (size-dependent) relative velocities that can be caused by a variety of mechanisms (*Völk et al.,* 1980; *Weidenschilling,* 1984; *Nakagawa et al.,* 1986; *Weidenschilling and Cuzzi,* 1993; *Ossenkopf,* 1993; *Cuzzi and Hogan,* 2003; *Ormel and Cuzzi,* 2007). Low impact velocities (and thus low impact energies) tend to allow small particles to grow more efficiently because collisions can be completely inelastic (e.g., *Wurm and Blum,* 1998). However, as grains collide to form larger and larger agglomerates, the level of coupling between growing particles and the gas decreases, and particle velocities relative to both other particles and the gas increases. As a result, larger particles begin to experience higher impact energies during interparticle collisions that can lead to fragmentation or erosion, rather than growth.

When the level of coupling between particles and the gas decreases, dust grains, which can be initially suspended at distances well above and below the solar nebula midplane, begin to gravitationally settle toward the midplane. As particles settle, they tend to grow. The time it takes for a particle to settle is a function of its size, and may be hundreds, thousands, or more orbital periods depending on ambient nebula conditions. Very small particles (e.g., submicrometer grains) have such long settling times relative to the lifetime of the gas disk that they are considered to be fully "entrained" in the gas. On the other end, some particles may become large enough that their settling time is comparable to their orbital period. Although technically these particles "couple" with the gas once per orbit, they are essentially viewed as the transition size from a regime in which particles collectively move with the gas, to a regime in which particles would prefer to move at the local Kepler orbital speed. These transitional particles are referred to as the "decoupling size," and as they settle to the nebular midplane they continue to grow by sweeping up dust and rubble (*Cuzzi et al.,* 1993; *Weidenschilling,* 1997).

The fact that particles tend to drift relative to the surrounding gas indicates that the gas component itself does not rotate at the local Kepler orbital velocity, v_K. This is because the nebula gas is not supported against the Sun's gravity by rotation alone. In a rotating frame there are three forces at work on a gas parcel in the nebula disk: a gravitational force radially directed toward the Sun; a centrifugal force directed radially outward from the nebula's rotation axis; and (because the gas is not pressureless) an outward-directed pressure gradient force that works to counter the effective gravity. The condition for equilibrium then requires that the nebula gas orbit at a velocity slightly less than v_K. Solid objects, which might otherwise orbit at the local Keplerian orbital speed, are too dense to be supported by the pressure gradient, and are subject to a drag force that can systematically remove their angular momentum, leading to orbital decay (*Weidenschilling,* 1977a).

Thus, a more formal definition of the decoupling size is the size at which the time needed for the gas drag force to dissipate a particles momentum relative to the gas (known as the "stopping time") is similar to its orbital period. At Jupiter, this size is $R \sim 1$ m for the canonical MMSN model (e.g., *Cuzzi et al.,* 1993). Since decoupling-size particles encounter the strongest drag force, they tend to have the most rapid inward orbital migration. For example, such a particle (assuming no growth) would eventually spiral in to the Sun in $\sim 10^4$–10^5 yr. This time is short compared to planetary accretion timescales ($\gtrsim 10^6$–10^7 yr), but, as objects grow larger, the drag force on them decreases and the stopping times can become quite long. Kilometer-sized planetesimals are mostly unaffected by the gas due to their greater mass-to-surface-area ratio. Thus, getting from decoupling-size objects to the safety of relatively immobile ($\gtrsim 1$ km) planetesimals is a key issue in planet formation.

Whether growth beyond the decoupling size happens depends on the turbulent viscosity ν in the disk. In general, the dissipation of viscous energy leads to the transport of mass inwards, facilitating further accretion onto the central object. Angular momentum is transported outward, causing the disk to spread (section 4.1). The efficiency of the mechanism of angular momentum transport is most commonly characterized by a turbulent parameter $\alpha \propto \nu$ (*Shakura and Sunyaev,* 1973). The level of turbulence is important because collisional velocities for decoupling-size objects can reach tens of meters per second in turbulence, which has more of a tendency to fragment them than to promote growth. This suggests that, while the incremental growth of sufficiently small grains and dust may take place irrespective of the level of nebula turbulence, successful growth past the decoupling size may require very low levels of turbulence (e.g., *Youdin and Shu,* 2002; *Cuzzi and Weidenschilling,* 2004), which helps to overcome the two main obstacles for this critical size: collisional disruption and rapid orbital migration (*Brauer et al.,* 2008; *Birnstiel et al.,* 2009). Several mechanisms have been proposed in the literature to explain planetesimal growth under nonturbulent (laminar) conditions (*Safronov,* 1960; *Goldreich and Ward,* 1973; *Weidenschilling,* 1997; *Sekiya,* 1998; *Youdin and Chiang,* 2004). Yet, in spite of the difficulties associated with potentially destructive collisions, mechanisms that attempt to explain various stages of planetesimal formation under turbulent conditions have also recently been advanced (see *Cuzzi and Weidenschilling,* 2006; *Cuzzi et al.,* 2007; *Johansen et al.,* 2007). One can argue on the basis of these works that growth under nonturbulent conditions provides a relatively straightforward pathway to planetesimal formation, while growth in the turbulent regime, if feasible, is a more complicated process with many stages. Indeed, the transition from agglomerates to planetesimals continues to provide the major stumbling block in planetary origins (*Weidenschilling,* 1997, 2002, 2004; *Weidenschilling and Cuzzi,* 1993; *Stepinski and Valageas,* 1997; *Dullemond and Dominik,* 2005; *Cuzzi and Weidenschilling,* 2006; *Dominik et al.,* 2007).

2.2. Core Accretion

The initial stage of Jupiter's formation entails the accretion of its solid core from the available reservoir of heliocentric planetesimals. Once planetesimals grow large enough, gravitational interactions and physical collisions between pairs of solid planetesimals provide the dominant perturbation of their basic Keplerian orbits. At this stage, effects that were more influential in the earlier stages of growth such as electromagnetic forces, collective gravitational effects, and in most circumstances gas drag, play minor roles. These planetesimals continue to agglomerate via pairwise mergers, with the rate of solid-body accretion by a planetesimal or planetary embryo (basically a very large planetesimal) being determined by the size and mass of the planetesimal/planetary embryo, the surface density of planetesimals, and the distribution of planetesimal velocities relative to the accreting body.

The planetesimal velocity distribution is probably the most important factor that controls the growth rate of planetary embryos into the core of a giant planet. As larger objects accrete, gravitational scatterings and elastic collisions can convert the ordered relative motions of orbiting planetesimals (i.e., Keplerian shear) into random motions, and can "stir up" the planetesimal random velocities up to the escape speed from the largest planetesimals in the swarm (*Safronov,* 1969). The effects of this gravitational stirring, however, tend to be balanced by collisional damping, because inelastic collisions (and, for smaller objects, gas drag) can damp eccentricities and inclinations. If one assumes that planetesimal pairwise collisions lead to perfect accretion, i.e., that all physical collisions are completely inelastic (fragmentation, erosion, and bouncing do not occur), this stage of growth can be initially quite rapid. With this assumption, the planetesimal accretion rate, $\dot{M}_Z$, is

$$\dot{M}_Z = \pi R^2 \sigma_Z \Omega F_g \quad (1)$$

where R is the radius of the accreting body, σ_Z is the surface density of solid planetesimals in the solar nebula, Ω is the orbital frequency at the location of the growing body, and F_g is the gravitational enhancement factor. The gravitational enhancement factor F_g arises from the ratio of the distance of close approach to the asymptotic unperturbed impact parameter, and in the two-body approximation (ignoring the tidal effects of the Sun's gravity) it is given by

$$F_g = 1 + \left(\frac{v_e}{v}\right)^2 \quad (2)$$

Here, v_e is the escape velocity from the surface of the planetary embryo, and v is the velocity dispersion of the planetesimals being accreted. The evolution of the planetesimal size distribution is determined by F_g, which favors collisions between bodies having larger masses and smaller relative velocities. Moreover, they can accrete almost every planetesimal they collide with (i.e., the perfect accretion approximation works best for the largest bodies).

As the planetesimal size distribution evolves, planetesimals (and planetary embryos) may pass through different growth regimes. These growth regimes are sometimes characterized as either orderly or runaway. When the relative velocities between planetesimals is comparable to or larger than the escape velocity of the largest body, $v \gtrsim v_e$, the growth rate $\dot{M}_Z$ is approximately proportional to R^2. This implies that the growth in radius is roughly constant (as can be easily derived from equation (1)). Thus the evolutionary path of the planetesimals exhibits an orderly growth across the entire size distribution so that planetesimals containing most of the mass double their masses at least as rapidly as the largest particle. When the relative velocity is small, $v \ll v_e$, the gravitational enhancement factor $F_g \propto R^2$, and so the growth rate $\dot{M}_Z$ is proportional to R^4. By virtue of its large, gravitationally enhanced cross-section, the most massive embryo doubles its mass faster than the smaller bodies do, and detaches itself from the mass distribution (*Levin,* 1978; *Greenberg et al.,* 1978; *Wetherill and Stewart,* 1989; *Ohtsuki et al.,* 2002).

Eventually the runaway body can grow so large (its F_g can exceed ~1000) that it transitions from dispersion-dominated growth to shear-dominated growth (*Lissauer,* 1987). This means that for these extremely low random velocities ($v \ll v_e$), the rate at which planetesimals encounter the growing planetary embryo is determined by the Keplerian shear in the planetesimal disk, and not by the random motions of the planetesimals. At this stage, larger embryos take longer to double in mass than do smaller ones, although embryos of all masses continue their runaway growth relative to surrounding planetesimals. This phase of rapid accretion of planetary embryos is known as oligarchic growth (*Kokubo and Ida,* 1998).

Rapid runaway or oligarchic accretion requires low random velocities, and thus small radial excursions of planetesimals. The planetary embryo's feeding zone is therefore limited to the annulus of planetesimals that it can gravitationally perturb into embryo-crossing orbits. Rapid growth stops when a planetary embryo has accreted most of the planetesimals within its feeding zone. Thus, runaway/oligarchic growth is self-limiting in nature, which implies that massive planetary embryos form at regular intervals in semimajor axis. The agglomeration of these embryos into a small number of widely spaced bodies necessarily requires a stage characterized by large orbital eccentricities. The large velocities implied by these large eccentricities imply small collision cross-sections (equation (2)) and hence long accretion times. Growth via binary (pairwise) collisions proceeds until the spacing of planetary orbits become dynamically isolated from one another, i.e., spacing sufficient for the configuration to be stable to gravitational interactions among the planets for the lifetime of the system (*Safronov,* 1969; *Wetherill,* 1990; *Lissauer,* 1993, 1995; *Agnor et al.,* 1999; *Laskar,* 2000; *Chambers,* 2001).

For shear-dominated accretion, the mass at which such a planetary embryo becomes isolated from the surrounding circumsolar disk via runaway accretion is given by (*Lissauer,* 1993)

$$M_{iso} = \frac{(8\pi\sqrt{3}r^2\sigma_Z)^{3/2}}{(3\,M_\odot)^{1/2}} \approx 1.6 \times 10^{25}(\sigma_Z^{3/2} r_{AU}^3)\ \mathrm{g} \quad (3)$$

where r_{AU} is the distance from the Sun in astronomical units (AU = 1.496×10^{13} cm). For the MMSN in which only local growth is considered, the mass at which runaway accretion would have ceased in Jupiter's accretion zone is ~1 $M_\oplus$ (*Lissauer,* 1987), which is likely too small to explain Jupiter's formation (see below). Rapid accretion can persist beyond the isolation mass if additional solids can diffuse into its feeding zone (*Kary et al.,* 1993; *Kary and Lissauer,* 1995). There are three plausible mechanisms for such diffusion: scattering between planetesimals, perturbations by neighboring planetary embryos, and migration of smaller planetesimals due to gas drag (section 2.1).

Other mechanisms that can lead to the migration of embryos from regions that are depleted in planetesimals into regions that are not depleted include gravitational torques resulting from the excitation of spiral density waves in the gaseous component of the disk (see section 4.3.3), and dynamical friction [if significant energy is transferred from the planetary embryo to the protoplanetary disk; e.g., see *Stewart and Wetherill* (1988)].

In the inner part of protoplanetary disks, Kepler shear is too great to allow the accretion of solid planets larger than a few $M_\oplus$ on any timescale unless the surface densities are considerably above that of the MMSN or a large amount of radial migration occurs. However, the fact that the relatively small terrestrial planets orbit deep within the Sun's potential well suggests that they likely were unable to eject substantial amounts of material from the inner solar system. Thus, the total amount of mass present in the terrestrial region during the runaway accretion epoch was probably not much more than the current mass of the terrestrial planets, implying that a high-velocity growth phase subsequent to runaway accretion was necessary in order to explain their present configuration (*Lissauer,* 1995).

This high-velocity final growth stage takes $O(10^8)$ yr in the terrestrial planet zone (*Safronov,* 1969; *Wetherill,* 1980; *Agnor et al.,* 1999; *Chambers,* 2001), but would require $O(10^9–10^{10})$ yr in the giant planet zone (*Safronov,* 1969) if one assumes local growth in a MMSN disk. These timescales reflect a slowing down of the accretion rate during the late stages of planetary growth due to a drop in the planetesimals surface density (*Wetherill,* 1980, 1986). The growth timescales quoted above are far longer than any modern estimates of the lifetimes of gas within protoplanetary disks ($\lesssim 10^7$ yr) (*Meyer et al.,* 2007), implying that Jupiter's core must grow large enough during the rapid runaway/oligarchic growth. The epoch of runaway/rapid oligarchic growth lasts only millions of years or less near the location of Jupiter (r = 5.2 AU), but can produce ~10 $M_\oplus$ cores if the circumsolar disk is enhanced in solids by only a few times the MMSN (*Lissauer,* 1987). The limits on the initial surface density of the disk are less restrictive in the giant planet region, because excess solid material can be ejected to the Oort cloud, or out of the solar system altogether.

The masses at which planets become isolated from the disk, thereby terminating the runaway/rapid oligarchic growth epoch, are likely to be comparably large at greater distances from the Sun. However, at these large distances, random velocities of planetesimals must remain quite small for accretion rates to be sufficiently rapid for planetary embryos to approach M_{iso} within the lifetimes of gaseous disks (*Pollack et al.,* 1996). Indeed, if planetesimal velocities become too large, material is more likely to be ejected to interstellar space than accreted by the planetary embryos.

2.3. Gas Accretion

2.3.1. Tenuous extended envelope phase. As the core grows, its gravitational potential well gets deeper, allowing its gravity to pull gas from the surrounding nebula toward it. At this stage, the core may begin to accumulate a gaseous envelope. A planet on the order of 1 $M_\oplus$ [the value for a specific planet depends upon the mean molecular weight and opacity of the atmosphere (*Stevenson,* 1982)] is able to capture an atmosphere because the escape speed from its physical surface is large compared to the thermal velocity (or sound speed) c_s of the surrounding gaseous protoplanetary disk. Initially, all gas with gravitational binding energy to the planet larger than the thermal energy is retained as part of the planet (*Cameron et al.,* 1982; *Bodenheimer and Pollack,* 1986). The radial extent of this region is ~GM_p/c_s^2, where G is the gravitational constant. If the protoplanet is small, this bound region can be significantly smaller than the protoplanet's gravitational domain, whose size is typically a significant fraction of the protoplanet's Hill sphere, R_H, which is given by

$$R_H = \left(\frac{M_p}{3\,M_\odot}\right)^{1/3} r \tag{4}$$

Here, M_p is the mass of the protoplanet, and r is the distance between the protoplanet and the Sun. The Hill radius denotes the distance from a planet's center along the planet-Sun line at which the planet's gravity equals the tidal force of the solar gravity relative to the planet's center. As the protoplanet increases in mass, the region in which gas can be bound increases and may become a significant fraction of R_H.

An embryo begins to accrete gas slowly, so its gaseous envelope is initially optically thin and isothermal with the surrounding protoplanetary disk. As the envelope gains mass it becomes optically thick to outgoing thermal radiation, and its lower reaches can get much warmer and denser than the gas in the surrounding protoplanetary disk. As the protoplanet's gravity continues to pull in gas from the surrounding disk toward it, thermal pressure from the existing envelope limits further accretion. This is because, for much of its gas accretion stage, the key factor limiting the protoplanet's accumulation of gas is its ability to radiate away the gravitational energy provided by the continued accretion of planetesimals and the contraction of the envelope; this energy loss is necessary for the envelope to further contract and allow more gas to enter the protoplanet's gravitational domain.

As the energy released by the accretion of planetesimals and gas is radiated away at the protoplanet's photosphere, the photosphere cools and a subsequent pressure drop causes the envelope to contract. This is referred to as a Kelvin-Helmholtz contraction. Compression heats the envelope and regulates the rate of contraction, which in turn controls how rapidly additional gas can enter the planet's gravitational domain and be accreted.

This suggests that the rate and manner in which a giant planet accretes solids can substantially affect its ability to attract gas. Initially accreted solids form the planet's core (section 2.2), around which gas is able to accumulate. Calculated gas accretion rates are very strongly increasing func-

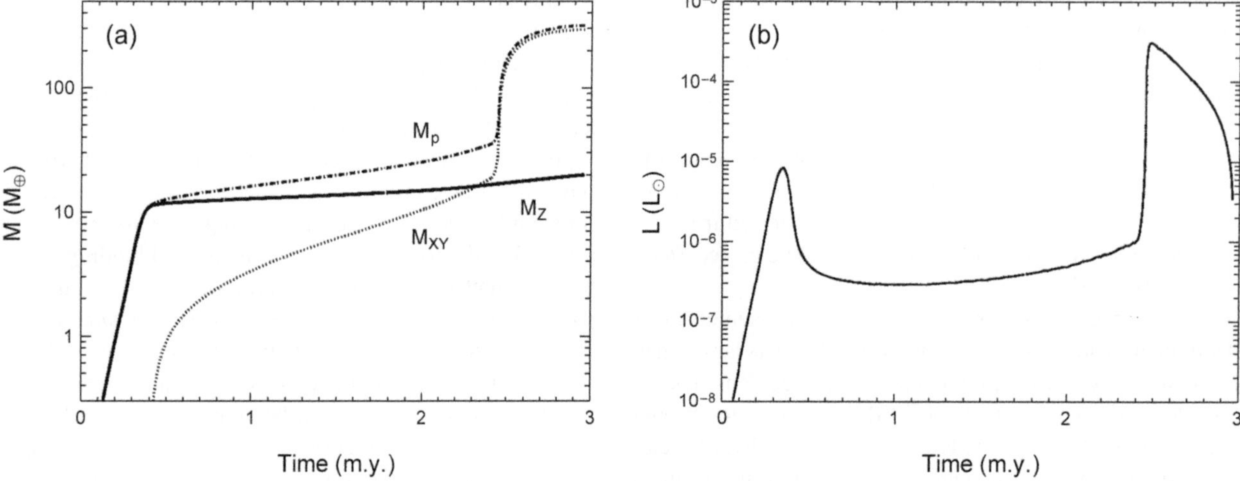

Fig. 1. Evolution of a proto-Jupiter within a protoplanetary disk with surface density of solids $\sigma_Z = 10$ g cm^{-2} and grain opacity in the protoplanet's envelope assumed to be at 2% of the interstellar value. Details of the calculation are presented in *Lissauer et al.* (2009). **(a)** The mass of solids in the planet (solid curve), gas in the planet (dotted curve), and the total mass of the planet (dot-dashed curve) are shown as functions of time. Note the slow, gradually increasing, buildup of gas, leading to a rapid growth spurt, and finally a slow tail off in accretion. Figure courtesy of O. Hubickyj. **(b)** The planet's luminosity is shown as a function of time. The rapid contraction of the planet just before t = 2.5 m.y. coincides with the highest luminosity and the epoch of most rapid gas accretion. From *Lissauer et al.* (2009).

tions of the total mass of the protoplanet, implying that rapid growth of the core is a key factor in enabling a protoplanet to accumulate substantial quantities of gas. Continued accretion of solids acts to reduce the protoplanet's growth time by increasing the depth of its gravitational potential well, but also counters growth by providing additional thermal energy to the envelope from solids that sink to the core. Another hurdle to rapid growth that planetesimal accretion provides is the increased atmospheric opacity from dust grains that are released (ablated) in the upper parts of the envelope. If the opacity is sufficiently high, much of the growing planet's envelope transports energy via convection. However, the distended very low density outer region of the envelope has thermal gradients that are too small for convection, but are large enough that they can act as an efficient thermal blanket if it is sufficiently dusty to be moderately opaque to outgoing radiation. This has the effect of slowing contraction and frustrating further accretion of gas, and lengthening the timescale for planet accretion.

Figure 1 shows the evolution of the mass and luminosity from a recent model of Jupiter's formation (*Lissauer et al.,* 2009). During the runaway planetesimal accretion epoch (when the core is predominantly being formed), the protoplanet's mass increases rapidly. Although at this point the gaseous atmosphere is quite tenuous, the internal temperature and thermal pressure of the envelope increases, which prevents substantial amounts of nebular gas from falling onto the protoplanet. When the rate of planetesimal accretion decreases (roughly around $M_p \gtrsim 10\ M_\oplus$), gas falls onto the protoplanet more rapidly as the additional component of thermal energy contributed by the accreting planetesimals decreases. At this stage the envelope mass is a sensitive function of the total mass, with the gaseous fraction increasing rapidly as the planet accretes (*Pollack et al.,* 1996). When the envelope reaches a mass comparable to that of the core, the self-gravity of the gas becomes substantial, and the envelope contracts when more gas is added. Eventually, increases in the planet's mass and the radiation of energy allow the envelope to shrink rapidly. Further accretion is then governed by the availability of gas rather than thermal considerations. At this point, the factor limiting the planet's growth rate is the flow of gas from the surrounding protoplanetary disk (*Lissauer et al.,* 2009).

The time required to reach this stage of rapid gas accretion is governed primarily by three factors: the mass of the solid core; the rate of energy input from continued accretion of solids; and the opacity of the envelope. These three factors appear to be key in determining whether giant planets are able to form within the lifetimes of protoplanetary disks ($\lesssim 10^7$ yr). For example, in a disk with initial $\sigma_Z =$ 10 g cm^{-2} at 5.2 AU from a 1 $M_\odot$ star, a planet whose atmosphere has 2% interstellar opacity forms with a 16 $M_\oplus$ core in 2.3 m.y.; in the same disk, a planet whose atmosphere has full interstellar opacity (~1 cm^2 g^{-1}) forms with a 17 $M_\oplus$ core in 6.3 m.y.; a planet whose atmosphere has 2% interstellar opacity but stops accreting solids at 10 $M_\oplus$ forms in 0.9 m.y., whereas if solids accretion is halted at 3 $M_\oplus$ accretion of a massive envelope requires 12 m.y. (*Hubickyj et al.,* 2005). These results suggest that if Jupiter's core mass is significantly less than 10 $M_\oplus$, then it presents a problem for formation models mainly because disk dispersal times are observed to be shorter than the time it takes for a smaller core mass to accrete a massive enough envelope (unless the opacity of the envelope to outgoing radiation is significantly less than 2% of the interstellar medium).

Thus, the key to forming Jupiter prior to the dispersal of the nebula is the rapid formation of a massive core coupled with a combination of a decreased solids accretion rate and/

or the outer regions of the giant planet envelope being transparent to outgoing radiation. However, since there is little in the way of observational constraints, our understanding continues to be handicapped by uncertainties in quantities such as the opacity and solids accretion rate that are derived from planet formation models. The compositions of the atmospheres of the giant planets may provide some insight. As the envelope becomes more massive, late-accreting planetesimals (but, early-arriving in the context of satellite formation, see section 4) sublimate before they can reach the core, thereby enhancing the heavy element content of the envelope considerably. In section 2.4, we discuss more on the composition of Jupiter's envelope.

2.3.2. Hydrodynamic phase. As demonstrated in Fig. 1, a protoplanet accumulates gas at a gradually increasing rate until its gas component is comparable to its heavy element mass (i.e., the envelope and core are of comparable mass). At this point, the protoplanet has enough mass for its self-gravity to compress the envelope substantially. The rate of gas accretion then accelerates rapidly, and a gas runaway occurs (*Pollack et al.,* 1996; *Hubickyj et al.,* 2005). This accretion continues as long as there is gas in the vicinity of the protoplanet's orbit. The ability of the protoplanet to accrete gas does not depend strongly on the outer boundary conditions (temperature and pressure) of the surrounding nebula, if there is adequate gas around to be accreted (*Mizuno,* 1980; *Stevenson,* 1982; *Pollack et al.,* 1996). Hydrodynamic limits allow quite rapid gas flow to the planet in an unperturbed disk. But in realistic scenarios, the protoplanet not only alters the disk by accreting material from it, but also by exerting gravitational torques on it (see section 4). Both of these processes can lead to a formation of a gap in the circumsolar disk (e.g., *Lin and Papaloizou,* 1979) and isolation of the planet from the surrounding gas, thus providing a means of limiting the final mass of the giant planet.

Observationally, such gravitationally induced gaps have been observed around small moons within Saturn's rings (*Showalter,* 1991; *Porco et al.,* 2005). Numerically, gas gap formation has been studied extensively. For example, *D'Angelo et al.,* (2003b) used a three-dimensional adaptive mesh refinement code to follow the flow of gas onto accreting giant planets of various masses embedded within a gaseous protoplanetary disk. *Bate et al.* (2003) have performed three-dimensional simulations of this problem using the ZEUS hydrodynamics code. Using parameters appropriate for a moderately viscous MMSN protoplanetary disk at 5 AU ($\alpha \sim 4 \times 10^{-3}$, see section 3.3), both groups found that <10 $M_\oplus$ planets do not perturb the protoplanetary disk enough to significantly affect the amount of gas that flows toward them. Gravitational torques on the disk by larger planets under these disk conditions drive away gas. In a moderately viscous disk, hydrodynamic limits on gas accretion reach to a few $\times 10^{-2}$ $M_\oplus$ per year for planets in the ~50–100 M range, and then decline as the planet continues to grow. An example of gas flow around/to a 1 M_J planet is shown in Fig. 2. In general, caution must be exercised in the interpretation of these types of calculations when attempting to connect them with the formation of the giant planet itself. Thus, for example, these calculations do

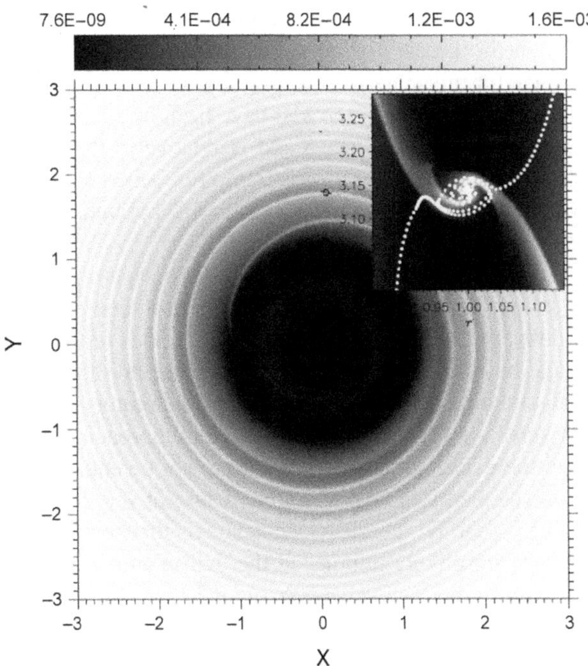

Fig. 2. The surface mass density of a gaseous disk containing a Jupiter-mass planet on a circular orbit located 5.2 AU from a 1 $M_\odot$ star. The ratio of the scale height of the disk to the distance from the star is 1/20, and the dimensionless viscosity at the location of the planet is $\alpha = 4 \times 10^{-3}$. The distance scale is in units of the planet's orbital distance, and surface density of 10^{-4} corresponds to 33 g cm^{-2}. The inset at the upper right shows a close-up of the disk region around the planet, plotted in cylindrical coordinates. The two series of white dots indicate actual trajectories (real particle paths, not streamlines) of material that is captured in the gravitational well of the planet and eventually accreted by the planet. See *D'Angelo et al.* (2005) for a description of the code used.

not include the thermal pressure on the nebula from the hot planet, which is found to be the major accretion-limiting factor for planets up to a few tens of $M_\oplus$ by the simulations discussed in section 2.3.1 (*Hubickyj et al.,* 2005; *Lissauer and Stevenson,* 2007; *Lissauer et al.,* 2009). It should be noted that ability for a protoplanet to open a gap is dependent on the viscosity of the disk. In nearly inviscid disks, for example, a ~10 $M_\oplus$ protoplanet may be capable of opening a gap (*Rafikov,* 2002a,b) (section 4.3.3).

If the planet successfully cuts off its supply of gas by the opening of a gap in the circumsolar disk, the planet effectively enters the isolation stage. Jupiter then contracts and cools to its present state at constant mass.

2.4. The Composition of Jupiter's Envelope

2.4.1. Enhancement in heavy elements. The elemental abundances of gases in Jupiter's atmosphere that are quite volatile, but unlike H and He still condensible within the giant planet region of the solar nebula, are about ~3–4 times solar (*Atreya et al.,* 1999; *Mahaffy et al.,* 2000; *Young,* 2003). If the relative abundances of all condensible elements in Jupiter's envelope are the same as in the Sun, then such material must account for ~18 $M_\oplus$ (*Owen and Encrenaz,*

2003). This suggests that solar ratio solids must have been abundant in the early solar system. However, present-day evidence for this material remains elusive, because no solid objects (e.g., comets, asteroids) have been found that have solar ratios of Ar, Kr, Xe, S and N relative to C, as does Jupiter (although solar S/C was found for Comet Halley by Giotto, the detection of noble gas and N_2 abundances in cometary comae is quite challenging).

Explanation for this enhancement has been attributed to one of two different mechanisms. The first idea relies on the delivery of volatiles from the outer regions of the solar nebula that were trapped in amorphous ice and then incorporated into planetesimals. Based on the laboratory work of *Bar-Nun et al.* (1988; see also 2007), this approach initially led to the expectation that Jupiter may not be enhanced in volatile elements like Ar, which condenses at very low temperatures (<30 K), because planetesimals that formed near the snow line likely dominated the delivery of heavy elements to Jupiter's envelope. A second view for Jupiter's enhancement proposed by *Gautier et al.* (2001a,b; also cf. *Hersant et al.,* 2004) is that volatiles were trapped in crystalline ices in the form of clathrate-hydrates (*Lunine and Stevenson,* 1985) at different temperatures in Jupiter's feeding zone and were then incorporated into the planetesimals that went into Jupiter. But even though Ar can be trapped in clathrates at temperatures above its condensation temperature, clathration of Ar still requires very low solar nebula temperatures T ~ 36 K, which is inconsistent with temperatures at Jupiter's location even using cool passive disk models (*Chiang and Goldreich,* 1997; *Sasselov and Lecar,* 2000; *Chiang et al.,* 2001).

Given that Ar does not condense at temperatures higher than ≳30 K, one might expect that the ratio of Ar to H in Jupiter should be the same in Jupiter as the Sun, if indeed Jupiter's formation were characterized by local growth (see discussion at beginning of section 2). However, there likely was considerable migration of solids due to gas drag in the outer solar nebula (see Fig. 3), so it cannot be assumed that material (of solar proportions) remained at the location in the protoplanetary disk where it condensed. This argument is bolstered by high-resolution submillimeter continuum observations that indicate the average dust disk sizes around T Tauri stars are ≈200 AU (*Andrews and Williams,* 2007), with similar results being obtained via millimeter interferometry (*Kitamura et al.,* 2002). This disk of solids eventually shrinks (even if the gas disk spreads outwards) presumably due to coagulation with objects eventually growing large enough to decouple from the gas and migrate inward (section 2.1). Some planetesimal formation takes place at sufficient distances that the circumsolar disk is very cold (see Fig. 3), perhaps cold enough to allow for the trapping of volatiles within the interiors of planetesimals in either amorphous or crystalline (in the form of clathrates) ice, depending on the trapping efficiency and kinetics. As these planetesimals drift in, they likely encounter warmer regions. Further growth to comet sizes ~1 km occurs at some point (*Weidenschilling,* 1997; *Kornet et al.,* 2004), at which time they attain drag times comparable to the lifetime of the circumsolar gas disk (≲10^7 yr).

The success of this solids migration mechanism depends on whether inwardly migrating planetesimals that possibly included amorphous ice within their interiors at the outset (*Mekler and Podolak,* 1994) were altered by the higher temperatures of the inner nebula and became mostly crystalline, losing the volatiles that would have been trapped in them (*Bar-Nun et al.,* 1985, 1987). Amorphous ice undergoes a transformation to crystalline ice at a rate that depends strongly on temperature (*Schmitt et al.,* 1989; *Kouchi et al.,* 1994; *Mekler and Podolak,* 1994). An ice grain may retain its amorphous state for the lifetime of the nebula (≲10^7 yr) provided the temperature is <85 K. That is, low temperatures favor the preservation of amorphous ice, while long time spans and even temporarily elevated temperatures drive the ice toward crystallization.

Thus, while the temperature constraint for incorporation of Ar in grains is quite stringent, the temperature constraint for planetesimals to preserve volatiles they acquired in cold portions of the disk may be much less so. Indeed, a temperature of <85 K is quite consistent with cool passive disk models at the location of Jupiter (*Chiang and Goldreich,* 1997; *Chiang et al.,* 2001) (see Fig. 3). However, if such planetesimals incorporated significant amounts of short-lived radionuclides such as ^{26}Al, radioactive decay would provide heating that would further complicate the ability for planetesimals to preserve their amorphous state (*Prialnik and Podolak,* 1995). Nevertheless, in the outer disk one would naturally expect longer accretion times, which would result in weaker radioactive heating and lower temperatures. It remains to be shown whether one can expect amorphous ice to be preserved within migrating planetesimals over the lifetime of the solar nebula, making it possible to deliver argon and other volatiles to Jupiter.

2.4.2. The snow line and planetesimal delivery. A likely consequence of this picture is that a shrinking (dust) disk would lead to a higher solids fraction in the planetary region than given by a MMSN. As we noted in section 2.2, this is consistent with the requirement that the circumsolar disk be enhanced in solids by at least a factor of a few in order for ~10 $M_\oplus$ cores to be formed during the runaway/oligarchic growth phase. Subsequently, some of the "excess" material, variously estimated between 50 and 100 $M_\oplus$ (*Stern,* 2003; *Goldreich et al.,* 2004), may wind up in the Oort cloud. Since Earth-sized objects may migrate (not via gas drag, but by gravitational interaction with the gaseous disk, see section 4.3.3) in a time shorter than or comparable to the nebula dissipation time, some of this solid material may also be lost to the Sun.

On the other hand, meter-sized objects have relatively short gas drag migration times (~10^4–10^5 yr at Jupiter), so it is possible that some fraction of the solid content of the disk drifted in until it encountered the snow line. Inwardly drifting planetesimals might then sublimate at this (water-ice) evaporation front (e.g., *Stevenson and Lunine,* 1988; *Ciesla and Cuzzi,* 2006). Within the context of the model

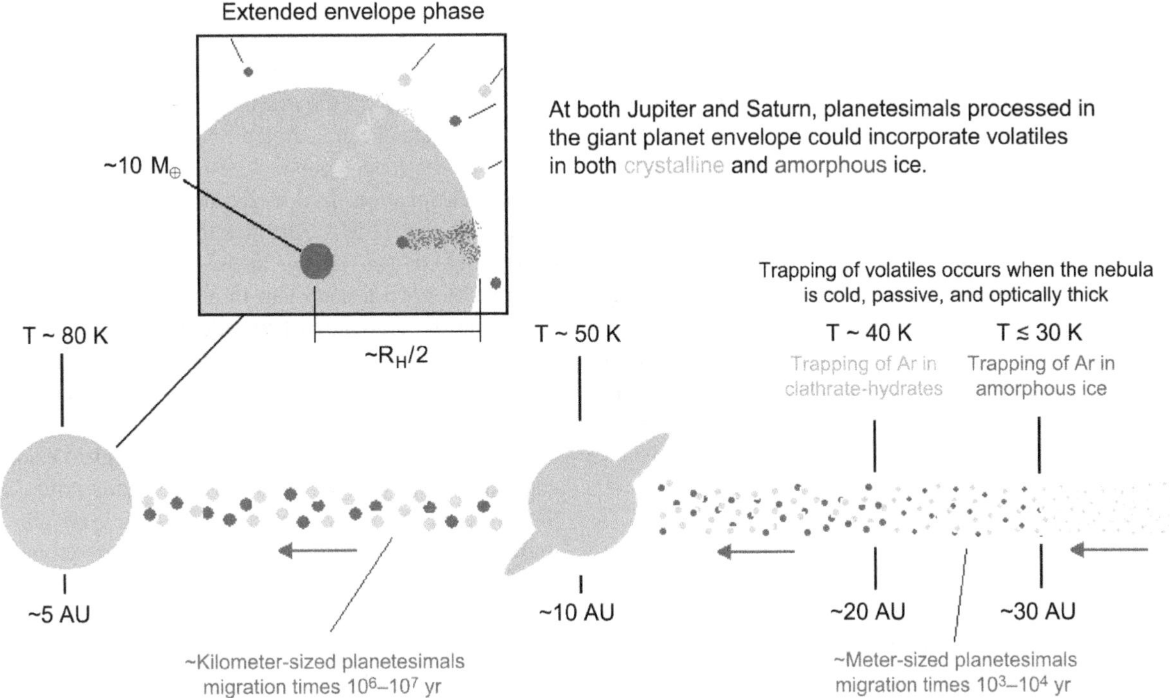

Fig. 3. Sufficiently far from the Sun (~30 AU), amorphous ice forming at low temperatures (≲30 K) can trap volatiles (*Owen et al.,* 1999). In warmer regions of the nebula closer to the Sun, ice is crystalline and volatiles may be trapped in clathrate hydrates (*Lunine and Stevenson,* 1985; *Gautier et al.,* 2001a,b). Planetesimals that form in cold regions and cross into warmer regions suffer a transition from amorphous to crystalline ice at a rate that depends on temperature (*Schmitt et al.,* 1989; *Mekler and Podolak,* 1994). The condition that this transition takes longer than the lifetime of the nebula defines a location outside of which amorphous ice can mix in provided cold planetesimals are delivered by gas drag migration in a similar timescale (which applies to planetesimals <1 km for a MMSN). An ice grain may retain its amorphous state for the age of the solar nebula (~10^7 yr) provided that the temperature is <85 K. Thus, it may be possible to deliver volatiles to the atmospheres of the forming giant planets either in amorphous ice or crystalline ice, depending on trapping efficiency in the two ice-phases, the initial distribution of mass in the primordial nebula, and the specifics of the growth, migration, and thermal evolution of planetesimals. This point of view provides a natural fit for the existence of the outer edge of the classical Kuiper belt — i.e., primordial planetesimals located outside ~30 AU may have migrated to the inner solar system by gas drag, delivering volatiles and enhancing the solid fraction in the planetary region (this outer edge location refers to the time prior to the outward migration of Neptune, which could have pushed the Kuiper belt out) (*Levison and Morbidelli,* 2003). Note that the temperature chosen at the location of Jupiter and Saturn is consistent with passive disk models (e.g., *Chiang et al.,* 2001).

proposed by these workers, the snow line might receive most of the volatile enhancement, even for heavy elements more volatile than water. If the solar nebula were turbulent, then diffusion due to turbulence might spread the effects of this evaporation front over a larger region. If the snow line is somewhere between 3 and 5 AU (*Morfill and Völk,* 1984; *Stevenson and Lunine,* 1988; *Cuzzi and Zahnle,* 2004), one might expect that the volatile heavy element enrichment in Jupiter's envelope is due to the high-volatile content of the nebula at its location. We should emphasize that this scenario is complicated by the likely temporal variation in the location of the snow line. The snow line may have been much closer to the Sun as would be predicted by more recent thermal structure models (e.g., *Sasselov and Lecar,* 2000; *Chiang et al.,* 2001).

Regardless of the snow line's temporal variation, the condensation front scenario would imply that Jupiter's enrichment should be more pronounced than Saturn's. Yet this conclusion appears to be in conflict with observations of Saturn's elemental ratio with respect to solar of C/H (~2–6) (*Courtin et al.,* 1984; *Buriez and de Bergh,* 1981) and N/H (~2–4) (*Marten et al.,* 1980; *de Pater and Massie,* 1985). The enhancement at the water-ice evaporation front would be dependent on how much of this material is delivered and how this material is incorporated into the Jupiter region. If most of the highly volatile content of planetesimals (e.g., Ar) vaporizes prior to either being mixed into the planet or somehow "embedded" in the circumplanetary subdisk, the enhancement in Jupiter's atmosphere may still be explained. The delivery of volatile-laden, pristine planetesimals may be a more efficient process, especially if the temperature at Jupiter's location is more consistent with a passive disk (Fig. 3). Thus, it is presently unclear whether it is possible to deliver and enhance heavy elements more volatile than water simultaneously at Jupiter (and Saturn) by enriching the nebula gas instead of direct planetesimal delivery. The latter would likely predict a higher heavy element enhancement for planets that accreted less gas from the nebula.

3. FORMATION OF THE CIRCUMJOVIAN DISK

As Jupiter's core mass grows, it obtains a substantial atmosphere through the collection of the surrounding solar nebula gas. Early on (still in the nebular stage), this gas falls onto a distended envelope that extends out to a significant fraction of the Hill sphere. Once the protoplanet reaches a mass of ~50–100 $M_\oplus$, the envelope contracts rapidly (transition stage). Eventually, Jupiter becomes massive enough to truncate the gas disk by the opening of a deep gas gap in the solar nebula and/or because all the gas in its feeding zone is depleted. Once all the gas within reach of the planet is depleted, accretion ends (isolation stage). We now address how the circumplanetary gas disk fits into this multistage process.

3.1. Introduction

A very basic characteristic of the circumjovian disk is its radial extent: How far does it extend from the planet? The size of the subnebula upon inflow depends on the specific angular momentum of gas flowing into the giant planet's gravitational domain (*Lissauer,* 1995; *Mosqueira and Estrada,* 2003a). Before and during much of the runaway gas accretion phase of Jupiter's formation (which spans a period prior to and after envelope contraction), the gas that enters the protoplanet's Roche lobe (equivalently, its Hill sphere, see equation (4)) has low specific angular momentum. This is because the mass of the protoplanet for much of this period is not sufficient to open a significant gas gap in the solar nebula for typical circumsolar disk model parameter choices (see section 3.4, and below). As a result, most of the gas being accreted during this period originates within the vicinity of the protoplanet (i.e., at heliocentric distances $\lesssim R_H$ from the planet). Prior to envelope contraction, gas accretes onto the giant planet's distended atmosphere. The contraction of the distended envelope happens early during the runaway gas accretion phase. Current models have Jupiter contracting prior to reaching approximately one-third of its final mass. Moreover, the timescale for envelope collapse (down to a few planetary radii) is relatively quick compared to the runaway gas accretion epoch (see *Lissauer et al.,* 2009). This indicates that the subnebula (circumplanetary disk) begins to form relatively early in the planet formation process when the planet is not sufficiently large to truncate the gas disk. As a result, the gas that continues to flow into the Roche lobe has mostly low specific angular momentum. Since the proto-Jupiter must still accrete $\gtrsim$200 $M_\oplus$ after envelope collapse, this suggests that the gas mass deposited over time in the relatively compact subnebula could be substantial. Furthermore, this still preliminary picture may be consistent with the view that the planet and disk may receive similar amounts of angular momentum (e.g., *Stevenson et al.,* 1986) (section 4).

As the protoplanet grows more massive, it clears the surrounding nebula gas through the actions of gas accretion and tidal interaction, leading to the formation of a gas gap in the solar nebula (section 2.3.2). Sufficiently massive objects may actually truncate the disk, which we take to mean that almost all the available gas in the vicinity of the giant planet has been accreted or shoved aside, so that the density in the (deep) gap is $\lesssim 10^{-2}$ of the unperturbed density at that location. In a moderately viscous disk, a Jupiter-mass planet (1 M_J) truncates the disk in a few hundred orbital periods (or ~10^3 yr at 5.2 AU). Larger masses are required to open a deep gap in a viscous disk in order to overcome the tendency of turbulent diffusion to refill the gap. This picture applies to an accretion scenario for Jupiter that assumes that no other planets influence the evolution of the nebula gas. In a scenario in which the giant planets of the solar system start out much closer to each other and subsequently migrate outward (*Tsiganis et al.,* 2005), the two planets jointly open a gap (if they grow almost simultaneously). In such a scenario, Saturn's local influence would dominate over that of turbulence in a moderately viscous disk (*Morbidelli and Crida,* 2007).

The process of gap-opening in the circumsolar gas disk has direct bearing on the satellite formation environment. As the gap becomes deeper, the continued gas inflow through this gap can significantly alter the properties of the subnebula. In particular, as gas in the protoplanet's vicinity is depleted, the inflow begins to be dominated by gas with specific angular momentum that is much higher than what previously accreted onto the distended envelope and/or compact disk. This is because the gas must now come from farther away (heliocentric distances $\gtrsim R_H$). Because satellite formation is expected to begin at this time (see section 4), the character of the gas inflow during the waning stage of Jupiter's accretion is a key issue in determining the environment in which the Galilean satellites form.

It is worth noting that Europa and the other Galilean satellites occupy a compact region (roughly ~4% of the Hill radius). The mass (and angular momentum) of outermost Callisto is comparable to that of Ganymede. This similarity in mass of these two Galilean satellites is puzzling since one might expect the outermost satellite to have significantly less mass. This is because it is *a priori* difficult to envision how the surface density of the satellite disk could have been large enough to make a massive satellite such as Callisto at its location, but insufficient to form other, smaller satellites outside its orbit (*Mosqueira and Estrada,* 2003a). Callisto's large mass most likely indicates that the circumjovian disk extended significantly beyond Callisto's orbit; thus, the lack of regular satellites outside Callisto requires explanation.

Until recently, numerical models that simulate gas accretion onto a "giant planet" embedded in a circumstellar disk could not resolve scales smaller than ~0.1 R_H (e.g., *Lubow et al.,* 1999; *Bate et al.,* 2003), a region several times larger than is populated by the regular satellites. It is only with the advent of higher resolution two- and three-dimensional simulations (described in section 3.3) that it became possible to resolve structure on the scale of the radial extent of the Galilean satellites. These recent simulations indicate that the circumplanetary disk formed by the gas inflow through the gap — irrespective of any subsequent viscous evolution and spreading — likely extended as much as ~5 times the size of the Galilean system (*D'Angelo et al.,* 2003b; *Ayliffe and Bate,* 2009). Thus, the specific angular momentum of gas inflow through a gap is significantly larger than that of the satellites themselves.

3.2. Analytical Estimates of Disk Sizes

We can obtain an estimate of the characteristic disk size formed by the accretion of low specific angular momentum gas before gap-opening, using angular momentum conservation. Assuming that Jupiter travels on a circular orbit, that the solar nebula gas moves in Keplerian orbits, and that prior to gap-opening Jupiter accretes gas parcels with semimajor axes originating from up to RH of its location, then the specific angular momentum ℓ of the accreted gas is approximately given by (*Lissauer,* 1995)

$$\ell \approx -\Omega \frac{\int_0^{R_H} \frac{3}{2} x^3 dx}{\int_0^{R_H} x dx} + \Omega R_H^2 \approx \frac{1}{4} \Omega R_H^2 \qquad (5)$$

The expression for the specific angular momentum estimate given above has two contributions. The first term is the specific angular momentum flux flowing into the planet due to Keplerian shear computed in the frame rotating at the planet's angular velocity. The second contribution is a correction to translate back to an inertial frame (see *Lissauer,* 1995, and references therein). Equation (5) neglects the gravitational effect of the planet, and assumes that the angular momentum of the inflowing gas is delivered to the Roche lobe of the giant planet. Using conservation of angular momentum, balancing centrifugal and gravitational forces $\ell^2/r_c^3 \approx GM_J/r_c^2$ we obtain the centrifugal radius (*Cassen and Pettibone,* 1976; *Stevenson et al.,* 1986; *Lissauer,* 1995; *Mosqueira and Estrada,* 2003a)

$$r_c \approx R_H/48 \qquad (6)$$

For a fully grown Jupiter, the centrifugal radius is located at $r_c \approx 15\ R_J$ (where $R_J = 71492$ km is a Jupiter radius) just outside the position of Ganymede [for Saturn, r_c lies just outside of Titan; see Fig. 1 of *Mosqueira and Estrada* (2003a)]. While this calculation would seem to employ unrealistic assumptions, recent three-dimensional simulations indicate that it provides a meaningful estimate (*Machida et al.,* 2008). The resulting disk size is consistent with the radial extent of the Galilean satellites.

After gap-opening, accretion may continue through the planetary Lagrange points, as seen in some simulations (e.g., *Artymowicz and Lubow,* 1996; *Lubow et al.,* 1999). Specifically, accretion occurs through the L_1 and L_2 points, which are located at roughly a distance R_H from the planet, and along the line connecting protoplanet and Sun (see Fig. 4). At these Lagrange points, the gravitational fields of both the protoplanet and the Sun combined with the centrifugal force are in balance. As seen in the rotating frame, a massless body placed at this location with zero relative velocity would remain stationary. We can obtain an estimate of the specific angular momentum of the gas as it passes through the Lagrange points by assuming that the inflow takes place at a low velocity in the rotating frame and it is directed nearly toward the planet. This may be done by keeping only the change of frame contribution of equation (5) or $\ell \sim \Omega R_H^2$. Again, using conservation of angular momentum, the estimated characteristic disk size formed by the inflow is significantly larger than before, roughly $\sim R_H/3 \sim 260\ R_J$ (e.g., *Quillen and Trilling,* 1998; *Mosqueira and Estrada,* 2003a).

These estimates indicate that gas flowing through a gap in the circumsolar disk brings with it significantly higher specific angular momentum, which produces a larger characteristic circumplanetary disk size, than does incoming gas when no gap is present. This information combined with the observed mass distribution of the regular satellites of Jupiter (and Saturn) can be used to argue in favor of a two-component circumplanetary disk: (1) a compact, relatively massive disk that forms over a period of time post envelope collapse and prior to disk truncation, and (2) a more extended, less massive outer disk that forms from gas flowing through a gap and at a lower inflow rate (e.g., *Bryden et al.,* 1999; *D'Angelo et al.,* 2003b; *Ayliffe and Bate,* 2009). An idealization of this two-component subnebula is shown in Fig. 8 (see Fig. 8a and caption). However, the details of the formation of the giant planet from the envelope collapse phase, to gap-opening and isolation, remains to be shown using hydrodynamical simulations. Numerical simulations of giant planet formation tend either to treat the growth of the protoplanet in isolation (e.g., *Pollack et al.,* 1996; *Hubickyj et al.,* 2005) (see section 2), or to treat giant planets (~1 M_J) embedded in circumstellar disks in the presence of a well-defined, deep gap (e.g., *Lubow et al.,* 1999; *Kley,* 1999; *Bate et al.,* 2003; *D'Angelo et al.,* 2003b). Mainly because of the computational demands, the latter simulations do not model changing planetary or nebula conditions. Not surprisingly then, the transition between a distended planetary envelope and a subnebula disk has received scant attention.

A consequence of continued gas inflow through the gap is that the subnebula will continue to evolve due to the turbulent viscosity generated by gas accretion onto the circumplanetary disk. However, even weak turbulence can pose a problem for satellitesimal formation (section 2.1). The turbulent circumplanetary disk environment generated by the inflow likely means that satellite formation does not begin until late in the planetary formation sequence when the gas inflow (through the gap) wanes, at which point turbulence in the subnebula may decay. The formation of the satellites at the stage where the planet approaches its final mass is

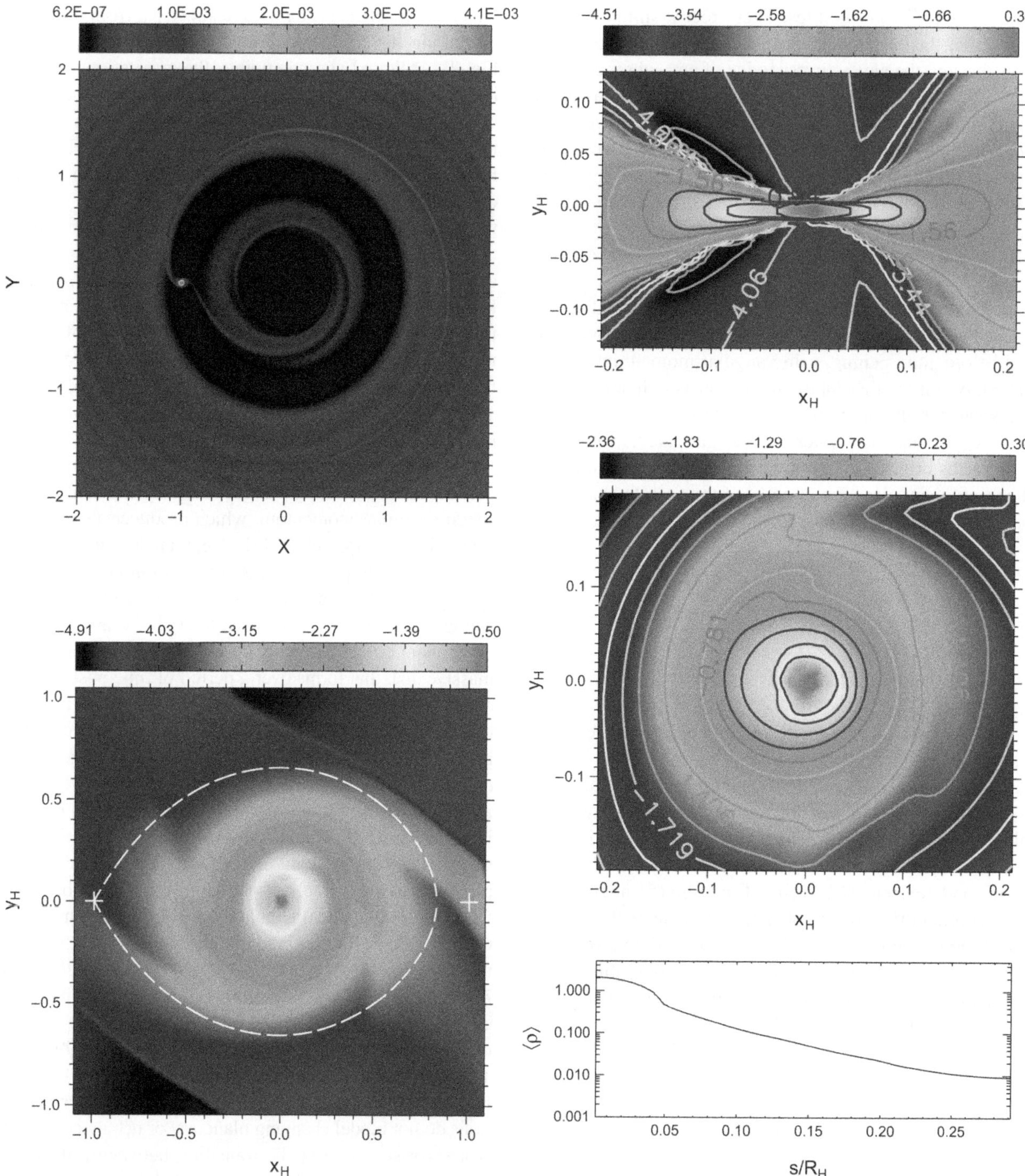

Fig. 4. See Plate 1. Formation of a circumplanetary disk around a Jupiter-mass planet in three dimensions. The top panel on the left shows the mass density distribution, ρ, in the circumstellar disk's midplane. The bottom panel on the left as well as the top and center panels on the right show density distributions, in logarithmic scale, within the planet's Roche lobe (the teardrop-shaped region marked by the dashed line). Iso-density contours are also indicated in two panels on the right. The logarithm (base 10) of the azimuthally averaged density in the disk's midplane is shown in the bottom panel on the right, where s represents the distance from the planet. The units on the axes are either the planet's orbital radius, r_p (X and Y coordinates), or the Hill radius, R_H (x_H, y_H, and z_H coordinates). The units of ρ are such that 10^{-3} corresponds to 10^{-12} g cm^{-3}.

further supported by the fact that even weak, ongoing inflow through the gap can generate a substantial amount of heating due to turbulent viscosity, which would generally result in a circumplanetary disk that is too hot for ice to condense and satellites to form and survive (e.g., *Coradini et al.,* 1989; *Makalkin et al.,* 1999; *Klahr and Kley,* 2006). As a result, a very low gas inflow rate — orders of magnitude lower than the accretion rates through gas gaps in numerical simulations (~10^{-2} $M_\oplus$ yr^{-1}) (e.g., *Lubow et al.,* 1999) — is probably a requirement of any satellite formation model. Even so, some applicable conclusions can be drawn from existing simulations.

3.3. Numerical Results in Two Dimensions/ Three Dimensions in the Presence of a Gap

Two-dimensional hydrodynamics calculations of a Jupiter-mass planet embedded in a circumstellar disk show prograde circulation of material within the planet's Roche lobe that is reminiscent of a circumplanetary disk (*Lubow et al.,* 1999, *Kley,* 1999). These simulations show that gas can flow through the gap formed by the giant planet, depending on the value for the nebula turbulence parameter ($\alpha \gtrsim 10^{-4}$) (*Bryden et al.,* 1999). A prominent feature exhibited in these Roche-lobe flows or streams is a two-arm spiral wave structure (see left panels of Fig. 4, which is a three-dimensional simulation).

As gas flow enters the Roche lobe near the planetary Lagrange points, these streams encircle the planet and impact one another on the opposite side (from which they entered). The resulting collision shocks the material, and deflects the flow toward the planet (e.g., *D'Angelo et al.,* 2002). In two dimensions, the spiral wave structure is weaker (not as tightly wound) for decreasing protoplanet mass, and disappears altogether for ~1 $M_\oplus$ protoplanetary masses. In three-dimensional simulations, these spiral waves are also less marked than in two dimensions as a consequence of the flow no longer being restricted to a plane (see, e.g., *D'Angelo et al.,* 2003a; *Klahr and Kley,* 2006). Despite the differences of the accretion flow, gas accretion rates in two and three dimensions are similar.

Detailed simulations of such systems pose a significant challenge from a numerical point of view since they demand that both the circumstellar disk and the hydrodynamics deep inside the planet's Roche lobe must be resolved. This requirement means that length scales must be resolved over more than two orders of magnitude, from the planet's orbital radius, r_p, down to a few percent of the Hill radius, R_H. *D'Angelo et al.* (2002, 2003b; see also *Ayliffe and Bate,* 2009) carry out a quantitative analysis of the properties of circumplanetary disks around jovian- and subjovian-mass planets. By treating the circumstellar disk as a locally isothermal and viscous fluid, and using a grid refinement technique known as "nested grids" that allow them to resolve length scales around the planet on the order of 0.01 R_H (~7 R_J for 1 M_J), these authors are able to show that the dynamical properties of the material orbiting within a few tenths of R_H from the planet are indeed consistent with a disk in Keplerian rotation.

Figure 4 shows the mass density distributions from a three-dimensional, local isothermal model (see *D'Angelo et al.,* 2003b for details). The temperature at 5.2 AU is assumed to be $T \simeq 110$ K (if the mean molecular weight of the gas is about 2.2) and the kinematic viscosity, ν, in this case is assumed to be constant in space and time (see discussion at the end of section 3.2). The aspect ratio of the circumstellar disk, which is given by the ratio of the disk's semithickness (generally denoted by the pressure or nebula scale height H) at the location of the planet to the planet's semimajor axis, is $H/r_p \sim 0.05$. For the disk parameters chosen, ν is comparable to a turbulent viscosity with an α-parameter of 4×10^{-3} at r_p. The top panel on the left illustrates the circumstellar disk and the density gap produced by the planet that exerts gravitational torques on the disk material. The other panels show the mass density, in logarithmic scale, over lengths ~R_H (left) and ~0.1 R_H (right). The bottom panel on the right displays the density in the disk's midplane, azimuthally averaged around the planet, as a function of the distance from the planet, s. The models shown in Fig. 4 can be rescaled by the unperturbed value of the mass density ρ (i.e., that of the circumstellar disk when the planet is not present) at r_p, which is a consequence of the locally isothermal approach. Therefore, the calculated density structure in the disk is independent of the unperturbed value of ρ. For the value of ρ chosen in Fig. 4, the disk mass within 0.2 R_H is 10^{-4} M_J.

D'Angelo et al. (2003a) present thermodynamical models of circumjovian disks in two dimensions. In these calculations, the energy budget of the disk accounts for advection and compressional work, viscous dissipation and local radiative dissipation. Characteristic temperatures and densities in these models depend mainly upon viscosity, opacity tables, and initial mass of the circumstellar disk. In this case, the results are not scalable by the unperturbed mass density, because the opacity depends on the value of ρ chosen.

Figure 5 displays surface density (Fig. 5a) and temperature profiles (Fig. 5b) obtained from calculations with different prescriptions and magnitude of the kinematic viscosity. These models rely on the opacity tables of *Bell and Lin* (1994) and assume that the initial unperturbed surface density, Σ, at 5.2 AU is roughly 100 g cm^{-2}. The circumstellar disk contains about 4.8 M_J within 13 AU of the star. The model with highest density and temperature (black curves) has a constant (in space and time) kinematic viscosity $\nu = 10^{15}$ cm^2 s^{-1}. The other models assume an α-viscosity $\nu = \alpha c_s H$, where the sound speed c_s and the pressure scale height H are a function of space and time while α is a constant. In Fig. 5, for increasing density and temperature, models have $\alpha = 10^{-4}$ (short-dashed curves), 10^{-3} (long-dashed curves), and 10^{-2} (dot-dashed curves), respectively. The value of α applies to both the circumsolar and circumplanetary disks. In the cases shown in the figure, the amount of mass within about 0.2 R_H of the planet is in the range between ~10^{-5} and 10^{-4} M_J.

The specific angular momentum from three-dimensional as well as two-dimensional models for a 1 M_J mass planet is plotted in Fig. 6. Within ~0.15 R_H of the planet, the rotation of the disk follows the rotation curve of a Keplerian disk (dashed curve). The curves correspond to the models in Fig. 5, while the multiple dot-dashed curve represents the three-dimensional case. It is interesting to note in the latter

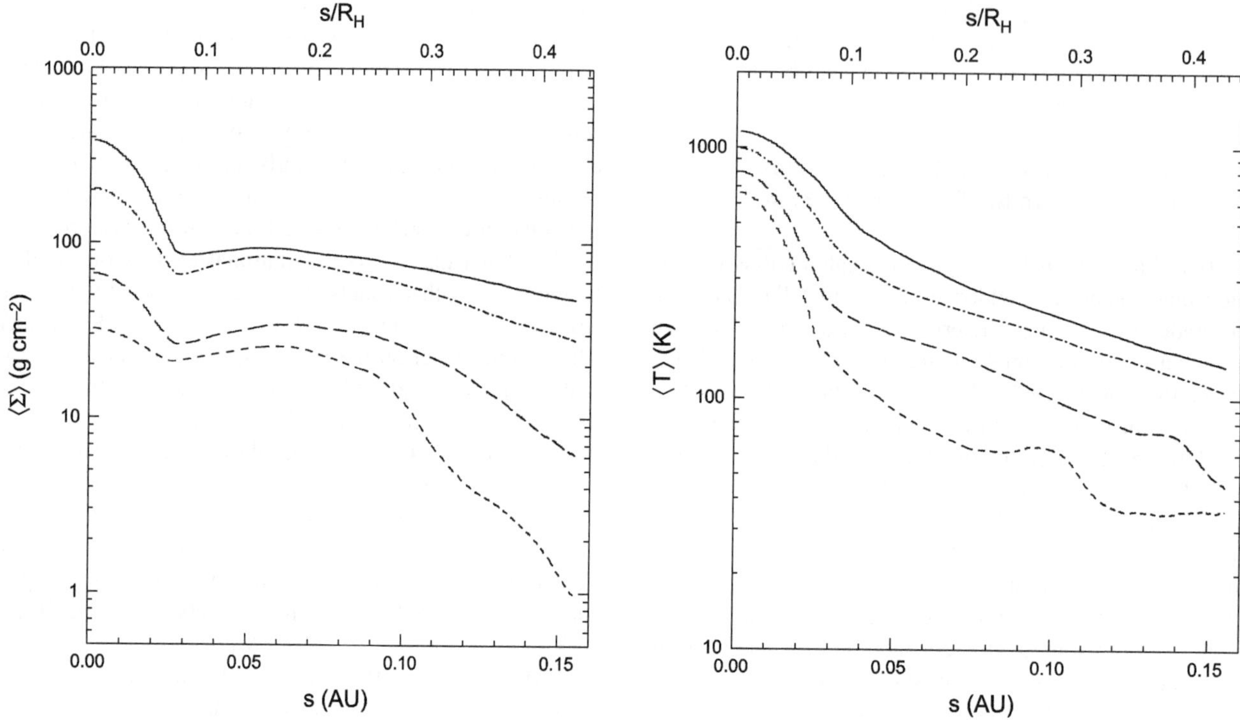

Fig. 5. **(a)** Surface density and **(b)** temperature of two-dimensional circumjovian disk models with viscous heating and radiative cooling (see text for further details). The quantities represent azimuthal averages around the planet. The models differ in the adopted viscosity prescription. The calculation that produces the highest density and temperature (solid curve) assumes a constant $\nu = 10^{15}$ cm^2 s^{-1}. The other calculations assume an α-type viscosity, $\nu = \alpha c_s H$ (the same value of α applies to nebula and subnebula), so ν is therefore is space- and time-dependent. For increasing density and temperature, models have $\alpha = 10^{-4}$ (short dash), 10^{-3} (long dash), and 10^{-2} (dot-dash), respectively.

case that, even though the subnebula temperature is assumed to be constant, the specific angular momentum distribution is consistent with that of the two-dimensional models, which are determined by means of calculations that allow for heating and cooling processes. Thus, it appears that for large planetary masses in which a deep gap is present in the circumstellar gas disk, two- and three-dimensional simulations give comparable results for the specific angular momentum. This is because the gas flowing across such a deep gap into the Roche lobe is coming from much farther away than R_H (see upper left panel of Fig. 4). Although the flow pattern in two and three dimensions differ, by the time the gas reaches the planet the specific angular momentum delivered with the inflow in both cases is qualitatively similar.

Finally, for comparison, the specific angular momentum of the Galilean satellites is also indicated (solid horizontal line). These simulations indicate that gaps correspond to higher specific angular momentum, and a larger characteristic disk size formed by the inflow, than that of the regular satellites themselves.

3.4. Connecting the Planet, Its Disk, and the Satellites

As was pointed out in section 3.2, planetary formation models tend to focus on either the growth of the planet in isolation (section 2) or a protoplanet of fixed mass embedded in a circumstellar disk in the presence of a well-formed gap (section 3.3). As a result, our understanding of the formation of the circumjovian gas disk remains incomplete, as no simulations have yet been done that model disk formation starting after the giant planet's envelope contraction to the time at which inflow from the circumsolar disk ceases.

During the period over which circumjovian disk formation occurs, the characteristic disk size may be roughly estimated by a balance between gravitational and centrifugal forces (section 3.2). Because the protoplanet likely accretes most of its gas mass after envelope collapse, a significant fraction of this gas may end up in the circumjovian gas disk leading to an initially massive subnebula. A legitimate concern that arises from this scenario is why Jupiter is not rotating near breakup velocity. The origin of Jupiter's current spin angular momentum remains poorly understood. A massive circumplanetary disk may be a requirement for despinning the planet (see, e.g., *Korycansky et al.,* 1991; *Takata and Stevenson,* 1996), although Jupiter's spin may require consideration of the role of magnetic fields. The question of Jupiter's spin thus represents a key piece of the jovian system puzzle that is in need of further investigation.

A compact massive disk will have important implications for satellite formation models. The characterization of this disk component will require the marriage of isolated growth models and embedded planet simulations, a milestone that is just beginning to be explored. Indeed, recent simulations of gas accretion onto a low-mass protoplanet (i.e., a deep

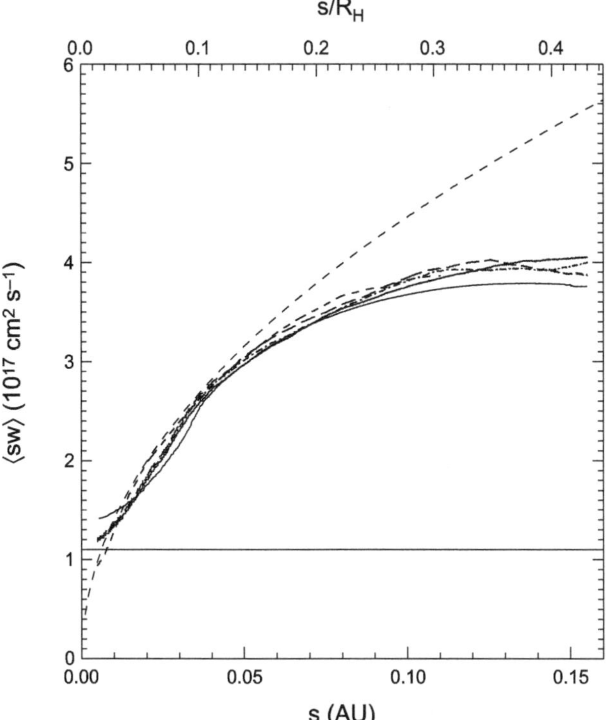

Fig. 6. Specific angular momentum of circumjovian disk models, in two and three dimensions, azimuthally averaged around the planet. The quantity w is the azimuthal velocity around the planet. Results from models in Figs. 4 and 5 are displayed. The multiple dot-dash curve represents the three-dimensional isothermal case. The less bold dashed line represents the Keplerian angular momentum. The horizontal line is the specific angular momentum of the Galilean satellites.

gap is not present) embedded in a circumstellar disk indicate that gas may not be bound to the planet outside of ~0.25 R_H (*Lissauer et al.,* 2009). Measurements of the specific angular momentum contained within the bound region for different pre-gap-opening protoplanet masses are consistent with equation (5) yielding $\ell \lesssim \Omega R_H^2/4$ (it is important to note that a parcel of gas that crosses within 0.25 R_H of the planet may originate from a radial distance as far as R_H).

Recently, *Machida et al.* (2008) used three-dimensional hydrodynamical simulations to model the angular momentum accretion into the giant planet Hill sphere when a partially depleted gap is present. These authors find for a range of planetary masses that a significant fraction of the total angular momentum may contribute to the formation of a compact circumplanetary disk. These results are inapplicable for masses comparable to 1 M_J. This is because the local simulation used by these workers is inappropriate to treat gap formation because of the radial boundary (*Miyoshi et al.,* 1999), so that the depth and width of the gap depends on the size of the simulation box used. However, as we pointed out earlier, an important conclusion that may be drawn from the results of Machida et al. is that a circumplanetary disk formed when a partial gap is present is compact due to the lower specific angular momentum of the inflow.

For large planetary masses in the presence of a deep gap, the circumplanetary disk size is qualitatively insensitive to the inflow rate, or whether the flow is treated in two dimensions or three. However, three-dimensional simulations are required early on in the accretion of the protoplanet when a deep gap is not present. We can understand this by noting that material in the nebula midplane whose semimajor axis is close to the planet ($\lesssim R_H$) actually does not accrete onto the planet, but instead undergoes horseshoe orbits (e.g., *Lubow et al.,* 1999; *Tanigawa and Watanabe,* 2002). Three-dimensional flows are then necessary to allow for low-angular-momentum gas from radial distances $\lesssim R_H$ and away from the midplane to be accreted directly onto the planet or into a compact disk (*Bate et al.,* 2003; *D'Angelo et al.,* 2003a). Moreover, *D'Angelo et al.* (2003b) point out that because of the flow circulation away from the disk midplane, less angular momentum is carried inside the Roche lobe by the midplane flow, so that in general three-dimensional simulations are required in order to properly account for the angular momentum.

What does this all mean for the formation of the Galilean satellites, which lie in a quite compact region close to the planet? When a well-formed gap is present, the characteristic circumjovian disk size formed by the inflow is >0.1 R_H, or >70 R_J (by comparison Ganymede is located at ~15 R_J). From Fig. 6, it can be seen that the specific angular momentum of the inflow is about a factor of 3–4 larger than that of the Galilean satellites (~1.1×10^{17} cm^2 s^{-1}), which means that the gas will achieve centrifugal balance at a radial location of ~200 R_J. While this disk is compact compared to RH, it is extended in terms of the locations of the Galilean satellites, and possibly linked to the location of the irregulars (the innermost ones of which lie near ~150 R_J). Thus, these results indicate that the inflow through a gap in the circumsolar disk results in an extended circumplanetary disk of characteristic size ~70–200 R_J, implying a mismatch between the size of the circumplanetary disk formed by gas inflow through the giant planet's gap and the compact region where the regular satellites are found.

This mismatch has important consequences for satellite formation. Taken at face value, it may mean one of two things: (1) the solid material coupled to the gas coming through the gap did not provide the bulk of the material that formed the regular satellites — planetesimals that were not coupled to the gas provided this source instead (see section 4.2); or (2) the regular satellites migrated distances considerably larger than their current distances from Jupiter. The first option would effectively preclude gas inflow through the gap as the source of solids for regular satellite formation. The second option would make questionable any model that explains the Galilean satellite compositional gradient with subnebula "snow line" arguments; i.e., all satellites would presumably start out far from the planet and outside the snow line, and receive their full complement of ices.

These options assume that after gap-opening the disk becomes cool enough, and thus the inflow is weak enough, that icy objects can form. Prior to gap-opening, the inflow is likely to be fast, so that the circumplanetary disk would be too hot (and turbulent) for the concurrent formation of ice-rich satellites like Callisto and Ganymede. Therefore, it is not possible to form a compact disk concurrently with the accretion of ice-rich, close-in satellites, indicating that any satellite formation in a rotationally supported circumplanetary disk likely doesn't begin until the gas inflow wanes and turbulence decays, or the subnebula gas disk has dissipated.

4. CONDITIONS FOR SATELLITE FORMATION

In the core nucleated accretion model, Jupiter's formation is thought to occur in three stages (section 2): nebular, transition, and isolation. The circumjovian gas disk forms during the transition stage of Jupiter's accretion and should be viewed as a drawn-out process that begins after the contraction of the envelope and ends when Jupiter is isolated from the solar nebula: (1) During the nebular stage, most of the solids reside in large planetesimals, some of which may dissolve in the growing envelope during the latter part of this stage. Most of the high-Z mass delivery takes place before the "cross-over" time when the mass of the gaseous envelope grows larger than the core. This stage of growth may be followed by a dilution as the gas accretion rate accelerates. (2) The envelope eventually becomes sufficiently massive to contract and accrete gas from the circumsolar disk hydrodynamically (transition stage). Contraction down to a few planetary radii happens quickly relative to the runaway gas accretion timescale. (3) After envelope collapse, the protoplanet's mass is still too small for it to clear a significant gap in the surrounding solar nebula, so most of the gas flowing into the Roche lobe accretes onto both the planet and a rotationally-supported compact disk. As the protoplanet becomes more massive, it depletes the gas in its vicinity of the solar nebula, truncating the disk. (4) The nebula gas continues to flow through the Lagrange points (see Fig. 4), leading to the formation of an extended disk. Meanwhile, planetesimals in Jupiter's feeding zone undergo an intense period of collisional grinding, leading to a fragmented population (section 4.2). Continued solids enhancement of the entire circumjovian disk occurs due to ablation of disk-crossing planetesimal fragments (see Fig. 7). The accretion of the satellites is expected to occur toward the tail end of Jupiter's formation, but how satellite formation proceeds largely depends on the level and persistence of turbulence both in the circumsolar and circumplanetary nebulae.

4.1. Turbulence and Its Implications for Satellite Accretion

For disks to accrete onto the central object, angular momentum must be transported outward. For the case of disks around young stellar objects (YSOs), magnetohydrodynamic (MHD) and self-gravitating mechanisms have been investigated in some detail (e.g., *Gammie and Johnson,* 2005). The self-gravitating mechanism eventually turns itself off as the gas disk dissipates, at which point MHD may gain relevance. Differentially rotating disks are subject to a local instability referred to as a magnetorotational instability, or MRI (*Balbus and Hawley,* 1991). Significant portions of the disk (specifically the planet formation regions) may be insufficiently ionized for MRI to be effective, creating a "dead zone" (*Gammie,* 1996; *Turner et al.,* 2007), or region of inactivity.

In the absence of an MHD mechanism, one would have to resort to a purely hydrodynamic mechanism. However, it is now known that hydrodynamic Keplerian disks are stable to linear perturbations (e.g., *Ryu and Goodman,* 1992; *Balbus et al.,* 1996). Possible sources of turbulence, such as convection (*Lin and Papaloizou,* 1980) and baroclinic effects (*Li et al.,* 2000; *Klahr and Bodenheimer,* 2003), may provide inadequate, decaying transport in three-dimensional disks (*Barranco and Marcus,* 2005; *Shen et al.,* 2006), subside as the disk becomes optically thin, and may fail to apply to isothermal portions of the disk. A number of analytical studies have suggested transient growth mechanisms for purely hydrodynamic turbulence that would lead to the excitation of nonlinear behavior (*Chagelishvili et al.,* 2003; *Umurhan and Regev,* 2004; *Afshordi et al.,* 2005).

However, numerical simulations (*Hawley et al.,* 1999; *Shen et al.,* 2006) and laboratory experiments (*Ji et al.,* 2006) cast doubt on the ability of purely hydrodynamic turbulence to transport angular momentum efficiently in Keplerian disks, even for high Reynolds number (*Lesur and Longaretti,* 2005). Although the evidence that Keplerian disks are laminar is not conclusive because the Reynolds numbers in disks are much larger than those accessible to computers or experiments, it is fair to say that what makes disks around YSOs accrete remains an open problem.

While there is a consensus that both the nebula and the subnebula undergo turbulent early phases, we presently lack a mechanism that can sustain turbulence in a dense, mostly isothermal subnebula (*Mosqueira and Estrada,* 2003a,b). In order to sidestep this problem, one is then forced to postulate a low-density gas disk (*Estrada and Mosqueira,* 2006), and invoke MRI turbulence to sustain turbulence in such a disk. But here again the likely presence of dust complicates the situation, even in the low-density case. Hence, a sufficiently general mechanism for sustaining turbulence in poorly ionized disks has yet to be found, suggesting that alternative mechanisms of disk removal need to be explored. In particular, the role that the planets and satellites themselves play in driving disk evolution has only begun to be explored (*Goodman and Rafikov,* 2001; *Mosqueira and Estrada,* 2003b; *Sari and Goldreich,* 2004).

4.2. Methods of Solids Delivery

There are several ways in which solids can be delivered to the circumjovian disk. Although all these mechanisms likely played a role in satellite formation, it should be emphasized that at the time of giant planet formation, most of the available mass of solids are in planetesimals in sizes ≥1 km (*Wetherill and Stewart,* 1993; *Weidenschilling,* 1997; *Ken-*

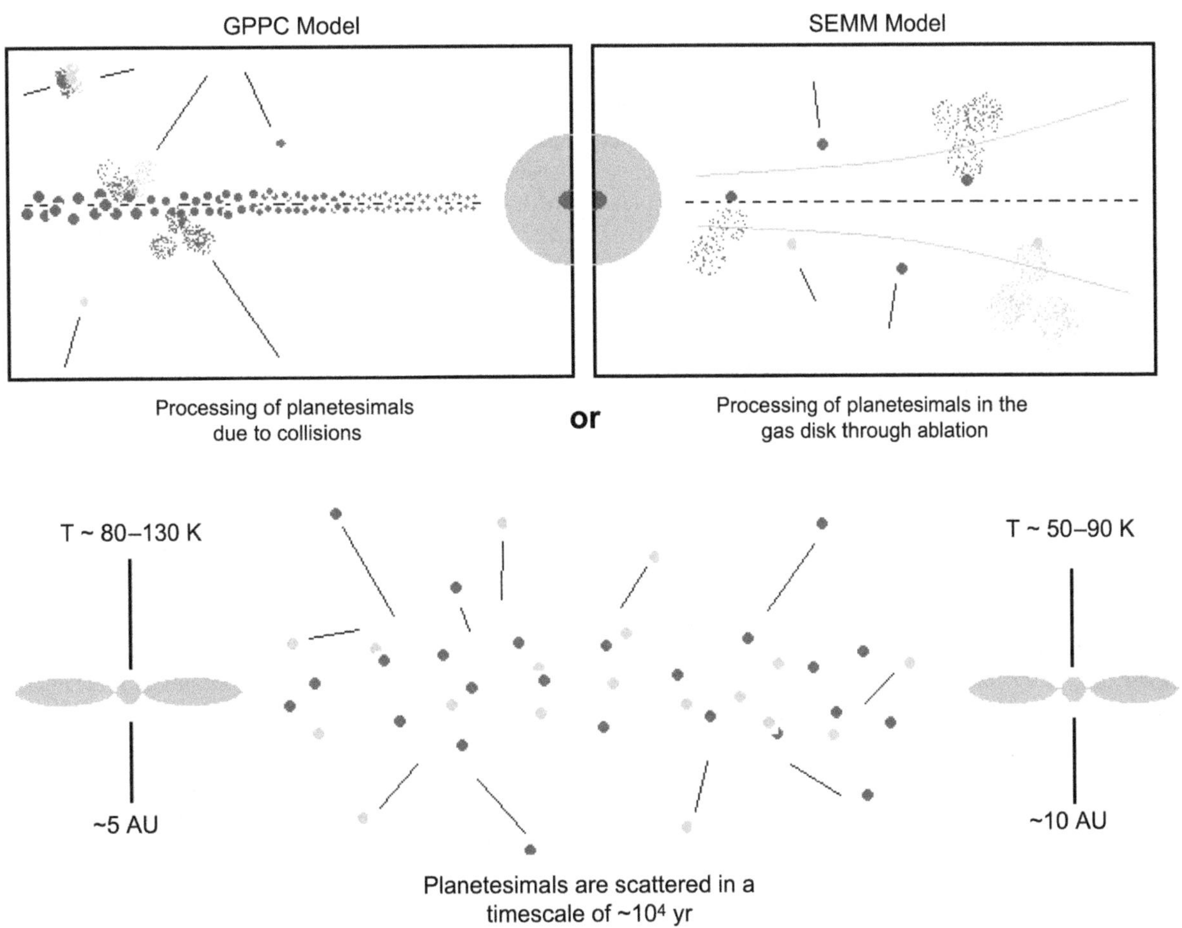

Fig. 7. Vignette of the dominant planetesimal delivery mechanisms (section 4.2) for the two satellite formation environments discussed. In the left window is the gas-poor planetesimal capture (GPPC) model for the circumjovian disk in which processing of planetesimals occurs through planetesimal-planetesimal, and planetesimal-satellitesimal collisions. In the right window is the solids-enhanced minimum mass (SEMM) model for the circumjovian disk in which processing of heliocentric planetesimals occurs through their ablation as they pass through the jovian subnebula. The different scenarios have implications for the compositional evolution of the Galilean satellites. In either case, it is expected that the planetesimal population in the feeding zone of Jupiter (or between Jupiter and Saturn, if Saturn is present) will mostly be scattered away in ~10^4 yr. Some of this material ends up in the circumplanetary disk through collisional capture and/or through direct passage across the disk plane prior to being scattered.

yon and Luu, 1999; *Charnoz and Morbidelli,* 2003). In the Jupiter-Saturn region the collisional timescale for kilometer-sized objects is similar to the ejection timescale (≲10^5 yr) (*Goldreich et al.,* 2004), so that a significant fraction of the mass of solids are fragmented into objects smaller than ~1 km (*Stern and Weissman,* 2001; *Charnoz and Morbidelli,* 2003). Sufficiently small planetesimals (~1 m) are protected from further collisional erosion by gas-drag and by collisional eccentricity and inclination damping. Given that fragmentation likely plays a significant role in the continued evolution of the heliocentric planetesimal population following the formation of Jupiter (*Stern and Weissman,* 2001; *Charnoz and Morbidelli,* 2003), the 1 m–1 km size range of planetesimals likely plays a key role in satellite formation.

1. *Break-up and dissolution of planetesimals in the extended envelope.* Since the giant planet envelope probably filled a fair fraction of its Roche lobe during a significant fraction of its gas accretion phase, its cross section would have been greatly enlarged (*Bodenheimer and Pollack,* 1986; *Pollack et al.,* 1996), so that early arriving planetesimals would break up and/or dissolve in the envelope, enhancing its metallicity. In the earliest stages of growth when the envelope mass is low, most planetesimals may reach the core intact. As the gaseous envelope becomes more massive, planetesimals begin to deposit significant amounts of their mass in the distended envelope (e.g., *Podolak et al.,* 1988). Some dust and debris deposited during the extended envelope and relatively rapid envelope collapse phases would have been left behind in any the subnebula.

2. *Ablation and gas drag capture of planetesimals through the circumjovian gas disk.* Depending on the gas surface density of the subnebula, disk crossers can either ablate, melt, vaporize, or be captured as they pass through the disk. Planetesimal fragments in the size range meter-

to-kilometer may ablate and be delivered to the subdisk. The total mass budget depends on the surface density of solids in the solar nebula, as well as the deposition efficiency, but it may be possible to deliver more than the mass of the Galilean satellites, a fraction of which could have been lost later due to the inefficiencies of satellite accretion.

3. *Collisional capture of planetesimals.* Once the planetesimal population has been perturbed into planet-crossing orbits (*Gladman and Duncan,* 1990), both gravitational and inelastic collisions between planetesimals within Jupiter's Hill sphere occur. If inelastic collisions occur between planetesimals of similar size, the loss of kinetic energy through their collision may lead to capture, and eventually to the formation of a circumplanetary accretion disk. Gravitational collisions between planetesimals also leads to mass capture.

4. *Dust coupled to the gas inflow through the gap.* Essentially all numerical models of giant planet formation indicate that there is flow of gas through the gap, with the strength of the flow dependent on the assumed value of the solar nebula turbulence (e.g., *Artymowicz and Lubow,* 1996; *Lubow et al.,* 1999; *Bryden et al.,* 2000; *Bate et al.,* 2003; *D'Angelo et al.,* 2003b). None of these studies incorporated dust in their simulations; however, there are numerous arguments as to why dust inflow cannot be the dominant source of solid material (for more discussion, see *Mosqueira and Estrada,* 2003a; *Estrada and Mosqueira,* 2006). First, little mass remains in dust at the time of planet formation (*Mizuno et al.,* 1978; *Weidenschilling,* 1997; *Charnoz and Morbidelli,* 2003). Second, as was discussed in section 3.3, the specific angular momentum of the inflow through a gap forms an extended disk, which would lead to the formation of satellites far from the planet where they are not observed. Third, the edges of gaps opened by giant planets act as effective filters, restricting the size and amount of dust that can be delivered this way (*Paardekooper and Mellema,* 2006; *Rice et al.,* 2006). Entrained particle sizes would be orders of magnitude smaller in radius than the local decoupling size (~1 m at Jupiter). Fourth, the dust content of the inflowing gas may be substantially decreased with respect to solar nebula gas. Embedded planet simulations in three dimensions show that the gas inflow comes from lower-density regions (e.g., see Fig. 4) above and below the midplane (e.g., *Bate et al.,* 2003; *D'Angelo et al.,* 2003b). The typical maximum flow velocity occurs at a pressure scale height. This further restricts particle sizes entrained in the gas inflow.

4.3. Gas-rich Environment

In section 3, we discussed the formation of a massive circumplanetary disk. Traditionally, the approach has been to calculate (akin to the MMSN) a minimum mass model for the circumjovian disk. In such a model, the total disk mass (gas + solids) is set by spreading the reconstituted mass (accounting for lost volatiles) of the Galilean satellites over the disk and adding enough gas to make the subnebula solar in composition [typically a factor of ~100 (e.g., *Pollack et al.,* 1994)]. The total mass of the circumplanetary disk that results from the approach described above is ~10^{-2} M_J (note that the disk-to-primary mass ratio for Jupiter is similar to Sun). Interestingly, the total angular momentum of this gaseous disk is comparable to the spin angular momentum of Jupiter (*Stevenson et al.,* 1986; see Table 3 of *Mosqueira and Estrada,* 2003a). Equipartition of angular momentum between planet and disk would result in a massive subnebula.

4.3.1. Decaying turbulence satellite formation scenario. As long as gas inflow through the gap continues, the circumplanetary disk should remain turbulent (driven by the inflow itself) and continue to viscously evolve. The gas inflow from the circumsolar disk wanes in the gap-opening timescale of <10^5 yr in a weakly turbulent solar nebula (*Bryden et al.,* 1999, 2000; *Mosqueira and Estrada,* 2006; *Morbidelli and Crida,* 2007). As the subnebula evolves, the gas surface density may decrease. Once the gas inflow ceases, a different driving mechanism is needed to facilitate further disk evolution.

This circumplanetary disk environment in which the regular satellites form has been dubbed the solids enhanced minimum mass (SEMM) disk (*Mosqueira and Estrada,* 2003a,b), because satellite formation occurs once sufficient gas has been removed from an initially massive subnebula and turbulence in the circumplanetary disk subsides. There are a number of processes that may raise the solids-to-gas ratio and lead to a SEMM disk. In particular, preferential removal of gas (e.g., *Takeuchi and Lin,* 2002) during the inflow-driven subnebula evolution phase may lead to enhancement of solids in the circumplanetary disk. Ablation of heliocentric planetesimals crossing the disk may result in further enhancement of solids. Such a disk may then be enhanced in solids by a factor of ~10 above solar consistent with the solids enhancement observed in the jovian atmosphere. This factor is also consistent with theoretical constraints based on the condition for gap-opening and satellite formation and migration in such a disk (section 4.3.3). We stress that the properties of a SEMM model are distinct from those of a minimum mass model in terms of disk cooling and satellite formation and migration timescales.

4.3.2. Satellite growth in the circumjovian disk. The growth of satellitesimals and embryos in the circumplanetary disk is controlled first by sweepup of dust and rubble (e.g., *Cuzzi et al.,* 1993; *Weidenschilling,* 1997). As the inflow from the circumsolar disk wanes and turbulence decays, the inner more massive region becomes weakly turbulent (while the outer extended disk becomes isothermal and quiescent). Once this occurs, dust coagulation and settling are assumed to proceed rapidly, given that dynamical times are ≳10^3 times faster in the inner disk of Jupiter than in the local solar nebula.

Once a significant fraction of the solids in the disk have aggregated into objects large enough to decouple from the gas (radii $R \gtrsim 10$–100 m for $\Sigma \gtrsim 10^4$ g) and settle to the disk midplane, they drift inward at different rates due to gas drag, leading to further "drift-augmented" accretion. In the weak turbulence regime, the ratio of the sweepup time (which assumes most of mass of the disk is initially in small particles entrained in the gas) to gas drag is

$$\frac{\tau_{sweep}}{\tau_{gas}} \approx \frac{4\rho_s R}{3\bar{\rho}_p \Delta v} \Big/ \frac{4\rho_s R v_K}{3C_D \rho (\Delta v)^2} \sim C_D \eta \frac{\rho}{\bar{\rho}_p} < 1 \qquad (7)$$

in the inner (and outer) disk. Here, $\bar{\rho}_p$ is the average solids density in the midplane, ρ_s is the satellitesimal density, C_D the gas drag coefficient, and $\eta = \Delta v/v_K$ (e.g., $\eta \sim 10^{-2}$ at Ganymede) measures the difference between v_K and the pressure supported gas velocity. Equation (7) implies that it is possible to form satellites/embryos of any size <1000 km at any radial location on a faster timescale then their inward migration due to gas drag.

Although perfect sticking is assumed in the sweepup model described above, it is likely that some fragmentation and erosion occurred. In particular, the relative velocities between decoupling particles and larger, relatively "immobile" satellitesimals are considerably greater in the circumplanetary disk compared to the solar nebula due to much faster dynamical times, which likely resulted in erosive impacts (turbulence exacerbates this problem; section 2.1). In addition, other factors may have contributed to less favorable growth (see *Mosqueira and Estrada,* 2003a, for more discussion). However, in the SEMM model $\rho/\bar{\rho}_p \lesssim 10$ due to particle settling, so that equation (7) may remain satisfied even for inefficient growth. But, it should be noted that a detailed simulation of the accretion of satellites from satellitesimals in circumplanetary nebulae remains to be done.

Eventually, as the reservoir of dust and rubble is depleted, sweepup is less efficient, and growth of the embryo begins to be controlled by gas drag drift-augmented accretion of satellitesimals and smaller embryos. Satellite embryos pose an effective barrier for inwardly drifting satellitesimals due to high impact probabilities (*Kary et al.,* 1993; *Mosqueira and Estrada,* 2003a). Thus embryos choke off the supply of material to the inner satellites. The protosatellite formation timescale is determined by the inward drift of the characteristic size of satellitesimals (or embryos) it accretes. In this picture Ganymede (as well as Io and Europa) forms in ~10^3–10^4 yr, whereas Callisto takes significantly longer (~10^6 yr) because it derives solids from the extended low-density outer disk. Nevertheless, the processes that lead to satellite formation are essentially the same in the outer and inner disks. It is important to keep in mind that in the SEMM model, the delivery of material to the circumplanetary disk (either gas or solids) takes place in a ~10^4-yr timescale, which is comparable to Ganymede's formation time but shorter than that of Callisto. We emphasize that while satellite embryos form quickly, the SEMM model has full-sized satellites forming on a longer timescale. This is because, unlike traditional minimum mass models, the SEMM model is not a local growth model; i.e., a full-sized satellite formation timescale is controlled by the timescale over which the feeding zone of the embryo is replenished by other embryos or satellitesimals. Thus, the mass of the satellites is spread out over the entire disk, and full-sized satellites must accrete material from well outside their feeding zones. In fact, most of the present-day satellite disk is empty, presumably due to gas drag clearing of satellitesimals.

4.3.3. Satellite survival in a gas-rich disk. A gas-rich disk promotes accretion of satellites; however, such a disk can also lead to orbital decay or even loss of satellites on timescales much faster than it would take the circumplanetary gas disk to dissipate. On the one end, gas drag migration (section 2.1) dominates for smaller objects (e.g., *Weidenschilling,* 1988), and is the primary mechanism for drift-augmented accretion. On the other end of the size scale, the migration rate of larger objects is determined by the gravitational interaction with a gaseous disk at Lindblad resonances, or gas tidal torque (see *Goldreich and Tremaine,* 1979). Intermediate-sized protosatellites (~1000 km) must contend both with gas drag migration and gas tidal torques that may lead to catastrophically fast (generally inward) migration rates.

As a satellite grows in size, its migration speed increases (the torque is proportional to the mass squared). However, sufficiently large satellites (mass ratio of satellite to planet of $\mu \sim 10^{-4}$) may stall and begin to open a gap (*Ward,* 1997; *Rafikov,* 2002b; *Mosqueira and Estrada,* 2003b). As a consequence of the satellite's tidal interaction with the disk (and concomitant angular momentum transfer), the satellite can actually drive the evolution of the disk (e.g., *Sari and Goldreich,* 2004) by producing a local effective viscosity (*Goodman and Rafikov,* 2001). Admittedly, the physics of disk-satellite interactions is complex. We simply note that other satellite formation models do not rely on gap-opening for satellite survival because the gas disk dissipates on a timescale comparable to the satellite formation timescale (*Canup and Ward,* 2002; *Alibert et al.,* 2005a; *Estrada and Mosqueira,* 2006).

Mosqueira and Estrada (2003b) explored static models of the jovian subnebula and determined the conditions under which the largest satellites may stall, open gaps, and survive under gas-rich conditions. Figure 8b shows results from these calculations. First, these results indicate that a "minimum mass disk" (solid curve) is likely too massive to allow for the survival of any of the inner disk Galilean satellites, and that a significant decrease in the gas surface density is required for satellite migration to stall (dotted and bold long-dashed curves, SEMM disk). Thus, the gap-opening condition itself argues in favor of a subnebula solids enhancement of a factor of ~10 with respect to solar composition. Second, the critical mass for a satellite to stall increases with distance from Jupiter. This allows a satellite to drift in until it finds an equilibrium position, so long as it is sufficiently massive to stall somewhere in the disk. In a regime of limited satellite migration, the largest satellites would tend to be located near the centrifugal radius. Finally, due to the gradient in the disk temperature (taken to be controlled by Jupiter's luminosity, see Fig. 1), the slopes of the

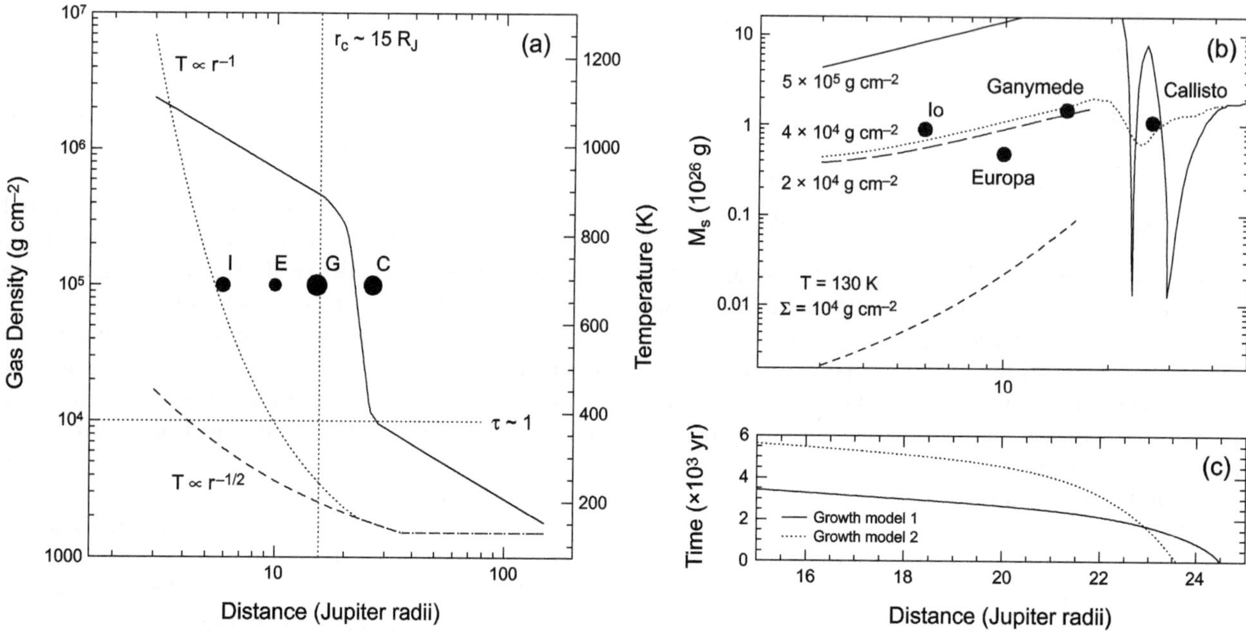

Fig. 8. (**a**) Idealization of the initial minimum mass Σ and assumed photospheric T profiles for the circumjovian subnebula. The reconstituted mass of Io, Europa, and Ganymede determine the mass of the optically thick inner disk, while the mass of Callisto is spread out over the optically thin outer disk. Ganymede lies just inside the centrifugal radius r_c, while Callisto lies outside a transition region that separates the inner and outer disks. The temperature is set to agree with the compositional constraints of the Galilean satellites (e.g., *Lunine and Stevenson,* 1982; *Mosqueira and Estrada,* 2003a,b), which implies a jovian luminosity of ~10^{-5} $L_{\odot}$, for a planetary radius of ~1.5–2 R_J consistent with planet formation models (*Hubickyj et al.,* 2005). (**b**) Critical mass at which migration stalls as a function of Jupiter radii for various Σ profiles using both vertically thermally stratified (solid and dotted curves) and vertically isothermal models (bold dashed curve). Gas drag is included. The solid curve corresponds to the minimum mass model, while the dotted and dashed curves correspond to the SEMM model. The short-dashed curve is a constant Σ and T model. (**c**) Migration and growth models for proto-Ganymede. The full-sized Ganymede is evolved backward in time from the location where it opens a gap to the point where it reaches embryo size (~1000 km) for a SEMM disk. Two models for growth are used. Solid curve: linear growth model; dotted curve: growth rate proportional to the disk surface density. Growth is consistent with the limited migration of Ganymede. See *Mosqueira and Estrada* (2003b) for detailed descriptions.

curves in Fig. 8 are shallow, limiting the range of masses that may stall to mass ratios of $\mu \sim 10^{-4}$. Additionally, the density waves launched by objects of this size shock-dissipate in a length scale smaller than their semimajor axis (*Goodman and Rafikov,* 2001), which allows the largest satellites to drive the evolution of the disk, open a cavity, and stall.

It must be stressed that there are a number of assumptions involved in the results described above. The initial gas surface density has been obtained by placing Callisto in the outer disk, and Ganymede, Europa, and Io in the inner disk, separated by an assumed transition between the outer and the inner disks. However, there are presently no detailed simulations of the formation of such a disk. Furthermore, the temperature structure of the disk is heuristic. Nevertheless, we stress that the overall survival mechanism need not be dependent on the specifics of the model. For instance, Callisto may stall because its migration is halted by the change in the surface density due to the presence of Ganymede. Alternatively, satellites may open gaps collectively. This latter option would be particularly relevant if satellite-disk interactions can lead to the excitation and growth of satellite eccentricities, which would result in spatially extended gap formation. Another process to consider is photoevaporation, which may take place in a timescale comparable to that of the formation and migration of Callisto (and Iapetus for Saturn). The key point here is that the largest satellites of each satellite system, and not disk dissipation due to turbulence, may be responsible for giant planet satellite survival.

Finally, it is useful to evaluate whether migration times are consistent with the growth timescales of the Galilean satellites. This calculation serves to provide an estimate of how far a satellite may have migrated. By associating the current locations of the satellites with their stalling location (Fig. 8b), *Mosqueira and Estrada* (2003b) integrated the migration of the full-sized Ganymede back in time to embryo size using different growth models (Fig. 8c). Here again, migration and formation of full-sized satellites is taken to occur in a subnebula enhanced in solids by a factor of ~10 with respect to solar composition mixtures. These results imply that SEMM disks are consistent with the limited migration of at least Ganymede, so that Ganymede likely formed close to the location of the centrifugal radius, r_c.

4.4. Gas-poor Environment

In the gas-poor scenario sustained turbulence (possibly hydrodynamic turbulence) or some other mechanism removes the gaseous circumplanetary disk quickly compared to the accretion timescale of the satellites (which in this scenario is tied to the timescale for clearing heliocentric planetesimals from the giant planet's feeding zone). By construction, in this model this timescale is ~10^5–10^6 yr. The issue arises whether delivery of planetesimals can last for this long. In section 4.5.2, we discuss possible ways to lengthen the planetesimal delivery timescale. Thus, satellite formation is taken to be somewhat akin to the formation of the terrestrial planets as the gas surface density is taken to be low but is left unspecified.

The gas-poor environment does not face the survival issues associated with the presence of significant amounts of gas, yet the remnant circumplanetary gas disk may still circularize orbits, or clear the disk of collisional debris. The way in which the solid material that makes up the satellites is delivered to the circumplanetary disk must differ significantly from the gas-rich case. This scenario relies on the formation of an accretion disk resulting from the capture into circumplanetary orbit of heliocentric planetesimals undergoing inelastic collisions (*Safronov et al.,* 1986; *Estrada and Mosqueira,* 2006; *Sari and Goldreich,* 2006) or gravitational scatterings (*Goldreich et al.,* 2002; *Agnor and Hamilton,* 2006) within the jovian Hill sphere (see section 4.2). Furthermore, here the angular momentum of the satellite system is largely determined by circumsolar planetesimal capture dynamics.

4.4.1. The circumplanetary swarm. The idea that the regular satellites could form out of a collisionally captured, gravitationally bound swarm of circumplanetary satellitesimals has been suggested and explored in a number of classic publications (*Schmidt,* 1957; *Safronov and Ruskol,* 1977; *Ruskol,* 1981, 1982; *Safronov et al.,* 1986). However, it has only been recently that a gas-poor planetesimal capture (GPPC) model has been advanced (*Estrada and Mosqueira,* 2006). Collisional capture mechanisms have been explored in terms of general accretion (*Sari and Goldreich,* 2006) and applied to the formation of Kuiper belt binaries (*Schlichting and Sari,* 2008).

The GPPC formation scenario is separated into two stages: an early stage in which the circumplanetary disk is initially formed, and a late stage in which a quasi-steady-state accretion disk is in place around the giant planet. The basic idea is that in the early stage, inelastic and gravitational collisions within the Hill sphere of the giant planet lead to the creation of a protosatellite "swarm" of both retrograde and prograde satellitesimals extending out as far as circumplanetary orbits are stable, ~$R_H/2$. At present, the complicated process of circumplanetary swarm generation remains to be modeled.

Estrada and Mosqueira (2006) treat the late stage of satellite formation in a gas-poor environment in which a circumplanetary accretion disk is assumed to be already present, focusing on inelastic collisions between incoming planetesimals with *larger* satellitesimals within the accretion disk as the mass capture mechanism. Planetesimal-satellitesimal collisions can lead to the capture of solids if the incoming planetesimal encounters a mass comparable to, or larger than, itself (*Safronov et al.,* 1986).

Since the circumplanetary disk has a significant surface area, the total mass of planetesimals that crosses the subdisk may be substantial. A reasonable estimate of the amount of mass in residual planetesimals in Jupiter's feeding zone is ~10 $M_\oplus$ ~ 10^{29} g. These planetesimals may cross the circumplanetary disk a number of times before being scattered away by Jupiter, and thus may be subject to capture. The amount of mass captured depends on the size distribution, and total mass of satellitesimals in the accretion disk, as well as the timescale over which heliocentric planetesimals are fed into the system.

Not all the mass delivered to the circumplanetary disk is accreted by the regular satellites. Bound objects may be dislodged by passing planetesimals, or during close encounters between the giant planets during the excitation of the Kuiper belt, as in the Nice model (*Tsiganis et al.,* 2005). Also, satellitesimals may be accreted onto the planet or lost by an evection resonance (*Nesvorny et al.,* 2003). The timescale constraint imposed by Callisto's partially differentiated state would require that Callisto must form in ≳10^5 yr, so that the feeding timescale of planetesimals should be at least this long. A condition for the GPPC model to satisfy the Callisto constraint is that the mass of the extended disk of solids at any one time (excluding satellite embryos) is typically a small fraction of a Galilean satellite (*Estrada and Mosqueira,* 2006).

Collisions in the circumplanetary disk can lead to fragmentation, accretion, or the removal of material from the outer to the inner portions of the disk, where the satellites form. This collisional removal of material (to the inner disk) is assumed to be replenished by the collisional capture of heliocentric planetesimal fragments. Removal of material from the outer regions to the inner regions can occur because the net specific angular momentum of the satellitesimal swarm (which consists of both retrograde and prograde satellitesimals) results in a much more compact prograde disk.

4.4.2. Angular momentum delivery. Initially, the specific angular momentum ℓ_z is small, but as the planet begins to clear its feeding zone, sufficient planetesimals are fed from the outermost regions of the feeding zone where inhomogeneities in the circumsolar planetesimal disk significantly increase ℓ_z of the circumplanetary swarm (*Lissauer and Kary,* 1991; *Dones and Tremaine,* 1993; *Estrada and Mosqueira,* 2006), whether they may or may not be captured (see below). *Estrada and Mosqueira* (2006) posited that these collisional processes may deliver enough ℓ_z to account for the total mass and angular momentum contained in the Galilean satellites. Solutions in Fig. 9 were found for a range of gap sizes in the heliocentric planetesimal population, R_{gap} ~ 0.5–1.5 R_H. In this case, a gap refers to a

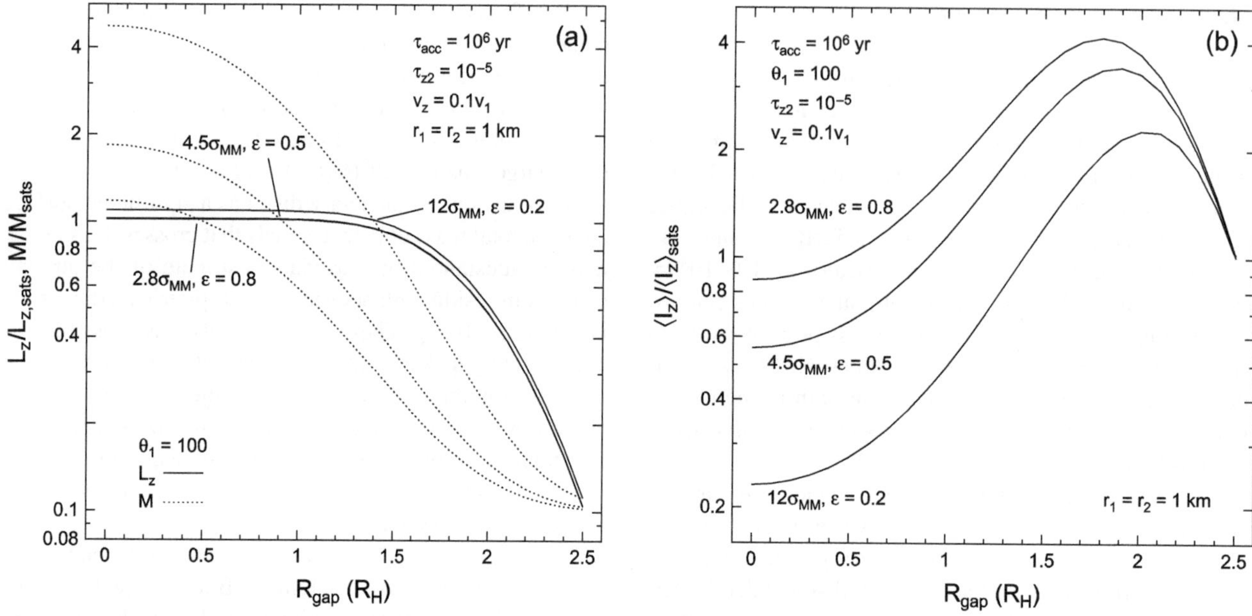

Fig. 9. **(a)** Mass (dotted lines) and angular momentum (solid lines) normalized to the values characteristic of the Galilean satellites ($M_{sats} \sim 4 \times 10^{26}$ g, $L_{z,sats} \sim 4 \times 10^{43}$ g cm^2 s^{-1}) delivered to the circumplanetary disk in $\tau_{acc} = 10^6$ yr as a function of gap size for several values of the solids surface density of the solar nebula. Surface density is expressed in terms of the MMSN value, $\sigma_{MM} = 3.3$ g cm^{-2} (cf. Fig. 1). A cold planetesimal population [$\theta_1 \sim 100$ where $\theta_1 = 0.5(v_e/v_1)^2$ is the Safronov parameter; cf. equation (2)] is assumed. Solutions for the given parameters are indicated by a pointer to where corresponding lines of M and L_z intersect. **(b)** Corresponding solutions for the specific angular momentum. In these plots, the "optical depth" τ_{z2} is the collision probability between the largest satellitesimals in the circumplanetary disk, and the planetesimal scale height is $\sim v_z/\Omega$. See *Estrada and Mosqueira* (2006) for details.

depletion of solids in the circumsolar disk, analogous to a gas gap. As can be seen from Fig. 9b, the specific angular momentum contribution increases as the solids gap grows larger. The total amount of mass and angular momentum that can be delivered is limited by the size of the planet's feeding zone, roughly $R_{gap} \sim 2.5\ R_H$. Thus, the angular momentum is seen to reach a maximum before beginning to decrease sharply as the gap size chokes off the mass inflow. The mass inflow will drop to nearly zero unless there are mechanisms that can replenish the solids in the planet's feeding zone.

4.5. Satisfying the Constraints

4.5.1. The compositional gradient. Io is rocky, Europa is ~90% rock, ~10% water-ice, whereas Ganymede and Callisto may be only ~50% rock, and ~50% water-ice by mass (*Sohl et al.,* 2002). There are three main explanations for this observation. The first ascribes the high-silicate fraction of Europa relative to Ganymede and Callisto to the subnebula temperature gradient due to Jupiter's luminosity at the time of satellite formation (e.g., *Pollack et al.,* 1976; *Lunine and Stevenson,* 1982). In this view, Ganymede's temperature is typically set at T ~ 250 K to allow for the condensation of ice at its location. Closer in, a persistently hot subnebula would prevent the condensation of volatiles close to Jupiter (*Pollack and Reynolds,* 1974; *Lunine and Stevenson,* 1982). In the optically thick case, this may imply a planetary luminosity $\sim 10^{-5}\ L_\odot$ (where $L_\odot = 3.827 \times 10^{33}$ ergs s^{-1} is the solar luminosity) for a planetary radius of 1.5–2 R_J. However, this picture runs into trouble because the inner small satellite Amalthea (located at ~2.5 R_J; Io is at ~6 R_J) has such a low mean density (0.86 g cm^{-3}) that models for its composition require that water-ice be a major constituent, even for improbably high values of porosity (*Anderson et al.,* 2005).

An alternative explanation argues that the compositional gradient may be due to the increase of impact velocities and impactor flux of Roche-lobe interlopers deep in the planetary potential well, leading to preferential volatile depletion in the cases of Io and Europa (*Shoemaker,* 1984). In this view, all the Galilean satellites would start out ice-rich, but some would lose more volatiles than others. One might expect a stochastic compositional component deep in the planetary potential well due to high-speed ≥10 km s^{-1} impacts with large (perhaps ~10–100 km) Roche-lobe interlopers. Such hyperbolic collisions might conceivably remove volatiles from the mantle of a differentiated satellite and place them on neighboring satellites [possibly analogous to the impact that may have stripped Mercury's mantle (*Benz et al.,* 1988)]. This might then explain the volatile depletion in Europa relative to Ganymede and Calliso, but needs to be quantitatively evaluated. Amalthea then may be a remnant of such a collisional process. The gas-poor satellite accretion environment (section 4.4) fits with this scenario, although a stochastic component may also apply to the gas-rich case. A third possibility is that Io, and possibly Europa, can lose their volatiles due to the Laplace resonance alone. A similar argument may apply to Enceladus and other mid-sized saturnian satellites.

4.5.2. The Callisto constraint. If Callisto is partially differentiated, it argues that Callisto's formation took place over a timescale $\gtrsim 10^5$ yr (*Stevenson et al.,* 1986; *Mosqueira et al.,* 2001; *Mosqueira and Estrada,* 2003a,b; *Canup and Ward,* 2002; *Alibert et al.,* 2005a; *Estrada and Mosqueira,* 2006). A long formation timescale may be required because the energy of accretion must be radiated away (*Safronov,* 1969; *Stevenson et al.,* 1986) to avoid melting the interior. Proper treatment of this problem should include the effects of an atmosphere (*Kuramoto and Matsui,* 1994), which allows for hydrodynamic and collisional blow-off, or convective transport to the subnebula (*Lunine and Stevenson,* 1982). Yet nonhydrostatic mass anomalies at the boundary of the rocky core could still imply a differentiated state (*McKinnon,* 1997; *Stevenson et al.,* 2003).

Assuming hydrostatic equilibrium, the simplest interpretation for Callisto's moment of inertia is a satellite structure consisting of a 300-km rock-free ice shell over a homogeneous rock and ice interior (*Anderson et al.,* 2001; *Schubert et al,*. 2004). The magnetic induction results may imply that an ocean occupies the lowermost part of Callisto's icy shell (*Zimmer et al.,* 2000). If Callisto's internal structure is primordial, and it was accreted homogeneously, impacts during the late stages of its growth can be used to raise the temperature of the surface regions of the satellite to the melting point and to supply the latent heat of melting (and vaporization), possibly leading to partial differentiation of just the surface layers.

On the other hand, the energy liberated by the sinking rocky component can eventually lead to a runaway process (*Friedson and Stevenson,* 1983). Moreover, both radiogenic heating and the presence of ammonia in the interior can result in a more stringent constraint on Callisto's thermal history. The relevance of these factors is uncertain: Although ammonia may help to sustain an ocean in Callisto [inferred to be present from Galileo magnetometer data (*Zimmer et al.,* 2000)], salts concentrated in the liquid layer and/or a satellite surface regolith can also help in this regard (*Spohn and Schubert,* 2003), and the addition of ^{26}Al, as specified by CAIs, depends on the assumption of spatial and temporal homogeneity (see *Wadhwa et al.,* 2007, and references therein; *Castillo-Rogez et al.,* 2007). We stress that the likely presence of an ocean means that melting did take place, presumably during satellite accretion. Furthermore, the possible presence of ammonia need not result in full differentiation (*Ellsworth and Schubert,* 1981).

A number of explanations for the Callisto-Ganymede dichotomy have been offered that rely on fine-tuning uncertain parameters (*Schubert et al.,* 1981; *Lunine and Stevenson,* 1982; *Friedson and Stevenson,* 1983; *Stevenson et al.,* 1986). *Showman and Malhotra* (1997) proposed an explanation based on the Laplace resonance, yet it is unclear that Ganymede suffered sufficient tidal heating to explain the differences between the two satellites.

The differences between Ganymede and Callisto can instead be a consequence of the disk formation and evolution scenario described in section 4.3, in which Callisto's formation time is long compared to that of Ganymede. In this SEMM model, Callisto's accretion timescale is tied to the disk clearing timescale (~10^6 yr) (*Mosqueira and Estrada,* 2003a). That is, in this view Callisto derives its full mass from satellitesimals in the extended outer disk that are brought into its feeding zone by gas drag migration. On the other hand, because Callisto poses an effective barrier for inwardly migrating objects, Ganymede must derive most of its mass from a more compact, denser region inside of Callisto's orbit, resulting in a much shorter accretion timescale ($\lesssim 10^4$ yr). This means that while Callisto may have enough time to radiate away its energy of accretion, Ganymede does not. This explanation for the Ganymede-Callisto dichotomy does not require special pleading for Callisto, but rather relies on its outer location to explain its accretional and thermal history. Note that a similar formation timescale would also apply to Iapetus in the saturnian system for the same reason, i.e., its outermost location would lead to long disk clearing times, which is directly tied to the satellite accretion time.

We stress that in the SEMM model, a natural outcome of a gas surface density distribution with a long tail out to the location of the irregular satellites is the formation of some satellites in dense inner portions, and others in extended regions of the disk. In this context, the large separation between Titan and Iapetus provides strong indirect evidence for a two-component subnebula. Therefore, it is not surprising that the same can be said for both Jupiter's and Saturn's regular satellites; i.e., Ganymede and Titan formed in more compact, higher-density regions of the disk than did Callisto and Iapetus. Finally, we point out that it is implausible to argue that the region between Callisto and the irregular satellites was empty. The solids that must have resided there had to be cleared and most ended up accreting onto Callisto, which accounts for its longer formation timescale.

An additional concern is that impacts with large embryos could result in Callisto's differentiation (*McKinnon,* 2006). In the SEMM model, typical radii of inwardly migrating embryos that form in the outer disk are ~200–500 km. However, these embryo sizes should not be confused with the typical sizes of the objects that accrete onto Callisto. Characterizing typical impactor sizes requires treatment of the interaction between a late-stage protosatellite and a swarm of satellitesimals, including the effects of collisional fragmentation.

Work by *Weidenschilling and Davis* (1985) shows that a combination of gas drag and perturbations due to mean-motion resonances with a planet (or satellite) has important consequences for the evolution of planetesimals (or satellitesimals/embryos). Inwardly migrating embryos can be readily captured into resonance before reaching Callisto (note that Hyperion is in such a resonance with Titan). *Malhotra* (1993) points out, though, that resonance trapping is vulnerable to mutual planetesimal (or satellitesimal) interactions, so that as the embryos approach a satellite, collisions among the embryos are expected to be destructive [see

Agnor and Asphaug (2004) for an analogous argument for planets], grinding them down and knocking them out of resonance. In this view, Hyperion (radius ~150 km) might represent a collisional remnant or survivor of such a collisional cascade. It is reasonable to expect that Hyperion-like or smaller objects might be typical impactors in the late-stage formation of Callisto (and Titan). Furthermore, one might expect such impactors to be porous, of low density (~0.5 g cm^{-3}), and likely undifferentiated, all of which tend to favor shallower energy deposition during satellite accretion. The question then becomes whether Callisto's ~10^6-yr accretion from Hyperion-like (or smaller), porous impactors striking the satellite at its escape speed would result in a partially differentiated state. This problem remains to be tackled in detail.

Alternatively, models that make Ganymede and Callisto in the same timescale rely on fine-tuning unknown parameters to explain the differences between these two satellites. This also applies to the gas-poor GPPC model, which forms both Callisto and Ganymede by the delivery of small collisional fragments and debris to the circumplanetary disk over a long timescale. Yet the planetesimal-clearing timescale in the Jupiter-Saturn region is ~10^4 yr (*Charnoz and Morbidelli,* 2003). Note that in the Nice model (*Tsiganis et al.,* 2005), the timescale for planetesimal delivery could be even shorter since Jupiter and Saturn are initially closer together. Therefore, a challenge in the GPPC model is to deliver enough mass at later times to lead to the formation of a partially differentiated Callisto. Possible ways to lengthen the planetesimal delivery timescale include stirring of the circumsolar disk by Uranus and Neptune, and planetesimal replenishment by gas-drag inward migration of meter-sized bodies.

In conclusion, it is important to point out that neither of the disk models discussed in sections 4.3 and 4.4 hinges on Callisto's internal state. If Callisto turns out to be differentiated, both models remain viable even if Callisto forms in a shorter timescale.

5. SUMMARY

We have attempted to provide a broad picture of the origin of the jovian system from the accumulation of the first generation of planetesimals to the birth of the magnificently complex system we see today. We have summarized our current understanding of the various stages of Jupiter's formation with the support of the most current numerical models of its accretion. There are several issues that still need to be addressed in models that treat Jupiter's growth (see *Lissauer and Stevenson,* 2007); most notably, more realistic opacity models for the giant planet envelope need to be implemented (e.g., *Movshovitz and Podolak,* 2007). This issue is tied to the requirement that Jupiter form faster than the nebula dispersal time. Yet, the level of sophistication in these growth models continues to improve.

We have identified where the circumplanetary disk, a byproduct of Jupiter's later stages of accretion, fits into Jupiter's formation history. There are two components to this disk: a more radially compact disk that forms as a result of the accretion of low specific angular momentum gas during a period after envelope contraction and prior to opening a deep gap in the solar nebula; and a more radially extended disk that arises as a result of continued gas inflow through a deep gap as Jupiter approaches its final mass. The resulting disk may be initially massive, as one might expect from a rough equipartition of angular momentum between the planet and disk (e.g., *Stevenson et al.,* 1986), but is expected to evolve at least until the gas inflow wanes. The subnebula may thus play a key role in determining the final spin angular momentum of Jupiter.

Presently, planetary formation models tend to focus on either the growth of the planet in the spherically symmetric approximation, or a protoplanet embedded in a circumstellar disk in the presence of a well-formed gap. As a result, our understanding of the formation of the circumplanetary disk remains incomplete. Furthermore, no simulations yet exist that model disk formation from envelope contraction to the isolation of the giant planet, when accretion ends. Thus, caution must be exercised in interpreting current numerical results. One important result from recent numerical simulations, however, is that the size of the circumplanetary disk formed is dependent on the specific angular momentum of the gas flowing into the planet's Roche lobe, which in turn depends on whether a deep gap is present or not. Gap-opening may take place toward the tail end of the runaway gas accretion phase as Jupiter approaches its final mass. If the architecture of the solar system were such that Saturn began much closer to Jupiter [as in the Nice model (*Tsiganis et al.,* 2005)], disk truncation would be accomplished jointly.

How the circumjovian gas disk evolves over time is dependent on the level of both solar nebula and subnebula turbulence. It is expected that as long as there is inflow through the giant planet gap, it will drive subnebula evolution. However, whether or not there is a sustained, intrinsic source of turbulence in the circumplanetary disk has a profound effect on the environment in which the Galilean satellites formed. We have offered examples of different pathways to satellite growth and survival dependent on if one posits sustained turbulence, or turbulence decays once the gas inflow subsides. The assumption leads to two qualitatively different formation environments that can account for the angular momentum of the regular satellites. Either model can in principle allow for a differentiated or partially differentiated Callisto.

In addition, both models rely on planetesimal delivery mechanisms to provide the mass necessary to form the satellites, and do not rely on dust entrained in the gas inflow to deliver solids. The compositional diversity and potential similarities in the primordial compositions of the satellites of Jupiter and Saturn (*Hibbitts,* 2006) hint that Jupiter and its satellite system may be derived from planetesimals formed locally as well as in more distant regions of the solar nebula. The overall implication then is that planetesimal delivery mechanisms likely provide the bulk of material for satellite accretion, regardless of the gas mass contained in the circumplanetary disk.

Finally, as the inventory of discovered extrasolar planets increases, the subject herein gains relevance. In most cases, there are dynamical differences between these newly discovered planetary systems and our own. Yet, having an understanding of Jupiter's formation from its beginning to its late stages serves as a benchmark to our general understanding of giant planet formation, and, by analogy, of the formation of the satellite systems that likely await as secondary discoveries around these extrasolar giants.

Acknowledgments. We would like to thank D. Stevenson and S.Weidenschilling for their reviews of this work. We also thank W. McKinnon for a thorough reading of the manuscript and suggestions on how to improve its exposition. P.R.E. acknowledges the support of a grant from NASA's Origins of Solar Systems (SSO) program.

REFERENCES

Afshordi N., Mukhopadhyay B., and Narayan R. (2005) Bypass to turbulence in hydrodynamic accretion: Lagrangian analysis of energy growth. *Astrophys. J., 629,* 373–382.

Agnor C. B. and Asphaug E. (2004) Accretion efficiency during planetary collisions. *Astrophys. J. Lett., 613,* L157–L160.

Agnor C. B. and Hamilton D. P. (2006) Neptune's capture of its moon Triton in a binary-planet gravitational encounter. *Nature, 441,* 192–194.

Agnor C. B., Canup R. M., and Levison H. F. (1999) On the character and consequences of large impacts in the late stage of terrestrial planet formation. *Icarus, 142,* 219–237.

Alibert Y., Mousis O., and Benz W. (2005a) Modeling the jovian subnebula. I. Thermodynamical conditions and migration of proto-satellites. *Astron. Astrophys., 439,* 1205–1213.

Alibert Y., Mousis O., Mordasini C., and Benz W. (2005b) New Jupiter and Saturn formation models meet observations. *Astrophys. J. Lett., 626,* L57–L60.

Anderson J. D., Schubert G., Jacobson R. A., Lau E. L., Moore W. B., and Sjogren W. L. (1998) Distribution of rock, metals, and ices in Callisto. *Science, 280,* 1573–1576.

Anderson J. D., Jacobson R. A., McElrath T. P., Moore W. B., Schubert G., and Thomas P. C. (2001) Shape, mean radius, gravity field, and interior structure of Callisto. *Icarus, 153,* 157–161.

Anderson J. D. and 11 colleagues (2005) Amalthea's density is less than that of water. *Science, 308,* 1291–1293.

Andrews S. M. and Williams J. P. (2007) High-resolution submillimeter constraints on circumstellar disk structure. *Astrophys. J., 659,* 705–728.

Artymowicz P. and Lubow S. H. (1996) Mass flow through gaps in circumbinary disks. *Astrophys. J. Lett., 467,* L77–L80.

Atreya S. K., Wong M. H., Owen T. C., Mahaffy P. R., Niemann H. B., de Pater I., Drossart P., and Encrenaz T. (1999) A comparison of the atmospheres of Jupiter and Saturn: Deep atmospheric composition, cloud structure, vertical mixing, and origin. *Planet. Space. Sci., 47,* 1243–1262.

Ayliffe B. A. and Bate M. R. (2009) Circumplanetary disc properties obtained from radiation hydrodynamical simulations of gas accretion by protoplanets. *Mon Not R. Astron. Soc.,* DOI: 10.1111/j.1365-2966.2009.15002.X.

Balbus S. A. and Hawley J. F. (1991) A powerful local shear instability in weakly magnetized disks. I. Linear analysis. *Astrophys. J., 376,* 214–222.

Balbus S. A., Hawley J. F., and Stone J. M. (1996) Nonlinear stability, hydrodynamical turbulence, and transport in disks. *Astrophys. J., 467,* 76–86.

Bar-Nun A., Herman G., Laufer D., and Rappaport M. L. (1985) Trapping and release of gases by water ice and implications for icy bodies. *Icarus, 63,* 317–332.

Bar-Nun A., Dror J., Kochavi E., and Laufer D. (1987) Amorphous water ice and its ability to trap gases. *Phys. Rev. B, 35,* 2427–2435.

Bar-Nun A., Kleinfeld I., and Kochavi E. (1988) Trapping of gas mixtures by amorphous water ice. *Phys. Rev. B, 38,* 7749–7754.

Bar-Nun A., Notesco G., and Owen T. (2007) Trapping of N_2, CO and Ar in amorphous ice. Application to comets. *Icarus, 190,* 655–659.

Barranco J. A. and Marcus P. S. (2005) Three-dimensional vortices in stratified protoplanetary disks. *Astrophys. J., 623,* 1157–1170.

Bate M. R., Lubow S. H., Ogilvie G. I., and Miller K. A. (2003) Three-dimensional calculations of high- and low-mass planets embedded in protoplanetary disks. *Rev. Mod. Phys., 70,* 1–53.

Bell K. R. and Lin D. N. C. (1994) Using FU Orionis outbursts to constrain self-regulated protostellar disk models. *Astrophys. J., 427,* 987–1004.

Benz W., Slattery W. L., and Cameron A. G. W. (1988) Collisional stripping of Mercury's mantle. *Icarus, 74,* 516–528.

Birnstiel T., Dullemond C. P., and Brauer F. (2009) Dust retention in protoplanetary disks. *Astron. Astrophys.,* in press, DOI: 10.1051/0004-6361/200912452.

Bodenheimer P. and Pollack J. B. (1986) Calculations of the accretion and evolution of the giant planets: The effects of solid cores. *Icarus, 67,* 391–408.

Bodenheimer P., Burket A., Klein R., and Boss A. P. (2000a) Multiple fragmentation of protostars. In *Protostars and Planets IV* (V. Mannings et al., eds.), pp. 675–701. Univ. of Arizona, Tucson.

Bodenheimer P., Hubickyj O., and Lissauer J. J. (2000b) Models of the in situ formation of detected extrasolar giant planets. *Icarus, 143,* 2–14.

Boss A. P. (2000) Gas giant protoplanet formation: Disk instability models with thermodynamics and radiative transfer. *Astrophys. J. Lett., 536,* L101–L104.

Brauer F., Dullemond C. P., and Henning Th. (2008) Coagulation, fragmentation and radial motion of solid particles in protoplanetary disks. *Astron Astrophys., 480,* 859–877.

Brownlee D. and 182 colleagues (2006) Comet 81P/Wild 2 under a microscope. *Science, 314,* 1711–1716.

Bryden G., Chen X., Lin D. N. C., Nelson R. P., and Papaloizou C. B. (1999) Tidally induced gap formation in protostellar disks: Gap clearing and suppression of protoplanetary growth. *Astrophys. J., 514,* 344–367.

Bryden G., Rozyczka M., Lin D. N. C., and Bodenheimer P. (2000) On the interaction between protoplanets and protostellar disks. *Astrophys. J., 540,* 1091–1101.

Buriez J. C. and de Bergh C. (1981) A study of the atmosphere of Saturn based on methane line profiles near 1.1 microns. *Astron. Astrophys., 94,* 382–390.

Cai K., Durisen R. H., Michael S., Boley A. C., Meja A. C., Pickett

M. K., and D'Alessio P. (2006) The effects of metallicity and grain size on gravitational instabilities in protoplanetary disks. *Astrophys. J. Lett., 636,* L149–L152.

Calvin W. M., Clark R. N., Brown R. H., and Spencer J. R. (1995) Spectra of the icy Galilean satellites from 0.2 to 5 μm: A compilation, new observations, and a recent summary. *J. Geophys. Res.–Planets, 100,* 19041–19048.

Cameron A. G. W., Decampli W. M., and Bodenheimer P. (1982) Evolution of giant gaseous protoplanets embedded in the primitive solar nebula. *Icarus, 49,* 298–312.

Canup R. and Ward W. R. (2002) Formation of the Galilean satellites: Conditions for accretion. *Astron. J., 124,* 3404–3423.

Cassen P. and Pettibone D. (1976) Steady accretion of a rotating fluid. *Astrophys. J., 208,* 500–511.

Castillo-Rogez J. C., Matson D. L., Sotin C., Johnson T. V., Lunine J. I., and Thomas P. C. (2007) Iapetus' geophysics: Rotation rate, shape, and equatorial ridge. *Icarus, 190,* 179–202.

Chaban G. M., Bernstein M., and Cruikshank D. P. (2006) Carbon dioxide on planetary bodies: Theoretical and experimental studies of molecular complexes. *Icarus, 187,* 592–599.

Chagelishvili G. D., Zahn J.-P., Tevzadze A. G., and Lominadze J. G. (2003) On hydrodynamic shear turbulence in Keplerian disks: Via transient growth to bypass transition. *Astron. Astrophys., 402,* 401–407.

Chambers J. E. (2001) Making more terrestrial planets. *Icarus, 152,* 205–224.

Charnoz S. and Morbidelli A. (2003) Coupling dynamical and collisional evolution of small bodies: An application to the early ejection of planetesimals from the Jupiter-Saturn region. *Icarus, 166,* 141–156.

Chiang E. I. and Goldreich P. (1997) Spectral energy distributions of T Tauri stars with passive circumstellar disks. *Astrophys. J., 490,* 368–376.

Chiang E. I., Joung M. K., Creech-Eakman M. J., Qi C., Kessler J. E., Blake G. A., and van Dishoeck E. F. (2001) Spectral energy distributions of passive T Tauri and Herbig Ae disks: Grain minerology, parameter dependences, and comparison with Infrared Space Observatory LWS observations. *Astrophys. J., 547,* 1077–1089.

Ciesla F. J. and Cuzzi J. N. (2006) The evolution of the water distribution in a viscous protoplanetary disk. *Icarus, 81,* 178–204.

Coradini A., Magni G., and Federico C. (1981) Gravitational instabilities in satellite disks and formation of regular satellites. *Astron. Astrophys., 99,* 255–261.

Coradini A., Federico C., and Luciano P. (1982) Ganymede and Callisto: Accumulation heat content. In *The Comparative Study of Planets* (A. Coradini and N. Fulchignoni, eds.), pp. 61–70. Reidel, Dordrecht.

Coradini A., Cerroni P., Magni G., and Federico C. (1989) Formation of the satellites of the outer solar system — Sources of their atmospheres. In *Origin and Evolution of Planetary and Satellite Atmospheres* (S. K. Atreya et al., eds.), pp. 723–762. Univ. of Arizona, Tucson.

Courtin R., Gautier D., Marten A., Bezard B., and Hanel R. (1984) The composition of Saturn's atmosphere at northern temperate latitudes from Voyager IRIS spectra — NH_3, PH_3, C_2H_2, C_2H_6, CH_3D, CH_4, and the saturnian D/H isotopic ratio. *Astrophys. J., 287,* 899–916.

Cruikshank D. P. and 30 colleagues (2007) Surface composition of Hyperion. *Nature, 448,* 54–56.

Cruikshank D. P. and 27 colleagues (2008) Hydrocarbons on Saturn's satellites Iapetus and Phoebe. *Icarus, 193,* 334–343.

Cuzzi J. N. and Hogan R. C. (2003) Blowing in the wind I. Velocities of chondrule-sized particles in a turbulent protoplanetary nebula. *Icarus, 164,* 127–138.

Cuzzi J. N. and Weidenschilling S. J. (2004) Formation of planetesimals in the solar nebula. In *Protostars and Planets III* (E. H. Levy and J. I. Lunine, eds.), pp. 1031–1060. Univ. of Arizona, Tucson.

Cuzzi J. N. and Weidenschilling S. J. (2006) Particle-gas dynamics and primary accretion. In *Meteorites and the Early Solar System II* (D. S. Lauretta and H. Y. McSween Jr., eds.), pp. 353–381. Univ. of Arizona, Tucson.

Cuzzi J. N. and Zahnle K. J. (2004) Material enhancement in protoplanetary nebulae by particle drift through evaporation fronts. *Astrophys. J., 614,* 490–496.

Cuzzi J. N., Dobrovolskis A. R., and Champney J. M. (1993) Particle-gas dynamics in the midplane of the protoplanetary nebula. *Icarus, 106,* 102–134.

Cuzzi J. N., Hogan R. C., and Shariff K. (2007) Towards a scenario for primary accretion of primitive bodies (abstract). In *Lunar and Planetary Science XXXVIII,* Abstract #1338. Lunar and Planetary Institute, Houston (CD-ROM).

D'Angelo G., Henning T., and Kley W. (2002) Nested-grid calculations of disk-planet interaction. *Astron. Astrophys., 385,* 647–670.

D'Angelo G., Kley W., and Henning T. (2003a) Orbital migration and mass accretion of protoplanets in three-dimensional global computations with nested grids. *Astrophys. J., 586,* 540–561.

D'Angelo G., Henning T., and Kley W. (2003b) Thermohydrodynamics of circumstellar disks with high-mass planets. *Astrophys. J., 599,* 548–576.

D'Angelo G., Bate M. R., and Lubow S. H. (2005) The dependence of protoplanet migration rates on co-orbital torques. *Mon. Not. R. Astron. Soc., 358,* 316–332.

de Pater I. and Massie S. T. (1985) Models of the millimeter-centimeter spectra of the giant planets. *Icarus, 62,* 143–171.

Dominik C., Blum J., Cuzzi J. N., and Wurm G. (2007) Growth of dust as the initial step toward planet formation. In *Protostars and Planets V* (B. Reipurth et al., eds.), pp. 783–800. Univ. of Arizona, Tucson.

Dones L. and Tremaine S. (1993) On the origin of planetary spins. *Icarus, 103,* 67–92.

Dullemond C. P. and Dominik C. (2005) Dust coagulation in protoplanetary disks: A rapid depletion of small grains. *Astron. Astrophys., 434,* 971–986.

Dullemond C. P., Hollenbach D., Kamp I., and D'Alessio P. (2007) Models of the structure and evolution of protoplanetary disks. In *Protostars and Planets V* (B. Reipurth et al., eds.), pp. 559–572. Univ. of Arizona, Tucson.

Durisen R. H., Boss A. P., Mayer L., Nelson A. F., Quinn T., and Rice W. K. M. (2007) Gravitational instabilities in gaseous protoplanetary disks and implications for giant planet formation. In *Protostars and Planets V* (B. Reipurth et al., eds.), pp. 607–622. Univ. of Arizona, Tucson.

Ellsworth K. and Schubert G. (1983) Saturn's icy satellites — Thermal and structural models. *Icarus, 54,* 490–510.

Estrada P. R. (2002) Formation of satellites around gas giant planets. Ph.D. thesis, Cornell University, Ithaca. 326 pp.

Estrada P. R. and Mosqueira I. (2006) A gas-poor planetesimal capture model for the formation of giant planet satellite systems. *Icarus, 181,* 486–509.

Fegley B. Jr. (1993) Chemistry of the solar nebula. In *The Chemistry of Life's Origins* (J. M. Greenberg et al., eds.), pp. 75–

147. Kluwer, Dordrecht.

Friedson A. J. and Stevenson D. J. (1983) Viscosity of rock-ice mixtures and applications to the evolution of icy satellites. *Icarus, 56,* 1–14.

Gammie C. F. (1996) Linear theory of magnetized, viscous, self-gravitating gas disks. *Astrophys. J., 463,* 725.

Gammie C. F. and Johnson B. M. (2005) Theoretical studies of gaseous disk evolution around solar mass stars. In *Chondrites and the Protoplanetary Disk* (A. N. Krot et al., eds.), p. 145. ASP Conference Series 341, Astronomical Society of the Pacific, San Francisco.

Garaud P. and Lin D. N. C. (2007) The effect of internal dissipation and surface irradiation on the structure of disks and the location of the snow line around Sun-like stars. *Astrophys. J., 654,* 606–624.

Gautier D., Hersant F., Mousis O., and Lunine J. I. (2001a) Enrichments in volatiles in Jupiter: A new interpretation of the Galileo measurements. *Astrophys. J. Lett., 550,* L227–L230.

Gautier D., Hersant F., Mousis O., and Lunine J. I. (2001b) Erratum: Enrichments in volatiles in Jupiter: A new interpretation of the Galileo measurements. *Astrophys. J. Lett., 559,* L183.

Gladman B. and Duncan M. (1990) On the fates of minor bodies in the outer solar system. *Astron. J., 100,* 1680–1693.

Goldreich P. and Tremaine S. (1979) The excitation of density waves at the Lindblad and corotation resonances by an external potential. *Astrophys. J., 233,* 857–871.

Goldreich P. and Ward W. R. (1973) The formation of planetesimals. *Astrophys. J., 183,* 1051–1062.

Goldreich. P., Lithwick Y., and Sari R. (2002) Formation of Kuiper-belt binaries by dynamical friction and three-body encounters. *Nature, 420,* 643–646.

Goldreich P., Lithwick Y., and Sari R. (2004) Final stages of planet formation. *Astrophys. J., 614,* 497–507.

Goodman J. and Rafikov R. R. (2001) Planetary torques as the viscosity of protoplanetary disks. *Astrophys. J., 552,* 793–802.

Greenberg R., Hartmann W. K., Chapman C. R., and Wacker J. F. (1978) Planetesimals to planets — Numerical siumulation of collisional evolution. *Icarus, 35,* 1–26.

Hawley J. F., Balbus S. A., and Winters W. F. (1999) Local hydrodynamic stability of accretion disks. *Astrophys. J., 518,* 394–404.

Hayashi C. (1981) Structure of the solar nebula, growth and decay of magnetic fields, and effects of magnetic and turbulent viscosities on the nebula. *Prog. Theor. Phys. Suppl., 70,* 35–53.

Hersant F., Gautier D., and Lunine J. I. (2004) Enrichment in volatiles in the giant planets of the solar system. *Planet. Space Sci., 52,* 623–641.

Hibbitts C. A. (2006) Intriguing differences and similarities in the surface compositions of the icy saturnian and galilean satellites (abstract). *American Geophysical Union, Fall Meeting 2006,* Abstract #P32A-08.

Hibbitts C. A., McCord T. B., and Hansen G. B. (2000) Distributions of CO_2 and SO_2 on the surface of Callisto. *J. Geophys. Res., 105,* 22541–22558.

Hibbitts C. A., Pappalardo R., Klemaszewski J., McCord T. B., and Hansen G. B. (2001) Comparing carbon dioxide distributions on Ganymede and Callisto (abstract). In *Lunar and Planetary Science XXXII,* Abstract #1263. Lunar and Planetary Institute, Houston (CD-ROM).

Hibbitts C. A., Pappalardo R. T., Hansen G. B., and McCord T. B. (2003) Carbon dioxide on Ganymede. *J. Geophys. Res., 108(2),* 1–22.

Hubbard W. B., Burrows A., and Lunine J. I. (2002) Theory of giant planets. *Annu. Rev. Astron. Astrophys., 40,* 103–136.

Hubickyj O., Bodenheimer P., and Lissauer J. J. (2005) Accretion of the gaseous envelope of Jupiter around a 5–10 Earth-mass core. *Icarus, 179,* 415–431.

Ji H., Burin M., Schartman E., and Goodman J. (2006) Hydrodynamic turbulence cannot transport angular momentum effectively in astrophysical disks. *Nature, 444,* 343–346.

Johansen A., Oishi J. S., Low M.-M., Klahr H., Henning T., and Youdin A. N. (2007) Rapid planetesimal formation in turbulent circumstellar disks. *Nature, 448,* 1022–1025.

Kary D. M. and Lissauer J. J. (1995) Nebular gas drag and planetary accretion. II. Planet on an eccentric orbit. *Icarus, 117,* 1–24.

Kary D. M., Lissauer J. J., and Greenzweig Y. (1993) Nebular gas drag and planetary accretion. *Icarus, 106,* 288.

Kenyon S. J. and Hartmann L. (1987) Spectral energy distributions of T Tauri stars — Disk flaring and limits on accretion. *Astrophys. J., 323,* 714–733.

Kenyon S. J. and Luu J. X. (1999) Accretion in the early outer solar system. *Astrophys. J., 526,* 465–470.

Kitamura Y., Momose M., Yokogawa S., Kawabe R., Tamura M., and Ida S. (2002) Investigation of the physical properties of protoplanetary disks around T Tauri stars by a 1 arcsecond imaging survey: Evolution and diversity of the disks in their accretion stage. *Astrophys. J., 581,* 357–380.

Klahr H. H. and Bodenheimer P. (2003) Turbulence in accretion disks: Vorticity generation and angular momentum transport via the global baroclinic instability. *Astrophys. J., 582,* 869–892.

Klahr H. and Kley W. (2006) 3D-radiation hydro simulations of disk-planet interactions. I. Numerical algorithm and test cases. *Astron. Astrophys., 445,* 747–758.

Kley W. (1999) Mass flow and accretion through gaps in accretion discs. *Mon. Not. R. Astron. Soc., 303,* 696–710.

Kokubo E. and Ida S. (1998) Oligarchic growth of protoplanets. *Icarus, 131,* 171–178.

Kornet K., Różyczka M., and Stepinski T. F. (2004) An alternative look at the snowline in protoplanetary disks. *Astron. Astrophys., 417,* 151–158.

Korycansky D. G., Pollack J. B., and Bodenheimer P. (1991) Numerical models of giant planet formation with rotation. *Icarus, 92,* 234–251.

Kouchi A., Yamamoto T., Kozasa T., Kuroda T., and Greenberg J. M. (1994) Conditions for condensation and preservation of amorphous ice and crystallinity of astrophysical ices. *Astron. Astrophys., 290,* 1009–1018.

Kuramoto K. and Matsui T. (1994) Formation of a hot protoatmosphere on the accreting giant icy satellite: Implications for the origin and evolution of Titan, Ganymede, and Callisto. *J. Geophys. Res., 99,* 21183–21200.

Laskar J. (2000) On the spacing of planetary systems. *Phys. Rev. Lett., 84,* 3240–3243.

Lesur G. and Longaretti P.-Y. (2005) On the relevance of subcritical hydrodynamic turbulence to accretion disk transport. *Astron. Astrophys., 444,* 25–44.

Levin B. J. (1978) Relative velocities of planetesimals and the early accumulation of planets. *Moon Planets, 19,* 289–296.

Levison H. F. and Morbidelli A. (2003) The formation of the Kuiper belt by the outward transport of bodies during Neptune's migration. *Nature, 426,* 419–421.

Li H., Finn J. M., Lovelace R. V. E., and Colgate S. A. (2000) Rossby wave instability of thin accretion disks. II. Detailed linear theory. *Astrophys. J., 533,* 1023–1034.

Lin D. N. C. and Papaloizou J. (1979) Tidal torques on accretion discs in binary systems with extreme mass ratios. *Mon. Not. R. Astron. Soc., 186,* 799–812.

Lin D. N. C. and Papaloizou J. (1980) On the structure and evolution of the primordial solar nebula. *Mon. Not. R. Astron. Soc., 191,* 37–48.

Lissauer J. J. (1987) Timescales for planetary accretion and the structure of the protoplanetary disk. *Icarus, 69,* 249–265.

Lissauer J. J. (1993) Planet formation. *Annu. Rev. Astron. Astrophys., 31,* 129–174.

Lissauer J. J. (1995) Urey prize lecture: On the diversity of plausible planetary systems. *Icarus, 114,* 217–236.

Lissauer J. J. (2007) Planets formed in habitable zones of M dwarf stars probably are deficient in volatiles. *Astrophys. J. Lett., 660,* L149–L152.

Lissauer J. J. and Kary D. M. (1991) The origin of the systematic component of planetary rotation. I. Planet on a circular orbit. *Icarus, 94,* 126–159.

Lissauer J. J. and Stevenson D. J. (2007) Formation of giant planets. In *Protostars and Planets V* (B. Reipurth et al., eds.), pp. 591–606. Univ. of Arizona, Tucson.

Lissauer J. J., Hubickyj O., D'Angelo G., and Bodenheimer P. (2009) Models of Jupiter's growth incorporating thermal and hydrodynamical constraints. *Icarus, 199,* 338–350.

Lubow S. H., Seibert M., and Artymowicz P. (1999) Disk accretion onto high mass planets. *Astrophys. J., 526,* 1001.

Lunine J. I. and Stevenson D. J. (1982) Formation of the Galilean satellites in a gaseous nebula. *Icarus, 52,* 14–39.

Lunine J. I. and Stevenson D. J. (1985) Thermodynamics of clathrate hydrate at low and high pressures with application to the outer solar system. *Astrophys. J. Suppl. Ser., 58,* 493–531.

Machida M. N., Kokubo E., Inutsuka S.-I., and Matsumoto T. (2008) Angular momentum accretion onto a gas giant planet. *Astrophys. J., 685,* 1220–1236.

Mahaffy P. R., Niemann H. B., Alpert A., Atreya S. K., Demick J., Donahue T. M., Harpold D. N., and Owen T. C. (2000) Noble gas abundance and isotope ratios in the atmosphere of Jupiter from the Galileo Probe Mass Spectrometer. *J. Geophys. Res., 105,* 15061–15072.

Makalkin A. B., Dorofeeva V. A., and Ruskol E. L. (1999) Modeling the protosatellite circumjovian accretion disk: An estimate of the basic parameters. *Solar Sys. Res., 33,* 456.

Malhotra R. (1993) Orbital resonances in the solar nebula — Strengths and weaknesses. *Icarus, 106,* 264–273.

Marten A., Courtin R., Gautier D., and Lacombe A. (1980) Ammonia vertical density profiles in Jupiter and Saturn from their radioelectric and infrared emissivities. *Icarus, 41,* 410–422.

Mayer L., Quinn T., Wadsley J., and Standel J. (2002) Formation of giant planets by fragmentation of protoplanetary disks. *Science, 298,* 1756–1759.

McCord T. B. and 11 colleagues (1997) Organics and other molecules in the surfaces of Ganymede and Callisto. *Science, 278,* 271–275.

McKinnon W. B. (1997) NOTE: Mystery of Callisto: Is it undifferentiated? *Icarus, 130,* 540–543.

McKinnon W. B. (2006) Differentiation of the Galilean satellites: It's different out there (abstract). In *Workshop on Early Planetary Differentiation,* pp. 66–67. LPI Contribution No. 1335, Lunar and Planetary Institute, Houston.

Mekler Y. and Podolak M. (1994) Formation of amorphous ice in the protoplanetary nebula. *Planet. Space Sci., 42,* 865–870.

Meyer M. R., Backman D. E., Weinberger A. J., and Wyatt M. C. (2007) Evolution of circumstellar disks around normal stars: Placing our solar system in context. In *Protostars and Planets V* (B. Reipurth et al., eds.), pp. 573–590. Univ. of Arizona, Tucson.

Miyoshi K., Takeuchi T., Tanaka H., and Ida S. (1999) Gravitational interaction between a protoplanet and a protoplanetary disk. I. Local three-dimensional simulations. *Astrophys. J., 516,* 451–464.

Mizuno H. (1980) Formation of the giant planets. *Prog. Theor. Phys., 64,* 544–557.

Mizuno H., Nakazawa K., and Hayashi C. (1978) Instability of a gaseous envelope surrounding a planetary core and formation of giant planets. *Prog. Theor. Phys., 60,* 699–710.

Morbidelli A. and Crida A. (2007) The dynamics of Jupiter and Saturn in the gaseous protoplanetary disk. *Icarus, 191,* 158–171.

Morfill G. E. and Völk H. J. (1984) Transport of dust and vapor and chemical fractionation in the early protosolar cloud. *Astrophys. J., 287,* 371–395.

Mosqueira I. and Estrada P. R. (2000) *Formation of Large Regular Satellites of Giant Planets in an Extended Gaseous Nebula: Subnebula Model and Accretion of Satellites.* Technical Report, NASA Ames Research Center, Moffett Field, California.

Mosqueira I. and Estrada P. R. (2003a) Formation of the regular satellites of giant planets in an extended gaseous nebula I: Subnebula model and accretion of satellites. *Icarus, 163,* 198–231.

Mosqueira I. and Estrada P. R. (2003b) Formation of the regular satellites of giant planets in an extended gaseous nebula II: Satellite migration and survival. *Icarus, 163,* 232–255.

Mosqueira I. and Estrada P. R. (2006) Jupiter's obliquity and a long-lived circumplanetary disk. *Icarus, 180,* 93–97.

Mosqueira I., Estrada P. R., Cuzzi J. N., and Squyres S. W. (2001) Circumjovian disk clearing after gap-opening and the formation of a partially differentiated Callisto (abstract). In *Lunar and Planetary Science XXXII,* Abstract #1989, Lunar and Planetary Institute, Houston (CD-ROM).

Mousis O. and Gautier D. (2004) Constraints on the presence of volatiles in Ganymede and Callisto from an evolutionary turbulent model of the jovian subnebula. *Planet. Space Sci., 52,* 361–370.

Movshovitz N. and Podolak M. (2007) The opacity of grains in protoplanetary atmospheres. *Icarus, 194,* 368–378.

Nakagawa Y., Sekiya M., and Hayashi C. (1986) Settling and growth of dust particles in a laminar phase of a low-mass solar nebula. *Icarus, 67,* 375–390.

Natta A., Testi L., Calvet N., Henning T., Waters R., and Wilner D. (2007) Dust in protoplanetary disks: Properties and evolution. In *Protostars and Planets V* (B. Reipurth et al., eds.), pp. 767–782. Univ. of Arizona, Tucson.

Nesvorný D., Alvarellos J. L. A., Dones L., and Levison H. F. (2003) Orbital and collisional evolution of the irregular satellites. *Astron. J., 126,* 398–429.

Ohtsuki K., Stewart G. R., and Ida S. (2002) Evolution of planetesimal velocities based on threebody orbital integrations and growth of protoplanets. *Icarus, 155,* 436–453.

Ormel C. W. and Cuzzi J. N. (2007) Closed-form expressions for particle relative velocities induced by turbulence. *Astron. Astrophys., 466,* 413–420.

Ossenkopf V. (1993) Dust coagulation in dense molecular clouds: The formation of fluffy aggregates. *Astron. Astrophys., 280,* 617–646.

Owen T. and Encrenaz T. (2003) Element abundances and isotope ratios in the giant planets and Titan. *Space Sci. Rev., 106,* 121–138.

Owen T., Mahaffy P., Niemann H. B., Atreya S., Donahue T., Bar-Nun A., and de Pater I. (1999) A low-temperature origin for the planetesimals that formed Jupiter. *Nature, 402,* 269–270.

Paardekooper S.-J. and Mellema G. (2006) Dust flow in gas disks in the presence of embedded planets. *Astron. Astrophys., 453,* 1129–1140.

Podolak M., Pollack J. B., and Reynolds R. T. (1988) Interactions of planetesimals with protoplanetary atmospheres. *Icarus, 73,* 163–179.

Pollack J. B. and Reynolds R. T. (1974) Implications of Jupiter's early contraction history for the composition of the Galilean satellites. *Icarus, 21,* 248–253.

Pollack J. B., Grossman A. S., Moore R., and Graboske H. C. Jr. (1976) The formation of Saturn's satellites and rings as influenced by Saturn's contraction history. *Icarus, 29,* 35–48.

Pollack J. B., Grossman A. S., Moore R., and Graboske H. C. Jr. (1977) A calculation of Saturn's gravitational contraction history. *Icarus, 30,* 111–128.

Pollack J. B., Hollenbach D., Beckwith S., Simonelli D. P., Roush T., and Fong W. (1994) Composition and radiative properties of grains in molecular clouds and accretion disks. *Astrophys. J., 421,* 615–639.

Pollack J. B., Hubickyj O., Bodenheimer P., Lissauer J. J., Podolak M., and Greenszeig Y. (1996) Formation of the giant planets by concurrent accretion of solids and gas. *Icarus, 124,* 62–85.

Porco C. C. and 34 colleagues (2005) Cassini imaging science: Initial results on Saturn's rings and small satellites. *Science, 307,* 1226–1236.

Prialnik D. and Podolak M. (1995) Radioactive heating of porous comet nuclei. *Icarus, 117,* 420–430.

Quillen A. C. and Trilling D. E. (1998) Do proto-jovian planets drive outflows? *Astrophys. J., 508,* 707–713.

Rafikov R. R. (2002a) Nonlinear propagation of planet-generated tidal waves. *Astrophys. J., 569,* 997–1008.

Rafikov R. R. (2002b) Planet migration and gap formation by tidally induced shocks. *Astrophys. J., 572,* 566–579.

Rafikov R. R. (2005) Can giant planets form by direct gravitational instability? *Astrophys. J. Lett., 621,* L69–L69.

Rice W. K. M., Armitage P. J., Wood K., and Lodato G. (2006) Dust filtration at gap edges: Implications for the spectral energy distributions of discs with embedded planets. *Mon. Not. R. Astron. Soc., 379,* 1619–1626.

Ruskol E. L. (1981) Formation of planets. In *The Solar System and Its Exploration,* pp. 107–113. ESA, Noordwijk, The Netherlands (SEE N82-26087 1688).

Ruskol E. L. (1982) Origin of planetary satellites. *Izv. Earth Phys., 18,* 425–433.

Ryu D. and Goodman J. (1992) Convective instability in differentially rotating disks. *Astrophys. J., 308,* 438–450.

Safronov V. S. (1960) On the gravitational instability in flattened systems with axial symmetry and non-uniform rotation. *Ann. Astrophys., 23,* 979–982.

Safronov V. S. (1969) *Evolution of the Protoplanetary Cloud and Formation of the Earth and Planets.* Nauka, Moscow. (Translated in NASA TTF-667, Israel Program for Scientific Translations, 1972.)

Safronov V. S. and Ruskol E. L. (1977) The accumulation of satellites. In *Planetary Satellites* (J. A. Burns, ed.), pp. 501–512. Univ. of Arizona, Tucson.

Safronov V. S., Pechernikova G. V., Ruskol E. L., and Vitiazev A. V. (1986) Protosatellite swarms. In *Satellites* (J. A. Burns and M. S. Matthews, eds.), pp. 89–116. Univ. of Arizona, Tucson.

Sandford S. A. and 54 colleagues (2006) Organics captured from comet 81P/Wild 2 by the Stardust spacecraft. *Science, 314,* 1720–1725.

Sari R. and Goldreich P. (2004) Planet-disk symbiosis. *Astrophys. J. Lett., 606,* L77–L80.

Sari R. and Goldreich P. (2006) Spherical accretion. *Astrophys. J. Lett., 642,* L65–L67.

Sasselov D. D. and Lecar M. (2000) On the snow line in dusty protoplanetary disks. *Astrophys. J., 528,* 995–998.

Schlichting H. E. and Sari R. (2008) Formation of Kupier belt binaries. *Astrophys. J., 673,* 1218–1224.

Schmidt O. Yu. (1957) *Four Lectures on the Theory of the Origin of the Earth, 3rd edition.* NA SSSR, Moscow.

Schmitt B., Espinasse S., Grim R. J. A., Greenberg J. M., and Klinger J. (1989) Laboratory studies of cometary ice analogues. In *Proceedings of an International Workshop on Physics and Mechanics of Cometary Materials* (J. Hunt and T. D. Guyeme, eds.), pp. 65–69. ESA SP-302, Noordwijk, The Netherlands.

Schubert G., Stevenson D. J., and Ellsworth K. (1981) Internal structures of the Galilean satellites. *Icarus, 47,* 46–59.

Schubert G., Anderson J. D., Spohn T., and McKinnon W. B. (2004) Interior composition, structure and dynamics of the Galilean satellites. In *Jupiter: The Planet, Satellites and Magnetosphere* (F. Bagenal et al., eds.), pp. 281–306. Cambridge Univ., Cambridge.

Scott E. R. D. and Krot A. N. (2005) Thermal processing of silicate dust in the solar nebula: Clues from primitive chondrite matrices. *Astrophys. J., 623,* 571–578.

Sekiya M. (1998) Quasi-equilibrium density distributions of small dust aggregations in the solar nebula. *Icarus, 133,* 298–309.

Shakura N. I. and Sunyaev R. A. (1973) Black holes in binary systems. Observational appearance. *Astron. Astrophys., 24,* 337–355.

Shen Y., Stone J. M., and Gardiner T. A. (2006) Three-dimensional compressible hydrodynamic simulations of vortices in disks. *Astrophys. J., 653,* 513–524.

Shoemaker E. M. (1984) Kuiper Prize Lecture. *16th Annual Meeting of the Division for Planetary Sciences,* Kona, Hawaii.

Showalter M. R. (1991) Visual detection of 1981S13, Saturn's eighteenth satellite, and its role in the Encke gap. *Nature, 351,* 709–713.

Showman A. P. and Malhotra R. (1997) Tidal evolution into the Laplace resonance and the resurfacing of Ganymede. *Icarus, 127,* 93–111.

Shu F. H., Johnstone D., and Hollenbach D. (1993) Photoevaporation of the solar nebula and the formation of the giant planets. *Icarus, 106,* 92–101.

Shu F. H., Shang H., Gounelle M., Glassgold A. E., and Lee T. (2001) The origin of chondrules and refractory inclusions in chondritic meteorites. *Astrophys. J., 548,* 1029–1050.

Sohl F., Spohn T., Breuer D., and Nagel K. (2002) Implications from Galileo observations on the interior structure and chemistry of the Galilean satellites. *Icarus, 157,* 104–119.

Spohn T. and Schubert G. (2003) Oceans in the icy Galilean satellites of Jupiter? *Icarus, 161,* 456–467.

Stepinski T. F. and Valageas P. (1997) Global evolution of solid matter in turbulent protoplanetary disks. II. Development of icy planetesimals. *Astron. Astrophys., 319,* 1007–1019.

Stern S. A. (2003) The evolution of comets in the Oort cloud and Kuiper belt. *Nature, 424,* 639–642.

Stern S. A. and Weissman P. R. (2001) Rapid collisional evolution of comets during the formation of the Oort cloud. *Nature, 409,* 589–591.

Stevenson D. J. (1982) Formation of the giant planets. *Planet. Space Sci., 30,* 755–764.

Stevenson D. J. and Lunine J. I. (1988) Rapid formation of Jupiter by diffuse redistribution of water vapor in the solar nebula. *Icarus, 75,* 146–155.

Stevenson D. J., Harris A. W., and Lunine J. I. (1986) Origins of satellites. In *Satellites* (J. A. Burns and M. S. Matthews, eds.), pp. 39–88. Univ. of Arizona, Tucson.

Stevenson D. J., McKinnon W. B., Canup R., Schubert G., and Zuber M. (2003) The power of JIMO for determining Galilean satellite internal structure and origin (abstract). *American Geophysical Union, Fall Meeting 2003,* Abstract #P11C-05.

Stewart G. R. andWetherill G. W. (1988) Evolution of planetesimal velocities. *Icarus, 74,* 542–553.

Takata T. and Stevenson D. J. (1996) Despin mechanism for protogiant planets and ionization state of protogiant planetary disks. *Icarus, 123,* 404–421.

Takeuchi T. and Lin D. N. C. (2002) Radial flow of dust particles in accretion disks. *Astrophys. J., 581,* 1344–1355.

Tanigawa T. and Watanabe S. (2002) Gas accretion flows onto giant protoplanets: High-resolution two-dimensional simulations. *Astrophys. J., 580,* 506–518.

Tsiganis K., Gomes R., Morbidelli A., and Levison H. F. (2005) Origin of the orbital architecture of the giant planets of the solar system. *Nature, 435,* 459–461.

Turner N. J., Sano T., and Dziourkevitch N. (2007) Turbulent mixing and the dead zone in protostellar disks. *Astrophys. J., 659,* 729–737.

Umurhan O. M. and Regev O. (2004) Hydrodynamic stability of rotationally supported flows: Linear and nonlinear 2D shearing box results. *Astron. Astrophys., 427,* 855–872.

Völk H. J., Morfill G. E., Roeser S., and Jones F. C. (1980) Collisions between grains in a turbulent gas. *Astron. Astrophys., 85,* 316–325.

Wadhwa M., Amelin Y., Davis A. M., Lugmair G. W., Meyer B., Gounelle M., and Desch S. J. (2007) From dust to planetesimals: Implications for the solar protoplanetary disk from short-lived radionuclides. In *Protostars and Planets V* (B. Reipurth et al., eds.), pp. 835–848. Univ. of Arizona, Tucson.

Ward W. R. (1997) Protoplanet migration by nebula tides. *Icarus, 126,* 261–281.

Weidenschilling S. J. (1977a) Aerodynamics of solid bodies in the solar nebula. *Mon. Not. R. Astron. Soc., 180,* 57–70.

Weidenschilling S. J. (1977b) The distribution of mass in the planetary system and solar nebula. *Astrophys. Space Sci., 51,* 153–158.

Weidenschilling S. J. (1984) Evolution of grains in a turbulent solar nebula. *Icarus, 60,* 553–567.

Weidenschilling S. J. (1988) Formation processes and timescales for meteorite parent bodies. In *Meteorites and the Early Solar System* (J. F. Kerridge et al., eds.), pp. 348–371. Univ. of Arizona, Tucson.

Weidenschilling S. J. (1997) The origin of comets in the solar nebula: A unified model. *Icarus, 127,* 290–306.

Weidenschilling S. J. (2002) Structure of a particle layer in the midplane of the solar nebula: Constraints on mechanisms for chondrule formation. *Meteoritics & Planet. Sci., 37,* A148.

Weidenschilling S. J. (2004) From icy grains to comets. In *Comets II* (M. C. Festou et al., eds.), pp. 97–104. Univ. of Arizona, Tucson.

Weidenschilling S. J. and Cuzzi J. N. (1993) Formation of planetesimals in the solar nebula. In *Protostars and Planets III* (E. H. Levy and J. I. Lunine, eds.), pp. 1031–1060. Univ. of Arizona, Tucson.

Weidenschilling S. J. and Davis D. R. (1985) Orbital resonances in the solar nebula — Implications for planetary accretion. *Icarus, 62,* 16–29.

Wetherill G. W. (1980) Formation of the terrestrial planets. *Annu. Rev. Astron. Astrophys., 18,* 77–113.

Wetherill G. W. (1986) Accumulation of the terrestrial planets and implications concerning lunar origin. In *Origin of the Moon* (W. K. Hartmann et al., eds.), pp. 519–550. Lunar and Planetary Institute, Houston.

Wetherill G. W. (1990) Formation of the Earth. *Annu. Rev. Earth Planet. Sci., 18,* 205–256.

Wetherill G. W. and Stewart G. R. (1989) Accumulation of a swarm of small planetesimals. *Icarus, 77,* 330–357.

Wetherill G. W. and Stewart G. R. (1993) Formation of planetary embryos — Effects of fragmentation, low relative velocity, and independent variation of eccentricity and inclination. *Icarus, 106,* 190.

Wong M. H., Lunine. J. I., Atreya S. K., Johnson T., Mahaffy P. R., Owen T. C., and Encrenaz T. (2008) Oxygen and other volatiles in the giant planets and their satellites. In *Oxygen in the Solar System* (G. MacPherson et al., eds.) pp. 219–246. Reviews in Mineralogy and Geochemistry, Vol. 68, Mineralogical Society of America.

Wuchterl G., Guillot T., and Lissauer J. J. (2000) Giant planet formation. In *Protostars and Planets IV* (V. Mannings et al., eds.), pp. 1081–1109. Univ. of Arizona, Tucson.

Wurm G. and Blum J. (1998) Experiments on preplanetary dust aggregation. *Icarus, 132,* 125–136.

Youdin A. N. and Chiang E. I. (2004) Particle pileups and planetesimal formation. *Astrophys. J., 601,* 1109–1119.

Youdin A. N. and Shu F. H. (2002) Planetesimal formation by gravitational instability. *Astrophys. J., 580,* 494–505.

Young R. E. (2003) The Galileo probe: How it has changed our understanding of Jupiter. *New Astron. Rev., 47,* 1–51.

Zimmer C., Khurana K. K., and Kivelson M. G. (2000) Subsurface oceans on Europa and Callisto: Constraints from Galileo magnetometer observations. *Icarus, 147,* 329–347.

Zolensky M. E. and 74 colleagues (2006) Minerology and petrology of comet 81P/Wild 2 nucleus samples. *Science, 314,* 1735–1741.

Origin of Europa and the Galilean Satellites

Robin M. Canup and William R. Ward
Southwest Research Institute

Europa is believed to have formed near the very end of Jupiter's own accretion, within a circumplanetary disk of gas and solid particles. We review the formation of the Galilean satellites in the context of current constraints and understanding of giant planet formation, focusing on recent models of satellite growth within a circumjovian accretion disk produced during the final stages of gas inflow to Jupiter. In such a disk, the Galilean satellites would have accreted slowly, in more than 10^5 yr, and in a low-pressure, low-gas-density environment. Gravitational interactions between the satellites and the gas disk lead to inward orbital migration and loss of satellites to Jupiter. Such effects tend to select for a maximum satellite mass and a common total satellite system mass compared to the planet's mass. One implication is that multiple satellite systems may have formed and been lost during the final stages of Jupiter's growth, with the Galilean satellites being the last generation that survived as gas inflow to Jupiter ended. We conclude by discussing open issues and implications for Europa's conditions of formation.

1. INTRODUCTION

Europa belongs to the family of four Galilean satellites, which have similar masses and orbits that are prograde, nearly circular, and approximately co-planar. These basic properties suggest that the Galilean satellites formed contemporaneously within a shared circumplanetary disk orbiting within Jupiter's equatorial plane. The conditions in this early disk established the satellites' initial masses, compositions, and thermal and orbital states. Observed properties of the Galilean satellites can provide constraints on this formation environment, which in turn can provide clues to the conditions of Jupiter's formation.

Regular satellites originate from material in circumplanetary orbit, which can arise from a number of diverse processes (e.g., *Pollack et al.,* 1991). A large collision into a solid planet can produce an impact-generated disk or in some cases an intact satellite (e.g., *Canup,* 2004, 2005). Although an impact origin is typically implicated for Earth's Moon and Pluto's Charon, this seems an unlikely explanation for the Galilean satellites given Jupiter's small ~3° obliquity. Collisions between objects in solar orbit that occur relatively close to a planet can also leave material in planetary orbit, in a process known as coaccretion (e.g., *Safronov et al.,* 1986). Forming the Galilean satellites by coaccretion is quite difficult, primarily because the rate of such collisions is generally low and captured material tends to not have enough net angular momentum to account for the current satellite orbits (*Estrada and Mosqueira,* 2006).

For gas planets, two processes are thought to be effective means for producing circumplanetary disks of gas (H + He) and solids (rock + ice). When a gas giant undergoes runaway gas accretion, it nearly fills the entire region within which its gravity dominates over that of the Sun, known as the planet's Hill sphere. For Jupiter's distance and current mean density, the Hill sphere of a planet is more than 700 times the planet's radius. A forming gas planet will begin to contract within its Hill sphere once the rate of its gas accretion can no longer compensate for the increasing rate of its gravitational contraction due to the planet's growing mass and luminosity. Accreted gas delivers angular momentum to the planet, and as the spinning planet contracts, conservation of angular momentum dictates that its rate of rotation increases. When this rate reaches the critical rate for rotational instability, the planet's outer equa-torial layers begin to shed, forming a so-called "spin-out" disk. A second means for forming a disk occurs if the solar nebula is still present during the contraction phase, so that while the planet contracts, it is also continuing to accrete gas. As the planet becomes smaller and smaller, at some point the gas inflowing from solar orbit contains too much angular momentum to allow it to fall directly onto the planet. Instead, gas flows into orbit around the planet and directly creates a disk (e.g., Fig. 1). Such a structure is known in astrophysical contexts as an "accretion" disk.

Whether the Galilean satellites formed in a spin-out or an accretion disk depends on the relative timing of Jupiter's contraction and the dispersal of the solar nebula. Recent models (e.g., *Papaloizou and Nelson,* 2005) predict that a Jupiter-mass planet will contract to within the current orbit of Io in less than a million years, while estimates of the lifetime of the solar nebula are longer, ~10^6–10^7 yr (e.g., *Haisch et al.,* 2001). Thus it appears probable that an accretion disk would have predominated during the final stages of Jupiter's formation.

The implication is that the Galilean satellites formed in a primordial gas disk surrounding Jupiter, some millions to perhaps 10 m.y. after the origin of the first solar system solids. The jovian circumplanetary disk has often been viewed as analogous to the circumsolar disk, and in many respects the growth of satellites within this "mini solar system" would proceed similarly to the well-studied growth of solid planets from the nebular disk. Collisions between orbiting objects would lead to growth of progressively larger

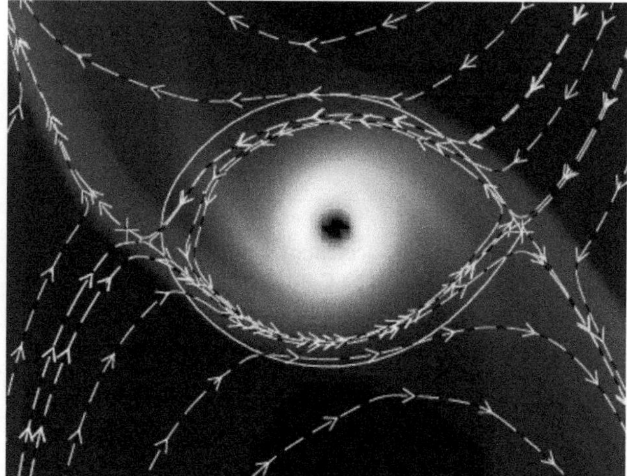

Fig. 1. Example results of a three-dimensional hydrodynamical simulation of gas inflow to a Jupiter-mass planet orbiting a solar mass star at 5.2 AU, assuming $\alpha = 4 \times 10^{-3}$. Shown is a close-up on the planet's Hill sphere (solid curve), with the central star located to the right, and the planet orbiting the star in the counterclockwise direction. Brightness scales with gas density, while dashed lines and arrows indicate flow streamlines in the disk midplane. Material in the numerical grid zones surrounding the "planet" is assumed to be accreted, and is removed from the simulation (dark area at the center of the Hill sphere). In the midplane, gas flows into the Hill sphere primarily near the L_1 and L_2 Lagrange points (crosses), undergoes shocks, and settles into a prograde circumplanetary disk. From *Bate et al.* (2003).

bodies. Interactions between growing satellites and the gas disk would modify their orbits, generally causing them to become more circular and co-planar, and then on longer timescales to spiral inward toward the planet. However, a fundamental difference between satellite and planet growth is that the satellite formation environment around a gas planet is coupled to and regulated by the evolution of the planet and the surrounding solar nebula. This leads to disk conditions that are not adequately described as simply a scaled-down version of the solar nebula, a theme that will be emphasized here.

In this chapter, we focus on a relatively recent model (*Canup and Ward,* 2002, 2006; Ward and Canup, in preparation) in which the Galilean satellites form within a gaseous accretion disk produced at the very end of Jupiter's own accretion (e.g., Fig. 2). In this construct, the disk is actively supplied by an ongoing inflow of gas and small solids to circumjovian orbit as satellites are forming. Disk conditions evolve in a quasi-steady-state fashion in response to the evolution of Jupiter's own state and the rate of its gas accretion. This leads to lower disk gas densities and/or temperatures at the time of Galilean satellite formation than assumed in prior works. Within the disk, orbiting solids accumulate into satellites. Satellite growth rates are governed by the rate of delivery of solid particles from solar orbit, which leads to orders-of-magnitude slower satellite accretion than typically predicted by models that consider a static rather than an actively supplied disk. We envision that Jupiter had a circumplanetary disk for an extended time period, and that it is probable that a total mass of nebular gas and solids much larger than the amount needed to form the Galilean satellites would have been processed through this disk. As satellites grow to masses comparable to those of the Galilean satellites, their gravitational interactions with the gas disk can cause their orbits to spiral inward toward the planet. Earlier generations of satellites at Jupiter may have formed and been lost to collision with Jupiter, with the Galilean satellites we see today representing the last surviving generation of jovian satellites that formed as gas inflow to the planet ended.

In section 2, we begin by outlining key constraints on Galilean satellite origin that arise from the properties of the satellites themselves. We then describe the development and expected characteristics of a circumplanetary disk in the context of Jupiter's mass accretion and contraction (sec-

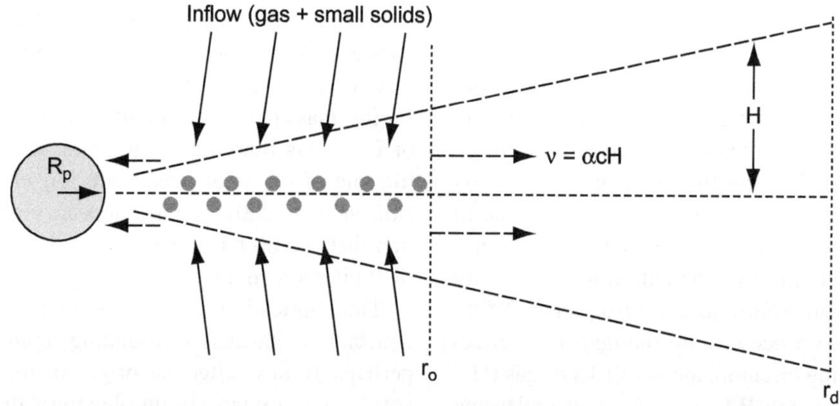

Fig. 2. An inflow-produced accretion disk. Inflowing gas and solids initially achieve centrifugal force balance across a region extending out to distance r_0, and solids accrete into satellites throughout this region. The gas spreads viscously onto the planet and outward to a removal distance, r_d, with $r_d \gg r_0$. The half-thickness of the gas disk is $H \sim c/\Omega$, where c is gas sound speed and Ω is orbital frequency, with $H/r \sim 0.1$. After *Canup and Ward* (2002).

TABLE 1. Satellite properties.

	a/R_P	M_S (10^{25} g)	R_S (km)	ρ_S (g cm^{-3})	$C/(M_S R_S^2)$
Io	5.9	8.93	1822	3.53	0.378
Europa	9.4	4.80	1562	3.01	0.346
Ganymede	15.0	14.8	2631	1.94	0.312
Callisto	26.4	10.8	2410	1.83	0.355

From *Schubert et al.* (2004), and chapter by Schubert et al.

tion 3). In section 4 we discuss the expected conditions within the disk at the time of the Galilean satellites' formation, as well as models of the growth (and loss) of large satellites within such a disk. Section 5 reviews primary outstanding issues and areas requiring future research, and in section 6 we outline potential implications for Europa's origin.

The formation of the Galilean satellites has been discussed in many other review chapters, including *Stevenson et al.* (1986), *Coradini et al.* (1989), *Pollack et al.* (1991), *Peale* (1999), *Lunine et al.* (2004), *McKinnon* (2007), *Peale* (2008), and the chapter in this volume by Estrada et al. The reader is referred to these reviews for other viewpoints on this topic.

2. CONSTRAINTS

A successful origin model must be consistent with basic properties of the Galilean satellites (see Table 1).

2.1. Satellite Masses

The Galilean satellites have similar masses, increasing by only about a factor of 3 between Europa and Ganymede. Together, the four satellites comprise a total satellite system mass, $M_T = 3.9 \times 10^{26}$ g, which is a small fraction of Jupiter's mass, $M_J = 1.9 \times 10^{30}$ g, with $(M_T/M_J) = 2.1 \times 10^{-4}$. This fraction is nearly identical to the analogous quantity for the saturnian satellite system (2.5×10^{-4}). Current interior models estimate that Jupiter and Saturn each contain substantial quantities of heavy elements other than H and He, with Jupiter having ~10–40 Earth masses ($M_\oplus$) in heavy elements and Saturn having ~2–8 $M_\oplus$ (e.g., *Guillot,* 2005). Yet their largest satellites are all relatively small, containing within a factor of 2 of the mass of the Moon, which contains a much larger fraction of Earth's mass ($\approx 0.01\ M_\oplus$). This suggests that processes that formed the gas planet satellites limited — and possibly even selected for — a maximum size to which the satellites could grow.

Assuming that the Galilean satellites formed from solar composition material (i.e., containing a gas-to-solids ratio of about 10^2), a minimum mass of ~0.02 M_J in gas and solids is required to produce them in the limit of perfectly efficient accretion. Historically, it is assumed that the satellites formed in a disk containing a comparable amount of material spread over an area extending from approximately the planet's surface to the orbit of the outermost satellite (e.g., *Lunine and Stevenson,* 1982; *Coradini and Magni,* 1984; *Coradini et al.,* 1989; *Pollack et al.,* 1991). Such a disk is referred to as the "minimum mass subnebula," or MMsN, in direct analogy to the "minimum mass solar nebula," or MMSN, defined for the circumsolar disk (e.g., *Hayashi et al.,* 1985). The corresponding MMsN gas surface density is on the order of $\sigma_{g,MMSN} \sim 0.02\ M_J/[\pi(30R_J)^2] \sim$ few $\times$ 10^5 g/cm^2, where $R_J = 7.15 \times 10^4$ km is Jupiter's radius. A variant on the MMsN has recently been advocated by *Mosqueira and Estrada* (2003) (see also the chapter by Estrada et al.), who invoke a similarly high gas density disk interior to ~15 R_J, accompanied by an outer, lower-density disk (with $\sigma_g \sim 10^3$–10^4 g/cm^2) in the region exterior to Ganymede's orbit. In general, the MMsN construct implies a gas-rich environment for Galilean satellite formation, in which circumplanetary disk gas surface densities are many orders of magnitude higher than those in the surrounding solar nebula.

Considering a preexisting disk as an initial condition reflects an implicit assumption that the disk forms quickly relative to the time it takes the objects within it to grow. For the circumsolar disk this assumption may be valid: The timescale over which the disk forms due to molecular cloud core collapse [~10^5–10^6 yr (e.g., *André et al.,* 2000)] may be shorter than the timescale for planets to accrete [~10^6–10^8 yr (e.g., *Nagasawa et al.,* 2007; *Lissauer and Stevenson,* 2007)]. However, for a circumplanetary disk, the timescale over which the disk forms is likely comparable to the formation time of the giant planet itself [~10^5–10^7 yr (e.g., *Lissauer and Stevenson,* 2007)], and this is much longer than the timescale for satellite growth predicted within an MMsN-type disk [$\lesssim 10^3$ yr (e.g., *Canup and Ward,* 2002)]. In such a situation, invoking a preexisting MMsN disk is not appropriate, because the disk's formation would be ongoing as the satellites grow. While it is then still the case that a minimum mass of 0.02 M_J in solar composition material must be delivered to the disk to produce the Galilean satellites, it need not be present in the disk all at one time (*Canup and Ward,* 2002). In a dynamically evolving disk, the instantaneous disk mass and surface density could be orders of magnitude less than that of the MMsN, so long as an appropriate total mass is processed through the disk during its lifetime.

To summarize, a minimum total mass in solar composition material ~0.02 M_J (corresponding to a minimum solid mass of ~2×10^{-4} M_J) must be delivered to circumjovian orbit to produce the Galilean satellites. The corresponding protosatellite disk surface density then depends on whether the disk is considered to be a closed system with a fixed total mass (as assumed in the MMsN models) or one that is actively supplied as the satellites grow (as we advocate here).

2.2. Compact Radial Scale

The Galilean satellites are located between about 6 and 26 R_J from the center of the planet. This is a quite compact configuration compared to the maximum orbital radius for gravitationally bound circumplanetary material, which is comparable to Jupiter's Hill radius, $R_{H,J} \equiv a_J(M_J/3\ M_\odot)^{1/3} \approx 744\ R_J$, where a_J = 5.2 AU = 7.78 × 10^{13} cm is Jupiter's semimajor axis and $M_\odot$ = 1.99 × 10^{33} g is the Sun's mass. While Jupiter's irregular satellites have much more extended orbits, their total mass is negligible compared to that of the four large satellites and their capture may have occurred much later (e.g., *Jewitt and Haghighipour,* 2007). Primordial process(es) that supplied the bulk of protosatellite material only resulted in large satellites on orbits relatively close to Jupiter, and this trend is also seen in the regular satellites of Saturn and Uranus.

2.3. Compositional Gradient

The decrease in mean satellite density with increasing distance from Jupiter (Table 1) is associated with a progressive increase in satellite ice mass fraction. Innermost Io is presently anhydrous, and its 3.53 g/cm^3 density suggests an interior composed of rock and metal. Europa, with a density of 3.01 g/cm^3, is thought to contain about 8% ice and water by mass, with the remainder rock (e.g., *Schubert et al.,* 2004). Outer Ganymede and Callisto have similar densities (1.94 and 1.83 g/cm^3, respectively), implying mixed ice/rock compositions and rock mass fractions ~0.55 for Ganymede and ~0.44 for Callisto (*McKinnon,* 1997). The compositional gradient seen in the Galilean satellites has traditionally been viewed as evidence that the satellites formed in a disk whose temperature decreased with orbital distance from the planet, with the ice-rich compositions of Ganymede and Callisto implying that the stability line for ice during the bulk of Galilean satellite growth was near the distance at which Ganymede formed (e.g., *Lunine and Stevenson,* 1982). This distance could be somewhat exterior to Ganymede's current orbital radius due to inward satellite migration (see section 4.3.4). Europa's small ice content would then imply that the ice stability line moved closer to Jupiter during the final stages of satellite growth, allowing limited ice accretion onto this satellite.

The above interpretation assumes that the current satellite densities reflect their initial compositions. However, it is possible — at least based on purely energetic arguments — that Europa could have begun with a more ice-rich composition, due to the potential for volatile loss associated with tidal dissipation over its history in the Laplace resonance. The current total heat flow from Europa has not been measured, but models predict $F_t \sim 6 \times 10^{18}$ erg/s to 3 × 10^{19} erg/s (e.g., *Hussmann et al.,* 2002; *Moore,* 2006). If heat generated by tidal dissipation were balanced by radiation from a uniform temperature surface, the implied surface temperature is only $T \sim [F_t/4\pi R_S^2\sigma_{SB}]^{1/4} \sim 30\ K(F_t/10^{19}\ erg/s)^{1/4}$, where R_S = 1561 km is Europa's radius and σ_{SB} is the Stephan-Boltzman constant, implying minimal sublimation loss of ice. Alternatively, warm tidally melted material could periodically erupt to the satellite's surface, with most of the energy lost through localized "hot-spots." In the limits that all tidally generated heat went into vaporizing water and that Europa has occupied the Laplace resonance for 4.5 × 10^9 yr [as predicted by some models, including *Peale and Lee* (2002)], a mass in ice equal to Europa's current mass (M_S = 4.8 × 10^{25} g) could have been lost over its history, because (10^{19} erg/s)(1.4 × 10^{17} s)/M_S ~ 3 × 10^{10} erg/g is comparable to both the latent heat of vaporization of water, ~2 × 10^{10} erg/g, and to the specific kinetic energy needed to escape Europa's gravity, which is also ~2 × 10^{10} erg/g. If Europa was initially ice-rich, Io could have been as well by the above arguments, given its much higher observed heat flow [$F_t \sim 10^{21}$ erg/s (e.g., *Moore et al.,* 2007)].

It is not clear, however, that conversion of tidal heat into vaporization and water vapor loss could have been efficient in an early, ice-rich Europa. Once its rocky core was surrounded by a thick layer of water and an ice shell, transport of hot material to the surface would likely have been limited (W. McKinnon, personal communication). If conversion of tidal heat into water vaporization was substantially less than 100% efficient at Europa, Europa's ice deficit is likely primordial, supporting the standard view that it formed predominantly in an environment too warm for ice. In this case Io would have also formed without substantial ice, because by virtue of being interior to Europa it would have been subject to higher disk temperatures.

2.4. Callisto's Interior

Galileo gravity data allow for estimates of the satellites' axial moments of inertia, C (Table 1), which can be used to constrain the degree of differentiation under the assumption that the satellites are in hydrostatic equilibrium. Such models imply that Io, Europa, and Ganymede are differentiated. For example, three-layer interior models for Ganymede that can account for its bulk density and low $C/M_SR_S^2$ = 0.3115 ± 0.0028 value (where M_S and R_S are the satellite's mass and radius) have a metallic core overlain by a rock mantle and an ice shell (e.g., *Schubert et al.,* 2004). It is unknown whether differentiation in the inner three Galilean satellites occurred during their accretion or long after, with the latter possible due to a combination of tidal heating, resonance passage(s) associated with their evolution into the Laplace resonance, and/or radiogenic heating from their rock components (e.g., *Friedson and Stevenson,* 1983; *Malhotra,* 1991; *Showman et al.,* 1997).

Unlike the inner Galilean satellites, outermost Callisto appears to be largely undifferentiated. Callisto's $C/M_SR_S^2$ =

0.3549 ± 0.0042 (*Anderson et al.,* 2001) is higher than that of differentiated Ganymede but lower than that predicted for a completely undifferentiated state [$C/M_SR_S^2$ ~ 0.38 including compression effects (*McKinnon,* 1997)]. An example two-layer model for Callisto (*Anderson et al.,* 2001) invokes an outer ice shell ~300 km thick overlying an undifferentiated, ~2100-km mixed rock-ice interior. While such solutions are nonunique, all involve some undifferentiated component in order to account for Callisto's inferred $C/M_SR_S^2$ (*Schubert et al.,* 2004). Some uncertainty in this interpretation persists, because the Galileo flybys were nearly equatorial and because Callisto rotates slowly. As a result, available data do not independently constrain J_2 and C_{22}, and so it has not been possible to determine the (J_2/C_{22}) ratio as an independent check on whether Callisto is well described by a hydrostatic state. If Callisto is nonhydrostatic, it is possible that it could be differentiated and still have a $C/M_SR_S^2$ consistent with the limited Galileo data, for example, due to structure on the surface of its rocky core (e.g., *McKinnon,* 1997). However, hydrostatic equilibrium is generally believed to be a good assumption given Callisto's size (*Schubert et al.,* 2004).

At face value, Galileo data imply that Callisto is only partially differentiated. Because differentiation is an irreversible event, the presence of a mixed ice/rock core or mantle in Callisto requires that the satellite avoided widescale melting of its ice throughout its entire history, including during its formation. This requirement in turn provides constraints on the rate and timing of Callisto's accretion, and by extension on the accretion rate and formation time of the other satellites as well.

The gravitational binding energy per unit mass associated with assembling Callisto is ~$(3/5)GM_S/R_S$ ~ 1.8 × 10^{10} erg/g, which is much larger than the latent heat of fusion for water ice, l ~ 3 × 10^9 erg/g. If a Callisto-sized satellite formed rapidly in ≲10^4 yr, its ice would melt throughout nearly its entire volume (e.g., *Stevenson et al.,* 1986), leading to complete differentiation (*Friedson and Stevenson,* 1983). Avoiding this outcome requires that Callisto formed slowly, so that radiative cooling counteracts accretional heating. The coldest satellite is achieved in the limit that Callisto forms by the accretion of small, low-velocity objects that deposit their impact energy at or very near its surface where it can be radiated away. In this limit, the temperature T_s at Callisto's surface is determined by the difference between the rates of accretional energy delivery and radiative cooling (e.g., *Pritchard and Stevenson,* 2000; *Barr and Canup,* 2008)

$$\frac{1}{2}\dot{M}_s v_{imp}^2 - \sigma_{SB}(T_s^4 - T^4)4\pi R_S^2 = \dot{M}_S C_p (T_s - T_0) \quad (1)$$

where M_S and R_S are the time-dependent satellite mass and radius, v_{imp} is the characteristic impact velocity of accreting material, σ_{SB} is the Stefan-Boltzmann constant, T_0 is the initial temperature of accreting matter, T is the temperature in the circumplanetary disk, and C_p ~ 1.5 × 10^7 erg/g/K is specific heat. Equation (1) assumes that the satellite radiates into an optically thick disk, which is probable for standard dust opacities (see section 4.1.2). In the region where Callisto formed, T was likely ~100 K (e.g., section 4.1.2 and Fig. 5b), while the minimum impact velocity is the satellite's escape velocity, $v_{esc} \equiv (2GM_S/R_S)^{1/2}$. Defining the accretion timescale $\tau_{acc} \equiv M_S/\dot{M}_S$, setting $v_{imp} = v_{esc}$ and $T = T_0$ = 100 K, and solving for the minimum τ_{acc} necessary to avoid melting of water ice in Callisto gives $\tau_{acc} \geq 5 \times 10^5$ yr (*Barr and Canup,* 2008; cf. *Stevenson et al.,* 1986).

Another potential heat source to a growing Callisto is radiogenic heating due to the decay of short-lived radioisotopes in the satellite's rocky components. The amount of radiogenic heating depends strongly on the absolute time that the satellite forms, because the primary heat source, ^{26}Al, decays with a half-life of 0.716 m.y. For Callisto to avoid melting due to radiogenic heating alone (assuming no retention of accretional energy), it must have completed its accretion no earlier than ~2.6–3 m.y. after the origin of the calcium-aluminum rich inclusions (CAIs) (*McKinnon,* 2006). Including both accretional and radiogenic heating implies that Callisto's formation was completed no earlier than about 4 m.y. after CAIs for a disk temperature T ~ 100 K (*Barr and Canup,* 2008).

An incompletely differentiated Callisto thus provides constraints on the start time of accretion, its accretion timescale, and therefore on the end time of its accretion. To avoid melting, the satellite must have (1) started accreting at least several million years after CAIs; (2) accreted its mass slowly, in ~5 × 10^5 yr or more; and (3) completed its accretion no earlier than ~4 m.y. after CAIs, given the total time implied by requirements (1) and (2). The completion time in requirement (3) is similar to mean observed circumsolar disk lifetimes [~3 m.y. (*Haisch et al.,* 2001)] and the time needed to grow Jupiter based on current core accretion models [~5 m.y. (*Hubickyj et al.,* 2005)]. Requirement (2), that τ_{acc} must be ≥5 × 10^5 yr, provides the most restrictive constraint on disk conditions, in particular on the supply rate of material to the disk. If all the material needed to form Callisto were orbiting Jupiter near the satellite's current orbit (the presumed starting condition of an MMsN disk), the satellite would accrete in ~10^3 yr (e.g., *Canup and Ward,* 2002) because of Callisto's short orbital period of only ~2 weeks. Yet Callisto must have formed orders of magnitude more slowly to avoid melting. Assuming that the Galilean satellites formed together through common overall processes, similarly long formation times are then implied for Io, Europa, and Ganymede as well.

2.5. Summary

Constraints that must be satisfied by any Galilean satellite origin model thus include (1) creation of a protosatellite disk to which a minimum of ~2 × 10^{-4} M_J in solids is supplied (corresponding to ~0.02 M_J in solar composition material); (2) final satellites with similar individual masses comparable to that of Earth's Moon; (3) an outer edge to the regular satellite region of approximately a few tens of jovian radii; (4) temperatures low enough for incorporation of substantial water ice into Ganymede and Callisto; and (5) protracted and relatively late satellite accretion, assuming that Callisto is indeed only partially differentiated. We now turn to a discussion of how such constraints might naturally be satisfied during the late stages of gas accretion onto Jupiter.

3. CIRCUMPLANETARY DISK DEVELOPMENT

As discussed in the introduction, a gas planet can develop a disk when either (1) its rotation rate increases to the point of rotational instability, so that the planet starts to shed material creating a so-called spin-out disk, or (2) the radius of the planet decreases below the outer radial extent of the pattern of inflowing gas, so that material flows directly into an accretion disk rather than directly onto the planet. Whether a spin-out or an accretion disk predominated for Jupiter is a function of the planet's gas accretion and contraction history.

Contraction models often assume that Jupiter accreted all its mass prior to its contraction (e.g., *Bodenheimer and Pollack,* 1986). However, it appears likely that Jupiter would have still been accreting gas while it contracted. *Papaloizou and Nelson* (2005) tracked a jovian planet's evolution including ongoing gas accretion, modeling the planet's state by numerically integrating stellar structure equations modified by the existence of a solid core [an approach also used by, e.g., *Bodenheimer and Pollack* (1986) and *Pollack et al.* (1996)]. Two classes of models were considered: Model A assumed the planet extends to the boundary of the Hill sphere and is relevant to the early phase of accretion, while Model B assumed a free boundary of a contracting planet that accretes mass from a circumplanetary disk, appropriate for Jupiter's final growth stages. The *Papaloizou and Nelson* (2005) models do not track the accreted angular momentum. Yet the existence of a circumplanetary disk at all implies a net angular momentum content of the inflowing gas, and for self-consistency, one must account for the removal of any excess angular momentum that would otherwise inhibit the planet's contraction. Ward and Canup (in preparation) have developed a model of disk formation around an accreting and contracting gas planet that focuses in particular on the angular momentum budget and the circumplanetary disk properties.

In this section we begin by reviewing estimates of the net angular momentum content expected for gas inflowing to Jupiter and the implied point at which rotational instability will occur as the planet contracts. We use this to first consider the simplest case of the contraction of a full-mass Jupiter in the limit of no gas accretion and no gas viscosity, and estimate the properties of a resulting spin-out disk. We then consider what we believe to be the more realistic case of a Jupiter that accretes gas as it contracts and is accompanied by a viscous circumplanetary disk. In this case, although a spin-out disk forms during the earliest stages of the planet's contraction, an accretion disk dominates once the planet has contracted to within the region of the current satellites.

3.1 Angular Momentum of Inflowing Gas

The net inflow of gas to Jupiter contained some average specific angular momentum, $j_c(t)$. This is difficult to estimate precisely without detailed hydrodynamic modeling, but some dimensional arguments can be employed to provide rough estimates (e.g., *Lissauer and Kary,* 1991). The planet will accrete gas parcels with both positive and negative angular momentum with respect to its center of mass. If we assume that the planet accretes all material that passes within some distance R' from its center, and that the gas flows on approximately Keplerian orbits, the net angular momentum delivery rate to the planet, $\dot{L} \sim \ell \dot{M} R'^2 \Omega_J$, can be simply obtained by integrating over the accreting region. Here $\dot{M}$ and Ω_J are the planet's mass accretion rate and orbital frequency (so that $\Omega_J \equiv \sqrt{GM_\odot/a_J^3}$ for Jupiter), and ℓ is the angular momentum bias, with $\ell > 0$ corresponding to a prograde rotating planet. For a two-dimensional inflow, appropriate if gas is accreted primarily within the planet's orbital plane, $\ell \approx 1/4$ (*Ruskol,* 1982; *Lissauer and Kary,* 1991; *Mosquiera and Estrada,* 2003). If we set the radius of the accreting region, R', to be comparable to the planet's Hill radius, $R_H \equiv a_J(M/3M_\odot)^{1/3}$, then

$$j_c \approx \ell R_H^2 \Omega_J = \frac{a_J^2 \Omega_J}{4}\left(\frac{M}{3M_\odot}\right)^{2/3} \tag{2}$$

where M is the planet's time-dependent mass. Other rationales for choosing R' are discussed by Ward and Canup (in preparation), but for simplicity we use equation (2) in the discussion here. We define a centrifugal radius for the inflowing material, r_c, as the distance from the planet's center at which the specific angular momentum of a circular orbit, $(GMr_c)^{1/2}$, equals that of the inflow average, j_c (e.g., *Cassen and Summers,* 1983). For equation (2), $r_c = \ell^2 R_H/3 \sim 0.02 R_H$, so that inflow to a full-mass Jupiter would have a centrifugal radius $r_c \sim 15\ R_J$.

Of course it is unlikely that at any given time, all the inflowing material has a specific angular momentum, j, equal to that of the average, j_c. Hydrodynamical simulations have not yet determined this angular momentum distribution. In the absence of such information, we adopt a heuristic model in which incoming mass ΔM is distributed evenly across a range of specific angular momenta $(j_c - \Delta j/2) < j < (j_c + \Delta j/2)$, so that there is a mass $dM \sim (\Delta M/\Delta j)dj$ in any interval dj. When this material achieves centrifugal balance at an orbital distance r from the planet's center, the corresponding range of specific angular momenta is $dj = (GMr)^{1/2} dr/2r$. For a mass inflow rate $\mathscr{F}$ this implies an inflow rate per unit area, f_{inflow}, to the planet's and/or disk's midplane of

$$f_{inflow} = \frac{\mathscr{F}}{4\pi r_c^2}\left(\frac{j_c}{\Delta j}\right)\left(\frac{r_c}{r}\right)^{3/2} \tag{3}$$

Here we will adopt $\Delta j \sim j_c$ for specificity, so that the inner and outer radial boundaries of the inflow region are $r_i \sim r_c/4$ and $r_o \sim 9r_c/4$. For a full-mass Jupiter, this corresponds to a region extending from about 4 R_J to 34 R_J. Although based on a simplified estimate of the angular momentum content of the inflowing gas, this range is roughly comparable to the Galilean satellite region. For comparison, three-dimensional hydrodynamic simulations with high spatial resolution within the planet's Hill sphere find an average specific angular momentum for gas inflowing to a Jupiter-mass planet of $j_c \sim 0.3\ R_H^2\Omega_J$, or $r_c \sim 22\ R_J$ (*Machida et al.,* 2008; *Machida,* 2008).

3.2. Rotational Instability

As Jupiter contracts, its spin frequency, $\omega \sim L/C$, increases, where L is its total angular momentum, $C = \lambda MR^2$ is its moment of inertia, and we will assume for simplicity that convection maintains nearly uniform rotation within the planet. A moment of inertia parameter can be roughly estimated from a polytropic model (e.g., *Chandrasekhar,* 1958). For example, for a polytropic index n = 2.5, appropriate for molecular hydrogen and the early stages of Jupiter's contraction, $\lambda \approx 0.17$. For comparison, Jupiter's current value is $\lambda = 0.26$.

When the planet's spin frequency reaches the critical value for rotational instability, $\omega_c \equiv (GM/R^3)^{1/2}$, the planet starts to shed material from its equator into the disk. The planet's cumulative angular momentum, L, is found by integrating equation (2), with $L = \int j_c dM = (3\ell/5)MR_H^2\Omega_J$. For a full mass Jupiter, this gives $L = \ell(5.4 \times 10^{47})$ g cm^2/s [much larger than the current angular momentum of Jupiter and its satellites augmented to solar composition, ~8.6×10^{45} g cm^2/s (e.g., *Korycansky et al.,* 1991)]. As Jupiter contracts, rotational instability sets in once L equals the critical value, $L_c = \lambda MR^2\omega_c$, which determines the critical planet radius, R_{rot}, for commencement of spin-out from the planet

$$R_{rot} = \frac{1}{3}\left(\frac{3\ell}{5\lambda}\right)^2 R_H \tag{4}$$

With $\ell = 1/4$, $\lambda = 0.17$, and $R_H = 744\ R_J$, this gives $R_{rot} \sim 200\ R_J$.

3.3. Contraction of a Full-Mass Jupiter: Properties of an Inviscid Spin-Out Disk

We first examine a case in which the bulk of the planet's mass has already been accreted before rotational instability sets in, so that there is no additional mass inflow to the planet as it contracts (i.e., $\mathcal{F} \approx 0$). Once $R \leq R_{rot}$, the planet begins to shed mass into a disk.

The behavior of the disk material depends on its ability to transport angular momentum via viscous stresses. If we consider the extreme of a completely inviscid disk, shed material remains in orbit where it first leaves the planet. In the absence of inflow, the change in the planet's mass is equal to that shed to the disk, $dM = 2\pi r\sigma_g dR$, where σ_g is the disk surface density. As the planet sheds mass at rate $\dot{M}$, its angular momentum decreases at a rate $\dot{L} = \dot{M}(GMR)^{1/2}$. Requiring the planet to maintain a critical rotation rate with $L = L_c = \lambda M(GMR)^{1/2}$ gives

$$\dot{L} = L\left(\frac{3}{2}\frac{\dot{M}}{M} + \frac{1}{2}\frac{\dot{R}}{R}\right) \tag{5}$$

which together with $\dot{L} = \dot{M}(GMR)^{1/2}$ implies

$$\frac{R}{M}\frac{dM}{dR} = \frac{\lambda}{2-3\lambda} \equiv q \tag{6}$$

This is easily integrated to find the mass retained in the contracting planet as a function of its radius, $M = M_0(R/R_{rot})^q$, where q = 0.11 for $\lambda = 0.17$, and M_0 is the planet's initial mass before its contraction. The total mass deposited in the spin-out disk is then $M_d = M_0\,[1 - (R/R_{rot})^q]$, with the final disk mass obtained by setting $R \approx R_J \ll R_{rot}$. If $R_{rot} \approx 10^2\ R_J$, $M_{final} \sim 0.6\ M_0$ so that requiring $M_{final} = M_J$ implies $M_0 = 1.66\ M_J$. The resulting inviscid spin-out disk would then contain ~$0.4\ M_0$, or ~66% of a Jupiter mass. If R_{rot} were $\approx 25\ R_J$, due, for example, to a factor of 2 lower value for ℓ, the disk mass decreases somewhat but is still ~$0.4\ M_J$. These disks are vastly more massive than the MMsN disk, which contains ~$0.02\ M_J$. Indeed, the mass is so large that a self-consistent treatment should include the disk's self-gravity. In addition, such massive disks would likely be gravitationally unstable, which would generate a viscosity and invalidate the starting assumption of an inviscid disk.

The discussion above assumes a largely convective and uniformly rotating planet. An alternative model is described by *Korycansky et al.* (1991), in which the planet's structure includes a differentially rotating radiative region of constant specific angular momentum between two uniformly rotating convective regions. This results in a more centrally condensed object with a larger critical radius for rotational instability. In this case, a smaller total disk mass is shed, but the disk has a much larger outer edge, ~300 planetary radii for the case of a Saturn-mass planet.

Thus given the estimated angular momentum budget associated with accreting Jupiter, the angular momentum that would need to be contained in an inviscid spin-out disk is so large that the resulting disk is too massive and/or too radially extended compared to the MMsN construct. To produce an appropriate MMsN-type disk would require instead that $\ell \sim 0.02$, or that R' be significantly less than the planet's Hill radius, which seems difficult to justify given results of hydrodynamic simulations (e.g., Fig. 1).

3.4. Viscosity in the Circumplanetary Disk

The starting condition of a static, nonviscous MMsN disk invoked in much early work on satellite formation appears difficult to reconcile with current angular momentum estimates and the subsequent contraction process for a Jupiter-like planet. However, a number of factors suggest that once gas is in circumplanetary orbit, it will not remain static but will instead radially spread due to viscous angular momentum transport. Perhaps the most compelling argument for viscosity is that viscous diffusion allows most of the disk's mass to evolve inward and be accreted by the planet, while most of the disk's angular momentum (contained in a small fraction of its mass) is transported outward where it can eventually be removed from circumplanetary orbit. Thus a much smaller disk mass can be consistent with the expected angular momentum budget than in the inviscid case. A circumplanetary disk also shares basic traits with the circumsolar disk, in which viscosity is commonly invoked to explain, e.g., observed mass accretion rates onto stars in extrasolar systems (e.g., *Stone et al.,* 2000).

The source and magnitude of viscosity in cold disks is actively debated, and while a number of candidate mechanisms have been proposed (see section 5), the mode and rate of angular momentum transport are quite uncertain. Given this uncertainty, a simplified parameterization is often used. The so-called "α model" (*Shakura and Sunyaev,* 1973) defines a viscosity $\nu = \alpha c H \approx \alpha c^2/\Omega$, where α is a constant, c is the gas sound speed in the disk, and $H \approx c/\Omega$ its vertical scale height (i.e., the disk's half thickness), with $\Omega \equiv (GM/r^3)^{1/2}$ the orbital frequency at radius r in the disk. The α model envisions angular momentum transport due to communication between turbulent disk eddies. The α-viscosity expression is similar to that for molecular viscosity, only with the molecular velocity replaced by an eddy velocity, $\alpha^{1/2}c$, and the molecular mean free path replaced by an eddy length scale, $\alpha^{1/2}H$ (e.g., *Dubrulle et al.,* 1995). Because the characteristic eddy velocity and length scale would be no larger than the disk sound speed and the disk thickness, respectively, the α parameter is a constant less than unity. Circumstellar disk models typically consider $10^{-4} \le \alpha \le 0.1$ (e.g., *Stone et al.,* 2000).

Angular momentum transport is characterized by a torque due to viscous shear, $g = 3\pi\sigma_g \nu j$, exerted by the inner disk on the outer disk across the circumference 2πr, where σ_g is the gas surface density, and $j = (GMr)^{1/2}$ is the specific angular momentum at orbital radius r; g is often referred to as the viscous couple (*Lynden-Bell and Pringle,* 1974). The viscous torque causes the angular momentum of an annular ring of mass $\delta m = 2\pi r\sigma_g \delta r$ to change at a rate

$$\delta m \frac{d}{dt}(j) = g(r) - g(r + \delta r) = -\frac{\partial g}{\partial r}\delta r \qquad (7)$$

With $(d/dt)j = (\partial j/\partial r)u$, where $u = (dr/dt)$ is the radial velocity of disk material with $u > 0$ defined as an outward velocity, equation (7) gives

$$2\pi r\sigma_g \delta r \frac{\partial j}{\partial r} u = -\frac{\partial g}{\partial r}\delta r \qquad (8)$$

Thus the couple drives an in-plane radial mass flux in the disk, $F = 2\pi r\sigma_g u$, and F satisfies (*Lynden-Bell and Pringle,* 1974)

$$F\frac{\partial j}{\partial r} = -\frac{\partial g}{\partial r} \qquad (9)$$

When the outer edge of the disk expands to a distance comparable to the planet's Hill radius, material escapes from the gravity of the planet and is returned to circumstellar orbit. Accordingly a viscous circumplanetary disk will have an outer edge $r_d = \gamma R_H$, where γ is a moderate fraction of unity, and a vanishing couple there, with $g(r_d) = 0$. Provided the viscous spreading time, $\tau_\nu \sim r_d^2/\nu$, is less than the characteristic timescale over which processes supplying material to the disk change, a quasi-steady-state is attained wherein the time-variation in the flux F can be ignored when integrating equation (9), so that $F \approx F(r)$ (see also section 3.5.5).

3.5. Contraction of Jupiter with Ongoing Gas Inflow: Evolution from a Viscous Spin-Out to an Accretion Disk

In section 3.2 we considered a planet that had acquired its full mass prior to its contraction. However, estimates of maximum gas accretion rates [$\sim 10^{-2}\ M_\oplus$/yr (e.g., *Hayashi et al.,* 1985)] imply that Jupiter would begin to contract to a size smaller than its Hill sphere once it exceeded ~0.2 to 0.3 M_J (e.g., *Tajima and Nakagawa,* 1997). In addition, predicted contraction timescales for Jupiter to achieve a size comparable to or less than the regular satellite region are shorter than the expected nebular lifetime and the duration of the inflow (e.g., *Papaloizou and Nelson,* 2005). Thus it seems most likely that for Jupiter, contraction and accretion occurred simultaneously. In this case, the system passes through stages in which the circumplanetary disk transitions from a spin-out disk to an accretion disk, as we next describe.

A comparison of the outer radial boundary for inflowing gas, $r_o \sim (9/4)(\ell^2/3)R_H$ (section 3.1), with the critical radius for instability R_{rot} from equation (4), reveals that $R_{rot}/r_o \sim (2/5\lambda)^2 \approx 5.5$ for $\lambda = 0.17$, with (R_{rot}/r_o) independent of the planet's mass. Thus for nominal parameter choices, Jupiter will begin to shed material into a spin-out disk first, while it is still larger than the outer boundary of the effective pattern of inflowing gas.

Figure 3 illustrates the three stages that Jupiter and its disk pass through as a function of the planet's time-dependent radius, R. A disk first appears when $r_o < R < R_{rot}$ (stage 1). During stage 1, inflowing material falls directly onto the planet, but because the planet is rotating at the critical rate it must shed material to the disk as it contracts. Thus stage 1 involves a purely spin-out disk. As the planet contracts further so that $r_i < R < r_o$ (stage 2), part of the inflowing gas falls directly onto the planet while part falls onto the disk, and the resulting disk can either be predominately a spin-out or an accretion disk. By the time Jupiter contracts to a radius smaller than that of the inner edge of the inflow pattern ($R < r_i$, stage 3), all the inflowing gas falls onto the disk, which approaches a pure accretion disk. We now discuss the properties of the circumplanetary disk as a function of time and position within the disk, assuming that the disk maintains a quasi-steady-state, following the

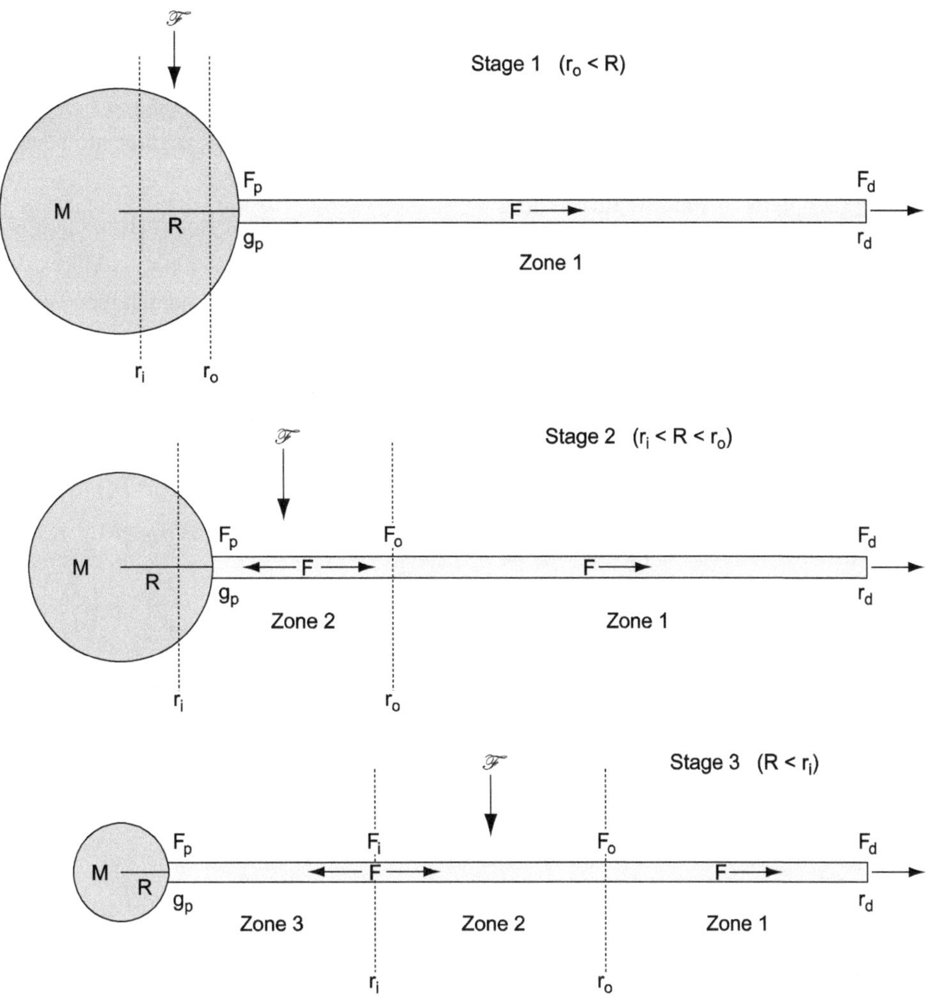

Fig. 3. Possible stages of the planet-disk configuration during a jovian planet's formation. Inflow $\mathcal{F}$ from the solar nebula occurs between r_i and r_o. These distances grow as the planet's mass M increases, while at the same time the planet's radius R contracts. In stage 1, the inflow is entirely on the planet; in stage 2 it is partially on the planet and partially on the disk; in stage 3 it is entirely on the disk. The disk inplane flux is F. The flux leaving the disk at its outer radius, r_d, is F_d, and is always positive; the flux at the planet, F_p, is the mass exchange rate between planet and disk and can be either positive (a spin-out disk) or negative (an accretion disk). In zone 2, the inplane flux changes with radius due to the infall, while in zones 1 and 3, there is no infall and in quasi-steady-state the zone fluxes remain constant.

framework of Ward and Canup (in preparation). At the end of this section we discuss when the steady-state assumption is likely to be valid.

3.5.1. In-plane mass flux. Wherever inflow impacts the disk, the continuity equation gives (*Canup and Ward,* 2002)

$$\frac{\partial \sigma_g}{\partial t} = f_{\text{inflow}} - \frac{1}{2\pi r}\frac{dF}{dr} \tag{10}$$

which for a disk in steady-state (i.e., $\partial\sigma_g/\partial t) = 0$) implies

$$\frac{dF}{dr} = 2\pi r f_{\text{inflow}} = \frac{\mathcal{F}}{2r_c^{1/2}r^{1/2}} \tag{11}$$

where equation (3) has been used to relate the inflow rate per disk area, f_{inflow}, to the global mass inflow rate, $\mathcal{F}$. Note that while (dF/dr) is positive definite, F itself can be positive or negative, depending on the position within the disk. Integration of equation (11) gives the flux variation with r for the various radial zones in the disk. We define two distance markers: The first, r_s, changes with the stage of the planet's contraction, while the second, r_z, changes with the position (zone) within the disk, viz.

$$r_s(R) \equiv \begin{cases} r_o & \text{for } r_o < R \text{ (stage 1)} \\ R & \text{for } r_i < R < r_o \text{ (stage 2)} \\ r_i & \text{for } R < r_i \text{ (stage 3)} \end{cases}$$

and (12)

$$r_z(r) \equiv \begin{cases} r_o & \text{for } r_o < r \text{ (zone 1)} \\ r & \text{for } r_i < r < r_o \text{ (zone 2)} \\ r_i & \text{for } r < r_i \text{ (zone 3)} \end{cases}$$

Using these, the flux can then be written compactly as

$$F(r) = F_p + \mathcal{F}\left[\left(\frac{r_z}{r_c}\right)^{1/2} - \left(\frac{r_s}{r_c}\right)^{1/2}\right] \quad (13)$$

where F_p is the flux at the inner edge of the disk denoting the mass exchange rate between the disk and the planet. Evaluating equation (13) at r_d tells us that the flux exiting the disk at the outer edge is $F_d = F_p + \mathcal{F}[(r_o/r_c)^{1/2} - (r_s/r_c)^{1/2}]$. With this, a useful alternative expression for F in terms of F_d can be found

$$F(r) = F_d - \mathcal{F}\left[\left(\frac{r_o}{r_c}\right)^{1/2} - \left(\frac{r_z}{r_c}\right)^{1/2}\right] \quad (14)$$

3.5.2. The viscous couple. To determine the structure of the disk, we integrate equation (9) with the appropriate boundary conditions for each zone. In zones 1 and 3 where the flux is constant in r, the combination Fj + g will be constant as well. However, in zone 2 where there is inflow onto the disk, we use Fdj/dr = d(Fj)/dr – jdF/dr to rewrite equation (9) as

$$\frac{d}{dr}(Fj + g) = j\frac{dF}{dr} \quad (15)$$

and integrate with the help of equation (11) (*Canup and Ward,* 2002). This gives

$$g = g_p + F_p j_p - Fj + \frac{\mathcal{F}}{2}\left[\left(\frac{r_z}{r_c}\right) - \left(\frac{r_s}{r_c}\right)\right] j_c \quad (16)$$

where $j_p = j(R)$, $j_c = j(r_c)$, and $g_p = g(R)$. Next, the boundary condition that the couple vanishes at the outer edge is applied to constrain its value at the planet, i.e.

$$g_p = F_d j_d - F_p j_p - \frac{\mathcal{F}}{2}\left[\left(\frac{r_o}{r_c}\right) - \left(\frac{r_s}{r_c}\right)\right] j_c \quad (17)$$

where $j_d = j(r_d)$. Equations (14) and (17) can now be used to eliminate F and g_p in equation (16) to yield (Ward and Canup, in preparation)

$$g = F_d(j_d - j) + \mathcal{F}\left[\left(\frac{r_o}{r_c}\right)^{1/2} - \left(\frac{r_z}{r_c}\right)^{1/2}\right] j - \frac{\mathcal{F}}{2}\left[\frac{r_o}{r_c} - \frac{r_z}{r_c}\right] j_c \quad (18)$$

In a quasi-steady-state, conservation of mass dictates that the rate of inflow to the system must equal the rate of change of the planet's mass plus any mass loss at the outer edge, i.e., $\mathcal{F} = \dot{M} + F_d$. Substituting this into equation (18), setting $r_o = 9r_c/4$, $g = 3\pi\sigma_g \nu j$, and dividing throughout by j gives

$$3\pi\sigma_g\nu = (\mathcal{F} - \dot{M})\left[\left(\frac{r_d}{r}\right)^{1/2} - 1\right] + \mathcal{F}\left[\frac{3}{2} - \left(\frac{r_z}{r_c}\right)^{1/2}\right] - \frac{\mathcal{F}}{2}\left[\frac{9}{4} - \frac{r_z}{r_c}\right]\left(\frac{r_c}{r}\right)^{1/2} \quad (19)$$

The remaining unknown is $\dot{M}$, and to find this, the planet's evolution including the disk interaction must now be considered.

3.5.3. Evolution of the planet. A constraint on $\dot{M}$ is provided by the planet's angular momentum evolution

$$\dot{L} = \frac{\mathcal{F}}{2}\left[\frac{r_s}{r_c} - \frac{r_i}{r_c}\right] j_c - (F_p j_p + g_p) \quad (20)$$

where the first term is the angular momentum falling directly on the planet. Using equation (17) to eliminate g_p, this becomes $\dot{L} = \mathcal{F} j_c - F_d j_d = \mathcal{F} j_c + (\dot{M} - \mathcal{F}) j_d$, which is just conservation of angular momentum and when equated to equation (5) implies

$$\left[3 - \frac{2}{\lambda}\left(\frac{r_d}{R}\right)^{1/2}\right]\frac{\dot{M}}{M} + \frac{\dot{R}}{R} = -\frac{2}{\lambda}\left(\frac{r_d}{R}\right)^{1/2}\frac{\mathcal{F}}{M}\left[1 - \left(\frac{r_c}{r_d}\right)^{1/2}\right] \quad (21)$$

In the limit $(r_d/R)^{1/2} \gg 1$, this expression reduces to

$$\dot{M} \approx -\frac{\lambda}{2}\left(\frac{R}{r_d}\right)^{1/2}\left|\frac{\dot{R}}{R}\right| M + \mathcal{F}\left[1 - \left(\frac{r_c}{r_d}\right)^{1/2}\right] \quad (22)$$

Note that the disk outflow is

$$F_d = \mathcal{F} - \dot{M} \approx \frac{\lambda}{2}\left(\frac{R}{r_d}\right)^{1/2}\frac{M}{\tau_{KH}} + \mathcal{F}\left(\frac{r_c}{r_d}\right)^{1/2} \quad (23)$$

where we define an instantaneous planet contraction timescale, $\tau_{KH} \equiv |R/\dot{R}|$. The flux across the disk's outer edge, F_d, is always positive and contains a contribution generated by the contraction and one by the inflow. However, the flux at the planet is

$$F_p = F_d - \mathcal{F}\left[\frac{3}{2} - \left(\frac{r_s}{r_c}\right)^{1/2}\right] \quad (24)$$

which can be either positive or negative. In stage 1, $F_p = F_d > 0$, and the disk is due to spin-out. However, F_p progressively decreases from F_d in stage 2, until in stage 3, $F_p = F_d - \mathcal{F}$, and the planet accretes material from the disk ($F_p < 0$) if

$$\frac{\lambda}{2}\left(\frac{R}{r_d}\right)^{1/2}\frac{M}{\tau_{KH}} < \mathcal{F}\left[1 - \left(\frac{r_c}{r_d}\right)^{1/2}\right] \quad (25)$$

3.5.4. Example planet contraction and disk formation history. Equations (19) and (22) can be used to determine the evolution of the planet mass and the disk structure as a function of the inflow and contraction rates. Alternatively, if $\dot{M}(t)$ and $\dot{R}(t)$ are known from an accreting giant planet contraction model, such as given by *Papaloizou and Nelson* (2005, their Fig. 11), shown here in Figs. 4a,b, equation (22) can be used to eliminate $\mathcal{F}$ in equation (19). Figure 4c shows

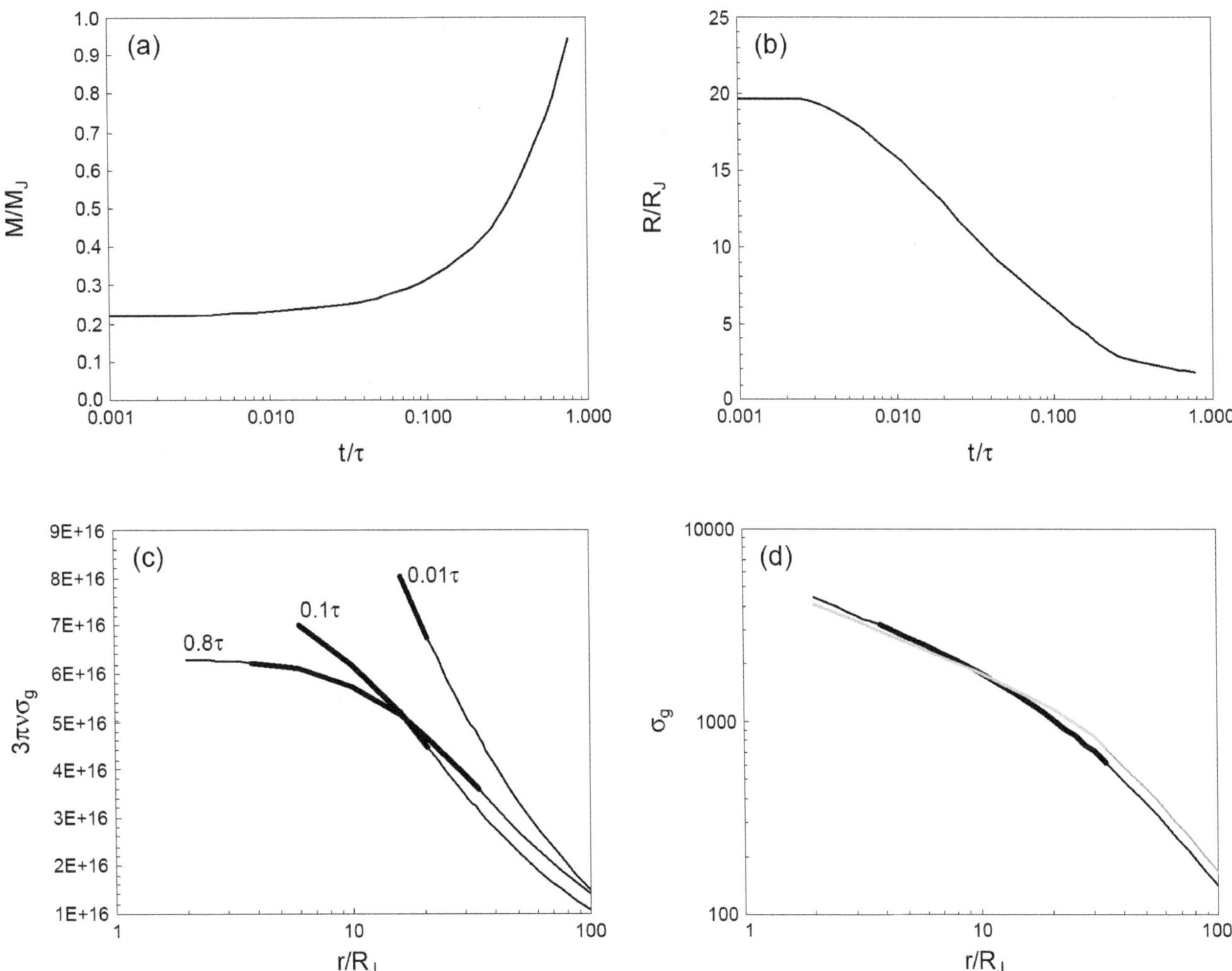

Fig. 4. Circumplanetary disk properties produced during an example Jupiter growth and contraction history. **(a),(b)** Planet mass in Jupiter masses and planet radius in jovian radii shown as a function of time in units of the planet accretion timescale, τ. This accretion and planet contraction history is from *Papaloizou and Nelson* (2005), and assumes a 5 $M_\oplus$ core, $\tau = 9 \times 10^5$ yr, and the opacity model of *Bell and Lin* (1994). **(c)** Black lines show $3\pi\sigma_g\nu$ (in cgs units) for the resulting disk produced by both inflow and spin-out, per equation (19), as a function of radial distance from the planet's center in Jupiter radii. The three curves correspond to three times in the planet history shown in **(a)** and **(b)**: $t = 0.01\tau$, 0.1τ, and 0.8τ. Bold sections indicate the inflow-supplied region (i.e., where $r_i \leq r \leq r_0$), assuming $\lambda = 0.17$, $l = 1/4$, and $r_d = 0.5R_H(M)$. **(d)** Sample predicted disk gas surface densities as a function of radial distance in planet radii for the $t = 0.8\tau$ curve in **(c)**, generated using both the Ward and Canup model detailed in section 3.5 (black) and the *Canup and Ward* (2002) model (gray). The latter is a simpler model that considers a uniform inflow per disk area and that the disk is a pure accretion disk. Both curves assume $\alpha = 10^{-3}$, a disk temperature of 1000 K at $r = 5$ R_J, and that the disk temperature decreases with $1/r$, so that $\nu \propto r^{1/2}$. Bold sections indicate the predicted inflow delivery region.

the resulting behavior of $3\pi\sigma_g\nu$ as a function of r at various times during the evolution. With time the inflow slows and the planet contracts, causing the inner disk edge to move inward and the peak value for $3\pi\sigma_g\nu$ to decrease. For the case shown here, once the planet contracts to a radius smaller than about 13 R_J, the flux at the planet becomes negative ($F_p < 0$) and the disk is increasingly well approximated as an accretion disk. Figure 4d shows the predicted gas surface density profile, $\sigma_g(r)$, at the time when the planet has accreted more than 90% of its final mass, assuming a simple model for the viscosity with $\nu \propto r^{1/2}$ (see caption for Fig. 4). Shown for comparison is the analogous surface density profile using the simpler model of *Canup and Ward* (2002) (see section 4), which considers a pure accretion disk and ignores possible contributions from spin-out. At late times in Jupiter's growth, the full disk model described in this section approaches that of a pure accretion disk.

3.5.5. Discussion. The coupled planet-disk evolution model described here assumes that the viscous torque between the disk and the surface of the planet will maintain the planet's rotation rate at the limit of rotational stability as the planet contracts. Yet ultimately, Jupiter must be left with less angular momentum than this to account for its current ~10-hour rotational day, which is about a factor of 3 longer than the critical rotation period for an $M = M_J$ and $R = R_J$ planet.

Accounting for subcritical rotation is a well-known issue for Jupiter and Saturn, as well as for protostars. Protostars, also believed to grow through mass delivered through a viscous accretion disk, have observed rotation rates that are often much slower than breakup (e.g., *Herbst et al.,* 2002). Proposed solutions to account for protostar angular momentum loss may also apply to gas giant planets, including (1) "disk-locking," or magnetic coupling between the central object and its disk that results in angular momentum transferred to the disk beyond the corotation radius [where the Kepler angular velocity equals the primary rotational velocity (e.g., *Konigl,* 1991; *Takata and Stevenson,* 1996)]. and (2) magnetocentrifugally driven "X-winds" that originate from the inner accretion disk, diverting mass and angular momentum that would otherwise be delivered to the primary (e.g., *Shu et al.,* 2000).

The first alternative has been explored in the context of a circumjovian disk by *Takata and Stevenson* (1996). From their work, one can derive a magnetic torque density of the form $dT_M/dr = -2B_J^2R^2Re(R/r)^3(\Omega - \omega)/\omega_c$, where B_J is the midplane magnetic field strength at R, and Re is the magnetic Reynolds number that depends in part on the conductivity of the disk material. Integrating the torque density throughout the disk gives $T_M = B_J^2R^3Re[\omega/\omega_c - (4/7)]$. The magnetic torque on the planet, $-T_M$, is thus negative for $\omega > 4\omega_c/7$, which could cause its rotation to become subcritical at some point as the inflow and disk mass decrease. The current spin rate of Jupiter is about 0.3 times its critical value. However, if the disk surface density becomes low enough that the magnetic torque density can force it to corotate with the planet (as a magnetosphere) out to some distance $r_M > R$, the total magnetic torque on the planet can remain negative for rotation rates below $4\omega_c/7$. A full treatment of this issue will involve modifying the disk profile to include the magnetic torque and assessing the expected degree of ionization in a gas-starved disk, which will be a topic of future work.

The model described in this section also assumes that the inplane flux can be approximated as a quasi-steady-state. This is a valid approximation when the viscous spreading time of the circumplanetary disk, τ_v, is short compared to both the planet's contraction timescale and the timescale over which the inflow changes. The viscous timescale is $\tau_v = r_d^2/\nu \approx 3 \times 10^4 \text{ yr}(10^{-3}/\alpha)(r_d/300\,R_J)^{3/2}(0.1/\{H/r\})^2$. In the planet's early contraction phase, the contraction timescale τ_{KH} was likely shorter than τ_v (e.g., Fig. 4b), so that the resulting disk would be more massive than that predicted with a steady-state analysis, although this may be mitigated by gravitational instability, which could increase the effective viscosity. However, once the planet's contraction rate slows so that $\tau_v < \tau_{KH}$, disk material left in orbit earlier has time to viscously spread out to r_d, and a quasi-steady-state is achieved. Similarly, an early, potentially rapidly varying gas inflow to the planet would produce a non-steady-state disk, if a disk existed at all at that time. However during the planet's final growth, changes in the gas inflow rate were most likely regulated by the removal of the solar nebula itself with a characteristic timescale of ~10^6–10^7 yr (e.g., *Haisch et al.,* 2001), which is much longer than τ_v. Thus as Jupiter approached its final mass, a surrounding viscous disk would have been well described as a quasi-steady-state.

4. DISK CONDITIONS AND GROWTH OF SATELLITES

We argued in section 3 that in the late stages of its formation, Jupiter was surrounded by an inflow-supplied accretion disk. This contrasts with models that consider the protosatellite disk as an isolated system with a fixed total mass and no ongoing gas or solid supply (e.g., *Lunine and Stevenson,* 1982; *Mosqueira and Estrada,* 2003). An actively supplied accretion disk has been considered by multiple works. *Coradini et al.* (1989) computed steady-state conditions within a viscous accretion disk, assuming that the instantaneous disk mass would need to be equivalent to that of the MMsN in order to produce the Galilean satellites. Creating a disk this massive required rapid inflow rates, which in turn implied high disk temperatures inconsistent with icy satellites. As such, Coradini et al. proposed a two-stage evolution, with an accretion disk forming during the first stage, followed by a second stage of satellite growth after gas inflow to the disk stopped and the disk cooled. *Makalkin et al.* (1999) observed that appropriately low disk temperatures for producing icy satellites could result during gas inflow if the inflow rate was slow, but found that in this case the disk mass at any one time would be much less than that of the MMsN. Makalkin et al. concluded that the total mass and temperature constraints needed to produce the Galilean satellites could not be simultaneously satisfied in an accretion disk. *Canup and Ward* (2001, 2002) showed that a disk produced by a slow-inflow would not only have low temperatures, but could also (1) allow for survival of Galilean-sized satellites against orbital decay due to density wave interactions with the gas disk, and (2) lead to slow satellite accretion, as needed for consistency with Callisto's apparent interior state. While the Canup and Ward disk density was much lower than the MMsN, they argued that an appropriate total mass in solids could nonetheless build up in the disk over time to form perhaps multiple Galilean-like systems, with earlier satellites lost to collision with the planet. *Stevenson* (2001a,b) also advocated an actively supplied circumjovian disk, which he referred to as a "gas-starved" disk. *Alibert et al.* (2005) coupled an accretion disk model to a time-dependent prescription for gas delivery to Jupiter, and predicted satellite compositional and loss patterns assuming a linear satellite mass growth rate. However, in their models, inflowing mass was delivered only at the outer edge of the circumplanetary disk, and angular momentum was not carried away by outflow from this point. Hence to maintain quasi-equilibrium, the assumed inflow would have to add almost no net angular momentum, and it is unclear why such a flow would form a disk. *Canup and Ward* (2006) performed direct simulations of satellite accretion within a disk supplied by an ongoing, time-dependent inflow and found that systems similar to the Galilean satellites could be created over a relatively broad range of disk properties.

In this section, we begin by evaluating conditions expected within an accretion disk at the time the Galilean satellites formed, including disk density and temperature profiles. We then discuss possible sources of solid material (rock and ice) to the disk, the conditions in which such material will tend to accumulate into satellites, and the expected satellite growth timescales. The effects of density wave interactions between the growing satellites and the background gas disk are then described, including the tendency for massive satellites to spiral into the planet through inward migration. We demonstrate that migration selects for a maximum satellite mass and a common total satellite system mass, and conclude by showing example satellite systems produced by N-body accretion simulations.

4.1. Properties of the Gas Disk

To characterize the protosatellite environment, we wish to estimate how the disk gas density and temperature evolve. In the discussion to follow, we adopt the disk model of *Canup and Ward* (2002), which assumes that Jupiter's disk was a pure accretion disk that derived its mass solely from gas inflowing from solar orbit. As we have shown above (e.g., Fig. 4d), this should be a good approximation during the late stages of Jupiter's growth when spin-out from the planet can be neglected. However future work should incorporate models (such as that described in section 3) that treat the inflow and the planet's state self-consistently.

4.1.1. Gas surface density. In a pure accretion disk, the gas surface density reflects a balance between the inflow supply of gas, and gas removal as it viscously spreads onto the planet or beyond the disk's outer edge, r_d (e.g., Fig. 2). Using methods similar to those employed in section 3, the steady-state gas surface density, σ_g, can be determined as a function of $\mathcal{F}$, ν, r_d, and the outer edge of the inflow region, r_o (*Lynden-Bell and Pringle,* 1974; *Canup and Ward,* 2002). For the simple case of a uniform inflow (i.e., f_{inflow} = constant for $r < r_0$), this gives

$$\sigma_g(r) \approx \frac{4\mathcal{F}}{15\pi\nu}\begin{cases} \frac{5}{4} - \left(\frac{r_o}{r_d}\right)^{1/2} - \frac{1}{4}\left(\frac{r}{r_o}\right)^2 & \text{for } r < r_o \\ \left(\frac{r_o}{r}\right)^{1/2} - \left(\frac{r_o}{r_d}\right)^{1/2} & \text{for } r \geq r_o \end{cases} \tag{26}$$

(*Canup and Ward,* 2002). The surface density is proportional to the mass inflow rate, $\mathcal{F}$, so a more massive circumplanetary disk is expected during rapid inflow and σ_g decreases as the inflow wanes. The surface density is inversely proportional to ν because as the viscosity increases, the disk spreads more rapidly, which reduces σ_g for a fixed inflow rate.

4.1.2. Disk temperature. The disk temperature affects the composition of accreting material as well as the disk's viscosity in the α-model, in which ν depends on the square of the gas sound speed, c^2, and therefore on the disk midplane temperature, T, with $c^2 = \gamma\Re T/\mu_{mol}$ [where $\gamma = 1.4$ is the adiabatic index and $\mu_{mol} \sim 2$ g/mol for molecular hydrogen, and $\Re = 8.31 \times 10^7$ erg/(mol K) is the gas constant]. The disk is heated by luminosity from Jupiter (which contributes even for a radially optically thick disk if the disk's vertical scale height increases with distance from the planet), viscous dissipation, and energy dissipation associated with the difference between the free-fall energy of the incoming gas and that of a Keplerian orbit. These energy sources are balanced by radiative cooling, which occurs predominantly in the vertical direction for a disk that is thin compared to its radial extent.

Jupiter's luminosity provides an energy rate per unit area to the region between orbital radii r to (r + Δr)

$$\begin{aligned} \dot{E}_J &= 2\sigma_{SB}T_J^4\left(\frac{R_J}{r}\right)^2 2\pi r[H(r+\Delta r) - H(r)]/2\pi r\Delta r \\ &\approx \frac{9}{4}\sigma_{SB}T_J^4\left(\frac{R_J}{r}\right)^2\left(\frac{c}{r\Omega}\right) \end{aligned} \tag{27}$$

where T_J is Jupiter's temperature, and a $T \propto 1/r^{3/4}$ dependence has been assumed in calculating H(r) (*Canup and Ward,* 2002). The energy production rate per area due to viscous dissipation is $\dot{E}_\nu = (9/4)\nu\Omega^2\sigma_g$, while that due to the infalling gas is $\dot{E}_{in} \approx (GM_J/2r)(\mathcal{F}/\pi r_o^2)$. The equation for thermal balance is (e.g., *Canup and Ward,* 2002)

$$\begin{aligned} &\frac{9}{4}\left[\sigma_{SB}T_J^4\left(\frac{R_J}{r}\right)^2\frac{c}{r\Omega} + \zeta\alpha c^2\Omega\sigma_g\right] = \\ &2\sigma_{SB}T_{eff}^4[1 - (T_{neb}/T_{eff})^4] \end{aligned} \tag{28}$$

where $\zeta \equiv 1 + 3/2[(r_o/r)^2 - 1/5]^{-1}$ accounts for the $\dot{E}_{in}$ term (generally small compared to the other energy sources), T_{neb} is the ambient temperature into which the circumplanetary disk radiates, T_{eff} is the disk's effective temperature, and the factor of 2 on the righthand side is because the disk has both upper and lower surfaces. Equation (28) assumes a radiative disk; convective energy transport is generally a small contribution, even in a turbulent disk (e.g., *Ruden and Lin,* 1986).

Prior Galilean satellite formation models have set T_{neb} equal to the solar nebula temperature at Jupiter's orbit, which recent works estimate to be ~100 K (e.g., *Garaud and Lin,* 2007). However a Jupiter-mass planet will likely open a gap in the circumsolar disk (e.g., *Bate et al.,* 2003), allowing the circumplanetary disk and planet to radiate to a reduced density and colder background over a substantial fraction of the available solid angle. Numerical simulations by *D'Angelo et al.* (2003) consider the thermohydrodynamic evolution of circumstellar disks containing Jupiter-sized protoplanets. Assuming standard dust opacities, they find optical depths within the gap region are typically on the order of unity (depending on the choice of viscosity), with gap temperatures of $T_{neb} \sim 20$ to 50 K. Note, however, that because the term in equation (28) that accounts for the

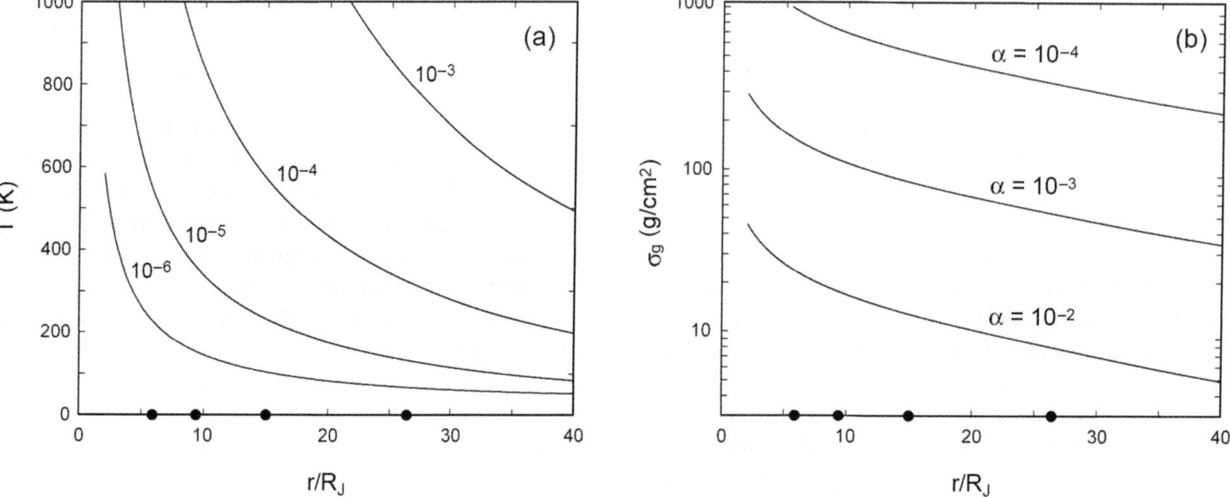

Fig. 5. Properties in a protosatellite accretion disk. Current positions of the Galilean satellites are shown as circles along the x-axes. **(a)** Radial midplane temperature profile shown as a function of inflow rate in Earth masses per year. All four curves assume $\alpha = 0.005$, $r_0 = 30R_P$, $r_d = 300R_P$, $T_{neb} = 20$ K, $T_J = 500$ K, and $\kappa_R \sim \kappa_P = 1$ cm^2/g. As the inflow to the planet progressively slows, disk temperatures fall, with inflow rates $<10^{-5}$ $M_\oplus$ yr^{-1} yielding water ice stability in the Callisto-Ganymede region. Varying the viscosity and/or opacity for a given inflow rate affects the midplane temperature as $T \propto (\kappa_R/\alpha)^{1/5}$. Conditions during Galilean satellite accretion were likely intermediate to the 10^{-5} $M_\oplus$ yr^{-1} and the 10^{-6} $M_\oplus$ yr^{-1} curves. **(b)** Radial gas surface density profile for $\mathcal{F} = 10^{-5}$ $M_\oplus$ yr^{-1}, shown as a function of the assumed disk viscosity α-parameter.

ambient temperature is proportional to $(T_{neb}/T_{eff})^4$, it becomes unimportant when the effective temperature substantially exceeds T_{neb}, which would be the case during most of the disk's evolution in the region of the regular satellites (Fig. 5a).

The temperature T_{eff} is that of an equivalent blackbody emitting the same radiative flux as the disk. For an optically thick disk, T_{eff} is the disk's surface temperature. For an optically thin disk, the disk is isothermal, but T_{eff} is less than the disk's temperature, reflecting a reduced emissivity. Given equation (28), an additional relationship is needed to determine the disk midplane temperature, T. A generalized calculation of a disk's vertical temperature profile is a complex radiative transfer problem, although analytical approximations exist for many cases. In the limit that viscous dissipation occurs primarily in the disk midplane, while heating due to Jupiter's luminosity and infalling mass are concentrated at the disk's surface, an approximate expression for the midplane temperature appropriate in both high and low optical depth regimes is (*Nakamoto and Nakagawa,* 1994)

$$2\sigma_{SB}T^4 = \left(1 + \frac{3}{8}\kappa_R\sigma_g + \frac{1}{2\kappa_P\sigma_g}\right)\dot{E}_\nu + \left(1 + \frac{1}{2\kappa_P\sigma_g}\right)(\dot{E}_J + \dot{E}_{in}) + 2\sigma_{SB}T^4_{neb} \tag{29}$$

where κ_R and κ_P are the disk's frequency-averaged Rosseland and Planck opacities. The Rosseland mean opacity is relevant for optically thick regions, while the Planck mean opacity is relevant for optically thin regions in which scattering can be ignored and cooling occurs predominantly in resonance lines of abundant species (e.g., *Hubeny,* 1990). For a solar composition gas, κ_P is always much larger than κ_R at low temperatures; e.g., at 500 K, $\kappa_P \sim O(10^{-1})$ cm^2/g while $\kappa_R \sim O(10^{-5})$ cm^2/g (*Helling et al.,* 2000). If dust grains are present, opacities are much higher, with $\kappa_R \sim \kappa_P \sim$ a few cm^2/g (e.g., *Semenov et al.,* 2003). In the next section, we will argue that small particles appear the most promising source of solid material to an accretion disk, and in this case it seems reasonable to imagine an accompanying population of dust grains, although their size distribution is unknown.

Equations (28) and (29) simplify when viscous dissipation is the dominant energy source (see also *Hubeny,* 1990), as is typically the case in most of the regular satellite region (*Canup and Ward,* 2002). For $T^4_{eff} \gg T^4_{neb}$

$$T^4_{eff} \approx \frac{9\Omega^2}{8\sigma_{SB}}\nu\sigma_g = \frac{3\Omega^2\mathcal{F}}{8\pi\sigma_{SB}}\left[1 - \frac{4}{5}\sqrt{\frac{r_o}{r_d}} - \frac{1}{5}\left(\frac{r}{r_o}\right)^2\right] \tag{30}$$

$$T^4 \approx \left(1 + \frac{3}{8}\kappa_R\sigma_g + \frac{1}{2\kappa_P\sigma_g}\right)T^4_{eff} \tag{31}$$

Equation (30) shows that the effective disk temperature is a function of the inflow rate, $\mathcal{F}$, but is *independent* of the gas surface density and viscosity. This is because from equation (26), $\sigma_g \propto 1/\nu$, so that the combination $(\nu\sigma_g)$ in equation (30) does not depend on viscosity. The behavior of the midplane temperature, T, depends on whether the disk is optically thick or thin. In an optically thick disk (with optical depth $\tau_R \equiv \kappa_R\sigma_g > 1$), the second term on the right-hand side of equation (31) dominates. Because $\sigma_g \propto \mathcal{F}/\nu \propto$

$\mathscr{F}/(\alpha T)$, this gives $T \propto (\mathscr{F}\kappa_R/\alpha)^{1/5}T_{eff}^{4/5} \propto \mathscr{F}^{2/5}(\kappa_R/\alpha)^{1/5}$, so that the midplane temperature depends rather weakly on the opacity and viscosity, and is also more strongly affected by the inflow rate. For a dusty disk with $\kappa_R \sim$ unity, the optically thick regime applies for $\sigma_g > O(1)$ g/cm^2, or for gas surface densities as small as 10^{-5} times that of the MMsN. For lower opacities, the optically thin regime applies for higher gas surface densities, and in this case the Planck term dominates, the disk is isothermal, and the viscosity and opacity are important in setting T.

For an optically thick disk supplied by an inflow, the inflow rate is a key control on the disk midplane temperature, unlike the frequently studied case of a disk with a fixed total mass, in which the viscosity is the primary variable that controls the disk temperature (e.g., *Lunine and Stevenson,* 1982). Equations (30) and (31) provide an important constraint on the conditions needed to produce the Galilean satellites, because they define a maximum inflow rate consistent with temperatures low enough for forming ice-rich satellites.

Figure 5a shows the midplane temperature as a function of the assumed inflow rate from a solution of equations (26), (28), and (29). Disk temperatures low enough for water ice [which is stable below ~150 to 200 K, depending on pressure (e.g., *Prinn and Fegley,* 1989; see also *Canup and Ward,* 2002)] in the $r \geq 15\ R_J$ region require $\mathscr{F} \leq 6 \times 10^{-6}\ M_\oplus$ yr^{-1} for the parameter choices in Fig. 5a. A solar composition inflow having this inflow rate would take $\geq 10^6$ yr to supply a mass in solids comparable to that of the Galilean satellites. Figure 5a shows that temperatures remain too high for ices in the innermost circumjovian disk even for very slow inflow rates, consistent with Io's anhydrous composition. While T does not vary strongly with the choice of disk viscosity, the resulting gas surface density varies as $1/\nu$, so that an inflow with $\mathscr{F} = 10^{-5}\ M_\oplus$ yr^{-1} would for $10^{-2} \leq \alpha \leq 10^{-4}$ produce a disk having $\sigma_g \sim 10$–10^3 g/cm^2 in the satellite region (Fig. 5b). A dusty disk having these surface densities would be optically thick. The corresponding disk pressures for the Fig. 5b curves range from ~10^{-5} to 10^{-2} bars in the $r < 30\ R_P$ region. These are much lower surface densities and pressures than that of an MMsN disk, with $\sigma_{g,MMsN}$ ~few × 10^5 g/cm^2, and characteristic pressures ~1–10 bars (e.g., *Prinn and Fegley,* 1989). A disk with a low, gaseous opacity and a lower viscosity ($\alpha = 10^{-4}$) has ice stability in the $r \geq 15\ R_J$ region for $\mathscr{F} < 5 \times 10^{-5}\ M_\oplus$ yr^{-1}, implying that $>10^5$ yr would be needed to supply a mass in solids equal to that of the Galilean satellites, assuming a solar composition inflow.

Because the satellite growth rate is regulated by the rate that solids are supplied to the disk (see section 4.2.3), a maximum inflow rate consistent with icy Ganymede and Callisto corresponds to a minimum satellite accretion timescale. Across a wide range of disk parameters, satellite formation timescales on the order of 10^5 to 10^6 yr are then implied. Such timescales seem reasonable if the final stages of inflow were regulated by the dispersal of the solar nebula.

4.1.3. Discussion. In an actively supplied accretion disk, the inflow rate is the primary regulator of disk temperature. Temperatures in the region of the outer Galilean satellites would have become cold enough for ices only once the rate of inflow slowed substantially compared to its peak. This implies that the Galilean satellites (1) formed very late in Jupiter's gas accretion, (2) originated within an orders-of-magnitude lower gas density disk than suggested by MMsN models, and (3) accreted very slowly (in $>10^5$ yr for a solar composition inflow; see also section 4.2.3). While the third condition is potentially consistent with an undifferentiated Callisto (per discussion in section 2.4), we emphasize that in the *Canup and Ward* (2002) model, the requirement of slow accretion emerges primarily from the ice-rich compositions of Ganymede and Callisto, and is not directly dependent on Callisto's interior state.

As we discuss in section 4.3, it is certainly possible that earlier generations of satellites accreted (and were lost) during periods of faster gas inflow. Higher disk gas densities and temperatures during those periods suggest that such satellites would have been rock-rich, rather than icy (*Canup and Ward,* 2002; *Alibert et al.,* 2005).

4.2. Delivery and Evolution of Disk Solids

4.2.1. Sources of solids. To form satellites, rock and ice must be supplied to the accretion disk in addition to the gas. Possible sources of solid material include (1) direct transport of small particles into the disk with the inflowing gas, (2) capture of Sun-orbiting planetesimals as they pass through the disk and lose energy due to gas drag, and (3) capture of collisional debris from collisions between planetesimals within the planet's Hill sphere, or between a planetesimal and a planet-orbiting object.

Mechanism (1) applies for particles small enough to be aerodynamically coupled to the gas. The behavior of a Sun-orbiting particle of radius r_p and density ρ_p is controlled by the motion of the gas rather than by solar gravity if its so-called "stopping time" due to interaction with the gas, $t_e \equiv r_p\rho_p/(c\rho_g)$, is shorter than its orbital timescale, $\Omega^{-1}{}_p \equiv (GM_\odot/r^3)^{-1/2}$ (e.g., *Weidenschilling and Cuzzi,* 1993). This is true for particles with radii

$$r_p \ll \frac{\rho_{g,n}c}{\rho_p\Omega_p} \approx \frac{\sigma_{g,n}}{2\rho_p} \sim O(1)\text{m}\left(\frac{\sigma_{g,n}}{400\ \text{g/cm}^2}\right)\left(\frac{1.5\ \text{g/cm}^3}{\rho_p}\right) \quad (32)$$

where vertical hydrostatic equilibrium (so that $H \approx c/\Omega$ and $\sigma_{g,n} = 2H\rho_{g,n}$) has been assumed, and $\rho_{g,n}$ and $\sigma_{g,n}$ are the solar nebula density and surface density. Particles smaller than meter-sized would be delivered to the disk by the inflowing gas with a specific angular momentum similar to the gas itself, so that their initial orbits would have $r \leq r_o$. For $r_o \sim 30\ R_J$ (as estimated in section 3.1), this region would be comparable to the radial scale of the Galilean system.

To be captured via mechanisms (2) or (3), a particle on an initially heliocentric orbit must lose sufficient energy while in the presence of the planet to become planet-bound. One potential source of dissipation is gas drag due to passage through the circumplanetary disk. To be captured during a single passage through the disk, a particle must encounter a mass in gas comparable to or greater than its own; this would occur for $4\pi r_p^2\sigma_g \geq 4\pi r_p^3\rho_p/3$, or for particle radii

$$r_p \leq 2\text{m}\left(\frac{\sigma_g}{10^2\ \text{g/cm}^2}\right)\left(\frac{1.5\ \text{g/cm}^3}{\rho_p}\right) \tag{33}$$

Repeat passages would allow larger objects to be captured. However, while meter-sized and smaller particles would be continually resupplied to the region near Jupiter by transport with the gas, larger objects could be relatively less plentiful, having already been depleted by accretion onto Jupiter's core and not as easily transported to the planet's vicinity. Gas-drag capture would supply solids across the full radial width of the accretion disk (i.e., for $r < r_d$), and so if this had been a dominant source of solid material for regular satellite accretion it is difficult to explain the compactness of the Galilean system.

An alternative mode of capture that is independent of the presence of gas involves mutual collisions, either between heliocentric orbiting objects colliding within the planet's Hill sphere ("free-free" collisions), or between a free object and a preexisting satellite (a "free-bound" collision). For discussion of both processes, the reader is referred to *Safronov et al.* (1986), *Estrada and Mosqueira* (2006), and the chapter by Estrada et al. In the context of the accretion disk model we address here, mass delivery from free-free collisions, which is quite inefficient, is likely to be less than that due to small particles delivered with the inflowing gas. *Estrada and Mosqueira* (2006) estimate a free-free delivery rate of ~2×10^{18} g/yr for a minimum-mass nebula whose solids are contained in 1-m to 10-km objects, while direct inflow of meter-sized and smaller particles with the gas can provide ~6×10^{20} g/yr ($\mathcal{F}/10^{-5}\ M_\oplus\ \text{yr}^{-1}$)($10^2$/f). However, free-bound collisions could provide an additional source of solid material to the $r < r_o$ region once satellites have begun to form there.

Although both direct inflow and capture processes would have been ongoing as Jupiter's satellites formed, direct inflow of small particles with the gas appears the mode of solid delivery most naturally able to account for the radial scale of the satellite system. In general, this mechanism becomes more effective as the assumed number density of small objects in the solar nebula is increased. Small bodies may be continually generated in late-stage planet formation, because the growing planets stir neighboring planetesimals to sufficient velocities so that collisions between planetesimals become fragmentary (e.g., *Goldreich et al.,* 2004; *Rafikov,* 2004; *Chiang et al.,* 2007; *Schlichting and Sari,* 2007).

4.2.2. Retention of small particles. The initial evolution of small disk particles will be governed by a balance between accretional growth and orbital decay due to aerodynamic gas drag. The rate of gas drag orbital decay is inversely proportional to particle radius for particles with $t_e \geq \Omega^{-1}$ (i.e., for sizes comparable to and larger than the particle size most strongly coupled to the gas), so that if particles grow larger quickly enough, they can avoid decay. The decay timescale, $\tau_{gd} \equiv r/\dot{r}$, for a radius r_p and density ρ_p particle at orbital distance r is $\tau_{gd} \sim (8/3C_D)(\rho_p r_p/\sigma_g)(r\Omega/c)^3\Omega^{-1}$, where C_D is the drag constant on the order of unity for particles too small to gravitationally perturb the gas. The timescale for a mass m_p and radius r_p particle to grow by accretion, $\tau_{acc} \equiv m_p/\dot{m}_p$ (or alternatively, the time spent as a mass m_p particle), is $\tau_{acc} \sim \rho_p r_p/(\sigma_p\Omega F_G)$. Here σ_p is the surface density of disk particles, and $F_G \equiv (1 + v_{esc}^2/v^2)$ is the gravitational focusing factor, a function of the ratio between the mutual escape velocity of colliding particles, v_{esc}, and their relative velocities at large separation, v, and estimated to be $F_G \sim 5$ (*Barr and Canup,* 2008).

Comparing τ_{acc} to τ_{gd} gives (e.g., *Makalkin et al.,* 1999; *Canup and Ward,* 2002)

$$\frac{\tau_{acc}}{\tau_{gd}} \sim 10^{-3}\left(\frac{5}{F_G}\right)\left(\frac{\sigma_g/\sigma_p}{100}\right)\left(\frac{c/r\Omega}{0.1}\right)^3 \tag{34}$$

independent of particle size. When this ratio is less than unity, particles can grow through mutual collisions faster than they are lost due to gas drag. The term in equation (34) that is least certain is the disk gas-to-solid surface density ratio. For a solar composition mixture, $(\sigma_g/\sigma_p) \sim 10^2$, and even a disk that was 1 to 2 orders of magnitude more gas-rich would still have $(\tau_{acc}/\tau_{gd}) \ll 1$. When $(\tau_{acc}/\tau_{gd}) \ll 1$, solids are expected to accrete across the $r < r_o$ region where they are delivered by the inflow.

While the inflowing gas viscously spreads and maintains a quasi-steady-state surface density, the inflowing solids will effectively be decoupled from the gas once they begin to accrete. This means that solids will build up in the circumplanetary disk and that the gas-to-solids ratio in the disk will be lower than in the inflowing material. For example, spreading the mass of the Galilean satellites into a disk extending to 30 R_P gives an average solid surface density $\sigma_p \sim 3 \times 10^3$ g/cm^2. Requiring that the inflow rate was slow enough for ices in the Callisto-Ganymede region implies a disk gas surface density <10^3 g/cm^2 for $\alpha > 10^{-4}$ (Fig. 5b), implying a circumplanetary disk gas-to-solids ratio $(\sigma_g/\sigma_p) \leq 1$, and that the Galilean satellites formed in gas-starved conditions.

4.2.3. Rate of satellite growth. The satellite accretion rate is a control on the initial thermal states of the satellites and is key to explaining an incompletely differentiated Callisto. In an MMsN disk, all the material needed to form the satellites is presumed to be in the disk as a starting condition. In this case, the time needed to accrete an $R_S \sim$ 2500 km, $\rho_S \sim 1.8$ g/cm^3 Callisto-like satellite is $\tau_{acc} \sim \rho_S R_S/(\sigma_p\Omega F_G) \sim 600$ yr (10^3 g/cm$^2/\sigma_p$)($r/26\ R_J)^{3/2}(5/F_G)$. As discussed in section 2.4, with such rapid accretion it appears impossible to avoid melting Callisto's ice, because peak temperatures in the satellite — even in the limit of maximally effective cooling — reach ~10^3 K (e.g., *Stevenson et al.,* 1986).

However, in an accretion disk the formation time of large satellites can be much longer, because it is regulated by the supply rate of particles to the disk. For a gas inflow rate slow enough to allow ices in the Ganymede-Callisto region (i.e., $\mathscr{F} \sim 10^{-5}\,M_\oplus\,yr^{-1}$ from section 4.1.2), the time required for the inflow to deliver a mass in solids equal to that of the Galilean satellites, $M_T = 2.1 \times 10^{-4}\,M_J$, is $\tau_{in} = (f/\mathscr{F})M_T \sim 7 \times 10^5\,yr\,(f/10^2)(10^{-5}\,M_\oplus\,yr^{-1}/\mathscr{F})$, where f is the gas-to-solids ratio in the inflowing material. Such long formation times are comparable to those required to explain Callisto [$\tau_{acc} > 5 \times 10^5$ yr (*Barr and Canup,* 2008)]. In general, in an inflow-sustained disk slow satellite growth arises directly from the same conditions necessary to produce ice-rich satellites, and accounting for Callisto's apparent interior state does not require the imposition of additional constraints.

4.3. Growth and Loss of Large Satellites

Just 2% of Jupiter's mass in solar composition material is needed to produce the Galilean satellites, and yet a much greater total mass — as much as tens of percent of Jupiter's mass — may well have been processed through the jovian accretion disk (e.g., section 3). The latter would imply a total mass in inflowing solid material that is many times larger than that contained in the Galilean satellites. Once inflowing particles begin to accrete, they are likely to grow faster than they would be lost to aerodynamic gas drag (equation (34)), implying a high efficiency for their incorporation into satellites. Why then aren't the Galilean satellites Mercury- or even Mars-sized? *Canup and Ward* (2006) proposed that Type I migration precludes the formation of satellites larger than a critical mass, and that migration will effectively select for the formation of satellite systems around gas planets containing a total mass ~ $O(10^{-4})$ times the planet's mass, as we now describe.

4.3.1. Satellite orbital migration: Type I and Type II. The effects of disk torques that arise from the interaction of an orbiting object with a companion gas or particle disk have received increasing attention in recent decades. The discovery of "hot Jupiters" in extrasolar systems — i.e., Jupiter-like planets that occupy orbits extremely close to their central stars — seems to require that gas giant planets were able to migrate inward over large distances. Disk torques provide a natural means for inward planet migration and even the loss of planets due to collision with the central star. Analogous satellite-disk torques have important implications for Galilean satellite formation.

As a satellite grows, its gravitational interactions with a background gas disk become increasingly important, and these interactions lead to torques on the satellite. The satellite's gravity induces spiral density waves in the gas disk, which are initiated at Lindblad resonant locations both interior and exterior to the satellite and propagate as pressure waves (e.g., *Goldreich and Tremaine,* 1980). The torque on the satellite due to the exterior waves leads to a decrease in the satellite's orbital angular momentum, while the interior waves lead to a positive torque on the satellite. Because the total torque due to the exterior waves is on the order of 10% larger than that of the interior waves, the net torque on the satellite is negative and acts as a drag on the satellite's orbit, causing it to slowly spiral inward toward the planet through a process known as Type I migration (e.g., *Ward,* 1986, 1997). The density wave torque, $\dot{L}_S$, is proportional to M_S^2, while the satellite's orbital angular momentum, L_S, is proportional to M_S, so that the timescale over which L_S changes is $L_S/\dot{L}_S \propto 1/M_S$, and migration becomes more rapid as a satellite grows.

The Type I migration timescale is

$$\tau_I \equiv \frac{r}{\dot{r}} = \frac{1}{C_a\Omega}\left(\frac{M_P}{M_S}\right)\left(\frac{M_P}{r^2\sigma_g}\right)\left(\frac{c}{r\Omega}\right)^2 \tag{35}$$
$$\sim 4 \times 10^5\,yr\left(\frac{3.5}{C_a}\right)\left(\frac{50\,g/cm^2}{\sigma_g}\right)\left(\frac{10^{26}\,g}{M_S}\right)\left(\frac{30}{r/R_J}\right)^{1/2}\left(\frac{c/r\Omega}{0.1}\right)^2$$

where M_P is the planet's mass, $\Omega \equiv (GM_p/r^3)^{1/2}$, $(c/r\Omega) \sim 0.1$, and C_a is a constant (*Tanaka et al.,* 2002). Approximate values appropriate for Galilean satellite accretion within a slow-inflow-produced disk (e.g., Fig. 5b) are given in the second expression. Density wave interactions also circularize satellite orbits, with an eccentricity (e) damping timescale, $\tau_e = e/\dot{e} \sim \tau_I(c/r\Omega)^2$, which is much shorter than the Type I timescale (e.g., *Ward,* 1988; *Artymowicz,* 1993; *Papaloizou and Larwood,* 2000; *Tanaka and Ward,* 2004).

For a sufficiently massive satellite, the response of the disk to its density wave perturbations becomes nonlinear, and the satellite clears an annular region of the disk centered at its orbital radius, producing a "gap" of reduced surface density compared to that of the background disk. A satellite massive enough to open a gap will be subject to Type II migration, in which its orbital evolution is coupled to the viscous evolution of the disk (e.g., *Lin and Papaloizou,* 1986). A standard criterion for gap opening and transition to Type II migration is that the cumulative torque on the disk by the satellite must exceed the rate of angular momentum transport due to the disk's viscosity, so that the satellite repels local disk material faster than it can be resupplied by viscous diffusion (e.g., *Lin and Papaloizou,* 1986; *Ward,* 1986). If the waves are damped locally, a gap-opening criterion that considers an α viscosity model is (e.g., *Ward and Hahn,* 2000)

$$\frac{M_S}{M_P} > c_v\sqrt{\alpha}\left(\frac{c}{r\Omega}\right)^{5/2} \tag{36}$$
$$\sim 2 \times 10^{-4}\,c_v\left(\frac{\alpha}{3 \times 10^{-3}}\right)^{1/2}\left(\frac{c/r\Omega}{0.1}\right)^{5/2}$$

where c_v is a constant on the order of unity to 10. For $c_v = 3$ and $(c/r\Omega) = 0.1$, the mass cutoff in equation (36) is larger than Ganymede's mass for $\alpha > 7 \times 10^{-5}$. If a satellite grows large enough to open a gap, Type II migration will generally also lead to orbital decay with a timescale comparable to the local disk's viscous spreading time

$$\tau_{II} \approx \tau_v = \frac{r^2}{\nu} \approx 300\,yr\left(\frac{3 \times 10^{-3}}{\alpha}\right)\left(\frac{r}{30R_J}\right)^{3/2}\left(\frac{0.1}{c/r\Omega}\right)^2 \tag{37}$$

4.3.2. Maximum satellite mass. A satellite will grow through the sweep-up of inflowing solids until it reaches a critical mass for which the characteristic time for its further growth is comparable to its Type I orbital decay timescale. Satellites cannot grow substantially larger than this critical mass before they are lost to collision with the planet, because Type I migration becomes more rapid as a satellite's mass increases.

Consider a satellite at orbital distance r that accretes material across an annulus of width Δr in an inflow-supplied accretion disk. The satellite's growth timescale is $\tau_{acc} \sim fM_S/(2\pi r \Delta r \mathcal{f}_{inflow})$, where $\mathcal{f}_{inflow}$ is the total mass inflow rate per area (gas and solids), and f is the mass ratio of gas-to-solids in the inflow. The annulus width Δr is set by the larger of two quantities: the distance over which a satellite can gravitationally perturb particles into a crossing orbit with itself (which is proportional to the satellite's Hill radius), or the radial excursion executed by an eccentric satellite during a single orbit, with $\Delta r \sim 2er$. From numerical simulations (see section 4.4), *Canup and Ward* (2006) found that Δr is set by the characteristic maximum eccentricity, e, that results from a balance between eccentricity damping by density waves and excitation via gravitational scatterings with similarly sized objects (e.g., *Ward,* 1993), with $e \sim (c/r\Omega)(M_S/4\pi r H\sigma_g)^{1/5}$. Using this estimate together with $\Delta r \sim 2er$ gives the satellite accretion timescale, $\tau_{acc} \sim (f\sigma_g/\mathcal{f}_{inflow})(M_S/4\pi r H\sigma_g)^{4/5}$.

The critical maximum satellite mass, M_{crit}, is then found by setting $\tau_{acc} \sim \tau_1$ from equation (35), using σ_g from equation (26), and solving for $M_S = M_{crit}$. Defining $\tau_G \equiv M_p/\mathcal{F}$ as the time in which a mass equal to the planet's mass is delivered by the inflow, the critical satellite mass in planet masses is (*Canup and Ward,* 2006)

$$\left(\frac{M_{crit}}{M_P}\right) \approx 5.4\left(\frac{\pi}{C_a}\right)^{5/9}\left(\frac{c}{r\Omega}\right)^{26/9}\left(\frac{r}{r_o}\right)^{10/9}\left(\frac{\alpha}{f}\right)^{2/3}[\Omega\tau_G f]^{1/9} \tag{38}$$

$$\sim 5.6\times10^{-5}\chi\left(\frac{3.5}{C_a}\right)^{5/9}\left(\frac{c/r\Omega}{0.1}\right)^{26/9}\left(\frac{r/r_o}{0.5}\right)^{10/9}\left(\frac{\alpha/f}{3\times10^{-5}}\right)^{2/3}$$

where $\chi \equiv [(1\text{ week}/\{2\pi/\Omega\})(f/10^2)(\tau_G/10^7\text{ yr})]^{1/9}$ is a factor close to unity. The Galilean satellites have (M_S/M_P) values ranging from 2.5×10^{-5} to 7.8×10^{-5}, comparable to the critical mass in equation (38). The predicted (M_{crit}/M_P) ratio depends very weakly on the inflow rate through the $(\tau_G)^{1/9}$ term in χ, which implies that a similar maximum satellite mass would be expected across a wide range of inflow rates. The quantity $(c/r\Omega)$ varies weakly with r for most disks, e.g., varying from ~0.09 to 0.11 across the $R_P < r < 30\,R_P$ region for the inflow conditions shown in Fig. 5b. If satellites form throughout the inflow region, the ratio (r/r_o) will be on the order of 0.1 to unity. The last term in equation (38) contains the ratio of two uncertain parameters: the α viscosity parameter and f. For a given inflow rate, a higher viscosity yields lower disk gas surface densities, and thus allows larger mass satellites to survive against Type I decay, while a lower f implies a more solid-rich inflow, which speeds up satellite growth allowing objects to grow larger before they are lost.

The above assumes that satellites undergo Type I migration and do not grow large enough to open gaps. Comparing the gap opening mass, M_{gap}, from equation (36) to M_{crit} from equation (38) gives

$$\left(\frac{M_{gap}}{M_{crit}}\right) \sim 9\left(\frac{c_v}{3}\right)\left(\frac{C_a}{3.5}\right)^{5/9}\left(\frac{3\times10^{-3}}{\alpha}\right)^{1/6} \times\left(\frac{0.1}{c/r\Omega}\right)^{7/18}\left(\frac{0.5}{r/r_o}\right)^{10/9}\left(\frac{f}{10^2}\right)^{2/3} \tag{39}$$

Unless the disk is extremely viscous and/or the inflow is substantially more solid-rich than solar composition, $M_{gap}/M_{crit} > 1$, and satellites will be lost to Type I decay before they grow large enough to open gaps and transition to Type II migration. In the satellite systems produced by the *Canup and Ward* (2006) accretion simulations (section 4.4), the final satellites were all less massive than M_{gap}, with the larg-est satellite in each system having an average mass $\sim 0.2 \pm 0.1\,M_{gap}$ for $c_v = 3$.

4.3.3. Creation of common (M_T/M_P). The limiting satellite mass in equation (38) also implies a limit on the total mass of a satellite system. Consider an inflow that persists for a time exceeding that needed for satellites of mass M_{crit} to form. Within a given annulus in the disk, a satellite grows to a mass $\sim M_{crit}$ before being lost to Type I decay, but in a comparable timescale to its demise another similarly massive satellite grows in its place, because $\tau_1 \sim \tau_{acc}$. In this way, the disk is regulated to contain a total mass in satellites, M_T, comparable to a distribution of mass M_{crit} objects across the inflow region. For $(c/r\Omega)$ and f that are approximately constant across the disk, the predicted satellite system mass fraction, (M_T/M_P), is (*Canup and Ward,* 2006)

$$\left(\frac{M_T}{M_P}\right) = \int_{r_i\sim R_P}^{r_o}\frac{(m_{crit}/M_p)}{\Delta r}dr \approx 2.1\left(\frac{\pi}{C_a}\right)^{4/9}\left(\frac{c}{r\Omega}\right)^{10/9}\left(\frac{\alpha}{f}\right)^{1/3}\frac{1}{[\Omega\tau_G f]^{1/9}} \tag{40}$$

$$\sim 2.5\times10^{-4}\frac{1}{\chi}\left(\frac{3.5}{C_a}\right)^{4/9}\left(\frac{c/r\Omega}{0.1}\right)^{10/9}\left(\frac{\alpha/f}{3\times10^{-5}}\right)^{1/3}$$

similar to (M_T/M_P) for both the jovian and saturnian systems. The fraction (M_T/M_p) is insensitive to inflow rate through χ, lacks any dependence on r_o, and depends quite weakly on (α/f) to the one-third power. The implication is that satellite systems formed within accretion disks of gas planets will have $(M_T/M_p) \sim O(10^{-4})$ across a wide range of inflow and disk parameters, so long as the disk is viscous and Type I migration is active. This offers an explanation for why the satellite systems of Jupiter and Saturn would contain a similar fraction of their central planet's mass, even though the individual accretion histories of these two planets (e.g., gas inflow rates, total mass processed through their accretion disks, etc.) may well have differed.

4.3.4. Extent of migration of Galilean satellites. Initially a satellite's mass accretion timescale is short compared to its Type I migration timescale. However, once its mass exceeds M_{crit} given in equation (38), these timescales invert. For a disk produced by a uniform inflow per area having $\sigma_g \propto 1/r^{3/4}$ and $(c/r\Omega) \propto r^{1/8}$ (*Canup and Ward,* 2002), equation (35) gives $\tau_I \propto r^{1/2}/M_S$. Once a satellite mass exceeds M_{crit} it migrates inward faster than it grows, and its characteristic migration timescale shortens as r decreases and M_S increases.

It seems probable that earlier generations of satellites formed and were then lost to collision with Jupiter by this process. The final Galilean satellites must have avoided this runaway migration phase, implying that their final masses were comparable to or smaller than M_{crit} in their precursor disk at the time the inflow ended and the disk dissipated. However, they would still have migrated inward somewhat as they formed. The satellite accretion rate (assuming a uniform inflow per area) (*Canup and Ward,* 2006) (see also section 4.3.2) and the Type I migration rate are

$$\frac{dM_S}{dt} \sim M_S\left(\frac{f_{inflow}}{f\sigma_g}\right)\left(\frac{4\pi rH\sigma_g}{M_S}\right)^{4/5}$$
$$\frac{dr}{dt} \sim -r\Omega C_a\left(\frac{M_S}{M_P}\right)\left(\frac{r^2\sigma_g}{M_P}\right)\left(\frac{r\Omega}{c}\right)^2 \quad (41)$$

Together these give the change in satellite mass as a function of radius, $dM_S/dr = (dM_S/dt)/(dr/dt)$, which can be directly integrated under the assumption that the gas surface density and the normalized disk scale height can be approximated as $\sigma_g = \sigma_{g,0}(R_P/r)^{\gamma_g}$ and $(c/r\Omega) = (c/r\Omega)_0(r/R_P)^{\gamma_c}$, where the terms subscripted zero refer to values at $r = R_P$. For $f_{inflow} \equiv \mathcal{F}/(\pi r_0^2)$, and $\sigma_{g,0}$ and $(c/r\Omega)_0$ determined from equation (26), the initial satellite orbital radius r_0 as a function of the final satellite mass (M_S) and final position (r_f) is

$$\frac{r_o}{R_P} = \left[(\beta+1)K\left(\frac{M_S}{M_P}\right)^{9/5} + \left(\frac{r_f}{R_P}\right)^{\beta+1}\right]^{1/(\beta+1)} \quad (42)$$

where $\beta \equiv (1/10) + (6/5)\gamma_g + (14/5)\gamma_c$, and

$$K \cong 2.16\times10^9\left(\frac{C_a}{3.5}\right)\left(\frac{1.25-\sqrt{r_o/r_d}}{0.8}\right)^{6/5}\left(\frac{r_o/R_P}{30}\right)^2 \quad (43)$$
$$\left(\frac{3\times10^{-5}}{\alpha/f}\right)\left(\frac{3\times10^{-3}}{\alpha}\right)^{1/5}\left(\frac{\mathcal{F}}{10^{-5}\,M_\oplus\,yr^{-1}}\right)^{1/5}\left(\frac{0.1}{(c/r\Omega)_o}\right)^{26/5}$$

For a uniform inflow per area, *Canup and Ward* (2002) find $\gamma_g \sim 3/4$ and $\gamma_c \sim 1/8$ in the inner disk, or $\beta = 27/20$. With this β, the nominal values given in equation (43), and the current satellite masses and orbital radii, the predicted starting radii for Io, Europa, Ganymede, and Callisto from equation (42) are then ~8, 10, 17, and 27 R_J, respectively.

Disk temperatures exterior to ~20 R_J drop below the ice stability line once the inflow rate has declined to about $\mathcal{F} < 10^{-5}\,M_\oplus\,yr^{-1}$ for the parameter choices in Fig. 5a, consistent with Ganymede and Callisto's ice-rich compositions, including effects of inward migration during their formation. The simple migration estimate in equation (42) neglects interactions between the growing satellites, including scattering, resonances, and/or mutual collisions and mergers. Accretion simulations including such effects find that the outer large satellites migrate inward by up to about 5 planetary radii in those systems that most closely resemble the Galilean satellites (*Canup and Ward,* 2006, e.g., their Fig. 3b).

4.4. Direct Accretion Simulations

Relatively few works have modeled Galilean satellite accretion. *Coradini et al.* (1989) analytically estimated satellite masses and orbital separations assuming a variety of values for the gravitational focusing factor. *Richardson et al.* (2000) used direct N-body simulations to model accretion within an MMsN disk, tracking the growth of satellites containing up to ~10% of the mass of Io. *Canup and Ward* (2002) and *Alibert et al.* (2005) considered satellite growth and loss due to Type I migration within an accretion disk by assuming a linear growth in satellite mass with time.

Canup and Ward (2006) presented the first direct simulations of the accretion of Galilean-sized satellites. They used an N-body integrator (*Duncan et al.,* 1998) modified to include the effects of ongoing inflow to the disk and interactions between the growing satellites and the gas disk. To mimic the inflow of solids, orbiting objects were added to the disk with random position within $r < r_o$ at a rate proportional to $\mathcal{F}/f$. Computational limitations on the number of objects dictated that the added objects were large, typically ~10^2 km. All objects were tracked with direct orbit integration, and collisions were assumed to result in the inelastic merger.

Figure 6 shows results of three accretion simulations that assumed a constant rate of gas inflow and different values for (α/f). Initially the inflow of solids causes the total mass in orbiting satellites, M_T, to increase with time until objects of mass ~M_{crit} form (equation (38)). The orbits of the largest satellites then begin to decay inward due to Type I migration, and M_T decreases in discrete steps as satellites are removed due to collision with Jupiter. Continuing solid inflow to the disk leads to the growth of another generation of mass ~M_{crit} satellites on a timescale comparable to that of the prior satellites' orbital decay. The net result is a repeating cycle of satellite formation and loss. At any time beyond that needed to initially form the first mass M_{crit} satellites, the total mass contained in satellites relative to that in the planet, (M_T/M_P), oscillates about a nearly constant value similar to that in equation (40). The analytical

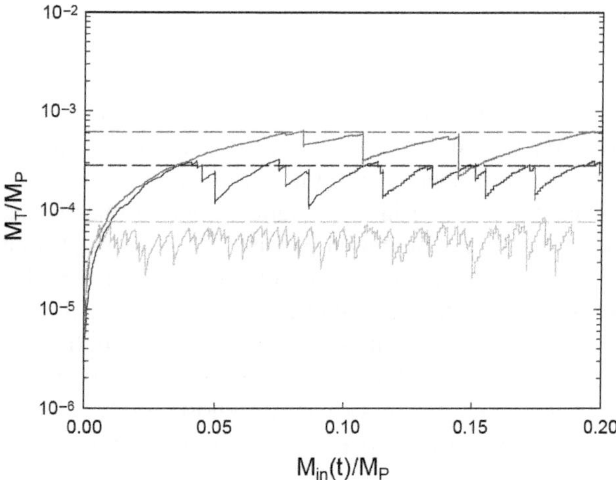

Fig. 6. Results of three satellite accretion simulations that consider a time-constant rate of gas inflow to the circumjovian disk, with $\mathscr{F} = 6 \times 10^{-5}\ M_\oplus\ yr^{-1}$. The total mass in satellites (M_T) scaled to the planet's final mass (M_P) is shown vs. the cumulative total mass delivered to the disk ($M_{in}\{t\}$) in planet masses. The light gray (bottom), black (middle), and gray (top) curves correspond respectively to simulations with (α/f) = 10^{-6}, 5×10^{-5}, and 5×10^{-4}. A satellite grows until it reaches a critical mass at which the timescale for its further growth is comparable to its Type I decay timescale. The satellite then spirals inward until it collides with Jupiter and is removed from the simulation, while in a comparable time a new, similarly sized satellite grows in its place. This process causes (M_T/M_P) to oscillate about a constant value. Dashed lines show analytically predicted (M_T/M_P) values from equation (40). From *Canup and Ward* (2006).

estimates in equations (38) and (40) consider disk annuli independently. But in the simulations, as satellites formed in the outer disk migrate inward they pass through interior zones and accrete additional material along the way. Migration-driven growth hastens their orbital decay, so that they are lost somewhat more quickly than the time to replenish their mass in their original radial zone. This causes the (M_T/M_p) value from equation (40) to be an approximate upper limit, as seen in Fig. 6. Because this predicted satellite system mass fraction is proportional to $(\alpha/f)^{1/3}$, the factor of 500 variation in (α/f) considered across the Fig. 6 simulations yields only a factor of ~10 spread in (M_T/M_p).

The actual inflow to Jupiter would have been time-dependent. In a system with an exponentially decaying inflow, the final system of satellites arises when the satellite growth timescale becomes comparable to the inflow decay timescale. As the inflow wanes and the gas disk disperses, Type I migration ends and a final system of satellites stabilizes. Figure 7 shows results of simulations with time-dependent inflows that were assumed to decay exponentially with time constants between 2×10^5 and 2×10^6 yr, for a variety of inflow and disk parameters. Many cases produce systems whose number of large satellites and satellite masses are similar to those of the Galilean system. The median number of large, (M_S/M_P) ≥ 10^{-5} final satellites in the *Canup and Ward* (2006) simulations was four, and the simulated satellites had average orbital separations comparable to those of the Galilean satellites. The radial extent of the system is set predominantly by the choice of r_o, with the most distant large satellite typically having a semimajor axis a_{max} ~ r_o, with this value varying by about a factor of 2 for a fixed

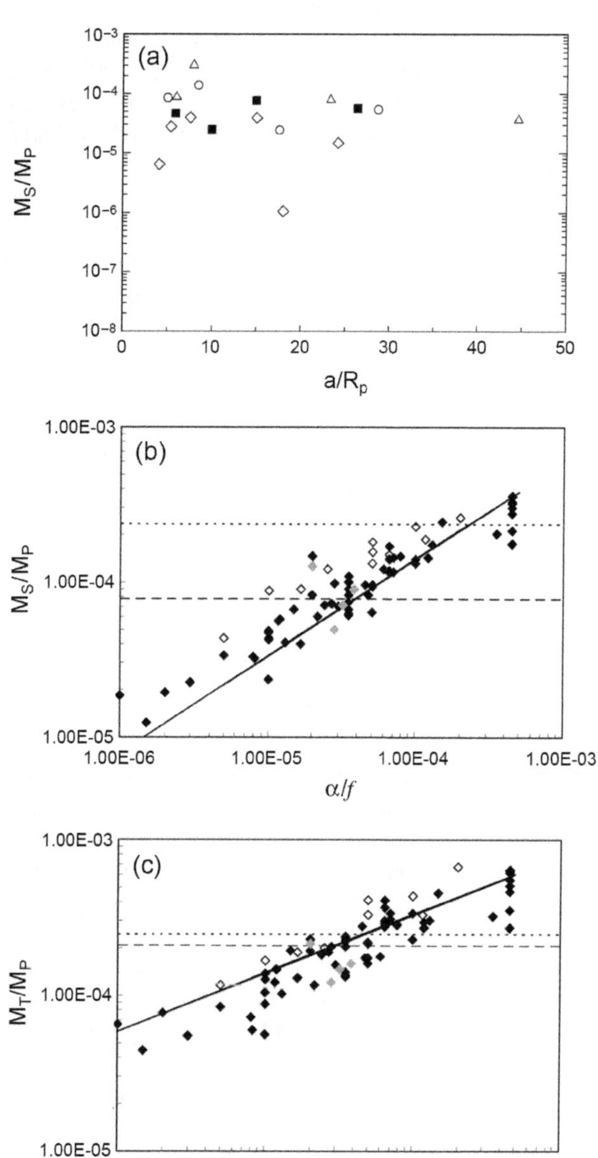

Fig. 7. Results of satellite accretion simulations that considered time-dependent rates of gas and solid inflow. **(a)** Three final simulated satellite systems, with the Galilean satellites shown for comparison (solid squares). All three runs had (r_o/R_P) = 30, with open diamonds, circles, and triangles corresponding to runs with (α/f) = 1.7×10^{-5}, 6.5×10^{-5}, and 5×10^{-4}, respectively. **(b)** The largest final satellite mass in planet masses shown as a function of (α/f) for 75 simulations. Gray, black, and open diamonds correspond to cases with (r_o/R_P) = 25, 30, and 44, respectively. Dashed and dotted lines show observed values for the Galilean and saturnian systems, while the solid line is the analytical prediction from equation (38). **(c)** The final satellite system mass in planet masses vs. (α/f), with symbols and lines having the same meanings as in **(b)** and the solid line derived from equation (40). From *Canup and Ward* (2006).

value of r_0 due to dynamical scattering and satellite migration (e.g., Fig. 7a, triangles vs. diamonds).

Canup and Ward (2006) showed that similar accretion conditions can also yield systems in which a single satellite contains most of the final satellite system mass, such as is the case for Titan and the saturnian satellites. This outcome occurs in an initially Galilean-like system when the inner satellites suffer a shorter characteristic migration timescale than the outermost satellite. If the gas inflow to the disk ends after the inner large satellites have been lost to collision with the planet but before a new complete generation of satellites is formed, a single large satellite can be the sole survivor. For the saturnian system, such a process would imply that the small and medium-sized saturnian satellites interior to Titan formed after all inner massive companions to Titan had been lost. Inflow rates during this final era could have been low enough to allow ice even in the inner saturnian disk, consistent with the ice-rich compositions of Saturn's inner satellites (*Canup and Ward,* 2006; *Barr and Canup,* 2008).

In summary, the balance between inflow-regulated satellite growth and gas-driven satellite loss causes the ratio between the total satellite system mass and the planet's mass to maintain a roughly constant value, which is nearly independent of both the inflow rate and its characteristic angular momentum (which sets r_0). This common fractional value is weakly dependent on the ratio of the disk viscosity α parameter and the inflow's gas-to-solid composition (Figs. 6 and 7c), with $(M_T/M_P) \propto (\alpha/f)^{1/3}$, which implies that satellite system masses similar to those of Jupiter and Saturn result for $10^{-6} < (\alpha/f) < 5 \times 10^{-4}$, i.e., a span of nearly 3 orders of magnitude. While quantitative predictions for α and f are lacking, (α/f) ratios throughout this range are consistent with commonly assumed values for α and f, e.g., $10^{-4} < \alpha < 0.1$ and $10 < f < 10^2$.

5. DISCUSSION

The accretion disk model reviewed here has a number of strengths. It describes satellite formation as a natural consequence of Jupiter's formation. Provided the planet contracted before nebular dissipation was complete, a circumplanetary accretion disk is expected, based on analytical estimates of the expected angular momentum of inflowing gas and on results of hydrodynamic simulations. If the disk is viscous, the outward angular momentum transport required to allow gas to be accreted by the planet is achieved. Between the period of peak runaway gas inflow expected for a gas giant planet and the cessation of gas accretion, inflow rates to the circumjovian disk would slow, allowing for disk temperatures low enough for ice stability in the general region of Ganymede and Callisto, with ice accretion at Europa's location possible late in the satellite's growth. The slow inflow rates needed for ice-rich satellites in turn imply that the Galilean satellites would have accreted slowly, in keeping with conditions needed for a Callisto-sized satellite to avoid melting and differentiation as it forms. Estimates of the average specific angular momentum of inflowing gas are broadly consistent with the radial scale of the Galilean system. Direct accretion simulations show the formation of Galilean-like satellite systems for a range of disk and inflow parameters, and suggest that Type I migration would regulate gas planet satellite systems to contain a common fraction of the planet's mass, independent of the total mass processed through the protosatellite disk. This offers an attractive explanation for the similarity of the Galilean and saturnian satellite system masses compared to their planet's mass.

However, some aspects of the accretion disk model are not well understood. Foremost is the origin of viscosity in the protosatellite disk, and indeed the origin of viscosity in the cool and/or dusty regions of the surrounding circumsolar disk as well. It is generally accepted that magnetorotational instability (MRI) in weakly magnetized disks drives turbulence and outward angular momentum transport, with effective α-values ~10^{-2} to 0.1 (*Balbus and Hawley,* 1991). To be operative, MRI requires a minimum ionization fraction, which can be produced thermally (for temperatures in excess of ~10^3 K), by galactic cosmic-rays [for regions of low gas surface density with $\sigma_g < 10^2$ g/cm^2 (e.g., *Dullemond et al.,* 2007)], or by stellar X-rays (e.g., *Glassgold et al.,* 1997). However, the presence of dust grains can deactivate MRI because they are effective charge absorbers and can cause the ionization fraction to plummet (*Sano et al.,* 2000). It is thus plausible that both the protosatellite disk and the circumsolar disk at Jupiter's distance would have been "layered": comprised of MRI "active" regions at the disk surfaces, together with a neutral, dust-rich "dead zone" surrounding the disk midplane (e.g., *Gammie,* 1996). It is challenging to determine the exact vertical and radial extent of the inner dead zone, because this depends on the disk chemistry and the dust grain size distribution (e.g., *Semenov et al.,* 2004); however, it is possible that in a gas-starved protosatellite disk the active region could be substantial, and this merits investigation.

Nonmagnetic hydrodynamic mechanisms have also been proposed as sources of turbulence and viscous transport in neutral disks, including notably those that rely on nonaxisymmetric effects, including decaying vorticies (e.g., *Lithwick,* 2007), baroclinic instability (*Klahr and Bodenheimer,* 2003), and shear instability (*Dubrulle et al.,* 2005). Recent experimental results claim to rule out angular momentum transport via axisymmetric hydrodynamical turbulence at levels corresponding to $\alpha > 10^{-5}$ (*Hantao et al.,* 2006). Gravitational torques from objects accreting within the disk may provide an effective viscosity (*Goodman and Rafikov,* 2001; see also *Canup and Ward,* 2002). In the specific context of an actively supplied accretion disk, the shock front resulting from the difference in the free-fall velocity of mass infalling to the disk and that of the orbiting gas is another potential source of turbulence and therefore viscosity (*Cassen and Moosman,* 1981). Such an inflow-driven viscosity would cease once the inflow had stopped completely, but it is possible that the last remaining orbiting gas could be then removed either by other sources of viscosity, e.g., by MRI once small grains have been removed through accretion onto the satellites. Spiral shocks generated by the inflowing gas may also drive angular momentum transport (*Lubow et al.,* 1999), although such shocks appear weaker in three-dimensional simulations than in two-dimensional simulations (*Bate et al.,* 2003).

Another area of uncertainty is the nature of the inflowing material. Hydrodynamical simulations clearly show the formation of a viscous circumplanetary disk (e.g., Fig. 1 from *Bate et al.,* 2003). A key open issue is the average specific angular momentum of the inflowing material (which controls r_0), as well as how the inflow varies with distance from the planet. *D'Angelo et al.* (2003) find a peak specific angular momentum of circumplanetary material about 2.6 times larger than that of material orbiting Jupiter at Callisto's distance. In contrast, simulations by *Machida et al.* (2008) and *Machida* (2008) that resolve Jupiter's Hill sphere with much higher resolution find an average inflow specific angular momentum comparable to that at Callisto's orbit, or $r_c \sim 22\ R_J$. *Machida* (2008) argues that this result is unaffected by the more local nature of the Machida et al. simulations. Further work is needed to address this important issue. The mass fraction of gas vs. solid particles in the inflow (the "*f*" parameter) is highly uncertain, and is likely to remain so. The predicted fraction of nebular solids contained in small particles as a function of time is inherently model-dependent, as is the rate of nebular dispersal and therefore the nebular gas surface density as a function of time. This means that both the numerator and denominator needed to estimate the ratio *f* are poorly constrained, with a wide range of values possible.

It has been suggested that a region at the outer edge of Jupiter's gap — within which there would be a positive gradient in gas density with orbital radius — may frustrate the delivery of small particles to Jupiter. *Rice et al.* (2006) argue that outward gas drag of small particles across this region would prevent all but very small dust particles from reaching the planet. However, this implies that small particles will build up at a preferred location at the gap's outer edge. Once the local particle density exceeds that of the gas, the particle drift rate will drop rapidly (e.g., *Nakagawa et al.,* 1986), making it more difficult to prevent particle diffusion into the planet's gap. This can result in a quasi-steady-state flow interior to the region of particle buildup that has a similar solids content as that of the global disk flow, alleviating the problem (*Ward,* 2009). Assessing whether this occurs will require models that incorporate the effects of particle concentration. It is also possible that strong turbulence con-centrated at the gap edges (e.g., Ward, in preparation) will weaken such efficient particle filtering by gas-drag.

The accretion simulations of *Canup and Ward* (2006) found that overall outcomes were rather weakly dependent of the ratio of (α/f), a fortuitous result given the uncertainties in both of these quantities discussed above. However, the smallest-sized particles in the Canup and Ward simulations were $\sim 10^2$-km-scale due to computational limitations, so that the regime of small particle growth was not resolved. Future work should address this regime, and in particular the conditions in which particle accretion can occur faster than particle loss due to aerodynamic gas drag. The potential role of other solid delivery mechanisms in the context of an inflow-supplied accretion disk, including gas drag and collisional capture, should also be assessed.

6. IMPLICATIONS

In the past two decades there have been important advances in our understanding of satellite properties, gas giant planet formation, and planet and satellite migration processes. This chapter has focused on a model motivated in part by such advances, in which Europa and the other Galilean satellites form within a gas-starved accretion disk produced during the final stages of Jupiter's own gas accretion. Compared to the pioneering models developed in the 1980s, including most notably the work of *Lunine and Stevenson* (1982), the slow-inflow accretion disk construct implies very different conditions for the disk environment in which Europa formed and for the nature of satellite accretion around gas planets. We conclude here by outlining overall implications of the *Canup and Ward* (2002, 2006) model.

1. *Satellite migration and loss.* If more than ~2% of Jupiter's mass in solar-composition material was processed through its accretion disk — which seems likely — multiple generations of satellites may have formed and ultimately been lost to Type I migration and collision with Jupiter. The Galilean satellites are then the last surviving generation, and their properties reflect conditions that existed as gas inflow to Jupiter was ending.

2. *Maximum satellite mass and a common total satellite system mass.* Type I migration precludes the formation of satellites larger than a critical mass, and causes the total satellite system mass to maintain a relatively constant value. While detailed characteristics of satellite systems of gas giants would vary, those that formed from accretion disks would be predicted to display a similar largest satellite mass and satellite system mass compared to their host planet's mass.

3. *Late formation times.* The growth of Europa and the other Galilean satellites is predicted to be concurrent with the dissipation of the solar nebula. This is likely to have occurred several million years or more after the formation of the earliest solar system solids, based on several lines of reasoning. Forming Jupiter via core accretion appears to require a few to ~5 m.y. (*Hubickyj et al.,* 2005), and observations of disks around other stars find a mean nebular lifetime of 3 m.y. (*Haisch et al.,* 2001). The requirement that Callisto did not melt as it formed implies that the satellites finished their accretion no earlier than about 4 m.y. after CAIs (*Barr and Canup,* 2008). These age estimates suggest that short-lived radioisotopes would have been relatively unimportant by the time the Galilean satellites accreted.

4. *Low disk gas densities, low disk pressures.* Accretion disks produced during the end stages of Jupiter's growth have low gas densities and pressures not too different from that of the background solar nebula. Infalling solid particles would therefore undergo little additional chemical processing, and their composition would be expected to generally reflect solar nebula conditions (e.g., *Mousis and*

Alibert, 2006). This contrasts to the dense and high-pressure disks implied by MMsN models, which implied substantial chemical alteration of disk material (e.g., *Prinn and Fegley,* 1981).

5. *Protracted satellite accretion.* The slow inflow rates required for temperatures low enough for ices in the Ganymede and Callisto region imply that the Galilean satellites accreted slowly, taking 10^5–10^6 yr to acquire their final masses for a solar composition inflow. Such long accretion times are also generally consistent with those needed for Callisto to avoid melting as it accreted. In contrast, traditional MMsN models predict much more rapid satellite accretion in $\leq 10^4$ yr.

6. *Accretion by predominantly small impactors.* Small particles delivered to the disk with the inflow will grow until their mutual collision timescale is comparable to their collision timescale with the dominant large satellite in their region of the disk. This balance yields a predicted characteristic impactor size of about a kilometer or less for the Galilean satellites (*Barr and Canup,* 2008). However, episodic collisions between large migrating satellites are possible.

7. *Possible primordial establishment of Laplace resonance.* The Galilean satellites would have likely migrated inward somewhat during their formation. Ganymede's larger mass means that it would have migrated inward at a faster rate than Europa and Io. *Peale and Lee* (2002) have shown that it is possible in such a situation to establish the Laplace resonance from outside-in during the satellite formation era.

8. *Heterogeneous accretion of Europa.* Disk temperatures near Europa's orbit are predicted to have been too warm for ices during much of the satellite's growth, so that most of Europa would have accreted as a purely rocky object. But during the final stages of Europa's accretion, the ice line likely moved within the satellite's orbit, so that the final ~10% of its accreted mass would have been a mixture of rock and ice.

REFERENCES

Alibert Y., Mousis O., and Benz W. (2005) Modeling the jovian subnebula. I. Thermodynamic conditions and migration of proto-satellites. *Astron. Astrophys., 439,* 1205–1213.

Anderson J. D., Jacobson R. A., McElrath T. P., Moore W. B., Schubert G., and Thomas P. C. (2001) Shape, mean radius, gravity field, and interior structure of Callisto. *Icarus, 153,* 157.

André R., Ward-Thompson D., and Barsony M. (2000) From prestellar cores to protostars: The initial conditions of star formation. In *Protostars and Planets IV* (V. Mannings et al., eds.), pp. 59–96. Univ. of Arizona, Tucson.

Artymowicz P. (1993) Disk-satellite interaction via density waves and the eccentricity evolution of bodies embedded in disks. *Astrophys. J., 419,* 166–180.

Balbus S. A. and Hawley J. F. (1991) A powerful local shear instability in weakly magnetized disks. I. Linear analysis. *Astrophys. J., 376,* 214–233.

Barr A. C. and Canup R. M. (2008) Constraints on gas giant satellite formation from the interior states of partially differentiated satellites. *Icarus, 198,* 163–177.

Bate M. R., Lubow S. H., Ogilvie G. I., and Miller K. A. (2003) Three-dimensional calculations of high- and low-mass planets embedded in protoplanetary discs. *Mon. Not. R. Astron. Soc., 341,* 213–229.

Bell K. R. and Lin D. N. C. (1994) Using FU Orionis outbursts to constrain self-regulated protostellar disk models. *Astrophys. J., 427,* 987–1004.

Bodenheimer P. and Pollack J. B. (1986) Calculations of the accretion and evolution of giant planets: The effects of solid cores. *Icarus, 67,* 391–408.

Canup R. M. (2004) Formation of the Moon. *Annu. Rev. Astron. Astrophys., 42,* 441–475.

Canup R. M. (2005) A giant impact origin of Pluto-Charon. *Science, 307,* 546–550.

Canup R. M. and Ward W. R. (2001) On the steady state conditions in a circum-jovian satellite accretion disk (abstract). In *Jupiter: Planets, Satellites and Magnetosphere,* June 25–30, 2001, Boulder, Colorado.

Canup R. M. and Ward W. R. (2002) Formation of the Galilean satellites: Conditions of accretion. *Astron. J., 124,* 3404–3423.

Canup R. M. and Ward W. R. (2006) A common mass scaling for satellites of gaseous planets. *Nature, 441,* 834–839.

Cassen P. and Moosman A. (1981) On the formation of protostellar disks. *Icarus, 48,* 353–376.

Cassen P. and Summers A. (1983) Models of the formation of the solar nebula. *Icarus, 53,* 26–40.

Chandrasekar S. (1958) *An Introduction to the Study of Stellar Structure.* Dover, New York.

Chiang E., Lithwick Y., Murray-Clay R., Buie M., Grundy W., and Holman M. (2007) A brief history of transneptunian space. In *Protostars and Planets V* (B. Reipurth et al., eds.), pp. 895–911. Univ. of Arizona, Tucson.

Coradini A. and Magni G. (1984) Structure of the satellitary accretion disk of Saturn. *Icarus, 59,* 376–391.

Coradini A., Cerroni P., Magni G., and Federico C. (1989) Formation of the satellites of the outer solar system: Sources of their atmospheres. In *Origin and Evolution of Planetary and Satellite Atmospheres* (S. K. Atreya et al., eds.), pp. 723–762. Univ. of Arizona, Tucson.

D'Angelo G. D., Henning T., and Kley W. (2003) Thermohydrodynamics of circumstellar disks with high-mass planets. *Astrophys. J., 599,* 548–576.

Dubrulle G., Morfill G., and Sterzik M. (1995) The dust subdisk in the protoplanetary nebula. *Icarus, 114,* 237–246.

Dubrulle G., Marié L., Normand Ch., Richard D., Hersant F., and Zahn J.-P. (2005) A hydrodynamic shear instability in stratified disks. *Astron. Astrophys., 429,* 1–13.

Dullemond C. P., Hollenbach D., Kamp I., and D'Alessio P. (2007) Models of the structure and evolution of protoplanetary disks. In *Protostars and Planets V* (B. Reipurth et al., eds.), pp. 555–572. Univ. of Arizona, Tucson.

Duncan M. J., Levison H. F., and Lee M. H. (1998) A multiple time step symplectic algorithm for integrating close encounters. *Astron. J., 116,* 2067–2077.

Estrada P. R. and Mosqueira I. (2006) A gas-poor planetesimal capture model for the formation of giant planet satellite systems. *Icarus, 181,* 486–509.

Friedson A. J. and Stevenson D. J. (1983) Viscosity of ice-rock mixtures and applications to the evolution of icy satellites. *Icarus, 56,* 1–14.

Gammie C. F. (1996) Layered accretion in T Tauri disks. *Astrophys. J., 457,* 355–362.

Garaud P. and Lin D. N. C. (2007) The effect of internal dissipation and surface irradiation on the structure of disks and the location of the snow line around Sun-like stars. *Astrophys. J., 654,* 606–624.

Glassgold A. E., Najita J. R., and Igea J. (1997) Neon fine-structure line emission by X-ray irradiated protoplanetary disks. *Astrophys. J., 656,* 515–523.

Goldreich P. and Tremaine S. (1980) Disk-satellite interactions. *Astrophys. J., 241,* 425–441.

Goldreich P., Lithwick Y., and Sari R. (2004) Planet formation by coagulation: A focus on Uranus and Neptune. *Annu. Rev. Astron. Astrophys., 42,* 549–601.

Goodman J. and Rafikov R. R. (2001) Planetary torques as the viscosity of protoplanetary disks. *Astrophys. J., 552,* 793–802.

Guillot T. (2005) The interiors of giant planets: Models and outstanding questions. *Annu. Rev. Earth Planet. Sci., 33,* 493–530.

Haisch K. E. Jr., Lada E. A., and Lada C. J. (2001) Disk frequencies and lifetimes in young clusters. *Astrophys. J. Lett., 553,* L153–L156.

Hantao J., Burin M., Schartman E., and Goodman J. (2006) Hydrodynamic turbulence cannot transport angular momentum effectively in astrophysical disks. *Nature, 444,* 343–346.

Hayashi C., Nakazawa K., and Nakagawa Y. (1985) Formation of the solar system. In *Protostars and Planets II* (D. C. Black and M. S. Matthews, eds.), pp. 1100–1153. Univ. of Arizona, Tucson.

Helling Ch., Winters J. M., and Sedlmayr E. (2000) Circumstellar dust shells around long-period variables. *Astron. Astrophys., 358,* 651–664.

Herbst W., Bailer-Jones C. A. L., Mundt R., Meisenheimer K., and Wackermann R. (2002) Stellar rotation and variability in the Orion Nebula Cluster. *Astron. Astrophys., 396,* 513–532.

Hubeny I. (1990) Vertical structure of accretion disks — A simplified analytical model. *Astrophys. J., 351,* 632–641.

Hubickyj O., Bodenheimer P., and Lissauer J. J. (2005) Accretion of the gaseous envelope of Jupiter around a 5 to 10 Earth-mass core. *Icarus, 179,* 415–431.

Hussmann H., Spohn T., and Wieczerkowski K. (2002) Thermal equilibrium states of Europa's ice shell: Implications for internal ocean thickness and surface heat flow. *Icarus, 156,* 143–151.

Jewitt D. and Haghighipour N. (2007) Irregular satellites of the planets: Products of capture in the early solar system. *Annu. Rev. Astron. Astrophys., 45,* 261–295.

Klahr H. H. and Bodenheimer P. (2003) Turbulence in accretion disks: Vorticity generation and angular momentum transport via the global baroclinic instability. *Astrophys. J., 582,* 869–892.

Koenigl A. (1991) Disk accretion onto magnetic T Tauri stars. *Astrophys. J. Lett., 370,* L39–L43.

Korycansky D. G., Bodenheimer P. and Pollack J. B. (1991) Numerical models of giant planet formation with rotation. *Icarus, 92,* 234–251.

Lin D. N. C. and Papaloizou J. (1986) On the tidal interaction between protoplanets and the protoplanetary disk III — Orbital migration of protoplanets. *Astrophys. J., 309,* 846–857.

Lissauer J. J. and Kary D. M. (1991) The origin of the systematic component of planetary rotation. I. Planet on a circular orbit. *Icarus, 94,* 126–159.

Lissauer J. J. and Stevenson D. J. (2007) Formation of giant planets. In *Protostars and Planets V* (B. Reipurth et al., eds.), pp. 591–601. Univ. of Arizona, Tucson.

Lithwick Y. (2007) Nonlinear evolution of hydrodynamical shear flows in two dimensions. *Astrophys J., 670,* 789–804.

Lubow S. H., Seibert M., and Artymowicz P. (1999) Disk accretion onto high-mass planets. *Astrophys. J., 526,* 1001–1012.

Lunine J. I. and Stevenson D. J. (1982) Formation of the Galilean satellites in a gaseous nebula. *Icarus, 52,* 14–39.

Lunine J. I., Coradini A., Gautier D., Owen T. C., and Wuchterl G. (2004) The origin of Jupiter. In *Jupiter: The Planet, Satellites and Magnetosphere* (F. Bagenal et al., eds.), pp. 19–34. Cambridge Univ., Cambridge.

Lynden-Bell D. and Pringle J. E. (1974) The evolution of viscous discs and the origin of the nebular variables. *Mon. Not. R. Astron. Soc., 168,* 603–637.

Machida M. N. (2008) Thermal effects of circumplanetary disk formation around proto-gas giant planets. *Mon. Not. R. Astron. Soc.,* in press.

Machida M. N., Kokubo E., Inutsuka S., and Matsumoto T. (2008) Angular momentum accretion onto a gas giant planet. *Astrophys. J., 685,* 1220–1236.

Makalkin A. B., Dorofeeva V. A., and Ruskol E. L. (1999) Modeling the protosatellite circum-jovian accretion disk: An estimate of the basic parameters. *Solar Sys. Res., 33,* 578.

Malhotra R. (1991) Tidal origin of the Laplace resonance and the resurfacing of Ganymede. *Icarus, 94,* 399–412.

McKinnon W. B. (1997) Mystery of Callisto: Is it undifferentiated? *Icarus, 130,* 540–543.

McKinnon W. B. (2006) Formation time of the Galilean satellites from Callisto's state of partial differentiation (abstract). In *Lunar and Planetary Science XXXVII,* Abstract #2444. Lunar and Planetary Institute, Houston (CD-ROM).

McKinnon W. B. (2007) Formation and early evolution of Io. In *Io after Galileo* (R. M. C. Lopes and J. R. Spencer, eds.), pp. 61–88. Springer Verlag, Berlin.

Moore W. B. (2006) Thermal equilibrium in Europa's ice shell. *Icarus, 180,* 141–146.

Moore W. B., Schubert G., Anderson J. D., and Spencer J. R. (2007) The interior of Io. In *Io after Galileo* (R. M. C. Lopes and J. R. Spencer, eds.), pp. 89–108. Springer Verlag, Berlin.

Mosqueira I. and Estrada P. R. (2003) Formation of regular satellites of giant planets in an extended gaseous nebula I: Subnebula model and accretion of satellites. *Icarus, 163,* 198–231.

Mousis O. and Alibert Y. (2006) Modeling the jovian subnebula: II. Composition of regular satellite ices. *Astron. Astrophys., 448,* 771–778.

Nagasawa M., Thommes E. W., Kenyon S. J., Bromley B. C., and Lin D. N. C. (2007) The diverse origins of terrestrial-planet systems. In *Protostars and Planets V* (B. Reipurth et al., eds.), pp. 639–654. Univ. of Arizona, Tucson.

Nakagawa Y., Sekiya M., and Hayashi C. (1986) Settling and growth of dust particles in a laminar phase of a low-mass solar nebula. *Icarus, 67,* 375–390.

Nakamoto T. and Nakagawa Y. (1994) Formation, early evolution, and gravitational stability of protoplanetary disks. *Astrophys. J., 421,* 640–650.

Papaloizou J. C. B. and Larwood J. D. (2000) On the orbital evolution and growth of protoplanets embedded in a gaseous disc. *Mon. Not. R. Astron. Soc., 315,* 823–833.

Papaloizou J. C. B. and Nelson R. P. (2005) Models of accreting gas giant protoplanets in protostellar disks. *Astron. Astrophys., 433,* 247–265.

Peale S. J. (1999) Origin and evolution of the natural satellites. *Annu. Rev. Astron. Astrophys., 37,* 533–602.

Peale S. J. (2008) The origin of the natural satellites. In *Treatise on Geophysics, Volume 10: Planets and Moons* (T. Spohn, ed.), pp. 465–508. Elsevier, New York.

Peale S. J. and Lee M. H. (2002) A primordial origin of the

Laplace relation among the Galilean satellites. *Science, 298,* 593–597.
Pollack J. B., Lunine J. I., and Tittemore W. C. (1991) Origin of the Uranian satellites. In *Uranus* (J. T. Bergstralh et al., eds.), pp. 469–512. Univ. of Arizona, Tucson.
Pollack J. B., Hubickyj O., Bodenheimer P. Lissauer J. J., Podolak M., and Greenszweig Y. (1996) Formation of giant planets by concurrent accretion of solids and gas. *Icarus, 124,* 62–85.
Prinn R. G. and Fegley B. Jr. (1981) Kinetic inhibition of CO and N_2 reduction in circumplanetary nebulae — Implications for satellite composition. *Astrophys. J., 249,* 308–317.
Prinn R. G. and Fegley B. Jr. (1989) Solar nebula chemistry: Origins of planetary, satellite and cometary volatiles. In *Origin and Evolution of Planetary and Satellite Atmospheres* (S. K. Atreya et al., eds.), pp. 78–136. Univ. of Arizona, Tucson.
Pritchard M. E. and Stevenson D. J. (2000) Thermal aspects of a lunar origin by giant impact. In *Origin of the Earth and Moon* (R. M. Canup and K. Righter, eds.), pp. 179–196. Univ. of Arizona, Tucson.
Rafikov R. R. (2004) Fast accretion of small planetesimals by protoplanetary cores. *Astron. J., 128,* 1348–1363.
Rice W. K. M., Armitage P. J., Wood K., and Lodato G. (2006) Dust filtration at gap edges: Implications for the spectral energy distributions of discs with embedded planets. *Mon. Not. R. Astron. Soc., 373,* 1619–1626.
Richardson D. C., Quinn T., Stadel J., and Lake G. (2000) Direct large-scale N-body simulations of planetesimal dynamics. *Icarus, 143,* 45–59.
Ruden S. P. and Lin D. N. C. (1986) The global evolution of the primordial solar nebula. *Astrophys. J., 308,* 883–901.
Ruskol E. L. (1982) Origin of planetary satellites. *Izvestiva Earth Phys., 18,* 425–433 (in Russian).
Safronov V. S., Pechernikova G. V., Ruskol E. L., and Vitiazev A. V. (1986) Protosatellite swarms. In *Satellites* (J. Burns and M. Matthews, eds.), pp. 89–116. Univ. of Arizona, Tucson.
Sano T., Miyama S. M., Umebayasi T., and Nakano T. (2000) Magnetorotational instabiltiy in protoplanetary disks II. Ionization state and unstable regions. *Astrophys. J., 543,* 486–501.
Schlichting H. E. and Sari R. (2007) The effect of semicollisional accretion on planetary spins. *Astrophys. J., 658,* 593–597.
Schubert G., Anderson J. D., Spohn T., and McKinnon W. B. (2004) Interior composition, structure and dynamics of the Galilean satellites. In *Jupiter: The Planet, Satellites and Magnetosphere* (F. Bagenal et al., eds.), pp. 281–306. Cambridge Univ., Cambridge.
Semenov D., Henning Th., Helling Ch., Ilgner M., and Sedlmayr E. (2003) Rosseland and Planck mean opacities for protoplanetary discs. *Astron. Astrophys., 410,* 611–621.
Semenov D., Wiebe D., and Henning Th. (2004) Reduction of chemical networks II. Analysis of the fractional ionisation in protoplanetary discs. *Astron. Astrophys., 417,* 93–106.
Shakura N. I. and Sunyaev R. A. (1973) Black holes in binary systems. Observational appearance. *Astron. Astrophys., 24,* 337–355.
Showman A. P., Stevenson D. J., and Malhotra R. (1997) Coupled orbital and thermal evolution of Ganymede. *Icarus, 129,* 367–383.
Shu F. H., Najita J. R., Shang H., and Zhi-Yun L. (2000) X-winds: Theory and observations. In *Protostars and Planets IV* (V. Mannings et al., eds.), pp. 789–813. Univ. of Arizona, Tucson.
Stevenson D. J. (2001a) Origins of the Galilean satellites (abstract). In *Jupiter: Planets, Satellites and Magnetosphere,* June 25–30, 2001, Boulder, Colorado.
Stevenson D. J. (2001b) Jupiter and its moons. *Science, 294,* 71–72.
Stevenson D. J., Harris A. W., and Lunine J. I. (1986) Origins of satellites. In *Satellites* (J. A. Burns and M. S. Matthews, eds.), pp. 39–88. Univ. of Arizona, Tucson.
Stone J. M., Gammie C. F., Balbus S. A., and Hawley J. F. (2000) Transport processes in protostellar disks. In *Protostars and Planets IV* (V. Mannings et al., eds.), pp. 589–612. Univ. of Arizona, Tucson.
Tajima N. and Nakagawa Y. (1997) Evolution and dynamical stability of the proto-giant-planet envelope. *Icarus, 126,* 282–292.
Takata T. and Stevenson D. J. (1996) Despin mechanism for protogiant planets and ionization state of protogiant planetary disks. *Icarus, 123,* 404–421.
Tanaka H. and Ward W. R. (2004) Three-dimensional interaction between a planet and an isothermal gaseous disk. II. Eccentricity waves and bending waves. *Astrophys. J., 602,* 388–395.
Tanaka H., Takeuchi T., and Ward W. R. (2002) Three-dimensional interaction between a planet and an isothermal gaseous disk. I. Corotation and Lindblad torques and planet migration. *Astrophys. J., 565,* 1257–1274.
Ward W. R. (1986) Density waves in the solar nebula — Differential Lindblad torque. *Icarus, 67,* 164–180.
Ward W. R. (1988) On disk-planet interactions and orbital eccentricities. *Icarus, 73,* 330–348.
Ward W. R. (1993) Density waves in the solar nebula — Planetesimal velocities. *Icarus, 106,* 274–287.
Ward W. R. (1997) Protoplanet migration by nebula tides. *Icarus, 126,* 261–281.
Ward W. R. (2009) Particle filtering by a planetary gap. In *Lunar and Planetary Science XL,* Abstract #1477. Lunar and Planetary Institute, Houston (CD-ROM).
Ward W. R. and Hahn J. (2000) Disk-planet interactions and the formation of planetary systems. In *Protostars and Planets IV* (V. Mannings et al., eds.), pp. 1135–1155. Univ. of Arizona, Tucson.
Weidenschilling S. J. and Cuzzi J. N. (1993) Formation of planetesimals in the solar nebula. In *Protostars and Planets III* (E. H. Levy and J. Lunine, eds.), pp. 1031–1060. Univ. of Arizona, Tucson.

Tides and Tidal Heating on Europa

Christophe Sotin
NASA Jet Propulsion Laboratory/California Institute of Technology

Gabriel Tobie
Centre National de la Recherche Scientifique, Université Nantes Atlantique

John Wahr
University of Colorado

William B. McKinnon
Washington University in St. Louis

Because of its close distance to Jupiter and its elliptic orbit, Europa experiences strong tidal forces. This chapter describes the tidal forces, the manner in which icy and silicate materials deform, the deformation and stress induced by tidal forces, the resulting generation of tidal heating, and its consequence for the thermal and orbital evolution of Europa. It is shown that most of the tidal heating is produced in the icy shell. The amount of tidal heating in the icy shell may be several times larger than radiogenic heating and may explain why Europa did not completely freeze after the accretion period. Europa's orbital evolution is coupled with those of Ganymede and Io, which eventually ended into a Laplace resonance. Some models suggest that Europa had a higher eccentricity in the past. They imply that tidal heating would have provided an energy source significantly larger than the present one. It is often thought that the silicate mantle could be as active as that on Io, where active volcanism is caused by tidal dissipation. Although one cannot rule out that tidal heating may have affected Europa's silicate mantle during its early history, current models predict that most of the tidal forces are dissipated into the outer icy shell and little in the silicate part. For reasonable values of viscosity for silicates, tidal heating in the silicate mantle is considerably less than radiogenic heating, which is much too small to yield partial melting at the present time. The presence of active volcanism at the ocean/silicate interface is therefore questionable.

1. INTRODUCTION

The global characteristics of Europa (Table 1) were determined during the Galileo mission. The few flybys provided information about the moment of inertia, the magnetic field, and the surface composition and morphology. The gravity data strongly suggest that Europa is fully differentiated into an iron core, a silicate mantle, and a H_2O layer (*Anderson et al.,* 1998). Europa's average density is close to that of silicates (Table 1). As a comparison, its density is only 15% smaller than that of Io (ρ_{Io} = 3528 kg/m^3), its closest neighbor in the jovian system. This indicates that Europa contains a relatively small amount of H_2O compared to the large icy satellites Ganymede and Callisto. The total thickness of the H_2O layer is estimated at between 100 and 200 km (*Anderson et al.,* 1998; *Sohl et al.,* 2002).

The magnetic data revealed the presence of an induced magnetic field, which is best understood with the presence of a liquid layer several tens of kilometers below the outer icy shell (*Kivelson et al.,* 1997; *Khurana et al.,* 1998, see chapter by Khurana et al.). Compared to the other icy satellites, the relatively small amount of H_2O does not allow for the presence of a high-pressure ice layer between the ocean and the silicate mantle. The water is therefore in direct contact with silicates. Interestingly, the pressure and temperature conditions at the silicate-ocean boundary are quite similar to those existing on Earth's seafloor, where organisms live without solar energy (e.g., *Kelley et al.,* 2001).

The question of a liquid water layer within Europa had already been discussed at the time of the Voyager mission (e.g., *Cassen et al.,* 1979). The observation of various re-

TABLE 1. Characteristics of Europa.

Parameter	Quantity and Dimension
Radius, R_S	1560.8 ± 0.5 km
Average density, $\bar{\rho}$	3013 ± 5 kg m^{-3}
GM	3202.72 ± 0.02 km^3 m^{-2}
Surface gravity, g(R_S)	1.32 m s^{-2}
J_2	435.5 ± 8.2 × 10^{-6}
C/MR^2	0.347
Period of rotation, T_E	3.55 days
Distance to Jupiter, a	670.9 × 10^6 m
Eccentricity, e	0.9–1%

GM and J_2 from chapter by Schubert et al.; radius from *Nimmo et al.* (2007).

cent geological features by Voyager, and later by Galileo, suggested that warm, mobile material, possibly liquid water, exists below the cold icy crust (*Pappalardo et al.*, 1999). The presence of liquid water is also supported by several thermal models. Various theoretical models of heat transfer by solid state convection in the icy shell suggest that Europa should be able to freeze completely if the main heat source is radiogenic and/or if the liquid layer does not contain any antifreezing compounds such as ammonia (*Deschamps and Sotin*, 2001; *Spohn and Schubert*, 2003). Indeed, if the only internal heating source is the decay of radiogenic elements contained in the silicate mantle, the heat flux at the rock-water interface would have reached maximum values on the order of 35 mW m^{-2} in the past and would have decayed to a value of 10 mW m^{-2} at the present time (see chapter by Moore and Hussmann). In this condition, the icy shell in equilibrium with the radiogenic heat flux would rapidly exceed a few tens of kilometers. For icy shell thicknesses larger than 20–30 km, numerous studies (*McKinnon*, 1999; *Deschamps and Sotin*, 2001; *Hussmann et al.*, 2002; *Tobie et al.*, 2003; *Mitri and Showmann*, 2005; *Barr and Pappalardo*, 2005; *Barr and McKinnon*, 2007; see chapter by Barr and Showman) have suggested that subsolidus convection would occur, increasing the efficiency of heat transfer and hence accelerating the crystallization rate of the liquid layer. One possible way to limit the crystallization of the liquid water layer is to have an additional heat source in the silicate mantle, in the water layer, or in the icy shell. The missing heat source required to maintain a liquid water ocean on Europa could be provided by the dissipation of energy due to solid tides generated by Jupiter's gravitational forces.

The tremendous surface heat flux [≥2 W m^{-2} (*Veeder et al.*, 1994; *Spencer et al.*, 2000)] and the intense volcanism of Io suggest that tidal dissipation constitutes a significant heat source inside some satellites, and that it can strongly influence their thermal state (e.g., *Segatz et al.*, 1988). On Europa, although tidal heating is probably lower due to the longer orbital period and the smaller radius of the satellite, it is still expected to contribute to the energy balance, and to favor the existence of a deep ocean below an icy shell thinner than 30 km (*Ojakangas and Stevenson*, 1989; *Hussmann et al.*, 2002; *Tobie et al.*, 2003). The complex tectonic activities observed on Europa's surface also indicate that tidal deformation and nonsynchronous rotation (NSR) play a key role in the geodynamical evolution of the satellite (e.g., *Greenberg et al.*, 1998).

In the present chapter, after introducing the mechanical properties of the materials that compose Europa's interior in section 2, we describe the origin and derivation of the time-varying potential induced by tidal interaction and NSR in section 3. Section 4 briefly recalls the classical formulation of tidal deformation for elastic and viscoelastic bodies. In section 5, we discuss some results on the global deformation of the satellite by exploring how the interior structure and rheology affect the determination of the Love numbers, quantities that define the global response of the satellite to periodic forcing. The calculations of tidal deformation are then used to derive the stress field induced by tidal forces and NSR in section 6. In section 7, different methods of computing the tidal dissipation rate within Europa's interior are presented and are used to derive the distribution of tidal heating within each internal layer. The implications of tides and tidal heating for the internal evolution of Europa are discussed in section 8. Given the intimate coupling between Europa's internal evolution and the tidal evolution of its orbit, we also summarize, in section 9, extant hypotheses for the formation of the Laplace resonance among Io, Europa, and Ganymede and Europa's long-term tidal evolution.

2. MECHANICAL PROPERTIES OF SATELLITE INTERIOR

2.1. General Principles

The gravitational forces along Europa's orbit vary periodically owing to the orbital eccentricity (T_E = 3.55 d; cf. Table 1). The materials that compose Europa will deform according to these periodic force fluctuations. The response of each layer in Europa's interior will be determined by the thermomechanical properties of the dominating material (liquid iron for the core, olivine for the silicate mantle, liquid water for the ocean, hexagonal ice I for the icy shell). Each of these materials behaves differently at the tidal forcing frequency. Part of the deformation will be purely elastic and is determined by the elastic moduli of the materials (shear and bulk modulii, μ and K respectively). The other part is anelastic and is determined by the different relaxation or attenuation mechanisms that can occur in the material. (Relaxation is usually used in the time domain, and attenuation in the frequency domain.) The relaxation processes are strongly frequency- and temperature-dependent, and they are determined by the microscopic structure of the material.

In polycrystals, such as hexagonal ice I or olivine, the attenuation mechanisms are mainly related to the motions of defects and their interaction with the crystal lattice, resulting in friction. During the first quarter of a tidal cycle, strain energy is progressively stored as compressive (or tensile) stress increases. During the second quarter, strain energy is restored as stress decreases. The same process operates during the other half period when stress is tensile (or compressive). The back and forth motions of crystal defects induced by alternating compression and tension convert part of the strain energy into heat. The efficiency of such dissipative processes depends on the population of existing defects (or dislocations) and therefore is very sensitive to the density of existing dislocations within the crystal lattice (e.g., *Cole and Durell*, 1995; *Jackson et al.*, 2000). From a macroscopic point of view, the anelastic behavior of a material sample results in a time delay of the material response relative to tidal forcing, in a smaller effective shear modulus and in the production of heat.

Formally, material attenuation is expressed through the phase lag between the application of the tidal stress and the actual deformation of the object. If a periodic stress $\tau = \tau_0$

$\exp[i\omega t]$ of angular frequency ω is applied to a material sample, the resulting strain at time t is given by $\varepsilon = \varepsilon_0 \exp[i(\omega t - \delta_m)]$, where δ_m is called the phase lag angle for mechanical relaxation, and τ_0 and ε_0 are the maximal amplitude of stress and strain respectively. The decrease ΔE_{diss} of energy per cycle due to dissipation is given by (*Hobbs,* 1974)

$$\Delta E_{diss} = \int_0^{2\pi/\omega} \tau \frac{\partial \varepsilon}{\partial t} dt = \pi \tau_0 \varepsilon_0 \sin(\delta_m) \quad (1)$$

The peak energy E reached during one cycle is equal to $\tau_0\varepsilon_0/2$, which implies $\frac{\Delta E_{diss}}{E} = 2\pi \sin(\delta_m)$. By definition, the quality factor Q, also called the specific dissipation function, is related to the energy loss and the phase lag: $Q^{-1} = \frac{\Delta E_{diss}}{2\pi E} = \sin(\delta_m)$.

The tidal phase lag can be measured from satellite tracking and altimetry. This has been done, e.g., to assess the tidal dissipation function of Earth's mantle from TOPEX/POSEIDON data (e.g., *Ray et al.,* 2001). In the absence of such data, the tidal dissipation function can in principle be estimated from secular changes of a satellite's mean motion. But the separation of the effects of the planet's dissipation from those of the satellite's dissipation is not straightforward (e.g., *Lainey and Tobie,* 2005). All these measurements yield constraints on the viscoelastic properties of the interior. The viscoelastic response of the satellite can also be approximated by using linear viscoelastic models, such as Maxwell or Burgers models, or directly from laboratory measurements on ice and rock samples and by integrating the attenuation properties for multilayered models using existing software (e.g., *Tobie et al.,* 2005).

For long-period forcing, the simplest and most applicable model of an elasticity is that of a Maxwell viscoelastic solid, which can be represented pictorially as an elastic spring and viscous dashpot in series. If a load is placed on a Maxwell solid, shear stresses will initially be induced in the solid, and the solid will instantaneously deform (elastic deformation). If the load is held in place, the shear stresses will gradually relax to zero, causing the deformation to grow (viscous deformation) with a time constant, called the Maxwell time t_M, that is defined by the ratio between the viscosity and the shear modulus, $t_M = \eta/\mu$. The long-term final state of the solid ($t \gg t_M$) will be the same as if the material were a fluid in hydrostatic equilibrium.

In the frequency domain, tidal deformation can be computed using the classical elastic formulation by defining a complex effective shear modulus μ^c (e.g., *Tobie et al.,* 2005). The effective complex shear modulus can be calculated from the dynamic compliance $D^c = D_1 + iD_2$, which relates strain to stress in the frequency domain (e.g., *Jackson et al.,* 2000)

$$\Re(\mu^c) = \frac{D_1}{D_1^2 + D_2^2} \quad \text{and} \quad \Im(\mu^c) = \frac{D_2}{D_1^2 + D_2^2} \quad (2)$$

For a Maxwell rheology

$$D_1 = \frac{1}{\mu} \quad \text{and} \quad D_2 = \frac{1}{\eta\omega} \quad (3)$$

where μ is the elastic shear modulus, η is the long-term viscosity, and ω is the angular frequency. For a Burgers rheology

$$D_1 = \frac{1}{\mu} + \frac{\frac{1}{\mu_B}}{1 + \omega^2\tau_B^2} \quad \text{and} \quad D_2 = \frac{1}{\eta\omega} + \frac{\frac{\omega\tau_B}{\mu_B}}{1 + \omega^2\tau_B^2} \quad (4)$$

where the two additional parameters, μ_B and $\tau_B = \mu_B/\eta_B$, describe a second relaxation process that occurs before the Maxwell relaxation becomes important. For experimentally-derived models, the compliance function can also depend on other parameters such as the dislocation density, salinity, porosity. etc. [see *Cole and Durell* (1995) for a complete expression of the dynamical compliance], but the principle remains the same: As long as the real and imaginary part of the dynamical compliance function can be determined, a complex shear modulus can be defined and the elastic formalism can be applied.

2.2. Viscoelastic Models for Ice

The Maxwell model is usually used for computing tidal heating in the icy shell (e.g., *Ojakangas and Stevenson,* 1989; *Tobie et al.,* 2005). It is adapted to describe material relaxation properties when the forcing period is close to their Maxwell time. However, it fails to quantify material attenuation over a wide range of frequencies and temperatures. Other models, based on laboratory measurements (*Tatibouet et al.,* 1986; *Cole and Durell,* 1995) or on analysis of tidal bending of polar glaciers (*Reeh et al.,* 2003) are probably more appropriate for describing the viscoelastic response across a wide range of temperatures. Note, however, that cyclic loading in the laboratory (*Tatibouet et al.,* 1986; *Cole and Durell,* 1995) is usually performed at frequencies (10^{-4}–10^{-3} Hz) higher than tidal frequencies (10^{-6}–10^{-5} Hz), so that mathematical extrapolation of experimentally-derived viscoelastic parameters to lower frequencies are required.

Figure 1 illustrates the dissipation function Q^{-1}, and the amplitude of the effective shear modulus μ_{eff} calculated for different viscoelastic models, as a function of forcing period and temperature. For a forcing period of the order of few days and temperatures larger than 240 K, the three viscoelastic models predict similar dissipation function values (Figs. 1a,b). The largest discrepancy is between the Maxwell and Burgers models, but that discrepancy is still smaller than a factor of 2. At lower temperatures, T < 240 K, the Maxwell model strongly underestimates the dissipation function compared to the empirical, experimentally-derived models of *Cole and Durell* (1995). Conversely, the Burgers model probably overestimates the dissipation function for temperature between 180 and 220 K. Each of the three mod-

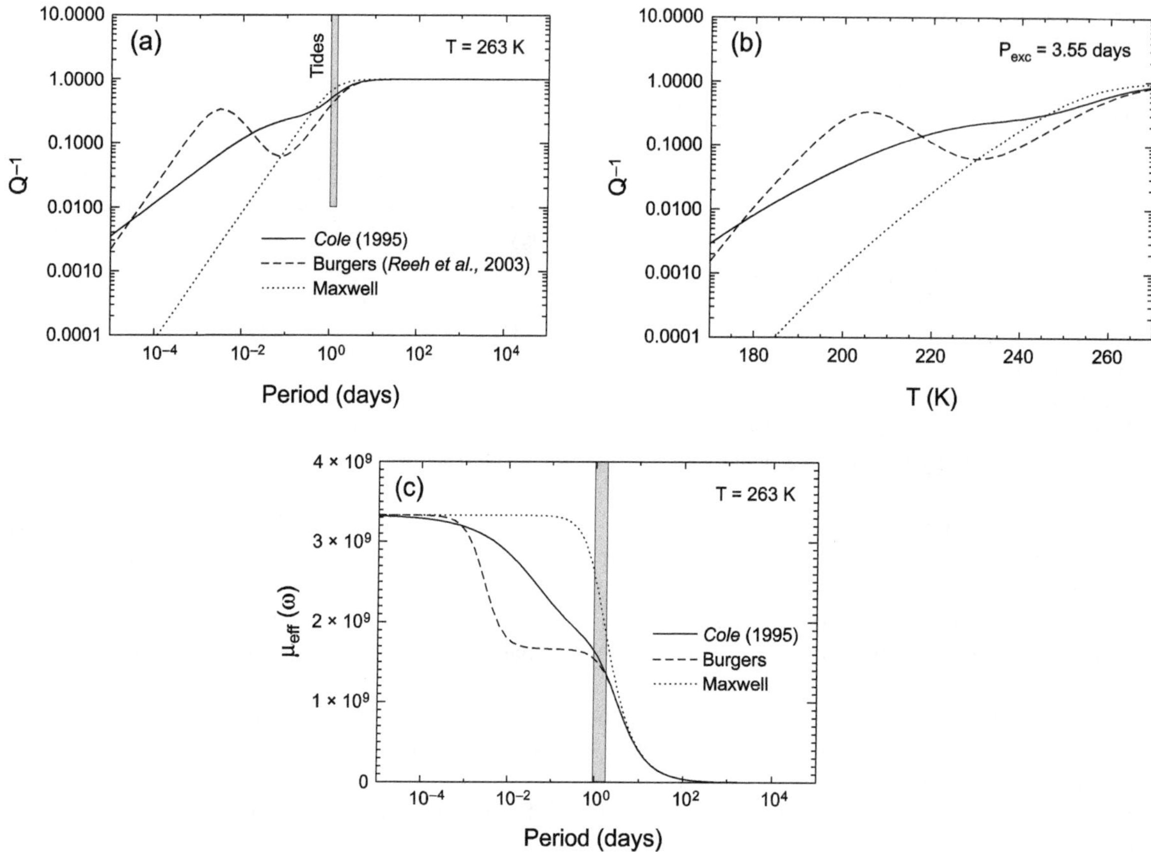

Fig. 1. Specific dissipation function Q^{-1} in ice I, for three different viscoelastic models, as a function of **(a)** forcing period and **(b)** temperature. **(c)** The corresponding effective shear modulus as a function of forcing period.

els predicts a decrease of the effective shear modulus for forcing periods approaching few days (Fig. 1c). The Maxwell model, however, predicts a smaller decrease than the two other models. For the two other models, the effective shear modulus starts decreasing before reaching periods close to the Maxwell time owing to the existence of a transient response related to other relaxation mechanisms with a time constant shorter than the Maxwell time constant. Even though the simple Maxwell model does not perfectly reproduce the viscoelastic response of water ice, it gives a reasonable first-order description. It has the advantage of depending on only three parameters that can be experimentally-derived: the elastic shear modulus, the elastic bulk modulus, and the viscosity.

2.3. Viscoelastic Models for Silicate

The Maxwell model is also classically used when modeling the tidal response of Europa's silicate interior (e.g., *Moore,* 2003; *Hussmann and Spohn,* 2004; *Tobie et al.,* 2005). On Earth, this model has been successfully used to describe the relaxation process associated with postglacial rebound, which occurs on timescales of a few thousand years (e.g., *Peltier,* 1974). The postglacial rebound data combined with constraints on geoid anomalies and plate motions thus indicate that the long-term viscosity of Earth's upper mantle, below the asthenosphere, is on the order of 10^{20}–10^{21} Pa s (e.g., *Cadek and Fleitout,* 1999; *Bills et al.,* 2007). As illustrated in Fig. 2, in the case of Earth, the extrapolation of the Maxwell model to tidal periods leads to dissipation values smaller than the global value derived from observations, when using the reference long-term viscosity of 10^{20}–10^{21} Pa s. Similar dissipation values can be obtained only if viscosity values one to two orders of magnitude smaller than the reference Earth's mantle viscosity are assumed. In addition, the Maxwell model predicts a steeper variation as a function of period than the variations inferred from the combination of tidal and rotational data (*Benjamin et al.,* 2006).

The observed period dependence is more consistent with the anelastic models proposed in the seismic period range (1–1000 s) based on the attenuation data of seismic waves in Earth's mantle (e.g., *Anderson and Given,* 1982; *Romanowicz and Durek,* 2000) and mechanical tests on rock samples in the laboratory (e.g., *Jackson,* 2000). However, the seismic model predicts a decrease of the dissipation function for periods larger than 10^{-3} d (100 s). For Earth's lower mantle, a similar decrease should occur for periods larger than 10 d, which is inconsistent with the rotational data (*Benjamin et al.,* 2006). Therefore, the extrapolation of the seismic anelastic model, which is well constrained in a given range of forcing period, also seems problematic.

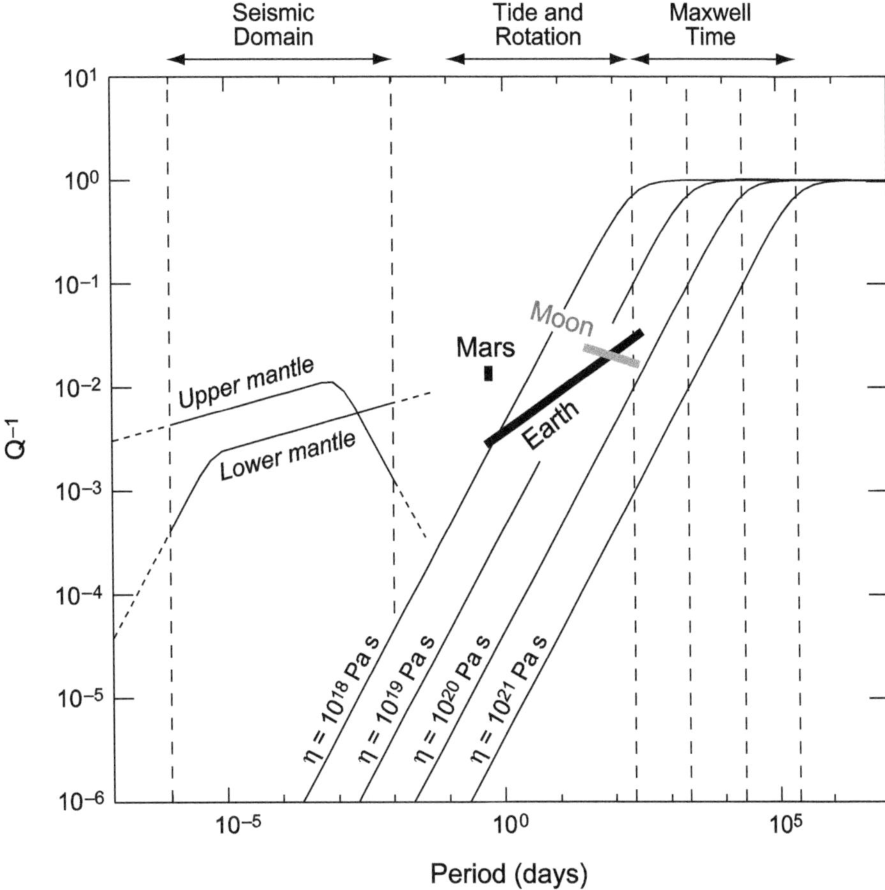

Fig. 2. Specific dissipation function Q^{-1} in the silicate mantle as a function of forcing period, constrained from an anelastic seismic model developed for Earth's mantle (*Anderson and Given,* 1982) from observations of tidal deformation and rotation changes on Earth (*Ray et al.,* 2001; *Benjamin et al.,* 2006), on the Moon (*Williams et al.,* 2001), and on Mars (*Yoder et al.,* 2003; *Bills et al.,* 2005) from a Maxwell model. In the seismic domain, the typical dissipation functions corresponding to Earth's upper mantle and lower mantle are shown, and extrapolated outside the seismic band with dashed lines. For the Maxwell model, an elastic modulus of 70 GPa and viscosity values ranging from 10^{18} to 10^{21} Pa s, corresponding to different Maxwell times indicated by gray vertical dashed lines, have been used.

The fact that Mars has an apparent dissipation function larger than that of Earth and that the period dependence for the Moon's dissipation function is inverted compared to Earth's is also puzzling (Fig. 2). In the absence of reliable data over a wide range of periods, it is difficult to propose a viscoelastic model capable of explaining the dissipation function from seismic periods to the Maxwell time.

As long as the forcing period is much shorter than the Maxwell time, the Maxwell model strongly underestimates the dissipation function. Therefore, for cold silicate interiors, the use of a Maxwell model is surely not relevant. On the other hand, for viscosity values on the order of 10^{18} Pa s, which is the estimated value near the melting point of olivine (*Karato and Wu,* 1993), the Maxwell time is equal to 150 d. For a tidal period of a few days, the Maxwell model probably gives a reasonable estimation of the dissipation rate. For hot silicate interiors, the Maxwell model is more relevant, even if it does not fully describe the dissipation processes. The main advantage of the Maxwell model is that it couples the temperature evolution and the dissipation via the viscosity in a very simple manner, and that it permits the study of the role of tidal dissipation on the thermal evolution of a rocky mantle close to its melting point. Note, however, that when the interior reaches the melting point, the Maxwell model probably does not correctly describe the dissipation rate, as the presence of melt strongly affects the mechanical properties of rock (e.g., *Schmeling,* 1985). In the literature, the change in the viscoelastic response of silicate materials due to melting is classically included by assuming a Maxwell rheology and by considering an abrupt change of the effective viscosity and of the effective shear modulus (*Fisher and Spohn,* 1990; *Moore,* 2003). In reality, the molten rock behaves as a two-phase media, where dissipation is probably related to complex interactions between the liquid phase and the solid matrix. More complex two-phase media models would probably be relevant. However, in the absence of reliable data to constrain it, a Maxwell-type model that includes viscosity and a shear modulus corrected for the presence of melt is probably a reasonable assumption as long as the melt fraction remains small.

3. THE TIDAL FORCE

Tides are present on a planet's moon because the gravitational force from the planet is not the same at every point in the moon. The spatially varying component of the force, denoted as the tidal force, causes deformation of the satellite. That deformation is the tide.

To understand the spatial pattern of the tide, consider Fig. 3a. Europa's orbital plane coincides with the plane of the paper, and we assume that Europa's rotation axis is perpendicular to that plane (i.e., that Europa's obliquity vanishes). The arrows indicate the direction and magnitude of the gravitational acceleration (the force per mass) caused by Jupiter, at a few points in Europa. The acceleration at every point is directed toward Jupiter. It is largest at points closest to Jupiter because of the $1/r^2$ dependence of the gravitational force. (The differences in arrow length and direction are greatly exaggerated.) The average of all these acceleration vectors is the net acceleration of Europa toward Jupiter. This net acceleration leads to Europa's orbital motion about Jupiter, and is approximately equal to the gravitational acceleration at the center of Europa. If we subtract the arrow at the center of Europa from the other arrows, we are left with the residual arrows shown in Fig. 3b.

Similarly, Fig. 3c shows the pattern of residual arrows as seen by an observer in the orbital plane. These residual arrows, which illustrate the component of the gravitational acceleration field that causes Europa to deform, represent the tidal force. Note from Figs. 3b and 3c that the force is radially outward on the sides toward and away from Jupiter, and is radially inward on all other sides. This forcing pattern remains fixed with respect to Jupiter. Motion of Europa relative to Jupiter causes the tidal force at fixed points in Europa to vary with time, and this leads to time-variable tidal displacements and stresses within Europa.

The temporal and spatial characteristics of tides on Europa are thus determined by Europa's motion relative to Jupiter, which leads to a modulation of the position vector in both amplitude and direction. The most rapidly varying part of that motion is caused by Europa's orbital motion. But there could also be contributions from NSR of Europa's outer icy shell. Here, we discuss both types of motion.

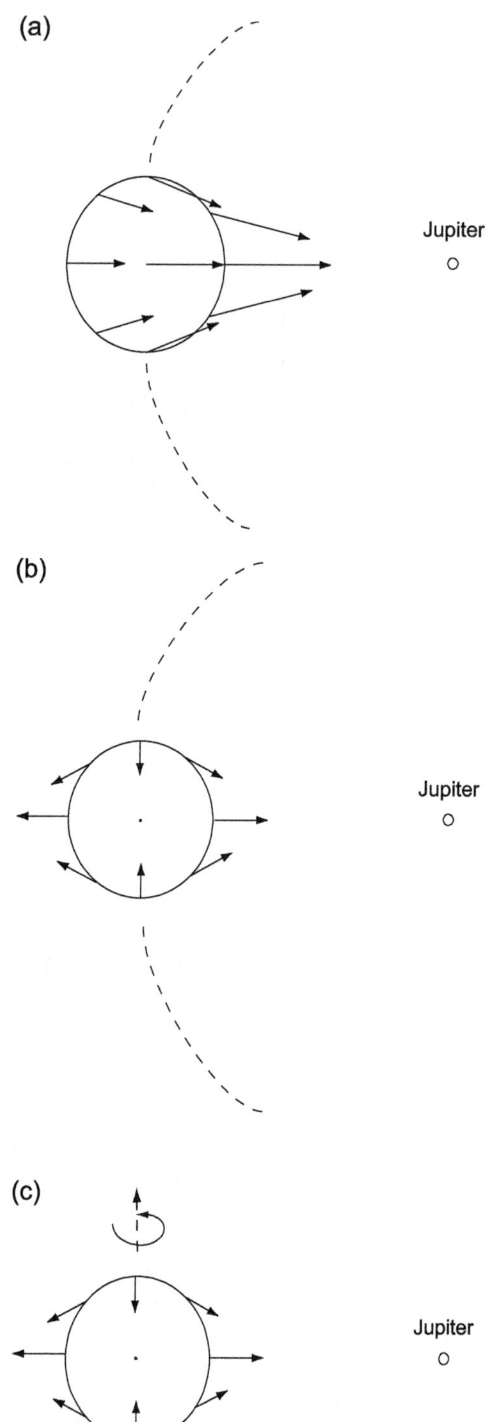

Fig. 3. Principles of tidal interaction. **(a)** Arrows illustrate the total gravitational acceleration vectors at points in Europa, caused by Jupiter. Europa's orbital plane is the plane of the paper (the orbit is shown as a dashed line.) All vectors point toward Jupiter. The acceleration is larger at points closer to Jupiter than at points farther away. **(b)** Similar to **(a)**, except the acceleration at the center of Europa has been subtracted from all the other vectors. The acceleration vector at Europa's center shown in **(a)** is equal to the total gravitational acceleration of Europa, and so the vectors shown here represent the tidal force, i.e., the part of Jupiter's gravitational force that causes deformation of Europa. **(c)** Similar to **(b)**, but now Europa is seen from a point in its equatorial plane. Europa's rotation vector is the heavy arrow pointing upward.

3.1. Orbital (Diurnal) Tides

Suppose Europa is rotating synchronously with its orbital motion. Thus, both the rotation and orbital periods are equal to T = 3.55 (Earth) days. If Europa's eccentricity and obliquity were both zero, then Europa would keep the same face toward Jupiter, the distance between Europa and Jupiter would not change, and so every point in Europa would remain fixed relative to Jupiter. The tidal force illustrated in Figs. 3b and 3c would cause a static tidal displacement field in Europa, with uplift at points on the sides that face toward and away from Jupiter, and subsidence at points on all other sides. Because this static tidal force would have been in place ever since Europa began synchronously rotating, all tidal shear stresses induced within Europa would probably have relaxed away long ago, resulting in a hydrostatic tidal response.

Orbital resonances between Europa and Io, and between Europa and Ganymede, circumvent Europa from residing in a circular (e = 0) orbit (section 9 presents details of these resonances). Instead, Europa is in an eccentric orbit, with eccentricity e ~ 0.009–0.010 [periodically fluctuating on short timescales (e.g., *Lainey,* 2002)]. This causes fixed points in Europa to move relative to Jupiter, for two reasons. First, for an eccentric orbit the distance between the satellite and planet changes with time. This changes the amplitude of the planet's gravitational force on the satellite, causing what is often referred to as the "radial tide." At perijove the tidal bulge is larger than at apojove, and this daily deformation results in diurnally varying displacements and stresses. The spatial pattern is identical to the tidal pattern shown in Figs. 3b and 3c, since the amplitude increases and decreases by the same factor at every point, over one tidal cycle.

Second, a synchronously orbiting satellite in an eccentric orbit does not always keep the same face toward the planet. A satellite at perijove is orbiting slightly faster than its (constant) rotation rate, and at apojove is orbiting slightly slower. This causes the tidal force pattern shown in Fig. 3b to rock back and forth relative to fixed points in the satellite, inducing a "librational tide" (see also chapter by Bills et al.).

Both the radial and librational tides occur at the orbital (diurnal) period of 3.55 d. These tides cause displacements that are much smaller (by a factor of 1/e) than the displacements caused by the static tide. The shear stresses in the solid parts of Europa (the icy shell and the rocky mantle) caused by the radial and librational tides are likely to be much larger than the shear stresses caused by the static tide. The shear stresses caused by the static tide have had time to relax away. But, as long as the viscous relaxation times of the icy shell and rocky mantle are on the order of 3.55 d or longer, the radial and librational tides would introduce nonzero shear stresses in those regions.

We will not consider the possible tidal effects of a nonzero obliquity in any detail. However, for a brief qualitative picture, suppose Europa is synchronously rotating with zero eccentricity, but with a nonzero obliquity = δ (see, e.g., *Kaula,* 1964). To a europan-fixed observer, Jupiter would then appear to be stationary in the east-west direction, but move from north of the equator to south of the equator with an angular amplitude of δ, at a 3.55-d period. The result is that the Fig. 3c force pattern would rock back-and-forth from north to south during the tidal cycle.

3.2. Tides Caused by Nonsynchronous Rotation

The presence of a global ocean on Europa would decouple the motion of the icy shell from that of the interior, unless coupling mechanisms such as internal gravitational, viscous, or pressure couplings counteract its effect (see chapter by Bills et al.). A decoupled shell could experience a net tidal torque, and so could rotate slightly faster than synchronously. If that was the case, then from the rotating shell's point of view the apparent location of Jupiter would move slowly across the sky. The tidal force pattern shown in Fig. 3b would thus appear to migrate westward: the direction opposing the NSR. The period of the tidal forcing would be one-half the NSR period, because the shell would pass through two outward tidal bulges during one NSR revolution.

3.3. The Tidal Potential

It is usual to describe the tidal force as the gradient of a scalar field, called the tidal potential. To derive this relationship, note that the total gravitational force per mass from Jupiter at a point $\vec{x}$ in Europa can be written as the gradient of the total gravitational potential

$$V(\vec{r}) = \frac{GM_J}{\left|\vec{r} - \vec{R}_J\right|} \tag{5}$$

Here $\vec{R}_J$ is the position vector of Jupiter (Jupiter is assumed to be a point mass), M_J = Jupiter's mass, and G is Newton's gravitational constant. Assume our coordinate system is fixed to Europa's outer shell, with an origin located at Europa's center. Expanding $1/|\vec{r} - \vec{R}_J|$ in terms of Legendre polynomials $P_l(\cos\gamma)$, where γ is the angle between $\vec{r}$ and $\vec{R}_J$, gives [see equation (3.38) of *Jackson* (1999)]

$$V(\vec{r}) = GM_J \sum_{l=0}^{\infty} \frac{r^l}{R_J^{l+1}} P_l(\cos\gamma) \tag{6}$$

where $r = |\vec{r}|$ is the radial coordinate of the point in Europa, and $R_J = |\vec{R}_J|$ is the radial coordinate of Jupiter. The l = 0 term in equation (6) is a constant and has no physical significance (it does not contribute to the force, since its gradient vanishes). The gradient of the l = 1 term is a constant vector. In fact, the l = 1 term represents the gravitational acceleration from Jupiter averaged over all Europa. Thus, the tidal force per mass, defined as the total force per mass minus the average acceleration, is the gradient of

$$V_T(\vec{r}) = GM_J \sum_{l=2}^{\infty} \frac{r^l}{R_J^{l+1}} P_l(\cos\gamma) \tag{7}$$

defined as the tidal potential, where now the sum begins at l = 2. Since r/R_J is small (r is no larger than the radius of Europa ≈1500 km, and $R_J \approx$ 670,000 km), it is usual to keep only the l = 2 term in equation (7), so that the tidal potential reduces to

$$V_T(\vec{x}) = GM_J \frac{r^2}{R_J^3} P_2(\cos\gamma) \tag{8}$$

The angle γ can be written in terms of the co-latitude and eastward longitude (θ, ϕ) of the europan point, and of similar angles (θ', ϕ') describing Jupiter's position relative to our europan-fixed coordinate system $(\cos\gamma = \cos\theta \cos\theta' + \sin\theta \sin\theta' \cos(\phi - \phi'))$. For Jupiter, the two angles and the radial coordinate R depend on time because of Europa's orbital motion and, possibly, its NSR. After including that time dependence, and keeping terms to first order in eccentricity, e, the tidal potential reduces to

$$V_T(r, \theta, \phi, t) = \frac{3GM_J R_s^2}{2a^3}\left(\frac{r}{R_s}\right)^2 [T_* + T_0 + T_1 + T_2] \quad (9)$$

where

$$T_* = \frac{1}{6}(1 - 3\cos^2\theta) \quad (10)$$

$$T_0 = \frac{1}{2}\sin^2\theta \cos(2\phi + 2bt) \quad (11)$$

$$T_1 = \frac{e}{2}(1 - 3\cos^2\theta)\cos(nt) \quad (12)$$

$$T_2 = \frac{e}{2}\sin^2\theta[3\cos(2\phi)\cos(nt) + 4\sin(2\phi)\sin(nt)] \quad (13)$$

Here, a and n are the semimajor axis and mean motion of Europa's orbit ($n = 2\pi/3.55$ radians/d), R_s is Europa's radius, b is the angular rate of NSR (radians per time), and t is time relative to perijove. To derive this result we have assumed the NSR rate is much slower than the diurnal rotation, and we have taken Europa's obliquity to be zero. [For a derivation of this result based on a more general representation given by *Kaula* (1964), see Appendix A of *Wahr et al.* (2009).]

The T_* term in equation (9) is static, and therefore causes time-independent tidal displacements. The rate of change of the T_0 term depends on the rate of NSR. Note that if there is no NSR (i.e., b = 0), then the T_0 term is also static. The T_1 and T_2 terms represent the diurnal tidal potential. Those terms are smaller than the NSR tidal potential (T_0) by a factor of eccentricity. T_1 and the cos(nt) term in T_2 cause the radial tide, and the sin(nt) term in T_2 causes the librational tide.

4. TIDAL DISPLACEMENTS AND INDUCED POTENTIAL

The tidal force per mass, equal to the gradient of the tidal potential (equation (9)), causes tidal displacements in Europa. Let the tidal displacement vector at any point (r, θ, ϕ) within Europa be $\vec{s}(r, \theta, \phi)$. The vector $\vec{s}$ is related to the applied tidal potential V_T through the differential equation describing conservation of linear momentum [cf. equation (2.2) of *Wahr* (1981), ignoring centrifugal and Coriolis forces]

$$\rho\partial_t^2\vec{s} = -\rho\nabla\phi_1^E - \rho\vec{s}\cdot\nabla(\nabla\phi_0) + \nabla\cdot\tau + \rho\nabla\phi_0\cdot[(\nabla\cdot\vec{s})\mathbf{I} - (\nabla\vec{s})^T] + \rho\nabla V_T \quad (14)$$

where τ is the stress tensor, ϕ_1^E is the gravitational potential arising from tidal deformation of Europa, ρ is the density, ϕ_0 is Europa's equilibrium (nontidal) gravitational potential, **I** is the identity matrix, and the superscript T denotes transpose. Poisson's equation, relating ϕ_1^E to the tidal displacement field, is

$$\nabla^2\phi_1^E = -4\pi G\nabla\cdot(\rho\vec{s}) \quad (15)$$

The stress-displacement relation is

$$\tau = \lambda(\nabla\cdot\vec{s}) + \mu[\nabla\vec{s} + (\nabla\vec{s})^T] \quad (16)$$

where τ is the stress tensor, and μ and λ are the Lamé parameters (μ is the shear modulus). These differential equations (14)–(16) assume Europa is compressible and self-gravitating. The material properties, described by ρ, μ, and λ, can vary spatially. The unknown variables, $\vec{s}$, τ, and ϕ_1^E, are determined by solving these differential equations subject to continuity conditions across internal boundaries and at the outer surface.

The tidal potential, V_T (equation (9)), is composed solely of second-degree spherical harmonics in (θ, ϕ) [T_* and T_1 have (degree, order) = (2,0), and T_0 and T_2 have (degree, order) = (2,2)]. We assume that Europa's equilibrium state is spherically symmetric, so that all boundaries are spheres and ρ, μ, and λ depend only on r (i.e., they are independent of θ and ϕ). In this case, spherical harmonics separate the differential equations and boundary conditions, and so $\vec{s}$ is composed of those same second-order harmonics. Specifically, $\vec{s} = s_r\hat{r} + s_\theta\hat{\theta} + s_\phi\hat{\phi}$ at the outer surface ($r = R_s$) can be related to V_T at the outer surface, using the two second-degree dimensionless Love numbers, h and ℓ (see, e.g., *Munk and MacDonald,* 1960; *Lambeck,* 1990) as

$$s_r(r = R_s, \theta, \phi, t) = \left(\frac{h}{g}\right)V_T(r = R_s, \theta, \phi, t) \quad (17)$$

$$s_\theta(r = R_s, \theta, \phi, t) = \left(\frac{\ell}{g}\right)\frac{\partial V_T}{\partial\theta}(r = R_s, \theta, \phi, t) \quad (18)$$

$$s_\phi(r = R_s, \theta, \phi, t) = \left(\frac{\ell}{g\sin\theta}\right)\frac{\partial V_T}{\partial\theta}(r = R_s, \theta, \phi, t) \quad (19)$$

where g is the gravitational acceleration at Europa's surface. The Love number h describes the amplitude of the radial displacement (s_r), and ℓ describes both the southward (s_θ) and eastward (s_ϕ) lateral displacements. Similarly, the gravitational potential at the outer surface caused by Europa's tidal deformation can be described with a third dimensionless Love number, k, as

$$\phi_1^E(r = R_s, \theta, \phi, t) = kV_T(r = R_s, \theta, \phi, t) \quad (20)$$

4.1. Viscoelastic Effects

The stress-displacement relation given in equation (16) assumes the material is elastic, i.e., that the material deforms fully the instant a stress is applied. As discussed in section 2, all real materials are at least slightly anelastic; energy is dissipated during a deformation cycle. The stress-displacement law for a Maxwell solid relates the stress tensor to time derivatives of both the displacement field and the stress tensor itself. The relation is considerably more complicated than that shown in equation (16). The result, however, can be simplified considerably by using the correspondence principle (e.g., *Peltier,* 1974).

The correspondence principle implies that if a periodic force with time dependence exp[iωt] is applied to a Maxwell solid (ω is angular frequency of the motion and $i = \sqrt{-1}$), then the stress-strain relation is still given by equation (16) but with the elastic Lamé parameters μ and λ replaced by

$$\tilde{\mu}(\omega) = \mu\left(\frac{i\omega}{i\omega + \frac{\mu}{\eta}}\right) \quad (21)$$

$$\tilde{\lambda}(\omega) = \lambda\left(\frac{i\omega + \frac{\mu}{\eta}\left(\frac{2\mu + 3\lambda}{3\lambda}\right)}{i\omega + \frac{\mu}{\eta}}\right) \quad (22)$$

where η is the viscosity and μ and λ are now the values of the Lamé parameters in the elastic (i.e., $\omega = \infty$) limit (see, e.g., *Wahr et al.,* 2009). These choices for $\tilde{\mu}$ and $\tilde{\lambda}$ assume that viscoelastic relaxation occurs for shear stresses, but not for bulk stresses.

The Maxwell relaxation time is defined as $t_M = \eta/\mu$. The forcing period is $T = 2\pi/\omega$. Defining the dimensionless parameter δ as

$$\delta \equiv \frac{T}{2\pi t_M} = \frac{\mu}{\eta\omega} \quad (23)$$

equations (21) and (22) become

$$\tilde{\mu}(\omega) = \mu\left(\frac{1}{1 - i\delta}\right) \quad (24)$$

$$\tilde{\lambda}(\omega) = \lambda\left(\frac{1 - i\delta\left(\frac{2\mu + 3\lambda}{3\lambda}\right)}{1 - i\delta}\right) \quad (25)$$

Note that $\tilde{\mu}$ and $\tilde{\lambda}$ are complex and depend on frequency. Thus, all unknown variables ($\vec{s}$, τ, and ϕ_1^E) in the equations of motion (14)–(16) are also complex and frequency dependent, as are the Love numbers (still defined by equations (17)–(20)). The complex Love numbers have the general form

$$\tilde{h}(\omega) = h_{re}(\omega) + ih_{im}(\omega) \quad (26)$$

$$\tilde{\ell}(\omega) = \ell_{re}(\omega) + i\ell_{im}(\omega) \quad (27)$$

$$\tilde{k}(\omega) = k_{re}(\omega) + ik_{im}(\omega) \quad (28)$$

where h_{re}, h_{im}, ℓ_{re}, ℓ_{im}, k_{re}, and k_{im} are real functions of frequency.

The imaginary parts of the Love numbers cause phase shifts in the body's response to the periodic forcing. Using $V_T(r = R_s, \theta, \phi, t) = V_0(\theta, \phi)e^{i\omega t}$ in equation (17), and using equation (26) for a Maxwell solid, and then taking the real part of the result for s_r, gives $s_r = V_0(\theta, \phi)\,[h_{re}(\omega)\cos(\omega t) - h_{re}(\omega)\sin(\omega t)]$. The presence of the sin(ωt) term means there is a phase difference between the forcing and the response. The imaginary parts of the Love numbers turn out to be negative and the real parts are positive (see below), which ends up implying that the maximum displacement occurs after the maximum force.

The importance of viscoelasticity depends on the value of δ, which in turn depends on the ratio of the forcing period to the Maxwell time. If the forcing period is much shorter than the Maxwell time, then $\delta \ll 1$, and the imaginary terms in equations (24) and (25) become negligible. In that case, $\tilde{\mu}(\omega) \approx \mu$ and $\tilde{\lambda}(\omega) \approx \lambda$, so that the effects of viscoelasticity are unimportant, and the satellite behaves elastically. In the other extreme, where the forcing period is much longer than the Maxwell time, then $\delta \gg 1$, and so $\tilde{\mu}(\omega) \approx 0$ and $\tilde{\lambda}(\omega) \approx \lambda + 2\mu/3$ = the elastic bulk modulus. Thus, at very long periods the material cannot support shear stresses, and so behaves as a fluid.

5. LOVE NUMBER RESULTS

The values of the Love numbers depend on Europa's internal structure. They are sensitive to such things as whether Europa has an ocean or not, and to the thickness and viscosity of the outer icy shell. Observations of surface displacements or of the external gravitational field could provide constraints on the Love numbers, and so on the internal structure. Because the Love numbers are particularly sensitive to the presence of an ocean (e.g., *Chyba et al.,* 1998; *Castillo et al.,* 2000; *Moore and Schubert,* 2000; *Wu et al.,* 2001; *Wahr et al.,* 2006), measuring tidal amplitudes has long been one of the important goals of a proposed Europa orbiter mission. In this section, we illustrate the sensitivity to internal structure by providing Love number values for various structural models and parameters. The numerical method we use to generate these values is based on standard algorithms used by geophysicists to compute

TABLE 2. Reference model parameter values (subscripts i, o, m, and c refer to the icy layer, liquid ocean, rocky mantle, and liquid metallic core, respectively).

$\rho_i = 920$ kg/m^3	$\rho_o = 1000$ kg/m^3	$\rho_c = 5150$ kg/m^3	$\rho_{ave} = 2989$ kg/m^3
$\mu_i = 3.3 \times 10^9$ Pa	$\mu_o = 0$	$\mu_c = 0$	$\mu_m = 4.0 \times 10^{10}$ Pa
$K_i = 9.3 \times 10^9$ Pa	$K_o = 2.0 \times 10^9$ Pa	$K_c = 5.0 \times 10^{11}$ Pa	$K_m = 6.7 \times 10^{10}$ Pa
$d_i = 20$ km	$d_o = 100$ km	$R_c = 700$ km	$R_E = 1561$ km

Love numbers for a stratified, compressible, and self-gravitating Earth. In fact, our numerical code is a modified version of the code employed by *Dahlen* (1976) to compute terrestrial Love numbers. We begin by assuming Europa is perfectly elastic. We assign uniform material properties (i.e., the density ρ, shear modulus μ, and bulk modulus $K = \lambda + 2/3\mu$) to each region: icy shell, liquid ocean (if present), rocky mantle, and liquid metallic core (if present). The model would allow us to include material properties that vary smoothly with radius; however, for simplicity, we elect here instead to use uniform material properties within each layer.

Our reference numerical values for all elastic parameters are listed in Table 2. Some of these reference values will be perturbed (below) to illustrate their impact on the Love numbers. In this table d_i and d_o are the thicknesses of the icy shell and liquid ocean, R_c is the radius of the fluid core, and R_E and ρ_{ave} are the radius and average density of the entire satellite. As we vary the thickness of the icy shell, the radius of the mantle, and the density and radius (and, for that matter, the existence) of the liquid core, the mantle density is always adjusted so that the average density of the satellite equals $\rho_{ave} = 2989$ kg/m^3.

Any satellite constraint on Love numbers would be based on observations of the diurnal tide. Nonsynchronous rotation tides occur slowly enough [observations presently constrain the NSR period to be $>10^4$ yr (*Hoppa et al.*, 1999)] that NSR tidal deformation would be indistinguishable from the static topography or gravity field over a satellite lifetime. The diurnal tides could presumably be detected either through altimeter observations of changing elevations at points on Europa's surface, or by tracking the trajectories of spacecraft as they orbit or otherwise pass through Europa's tidally changing gravity field. Altimeter observations would constrain h, and spacecraft tracking measurements would constrain k.

Model results (below) suggest that a measurement of either h or k could provide strong evidence of the presence (or absence) of a fluid ocean underlying the icy shell. In general, the Love numbers would be much larger if there is an ocean than if there is not. This is because the icy shell is thin compared with Europa's 1561-km radius. It is thin enough that unless its shear modulus is either unusually large or unusually small, it basically just rides up and down on whatever lies directly beneath it, with its outer surface deforming pretty much like the underlying surface. The rocky mantle does not deform much during a tidal cycle. Thus if there is no ocean, and therefore the icy shell sits directly on the mantle, the icy shell won't deform much either. The Love numbers would be small.

An ocean sitting on the mantle, though, would deform a good deal, because it cannot support the shear stresses necessary to keep the top of the ocean tied to the bottom of the ocean. Instead, the upper surface of the ocean tries to stay in a state of hydrostatic equilibrium with the tidal force. This is a state where the surface is continually adjusting its shape to coincide with a surface of constant gravitational potential, which is changing periodically because of the tidal force. Thus, if the icy shell lies above an ocean, then the shell's outer surface will also remain close to a state of hydrostatic equilibrium — because its lower surface does. In this case, the Love numbers are large.

Computations using the default structural parameters given in Table 2 (representing an elastic satellite with a 20-km-thick icy layer overlying a 100-km-thick ocean) give Love number values of h = 1.209, l = 0.299, and k = 0.261. Using the same parameter values, but removing the liquid ocean so that the icy shell extends all the way down to the rocky mantle, gives h = 0.069, l = 0.012, and k = 0.037. Thus, the presence of a liquid ocean increases the displacement Love numbers (h and l) by factors of about 20, and increases the gravitational Love number (k) by a factor of 7. It is these large differences that offer the possibility of using Love number observations to decide whether Europa has an ocean. For example, these values of h translate into a maximum radial diurnal tidal displacement of about 26 m when there is an ocean, and only 1.5 m when there is not.

Suppose, as is generally believed, there is a fluid ocean. The thickness of the icy shell is not well known, which raises the possibility of constraining that thickness using tidal observations. Figure 4 shows the Love numbers h (Fig. 4a) and k (Fig. 4b) as functions of ice thickness, using the reference values for the other parameters, but also after modifying the reference values in various ways (specified on the figure). Note that the Love numbers appear to decrease linearly with increasing ice thickness, as has been noted by *Moore and Schubert* (2000) and *Wahr et al.* (2006). This reflects the fact that a thick shell is more effective at holding back the underlying ocean than a thin shell.

A problem with using h or k alone to constrain ice thickness is that there are other poorly known parameters that affect the Love numbers. Figures 4a,b show what happens when the default model is perturbed by varying the mantle's shear modulus, varying the thickness of the combined ocean + ice layer, varying the density of the ocean/ice system, and removing the fluid core. Of these perturbations, the change in ocean/ice density has the largest impact (note that changing the thickness of the combined ocean + ice layer has almost no effect). For the reference model we take values of ρ_o and ρ_i equal to those of pure water and pure

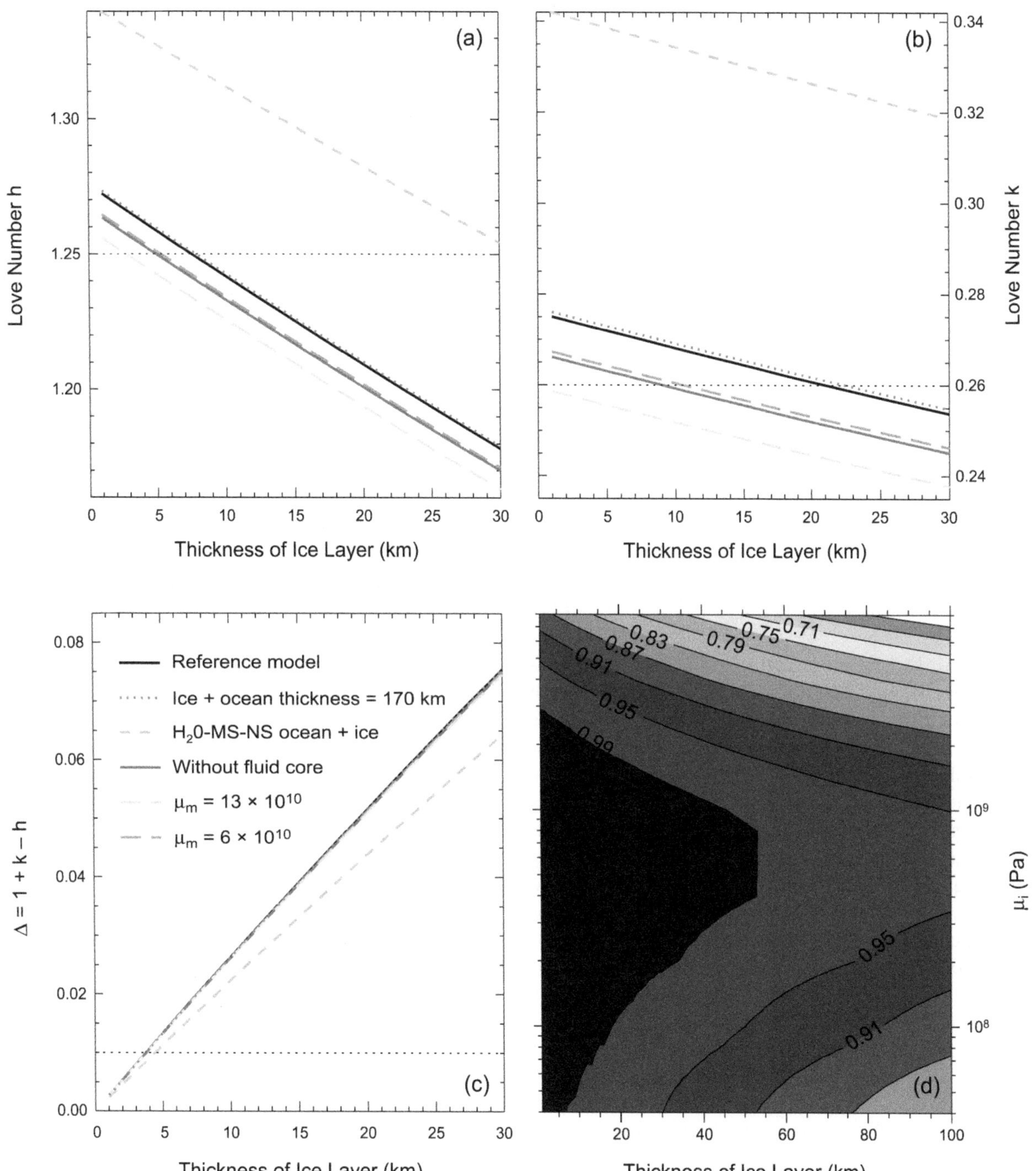

Fig. 4. See Plate 2. Elastic results for **(a)** Love number h, **(b)** k, and **(c)** $\Delta = 1 + k - h$. All results are plotted vs. the thickness of the icy shell. The different lines correspond to different assumptions about Europa's structural parameters. If not specified in the legend in **(c)**, all parameters are equal to the default values given in Table 2. **(d)** Ratio between our exact numerical results for Δ and the approximation in equation (31), as a function of d and μ_i. The reference parameter values are used for all other parameters. The results show that equation (31) serves as a reasonable approximation to Δ, especially for small values of d.

water ice. The H_2O-MS-NS case noted in Fig. 4 refers to a composition consistent with the eutectic system H_2O-$MgSO_4$-Na_2SO_4 brine, one of the possibilities proposed by *Kargel et al.* (2000) (see also chapter by Zolotov and Kargel). For that case the ocean and ice densities are taken to be $\rho_o = 1208$ kg/m^3 and $\rho_i = 1144$ kg/m^3. The larger densities in this case cause a larger gravitational signal (thus increasing k) when the ocean/ice layer is tidally deformed, and that signal acts back on the ocean/ice layer itself to cause a larger displacement (thus increasing h).

The dependence of h and k on other parameters complicates the problem of constraining ice thickness. For example, suppose observations give h = 1.25 or k = 0.26, values represented by horizontal dotted lines in Figs. 4a,b.

Those lines illustrate the range of possible values of ice thickness that could be inferred from observations of h or k alone, depending on what is assumed about the other structural parameters. The presence or absence of a fluid core could make a difference in the inferred thickness of about 3 km when interpreting this value of h and about 12 km when interpreting k. Varying μ_m between 4×10^{10} Pa and 13×10^{10} Pa, a somewhat arbitrarily chosen but not implausible range of values, could make a difference of about 5 km when interpreting h and about 23 km when interpreting k. The difference between the results computed assuming a pure H_2O ice + ocean layer and a eutectic H_2O-MS-NS layer corresponds to a difference in inferred ice thickness of about 23 km when interpreting h and about 95 km when interpreting k.

The basic problem is this. Let d be the thickness of the ice layer, and R be Europa's radius. The Love numbers h and k are functions of d. Expand those Love numbers in a Taylor series in powers of d/R. For a thin ice layer, $d/R \ll 1$, and so we truncate that power series to first order in d/R to get

$$h \approx h_0 + h_1 \frac{d}{R}, \quad k \approx k_0 + k_1 \frac{d}{R} \tag{29}$$

where h_0, h_1, k_0, and k_1 depend on various structural parameters, but not on d. The d/R terms in equation (29) cause the slopes of the lines shown in Figs. 4a,b, and are the terms that would need to be exploited to determine d from observations of h or k. Note from Fig. 4a that the slopes of the various lines for h are all about the same, as are the slopes of the lines for k shown in Fig. 4b. Mostly what's different about those lines is their y-intercepts; i.e., the h_0 and k_0 terms in equation (29). Those terms are the values the Love numbers would assume if there was no ice layer, so that the ocean extended right to the outer surface. h_0 and k_0 are much larger than the contributions from the d/R terms, and so relatively small errors in the predictions of those terms could cause large uncertainties in inferences of d.

Much of this ambiguity could be removed if both h and k could be measured. The combination of Love numbers $\Delta = 1 + k - h$ would vanish if there was no outer icy layer: i.e., $1 + k_0 - h_0 = 0$. This is simply the condition that the outer surface of an unconstrained fluid in hydrostatic equilibrium coincides with a surface of constant potential. Thus, when a thin outer ice layer is included and Δ is expanded to first order in d/R, the zero-order term vanishes, giving

$$\Delta \approx \Delta_1 \frac{d}{R} \tag{30}$$

where $\Delta_1 = k_1 - h_1$. Since a zero-order term is not present, there would be no uncertainty in that term to contaminate the estimate of d. Figure 4c, for example, shows results for Δ as a function of ice layer thickness (d). The lines now have a common y-intercept (zero), and the only line with a different slope is the one computed for the eutectic density distribution. If the observed values were still h = 1.25 and k = 0.26, so that $\Delta = 0.01$, then a mistaken assumption about whether the density distribution was that of pure water or that of H_2O-$MgSO_4$-Na_2SO_4 brine would cause an error of less than 1 km in the ice thickness inferred from Δ.

The difference in slope shown in Fig. 4c between the default model and the model using the eutectic brine density suggests that Δ_1 in equation (30) must depend on density. *Wahr et al.* (2006) derived a first-order (in d/r) analytical approximation for Δ of

$$\Delta \approx \left[\frac{15R^2G\pi(16\bar{\rho} - 11\rho_0)(\rho_0 - \rho_i) + 90\mu_i}{11R^2G\pi(5\bar{\rho} - 3\rho_0)\rho_0} - 1.4\frac{\mu_1}{K_i} \right] \frac{d}{R} \tag{31}$$

This result shows that the slope of Δ vs. d depends not only on the ocean and ice densities, ρ_o and ρ_i, but also on the shear and bulk modulii of the ice, μ_i and K_i. Figure 4c shows the exact numerical result for Δ divided by the approximation in equation (31) as a function of d and μ_i, computed using the Table 2 reference model values for all other variables. The results show the approximation in equation (31) is reasonably accurate for small values of d (x-axis), although it begins to break down as d increases, especially when μ_i is large. This disagreement is caused by the neglect of second- and higher-order terms in d/R when deriving equation (31).

By far the worst-known parameter in equation (31), other than d, is μ_i. More than that, using the reference parameter values on equation (31) shows that Δ is an order-of-magnitude more sensitive to errors in μ_i than to errors in $\rho_o - \rho_i$, another poorly known parameter. Thus, it is fair to conclude that the dominant error in any estimate of d caused by inadequacies in model parameters would be caused by errors in μ_i. In fact, a reasonable summary is that observations of Δ would constrain the product $\mu_i d$, so that the uncertainty in d would be approximately proportional to the uncertainty in $1/\mu_i$.

The general conclusion that tidal observations could be used to determine the presence or absence of an ocean is based on the result that Love numbers are large when there is an ocean and small when there is not. Viscoelasticity in the icy shell would reduce the shear modulus, and so cause the shell to behave more like a fluid. This would increase the tidal amplitude in the oceanless case. It raises the question of whether viscoelastic effects could be large enough that tides for a grounded, icy shell might be indistinguishable from those for a shell floating on a truly fluid ocean.

The importance of viscoelasticity depends on the value of δ, defined by equation (23). Viscoelastic effects could start to become significant if δ is, say, on the order of 0.2 or larger. Since the period T = 3.55 d for diurnal tides, this suggests (using the default value of μ_i given in Table 2) that viscoelasticity could begin to be important if the ice viscosity is on the order of 5×10^{14} Pa s or smaller. The thin, highly fractured outer surface layer of Europa might re-

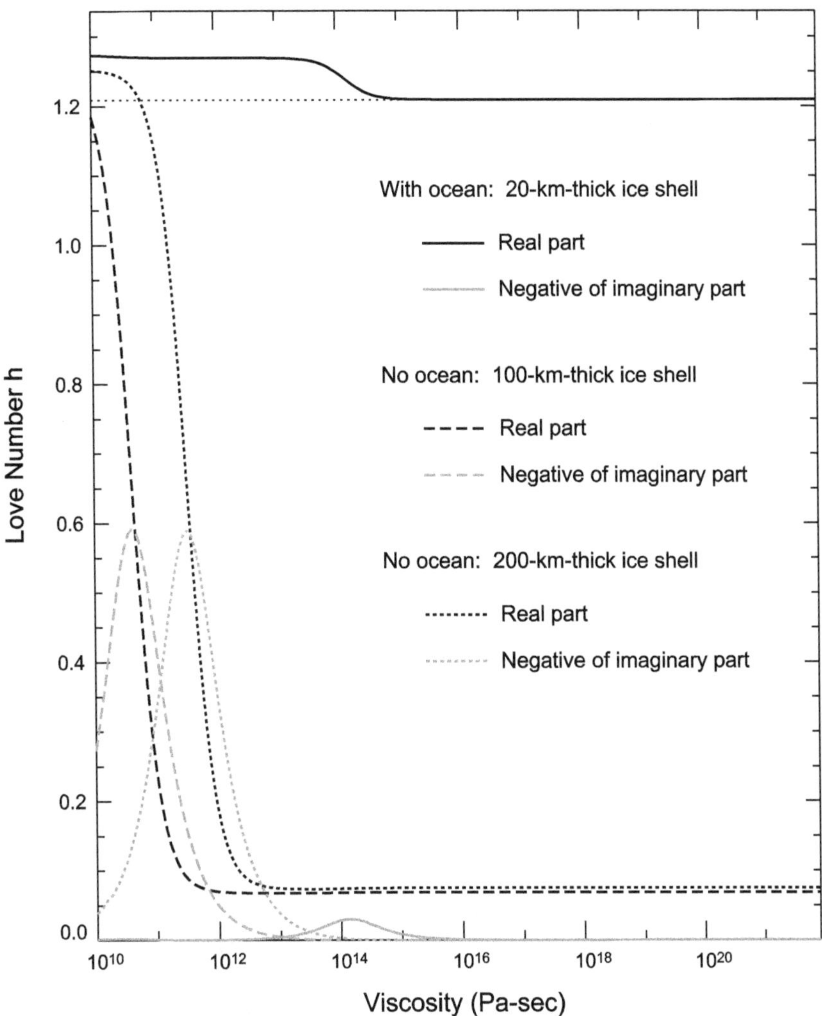

Fig. 5. Results for the Love number h for a viscous icy shell, as a function of shell viscosity. Shown are results for a 20-km-thick icy shell floating on a fluid ocean, as well as for two oceanless cases with different shell thicknesses. The reference values shown in Table 2 are used for all other structural parameters. In each case, results are shown for both the real part and the negative of the imaginary part. The horizontal dotted line shows the value for an elastic 20-km icy shell overlying a fluid ocean.

spond to tidal forcing as though it were a viscoelastic solid with very low viscosity. But the underlying cold ice layers in the rest of the upper shell are likely to have a large viscosity, perhaps on the order of 10^{22} Pa. Viscosities smaller than 10^{15} Pa s are conceivably possible for deeper layers in the ice, however.

Viscoelastic effects in the icy shell would not have a significant impact on the diurnal Love numbers when there is an ocean. In that case, the outer surface of the icy shell is already nearly consistent with hydrostatic equilibrium even for an elastic rheology, because a thin shell is not strong enough to hold back the underlying ocean (see above). Introducing viscoelasticity would further weaken the shell, but an elastic shell is already too weak to have much of an impact. This can be seen from Fig. 5 for the Love number h, where the viscoelastic results for a 20-km-thick shell overlying a liquid ocean differ from the elastic results by no more than 5%, no matter how small the viscosity. Note that viscoelastic effects begin to be evident only when the viscosity has decreased to about 5×10^{14} Pa s, as anticipated above.

Viscoelastic effects could, in principle, be more important if there is no ocean. Figure 5 shows that as the viscosity converges to zero, the real part of the Love number h in the oceanless case converges to the Love number value for a fluid ocean. The imaginary part of the Love number, which causes a phase lag between the tidal displacements and the tidal force, rises to about half the real value across the range of viscosities where the slope of h_{re} vs. viscosity is largest. Thus, observations of the tidal phase lag could perhaps be used to distinguish the ocean and oceanless cases, if there was ever a concern that a small ice viscosity might be confusing the oceanless vs. ocean interpretation as proposed for Titan (*Rappaport et al.,* 2008).

Such a concern, however, seems unlikely for Europa. In the oceanless case, the viscoelastic effects do not become important until the viscosity is as small as 10^{12} Pa s for a 200-km outer ice layer, and 10^{11} Pa s for a 100-km layer. The fact that these viscosity values are smaller than those expected based on the argument above that the value of δ need only be on the order of 0.2 or larger suggests that the true relaxation time of an icy layer lying on top of a rocky mantle is significantly longer than the Maxwell time. This, in fact, is a well-known characteristic of viscoelastic relaxation on Earth (*Peltier,* 1985). It is unlikely that the ice viscosity would be nearly as small as 10^{11}–10^{12} Pa s, especially through the entire 100–200-km icy layer. A viscosity value this small might conceivably provide a reasonable representation of densely broken ice at the outer surface. But, as Fig. 5 suggests, the thinner the viscous layer, the smaller the viscosity would need to be to have an impact. For example, if the low-viscosity layer was restricted to the upper 1 km of the icy shell, the viscosity would have to be as small as 5×10^4 Pa s to produce a Love number as large as 0.6.

6. TIDAL STRESSES

The diurnal and NSR tides affect Europa's stress field. Tidal stresses in the icy shell are likely to be an important factor in determining the evolution of surface tectonics, causing cracking and the formation of lineaments on the surface (*Greenberg et al.,* 1998). The stress field at Europa's surface at any time during the tidal cycle can always be diagonalized so that there is compression along one horizontal axis and tension along the other. If the tension is strong enough and/or the ice weak enough, the ice will fail by pulling apart in the direction of the tensional axis. Thus, the surface lineament pattern is apt to be a partial indicator of past orientations of the icy shell relative to Jupiter. Unraveling that information depends on understanding the relative importance of diurnal and NSR tides, since those tides introduce different stress patterns. Their relative importance depends on the ice viscosity value and on the NSR rate. Specifically, it depends on the value of the parameter δ (equation (23)), which in the NSR case is proportional to the ratio of the NSR period to the viscoelastic relaxation time of the ice. Since NSR is highly unlikely if there is no liquid ocean (see chapter by Bills et al.), so that the icy shell is not decoupled from the rocky mantle, we will assume in the remainder of this section that a liquid ocean does exist.

Tidal stresses on Europa have traditionally been computed using a "flattening" approximation (*Melosh,* 1977; *Helfenstein and Parmentier,* 1985; *Leith and McKinnon,* 1996; *Hoppa,* 1998; *Greenberg et al.,* 1998; *Hurford et al.,* 2007). The flattening model is inherently elastic, but viscoelasticity is almost certainly critical for accurate modeling of NSR stresses (see below). To simulate viscoelastic effects in the NSR case, applications of the flattening model use the difference between two elastic stress fields, with one field rotated about the $\hat{z}$ axis relative to the other. In effect, one stress field is computed for bt = 0 in equation (11), and the other for bt = some specified angle.

Stress models that include viscoelasticity directly into the equations of motion have recently been developed (*Harada and Kurita,* 2007; *Preblich et al.,* 2007; *Wahr et al.,* 2009). Here we describe results based on the model of *Wahr et al.* (2009).

The stress, τ, is a tensor field. Its relationship to the displacement field, $\vec{s}$, is shown in equation (16). The tidal stress at any point depends on the tidal displacements at that point, as well as on the Lamé parameters μ and λ at that point. The Lamé parameters convert strain, described by the derivatives of $\vec{s}$ in equation (16), into stress. Tidal displacements at the outer surface of the icy shell can be represented in terms of the Love numbers, h and l, and the tidal potential at the outer surface, V_T (r = R_s, θ, ϕ, t) (see equations (17)–(19)). Thus, expressions for the stress components at the outer surface reduce to a sum of latitude- and longitude-dependent terms, with each term proportional to a linear combination of the Love numbers h and l, multiplied by a linear combination of the outer surface μ and λ. The latitude and longitude dependence depends on V_T (r = R_s, θ, ϕ, t), and is different for NSR tides (equation (11)) than for diurnal tides (equations (12)–(13)).

Viscoelasticity affects the outer surface stresses because of its impact both on the Love numbers and on the surface Lamé parameters. The impact on the Love numbers, though, is small when there is a liquid ocean (see section 5). Whether the shell is viscoelastic or not, it is not strong enough to hold back the underlying ocean tides. But the effects of viscoelasticity on the outer surface Lamé parameters could be large, and those effects get mapped directly into the tidal stress.

Viscoelasticity affects the Lamé parameters through the parameter δ, as shown in equations (24)–(25). δ (equation (23)) is proportional to the tidal forcing period divided by the Maxwell relaxation time, with the latter being proportional to viscosity. The forcing period for diurnal tides is 3.55 d. The forcing period for the NSR tides is half the NSR period (the icy shell passes through two maxima in the tidal force during one complete revolution). The NSR period is unknown, but is probably larger than 10^4 yr (*Hoppa et al.,* 1999). Thus, δ is at least $(0.5 \times 10^4 \times 365)/3.55 = 5 \times 10^5$ times larger for NSR tides than for diurnal tides.

The general description of how the stress depends on δ is the same in the diurnal and NSR cases. In each case, there are two extremes. The maximum stress occurs when the viscoelastic relaxation time is much longer than the forcing period (so that $\delta \ll 1$). There is then not enough time during a tidal cycle for the shear stress to relax appreciably. It begins to relax as the tides deform the ice in one direction. But before it can relax by any significant amount, the tides start deforming the ice in the opposite direction. As a result, the stress always remains close to its elastic value. This

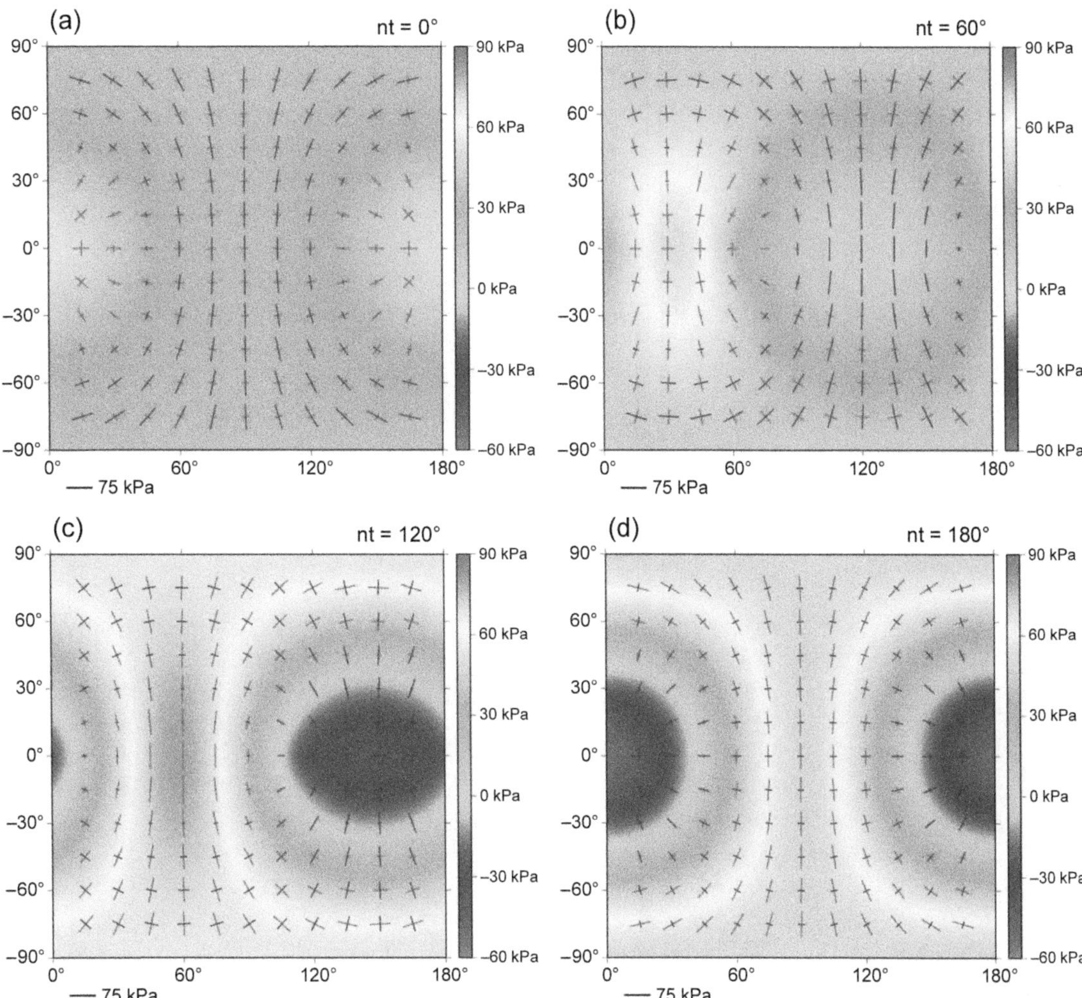

Fig. 6. See Plate 3. Diurnal stress results from the viscoelastic model, at different values of the angle after perijove (nt in equations (12)–(13)), computed using parameters from Table 2. The subjovian point at perijove is at latitude (y-axis) 0°, longitude (ϕ, x-axis) 0°. East is positive. Results for $180° < \phi < 360°$ are the same as those between 0° and 180°: $\tau(\phi + 180°) = \tau(\phi)$. Stresses in the second half of the orbit ($180° < nt < 360°$) are east-west reflections of those in the first half. Tics show the magnitude and orientation of the principal components of the stresses on the surface of the satellite. Compression (blue) is negative and tension (red) is positive. Background color indicates the magnitude of the most tensile of the two principal components. From *Wahr et al.* (2009); figure provided by Zane Selvans.

can be seen from equations (24) and (25), by noting that when $\delta \ll 1$, $\tilde{\mu}$ and $\tilde{\lambda}$ are approximately equal to their elastic values.

The other extreme occurs when the relaxation time is much shorter than the forcing period ($\delta \gg 1$). In that case, the shear stress has time to almost completely relax during a tidal cycle, and so it nearly vanishes. This can be seen from equations (24) and (25), by noting that when $\delta \gg 1$, $\tilde{\mu} \rightarrow 0$ and $\tilde{\lambda} \rightarrow K$.

At intermediate periods, where δ is on the order of unity, the stress amplitudes fall somewhere between these two extremes. Mathematical expressions for the outer surface stress components in the fully viscoelastic case, given in terms of the viscoelastic Love numbers, are given in *Wahr et al.* (2009).

6.1. Diurnal Tidal Stresses

Numerical results for the diurnal stress are shown in Fig. 6 at different points in the orbit, i.e., at different values of nt in equations (12)–(13). The effects of viscoelasticity are included in these results, with an assumed outer layer viscosity of 10^{22} Pa s. This corresponds to a value of $\delta = 1.6 \times 10^{-8}$, which is small enough that the stress results are virtually indistinguishable from those in the elastic case.

In these plots, the subjovian point at perijove is at latitude (y-axis) 0°, longitude (x-axis) 0°. Note that during a tidal cycle the stress pattern does not just rock back-and-forth about the subjovian point, but its shape actually changes. This is because the diurnal stress is driven by two terms, T_1 (equation (12)) and T_2 (equation (13)), that have

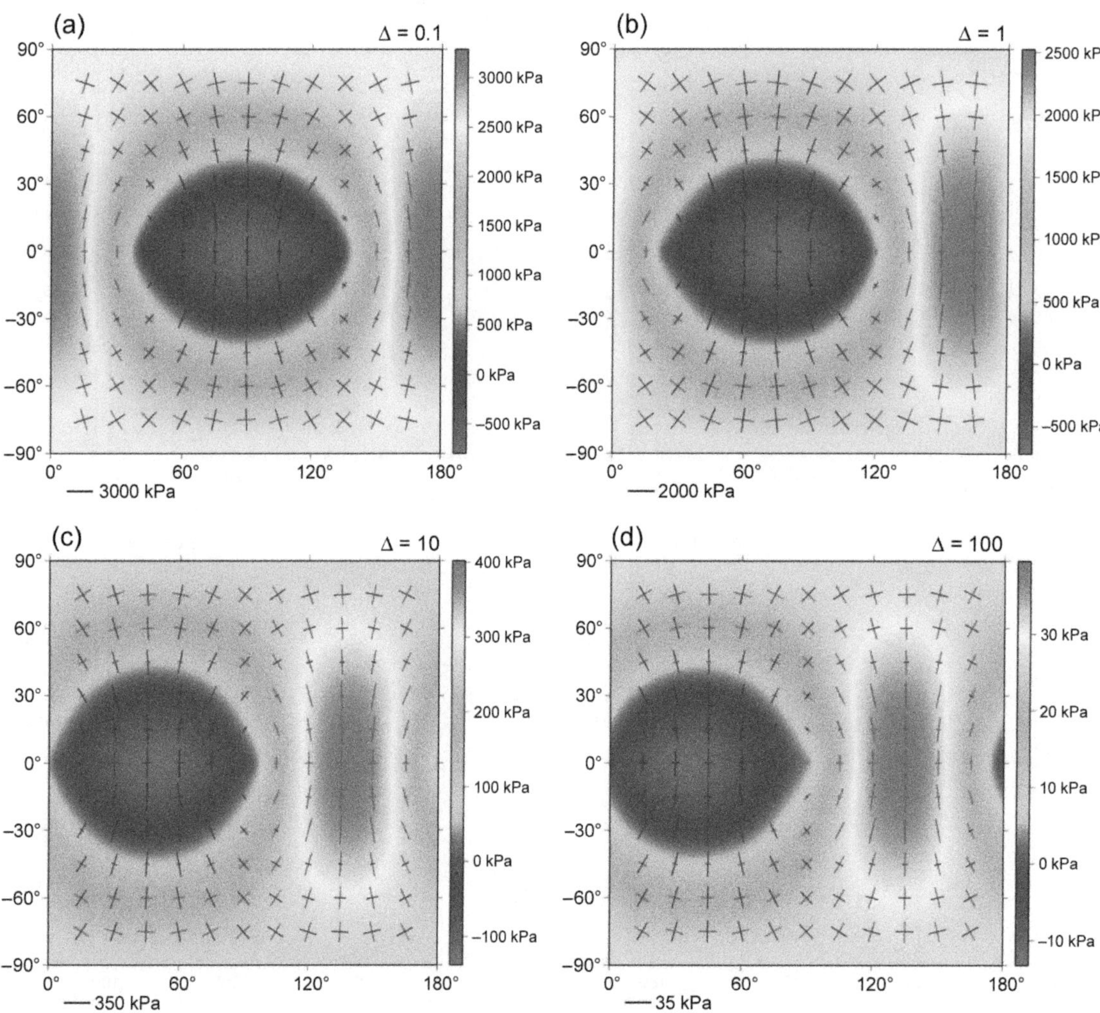

Fig. 7. See Plate 3. NSR stress results from our viscoelastic model for different values of δ. Plots are similar to those in Fig. 6. The subjovian point is at latitude = longitude = 0°. And $\tau(\phi + 180°) = \tau(\phi)$. Results are computed using the model parameters shown in Table 2, and using a viscosity of 10^{22} Pa s in the outer ice layer. Note that the arrow length and color scales are different in the different panels. From *Wahr et al.* (2009); figure provided by Zane Selvans.

different (θ, ϕ) dependencies, and the relative importance of those terms varies during the tidal cycle.

The red tic marks in Fig. 6 denote axes of tension. The stress tries to pull the surface apart in those directions, and so the preferred orientation of new cracks is perpendicular to those axes. The magnitude of the most tensile stress is shown by the background color. In any one of these plots, all else being equal, cracks are most likely to open where the underlying color is reddest.

6.2. Nonsynchronous Rotation Stresses

Numerical results for the NSR stresses are shown in Fig. 7, for different values of δ in the outer ice layer. In each panel, the subjovian point at perijove is at latitude 0°, longitude 0°, and the tic marks and background color scale have the same interpretation as in Fig. 6. In each case, the stress is evaluated at time t = 0. But the maps have a more general interpretation. Nonsynchronous rotation causes the icy shell to rotate slowly eastward relative to Jupiter. From the shell's point of view, the stress and displacement patterns, which are fixed relative to Jupiter, appear to drift westward without changing their shape or amplitude. At any time t, NSR would cause the subjovian point to be at longitude $\phi = -bt$ relative to the europan surface (see equation (11)). Thus, the Fig. 7 maps can be interpreted as showing the surface stresses at any time, relative to a subjovian point at latitude 0° and longitude 0°.

Note that as δ increases, two things happen. One is that the stress amplitudes get smaller (the color scales are different for the different panels). This is as expected, because a larger δ means that the outer ice layer behaves more like a fluid and so is less able to support shear stresses at the tidal period. The other thing that happens as δ increases is that the stress pattern drifts westward (leftward in the plots), opposite to the direction of the shell's rotation. This somewhat counterintuitive result occurs because viscoelasticity causes the maximum displacement to occur after the maximum stress. Thus since the ocean-imposed displacement field in the shell is oriented toward the parent planet, the stress pattern is shifted in the direction from which the shell has rotated.

Figure 7 shows results for values of δ between 0.1 and 100. These numbers are well within the possible range of true values, given the uncertainties in the NSR rate and the outer layer viscosity. If, for example, the outer layer viscosity was 10^{22} Pa s, a somewhat arbitrarily chosen but not unreasonable value, δ of 0.1 and 100 would correspond to NSR rotation periods of 6×10^4 yr and 6×10^7 yr, respectively; both of which are feasible.

If Europa was elastic, the NSR stresses would be larger than the diurnal stresses by a factor on the order of $1/e \approx 100$ (compare equation (11) with equations (12)–(13)). Viscoelasticity is far more effective at reducing the NSR stresses than the diurnal stresses, because of the longer NSR period (i.e., the larger value of δ). Comparing the NSR amplitudes in Fig. 7 with the diurnal amplitudes in Fig. 6 suggests that the NSR stresses tend to be larger than the diurnal stresses for NSR values of $\delta < \sim 50$, which, for an upper ice viscosity of 10^{22} Pa s, corresponds to an NSR period shorter than $\sim 3 \times 10^7$ yr. The NSR stresses dominate the diurnal stresses when δ is much smaller than this, and the diurnal stresses dominate when δ is much larger than this. Note, however, that if the NSR stresses become too high, the yield strength of the lithosphere could be reached, leading to brittle failure and stress release. The occurrence of lithospheric rupture would affect its mechanical properties and the stress field derived for an elastic or viscoelastic case would no longer be valid. All the calculations shown here are valid only as long as the material behaves as a cohesive and continuous media. The elastic and viscoelastic stress pattern can however be used to predict where maximal stresses take place and where faults are more likely to develop (see, e.g., *Greenberg et al.,* 1998; chapter by Nimmo and Manga; chapter by Prockter and Patterson).

7. TIDAL HEATING

A significant part of the strain energy involved during tidal deformation can be converted into heat by viscous friction within the viscoelastic interior or friction along faults at the surface. The amount and distribution of energy dissipated within the interior depend on the distribution of tidal strain energy and on the local dissipation processes (e.g., *Ross and Schubert,* 1987; *Ojakangas and Stevenson,* 1989). After describing the techniques to compute the tidal dissipation rate within the interior, we discuss the distribution of tidal heating within each internal layer. For simplicity, the interior is assumed to behave as a Maxwell body, but all the calculations presented here could be performed with any other rheological model.

7.1. Computation of Tidal Heating

7.1.1. Energy balance. The first law of thermodynamics states that any body possesses an internal energy, which varies with its deformation. The rate of doing mechanical work $\dot{E}_{mech}$ plus the rate of dissipation $\dot{E}_{diss}$ are equal to the rate of increase of the kinetic and internal energies of the body, $\dot{E}_{kin}$ and $\dot{E}_{int}$, respectively

$$\dot{E}_{mech} + \dot{E}_{diss} = \dot{E}_{kin} + \dot{E}_{int} \quad (32)$$

The rate of doing mechanical work $\dot{E}_{mech}$ is given by

$$\dot{E}_{mech} = \iiint_V \rho \frac{\partial \vec{s}}{\partial t} \vec{grad}(\phi_T^E) dV + \iint_S \frac{\partial \vec{s}}{\partial t} \vec{\tau}_s dS \quad (33)$$

where $\vec{\tau}_s$ is the surface traction, and V and S are the volume and the outer surface of the deformed body, respectively. Using the equations of motion and Gauss's divergence theorem, equation (33) can be written as

$$\dot{E}_{mech} = \frac{\partial}{\partial t} \iiint_V \frac{1}{2} \rho \dot{s}_i^2 dV + \iiint_V \tau_{ij} \dot{\varepsilon}_{ij} dV \quad (34)$$

Assuming that all the dissipated energy is converted into heat, the rate of heating $\dot{E}_{diss}$ is given by

$$\dot{E}_{diss} = \iiint_V \dot{Q} dV \quad (35)$$

where $\dot{Q}$ is the heating rate per unit volume, and the rate of kinetic energy $\dot{E}_{kin}$ is defined by

$$\dot{E}_{kin} = \frac{\partial}{\partial t} \iiint_V \frac{1}{2} \rho \dot{s}_i^2 dV \quad (36)$$

Thus, the rate of increase of internal energy per unit volume $\dot{U}$ is expressed as

$$\dot{U} = \dot{Q} + \tau_{ij} \dot{\varepsilon}_{ij} \quad (37)$$

For a purely elastic deformation, the heat content of any unit volume remains constant, and

$$\dot{U}_{elas} = \tau_{ij} \dot{\varepsilon}_{ij} \quad (38)$$

For a viscoelastic deformation, $\dot{Q} > 0$.

7.1.2. Dissipation rate per unit volume. By using the Fourier transformation of the stress and strain rate tensors, and assuming that the body responds at the frequency ω_f imposed by the forcing, the stress tensor τ_{ij} and the strain rate tensor $\dot{\varepsilon}_{ij}$ are written as functions of time

$$\tau_{ij}(t) = \int_0^\infty \tilde{\tau}_{ij}(\omega) e^{i\omega t} \delta(\omega - \omega_f) d\omega = \tilde{\tau}_{ij}(\omega_f) e^{i\omega_f t} \quad (39)$$

$$\dot{\varepsilon}_{ij}(t) = \frac{\partial}{\partial t} \int_0^\infty \tilde{\varepsilon}_{ij}(\omega) e^{i\omega t} \delta(\omega - \omega_f) d\omega = i\omega_f \tilde{\varepsilon}_{ij}(\omega_f) e^{i\omega_f t} \quad (40)$$

where $\tilde{\tau}_{ij}(\omega_f)$ and $\tilde{\varepsilon}_{ij}(\omega_f)$ are the complex amplitudes of the stress and strain tensors.

In the frequency domain, the complex power P^c per unit volume is defined from the complex components of the stress and strain tensors

$$P^c = i\frac{\omega_f}{2}\tilde{\tau}_{ij}\tilde{\varepsilon}^*_{ij} = -\frac{\omega_f}{2}\underbrace{[\mathrm{Im}(\tilde{\tau}_{ij})\mathrm{Re}(\tilde{\varepsilon}_{ij}) - \mathrm{Re}(\tilde{\tau}_{ij})\mathrm{Im}(\tilde{\varepsilon}_{ij})]}_{\text{real}} - i\frac{\omega_f}{2}\underbrace{[\mathrm{Re}(\tilde{\tau}_{ij})\mathrm{Re}(\tilde{\varepsilon}_{ij}) + \mathrm{Im}(\tilde{\tau}_{ij})\mathrm{Im}(\tilde{\varepsilon}_{ij})]}_{\text{imaginary}} \quad (41)$$

the star indicating the complex conjugate. The factor $\frac{1}{2}$ corresponds to the average value of the square of the sinusoidal function over one cycle. The real part of the complex power represents the dissipated power averaged over one forcing cycle, whereas the imaginary part corresponds to the elastic energy instantaneously stored by the body deformation during one quarter of a cycle and restored during the next quarter. The tidal heating rate per unit volume averaged over one cycle is then equal to

$$\bar{h}_{tide} = \frac{\omega_f}{2}\left|[\mathrm{Im}(\tilde{\tau}_{ij})\mathrm{Re}(\tilde{\varepsilon}_{ij}) - \mathrm{Re}(\tilde{\tau}_{ij})\mathrm{Im}(\tilde{\varepsilon}_{ij})]\right| \quad (42)$$

7.2. Variational Equations and Tidal Dissipation

Following the classical approach used in free oscillation problems (*Backus and Gilbert,* 1968; *Takeushi and Saito,* 1972; for a review, see *Romanowicz and Durek,* 2000), the radial distribution of the dissipation rate can be computed by using the variational principle. The application of the variational technique to the system of differential equations that describes the tidal deformation (equations (14)–(16)) provides explicit expressions of the kinetic and strain energy-density integrals associated with the motion. In the frequency domain, and using Green's first transformation, the set of equations (32) to (36) reduces to

$$\iiint_V \tau_{ij}\dot{\varepsilon}_{ij}dV - \iiint_V \rho\frac{\partial\vec{s}}{\partial t}\vec{\mathrm{grad}}(\Phi_1^E)dV = \omega I_2\int_0^{\pi}\int_0^{2\pi}[V_T(\theta,\phi)]^2\sin\theta d\theta d\phi \quad (43)$$

$$\dot{E}_{mech} = \iiint_V \rho\frac{\partial\vec{s}}{\partial t}\vec{\mathrm{grad}}(\Phi_1^E)dV + \omega I_3\int_0^{\pi}\int_0^{2\pi}[V_T(\theta,\phi)]^2\sin\theta d\theta d\phi \quad (44)$$

$$\dot{E}_{kin} = -\omega^3 I_1\int_0^{\pi}\int_0^{2\pi}[V_T(\theta,\phi)]^2\sin\theta d\theta d\phi \quad (45)$$

where I_1, I_2, and I_3 are the radial energy integrals related to the kinetic, strain, and potential energies, respectively (*Takeushi and Saito,* 1972). The conservation of energy (equation (32)) requires

$$I_3 = -\omega^2 I_1 + I_2 \quad (46)$$

The energy integrals can be determined from the radial functions of displacement, stress and potential (see Appendix A for their complete expression). These radial functions can be calculated from the integration of a set of six differential equations resulting from equations (14)–(16) (for more details, see *Takeushi and Saito,* 1972; *Tobie et al.,* 2005). At the surface, the radial functions, commonly termed y_i, are related to the Love numbers defined in section 4. The I_2 integral can be rearranged as

$$I_2 = \int_0^{R_s}[H_K K + H_\mu \mu + H_\rho \rho + H_0]dr \quad (47)$$

where H_K, H_μ, and H_ρ represent the radial sensitivity to the bulk modulus K, to the shear modulus μ, and to the density ρ, respectively. These parameters depend on the radial structure of the satellite and on the radial functions of displacement, stress and potential (see Appendix A). These three parameters (H_K, H_μ, and H_ρ) are purely real, so that all the information on dissipation is only contained in the imaginary part of the bulk and shear modulus. The bulk dissipation $I_m(K)$ of planetary materials is badly constrained and is *a priori* much smaller than $I_m(\mu)$. Therefore, only the dissipation associated with shear motion is considered and the dissipation rate is proportional to the imaginary part of the complex effective shear modulus μ(ω).

By rearranging the expression of the energy density integrals, and using the surface boundary conditions and the second-degree tide-generating potential, it can be shown that the radial distribution of the dissipation rate per unit volume is

$$\bar{h}_{tide}(r) = -\frac{21}{10}\frac{\omega^5 R_s^4 e^2}{r^2}H_\mu \mathrm{Im}(\mu) \quad (48)$$

Equation (48) permits the computation of the radial distribution of dissipation rate within any planetary interior using the profiles of $I_m(\mu)$ and H_μ, determined in each internal layer from its density and viscoelastic rheological properties (μ, K, and η).

7.3. Global Dissipation

The global dissipation can be obtained by integrating $\bar{h}_{tide}(r)$ over the radius or $\bar{h}_{tide}(r, \theta, \phi)$ over the volume of the sphere. Furthermore, *Zschau* (1978) showed that the volume integral can be transformed into a surface integral with the help of Green's first transformation and can be expressed as a function of the imaginary part of the secondary Love number k (*Segatz et al.,* 1988)

$$\dot{E}_{diss} = \frac{5\mathrm{Im}(k)}{8\pi^2 G R_p}\iint_S\int_0^T\left(\frac{\partial V_T}{\partial T}\right)^2 dtdS \quad (49)$$

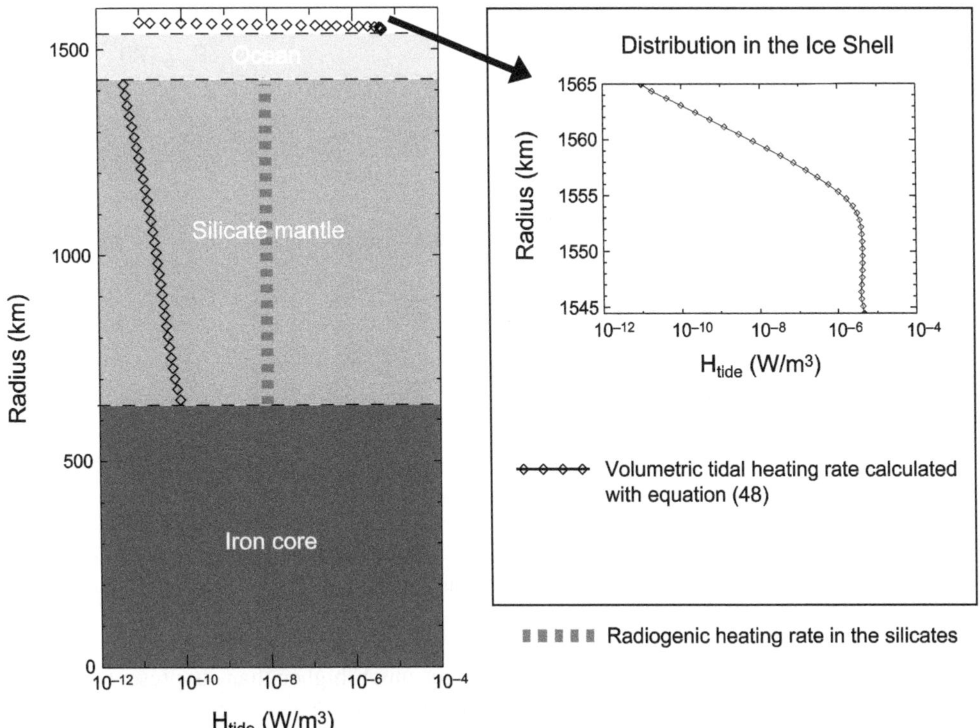

Fig. 8. Radial distribution of tidal heating within Europa and detailed distribution in the ice layer, computed from a given viscosity profile. The radiogenic heating rate per unit volume is also displayed for comparison. From *Tobie et al.* (2003).

where V_T is the second-degree tide-generating potential. If one considers only the diurnal tidal potential (T_1 and T_2, equation (9)), it reduces to the following expression

$$\dot{E}_{diss} = -\frac{21}{2}\text{Im}(k_2)\frac{(\omega R_s)^5}{G}e^2 \quad (50)$$

This can also be expressed as a function of the global dissipation factor $Q = -|k_2|/\text{Im}(k_2)$, which is the ratio between the stored peak elastic energy and the dissipated energy during one cycle.

$$\dot{E}_{diss} = \frac{21}{2}\frac{|k_2|}{Q}\frac{(\omega R_s)^5}{G}e^2 \quad (51)$$

Although tidal heating is known to be heterogeneously distributed within the interior (*Segatz et al.,* 1988; *Ojakangas and Stevenson,* 1989; *Tobie et al.,* 2003), the latter formalism is commonly used to assess the global dissipation rate and is incorporated in thermal evolution models as uniform internal heating from the global value (equation (50)) (e.g., *Peale et al.,* 1979; *Fischer and Spohn,* 1990; *Showman et al.,* 1997; *Hussmann et al.,* 2002; *Hussmann and Spohn,* 2004). As illustrated later, the strong coupling between internal dynamics and tidal dissipation (*Tobie et al.,* 2003, 2008; *Mitri and Showman,* 2008) requires the determination of both radial and lateral distributions of the tidal heating rate in order to correctly assess its effect on thermal evolution.

7.4. Distribution of Dissipation within Europa's Interior

7.4.1. Radial distribution. Figure 8 shows the radial distribution of the tidal dissipation rate per unit volume determined from equation (48) as a function of depth in the icy shell and in the deeper interior. In this calculation, the deeper interior is assumed to be differentiated into a liquid iron core and a silicate mantle with a constant viscosity of 10^{21} Pa s (close to the viscosity of Earth's mantle). The icy shell is separated from the rocky mantle by a liquid water layer and is assumed to be convective. The viscosity profile used to compute the dissipation rate in the icy shell has been determined from thermal convection experiments performed in a two-dimensional Cartesian domain with reflecting side boundaries and a free-slip condition on the top and bottom boundaries (*Tobie et al.,* 2003). The whole interior is assumed to behave as a compressible Maxwell body (equations (21) and (22)).

The liquid layer, by decoupling the deep interior from the surface, increases the tidal strain rate in the outer icy shell. As the ice is highly dissipative in the convective part, this leads to heating rates larger than 10^{-6} W m^{-3}, which is about 100 times larger than the volumetric radiogenic heat-

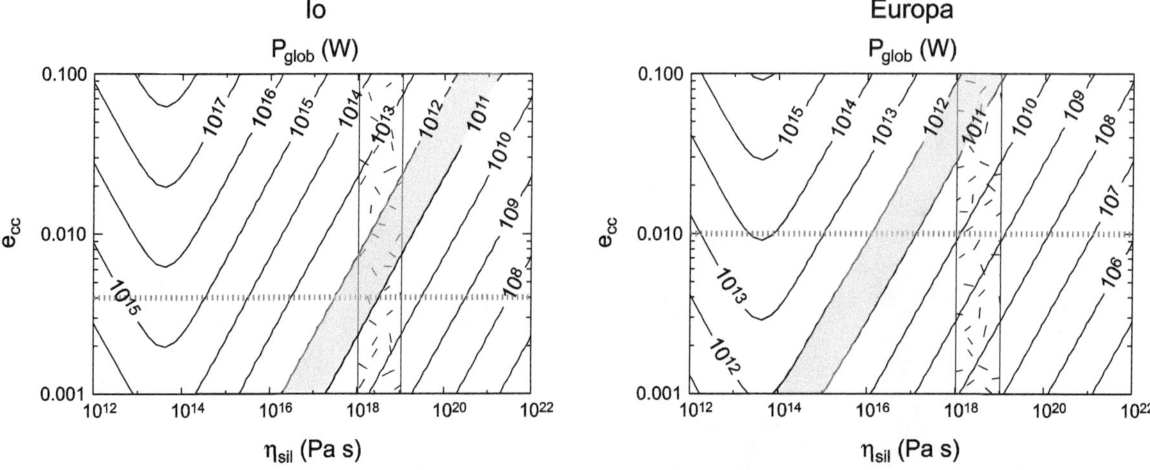

Fig. 9. Global dissipated power, $\dot{E}_{diss}$, in the silicate mantles of Io and Europa as a function of silicate viscosity η_{sil} and orbital eccentricity e. The dotted horizontal line indicates the current eccentricity. The dashed area indicates the possible viscosity range of silicate near the melting point. The gray band represents the typical value of radiogenic heating power. From *Tobie* (2003).

ing rate in the silicate mantle. In contrast, tidal dissipation in the silicate mantle remains small and is negligible compared to the radiogenic heating rate for viscosity values similar to Earth's mantle.

7.4.2. Dissipation in the silicate mantle. The only way to obtain significant dissipation in the silicate mantle is to strongly reduce the viscosity relative to typical terrestrial values [10^{20}–10^{21} Pa s (*Karato and Wu,* 1993)]. Figure 9 compares the global dissipated power that can be generated in the mantles of Io and of Europa as a function of viscosity and eccentricity, assuming a Maxwell rheology. It can be seen that for viscosity values on the order of 10^{18}–10^{19} Pa s, which correspond to the expected value near the melting point (*Karato and Wu,* 1993), and for the current eccentricity, tidal heating is similar to radiogenic heating on Io, whereas it is two orders of magnitude smaller than radiogenic heating on Europa. In reality, the global dissipated power on Io is as large as 100 TW (*Spencer et al.,* 2000). This large dissipation rate can be reproduced only if the mean viscosity of the silicate mantle is on the order of 10^{15} Pa s, which is an unrealistic value. For such a large dissipation rate, a large amount of melt would probably be produced in the mantle, so that the Maxwell model would not be valid for determining the magnitude of tidal dissipation.

Although the Maxwell model is not appropriate for determining dissipation rates for a satellite, such as Io, with such a highly dissipative interior, the Maxwell model can be used to assess the likelihood of the interior becoming this dissipative to begin with. On Io, the dissipation rate already exceeds the radiogenic heating rate for a viscosity of 10^{18} Pa s, which is a possible value for temperature close to the melting point. This condition is likely to lead to a thermal runaway where the interior gets more and more dissipative as the temperature increases [up to the limit of substantially molten (*Moore,* 2003)]. On Europa, the only way to have a similar thermal runaway is to have an eccentricity much higher than its present-day value. For a viscosity of 10^{18} Pa s, an eccentricity of 0.05 is required to induce a global dissipated power similar to the radiogenic power. In that case, a significant increase in mantle temperature and perhaps an Io-like thermal runaway might have occurred in the past. If such an event occurred, it could have been possible to sustain a highly dissipative state before the eccentricity decreased to its present-day value.

The simplest way to estimate the maximum dissipation rate within the silicate mantle at the present time is to scale the dissipation rate of Io, estimated from the observed thermal emission, to Europa (e.g., *Thomson and Delaney,* 2001). The global dissipation power is a function of the orbital eccentricity and orbital frequency, and of the radius of the rocky mantle. Assuming that Europa's interior behaves like Io's [Europa's Im(k) is assumed to be similar to Io's], the tidally generated power in Europa's silicate mantle can be estimated by

$$P_E = P_I \times \left(\frac{\omega_E R_E}{\omega_I R_I}\right)^5 \left(\frac{e_E}{e_I}\right)^2 \qquad (52)$$

where the subscripts $_E$ and $_I$ refer to Europa and Io, respectively. By taking R_E 1441 km, R_I = 1821 km, e_E = 0.0096, e_I = 0.004, $\omega_E = 2.05 \times 10^{-5}$ rad s^{-1}, $\omega_I = 4.11 \times 10^{-5}$ rad s^{-1}, and P_I ~ 100 TW (*Spencer et al.,* 2000), this scaling gives P_E ~ 5.5 TW.

As the produced energy in Europa's mantle is about 20 times less than in Io, its interior temperature is probably smaller, and therefore Europa's mantle is probably less dissipative than Io. Furthermore, the existence of a liquid water layer that decouples the crust from the deep interior make Europa's silicate mantle less sensitive to tidal forcing. The application of the variational technique described before shows that Europa's mantle is three times less sensitive to tidal fluctuations owing to the decoupling effect of the

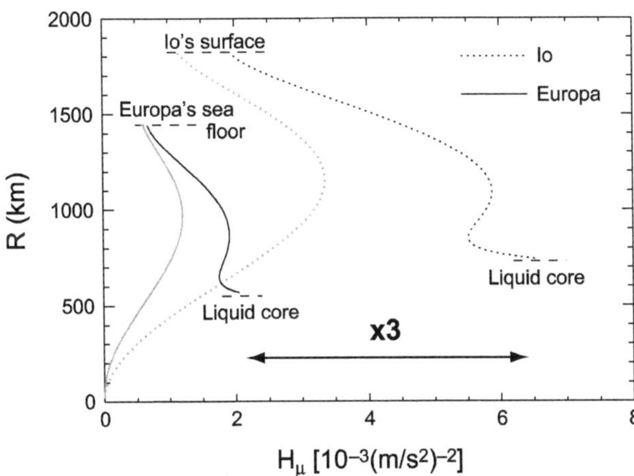

Fig. 10. Sensitivity parameter H_μ within Europa's and Io's silicate interior, before differentiation (ρ = 3500 kgm^{-3}, gray curves) and after differentiation into a silicate mantle and a liquid iron core (ρ_C = 7000 kgm^{-3} and ρ_M = 3300 kgm^{-3}, black curves). From *Tobie et al.* (2005).

ocean and to its smaller size (Fig. 10) (*Tobie et al.,* 2005). Indeed, when no decoupling liquid layer is present, the icy shell, which is much more deformable that the rocky interior, exerts a traction on the silicate mantle. When a decoupling layer is added, this traction disappears and the tidal deformation of the silicate mantle is reduced. Finally, tidal dissipation in the rocky mantle is limited by the amount of dissipation in the icy shell. The requirement of orbital equilibrium limits the total amount of energy that can be dissipated within the interior. If tidal dissipation is too large, the eccentricity would rapidly decay, and the tidal dissipation would cease. High dissipation in the silicate interior and in the icy shell are mutually exclusive [see, e.g., *Bland et al.* (2008) for an illustration of this effect in the case of Ganymede].

All these effects reduce the capability of Europa's rocky mantle to produce large amounts of energy by tidal friction, so that the current eccentricity is perhaps too low for Europa's mantle to be in a highly dissipative state. In this condition, the total tidally produced power at present would not exceed the radiogenic power (0.5 TW). It is, however, conceivable that Europa experienced periods with elevated orbital eccentricities in the past (*Showmann et al.,* 1997; *Hussmann and Spohn,* 2004), resulting in higher dissipation rates and higher probabilities of initiating thermal runaway events. Even if large spikes in eccentricity and dissipation rate did occur, damping due to tidal friction would limit the duration of such events. As Europa's eccentricity is continually forced by Io and to a lesser extent by Ganymede owing to the Laplace resonance, it is difficult to evaluate the timescale of eccentricity damping. Future modelling efforts are required to assess the likelihood and duration of Io-like dissipation events in Europa.

7.4.3. Distribution of dissipation in the icy shell. In contrast to the rocky mantle, the presence of a decoupling liquid layer strongly enhances the dissipation rate in the icy shell. Works by *Cassen et al.* (1979), *Ojakangas and Stevenson* (1989), and more recently by *Tobie et al.* (2003, 2005) showed that the tidal deformation of the icy shell is mainly controlled by the radial displacement of the subsurface ocean, imposing a constant tidal strain rate with depth. Moreover, the ocean response to tidal fluctuations induces lateral variations of shear strain rate in the icy shell, reaching maximum values at the poles and minimum values near the equator. Therefore, for an icy shell with a constant viscosity above an ocean, the tidal dissipation rate is maximum at the poles and minimum in the equatorial regions, and remains constant throughout the icy shell (Figs. 11c,d). This strongly contrasts with the tidal dissipation patterns obtained for models with no liquid layer. In the absence of decoupling with the rocky mantle, the displacement of the icy shell is limited and the distribution of tidal strain rate strongly varies between the base of the ice layer and the surface (Figs. 11a,b).

The results presented above assume an isoviscous ice layer. In reality, temperature and hence viscosity strongly vary as a function of depth in the icy shell, having minimum values of 60–110 K at the surface (depending on latitude), and maximum values of 260–270 K at the interface between the icy shell and the ocean. As the tidal dissipation rate depends on the imaginary part of the effective shear modulus, which is a function of viscosity and frequency, the tidal dissipation rate at a given tidal strain rate is expected to vary from the surface to the bottom of the layer. Furthermore, if the icy shell is convective, lateral variations of dissipation owing to lateral variations of ice viscosity are also expected (*Tobie et al.,* 2003; *Mitri and Showman,* 2008).

The viscosity (temperature) dependence of the time-averaged volumetric dissipation rate H can be approximated by assuming an incompressible Maxwell medium, subjected to sinusoidal variations of strain rate $\dot{\varepsilon}$

$$H = \frac{2\eta\bar{\dot{\varepsilon}}^2}{1 + (\omega\eta/\mu_E)^2} \tag{53}$$

A dissipation peak occurs when $\omega\eta/\mu_E = 1$, i.e., when η = 1.5×10^{14} Pa s. Figure 12 illustrates the viscosity dependence of the tidal dissipation rate for two extreme values of the tidal strain rate expected within Europa's icy shell. For an ice viscosity lower than 10^{17} Pa s, the tidal heating rate is larger than the radiogenic heating rate in the silicate mantle. The viscosity for temperatures close to the melting point is expected to range between 10^{13} and 10^{15} Pa s, depending on the grain size (*Montagnat and Duval,* 2000; *Durham et al.,* 2001; *Goldsby and Kolhstedt,* 2001; *Montagnat,* 2004), which leads to large dissipation in the bottom part of the icy shell (e.g., *McKinnon,* 1999).

If the icy shell is convecting (see also chapter by Barr and Showman), at least half of the icy shell has a temperature close to the melting point (>240–250 K) and large amounts of energy can be dissipated (Fig. 8). If one assumes that convective heterogeneities have a small effect on the

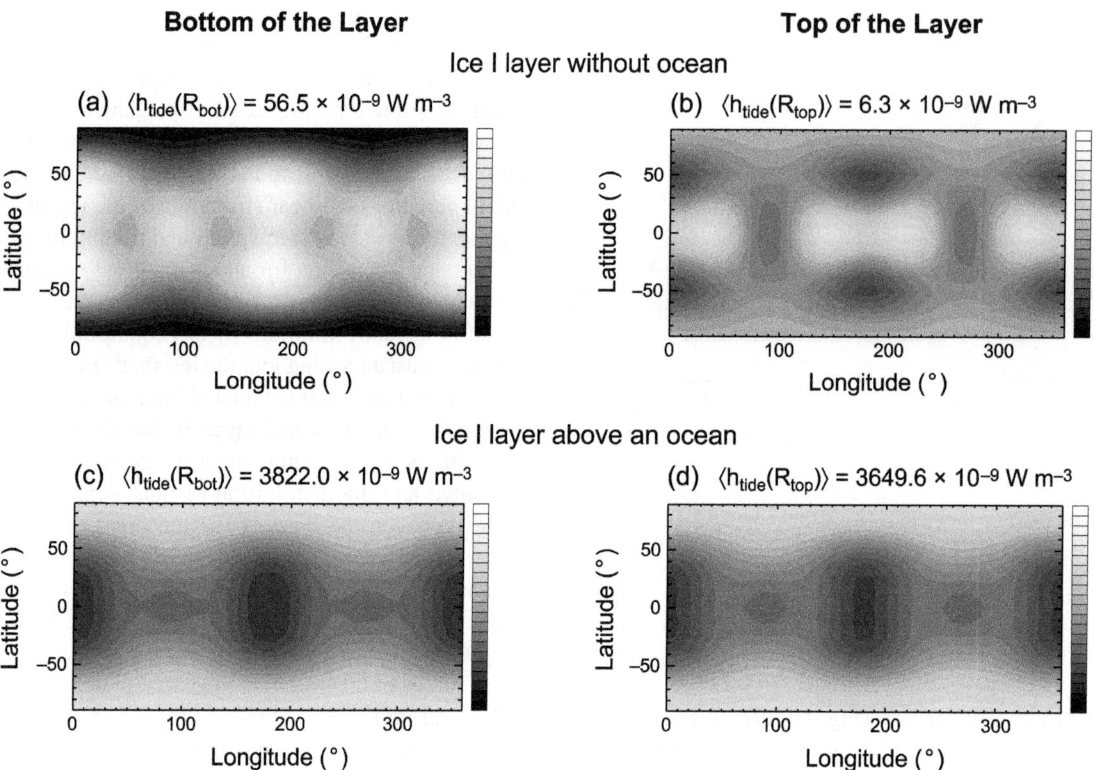

Fig. 11. Lateral distribution, in percent (%) of the mean value indicated above each panel, of the tidal dissipation rate at the bottom and at the top of the ice I layer for a model with **(a,b)** no liquid water layer and **(c,d)** a model with a global ocean. From *Tobie et al.* (2005).

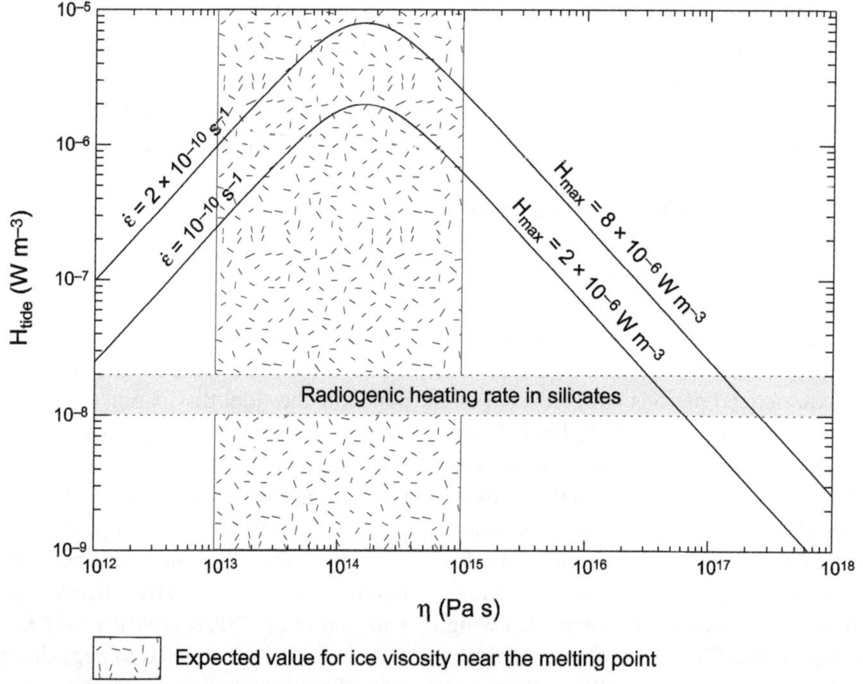

Fig. 12. Volumetric tidal heating rate within Europa's icy shell as a function of ice viscosity, calculated with equation (53) for two extreme values of tidal strain rate: $\dot{\varepsilon} = 2 \times 10^{-10}$ s^{-1} near the poles and $\dot{\varepsilon} = 10^{-10}$ s^{-1} at the equator. For viscosities lower than 5×10^{16}–10^{17} Pa s, the tidal heating rate per unit volume is larger than the radiogenic heating rate per unit volume of the silicate mantle. From *Tobie et al.* (2003).

tidal strain rate, equation (53) can be used to estimate the local heat production rate. This assumption is probably correct because as explained above the tidal strain rate is mainly driven by the degree-two ocean displacement. As long as an ocean is present and the lateral viscosity variations are characterized by much higher degrees, they probably do not affect the tidal strain field and the above assumption is valid. When the viscosity structure is described by lower-degree harmonics, more complex theories capable of describing the variation of strain rate as a function of viscosity are required. This is, for instance, the case on Enceladus, where a large-scale viscosity anomaly is expected at the south pole (*Tobie et al.,* 2008; *Nimmo and Pappalardo,* 2006).

Figure 13 presents the tidal dissipation rate obtained in the convective icy shell by assuming a Maxwell rheology. Most of the dissipation occurs in the convecting region of the icy shell, and tidal dissipation rates remain very small in the cold conductive lid. Depending on the viscosity value near the melting point, maximum dissipation occurs in the center of hot plumes, leading to partial melting, or in the cold downwellings. These calculations assume a simple Maxwell rheology and a temperature-dependent Newtonian viscosity, whereas in reality the rheology of ice might be much more complex. It depends on several parameters, including dislocation density, grain size, impurity content, applied stress, loading history, etc. The assumption of a more complex rheology would probably not modify the conclusions about the influence of an internal liquid layer on the tidal strain field, but it could change the distribution of dissipation inside the icy shell. In particular, the Maxwell rheology usually underestimates the dissipation rate at cold temperatures. Future experimental and theoritical work is required to better understand the impact of tidal dissipation on the internal dynamics of Europa's icy shell.

7.4.4. Shear heating along faults. Localized deformation along faults within the cold lithosphere may also provide a significant source of dissipation. By applying a shear deformation model initially developed for Earth's application (*Yuen et al.,* 1978), *Nimmo and Gaidos* (2002) proposed that shear heating owing to tidally induced strike-slip motion along fractures could locally raise the temperature, possibly up to the melting point at shallow depths (see chapters by Kattenhorn and Hurford and by Prockter and Patterson). The large heating reported by *Nimmo and Gaidos* (2002) requires, however, relatively large shear velocities. Furthermore, in their model, *Nimmo and Gaidos* (2002) prescribed the displacement rate, independently of the amount of work that might be needed to create the shear motions. Later, *Preblich et al.* (2007) showed that tidal displacements

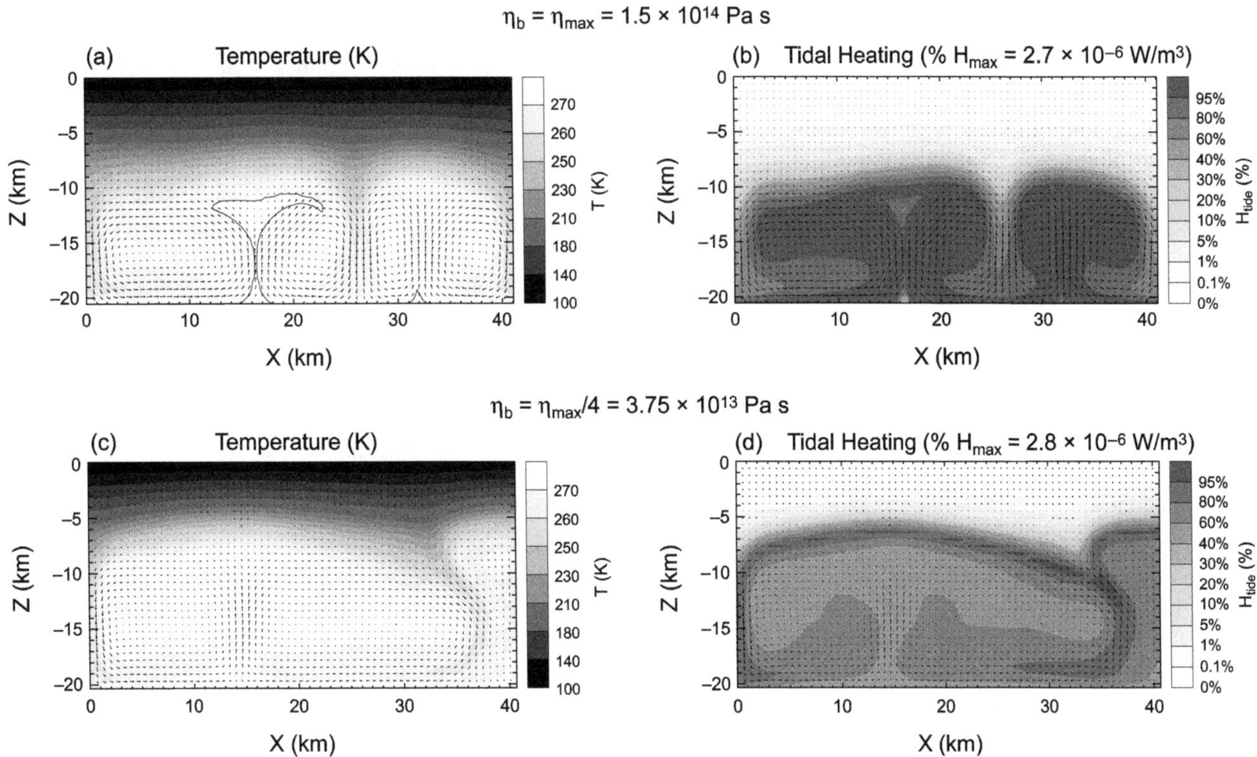

Fig. 13. **(a,c)** Temperature and **(b,d)** tidal heating rate per unit volume fields obtained for an icy shell thickness of b = 20 km, a bottom temperature of 270 K, a maximum tidal heating rate of 2.7×10^{-6} W m^{-3} and two different possible values for the ice viscosity at the melting point of **(a,b)** 1.5×10^{14} Pa s and **(c,d)** 3.75×10^{14} Pa s. The first value leads to maximum dissipation in the hot plumes, whereas the second one leads to maximum dissipation in the cold downwellings. The black isocontour delimits the area where temperature equals to the melting temperature of pure water ice. From *Tobie et al.* (2003).

along tectonic faults are smaller than initially anticipated by *Nimmo and Gaidos* (2002). This mainly follows from the fact that the shear velocity rates prescribed by Nimmo and Gaidos are valid for a rigid lithosphere above a totally fluid layer. It does not take into account the resistance of the viscous ice layer below, which would tend to reduce the shear motion rate. Despite these limitations, shear heating along tidally forced faults remains a viable mechanism, but it is probably overestimated in existing models. Shear heating may contribute locally to the energy budget but cannot be the sole source of energy in Europa's icy shell.

8. IMPACT OF TIDAL HEATING ON INTERNAL DYNAMICS AND HEAT TRANSFER

The previous sections have shown that tidal heating due to viscous friction is mainly localized in the icy shell and is strongly dependent on the viscosity of water ice. In this section, we briefly discuss some consequences of this strong heating on the thermal state and the geodynamics of the icy shell.

8.1. Equilibrium Thickness of a Tidally Heated Icy Shell

The equilibrium thickness of a tidally heated icy shell was first modeled assuming that heat was transferred by conduction (*Ojakangas and Stevenson,* 1989). Tidal heating is mainly produced at the bottom of the icy shell where the temperature is closest to the melting point, which gives a viscosity close to the optimum viscosity for tidal dissipation (Fig. 12). Taking into account the strong temperature dependence of thermal conductivity and the variations in tidal heating due to longitude and latitude, *Ojakangas and Stevenson* (1989) predict an equilibrium thickness on the order of 25 km. This value depends on the amount of heat that comes from the silicate mantle; the smaller the heat from the silicate mantle, the larger the thickness. A heat flux of 10 mW m^{-2} decreases the mean thickness by about 4 km (see chapter by Barr and Showman). It must be noted that large thickness variations on the order of 10 km are predicted as a function of latitude and longitude: The maximum thickness is at the equator at the jovian and antijovian longitude.

Later studies used scaling laws and numerical simulations to investigate the possibility of subsolidus convection and its implications for temporal and spatial variations of icy shell thickness (e.g., *Hussmann et al.,* 2002; *Tobie et al.,* 2003; *Moore,* 2003; *Mitri and Showman,* 2008; *Nimmo et al.,* 2007; see also chapter by Barr and Showman). The utility of scaling laws is limited by the fact that tidal heating depends strongly on viscosity. The viscosity difference between the material in an upwelling plume and the bulk ice is on the order of one order of magnitude (e.g., *Deschamps and Sotin,* 2001), which translates into a one order of magnitude variation in tidal heating. Using the correspondence principle, *Tobie et al.* (2003) simulated convection processes using a two-dimensional convection code with strongly temperature dependent viscosity (Fig. 13). Hot plumes form at the interface between the ice crust and the ocean and cold plumes form below the thermal boundary layer. The choice of viscosity law controls the location of tidal heating and the pattern of convection.

For a viscosity of 1.5×10^{14} Pa s at the melting point, tidal heating is maximum in the hot plume (Figs. 13a,b). The temperature increases in the hot plume and the latent heat of solid/liquid transition controls the temperature at the melting point. The amount of tidal heating in the cold plume is 10 times smaller than that in the hot plume.

For a melting point viscosity that is four times smaller (Figs. 13c,d), the distribution of tidal heating is quite different, with the maximum tidal heating occurring in the downwelling plumes. It must be noted that the convection pattern is dominated by cold plumes and that no hot plume forms at the ocean/icy crust interface. This pattern resembles that of volumetrically heated fluids.

In the examples described in Fig. 13, the thickness of the icy shell is about 20 km, a value lower than the equilibrium thickness obtained in the conductive case. It suggests that convection is possible in Europa's icy crust. As in the conductive case, the amount of tidal heating also depends on latitude and longitude (Figs. 11c,d), with minimum values at the jovian and antijovian points along the equator and maximum values at the poles. To compare with the conductive case described above, a calculation of the equilibrium thickness with 10 mW m^{-2} coming from the silicate mantle is shown in Fig. 14. Variations on the order of 8 km are observed. The surface heat flux is larger where the icy crust thickness is thinner. Whereas the variations in thickness are about 30% (Fig. 14a), variations of heat flux (Fig. 14c) are only 15% due to the effect of a conductive lid that has little lateral thickness variations (Fig. 14b).

As described above, tides and tidal heating strongly affect the internal dynamics and the thermal evolution of Europa's icy crust. They prevent Europa's ocean from completely freezing. However, it must be noted that several parameters could change the results of these models. For example, the presence of antifreeze material would lower the temperature of the ocean, resulting in a colder ice crust and less tidal heating. Antifreeze material would also prevent the complete freezing of the ocean, but tidal heating would not be as important because the ice viscosity would be much higher. One critical parameter is viscosity because small viscosity changes strongly modify the amount of tidal heating. Also, it is not clear that the deformation mechanisms involved in the convection process are the same as those involved in tidal dissipation because the timescales are quite different. Viscosities could be different, and the implications must still be investigated by both laboratory experiments and numerical simulations. Most simulations assume Maxwell bodies. More complex models may be necessary to describe tidal processes.

8.2. Tidal Heating and Formation of Chaotic Terrains

The presence of chaotic terrains on Europa (e.g., *Pappalardo et al.,* 1999) has been interpreted as indicating either

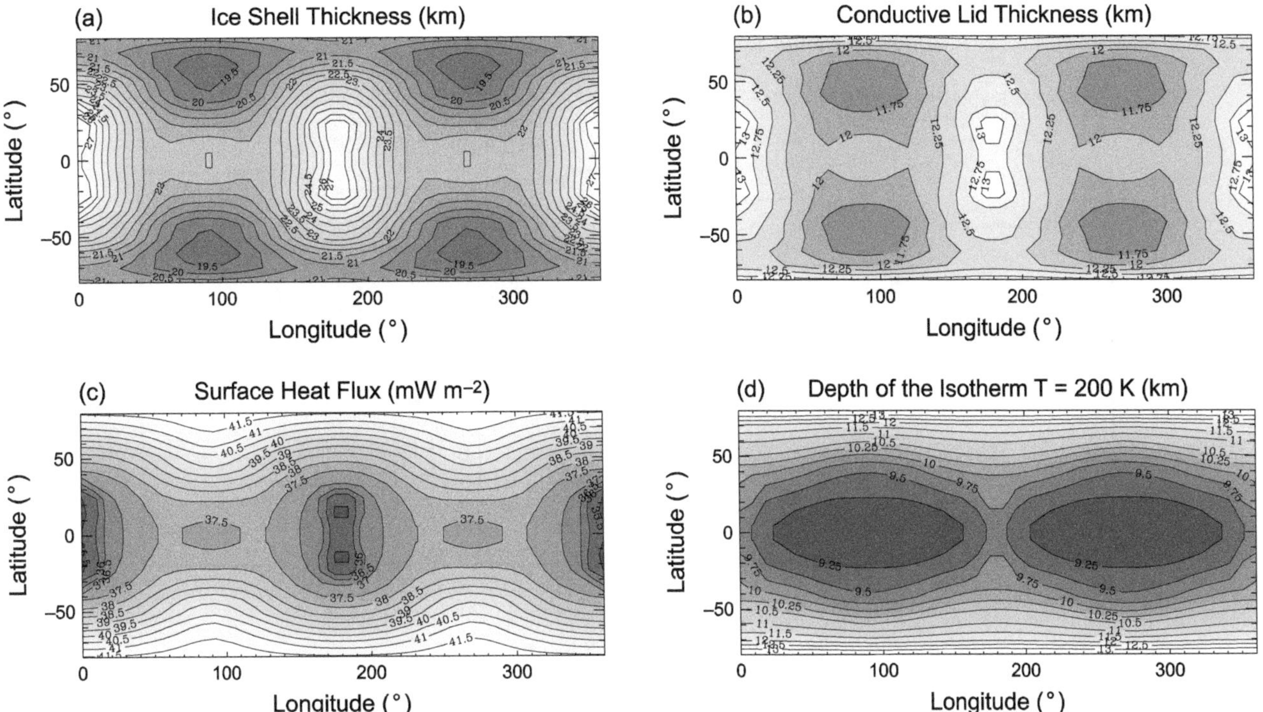

Fig. 14. **(a)** Equilibrium icy shell thickness for a maximal dissipation in hot plumes ($\eta_b = 1.5 \times 10^{14}$ Pa s) with a heat flux from the silicate mantle of 10 mW m^{-2}, and the corresponding **(b)** conductive lid thickness, **(c)** surface heat flux, and **(d)** depth of the isotherm T = 200K. From *Tobie et al.* (2003).

the presence of the ocean at shallow depth (*Thomson and Delaney,* 2001) or partial melting of ice at shallow depth (*Collins et al.,* 2000; *Sotin et al.,* 2002) (Fig. 15) (see chapters by Collins and Nimmo and by Vance and Goodman).

In the model proposed by *Thomson and Delaney* (2001), chaos-type features on the surface of Europa result from melt-through structures formed by rotationally confined, steady and/or episodic oceanic plumes. These plumes form at places of the seafloor where tidal dissipation in the rocky interior generates local heated spots (Fig. 15b). This model is difficult to reconcile with the fact that more tidal heating is probably dissipated in the icy shell than in the silicate shell (see above). Furthermore, *Goodman et al.* (2004) showed that the heat flux transported by ascending hot water plumes is probably too weak to allow complete melt-through of the ice layer.

Alternatively, *Sotin et al.* (2002) proposed that tidal heating in upwelling plumes in the solid icy crust could explain the formation of chaotic regions. Because the ascent of a plume occurs quickly compared to thermal conduction, tidal heating heats the ice to its melting point. The amount of partial melt that can be produced is the ratio of latent heat to tidal heating. The partial melt that occurs in upwelling plumes has interesting implications for Europa's surface features. First, because water is denser than ice, the melting of ice results in a volume decrease that can cause subsidence of the overlying material. Second, the liquid sinks back to the liquid layer, replenishing the internal ocean. Third, the excess density caused by the presence of the liquid creates a negative buoyancy that balances the thermal buoyancy responsible for the ascent of the plume. Fourth, the presence of partial melt may affect the viscosity (*De la Chapelle et al.,* 1999) and therefore the amount of tidal heating and the efficiency of convection. The hot plume dies away and new hot plumes form at different locations at the ice/ocean interface (*Tobie et al.,* 2003). It is interesting to note that a similar mechanism can be invoked for Enceladus (*Tobie et al.,* 2008), another body in the solar system where tidal heating plays a major role in its thermal evolution and surface dynamics.

9. ECCENTRICITY EVOLUTION AND ORIGIN OF THE IO–EUROPA–GANYMEDE LAPLACE RESONANCE

The tides and tidal effects that are so important to Europa are a function of the remarkable set of orbital resonances among the three Galilean satellites Io, Europa, and Ganymede. The strongest resonances are those between Io and Europa and between Europa and Ganymede. The ratios of the mean motions (average orbital angular velocity) of each satellite pair are very close to (but not identical to) 2:1. This commensurability excites the orbital eccentricities of all three satellites, but principally Io and Europa.

Conjunctions between each satellite pair drift at mean angular velocities

$$2n_2 - n_1 = \omega_1 \tag{54}$$

$$2n_3 - n_2 = \omega_2 \tag{55}$$

where n_1, n_2, and n_3 are the mean motions of Io, Europa, and Ganymede, respectively.

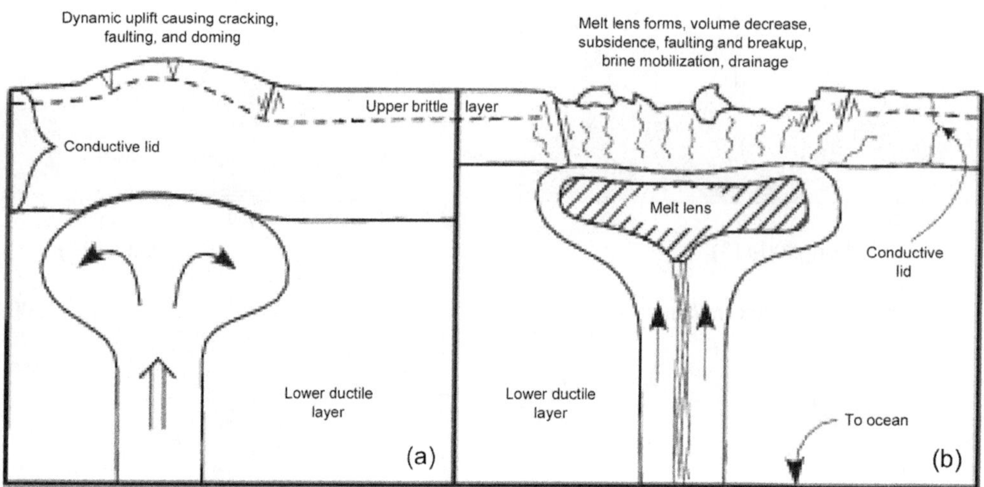

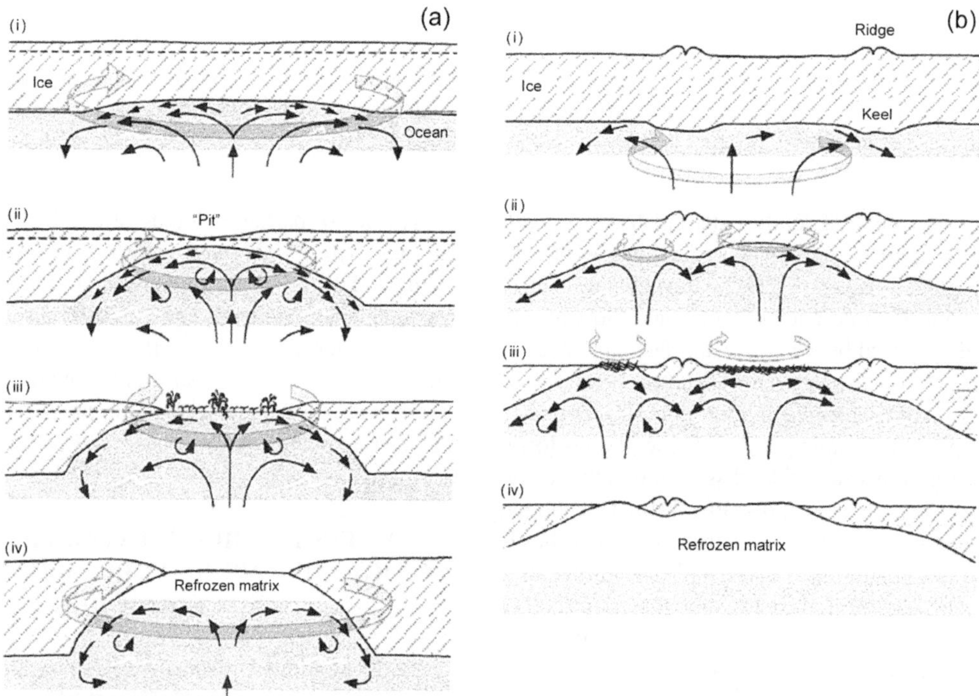

Fig. 15. Two models for the formation of chaotic terrains on Europa. *Model 1:* Interpretation of the effects of tidal heating on a plume. **(a)** Without tidal heating, plume rises buoyantly, impinges on overlying conductive lid, and causes doming. **(b)** Plume is tidally heated and melt lens builds up in the upper part and leads to the collapse and modification seen in chaos and lenticulae. From *Sotin et al.* (2002). *Model 2:* Possible steps in the formation of plume-derived melt-through regions on the surface of Europa. **(a)** Melting in regions several plume radii removed from under-ice topographic structures and **(b)** melting in proximity to ice ridges. Buoyant plumes transport heat to the base of the icy shell where melting begins to take place. As the fluids spread laterally outward from the center of the plume, the Coriolis force causes the fluid to turn in an anticyclonic sense (to the right of the direction of motion in the northern hemisphere) so that heat fluid remains in a lens at the base of the ice cover. From *Thomson and Delanay* (2001).

Presently, ω_1 and ω_2 are small and equal (in a time-averaged sense), and equal to the orbital precession rate of either Io or Europa (0.74° d^{-1}). The equality of ω_1 and ω_2, a 1:1 commensurability between the angular drift rates of the Io-Europa and Europa-Ganymede conjunctions, essentially defines the Laplace resonance between all three satellites. It is the Laplace resonance that supplies the orbital eccentricity that drives Europa's tectonics and heats its interior.

Given the example of Earth's Moon, in which tides raised on Earth by the Moon supply angular momentum to the Moon's orbit, driving it outward over geologic time, it has long been thought that the many mean-motion resonances among the satellites of the outer planets are not primordial, but rather the product of differential tidal evolution in which an inner satellite evolves into resonance with an outer one (*Goldreich,* 1965). The rate of expansion of a satellite orbit due to the tide it raises on its primary is proportional to $m_s a^{-11/2}$, where m_s is the satellite mass and a its semimajor axis (e.g., *Goldreich and Soter,* 1966). Thus, with regard to the Galilean satellites, Io will be driven out faster than Europa, and Europa faster than Ganymede. It is natural to assume that, in the past, differential tidal expansion first caused Io to move into the 2:1 mean-motion resonance with Europa, after which the pair was tidally driven outward until Ganymede was captured into the 2:1 resonance with Europa as well (this orbital evolution path is illustrated in Fig. 16).

The first successful model of this resonance capture and inside-out assembly of the Laplace resonance was accomplished by *Yoder* (1979) and *Yoder and Peale* (1981). In this model, Europa's eccentricity was much smaller (~0.0014) when it was solely in the 2:1 resonance with Io, and it is only when the Laplace resonance formed that modern values of Europa's eccentricity e ~ 0.01 were achieved. If so, then the strong tidal heating that Europa experiences is a latter-stage development in the satellite's geological history.

A key aspect of the Yoder and Peale model, and the secret of its success in explaining the present-day Laplace resonance, is the inclusion of the dissipative properties of the satellites. Tidal torques from Jupiter continue to act to drive Io closer to Europa, reducing the absolute value of ω_1 (i.e., acting to move it toward zero), which drives the satellite pair deeper into their 2:1 mean-motion resonance. This in turn acts to increase the eccentricity of both satellites, but the increased tidal dissipation causes Io's orbit to shrink (back away from Europa's orbit and thus away from commensurability). It is the balance of these two effects that stabilizes the resonant configuration. The importance of satellite interior properties for orbital evolution is now an established "truth" in planetary science. Orbital and interior thermal evolution are intimately coupled, and one cannot be understood in the absence of the other.

Examining the tidal assembly hypothesis for the formation of the Laplace resonance in greater depth, *Malhotra* (1991) and Showman and *Malhotra* (1997) subsequently found that before Io and Europa entered the 2:1 mean-motion commensurability, all three satellites (Io, Europa, and Ganymede) may have been temporarily captured into low-order Laplace-like resonances characterized by ω_1/ω_2 = 1:2, 3:2, or 2, after which the satellites evolved into the present $\omega_1/\omega_2 = 1$ (Figs. 16 and 17a). Eccentricities are excited in these temporary resonances for both Europa and Ganymede, with Europa attaining e values that varied but could have averaged up to ~0.01 (somewhat greater as the ω_1/ω_2 = 3:2 and 2 resonances were exited) (Fig. 17b). Thus Europa's modern level of tidal heating could date from this intermediate era in solar system history.

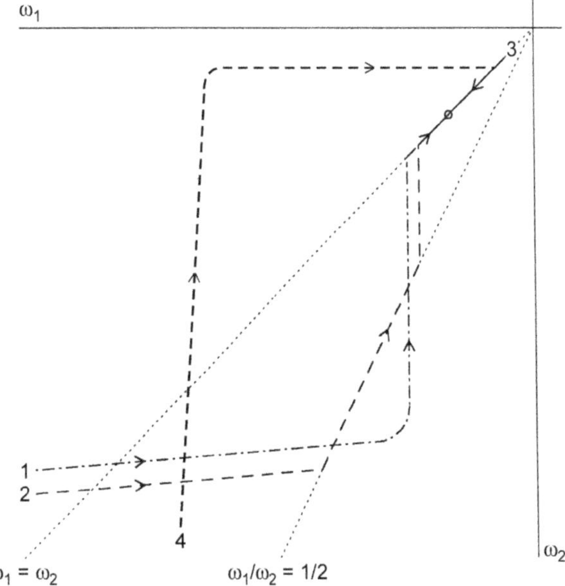

Fig. 16. Schematic paths to the current Laplace resonance state (circle), as shown in $\omega_1\omega_2$ space. The $\omega_1/\omega_2 = 1$ refers to the Laplace relation in general, and $\omega_1/\omega_2 = 1/2$ is a Laplace-like resonance. The initial (t = 0) position of satellites in this figure is unknown, but jovian tides will drive the paths sharply to the right, as ω_2 is much less affected by comparison. In the *Yoder* (1979) model (line 1, dot-dashed), the path turns vertical when Io and Europa reach the 2:1 mean-motion resonance, with eventual capture into the Laplace resonance. In the *Malhotra* (1991) model (line 2, dashed), the satellites first occupy the $\omega_1/\omega_2 = 1/2$ (for example), and then after tidally evolving deeper into this resonance, break free and are captured into the Laplace resonance. Greenberg's original 1982 hypothesis for a primordial origin had the satellites originate (accrete) deeper in resonance (line 3, solid line, upper right). A primordial origin in the model of *Peale and Lee* (2002), in contrast (line 4, thick dashes), moves steeply upward until the Europa-Ganymede 2:1 mean-motion resonance is established, at which point the path bends to the right, ultimately resulting in capture deep in the Laplace resonance. After the protojovian nebula dissipates, the resonance relaxes somewhat. Figure modified from *Showman and Malhotra* (1997).

The fundamental motivation of the *Showman and Malhotra* (1997) model lay in the geophysically significant excitation of Ganymede's eccentricity at an intermediate stage in solar system history, something that does not happen at all in the *Yoder and Peale* (1981) model. In principle, then, such a history could explain latter-stage internal ac-

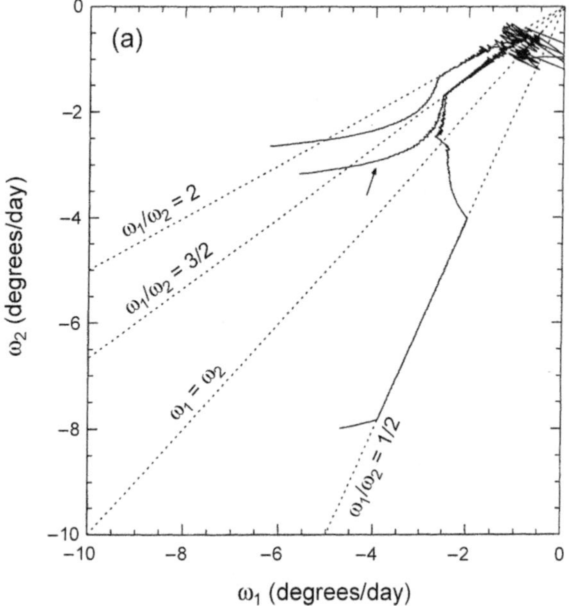

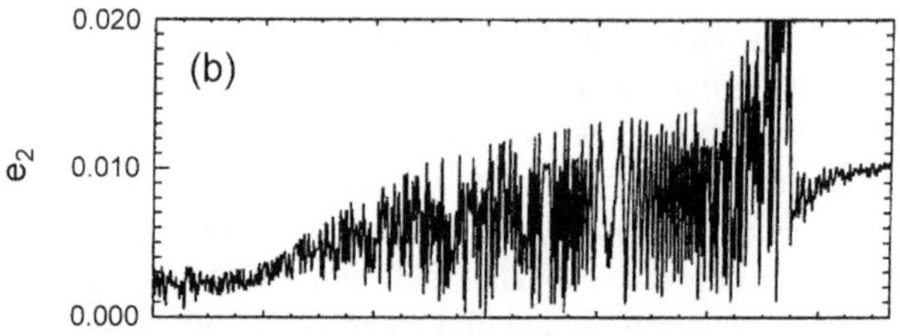

Fig. 17. **(a)** Example tidal evolutionary paths for Io, Europa, and Ganymede displayed in a ω_1–ω_2 diagram. The initial slopes are flatter because Io's outward evolution is initially the greatest of the three. **(b)** Time evolution of Europa's eccentricity for the example of temporary capture into the $\omega_1/\omega_2 = 3/2$ resonance [arrowed path in **(a)**]; Q_J is Jupiter's dissipation factor, generally taken to be ~10^5 (*Peale,* 2003). In this numerical calculation Io and Europa were relatively dissipative, while Ganymede was not, and the system jumped into the Laplace resonance ($\omega_1 = \omega_2$) when Io's dissipation was arbitrarily changed (lowered) by a factor of ~3. The state of the system at the end of the integration is close to that presently observed. Adapted from *Showman and Malhotra* (1997).

tivity or rejuvenation within Ganymede, specifically the formation of its grooved and smooth terrains and operation of its magnetic dynamo (and, of course, the difference between Ganymede and Callisto) (see *Pappalardo et al.,* 2004, and references therein). Initial models of Ganymede's internal evolution that included the resonance assembly scenario of *Showman and Malhotra* (1997) were not, however, able to show that global thermal runaways and resurfacing could occur (*Showman et al.,* 1997), nor have similar models to restart a magnetic dynamo in Ganymede's iron core been successful (*Bland et al.,* 2008). This does not mean that this otherwise appealing explanation for the Ganymede-Callisto dichotomy is disproven, only that we may not yet understand how tidal dissipation and dynamos really work. Additional discussion of the *Showman and Malhotra* (1997) model can be found in *Peale* (1999), and for further, encouraging progress in explaining Ganymede's resurfacing in this scenario, see *Bland et al.* (2009).

A much different solution to the origin of the Laplace resonance was offered by *Greenberg* (1982, 1987). Arguing that the upper limit on Jupiter's specific dissipation factor Q_J (<~10^6) necessary to assemble the resonance from the outside in was physically unrealistic (too low for a gas giant), he proposed that the Laplace resonance is instead primordial. The satellites formed (accreted) originally deeper in the resonance, and have been evolving slowly out from the resonance ever since (see Fig. 16). Such a slow evolution depends on a $Q_J \gg 10^6$, although *Peale* (1999, 2003) has pointed out that such low jovian dissipation is not consistent with Io's prodigious heat flow. In other words, without sufficient torque from jovian tides to counteract the dissipative loses within Io, the Io-Europa 2:1 mean-motion resonance would disassemble on a timescale short compared with the age of the solar system. The situation would have been even less tenable earlier in solar system history, because when deeper in resonance, forced eccentricities (and heating rates) would have been even higher.

Despite this bleak assessment, the Laplace resonance may indeed be primordial. To understand this, we need to consider the conditions during the primordial epoch, when the Galilean satellites were accreting in the protojovian nebula. If the satellites accreted in a slow-inflow, gas-starved disk similar to that described in the chapter by Canup and Ward, as the satellites grow to larger and larger sizes they will undergo inward type I migration (*Canup and Ward,* 2002, 2006). Ganymede is by far the most massive of the three, and as type I drift is proportional to satellite mass and nebular surface density, Ganymede can in principle migrate faster (see Fig. 16). *Peale and Lee* (2002) demonstrate numerically that in doing so Ganymede can capture Europa into the 2:1 mean-motion resonance, and then the pair can migrate fast enough to capture Io as well (Fig. 18). Given Io's greater mass than Europa, and the arguably greater disk surface density closer to Jupiter, Europa's capture into the 2:1 mean-motion resonance with Io depends on Io not evolving faster than Europa after Europa is captured into resonance with Ganymede.

Europa's eccentricity in Fig. 18, and that of Io, are well above current values, as the system is deeper in resonance than today. Orbital eccentricities during migration are limited (damped) by interactions with the gas disk. As the disk dissipates (i.e., as gas and dust infall from the solar nebula abates), these eccentricities should begin to decline due to tidal dissipation within the satellites. Remarkably, *Peale and Lee* (2002) show that the entire system (Io-Europa-Ganymede) naturally relaxes to its current Laplace configuration.

The model of *Peale and Lee* (2002) might be thought of as distressing, in that it obviates the solution to the Ganymede-Callisto dichotomy described above. Yet this primordial, outside-in assembly of the Laplace resonance is both elegant and comprehensive in its predictions. One of these predictions is that Europa's history of tidal heating, and ocean formation, is likely to have begun very early. In addition, the decay of Europa's primordial $e_I \simeq 0.15$ in Fig. 18

to the present-day value of 0.01 implies substantial dissipation within Europa, at least while this primordial eccentricity survives (which was paced by the decay of Io's eccentricity). For 1 to a few × 10^6 yr, tidal heating within Europa would have been ~100 times that today, all other things being equal.

While Europa's early history is not preserved in its geological record, perhaps the most important implication of the model of *Peale and Lee* (2002) is that the geophysical conditions conducive (if not necessary) for the origin of life (water, volcanism, hydrothermal mass transfer of biologically important elements) may have been met early on. Once the Laplace resonance is established, by any of the mechanisms described above, the eccentricity and tidal heating within Europa may evolve with time as well, and need not be the same as current-day values. *Hussmann and Spohn* (2004) explore such time-evolutionary models, and some examples have substantially more tidal heating earlier in Europa's history but within the context the Laplace configuration. Those particular examples, however, do not match Io's presently observed tidal heat output (the evolutions of Io, Europa, and Ganymede are coupled). The example that does (and see the chapter by Moore and Hussmann) is of the oscillatory type (*Ojakangas and Stevenson,* 1986), and Europa's average eccentricity over time is similar to today's value in this case. Future exploration of Europa (and of Io and Ganymede), and theoretical modeling of their tidal evolution, may ultimately yield further clues as to how and when Europa entered the Laplace resonance, and how this resonance evolved through time.

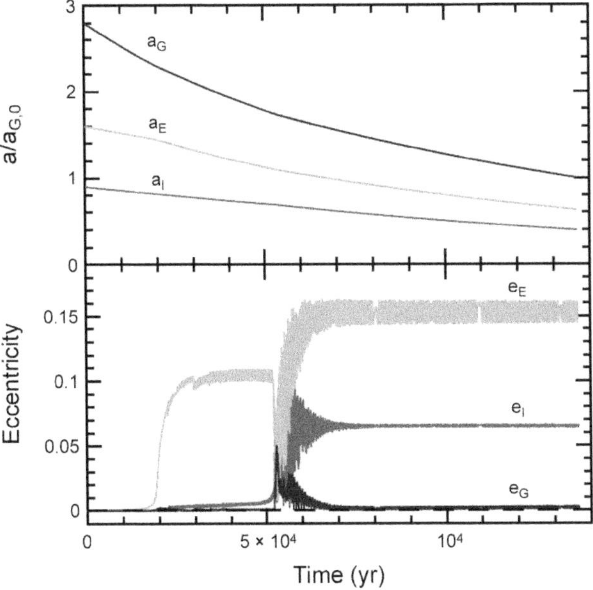

Fig. 18. Circumjovian-nebula-induced evolution of the Galilean satellites into the Laplace resonance. The semimajor axes of Io (a_I), Europa (a_E), and Ganymede (a_G) are normalized to Ganymede's current distance from Jupiter, $a_{G,0}$; e_I, e_E, and e_G are the eccentricities of Io, Europa, and Ganymede, respectively. Inward migration and eccentricity damping timescales are imposed, and the initial orbits are circular and coplanar. After circumjovian nebula dispersal, eccentricities are damped by solid-body tides within each satellite. Modified from *Peale* (2003).

10. CONCLUDING REMARKS

Exploration of the solar system has shown that tides on Io and Enceladus are so large that tidal heating is the main process that drives their internal dynamics. The studies of these two satellites of Jupiter and Saturn have fostered theoretical developments linking the internal dynamics and the orbital evolution of satellites. The present chapter has applied these calculations to Europa.

First, the equations that describe the viscoelastic deformation of a satellite during its eccentric orbit around the planet have been described. The limitations of the Maxwell model, which is widely used, were discussed. It was shown that this simple description holds as long as the forcing period is close to the Maxwell time, which depends on both the viscosity and the elastic parameters of the body. Outside of the Maxwell time, the calculation of dissipation can be quite far from reality as suggested by the comparison with observations of the dissipation in Earth-like planets.

In this chapter we find that Europa's eccentricity is probably large enough for tidal dissipation to play a key role in its thermal evolution. Two principal consequences are that (1) tidal deformation is a strong driver of surface tectonics and (2) tidal dissipation in the icy crust affects surface geology through the formation of chaotic terrains and probably other features involving melting of icy material. The presence of an ocean at some depths has been inferred from the presence of an induced magnetic field. Thermal evolution models suggest that such an ocean could still be present 4.5 b.y. after Europa formed because tidal heating in the icy crust prevented a complete freezing. Models describing the relationships between internal dynamics and tidal dissipation have been presented. The question of tidal heating in the silicate mantle has been discussed. Such heating in the silicate mantle has potential astrobiological implications if tidal heating is sufficient to melt silicates and to lead to volcanism at the silicate/ocean interface. The conditions would be very similar to those existing at the terrestrial seafloor. As an interesting consequence, section 7.4.2 has discussed this possibility and has given the conditions required for such a process to happen.

One limitation on the calculations of tidal dissipation is our knowledge of the viscoelastic behavior of ices and silicates at the forcing period relevant to Europa. Laboratory experiments describing the behavior of ices when submitted to cyclic stresses are needed in order to better constrain the amount of tidal dissipation in icy bodies. Such experiments should allow us to model the deformation of ice at laboratory conditions and to extrapolate these models to conditions relevant to the icy moons of the giant planets.

In addition to improving models of tidal heating, observations of Europa by the future Europa and Jupiter System Mission (EJSM) would be critical to validate models. Mapping of geological features, observations of surface deformation, determination of the ocean depth, intensity of the induced magnetic field, and determination of the surface composition will provide constraints on the role of tidal heating on the internal dynamics and overall evolution of this satellite. Such measurements will enable us to better assess the habitability of the subsurface ocean.

APPENDIX A

The three energy integrals I_1, I_2, and I_3, defined in section 7.2, are expressed in term of density ρ, shear and bulk moduli, μ and K, and six complex radial functions $y_i(r)$ that are classically used to characterize the radial dependence of shear stress, normal stress, displacement, and potential in the frequency domain (Table A1) (see *Tobie et al.*, 2005, for more details). At the surface, the radial functions, commonly termed y_i, are related to the Love numbers defined in section 4.

The I_2 integral can be rearranged as

$$I_2 = \int_0^{R_s} [H_K K + H_\mu \mu + H_\rho \rho + H_0] dr \qquad \text{(A1)}$$

where H_K, H_μ, and H_ρ represent the radial sensitivity to the bulk modulus K, to the shear modulus μ, and to the density ρ, respectively

$$\begin{aligned} H_K &= \left| y_2 - \frac{K - 2/3\mu}{r}[2y_1 - l(l+1)y_3] \right|^2 + \\ &\quad 2\mathrm{Re}\left\{ \frac{dy_1}{dr}[2y_1 - l(l+1)y_3] \right\} \left| 2y_1 - l(l+1)y_3 \right|^2 \\ H_\mu &= \frac{4}{3}\left| y_2 - \frac{K - 2/3\mu}{r}[2y_1 - l(l+1)y_3] \right|^2 - \\ &\quad \frac{4}{3}\mathrm{Re}\left\{ \frac{dy_1}{dr}[2y_1 - l(l+1)y_3] \right\} + \\ &\quad \frac{1}{3}\left| 2y_1 - l(l+1)y_3 \right|^2 + l(l+1)r^2|y_4|^2/|\mu|^2 \\ H_\rho &= 2(l+1)r\mathrm{Re}\{y_1^* y^5\} - 2l(l+1)r\mathrm{Re}\{y_3^* y^5\} - \\ &\quad 2gr\mathrm{Re}\{y_1^*[2y_1 - l(l+1)y_3]\} \\ H_0 &= (4\pi G)^{-1} r |y_6|^2 \end{aligned} \qquad \text{(A2)}$$

TABLE A1. Radial energy integrals.

	x_i any Complex Function	$x_i = y_i^*$
I_1	$\int_0^{R_s} \rho[y_1 x_1 + l(l+1)y_3 x_3] r^2 dr$	$\int_0^{R_s} \rho[|y_1|^2 + l(l+1)|y_3|^2] r^2 dr$
I_2	$\int_0^{R_s} \sum_{k=1}^{9} I_{2,k} dr$	$\int_0^{R_s} \sum_{k=1}^{9} I_{2,k} dr$
I_{21}	$(K + \frac{4}{3}\mu) r^2 \frac{dx_1}{dr}\frac{dy_1}{dr}$	$\frac{4}{3}\frac{r^2}{|K + 4/3\mu|^2}\left| y_2 - \frac{K - 2/3\mu}{r}[2y_1 - l(l+1)y_3] \right|^2 \times \mu$ $+ \frac{r^2}{|K + 4/3\mu|^2}\left| y_2 - \frac{K - 2/3\mu}{r}[2y_1 - l(l+1)y_3] \right|^2 \times K$
I_{22}	$(K - \frac{2}{3}\mu) r \left(\frac{dx_1}{dr}[2y_1 - l(l+1)y_3] + \frac{dy_1}{dr}[2x_1 - l(l+1)x_3] \right)$	$-\frac{4}{3} r\mathrm{Re}\left\{ \frac{dy_1^*}{dr}[2y_1 - l(l+1)y_3] \right\} \times \mu$ $2r\mathrm{Re}\left\{ \frac{dy_1^*}{dr}[2y_1 - l(l+1)y_3] \right\} \times K$
I_{23}	$(K + \frac{1}{3}\mu) r [2y_1 - l(l+1)y_3][2x_1 - l(l+1)x_3]$	$\frac{1}{3}|2y_1 - l(l+1)y_3|^2 \times \mu + |2y_1 - l(l+1)y_3|^2 \times K$
I_{24}	$l(l+1) r y_4 \left(r\frac{dx_3}{dr} + x_1 - x_3 \right)$	$l(l+1) r^2 |y_4|^2/|\mu|^2 \times \mu$
I_{25}	$l(l^2 - 1)(l+2)\mu x_3 y_3$	$l(l^2 - 1)(l+2)|y_3|^2 \times \mu$
I_{26}	$(l+1)\rho r(x_1 y_5 + x_5 y_1)$	$2(l+1) r\mathrm{Re}\{y_1^* y_5\} \times \rho$
I_{27}	$-l(l+1)\rho r(x_3 y_5 + x_5 y_3)$	$-2l(l+1) r\mathrm{Re}\{y_3^* y_5\} \times \rho$
I_{28}	$-\rho g r(x_1[2y_1 - l(l+1)y_3] + y_1[2x_1 - l(l+1)x_3]$	$-2gr\mathrm{Re}\{y_1^*[2y_1 - l(l+1)y_3]\} \times \rho$
I_{29}	$(4\pi G)^{-1} r y_6 \left[r\frac{dx_5}{dr} - (4\pi G)\rho r x_1 + (l+1)x_5 \right]$	$(4\pi G)^{-1} r |y_6|^2$
I_3	$\left\{ r^2 \left[x_1 y_2 + l(l+1) x_3 y_4 + \frac{x_5 y_6}{4\pi G} \right] \right\}_{r=R_s}$	$\left\{ r^2 \left[y_1^* y_2 + l(l+1) y_3^* y_4 + \frac{y_5^* y_6}{4\pi G} \right] \right\}_{r=R_s}$

Acknowledgments. Detailed reviews of this chapter by O. Karatekin and J. Roberts have significantly improved this chapter. Part of this work was carried out at the Jet Propulsion Laboratory, California Institute of Technology, under contract with NASA. C.S. acknowledges support by the JPL R&TD program. G.T. and C.S. were partly funded by INSU-Programme National de Planetologie and Agence National de Recherche (ETHER, no. 06-1-137653), France.

REFERENCES

Anderson D. L. and Given J. S. (1982) Absorption band Q model for the Earth. *J. Geophys. Res., 87,* 3893–3904.

Anderson J. D., Schubert G., Jacobson R. A., Lau E. L., Moore W. B., Sjogren W. L. (1998) Europa's differentiated internal structure: Inferences from four Galileo encounters. *Science, 281,* 2019–2022.

Backus G. and Gilbert F. (1968) The resolving power of gross Earth data. *Geophys. J. R. Astron. Soc., 16,* 169–205.

Barr A. C. and McKinnon W. B. (2007) Convection in ice I shells and mantles with self-consistent grain size. *J. Geophys. Res., 112(E2),* CiteID E02012.

Barr A. C. and Pappalardo R. T. (2005) Onset of convection in the icy Galilean satellites: Influence of rheology. *J. Geophys. Res., 110(E12),* CiteID E1(2005).

Benjamin D., Wahr J., Ray R. D., Egbert G. D., and Desai S. D. (2006) Constraints on mantle anelasticity from geodetic observations, and implications for the J2 anomaly. *Geophys. J. Intl., 165,* 3–16.

Bills B. G., Neumann G. A., D. E. Smith D. E., and Zuber M. T. (2005) Improved estimate of tidal dissipation within Mars from MOLA observations of the shadow of Phobos. *J. Geophys. Res., 110,* E07004, DOI: 10.1029/2004JE002376.

Bills B. G., Adams K. D., and Wesnousky S. G. (2007) Viscosity structure of the crust and upper mantle in western Nevada from isostatic rebound patterns of the late Pleistocene Lake Lahontan high shoreline. *J. Geophys. Res.–Solid Earth, 112,* B06405, DOI: 10.1029/2005JB003941.

Bland M. T., Showman A. P., and Tobie G. (2008) The production of Ganymede's magnetic field. *Icarus, 198,* 384–399.

Bland M. T., Showman A. P., and Tobie G. (2009) The orbital thermal evolution and global expansion of Ganymede. *Icarus, 200,* 207–221.

Cadek O. and Fleitout L. (1999) A global geoid model with imposed plate velocities and partial layering. *J. Geophys. Res., 104,* 29055–29075.

Canup R. M. and Ward W. R. (2002) Formation of the Galilean satellites: Conditions of accretion. *Astron. J., 124,* 3404–3423.

Canup R. M. and Ward W. R.(2006) A common mass scaling for satellite systems of gaseous planets. *Nature, 441,* 834–839.

Cassen P., Reynolds R. T., and Peale S. J. (1979) Is there liquid water on Europa? *Geophys. Res. Lett., 6,* 731–734.

Castillo J., Mocquet A., and Sotin C. (2000) Detecter la presence d'un ocean dans Europe partir de mesures altimtriques et gravimetriques. *C. R. Acad. Sci., 330,* 659–666.

Chyba C. J., Edwards B. C., England T., Geissler P. E., Johnson T. V., Ostro S. J., Pappalardo R. T., Peale S. J., Squyres S. W., Terrile R., Yoder C. F., and Zuber M. T. (1998) *Report of the Europa Science Definition Team,* pp. 2019–2022. National Aeronautics and Space Administration, Washington, DC.

Cole D. M. and Durell G. D. (1995) The cyclic loading of saline ice. *Philos. Mag., 72,* 209–229.

Collins G. C., Head J. W., Pappalardo R. T., and Spaun N. A. (2000) Evaluation of models for the formation of chaotic terrain on Europa. *J. Geophys. Res., 105,* 1709–1716.

Dahlen F. A. (1976) The passive influence of the oceans upon the rotation of the Earth. *Geophys. J. Intl., 46,* 363–406.

De La Chapelle S., Milsch H., Castelnau O., and Duval P. (1999) Compressive creep of ice containing a liquid intergranular phase: Rate-controlling processes in the dislocation creep regime. *Geophys. Res. Lett., 26,* 251–254.

Deschamps F. and Sotin C. (2001) Thermal convection in the outer shell of large icy satellites. *J. Geophys. Res., 106,* 5107–5121.

Durham W. B., Stern L. A., and Kirby S. H. (2001) Rheology of ice I at low stress and elevated confining pressure. *J. Geophys. Res., 106,* 11031–11042.

Fischer H.-J. and Spohn T. (1990) Thermal-orbital histories of viscoelastic models of Io (J1). *Icarus, 83,* 39–65.

Goldreich P. (1965) An explanation of the frequent occurrence of commensurable mean motions in the Solar System. *Mon. Not. R. Astron. Soc., 130,* 159–181.

Goldreich P. and Soter S. (1966) Q in the solar system. *Icarus, 5,* 375–389.

Goldsby D. L. and Kohlstedt D. L. (2001) Superplastic deformation of ice: Experimental observations. *J. Geophys. Res., 106,* 11017–11030.

Goodman J. C., Collins G. C., Marshall J., and Pierrehumbert R. T. (2004) Hydrothermal plume dynamics on Europa: Implications for chaos formation. *J. Geophys. Res., 109,* E03008.

Greenberg R. (1982) Orbital evolution of the Galilean satellites. In *Satellites of Jupiter* (D. Morrison, ed.), pp. 65–92. Univ. of Arizona, Tucson.

Greenberg R. (1987) Galilean satellites: Evolutionary paths in deep resonance. *Icarus, 70,* 334–347.

Greenberg R., Geissler P., Hoppa G., Tufts B. R., Durda D. D., Pappalardo R., Head J. W., Greeley R., Sullivan R., and Carr M. H. (1998) Tectonic processes on Europa: Tidal stresses, mechanical response, and visible features. *Icarus, 135,* 64–78.

Harada Y. and Kurita K. (2007) Effect of non-synchronous rotation on surface stress upon Europa: Constraints on surface rheology. *Geophys. Res. Lett., 34,* L11204.

Helfenstein P. and Parmentier E. M. (1985) Patterns of fracture and tidal stresses due to nonsynchronous rotation — Implications for fracturing on Europa. *Icarus, 61,* 175–184.

Hobbs P. V. (1974) *Ice Physics.* Clarendon, Oxford.

Hoppa G. V. (1998) Europa: Effects of rotation and tides on tectonic processes. Ph.D. thesis, Univ. of Arizona, Tucson. 227 pp.

Hoppa G. V., Greenberg R., Geissler P., Tufts B. R., Plassmann J., and Durda D. D. (1999) Strike-slip faults on Europa: Global shear patterns driven by tidal stress. *Icarus, 137,* 341–347, DOI: 10.1006/icar.1998.6065.

Hurford T. A., Sarid A. R., and Greenberg R. (2007) Cycloidal cracks on Europa: Improved modeling and non-synchronous rotation implications. *Icarus, 186,* 218–233.

Hussmann H. and Spohn T. (2004) Thermal-orbital evolution of Io and Europa. *Icarus, 171,* 391–410.

Hussmann H., Spohn T., and Wieczerkowski K. (2002) Thermal equilibrium states of Europa's ice shell: Implications for internal ocean thickness and surface heat flow. *Icarus, 156,* 143–151.

Jackson I. (2000) Laboratory measurement of seismic wave dispersion and attenuation: Recent progress. In *Earth's Deep Interior, Mineral Physics and Tomography, from the Atomic to Global Scale* (S. Karato et al., eds.), pp. 265–289. AGU Geophysical Monograph 117, Washington, DC.

Jackson J. D. (1999) *Classical Electrodynamics,* 3rd edition. Wiley & Sons, New York. 808 pp.

Karato S. I. and Wu P. (1993) Rheology of the upper mantle: A synthesis. *Science, 260,* 771–778.

Kargel J. S., Kaye J. Z., Head J. W., Marion G. M., Sassen R., et al. (2000) Europa's crust and ocean: Origin, composition, and prospects for life. *Icarus, 148,* 226–265.

Kaula W. M. (1964) Tidal dissipation by solid friction and the resulting orbital evolution. *Rev. Geophys., 2,* 661–685.

Kelley D. S., Karson J. A., Blackman D. K., Fruh-Green G. L., Butterfield D. A., Lilley M. D., Olson E. J., Schrenk M. O., Roe K. K., Lebon G. T., and Rivizzigno P. (2001) An off-axis hydrothermal vent field near the Mid-Atlantic Ridge at 30 degrees N. *Nature, 412,* 145–149.

Khurana K. K., Kivelson M. G., Stevenson D. J., Schubert G., Russel C. T., Walker R. J., and Polanskey C. (1998) Induced magnetic fields as evidence for subsurface oceans in Europa and Callisto. *Nature, 395,* 777–780.

Kivelson M. G., Khurana K. K., Joy S., Russell C. T., Southwood D. J., Walker R. J., and Polanskey C. (1997) Europa's magnetic signature: Report from Galileo's pass on 19 December (1996). *Science, 276,* 1239–1241.

Lainey V. (2002) Dynamical theory of the Galilean satellites (in French). Ph.D. thesis, Observatoire de Paris, Paris. 130 pp.

Lainey V. and Tobie G. (2005) New constraints on Io's and Jupiter's tidal dissipation. *Icarus, 179,* 485–489.

Lambeck K. (1990) *Geophysical Geodesy.* Cambridge Univ., New York.

Leith A. C. and McKinnon W. B. (1996) Is there evidence for polar wander on Europa? *Icarus, 120,* 387–398.

Malhotra R. (1991) Tidal origin of the Laplace resonance and the resurfacing of Ganymede. *Icarus, 94,* 399–412.

McKinnon W. B. (1999) Convective instability in Europa's floating ice shell. *Geophys. Res. Lett., 26,* 951–954.

Melosh H. J. (1997) Global tectonics of a despun planet. *Icarus, 31,* 221–243.

Mitri G. and Showman A. P. (2005) Convective conductive transitions and sensitivity of a convecting ice shell to perturbations in heat flux and tidal-heating rate: Implications for Europa. *Icarus, 177,* 447–460.

Mitri G. and Showman A. P. (2008) A model for the temperature-dependence of tidal dissipation in convective plumes on icy satellites: Implications for Europa and Enceladus. *Icarus, 195,* 758–764.

Montagnat M. (2004) The viscoplastic behaviour of ice in polar ice sheets: Experimental results and modelling. *C. R. Phys., 5,* 699–708.

Montagnat M. and Duval P. (2000) Rate controlling processes in the creep of polar ice, influence of grain boundary migration associated with recrystallization. *Earth Planet. Sci. Lett., 183,* 179–186.

Moore W. B. (2003) Tidal heating and convection in Io. *J. Geophys. Res., 108,* 5096.

Moore W. B. and Schubert G. (2000) The tidal response of Europa. *Icarus, 147,* 317–319.

Munk W. H. and MacDonald G. J. F. (1960) *The Rotation of the Earth.* Cambridge Univ., New York. 323 pp.

Nimmo F. and Gaidos E. (2002) Strike-slip motion and double ridge formation on Europa. *J. Geophys. Res., 107(E4),* DOI: 10.1029/2000JE001476.5021.

Nimmo F. and Pappalardo R. T. (2006) Diapir-induced reorientation of Saturn's moon Enceladus. *Nature, 441,* 614–616.

Nimmo F., Thomas P. C., Pappalardo R. T., and Moore W. B. (2007) The global shape of Europa: Constraints on lateral shell thickness variations. *Icarus, 191,* 183–192.

Ojakangas G. W. and Stevenson D. J. (1986) Episodic volcanism of tidally heated satellites with application to Io. *Icarus, 66,* 341–358.

Ojakangas G. W. and Stevenson D. J. (1989) Thermal state of an ice shell on Europa. *Icarus, 81,* 220–241.

Pappalardo R. T., Belton M. J. S., Breneman H. H., Carr M. H., Chapman C. R., Collins G. C., Denk T., Fagents S., Geissler P. E., Giese B., Greeley R., Greenberg R., Head J. W., Helfenstein P., Hoppa G., Kadel S. D., Klaasen K. P., Klemaszewski J. E., Magee K., McEven A. S., Moore J. M., Moore W. B., Neukum G., Phillips C. B., Prokter L., Schubert G., Senske D. A., Sullivan R. J., Tufts B. R., Turtle E. P., Wagner R., and Williams K. K. (1999) Does Europa have a subsurface ocean? Evaluation of the geological evidence. *J. Geophys. Res., 104,* 24015–24056.

Pappalardo R. T., Collins G. C., Head J. W., Helfenstein P., McCord T. B., Moore J. M., Prockter L. M., Schenk P. M., and Spencer J. R. (2004) Geology of Ganymede. In *Jupiter: The Planet, Satellites and Magnetosphere* (F. Bagenal et al., eds.), pp. 363–396. Cambridge Univ., Cambridge.

Peale S. J. (1999) Origin and evolution of the natural satellites. *Annu. Rev. Astron. Astrophys., 37,* 533–602.

Peale S. J. (2003) Tidally induced volcanism. *Cel. Mech. Dyn. Astron., 87,* 129–155.

Peale S. J. and Lee M. H. (2002) A primordial origin of the Laplace relation among the Galilean satellites. *Science, 298,* 593–597.

Peale S. J., Cassen P., and Reynolds R. T. (1979) Melting of Io by tidal dissipation. *Science, 203,* 892–894.

Peltier W. R. (1974) The impulse response of a Maxwell earth. *Rev. Geophys. Space Phys., 12,* 649–669.

Peltier W. R. (1985) The LAGEOS constraint on deep mantle viscosity: Result from a new normal mode method for the inversion of viscoelastic relaxation spectra. *J. Geophys. Res., 90,* 9411–9421.

Preblich B., Greenberg R., Riley J., and O'Brien D. (2007) Tidally driven strike-slip displacement on Europa: Viscoelastic modeling. *Planet. Space Sci., 55,* 1225–1245.

Rappaport N. J., Iess L., Wahr J., Lunine J. I., Armstrong J. W., Asmar S. W., Tortora P., Di Benedetto M., and Racioppa P. (2008) Can Cassini detect a subsurface ocean in Titan from gravity measurements? *Icarus, 194,* 711–720.

Ray R. D., Eanes R. J., and Lemoine F. G. (2001) Constraints on energy dissipation in the Earth's body tide from satellite tracking and altimetry. *Geophys. J. Intl., 144,* 471–480.

Reeh N., Christensen E. L., Mayer C., and Olesen O. B. (2003) Tidal bending of glaciers: A linear viscoelastic approach. *Ann. Glaciol., 37,* 83–89.

Romanowicz B. and Durek J. J. (2000) Seismological constraints on attenuation in the Earth: A review. In *Earth's Deep Interior, Mineral Physics and Tomography, from the Atomic to Global Scale* (S. Karato et al., eds.), pp. 217–295. AGU Geophysical Monograph 117, Washington, DC.

Ross M. N. and Schubert G. (1987) Tidal heating in an internal ocean model of Europa. *Nature, 325,* 133–134.

Schmeling H. (1985) Numerical models on the influence of partial melt on elastic, anelastic and electric properties of rocks. Part I: Elasticity and anelasticity. *Phys. Earth Planet. Inter., 41,* 34–57.

Segatz M., Spohn T., Ross M. N., and Schubert G. (1988) Tidal dissipation, surface heat flow, and figure of viscoelastic models of Io. *Icarus, 75,* 187–206.

Showman A. P. and Malhotra R. (1997) Tidal evolution into the Laplace resonance and the resurfacing of Ganymede. *Icarus, 127,* 93–111.

Showman A.P., Stevenson D. J., and Malhotra R. (1997) Coupled orbital and thermal evolution of Ganymede. *Icarus, 129,* 367–383.

Sohl F., Spohn T., Breuer D., and Nagel K. (2002) Implications from Galileo observations on the interior structure and chemistry of the Galilean satellites. *Icarus, 157,* 104–119.

Sotin C., Head J. W., and Tobie G. (2002) Europa: Tidal heating of upwelling thermal plumes and the origin of lenticulae and chaos melting. *Geophys. Res. Lett., 29,* DOI: 10.1029/2001GL013844.1233.

Spencer J. R., Rathburn J. A., Travis L. D., Tamppari L. K., Barnard L., Martin T. Z., and McEwen A. S. (2000) Io's thermal emission from the Galileo photopolarimeter-radiometer. *Science, 288,* 1198–1201.

Spohn T. and Schubert G. (2003) Oceans in the icy Galilean satellites of Jupiter. *Icarus, 161,* 456–467.

Takeushi H. and Saito M. (1972) Seismic surfaces waves. In *Methods in Computational Physics, Vol. 1* (B. A. Bolt, ed.), pp. 217–295. Academic, New York.

Tatibouet J., Perez J. ,and Vassoille R. (1986) High-temperature internal friction and dislocations in ice I_h. *J. Phys., 47,* 51–60.

Thomson R. E. and Delaney J. R. (2001) Evidence for a weakly stratified europan ocean sustained by seafloor heat flux. *J. Geophys. Res., 106,* 12355–12366.

Tobie G. (2003) Impact of tidal heating on the geodynamical evolution of Europa and Titan (in French). Ph.D. thesis, University of Nantes, Nantes. 267 pp.

Tobie G., Choblet G., and Sotin C. (2003) Tidally heated convection: Constraints on Europa's ice shell thickness. *J. Geophys. Res., 108(E11),* 10.1029/2003JE002099, CiteID 5124.

Tobie G., Mocquet A., and Sotin C. (2005) Tidal dissipation within large icy satellites: Applications to Europa and Titan. *Icarus, 177,* 534–549.

Tobie G., Cadek O., and Sotin C. (2008) Solid tidal friction above a liquid water reservoir as the origin of the south pole hotspot on Enceladus. *Icarus,* DOI: 10.1016/j.icarus.2008.03.008.

Veeder G. J., Matson D. L., Johnson T. V., Blaney D. L., and Goguen J. D. (1994) Io's heat flow from infrared radiometry: 1983–1993. *J. Geophys. Res., 99,* 17095–17162.

Wahr J. M. (1981) Body tides on an elliptical, rotating, elastic and oceanless Earth. Geophys. *J. R. Astron. Soc., 64,* 677–704.

Wahr J. M., Zuber M. T., Smith D. E., and Lunine J. I. (2006) Tides on Europa, and the thickness of Europa's icy shell. *J. Geophys. Res.–Planets, 111,* E12005, DOI: 10.1029/2006JE002729.

Wahr J., Selvans Z. A., Mullen M. E., Barr A. C., Collins G. C., Pappalardo R. T., and Selvans M. M. (2009) Modeling stresses on satellites due to non-synchronous rotation and orbital eccentricity using gravitational potential theory. *Icarus, 200,* 188–206.

Williams J. G., Boggs D. H., Yoder C. F., Ratcliff J. T., and Dickey J. O. (2001) Lunar rotational dissipation in solid body and molten core. *J. Geophys. Res., 106(E11),* 27933–27968.

Wu X. P., Bar-Sever Y. E., Folkner W. M., Williams J. G., and Zumberge J. F. (2001) Probing Europa's hidden ocean from tidal effects on orbital dynamics. *Geophys. Res. Lett., 28,* 2245–2248.

Yoder C. F. (1979) How tidal heating in Io drives the Galilean orbital resonance locks. *Nature, 279,* 767–770.

Yoder C. F. and Peale S. J. (1981) The tides of Io. *Icarus, 47,* 1–35.

Yoder C. F., Konopliv A. S., Yuan D. N., Standish E. M., and Folkner W. M. (2003) Fluid core size of Mars from detection of the solar tide. *Science, 300,* 299–303.

Yuen D. A., Fleitout L., Schubert G., and Froidevaux C. (1978) Shear deformation zones along major transform faults and subducting slabs. *Geophys. J. R. Astron. Soc., 54,* 93–119.

Zschau J. (1978) Tidal friction in the solid earth: Loading tides versus body tides. In *Tidal Friction and the Earth's Rotation* (P. Brosche and J. Sündermann, eds.), pp. 62–94. Springer-Verlag, New York.

Rotational Dynamics of Europa

Bruce G. Bills
NASA Goddard Space Flight Center and Scripps Institution of Oceanography

Francis Nimmo
University of California, Santa Cruz

Özgür Karatekin, Tim Van Hoolst, and Nicolas Rambaux
Royal Observatory of Belgium

Benjamin Levrard
Institut de Mécanique Céleste et de Calcul des Ephémérides and Ecole Normale Superieure de Lyon

Jacques Laskar
Institut de Mécanique Céleste et de Calcul des Ephémérides

The rotational state of Europa is only rather poorly constrained at present. It is known to rotate about an axis that is nearly perpendicular to the orbit plane, at a rate that is nearly constant and approximates the mean orbital rate. Small departures from a constant rotation rate and oscillations of the rotation axis both lead to stresses that may influence the location and orientation of surface tectonic features. However, at present geological evidence for either of these processes is disputed. We describe a variety of issues that future geodetic observations will likely resolve, including variations in the rate and direction of rotation, on a wide range of timescales. Since the external perturbations causing these changes are generally well known, observations of the amplitude and phase of the responses will provide important information about the internal structure of Europa. We focus on three aspects of the rotational dynamics: obliquity, forced librations, and possible small departures from a synchronous rotation rate. Europa's obliquity should be nonzero, while the rotation rate is likely to be synchronous unless lateral shell thickness variations occur. The tectonic consequences of a nonzero obliquity and true polar wander have yet to be thoroughly investigated.

1. INTRODUCTION

The primary objective of this chapter is to describe a variety of rotational phenomena that Europa is expected to exhibit, and that will, when properly observed, provide important diagnostic information about the internal structure. The rotational state of any planet or satellite is important to understand for at least three reasons. First, proper collation of observations at various epochs and locations requires a good understanding of the rotation rate and direction of the rotation pole. Second, some aspects of the mean rotation state, and all plausible variations in the rotation state, provide information about the structure of the interior. Third, variations in rotation rate or rotation axis orientation lead to global stresses, thus surface tectonic features may constrain the existence of such processes. All that is presently known, from observations, about the rotation state of Europa is that its rotation rate is very close to synchronous, and the direction of its spin pole is very close to that of its orbit pole, so the obliquity is very small.

As we will discuss below, dynamical arguments suggest that the mean rotation rate is indeed equal to the mean orbital rate, but that the obliquity (the angle between orbit pole and spin pole), while small, is nonzero. These assertions are at odds with what is usually assumed about the rotation of Europa in studies attempting to interpret the tectonics of the body, and thus need to be carefully explored and supported. In addition, gravitational torques from Jupiter are expected to cause librations, which are periodic variations in the direction and rate of rotation. As the position and mass of Jupiter are well known, the amplitude and phase of these variations are diagnostic of internal structure.

This chapter will consist of three main parts. (1) The first part will discuss the obliquity history of Europa, and explain how observations of the current orientation of the spin pole will constrain the moments of inertia of the body. (2) The second part will discuss forced librations, with primary emphasis on longitudinal librations. It will also be discussed how observations of the amplitude and phase of the periodic variations in rotation rate will constrain internal structure. (3) The third part will discuss arguments for and against nonsynchronous rotation (NSR). Most models of tidal dissipation predict that a body like Europa will be close to a synchronous rotation state, but that the rotation rate at which the tidal torque vanishes differs slightly from exact synchronism. From a dynamical perspective, the question is whether gravitational torques on a permanent asymmetry are large enough to "finish the job." We will also briefly discuss the associated issue of true polar wander (TPW) of the ice shell.

2. OBLIQUITY

In this section we discuss the obliquity of Europa. The obliquity of a planet or satellite is the angular separation between its spin pole and orbit pole, or equivalently, the angle between the equator plane and orbit plane. For Earth, the current obliquity is 23.439° (*Lieske et al.,* 1977), which sets the locations of the tropics of Cancer and Capricorn, which are the northern and southern limits at which the Sun appears directly overhead, and the Arctic and Antarctic circles, which are the equatorward limits beyond which the Sun does not rise on the days of the corresponding solstices. Earth's obliquity is presently decreasing (*Rubincam et al.,* 1998), and oscillates between 22.1° and 24.5° with a 41-k.y. period (*Berger et al.,* 1992; *Laskar et al.,* 1993), due to lunar and solar torques on Earth's oblate figure. The associated changes in seasonal and latitudinal patterns of insolation have a significant impact upon global climate (*Milankovitch,* 1941; *Hays et al.,* 1976; *Hinnov and Ogg,* 2007).

For planetary satellites, the solar radiation cycles can be more complex, depending as they do upon the obliquity of the planet, inclination of the satellite orbit, and obliquity of the satellite itself. Several recent studies of the radiative environment of Titan (*Flasar,* 1998; *Roos-Serote,* 2005; *Tokano and Neubauer,* 2005) have referred to the angular separation between the spin pole of Titan and the orbit pole of Saturn as Titan's obliquity. This angle, approximately equal to the 26.73° dynamical obliquity of Saturn (*Ward and Hamilton,* 2004; *Hamilton and Ward,* 2004), is certainly the relevant angle for consideration of radiative input to the atmosphere of Titan. However, from an orbital and rotational dynamics perspective, the important angle is the much smaller separation between the spin pole of Titan and the pole of its own orbit about Saturn. Likewise for Europa, we are interested in its own dynamical obliquity. As the obliquity of Jupiter is only about 3.1° (*Ward and Canup,* 2006), the solar radiative pattern at Europa is simpler than for Titan.

The obliquity of Europa is not currently known, other than that it is certainly quite small (*Lieske,* 1979). However, when measurement accuracies increase sufficiently to allow a determination of that value, it will provide information about the internal structure of the body. Most of the remainder of this section will attempt to explain that connection.

2.1. Moments and Precession

Measurements of the mass M and mean radius R of a satellite yield a mean density estimate, which for Europa is already rather well known (*Anderson et al.,* 1998a)

$$\langle \rho \rangle = (2989 \pm 46)\ \mathrm{kg\ m^{-3}} \qquad (1)$$

The Galilean satellites show an interesting progression of decreasing density with increasing distance from Jupiter (*Johnson,* 2005), but density only rather weakly constrains internal structure (*Consolmagno and Lewis,* 1978). However, the moments of inertia provide additional constraints on the radial density structure (*Bills and Rubincam,* 1995; *Sotin and Tobie,* 2004). There are several ways to estimate the moments of inertia, and the rotational dynamics provide several options.

Perturbations of spacecraft trajectories, either on captured orbits or during a close flyby, can be used to infer the low-degree terms in the gravitational potential. The coefficients of harmonic degree 2 in the gravitational potential of a body are related to the principal moments of inertia (A < B < C) via (*Soler,* 1984)

$$\begin{aligned} J_2MR^2 &= C - (A + B)/2 \\ C_{2,2}MR^2 &= (B - A)/4 \end{aligned} \qquad (2)$$

There are, in general, five terms of harmonic degree 2, and six independent terms in the inertia tensor. However, if the coordinate axes are chosen to coincide with the principal axes of the inertial ellipsoid, then only these two potential terms remain. Measurements of the gravitational field alone do not suffice to determine the moments of inertia, as the system of equations is underdetermined by 1.

One approach to estimating those moments, in the absence of further constraints, is to assume that the body is in hydrostatic equilibrium, and that the degree-2 harmonics of the gravity field reflect a response to the well-known tidal and rotational potentials. This approach was developed by *Hubbard and Anderson* (1978) and applied to Europa by *Anderson et al.* (1998a) (see chapter by Schubert et al.). That method provides the current best estimates of the moments of inertia of the Galilean satellites (*Anderson et al.,* 1996a,b, 1998a,b). In terms of the dimensionless polar moment

$$c = \frac{C}{MR^2} \qquad (3)$$

the Galilean satellite values are {0.379, 0.346, 0.311, 0.355}, for Io, Europa, Ganymede and Callisto, respectively (*Schubert et al.,* 2004). Recall that a homogeneous sphere has c = 2/5, and smaller values indicate a more centrally condensed structure.

The hydrostatic assumption can be verified if both J_2 and $C_{2,2}$ can be measured independently, since for a hydrostatic body the ratio of these two quantities is 10/3 (e.g., *Murray and Dermott,* 1999). However, determination of J_2 requires polar or near-polar flybys, while $C_{2,2}$ requires equatorial or near-equatorial trajectories, so that it is not always possible (as at Callisto) (*Anderson et al.,* 1998a) to verify the hydrostatic assumption.

Other approaches to determining internal structure rely upon the fact that the rotational dynamics of the body are controlled by the moments of inertia. It is often the case that the applied torques are well known, and that observations of the rotational response thus constrain the moments. For a rapidly rotating body, like Earth or Mars, the solar gravitational torque acting on the oblate figure of the body causes it to precess about its orbit pole. If we ignore effects of an eccentric orbit, the precessional motion of the unit vector $\hat{s}$, aligned with the spin pole, is governed by

$$\frac{d\hat{s}}{dt} = \alpha\,(\hat{n}\cdot\hat{s})\,(\hat{s}\times\hat{n}) \tag{4}$$

where $\hat{n}$ is the orbit pole unit vector, and α is a spin precession rate parameter given by (*Kinoshita,* 1977; *Ward,* 1973)

$$\begin{aligned}\alpha &= \frac{3}{2}\frac{n^2}{\omega}\left(\frac{C-(A+B)/2}{C}\right)\\ &= \frac{3}{2}\frac{n^2}{\omega}\left(\frac{J_2}{c}\right)\end{aligned} \tag{5}$$

with n the orbital mean motion, and ω the spin rate.

If both of the degree-2 gravity coefficients and the spin pole precession rate α can be measured, as has been done for Earth (*Hilton et al.,* 2006) and Mars (*Folkner et al.,* 1997), then the polar moment C can be estimated, without requiring the hydrostatic assumption. That is, in fact, how the moments of inertia of those two bodies were determined. A difficulty with this approach, of directly observing the spin pole precession rate, is that typical rates are very low. For Earth and Mars, the spin pole precession rates are 50 and 10 arcsec/yr, respectively. The challenge of seeing the spin pole of Europa precess, without a relatively long-lived lander, would be formidable. Fortunately, there are better ways to accomplish the same objective.

2.2. Spin Pole Trajectories

We now consider briefly how the spin pole precession trajectory depends on the motion of the orbit pole. In the simplest case, where the orbit plane orientation remains constant, the spin pole trajectory is along a circular cone centered on the orbit pole. In that case, the spin pole maintains a constant obliquity as it precesses. If the orbit pole is itself precessing, as is generally the case, the spin pole trajectory can be quite complex. If the orbit pole is precessing much faster than the spin pole can move, then the spin pole essentially sees a spin-averaged orbit pole, and precesses at nearly constant inclination to the invariable pole, which is the pole about which the orbit is precessing.

The most complex spin pole motion occurs when the orbit pole rates and spin pole rate are comparable. In that case, the motion of the spin pole is resonantly enhanced. These features are extensively discussed in the literature on Mars obliquity variations (*Ward,* 1973, 1992; *Bills,* 1990). The orbital precession amplitudes for Earth and Mars are similar, and the periods are identical, but Mars has obliquity variations that are substantially larger than those for Earth because the spin pole precession rate of Earth is too fast for resonance enhancement, whereas Mars does see resonant effects. In fact, it has been claimed that the obliquity variations for Mars are chaotic (*Touma and Wisdom,* 1993; *Laskar and Robutel,* 1993). However, even relatively small amounts of dissipation will suppress the chaotic variations (*Bills and Comstock,* 2005; *Bills,* 1994, 1999, 2005).

A resonant enhancement of spin pole motion requires orbital precession rates comparable to the spin pole precession rate. However, for most solar system bodies, the difference between polar and equatorial moments is a small fraction of either value, and thus the spin pole precession rates are much slower than the spin or orbital rates. However, there are often orbit-orbit interactions, so-called secular perturbations, that have periods much longer than the orbital periods. It is a near commensurability between the spin pole precession rate of Mars, and some of its secular orbital variations, which give rise to the large obliquity variations.

The situation at Europa is not particularly well approximated by either Earth or Mars. A somewhat more relevant analog is provided by the Moon. The rotational state of the Moon is well approximated by three features, first enunciated by G. D. Cassini in 1693, which can be paraphrased as (1) the spin period and orbit period are identical; (2) the spin axis maintains a constant inclination to the ecliptic pole; and (3) the spin axis, orbit pole, and ecliptic pole remain coplanar.

The first of these had, of course, been known much earlier, and the dynamical importance of the second and third laws was not fully appreciated until much later. It is now understood (*Colombo,* 1966; *Peale,* 1969; *Ward,* 1975a; *Gladman et al.,* 1996) that adjustment of the obliquity to achieve co-precession of the spin and orbit poles about an invariable pole can occur without synchronous locking of the spin and orbit periods. That is, Cassini's first law is at least partially decoupled from the other two. In fact, most features of the lunar spin pole motion are reproduced in a model where the lunar gravity field is approximated as axisymmetric (*Wisdom,* 2006).

The condition for this coplanar precession, in nearly circular orbits, can be written as (*Ward,* 1975a)

$$(v + (u - v)\cos[\varepsilon])\sin[\varepsilon] = \sin(i - \varepsilon) \tag{6}$$

where i is the inclination of the orbit pole to the invariable pole, and ε is the obliquity or separation of spin and orbit poles. The parameters u and v are related to the moments of inertia of the body, and the relative rates of orbital motion and orbital precession.

The first of these parameters has the form

$$u = U\,p \tag{7}$$

where the moment dependent factor is

$$U = \frac{3}{2}\left(\frac{C-A}{C}\right) = \frac{3}{2}\left(\frac{J_2 + 2C_{2,2}}{c}\right) \tag{8}$$

The relative rates of orbital motion and orbit plane precession is

$$p = \frac{n}{d\Omega/dt} \tag{9}$$

where n is the mean motion, and Ω is the longitude of the ascending node of the orbit. For most bodies, the node regresses and this ratio is thus negative. The second parameter has a similar factorization

$$v = V\,p \tag{10}$$

with

$$V = \frac{3}{8}\left(\frac{B-A}{C}\right) = \frac{3}{2}\left(\frac{C_{2,2}}{c}\right) \tag{11}$$

When these substitutions are made, the constraint equation (6) can be written in either of the alternative forms

$$2c\,\sin(i-\varepsilon) = 3p\,(C_{2,2} + (J_2 + C_{2,2})\cos[\varepsilon])\sin[\varepsilon] \tag{12}$$

$$8C\,\sin(i-\varepsilon) = 3p\,(B - A + (4C - B - 3A)\cos[\varepsilon])\sin[\varepsilon] \tag{13}$$

These constraint equations are linear in polar moment, but nonlinear in obliquity. Thus, if both gravitational coefficients and the inclination and obliquity can be measured, we could rather trivially solve for the polar moment as

$$c = \frac{3p}{2}\left(\frac{(C_{2,2} + (J_2 + C_{2,2})\cos[\varepsilon])\sin[\varepsilon]}{\sin[i-\varepsilon]}\right) \tag{14}$$

$$C = \frac{3p}{2}\frac{(B - A - (3A + B)\cos[\varepsilon])\sin[\varepsilon]}{4\sin[i-\varepsilon] - 3p\,\sin[2\varepsilon]} \tag{15}$$

If only one of J_2 and $C_{2,2}$ are known, the hydrostatic assumption can be used (see section 2.1). When solving these constraint equations for obliquity, the situation is somewhat more subtle. In general, there are either two or four distinct real solutions for obliquity, depending upon the values of the input parameters. In all cases, the spin pole $\hat{s}$, orbit pole $\hat{n}$, and invariable pole $\hat{k}$ are coplanar. It is also convenient to define a signed obliquity, with positive values corresponding to $\hat{s}$ and $\hat{n}$ on opposite sides of $\hat{k}$. Following *Peale* (1969), the usual numbering of these separate Cassini states $\{S_1, S_2, S_3, S_4\}$ is that S_1 is $\hat{s}$ near to $\hat{k}$ and on the same side as $\hat{n}$; S_2 is somewhat farther from $\hat{k}$, and on the opposite side from $\hat{n}$; S_3 is retrograde, and thus nearly antiparallel to $\hat{n}$; and S_4 is on the same side of $\hat{k}$ as S_1, but farther from $\hat{n}$ and $\hat{k}$. These spin states represent tangential intersections of a sphere (possible orientations of the spin pole) and a parabolic cylinder representing the Hamiltonian.

If the radius of curvature of the parabola is too large, there are only two possible spin states, otherwise there are four. At the transition point, states 1 and 4 coalesce and vanish. In the axisymmetric case, for which v = 0, the transition occurs at (*Henrard and Murigande,* 1987; *Ward and Hamilton,* 2004)

$$u = -(\sin[i]^{2/3} + \cos[i]^{2/3})^{3/2} \tag{16}$$

and

$$\tan[\varepsilon] = -\tan[i]^{1/3} \tag{17}$$

If the magnitude of the parameter u is larger than the value given by equation (16), then all four Cassini states exist.

All four of the Cassini states represent equilibrium configurations. That is, if the spin pole $\hat{s}$ is placed in such a state, it will precess in such a way as to maintain a fixed orientation relative to $\hat{n}$ and $\hat{k}$. The states S_1, S_2, and S_3 are stable, in the sense that small departures from equilibrium will lead to finite amplitude librations. Each of these states is the dynamical center of a domain of stable librations, and these three domains cover the entire sphere. In contrast, S_4 is unstable. On longer timescales, when tidal effects are included, only states S_1 and S_2 appear as secularly stable (*Peale,* 1974), and *Gladman et al.* (1996) have further argued that whenever $S_1$and S_2 both exist, S_1 will be favored.

Returning briefly to consideration of the Moon, it is the only body in the solar system known to occupy Cassini state S_2. *Ward* (1975) has argued that the Moon initially occupied S_1, but during its orbit evolution outward to the present distance from Earth, the states S_1 and S_4 merged and disappeared, forcing the Moon to transition to state S_2.

2.3. Application to Europa

From the perspective of obliquity dynamics, there are two important ways in which Europa differs from the Moon. The orbit precession for Europa is not steady, because its orbit is significantly perturbed by Io, Ganymede, and Callisto (*Lieske,* 1998; *Lainey et al.,* 2004a,b) and the presumed presence of an icy shell decoupled from the underlying material implies that the moments of inertia of the shell itself need to be considered. The first effect, as discussed below, can be included by considering precession effects on a mode-by-mode basis. The latter effect is considered in some detail in the following discussion on forced librations (section 3).

Obliquity variations for dissipative bodies in nonuniformly precessing orbits can be easily accommodated via a linearized analysis of the torque balance. Similar linear analyses of spin pole precession have been constructed previously, in the context of studying obliquity variations of the Earth (*Miskovitch,* 1931; *Sharaf and Boudnikova,* 1967; *Vernekar,* 1972; *Berger,* 1976), Mars (*Ward,* 1973, 1992), Venus (*Ward and de Campli,* 1979; *Yoder and Ward,* 1979; *Yoder,* 1995, 1997), Mercury (*Bills and Comstock,* 2005), and the Galilean satellites (*Bills,* 2005).

The first step in that process is to represent the unit vectors $\hat{s}$ and $\hat{n}$, which point along the spin pole and orbit pole, in terms of complex scalars S and N, by projecting each of them onto the invariable plane. In the present context, that will be approximated by Jupiter's equator plane. That is, we are ignoring the slow precession of Jupiter's spin pole, since it is much slower than the Galilean satellite precession rates. If we also ignore the variations in satellite orbital eccentricity values, the governing equation for spin pole precession can now be written in the simple linear form

$$\frac{dS}{dt} = -i\alpha\,(N - S) \tag{18}$$

where α is the spin precession rate parameter (equation (5)). If the orbit pole evolution is represented via the series

$$N[t] = \sum_j n_j \exp[i\,(f_j t + \gamma_j)] \tag{19}$$

where f_j is an orbit pole precession rate and γ_j represents a phase offset, then the corresponding solution for the spin pole can be written simply as

$$S[t] = S_{free} + S_{forced} \tag{20}$$

where the free pole motion, which depends only on the initial condition, is

$$S_{free} = S[0]\exp(i\alpha t) \tag{21}$$

and the forced motion is

$$S_{forced} = \sum_j s_j\,[\exp[if_j t] - \exp[i\alpha t]]\exp[i\gamma_j] \tag{22}$$

with amplitudes given by

$$s_j = \left(\frac{\alpha}{\alpha + f_j}\right) n_j \tag{23}$$

Each term in the series describing the orbit pole has a corresponding term in the forced spin pole series. The spin rate parameter α is positive, and all the orbit pole rates f_j are negative. If one of the sums $\alpha + f_j$ is close to zero, then the corresponding amplitude in the spin trajectory will be amplified.

Dissipation can be easily introduced by simply making the spin precession parameter complex: $\alpha \rightarrow \alpha + i\,\beta$. When included this way, the dissipation completely damps the free term and somewhat modifies the forced terms. Assuming that the damping term is small, the resulting model for damped forced spin evolution takes the form

$$S[t] = \sum_j s_j\,(\exp[if_j t])\exp[i\gamma_j] \tag{24}$$

The second of the terms in square brackets in the original equation for forced response is removed by dissipation. To obtain this expression, we allow a finite value of β, take the limit as $t \rightarrow \infty$, and then set β back to zero. It is evident that the orbit pole and spin pole trajectories are characterized by identical frequencies and phases, but different amplitudes.

This solution can be viewed as a rough generalization of the Cassini state for the case of nonuniform orbit precession. In the case of a single orbit precession frequency, the expected end-state for dissipative spin evolution is a special situation in which the obliquity has adjusted to a value at which the system maintains a constant relative geometry. That is, the spin pole and orbit pole remain coplanar with the invariable pole as the spin pole precesses about the orbit pole and the orbit pole precesses about the invariable pole (*Colombo,* 1966; *Peale,* 1969; *Ward,* 1975b; *Henrard and Murigande,* 1987).

If the orbit pole precession is not steady, no such coplanar configuration is attainable. However, the motions of the orbit and spin poles can achieve a mode-by-mode equivalent of the Cassini state. The solution above is such that each mode of the orbit pole precession, with amplitude n_j, rate f_j, and phase γ_j, has a corresponding mode of spin pole precession with rate and phase identical to the orbit mode values, and with an amplitude proportional to the orbit amplitude. The constant of proportionality is just the ratio $\alpha/(\alpha + f_j)$ of the spin precession rate to the relative spin-orbit precession rate.

The angular separation between spin pole and orbit pole has a simple expression

$$\Delta S[t] \equiv S[t] - N[t] = \sum_j \Delta s_j \exp[i(f_j t + \gamma_j)] \tag{25}$$

The amplitude of each term is just the difference in amplitudes of the spin and orbit solutions

$$\Delta s_j = s_j - n_j = \left(\frac{\alpha}{\alpha + f_j} - 1\right) n_j = \left(\frac{-f_j}{\alpha + f_j}\right) n_j \tag{26}$$

The magnitude of this complex quantity is the obliquity. It has the same frequencies as the orbital inclination, but different amplitudes.

Application of this theory to Europa has been made (*Bills,* 2005), using the orbit model of *Lieske* (1998) and moment estimates from *Anderson et al.* (1998a). The predicted spin pole precession rate parameter α and main orbital inclination rates are given in Table 1. The average inclination of Europa's orbit to Jupiter's equator plane is ~0.5° and, based on the measured gravitational moments, the predicted average obliquity is ~0.1°.

If the ice shell is fully decoupled from the deeper interior, then the moments of inertia of the shell will determine

TABLE 1. Precession rates.

Parameter	Rate (10^{-3}°/day)	Period (yr)
α	191	5.16
f_1	–133	7.42
f_2	–32.6	30.2
f_3	–7.18	137
f_4	–1.75	560

the spin pole precession rate. The ice shell thickness estimates from *Ojakangas and Stevenson* (1989a) imply moments of inertia of

$$\begin{bmatrix} A \\ B \\ C \end{bmatrix} = 0.610 M_s R^2 \left(\begin{bmatrix} 1 \\ 1 \\ 1 \end{bmatrix} + \begin{bmatrix} -58.24 \\ +48.76 \\ +9.78 \end{bmatrix} \times 10^{-3} \right) \tag{27}$$

where M_s is the satellite mass, and a corresponding spin pole precession rate parameter of

$$\alpha_s = 0.0218 \text{ n} = 2.21°/\text{day} \tag{28}$$

or roughly 12 times the solid body result. This is much higher than any of the secular orbital rates, but a value in that range could produce a resonant interaction with short-period terms, and thereby produce a larger obliquity.

Two aspects of these moment estimates deserve comment. First is that B > C, contrary to the stated convention (just above equation (2)). This implies an unstable rotational configuration, as is discussed at length by *Ojakangas and Stevenson* (1989b). Second is that the fractional departures from spherical symmetry are substantially larger than the values cited by *Ojakangas and Stevenson* (1989a,b). However, their estimates include isostatic compensation of the ice shell by the underlying fluid layer, which is certainly relevant to the body as a whole, but we are interested in the moments of the shell itself. How appropriate these moment estimates are for Europa's spin state is difficult to determine, but we note that a recent determination of the obliquity of Titan (*Stiles et al.,* 2007), suggests that it is larger than expected for a solid body, and can be interpreted as indicative of a decoupled shell (*Bills and Nimmo,* 2008).

In the absence of direct detection of the obliquity of Europa, we might hope to place constraints on that parameter via its influence on the tidal stress field and resulting tectonic patterns on the surface. Some progress has been made in that direction (e.g., *Hurford et al.,* 2006) (see also chapter by Kattenhorn and Hurford).

2.4. Summary

In simple cases, a satellite's spin precession rate depends on its gravitational moments, its orbital inclination and obliquity, and its polar moment of inertia (equation (14)). If all the other factors are known, the polar moment of inertia can be determined directly. The analysis is similar but more complicated if, as at Europa, dissipation or non-uniform precession occur. Although Europa's current obliquity is not known, it is likely to be ~0.1° unless the ice shell is decoupled from the interior, in which case it is likely to be somewhat larger.

3. LIBRATIONS

We consider the physical librations (i.e., periodic variations in the rotation of Europa) due to the gravitational torque of Jupiter on Europa's nonspherical shape. The other solar system bodies also exert torques on Europa but the Jupiter torque is at least three orders of magnitude larger than the torque by any other body because of Jupiter's mass and distance. Physical librations can be in longitude (in the equatorial plane) as well as in latitude (normal to the equatorial plane). The former corresponds to changes in rotation rate (see section 4 below) and the latter to variations in polar orientation. Among the primary parameters of interest are the deviation from spherical symmetry given by the moments of inertia difference B – A [or $C_{2,2} = (B - A)/4MR^2$] for the longitudinal librations and C – (A + B)/2 ($=MR^2J_2$) for the latitudinal librations. Orbital parameters such as the eccentricity, e, and the obliquity, ε, are also important for the longitudinal and latitudinal librations, respectively. Because of Europa's likely small obliquity (see section 2 above), the latitudinal librations are at least one order of magnitude smaller (*Henrard,* 2005). Secular changes in latitude (true polar wander) may arise under some circumstances, and have important geological effects (*Ojakangas and Stevenson,* 1989b) (see also section 4), but here we consider only the longitudinal libration (*Comstock and Bills,* 2003).

3.1. Rigid Librations

In the absence of a subsurface ocean, Europa's response can be approximated as that of a rigid triaxial body deformed from sphericity by centrifugal forces as well as the permanent tides of Jupiter. In this case, the spin angular momentum H of Europa can be given as

$$H = C\frac{d\theta}{dt} \tag{29}$$

where θ is the angle of rotation of Europa (see Fig. 1). By neglecting tidal deformations over the period of interest (i.e., the orbital period), the change in rotation rate is proportional to the gravitational torque L of Jupiter on Europa.

$$\frac{dH}{dt} = C\frac{d^2\theta}{dt^2} = L \tag{30}$$

Europa, assumed to be in synchronous rotation, does not have a uniform orbital motion due to its noncircular orbit

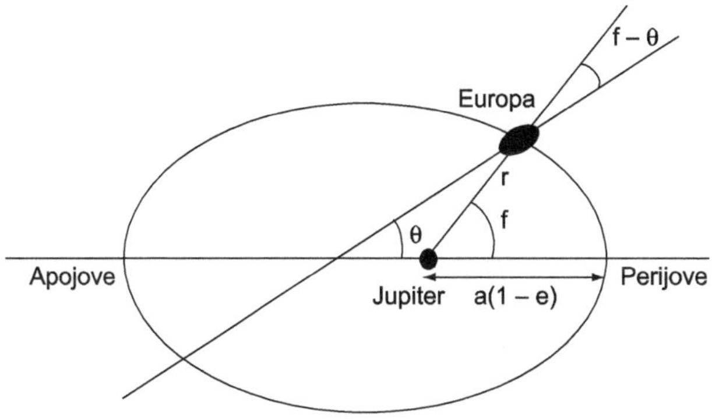

Fig. 1. Geometry of Europa's spin-orbit motion.

(e = 0.0094). Europa's long axis hence is not always directed exactly toward Jupiter. The gravitational torque of Jupiter on Europa's nonspherical figure tends to modify the satellite's rotation, resulting in the longitudinal librations. This torque depends on the differences of the equatorial moments of inertia B – A (e.g., *Goldreich and Peale,* 1966)

$$L = -\frac{3}{2}n^2(B - A)\left(\frac{\alpha}{r}\right)^3 \sin[2(\theta - f)] \qquad (31)$$

where n is the mean motion of Europa, a the semimajor axis, f the true anomaly, and r the distance between the mass centers of Jupiter and Europa (Fig. 1).

The governing equation describing the rotation of Europa becomes

$$C\frac{d^2\theta}{dt^2} + \frac{3}{2}n^2(B - A)\left(\frac{\alpha}{r}\right)^3 \sin[2(\theta - f)] = 0 \qquad (32)$$

Equation (32) is a differential equation of the second order in time t analogous to the classical pendulum equation with the exception that the second term on the lefthand side contains time-dependent variables r and f due to the eccentric orbit. Europa can librate, circulate, or tumble chaotically depending on the energy of the system, or equivalently on the initial conditions (θ_0, $\dot{\theta}_0$). The qualitative behavior of the librational motion of Europa can be best described by using the Poincaré surface of sections (*Poincaré,* 1892). Figure 2 shows a surface of section plot resulting from the numerical integration of equation (32) where θ and $\dot{\theta}$/n are plotted at each periapse passage. The global behavior of the rotation of Europa is regular and the chaotic zones are not visible at this scale. The central area of the graph with closed curves represents the libration zone where the mean rotational velocity is equal to n and Europa is in spin-orbit resonance. Outside this region, Europa is out of resonance, the system has more energy than $3n^2(B - A)/4$, and θ is unbounded. The width of the libration area is equal to $\dot{\theta}$/n = 0.134 and represents twice the maximum increment in the instantaneous velocity of Europa for which Europa remains in spin-orbit resonance. The separatrix, which separates librational from circulatory behavior, is shown with a bold line. Figure 2 shows three equilibrium points. The points with coordinates (+π/2,1) and (–π/2,1) are unstable hyperbolic points, whereas the point at the center (0,1) corresponds to an elliptical point that is stable. In the following, we describe the librational motion close to the elliptical point where the amplitude of libration is small. In this case, the time taken for θ to move along one librational curve is equal to $2\pi/\omega_f$ where $\omega_f = n\sqrt{3(B - A)/C}$ is the free libration frequency, which is determined by solving equation (32) using the average value of the second term.

For small angular difference θ – f, equation (32) can be solved by expressing the time-dependent parameters (f, r) in terms of Fourier series in eccentricity e and mean anomaly M (see *Murray and Dermott,* 1999). The main periodic forcing term has a period equal to the orbital period (3.55 days). The first order solution in e yields the libration amplitude as

$$A_\theta = -\frac{2\omega_f^2 e}{n^2 - \omega_f^2 e} \approx 6\frac{(B - A)}{C}e \qquad (33)$$

Since the moment of inertia ratio (B – A)/C is usually much smaller than unity, the free libration frequency ω_f compared to the orbital frequency n is small ($\omega_f \approx 0.067n$).

The above equation shows that the amplitude of libration depends linearly on the difference of equatorial moments of inertia B – A and the eccentricity e and is inversely proportional to the polar moment of inertia C. Values for the moment of inertia ratio (B – A)/C = 0.0015 and the polar moment of inertia factor $C/MR^2 = 0.3479$ can be deduced by using the measured $C_{2,2} = 131.5 \pm 2.5\ 10^{-6}$ from the Galileo flyby (*Anderson et al.,* 1998b) and by assuming that Europa is in hydrostatic equilibrium (*Hubbard and Anderson,* 1978). The resulting amplitude of forced librations is then 8.52×10^{-5} rad, which corresponds to an equatorial surface displacement of ±133 m over an orbital period.

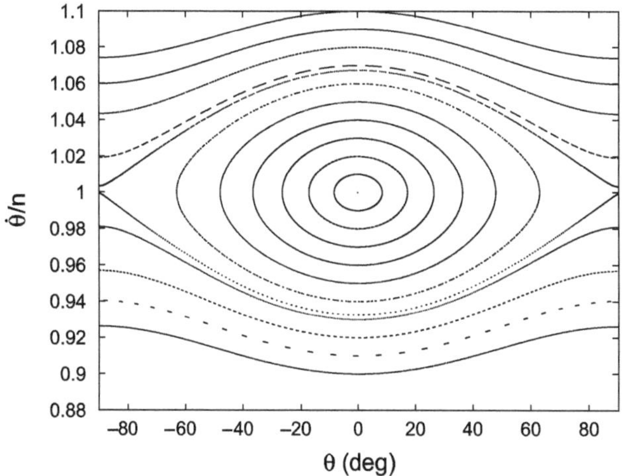

Fig. 2. Phase diagram for spin-orbit dynamics of a rigid Europa.

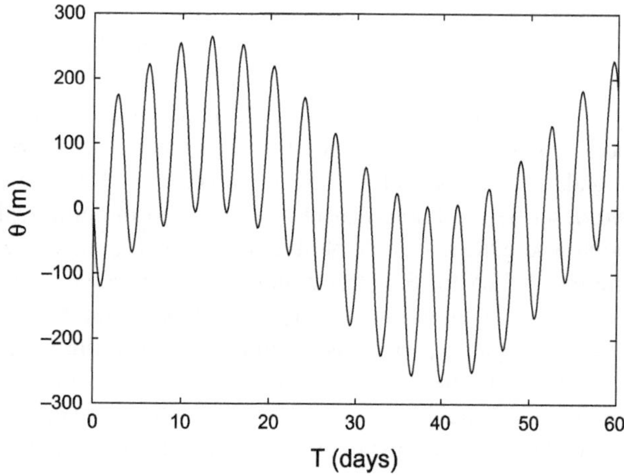

Fig. 3. Librational response of a rigidly rotatating Europa, with both free and forced components.

Figure 3 illustrates this forced motion and also shows the presence of a long-period oscillation that corresponds to the free libration frequency of the dynamical system. The librational response can thus be decomposed into an angular motion at the orbital period, i.e., forced libration, and a rotation rate variation at a longer period, i.e, free libration. It is a characteristic of the free libration period that the associated amplitude is arbitrary. However, it is expected to be close to zero since tidal dissipation within the satellite damps the free librations (as with the free obliquity term in the preceding section). The free libration frequency depends on the internal structure, whereas the period of the forced libration is determined by the period of the Jupiter torque, i.e., 3.55 days. Note that for a sphere or an oblate spheroid with $B - A = 0$, the free libration period goes to infinity. With the estimated value of $(B - A)/C = 0.015$, the free libration period of Europa is 52.7 days.

Equation (33) shows that if the libration amplitude can be measured along with $C_{2,2}$, then the polar moment of inertia can be determined directly. Investigations of Mercury using radar speckle interferometry have detected a forced libration component (*Margot et al.,* 2007). The amplitude of that libration is indicative of decoupling of the core from the mantle (*Peale,* 1976; *Peale et al.,* 2002; *Rambaux et al.,* 2007). Future spacecraft missions to Europa will certainly be able to detect radial tidal motions (*Wahr et al.,* 2006), and may also resolve libration effects.

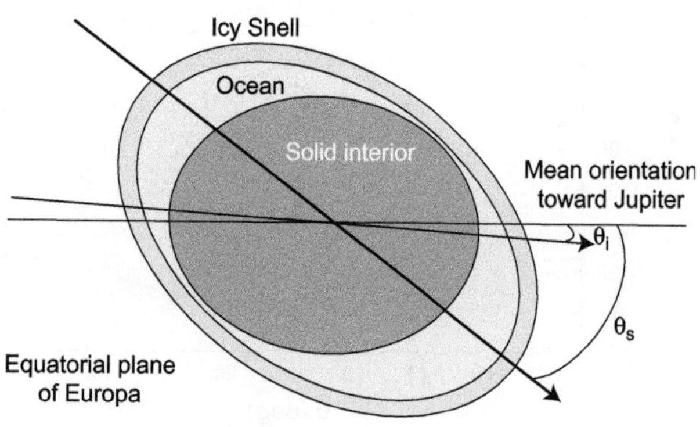

Fig. 4. Schematic geometry of differential librations of Europa.

3.2. Differential Librations

If an ocean exists under the icy shell, Europa can no longer be assumed to perform rigid rotations. The interior [likely composed of a metallic core and a silicate mantle (*Anderson et al.,* 1998b)] and the icy shell perform differential librations under the gravitational torque of Jupiter, with amplitudes proportional to their respective equatorial moment of inertia differences. If their motions are decoupled, the libration amplitudes would be inversely proportional to their respective polar moments of inertia (C_i and C_s where the subscripts i and s stand for the interior and the shell, respectively). The amplitude of the libration on Europa's surface would then be amplified by a factor C/C_s in comparison with the rigid librations, provided that the equatorial moment of inertia difference $(B - A)$ of the shell and rigid satellite are similar. Depending on the thickness of the icy shell ($1 \text{ km} < h < 100 \text{ km}$), C/C_s varies approximately between 10 and 1000, suggesting periodic equatorial displacements up to hundreds of kilometers for a decoupled thin icy shell. Such displacements would generate large surface stresses, likely visible as tectonic features, as well as significant dissipation within the icy shell.

In reality, it is more likely that the motions of the icy shell and the interior are coupled, through either the gravitational force or through pressure components in the ocean caused by the rotational potential [so-called Poincaré torques; see *Ojakangas and Stevenson* (1989b)]. For longitudinal librations, only the former are likely to be important, and we focus on them here.

An internal gravitational torque between the solid interior and the outer icy shell arises when the principal axes of the icy shell and the interior are not aligned (see Fig. 4). The gravitational coupling has been studied in the context of the rotation variations of the Earth, in addition to other couplings due to pressure on topography, and viscous and electromagnetic torques (see, e.g., *Buffett,* 1996; *Szeto and Xu,* 1997). For terrestrial planets, the principal gravitational coupling occurs between the large mantle and the small solid inner core. In the case of Europa, the massive interior with a much larger moment of inertia than the thin outer icy shell has a very significant effect on the rotation of the surface layer. Taking into account gravitation as the principal internal coupling, the libration of Europa can be expressed in terms of two angular momentum equations for the two solid layers, the icy shell and the interior (*Van Hoolst et al.,* 2008)

$$C_s \frac{d^2\theta_s}{dt^2} + \frac{3}{2}(B_s - A_s)\, n^2 \left(\frac{\alpha}{r}\right)^3 \sin[2(\theta_s - f)] = -K \sin[2(\theta_s - \theta_i)] \tag{34}$$

$$C_i \frac{d^2\theta_i}{dt^2} + \frac{3}{2}(B_i - A_i)\, n^2 \left(\frac{\alpha}{r}\right)^3 \sin[2(\theta_i - f)] = +K \sin[2(\theta_s - \theta_i)] \quad (35)$$

These equations are coupled by the gravitational coupling parameter

$$K = \frac{4\pi G}{5} \frac{8\pi}{15} \rho_s \beta_s ((\rho_m - \rho_s)\beta_m r_m^5 + (\rho_c - \rho_m)\beta_c r_c^5) \quad (36)$$

where β is the equatorial flattening; ρ is the density; r is the radius; and the subscripts s, m, and c stand for the icy shell, silicate mantle, and iron core, respectively. Here, the internal ocean is assumed to have the same density as the icy shell. In the above equations the dynamics of the ocean are neglected; in particular, the effect of pressure on the ellipsoidal boundaries is not included (cf. *Ojakangas and Stevenson,* 1989b).

The polar moments of inertia and the equatorial flattening of Europa's inner layers can be calculated from Clairaut's theory by taking into account both rotational and tidal deformations (e.g., *Murray and Dermott,* 1999). Note that gravitational coupling would be small if Europa were not completely differentiated ($\rho_s \approx \rho_m \approx \rho_c$), or if internal flattening is negligible ($\beta_m \approx \beta_c \approx 0$).

The internal gravitational coupling strongly reduces the libration amplitude with respect to decoupled shells to maximum surface displacement of about 140–150 m (*Van Hoolst et al.,* 2008). The libration amplitude increases linearly proportional to the ice shell thickness for h > 10 km. This is not surprising considering the linear relationship between C_s and h. On the other hand, the equatorial displacement of the interior remains constant at about 140 m, independent of the thickness of the icy shell, because the influence of the thin outer layer on the massive interior is small. A particular consequence of the inclusion of the gravitational coupling between the shell and the interior is that the libration amplitude is resonantly amplified at h ≈ 1 km. The resonance occurs when one of the two free libration periods of the angular momentum equations (34) and (35) approaches the orbital period.

3.3. Discussion

The libration amplitude of Europa depends on the presence of a global ocean between the icy shell and the silicate mantle. In its absence, Europa performs rigid librations with an amplitude of ≈133 m. In the presence of a subsurface ocean, the icy shell and the interior (likely composed of a silicate mantle and metallic core) perform differential librations. If their motions are decoupled, the amplitude of the libration would be amplified by a factor of 10 to 1000 with respect to the rigid libration, resulting in periodic equatorial surface displacements up to hundreds of kilometers. However, the rotation of the icy shell is likely coupled with that of the interior. The gravitational coupling between the massive interior and the thin outer shell does not allow the surface to librate with an amplitude larger than 140–150 m for reasonable ice shell thicknesses (several tens of kilometers). This is only ≈10% larger than the rigid body libration amplitude.

Because the libration amplitude is a function of the internal structure, its observations can yield information about the interior of Europa. If Europa reacts as a rigid body to the gravitational torque of Jupiter, the measurements provide information principally on the moment of inertia ratio (B – A)/C of the satellite. Europa's surface performs larger librations in the presence of a subsurface ocean. If the gravitational torque between the shell and the interior is the dominant coupling mechanism, the existence of a subsurface ocean can be determined if a libration measurement precision on the order of several meters can be achieved. With a better accuracy, information on the thickness of the icy shell can be obtained. As the position of a spacecraft in orbit will be perturbed mostly by the librations of the solid interior (*Wu et al.,* 2000), simultaneous observations of the librations of both the interior and the surface could enhance the knowledge of Europa's internal structure, in particular the density and radius of the solid interior.

4. NONSYNCHRONOUS ROTATION

We now consider the mean rotation rate of Europa. It is clear that Europa is very nearly in a synchronous state, similar to that of the Moon, in which the rotation period and orbit period are exactly equal. However, both theoretical and observational arguments have been presented for Europa deviating slightly from a state of exact synchronous rotation. This question is important because nonsynchronous rotation (NSR) results in tectonic stresses that can be up to 2 orders of magnitude larger than the diurnal tidal stresses thought to be responsible for many of the surface tectonic features (*Greenberg et al.,* 1998) (see also chapters by Sotin et al. and Kattenhorn and Hurford). However, we will differ from most previous works and argue below that NSR is unlikely on dynamical grounds, unless lateral variations in icy shell thickness occur, and that a small but finite obliquity might instead be responsible for the geological observations.

4.1. Geological Evidence

Direct determination of the rotation rate of Europa's ice shell should be possible by comparing the position of various surface features as imaged at different times, but up to now, no image has combined a sufficient resolution and/or a precise location to be used in such a way. *Hoppa et al.* (1999a) identified a pair of images each taken at the same resolution (1.6 km/pixel) by the Voyager 2 and Galileo spacecrafts 17 years apart, which show the same region relative to the terminator. They found that the longitudinal variation of the terminator is within the error bar (estimated to be ~0.5°), suggesting that the period of rotation relative to the direction of Jupiter must be greater than 12,000 years, with no compelling evidence of NSR.

Observations and interpretations of tectonic features in the icy shell have furnished other lines of evidence for a NSR. Some classes of tectonic features on Europa are thought to be the result of diurnal tidal stresses of around 100-kPa amplitude caused by Europa's eccentric orbit (e.g., *Hoppa et al.,* 1999c). The evolution and propagation of some tectonic features is thought to follow the ever-changing stress field, in both amplitude and orientation. The addition of a long-period component associated with a small supersynchronous rotation of the shell modifies the amplitude of the tidal stress and induces an increasingly clockwise reorientation of tectonic features as the surface reorients relative to fixed global patterns of tidal stress. The first data-based argument for NSR was the observation of systematic longitudinal and azimuthal shifts in the crack locations from those predicted by an infinitesimal amount of NSR (e.g., *McEwen,* 1986). Three types of tectonic features are generally interpreted to constrain the rotation rate (*Greenberg et al.,* 2002):

1. Orientations of tectonic lineaments: The cross-cutting sequences of global and regional lineaments exhibit a rotation of azimuthal orientation over time, which is consistent with predictions based on tidal stress calculations (e.g., *Geissler et al.,* 1998). More detailed analyses have suggested that correlations are better if both diurnal and nonsynchronous tidal stresses are taken into account (*Greenberg et al.,* 1998). The observation that lineaments in the southern hemisphere show similar cross-cutting sequences but an opposite sense of azimuth rotation to those in the northern hemisphere is also consistent with the theory. However, it has been shown that NSR in the opposite sense to that predicted provides an equally good fit to the observations (*Sarid et al.,* 2004). This work, and subsequent work by the same authors (e.g., *Sarid et al.,* 2006), suggest that the observational evidence for NSR from lineament orientations is weak, at best. These investigations do not provide constraints on Europa's rotation rate. One reason is that several sets of older lineaments do not exhibit a monotonic change in azimuthal orientation, suggesting that they may have formed over several rotational periods and indicating that there are various scales over which fracture interpretation can be carried out (e.g., *Kattenhorn,* 2002). In this context, it was concluded that the rotation period must be significantly lower than the age of the surface (e.g., *Greenberg et al.,* 2002), estimated to be ~10^8 yr (*Zahnle et al.,* 2003).

2. Strike-slip faults: Strike-slip displacements, in which one part of the surface has sheared past another along a fault, are common at Europa's surface and have been explained as a result of "tidal walking" driven by diurnal tides (e.g., *Hoppa et al.,* 1999b). According to the theory, the sense of the shear depends on location and crack azimuth but the fit to tidal stress patterns is better if one allows an additional slow NSR of several tens of degrees (*Hoppa et al.,* 1999b).

3. Cycloidal cracks: Many of the lineaments on Europa appear in the form of arcuate cracks called cycloids. The formation and propagation of such features have been explained by diurnally varying tidal stresses (*Hoppa et al.,* 1999c). Again, accounting for the additional tidal stress due to NSR allows a better fit between observed and modeled cycloid orientations (*Hoppa et al.,* 1999c; *Hurford et al.,* 2007). Additional constraints on Europa's rotation rate have been investigated by *Hoppa et al.* (2001). They analyzed a set of cycloidal ridges having cross-cutting relationships and located in the Astypalaea region. *Hoppa et al.* (2001) concluded that the cycloids probably formed at different longitudes from their current location and during different NSR cycles, lending credence to the idea that only a few such ridges form over a single NSR period. If only a few such ridges form over one such period, determining the amount or timing of reorientation from lineament orientations is very challenging (see above).

Another observational argument suggesting that Europa's rotation is nonsynchronous is implicit in the work of *Shoemaker and Wolke* (1982) and *Passey and Shoemaker* (1982). These authors predicted that leading-trailing hemisphere crater densities would be found on synchronously rotating satellites. Galileo images show a weak apexantapex crater density variation on Ganymede (*Zahnle et al.,* 2001; *Schenk et al.,* 2004), suggesting that NSR has smeared out much of the expected signal. Europa's surface is much less heavily cratered than that of Ganymede, but the global distribution of 1-km craters on Europa shows no sign of a statistically significant asymmetry (*Schenk et al.,* 2004). This result suggests a NSR period much less than the surface's age.

In the next section, we discuss the dynamical arguments concerning the possibility of a NSR of Europa.

4.2. Dynamical Considerations

As for many satellites close to their primary, tidal torques tend to slow down the rotation of an initially fast rotating satellite. Using the reasonable value Q ~ 100 for the tidal quality factor leads to a rough estimate of Europa's despinning timescale of about 10^5 yr (e.g., *Goldreich and Peale,* 1966; *Murray and Dermott,* 1999), suggesting that Europa has currently reached its equilibrium mean rotation rate. The corresponding timescale for the eccentricity damping (~10^8 yr) is also lower than the age of the system, but the Laplace resonance between Europa, Io, and Ganymede maintains a nonzero orbital eccentricity e close to 0.01 (e.g., *Peale and Lee,* 2002).

As a consequence of its nonzero eccentricity, Europa will experience periodically reversing torques due to the gravitational influence of Jupiter if its figure (as measured by B – A) contains permanent asymmetries. These torques give rise to forced librations (see section 3), and the averaged torque favors stable librations about spin rates of p × n where p is an integer or an half-integer corresponding to spin-orbit resonances. However, if tidal dissipation occurs in the satellite, there are additional torques that arise due to its tidally deformed figure. These tidal torques will tend to drive the satellite to a rotation state slightly faster than synchronous (*Greenberg and Weidenschilling,* 1984). It is the balance between the tidal torques and the permanent torques that determines the rotational state of the satellite. Since the orbital eccentricity of Europa is small, only the possibility of capture into a synchronous state is discussed here (*Goldreich and Peale,* 1966; *Correia and Laskar,* 2004). In all the following discussion, we also suppose that Europa's obliquity is zero (although see section 2 above). We will consider additional effects that can arise when Europa's shell thickness varies laterally at the end of this section.

Around the resonance, it is convenient to introduce the angular variable γ such as $d\gamma/d_t = \omega - n$, where ω is Europa's rotational velocity and n is the orbital mean motion. The time-averaged evolution of the satellite's rotation over an orbital period is then given by (*Goldreich and Peale,* 1966)

$$C\frac{d^2\gamma}{dt^2} = -\frac{3}{2}H[1,e](B - A)\,n^2 \sin[2\gamma] + \Gamma_{tidal} \quad (37)$$

where Γ_{tidal} is the mean tidal torque acting to brake the spin of the satellite, H[1,e] is the Hansen coefficient for the synchronous resonance, and $A \leq B \leq C$ are the satellite's principal moments of inertia. For very small eccentricity, it is reasonable to consider that $H(1,e) \simeq 1 - 5e^2/2 \simeq 1$. The synchronous resonance has a finite angular frequency of libration and the width of the resonance is

$$\Delta\omega = n\sqrt{3\frac{B - A}{C}H[1,e]} \quad (38)$$

The oblateness parameter $\beta = (B - A)/C$ is related to the second-degree gravitational coefficient $C_{2,2}$ by

$$\beta = \left(\frac{B - A}{4MR^2}\right)\left(\frac{4MR^2}{C}\right) = C_{2,2}\left(\frac{4MR^2}{C}\right) \quad (39)$$

Using $C_{2,2} = (131.5 \pm 2.5)\,10^{-6}$ and $C/MR^2 = 0.346 \pm 0.005$, which are values determined from Doppler measurements of the Galileo spacecraft (*Anderson et al.,* 1998b), we have $\beta = (B - A)/C \simeq 1.52 \times 10^{-3}$ so that $\Delta\omega \simeq 0.0675$ n. As a comparison, $\beta \simeq 10^{-4}$ for Mercury and the Moon.

To determine whether Europa has some chance of being trapped into this resonance requires an estimate of the magnitude of the tidal torque Γ_{tidal}. Synchronous rotation will be achieved only if the maximum restoring torque exceeds the net tidal torque at the synchronous rate. If the tidal torque yields a stable equilibrium rotational state ω_{eq}, in which the tidal torque vanishes, it is easy to show that an equivalent condition for the capture into synchronous resonance (i.e., $\omega = n$) is that the equilibrium rotation rate is located within the resonance, that is

$$|\omega_{eq} - n| < \Delta\omega \quad (40)$$

Conversely, if the latter condition is not satisfied, the final rotation state is given by ω_{eq}.

Mechanisms of tidal dissipation are poorly constrained in the solar system and current models of tides result more from mathematical simplifications than reliable physical arguments (see *Greenberg and Weidenschilling,* 1984). The simplest model of tidal response is generally called the "viscous" model as described in *Mignard* (1980). He assumed a constant time lag for any frequency component of the tidal perturbation. In other words, the tidally deformed surface of the satellite always assumes the equipotential surface it would have formed a constant time lag Δt ago, in the absence of dissipation. In this case, the ratio 1/Q is proportional to the frequency of the tides. The expression for the average tidal torque can be thus written as (*Goldreich and Peale,* 1966; *Correia and Laskar,* 2004)

$$\Gamma_{tidal} = -2\frac{K}{n}\left(\left(\frac{1 + 3e^2 + 3e^4/8}{(1 - e^2)^{9/2}}\right)\frac{\omega}{n} - \left(\frac{1 + 15e^2/2 + 45e^4/8}{(1 - e^2)^6}\right)\right) \quad (41)$$

with

$$K = \frac{3}{2}\frac{k_2}{Q}\left(\frac{GM_s^2}{R_s}\right)\left(\frac{M_J}{M_s}\right)^2\left(\frac{R_s}{a}\right)^6 n \quad (42)$$

where k_2 is Europa's Love number of degree 2, R_s is Europa's radius, M_J and M_s the respective masses of Jupiter and Europa, and a is the orbit semimajor axis. The equilibrium rotational state in which the tidal torque vanishes is then given by

$$\omega_{eq} = n\left(\frac{1 + 15e^2/2 + 45e^4/8}{(1 - e^2)^{3/2}(1 + 3e^2 + 3e^4/8)}\right) \quad (43)$$

which can be reasonably approximated by $\omega_{eq} \simeq n(1 + 6e^2)$ for $e \leq 0.4$. Although the equilibrium spin rate is slightly faster than an exactly synchronous rate, this particular model for Q does not support the rotational properties inferred from the interpretation of tectonic features:

1. The small but nonzero value of Europa's eccentricity provides a pseudo-synchronous rotation rate of $\omega_{eq} \simeq 1.0006$ n, implying that the subjovian point would travel around Europa's equator with a synodic period of approximately 15 years. This clearly disagrees with the minimum ~10^4-yr synodic period estimated by *Hoppa et al.* (1999a).

2. The deviation from the synchronous state (~6 e^2n ~ 0.0006 n) is lower than the resonance width (0.0675 n), indicating that the tidal torque is not large enough to offset the rigid torque due to the permanent asymmetry of Europa's mass distribution, as described by $C_{2,2}$. Equations (38) and (40) shows that the "viscous" model would require an eccentricity larger than

$$e_c = \left(\frac{1}{12}\left(\frac{B - A}{C}\right)\right)^{1/4} \simeq 0.1 \quad (44)$$

to escape from synchroneity, which is not possible in Europa's current orbital configuration.

Another tidal model that is commonly used is the constant-Q model where the tidal quality factor Q is assumed to be independent of frequency (cf. *Greenberg and Weidenschilling,* 1984). This model is not appropriate close to spin-orbit resonances because it gives rise to nonphysical discontinuities in the expression of the tidal torque (*Kaula,* 1964). Furthermore, for any eccentricity lower than 0.235, the equilibrium spin rate is the synchronous state (e.g., *Kaula,* 1964; *Goldreich and Peale,* 1966) so that tidal dissipation will ultimately drive Europa to this state independent of the asymmetry of its mass distribution.

More realistic assumptions can be made concerning the deformation of Europa's interior due to the tidal perturbing potential. A common assumption for planetary or satellite interiors is that they behave like viscoelastic bodies, having a Maxwell rheology. This implies that elastic and viscous effects are combined in series such that a body's response is that of a purely elastic material at short periods but that of a Newtonian viscous material at long periods. A key parameter of the viscoelastic response is the Maxwell relaxation time τ, which is defined as the ratio between the effective Newtonian viscosity η and the elastic shear modulus μ of the material: $\tau = \eta/\mu$. Here we examine configurations in which the net average tidal torque could vanish assuming the simplest hypotheses for Europa's internal structure. Extended work and discussions are in progress on that topic (see *Levrard et al.,* 2007; *Levrard,* 2008; *Wisdom,* 2008). If an internal ocean decouples the outer icy shell from an inner mantle, the icy shell is expected to follow the fluid deformation of the ocean so that its deformation is significantly greater than if an ocean were not present (*Moore and Schubert,* 2000). In that case, the tidal response of Europa's surface depends mainly on the properties of the ice shell. Rheological and thermal properties of ice at low stress levels are not well constrained and we adopt the values $\mu = 3 \times 10^9$ Pa and $\eta = 10^{14}$ Pa s as is currently assumed for internal structure models of Europa and Titan (e.g., *Tobie et al.,* 2005). These parameter choices correspond to a Maxwell time $\tau \sim 9$ h, about 1 order of magnitude lower than Europa's orbital period (~3.55 days); higher viscosities or lower rigidities would result in Maxwell times comparable to the orbital period. For τ much less than the orbital period, Europa would behave almost as a viscous fluid. In this case, the viscoelastic model matches the "viscous" model previously described (e.g., *Darwin,* 1908), leading to the same conclusions. The tidal torque would vanish at a slightly supersynchronous rotation rate, but Jupiter's torque on the permanent departure from spherical symmetry (presumably due to irregularities in the silicate interior) is large enough to maintain synchronicity.

Alternatively, if an internal ocean is not present, Europa's tidal deformation is expected to follow the contribution of the deep rocky interior. Europa is most likely differentiated into a metallic core surrounded by a rock mantle (e.g., *Schubert et al.,* 2004). We consider a silicate mantle with physical properties ($\mu = 7 \times 10^{10}$ Pa and $\eta = 10^{20}$ Pa s) close to Earth's values (e.g., *Tobie et al.,* 2005) such that the Maxwell relaxation time is $\sim 10^4$ yr, much longer than the orbital period. The tidal torque is expected to slow down Europa's rotation when its rotational velocity ω is slightly larger than n ($\Gamma_{tidal} \rightarrow 0^-$) and speed up Europa's rotation when ω is slightly smaller than n ($\Gamma_{tidal} \rightarrow 0^+$). Conversely, at the approach of the resonance, the tidal frequency approaches zero when $\omega \rightarrow n$, corresponding to very long forcing periods for which the mantle's response is close to that of a perfect fluid. In that case, no tidal dissipation occurs and the average tidal torque also tends to vanish. Such an analysis suggests that the sign of the tidal torque must change at the synchronous rotation and that this point corresponds to a stable equilibrium state for the viscoelastic tidal torque.

All the above models of tidal dissipation operate such that they will drive a body like Europa into the synchronous resonance, either because the gravitational torque on Europa's mass asymmetry distribution is large enough to overcome the tidal torque, or because the tidal endpoint is exact synchronism.

However, there is at least one possible mechanism by which NSR can still occur. It was pointed out by *Ojakangas and Stevenson* (1989a) that if the floating icy shell is conductive, lateral variations in shell thickness, and thus differences in A, B, and C for the shell, will arise. If $B \neq C$ then the gravitational influence of Jupiter acting on the shell will rotate the shell slightly until this torque is balanced by the tidal torque. However, in this configuration the shell will no longer be in thermal equilibrium, because tidal heating is symmetrical about the subjovian point. The icy shell thickness distribution will thus change with time, leading to further rotation, and so on. In this manner, a steady-state NSR of the shell may still arise.

Of course, it is not currently clear whether the required lateral shell thickness variations exist. Various surface features on Europa have been attributed to convection in the ice shell (see chapter by Barr and Showman), in which case the shell thickness is certainly uniform. Even in the absence of convection, lateral thickness variations can be smoothed out over timescales depending primarily on the background shell thickness (e.g., *Nimmo,* 2004). Existing topographic measurements are insufficient to determine whether or not shell thickness variations exist (*Nimmo et al.,* 2007). Nonetheless, the mechanism proposed by *Ojakangas and Stevenson* (1989a) remains a viable way of generating NSR for a floating, conductive icy shell.

4.3. Secular Polar Motion

Nonsynchronous rotation, as proposed by *Greenberg and Weidenschilling* (1984), and further elaborated by *Ojakangas and Stevenson* (1989a), is rather similar in many regards to secular polar motion or true polar wander (TPW). The slightly peculiar phrase "true polar wander" indicates that the geographic location of the rotation axis is changing, and was needed in the terrestrial literature to distinguish from "apparent polar wander," which is simply due to changes in the orientation of the magnetic field.

In both NSR and TPW the concern is not with the rotational dynamics of a rigid body, but rather the slow deformation and reorientation of the material within the body, while the departure from spherical symmetry of the mass distribution remains essentially constant. True polar wander is driven by the application of internal or surface loads (e.g., shell thickness variations, impact basins, convective diapirs) and resisted by the elastic portion of the tidal and rotational bulges. The most important effect of TPW is that, as with NSR, it generates large (up to several MPa) stresses and can thus give rise to observable global tectonic patterns (e.g., *Leith and McKinnon,* 1996).

The circumstances under which TPW may occur on satellites, and the resulting stresses, have been recently reviewed by *Matsuyama and Nimmo* (2007, 2008). If Europa's icy shell does show lateral thickness variations, it may be rotationally unstable and undergo large TPW events, unless stabilized by dissipation within the ice shell (*Ojakangas and Stevenson,* 1989b).

Leith and McKinnon (1996) found only minor evidence for TPW from a global survey of lineations as seen by Voyager. More recently, *Sarid et al.* (2002) argued that the style of strike-slip offsets could be consistent with a TPW of 30°, while *Schenk et al.* (2008) suggested that two large, antipodal cir-cular features were caused by roughly 90° of TPW. There is currently no consensus on whether significant TPW has occurred, but it is a dynamically plausible mechanism that needs to be taken into account when considering how global lineament patterns may have formed.

If Europa does exhibit TPW, as has been recently claimed (*Schenk et al.,* 2008), then it would seem likely that it could also exhibit NSR. In both cases, the critical issue is whether the rheology of the icy shell allows it to maintain long-term elastic strength. If the deformation required for NSR or TPW were to occur on Europa, but were mainly accommodated by viscous flow, rather than brittle failure, it might have only subtle expression in surface features (*Harada and Kurita,* 2006, 2007).

4.4. Discussion

The rotational motion of a solid shell decoupled from a liquid one has been widely investigated to model the dynamics of planets having a mantle overlying a liquid core, like Mercury or Earth (e.g., *Goldreich and Peale,* 1967; *Correia,* 2006), and most of the corresponding results can be used to provide some insights about the rotation of the external icy shell of Europa. In particular, because of their different shape and densities, a water ocean and the icy shell do not have the same permanent departure from the spherical symmetry and the two parts tend to precess and librate at different rates.

This tendency is probably counteracted by different interactions acting at the interface: the torque of nonradial inertial pressure forces of the solid shell over the ocean provoked by the nonspherical shape of the interface and the torque of the viscous (or turbulent) friction between the solid and liquid layers (e.g., *Ojakangas and Stevenson,* 1989b). Note that in the specific case of Europa, significant variations or irregularities of the shell thickness would add a "topographic" torque, but its contribution is poorly constrained.

Assuming that the effective viscous torque is proportional to the differential rotation between the core and the mantle, *Correia* (2006) provided complete analytical solutions of the average motion of the core and mantle spins. He found that the general formulation of the mantle's spin evolution noticeably depends on the efficiency of the viscous coupling. If the coupling between the two layers is strong (compared to the amplitude of the gravitational torque), the entire planet (or satellite) is expected to participate in the librations so that the moments of inertia used in equation (37) are still appropriate and the width of the resonance is not affected. If the coupling is weak, *Goldreich and Peale* (1967) and *Correia* (2006) pointed out that near a spin-orbit resonance, the liquid layer will not respond instantaneously to the periodic librations of the solid shell, resulting in an additional source of dissipation. As a consequence, this mechanism greatly increases the chances of capturing a planet or satellite into resonance. One reason is that the appropriate moment of inertia in equation (37) would be now that of the icy shell alone C_s, leading to an increase of the width of the resonance by a factor $\sqrt{C/C_s}$ as indicated by equation (38). Considering a reasonable 100-km-thick icy shell and using the ratio $C/MR^2 = 0.346$, we estimate the width of the resonance to be about three times larger than previously, i.e., $\Delta\omega \sim 0.2$ n. Hence, this clearly makes synchronous rotation even more likely in the present Europa configuration.

We have only considered the coupling between the icy shell and a hypothetical subsurface ocean. It is important to note that all the previous effects are enhanced if Europa has a metallic core that is also decoupled from the silicate mantle.

The arguments presented here show that dynamical models neglecting shell thickness variations predict quite different scenarios for the current rotational state of Europa to those based on geological analyses. Although tidal models are still not sufficiently well developed to furnish conclusive information on Europa's rotation, we stress that there may also be other ways to interpret the formation and evolution of surface tectonic features. In particular, most models of tidal stresses currently suppose that the tidal forcing is only due to the nonzero eccentricity of Europa and that this latter is a constant value.

Variations of Europa's orbital and rotational parameters may also affect the geometry of tidal stress. First, the influence of longitudinal and latitudinal librations should be tested in current models of external tidal forcing. Second, the eccentricity of Europa varies significantly (e.g., *Lainey et al.,* 2006), but these variations can occur over a wide range of timescales: from a few days (corresponding to orbital periods) to 10^5–10^6 yr (corresponding to secular changes in Jupiter's orbit). Intermediate periods are also expected to result from mutual gravitational perturbations between Galilean satellites (e.g., *Hussmann and Spohn,* 2004). Third, although Europa's forced obliquity is small, time-dependent evolution of the precessional motion and of the obliquity of Europe occur on similar timescales (e.g., *Bills,* 2005), leading to a complex history of the tidal geometry at the surface of Europa (see section 2 above). A promising investigation has been recently made by *Hurford et al.* (2006). They have shown that tidal stress patterns similar to those often attributed to NSR can be generated by librations due to a small (but nonzero) obliquity. Fourth, TPW is another mechanism closely related to NSR that may have important geological effects (see above). More generally, the possibility that Europa's tectonic features could be generated by cycles of stress and strain related to orbital forcing or pole reorientation, without requiring NSR, will require further investigation.

5. SUMMARY AND CONCLUSIONS

Variations in the rotation rate or spin axis orientation of Europa can arise, primarily due to tidal interactions with Jupiter, and can result in significant surface stresses. Europa's obliquity is predicted to vary on a 10–10^3-yr timescale, and to have a mean value of ~0.1° or perhaps several times larger, for cases without and with a subsurface ocean, respectively. Librations in longitude will have an amplitude of ≈130 m in the case of a rigid Europa; the amplitude will only increase by ≈10% in the case of a subsurface ocean, due to the strong gravitational coupling between shell and interior. Either the mean obliquity or the libration amplitude can be used to determine Europa's polar moment of inertia, if the gravitational moments are known. Nonsynchronous rotation occurs if the tidal torques due to Europa's eccentric orbit exceed the torques arising due to Europa's permanent shape asymmetry. We conclude that the latter are likely to dominate and thus that NSR is not likely to occur, unless lateral variations in shell thickness exist. Tidal stresses due to Europa's finite obliquity, or polar wander, may be a viable alternative to NSR in explaining the observed global tectonic patterns.

Acknowledgments. B.L. and J.L. thank A. Morgado Correia for useful and valuable discussions. Helpful reviews by G. Ojakangas, I. Matsuyama, and W. B. McKinnon are also gratefully acknowledged. Parts of this work were supported by the NASA Planetary Geology and Geophysics and Outer Planets Research programs.

REFERENCES

Anderson J. D., Lau E. L., Sjogren W. L., Schubert G., and Moore W. B. (1996a) Gravitational constraints on the internal structure of Ganymede. *Nature, 384,* 541–543.

Anderson J. D., Sjogren W. L., and Schubert G. (1996b) Galileo gravity results and the internal structure of Io. *Science, 272,* 709–712.

Anderson J. D., Schubert G., Jacobson R. A., Lau E. L., Moore W. B., and Sjogren W. L. (1998a) Distribution of rock, metals, and ices in Callisto. *Science, 280,* 1573–1576.

Anderson J. D., Schubert G., Jacobson R. A., Lau E. L., Moore W. B., and Sjogren W. L. (1998b) Europa's differentiated internal structure: Inferences from four Galileo encounters. *Science, 281,* 2019–2022.

Berger A. L. (1976) Obliquity and precession for the last 5 million years. *Astron. Astrophys., 51,* 127–135.

Berger A., Loutre M. F., and Laskar J. (1992) Stability of the astronomical frequencies over the Earth's history. *Science, 255,* 560–566.

Bills B. G. (1990) The rigid body obliquity history of Mars. *J. Geophys. Res.-Solid Earth Planets, 95,* 14137–14153.

Bills B. G. (1994) Obliquity-oblateness feedback: Are climatically sensitive values of obliquity dynamically unstable. *Geophys. Res. Lett., 21,* 177–180.

Bills B. G. (1999) Obliquity-oblateness feedback on Mars. *J. Geophys. Res.-Planets, 104,* 30773–30797.

Bills B. G. (2005) Free and forced obliquities of the Galilean satellites of Jupiter. *Icarus, 175,* 233–247.

Bills B. G. and Comstock R. L. (2005) Forced obliquity variations of Mercury. *J. Geophys. Res.-Planets, 110.*

Bills B. G. and Nimmo F. (2008), Forced obliquity and moments of inertia of Titan. *Icarus, 196,* 293–297.

Bills B. G. and Rubincam D. P. (1995) Constraints on density models from radial moments: Applications to Earth, Moon, and Mars. *J. Geophys. Res.-Planets, 100,* 26305–26315.

Buffett B. A. (1996) Gravitational oscillations in the length of day. *Geophys. Res. Lett., 23(17),* 2279–2282.

Colombo G. (1966) Cassini's second and third laws. *Astron. J., 71,* 891–896.

Comstock R. L. and Bills B. G. (2003) A solar system survey of forced librations in longitude. *J. Geophys. Res. 108(E9),* 5100, DOI: 10.1029/2003JE00210.

Consolmagno G. J. and Lewis J. S. (1978) Evolution of icy satellite interiors and surfaces. *Icarus, 34,* 280–293.

Correia A. C. M. (2006) The core-mantle friction effect on the secular spin evolution of terrestrial planets. *Earth. Planet. Sci. Lett., 252,* 398–412.

Correia A. C. M. and Laskar J. (2004) Mercury's capture into the 3/2 spin-orbit resonance as a result of its chaotic dynamics. *Nature, 429,* 848–852.

Darwin G. (1908) Tidal friction and Cosmogony. In *Scientific Paper Vol. 2,* Cambridge Univ., New York.

Flasar F. M. (1998) The dynamic meteorology of Titan. *Planet. Space Sci., 46,* 1125–1147.

Folkner W. M., Yoder C. F., Yuan D. N., Standish E. M., and Preston R. A. (1997) Interior structure and seasonal mass redistribution of Mars from radio tracking of Mars Pathfinder. *Science, 278,* 1749–1752.

Geissler P., Greenberg R., Hoppa G., Helfenstein P., McEwen A., et al. (1998) Evidence for non-synchronous rotation of Europa. *Nature, 391,* 368–370.

Gladman B., Dane Quinn D., Nicholson P., and Rand R. (1996) Synchronous locking of tidally evolving satellites. *Icarus, 122,* 166–192.

Goldreich P. and Peale S. (1966) Spin-orbit coupling in the solar system. *Astron. J., 71,* 425–438.

Goldreich P. and Peale S. (1967) Spin-orbit coupling in the solar system. II. The resonant rotation of Venus. *Astron. J., 72,* 662–668.

Greenberg R. and Weidenschilling S. J. (1984) How fast do Galilean satellites spin? *Icarus, 58,* 186–196.

Greenberg R., Geissler P., Hoppa G., Tufts B. R., Durda D. D., et al. (1998) Tectonic processes on Europa: Tidal stresses, mechanical response, and visible features. *Icarus, 135,* 64–78.

Greenberg R., Hoppa G. V., Geissler P., Sarid A., and Tufts B. R. (2002) The rotation of Europa. *Cel. Mech. Dyn. Astron., 83,* 35–47.

Hamilton D. P. and Ward W. R. (2004) Tilting Saturn. II. Numerical model. *Astron. J., 128,* 2510–2517.

Harada Y. and Kurita K. (2006) The dependence of surface tidal stress on the internal structure of Europa: The possibility of cracking of the icy shell. *Planet. Space Sci., 54,* 170–180.

Harada Y. and Kurita K. (2007) Effect of non-synchronous rotation on surface stress upon Europa: Constraints on surface rheology. *Geophys. Res. Lett., 34,* L11204, DOI: 10.1029/2007GL029554.

Hays J. D., Imbrie J., and Shackleton N. J. (1976) Variations in Earth's orbit — Pacemaker of the Ice Ages. *Science, 194,* 1121–1132.

Henrard J. (2005) The rotation of Europa. *Cel. Mech. Dyn. Astron., 91,* 131–149.

Henrard J. and Murigande C. (1987) Colombo's top. *Cel. Mech., 40,* 345–366.

Hilton J. L., Capitaine N., Chapront J., Ferrandiz J. M., Fienga A., et al. (2006) Report of the International Astronomical Union Division I Working Group on precession and the ecliptic. *Cel. Mech. Dyn. Astron., 94,* 351–367.

Hinnov L. A. and Ogg J. C. (2007) Cyclostratigraphy and the astronomical time scale. *Stratigraphy, 4,* 239–251.

Hoppa G., Greenberg R., Geissler P., Tufts B. R., Plassmann J., and Durda D. D. (1999a) Rotation of Europa: Constraints from terminator and limb positions. *Icarus, 137,* 341–347.

Hoppa G., Tufts B. R., Greenberg R., and Geissler P. E. (1999b) Strike-slip faults on Europa: Global shear patterns driven by tidal stress. *Icarus, 141,* 287–298.

Hoppa G., Tufts B. R., Greenberg R., and Geissler P. E. (1999c) Formation of cycloidal features on Europa. *Science, 285,* 1899–1902.

Hoppa G., Tufts B. R., Greenberg R., Hurford T. A., O'Brien D. P., and Geissler P. E. (2001) Europa's rate of rotation derived from the tectonic sequences in the Astypalaea region. *Icarus, 153,* 208–213.

Hubbard W. B. and Anderson J. D. (1978) Possible flyby measurements of Galilean satellite interior structure. *Icarus, 33,* 336–341.

Hurford T. A., Bills B. G., Sarid A. R., and Greenberg R. (2006) Unraveling Europa's tectonic history: Evidence for a finite obliquity? (abstract). In *Lunar and Planetary Science XXXVII,* Abstract #1303. Lunar and Planetary Institute, Houston (CD-ROM).

Hurford T. A., Sarid A. R., and Greenberg R. (2007) Cycloidal cracks on Europa: Improved modeling and nonsynchronous rotation implications. *Icarus, 186,* 218–233.

Hussmann H. and Spohn T. (2004) Thermal-orbital evolution of Io and Europa. *Icarus, 171,* 391–410.

Johnson T. V. (2005) Geology of the icy satellites. *Space Sci. Rev., 116,* 401–420.

Kattenhorn S. A. (2002) Nonsynchronous rotation evidence and fracture history in the Bright Plains region, Europa. *Icarus, 157,* 490–506.

Kaula W. M. (1964) Tidal dissipation by solid friction and the re-sulting orbital evolution. *Rev. Geophys., 2,* 661–685.

Kinoshita H. (1977) Theory of rotation of rigid Earth. *Cel. Mech., 15,* 277–326.

Lainey V., Arlot J. E., and Vienne A. (2004a) New accurate ephemerides for the Galilean satellites of Jupiter — II. Fitting the observations. *Astron. Astrophys., 427,* 371–376.

Lainey V., Duriez L., and Vienne A. (2004b) New accurate ephemerides for the Galilean satellites of Jupiter — I. Numerical integration of elaborated equations of motion. *Astron. Astrophys., 420,* 1171–1183.

Lainey V., Duriez L., and Vienne A. (2006) Synthetic representation of the Galilean satellites' orbital motions from L1 ephemerides. *Astron. Astrophys., 456,* 783–788.

Laskar J. and Robutel P. (1993) The chaotic obliquity of the planets. *Nature, 361,* 608–612.

Laskar J., Joutel F., and Boudin F. (1993) Orbital, precessional, and insolation quantities for the Earth from –20 MYR to +10 MYR. *Astron. Astrophys., 270,* 522–533.

Leith A. C. and McKinnon W. B. (1996) Is there evidence for polar wander on Europa? *Icarus, 120,* 387–398.

Levrard B. (2008) A proof that tidal heating in a synchronous rotation is always larger than in an asymptotic nonsynchronous ro-tation state. *Icarus, 193,* 641–643.

Levrard B., Correia A. C. M., Chabrier G., Baraffe I., Selsis F., and Laskar J. (2007) Tidal dissipation within hot Jupiter: A new appraisal. *Astron. Astrophys., 462,* L5–L8.

Lieske J. H. (1979) Poles of the Galilean satellites. *Astron. Astrophys., 75,* 158–163.

Lieske J. H. (1998) Galilean satellite ephemerides E5. *Astron. Astrophys. Suppl. Ser., 129,* 205–217.

Lieske J. H., Lederle T., Fricke W., and Morando B. (1977) Expressions for precession quantities based upon IAU (1976) system of astronomical constants. *Astron. Astrophys., 58,* 1–16.

Margot J. L., Peale S. J., Jurgens R. F., Slade M. A., and Holin I. V. (2007) Large longitude libration of Mercury reveals a molten core. *Science, 316,* 710–714.

Matsuyama I. and Nimmo F. (2007) Rotational stability of tidally-deformed planetary bodies. *J. Geophys. Res., 112,* E11003.

Matsuyama I. and Nimmo F. (2008) Tectonic patterns on reoriented and despun planetary bodies. *Icarus, 195,* 459–473.

McEwen A. S. (1986) Tidal reorientation and the fracturing of Jupiter's moon Europa. *Nature, 321,* 49–51.

Mignard F. (1980) The evolution of the lunar orbit revisited, II. *Moon Planets, 23,* 185–201.

Milankovitch M. (1941) *Kanon der Erdebestralung und seine Anvendung auf des Eiszeitproblem.* Royal Serbian Academy, Belgrade.

Miskovitch V. V. (1931) Variations seculaires de elementss astronomiques de l'orbite terrestre. *Glas. Spr. Kalyevske Acad., 143.*

Moore W. B. and Schubert G. (2000) The tidal response of Europa. *Icarus, 147,* 317–319.

Murray C. D. and Dermott S. F. (1999) *Solar System Dynamics.* Cambridge Univ., Cambridge.

Nimmo F. (2004) Non-Newtonian topographic relaxation on Europa. *Icarus, 168,* 205–208.

Nimmo F., Thomas P. C., Pappalardo R. T., and Moore W. B. (2007) The global shape of Europa: Constraints on lateral shell thickness variations. *Icarus, 191,* 183–192.

Ojakangas G. W. and Stevenson D. J. (1989a) Thermal state of an ice shell on Europa. *Icarus, 81,* 220–241.

Ojakangas G. W. and Stevenson D. J. (1989b) Polar wander of an ice shell on Europa. *Icarus, 81,* 242–270.

Passey Q. R. and Shoemaker E. M. (1982) Craters and basins on Ganymede and Callisto: Morphological indicators of crustal evolution. In *Satellites of Jupiter* (D. Morrison, ed.), pp. 379–434. Univ. of Arizona, Tucson.

Peale S. J. (1969) Generalized Cassinis Laws. *Astron. J., 74,* 483–490.

Peale S. J. (1974) Possible histories of the obliquity of Mercury. *Astron. J., 79,* 722–744.

Peale S. J. (1976) Does Mercury have a molten core? *Nature, 262,* 765–766.

Peale S. J. and Lee M. H. (2002) A primordial origin of the Laplace relation among the Galilean satellites. *Science, 298,* 593–597.

Peale S. J., Phillips R. J., Solomon S. C., Smith D. E., and Zuber M. T. (2002) A procedure for determining the nature of Mercury's core. *Meteoritics & Planet. Sci., 37,* 1269–1283.

Poincaré H. (1892) *Les Méthodes nouvelles de la Mécanique Céleste.* Gauthier-Villars, Paris.

Rambaux N., Van Hoolst T., Dehant V., and Bois E. (2007) Inertial core-mantle coupling and libration of Mercury. *Astron. Astrophys., 468,* 711–719.

Roos-Serote M. (2005) The changing face of Titan's haze: Is it all dynamics? *Space Sci. Rev., 116,* 201–210.

Rubincam D. P., Chao B. F., and Bills B. G. (1998) The incredible shrinking tropics. *Sky Telescope, 95,* 36–37.

Sarid A. R., Greenberg R., Hoppa G. V., Hurford T. A., Tufts B. R., and Geissler P. (2002) Polar wander and surface convergence of Europa's ice shell: Evidence from a survey of strike-slip displacement. *Icarus, 158,* 24–41.

Sarid A. R., Greenberg R., Hoppa G. V., Geissler P., and Preblich B. (2004) Crack azimuths on Europa: Time sequence in the southern leading face. *Icarus, 168,* 144–157.

Sarid A. R., Greenberg R., and Hurford T. A. (2006) Crack azimuths on Europa: Sequencing of the northern leading hemisphere. *J. Geophys. Res., 111,* E08004.

Schenk P. M., Chapman C. R., Zahnle K., and Moore J. M. (2004) Ages and interiors: The cratering record of the Galilean satellites. In *Jupiter: The Planets, Satellites and Magnetosphere* (F. Bagenal et al., eds.), pp. 427–456. Cambridge Univ., New York.

Schenk P., Matsuyama I., and Nimmo F. (2008) True polar wander on Europa from global-scale small-circle depressions. *Nature, 453,* 368–371.

Schubert G., Anderson J. D., Spohn T., and McKinnon W. B. (2004) Interior composition, structure and dynamics of the Galilean satellites. In *Jupiter: The Planets, Satellites and Magnetosphere* (F. Bagenal et al., eds.), pp. 281–306. Cambridge Univ., New York.

Sharaf S. G. and Boudnikova N. A. (1967) Secular variations of elements of the Earth's orbit which influence climates of the geological past. *Bull. Inst. Theor. Astron., 11,* 231–261.

Shoemaker E. M. and Wolfe R. F. (1982) Cratering time scales for the Galilean Satellites. In *Satellites of Jupiter* (D. Morrison, ed.), pp. 277–339. Univ. of Arizona, Tucson.

Soler T. (1984) A new matrix development of the potential and attraction at exterior points as a function of the inertia tensors. *Cel. Mech., 32,* 257–296.

Sotin C. and Tobie G. (2004) Internal structure and dynamics of the large icy satellites. *Compt. Rend. Phys., 5,* 769–780.

Stiles B. W., Lorenz R. D., Kirk R. L., Hensley S., Lee E. M., et al. (2007) Titan's spin state from Cassini SAR data: Evidence for an internal ocean. *Eos Trans AGU, 88(52),* Fall Meeting Suppl., Abstract P21D-07.

Szeto A. M. K. and Zu S. (1997) Gravitational coupling in a triaxial ellipsoidal Earth. *J. Geophys. Res., 102,* 27651–27657.

Tobie G., Mocquet A., and Sotin C. (2005) Tidal dissipation within large icy satellites: Applications to Europa and Titan. *Icarus, 177,* 534–549.

Tokano T. and Neubauer F. M. (2005) Wind-induced seasonal angular momentum exchange at Titan's surface and its influence on Titan's length-of-day. *Geophys. Res. Lett., 32,* L24203.

Touma J. and Wisdom J. (1993) The chaotic obliquity of Mars. *Science, 259,* 1294–1296.

Van Hoolst T., Rambaux N., Karatekin Ö., Dehant V., and Rivoldini A. (2008) The librations, shape, and icy shell of Europa. *Icarus, 195,* 386–399.

Vernekar A. D. (1972) Long period global variations of incoming solar radiation. *Meteorol. Monogr., 51,* 1–22.

Wahr J. M., Zuber M. T., Smith D. E., and Lunine J. I. (2006) Tides on Europa and the thickness of Europa's icy shell. *J. Geophys. Res., 111,* E12005.

Ward W. R. (1973) Large-scale variations in obliquity of Mars. *Science, 181,* 260–262.

Ward W. R. (1975a) Tidal friction and generalized Cassini's laws in the solar system. *Astron. J., 80,* 64–70.

Ward W. R. (1975b) Past orientation of the lunar spin axis. *Science, 189,* 377–379.

Ward W. R. (1992) Long term orbital and spin dynamics of Mars. In *Mars* (H. H. Kieffer et al., eds.), pp. 298–320. Univ. of Arizona, Tucson.

Ward W. R. and Canup R. M. (2006) The obliquity of Jupiter. *Astrophys. J. Lett., 640,* L91–L94.

Ward W. R. and Decampli W. M. (1979) Comments on the Venus rotation pole. *Astrophy. J. Lett., 230,* L117–L121.

Ward W. R. and Hamilton D. P. (2004) Tilting Saturn. I. Analytic model. *Astron. J., 128,* 2501–2509.

Wisdom J. (2006) Dynamics of the lunar spin axis. *Astron. J., 131,* 1864–1871.

Wisdom J. (2008) Tidal dissipation at arbitrary eccentricity and obliquity. *Icarus, 193,* 637–640.

Wu X., Bender P. L., Peale S. J., and Rosborough G. W. (1997) Determination of Mercury's 88 day libration and fluid core size from orbit. *Planet. Space Sci., 45,* 15–19.

Yoder C. F. (1995) Venus free obliquity. *Icarus, 117,* 250–286.

Yoder C. F. (1997) Venusian spin dynamics. In *Venus II — Geology, Geophysics, Atmosphere, and Solar Wind Environment* (S. W. Bougher et al., eds.), pp. 1087–1124. Univ. of Arizona, Tucson.

Yoder C. F. and Ward W. R. (1979) Does Venus wobble? *Astrophys. J. Lett., 233,* L33–L37.

Zahnle K., Schenk P., Sobieszczyk S., Dones L., and Levison H. F. (2001) Differential cratering of synchronously rotating satellites by ecliptic comets. *Icarus, 153,* 111–129.

Zahnle K., Schenk P., Levison H., and Dones L. (2003) Cratering rates in the outer solar system. *Icarus, 163,* 263–289.

Part II:

Geology and Surface

Geologic Stratigraphy and Evolution of Europa's Surface

Thomas Doggett, Ronald Greeley, and Patricio Figueredo
Arizona State University

Ken Tanaka
United States Geological Survey

The icy surface of Europa is a window into its interior, including a suspected subsurface ocean that could be habitable. Europa's surface is remarkably young, recording only the last ~40–90 m.y., and shows a visible geological history that begins with a period of tectonic resurfacing, followed by localized chaos formation, formation of bands from extensional tectonics, emplacement of globally extensive ridges, and development of large regions of chaos. The formation models for many of these features, combined with their geologically young age, suggest that the ocean or warm ice layer is currently extant beneath a ~20-km-thick icy shell.

1. INTRODUCTION

Since its discovery in 1610 by Galileo Galilei, Europa has attracted observation by increasingly sophisticated Earth-based investigations and spacecraft (see chapter by Alexander et al.). With a radius of 1562 km, Europa is about the same size as Earth's Moon, while gravity data from the Galileo spacecraft indicate an overall density of 3006.2 kg m^{-3} and a silicate mantle surrounding an iron core (see chapter by Schubert et al.). Its high surface albedo was explained by spectroscopic detections of water ice (*Kuiper,* 1957; *Moroz,* 1965; *Pilcher et al.,* 1972), reflecting the composition of a ~100-km-thick outer shell (*Anderson et al.,* 1998).

Europa has a notable paucity of large impact craters (*Smith et al.,* 1979a,b; *Lucchitta and Soderblom,* 1982; see chapters by Bierhaus et al. and Schenk and Turtle) compared to the heavily cratered terrains seen on other bodies, including neighboring Ganymede and Callisto (*Schenk,* 1991; *Chapman et al.,* 1998). Based on size frequency distributions of impact craters, the surface age of Europa is estimated at ~40–90 m.y. (see chapter by Bierhaus et al.). Thus, the visible geological record on Europa reflects only a fraction of solar system history. Europa is presumed to have formed in a flattened, rotating circumplanetary disk of gas and solid particles around a proto-Jupiter ~4.6 G.y. ago (*Schubert et al.,* 2004; see chapters by Estrada et al. and Canup and Ward). Europa then underwent a process of thermal evolution (see chapter by Moore and Hussman) that included differentiation of a metallic core and rocky mantle (see chapter by Schubert et al.). At some point the modern 4:2:1 orbital resonance (*Laplace,* 1805) developed among Io, Europa, and Ganymede, with competing models proposing different evolutionary paths in the orbital relationships of the Galilean satellites (see chapter by Sotin et al.).

The Laplace resonance produces forced eccentricities in the orbits of the three moons, leading to a diurnal variation in the gravitational pull of Jupiter on each moon. *Peale et al.* (1979) proposed dissipation of the resulting tidal energy as a mechanism for internal heating. For Io, this is sufficient to fuel the active volcanism detected by Voyager 1 (*Morabito et al.,* 1979) and was suggested as a means to melt the base of Europa's icy crust into a subsurface ocean (*Cassen et al.,* 1979, 1982). This ocean could be an abode for life (*Reynolds et al.,* 1983; see chapter by Hand et al.) and a source of liquid water as a resurfacing agent (*Squyres et al.,* 1983).

Data returned by the Galileo spacecraft strongly supported, but did not definitively prove, the existence of a subsurface ocean (*Pappalardo et al.,* 1999; *Greenberg et al.,* 2000, *Greeley et al.,* 2004; and other review chapters in *Bagenal et al.,* 2004; see also chapter by McKinnon et al.). The principal evidence for an ocean comes from the analysis of surface geology (*Belton et al.,* 1996; *Greeley,* 1997; *Carr et al.,* 1998; *Greeley et al.,* 1998a,b; *Pappalardo et al.,* 1999) derived from Galileo images, and from the detection of an induced magnetic field consistent with a shallow conducting layer (*Khurana et al.,* 1998; *Kivelson et al.,* 1999, 2000; *Zimmer et al.,* 2000; see chapter by Khurana et al.). A liquid ocean would decouple the icy shell from the rocky interior and potentially lead to nonsynchronous rotation of the icy shell (*Goldreich,* 1966; *Greenberg and Weidenschilling,* 1984; *Ojakangas and Stevenson,* 1989; *Geissler et al.,* 1998a).

The presence of a non-ice surface component, tentatively identified from telescopic studies, was confirmed by Voyager data (*McEwen,* 1986a) and further studied with data from Galileo's Near Infrared Mapping Spectrometer (NIMS) instrument (*Carlson et al.,* 1996). The exact composition of the non-ice component is uncertain (see chapter by Carlson et al.), but is generally categorized as hydrated salts and sulfates (*McCord et al.,* 1998b, 1999). Another possibility is hydrated sulfuric acid (*Carlson et al.,* 1999b) produced by radiolysis of sulfur or sulfate salts. The non-ice component

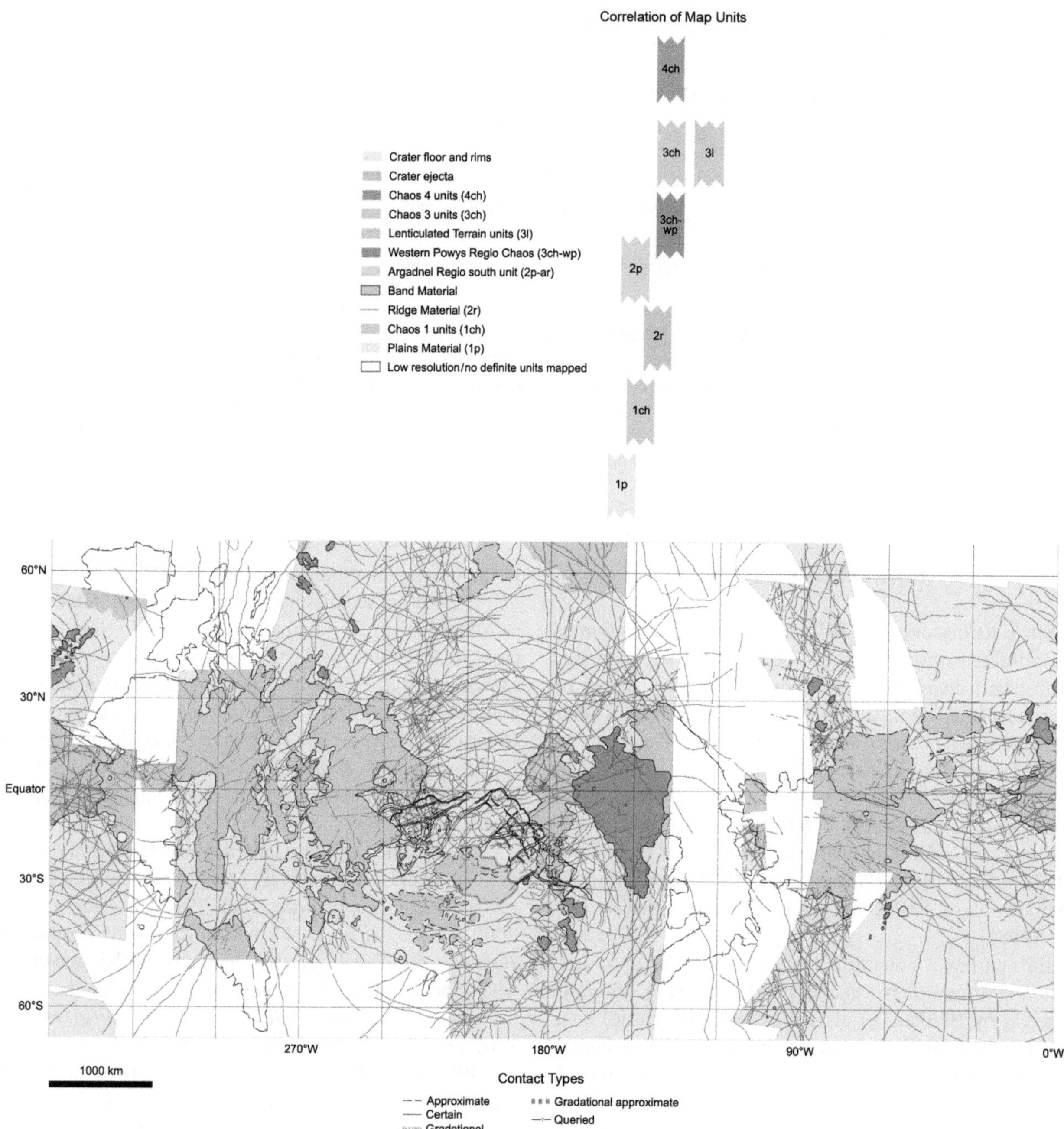

Fig. 1. See Plate 5. Global geological map of Europa, including stratigraphic column of map units.

is associated with ridges and chaos, suggesting that it is emplaced by endogenic processes (*Fanale et al.,* 2000; *McCord et al.,* 1998b, 1999), but deposition of sulfur emitted from Io's volcanos may also be occurring.

Starting with the pioneering work on early lunar missions, the field-based techniques of terrestrial geological mapping have been adapted to photogeological mapping of other worlds (see review by *Batson et al.,* 1990). In both terrestrial and planetary mapping, the spatial extent of three-dimensional rock units (or ice units, in the case of outer solar system bodies such as Europa) are mapped, and the temporal relations, or stratigraphy, among those geologic units are determined. Geological maps have been made for portions of Europa (*Head et al.,* 1998; *Senske et al.,* 1998; *Spaun et al.,* 1998a, b; *Klemaszewski et al.,* 1999; *Prockter et al.,* 1999; *Sullivan et al.,* 1999b; *Williams et al.,* 1999; *Figueredo and Greeley,* 2000, 2004; *Kadel et al.,* 2000; *Figueredo et al.,* 2002; *Prockter and Schenk,* 2002; *Kattenhorn,* 2002), and recently a global map has been produced (Fig. 1).

This chapter reviews the overall physiography of Europa, the mapped material units in a global geological map, and the geological history determined from the stratigraphic relations, as well as the interpreted formation mechanisms and modification histories. The global geological map (Fig. 1) uses a controlled photomosaic of Europa (*USGS,* 2002) that combines the best coverage from the Galileo Solid State Imager (SSI) and Voyager 1 and 2 cameras.

2. PHYSIOGRAPHY

Europa's orbital and tidal mechanics provide a convenient basis for subdividing the satellite (Fig. 2). Because Europa's rotation is essentially synchronous with its orbit around Jupiter, Europa can be divided into leading and trailing hemispheres, based on the direction of Europa's orbit (counterclockwise from the perspective of Jupiter's north pole), and subjovian and antijovian hemispheres, based on the orientation of the surface with respect to Jupiter. Here we use those divisions to define quadrants (Fig. 3) for ease of discussion of the regional geography (section 2.3). By convention, the coordinate system for Europa is defined by the crater Cilix at 182°W, and longitude is measured in a west positive system (*Davies et al.,* 1998; *USGS,* 2002).

2.1. Topography

Early topographic data for Europa are limited to limb observations collected during the Voyager flybys that indicated a relatively flat surface, with knobs and ridges less than a few hundred meters high (*Lucchitta and Soderblom,* 1982). The failure of Galileo's high-gain antenna meant that only 5% of the surface was imaged stereoscopically, although complementary photoclinometry ("shape-from-shading") techniques have been used to assess the topography of an additional 20% of the surface (*Schenk,* 2006). Photoclinometric data are inherently uncertain because of albedo differences of the surface, which in turn are related variations in composition, particle size, and texture of the surface material. Interpretation of existing stereo data by *Schenk* (2004, 2006) indicates global relief of some 2 km. This is expressed in high-standing plateaus, fault blocks, irregular depressions with hundreds of meters of relief, and arcuate troughs (*Schenk et al.,* 2008).

Other topographic studies indicate that chaos terrain, including Conamara and Murias Chaos, is up to 250 m higher in elevation than the surrounding terrain (*Figueredo et al.,* 2002; *Schenk and Pappalardo,* 2004; *Nimmo and Giese,* 2005), supporting a diapiric model for chaos formation. However, a uniform, global characterization of Europa's topography awaits further data from a future mission.

2.2. Surface Feature Types and Nomenclature

A variety of surface features are apparent on Europa. The International Astronomical Union (IAU) has assigned official names (Table 1) derived from Celtic mythology and the Greek myth of Europa to seven types of features: regiones, chaos, craters, large ringed features, flexūs, lineae, and

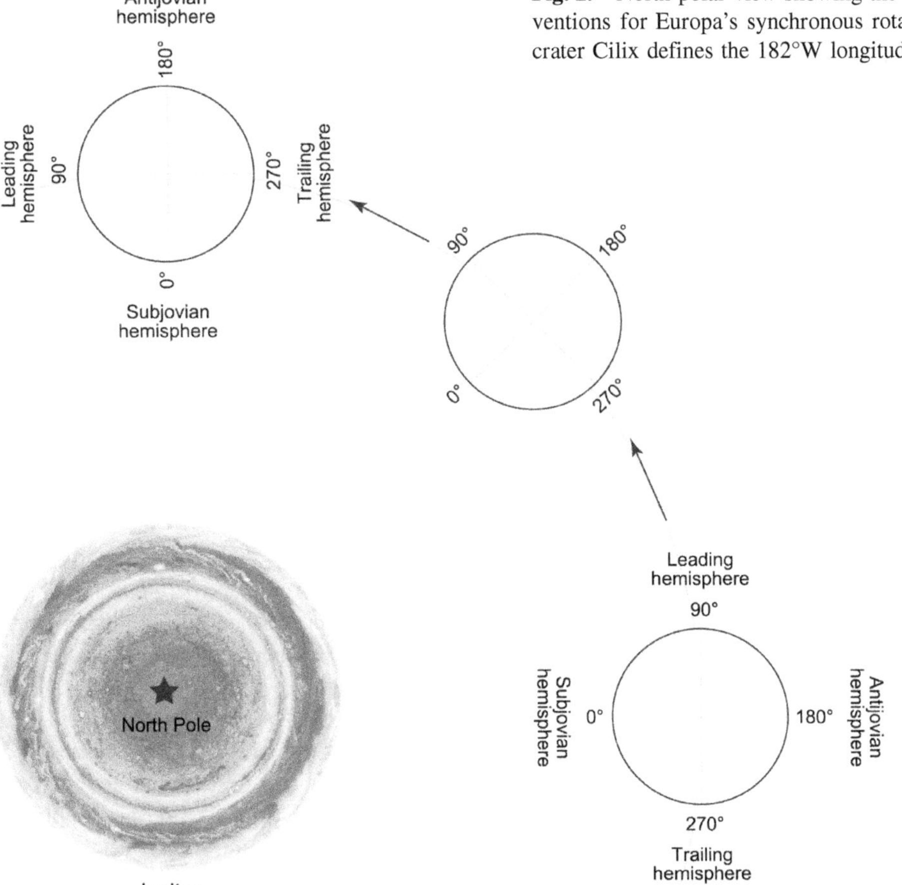

Fig. 2. North polar view showing the hemisphere naming conventions for Europa's synchronous rotation around Jupiter. The crater Cilix defines the 182°W longitude line.

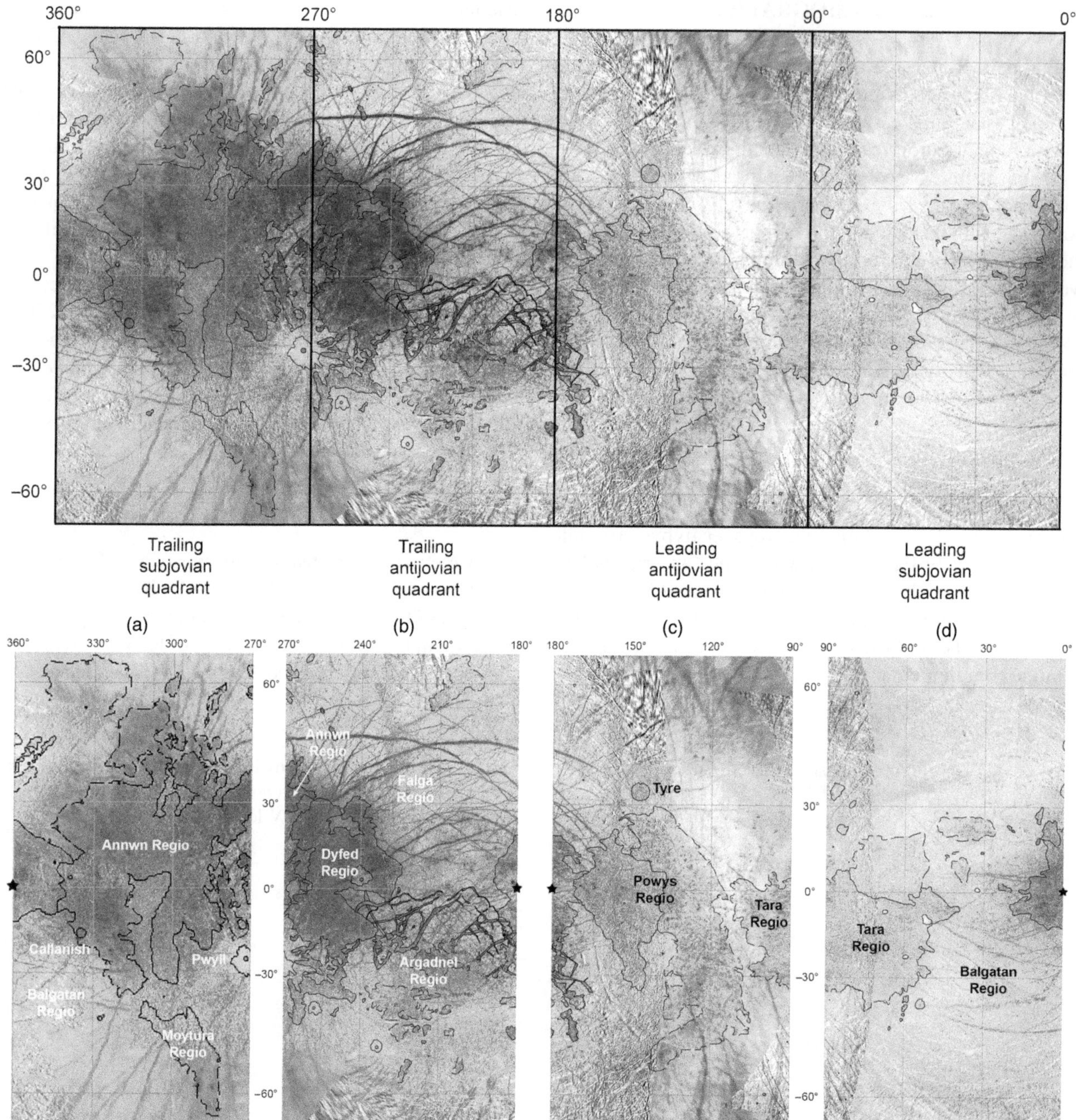

Fig. 3. Mercator projection of Galileo/Voyager mosaic from 63°N to 63°S, showing the four quadrants (90° of longitude each) discussed in the text. **(a)** Trailing subjovian quadrant; star indicates subjovian point. **(b)** Trailing antijovian quadrant; star indicates antijovian point. **(c)** Leading antijovian quadrant; star indicates antijovian point. **(d)** Leading subjovian quadrant; star indicates subjovian point.

maculae; an additional feature type, lenticulae, has been defined, but no official names have yet been given for specific lenticulae.

2.2.1. Regiones. Regiones (singular: regio) are defined as large areas marked by reflectivity or colors distinct from adjacent areas, or as broad geographic regions (Figs. 3a–d). Regiones on Europa fall into two broad categories: bright plains that are criss-crossed by ridges, such as Falga Regio and Balgatan Regio, and darker regions akin to chaos (discussed below; see also chapter by Collins and Nimmo), such as Annwn Regio and Dyfed Regio. This dichtotomy was first noted in Voyager data, where *Lucchitta and Solderblom* (1982) divided the surface into brighter plains and darker mottled terrain. Europan regiones are named for places from Celtic mythology.

2.2.2. Chaos. The term chaos (see chapter by Collins and Nimmo) is defined by the IAU as distinctive areas of broken terrain and was first applied to areas of broken terrain on Mars. For Europa, the term is applied to a terrain type seen in higher-resolution Galileo images, exemplified by Conamara Chaos (Fig. 4). Europan chaos is characterized by polygonal blocks of ridged plains within a matrix

TABLE 1. Europa nomenclature: Official names for features on Europa approved by the International Astronomical Union Working Group on Planetary System Nomenclature (IAU WGPSN) as of October 2009; up-to-date information is available at the Gazeteer of Planetary System Nomenclature website (*planetarynames.wr.usgs.gov*).

Name	Central Latitude	Central Longitude	Diameter (km)	Origin of Name
Chaos				
Arran Chaos	13.4	80.5	26.0	Island where Manannán had a palace.
Conamara Chaos	9.7	272.7	143.7	Rugged part of western Ireland named for Conmac.
Murias Chaos	22.4	83.8	116.0	One of the four great cities of the Tuatha Dé Danann in Irish Celtic myths.
Narberth Chaos	–26.0	273.0	20.0	Chief court of Pwyll; he first saw his future wife Rhiannon at a nearby mound.
Rathmore Chaos	25.4	75.0	57.0	Seat of Mongan, a son of the sea god Manannán.
Craters				
Áine	–43.0	177.5	5.0	Celtic goddess of love and fertility.
Amaethon	13.82	177.47	1.7	Celtic god of agriculture.
Amergin	–14.7	230.6	17.0	Legendary Irish druid and poet.
Angus	–12.6	75.1	4.5	Beautiful Celtic god of love.
Avagddu	1.4	169.5	10.0	Celtic storm deity, ill-fated son of Tegid the Bald.
Balor	–52.8	97.8	4.8	Celtic god of the night whose evil eye caused death.
Bress	37.64	98.66	10.0	Beautiful son of Elatha in Celtic mythology.
Brigid	10.8	81.3	9.5	Celtic goddess of healing, smiths, fertility, and poetry.
Camulus	–26.5	81.1	4.5	Gaelic war god.
Cilix	2.6	181.9	15.0	Brother of Europa.
Cliodhna	–2.5	76.4	3.0	Celtic goddess of beauty.
Cormac	–36.9	88.1	4.0	Cormac Mac Art, High King of Ulster in Irish myths.
Dagda	37.35	168.74	9.8	One of the chief deities of the Tuatha de Danann in Irish mythology.
Deirdre	–65.4	207.3	4.5	The most beautiful woman in Irish myths.
Diarmuid	–61.3	102.0	8.2	Handsome Irish mythological warrior, husband of Gráinne.
Dylan	–55.3	84.4	5.3	Celtic sea god.
Elathan	–31.9	79.8	2.5	Handsome Celtic king, father of Sun god Bres.
Eochaid	–50.48	233.33	10.6	King of the Fir Bolgs in Celtic mythology.
Govannan	–37.3	302.8	11.5	One of the Children of Don, a smith and brewer.
Gráinne	–59.7	99.4	13.5	Daughter of Cormac Mac Art and wife of Diarmuid.
Gwern	9.14	344.54	22.2	Son of Branwen in Celtic mythology.
Gwydion	–60.5	81.6	5.0	Celtic poet, one of the children of the mother goddess Don.
Llyr	–1.8	221.8	1.1	Celtic sea god.
Luchtar	–40.2	257.57	19.9	Celtic god of carpentry.
Lug	27.99	44.31	11.0	Irish omnicompetent god.
Mael Dúin	–16.8	197.9	2.0	Celtic hero.
Maeve	58.8	78.9	21.3	Mythological Irish queen of Connacht province.
Manannán	3.1	239.7	30.0	Irish sea and fertility god.
Math	–25.6	183.7	10.8	Celtic god of wealth and treasure.
Midir	3.65	338.75	37.4	Gaelic fate and underworld deity.
Morvran	–4.9	152.6	15.0	Celtic; ugly son of Tegid.
Niamh	21.1	216.9	5.0	Golden-haired daughter of the Celtic sea and fertility god Manannán.
Ogma	87.45	287.86	5.0	Celtic god of eloquence and literature, a son of Dagda.
Oisín	–52.3	213.4	6.2	Mythical Irish warrior, son of Fionn Mac Cumhail and Sadb.
Pryderi	–66.1	159.1	1.7	Son of Pwyll, Celtic god of the underworld.
Pwyll	–25.2	271.4	45.0	Celtic god of the underworld.
Rhiannon	–80.9	194.9	15.9	Celtic heroine.
Taliesin	–22.8	138.0	50.0	Celtic, son of Bran; magician.
Tegid	0.8	164.4	29.7	Celtic hero who lived in Bula Lake.
Tuag	59.92	172.36	15.2	Irish dawn goddess.
Uaithne	–48.5	90.7	6.5	The harpist for Dagda, the father of all gods in Celtic myths.
Flexūs				
Cilicia Flexus	–59.5	171.7	1312.0	Land named for Cilix on his search for Europa.
Delphi Flexus	–68.2	174.1	793.0	Where the cow led Cadmus before it stopped at the site of Thebes.
Gortyna Flexus	–42.1	144.6	940.0	Place on Crete where Zeus brought Europa.
Phocis Flexus	–44.5	198.4	242.0	Where the cow lead Cadmus before it stopped at the site of Thebes.
Sidon Flexus	–66.4	183.4	1133.0	Another name for Tyre; where Europa was born.
Large Ringed Features				
Callanish	–16.7	334.5	107.0	Stone circle in the Outer Hebrides, Scotland.
Tyre	33.6	146.6	149.0	Greek; the seashore where Zeus abducted Europa. Changed from Tyre Macula.

TABLE 1. (continued).

Name	Central Latitude	Central Longitude	Diameter (km)	Origin of Name
Lineae				
Adonis Linea	–61.0	122.6	1560.0	Greek; son of Phoenix, nephew of Europa.
Agave Linea	12.8	273.1	1440.0	Daughter of Harmonia and Cadmus.
Agenor Linea	–43.8	213.5	1496.0	Greek; Europa's father.
Alphesiboea Linea	–25.1	175.9	1438.0	Son of Phoenix, nephew of Europa.
Androgeos Linea	11.7	279.3	723.0	Son of Minos in Greek mythology.
Argiope Linea	–1.7	195.6	689.0	Greek; another name for Telephassa.
Asterius Linea	14.9	270.8	1943.0	Greek; Europa's husband after Zeus.
Astypalaea Linea	–75.8	212.1	817.0	Sister of Europa.
Autonoë Linea	18.2	165.1	760.0	Daughter of Harmonia and Cadmus in Greek mythology.
Belus Linea	9.3	231.4	2437.0	Greek; Agenor's twin brother.
Butterdon Linea	–44.7	0.1	1900.0	Stone row in England.
Cadmus Linea	38.7	191.7	3548.0	Greek; brother of Europa.
Chthonius Linea	–1.4	304.2	2180.0	Survivor of the men Cadmus sowed with dragon's teeth, a founder of Thebes.
Corick Linea	17.8	18.3	1300.0	Stone row in Ireland.
Drizzlecomb Linea	7.7	111.7	1500.0	Stone row in England.
Drumskinny Linea	48.3	161.0	1375.0	Stone row in Ireland.
Echion Linea	–11.6	185.2	1026.0	Survivor of the men Cadmus sowed with the dragon's teeth; a founder of Thebes.
Euphemus Linea	–11.4	45.7	1250.0	In Greek mythology, son of Europa and Poseidon who could walk on water.
Glaukos Linea	57.8	230.9	1400.0	Son of Minos in Greek mythology.
Harmonia Linea	28.0	171.7	1154.0	Wife of Cadmus.
Hyperenor Linea	–12.1	324.4	2996.0	Survivor of the men Cadmus sowed with dragon's teeth, a founder of Thebes.
Ino Linea	–1.7	174.6	1515.0	Daughter of Harmonia and Cadmus.
Katreus Linea	–38.8	213.3	195.0	Son of Minos in Greek mythology.
Kennet Linea	–41.0	312.0	3200.0	Stone row in England.
Libya Linea	–54.0	181.0	366.0	Greek; Agenor's mother.
Mehen Linea	56.0	236.7	1500.0	Stone row in Brittany, France.
Merrivale Linea	–41.0	299.5	1600.0	Stone row in England.
Minos Linea	47.2	195.2	2170.0	Greek; son of Europa and Zeus.
Onga Linea	–38.7	211.3	870.0	Phoenician name for Athene.
Pelagon Linea	35.5	173.6	616.7	King who sold Cadmus the cow with a white full moon on each flank.
Pelorus Linea	–19.8	188.3	1535.0	Greek; survivor of the men Cadmus sowed with the dragon's teeth.
Phineus Linea	–29.8	319.9	2004.0	Greek; brother of Europa.
Phoenix Linea	16.6	188.8	1621.0	Brother of Europa.
Rhadamanthys Linea	19.3	200.5	1747.0	Son of Europa and Zeus.
Sarpedon Linea	–49.5	92.9	900.0	Greek; son of Europa and Zeus.
Sharpitor Linea	65.4	171.7	1650.0	Stone row in England.
Sparti Linea	59.3	245.5	1600.0	In Greek mythology, warriors who sprouted from dragon's teeth.
Staldon Linea	–0.8	27.4	1525.0	Stone row in England.
Tectamus Linea	26.9	199.2	2096.0	Father of Asterius.
Telephassa Linea	–0.8	177.2	777.0	Europa's mother.
Thasus Linea	–66.1	184.0	669.3	Greek; brother of Europa.
Thynia Linea	–59.2	154.5	412.6	Peninsula between Black and Marmara Seas, where Phineus sought Europa.
Tormsdale Linea	47.7	258.0	875.0	Stone row in Ireland.
Udaeus Linea	48.6	239.4	2050.0	In Greek mythology, survivors of the men Cadmus sowed with dragon's teeth.
Yelland Linea	–16.7	196.0	186.0	Stone row in England.
Maculae				
Boeotia Macula	–53.6	166.8	30.0	Place where Cadmus led cow before it stopped at site of Thebes.
Castalia Macula	–1.6	225.7	35.0	Greek; spring where Cadmus, brother of Europa, killed the dragon.
Cyclades Macula	–62.5	191.3	107.0	Islands where Rhadamanthys reigned.
Thera Macula	–46.7	181.2	95.0	Greek; place where Cadmus stopped in his search for Europa.
Thrace Macula	–45.9	172.1	180.2	Place in northern Greece where Cadmus stopped in his search for Europa.
Regiones				
Annwn Regio	20.0	320.0	2300.0	Traditional name of the Welsh Otherworld.
Argadnel Regio	–14.6	208.5	1900.0	In Celtic mythology, one of the islands of Earthly paradise.
Balgatan Regio	–50.0	30.0	2500.0	Celtic; pass to which Tuatha Dé Dannan retreated before battle with Fir Bolgs.
Dyfed Regio	10.0	250.0	1750.0	In Welsh mythology; southwestern kingdom just east of Annwn.
Falga Regio	30.0	210.0	2500.0	In Celtic mythology, island where Midir had a stronghold.
Moytura Regio	–50.0	294.3	483.0	Location of battles between the Fomorians and the Tautha de Danann.
Powys Regio	0.0	145.0	2000.0	In Celtic mythology, ancient kingdom of mid-Wales.
Tara Regio	–10.0	75.0	1780.0	In Celtic mythology, the main royal residence of the High Kings.

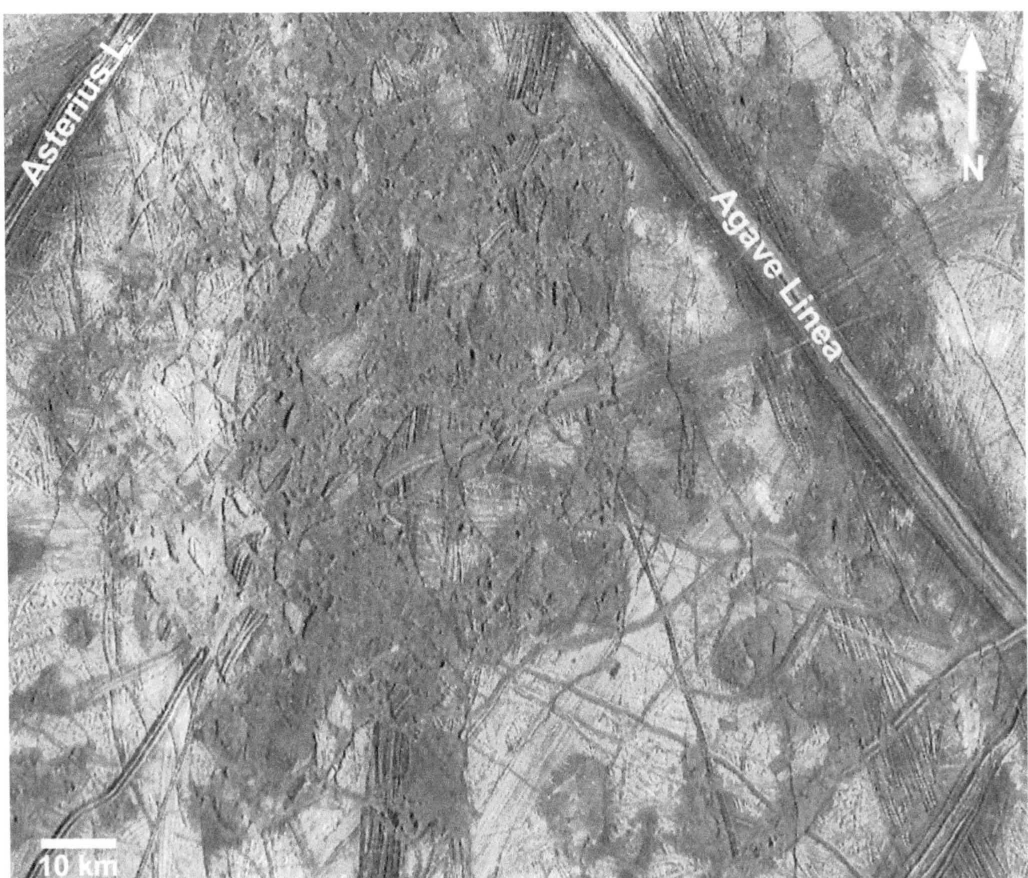

Fig. 4. Conamara Chaos, a 75 × 100-km zone south between Agave Linea and Asterius Linea (labeled). In detail (Fig. 18), Conamara contains blocks of ice preserving the previous ridged plains surface in a matrix of knobby material. Green filter image from Galileo SSI observation 12ESDRKLN_01 at 169 m/pixel.

of hummocky material (*Carr et al.,* 1998; *Spaun et al.,* 1998a). Chaos matrix material can be either low-lying or high-standing relative to the surrounding plains (*Collins et al.,* 2000, *Greeley et al.,* 2000). Chaos generally corresponds to the darker mottled terrain in Voyager images; however, albedo does not always correspond to the morphology seen in higher-resolution images (*Pappalardo et al.,* 1998a). It should be noted that mottled terrain is also generally characterized by lenticulae. Like regiones, chaos on Europa are named for places from Celtic mythology.

2.2.3. Craters. Impact craters on Europa (see chapter by Schenk and Turtle) were first observed in Voyager images (*Smith et al.,* 1979a,b). Just as on rocky bodies, smaller craters are bowl-shaped while larger craters have more complex morphology, to include flat floors, central peaks, and terraces (Fig. 5) (*Moore et al.,* 1998, 2001; *Schenk,* 2002). However, the minimum diameter for the transition from simple to complex morphology is smaller on Europa, 5–6 km, compared to 10–20 km for the terrestrial planets. Based on this onset diameter and also estimates of transient crater depths, *Moore et al.* (2001) concluded that craters ~10–18 km in diameter (3–6-km-deep transient craters) did not penetrate to a liquid layer, and thus the ice must have been at least several kilometers thick at the time of crater formation. The crater Pwyll is notable for its extensive bright rays and pedestal-like proximal ejecta. *Moore et al.*

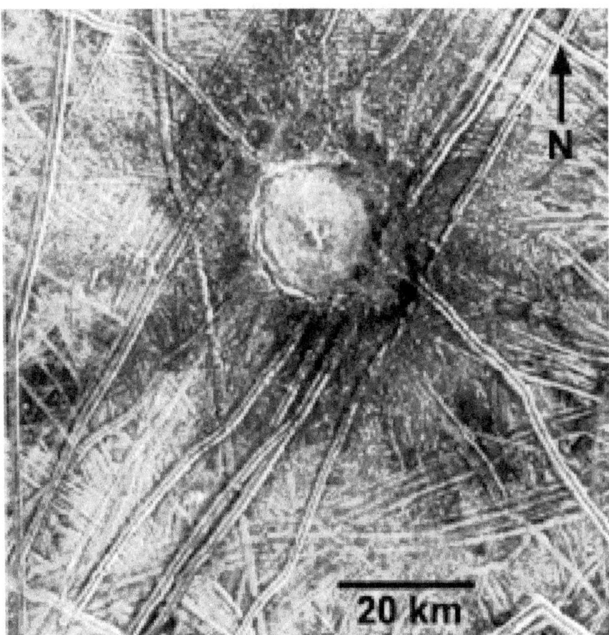

Fig. 5. Cilix is an intermediate-aged crater on Europa, superimposed on older ridged plains. It displays a central peak and incomplete terraces on the crater walls. The dark ejecta is thought to include non-ice materials excavated from the subsurface and/or brought by the impactor. From Galileo SSI observation 16ESCILIXS01 at 110 m/pixel.

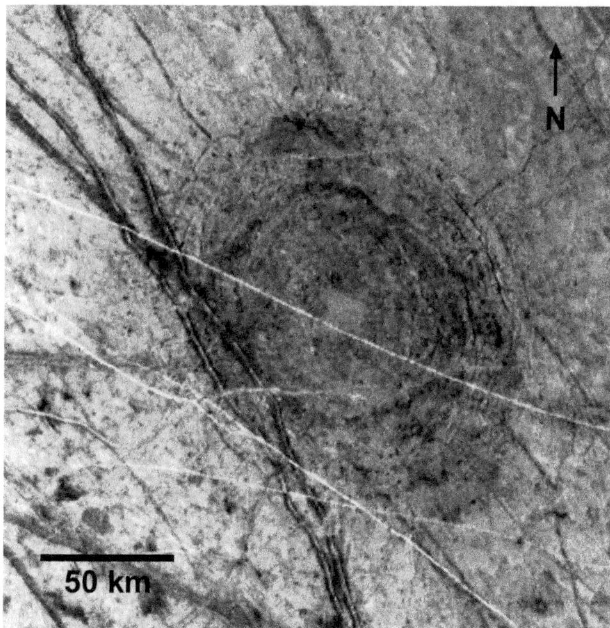

Fig. 6. Tyre is one of Europa's two large multiringed features. From stratigraphic relationships, there are both older ridges that it superimposes and younger ridges that cut across it, and it is the oldest known impact feature on Europa. From Galileo SSI observation G7ESTYRMAC01 at 570 m/pixel.

(1998) concluded that the pedestal is due to a convex upward scarp, possibly formed by outward creep of warm, plastically deforming ice. Craters on Europa are named for Celtic gods and heroes.

2.2.4. Large ringed features. Large ringed features appeared in Voyager images as dark, low-topography circular spots (*Lucchitta and Soderblom,* 1982; *Malin and Pieri,* 1986). In higher-resolution Galileo images (Fig. 6) they are seen as a series of concentric rings and are interpreted to be relic impacts (*Moore et al.,* 1998; *Schenk,* 2002) that have penetrated into liquid water or warm ice. If this interpretation is correct, large ringed features like Callanish and Tyre are among only six known impact structures on Europa >30 km in diameter (see chapter by Bierhaus et al.). They are officially named for Celtic stone circles, with the exception of Tyre (which was originally classified as a macula from Voyager data).

2.2.5. Flexūs. Cycloidal ridges, or flexūs (singular: flexus), consist of connected arcuate segments of similar length that show planforms that are convex in the same direction, forming congruent curved segments joined at sharp cusps (Fig. 7). Their shape is suggested to be controlled by a changing direction of crack propagation under the control of Europa's diurnally rotating stress field at propagation speeds slower than Europa's orbit around Jupiter (*Hoppa et al.,* 1999b). Flexūs are named for places associated with the Europa myth, or Celtic stone rows.

2.2.6. Lineae. Lineae (singular: linea) are defined as elongate markings, and may be either curved or straight, dark or bright, and negative (troughs) or positive (ridges) in relief.

Troughs ("fractures" of the chapter by Kattenhorn and Hurford) are generally V- or U-shaped in cross section and commonly have slightly raised rims. Flat-floored and/or curvilinear troughs occur less frequently. Troughs can transect all terrain types. Narrow (100–300 m wide) troughs are not generally recognized at global scales but appear as hairline lineaments. In addition, there is also a class of wider (25–40 km), arcuate troughs (*Schenk,* 2005) that form incomplete global scale (>2000 km diameter) circles that are antipodal to one another (*Schenk et al.,* 2008). Some sections of these troughs are only viewable in images of Europa from the New Horizons Jupiter flyby (*Grundy et al.,* 2007), which, while lower in resolution than Galileo, had better viewing angles for observing topography. *Schenk et al.* (2008) interpret these partial circles to be evidence of polar wander by Europa's ice shell.

Ridges are the dominant surface features on Europa (see chapters by Kattenhorn and Hurford and by Prockter and Patterson). They range in width from ~200 m to >4 km, commonly exceed 1000 km in length (Fig. 8), and potentially are as high as 200–350 m (*Kadel et al.,* 1998). Ridges include features that are straight, curvilinear, or cycloidal (see section 2.2.5). Ridge material forms single ridges, double ridges (two ridges separated by a trough), or ridge complexes consisting of more than two ridges (*Pappalardo et al.,* 1998b). In high-resolution images (Fig. 9) the cross section of some ridges depicts relatively steep outer flanks, rising to a flat-topped summit. The steep flanks may consist of mass-wasted material, apparently emplaced at its

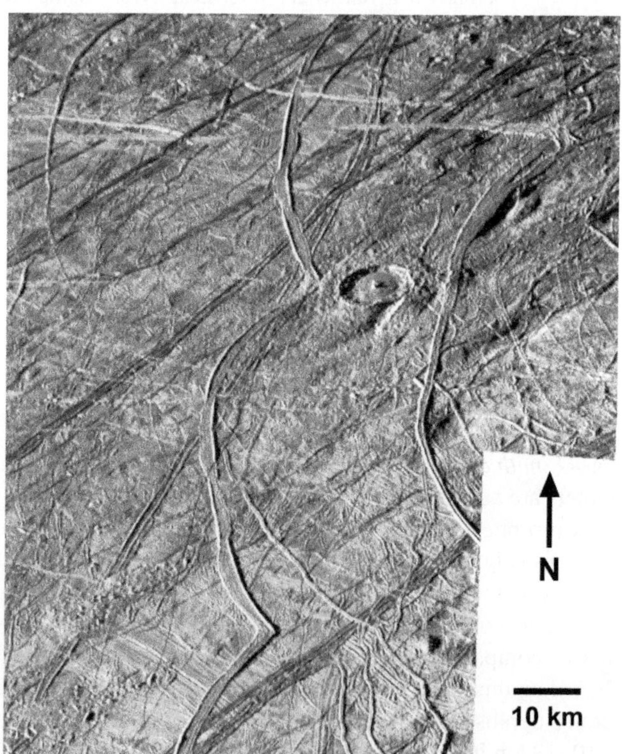

Fig. 7. Example of a cycloidal ridge in the northern leading subjovian quadrant. From Galileo SSI observation 15ESREGMAP02 at 234 m/pixel.

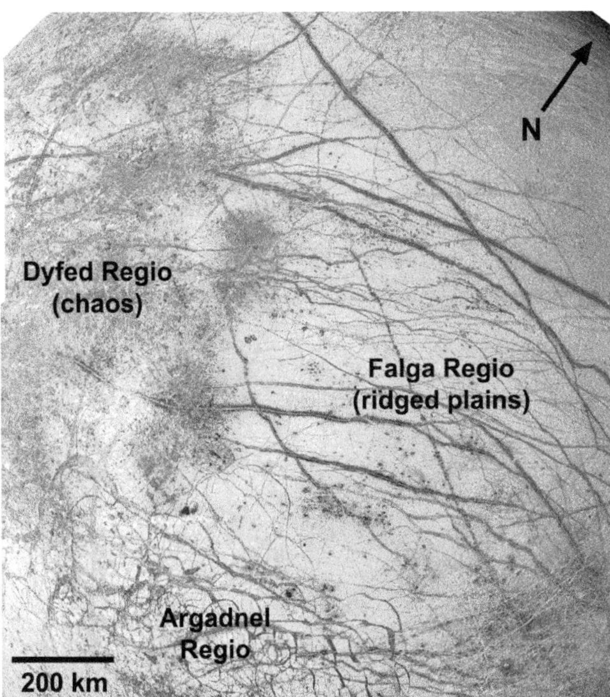

Fig. 8. Trailing antijovian quadrant showing ridged plains (Falga Regio) with scattered occurrences of lenticulae (small dark spots). To the west is a large region of chaos (Dyfed Regio), and to the south is the "wedges" (Argadnel Regio) region. From Galileo SSI observation G1ESGLOBAL01 at 1.6 km/pixel.

angle of repose (*Sullivan et al.,* 1999a). Stereoscopic images of other ridges reveal that the flanking terrain may be elevated (*Giese et al.,* 1999a, b) while other ridges may be flanked by depressions, and possibly marked by parallel fractures. Ridge complexes can form anastomosing and discontinuous sets of ridges and also sets of ridges that "split" away from the main trend of the complex (*Figueredo and Greeley,* 2004). The topographically high morphology of ridges produces sharp contacts with surrounding units. Triple bands were described from Voyager images as lineaments consisting of a bright stripe flanked on each side by a dark band (*Lucchitta and Soderblom,* 1982). Galileo images show that most triple bands are double ridges (comprising the bright part seen on Voyager images) flanked by dark, diffuse material lacking a sharp outer margin (*Belton et al.,* 1996; *Greeley,* 1997; *Fagents et al.,* 2000). Ridges are named for people associated with the Europa myth, or Celtic stone rows.

2.2.7. Maculae. Maculae (singular: macula) are dark spots of irregular shape (Fig. 10) that are named for places associated with the Europa myth. A study of Castalia Macula (*Prockter and Schenk,* 2005) interpreted it to be dark, endogenic material that flooded a small basin. The largest and most prominent maculae on Europa, Thera and Thrace Maculae, were originally interpreted as possible cryovolcanic flows (*Wilson et al.,* 1997), but subsequent Galileo images depict them to be darkened and degraded versions of the surrounding ridged plains, resembling chaos (*Fagents,* 2003).

2.2.8. Lenticulae. Lenticulae (singular: lenticula, from the Latin for freckles) are defined by the IAU as small dark spots on Europa (Fig. 11). Lenticulae are commonly defined as circular to elliptical features that are tens to hundreds of meters in relief and are either positive (domes) or negative (pits) (*Pappalardo et al.,* 1998a), while lenticulae with no apparent relief but lower albedo are informally termed "spots." Most pit lenticulae have lower albedo than the surrounding terrain, disrupting preexisting terrain with chaos-like materials (*Spaun et al.,* 1998a), whereas dome lenticulae generally have the same albedo as the surrounding terrain and barely disrupt the surface (*Pappalardo and Barr,* 2004). Lenticulae are most apparent in images with low-angle illumination, although the texture of lenticulae is apparent in most illumination geometries. Some taxonomies group some or all lenticulae with chaos regions (*Greenberg et al.,* 2003; see chapter by Collins and Nimmo).

2.3. Regional Geography

Ridged plains dominate ~60% of the surface. The lower albedo in the remaining ~40% is strongly correlated to chaotic terrain (*McCord et al.,* 1998a, 1999; *Fanale et al.,* 2000) and the non-ice component seen in spectral data. Overlapping this geomorphology-correlated distinction in albedo and composition is a geographic variation where the trailing hemisphere has an overall lower albedo and deeper sulfur absorption features in violet and ultraviolet wavelengths (*McEwen,* 1986a; *Hendrix et al.,* 1998). This is at-

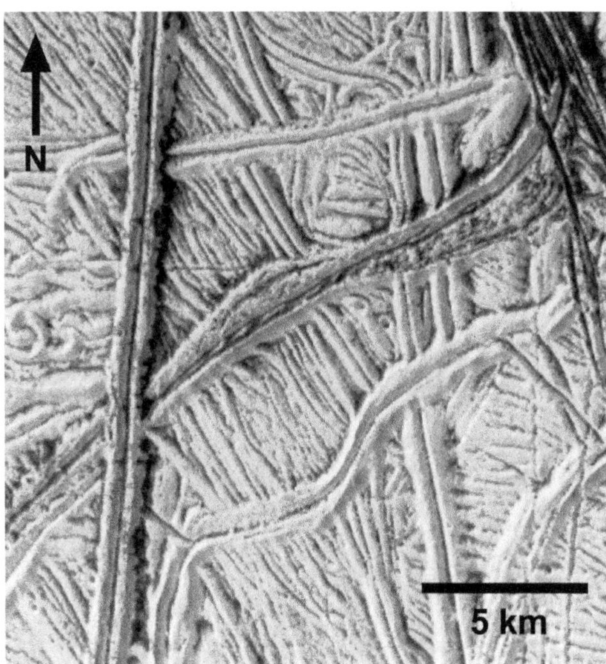

Fig. 9. Closeup view of ridged plains material, showing intricate overlapping and cross-cutting structures. Note that the matrix consists of smaller ridges inferred to be the oldest features observed on Europa. From Galileo SSI observation 12ESWEDGE_02 at 27 m/pixel.

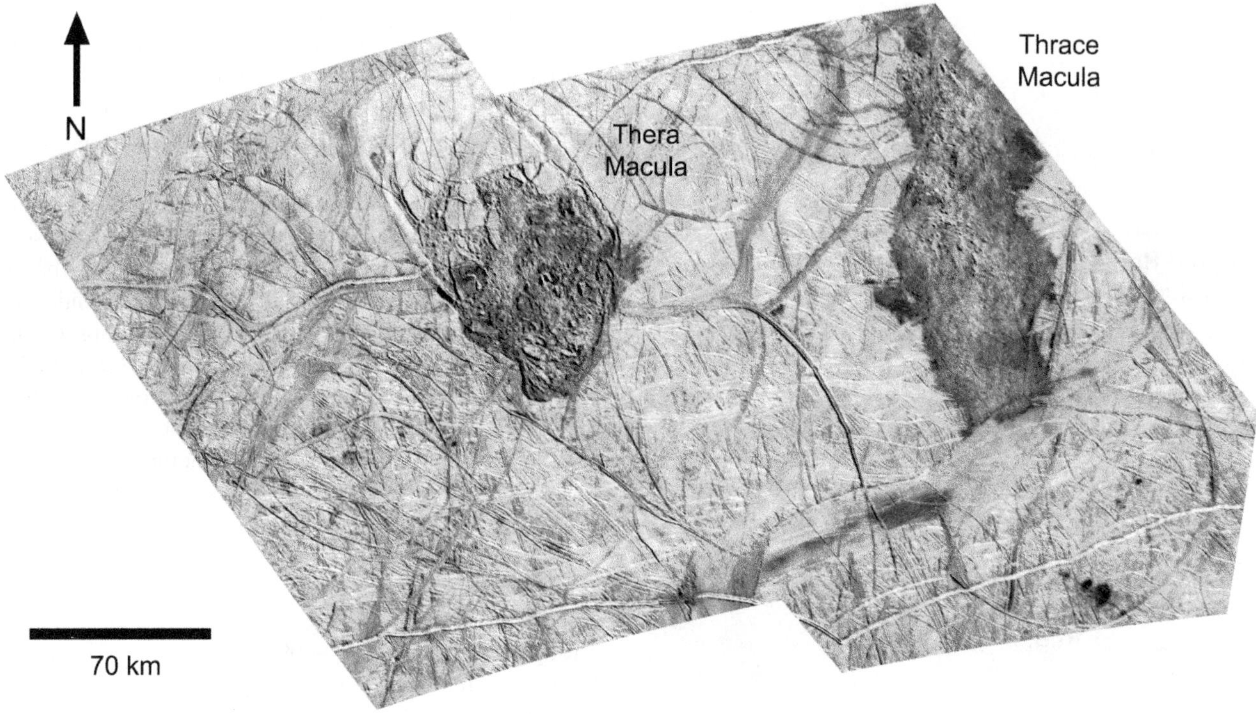

Fig. 10. Thera Macula and Thrace Macula in the antijovian hemisphere. While similar to other chaos in some respects (e.g., incorporating "rafts" of ridged plains material), the maculae are distinctive as the apparent source for dark material that floods or overlays adjacent ridged plains and lineae. From Galileo SSI observation 17ESREGMAP01 at 220 m/pixel.

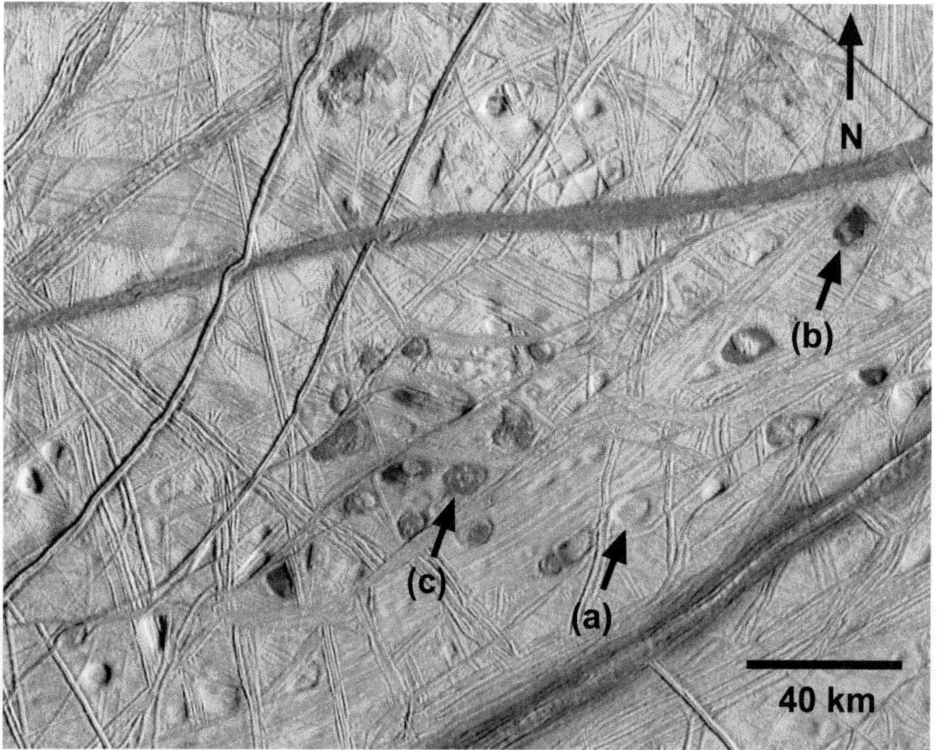

Fig. 11. Examples of lenticulae that formed in the ridged plains in Falga Regio. Lenticulae include **(a)** pits, **(b)** spots, and **(c)** domes. From Galileo SSI observation 15ESREGMAP01 at 228 m/pixel.

tributed to heavier bombardment by charged particles in the jovian magnetosphere (see chapters by Carlson et al. and Paranicas et al.). The geography of individual quadrants (Fig. 3) is discussed below.

2.3.1. Trailing subjovian quadrant (270° to 360°W). This quadrant (Fig. 3a) is dominated by Annwn Regio, a low-albedo region stretching from 50°N to 40°S and from 270°W to 340°W. Lying between Annwn Regio and Dyfed Regio, and south of the prominent intersection of Agave and Asterius Lineae, is Conamara Chaos. Conamara is the archetype for "platy" chaos (*Greeley et al.,* 2000). South of Conamara Chaos is Pwyll impact crater, which has an extensive ejecta blanket and rays that cross both Annwn Regio and Dyfed Regio. Moytura Regio is another low-albedo region in the southern part of this quadrant.

2.3.2. Trailing antijovian quadrant (180° to 270°W). In this quadrant (Fig. 3b), the eastern area of the antijovian dark terrain is separated by Argadnel Regio from the equatorial low-albedo region, Dyfed Regio. Argadnel is also known as the "wedges" region (Fig. 8) and is characterized by curved, tapering dark bands (*Prockter et al.,* 1999). North of the wedges is Falga Regio (Fig. 8), a region crisscrossed with arcuate lineae that is the archetype for ridged plains (*Greeley et al.,* 2000). To the south of Argadnel Regio are occurrences of lenticulated terrain. Southeast of Argadnel Regio is the distinctive low-albedo feature, Thera Macula, which is adjacent to Thrace Macula. This area is also notable for the occurrence of cycloidal ridges (flexūs).

2.3.3. Leading antijovian quadrant (90° to 180°W). This quadrant (Fig. 3c) contains the eastern part of Tara Regio, which is separated by a narrow zone of plains from another dark albedo region named Powys Regio. Powys Regio stretches from 30°N to nearly 60°S. The western and eastern parts of Powys Regio are distinct, where they were first noticed in Voyager data by color as gray and brown mottled terrain, respectively (*Lucchitta and Soderblom,* 1982). The western portion of Powys Regio borders a dark terrain, centered on the antijovian point, that is similar to the subjovian dark terrain previously discussed. In the plains just north of Powys Regio is the large ringed feature Tyre. Tyre is generally interpreted as a relatively old impact feature.

2.3.4. Leading subjovian quadrant (0° to 90°W). This quadrant (Fig. 3d) contains portions of two distinctly different low-albedo regions. The first, centered near the subjovian point, is criss-crossed with lineae that extend from the adjacent plains. These plains are exemplified by Balgatan Regio in the southern portion of the quadrant. The second low-albedo region is largely south of the equator in eastern Tara Regio, where higher-resolution images show chaos (*Figueredo and Greeley,* 2004). Adjacent parts of the northern hemisphere were mapped as lenticulated terrain, while the plains separating Tara Regio and the subjovian dark terrain contain more widely spaced and generally smaller lenticulae. Also of note are isolated patches of chaos such as Murias Chaos, the archetypical example of elevated chaos (*Figueredo et al.,* 2002).

3. MATERIAL UNITS

While Pioneer 10 returned a single low-resolution (161 km/pixel) image of Europa (*Fimmel et al.,* 1974, 1980), which showed light and dark patches that hinted at surface heterogeneity, the Voyager flybys of the jovian system were the first to image Europa at sufficient resolution to discern surface features (*Smith et al.,* 1979a,b). Two distinct surface types, plains and "mottled" terrain, were identified, as well as globally extensive (>1000 km) lineaments and a paucity of large impact craters suggestive of a young surface (*Pieri,* 1981; *Lucchitta and Soderblom,* 1982; *Malin and Pieri,* 1986).

The first geological sketch map of Europa was produced by *Lucchitta and Soderblom* (1982) and covered ~25% of Europa. It was based on Voyager 2 images (~2 km/pixel) and was not improved upon until the Galileo mission (*Belton et al.,* 1996; *Carr et al.,* 1998; *Greeley et al.,* 1998b). However, only 36.2% of the surface was imaged at 1 km/pixel or better resolution (Fig. 12) in Galileo data. Consequently, much science remains to be done in the global context (Fig. 13) when new imaging data of uniform resolution are obtained from a future mission (see chapter by Greeley et al.).

Greeley et al. (2000) summarized initial Galileo-based observations and mapping (*Greeley et al.,* 1998b, *Head et al.,* 1998; *Senske et al.,* 1998; *Spaun et al.,* 1998a; *Klemaszewski et al.,* 1999; *Prockter et al.,* 1999; *Sullivan et al.,* 1999b; *Williams et al.,* 1999; *Figueredo and Greeley,* 2000; *Kadel et al.,* 2000), and they have defined five main surface units: plains, chaos, band, ridge, and crater-related materials, discussed below. This taxonomy was sustained in

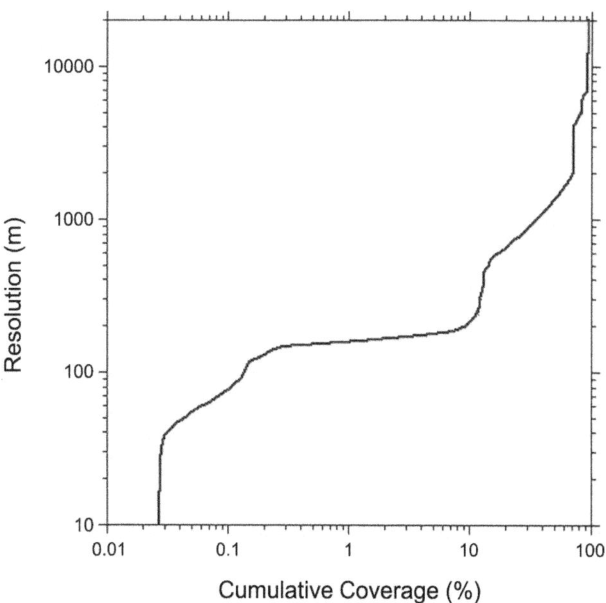

Fig. 12. Existing imaging coverage of Europa, indicating the cumulative percent of the surface area imaged as a function of resolution.

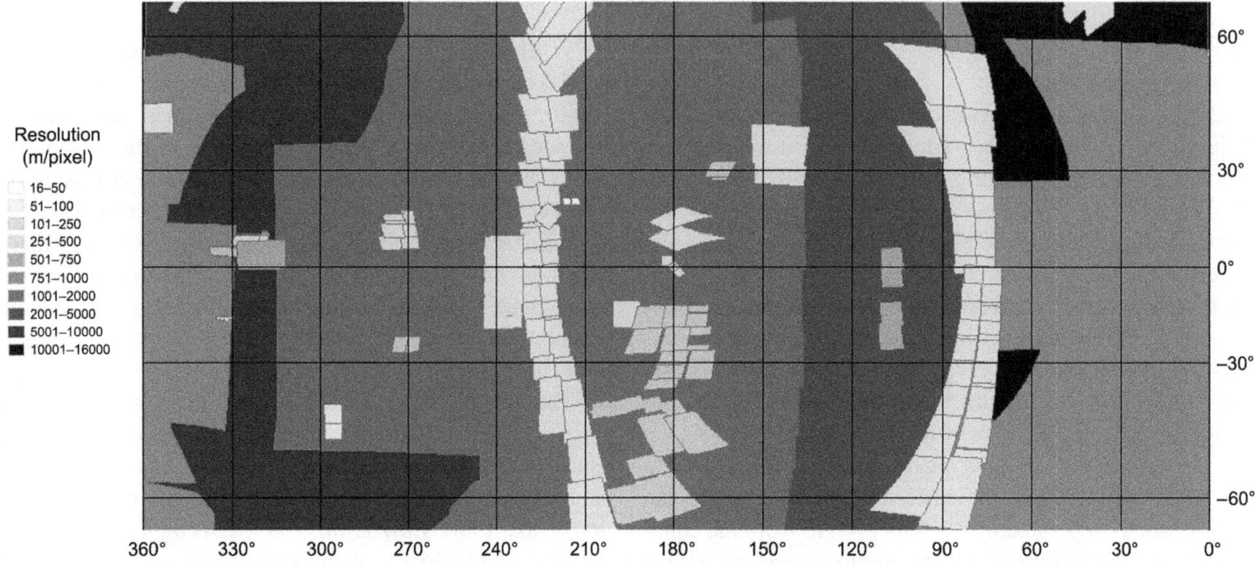

Fig. 13. Map showing the resolution of images that compose the mosaic used as a base for the global geological map. The mosaic is based almost entirely on Galileo images, except for a strip centered near 150°W that uses data from Voyager 2.

subsequent local- and regional-scale mapping using higher-resolution Galileo images (*Figueredo et al.,* 2002; *Prockter and Schenk,* 2002; *Kattenhorn,* 2002), including two pole-to-pole swaths on roughly opposite sides of Europa (*Figueredo and Greeley,* 2000, 2004). A global geological map (Fig. 1) was produced based on Voyager and Galileo images (*USGS,* 2002), defining an additional unit of lenticulated terrain. Although the imaging campaigns during Galileo's multiple Europa flybys did attempt to sample the diversity of terrains, there remains an important caveat that the lack of global coverage at higher (<1 km/pixel) resolution means the resulting map is necessarily tentative and an interim step awaiting a future mission (see chapter by Greeley et al.).

3.1. Plains Material

The oldest unit visible on the surface of Europa is the ridged plains material (*Chapman et al.,* 1998; *Greeley et al.,* 1998b, 2000; *Senske et al.,* 1998; *Head et al.,* 1999b). It is a globally extensive unit (Fig. 8) and was mapped as smooth plains in Voyager images (*Lucchitta and Soderblom,* 1982). The highest-resolution Galileo images show it to consist of ridges and troughs (Fig. 9) (*Head et al.,* 1998, 1999a; *Spaun et al.,* 1998b; *Sullivan et al.,* 1999b; *Kattenhorn,* 2002). The ridged plains correspond to the brightest areas of Europa, such as Falga Regio in the northern antijovian hemisphere and Balgatan Regio in the southern leading hemisphere.

Overall the ridged plains unit forms a relatively flat to gently undulating surface (*Figueredo and Greeley,* 2004). The ridges and troughs cross-cut one another and have a wide range of orientations, lengths, widths, and states of preservation (*Sullivan et al.,* 1999a,b; *Spaun et al.,* 2003). Sets of parallel and subparallel ridges are evident, along with less organized sets of various orientations that are typically shorter in length (*Greeley et al.,* 1998b; *Patel et al.,* 1999). In some small areas (~80 km^2 or less), the topographic relief of ridged plains is subdued and brighter than surrounding terrain, which may result from local heating that viscously relaxes the icy crust. The ridged plains are disrupted in some locations by pits and domes, and cross-cut by wide ridge complexes and bands. Differences in topography and albedo typically cause sharp contacts between ridged plains and other units. Overall, the unit represents a history of crustal deformation that "erases" (resurfaces) traces of previous surface features. This history is not simple, or necessarily identical in all locations (*Figueredo and Greeley,* 2004), but cannot be further subdivided on the basis of existing images.

The ridged plains are interpreted to result from either multiple episodes of ridge building (*Geissler et al.,* 1998b; *Patel et al.,* 1999; *Sullivan et al.,* 1999b; *Figueredo and Greeley,* 2004) or from a continuous process of ridge building (*Greenberg et al.,* 1998); in either case, the process represents a style of tectonic resurfacing (*Head et al.,* 1997). *Greenberg et al.* (1998) argue that tectonic resurfacing and ridge-building were contemporaneous with the formation of other units, while others (*Head et al.,* 1999b; *Pappalardo et al.,* 1999; *Kadel et al.,* 2000; *Spaun et al.,* 1999a) propose that it was most intensive in the earliest parts of Europa's visible history, with a subsequent transition toward other styles of resurfacing. Many of the albedo-based subdivisions of the plains suggested by *Lucchitta and Soderblom* (1982) are not supported in Galileo-based mapping (*Greeley et al.,* 2000) at improved image resolution.

3.2. Bands Material

Bands are linear zones distinguished by their contrast in albedo and/or surface texture compared with the surrounding terrain (Fig. 14) (*Greeley et al.,* 2000; see chapter by Prockter and Patterson). Band margins are generally sharp, and some show bounding ridges (Fig. 15).

They have an internal structure of ridges and troughs trending subparallel to each other and the boundaries of the band, and in some cases there is a single narrow central trough surrounded by a hummocky textured zone on both sides. Outside the hummocky zone are subparallel ridges and troughs, ~300–400 m wide, that are regularly spaced and somewhat triangular in cross-section.

Some bands have also been termed "pull-aparts" (*Sullivan et al.,* 1998; *Greeley et al.,* 1998b), and these are most clearly seen in a 2000 × 600-km zone southwest of the antijovian point, named Argadnel Regio, where they separate the crust into plates (*Schenk and McKinnon,* 1989; *Belton et al.,* 1996; *Sullivan et al.,* 1998; *Prockter et al.,* 1999; *Tufts et al.,* 2000). Their geometries enable reconstruction of the original configuration, restoring structures that were apparently displaced as the bands opened along fractures. This implies the icy surface layer behaved brittlely, separating and translating over a low-viscosity subsurface, making room for darker, mobile material to fill the gap (either the same layer that the plates moved on, or from deeper sources).

Analogies have been made between pull-apart formation on Europa and the formation of leads in terrestrial sea ice (*Pappalardo and Coon,* 1996; *Greeley et al.,* 1998a,b) or to spreading centers in the terrestrial lithosphere (*Sullivan et al.,* 1998; *Prockter et al.,* 2002), in that new material (liquid and/or solid-state) is intruded and extruded as the opposing plates are separated. *Greenberg et al.* (1998) and *Tufts et al.* (2000) consider that cyclical tension and compression due to Europa's diurnal tidal flexing might create bands through a "ratcheting" process (see also *Manga and Sinton,* 2004). In this view, cracks open during the tensile phase of the diurnal cycle, allowing water to rise and freeze. The cracks are unable to close completely during the compressional phase due to the addition of new material; hence, the band widens with time as new material is added.

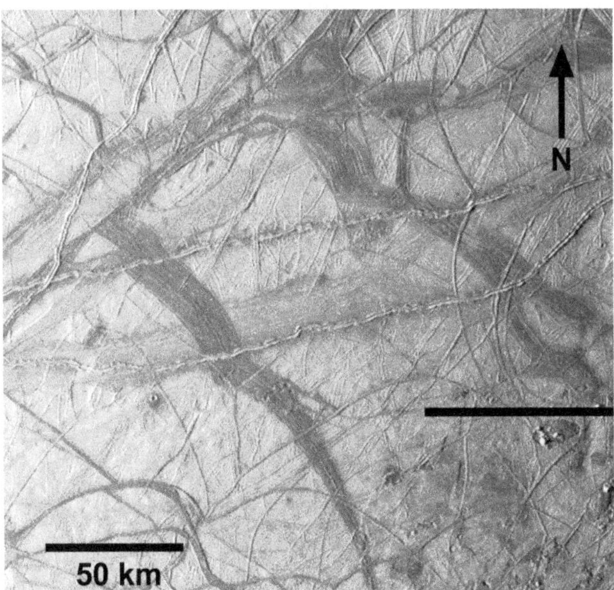

Fig. 14. Wide pull-apart bands in Argadnel Regio. These bands are inferred to be "spreading zones" in the ridged plains that are infilled with darker material derived from the interior. With time, the dark materials brighten. From Galileo SSI observation C3ESWEDGES01 at 420 m/pixel.

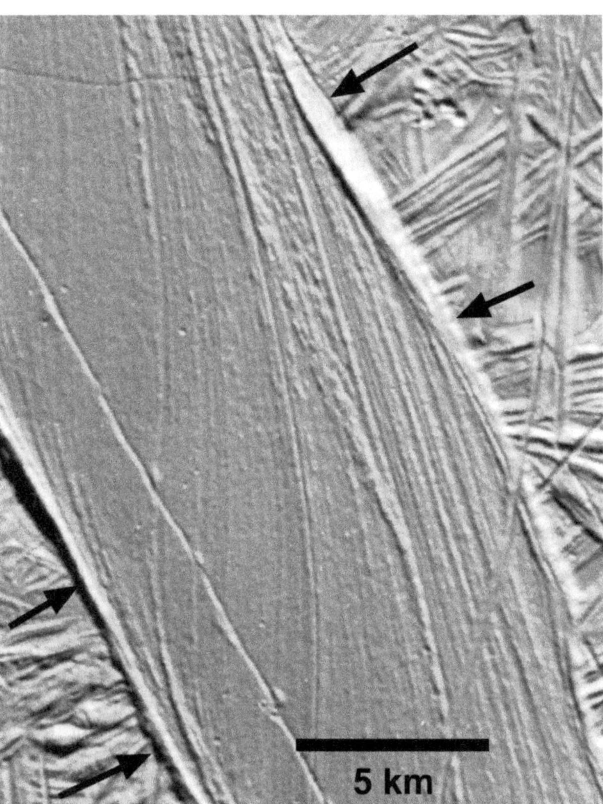

Fig. 15. A portion of the band Astypalaea Linea, showing bounding ridges (arrows). From Galileo SSI observation 17ESSTRSLP01 at 39 m/pixel.

Stratigraphic evidence indicates that dark bands brighten with time (*Pappalardo and Sullivan,* 1996; *Greeley et al.,* 1998a,b; *Geissler et al.,* 1998a), perhaps related to chemical changes and/or sputtering processes that redistribute surface frost (*Geissler et al.,* 1998b). Some bands are noted to be bright in near-infrared data and darker at shorter wavelength, but indistinguishable in clear filter images (*Geissler et al.,* 1998b); these bands are also the oldest in observed stratigraphic sequences of lineaments.

3.3. Ridge Material

Ridges are Europa's most ubiquitous landform (see chapters by Kattenhorn and Hurford and by Prockter and Patterson). Although ridge material can be illustrated on detailed maps, the global map of Fig. 1 is too small to show it as individual units. We do, however, discuss ridge material for completeness. Globally-extensive ridges are shown on Fig. 1 with dark lines. Ridge material most commonly occurs in the form of double ridges, i.e., a ridge pair separated by a medial trough (Fig. 9). Ridges range in length from a few kilometers to >1000 km. Double ridges are 0.2 to >4 km wide; have flank slopes near the angle of repose for loose, blocky material; and are characterized by a continuous axial trough. In cross section, some double ridges are slightly convex to trapezoidal, with a central depression. Preexisting topography has been identified on some ridge flanks (*Head et al.*, 1999a). Mass wasting is prevalent along the ridges, with the debris apparently draping over preexisting terrain (*Sullivan et al.*, 1999a; see chapter by Moore et al.).

Many ridges show evidence for strike-slip motion, a characteristic not seen in isolated troughs (*Hoppa et al.*, 1999a; see chapters by Kattenhorn and Hurford and by Prockter and Patterson). Apparent ridge morphology is sensitive to lighting geometry. At low-solar-incidence angles, some complex ridges appear as triple bands (*Greeley et al.*, 1997) with diffuse dark material on the flanks, while topographic details are better seen at higher-solar-incidence angles, where the dark material is seen as infilling topographic lows. In high-resolution stereo images, it is apparent that dark material occurs in local topographic lows, such as the floors of axial troughs and on wall terraces. Some ridges are flanked by topographic depressions and/or fine-scale fractures, suggesting loading of the lithosphere either from above (due to the weight of the ridge material) or from below (e.g., due to withdrawal of subsurface material) during ridge formation (*Head et al.*, 1999a; *Sullivan et al.*, 1999b; *Tufts et al.*, 2000).

A variety of mechanisms for ridge formation has been proposed (see chapter by Prockter and Patterson; also reviewed in *Pappalardo et al.*, 1999; *Greeley et al.*, 2004) with different implications for the presence and distribution of liquid water, including tidal "pumping" (*Greenberg et al.*, 1998), volcanism (*Fagents et al.*, 1998; *Kadel et al.*, 1998), dike intrusion (*Turtle et al.*, 1998), tectonic compression (*Sullivan et al.*, 1998; *Patterson et al.*, 2006), linear diapirism (*Head et al.*, 1999a), shear heating (*Gaidos and Nimmo*, 2000; *Nimmo and Gaidos*, 2002), and localized zones of volumetric strain as compaction, dilation, and/or shear bands (*Aydin*, 2006).

Stresses in the crust from diurnal tides and nonsynchronous rotation can provide the extensional forces (*Helfenstein and Parmentier*, 1980, 1983, 1985; *McEwen*, 1986b; *Leith and McKinnon*, 1996; *Greenberg et al.*, 1998, 2002; *Hoppa et al.*, 1999a,b) needed for most of these formation models. The cycloidal pattern of some ridges is best explained by propagation at rates slower than Europa's orbit around Jupiter, such that the direction of crack propagation varies with the diurnally rotating stress field (*Hoppa et al.*, 1999b), while straight ridges could have formed from relatively straight cracks, which perhaps had greater initial propagation speeds. Further complexity arises if the development of cycloid cusps is governed by shear stresses at the tips of cyloid cracks, even as the growth of the cycloidal cracks are governed by diurnal stresses (*Marshall and Kattenhorn*, 2005; *Groenleer and Kattenhorn*, 2008).

Greenberg et al. (1998) suggest a ridge load could depress a thin ice lithosphere below a water line to cause surface flooding; *Pappalardo and Coon* (1996) argue that this is possible only if the lithosphere is <2 km thick. *Greeley* (1997) and *Fagents et al.* (2000) considered the possibility that the dark flanks of triple bands were created by ballistic emplacement of dark materials entrained in gas-driven cryovolcanic eruptions, or that the materials are thin lag deposits, formed adjacent to a water or solid-state ice intrusion, due to sublimation of surface frosts and local concentration of refractory materials. This is along the lines of a model suggested by *Head and Pappalardo* (1999), in which ridge-related heating might trigger partial melting, perhaps mobilizing brines that contribute to formation of the dark flanks.

3.4. Lenticulated Terrain Material

Lenticulae (see section 2.2.8) exist both as relatively widely spaced occurrences of typically smaller lenticulae in ridged plains areas (Fig. 16) and in denser associations of typically larger lenticulae. The latter are sometimes termed "microchaos" and may arise from the merger of several individual lenticulae (*Spaun et al.*, 1999a,b).

In the global map (Fig. 1), widely spaced lenticulae are marked as structural features where there is sufficient resolution and appropriate illumination to identify them. However, where they are of sufficient density and size to be considered microchaos, they are mapped as a separate material unit: lenticulated terrain. Such lenticulated terrain is mapped as such even though the older ridged plains terrain that has been disrupted is still discernable surrounding the lenticulae and microchaos within the unit.

The morphological transition from lenticulae to larger chaos areas (Fig. 17) suggests related formational processes (*Spaun et al.*, 1999a; *Riley et al.*, 2000; *Sotin et al.*, 2002).

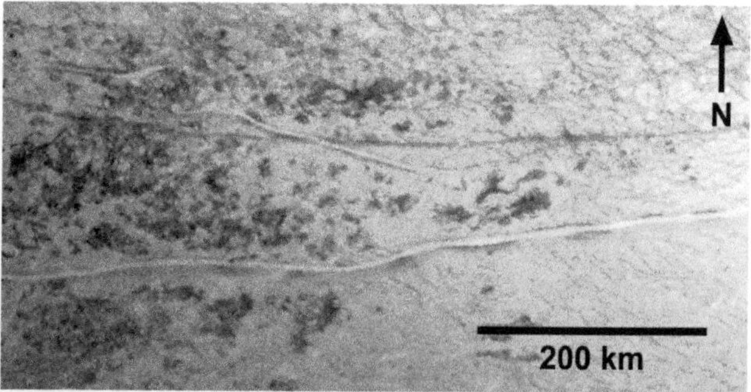

Fig. 16. Area south of Argadnel Regio, showing a transition from ridged plains (right side) with sparse lenticulae, to increasing density and size of lenticulae and "microchaos" to the west. The western region has been mapped as lenticulaed terrain on the global geological map. From Galileo SSI observation E6ESGLOBAL01 at 1.6 km/pixel.

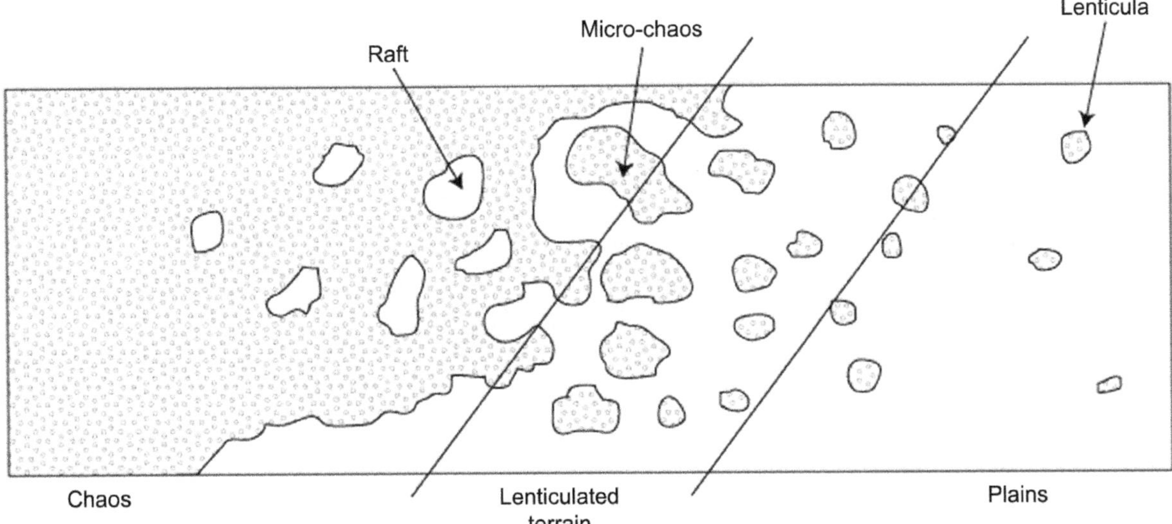

Fig. 17. Schematic representation of the inferred gradation in crustal disruption by endogenic processes, and equivalent map units. Plains are disrupted by widely spaced lenticulae (right side), which merge into closer-packed and larger "microchaos" in lenticulated terrain, into terrain that is completely disrupted into chaos. Although a time sequence or evolution from lenticulae to full-on chaos might be inferred, no *a priori* age relations can be assumed among the units, because they may have formed simultaneously, differing only in the degree of modification.

Proposed models of lenticula formation include (1) ice volcanism (*Fagents et al.,* 1998), (2) diapirism (*Pappalardo et al.,* 1998a; *Rathbun et al.,* 1998) possibly accompanied by partial melting of a salt-rich icy lithosphere (*Head and Pappalardo,* 1999), and (3) melt-through of the icy shell (*Carr et al.,* 1998; *Greenberg et al.,* 1999; *O'Brien et al.,* 2002; *Greenberg et al.,* 2003).

Pappalardo et al. (1998a) argue that most lenticulae have diameters of ~10 km (see also *Spaun et al.,* 1999a,b, 2001), and that the size similarity and morphological gradation among pits, domes, and spots suggest that they are genetically related (*Carr et al.,* 1998; *Greeley et al.,* 1998b; *Pappalardo et al.,* 1998a; *Spaun et al.,* 1998a). This suggestion is consistent with an origin through diapirism as the manifestation of solid-state convection of Europa's icy shell (*Rathbun et al.,* 1998). However, *Greenberg et al.* (1999) argue that pits and spots are only small members of a continuous size distribution of chaos terrain. New insights into this issue are presented in the chapter by Collins and Nimmo.

Carr et al. (1998) and then *Pappalardo and Barr* (2004) subdivided domes into two endmember types based on their inferred genetic relationship to the preexisting surface. Type 1 domes consist of darker material that either replaced or covered over the preexisting terrain. In contrast, preexisting terrain is preserved on the surfaces of Type 2 domes, suggesting that the surface was upwarped but not destroyed, although fractures can occur along their crests. Type 2 dome boundaries can be continuous with respect to the surrounding terrain with no discrete scarp, producing a gentle convex dome, consistent with flexure of an elastic plate warped upward from below. Other Type 2 domes are bounded by abrupt scarps of apparent tectonic origin, with associated dome tops that are tilted or relatively flat, consistent with laccolith-like brittle failure and punching upward of the dome along bounding fractures (*Pappalardo and Barr,* 2004).

3.5. Chaos Material

Chaos (Figs. 4 and 18) has a variety of expressions on Europa. However, because of the small size of Fig. 1, and variations in appearance due to uneven image resolution and lighting, all chaos of the same general age is mapped as a combined unit. We note that various classifications of chaos have been used in other detailed mapping: (1) an albedo-based distinction between central, dark mottled terrain and peripheral bright mottled terrain (*Senske et al.,* 1998); (2) a three-fold division among (a) chaos interpreted to be "fresh" (unmodified), (b) chaos that that has been modified in some way, and (c) chaos consisting of isolated massifs in the middle of ridged plains interpreted to be remnants of old chaos terrain (*Greenberg et al.,* 1999); (3) a two-fold division between knobby and platy chaos (*Greeley et al.,* 2000); and (4) a three-fold division for younger, elevated chaos (Fig. 19), chaos that is intermediate in topography and age, and older, subdued chaos (*Figueredo and Greeley,* 2004). Other terms that have been used in chaos mapping are speckled material or speckled chaos (*Prockter et al.,* 1999; *Prockter and Schenk,* 2002) to describe chaos that has no discernable plates at the resolution of the image. Moreover, merged or microchaos are interpreted as transitional from lenticulae to chaos (*Spaun et al.,* 1999a,b).

Chaos regions are interpreted as areas of focused heat flow and perhaps local melting (see summaries in *Collins et al.,* 2000, and in the chapter by Collins and Nimmo). In

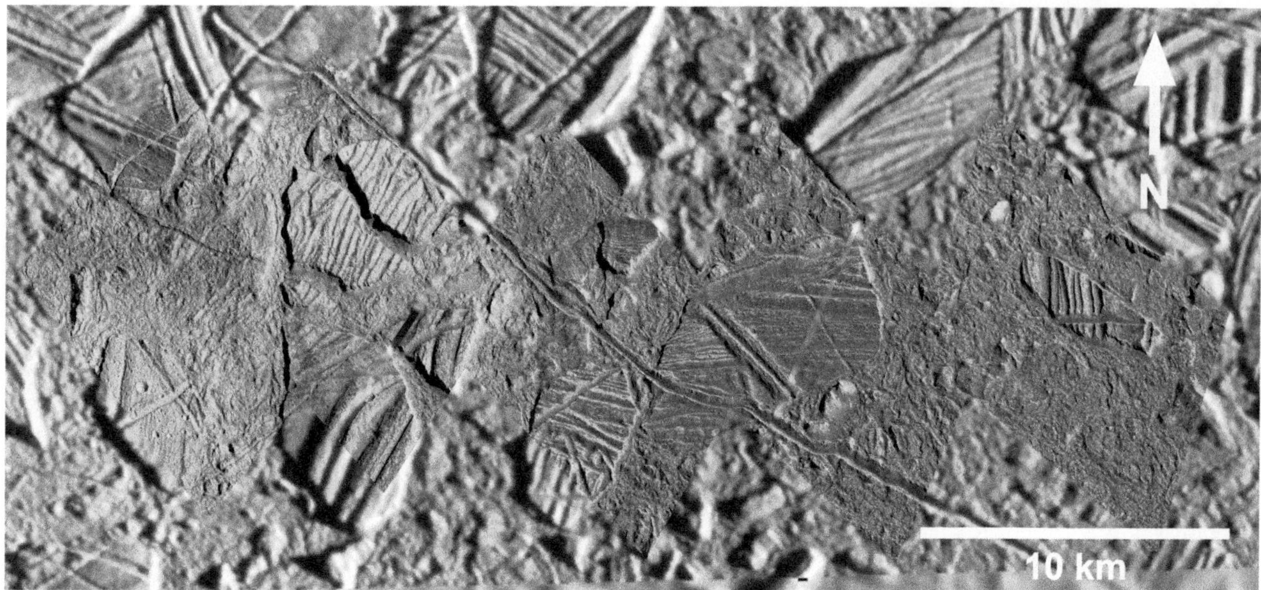

Fig. 18. Closeup view of icy blocks of ridged plains within Conamara Chaos (Fig. 4). From Galileo SSI image E6ESDRKLN_01 at 180 m/pixel, with insets from Galileo SSI observation 12ESCHAOS_01 at 9 m/pixel, composite by Deutsches Zentrum für Luft- und Raumfahrt e.V. (DLR).

a melting model, blocks are analogous to buoyant icebergs (e.g., *Greenberg et al.,* 1999). Alternatively, solid-state ice might rise diapirically to the surface, a plausible scenario if partial melting takes place (*Collins et al.,* 2000), disrupting the relatively cold and rigid lithosphere similar to the means proposed for lenticulae and moving fragmented slabs of colder lithospheric material (*Pappalardo et al.,* 1998a).

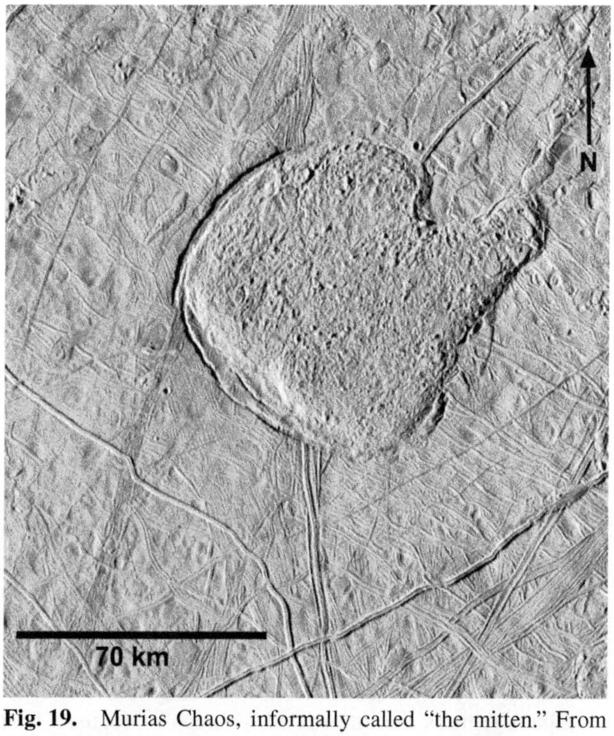

Fig. 19. Murias Chaos, informally called "the mitten." From Galileo SSI observation E15REGMAP02 at 234 m/pixel.

Topographic data across Conamara Chaos show that much of it is ~300 m higher than the surrounding plains (*Williams and Greeley,* 1998; *Schenk and Pappalardo,* 2002). Similar results are found for Murias Chaos ("the mitten") in the leading hemisphere (*Figueredo et al.,* 2002). In Conamara Chaos, ~60% of the preexisting terrain was replaced or converted into matrix material, and many of the surviving blocks can be restored to their original positions through translation and rotation (*Spaun et al.,* 1998a).

3.6. Crater Materials

Crater materials consist of the floor, crater rim, and ejecta deposits formed by impact events (Fig. 5), as well as units associated with the known multiring impact structures, Tyre and Callanish. Impact crater morphologies appear to place the base of a solid ice layer at >20 km depth when the impacts occurred (*Schenk,* 2002).

The relatively small number of primary impact craters on Europa (e.g., *Chapman et al.,* 1998; *Moore et al.,* 2001) suggest a relatively young age for its surface of ~40–90 m.y. (see chapter by Bierhaus et al.), which refines earlier estimates in the range ~10–60 m.y. (*Zahnle et al.,* 1998, 2003; *Schenk et al.,* 2004). Alternatively, *Neukum* (1997) assumed that large basin-forming impacts on the Galilean satellites date from the period of late heavy bombardment, resulting in an older inferred age of 1–3 G.y. for the europan surface. Further evidence of young age comes from crater size-frequency distributions of Pwyll's secondaries (Fig. 20), which suggest that most of the small craters on Europa are secondaries rather than small primaries (*Bierhaus et al.,* 2001; *Moore et al.,* 2001; *Schenk et al.,* 2004; see chapter by Bierhaus et al.).

4. GEOLOGICAL HISTORY

A fundamental application of geological mapping is to determine the sequence of geological events represented by the mapped units and the derivation of the geological history expressed on the surface. Mapped units are considered to be composed of three-dimensional materials (i.e., they have thickness that extends below the surface). These materials can be placed in a stratigraphic position relative to other units and structures on the basis of cross-cutting, superposition, and embayment relationships (*Wilhelms,* 1990). Consequently, geological mapping is a tool for understanding the evolution of a planetary surface. In the case of Europa, this surface only reflects a small fraction of the age of the solar system (Fig. 21), while the geological "prehistory" of Europa from accretion (see chapters by Estrada et al. and Canup and Ward) to the oldest surface unit cannot be directly addressed by geological mapping.

Although a formal europan geological timescale must await more complete imaging coverage of the surface, mapping using available data suggests a distinct sequence in its history. Both the pole-to-pole mapping (*Figueredo and Greeley,* 2004) and global mapping (Fig. 1) divide the chronology into numbered episodes, from "1" (oldest) to "4" (youngest). This is conceptually similar to the initial stratigraphy proposed for Earth of primary, secondary, tertiary, and quaternary (*Lehmann,* 1756).

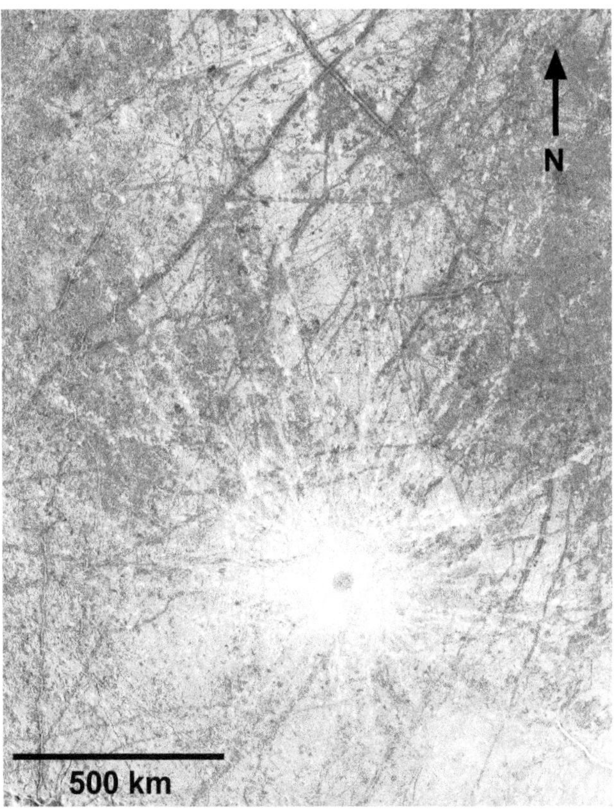

Fig. 20. Pwyll is the type example for the youngest impact craters on Europa. Rays formed by the impact overlap adjacent chaos material in Dyfed Regio (upper right) and Annwn Regio (upper left). From Galileo SSI observation E4ESGLOMAP01 at 1.2 km/pixel.

4.1. First Episode

The earliest episode recorded on Europa's surface is represented by the ridged plains unit, indicated as 1p in Fig. 1. A constraint on the age of this episode comes from crater density and impact flux models that estimate the global age of the surface as ~40–90 m.y. (see chapter by Bierhaus et al.). This is an average age, implying that some surfaces may be older, and it places most of the span of Europa's visible record as roughly coincident with the Cenozoic or late Mesozoic Eras on Earth (Fig. 21). The general interpretation is that earlier surface features, including most of the impact crater record, were obliterated by resurfacing. This entails ridge building as the dominant process that then led to the formation of densely spaced ridges covering the surface. It is possible that some massifs are remnant of previous terrains (e.g., *Greenberg et al.,* 1999), rather than blocks that were subsequently uplifted. Just as it is inferred that evidence of previous impact cratering was obliterated, earlier terrain disruption also cannot be ruled out. As the intensity of tectonic resurfacing waned with time, local endogenic deformation occurred that produced the stratigraphically lower chaos regions (1ch units).

4.2. Second Episode

The transition from the first to second episode is characterized by the formation of more prominent ridges that are spaced farther apart. These ridges superpose the ridged plains and chaos units in the antijovian and subjovian areas of the first period (units 1p and 1ch). The earliest stages in this period include the formation of the pull-apart bands in Argadnel Regio that are, in turn, overprinted by the dense accumulation of ridges in the Argandel Regio south unit (2ar). The impacts of Tyre and Callinish (unit 2c) occurred in this period, as they disrupt some ridges but are superposed by other ridges.

4.3. Third Episode

The third episode is characterized by the widespread formation of chaos units (3ch). Large regions of chaos were emplaced, including Powys Regio (units 3chwp and 3chep) and Tara Regio (3cht) in the leading hemisphere, and Annwn Regio (3cha) and Dyfed Regio (3chd) in the trailing hemisphere. In both hemispheres there is a stratigraphic sequence from ridged plains to early chaos in the first episode, to lineae in the second episode, to chaos in the third episode (*Figueredo and Greeley,* 2004). The leading hemisphere includes the Powys west unit (3chwp), which is intermediate between plains and chaos both morphologically and stratigraphically. There is also widespread emplacement of len-

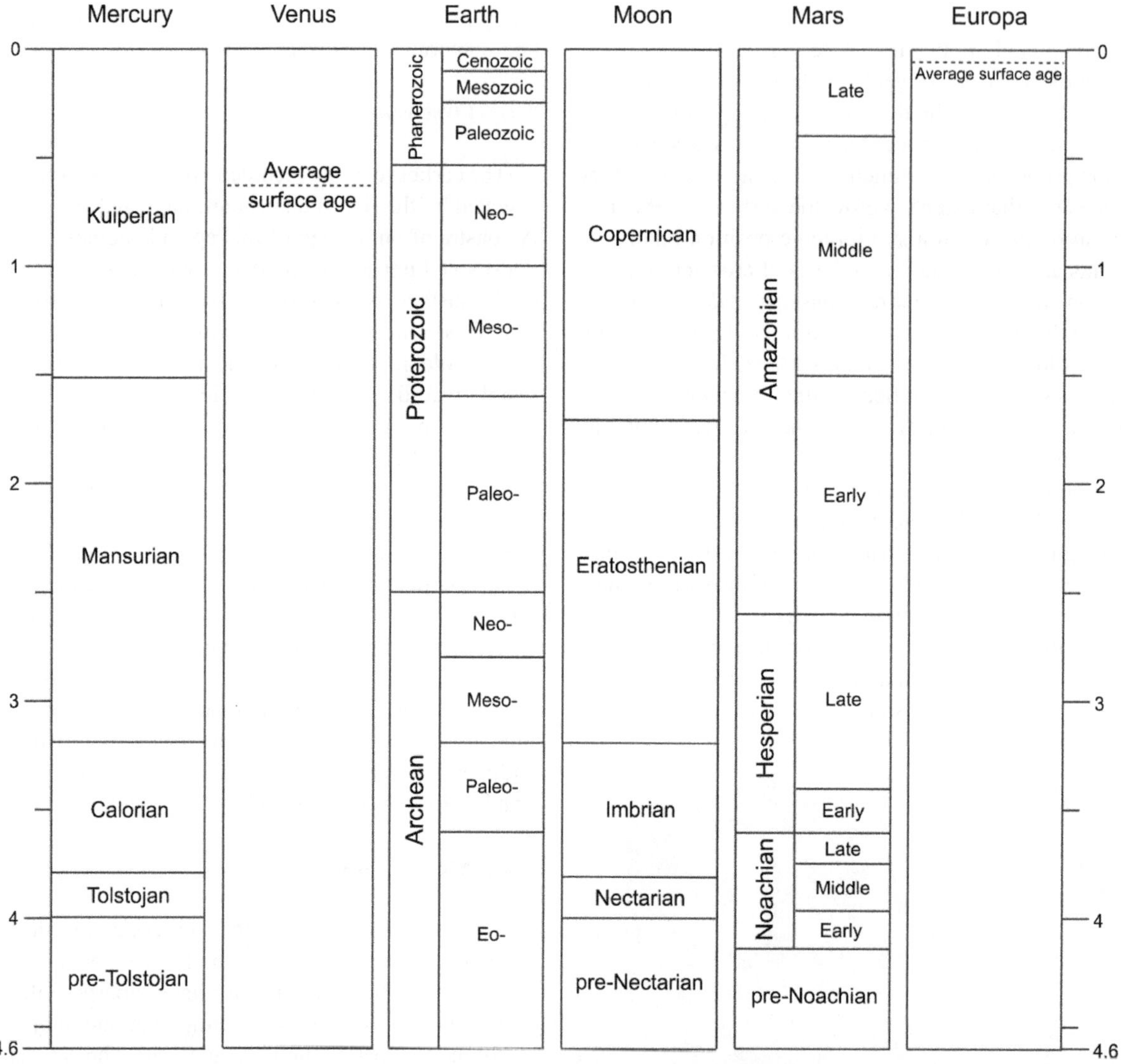

Fig. 21. Comparison of Europa's 40–90-m.y. average surface age (see chapter by Bierhaus et al.) to the geological record of terrestrial bodies. Europa's young surface might be most directly compared to Venus, which also has a sparse impact record and is inferred to have undergone geologically recent global resurfacing.

ticulated terrain units (3l), superposed on second-episode lineae. Some lineae, notably Agenor, show evidence of reactivation during this time (*Klemaszewski et al.,* 1999). Most preserved impact craters (3c) formed in this episode.

4.4. Fourth Episode

Superposed on third-episode chaos units are the youngest craters (4c) on Europa, the most prominent of which is Pwyll with its extensive ejecta blanket and rays. This would be conceptually similar to the Copernican episode in lunar stratigraphy (*Shoemaker and Hackman,* 1962), which is marked by impact craters with fresh rays. Some small exposures of chaos, notably the knobby type exemplified by Murias Chaos (*Greeley et al.,* 2000) and Castalia Macula, are mapped in this episode. Young fractures can be seen that cut across chaos terrains (Fig. 18) and band material (Fig. 15).

5. SUMMARY

The existence, stability, and evolution of a subsurface ocean through time are believed to be intricately tied to Europa's surface age and geological activity. Most chaos regions and dark plains materials are among Europa's youngest units, cross-cutting older bands and ridged plains, while ridged plains materials are commonly inferred to be the oldest units (*Head et al.,* 1999b; *Sullivan et al.,* 1999b; *Prockter et al.,* 1999; *Figueredo and Greeley,* 2000; *Kadel et al.,* 2000; *Greeley et al.,* 2000). *Greenberg et al.* (1999, 2002) and *Riley et al.* (2000) argue that this interpretation is an artifact of the difficulty of recognizing older terrains that have been disrupted by fractures, ridges, and bands. If the former interpretation is correct, as we infer here from the visible stratigraphy, then Europa would appear to have changed in geological style through time from ridged plains formation to mottled terrain formation. It has been suggested that mottled terrain units formed by diapirism triggered as an ice shell cooled and thickened to the point at which it reached a critical thickness and solid state convection was initiated (*Pappalardo et al.,* 1998a; *McKinnon,* 1999; *Figueredo and Greeley,* 2004).

The current surface only reflects a small fraction of Europa's history, and the nature of Europa's prior geological history cannot be addressed directly by stratigraphic mapping of the current surface of the icy shell; therefore, we can only cautiously address the "missing" geological history. Because the surface is so young, and some units appear to be manifestations of the activity of a ductile or liquid sublayer, it is considered to be unlikely that the putative ocean or warm ice would have frozen completely in the last tens of million years (*Greeley et al.,* 2004). *Pappalardo et al.* (1999) lists the most likely scenarios: either steady-state, episodic, or sporadic resurfacing. Support for the theory of sporadic resurfacing comes from evolution models that consider the effect and history of Europa's tidal stresses, which could wax and wane cyclically (*Ojakangas and Stevenson,* 1986; *Hussman and Spohn,* 2004; see chapter by Nimmo and Manga).

Determining the location and distribution of units associated with tectonic and cryovolcanic processes that may have emplaced materials from the underlying ice or liquid water ocean on the surface, and vice versa, have direct implications for the ongoing investigation of the astrobiological potential of Europa (*Reynolds et al.,* 1983; *Greenberg et al.,* 2000; *Chyba and Hand,* 2001; *Figueredo et al.,* 2003; see chapter by Hand et al.). Surface-subsurface exchange could transport organic compounds and oxidants from the surface to the subsurface, where conditions for life sustainment are more favorable (e.g., *Hall et al.,* 1995; *McCord et al.,* 1998a,b; *Carlson et al.,* 1999a; *Prockter and Schenk,* 2002).

The existing imaging coverage of Europa is incomplete: only a small fraction of the planet has been imaged at the same resolutions as comparably better understood silicate bodies in the inner solar system. This, along with the exotic composition of the crust (by our standards — compared to the baseline established by terrestrial geology) complicates definitive assessment of the terrains including their geological diversity, relative ages, and underlying processes. Higher-resolution and complete imaging by a future Europa mission (see chapter by Greeley et al.; *Space Studies Board,* 1999, 2003, 2007; *Clark et al.,* 2007), ideally with global coverage at uniform illumination and viewing geometries, will allow further understanding of Europa's geology and stratigraphy. In the meantime, our current knowledge of the geologic stratigraphy provides a framework for other aspects of Europa science, as well as focusing questions for future exploration.

Acknowledgments. This work was supported by NASA through the Planetary Geology and Geophysics Program. We thank D. Ball of the ASU Space Photography Laboratory and T. Hare of the U. S. Geological Survey for access to images and cartographic materials, M. Furst for assistance in the preparation of figures, C. Bradbury for computer support, and D. Williams, D. Senske, J. Blue, N. Spaun, and R. Pappalardo for reviews of the manuscript.

REFERENCES

Anderson J. D., Schubert G., Jacobson R. A., Lau E. L., Moore W. B., and Sjogren W. L. (1998) Europa's differential internal structure: Inferences from four Galileo encounters. *Science, 281,* 2019–2022.

Aydin A. (2006) Failure modes of the lineaments on Jupiter's moon, Europa: Implications for the evolution of its icy crust. *J. Struct. Geol., 28,* 2222–2236.

Bagenal F., Dowling T. and McKinnon W., eds. (2004) *Jupiter: The Planet, Satellites and Magnetosphere.* Cambridge Univ., Cambridge. 719 pp.

Batson R. M., Whitaker E. A., and Wilhelms D. E. (1990) History of planetary cartography. In *Planetary Mapping* (R. Greeley and R. M. Batson, eds.), pp. 12–59. Cambridge Univ., Cambridge.

Belton M. J. S. and 33 colleagues (1996) Galileo's first images of Jupiter and the Galilean satellites. *Science, 274,* 377–385.

Bierhaus E., Chapman C., Merline W., Brooks S., and Asphaug E. (2001) Pwyll secondaries and other small craters on Europa. *Icarus, 153,* 264–276.

Carlson R. W. and 39 colleagues (1996) Near-infrared spectroscopy and spectral mapping of Jupiter and the Galilean satellites: Results from Galileo's initial orbit. *Science, 274,* 385–388.

Carlson R. W. and 13 colleagues (1999a) Hydrogen peroxide on the surface of Europa. *Science, 283,* 2062–2064.

Carlson R. W., Johnson R. E., and Anderson M. S. (1999b) Sulfuric acid on Europa and the radiolytic sulfur cycle. *Science, 286,* 97–99.

Carr M. H., Belton M. J. S., Chapman C. R., Davies M. E., Geissler P., Greenberg R., McEwen A. S., Tufts B. R., Greeley R., and Sullivan R. (1998) Evidence for a subsurface ocean on Europa. *Nature, 391,* 363–365.

Cassen P., Reynolds R. T., and Peale S. J. (1979) Is there liquid water on Europa. *Geophys. Res. Lett., 6,* 731–734.

Cassen P. M., Peale S., and Reynolds R. T. (1982) Structure and thermal evolution of the Galilean satellites. In *Satellites of Jupiter* (D. Morrison, ed.), pp. 93–128. Univ. of Arizona, Tucson.

Chapman C. R., Merline W. J., Bierhaus B., and Brooks S. (1998) Cratering in the jovian system: Intersatellite comparisons (abstract). In *Lunar and Planetary Science XXIV,* Abstract #1927. Lunar and Planetary Institute, Houston (CD-ROM).

Chyba C. F. and Hand K. P. (2001) Life without photosynthesis. *Science, 292,* 2026–2027.

Clark K., Greeley R., Pappalardo R., and Jones C. (2007) *2007 Europa Explorer Mission Study: Final Report,* JPL D-38502. Jet Propulsion Laboratory, Pasadena. 282 pp.

Collins G. C., Head J. W., Pappalardo R. T., and Spaun N. A. (2000) Evaluation of models for the formation of chaotic terrain on Europa. *J. Geophys. Res., 105,* 1709–1716.

Davies M. E. and 11 colleagues (1998) The control networks of the Galilean satellites and implications for global shape. *Icarus, 135,* 372–376.

Fagents S. A. (2003) Considerations for effusive cryovolcanism on Europa: The post-Galileo perspective. *J. Geophys. Res., 108,* DOI: 10.1029/2003JE002128.

Fagents S. A., Kadel S. D., Greeley R., Kirk R. L., and the Galileo SSI Team (1998) Styles of cryovolcanism on Europa: Summary of evidence from the Galileo nominal mission (abstract). In *Lunar and Planetary Science XXIX,* Abstract #1721. Lunar and Planetary Institute, Houston (CD-ROM).

Fagents S. A., Greeley R., Sullivan R. J., Pappalardo R. T., Prockter L. M., and the Galileo SSI Team (2000) Cryomagmatic mechanisms for the formation of Rhadamanthys Linea, triple band margins, and other low-albedo features on Europa. *Icarus, 144,* 54–88.

Fanale F. P. and 10 colleagues (2000) Tyre and Pwyll: Galileo orbital remote sensing of mineralogy versus morphology at two selected sites on Europa. *J. Geophys. Res., 105,* 22647–22657.

Figueredo P. H. and Greeley R. (2000) Geologic mapping of the northern leading hemisphere of Europa from Galileo solid-state imaging data. *J. Geophys. Res., 105,* 22629–22646.

Figueredo P. H. and Greeley R. (2004) Resurfacing history of Europa from pole-to-pole geological mapping. *Icarus, 167,* 287–312.

Figueredo P. H., Chuang F. C., Rathbun J., Kirk R. L., and Greeley R. (2002) Geology and origin of Europa's mitten feature (Murias Chaos). *J. Geophys. Res., 107,* DOI: 10.1029/2001JE001591.

Figueredo P. H., Greeley R., Neuer S., Irwin L., and Schulze-Makuch D. (2003) Locating potential biosignatures on Europa from surface geology observations. *Astrobiology, 3,* 851–861.

Fimmel R. O., Swindell W., and Burgess E. (1974) *Pioneer Odyssey: Encounter with a Giant* NASA SP-349, Washington, DC.

Fimmel R. O., Van Allen J. A., and Burgess E. (1980) *Pioneer: First to Jupiter, Saturn and Beyond.* NASA SP-446, Washington, DC.

Gaidos E. J. and Nimmo F. (2000) Tectonics and water on Europa. *Nature, 405,* 637.

Geissler P., Greenberg R., Hoppa G., Helfenstein P., McEwen A., Pappalardo R., Tufts R., Ockert-Bell M., Sullivan R., and Greeley R. (1998a) Evidence for non-synchronous rotation of Europa. *Nature, 391,* 368–370.

Geissler P. and 16 colleagues (1998b) Evolution of lineaments on Europa: Clues from Galileo multispectral imaging observations. *Icarus, 135,* 107–126.

Giese B., Wagner R., Neukum G., Sullivan R., and the SSI Team (1999a) Doublet ridge formation on Europa: Evidence from topographic data. *Bull. Am. Astron. Soc., 31(4),* Abstract #62.08.

Giese B., Wagner R., and Neukum G. (1999b) The local topography of Europa: Stereo analysis of Galileo SSI images and implications for geology. In *Geophys. Res. Abstr., 1,* Abstract #6208. European Geosciences Union (CD-ROM).

Goldreich P. (1966) Final spin states of planets and satellites. *Astrophys. J., 71,* 1–7.

Greeley R. (1997) Geology of Europa: Galileo update. In *The Three Galileos: The Man, the Spacecraft, the Telescope* (C. Barbieri et al., eds.), pp. 191–200. Kluwer, Dordrecht.

Greeley R., Sullivan R., Bender K. C., Homan K. S., Fagents S. A., Pappalardo R. T., and Head J. W. (1997) Europa triple bands — Galileo images. In *Lunar and Planetary Science XXVIII,* pp. 455–456. Lunar and Planetary Institute, Houston.

Greeley R., Sullivan R., Coon M. D., Geissler P. E., Tufts B. R., Head J. W., Pappalardo R. T., and Moore J. M. (1998a) Terrestrial sea ice morphology: Considerations for Europa. *Icarus, 135,* 25–40.

Greeley R. and 20 colleagues (1998b) Europa: Initial Galileo geological observations. *Icarus, 135,* 4–24.

Greeley R. and 17 colleagues (2000) Geologic mapping of Europa. *J. Geophys. Res., 105,* 22559–22578.

Greeley R., Chyba C. F., Head J. W., McCord T. B., McKinnon W. B., Pappalardo R. T., and Figueredo P. (2004) Geology of Europa. In *Jupiter: The Planet, Satellites and Magnetosphere* (F. Bagenal et al., eds.), pp. 329–362. Cambridge Univ., Cambridge.

Greenberg R. and Weidenschilling S. J. (1984) How fast do Galilean satellites spin? *Icarus, 58,* 186–196.

Greenberg R., Geissler P., Hoppa G., Tufts B. R., Durda D. D., Pappalardo R., Head J. W., Greeley R., Sullivan R., and Carr M. H. (1998) Tectonic processes on Europa: Tidal stresses, mechanical response, and visible features. *Icarus, 135,* 64–78.

Greenberg R., Hoppa G. V., Tufts B. R., Geissler P., Riley J., and Kadel S. (1999) Chaos on Europa. *Icarus, 141,* 263–286.

Greenberg R., Geissler P., Tufts B. R., and Hoppa G. V. (2000) Habitability of Europa's crust: The role of tidal-tectonic processes. *J. Geophys. Res., 105,* 17551–17562.

Greenberg R., Geissler P., Hoppa G., and Tufts B. R. (2002) Tidal tectonic processes and their implications for the character of Europa's icy crust. *Rev. Geophys., 40,* DOI: 10.1029/2000RG000096.

Greenberg R., Leake M. A., Hoppa G. V., and Tufts B. R. (2003) Pits and uplifts on Europa. *Icarus, 161,* 102–126.

Groenleer J. M. and Kattenhorn S. A. (2008) Cycloid crack sequences on Europa: Relationship to stress history and constraints on growth mechanics based on cusp angles. *Icarus, 193,* 158–181.

Grundy W. M. and 15 colleagues (2007) New Horizons mapping of Europa and Ganymede. *Science, 318,* 234–237.

Hall D. T., Strobel D. F., Feldman P. D., McGrath M. A., and Weaver H. A. (1995) Detection of an oxygen atmosphere on Jupiter's moon Europa. *Nature, 373,* 677–679.

Head J. W. and Pappalardo R. T. (1999) Brine mobilization during lithospheric heating on Europa: Implications for formation of chaos terrain. *J. Geophys. Res., 104,* 27143–27156.

Head J. W., Pappalardo R., Collins G., Greeley R., and the Galileo Imaging Team (1997) Tectonic resurfacing on Ganymede and its role in the formation of grooved terrain. In *Lunar and Planetary Science XXVIII,* pp. 535–536. Lunar and Planetary Institute, Houston.

Head J. W., Sherman N. D., Pappalardo R. T., Greeley R., Sullivan R., Senske D. A., McEwen A., and the Galileo Imaging Team (1998) Geologic history of the E4 region of Europa: Implications for ridge formation, cryovolcanism, and chaos formation. In *Lunar and Planetary Science XXIX,* Abstract #1412. Lunar and Planetary Institute, Houston (CD-ROM).

Head J. W., Pappalardo R., and Sullivan R. (1999a) Europa: Morphological characteristics of ridges and triple bands from Galileo data (E4 and E6) and assessment of a linear diapirism model. *J. Geophys. Res., 104,* 24223–24235.

Head J. W., Pappalardo R. T., Prockter L. M., Spaun N. A., Collins G. C., Greeley R., Klemaszewski J., Sullivan R., Chapman C., and the Galileo SSI Team (1999b) Europa: Recent geological history from Galileo observations. In *Lunar and Planetary Science XXX,* Abstract #1404. Lunar and Planetary Institute, Houston (CD-ROM).

Helfenstein P. and Parmentier E. M. (1980) Fractures on Europa — Possible response of an ice crust to tidal deformation. *Proc. Lunar Planet. Sci. Conf. 11th,* pp. 1987–1998.

Helfenstein P. and Parmentier E. M. (1983) Patterns of fracture and tidal stresses on Europa. *Icarus, 53,* 415–430.

Helfenstein P. and Parmentier E. M. (1985) Patterns of fracture and tidal stresses due to nonsynchronous rotation — Implications for fracturing on Europa. *Icarus, 61,* 175–184.

Hendrix A. R., Barth C. A., and Hord C. W. (1998) Europa: Disk-resolved ultraviolet measurements using the Galileo Ultraviolet Spectrometer. *Icarus, 135,* 72–94.

Hoppa G. V., Tufts B. R., Greenberg R., and Geissler P. (1999a) Strikeslip faults on Europa: Global shear patterns driven by tidal stress. *Icarus, 141,* 287–298.

Hoppa G. V., Tufts B. R., Greenberg R., and Geissler P. E. (1999b) Formation of cycloidal features on Europa. *Science, 285,* 1899–1902.

Hussmann H. and Spohn T. (2004) Thermal-orbital evolution of Io and Europa. *Icarus, 171,* 391–410.

Kadel S. D., Fagents S. A., Greeley R., and the Galileo SSI Team (1998) Trough-bounding ridge pairs on Europa — Considerations for an endogenic model of formation. In *Lunar and Planetary Science XXIX,* Abstract #1078. Lunar and Planetary Institute, Houston (CD-ROM).

Kadel S. D., Chuang F. C., Greeley R., Moore J. M., and the Galileo SSI Team (2000) Geological history of the Tyre region of Europa: A regional perspective on europan surface features and ice thickness. *J. Geophys. Res., 105,* 22656–22669.

Kattenhorn S. A. (2002) Nonsynchronous rotation evidence and fracture history in the bright plains region, Europa. *Icarus, 157,* 490–506.

Khurana K. K., Kivelson M. G., Stevenson D. J., Schubert G., Russell C. T., Walker R. J., and Polanskey C. (1998) Induced magnetic fields as evidence for subsurface oceans on Europa and Callisto. *Nature, 395,* 777–780.

Kivelson M. G., Khurana K. K., Stevenson D. J., Bennett L., Joy S., Russell C. T., Walker R. J., Zimmer C., and Polanskey C. (1999) Europa and Callisto: Induced or intrinsic fields in a periodically varying plasma environment. *J. Geophys. Res., 104,* 4609–4626.

Kivelson M. G., Khurana K. K., Russell C. T., Volwerk M., Walker R. J., and Zimmer C. (2000) Galileo magnetometer measurements: A stronger case for a subsurface ocean at Europa. *Science, 289,* 1340–1343.

Klemaszewski J. E., Greeley R., Prockter L. M., Geissler P. E., and the Galileo SSI Team (1999) Geologic mapping of Eastern Agenor Linea, Europa. In *Lunar and Planetary Science XXX,* Abstract #1680. Lunar and Planetary Institute, Houston (CD-ROM).

Kuiper G. P. (1957) Infrared observations of planets and satellites. *Astron. J., 62,* 291–306.

Laplace P. S. (1805) *Traité de Méchanique Céleste, Vol. 4.* De Crapelet, Paris. 548 pp.

Lehmann J. G. (1756) *Versuch einer Geschichte von Flötz-Gebürgen betreffend deren Entstehung, Lage, darinne befindliche Metallen, Mineralien und Foßilien größtentheils aus eigenen Wahrnehmungen und aus denen Grundsätzen der Natur-Lehre hergeleitet, und mit nöthigen Kupfern versehen.* F. A. Lange, Berlin. 329 pp.

Leith A. C. and McKinnon W. B. (1996) Is there evidence for polar wander on Europa? *Icarus, 120,* 387–398.

Lucchitta B. K. and Soderblom L. A. (1982) The geology of Europa. In *Satellites of Jupiter* (D. Morrison, ed.), pp. 521–555. Univ. of Arizona, Tucson.

Malin M. C. and Pieri D. C. (1986) Europa. In *Satellites* (J. A. Burns and M. S. Matthews, eds.), pp. 689–717. Univ. of Arizona, Tucson.

Manga M. and Sinton A. (2004) Formation of bands and ridges on Europa by cyclic deformation: Insights from analogue wax experiments. *J. Geophys. Res., 109,* E09001, DOI: 10.1029/2004JE002249.

Marshall S. T. and Kattenhorn S. A. (2005) A revised model for cycloid growth mechanics on Europa: Evidence from surface morphologies and geometries. *Icarus, 177,* 341–366.

McCord T. B. and 12 colleagues (1998a) Non-water-ice constituents in the surface material of the icy Galilean satellites from the Galileo near-infrared mapping spectrometer investigation. *J. Geophys. Res., 103,* 8603–8626.

McCord T. B. and 11 colleagues (1998b) Salts on Europa's surface detected by Galileo's Near Infrared Mapping Spectrometer. *Science, 280,* 1242–1245.

McCord T. B. and 11 colleagues (1999) Hydrated salt minerals on Europa's surface from the Galileo Near-Infrared Mapping Spectrometer (NIMS) investigation. *J. Geophys. Res., 104,* 11827–11852.

McEwen A. S. (1986a) Exogenic and endogenic albedo and color patterns on Europa. *J. Geophys. Res., 91,* 8077–8097.

McEwen A. S. (1986b) Tidal reorientation and the fracturing of Jupiter's moon Europa. *Nature, 321,* 49–51.

McKinnon W. B. (1999) Convective instability in Europa's floating ice shell. *Geophys. Res. Lett., 26,* 951–954.

Moore J. M. and 17 colleagues (1998) Large impact features on Europa: Results of the Galileo nominal mission. *Icarus, 135,* 127–145.

Moore J. M. and 24 colleagues (2001) Impact features on Europa: Results of the Galileo Europa Mission (GEM). *Icarus, 151,* 93–111.

Morabito L. A., Synnott S. P., Kupferman P. N., and Collins S. A. (1979) Discovery of currently active extraterrestrial volcanism. *Science, 204,* 972.

Moroz V. L. (1965) Infrared spectrophotometry of the Moon and the Galilean satellites of Jupiter. *Astron. Zh., 42,* 1287.

Neukum G. (1997) Bombardment history of the jovian system. In *The Three Galileos: The Man, the Spacecraft, the Telescope* (C. Barbieri et al., eds.), pp. 201–212. Kluwer, Dordrecht.

Nimmo F. and Gaidos E. (2002) Strike-slip motion and double ridge formation on Europa. *J. Geophys. Res., 107,* DOI: 10.1029/2000JE001476.

Nimmo F. and Giese B. (2005) Thermal and topographic tests of Europa chaos formation models from Galileo E15 observations. *Icarus, 177,* 327–340.

O'Brien D. P., Geissler P., and Greenberg R. (2002) A melt-through model for chaos formation on Europa. *Icarus, 156,* 152–161.

Ojakangas G. W. and Stevenson D. J. (1986) Episodic volcanism of tidally heated satellites with application to Io. *Icarus, 66,* 341–358.

Ojakangas G. W. and Stevenson D. J. (1989) Polar wander of an ice shell on Europa. *Icarus, 81,* 242–270.

Pappalardo R. T. and Barr A. C. (2004) Origin of domes on Europa: The role of thermally induced compositional buoyancy. *Geophys. Res. Lett., 31,* L01701, DOI: 10.1029/2003GL019202.

Pappalardo R. and Coon M. D. (1996) A sea ice analog for the surface of Europa. In *Lunar and Planetary Science Conference*

XXVII, pp. 997–998. Lunar and Planetary Institute, Houston.

Pappalardo R. T. and Sullivan R. J. (1996) Evidence for separation across a gray band on Europa. *Icarus, 123,* 557–567.

Pappalardo R., Head J. W., Greeley R., Sullivan R. J., Pilcher C., Schubert G., Moore W. B., Carr M. H., Moore J. M., and Belton M. J. S. (1998a) Geological evidence for solid-state con-vection in Europa's ice shell. *Nature, 391,* 365–368.

Pappalardo R., Head J. W., Sherman N. D., Greeley R., Sullivan R. J., and the Galileo Imaging Team (1998b) Classification of europan ridges and troughs and a possible genetic sequence. In *Lunar and Planetary Science XXIX,* Abstract #1859. Lunar and Planetary Institute, Houston (CD-ROM).

Pappalardo R. T. and 31 colleagues (1999) Does Europa have a subsurface ocean? Evaluation of the geological evidence. *J. Geophys. Res., 104,* 24015–24056.

Patel J. G., Pappalardo R. T., Prockter L. M., Collins G. C., Head J. W., and the Galileo SSI Team (1999) Morphology of ridge and trough terrain on Europa: Fourier analysis and comparison to Ganymede. *Eos Trans. AGU, 80(17),* Spring Meet. Suppl., S210.

Patterson G. W., Head J. W., and Pappalardo R. T. (2006) Plate motion on Europa and nonrigid behavior of the icy lithosphere: The Castalia Macula region. *J. Struct. Geol., 28,* 2237–2258.

Peale S. J., Cassen P., and Reynolds R. T. (1979) Melting of Io by tidal dissipation. *Science, 203,* 892–894.

Pieri D. C. (1981) Lineament and polygon patterns on Europa. *Nature, 289,* 17–21.

Pilcher C. B., Ridgeway S. T., and McCord T. B. (1972) Galilean satellites: Identification of water frost. *Science, 178,* 1087–1089.

Prockter L. M. and Schenk P. M. (2002) Mapping of Europa's youthful "dark spot" — A potential landing site. In *Lunar and Planetary Science XXXIII,* Abstract #1732. Lunar and Planetary Institute, Houston (CD-ROM).

Prockter L. and Schenk P. (2005) Origin and evolution of Castalia Macula, an anomalous young depression on Europa. *Icarus, 177,* 305–326.

Prockter L. M., Antman A. M., Pappalardo R. T., Head J. W., and Collins G. C. (1999) Europa: Stratigraphy and geological history of the anti-jovian region from Galileo E14 Solid State Imaging data. *J. Geophys. Res., 104,* 16531–16540.

Prockter L. M., Head J. W., Pappalardo R. T., Sullivan R. J., Clifton A. E., Giese B., Wagner R., and Neukum G. (2002) Morphology of europan bands at high resolution: A mid-ocean ridge-type rift mechanism. *J. Geophys. Res., 107,* DOI: 10.1029/2000JE001458.

Rathbun J. A., Musser G. S., and Squyres S. W. (1998) Ice diapirs on Europa: Implications for liquid water. *Geophys. Res. Lett., 25,* 4157–4160.

Reynolds R. T., Squyres S. W., Colburn D. S., and McKay C. P. (1983) On the habitability of Europa. *Icarus, 56,* 246–254.

Riley J., Hoppa G. V., Greenberg R., Tufts B. R., and Geissler P. (2000) Distribution of chaotic terrain on Europa. *J. Geophys. Res., 105,* 22599–22616.

Schenk P. M. (1991) Ganymede and Callisto: Complex crater morphology and formation processes on icy satellites. *J. Geophys. Res., 96,* 15635–15664.

Schenk P. M. (2002) Thickness constraints on the icy shells of the Galilean satellites from a comparison of crater shapes. *Nature, 417,* 419–421.

Schenk P. M. (2004) Sinking to new lows and rising to new heights: The topography of Europa. In *Workshop on Europa's Icy Shell: Past, Present and Future,* pp. 82–83. LPI Contribution 1195, Lunar and Planetary Institute, Houston.

Schenk P. M. (2005) The crop circles of Europa. In *Lunar and Planetary Science XXXVI,* Abstract #2081. Lunar and Planetary Institute, Houston (CD-ROM).

Schenk P. M. (2006) Europa's topographic story (abstract). In *Europa Focus Group Workshop 5,* p. 119.

Schenk P. M. and McKinnon W. B. (1989) Fault offsets and lateral crustal movement on Europa — Evidence for a mobile ice shell. *Icarus, 79,* 75–100.

Schenk P. M. and Pappalardo R. T. (2002) Stereo and photoclinometric topography of chaos and anarchy on Europa: Evidence for diapiric origins. In *Lunar and Planetary Science XXXIII,* Abstract #2035. Lunar and Planetary Institute, Houston (CD-ROM).

Schenk P. M. and Pappalardo P. (2004) Topographic variations in chaos on Europa: Implications for diapiric formation. *Geophys. Res. Lett., 31,* L1703, DOI: 10.1029/2004GL019978.

Schenk P. M., Chapman C. R., Zahnle K., and Moore J. M. (2004) Ages and interiors: The crater record of the Galilean satellites. In *Jupiter: The Planet, Satellites and Magnetosphere* (F. Bagenal et al., eds.), pp. 427–456. Cambridge Univ., Cambridge.

Schenk P. M., Matsuyama I., and Nimmo F. (2008) True polar wander on Europa from global-scale small-circle depressions. *Nature, 453,* 368–371.

Schubert G., Anderson J. D., Spohn T., and McKinnon W. B. (2004) Interior composition, structure and dynamics of the Galilean satellites. In *Jupiter: The Planet, Satellites and Magnetosphere* (F. Bagenal et al., eds.), pp. 281–306. Cambridge Univ., Cambridge.

Senske D. A., Greeley R., Head J., Pappalardo R., Sullivan R., Carr M., Geissler P., Moore J., and the Galileo SSI Team (1998) Geologic mapping of Europa: Unit identification and stratigraphy at global and local scales. In *Lunar and Planetary Science XXIX,* Abstract #1743. Lunar and Planetary Institute, Houston (CD-ROM).

Shoemaker E. M. and Hackman R. J. (1962) Stratigraphic basis for a lunar time scale. In *The Moon* (Z. Kopal and Z. K. Mikhailov, eds.), p. 289–300. Academic, London.

Smith B. A. and 21 colleagues (1979a) The Galilean satellites and Jupiter — Voyager 2 imaging science results. *Science, 206,* 927–950.

Smith B. A. and 21 colleagues (1979b) The Jupiter system through the eyes of Voyager 1. *Science, 204,* 951–957.

Sotin C., Head J. W., and Tobie G. (2002) Europa: Tidal heating of upwelling thermal plumes and the origin of lenticulae and chaos melting. *Geophys. Res. Lett., 29,* DOI: 10.1029/2001GL013844.

Space Studies Board (1999) *A Science Strategy for the Exploration of Europa.* National Academies, Washington, DC. 68 pp.

Space Studies Board (2003) *New Frontiers in the Solar System: An Integrated Exploration Strategy.* National Academies, Washington, DC. 231 pp.

Space Studies Board (Committee on Assessing the Solar System Exploration Program) (2007) *Grading NASA's Solar System Exploration Program: A Midterm Review.* National Academies, Washington, DC. 100 pp.

Spaun N. A., Head J. W., Collins G. C., Prockter L. M., and Pappalardo R. T. (1998a) Conamara Chaos region, Europa: Reconstruction of mobile polygonal ice blocks. *Geophys. Res. Lett., 25,* 4277–4280.

Spaun N. A., Head J. W., Pappalardo R. T., and the Galileo Imaging Team (1998b) Geologic history, surface morphology and deformation sequence in an area near Conamara Chaos, Europa. In *Lunar and Planetary Science XXIX,* Abstract #1899. Lunar and Planetary Institute, Houston (CD-ROM).

Spaun N. A., Head J. W., Pappalardo R. T., and the Galileo SSI Team (1999a) Chaos and lenticulae on Europa: Structure, morphology and comparative analysis. In *Lunar and Planetary Science XXX,* Abstract #1276. Lunar and Planetary Institute, Houston (CD-ROM).

Spaun N. A., Prockter L. M., Pappalardo R. T., Head J. W., Collins G. C., Antman A., Greeley R., and the Galileo SSI Team (1999b) Spatial distribution of lenticulae and chaos on Europa. In *Lunar and Planetary Science XXX,* Abstract #1847. Lunar and Planetary Institute, Houston (CD-ROM).

Spaun N. A., Pappalardo R. T., and Head J. W. (2001) Equatorial distribution of chaos and lenticulae on Europa. In *Lunar and Planetary Science XXXII,* Abstract #2132. Lunar and Planetary Institute, Houston (CD-ROM).

Spaun N. A., Pappalardo R. T., and Head J. W. (2003) Evidence for shear failure in forming near-equatorial lineae on Europa. *J. Geophys. Res., 108,* DOI: 10.1029/2001JE001499.

Squyres S. W., Reynolds R. T., and Cassen P. M. (1983) Liquid water and active resurfacing on Europa. *Nature, 301,* 225–226.

Sullivan R., Greeley R., Homan K., Klemaszewski J., Belton M. J. S., Carr M. H., Chapman C. R., Tufts R., Head J. W., and Pappalardo R. (1998) Episodic plate separation and fracture infill on the surface of Europa. *Nature, 391,* 371–373.

Sullivan R., Moore J., and Pappalardo R. (1999a) Mass-wasting and slope evolution on Europa. In *Lunar and Planetary Science XXX,* Abstract #1747. Lunar and Planetary Institute, Houston (CD-ROM).

Sullivan R., Greeley R., Klemaszewski J., Moreau J., Head J. W., Pappalardo R., Moore J., and Tufts B. R. (1999b) High resolution Galileo geological mapping of ridged plains on Europa. In *Lunar and Planetary Science XXX,* Abstract #1925. Lunar and Planetary Institute, Houston (CD-ROM).

Tufts B. R., Greenberg R., Hoppa G., and Geissler P. (2000) Lithospheric dilation on Europa, *Icarus, 146,* 75–97.

Turtle E. P., Melosh H. J., and Phillips C. B. (1998) Tectonic modeling of the formation of europan ridges. *Eos Trans. AGU, 79(45),* Fall Meet. Suppl., F541.

U.S. Geological Survey (2002) *Controlled Photomosaic Map of Europa.* Geological Investigations Series I-2757, 1:15,000,000. U.S. Geological Survey, Washington, DC.

Wilhelms D. E. (1990) Geologic mapping. In *Planetary Mapping* (R. Greeley and R. M. Batson, eds.), pp. 208–260. Cambridge Univ., Cambridge.

Williams D. A. and 13 colleagues (1999) Terrain variation on Europa: Overview of Galileo orbit E17 imaging results. In *Lunar and Planetary Science XXX,* Abstract #1396. Lunar and Planetary Institute, Houston (CD-ROM).

Williams K. K. and Greeley R. (1998) Estimates of ice thickness in the Conamara Chaos region of Europa. *Geophys. Res. Lett., 25,* 4273–4276.

Wilson L., Head J. W. and Pappalardo R. T. (1997) Eruption of lava flows on Europa: Theory and application to Thrace Macula. *J. Geophys. Res., 102,* 9263–9272.

Zahnle K., Dones L., and Levison H. F. (1998) Cratering rates on the Galilean satellites. *Icarus, 136,* 202–222.

Zahnle K., Schenk P., Levison H., and Dones L. (2003) Cratering rates in the outer solar system. *Icarus, 163,* 263–289.

Zimmer C., Khurana K. K., and Kivelson M. G. (2000) Subsurface oceans on Europa and Callisto: Constraints from Galileo magnetometer observations. *Icarus, 147,* 329–347.

Europa's Crater Distributions and Surface Ages

Edward B. Bierhaus
Lockheed Martin

Kevin Zahnle
NASA Ames Research Center

Clark R. Chapman
Southwest Research Institute

We review the state of knowledge regarding Europa's impact-crater distributions, the projectile populations in the jovian system and their relative impact rates, and the resulting derived surface ages. In particular, we estimate Europa's best-fit average surface age to be between 40 and 90 m.y. (error bars render the spread 20 to 200 m.y.), consistent with previous estimates and quite young within the timescale of the solar system; we confirm previous observations that the size-frequency distribution of Europa's primary craters does not match that of the Moon; we discuss the contributions of nonprimary craters to the small crater record, and the associated possible error in derived age; and we follow up on earlier observations of the near absence of any crater degradation. In reviewing the current knowledge, we provide some additional observations: We find possible evidence for a slight increase in crater density on Europa's leading hemisphere, which is expected based on the dynamics of heliocentric impactors; within the approximately 10% of the surface imaged at regional resolution (about 250 m/pixel) we find inconsistencies between stratigraphic ages and crater densities; and we note that the scarcity of rays from large craters, even on such a youthful surface, is evidence for the rays' rapid disappearance, perhaps in part due to the intense radiation seen by Europa's surface, and perhaps in part due to lack of ejecta products (melt glass) seen in rocky ejecta.

1. INTRODUCTION

Europa is a world of intrigue, from geologic, biologic, and chronologic perspectives. In this chapter we focus on chronology as revealed by crater populations. Cataloguing crater size-frequency distributions (SFDs), areal densities, and their spatial variations in the context of projectile infall rates is the best method to determine age(s) in the absence of material samples. Simply put, Europa's cratering record indicates a youthful surface. Some of Europa's surface may express geological activity of comparable age to recent geology on Earth. The chapter by Schenk and Turtle discusses Europa crater morphology, although we briefly mention the topic in this chapter as it has relevance for interpreting the crater SFDs and surface ages (section 5).

Europa's icy shell exhibits ridge systems that are local to global in extent, whose complex stratigraphy and relative orientations make heavily ridged areas look like a "ball of string" (see chapter by Kattenhorn and Hurford). "Chaotic" regions, areas in which the preexisting surface was partly or completely disrupted, are a few kilometers to a few hundred kilometers in size and appear across the surface (see chapter by Collins and Nimmo). Craters, however, are rare at all scales, and although more numerous at small sizes, are still underabundant relative to the populations on Ganymede and Callisto. All these features reside in a global icy shell (see chapter by Nimmo and Manga) that almost certainly hides a global ocean.

The astonishing variety and stratigraphy of surface features (see chapter by Doggett et al.), coupled with the low surface density of craters, implies ongoing surface activity. Unfortunately, despite efforts on the part of theorists and modelers, we do not yet have strong constraints on the formation *rate* of Europa's native surface features, and therefore they can tell us only that Europa is active. Despite sparse numbers, impact craters are the best means to evaluate absolute surface ages for the ocean moon.

1.1. Crater Size-Frequency Distributions

To establish context for subsequent discussions, we define our default format for SFDs, as there are several forms used in the literature to describe crater distributions.

The number of craters within a particular diameter range tends to follow a power law, such that the number of craters dN in the diameter range D to D + dD follows the relation $dN = kD^b dD$, where k and b are constants (usually found by a nonlinear least-squares fit to the data). The parameter k corresponds to crater surface density, and thus presumably age; a higher surface density results in larger k, and usually implies greater age. The parameter b is frequently referred to as the SFD "slope" (because in log-log plots a power law appears linear with slope b), and we will use that nomenclature here. Note that we are using the differential form of the power law, while some use the cumulative form of the power law [N (≥D) = cD^a], in which the number of craters N of diameter greater than or equal to D is cD^a. In the world of pure mathematics, the differential is the derivative of the cumulative, and thus the differential slope is one unit steeper than the cumulative slope (i.e., if a = –3 cumulative, then b = –4 differential). However, natural variation within crater SFDs causes departure from the ideal such that a least-squares fit to the cumulative and differential forms of the same dataset may not hew exactly to this relationship. Finally, we plot crater SFDs using the relative form, or R plot, which is the differential distribution normalized by D^{-3}; thus, a differential slope of –3 will plot as a horizontal line in the R plot, and differential slopes of >–3 will have positive slopes. We use the R plot because it accentuates deviations from the pure power-law form, the height of data within the R plot is more easily understood in terms of crater surface density, and for pure practicality, it is a more compact means to display the data.

1.2. Voyager Perspective on Europa's Craters

Europa was incompletely imaged during the Voyager flybys, with no better than ~2 km/pixel image scale (*Smith et al.,* 1979). However, these image data were sufficient to show that, despite the lack of obvious activity such as that displayed by Io, Europa was young relative to the other icy Galilean satellites. *Lucchitta and Soderblom* (1982) identified a maximum of eight candidate craters from Voyager 2 imagery, compared with Callisto's nearly saturated surface and Ganymede's densely cratered surface. *Malin and Pieri* (1986) later commented on additional possible craters and irregular pits visible near the terminator; they also noted that large "maculae" (dark spots) and smaller spots might be related to impact craters. Some of the features recognized in Voyager images are now understood to be part of the larger, more prominent examples of a population of endogenic pits or depressions with typical diameters ~8–10 km (*Chapman et al.,* 1997).

1.3. Galileo Perspective on Europa's Craters

Despite the constricted data return, caused by the failed deployment of the main antenna, the Galileo spacecraft managed to return new views of Europa at global, regional, and high-resolution scales. These images confirmed the sparse large-crater population suggested by the Voyager images, but Galileo imaging brought some surprises. In particular, imaging enabled the following new observations of cratering on Europa: (1) the multiring structures Tyre (~41 km) and Callanish (~33 km) are in fact impact features and not the result of surface activity (*Moore et al.,* 1998); (2) the transition in crater morphologies from complex to multiring at such surprisingly small sizes (relative to the Moon and Callisto) is among the best lines of evidence for an ocean beneath the ice (*Moore et al.,* 1998, 2001; *Schenk,* 2002); (3) the SFD of europan primary craters is more shallow than the SFD of lunar craters of similar size (*Schenk et al.,* 2004); (4) because most small-crater populations (<1 km) have steep SFDs and appear in clusters, they are secondary craters rather than primary craters (*Bierhaus et al.,* 2005).

1.4. Cratering in Ice

Impact cratering in ice is similar to that in rock (*Melosh,* 1989), but sufficient differences exist in crater formation and evolution that one must be aware of how target properties affect the production and preservation of crater distributions. For Europa in particular, a global ocean underneath the icy shell introduces a layered target unlike the terrestrial planets, and ongoing modeling and experiments seek to quantify how Europa's ice-over-water structure affects crater formation and subsequent morphology. Ice exhibits temperature-dependent effects that are not completely characterized in terms of crater formation and evolution, but colder ice tends to be stronger and flow more slowly in time. This adds an additional subtlety to interpreting craters on Europa, as Europa's surface is near 100 K (*Spencer et al.,* 1999), while the ice-water interface is within a few degrees of 273 K (0°C), depending on non-H_2O components.

1.4.1. Crater formation. Ice's weaker yield strength and lower temperature limits for its liquid and vapor states suggest that a given energy impact will make a larger crater in ice than in rock. Ice's weaker strength also means craters on icy surfaces will transition to gravity-dominated crater forms at smaller diameters than on a rocky surface of equivalent surface gravity. [See *Holsapple* (1993) for a comprehensive overview of crater-scaling laws.] In section 3.2 we take the effective strength of ice to be about 20 times less than rock, but this is an estimate based on measurements of polycrystalline ice; details of Europa's ice structure (such as near-surface porosity) that contribute to effective strength are still uncertain. Comparing Europa's crater SFD with those of rocky worlds (such as the Moon) requires careful consideration of differences in the target and the impactor velocity, and the resulting differences in crater size and morphology.

1.4.2. Crater evolution. On icy satellites with sufficient gravity, ice adjusts to topographic equilibrium over geological timescales, smoothing high and low elevations to a similar height (*Dombard and McKinnon,* 2000, 2006). Whereas large craters (even giant basins) on rocky surfaces maintain significant topography over long timescales, such as the South Pole Aitken Basin on the Moon or Hellas on Mars, topographic expression of large craters on icy surfaces can relax to more shallow profiles. Europa's surface is sufficiently young that its largest recognized craters are either too small or too youthful (or both) to have experienced topographic relaxation. Europa's larger craters show unusual topography despite their youth, see the chapter by Schenk and Turtle for more details.

1.4.3. Layered target materials. Although still not a "smoking gun," evidence for a global ocean on Europa is titanic. Smaller impacts form entirely within the icy shell, but the transient-crater depths of progressively larger impacts approach, and eventually penetrate, the ice/water boundary. The multiring structures of Callanish and Tyre are the two largest recognized impacts on Europa, yet these are still not large relative to large craters and basins on Ganymede and Callisto (hundreds to thousands of kilometers in diameter). The appearance of multiring structures on Europa at sizes at least an order of magnitude smaller than on other solid surfaces is a key indicator of the subsurface ocean. There is a growing realization that impacts larger than Tyre may create features utterly foreign to traditional impact morphologies, so the layered ice-over-water target has important implications for the completeness and interpretation of the large-diameter crater SFD.

1.5. Inner Versus Outer-Solar System Impactors

The Moon, because it is the only non-Earth surface for which we have samples of known provenance, is the reference standard for correlating crater density to surface age, but applicability of a lunar-derived chronology to other planets and satellites is not warranted without a complete understanding of the impactor populations throughout the solar system. Hundreds of years of asteroid observations, combined with recent understanding of non-Keplerian effects [e.g., the Yarkovksy and YORP effects (*Bottke et al.,* 2002a)] and numerical modeling of the dynamical evolution of small-body populations, confirm that asteroids are the dominant impactors of the inner solar system. *Shoemaker and Wolfe* (1982) showed convincingly that Jupiter-family comets (JFCs, then known as short-period comets) were by far the major source of projectiles in the jovian system. *Zahnle et al.* (1998, 2003), *Levison et al.* (2000), and *Levison and Duncan* (1997) revisited Shoemaker's idea in the light of a modern understanding of the origin of JFCs from the Kuiper belt and scattered disk (see section 4.1). The newer work confirms the assumptions made in the earlier work, and differences between the cratering rates are well within the uncertainties. Observations and solar system dynamics all point to comets dominating outer planet impact flux. Accurate estimates of the impact flux are essential to turn crater densities into absolute ages. As precise estimates are currently impossible, we will make do with what we have.

1.6. The Structure of this Chapter

The remainder of this chapter proceeds as follows: In section 2 we review the known and estimated primary crater populations on Europa; in section 3 we discuss nonprimary craters, and their effect on crater SFDs and age interpretation; in section 4 we review the current state of knowledge of the jovian-system impacting flux, and from that flux we discuss relative and absolute ages for Europa; in section 5 we discuss other markers of age; in section 6 we briefly discuss Europa in the context of the jovian system, particularly what the cratering records of Callisto and Ganymede imply for the history of Europa; and in section 7 we summarize our state of knowledge for Europa's crater populations and surface age.

2. PRIMARY CRATERS

Europa is one of the few solar system surfaces on which counting the primary craters is not an exercise in tedium. There are only 24 craters ≥10 km diameter (*Schenk et al.,* 2004) identified in Galileo images, although global imagery is incomplete except at low resolutions. Nevertheless, there are sufficient numbers to make meaningful comparisons with Ganymede and Callisto (and Io), and the sparseness of large primaries is the strongest evidence for ongoing activity within Europa's icy shell.

2.1. Craters ≥30 km

There are six known craters with diameters ≥30 km: Tyre, Callanish, Taliesin, Tegid, and two unnamed craters. Craters of these sizes are likely the most completely catalogued, with an exception for nontraditional impact morphologies (discussed below). The combination of Voyager and Galileo imagery creates a global map of about 2 km image scale, presumably sufficient to completely identify all craters larger than about 10 km. However, the phase angles (lighting geometry) are nonuniform for these image data; portions of the global map are high-Sun images, in which craters can be notoriously difficult to identify. (For example, low-resolution, high-Sun Galileo images at 1.6 km/pixel of 19-km Cilix could not confirm it as an impact structure; confirmation required subsequent, low-Sun higher-resolution images.) As a result, the 2-km global map at best enables complete mapping of craters ≥30 km.

However, a number of experiments, starting with *Greeley et al.* (1982), and most recently by *Ong* (2004), demonstrate that in certain conditions impacts into layered ice-over-water targets can generate features utterly foreign to traditional impact craters. When an impactor completely penetrates the icy shell, it creates a turbulent ice-water interface for several crater radii around the impact site. The feature that results is a merger of the traditional excavation processes of the impact with extensive fracturing and adjustment of the ice crust around the impact site. The latter could obliterate any of the circularity that typifies an impact feature. Such an impact feature would result from an impactor whose diameter is roughly half the icy shell depth at the point of impact. Extrapolation from the known primary craters suggests that one such impact may exist. Identifying such a feature would be a tremendous boon for constraining the ice thickness (at least in that spot at the time of formation),

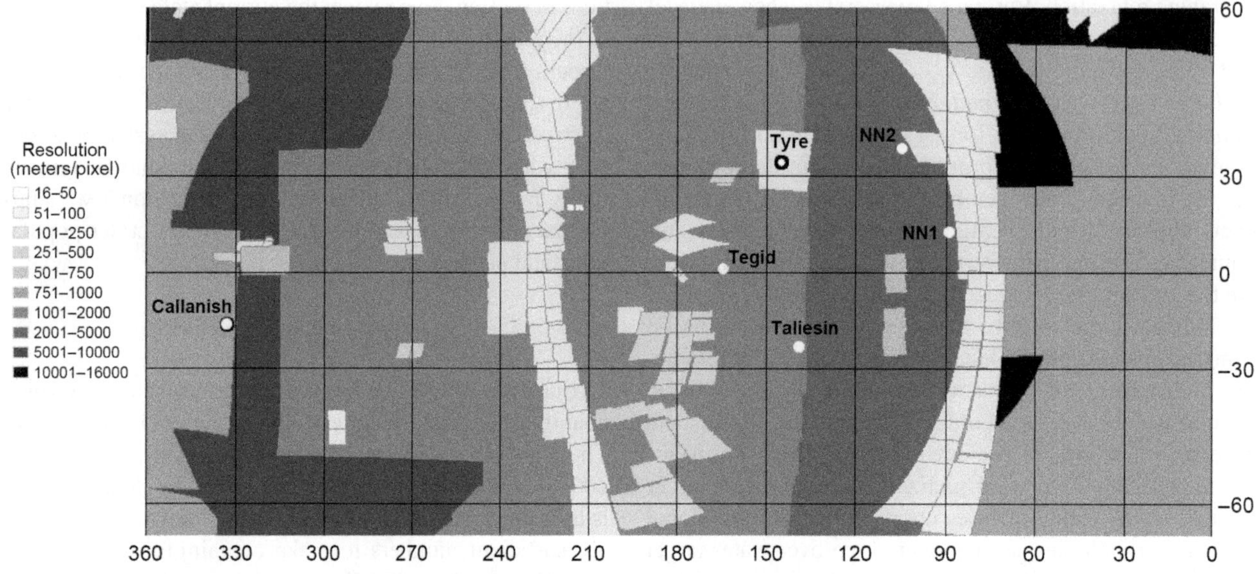

Fig. 1. The six europan craters ≥30 km, superimposed on a resolution map showing the highest-resolution coverage using Galileo and Voyager image data. The craters appear concentrated between longitudes 100° and 170°, a relationship that may not be a product of the varying resolution over the surface since they are not coincident with the highest-resolution coverage. Comparison with a simple simulation (see Fig. 2) that places six craters at random locations on Europa suggests that this spatial distribution is perhaps nonrandom, either due to preferential destruction of large craters within certain regions on Europa, and/or the existence of currently unrecognized impact structures. The two labels "NN1" and "NN2" are for unnamed craters 1 and 2. Resolution map courtesy of T. Doggett.

but would not significantly affect our understanding of Europa's age (due to statistics of small numbers for that impactor size).

Figure 1 shows the spatial distribution of the known craters ≥30 km on Europa. The craters seem to be concentrated within a longitude band between 100° and 170°, although statistics of small numbers can suggest patterns where there are none. To investigate the spatial distribution of these six craters, we ran 100,000 simple simulations in which we randomly placed 6 craters on Europa's surface, and compared nearest-neighbor statistics between the simulation and the data values. Figure 2 shows the results for the first through fifth nearest neighbor (each of the six craters has five "nearest neighbors"); the black points and error bars represent the mean and standard deviation for the simulated nearest-neighbor distances, and the gray points are the Europa data. The actual fifth nearest neighbor distance is greater than the simulated value, caused by the large distance between Callanish and the other five craters. While not especially rigorous, this simple comparison hints (but no more than hints) that perhaps the region between 100° and 170° longitude is one of the older regions of Europa, and/or there are large impact structures, perhaps with nontraditional morphologies, that are not yet identified. Better global imaging at uniform illumination and resolution is required to resolve this ambiguity, which may be the result of biased observations from inconsistent global coverage.

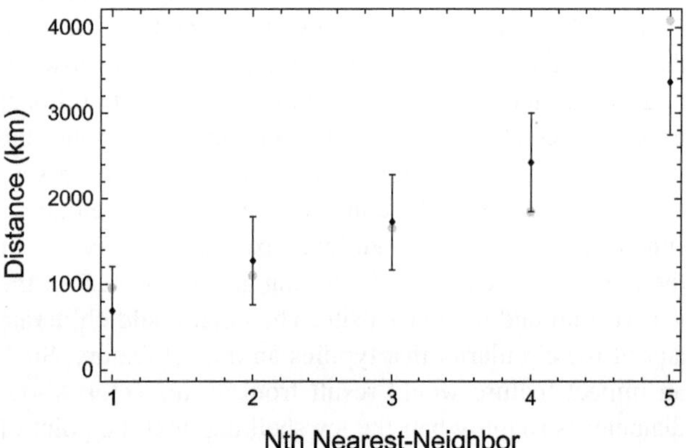

Fig. 2. The black points and error bars are the mean and standard deviation of the nearest-neighbor distances for 100,000 simulations of six randomly placed craters. The gray points are the Europa data.

2.2. Craters with 3 km ≤ D < 30 km

During orbits 15 and 17, Galileo obtained near pole-to-pole swaths of regional resolution (~230–450 m/pixel) images of Europa's leading and trailing hemispheres. These image data, hereafter referred to as "REGMAP," represent the best opportunity to measure and establish the primary crater SFD. Table 1 summarizes the areas of these regions. The data listed cover about 9% of Europa's surface.

Within the REGMAP data, the low-latitude images have the best resolution and are the least foreshortened; the high-latitude images are the poorest resolution and the most foreshortened. The "completeness diameter" (the smallest diameter for which all craters of that size are completely identified, without losses due to resolution) varies across the mosaics, from ~1 km at low latitudes to ≥3 km at high latitudes. Because both swaths sample nearly the entire

TABLE 1. Measured regional resolution mosaics.

Image Sequence	Minimum Pixel Scale (m/pix)	Maximum Pixel Scale (m/pix)	Area (km^2)
Leading Hemisphere			
15ESREGMAP02	230	607	829,484
17ESREGMAP02	216	434	607,380
Total Area			1,436,864
Trailing Hemisphere			
15ESREGMAP01	230	416	358,207
17ESREGMAP01	223	443	729,602
17ESNERTRM01	211	260	246,629
Total Area			1,334,438

range of latitudes, we do not believe a latitudinal sampling bias exists. However, the mosaics are located on the leading and trailing hemispheres of Europa; because the tidal-stress induced fracturing is a function of surface position, including longitude, we cannot be certain that the crater populations seen on the leading and trailing hemispheres are representative of the populations on the jovian or antijovian hemispheres. (Consistency with other data suggests this may not be a concern; see section 4.4.)

Figure 3 shows an R plot for the REGMAP data, a point for the global count of craters ≥30 km, and measurements for craters <3 km, which we will discuss in the next section. The error bars are 1σ for Poisson statistics ($\sqrt{N}$). The slope for data >3 km is about –2 differential, or about –1 cumulative. These measurements match those of *Schenk et al.* (2004); two independent measurements that agree so well

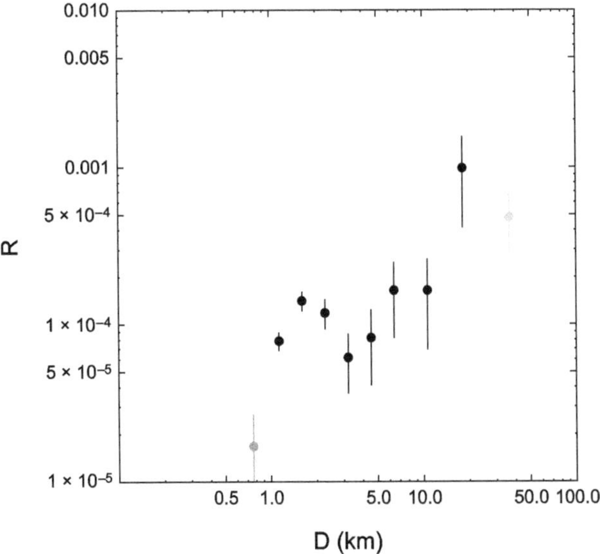

Fig. 3. R plot for the combined measurements, both leading side and trailing side, of the REGMAP data (black points), a point for the global count of all known craters ≥30 km (light gray point), and an estimate of the small primary population from high-resolution images (dark gray point).

(also see Fig. 10) lend confidence that this is indeed the primary population visible within the REGMAP areas. Our measurements do not include primaries specifically targeted for higher-resolution imaging (e.g. Pwyll, Cilix, Rhiannon, and others), as to not artificially inflate the crater density.

2.3. Craters with D < 3 km

Identifying primary craters in this diameter range is tricky business due to serious contamination from secondary, and sesquinary, sources. Secondary craters, formed by the reimpact of ejecta launched during primary crater formation, tend to form in spatial clusters, although they also contribute to the spatially random population as well. "Sesquinary" is a term coined by *Zahnle et al.* (2008) to describe ejecta that escape one body (e.g, Io) and subsequently either return to the parent body, or strike another (e.g., Europa). Using spatial statistics and SFDs of small craters, *Bierhaus et al.* (2005) estimated that <5% of craters <1 km are primaries; this was without consideration of sesquinaries. Depending on the delivery rate of Io ejecta to Europa, the actual percentage of small craters that are primaries could be less than 1%. We use two types of image data to evaluate primaries <3 km: the REGMAP data, and high-resolution (scale ≤50 m/pixel) images. We discuss each of these in more detail below.

2.3.1. REGMAP. The REGMAP data are at their limit of usefulness at these scales; the measurements suffer incompleteness at diameters smaller than ~1 km at low latitudes, and at diameters smaller than 3 km at high latitudes. Nevertheless, the data are a useful bound on the crater density. Figure 3 shows a "bump" in the SFD curve between 1 and 3 km. This feature may result from secondary and sesquinary craters, or reflect the primary SFD, or both. The data and modeling favor the role of nonprimaries (secondaries, and possible sesquinaries). The 15ESREGMAP02 data clearly exhibit two cases of adjacent secondary fields (secondaries within a few radii of their parent primary) whose source primary is just beyond the boundaries of the image. The global map has insufficient resolution to definitively locate these craters. Although we did not include any measurements from within the two adjacent fields, it is possible that more distant 1-km craters are secondaries of the two "off-screen" primaries; we know that Pwyll generated secondaries up to 500 m diameter at almost 1000 km distance (*Bierhaus et al.,* 2001), so it is conceivable that 1-km secondaries appear within a few hundred kilometers of their source primary. Figure 4 shows one of these cases. A 1-km crater may be part of a crater cluster whose smaller members are unresolved at the resolution of the REGMAP image data.

2.3.2. High-resolution images. Incompleteness concerns within the REGMAP data, or ambiguities such as those discussed in the previous section, mean that the high-resolution data (≤50 m/pixel) are ostensibly the best resource to identify the small primary population. However, these data suffer serious contamination from nonprimary sources, and so are not without their limitations. *Bierhaus* (2006)

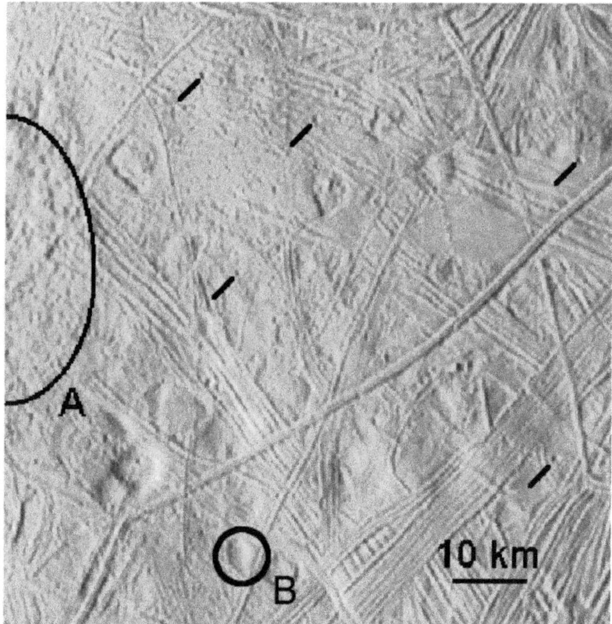

Fig. 4. Portion of 15ESREGMAP02 frame 74265; average image scale is about 250 m/pixel, north is approximately up, and the Sun is from the left. The boundary marked by "A" is the possible edge of a continuous ejecta blanket for a primary crater that is just beyond the image edge. Adjacent secondaries appear throughout the subimage; several examples appear at the righthand edges of the black diagonal slashes. Also visible are several endogenic depressions that, at lower resolutions, could be mistakenly considered as primary craters. The circle, marked by "B", is one such example.

searched for small primaries using the following criteria: (1) Set the largest possible contamination diameter from secondaries using the adjacent secondary field around the largest primary, Tyre. Tyre's largest distant secondary is ~1.5 km, so any crater larger, and far from a large impact structure, must be a primary. There are only two craters with D > 1.5 km (1.7 km, 1.8 km) in all the high-resolution images. (2) Find individual, small craters with rays (as opposed to within the ray of a larger crater). Only one small crater has a ray system within the high-resolution data.

Thus there are three craters in the roughly 1-km range, in the ~52,000 km^2 covered in the high-resolution images, that are (almost certainly) primaries. The dark gray point in Fig. 3 is for these craters. (The horizontal location of the point within the plot is the weighted median of the diameter bin width, about 0.8 km.) This is a lower limit, because (on Europa) rays fade quickly and a small primary that forms near a cluster of secondaries could be mistaken for a secondary. The point falls along the trend established by the larger diameters; a possible interpretation is that the –2 differential slope seen at larger diameters continues to subkilometer sizes. However, we caution that coincidence may be an equally robust explanation.

Enceladus may offer some clues to the relative abundances of primary and nonprimary craters at small diameters. The lightly cratered, south-polar terrain on Enceladus may be sufficiently youthful that it is free from nonprimary craters. *Kirchoff et al.* (2007) estimate a cumulative slope of –1.31 (roughly –2.3 differential), which is similar to the slope seen on Europa at larger diameters. There are scaling considerations and impactor sources that prevent a direct comparison between a jovian satellite and a saturnian satellite, but we may be seeing the emergence of a relatively shallow trend in the outer solar system for primaries to diameters as small as 1 km.

In sum, our knowledge of europan primary craters smaller than a few kilometers diameter remains unsatisfactory. The limited high-resolution data, covering less than 1% of Europa's surface, are too contaminated by nonprimary sources (see next section), and the more extensive but lower-resolution REGMAP data cannot resolve the ambiguity between primary and nonprimary at these diameters. The SFD slope for small primaries could continue the trend seen at larger diameters, but we cannot dismiss other functional forms, including transitioning to steeper, or more shallow, slopes. More high-resolution data of Europa (or Io) are necessary to unambiguously resolve this question for the jovian system.

3. SECONDARY AND SESQUINARY CRATERS

The seemingly straightforward correlation between crater density and surface age is confounded by a variety of effects, including secondary craters. Millions, maybe billions, of secondaries form from the fallback of ejecta excavated by a primary impact. The secondaries generated by a single primary can be grouped into three crude categories: (1) The *adjacent secondary* population, which appears at the edge of the continuous ejecta blanket and continues for several parent-crater radii. Adjacent secondaries usually overwhelm the preexisting local crater population. (2) *Distant secondaries*, which can be regional to global in extent. These are usually clustered and can appear inside the bright rays of the parent primary. Within a cluster, the distant secondaries also dominate the local crater population. (3) *Background secondaries*, which are spatially random and can be essentially indistinguishable from small primary craters. Because some ejecta leave the primary impact site at speeds up to and exceeding the escape speed of the parent body (in Europa's case, only ~2 km/s), a single primary can broadcast globally distributed secondaries. In multibody systems, such as the Galilean satellites, ejecta that escape one body can hit another body. *Zahnle et al.* (2008) called these interobject craters "sesquinaries." In short, considering all small impact craters as primary craters could easily lead to gross misconceptions of surface age. This is of particular interest on Europa, as the sparse large crater population makes it hard to resist the temptation to use the far more numerous, and thus ostensibly more statistically significant, small craters to set relative and absolute ages for Europa.

3.1. Secondary Craters

Small craters (<1 km) on Europa display two traits that distinguish them from their larger-diameter cousins: They exist in greater numbers than expected from extrapolating the large-crater SFD, and they display distinct, nonrandom spatial patters. Clusters of small craters with steep SFDs appear in every high-resolution image. The clusters cross geologic boundaries, and in most cases the clustering is not caused by preferential destruction or preservation of a particular area with superimposed craters. *Bierhaus* (2004) and *Bierhaus et al.* (2005) concluded that at least 95% of these craters are secondaries, based on clustering analysis and comparison of the SFDs between high-resolution measurements; Figure 5 shows a few examples of crater clusters. *Bierhaus et al.* (2005) also demonstrated most of the unclustered craters are also secondaries, primarily due to lack of convergence of the SFD for the spatially random populations between regions; the actual percentage of true primaries is likely only a few percent. *Preblich et al.* (2007) find that secondaries from the martian crater Zunil have different SFDs depending on whether the secondaries are inside or outside a ray (where ray secondaries have a steeper SFD than those outside a ray). The variability of the secondary-crater SFDs, both for those in clusters (rays) and for those outside rays, impose significant barriers to identifying the primary-crater SFD at these diameters.

The relative scarcity of large primaries but numerous secondaries suggests that a single primary generates a large population of secondaries. Mass-balance scaling laws (mass ejected from the primaries vs. mass required to create the secondaries) indicate the relationship is physically plausible (*Bierhaus et al.,* 2001), and the connection between primaries and secondaries is borne out by several lines of recent

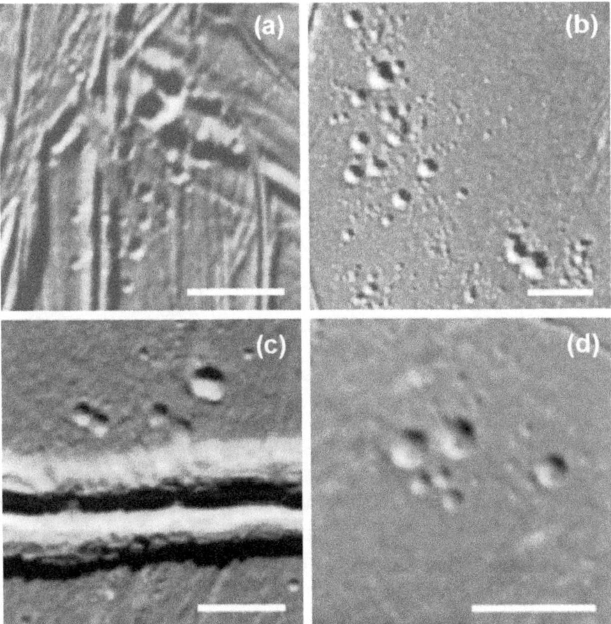

Fig. 5. Examples of secondary crater clusters visible in high-resolution images. The scale bar is 2 km for each subimage. The subimages are from **(a)** 17ESSOUTHP01 (Sun from the left), **(b)** 17ESSTRSLP01 (Sun from the upper right), **(c)** 17ESTHYLIN01 (Sun from the top), and **(d)** 17ESLIBLIN01 (Sun from the upper right).

evidence from other surfaces, computer models, and analytic arguments. We review some of the major contributions in the following.

Although the more densely cratered surfaces of the Moon, Mars, and Mercury mask the cleanly resolved clustering of secondaries seen on Europa's surfaces, there are several recent observations on those surfaces that support a large secondary-crater population. *McEwen et al.* (2005) estimate that the 10-km martian crater Zunil created 10^7 secondaries between 10 m and 200 m diameter, extending up to 1600 km from Zunil. *Preblich et al.* (2007) update the McEwen et al. results, and find Zunil created more than 7×10^7 secondaries larger than 15 m diameter, over 10^8 secondaries larger than 10 m diameter, and they suggest these are lower limits due to observational constraints. Some Zunil secondary populations are 1700 km distant. *Dundas et al.* (2007) estimate the lunar crater Tycho created almost 10^6 secondaries larger than 63 m diameter; this estimate is also a lower limit on the secondaries due to observational constraints. Most recently, the first MESSENGER flyby of Mercury revealed prominent ray systems and extensive fields of secondaries; *Strom et al.* (2008) estimate that secondaries are more abundant than primaries at diameters less than 8–10 km.

Numerical modeling suggests an individual impact generates numerous secondaries. Modeling of vertical impacts by *Head et al.* (2002) suggests that a 3-km martian crater creates 10^7 decimeter-sized fragments that *escape* Mars. Such fragments are on the higher end of the ejection-velocity curve and the more numerous lower-velocity fragments fall back to make secondaries. *Artemieva and Ivanov* (2004) model oblique impacts into the martian surface and include atmospheric effects. They find similar results to those of *Head et al.* (2002) even with more sophisticated treatment of atmospheric conditions.

Comparisons between the rocky surfaces of the terrestrial planets and the icy surface of Europa indicates secondaries should be at least as numerous as on terrestrial surfaces. For example, primary impacts on Europa occur at speeds on the order of 26 km/s (*Zahnle et al.,* 1998), compared with the ~10 km/s (*Bottke et al.,* 2002b, 1994) on Mars. Europa's surface gravity is 1.3 m/s^2, compared with 3.7 m/s^2 on Mars. The combined effects of higher primary impact speeds [and thus higher fragment ejection speeds, assuming ejection speeds are a function of impact speeds (*Melosh,* 1989)], and the lower surface gravity of Europa relative to Mars indicate that a single, large primary crater could create a global network of secondaries even more easily on Europa than on Mars. (A 10-km crater on Europa should be able to broadcast secondaries at least as well as Zunil.)

Multiple observations of the Europa crater population suggest secondaries are the dominant small crater population. *Bierhaus et al.* (2001) measurements of Pwyll secondaries in two high-resolution mosaics demonstrate that Pwyll deposited over 3300 craters larger than 50 m diameter in these two small areas (about 1250 m^2 total), which are 1000+ km distant from Pwyll. Rough approximation of

Pwyll's total ray area suggests Pwyll created 10^6 secondaries. This estimate is also a lower limit given that secondaries, while concentrated in rays, also appear outside rays. *Bierhaus* (2004) measured almost 7000 Tyre secondaries within a few Tyre radii ("adjacent secondaries") that are larger than 1 km, the smallest diameter robustly resolved in the Galileo data. Given the steep SFD of these craters, many more must exist at smaller diameters. Tyre may have generated 10^5 secondaries larger than a few hundred meters diameter in its adjacent field alone. There was ample material for spalled ejecta that makes distant secondaries. Modeling by *Turtle and Ivanov* (2002) indicates the shell thickness at the Tyre impact site had to be at least ~12 km thick, and perhaps 18 km thick. Therefore Tyre excavated and ejected an ice layer several kilometers thick, creating a significant *distant* secondary population. Tyre had to eject more material than Pwyll, suggesting Tyre created *at least* 10^6 secondaries.

McEwen and Bierhaus (2006) demonstrated that for equivalent impactor SFDs, the secondary-crater SFDs will be steeper than primary-crater SFDs because of the inverse mass-velocity relationship for ejecta fragments. Thus, for known impactor populations, there will be some diameter below which secondaries are the dominant crater form.

Secondaries are a major component of Europa's small crater population, and have been a frustratingly effective mask of Europa's small primary-crater SFD. Their ubiquity introduces ambiguity into lower-resolution images in which a 1-km crater is the smallest resolved crater — an isolated 1-km crater could be a single crater, or could be the largest member of an otherwise unresolved secondary cluster of craters.

3.2. Sesquinary Craters

The latest development for interpreting craters on Europa is a model predicting the formation of sesquinary craters. *Alvarellos et al.* (2008) and *Zahnle et al.* (2008) developed a description of secondary and sesquinary cratering based on theoretical ideas proposed by *Melosh* (1984, 1985a,b, 1989) and first implemented by *Vickery* (1986). Melosh split impact ejecta into two kinds: "Grady-Kipp fragments," to describe rocks from below the surface embedded in the main excavation flow, and a smaller mass of larger, faster "spalls" that originate from constructive wave interference at a competent surface. The models predict the size-velocity distribution of ejecta, which can be used to develop size-number distributions of secondary and sesquinary impact craters. The model was applied to the well-studied young martian crater Zunil for ground truth (*McEwen et al.,* 2005; *McEwen and Bierhaus,* 2006; *Preblich et al.,* 2007). Zunil is set on young volcanic plains that should present a competent surface and a good analog to Io. *Zahnle et al.* (2008) found that the predicted size of spall plates provides a good fit to the sizes of Zunil's biggest secondary craters. But in general spall plates are expected to break up into fragments of a scale comparable to the thickness of the plates, and at the high ejection velocities appropriate to escape from Io, are liable to break up further into Grady-Kipp-sized fragments (*Melosh,* 1989).

To simulate the orbital evolution of impact ejecta from Io, *Alvarellos et al.* (2008) used Levison and Duncan's Swift numerical package (*Levison and Duncan,* 1994). Swift is a regularized, mixed-variable symplectic integrator (*Wisdom and Holman,* 1991). Test particles were integrated for 10^4 yr. Most particles that escape from Io ultimately return to Io. About 9% of the particles hit Europa, 5% hit Ganymede, and 0.4% hit Callisto. The median time for transfer of ejecta from Io to Europa is 56 yr. These results are consistent with results that *Alvarellos et al.* (2002) obtained in a previous study of impact ejecta launched from Ganymede. They found that ~10% of what escaped Ganymede hit Europa, and another ~13% hit Callisto. They expect that roughly 70% of ejecta that escape from Europa will hit Europa, roughly 13% hit Io, ~16% hit Ganymede, and the remaining 1% hit Callisto. Ionian sesquinaries impact Europa in a nearly random spatial distribution, with a tendency to avoid Europa's trailing pole.

Zahnle et al.'s (2008) model predicts that sesquinary craters made by ejecta from Io are roughly as important on Europa as conventional secondary craters at sizes 200–1000 m diameter, counted by *Bierhaus et al.* (2005). Figure 6 shows the predicted populations for spalls and Grady-Kipp fragments. However, Bierhaus et al. concluded, primarily based

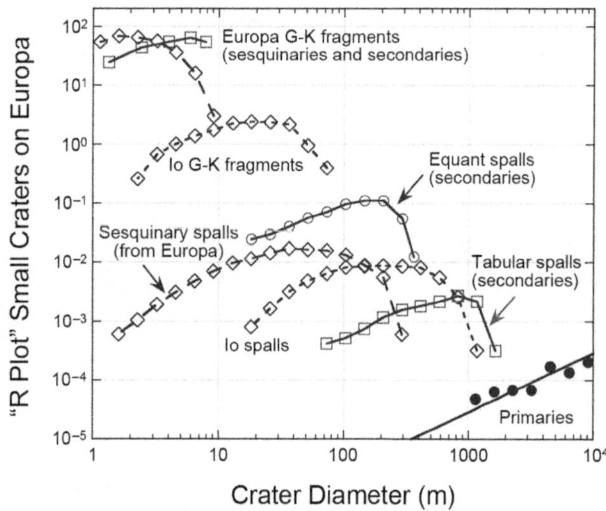

Fig. 6. Simulated populations of secondary and sesquinary impact craters on Europa generated using equations described by *Zahnle et al.* (2008) and expressed as an R plot. The simulations sum up the contributions of a range of primary craters on Io and Europa. Genetically distinct populations are labeled. Sesquinary craters refer to impacts by debris that was in orbit about Jupiter and secondary craters refer to craters made by ordinary suborbital ejecta. The ejecta themselves are broken down into spalls (equant spalls of dimension $Z_{sp} \times Z_{sp} \times Z_{sp}$ and tabular spalls of dimension $10Z_{sp} \times 10Z_{sp} \times Z_{sp}$) and Grady-Kipp fragments. The models predicts that sesquinary craters made by ejecta from Io should be abundant in the 500-m-diameter range. However, it is uncertain whether spalls from Io could actually remain intact at the high launch velocities (v > 2.4 km/s) required to escape Io.

on clustering analysis, that most craters on Europa in this size range are conventional secondary craters. If most small europan craters were ionian sesquinaries, they should not be clustered. This discrepancy deserves some discussion, especially given the intuitive expectation that impact ejecta from craters on Europa ought to be vastly more important than impact ejecta from craters on Io.

By construction, craters in the 100–1000-m diameter range are mostly made by spalls, either from Io or from Europa. *Melosh* (1989) approximates the thickness of the spall plate by

$$Z_{sp} = \frac{Yd}{\rho_t C_L v} \quad (1)$$

where Y is the dynamic strength of the material, d the diameter of the impactor, ρ_t the density of the target material, C_L the longitudinal sound speed in the target, and v the ejection velocity of the spall. The length and breadth of the spall plates are at genesis typically an order of magnitude larger than their thickness, although we expect the tiles to break up into more equant bodies with dimensions on the order of Z_{sp}. Comparison between Io and Europa is in most respects a comparison between rock and ice. We take Y = 2×10^9 dynes/cm^2 for rock and 1×10^8 dynes/cm^2 for ice; ρ_t = 3 g/cm^3 for rock and 0.9 g/cm^3 for ice; and C_L = 5 km/s for rock and 2 km/s for ice.

A plausible flaw in the argument is in the presumption that most of the big spalls from Io get launched intact. Melosh suggests that equation (1) is questionable for ejection velocities higher than 1 km/s, and that what is more likely is that these abused spalls break up into the smaller Grady-Kipp fragments during launch. Another alternative is that ionian lava-flow thickness may contribute to spall thickness (e.g., the boundaries between flows are structural weaknesses that the high-pressure excavation wave exploits); if Io lava flows are on the order of 1 m thick (*Davies et al.*, 2000), ionian spalls may rarely be larger than 1 m. Of course it should not be forgotten that these predicted numbers of sesquinary and secondary craters depend rather sensitively on small number statistics, since it is the biggest cometary impacts to occur on Io or Europa that are expected to dominate the production of globally dispersed small craters on Europa.

Predicted sesquinaries from ionian Grady-Kipp fragments are also shown in Fig. 6. These are expected to be in the 10–100-m diameter range. It is apparent that these too are predicted to be very numerous on Europa. Moreover, it is much less plausible to argue that they would be made less visible by being broken down further into smaller fragments, since by construction the Grady-Kipp fragments *are* just those smaller fragments. Thus the prediction remains that there should be a huge number (on the order of 10^{10}) of 10–100-m iogenic sequinary craters on Europa. However, we would not necessarily expect these to be more abundant than comparable europan secondaries, because the Grady-Kipp fragments sample the excavated mass as a whole, and not the mass of the spall zone (which because of rock's greater strength makes up a bigger fraction of Io's crater ejecta than Europa's).

The ejection model (spallation and production of Grady-Kipp fragments) that works so well for Zunil on Mars overpredicts the importance of ionian sesquinaries on Europa due to uncertainty in the relative strengths of the ionian and europan surfaces, as well as uncertainty in the maximum dimension of the fragments that make sesquinaries. While early versions of the model do not match the observations, the underlying process (material exchange) exists and occurs elsewhere (e.g., martian and lunar meteorites found on Earth), and so while the magnitude is currently uncertain, material exchange between the Galilean satellites must occur.

3.3. Effect on Crater Size-Frequency Distributions and Age Interpretation

Secondaries and sesquinaries alter the crater SFD, in both slope and surface density, from what the purely primary SFD would be alone. Decades of observations [starting with *Shoemaker* (1965)] demonstrate consistent trends for the effect of secondaries, namely that they contribute a generally steep (<-4) component to the overall SFD. The slope can vary (for example, secondaries on Europa have slopes between –4 and –6.5), but are generally steep. The steep slope cannot continue down to arbitrarily small sizes, so within a given cluster the SFD eventually "rolls over," and the number of smaller craters goes to zero. The superposition of several clusters, from primaries of different sizes and various ranges, could extend the steep slope over a larger diameter range. Figure 10 shows representative measurements of small-crater SFDs on Europa, as well as other features we discuss in later sections.

Folding the effect of sesquinaries into crater SFDs is still an emerging field, and requires understanding of mass and velocity distributions of ejected material, how that material dynamically evolves in multibody gravitational environments, and the appropriate scaling law that translates a fragment into a crater on the surface of the recipient object. Certainly the presence of lunar and martian meteorites on Earth unequivocally demonstrates that material exchange can occur in systems that are far less dynamically coupled than the Galilean satellites, which are mere handshakes away in comparison. In theory, larger primary impacts on sibling moons could create larger sesquinaries on Europa, but because larger craters are progressively older (usually), Europa's youth sets a relatively firm limit on the maximum sesquinary expected from Io and other satellites; *Zahnle et al.* (2008) estimate that in a 60-m.y. window, the largest ionian sesquinary on Europa is 1.1 km.

Primary craters accumulate randomly in space and in time; in contrast, millions of secondary craters from a single primary impact form in minutes to tens of minutes, adding both clustered and spatially random populations at regional, and probably global, scales. Sesquinaries are intermediate between primaries and secondaries — they do not form "at once" as secondaries do, but *Alvarellos et al.* (2008) find the median transfer time between Io and Europa to be 56 yr, so an ionian impact of sufficient size could deliver hundreds of fragments to Europa in a few decades, creating craters between tens of meters and 1 km in size. The sesquinaries will generally be spatially random, although they tend to avoid the trailing hemisphere.

Primaries, secondaries, and sesquinaries do not function as equivalent chronological markers. *Hartmann* (2007), using the *Malin et al.* (2006) MOC data, argues that on Mars the accumulation of background secondaries (those that are not in obvious clusters) approximates the accumulation of small primaries, such that the superposition of the two populations still provides a reliable (within a factor of 3 to 10) estimate of age. *McEwen and Tornabene* (2007) follow up Hartmann's analysis using MRO HiRISE data to constrain whether or not the craters identified by *Malin et al.* (2006) were, in fact, formed within the past decade. While the HiRISE data confirm that some of the craters could be recent, *McEwen and Tornabene* (2007) also observe that primaries smaller than 100 m generate secondaries, and that secondaries may form even in ancient terrains in which the megaregolith is indurated and the loose regolith is only meters thick. They also point out there is still a discrepancy of about 100 between the estimated impact rates from the *Malin et al.* (2006) results and from the lack of superimposed small craters on fresh, rayed craters. So the chronology of small martian craters has not yet converged.

There is an emerging discussion that the impact rate may vary from the typically assumed constant flux. The creation of asteroid families near orbital resonances could deliver a short-term spike of impactors to the inner solar system. Asteroid falls on Earth with similar ages to small spikes in the lunar impact flux [as recorded in Apollo samples; see *Hartmann et al.* (2007) for discussion] provide initial demonstration of plausibility for this hypothesis, but analysis is still preliminary and one has to be careful not to invoke *ad hoc* arguments to explain observations.

Ultimately, Mars and Europa may be sufficiently different to render comparisons crudely instructive at best. Different target surfaces (rock vs. ice), surface gravity (Mars' is almost three times greater than Europa's), impacting populations (asteroids vs. comets), and dynamical environments (sources of sesquinaries) impose several layers of scaling and modeling, and thus large uncertainties, to back out a direct comparison between the surfaces. There is ample evidence from the Europa data that the small comet population (the things that hit Europa) is different from the small asteroid population (the things that hit Mars, the Moon, and Earth). While small asteroids may have a slope as steep as –3.5, small comets may have a slope as shallow as –2 (both slopes are differential). The rate at which small comets accumulate on outer satellites (such as Europa) could be much lower than the impact rate of small asteroids on terrestrial planets (such as Mars); thus, small europan primaries would be overwhelmed by the production of secondaries from larger primaries. (This is a description that must approximate reality due to the dominance of crater clusters within Europa's small crater population.) Since the distribution of ejecta is roughly ballistic, and therefore inversely proportional to surface gravity, a given velocity ejecta fragment will travel three times farther on Europa than on Mars. And although still not fully quantified, Europa will certainly suffer more contamination from sesquinaries than Mars.

If sesquinaries are insignificant compared with the production of secondaries and primaries, it may be possible to use relative densities of secondaries to evaluate relative ages *if* it is possible to tie individual clusters to their parent primary. (That is, a particular region has clusters from three different primary craters of size D, and so is an equivalent age to having accumulated the three primaries.) However, making such connections between clusters and their parent primary has been difficult, in part because of incomplete coverage and uncertainty in the minimum-sized primary that generates the measured secondaries. Moreover, it is not clear that sesquinaries are insignificant relative to primaries or secondaries; this depends on the flux of small objects that hits Europa (a topic with significant uncertainty, as just discussed), and on the flux of larger objects hitting both Europa and Io.

Because of the unfortunate, coincidental convergence of the "onset diameter" of secondaries and sesquinaries at about 1 km, we urge extreme caution when using craters smaller than 1 km for any aspect of age determination. Stratigraphic relationships are likely the best means to evaluate relative ages on a local scale.

4. RELATIVE AND ABSOLUTE SURFACE AGES

4.1. Comet Flux and Average Age

Jupiter-family comets are the chief source of primary impact craters on the Galilean satellites (*Shoemaker and Wolfe,* 1982; *Zahnle et al.,* 1998, 2003; *Levison and Duncan,* 1997; *Levison et al.,* 2000). Other plausible sources — long-period comets, trojan asteroids, and main-belt asteroids — are easily shown to be unimportant by comparison (*Shoemaker and Wolfe,* 1982; *Zahnle et al.,* 1998).

Jupiter-family comets comprise at least two distinct populations that extend from source regions in the Kuiper belt to the inner solar system. The two important populations are the traditional Kuiper belt (generally low-eccentricity, low-inclination objects that could have formed in place) and the scattered disk (generally higher-eccentricity and higher-inclination objects that appear to have been scattered to their present distant orbits by interactions with planets). Numerical simulations of how these orbits evolve when they become unstable provide a theoretical statistical description of the steady-state distribution of comet orbits in the solar system today. Observed comets sample parts of this distribution. At Earth's orbit the survivors are sampled more or less completely, but most of these comets are short lived owing to thermal disruption, and thus this sample must give an underestimate. Big comets can be counted directly in the Kuiper belt, and some (known as Centaurs) can be seen between the outer planets, but these big comets produce basin-sized impacts, akin to Gilgamesh on Ganymede, and are therefore too few to use for cosmic dating. To relate the big comets of the outer solar system to the small comets that make the countable craters requires an SFD that is most readily determined from the SFD of the craters themselves. A third approach makes use of the historic record of close encounters of comets with Jupiter. Close encounters with Jupiter can make dark comets more visible, either by scattering them into the inner solar system where we can see them better, or by jostling them enough to awaken them to activity, or both. Many of these comets are kilometer-sized and thus make small, countworthy craters.

Zahnle et al. (2003) used craters on Europa (for small comets), Ganymede and Callisto (for midsized comets), and the observed SFD of Kuiper belt objects (big comets) to piece together a united SFD of comets encountering Jupiter and the Galilean satellites. Comet size is related to crater size by assuming that comets have a density of 0.6 g/cm^3 and using impact velocities listed by *Zahnle et al.* (2003). The resulting rate $\dot{N}(>d)$ that comets strike Europa can then be described by a cumulative SFD assembled from four power laws (*Zahnle et al.*, 2003)

$$\begin{aligned}\dot{N}(>d) &= \frac{1}{\tau_0}\left(\frac{d_0}{d}\right)^{b_0} && d < 1.5 \\ &= \frac{1}{\tau_0}\left(\frac{d_0}{d}\right)^{b_1} && 1.5 < d < 5.0 \\ &= \frac{1}{\tau_0}\left(\frac{d_0}{d_1}\right)^{b_1}\left(\frac{d_1}{d}\right)^{b_2} && 5.0 < d < 30 \\ &= \frac{1}{\tau_0}\left(\frac{d_0}{d_1}\right)^{b_1}\left(\frac{d_1}{d_2}\right)^{b_2}\left(\frac{d_2}{d}\right)^{b_3} && d > 30\end{aligned} \quad (2)$$

where τ_0 = 3.1 m.y.; the diameters are d_0 = 1.5 km, d_1 = 5 km, and d_2 = 30 km; and the exponents are b_0 = 1, b_1 = 1.7, b_2 = 2.5, and b_3 = 3.2. This distribution is deficient in both large (>30 km diameter) and small (<1.5 km diameter) bodies compared with near-Earth asteroids. Uncertainty in the number and sizes of comets at Jupiter introduces a factor of 3 uncertainty in τ_0. The size-frequency distribution of the smallest comets is the most uncertain because primary kilometer-sized craters on Europa are subject to confusion with secondary and sesquinary craters. Also, it should be stressed that subkilometer-sized comets cannot yet be used to date Europa's surface, because the number of these comets has been determined by Europa's craters.

We use Monte Carlo simulations of comet fluxes with different exponents in equation (2). Figure 7 illustrates that there is not much cause to choose between models with 40 m.y. and 90 m.y. surfaces, although if we take a cue from early Enceladus measurements (*Kirchoff et al.*, 2007), the shallow production curve associated with 60 m.y. may be "first among equals." This range of ages presumes that we know the comet impact rate exactly; in fact the comet impact rate is at best known to a factor of 3. Presuming uncorrelated log-normal uncertainites in the Monte Carlo model fits and in the comet flux lead to error bars described by

$$\log_{10}(\text{age}) = 1.78 \pm 0.18 \text{ (Monte Carlo spread)}$$
$$\log_{10}(\text{age}) = 1.78 \pm 0.47 \text{ (comet flux uncertainty)}$$

then,

$$\log_{10}(\text{age}) = 1.78 \pm \sqrt{0.18^2 + 0.47^2} = 1.78 \pm 0.5 \quad (3)$$

leading to the conclusion that, with uncertainties, the global average surface age of Europa is between 19 and 190 m.y. (or 20 and 200 m.y., depending on rounding preferences).

Zahnle et al. (2003) estimate a best-fit average age of between 30 and 70 m.y.; the error bars between their estimate and our estimate render the two essentially identical. Our best-fit age (40–90 m.y.) is slightly shifted to older values because we decrease the number of small primary craters. In other words, as the flux goes down, the age goes up. Nevertheless, Europa remains young.

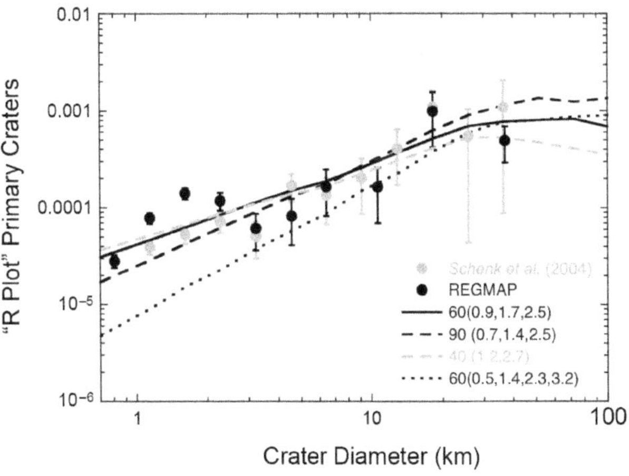

Fig. 7. Four Monte Carlo simulations of primary impact craters on Europa compared with crater counts from *Schenk et al.* (2004) and our REGMAP data. The simulations use algorithms described fully in *Zahnle et al.* (2008). Resurfacing is neglected. The four models are labeled by surface age in million years and by the exponents [b_0, b_1, b_2, (b_3)] describing the size-number distributions of small comets in equation (2). Small craters cannot be used to determine surface ages because (1) some or most are secondaries or sesquinaries and (2) the size-number distribution of the small comets is determined by fitting to just these same craters; the argument would be circular.

4.2. Terrain Comparisons

Figueredo and Greeley (2004) use the REGMAP data to develop a stratigraphic column for Europa's surface features, and present an interesting case for a transition in resurfacing style from ridge formation to chaos formation. Chaos terrain seems to be the most recent form of surface activity, and so they support a cyclic transition between

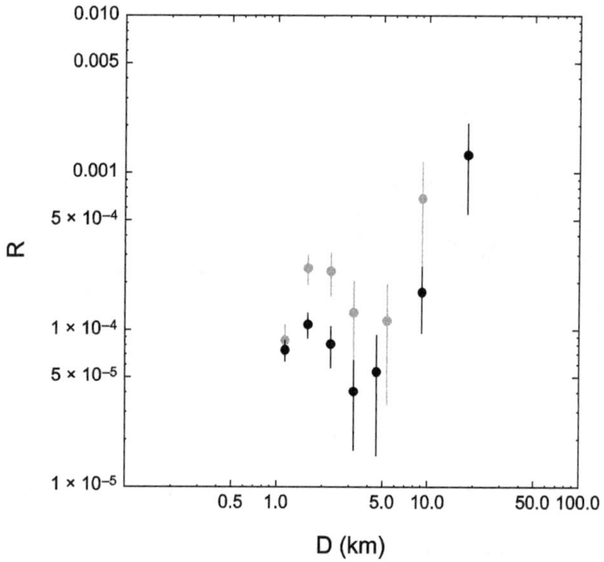

Fig. 8. R-plot comparison between primary crater measurements on ridged terrain (black points) and chaos terrain (gray points). Stratigraphic relationships (*Figueredo and Greeley,* 2004) suggest that chaos regions are generally younger than ridged terrains, yet the density on chaos terrains is generally higher.

ridge-dominated surface activity and chaos-dominated surface activity. [Although they cannot rule out a secular change in time, they consider such a scenario as unlikely since it requires that we are observing Europa at a unique period in its geological history; see also *Pappalardo et al.* (1999).] Small cracks and/or immature ridges often are superimposed on chaos; *Greenberg et al.* (1999) present several examples of immature cracks and double ridges that appear within chaos structures. They may have formed after the chaos, or they may have been preexisting and survived the creation of chaos. While the details of particular cases are unclear, the general relationship seems to be that chaos terrains generally are younger than ridged terrains (see chapter by Doggett et al.).

We use the REGMAP crater data to evaluate crater densities on Europa's terrains. Although *Figueredo and Greeley* (2004) define multiple subclasses for ridged and chaos terrains, there are insufficient numbers of primary craters on each subclass to make a meaningful comparison, so we limit our analysis to ridged vs. chaos, shown in Fig. 8. One might reasonably expect that the chaos crater density would be lower than that of ridged terrain, given that chaos terrain is generally younger (by stratigraphic analysis). However, we find that the densities on chaos terrains can be higher than those on ridged terrains.

Because the density contrast exists at most diameters, we rule out observational effects (e.g., craters are more easily identified on chaos terrain). Because these measurements do not include diameters less than 1 km, secondaries should not contribute any confusion. Part of the difference may stem from our oversimplification of terrain types, in which we grouped all chaos measurements together and all ridged-terrain measurements together, when in fact there are multiple subcategories for each terrain type that appear at different points in the stratigraphic column. Moreover, the error bars overlap in several diameter bins. But the chaos terrain has a systematically higher density, and in two diameter bins this difference is factors of several greater than the error bars.

Mars also has examples of young terrains with higher crater density, due to suspected but complex (and ultimately not entirely understood) interactions between deposition and erosion. But Mars is not Europa, and the obliquity cycles, periglacial processes, wind/dust interactions, etc., that drive the complex martian geology cannot serve as analogs for Europa. The lack of clear consistency between stratigraphy and crater density of ridged and chaotic terrains is a topic that deserves more attention, and would benefit from additional observations from future missions.

4.3. Hemispherical Comparisons

The REGMAP data, used for our crater measurements, are two north-south swaths that are approximately the leading and trailing hemispheres. *Zahnle et al.* (2003) demonstrate that heliocentric impactors striking sychronously rotating satellites should create a hemispherical crater asymmetry: because the leading hemisphere is more often hit, it should have a higher crater density than the trailing side. The dichotemy is not yet observed on Callisto, perhaps in part due to the equilibrium density (or near-equilibrium density) crater population. Ganymede exhibits the asymmetry, although at a level less than predicted. *Zahnle et al.* (2003) propose that nonsynchronous rotation (e.g., while the vast majority of the moon's mass is tidally locked, the relatively thin icy shell is not) may account for the lower-than-expected asymmetry on Ganymede.

Figure 9 plots the crater populations for the leading (black) and trailing (gray) hemispheres of Europa. The leading hemisphere, on average, has a higher density than the trailing hemisphere, consistent with the *Zahnle et al.* (2003) models. While Europa also is expected to experience nonsynchronous rotation, the rate is unknown (see the chapters by Kattenhorn and Hurford, Nimmo and Manga, and Sotin et al.), and could be sufficiently slow that a hemispherical asymmetry accumulates.

4.4. Comparison with Other Data and Models

4.4.1. Other cratering data. *Schenk et al.* (2004) present measurements of Europa's primary crater population. Their data include measurements from the REGMAP mosaics we examined, as well as other mosaics of similar resolutions but in different locations. The two datasets match well, easily within the error bars of one another. Figure 10 recreates Fig. 3 (our REGMAP data), with the addition of the primary crater measurements from *Schenk et al.* (2004), small-crater measurements seen in the high-resolution data (*Bierhaus et al.,* 2001, 2005), and the Neukum production

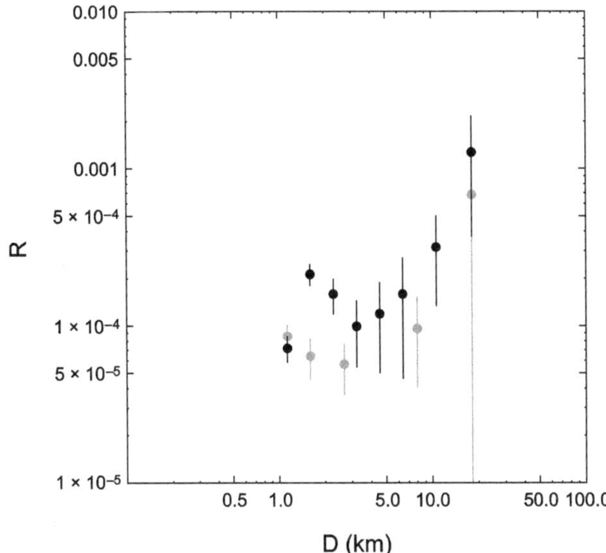

Fig. 9. R plot of the crater measurements on the leading side (black points) and trailing side (gray points) of Europa. The leading side, on average, has a slightly higher density than the trailing side.

function (NPF) as defined by *Ivanov et al.* (2002) (more about the NPF is given below). There are two satisfactory conclusions to reach when comparing our data with that of *Schenk et al.* (2004): (1) The close agreement between the two independent datasets lends significant confidence that these data reflect the primary SFD seen within the REGMAP data, and (2) because the Schenk et al. data include measurements from locations outside the leading/trailing hemispheres, we have greater confidence that the REGMAP data do indeed reflect the global trend for the total primary population (and thus it is less likely there are strong longitudinal variations in crater density, other than the possible leading/trailing asymmetry).

4.4.2. Other cratering models. There have been two models to estimate surface ages of outer solar system objects: The first (which is our preferred model) finds, by observation and by modeling of small-body orbital evolution, that comet populations are the dominant impactors. The second model (*Neukum et al.,* 1999) assumes, despite dynamical implausibility, that after capture into orbit around an outer planet (in the case of this discussion, Jupiter), asteroids are the dominant impactors. The second model requires some unexplained dynamical mechanism that delivers asteroids from the main belt into circumjovian orbits throughout solar system history, including current times.

Neukum derived a 13th-order polynomial to describe crater SFDs from inner-solar system surfaces, known as the Neukum production function (NPF). We compare the Europa data with the NPF. While *Neukum et al.* (1999) suggest that the NPF provides a good fit to Europa's crater population, we do not find good agreement between the data and the NPF. The black curve in Fig. 10 is the NPF for a 1-G.y. surface (one of several average surface ages for Europa proposed by *Neukum et al.*). The figure clearly demonstrates two important points: (1) There is a disconnect between the primary-crater SFD and the small-crater SFD visible in the high-resolution data. The high-resolution data are unlikely to sample atypical regions because (a) they are from disperse locations across Europa and yet have similar crater densities; and (b) a random sampling of a small fraction of the total surface area has a better chance of selecting *typical* regions than *atypical* regions. Therefore we assume the high-resolution measurements are representative of small craters everywhere on Europa. In order for the small craters to be part of the same population as the primary data, the SFD would have to be very steep in the diameter range between a few kilometers and 1 km. However, one does not have to invoke an unusually (and perhaps nonphysically) steep production in that diameter range if, as spatial statistics strongly indicate, most of the small craters are in fact secondaries, and not primaries. Therefore the high-resolution and REGMAP measurements are different populations that do not need to be unified in a single production function (be it the NPF or otherwise). (2) While it is possible to shift the NPF vertically and/or horizontally to achieve a "fit" to the measured data over a certain diameter range, the NPF does not represent the data over the full range of diameters. The NPF curve is not scaled to account for differences in target properties or impact velocities, but in this case no amount of scaling can repair the fundamental mismatch in the functional form of the data and curve.

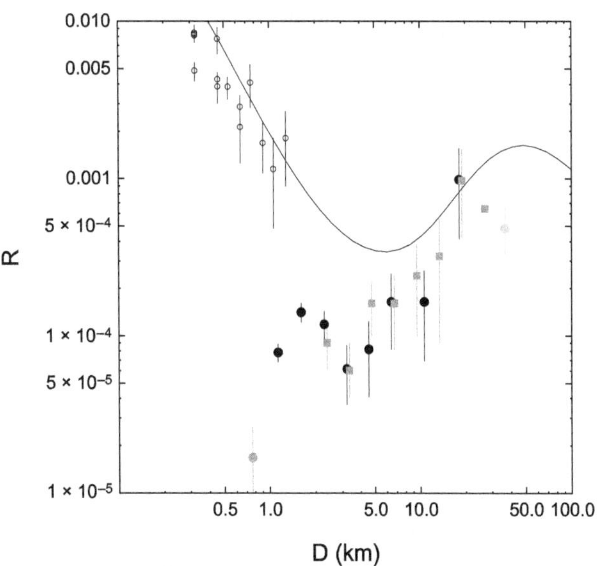

Fig. 10. In addition to the data from Fig. 3, this figure displays three additional datasets: The gray squares are primary crater measurements from *Schenk et al.* (2004); the small, open circles are representative counts of craters seen in high-resolution images; and the black curve is the Neukum production function (NPF), as defined by *Ivanov et al.* (2002). We interpret the NPF fit is poor, and that the measured crater distribution represents two populations: Most craters below ~1 km diameter are secondaries, while those ≥1 km are primaries.

At its current location, the NPF fits the steep branch of small craters, but differs by factors of several to almost an order of magnitude at other diameters. Adding appropriate scaling for higher impact velocities and weaker target surface would shift the curve to the right, potentially improving the fit at larger diameters but degrading the fit at smaller diameters. The gross similarities to the NPF that were inferred during initial examination of the Galileo data are not evident with a more complete understanding of the crater population.

5. OTHER PROXIES FOR SURFACE AGE

5.1. Crater Rays

Crater rays have been used as indicators of crater age on the Moon (*McEwen et al.,* 1997; *Grier et al.,* 2001). Although ray formation is still not completely understood (*Hawke et al.,* 2004), they are associated with younger craters, but "young" requires context. Ray preservation depends upon the presence or absence of an atmosphere (martian rays seem particularly short-lived due to atmospheric erosion and/or dust deposition) and the component materials of the ray relative to the substrate. *Hawke et al.* (2004) estimate some lunar rayed craters are older than 1.1 G.y.; *McEwen et al.* (1997) estimate that some mature lunar rays are over 3 G.y. In contrast, visible rays do not last very long on Mars, although rays persist longer in the infrared (IR) since the daily thermal wave penetrates a thin coating of dust. Using thermal IR images, *Tornabene et al.* (2006) identified four definite and three probable rayed martian craters. Where active processes rapidly affect the optical properties of a surface, such as dust migration on Mars, the visible-wavelength evidence for rays fade relatively rapidly, while nonvisible evidence (thermal signature due to particle-size contrast with surrounding terrain) can remain for longer periods. On more dormant surfaces, such as the Moon, visible-wavelength ray features can remain for much longer periods (over a billion years).

Like the Moon, Europa has no appreciable atmosphere, but like Mars, Europa has active surface processes, both from endogenic processes (ridge and chaos formation) and the routine pelting of high-energy (MeV) charged particles from Jupiter's radiation environment (see chapters by Paranicas et al. and Johnson et al.). Unlike both the Moon and Mars, Europa's surface is mostly water ice. Impacts into rocky surfaces generate melt glass, and other materials, that exist in mechanically and chemically different states from the original rock, transformed by the high pressures and temperatures (PT) of an primary impact. These altered forms are stable over time. While ice also experiences mechanical deformation during impact (*Stewart and Ahrens,* 2005), transforming into various high-PT phases distinct from ice Ih, we do not know what states are stable after the transient high-PT from an impact dissipate.

While we know little about the ejecta products and ray creation on icy surfaces, the lack of visible rays around most of Europa's large craters, despite Europa's youthful surface, suggests that visible rays disappear relatively rapidly. The ejecta products may "relax" back to ice Ih, and/or sputtering from the radiation environment could affect ray visibility. In their chapter, Paranicas et al. estimate ion sputtering erodes about 10^{27} H_2O molecules per second for the whole surface, which is approximately 6×10^4 yr to erode 1 mm of ice everywhere, or 6×10^5 yr to erode (or at least redistribute) 1 cm.

Fig. 11. A high-Sun image of the Pwyll crater and its extensive ray system. Pwyll is the only large primary crater on Europa that exhibits an extensive ray system, which extends over 1000 km from the crater. Other craters suggest that rays fade quickly on Europa, so Pwyll must be young.

Europa's roughly 25-km crater Pwyll is the only large crater still to possess an extensive ray system (Fig. 11). Other europan primaries, such as Cilix and Manannán, express limited rays (*Moore et al.,* 2001), but they are far more limited in extent than the hemisphere-crossing Pwyll rays, and may be the remnants of what were once larger ray systems. Considering the Paranicas et al. results, this suggests Pwyll is less than 10^6 yr old. *Zahnle et al.* (2003) estimate a 20-km crater forms once every 2 m.y., within roughly a factor of 2 (which is quite good agreement for this line of work).

The presence of rays can serve as another check on the Neukum model for europan cratering; *Neukum et al.* (1999) estimate that Pwyll is ~20 m.y. old. If Europa's surface erodes, or is redistributed, at the rate of 1 cm per 6×10^5 yr, then the upper 1 cm could have been reworked 33 times since Pwyll's formation. This should be more than enough

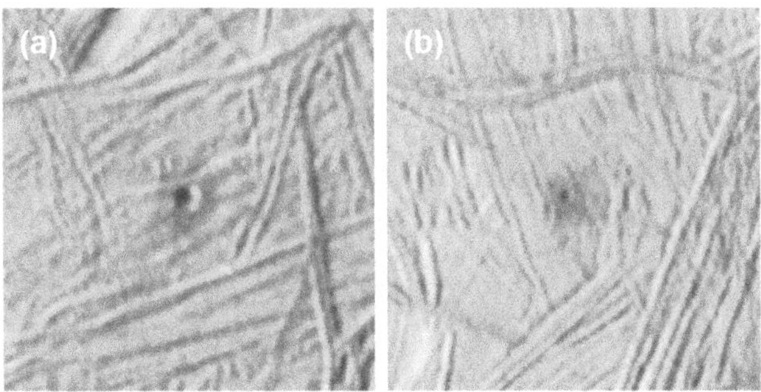

Fig. 12. Two small craters with dark halos. The Sun is from the left in both images. **(a)** Portion of 17ESREGMAP01 frame 64200; the crater is about 1.7 km diameter. **(b)** Portion of 15ESREGMAP01 frame 61878; the crater is about 0.9 km diameter.

to obliterate any evidence of the ray, assuming optical properties (specifically, the albedo contrast between the ray and background surface) of the surface's upper centimeter control ray appearance in visible wavelengths. The error bars on the Neukum et al. estimate permit younger (and slightly older) ages for Pwyll, but the best-fit age of 20 m.y. seems incompatible with the age constraint provided by sputtering rates.

Some small craters exhibit rays. We find several examples within the REGMAP data of <2-km craters with dark and bright ejecta that extend a few crater radii from the rim. The dark ejecta craters are particularly interesting, as some of them may represent ionian spalls. Figure 12 shows two such examples of small craters with dark ejecta. Since *Zahnle et al.* (2003) predict that Io's impact rate is roughly twice that of Europa, it is reasonable to expect that two "Pwylls-on-Io" have formed in the past 2 m.y. (or one has formed in the past 1 m.y.), which deliver dark, basaltic fragments to Europa's surface.

We do not have sufficient laboratory data or model results to describe the production of icy ejecta, but the scarcity of rays on a surface as young as Europa's suggests icy rays are more ephemeral than their rocky cousins. A high-energy radiation environment that continuously redistributes the optical layer of Europa's surface may provide an independent time constraint (an upper limit) on the age of a crater that still expresses rays, and possibly an upper limit on the age of dark ridges and bands (*Geissler et al.*, 1998). But more work is necessary to quantify the relationship between color and age beyond estimates of the sputtering rate (see the Paranicas et al. chapter), and the simple assumption that a ray is eliminated after the upper 1 cm is reworked by the sputtering.

5.2. Crater Degradation States

A crater's degradation state can be a useful indicator of its age, although that depends on knowledge of the processes and their rates that degrade the crater. On a geologically inactive, airless world such as the Moon, craters are degraded mainly by the superposition of later craters, which — depending on their size — erode earlier crater topography, overlap earlier craters, or fill craters with ejecta, resulting in crater erosion that is either gradual or instantaneous depending on proximity to the newly formed crater(s). For multikilometer craters on the Moon, crater degradation can be a multi-billion-year process. On geologically active worlds, such as Europa, the distribution of degradation states can set constraints on how the surface changes with time: A "continuous" spectrum of degradation states, from pristine to barely recognizable, suggests that the surface activity also occurs continuously; discrete degradation states suggest that surface activity takes place episodically.

Because of the limited Galileo coverage, we necessarily focus on the degradation states of the large craters imaged at sufficient resolution to evaluate crater morphology. See the chapter by Schenk and Turtle or *Moore et al.* (1998, 2001) for detailed discussions of crater morphology; here we focus on morphology and degradation exclusively as a means to constrain Europa's surface age.

5.2.1. Tyre. As the largest recognized impact structure on Europa, Tyre is likely to be one of the oldest. During the E14 orbit, Galileo obtained a nine-image mosaic of Tyre at about 200 m/pixel, so features >1 km are resolved well enough to determine stratigraphy and morphology (Fig. 13). Two well-developed double ridges cross-cut the southwest

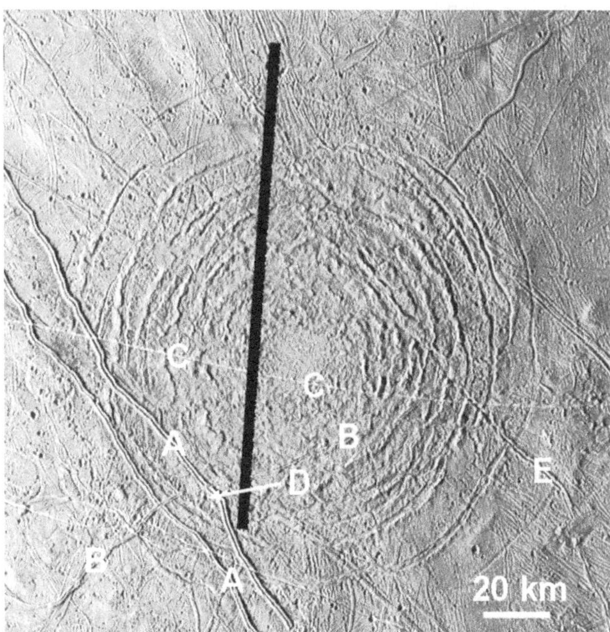

Fig. 13. The Tyre multiring impact structure, and part of the surrounding field of secondary craters. The Sun is from the left. Despite its relatively large size (and thus presumably greater age), only a few features clearly postdate the central feature or the concentric rings: **(a)** two well-developed double ridges; **(b)** one "immature" double ridge, marked in two places; **(c)** a simple fracture, marked in two places; and **(d)** one 2-km crater. The double ridge marked (E) appears to have been active before Tyre (contains superimposed Tyre secondaries), and may have been active after Tyre (the double ridge reformed over parts of Tyre). The black bar is a data gap.

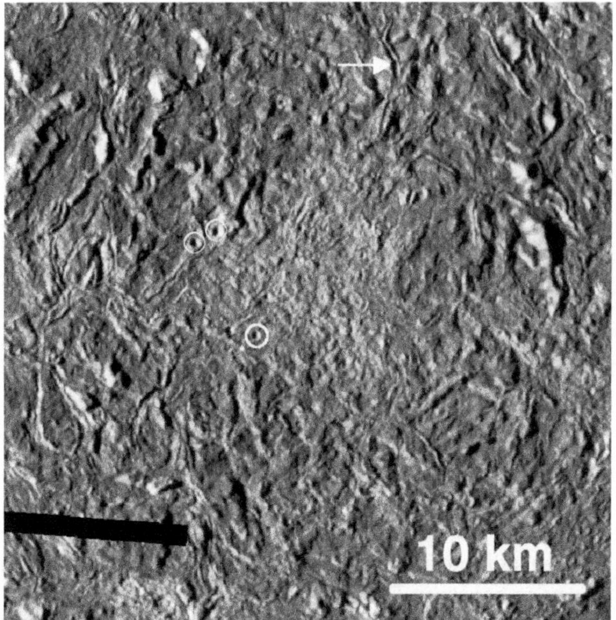

Fig. 14. The central unit of the Callanish multiring structure (the rings are beyond the boundaries of the image). The Sun is from the left. The circles highlight small craters, maybe secondaries from another primary, that formed after Callanish. The arrow points to one of a few small, sinuous fractures within the central unit. Other than these small features, Callanish is essentially unmodified by ridges or chaos. (The tilted black bar in the lower left is a data gap.)

portion of Tyre's ring structure, one less-developed double ridge crosses Tyre's southern ring structure (and subsequently transects the two well-developed double-ridges), and one simple fracture crosses tangentially to Tyre's central smooth region (as well as overprinting the three double ridges). Only one resolved crater appears to postdate Tyre; a 2-km crater sits within the southwest portion of Tyre's ring structure. One double ridge seems to be unique in that it appears to have been active both before and after Tyre's for-mation (see Fig. 13). The region around Tyre is covered with Tyre secondaries and chaos features. Some of the chaos features are surrounded by secondaries yet do not have superimposed secondaries themselves, suggesting they appeared (or at least continued to evolve) after Tyre formed.

5.2.2. Callanish. Galileo imaged portions of Callanish at about 120 m/pixel during orbit E4, and during orbit E26 aqcuired a roughly 50-m/pixel sequence of images that transect a portion of the eastern continuous ejecta zone, the eastern ring structure, and through the central impact zone. The E26 data reveal some small (400–600 m) craters and small (~200 m width) sinuous fractures that are superimposed over the central Callanish structure (Fig. 14), and which are not visible in the lower-resolution E4 data. The area occupied by these features is a few percent or less than the total area of Callanish.

Larger craters tend to be older (they occur less frequently, and thus have a lower likelihood that one formed recently), and certainly Tyre and Callanish show some signs of age in the form of superimposed ridges and craters on their central impact site and concentric ring structures. But no chaos formation blemishes their visage, and in terms of area-preserved vs. area-destroyed (of the impact feature), they are essentially unmodified.

5.2.3. Craters 1 km $\leq D \leq$ 30 km. Galileo imaged only a few craters in this diameter range with sufficient resolution to evaluate morphology and degradation, summarized by *Moore et al.* (1998, 2001). None of these craters possesses postimpact modification by ridge or chaos formation. Without knowing their appearance when freshly formed, we cannot definitively state that they are completely unmodified (especially since aspects of large-crater morphology on Europa are unique to that surface). However, we can assert that the craters are essentially unaffected by the two dominant europan resurfacing mechanisms, namely ridge and chaos formation.

5.2.4. Craters ≤1 km. Most craters of this size are secondaries, and are thus not markers of continuous accumulation but rather the more infrequent formation of large primaries. Comparing degradation states of secondaries from the same primary can reveal variations in surface activity. There is only one identified example of this, primarily because of a fortuitously placed 10-m/pixel image sequence: *Bierhaus et al.* (2001) report that secondaries from Pwyll in Conamara Chaos show nonuniform morphology, a possible indication that Conamara Chaos experienced surface activity since Pwyll's formation. Otherwise, the typical ~50-m/pixel scale of the high-resolution images was not sufficient to evaluate relative degradation states of craters ≤1 km.

5.2.5. Summary of crater degradation. The "big picture" of crater erosion on Europa is striking in that the craters are, with small exceptions for the largest craters, relatively untouched by the seemingly ubiquitous and ongoing ridge and chaos formation. This suggests that whatever erases the craters is an abrupt event, rather than a slow, continuous process that would, at any time, create a suite of degradation states (dependent on age) within a given crater diameter range. Craters within heavily cratered terrains on the Moon and Mars can be significantly obscured by superimposed craters, but still recognizable. The experience of crater mapping on other surfaces provides some confidence that even heavily degraded craters should be recognizable. The apparent freshness of europan craters is consisent with the regional-scale mapping by *Figueredo and Greeley* (2004), who map craters to the youngest period (the top of the stratigraphic column) on Europa.

6. EUROPA IN TIME

6.1. Europa in the Present

Europa's average surface age is young, a Voyager-era suspicion that Galileo data unequivocally confirm. But for Europa, the simplicity of youth is hardly simple. The primary crater SFD presents two questions that seem at odds with other observations: (1) Why do chaos terrains, which are likely among the youngest features on Europa, appear to have generally higher crater density? (2) Why do large primary craters show little or no sign of degradation, either from ridge or chaos formation?

We can make a tentative appeal to Venus as a (relatively) youthful surface with a deceptively simple crater record. Initial analysis (*Phillips et al.,* 1992) of Magellan data revealed a crater distribution thought to be spatially random, and perhaps reflecting a global resurfacing that ended, or took place, about 500 m.y. ago. Subsequent analysis revealed a more complex story. *Hauck et al.* (1998) demonstrate spatial randomness cannot constrain resurfacing models, and show that the venusian plains could represent a spread of ages. *Campbell* (1999) presents a fascinating analysis of small number statistics, illustrating that a single measurement of a small number of craters can at best represent an estimate of the mean density, and the associated Poisson error bars reflect the spread in the mean; he also demonstrates that multiple resurfacing histories are consistent with the venusian cratering record. *Rumpf et al.* (2005) show that some craters do indeed exhibit postformation modification by tectonism and volcanic flows.

Keep in mind that the "small number statistics" for Venus is nearly 1000 craters, compared with the mere tens of large craters on Europa. The venusian analyses are also near-global in extent, while for all except the very largest diameters, europan analysis involves 9–11% of the moon's surface. We may be faced with the happy conundrum that Europa's compelling youth, and associated sparse crater population, means that we can use the crater population to do little more than confirm Europa is young.

Pappalardo et al. (1999) proposed a cyclic history for Europa's endogenic activity, transitioning between ridge building and chaos formation. Further analysis by *Figueredo and Greeley* (2004) supports this hypothesis. The similar crater densities between the two terrains carry the condition that the ridge building and chaos formation, and the transition between the two, happened quickly enough that their combined effect was essentially a global resurfacing, creating a clean surface that then accumulated impact craters. To preserve fresh crater morphology, this also requires that whatever activity that has occurred since the last major episode of chaos formation has not affected the large craters.

The apparent lack of crater degradation is consistent with estimates of resurfacing rates. *Phillips et al.* (2000) searched for surface activity by comparing Voyager and Galileo images of the same region to identify changes. They found none, and set an upper limit on the steady-state surface alteration rate of 1 km^2 y^{-1}. The known and estimated crater population account for less than 0.1% of Europa's surface area, so there could have been small- or regional-scale resurfacing on parts of Europa that are not near any of the large primaries. (Nor can we rule out the total, and perhaps rapid, destruction of a ≤10-km primary by the numerous ≥10-km chaos regions.)

Hussman and Spohn (2004) present a model in which tidal dissipation of orbital energy generates internal heat within Europa and Io. In one of the cases they consider, the oscillation period of the satellites' eccentricity, and associated variation in heating rate, is about 150 m.y. Their model is consistent with periods of high surface activity during the high-heat phase, followed by periods of low surface activity during the low-heat phase. It is plausible that the most recent phase of resurfacing created ridges and chaos terrains in close succession, erasing the then-existing crater population. The only craters currently present are those formed after the rapid-resurfacing phase ended, and although ridges and chaos formed approximately sequentially, the time between the two resurfacing phases was short enough that the terrain types did not accumulate significantly different crater densities.

The consistency between our REGMAP-only measurements and the *Schenk et al.* (2004) measurements, which include data from elsewhere on Europa, imply that REGMAP is close to the global average. Global coverage at regional resolution would resolve both the uncertainty in the average crater SFD, and the possible discrepancy between stratigraphic relationships and crater density for the ridged and chaos terrains.

6.2. Europa in the Past

The usefulness of Europa's craters for chronology extend only to the time of their formation, which is recent relative to the history of the Galilean satellites and jovian system. But there are important discussions that consider Europa as a potentially habitable environment (see chapter by Hand et al.) that require a longer time history than Europa's craters provide. Ganymede and Callisto provide clues to the early, and probably violent, history of Europa, with consequences for early geology and habitability. Both Ganymede and Callisto possess numerous craters several hundred kilometers across, and a few thousand-kilometer-scale multiring basins. Notable are Callisto's 4000-km Valhalla and 1640-km Asgard (both dimensions are the outer-ring diameter), and Ganymede's 600-km Gilgamesh. It is reasonable to conclude that early Europa was pummeled with multiple enormous impacts. We don't know the early icy shell/ocean structure of Europa (see the chapters by Schubert et al. and Moore and Hussmann), but for the following discussion assume that the current ~100-km H_2O layer (ice over liquid) was in place early enough that it still experienced several enormous impacts.

At these crater sizes (hundreds to a few thousand kilometers diameter for final basin size), an icy shell tens of kilometers thick is effectively transparent to the impactor (which itself is tens to a few hundred kilometers in size), and the resulting transient crater could penetrate to the bottom of the ocean layer. The impact will generate a series of waves as the transient crater collapses; the initial-wave amplitude will be almost as high as the transient crater depth.

The presence of the ice layer over the ocean is a damping mechanism for impact waves, but the *Ong* (2004) and *Greeley et al.* (1982) experiments demonstrate that turbulent interaction between waves generated by an ice-penetrating impact will disrupt the ice for several crater radii beyond the transient crater. A Valhalla-scale impact on Europa could catastrophically disrupt the icy shell over an entire hemisphere, and perhaps cause less severe disruption on a global scale. *Ojakangas and Stevenson* (1989) and *Nimmo and Pappalardo* (2006) demonstrate that large mass imbalances of a decoupled icy shell can cause the icy shell to reorient the lowest-mass portion of the shell (in the case of a large impact, the location of the impact) over a pole. Other factors that affect the environment include the mass of water lost during ejection, the dynamics of water redistribution to a new equilibrium state, and the reaccumulation of some ejected material.

Certainly a Gilgamesh- or Valhalla-scale impact would cause global disruption of any ecosystem that might have existed at the time of the impact. Smaller impacts would have not been as severe but could still seriously affect europan (pre-) biology. Early life on Europa (if it existed) would have experienced one of the same challenges experienced by life on early Earth, namely the survival of enormous impacts that caused rapid and severe disruption to the global environment.

7. SUMMARY

Certainly we now know more about Europa's craters and age than we did before the Galileo spacecraft, but uncertainties remain. We summarize by parsing our previous discussions into (1) what we know, (2) what we may know, and (3) what remains unknown.

What we know:

1. Primary craters with diameters larger than 3 km display a differential slope near –2.

2. Secondaries dominate the crater population at diameters smaller than 1 km.

3. The overall shape of Europa's crater SFD is inconsistent with the shape of the crater SFD seen in the inner solar system (e.g., the Moon).

4. Because the impacting population and dynamical environment of the Galilean satellites are sufficiently different from those of the Moon and Mars, crater production functions derived from the Moon or Mars are not useful as templates for Europa.

5. If there were no uncertainty in the comet impact rate or cratering mechanics, Europa's average surface age would be 40–90 m.y. Uncertainties in the comet impact rate and cratering mechanics widens the plausible range to 20–200 m.y. The upper-limit age is still young on solar system timescales and entirely consistent with current geological activity and the presence of an interior ocean.

What we may know:

1. There is slight evidence, both in the number and spatial distribution of craters ≥30 km, that there may be an as-yet unrecognized crater larger than Tyre; if true, this would not change our conclusions regarding Europa's average age.

2. There may be a slight enhancement of craters on Europa's leading side, not unexpected due to the dynamics of heliocentric impactors. However, nonsynchronous rotation of Europa's icy shell (see the chapter by Nimmo and Manga) should damp hemispherical cratering asymmetries (*Zahnle et al.,* 2003), so the difference is not expected to be great.

3. Melosh's theory (*Melosh,* 1984, 1985a,b, 1989) of spalls and Grady-Kipp fragments provides a good first-order fit to the properties of secondary craters, both in number and slope. But the theory overestimates ionian sesquinaries.

4. Sesquinaries from Io contribute to the crater population for diameters ≤1 km. But because of uncertainty in both the rate at which primary craters form on Io that deliver ≤1 km sesquinaries, and in the scaling laws used to estimate sesquinary production from a given impact, we do not know what fraction of the spatially random small crater population is sesquinary in origin.

5. Crater rays may provide an independent means to evaluate crater age if the sputtering rate is directly related to the destruction of ray visibility.

6. Because Europa's surface is so young, it can tell us little about times long ago. However, the giant impact scars on Callisto and Ganymede imply that Europa was similarly pummelled in the past, with possible severe consequences to early habitability and biology.

What remains unknown:

1. The analysis of crater density on a crude grouping of two basic terrains, ridges and chaos, creates the unexpected result that the generally younger chaos terrains possibly have a higher crater density. This remains unexplained, but could be due to small number statistics, and awaits more data and analysis from a future mission.

2. As a result of the significant uncertainty in the relative numbers of primaries, sesquinaries, and spatially random secondaries smaller than 1 km, the shape and density of the primary crater SFD for diameters ≤1 km is still poorly constrained.

Acknowledgments. We thank A. McEwen and C. Phillips for helpful reviews that improved the clarity of the chapter. E.B.B. and K.Z. acknowledge the NASA Outer Planets Research Program for support. All images in this chapter are from the Galileo mission to Jupiter, and are available online at *pds.nasa.gov* (page used with permission of NASA and the Jet Propulsion Laboratory/NASA California Institute of Technology).

REFERENCES

Alvarellos J. L., Zahnle K. J., Dobrovolskis A. R., and Hamill P. (2002) Orbital evolution of impact ejecta from Ganymede. *Icarus, 160,* 108–123.

Alvarellos J. L., Zahnle K. J., Dobrovolskis A. R., and Hamill P. (2008) Transfer of mass from Io to Europa and beyond due to cometary impacts. *Icarus, 194,* 636–646.

Artemieva N. and Ivanov B. (2004) Launch of martian meteorites in oblique impacts. *Icarus, 171(1),* 84–101.

Bierhaus E. B. (2004) Discovery that secondary craters dominate Europa's small crater population. Ph.D. thesis, University of Colorado, Boulder. 293 pp.

Bierhaus E. B. (2006) Crater signal-to-noise: Extracting Europa's small primary crater population. In *Workshop on Surface Ages and Histories: Issues in Planetary Chronology,* pp. 14–15. LPI Contribution No. 1320, Lunar and Planetary Institute, Houston.

Bierhaus E. B., Chapman C. R., Merline W. J., Brooks S. M., and Asphaug E. (2001) Pwyll secondaries and other small craters on Europa. *Icarus, 153,* 264–276.

Bierhaus E. B., Chapman C. R., and Merline W. J. (2005) Secondary craters on Europa and implications for cratered surfaces. *Nature, 437,* 1125–1127.

Bottke W. F., Nolan M. C., Kolvoord R. A., and Greenberg R. (1994) Velocity distribution among colliding asteroids. *Icarus, 107,* 255–268.

Bottke W. F., Vokrouhlický D., Rubincam D. P., and Brož M. (2002a) The effect of Yarkovsky thermal forces on the dynamical evolution of asteroids and meteoroids. In *Asteroids III* (W. F. Bottke Jr. et al., eds.), pp. 395–408. Univ. of Arizona, Tucson.

Bottke W. F., Morbidelli A., Jedicke R., Petit J.-M., Levison H., Michel P., and Metcalfe T. S. (2002b) Debiased orbital and size distributions of the near-Earth objects. *Icarus, 156,* 399–433.

Campbell B. A. (1999) Surface formation rates and impact crater densities on Venus. *J. Geophys. Res., 104(E9),* 21951–21956.

Chapman C. R., Merline W. J., Bierhaus B., Keller J., and Brooks S. (1997) Impactor populations on the galilean satellites. *Bull. Am. Astron. Soc., 29,* 984, Abstract #12.10.

Davies A. G., Lopes-Gautier R., Smythe W. D., and Carlson R. W. (2000) Silicate cooling model fits to Galileo NIMS data of volcanism on Io. *Icarus, 148(1),* 211–225.

Dombard A. J. and McKinnon W. B. (2000) Long-term retention of impact crater topography on Ganymede. *Geophys. Res. Lett., 27,* 3663–3666.

Dombard A. J. and McKinnon W. B. (2006) Elastoviscoplastic relaxation of impact crater topography with application to Ganymede and Callisto. *J. Geophys. Res., 111,* E01001.

Dundas C. M. and McEwen A. S. (2007) Rays and secondary craters of Tycho. *Icarus, 186,* 31–40.

Figueredo P. H. and Greeley R. (2004) Resurfacing history of Europa from pole-to-pole geologic mapping. *Icarus, 167,* 287–312.

Geissler P. E. and 16 colleagues (1998) Evolution of lineaments on Europa: Clues from Galileo multispectral imaging observations. *Icarus, 135,* 107–126.

Greeley R., Fink J. H., Gault D. E., and Guest J. E. (1982) Experimental simulation of impact cratering on icy satellites. In *Satellites of Jupiter* (D. Morrison, ed.), pp. 340–378. Univ. of Arizona, Tucson.

Greenberg R., Hoppa G. V., Tufts B. R., Geissler P., and Riley J. (1999) Chaos on Europa. *Icarus, 141,* 263–286.

Grier J. A., McEwen A. S. Lucey P. G., Milazzo M., and Strom R. G. (2001) Optical maturity of ejecta from large rayed lunar craters. *J. Geophys. Res., 106(E12),* 32847–32862.

Hartmann W. K., Quantin C., and Mangold N. (2007) Possible long-term decline in impact rates 2. Lunar impact-melt data regarding impact history. *Icarus, 186,* 11–23.

Hauck S. A., Phillips R. J., and Price M. H. (1998) Venus: Crater distribution and plains resurfacing models. *J. Geophys. Res., 103(E6),* 13635–13642.

Hawke B. R., Blewett D. T., Lucey P. G., Smith G. A., Bell J. F., Campbell B. A., and Robinson M. S. (2004) The origin of lunar crater rays. *Icarus, 170,* 1–16.

Head J. N., Melosh H. J., and Ivanov B. A. (2002) Martian meteorite launch: High-speed ejecta from small craters. *Science, 298,* 1752–1756.

Holsapple K. A. (1993) The scaling of impact processes in planetary sciences. *Annu. Rev. Earth Planet. Sci., 21,* 333–373.

Hussmann H. and Spohn T. (2004) Thermal-orbital evolution of Io and Europa. *Icarus, 171,* 391–410.

Ivanov B. A., Neukum G., Bottke W. F., and Hartmann W. K. (2002) The comparison of size-frequency distributions of impact craters and asteroids and the planetary cratering rate. In *Asteroids III* (W. F. Bottke Jr. et. al, eds.), pp. 89–101. Univ. of Arizona, Tucson.

Kirchoff M. R., Schenk P., and Seddio S. (2007) Cratering records of Enceladus, Dione and Rhea — Results from Cassini ISS imaging. In *Lunar and Planetary Science XXXVIII,* Abstract #1338. Lunar and Planetary Institute, Houston (CD-ROM).

Levison H. F. and Duncan M. J. (1994) The long-term dynamical behavior of short-period comets. *Icarus, 108(1),* 18–36.

Levison H. F. and Duncan M. J. (1997) From the Kuiper belt to Jupiter-family comets: The spatial distribution of ecliptic comets. *Icarus, 127,* 13–32.

Levison H. F., Duncan M. J., Zahnle K., Holman M., and Dones L. (2000) NOTE: Planetary impact rates from ecliptic comets. *Icarus, 143,* 415–420.

Lucchitta B. K. and Soderblom L. A. (1982) The geology of Europa. In *Satellites of Jupiter* (D. Morrison, ed.), pp. 521–555. Univ. of Arizona, Tucson.

Malin M. C. and Pieri D. C. (1986) Europa. In *Satellites* (J. A. Burns and M. S. Matthews, eds.), pp. 689–717. Univ. of Arizona, Tucson.

Malin M. C., Edgett K. S., Posiolova L. V., McColley S. M., and Dobrea E. Z. N. (2006) Present-day impact cratering rate and contemporary gully activity on Mars. *Science, 314,* 1573–1577.

McEwen A. S. and Bierhaus E. B. (2006) the importance of secondary cratering to age constraints on planetary surfaces. *Annu. Rev. Earth Planet. Sci., 34,* 535–567.

McEwen A. S. and Tornabene L. L. (2007) Modern Mars: HiRISE observations of small, recent impact craters. In *Seventh International Conference on Mars,* Abstract #3086. Lunar and Planetary Institute, Houston.

McEwen A. S., Moore J. M., and Shoemaker E. M. (1997) The Phanerozoic impact cratering rate: Evidence from the farside of the Moon. *J. Geophys. Res., 102(E4),* 9231–9242.

McEwen A. S., Preblich B. S., Turtle E. P., Artemieva N. A., Golombek M. P., Hurst M., Kirk R. L., Burr D. M., and Christensen P. R. (2005) The rayed crater Zunil and interpretations of small impact craters on Mars. *Icarus, 176,* 351–381.

Melosh H. J. (1984) Impact ejection, spallation, and the origin of meteorites. *Icarus, 59,* 234–260.

Melosh H. J. (1985a) Impact cratering mechanics — Relationship between the shock wave and excavation flow. *Icarus, 62,* 339–343.

Melosh H. J. (1985b) Ejection of rock fragments from planetary bodies. *Geology, 13,* 144–148.

Melosh H. J. (1989) *Impact Cratering: A Geologic Process.* Oxford Univ., New York. 245 pp.

Moore J. M. and 17 colleagues (1998) Large impact features on Europa: Results of the Galileo nominal mission. *Icarus, 135,* 127–145.

Moore J. M. and 25 colleagues (2001) Impact features on Europa: Results of the Galileo Europa mission (GEM). *Icarus, 151,* 93–111.

Neukum G., Wagner R., Wolf U., and the Galileo SSI Team (1999) Cratering record of Europa and implications for time-scale and crustal development. In *Lunar and Planetary Science XXX,* Abstract #1992. Lunar and Planetary Institute, Houston (CD-ROM).

Nimmo F. and Pappalardo R. T. (2006) Diapir-induced reorientation of Saturn's moon Enceladus. *Nature, 441,* 614–616.

Ojakangas G. W. and Stevenson D. J. (1989) Polar wander of an ice shell on Europa. *Icarus, 81,* 242–270.

Ong L. C. F. (2004) What lies beneath the surface? Europa's icy enigma. B.A. thesis, Williams College, Williamstown, Massachusetts. 91 pp.

Pappalardo R. T. and Collins G. C. (1999) Extensionally strained craters on Ganymede. In *Lunar and Planetary Science XXX,* Abstract #1773. Lunar and Planetary Institute, Houston (CD-ROM).

Pappalardo R. T. and 31 colleagues (1999) Does Europa have a subsurface ocean? Evaluation of the geological evidence. *J. Geophys. Res., 104,* 24015–24055.

Phillips C. B., McEwen A. S., Hoppa G. V., Fagents S. A., Greeley R., Klemaszewski J. E., Pappalardo R. T., Klaasen K. P., and Breneman H. H. (2000) The search for current geologic activity on Europa. *J. Geophys. Res., 105(E9),* 22579–22598.

Phillips R. J., Raubertas R. F., Arvidson R. E., Sarkar I. C., Herrick R. R., Izenberg N., and Grimm R. E. (1992) Impact craters and Venus resurfacing history. *J. Geophys. Res., 97(E10),* 15923–15948.

Preblich B. S., McEwen A. S., and Studer D. M. (2007) Mapping rays and secondary craters from the martian crater Zunil. *J. Geophys. Res., 112,* E05006.

Rumpf M. E., Herrick R., Gregg T. K. (2005) The complicated geologic histories of large venusian impact craters. *Eos Trans. AGU, 86(52),* Fall Meet. Suppl., Abstract #P33A-0233.

Schenk P. M. (2002) Thickness constraints on the icy shells of the Galilean satellites from a comparison of crater shapes. *Nature, 417,* 419–421.

Schenk P. M., Chapman C. R., Zahnle K., and Moore J. M. (2004) Ages and interiors: The cratering record of the Galilean satellites. In *Jupiter: The Planet, Satellites and Magnetosphere* (F. Bagenal et al., eds.), pp. 427–456. Cambridge Univ., Cambridge.

Smith B. A. and 21 colleagues (1979) The Galilean satellites and Jupiter: Voyager 2 imaging science results. *Science, 206,* 927–950.

Shoemaker E. M. (1965) Preliminary analysis of the fine structure of the lunar surface in Mare Cognitum. In *The Nature of the Lunar Surface* (W. N. Hess et al., eds.), pp. 23–77. JPL Tech. Report No. 32–700, Johns Hopkins Press, Baltimore.

Shoemaker E. M. and Wolfe R. F. (1982) Cratering time scales for the Galilean satellites. In *Satellites of Jupiter* (D. Morrison, ed.), pp. 277–339. Univ. of Arizona, Tucson.

Spencer J. R., Tamppari L. K., Martin T. Z., and Travis L. D. (1999) Temperatures on Europa from Galileo PPR: Nighttime thermal anomalies. *Science, 284,* 1514–1516.

Stewart S. T. and Ahrens T. J. (2005) Shock properties of H_2O ice. *J. Geophys. Res.–Planets, 110,* E03005.

Strom R. G., Chapman C. R., Merline W. J., Solomon S. C., and Head J. W. (2008) Mercury cratering record viewed from MESSENGER's first flyby. *Science, 321,* 79.

Tornabene L. L., Moersch J. E., McSween H. Y., McEwen A. S., Piatek J. L., Milam K. A., and Christensen P. R. (2006) Identification of large (2–10 km) rayed craters on Mars in THEMIS thermal infrared images: Implications for possible martian meteorite source regions. *J. Geophys. Res., 111,* E10006.

Turtle E. P. and Ivanov B. A. (2002) Numerical simulations of impact crater excavation and collapse on Europa: Implications for ice thickness. In *Lunar and Planetary Science XXXIII,* Abstract #1431. Lunar and Planetary Institute, Houston (CD-ROM).

Vickery A. M. (1986) Size-velocity distribution of large ejecta fragments. *Icarus, 67,* 224–236.

Wisdom J. and Holman M. (1991) Symplectic maps for the n-body problem. *Astron. J., 102,* 1528–1538.

Yamamoto S., Kadono T., Sugita S., and Matsui T. (2005) Velocity distributions of high-velocity ejecta from regolith targets. *Icarus, 178,* 264–273.

Zahnle K., Dones L., and Levison H. F. (1998) Cratering rates on the Galilean satellites. *Icarus, 136,* 202–222.

Zahnle K., Schenk P., Levison H. F., Dones L. (2003) Cratering rates in the outer solar system. *Icarus, 163,* 263–289.

Zahnle K., Alvarellos J. L., Dobrovolskis A., and Hamill P. (2008) Secondary and sesquinary craters on Europa. *Icarus, 194,* 660–674.

Europa's Impact Craters: Probes of the Icy Shell

Paul M. Schenk
Lunar and Planetary Institute

Elizabeth P. Turtle
Johns Hopkins University Applied Physics Laboratory

Impact craters offer unique probes into Europa's substructure. Despite limited global mapping, crater morphology and topography reveal several craterforms unique to Europa. Comparison to Ganymede is especially revealing, given that the interior of Europa is believed to be dominated by significantly higher temperatures, leading to a water ocean potentially within reach of the largest craters. Morphologies of craters <18 km across are very similar to those on Ganymede (although those between 8 and 18 km across are comparatively shallow). Only a few craters large than 20 km across have been observed. These include the modified central peak craters Pwyll and Manannán, whose central peaks and rim scarps are irregular or disrupted and whose crater depths are very low. The largest features are the multiring basins Callanish and Tyre, with estimated rim diameters of 33 and 38 km, respectively. The pronounced central peak in Pwyll and the rollover in depth/diameter curve both appear to constrain the icy shell to be at least 10–19 km thick. Widths of ring graben surrounding Callanish and Tyre suggests that heat flows were quite high, although considerable uncertainty remains in interpretation of these results. Only one crater has been identified as being possibly modified by postimpact processes. This raises but does not yet constrain or require the possibility that Europa was resurfaced catastrophically or has been quiescent since most of the resurfacing was accomplished.

1. INTRODUCTION

Europa is believed to possess a liquid water ocean, possibly within a few tens of kilometers of the icy surface (e.g., *Kivelson et al.,* 2000; *Pappalardo et al.,* 1999). Because impact crater formation encompasses and is influenced by a broad hemispheric zone in the target surrounding the point of impact (e.g., *Melosh,* 1989), study of Europa's few impact craters has focused on elucidating the nature of its icy surface and searching for evidence of an underlying ocean. Europa may provide key demonstrations of the influence of target properties on crater morphology. Its impact craters have been described in detail elsewhere (*Moore et al.,* 1998, 2001; *Schenk,* 2002; *Schenk et al.,* 2004); here we summarize those findings and describe new details that have emerged from ongoing analyses. We focus on the constraints they place on the crater formation process and the properties of the icy shell, in particular how differences between Europa's craters and their counterparts on other icy satellites inform these constraints. In this context, we discuss work to model the observed craters, efforts that offer some of our best prospects for probing the deeper structure and character of Europa's icy shell.

2. CRATER MORPHOLOGIES

Voyager provided the first observations of natural impact craters formed in icy, as opposed to silicate, targets. Voyager imaging revealed the paucity of Europa's impact craters (*Smith et al.,* 1979; *Lucchitta and Soderblom,* 1982), indicating the youthful state of the surface, but revealed little detail of impact morphologies: Only five or so definitive impact craters were detected on the ~20% of the surface observed at ~2 km/pixel. A large, apparently ringed, dark spot was also imaged (Tyre Macula, as it was known then), but its impact origin was uncertain at these resolutions. Voyager (supplemented by Galileo) achieved much greater success at Ganymede, where impact craters are both plentiful and well preserved on younger bright terrain. Large craters on Ganymede's older dark terrain are usually highly relaxed, while those on Callisto are typically very similar to Ganymede's but are often eroded or relaxed (*Moore et al.,* 2004). Hence Ganymede's bright-terrain craters provide the best examples for comparative planetology with Europa.

Galileo's medium-resolution survey imaging (500–2000 m/pixel) of ~75% of Europa's surface revealed additional impact craters, some in great detail, despite the antenna failure that prevented acquisition of a global image map and thus a true global inventory of impact craters. Galileo mapping coverage (100–400 m/pixel) includes only two pole-to-pole swaths ~15° wide, plus several isolated targets of particular interest (see chapter by Doggett et al.). These data provide a sample of the impact crater population for statistical analysis down to diameters of ~1 km over nearly 10% of the surface in two swaths, one each on the leading and trailing hemispheres. These data are summarized in *Schenk et al.* (2004) and Bierhaus et al. (this volume). This sample can be extrapolated globally; however, mosaics targeted on large impact features Pwyll, Callanish, and Tyre were not used in this survey because of the large number of known secondaries associated with each.

Galileo was able to focus its limited resources on a few key larger impact craters. In addition to high-resolution imaging, Galileo also provided important topographic data through stereo imaging and shape-from-shading (photoclinometry) from low-Sun illumination imaging (e.g., *Schenk,* 2002) for many of the larger craters. These data, together with the partial global survey, revealed several surprises. In order to understand what the anomalous shapes and structures imply for Europa and its interior we must discuss what the nominal expectations were.

2.1. Craters on Ganymede and Callisto: Benchmarks for Europa

The transition to complex crater morphologies (i.e., from simple bowl-shaped to flat-floored craters with central peaks and/or rim terraces) is most likely controlled by a combination of surface gravity and the composition and strength of the lithosphere (e.g., *Pike,* 1980; *Melosh,* 1989; *Schenk et al.,* 2004). Indeed, icy craters do follow an inverse correlation between transition diameters and surface gravity, albeit at lower overall diameters than on rocky worlds (*Schenk et al.,* 2004). The difference between rocky and icy worlds is most commonly attributed to the inherent mechanical weakness of water ice (e.g., *Melosh,* 1982; *Passey and Shoemaker,* 1982).

The three icy Galilean satellites offer a curious but highly useful natural coincidence in that surface gravity is nearly identical on all three bodies. All three satellites also have surface gravity only slightly weaker than the Moon, a solar system crater morphology benchmark. When surface gravity is similar, then impact crater morphologies and the diameters at which transitions occur should also be similar (e.g., *Melosh,* 1982; *Pike,* 1980), as is the case for Ganymede and Callisto (surface modification histories notwithstanding). Such cases offer the potential to compare lithospheric properties directly using morphology [as in the case of Ganymede and the Moon, which also have similar surface gravities but different material properties and thus exhibit differences in crater morphology and transition diameters (*Schenk,* 1991)]. As Galileo approached Europa in 1996, it was already known that Europa was different from Ganymede and Callisto, and an ocean was suspected based on indirect geological arguments, but how such differences would be manifested in the morphologies of Europa's craters was only guessed (e.g., *McKinnon and Melosh,* 1980).

On Ganymede and Callisto, as on the terrestrial planets and satellites, we see a well-defined morphologic continuum from smaller to larger craters (Fig. 1) (e.g., *Passey and Shoemaker,* 1982; *Schenk,* 1991; *Schenk et al.,* 2004). Post-Galileo study confirms that central peak craters are the dominant landform between ~3 to ~23 km diameter. Their morphology is not unlike that seen on the Moon or Mercury, except for the narrower rimwalls of Ganymede craters (*Passey and Shoemaker,* 1982; *Schenk,* 1991). (We note that there are always a few kilometers of overlap between the occurrences of the various classifications due to natural variations in impact and target parameters.) Flat-floored complex craters (Fig. 1) are approximately as abundant as central-peak craters between ~17 and 23 km diameter. Between ~25 and 72 km, however, central-pit craters dominate (*Passey and Shoemaker,* 1982; *Schenk,* 1993). Central-dome craters dominate between 72 and ~160 km diameters (*Schenk,* 1993). Both central-pit and central-dome landforms are rare to nonexistent on terrestrial planets [with the exception of occasional central pit craters on Mars (*Barlow,* 2006)].

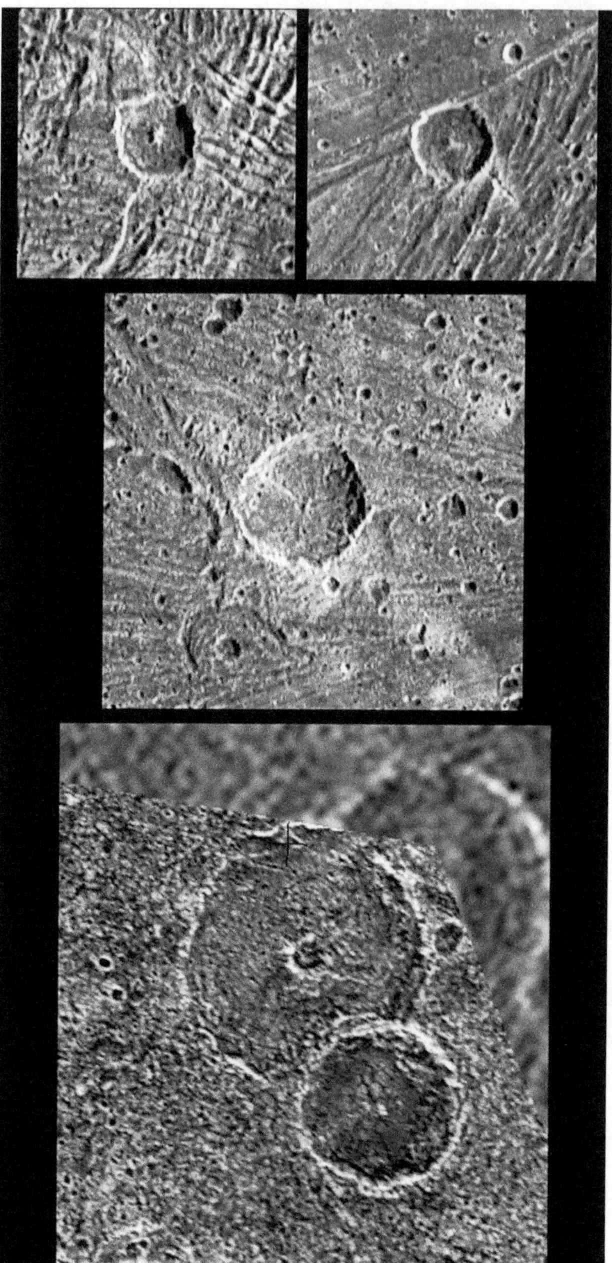

Fig. 1. Complex craters on Ganymede. The two central-peak craters at top are 11 km across; the flat-floored crater at center is 19 km across; and the central-pit crater at bottom is 50 km across. Given Europa's similar gravity and surface composition, its craters would be expected to resemble these examples.

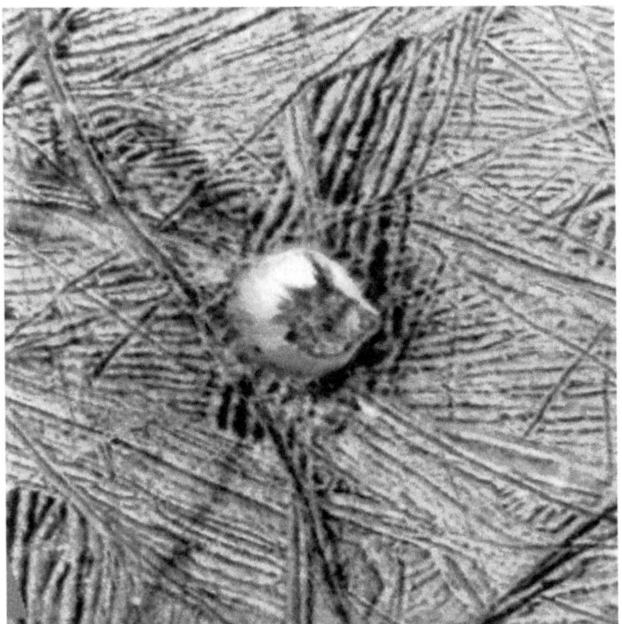

Fig. 2. Niamh, an ~4-km-diameter simple crater on Europa imaged at 30 m/pixel. Note hummocky debris on crater floor.

2.2. Craters on Europa

2.2.1. Simple craters. Extrapolation of the observed Voyager crater population suggested that Europa should have a significant population of kilometer-sized craters. In the 20% or so of the surface mapped at better than 500 m resolution, Galileo observed several thousand craters. Many of these are smaller than 1 km across and appear to be dispersed secondary craters (*Bierhaus et al.,* 2001). We confine ourselves to craters larger than 1 km in diameter. Simple bowl-shaped craters are common at diameters <4 km but few were observed at resolutions great enough to make out morphologic details. Niamh (D ~ 4.5 km) is one of the best-resolved examples (Fig. 2). Simple craters on Europa have mean depth/diameter ratios of ~0.17 (*Schenk et al.,* 2004). They are the only simple craters on the icy satellites, including Enceladus, which have depth/diameter ratios of less than 0.2, the lunar value (*Pike,* 1977). The origin of this anomaly is not understood.

2.2.2. "Normal" complex craters. The population of known impact craters larger than 10 km across on Europa is very limited (Fig. 3; Table 1). Nonetheless, we can see that those craters <15 km in diameter or so have morphologies consistent with those on the other Galilean satellites. The dominant morphology at diameters between 4 and 20 km is that of central peak craters. Examples (Figs. 4 and 5) include Grainne (D ~ 14 km), Cilix (D ~ 19 km), and Maeve (D ~ 21 km). These craters are indistinguishable from similarly sized craters on Ganymede and Callisto. Complex craters on Europa feature narrow rimwalls similar to those on Ganymede [roughly half the width of rim-walls in lunar craters (*Schenk,* 1991)], flat floors, and small central peaks or peak complexes. A few flat-floored craters without central peaks also exist in the size range 13–16 km. These include Rhiannon (Fig. 6), first observed by Voyager. Although a terrace or two can be seen in some of these craters (Figs. 4–6), most notably in Cilix and Rhiannon, the narrow rimwalls correlate with a general lack of coherent terrace wreath development in complex craters on Ganymede (*Schenk,* 1991). Hummocky rounded deposits are often seen at the bases of these steep rimwall scarps at high resolution (Fig. 5), suggesting that rimwall collapse may involve less-coherent material than for lunar or mercurian craters.

2.2.3. Complex craters: Pwyll and Manannán. On Ganymede and Callisto, central pit craters (Fig. 1) dominate at diameters of ~40 km and larger, although examples occur at diameters as small as 25 km. However, only three craters larger than 25 km have been identified on Europa (Table 1), and all have morphologies different from their Ganymede counterparts. Indeed, at diameters greater than ~20 km on Europa, the standard complex crater landform

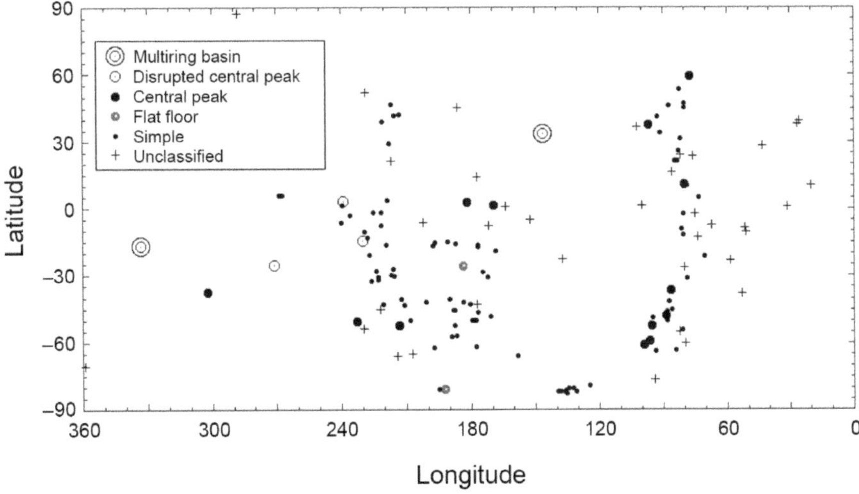

Fig. 3. Global map of known craters on Europa larger than 1 km. The extreme variation in crater density is due to the large resolution range (<200 m to 4 km per pixel) inherent in the available images used in global mapping. This nonuniformity makes determination of possible global asymmetries and age dating impossible.

TABLE 1. Impact craters on Europa larger than 10 km.

Name	Morphologic Type	Latitude	Longitude	Diameter (km)	Best Resolution (km/pxl)	Comment
Tyre	Multiring basin	33.7	146.3	37.8	0.16	
Taliesin	—	–22.7	137.5	(~36)	2.00	Diameter may be 14 km
Callanish	Multiring basin	–16.7	333.1	32.7	0.12	
Tegid	—	0.8	163.9	(30)	1.44	May be smaller
Pwyll	Disrupted central peak	–25.3	270.9	26.6	0.13	Disrupted central peak/pit
ZN02:100	—	1.5	99.9	24.0	3.20	Bright spot
Manannán	Disrupted central peak	3.3	239.2	22.6	0.02	Disrupted central peak/pit
Maeve	Central peak	59.3	77.2	21.1	0.23	Double central peak
ZN16:086	—	16.3	85.6	(~20)	3.20	Identified by nearby secondary field
Cilix	Central peak	2.7	181.6	19.0	0.10	Elongated central peak
Amergin	Disrupted central peak	–14.6	230.1	18.8	0.21	Complex central structure
Morvran	—	–4.8	152.3	16.8	2.00	
Eochaird	Central peak	–50.4	232.6	16.8	1.22	Dark ring
ZN37:102	—	36.6	102.6	(~16)	3.20	Bright spot, secondary craters
Rhiannon	Flat floored	–80.9	192.1	15.5	0.04	
ZS70:359	—	–70.6	359.1	15.0	1.00	
Math	Flat floored	–25.6	183.3	14.6	0.23	
Lug	—	28.1	43.6	14.0	1.00	Dark ring
Grainne	Disrupted central peak	–59.3	96.5	13.8	0.22	Horseshoe-shaped central peak
ZN39:026	—	39.4	25.6	13.0	1.00	
ZS54:230	—	–53.6	229.6	12.5	1.66	Bright spot with rim shadows
Avagddu	Central peak	1.5	169.2	11.3	1.44	
Govannan	Central peak	–37.4	302.3	10.6	1.22	

seen on Ganymede begins to change into progressively more unusual landforms. An example of this transitional landform is Amergin, a 19-km-wide crater partially imaged by Galileo (Fig. 7). While the rim seems coherent, the central uplift appears chaotic. An alternative possibility is that postimpact disruption related to chaos formation (e.g., *Greeley et al.,* 2004) may have broken up the crater floor, although the image resolution is insufficient to evaluate this possibility. Another example is 17-km-wide Grainne (Fig. 4), in which the central massif has a distinct U-shape.

Two complex craters between 20 and 30 km were observed at close range by Galileo (*Moore et al.,* 2001): the extremely young bright-rayed Pwyll (D ~ 27 km; Fig. 8) and the somewhat older ~23-km-wide Manannán crater (Fig. 8). Although both have central structures, their morphology is significantly different from that of impact craters of similar size on Ganymede [Maeve, 21 km across, was also observed but displays a classic complex morphology (Fig. 4)]. Both were Galileo targets of interest and were imaged in stereo, allowing the generation of digital elevation models (DEMs). Although the planned Pwyll high-resolution mosaic just missed the crater rim, Manannán was successfully imaged at 20 m/pixel under high Sun. Pwyll and Manannán share many properties, but we focus on Pwyll first.

Pwyll (Fig. 8) features a crater rim scarp and a central peak complex, but on close examination they can be seen to have anomalous properties. A 27-km crater on Ganymede is close to the transition from central peak to central pit morphology (*Schenk,* 1991) and could have either on Europa. Although there is some suggestion of a central pit at Pwyll, the central peak complex is characterized by a prominent fragmented asymmetric massif. The largest of these massifs stands nearly 500 m high and encloses a small semicircular amphitheater. The rimwall is not uniformly elevated as in other craters (Figs. 4–6), but is highly variable in elevation, and in some areas nearly disappears altogether. Finally, the crater floor is not depressed but rises to at least the level of the original surface (Fig. 8b,c). Pwyll is likely the youngest complex crater on Europa and features a prominent bright ray system (Fig. 8d) similar to that of Tycho on the Moon. It also features a prominent dark ring, which at higher resolution corresponds with the inner (thicker) portion of the continuous ejecta deposits, as well as the crater floor itself.

Manannán (Fig. 9a) is broadly similar to Pwyll. A crater rim is apparent along part of the circumference but is generally weakly expressed or absent. A small offset central peak complex is also partly preserved in the eastern floor. Here, however, a higher-resolution mosaic and topographic map (Fig. 9b) afford additional insights. As at Pwyll, the crater floor lies essentially at the level of the original surface (Figs. 9b,c). A peculiarity of the floor structure of Manannán is the small, approximately central depression ~150 m deep (Fig. 9b) (*Moore et al.,* 2001). (If Pwyll has such a depression, it was not detected.) This feature has no raised rim or evidence of upraised blocky material (e.g., *Schenk,* 1991, 1993). Rather the feature appears to be a depression within the mottled crater floor material itself. Radial lineaments extend from the center of the depression,

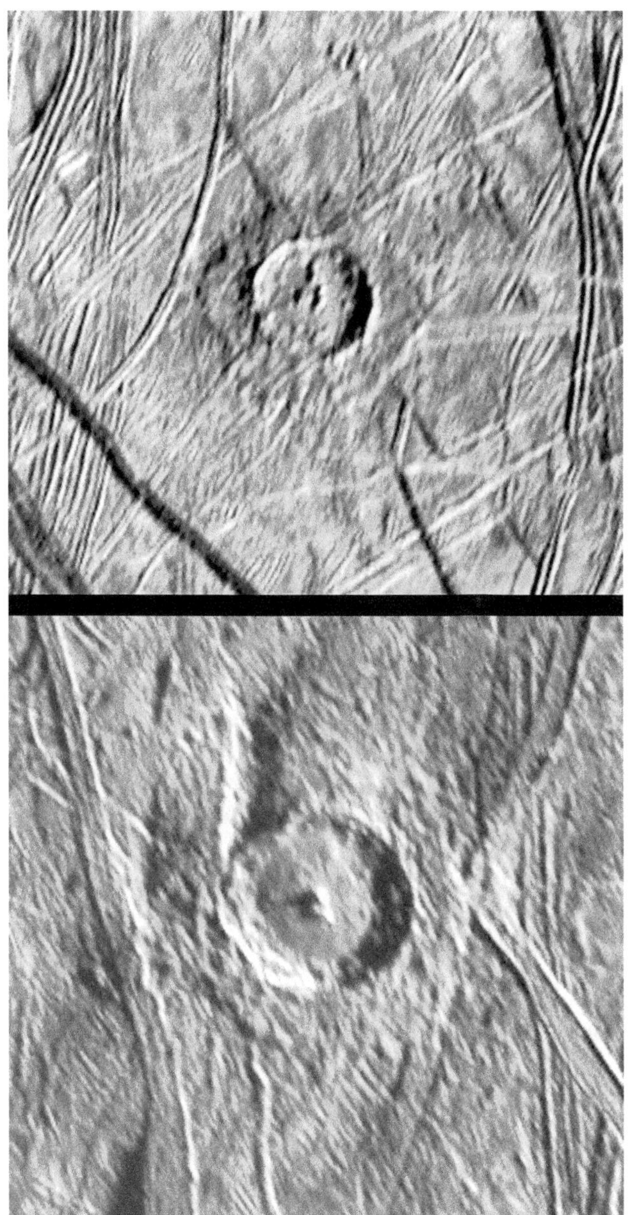

Fig. 4. Examples of classic complex craters on Europa. From top, Maeve (D ~ 21 km) and Grainne (D ~ 17 km). One or two terrace scarps are evident in craters of this size, but the rimwalls are considerably narrower than for lunar craters, where they constitute roughly half the crater diameter (*Pike,* 1980). The annular "pancake" ejecta deposits first observed on Ganymede (*Horner and Greeley,* 1982; *Schenk et al.,* 2004) are well developed in these craters.

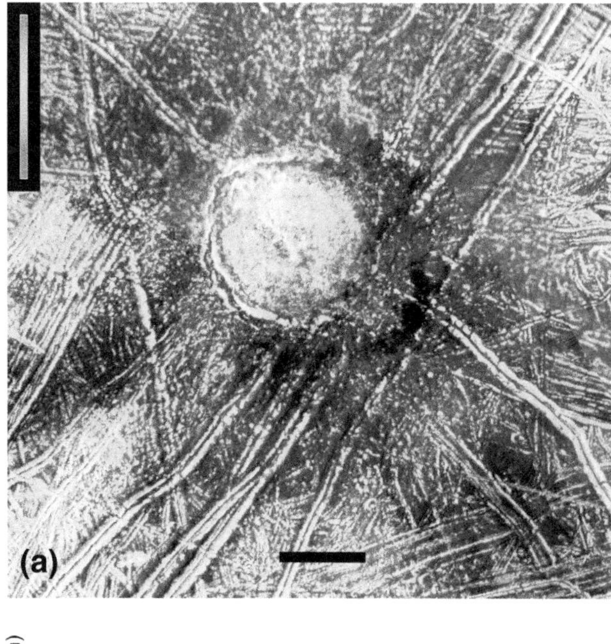

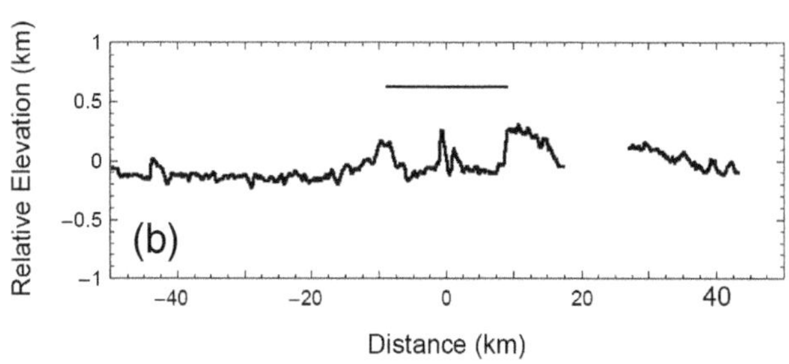

Fig. 5. Cilix (D ~ 19 km) is one of the largest, preserved, and best-imaged complex craters on Europa. **(a)** See Plate 6. High-resolution mosaic shows the basic features including a small and slightly offset central peak complex and one or two terrace scarps. The rimwall is otherwise much narrower than lunar craters of similar size. Note the dark annulus, which may be ejecta or the disruption of surface frosts. Some of this dark material ponds in topographic lows, although this could be due to mass wasting or drainage of mobilized ejecta material. Portions of a pancake ejecta deposit can be seen onlapping ridged plains at upper right and center right. Steep local ridge topography gives this unit an artificial lobate or feathered appearance. Mosaic has been color-coded to show stereo-derived topographic information, showing the classic flat depressed floor and raised rim. Horizontal scale bar is 10 km, vertical topography scale is 1 km. **(b)** Topographic profile running west to east across crater center. We note the variable topography of the surrounding plains and across the floor of Cilix itself, suggesting Cilix may have formed on a local rise, making determination of the absolute elevation of the crater floor difficult.

suggestive of fracture during subsidence of the crater center (Fig. 9e).

Color imaging shows the floor to be distinctly bluer than either surrounding materials or the dark red ejecta ring (Fig. 9e). This is interpreted as evidence of a distinct impact deposit of some kind. Whether this deposit is an impact melt sheet (i.e., refrozen water and icy breccia) or a deposit with different ice grain sizes is unknown, but the deposit appears to be confined mostly within the crater rim. The ruggedness of the crater floor and correlation of the bluish units with lower-lying and smoother terrains indicates that this deposit did not cover the entire crater floor and ponded in low areas. Furthermore, lobate flow units on the rim of Manannán (Fig. 9d) clearly show that mobile impact deposits (possibly a mixture of fragmented bedrock and impact melt, such as suevite) can form on the surfaces of icy bodies. The DEM shows that in at least one case these flows traversed uphill and over the rim crest (Fig. 9d), with an

Fig. 6. Flat-floored ~15-km-diameter complex crater Rhiannon, observed at low Sun by Galileo at 45 m/pixel. Several rounded scalloped deposits and a few partially developed arcuate terrace blocks can be seen at the base of the rim scarp but coherent rim failure has not been otherwise identified on the icy satellites. Beyond the rim can be seen the plateau-like topographic bench referred to as a pancake ejecta deposit (arrows).

elevation difference of 100–200 m. Flow uphill over the rim crest, a phenomenon also observed in volcanic pyroclastic flows and landslides, is most likely due to residual inertia imparted during or immediately after impact excavation.

Although Manannán lacks the bright rays of Pwyll, it is evidently relatively young. At medium resolution (Figs. 9a,e), a partially preserved dark ring can be seen outside the rim. At high resolution, this dark ring corresponds with the inner ejecta deposit (Fig. 9b) associated with pancake deposits on Ganymede (*Horner and Greeley,* 1982; *Schenk and Ridolfi,* 2002). Manannán postdates all terrains in the immediate vicinity although partial postimpact disruption of the eastern crater floor cannot be ruled out. Topography shows several block-like features associated with the central uplift and a depression along the eastern rimwall (Fig. 9b). Both may simply be the influence of rugged preexisting topography on the modification stage of crater formation.

We observe that Maeve (Fig. 4) is only 1 km smaller than Manannán but strikingly different in morphology, including the formation of a fully preserved complex (central peak and rimwall) morphology. Unless the transition to anomalous morphology at ~20 km is extremely abrupt, we must consider possible variations in shell properties with time or with location to explain the different morphologies. Temporal variations in shell thickness are suggested by thermal models of the water and ice layers (*Hussmann and Spohn,* 2004). However, until these craters can be reliably dated and the rest of Europa mapped (in the hopes of finding more transitional craters), evidence for temporal variations recorded in crater morphology will remain conjectural. Latitudinal variations in ice-shell thickness are also suggested by thermal equilibrium models (*Ojakangas and Stevenson,* 1989) that also predict possible true polar wander of the

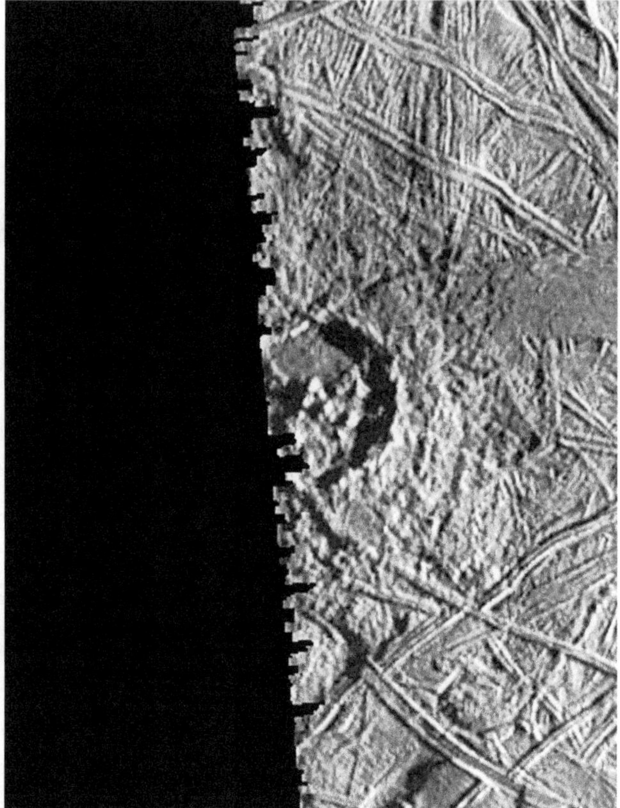

Fig. 7. Impact crater Amergin (D ~ 19 km), as observed by Galileo at 230 m/pixel. Although the crater rim scarp is relatively well preserved, the central complex is more chaotic than in other craters in this diameter range (cf. Fig. 4). This could represent a transition to more disrupted crater landforms such as Manannán, local differences in target properties, or a rare example of postimpact modification by other geological processes.

ice shell. New observational evidence for polar wander in turn supports possible thickness variations (*Sarid et al.,* 2002; *Schenk et al.,* 2008). If Maeve formed near the (pre-polar-wander) paleopole, its morphology may reflect a thicker ice shell, as may high-latitude (81°S) Rhiannon. Again, we are severely handicapped by the lack of a global crater inventory at all latitudes and longitudes with which to test such concepts. Finally, it is worth noting that morphological transitions are rarely instantaneous but spread over a finite diameter range, even on silicate planets like the Moon (e.g., *Pike,* 1977, 1980). This is usually attributed to subtle local variations in impact conditions or crustal rheology.

2.2.4. Multiring features: Tyre and Callanish. Ganymede and Callisto are dominated by large ancient multiring structures, the type example of which is the 2000-km-radius Valhalla ring system on Callisto (e.g., *McKinnon and Melosh,* 1980; *Schenk et al.,* 2004). Multiring structures on both Ganymede and Callisto are hundreds to thousands of kilometers across and are characterized by numerous closely spaced concentric ridges and scarps and by low relief (*Passey and Shoemaker,* 1982; *McKinnon and Melosh,* 1980; *Schenk and McKinnon,* 1987). That multiring basins occur on Europa was not a great surprise. The surprise was

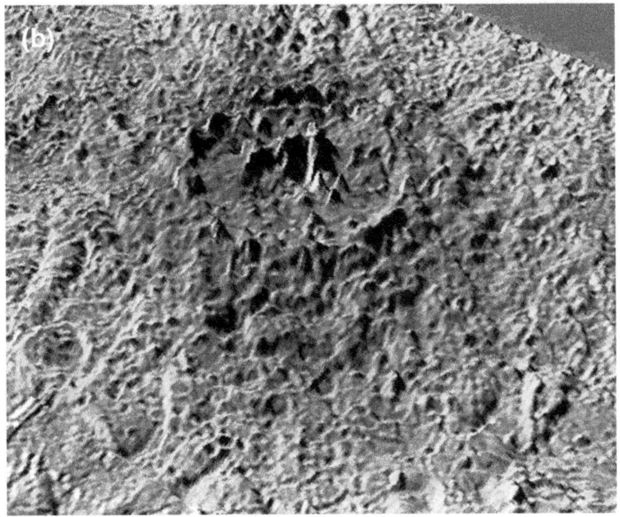

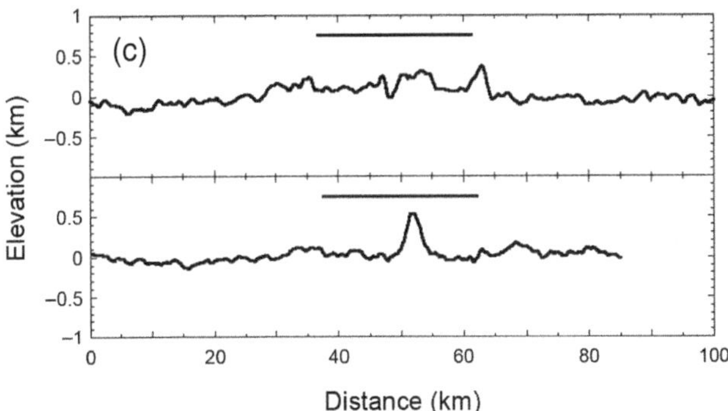

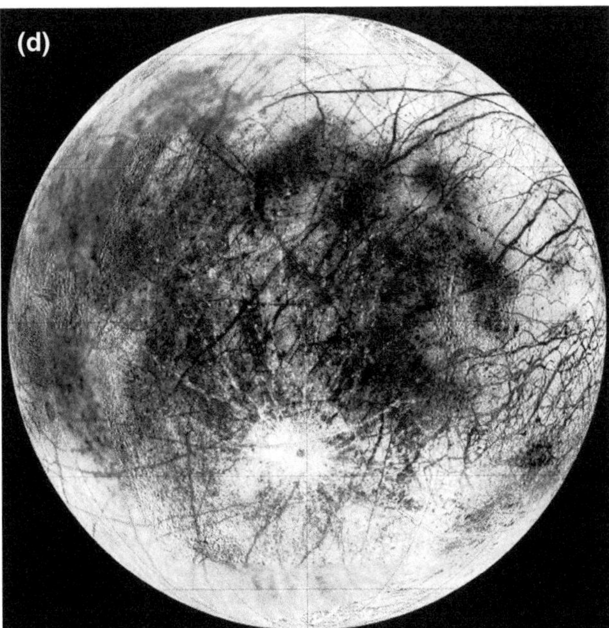

the small size of the two examples discovered on Europa, Tyre (first seen by Voyager), and Callanish (*Moore et al.,* 1998, 2001). We begin by discussing Tyre.

Tyre was observed to be a dark spot ~150 km across, coinciding roughly with a concentric set of ridges and graben ~160 km wide (Fig. 10). The dark spot has a distinct color difference compared to surrounding terrains (e.g., *Kadel et al.,* 2000). Tyre is surrounded by an extensive field of secondary craters, confirming its impact origins. The secondaries lie just beyond the edge of the dark spot, confirming that this low-albedo material corresponds roughly with the continuous ejecta deposit. Based on scaling arguments, these features suggest an equivalent final crater diameter of 38 km, just inside the radius of the prominent scarps and graben. [This value is revised slightly from previous reports (*Schenk and Ridolfi,* 2002; see also *Kadel et al.,* 2000) based on updated mapping of the ejecta deposit.] Three major lineaments appear to cross and postdate Tyre (Fig. 10b): the long narrow, approximately east-west fracture crosscutting ejecta and floor material in the center of the basin, and two sets of prominent double ridges to the southeast crosscutting the outer circumferential fractures. No other units cut Tyre with certainty, so the relative stratigraphic age of this basin remains uncertain. No high-resolution views of Tyre were obtained.

Callanish (Fig. 11) is broadly similar in morphology but somewhat smaller, with an observed ring system diameter of 95 km. The final crater diameter estimated from ejecta scaling is only 33 km, however, and a classical crater rim is not apparent. As at Tyre, the most prominent set of concentric ridges formed outside the nominal crater itself. This interpretation is consistent with the fact that preexisting surface features are preserved between the ring graben. A high-resolution mosaic (47 m/pixel) was obtained across the center of Callanish (Fig. 11c). Crenulated surface textures are evident across most of the basin floor and beyond the

Fig. 8. **(a)** Galileo images of 27-km-diameter Pwyll, Europa. Prominent features include the dark annulus corresponding to the crater floor and the inner (pancake) ejecta deposit and best seen in the high-Sun image at left, and the rugged asymmetric central uplift and the irregular rim elevations best seen in the lower-Sun image at right. Resolution in left image is 120 m/pixel, on right is 190 m/pixel. **(b)** Perspective view of Pwyll (D ~ 27 km). View rendered using DEM based on combined stereo-controlled photoclinometric data analysis (total dynamic range in topography is 1 km). **(c)** Two topographic profiles across Pwyll crater. Top profile is north-south, bottom profile is west-east. Notable are the inconsistent presence of a crater rim (most evident in the top profile), the lack of any floor depression, and the prominent offset central peak almost 500 m high in the bottom profile. Crater floor (rim-to-rim) is marked by the horizontal bars. Profiles were generated from data shown in **(b)**. Vertical precision in DEM and profiles is ~40 m. **(d)** Global view of the leading hemisphere of Europa, highlighting the prominent bright ray system radiating from the bright crater Pwyll just below center. At this resolution Pwyll itself is a small dark spot in the center of the bright ray system, which corresponds to the crater floor and inner ejecta unit.

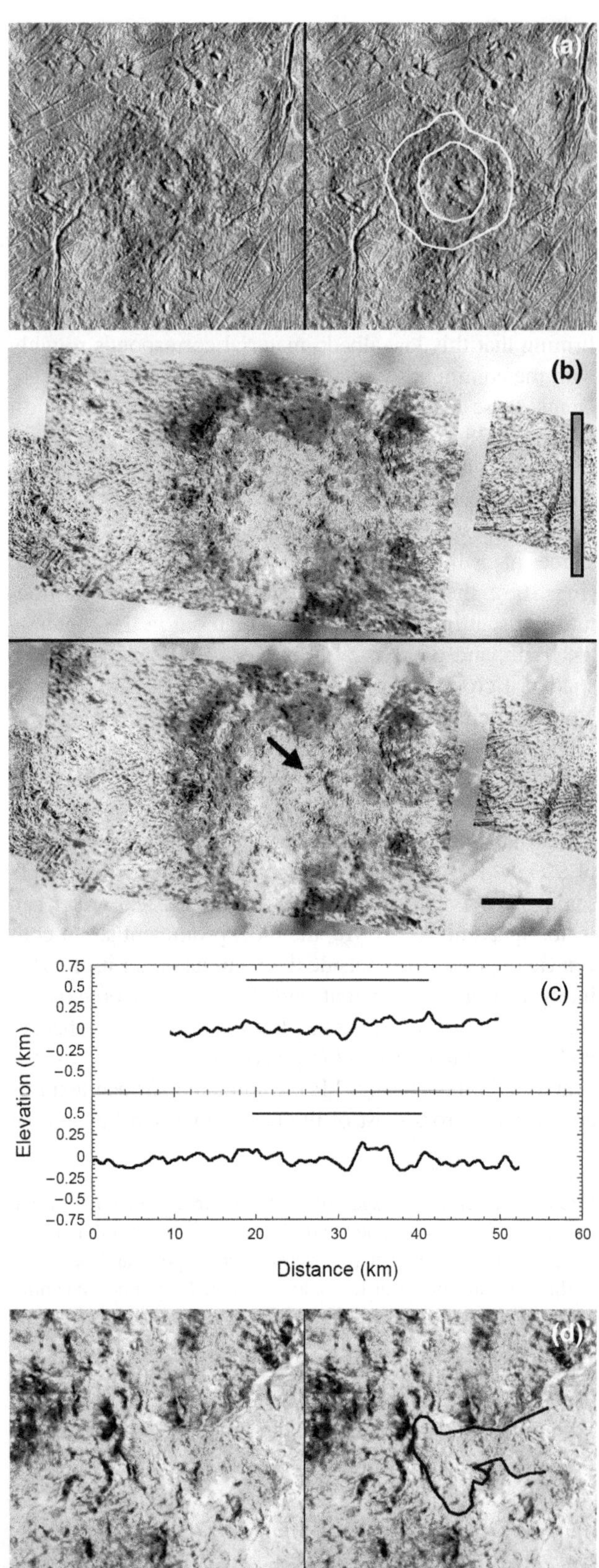

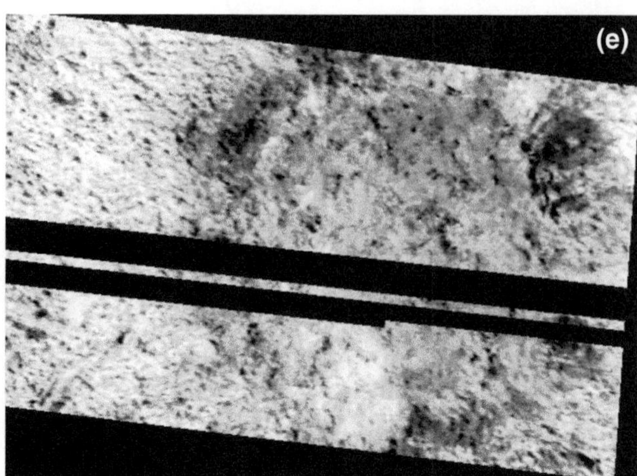

Fig. 9. **(a)** Regional view of Manannán (D ~ 23 km). Right view outlines location of rim (inner circle) and pancake ejecta deposit (outer circle), also apparent as a dark annulus and an irregular outward-facing topographic scarp. Image resolution ~230 m. North is up. **(b)** See Plate 7. High-resolution mosaics and topographic maps of Manannán crater. The mosaic consists of a high-resolution (20 m/pixel) strip and an 80 m/pixel context image, here overlain on a much-lower-resolution base mosaic. Two versions of the mosaic are shown, each color-coded with topographic data (purple and blue, low; red, high). Top view is based on stereo-derived, bottom on photoclinometry-derived DEM data. The uncertainty in the stereo data is ~10 m. Small central depression (arrow) is located west of asymmetric central peak. Horizontal scale bar is 10 km; vertical topography scale is 500 m. **(c)** Topographic profiles across Manannán crater. Top profile is north-south, bottom profile west-east. A small central depression is evident in both, although the top profile misses the small central peak complexes to the east (see bottom profile). As in Pwyll, the actual rim scarp and crater is difficult to identify without a marker due to the rugged topography. Manannán evidently formed in a region dominated by rugged chaos. Crater floor (rim-to-rim) is marked by the horizontal bars. Profiles were generated from data shown in **(b)**. Vertical precision is ~10 m. **(d)** See Plate 8. High-resolution view of rim of Manannán showing lobate flow material. Deposit is interpreted as impact-melt-rich material flowing over the rim scarp. View is enlargement of **(b)** (top). Scene is ~10 km across. **(e)** See Plate 9. Three-color mosaic of Manannán obtained by Galileo. The crater floor (deposit inside the dark ring) is covered by bluish and brownish units, some of which appear to correlate with morphological evidence of impact melt. Color filters are 0.9, 0.6, and 0.4 μm. Image resolution is 80 m.

nominal crater rim, suggesting that if a competent crater rim ever formed, basin floor materials buried or obliterated it. Some of these materials abut concentric ridges and appear to be deformed impact-melt-rich or other low-viscosity deposits. Despite the abundant evidence for impact melt, underlying structures (including both angular ridges and graben-like fractures) are well preserved, suggesting that the fluid-like deposits are not very thick. They also show that the emplacement of dark ejecta was controlled by local topography (Fig. 11c).

As on Ganymede and Callisto, no obvious crater rim scarp can be identified or confirmed among the numerous concentric ridges. The final crater diameters [see *Turtle et al.* (2005) for a discussion clarifying the terminology of crater diameters], 33 km for Callanish and 38 km for Tyre, were estimated by scaling from the inner limit of the ob-

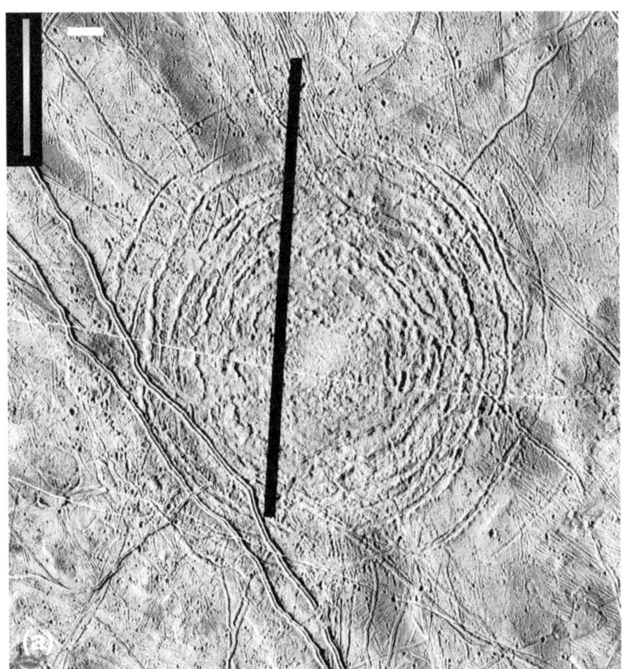

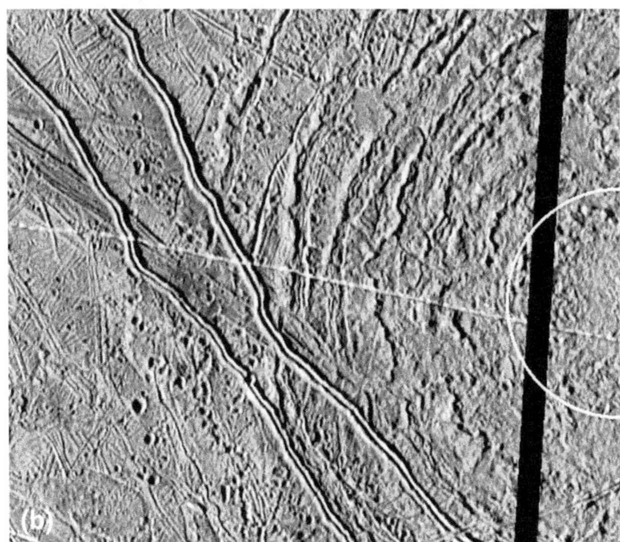

Fig. 10. **(a)** See Plate 10. Tyre multiring impact feature, Europa. Color-coding displays topographic data obtained from photoclinometry (purple, low; red, high). Many of the domed and depressed features are real, based on shading information, but the general reliability of longer-wavelength information is unknown. Whether the topographic rise in the southeast portion of Tyre is real, for example, is uncertain, but similar features have been identified in stereo DEMs, where long-wavelength topography is controlled. White horizontal scale bar is 10 km; vertical topography scale is 750 m. **(b)** Close-up of western section of Tyre impact basin, Europa (image from **(a)**, without topography layer). Basin center is to the right. Position of the final crater diameter estimated from scaling is shown by the white circle. Crenulated floor materials grade outward into ridged plains populated by secondary craters. Note the narrow fracture (bright in this view due to oblique viewing geometry) crossing the basin. The relative ages of the two prominent double ridges are indeterminate.

served secondary crater fields (*Moore et al.,* 2001; *Schenk and Ridolfi,* 2002; *Schenk,* 2002). These diameters would put the location of a preserved final crater rim inside most of the observed concentric ridges and scarps.

The visual impression of both basins is of low relief. Topography of Tyre is estimated from photoclinometry only. No true stereo observations were successfully acquired of either basin. However, high-resolution imaging across the center of Callanish was acquired with opposite Sun illumination directions during orbits E4 and E26. By reversing the contrast in one of the images, an effective stereo pair can be created, which allows us to create a partial stereo DEM across the feature. The nature of the shadowing in this case precludes stereo mapping of tall steep-sided ridges and photoclinometry (PC) has been used to fill in these data. The combination of the stereo and PC-DEM maps provides the first high-resolution controlled elevation across one of these structures (Fig. 11a). These data confirm that Callanish has relief of no more than ~100 m (Fig. 11b), with the exception of the steep-sided ridges, and possesses a small central depression similar to that observed at Manannán.

2.2.5. Transition craters. The transition on Europa from modified central peak craters (Pwyll and Manannán) to structures with external ring fractures (Callanish and Tyre) occurs within the very narrow diameter range of 27 to 33 km, which is abrupt by planetary standards. Typically other morphologic types occur between these central peak and multiring morphologies, e.g., central pit and peak ring or central dome structures (see discussion in section 3). However, the statistics are quite poor, and with only a few examples of each type of crater, variations in target structure or properties that control final crater morphology cannot be ruled out. Galileo observed additional large craters that may fit in this diameter range (Table 1), but low resolution (>1 km) precludes reliable classification. These include Tegid and Taliesin (Fig. 12), as well as several large features on the leading hemisphere observed late in the mission during the twenty-fifth orbit (designated I25) under high-Sun illumination. Two of the I25 craters occur just outside the leading hemisphere north-south mapping mosaic, and are identified as impact structures by the fields of secondary craters that extend into the mosaic (e.g., Fig. 13). Tegid consists of a large relatively bright central feature (resembling central domes on Ganymede) surrounded by a zone of concentric features. The inner feature has a diameter of ~30 km and the outermost limit of topographic relief is ~60 km in diameter. Similarly sized Taliesin (first observed by Voyager) has a small central ring and peak surrounded by irregular outer rings, although the diameter cannot be confidently identified and could be either 14 or 36 km.

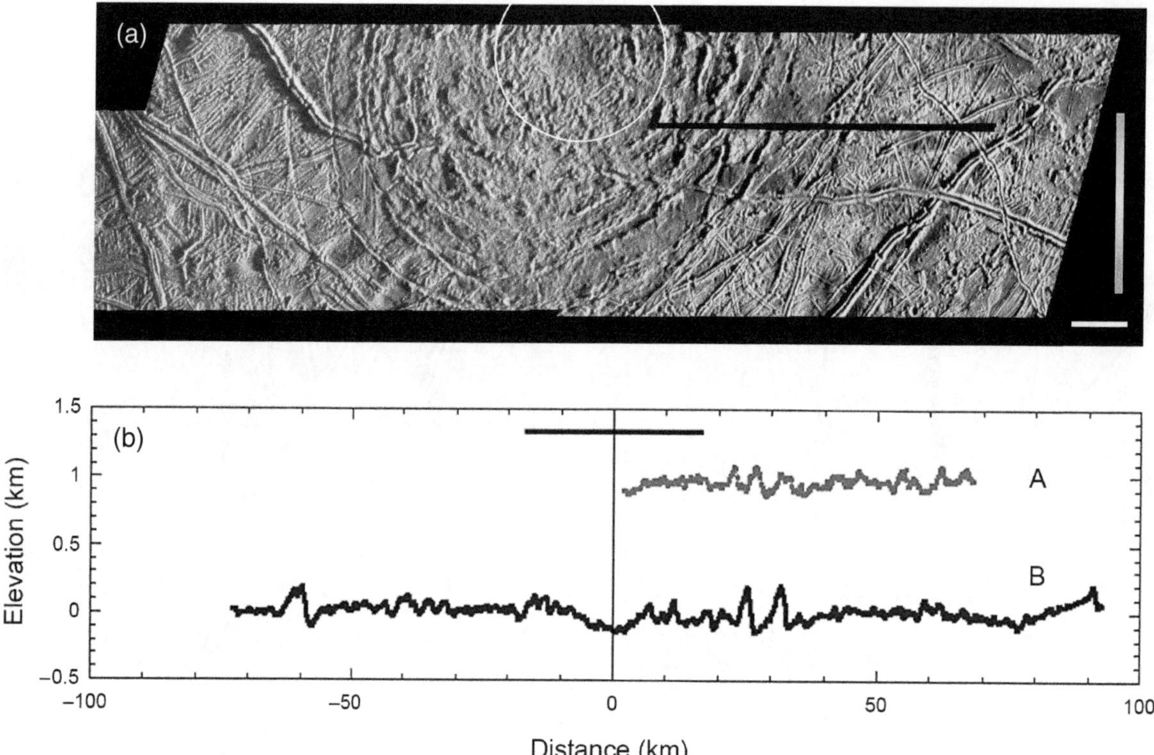

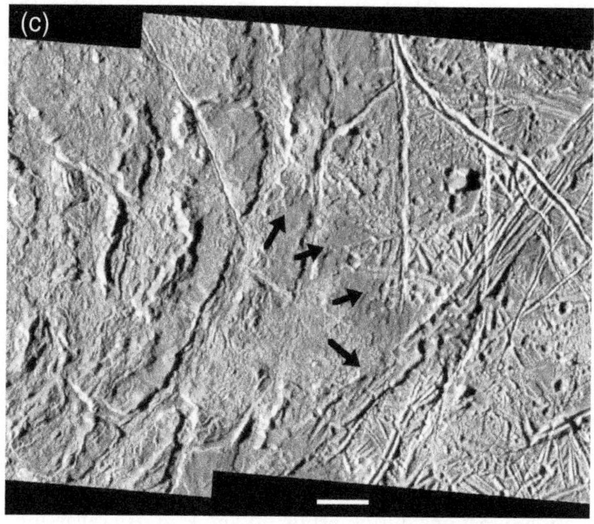

Fig. 11. **(a)** See Plate 11. Callanish multiring impact feature, Europa. Position of the final crater diameter estimated from scaling is shown by the white circle. Color-coding displays topographic data obtained from stereo-controlled photoclinometry (in which the stereo component controls long-wavelength topography). Mosaic resolution 120 m per pixel. Horizontal scale bar is 10 km; vertical topography scale bar is 750 m. Vertical DEM precision is ~25 m. **(b)** Topographic profiles across Callanish impact basin. Extent of final crater diameter estimated from the range to secondary craters is shown by horizontal bar. The outer limit of concentric rings occurs at radii of ~48 km. Profile A is due south from crater center to map edge, profile B is from west to east across basin center to map edge. Data from **(a)**. **(c)** Portion of highest-resolution view of Callanish. Basin center is to left. Crenulated surface textures within the basin and abutting concentric ridges are consistent with flow and settling of impact melt. Dark ejecta material is seen embaying ridges and plains (arrows). Ejecta grades into dispersed secondary craters outward from the dark zone. Mosaic resolution is 47 m/pixel, scale bar is 5 km.

Although little can be said about these structures, they are likely to be key to understanding the transition to multiring morphologies and are potentially indicative of regional spatial or temporal variations in ice shell properties. As such they should be a priority for any global mapping mission.

2.2.6. Impact melt and ejecta deposits. Models generally predict the production of large quantities of impact melt (water) in larger impact craters on the icy satellites (*Turtle and Pierazzo,* 2001). With few exceptions, only isolated occurrences of what appears to be ponded impact melt have been identified. The floors of Pwyll and Manannán have considerable relief (Figs. 8 and 9), despite abundant evidence of impact melt or flow-like features (Fig. 9d). These observations suggest that impact-melt ponding may be less extensive than in large lunar craters. It has been noted that if the density of liquid water is truly higher than that of europan or ganymedean crust, then melt would presumably prefer to drain downward under gravity through the countless fissures opened during impact.

Voyager observed what were termed pedestal ejecta deposits surrounding many craters on Ganymede (*Horner and Greeley,* 1982). These appear in medium-resolution (350–1500 m/pixel) images as circular mesa-like plateaus just outside the rims of most complex craters. Similar deposits are noted around complex craters on Europa as well (e.g., Figs. 4 and 5), although the extremely rugged nature of Eu-

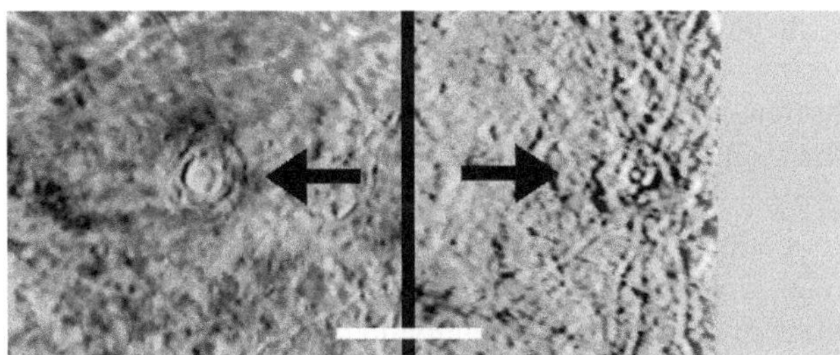

Fig. 12. Possible transition craters Tegid (left) and Taliesin (right). It is not clear which concentric feature corresponds to the nominal crater rim in either case. White scale bar is 100 km.

Fig. 13. An unusually high concentration of <1- to 2-km-diameter impact craters (black circles) on Europa. These craters resemble the outer reaches of a secondary crater field, although they are too large to be attributable to Bress, the 10-km-wide crater at upper right. The field is centered in the approximate vicinity of the white circle, which also corresponds to the location and approximate size of an ~30-km-diameter white spot visible in low-resolution (4 km/pixel) images of this area. The actual crater size remains unknown.

ropa's surface causes the outer edges of these deposits to follow local contours, distorting the shapes of these deposits in planform. *Schenk et al.* (2004) have suggested renaming these deposits as pancake ejecta, the revised name of the relevant martian morphological equivalent (*Barlow and Perez,* 2003), and we adopt that terminology here. Pancake deposits are only the thicker inner portion of the continuous ejecta deposit (e.g., *Moore et al.,* 1974). Galileo observations of pancake deposits on Ganymede at higher resolution revealed no indication of the involvement of fluidized ejecta of the type conjectured for Mars (e.g., *Schenk et al.,* 2004), suggesting that these deposits are emplaced ballistically. No credible evidence has been found at higher resolutions on either Ganymede or Europa for true lobate ejecta deposits of the type found on Mars and attributed to fluidized ejecta (*Barlow et al.,* 2003).

Beyond the continuous ejecta units lie the discontinuous secondary crater fields, which are most concentrated just beyond the outer edge of the ejecta deposit. *Schenk and Ridolfi* (2002) found that both the pancake and continuous ejecta deposits on Ganymede have a systematic relationship in diameter to that of the crater rim. Thus, the inner limit of the secondary crater field can be used as a proxy for the continuous ejecta margin (assuming similiarites in cratering mechanics but possibly subject to variations in target properties), and we can use the inner range of secondary craters to estimate original crater diameters where no rim scarp is apparent. This method was employed to determine the effective or nominal final diameters of the multiring features Tyre and Callanish. This method would also provide insight into Tegid and Taliesin, were images with resolutions high enough to observe the ejecta and secondary crater fields available.

A peculiarity of europan ejecta is the relatively low albedo of the pancake deposits in many instances. This effect is most evident in larger craters, including Cilix (Fig. 5), Pwyll (Fig. 8), and Manannán (Fig. 9d). High-resolution images show that the craters themselves fit within the dark units, which correspond to the inner ejecta deposits. These relatively dark circular deposits can be used to identify potential impact craters in poorly resolved areas (Fig. 14). Indeed, most craters in areas imaged at low resolution might not be recognized without the presence of albedo features. The youngest craters are recognized by small bright ray patterns, nearly all of which are observed on the leading hemisphere. [Whether this constitutes proof of the factor of 30–40 leading-trailing asymmetry in impact crater accumulation predicted by *Zahnle et al.* (2003) remains to be seen due to the variable resolution and illumination geometry of the different image sets currently available to map Europa.]

Although the global crater catalog is very irregular, craters smaller than ~15 km across appear to lack these dark rings. This observation suggests that perhaps only craters larger than a critical size are excavating into unknown dark material at depth; scaling relations (*Melosh,* 1989; *McKinnon and Schenk,* 1995) yield an excavation depth of ~1 km for a crater with a final diameter of 15 km. At Pwyll, Tyre, and Callanish, where excavation depths may exceed 2 km, the dark spot corresponds to both the crater floor and ejecta deposits, a correlation that is consistent with the prior conclusion that the larger craters are excavating a dark subsurface unit.

2.3. Postimpact Deformation and Catastrophic Resurfacing

The visual impression from Galileo's partial global survey is that few impact craters have experienced tectonic or other geological modification. Deformation of Europa's impact craters could reveal whether or not resurfacing has been episodic or (more-or-less) continuous over the 40–90-m.y. age of the observed surface (see chapter by Bierhaus et al.). A similar question was raised in the case of Venus based on the observation that few craters on that planet had been modified after formation. The resulting catastrophic

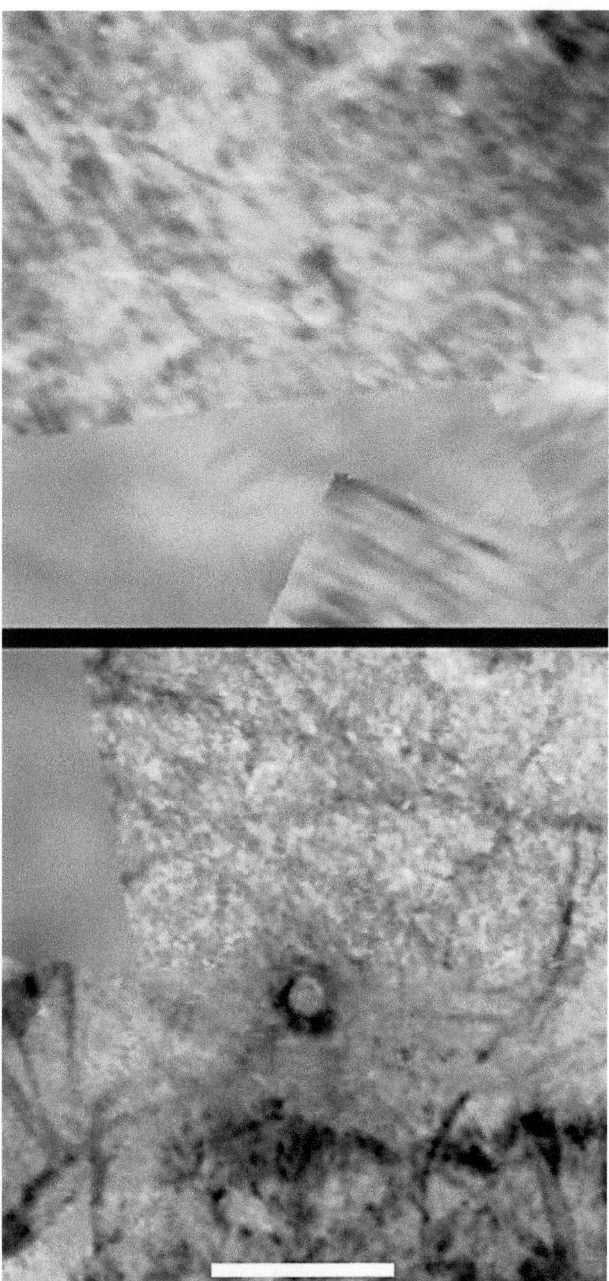

Fig. 14. Examples of poorly resolved dark-ring craters. These dark rings are similar to those observed at Cilix and Manannán, and appear to be characteristic of craters at these diameters. From top to bottom: Eochaird (D ~ 17 km) and Lug (D ~ 15 km). Eochaird appears to be a central peak crater, Lug a flat-floored or other type. The floor of Eochaird is distinctly bluish in color, similar to Manannán (Fig. 9). White scale bar is 100 km.

resurfacing model suggested most of Venus' resurfacing occurred in a major pulse followed by quiescence. Detailed examination of crater morphometry (e.g., *Herrick and Sharpton,* 2000) revealed that many venusian craters have been partly flooded by dark lavas, reducing their depths, countering arguments for volcanic catastrophism on Venus.

In the case of Europa, an evaluation of catastrophic resurfacing is crippled by the lack of a global survey at a resolu-tion of ~200 m or better. We can nonetheless examine the available record in lieu of such a survey. The dominant geological processes on Europa are tectonic faulting, band formation, and (arguably) diapir-driven surface disruption (e.g., *Greeley et al.,* 2004). Only a dozen or so craters were mapped at sufficient resolution to discriminate such deformation. Tyre and possibly Amergin are the only craters known to have been modified after impact. In the case of Tyre deformation is limited to three global-scale cross-cutting lineaments. Whether this is related to Tyre's large size and increased likelihood of cross-cutting, or to formation in a geologically active area, is unknown. Viscous relaxation is expected, but there is no direct evidence for this process occurring on Europa. Amergin (Fig. 7) remains the only case in which the crater floor morphology is suggestive of possible postimpact disruption, although this evidence is ambiguous. At least another dozen impact craters larger than a few kilometers across exist in less favorably imaged areas where postimpact modification of this type would go undetected. Thus the possibility of catastrophic (or rapid) resurfacing of Europa remains open because of the lack of a true global crater survey (including deformation state), a goal that should be a prime focus of any future mission.

3. CRATER MORPHOLOGY AND THE INTERIOR: MODELING

The morphology of a crater is influenced by the energy and momentum of the impact event (related to crater size) as well as by the gravity, material properties, and near-surface structure of the target. The characteristics of impact craters can therefore be used to infer information about the target. Europa's impact craters are peculiar in many respects, a number of which stand out as diagnostic and can potentially be used to constrain the environment in which, and the mechanics by which, these craters formed. Such characteristics include the preservation of central peaks in the anomalous craters Pwyll and Manannán (Figs. 8 and 9); the very rapid transition from fairly typical complex craters to multiring structures and the Valhalla-type morphology thereof (Figs. 10 and 11); and the abrupt rollover in the depth/diameter curve (described below). For example, transition diameters depend upon the strength of the target material (e.g., *Melosh,* 1982). The concentric fractures around the two largest impact structures may have formed as a result of Europa's near-surface structure (e.g., *Melosh and McKinnon,* 1978; *Turtle,* 1998); and Europa's anomalously shallow craters (e.g., *Moore et al.,* 2001; *Schenk,* 2002) may be a result of the collapse process itself under europan conditions (*Turtle and Ivanov,* 2002). Thus, an effective approach to investigating the constraints that the observed crater morphologies put on Europa's near-surface structure is to use numerical methods (e.g., hydrocode and finite-

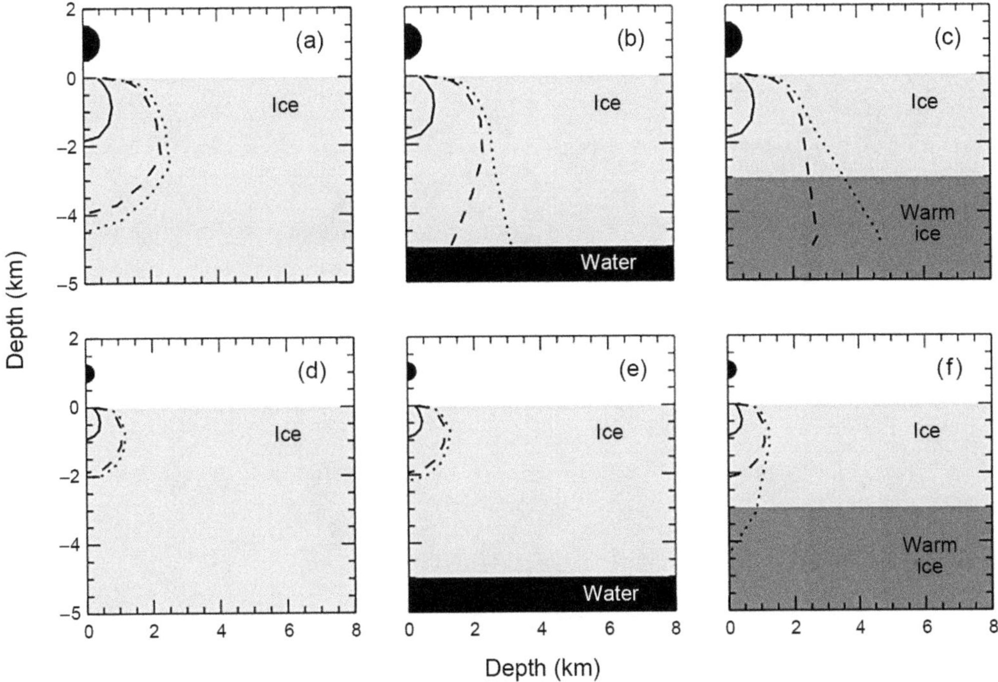

Fig. 15. Results of numerical simulations of cratering on Europa (after Fig. 3 in *Turtle and Pierazzo,* 2001). **(a)**–**(c)** Impacts of large projectiles; **(d)**–**(f)** small projectiles. The target consists of 9-km-thick ice (gray) over liquid water in **(a)** and **(d)**; 5-km-thick ice (gray) over liquid water (black) in **(b)** and **(e)**; and a 3-km-thick conductive lid (light gray) over convecting ice (dark gray) in **(c)** and **(f)**. Solid, dashed, and dotted lines delineate the regions of complete vaporization, complete melting, and 50% melting, respectively. Ice strength properties are assumed to be temperature dependent, but other weakening mechanisms (such as acoustic fluidization) are not included.

element modeling) to simulate impact cratering under europan conditions. By reproducing the observed crater morphologies it is possible to constrain the near-surface structure and conditions.

3.1. Central Peaks

Although europan complex craters are shallow, they otherwise resemble complex craters found elsewhere in the solar system. Their central peaks are morphologically similar to those observed on other planets, so the simplest explanation is that they formed by the same mechanism. In lunar and terrestrial craters, central peaks consist of deeply buried material that was uplifted during crater formation (e.g., *Grieve et al.,* 1981; *Pieters et al.,* 1982). Therefore, their occurrence on Europa presumably requires that an ice shell not be breached during crater formation; moreover, it requires that the material uplifted from the deeper crust has the strength to remain so over europan geological timescales (up to several 10^7 yr) (*Zahnle et al.,* 2003).

To investigate the implications of Europa's central-peak craters, *Turtle and Pierazzo* (2001) conducted hydrocode simulations of vapor and melt production during impacts into ice layers with various thicknesses and thermal gradients overlying liquid water. Zones of complete vaporization or melting that penetrate to liquid water or warm ice would presumably preclude the formation and preservation of central peaks, therefore these models can constrain the thickness of Europa's crust. *Turtle and Pierazzo* (2001) simulated impacts that would produce transient craters with diameters of ~12 and ~21 km, approximate transient crater diameters predicted for Europa's largest central-peak crater (Cilix) and the multiple-ring crater Callanish respectively [based on *McKinnon and Schenk* (1995) scaling]. (The estimated transient diameter for Pwyll is ~16 km.) These simulations (Fig. 15) demonstrated that at the times and locations of complex crater formation on Europa, the thickness of the cold ice layer had to exceed 3–4 km. Since impacts disrupt target material well beyond the zone of partial melting (e.g., *Melosh,* 1982), these simulations put a stringent lower limit on the thickness of an icy shell.

To address the issue of central-peak formation more directly, *Turtle and Ivanov* (2002) conducted hydrocode simulations of the excavation and collapse of 10-km-diameter transient craters [within the size range expected for Europa's central peak craters (*Moore et al.,* 2001)] in ice layers ranging from 5 to 11 km thick (Fig. 16), assuming linear thermal gradients. They found a strong dependence of final crater morphology on ice thickness: For ice less than ~8 km thick, liquid water penetrated to the surface dur-

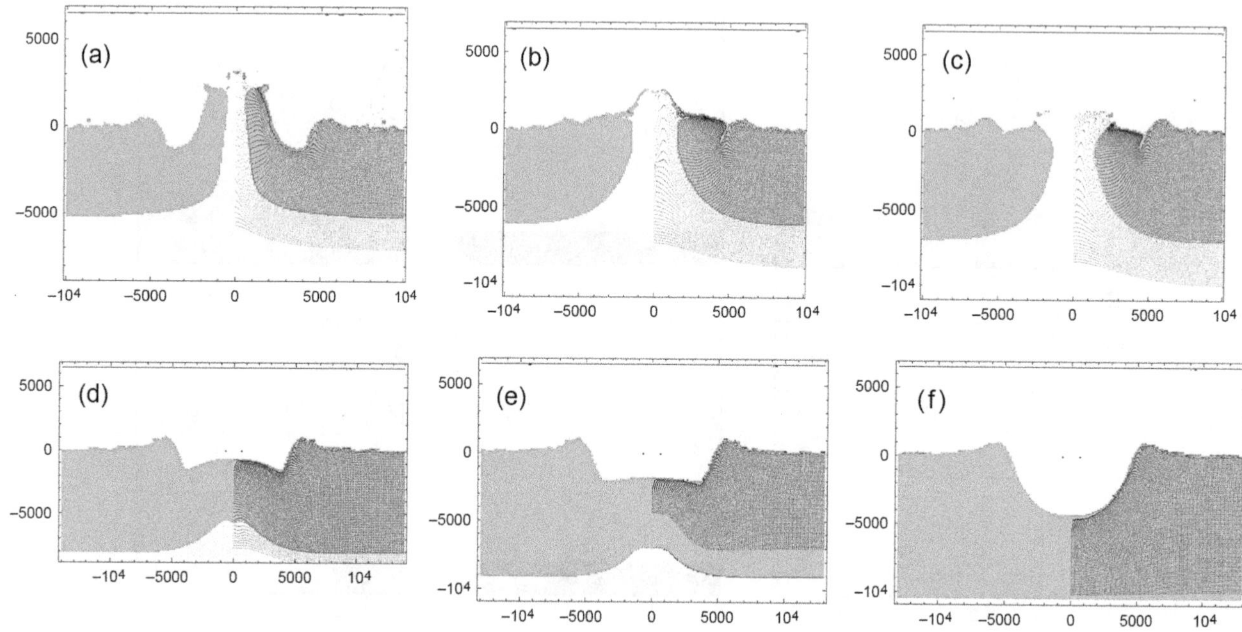

Fig. 16. Cross-sections of the final stages of crater collapse from numerical simulations of impacts into ice overlying water (after *Turtle and Ivanov,* 2002). **(a)–(f)** Illustrations of ice (gray) layers 5, 6, 7, 8, 9, and 11 km thick, respectively. The figures also demonstrate penetration of water (white) to the surface during crater collapse in ice ≤7 km thick. (Black points on the right side of each figure are the locations of tracer particles in the simulations.)

ing the collapse of the transient crater, thus precluding central-peak (as opposed to central-pit) formation. The simulations also demonstrated that, under europan conditions, crater collapse may produce the extremely low-topography structures observed, thereby obviating the need for postcollapse modification to reduce the rim-to-floor depths. These results provide a lower limit, contingent on the strength model employed, for the thickness of the europan ice at the times and locations of central peak crater formation: To prevent craters from penetrating to an ocean, the icy-shell thickness must be comparable to or greater than the transient crater diameter. In the case of Pwyll ($D_{transient}$ ~ 16 km), this implies an icy shell at least ~15 km thick. This range is consistent with the results of *Schenk* (2002) based on the diameters at which Europa's craters are observed to undergo morphologic transitions (see discussion below). Further investigations into icy crater collapse should provide more definitive constraints.

3.2. Multiring Basins

Substantial research has been conducted to explain the formation of multiring impact structures. The ring-tectonic hypothesis (*Melosh and McKinnon,* 1978; *McKinnon and Melosh,* 1980) relies on crater collapse in a stratified target with a surface layer that behaves elastically on the timescale of impact crater formation that overlies a layer that behaves like a fluid on the same timescale. If the crater, or the disrupted region around it, penetrates to the fluid-like layer, as the crater collapses the low-viscosity material will flow inward, exerting a drag force on the upper elastic layer, and if the resulting radial extensional stress is sufficient to fracture the upper layer, ring faults will form. This model predicts four possible morphologies, depending on the thickness of the overlying rigid layer relative to the impact (Fig. 3 of *McKinnon and Melosh,* 1980), including both Valhalla-type (seen on Callisto) and Orientale-type (seen on the Moon and terrestrial planets) structures. For very thin rigid layers, this model predicts both circumferential and radial fracturing.

Finite-element modeling by *Turtle* (1998) simulating this scenario with target parameters appropriate to Europa demonstrated that ice thicknesses of at least 12–16 km were consistent with the formation of ring faults around craters the size of Callanish and Tyre, but not around Pwyll-sized or smaller craters. Furthermore, in such cases the ice was not thin enough for circumferential extensional stresses to be sufficient to predict radial fracturing, also consistent with the observed crater forms on Europa (*Turtle,* 1998). Hydrocode modeling by *Moore et al.* (1998) also demonstrated that not only ring formation but also a relatively low density of secondaries around Callanish implies formation in ice overlying liquid, a situation that leads to fewer, and slower, solid ejecta fragments. However, the fact that radial faulting did not occur during the formation of these structures strongly suggests that the icy shell was not breached (*Turtle,* 1998; *Kadel et al.,* 2000).

Kadel et al. (2000) also used scaling relationships to estimate the thickness of the brittle lithosphere, arriving at a lower limit of 3.5 ± 0.5 km. More recently, *Lichtenberg et al.* (2006) used the widths of the ring graben in Tyre and Callanish in an attempt to constrain icy shell properties. They estimated the depth to the brittle-ductile transition to be 1200–1700 m and heat flows to be in excess of 100 (and possibly 200) mW/m^2. Whether these estimates are realistic remains to be seen; for comparison, estimates of elastic lithospheric thickness based on loading over much longer timescales by europan ridges average ~200 m (*Hurford et al.,* 2005).

3.3. Depth Versus Diameter

The rim-to-floor crater depths of larger craters on Europa are shallower than smaller complex craters, in contrast with the canonical trend of increasing depth with diameter observed on all other planets (e.g., *Pike,* 1974, 1988; *Schenk,* 1989, 2002). This rollover in the depth/diameter curve (Fig. 17) occurs within a narrow diameter range (8–12 km). It also occurs at diameters smaller than those for which morphologic disruption occurs. Cilix, for example, is shallower than would be predicted for its diameter, although it displays no obvious anomalous morphologies. The multiring features have relief of 100 m or less, as confirmed by the new stereo results of Callanish (Figs. 11a,b).

As discussed in section 2.3, the shallow topography of complex craters is probably not a viscous relaxation effect, but rather a direct result of crater collapse, consistent with numerical simulations (*Turtle and Ivanov,* 2002) and morphological interpretation (*Kadel et al.,* 2000). Pwyll is the youngest complex impact crater known on Europa, with a bright ray system that extends several thousand kilometers (Fig. 8d) (e.g., *Bierhaus et al.,* 2001). Viscous relaxation could proceed on relatively short geological timescales in sufficiently warm ice, but relaxation preferentially preserves short-wavelength morphologic features despite reduction in, or complete removal of, long-wavelength topographic amplitude (e.g., *Parmentier and Head,* 1981) and when seen elsewhere has not been observed to disrupt central uplifts or destroy rim coherence (e.g., *Passey,* 1983; *Dombard and McKinnon,* 2006; *Schenk,* 2006). Therefore, the anomalous appearances of the rims and central structures in these craters, and by association crater depth, must be the result of collapse and modification processes, which occur on timescales of minutes (e.g., *Melosh,* 1989), rather than millennia.

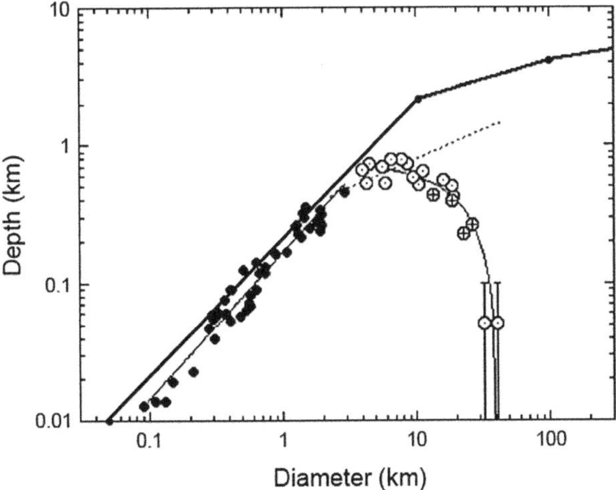

Fig. 17. Depth vs. diameter curve for Europa, showing rollover at diameters >8 km. Simple craters (solid dots), central peak craters (open circles), distorted central peak craters (crossed circles), and multiring basins (circles with error bars) are shown. Figure modified from *Schenk* (2002).

The most likely origin of the depth/diameter rollover is that, as crater size increases, the excavation cavity and the hemispheric zone surrounding it (which is involved in the modification process) penetrate more deeply into a highly mobile layer at depth (*Schenk,* 2002). Thus larger craters are more sensitive to the properties of the deeper icy shell. *Schenk* (2002) argues that craters in the 8–25-km size range exhibit an enhanced degree of crater modification due to the very weak rheology of the warm lower part of the icy shell, consistent with the simulations of *Turtle and Ivanov* (2002). The largest impact features, Tyre and Callanish, are formed and modified in a shell that is essentially without strength as far as the impact process is concerned. Ring tectonic theory (e.g., *McKinnon and Melosh,* 1980) requires that these craters effectively form in a weak asthenosphere with only a thin brittle cold surface layer that is structurally negligible compared to the thicker layer underneath. The transitions to these modified crater landforms begin with the rollover in the depth/diameter curve at ~8 km, related to penetration into the warm ductile center of the shell. The transition from modified central peak craters to multiring structures at ~30 km is related to excavation into the lower part of the icy shell, indicating that the icy layer may be on the order of 19 km thick (*Schenk,* 2002), with an effective uncertainty of 5 or 6 km. This estimate is based on the rule of thumb that the rheology at a given depth affects transient craters of that scale, and is consistent with the lower limits determined by *Turtle and Pierrazo* (2001) and *Turtle and Ivanov* (2002) and the morphological interpretation of *Kadel et al.* (2000).

Ganymede and Callisto are also believed to possess liquid water oceans at depth (e.g., *Kivelson et al.,* 2002, and *Khurana et al.,* 1998, respectively). The difference in crater morphology between Europa and the other icy Galilean satellites may be related to the depths of these oceans (*Schenk,* 2002). That on Ganymede is believed to be considerably deeper, >100 km (*Kivelson et al.,* 2002). Thus, under present conditions (and in the past), much larger craters would be required on Ganymede to sense warmer ice layers or the ocean itself. Central dome craters on Ganymede and Callisto only manifest transitional morphologies at diameters >175 km, at which point, like Pwyll and Manannán, poorly developed rim structures are observed (*Schenk et al.,* 2004). Perhaps these large craters on Ganymede and Callisto also sensed the deeper warmer layers beneath.

An important caveat to the modeling results is that although the morphologies of Europa's craters can put a lower limit on the depth to a subsurface layer of weak material, they do not prove conclusively that the mobile layer is a liquid ocean. Another limiting factor is that the morphological interpretations only apply at the location and time of crater formation. Although central-peak craters are fairly well distributed over the globe and Callanish and Tyre are ~180° apart in longitude, given the dearth of craters inferring general europan conditions from these results is rather risky and temporal and spatial variations in shell properties cannot be ruled out.

4. DISCUSSION

Europa's impact craters are few but the similarities and differences from their counterparts on other icy satellites are telling. High-resolution images reveal impact-melt-rich deposits, central peak structures, multiring structures, and a variety of ejecta morphologies including extensive secondary cratering.

One of the key motivations for interest in europan impact morphologies is that differences compared to their Ganymedean/Callistoan counterparts offer the hope that they are influenced by, and thus will provide insight into, the conditions beneath the surface of Europa. The endogenic geological processes acting on Europa are unique to Europa, and therefore impact cratering, being a ubiquitous and comparatively well understood process in the solar system, offers an essential perspective for study of the icy shell, leading to the natural if not inevitable question: Can Europa's craters divulge the thickness of the ice layer?

The observables are well enough defined for us to proceed with the attempt, although there remain critical questions regarding temporal and spatial variability that only a global survey can answer (even with a completely global dataset, such an investigation will be impeded to a certain extent by Europa's low number of impact craters). The key observables are that impact crater morphology and shape (i.e., d/D) change on Europa at diameters greater than ~8 km or so, when compared with craters on similar-sized icy bodies. This is manifest as a rollover or decline in d/D ratios at larger diameters (Fig. 17) and a disruption in classical impact crater morphologies, namely a lack of coherent central structures, restricted or partial rim development, elevated floor elevation, and in the largest examples, multiring development and the lack of a crater landform. The interpretation of these transitions have been that forces unrelated to normal impact processes are involved in crater formation and modification stages on Europa, and increasingly so with larger diameters (e.g., *Moore et al.,* 1998, 2001; *Turtle and Pierrazo,* 2001; *Schenk,* 2002). The inference has been that these larger craters are sensing Europa's shallow water ocean, and this liquid layer is altering impact formation processes.

As outlined above, several groups have attempted to use crater morphologies and transitions to constrain icy shell thickness, using the above interpretation as a framework. Modeling of any impact process is a challenge as the answers depend on our understanding of the impact process itself (under any conditions) as well as of the physical properties of the icy shell, which are poorly known, e.g., the thermal gradient through the icy shell may not be linear (see chapter by Nimmo and Manga and chapter by Barr and Showman). However, the violence of impact cratering is a great neutralizer in that it vaporizes, melts, and comminutes any target extensively in the zone surrounding the crater. The initial conditions that are of the most importance are temperature, composition, and strength of the target material. There are, of course, uncertainties in these parameters, but reasonable estimates can be derived for all of them.

The various models of impact crater morphology on Europa give somewhat different answers for the thickness of the icy shell, but all provide minimum thicknesses in the range of 4–20 km at the times and locations of impact, converging on the range of 15–19 km. The assumptions built into these models are based in good part on experience with terrestrial cratering and are unlikely to be egregiously wrong. Certainly, given the extensive research on terrestrial and lunar impact cratering, it is extremely difficult to conceive that a crater of Pwyll's size, with its towering 500-m-high central peak complex, could have formed in an icy shell only a few kilometers thick. Thus, Europa's impact crater record would appear to effectively eliminate thin shell models from consideration (see chapter by McKinnon et al.), at least at the times and locations of cratering; spatially or temporally localized occurrences of thin ice, as well as cyclical changes in global shell thickness (*Hussmann and Spohn,* 2004), cannot be ruled out due to the sparse crater population. Further modeling is strongly recommended in order to test different internal conditions and to explore the roles of ice properties in controlling crater morphologies. Complete global mapping of the impact crater population and other features that place constraints on ice-shell properties on Europa is paramount.

Acknowledgments. Thanks are extended to N. Barlow, K. Zahnle, H. J. Melosh, and W. B. McKinnon for their excellent and provocative reviews, and to B. Ivanov for his modeling contributions. P.S. and E.T. acknowledge funding support from NASA's Planetary Geology and Geophysics Program. This paper is LPI Contribution No. 1456.

REFERENCES

Barlow N. G. (2006) Impact craters in the northern hemisphere of Mars: Layered ejecta and central pit characteristics. *Meteoritics & Planet. Sci., 41,* 1425–1436.

Barlow N. G. and Perez C. B. (2003) Martian impact crater ejecta morphologies as indicators of the distribution of subsurface volatiles. *J. Geophys. Res., 108,* 4–1, DOI: 10.1029/2002JE002036.

Bierhaus E. B., Chapman C. R., Merline W. J., Brooks S. M., and Asphaug E. (2001) Pwyll secondaries and other small craters on Europa. *Icarus, 153,* 264–276.

Dombard A. J. and McKinnon W. B. (2006) Elastoviscoplastic relaxation of impact crater topography with application to Ganymede and Callisto. *J. Geophys. Res., 111,* E01001.

Greeley R., Chyba C. F., Head J. W. III; McCord T. B., McKinnon

W. B., Pappalardo R. T., and Figueredo P. H. (2004) Geology of Europa. In *Jupiter: The Planet, Satellites and Magnetosphere* (F. Bagenal et al., eds.), pp. 329–362. Cambridge Univ., Cambridge.

Grieve R. A. F., Robertson P. B., and Dence M. R. (1981) Constraints on the formation of ring impact structures, based on terrestrial data. In *Multiring Basins* (P. H. Schultz and R. B. Merrill, eds.), *Proc. Lunar Planet. Sci. Conf. 12A,* pp. 37–57.

Herrick R. R. and Sharpton V. L. (2000) Implications from stereo-derived topography of venusian impact craters. *J. Geophys. Res., 105,* 20245–20262.

Horner V. M. and Greeley R. (1982) Pedestal craters on Ganymede. *Icarus, 51,* 549–562.

Hurford T. A., Beyer R. A., Schmidt B., Preblich B., Sarid A. R., and Greenberg R. (2005) Flexure of Europa's lithosphere due to ridge-loading. *Icarus, 177,* 380–396.

Hussmann H. and Spohn T. (2004) Thermal-orbital evolution of Io and Europa. *Icarus, 171,* 391–410.

Kadel S. D., Chuang F. C., Greeley R., and Moore J. M. (2000) Geological history of the Tyre region of Europa: A regional perspective on europan surface features and ice thickness. *J. Geophys. Res., 105,* 22656–22669.

Khurana K. K., Kivelson M. G., Stevenson D. J., Schubert G., Russell C. T., Walker R. J., and Polanskey C. (1998) Induced magnetic fields as evidence for subsurface oceans in Europa and Callisto. *Nature, 395,* 777–780.

Kivelson M. G., Khurana K. K., Russell C. T., Volwerk M., Walker R. J., and Zimmer C. (2000) Galileo magnetometer measurements: A stronger case for a subsurface ocean at Europa. *Science, 289,* 1340–1343.

Kivelson M. G., Khurana K. K., and Volwerk M. (2002) The permanent and inductive magnetic moments of Ganymede. *Icarus, 157,* 507–522.

Lichtenberg K. A., McKinnon W. B., and Barr A. C. (2006) Heat flux from impact ring graben on Europa. In *Lunar and Planetary Science XXXVII,* Abstract #2399. Lunar and Planetary Institute, Houston (CD-ROM).

Lucchitta B. K. and Soderblom L. A. (1982) The geology of Europa. In *Satellites of Jupiter* (D. Morrison and M. S. Matthews, eds.), pp. 521–555. Univ. of Arizona, Tucson.

McKinnon W. B. and Melosh H. J. (1980) Evolution of planetary lithospheres — Evidence from multiringed structures on Ganymede and Callisto. *Icarus, 44,* 454–471.

McKinnon W. B. and Schenk P. M. (1995) Estimates of comet fragment masses from impact crater chains on Callisto and Ganymede. *Geophys. Res. Lett., 22,* 1829–1832.

Melosh H. J. (1982) A schematic model of crater modification by gravity. *J. Geophys. Res., 87,* 371–380.

Melosh H. J. (1989) *Impact Cratering: A Geologic Process*. Oxford Univ., New York. 253 pp.

Melosh H. J. and McKinnon W. B. (1978) The mechanics of ringed basin formation. *Geophys. Res. Lett., 5,* 985–988.

Moore H. J., Hodges C. A., and Scott D. H. (1974) Multiringed basins — Illustrated by Orientale and associated features. *Proc. Lunar Sci. Conf. 5th,* pp. 71–100.

Moore J. M., Asphaug E., Sullivan R. J., Klemaszewski J. E., Bender K. C., Greeley R., Geissler P. E., McEwen A. S., Turtle E. P., Phillips C. B., Tufts B. R., Head J. W., Pappalardo R. T., Jones K. B., Chapman C. R., Belton M. J. S., Kirk R. L., and Morrison D. (1998) Large impact features on Europa: Results of the Galileo nominal mission. *Icarus, 135,* 127–145.

Moore J. M., Asphaug E. B., Michael J. S., Bierhaus B., Breneman H. H., Brooks S. M., Chapman C. R., Chuang F. C., Collins G. C., Giese B., Greeley R., Head J. W., Kadel S., Klaasen K. P., Klemaszewski J. E., Magee K. P., Moreau J., Morrison D., Neukum G., Pappalardo R. T., Phillips C. B., Schenk P. M., Senske D. A., Sullivan R. J., Turtle E. P., and Williams K. K. (2001) Impact features on Europa: Results of the Galileo Europa Mission (GEM). *Icarus, 151,* 93–111.

Moore J. M., Schenk P. M., Bruesch L. S., Asphaug E., and McKinnon W. B. (2004) Large impact features on middle-sized icy satellites. *Icarus, 171,* 421–443.

Ojakangas G. W. and Stevenson D. J. (1989) Polar wander of an ice shell on Europa. *Icarus, 81,* 242–270.

Pappalardo R. T., Belton M. J. S., Breneman H. H., Carr M. H., Chapman C. R., Collins G. C., Denk T., Fagents S., Geissler P. E., Giese B., Greeley R., Greenberg R., Head J. W., Helfenstein P., Hoppa G., Kadel S. D., Klaasen K. P., Klemaszewski J. E., Magee K., McEwen A. S., Moore J. M., Moore W. B., Neukum G., Phillips C. B., Prockter L. M., Schubert G., Senske D. A., Sullivan R. J., Tufts B. R., Turtle E. P., Wagner R., and Williams K. K. (1999) Does Europa have a subsurface ocean? Evaluation of the geological evidence. *J. Geophys. Res., 104,* 24015–24056.

Parmentier E. M. and Head J. W. (1981) Viscous relaxation of impact craters on icy planetary surfaces — Determination of viscosity variation with depth. *Icarus, 47,* 100–111.

Passey Q. R. (1983) Viscosity of the lithosphere of Enceladus. *Icarus, 53,* 105–120.

Passey Q. R. and Shoemaker E. M. (1982) Craters and basins on Ganymede and Callisto: Morphological indicators of crustal evolution. In *Satellites of Jupiter* (D. Morrison and M. S. Matthews, eds.), pp. 379–434. Univ. of Arizona, Tucson.

Pieters C. M. (1982) Copernicus crater central peak — Lunar mountain of unique composition. *Science, 215,* 59–61.

Pike R. J. (1974) Depth/diameter relations of fresh lunar craters — Revision from spacecraft data. *Geophys. Res. Lett., 1,* 291–294.

Pike R. J. (1977) Size-dependence in the shape of fresh impact craters on the moon. In *Impact and Explosion Cratering: Planetary and Terrestrial Implications* (D. J. Roddy et al., eds.), pp. 489–509. Pergamon, New York.

Pike R. J. (1980) Formation of complex impact craters — Evidence from Mars and other planets. *Icarus, 43,* 1–19.

Pike R. J. (1988) Geomorphology of impact craters on Mercury. In *Mercury* (C. R. Chapman et al., eds.), pp. 165–273. Univ. of Arizona, Tucson.

Sarid A. R., Greenberg R., Hoppa G. V., Hurford T. A., Tufts B. R., and Geissler P. (2002) Polar wander and surface convergence of Europa's ice shell: Evidence from a survey of strike-slip displacement. *Icarus, 158,* 24–41.

Schenk P. M. (1989) Crater formation and modification on the icy satellites of Uranus and Saturn — Depth/diameter and central peak occurrence. *J. Geophys. Res., 94,* 3813–3832.

Schenk P. M. (1991) Ganymede and Callisto — Complex crater formation and planetary crusts. *J. Geophys. Res., 96,* 15635–15664.

Schenk P. M. (1993) Central pit and dome craters — Exposing the interiors of Ganymede and Callisto. *J. Geophys. Res., 98,* 7475–7498.

Schenk P. M. (2002) Thickness constraints on the icy shells of the galilean satellites from a comparison of crater shapes. *Nature, 417,* 419–421.

Schenk P. M. (2006) Impact crater morphology on saturnian satellites — First results. In *Lunar and Planetary Science XXXVII,* Abstract #2339. Lunar and Planetary Institute, Houston (CD-ROM).

Schenk P. M. and McKinnon W. B. (1987) Ring geometry on Ganymede and Callisto. *Icarus, 72,* 209–234.

Schenk P. and Ridolfi F. (2002) Morphology and scaling of ejecta deposits on icy satellites. *Geophys. Res. Lett., 29,* 31(1)–31(4).

Schenk P. M., Chapman C. R., Zahnle K. M., and Jeffrey M. (2004) Ages and interiors: The cratering record of the Galilean satellites. In *Jupiter: The Planet, Satellites and Magnetosphere* (F. Bagenal et al., eds.), pp. 427–456. Cambridge Univ., Cambridge.

Schenk P., Matsuyama I., and Nimmo F. (2008) True polar wander on Europa from global-scale small-circle depressions. *Nature, 453,* 368–371.

Smith B. A., Soderblom L. A., Beebe R., Boyce J., Briggs G., Carr M., Collins S. A., Johnson T. V., Cook A. F. II, Danielson G. E., and Morrison D. (1979) The Galilean satellites and Jupiter — Voyager 2 imaging science results. *Science, 206,* 927–950.

Turtle E. P. (1998) Finite-element modeling of large impact craters: Implications for the size of the Vredefort structure and the formation of multiple ring craters. Ph.D. dissertation, Univ. of Arizona, Tucson. 176 pp.

Turtle E. P. and Ivanov B. A. (2002) Numerical simulations of impact crater excavation and collapse on Europa: Implications for ice thickness. In *Lunar and Planetary Science XXXIII,* Abstract #1431. Lunar and Planetary Institute, Houston (CD-ROM).

Turtle E. P. and Pierazzo E. (2001) Thickness of a europan ice shell from impact crater simulations. *Science, 294,* 1326–1328.

Turtle E. P., Pierazzo E., Collins G. S., Osinski G. R., Melosh H. J., Morgan J. V., Reimold W. U. and Spray J. G. (2005) Impact structures: What does crater diameter mean? In *Large Meteorite Impacts III* (T. Kenkmann et al., eds.), pp. 1–24. Geol. Soc. Am. Spec. Paper 384, Geological Society of America.

Zahnle K., Schenk P., Levison H., and Dones L. (2003) Cratering rates in the outer solar system. *Icarus, 163,* 263–289.

Tectonics of Europa

Simon A. Kattenhorn
University of Idaho

Terry Hurford
NASA Goddard Space Flight Center

Europa has experienced significant tectonic disruption over its visible history. The description, interpretation, and modeling of tectonic features imaged by the Voyager and Galileo missions have resulted in significant developments in four key areas addressed in this chapter: (1) The characteristics and formation mechanisms of the various types of tectonic features; (2) the driving force behind the tectonics; (3) the geological evolution of its surface; and (4) the question of ongoing tectonics. We elaborate upon these themes, focusing on the following elements: (1) The prevalence of global tension, combined with the inherent weakness of ice, has resulted in a wealth of extensional tectonic features. Crustal convergence features are less obvious but are seemingly necessary for a balanced surface area budget in light of the large amount of extension. Strike-slip faults are relatively common but may not imply primary compressive shear failure, as the constantly changing nature of the tidal stress field likely promotes shearing reactivation of preexisting cracks. Frictional shearing and heating thus contributed to the morphologic and mechanical evolution of tectonic features. (2) Many fracture patterns can be correlated with theoretical stress fields induced by diurnal tidal forcing and long-term effects of nonsynchronous rotation of the icy shell; however, these driving mechanisms alone probably cannot explain all fracturing. Additional sources of stress may have been associated with orbital evolution, polar wander, finite obliquity, ice shell thickening, endogenic forcing by convection and diapirism, and secondary effects driven by strike-slip faulting and plate flexure. (3) Tectonic resurfacing has dominated the ~40–90 m.y. of visible geological history. A gradual decrease in tectonic activity through time coincided with an increase in cryomagmatism and thermal convection in the icy shell, implying shell thickening. Hence, tectonic resurfacing gave way to cryomagmatic resurfacing through the development of broad areas of crustal disruption called chaos. (4) There is no definitive evidence for active tectonics; however, some tectonic features have been noted to postdate chaos. A thickening icy shell equates to a decreased tidal response in the underlying ocean, but stresses associated with icy shell expansion may still sufficiently augment the contemporary tidal stress state to allow active tectonics.

1. INTRODUCTION AND HISTORICAL PERSPECTIVE

It took 369 years after the discovery of Europa, the smallest of the Galilean moons, before humans finally managed a close look at this icy world as the Voyager spacecraft sped by in 1979. The analysis of Voyager images of Europa (e.g., *Finnerty et al.,* 1981; *Pieri,* 1981; *Helfenstein and Parmentier,* 1980, 1983, 1985; *Lucchitta and Soderblom,* 1982; *McEwen,* 1986; *Schenk and McKinnon,* 1989), which had resolutions of 2 km to tens of kilometers per pixel, resulted in the identification of a multitude of superposed crosscutting lineaments. The overall appearance, much akin to a "ball of string" (*Smith et al.,* 1979), spoke of a history of intense tectonic activity (Fig. 1), but a paucity of large impact craters suggested a geologically young surface. The surface deformation suggested an efficient tectonic resurfacing process, perhaps accompanied by cryovolcanism, resulting in a broad classification of Europa's surface into tectonic terrain and chaotic or mottled (cryomagmatically disaggregated) terrain (*Smith et al.,* 1979; *Lucchitta and Soderblom,* 1982). Hence, competing tectonic and endogenic processes have both been important in shaping Europa's geology. The notion of a tectonically active world implied an effective tidal forcing of the icy shell, leading researchers to hypothesize about the presence of a liquid ocean beneath the icy exterior of the moon (see chapter by Alexander et al.).

Detailed analyses of higher-resolution (tens to hundreds of meters per pixel) Galileo spacecraft images 20 years after Voyager resulted in a more precise classification of the many types of lineaments (e.g., troughs, ridges, bands, cycloids, strike-slip faults) and a reexamination of lineament-formation mechanisms (e.g., summary in *Pappalardo et al.,* 1999). In many studies, crosscutting relationships among multiple episodes of lineaments allowed a complex tectonic history to be unraveled. Cryomagmatism and chaos formation (see chapter by Collins and Nimmo) generally postdated tectonic resurfacing, although there was broad overlap between tectonic and cryomagmatic processes and some tectonic features are clearly geologically young (see chapter by Doggett et al.). Most of the emphasis in tectonic analyses of Galileo data was given to determining what caused individual features to form in the first place, and what mech-

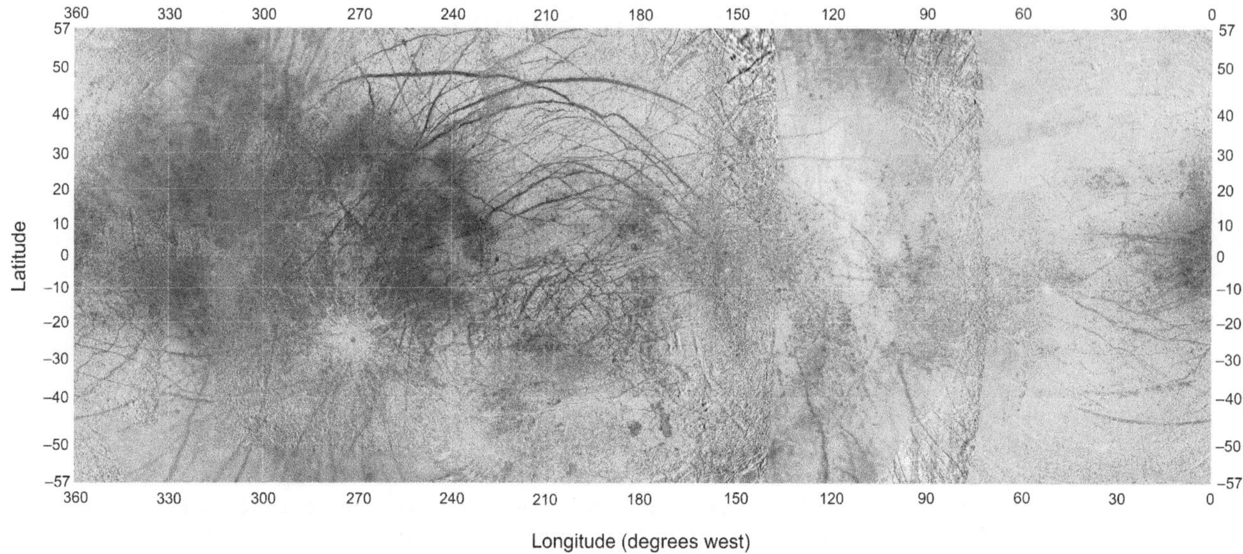

Fig. 1. Global mosaic of Europa, in Mercator projection, based on Voyager and Galileo imagery (courtesy of USGS, Flagstaff). The center is at longitude 180°W and the latitude range is +57° to –57°.

anisms resulted in their morphologic and geometric differences. Over the past dozen or so years, this work has produced a convincing framework for the role of tidal stresses in deforming the icy shell to produce fracturing, assisted by some amount of buildup of stress due to nonsynchronous rotation (see chapter by Sotin et al.). These stresses can be quantified and show a remarkable correlation to many lineament orientations after accounting for the effects of longitudinal migration of the icy shell by nonsynchronous rotation. The magnitude of the tidal distortion required for pervasive fracturing of the icy shell strongly suggests the presence of a tidally responding ocean beneath a relatively thin shell that was repeatedly stressed and strained. The presence of such an ocean is also strongly implied by Galileo magnetometer measurements (*Kivelson et al.,* 2000; see chapter by Khurana et al.).

The goal of this chapter is to provide a thorough examination of the wide range of tectonic features that pervasively damaged the icy shell of Europa during its visible geological history and their implications for characterizing the nature of the icy shell, the interior dynamics, the tidal deformation history, and the prospects for active tectonics. To accomplish this, the chapter is divided into three main components. First, we discuss the tectonic features themselves, summarizing their likely formation mechanisms in terms of being extensional, compressive, or lateral shearing structures. Second, we provide an overview of the causal factors that drive tectonic deformation at all scales, focusing on the production of stresses in the icy shell through tidal forcing and other means. Finally, we examine the tectonic history of Europa interpreted from the various types of tectonic features placed in the context of the tidal stress history. In so doing, we also address a number of topical issues such as the evidence for diminishing tectonic activity through time, reasons for the disparate geometries of lineaments, and the prospect of active tectonics. We do not speculate on the exact thickness of the icy shell in this chapter other than to infer that it is sufficiently thin (likely <30 km) to enable a strong tidal response in the underlying ocean, and consequently inducing tectonic deformation of the shell. One reason for this omission is that any constraints on icy shell thickness based on tectonic features are inconclusive. Second, fractures that have been used to deduce brittle or elastic thicknesses do not necessarily capture the entire thickness of the icy shell (e.g., *Billings and Kattenhorn,* 2005), much of which may be behaving inelastically depending on the timescale of deformation (see chapter by Nimmo and Manga). Various terms have been used to describe tectonic features on Europa. We aim to define and standardize the relevant terms adopted in this chapter and that of Prockter and Patterson with the glossary found in Table 1.

2. DEFORMATION STYLES

2.1. Extensional Tectonics

Analyses of low-resolution Voyager images followed by higher-resolution Galileo images resulted in the recognition and characterization of ubiquitous fractures in Europa's icy shell (Fig. 2). The prevalence of linear or curvilinear lineaments (e.g., *Lucchitta and Soderblom,* 1982) led to comparisons to terrestrial tension cracks, resulting in a body of published works that interpreted the majority of europan lineaments as tensile features (e.g., *McEwen,* 1986; *Leith and McKinnon,* 1996). Exceptions included the identification of lateral offset lineaments, interpreted to be strike-slip faults (see section 2.3). The assumption that tensile failure predominates in the icy shell is supported by numerous lines of evidence. First, ice has been shown to be particularly weak in tension under terrestrial conditions and in low-tem-

TABLE 1. Glossary of terms used to describe tectonic features on Europa.

band — A general term used to describe a tabular feature that formed by dilation (*dilational band*), contraction (*convergence band*), and/or strike-slip motions (*band-like strike-slip fault*). A *dilational band* is *lineated* (faulted or ridged varieties) or *smooth*. In low-resolution images, relative albedo may be used to distinguish between a *bright band*, *gray band*, or *dark band*.

band-like strike-slip fault — A type of *strike-slip fault* that morphologically resembles a *dilational band* and implies an oblique cumulative opening vector.

bright band — A term used to refer to high-relative-albedo, tabular features in low-resolution imagery. These bands may have experienced some combination of dilation, strike-slip, and/or convergence.

chaos — Regions of crustal disruption or disaggregation related to an underlying endogenic process such as cryomagmatism.

complex ridge — See *ridge complex*. Use of this term is discouraged.

convergence band — a tabular zone that appears to represent missing crust after tectonic reconstructions are undertaken, implying a localized zone of contraction.

crack — See *trough*.

cycloid — Curved, cuspate structures that form in chains of arcuate segments linked at sharp cusps. If a central *trough* is flanked by ridge edifices, the feature is called a *cycloidal ridge*. Also called a *flexus* (plural *flexūs*).

cycloidal fracture — A type of *trough* that forms arcuate segments linked at sharp cusps. They are the ridgeless progenitor of a *cycloidal ridge*.

cycloidal ridge — A *cycloid* that has developed ridges to either side of a central *trough*.

dark band — A term sometimes used to refer to a low-albedo *dilational band*.

dilational band — A tabular zone of dilation in the icy shell where intrusion of new crustal material occurred between the walls of a crack. Also called a *pull-apart band*. If fine, internal lineations are observed, the term *lineated band* may be used. The lineations may be defined by *normal faults* (hence, a *faulted band*) or by parallel *ridges* (a *ridged band*). If no lineations are observable (commonly a resolution effect), the term *smooth band* may be used.

diurnal tidal stress — The stress produced globally in the icy shell in response to the oscillating tidal response of the satellite during its eccentric orbit.

double ridge — See *ridge*.

endogenic fracture — A type of *trough* that forms above or adjacent to a zone of endogenic activity in the icy shell, thus commonly occurring adjacent to regions of *chaos*.

faulted band — A type of *lineated band* in which the lineations are caused by *normal faults* that have dissected the surface of the *dilational band*.

flexure fracture — A type of *trough* that forms alongside a *ridge* in response to flexing of the icy shell beside a *ridge*.

flexus — See *cycloid*.

fold — A rare form of contractional deformation in which the icy shell warps into anticlinal and synclinal undulations, such as within *dilational bands*.

fold hinge fracture — A type of *trough* that forms along the crest of an anticline.

gray band — A term sometimes used to refer to an intermediate relative albedo *dilational band*.

lineated band — A type of *dilational band* characterized by a fine lineated internal texture. This term may be used generically regardless of the inferred cause of the lineation. Two varieties are *faulted bands* and *ridged bands*.

nonsynchronous rotation — The proposed process by which the icy shell gradually migrates longitudinally eastward (i.e., about the rotational poles) in response to rotational torques. As a result, all locations on the surface migrate across the tidal bulges, resulting in a global component of stress that may contribute to the tectonics.

normal fault — An extensional shear fracture across which a vertical component of motion is inferred. Interpreted to define the lineations within a *faulted band*.

protoridge — The progenitor of a *ridge*, composed of a central trough flanked by poorly developed edifices. Also called a *raised-flank trough*.

pull-apart band — See *dilational band*.

raised-flank trough — See *protoridge*.

ridge — The most common tectonically related feature on Europa, comprising a central crack or trough flanked by two raised edifices, up to a few hundred meters high and less than 5 km wide. Also called a *double ridge*.

ridge complex — Several adjacent ridges that can be mutually parallel or commonly sinuous and anastomosing. Individual ridges in the complex are readily identifiable.

ridged band — A type of *lineated band* in which the lineations are created by numerous parallel ridges that define the *dilational band*.

ridged plains — The oldest and most expansive portions of the surface of Europa, composed of a multitude of low, generally high-albedo *ridges, bands,* and other structures that were repeatedly overprinted by younger features. Also called *subdued plains*.

ridge-like strike-slip fault — A type of *strike-slip fault* that morphologically resembles a *ridge*.

small-circle depressions — Up to 1.5-km-deep depressions in the icy shell that form broad, circular map patterns centered ~25° from the equator in an antipodal relationship on the leading and trailing hemispheres.

smooth band — A type of *dilational band* lacking an observable internal lineated texture, commonly in response to image resolution constraints, that may include small-scale hummocks.

strike-slip fault — A lineament along which older crosscut features were translated laterally during shearing.

subdued plains — See *ridged plains*.

tailcrack — A type of *trough* that forms where tension occurs at the tip of a *strike-slip fault* in response to fault motion.

tectonic fracture — A type of linear or broadly curving *trough* that forms locally or regionally, probably in response to tensile tidal stress.

triple band — A now-discouraged term originally used to describe lineaments in low-resolution imagery that appeared as a bright central stripe flanked by dark edges. Such features were ultimately identified in higher-resolution images to be predominantly *ridge complexes* flanked by dark materials, but may also be *double ridges* or *bright bands* flanked by dark materials.

trough — A ridgeless fracture with a visible width that results in a linear indentation at the surface of the icy shell. Also called a *crack*.

perature experiments, with tensile strengths of perhaps hundreds of kilopascals to as much as ~2 MPa (cf. *Schulson,* 1987, 2001; *Rist and Murrell,* 1994; *Lee et al.,* 2005). Second, unlike Earth, Europa's surface experiences absolute tension on a regular basis due to oscillating tidal bulges during the diurnal cycle (see section 4.1.5 and chapter by Sotin et al.). Third, some lineaments of particular ages typically form as multiple, parallel sets analogous to terrestrial tension joint sets. Fourth, the majority of lineaments on Europa do not appear to show lateral offsets, implying only crack-orthogonal motions (mode I cracks, in fracture mechanics terminology). Finally, there is clear evidence of complete dilational separation of parts of the icy shell, with infill of material from below to create new surface area within the dilated crack. A number of tensile tectonic features have been identified on the basis of these lines of evidence, including ridges, cycloids, dilational bands, and troughs. Additionally, normal faults on Europa indicate that extensional tectonics can occur through shear failure where there is deviatoric tension in a compressive stress field. Inferred extensional features, called small-circle depressions (*Schenk et al.,* 2008), form an antipodal pattern in the leading and trailing hemispheres, but an explicit formation mechanism has not been identified (see section 4.1.7).

2.1.1. Ridges. The most common lineaments on Europa are ridges, also referred to as *double ridges* due to the typical morphology in which a central crack or trough is flanked by two raised edifices (e.g., Androgeos Linea in Fig. 2b). As a result, ridges typically appear in medium- to high-resolution surface images as having a central dark stripe (the trough) flanked by two bright thicker stripes (the raised edifices; Fig. 3c), and may extend from a few kilometers to in excess of 1000 km across the surface. Ridges are analyzed in great detail elsewhere in this book (see chapter by Prockter and Patterson) but are also described here due to their prominence among Europa's tectonic features. Not only are ridges the most common type of structural lineament on Europa, they are also the most persistent, being the primary component of the very oldest *ridged plains* of the icy surface (*Figueredo and Greeley,* 2000; *Kattenhorn,* 2002; *Doggett et al.,* 2007) as well as constituting some of the youngest geological features. Hence, whatever process is responsible for their development must have been an ongoing process throughout Europa's visible geological history. Ridges of different ages commonly occur in disparate orientations, resulting in a complex network of multiply crosscutting generations of ridge sets and indicating significant changes in stress fields through time (see sections 5.1 and 5.2). Topographically, ridges range from barely perceptible in parts of the background plains to as high as several hundred meters (*Malin and Pieri,* 1986; *Greenberg et al.,* 1998; *Greeley et al.,* 2000). Individual ridge widths are typically around a few hundreds of meters (<400 m) but higher ridges tend to be wider, with maximum

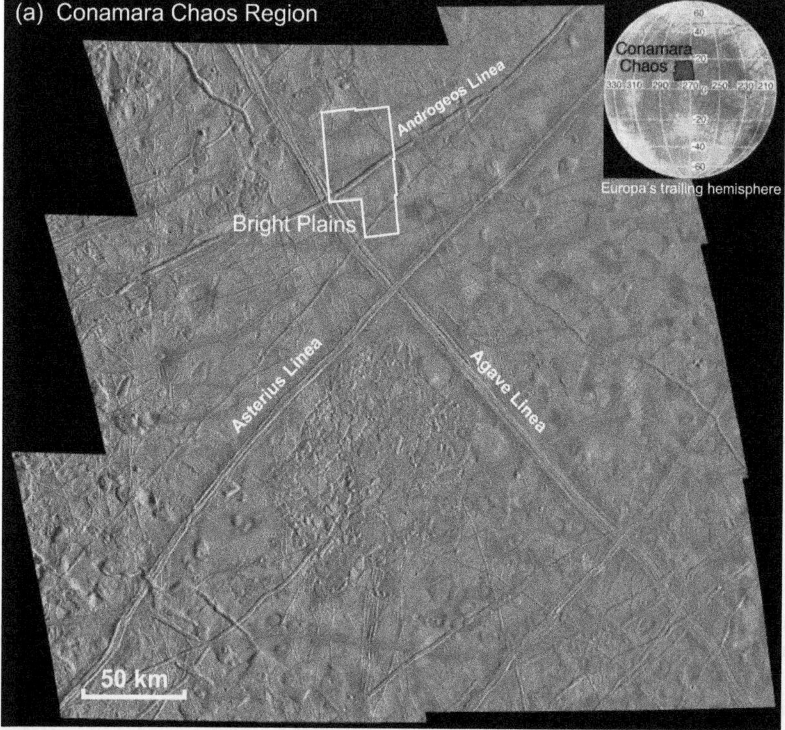

Fig. 2. **(a)** Conamara Chaos in the northern trailing hemisphere (orthographic mosaic E6ESDRKLIN01). The region is dominated by the prominent tectonic lineaments Asterius and Agave Lineae and the cryomagmatically disrupted region of chaos south of where the two lineaments cross. **(b)** The Bright Plains region (Galileo mosaic E6ESBRTPLN02) highlights the wide range of tectonic features in different orientations that characterizes much of the europan surface. This region shows ridges, dilational bands, and troughs, with the prominent Androgeos Linea cutting across the center of the image. Images courtesy of the NASA Space Photography Laboratory at Arizona State University.

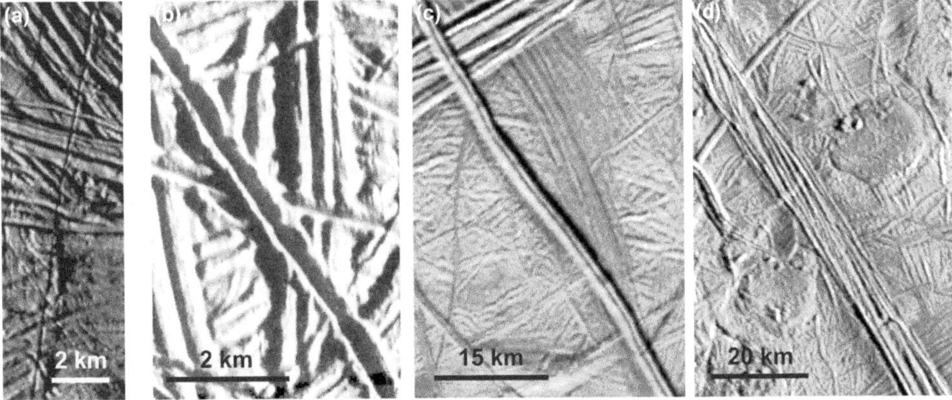

Fig. 3. Progressive evolution of fractures from **(a)** simple troughs to the development of **(b)** nascent ridges, then **(c)** fully developed ridges, and finally **(d)** a ridge complex. The image in **(a)** was taken during Galileo's E11orbit, image 420626739; those in **(b)**–**(d)** are all from the Conamara Chaos and Bright Plains region shown in Fig. 2, taken during Galileo's E6 orbit. The ridge in **(c)** is Androgeos Linea. The ridge complex in **(d)** is Agave Linea.

surface widths of <5 km (*Coulter et al.,* 2009). *Kattenhorn* (2002) examined the 20 m/pixel "Bright Plains" images (Fig. 2b) and concluded that ridges became higher, wider, and fewer in number during the geological sequence in that region.

Ridges do not necessarily occur in isolation but may develop prominent lineaments comprising several, commonly braided and inosculating, superposed ridges up to 15 km wide (Fig. 3d). Such features are called *ridge complexes* (e.g., *Greenberg et al.,* 1998) but are sometimes referred to as *complex ridges* (*Figueredo and Greeley,* 2000, 2004; *Greeley et al.,* 2000) and imply that an early formed ridge may create a zone of weakness that localizes and superposes later ridge development (Patterson and Head, in preparation), possibly accompanied by shearing (*Aydin,* 2006). Prominent examples include Agave and Asterius Lineae in the Conamara Chaos region (Fig. 2a). In low-resolution images, ridge complexes commonly appear bright with flanking dark stripes, and were originally referred to as triple bands (*Lucchitta and Soderblom,* 1982), now a defunct term as it does not imply any explicit geological feature type. Ridges and ridge complexes can produce a surface line-load that impinges on the elastic portion of the icy shell, resulting in downward flexing to either side of the ridge (*Tufts,* 1998; *Head et al.,* 1999; *Kattenhorn,* 2002; *Billings and Kattenhorn,* 2005; *Hurford et al.,* 2005). The associated bending stresses may induce flanking lines of tension fractures to either side of the ridge (see section 2.1.4).

A key observation about ridges is that they appear to represent an advanced stage in a genetic sequence of lineament development (*Geissler et al.,* 1998a; *Head et al.,* 1999) (Fig. 3). This idea stemmed from the fact that some ridges change in morphology along their trend, becoming less prominent along their lengths with a central crack flanked by underdeveloped edifices (Fig. 3b). This less-evolved stage has been called a *protoridge* (*Kattenhorn,* 2002) or a *raised-flank trough* (*Head et al.,* 1999). Ultimately, the protoridge may disappear along its trend to reveal its progenitor: a ridgeless crack called a *trough* (Fig. 3a). This sequence of development has been interpreted to imply that all ridges originate as troughs that form as tension cracks, evolving into protoridges and then finally ridges (Fig. 3c) or sometimes ridge complexes (Fig. 3d), although the nature of this evolution (i.e., the ridge growth process) is neither straightforward nor generally agreed upon.

The majority of published models of ridge development assume that they are fundamentally tensile features, but with differing mechanisms to explain the manner in which the ridge edifices are constructed. *Kadel et al.* (1998) attributed ridge growth to the effects of gas-driven cryovolcanic fissure eruptions. Weaknesses of this model include no explanation for a subsurface source of cryovolcanic material and the inherent difficulty in attempting to account for the remarkably consistent morphology of ridges, commonly for hundreds of kilometers along their lengths. A linear diapirism model (*Head et al.,* 1999) invokes buoyant upwelling of ice beneath a tension crack near the surface, resulting in an upward bending of the brittle carapace to either side of the central crack, thus forming ridges. Problems with this model include no explicit mechanism to explain why such upwelling would occur beneath a surface crack (although see the shear heating discussion below, for one possibility) and it necessitates that the preexisting terrain is preserved within the upwarped ridge slopes. This final point does not seem to be characteristic of europan ridges even though some examples have been described (*Giese et al.,* 1999; *Head et al.,* 1999; *Cordero and Mendoza,* 2004). The along-trend morphology of a well-developed ridge is invariably unaffected by the nature of the ridged plains it crosses, which may imply that upward warping of these plains is not the primary process by which ridges are created. Mass wasting could conceivably have removed the original surface roughness along ridges; however, ridge slopes are typically less than 30° and on average only about 10° (*Coulter et al.,* 2009), which seems to be too low to be consistent with a mass wasting process. An incremental wedging model (*Turtle et al.,* 1998; *Melosh and Turtle,* 2004) explains ridge development as being due to a gradual forcing apart of the walls of a crack intruded by material from below that freezes within the crack, causing upward plastic deformation at the surface to construct the ridge. In this model, it is unclear what the source of the intruded material would be, or why it is so readily accessible by surface cracks.

The ridge model of *Greenberg et al.* (1998) attributes their formation to the cyclical extrusion of a slurry of lead ice created by exposure and instantaneous near-surface freezing of an underlying ocean during the opening of a crack by diurnal tidal stresses. During subsequent closure of the crack, the ice is squeezed onto the surface, gradually building piles of ice debris to either side of the central crack. A caveat to this model is the need for a very thin icy shell in order to reconcile the requirement of complete separation of the ice layer to expose the underlying ocean (e.g., *Rudolph and Manga,* 2009) with the fact that tidal stresses are typically very small (likely a few tens of kilopascals) and thus would quickly be overcome by overburden pressure with increasing depth. Furthermore, because of the density contrast, ocean water can only rise to a level about 10% of the total icy shell thickness down from the surface. It would therefore have become increasingly difficult for material to be squeezed up to the surface if the icy shell thickened through time, as is suggested by thermodynamic models and the evolution of surface geological features (see section 5.2), unless the underlying ocean is sufficiently pressurized to allow cryovolcanic eruption, which is unlikely to be the case on Europa (*Manga and Wang,* 2007).

Gaidos and Nimmo (2000) and *Nimmo and Gaidos* (2002) infer that ridge development is ultimately driven by frictional shear heating along an existing crack (see section 2.3.1), even for friction coefficients as low as 0.1. Shearing is driven by the ever-rotating diurnal tidal stress field (see section 4.1.5) with inferred diurnal timescale shear velocities of 10^{-6}–10^{-7} m s^{-1}. Subsequent heating results in an almost sevenfold increase in the local surface heat flux causing buoyant rising of warm ice toward the surface and resultant upwarping of a near-surface brittle layer to form a ridge. For shear velocities $>\sim 10^{-6}$ m s^{-1}, local melting along the shear zone will cause downward draining of any melt products and the production of void space that may promote sagging or lateral contraction. This model places only mild depth constraints on the preexisting cracks that get sheared (i.e., the cracks do not need to penetrate the icy shell to an underlying ocean), it is based on a proven deformation mechanism on Europa (i.e., shearing; see section 2.3), and is a process that has likely been occurring on Europa as long as there has been a tidally responding ocean beneath the icy shell. Caveats to the model include the lack of observable lateral offsets along the majority of ridges and the lack of a clear mechanism for how the buoyant upwelling ultimately constructs the ridges. The mechanism appears to resemble the linear diapirism model described above and the two ideas are probably not mutually exclusive and thus share similar caveats regarding the ridge construction process. However, a source of heating may also perhaps explain the relatively low outer slopes of ridges if viscoplastic flow is the dominant form of ridge slope modification (*Coulter et al.,* 2009). *Han and Showman* (2008) explicitly model a linear zone of thermal upwelling beneath a frictionally shear heated zone, producing narrow, laterally continuous, ridge-like features with heights of up to 120 m. Ridges are known to attain heights of more than twice this amount; however, this small disparity may perhaps be circumvented if there is a component of fault-perpendicular contraction during the shear process that contributes to the construction of ridges (*Nimmo and Gaidos,* 2002; *Vetter,* 2005; *Aydin,* 2006; *Patterson et al.,* 2006; *Kattenhorn et al.,* 2007; *Bader and Kattenhorn,* 2007) or additional buoyancy mechanisms such as depletion of salts during heating (cf. *Nimmo et al.,* 2003). Possible evidence for contraction across ridges is described in more detail in section 2.2.3, but essentially allows ridge development to accommodate convergence across weak zones caused by shear heating along cracks. The associated contraction of the brittle icy shell at the surface may develop a permanent positive relief structure, perhaps negating the problem of expected relaxation of buoyant-upwelling-induced ridges over the thermal diffusion timescale of $\sim 10^7$ yr (*Han and Showman,* 2008).

Regardless of the precise formation mechanism for ridges, the fact remains that ridge building has been an effective geological process on Europa. Considering the relatively young surface age based on crater densities (40–90 m.y.) (*Zahnle et al.,* 2003; see chapter by Bierhaus et al.) and the great number of ridges of different ages and orientations, each individual ridge must form in a relatively short period of time. *Greenberg et al.* (1998) suggest that ridges may form within 30,000 yr based on their tidal pumping model; however, that estimate is necessarily based on a number of uncertain assumptions regarding crack dilation and spacing, as well as the effectiveness of the ice extrusion process. *Melosh and Turtle* (2004) estimate 10,000 yr to form a ridge through an incremental ice-wedging process. *Gaidos and Nimmo* (2000) estimate that shear heating may result in buoyant uprising of warm ice at a rate that could conceivably construct a ridge in only ~10 yr. Presumably, the amount of time during which a crack may remain active (whether in dilation, shearing, or contraction) with the capability of developing a ridge is ultimately controlled by the amount of time over which the global tidal stresses drive crack activity. This timing may be controlled by the rate of nonsynchronous rotation, if present, relative to a tidally locked interior (see sections 4.1.6 and 5.1), which would eventually rotate an active crack away from a stress field conducive to crack activity. One cycle of nonsynchronous rotation takes in excess of 12,000 yr (*Hoppa et al.,* 1999b) and perhaps as much as 1.3 m.y. [cf. *Hoppa et al.* (2001), accounting for the revised surface age of Europa in the chapter by Bierhaus et al., and a factor of 3 uncertainty], suggesting that an estimate of a few tens of thousands of years during which a crack can remain active and thus form a ridge may be reasonable.

2.1.2. Cycloids. Europa also exhibits unique features morphologically similar to ridges called cycloids, also referred to as *cycloidal ridges* or *flexūs*. Cycloids, first described from Voyager data (*Pieri,* 1981; *Lucchitta and Soderblom,* 1982; *Helfenstein and Parmentier,* 1983), are curved, cuspate cracks that form chains of multiple, concatenated segments extending hundreds to thousands of kilometers across the surface (Fig. 4). Each curved segment of a cycloid chain is linked to the adjacent one at an abrupt kink called a cusp. Individual segments are typically tens of kilometers long, measured in a direct line from cusp to cusp, but are locally up to several hundred kilometers long. The central crack is commonly, but not necessarily, flanked by ridges, implying that progressive ridge development occurs to either side of an initial crack, analogous to linear ridges described in section 2.1.1. In some cases, ultimate dilation of the cracks occurs to form cycloidal bands (see chapter by Prockter and Patterson). Cycloids are distinctly different from other tectonic cracks on Europa and, other than one possibly analogous feature in the south polar region of Enceladus, are unique in the solar system. Therefore, their mode of formation must also differ from other tectonic lineaments on Europa, implying significant variability in crack-driving processes during the long-term tectonic history.

Hoppa and Tufts (1999) and *Hoppa et al.* (1999a) proposed that cycloidal cracks form as a result of tensile cracking in response to diurnally varying tidal stresses produced by Europa's orbital eccentricity (these stresses are described in detail in section 4.1.5; also see chapter by Sotin et al.). As Europa orbits Jupiter, it is constantly being reshaped as the size and location of the tidal bulges change, producing a time-dependent diurnal tidal stress field. At any location on the surface during the orbital period, the orientations of the principal stresses rotate and the magnitudes oscillate. The stresses rotate counterclockwise in the northern hemisphere and clockwise in the southern hemisphere, 180° each orbit (while only magnitudes change along the equator). The individual cycloid segments are hypothesized to grow as tension fractures that propagate perpendicular to the rotating direction of maximum tensile stress, resulting in curved segments. Cracking begins when the tensile strength of europan ice is overcome and ceases when the tensile stress drops below the crack propagation strength, assumed to be less than the tensile strength. These parameters are unknowns on Europa; therefore, *Hoppa et al.* (1999a, 2001) estimated them in such a way so as to provide the best match between observed cycloid shapes and theoretical stress fields. This growth model for cycloids predicts that all northern hemisphere cycloids that are concave toward the equator grew from east to west, whereas all cycloids concave toward the poles grew from west to east. Similarly, northern hemisphere cycloids that are concave to the east are predicted to have grown from north to south, whereas cycloids that are concave to the west grew from south to north. The opposite sense of growth is true in all cases in the southern hemisphere. Also, because the diurnal stress characteristics are dependent on latitude and longitude, expected cycloid shapes are correspondingly variable across the surface of Europa, ranging from broadly curved to squarish (*Bart et*

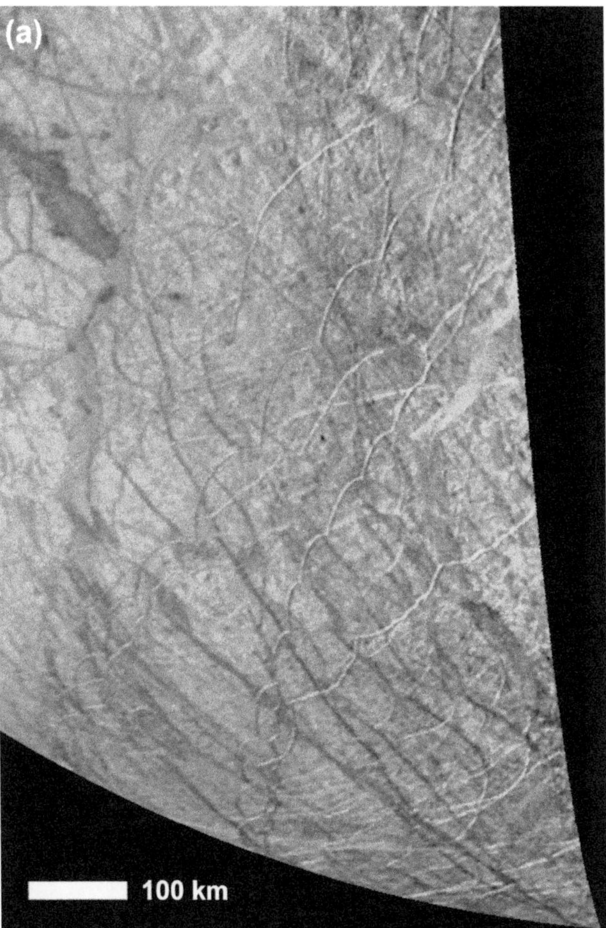

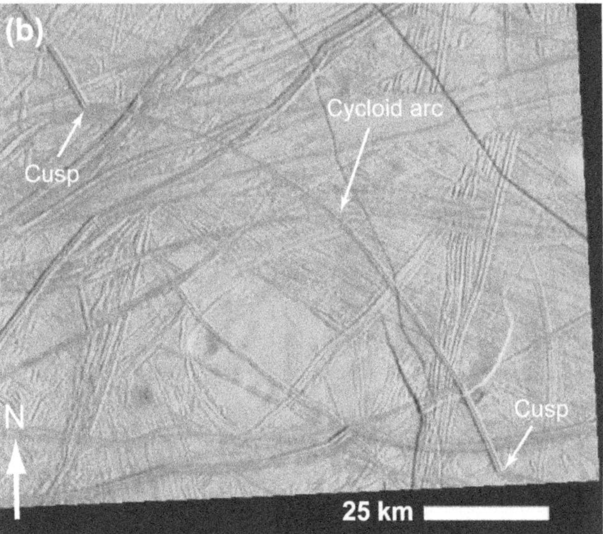

Fig. 4. **(a)** Voyager 2 image (c2065219) showing several chains of cycloidal ridges. **(b)** Example of a cycloidal fracture in the northern trailing hemisphere with ridge development along a small part of the arc (Galileo image 449961879, E15 orbit).

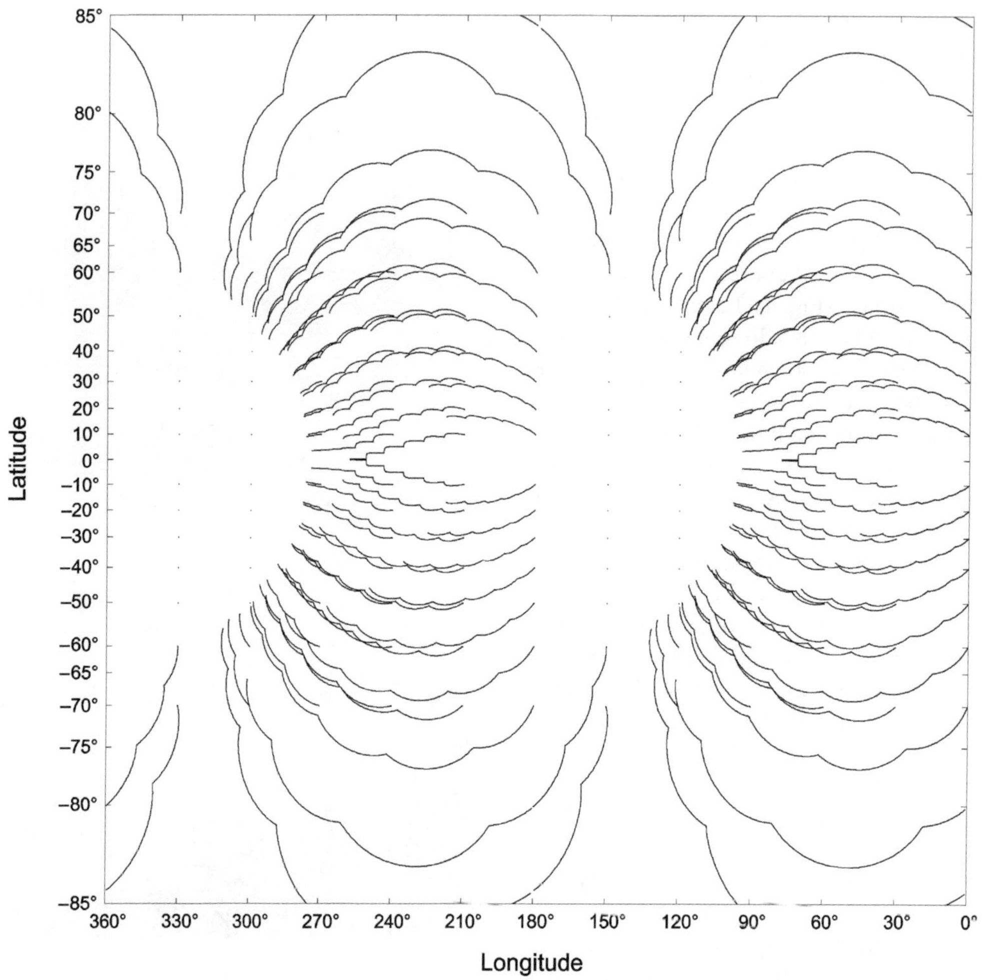

Fig. 5. Theoretical distribution of cycloids that grow from east to west (*Hurford et al.,* 2007a) in a stress field composed of a diurnal component plus stress due to 1° of nonsynchronous rotation. East-growing cycloids will have an opposite curvature in each hemisphere but a similar overall distribution.

al., 2003; *Hurford et al.,* 2007a). Cycloid chains commonly exhibit a large-scale curvature due to regional variability in global tidal stress fields. The growth of these broadly curved cycloid chains ultimately ceases when they propagate into an area where critical stress values are not attained, resulting in some regions where cycloids theoretically never form (Fig. 5).

The growth of a single cycloid arc is assumed to be a continuous process during the course of a europan day and must occur at a low crack propagation speed (~3–5 km/h) in order for curved segments to produce a match to the predicted stresses. Actual growth speeds may be much faster, albeit occurring in short, discrete spurts to produce a much lower effective propagation speed (*Lee et al.,* 2005). After cessation of growth of a cycloid arc, the next arc starts growing when the tensile strength of the ice is again overcome in the subsequent orbit; however, because the stresses rotate during the period of no crack growth, the new cycloid arc propagates at an angle away from the tip of the previous one, forming a sharp cusp. These rotated stresses resolve a component of shearing along the arrested cycloid segment immediately prior to the development of the cusp, inducing a concentration of stress at its tip that drives the development of the new cycloid arc (*Marshall and Kattenhorn,* 2005; *Groenleer and Kattenhorn,* 2008). In this way, cycloid cusps form identically to features called tailcracks that develop at the tips of strike-slip faults on Earth and elsewhere throughout the solar system, including Europa (see section 2.3.3). The tailcrack then continues to grow in tension to form a new cycloid arc, driven by the diurnal stresses. Ongoing shearing near cycloid cusps has also been suggested to be the cause of multiple tailcrack-like splays of fractures emanating from cusp regions, producing complex cusps (*Marshall and Kattenhorn,* 2005) that resemble horsetail fracture splays along terrestrial strike-slip faults.

2.1.3. Dilational bands. Also referred to as *pull-apart bands* (see chapter by Prockter and Patterson), dilational bands represent clear evidence of prolonged dilation in the icy shell and hence a resurfacing process on the icy moon. A dilational band is a tabular zone of new crustal material that intruded between the progressively dilating walls of a tension fracture (Fig. 6a). The surface of this material ap-

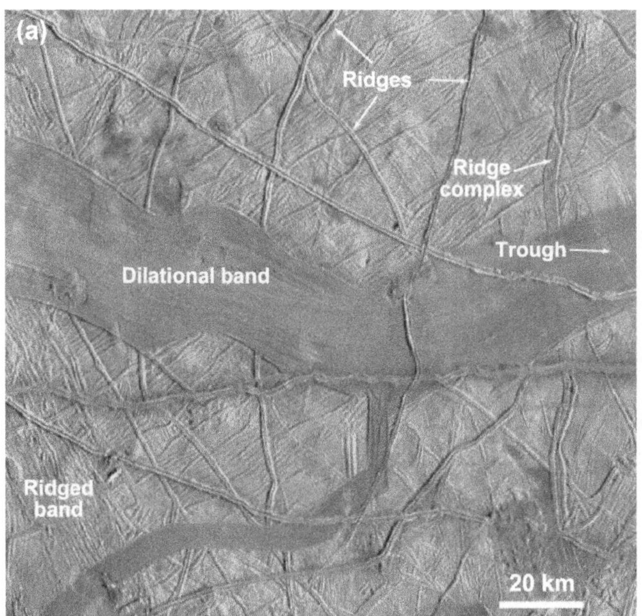

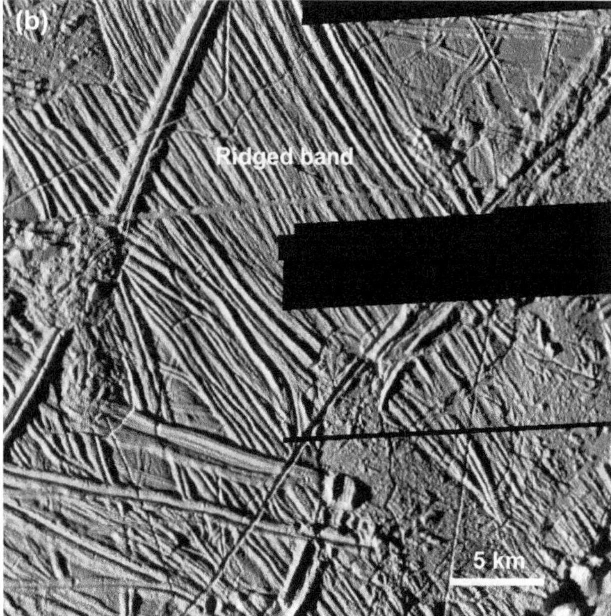

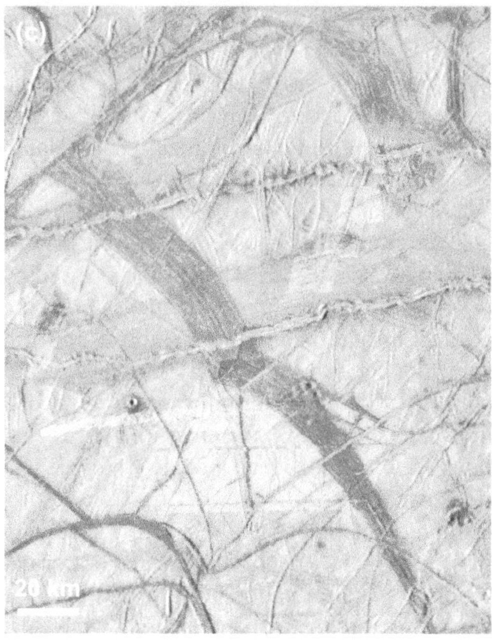

Fig. 6. **(a)** A dilational band (provisionally named Phaidra Linea) in the equatorial trailing hemisphere. Note how features can be matched up to either side of the band. Although the dilational band is one of the younger features in this region, a number of younger ridges and troughs can be seen to crosscut the band (from Galileo observation 11ESREGMAP01). **(b)** A 17-km-wide ridged band in the northern leading hemisphere (from Galileo image mosaic 11ESMORPHY01). **(c)** The dark dilational band Yelland Linea in the "Wedges" region, Argadnel Regio (from Galileo image mosaics 12ESWEDGE_01/02/03 superimposed on C3ESWEDGES01).

pears featureless in lower-resolution images, in which case the term *smooth band* can be used as a descriptor; however, if an internal geometry of fine lineations is observable (usually at medium to high resolution, as in Fig. 6a), the term *lineated band* may be used. Complete separation and infill of the surface is evidenced by the fact that the ridged plains to either side of a dilational band typically match up, implying that the dilational band material represents new surface area (*Schenk and McKinnon,* 1989; *Sullivan et al.,* 1998). The exact source of the dilational band material is unclear. Some have suggested that the dilation occurred across cracks that fully penetrated the icy shell, with the smooth material representing frozen portions of an exposed, underlying ocean (*Tufts et al.,* 2000). Alternatively, dilational bands may represent regions where the brittle portion of the icy shell dilated slowly enough that ductile ice underwent buoyant upwelling from deeper portions of the icy shell to fill in the dilational gap (*Pappalardo and Sullivan,* 1996; *Sullivan et al.,* 1998; *Prockter et al.,* 1999, 2002). The low albedo of most young dilational band material has been suggested to represent the effects of magnetospheric particle bombardment of endogenic sulfur-bearing compounds (*Kargel,* 1991; *Noll et al.,* 1995); however, continued exposure at the surface ultimately results in a brightening of the dilational band such that the oldest dilational bands have a similar albedo to the surrounding ridged plains (*Pappalardo and Sullivan,* 1996; *Geissler et al.,* 1998a; *Prockter et al.,* 1999, 2002), which historically resulted in use of the descriptor *bright bands.* Although tectonic plates (in the terrestrial sense) do not exist in the icy shell (see section 3), dilational bands represent a spreading phenomenon analogous to mid-ocean ridge spreading centers, making europan dilational bands the only other known feature in the solar system where complete lithospheric separation has occurred. Similar to mid-ocean ridges, dilational bands stand higher than the surrounding plains by up to 200 m (*Prockter et al.,* 1999, 2002; *Nimmo et al.,* 2003) and may have a relatively reduced elastic thickness (in the range of a few hundred meters to a few kilometers) (*Prockter et al.,* 2002; *Billings and Kattenhorn,* 2005; *Stempel et al.,* 2005). Opening vectors across dilational bands indicate that both orthogonal and oblique dilations are common.

Similar to ridges, dilational bands can be hundreds of kilometers in length, implying a regional process responsible for the recorded period of spreading. Dilational band widths generally do not exceed ~30 km and are commonly only a few kilometers, suggesting that the process responsible for driving the dilation is unable to sustain spreading beyond a certain time and/or width limit. Nonetheless, a dilational band formation event likely represents a prolonged period of uninterrupted tectonic extension. Discrete episodes of dilation are evidenced by an internal fabric of fine lineaments within lineated bands (Fig. 6a) that commonly form a bilateral symmetry about a central axis and exhibit a consistent spacing on the order of ~500 m (*Sullivan et al.,* 1998; *Prockter et al.,* 1999, 2002; *Stempel et al.,* 2005). If these lineaments are normal faults, creating a type of lineated band called a *faulted band,* they may perhaps form a graben-like system centered about the middle of the dilational band (see section 2.1.5), with older faults being translated successively further away from the actively spreading central axis over time. At slow spreading rates, the faulted surface of the dilational band may become rugged, with tilted fault blocks creating repeating valleys and ramparts (*Prockter et al.,* 2002; *Stempel et al.,* 2005).

Some dilational bands have been noted to exhibit a maximum dilation near the center of the length of the band, with dilation decreasing toward either tip, such as Thynia Linea (*Pappalardo and Sullivan,* 1996) and Yelland Linea (Fig. 6c) (*Stempel et al.,* 2005). Such dilational bands resemble typical cracks in an elastic layer that are dilated by a regional tensile stress acting perpendicular to the feature. Nonetheless, some dilational bands are more rhomboidal where they occur in extensional stepover zones (i.e., pull-aparts) along strike-slip faults (see section 2.3.3) such as Astypalaea Linea in the southern antijovian region (*Tufts et al.,* 1999, 2000; *Kattenhorn,* 2004a) and wedge-shaped bands in Argadnel Regio (*Schulson,* 2002; *Kattenhorn and Marshall,* 2006), implying a localized tectonic phenomenon.

Although it is possible that the driving stress responsible for a dilational band also created the original crack across which dilation ensued, some dilational bands show evidence of raised flanks that appear to be the two halves of an erstwhile ridge. Hence, some dilational bands dilated preexisting cracks that had already developed ridges along their rims. Considering that ridge formation is not instantaneous but may take tens of thousands of years and involve a range of mechanisms (dilational, contractional, and shearing), the process responsible for dilational band formation is unlikely to be a natural endmember of the ridge formation process. Ridges simply provide weakness zones within the icy shell that may be utilized when the conditions conducive to dilational band formation occur. The actual spreading across dilational bands may be driven by underlying convection in a region of locally high heat flow (*Prockter et al.,* 2002), perhaps originally initiated by the cracking of the icy shell (cf. *Han and Showman,* 2008). This upwelling may explain why portions of ridged plains immediately adjacent to some dilational bands are also slightly elevated relative to the surroundings (*Nimmo et al.,* 2003). Comparisons to terrestrial midocean-ridge spreading models suggest strain rates in the range 10^{-15}–10^{-12} s^{-1} at europan dilational bands (*Nimmo,* 2004a,b; *Stempel et al.,* 2005), implying spreading rates of 0.2–40 mm yr^{-1}, similar to terrestrial mid-ocean ridge spreading rates. *Nimmo* (2004b) estimates a maximum duration of 10 m.y. for dilational band formation, with the observed narrow rift geometry of dilational bands implying a high strain rate or a relatively thick shell (but probably <15 km) at the time of dilational band formation. *Stempel et al.* (2005) estimate the duration of spreading to be in the range 0.1–30 m.y. (with lower estimates representing lower coefficients of friction of the ice), which is well within the limits of surface ages deduced from the cratering record (*Zahnle et al.,* 2003; see chapter by Bierhaus et al.). Nonetheless, the great number and different ages of dilational bands and other tectonic features on Europa suggest that dilational band activity is likely to be at the lower end of this duration estimate. This deduction is also supported by the requirement for self-similar stress conditions during the development of dilational bands that exhibit a uniform lineated texture, although some dilational bands may record a more complicated dilational/stress history. The needed stresses to drive spreading are on the order of a few megapascals (*Stempel et al.,* 2005), which are likely provided by nonsynchronous rotation (see section 4.1.6). Considering that nonsynchronous rotation would result in a gradually changing stress field in any area of dilational band formation, the duration of band formation must be less than the time it would take for nonsynchronous rotation to move the developing dilational band into a new stress state.

Some cycloidal cracks also show evidence of having dilated to form cycloidal dilational bands (*Marshall and Kattenhorn,* 2005), including Thynia Linea (*Pappalardo and Sullivan,* 1996; *Tufts et al.,* 2000), wedge-shaped bands in Argadnel Regio (*Prockter et al.,* 2002), and the prominent example of the "Sickle" (provisionally named Phaidra Linea) in the equatorial trailing hemisphere (*Tufts et al.,* 2000; *Prockter et al.,* 2002) (Fig. 6a). In these examples, the opening vector across each dilational band is constant, resulting in portions of the band having undergone oblique dilation relative to the margins.

An apparent type of lineated dilational band morphologically similar to a faulted band, but composed of ridges, is a tabular spreading zone referred to here as a *ridged band* (cf. *Figueredo and Greeley,* 2000, 2004) (Fig. 6b). *Stempel et al.* (2005) used this same terminology to refer to faulted bands; however, we abandon that use of the nomenclature because ridges (which have an explicit meaning on Europa) are not the dominant feature in faulted bands. In low-resolution imagery, ridged bands may be mistaken for smooth bands if the internal lineated texture is not observable; however, ridged bands are essentially up to 60-km-wide complexes of multiple adjacent ridges and furrows (i.e., essentially dilational bands composed of ridges). Ridged bands appear to be analogous to the "Class 2 ridges" of *Greenberg et al.* (1998) and *Tufts et al.* (2000) and have also been classified as "complex ridges" (*Kattenhorn,* 2002; *Spaun et al.,* 2003). This latter terminology causes confusion and should be avoided in light of previous usages and to avoid confusion with the definition of a ridge complex as a zone of multiple superposed and interweaving ridges (see section 2.1.1). Ridged bands also represent zones of spreading in the icy shell but the infill process differs from smooth bands and faulted bands, although hybrid features may exist that take on the appearance of both smooth bands and ridged bands (*Tufts et al.,* 2000). Instead of upwelling of ductile material from below, ridged bands may represent a different form of spreading. One model invokes injection of material from below into discrete fractures, analogous to a terrestrial dike swarm. At the surface, each of these cracks may then take on the form of a ridge through some ridge-forming mechanism (see section 2.1.1). Like a dike swarm, there is no necessity for symmetry about a central crack, as the sequence of intrusion through vertical dike-like features may be somewhat random across the spreading zone (*Tufts et al.,* 2000). Nonetheless, bilateral symmetry is possible (e.g., *Kattenhorn,* 2002).

2.1.4. Troughs. Considering the lack of raised edifices along the crack margins, as typifies ridges, troughs are easily overlooked in images of the surface. They likely represent tension fractures in the ice shell that never developed into a more evolved landform, such as a ridge, and have been referred to simply as *cracks*. They have been acknowledged (*Figueredo and Greeley,* 2000, 2004; *Kadel et al.,* 2000; *Prockter et al.,* 2000; *Kattenhorn,* 2002) but have received little attention, despite being relatively common in high-resolution images of the surface. The majority of the tectonic features on Europa likely owe their beginnings to tension fracturing at the surface of the icy shell, starting as troughs.

Confining stress considerations dictate that most fractures are likely tension fractures that are initiated at the surface and then propagate downward. Whether or not the fractures completely penetrate the icy shell is dependent on the availability of stresses to overcome the overburden, which becomes increasingly difficult for thicker icy shells. Diurnal tidal stresses (see section 4.1.5) are rapidly overwhelmed by the overburden in the icy shell, which, for an ice density of 0.91 g/cm^3, increases at a rate of about 1.2 MPa/km. *Hoppa et al.* (1999a) suggest that fracture depths should thus not exceed ~65 m; however, consideration of crack-tip stresses increase this depth to a few hundred meters (*Lee et al.,* 2005; *Qin et al.,* 2007) and perhaps up to several kilometers when ice porosity effects are considered, allowing complete penetration of the icy shells if its thickness does not exceed 3 km. The addition of stress due to nonsynchronous rotation (see section 4.1.6) allows penetration depths of several kilometers (*Panning et al.,* 2006) to approaching 10 km (*Golombek and Banerdt,* 1990; *Leith and McKinnon,* 1996), perhaps allowing complete penetration of an icy shell in the thickness range 6–13 km for a plausible range of ice porosity (*Lee et al.,* 2005). If there has been more than 10 km of thickening of the icy shell through time (see section 4.1.8), sufficient stress may have been produced at the base of the icy shell to initiate cracking and permit complete penetration through the shell (*Manga and Wang,* 2007), with upward propagation aided by fluid pressure (cf. *Crawford and Stevenson,* 1988), in which case it would not be unreasonable to assume that tension fractures provide potential connection pathways between the surface and an underlying ocean. Nonetheless, viscoelastic relaxation in the lower part of the icy shell over the shell thickening timescale likely limits complete penetration by tension fractures to icy shells ≤2.5 km thick (*Rudolph and Manga,* 2009).

Troughs constitute the youngest tectonic features on Europa based on crosscutting relationships. Hence, they provide the most promising indicator of recent to current tectonic activity (see section 5.3). *Kattenhorn* (2002) showed that troughs make up a significant portion (the youngest 20%) of the tectonic history of the Bright Plains region. Although the surface geological history appears to record a transition from principally tectonic to predominantly cryomagmatic activity, with associated formation of regions of chaos and lenticulae (see section 5.2 and chapter by Doggett et al.), troughs have been noted to cross regions of chaos and thus may be very recent features.

Troughs vary greatly in geometry, scale, orientation, and location, reflecting differences in causal mechanisms driving fracturing (e.g., *Figueredo and Greeley,* 2004). A classification scheme for troughs is suggested here to address these differences (Fig. 7): (1) *Tectonic fractures* range in length from <10 km to hundreds of kilometers (Fig. 7a). They are presumably the result of global tidal stresses plus any other global, regional, or local contributing stress components that may drive global tectonics (see section 4). They exhibit a range of orientations and are characterized by linear or broadly curving geometries. They are sometimes segmented along their lengths or may show evidence of having formed by the coalescence of numerous segments, analogous to terrestrial joints. (2) *Cycloidal fractures* (Figs. 4b and 7b) are also tectonic fractures but are specifically the ridgeless progenitors to cycloidal ridges. They are distinct in their curved and cuspate nature, probably reflecting the dominant control of the diurnal stress field in their formation. Although cycloidal ridges are more common, relatively younger cycloidal fractures have also been identified (e.g., *Marshall and Kattenhorn,* 2005), indicating that cycloid development has persisted until at least geologically recent times. (3) *Tailcracks* emanate from the tips of strike-slip faults (Fig. 7b) and represent brittle accommodation of fault motion in the

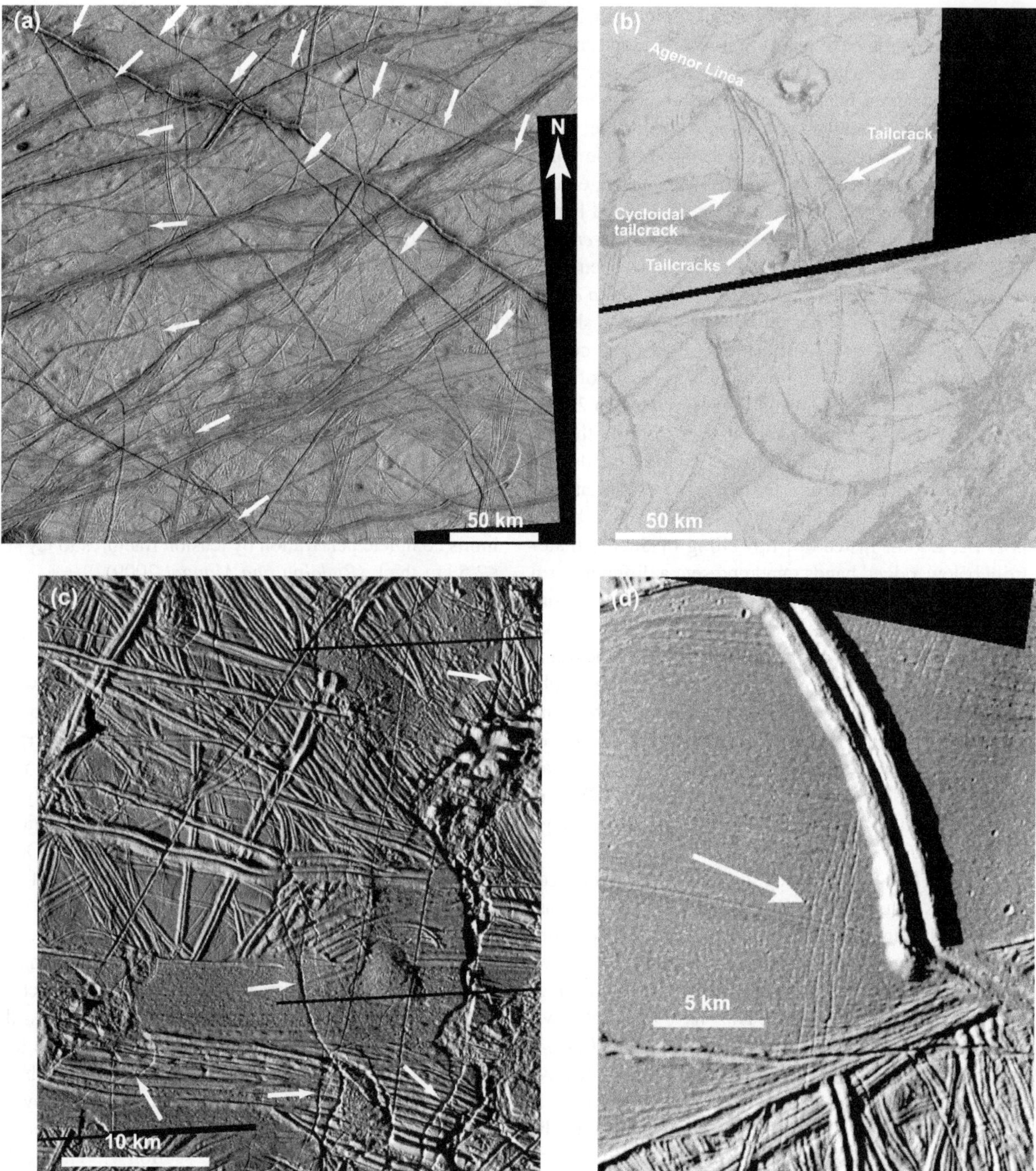

Fig. 7. Troughs (indicated by white arrows). **(a)** *Tectonic fractures* refer to any generic troughs induced by global deformation of the icy shell, such as due to tidal forcing (from Galileo mosaic 15ESREGMAP01). **(b)** *Tailcracks* form at the tip of a strike-slip fault (in this case, Agenor Linea) where localized tension is induced by fault motion. One tailcrack here is also a *cycloidal fracture* (see Fig. 4). From Galileo images 466664413 and 466665378, E17 orbit. **(c)** *Endogenic fractures* may be irregular and typically form adjacent to chaos, as occurs here in Galileo image mosaic 11ESMORPHY01. **(d)** *Fold hinge fractures* are caused by tension along anticlinal fold hinges. This example is within Astypalaea Linea, from Galileo image 466670113, E17 orbit. A fifth fracture type is a *flexure fracture* as occurs to either side of Androgeos Linea (see Fig. 2b) in response to elastic bending of the icy shell.

tensile quadrant of a fault tip (*Schulson,* 2002; *Kattenhorn,* 2004a; *Kattenhorn and Marshall,* 2006). Also referred to as horsetail splays or wing cracks in terrestrial analogs, they have been recognized in many locations on Europa such as the impressive example at the southeast end of Agenor Linea (*Prockter et al.,* 2000; *Kattenhorn,* 2004a). Tailcracks form in response to the locally perturbed stress field at the tip of a fault and thus do not provide a direct indicator of the global stress field at the time of their formation, except where the tailcracks extend far enough from the fault tip that their orientations are controlled by regional stresses. (4) *Endogenic fractures* are most commonly associated with regions of chaos (Fig. 7c) and thus reflect a local process likely driven by thermal upwelling or diapirism in a convecting icy shell, which in turn disrupts the brittle carapace (*Collins et al.,* 2000; *Figueredo and Greeley,* 2004; *Schenk and Pappalardo,* 2004; *Mitri and Showman,* 2008). These fractures may also be associated with broad scale warping at the surface of the icy shell (e.g., Fig. 3b in *Prockter and Pappalardo,* 2000), which is also likely driven by an endogenic process. Thus, the orientations of endogenic fractures are not controlled by a global stress field. (5) *Flexure fractures* occur along the flanks of many large ridges (e.g., Fig. 2b) that have caused elastic flexing of the adjacent icy shell, probably due to the loading caused by their weight (*Tufts et al.,* 2000; *Hurford et al.,* 2005) and/or by withdrawal of material from beneath the ridge flanks (*Head et al.,* 1999). Flexing of the elastic portion of the icy shell alongside a ridge induces bending stresses that can exceed the tensile strength of the ice, at which point one or more fractures develops (*Billings and Kattenhorn,* 2005). (6) *Fold hinge fractures* (Fig. 7d) form in response to tensile bending stresses along the hinge lines of surface anticlines (see section 2.2.1), although such features appear to be relatively rare (*Prockter and Pappalardo,* 2000).

2.1.5. Normal faults. Although tensile fracturing is dominant, some of the extension of the icy shell occurred by normal faulting, evidenced by fault scarps that show a component of vertical motion. Tensile stresses are common in the shell (see section 4.1); however, normal faulting does not occur under conditions of absolute tension, only deviatoric tension. Normal faulting is possible when the minimum horizontal compressive stress is less than the overburden compressive stress by a sufficient amount to overcome the internal friction (*Beeman et al.,* 1988; *Pappalardo and Davis,* 2007). Compressive stresses are common in the icy shell during the tidal cycle, therefore adequate conditions for extensional shear failure are likely to be common.

The great majority of normal faults are inferred to constitute the fine striations within lineated bands (see section 2.1.3) that parallel the band boundaries, contributing to the similarity between dilational bands and terrestrial mid-ocean ridge spreading centers. Rare normal faults with significant vertical displacements (~300 m) have also been noted to crosscut ridged plains (*Nimmo and Schenk,* 2006), and it is possible that up to 1.5-km-deep troughs in the icy shell that define antipodal small-circle depressions in the leading and trailing hemispheres are also normal fault related (*Schenk et al.,* 2008). In lineated bands, bilateral symmetry about a central trough is reminiscent of repeating pairs of inward-dipping, graben-bounding faults that move progressively outward from the spreading axis with time. In high-resolution images (less than a few tens of meters per pixel), the striations in the lineated bands are clearly normal faults, showing smooth and highly reflective fault planes bounding rotated blocks of band material with low-albedo tilted upper surfaces. Mechanical interactions are evident in fault trace patterns, as well as relay ramps in overlap zones, analogous to terrestrial normal fault systems (*Kattenhorn,* 2002). Normal faults in lineated bands do not necessarily dip inward towards an axial trough but may dip consistently in one direction across the entire width of a band or be restricted to certain parts of the band only (*Kattenhorn,* 2002; *Prockter et al.,* 2002). This pattern of faulting is distinctly different to mid-ocean ridge rift zones and may imply that some dilational bands were extended by normal faulting at some time after initial band formation, perhaps due to the reduced brittle thickness of the dilational band relative to the surrounding ridged plains, causing the bands to act as necking instabilities that focused extension. It is unclear, however, why some dilational bands localize normal faulting (forming faulted bands) rather than localizing tension fractures (in which case, ridged bands may ultimately develop).

2.2. Compressive Tectonics

An apparent consequence of the great amount of extension on Europa is the need for corresponding contraction to provide a balanced surface area budget. Although Europa has likely undergone some amount of expansion due to cooling and thickening of the icy shell, the maximum likely extensional strain after even 100 m.y. of cooling would be ~0.35% (*Nimmo,* 2004a), which is insufficient to account for the amount of new surface area created at spreading bands, which occupy ~5% of the surface area in the pole-to-pole regional maps of the leading and trailing hemispheres (*Figueredo and Greeley,* 2004). Hence, some amount of contraction must have occurred during the visible tectonic history in order to create a balanced surface area budget. Given this need, contractional features might be expected to be as common on the surface as extensional features. In actuality, contractional deformation on Europa is visibly sparse and by no means obvious. *Greenberg et al.* (1999) contemplated whether much of the contraction may have been accommodated within areas of chaos terrain, which covers ~10% of the surface; however, no ultimate geological evidence arose to imply that chaos represents sites of contraction.

Part of the reason for an apparent lack of contractional features may be a historical emphasis on the many forms of extensional structures such as ridges and dilational bands, which are morphologically dominant; however, a potentially greater hindrance to the identification of contraction has been a failure to clearly recognize the manifestation of such deformation. In terrestrial settings, contraction may be accommodated variably over a range of scales. At the outcrop scale, small amounts of contraction may occur along deformation bands (tabular zones of cataclasis and porosity reduction in granular rocks) or along pressure-solution surfaces (sometimes called anticracks, where part of the rock is removed in solution resulting in a volume loss). At the regional scale, contraction is accommodated in brittle materials through the development of thrust faults (with associated mountain building) or elastic warping, and in ductile materials through folding. At the planetary scale, broad-scale convergence occurs at subduction zones to accommodate the creation of new oceanic lithosphere at spreading ridges. Hence, any contraction on Europa is likely manifested through one or more of these mechanisms. There is no evidence for global plate tectonics on Europa (see section 3) and thus no removal of surface area along subduction zones; however, all other forms of contraction described above may be viable on Europa.

2.2.1. Folding. The first documented evidence for contraction on Europa was the identification of several parallel folds within a dilational band associated with the right-lateral strike-slip fault Astypalaea Linea (*Prockter and Pappalardo,* 2000). The folds are visible due to subtle differences in surface brightness and have a wavelength of ~25 km and crest-to-trough amplitudes of 250 ± 50 m (*Dombard and McKinnon,* 2006). In adjacent ridged terrain, the folds gradually disappear or are not resolvable, suggesting that dilational band material is more easily folded, perhaps due to a localized high heat flow (*Prockter,* 2001). Other telltale clues for the presence of the folds include clusters of bending-induced tensile cracks along the hinge lines of anticlines and compressive crenulations along the hinge lines of synclines (Fig. 7d). The high heat flow in a dilating band (see section 2.1.3) relative to adjacent ridged terrain, combined with a reduced ice thickness (*Billings and Kattenhorn,* 2005), causes dilational bands to become localized zones of crustal weakness that may be able accommodate contraction by folding. *Prockter and Pappalardo* (2000) suggest that sufficient compressive stress may accumulate to drive folding over <40° of nonsynchronous rotation of the icy shell, and also suggest that fold axis orientations are consistent with predicted stress fields. In contrast, *Mével and Mercier* (2002) suggest that late transpressive motion along Astypalaea Linea may have caused the localized folding. Over time, folds may relax away due to ductile flow of deeper, warm ice; however, *Dombard and McKinnon* (2006) suggest that such relaxation would occur so slowly relative to the age of Europa's surface (e.g., as little as 4% relaxation over 100 m.y.) that all folds that formed during the geologically visible past should still be apparent. Hence, the distinct scarcity of visible folding on Europa suggests that it is not a significant accommodator of crustal contraction and/or is difficult to recognize in existing images, and is certainly insufficient to balance out the amount of extension.

2.2.2. Convergence bands. A second candidate feature for localized contraction, first identified by *Sarid et al.* (2002), is a convergence band, across which a tectonic reconstruction reveals a zone of "missing" crust. These bands superficially resemble dilational bands caused by plate spreading in that they form broad zones that may be many kilometers wide and perhaps tens of kilometers long, and appear to disrupt the surrounding ridged terrain (*Greenberg,* 2004) (Fig. 8). They differ from dilational bands, however, in that their edges may be nonlinear or even inosculating, and one edge of the band is not necessarily a mirror image of the other side, unlike dilational bands that form by icy shell separation and spreading. Convergence bands are among the least-studied features on Europa; therefore, much work is still needed to fully characterize their broad-scale geometries, internal morphologies, and mechanical evolution. Nonetheless, existing studies imply two varieties of convergence bands: (1) those driven by motion along strike-slip faults, with resultant convergence adjacent to the fault in the tip-region compressional quadrants; and (2) those that develop along zones of preexisting weakness in the icy shell, such as dilational bands and dilational strike-slip faults.

Sarid et al. (2002) describe two sites in the trailing hemisphere where convergence is driven by motions along adjacent strike-slip faults (type 1 above). One example in the southern trailing hemisphere (in the Galileo regional image mosaic 17ESREGMAP01), between Argadnel Regio and Castalia Macula (Fig. 8a), is suggested to be a zone of convergence related to right-lateral motion along an approximately north-south-oriented strike-slip fault north of the zone of convergence. *Patterson et al.* (2006) describe evidence of dilational band development in the same vicinity. We infer that both processes may have occurred here at different times (early convergence with later dilation). Strike-slip fault driven convergence is suggested to have occurred in the marginal blocks alongside Astypalaea Linea (*Mével and Mercier,* 2002), resulting in up to 55% contraction along numerous distributed ridge-like crenulations (perhaps a related contractional mechanism to convergence bands). Evidence for strike-slip fault-related convergence bands is also presented by *Kattenhorn and Marshall* (2006), who characterize compressive stress concentrations in the tip regions of strike-slip faults in Argadnel Regio (see section 2.3.3).

Convergence bands that localize along sites of crustal weakness such as preexisting dilational bands (type 2 above) may also be driven by nearby strike-slip fault motions. This phenomenon appears to have occurred in the northern trailing hemisphere (Fig. 8b) example described by *Sarid et al.* (2002), although they did not recognize it as such. Several major strike-slip faults on Europa dilated during their development to become band-like (see section 2.3.3) and may have subsequently acted as loci for a component of convergence if conditions changed from transtensional to transpressional. Candidate site examples include the bright bands Corick Linea, Katreus Linea, and Agenor Linea (see chapter by Prockter and Patterson). Agenor Linea was interpreted by *Prockter et al.* (2000) to contain contractional features related to shear-related duplexing, suggesting transpression. *Greenberg* (2004) inferred 20 km of convergence across Corick Linea in order to match up offset features, resulting in the formation of a 5-km-wide convergence band, implying 75% shortening.

It is unclear how inferred contraction is physically accommodated within a convergence band. *Sarid et al.* (2002) describe an internal fabric of numerous subtle parallel striations. Conceivably, these could be the traces of thrust faults. In the 15ESREGMAP01 example (Fig. 8b), 8 km of convergence is accommodated within a band that is only 2.7 km wide, indicating 66% shortening. Even with a perhaps un-

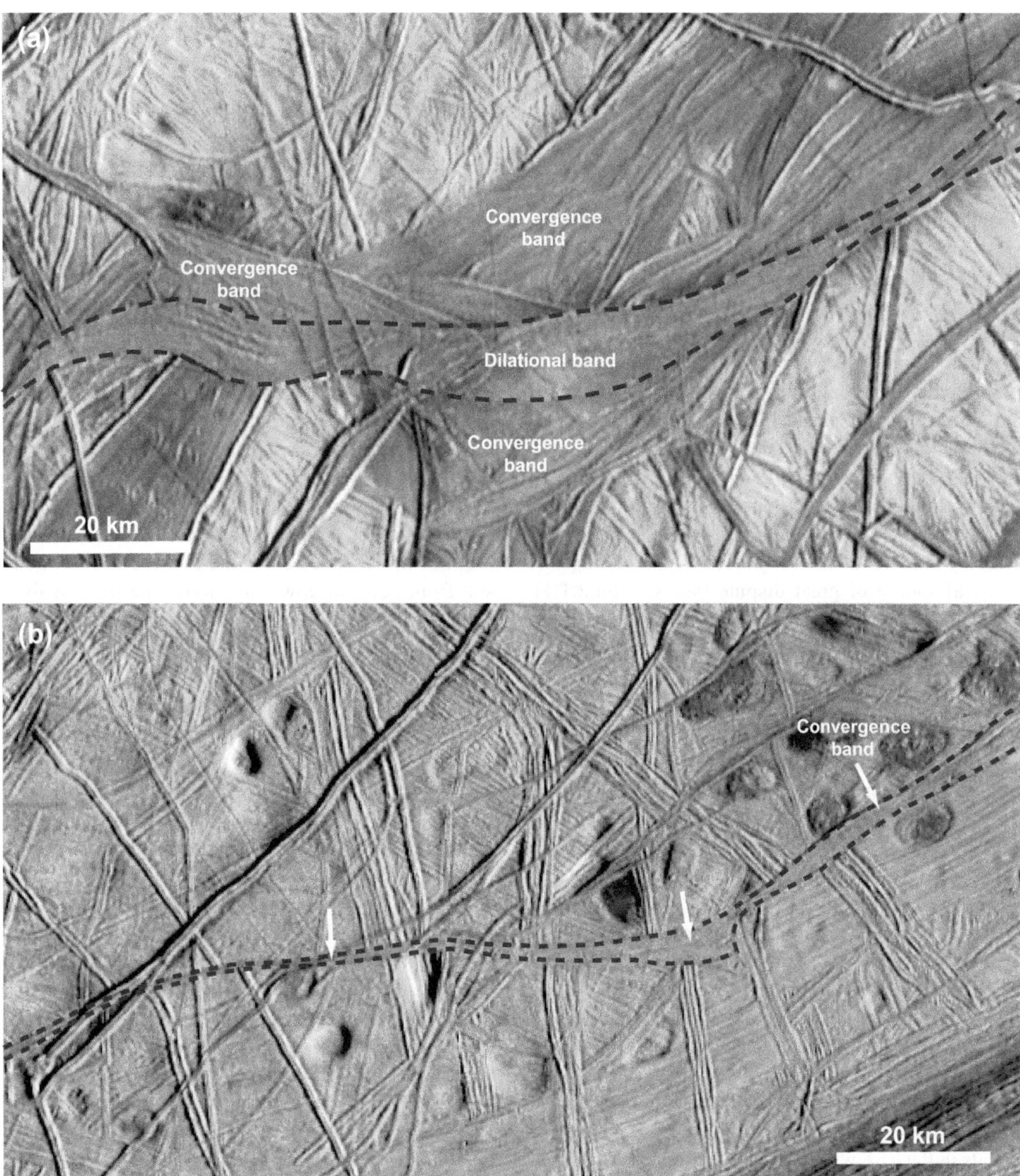

Fig. 8. Two sites of potential contraction in the icy shell, manifested as convergence bands. **(a)** Possible convergence bands where ~15 km of convergence may have occurred in the equatorial trailing hemisphere, near Castalia Macula. The inferred convergence bands resemble dilational bands but have irregular margins and do not have matching geology to either side of the band. This particular example is superimposed with a later dilational band that passes across the center of the convergence band (*Patterson et al.,* 2006). From Galileo mosaic 17ESREGMAP01. **(b)** An irregular convergence band (white arrowed zone between dashed lines) where ~8 km of crust appears to be missing, resulting in mismatches of older, crosscut features (from Galileo mosaic 15ESREGMAP01).

realistically small fault spacing of 100 m, each fault would need to accommodate 300 m of horizontal shortening, resulting in 170–300 m of vertical relief across each fault (depending on the fault dip, assumed to be in the range 30°–45°). These numbers would increase as the fault spacing increases. The convergence bands do not appear to display such rugged relief, perhaps casting doubt on thrust faulting as a mechanism for contraction. A related complicating factor is the relatively large amount of differential stress (up to 10 MPa by a depth of 3 km) required for thrust faults to develop on Europa (*Pappalardo and Davis,* 2007), which would hinder thrust fault development in the icy shell except at very shallow depths (perhaps less than a few hundred meters) due to the typical magnitudes of tidal stresses (unless some other source of compressive stress exists). Other possible contractional mechanisms include actual volume loss within convergence bands due to remobilization of warm ice toward deeper portions of the icy shell, or compaction through porosity reduction (e.g., *Aydin,* 2006).

The documented large values of contraction at convergence bands should make them easy to identify through disruption of the continuity of the ridged plains, except perhaps where convergence occurs within preexisting dilational bands and the convergence is less than the original dilation. Convergence band-like features might be common in Europa images but may have been previously misidentified as dilational bands; therefore, they are potentially important tectonic features for balancing Europa's surface area budget. Additional work will be needed to test this hypothesis.

2.2.3. Contraction across ridges. A third possibility for sites of convergence on Europa is the ubiquitous double ridges that dominate the tectonic fabric of the icy shell, first suggested by *Sullivan et al.* (1997). The origin of ridges is an equivocal source of great dispute (see section 2.1.1). Some ridges show evidence for lateral offsets along them but the majority of ridges do not. Nonetheless, recent models (*Nimmo and Gaidos,* 2002) hypothesize that ridge formation may be driven by lateral shearing and frictional heating along cracks. An analysis of ridges containing relatively offset features (i.e., where relatively older lineaments have undergone apparent lateral displacements along a ridge) reveals that the offsets cannot be related to lateral shearing alone (*Patterson and Pappalardo,* 2002; *McBee et al.,* 2003; *Vetter,* 2005; *Patterson et al.,* 2006; *Bader and Kattenhorn,* 2007; *Kattenhorn et al.,* 2007). Although pure dilation across any lineament (e.g., a dilational band) cannot produce lateral offsets of features split apart by the dilation, convergence across a lineament will result in the production of apparent lateral offsets of all nonorthogonal crosscut features along the lineament that have nothing to do with lateral shearing.

Aydin (2006) identified ridge-like contraction lineaments, which he referred to as compaction bands or compactive shear bands in the case where lateral shearing also occurred. As used by *Aydin* (2006), the word "band" has no association or morphological resemblance to either dilational bands or convergence bands described earlier, but rather originates from terrestrial analogs called deformation bands, across which compaction can occur through the comminution of granular materials. We therefore avoid use of "compaction band" to describe these features, which morphologically resemble ridges. The amount of apparent lateral offset along a ridge is controlled by the amount of convergence as well as the relative angle α measured clockwise from the ridge toward any offset feature (Fig. 9). It is possible to differentiate between apparent offsets caused by convergence and those caused by true lateral shearing by examining the distribution of normalized *separation* as a function of α, where separation is the orthogonal distance between a linear feature on one side of a ridge and the projection of that same feature on the other side of the ridge (i.e., the conventional structural geologic definition of separation). Separation is uniquely related to α and the amount of convergence relative to the amount of true lateral shear motion (*Vetter,* 2005; *Bader and Kattenhorn,* 2008).

Apparent lateral offsets along ridges related to convergent motions raise the possibility that ridge development may, in some cases, be partially driven by contraction across sheared lineaments. Such a notion is congruent with the *Nimmo and Gaidos* (2002) frictional shear heating model for ridge development. Such heating may result in a more mobile wedge of ice alongside a shearing lineament, causing a localized zone of weakness in the icy shell where contraction may be accommodated. In this model, part of the loss of volume along the sheared lineament occurs by remobilization of warm ductile ice into the deeper parts of the shell or by localized melting and downward draining within the frictionally heated zone. Part of the near-surface convergence, however, is manifested through the construction of ridge edifices along the crack. The ridges could result from buoyant upwelling above the frictionally heated zone, buckling, or porosity reduction in near-surface ices, assuming an initially high enough porosity to allow volumetric compaction. Thus, the possibility exists that this is a plausible mechanism for ridge development, particularly because it is likely that all active lineaments on Europa are subject to shearing and thus heating during the europan day in response to the constantly rotating diurnal tidal stresses (see section 2.3.4). Those ridges that have been shown to exhibit a component of contraction across them invariably show a larger amount of strike-slip motion accompanying the contraction (*Vetter,* 2005), providing a strong argument that the contraction is related to the process of shearing. The implication of this mechanism for ridge development is that the majority of the elusive contraction in Europa's icy shell may, in fact, be accommodated across its most common type of feature (i.e., ridges), with only a minimal amount of convergence needed across any one ridge to allow for a balanced surface area budget. Unfortunately, other than a few prominent ridge examples, this hypothesis has not been convincingly demonstrated, plausibly because apparent offsets along ridges are below existing resolution limits, perhaps due to minimal amounts of convergence and/or lateral offsets. Therefore, any inferences about ridges as important accommodators of convergence remain mostly conceptual for the time being. An added complication is that many sheared ridges have been shown to exhibit evidence of dilation using the same technique (α vs. separation graphs) described above (*Bader and Kattenhorn,* 2008), indicating that ridges may be subject to a combination of dilational, convergent, and shearing motions during their history.

2.3. Lateral Shearing

2.3.1. Shear failure of ice. Lateral shearing refers to strike-slip motions along lineaments. Voyager images indi-

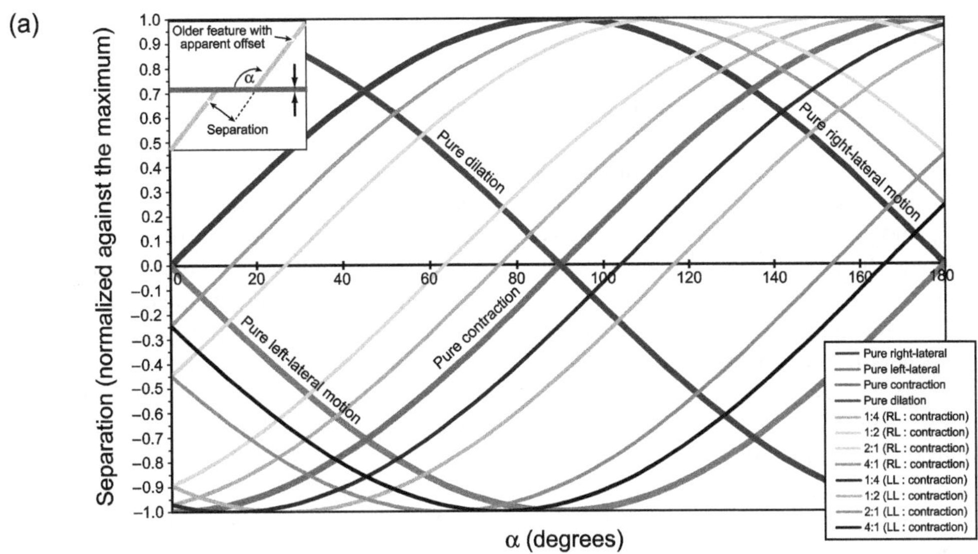

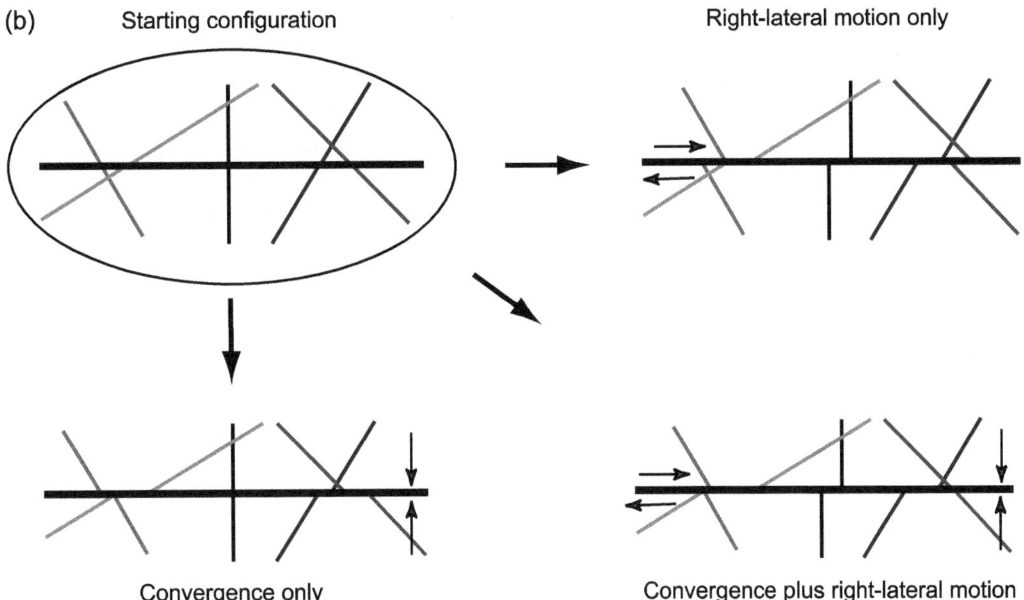

Fig. 9. See Plate 12. **(a)** Analytical curves can be used to differentiate the relative amounts of strike-slip motion (whether right-lateral or left-lateral) and contraction across a ridge/strike-slip fault based on the amount of separation between two halves of an apparently offset feature (inset, upper left). α is the angle measured from the ridge/strike-slip fault to the offset feature in a clockwise sense. A separate set of curves exists for the case of dilation plus strike-slip motion (*Vetter,* 2005). **(b)** Illustration of various configurations of mismatches of crosscut features in response to pure right-lateral motion, convergence only, or a combination (motion is along the thicker black line, representing a ridge or strike-slip fault). If α is 90°, offsets only occur if there is lateral motion. Pure convergence can produce a mixture of left- and right-lateral offsets, depending on α.

cated the presence of lateral offsets in the europan icy shell at scales of tens of kilometers (*Schenk and McKinnon,* 1989). On Earth, such motions may occur for two reasons: (1) through primary shear failure of rock in a stress field where the intermediate compressive principal stress (σ_2) is vertical and the horizontal differential stress ($\sigma_1 - \sigma_3$) exceeds the rock frictional strength; and (2) through reactivation of existing structures (faults and fractures) in response to a temporal change in the stress field relative to when the existing structures first formed. In the first case, approximately vertical strike-slip faults are produced on Earth and include transform plate boundary faults (such as the San Andreas fault) as well as intraplate faults. In the second case, any of the principal stresses could be vertical, resulting in oblique-slip motion in cases where the faults are not vertical.

On Europa, the second of these situations most certainly exists. The stress field changes constantly due to variations in the tidal figure during each orbit (see section 4.1); therefore, any active lineament on Europa necessarily experiences shear stresses that could induce lateral shearing. If a crack is dilated during shearing (i.e., the normal stress is tensile), there is no frictional constraint on shear motion. For the case of sliding of a closed crack, the limiting factor is the coefficient of friction, μ, based on the Coulomb failure criterion: $\tau = S_0 + \mu\sigma_n$ at failure, where τ is shear stress, σ_n is normal stress, and S_0 is the inherent shear strength (cohesion) in the absence of normal stress.

The value of μ for europan ice is unknown but is almost certainly less than occurs in rocky materials and is commonly approximated on the basis of comparisons to terrestrial observations and laboratory experiments (*Beeman et al.,* 1988; *Rist and Murrell,* 1994; *Rist,* 1997; *Schulson,* 2009). Complications exist in that μ is likely affected by ice chemistry, sliding speed, and temperature, with the friction increasing as temperature decreases. Based on low-temperature (77–115 K) experiments on sawcut pure water ice, *Beeman et al.* (1988) provide perhaps the most appropriate estimate of μ applicable to europan conditions in the upper part of the icy shell, but even so, μ is variable depending on loading conditions. At low confining pressures ($P \leq 5$ MPa; or down to a depth of ~4 km on Europa), $\mu = 0.55$ and the inherent shear strength, $S_0 = 1$ MPa. Where $P \geq 10$ MPa (deeper than ~8 km), $\mu = 0.2$ and the inherent shear strength, $S_0 = 8.3$ MPa. Potential caveats to the laboratory technique include the 2–5 orders-of-magnitude-faster sliding velocities used than may occur on europan faults, which underestimates the friction coefficient (e.g., *Kennedy et al.,* 2000; *Fortt and Schulson,* 2004); a time-dependency to crack strength, increasing with the number of slip events or due to development of fault gouge; the usage of pure water ice; the possibility that frictionally heated ice along a shearing crack may experience a decreased friction coefficient (as has been suggested to occur on Enceladus) (*Smith-Konter and Pappalardo,* 2008); and the assumption of a preexisting fault.

The assumption of a preexisting fault is only problematic in attempting to make predictions about the orientations of primary shear fractures relative to the principal stresses. In laboratory experiments on water ice, this angle is commonly reported to be ~45° (*Durham et al.,* 1983; *Sammonds et al.,* 1989; *Rist and Murrell,* 1994; *Rist et al.,* 1994; *Rist,* 1997). Considering that the angle, θ, between a developing shear fracture and the maximum compressive principal stress, σ_1, is given by $\theta = 0.5 \tan^{-1} (1/\mu)$, an angle of 45° implies $\mu = 0$, which disagrees with deduced values of μ described above. *Schulson et al.* (1999) suggest that the 45° shear fracture angles were an artifact of the loading apparatus. They produced shear fractures in ice where $\theta = 30°$ (i.e., $\mu \approx 0.6$), more congruous with the low confining stress results of *Beeman et al.* (1988). If low sliding velocities (10^{-6}–10^{-7} m s^{-1}) (*Nimmo and Gaidos,* 2002) occur on Europa, μ may actually be in the range 0.6–0.8 (*Kennedy et al.,* 2000), creating shear fractures at $\theta = 25°$–$30°$ to the orientation of σ_1, unless the ice is sufficiently warm (e.g., due to shear heating) that μ is reduced.

2.3.2. Lateral shear failure on Europa. Normal faults have already been shown to be a common form of extensional deformation (see section 2.1.5), therefore lateral shearing would tend to produce oblique-slip motions along them. *Kattenhorn* (2002) describes evidence of *en echelon* breakdown zones at the tips of normal faults within dilational bands, implying oblique motions during fault tip propagation. Nonetheless, vertical fractures predominate on Europa due to tensile failure perpendicular to the horizontal maximum extension direction. As a result, troughs, ridges, and dilational band margins are all likely to be vertical and prone to reactivation through lateral shearing as long as the fractures have not yet healed. Strike-slip motions along these features thus also fall into category 2 faults described in section 2.3.1. Whether or not category 1 strike-slip faults exist on Europa (i.e., due to primary shear failure) is still unclear, especially seeing as they may be impossible to decipher from the myriad other vertical cracks that underwent lateral shearing due to reactivation. Nonetheless, primary shear failure is ultimately dependent on whether or not the stress conditions favor strike-slip fault formation, exactly as it would on Earth. It should also be noted that lateral shearing has been noted in terrestrial ice shelfs (*Wilson,* 1960) and glaciers (e.g., during the rupture of the Denali fault in Alaska in 2002), but typically takes on the form of *en echelon* fracture arrays, with individual fractures oriented obliquely to the trend of the fault zone (*Haeussler et al.,* 2004). Very few analogous *en echelon* crack geometries have been described in europan strike-slip fault zones (e.g., *Prockter et al.,* 2000; *Michalski and Greeley,* 2002), and appear to be more commonplace within shear-reactivated dilational bands, perhaps implying that strike-slip motions typically reactivate existing cracks that initially formed in tension (cf. *Greenberg et al.,* 1998; *Tufts et al.,* 1999). In some instances, shearing across regions of closely spaced tension cracks may result in fragmentation along the developing shear zone (*Aydin,* 2006).

Any evaluation of the formation mechanisms for strike-slip faults on Europa must be placed within the context of the mechanics of shear failure of ice and must incorporate a candid analysis of the pros and cons of shear failure vs. tensile failure interpretations for europan lineaments showing apparent offsets (e.g., *Kattenhorn,* 2004b). For example, *Spaun et al.* (2003) suggest that northeast- and northwest-oriented ridges with strike-slip offsets in the equatorial trailing hemisphere formed as primary shear fractures because they form X-shaped patterns that superficially resemble conjugate shear sets, and because of their locations relative to expected nonsynchronous rotation stresses. Although the X-shapes are reminiscent of conjugate shear fractures, shear offsets are either absent or inconsistent along ridges of a particular orientation and no explicit crosscutting relationships exist, probably implying that the ridges are distinctly different in age. Also, potentially conjugate ridge sets (i.e., X-shapes for ridges of similar stratigraphic age) show no consistent conjugate angle, 2θ, between them. This would not be expected for true conjugate sets as 2θ is explicitly controlled by the coefficient of friction of the ice, μ (see section 2.3.1), which should be somewhat consistent. Hence, there is currently no convincing geological evidence that primary shear failure of the icy shell, analogous to strike-slip faulting, produced global lineaments. Nonetheless, given that extensional shear failure is likely responsible for normal faults in the icy shell, the possibility that some lineaments formed through strike-slip shear fracturing cannot be dismissed.

2.3.3. Strike-slip fault morphologies. Lateral offsets can potentially occur along all lineament types (i.e., troughs, ridges, cycloids, and dilational and convergent bands), regardless of how they initially formed. Nonetheless, cumulative offsets that are large enough to be resolvable in Galileo images (generally hundreds of meters or more) typically occur along ridges and dilational bands. Accordingly, *Kattenhorn* (2004a) distinguishes two predominant types of strike-slip faults on Europa: ridge-like and band-like, based on their morphologic similarity to ridges and dilational bands (Fig. 10). The mechanical evolution of the two types of strike-slip faults is distinctly different. Ridges likely initiated as troughs (see section 2.1.1); therefore, any lateral shear motions along them probably happened later, potentially contributing to the process of ridge development. In the case of dilational bands (see section 2.1.3), the question arises as to whether strike-slip offsets along them happen prior, during, or after dilation. Strike-slip offsets across dilational bands are evidenced by the need for oblique closing to reconstruct older features affected by the dilation. Sigmoidal lineations within dilational band material may imply oblique dilation (i.e., concurrent dilation and strike-slip motion), as occurs along Astypalaea Linea; however, the timing of strike-slip motions along dilational bands cannot necessarily be determined based on band morphology alone (*Prockter et al.,* 2002).

Kattenhorn (2004a) and *Kattenhorn and Marshall* (2006) suggest that secondary cracks at fault tips, called tailcracks, provide insights into the mechanics of ridge-like vs. band-like strike-slip fault development. Tailcracks are secondary tension fractures commonly observed at the tips of strike-slip faults on Earth, and which have also been documented on Europa (see sections 2.1.4 and 4.2). The intersection of the fault and its associated tailcrack is manifested by a sharp kink, with an angle described by linear elastic fracture mechanics theory as being controlled by the ratio of shear stress to normal stress at the instant of tailcrack development. Hence, it is possible to determine whether or not a fault is dilating at the instant it is shearing laterally based on the geometry of its tailcracks. Using this line of reasoning, *Kat-*

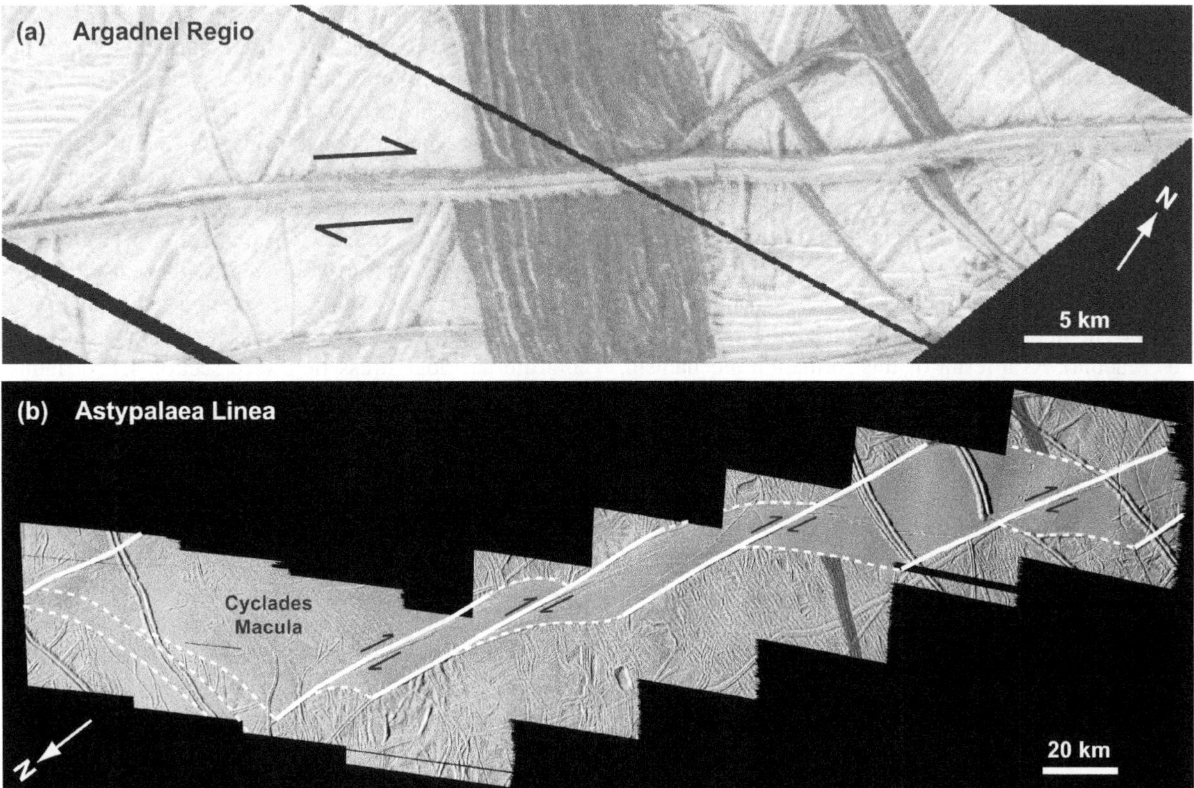

Fig. 10. Examples of strike-slip faults. **(a)** Unnamed right-lateral, ridge-like strike-slip fault cutting Yelland Linea in Argadnel Regio, from Galileo mosaic 12ESWEDGE_03. **(b)** A portion of the band-like strike-slip fault Astypalaea Linea in the southern antijovian region. White solid lines are fault segments that slipped right-laterally, producing dilational pull-aparts along linking cracks (white dashed lines). A ridge that was offset by 77 km (gray) illustrates the fault kinematics. From Galileo mosaic 17ESSTRSLP02.

tenhorn (2004a) showed that ridge-like faults (Fig. 10a) commonly undergo lateral shearing while slightly dilated or frictionally closed (see also *Aydin,* 2006), whereas the band-like faults they examined (Fig. 10b) accrue strike-slip offsets concomitant with significant dilation. Hence, many band-like strike-slip faults are essentially transtensional, oblique spreading zones. They exhibit typical lateral offsets of tens of kilometers (e.g., ~77 km along Astypalaea Linea) (*Kattenhorn,* 2004a) and lengths of up to ~1500 km (e.g., Agenor Linea) (*Prockter et al.,* 2000).

Despite these interpretations, band-like strike-slip faults should not be considered to be simply types of dilational bands, as some strike-slip motions almost certainly occur after dilation has ceased and may even be associated with some amount of later contraction. For example, the bright bands originally described from Voyager imagery, such as Agenor Linea (Fig. 7b) (*Schenk and McKinnon,* 1989), may reflect geologically recent reactivation of older dilational bands to create transpressional strike-slip faults. The interior structure of Agenor Linea is convoluted (unlike simple dilational bands) with some portions being reminiscent of transpressional duplexing along terrestrial fault zones (*Prockter et al.,* 2000). This duplexing probably occurred during shearing of the original band, with restraining bends developing because of the irregular geometry of the band margins. Tectonic annihilation of the margins of Agenor Linea during shearing may have resulted in ridged plains that cannot be matched from one side of the fault zone to the other, unlike dilational bands.

We infer that some band-like strike-slip faults are essentially ridge-like/band-like hybrids, in that the band portions of the fault formed as dilational pull-aparts along segmented ridge-like faults, where the sense of step between adjacent fault segments was the same as the sense of slip along them. A prominent example is the right-lateral fault Astypalaea Linea in the south-polar region (*Tufts et al.,* 1999; *Kattenhorn,* 2004a), where rhomboidal pull-apart bands formed between right-stepping ridge-like fault segments (e.g., Cyclades Macula). As a result of this formation mechanism, the interior striations of these bands (which trend perpendicular to the inferred spreading direction) are highly oblique to the overall trend of Astypalaea Linea.

As with distinctions between dilational bands and band-like faults, ridge-like strike-slip faults should not be considered to be simply types of ridges. Crosscut features that can be matched to either side of ridges (producing so-called piercing points that used to be together) typically show zero lateral offsets, indicating that measurable cumulative lateral offsets are not a requirement for ridge development. Nonetheless, some ridges do show appreciable lateral offsets, implying a long-term process of shearing and strike-slip accumulation, whereas troughs never show lateral offsets (*Hoppa et al.,* 1999c), suggesting that lateral motions may be an important aspect of the ridge building process. Ridge-like fault offsets are typically in the range of hundreds of meters to several kilometers (*Hoppa et al.,* 2000) but have been measured as high as 83 km (*Sarid et al.,* 2002). An important observation is that the lateral offset along a ridge-like fault does not scale with the length of the ridge, as would be true of terrestrial strike-slip faults and theoretical predictions for shear fractures based on linear elastic fracture mechanics (*Pollard and Segall,* 1987). For example, Agave Linea, which passes north of Conamara Chaos (Fig. 2a), is at least 2000 km long, yet shows lateral offsets of ~5 km, implying that it did not initiate or grow as a primary shear fracture but rather accrued a minor amount of lateral offset during subsequent movement. Therefore, models that favor ridge formation through primary shear failure of the icy shell do not seem justified in these cases. Nonetheless, frictional shearing may very well contribute to the ridge developmental process (*Nimmo and Gaidos,* 2002) (see section 2.1.1). Accordingly, tailcrack geometries along ridge-like faults confirm that frictional sliding of closed cracks commonly did occur to form these features (*Kattenhorn,* 2004a). The caveat to this model is that not all ridges show lateral offsets, implying two possible scenarios: (1) repeated forward and backward frictional shear motion along cracks during the diurnal cycle resulted in zero or unresolvable cumulative strike-slip offset but still produced enough total heat to drive ridge construction; or (2) frictional shearing is just one of perhaps several contributing factors to ridge development (see section 2.1.1), such that the absence of strike-slip motions does not preclude ridge formation. Neither scenario can be proven; however, the creation of apparent lateral offsets through a component of convergence, if present (see section 2.2.3), implies other processes may also contribute to ridge development. The upshot is that strike-slip motions are relatively common on Europa and that the lateral motions commonly utilize ridges and dilational bands, which must therefore act as planar weaknesses in the icy shell along which lateral shearing is accommodated.

2.3.4. Tidal walking. True strike-slip motions along faults are suggested to be driven by diurnal tidal stresses through a process informally referred to as tidal walking (*Hoppa et al.,* 1999c). In response to the constantly changing diurnal tidal stress field (see section 4.1.5), faults repeatedly experience tension then compression out of phase with left- and right-lateral shear stresses, because of the manner in which the tidal stress tensor resolves normal and shear stresses onto the fault surfaces. As with any fault or fracture, certain failure criteria rules apply. Where tension opens a crack, there would be no frictional resistance to shear motion; therefore, concomitant shearing should produce lateral offset. If a crack is closed, shear motion is limited by the frictional strength; therefore, lateral offsets can only be produced if the shear stress exceeds the normal stress multiplied by the coefficient of static friction, μ (i.e., the Coulomb failure criterion). Hence, the tidal walking theory implies that when the stress normal to a fault is tensile, the fault opens at the surface, allowing shear stress to produce a small amount of offset. The net sense of shear during this dilational phase, which depends on the extent to which the normal and shear stress curves are out of phase with each other (*Hoppa et al.,* 1999c; *Groenleer and Kattenhorn,* 2008), controls the sense of strike-slip offset. As

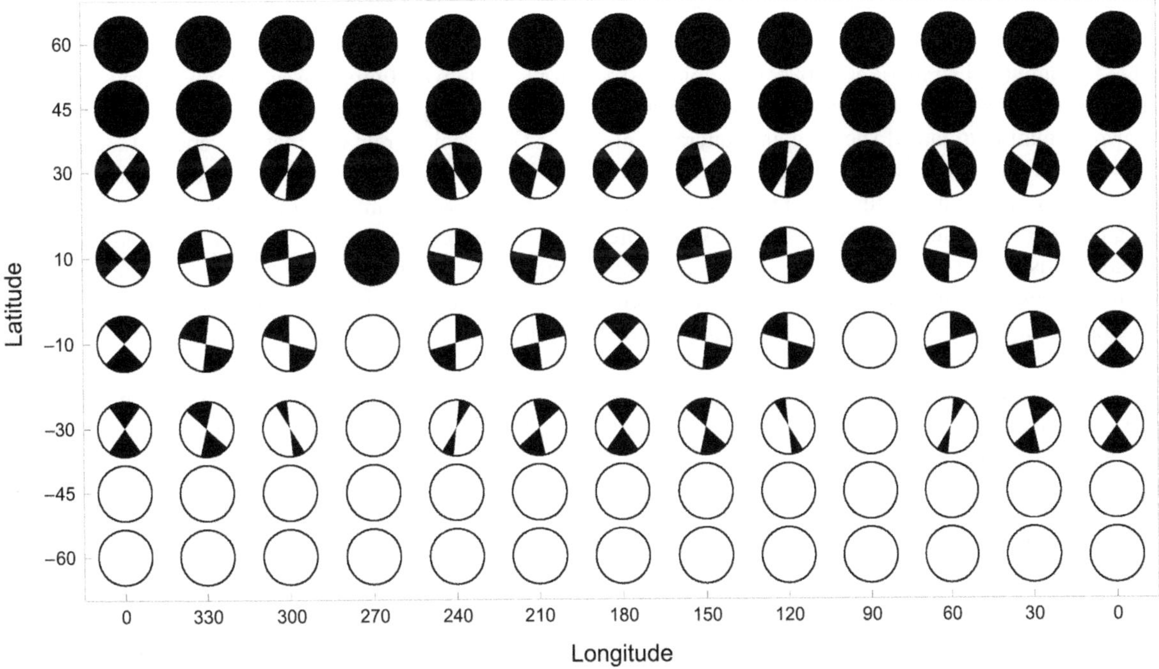

Fig. 11. Global pattern of strike-slip motion sense of faults of all orientations as a result of diurnal tidal stresses and the process of tidal walking. Fault orientations are indicated by the circular rose diagrams where the top of each circle is the 0° or north-south orientation. There is a prevalence of left-lateral motions (black) in the northern hemisphere and right-lateral motion (white) in the southern hemisphere.

the normal stress acting across the fault changes from tension to compression about half an orbit later, the fault surfaces must be in frictional contact. The friction along the fault then limits the ability of a changing sense of shear stress to completely remove the recently accrued offset. Consequently, after each diurnal cycle, the fault accumulates a small net sense of strike-slip offset along its length. Over many successive cycles of this process, the accumulation of a multitude of small strike-slip offsets produces a visible amount of strike-slip displacement.

The sense (left- or right-lateral) of strike-slip displacement along a fault depends on the orientation and location of the fault (Fig. 11). Poleward of 45° latitude, regardless of fault orientation, tidal walking must result in only left-lateral faults in the northern hemisphere and right-lateral faults in the southern hemisphere. A mixture of left- and right-lateral displacements is predicted in the midlatitudes, depending on fault orientation. Observations of many strike-slip fault offsets appear to support this theoretical expectation (*Hoppa et al.,* 2000). In the leading and trailing hemisphere Galileo regional mapping observations, *Sarid et al.* (2002) observed left-lateral faults in the high northern latitudes and right-lateral faults in the high southern latitudes; however, the transition to orientation-controlled offset sense was not observed to occur at the predicted 45° latitude. They posited that a small amount of polar wander may be responsible for the mismatch between tidal walking theory and their strike-slip fault observations.

Tidal walking provides a simple mechanism for driving strike-slip motions on Europa, with a similar phenomenon now suggested to be driving fault activity in the south polar region of Enceladus (*Hurford et al.,* 2007b; *Smith-Konter and Pappalardo,* 2008), possibly responsible for plume eruptions. Nonetheless, while band-like faults show consistent agreement with the predictions of tidal walking, there are exceptions for ridge-like faults (such as left-lateral ridge-like faults in the south-polar region) (*Kattenhorn,* 2004a). A potential caveat is that the theory does not incorporate the longer-term effects of nonsynchronous rotation stresses on strike-slip motions, which are expected to be important. For example, *Schulson* (2002) uses a frictional sliding calculation to infer that a significant amount of nonsynchronous rotation stress buildup would be needed to drive motion along a strike-slip fault with sufficient force to allow a developing tailcrack to penetrate down to a depth of 1 km into the icy shell. Considering that some tailcracks have developed into dilational bands, and thus must have penetrated sufficiently deep to reach viscoplastically deformable ice, nonsynchronous rotation stresses may indeed have been an important contributor to this process. A final caveat is that the tidal walking theory is currently conceptual, as it does not incorporate a quantitative test of shear motions using frictional failure criteria. In fact, the recent work involving tidal stresses acting on faults on Enceladus has produced a conceptual model for slip along faults that is based on the mechanics of strike-slip faults on Earth (*Smith-Konter and Pappalardo,* 2008). This newer model of tidally driven slip along faults might still be considered a form of tidal walking but its predictions of the sense of slip may be different in detail from the *Hoppa et al.* (1999c) model shown in Fig. 11.

2.3.5. Lateral shearing and cycloid growth. Although cycloids are probably tension fractures (see section 2.1.2), lateral shearing is likely to play an important part in their developmental mechanics. *Marshall and Kattenhorn* (2005) highlighted geometrical similarities between cycloid cusps and tailcracks and developed a conceptual model that ascribed cusp formation to lateral shearing and tailcrack initiation in the tip region of a recently formed cycloid arc in response to the rotating diurnal stresses. The tailcrack then continues to grow in tension, forming the next cycloid arc in the chain. This model was bolstered by the demonstration of an excellent match between cusp angles and theoretical tailcrack angles predicted using linear elastic fracture mechanics equations and calculated tidal stress fields (*Groenleer and Kattenhorn,* 2008). The point in the orbit at which cusp development and new arc initiation occurs is therefore controlled by the interplay between resolved shear stress and normal stress in the tip region of the previously formed arc, not the timing of the maximum tensile principal stress during the orbit. This point in the orbit changes for each cycloid arc, depending on the orientation of the end region of the most recently formed arc.

3. LATERAL MOTIONS WITHOUT GLOBAL PLATE TECTONICS

Plate tectonics, in the terrestrial sense, implies that the entire brittle outer layer (lithosphere) is broken into numerous discrete fragments (plates) that move across the surface relative to each other in response to a combination of three types of motions along their boundaries: divergent, convergent, and transform. Europa contains a highly fractured outer layer (cryosphere), raising the issue of whether it too is broken into discrete plates with discernable boundaries. The existence of dilational bands on Europa (see section 2.1.3) indicates that new surface area was created through the process of complete separation of the icy shell and concurrent filling of the resultant lithospheric gap by material from below, analogous to mid-ocean ridge spreading centers. However, these tabular dilational gashes in the icy shell tend to vary in width along their lengths and may have discrete ends, or tips, indicating that they do not facilitate rigid motions of the two opposing sides away from the spreading site as typifies tectonic plates on Earth. Terrestrial plate tectonics has created a balance between surface area creation at mid-ocean ridges and surface area removal at subduction zones. In contrast, there are no known subduction zones on Europa (*Sullivan et al.,* 1998; *Prockter et al.,* 2002) and no obvious global pattern of tectonic plate boundaries that localize deformation. Rather, brittle deformation of the icy shell is globally pervasive, although the intensity of deformation may be localized (e.g., at dilational bands and major strike-slip faults). The locations of tensile fracturing or lateral shearing at any point in time is likely driven by the characteristics of the global tidal stress field, which is geographically variable (see section 4). However, any long-term reorientation of the icy shell, such as by nonsynchronous rotation (section 4.1.6) or polar wander (section 4.1.7), could move all locations on the moon's surface into optimal stress zones for brittle deformation at some point, allowing for the global pervasiveness of brittle fracturing. The upshot is that there is no indication of a terrestrial-like, global plate tectonic system on Europa. Accordingly, there is no known (or likely) driving mechanism for global plate tectonics either.

The problem remains that the repeated creation of tens-of-kilometers-wide dilational gashes in the icy shell must be somehow accommodated without global plate tectonics. Localized contractional features apparently do exist on Europa to accommodate some of the spreading occurring at dilational bands (see section 2.2), including convergence bands, ridges, and to a much lesser extent, folds. It is also possible that much of the opening across a band is accommodated elastically in the icy shell at timescales less than the Maxwell relaxation time. In a linear elastic half-space, a tension fracture causes significant elastic deformation within the material within which it is embedded out to a distance that scales with the smallest dimension of the crack (e.g., *Pollard and Segall,* 1987). On Europa, this dimension is likely to be the fracture depth for the case of fractures that do not fully penetrate the icy shell. For a very long tension fracture or dilational band, elastic deformation will mostly dissipate within about three crack depths perpendicularly away from the structure, which should be in the range of a few kilometers to a few tens of kilometers. If the crack fully penetrates the shell (a situation that has been proposed but never confirmed on Europa), elastic deformations will extend away from the crack out to a distance that scales with the crack length, as long as this length exceeds the shell thickness (also see *Sandwell et al.,* 2004). Hence, shell-penetrating cracks that extend hundreds of kilometers across the surface would cause elastic deformations in the adjacent icy shell out to distances of hundreds of kilometers. The point here is that dilations could conceivably be accommodated by elastic contractional strains alongside a dilating crack without the need for visible contractional deformation. Because dilational bands maintain a permanent dilation, the elastic deformation of the icy shell can only be dissipated by long-term viscous relaxation within the icy shell. Considering that the relaxation time for the cold outer brittle portion of the icy shell (~80 G.y.) (*Nimmo,* 2004a) exceeds the age of the solar system, ongoing opening of a dilational band should ultimately cause stored elastic energy within any elastically perturbed zone around the dilational band to be released through brittle deformation, such as shear failure (e.g., strike-slip faulting) or convergence along existing structures. Hence, some motions along structures may be driven by locally perturbed stresses alongside dilating bands rather than global tidal stresses.

Terrestrial strike-slip faults can be isolated or transcurrent intraplate faults or transform plate boundary faults. Isolated faults have distinct tips at the end of the fault trace, toward which the lateral offset decreases to zero from some maximum along the fault trace (commonly near the geometric center of the fault). In contrast, transcurrent and transform faults may be unconstrained by fault tips as a result of the presence of deformation belts, spreading centers, or subduction zones at either tip, allowing the fault to behave analogous to a rigid-body translation. In these cases, the total offset along the fault may be approximately constant along the fault length. In the absence of global plate tectonics on Europa, one might expect only nontransform faults; however, palinspastic reconstruction of fault offsets can commonly be achieved via a rigid-body technique (e.g., Astypalaea Linea) (*Tufts et al.,* 1999; *Kattenhorn,* 2004a). This is possible because even in the absence of plate tectonics, fault-tip motions can potentially be accommodated by dilational bands where extension is needed and by convergence bands where contraction is needed (*Sarid et al.,* 2002; *Schulson,* 2002; *Kattenhorn and Marshall,* 2006). Nonetheless, it is not always possible to match up features across a palinspastically restored strike-slip fault using rigid translations because of variability in the amount of strike-slip displacement associated with elastic displacement gradients (e.g., *Vetter,* 2005). This behavior may suggest that isolated strike-slip faults exist on Europa with slip gradients along their lengths and should not be treated as rigid-body transforms in tectonic reconstructions.

Although the preceding discussion highlights the lack of global plate tectonics on Europa, rigid translations and rotations appear to have occurred locally along circumferentially detached portions of the icy shell (i.e., areas that are completely surrounded by prominent, active cracks with variable motion behaviors, analogous to plate boundary processes). These localized plate-like fragments of the icy shell have been referred to as microplates (*Schenk and McKinnon,* 1989; *Rothery,* 1992; *Sullivan et al.,* 1998; *Sarid et al.,* 2002) and sometimes form triple junctions where three microplates meet (*Patterson and Head,* 2007). The existence of these mostly intact portions of the icy shell that undergo lateral shifts or rotations justifies local usage of plate tectonic analysis techniques, similar to terrestrial analogs, despite the lack of a global plate tectonic system. Accordingly, *Patterson et al.* (2006) examined fault- and dilational band-bounded microplates, tens to hundreds of kilometers wide, in the equatorial region near Castalia Macula, northwest of Argadnel Regio, and were able to determine Euler poles of rotation. Nonetheless, the nonuniqueness of derived Euler poles using several adjacent microplates and a deduced deformation sequence implies nonrigid deformation within the microplates. In the absence of a known driving mechanism for any type of large-scale plate motion on Europa, small rotations of inferred microplates are ostensibly driven by differential motions induced by local tectonic activity along the microplate boundary (such as the opening of a dilational band), or in adjacent deforming regions. This process may be particularly common where the icy shell is heavily dissected by multiply oriented dilational bands (such as in Argadnel Regio) because dilational bands potentially form zones of weakness within the icy shell across which later deformation may be localized.

4. CAUSES OF TECTONIC DEFORMATION

Fractures on Europa are produced in response to stress in the icy shell. As described in previous sections of this chapter, a diverse range of tectonic features has been produced in response to these stresses, causing initial brittle failure of the ice and subsequent modification of these structures over time. Thus, the tectonic record preserves a record of the stresses experienced by the icy shell and holds the key to identifying the processes that imparted those stresses. We now highlight the processes that can create the stresses responsible for fractures in the europan shell at global, regional, and local scales (also see chapters by Nimmo and Manga, Sotin et al., and Bills et al.).

4.1. Global-Scale Stress

Some global-scale ridges span more than 50% of the circumference of Europa, so the stress conditions needed to form the fractures along which these ridges formed must have been global in scale. We therefore examine the range of plausible factors that may induce global-scale stress fields, which, at their essence, require a global change in shape of the satellite.

4.1.1. Thin shell approximation. Using Mercury as an example, *Melosh* (1977) showed that a global change in shape caused by a change in rotation rate produces stress that could drive the formation of tectonic features. In its simplest form, the model approximates the surface stress as occurring in a thin elastic shell that is decoupled from a fluid interior as it deforms. The horizontal strain in the shell induced by the tidally distorted interior results in stress on the surface, given by

$$\sigma_{\theta\theta} = -\frac{1}{3}\mu f\left(\frac{1+\nu}{5+\nu}\right)(5 + 3\cos 2\theta) \qquad (1)$$

and

$$\sigma_{\phi\phi} = \frac{1}{3}\mu f\left(\frac{1+\nu}{5+\nu}\right)(1 - 9\cos 2\theta) \qquad (2)$$

The quantity $\sigma_{\theta\theta}$ is the stress along the surface in the direction radial to the axis of symmetry, while $\sigma_{\varphi\varphi}$ is the stress along the surface in a direction orthogonal to the $\sigma_{\theta\theta}$ stress. Also, θ is the angular distance to any point on the surface measured with respect to the tidal distortion's axis of symmetry, f is the flattening, μ is the rigidity (shear modulus) of the shell, and ν is Poisson's ratio. Compressive stresses are defined here to be positive and tensional stresses negative. Flattening, f, is defined as $f = (r_{sym} - r_{orth})/r$ where r is the mean radius of the body, r_{sym} is the radius along the axis of deformation symmetry, and r_{orth} is the radius along an orthogonal axis. The flattening is positive when $r_{sym} > r_{orth}$, otherwise it is negative.

The thin shell approximation of surface stress is also applicable to Europa, where conservative estimates of tidal heating predict a H_2O layer that might be liquid underneath an icy shell (*Peale and Cassen,* 1978; *Moore,* 2006). Moreover, magnetometer measurements from Galileo strongly imply the existence of a liquid layer (*Kivelson et al.,* 2000).

A liquid ocean would decouple the icy shell from the interior of Europa. Although the thickness of the icy shell is not known, a range of techniques have constrained it to be probably less than 30 km (*Billings and Kattenhorn,* 2005, and references therein; see chapter by McKinnon et al.). In the absence of a good physical constraint on the thickness of the icy shell, the assumption that it behaves as a thin elastic layer is assumed to be valid. If this icy shell is on the order of 10 km thick, the stresses on its surface may still approximate a thin elastic shell (*Hurford,* 2005). A thicker icy shell is expected to have its upper portion acting elastically (*Williams and Greeley,* 1998; *Billings and Kattenhorn,* 2005; *Hurford et al.,* 2005) with deeper, warmer ice behaving viscously at low strain rates (*Rudolph and Manga,* 2009). Even so, the entire icy shell may still behave in a manner analogous to a thin elastic layer if deformed rapidly, such as at the timescale of diurnal stressing (*Wahr et al.,* 2009).

Although we describe the results of a thin shell approximation for the tidal stress, an alternative model that can account for thicker icy shells has been recently developed (*Wahr et al.,* 2009). Models of tidal stress for arbitrary icy shell thicknesses are based on standard techniques used to calculate tides on Earth (*Dahlen,* 1976). These models define the stress in terms of the material properties of the satellite at it surface and a description of the satellite's tidal deformation, which is given by tidal Love numbers h_2 and l_2. The models also assume that the satellite is composed of distinct layers, each defined by its own material properties: density ρ, viscosity η, rigidity μ, and compressibility λ. Measurements of the moment of inertia allow some constraints on the distribution of mass within the body (*Anderson et al.,* 1997). However, the structure of Europa's interior is poorly constrained.

4.1.2. Despinning. After the Voyager flybys, the pattern of global ridges on Europa (Fig. 1) was described as consisting of radial and concentric fractures resembling tension cracks near the sub- and antijovian points, while elsewhere on the surface, 60° intersections of fractures were interpreted as conjugate shear fractures (*Helfenstein and Parmentier,* 1980). This pattern suggested that cracks were produced by a change in shape of Europa by a tidal process and not by a change in shape controlled by the rotation of Europa. Early analysis favored a change in shape induced by a tidal response driven by Europa's changing distance from Jupiter because of its finite eccentricity (*Helfenstein and Parmentier,* 1980). In Europa's early history, it is expected that the entire satellite spun nonsynchronously; however, tidal torques would have changed its spin rate, despinning it on a timescale much shorter than the age of the solar system such that the satellite's solid interior now rotates synchronously (*Peale,* 1977; *Squyres and Croft,* 1986), although the decoupled icy shell may rotate nonsynchronously (see section 4.1.6). Nonetheless, even after reaching a synchronously rotating state, the spin rate will continue to change as Europa's orbit evolves. Outward orbital migration (*Yoder,* 1979) increases the semimajor axis length, forcing Europa's spin rate to decrease as tidal torques further slow its rotation rate in order to maintain a synchronous rotation state.

A spinning satellite in hydrostatic equilibrium deforms into a shape in which the force of gravity balances the centripetal force produced by its rotation. This deformation produces an oblate spheroid whose radius is depressed along an axis defined by the pole of rotation and enhanced along any orthogonal axis, extending through the equator. If the spin rate changes, then the oblateness of the spheroid will change in response and produce stress on the surface. Despinning reduces its oblateness ($\Delta f > 0$), producing a rebound of the radius along the axis of symmetry (through the rotational poles) and all-around tensile stress (i.e. $\sigma_{\theta\theta} < 0$ and $\sigma_{\varphi\varphi} < 0$) at latitudes poleward of ±48.2° (Fig. 12a). In the midlatitudes, meridional stresses (i.e., radial to the poles) are tensile while stresses in the orthogonal direction are compressive (i.e. $\sigma_{\theta\theta} < 0$ and $\sigma_{\varphi\varphi} > 0$). Tectonic features produced by this type of global stress pattern would most likely consist of radial and concentric tensile fractures centered on the polar regions of Europa. These have not been observed, suggesting that despinning is not responsible for the fracture patterns on Europa.

4.1.3. Orbital recession. The pattern of stress due to tidal deformation (as opposed to despinning) predicts a symmetry more consistent with the pattern of global fractures observed. Tidal deformation of Europa produces a prolate spheroid elongated along an axis connecting the center of Europa and Jupiter. Along any orthogonal axis the radius of the deformed body is depressed. The tide-raising potential, W, and the response of the body to that potential, denoted by the Love number h_2, determines the height of the tidal deformation given by $H = -h_2W/g$, where g is the acceleration of gravity. The value of h_2 depends on the material properties and configuration of mass within Europa's interior. The tide-raising potential W depends strongly on the distance between Europa and Jupiter: $W = -GMa^{-3} R^2 (1.5 \cos^2 \theta - 0.5)$, where M is the mass of Jupiter, a is the semimajor axis of the orbit, R is the radius of Europa, and θ is the angular distance between any point on Europa's surface and the axis between the centers of Jupiter and Europa.

Since Jupiter rotates faster than Europa orbits, the tidal bulge raised on Jupiter by Europa is oriented slightly ahead of its position, providing orbital energy to Europa, which causes the orbit to migrate outward. As Europa's orbit migrates outward, the tide-raising potential deforming its surface decreases and the height of the tide raised on its surface is gradually reduced (*Squyres and Croft,* 1986). This change in shape ($\Delta f < 0$) should produce an all-around compressive stress on its surface ($\sigma_{\theta\theta} > 0$ and $\sigma_{\varphi\varphi} > 0$) in the region within 41.8° of the axis of symmetry, defined by the line between the centers of Europa and Jupiter (Fig. 12b). In the region between the compressive zones, the stresses are compressive radial to the axis of symmetry and tensile in the orthogonal direction ($\sigma_{\theta\theta} > 0$ and $\sigma_{\varphi\varphi} < 0$). If frac-

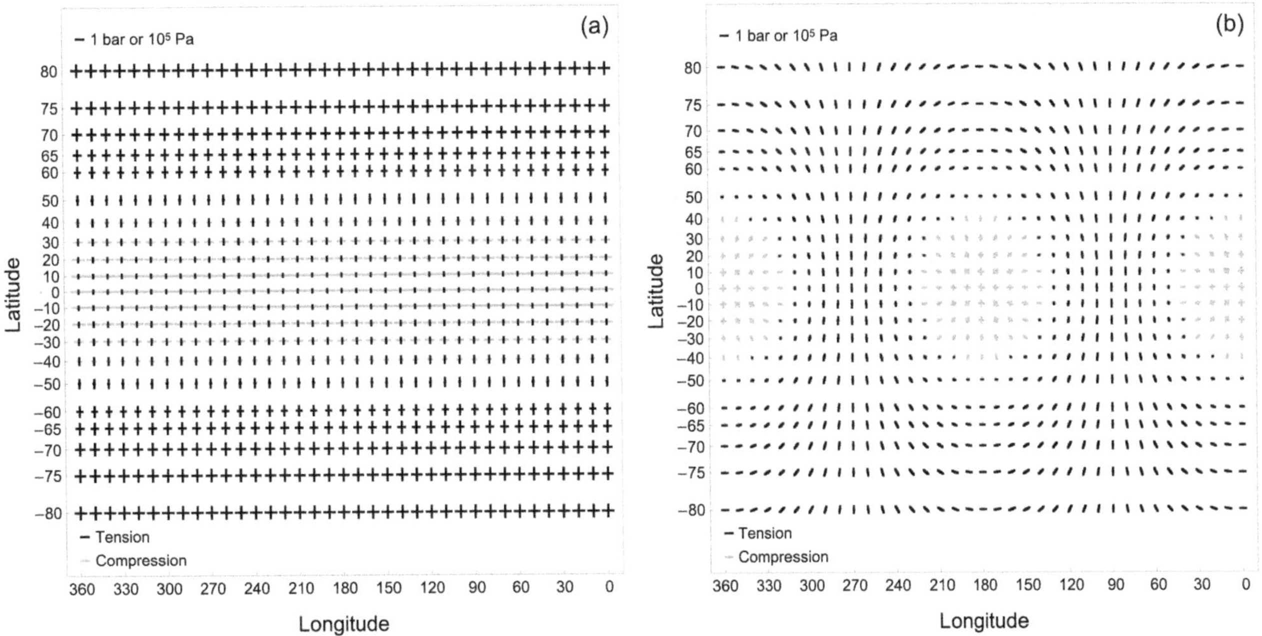

Fig. 12. **(a)** The stress field produced by a 5% decrease in Europa's rotation rate due to orbital recession and subsequent despinning. **(b)** The stress field produced by outward orbital migration, resulting in a 1% semimajor axis lengthening.

tures on Europa form mainly in tension, the stress from orbital recession would not produce fractures in the sub- and antijovian regions. Moreover, the cracks that would form between the sub- and antijovian regions would be radial to those regions. Hence, the stress field due to orbital recession also fails to produce tensile fractures in a pattern that matches Europa observations (Fig. 1).

4.1.4. Internal differentiation. During differentiation, mass within a body such as Europa is redistributed, with heavier materials moving toward the center. The redistribution of mass, along with any changes in material properties that occur, will change Europa's response to a tide-raising potential (cf. *Squyres and Croft,* 1986). This change manifests itself as a change in Europa's Love number h_2. As Europa's response to Jupiter's tide-raising potential changes, the prolateness of Europa is affected and the resulting change in shape produces stress on its surface.

Differentiation reduces Europa's response to a tide-raising potential, reducing the prolateness of the shape ($\Delta f < 0$). The resultant stress on the surface is similar to the stress produced by outward orbital migration: compressive in the region within 41.8° of the axis of symmetry, defined by the line between the centers of Europa and Jupiter. In the region between the compressive zones centered about the axis of symmetry, the stresses are compressive radial to the axis of symmetry and tensile in the orthogonal direction (i.e., analogous to Fig. 12b). Again, this stress field would not produce tension fractures in the sub- and antijovian regions but would produce cracks radial to those regions, which does not agree with the pattern of lineaments observed (Fig. 1). Moreover, the initial differentiation of Europa occurred early in its history, not long after its initial formation. Because Europa's surface is young (~40–90 m.y.), any tectonic features that formed as a result of the stress produced by differentiation have been erased by the formation of subsequent terrains.

4.1.5. Eccentricity and diurnal tidal stresses. If the semimajor axis were to migrate inward, the stress field would be the exact opposite of that described above for orbital recession, with a region of tension within 41.8° of the axis of symmetry. This stress field could produce tensile fractures in the sub- and antijovian regions as observed (Fig. 1). For this reason, *Helfenstein and Parmentier* (1983) proposed that global-scale lineaments formed in response to orbital eccentricity, e, which causes Europa's distance from Jupiter to change throughout the orbit, resulting in an oscillating diurnal tidal height response. At perijove, Europa is closer than average and the stresses near the sub- and antijovian points are tensile, allowing fractures to form. At apojove, Europa is farther than average and the stresses in this region are compressive. We now understand that this characterization of the stress field is incomplete. It neglects other important effects that orbital eccentricity has on tidal deformation, such as an oscillation in the longitudinal location of the tidal bulge, resulting in a poor match to tectonic patterns.

Because of Europa's finite eccentricity, as Europa moves from perijove to apojove, there is a small variation in the radial tide with an amplitude of $\Delta H = (9eh_2MR)/(4\pi\rho_{av}a^3)$ that affects f. In addition, Jupiter's angular position with respect to a fixed location above Europa's surface oscillates with an amplitude of 2e radians in longitude (and ε radians in latitude if there is a finite obliquity), changing slightly the angular distance θ of any point on the surface relative to the center of the tidal bulge. These two effects combine, yielding a diurnally varying component of the tide (*Greenberg et al.,* 1998; *Hurford et al.,* 2009).

The diurnal stress produced by Europa's orbital eccentricity thus changes throughout the orbit (Fig. 13). Zones of tension and compression along the equator migrate eastward throughout the day. The orientations of the principal stress axes rotate clockwise in the southern hemisphere and counterclockwise in the northern hemisphere, with 180° of rotation in principal stress orientations each orbit. This changing stress field provided the context for the development of the tidal walking hypothesis for strike-slip fault motions (see section 2.3.4) as well as for unraveling the patterns of cycloidal cracks on Europa (see section 2.1.2). Nonetheless, there are caveats to the use of the diurnal stress field pattern for unraveling all tectonic features. The constantly changing nature of the stress field makes it difficult to account for the extremely linear nature of most cracks, and the stress magnitudes are extremely small (on the order of a few tens of kilopascals) (*Hoppa et al.,* 1999a) such

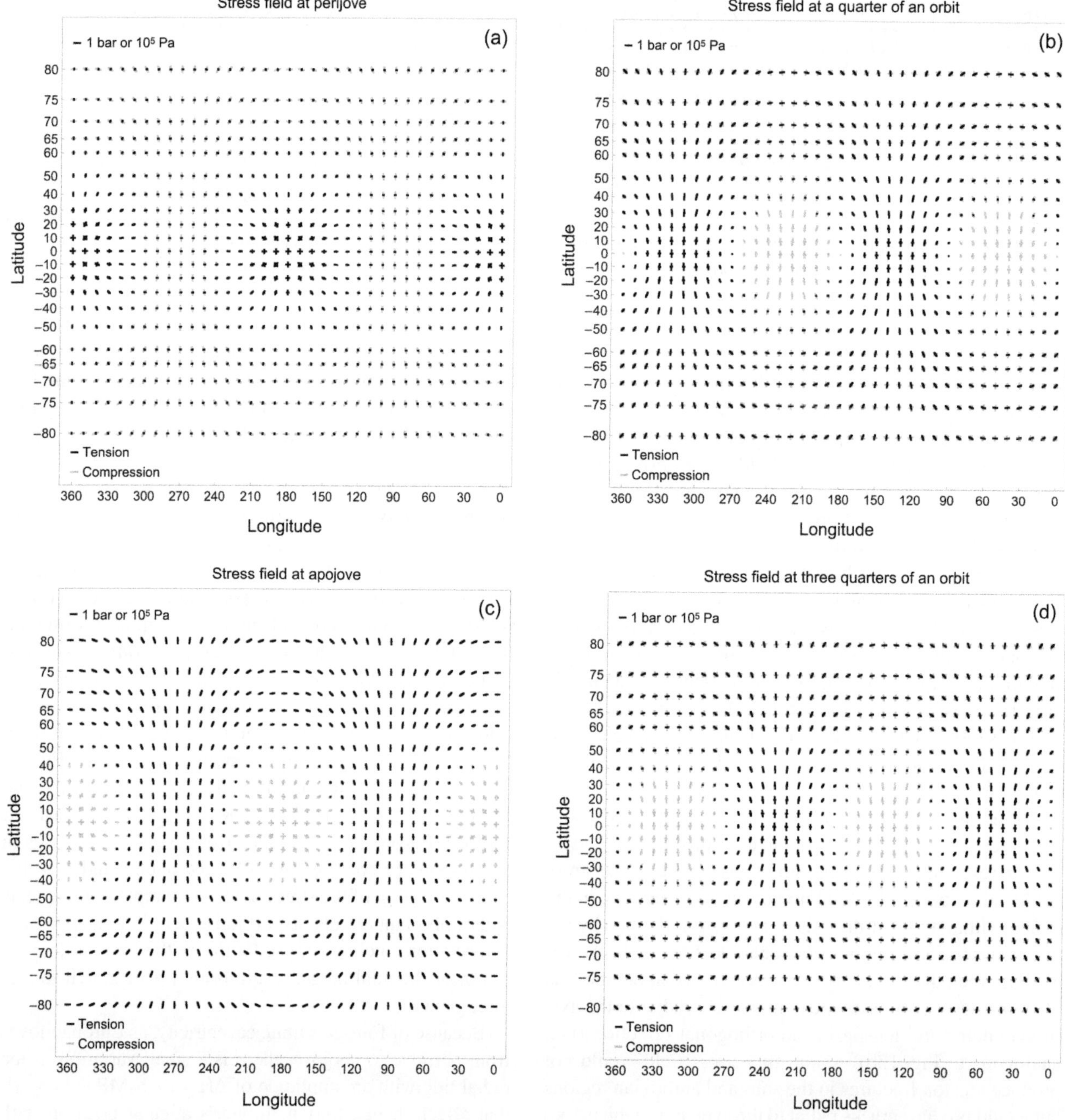

Fig. 13. Stresses produced by the diurnal oscillations of Europa's tidal bulge are shown in one-quarter orbital increments beginning at perijove. Zones of tension and compression along the equator migrate eastward throughout the day. The orientation of the principal stress axes rotate clockwise in the southern hemisphere and counterclockwise in the northern hemisphere, changing by 180° each orbit.

that it is has been questioned whether they are sufficient to overcome the likely tensile strength of ice at the low surface temperatures of Europa (*Harada and Kurita,* 2006). Also, diurnally varying tidal stresses cannot account for cycloids that cross the equator because of the mutually opposite rotation sense of stresses in each hemisphere. Such cycloids may provide evidence that there also exists a component of stress due to a small amount (perhaps ~0.1°) of obliquity that affected their formation patterns (*Hurford et al.,* 2006, 2009; *Sarid-Rhoden et al.,* 2009; see chapter by Bills et al.). The presence of a small forced libration (perhaps 150 m or so) of the decoupled icy shell over the diurnal period may also contribute a component of global stress (e.g., *Rambaux et al.,* 2007; *Hurford et al.,* 2008; *Van Hoolst et al.,* 2008; see chapter by Bills et al.). It is also possible that Europa's eccentricity has changed through time (see chapter by Sotin et al.), resulting in different diurnal stress field characteristics during the geological history. Nonetheless, pervasive fracturing of the icy shell and the existence of global-scale lineaments that are linear or broadly curved over great distances implies the existence of a higher-magnitude, static state of stress that dominates in controlling fracture geometries. Such a stress field may have been created in the icy shell through the process of nonsynchronous rotation.

4.1.6. Nonsynchronous rotation. Even though tidal torques work to force Europa to rotate synchronously, that rotation state can only be maintained if the orbit is circular or there exists a permanent and significant mass asymmetry within Europa (*Greenberg and Weidenschilling,* 1984). However, the Laplace resonance between Io, Europa, and Ganymede prevents tidal torques from circularizing Europa's orbit, forcing a small but finite orbital eccentricity. In a noncircular orbit, torques on Europa will tend to force it to rotate slightly faster than synchronous. This process would be accentuated by the potential development of thickness changes in an icy shell floating on a liquid ocean as tidal dissipation moves it toward thermal equilibrium (*Ojakangas and Stevenson,* 1989a). Gravitational torques from Jupiter acting on a variable thickness icy shell could induce nonsynchronous rotation on the thermal diffusion timescale (<10 m.y. per rotation), resulting in the shell attaining a state of dynamic equilibrium. Hence, as long as tidal heating within Europa prevented any permanent mass asymmetries from forming within the icy shell, the tidal torques may have forced Europa to rotate nonsynchronously (*Greenberg and Weidenschilling,* 1984), producing a significant contribution to the global stress state (*Helfenstein and Parmentier,* 1985).

Stresses from nonsynchronous rotation are caused by the reshaping of Europa as the location of the tidal bulge moves with respect to a fixed location on the surface. The actual amount of tidal deformation does not change, only the geographic location relative to the tidal bulges. The stress produced by 1° of nonsynchronous rotation produces an equatorial zone of tension to the west of the sub- and antijovian points and a zone of compression to the east of the sub-

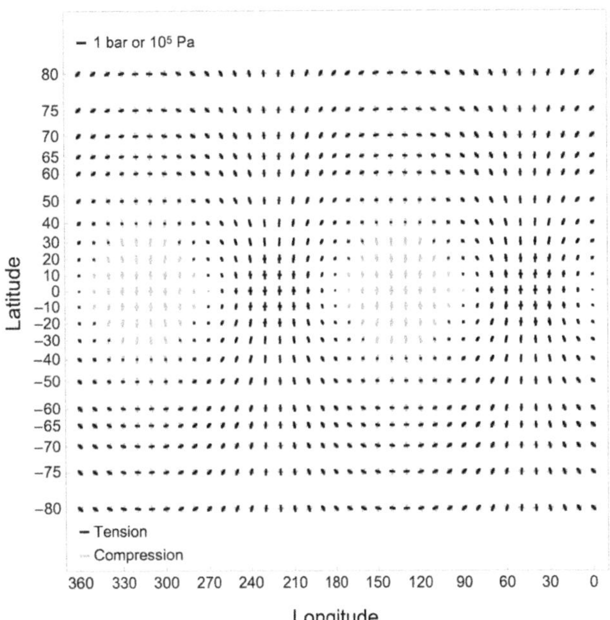

Fig. 14. The stress field produced by 1° of nonsynchronous rotation produces zones of tension just west of 0° and 180° longitude.

and antijovian points (Fig. 14). At high latitudes, the principal stresses ($\sigma_{\theta\theta}$ and $\sigma_{\phi\phi}$) have a range of magnitudes and signs. This stress field is broadly consistent with the pattern of ridges observed on the surface (Figs. 1 and 15). Therefore, although other sources of stress may contribute to global-scale crack formation, the stress from nonsynchronous rotation may be an important contributor to the global tectonics. Moreover, the amount of nonsynchronous rotation needed to produce stress of sufficient magnitude to allow tensile failure can be obtained within a few degrees of reorientation of the icy shell, assuming the rate at which the shell reorients is sufficiently rapid that stresses can accumulate elastically (*Harada and Kurita,* 2007; *Wahr et al.,* 2009). However, a far greater amount of cumulative nonsynchronous rotation is needed to account for the current longitudes of many fractures relative to their probable formation locations, if they indeed formed in response to nonsynchronous rotation (*McEwen,* 1986) (Fig. 15; see section 5.1).

Geissler et al. (1998a,b) identified what appeared to be a systematic change in the azimuth of cracks over time, suggesting that Europa's surface has migrated eastward relative to the direction of Jupiter. Hence, at different times in the past, regions where cracks formed in a range of orientations may have been located within a zone of tension within the nonsynchronous rotation stress field, allowing these features to have formed as unrelated, superposed episodes of tension cracks that were subsequently translated eastward with respect to the zone of tension. As the icy shell migrated eastward, the changing orientation of the stress field at these locations resulted in new episodes of tension fracturing being superposed on older episodes, sometimes resulting in a seemingly conjugate pattern. Similar studies, utiliz-

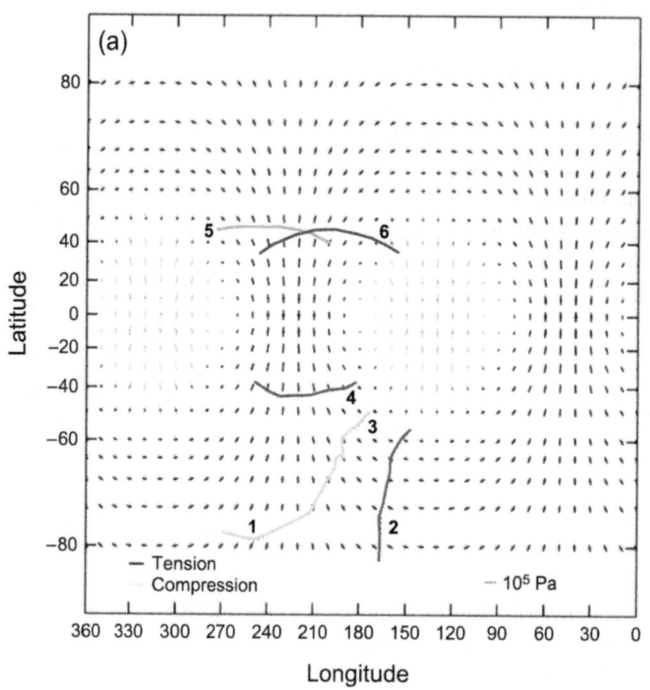

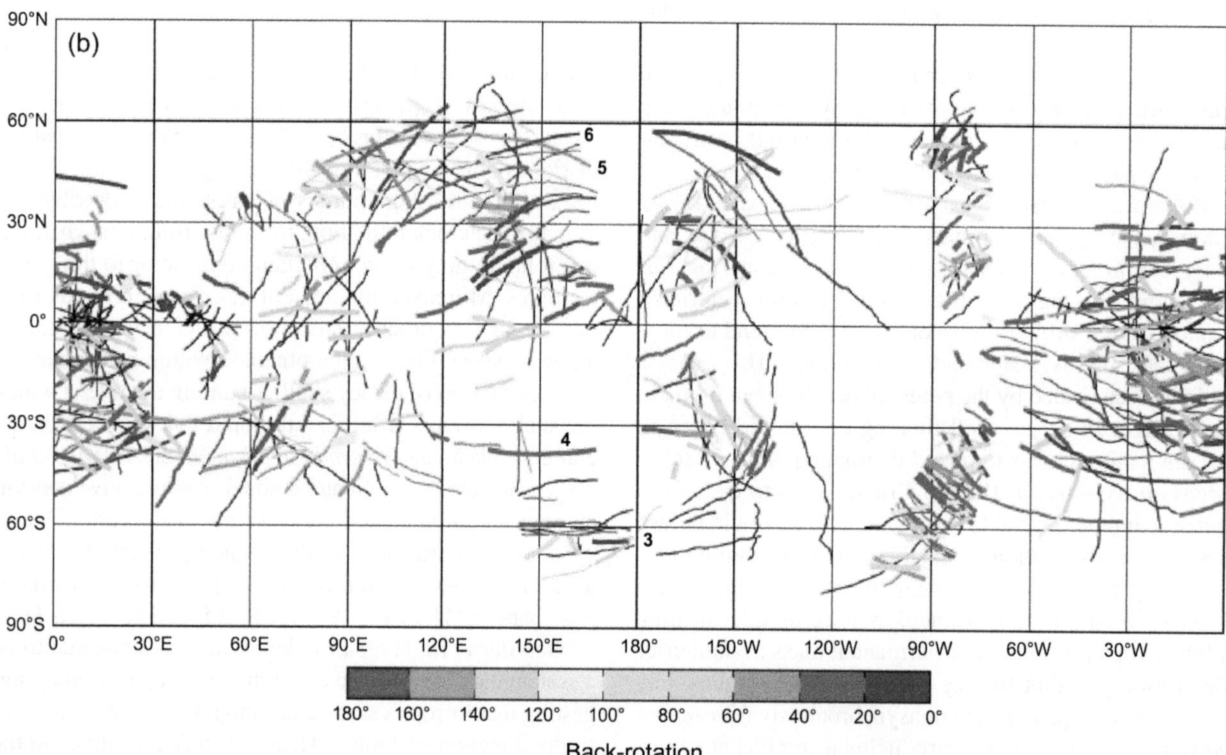

Fig. 15. See Plate 13. **(a)** The tidal stress-field produced by 1° of nonsynchronous rotation (*Greenberg et al.,* 1998) shows a good fit to the locations and orientations of several lineaments. The lineaments are numbered (1) Astypalaea, (2) Thynia, (3) Libya, (4) Agenor, (5) Udaeus, and (6) Minos Lineae. A better fit is produced if fractures are back-rotated westward in longitude relative to the stress field [by an amount given by their color coding, as defined in **(b)**] such that their orientations are perpendicular to the direction of maximum tension, providing a possible original longitude for the creation of the crack. **(b)** The global pattern of lineaments based on Galileo observations illustrates a range of orientations that cannot all be fitted to the same stress template, such as that shown in **(a)**, because they formed at different times. The lineaments are color coded to indicate how far westward they must be back-rotated in order to form perpendicular to the maximum tension produced by the stress of nonsynchronous rotation [map courtesy of Z. Selvans (Selvans et al., in preparation)]. Some of the numbered lineae in **(a)** are numbered accordingly in **(b)**.

ing higher-resolution data, have confirmed that a correlation between age and orientation exists (see section 5.1), and have been related to various interpretations of the effects of nonsynchronous rotation of the icy shell (*Figueredo and Greeley,* 2000; *Kattenhorn,* 2002; *Sarid et al.,* 2004, 2005, 2006; *Groenleer and Kattenhorn,* 2008).

4.1.7. Polar wander. Another potential contributor to the global stress state is provided by long-term latitudinal reorientation of the icy shell, or polar wander, in response to latitudinal changes in ice thickness as the icy shell thermally evolves (*Ojakangas and Stevenson,* 1989a,b). *Leith and McKinnon* (1996) explored the theoretical stress produced by this phenomenon and found only limited evidence for polar wander within observable crack patterns. *Sarid et al.* (2002) noted that the slip sense along some near-equatorial strike-slip faults did not agree with the predictions of tidal walking (see section 2.3.4) and that a small amount of polar wander since fault motions accrued may explain the distribution of the observed slip sense. However, finite obliquity is another potential mechanism for such aberrations (*Hurford et al.,* 2009) as well as the effects of convergence across ridges, which may create an apparent sense of offset in the absence of any actual strike-slip motions (see section 2.2.3). *Schenk et al.* (2008) interpret the locations and geometries of inferred extensional zones called small-circle depressions (SCDs) as being due to true polar wander. These curved depressions are up to 1.5 km deep and trace out segments of the arcs of almost perfect circles centered ~25° from the equator, forming an antipodal pattern in the leading and trailing hemispheres. The shapes and locations of the troughs are suggested to be consistent with the stress field that would result from 80° of polar wander. It is unclear what the timing of this hypothesized shell reorientation event may have been because other surface features are seemingly unaffected by the locations of the SCDs. The associated polar wander stress field (e.g., *Matsuyama and Nimmo,* 2008) has also not been compared to the orientations of the vast number of other tectonic features on Europa. Therefore, polar wander has yet to be shown to be a likely source of global stresses on Europa in terms of its potential contribution to the pattern of global tectonics.

4.1.8. Ice shell thickening. If Europa's icy shell increased in thickness relatively recently, the related stress may have contributed to tectonic activity (see chapter by Nimmo and Manga). For example, *Nimmo* (2004a) showed that for a nonconvecting icy shell that thickens in excess of 20 km, isotropic tensile stresses are produced within the icy shell, attaining magnitudes of perhaps 10 MPa at the surface, which are sufficient to induce tension fracturing and perhaps even shell-penetrating cracks (see also *Manga and Wang,* 2007). Because these stresses are isotropic, the principal stress orientations would still be controlled by the nonisotropic (e.g., tidal) components, which would thus control fracture orientations. This effect provides a potential solution to the mystery of tensile fracturing in equatorial compressive zones within the nonsynchronous rotation stress field, as described by *Spaun et al.* (2003) (see section 2.3.2). Another effect of an increasing icy shell thickness is a decreasing tidal response (e.g., a ~1% reduction if the shell thickens from 1 km to 10 km) (*Moore and Schubert,* 2000; *Hurford,* 2005), resulting in slightly reduced tidal stress magnitudes. Also, a change in ice thickness results in a change in the Love number, h_2, creating an additional component to the stress field similar to that induced by internal differentiation (see section 4.1.4).

4.2. Regional and Local Stress

Although global-scale stresses are responsible for the formation of the majority of the cracks on Europa's surface, regional and local scale stresses have also played an important role in the tectonic history. These stresses essentially represent regional and local perturbations to the global stress state and are significant for the tectonics if their magnitudes are on the order of the global stresses or higher. Although *regional* and *local* have somewhat arbitrary definitions, the important distinction is that they do not result from the global effects of tidal deformation or reorientation of the icy shell. Regional effects may encompass thousands of kilometers across the surface, depending on the causal factor. For example, endogenic processes such as thermal or compositional diapirism (see chapter by Barr and Showman) create upwelling that can impart regional stress conditions on an overlying stagnant lid. The formation of chaotic terrain, which represents the destruction of an earlier surface, is likely related to this process (*Collins et al.,* 2000; *Sotin et al.,* 2002; *Mitri and Showman,* 2008; see chapter by Collins and Nimmo). Chaos areas can only become disaggregated through the initial development of fractures that create loose blocks of brittle ice that are subsequently rafted around by the underlying motion of warmer ice and melt. Indeed, one of the types of tension fractures described in section 2.1.3, endogenic fractures, is specifically related to this phenomenon. The characteristics of the stress field caused by a diapir impinging on the brittle ice lid from below are not easily quantifiable, as they would likely depend on the shape and size of the diapir. Although the stress magnitudes are likely to be small (tens of kilopascals), they may be sufficient to drive deformation in the brittle lid (*Showman and Han,* 2005). A common observation is that endogenic fractures form concentric patterns about regions of chaos, suggesting that the causal stress field may be somewhat similar to that of an ascending bulge, analogous to upwarps above laccolith intrusions on Earth (*Jackson and Pollard,* 1988). Regional perturbations to the global stress state may also be imparted by large meteorite impacts, which can result in both radial and concentric fracture patterns for the case of complete shell penetration (*Melosh,* 1989), although radial fractures are not observed around the large impact site Tyre (*Kadel et al.,* 2000).

Local stress effects occur at scales on the order of hundreds of meters to perhaps hundreds of kilometers. The causal mechanism is typically apparent, usually related to an existing ridge, strike-slip fault, or fold, or to regions of lenticulae (small disruptions related to underlying diapiric activity). For example, ridge constructs are up to several hundreds of meters high and their weight is expected to push down on the icy shell. As a result, the elastic portion of the shell flexes downward beneath the ridge, apparently producing secondary uplifts, or forebulges, to either side of the ridge, evident in subtle changes in shading and shadow patterns (*Tufts,* 1998; *Hurford et al.,* 2005). The resultant stresses induced by bending of the icy shell are sufficient to fracture the icy shell to either side of the ridge (Fig. 2b) and can be quantified using the equations describing the flexure of an elastic plate beneath a line-load (*Turcotte and Schubert,* 2002). *Billings and Kattenhorn* (2005) used these equations to formulate a relationship between elastic ice thickness and the distance from a ridge load to its flanking secondary cracks, allowing the thickness of the elastic portion of the icy shell to be constrained to between ~200 m and ~2400 m. Bending stresses are also responsible for the local development of surface cracks along the hinge lines of anticlinal folds, such as occurs within Astypalaea Linea (*Prockter and Pappalardo,* 2000).

Strike-slip faults produce local perturbations to the global stress field in response to lateral motions that create stress concentrations at the fault tips (*Kattenhorn,* 2004a). These stress fields are quantifiable (see section 2.3.3) and have been shown to be responsible for the development of secondary fractures at fault tips called tailcracks (see section 2.1.4) (*Prockter et al.,* 2000; *Schulson,* 2002; *Kattenhorn,* 2004a; *Kattenhorn and Marshall,* 2006). Tailcracks form in the locally perturbed stress field in the vicinity of a tip of a shearing fault or fracture. The remote (far-field) stresses govern the sense of slip on the fault itself while shear motion along the fault locally perturbs the stress field, creating near-tip quadrants of extension and compression, with an antisymmetric pattern at opposing tips. Tailcracks form in extensional quadrants and propagate away from the fault tip perpendicular to the direction of maximum local tension. A similar phenomenon is responsible for the development of cusps along cycloidal chains (*Groenleer and Kattenhorn,* 2008) in response to lateral shearing effects during the diurnal cycle. Compressive quadrants may also experience local deformation in the form of folds or anticracks, the latter having been suggested to exist in the "Wedges" region, Argadnel Regio (*Kattenhorn and Marshall,* 2006).

5. TECTONIC EVOLUTION OF EUROPA

5.1. Tectonic Evidence for Nonsynchronous Rotation

Theoretical considerations of nonsynchronous rotation stress fields (see section 4.1.6) and their match to the orientations of geologically recent large-scale lineaments prompts the search for direct evidence for extensive nonsynchronous rotation to determine its relative importance throughout Europa's visible tectonic history. Comparisons between lineament orientations on Europa and theoretical global stress fields suggest that many geologically recent lineaments did not form in their current longitudes (*McEwen,* 1986; *Greenberg et al.,* 1998; *Hoppa et al.,* 1999b). Instead, these lineaments are inferred to have translated longitudinally due to nonsynchronous rotation by a few tens of degrees (Fig. 15). Many locations on Europa show a complex sequence of multiply superposed lineaments with differing orientations, reflecting the dominant tensile stress orientation at that location at the time of lineament development. This is plausibly because the regional stress orientation changed as the icy shell migrated steadily eastward through time, rotating through a full 360° per nonsynchronous rotation cycle with a clockwise sense in the northern hemisphere and a counterclockwise sense in the southern hemisphere. Accordingly, lineament orientations have been suggested to display this rotation sense through time (*McEwen,* 1986; *Geissler et al.,* 1998a; *Figueredo and Greeley,* 2000; *Kadel et al.,* 2000; *Kattenhorn,* 2002). The amount of reorientation of the icy shell inferred from the tectonic history based on these studies is as little as 50° and up to 1000°, but may be much more depending on the amount of time between individual lineament formation events, which is unknown. A problem with these inferences is that changing fracture orientations do not directly indicate that nonsynchronous rotation actually occurred. Considering that nonsynchronous stress orientations sweep through all angles, it is not surprising that interpretations are equivocal. In fact, if one arbitrarily assumes that the sense of rotation of lineaments through time is the opposite of what would be expected in response to nonsynchronous rotation, the number of complete shell rotations needed to account for the full range of lineament orientations through time does not differ greatly to the nonsynchronous rotation result (*Sarid et al.,* 2004, 2005, 2006). Hence, although changing lineament rotations through time are certainly consistent with nonsynchronous rotation, this phenomenon does not provide direct proof that nonsynchronous rotation occurs.

Evidence for nonsynchronous rotation was suggested by *Hoppa et al.* (1999c) to be evident in the slip sense of strike-slip faults in equatorial areas, where the sense of cumulative slip in the context of the tidal walking model (see section 2.3.4) is completely dependent on the fault orientation. Some faults do not have the correct sense of slip for their orientation and location; however, back-rotation of a nonsynchronously rotated shell by up to 90° can place each fault at a longitude where the slip sense is compatible with expectations.

Additional evidence for icy shell reorientation, probably through nonsynchronous rotation, has emerged from studies of cycloids. *Hoppa et al.* (2001) show examples of southern hemisphere cycloids that could not have formed in their current locations but rather at locations up to 80° further to the west, after which nonsynchronous reorientation of the icy shell translated them eastward. *Hurford et al.* (2007a) reduced this estimate to 24° for these particular cycloids by accounting for the presence of a small amount of nonsynchronous rotation stress during cycloid growth, indicating that reorientation of the icy shell is still necessary to fit the cycloid shapes. However, the inclusion of a small amount of obliquity may remove the need for a component of nonsynchronous rotation stress to account for the shapes of some cycloid examples (*Sarid-Rhoden et al.,* 2009). *Groenleer and Kattenhorn* (2008) document a history of numerous cycloidal episodes in the northern trailing hemisphere that necessitate at least 600° of nonsynchronous reorientation of the icy shell during the period of cycloid development alone (i.e., excluding all other lineament types). Crosscutting relationships among these cycloids imply a definitive age sequence that can best be reconciled with cycloid chains having formed during different nonsynchronous rotation cycles, lending credence to the idea that only a few cycloid chains form during each nonsynchronous rotation cycle.

The wide range of estimates of nonsynchronous rotation amounts may seem contradictory; however, disparate results likely reflect variances in both the types of features used to make the estimates and the differing resolutions of images studied in different regions. The combined studies suggest that the decoupled icy shell of Europa may have undergone almost three complete rotations relative to the rocky interior during the visible tectonic history, although this number may be much higher in reality. In support of this notion, given the 40–90-m.y. surface age of Europa (see chapter by Bierhaus et al.), nonsynchronous rotation must have occurred rapidly enough to accrue elastic stresses before they could be relieved viscously beyond the Maxwell relaxation time (*Harada and Kurita,* 2007; *Wahr et al.,* 2009). Given this rate of rotation, there have likely been numerous rotations of the icy shell relative to the interior during Europa's visible history. Nonetheless, if nonsynchronous rotation effects were relatively constant over the entire visible geological history, one might expect there to be a somewhat unchanging tectonic response throughout this time. On the contrary, Europa exhibits a diverse geological history with distinct temporal variability in the tectonic features that developed, implying that other factors need to be considered when examining the tectonic evolution of Europa, such as variations in the rate of nonsynchronous rotation (*Nimmo et al.,* 2005; *Hayne et al.,* 2006).

5.2. Temporal Changes in Tectonic Style

Previous sections of this chapter considered the range of tectonic features that pervasively deform the icy shell of Europa and the possible sources of stress responsible for those features. We now turn to the geological development of Europa inferred from the many episodes of tectonic events evident on the surface. Over the visible surface history of the icy shell, repeated fracturing events have left a remarkably coherent story regarding the tectonic evolution in the form of clearly identifiable crosscutting relationships. In so doing, tectonic features reveal that there has been a gradual change in the nature of deformation that points toward temporal variations in the thickness of the icy shell, processes occurring within the shell, and the response of the shell to tidal deformation.

Numerous studies aimed at unraveling the geological history of Europa have revealed significant changes through time (*Lucchitta and Soderblom,* 1982; *Prockter et al.,* 1999, 2002; *Figueredo and Greeley,* 2000; 2004; *Greeley et al.,* 2000; *Kadel et al.,* 2000; *Kattenhorn,* 2002; *Spaun et al.,* 2003; *Riley et al.,* 2006; *Doggett et al.,* 2007). A simple summary of these changes has early development of ridged plains (a multitude of closely spaced ridges) being followed by periods of ridge and band development, endogenic cryomagmatic disruption and cryovolcanism (chaos, lenticulae, and smooth plains) with some contemporaneous and subsequent ridge formation, and a late stage of tectonic fracturing (see chapter by Doggett et al.). Many lineaments were reactivated as strike-slip faults during the periods of tectonic activity. The following synthesis considers the individual evolution of the most prominent types of tectonic features that have been placed into a stratigraphic context.

5.2.1. Ridges. The oldest portions of the surface appear to consist of finely crenulated *ridged plains* or *subdued plains* that may represent the earliest visible ridges, with spacings on the order of hundreds of meters. Most of the ridged plains were subsequently destroyed by the repeated development of successively younger double ridges (i.e., a prominent central trough flanked by two raised rims) and dilational bands in a range of orientations. Older ridges appear to be more numerous, narrower, lower, and more closely spaced (tens to hundreds of kilometers) than relatively younger ridges (*Kadel et al.,* 2000; *Kattenhorn,* 2002; *Figueredo and Greeley,* 2004). Many of the prominent (and typically geologically young) global-scale ridges that cover broad portions of the europan surface have evolved into ridge complexes, forming the highest and widest ridge structures. Ridge development thus spans most of the geological history of the surface, although the youngest ridges could have formed many tens of degrees of nonsynchronous rotation in the past. Hence, it is unclear if ridges continue to form or if the processes responsible for their development (see section 2.1.1) still occur in the icy shell.

5.2.2. Cycloids. There has been some uncertainty as to whether cycloid development represents a geologically recent phenomenon on Europa. *Groenleer and Kattenhorn* (2008) show that there has been an extended period of cycloid development in the trailing hemisphere encompassing at least 600° of nonsynchronous rotation, with the most recent event occurring at least 30° of nonsynchronous rotation in the past. These cycloid-forming events were punctuated by periods of intervening linear ridge development, suggesting that there was an oscillation between the conditions conducive to cycloid and linear fracture formation, respectively. Although these oscillations are observed in specific areas, it is unknown whether or not a similar global temporal pattern of fracturing occurs or whether cycloid development in one region of Europa could have occurred contemporaneously with linear ridge development elsewhere. Regardless, whatever processes were responsible for ridge construction continued regardless of whether cracks were linear or cycloidal, indicated by ridge development along the flanks of cycloids as well as linear fractures. Unlike linear ridges, however, cycloids have not been explicitly described from the earliest part of the geological history, raising the possibility that the initiation of the conditions needed to form cycloids occurred at some critical juncture during the tectonic history.

5.2.3. Dilational bands. There is also a wide range in the ages of dilational bands, some of which formed around the time of the earliest ridges (*Greeley et al.,* 2000; *Kattenhorn,* 2002); however, it is possible that the driving process behind dilational band development ceased on Europa even while ridge development continued. *Prockter et al.* (2002) indicate that there are always features younger than dilational bands, including troughs, ridges, and lenticulae. Nonetheless, dilational bands change in albedo from dark to light through time and the youngest dilational bands still have dark albedos. It is not known how quickly this lightening process occurs, so dilational bands could be many millions of years old. Dilational band widths are inferred to have decreased through time (*Figueredo and Greeley,* 2004), perhaps indicating a thinner icy shell and higher thermal gradients earlier in Europa's visible geological history. *Nimmo* (2004b) modeled the extension of icy shells and demonstrated a trade-off between strain rate and shell thickness in terms of whether wide or narrow rifts will develop. Narrow rifts are analogous to europan dilational bands and imply high strain rates or relatively thick icy shells. It is possible that changing shell thickness and strain rate conditions on Europa resulted in a transition in the mode of extension to no longer favor dilational band development. Hence, dilational bands rarely postdate chaos (see chapter by Prockter and Patterson). Those young dilational bands that appear to postdate chaos formation have dilated preexisting cycloids (*Figueredo and Greeley,* 2000); therefore, the initiation of the process responsible for cycloid development did not correspond to a termination of the process responsible for dilational band development, with both processes possibly overlapping with the period of chaos formation.

Figueredo and Greeley (2004) suggest that a long period of tectonic resurfacing by ridge and band development was followed by a rapid decrease in tectonic activity through time, concomitant with a steady increase in "cryovolcanic" activity that resulted in the formation of broad areas of chaos and lenticulae. These changes were attributed to the effects of a thickening icy shell that ultimately reached the threshold thickness required to induce solid-state convection, diapirism, and cryovolcanic resurfacing (*Pappalardo et al.,* 1999; *Barr et al.,* 2004; *Barr and Pappalardo,* 2005; see chapter by Nimmo and Manga). Nonetheless, an even younger period of trough development occurred (both linear and cycloidal), concomitant with recent cratering events. Considering that the visible geological history of Europa only spans around 2% of the total age of the satellite, it is not known if tectonic resurfacing and cryovolcanic processes occur in an oscillating cycle related to repeated thinning and thickening of the icy shell, perhaps associated with changes in Europa's orbital eccentricity (see chapters by Nimmo and Manga, Moore and Hussman, and Sotin et al.). A further complication is provided by the possibility of lateral variations in icy shell thickness due to latitudinal and longitudinal variations in tidal heating (*Ojakangas and Stevenson,* 1989a) and past tectonic modification of the icy shell (*Billings and Kattenhorn,* 2005). Regardless, the inferred thickening of the icy shell over the past 40–90 m.y. clearly influenced the tectonics.

5.3. Active Tectonics on Europa?

Although many bodies in the solar system show evidence of a remarkable geological history of tectonic activity (e.g., Mars, Venus, icy moons of Jupiter, Saturn, Uranus, and Neptune), evidence of current tectonic activity outside of Earth has been very elusive. With the recent discovery of eruptive water-ice plumes emanating from cracks (called "tiger stripes") in the south polar area of Saturn's icy moon Enceladus (*Porco et al.,* 2006; *Spencer et al.,* 2006), a relationship to active tectonics has been inferred, based on models of crack motions within a tidal stress field (*Hurford et al.,* 2007b; *Nimmo et al.,* 2007; *Smith-Konter and Pappalardo,* 2008). Any icy satellite with a liquid ocean beneath an icy shell that has apparently deformed due to the tidal response of the underlying ocean therefore provides a good candidate study for active tectonics. Hence, Europa is a promising candidate for active tectonics. It is generally accepted that a relatively thin icy shell (≤30 km) overlies a liquid ocean on Europa (see section 4.1.1) that experiences an oscillating tidal response due to its orbital eccentricity.

Despite the extensive body of work on the characterization and interpretation of europan tectonic features discussed in this chapter, this work could not directly answer the question of whether there is tectonic activity on Europa today. Past analyses demonstrated a good correlation between tidal stresses and tectonic lineaments, but focused on relatively old geological features such as ridges and dilational bands. Although some ridges can be placed into the youngest portions of the geological history (*Figueredo and Greeley,* 2000, 2004), especially cycloidal ridges, an analysis of ridge orientations or morphologies cannot be used to answer the question about active tectonics for three important reasons.

First, even geologically young cycloidal ridges currently occur in longitudinal locations (relative to the tidal bulges) that do not match the current tidal stress fields, suggesting that some amount of nonsynchronous rotation has occurred, implying at least several tens of thousands of years since those ridges formed (*Hoppa et al.,* 2001; *Hurford et al.,* 2007a; *Groenleer and Kattenhorn,* 2008). Although this is not a great period of time, geologically speaking, it must be remembered that the nonsynchronous rotation rate is still not well constrained and that these ridges are not the youngest tectonic features. Also, they could have formed 30° modulo 180° in the past (based on the global symmetry of the tidal stress field). The upshot is that cycloidal ridges do not provide proof of current tectonic activity.

Second, ridge-producing cracks may take a long time to construct their ramparts to either side of the crack (see section 2.1.1). The construction of a ridge is a late stage in an evolutionary process that begins with the development of a trough. It is not known how long this ridge-building process takes, mostly because many years of ridge analyses have still not conclusively determined their mode of development. Typical ridges may take tens of thousands of years to form and can only do so when optimally oriented with respect to the changing stress field during the nonsynchronous rotation process. Hence, ridges cannot be used to make inferences about current tectonic activity on Europa, even if they appear to be geologically young.

Third, the still-present topography of up to 200 m still observed along dilational bands and even higher elevations along ridges that date back through the ~40–90 m.y. of geological history indicates that topography can survive for long time periods and is thus unlikely to be an indicator of recent tectonics. *Dombard and McKinnon* (2006) suggest that appreciable topography may actually survive for >100 m.y. considering the low rate of viscous relaxation, although their results vary as a function of heat flow and surface temperature. *Nimmo et al.* (2003) show that compositional buoyancy due to lateral density differences in a cold, near-surface elastic layer (perhaps related to variability in either ice porosity or salt concentration) could explain prolonged topography that might otherwise be eradicated in less than 0.1 m.y. by ductile flow of ice from below an uncompensated surface load. *Luttrell and Sandwell* (2006) indicate that short-wavelength topography (~20 km), as would characterize ridges, can in fact be supported by the strength of the brittle portion of icy shell itself without the need for underlying support.

At least three factors have complicated attempts to make inferences about active tectonics on Europa:

1. There is a distinct lack of observational evidence for tectonic activity in the time interval between the imaging periods of the Voyager and Galileo spacecraft missions. A comparison of Voyager and Galileo images based on an iterative coregistration technique detected no evidence for deposition by plume activity (such as that which accompanies tectonic activity on Enceladus), nor any evidence for new fracturing during the 20-year interval between missions (*Phillips et al.,* 2000). The lack of inferred plume activity could simply be the result of negative buoyancy effects that hamper surface eruptions (*Crawford and Stevenson,* 1988), regardless of the fact that cracks might fully penetrate an icy shell that has substantially cooled and thickened (*Manga and Wang,* 2007). Hence, absent plumes do not necessarily imply no active tectonics. Additionally, no evidence for any geological (tectonic or otherwise) changes was discovered, so inferences regarding active tectonics are inconclusive. Part of the difficulty introduced into attempts at comparing these datasets is the low resolution of Voyager data and differences in photometric angles, which can dramatically alter the appearance and albedo of surface features. Future missions could be targeted to match Galileo coverage at high resolutions, which may increase the chances of measuring geological changes (*Phillips and Chyba,* 2003).

2. Inferences about tectonic activity based on the geological history in general have been inconclusive. Based on the low density of craters relative to the expected impact flux rate, Europa's surface appears to be geologically young (average age of 40–90 m.y.). This raises the likelihood that tectonic resurfacing has been a geologically recent process that may continue today. Although references have been made to "recent" tectonic lineaments on Europa (e.g., *Geissler et al.,* 1998a), inferences about what features are geologically recent are commonly thwarted by analyses of higher-resolution images that reveal comparatively younger geological features (tectonic fractures and endogenically driven chaos and lenticulae). Europa has experienced numerous changes in tectonic style over its visible history, with a prevalence of endogenically driven surface disruption in the recent geological history (see section 5.2). Nonetheless, a younger period of postchaos fracturing has been noted in several studies (*Kadel et al.,* 2000; *Figueredo and Greeley,* 2000, 2004; *Doggett et al.,* 2007), implying that the development of chaos did not equate to the cessation of tectonic activity on Europa. Although these geological studies point to a high likelihood that there was a geologically young period of tectonic activity, they have not unequivocally determined whether such activity is ongoing.

3. Theoretical stresses in a thickening icy shell allow ongoing fracturing but imply a reduction or cessation of tidally driven tectonic activity. Geological inferences of a thickening icy shell based on convincing evidence that the threshold thickness for solid-state convection was attained (*Barr et al.,* 2004; *Barr and Pappalardo,* 2005; see chapter by Barr and Showman) has led to the development of theoretical models to address the impact of a thickening icy shell on the tidal stress field. As a result of the decrease in tidal deformation as an icy shell thickens, and the fact that tidal stresses are relatively small even in a thin icy shell, there may be a decreasing likelihood of tidally driven tectonic activity through time on Europa (*Hurford,* 2005), reflected in the apparent transition from tectonic resurfacing to cryovolcanic resurfacing (see section 5.2). Hence, processes such as cycloid development and tidal walking along strike-slip faults, which are driven primarily by diurnal tidal stresses, may decline. Nonetheless, tensile stresses produced by expansion of a thickening icy shell (in the 10 MPa range) can far exceed those induced by diurnal or nonsynchronous rotation effects and could promote surface fracturing (see section 4.1.8). Because the stresses are isotropic, orientations of tectonic features will still be controlled by the nonisotropic components of the total stress field (e.g., tidal stresses), even if they are small. Thus, theoretical considerations show that the potential for sufficiently high stresses exists to promote active tectonics that can be reconciled with the current tidal stress state. Nonetheless, it cannot be

proven whether or not tectonic activity actually exists. Although thickening of the icy shell may have indeed induced changes in the tectonic style on Europa through time, as well as plausibly initiating convective overturn and cryovolcanism, it did not necessarily herald the death knell for tectonics.

The question of active tectonics on Europa is open ended. Its importance for understanding the interior dynamics of this icy world, as well as providing potential access pathways to the subsurface ocean, will undoubtedly motivate future studies in this direction, particularly in the context of future mission goals and design (e.g., *Sandwell et al.,* 2004; *Panning et al.,* 2006; see chapter by Greeley et al.).

6. CONCLUDING REMARKS

The icy shell of Europa has experienced pervasive deformation that has resulted in a wide range of tectonic features over the 40–90-m.y. visible history. This deformation was driven by a number of factors, whether global, regional, or local. The two most important global contributors to stresses were probably diurnal tidal deformation induced by a tidally responding subsurface ocean and a longer timescale longitudinal reorientation of the icy shell relative to its tidal bulges by nonsynchronous rotation.

Such stress fields commonly produced tensile components, resulting in ubiquitous extensional deformation. Tension fractures at the surface were reworked over many tens of thousands of years to form ridges. These ridges, composed of raised rims to either side of a central fracture, likely formed in response to a combination of extensional, compressive, and shear-related processes. Complete separation of the icy shell is evidenced by the development of dilational bands, which formed in response to tidally driven extension and endogenic processes such as local upwelling of ductile portions of the shell. Cycloidal fractures provide a strong indication that the diurnal stress field dominated in the creation of tectonic features at times, although this may have been more common later in the geological history, potentially as a result of a decrease in the rate of nonsynchronous rotation stress build-up as the icy shell gradually thickened. Long, linear, or curvilinear fractures likely formed in response to dominant nonsynchronous rotation stresses. Extension through normal faulting is seemingly rare within the ridged plains but commonplace within dilational bands, which resemble terrestrial mid-ocean ridges in many respects. Long-term changes in the orientations of multiply superposed extensional lineaments strongly suggest a temporally variable stress field, ostensibly in response to nonsynchronous rotation of the icy shell.

Contractional deformation is required to balance the surface area budget in light of the new surface area created at dilational bands, which exceeds the amount that would be expected purely as a result of icy shell thickening and expansion. Initially elusive in Galileo images, mounting evidence suggests that contraction may be common on Europa. Although folding is not likely to be globally significant, convergence bands and ubiquitous ridges are plausibly prominent accommodators of contraction. Unanswered questions remain regarding where convergence bands preferentially develop and whether contraction across ridges contributes to ridge construction in tandem with shearing and frictional heating.

Lateral shearing is common as a result of the constantly changing orientations of principal stresses predicted on the diurnal timescale as well as over longer time periods due to nonsynchronous rotation. As long as a fracture has not healed, it can potentially be reactivated by shear stresses that resolve onto the fracture plane from the tidal stress field. These shear stresses may induce strike-slip motions, which may or may not be frictionally resisted along the crack depending on whether the concomitant normal stresses are compressive or tensile (i.e., whether the crack is closed or dilated during the shearing). Active fractures should undergo cycles of tension and compression, as well as backward and forward motion, over the course of the europan day. These motions may result in a slow ratcheting with a preferred motion sense depending on location and crack orientation, ultimately accumulating visible strike-slip offsets; this process is referred to as tidal walking. Convergence across ridges may also produce apparent lateral offsets, resembling those created by strike-slip motions.

Europa is clearly one of the most tectonically diverse bodies in the solar system. Its crack-riddled surface has left a trail of clues for unraveling its deformation and stress history. The moon likely reached the pinnacle of its tectonic activity a long time ago (perhaps millions of years). Gradual thickening of the icy shell may have ultimately resulted in the attainment of a threshold thickness for convectional overturn. This thickening may have resulted in a reduced tidal response, a slower rate of nonsynchronous rotation, and thus smaller global stresses, causing a decrease in the amount of tectonic activity. Nonetheless, tectonic features have been noted to postdate endogenically driven disruptions of the surface that formed regions of chaos. It is probable that tectonic activity has continued to the current day on Europa, with driving stresses plausibly being composed partly of small magnitude tidal components in addition to a large isotropic tension created by the expansion of the icy shell. As with Cassini's encounter with an obviously active Enceladus, it is seems likely that future Europa missions may very well be met with a few surprises.

Acknowledgments. The authors thank G. W. Patterson and R. Pappalardo for their insightful and constructive comments on the original manuscript and acknowledge NASA for its financial support of the authors' research efforts. Useful discussions with L. Prockter, W. Patterson, P. Schenk, and C. Coulter helped improve this work.

REFERENCES

Anderson J. D., Lau E. L., Sjogren W. L., Schubert G., and Moore W. B. (1997) Europa's differentiated internal structure: Inference from two Galileo encounters. *Science, 276,* 1236–1239.

Aydin A. (2006) Failure modes of the lineaments on Jupiter's moon, Europa: Implications for the evolution of its icy crust.

J. Struct. Geol., 28, 2222–2236.

Bader C. E. and Kattenhorn S. A. (2007) Formation of ridge-type strike-slip faults on Europa. *Eos Trans. AGU, 88,* Fall Meet. Suppl., Abstract P53B–1243.

Bader C. E. and Kattenhorn S. A. (2008) Formation mechanisms of europan ridges with apparent lateral offsets. In *Lunar and Planetary Science XXIX,* Abstract #2036. Lunar and Planetary Institute, Houston (CD-ROM).

Barr A. C. and Pappalardo R. T. (2005) Onset of convection in the icy Galilean satellites: Influence of rheology. *J. Geophys. Res., 110,* E12005, DOI: 10.1029/2004JE002371.

Barr A. C., Pappalardo R. T., and Zhong S. (2004) Convective instability in ice I with non-Newtonian rheology: Application to the icy Galilean satellites. *J. Geophys. Res., 109,* E12008, DOI: 10.1029/2004JE002296.

Bart G. D., Greenberg R., and Hoppa G. V. (2003) Cycloids and wedges: Global patterns from tidal stress on Europa. In *Lunar and Planetary Science XXIV,* Abstract #1396. Lunar and Planetary Institute, Houston (CD-ROM).

Beeman M., Durham W. B., and Kirby S. H. (1988) Friction of ice. *J. Geophys. Res., 93,* 7625–7633.

Billings S. E. and Kattenhorn S. A. (2005) The great thickness debate: Ice shell thickness models for Europa and comparisons with estimates based on flexure at ridges. *Icarus, 177,* 397–412.

Collins G. C., Head J. W. III, Pappalardo R. T., and Spaun N. A. (2000) Evaluation of models for the formation of chaotic terrain on Europa. *J. Geophys. Res., 105(E1),* 1709–1716.

Cordero G. and Mendoza B. (2004) Evidence for the origin of ridges on Europa by means of photoclinometric data from the E4 Galileo orbit. *Geofís. Intl., 43,* 301–306.

Coulter C. E., Kattenhorn S. A., and Schenk P. M. (2009) Topographic profile analysis and morphologic characterization of Europa's double ridges. In *Lunar and Planetary Science XL,* Abstract #1960. Lunar and Planetary Institute, Houston (CD-ROM).

Crawford G. D. and Stevenson D. J. (1988) Gas-driven water volcanism and the resurfacing of Europa. *Icarus, 73,* 66–79.

Dahlen F. (1976) The passive influence of the oceans upon the rotation of the earth. *Geophys. J. R. Astron. Soc., 46,* 363–406.

Doggett T., Figueredo P., Greeley R., Hare T., Kolb E., Mullins K., Senske D., Tanaka K., and Weiser S. (2007) Global geologic map of Europa. In *Lunar and Planetary Science XXXVIII,* Abstract #2296. Lunar and Planetary Institute, Houston (CD-ROM).

Dombard A. J. and McKinnon W. B. (2006) Folding of Europa's icy lithosphere: An analysis of viscous-plastic buckling and subsequent topographic relaxation. *J. Struct. Geol., 28,* 2259–2269.

Durham W. B., Heard H. C., and Kirby S. H. (1983) Experimental deformation of polycrystalline H_2O ice at high pressure and low temperature: Preliminary results. *Proc. Lunar Planet. Sci. Conf. 14th,* in *J. Geophys. Res., 88,* B377–B392.

Figueredo P. H. and Greeley R. (2000) Geologic mapping of the northern leading hemisphere of Europa from Galileo solid-state imaging data. *J. Geophys. Res., 105,* 22629–22646.

Figueredo P. H. and Greeley R. (2004) Resurfacing history of Europa from pole-to-pole geological mapping. *Icarus, 167,* 287–312.

Finnerty A. A., Ransford G. A., Pieri D. C., and Collerson K. D. (1981) Is Europa surface cracking due to thermal evolution? *Nature, 289,* 24–27.

Fortt A. L. and Schulson E. M. (2004) Post-terminal compressive deformation of ice: Friction along Coulombic shear faults. *Eos Trans. AGU, 85,* T13A–05.

Gaidos E. J. and Nimmo F. (2000) Tectonics and water on Europa. *Nature, 405,* 637.

Geissler P. E. and 16 colleagues (1998a) Evolution of lineaments on Europa: Clues from Galileo multispectral imaging observations. *Icarus, 135,* 107–126.

Geissler P. E. and 15 colleagues (1998b) Evidence for non-synchronous rotation of Europa. *Nature, 391,* 368–370.

Giese B., Wagner R., Neukum G., Sullivan R., and the Galileo SSI Team (1999) Doublet ridge formation on Europa: Evidence from topographic data. 31st DPS Meeting Abstracts, *Bull. Am. Astron. Soc., 31(4),* 62.08.

Golombek M. P. and Banerdt W. B. (1990) Constraints on the subsurface structure of Europa. *Icarus, 83,* 441–452.

Greeley R. and 17 colleagues (2000) Geologic mapping of Europa. *J. Geophys. Res., 105,* 22559–22578.

Greenberg R. (2004) The evil twin of Agenor: Tectonic convergence on Europa. *Icarus, 167,* 313–319.

Greenberg R. and Weidenschilling S. J. (1984) How fast do Galilean satellites spin? *Icarus, 58,* 186–196.

Greenberg R., Geissler P., Hoppa G., Tufts B. R., Durda D. D., Pappalardo R., Head J. W., Greeley R., Sullivan R., and Carr M. H. (1998) Tectonic processes on Europa: Tidal stresses, mechanical response, and visible features. *Icarus, 135,* 64–78.

Greenberg R., Hoppa G. V., Tufts B. R., Geissler P., and Riley J. (1999) Chaos on Europa. *Icarus, 141,* 263–286.

Groenleer J. M. and Kattenhorn S. A. (2008) Cycloid crack sequences on Europa: Relationship to stress history and constraints on growth mechanics based on cusp angles. *Icarus, 193,* 158–181.

Han L. and Showman A. P. (2008) Implications of shear heating and fracture zones for ridge formation on Europa. *Geophys. Res. Lett., 35,* L03202, DOI: 10.1029/2007GL031957.

Haeussler P. J., Schwartz D. P., Dawson T. E., Stenner H. D., Lienkaemper J. J., Sherrod B., Cinti F. R., Montone P., Craw P. A., Crone A. J., and Personius S. F. (2004) Surface rupture and slip distribution of the Denali and Totschunda faults in the 3 November 2002 M 7.9 earthquake, Alaska. *Bull. Seismol. Soc. Am., 94(6B),* S23–S52.

Harada Y. and Kurita K. (2006) The dependence of surface tidal stress on the internal structure of Europa: The possibility of cracking of the icy shell. *Planet. Space Sci., 54,* 170–180.

Harada Y. and Kurita K. (2007) Effect of non-synchronous rotation on surface stress upon Europa: Constraints on surface rheology. *Geophys. Res. Lett., 34,* L11204, DOI: 10.1029/2007GL029554.

Hayne P. O., Sleep N. H., and Lissauer J. J. (2006) Thickness variations in Europa's icy shell: Stress and rotation effects. In *Europa Focus Group Workshop Abstracts, Vol. 5* (R. Greeley, ed.), pp. 53–54. Arizona State Univ., Tempe.

Head J. W., Pappalardo R. T., and Sullivan R. (1999) Europa: Morphological characteristics of ridges and triple bands from Galileo data (E4 and E6) and assessment of a linear diapirism model. *J. Geophys. Res., 104,* 24223–24236.

Helfenstein P. and Parmentier E. M. (1980) Fractures on Europa: Possible response of an ice crust to tidal deformation. *Proc. Lunar Planet. Sci. Conf. 11th,* pp. 1987–1998.

Helfenstein P. and Parmentier E. M. (1983) Patterns of fracture and tidal stresses on Europa. *Icarus, 53,* 415–430.

Helfenstein P. and Parmentier E. M. (1985) Patterns of fracture and tidal stresses due to nonsynchronous rotation: Implications for fracturing on Europa. *Icarus, 61,* 175–184.

Hoppa G. V. and Tufts B. R. (1999) Formation of cycloidal features on Europa. In *Lunar and Planetary Science XXX,* Abstract #1599. Lunar and Planetary Institute, Houston (CD-ROM).

Hoppa G. V., Tufts B. R., Greenberg R., and Geissler P. E. (1999a) Formation of cycloidal features on Europa. *Science, 285,* 1899–1902.

Hoppa G., Greenberg R., Geissler P., Tufts B. R., Plassmann J., and Durda D. D. (1999b) Rotation of Europa: Constraints from terminator and limb positions. *Icarus, 137,* 341–347.

Hoppa G., Tufts B. R., Greenberg R., and Geissler P. (1999c) Strike-slip faults on Europa: Global shear patterns driven by tidal stress. *Icarus, 141,* 287–298.

Hoppa G., Greenberg R., Tufts B. R., Geissler P., Phillips C., and Milazzo M. (2000) Distribution of strike-slip faults on Europa. *J. Geophys. Res., 105,* 22617–22627.

Hoppa G. V., Tufts B. R., Greenberg R., Hurford T. A., O'Brien D. P., and Geissler P. E. (2001) Europa's rate of rotation derived from the tectonic sequence in the Astypalaea region. *Icarus, 153,* 208–213.

Hurford T. A. (2005) Tides and tidal stress: Applications to Europa. Ph.D. thesis, Univ. of Arizona, Tucson.

Hurford T. A., Beyer R. A., Schmidt B., Preblich B., Sarid A. R., and Greenberg R. (2005) Flexure of Europa's lithosphere due to ridge-loading. *Icarus, 177,* 380–396.

Hurford T. A., Bills B. G., Sarid A. R., and Greenberg R. (2006) Unraveling Europa's tectonic history: Evidence for a finite obliquity? In *Lunar and Planetary Science XXXVII,* Abstract #1303. Lunar and Planetary Institute, Houston (CD-ROM).

Hurford T. A., Sarid A. R., and Greenberg R. (2007a) Cycloidal cracks on Europa: Improved modeling and non-synchronous rotation implications. *Icarus, 186,* 218–233.

Hurford T. A., Helfenstein P., Hoppa G. V., Greenberg R., and Bills B. G. (2007b) Eruptions arising from tidally controlled periodic openings of rifts on Enceladus. *Nature, 447,* 292–294.

Hurford T. A., Bills B. G., Greenberg R., Hoppa G. V., and Helfenstein P. (2008) How libration affects strike-slip displacement on Enceladus. In *Lunar and Planetary Science XXXIX,* Abstract #1826. Lunar and Planetary Institute, Houston (CD-ROM).

Hurford T. A., Greenberg R., Bills B. G., and Sarid A. R. (2009) The influence of obliquity on europan cycloid formation. *Icarus,* DOI: 10.1016/j.icarus.2009.02.036.

Jackson M. D. and Pollard D. D. (1988) The laccolith-stock controversy: New results from the southern Henry Mountains, Utah. *Geol. Soc. Am. Bull., 100,* 117–139.

Kadel S. D., Fagents S. A., and Greeley R. (1998) Trough-bounding ridge pairs on Europa: Considerations for an endogenic model of formation. In *Lunar and Planetary Science XXIX,* Abstract #1078. Lunar and Planetary Institute, Houston (CD-ROM).

Kadel S. D., Chuang F. C., Greeley R., Moore J. M., and the Galileo SSI Team (2000) Geological history of the Tyre region of Europa: A regional perspective on europan surface features and ice thickness. *J. Geophys. Res., 105,* 22657–22669.

Kargel J. S. (1991) Brine volcanism and the interior structures of asteroids and icy satellites. *Icarus, 94,* 368–390.

Kattenhorn S. A. (2002) Nonsynchronous rotation evidence and fracture history in the Bright Plains region, Europa. *Icarus, 157,* 490–506.

Kattenhorn S. A. (2004a) Strike-slip fault evolution on Europa: Evidence from tail-crack geometries. *Icarus, 172,* 582–602.

Kattenhorn S. A. (2004b) What is (and isn't) wrong with both the tension and shear failure models for the formation of lineae on Europa. In *Workshop on Europa's Icy Shell: Past, Present, and Future,* pp. 38–39. LPI Contribution No. 1195, Lunar and Planetary Institute, Houston.

Kattenhorn S. A. and Marshall S. T. (2006) Fault-induced perturbed stress fields and associated tensile and compressive deformation at fault tips in the ice shell of Europa: Implications for fault mechanics. *J. Struct. Geol., 28,* 2204–2221.

Kattenhorn S. A., Groenleer J. M., Marshall S. T., and Vetter J. C. (2007) Shearing-induced tectonic deformation on icy satellites: Europa as a case study. In *Workshop on Ices, Oceans, and Fire: Satellites of the Outer Solar System,* pp. 74–75. LPI Contribution No. 1357, Lunar and Planetary Institute, Houston.

Kennedy F. E., Schulson E. M., and Jones D. E. (2000) The friction of ice on ice at low sliding velocities. *Philos. Mag. A., 80,* 1093–1110.

Kivelson M. G., Khurana K. K., Russell C. T., Volwerk M., Walker R. J., and Zimmer C. (2000) Galileo magnetometer measurements: A stronger case for a subsurface ocean at Europa. *Science, 289,* 1340–1343.

Lee S., Pappalardo R. T., and Makris N. C. (2005) Mechanics of tidally driven fractures in Europa's ice shell. *Icarus, 177,* 367–379.

Leith A. C. and McKinnon W. B. (1996) Is there evidence for polar wander on Europa? *Icarus, 120,* 387–398.

Lucchitta B. K. and Soderblom L. A. (1982) The geology of Europa. In *The Satellites of Jupiter* (D. Morrison, ed.), pp. 521–555. Univ. of Arizona, Tucson, Arizona.

Luttrell K. and Sandwell D. (2006) Strength of the lithosphere of the Galilean satellites. *Icarus, 183,* 159–167.

Malin M. C. and Pieri D. C. (1986) Europa. In *Satellites* (J. A. Burns and M. S. Matthews, eds.), pp. 689–717. Univ. of Arizona, Tucson, Arizona.

Manga M. and Wang C.-Y. (2007) Pressurized oceans and the eruption of liquid water on Europa and Enceladus. *Geophys. Res. Lett., 34(7),* L07202, DOI: 10.1029/2007GL029297.

Marshall S. T. and Kattenhorn S. A. (2005) A revised model for cycloid growth mechanics on Europa: Evidence from surface morphologies and geometries. *Icarus, 177,* 341–366.

Matsuyama I. and Nimmo F. (2008) Tectonic patterns on reoriented and despun planetary bodies. *Icarus, 195,* 459–473.

McBee J. H., Hartmann D., and Collins G. C. (2003) Strain across ridges on Europa. In *Lunar and Planetary Science XXXIV,* Abstract #1783. Lunar and Planetary Institute, Houston (CD-ROM).

McEwen A. S. (1986) Tidal reorientation and the fracturing of Jupiter's moon Europa. *Nature, 321,* 49–51.

Melosh H. J. (1977) Global tectonics of a despun planet. *Icarus, 31,* 221–243.

Melosh H. J. (1989) *Impact Cratering: A Geologic Process.* Oxford Univ., New York. 245 pp.

Melosh H. J. and Turtle E. P. (2004) Ridges on Europa: Origin by incremental ice wedging. In *Lunar and Planetary Science XXXV,* Abstract #2029. Lunar and Planetary Institute, Houston (CD-ROM).

Mével L. and Mercier E. (2002) Geodynamics on Europa: Evidence for a crustal resorption process. In *Lunar and Planetary Science XXXIII,* Abstract #1476. Lunar and Planetary Institute, Houston (CD-ROM).

Michalski J. R. and Greeley R. (2002) En echelon ridge and trough structures on Europa. *Geophys. Res. Lett., 29,* 1498, DOI: 10.1029/2002GL014956.

Mitri G. and Showman A. P. (2008) A model for the temperature-dependence of tidal dissipation in convective plumes on icy satellites: Implications for Europa and Enceladus. *Icarus, 195,* 758–764.

Moore W. B. (2006) Thermal equilibrium in Europa's ice shell. *Icarus, 180,* 141–146.

Moore W. B. and Schubert G. (2000) The tidal response of Europa. *Icarus, 147,* 317–319.

Nimmo F. (2004a) Stresses generated in cooling viscoelastic ice shells: Application to Europa. *J. Geophys. Res., 109,* E12001, DOI: 10.1029/2004JE002347.

Nimmo F. (2004b) Dynamics of rifting and modes of extension on icy satellites. *J. Geophys. Res., 109,* E01003, DOI: 10.1029/2003JE002168.

Nimmo F. and Gaidos E. (2002) Strike-slip motion and double ridge formation on Europa. *J. Geophys. Res., 107,* 1–8.

Nimmo F. and Schenk P. (2006) Normal faulting on Europa: Implications for ice shell properties. *J. Struct. Geol., 28,* 2194–2203.

Nimmo F., Pappalardo R. T., and Giese B. (2003) On the origins of band topography, Europa. *Icarus, 166,* 21–32.

Nimmo F., Pappalardo R. T., and Moore W. M. (2005) Icy satellite shell thickening: Consequences for non-synchronous rotation rates and stresses. *Eos Trans. AGU, 86(52),* Fall Meet. Suppl., Abstract P22A-05.

Nimmo F., Spencer J. R., Pappalardo R. T., and Mullen M. E. (2007) Shear heating as the origin of the plumes and heat flux on Enceladus. *Nature, 447,* 289–291.

Noll K. S., Weaver H. A., and Gonnella A. M. (1995) The albedo spectrum of Europa from 2200 Å to 3300 Å. *J. Geophys. Res., 100,* 19057–19059.

Ojakangas G. W. and Stevenson D. J. (1989a) Thermal state of an ice shell on Europa. *Icarus, 81,* 220–241.

Ojakangas G. W. and Stevenson D. J. (1989b) Polar wander of an ice shell on Europa. *Icarus, 81,* 242–270.

Panning M., Lekic V., Manga M., Cammarano F., and Romanowicz B. (2006) Long-period seismology on Europa: 2. Predicted seismic response. *J. Geophys. Res., 111,* E12008, DOI: 10.1029/2006JE002712.

Pappalardo R.T. and Davis D. M. (2007) Where's the compression? Explaining the lack of contractional structures on icy satellites. In *Workshop on Ices, Oceans, and Fire: Satellites of the Outer Solar System,* pp. 108–109. LPI Contribution No. 1357, Lunar and Planetary Institute, Houston.

Pappalardo R. T. and Sullivan R. J. (1996) Evidence for separation across a gray band on Europa. *Icarus, 123,* 557–567.

Pappalardo R.T. and 31 colleagues (1999) Does Europa have a subsurface ocean? Evaluation of the geological evidence. *J. Geophys. Res., 104,* E10, 24015–24055.

Patterson G. W. and Head J. W. (2007) Kinematic analysis of triple junctions on Europa. In *Workshop on Ices, Oceans, and Fire: Satellites of the Outer Solar System,* pp. 110–111. LPI Contribution No. 1357, Lunar and Planetary Institute, Houston.

Patterson G. W. and Pappalardo R. T. (2002) Compression across ridges on Europa. In *Lunar and Planetary Science XXXIII,* Abstract #1681. Lunar and Planetary Institute, Houston (CD-ROM).

Patterson G. W., Head J. W., and Pappalardo R. T. (2006) Plate motion on Europa and nonrigid behavior of the icy lithosphere: The Castalia Macula region. *J. Struct. Geol., 28,* 2237–2258.

Peale S. J. (1977) Rotation histories of the natural satellites. In *Planetary Satellites* (J. Burns, ed.), pp. 87–112. Univ. of Arizona, Tucson.

Peale S. J. and Cassen P. (1978) Contribution of tidal dissipation to lunar thermal history. *Icarus, 36,* 245–269.

Phillips C. B. and Chyba C. F. (2003) Methods for detecting current geological activity on Europa. *Forum on Concepts and Approaches for Jupiter Icy Moon Orbiter,* Abstract #9018. LPI Contribution No. 1163, Lunar and Planetary Institute, Houston.

Phillips C. B., McEwen A. S., Hoppa G. V., Fagents S. A., Greeley R., Klemaszewski J. E., Pappalardo R. T., Klaasen K. P., and Breneman H. H. (2000) The search for current geologic activity on Europa. *J. Geophys. Res., 105,* 22579–22597.

Pieri D. C. (1981) Lineament and polygon patterns on Europa. *Nature, 289,* 17–21.

Pollard D. D. and Segall P. (1987) Theoretical displacements and stresses near fractures in rock: With applications to faults, joints, veins, dikes, and solution surfaces. In *Fracture Mechanics of Rock* (B. K. Atkinson, ed.), pp. 277–349. Academic, London.

Porco C. C. and 24 colleagues (2006) Cassini observes the active south pole of Enceladus. *Science, 311,* 1393–1401.

Prockter L. M. (2001) Creation and destruction of lithosphere on Europa: From bands to folds. In *Lunar and Planetary Science XXXII,* Abstract #1452. Lunar and Planetary Institute, Houston (CD-ROM).

Prockter L. M. and Pappalardo R. T. (2000) Folds on Europa: Implications for crustal cycling and accommodation of extension. *Science, 289,* 941–943.

Prockter L. M., Pappalardo R. T., Sullivan R., Head J. W., Patel J. G., Giese B., Wagner R., Neukum G., and Greeley R. (1999) Morphology and evolution of europan bands: Investigation of a seafloor spreading analog. In *Lunar and Planetary Science XXX,* Abstract #1900. Lunar and Planetary Institute, Houston (CD-ROM).

Prockter L. M., Pappalardo R. T., and Head III J. W. (2000) Strike-slip duplexing on Jupiter's icy moon Europa. *J. Geophys. Res., 105,* 9483–9488.

Prockter L. M., Head J., Pappalardo R., Sullivan R., Clifton A. E., Giese B., Wagner R., and Neukum G. (2002) Morphology of europan bands at high resolution: A mid-ocean ridge-type rift mechanism. *J. Geophys. Res., 107,* 1–26.

Qin R., Buck W. R., and Germanovich L. (2007) Comment on "Mechanics of tidally driven fractures in Europa's ice shell" by S. Lee, R. T. Pappalardo, and N. C. Makris (2005, *Icarus 177,* 367–379). *Icarus, 189,* 595–597.

Rambaux N., Karatekin Ö., and Van Hoolst T. (2007) Europa's librations and ice shell thickness. *Soc. Franc. Astron. Astrophys., SF2A,* 1–6.

Riley J., Greenberg R., and Sarid A. (2006) Europa's South Pole region: A sequential reconstruction of surface modification processes. *Earth Planet. Sci. Lett., 248,* 808–821.

Rist M. A. (1997) High-stress ice fracture and friction. *J. Phys. Chem., 101,* 6263–6266.

Rist M. A. and Murrell S. A. F. (1994) Ice triaxial deformation and fracture. *J. Glaciol., 40,* 305–318.

Rist M. A., Jones S. J., and Slade T. D. (1994) Microcracking and shear fracture in ice. *Ann. Glaciol., 19,* 131–137.

Rothery D. A. (1992) *Satellites of the Outer Planets.* Clarendon, Oxford. 208 pp.

Rudolph M. L. and Manga M. (2009) Fracture penetration in planetary ice shells. *Icarus, 199,* 536–541.

Sammonds P. R., Murrell S. A. F., and Rist M. A. (1989) Fracture of multi-year sea ice under triaxial stresses: Apparatus description and preliminary results. *J. Offshore Mech. Arct.*

Eng., 111, 258–263.

Sandwell D., Rosen P., Moore W., and Gurrola E. (2004) Radar interferometry for measuring tidal strains across cracks on Europa. *J. Geophys. Res., 109,* E11003, DOI: 10.1029/2004JE002276.

Sarid A. R., Greenberg R., Hoppa G., Hurford T. A., Tufts R., and Geissler P. (2002) Polar wander and surface convergence of Europa's ice shell: Evidence from a survey of strike-slip displacement. *Icarus, 158,* 24–41.

Sarid A. R., Greenberg R., Hoppa G. V., Geissler P., and Preblich B. (2004) Crack azimuths on Europa: Time sequence in the southern leading face. *Icarus, 168,* 144–157.

Sarid A. R., Greenberg R., Hoppa G. V., Brown D. M., and Geissler P. (2005) Crack azimuths on Europa: The G1 lineament sequence revisited. *Icarus, 173,* 469–479.

Sarid A. R., Greenberg R., and Hurford T. A. (2006) Crack azimuths on Europa: Sequencing of the northern leading hemisphere. *J. Geophys. Res., 111,* E08004, DOI: 10.1029/2005JE002524.

Sarid-Rhoden A. R., Militzer B., Huff E. M., Hurford T. A., Manga M., and Richards M. (2009) Implications for Europa's obliquity from cycloid modeling. In *Lunar and Planetary Science XL,* Abstract #1891. Lunar and Planetary Institute, Houston (CD-ROM).

Schenk P. and McKinnon W. B. (1989) Fault offsets and lateral crustal movement on Europa: Evidence for a mobile ice shell. *Icarus, 79,* 75–100.

Schenk P. M. and Pappalardo R. T. (2004) Topographic variations in chaos on Europa: Implications for diapiric formation. *Geophys. Res. Lett., 31,* L16703, DOI: 10.1029/2004GL019978.

Schenk P., Matsuyama I., and Nimmo F. (2008) True polar wander on Europa from global-scale small-circle depressions. *Nature, 453,* 368–371.

Schulson E. M. (1987) The fracture of ice I_h. *J. Phys. (Paris), Colloque C1, 48,* 207–218.

Schulson E. M. (2001) Brittle failure of ice. *Engrg. Fract. Mech., 68,* 1839–1887.

Schulson E. M. (2002) On the origin of a wedge crack within the icy crust of Europa. *J. Geophys. Res., 107,* 5107, DOI: 10.1029/2001JE001586.

Schulson E. M. (2009) Frictional sliding of cold ice. In *Lunar and Planetary Science XL,* Abstract #1795. Lunar and Planetary Institute, Houston (CD-ROM).

Schulson E. M., Iliescu D., and Renshaw C. E. (1999) On the initiation of shear faults during brittle compressive failure: A new mechanism. *J. Geophys. Res., 104,* 695–705.

Showman A. P. and Han L. (2005) Effects of plasticity on convection in an ice shell: Implications for Europa. *Icarus, 177,* 425–437.

Smith B. A. and the Voyager Imaging Team (1979) The Galilean satellites of Jupiter: Voyager 2 imaging science results. *Science, 206,* 951–972.

Smith-Konter B. and Pappalardo R. T. (2008) Tidally driven stress accumulation and shear failure of Enceladus's tiger stripes. *Icarus, 198,* 435–451.

Sotin C., Head J. W. III, and Tobie G. (2002) Europa: Tidal heating of upwelling thermal plumes and the origin of lenticulae and chaos melting. *Geophys. Res. Lett., 29,* DOI: 10.1029/2001GL013844.

Spaun N. A., Pappalardo R. T., and Head J. W. III (2003) Evidence for shear failure in forming near-equatorial lineae on Europa. *J. Geophys. Res., 108(E6),* 5060, DOI: 10.1029/2001JE001499.

Spencer J. R., Pearl J. C., Segura M., Flasar F. M., Mamoutkine A., Romani P., Buratti B. J., Hendrix A. R., Spilker L. J., and Lopes R. M. C. (2006) Cassini encounters Enceladus: Background and the discovery of a south polar hot spot. *Science, 311,* 1401–1405.

Squyres S. W. and Croft S. K. (1986) The tectonics of icy satellites. In *Satellites* (J. A. Burns and M. S. Matthews, eds.), pp. 293–341. Univ. of Arizona, Tucson.

Stempel M. M., Barr A. C., and Pappalardo R. T. (2005) Model constraints on the opening rates of bands on Europa. *Icarus, 177,* 297–304.

Sullivan R. and 15 colleagues (1997) Ridge formation on Europa: Examples from Galileo high resolution images. *GSA Annual Convention, Abstr. with Progr., 29,* A-312.

Sullivan R. and 12 colleagues (1998) Episodic plate separation and fracture infill on the surface of Europa. *Nature, 391,* 371–373.

Tufts B. R. (1998) Lithospheric displacement features on Europa and their interpretation. Ph.D. thesis, Univ. of Arizona, Tucson. 288 pp.

Tufts B. R., Greenberg R., Hoppa G., and Geissler P. (1999) Astypalaea Linea: A large-scale strike-slip fault on Europa. *Icarus, 141,* 53–64.

Tufts B. R., Greenberg R., Hoppa G., and Geissler P. (2000) Lithospheric dilation on Europa. *Icarus, 146,* 75–97.

Turcotte D. L. and Schubert G. (2002) *Geodynamics,* 2nd edition. Cambridge Univ., New York. 456 pp.

Turtle E. P., Melosh H. J., and Phillips C. B. (1998) Tectonic modeling of the formation of europan ridges. *Eos Trans. AGU, 79,* F541.

Van Hoolst T., Rambaux N., Karatekin Ö., Dehant V., and Rivoldini A. (2008) The librations, shape, and icy shell of Europa. *Icarus, 195,* 386–399.

Vetter J. C. (2005) Evaluating displacements along europan ridges. M.S. thesis, Univ. of Idaho, Moscow.

Wahr J., Selvans Z. A., Mullen M. E., Barr A. C., Collins G. C., Selvans M. M., and Pappalardo R. T. (2009) Modeling stresses on satellites due to nonsynchronous rotation and orbital eccentricity using gravitational potential theory. *Icarus, 200,* 188–206.

Williams K. K. and Greeley R. (1998) Estimates of ice thickness in the Conamara Chaos region of Europa. *Geophys. Res. Lett., 25,* 4273–4276.

Wilson G. (1960) The tectonics of the 'Great Ice Chasm', Filchner Ice Shelf, Antarctica. *Proc. Geol. Assoc., 71,* 130–138.

Yoder C. F. (1979) How tidal heating on Io drives the Galilean resonance locks. *Nature, 279,* 767–770.

Zahnle K., Schenk P., Levison H., and Dones L. (2003) Cratering rates in the outer solar system. *Icarus, 163,* 263–289.

Morphology and Evolution of Europa's Ridges and Bands

Louise M. Prockter and G. Wesley Patterson
Johns Hopkins University Applied Physics Laboratory

Europa's surface exhibits a variety of landforms, both familiar and unique. Spacecraft exploration has enabled the role of these surface features to be better understood in the context of Europa's history and evolution. Two of the most prominent geological phenomena on Europa are ridges and bands, linear features that are strikingly young by planetary standards. In this chapter we investigate their detailed morphological characteristics, stratigraphic relationships, and formation mechanisms. Several different morphological types of ridge are present on the surface at a variety of scales and most likely formed through shear heating. Pull-apart bands may have formed in a manner analogous to that of mid-ocean ridges on Earth, in response to extensional stresses, and they can stand significantly higher than their surroundings, suggesting a solid-state origin. Other types of bands have undergone considerable shear deformation and possibly contraction. Many linear features have dark diffuse deposits along their margins that may represent "cryoclastic" eruptions or local thermal effects. The origins of ridges and bands may not be directly related, although bands may exploit ridges as zones of weakness.

1. INTRODUCTION

The surface of Europa was first imaged by the Voyager spacecraft ~30 years ago at relatively low resolution (up to ~2 km/pixel), and it was immediately clear that linear features played a prominent role in the visible surface history of the satellite (e.g., *Smith et al.,* 1979; *Pieri,* 1981; *Lucchitta and Soderblom,* 1982). These features were mapped and classified based on shape and albedo, and it was suggested that their formation was controlled by fracturing processes and tidal deformation (*Pieri,* 1981; *Lucchitta and Soderblom,* 1982).

Two decades later, the Galileo spacecraft provided images of the surface at resolutions of one to two orders of magnitude better than those obtained by Voyager (*Belton et al.,* 1996; *Greeley et al.,* 2000, 2004). Here, we highlight the two structural feature types that most excite us about this enigmatic satellite: ridges and bands. Both features appear to initiate as fractures in Europa's brittle lithosphere, and are subsequently modified to produce the morphology visible on the surface. The formation and modification of Europa's linear features has been attributed to deformation of the icy shell as the result of stresses induced by diurnal tides, nonsynchronous rotation, and/or polar wander (*Greeley et al.,* 2004; see chapter by Kattenhorn and Hurford).

Evidence suggests that linear and cycloidal fracture patterns observed on Europa's surface can be explained by diurnal tidal stresses that vary on the 3.55-d timescale of the satellite's orbit and result from its orbital eccentricity and proximity to Jupiter (*Helfenstein and Parmentier,* 1985; *Greenberg et al.,* 1998; *Hoppa et al.,* 1999b). It has also been suggested that torques imposed on Europa by tidal forces could lead to nonsynchronous rotation of the satellite's decoupled outer shell with respect to its interior (*Greenberg and Weidenshilling,* 1984; *Ojakangas and Stevenson,* 1989a). Action of this latter mechanism is supported by the orientations of many prominent global-scale surface features (*McEwen,* 1986; *Leith and McKinnon,* 1996; *Geissler et al.,* 1998a) and by regional-scale mapping efforts, which have inferred systematic variations in the orientations of co-located features with respect to relative age (*Figueredo and Greeley,* 2000; *Kattenhorn,* 2002; *Figueredo and Greeley,* 2004). The stresses involved with the nonsynchronous mechanism are predicted to migrate across the surface on timescales $>10^3$ yr (*Greenberg et al.,* 1998; *Hoppa et al.,* 1999b, 2001). It has also been suggested that significant variation in the thickness of Europa's icy shell from equator to pole could lead to true polar wander, a process that would impart significant stresses into the satellite's lithosphere (*Ojakangas and Stevenson,* 1989b; *Leith and McKinnon,* 1996). Based on the orientation and distribution of various features mapped from Galileo spacecraft images, it has been proposed that ~30° (*Sarid et al.,* 2002) to ~80° (*Schenk et al.,* 2008) of true polar wander may have occurred in Europa's relatively recent past. These driving forces for the formation of surface features on Europa are mentioned throughout this chapter as they relate to the formation and evolution of ridges and bands. More detailed discussions of these topics are provided in the several chapters by Sotin et al., Bills et al., Kattenhorn and Hurford, and Nimmo and Manga.

Every part of Europa that has been imaged to date contains ridges, the most ubiquitous of which is the double ridge, consisting of two raised crests flanking a central V-shaped trough. These ridges may be part of a continuum of features that form when a simple fracture gets worked by tidal and other global stresses, resulting in a distinctive morphology (Fig. 1). Further reworking may result in more complex ridge types, including some that are composed of numerous anastomosing double ridges. Several models have

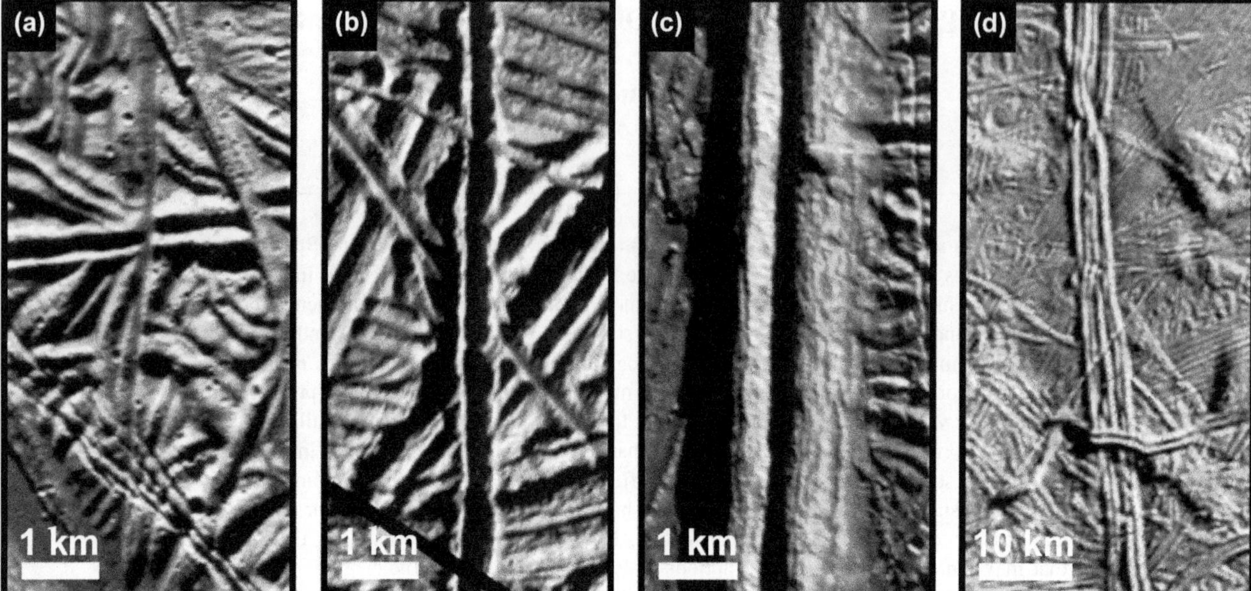

Fig. 1. A classification sequence for ridges on Europa: **(a)** troughs; **(b)** raised-flank troughs; **(c)** double ridges; and **(d)** ridge complexes (cf. *Pappalardo et al.,* 1998). Panels **(a)** and **(b)** are found in Galileo observation E6ESBRTPLN02 at a resolution of 30 m/pixel. **(c)** Prominent double ridge Androgeous Linea (see Fig. 2) is also found in Galileo observation E6ESBRTPLN02. **(d)** A portion of Asterius Linea, found in Galileo observation E6ESDRKLIN01 at a resolution of 220 m/pixel [note change in scale for **(d)** relative to the others].

been proposed for the formation of ridge morphology, but as we will discuss, it seems likely that shear deformation plays a primary role. Most ridge types found on Europa appear to be unique to that moon, but some double-ridge structures have also been identified on Triton (*Prockter et al.,* 2005), and on Enceladus (*Hurford et al.,* 2007; *Nimmo et al.,* 2007) where they appear to be the locus of heat production and source regions for plume activity (*Abramov and Spencer,* 2009).

A significant portion of Europa's surface has been replaced by the formation of bands — linear, wedge-shaped, or sickle-shaped features that are clearly visible cutting swaths through background ridged plains at a variety of different scales, and which themselves are disrupted and overprinted by later chaos and ridge formation. The significance of "pull-apart" or "dilational" bands (also historically known as "dark" or "gray" bands) and their role in the resurfacing of Europa was first recognized by *Schenk and McKinnon* (1989) from Voyager data. These authors proposed that the icy crust containing bands was probably mechanically decoupled from the silicate interior during their formation, allowing bands to form by opening atop a ductile substratum. Studies of the prominent band Thynia Linea by *Pappalardo and Sullivan* (1996) led those authors to suggest a possible volcano-tectonic scenario for the resurfacing of Europa. These authors also noted that bands appear to brighten with time, as young bands are relatively dark, while older bands are as bright as Europa's background plains.

Detailed morphological studies of bands were made possible with the advent of high-resolution Galileo images, and showed that a variety of features may exist within the interiors of bands. We denote two major classes of bands: pull-apart (or dilational) bands and bright bands. Pull-apart bands are extensional, commonly with some associated shear: They formed when the surface cracked apart and new material was emplaced into the gap from below. The margins of these bands can be reconstructed to show how the terrain looked before the episode of band formation. Pull-apart bands can be further subdivided based on interior morphology. Bright bands are not so well understood; they do not reconstruct perfectly, and may have undergone different types of deformation including shear and/or compression.

This chapter is intended to be complementary to that of Kattenhorn and Hurford. Those authors discuss the underlying causes of tectonic features on Europa; here we examine in detail the surface expressions of ridges and bands, which are highly unusual even among icy satellites, but are ubiquitous across the surface of Europa. We first discuss the different ridge types including their morphology, topography, and stratigraphy, and models for how they may have formed. We then present the two major types of bands, including their morphology and other characteristics, and discuss their potential formation mechanisms. We finish with a discussion of the outstanding questions about both types of features.

2. RIDGES

Ridges are the most morphologically common feature on Europa, but they remain incompletely understood. Characterizing ridges can give insight into their modes of formation and provide valuable information regarding the satellite's geophysical state, including the nature of its tidal processes and the presence and distribution of liquid water. In this section, we discuss Europa's ridges in terms of morphology, topography, superposition relationships, and formation mechanisms (additional discussion involving formation mechanisms can be found in the chapter by Kattenhorn and Hurford).

2.1. Morphology

Ridge features on Europa come in a variety of morphological forms including raised-flank troughs, double ridges, and morphologically complex ridges that consist of a series of anastomosing and inosculating component ridges (Fig. 1). Morphological classification schemes have been proposed that suggest that these classes are related, with initial cracks transitioning into raised-flank troughs, which then transition into double ridges, and then into more morphologically complex features (*Greenberg et al.,* 1998; *Head et al.,* 1999). Ridges and troughs are observed on length scales of up to thousands of kilometers and can range from linear to cycloidal in planform. Double ridges are by far the most common ridge type on Europa and appear to be a fundamental building block that has shaped the landscape we observe today. It is not clear whether ridges and bands are related; it has been suggested that bands are an endmember of the process by which ridges are formed (section 3.1) (*Tufts et al.,* 2000), but it has also been noted that some ridges appear to have been exploited by subsequent band formation (section 3.1) (*Prockter et al.,* 2002).

2.1.1. Double ridges. The "double ridge" morphology is ubiquitous and is characterized by the presence of an axial trough flanked by twinned ridge crests (Figs. 1c and 2). Double ridges can range in width from ~200 m to >4 km and have lengths from tens of kilometers to many hundreds of kilometers (*Greeley et al.,* 2000). The morphological characteristics of double ridges are generally uniform over long distances and the ridges themselves can be remarkably straight for significant portions of their length.

The small-scale morphological characteristics of double ridges have been described in detail by *Head et al.* (1999), who noted that the ridge crests may contain narrow apical zones of small-scale, ridge-parallel faulting (Figs. 2b,c). The paired ridges themselves each have an inner and outer slope. The inner slopes are relatively straight, typically producing a V-shaped axial trough. They can be striated perpendicular to the ridge length, leading *Head et al.* (1999) to suggest that slumping and local faulting parallel to the ridge, followed by mass wasting, have played an important role in the development of the inner wall (Fig. 2). Some double ridges are characterized by a smaller inner ridge or ridges that typically lie at a lower elevation than the flanking ridge pair. Preexisting structures from the surrounding terrain can sometimes be traced up the outer slopes of the ridge flanks, while in other cases they appear to be partially buried (arrows, Fig. 2b). The flanks of ridges are commonly mantled by continuous deposits of dark, reddish material (Fig. 2c) that subdues the topography, and they can show signatures of mass wasted debris, including boulders (also see chapter by Moore et al.). Discontinuous lineations and irregular crenulations are common along the lengths of the outer slopes, suggestive of stepped terraces.

Subtle and broad marginal troughs a few tens of meters deep have also been observed in association with double ridges (*Pappalardo and Coon,* 1996; *Tufts,* 1998; *Billings and Kattenhorn,* 2005; *Hurford et al.,* 2005), and subparallel flanking fractures (presumably tensile) may be found near the outer reaches of these troughs (Fig. 2a). It has been suggested that these troughs and associated fractures result from either loading of the lithosphere (*Tufts et al.,* 2000; *Billings and Kattenhorn,* 2005) or uplift plausibly due to shear heating and intrusion along the underlying fault (*Gaidos and Nimmo,* 2000). Additionally, these marginal troughs can possess diffuse, discontinuous deposits of lower relative albedo material that appear to embay and cover preexisting terrain (Fig. 2a).

2.1.2. Ridge complexes. Our morphological classification of ridge complexes encompasses features that have been previously described as "ridge complexes" and "complex ridges" as well as some features described as "triple bands" (*Lucchitta and Soderblom,* 1982; *Greenberg et al.,* 1998; *Head et al.,* 1999; *Figueredo and Greeley,* 2000, 2004). Ridge complexes (Figs. 1d and 3) are defined here as features up to tens of kilometers wide consisting of a collection of subparallel anastamosing and inosculating single or double ridges. Ridge complexes have trends that are linear to curvilinear, no obvious bilateral symmetry, and margins that are sinuous or undulating. Individual ridges within a complex occasionally divert away from the main set and otherwise isolated ridges appear to occasionally divert toward and become incorporated into a throughgoing complex. Some of the most prominent linear-to-curvilinear features observable in global-scale images of the satellite are of this ridge type, and much of Europa's ridged plains (section 2.1.3) appear to be composed of ridge complexes.

We present this definition of ridge complexes in the hope that it will be adopted by future workers to avoid confusion of categorization over similar features in the literature. For example, one prominent feature (see Fig. 8c) has previously been termed a "ridge complex" (Fig. 10 of *Greeley et al.,* 2000), a "complex ridge" (Fig. 6 of *Head et al.,* 1999), and a "Class 2" ridge (Fig. 2 of *Tufts et al.,* 2000). Using the present classification, this feature is not a ridge complex at all, but is better characterized as a faulted band (section 3.1.1.2), based on specific morphological characteristics. Primary among these are that it is not composed of a set of anastomosing ridges and that the preexisting terrain it interrupts is reconstructable, something that is characteristic of band-type features (section 3.1), but generally not of ridge-type features.

2.1.3. Ridged plains. The majority of Europa's surface is composed of terrain defined as "undifferentiated plains material" (*Greeley et al.,* 2000), or more popularly as "ridged plains" (see chapter by Doggett et al.). These plains (Fig. 4) are composed of relatively old ridged structures so heavily overprinted by numerous younger features that it is

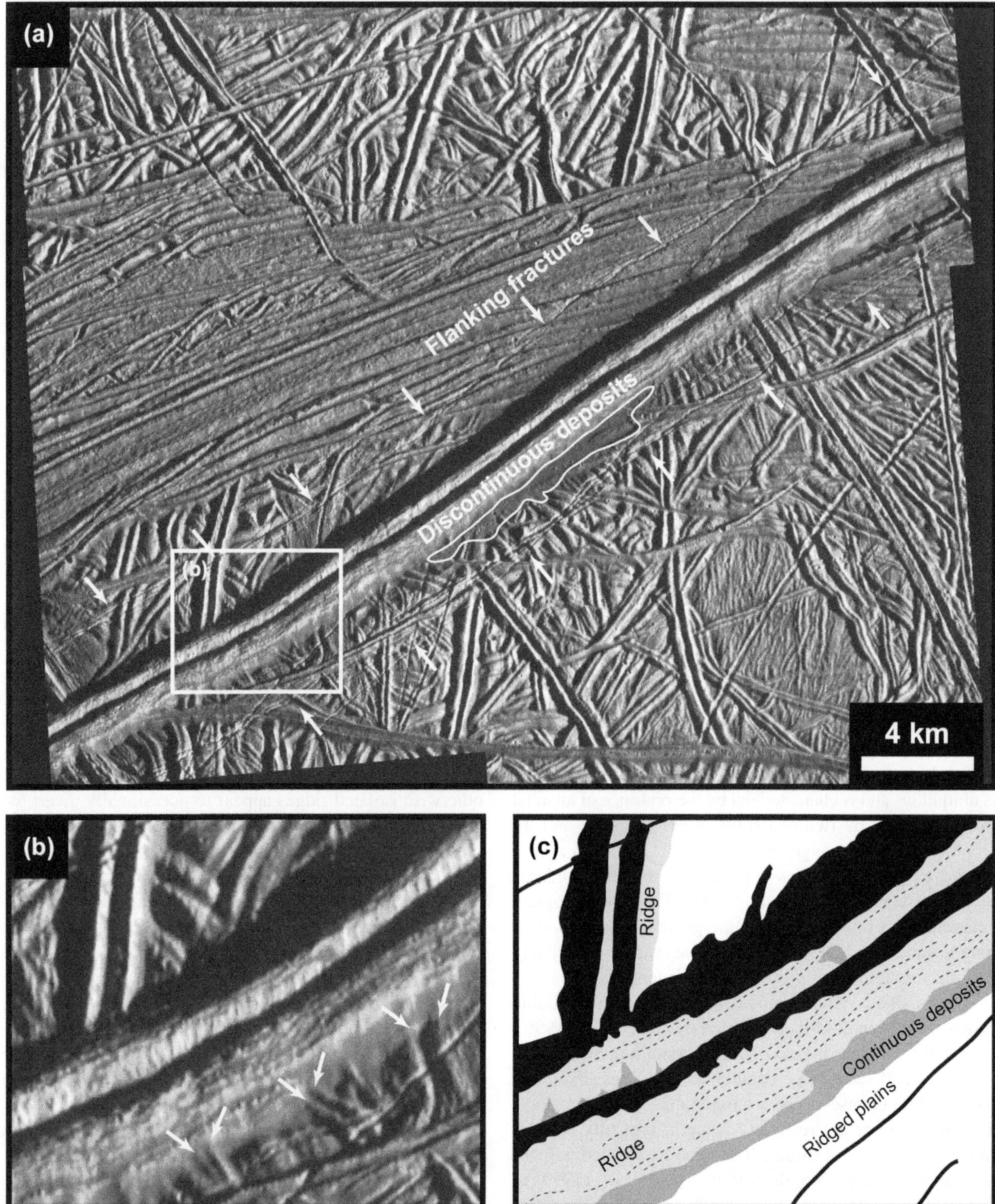

Fig. 2. High-resolution view (~30 m/pixel) of a relatively young and prominent double ridge, Androgeos Linea (14.7°N, 273.4°W, E6ESBRTPLN02). **(a)** Flanking fractures associated with the double ridge are indicated by white arrows and discontinuous deposits are outlined with a solid white line. White box indicates location of inset **(b)**. **(b)** Arrows indicate preexisting structures that can be traced into the outer slopes of the flanks of Androgeos. **(c)** Sketch map indicating the boundaries of the double ridge with its associated continuous deposits and the surrounding ridged plains. Dashed lines indicate linear striations found on the inner and outer slopes of the ridge flanks. Slumps associated with mass wasting and possible terraces are also shown in a darker shade of gray on the inner slope of the ridge. Solid black lines represent locations of flanking fractures.

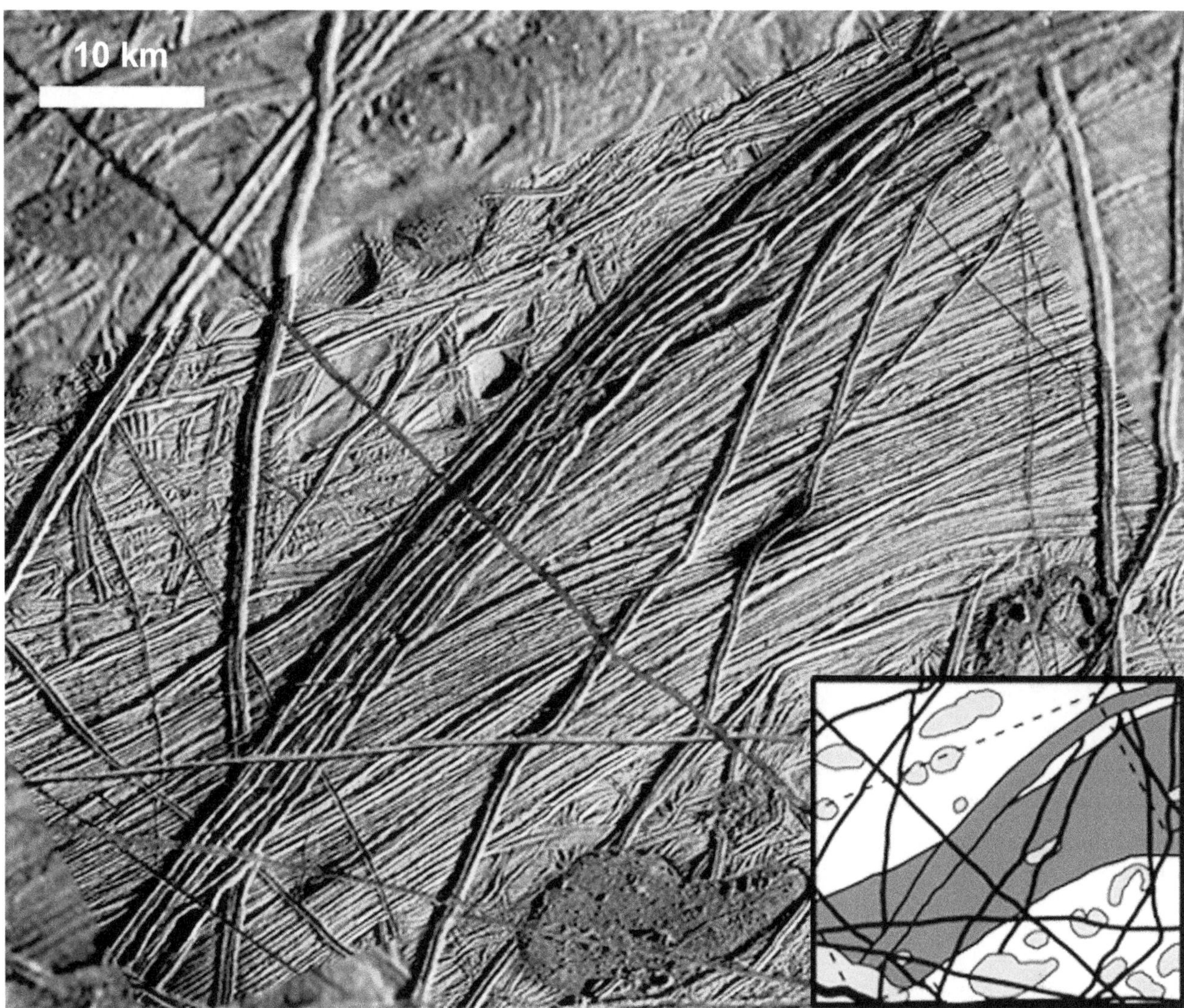

Fig. 3. Image mosaic of a portion of Europa's northern trailing hemisphere (~35°N, 225°W) combining data taken from the Galileo 15ESREGMAP01 and 19ESRHADAM01 observations at 220 m/pixel and 65 m/pixel respectively. An unnamed ridge complex extends from lower left to upper right across the image and superposes a wider ridged band with a similar trend. Inset sketch map shows their boundaries along with the locations of prominent cross-cutting troughs/ridges (black lines) and relatively younger disrupted terrain (gray polygons).

prohibitively difficult try to reconstruct their history. Ridged plains contain a tremendous diversity of interlaced and interwoven ridges exhibiting a wide range of azimuthal orientations (Fig. 2 and 4). Studies of some of the highest-resolution images obtained by the Galileo spacecraft (such as those in Fig. 4) allowed *Patel et al.* (1999) to recognize seven subtypes of "ridge and trough terrain" within the ridged plains, each with distinct morphological characteristics. They used Fourier analysis to derive the approximate spacing of the constituent ridges and troughs within each terrain type, finding a gradation among them that they interpreted to represent multiple formation processes and/or a morphological evolution.

2.1.4. Dark deposits. Some double ridges and ridge complexes are flanked by deposits that are dark at visible wavelengths (*Lucchitta and Soderblom,* 1982; *Belton et al.,* 1996) (Fig. 5). These relatively low-albedo deposits are generally patchy and diffuse, extending for up to 10 km on either side of the feature with which they are associated (*Belton et al.,* 1996). The dark material is likely a surficial deposit (*Geissler et al.,* 1998b), as it generally drapes over the surrounding terrain (Fig. 5b). In high-resolution images of Rhadamanthys Linea (Fig. 5c), preexisting structural features are clearly visible through patchy diffuse deposits, suggesting that the deposits are relatively thin (*Fagents et al.,* 2000).

Dark deposits may be continuous along the flanks of a ridge (e.g., Belus Linea in Figs. 5a,b), or spaced in discrete subcircular regions along the margins of a ridge (e.g., Rhadamanthys Linea in Fig. 5c). This striking combination of a bright central lineament flanked by dark deposits on either side led to these features being described as "triple bands" in early Voyager studies (e.g., *Lucchitta and Soderblom,* 1982). These deposits are predominately, but not ex-

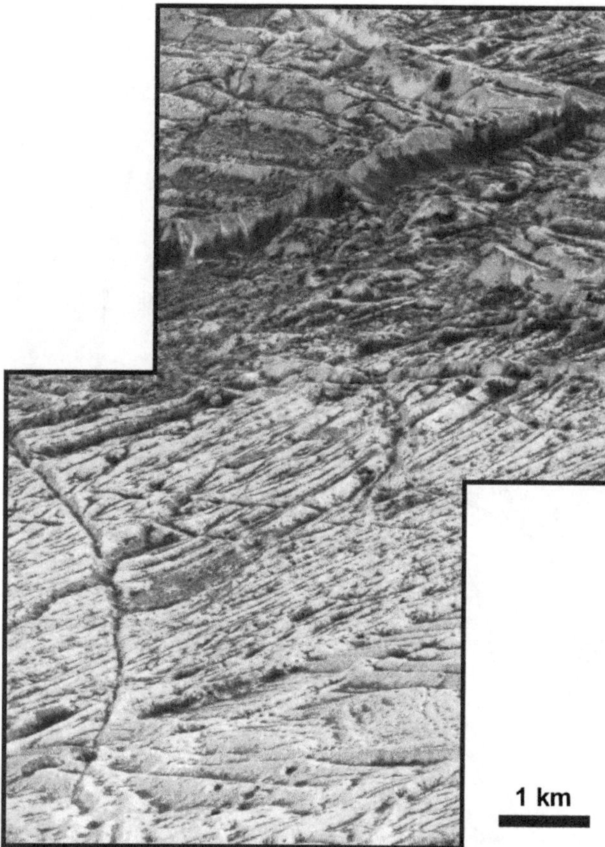

Fig. 4. Image mosaic acquired obliquely at ~12 m/pixel (from Galileo observation 12ESMOTTLE02), illustrating the complex nature of Europa's ridged plains (foreground).

clusively, observed in association with ridges on Europa and may serve as a proxy for relative age (section 2.3).

2.2. Topography

As suggested by their consistent morphology over long distance, the crests of double ridges have generally uniform heights that range from just tens of meters to as much as 350 m above the terrain that surrounds them (Fig. 6). Photoclinometric profiles of ridge flanks (*Kadel et al.,* 1998) suggest that they have inward-facing slopes averaging ~36° and outward-facing slopes averaging ~38°. These values are consistent with the expected 34° angle of repose, but measured exterior slopes may be up to 55° locally, implying that the constituent materials are at least partially consolidated (*Kadel et al.,* 1998).

2.3. Stratigraphic Relationships

Ridges are observed over most of Europa's visible surface history (e.g., *Figueredo and Greeley,* 2000, 2004; *Kattenhorn,* 2002). Cross-cutting relationships and relative albedo suggest that ridged plains are the oldest mappable terrain on Europa (e.g., *Greeley et al.,* 2000, 2004). Double ridges and ridge complexes cross-cut and are superposed on ridged plains.

Well-observed ridge termini are remarkably rare, as ridges are commonly truncated by a younger feature before terminating, or they continue beyond the extent of the limited Galileo high-resolution imaging coverage. The youngest and widest ridges are also topographically most prominent, while stratigraphically older ridges tend to be less prominent in height and display more subdued relative topography (*Figueredo and Greeley,* 2004). Ridges typically cross-cut older terrain without deflection of the younger ridge, and they can be cut by subsequent fractures and ridges. Many examples exist of older structures on either side of a ridge that have been displaced across the ridge by strike-slip motion, and/or minor extension or contraction (see chapter by Kattenhorn and Hurford).

It has been suggested that the distribution and relative albedo of deposits flanking ridges can serve as an indicator of relative age (*Belton et al.,* 1996; *Geissler et al.,* 1998b). Geissler et al. observed that dark flanking deposits are rarely associated with the youngest ridges and concluded that such deposits likely form only after ridge construction has ceased. *Belton et al.* (1996) and *Geissler et al.* (1998b) both suggested that features with discontinuous dark flanking deposits are relatively younger than those with continuous dark deposits along their flanks. Using cross-cutting relationships involving these features, they were also able to demonstrate that these flanking deposits brighten with time, approaching or even surpassing the albedo of the ridged plains they superpose. Potential causes and consequences of this process of surface brightening are discussed in more detail in the chapter by Carlson et al.

2.4. Formation Mechanisms

2.4.1. Double ridges. A number of models have been proposed for the formation of double ridges on Europa. These include accumulation through cryovolcanic processes (*Kadel et al.,* 1998), incremental buildup by successive diking events or "wedging" (*Turtle et al.,* 1998), upwarping through linear diapirism (*Head et al.,* 1999), or compression along fractures (*Sullivan et al.,* 1998). The viability of each of these models, however, has faced challenges when tested against observations. The cryovolcanism model does not easily explain the remarkable uniformity of ridge crests over large distances, or the distinct V-shaped trough between them. The wedging model, in which cracks penetrate upward from the ocean into the icy shell, and then liquid is injected into cracks from beneath to upwarp the surface, could explain the general morphology of ridges but still has difficulty explaining the uniformity of the crests and the morphologies of more complex ridge forms. Upwarp through linear diapirism could explain some observations, such as why preexisting terrain is occasionally seen running up ridge flanks; however, it does not explain how ridges with multiple sets might form. The model of compression was originally thought unlikely, but recent studies (*Patterson et al.,* 2006; see chapter by Kattenhorn and Hurford) suggest that perhaps ridges do play a role in accommodating some of Europa's apparently widespread ex-

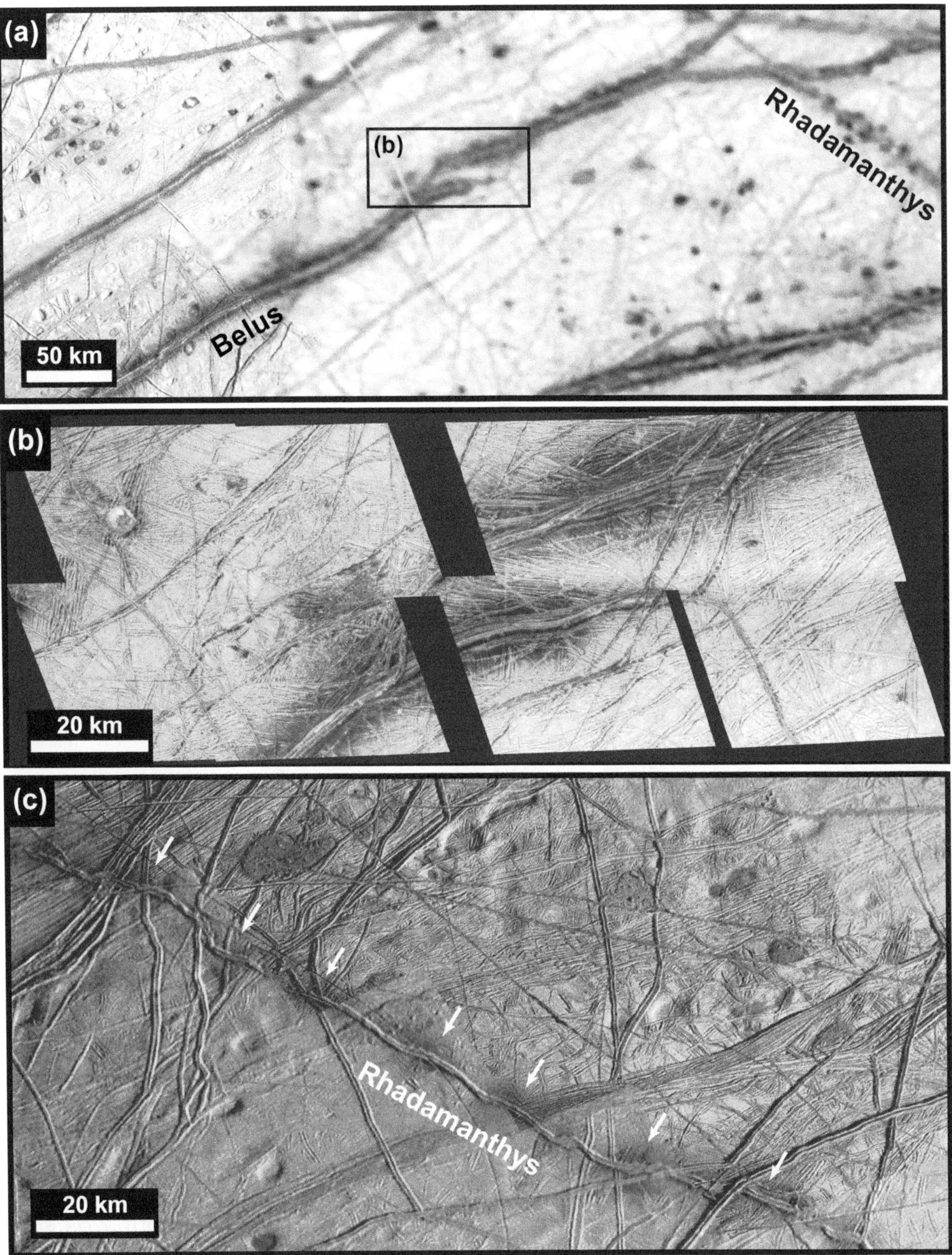

Fig. 5. Image taken from USGS controlled photomosaic of Europa (map I-2757) that includes portions of the prominent ridge complex Belus Linea and double ridge Rhadamanthys Linea. Dark deposits flank both features, but while continuous along Belus Linea, they are discontinuous along Rhadamanthys. **(b)** Mosaic of images obtained during the 14ESTRPBND01 Galileo observation of Belus Linea at a resolution of 75 m/pixel, illustrating the diffuse nature of dark deposits and how they drape over the adjacent terrain. **(c)** Mosaic of images obtained during the 19ESRHADAM01 Galileo observation at a resolution of 65 m/pixel superposed on lower-resolution data (220 m/pixel) acquired during the 15ESREGMAP01 observation. Dark deposits along Rhadamanthys Linea appear as discrete subcircular regions discontinuously distributed along its length (at arrows).

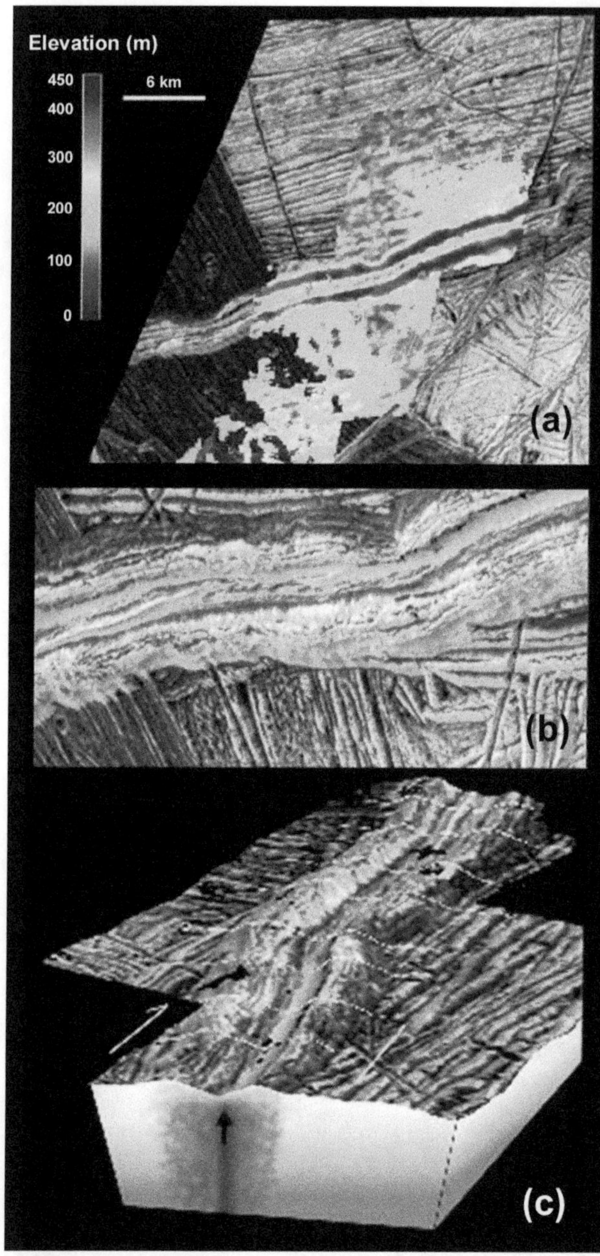

Fig. 6. See Plate 14. **(a)** Topography derived from stereo imaging across a prominent double ridge observed during the Galileo 12ESWEDGES01 and 02 observations shows that the ridge flanks stand several hundred meters above the surrounding terrain (courtesy of B. Giese, DLR). **(b)** High-resolution image (~12 m/pixel) of the same ridge showing many of the distinctive morphological features indicative of double ridges (image is ~15 km wide). **(c)** A perspective view (courtesy of B. Giese and R. Pappalardo) of the double ridge that illustrates key aspects of the shear heating model for ridge formation (*Gaidos and Nimmo,* 2000; *Nimmo and Gaidos,* 2002). Here, the ridge forms as a result of strike-slip motion along a tensile fracture, causing the ice to warm in a zone surrounding the original trough, with the maximum heating near the brittle-ductile transition depth. The warmed ice is more buoyant than the colder surroundings, leading it to rise up, and/or focus contractional deformation, forming ridge crests that flank a distinct trough.

tension. However, exactly what that role is and its significance in the formation of the ridge itself remains uncertain. Two other models of double ridge formation have fared better when tested against observations; the first is based on a terrestrial sea ice analog, and the second relies on thermal buoyancy induced via shear heating along slipping fractures.

2.4.1.1. Tidal pumping (ridges): A sea ice analog model, known as "tidal pumping," was originally proposed by *Pappalardo and Coon* (1996) and later developed by *Greenberg et al.* (1998) and *Tufts et al.* (2000). It suggests that ridges on Europa are analogous to pressure ridges found in terrestrial arctic sea ice and that their formation is controlled by diurnally varying tidal stresses. In the tidal pumping model, a trough is opened through the cyclically varying diurnal tidal stresses experienced by Europa during its 3.5-d orbit around Jupiter, allowing water to seep up from the underlying ocean and freezing to form a slurry. Later in the same cycle, diurnally applied compressive stresses along the same trough push the margins back together, forcing the partially frozen ice out of the trough to form the initial crests of a double ridge. This process is repeated over successive tidal cycles and, it is proposed, either culminates in the formation of a double ridge or continues to operate resulting in more complex morphologies. *Tufts et al.* (2000) developed the model further, and proposed that it could also explain band formation on Europa (see section 3.1.5.1). This led them to propose that double ridges and bands are endmember morphologies within a continuum of tectonic features that all form through the same process (section 4).

The tidal pumping model is attractive in that somewhat analogous mechanisms operate on Earth, and Europa's ridges superficially resemble terrestrial lead ice (*Pappalardo and Coon,* 1996; *Greeley et al.,* 1998), although they are of vastly different scales, and require the presence of liquid water very close to the surface over a wide area. This model can explain the linear nature of ridges over long distances, and many observed local-scale characteristics (e.g., the presence of small subparallel ridges, and fractures atop some double ridge crests). However, geophysical and stratigraphic arguments create challenges that are difficult to overcome for the tidal pumping mechanism. One geophysical argument involves the shallow depths to which surface fractures can penetrate ice (~200 m) given the magnitude of stresses (approximately tens of kPa) induced by diurnal tidal forces (*Crawford and Stevenson,* 1988; *Hoppa et al.,* 1999a; *Lee et al.,* 2003). This falls far short of even the current minimum estimate for the total thickness of the brittle portion of the satellite's icy shell [1–2 km (*Greenberg et al.,* 1998)]. More recent work by *Lee et al.* (2005) has suggested that, if reasonable values of porosity are taken into account, it may be possible for fractures to penetrate a brittle ice layer up to 3 km in thickness. However, even for that optimistic case, another geophysical argument against the tidal pumping model arises: The instantaneous opening widths of fractures controlled by stresses of the magnitudes expected in association with diurnal tidal forces would be on the order of centimeters (*Lee et al.,* 2005). This creates two problems. The first involves the freezing time of water within the fracture by conduction of heat to the walls and the second involves the magnitude of force necessary to extrude ice from so narrow an orifice. Neither would appear to be sufficient to lead to the formation of ridges as described in the tidal pumping mechanism (*Gaidos and Nimmo,* 2000).

Finally, as also discussed in the chapter by Kattenhorn and Hurford, geological evidence suggests that the style of tectonic deformation observed at Europa's surface appears to have changed with time (*Lucchitta and Soderblom,* 1982; *Prockter et al.,* 1999, 2002; *Figueredo and Greeley,* 2000, 2004; *Greeley et al.,* 2000; *Kattenhorn,* 2002). Based on this evidence, it has been proposed that Europa's icy shell has gradually thickened through its visible surface history (e.g., *Figueredo and Greeley,* 2004; *Greeley et al.,* 2004; see chapter by Kattenhorn and Hurford). As we have already discussed (section 2.3), stratigraphic evidence suggests that ridges have formed over much of Europa's surface history. However, their morphology and morphometry do not appear to vary appreciably with time even though a thickening icy shell should progressively inhibit the formation of ridges in the tidal pumping model, given the requirement for a very thin shell.

2.4.1.2. Shear heating: The shear heating model suggests that double ridge crests form through thermal buoyancy that occurs when a trough undergoes strike-slip motion resulting from shear stresses induced by Europa's diurnal tides (*Gaidos and Nimmo,* 2000; *Nimmo and Gaidos,* 2002). In this model, shear associated with a trough leads to frictional heating along its walls and warms the subsurface ice (Fig. 6c). This warm ice then rises beneath the trough and forms the crests of a double ridge. This process could also generate subsurface melting along the walls of the trough and the draining of that melt would explain the distinct V-shaped central trough associated with double ridges.

The shear-heating model appears to fit observations of ridge morphology. The uniform width and height of ridge flanks would be a natural consequence of thermal buoyancy induced by shear along a trough, and subsequent episodes could form additional ridges, forming more complex ridge types. This model predicts that ridge crests could be built to a few hundred meters in only a decade or so. Furthermore, shear heating would be expected to soften the ice along the ridge, which might enable contraction to more easily be accommodated. This model is also bolstered by the morphological similarity of Europa's ridges to those on Triton and Enceladus, where icy shells may have been thicker than Europa's at the time of ridge formation (inhibiting deep cracking), yet diurnally driven shear heating is potentially pertinent to ridge formation (*Prockter et al.,* 2005; *Nimmo et al.,* 2007). Because shear heating does not require that fractures penetrate the entire icy shell of Europa, it does not suffer from the same challenges that the tidal pumping model must overcome.

2.4.2. Ridge complexes. The morphological similarity of component ridges within a ridge complex to that of individual double ridges on the surface of Europa suggests that the two likely share a similar formation mechanism. Indeed, this has led to the common assumption that these features form by the successive buildup of individual double ridges within a confined region and over a period of time (*Geissler et al.,* 1998b; *Greenberg et al.,* 1998; *Figueredo and Greeley,* 2000), an inference supported by both the tidal pumping and shear heating models for the formation of double ridges. Transitional morphologies between double ridges and ridge complexes have been described (*Belton et al.,* 1996; *Geissler et al.,* 1998b) but it remains unclear whether, given enough time, all active ridges would transition into ridge complexes or if a unique set of circumstances leads to the formation of these features. For instance, based on mapping of selected areas of Europa's surface and analogous terrestrial morphologies, *Aydin* (2006) suggests that ridge complexes can be best described as volumetric zones of deformation, rather than successive build-up of multiple double ridges over time. This implies that ridge complexes may form under different conditions than double ridges, which would challenge the notion that the two are related. In contrast, experimental work done by *Manga and Sinton* (2004) found a possible connection between how double ridges and ridge complexes might form based on the tidal pumping model.

Manga and Sinton (2004) used analog wax experiments in an attempt to better understand the processes and conditions that lead to lineament formation on Europa. In these experiments, they looked at how the ratios of net opening in a tidal cycle to the amplitude of crack opening in that cycle [γ— equivalent to the dilation quotient defined in *Tufts et al.* (2000)] and lithospheric cooling to deformation timescales (Ψ) affected the formation of surface features. When the lithospheric cooling and deformation timescale were equivalent (i.e., $\Psi = 1$) and the ratio of secular dilation to the amplitude of crack opening was low (i.e., $\gamma < 0.5$), they observed the formation of deformation belts within the wax that had interior morphologies resembling the interweaving and discontinuous ridges in ridge complexes. The conditions that led to the formation of these features were analogous to those described in the tidal pumping model for the formation of ridges (*Tufts et al.,* 2000), but *Manga and Sinton* (2004) suggested that the periodic strains involved were the result of nonsynchronous rotation instead of diurnal tides.

2.4.3. Ridged plains. As described in section 2.1.3, *Patel et al.* (1999) describe seven subtypes of "ridge and trough terrain" within the ridged plains. They identified one morphological subtype, termed "parallel-linear ridge and trough terrain," with a characteristic ridge and trough spacing of 250–550 m and suggested that it was morphologically similar to parallel ridged terrain on Ganymede, interpreted to have formed by tilt-block-style normal faulting (*Pappalardo et al.,* 1998a). The ridge and trough spacings on Europa were found to be about 2–3 times smaller than those on Ganymede. *Patel et al.* (1999) suggested that if the parallel-linear ridge and trough terrain did originate through tilt-block style normal faulting, it would imply a significantly thinner brittle lithosphere than on Ganymede at the time of formation. The many other "ridge and trough terrains" subtypes within the ridged plains may yet reveal similarly interesting clues about the evolution of Europa. Here too, *Aydin* (2006) suggests that the morphology of ridged plains could be best described by multiple overprinting volumetric deformation zones.

2.4.4. Associated dark deposits. It has been suggested that the dark deposits observed alongside many ridges formed from the eruption of icy cryovolcanic material that seeped out along the ridge flanks, embaying the surrounding terrain, or by the deposition of material from vapor plumes formed as the result of shear heating (*Nimmo and Gaidos,* 2002; *Nimmo et al.,* 2007). Such eruptions could also have resulted from particulates entrained in exsolved gases brought to the surface through fractures in the icy lithosphere (*Crawford and Stevenson,* 1988; *Kadel et al.,* 1998; *Fagents et al.,* 2000). Fagents et al. studied 70 dark spots along Rhadamanthys Linea, and found that their diameters were consistent with what would be expected from cryoclastic eruptions, given plausible eruption velocities and volatile contents. However, because dark deposits are inferred to brighten with age, the fact that the youngest ridges observed by *Geissler et al.* (1998b) do not appear to be associated with dark deposits (section 2.3) represents a challenge to this model.

Fagents et al. (2000) also investigated whether dark deposits could form as the result of local sublimation of icy surface materials leaving a layer enriched in refractory dark deposits, analogous to thermal segregation processes suggested to occur on Ganymede (*Spencer,* 1987). They found this mechanism satisfactorily explains dark deposits with the dimensions of a few kilometers, but that for larger features, measuring up to 25 km in width, additional heat sources or multiple intrusions are necessary to produce the observed dark flanking materials. This model could explain the lack of dark deposits associated with the youngest ridges (*Geissler et al.,* 1998b) because it implicitly requires a time delay while the thermal front propagates through the cold crustal ice and initiates changes in the surface ice. However, a drawback with the sublimation model is that it does not explain why some bands are uniformly dark, yet do not have any diffuse deposits alongside them, or why patchy deposits are seen alongside some lineae, rather than continuous deposits.

3. BANDS

Bands are a common feature on Europa. In this section, we will describe the morphology, topography, relative albedo, boundaries, and probable formation mechanisms of several distinct categories of band. The most commonly observed are informally known as "pull-apart" bands, or dilational bands (*Tufts et al.,* 2000). In general, the margins of this band type can be reconstructed extremely well (that is, they fit back together, like pieces of a jigsaw puzzle) (*Schenk and McKinnon,* 1989; *Pappalardo and Sullivan,* 1996; *Sullivan et al.,* 1998) (Fig. 7), indicating that the interiors of pull-apart bands consist of subsurface material that has been emplaced at the surface of Europa. This band type can be further subdivided on the basis of interior morphology. Opening across dilational bands is primarily extensional (i.e., perpendicular to the band's long axis), although there is commonly also a component of lateral motion resulting in oblique opening (*Tufts et al.,* 2000; *Prockter et al.,* 2002). Maximum band widths are about 30 km, suggesting that band formation is limited by some geophysical process. Band formation clearly represents a significant process by which Europa's crust has been resurfaced (*Schenk and McKinnon,* 1989; *Pappalardo and Sullivan,* 1996); in one region of the subjovian hemisphere it has been estimated that over half of the surface is formed from bands alone (*Prockter et al.,* 2002).

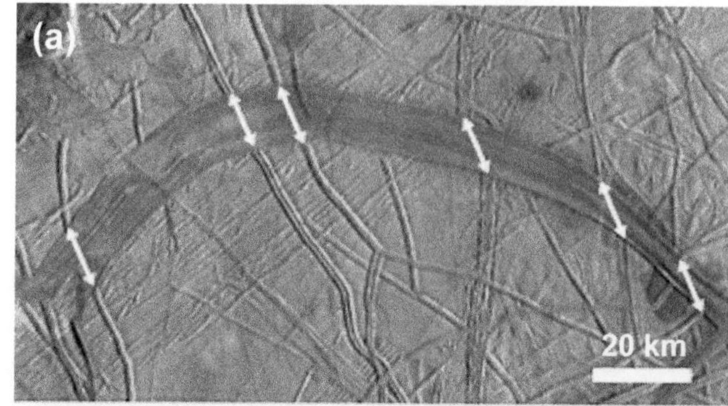

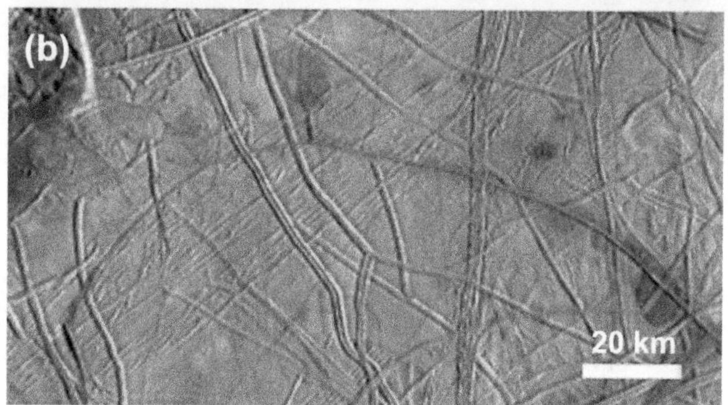

Fig. 7. Reconstruction of an ~12-km-wide sickle-shaped pull-apart band in the Castalia Macula region of Europa (Galileo observation 17ESREGMAP01 at 220 m/pixel). **(a)** Current surface configuration. **(b)** Reconstruction obtained by removing band material and moving north and south sides of the band as coherent plates. Note that preexisting structures with a wide range of trends reconstruct. The band appears to have opened by exploiting preexisting weaknesses along a cycloidal ridge.

A separate category of bands are termed bright bands. These are linear features that disrupt preexisting terrain and have internal textures reminiscent of pull-apart bands. However, unlike pull-apart bands, these enigmatic features defy reconstruction and have a higher relative albedo than the terrain they disrupt. Formation mechanisms relying on dilational, contractional, and/or lateral deformation have all been proposed to explain the unique characteristics of this type of band (*Schenk and McKinnon,* 1989; *Prockter et al.,*

2000; *Greenberg,* 2004; *Kattenhorn,* 2004). Examples of bright bands include Agenor, Corick, and Katreus Linea.

3.1. Pull-Apart Bands

3.1.1. Morphology. While the basic process of pull-apart band formation appears to be consistent across Europa's surface, detailed band morphology differs (Fig. 8). Most workers have classified them as generally falling into two types, known as smooth bands and lineated bands (e.g., *Greeley et al.,* 2000, 2004; *Prockter et al.,* 2002), although *Tufts et al.* (2000) use a slightly different classification (as described in section 4). Smooth bands appear to have a uniform texture at regional resolutions, but at high resolution are predominantly composed of small, closely spaced hummocks with little other original structure (Fig. 8a). Lineated bands, as their name suggests, exhibit internal lineations oriented subparallel to their central axes. These lineations may either be normal faults, or closely spaced linear ridges and troughs, allowing lineated bands to be further subdivided into faulted bands and ridged bands, respectively (Figs. 8b,c). Many lineated bands appear to be smooth at lower resolutions. One type of band has been referred to in the literature as "ridged" bands (*Figueredo and Greeley,* 2000; *Stempel et al.,* 2005), but these may be more analogous to ridge complexes (see chapter by Kattenhorn and Hurford).

Most bands exhibit bilateral symmetry around a central trough oriented parallel to the band margins (*Sullivan et al.,* 1998; *Prockter et al.,* 2002). Some bands appear to have undergone more than one episode of deformation, in which a preexisting band has apparently been reactivated along the central spreading axis, and a younger central band with a distinct morphology has formed.

3.1.1.1. Smooth bands: Most smooth bands have distinct central troughs, although some have no obvious central trough, or a very inconspicuous trough. Smooth bands are characterized by numerous small hummocks (e.g., Figs. 8a, 9), very closely spaced together, forming a texture that appears smooth at low resolutions but rough at higher resolutions (10–100 m/pixel). The hummocks tend to be either uniformly whale-backed or hemispherical in shape within a single band, and generally show no systematic variation in size or height either toward the margins of the band, or along its length, suggesting a formation mechanism that was consistent throughout band formation. The hummocks may be subtly aligned with the central axis of the band, a likely result of its formation by symmetrical spreading (Fig. 8a). One interesting variation on this morphology is found within the band Thynia Linea (Fig. 9). Thynia has a generally smooth appearance with a small-scale hummocky texture; however, superimposed on this texture are a number of distinct, larger chains of hummocks typically ~500 m in length, and arranged in chevron patterns that are largely bilaterally symmetric.

Astypalaea Linea (Fig. 10) is a prominent smooth band in the southern antijovian region of Europa that extends for over ~800 km from ~79°S, 268°W to 60°S, 191°W. Its interior morphology is consistent with that described for smooth pull-apart bands, but it has other unique characteristics that set it apart from typical pull-apart bands, prompting *Kattenhorn* (2004) to classify it as a band-like strike-slip fault. Reconstructions of the feature (*Tufts et al.,* 1999; *Kattenhorn,* 2004) suggest it consists of several north-south-

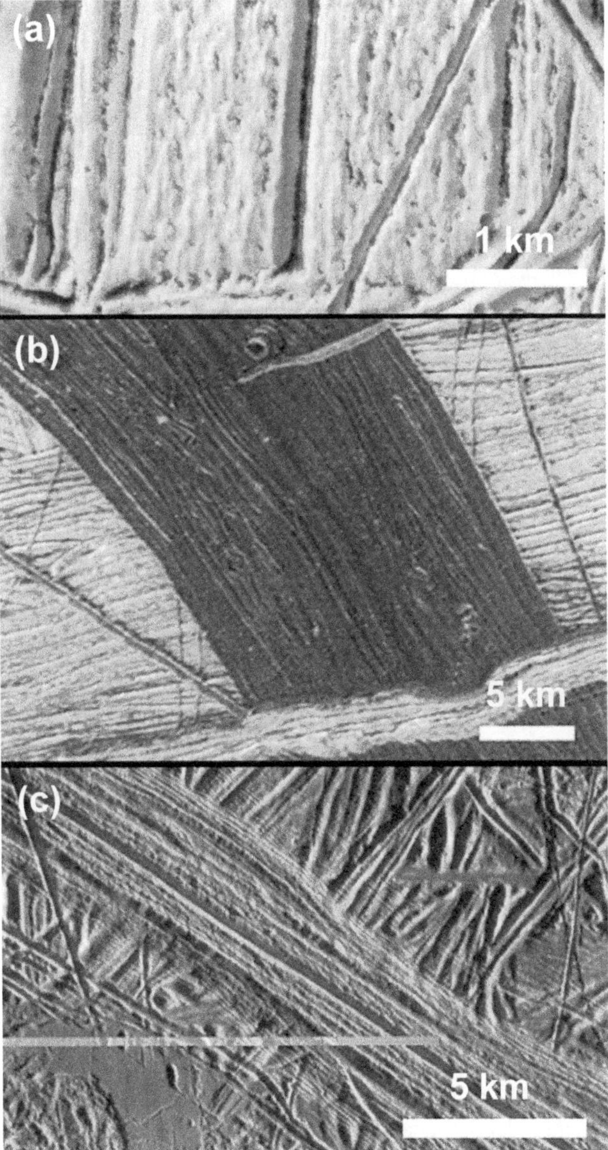

Fig. 8. Smooth and lineated bands on Europa. **(a)** An older smooth pull-apart band seen at high resolution, in Argadnel Regio, the antijovian "wedges" region (Galileo observation 12ESWEDGES01 at 15 m/pixel); this "smooth" band is quite hummocky in detail. **(b)** Yelland Linea, a dark lineated pull-apart band in the same region (12ESWEDGES03 — 90 m/pixel); a higher-resolution strip across Yelland is shown in Fig. 11. Yelland Linea overprints the intermediate-aged gray band Ino Linea, trending east-northeast/west-southwest, which shows similar morphological details. **(c)** A linear feature seen in the E4ESDRKMAT02 Galileo observation (30 m/pixel) has been variously categorized, but is here classified as a faulted pull-apart band.

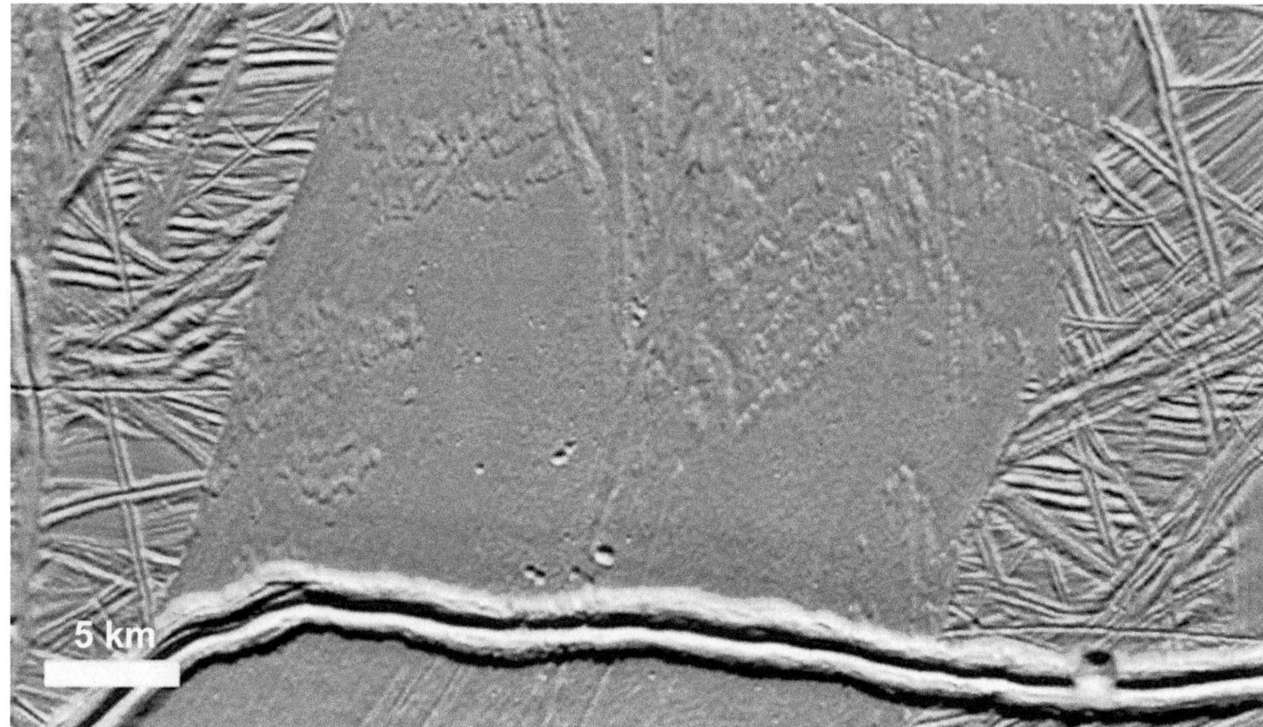

Fig. 9. Portion of Thynia Linea imaged during the 17ESTHYLIN01 Galileo observation at a resolution of 40 m/pixel. This smooth band is characterized by numerous closely spaced and small hummocks that tend to be either whalebacked or hemispherical in shape. This particular band also shows an interesting variation on this morphology: It has a superimposed texture consisting of larger chains of hummocks, typically ~500 m in length, arranged in chevron patterns that are largely bilaterally symmetric.

trending ridge segments measuring from 46 to 88 km long that are aligned in a right-stepping *en echelon* pattern with a north-northeast/south-southwest trend. Oblique displacement along Astypalaea Linea is right-lateral with magnitudes of at least 42 to 56 km (*Tufts et al.,* 1999; *Kattenhorn,* 2004). The ridge segments define the boundaries of several rhomboidal pull-apart segments, including Cyclades Macula (Fig. 10a). The orientations of parallel lineations within these pull-aparts suggest that Astypalaea opened at an angle highly oblique to its trend. *Greenberg et al.* (1998) have suggested that Astypalaea was ideally oriented to have experienced the combined dilation and right-lateral motion, indicated by its morphological characteristics if it opened when situated about 20° of longitude west of its current location on the surface, and subsequently translated by nonsynchronous rotation.

3.1.1.2. Lineated bands: These bands are so named because they show clear evidence of linear structure parallel to their margins whether viewed at regional or local scales. Two types of lineated band exist: (1) faulted and (2) ridged.

Faulted bands appear to be smooth bands that have undergone subsequent tectonic deformation, creating a linear texture visible at lower resolutions (e.g., Fig. 8c). The small hummocks visible at high resolution on smooth bands form the background surface texture of lineated bands. Like smooth bands, lineated bands have distinct troughs along their centers, which is an axis of morphological symmetry, inferred to be the spreading axis. High-resolution imaging (Fig. 11) has shown several different morphological zones within some bands (*Prockter et al.,* 2002). Some lineated bands have a zone of small hummocks surrounding the central trough to an equal distance on either side, persisting along the length of the band. Further away from the central trough, the terrain becomes much more rugged, with closely spaced subparallel ridges and troughs reaching to the band margins. These lineaments may be both fissures and clearly defined fault blocks with distinct striated scarps and associated talus slopes. The hummocky textured terrain is visible on some of the backtilted faces of these fault blocks. Other lineated bands show evidence of faulting, but do not show a distinct zone of hummocks surrounding the central trough. While in most lineated bands the faults appear symmetrical on either side of the inferred spreading axis, some exhibit faulting on one side of the band only.

Ridged bands are composed of generally uniformly spaced ridges and troughs that make up the fabric of the band (see wide band in Fig. 3). Commonly no distinct central trough is present, and these ridges run parallel to the band margins. These bands appear to be a major component of the oldest ridged plains, although it is difficult to construct their histories, because they are so disrupted.

Lineaments within a band typically run parallel to the band margins. However, in at least one case where the margins of a band are distinctly curved (Fig. 13 of *Prockter et al.,* 2002), the lineaments become straighter toward the center of the band. This suggests that the spreading process

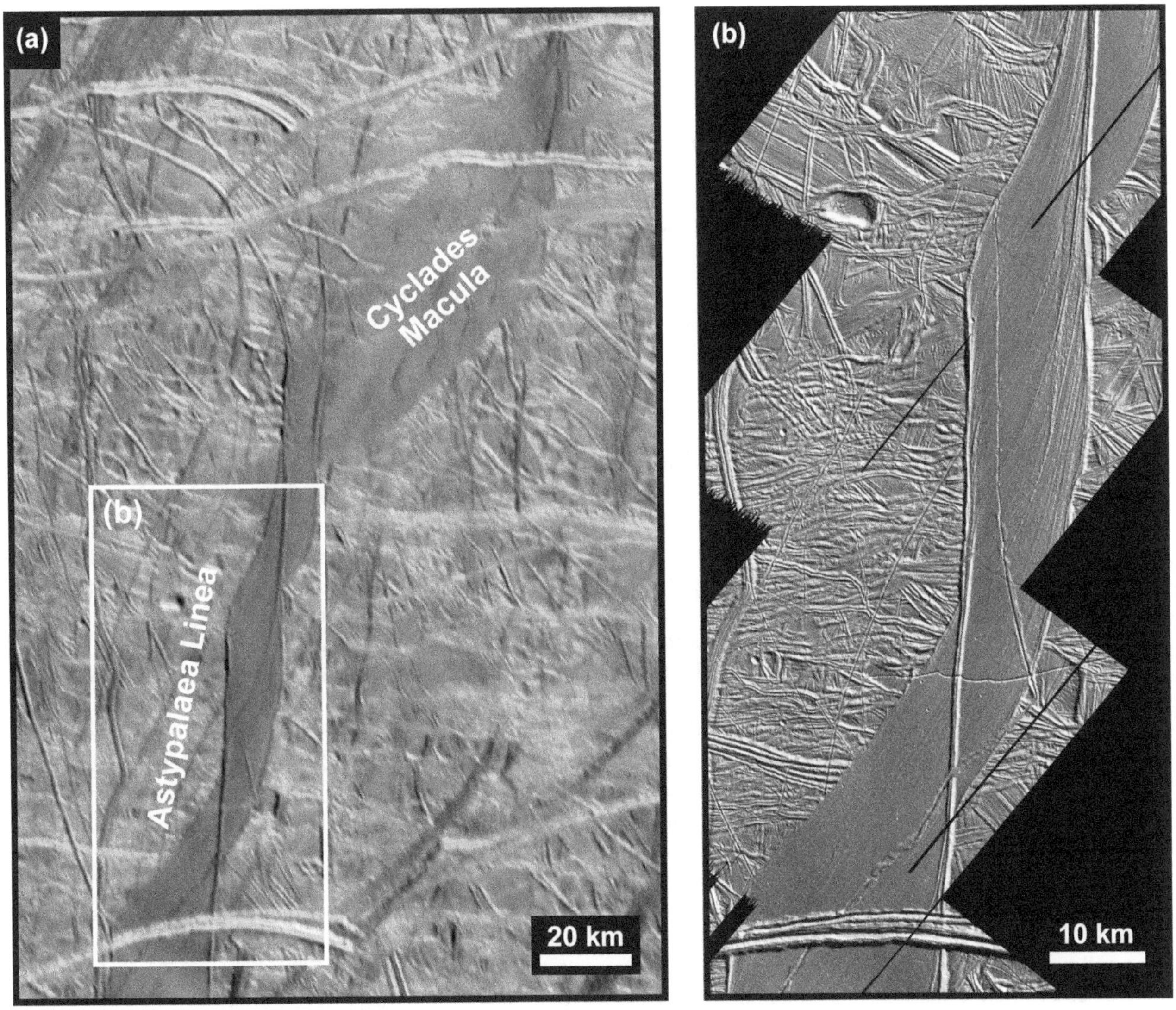

Fig. 10. Images of the smooth band-like strike-slip fault Astypalaea Linea and the associated rhomboidal pull-apart zone Cyclades Macula. **(a)** Portion of a mosaic of images acquired during the 17ESREGMAP01 Galileo observation at a resolution of 220 m/pixel; white box shows location of the subsequent figure. **(b)** Portion of Astypalaea Linea imaged at 50 m/pixel during the 17ESSTRSLP01 Galileo observation.

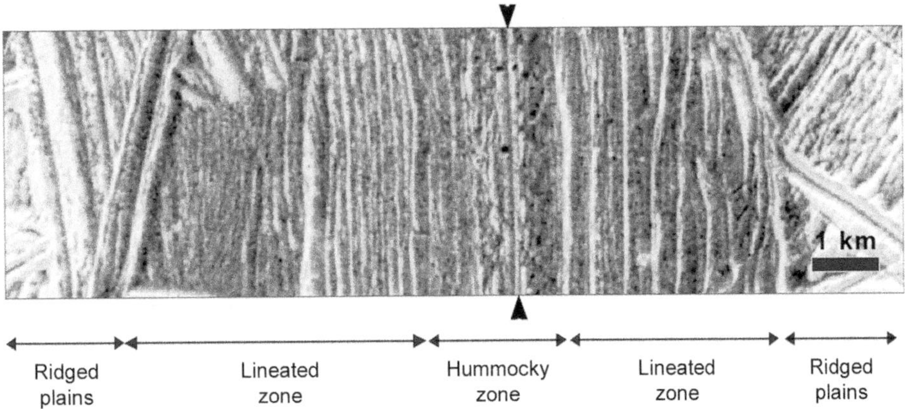

Fig. 11. Major geological units within the lineated pull-apart band Yelland Linea, imaged during the 12ESWEDGES02 Galileo observation at 35 m/pixel. Note the broad similarities in width and style on each side of the central trough (vertical arrows), which is the inferred spreading axis, although there are variations in small-scale details such as hummock morphology and fracture spacing.

tended toward the least energetic configuration (i.e., straight) during band formation.

3.1.2. Boundaries. High-resolution images of bands show that they have a distinct boundary with the preexisting terrain. Some are bounded on each margin by a single ridge (*Tufts et al.,* 2000; *Prockter et al.,* 2002). In such cases, when the band margins are reconstructed, a double ridge is clearly visible (e.g., Fig. 7), implying that the double ridge provided a zone of weakness that was exploited during band formation. *Prockter et al.* (2002) showed that some bands have opened along more than one ridge along different portions of their lengths. If no ridge is present along the boundary with the surrounding terrain, the margins reconstruct simply to a crack or trough.

Whether a ridge is present or not, the boundary with the surrounding terrain generally appears consistent along the length of the band, and is remarkably distinct. Elsewhere on Europa, many sites of chaos and lenticulae (see chapter by Doggett et al.) are associated with low-albedo plains material that has embayed the surrounding terrain (e.g., *Pappalardo et al.,* 1998b; *Greeley et al.,* 2000). In at least one disrupted region, Murias Chaos, material appears to have extruded onto the surface and flowed over the boundary with the surrounding terrain for several tens of kilometers (*Figuredo et al.,* 2002). In contrast, although bands generally stand higher than the surrounding terrain (see section 3.1.3), there is always a distinct boundary between the band and the preexisting plains, with no observed spillover of band material or any associated low-albedo material. This implies that band material was emplaced in a solid state rather than as a fluid.

3.1.3. Topography. *Nimmo et al.* (2003) investigated three pull-apart bands in the antijovian region imaged during Galileo orbit E14, and concluded that all three bands, whether young or old, stand about 100–150 m higher than their surroundings, with ridges within the band standing higher still. A combination of stereo and photoclinometry using Galileo and Voyager data (*Schenk et al.,* 2001) has enabled bands to be studied at several other locations on Europa, and similar height estimates have been derived (Patterson et al., in preparation). Patterson et al. studied 14 bands, 11 of which stand higher than their surroundings. They report that there may be a slight bias toward younger bands standing higher than older bands, and that there is no apparent correlation between morphology and topography, but the study is limited by the lack of high-resolution imaging and topographic data. Furthermore, there does not seem to be any marked variation in topography along the length of each band studied; they appear to be grossly uniform in elevation in both height and width. This uniformity lends further support to the idea that bands formed relatively rapidly, without abrupt changes in stress regime either along or across the band during opening.

3.1.4. Stratigraphy. Pull-apart bands have a range of relative albedos. As with other features on Europa, there appears to be a distinct correlation between albedo and age, with the oldest pull-apart bands exhibiting the highest albedo — they are as bright as the oldest background plains at visible wavelengths, and can even be brighter than the background plains at near-infrared wavelengths (*Geissler et al.,* 1998b). The IR-bright bands have never been satisfactorily explained. Because most of the background plains appear to be composed of dissected band and ridge material, it is not clear why younger bands would be even brighter than their older counterparts. Higher-resolution images of the IR-bright bands, while limited, imply that they are smooth or ridged bands, so they do not appear to be of a distinct morphological type. Thus, it seems likely that this phenomenon is the result of grain size or compositional variations over time. In contrast, the most recent bands appear almost as dark as some of the youngest dark plains units (*Geissler et al.,* 1998b; *Helfenstein et al.,* 1998; *Greeley et al.,* 2000). Most bands fall in between the extremes of dark and bright and are intermediate in albedo, leading to their popular designation as "gray bands." Although the absolute age of bands relative to other europan surface features cannot be assessed because of the dearth of impact craters on Europa (see chapter by Bierhaus et al.), this change in brightness of pull-apart bands over time can be a useful tool for estimating a band's relative age in imaged regions.

Geological mapping shows that bands generally predate Europa's other widespread surface features, chaos and lenticulae (see chapters by Collins and Nimmo and by Doggett et al.). Bands are commonly disrupted by chaos and lenticulae as well as by subsequent ridge and crack formation, but very few bands overprint chaos or lenticulae (*Prockter et al.,* 1999, 2000, 2002; *Greeley et al.,* 2000, 2004). These observations suggest that the conditions necessary for band formation have not been available in Europa's recent history and that the style of deformation on Europa has broadly changed from lateral (bands) to more vertical (chaos) tectonics (see chapter by Doggett et al.).

3.1.5. Formation mechanisms. It was first demonstrated by *Schenk and McKinnon* (1989) that if the material composing some bands were removed, the margins of those bands could be fit back together so that the surrounding preexisting lineaments were reconstructed. Based on these reconstructions, they suggested that pull-apart bands form by fracturing of a brittle surface layer overlying a more ductile subsurface. In this scenario, failure of the brittle layer would occur along near-vertical tension fractures (as opposed to along normal faults). This would lead to complete separation (or rifting) of the brittle layer, and infilling of lower albedo material in the newly created gap. The reconstruction of numerous additional bands in a similar manner has reinforced this interpretation (e.g., *Pappalardo and Sullivan,* 1996; *Tufts et al.,* 1999, 2000; *Prockter et al.,* 2002). It is apparent from the morphology of some pull-apart bands (e.g., Astypalaea Linea), and the presence of *en echelon* structures (*Michalski and Greeley,* 2002) within others, that some of these features opened at an oblique angle to the maximum tensile stress directions.

Manga and Sinton (2004) performed analog wax experiments in an attempt to better understand the processes and conditions that lead to ridge and band formation on Europa, as described in section 2.4.2. They observed the formation of deformation belts within the wax that had interior morphologies resembling both ridge complexes and ridged bands. As γ was increased, they observed that ridges within the deforming region formed in a more parallel sense to each other and that the deformation belt was more bilaterally symmetric, an observed characteristic of many bands on Europa. They also found that as γ increased the relief of ridges within the deformation belt decreased. Manga and Sinton also indicated that oblique opening and shear were common in their experiments, as is also inferred to be the case on Europa, and that the formation of bands requires opening atop ductile ice rather than liquid water.

Two endmember models have been proposed for the formation of pull-apart bands, and each has significant implications for the state of Europa's icy shell during band formation. One is the tidal pumping model proposed by *Tufts et al.* (2000) (described in section 2.4.1.1 in relation to ridge formation), which suggests that bands are part of a continuum process that begins with the formation of a crack, progresses through increasingly more complicated ridge forms, and ultimately ends in the formation of a pull-apart band. The second model, by *Prockter et al.* (2002), proposes that bands and ridges are distinct in their formation mechanisms, and that bands instead form in a manner analogous to terrestrial mid-ocean ridges and exploit preexisting cracks or ridges. The differences between these two models are significant: The tidal pumping model relies on cyclical stressing and bands are constructed from liquid freezing at the surface, while the mid-ocean ridge model does not rely on cyclical stresses and requires solid-state material.

3.1.5.1. Tidal pumping (bands): In the Tufts et al. model, described in detail in section 2.4.1.1, the cyclically varying diurnal tidal stresses result in cracks dilating and closing (see chapter by Kattenhorn and Hurford). In this model the icy shell is thin (only a few kilometers), and as the walls of the crack move fully apart, liquid water is exposed to the surface, and a layer of ice forms. As the crack walls are subsequently forced back together, the newly formed ice is trapped and forced upward, creating a pair of linear ridges with a trough between them. Subsequent tidal cycles result in the crack within the double ridge opening up further, forming a new set of paired ridges each time, and continuing until a band has formed.

It is highly unlikely, however, that Europa's ocean is close enough to the surface for this mechanism to operate: In order to support the topography observed on the surface, ranging from troughs 1 km deep (*Schenk,* 2005; *Schenk et al.,* 2008) to chaos regions and domes standing several hundreds of meters high (see chapter by Collins and Nimmo), the ice would need to be at least 10–20 km thick (*Figueredo et al.,* 2002; *Schenk,* 2002; *Schenk and Pappalardo,* 2004; *Prockter and Schenk,* 2005). It is possible that Europa's icy shell was thinner during the time of band formation, as there appears to have been a change in the style of morphological deformation over most of the surface from bands to chaos and lenticulae (*Prockter et al.,* 1999; *Greeley et al.,* 2000, 2004; see chapter by Doggett et al.), but band topography still requires a relatively thick shell at the time of their formation. If bands are an endmember of a continuum that begins with crack and ridge formation, then we must explain why recent conditions are sufficiently different from those earlier in Europa's visible history (still only ~40–90 m.y. on average), such that double ridges still formed in recent times but bands did not.

Another issue with the tidal pumping model is that it does not adequately explain the morphology found within the bands themselves. The model predicts that all bands should all be lineated to some degree, and that smoother bands are the result of faster opening. In particular, several bands do not show much evidence for lineation, but instead have numerous small hummocks across them. It has been suggested by *Prockter et al.* (2002) that the chevron-shaped hummocks in Thynia Linea (Fig. 9) formed at the spreading axis from pulses of ice or partial melt that were relatively warm with respect to the surrounding band material. Small asymmetries in the distribution of the hummocks across the band may imply a reorientation of the spreading axis by different amounts. It is possible that these hummocks could form as the result of crushed ice, but it is hard to reconcile how and why they would be so uniform in size along and across the band.

Even if Europa's bands could form by tidal pumping, it is difficult to explain how they would stand over 100 m above the surrounding terrain, as most of them do. Topography may be key to understanding the formation mechanisms of bands. The viability of several mechanisms for isostatically supporting topography associated with dilational features was examined by *Nimmo et al.* (2003). They investigated thermal and compositional mechanisms that could lead to lateral variations in the density of the satellite's icy shell, which they assessed using a thin (~2 km) and thick (~20 km) shell approximation.

For the thin-shell approximation, *Nimmo et al.* (2003) found that the most feasible way to produce and maintain Europa's topography would be by variations in porosity (a compositional support mechanism) of ~20% between a band and the surrounding terrain. However, it is difficult to envisage a scenario by which such porosity variations might occur. For a thicker shell, *Nimmo et al.* (2003) found that observed topography could be produced by density variations driven by thermal and/or compositional mechanisms. Thermal mechanisms could include shear heating, while compositional mechanisms could consist of lateral porosity variations of ~2% in the ice or higher density contaminants within the icy shell compared to contaminant-depleted bands. Based on the likely detection of hydrated sulfates by the Galileo NIMS instrument (*McCord et al.,* 1999; *Carlson et al.,* 1999; see chapter by Carlson et al.), and cosmochemical arguments (*Kargel et al.,* 2000; *Fanale et al.,* 2000; *Zolotov and Shock,* 2001), high-density salt contaminants are probably present with Europa's icy shell. *Nimmo et al.* (2003) estimated that for a ~20-km-thick shell, only ~2.5% of the contaminants present within a given region would need to be removed in order to produce the observed topography. Topography resulting from lateral variations in the amount of contaminants in the icy shell or from high band porosity could be maintained for long periods of time, while topography resulting from active thermal mechanisms would quickly decay once the thermal source was removed.

3.1.5.2. Mid-ocean ridge (buoyancy) model: The second model originated with observations of bilateral symmetry within the bands (*Sullivan et al.,* 1998), the presence of specific types of morphological and structural features within them, and the progression of these morphological features on either side of the central trough. *Prockter et al.* (2002) noted that morphological features of europan bands are similar to those found at terrestrial mid-ocean ridges. On Earth, terrestrial seafloor spreading centers exhibit a central trough, flanked on either side by small volcanic mounds, formed as new material erupts at the central spreading axis. As the new seafloor material moves away from the spreading center, it cools and thickens, and becomes faulted once it reaches a critical depth. The central trough, small hillocks, and faults and fractures appear morphologically similar to features within europan bands, prompting Prockter et al. to propose that the formation mechanisms are analogous, despite the fact that bands form in icy material at extremely low temperatures. They suggested that band morphology could be used as a proxy for opening rate, because terrestrial mid-ocean ridge morphology is different at different spreading rates. Prockter et al. proposed that smooth bands may be analogous to fast-spreading terrestrial ridges, while faulted bands are created as smooth band material moves slowly enough to cool and fracture as it moves away from the central trough, in the same manner as terrestrial seafloor lithosphere on slow-spreading ridges.

Adopting the mid-ocean-ridge analog for the formation of pull-apart bands, *Stempel et al.* (2005) constrained spreading rates, active lifetimes, and formational strain rates for two bands (Yelland and Ino Lineae) in Argadnel Regio, the "wedges region" of Europa. They suggested that the hummocky zone within these features is the surface expression of lithosphere too thin to permit lithostatic stress sufficient to induce normal faulting and that it was instead dominated by shallow tension fractures. They found that opening rates could range from 0.2 to 40 mm yr^{-1}, active lifetimes from 0.1 to 30 m.y., and local strain rates from $<10^{-15}$ to 10^{-12} s^{-1}. They also suggested that, if the observed structures in the outer portions of ridged bands are indeed normal faults, then the tensile strength of ice in the bands could range from 0.4 to 2 MPa, consistent with nonsynchronous rotation as the dominant driving mechanism for band opening. Stempel et al. note that the inferred spreading and strain rates based on this analysis are similar to terrestrial continental rifting and mid-ocean-ridge strain rates.

While bands and mid-ocean ridges show remarkable similarities, the driving mechanism for band formation may be very different from that of oceanic crust formation on Earth. Seafloor forms as part of a constant cycling of oceanic crust back into the mantle, whereas it is still not well understood where or how contraction resulting from band formation is accommodated on Europa (see chapter by Kattenhorn and Hurford for additional discussion). Furthermore, bands appear to only reach a finite width, unlike terrestrial seafloor, which continues to spread and grow until it is subducted.

The presence of one crest of a double ridge bounding each margin of a band gives clues to the process of band formation. These boundaries have been interpreted in two ways. *Tufts et al.* (2000) interpret the boundaries as an integral part of band formation. In their tidal pumping model, the band forms as part of a continuum, initiating as a crack. If band formation ensues immediately after the crack event, then a band will form with no margins. Alternatively, the crack may first develop into the surface expression of a ridge, which then continues to form by extension into a band, thus exhibiting a single bounding ridge on each margin. Thus, bands may or may not have bounding ridges, but in the tidal pumping model, the band forms immediately following the cracking event. In contrast, *Prockter et al.* (2002) showed that at least one band, the "sickle" observed on Galileo orbit E11 (provisionally named Phaidra Linea), opened along segments of two distinct cycloidal ridges. They suggested that bands do not necessarily form contemporaneously with a cracking event (although they do not preclude this possibility), but rather form wherever a preexisting weakness or weaknesses can be exploited.

3.2. Bright Bands

Not all bands on Europa can be reconstructed in the manner of pull-apart bands. Three known "bright bands" are identified on the basis of their high relative albedo, and may share some of the same characteristics as lineated features; however, they appear to have very different formation mechanisms.

3.2.1. Morphology. Agenor Linea is Europa's best-imaged bright band. This prominent feature trends approximately east-west across Europa's antijovian southern hemisphere, extending for over 1500 km (Fig. 12). The high relative albedo of the band has led to the suggestion that it might be covered with relatively fine-grained material interpreted to be recent, bright frost (*Geissler et al.,* 1998b), or that the fault may contain relatively young, bright ice (*Lucchitta et al.,* 1981; *Schenk and McKinnon,* 1989). Portions of the band were imaged at up to 50 m/pixel resolution by Galileo (Figs. 12b,c), enabling *Prockter et al.* (2000) to map three subparallel swaths of material that comprise Agenor Linea on the basis of relative albedo and morphology (Fig. 12d). The northernmost zone is unusually bright, the central zone is intermediate in albedo, and the southernmost swath exhibits the lowest relative albedo. Short, left-stepping *en echelon* structures are identified close to the margins of each swath, and structures indicative of transpression and transtension are inferred within the curvilinear portions of the band, interpreted as restraining or releasing bends resulting from dextral shear. *Prockter et al.* (2000) also identified a small block of preexisting terrain (a "horse,"

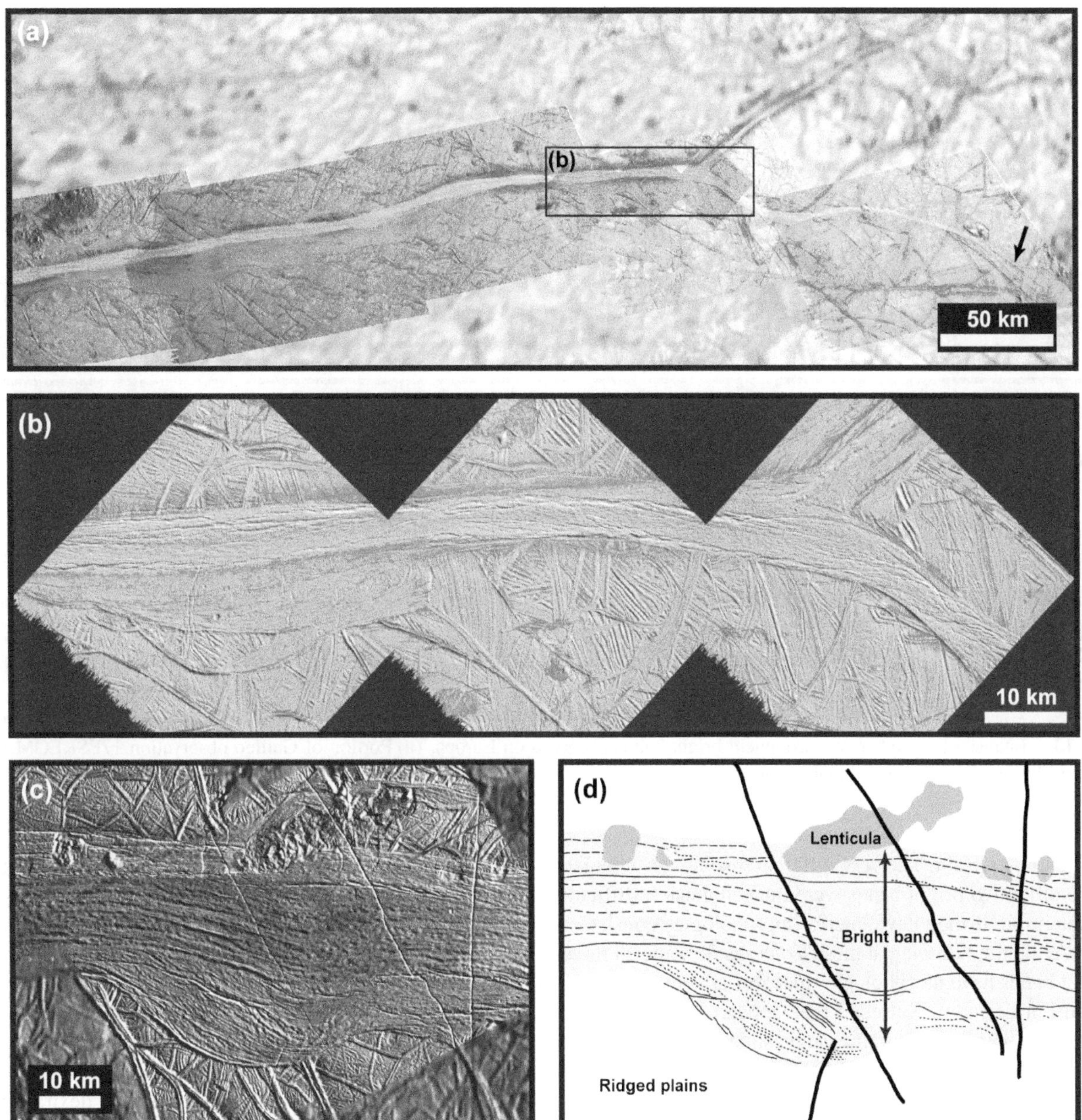

Fig. 12. Images illustrating the unique morphology of Agenor Linea. **(a)** Mosaic of images covering approximately half the ~1500-km extent of Agenor, including tailcrack structures associated with its eastern terminus (includes Galileo observations 14ESGLOCOL01, 17ESAGENOR01, and 17ESAGENOR03 at 1.5 km/pixel, 200 m/pixel, and 50 m/pixel respectively). **(b)** Image mosaic (17ESAGENOR03) acquired during the E17 encounter of the Galileo spacecraft at a resolution of ~50 m/pixel, with location indicated by the black box on **(a)**. **(c)** Portion of a separate image mosaic also acquired during the Galileo E17 encounter (17ESAGENOR03) at a resolution of ~60 m/pixel but along the western portion of Agenor Linea [to the west of the region shown in **(a)**]. **(d)** Sketch map indicating distinct interior morphological characteristics associated with Agenor Linea. Solid thin lines within the margins of the band indicate the boundaries of distinct constituent swaths proposed by *Prockter et al.* (2002). Dashed lines indicate trends of interior lineations, and bold lines are subsequent troughs. Heavy black lines indicate cross-cutting structures.

in structural geology terms) that was apparently decoupled from its surroundings resulting from right-lateral strike-slip along its northern and southern margins. Further evidence that shear played a significant role in Agenor Linea's formation comes from a splay of cracks (Fig. 12a) at the eastern end of the band, interpreted to be a trailing imbricate extensional fan, or "horsetail" complex (*Prockter et al.*, 2000; *Kattenhorn,* 2004).

Katreus and Corick Lineae (Fig. 13) are two other prominent bright bands that have been imaged at lower resolutions (~220 and 1000 m/pixel, respectively). Katreus approaches 800 km in length and is located just north of Agenor. Corick is somewhat shorter than Agenor Linea and is located almost diametrically opposite in longitude, although rather farther south. Like Agenor, Corick has sinuous margins and a relatively high albedo with respect to the

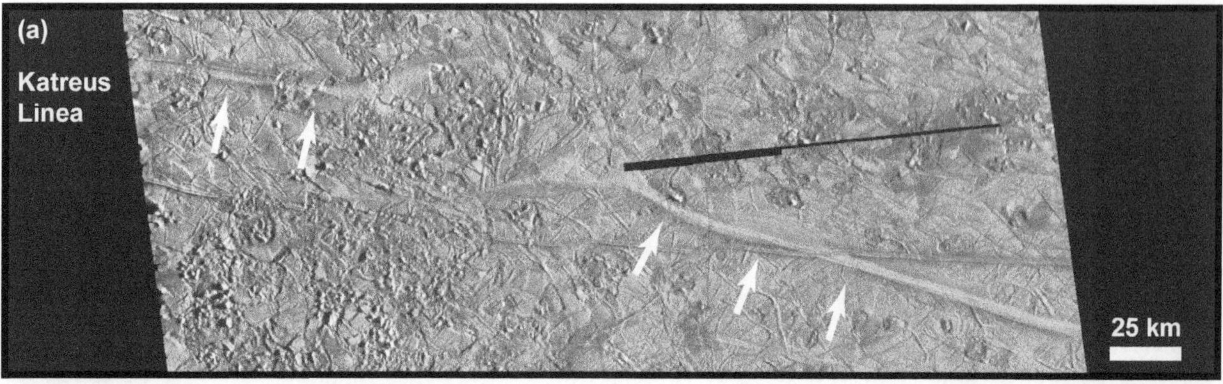

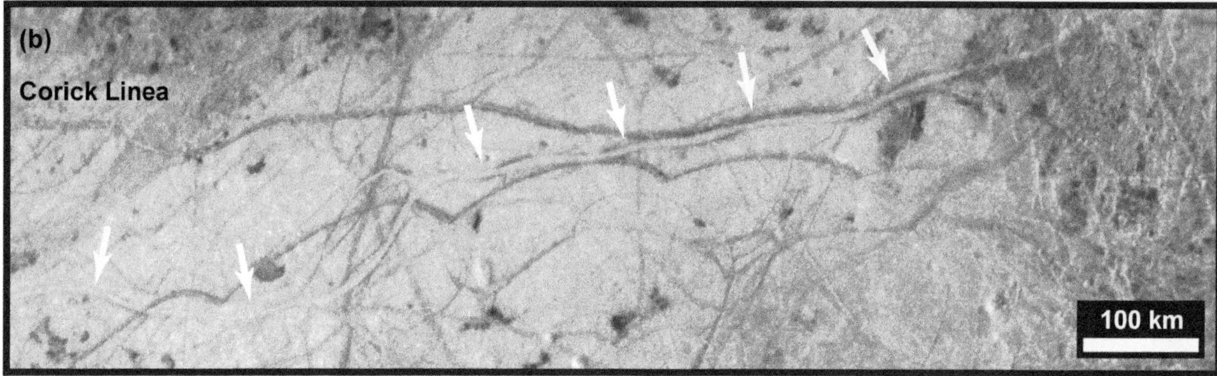

Fig. 13. Images of the other two prominent bright bands observed on Europa. **(a)** Portion of Galileo observation 17ESREGMAP01 at a resolution of ~220 m/pixel that includes a portion of the ~800-km-long Katreus Linea (arrows). **(b)** Corick Linea (arrows) shown at a resolution of ~1 km/pixel (Galileo observation 25ESGLOBAL01).

surrounding terrain. The very different resolutions at which these other two bright bands were imaged make it difficult to determine if they have interior morphologies consistent with Agenor's; however, at these lower resolutions, all three bright bands have morphologies that appear subtle and subdued. Furthermore, Katreus Linea appears to have albedo variations within its margins, suggesting that it too might be divided into distinct swaths (Fig. 13a).

3.2.2. Formation mechanism. The high albedo of Agenor and Katreus Lineae in low-resolution Voyager images led *Lucchitta et al.* (1981) to propose that bright bands might have formed as the result of a dike intrusion of clean, icy material. Based on the enigmatic appearance of Agenor Linea, coupled with its orientation relative to inferred opening directions of dilational bands within Argadnel Regio, *Schenk and McKinnon* (1989) suggested that bright bands might be contractional in nature. *Greenberg* (2004) used these arguments, along with a possible reconstruction of Corick Linea, to suggest that the formation of bright bands includes a significant component of contraction. He noted that if his reconstruction were correct, a zone of crust 25 km wide would have been compressed into a band only a few kilometers wide. He also proposed that, given Corick's apparent youth, the ice must have either relaxed away very quickly, or it was very thin to begin with, as would be the case if it originated as a band formed by the "tidal pumping" model of *Greenberg et al.* (1998). It is not clear how such a wide swath of terrain as Agenor could be smoothed away without leaving some remnant evidence of preexisting terrain or topographic expression, if it were merely background ridged plains that had been compressed to form the bright band.

In contrast, *Prockter et al.* (2002) found no evidence that Agenor Linea is a contractional structure, but instead proposed that Agenor's three constituent swaths formed sequentially from south to north, as the result of dextral shear, resulting in a series of strike-slip duplexes possibly with some associated extension. They noted offsets of up to ~20 km along the band, although *Kattenhorn* (2004) showed that the offset must decrease dramatically toward the lateral tip, suggesting a complex history. Extension would be consistent with its orientation in the tensile stress field induced by combined nonsynchronous and diurnal stresses (*Geissler et al.,* 1998a). *Prockter et al.* (2002) suggested that the band initiated along an arcuate trace, with each successive swath evolving into a more linear configuration, as expected of a shear zone. They proposed that the brightest, northernmost swath (Fig. 12) could have resulted from a greater amount of shear than had occurred across the rest of the band, or that it may have experienced resurfacing through cryovolcanic extrusion.

Studies of Agenor's tailcrack system provide strong evidence that at some time during its formation, the band underwent right-lateral shear failure (*Kattenhorn,* 2004), and the morphological evidence of *Prockter et al.* (2002) also supports such an origin. *Schenk et al.* (2008) have proposed that Europa's shell has undergone an episode of ~80° of true polar wander. They note that if Agenor and Corick Lineae are strike-slip faults, their locations and possibly orientations are consistent with predicted stresses in such a case. Some contraction may have subsequently occurred along Agenor Linea, perhaps localized along the fault zone (*Greenberg,* 2004), although this has not been conclusively demonstrated. However, it is clear that Agenor has experienced a complex history, and has undergone successive periods of deformation.

4. DISCUSSION

Despite numerous studies, the relationship between ridges and bands is still unclear, although much is now known about their morphology and distribution. It is apparent that bands formed from upwelling of solid-state material during lithospheric separation of the surface, and that brines within the icy shell played a significant part in their formation. By contrast, ridges appear to have formed primarily through shearing along cracks as a result of diurnal stresses.

Bands appear to comprise a significant fraction of Europa's visible surface, and dissected older bands are commonly found within the ridged plains. Insufficient data exists to establish whether there is a stratigraphic sequence among different band types, although it is worth noting that many of the bands found within the older ridged plains appear to be lineated ridged bands, while bright bands such as Agenor and Corick may exhibit unusual photometric properties because of frost deposition, implying they may have undergone relatively recent activity. Furthermore, band formation appears to have given way to the formation of chaos features. Even relatively young bands (as evidenced by their low albedo and stratigraphic position) are commonly cross-cut by chaos and lenticulae, but only rare examples have been found of the reverse relationship. This stratigraphic sequence has been proposed to imply a change in resurfacing style from horizontal to vertical tectonics. Thus, the process that formed bands is likely no longer active on the surface.

In contrast, ridge formation has apparently continued throughout Europa's visible history. Some ridges were exploited as zones of weakness during band formation, and many are visible in the ridged plains. The very youngest mapped features are troughs, suggesting that cracking is ongoing on Europa's surface, and that reworking of these fractures into ridges could be ongoing even today. Thus, the process of ridge formation likely differs from that of band formation; one only appears to require ongoing tidal stresses, while the other requires conditions that allow significant dilation of the surface, with upwelling of relatively mobile material from the subsurface. Furthermore, it is not yet known why some fractures form as linear features while others form as cycloids, although this may be related to the relative magnitude of diurnal and nonsynchronous stresses, among other factors (see chapter by Kattenhorn and Hurford). It appears that the surface morphology of the cycloidal ridges at local scales is similar in scale and transitional to that of other double ridges, so whatever mechanism has reworked the crack into a ridge appears to have formed both cycloidal ridges and other double ridges.

Two other icy satellites, Triton and Enceladus, each exhibit structures that are remarkably similar to double ridges on Europa. The Enceladus "tiger stripes" appear to be the locus of plume activity on that moon, and it seems likely that current activity on Europa could also be focused on double ridges in favorable stress regimes (*Nimmo et al.,* 2007; *Smith-Konter and Pappalardo,* 2008). This argues for a commonality of processes, probably related to diurnal stresses and consequent shear heating.

The issue of how, or whether, the ocean communicates with the surface of Europa is still unknown. Most workers agree that the icy shell is currently too thick to enable direct contact between the ocean and the surface, and that some intermediate process is needed to deliver ocean material to the surface. Geological observations of band topography, morphology, and margins imply that bands formed from solid-state material, rather than from liquid. *Nimmo et al.* (2003) note that observed band topography could only be possible through passive thermal support resulting from rifting if bands are extremely young, less than ~1 m.y., an age that is inconsistent with their position in the stratigraphic column. They propose that the most likely mechanism for band topography is lateral variations in brine content, caused by shear heating or passive advection, in turn causing upwelling of warm ice. In either case, bands are unlikely to represent places where the subsurface has communicated directly with the ocean, although an icy shell rich in brines could be the remnants of an ocean that was once closer to the surface but has subsequently frozen out. It is also possible that bands have formed over linear convection cells, something that could occur if Europa's shell switched rapidly from a conducting state to a convecting state as a result of changes in the basal heat flux and tidal-heating rate (*Mitri and Showman,* 2005). In this case, the upwelling material once could have been in direct communication with the ocean, with ocean material plausibly trapped within.

Another key question is how compressional stresses are accommodated on Europa, particularly those resulting from band formation. Although most bands appear primarily extensional, it has been suggested that some bands ["convergence bands" of *Sarid et al.* (2002)] are contractional in nature (see chapter by Kattenhorn and Hurford). If so, it would be expected that small-scale structures diagnostic of contraction would be observed, such as folds or thrust faults, but these have not been specifically identified within candidate convergence bands. Many bands are difficult to reconstruct, due to multiple overprinting by later linear features and disruption by younger chaos and lenticulae, and because of the limited resolutions and spatial extent of available imaging data. Certainly it appears that many bands may have undergone more than one type of deformation during their formation. It is more likely that some of Europa's extension has been accommodated by ridges (*Patterson et al.,* 2006; see chapter by Kattenhorn and Hurford). However, it is similarly challenging to show evidence for convergence of preexisting features across ridges, given the limited image datasets and the history of a large number of linear features generally present at any place on the surface that need to be individually unraveled. Nevertheless, this is a promising avenue of inquiry into Europa's "missing" surface, and studies continue.

In summary, many questions remain about how ridges and bands have formed and what their formation mechanisms imply for Europa's evolution. The formation mechanism of ridged dilational bands is not well understood, and that of bright bands and complex ridges remains enigmatic. To fully understand the difference between bands and ridges, as well as the variations in morphology among their subtypes, globally comprehensive imaging at a variety of resolutions is needed before the relative roles of shear deformation, extension, and potential contraction can be properly understood.

At the time of writing, NASA has just prioritized a flagship-class dual-spacecraft mission (with ESA) to Europa, Ganymede, and the Jupiter system (see chapter by Greeley et al.). Key questions for this mission to address regarding ridges and bands include the following: (1) What is the extent of the role of shear heating in ridge formation? (2) How are dark deposits related to ridge formation? (3) Do double ridges and ridge complexes form in the same way? (4) Are ridges and bands genetically related? (5) How deep do ridge fractures penetrate into the subsurface? (6) Have ridges communicated with the ocean? (7) Do bands accommodate contraction? (8) Are pull-apart bands and bright bands related? (9) What is the composition of bands and does it vary with morphology and/or age? (10) Why are some bands brighter in the infrared than the ridged plains? (11) Did bands communicate with the ocean? (12) When and why did bands cease forming?

These questions and others can be addressed by a combination of techniques that could include high-resolution imaging, regional color imaging, compositional mapping, topographic mapping, and radar sounding (see chapter by Greeley et al.). Europa is an archetype for icy satellite formation and evolution, and is surely one of the most compelling places to explore in the solar system.

Acknowledgments. We dedicate this chapter to the memory of Jiganesh Patel, a promising young scientist and wonderful friend whose life ended abruptly and unexpectedly in 2006, when he was only 28. While an undergraduate at Brown University, Jig studied Europa ridged plains and Ganymede grooved terrain using newly acquired Galileo images. His enthusiasm for everything he did was irrepressible and infectious, and we know he would have been thrilled to learn that NASA and ESA plan to send a pair of spacecraft back to Europa and Ganymede in 2020.

REFERENCES

Abramov O. and Spencer J. R. (2009) Endogenic heat from Enceladus' south polar fractures: New observations, and models of conductive surface heating. *Icarus, 199,* 189–196.

Aydin A. (2006) Failure modes of the lineaments on Jupiter's moon, Europa: Implications for the evolution of its icy crust. *J. Struct. Geol., 28,* 2222–2236.

Belton M. J. S., Head J. W. III, Ingersoll A. P., Greeley R., McEwen A. S., Klaasen K. P., Senske D., Pappalardo R., Collins G., Vasavada A. R., Sullivan R., Simonelli D., Geissler P., Carr M. H., Davies M. E., Veverka J., Gierasch P. J., Banfield D., Bell M., Chapman C. R., Anger C., Greenberg R., Neukum G., Pilcher C. B., Beebe R. F., Burns J. A., Fanale F., Ip W., Johnson T. V., Morrison D., Moore J., Orton G. S., Thomas P., and West R. A. (1996) Galileo's first images of Jupiter and the Galilean satellites. *Science, 274,* 377–385.

Billings S. E. and Kattenhorn S. A. (2005) The great thickness debate: Ice shell thickness models for Europa and comparisons with estimates based on flexure at ridges. *Icarus, 177,* 397–412.

Carlson R. W., Johnson R. E., and Anderson M. S. (1999) Sulfuric acid on Europa and the radiolytic sulfur cycle. *Science, 286,* 97–99.

Crawford G. D. and Stevenson D. J. (1998) Gas-driven water volcanism and the resurfacing of Europa. *Icarus, 73,* 66–79.

Fagents S. A., Greeley R., Sullivan R. J., Pappalardo R. T., Prockter L. M., and the Galileo SSI Team (2000) Cryomagmatic mechanisms for the formation of Rhadamanthys Linea, triple band margins, and other low-albedo features on Europa. *Icarus, 144,* 54–88.

Fanale F., Li Y.-H, De Carlo E, Farley C., Sharma S. K., Horton K., and Granahan J. (2000) An experimental investigation of Europa's "ocean" composition independent of Galileo orbital remote sensing. *J. Geophys. Res., 106,* 14595–14600.

Figueredo P. H. and Greeley R. (2000) Geologic mapping of the northern leading hemisphere of Europa from Galileo solid-state imaging data. *J. Geophys. Res., 105,* 22629–22646.

Figueredo P. H. and Greeley R. (2004) Resurfacing history of Europa from pole-to-pole geological mapping. *Icarus, 167,* 287–312.

Figueredo P. H., Chuang F. C., Rathbun J., Kirk R. L., and Greeley R. (2002) Geology and origin of Europa's "Mitten" feature (Murias Chaos). *J. Geophys. Res., 107,* DOI: 10.1029/2001JE001591.

Gaidos E. J. and Nimmo F. (2000) Tectonics and water on Europa. *Nature, 405,* 637.

Giese B., Wagner R. and Neukum G. (1999) The local topography of Europa: Stereo analysis of Galileo SSI images and implications for geology. *Geophys. Res. Abstr., 1,* 742.

Geissler P. E., Greenberg R., Hoppa G., Helfenstein P., McEwen A., Pappalardo R., Tufts R., Ockert-Bell M., Sullivan R., Greeley R., Belton M. J. S., Denk T., Clark B., Burns J., and Veverka J. (1998a) Evidence for non-synchronous rotation of Europa. *Nature, 391,* 368–370.

Geissler P. E., Greenberg R., Hoppa G., McEwen A., Tufts R., Phillips C., Clark B., Ockert-Bell M., Helfenstein P., Burns J., Veverka J., Sullivan R., Greeley R., Pappalardo R. T., Head J. W., Belton M. J. S., and Denk T. (1998b) Evolution of lineaments on Europa: Clues from Galileo multispectral imaging observations. *Icarus, 135,* 107–126.

Greeley R., Sullivan R., Coon M. D., Geissler P. E., Tufts B. R., Head J. W. III, Pappalardo R. T., and Moore J. M. (1998) Terrestrial sea ice morphology: Considerations for Europa. *Icarus, 135,* 25–40.

Greeley R., Figueredo P. H., Williams D. A., Chuang F. C., Klemaszewski J. E., Kadel S. D., Prockter L. M., Pappalardo R. T., Head J. W. III, Collins G. C., Spaun N. A., Sullivan R. J., Moore J. M., Senske D. A., Tufts B. R., Johnson T. V., Belton M. J. S., and Tanaka K. L. (2000) Geologic mapping of Europa. *J. Geophys. Res., 105,* 22559–22578.

Greeley R., Chyba C., Head J. W., McCord T., McKinnon W. B., and Pappalardo R. T. (2004) Geology of Europa. In *Jupiter: The Planet, Satellites and Magnetosphere* (F. Bagenal et al., eds.), pp. 329–362. Cambridge Univ., Cambridge.

Greenberg R. (2004) The evil twin of Agenor: Tectonic convergence on Europa. *Icarus, 167,* 313–319.

Greenberg R. and Weidenshilling S. J. (1984) How fast do Galilean satellites spin? *Icarus, 58,* 186–196.

Greenberg R., Geissler P. E., Hoppa G., Tufts B. R., Durda D. D., Pappalardo R., Head J. W., Greeley R., Sullivan R., and Carr M. H. (1998) Tectonic processes on Europa: Tidal stresses, mechanical response, and visible features. *Icarus, 135,* 64–78.

Head J. W., Pappalardo R. T., and Sullivan R. (1999) Europa: Morphological characteristics of ridges and triple bands from Galileo data (E4 and E6) and assessment of a linear diapirism model. *J. Geophys. Res., 104,* 24223–24236.

Helfenstein P. and Parmentier E. M. (1985) Patterns of fracture and tidal stresses on Europa. *Icarus, 53,* 415–430.

Helfenstein P., Currier N., Clark B. E., Veverka J., Bell M., Sullivan R., Klemaszewski J., Greeley R., Pappalardo R. T., Head J. W. III, Jones T., Klaasen K., Magee K., Geissler P., Greenberg R., McEwen A., Phillips C., Colvin T., Davies M., Denk T., Neukum G., and Belton M. J. S. (1998) Galileo observations of Europa's opposition effect. *Icarus, 135,* 41–63.

Hoppa G. V., Tufts B. R., Greenberg R., and Geissler P. E. (1999a) Formation of cycloidal features on Europa. *Science, 285,* 1899–1902.

Hoppa G. V., Tufts B. R., Greenberg R., and Geissler P. E. (1999b) Strike-slip faults on Europa: Global shear patterns driven by tidal stress. *Icarus, 141,* 287–298.

Hoppa G. V., Tufts B. R., Greenberg R., Hurford T. A., O'Brien D. P., and Geissler P. E. (2001) Europa's rate of rotation derived from the tectonic sequence in the Astypalaea region. *Icarus, 153,* 208–213.

Hurford T. A., Beyer R. A., Schmidt B., Preblich B., Sarid A. R., and Greenberg R. (2005) Flexure of Europa's lithosphere due to ridge-loading. *Icarus, 177,* 380–396.

Hurford T. A., Helfenstein P., Hoppa G. V., Greenberg R., and Bills B. G. (2007) Eruptions arising from tidally controlled periodic openings of rifts on Enceladus. *Nature, 447,* 292–294.

Kadel S. D., Fagents S. A., Greeley R., and the Galileo SSI Team (1998) Trough-bounding ridge pairs on Europa — Considerations for an endogenic model of formation. In *Lunar and Planetary Science XXIX,* Abstract #1078. Lunar and Planetary Institute, Houston (CD-ROM).

Kattenhorn S. A. (2002) Nonsynchronous rotation evidence and fracture history in the Bright Plains region, Europa. *Icarus, 157,* 490–506.

Kattenhorn S. A. (2004) Strike-slip fault evolution on Europa: Evidence from tailcrack geometries. *Icarus, 172,* 582–602.

Kargel J. S., Kaye J. Z., Head J. W., Marion G., Sassen R., Crowley J. K., Ballesteros O. P., Grant S. A., and Hogenboom D. L. (2000) Europa's crust and ocean: Origin, composition and the prospects for life. *Icarus, 148,* 226–265.

Lee S., Zanolin M., Thode A. M., Pappalardo R. T., and Makris N. C. (2003) Probing Europa's interior with natural sound sources. *Icarus, 165,* 144–167.

Lee S., Pappalardo R. T., and Makris N. C. (2005) Mechanics of tidally driven fractures in Europa's ice shell. *Icarus, 177,* 367–379.

Leith A. C. and McKinnon W. B. (1996) Is there evidence for polar wander on Europa? *Icarus, 120,* 387–398.

Lucchitta B. K. and Soderblom L. A. (1982) The geology of Europa. In *The Satellites of Jupiter* (D. Morrison, ed.), pp. 521–555. Univ. of Arizona, Tucson.

Lucchitta B. K., Soderblom L. A., and Ferguson H. M. (1981) Structures on Europa. In *Lunar and Planetary Science XII,* pp. 1555–1567. Lunar and Planetary Institute, Houston.

Manga M. and Sinton A. (2004) Formation of bands and ridges on Europa by cyclic deformation: Insights from analogue wax experiments. *J. Geophys. Res., 109,* E09001, DOI: 10.1029/2004JE002249.

McCord T. B., Hansen G. B., Matson D. L., Johnson T. V., Crowley J. K., Fanale F. P., Carlson R. W., Smythe W. D., Martin P. D., Hibbitts C. A., Granahan J. C., and Ocampo A. (1999) Hydrated salt minerals on Europa's surface from the Galileo near-infrared mapping spectrometer (NIMS) investigation. *J. Geophys. Res., 104,* 11827–11851.

McEwen A. S. (1986) Tidal reorientation and fracturing of Jupiter's moon Europa. *Nature, 321,* 49–51.

Michalski J. R. and Greeley R. (2002) En echelon ridge and trough structures on Europa. *Geophys. Res. Lett., 29,* DOI: 10.1029/2002GL014956.

Mitri G. and Showman A. P. (2005) Convective-conductive transitions and sensitivity of a convecting ice shell to perturbations in heat flux and tidal-heating rate: Implications for Europa, *Icarus, 177,* 447–460.

Nimmo F. and Gaidos E. (2002) Strike-slip motion and double ridge formation on Europa. *J. Geophys. Res., 107,* DOI: 10.1029/2000JE001476.

Nimmo F., Pappalardo R. T., and Giese B. (2003) On the origins of band topography, Europa. *Icarus, 166,* 21–32.

Nimmo F., Spencer J. R, Pappalardo R. T., and Mullen M. E. (2007) A shear-heating mechanism for the generation of vapour plumes and high heat fluxes on Saturn's moon Enceladus. *Nature, 447,* 289–291.

Ojakangas G. W. and Stevenson D. J. (1989a) Thermal state of an ice shell on Europa. *Icarus, 81,* 220–241.

Ojakangas G. W. and Stevenson D. J. (1989b) Polar wander of an ice shell on Europa. *Icarus, 81,* 242–270.

Pappalardo R. T. and Coon M. D. (1996) A sea ice analog for the surface of Europa. In *Lunar and Planetary Science XXVIII,* pp. 997–998. Lunar and Planetary Institute, Houston.

Pappalardo R. T. and Sullivan R. J. (1996) Evidence for separation across a gray band on Europa. *Icarus, 123,* 557–567.

Pappalardo R. T., Head J. W., Collins G. C., Kirk R. L., Neukum G., Oberst J., Giese B., Greeley R., Chapman C. R., Helfenstein P., Moore J. M., McEwen A., Tufts B. R., Senske D. A., Breneman H. H., and Klaasen K. (1998a) Grooved terrain on Ganymede: First results from Galileo high-resolution imaging. *Icarus, 135,* 276–302.

Pappalardo R. T., Head J. W., Greeley R., Sullivan R. J., Pilcher C., Schubert G., Moore W. B., Carr M. H., Moore J. M., Belton M. J. S., and Goldsby D. L. (1998b) Geological evidence for solid-state convection in Europa's ice shell. *Nature, 391,* 365–368.

Patel J. G., Pappalardo R. T., Prockter L. M., Collins G. C., and Head J. W. (1999) Morphology of ridge and trough terrain on

Europa: Fourier analysis and comparison to Ganymede. *Eos Trans. AGU, Spring Meeting Supplement,* #P42A–12.

Patterson G. W., Head J. W., and Pappalardo R. T. (2006) Plate motion on Europa and nonrigid behavior of the icy lithosphere: The Castalia Macula region. *J. Struct. Geol., 28(12),* 2237–2258.

Pieri D. C. (1981) Lineament and polygon patterns on Europa. *Nature, 289,* 17–21.

Prockter L. M. and Schenk P. M. (2005) Origin and evolution of Castalia Macula, an anomalous young depression on Europa. *Icarus, 177,* 305–326.

Prockter L. M., Antman A. M., Pappalardo R. T., Head J. W., and Collins G. C. (1999) Europa: Stratigraphy and geological history of the anti-jovian region from Galileo E14 solid-state imaging data. *J. Geophys. Res., 104,* 16531–16540.

Prockter L. M., Pappalardo R. T., and Head J. W. (2000) Strike-slip duplexing on Jupiter's icy moon Europa. *J. Geophys. Res., 105,* 9483–9488.

Prockter L. M., Head J. W. III, Pappalardo R. T., Patel J. G., Sullivan R. J., Clifton A. E., Giese B., Wagner R., and Neukum G. (2002) Morphology of europan bands at high resolution: A mid-ocean ridge-type rift mechanism. *J. Geophys. Res., 107,* DOI: 10.1029/2000JE001458.

Prockter L. M., Nimmo F., and Pappalardo R. T. (2005) A shear heating origin for ridges on Triton. *Geophys. Res. Lett., 32,* L14202.

Sarid A. R., Greenberg R., Hoppa G. V., Hurford T. A., Tufts B. R., and Geissler P. (2002) Polar wander and surface convergence of Europa's ice shell: Evidence from a survey of strike-slip displacement. *Icarus, 158,* 24–41.

Schenk P. M. (2002) Thickness constraints on the icy shells of the Galilean satellites from a comparison of crater shapes. *Nature, 417,* 419–421.

Schenk P. M. (2005) The crop circles of Europa. In *Lunar and Planetary Science XXXVI,* Abstract #2081. Lunar and Planetary Institute, Houston (CD-ROM).

Schenk P. M. and McKinnon W. B. (1989) Fault offsets and lateral crustal movement on Europa — Evidence for a mobile ice shell. *Icarus, 79,* 75–100.

Schenk P. M. and Pappalardo R. T. (2004) Topographic variations in chaos on Europa: Implications for diapiric formation. *Geophys. Res. Lett., 31,* L16703, DOI: 10.1029/2004GL019978.

Schenk P., Gwynn D., McKinnon W., and Moore J. (2001) Topography of Ganymede's resurfaced terrains: Evidence for flooding by low-viscosity aqueous lavas. *Nature, 410,* 57–60.

Schenk P. M., Matsuyama I., and Nimmo F. (2008) True polar wander on Europa from global-scale small-circle depressions. *Nature, 453,* 368–371.

Smith B. A. and the Voyager Imaging Team (1979) The Galilean satellites of Jupiter: Voyager 2 imaging science results. *Science, 206,* 951–972.

Smith-Konter B. and Pappalardo R. T. (2008) Tidally driven stress accumulation and shear failure of Enceladus's tiger stripes. *Icarus, 451,* 435–451.

Spencer J. R. (1987). Thermal segregation of water ice on the Galilean satellites. *Icarus, 69,* 297–313.

Stempel M. M., Barr A. C., and Pappalardo R. T. (2005) Model constraints on the opening rates of bands on Europa. *Icarus, 177,* 297–304.

Sullivan R., Greeley R., Homan K., Klemaszewski J., Belton M. J. S., Carr M. H., Chapman C. R., Tufts R., Head J. W. III, Pappalardo R., Moore J., Thomas P., and the Galileo Imaging Team (1998) Episodic plate separation and fracture infill on the surface of Europa. *Nature, 391,* 371–373.

Tufts B. R. (1998) Lithospheric displacement features on Europa and their interpretation. Ph.D. thesis, University of Arizona, Tucson. 288 pp.

Tufts B. R, Greenberg R., Hoppa G., and Geissler P. (1999) Astypalaea Linea, a large-scale strike-slip fault on Europa. *Icarus, 141,* 53–64.

Tufts B. R., Greenberg R., Hoppa G., and Geissler P. (2000) Lithospheric dilation on Europa. *Icarus, 146,* 75–97.

Turtle E. P., Melosh H. J., and Phillips C. B. (1998) Tectonic modeling of the formation of europan ridges. *Eos Trans. AGU, 79(45), Fall Meeting Supplement,* F541.

Zolotov M. Y. and Shock E. (2001) Composition and stability of salts on the surface of Europa and their oceanic origin. *J. Geophys. Res., 106,* 32815–32827.

Chaotic Terrain on Europa

Geoffrey Collins
Wheaton College

Francis Nimmo
University of California Santa Cruz

Chaotic terrain covers approximately one-quarter of Europa's surface, and is formed by disruption of the preexisting surface into isolated plates, and formation of lumpy matrix material between the plates. Key observations include the motion of plates of preexisting terrain within chaos areas, the matrix material commonly elevated above the background terrain, associated dark hydrated materials exposed at the surface, concentration into two antipodal areas around the equator, and a large and continuous range of different sizes and morphologies. Of the models that have been proposed to explain chaotic terrain formation, a melt-through model and a brine-mobilization model best fit the observations. The melt-through model faces serious energy problems, while some details of the brine-mobilization model lack sufficient quantification. None of the existing models offer a completely satisfactory explanation for chaos formation yet, but future data from an orbiting spacecraft could decisively test the hypotheses.

1. INTRODUCTION

It was recognized from Voyager images that isolated patches of Europa's surface have been disrupted, exhibiting a hummocky surface, breakup of prominent ridges, and emplacement of dark reddish-brown material (*Lucchitta and Soderblom,* 1982). These collections of uneven splotches on Europa's smooth face were dubbed "mottled terrain" in Voyager-based analyses; later small, isolated spots were dubbed "lenticulae" after the Latin word for freckles. Early ideas for the formation of mottled terrain involved some kind of tectonic disruption of preexisting terrain, accompanied by emplacement of dark material from below (*Lucchitta and Soderblom,* 1982). An alternative idea was that mottled terrain is older and has been subjected to more exogenic (impact and charged particle) modification (*Malin and Pieri,* 1986). High-resolution Galileo images of Europa have revealed mottled terrain to be a collection of highly disrupted areas on Europa's surface, now dubbed "chaotic terrain." With more detailed data, models for chaotic terrain formation can be more tightly constrained, but none of them yet provide an entirely satisfactory explanation. Our intent is that the synthesis of observations and models outlined in this chapter will inspire further efforts to understand this widespread and enigmatic type of terrain, which appears to be unique to Europa.

Conamara Chaos (Fig. 1) has become one of the iconic images of Europa. Because it is the most well-defined and well-imaged chaos area on Europa, it has also become the most well-studied (e.g., *Carr et al.,* 1998; *Spaun et al.,* 1998; *Greenberg et al.,* 1999; *Head and Pappalardo,* 1999; *Schenk and Pappalardo,* 2004). Figure 2 is a higher-resolution closeup of the center of Conamara, and Fig. 3 is from an even higher-resolution image sequence near the center of Fig. 2. Figures 1 and 2 show many of the typical characteristics of chaotic terrain: an area where the preexisting surface has been disrupted, commonly broken into separate plates, with a lumpy matrix filling in between the plates (*Carr et al.,* 1998). Shadows in these low-Sun images give the impression that the matrix is lower than the plates and the surrounding terrain, because a scarp commonly separates the original surface from the matrix below. Some of the plates appear to have moved from their original positions, and some of them are tilted. Are these observations truly typical of chaotic terrain? In this chapter we will show morphological, topographic, and size characteristics of a variety of chaotic terrain areas to gain a wider perspective.

One point of contention has been the definition of chaotic terrain, and which features should be lumped together, and which ones should be split apart. There is agreement that all features showing the type of lumpy, disrupted matrix such as that seen between the plates in Conamara should be considered chaotic terrain. However, there is disagreement (e.g., *Pappalardo et al.,* 1999; *Greenberg et al.,* 1999) about whether other features, termed "pits, spots, and domes" should be considered in the same class as chaotic terrain. Figure 4 shows a provocative example of such features mixed with chaotic terrain. In this image, there are several small features that appear to have disrupted chaotic terrain in their centers, usually with associated dark mate-

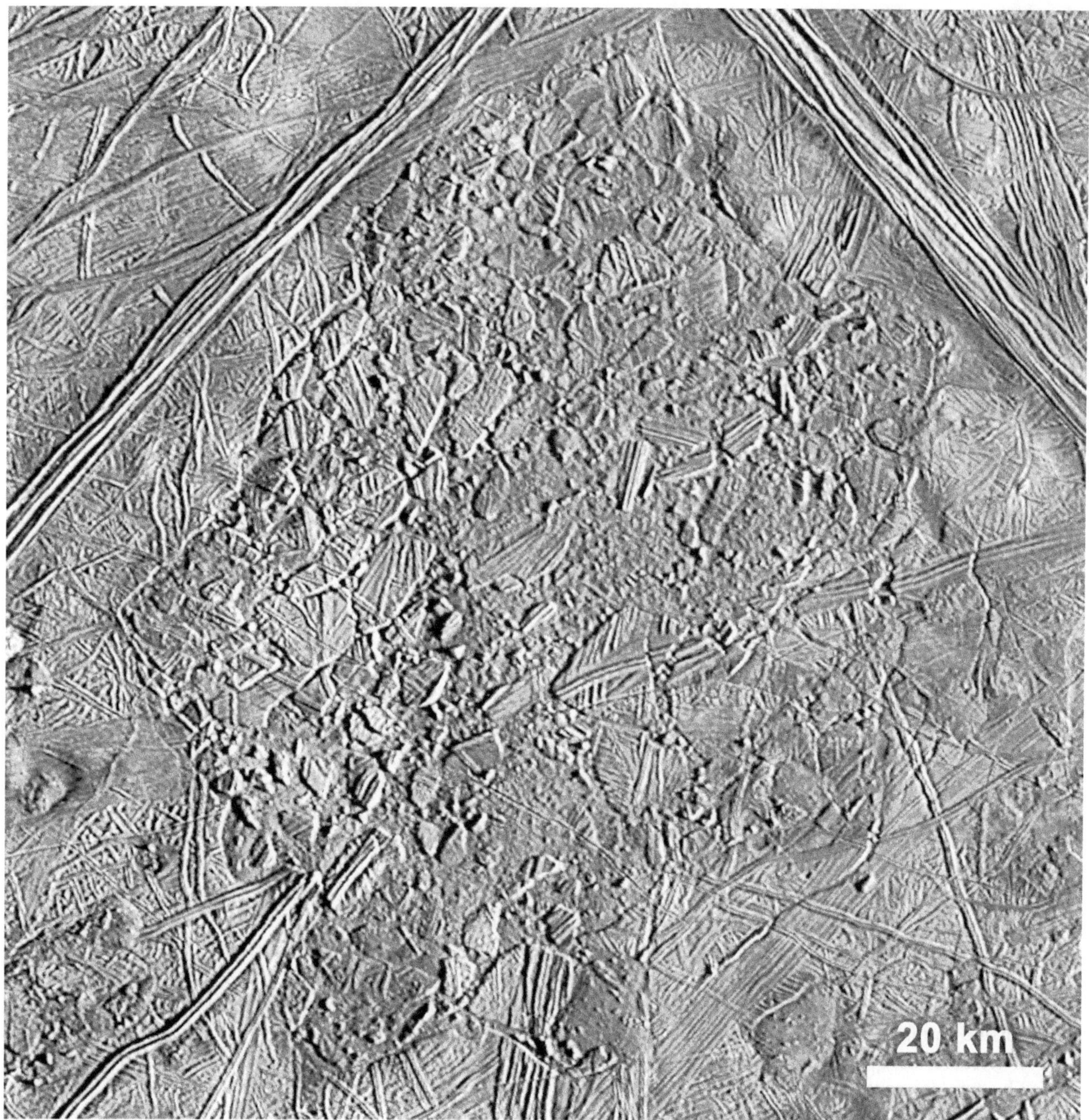

Fig. 1. As one of the most well-studied regions on Europa, Conamara Chaos is the archetype for chaotic terrain. Broken plates of pre-existing terrain are surrounded by a disorganized lumpy matrix. Galileo E6ESDRKLIN01 observation, with illumination from the right. North is to the top.

rial around the edge. Intermixed with these chaotic terrain features are both negative-relief pits and positive-relief domes where the original surface has not been disrupted. There are also patches covered in dark material (spots). There is even one clear example (Fig. 4, arrow) of a pit, filled with dark material, with a dome of disrupted material rising from the center. Similar relationships can be observed in the area surrounding Conamara Chaos (*Pappalardo et al.,* 1998a). The principle of parsimony would guide us to seek a simple explanation for all of these features together, rather than separate explanations for each type of feature.

In this chapter we use "chaotic terrain" as an umbrella term to include features referred to in other works as chaos, lenticulae, micro-chaos, or "pits, spots, and domes." There are combined and transitional examples among all of these features, and we seek a common explanation for all of them, so we do not wish to give them separate names and draw what may be arbitrary boundaries within this class of features. We begin by considering the morphology, topography, spectral properties, size distribution, and geographic distribution of chaotic terrain, and then we compare all the observational constraints to various theoretical models that

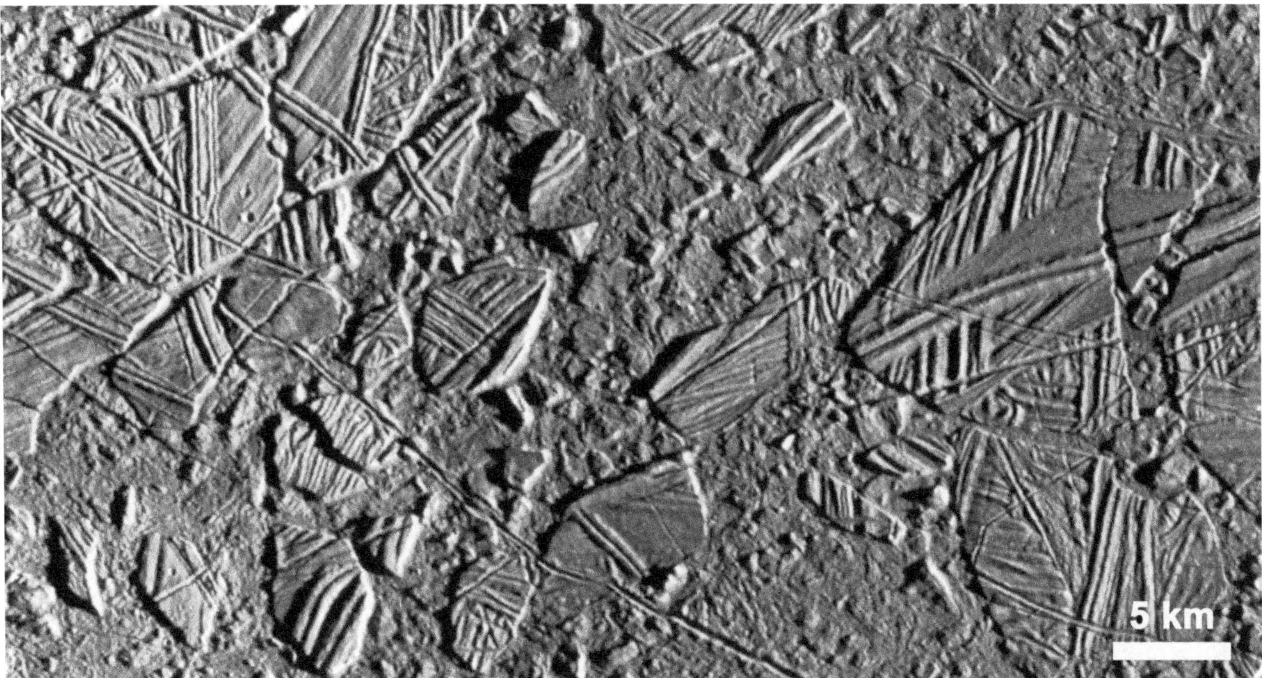

Fig. 2. Closeup of the center of Conamara Chaos, showing tilted plates and jumbled blocks within the matrix. Galileo E6ESBRTPLN01 observation, with illumination from the right.

have been proposed to explain their formation. Because chaotic terrain formation is a widespread process on Europa, finding an appropriate geophysical mechanism is important for understanding the implications of chaos for icy shell thickness, sources of heat in Europa, and material exchange between the icy shell and the underlying ocean.

2. MORPHOLOGY

Several studies have taken the approach of mapping chaotic terrain on a feature-by-feature classification basis (see chapter by Doggett et al.). *Greeley et al.* (2000) described a set of basic mapping conventions for Europa, and subdivided chaotic terrain into "platy chaos material" that has preserved plates of preexisting terrain within a hummocky matrix and "knobby chaos material" that appears to be entirely composed of rugged matrix material. *Figueredo and Greeley* (2004) added "elevated chaos" that appears to be elevated above the surrounding plains, and "subdued chaos" where the plates are present but less distinct from the matrix. *Prockter et al.* (1999) produced a geological map of an area imaged only at high-Sun angles, and derived a different classification scheme more heavily based on albedo patterns than the near-terminator mapping by *Figueredo and Greeley* (2004).

While the classification approach is expedient for geological mapping, it can sometimes draw arbitrary distinctions between types of chaotic terrain when there is a continuum of morphology observed. Here we take an approach of describing each type of morphological element commonly observed in chaotic terrain, and then discuss how these elements occur together to form different overall morphologies of chaotic terrain. Because the goal of this chapter is to discuss the different geophysical models for forming chaotic terrain, concentrating on the morphological elements that such models must produce is the most illuminating path.

2.1. Matrix

The irregular lumpy material that comprises the matrix between plates is part of what gives chaotic terrain its name. Figure 3 is a high-resolution image from the center of Conamara Chaos, showing a small plate surrounded by matrix material. There is no apparent pattern in this area; subkilometer-sized ice blocks poke up in a jumble. Yet the pattern in the background between the larger blocks is not entirely random, as many areas show similarly sized small lumps, spaced out across the surface, giving a sponge-like texture. *Head et al.* (1999) also pointed out examples in this area of groups of discontinuous ridges, and "subdued lineated polygons" that appear to be transitional between the plates of preexisting terrain and the lumpy matrix. An example is marked with an arrow in Fig. 3, where a small square of slightly elevated matrix is crossed by several parallel lineaments.

A more dramatic and unusual example of the preservation of order within chaotic matrix is in Thrace Macula (Fig. 5), where many of the ridges in the exterior terrain can be traced through the otherwise irregular, lumpy chaotic

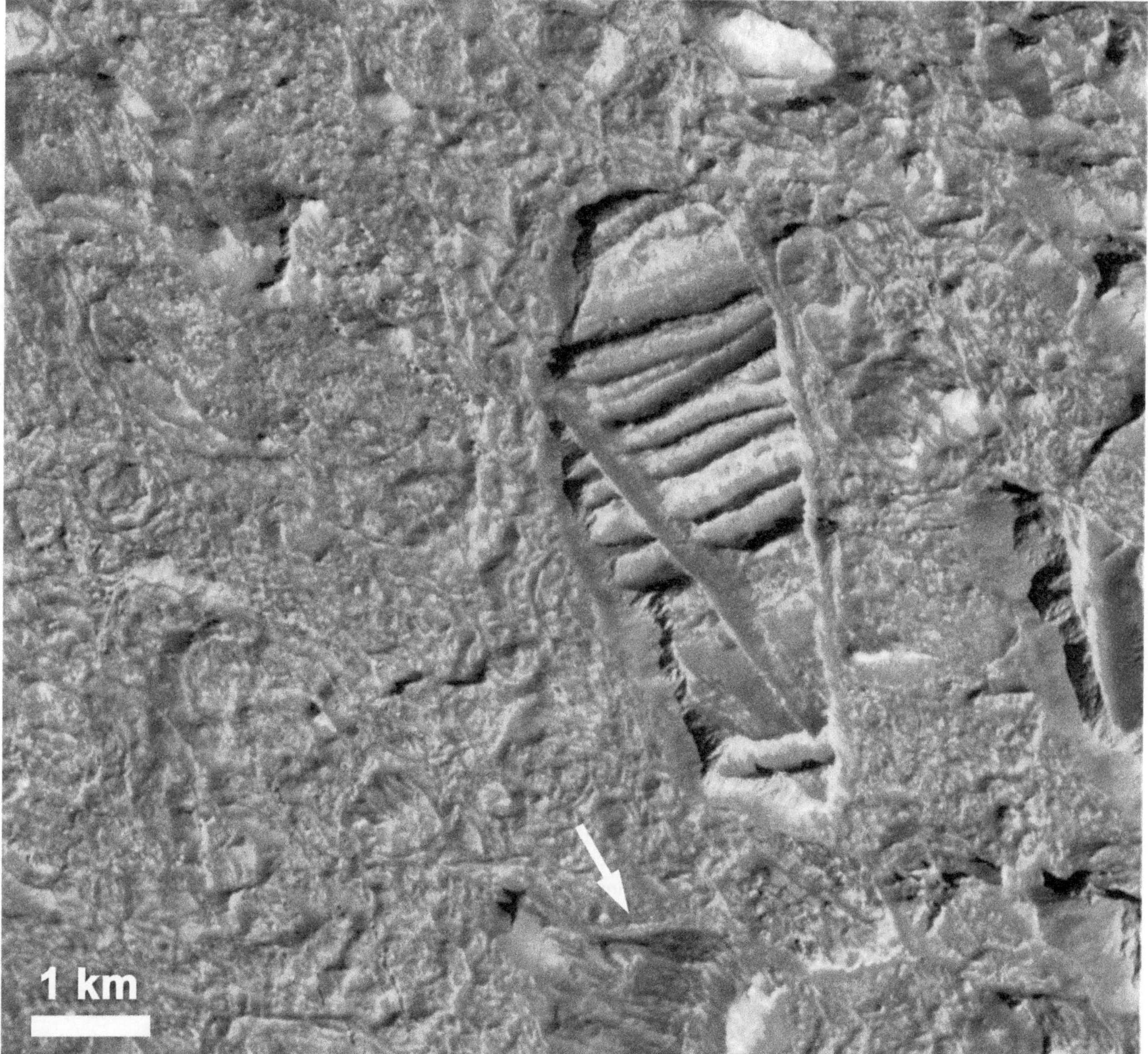

Fig. 3. A single plate of preexisting terrain surrounded by matrix material in the center of Conamara Chaos. Note the cliffs and talus slopes along the bottom and left edges of the plate, and the distributed low hills and discontinuous ridges in the matrix. Arrow points out a subdued lineated polygon (see text). Galileo 12ESCHAOS_01 observation, with illumination from the lower right. North is to the top right.

matrix. It appears that the material that makes up the chaotic terrain matrix in Thrace has been disrupted or degraded *in situ* rather than being entirely new material emplaced cryovolcanically (cf. *Fagents,* 2003). *Head et al.* (1999) came to a similar conclusion based on the high-resolution images of Conamara, arguing that the matrix is largely preexisting terrain that has been disrupted in place. More high-resolution images of chaotic terrain are necessary to resolve if this phenomenon is widespread.

2.2. Plates

Plates of preexisting terrain within chaos areas span about an order of magnitude in size. The largest plates in Fig. 1 are about 20 km across, while the smallest recognizable plates in Fig. 2 are approximately 1 km across. Below this size, only locally high-standing peaks are seen within the matrix, which may be remnants of smaller plates that have either tilted or undergone mass wasting to obscure their original surface texture (*Collins et al.,* 2000).

The shadows in Figs. 1 and 2 show that the plates in Conamara Chaos are locally higher than the matrix material. *Williams and Greeley* (1998) used these shadows to estimate that the blocks lie 40–150 m above the matrix. Figure 3 shows a typical margin of one of these high-standing blocks. Along the bottom and left sides of the block, it is apparent that the block terminates in a steep cliff, and that a talus slope of darker loose material has accumulated at the base of the slope (see chapter by Moore et al.), possibly aided by sublimation degradation (*Moore et al.,* 1999).

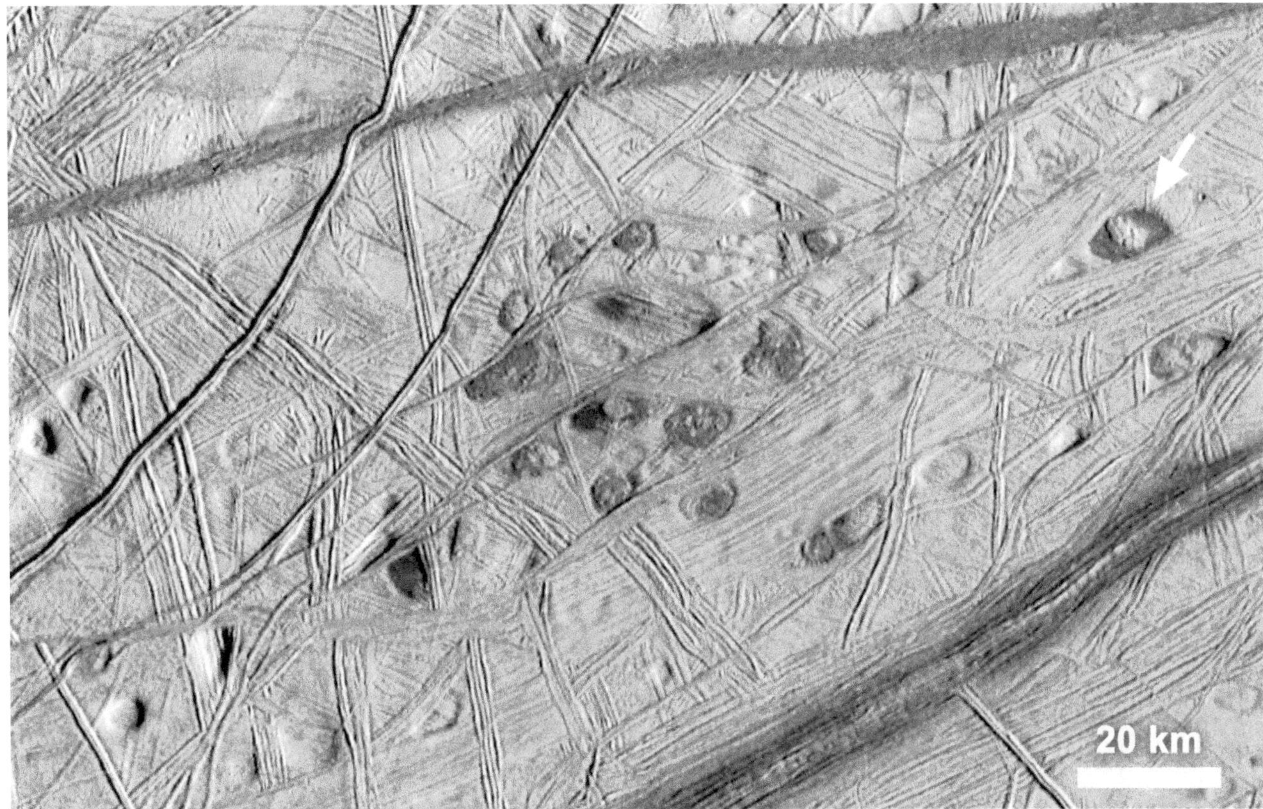

Fig. 4. A collection of small chaotic terrain features intermixed with undisrupted pits and domes, many with associated dark material. At least one feature (top right, arrow) prominently displays pit, dark spot, dome, and chaotic terrain characteristics, supporting the consideration of all such features as a group. Galileo 15ESREGMAP01 observation, with illumination from the right.

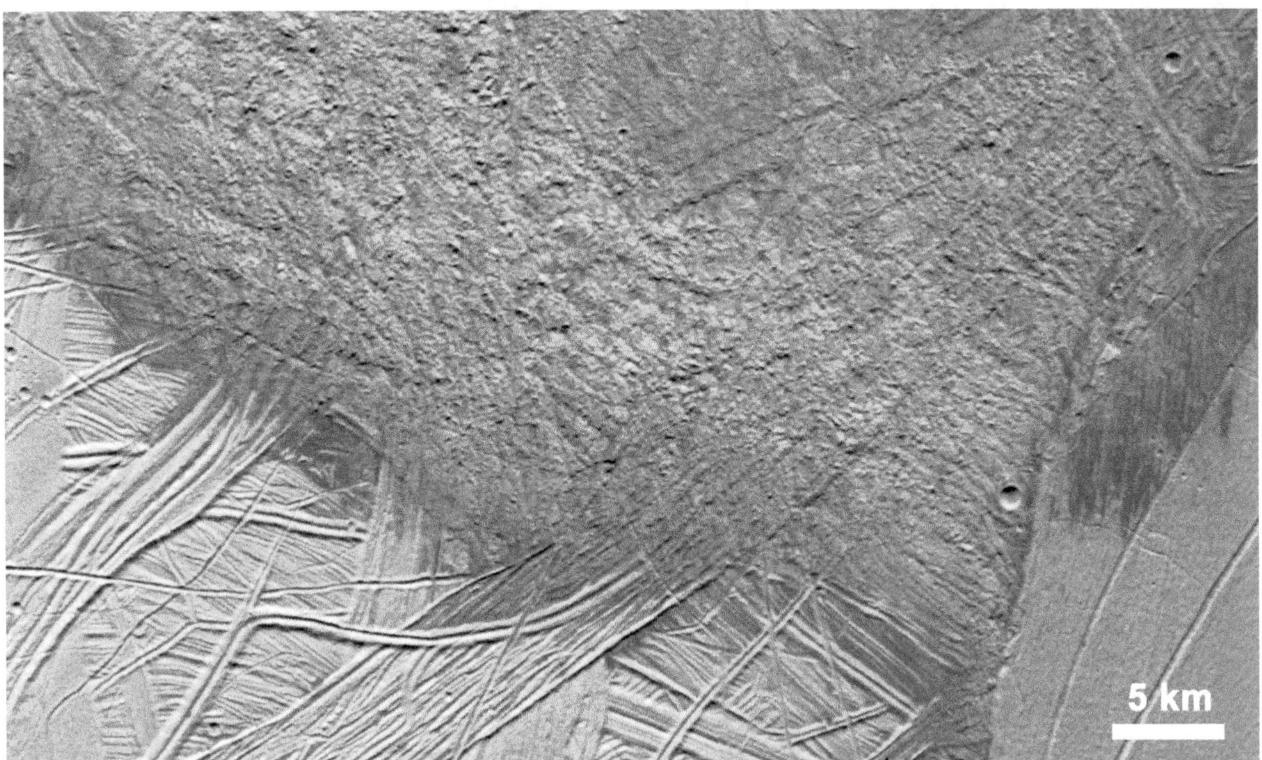

Fig. 5. The matrix material in Thrace Macula shows clear remnants of the preexisting ridged terrain, implying that some process has degraded the surface material *in situ*. Galileo 17ESTHRACE01 observation, with illumination from the top.

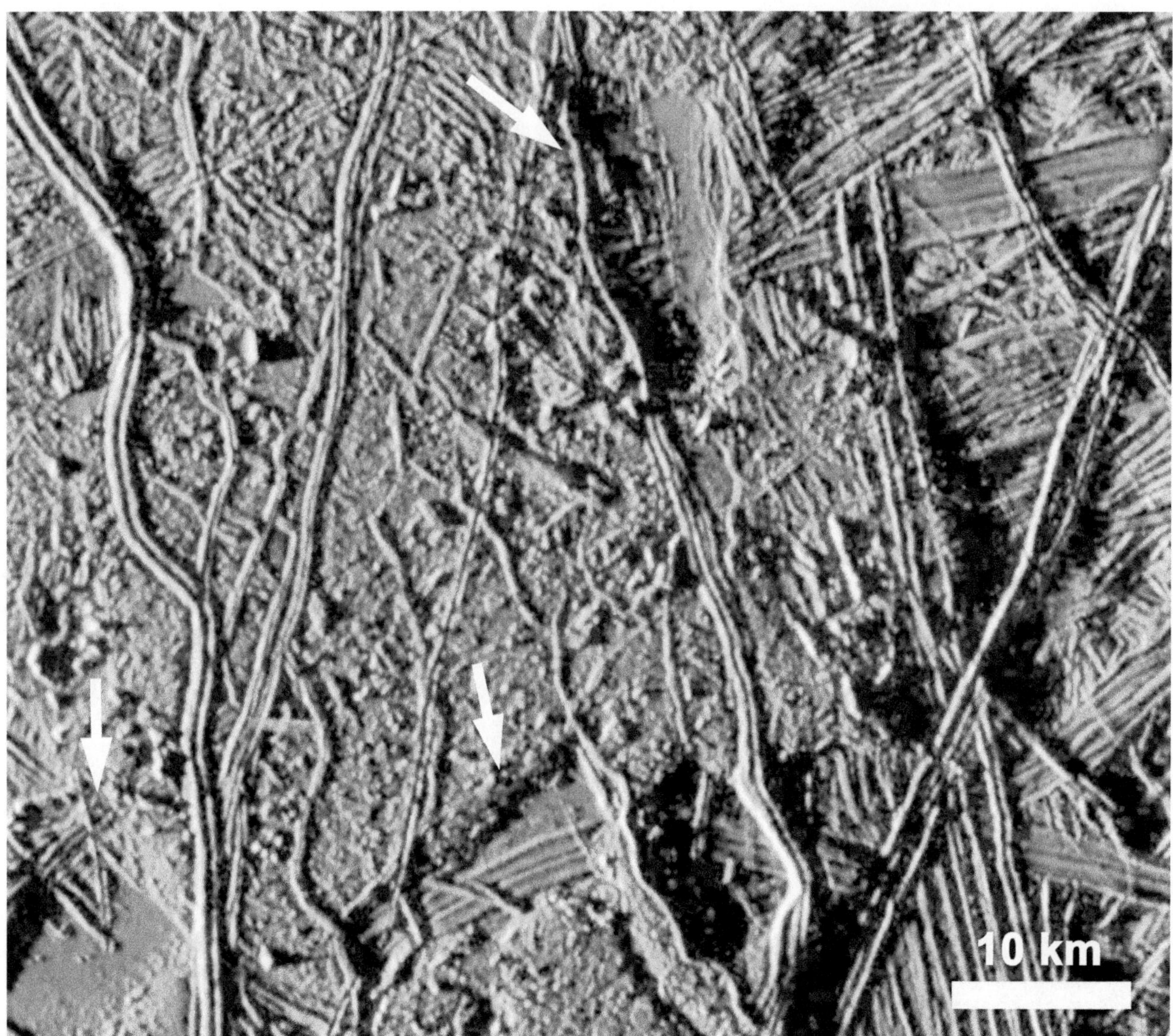

Fig. 6. An area of chaotic terrain in which the isolated "plates" or preexisting material are lower than the matrix (arrows). Many of them appear to be embayed by smooth material. Several younger ridges cut across this area of chaotic terrain. Galileo 25ESNPOLE01 observation, with illumination from the left; the north pole of Europa is just outside the scene to the lower left.

There are also several examples within Conamara of tilted plates that have a cliff on one edge, and slope gently into the matrix along another edge (Fig. 2).

Figure 6 shows a possible counterexample to the locally high-standing plates common to many areas. In this area of chaotic terrain near the north pole of Europa, the matrix material is locally higher standing than the isolated patches of preexisting material within it. In this case, it appears that the matrix has a rounded edge that slopes down onto the surface of the plates. The plates in this chaos area are unusual in that the preexisting ridges appear to be embayed by smooth material (possibly similar to embayed ridges at some chaos margins; see next section). Whether these isolated patches of preexisting terrain within the matrix of this chaos area meet the normal definition of "plates" is debatable, but it serves as an interesting example to contrast with Conamara.

One of the most intriguing observations about plates is that many of them have moved from their original positions. In numerous areas of chaotic terrain, one can observe pieces of preexisting ridges on plates that have moved from their original locations, which are interpolated from ridges in the surrounding undisrupted terrain. In most areas, only a handful of plates can be reconstructed in such a fashion because of resolution or area limitations, but in Conamara Chaos, *Spaun et al.* (1998) used several throughgoing ridges to reconstruct the inferred original positions of dozens of plates (cf. Fig. 10 of *Greenberg et al.,* 1999). The reconstruction showed that at least 78% of the blocks shifted from their initial position, by an average of 2 km of lateral translation. The largest translation was 8 km, and 22% of the plates moved over 5 km. Plates appeared to move inward from the boundaries, and there was no apparent pattern of movement toward or away from matrix-rich areas. There appeared to be an overall clockwise movement of plates around the chaos, but we should point out that the study lacked a formal error analysis in the reconstructions, instead giving an estimate of 0.5–1 km confidence in the initial positions. The method of reconstruction by throughgoing ridges is more accurate perpendicular to the ridge than parallel to the ridge, so the exact pattern of movement may not be entirely reliable.

In addition to lateral translation, *Spaun et al.* (1998) measured rotation of the plates around their vertical axes. The inferred rotation was evenly distributed between clockwise and counterclockwise. The magnitude of rotation was generally small, with 75% of the plates having rotated less than 15° relative to their initial orientations.

2.3. Margins

Like many areas of chaotic terrain, most of the boundary of Conamara is defined by an inward-facing scarp (Fig. 1). Closer inspection of this image shows that part of the boundary (especially in the southeastern portion) is formed by areas where the chaotic matrix is level with, or even appears to dome up above, the surrounding terrain. Such a relationship is shown more clearly in an area of chaotic terrain near Agenor Linea (Fig. 7). Near the top of the image, there is a clear scarp leading down into the chaotic matrix, while near the bottom, the matrix appears to grade directly into the surrounding terrain. It is interesting to note on this image that right next to the inward-facing scarp, some of the plates are clearly higher than the undisrupted terrain nearby, as they cast shadows all the way across the matrix and scarp and onto the surrounding terrain.

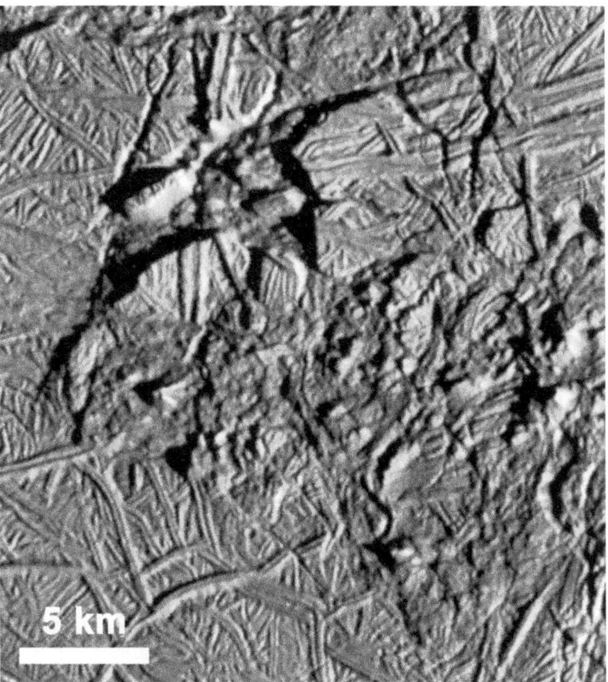

Fig. 7. Chaos region north of Agenor Linea shows both an inward-facing scarp boundary (top) and a level boundary (bottom) between matrix and surrounding terrain. Note that some plates within the chaos cast shadows beyond the edge of the chaos region, showing that the plates have been uplifted relative to the surrounding terrain. Galileo 17ESAGENOR03 observation, with illumination from the right.

In contrast to the inward-facing scarps, there are also many chaotic terrain boundaries formed by matrix domed up over the surrounding terrain. In some cases it appears as if the matrix material has viscously flowed onto its surroundings. For example, the boundaries of Murias Chaos (Fig. 8a) are formed of matrix doming above the surroundings, and circumferential fractures just outside the matrix edge may be due to flexure as a load is emplaced on the crust (*Figueredo et al.,* 2002). Troughs around smaller areas of chaotic terrain with such margins (sometimes referred to as disrupted domes) have been estimated to be hundreds of meters deep, and may also be due to flexure (*Williams and Greeley,* 1998). In another example, a domed region of chaotic matrix near Europa's south pole appears to have

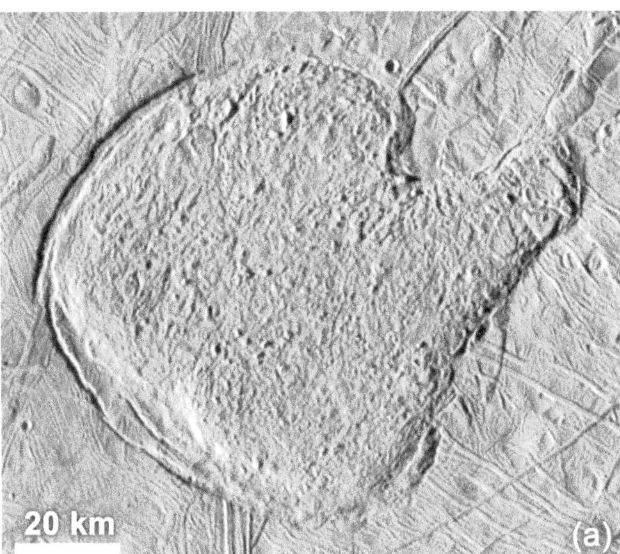

Fig. 8. Chaotic terrain on Europa in which the matrix material slopes upward from the surrounding terrain. **(a)** Murias Chaos stands 400–600 m above its base, and the fractured trough along the west side may be due to flexure from loading (*Figueredo et al.,* 2002). Galileo 15ESREGMAP02 observation, with illumination from the left. **(b)** Chaotic matrix appears to flow into the ridge running left-to-right across the image. Galileo 17ESSOUTHP01 observation, with illumination from the top left.

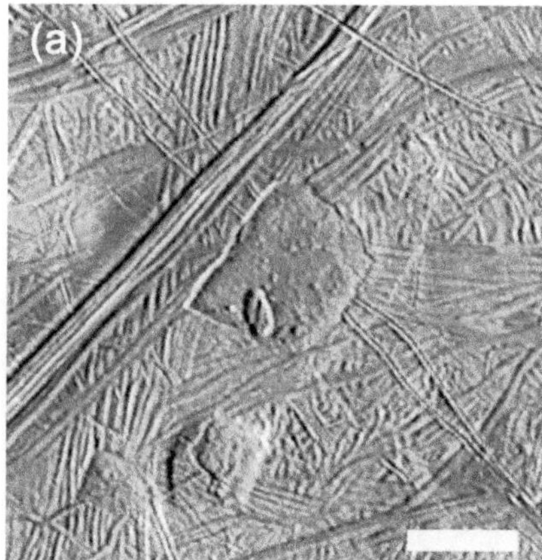

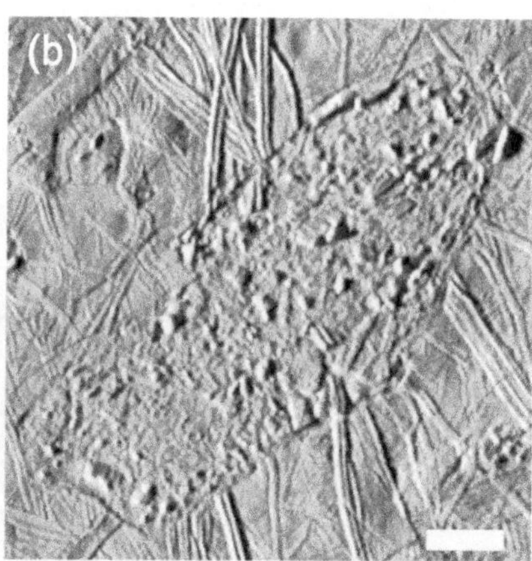

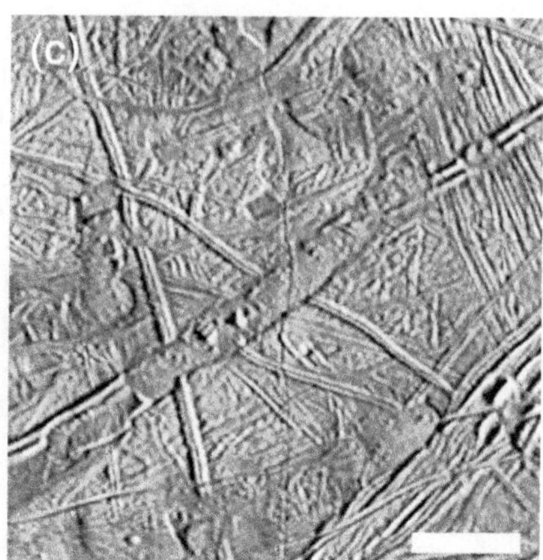

Fig. 9. There is no single mode of interaction between chaotic terrain and prominent preexisting ridges. **(a)** Chaos appearing to avoid a ridge. Galileo E6ESDRKLIN01 observation, with illumination from the right. **(b)** Chaos ignoring the presence of ridges. Galileo 15ESREGMAP02 observation, with illumination from the left. **(c)** Chaos forming largely within the confines of a ridge. Galileo E6ESDRKLIN01 observation, with illumination from the right. All scale bars are 10 km.

pushed up against the side of a preexisting ridge, suggesting horizontal flow (Fig. 8b).

Greenberg et al. (1999) observed that the margins of some chaotic terrain areas tend to avoid large preexisting ridges. The margins on Conamara avoid the two large crossing ridges to the north (Fig. 1), and smaller chaos areas (Fig. 9a) sometimes behave similarly. However, there are also many counterexamples in which chaos margins ignore ridges (Fig. 9b), or even seem to be confined within ridges (Fig. 9c). A more rigorous analysis of the interaction between chaotic terrain and large ridges is needed to see whether ridges affect chaos terrain margins.

Boundaries of chaotic terrain areas are commonly accompanied by deposits of smooth dark material. Figure 5 shows an excellent example of this at Thrace Macula, with smooth dark material diffusing out from the morphological chaos boundary into topographic lows. Whether this is due to liquid coming from the subsurface (*Fagents,* 2003) or from sublimation of bright surface frosts (*Fagents et al.,* 2000) is unclear. Similar dark deposits infilling between ridges are seen around Castalia Macula, a large dark area next to chaotic terrain. Stereo data of that area shows that while the dark material is confined to topographic lows, it is far from being flat (*Prockter and Schenk,* 2005), so a simple explanation of pooling liquids does not fit the observations, unless substantial surface motion has taken place afterward.

2.4. Combinations of Morphological Elements

Areas of chaotic terrain can range from being comprised mostly of plates with small intervening lanes of matrix, to being mostly matrix with few or no plates. Conamara Chaos lies in the middle, with 59% matrix and 41% plates by area (*Spaun et al.,* 1998). *Spaun* (2002) mapped the interiors of chaotic terrain in eight additional regions and found that a roughly half-and-half mix of plates and matrix was typical for the majority of chaos areas.

One correlation that is observed with chaos margins is that areas with a prominent inward-facing scarp tend to have abundant plates (e.g., Figs. 1 and 7), while areas with matrix doming above the surrounding terrain tend to be mostly plate-free (e.g., Fig. 8) (*Figueredo and Greeley,* 2004). As stated at the beginning of this section, most mapping to date has been on a chaos-by-chaos basis instead of mapping the morphological features inside areas of chaotic terrain. *Spaun* (2002) mapped the interior morphology within several regions of chaotic terrain but there was little quantification of the proportions of plates and matrix for areas other than Conamara. More work needs to be done to quantify the relationships between matrix types, abundance and motion of blocks, and the nature of chaos margins.

3. TOPOGRAPHY

In some areas of chaotic terrain, the matrix appears to be domed above the adjacent terrain. However, there is usually a trough surrounding this type of chaos, so it is not immediately apparent how high the chaos matrix stands with respect to distant background terrain. Photoclinometry measurements of Murias Chaos (Fig. 8a) show that even though the surrounding trough is depressed by 300–400 m from the background terrain, the chaos matrix itself rises 400–600 m above its base, placing it about 100 m above the background terrain (*Figueredo et al.,* 2002). Stereo topography of a large dome-shaped feature next to Castalia Macula, with chaotic terrain exposed on its crest, shows that the top of the feature rises more than 1 km from its base (*Prockter and Schenk,* 2005). An unusual chaos area (Fig. 10) is cut by a block of crust that seems to have been "punched up," raising part of the chaos onto a flat mesa 900 m high (*Figueredo and Greeley,* 2000).

If chaos areas that appear to have domical topography are in fact high standing, what about the topography of chaos areas with inward-facing scarps, where the matrix appears to be low? Stereo topography of two small (~10 km diameter) chaos areas near Tyre show that even though they have an inward-facing scarp, the matrix in the center of the chaos areas stands 100–200 m above the background terrain (*Nimmo and Giese,* 2005). Stereo coverage of Cona-

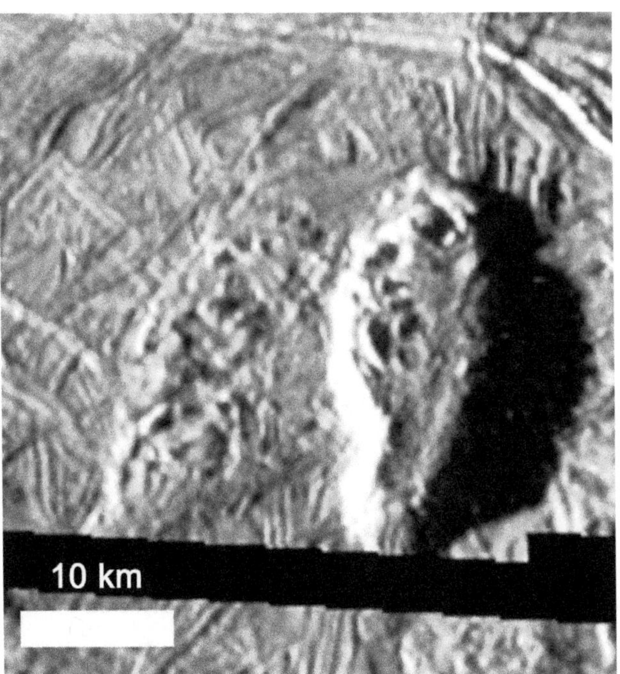

Fig. 10. A plateau nearly 1 km high cuts across this chaotic terrain area. Galileo 15ESREGMAP02 observation, with illumination from the left.

mara Chaos shows that the matrix in the interior is also uplifted in several dome-shaped regions, standing up to 250 m above the surrounding terrain (*Schenk and Pappalardo,* 2004). The uplifts in Conamara appear to be associated with the terrain that is composed of matrix without plates. In fact, even though most plates are clearly locally higher than the matrix, many of the surfaces of the plates are actually topographically lower than the more distant matrix-rich areas (see Fig. 15 for an illustration).

Even with the limited reliable topographic coverage from the Galileo mission, the message seems clear that the formation of chaotic terrain involves doming of the surface at some stage in the process. It is interesting to note that Murias Chaos, an almost plate-free chaos, appears to be one large dome, while Conamara Chaos appears to be composed of multiple domes, with the abundant plates existing in the interstices between the domes.

4. SPECTRAL PROPERTIES

Areas of chaotic terrain are almost always associated with dark reddish-brown material on the surface. The exact composition of this material is unknown (see chapter by Carlson et al.), but the color is probably due to radiolysis of surface materials producing reddish sulfur compounds (*Carlson et al.,* 2002). Infrared spectra of this material shows distorted water absorption bands, interpreted to be a hydrated material such as heavily hydrated sulfate salts (*McCord et al.,* 2002; *Dalton et al.,* 2005), or sulfuric acid hydrate (*Carlson et al.,* 2005). Knowledge of the composition of the contaminant material mixed into the water ice in chaotic terrain is important, because hydrated salts or sulfuric acid hydrate will melt at lower temperatures than pure water ice, and could possibly play an important role in the formation of chaotic terrain (see section 8.3).

As noted in section 2 above, some areas of chaotic terrain are surrounded by a deposit of dark material that seems to fill topographic lows. On orbit E12, Galileo collected a moderately high-resolution color observation of Conamara Chaos. *Head and Pappalardo* (1999) showed from this observation that both the matrix and most of the smaller blocks are covered in the reddish-brown material, while some large blocks with prominent ridge segments tend to be closer to the color of the undisrupted background terrain. Some of the chaos boundaries are also sharp color transitions, while others show small amounts of the dark material fingering in between ridges (like the boundary of Thrace Macula; see Fig. 5). The large blocks within the chaos also show an enhanced amount of dark material lying in between the ridges, as compared to the background terrain outside the chaos.

5. SIZE DISTRIBUTION

Features that match the description of chaotic terrain range over at least a 3-order-of-magnitude size range, from kilometer-scale features all the way up to regions more than 1300 km across. For some regions of chaotic terrain, it is relatively simple to define the boundary and map the surface area. However, there are many cases where fractures extend out from large patches of chaotic terrain to link up with nearby smaller patches. Figure 11 shows one such example, where larger chaos areas at the top and right edges of the frame are connected via narrow fractures to a small chaos area near the center. Should all this be counted as one chaos area? Are the regions between the fractures counted as undisrupted preexisting terrain or as chaos plates? Even Conamara Chaos exhibits similar boundary definition problems on the eastern side, with fractures leading to small neighboring chaos areas. Faced with a similar situation with fractures connecting two neighboring domes, *Riley et al.* (2000) decided to count them as separate chaos areas. In contrast, *Spaun et al.* (1999) and *Figueredo and Greeley* (2004) argued that these observations show the process of chaotic terrain growing by merging smaller chaos areas together (see chapter by Doggett et al.). Seen in this light, the topographic observations within Conamara where the matrix forms several domes with lower broken plates in between (*Schenk and Pappalardo,* 2004) appears to be a more advanced stage of the merging process that may be observed in Fig. 11.

Much ink has been expended discussing the exact nature of the size distribution of chaotic terrain, an issue of contention because the apparent existence of a preferred size and spacing for chaotic terrain features was used to advocate a diapiric model of formation (e.g., *Pappalardo et al.,* 1998a, 1999; *Spaun,* 2002). Advocates for the alternative melt-through model (e.g., *Greenberg et al.,* 1999; *Riley et al.,* 2000) instead argued that there is no preferred size for chaotic terrain, and that the number of features rises continuously with decreasing size. At the worst, the reader was confronted with data from two research groups plot-

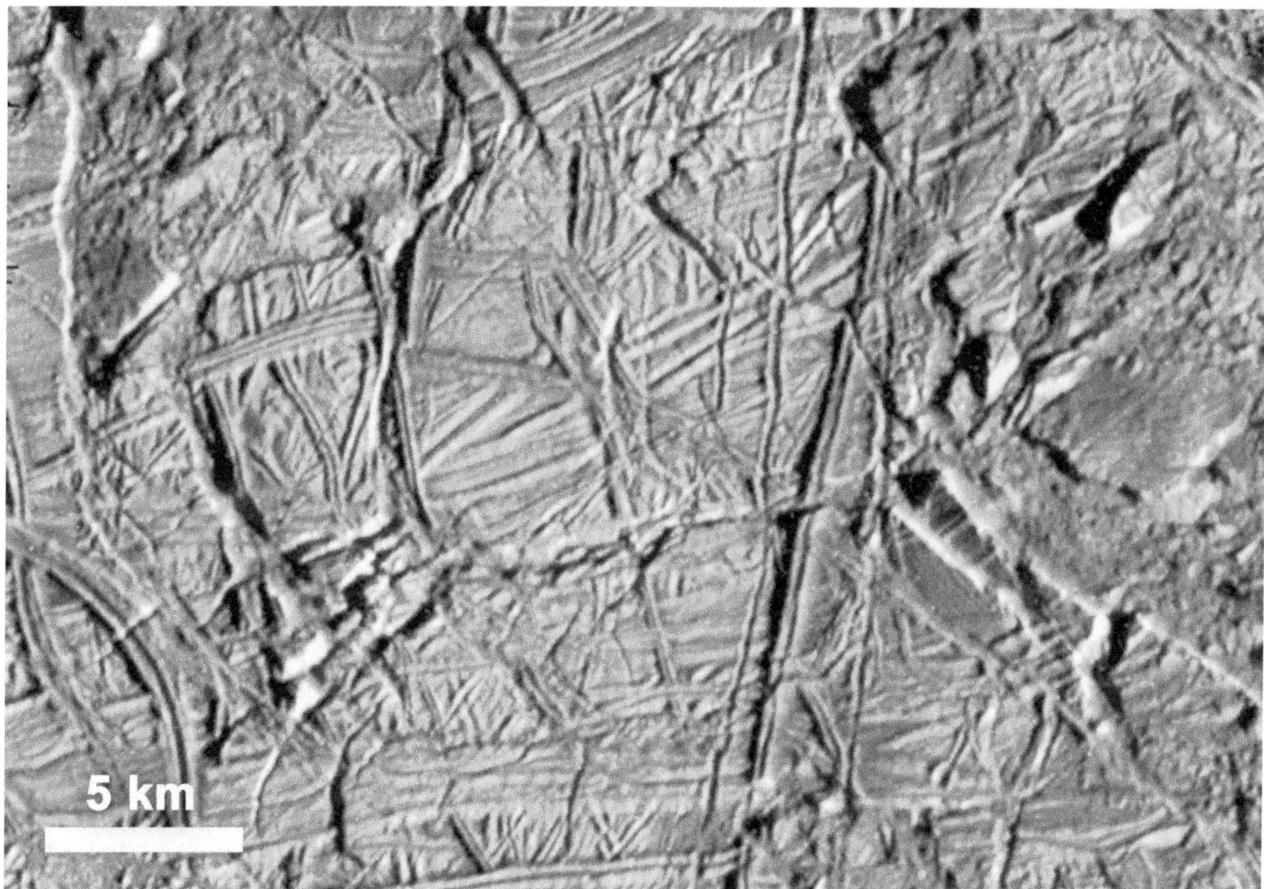

Fig. 11. Two large chaos regions at the top and right are connected to a smaller chaos region in the lower left by narrow troughs or fractures. Should this be considered as one chaos area? In the view of *Spaun et al.* (1999) and *Figueredo and Greeley* (2004), this may be an example of chaotic terrain growth through absorbing neighboring chaos regions. Galileo 17ESDISSTR01 observation, with illumination from the right.

ted in such a way as to make it impossible to visually compare them, e.g., a histogram of feature diameters on a linear scale (*Spaun,* 2002) vs. a cumulative plot of feature areas on a logarithmic scale (*Greenberg et al.,* 1999). In an attempt to settle this confusion, we have rebinned all the chaotic terrain size data published in the literature and plotted it on one graph (Fig. 12). The two main opposing studies on the subject, *Riley et al.* (2000) and *Spaun* (2002), both mapped large regions of Europa at regional resolution (~200 m/pixel). Replotting their data in a common format on Fig. 12a shows that their regional results are almost identical. It is ironic that the artifacts of data display led Riley et al. to argue for the existence of more small chaos areas than Spaun, as the most significant divergence between the two data sets is the larger number of small chaos areas measured by Spaun.

Hoppa et al. (2001) used a region of Europa imaged at two different resolutions to show that more small chaos areas will be recognized at higher resolution, the implication being that some, if not all, of the downturn in chaos abundance at small sizes in the regional data is a resolution effect. Both *Riley et al.* (2000) and *Spaun* (2002) mapped a few small areas imaged at higher resolution, and these data are also plotted on Fig. 12a. The flattening of the distribution at small sizes in the high-resolution data for both mapping groups demonstrates that Hoppa et al.'s basic point is correct. However, this high-resolution mapping took place in areas of mottled terrain, so the absolute abundance of small chaos areas (the vertical position on the graph) at small sizes may not be applicable globally. Riley et al. claim that the chaos abundance from the high-resolution data can be used on the rest of the global database to "correct for recognizability" and that doing so erases any peak in the size-frequency distribution of chaotic terrain. However, the distribution must roll over somewhere; chaotic terrain cannot become infinitesimally small. Indeed, the highest-resolution images of Europa at 6 m/pixel within mottled terrain did not reveal a new class of even smaller chaotic terrain areas. It appears that at the tens-of-meters scale, the primary processes modifying Europa's surface are surface regolith processes, and not endogenic processes like chaos formation. Perhaps the noted dropoff in chaos abundance below an area of 3 km^2 (~2 km diameter) is real, and not a resolution artifact. Only the collection of a large sample of high-resolution imaging data of Europa's surface can confidently resolve this issue.

Two groups have also reported the size distribution of pits and domes separately from chaotic terrain in general. Figure 12b shows the size distribution of pits and domes from *Greenberg et al.* (2003) compared to domes measured by *Rathbun et al.* (1998). The Greenberg et al. measurements are from several observations, while the Rathbun et al. measurements are only from the regional resolution Galileo E6ESDRKLIN01 mosaic, in the area of Conamara Chaos. The dropoff at small sizes in the Rathbun data therefore may be a resolution effect, much like the resolution effect for chaos. In fact, the Rathbun et al. line lies almost directly on top of the peak of the *Riley et al.* (2000) and *Spaun* (2002) regional size distribution lines for chaotic terrain shown in Fig. 12a. Compared to the size distribution of small chaos regions mapped at high resolution by Riley et al. and Spaun, the Greenberg et al. data for pits and domes appears to be more strongly peaked near an area of 10 km^2 (3–4 km diameter), but this may be a resolution effect, because Greenberg et al. did not include the same high-resolution images included by Riley et al. It is interesting to note that the size distribution of pits and domes is very similar, further reinforcing the impression from Fig. 4 that they are somehow related.

6. GEOGRAPHIC AND TEMPORAL DISTRIBUTION

After mapping the 10% of the surface seen in regional-scale (~200 m pixel^{-1}) Galileo imaging of Europa, *Riley et al.* (2000) concluded that 28% of the surface is covered by chaotic terrain. Sampling from high-resolution data covering 7400 km^2 revealed a larger population of small chaos areas that were not observed in regional imaging (*Riley et al.*, 2000; *Hoppa et al.*, 2001) (see previous section). Extrapolating from high-resolution sampling to the rest of the globe, Riley et al. added another 12% to the chaos surface coverage estimate for a total of ~40%. In separate geological mapping work by *Figueredo and Greeley* (2003), they reported that approximately 20% of the surface is covered by chaotic terrain. This leaves a significant discrepancy in the reported surface coverage of chaotic terrain.

There is a direct correlation between areas of chaotic terrain seen at moderate and high resolution, and the "mottled terrain" seen in lower-resolution global-scale images. *Pappalardo et al.* (1998b) mapped mottled terrain distribution across Europa from global-scale images. They showed that the mottled terrain is concentrated in two roughly oval-shaped and antipodal areas filling areas within 40° of the equator, centered at approximately 120° and 300°W (west longitude). This pattern may also be seen in the global geological map of Europa presented in the chapter by Doggett et al. Geological mapping of regional-scale images by *Figueredo and Greeley* (2004) shows a similar concentration of chaotic terrain near the equator. *Pappalardo et al.* (1998b) argued that the observed mottled terrain distribution is best fit by the compressional zones predicted from the nonsynchronous rotation stress model, swept out through the last few tens of degrees of rotation. *Figueredo and Greeley* (2000), with a more spatially limited but more highly detailed dataset, argued instead that the pattern of chaos dis-

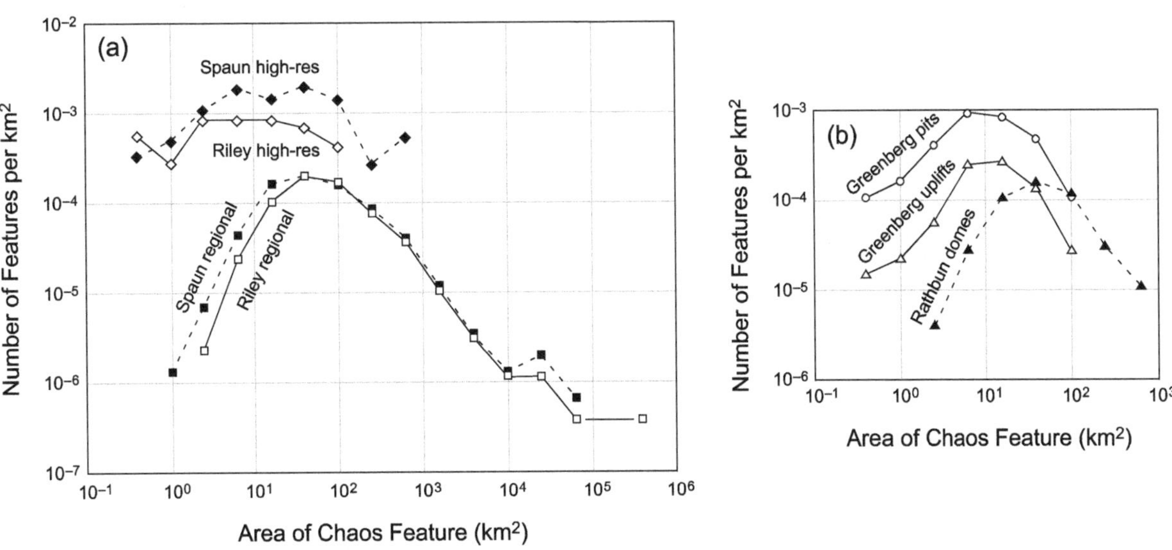

Fig. 12. **(a)** Size distributions of chaotic terrain measured by *Riley et al.* (2000) and *Spaun* (2002), rebinned and plotted on the same axes. The two studies, which came to different conclusions about the size distribution, show remarkably good agreement. The "regional" lines come from a large amount of Galileo regional image data obtained at approximately 200 m pixel^{-1}, while the "high-res" lines are from Galileo observations at a few tens of meters per pixel. More chaotic terrain is recognized at higher resolution, in agreement with *Hoppa et al.* (2001), although it is unclear whether the dropoff in chaos abundance at ~3 km^2 (~1 km diameter) is real. **(b)** Size distribution of pits and domes from *Greenberg et al.* (2003) compared to domes measured by *Rathbun et al.* (1998). For comparison to **(a)**, the Rathbun et al. line lies almost directly on top of the peak of the *Riley et al.* (2000) and *Spaun* (2002) regional size distribution lines for chaotic terrain.

tribution is a good fit to local icy shell thickness minima predicted by tidal heating models near the leading and trailing points (*Ojakangas and Stevenson,* 1989). Based on mapping of high-resolution areas, *Spaun et al.* (2004) reported that chaotic terrain occurs in smaller, more abundant and closely spaced patches near the leading and trailing points and at 330°W, rather than near the antijovian point, but they concluded that the interpretation of geographic patterns at high resolution is hampered by the limited imaging data coverage. It is also interesting to note that the concentrations of chaos regions near 120°W and 300°W closely correspond to the positions of the paleopoles proposed by *Schenk et al.* (2008) in their polar wander scenario. Because tidal heating is most strongly concentrated near the poles, this possible correlation hints at tidal heating being a strongly controlling factor in chaotic terrain formation. In this hypothesis, it also implies that chaos formation would have predated the reorientation event, although the timing of that event is currently unknown.

Geological mapping of Europa has generally found that chaotic terrain crosscuts almost all other features (e.g., *Prockter et al.,* 1999; *Figueredo and Greeley,* 2000; see chapter by Doggett et al.) and thus puts it at or near the top of the time sequence of feature formation. This suggests that the formation of chaotic terrain may be a relatively recent phenomenon. There are also interesting examples of chaotic terrain cut by younger features. *Prockter and Schenk* (2005) showed older chaotic terrain cut by the formation of a nearly 1-km-tall dome at Castalia Macula (Fig. 13), while *Riley et al.* (2006) pointed out an area of chaotic terrain near the south pole crosscut by ridges and younger chaos (Fig. 14). *Greenberg et al.* (1999) mapped chaotic

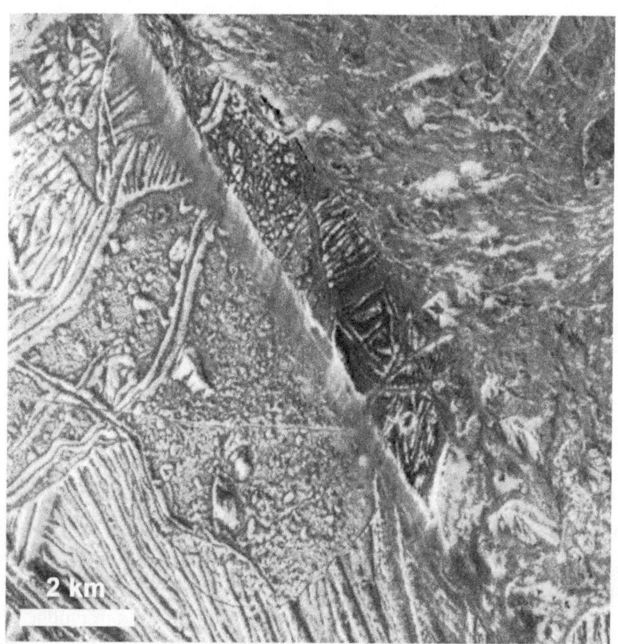

Fig. 13. An older patch of chaotic terrain is cut by a large cliff resulting from surface uplift north of Castalia Macula. Galileo 14ESDRKSPT01 observation, with illumination from overhead.

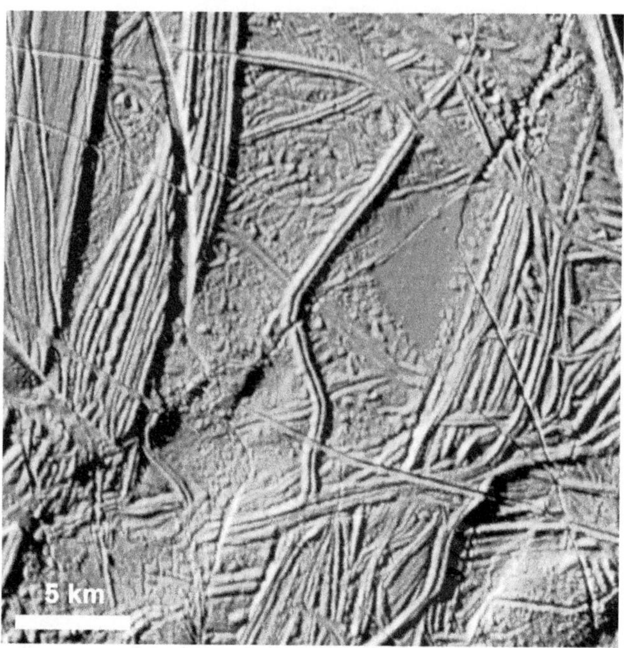

Fig. 14. Possible example of old chaotic terrain near the south pole of Europa, mapped by *Riley et al.* (2006). Lumpy matrix-like material is exposed between several generations of ridges. Galileo 17ESSOUTHP01 observation, with illumination from the left.

terrain on Europa, splitting it into two categories: "fresh" chaos, which is largely unmodified by any other features; and "modified" chaos, which is distinguished by a smoother matrix and more subdued edges. They reported that modified chaos appears to be more commonly crosscut by younger tectonic features, although the extent to which this occurs has not been quantified. *Figueredo and Greeley* (2004) also included a "subdued" chaos unit in their geological mapping, with similar characteristics to the "modified" chaos of Greenberg et al., and found that it was slightly older than other chaotic terrain units in the time sequence.

Greenberg et al. (1999) also identified features such as isolated polygonal blocks and "fins" inferred to be tilted crustal blocks, unassociated with any surrounding chaotic terrain. They interpreted these features as evidence of older chaotic terrain that has been mostly destroyed by later resurfacing, and *Hoppa et al.* (2001) used this to argue that chaos formation and ridge formation are both continuous and ongoing processes throughout europan history. A thorough statistical analysis of the frequency of crosscutting relationships would more thoroughly address whether ridges and chaos could be forming at a constant rate through time, or whether chaos formation is more common in recent times.

A possible constraint on the duration of the chaos formation process comes from studying the relationship between chaotic terrain and secondary craters. *Bierhaus et al.* (2001) studied the distribution of small craters in high-resolution images of Conamara Chaos, and found that the vast majority of the craters were related to a ray from the impact crater Pwyll to the south. These secondaries all formed in a

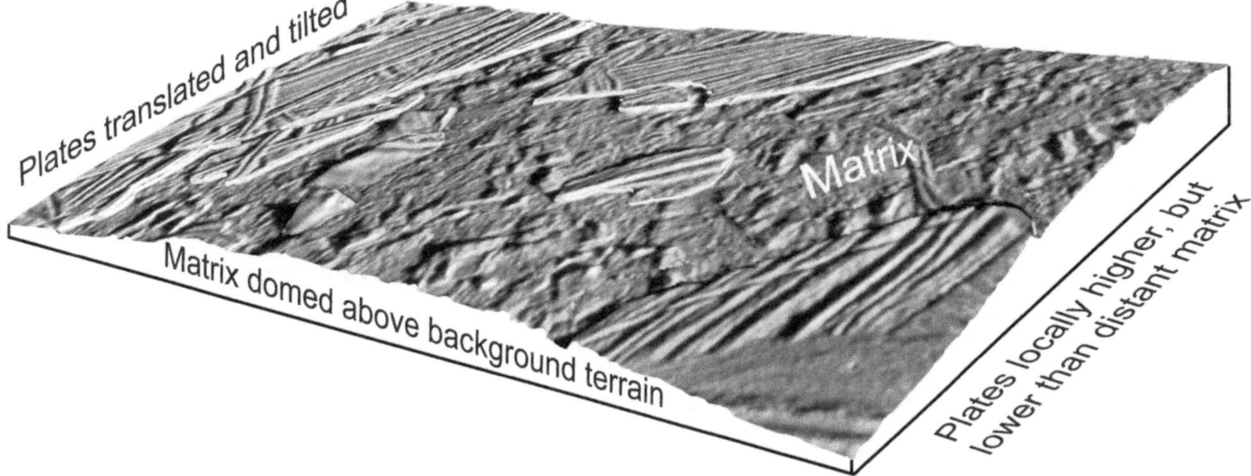

Fig. 15. Illustration of four of the eight "hard" observational constraints on chaotic terrain formation.

geological instant, and yet there are more Pwyll secondary craters preserved on the blocks of preexisting terrain in the chaos than there are in the matrix. Bierhaus et al. argued that this is not an effect of lack of recognition in the matrix, but rather there has been post-Pwyll activity in Conamara that has modified the matrix and not the blocks. An alternative explanation would be that the matrix material is less cohesive and may not preserve impact craters as well (similar to the loss of small craters in the lunar highlands), but such an effect would have to not be size selective in order to explain the crater distribution data. Even more perplexingly, the density of craters in all chaotic terrain is higher than in the ridged terrain (with overlapping error bars; see chapter by Bierhaus et al.). If this crater density difference is borne out by future observations, this would contradict the cross-cutting relationships.

7. SUMMARY OF OBSERVATIONAL CONSTRAINTS FOR CHAOS FORMATION MODELS

Before we consider theoretical models for chaotic terrain formation, let us first summarize the observational constraints from the previous five sections. A few of these constraints are illustrated in Fig. 15. Any viable model for chaotic terrain formation must be able to explain the following "hard" observational constraints:

H1: Formation of irregular, lumpy matrix material

H2: Plates of preexisting terrain usually higher than adjacent matrix (~100 m)

H3: Plates tilted, rotated, and translated horizontally (a few kilometers distance)

H4: Chaos with abundant plates are usually bounded by inward-facing scarps; chaos with few plates are usually bounded by matrix doming above the surroundings

H5: Matrix material is usually topographically higher than background terrain by 100–250 m, and can stand higher than the plates in the same chaos region

H6: Exposure of dark hydrated sulfurous salt/acid material in matrix, on small plates, and in surrounding topographic lows

H7: Areas of chaotic terrain range from approximately 1 km to over 1000 km in diameter

H8: Chaotic terrain is more concentrated in antipodal regions near the equator

In addition, there are some "soft" observational constraints that may be real, or they may be observational biases, misinterpretations, or misclassifications of feature types presently assumed to be part of chaotic terrain. Models that can also explain these observations will be considered more successful than those models that cannot explain the following:

S1: Undisrupted pits and domes exist in association with chaotic terrain

S2: Preservation of preexisting structure in degraded form within the chaos matrix

S3: Plates of preexisting terrain have a minimum size of approximately 1 km

S4: Ridges are preferentially preserved during chaos formation

S5: Dark matrix material can flow out onto preexisting surface

S6: Domes cutting across chaotic terrain can stand up to 1 km high

S7: Large chaotic terrain regions grow by a process of merging smaller regions together

S8: Matrix formation may be long-lived, wiping out small craters

8. MODELS FOR CHAOTIC TERRAIN FORMATION

In this section, we discuss five models for the formation of chaotic terrain that have been proposed in the literature and compare each model to the observational constraints above. The results of this section are summarized in Table 1,

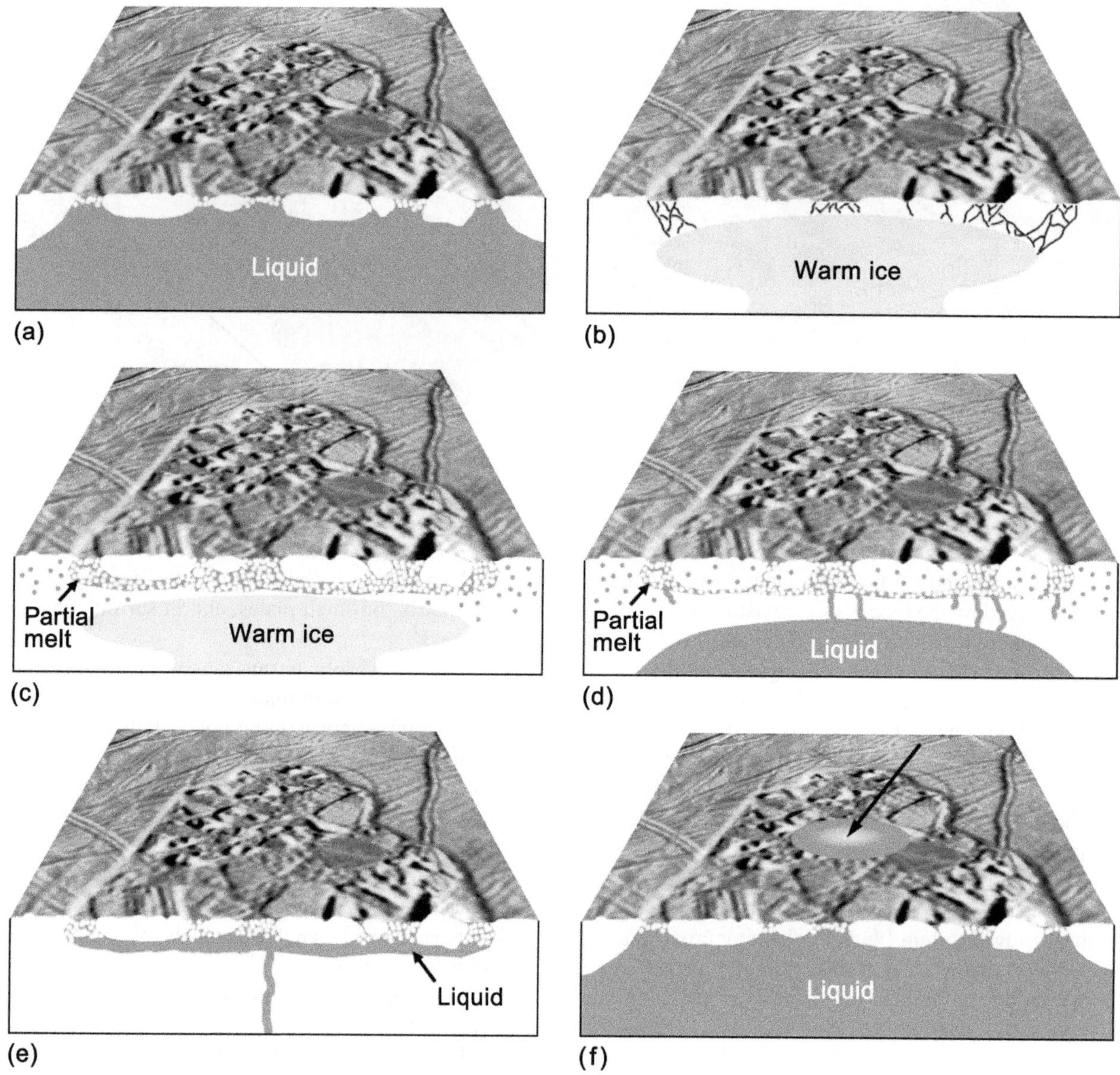

Fig. 16. Illustration of candidate chaotic terrain formation models: **(a)** melt-through; **(b)** diapirism; **(c)** brine mobilization driven by diapirism; **(d)** brine mobilization driven by partial melt-through; **(e)** sill formation; **(f)** impact. See text for explanation of the models.

which shows for each combination of model and observation whether there is a natural explanation, whether special circumstances are required, or whether the model does not plausibly explain that observation.

8.1. Melting Through the Icy Shell

The visual similarity between chaotic terrain and terrestrial pack ice inspired some to consider models in which the plates are the equivalent of tabular icebergs floating in an underlying ocean, and the matrix is composed of lower-lying sea ice formed as the top of that ocean froze (Fig. 16a) (*Carr et al.,* 1998; *Greeley et al.,* 1998). This is envisioned to happen as a result of a heat source at the base of the icy shell melting the overlying ice, creating a hole melted through the ice and exposing the ocean below (*Greenberg et al.,* 1999; *Thomson and Delaney,* 2001). If the plates observed in Conamara Chaos were free-floating icebergs, their heights would indicate that the ice in those blocks was 0.5–2 km thick at the time they froze into the surrounding matrix (*Williams and Greeley,* 1998).

The model of *Greenberg et al.* (1999) begins with the first stage of melting, thinning the ice crust and producing a pit. If melting progresses at a constant rate as the melting front reaches the surface, then topographic highs such as ridges, which may be isostatically supported regions of thicker ice, would be the last areas to melt and may be preserved as the floating plates within the chaotic terrain. However, as the melting front nears the surface, its propagation speed will be inversely proportional to its proximity to the surface, due to thermal conduction through the ice and radiation of the energy into space, so in actuality the bases of ridges will melt faster and the topography should flatten out instead of producing ridged plates (*Goodman et al.,* 2004).

Initial objections to the melt-through model were raised because of the tremendous amount of energy required to produce and maintain melt at Europa's surface (*Pappalardo et al.,* 1999; *Collins et al.,* 2000), and the tendency for the warm ice at the base of the shell to flow in from the sides to fill the hole (*Stevenson,* 2000). Of the two objections, the energy argument is more difficult to surmount. Maintaining a surface temperature at 270 K would require a heat flux of roughly 300 W m^{-2} (*Goodman et al.,* 2004), about 25 times the current solar insolation, and more than 10^3 times the likely globally averaged tidal heat production, even assuming that Europa's silicate interior is as dissipative as Io's (e.g., *Nimmo and Giese,* 2005). Complete melt-through of the icy shell may not actually be necessary to form chaotic terrain, and thinning the icy shell may have interesting effects, but this remains to be quantified (see the end of section 8.3). Regarding the ice flow objection, more careful consideration of ice rheology shows that the ice infill process is somewhat slower than envisioned by Stevenson (*Nimmo,* 2004), but the main unknown is the thickness of the icy shell. If Europa's silicate mantle is Io-like in its tidal dissipation, the equilibrium icy shell thickness is about 2 km, while in the absence of silicate dissipation the equilibrium thickness is tens of kilometers (e.g., *Ojakangas and Stevenson,* 1989). Thicker icy shells not only have more total ice to melt through, but they also have a deeper channel of warm ice at the bottom to promote inflow. There will be a critical icy shell thickness (probably on the order of 10 km) above which inflow will be faster than melting, preventing melt-through, so an independent constraint on the thickness of the icy shell is crucial to the validity of this model.

O'Brien et al. (2002) modeled the melting process and found that complete melt-through of an icy shell 6 km thick could be achieved in ~10^4 yr by concentrating a few percent of Europa's total tidal heat output (roughly 7 TW, assuming tidal dissipation in an Io-like mantle) within a region 200 km across. They also demonstrated that the melting process was orders of magnitude faster than ice inflow for the shell thicknesses they assumed, limiting the thickness of the warm ice channel that could flow at the base of the shell. The complete melt-through found in this model is at odds with a simple energy-balance model, because the heat supplied in the model to melt the ice is insufficient to maintain open water at the surface (*Goodman et al.,* 2004). It turns out that the numerical scheme of O'Brien et al. lacked the resolution to accurately treat near-surface effects; the constant grid spacing of 100 m lost resolution of the heat conduction problem and became numerically unstable as the melt approached the surface. An improved model with adaptive grid spacing (*Goodman et al.,* 2004) showed that the ice will asymptotically approach an equilibrium thickness, on the order of 100 m for the energy and shell thickness values assumed by O'Brien et al. It is thus unclear whether melt-through of the kind originally envisaged (*Carr et al.,* 1998; *Greenberg et al.,* 1999; *Thomson and Delaney,* 2001), with open water at Europa's surface, is physically possible.

What is a plausible source of heat supplied to the base of the icy shell? As discussed above, it is possible that Europa's silicate mantle is highly dissipative and generates ~7 TW of heat. A direct path from seafloor heat sources to the base of the icy shell is hydrothermal plumes rising through the ocean (see chapter by Vance and Goodman). The hot water will rise until it reaches neutral buoyancy, so the density stratification of the ocean is critically important for determining if a plume will reach the base of the icy shell. *Thomson and Delaney* (2001) investigated hydrothermal plumes that would rise and then stall at the base of the icy shell, delivering enough heat for melt-through events. They envisioned an ocean that undergoes stratification and destratification events, with chaos formation coinciding with oceanic conditions that allow hydrothermal plumes to deliver heat to the icy shell. *Goodman et al.* (2004) argued for an essentially unstratified ocean and conducted scaled physical experiments to determine the dynamics of hydrothermal plumes in an unstratified, rotating environment. The plumes start as narrow, rotationally confined cylinders, but when they run up against a solid boundary such as the base of the icy shell, they spread laterally into inverted cones with a width of 25–50 km. This will spread the heat from a point source on the ocean floor over a large area of the icy shell. The resulting heat flux (0.1–10 W m^{-2}) is insufficient by more than an order of magnitude to cause complete melt-through events. Melt-through requires a near-surface heat flux of ~300 W m^{-2}, or 3% of the total heat flow of Europa assumed by *O'Brien et al.* (2002), concentrated into a region 25 km wide. Putting more heat into the bottom of the plume at the seafloor heat source serves to spread the heat over a larger area of the icy shell as the plume grows wider. *Lowell and DuBose* (2005) investigated the nature of possible hydrothermal source areas, and found that the reduced gravity on Europa leads to weaker hydrothermal plume sources as compared to sources on the Earth's seafloor. The heating patch size found by *Goodman et al.* (2004) is smaller than chaos areas such as Conamara, but much larger than the more numerous small chaos areas.

Given the constraints on hydrothermal plume size, *Goodman et al.* (2004, appendix A) used the *O'Brien et al.* (2002) melting model to investigate the expected size distribution of chaotic terrain formed from these plumes. A power-law distribution of plume lifetimes could explain the observed distribution of chaos areas much larger than 10 km diameter. However, a power-law distribution of plume lifetimes cannot explain the observed size distribution of chaos areas smaller than 10 km. In order to produce the observed numerous small chaos areas, one would need to posit an infinite spike in the plume distribution at a total power equal to the amount needed to melt through only at the center of the plume. This is equivalent to turning off the plume at its base as soon as the first melt reaches the surface, but there is no readily apparent feedback mechanism between melt at the surface and a seafloor heat source supplying it.

The question of stratification in Europa's ocean brings up another interesting idea for melting through the icy shell of Europa. *Melosh et al.* (2004) pointed out that if the ocean is composed of relatively fresh water, a "stratosphere" of near-freezing water can stably stratify on top of water a few degrees warmer. As the warmer bottom water heats up from the rocky core below, the stratification can overturn, bringing warm water in contact with the icy shell and leading to episodic, widespread melt through events. Salt dissolved in the water eliminates the dip in the temperature-density relationship, and induced magnetic field evidence from Europa indicates that the ocean must be somewhat salty (*Zimmer et al.,* 2000; *Hand and Chyba,* 2007). However, there are also pressure effects on the solution of salt that can operate deep in Europa's ocean and lead to stratification and inhibition of hydrothermal plume rise (*Vance and Brown,* 2005; see chapter by Vance and Goodman).

The melt-through model explains the motion of the chaos plates by currents in the water pushing the floating plates through the liquid. *Thomson and Delaney* (2001) noted that the motion of the plates in Conamara Chaos measured by *Spaun et al.* (1998) is consistent with Coriolis-driven motion in the top of an ascending hydrothermal plume. However, *Goodman et al.* (2004) calculated the forces on the plates due to such fluid motion, and found that the stresses exerted by currents on the plates (~1 Pa) were too small to move the plates even through a thin layer of weak slush.

Once the heat source is turned off, the hole in the icy shell will begin to refreeze back to its equilibrium thickness, staying relatively thin for ~1 m.y. (*Buck et al.,* 2002). As the ice thickens, the surface will return to its premelting elevation and the original pit will disappear. If the matrix ice is welded to the margins, it will appear to dome up in the center as refreezing progresses, but it will not rise above the elevation of the background terrain unless there are compositional contrasts between the refreezing ice and the original unmelted ice (*Nimmo and Giese,* 2005). If there is a mechanism other than refreezing that can trap salts within the icy shell, it may be possible to produce such compositional contrasts, but this issue has not been well explored.

Melt-Through Model:
Comparison to Observational Constraints

The melt-through model can be compared to the hard and soft observational constraints on chaos formation as follows.

H1: Matrix material is formed by melting and disruption of parts of the surface ice.

H2: The plates are floating in the ocean below and stand high because they are thicker remnants of the icy shell than the material in the matrix.

H3: Plates can easily translate, rotate, and tilt if the icy shell is melted through, but the driving force for plate motion is unclear.

H4: Inward-facing scarps form by calving icebergs from the margin of the melted area; matrix could slope onto pieces of surrounding terrain that originally tilted into the molten matrix.

H5: The surface will not rise above the background terrain unless there is a mechanism to produce significant density variations between the refrozen ice and the original icy shell.

H6: Exposure of the ocean at the surface provides the hydrated sulfurous contaminants.

H7: The range of possible sizes of melt-through areas depends on the mechanism invoked to heat the base of the icy shell. Some mechanisms, such as hydrothermal plumes in an unstratified ocean, would predict a paucity of small (<~5 km diameter) chaos areas (see appendix A of *Goodman et al.,* 2004).

H8: The geographic distribution of chaotic terrain may be due to enhanced tidal heating, because melt-through events may be more common where tidal heating is more intense. If this was the controlling factor, chaos should be more common at the poles, but perhaps the cold surface temperature at the poles inhibits surface melting, or polar wander has subsequently occurred.

The melt-through model can also easily explain two of the soft constraints. Pits (but not domes) are a natural consequence of melting. The minimum size of plates may be a consequence of tilting when the plate width is smaller than the plate thickness.

8.2. Diapirism

It was recognized early in the Galileo tour that pits, spots, and domes on Europa (Fig. 4), including some with chaos characteristics, might be the surface expression of rising diapirs (Fig. 16b) (*Pappalardo et al.,* 1998a), analogous to salt diapirs on Earth. Diapirs have been used to explain collections of similar-sized small domes (*Rathbun et al.,* 1998) as well as large dome features such as Murias Chaos (*Figueredo et al.,* 2002) and Thera Macula (*Mével and Mercier,* 2007). As discussed below, such diapirs typically arise due to either thermal or compositional buoyancy. On Europa, there is also the potential complication of locally enhanced tidal heating and generation of melt (discussed below). Theoretical aspects of diapirism are discussed in more detail in the chapter by Barr and Showman.

One significant problem for the diapirism hypothesis is that in conventional thermal convection, the size of individual diapirs is rather uniform, because it is set by the thickness of the thermal boundary layer. This behavior is clearly different from the observed wide size-spectrum of chaos regions. While it is possible that some large chaos regions are generated by merging diapirs (*Spaun et al.,* 1999; *Figueredo and Greeley,* 2004; *Schenk and Pappa-

lardo, 2004; *Mével and Mercier,* 2007), the wide size range of individual, roughly axisymmetric features is difficult to explain by such merging.

The behavior of thermal convection in strongly temperature-dependent materials, such as ice, is fairly well understood (e.g., *Solomatov,* 1995; chapter by Barr and Showman). The temperature contrasts are set by the rheological properties of the ice, and are typically small (a few tens of Kelvin) (*Nimmo and Manga,* 2002). As a result of the low-temperature contrast and the presence of a thick stagnant lid, the topography generated by simple thermal diapirs is also small (typically a few tens of meters) (*Showman and Han,* 2004). Because the lid is cold, the convective strains are small and incompatible with the high strains associated with chaos block rotation or lateral matrix flow. Thus, simple thermal convection is not a satisfactory mechanism for generating chaos, and as a result several additional physical processes have been proposed.

One potentially important process is the yielding of the near-surface material. If stresses are large enough to cause yielding (plastic or brittle failure) of the near-surface ice then both surface velocity and surface topography can increase significantly. For instance, *Showman and Han* (2005) demonstrated that if the plastic yield stresses are sufficiently small (a few tens of kilopascals), significant surface strains and topographic amplitudes of ~100 m can result. For reference, a plastic yield stress on the order of 10^4 Pa is similar to the yield stress required to move giant icebergs through polar pack ice on Earth (*Lichey and Hellmer,* 2001). It should be noted that the elegant model of *Hoppa et al.* (1999) for the formation of cycloidal fractures on Europa requires a similarly low fracture propagation stress. The *Showman and Han* (2005) model demonstrated complete surface recycling, which is bad for preserving plates, but could plausibly provide a source for matrix-rich features that appear to flow over the background terrain, such as Murias Chaos (*Figueredo et al.,* 2002). *Miyamoto et al.* (2005) showed that such surface flows of warm ice are possible if the effusion rate is on the order of 0.01–0.1 $km^3\ yr^{-1}$, which would require a wide supply conduit. One problem with the *Showman and Han* (2005) model is that it is unable to produce features smaller than 30–100 km diameter (a few times the shell thickness), which is much larger than many of the observed domes (e.g., *Rathbun et al.,* 1998; *Greenberg et al.,* 2003) that have been proposed as diapiric features.

Another way of increasing surface topography is to appeal to compositional buoyancy. *Han and Showman* (2005) demonstrated that, as expected, larger surface topography can be generated if compositional density contrasts are present in addition to thermal contrasts. However, as with pure thermal convection, the generation of surface topography comparable to that observed in some chaos regions requires a weak near-surface ice layer, again suggesting that some kind of yielding process might be important, while the length-scale of features generated is again significantly larger than many chaos regions. *Pappalardo and Barr* (2004) specifically appealed to compositional density contrasts arising from the preferential removal of a low-melting-temperature, dense component (brine) from the icy shell in warm upwellings. This idea of the linkage between convection and brine production will be discussed in more detail in the next section.

Another important way in which Europa's icy shell differs from that of standard convective settings is that the heat production is expected to vary with the local viscosity (e.g., *Ojakangas and Stevenson,* 1989). This effect is potentially extremely important, because it can lead to a thermal runaway and the production of local melt even in pure ice (*McKinnon,* 1999; *Wang and Stevenson,* 2000; *Sotin et al.,* 2002; *Tobie et al.,* 2003; *Mitri and Showman,* 2008; chapter by Barr and Showman). In the absence of other effects, however, this melting is still relatively deep because of the presence of a stagnant lid at the top of the icy shell (e.g., *Sotin et al.,* 2002; *Nimmo and Giese,* 2005). Because tidal strain rates are higher toward the poles, a prediction of any model in which tidal heating is important would be that chaotic terrain is formed more frequently nearer the poles.

Diapirism Model: Comparison to Observational Constraints

The diapirism model can be compared to the hard and soft observational constraints on chaos formation as follows.

H1: Matrix material could be formed by pervasive fracturing and plastic yielding of surface ice, although the details of this process have not been investigated.

H2: It is difficult to explain why the matrix material would subside relative to the plates without calling for material removal through tidally induced melting right near the surface.

H3: Unless yielding occurs, the diapirism model generates surface strains close to zero. Topographic gradients and a sufficiently weak lid might explain the formation and mobility of the plates (*Showman and Han,* 2005).

H4: Inward-facing scarps are difficult to explain for the same reason as H2. Doming of material above surroundings in plate-free chaos may be similar to the ice extrusion mechanism outlined by *Figueredo et al.* (2002).

H5: Matrix material will stand significantly higher than the background and plates if yielding and/or compositional convection are important.

H6: Exposure of dark material may be due to sublimation of warm surface ice.

H7: The diapirism model has a strong preferential size distribution, which is not simple to reconcile with the large range of sizes observed. The creation of larger features through merging diapirs or thermal runaways may partially alleviate this problem.

H8: The geographic distribution of chaotic terrain may be a result of enhanced tidal heating, which would drive more intense convection and possibly more melting within the diapirs. If this was the controlling factor, chaos should be more common at the poles, but perhaps the cold surface temperature thickens the stagnant lid or the pole has shifted.

The diapirism model also offers easy explanations for the soft constraints regarding the formation of undisrupted pits and domes, preservation of preexisting structure in the matrix, and growing chaotic terrain through mergers. This relatively slow process may be the best one to explain the apparently long formation time for matrix material.

8.3. Brine Mobilization

While there is certainly compositional heterogeneity on the surface of Europa, compositional variations within the icy shell are relatively unconstrained. Interesting behaviors can arise from consideration of non-water-ice materials within the shell, some of which may be relevant to the formation of chaotic terrain. *Head and Pappalardo* (1999) and *Collins et al.* (2000) considered the effect of low-melting-point materials mixed within the ice as the ice was being heated. Heating induces partial melting within the icy shell, dramatically lowering the ice viscosity and allowing percolation of liquids through the shell.

Head and Pappalardo (1999) specifically examined salts trapped within the ice, comparing potential processes on Europa to processes observed to happen within salty terrestrial sea ice. If sea ice freezes rapidly, it can trap significant amounts of salt within its structure, while slower-freezing ice rejects brines from its growing crystal structure. A similar process may occur on Europa whenever ocean water is in the upper part of the icy shell, trapping brines near the surface and freezing cleaner ice beneath. Such excursions of the ocean to the near-surface may occur due to overall variations in icy shell thickness, complete or partial melt-through events, or intrusion of ocean water into the icy shell as dikes or sills.

In terrestrial sea ice, frozen brines tend to form interconnected networks, allowing them to mobilize when a critical temperature (and thus a critical partial melt fraction) is reached. This brine mobilization could lead to disintegration of surface features in the matrix, effusion of salty fluids into low areas within and surrounding the chaos, and a detachment layer allowing for plate motion (*Head and Pappalardo,* 1999). Drainage of brine from the icy shell could also compositionally enhance the buoyancy of the remaining ice (e.g., *Pappalardo and Barr,* 2004). For example, a density contrast between briny ice and clean ice of 100 kg m^{-3} could produce the observed 100–200-m elevations if a 1-km thickness of ice was "cleaned out." Basal flow of ice would be required to balance volume loss from brine drainage.

Another effect of impurities in the icy shell is that they will change the thermal conductivity. For example, the hydrated salts that could form the frozen brine channels near Europa's surface have a low thermal conductivity and thus could enhance the trapping of subsurface heat (*Prieto-Ballesteros and Kargel,* 2005), if they are present in sufficient quantities. *Kargel et al.* (2007) used a similar argument of heat trapping by hydrated salts and clathrates to explain chaotic terrain on Mars (which is only superficially similar to europan chaos).

Both *Head and Pappalardo* (1999) and *Collins et al.* (2000) envisioned a diapiric heat source where clean ice in its solid state could mobilize lower-melting-temperature brines in the upper icy shell (Fig. 16c). A problematic aspect of conventional thermal convection is the presence of a thick stagnant lid, as outlined in the previous section. This lid, combined with the low surface temperatures, ensures that it is very difficult to generate melting within a few kilometers of the surface of the satellite (e.g., *Tobie et al.,* 2003), which makes it hard to cause plate motion or matrix degradation. *Nimmo and Giese* (2005) examined the thermal equilibration of a warm ice diapir with cold briny near-surface ice and could not generate significant melt within 7 km of the surface. It is possible that some combination of brine production plus surface yielding (e.g., *Han and Showman,* 2005) could allow near-surface melting and generate the observed features. Feedbacks between convection and melt production are likely to be complex. *Showman and Han* (2005) proposed that plastic yielding near the surface may occur preferentially where partial melt is formed, providing a way to produce a combination of plates and matrix.

Although the brine-mobilization model has previously been tied to the diapirism model in the literature, the essential aspect of the model is that subsurface heating mobilizes brines within the icy shell. Ice shell thinning from a heat source applied to the base of the shell could also cause brine mobilization within the shell as the partial-melting isotherms intersect the near-surface ice (Fig. 16d). Such a hybrid model would allow near-surface melting (a problem with most diapirism models), while not requiring the enormous magnitude of energy demanded by the melt-through model. Localization of brine mobilization and drainage beneath areas of enhanced matrix formation could also leave them topographically elevated via Pratt isostasy once the icy shell returned to its equilibrium thickness, addressing another difficulty of the melt-through model. A basal heat source could be locally enhanced by tidal heating arising from the reduced ice viscosity related to warming and melt production. Although we propose such a hybrid model as a promising avenue for further investigation, this model has not been fully developed, so our evaluation of the brine-mobilization model below is based on the diapir-driven version previously proposed in the literature.

Brine-Mobilization Model: Comparison to Observational Constraints

The brine-mobilization model can be compared to the hard and soft observational constraints on chaos formation as follows.

H1: Matrix material is formed by disaggregation of near-surface ice as the brine within it mobilizes.

H2: If the brine drains from the matrix material to other

locations, it will lose volume compared to the nearby plates, leaving it locally lower standing.

H3: The significant drop in viscosity from partial melting may enable motion of plates, although the volume and depth requirements on this melt have not been quantified. Thermal equilibration of diapirs with near-surface ice may not produce enough melt to significantly mobilize the surface (*Nimmo and Giese,* 2005), although it could enhance the yielding process (see the diapir model).

H4: Inward-facing scarps are related to loss of volume in the matrix. Doming of material above surroundings could be related to ice upwelling.

H5: Drainage of brine from the area of chaotic terrain leaves the ice compositionally buoyant. Reasonable density differences will produce the required topography (*Pappalardo and Barr,* 2004), but the detailed mechanisms of initial brine emplacement and brine drainage could be better quantified.

H6: Brines may effuse onto low areas of the surface on and around the chaos.

H7: The size distribution of diapirs is a critical challenge to compare to the size distribution of chaotic terrain. However, the possibility of isolated brine "pockets" and the lateral mobility of brine give additional free parameters to help explain the full range of chaotic terrain sizes.

H8: Enhanced tidal heating could help to explain the geographic distribution of chaotic terrain. If this was the controlling factor, chaos should be more common at the poles, but perhaps the cold surface temperature at the poles inhibits chaos formation or polar wander has occurred.

The brine-mobilization model inherits many of the soft constraint compatibilities from the diapirism model, but pockets of mobilized brine may do a better job of explaining the sizes of plates.

8.4. Injection of Sills

Another way to deliver liquid into the icy shell of Europa is to inject it directly from the ocean. *Crawford and Stevenson* (1988) investigated the propagation of fluid-filled cracks from the base of the icy shell, and found that the process is difficult unless exsolved gases help to open the crack. *Collins et al.* (2000) suggested that sills of melt could form within Europa's icy shell from pressurized water injected from the ocean (Fig. 16e). *Manga and Wang* (2007) investigated the pressure that would build up in the ocean as Europa's icy shell thickens. They found that, although it is difficult to get water all the way to erupting at the surface of Europa, it is possible to form sills within the icy shell from pressure buildup in the ocean. The maximum sill thickness they could generate through this mechanism was about 10 m, meaning that about 10 m of surface topography would be generated. Compared to the above mechanisms for chaotic terrain formation, sill injection has not been extensively studied. Hybrid sill/brine-mobilization models are not as promising as other hybrid models, because the injection of ocean water into the sill does not produce topographic uplift from decreasing the column density of the icy shell.

Sill-Injection Model: Comparison to Observational Constraints

The sill-injection model can be compared to the hard and soft observational constraints on chaos formation as follows.

H1 Matrix material may be formed by fracturing and yielding of the near-surface ice, although the exact mechanism is unclear.

H2: If the plates are floating in the liquid of the sill, the topographic difference between plates and matrix could be explained, but the expected thinness of the sill precludes free-floating plates.

H3: Plates will be decoupled by the sill and could easily translate horizontally, but tilting may be more difficult to explain in a thin liquid layer.

H4: Sills do not offer a natural explanation for chaos margins.

H5: Freezing of the sill can cause updoming above surrounding terrain, but not enough to explain the observations unless material is continually pumped into the same sill to freeze a thicker layer.

H6: Pressurization of injected ocean water as the sill freezes may drive small effusions of cryovolcanic materials.

H7: The horizontal extent of likely sills on Europa has not been modeled.

H8: Sills may form where it is easiest to propagate dikes into the icy shell from a pressurized ocean, but this does not bear any obvious relationship to the observed distribution of chaotic terrain.

Sills do not offer any particularly easy explanations for any of the soft observational constraints.

8.5. Impact

Because chaotic terrain formation involves disruption of surface materials, an exogenic impact mechanism has been posited for their formation, in contrast to endogenic heating mechanisms (Fig. 16f). *Billings and Kattenhorn* (2003) noted similar morphologies between chaotic terrain on Europa and terrestrial explosion craters on floating ice. In the explosion craters, floating plates of the original ice surface are preserved in a slushy matrix, filling an irregular hole in the ice. *Cox et al.* (2005) compared chaotic terrain to impact experiments into ice targets floating on a liquid substrate. Their impact experiments show catastrophic disruption of the floating ice over some threshold energy value, and Cox et al. argued that this could explain some large areas of chaotic terrain.

A difficulty for the impact hypothesis is that there are already two good examples on Europa of impacts that seem to have penetrated the icy shell to some extent, the features Tyre and Callanish (see chapter by Schenk and Turtle). Both of them are surrounded by secondary craters and by multiple rings of concentric fractures, much like large impact

TABLE 1. Comparison of models for chaotic terrain formation to the observational constraints enumerated in section 7.

Observational Constraint (H = hard constraint; S = soft constraint)	Melt Through	Diapirism	Brine Mobilization	Sill Injection	Impact
H1: Formation of matrix material	✓	✱	✓	✗	✓
H2: Plates locally higher than matrix	✓	✗	✓	✗	✓
H3: Plates tilt, rotate, and translate	✓	✱	✱	✱	✓
H4: Nature of chaotic terrain margins	✱	✱	✱	✗	✱
H5: Matrix topographically high	✱	✱	✓	✱	✱
H6: Dark hydrated salts/acid	✓	✱	✓	✓	✓
H7: Diameter range ~1–1000 km	✱	✗	✱	✓	✗
H8: Concentrated near the equator	✱	✱	✱	✗	✗
S1: Associated pits and domes	✗	✓	✓	✗	✗
S2: Preexisting structures preserved	✗	✓	✓	✓	✗
S3: Plate size >1 km	✓	✗	✱	✱	✓
S4: Ridges preferentially preserved	✓	✱	✱	✱	✗
S5: Matrix material forms viscous flows	✗	✓	✓	✱	✱
S6: Associated domes ~1 km high	✗	✱	✱	✱	✗
S7: Chaos regions grow by merging	✱	✓	✓	✓	✗
S8: Matrix formation long-lived?	✗	✱	✱	✱	✗

Symbols: ✓ = This model naturally explains this observation. ✱ = Special, but plausible, circumstances may be required to produce this observation from this model. ✗ = This model does not plausibly explain this observation.

basins on other bodies in the solar system. The impact hypothesis can only explain large areas of chaotic terrain and not small ones (which would not penetrate the ice), leading to two problems with the basic argument. First, there is a case of multiple explanations for the same thing: If impacts explain large chaos areas, then another explanation is needed for small ones. Second, there is a case of one explanation for multiple unrelated features: Tyre and Callanish being formed through the same mechanism as Conamara Chaos and Murias Chaos. Therefore, the principle of parsimony does not favor the impact hypothesis for chaotic terrain formation.

Impact Model: Comparison to Observational Constraints

The impact model can be compared to the hard and soft observational constraints on chaos formation as follows.

H1: Matrix material is formed by catastrophic disruption of the surface ice during the impact event.

H2: The plates are floating in the ocean below and are remnants of the original icy shell.

H3: Plates can easily translate, rotate, and tilt, and there are likely to be dynamic waves and currents in the water that would push the plates.

H4: Inward-facing scarps show the boundary of the disrupted area, but it is difficult to explain other chaos areas where matrix slopes down onto the adjacent terrain.

H5: There is no mechanism to dome the surface above the background terrain unless the refrozen ice has a different density than the original icy shell (see point H5 for the melt-through model).

H6: Exposure of the ocean provides the hydrated sulfurous contaminants.

H7: This model can only explain the largest end of the observed size range for chaotic terrain, unless radical variations in icy shell thickness are called upon.

H8: Impacts should be evenly distributed, so geographic preference would have to be explained through large differences in the target material from region to region.

Impacts can naturally explain the soft constraint of the minimum size of plates, because plates that are taller than they are wide will tip over in the ocean. Most of the other soft constraints are extremely difficult or impossible to explain in the impact scenario.

9. SUMMARY AND CONCLUSIONS

Table 1 summarizes how well the various models outlined in section 8 fit the observations. This table is not meant to be the end of the story, but rather is meant as a guide to where future work could be done to build a better model of chaotic terrain formation. While the simple diapir, sill-injection, and impact models fail to explain several key observations, the melt-through and brine-mobilization models can explain the key observations and many of the secondary observations.

As discussed above, a major issue with the melt-through model is the large energy fluxes required to completely melt the icy shell. A further problem is that, at least for the hydrothermal plume model, it appears to be difficult to generate a significant number of the most abundant features, which are smaller than 10 km across. On the other hand, because it is likely that the shell thickness exerts a strong control on the minimum lengthscale of surface features, the melt-through model is more capable of explaining small chaos features than models requiring shells tens of kilometers thick, if the melting patch can be of arbitrary size. One further aspect of the melt-through model that requires further quantification is the possible formation of lateral density contrasts after refreezing, leading to uplifted matrix material. The process that initially charges the upper icy shell with salts would have to be significantly different than the refreezing process to explain such variations, and the injection of ocean water into dikes and sills is a promising possibility in this respect.

Some aspects of the brine-mobilization model, especially the ability to generate surface motion and the details of draining brines out of the icy shell, require more quantitative verification. The generation of brine-rich melts at shallow depths is difficult, but possible if diapirism resulting in localized tidal heating and/or surface yielding is invoked. Diapirism of pure ice alone is unable to explain the wide size spectrum of chaos regions, but the presence of local pockets of brine could plausibly produce smaller features that would explain the observations. Similar to the melt-through model, the process by which the icy shell accumulates impurities needs further investigation.

We conclude that a hybrid model invoking some degree of shell thinning or surface yielding (thus generating near-surface melting and smaller-scale features) coupled with brine production (probably as a result of tidal dissipation and diapirism) is perhaps the most plausible mechanism to explain all the observations. As yet, no quantitative model incorporating all the relevant processes has been developed.

A particular issue for all the existing models is that it seems hard to generate both very large (~1000 km) and very small (~1 km) features with the same process. Current models of both hydrothermal plumes and diapirs produce sharply peaked size distributions; invoking variably sized brine pockets is currently *ad hoc* and shifts the question to what originally caused those pockets. One possibility is that feedbacks between convective stresses and yielding, or between temperature and tidal heating, yield a broader spectrum of convective length scales.

Despite the decade of study since Galileo returned stunning images of chaotic terrain, it is clear that its origin is still not fully understood. Further modeling efforts still may not resolve the issue without new observations. Fortunately, plans are currently under development to launch a Europa orbiter. Uniform high-resolution image coverage would provide a much more stable platform of observations on which to base chaos formation models. Moreover, the spacecraft would be equipped with instruments such as laser ranging and ice-penetrating radar, which will be able to definitively test the different model predictions for chaos formation. This current review will undoubtedly appear naïve to the authors of a post-Europa-orbiter review of chaos formation. Nonetheless, we hope to be alive to read it, and that the enigma of chaotic terrain will finally be resolved.

Acknowledgments. A. Dombard and D. O'Brien provided helpful and thorough reviews that improved this manuscript. We acknowledge support from the NASA Outer Planets Research and Planetary Geology and Geophysics programs.

REFERENCES

Bierhaus E. B., Chapman C. R., Merline W. J., Brooks S. M., and Asphaug E. (2001) Pwyll secondaries and other small craters on Europa. *Icarus, 153,* 264–276.

Billings S. E. and Kattenhorn S. A. (2003) Comparison between terrestrial explosion crater morphology in floating ice and Europan chaos. In *Lunar and Planetary Science XXXIV,* Abstract #1955. Lunar and Planetary Institute, Houston (CD-ROM).

Buck L., Chyba C. F., Goulet M., Smith A., and Thomas P. (2002) Persistence of thin ice regions in Europa's ice crust. *Geophys. Res. Lett., 29,* DOI: 10.1029/2002GL016171.

Carlson R. W., Anderson M. S., Johnson R. E., Schulman M. B., and Yavrouian A. H. (2002) Sulfuric acid production on Europa: The radiolysis of sulfur in water ice. *Icarus, 157,* 456–463.

Carlson R. W., Anderson M. S., Mehlman R., and Johnson R. E. (2005) Distribution of hydrate on Europa: Further evidence for sulfuric acid hydrate. *Icarus, 177,* 461–471.

Carr M. H., Belton M. J. S., Chapman C. R., Davies M. E., Geissler P., Greenberg R., McEwen A. S., Tufts B. R., Greeley R., Sullivan R., Head J. W., Pappalardo R. T., Klaasen K. P., Johnson T. V., Kaufman J., Senske D., Moore J., Neukum G., Schubert G., Burns J. A., Thomas P., and Veverka J. (1998) Evidence for a subsurface ocean on Europa. *Nature, 391,* 363–365.

Collins G. C., Head J. W., Pappalardo R. T., and Spaun N. A. (2000) Evaluation of models for the formation of chaotic terrain on Europa. *J. Geophys. Res., 105,* 1709–1716.

Cox R., Ong L. C. F., and Arakawa M. (2005) Is chaos on Europa caused by crust-penetrating impacts? In *Lunar and Planetary Science XXXVI,* Abstract #2101. Lunar and Planetary Institute, Houston (CD-ROM).

Crawford G. D. and Stevenson D. J. (1988) Gas-driven water volcanism in the resurfacing of Europa. *Icarus, 73,* 66–79.

Dalton J. B., Prieto-Ballesteros O., Kargel J. S., Jamieson C. S., Jolivet J., and Quinn R. (2005) Spectral comparison of heavily hydrated salts with disrupted terrains on Europa. *Icarus, 177,* 472–490.

Fagents S. A. (2003) Considerations for effusive cryovolcanism on Europa: The post-Galileo perspective. *J. Geophys. Res., 108,* DOI: 10.1029/2003JE002128.

Fagents S. A., Greeley R., Sullivan R. J., Pappalardo R. T., and Prockter L. M. (2000) Cryomagmatic mechanisms for the formation of Rhadamanthys Linea, triple band margins, and other low-albedo features on Europa. *Icarus, 144,* 54–88.

Figueredo P. H. and Greeley R. (2000) Geologic mapping of the northern leading hemisphere of Europa from Galileo solid-state imaging data. *J. Geophys. Res., 105,* 22629–22646.

Figueredo P. H. and Greeley R. (2003) The emerging resurfacing history of Europa from pole-to-pole geologic mapping. In *Lunar and Planetary Science XXXIV,* Abstract #1017. Lunar and Planetary Institute, Houston (CD-ROM).

Figueredo P. H. and Greeley R. (2004) Resurfacing history of Europa from pole-to-pole geological mapping. *Icarus, 167,* 287–312.

Figueredo P. H., Chuang F. C., Rathbun J., Kirk R. L., and Greeley R. (2002) Geology and origin of Europa's "mitten" feature (Murias Chaos). *J. Geophys. Res., 107,* DOI: 10.1029/2001JE001591.

Goodman J. C., Collins G. C., Marshall J., and Pierrehumbert R. T. (2004) Hydrothermal plume dynamics on Europa: Implications for chaos formation. *J. Geophys. Res., 109,* DOI: 10.1029/2003JE002073.

Greeley R., Sullivan R., Coon M. D., Geissler P. E., Tufts B. R., Head J. W., Pappalardo R. T., and Moore J. M. (1998) Terrestrial sea ice morphology: Considerations for Europa. *Icarus, 135,* 25–40.

Greeley R., Figueredo P. H., Williams D. A., Chuang F. C., Klemaszewski J. E., Kadel S. D., Prockter L. M., Pappalardo R. T., Head J. W., Collins G. C., Spaun N. A., Sullivan R. J., Moore J. M., Senske D. A., Tufts B. R., Johnson T. V., Belton M. J. S., and Tanaka K. L. (2000) Geologic mapping of Europa. *J. Geophys. Res., 105,* 22559–22578.

Greenberg R., Hoppa G. V., Tufts B. R., Geissler P., Riley J., and Kadel S. (1999) Chaos on Europa. *Icarus, 141,* 263–286.

Greenberg R., Leake M. A., Hoppa G. V., and Tufts B. R. (2003) Pits and uplifts on Europa. *Icarus, 161,* 102–126.

Han L. and Showman A. P. (2005) Thermo-compositional convection in Europa's icy shell with salinity. *Geophys. Res. Lett., 32,* DOI: 10.1029/2005GL023979.

Hand K. P. and Chyba C. F. (2007) Empirical constraints on the salinity of the europan ocean and implications for a thin ice shell. *Icarus, 189,* 424–438.

Head J. W. and Pappalardo R. T. (1999) Brine mobilization during lithospheric heating on Europa: Implications for formation of chaos terrain, lenticula texture, and color variations. *J. Geophys. Res., 104,* 27143–27155.

Head J. W., Pappalardo R. T., Spaun N. A., Prockter L. M., and Collins G. C. (1999) Chaos terrain on Europa: Characterization from Galileo E12 very high-resolution images of Conamara Chaos: 2. Matrix. In *Lunar and Planetary Science XXX,* Abstract #1587. Lunar and Planetary Institute, Houston (CD-ROM).

Hoppa G. V., Tufts B. R., Greenberg R., and Geissler P. E. (1999) Formation of cycloidal features on Europa. *Science, 285,* 1899–1902.

Hoppa G. V., Greenberg R., Riley J., and Tufts B. R. (2001) Observational selection effects in Europa image data: Identification of chaotic terrain. *Icarus, 151,* 181–189.

Kargel J. S., Furfaro R., Prieto-Ballesteros O., Rodriguez J. A. P., Montgomery D. R., Gillespie A. R., Marion G. M., and Wood S. E. (2007) Martian hydrogeology sustained by thermally insulating gas and salt hydrates. *Geology, 35,* 975–978.

Lichey C. and Hellmer H. H. (2001) Modeling giant-iceberg drift under the influence of sea ice in the Weddell Sea, Antarctica. *J. Glaciol., 47,* 452–460.

Lowell R. P. and DuBose M. (2005) Hydrothermal systems on Europa. *Geophys. Res. Lett., 32,* DOI: 10.1029/2005GL022375.

Lucchitta B. K. and Soderblom L. A. (1982) The geology of Europa. In *Satellites of Jupiter* (D. Morrison, ed.), pp. 521–555. Univ. of Arizona, Tucson.

Malin M. C. and Pieri D. C. (1986) Europa. In *Satellites* (J. A. Burns and M. S. Matthews, eds.), pp. 689–717. Univ. of Arizona, Tucson.

Manga M. and Wang C.-Y. (2007) Pressurized oceans and the eruption of liquid water on Europa and Enceladus. *Geophys. Res. Lett., 34,* DOI: 10.1029/2007GL029297.

McCord T. B., Teeter G., Hansen G. B., Sieger M. T., and Orlando T. M. (2002) Brines exposed to Europa surface conditions. *J. Geophys. Res., 107,* DOI: 10.1029/2000JE001453.

McKinnon W. B. (1999) Convective instability in Europa's floating ice shell. *Geophys. Res. Lett., 26,* 951–954.

Melosh H. J., Ekholm A. G., Showman A. P., and Lorenz R. D. (2004) The temperature of Europa's subsurface water ocean. *Icarus, 168,* 498–502.

Mével L., and Mercier E. (2007) Large-scale doming on Europa: A model of formation of Thera Macula. *Planet. Space Sci., 55,* 915–927.

Mitri G. and Showman A. P. (2008) A model for the temperature-dependence of tidal dissipation in convective plumes on icy satellites: Implications for Europa and Enceladus. *Icarus, 195,* 758–764.

Miyamoto H., Mitri G., Showman A. P., and Dohm J. M. (2005) Putative ice flows on Europa: Geometric patterns and relation to topography collectively constrain material properties and effusion rates. *Icarus, 177,* 413–424.

Moore J. M., Asphaug E., Morrison D., Spencer J. R., Chapman C. R., Bierhaus B., Sullivan R. J., Chuang F. C., Klemaszewski J. E., Greeley R., Bender K. C., Geissler P. E., Helfenstein P., and Pilcher C. B. (1999) Mass movement and landform degradation on the icy Galilean satellites: Results of the Galileo nominal mission. *Icarus, 140,* 294–312.

Nimmo F. (2004) Non-newtonian topographic relaxation on Europa. *Icarus, 168,* 205–208.

Nimmo F. and Giese B. (2005) Thermal and topographic tests of Europa chaos formation models from Galileo E15 observations. *Icarus, 177,* 327–340.

Nimmo F. and Manga M. (2002) Causes, characteristics, and consequences of convective diapirism on Europa. *Geophys. Res. Lett., 29,* DOI: 10.1029/2002GL015754.

O'Brien D. P., Geissler P., and Greenberg R. (2002) A melt-through model for chaos formation on Europa. *Icarus, 156,* 152–161.

Ojakangas G. W. and Stevenson D. J. (1989) Thermal state of an ice shell on Europa. *Icarus, 81,* 220–241.

Pappalardo R. T. and Barr A. C. (2004) The origin of domes on Europa: The role of thermally induced compositional diapirism. *Geophys. Res. Lett., 31,* DOI: 10.1029/2003GL019202.

Pappalardo R. T., Head J. W., Greeley R., Sullivan R. J., Pilcher C., Schubert G., Moore W. B., Carr M. H., Moore J. M., Belton M. J. S., and Goldsby D. L. (1998a) Geological evidence for solid-state convection in Europa's ice shell. *Nature, 391,* 365–368.

Pappalardo R. T., Sherman N. D., Head J. W., Collins G. C., Greeley R., Klemaszewski J., Sullivan R., Phillips C. B., McEwen A. S., and Geissler P. E. (1998b) Distribution of mottled terrain on Europa: A possible link to nonsynchronoous rotation stresses. In *Lunar and Planetary Science XXIX,* Abstract #1923. Lunar and Planetary Institute, Houston (CD-ROM).

Pappalardo R. T., Belton M. J. S., Breneman H. H., Carr M. H., Chapman C. R., Collins G. C., Denk T., Fagents S., Geissler P. E., Giese B., Greeley R., Greenberg R., Head J. W., Helfenstein P., Hoppa G., Kadel S. D., Klaasen K.P., Klemaszewski J. E., Magee K., McEwen A. S., Moore J. M., Moore W. B., Neukum G., Phillips C. B., Prockter L. M., Schubert G., Senske D. A., Sullivan R. J., Tufts B. R., Turtle E. P., Wagner R., and Williams K. K. (1999) Does Europa have a subsurface ocean? Evaluation of the geological evidence. *J. Geophys. Res., 104,* 24015–24055.

Prieto-Ballesteros O. and Kargel J. S. (2005) Thermal state and complex geology of a heterogeneous salty crust of Jupiter's satellite, Europa. *Icarus, 173,* 212–221.

Prockter L. M. and Schenk P. (2005) Origin and evolution of Castalia Macula, an anomalous young depression on Europa. *Icarus, 177,* 305–326.

Prockter L. M., Antman A. M., Pappalardo R. T., Head J. W., and Collins G. C. (1999) Europa: Stratigraphy and geological history of the anti-jovian region from Galileo E14 solid-state imaging data. *J. Geophys. Res., 104,* 16531–16540.

Rathbun J. A., Musser G. S., and Squyres S. W. (1998) Ice diapirs on Europa: Implications for liquid water. *Geophys. Res. Lett., 25,* 4157–4160.

Riley J., Hoppa G. V., Greenberg R., Tufts B. R., and Geissler P. (2000) Distribution of chaotic terrain on Europa. *J. Geophys. Res., 105,* 22599–22615.

Riley J., Greenberg R., and Sarid A. (2006) Europa's south pole region: A sequential reconstruction of surface modification processes. *Earth Planet. Sci. Lett., 248,* 808–821.

Schenk P. M. and Pappalardo R. T. (2004) Topographic variations in chaos on Europa: Implications for diapiric formation. *Geophys. Res. Lett., 31,* DOI: 10.1029/2004GL019978.

Schenk P., Matsuyama I., and Nimmo F. (2008) True polar wander on Europa from global-scale small-circle depressions. *Nature, 453,* 368–371.

Showman A. P. and Han L. (2004) Numerical simulations of convection in Europa's ice shell: Implications for surface features. *J. Geophys. Res., 109,* DOI: 10.1029/2003JE002103.

Showman A. P. and Han L. (2005) Effects of plasticity on convection in an ice shell: Implications for Europa. *Icarus, 177,* 425–437.

Solomatov V. S. (1995) Scaling of temperature- and stress-dependent viscosity convection. *Phys. Fluids, 7,* 266–274.

Sotin C., Head J. W., and Tobie G. (2002) Europa: Tidal heating of upwelling thermal plumes and the origin of lenticulae and chaos melting. *Geophys. Res. Lett., 29,* DOI: 10.1029/2001GL013844.

Spaun N. A. (2002) Chaos, lenticulae, and lineae on Europa: Implications for geological history, crustal thickness, and the presence of an ocean. Ph.D. thesis, Brown University, Providence, Rhode Island.

Spaun N. A., Head J. W., Collins G. C., Prockter L. M., and Pappalardo R. T. (1998) Conamara Chaos region, Europa: Reconstruction of mobile polygonal ice blocks. *Geophys. Res. Lett., 25,* 4277–4280.

Spaun N. A., Prockter L. M., Pappalardo R. T., Head J. W., Collins G. C., Antman A., and Greeley R. (1999) Spatial distribution of lenticulae and chaos on Europa. In *Lunar and Planetary Science XXX,* Abstract #1847. Lunar and Planetary Institute, Houston (CD-ROM).

Spaun N. A., Head J. W., and Pappalardo R. T. (2004) Europan chaos and lenticulae: A synthesis of size, spacing, and areal density analyses. In *Lunar and Planetary Science XXXV,* Abstract #1409. Lunar and Planetary Institute, Houston (CD-ROM).

Stevenson D. J. (2000) Limits on the variation of thickness of Europa's ice shell. In *Lunar and Planetary Science XXXI,* Abstract #1506. Lunar and Planetary Institute, Houston (CD-ROM).

Thomson R. E. and Delaney J. R. (2001) Evidence for a weakly stratified europan ocean sustained by seafloor heat flux. *J. Geoxphys. Res., 106,* 12355–12365.

Tobie G., Choblet G., and Sotin C. (2003) Tidally heated convection: Constraints on Europa's ice shell thickness. *J. Geophys. Res., 108,* DOI: 10.1029/2003JE002099.

Vance S. and Brown J. M. (2005) Layering and double-diffusion style convection in Europa's ocean. *Icarus, 177,* 506–514.

Wang H. and Stevenson D. J. (2000) Convection and internal melting of Europa's ice shell. In *Lunar and Planetary Science XXXI,* Abstract #1293. Lunar and Planetary Institute, Houston (CD-ROM).

Williams K. K. and Greeley R. (1998) Estimates of ice thickness in the Conamara Chaos region of Europa. *Geophys. Res. Lett., 25,* 4273–4276.

Zimmer C., Khurana K., and Kivelson M. G. (2000) Subsurface oceans on Europa and Callisto: Constraints from Galileo magnetometer observations. *Icarus, 147,* 329–347.

Europa's Surface Composition

R. W. Carlson
NASA Jet Propulsion Laboratory/California Institute of Technology

W. M. Calvin
University of Nevada

J. B. Dalton
NASA Jet Propulsion Laboratory/California Institute of Technology

G. B. Hansen
University of Washington

R. L. Hudson
Eckerd College

R. E. Johnson
University of Virginia

T. B. McCord
Bear Fight Center

M. H. Moore
NASA Goddard Space Flight Center

Europa is unique in the solar system, having a young, icy surface bombarded by high-energy radiation and possessing many possible sources of surface material. One possible source is Europa's putative subsurface ocean, from which material may be emplaced through cryovolcanic activity or effusive flows. Impact ejecta from Io and implantation of iogenic sulfur, oxygen, sodium, potassium, and chlorine ions on Europa's trailing hemisphere are likely sources, as well as direct meteoritic and cometary impacts and outer-satellite-derived impact ejecta that spiral toward Jupiter. While we cannot yet answer the central question of where the non-ice material on Europa's surface comes from, we can identify and quantify the species that are known or thought to be present: H_2O, a hydrate, SO_2, elemental sulfur, O_2, H_2O_2, CO_2, Na, and K. Europa, like many satellites, has a hemispherical dichotomy, in this case a reddish trailing hemisphere (in the sense of orbital motion) and a brighter, leading hemisphere. The purest H_2O is found on the leading hemisphere while the trailing hemisphere contains the highest concentration of the next most prevalent species, a hydrated material of unknown composition. The H_2O ice on the leading side is amorphous on the upper surface, with crystalline ice present at submillimeter depths. The trailing hemisphere contains ice plus a hydrated component that may be hydrated salts, derived from the ocean as brine, and/or hydrated sulfuric acid, the major equilibrium product from radiolysis of sulfurous material and H_2O. The source of sulfurous material could be endogenic or from implantation of iogenic sulfur, or both. Sulfur dioxide and sulfur are thought to be present, mainly on the trailing hemisphere. This is consistent with ion implantation, but the sulfur distribution and that of the hydrate show correlations with geological features, so there must be some endogenic control of these constituents, either as a source or modification process. All the species in the ~1-m regolith are affected by radiation, but the archetypal radiolytic species, observed on both hemispheres, are molecular oxygen and hydrogen peroxide. These are certainly radiolytic products since continuous production is required, with O_2 being volatile and escaping easily, while H_2O_2 is quickly destroyed by sunlight. Carbon dioxide is present and poses a mystery. It could be outgassing from the interior or a photolytic or radiolytic product of micrometeorite-derived carbonaceous material. Sodium and potassium atoms are found in the tenuous atmosphere and arise from sputtering of surface material. These atoms can be derived initially from the iogenic plasma and from endogenic salts, but the implantation flux rates are not known well enough to establish the source.

1. INTRODUCTION

1.1. Introduction

Europa is a fully differentiated planet-sized satellite with a Fe or Fe-FeS core, a silicate mantle, and a 100- to 160-km-thick outer layer consisting of an icy crust covering a probable ocean (see chapter by Schubert et al.). Like Io, Europa is tidally heated and is currently (or recently) geologically active, exhibiting a crater age of 40 to 90 m.y. (*Zahnle et al.,* 2003; see chapter by Bierhaus et al.). While the surface is smooth on a large scale, at higher spatial detail it is modified everywhere by cracking of the brittle surface and possible convective motions within its ductile icy shell, producing lineae (long ridge systems often with a bright center band and dark margins), lenticulae (termed pits, spots, and domes), and chaos regions suggestive of partial melting. An active Io-like body may be hidden below the ice cover with possible hydrothermal activity at the ocean-mantle interface (*Kargel et al.,* 2000) and oceanic material may be emplaced on the surface. Other sources of surface material include exogenic material from Io and possibly other satellites, from the jovian magnetosphere, and from comet and meteorite impacts. These impacts also churn and mix the surface, producing an icy regolith. In addition, Europa is irradiated by high-energy particles that radiolytically modify the surface material and produce a tenuous sputtered atmosphere with species that are indicative of the surface composition. Impact gardening and micrometeoric deposition occurs predominantly on the leading side (in the sense of orbital motion), while plasma implantation and bombardment by high-energy electrons are strongest in the trailing hemisphere. These processes can lead to the observed hemispheric albedo differences, producing the "white" and "red" hemispheres (*Johnson and Pilcher,* 1977) shown in Fig. 1. Nonsynchronous rotation of Europa's shell, if occurring, would moderate these influences. While the surface appears geologically younger, brighter, and much icier than the other icy Galilean satellites, it does show the presence of non-ice materials, and these materials can provide clues to Europa's formation and evolution. Distinguishing between exogenic or endogenic source(s) of Europa's non-ice material is of particular interest. This is a question that we cannot yet answer, but is a guiding theme of this chapter.

In this chapter we first discuss the possible sources of surface species (section 1.2) and then chemical and physical alteration processes that can change the composition or distribution (section 1.3). The observational methods are briefly described (section 1.4) and a listing of known and suggested species is presented. Supporting laboratory studies are briefly discussed (section 1.5), followed by detailed discussions of each species (section 2). Some recommendations for future observations and experiments conclude the chapter.

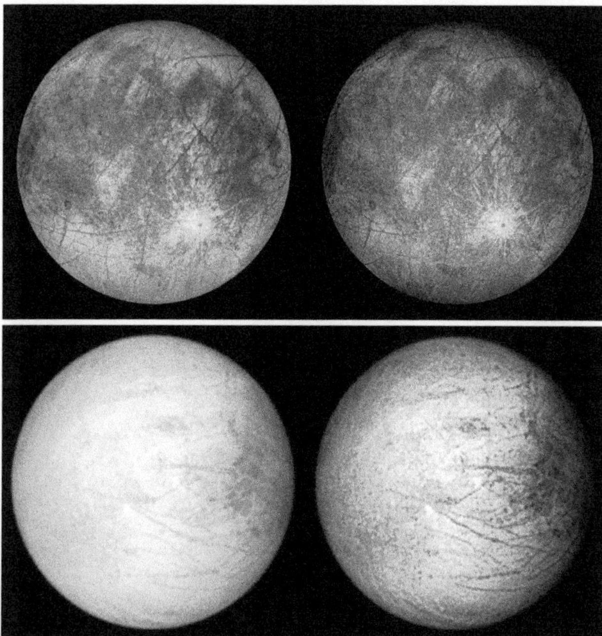

Fig. 1. See Plate 15. Europa's trailing and leading hemispheres (top and bottom panels, respectively), as imaged by Galileo's Solid State Imaging camera, illustrating the differences between the two sides and the association of the brownish material with geological features. The images at left are natural color, while the images at right are enhanced to show structural detail. The resolutions (and JPL Photojournal number) are 6.9 km/pixel (PIA00502) for the trailing sides and 12.7 km/pixel (PIA01295) for the leading side.

1.2. Sources of Surface Species

1.2.1. Endogenic sources. Jupiter's satellites formed from either a warm and dense protojovian nebula or a thin and cold disk (*Schubert et al.,* 2004; chapter by Canup and Ward). Thermochemical reactions in a warm nebula would have hydrated the silicates to serpentine, oxidized the iron to Fe_3O_4, and reduced the C and N compounds to CH_4, NH_3, and small amounts of HCN (*Prinn and Fegley,* 1981, 1989). Methane is too volatile to have condensed appreciably in the protojovian nebula, and ammonia would not have condensed at Europa's likely formation temperature. In the thin and cold nebula case, little thermochemical processing occurs before formation; unaltered silicates, Ni-Fe alloy, iron sulfide, organic matter, and water ice are retained and incorporated into the satellite. If the ice grains' compositions are similar to that of the interstellar medium (*Gibb et al.,* 2004), then CO_2, CH_3OH, OCS, H_2CO, HCOOH, NH_3, and OCN^- and perhaps CO and CH_4 may have been incorporated into the forming satellite, perhaps as clathrates. Some of the more volatile species may be outgassing at the present time, with the molecules diffusing to the surface and forming a tenuous atmosphere before being ionized and lost to the magnetosphere.

If Europa has an ocean (see chapter by McKinnon et al.), it is plausible that oceanic material could be emplaced on the surface. The young surface age and the absence of a dark meteoritic blanket, such as that covering Callisto, suggest recent replenishment of the surface. Whether or not the brownish material originates from the ocean is unknown, but there are tantalizing associations of non-ice material

TABLE 1. Estimated exogenic fluxes on Europa's surface, expressed in 10^6 atoms cm^{-2} s^{-1}; outer satellite ejecta sources are not included.

Element	Micrometeoroid (cometary)*	Io Plasma Torus (as ions)†	Io impact ejecta‡				
			Tholeiite basalt	Alkali basalt	Komatiite	Dunite	B1 CAI
H	8.2						
C	1.2	0.05, <0.14§,¶					
N	0.31						
O	2.7	~300**	9.6	9.4	9.8	9.4	9.4
Na	0.011	>5.3††	0.36	0.36	0.053	0.034	0.021
Mg	0.11		0.41	0.97	2.6	3.8	0.9
Al	0.0072	<0.05‡‡	1.0	0.96	0.28	0.056	2.1
Si	0.20	<0.6§	3.0	2.6	2.8	2.4	1.7
S	0.077	140					
Cl		1.8–5.6§,§§					
K	0.00021	~0.5¶¶	0.058	0.082	0.0067	0.0075	0.0075
Ca	0.0067	<0.006‡‡	0.56	0.64	0.34	0.51	1.81
Fe	0.056	<0.2‡‡	0.67	0.61	0.57	0.65	0.30

*Average mass influx from *Cooper et al.* (2001). Relative composition from *Anders and Grevasse* (1989).

†Sulfur flux value from *Johnson et al.* (2004) for trailing side apex and based on *Bagenal* (1994). No plasma deflection is included.

‡Equatorial mass influx rate averaged over 10-m.y. simulation, from *Zahnle et al.* (2008). The impactors' elemental composition is estimated for four models of Io basalts and an Allende-type calcium aluminum ceramic (B1 CAI), all from *Schaefer and Fegley* (2004).

§*Feldman et al.* (2004). Their C flux could be of solar wind origin.

¶Assumed a C/S ratio of 10^{-3} for Io plumes (*Schaefer and Fegley,* 2005), consistent with Voyager upper limits (*Pearl et al.,* 1979).

**O:S ~2:1 (*Hall et al.,* 1994) or higher at Europa (*Bagenal,* 1994).

††Lower limit for Na^+ from *Hall et al.* (1994). Note: comparable to Cl flux.

‡‡Neutral atom source limits relative to sodium, from *Na et al.* (1998).

§§*Kuppers and Schneider* (2000); *Feldman et al.* (2001).

¶¶*McGrath et al.* (2004) give a [Na]/[K] ratio of 10 ± 5 for atoms at 10–20 Io radii. We assume this ratio will be preserved in the plasma torus.

with geological features, suggestive of emplacement from below (Fig. 1). However, there is no direct evidence for surface-ocean exchange and other processes can produce geological and compositional associations (see below).

Europa's putative ocean overlies the silicate mantle, and may chemically react with it (see chapter by Zolotov and Kargel). Two different pathways are possible, depending on whether chemically evolved H_2 escapes or is trapped in the ocean by the icy shell. In the latter (closed) case, the presence of H_2 limits oxidation and species such as sulfates are not formed. If H_2 escapes, a likely scenario, the ocean initially evolves to an alkaline solution of (in order of concentration) OH^-, Na^+, $NaHSiO_3$, Cl^-, Ca^{2+}, NaOH, K^+, $HSiO_3^-$, H_2, $CaOH^+$, . . . (*Zolotov and Mironenko,* 2007). Cations (Na^+, Ca^{2+}, . . .) are supplied to the ocean through dissolution of the rock's silicates. Subsequent dissolution and long-term hydrothermal reactions may lead to a sulfate-bearing salty or acidic ocean (*Kargel et al.,* 2000; *Zolotov and Shock,* 2001; *Marion,* 2002; *Zolotov and Shock,* 2003; *Marion and Kargel,* 2008; chapter by Zolotov and Kargel). However, *McKinnon and Zolensky* (2003) have argued that sulfate is not easily formed in Europa's early ocean, which would have been sulfidic (e.g., *Zolotov and Mironenko,* 2007), but small amounts of sulfate introduced into the ocean from the mantle could overwhelm the initial sulfidic state (W. McKinnon, personal communication, 2008). It has also been suggested that SO_2 (*Kargel et al.,* 2000; *McKinnon and Zolensky,* 2003) could be vented into Europa's ocean. The SO_2 can form sulfurous acid or a clathrate (*Prieto-Ballesteros et al.,* 2005; *Hand et al.,* 2006) or, under oxidizing conditions, could form sulfuric acid. High-temperature decomposition of accreted organic compounds could also have supplied C, N, and S species to the ocean and icy shell (see chapter by Zolotov and Kargel). Finally, primary or modified organic matter may be transported toward the surface (*Kargel et al.,* 2000; chapter by Zolotov and Kargel).

1.2.2. Exogenic sources. Three sources are thought to be providing material to Europa's surface, the first being delivery of material from outside the jovian system through direct impacts of comets, asteroids, meteorites, and micrometeorites. Numerical estimates of the globally averaged elemental fluxes from micrometeoroid impacts are given in Table 1 using the flux from *Cooper et al.* (2001) and cometary abundances (*Anders and Grevasse,* 1989). The flux distribution on the surface will be nonuniform; orbital motion increases the flux striking the leading hemisphere and decreases the trailing-side flux (*Zahnle et al.,* 1998) (see section 1.3.2).

The second source of material for the Galilean satellites' surfaces is material ejected from the outer irregular satellites of Jupiter. This source was discussed by *Pollack et al.* (1978), inferring from the low albedo of these small bodies that they are composed of carbonaceous-chondritic-like material. These authors thought that this source is more potent than micrometeoroid bombardment due to the lower impact velocities, but no fluxes were estimated. It is also possible that ejecta from the outer Galilean satellites could contribute material to Europa. We can estimate this source strength as roughly equal to the iogenic impact ejecta source discussed below (K. Zahnle, personal communication, 2008); however, we do not further consider outer satellite contributions.

The third source is the neighboring inner satellite Io, from which material is brought to Europa by Io's thermal plasma torus, by higher-energy iogenic ions, and by impact ejection of crustal material from Io (*Alvarellos et al.,* 2008; *Zahnle et al.,* 2008, see also chapter by Zolotov and Kargel). Thermal plasma from Io deposits material mainly on Europa's trailing hemisphere. Implantation of more energetic ions occurs uniformly over the surface and, for sulfur, the flux is about 10% of the maximum flux of thermal S ions at the trailing antapex. An estimate of the undeflected trailing-side sulfur plasma flux from *Johnson et al.* (2004) is used with relative ion density measurements to generate the plasma input in Table 1. Plasma deflection could reduce the implantation flux. *Ip* (1996) predicted a 20% reduction, whereas modeling by *Saur et al.* (1998) suggested a factor of 5 reduction. Using Galileo data, *Paranicas et al.* (2002) and *Volwerk et al.* (2004) have estimated that only 10% of the impinging plasma is diverted around Europa. These plasma effects are discussed in the chapter by Kivelson et al., but no further estimates are currently available.

Io's impact ejecta, suggested to consist largely of basaltic spall fragments (*Alvarellos et al.,* 2008; *Zahnle et al.,* 2008), strike Europa at high velocities and will be largely vaporized and undergo reactions in the plume. The fluxes are given in Table 1 for five models of Io's magma composition. Io's volcanos are dust sources (*Postberg et al.,* 2006), but their contribution to Europa's surface is inconsequential (*Kruger et al.,* 2003).

Exogenic material will accumulate on the surface and be buried by micrometeoroid gardening. We can crudely estimate the expected concentrations by assuming that the resurfacing age of the icy crust is the same as the ~50-m.y. cratering age (see *Zahnle et al.,* 2003) and ignoring the asymmetric flux rates and gardening patterns. We then find that a flux of 10^6 atoms cm^{-2} s^{-1} in a nominal 1-m-deep regolith (see Fig. 2) will result in a longitudinally and vertically averaged volume mixing ratio relative to H_2O of

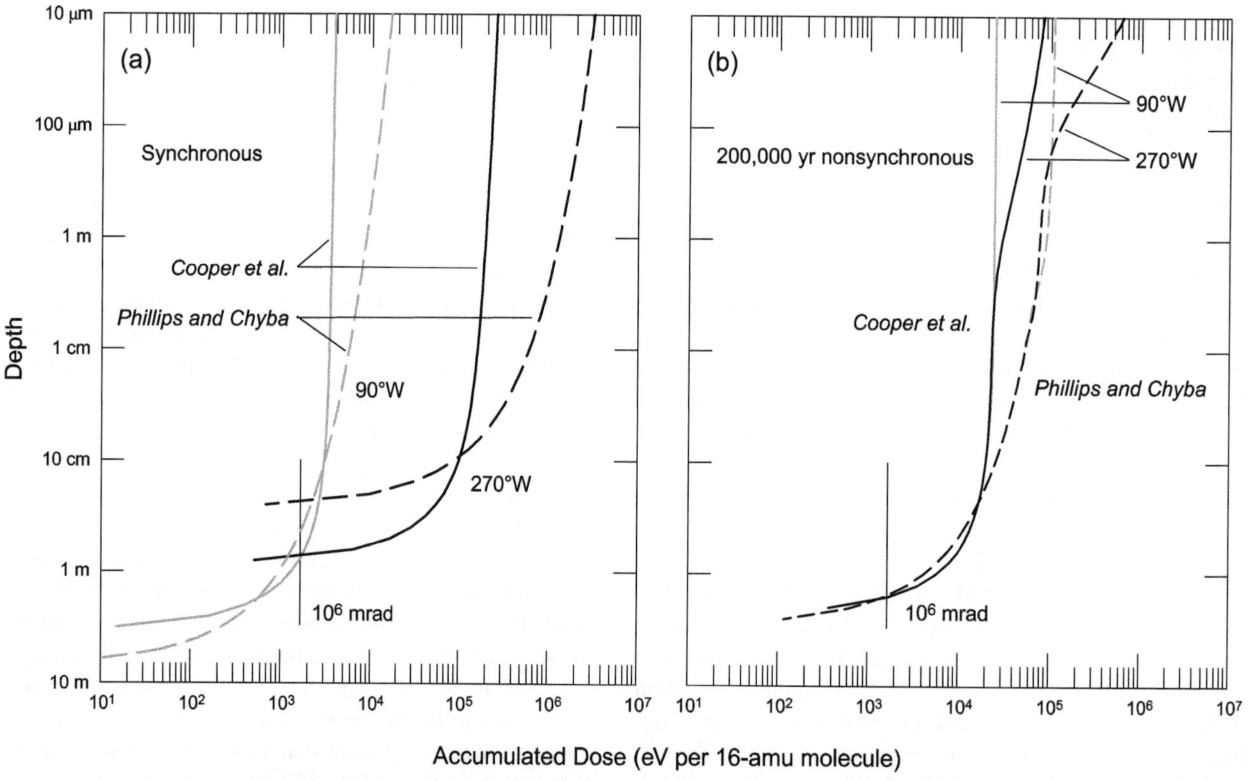

Fig. 2. Radiolytic dose for molecules as a function of their current depth for a 50-m.y. surface. The doses for the leading-side apex and trailing-side antapex are shown as gray and black lines, respectively. This vertical variation will also approximate the relative distribution of implanted iogenic sulfur, sodium, and other thermal plasma atoms since their longitudinal flux distrubutions are somewhat similar to the ionizing radiation distribution. Gardening models of *Cooper et al.* (2001) (solid lines) and *Phillips and Chyba* (2001) (dashed lines) are shown for **(a)** synchronous rotation and **(b)** a 200,000-yr nonsynchronous rotation period. The trailing-side enhancement of dose is due to the higher electron flux contribution and lower gardening rate for that hemisphere. Remote optical sensing samples approximately the top 100 μm to 1 mm. Porosity is neglected in these calculations.

roughly 500 ppm. If Io were the only source of non-ice material to Europa's surface and no loss occurred, then using the flux values from Table 1, sulfur compounds could be present on the surface at ~7% (molar abundance) relative to H_2O, while Na and Cl could reach 0.3%. Silicon and magnesium could be comparable or slightly less than Na and Cl. These estimates assume uniform mixing and ignore hemispherical flux and gardening rate differences, which can produce surface concentrations that are a factor of 10 or more different between the leading and trailing sides (see Fig. 2 and caption).

1.3. Chemical and Physical Processes

In the following subsections we briefly describe various processes that can chemically or physically alter the surface composition.

1.3.1. Radiolysis. The Galilean satellites are imbedded in an intense radiation environment, with their surfaces being constantly bombarded by energetic electrons, protons, and heavy ions (see chapter by Paranicas et al.), along with a lesser-energy flux of solar ultraviolet (UV) photons (the mean solar UV energy flux that can dissociate H_2O is ~2% of the magnetospheric flux). *Burns* (1968) and *Morrison and Burns* (1976) were the first to point out that the surfaces of the Galilean satellites could be modified by magnetospheric irradiation, causing albedo and color variations, but the magnitude of radiation effects was not fully appreciated until two definitive radiolytic species, molecular oxygen and hydrogen peroxide, were discovered on Europa (see sections 2.5 and 2.6).

A single keV to MeV electron or ion passing through an icy molecular solid produces a trail of ionizations and excitations as the original particle's energy is degraded. Each ionization event along the track will produce secondary electrons that, in turn, travel through the ice, creating separate tracks of yet more ionizations, excitations, and subsequent reactions and chemical changes. In this way, the direct chemical action of the incident particle is overshadowed by the chemistry produced by the secondary electrons. This implies that, to a first approximation, the chemical reactants produced by various ionizing radiations (e^-, H^+, He^+, X-rays, γ-rays) are identical, although product yields may depend on specific doses, dose rates, and the linear energy transfer rate of the primary particle. The average penetration depth for magnetospheric electrons at Europa is 0.6 mm (*Cooper et al.,* 2001) but high-energy electrons and bremmstrahlung can deposit energy to meter depths. Ions have a much shorter range. Protons at Europa's orbit have an average range of 0.01 mm (see chapter by Paranicas et al. for dose vs. depth curves). Range and stopping power data for electrons and ions are available from *ICRU* (1984) and *Ziegler et al.* (1985), respectively. The time-integrated particle or energy flux incident on a surface is termed the fluence, and the absorbed energy density in the material is the dose, expressed in various units including eV $molecule^{-1}$, eV $16\text{-}amu^{-1}$, rads (1 rad = 100 erg g^{-1}), and grays (1 Gy = 1 J kg^{-1}). Additional information can be found in standard references on general radiation chemistry (e.g., *O'Donnell and Sangster,* 1970; *Swallow,* 1973; *Spinks and Woods,* 1990; *Mozumder,* 1999) and charged-particle interactions with planetary surfaces (*Johnson,* 1990; *Johnson et al.,* 2004). A pre-Galilieo summary of photolysis and radiolysis on icy satellites is found in *Johnson and Quickenden* (1997) and a special section of the *Journal of Geophysical Research* is devoted to photolysis and radiolysis in the outer solar system (*Domingue and Allamandola,* 2001).

The chemical composition of Europa's regolith is profoundly influenced by jovian magnetospheric radiation. At the same time, this radiation can also alter the molecular environment, and therefore the positions and shapes of spectral bands (see section 2.2.3). It also can produce defects and disorder in the ice and thereby alter the thermophysical and optical scattering properties. The dominant ionizing particles at Europa are electrons and protons (and, to a lesser extent, multiply charged S and O ions) with energies ranging from <10 keV to >10 MeV and with average energies in the MeV range. Energetic particle observations from both flyby and orbiting spacecraft show that radiation doses of 1 eV per H_2O molecule (~600 Mrad) is achieved in three years or less at depths of ~100 μm, which is a typical remote-sensing depth (*Cooper et al.,* 2001). Extremely energetic electrons and bremmstrahlung X-rays will penetrate more deeply, while micrometeoroid-induced gardening simultaneously buries the radiation products and brings material from depth. Irradiated material will be vertically mixed throughout the regolith (Fig. 2). Significant changes occur for doses of only a few eV per molecule (the dose often expressed as eV per 16-amu), so most molecules in the regolith will have been altered many times over the age of the surface.

A measure of the initial radiolytic production or destruction rate is given by the G value, the number of molecules produced or destroyed per 100 eV of absorbed energy. As an example, from the compilation by *Johnson et al.* (2004), CO_2 in H_2O ice is destroyed at a rate $G(-CO_2) = 0.55$ per 100 eV. Therefore, at an accumulated dose of about 5×10^4 eV $16\text{-}amu^{-1}$ (a rough global average in the upper millimeter for the 200,000-yr nonsynchronous rotation case; see Fig. 2) a CO_2 molecule would have been destroyed 800-fold. The alkaline Earth sulfates and their hydrates are among the most radiolytically stable molecules. They can be dehydrated and decomposed by producing SO_2 with $G(SO_2) = 0.004$ (*Johnson et al.,* 2004), so hydrates such as epsomite ($MgSO_4 \cdot 7H_2O$) will have suffered ~30 destructive events for the conditions considered above. For newly emplaced or exposed material, it is of interest to know the time required to accumulate a dose of 1 eV $16\text{-}amu^{-1}$, a typical dose to establish new products. For a typical optical sampling depth of 100 μm, and using the dose values from the chapter by Paranicas et al., the time for a pristine sample to generate these new species and to reach radiolytic equilibrium is ~20 months on the trailing side (at the anta-pex) and 40 months on the leading side (apex).

Because Europa's ice is dominated by water molecules, it is appropriate to briefly consider their radiation chemistry (*Buxton,* 1987; *Spinks and Woods,* 1990). A keV to MeV electron encountering an H_2O molecule will cause either an excitation or ionization. These will yield, in turn, a set of primary products that include charged species, radicals, and closed-shell molecules, summarized as

$$H_2O \rightarrow H_2O^+ + e^- \rightarrow H + OH \rightarrow H_2 + H_2O_2$$

H_2 molecules rapidly escape from the surface and even the satellite, so the surface becomes oxidizing. Some reduction can occur from energetic magnetospheric proton implantation, as on Mercury, the Moon, and asteroids (*Hapke,* 2001) from the solar wind, but this will be a secondary effect due to the relatively low proton flux.

The incident radiation also decomposes peroxide, so equilibrium concentrations will be achieved where production equals destruction. This decomposition, and other secondary reactions, will make HO_2, HO_3, O_2, and O_3, and electron attachment to OH (hydroxyl) will produce OH^- (hydroxide). Given these radiation products from H_2O, subsequent reactions with other molecules in the original ice may include H^+ transfer (acid-base reaction), e^- transfer (oxidation-reduction), and free-radical reactions, such as radical combinations, H and OH addition, disproportionation, and atom abstraction by H and OH. In general, all these reactions are sufficient to explain many chemical species identified on Europa (e.g., H_2O_2, H_2SO_4 hydrates, O_2). The dissociation products of H_2O_2 are species such as OH, OH^-, HO_2, O_2, H_2, and H_2O. Since H_2 readily diffuses out of the ice at Europa's temperature, the ice surface is permanently modified and becomes more oxidized as the $H_2O \leftrightarrow H_2O_2$ reaction cycle continues and H_2 is lost.

Sulfur species thought to be present in Europa's ice, elemental sulfur, SO_2 and the SO_4^{2-} ion, are part of the dynamic sulfur cycle driven by interactions with the jovian magnetosphere on relatively short timescales. In Europa's water ice, sulfur, SO_2, and other sulfur species are oxidized to SO_4^{2-} with radiation processing. Some of the species formed are

$$H_2O + SO_2 \rightarrow SO_3, HSO_3^-, HSO_4^-, SO_4^{2-}, H_3O^+$$

Reactions of SO_2 with H_2O_2 can form sulfates. Irradiation of possible hydrated salts, e.g., $Mg_2SO_4 \bullet 7H_2O$ and $Na_2 \bullet 10H_2O$, yields metal oxides, the SO_4^{2-} ion in the form of sulfuric acid and sulfate salts, and SO_2. Based on laboratory experiments (*Moore et al.,* 2007), it is estimated that these radiolytic processes form the observed abundances of SO_4^{2-} on Europa in $<10^4$ yr. The radiolytic cycle on Europa continues as H_2SO_4 and its hydrates are dissociated, resulting in species such as SO_3, HSO_4^-, SO_2, H_2, and S. Thus the sulfur cycle, $S \rightarrow H_2SO_4 \rightarrow SO_2 \rightarrow S$, results in a dynamic equilibrium where the relative abundances are established by production and loss mechanisms. The largest reservoir of sulfur on Europa is thought to be in the more stable sulfate and other sulfur-containing ions, 93–98%, compared to 1.5–6.9% as SO_2 [abundances relative to total sulfur (*Moore et al.,* 2007)]. Note that the starting point for these cycles is immaterial; one could start with sulfur, sulfide, or a sulfate, eventually reaching equilibrium with sulfate (as acid and salts, if the metal cations are present), SO_2, and sulfur allotropes.

The carbon volatile, CO_2, identified in Europa's H_2O ice and discussed further in section 2.7, probably takes part in the radiolytic carbon cycle. Some of the species formed by irradiation are

$$H_2O + CO_2 \rightarrow CO_3, CO, O_2, O_3, H_2CO_3$$

On Europa, it is estimated that CO_2 will convert to CO and H_2CO_3 (carbonic acid) with a half-life <2000 yr. With further irradiation, H_2CO_3 is destroyed reforming CO_2

$$H_2CO_3 \rightarrow CO_2 + H_2O$$

A large destruction rate for H_2CO_3 has been measured in laboratory studies, so at equilibrium ($H_2O + CO_2 \leftrightarrow H_2CO_3$) most of the carbon will be as CO_2. Europa's surface is an open system and CO_2 can slowly enter the atmosphere and escape to space. Thus, a continuing source of carbon may be required. The carbon cycle is discussed in more detail in section 2.7.

1.3.2. Sputtering redistribution. When high-energy ions strike a surface, they eject atoms and molecules through both elastic collisions and through localized electronic excitation (see *Johnson,* 1990). While electrons and photons also sputter material, the most efficient particles are heavy high-energy ions. Jupiter's magnetosphere contains energetic protons, sulfur ions, and oxygen ions, with the latter two species producing most of the sputtering at Io. Because of their large gyroradii, these ions strike Europa fairly uniformly in longitude. Relevant features are sputtered Na and K atoms that form a tenuous ballistic atmosphere and a flux of escaping atoms. Europa also has an O_2 and presumably H_2O atmosphere (see chapters by Johnson et al. and McGrath et al.), formed by the sputtering products of water — H_2O, O_2, and the light molecule H_2 that directly escapes. Measurement of the sputtered atmosphere offers a means of exploring the surface composition. This process also redistributes material over the surface, as discussed in the following.

Tiscareno and Giessler (2003) computed the globally averaged erosion rate of 0.0147 μm yr, in agreement with prior calculations by *Cooper et al.* (2001). Of these molecules, 42–86% of the molecules restrike the surface. There is more sputtering on the trailing hemisphere and the net transfer, mainly of water molecules, from the trailing to leading side is <0.003 μm yr^{-1}. This rate is small compared to *Cooper et al.*'s (2001) vertical gardening rate of 1.2 μm yr^{-1}. In the 50-m.y. age of the surface, a maximum net of transfer of 15 cm of material will be mixed into the nominal 1-m leading-side regolith (see below). While unimportant for H_2O, this process may be important in redistributing the Na and K atoms to explain the orbital behavior of the Na and K clouds (*Leblanc et al.,* 2005; but see *Cipriani et al.,* 2008).

1.3.3. Impact-induced chemistry. Hypervelocity impacts on ice produce high-temperature shock waves, vaporization, electronic excitation, light emission, and ionization (*Eichhorn and Grun,* 1993; *Burchell et al.,* 1996; *Kissell and Krueger,* 1987) and there will be chemical changes produced in this highly energetic process. Although the average power in micrometeoroid impacts is about a factor of 10^5 less than that for energetic particles, and therefore not globally significant, large impacts may produce localized effects (*Cooper et al.,* 2001).

The chemical effects of impact heating and shock-wave generation may be simulated by a variety of processes (*Kissell and Krueger,* 1987; *Scattergood et al.,* 1989) including pulsed lasers. Laser ablation experiments on mixed ices to simulate Europa have been performed by *Nna-Mvondo et al.* (2008). For example, for H_2O:CO_2 ices they find H_2O_2, CO, and CH_3OH being produced, while H_2O ice with Na_2CO_3 generates CO and CO_2. Irradiation of water ice containing methanol yields more complex products, indicating that large impacts may produce a quantity of diverse and interesting compounds. Additional quantitative work is needed to assess the importance of this process for icy satellites.

1.3.4. Micrometeoroid gardening. The formation of regoliths, the fragmented, porous "rock blankets" produced by meteoritic impact, directly influences the surface composition. Meteoritic impacts introduce new material and produce craters, excavating the existing surface and covering the adjacent surface. As Europa's surface is being implanted with sulfur ions and bombarded by high-energy radiation that forms new molecules, gardening simultaneously buries these products, and brings fresh (and previously irradiated) material to the surface where the implantation and bombardment process continues. Regoliths can also present large areas for chemical and physical interaction with atmospheric species (*Cassidy et al.,* 2007).

The volume of material ejected by impacts is 10 to 100 times greater for ice targets than crystalline rocks (*Lange and Ahrens,* 1987), so a thick, porous regolith may expected on Europa, depending on the surface age. This mixing ("gardening") extends to a depth that increases with time but can vary significantly over the surface due to statistical variations of the impactor's mass, velocity, and flux as well as location and orbital geometry. One of the first estimates of regolith growth on Europa is by *Varnes and Jakosky* (1999), who predicted 1- to 10-cm-deep regoliths for a 10-m.y.-old surface. *Cooper et al.* (2001) developed a depth vs. time relationship using lunar examples and Ganymede impact rates derived by *Shoemaker et al.* (1982) and *Shoemaker and Wolfe* (1982). The initial rate is about 1.2 μm per year, slowing as the regolith forms and penetration to deeper levels becomes progressively more unlikely. *Phillips and Chyba* (2001, 2004) developed two regolith growth models, the first normalized to large craters (*Phillips and Chyba,* 2001) and the second to small craters (*Phillips and Chyba,* 2004). Average regolith depths for an assumed 10-m.y.-old surface are 1.3 m and 0.7 m for the *Cooper et al.* (2001) and *Phillips and Chyba* (2001) formulations, respectively. These depths are averages over the satellite surfaces.

The micrometeorite impact rate and their velocities are greatest on the leading hemispheres (*Zahnle et al.,* 1998), so there will be differences between regoliths on the leading and trailing sides. If Europa is not tidally locked, but undergoes nonsynchronous rotation, then the hemispherical differences can average out, depending upon the rotation rate.

Carlson (2003) used two different gardening formulations and computed the mean vertical dose profile that a molecule, now at that depth, has received in a given time, subject to asymmetric charged particle irradiation (*Paranicas et al.,* 2001), asymmetric meteoritic erosion rates, and various cases of nonsynchronous rotation. The dose is normalized to a 16-amu molecule and shown in Fig. 2 for a 50-m.y. exposure [the crater age of the surface, from *Zahnle et al.* (2003)]. The gardening depth on the leading side is about 1 m for both models in the tidally locked case, but the trailing-side depths differ by a factor of ~10, from ~5 cm in the Phillips-Chyba model to ~50 cm using Cooper et al.'s formulation. In the nonsynchronous rotation case, the gardening depths are about the same due to the spin averaging.

The optical surface is overturned rapidly. Using *Cooper et al.*'s (2001) formulation the optical surface is gardened at a rate of 1.2 μm yr^{-1}, so for a nominal 100-μm remote sensing sampling depth, gardening excavation and overturning occurs in about 80 yr.

1.3.5. Thermal processes. The surface of Europa can be heated by solar insolation and by internal heat sources. We first consider the solar heating. While H_2O absorbs only weakly in the visible region, it does absorb infrared (IR) radiation, and this must be considered when computing thermal effects. For impure ice and hydrated material, additional absorption can occur, depending on the absorption properties, and can produce heating rates and temperatures greater than for ice. The increased temperature will promote greater H_2O sublimation from the impure ice. The sublimation rates of hydrates depend on the particular hydrate and can be less than for water ice. The net effect is to thermally segregate H_2O molecules from darker materials (*Spencer,* 1987a) and possibly "garden" the surface (*Grundy and Stansberry,* 2000). Since sublimation and recondensation occurs diurnally, with sublimated molecules condensing on both dark and icy regions, the segregation is probably not carried to completion. The surface of Europa contains amorphous ice (section 2.1), and amorphous ice has a higher vapor pressure than crystalline ice, so the local sublimation and condensation fluxes may be as high as 10^{12} molecules cm^{-2} s^{-1} at 120 K (*Sack and Baragiola,* 1993, their Fig. 5), to be compared to the average sputtering rate of ~1.5×10^9 molecules cm^{-2} s^{-1}. This amorphous ice sublimation rate is greater than the crystalline ice sublimation rates used by *Shi et al.* (1995) in their comparison of icy satellite sublimation and sputtering rates. However, there are different forms

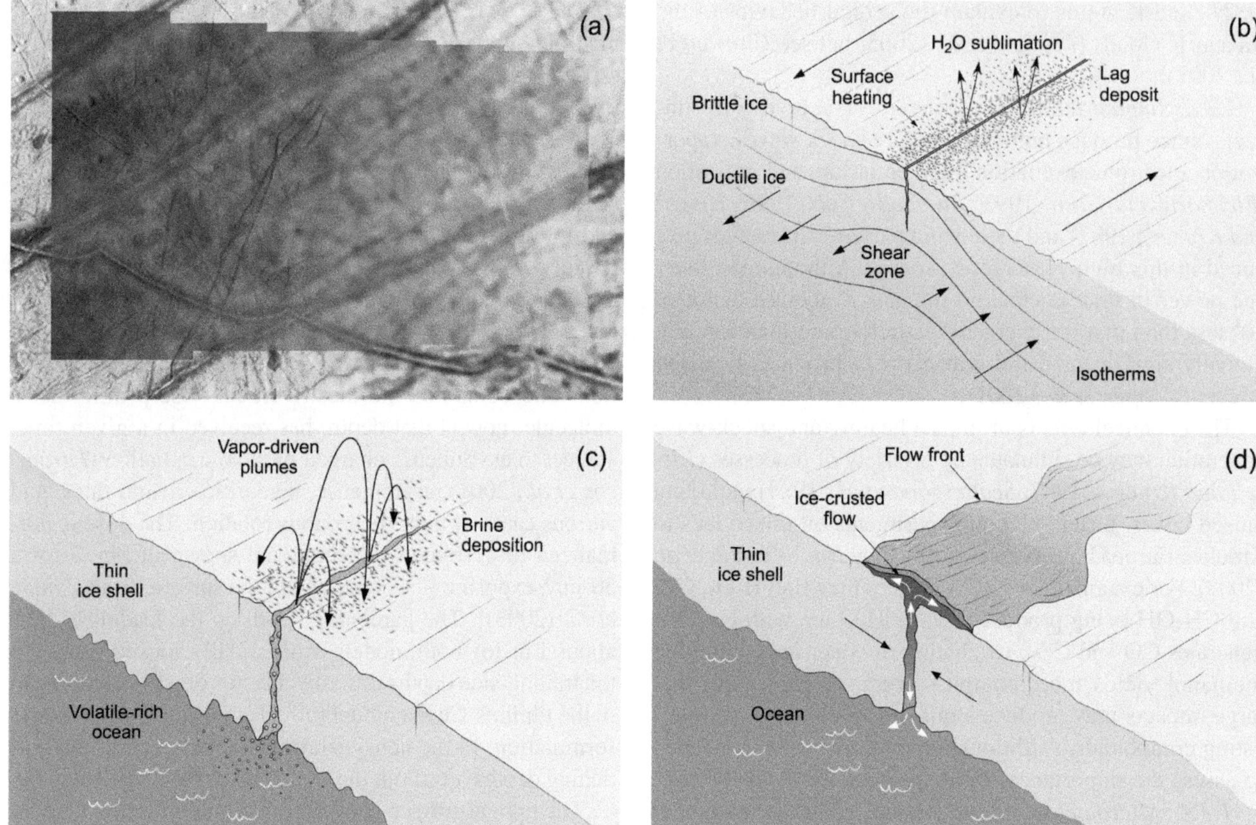

Fig. 3. See Plate 16. Association of Europa's dark material with lineae and possible geological mechanisms. **(a)** False-color Galileo NIMS observation 11ENCYCLOD01 overlaid upon visible imaging data. Red indicates hydrated materials while blue denotes more pure water ice and frost. Note correlation between red (hydrated) material and lineae. In higher-resolution spectral maps the bright central band of lineae are observed to be water ice (*Carlson et al.,* 2005a). **(b)** Thermal modification of the surface from shear heating of the surface in the region of the linea, sublimating H_2O and leaving enhanced concentrations of higher vapor pressure hydrate and darker, possibly sulfurous material. **(c)** Thin shell volatile-driven explosive cryovolcanism depositing oceanic brine. **(d)** Effusive emplacement of near-surface liquid by tide-induced opening and closing of a crack. Other mechanisms are possible.

of amorphous ice, so the sublimation rate of amorphous ice produced by irradiation needs to be measured.

Heating by geological processes can affect the surface through sublimation, by softening or melting of surficial material, and by the generation of plumes. The sources of heat can be rising diapirs and shear heating of cracks (*Nimmo and Gaidos,* 2002) (see Fig. 3b). Sublimation will produce local thermal segregation and lag deposits of more refractory material (*Head et al.,* 1999; *Fagents et al.,* 2000; *Fagents,* 2003). Diapiric heating can soften and mobilize the ice crust and even produce localized melting [brine mobilization (see *Head and Pappalardo,* 1999; *Fagents et al.,* 2000; *Fagents,* 2003)]. Europa exhibit some evidence for small-scale, low-viscosity flooding, which could be from liquid effusion to the surface (see below) or from heating and melting of surface material. While most sulfate salts (see section 2.2) depress the freezing point by less than 5°, one sulfate salt, ferric sulfate, has a low melting point (*Chevrier and Altheide,* 2008), although there is little evidence for iron compounds on the surface (section 2.9.5). Sulfuric acid-water solutions have a eutectic point as low as ~200 K (*Zeleznik,* 1991), forming a low-temperature liquid that may explain some observed features (*Collins et al.,* 2000; *Fagents,* 2003). Note, however, that significant quantities of sulfuric acid, ferric sulfate, or other low-melting-point material would have to be present in the upper icy shell to explain many of these geological features if they are not from effused liquids. The necessary quantities remain to be investigated.

Plumes created by shear heating will contain near-surface material that can be ejected at 450 m s^{-1} and rise to altitudes of 70 km (*Nimmo et al.,* 2007). The plumes can produce local resurfacing with a rate of ~50 μm yr^{-1} (*Nimmo et al.,* 2007).

1.3.6. Cryovolcanism. Cryovolcanism may be operating on Europa and introducing material from the icy shell and even from the ocean. There are two descriptions of the ice crust, thin or thick, and this parameter influences the cryovolcanic mechanisms. Thin crusts allow direct contact of the ocean to the surface by melt-through, explosive cryovolcanism, and through deep, tidally worked cracks. In the thick case, uprising diapirs can cause thermal modifications of the surface, as discussed above, and can break through the surface. The icy shell thickness can vary over the surface and evolve with time. Mechanisms and examples are given below.

1.3.6.1. Explosive cryovolcanism: A crack penetrating through a thin shell to the ocean can be an energetic source of material (Fig. 3c) if the ocean contains volatiles such as CO_2, CO, SO_2, NH_3, and CH_4 (see analysis by *Fagents et al.,* 2000). Volatiles released by exposure to low pressure can erupt and spray gases and particulates (ice crystals, liquid droplets) at velocities of 30 to 250 m s^{-1} and form 1- to 25-km-high plumes. If multiple plumes occur along a crack, the deposited oceanic material will form a margin a few kilometers to tens of kilometers wide. This mechanism is consistent with observations, but *Fagents et al.*'s (2000) preferred explanation for the dark margins along lineae was the production of sublimational lag deposits [of possibly sulfurous material (see *Fagents,* 2003)].

1.3.6.2. Effusive flow: The flow of liquid onto a low-temperature icy surface has been analyzed in general by *Allison and Clifford* (1987) and specifically for effusion on Europa by *Fagents* (2003). This process may occur by direct, localized melt-through of a thin crust when the surface is below the water line and may form the smooth low plains found on Europa. Effusive flow can also be produced if cracks penetrate the thin crust to the ocean and partially fill with denser oceanic liquid. Tidally forced opening and closing of the crack extrudes liquid to the surface, probably under a thin frozen crust (Fig. 3d).

For thick icy shells, internal convection produces rising diapirs of warm ice that, in general, reach a "stagnant lid" and do not penetrate the surface. However, some positive relief features such as Murias Chaos (the "mitten") appear to show high-viscosity flow, suggesting that a large thermal diaper of warm plume pierced the crust. Other features such as bands may be indicative of processes similar to seafloor spreading. These high-standing features may be produced from compositionally and thermally buoyant ice (*Prockter et al.,* 2002). The relationship to the ocean and the composition of diapirs is not established.

1.3.7. Other processes.

1.3.7.1. Mass wasting: High-resolution imagery has shown evidence for downslope motions of dark material that we will show later is associated with hydrate material. These observations suggest that the dark material forms a thin veneer of unknown thickness. This mass wasting uncovers brighter ice and may be an explanation, along with others (*Carlson et al.,* 2002), for the brightening of lineae found by *Geissler et al.* (1998). The accumulation of dark matter at the base of ridges concentrates the material. Mass wasting may be assisted by charging and electrostatic levitation.

1.3.7.2. Impact exhumation: Impacts excavate material forming bright icy rays, and some impacts may have penetrated to the ocean: *Fanale et al.* (2000) investigated the Tyre and Pwyll impact sites and suggested that Europa's subsurface was laced with numerous liquid intrusions lying below a superficial ice veneer. They suggest that this thin ice layer, prominent on the leading hemisphere, is produced by sputtering, eroding the trailing side and depositing the material on the opposite hemisphere. However, as discussed in section 1.3.2, sputtering redistribution is too slow to build up such a surface.

1.3.7.4. Clathrate disruption: Clathrates have been suggested to be present in the ocean (*Prieto-Ballesteros et al.,* 2005). Bouyant clathrates may be incorporated into the ice crust and, through convective or other motions, be introduced to the surface where they are unstable. Explosive gas release could mechanically disrupt the surface and perhaps alter its composition (*Prieto-Ballesteros et al.,* 2005).

1.4. Observational Methods and Europa Surface Species

The primary source of our compositional information is remote sensing, by groundbased and Earth-orbiting telescopes and from spacecraft measurements. Earth-based spectra often are recorded at visible and near-IR wavelengths where there is efficient reflection of solar radiation. For groundbased observations, absorption by the Earth's atmosphere limits the accessible spectrum and the object's distance limits the spatial resolution. The majority of these telescopic observations observed reflected sunlight for wavelengths less than 3 μm, but some measurements have been performed at longer wavelengths (*Lebofsky and Freieburg,* 1985; *Noll and Knacke,* 1993) and in the thermal IR (*Mills and Brown,* 2000). Earth-orbiting telescopes such as the International Ultraviolet Explorer (IUE) and Hubble Space Telescope (HST) have been used to probe Europa's surface at UV wavelengths obscured by Earth's ozone and oxygen (0.2 to 0.3 μm). Both HST and ground-based telescopes are used to study Europa's tenuous atmosphere, which is formed by sputtering from the surface and therefore serves as a useful indicator of surface composition. Voyager provided thermal IR spectra (*Spencer et al.,* 2004) and the Galileo mission obtained UV and IR spectroscopic measurements, the former in the 0.2- to 0.3-μm range and the latter in the 0.7- to 5.2-μm region. A major limitation of the Galileo measurements is radiation-induced noise, which affects the UV and the long-wave IR measurements (wavelengths $\lambda >$ 3 μm). Consequently, some of the features seen on Ganymede and Callisto could not be investigated at Europa with sufficient sensitivity. Some long-wave features were detected by taking measurements from afar, near Ganymede's orbit. The deficiency of high-quality spectra for wavelengths greater than 3 μm is a major detriment to understanding Europa's surface composition. The New Horizons Jupiter flyby augmented Galileo's spectral mapping coverage of Europa in the 1.25- to 2.5-μm wavelength range.

The IR region senses vibration-rotation transitions of molecules, with the strongest transitions, the "fundamental bands," generally occurring at longer wavelengths, while the shorter-wavelength near-IR region contains weaker overtones and combination bands of two or more vibrational modes. The exception is hydrides such as H_2O, where the low mass of the H atom moves the fundamental OH stretching vibration to ~3 μm. The visible and UV regions contain spectral features generally due to electronic transitions and there are some electronic transitions observed in the near-IR for Fe-containing compounds. In condensed matter, electronic transitions are often quite broad, so definitive identifications are difficult to make, but the absorptions can be strong, providing good sensitivity to minor species. There is a lack of laboratory UV reflectance spectra for materials of interest for Europa, particularly at relevant temperatures.

Ultraviolet, visible, and IR spectroscopy probes just the upper portion of the surface, and it is important to be able to estimate those depths (Z) and corresponding photon path lengths (L). These wavelength-dependent quantities are determined by the single scattering phase function, the grain diameter D and absorption coefficient α. We computed the reflectance, R, and the mean optical path length (MOPL) (*Clark and Roush,* 1984), for refractive scattering and various values of αD. Dividing the mean optical path length (MOPL $\equiv -\ln(R) = \alpha L$) by αD gives the path length-to-grain diameter ratio L/D as a function of reflectance. Assuming that the sampling depth is one-fourth of the mean photon path length, depths of a few grain diameters are probed in dark absorbing media and greater depths are sampled in bright materials (e.g., for reflectivity R = 0.1, 0.5, and 0.9, we find Z/D ~ 2, 5, and 30, respectively). These depths, uncorrected for porosity, are in good agreement with calculations by W. Grundy (personal communication, 2008), found using *Grundy et al.*'s (2000) Monte Carlo routine for irregularly shaped particles. They are also consistent with the MOPL calculations of *Clark and Lucey* (1984) for near-IR reflectance spectra of laboratory frost samples. Later we will use the above results with independently determined grain sizes to estimate absorption properties and abundances of minor species in the scattering grains.

Several approaches have been used to estimate concentration by comparing observed satellite spectra with laboratory-derived results such as (1) Spectral radiative transfer models employing optical constants for candidate materials. Linear mixing, granular mixing, and molecular mixing models are used in such models. (2) Laboratory reflectance spectra of pure or mixed species; the latter can be approximately simulated by numerical linear mixing of results for pure species. (3) Laboratory transmission spectra, inferring concentrations from band depths. (4) Comparing equivalent widths (integrated band areas expressed as the width times the continuum level) of Europa spectral features with experimentally determined integrated band strengths (e.g., *d'Hendecourt and Allamandola,* 1986; *Allamandola et al.,* 1988; *Gerakines et al.,* 1995, 2005; *Moore and Hudson,* 1998; *Kerkhof et al.,* 1999; *Moore et al.,* 2003; and others).

In addition to remote sensing methods, there are suggestive *in situ* plasma, plasma wave, and energetic neutral atom measurements (see chapter by Kivelson et al.). Probable and possible surface species from both remote sensing and *in situ* observations are collected in Table 2; however, this list must be considered incomplete, as there are likely numerous species awaiting discovery. Spectral features similar to those observed from Ganymede and Callisto by Galileo's Near-Infrared Mapping Spectrometer (NIMS) may be present on Europa. We also include the neutral Europa torus (Table 2, bottom) because remote observations of atomic and molecular emissions from this torus may be fruitful in determining minor species originating from Europa, as demonstrated by *Hansen et al.* (2005).

1.5. Laboratory Methods

It is often the case that new planetary observations provide unanticipated results, necessitating new laboratory measurements to aid interpretation. This situation is true for Europa's surface composition and particularly applicable for understanding radiolytic species and the reflectance spectroscopy of hydrates. These two efforts are briefly discussed below.

1.5.1. Radiation chemistry. Since samples retrieved from Europa's surface are not yet available, chemical compositions are determined most directly from spectral comparisons with laboratory analogs. Experimental samples of the volatile ices can be prepared by condensation of a vapor onto a precooled substrate inside a vacuum chamber. Substrate temperatures relevant to Europa are appropriate, but not always always used. In some cases, ices are made by flash cooling of a room-temperature liquid mixture or spraying from a nebulizer onto a cold plate. Following ice formation, the ice sample's transmission or reflection spectrum can be recorded at various temperatures and irradiation conditions.

Laboratory ice analogs also can be studied from the UV to IR, with each region carrying its own benefits and disadvantages. For example, UV and visible-light measurements often give only broadly sloping but otherwise generally featureless spectra, making unique chemical assignments difficult. Near-IR spectra can possess distinct absorptions, but usually only for the more abundant species since near-IR bands generally are due to combination and overtone transitions and are usually much weaker than the fundamental transitions. The most productive laboratory work, in terms of assigning molecular bands and unraveling chemical change, has been done with mid-IR spectroscopy (2.5–25 μm, 4000–400 cm^{-1}). Spectra in this region are from vibrations involving functional groups (groups of bonded atoms), with many functional groups having very diagnostic wavelengths. Relatively little work has been done in the far-IR, although this region can be useful for distinguishing the amorphous or crystalline phase and determining the clathrate nature of an ice. In general, all this suggests that using laboratory measurements to understand Europa's surface chemistry requires measurements over a wide spectral range.

1.5.2. Reflectance spectroscopy. Interpretation of spectral observations of icy satellites relies upon comparison to

TABLE 2. Known and suggested identifications of species on Europa (see individual sections for discussion).

Identification	Method	Wavelength or Region	Comments
H_2O ice	Solar reflectance	1.5, 2, 3 μm	Amorphous and crystalline
Hydrate or hydronium	"	1.3, 1.5, 2 μm	Salts and/or acid
S_μ, S_4, S_8	"	0.3–0.6, 0.53 μm	Trailing/leading-side differences
SO_2	"	0.25–0.32, 4 μm	Trailing-side enhancement
O_2	"	0.577, 0.628 μm	Radiolytic, surface and atmosphere
H_2O_2	"	3.5 μm, 0.2–0.3 μm	Radiolytic
CO_2	"	4.26 μm	
Possible transient NH_3 H_2O*	"	2.21, 2.32 μm	Possibly spurious*
Possible amide features $-NH_2$†	"	2.05, 2.17 μm	
Na, K	Atmospheric resonance scattering	0.589, 0.590, 0.766, 0.770 μm	In sputter atmosphere and escaping from Europa.
H_2O^+	Plasma mass spectra	M/Z = 18	Pickup ions‡
H_3O^+ or K^{2+}	"	M/Z = 19	§
O_2^+	Ion cyclotron waves		Possible trace pickup ions¶
Cl^+, Cl^-	"		"
Na^+ or Ca^+, Mg^+ or K^+	"		"
SO^+, Si^+	"		"
Water group atoms and molecules (inferred)	Energetic Neutral Atoms (H)	H^+ charge exchange	In gas torus around Jupiter**
H††	Emission spectra	0.12 μm	"

*Brown et al. (1988).
†Dalton et al. (2003).
‡Paterson et al. (1999a).
§McNutt (1993).
¶Volwerk et al. (2001).
**Mauk et al. (2003).
††Hansen et al. (2005).

reference spectra of candidate materials measured under controlled conditions. Reflectance spectra of water ice and ices of other volatiles at low temperatures have been performed since the 1970s and numerous minerals and ices and have been studied spectroscopically. While large spectral databases exist for many materials (*Clark et al.,* 1993, 2003, 2007; *Henning et al.,* 1999; *Christensen et al.,* 2000) and applications to Mars have prompted spectroscopic studies of sulfate minerals (*Cloutis et al.,* 2006), many measurements do not encompass Europa's entire solar reflection regime, which extends from the UV to the mid-IR, from 0.2 to about 7 μm (thermal emission will dominate at longer wavelengths). Furthermore, most of these measurements were not obtained at sample temperatures appropriate to the surface of Europa. At these low temperatures [~100–132 K for the dayside (*Spencer et al.,* 1999)] spectra of many materials can be quite different (*Pauling,* 1935; *Grundy and Schmitt,* 1998; *Hinrichs and Lucey,* 2002). Early reflectance spectroscopy measurements of frozen volatiles by *Kieffer* (1970), *Kieffer and Smythe* (1974), *Lebofsky and Fegley* (1976), *Clark* (1981a,b), and others established the field of planetary laboratory reflectance spectroscopy that continues today. Recent application of cryogenic reflectance spectroscopy to candidate Europa surface materials (*McCord et al.,* 2001; *Carlson et al.,* 2005a; *Dalton et al.,* 2005) has continued to improve our knowledge of possible hydrated compounds that may be on the surface. Hydrocarbon reflectance spectra, of potential use for Europa studies, are being obtained by *Clark et al.* (2008a).

2. SURFACE SPECIES

We begin this discussion with the two major constituents: water ice and a hydrated species. Sulfur compounds (sulfur dioxide and elemental sulfur) are then discussed. While all molecules on Europa are influenced by radiation, the two obvious radiolytic products, molecular oxygen and hydrogen peroxide, are presented in order of discovery. Carbon dioxide is then discussed, followed by sodium and potassium, and finally other suggested but as yet unobserved or unverified species.

2.1. Water Ice

2.1.1. Introduction. Water, present throughout the solar system (*Encrenaz,* 2008), is expressed on Europa's icy crust (see chapter by Schubert et al.) as relatively fine-grained frost, combined with the hydrated materials described in section 2.2. Other possible H_2O forms include clathrate hydrates, discussed in sections 2.4 and 2.6. Water ice is indicated on Europa by the appearance of prominent vibrational bands in spectra of Europa (Fig. 4). Early

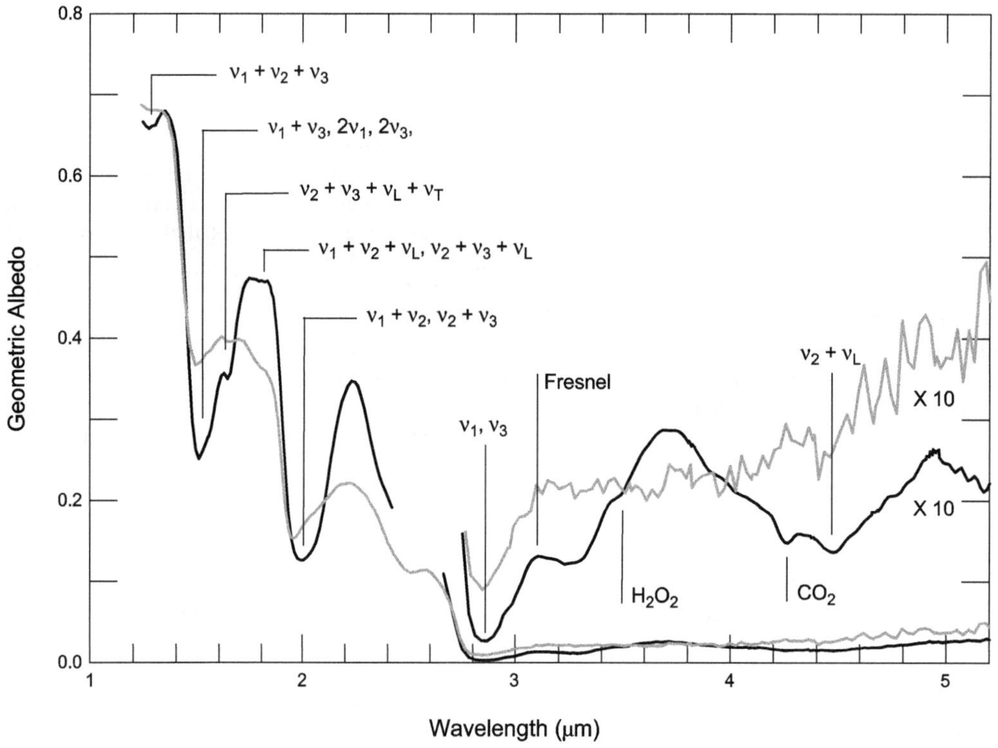

Fig. 4. Galileo NIMS near-IR spectra of Europa's leading (black line) and trailing (gray line) hemispheres. The leading-side spectrum indicates nearly pure H_2O ice with about ~30-μm grain size, along with H_2O_2 and CO_2. Band assignments are from *Ockman* (1958); ν_1 and ν_3 are the fundamental symmetric and asymmetric stetch modes and ν_2 is the fundamental bending mode. T denotes translational excitations, L denotes librational excitations. The HDO ν_3 fundamental occurs at 4.10 μm, but is not found yet in NIMS spectra. A Fresnel reflection feature (reststrahlen band) is found in at 3.1 μm where the ν_1 and ν_3 absorption is so strong that ice behaves like a metal (the reflection minima at 2.85 μm occurs where the real index crosses unity) The shape of Europa's diffuse reststrahlen feature indicates amorphous ice in the leading-side surface, but the feature at 1.65 μm that involves lattice excitations indicates crystalline ice at submillimeter sampling depths. The average trailing-side spectrum shows the asymmetric water bands of Europa's hydrated material. This spectrum, obtained during the G2 Galileo Europa flyby, is noisier than the leading-side spectrum, which was obtained during the E11 orbit at a large distance from Europa, outside its intense radiation environment.

telescopic observations of the Galilean satellites by *Kuiper* (1957) suggested the presence of water bands, confirmed by near-IR spectra obtained by *Moroz* (1965) and *Johnson and McCord* (1971) (see chapter by Alexander et al.). High-quality spectra were obtained by *Pilcher et al.* (1972) and *Fink et al.* (1973) and show Europa to have a predominantly H_2O covered surface with much more ice coverage than on Ganymede or Callisto. *Pollack et al.* (1978) obtained spectra of the Galilean satellites from airborne telescopic measurements and *Clark* (1980) obtained high-quality ground-based spectra. They both noted that the H_2O bands on the trailing hemisphere were distorted, and we now know that these distorted bands are due to a hydrate and not pure H_2O frost. Europa's trailing side contains hydrate and H_2O ice in variable proportions, whereas the leading hemisphere's surface is dominated by water ice.

2.1.2. Ice phases and their formation. The phases of ice are of interest because they indicate surface processes and likely play a role in trapping of volatile molecules such as O_2 and CO_2. The lowest-energy lattice arrangement of water ice at low pressures is hexagonal. Ice produced by freezing liquid water is this form except when flash frozen under very special conditions, forming amorphous ice (*Mayer,* 1985). Condensation from the gas phase produces different forms depending on the temperature (*Jenniskens et al.,* 1998; *Baragiola,* 2003). Water ice grown within about 100° of its 273 K melting point is hexagonal. A metastable cubic crystalline structure (in which the alternate planes of the hexagonal structure are shifted one position) is formed at temperatures of approximately 140 to 150 K, and disordered amorphous structures are made below 150 K. There are three known amorphous phases (see *Jenniskens et al.,* 1998), a high-density phase (I_ah) formed at temperatures <30 K, a low-density phase (I_al) formed at temperatures <100 K, and restrained amorphous ice (I_ar) that coexists with cubic ice and formed when either I_al is heated or irradiated, or when water molecules are condensed in the temperature range of 100 to 140 K.

Once formed, the various forms of ice will eventually become hexagonal ice, with the rates being very dependent on temperature. Observed europan temperatures range from <76 K at night to a dayside maximum of 132 K (*Spencer et al.,* 1999), but ice and darker materials may segregate by sublimation (*Spencer,* 1987a), leading to higher-albedo icy regions having low daytime temperatures of perhaps ~110 K (*Grundy et al.,* 1999). Even at these temperatures, amorphous ice will crystallize over short timescales, e.g., at 100 K it will crystallize within 10 yr to cubic ice (*Jenniskens et al.,* 1998; *Baragiola,* 2003) and perhaps within 20 yr to hexagonal ice [extrapolated from measurements by *Dowell and Rinfret* (1960)].

Amorphous ice on Europa can be continuously created from the condensation of previously sublimated or sputtered molecules (there may be deposition rate effects that favor the crystal formation (*Kouchi et al.,* 1994; but see *Baragiola,* 2003). Another likely amorphization mechanism is the disruption of crystalline ice by particle radiation. Ultraviolet, electron, and ion irradiation produce disorder in crystalline ice (*Kouchi and Kuroda,* 1990; *Baratta et al.,* 1991; *Moore and Hudson,* 1992; *Strazzulla et al.,* 1992; *Baragiola,* 2003; *Raut et al.,* 2004; *Baragiola et al.,* 2005; *Mastrapa and Brown,* 2006, *Leto and Baratta,* 2003; *Leto et al.,* 2005, and references therein). Note that a numerical cross-section error in *Kouchi and Kuroda* (1990) is corrected in *Leto et al.* (2003). Amorphization by irradiation is more effective at temperatures lower than those of Europa's surface. However, the decrease in efficiency with increasing temperature appears to depend on the type of incident particle. It may also depend on the dose rate and on the specific measurement, e.g., electron diffraction, the far-IR lattice bands, the 3.1-μm fundamental stretch band, or the 1.65-μm combination band, each of which may indicate different amorphous properties. *Moore and Hudson* (1992) first noted particle dependence when comparing the temperature dependence of the amorphization rate for proton irradiation with *Strazzulla et al.*'s (1992) results for helium ion bombardment. Similar conclusions are suggested by comparing *Leto and Baratta*'s (2003) and *Raut et al.*'s (2004) argon ion experiments conducted at 16 K and 70 K, respectively, for which there is only a factor of ~2 decrease in efficiency, compared to a factor of 200 for protons. (Note that the ion energies were somewhat different.) Amorphization by electrons exhibits a small rate decrease with temperature (*Strazzulla et al.,* 1992). It seems likely that electrons and the heavy energetic (keV to MeV) sulfur and oxygen ions produce the amorphous ice seen on Europa. We also note that irradiation of existing condensed amorphous ice increases its resistance to crystallization (*Baratta et al.,* 1994). The amorphization rate will decrease with depth and at some level below the surface the crystallization rate will exceed the amorphization rate and a transition from amorphous to crystalline will be formed.

2.1.3. Spectral properties. The water molecule has a fundamental H–O–H bending transition (ν_2) at 6 μm and O–H fundamental stretching bands (ν_1, ν_3) at 3 μm. Overtones and combination bands produce features at shorter wavelengths (Fig. 4). Ice exhibits lattice excitations (phonons) at ~45 μm and ~12 μm, corresponding to molecular translations (ν_T) or librations (ν_L), respectively. Many weak combination bands include these modes.

The spectrum of hexagonal and cubic ice are essentially indistinguishable from each other at near- and mid-IR wavelengths (*Bertie and Whalley,* 1964, 1967; *Bertie et al.,* 1969), and the difference of internal energy between cubic and hexagonal ice is small (*Handa et al.,* 1988). There is some evidence of absorptivity differences between cubic and hexagonal ice in the 60-μm region (*Bertie and Jacobs,* 1977; *Curtis et al.,* 2005). There are also possible differences between cubic and hexagonal ice absorption properties in the far UV (*Onaka and Takahashi,* 1968).

The amorphous forms have highly variable spectra, depending on their temperature history. "Annealed" forms of amorphous ice (that have been warmed above 100 K) have well-defined spectra that are distinct from the spectra of crystalline ice (*Hagen et al.,* 1981; *Schmitt et al.,* 1998). There are several features in the near-IR reflection spectrum of water ice that can be used to probe lattice order. These include the narrow combination band at 1.65 μm that involves lattice motions as well as molecular vibrations and exhibits a greatly reduced strength in warm crystalline or amorphous ice (*Hagen et al.,* 1981; *Grundy and Schmitt,* 1998; *Schmitt et al.,* 1998; *Mastrapa and Brown,* 2006). This band is temperature sensitive, and has been used as a thermometer for some outer solar system satellites, but has not been useful for determining Europa temperatures due to the presence of amorphous ice, hydrated mineral phases, and radiation-damaged crystalline ice (*Grundy et al.,* 1999). The fundamental absorption near 3.1 μm appears as a reflection peak in frost spectra. It is broad and weak for amorphous and warm ice (*Hagen et al.,* 1981; *Wood and Roux,* 1982), and narrower and stronger with a triplet structure for cold crystalline ice (*Bertie et al.,* 1969; *Bergren et al.,* 1978; *Hagen et al.,* 1981). Subtle band-center shifts and bandwidth changes are also apparent for all the IR bands as a function of temperature and crystallinity (*Hagen et al.,* 1981; *Grundy and Schmitt,* 1998; *Mastrapa et al.,* 2008; *Mastrapa and Sandford,* 2008), as well as other parameters such as grain size, purity, and illumination and observation geometry.

2.1.4. NIMS observations. Although several telescopic spectral studies of Europa were conducted earlier (see *Calvin et al.,* 1995), the spatial and spectral resolution needed to study the physical state of the ice and to separate the ice and hydrate components was not available until the Galileo Near Infrared Mapping Spectrometer (NIMS) (*Carlson et al.,* 1992) orbited Jupiter. *Hansen and McCord* (2004), using NIMS data, studied the balance between crystal disruption by radiation vs. crystal formation by thermal processes for the icy Galilean satellites' surfaces. They used the 3.1-μm reststrahlen or Fresnel reflection peak, which is formed by reflection off the facets of the water ice grains and is effectively from depths of approximately a wavelength. The strength and shape of the 3.1-μm peak does not vary significantly with grain size, as long as the grains are larger than about 10 μm.

Hansen and McCord (2004) found that the nearly pure ice on the uppermost surface of Europa's leading hemisphere appeared to be uniformly amorphous, implying that radiation processes dominated thermal processes in that hemisphere. The trailing side contains predominantly hydrated material rather than pure ice, so no definitive statements about phase can be made, other than that there is some crystalline H_2O ice present, as indicated by the presence of a weak 1.65-μm band. The 2.71-μm dangling bond feature, formed in porous amorphous ice, is not apparent in NIMS spectra and is not expected to be present due to rapid compaction and pore closure by ion irradiation (*Palumbo,* 2006; *Raut et al.,* 2007). From the presence of the 1.65-μm band on the leading side, *Hansen and McCord* (2004) inferred that ice at ~1 mm depth was crystalline, so the transition zone from amorphous to crystalline must take place somewhere above 1 mm depth. This depth value can be refined somewhat. Using the sampling depth calculations of section 1.4 with a reflectivity of R (1.65-μm) ~ 0.4 (see Fig. 4) and a grain diameter D ~ 30 to 40 μm, then the sampling depth Z is about 120 to 160 μm. It is plausible that the ice phase is the restrained amorphous form. Within that depth, there will be both amorphous and crystalline ice and the relative amounts can be studied using the ratio of the 1.65-μm and 1.5-μm band areas. From Fig. 4, an area ratio of 0.025 is found and is about one-half or less than that for crystalline ice at a nominal Europa temperature of 120 K (*Grundy and Schmitt,* 1998; *Leto et al.,* 2005, *Mastrapa and Brown,* 2006). Leto et al. have studied the amorphization of cubic ice irradiated at 90 K by 200-keV protons, as indicated by these band ratios, and find that a dose of ~5 eV/16-amu will produce the observed value. They found that complete amorphization is accomplished at ~10 eV/16-amu. At the midpoint of the sampled depth (Z/2 ~ 70 μm) the heavy ion dose rate is ~0.05 eV/H_2O-molecule/year (see chapter by Paranicas et al.), giving a very crude estimate of amorphization timescales of ~100 yr. If protons are included in the dose rate, the timescale will be less. Note that Leto et al.'s results are different than *Mastrapa and Brown*'s (2006) measurements for the same band and the same ionizing particle (protons); the latter authors suggest that differences in experimental film thicknesses may be important. The phase state of the surface may be a useful indicator of age, but more experimental work on amorphization and crystallization at Europa-like temperatures is needed.

Since the vapor pressure of amorphous ice is up to 100 times that of crystalline ice (*Kouchi,* 1987; *Sack and Baragiola,* 1993), the preponderance of amorphous ice on the surface could increase the role of sublimation relative to sputtering and micrometeoroid impacts on producing an atmosphere, H_2O redistribution, and resurfacing (*Shi et al.,* 1995; *Tiscareno and Geissler,* 2003) (see sections 1.3.2 and 1.3.5).

Water ice grain diameters for the leading, icy hemisphere have been determined to be ~30–40 μm using theoretical water ice spectra and comparing them to the NIMS data (*Hansen and McCord,* 2004). In the trailing hemisphere, where there is both water and hydrate contributing to the absorption, the two different average grain sizes were determined using intimate mixing of these two materials (*Carlson et al.,* 2005a). Hydrate grain sizes are given in section 2.2; for H_2O, there is an equator-to-pole increase in H_2O grain sizes, at least for the northern hemisphere. H_2O grains of about 20 μm in diameter are found for equatorial regions, increasing to 50 μm at midlatitudes (45°). Mean diameters are about 100 μm at 60°N. The particle sizes are probably controlled by micrometeoroid-induced gardening and comminution, balanced by the sputter destruction of small grains (*Clark et al.,* 1983).

2.2. Hydrates

2.2.1. Introduction. Europa exhibits H_2O absorption bands that are distorted and asymmetric compared to pure H_2O (Fig. 4). These bands were first noted by *Pollack et al.* (1978), who suggested that the bands were broadened by magnetospheric particle bombardment, as found in laboratory measurements of irradiated silicates (*Dybwad,* 1971). *Clark* (1980) found similar features in trailing-side spectra and remarked that they were most unusual spectra among the four objects he studied (Europa, Ganymede, Callisto, and Saturn's rings), and possibly due to ice mixed with other minerals as studied in his laboratory (*Clark,* 1981b). He presciently noted that no hydrated minerals could be ruled out. These distorted bands, noted in early NIMS spectra and found to be similar over much of the trailing side, were initially considered to be from water in clays, which can contain interlayer H_2O molecules, or in natural zeolites such as chabazite and copiapite that can trap H_2O and other molecules. In both cases distorted and asymmetric water bands are found (*Clark et al.,* 1990, 1993) but the band shapes did not provide good matches to Europa spectra (*McCord et al.,* 1998b). Spectra of OH-bearing minerals, particularly phyllosilicates, also bear resemblance to those of Europa's dark terrain. However, lacking the H–O–H stretching plus bending combination bands ($\nu_1 + \nu_2$, $\nu_2 + \nu_3$) at 2 μm, they do not exhibit all the combination features observed in Europa's 1- to 3-μm spectrum (*Hunt and Ashley,* 1979; *Clark et al.,* 1990). Attempts to reproduce the distorted Europa spectral features using varying grain sizes of water ice frosts and glazes, as well as neutral scattering elements such as bubbles or dust, achieved only limited success (*Carlson et al.,* 1996; *McCord et al.,* 1999; *Dalton,* 2000). The remaining possibility was hydrates.

Two explanations for the asymmetric bands were proposed, based on two different classes of hydrated molecules: hydrated salts (possibly sulfates) (*McCord et al.,* 1998b) and hydrated sulfuric acid (*Carlson et al.,* 1999b). In the first case, the initial source of the salts is considered to be upwelled material from the ocean, while the second, sulfuric acid, is the major equilibrium product of the radiolysis of sulfurous material and H_2O. In the latter case the initial source of sulfur is masked by radiolytic chemistry, and could be iogenic sulfur ions or endogenic sulfate salts, sulfides, or sulfoxides. It is important to note that the hydrate bands are from vibrations of water molecules, distorted in the hydrate structure, and this alteration can produce non-unique spectra. Thus, it is difficult to establish unambiguous identification with the current data, and there is presently no definitive, unequivocal evidence for either acids or salts on Europa.

2.2.2. Spectroscopy of hydrates. Hydrated compounds are molecules or ions with surrounding shells of water molecules that are held in place by polar bonds. (Hydrates should not be confused with clathrate hydrates, which are crystalline arrangements of H_2O molecules that form cages surrounding trapped "guest" molecules.) Hydrated molecules exist in aqueous solutions, in mixed amorphous ices, and as stoichoimetric crystals. In crystalline hydrates, water molecules are bound at specific sites (*Bauer,* 1964; *Hunt and Salisbury,* 1970; *Hunt et al.,* 1971a,b; *Hunt,* 1977; *Crowley,* 1991) and comprise part of the crystal lattice. This configuration allows for some but not all of the normal water vibrational modes (*Herzberg,* 1950; *Whalley and Bertie,* 1967; *Hunt and Salisbury,* 1970; *Hunt et al.,* 1973). The electrostatic influence of the other molecules leads to distortion of the crystal lattice and hydrogen bonds (*Pauling,* 1935; *Herzberg,* 1945) and consequently shifts the allowed vibrational frequencies and alters the corresponding features' shapes (*Whalley,* 1968; *Bertie et al.,* 1969; *Hunt and Salisbury,* 1970; *Hobbs,* 1974; *Hunt,* 1977). These shifts are illustrated by the near-IR spectra of hydrated magnesium sulfates ($MgSO_4$•nH_2O; n = 0, 1.5, 2, 3, 4, 5, 6, 7) shown in Fig. 5. The spectrum of anhydrous $MgSO_4$ (top) is nearly featureless, because $MgSO_4$ has no vibrational features in this wavelength range (*Gaffey et al.,* 1993; *Chaban et al.,* 2002; *Dalton,* 2003). The two small absorptions seen in the anhydrous $MgSO_4$ spectrum in Fig. 5 are actually caused by adsorbed water, which is virtually impossible to remove even under stringently controlled laboratory conditions because of the hygroscopic nature of the $MgSO_4$ (*Bauer,* 1964; *Crowley,* 1991; *Dalton,* 2003). Magnesium sulfate readily accepts water molecules, and $MgSO_4$•H_2O (kieserite) has features at ~1.0, 1.25, 1.5, and 2.0 μm whose positions correlate with those in the water ice (compare Fig. 4 and Fig. 5) and, indeed, arise from the bound water. As water content increases, more vibrational modes become possible, and several smaller features become apparent in the spectrum. As the number of waters of hydration increases, the magnesium sulfate hydrate spectra become more complex, while various small bands overlap and blend to give rise to broader absorption features. In a very general sense, as the number of water molecules of hydration increases, the spectrum of hydrated magnesium sulfate begins to more closely resemble the spectrum of the non-icy material on Europa.

At the low temperatures that prevail on icy satellites of the outer solar system, many hydrated compounds exhibit markedly different spectral behavior compared to spectra obtained at "room" temperature (*Dalton and Clark,* 1999; *McCord et al.,* 1999; *Dalton et al.,* 2005). In the 100–132-K range of dayside surface temperatures observed at Europa by Voyager and Galileo (*Spencer,* 1987b; *Spencer et al.,* 1999), the decreased thermal motion lowers the intermolecular coupling and causes the individual bands to narrow, reducing their overlap and producing several discrete, fine absorption features. This is illustrated in Fig. 6 for hexahydrite ($MgSO_4$•$6H_2O$) and bloedite [$Na_2Mg(SO_4)_2$•$4H_2O$]. The effects are most pronounced within the complexes that make up the 1.5- and 2.0-μm absorption features. At 300 K, these features are smooth and broad. At 120 K, however, these and other minor absorption features become much more pronounced. Several very fine absorption features, with widths ranging from 10 to 50 nm, can be observed within the 1.5-μm feature in both species.

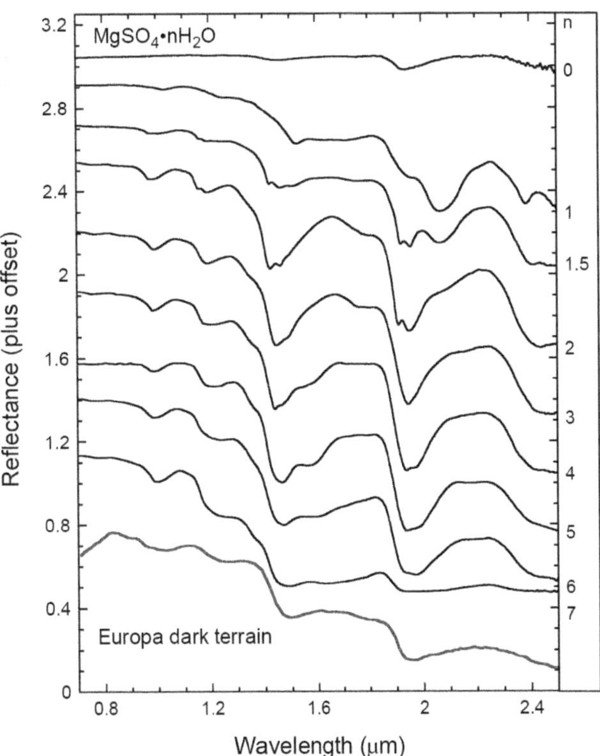

Fig. 5. Spectra of magnesium sulfate hydrates, $MgSO_4$•nH_2O, for n = 0–7. The addition of each water of hydration produces additional absorption features. As n reaches higher values, the absorptions overlap and blend together, producing broader features. These spectra were measured for samples at room temperature. Adapted from *Dalton* (2003); pentahydrite spectrum from *Crowley* (1991).

Near-IR reflectance spectra of numerous hydrated materials have been studied in the laboratory (*Hunt et al.,* 1971b; *Crowley,* 1991; *Carlson et al.,* 1999b, 2005a; *Dalton,* 2000, 2003, 2007; *McCord et al.,* 2001, 2002; *Crowley et al.,* 2003; *Dalton et al.,* 2003, 2005; *Orlando et al.,* 2005) and many hydrates show diagnostic absorptions in their spectra. A great number of these features are of sufficient strength and depth to be distinguishable at the Galileo NIMS resolution (25 nm), yet have not been detected in any examination of the NIMS observations to date, nor in high-resolution, high-signal-to-noise, spatially resolved telescopic spectra of Europa's trailing side obtained by *Spencer et al.* (2005). Experiments with flash-frozen brines (*Dalton and*

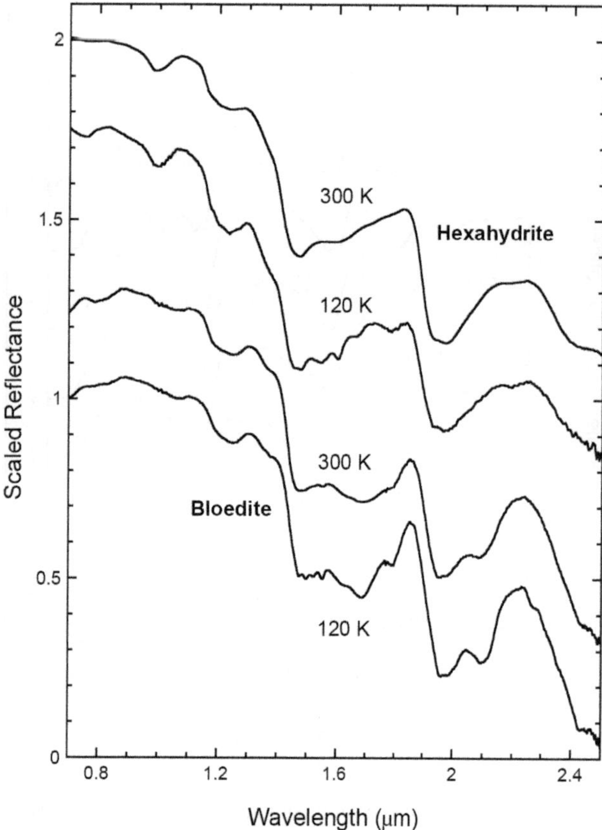

Fig. 6. Laboratory spectra of hexahydrite ($MgSO_4 \cdot 6H_2O$) and bloedite [$Na_2Mg(SO_4)_2 \cdot 4H_2O$] at 300 K and 120 K. At low temperature, the individual vibrational overtones and combinations that make up the broad water absorption features change in frequency, leading to differences in absorption feature position, strength, and width. Additional minor features become apparent and sharpen as the temperature decreases further. Adapted from *Dalton* (2003).

Clark, 1998; *McCord et al.,* 2002; *Dalton et al.,* 2005; *Orlando et al.,* 2005) combine the effects of hydration and small grain size to approximate several of the Europa spectral features. In these frozen brines, mixtures of pure water, anions, cations, and commingled molecules of varying levels of hydration match the overall character and slope of the Europa spectrum, but no perfect match to all the band shapes and positions has yet been obtained.

While magnesium and sodium sulfate hydrates have been studied intensively in the literature, sulfuric acid hydrates also give rise to distorted and asymmetric features in the 1.5- and 2.0-μm range that are strikingly similar to those seen in the spectrum of Europa (*Carlson et al.,* 1999b, 2002, 2005a). Close examination of the spectrum of sulfuric acid hydrate in Fig. 7 (from *Carlson et al.,* 1999b) reveals that the features correspond to those seen in sulfates and other hydrates, at 1.25, 1.5, 1.8, and 2.0 μm. The complex index of refraction of $H_2SO_4 \cdot 8H_2O$ was subsequently measured and used to generate comparison spectra (*Carlson et al.,* 2005a), showing that the relative strengths and shapes of the laboratory-derived features closely match those seen in Europa spectra. Shifts of the 1.5- and 2.0-μm bands may be due to radiation effects as discussed below.

Similar spectral qualities have also been produced by the addition of acids (HCl, HBr) to water ice, resulting in creation of hydronium (H_3O^+) and altering the ice structure (*Clark,* 2004). Irradiation of water ice may also create H_3O^+ (*Clark,* 2004). These changes in the ice structure may shift spectral absorption feature positions and alter spectral shapes and match Europa's profile (*Clark,* 2004). HCl and HBr also form numerous hydrates whose spectral properties resemble those of Europa (R. Clark, personal communication, 2008).

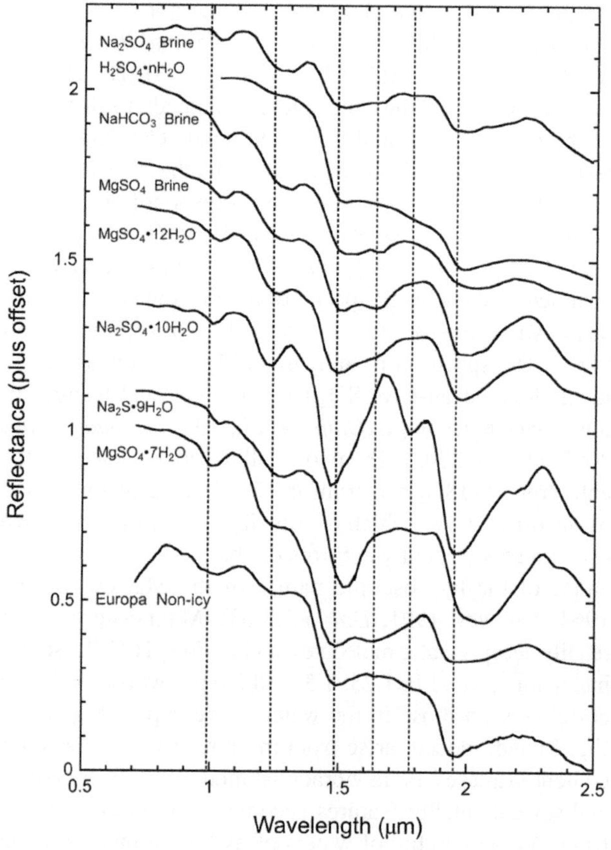

Fig. 7. Cryogenic near-IR reflectance spectra of highly hydrated compounds and frozen brines. Each spectrum has been offset vertically for clarity. All spectra in this figure were measured at 100 K, except sulfuric acid hydrate at 140 K. Spectra were convolved to the NIMS resolution for comparison with the Europa spectrum at bottom. Vertical bars denote band center positions in the Europa spectrum. From bottom, in order of increasing hydration state: epsomite, $MgSO_4 \cdot 7H_2O$; sodium sulfide nonahydrate, $Na_2S \cdot 9H_2O$; mirabilite, $Na_2SO_4 \cdot 10H_2O$; magnesium sulfate hendecahydrate, $MgSO_4 \cdot 11H_2O$; sulfuric acid octahydrate, $H_2SO_4 \cdot 8H_2O$; $MgSO_4$, $NaHCO_3$, and Na_2SO_4 saturated brines, flash-frozen at 77 K. [Sulfuric acid hydrate spectrum from *Carlson et al.,* (1999b); others from *Dalton et al.* (2005).] The hydration state of $MgSO_4 \cdot 11H_2O$ was established by *Peterson and Wang* (2006).

Biological materials also contain a number of hydrated compounds, and cryogenic spectroscopy of extremophilic organisms has demonstrated that microbes in low-temperature ice provide as close a spectral match to the Europa deposits as any individual material proposed thus far (*Dalton et al.,* 2003). Before such measurements can be considered as potential evidence for extant or extinct biological activity at Europa, however, spectral properties of other candidate materials, including many hydrates, must be investigated.

Midinfrared spectra within the reflected sunlight regime (i.e., for wavelengths greater than ~2.5 μm) may provide anion-specific information. While the fundamental stretching and bending transitions occur at longer wavelengths, combination bands will appear at shorter wavelengths, within the reflected sunlight regime (to ~7 μm for Europa). For example, the frequencies of the fundamental stretching transitions of sulfate anions are in the 1040 to 1210 cm^{-1} region (~8 to 12 μm), and the overtone occurs at about 4.5 μm. Water-of-hydration also can add additional combination bands that can be diagnostic of the anions, and perhaps the cations. However, the added H_2O molecules introduce more absorption and can mute the band structure. Examples are shown in Fig. 8, illustrating that diagnostic information may be gleaned from mid-IR spectra, but such spectra can vary with grain size, measurement geometry, and temperature. NIMS trailing-side measurements (Fig. 4) confirm the low and seemingly featureless reflectance found through groundbased spectrophotemetry by *Lebofsky and Fegely* (1985), consistent with the presence of an acid hydrate or a mixture of salt hydrates, but the amount of radiation noise in NIMS's long-wave channels has discouraged comprehensive investigations. More laboratory work and data analysis needs be performed to make full use of this spectral region.

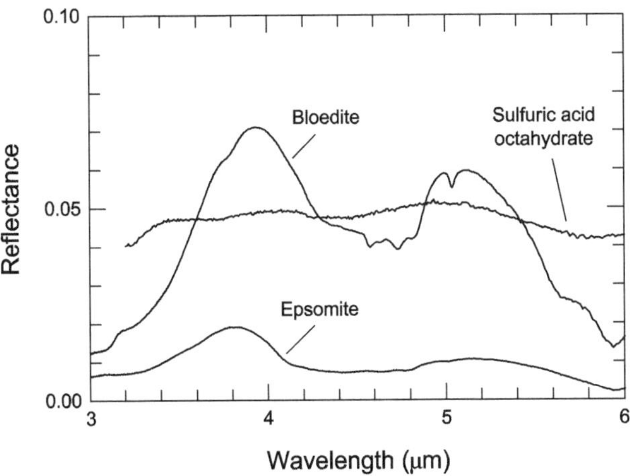

Fig. 8. Reflectance of example sulfate hydrates in the 3- to 6-μm region. Although the reflectance is low, some hydrates exhibit diagnostic structure that may be useful for hydrate identification. The bloedite [$Na_2Mg(SO_4)_2$•$4H_2O$] and epsomite ($MgSO_4$•$7H_2O$) spectra are from the USGS splib06a library (*Clark et al.,* 2007) and are from measurements by *Crowley et al.* (1991). The samples were at room temperature and the grain sizes are about 250 μm (*Crowley et al.,* 1991). Other hydrated salt spectra are presented by *McCord et al.* (1999). The acid hydrate spectrum is for a sample at 77 K (*Carlson et al.,* 1999b). Reflectance spectra are sensitive to grain size, so spectra of other samples may be brighter or darker and exhibit different shapes.

The UV region may be useful in discriminating between various candidate species, using existing spectra from IUE, Hubble, and Galileo (see section 2.3) and future observations, and comparing them with laboratory measurements. Water ice is relatively transparent for wavelengths greater than ~0.2 μm (*Dressler and Schnepp,* 1960; *Onaka and Takahashi,* 1968; *Pipes et al.,* 1974; *Warren,* 1984; *Warren and Brandt,* 2008; *Hapke et al.,* 1981), so features from absorbers such as hydrates may be present in UV spectra of Europa. As an example, two hydrated salts that have been suggested for Europa, mirabilite and epsomite (see section 2.2.5), show potentially diagnostic UV absorption features [see *Clark et al.*'s (2007) splib06a spectral library and *Crowley et al.* (1991)]. These reflectance spectra are compared with Europa measurements in Fig. 9. Mirabilite exhibits UV absorption with an onset at ~0.4 μm and grows in strength as the wavelength decreases. Epsomite has a weak absorption at 0.27 μm and a stronger band at 0.23 μm with a relative band depth of 25%. The position of weaker band coincides with Europa's 0.27-μm absorption band, but the strength may be too low to explain Europa's absorption, generally considered to be due to SO_2 (see section 2.3). The 0.23-μm feature of epsomite is not apparent in Galileo UVS measurements, but the Europa observations are noisy at these short wavelengths due to the low solar flux and the strength of the epsomite band may be muted by Europa's strong UV absorption, thought to arise from sulfur (see section 2.4). Future UV laboratory measurements and analyses such as those by *Hendrix et al.* (2008) can provide useful constraints for Europa's surface composition.

2.2.3. Spectral effects of radiation. Hydrates are susceptible to alteration by radiation, which can remove water molecules (dehydration), damage crystal structure (amorphization), and destroy either the water molecules or the host molecule (decomposition). One result of these processes is to change the H_2O vibrational modes and thus the corresponding spectrum. These spectral changes of irradiated hydrates are not well characterized, but results by *Nash and Fanale* (1977) illustrate the general effect on the spectrum (Fig. 10), which is to shift the positions of the hydrate bands and change their shape. The derived G value for H_2O decomposition of $MgSO_4$•$7H_2O$ and other sulfates is small [$G(H_2) = 0.0027$; see *Huang and Johnson* (1965)], but the results of *Nash and Fanale* (1977) indicate that the loss of H_2O (dehydration) is greater by a factor of about 100 or more. Band shifts of about 1–2% are also found for the fundamental bands of sulfates (*Spitsyn et al.,* 1969) and silicates (*Dybwad,* 1971). Radiation also darkens sulfates in the visible and UV (*Lebofsky and Fegley,* 1976; *Nash and Fanale,* 1977; see also Fig. 2 of *Carlson et al.,* 1999b), presumably by forming sulfur allotropes.

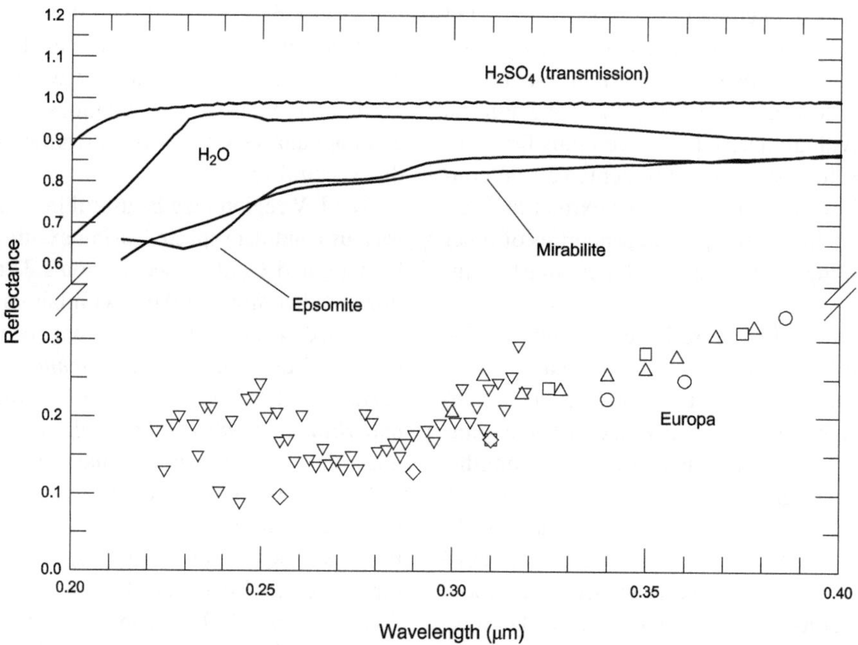

Fig. 9. Ultraviolet spectra of laboratory samples and Europa's trailing hemisphere. Mirabilite ($Na_2SO_4 \cdot 10H_2O$) and epsomite ($MgSO_4 \cdot 7H_2O$) reflectance spectra are from the USGS splib06a spectral library (*Clark et al.,* 2007) and are from measurements by *Crowley et al.* (1991). The samples were at room temperature and the grain sizes are about 250 μm (*Crowley et al.* 1991). The reflectance of H_2O frost at 77 K measured by *Hapke et al.* (1981) shows the onset of absorption at ~0.23 μm (see also *Pipes et al.,* 1974) as does the transmission spectrum of a 5-mm path-length of liquid sulfuric acid at 20°C, discussed by *Carlson et al.* (1999b). Europa data are from *Johnson* (1970) (○), *Wamsteker* (1972) (△), *McFadden et al.* (1980) (□), *Nelson et al.* (1987) (◇), and *Hendrix et al.* (1998) (▽). See Fig. 13 for extended spectral range and normalization information. Note the vertical scale change.

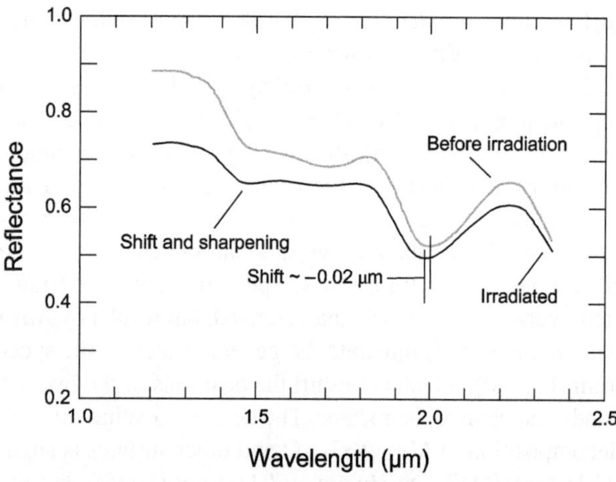

Fig. 10. Radiation-induced spectral shifts. Proton irradiation of bloedite [$Na_2Mg(SO_4)_2 \cdot 4H_2O$] shows hydrate band-center shifts to shorter wavelengths of about 20 nm. Similar shifts to shorter wavelengths are seen in ferric sulfate at doses of ~10 eV/16-amu (*Nash and Fanale,* 1977). Comparisons of laboratory and Europa spectra must allow for radiation-induced shifts of hydrate bands. Measurements from *Nash and Fanale* (1977). The spectrum of the irradiated sample was offset by –0.1 to improve clarity.

2.2.4. Abundance and distribution of Europa's hydrate. The global distribution of hydrated material was initially studied by *McCord et al.* (1999), who chose two endmembers (one mostly ice and the other mostly non-ice hydrated material) and mapped the relative fraction using linear mixing. They found that spectra for the ice-poor areas looked nearly identical for all regions investigated, implying that the hydrate was about the same composition everywhere studied. A subsequent analysis (*Carlson et al.,* 2005a) assumed intimate mixtures of hydrate grains and H_2O ice grains and, using measured values of the optical constants for sulfuric acid octahydrate and water ice, performed a radiative transfer fit for each spectrum in NIMS global observations. The results (Fig. 11b) are similar to those of the earlier work by *McCord et al.* (1999) where the two studies overlap, but the later work encompasses more of Europa and provides information about absolute abundance and grain sizes. The maximum hydrate concentration near the antapex, expressed as hydrate/(hydrate + H_2O) is about 80–90% by volume. For sulfuric acid hydrate, this would be about 1 sulfur atom (or sulfate anion) per 10 water molecules, roughly 2 of which are in H_2O ice grains and about 8 in the hydrate (*Carlson et al.,* 2005a). Hydrate grain sizes (diameters) are 6–14 μm on the equatorial trailing-side region, and 12–20 μm on the antijovian equatorial side. The grain sizes increase to about 25 μm at 45° to 60°N latitude. Observations during the recent New Horizons flyby produced a low-spatial resolution (180–250 km) hydrate map (Fig. 11a) that has more leading-side and southern hemisphere coverage than obtained with Galileo (*Grundy et al.,* 2007). The hydrate concentration was found to be very low throughout the observed leading hemisphere.

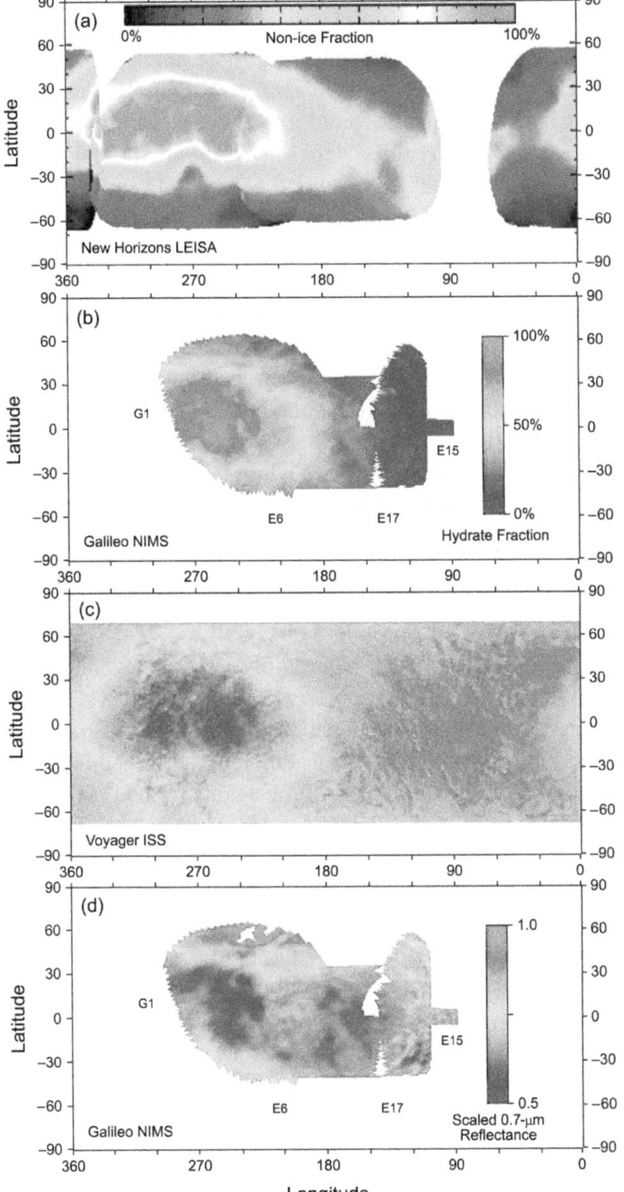

Fig. 11. See Plate 17. Distribution of hydrate and UV and near-IR albedos. **(a)** Non-ice distribution measured from New Horizons (*Grundy et al.,* 2007). **(b)** Molar distribution of hydrate assuming sulfuric acid hydrate, but representative of other hydrates (*Carlson et al.,* 2005a). **(c)** Voyager UV/V map, from *McEwen* (1986). Blue denotes high UV absorption and corresponding low UV reflectance. (d) Galileo NIMS 0.7 μm/1.2 μm ratio map, scaled (*Carlson et al.,* 2005a). Note the correlation of the non-ice hydrate distribution [**(a)** and **(b)**] with both the UV **(c)** and the near-IR absorber **(d)**, and the trailing-side enhancement of all three.

The hydrate distribution correlates with the UV absorber found in Voyager imagery (section 2.4) and with red albedo as measured by NIMS at 0.7 μm (Figs. 11c and 11d, respectively). The correlation of hydrate and decreased visual albedo means that either the hydrated material itself is dark, or that another species with low albedo is intimately associated with the hydrate. Since most hydrated salts and acids are colorless, the second possibility seems the more likely. One interpretation for the dark material is sulfur (S_8 and S_μ, cyclo-octal and polymeric sulfur), produced radiolytically from sulfate, SO_2, and possible sulfide compounds (see below). The hydrate distribution also correlates with Europa's SO_2 as measured in the UV by Galileo (section 2.3).

The hydrated material is least abundant in polar regions, where water ice dominates. The asymmetric spectral effects appear strongest in equatorial regions of the trailing hemisphere, where the iogenic plasma sulfur implantation flux is greatest. This argues for an exogenic origin. In the disrupted chaos regions, and in the immediate vicinity of lineae, the surface is darker and the spectra more asymmetric than in the surrounding terrains, suggesting an endogenic source or an endogenic modification process associated with these features. An endogenic source could be oceanic material introduced to the surface by any of several processes (section 1.3.6). An endogenic modification process could be localized shear or diapiric heating that produces a lag deposit of concentrated hydrate and associated dark material (section 1.3.5). Increasing the grain size can also enhance absorption and darken the surface. There is less hydrate in the leading hemisphere, although there are lineae and chaos regions there as well (*Riley et al.,* 2000). These leading-side features lack the color associated with the trailing "red" hemisphere. The lineae extending from the trailing fade in color in the leading hemisphere (*Nelson et al.,* 1986). Some examples of the hydrate distribution for geological features are shown earlier in Fig. 3 and in *McCord et al.* (1998b), *Fanale et al.* (2000), *McCord et al.* (1999), *Dalton* (2000), *Dalton et al.* (2003), and *Carlson et al.* (2005a).

The trailing-side enhancement suggests sulfur ion implantation as the source of hydrate, but a thin icy shell on the trailing side, allowing surface emplacement of brine, and a thick, impenetratable shell on the leading side might also explain the hemispheric dichotomy.

2.2.5. The hydrated salt hypothesis. *Clark* (1980) first suggested the possibility of hydrated minerals as the source of the asymmetric bands. The more specific hydrated salt explanation was advanced by *McCord et al.* (1998b), who noted a trailing-side enhancement of hydrated material and its association with dark material and geological features such as lineae. They found that sulfate and carbonate hydrates provided a better match to NIMS spectra than ice or hydrous silicate minerals, and that these compounds could be extruded on to the surface from the assumed ocean below. Flash evaporation, freezing, sublimation, and sputtering could concentrate the exposed brine, and leave behind crystallized salt hydrates. The possible existence of salts on the surface could provide evidence for an ocean. This work was expanded to include more mapping and to find possible combinations of minerals that provide good fits to the endmember non-ice spectrum (*McCord et al.,* 1999). Various combinations of natron ($Na_2CO_3 \cdot 10H_2O$), mirabilite ($Na_2SO_4 \cdot 10H_2O$), bloedite [$Na_2Mg(SO_4)_2 \cdot 4H_2O$], epsomite ($MgSO_4 \cdot 7H_2O$), and hexahydrate ($MgSO_4 \cdot 6H_2O$) were used to construct the linear mixing modeled reflectance. The hydrate spectra were from room temperature measurements.

The thermal and radiation stability of hydrated salt minerals epsomite, mirabilite, and natron were also investigated at temperatures relevant to Europa (*McCord et al.,* 2001). The thermal stability of epsomite was sufficient for it to remain hydrated at Europa temperatures well over geological timescales, whereas natron and mirabilite would dehydrate significantly in 10^8 and 10^3 yr, respectively. A G value for the destruction of $MgSO_4 \cdot 7H_2O$ by the decomposition of the sulfate anion, producing SO_2, was established as $G(SO_2) = 0.004$ and is consistent with other measurements [see *Johnson et al.* (2004) and section 1.3.1]. Flash freezing of brines that might occur on Europa if extruded brines condensed on high-thermal conductivity grains (*Baragiola,* 2003) was considered by *Dalton and Clark* (1998, 1999), *Dalton* (2000), *McCord et al.* (2002), and *Orlando et al.* (2005). The rapid freezing used in these experiments (10^4 K/min) results in disordered, glassy ices. Models of extrusion on the surfaces of icy satellites (*Allison and Clifford,* 1987; *Fagents,* 2003) do not predict such rapid cooling, but radiation can also reduce the order of crystalline samples (see section 2.2.3) and flash freezing experiments may simulate radiation-induced amorphization (*McCord et al.,* 2002). Since the crystalline order was reduced in the rapid freezing process, the resulting spectral structure was smoother than that found in crystalline hydrates and the frozen brine spectra, providing a better match to the NIMS spectra than obtained with crystalline salt hydrates. *Orlando et al.* (2005) used five combinations of magnesium sulfate, sodium sulfate, and sulfuric acid and found good fits for $MgSO_4$, Na_2SO_4, and H_2SO_4 in the ranges of 24–50%, 25–40%, and 25–35%, respectively (see section 2.2.7).

If salts were the original source, there will an assemblage of metal sulfates, hydrogen sulfates, and metal oxides and hydroxides (*Johnson,* 2001). Sulfate and other hydrates are destroyed in forming SO_2 and by removal of the metal atoms (e.g., Mg and Na), sometimes in the excited state (*Nash and Fanale,* 1977). Resulting products can be MgO, $Mg(OH)_2$, Na_2O, Na_2O_2, and NaOH, but production rates are not known, and there has been little work on oxide and hydroxide radiolysis in water ice. MgO in pure form is rapidly dissociated (*Wysocki,* 1986) with G = 1–4, but $Mg(OH)_2$ is quite stable with a destruction rate of $G < 0.03$ (*Glagolev et al.,* 1967). Of these two species, $Mg(OH)_2$ will likely be greater in abundance, but experiments on the initial oxide and hydroxide production rates have not been done, so one cannot predict the equilibrium value. Rough upper limits for the molar fractions of NaOH and $Mg(OH)_2$ have been established at 5% and 3% (*Carlson et al.,* 1999b; *Shirley et al.,* 1999), but it may be fruitful to reexamine spectra for these and other hydroxides (strong oxide features do not occur within the spectral range of current Europa data).

2.2.6. The sulfuric acid hypothesis and the radiolytic sulfur cycle. A different hypothesis was formulated by *Carlson et al.* (1999b), prompted by the efficient radiolytic production of sulfuric acid and the hydrophilic nature of H_2SO_4. In the frozen, crystalline state, the hydrates $H_2SO_4 \cdot nH_2O$ with n = 1,2,3,4,6.5, and 8 are formed while in the liquid, and presumably the amorphous solid state, the first hydration shell surrounding the sulfate anion contains 7–12 H_2O molecules, and there are four H_2O molecules in the first hydration shell around each proton [present as the hydronium ion, H_3O^+ (*Ohtaki and Radnai,* 1993)]. Spectra of the hemi-hexahydrate and octahydrate showed good agreement with Europa as measured by Galileo's NIMS (*Carlson et al.,* 1999b, 2005a) (see section 2.2.7). The position of the band minima for the crystalline samples is ~0.02 μm longer than observed on Europa and has been attributed to the amorphous nature of radiolytically produced sulfate compared to the ordered structure of presumed crystalline samples measured in the laboratory (*Carlson et al.,* 2005a).

The sulfate group, SO_4^{2-}, is a highly oxidized and stable complex and is the end product of numerous photolytic and radiolytic reactions. It is present on Venus as the main cloud particle constituent where it is formed by photolysis, it is present on Earth as acid rain from SO_2 oxidation, and perhaps was present on early Mars. Sulfuric acid is formed with high efficiency by the radiolysis of elemental sulfur in water ice at 77 K (*Carlson et al.,* 2002). H_2SO_4 is also made by sulfur ion implantation into water ice (*Strazzulla et al.,* 2007). When SO_2 in water ice is irradiated by energetic particles or photons, sulfuric acid is formed (*Schriver-Mazzuoli et al.,* 2003b; *Moore et al.,* 2007). The ions that are observed are sulfate, bisulfate (HSO_4^-), and bisulfite (HSO_3^-) (*Moore et al.,* 2007).

Sulfur dioxide and hydrogen sulfide are not produced in measurable quantities in either sulfur ion implantation or by radiolysis of elemental sulfur in water (*DellaGuardia and Johnston,* 1980; *Strazzulla et al.,* 2007). Instead, SO_2 is a product of sulfate destruction (see *Hochanadel et al.,* 1955; summary in *Johnson et al.,* 2004). Elemental sulfur is a minor decomposition product of both SO_2 (*Rothschild,* 1964; *Moore,* 1984) and sulfates (*Sasaki et al.,* 1978), but the efficiency or G value has not been obtained for SO_2 or sulfates in ice. Since sulfate is both produced and destroyed by ionizing radiation, the net result is that, whatever the starting point, an equilibrium mixture of sulfate, SO_2, elemental sulfur, and possibly some H_2S will be produced, with most of the sulfur atoms in the form of sulfate. The observed association of hydrate with dark, reddish material (presumably polymeric sulfur, S_μ; see section 2.4) and with SO_2 (see section 2.3) is consistent with the radiolytic sulfur cycle and supports this hypothesis. The timescale to establish equilibrium on Europa is two to four years (*Moore et al.,* 2007).

2.2.7. Spectral observations and fits. The two explanations discussed above are both plausible, and current analyses are unable to eliminate either of them. Indeed, it may be that both are currently operating on Europa. However, the possible existence of material derived from the ocean is an important question and definitive spectral evidence of endogenic material, such as salts or perhaps derived metal hydroxides, is needed to resolve this question. This is not possible with the current state of analysis, which we illustrate with three spectral fits in Fig. 12. Fits to Europa spectra using only sulfuric acid grains and water ice grains, intimately mixed, are shown (Fig. 12a) for a hydrated region an icy region, and an intermediate case (*Carlson et al.,* 2005a). Flash frozen acidic brines, from *Orlando et al.* (2005), are shown in Fig. 12b for their best match, consisting of 50% $MgSO_4$, 25% Na_2SO_4, and 25% H_2SO_4. An equivalent case using $NaHSO_4$ instead of equal mixtures of Na_2SO_4 and H_2SO_4 shows the same spectral behavior. Using cryogenic laboratory spectra, *Dalton* (2007) found a good match (Fig. 12b) to the Europa spectrum with a linear mixture of 62% sulfuric acid hydrate, 14% hexahydrite, 11% bloedite, and 12% mirabilite. It is not possible to distinguish between competing hypotheses at present.

2.3. Sulfur Dioxide

SO_2 was the second compound identified with any certainty on Europa. *Lane et al.* (1981), using IUE data, found an absorption band centered at 0.28 μm in Europa's trailing-side/leading-side ratio spectrum and attributed it to SO_2 in Europa's ice. These authors made the important suggestion that this compound was produced from implanted sul-

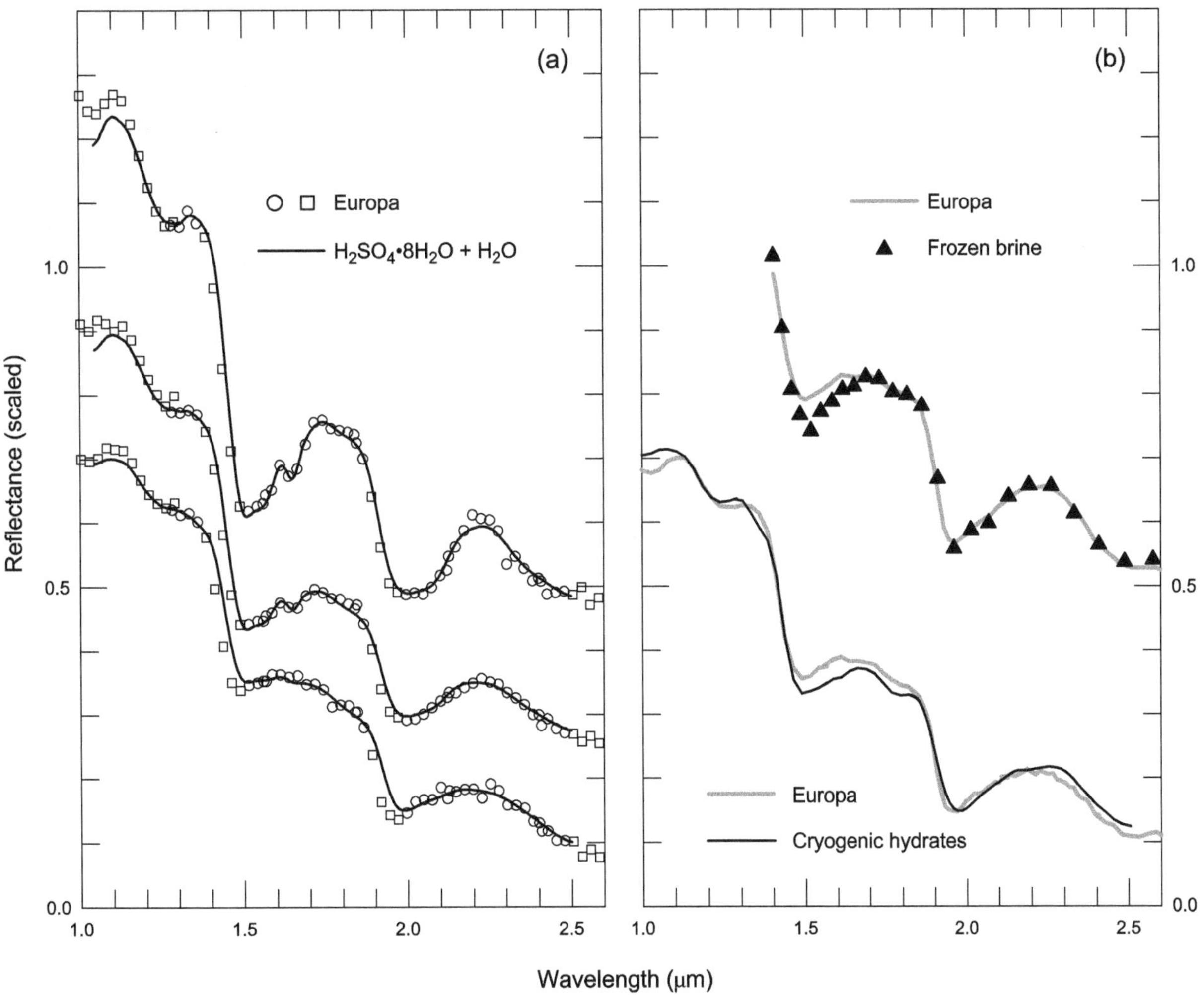

Fig. 12. Examples of the current state of spectral analysis of Europa's hydrate. **(a)** Spectra of intimately mixed sulfuric acid hydrate and water ice grains (black lines) fitted to individual NIMS spectra (symbols) using measured optical constants for H_2SO_4 hydrate and H_2O ice. The hydrate fractions for the three different pixels are, from top to bottom, 30%, 54%, and 89%. From *Carlson et al.* (2005a). **(b)** Hydrated salts and frozen brines compared to NIMS endmember spectra. The upper set compares the spectrum of a frozen brine solution (triangles) with NIMS spectra (gray line). The brine's non-H_2O composition was 50% $MgSO_4$, 25% Na_2SO_4, and 25% H_2SO_4. From *Orlando et al.* (2005). The lower set compares NIMS data (gray line) with spectrum of a numerically fitted linear mixture of individual cryogenic reflectance spectra, using 62% sulfuric acid hydrate, 14% hexahydrite, 11% bloedite, and 12% mirabilite. No H_2O was needed in this fit. From *Dalton* (2007).

fur ions from Io's plasma torus, an exogenic source contemporaneously recognized by *Eviatar et al.* (1981) (see Figs. 9 and 13 for HST and Galileo UV spectra of Europa).

The position and shape of Europa's feature closely matches spectra of condensed SO_2 (*Sack et al.,* 1992) and this molecule is the most stable sulfoxide. While sulfur monoxide, SO, exhibits absorption in the UV (*Jones,* 1950), this molecule and its dimer are extremely reactive and found only in the gas phase or within inert matrices at very low temperature (<31 K) (*Hopkins and Brown,* 1975). Disulfur monoxide, S_2O, is also unstable (see discussion in *Carlson et al.,* 2007; *Baklouti et al.,* 2008) and exhibits a UV band that peaks at 0.295 μm (*Phillips et al.,* 1969), inconsistent with Europa's band position. Sulfur trioxide, SO_3, rapidly reacts with H_2O to form H_2SO_4. Other substances absorb in the UV, including sulfur. In particular, S_8 in various solvents and S_8 in the gas phase exhibit absorption maximum at ~0.28 μm, as does polymeric sulfur (*Meyer et al.,* 1971; *Nishijima et al.,* 1976; *Nelson and Hapke,* 1978; *Sill and Clark,* 1982) and irradiated sulfur (*Hapke and Graham,* 1989). In contrast, orthorhombic α-S, the most stable form of cyclo-octal sulfur, has absorption minima at 0.28 μm (*Fuller et al.,* 1998). Therefore, while SO_2 is a plausible and likely candidate, its presence is not unequivocally established.

From IUE ratio spectra, *Lane et al.* (1981) found that the feature was stronger on the trailing side compared to the leading hemisphere (and may have been absent there; see below). The feature was strongest at 277° ± 3°W (*Ockert et al.,* 1987; *Nelson et al.,* 1987). *Sack et al.* (1992) simulated Europa's SO_2 in the laboratory by vapor deposition and found a good fit to observations from *Nelson and Lane* (1987) using reflection data for an SO_2 film about 0.12 μm thick (or about 2 × 10^{17} cm^{-2} within the sampling depth). *Noll et al.* (1995) obtained a Hubble Space Telescope UV reflectance spectrum of Europa's trailing side and also compared it to *Sack et al.*'s (1992) data, finding good agreement with Europa's absorption feature and laboratory spectra of a slightly thicker film (0.16 μm, or about 3 × 10^{17} cm^{-2}).

The UV absorption feature was mapped over Europa's surface by *Hendrix et al.* (1998) using Galileo UV data, finding abundances similar to those of *Noll et al.* (1995) and *Sack et al.* (1993). No SO_2 feature was observed on the leading side, as also shown in spectra presented by *Domingue and Lane* (1998). In contrast, *Spencer et al.* (1995) noted, in his in leading-side spectra, a possible absorption edge at ~0.38 μm that could be due to SO_2, but this seems unlikely considering the apparently low leading-

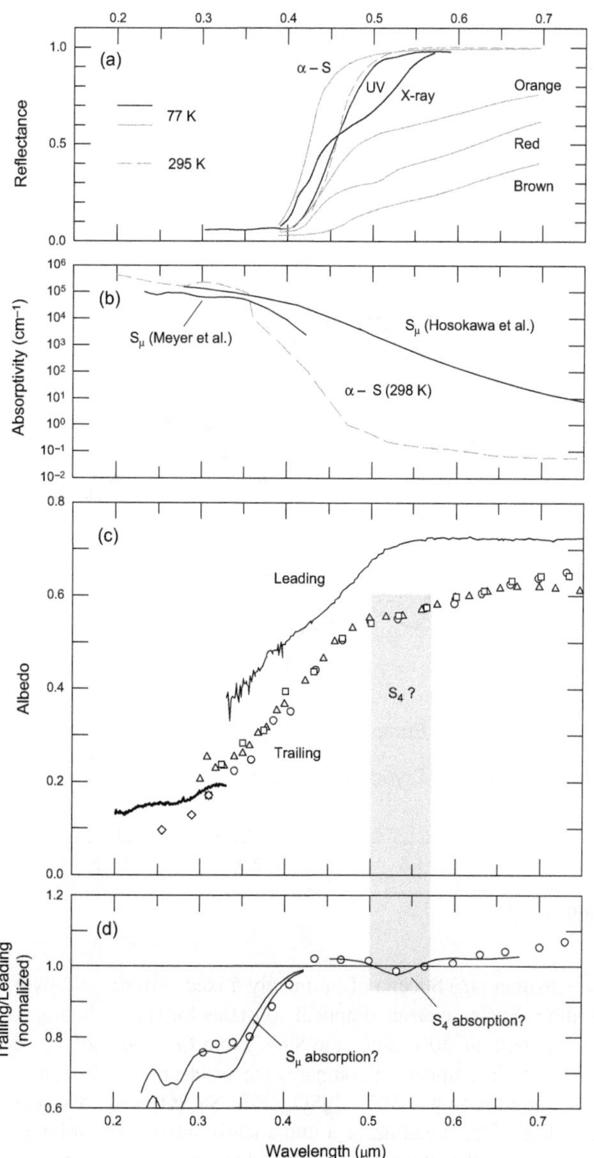

Fig. 13. Europa spectra and sulfur. **(a)** Laboratory reflectance for α-S_8 (orthorhombic cyclo-octal sulfur) at 295 K and 77 K, temperatures for which α-S_8 appears yellow or white, respectively. Spectra of 77 K quenched liquid samples (initially at 382 K, 47 5 K, and 718 K) forming orange, red, and brown sulfur with progressively increasing amounts of polymeric sulfur S_μ and S_4 (in "red sulfur"). These spectra are from *Gradie et al.* (1980). Ultraviolet- and X-ray-irradiated S_8 at 77 K (*Hapke and Graham,* 1989; *Nelson et al.,* 1990), showing S_μ and S_4 photolytic and radiolytic production. Note the shift in the band edge from 0.45 μm to 0.5 μm by UV irradiation and the presumed S_4 absorption from X-ray irradiation and the possible presence of S_3 at ~0.42 μm. **(b)** Absorption coefficients of α-S_8 at room temperature (*Fuller et al.,* 1998); liquid sulfur at 450°C showing S_μ absorption that extends to ~1.5 μm (*Hosokawa et al.,* 1994); and yellow polymeric sulfur from 250°C liquid quenched at 0°C (*Meyer et al.,* 1971) (arbitrary scaling). **(c)** Spectra of Europa. The leading-side spectrum is from *Spencer et al.* (1995). The trailing-side ground-based spectrophotometry measurements are from *Johnson* (1970) (○) and *Wamsteker* (1972) (△), both normalized at 0.56 μm to *McFadden et al.* (1980) (□) [as presented in *Calvin et al.* (1995)]. Note the possible S_4 absorption at ~0.53 μm. The UV spectrum is from *Noll et al.* (1995) (—) and the ◇ points are from *Nelson et al.* (1987). **(d)** The trailing-to-leading-side spectral albedo ratio, normalized at 0.56 μm, is from *Johnson* (1970) (○) and compared to S_μ and S_4 absorption using *Meyer et al.*'s (1971, 1972) absorption profiles for two optical depths.

side SO_2 abundance and the small absorption cross section at this wavelength.

Hendrix et al. (2002, 2008) found that SO_2 linearly correlates with Europa's hydrate, consistent with Europa's radiolytic sulfur cycle. At 80% hydrate concentration, approximately the maximum concentration of hydrate on the trailing side, they found that the SO_2 equivalent column density for a one-way path was 6×10^{17} cm^{-2} based on laboratory reflection measurements by *Sack et al.* (1992). We estimate the photon path length using the continuum reflectivity of R ~ 0.2 at 0.28 μm (Fig. 9) and an average hydrate grain size of 10 μm (section 2.2.4), giving L ~ 120 μm (see section 1.4). The column density of H_2O molecules along this path is ~3.6×10^{20} cm^{-2}, so the SO_2 molar density at high hydrate concentration is about 0.2%. The form of SO_2 is not known. It could exist as a dispersed molecular component in solid solution, as SO_2 inclusions, as hydrates or clusters of $(SO_2)_m(H_2O)_n$ (*Schriver et al.*, 1988; *Schriver-Mazzuoli et al.*, 2003a), or in a mixed clathrate (*Hand et al.*, 2006). Using IUE data, *Ockert et al.* (1987) and *Nelson et al.* (1987) found that the signature was unchanged within 20% over the eight-year span of observations. Combining IUE and Galileo results, *Domingue and Hendrix* (2005) showed that the feature was stable over the 1978–1996 period.

A strong feature in Io's spectra is the SO_2 $\nu_1 + \nu_3$ combination band at 4.07 μm, and has been suggested to be the absorber for a 4-μm band in Ganymede and Callisto's spectra (*McCord et al.*, 1997, 1998a; *Hibbitts et al.*, 2000) as well as for Europa's (*Smythe et al.*, 1998; *Hansen and McCord*, 2008). However, there is an inconsistency between Europa's UV and IR measurements if one assumes they are both due to SO_2. A simple radiative model shows that the relative reflectance decrease at low reflectance is $\Delta R/R = -(2/3)Df\alpha$ where f is the molar concentration of SO_2 and α is its absorption coefficient. Using the above UV derived SO_2 abundance of 0.2% and obtaining a using SO_2 optical constants (*Schmitt et al.*, 1994), then an IR band depth of only 0.1% is expected. If the 4-μm band depth of ~5–10% (*Hansen and McCord*, 2008) were due to SO_2 then the UV feature would be much stronger than observed. The same also applies to Ganymede, which does not exhibit an UV SO_2 feature (*Noll et al.*, 1996; *Hendrix et al.*, 1999), but does show a 4-μm band with a ~4% average absorption depth (*McCord et al.*, 1998a). It may be that the IR absorption strength of SO_2 is greatly enhanced in H_2O ice, or perhaps another absorber is responsible for this band.

Sulfur dioxide, while it could be an outgassing product from the interior (*Noll et al.*, 1996), is more likely a product of surficial chemical reactions. *Lane et al.* (1981) suggested SO_2 formation through implantation of energetic sulfur ions into the icy surface, but this mechanism does not directly produce SO_2 molecules with high efficiency (*Strazzulla et al.*, 2007). Irradiation of ice-coated sulfurous residues is also inefficient in producing SO_2 (*Gomis and Strazzulla*, 2008). SO_2 is produced by irradiation of H_2S:H_2O ices (*Moore et al.*, 2007); however, H_2S is not yet observed on Europa (see section 2.9.4). Sulfur dioxide can also be formed in the decomposition of sulfates [see review by *Johnson et al.* (2004) and the measurements by *Moore et al.* (2007)] and would be an equilibrium species in Europa's radiolytic sulfur cycle. The newly formed SO_2 is then itself photolytically and radiolytically decomposed (*Schriver-Mazzuoli et al.*, 2003b; *Moore et al.*, 2007) with a lifetime of a few years in Europa's top 100 μm (*Moore et al.*, 2007). The decomposition products reform sulfate in a repeating cycle, and a radiolytic equilibrium SO_2 abundance is formed that is sensitive to the total sulfur to water ratio (*Moore et al.*, 2007). For $[\Sigma S]/[H_2O]$ = 1/3, 1/10, and 1/30, the equilibrium SO_2 fractional abundance is found to be $[SO_2]/[H_2O]$ = 8%, 0.35%, and 0.003%, using $[SO_2]/[\Sigma S]$ values obtained by *Moore et al.* (2007) with proton irradiation. The intermediate case simultaneously mimics Europa's UV-derived SO_2 abundance and Europa's hydrate abundance {$[\Sigma S]/[H_2O]$ ~ 0.1 (*Carlson et al.*, 2005a)} within factors of ~3. The predominance of SO_2 on the trailing side suggests sulfur ion implantation as the source but SO_2 can also be produced by radiolysis of sulfate salts. Uniform outgassing of SO_2 over Europa's surface is ruled out by the paucity of SO_2 on the leading side.

2.4. Sulfur Allotropes

Native sulfur has long been a candidate for Europa's dark material. Multispectral photometry of the Galilean satellites (see Fig. 13c) showed an absorption band shortward of 0.5 μm that was suggestive of sulfur compounds. *Johnson and McCord* (1971) noted that polysulfides may be responsible for this UV downturn in the satellite spectra, although other candidates were noted, including radiation-damaged ice and iron compounds (see section 2.9.2). *Wamsteker* (1972) suggested that sulfur may be a common absorber on the icy Galilean satellites and particularly on Io (*Wamsteker et al.*, 1973). Recognition that the Io plasma torus contains sulfur ions that diffuse away from Io's orbit and can strike Europa prompted *Eviatar et al.* (1981) to suggest this exogenic source. Shortly thereafter, *Lane et al.* (1981) found evidence for sulfur implantation through IUE observations of SO_2 on Europa's trailing side.

Johnson et al. (1983) analyzed Voyager color maps and found three prominent spectral units, one being bright in orange light, and two darker regions, one with UV reflectance lower than the other. The low UV reflectance units were predominately on the trailing side and are responsible for the higher trailing-side UV absorption compared to the leading side discovered in groundbased spectrophotometry (Fig. 13c,d). These authors reiterated the suggestion of ion implantation of sulfur, suggesting that differential contamination by this element could produce the UV differences. *Nelson et al.* (1986) analyzed these data and found that all spectral units showed gradual changes with longitude. They suggested that all dark units are related, including the brown lineae, spots, and the two types of mottled terrain (UV dark and UV bright). The UV absorption feature was found to occur on both hemispheres with a distribution that followed a cosine dependence, with its minimum on the leading apex of orbital motion and maximum in the trailing antapex. *Nelson et al.* (1986) considered this to be a magnetospheric effect, and favored sulfur ion implantation on the trailing and leading sides, the latter due to high-energy ions, with modifications by sputtering redistribution and gardening. Grain size variations were also considered. *McEwen* (1986) similarly found a global pattern suggesting exogenic control, again possibly from sulfur-ion implantation, sputtering, and impact gardening, that could produce compositional and/or grain size gradients and the observed pattern. *Johnson et al.* (1988) and *Pospieszalska and Johnson* (1989) calculated the longitudinal implantation flux of plasma and high-energy sulfur ions and found good agreement with the Voyager UV images. *Spencer et al.* (1995) obtained spectra of Europa's leading side (see Fig. 13c) and found excellent fits, particularly at the 0.5 μm band edge, using α-sulfur, SO_2, and proton-irradiated NaSH. They also found a slope change at about 0.38 μm that may be part of the UV absorption feature prominent on the trailing side (Fig. 13c) and also present on the leading side. Since the 0.5-μm absorption edge is evident on both the trailing and leading sides, *Spencer et al.* (1995) suggested that the sulfur could be endogenic.

Early objections for sulfur being on Io were due to the temperature shift of sulfur's absorption band, shifting the band edge to shorter wavelengths and producing a white compound at low temperatures rather than yellow [see review of Io's surface composition and sulfur properties by *Carlson et al.* (2007)]. However, it was shown by *Steudel et al.* (1986) that UV radiation produced yellow polymeric sulfur at satellite temperatures, with the band edge shifting back to 0.5 μm (*Hapke and Graham,* 1989) and thus sulfur is consistent with Io's spectrum (and Europa's). Other radiation alteration effects may be operative. High-energy irradiation can open the S_8 ring and the products can recombine to produce longer chains and large rings, an allotrope known as polymeric sulfur and denoted S_μ or S_∞ (*Steudel and Eckert,* 2003). Solid polymeric sulfur is yellow or brown (*Meyer et al.,* 1972; *Steudel et al.,* 1986) and can be formed by irradiation and by quenching the liquid. It exhibits an absorption band starting at ~0.4 μm, with a band or shoulder at 0.36 μm seen in liquid sulfur, quenched sulfur, and annealed sulfur photolysis products (*Meyer et al.,* 1971; *Nishijima et al.,* 1976; *Eckert and Steudal,* 2003). It is possible that Europa's UV feature is due to absorption by polymeric sulfur. Figure 13d shows the absorption by this allotrope based on the relative absorption found by *Meyer et al.* (1971), and provides a good fit to *Johnson*'s (1970) trailing-to-leading-side spectral ratios. The preponderance of this feature on the trailing side may be due to enhanced irradiation on this hemisphere by energetic electrons (*Paranicas et al.,* 2001), in addition to there being simply more sulfur.

We estimate the differential amount of S_μ between the leading and trailing hemispheres using *Johnson*'s (1970) measurements (Fig. 13d) with typical S_μ absorption values from *Hosokawa et al.* (1994), finding that the concentration difference, relative to water, is $[S_\mu]/[H_2O] \sim 2 \times 10^{-4}$. This can be regarded as a lower limit for the differential and total sulfur abundance. The absolute abundance is difficult to estimate and will require radiative transfer calculations and spectral fitting with differing proportions of S_8 and S_μ.

Another sulfur feature, found in spectra of Io and perhaps present in Europa's spectrum, is the 0.53 μm band of tetrasulfur, S_4. This molecule is formed during radiolytic decomposition and has two absorption bands in the visible, the stronger one at ~0.53 μm and the weaker isomer band at ~0.63 μm. S_4 is produced in X-ray irradiation (equivalent to electron and proton irradiation) along with the less stable S_3 molecule (*Nelson et al.,* 1990). Tetrasulfur is produced photolytically (*Meyer and Stroyer-Hansen,* 1972) and in electric discharges (*Hopkins et al.,* 1973). The lifetime of photolytic S_4 is ~60 h at 171 K but the molecule may be more stable at europan temperatures (see below). Its possible presence on Europa is hinted at in Fig. 13c, which shows an inflection at ~0.53 μm in the trailing-side spectrophotometry, and in Fig. 13d, where a minimum is seen at the same wavelength. Although *Johnson and McCord* (1970) and *Johnson* (1971) noted no appreciable dip between 0.5 and 0.6 μm, *Wamsteker* (1972) did consider this a possible europan feature in his spectrophotometric data, as did *Nelson and Hapke* (1978) in theirs, and they identified the stronger ionian feature with S_4. *McFadden et al.* (1980) thought this a possible feature too, but not necessarily the same as Io's due to differing widths and positions. The absorption strength measurements of S_4 (*Meyer et al.,* 1972; *Krasnopolsky,* 1987; *Billmers and Smith,* 1991) are too discrepant to make a meaningful estimate of its abundance, but if this molecule is indeed present on Europa it must be continuously produced as its lifetime is probably less than 2 months, based on its apparent lifetime in Io plume ejecta (*Carlson et al.,* 2007). $S_3 + S_5$ are also produced by irradiation, and S_3 has a band at ~0.4 μm that is ~10 times stronger than the S_4 band, but such a feature is not apparent in Io or Europa spectra, probably due to the greater instability of S_3 (*Hopkins et al.,* 1973). There are other possibilities for the 0.53-μm feature, including ferric iron as suggested for Io by *Nash and Fanale* (1977), but the accompanying 0.8-μm Fe^{+++} band is not apparent in Europa's spectrum (see section 2.9.5).

As illustrated in sulfur reflectance and absorbance data (Figs. 13a,b), polymeric sulfur absorption can extend to near-IR wavelengths. An IR leading-trailing-side effect was observed by *Pollack et al.* (1978), who noted a lower albedo at ~1 μm for the trailing side compared to the leading hemisphere. This effect is seen in Galileo NIMS spectra, where the material that is dark in the visible region is absorbing for wavelengths up to 1 μm. This spectral region includes S_μ and sulfur dangling bond absorption (*Hosokawa et al.,* 1994). Figure 11d shows a map of this absorption at 0.7 μm and the strong correlation with the UV absorption feature (Fig. 11c).

Europa's sulfurous matter can arise from the S ion plasma influx, meteoritic and cometary infall, and from endogenic sources, such as SO_2 outgassing, emplacement of oceanic material containing sulfates and sulfides, or from existing sulfurous impurities in the ice crust. The amount brought in by ion impact can be estimated by assuming that the ice crust was emplaced 50 m.y. ago, and that the current plasma flux (Table 1) has been constant over that period. For the synchronous rotation case and with little plasma diversion, the antapex would receive enough sulfur (12 g) to form a 6-cm layer of elemental sulfur, or a 66-cm layer of sulfate hydrate. Gardening will reduce this to a concentration of several percent relative to H_2O, and nonsynchronous rotation will tend to average the distribution, just as it does for the accumulated ionizing dose (see Fig. 2). The implanted sulfur ions do not create elemental sulfur directly, but instead sulfuric acid (which is colorless) is preferentially produced (*Strazzulla et al.,* 2007), followed by decomposition of the sulfate to produce SO_2 (also colorless) and elemental sulfur (see *Johnson et al.,* 2004). Continued radiolysis reforms sulfate in timescales of a few years in a continuing radiolytic sulfur cycle (*Carlson et al.,* 1999b, 2002, 2005a; *Moore et al.,* 2007). Any endogenic sulfurous material will also be rapidly incorporated into the cycle. No matter what the source, the most stable and abundant form is expected to be sulfate, followed by SO_2 and elemental sulfur.

The distribution of the presumed sulfur is variegated and is controlled by the implantation pattern, the emplacement pattern of any endogenic sources, gardening, the random distribution of impacts, and geological processes such as tectonism, mass wasting, brine mobilization, and the production of lag deposits by near-surface heating events. Polymeric sulfur, if the cause of the UV and near-IR absorption, correlates with Europa's hydrated material (Fig. 11) and offers potential for deciphering surface history and source mechanisms.

2.5. Molecular Oxygen

In March and April of 1993 and 1994, telescopic observations uncovered two previously unseen absorption features on Jupiter's moon Ganymede at 0.6275 and 0.5773 μm (*Spencer et al.,* 1995). Subsequent searches identified weaker features on both Europa and Callisto (*Spencer and Calvin,* 2002) and the 0.5773-μm feature may be present on a Kuiper belt object (*Tegler et al.,* 2007). These bands were identified with condensed molecular oxygen based on the precise central wavelength match, the band asymmetry, and the relative strength of the two strongest visible absorptions (*Spencer et al.,* 1995; *Calvin et al.,* 1996). These absorptions arise from the simultaneous excitation of interacting pairs of O_2 molecules, producing the transitions $O_2(^1\Delta_g) + O_2(^1\Delta_g) \leftarrow O_2(^3\Sigma_g^-) + O_2(^3\Sigma_g^-)$, with the observed features being the first two members of the vibrational-excitation progression (*Landau et al.,* 1962). Higher excitation bands occur at shorter wavelengths in laboratory spectra but are too weak to have been observed from the satellites. The band shapes are similar to those of oxygen in the liquid or solid γ phase or a similar dense state.

The electronic absorption spectrum of condensed oxygen has been studied since at least the 1930s. General information on the absorption band positions, shape, and strength in both condensed and high-pressure oxygen can be found in the literature (*Landau et al.,* 1962; *Dianov-Klokov,* 1964, 1966; *Findlay,* 1970; *Greenblatt et al.,* 1990; see review by *Cooper et al.,* 2003a). There have been a number of attempts to model and explain the double electronic transition that results in the blue color of the condensed phase, which are now generally interpreted as collision-induced transitions (*Robinson,* 1967; *Blickensderfer and Ewing,* 1969a,b; *Tsai and Robinson,* 1969; *Long and Kearns,* 1973; *Long and Ewing,* 1973). The four strongest bands in the liquid are also observed in all three crystallographic phases of solid O_2 (*Landau et al.,* 1962; *Dianov-Klokov,* 1966). The central wavelengths of the visible features are similar in the high-pressure gas, the liquid, and the solid; however, there is a marked increase in the band asymmetry in the liquid and solid phases. Absorption band strengths of these collision-induced bands are strongly dependent on the density of O_2 as seen in transmission measurements of liquid O_2 at varying temperatures (W. Calvin et al., unpublished data). The absorption strength of the O_2 IR atmospheric system ($^1\Delta_g \leftarrow {}^3\Sigma_g^-$) is also enhanced by collisions and weak bands at 1.25 and 1.06 μm (v' = 0 and 1, respectively) may be expected to be present in Europa's spectrum, but H_2O has bands at these positions so the O_2 component has not been distinguished. *Cooper et al.* (2003a) has reviewed the spectroscopy of O_2 relevant to its presence and observation on the icy satellites.

This existence of surficial O_2 is surprising, since at the surface temperatures and pressures of the Galilean satellites, O_2 is expected to evaporate into the atmosphere immediately. The oxygen must be trapped in the ice and, since these bands are intrinsically weak, significant quantities of O_2 must be present in concentrated form. Assuming a linear dependence between the observed band strength and density, *Johnson et al.* (2003) estimated Europa's molar O_2 abundance, relative to H_2O, to be 0.017–0.17%. However, since the absorption occurs by excitation of two interacting molecules, a quadratic dependence may be appropriate if O_2 molecules are *randomly* distributed, for which a relative abundance of 1.2–4.6% is estimated (*Hand et al.,* 2006). If O_2 molecules are clustered, trapped within defects for example, then the same absorption could be obtained from a smaller average concentration. In contrast to Ganymede's O_2 features, which are strongest on the trailing hemisphere, Europa's bands are distributed over all longitudes with no significant longitudinal variation (*Spencer and Calvin,* 2002).

O_2 is a product of energetic plasma bombardment of the water ice surface (*Calvin et al.,* 1996; *Johnson and Jesser,* 1997; *Johnson and Quickenden,* 1997; *Sieger et al.,* 1998) and its existence on the surfaces of the icy Galilean satellites provided one of the first definitive indications that radiolysis is an important process on these bodies. Although O_2 formation in irradiated ice has been studied for almost a century, detailed reaction mechanisms have been difficult to determine. In a now-classic experiment, *Reimann et al.* (1984) showed that O_2 production in freshly deposited ice samples increased as the ice became increasing altered by the radiation and appeared to be correlated with the loss of H_2. Both the O_2 and H_2 were formed and trapped at depth in these samples, and not just at the surface layers, with an efficiency that increases with the ice temperature, as recently confirmed by *Teolis et al.* (2005b). They found, using 100-keV Ar^+ bombardment (approximately simulating Europa's oxygen and sulfur ion irradiation), that O_2 concentrations of up to 30% are produced in H_2O ice at 130 K. At 100 K the concentration is less, be-ing <5%. The fluence level to reach equilibrium O_2 concentrations was found by *Teolis et al.* (2005b) to be 1–3 × 10^{15} ions cm^{-2}, which is reached in 3–10 yr on Europa when considering only the energetic O and S ion fluxes (*Cooper et al.,* 2001). Electron and proton irradiation will shorten these timescales. Gardening and condensation of Europa's tenuous water vapor atmosphere will bury the O_2-rich ice, building up an oxidant-rich regolith.

O_2 production in bombarded water ice appears to require the formation of a precursor molecule that is stable at 120 K on timescales on the order of at least 1 h (*Sieger et al.,* 1998; *Orlando and Sieger,* 2003). The precursor is associated with the loss of H_2 (*Reimann et al.,* 1984) and has been suggested to be O trapped in a defect as O-H_2O (*Khriachtchev et al.,* 1997; *Johnson et al.,* 2003, 2005; *Orlando and Sieger,* 2003), a form that can also convert to peroxide under irradiation, or the precursor could be peroxide itself. *Cooper et al.* (2003b) have argued that production of O_2 dimers occurs by a different mechanism, through photolysis or radiolysis of peroxide aggregates, although *Loeffler and Baragiola* (2005) have concluded that H_2O_2 exists on Europa as a solution, rather than as discrete aggregates (see Fig. 15 and section 2.6).

The location and form of the trapped oxygen has been a subject of some debate (*Calvin and Spencer,* 1997; *Vidal et al.,* 1997; *Baragiola and Bahr,* 1998; *Baragiola et al.,* 1999b; *Johnson,* 1999; *Cooper et al.,* 2003b), but recent work (*Teolis et al.,* 2005a,b; *Loeffler et al.,* 2006d) supports the model suggested by *Johnson and Jesser* (1997) that O_2 trapping is facilitated by defect formation by radiation. The O_2 absorption bands are probably associated with multiple O_2 molecules trapped together in inclusions formed from multiple defects. These are often called microbubbles and are formed in a variety of irradiated materials. As discussed in the H_2O section, radiation amorphizes the ice, producing defects — pores and voids — within which the radiolytically produced O_2 can accumulate. Radiation also compacts the ice (*Palumbo,* 2006; *Raut et al.,* 2007), closing the pores and trapping the oxygen at densities of up to ~30% (*Teolis et al.,* 2005b). *Teolis et al.* (2006) simulated an icy satellite with a tenuous H_2O atmosphere that is recondensing on the surface as it is being irradiated. A column of radiolytic O_2 is built up as the compacted ice column replaces the sputtered H_2O, grows, and caps the O_2 below, protecting it from diffusive or sputtering loss. This interesting mechanism holds promise in understanding the O_2 content on the jovian satellites.

Radiation compaction occurs for both ion and electron irradiation and the trapping sites may be similar to clathrate structures (*Grieves and Orlando,* 2005; *Hand et al.,* 2006). Mixed clathrates have been suggested by *Hand et al.* (2006) for Europa's crust that could contain caged O_2 molecules in single and double occupancy, the latter potentially producing the observed O_2 collision pair features. Radiolysis of oxygen-rich ice will enhance the production of H_2O_2 and will produce ozone (O_3) and other reactive oxidants such as hydroperoxyl (HO_2) and hydrogen trioxide (HO_3) (*Cooper et al.,* 2006). Ozone, not yet observed on Europa, is discussed in section 2.9.3.

2.6. Hydrogen Peroxide

H_2O_2 was predicted to be on Europa as a radiolysis product of ice by *Johnson and Quickenden* (1997), and was soon thereafter discovered (*Carlson et al.,* 1999a) on the surface using Galileo IR and UV spectroscopy (Fig. 14). Peroxide is quickly dissociated by UV radiation so the existence and abundance of H_2O_2 dramatically indicated rapid radiolytic production and the importance of radiation effects on the Galilean satellites, especially Europa. This oxidizing molecule, along with O_2, is also of astrobiological interest (see chapter by Hand et al.).

The initial discovery made use of spectra of Europa's leading, and iciest, hemisphere, but subsequent analysis indicates that H_2O_2 is also present on the trailing hemisphere with comparable band depths (*Hansen and McCord,* 2008). The leading side's average molar abundance is about 0.13% (*Carlson et al.,* 1999a) but there are variations over the surface. The leading-side distribution of H_2O_2 (*Carlson,* 2004) seemed to correlate with the abundance of CO_2, possibly related to the production mechanism, discussed below. There may be temporal variations as well. *Domingue and Hendrix* (2005) have noted UV spectral slope decreases and darkening on the leading side, antijovian quadrant that occurred between the IUE era (1979–1984) and Galileo's (1995–1996). Similar changes were observed earlier on both the leading and trailing hemisphere (*Domingue and Lane,* 1998). It was suggested by *Domingue and Hendrix* (2005) that temporal variability of the space environment (e.g., Jupiter's magnetosphere or gardening rates) may have depleted the amount of H_2O_2.

An ionizing particle with energy in the keV to MeV range produces a track of lower-energy secondary electrons as it passes through the H_2O ice. Hydrogen peroxide forms along these tracks through electron-induced dissociation and ionization of H_2O molecules, producing H + OH radicals. The OH radicals can combine: $OH + OH \rightarrow H_2O_2$. Peroxide can be destroyed by electrons produced in a subsequent ionizing particle's avalanche and by UV and visible radia-

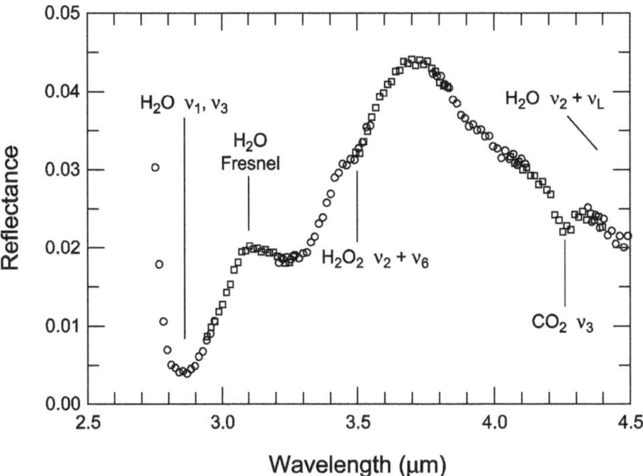

Fig. 14. Spectrum of Europa's leading side in the 2.5- to 4.5-μm region. The hydrogen peroxide combination band is identified at 3.5 μm and shown with continuum removed in Fig. 15. Ultraviolet spectra of Europa corroborate this identification. The CO_2 asymmetric stretch fundamental is at 4.25 ± 0.01 μm and consistent with the 4.258-μm band centers found for Ganymede and Callisto and many saturnian satellites (2.7). The broad, structureless Fresnel reflection peak of H_2O is indicative of amorphous ice. L denotes libration.

tion. The diurnally averaged H_2O_2 photolysis rate, ignoring any reduction from the cage effect, is about 10^{-6} s^{-1} when surface reflection is accounted for, giving a photodissociation lifetime of about 10 d. The cage effect (*Franck and Rabinowitsch,* 1934) can inhibit photodissociation rates by factors of 3 or more (*Schriever et al.,* 1991). The half-life of H_2O_2 in water ice under energetic proton irradiation is ~3 d in the top 100 μm (*Hudson and Moore,* 2006; see also *Loeffler et al.,* 2006c). Thus, to first order we can ignore photodissociation and can consider only particle-induced production and destruction processes. Equilibrium concentrations will be achieved when these rates are equal. Only ions, mainly protons, are thought to strike Europa's leading side (*Paranicas et al.,* 2001) and experimental ion irradiation results are consistent with the concentrations measured on that hemisphere. For example, ices irradiated by 100-keV protons show equilibrium peroxide concentrations of 0.14% and 0.1% for 80 K and 120 K ice samples, respectively (*Loeffler et al.,* 2006b). These results are also in general agreement with the ion irradiation measurements of *Moore and Hudson* (2000) and *Gomis et al.* (2004a,b), although *Moore and Hudson* (2000) found that the addition of electron scavengers such CO_2 increased the yield at Europa-like temperatures and may explain the apparent association of H_2O_2 with CO_2 as noted above.

H_2O_2 production by high-energy electron irradiation has been studied by *Hand* (2007), who found an inverse temperature dependency for the equilibrium values, similar to that for ion irradiation. He found equilibrium values that were about a factor of 3 smaller than for ion irradiation, being 0.04%, 0.03%, and 0.01% for 10-keV electron irradiation of 80 K, 100 K, and 120 K ice, respectively.

It is of interest to know the equilibration time. For both electron and ion irradiation, equilibrium concentrations are achieved at a fluence of about 10^{19} eV cm^{-2} (*Hand,* 2007; *Gomis et al.,* 2004a,b; *Loeffler et al.,* 2006b). Using *Cooper et al.*'s (2001) energy fluxes, an equilibration time on the trailing side is about 2 days, and roughly a week on the leading side. These times pertain to the mean penetration depths of the ionizing particle.

The position and shape of Europa's H_2O_2 band is indicative of the state of H_2O_2 in the ice matrix. *Loeffler and Baragiola* (2005) found different spectral properties for pure crystalline H_2O_2, aggregates of H_2O_2 within the ice, and H_2O_2 dispersed throughout the ice. The absorbance profiles for peroxide produced by proton and electron bombardment are similar to the shape and position observed for Europa's H_2O_2 (Fig. 15), and both are the same as spectra of H_2O_2 dispersed in ice (*Loeffler and Baragiola,* 2005). Using phase diagram information, *Loeffler and Baragiola* (2005) inferred that the dispersed H_2O_2 exists as individual trimers, H_2O-H_2O_2-H_2O, randomly dispersed throughout the H_2O matrix. At high temperatures (150 K), these trimers will precipitate as inclusions of the dihydrate $H_2O_2{\bullet}2H_2O$, rather than as aggregates of pure H_2O_2. This seems to contradict the O_2 production scheme proposed by *Cooper et al.* (2003b).

2.7. Carbon Dioxide

Carbon dioxide is a common constituent on icy satellites and was first observed on Ganymede and Callisto by Galileo's NIMS using the strong ν_3 absorption band at

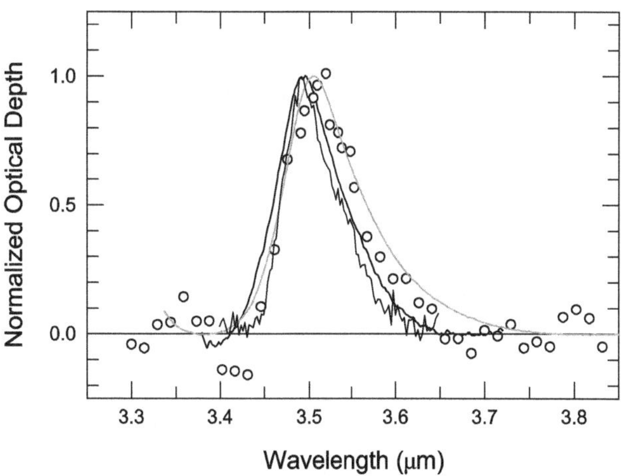

Fig. 15. Comparison of the absorbance profile of H_2O_2 on Europa with that produced in the laboratory by irradiation of H_2O ice. NIMS data are shown as circles and have the underlying continuum subtracted. The thick black line corresponds to H_2O_2 produced by irradiation of ice at 80 K [curve g of Fig. 1 in *Loeffler and Baragiola* (2005)]. Spectra of electron irradiated H_2O at 80 K (thin black line) show a similar profile (*Hand,* 2007). The NIMS profile and the two irradiation profiles are all similar to spectra of H_2O_2 dispersed in water ice (gray line; curve f of Fig. 1 in *Loeffler and Baragiola* (2005)].

4.3 μm (*Carlson et al.*, 1996; *McCord et al.*, 1997, 1998a; *Carlson,* 1999; *Hibbitts et al*., 2000, 2002, 2003). Evidence for the presence of CO_2 on Europa was subsequently obtained for the leading hemisphere using distant observations to avoid radiation noise (*Smythe et al.*, 1998); confirming previous hints of structure near 4.3 μm (*McCord et al.*, 1998a). A recent analysis of NIMS data provides evidence for CO_2 on the trailing side, with large band depths and a nonuniform distribution (*Hansen and McCord,* 2008).

Europa's 2.5–4.5-μm spectrum (Fig. 14) shows absorption due to the CO_2 ν_3 transition. The equivalent width of this feature was used by *Hand et al.* (2007) to estimate the leading-side CO_2 concentration at 360 ppmv. The position of the band is similar to that observed for the other icy Galilean satellites and for many CO_2-rich regions on the saturnian satellites (*Clark et al.*, 2005a,b, 2008b). The leading-side distribution, while obtained with poor spatial resolution (*Carlson,* 2001), shows a correlation with visibly dark material, similar to results for Ganymede and Callisto, and similar to the trailing-side analysis by *Hansen and McCord* (2008).

The existence of CO_2 on the icy satellites presents two puzzles. First, what is its origin — exogenic or endogenic — and second, how can it be stable at the temperatures of the Galilean satellites? The origin may be endogenic. It is now known that condensed CO_2 is widespread in the solar system, being present in icy satellites of Saturn (*Buratti et al.,* 2005; *Clark et al.,* 2005a,b, 2008b; *Brown et al.,* 2006a,b; *Waite et al.,* 2006), three uranian satellites (*Grundy et al.,* 2003, 2006), and Neptune's Triton (*Cruikshank et al.,* 1993). It is also present on Mars and in comets and interstellar grains (*Gibb et al.,* 2004). It is attractive, then, to assume a common, endogenic source for the icy satellites, with the CO_2 being a degassing product of primordial or internally produced volatiles. Since interstellar grains contain CO_2, and are likely to be a component of the initial solar nebula, a common and ubiquitous source of primordial CO_2 seems probable. There is observational evidence from the extent of sublimational erosion that Callisto's CO_2 is internally derived (*Moore et al.,* 1999). Furthermore, Cassini's mass spectrometer has directly measured CO_2 venting from Enceladus's interior through the south pole "tiger stripes" vents (*Waite et al.,* 2006), perhaps due to clathrate exsolvation. At Enceladus, and in other bodies, CO_2 could also be produced through high-temperature oxidation of organic compounds in the interior (see chapter by Zolotov and Kargel for details). Endogenic primordial or internally generated CO_2 seems plausible and likely for icy satellites.

On the other hand, surface photolysis and radiolysis may play a role in producing CO_2 from carbonaceous material. Mapping of Callisto's surficial CO_2 by *Hibbitts et al.* (2000) shows that CO_2 forms a trailing-side "bulls-eye," suggesting an influence by Jupiter's rotating magnetic field and magnetosphere. The correlation with dark material suggests that a carbon-containing material is involved. At Europa, oxidized carbon compounds (CO_3^{2-}, HCO_3^-, carbonates) or organic molecules are possibilities, and meteorites and micrometeoroids are certain sources of carbonaceous material (section 1.2, Table 1, section 2.9.1). Carbon ions are present in the jovian magnetosphere but the fluxes are too low to be of compositional significance (*Cohen et al.,* 2001).

When carbonaceous grains with water ice mantles are irradiated, CO and CO_2 molecules are formed at the interface (*Mennella et al.,* 2004; *Gomis and Strazzulla,* 2005; *Raut et al.,* 2005). Equilibrium surface densities at the interface are ~0.3–6 × 10^{15} CO_2 molecules cm^{-2}, equivalent to a few monolayers or less. *Gomis and Strazzulla* (2005) have argued that ices with small amounts of ion-irradiated submicrometer carbonaceous particles can contribute sufficient CO_2 to produce the absorptions observed on the Galilean satellites. The energy dose to reach equilibrium is ~100 eV/16-amu (*Gomis and Strazzulla,* 2005), corresponding to times of about 300 and 150 yr on Europa's leading and trailing hemisphere, respectively. Note that the equilibrium concentration is independent of the ionizing flux since production and destruction rates are equal. Consequently, similar processes may be occurring on other solar system bodies by cosmic-ray irradiation, albeit with longer timescales. An objection to this mechanism is that the resulting CO_2 band position [2339 cm^{-1} (*Mennella et al.,* 2004)] is inconsistent with that observed from the Galilean satellites and elsewhere (see discussion of CO_2 stability below).

If the carbonaceous material is mixed in ice on molecular scales, rather than as grains, access to carbon is greatly enhanced, and CO_2 production is increased. Examples of ion-irradiated mixed ices containing hydrocarbons are given in *Moore et al.* (1996), *Palumbo* (1997), *Moore and Hudson* (1998), and *Palumbo et al.* (1998), and for electron irradiation, by *Hand* (2007). Figure 16 shows a typical example of the production curves for the radiolysis products of a mixed ice of isobutane (C_4H_{10}) and H_2O. Carbon dioxide and carbon monoxide are generally produced, irrespective of the original C and O mixture (*Palumbo,* 1997).

Carbon dioxide on Europa's surface, whether it is endogenic, photolytic, or radiolytic, can be destroyed to yield carbonic acid (H_2CO_3) and carbon monoxide (*Moore et al.,* 1991; *Moore and Khanna,* 1991; *DelloRusso et al.,* 1993; *Brucato et al.,* 1997; *Gerakines et al.,* 2000) and these molecules will then be radiolyzed back to CO_2. The major species in this cycle is CO_2, with H_2CO_3 being present at relative molar density of $[H_2CO_3]/[CO_2]$ ~ 1.7% along with some CO (*Carlson et al.,* 2005b; *Hand,* 2007).

An O–H stretching band of H_2CO_3 was suggested by *Hage et al.* (1998) as a candidate for Ganymede and Callisto's 3.8-μm absorption feature. The relative intensities of the 3.8-μm and CO_2 ν_3 bands are about the same for the Ganymede and Callisto observations and the laboratory radiolysis measurements, lending support for Hage et al.'s suggestion. The H_2CO_3 band is not evident in Fig. 14 but may be just at the limit of detection. The structure of carbonic acid is $(HO)_2CO$ (*Moore and Khanna,* 1991), with the three oxygen atoms bonded to the carbon atom, so there are no C-H bonds. However, the C-H stretching band from alcohols should be present, but these bands are intrinsically

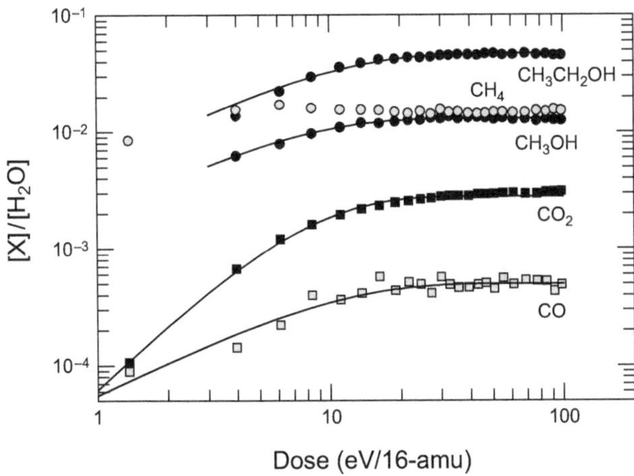

Fig. 16. Production curves for the radiolysis products of a 1:10 C_4H_{10}:H_2O ice at 80 K, illustrating radiolytic equilibrium. The major products are the alcohols methanol (CH_3OH) and ethanol (CH_3CH_2OH) and methane, with minor amounts of CO_2 and CO. The initial slope of the CO_2 curve indicates that this molecule is a secondary product, derived from a radiolytically-generated precursor. Ice containing methanol, when mildly heated in vacuum, forms clathrates that can also trap CO_2 and other molecules. This is one possible explanation for the existence and stability of trapped CO_2 on icy satellites.

weak compared to the CO_2 band, and can be a factor of 10 less intense than the CO_2 feature in spectra of radiolyzed ice containing hydrocarbons (*Moore et al.,* 1996; *Moore and Hudson,* 1998). Carbon monoxide (section 2.9.1) is also produced, but this volatile molecule will rapidly diffuse out of the ice and escape into the atmosphere. The apparent loss of CO_2 and possibly CO, the former indicated by the existence of a CO_2 atmosphere on Callisto (*Carlson,* 1999), implies that carbon is being lost from Europa. The widespread occurrence of CO_2 on icy satellites throughout the solar system suggests that surficial CO_2 is not a transient feature, so there must be continual replenishment of carbon atoms. The crudely estimated CO_2 loss rate from Callisto of 6×10^6 cm^{-2} s^{-1} is close to the meteoritic C atom input rate of 1.2×10^6 cm^{-2} s^{-1} (Table 1) and suggestive of meteoritic infall as the continuing supply.

Carbon dioxide ice is highly volatile, with a vapor pressure of 0.1 mbar at 125 K, so it cannot exist as CO_2 ice on the Galilean satellites. In dilute H_2O ice solutions, CO_2 is more stable, but much of it is lost when amorphous to cubic transitions take place (*Sandford and Allamandola,* 1990). The timescale for these phase changes on Europa is ~10 yr (see section 2.1). Some CO_2 still remains trapped on laboratory timescales, but stability over longer periods probably requires a more effective trapping mechanism. Early suggestions for trapping CO_2 were as clathrates, as fluid inclusions, or trapped in radiation-induced defects and voids (*Carlson et al.,* 1996; *McCord et al.,* 1998a). Many silicates effectively trap CO_2 and some show weaker bands at ~4 μm [see the *Clark et al.* (2007) spectral library] that could be related to the 4-μm feature present on Ganymede, Callisto, and perhaps Europa. The position of the CO_2 band is potentially diagnostic of the trapping mechanism. For example, the band position of CO_2 ice is 2343 cm^{-1}, and is 2340 cm^{-1} for CO_2 molecules in H_2O ice (*Sandford and Allamandola,* 1990), whereas the band lies at 2349 ± 2.2 cm^{-1} for the Galilean satellites (less accurately known for Europa). This position is very close to the 1-0 R-branch rotation-vibration line of the gas, arising from excitation of CO_2 in the nonrotating state (J" = 0). This suggests that the molecules' environment restricts their rotation but otherwise the molecules are nearly free. Hindered rotation of the J = 1 excited state probably broadens the line.

A specific trapping mechanism has not been identified and is the subject of ongoing work. Numerous prior studies of numerous CO_2-rich minerals and ices have shown band positions occurring over a 30-cm^{-1} span, but with few candidates in the interval of interest (2349 ± 2.2 cm^{-1}). Some zeolites provide adequate positions at room temperature, but few studies at relevant temperatures have been performed. *Hibbitts and Szanyi* (2007) have investigated physisorption on various minerals at icy satellite temperatures. Of the samples considered, Ca-montmorillonite, serpentine, goethite, and palagonite, CO_2 adsorbed only on Ca-montmorillonite where it remained for times longer than tens of minutes. The position of the asymmetric stretching band of the adsorbed CO_2 at 125 K was found to be 4.26 μm (2347 cm^{-1}), in excellent agreement with the Galilean satellites' CO_2 band position. While this remains a plausible candidate, the long-term stability needs to be studied. Alteration of the physisorbing medium by irradiation may be an important component in trapping CO_2 on the Galilean satellites (*Hibbitts and Szanyi,* 2007).

Trapping of CO_2 in clathrate structures is also a possibility (*Carlson and Hand,* 2006; *Hand et al.,* 2006). In particular, when hydrocarbons in ice are irradiated, methanol (CH_3OH) is produced, along with CO_2 and other species (Fig. 16). Water ice that contains methanol, when heated to about 120–125 K, transforms to a clathrate at vacuum (*Blake et al.,* 1991), and if other molecules are present, they can be incorporated into the clathrate cages, forming a mixed clathrate. The band position of enclathrated CO_2 shifts to 2346 cm^{-1} (*Blake et al.,* 1991; *Fleyfel and Devlin,* 1991), noticed earlier in mixed ice experiments (*Sandford and Allamandola,* 1990; *Ehrenfreund et al.,* 1999). Interactions between CO_2 and CH_3OH may also produce shifts in position (*Chaban et al.,* 2007). The band center for clathrate-trapped CO_2 is encouragingly close to observed values. Further work on clathrate production, physisorption on mineral grains, and other CO_2 production and trapping mechanisms is needed in order to understand the origin of CO_2 on Europa and icy satellites in general.

2.8. Sodium and Potassium Compounds

Although sodium and potassium compounds have not been directly detected on Europa's surface, these atoms have been observed in the atmosphere and are thought to be introduced there by sputtering (see chapters by McGrath et al. and Johnson et al.). The initial source of these atoms could be exogenic or endogenic, from Io's plasma torus or alkali-containing salts from Europa's putative ocean, therefore their detection is of great interest.

Potassium has been identified once (*Brown,* 2001), whereas sodium has been observed a few times (*Brown,* 1999, 2001; *Leblanc et al.,* 2002, 2005). An emission peak reported by *Porco et al.* (2003) when Europa was in Jupiter's shadow has been also associated with sodium (*Cassidy et al.,* 2008). Modeling of these observations has given insight into the transport and loss of sodium, as described in more detail in the chapter by McGrath et al. Such processes apply to other trace species ejected from Europa's surface, but not yet detected in the gas phase. There are also reports of alkali pickup ions observed near Europa (Table 2).

Alkali elements on Europa's surface could exist in the form of salts, hydroxides (*Johnson,* 2001), solvated ions, or neutral atoms (*Yakshinskiy and Madey,* 2001). Sputtering simulations showed that there may be considerable redistribution of sodium and potassium across the surface of Europa, which would also be the case for other trace species that are sputtered. Therefore, a significant fraction of the observed gas-phase alkalis are from previously sputtered atoms, adsorbed on the surface and again ejected into the atmosphere. Most are not directly ejected from an intrinsic salt mineral. Observations and modeling suggest that the dark region on the trailing hemisphere of Europa is likely the initial source (*Leblanc et al.,* 2005; *Cassidy et al.,* 2008).

Assuming sodium in an ice matrix is carried off with the sputtered water products, theoretical models indicated that the average molar surface sodium concentration, relative to H_2O, was ~0.5–1% (*Johnson et al.,* 2002; *Leblanc et al.,* 2002) and that the present escape rate is 5×10^6–12×10^6 Na atoms cm^{-2} s^{-1} (*Leblanc et al.,* 2002, 2005). *Cipriani et al.* (2008) have considered models that include orbital or temporal nonuniformities in the magnetospheric flux that sputters the Na, and infer Na escape rates of 3×10^6 atoms cm^{-2} s^{-1} or greater. If one assumes a plasma implantation reduction factor (*Saur et al.,* 1998) of $r_{Europa} = 0.1$ to 0.2 (i.e., implantation flux reduced by a factor of either 10 or 5), then the sodium implantation rate could not account for the loss rate implied by the observations (*Johnson,* 2000; *Leblanc et al.,* 2002). That implantation is not the principal source was also suggested by the different Na/K ratios at Europa (25 ± 3) and at Io (10 ± 3) (*Brown,* 2001; *Johnson et al.,* 2002; *Leblanc et al.,* 2005). A meteoroid source would have a Na/K ratio of ~13, whereas *Zolotov and Shock*'s (2001) model predicts that Europa's ocean will have a ratio of ~14–19. Since freezing of upwelling oceanic water may increase the Na/K ratio, material originating from Europa's ocean could produce a ratio at the surface consistent with the observations (*Zolotov and Shock,* 2001). Note that Na is predicted by geochemical models to be a major constituent of Europa's ocean, but that the K abundance depends strongly of the H_2 and CO_2 fugacities (*Zolotov,* 2008).

A critical assumption of the above discussion is that plasma is significantly deflected at Europa, implying that plasma implantation provides too little Na to satisfy the inferred ejection rate, and that an endogenic source is necessary. While there does seems to be diversion of flow on the jovian face of Europa (*Paterson et al.,* 1999b), the plasma reduction factor could be much less than considered above. *Ip* (1996) derives a reduction factor of $r_{Europa} = 0.8$, while analysis of combined magnetic field and energetic particle data (*Paranicas et al.,* 2002; *Volwerk et al.,* 2004) yields $r_{Europa} = 0.9$. In these high values of r_{Europa}, the rate of Na plasma input (>5.3×10^6 Na atoms cm^{-2} s^{-1}, Table 1) could balance Europa's loss rate noted above and an endogenic source is not be required. The source location, the trailing-side dark material, is also consistent with plasma implantation, which occurs primarily on the trailing hemisphere. In addition, the surface density of Na can be independently estimated by assuming S and Na ions are supplied in the flux ratio from Table 1 and using *Carlson et al.*'s (2005a) estimate of $[S]/[H_2O] \sim 0.1$ on Europa's trailing side (S as sulfate), giving a surface concentration of $[Na]/[H_2O] \geq 0.4\%$. This is comparable to the 0.5–1% estimate for Na from the model fits noted above. Better models and understanding of Europa's plasma interaction are being developed (see chapter by Kivelson et al.) and soon we may be able to better estimate endogenic and exogenic contributions.

The different values of [Na]/[K] for Io and Europa are often used to state that Europa's alkalis are endogenic. However, fractionation occurs in the various sputtering processes and in the escape process, and sputtering of ice mixtures has not been adequately studied. As an example, for low cascade density sputtering (*Johnson,* 1990), Europa's distant Na/K ratio can be estimated as follows: If gardening is more rapid than loss by sputtering, then Europa's Na and K surface densities would be expected to be in the same ratio as Io's escaping Na and K, consistently estimated to be 10 ± 3 (*Brown,* 2001) and 10 ± 5 (*McGrath et al.,* 2004), although higher values (20–30) have been reported (*Trafton,* 1981). The low cascade density sputtering yield Y for a minor species with concentration c in ice is $Y = c(U_{H_2O}/U) Y_{H_2O}$, where U is a characteristic cohesion energy (*Johnson,* 1990). The escape fraction is proportional to $U/(½Mv_{esc}^2)$, with M being the mass and v_{esc} the escape velocity. The escape fraction is therefore independent of the cohesion energy and fractionation will occur inversely proportional to mass. For an iogenic input with [Na]/[K] = 5–15, we find Europa's escape flux ratio to be 8.5 to 25.5. This upper value is consistent with Brown's observations. Other sputtering processes (e.g., thermal spikes) can exhibit different fractionation effects. We conclude that definitive conclusions about the endogenic or exogenic nature of Europa's alkalis cannot be made at the present.

2.9. Other Possible or Suggested Species

2.9.1. Carbon compounds.

2.9.1.1. Carbon-oxygen compounds: Carbonate minerals have been suggested as possible candidates for Ganymede and Callisto's 4-μm feature (*Johnson et al.,* 2004) but further analysis is needed to establish the potential of CO_3^{2-}

compounds to explain the icy satellite's feature. Since carbon dioxide is present on Europa, one might expect related compounds such as carbon monoxide to be present. CO could be an endogenic outgassing product, or produced by radiolysis of CO_2 (*Moore et al.,* 1991; *Moore and Khanna,* 1991; *Brucato et al.,* 1997; *Strazzulla et al.,* 2005) or hydrocarbons (*Strazzulla et al.,* 1995; *Moore et al.,* 1996; *Moore and Hudson,* 1998; *Strazzulla and Moroz,* 2005). CO_2 is a back reaction product of CO radiolysis, and at equilibrium, the ratio of CO to CO_2 for various initial compounds is about unity (*Strazzulla et al.,* 1995). CO is very volatile, so any outgassed or radiolytically produced CO may rapidly escape, resulting in a low surface concentration. We can place a conservative limit on the leading-side concentration by assuming a band depth of less than 5% for possible CO 1-0 band absorption at 4.67 μm, giving a limit to the equivalent width of ~0.6 cm^{-1} (using NIMS' 0.025-μm resolution). Using the band strength determined by *Gerakines et al.* (1995), we find $<5 \times 10^{16}$ CO molecules cm^{-2} in the photon path L. Using a reflectance of 2% to derive the MOPL = αL, and using the absorption coefficient for H_2O, we determine L and the corresponding column density of H_2O molecules. This procedure gives $[CO]/[H_2O] <$ 250 ppm. Limits on the trailing-side abundance are higher due to higher noise levels in the available spectra.

CO radiolysis produces carbon suboxide, C_3O_2, and formaldehyde, H_2CO. In light of the CO upper limit, neither is expected in great quantities, but H_2CO merits discussion as it can be derived from hydrocarbons (*Moore and Hudson,* 1998) and is of astrobiological interest (*Chyba,* 2000; chapter by Hand et al.). The positions of formaldehyde's absorption bands depend on the matrix and the H_2CO concentration; for low concentrations in H_2O the CH symmetric and asymmetric stretch bands occur at 2785 and 2853 cm^{-1} (3.59 and 3.51 μm), respectively (*va der Zwet et al.,* 1985). The asymmetric stretch band is close to Europa's H_2O_2 feature, but the symmetric stretch band is isolated and one can estimate an upper limit for the leading side. Assuming a detectable integrated band intensity that is one-third that for Europa's H_2O_2 band (Fig. 14), and using the H_2CO band strength (*va der Zwet et al.,* 1985) relative to that of CH_4 (*d'Hendecourt and Allamandola,* 1986) with *Loeffler et al.*'s (2006b) value for H_2O_2, we find a limit of $[H_2CO]/[H_2O] <$ 0.25% for the leading side.

2.9.1.2. Hydrocarbons: The aliphatic hydrocarbon CH stretching band has been reported in Callisto spectra (*McCord et al.,* 1997) but is not yet observed on Europa. An upper limit to the hydrocarbon content is given in the chapter by Hand et al. as the number of methylene groups per water molecule, $[CH_2]/[H_2O]$, being $<1.5 \times 10^{-3}$. The number expected for meteoritic infall and burial by gardening, if all the C atoms are associated with methylene, is about 500 ppm or about one-third of this upper limit (see Table 1).

2.9.2. Nitrogen compounds.

2.9.2.1. Nitriles: A feature at 4.57 μm is found in spectra of both Ganymede and Callisto (*McCord et al.,* 1997, 1998a) and was suggested to be due to absorption by molecules containing the nitrile, C≡N, group. The fundamental C≡N stretch band occurs in the 4.4- to 4.9-μm region (*Bernstein et al.,* 1997; *Lowenthal et al.,* 2002; *Raunier et al.,* 2003; *Gerakines et al.,* 2004). The well-known interstellar XCN feature [now known to be due to OCN^- (*Hudson et al.,* 2001)] occurs at 4.62 μm. Polymeric HCN and some tholins possess absorption features in the 4.53- to 4.61-μm region that may be consistent with the suggestion of cyanogens (*Cruikshank et al.,* 1991; *Khare et al.,* 1994). Whichever molecule(s) the Ganymede and Callisto features are due to, it is reasonable to expect it to also be present on Europa if the material emanates from meteoritic and cometary infall. However, Europa's surface age is young compared to Ganymede and Callisto, so the amount that has accumulated is less by perhaps by a factor of about 100. Therefore it is not surprising that the abundance is too low to be detected on Europa.

2.9.2.2. Ammonia: Noting possible features at 1.8, 2.1, 2.2, and 2.3 μm in spectra obtained by R. Clark in 1980, *Brown et al.* (1988) subsequently searched the 2.0- to 2.5-μm region using improved instrumentation. They found weak absorptions at 2.2 and 2.3 μm in 1985, but not in measurements a year later. They considered sources of systematic error and noted that, while transient $NH_3 \bullet H_2O$ features could explain the features, they were skeptical of invoking transient phenomena as an explanation. *Calvin et al.* (1995) obtained telescopic spectra in 1989 and found none of the reported 2.1-, 2.2-, and 2.3-μm features. They noted that incomplete cancellation of the stellar Brackett line, or interference by atmospheric gases, could introduce spurious features at these wavelengths. They also noted that the 1.8-μm feature in the 1980 measurements could be real, or could be an artifact caused by incomplete atmospheric water vapor removal. Ammonia is rapidly destroyed by radiation, forming N_2 and H_2 (*Loeffler et al.,* 2006a), both of which will rapidly escape the surface.

2.9.2.3. Amides: *Dalton et al.* (2003) examined IR methods for detection capabilities of biological molecules and found two possible features (at 2.05 and 2.17 μm) in NIMS spectra that could be due N–H-related combination bands of an amide. While intriguing, these features are close to the noise level, and there are other possibilities for lines at these positions. If amide functional groups can survive the radiation environment at Europa they could serve as potential biomarkers; however, more laboratory work is needed to assess their stability and to investigate other, abiotic sources for these features.

2.9.3. Ozone and oxygen compounds. There have been considerable experimental advances in understanding the production of ozone in ice. *Baragiola et al.* (1999a) used the Hartley UV absorption band to study ozone production in $H_2O{:}O_2$ ices and successfully described ozone production using the classic Chapman atmospheric reaction scheme. Recent laboratory work by *Cooper et al.* (2008), using thin films of oxygen aggregates in water ice, has confirmed the suggestion that ozone is created through the irradiation of O_2 inclusions, as suggested earlier by *Johnson and Jesser* (1997). Starting only with water ice at satellite-like temperatures, *Teolis et al.* (2006) formed O_2 and O_3 using heavy ion bombardment (section 2.5). The key element of this experiment was the simultaneous deposition of H_2O to simulate recondensation of a satellite's tenuous H_2O atmosphere. They found that the temperature, radiation, and recondensation conditions for O_3 production are met on the jovian and saturnian satellites. Therefore it is not surprising that O_3 is found on Ganymede (*Noll et al.,* 1996; *Hendrix et al.,* 1999) and might be expected on Europa and Callisto, where O_2 is also present [but in lesser amounts; see *Spencer and Calvin* (2002)]. This molecule is, surprisingly, also present on the surfaces of saturnian satellites (*Noll et al.,* 1997), where a less-intense magnetosphere would be expected to produce less precursor O_2.

Radiolysis of $H_2O + O_2$ produces H_2O_2, O_3, OH, HO_2, and HO_3 at exposures levels of only a few eV/16-amu. As the temperature of the radiolyzed $H_2O + O_2$ ices is raised, the amount of O_3 was found to decrease (*Cooper et al.,* 2008). This inverse temperature dependence may partly explain the existence of O_3 on the saturnian satellites even though their radiation environment is less intense than that of the jovian system. Ozone is also produced by irradiation of $H_2O + CO_2$ ice mixtures (*Strazzulla et al.,* 2005).

Ganymede's O_3 abundance in the observed column was measured by *Noll et al.* (1996) and *Hendrix et al.* (1999) as 4.5×10^{16} and 4.6×10^{16} cm^{-2}, respectively. Europa does not show an ozone-like feature (*Hendrix et al.,* 1998) in the UV, so the upper limit must be some fraction of Ganymede's abundance. Searches for a 4.8-μm O_3 band on the Galilean satellites with NIMS data have so far been unsuccessful, in part due to the low intrinsic absorption strength for this transition (P. Cooper, personal communication, 2007).

Oxides and hydroxides related to hydrated salts are discussed in section 2.2.5. The hydroxyl radical, OH, and other radicals are expected to be present in trace amounts (*Johnson and Quickenden,* 1997) but are below the detection limits of current datasets.

2.9.4. Sulfur compounds.

2.9.4.1. Hydrosulfides and hydrogen sulfide: A feature at 3.88 μm is observed on Ganymede and Callisto and suggested to arise from absorption by the SH stretch transition of a hydrosulfide (*McCord et al.,* 1997, 1998a). If this feature were present on Europa, then H_2S would be a candidate since the ocean, if nonoxidizing, could be rich in hydrogen sulfide. However, such an ocean arises if it remains reducing because of little hydrogen loss, implying weak communication to the surface and minimal surface emplacement of H_2S. Radiolytic H_2S was predicted in sulfate hydrate-rich regions at relative molar concentrations of 2×10^{-4} (*Carlson et al.,* 2002), neglecting UV destruction. However, Europa's oxidizing surface can effectively destroy H_2S. H_2S is very volatile and escapes from H_2O:H_2S ices at 132 K in laboratory timescales (*Moore et al.,* 2007). The band position for H_2S in H_2O ice is 3.90 μm. This species is not observed on Europa, but no limits have been established since the band strength in H_2O ice is not established. If it is present, it might be associated with sulfur and preferentially distributed on the trailing side.

2.9.4.2. Sulfanes: Sulfur chains with an H atom attached to the ends are termed sulfanes. Since polymeric sulfur containing long sulfur chains is possibly present on Europa's trailing hemisphere, then sulfanes might be present there as well. Sulfanes will exhibit the SH stretch band, but the exact positions and strengths are not known.

2.9.4.3. Polysulfur oxides: *Spencer et al.* (1995) noted the likelihood of sulfur compounds on Europa and followed *Sill and Clark*'s (1982) suggestion for Io, suggesting polymers of disulfur monoxide (S_2O) with SO_2 as coloring agents. Pure, concentrated S_2O can polymerize to form polysulfur oxides (PSO); however, the red color associated with S_2O and PSOs on Io is now thought to be due to the decomposition product S_4 (*Steudel and Steudel,* 2004). If S_2O or polysulfur oxides are present in significant amounts, they will produce overtone absorption bands at approximately 4.5 μm (*Baklouti et al.,* 2008). These bands are not seen in Europa's spectrum but upper limits to S_2O and PSO abundances cannot be calculated because the band strengths are not yet known.

2.9.4.4. Sulfurous acid: Aqueous solutions of SO_2 produces H^+ and HSO_3^-, an acidic solution termed sulfurous acid. In contrast to H_2SO_4 and H_2CO_3, the free molecule H_2SO_3 has not been found in nature. *Voegele et al.* (2004) calculated that, while pure H_2SO_3 molecules might have a long lifetime, the presence of H_2O greatly decreases the stability of the molecule. At 100 K, and with a 1:2 H_2SO_3:H_2O ratio, the computed lifetime is about 1 d. The molecule has not been found following proton implantation in frozen SO_2 (*Garozzo et al.,* 2008). H_2SO_3 molecules will persist on Europa in vanishingly small quantities, far below the detection limit.

2.9.5. Iron compounds. Ferric (Fe^{3+}) iron oxides and oxyhydroxides show strong absorption in the UV and blue, extending to about 0.55 μm, due to charge transfer between oxygen and Fe^{3+}. Electronic crystal field transitions in the Fe^{3+} ion produce a series of absorption bands. An excellent description of these electronic transitions is given in *Burns* (1993). For oxides and oxyhydroxides these band positions (and assignments) are 0.43 μm ($^6A_{1g} \rightarrow {}^4A_{1g}, {}^4E_g$), 0.63 μm ($^6A_{1g} \rightarrow {}^4T_{2g}$), and 0.87 μm ($^6A_{1g} \rightarrow {}^4T_{1g}$) (*Morris et al.,* 1985). Additionally, there is a double electron transition at 0.5 μm ($2{}^6A_{1g} \rightarrow 2{}^4T_{1g}$) (*Morris et al.,* 1997). The positions of these bands are independent of temperature, at least for hematite, but the bands sharpen somewhat as the temperature is reduced (*Morris et al.,* 1997). Even though these four bands correspond to spin-forbidden transitions, they all can be quite strong. The combined effect of these absorptions is to produce a very reddish spectrum for ferric oxides and oxyhydroxides. The positions and strengths of electronic transitions depend on the site symmetry and the mineral composition. For example, ferric sulfate nonahydrate, $Fe_2(SO_4)_3 \cdot 9H_2O$– coquimbite, exhibits bands at 0.42–0.43 μm ($^6A_{1g} \rightarrow {}^4A_{1g}, {}^4E_g$), 0.56 μm ($^6A_{1g} \rightarrow {}^4T_{2g}$), and 0.78 μm ($^6A_{1g} \rightarrow {}^4T_{1g}$), giving this mineral a gray-green tinge (*Rossman,* 1975).

Ferrous iron (Fe^{2+}) has spin-allowed crystal field transitions that produce the well known iron silicate absorption bands in the 1- and 2-μm regions, as well as spin-forbidden transitions at 0.45 μm ($^5T_{2g} \rightarrow {}^3T_{2g}$), 0.51 μm ($^5T_{2g} \rightarrow {}^1A_{1g}$), and 0.55 μm ($^5T_{2g} \rightarrow {}^3T_{1g}$) (*Hunt and Salisbury,* 1970; *Hunt et al.,* 1971b).

Iron compounds have been suggested for the Galilean satellites in general, based on the general red appearance of their spectra and those of iron compounds (*Johnson,* 1970; *Johnson and McCord,* 1970; *Sill and Clark,* 1982; *McEwen,* 1986), but no features are found in groundbased 0.9- to 1.1-μm spectra (*Johnson and Pilcher,* 1977) and none have been found in later telescopic or spacecraft spectra, indicating a lack of ferrous silicates.

Given the oxidizing nature of Europa's surface, ferric oxides and oxyhydroxides are possible candidates. *Clark* (1980) noted 0.87-μm absorption features in spectra of Europa, Ganymede, and Callisto that he suggested were due to absorption by ferric iron compounds. The absorption was weakest for Europa, for which the band was about 4% deep and had a full width at half maximum of ~0.1 μm. This position and width is consistent with Fe^{3+} absorption, although there were questions about the accuracy of the lunar reference spectrum in this spectral region (*Clark,* 1980). This band is not apparent in NIMS spectra at about the 2% level. Nanophase ferric oxides, and such minerals as ferrihydrite, goethite, hematite, maghemite, and lepidocrocite, have been extensively studied for Mars application and representative spectra are given in *Morris et al.* (1993, 2000). Ferric oxides show the 0.87-μm band, sometimes centered at longer wavelengths. Just shortward of this band, generally at about 0.7–0.8 μm, the reflectivity drops quickly due to the stronger 0.63-μm band, and at ~0.6 μm, another shoulder is reached, below which the reflectivity drops even more quickly. In many cases this second shoulder is absent, and the break-point in the spectrum is between 0.7 and 0.8 μm. At low temperature, the 0.6-μm shoulder becomes a local maximum, with the minimum at 0.63 μm. Thus, the turnover point in the spectrum occurs at a longer wavelength compared to either the leading or trailing-side spectra of Europa (Fig. 13c). We conclude that there is no strong evidence for ferric oxide or hydroxide compounds at observable amounts on Europa.

Ferric sulfate is worthy of consideration, since concentrated ferric sulfate aqueous solutions possess low freezing points, possibly as low as ~200 K (*Chevrier and Altheide,* 2008). *Nash and Fanale* (1977) obtained spectra of $Fe_2(SO_4)_3 \cdot xH_2O$ [see also *Rossman* (1975), *Crowley et al.* (2003), *Clark et al.*'s (2007) splib06 spectral library] and they suggested that the absorption bands at 0.5 and 0.8 μm might be responsible for two bands in Io's spectra. Although it is now thought that the 0.53-μm band in Io's spectrum is due to S_4 (*Carlson et al.,* 2007) and tetrasulfur is possible on Europa, it is appropriate to consider ferric sulfate. The depth of Europa's potential 0.53-μm band is about 4% (Fig. 13D) and *Rossman's* (1975) absorbance measurement suggests equal strengths for the 0.56- and 0.78-μm features, whereas *Nash and Fanale*'s (1977) measurements indicate that the 0.8-μm band should be about five times stronger in band depth than the 0.5-μm feature. However, NIMS spectra show less than a 2% feature at 0.8 μm. In addition, Europa's 0.4-μm reflectance (Fig. 13c) of ~40% is much higher than would be expected based on *Nash and Fanale*'s (1977) and *Crowley et al.*'s (2003) reflectance spectra, both of which show reflectance values at 0.4 μm of ~2%. While we cannot rule out some iron compounds on Europa's surface, any amount present is constrained to trace levels.

3. FUTURE MEASUREMENTS AND EXPERIMENTS

There are several avenues to pursue to improve understanding of Europa's surface composition in general and the hydrate question in particular. First, we need better understanding of the exogenic and endogenic sources. Source rates for material emanating from the small outer satellites and Ganymede and Callisto need quantification. We also need to understand Europa's plasma interaction better in order to verify (or refute) Europa as an endogenic source of the escaping Na and K. This understanding may be developed by further analysis theoretical modeling and analysis of Galileo particles and magnetic field data, as well as continued observations and analysis of Europa's extended Na and K atmosphere.

Additionally, the age and evolution of the surface needs to be understood. Does the ~50-m.y. cratering age imply that entire crust was replenished then, or has the ice that forms the crust remained in the crust since formation of the satellite? How pure was the pristine crust?

Telescopic measurements, both from the ground and by orbiting telescopes, can continue near-IR searches for hydrated salt features as well as searching for hydroxide features to identify any metals and nonmetals. Low light level luminescence emissions from Europa observed when in eclipse may allow detection of species such as magnesium, as emissions from neutral Mg are produced by ion bombardment (e.g., the Mg I lines $^3P^0$–^3S λλ5167 Å–5183 Å and ^1S–$^1P^0$ λ2852 Å as observed by irradiating $MgSO_4 \cdot 7H_2O$ (*Nash and Fanale,* 1977)). Telescopic observations of the atmosphere near Europa, as well as of the neutral torus, may provide identifications of new species that may have originated from the surface (e.g., Cl). Mapping of Europa's surficial O_2 may lead to better understanding of the oxygen production and trapping mechanism. For that matter, except for the observations by *Spencer et al.* (1995), there have been no spatially resolved imaging spectroscopy measurements of Europa in the visible region. Such measurements could provide better evidence for S_4 and possibly other absorbers and coloring agents. Further analysis of existing UV and IR data may be fruitful; a search for metal hydroxide features is one area of interest.

More laboratory spectra with extended wavelength coverage are needed for candidate materials. Many of the published spectra of candidate icy satellite surface materials were recorded during investigations of the properties and chemistry of the interstellar medium and comets. In many cases, such published spectra are from thin films (0.1–10 μm) measured in the mid-IR region at a few kelvins in temperature. However, the data needed for quantitative comparisons to spectra from Europa and other icy satellites often require different experimental conditions. Laboratory measurements for all candidate materials are needed, ideally with the following characteristics: First, spectra must be recorded in either diffuse reflectance or presented in terms of optical constants — quantities that enable quantitative abundance modeling. For diffuse reflectance, different grain sizes must be investigated. Second, laboratory measurements are needed across the full spectral range of typical and future spacecraft instruments (0.1–7 μm). Third, measurements must be conducted with samples sufficiently thick to produce useful absorption features, particularly the shapes and strengths of weak bands, such as in the near-IR region. Fourth, measurements must be temperature-appropriate for the bodies of interest, since many candidate compounds (especially ices) display marked spectral changes with temperature. For the specific case of Europa, laboratory measurements in the 80–130-K range are critical.

There have few laboratory studies of how radiation-induced dehydration, amorphization, and decomposition can alter hydrate band positions and profiles. Such measurements should be carried out for both ion and electron irradiation, as collision effects may be different for the two cases. The vapor pressure of radiation-produced amorphous ice is needed to accurately describe Europa's sublimational atmosphere and transport. Continued study of the radiolytic production and trapping of O_2, CO_2, and other volatiles will tell us how much O_2 is stored in Europa's regolith and if clathrates are important features of the icy shell.

The logical next stage of exploration would be an orbiter around Europa, instrumented with remote-sensing IR and UV spectrometers. The IR spectrometer should have a longer wavelength capability than NIMS had, in order to measure important functional groups such as C=O, C=C, N=O, NO_2, CO_2^-, and C-O, and the deformation transitions of N-H, NH_2, and O-H. Classes of molecules such as carbonates, carboxylic acid and their salts, nitrates, alcohols, esters, ketones, aldehydes, aromatic molecules, amines, and amides would be observable in this extended wavelength interval. Sufficient spatial resolution to acquire spectra of distinct geological regions will be critical to identify and discriminate surface materials and assessing their roles in evolution of the surface. Europa's sputter atmosphere provides another means of indirectly determining surface composition. An orbiting ion mass spectrometer or rotational line microwave spectrometer could respectively measure the ion or neutral component and provide new insights into Europa's surface composition. Concepts for a future landed mission to Europa require knowledge of the surface composition and structure. While many data sets are now available, there remains much work to be done in order to further explore this fascinating world and determine its habitability.

Acknowledgments. We thank R. Clark and C. Hibbits for their important and insightful reviews, and W. Grundy, K. Hand, M. Loeffler, W. McKinnon, U. Raut, J. Shirley, K. Zahnle, and M. Zolotov for their comments, data, and discussions. A portion of the research described in this chapter was carried out at the Jet Propulsion Laboratory, California Institute of Technology, and was done so under a contract with the National Aeronautics and Space Administration.

REFERENCES

Allamandola L. J., Sandford S. A., and Valero G. J. (1988) Photochemical and thermal evolution of interstellar/precometary ice analogs. *Icarus, 76,* 225–252.

Allison M. L. and Clifford S. M. (1987) Ice-covered water volcanism on Ganymede. *J. Geophys. Res., 92,* 7865–7876.

Alvarellos J. L., Zahnle K. J., Dobrovolskis A. R., and Hamill P. (2008) Transfer of mass from Io to Europa and beyond due to cometary impacts. *Icarus, 194,* 636–646, DOI: 10.1016/j.icarus.2007.09.025.

Anders E. and Grevasse N. (1989) Abundances of the elements: Meteoritic and solar. *Geochim. Cosmochim. Acta, 53,* 197–214.

Bagenal F. (1994) Empirical-model of the Io plasma torus — Voyager measurements. *J. Geophys. Res., 99,* 11043–11062.

Baklouti D., Schmitt B., and Brissaud O. (2008) S_2O, polysulfur oxide and sulfur polymer on Io's surface? *Icarus, 194,* 647–659.

Baragiola R. A. (2003) Water ice on outer solar system surfaces: Basic properties and radiation effects. *Planet. Space Sci., 51,* 953–961.

Baragiola R. A. and Bahr D. A. (1998) Laboratory studies of the optical properties and stability of oxygen on Ganymede. *J. Geophys. Res., 103,* 25865–25872.

Baragiola R. A., Atteberry C. L., Bahr D. A., and Jakas M. M. (1999a) Solid-state ozone synthesis by energetic ions. *Nucl. Instr. Meth. Phys. Res. B, 157,* 233–238.

Baragiola R. A., Atteberry C. L., Bahr D. A., and Peters M. (1999b) Comment on "Laboratory studies of the optical properties and stability of oxygen on Ganymede" by Raul A. Baragiola and David A. Bahr — Reply. *J. Geophys. Res., 104,* 14183–14187.

Baragiola R. A., Loeffler M. J., Raut U., Vidal R. A., and Wilson C. D. (2005) Laboratory studies of radiation effects in water ice in the outer solar system. *Rad. Phys. Chem., 72,* 187–191.

Baratta G. A., Leto G., Spinella F., Strazzulla G., and Foti G. (1991) The 3.1 μm feature in ion-irradiated ice. *Astron. Astrophys., 252,* 421–424.

Baratta G. A., Castorina A. C., Leto G., Palumbo M. E., Spinella F., and Strazzulla G. (1994) Ion irradiation experiments relevant to the physics of comets. *Planet. Space Sci., 42,* 759–766.

Bauer W. H. (1964) On the crystal chemistry of salt hydrates II: A neutron diffraction study of $MgSO_4 \cdot 4H_2O$. *Acta Crystallogr., 17,* 863–869.

Bergren M. S., Shuh D., Sceats M. G., and Rice S. A. (1978) The OH stretching region infrared spectra of low density amorphous solid water and polycrystalline ice I_h. *J. Chem. Phys., 69,* 3477–3482.

Bernstein M. P., Sandford S. A., and Allamandola L. J. (1997) The infrared spectra of nitriles and related compounds frozen in Ar and H_2O. *Astrophys. J., 476,* 932–942.

Bertie J. E. and Jacobs S. M. (1977) Far-infrared absorption by ices Ih and Ic at 4.3 K and the powder diffraction pattern of ice I_c. *J. Chem. Phys., 67,* 2445–2448.

Bertie J. E. and Whalley E. (1964) Infrared spectra of ices I_h and I_c in the range 4000 to 350 cm^{-1}. *J. Chem. Phys., 40,* 1637–1645.

Bertie J. E. and Whalley E. (1967) Optical spectra of orientationally disordered crystals. II. Infrared spectra of ice I_h and I_c from 360 to 50 cm^{-1}. *J. Chem Phys., 46,* 1271–1284.

Bertie J. E., Labbe H. J., and Whalley E. (1969) Absorptivity of ice I in the range 4000–30 cm^{-1}. *J. Chem. Phys., 50,* 1271–1284.

Billmers R. I. and Smith A. L. (1991) Ultraviolet-visible absorption spectra of equilibrium sulfur vapor: Molar absorptivity spectra of S_3 and S_4. *J. Phys. Chem., 95,* 4242–4245.

Blake D., Allamandola L., Sandford S., Hudgins D., and Freund F. (1991) Clathrate hydrate formation in amorphous cometary ice analogs in vacuo. *Science, 254,* 548–551.

Blickensderfer R. P. and Ewing G. E. (1969a) Collision-induced absorption spectrum of gaseous oxygen at low temperatures and pressures. II. Simultaneous transitions $^1\Delta_g + {}^1\Delta_g \leftarrow {}^3\Sigma_g^- + {}^3\Sigma_g^-$ and $^1\Delta_g + {}^3\Sigma_g^+ \leftarrow {}^3\Sigma_g^- + {}^3\Sigma_g^-$. *J. Chem. Phys., 51,* 5284–5289.

Blickensderfer R. P. and Ewing G. E. (1969b) Collision-induced absorption spectrum of gaseous oxygen at low temperatures and pressures. I. The $^1\Delta_g - {}^3\Sigma_g$ system. *J. Chem. Phys., 51,* 873–883.

Brown M. E. (1999) Trace elements in the atmosphere of Europa as a probe of surface composition (abstract). *Eos Trans. AGU, 80,* F604.

Brown M. E. (2001) Potassium in Europa's atmosphere. *Icarus, 151,* 190–195.

Brown R. H., Cruikshank D. P., Tokunaga A. T., Smith R. G., and Clark R. N. (1988) Search for volatiles on icy satellites I. Europa. *Icarus, 74,* 262–271.

Brown R. H., Baines K. H., Belluci G., Buratti B. J., Cappaccioni F., et al. (2006a) Observations in the Saturn system during approach and orbital insertion, with Cassini's visual and infrared mapping spectrometer (VIMS). *Astron. Astrophys., 446,* 707–716.

Brown R. H., Clark R. N., Buratti B. J., Cruikshank D. P., Barnes J. W., et al. (2006b) Composition and physical properties of Enceladus' surface. *Science, 311,* 1425–1428.

Brucato J. R., Palumbo M. E., and Strazzulla G. (1997) Carbonic acid by ion implantation in water/carbon dioxide ice mixtures. *Icarus, 125,* 135–144.

Buratti B. J., Cruikshank D. P., Brown R. H., Clark R. N., Bauer J. M., et al. (2005) Cassini infrared and visual mapping spectrometer observations of Iapetus: Detection of CO_2. *Astrophys. J., 622,* L149–L152.

Burchell M. J., Cole M. J., and Ratcliff P. R. (1996) Light flash and ionization from hypervelocity impacts on ice. *Icarus, 122,* 359–365.

Burns J. A. (1968) Jupiter's decametric radio emission and the radiation belts of its galilean satellites. *Science, 159,* 971–972.

Burns R. G. (1993) Origin of electronic spectra of minerals in the visible and near-infrared region. In *Remote Geochemical Analysis* (C. M. Pieters and P. A. J. Englert, eds.), pp. 3–29. Cambridge Univ., Cambridge.

Buxton G. V. (1987) Radiation chemistry of the liquid state, 1, Water and homogeneous aqueous solutions. In *Radiation Chemistry* (M. A. Farhataziz and J. Rodgers, eds.), pp. 321–349. Wiley, New York.

Calvin W. M. and Spencer J. R. (1997) Latitudinal distribution of O_2 on Ganymede: Observations with the Hubble Space Telescope. *Icarus, 130,* 505–516.

Calvin W. M., Clark R. N., Brown R. H., and Spencer J. R. (1995) Spectra of the icy Galilean satellites from 0.2 to 5 µm — A compilation, new observations, and a recent summary. *J. Geophys. Res., 100,* 19041–19048.

Calvin W. M., Johnson R. E., and Spencer J. R. (1996) O_2 on Ganymede: Spectral characteristics and plasma formation mechanisms. *Geophys. Res. Lett., 23,* 673–676.

Carlson R. W. (1999) A tenuous carbon dioxide atmosphere on Jupiter's moon Callisto. *Science, 283,* 820–821.

Carlson R. W. (2001) Spatial distribution of carbon dioxide, hydrogen peroxide, and sulfuric acid on Europa (abstract). *Bull. Am. Astron. Soc., 33,* 1125.

Carlson R. W. (2003) Europa's radiation processed regolith (abstract). *Bull. Am. Astron. Soc., 35,* 17.01.

Carlson R. W. (2004) Distribution of hydrogen peroxide, carbon dioxide, and sulfuric acid in Europa's icy crust. In *Workshop on Europa's Icy Shell: Past, Present and Future,* p. 15. LPI Contribution No. 1195, Lunar and Planetary Institute, Houston.

Carlson R. W. and Hand K. P. (2006) The mystery of carbon dioxide on icy satellites: A mixed clathrate hydrate? (abstract). *Bull. Am. Astron. Soc., 38,* 38.30.13.

Carlson R. W., Weissman P. R., Smythe W. D., Mahoney J. C., Aptaker I., Bailey G., Baines K., Burns R., Carpenter E., Curry K., Danielson G., Encrenaz T., Enmark H., Fanale F., Gram M., Hernandez M., Hickok R., Jenkins G., Johnson T., Jones S., Kieffer H., Labaw C., Lockhart R., Macenka S., Marino J., Masursky H., Matson D., McCord T., Mehaffey K., Ocampo A., Root G., Salazar R., Sevilla D., Sleigh W., Smythe W., Soderblom L., Steimle L., Steinkraus R., Taylor F., and Wilson D. (1992) Near–Infrared mapping spectrometer experiment on Galileo. *Space Sci. Rev., 60,* 457–502.

Carlson R. W., Smythe W., Baines K., Barbinis E., Becker K., Burns R., Calcutt S., Calvin W., Clark R., Danielson G., Davies A., Drossart P., Encrenaz T., Fanale F., Granahan J., Hansen G., Herrera P., Hibbitts C., Hui J., Irwin P., Johnson T., Kamp L., Kieffer H., Leader F., Lellouch E., Lopes-Gautier R., Matson D., McCord T., Mehlman R., Ocampo A., Orton G., Roos-Serote M., Segura M., Shirley J., Soderblom L., Stevenson A., Taylor F., Torson J., Weir A., and Weissman P. (1996) Near-infrared spectroscopy and spectral mapping of Jupiter and the Galilean satellites: Results from Galileo's initial orbit. *Science, 274,* 385–388.

Carlson R. W., Anderson M. S., Johnson R. E., Smythe W. D., Hendrix A. R., Barth C. A., Soderblom L. A., Hansen G. B., McCord T. B., Dalton J. B., Clark R. N., Shirley J. H., Ocampo A. C., and Matson D. L. (1999a) Hydrogen peroxide on the surface of Europa. *Science, 283,* 2062–2064.

Carlson R. W., Johnson R. E., and Anderson M. S. (1999b) Sulfuric acid on Europa and the radiolytic sulfur cycle. *Science, 286,* 97–99.

Carlson R. W., Anderson M. S., Johnson R. E., Schulman M. B., and Yavrouian A. H. (2002) Sulfuric acid production on Europa: The radiolysis of sulfur in water ice. *Icarus, 157,* 456–463.

Carlson R. W., Anderson M. S., Mehlman R., and Johnson R. E. (2005a) Distribution of hydrate on Europa: Further evidence for sulfuric acid hydrate. *Icarus, 177,* 461–471.

Carlson R. W., Hand K. P., Gerakines P. A., Moore M. H., and Hudson R. L. (2005b) Radiolytic production of carbonic acid and applications to Jupiter's icy satellites (abstract). *Bull. Am. Astron. Soc., 37,* 751–752.

Carlson R. W., Kargel J. S., Doute S., Soderblom L. A., and Dalton B. (2007) Io's surface composition. In *Io after Galileo* (R. M. C. Lopes and J. R. Spencer, eds.), pp. 193–229. Springer-Praxis, Chichester.

Cassidy T. A., Johnson R. E., McGrath M. A.., Wong M. C., and Cooper J. E. (2007) The spatial morphology of Europa's near-surface O_2 atmosphere. *Icarus, 191,* 755–764.

Cassidy T. A., Johnson R. E., Geissler P., and Leblanc F. (2008) Simulation of Na D emission near Europa during eclipse. *J. Geophys. Res., 113,* E02005, DOI: 10.1029/2007/je002955.

Chaban G. M., Huo W. M., and Lee T. J. (2002) Theoretical study of infrared and Raman spectra of hydrated magnesium sulfate salts. *J. Chem. Phys., 117,* 2532–2537.

Chaban G. M., Bernstein M. P., and Cruikshank D. P. (2007) Carbon dioxide on planetary bodies: Theoretical and experimental studies of molecular complexes. *Icarus, 187,* 592–599.

Chevrier V. F. and Altheide T. S. (2008) Low temperature aqueous ferric sulfate solutions on the surface of Mars. *Geophys. Res. Lett., 35,* L22101, DOI: 10.1029/2008GL035489.

Christensen P. R., Bandfield J. L., Hamilton V. E., Howard D. A., Lane M. D., Piatek J. L., Russ S. W., and Stefanov W. L. (2000) A thermal emission spectral library of rock-forming minerals. *J. Geophys. Res., 105,* 9735–9740.

Chyba C. F. (2000) Energy for microbial life on Europa. *Nature, 403,* 381–382.

Cipriani F., Leblanc F., Witasse O., and Johnson R. E. (2008) Sodium recycling at Europa: What do we learn from the sodium cloud variability? *Geophys. Res. Lett., 35,* L19201, DOI: 10.1029/2008gl035061.

Clark R. N. (1980) Ganymede, Europa, Callisto, and Saturn's rings — Compositional analysis from reflectance spectroscopy. *Icarus, 44,* 388–409.

Clark R. N. (1981a) Water frost and ice — The near-infrared spectral reflectance 0.65–2.5 microns. *J. Geophys. Res., 86,* 3087–3096.

Clark R. N. (1981b) The spectral reflectance of water-mineral mixtures at low-temperatures. *J. Geophys. Res., 86,* 3074–3086.

Clark R. N. (2004) The surface composition of Europa: Mixed water, hydronium, and hydrogen peroxide ice. In *Workshop on Europa's Icy Shell: Past, Present, and Future,* Abstract #7057. LPI Contribution 1195, Lunar and Planetary Institute, Houston.

Clark R. N. and Lucey P. G. (1984) Spectral properties of ice-particulate mixtures and implications for remote sensing 1. Intimate mixtures. *J. Geophys. Res., 89,* 6341–6348.

Clark R. N. and Roush T. L. (1984) Reflectance spectroscopy: Quantitative analysis techniques for remote sensing applications. *J. Geophys. Res., 89,* 6329–6340.

Clark R. N., Fanale F. P., and Zent A. P. (1983) Frost grain-size metamorphism: Implications for remote-sensing of planetary surfaces. *Icarus, 56,* 233–245.

Clark R. N., King T. V. V., Klejwa M., Swayze G. A., and Vergo N. (1990) High spectral resolution reflectance spectroscopy of minerals. *J. Geophys. Res., 95,* 12653–12680.

Clark R. N., Swayze G. A., Gallagher A., King T. V. V., and Calvin W. M. (1993) *USGS Digital Spectral Library, Version 1: 0.2 to 3.0 μm.* U.S. Geol. Surv. Open File Report 93 592, U.S. Geological Survey, Flagstaff, Arizona. 1340 pp. Available online at http://speclab.cr.usgs.gov.

Clark R. N., Swayze G. A., Wise R., Livo K. E., Hoefen T. M., Kokaly R. F., and Sutley S. J. (2003) *USGS Digital Spectral Library splib05a.* USGS Open File Report 03-395, U.S. Geological Survey, Flagstaff, Arizona.

Clark R. N., Brown R., Baines K., Belluci G., Bibring J.-P., et al. (2005a) Cassini VIMS compositional mapping of surfaces in the Saturn system and the role of water, cyanide compounds and carbon dioxide. *Bull. Am. Astron. Soc., 37,* 705.

Clark R. N., Brown R. H., Jaumann R., Cruikshank D. P., Nelson R. M., Buratti B. J., et al. (2005b) Compositional maps of Saturn's moon Phoebe from imaging spectroscopy. *Nature, 435,* 66–69.

Clark R. N., Swayze G. A., Wise R., Livo E., Hoefen T., Kokaly R., and Sutley S. J. (2007) *USGS Digital Spectral Library splib06a.* Digital Data Series 231, U.S. Geological Survey, Flagstaff, Arizona.

Clark R. N., Curchin J. M., Hoefen T. M., and Swayze G. A. (2008a) Reflectance spectroscopy of organic compounds I: Alkanes. *J. Geophys. Res.–Planets.,* in press.

Clark R. N., Curchin J. M., Jaumann R., Cruikshank D. P., Brown R. H., Hoefen T. M., Stephan K., Moore J. M., Buratti B. J., Baines K. H., Nicholson P. D., and Nelson R. M. (2008b) Compositional mapping of Saturn's satellite Dione with Cassini VIMS and implications of dark material in the Saturn system. *Icarus, 193,* 372–386.

Cloutis E. A., Hawthorne F. C., Mertzman S. A., Krenn K., Craig M. A., Marcino D., Methot M., Strong J., Mustard J. F., Blaney D. L., Bell J. F., and Vilas F. (2006) Detection and discrimination of sulfate minerals using reflectance spectroscopy. *Icarus, 184,* 121–157, DOI: 10.1016/j.icarus.(2006)04.003.

Cohen C. M. S., Stone E. C., and Selesnick R. S. (2001) Energetic ion observations in the middle jovian magnetosphere. *J. Geophys. Res., 106,* 29871–29881.

Collins G. C., Head J. W., Pappalardo R. T., and Spaun N. A. (2000) Evaluation of models for the formation of chaotic terrain on Europa. *J. Geophys. Res., 105,* 1709–1716.

Cooper J. F., Johnson R. E., Mauk B. H., Garrett H. B., and Gehrels N. (2001) Energetic ion and electron irradiation of the icy Galilean satellites. *Icarus, 149,* 133–159.

Cooper P. D., Johnson R. E., and Quickenden T. I. (2003a) Hydrogen peroxide dimers and the production of O_2 in icy satellite surfaces. *Icarus, 166,* 444–446.

Cooper P. D., Johnson R. E., and Quickenden T. I. (2003b) A review of possible optical absorption features of oxygen molecules in icy surfaces of outer solar system bodies. *Planet. Space Sci., 51,* 183–192.

Cooper P. D., Moore M. H., and Hudson R. L. (2006) Infrared detection of HO_2 and HO_3 radicals in water ice. *J. Phys. Chem. A., 110,* 7985–7988.

Cooper P. D., Moore M. H., and Hudson R. L. (2008) Radiation chemistry of $H_2O + O_2$ ices. *Icarus, 194,* 379–388.

Crowley J. K. (1991) Visible and near-infrared (0.4–2.5 μm) reflectance spectra of playa evaporite minerals. *J. Geophys. Res., 96,* 16231–16240.

Crowley J. K., Williams D. E., Hammarstrom J. M., Piatek N., Chou I.-M., and Mars J. C. (2003) Spectral reflectance properties (0.4–2.5 μm) of secondary Fe-oxide, Fe-hydroxide, and Fe-sulfate-hydrate minerals associated with sulphide-bearing mine wastes. *Geochemistry, 3,* 219–228.

Cruikshank D. P., Allamandola L. J., Hartmann W. K., Tholen D. J., Brown R. H., Mathews C. N., and Bell J. F. (1991) Solid C≡N bearing material on outer solar system bodies. *Icarus, 94,* 345–353.

Cruikshank D. P., Roush T. L., Owen T. C., Geballe T. R., Debergh C., Schmitt B., Brown R. H., and Bartholomew M. J. (1993) Ices on the surface of Triton. *Science, 261,* 742–745.

Curtis D. B., Rajaram B., Toon O. B., and Tolbert M. A. (2005) Measurement of the temperature-dependent optical constants of water ice in the 15–200 μm range. *Appl. Optics, 44,* 4102–4118.

Dalton J. B. and Clark R. N. (1998) Laboratory spectra of Europa candidate materials at cryogenic temperatures (abstract). *Bull. Am. Astron. Soc., 30,* 1081.

Dalton J. B. and Clark R. N. (1999) Observational constraints on Europa's surface composition from Galileo NIMS. In *Lunar and Planetary Science XXX,* Abstract #2064. Lunar and Planetary Institute, Houston (CD–ROM).

Dalton J. B. (2000) Constraints on the surface composition of Jupiter's moon Europa based on laboratory and spacecraft data. Ph.D. thesis, University of Colorado, Boulder, 253 pp.

Dalton J. B. (2003) Spectral behavior of hydrated sulfate salts: Implications for Europa mission spectrometer design. *Astrobiology, 3,* 771–784. DOI: 10.1089/153110703322736097.

Dalton J. B. (2007) Linear mixture modeling of Europa's non-ice material using cryogenic laboratory spectroscopy. *Geophys. Res. Lett., 34,* L21205, DOI: 10.1029/2007GL031497.

Dalton J. B., Mogul R., Kagawa H. K., Chan S. L., and Jamieson C. S. (2003) Near-infrared detection of potential evidence for microscopic organisms on Europa. *Astrobiology, 3,* 505–529, DOI: 10.1089/153110703322610618.

Dalton J. B., Prieto-Ballesteros O., Kargel J. S., Jamieson C. S., Jolivet J., and Quinn R. C. (2005) Spectral comparison of heavily hydrated salts with disrupted terrains on Europa. *Icarus, 177,* 472–490.

DellaGuardia R. A. and Johnston F. J. (1980) Radiation-induced reaction of sulfur and water. *Rad. Res., 84,* 259–264.

DelloRusso N., Khanna R. K., and Moore M. H. (1993) Identification and yield of carbonic acid and formaldehyde in irradiated ices. *J. Geophys. Res., 98,* 5505–5510.

d'Hendecourt L. and Allamandola L. (1986) Time dependent chemistry in dense molecular clouds. III. Infrared band cross sections of molecules in the solid state at 10 K. *Astron. Astrophys. Suppl., 64,* 453–467.

Dianov-Klokov V. I. (1964) Absorption spectrum of oxygen at pressures from 2 to 35 atmospheres in the region 12600–3600 Å. *Optika I Spektroskopiya, 16,* 409–416.

Dianov-Klokov V. I. (1966) Absorption spectrum of condensed oxygen in 1.26–0.3 μm region. *Optics and Spectroscopy-USSR, 20,* 530.

Domingue D. and Allamandola L. (2001) Introduction to the special section: Photolysis and radiolysis of outer solar system ices (PROSSI). *J. Geophys. Res.–Planets, 106,* 33273–33273.

Domingue D. and Hendrix A. (2005) A search for temporal variability in the surface chemistry of the icy Galilean satellites. *Icarus, 173,* 50–65.

Domingue D. L. and Lane A. L. (1998) IUE views Europa: Temporal variations in the UV. *Geophys. Res. Lett., 25,* 4421–4424.

Dowell L. G. and Rinfret A. P. (1960) Low-temperature forms of ice as studied by X-ray diffraction. *Nature, 188,* 1144–1148.

Dressler K. and Schnepp O. (1960) Absorption spectra of solid methane, ammonia, and ice in the vacuum ultraviolet. *J. Chem. Phys., 33,* 270–274.

Dybwad J. P. (1971) Radiation effects on silicates (5-keV H^+, D^+, He^+, H^{2+}). *J. Geophys. Res., 76,* 4023–4029.

Eckert B. and Steudal R. (2003) Molecular spectra of sulfur molecules and solid sulfur allotropes. *Topics Curr. Chem., 231,* 31–98.

Ehrenfreund P., Kerkhof O., Schutte W. A., Boogert A. C. A., Gerakines P. A., Dartois E., d'Hendecourt L., Tielens A. G. G. M., van Dishoeck E. F., and Whittet D. C. B. (1999) Laboratory studies of thermally processed H_2O-CH_3OH-CO_2 ice mixtures and their astrophysical implications. *Astron. Astrophys., 350,* 240–253.

Eichhorn K. and Grun E. (1993) High-velocity impacts of dust particles in low-temperature water ice. *Planet. Space Sci., 41,* 429–433.

Encrenaz T. (2008) Water in the solar system. *Annu. Rev. Astron. Astrophys., 46,* 57–87.

Eviatar A., Siscoe G. L., Johnson T. V., and Matson D. L. (1981) Effects of Io ejecta on Europa. *Icarus, 47,* 75–83.

Fagents S. A. (2003) Considerations for effusive cryovolcanism on Europa: The post-Galileo perspective. *J. Geophys. Res., 108(E12),* 5139.

Fagents S. A., Greeley R., Sullivan R. J., Pappalardo R. T., and Prockter L. M. (2000) Cryomagmatic mechanisms for the formation of Rhadamanthys linea, triple band margins, and other low-albedo features on Europa. *Icarus, 144,* 54–88.

Fanale F. P., Granahan J. C., Greeley R., Pappalardo R., Head J., Shirley J., Carlson R., Hendrix A., Moore J., McCord T. B., and Belton M. (2000) Tyre and Pwyll: Galileo orbital remote sensing of mineralogy versus morphology at two selected sites on Europa. *J. Geophys. Res., 105,* 22647–22655.

Feldman P. D., Ake T. B., Berman A. F., Moos H. W., Sahnow D. J., Strobel D. F., Weaver H. A., and Young P. R. (2001) Detection of chlorine ions in the Far Ultraviolet Explorer spectrum of the Io plasma torus. *Astrophys. J. Lett., 554,* L123–L126.

Feldman P. D., Strobel D. F., Moos H. W., and Weaver H. A. (2004) The far ultraviolet spectrum of the Io plasma torus. *Astrophys. J., 601,* 583–591.

Findlay F. D. (1970) Visible emission bands of molecular oxygen. *Can. J. Phys., 48,* 2107.

Fink U., Dekkers N. H., and Larson H. P. (1973) Infrared spectra of the Galilean satellites of Jupiter. *Astrophys. J. Lett., 179,* L155–L159.

Fleyfel F. and Devlin J. P. (1991) Carbon-dioxide clathrate hydrate epitaxial-growth — Spectroscopic evidence for formation of the simple Type-II CO_2 hydrate. *J. Phys. Chem., 95,* 3811–3815.

Franck J. and Rabinowitsch E. (1934) Some remarks about free radicals and the photochemistry of solutions. *Trans. Faraday Soc., 30,* 0120–0130.

Fuller K. A., Downing H. D., and Querry M. R. (1998) Orthorhombic sulfur (α-S). In *Handbook of Optical Constants of Solids III* (E. D. Palik, ed.), p. 899. Academic, San Diego.

Gaffey S. J., McFadden L. A., Nash D., and Pieters C. M. (1993) Ultraviolet, visible, and near-infrared reflectance spectroscopy: Laboratory spectra of geologic materials. In *Remote Geochemical Analysis* (C. M. Pieters and P. A. J. Englert, eds.), pp. 43–77. Cambridge Univ., Cambridge.

Garozzo M., Fulvio D., Gomis O., Palumbo M. E., and Strazzulla G. (2008) H-implantation in SO_2 and CO_2 ices. *Planet. Space Sci., 56,* 1300–1308.

Geissler P. E., Greenberg R., Hoppa G., McEwen A., Tufts R., Phillips C., Clark B., Ockert-Bell M., Helfenstein P., Burns J., Veverka J., Sullivan R., Greeley R., Pappalardo R. T., Head J. W., Belton M. J. S., and Denk T. (1998) Evolution of lineaments on Europa: Clues from Galileo multispectral imaging observations. *Icarus, 135,* 107–126.

Gerakines P. A., Schutte W. A., Greenberg J. M., and van Dishoeck E. F. (1995) The infrared band strengths of H_2O, CO, and CO_2 in laboratory simulations of astrophysical ice mixtures. *Astron. Astrophys., 296,* 810–818.

Gerakines P. A., Moore M. H., and Hudson R. L. (2000) Carbonic acid production in H_2O:CO_2 ices — UV photolysis vs. proton bombardment. *Astron. Astrophys., 357,* 793–800.

Gerakines P. A., Moore M. H., and Hudson R. L. (2004) Ultraviolet photolysis and proton irradiation of astrophysical ice analogs containing hydrogen cyanide. *Icarus, 170,* 202–213.

Gerakines P. A., Bray J. J., Davis A., and Richey C. R. (2005) The strengths of near-infrared absorption features relevant to interstellar and planetary ices. *Astrophys. J., 620,* 1140–1150.

Gibb E. L., Whittet D. C. B., Boogert A. C. A., and Tielens A. G. G. M. (2004) Interstellar ice: The Infrared Space Observatory Legacy. *Astrophys. J. Suppl. Ser., 151,* 35–73.

Glagolev V. L., Gordeeva V. A., Zhabrova G. M., and Kadenatsi B. M. (1967) On the radiation decomposition of aluminum and magnesium hydroxides. *High Energy Chem., 1,* 247–248.

Gomis O. and Strazzulla G. (2005) CO_2 production by ion irradiation of H_2O ice on top of carbonaceous material and its relevance to the Galilean satellites. *Icarus, 177,* 570–576.

Gomis O. and Strazzulla G. (2008) Ion irradiation of H_2O ice on top of sulfurous solid residue and its relevance to the Galilean satellites. *Icarus, 194,* 146–152.

Gomis O., Leto G., and Strazzulla G. (2004a) Hydrogen peroxide production by ion irradiation of thin water ice films. *Astron. Astrophys., 420,* 405–410.

Gomis O., Satorre M. A., Strazzulla G., and Leto G. (2004b) Hydrogen peroxide formation by ion implantation in water ice and its relevance to the Galilean satellites. *Planet. Space Sci., 52,* 371–378.

Gradie J., Thomas P., and Veverka J. (1980) The surface composition of Amalthea. *Icarus, 44,* 373–387.

Greenblatt G. D., Orlando J. J., Burkholder J. B., and Ravishankara A. R. (1990) Absorption-measurements of oxygen between 330 nm and 1140 nm. *J. Geophys. Res., 95,* 18577–18582.

Grieves G. A. and Orlando T. M. (2005) The importance of pores in the electron stimulated production of D_2 and O_2 in low temperature ice. *Surface Sci., 593,* 180–186.

Grundy W. M. and Schmitt B. (1998) The temperature-dependent near-infrared absorption spectrum of hexagonal ice. *J. Geophys. Res., 103,* 25809–25822.

Grundy W. M. and Stansberry J. A. (2000) Solar gardening and the seasonal evolution of nitrogen ice on Triton and Pluto. *Icarus, 148,* 340–346.

Grundy W. M., Buie M. W., Stansberry J. A., Spencer J. R., and Schmitt B. (1999) Near-infrared spectra of icy outer solar system surfaces: Remote determination of H_2O ice temperatures. *Icarus, 142,* 536–549.

Grundy W. M., Doute S., and Schmitt B. (2000) A Monte Carlo ray-tracing model for scattering and polarization by large particles with complex shapes. *J. Geophys. Res., 105,* 29291–29314.

Grundy W. M., Young L. A., and Young E. F. (2003) Discovery of CO_2 ice and leading-trailing spectral asymmetry on the uranian satellite Ariel. *Icarus, 162,* 222–229.

Grundy W. M., Young L. A., Spencer J. R., Johnson R. E., Young E. F., and Buie M. W. (2006) Distributions of H_2O and CO_2 ices on Ariel, Umbriel, Titania, and Oberon from IRTF/SpeX observations. *Icarus, 184,* 543–555.

Grundy W. M., Buratti B. J., Cheng A. F., Emery J. P., Lunsford A., McKinnon W. B., Moore J. M., Newman S. F., Olkin C. B., Reuter D. C., Schenk P. M., Spencer J. R., Stern S. A., Throop H. B., Weaver H. A., and the New Horizons Team (2007) New Horizons mapping of Europa and Ganymede. *Science, 318,* 234–237, DOI: 10.1126/science.1147623.

Hage W., Liedl K. R., Hallbrucker A., and Mayer E. (1998) Carbonic acid in the gas phase and its astrophysical relevance. *Science, 279,* 1332–1335.

Hagen W., Tielens A. G. G. M., and Greenberg J. M. (1981) The infrared spectra of amorphous solid water and ice I_c between 10 and 140 K. *Chem. Phys., 56,* 367–379.

Hall D. T., Gladstone G. R., Moos H. W., Bagenal F., Clarke J. T., Feldman P., McGrath M. A., Schneider N. M., Shemansky D. E., Strobel D. F., and Waite J. H. (1994) Extreme ultraviolet Explorer satellite observation of Jupiter's Io plasma torus. *Astrophys. J. Lett., 426,* L51–L54.

Hand K. P. (2007) On the physics and chemistry of the ice shell and sub-surface ocean of Europa. Ph.D. thesis, Stanford University, Stanford, California.

Hand K. P., Chyba C. F., Carlson R. W., and Cooper J. F. (2006) Clathrate hydrates of oxidants in the ice shell of Europa. *Astrobiology, 6,* 463–482, DOI: 10.1089/ast.(2006)6.463.

Hand K. P., Carlson R. W., and Chyba C. F. (2007) Energy, chemical disequilibrium, and geological constraints on Europa. *Astrobiology, 7,* 1006–1022.

Handa Y. P., Klug D. D., and Whalley E. (1988) Energies of phases of ice at low temperature and pressure relative to ice I_h. *Can. J. Chem., 66,* 919–924.

Hansen C. J., Shemansky D. E., and Hendrix A. R. (2005) Cassini UVIS observations of Europa's oxygen atmosphere and torus. *Icarus, 176,* 305–315.

Hansen G. B. and McCord T. B. (2004) Amorphous and crystalline ice on the Galilean satellites: A balance between thermal and radiolytic processes. *J. Geophys. Res., 109,* E01012, 1–19, DOI: 10.1029/2003JE002149.

Hansen G. B. and McCord T. B. (2008) Widespread CO_2 and other non-ice compounds on the anti-Jovian and trailing sides of Europa from Galileo/NIMS observations. *Geophys. Res. Lett., 35,* L01202.

Hapke B. (2001) Space weathering from Mercury to the asteroid belt. *J. Geophys. Res., 106,* 10039–10073.

Hapke B. and Graham F. (1989) Spectral properties of condensed phases of disulfur monoxide, polysulfur oxide, and irradiated sulfur. *Icarus, 79,* 47–55.

Hapke B., Wells E., Wagner J., and Partlow W. (1981) Far-UV, visible, and near-IR reflectance spectra of frosts of H_2O, CO_2, NH_3 and SO_2. *Icarus, 47,* 361–367.

Head J. W. and Pappalardo R. T. (1999) Brine mobilization during lithospheric heating on Europa: Implications for formation of chaos terrain, lenticula texture, and color variations. *J. Geophys. Res., 104,* 27143–27155.

Head J. W., Pappalardo R. T., and Sullivan R. (1999) Europa: Morphological characteristics of ridges and triple bands from Galileo data (E4 and E6) and assessment of a linear diapirism model. *J. Geophys. Res., 104,* 24223–24236.

Hendrix A. R., Barth C. A., Hord C. W., and Lane A. L. (1998) Europa: Disk-resolved ultraviolet measurements using the Galileo ultraviolet spectrometer. *Icarus, 135,* 79–94.

Hendrix A. R., Barth C. A., and Hord C. W. (1999) Ganymede's ozone-like absorber: Observations by the Galileo ultraviolet spectrometer. *J. Geophys. Res., 104,* 14169–14178.

Hendrix A. R., Carlson R. W., Mehlman R., and Smythe W. D. (2002) Europa as measured by Galileo NIMS and UVS (abstract). In *Jupiter after Galileo and Cassini,* p. 108. Observatorio Astronomico de Lisboa, Lisbon.

Hendrix A., Carlson R., and Johnson R. E. (2008) Europa's ultraviolet absorption feature: Correlation with endogenic surface features on the trailing hemisphere (abstract). *Bull. Am. Astron. Soc., 40,* Abstract No. 59.07.

Henning T., Il'in V. B., Krivova N. A., Michel B., and Voshchinnikov N. V. (1999) WWW database of optical constants for astronomy. *Astron. Astrophys. Suppl., 136,* 405–406.

Herzberg G. (1945) *Molecular Spectra and Molecular Structure II: Infrared and Raman Spectra of Polyatomic Molecules.* Van Nostrand, Princeton.

Herzberg G. (1950) *Molecular Spectra and Molecular Structure I: Spectra of Diatomic Molecules.* Van Nostrand, Princeton.

Hibbitts C. A. and Szanyi J. (2007) Physisorption of CO_2 on non-ice materials relevant to icy satellites. *Icarus, 191,* 371–380.

Hibbitts C. A., McCord T. B. and Hansen G. B. (2000) The distributions of CO_2 and SO_2 on the surface of Callisto. *J. Geophys. Res., 105,* 22541–22557.

Hibbitts C. A., Klemaszewski J. E., McCord T. B., Hansen G. B., and Greeley R. (2002) CO_2-rich impact craters on Callisto. *J. Geophys. Res., 107,* 14-1 to 14-12, 5084, DOI: 1029/2000JE001412.

Hibbits C. A., Pappalardo R. T., Hansen G. B., and McCord T. B. (2003) Carbon dioxide on Ganymede. *J. Geophys. Res., 108,* 2-1 to 2-21, 5036, DOI: 1029/2002JE001956.

Hinrichs J. I. and Lucey P. G. (2002) Temperature-dependent near-infrared spectral properties of minerals, meteorites, and lunar soil. *Icarus, 155,* 169–180.

Hobbs P. V. (1974) *Ice Physics.* Oxford Univ., London.

Hochanadel C. J., Ghormley J. A., and Sworski T. J. (1955) The decomposition of sulfuric acid by cobalt γ rays. *J. Am. Chem. Soc., 77,* 3215.

Hopkins A. G. and Brown C. W. (1975) Infrared spectrum of sulfur monoxide. *J. Chem. Phys., 62,* 2511–12.

Hopkins A. G., Tang S.-Y., and Brown C. W. (1973) Infrared and Raman spectra of the low-temperature products from discharged sulfur dioxide. *J. Am. Chem. Soc., 95,* 3486–3490.

Hosokawa S., Matsuoka T., and Tamura K. (1994) Optical absorption spectra of liquid sulphur over a wide absorption range. *J. Phys. Condens. Matter, 6,* 5273–5282.

Huang S. and Johnson E. R. (1965) The radiation-induced decomposition of some inorganic sulfates. In *Effects of High Energy Radiation on Inorganic Substances,* pp. 121–138. Special Tech. Publ. No. 400, ASTM, Seattle.

Hudson R. L. and Moore M. H. (2006) infrared spectra and radiation stability of H_2O_2 ices relevant to Europa. *Astrobiology, 6,* 483–489.

Hudson R. L., Moore M. H., and Gerakines P. A. (2001) The formation of cyanate ion (OCN–) in interstellar ice analogs. *Astrophys. J., 550,* 1140–1150.

Hunt G. R. (1977) Spectral signatures of particulate minerals in the visible and near infrared. *Geophysics, 42,* 501–533.

Hunt G. R. and Ashley R. P. (1979) Spectra of altered rocks in the visible and near infrared. *Econ. Geol., 74,* 1613–1629.

Hunt G. R. and Salisbury J. W. (1970) Visible and near-infrared spectra of minerals and rocks: I Silicate minerals. *Mod. Geol., 1,* 283–300.

Hunt G. R., Salisbury J. W., and Lenhoff C. J. (1971a) Visible and near-infrared spectra of minerals and rocks: III. Oxides and hydroxides. *Mod. Geol., 2,* 195–205.

Hunt G. R., Salisbury J. W., and Lenhoff C. J. (1971b) Visible and near-infrared spectra of minerals and rocks: IV. Sulphides and sulphates. *Mod. Geol., 3,* 1–14.

Hunt G. R., Salisbury J. W., and Lenhoff C. J. (1973) Visible and near-infrared spectroscopy of minerals and rocks: VI. Additional silicates. *Mod. Geol., 4,* 85–106.

ICRU (1984) *Stopping Powers for Electrons and Positrons.* ICRU Report 37, International Commission on Radiation Units and Measurements. Available online at http://physics.nist.gov/PhysRefData/Star/Text/ESTAR.html.

Ip W. H. (1996) Europa's oxygen exosphere and its magnetospheric interaction. *Icarus, 120,* 317–325.

Jenniskens P., Blake D. F., and Kouchi A. (1998) Amorphous water ice. In *Solar System Ices* (B. Schmitt et al., eds.), pp. 199–240. Kluwer, Boston.

Johnson R. E. (1990) *Energetic Charged-Particle Interactions with Atmospheres and Surfaces.* Springer-Verlag, Berlin.

Johnson R. E. (1999) Comment on "Laboratory studies of the optical properties and stability of oxygen on Ganymede" by Raul A. Baragiola and David A. Bahr. *J. Geophys. Res., 104,* 14179–14182.

Johnson R. E. (2000) Sodium at Europa. *Icarus, 143,* 429–433.

Johnson R. E. (2001) Surface chemistry in the jovian magnetosphere radiation environment. In *Chemical Dynamics in Extreme Environments* (R. A. Dressler, ed.), World Scientific, Singapore.

Johnson R. E. and Jesser W. A. (1997) O_2/O_3 microatmospheres in the surface of Ganymede. *Astrophys. J. Lett., 480,* L79–L82.

Johnson R. E. and Quickenden T. I. (1997) Photolysis and radiolysis of water ice on outer solar system bodies. *J. Geophys. Res., 102,* 10985–10996.

Johnson R. E., Nelson M. L., McCord T. B., and Gradie J. C. (1988) Analysis of Voyager images of Europa — Plasma bombardment. *Icarus, 75,* 423–436.

Johnson R. E., Leblanc F., Yakshinskiy B. V., and Madey T. E. (2002) Energy distributions for desorption of sodium and potassium from ice: The Na/K ratio at Europa. *Icarus, 156,* 136–142.

Johnson R. E., Quickenden T. I., Cooper P. D., McKinley A., and Freeman C. G. (2003) The production of oxidants in Europa's surface. *Astrobiology, 3,* 823–850.

Johnson R. E., Carlson R. W., Cooper J. F., Paranicas C., Moore M. H., and Wong M. (2004) Radiation effects on the surfaces of the Galilean satellites. In *Jupiter: The Planets, Satellites and Magnetosphere* (F. Bagenal et al., eds.), pp. 485–512. Cambridge Univ., Cambridge.

Johnson R. E., Cooper P. D., Quickenden T. I., Grieves G. A., and Orlando T. M. (2005) Production of oxygen by electronically induced dissociations in ice. *J. Chem. Phys., 123,* 184715-1-8.

Johnson T. V. (1970) Albedo and spectral reflectivity of the Galilean satellites of Jupiter. Ph.D. thesis, California Institute of Technology, Pasadena.

Johnson T. V. (1971) Galilean satellites: Narrowband photometry 0.3–1.1 microns. *Icarus, 14,* 94–111.

Johnson T. V. and McCord T. B. (1970) Galilean satellites: The spectral reflectivity 0.30–1.10 microns. *Icarus, 13,* 37–42.

Johnson T. V. and McCord T. B. (1971) Spectral geometric albedo of the Galilean satellites 0.3–2.5 microns. *Astrophys. J., 169,* 589–593.

Johnson T. V. and Pilcher C. B. (1977) Satellite spectrophotometry and surface compositions. In *Planetary Satellites* (J. A. Burns, ed.), pp. 232–268. Univ. of Arizona, Tucson.

Johnson T. V., Soderblom L. A., Mosher J. A., Danielson G. E., Cook A. F., and Kupferman P. (1983) Global multispectral mosaics of the icy galilean satellites. *J. Geophys. Res., 88,* 5789–5805.

Jones A. V. (1950) Infra-red and ultraviolet spectra of sulfur monoxide. *J. Chem. Phys., 18,* 1263–1268.

Kargel J. S., Kaye J., Head J. W. I., Marion G., Sassen R., Crowley J., Prieto O., and Hogenboom D. (2000) Europa's crust and ocean: Origin, composition, and prospects for life. *Icarus, 94,* 368–390.

Kerkhof O., Schutte W. A., and Ehrenfreund P. (1999) The infrared band strengths of CH_3OH, NH_3 and CH_4 in laboratory simulations of astrophysical ice mixtures. *Astron. Astrophys., 346,* 990–994.

Khare B. N., Sagan C., Thompson W. R., Arakawa E. T., Meisse C., and Tuminello P. S. (1994) Optical properties of poly-HCN and their astronomical applications. *Can. J. Chem., 72,* 678–694.

Khriachtchev L., Pettersson M., Tuominen S., and Rasanen M. (1997) Photochemistry of hydrogen peroxide in solid argon. *J. Chem. Phys., 107,* 7252–7259.

Kieffer H. H. (1970) Spectral reflectance of CO_2-H_2O frosts. *J. Geophys. Res., 75,* 501–509.

Kieffer H. H. and Smythe W. D. (1974) Frost spectra: Comparison with Jupiter's satellites. *Icarus, 21,* 506–512.

Kissel J. and Krueger F. R. (1987) Ion formation by impact of fast dust particles and comparison with related techniques. *Appl. Phys. A., 42,* 69–85.

Kouchi A. (1987) Vapor-pressure of amorphous H_2O ice and its astrophysical implications. *Nature, 330,* 550–552.

Kouchi A. and Kuroda T. (1990) Amorphization of cubic ice by ultraviolet-irradiation. *Nature, 344,* 134–135.

Kouchi A., Yamamoto T., Kozasa T., Kuroda T., and Greenberg J. M. (1994) Conditions for condensation and preservation of amorphous ice and crystallinity of astrophysical ices. *Astron. Astrophys., 290,* 1009–1018.

Krasnopolsky V. A. (1987) S_3 and S_4 absorption cross sections in the range of 340 to 600 nm and evaluation of the S_3 abundance in the lower atmosphere of Venus. *Adv. Space. Res., 7,* (12)25–(12)27.

Kruger H., Geissler P., Horanyi M., Graps A. L., Kempf S., Srama R., Moragas-Klostermeyer G., Moissl R., Johnson T. V., and Grun E. (2003) Jovian dust streams: A monitor of Io's volcanic plume activity. *Geophys. Res. Lett., 30,* 3-1 to 3-4.

Kuiper G. P. (1957) Infrared observations of planets and satellites (abstract). *Astron. J., 62,* 245.

Kuppers M. and Schneider N. M. (2000) Discovery of chlorine in the Io torus. *Geophys. Res. Lett., 27,* 513–516.

Landau A., Allin E. J., and Welsh H. L. (1962) The absorption spectrum of solid oxygen in the wavelength region from 12,000 Å to 3300 Å. *Spectrochim. Acta., 18,* 1–19.

Lane A. L., Nelson R. M., and Matson D. L. (1981) Evidence for sulfur implantation in Europa's UV absorption band. *Nature, 292,* 38–39.

Lange M. A. and Ahrens T. J. (1987) Impact experiments in low-temperature ice. *Icarus, 69,* 506–518.

Leblanc F., Johnson R. E., and Brown M. E. (2002) Europa's sodium atmosphere: An ocean source? *Icarus, 159,* 132–144.

Leblanc F., Potter A. E., Killen R. M., and Johnson R. E. (2005) Origins of Europa Na cloud and torus. *Icarus, 178,* 367–385.

Lebofsky L. A. and Fegley M. B. Jr. (1976) Laboratory reflection spectra for the determination of chemical composition of icy bodies. *Icarus, 28,* 379–387.

Lebofsky L. A. and Feierberg M. A. (1985) 2.7-μm to 4.1-μm spectrophotometry of icy satellites of Saturn and Jupiter. *Icarus, 63,* 237–242.

Leto G. and Baratta G. A. (2003) Ly-α photon induced amorphization of I_c water ice at 16 Kelvin. Effects and quantitative comparison with ion irradiation. *Astron. Astrophys., 397,* 7–13.

Leto G., Gomis O., and Strazzulla G. (2005) The reflectance spectrum of water ice: Is the 1.65 μm peak a good temperature probe? *Mem. Soc. Astron. Ital. Suppl., 6,* 57–62.

Loeffler M. J. and Baragiola R. A. (2005) The state of hydrogen peroxide on Europa. *Geophys. Res. Lett., 32,* L17202 1–4.

Loeffler M. J., Raut U., and Baragiola R. A. (2006a) Enceladus: A source of nitrogen and an explanation for the water vapor plume observed by Cassini. *Astrophys. J. Lett., 649,* L133–L136.

Loeffler M. J., Raut U., Vidal R. A., Baragiola R. A., and Carlson R. W. (2006b) Synthesis of hydrogen peroxide in water ice by ion irradiation. *Icarus, 180,* 265–273.

Loeffler M. J., Teolis B. D., and Baragiola R. A. (2006c) Decomposition of solid amorphous hydrogen peroxide by ion irradiation. *J. Chem. Phys., 124,* DOI: 10470210.1063/1.2171967.

Loeffler M. J., Teolis B. D., and Baragiola R. A. (2006d) A model study of the thermal evolution of astrophysical ices. *Astrophys. J. Lett., 639,* L103–L106.

Long C. A. and Ewing G. E. (1973) Spectroscopic investigation of van der waals molecules. 1. Infrared and visible spectra of $(O_2)_2$. *J. Chem. Phys., 58,* 4824–4834.

Long C. and Kearns D. R. (1973) Selection-rules for intermolecular enhancement of spin forbidden transitions in molecular-oxygen. *J. Chem. Phys., 59,* 5729–5736.

Lowenthal M. S., Khanna R. K., and Moore M. H. (2002) Infrared spectrum of solid isocyanic acid (HNCO): Vibrational assignments and integrated band intensities. *Spectrochim. Acta A., 58,* 73–78.

Marion G. M. (2002) A molal-based model for strong acid chemistry at low temperatures (187 to 298 K). *Geochim. Cosmochim. Acta, 66,* 2499–2516.

Marion G. M. and Kargel J. S. (2008) *Cold Aqueous Planetary Geochemistry with FREZCHEM.* Springer-Verlag, Berlin.

Mastrapa R. M. E. and Brown R. H. (2006) Ion irradiation of crystalline H_2O-ice: Effect on the 1.65-μm band. *Icarus, 183,* 207–214.

Mastrapa R. M. and Sandford S. A. (2008) New optical constants of amorphous and crystalline ice, 3–20 micrometers. *Eos Trans. AGU, 89,* Fall Meet. Suppl., Abstract MR13A-1700.

Mastrapa R. M., Bernstein M. P., Sandford S. A., Roush T. L., Cruikshank D. P., and Ore C. M. D. (2008) Optical constants of amorphous and crystalline H_2O-ice in the near infrared from 1.1 to 2.6 μm. *Icarus, 97,* 307–320.

Mauk B. H., Mitchell D. G., Krimigis S. M., Roelof E. C., and Paranicas C. P. (2003) Energetic neutral atoms from a trans-Europa gas torus at Jupiter. *Nature, 421,* 920–922.

Mayer E. (1985) New method for vitrifying water and other liquids by rapid cooling of their aerosols. *J. Appl. Phys., 58,* 663–667.

McCord T. B., Carlson R. W., Smythe W. D., Hansen G. B., Clark R. N., Hibbitts C. A., Fanale F. P., Granahan J. C., Segura M., Matson D. L., Johnson T. V., and Martin P. D. (1997) Organics and other molecules in the surfaces of Callisto and Ganymede. *Science, 278,* 271–275.

McCord T. B., Hansen G. B., Clark R. N., Martin P. D., Hibbits

C. A., Fanale F. P., Granahan J. C., Segura M., Matson D. L., Johnson T. V., Carlson R. W., Smythe W. D., Danielson G. E., and the NIMS Team (1998a) Non-water-ice constituents in the surface material of the icy Galilean satellites from the Galileo near infrared mapping spectrometer investigation. *J. Geophys. Res., 103,* 8603–8626.

McCord T. B., Hansen G. B., Fanale F. P., Carlson R. W., Matson D. L., Johnson T. V., Smythe W. D., Crowley J. K., Martin P. D., Ocampo A., Hibbitts C. A., and Granahan J. C. (1998b) Salts an Europa's surface detected by Galileo's Near Infrared Mapping Spectrometer. *Science, 280,* 1242–1245.

McCord T. B., Hansen G. B., Matson D. L., Johnson T. V., Crowley J. K., Fanale F. P., Carlson R. W., Smythe W. D., Martin P. D., Hibbitts C. A., Granahan J. C., Ocampo A., and the NIMS Team (1999) Hydrated salt minerals on Europa's surface from the Galileo NIMS investigation. *J. Geophys. Res., 104,* 11827–11851.

McCord T. B., Orlando T. M., Teeter G., Hansen G. B., Sieger M. T., Petrik N. K., and Van Keulen L. (2001) Thermal and radiation stability of the hydrated minerals epsomite, mirabilite, and natron under Europa environmental conditions. *J. Geophys. Res., 106,* 3311–3319.

McCord T. B., Teeter G., Hansen G. B., Sieger M. T., and Orlando T. M. (2002) Brines exposed to Europa surface conditions. *J. Geophys. Res., 107,* 4-1 to 4-6.

McEwen A. S. (1986) Exogenic and endogenic albedo and color patterns on Europa. *J. Geophys. Res., 91,* 8077–8097.

McFadden L. A., Bell J. F., and McCord T. B. (1980) Visible spectral reflectance measurements (0.33–1.1 μm) of the Galilean satellites at many orbital phase angles. *Icarus, 44,* 410–430.

McGrath M. A., Lellouch E., Strobel D. F., Feldman P. D., and Johnson R. E. (2004) Satellite atmospheres. In *Jupiter: The Planet, Satellites, and Magnetosphere* (F. Bagenal et al., eds.), pp. 457–483. Cambridge Univ., Cambridge.

McKinnon W. B. and Zolensky M. E. (2003) Sulfate content of Europa's ocean and shell: Evolutionary considerations and some geological and astrobiological implications. *Astrobiology, 3,* 879–897.

McNutt R. L. (1993) Possible in situ detection of K^{2+} in the jovian magnetosphere. *J. Geophys. Res., 98,* 21221–21229.

Mennella V., Palumbo M. E., and Baretta G. A. (2004) Formation of CO and CO_2 molecules by ion irradiation of water ice-covered hydrogenated carbon grains. *Astrophys. J., 615,* 1073–1080.

Meyer B. and Stroyer-Hansen T. (1972) Infrared spectrum of S_4. *J. Phys. Chem., 76,* 3968–3969.

Meyer B., Oommen T. V., and Jensen D. (1971) The color of liquid sulfur. *J. Phys. Chem., 75,* 912–917.

Meyer B., Stroyer-Hansen T., and Oommen T. V. (1972) The visible spectrum of S_3 and S_4. *J. Molec. Spectrosc., 42,* 335–343.

Mills F. P. and Brown M. E. (2000) Thermal infrared spectroscopy of Europa and Callisto. *J. Geophys. Res., 105,* 15051–15059.

Moore J. M., Asphaug E., Morrison D., Spencer J. R., Chapman C. R., Bierhaus B., Sullivan R. J., Chuang F. C., Klemaszewski J. E., Greeley R., Bender K. C., Geissler P. E., Helfenstein P., and Pilcher C. B. (1999) Mass movement and landform degradation on the icy Galilean satellites: Results of the Galileo nominal mission. *Icarus, 140,* 294–312.

Moore M. H. (1984) Studies of proton-irradiated SO_2 at low-temperatures — Implications for Io. *Icarus, 59,* 114–128.

Moore M. H. and Hudson R. L. (1992) Far-infrared spectral studies of phase changes in water ice induced by proton irradiation. *Astrophys. J., 401,* 353–360.

Moore M. H. and Hudson R. L. (1998) Infrared study of ion-irradiated water-ice mixtures with hydrocarbons relevant to comets. *Icarus, 135,* 518–527.

Moore M. H. and Hudson R. L. (2000) IR detection of H_2O_2 at 80 K in ion-irradiated laboratory ices relevant to Europa. *Icarus, 145,* 282–288.

Moore M. H. and Khanna R. K. (1991) Infrared and mass spectral studies of proton irradiated $H_2O + CO_2$ ice: Evidence for carbonic acid. *Spectrochim. Acta, 47A,* 255–262.

Moore M. H., Khanna R., and Donn B. (1991) Studies of proton irradiated $H_2O + CO_2$ and $H_2O + CO$ ices and analysis of synthesized molecules. *J. Geophys. Res., 96,* 17541–17545.

Moore M. H., Ferrante R. F., and Nuth J. A. (1996) Infrared spectra of proton irradiated ices containing methanol. *Planet. Space Sci., 44,* 927–935.

Moore M. H., Hudson R. L., and Ferrante R. F. (2003) Radiation products in processed ices relevant to Edgeworth-Kuiper-belt objects. *Earth Moon Planets, 92,* 291–306.

Moore M. H., Hudson R. L., and Carlson R. W. (2007) The radiolysis of SO_2 and H_2S in water ice: Implications for the icy jovian satellites. *Icarus, 189,* 409–423.

Moroz V. I. (1965) Infrared spectrophotometry of satellites: The Moon and the Galilean satellites of Jupiter. *Astron. Zh., 42,* 1287, trans. in *Soviet Astron.–AJ, 9,* 999–1006.

Morris R. V., Lauer H. V. Jr., Lawson C. A., Gibson E. K. Jr., Nace G. A., and Stewart C. (1985) Spectral and other physicochemical properties of submicron powders of hematite (α-Fe_2O_3), maghemite (γ-Fe_2O_3), magnetite (Fe_3O_4), goethite (α-FeOOH), and lepidocrocite (γ-FeOOH). *J. Geophys. Res., 90,* 3126–3144.

Morris R. V., Golden D. C., Bell J. F., Lauer H. V., and Adams J. B. (1993) Pigmenting agents in martian soils — inferences from spectral, Mossbauer, and magnetic-properties of nanophase and other iron-oxides in Hawaiian palagonitic soil PN-9. *Geochim. Cosmochim. Acta, 57,* 4597–4609.

Morris R. V., Golden D. C., and Bell J. F. III (1997) Low-temperature reflectivity spectra of red hematite and the color of Mars. *J. Geophys. Res., 102,* 9125–9133.

Morris R. V., Golden D. C., Bell J. F., Shelfer T. D., Scheinost A. C., Hinman N. W., Furniss G., Mertzman S. A., Bishop J. L., Ming D. W., Allen C. C., and Britt D. T. (2000) Mineralogy, composition, and alteration of Mars Pathfinder rocks and soils: Evidence from multispectral, elemental, and magnetic data on terrestrial analogue, SNC meteorite, and Pathfinder samples. *J. Geophys. Res., 105,* 1757–1817.

Morrison D. and Burns J. A. (1976) The jovian satellites. In *Jupiter* (T. Gehrels, ed.), pp. 991–1034. Univ. of Arizona, Tucson.

Mozumder A. (1999) *Fundamentals of Radiation Chemistry.* Academic, San Diego.

Na C. Y., Trafton L. M., Barker E. S., and Stern S. A. (1998) A search for new species in Io's extended atmosphere. *Icarus, 131,* 449–452.

Nash D. B. and Fanale F. P. (1977) Io's surface composition based on reflectance spectra of sulfur/salt mixtures and proton irradiation experiments. *Icarus, 31,* 40–80.

Nelson M. L., McCord T. B., Clark R. N., Johnson T. V., Matson D. L., Mosher J. A., and Soderblom L. A. (1986) Europa: Characterization and interpretation of global spectral surface units. *Icarus, 65,* 129–151.

Nelson R. M. and Hapke B. W. (1978) Spectral reflectivities of the Galilean satellites and Titan, 0.32 to 0.86 micrometers. *Icarus, 36,* 304–329.

Nelson R. M. and Lane A. L. (1987) Planetary satellites. In *Scientific Accomplishments of IUE* (Y. Kondo, ed.), pp. 67–99. Reidel, Dordrecht.

Nelson R. M., Lane A. L., Matson D. L., Veeder G. J., Buratti B. J., and Tedesco E. F. (1987) Spectral geometric albedos of the Galilean satellites from 0.24 to 0.34 micrometers: Observations with the International Ultraviolet Explorer. *Icarus, 72,* 358–380.

Nelson R. M., Smythe W. D., Hapke B. W. and Cohen A. J. (1990) On the effect of X-rays on the color of elemental sulfur: Implications for Jupiter's satellite Io. *Icarus, 85,* 326–334.

Nimmo F. and Gaidos E. (2002) Strike-slip motion and double ridge formation on Europa. *J. Geophys. Res., 107,* Article No. 5021.

Nimmo F., Pappalardo R., and Cuzzi J. (2007) Observational and theoretical constraints on plume activity at Europa. AGU Fall Meeting, Abstract #P51E-05.

Nishijima C., Kanamuru N., and Titmura K. (1976) Primary photochemical processes of sulfur in solution. *Bull. Chem. Soc. Japan., 49,* 1151–1152.

Nna-Mvondo D., Khare B., Ishihara T. and McKay C. P. (2008) Experimental impact shock chemistry on planetary icy satellites. *Icarus, 194,* 816–829. DOI: 10.1016/j.icarus.(2007)11.001.

Noll K. S. and Knacke R. F. (1993) Titan: 1–5 μm photometry and spectrophotometry and a search for variability. *Icarus, 101,* 272–281.

Noll K. S., Weaver H. A., and Gonnella A. M. (1995) The albedo spectrum of Europa from 2200 angstrom to 3300 angstrom. *J. Geophys. Res., 100,* 19057–19059.

Noll K. S., Johnson R. E., Lane A. L., Domingue D. L., and Weaver H. A. (1996) Detection of ozone on Ganymede. *Science, 273,* 341–343.

Noll K. S., Roush T. L., Cruikshank D. P., Johnson R. E., and Pendleton Y. J. (1997) Detection of ozone on Saturn's satellites Rhea and Dione. *Nature, 388,* 45–47.

O'Donnell J. H. and Sangster D. F. (1970) *Principles of Radiation Chemistry.* American Elsevier, New York.

Ockert M. E., Nelson R. M., Lane A. L., and Matson D. L. (1987) Europa's ultraviolet absorption band (260 to 320 nm): Temporal and spatial evidence from IUE. *Icarus, 70,* 499–505.

Ockman N. (1958) The infra-red and Raman spectra of ice. *Philos. Mag. Suppl., 7,* 199–220.

Ohtaki H. and Radnai T. (1993) Structure and dynamics of hydrated ions. *Chem. Rev., 93,* 1157–1204.

Onaka R. and Takahashi T. (1968) Vacuum UV absorption of liquid water and ice. *J. Phys. Soc. Japan., 24,* 548–550.

Orlando T. M. and Sieger M. T. (2003) The role of electron-stimulated production of O_2 from water ice in the radiation processing of outer solar system surfaces. *Surf. Sci., 528,* 1–7.

Orlando T. M., McCord T. B., and Grieves G. A. (2005) The chemical nature of Europa surface material and the relation to a subsurface ocean. *Icarus, 177,* 528–533.

Palumbo M. E. (1997) Production of CO and CO_2 after ion irradiation of ices. *Adv. Space. Res., 20,* 1637–1645.

Palumbo M. E. (2006) Formation of compact solid water after ion irradiation at 15 K. *Astron. Astrophys., 453,* 903–909.

Palumbo M. E., Baratta G. A., Brucato J. R., Castorina A. C., Satorre M. A., and Strazzulla G. (1998) Profile of the CO_2 bands produced after ion irradiation of ice mixtures. *Astron. Astrophys., 334,* 247–252.

Paranicas C., Carlson R. W., and Johnson R. E. (2001) Electron bombardment of Europa. *Geophys. Res. Lett., 28,* 673–676.

Paranicas C. P., Vollmer M., and Kivelson M. G. (2002) Flow diversion at Europa. AGU Spring Meeting, Abstract #P21B-07.

Paterson W. R., Frank L. A., and Ackerson K. L. (1999a) Galileo plasma observations at Europa: Ion energy spectra and moments. *J. Geophys. Res., 104,* 22779–22791.

Paterson W. R., Frank L. A., and Ackerson K. L. (1999b) Galileo plasma observations at Europa: Ion energy spectra and moments. *J. Geophys. Res., 104,* 22779–227791.

Pauling L. (1935) The structure and entropy of ice and other crystals with some randomness of atomic arrangement. *J. Am. Chem. Soc., 57,* 2680–2684.

Pearl J., Hanel R., Kunde V., Maguire W., Fox K., Gupta S., Ponnamperuma C., and Raulin F. (1979) Identification of gaseous SO_2 and new upper limits for other gases on Io. *Nature, 280,* 755–758.

Peterson R. C. and Wang R. Y. (2006) Crystal molds on Mars: Melting of a possible new mineral species to create martian chaotic terrain. *Geology, 34,* 957–960. 10.1130/g22678a.1

Phillips C. B. and Chyba C. F. (2001) Impact gardening rates on Europa: Comparison with sputtering. In *Lunar and Planetary Science XXXII,* Abstract #2111. Lunar and Planetary Institute, Houston (CD-ROM).

Phillips C. B. and Chyba C. F. (2004) Impact gardening, sputtering, mixing, and surface–subsurface exchange on Europa. In *Workshop on Europa's Icy Shell,* pp. 70–71. LPI Contribution No. 1195, Lunar and Planetary Institute, Houston.

Phillips L. F., Smith J. J., and Meyer B. (1969) The ultraviolet spectra of matrix isolated disulfur monoxide and sulfur dioxide. *J. Molec. Spectrosc., 29,* 230–243.

Pilcher C. B., Ridgeway S. T., and McCord T. B. (1972) Galilean satellites: Identification of water frost. *Science, 178,* 1087–1089.

Pipes J. G., Browell E. V., and Anderson R. C. (1974) Reflectance of amorphous-cubic NH_3 frosts and amorphous-hexagonal frosts at 77 K from 1400 to 3000 A. *Icarus, 21,* 283–291.

Pollack J. B., Witteborn F. C., Erickson E. F., Strecker D. W., Baldwin B. J., and Bunch T. E. (1978) Near-infrared spectra of the Galilean satellites: Observations and compositional implications. *Icarus, 36,* 271–303.

Porco C. C., West R. A., McEwen A., Genio A. D. D., Ingersoll, A. P., Thomas P., Squyres S., Dones L., Murray C. D., Johnson T. V., Burns J. A., Brahic A., Neukum G., Veverka J., Barbara J. M., Denk T., Evans M., Ferrier J. J., Geissler P., Helfenstein P., Roatsch T., Throop H., Tiscareno M., and Vasavada A. R. (2003) Cassini imaging of Jupiter's atmosphere, satellites, and rings. *Science, 299,* 1541–1547.

Pospieszalska M. K. and Johnson R. E. (1989) Magnetospheric ion-bombardment profiles of satellites — Europa and Dione. *Icarus, 78,* 1–13.

Postberg F., Kempf S., Srama R., Green S. F., Hillier J. K., McBride N., and Grun E. (2006) Composition of jovian dust stream particles. *Icarus, 183,* 122–134. DOI: 10.1016/j.icarus.(2006)02.001.

Prieto-Ballesteros O., Kargel J. S., Fernandez-Sampedro M., Selsis F., Martinez E. S., and Hogenboom D. L. (2005) Evaluation of the possible presence of clathrate hydrates in Europa's icy shell or seafloor. *Icarus,* 491–505.

Prinn R. G. and Fegley B. (1981) Kinetic inhibition of CO and N_2 reduction in circumplanetary nebulae — Implications for satellite compositions. *Astrophys. J., 249,* 308–317.

Prinn R. G. and Fegley B. Jr. (1989) Origin of planetary, satellite, and cometary volatiles. In *Origin and Evolution of Planetary and Satellite Atmospheres* (S. K. Atreya et al., eds.), pp. 8–136. Univ. of Arizona, Tucson.

Prockter L. M., Head J. W., Pappalardo R. T., Sullivan R. J., Clifton A. E., Giese B., Wagner R., and Neukum G. (2002) Morphology of europan bands at high resolution: A mid-ocean ridge-type rift mechanism. *J. Geophys. Res., 107,* DOI: 10.1029/2000JE001458.

Raunier S., Chiavassa T., Allouche A., Marinelli F., and Aycard J.-P. (2003) Thermal reactivity of HNCO with water ice: An infrared and theoretical study. *Chem. Phys., 288,* 197–210.

Raut U., Loeffler M. J., Vidal R. A., and Baragiola R. A. (2004) The OH stretch infrared band of water and its temperature and radiation dependence. In *Lunar and Planetary Science XXXV,* Abstract #1922. Lunar and Planetary Institute, Houston (CD-ROM).

Raut U., Loeffler M. J., Teolis B. D., Vidal R. A., and Baragiola R. A. (2005) Radiation synthesis of carbon dioxide in ice coated grains (abstract). *Bull. Am. Astron. Soc., 37,* 755.

Raut U., Teolis B. D., Loeffler M. J., Vidal R. A., Fama M., and Baragiola R. A. (2007) Compaction of microporous amorphous solid water by ion irradiation. *J. Chem. Phys., 126,* 244511.

Reimann C. T., Boring J. W., Johnson R. E., Garrett J. W., Farmer K. R., and Brown W. L. (1984) Ion induced molecular ejection from D_2O ice. *Surf. Sci., 147,* 227.

Riley J., Hoppa G. V., Greenberg R., and Tufts B. R. (2000) Distribution of chaotic terrain on Europa. *J. Geophys. Res., 105,* 22599–22615.

Robinson G. W. (1967) Intensity enhancement of forbidden electronic transitions by weak intermolecular interactions. *J. Chem. Phys., 46,* 572–585.

Rossman G. R. (1975) Spectroscopic and magnetic studies of ferric iron hydroxy sulfates: Intensification of color in ferric iron clusters bridged by a single hydroxide ion. *Am. Mineral., 60,* 698–704.

Rothschild W. G. (1964) γ-ray decomposition of pure liquid sulfur dioxide. *J. Am. Chem. Soc., 86,* 1307–1309.

Sack N. J. and Baragiola R. A. (1993) Sublimation of vapor-deposited water ice below 170 K and its dependence on growth conditions. *Phys. Rev. B, 48,* 9973–9978.

Sack N. J., Johnson R. E., Boring J. W., and Baragiola R. A. (1992) The effect of magnetospheric ion bombardment on the reflectance of Europa's surface. *Icarus, 100,* 534–540.

Sack N. J., Baragiola R. A., and Johnson R. E. (1993) Effect of plasma ion bombardment on the reflectance of Io's trailing and leading hemispheres. *Icarus, 104,* 152–154.

Sandford S. A. and Allamandola L. J. (1990) The physical and infrared spectral properties of CO_2 in astrophysical ice analogs. *Astrophys. J., 355,* 357–372.

Sasaki T., Williams R. S., Wong J. S., and Shirley D. A. (1978) Radiation damage studies by X-ray photoelectron spectroscopy. I. Electron irradiated $LiNO_3$ and Li_2SO_4. *J. Chem. Phys., 68,* 2718–2724.

Saur J., Strobel D. F., and Neubauer F. M. (1998) Interaction of the jovian magnetosphere with Europa: Constraints on the neutral atmosphere. *J. Geophys. Res., 103,* 19947–19962.

Scattergood T. W., McKay C. P., Borucki W. J., Giver L. P., Vanghyseghem H., Parris J. E., and Miller S. L. (1989) Production of organic-compounds in plasmas — A comparison among electric-sparks, laser-induced plasmas, and uv-light. *Icarus, 81,* 413–428.

Schaefer L. and Fegley B. (2004) A thermodynamic model of high temperature lava vaporization on Io. *Icarus, 169,* 216–241.

Schaefer L. and Fegley B. Jr. (2005) Predicted abundances of carbon compounds in volcanic gases on Io. *Icarus, 618,* 1079–1085.

Schmitt B., de Bergh C., Lellouch E., Maillard J.-P., Barbe A., and Doute S. (1994) Identification of three absorption bands in the 2-μm spectrum of Io. *Icarus, 111,* 79–105.

Schmitt B., Quirica E., Trotta F., and Grundy W. M. (1998) Optical properties of ices from the UV to infrared. In *Solar System Ices* (B. Schmitt et al., eds.), pp. 199–240. Kluwer, Dordrecht.

Schriever R., Chergui M., and Schwentner N. (1991) Cage effect on the photodissociation of H_2O in Xe matrices. *J. Phys. Chem., 95,* 6124–6128.

Schriver A., Schriver L., and Perchard J. P. (1988) Infrared matrix isolation studies of complexes between water and sulfur dioxide: Identification and structure of the 1:1, 1:2, and 2:1 species. *J. Molec. Spectrosc., 127,* 125–142.

Schriver-Mazzuoli L., Chaabouni H., and Schriver A. (2003a) Infrared spectra of SO_2 and SO_2: H_2O ices at low temperature. *J. Mol. Struct., 644,* 151–164.

Schriver-Mazzuoli L., Schriver A., and Chaabouni H. (2003b) Photo-oxidation of SO_2 and of SO_2 trapped in amorphous water ice studied by IR spectroscopy: Implications for Jupiter's satellite Europa. *Can. J. Phys., 81,* 301–309.

Schubert G., Anderson J. D., Spohn T., and McKinnon W. B. (2004) Interior composition, structure, and dynamics of the Galilean satellites. In *Jupiter: The Planet, Satellites and Magnetosphere* (F. Bagenal et al., eds.), pp. 281–306. Cambridge Univ., Cambridge.

Shi M., Baragiola R. A., Grosjean D. E., Johnson R. E., Jurac S., and Schou J. (1995) Sputtering of water ice surfaces and the production of extended neutral atmospheres. *J. Geophys. Res., 100,* 26387–26395.

Shirley J. H., Carlson R. W., and Anderson M. S. (1999) Upper limits for sodium and magnesium hydroxides on Europa (abstract). *Eos Trans. AGU, 80,* F604.

Shoemaker E. M. and Wolfe R. F. (1982) Cratering time scales for the Galilean satellites. In *Satellites of Jupiter* (D. Morrison, ed.), pp. 277–339. Univ. of Arizona, Tucson.

Shoemaker E. M., Lucchitta B. K., Wilhems D. E., Plescia J. B., and Squyres S. W. (1982) The geology of Ganymede. In *The Satellites of Jupiter* (D. Morrison, ed.), pp. 435–520. Univ. of Arizona, Tucson.

Sieger M. T., Simpson W. C. and Orlando T. M. (1998) Production of O_2 on icy satellites by electronic excitation of low-temperature water ice. *Nature, 394,* 554–556.

Sill G. T. and Clark R. N. (1982) Composition of the surfaces of the Galilean satellites. In *Satellites of Jupiter* (D. Morrison, ed.), pp. 174–212. Univ. of Arizona, Tucson.

Smythe W. D., Carlson R. W., Ocampo A., Matson D., Johnson T. V., McCord T. B., Hansen G. E., Soderblom L. A., and Clark R. N. (1998) Absorption bands in the spectrum of Europa detected by the Galileo NIMS instrument. In *Lunar and Planetary Science XXIX,* Abstract #1532. Lunar and Planetary Institute, Houston (CD-ROM).

Spencer J. R. (1987a) The surfaces of Europa, Ganymede, and Callisto: An investigation using Voyager IRIS thermal infrared spectra. Ph.D. thesis, University of Arizona, Tucson.

Spencer J. R. (1987b) Thermal segregation of water ice on the galilean satellites. *Icarus, 69,* 297–313.

Spencer J. R. and Calvin W. M. (2002) Condensed O_2 on Europa and Callisto. *Astron. J., 124,* 3400–3403.

Spencer J. R., Calvin W. M., and Person M. J. (1995) Charge-coupled-device spectra of the galilean satellites: Molecular-oxygen on Ganymede. *J. Geophys. Res., 100,* 19049–19056.

Spencer J. R., Tampari L. K., Martin T. Z., and Travis L. D. (1999) Temperatures on Europa from Galileo photopolarimeter-radiometer: Nighttime thermal anomolies. *Science, 284,* 1514–1516.

Spencer J. R., Carlson R. W., Becker T. L., Blue J. S. (2004) Maps and spectra of Jupiter and the Galilean satellites. In *Jupiter: The Planet, Satellites and Magnetosphere* (F. Bagenal et al., eds.), pp. 689–698. Cambridge Univ., Cambridge.

Spencer J. R., Grundy W. M., Dumas C., Carlson R. W., and McCord T. B. (2005) The nature of Europa's non-ice surface components: High spatial and spectral resolution spectroscopy from the Keck telescope. *Icarus, 182,* 202–210.

Spinks J. W. T. and Woods R. J. (1990) *An Introduction to Radiation Chemistry.* Wiley and Sons, New York.

Spitsyn V. I., Mikhailenko I. E., and Morozova T. V. (1969) Use of infrared spectroscopy for studying the change in the nature of the bond in the SO_4^{2-} ion of the radioactive sulfates of some group I and II elements. *Dokl. Phys. Chem., 186,* 358–361.

Steudel R. and Eckert B. (2003) Solid sulfur allotropes. *Topics Curr. Chem., 230,* 1–79.

Steudel R. and Steudel Y. (2004) The thermal decomposition of S_2O forming SO_2, S_3, S_4, and S_5O — An ab initio MO study. *Eur. J. Inorg. Chem., 2004,* 3513–3521.

Steudel R., Holdt G., and Young A. T. (1986) On the colors of Jupiter's satellite Io: Irradiation of solid sulfur at 77 K. *J. Geophys. Res., 91,* 4971–4977.

Strazzulla G. and Moroz L. (2005) Ion irradiation of asphaltite as an analogue of solid hydrocarbons in the interstellar medium. *Astron. Astrophys., 434,* 593–598.

Strazzulla G., Baratta G. A., Leto G., and Foti G. (1992) Ion-beam-induced amorphization of crystalline water ice. *Europhys. Lett., 18,* 517–522.

Strazzulla G., Castorina A. C., and Palumbo M. E. (1995) Ion irradiation of astrophysical ices. *Planet. Space Sci., 43,* 1247–1251.

Strazzulla G., Leto G., Spinella F., and Gomis O. (2005) Production of oxidants by ion irradiation of water/carbon dioxide frozen mixtures. *Astrobiology, 5,* 612–621.

Strazzulla G., Baratta G. A., and Gomis O. (2007) Hydrate sulfuric acid after ion implantation in water ice. *Icarus, 192,* 623–628.

Swallow A. J. (1973) *Radiation Chemistry.* Wiley and Sons, New York.

Tegler S. C., Grundy W. M., Romanishin W., Consolmagno G. J., Mogren K., Vilas F. (2007) Optical spectroscopy of the large Kuiper Belt objects 136472 (2005 FY_9) and 136108 (2003 EL_{61}). *Astron. J., 133,* 526–530.

Teolis B. D., Vidal R. A., Loeffler M. J., and Baragiola R. A. (2005a) Radiolysis and trapping of O_2 and O_3 in water ice. *Bull. Am. Astron. Soc., 37,* 774.

Teolis B. D., Vidal R. A., Shi J., and Baragiola R. A. (2005b) Mechanisms of O_2 sputtering from water ice by keV ions. *Phys. Rev. B, 72,* 245422-1-9.

Teolis B. D., Loeffler M. J., Raut U., Famá M., and Baragiola R. A. (2006) Ozone synthesis on the icy satellites. *Astrophys. J. Lett., 644,* L141–L144.

Tiscareno M. S. and Geissler P. E. (2003) Can redistribution of material by sputtering explain the hemispheric dichotomy of Europa? *Icarus, 161,* 90–101.

Trafton L. (1981) A survey of Io's potassium cloud. *Astrophys. J., 247,* 1125–1140.

Tsai S. C. and Robinson G. W. (1969) Why is condensed oxygen blue? *J. Chem. Phys., 51,* 3559–3568.

van der Zwet G. P., Allamandola L. J., Baas F., and Greenberg J. M. (1985) Laboratory identification of the emission features near 3.5 μm in the pre-main-sequence star HD97048. *Astron. Astrophys., 145,* 262–268.

Varnes E. S. and Jakosky B. M. (1999) Lifetime of organic molecules at the surface of Europa. In *Lunar and Planetary Science XXX,* Abstract #1082. Lunar and Planetary Institute, Houston (CD-ROM).

Vidal R. A., Bahr D., Baragiola R. A., and Peters M. (1997) Oxygen on Ganymede: Laboratory studies. *Science, 276,* 1839–1842.

Voegele A. F., Loerting T., Tautermann C. S., Hallbrucker A., Mayer E., and Liedl K. R. (2004) Sulfurous acid (H_2SO_3) on Io? *Icarus, 169,* 242–249.

Volwerk M., Kivelson M., and Khurana K. K. (2001) Wave activity in Europa's wake: Implications for ion pickup. *J. Geophys. Res., 106,* 26033–26048.

Volwerk M., Paranicas C., Kivelson M. G., and Khurana K. K. (2004) Europa's interaction with Jupiter's magnetosphere. *35th COSPAR Scientific Assembly,* Paris, p. 313.

Waite J. H., Combi M. R., Ip W.-H., Cravens T. E., McNutt R. L. J., and others (2006) Cassini ion and neutral mass spectrometer: Enceladus plume composition and structure. *Science, 311,* 1419–1422.

Wamsteker W. (1972) Narrowband photometry of the galilean satellites. *Comm. Lunar Planet. Lab., 167,* 171–177.

Wamsteker W., Kroes R. L., and Fountain J. A. (1973) On the surface composition of Io. *Icarus, 23,* 417–424.

Warren S. G. (1984) Optical constants of ice from the ultraviolet to the microwave. *Appl. Optics, 23,* 1206–1225.

Warren S. G. and Brandt R. E. (2008) Optical constants of ice from the ultraviolet to the microwave: A revised compilation. *J. Geophys. Res., 113,* D14220, DOI: 10.1029/2007JD009744.

Whalley E. (1968) Structures of ice and water as investigated by infrared spectroscopy. *Dev. Appl. Spectrosc., 6,* 277–296.

Whalley E. and Bertie J. E. (1967) Optical spectra of orientation-ally-disordered crystals I. Theory for translational lattice vibrations. *J. Chem. Phys., 46,* 1264–1270.

Wood B. E. and Roux J. A. (1982) Infrared optical properties of thin H_2O, NH_3, and CO_2 cryofilms. *J. Opt. Soc. Am., 72,* 720–728.

Wysocki S. (1986) γ-radiolysis of polycrystalline magnesium oxide. *J. Chem. Soc. Faraday Trans., 182,* 715–721.

Yakshinskiy B. V. and Madey T. E. (2001) Electron-and photon-stimulated desorption of K from ice surfaces. *J. Geophys. Res., 106,* 33303–33307.

Zahnle K., Dones L., and Levison H. F. (1998) Cratering rates on the galilean satellites. *Icarus, 136,* 202–222.

Zahnle K., Schenk P., Levison H., and Dones L. (2003) Cratering rates in the outer solar system. *Icarus, 163,* 263–289.

Zahnle K., Alvarellos J. L., Dobrovolskis A., and Hamill P. (2008)

Secondary and sesquinary craters on Europa. *Icarus, 194,* 660–674. DOI: 10.1016/j.icarus.(2007)10.024.

Zeleznik F. J. (1991) Thermodynamic properties of the aqueous sulfuric acid system to 350-K. *J. Phys. Chem. Ref. Data., 20,* 1157–1200.

Ziegler J. F., Biersack J. P., and U. Littmark (1985) *The Stopping and Range of Ions in Matter.* Pergamon, New York. Available online at http://www.srim.org/.

Zolotov M. (2008) Oceanic composition on Europa: Constraints from mineral solubilities. In Lunar and Planetary Science Conference XXXIX, Abstract #2349. Lunar and Planetary Institute, Houston (CD-ROM).

Zolotov M. Y. and Mironenko M. V. (2007) Chemical evolution of an early ocean on Europa: A kinetic-thermodynamic modeling. In *Workshop on Ices, Oceans, and Fire: Satellites of the Outer Solar System,* pp. 157–158. LPI Contribution No. 1357, Lunar and Planetary Institute, Houston.

Zolotov M. Y. and Shock E. L. (2001) Composition and stability of salts on the surface of Europa and their oceanic origin. *J. Geophys. Res., 106,* 32815–32827.

Zolotov M. Y. and Shock E. L. (2003) Energy for biologic sulfate reduction in a hydrothermally formed ocean on Europa. *J. Geophys. Res., 108,* 3-1 5022, DOI: 10.1029/2002JE001966.

Surface Properties, Regolith, and Landscape Degradation

Jeffrey M. Moore
NASA Ames Research Center

Greg Black
University of Virginia

Bonnie Buratti
NASA Jet Propulsion Laboratory/California Institute of Technology

Cynthia B. Phillips
SETI Institute

John Spencer
Southwest Research Institute

Robert Sullivan
Cornell University

Europa's surface at the smallest scales is covered by a surficial debris layer, most likely the product of tectonic fracturing and crushing, attendant downslope mass wasting, and accumulation of debris in topographic depressions. Impact-generated debris plays a minor role. High-velocity charged-particle sputtering of ice probably dominates over impact gardening. Thermal frost segregation deposits frost at relatively high elevations while more refractory material accumulates in adjacent topographic lows. Photometric and thermal properties suggest fine-grained porous material at centimeter-scale depths. Radar indicates that the surface layer is porous and highly fractured to a depth of at least a few meters.

1. INTRODUCTION

The europan surface layer is defined in this chapter as the unconsolidated material lying in the top few meters (or less) of the surface. Indeed, the measured photometric and spectroscopic properties involve that part of the surface that is a few optical depths thick; thus, for example, infrared spectroscopy is sensitive to the top millimeter of the surface. This layer is sometimes called the remote sensing layer. In this review we will focus on photometric, thermophysical, and radar properties. Additionally, we will review to what extent Europa has a charged-particle sputtered and impact-generated regolith, and the nature of debris generated and distributed by nonimpact mass wasting.

An understanding of this layer is important in several regards. First, any attempt using remote sensing to interpret the nature of the icy shell (or crust), its underlying probable ocean, and the silicate/metal subice interior will necessarily be "filtered" by the intervening uppermost surface layer. All visible light and infrared imaging and spectroscopy will never detect photons from deeper than a few centimeters (and often from only the few uppermost micrometers). Radar can penetrate perhaps as much as a few kilometers, but return signals would have to be deconvolved from those produced by surface layer scatterers. Endogenic landforms are modified by exogenic surface layer processes, which must be accounted for in inferring the actual nature of the internal activity primarily responsible for these landforms. Indeed, an argument could be made that, at the least, an understanding of the surface layer is essential if for no other reason than to understand what lies and operates beneath it. When the day comes that probes are dispatched to the surface, detailed surface layer knowledge will be an absolute prerequisite for probe design, landing safety, and surface operations.

In the first sections, we review techniques for probing the surface layer and their implications. We begin with the properties of the uppermost surface (photometry) and move to those that probe greater depths (thermal and radar), and then review the effects of many small hypervelocity impacts and the jovian charged-particle environment on materials that are or were exposed to the surface. Next, we consider thermal ice-segregation and redistribution of volatiles on the surface. Finally, we review erosion and mass wasting of surface materials, which operate to modify high-standing and slope topography and produce rubble in topographic lows.

2. PHOTOMETRY

2.1. Introduction

Photometry is the quantitative measurement of reflected or emitted radiation from a celestial body. Photometric measurements of icy surfaces are sensitive to the distribution of volatiles, as well as complex optical and mechanical attributes of the surface, including the compaction state of the upper regolith, the degree of macroscopic roughness, and the size and size distribution of particles. The most elementary photometric measurements express the brightness of a body as a function of the subobserver longitude or solar phase angle (the angle between the Sun, the surface, and the observer). Broadband albedos, including color maps that show the distribution of volatiles and minerals, can be reliably created by modeling how the intensity of the surface changes with viewing geometry. Finally, photometric measurements can yield disk-integrated and disk-resolved thermophysical quantities such as the Bond albedo. Physical parameters derived from photometric studies provide important clues to the geologic history of a planetary surface, including its collisional history, resurfacing events, deposition of exogenous dust, mass wasting, volcanic events, and volatile transport.

Prior to the exploration of Europa by spacecraft, ground-based observers measured the rotational brightness variations of Europa to reveal it is about 0.3 visual magnitudes (~30%) brighter on the leading side than on the trailing side (*Morrison et al.,* 1974). Furthermore, the darker trailing side is redder than the leading side (*Morrison and Morrison,* 1977). The solar phase curve of Europa was found to be less steep than those of the other Galilean satellites, a finding that could be attributed to its higher albedo (*Morrison et al.,* 1974; *Millis and Thompson,* 1975; *Veverka,* 1977).

2.2. Photometry and Physical Properties

The maximum excursion in solar phase angle from Earth for the jovian system is only 12°. Voyager 1 and Voyager 2 expanded this range to 32° in contiguous measurements, and the spacecraft also obtained isolated measurements at 94°, 103°–109°, and 143° (*Buratti and Veverka,* 1983). Later, the New Horizons spacecraft obtained additional measurements of Europa at 35°, 71°, and ~130° during its flyby of Jupiter en route to Pluto (Fig. 1) (*Grundy et al.,* 2007). The Galileo and Cassini cameras obtained observations at additional viewing geometries, but the data have not been photometrically studied in a systematic matter, except for disk-integrated opposition surge measurements of Europa by the Cassini Visual Infrared Mapping Spectrometer (VIMS) (*Simonelli and Buratti,* 2004) and disk-resolved opposition surge measurements by the Galileo camera (*Helfenstein et al.,* 1998).

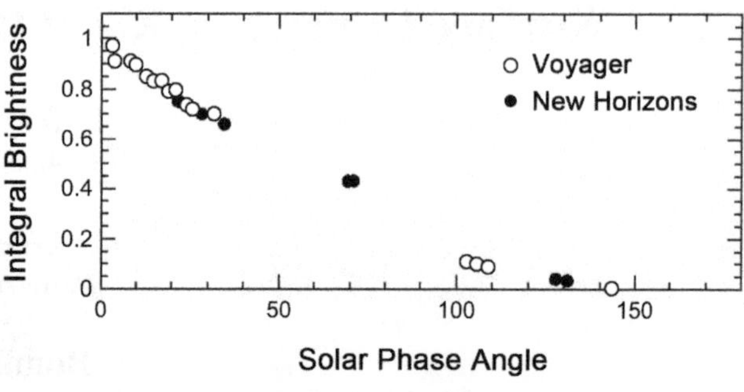

Fig. 1. Visible disk-integrated brightness of Europa derived from Voyager and New Horizons imaging experiments (*Grundy et al.,* 2007). Europa exhibits a more isotropic phase curve than other icy satellites: The brightness at 70° is only about 50% less than that near opposition.

Voyager observations showed that the steepness of Europa's solar phase curve was a strong inverse function of wavelength, perhaps because the albedo itself increases with wavelength (*Buratti and Veverka,* 1983). Higher albedos produce more multiply scattered photons to partly illuminate primary shadows. The disappearance of these shadows as a surface become fully illuminated to an observer causes a steeply rising phase curve (the so-called "opposition effect"). Europa was the first satellite for which it was realized that the simple lunar-like scattering law did not apply, again because of its high albedo and concomitant multiple scattering.

Two fundamental photometric parameters are the geometric albedo and the phase integral, both of which are wavelength dependent. The geometric albedo (p) is the integral brightness of a planet or satellite at a solar phase angle of 0° compared with a similarly sized disk that scatters isotropically. The phase integral (q) is the normalized integral brightness of a planetary body integrated over all solar phase angles. It is essentially the beam pattern of the object. The Bond albedo, or the spherical albedo, is the product of p and q. The bolometric Bond albedo is integrated over all wavelengths, and it expresses the amount of electromagnetic power coming from the body as a fraction of the incident power. The geometric albedo of Europa at 0.47 μm is 0.72 on the leading side and 0.62 on the trailing side (*Buratti and Veverka,* 1983). Based on the observations shown in Fig. 1, the Bond albedo at visible wavelengths is 1.01 ± 0.04 (*Grundy et al.,* 2007).

Photometric measurements can be fit to models that express the fraction of reflected radiation as a function of the physical characteristics of the surface. The most widely used model is that of *Hapke* (1981, 1984, 1986, 2002, 2008). The properties that can be modeled are the macroscopic roughness, which includes features ranging in size from clumps of particles to mountains and craters; the compaction state of the optically active portion of the upper regolith; the single-scattering albedo; and the single-particle phase function, which depends on the real and complex indices of refraction (and thus the composition) and the size, size distribution, and shape of the scatterers. The single-scattering albedo $\tilde{\omega}_0$ is the probability that a photon will be reflected *in any di-*

rection after it scatters. The single-particle phase function describes the probability that a photon undergoing a single scattering will exit in a given direction. To describe this function, we make use of the Henyey-Greenstein g (*Henyey and Greenstein,* 1941)

$$P(\cos\alpha, g) = (1 - g^2)/(1 + g^2 + 2g\cos\alpha)^{3/2} \quad (1)$$

where g is the asymmetry factor describing the directional scattering properties of individual particles and α is the solar phase angle. A g of –1 corresponds to pure backscattering, +1 corresponds to pure forward scattering, and 0 describes isotropic scattering.

The size of grains and their size distribution are important physical parameters that can also be derived from quantitative measurements of the light reflected from a planetary surface. In general, smaller particles tend to be more forward-scattering (at least in the geometric optics limit). The most accurate measurements of particle size can be obtained by studying the line positions of individual spectral components. Europa's spectrum (and those of the other icy Galilean satellites) are most closely matched by particle sizes in the range of 20 to several 100 μm (*Hansen et al.,* 2005). In contrast, the icy saturnian satellites have a component of submicrometer particles (*Hansen et al.,* 2005), which are possibly accreted E-ring particles.

The first photometric model fit to Europa (Fig. 2) implied that its surface was less rough than the Moon and the saturnian satellite Mimas (which is similar to Europa in albedo), with a surface mean slope angle of only 23° (*Buratti,* 1985). Further work was consistent with an even lower mean slope angle of 10° (*Domingue et al.,* 1991). Results from the International Ultraviolet Explorer (IUE) implied that the compaction state of Europa's regolith exhibited differences between its leading and trailing side: The fraction of void space on the leading side was inferred as 25%,

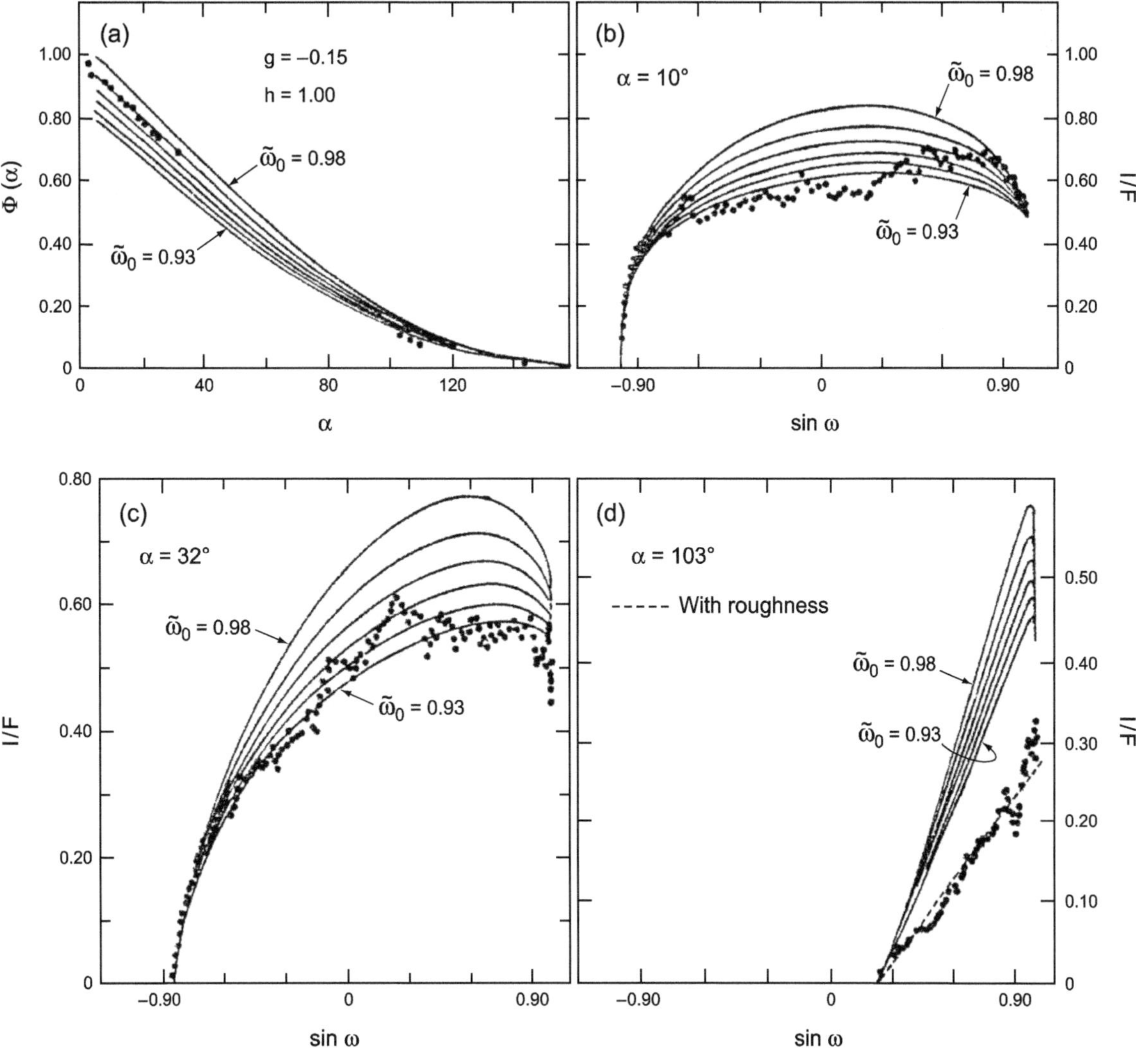

Fig. 2. Results of a photometric model for Europa. **(a)** The normalized disk-integrated solar phase curve, $\Phi(\alpha)$ where α is the solar phase angle, with models of single-scattering albedos ($\tilde{\omega}_0$) between 0.93 and 0.98. The roughness is defined by a mean slope angle of 23° while the parameter h is dependent on the compaction state of the regolith (*Hapke,* 1981). The Henyey-Greenstein g for this model is –0.15. **(b),(c),(d)** The model for disk-resolved scans of specific intensity (I/F) extracted from the photometric equator as a function of sin ω (where ω is the photometric longitude) for α = 10°, 32°, and 103°, respectively. Based on *Buratti* (1985).

TABLE 1. Summary of physical photometric parameters for Europa (numbers in brackets are references).

Single-Scattering Albedo, ϖ_0	Roughness (Mean Slope Angle, °)	Compaction State: Fraction of Void Space	Single-Particle Phase Function g
0.92 (leading)	10 [1]	25% [2]	–0.43; 0.079* [1]
0.90 (trailing)	10 [1]	79% [2]	–0.43; 0.287* [1]
0.97 (integral)	23 [2]		–0.15 [2]

*Two-term Henyey-Greenstein parameter including forward scattering.
References: [1] *Domingue et al.* (1991); [2] *Buratti* (1985).

but on the trailing side it was 79% (*Buratti et al.,* 1988). This difference could be due to recent activity on the high-albedo leading side, or to less intense magnetospheric bombardment there. Finally, the single-particle phase function is more isotropic than other icy satellites (*Buratti,* 1985), which suggests that Europa's surface particles are relatively transparent. In general, the phase functions of icy satellites are backscattering, while the phase function of terrestrial frost and snow is forward scattering (*Verbiscer and Veverka,* 1990). *Domingue and Verbiscer* (1997) attribute this difference to the complex nature of the structure and texture of surface particles, including nonspherical shapes and aggregated particles.

Compaction states implied by telescopic measurements of Europa at very small solar phase angles were not plausible when derived by the shadow-hiding model alone (*Domingue et al.,* 1991). One interpretation of this finding is that Europa exhibits coherent backscatter, a phenomenon in which multiply scattered photons traversing the same path exit coherently in the direction back to the scatterer (*Hapke,* 1990). Europa was earlier shown to exhibit coherent backscatter in radar data (*Ostro and Shoemaker,* 1990). Disk-resolved Galileo measurements of the opposition surge show substantial differences in the amplitude of the surge for specific terrains: The low-albedo units exhibit surges up to four times larger than the high-albedo, ice-rich plains (*Helfenstein et al.,* 1998). This result suggests the predominance of the shadow-hiding effect in the dark regions where primary shadows are not diluted by multiply scattered photons. A component of the surge due to coherent backscatter was predominant at very small (<0.2°) phase angles. This result was confirmed by Cassini observations at infrared wavelengths (*Simonelli and Buratti,* 2004). *Helfenstein et al.* (1998) attributed the textural character of the dark, red lineaments and ridges to the emission of particles from reactivated vents that expel material in an episodic fashion. In this scenario, younger, high-albedo surfaces surrounding the lineaments are originally composed of relatively compacted, large-grain material, but these regions mature into higher-porosity, darker material as ejecta from the vents accumulates in adjacent regions.

Although photometric models have been fit to a number of airless bodies, two major concerns surround this field of work: (1) How well can the models can be constrained? (2) How well do the idealized assumptions of the models reflect physical reality? The uniqueness of the models is strongly dependent on the amount of data collected, particularly on the total excursion and granularity of the solar phase angle. Particularly useful are measurements at small phase angles (less than approximately 12°), which can be isolated and fit to models of the opposition surge (*Irvine,* 1966; *Hapke,* 1984), and measurements at high solar phase angles, which are important for determining the degree of roughness, and the single-particle phase function. Also critical is the acquisition of disk-resolved measurements, where the degree of roughness can be determined independently of disk-integrated measurements by studying the functional form of the intensity of the surface near the limb of the satellites. Large solar phase angles (greater than 90°) are sensitive to both roughness and the forward scattering component of the single-scattering albedo. If observations of the opposition surge are used to define the compaction state of the upper regolith, and if disk-resolved measurements can be used to uniquely determine the roughness of the surface, the single-scattering albedo and the single-particle phase function can be well-constrained. Unfortunately, key data is missing for Europa, including published observations of the satellite at a full excursion of large solar phase angles. Table 1 summarizes our knowledge of photometric parameters for Europa.

The second common concern regarding photometric modeling is whether photometric parameters accurately describe the physical nature of planetary surfaces. Recent work has shown that Hapke's models do not correctly derive the known properties of a wide range of particulate samples that were measured in the laboratory at a wide range of viewing geometries (*Shepard and Helfenstein,* 2007; *Hapke,* 2008). Studies have shown that the failure to fully model multiple scattering renders roughness models inadequate for bright surfaces, and that a selection effect may prefer large features for bright surfaces because they more easily cast opaque shadows (*Buratti and Veverka,* 1985; *Helfenstein and Shepard,* 1999; *Shepard and Campbell,* 1998; *Shkuratov and Helfenstein,* 2001). However, much *comparative* information can be gleaned from photometric modeling, such as the comparison of macroscopic roughness among many objects of similar albedos, or the contrasting of a parameter for the leading and trailing sides of a body.

Detailed color and albedo measurements of specific terrains on Europa from Voyager images show that two fundamental types of material exist on Europa: bright ice comprising the plains, and darker, redder, silicate-rich material making up the brown spots and wedge-shaped bands (*Buratti and Golombek,* 1988). These results were refined considerably by measurements from Galileo, which show that the visible albedo of the darkest features is 0.17 at a solar phase angle of 5° (*Helfenstein et al.,* 1998). This value is comparable to low-albedo materials on Ganymede and Callisto, although the europan material is much redder. A comprehensive multispectral analysis of Europa based on Voyager images confirmed the result that all of Europa's surface can be modeled by two spectral end members: one represented by dark, red material comprising the dark spots, triple bands, and lineaments, and a second one represented by bright, icy plains material (*Clark et al.,* 1998). However, this study disagrees with the result of *Helfenstein et al.* (1998) that europan dark materials become brighter with time.

Domingue and Hapke (1992) analyzed the photometric properties of specific terrains on Europa. The single-scattering albedos in the Voyager clear filter (0.47 μm) range from 0.88 for the lowest albedo mottled terrain to 0.98 for the brightest plains. They found that the mean slope angle and the single-particle phase function are similar to disk-integrated values, except for the phase functions of two bright plains units. The brightest plains unit exhibit a strongly backscattering component, consistent with a high density of internal scatterers that may have been formed by ion bombardment and subsequent sputtering. A second plains unit, with a single-scattering albedo of 0.96, exhibit strongly forward-scattering behavior, suggesting clear, nearly spherical particles. Outside these exceptions, the global similarity of the single-scattering phase function is consistent with a relatively uniform particle size on Europa. Albedo differences are most likely caused by compositional differences — specifically different fractional additions of dark, red material — rather than particle size.

2.3. Albedo Maps of Europa

Most of the variation in intensity on a planetary surface is due to changing viewing geometry. A physical albedo, such as normal reflectance, can be derived if these changes are accounted for and modeled. Geophysical interpretations that incorporate albedo variegations can be made only when such corrections are done. A simple photometric function that expresses the specific intensity from a planetary surface and that is an approximation of the equation of radiative transfer is given by *Buratti and Veverka* (1983)

$$I/F = Af(\alpha)\cos i/(\cos i + \cos e) + (1 - A)\cos i \qquad (2)$$

where A is a parameter such that A = 1 is purely lunar-like scattering, in which multiple scattering is negligible, and A = 0 is Lambert scattering (which no planetary surface exhibits); α is the solar phase angle; i is the radiance incident angle; and e is the radiance emission angle. A is a function of wavelength and solar phase angle. Over the range of solar phase angles attained by Voyager, A is assumed a value between 0.54 and 0.77.

Another simple, but purely empirical, photometric function is given by Minnaert's equation

$$I/F = B_0(\alpha)\cos i^{k(\alpha)}\cos e^{k(\alpha) - 1} \qquad (3)$$

where B_0 is a brightness parameter that at $\alpha = 0°$ equals the normal reflectance, and $k(\alpha)$ is the limb darkening parameter.

Figure 3 shows a map of normal reflectance in the Voyager clear filter (0.47 μm) created from Voyager imaging data (*Johnson et al.,* 1983). Minnaert's equation was used to correct for the effects of limb darkening and solar phase variations. The results show that Europa is substantially darker on its trailing side, as suggested by its rotational lightcurve. Much of this dichotomy is due to the placement of darker, mottled terrain on its trailing hemisphere, but the effects of magnetospheric bombardment such as sputtering and implantation of sulfur from Io are also responsible for the leading-trailing pattern (see chapters by Carlson et al. and Paranicas et al.). The ultraviolet lightcurve of Europa underscores the importance of this effect: Sulfur is absorbing in the ultraviolet, and the amplitude of the rotational lightcurve of Europa is far larger in this spectral region (*Buratti and Veverka,* 1983; *McEwen,* 1986). Even with substantial magetospheric bombardment and darkening, Europa is the brightest of the icy Galilean satellites, an indication that its surface has far more fresh exposed water

Fig. 3. See Plate 18. A map of normal reflectances for Europa based on Voyager images (*Johnson et al.,* 1983). The values are green, 36–52%; yellow, 53–70%; red, >71%. The map is a global simple cylindrical projection centered on the antijovian point (0°, 180°W).

ice and fewer surface contaminants than Ganymede and Callisto. While most of Callisto's normal albedo at 0.47 μm is between 15% and 35%, and most of Ganymede's is between 35% and 52% (with a smattering of brighter fresh craters for both objects), Voyager data suggests Europa's normal albedos are mostly in the 53–70% range (*Johnson et al.,* 1983).

McEwen (1986) and *Nelson et al.* (1986) analyzed the separate contributions of endogenic and exogenic processes to Europa's albedo and color. They found that an exogenic pattern — caused primarily by ion bombardment, sputtering, and implantation of sulfur ions from Io — serve to darken and redden the surface, particularly in the ultraviolet (0.34 μm). These exogenic effects were maximized at the antapex of motion and gradually decreased in a sinusoidal pattern as a function of distance from this point (*McEwen,* 1986). The subtraction of this exogenic pattern from the underlying geologic units enables the estimation of albedos and colors that represent the underlying geology. In the Voyager orange filter (0.58 μm) the uncorrected normal reflectances range from 0.62 to 0.78, while the corrected values are 0.65 to 0.77. In the ultraviolet filter the corresponding values are 0.49 to 0.26 and 0.46 to 0.33. *McEwen* (1986) produced an albedo map of Europa in which the exogenic alterations of color and albedo were removed.

3. THERMOPHYSICAL PROPERTIES

3.1. Introduction

The thermophysical properties of Europa are derived from observations of surface temperatures (particularly those acquired from the same location) at different times of day. Surface temperatures, which are determined by diurnal heat flow within the top few centimeters of the surface and are sensitive to surface composition and morphology (for example, porosity and grain size and shape), can cast light on the various processes that shape Europa's uppermost surface: impact gardening, erosion (sputtering) by jovian magnetospheric ions (*Ip et al.,* 1998), insolation-controlled sublimation (*Spencer,* 1987a; *Clark et al.,* 1983), or endogenic processes. The best Voyager thermal measurements (*Spencer,* 1987a) had limited coverage and low spatial resolution (900 km).

The Galileo Photopolarimeter-Radiometer (PPR) (*Russell et al.,* 1992; *Orton et al.,* 1996) mapped thermal radiation from Europa's surface with a spatial resolution of 80–200 km. Near-blackbody radiation is expected from the surfaces of the icy Galilean satellites because of the high opacity of water ice (*Warren,* 1984), and Voyager thermal emission spectra of Europa were consistent with gray-body emission with a mean emissivity ε near 0.9 (*Spencer,* 1987a). Kinetic temperatures are slightly higher than the infrared brightness temperatures reported here, because $\varepsilon < 1$. Observations presented here (Figs. 4 and 5), with noise levels <1 K except for the noisier 17-μm data, were obtained during the day on Galileo orbits E6 (February 20, 1997) and G7 (April 5, 1997) at a wavelength of 27.5 μm, and orbit I25 (November 26, 1999) at a wavelength of 17 μm, and at night on orbit E17 (September 26, 1998), with a wide-open filter position sensitive to all radiation from 0.35 to ~100 μm.

3.2. Observations

Daytime brightness temperatures peak near 132 K for longitude 180°W (Fig. 4), consistent with ground-based disk-integrated radiometry (*Morrison,* 1977; *Blaney et al.,* 1999). Diurnal temperature variations (Fig. 5) constrain the surface bolometric albedo A and the effective thermal inertia Γ_e. If sunlight is absorbed at the surface, Γ_e is equal to the thermal inertia Γ, which is given by $\Gamma = \sqrt{(k\rho c)}$, where k is the surface thermal conductivity, ρ is the density, and c is the specific heat. Increasing Γ will increase nighttime temperatures. For a particulate surface, Γ variations tend to be dominated by variations in k, because of variations in compaction, grain size, or effectiveness of grain-to-grain heat flow (*Wechsler et al.,* 1972), although composition may also be important. In icy materials, however, sunlight may be absorbed below the surface (*Brown and Matson,* 1987), which also increases nighttime temperatures. Thermal inertia and sunlight penetration effects are difficult to distinguish on the basis of surface temperature measurements (*Urquhart and Jakosky,* 1996), and Γ_e as defined here may include contributions from Γ and from sunlight penetration. Because data from different times of the day come from different parts of the surface, Fig. 5 provides only an approximation to the diurnal temperature variation of any particular place on Europa's surface but gives some constraint on global A and Γ_e values. The solid curve on Fig. 5 assumes $\varepsilon = 0.9$, and a homogeneous thermal model with A = 0.55 and $\Gamma = 7 \times 10^4$ erg cm^{-2} s$^{-1/2}$ K^{-1}. This thermal inertia is 20 times lower than the value for solid ice, indicating a particulate surface, consistent with photometric and eclipse cooling studies (*Morrison,* 1977; *Helfenstein et al.,* 1998). These values match the observed diurnal brightness temperature range, but provide a poor fit to the data, which are more symmetrical around midday than the temperatures shown by the model. The difference may be largely because existing prenoon observations cover the darker and warmer trailing hemisphere of Europa (*Morrison,* 1977; *Blaney et al.,* 1999). However, unresolved vertical or lateral Γ_e inhomogeneities, required by eclipse observations of the icy Galilean satellites (*Spencer,* 1987a; *Morrison,* 1977), would also make the diurnal thermal profile more symmetrical (*Russsell et al.,* 1992). Our best-fit model has a bolometric albedo A, consistent with the photometrically derived value of 0.62 ± 0.14 (*Buratti and Veverka,* 1983), and gives an equatorial diurnal and seasonal mean surface temperature of 106 K, which serves as an important boundary condition for interior models.

In summary, the best constraints on surface thermal inertia and bolometric albedo are from Galileo PPR data, which has good spatial resolution during both day and night

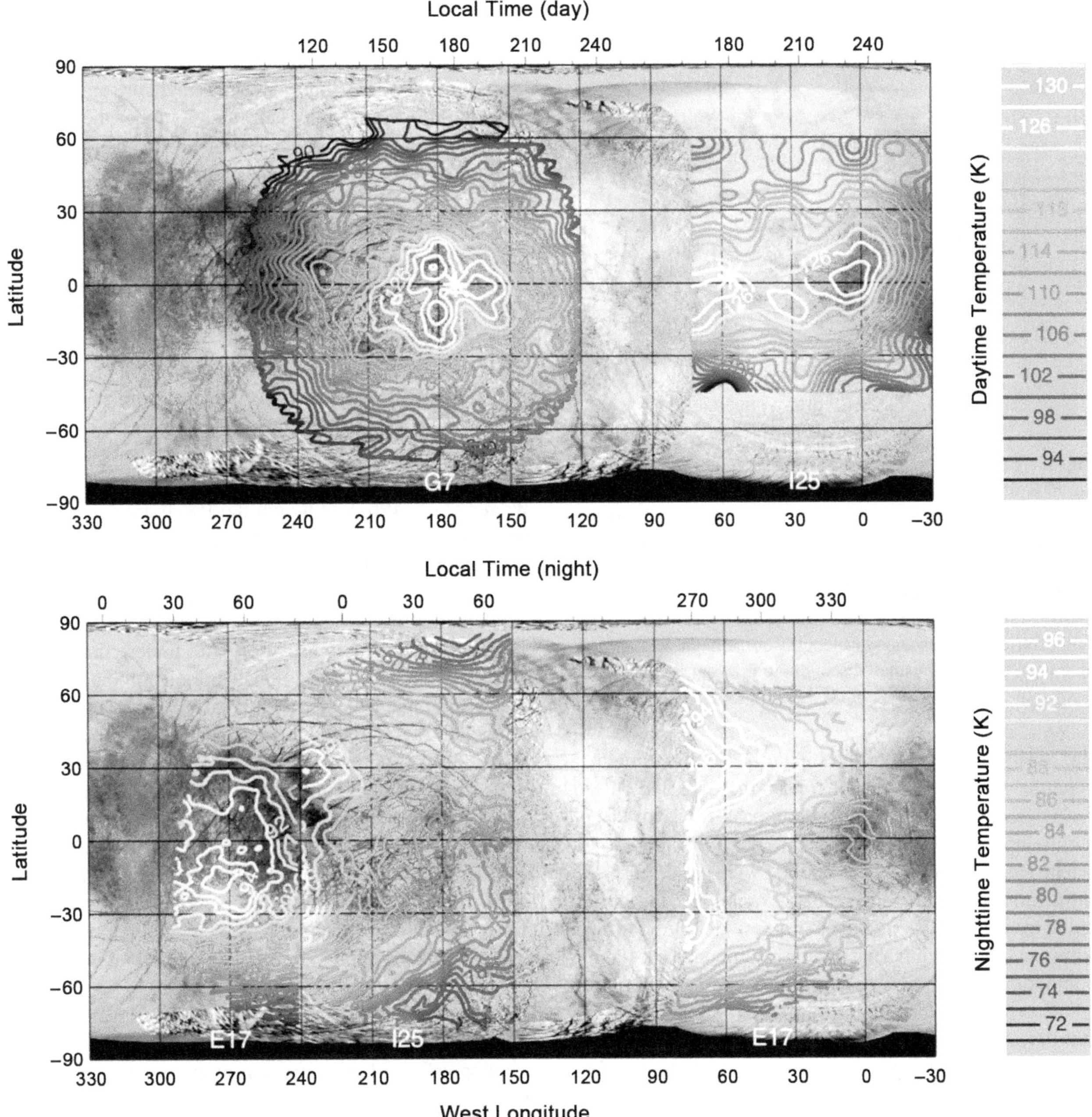

Fig. 4. See Plate 19. Contours of daytime (top) and nighttime (bottom) brightness temperature distributions on Europa inferred from Galileo PPR data (nighttime from Galileo orbits E17 and I25, and daytime from orbits G7 and I25). Contour interval is 1 K for the nighttime map and 2 K for the daytime map, and the contour color scheme is different for the two maps. Local time (top axis) is presented in degrees of rotation after midnight, and the subsolar points for the daytime data are shown by a white star. Base map is from Galileo and Voyager images.

(see *Spencer et al.,* 1999). Diurnal thermal inertia of about 5×10^4 erg cm^{-2} s$^{-1/2}$ K^{-1} is indicated, similar to the other Galilean satellites but higher than the saturnian satellites (e.g., *Abramov and Spencer,* 2008, 2009). The diurnal curve is not well matched by a simple homogeneous model, probably due to a combination of regional variations in thermal inertia and albedo, and local inhomogeneities. Groundbased observations of eclipse cooling indicate that some of the surface has lower thermal inertia still (*Morrison and Cruikshank,* 1973; *Hansen,* 1973), although to date there is no successful model of the entire eclipse cooling/heating curve.

Local daytime brightness temperatures correlate inversely with albedo, as expected from equilibrium with absorbed sunlight (Fig. 4). On a global scale, the same is true at night. Longitudes 0° through 75°W, observed after sunset, with a typical 0.48-μm normal albedo of 0.65 (*McEwen,* 1986) (see section 2.3), are generally colder than longitudes 240° through 300°W seen before dawn, where the typical albedo

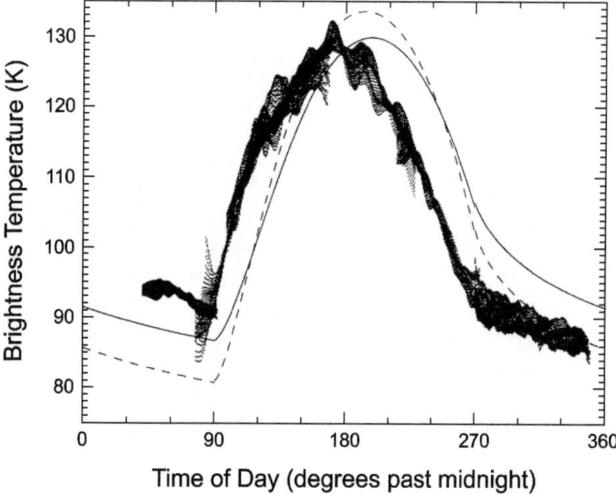

Fig. 5. Europa diurnal brightness temperature profile, combining PPR data from Galileo orbits E6 and G7 (daytime) and E17 (nighttime). Only regions within 15° of the equator, observed at less than 65° from vertical, are included. Shown are model diurnal temperature profiles for surfaces with A = 0.55 and $\Gamma = 7 \times 10^4$ erg $cm^{-2}\ s^{-1/2}\ K^{-1}$ (solid line) and $\Gamma = 4.5 \times 10^4$ erg $cm^{-2}\ s^{-1/2}\ K^{-1}$ (dashed line). The models were run separately for the different heliocentric distances and wavelengths of the daytime and nighttime data, resulting in small discontinuities at sunrise and sunset.

is 0.5. However, much of the smaller-scale nighttime temperature variability does not follow the expected rule of darker being warmer. For example, peak predawn brightness temperatures (95 K near 15°S, 280°W) include the bright ejecta blanket of the crater Pwyll (26°S, 271°W). Most remarkable is the systematic variation with latitude of the leading hemisphere temperatures: northern latitudes are warmer than equivalent southern latitudes, and there is a temperature minimum along the equator. At longitude 55°W, brightness temperatures are 94.5, 88.0, and 90.5 K at 30°N, 2°S, and 30°S, respectively. Voyager 0.48 μm images of these longitudes show a global-scale trend of increasing albedo away from the equator (*McEwen,* 1986), perhaps related to the thermal pattern, although opposite the trend expected if temperature was controlled directly by albedo. However, the Voyager images and the best Galileo images of this longitude range (with a resolution of 12 km $pixel^{-1}$, used as the base map in Fig. 4) also show albedo patterns on a 200-km scale that are comparable in magnitude to the equator-to-pole variations and are not reflected in the PPR temperature distribution.

Emissivity variations are unlikely to produce the observed large variations in broadband thermal emission, as decreased emissivity will increase surface kinetic temperature by inhibiting radiative heat loss. Difficulties with this and other alternative explanations for the observed nighttime temperature anomalies point to variation in Γ_e as the most likely cause. A change in Γ_e from 7×10^4 to 4.5×10^4 erg $cm^{-2}\ s^{-1/2}\ K^{-1}$ (produced, for instance, by a 2.4-fold drop in k) will reduce equatorial temperatures by 5 K at a local time 30° after sunset (Fig. 5), comparable to the magnitude of the observed thermal anomalies. Daytime observations of longitudes 0°–80°W obtained during Galileo orbit I25 (Fig. 4) confirm this hypothesis: The midlatitude regions that are anomalously warm at night are anomalously cold during the day, consistent with a higher thermal inertia than their surroundings. The difficulty then lies in explaining the spatial patterns of Γ_e variation inferred from the observations. Higher Γ_e values near the crater Pwyll might result from an increased abundance of large regolith particles in the ejecta blanket (of size comparable to or greater than the diurnal skin depth, which is 4 cm for our best-fit model and an assumed regolith density of 0.5 g cm^{-3}), or from 1 to 2 cm of sunlight penetration (*Urquhart and Jakosky,* 1996) in the ice-rich material of the ejecta. Lunar eclipse thermal data show a somewhat similar pattern of warm temperatures on crater ejecta blankets within 1 crater diameter of the rim, produced by the enhanced abundance of rocky blocks there (*Saari and Shorthill,* 1963; *Wildey et al.,* 1967), although the Pwyll anomaly is much larger, extending more than 250 km (10 crater diameters) from the rim, and is not perfectly centered on the crater. The leading-side nighttime temperatures are even more problematic, because they are correlated with latitude but there is no observed correlation with underlying geology. Indeed, simple theoretical considerations predict an opposite trend to that observed: The high daytime temperatures at low latitudes will tend to cause grain sintering (*Eluszkiewicz,* 1991), which will increase k and thus Γ_e at low latitudes. Large, sintered grains allow greater sunlight penetration, which also tends to increase Γ_e. Surface texture could also be affected by the impact of jovian magnetospheric ions, but large gradients in ion flux with latitude in the 0°–75°W region are not expected (*Pospieszalska and Johnson,* 1989), especially as Galileo has ruled out an intrinsic magnetic field strong enough to deflect bombarding ions (*Kivelson et al.,* 1997).

The nighttime temperature observations can also be used to search for small, local, endogenic thermal anomalies ("hot spots") on Europa, which might persist for decades after local geological activity has ceased (*Abramov and Spencer,* 2008). No such endogenic hot spots have been detected. Upper limits to hot spot circular-equivalent diameter and temperature in the 50% of Europa's surface covered by the most sensitive observations (the nighttime coverage shown in Fig. 4, lower panel) are 16.8 km at 130 K, 6.2 km at 200 K, 3.4 km at 273 K, or 2.0 km at 350 K. This is much fainter than a brief thermal event tentatively identified in 1981 groundbased observations (*Tittemore and Sinton,* 1989) but not reproduced.

4. RADAR STUDIES

4.1. Introduction

Radar provides a view of the surface and near subsurface, probing at least several wavelengths deep into typical non-ice targets but potentially much deeper into the low absorption layers of predominantly icy surfaces. The strength, polarization, and angular scattering characteristics of the returned echoes contain information on the scattering mechanism, and in turn the structure and composition of the layers from which those echoes originate. Here we review short, decimeter wavelength radar studies, which are sensitive to the upper regolith to depths of at most a few hundred wavelengths or tens of meters. Longer wavelength radars at wavelengths of several to tens of meters or more could penetrate proportionately deeper, sensing structures or liquid layers at kilometer depths (see chapter by Blankenship et al.).

4.2. Observations

The first radar detection of Europa was made in late 1975 using the 12.6-cm wavelength radar system at the Arecibo Observatory as part of a campaign to measure the radar properties of all Galilean satellites (*Campbell et al.,* 1977). Those observations were met with some surprise as they indicated Europa had an extremely high reflectivity, an inverted polarization signature, and entirely diffusive scattering behavior. Such properties had not previously been seen from any other object. Figure 6 shows a typical radar spectrum of Europa, which results from the spread of Doppler shifts induced by Europa's rotation on the single transmitted frequency. Such data can be thought of as scanning a slit across the disk of the target; more specific details of radar techniques and measured quantities can be found elsewhere (e.g., *Ostro,* 1993).

Those initial results would be confirmed in additional observations using Arecibo (*Campbell et al.,* 1978; *Ostro et al.,* 1980) as well as the Goldstone radar system, which operates at a wavelength of 3.5 cm (*Ostro et al.,* 1992). At both wavelengths Europa's radar reflectivity was about an order of magnitude larger than any other inner solar system target that had yet been observed. When the transmitted signal was circularly polarized, as is a standard experimental configuration, Europa's echo was inverted, in the sense that more power was returned in the same circular polarization as the transmitted signal, denoted σ_{SC}, vs. the opposite circular polarization, or σ_{OC}. Single reflections from a surface such as dominate the scattering from the Moon, Mars, Venus, and Mercury will switch the signal's polarization and return more power in the opposite circular sense instead. The echoes from those inner solar system targets in the stronger opposite circular polarization are dominated by this quasispecular surface reflection component near the center of the disk where the incidence angle is near normal, e.g., the center spike in the lunar-like spectrum in Fig. 6. At higher incidence angles a multiple scattering diffuse component contributes, which is typically well-described empirically by a $\cos^n\theta$ scattering law where θ is the incidence angle and n is usually in the range 1–2 (Table 2). Europa's echoes show no sign of the quasispecular component, with scattering at all angles behaving as a diffuse scattering component with an n of 1.5–1.8 (*Ostro et al.,* 1992).

TABLE 2. Europa's bulk radar properties.

	3.5 cm*	12.6 cm*	70 cm‡
σ_{OC}	0.91 ± 0.13	1.03 ± 0.08	≤0.17
σ_{SC}	1.40 ± 0.23	1.58 ± 0.14	≤0.17
μ_C	1.43 ± 0.24	1.53 ± 0.03	—
μ_L	—	0.43 ± 0.08†	—
n	1.7 ± 0.4	1.8 ± 0.2	—

*3.5-cm and 12-cm wavelength albedos, circular polarization ratio ($\mu_C = \sigma_{SC}/\sigma_{OC}$), and scattering law exponent n of $\cos^n\theta$ are from *Ostro et al.* (1992).

†12.6-cm wavelength linear polarization measurement ($\mu_L = \sigma_{SL}/\sigma_{OL}$) is from *Campbell et al.* (1977).

‡70-cm wavelength upper limits are from *Black et al.* (2001a).

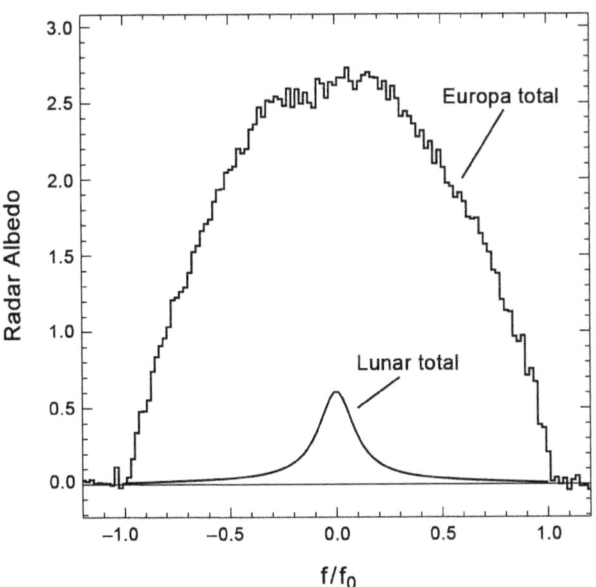

Fig. 6. A typical Doppler radar spectrum of Europa compared to a representative spectrum of Earth's Moon. The total radar albedo is shown, which is the total echo power normalized by the target's projected area. The echo is spread in frequency by a target's rotation, and the frequency axis has been normalized to each target's half Doppler bandwidth, f_0, which depends on target size and rotation rate.

Some variations in reflectivity and polarization properties with longitude are seen; for example, the little structure in the spectrum of Fig. 6. However, there has been only a minor effort to compare to other imaging data at other wavelengths to identify the radar features with any surface units. Such efforts are impeded by the nature of the one-dimensional radar spectra for which the constant proximity of the subradar point to Europa's equator makes it difficult to distinguish the latitude of any feature. *Ostro* (1982) reported a tentative alignment of a high radar polarization feature with dark regions at a longitude of 246° by using two radar spectra that overlap that region but from different rotation aspects. A more systematic mapping of radar properties with longitude was made by *Ostro et al.* (1992). From radar spectra centered at multiple longitudes, a crude inversion was made of the spectra into albedo maps and some latitude resolution was possible, although the maps were still north-south hemisphere ambiguous. The major features were lower radar reflective equatorial and polar regions and higher radar reflective midlatitude areas.

Observations made in 1988 and 1990 with the Arecibo 70-cm-wavelength radar (*Black et al.,* 2001a) failed to convincingly detect an echo from Europa, indicating that its reflectivity at 70 cm wavelength was roughly an order of magnitude less than at the shorter two wavelengths. Simultaneous 70-cm-wavelength observations of Ganymede and Callisto detected them but still showed a similar decline in radar reflectivity from shorter wavelengths. Their polarization remained inverted, suggesting that perhaps Europa's might also remain inverted at 70 cm; however, the echoes were below the detection threshold.

Use of the standard two-dimensional radar imaging technique where the echo is resolved in both frequency and time, also known as delay-Doppler mapping (cf. *Ostro,* 1993), to image Europa is complicated by its large size and relatively fast rotation, which makes it difficult to obtain sufficiently narrow resolutions in both dimensions simultaneously. A technique known as "long code" (cf. *Harmon et al.,* 1992) largely avoids this problem, although it provides reduced sensitivity. *Harmon et al.* (1994) were the first to successfully apply long code to Ganymede and Callisto, and *Harcke et al.* (2000) and *Harcke et al.* (2001a) applied the technique to Europa. Although that technique can produce a two-dimensional image of the radar properties of the surface, the reduced sensitivity has not permitted high enough contrast images to result in any useful correlation of radar properties with geologic or albedo features seen at other wavelengths. Another technique is to make a direct image of the target under radar illumination. *Harcke et al.* (2001b) have used this method to produce resolved images of Ganymede and Callisto with the Goldstone radar, and the Very Large Array as the imaging system. However, there have been no reported efforts to use this technique on Europa.

4.3. Models

It is clear that Europa's scattering properties are the result of the radio-wavelength transmission characteristics of cold water ice. Ganymede and Callisto have radar properties similar to those of Europa but with a decrease in overall albedos, and it was immediately recognized (*Campbell et al.,* 1977) that these properties were correlated with ice fraction on the surface. The radar reflectivities decrease in the following order: Europa, to Ganymede, to Callisto. This trend is in line with the decrease in icy purity on the surface (cf. *Calvin et al.,* 1995). Clean water ice at cold (~100 K) temperatures has an extremely low absorption coefficient at radio wavelengths (*Thompson and Squyres,* 1990; *Warren,* 1984), permitting penetration depths of hundreds or more wavelengths. The addition of contaminants such as meteoritic material or salts will increase the absorption. While pure ice permits long penetration depths into the surface, that alone is not sufficient to explain the echoes. Scattering centers within the ice are needed to reflect power back to the observer. Generically, these must be some contrast in dielectric constant, due to either changes in composition or density.

Several mechanisms have been explored to qualitatively explain the radar properties of Europa (and Ganymede and Callisto), although little has been done to interpret those models quantitatively in terms of physical subsurface parameters. All models must reproduce high reflectivity and an inverted polarization ($\mu_C = \sigma_{SC}/\sigma_{OC} > 1$), vs. what occurs for single scattering from a fairly smooth surface ($\mu_C \sim 0$). This is usually either accomplished by special geometries that include multiple reflections, or via a series of scattering events that are predominately forward scattering, thus tending to preserve the initial polarization. The lack of observable quasispecular surface reflection means the surface is either extremely rough, for which it may be physically difficult to realize a rough enough surface, or the impedance difference between it and free space is quite low, as may result from a very porous or finely powdered water ice regolith.

The first effort to explain this type of reflectivity and polarization behavior (actually from Ganymede) was made by *Goldstein and Green* (1980), who ran a Monte Carlo simulation to follow photons in a geometric optics calculation as they scattered between discrete fragments of solid water ice. They were able to obtain a reasonable match to the observations and found that, in this scenario, most of the scattering contributing to the echo was from internal scattering in the fragments off the ice-to-vacuum interfaces.

Double bounces off two interior walls of spherical surface craters were first invoked as an explanation by *Ostro and Pettengill* (1978). In this model, each bounce reverses the circular polarization, naturally resulting in the echo being dominated by the same polarization sense as the original signal. While simple in concept, this model may place fairly stringent requirements on the necessary crater geometry and requires a dielectric constant at least as high as solid water ice, but possibly higher. Thus, it may be implausible to have sufficient numbers of structures satisfying these exact conditions to produce the observations (*Ostro and Shoemaker,* 1990).

A model first proposed by *Eshleman* (1986a) improves on the spherical crater model by burying those craters under an overburden of fresher water ice, which is thus denser and has a higher dielectric constant than the older fractured ice layer containing the crater. This configuration can lead to total internal reflection from the crater walls within the overlying layer, favoring paths with reflections from multiple sides of the crater walls and preserving the initial polarization sense. Such a geometry might naturally result from older craters being covered by fresher eruptions of water, or from a process late in crater formation that results in a slight densification of the crater walls themselves. This model has been extended (*Gurrola and Eshleman,* 1990; *Gurrola,* 1995) and can reproduce the observed scattering law and wavelength dependencies. The necessary geometry is near, but not exactly, spherical. Deviations from symmetric craters are required, as perfect symmetry would actually produce a null in the exact backscatter direction as confirmed by numerical electromagnetic scattering computations (*Baron et al.,* 2003). As conceptually instructive as these geometric models are, however, actual europan topography has few craters, and they do not have these shapes (see chapter by Schenk and Turtle).

Hagfors et al. (1985, 1997) explored an explanation using refraction lenses, locations in the surface where a gradual change in the index of refraction slowly bends the radar signal around into the backscatter direction without reflection, thus preserving the polarization. Structures with a solid water ice index of refraction (~3.1), falling to near vacuum (~1) over the scale of a few wavelengths, are able to produce the high radar reflectivity and possibly the inverted polarization ratio (*Eshleman,* 1986b). The origin of such structures may be difficult to explain, but *Hagfors et al.* (1997) suggest, as one possible cause, that the introduction of high refractive index from meteoritic material could serve as the centers of such structures. Alternatively, they also note that somewhat similar structures are thought to be responsible for high radar reflectivity of portions of the Greenland ice sheet (*Rignot,* 1995); however, in that case seasonal melting and refreezing is their origin.

Hapke (1990) first suggested the explanation to be a coherent backscatter effect similar to what has been observed from planetary surfaces at optical and infrared wavelengths at times of opposition (see section 2.2). The effect arises from multiple scatterings within a low loss medium, whereby a coherent enhancement occurs around the exact backscatter direction. The polarization sense is largely preserved, as the individual scattering events that make up the full path are predominantly forward scattering, rather than the polarization-reversing scattering that would result from a single backscattering event. This mechanism does not place any specific requirements on the type or form of the scatterers. Although only exact backscatter is seen from groundbased observations, the magnitude and angular width of this effect depend on the ratio of the wavelength to the mean distance between scattering events, as well as the efficiency of those scattering events and the absorption in the medium between them. *Black et al.* (2001b) modeled the change in properties from 3.5 to 70 cm wavelength by considering a distribution of scatterer sizes within the coherent backscatter effect formulation of *Peters* (1992), which includes polarization, loss, and anisotropic scatterers. The wavelength dependence, most notably the drop in reflectivity between 12.6 and 70 cm wavelength, could be reproduced by a fairly steep distribution of scatterers such that few were available on the scale of the largest wavelength, essentially reducing the reflectivity by removing available scattering centers rather than by increasing any absorption losses. Models with ice scattering centers denser than the surrounding medium best match the data; more absorbing scatterers such as meteoritic or salt inclusions would not be efficient enough to produce the albedos in this scenario. Models with void spaces as the scattering centers within otherwise solid ice also match the data well. An alternative formulation of the coherent backscatter effect by *Mishchenko* (1992) that also includes polarization by using scattering matrices qualitatively produces similar albedo and polarization predictions, but it has not yet been matched to the observations in detail.

Correlation of radar data with geological and albedo maps at optical and infrared wavelengths is still a largely unexplored area that should prove fruitful. Future groundbased efforts to obtain higher-quality data will need improved sensitivity to make the long code imaging or direct imaging more productive. A huge improvement in the interpretation of radar scattering data is possible if scattering at other angles besides exact backscatter can be measured. This might be done, for example, by using a radar transmitter on a spacecraft in the Jupiter system with the receiving system on Earth, or vice versa.

5. CHARGED-PARTICLE AND IMPACT-GENERATED REGOLITH

5.1. Introduction

Most airless bodies in the solar system are observed or inferred to have an uppermost layer, usually referred to as regolith, of loose material, generated by the rain of mostly small but sometimes large impactors. The basaltic maria of Earth's Moon, for instance, are typically covered with several meters of regolith, which has formed over an impact exposure period in excess of 3 b.y. Moreover, many of the satellites of the other planets orbit within the intense charged-particle environment of their primaries' magnetosphere. Water-ice covered Europa certainly is such a satellite (see chapter by Paranicas et al.). The density and energy of these charged particles, in combination with H_2O ice's relatively weak bonds, results in the alteration of ice exposed at Europa's surface.

We define Europa's regolith as the uppermost loose layer that is produced by a combination of particle bombardment and impact overturn. Below we first summarize the processes of ultraviolet and charged-particle bombardment and sputtering (for further discussion, see chapters by Carlson et al., Paranicas et al., and Johnson et al.). Then we consider impact gardening and regolith generation, and how this competes with charged-particle bombardment (for complementary discussion, see chapter by Bierhaus et al.).

5.2. Ultraviolet and Charged-Particle Bombardment

Europa's charged-particle and ultraviolet (UV) radiation environment has been measured by the Galileo spacecraft and is summarized by *Cooper et al.* (2001) and in the chapter by Paranicas et al. The solar irradiation of the surface of Europa is smaller than that at Earth's distance from the Sun by a factor $(5.2)^2 \approx 27$ and totals 8×10^{12} keV cm^{-2} c^{-1}; the incident ultraviolet-C (UV-C) flux ($\lambda < 2800$ Å) is 4 ×

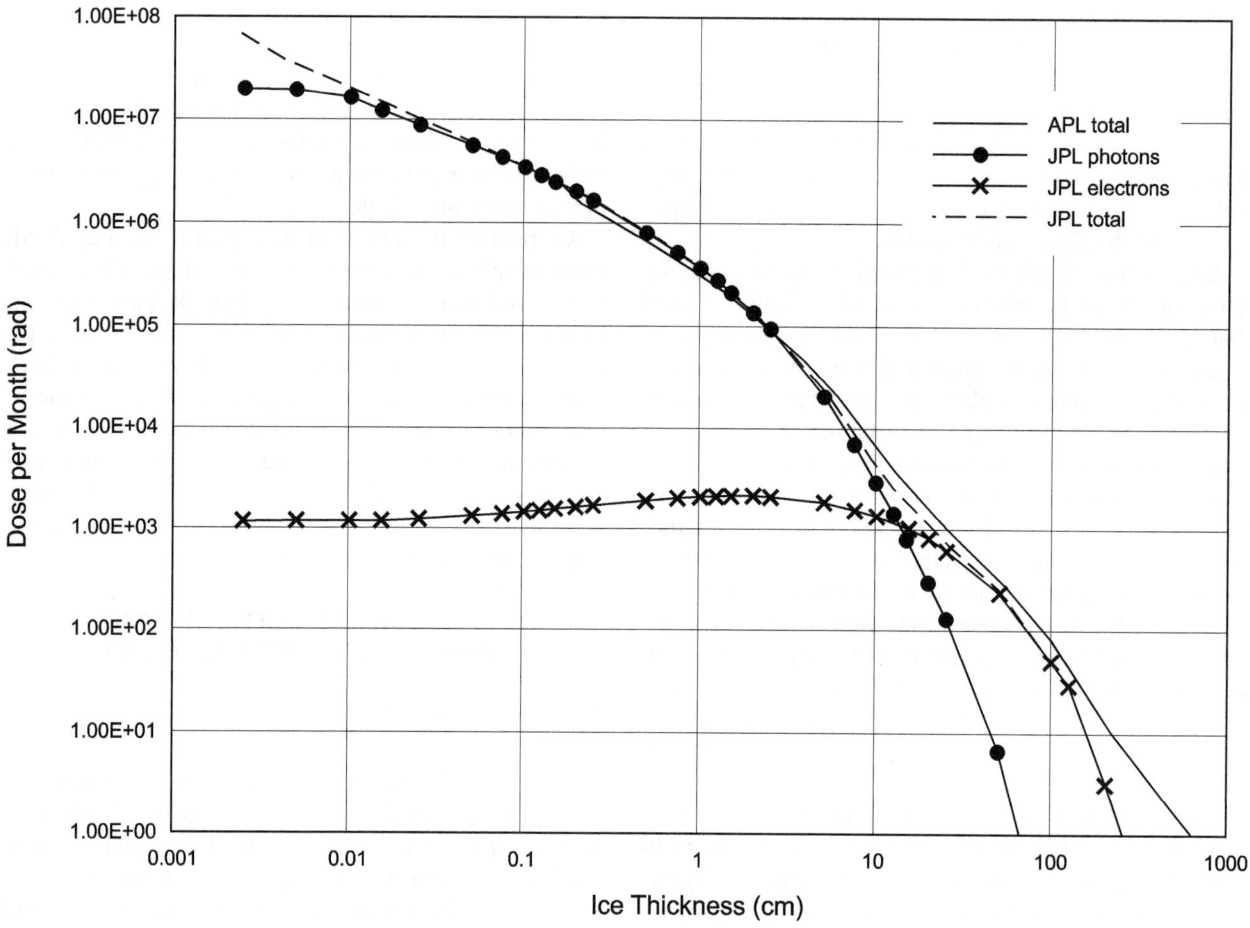

Fig. 7. Radiation dose in ice. The dose in units of rad in a water-ice surface per month, beneath a given thickness of ice. After *National Research Council Space Studies Board* (2000), with data and models provided by researchers at the Jet Propulsion Laboratory (JPL) and the Applied Physics Laboratory (APL).

10^{10} keV cm^{-2} c^{-1}. Energies greater than ~4.4 eV, corresponding to photon wavelengths below 2800 Å, are required to dissociate H_2O, so UV-C fluxes are those that are relevant to radiation chemistry.

Europa's harsh charged-particle radiation environment is primarily due to the acceleration of particles in Jupiter's magnetosphere. The charged-particle incident energy flux of about 8 × 10^{10} keV cm^{-2} s^{-1} is about twice that due to UV-C. It is dominated by energetic electrons in the keV to MeV range, but includes H^+, O^{n+}, and S^{n+} ions as well. Radiation processing depths depend on the density of the surface, which could range from that for solid ice (0.92 g cm^{-3}) to that for water frost (~0.1 g cm^{-3}). Because the density of Europa's uppermost surface is unknown, penetration depths are presented in terms of mass cross sections. The average stopping depth for the most penetrating charged particles, the electrons, is 0.62 g cm^{-2}, corresponding to a depth of ~0.7 mm in density 0.92 g cm^{-3} ice. These depths are much greater than those penetrated by UV photons, which affect only the top submicrometer layer; for example, Lyman-α photons (λ = 1216 Å) have a mean penetration depth of only 0.04 μm in ice.

A crude understanding of the implications of the incident radiation fluxes cited above for materials or biology can be obtained by calculating net volume radiation dose rates as a function of depth into Europa's ice (Fig. 7). Depths of 1 cm in Europa's ice (assuming an ice density of 1 g cm^{-3}) experience a volume dosage rate of ~0.3 Mrad per month (*Cooper et al.*, 2001). This dose is hundreds of times higher than the lethal dose for human beings — a lethal dose for 50% of humans is ~650 rad. The most radiation-resistant terrestrial organism known, the $R_{II}5$ strain of *Deinococcus radiodurans*, has 90% survival after 6 Mrad, dropping to 10^{-6} survival after 12 Mrad of ionizing radiation dose (*Auda and Emborg,* 1973). Radiation doses fall rapidly with depth in Europa's ice, so that at meter depths, monthly doses are ~100 rad; doses drop to values similar to those in Earth's biosphere at depths of 20–40 m.

5.3. Sputtering

Sputtering is erosion of a surface by charged-particle impacts (see chapter by Johnson et al.). The best estimate for the sputtering erosion rate on Europa due to energetic

H^+, O^{n+}, and S^{n+} ions is about 0.02 μm yr^{-1} for erosion of surface molecules of H_2O (*Cooper et al.,* 2001; see chapter by Johnson et al.), although some previous estimates have been as much as 100 times higher. Over Europa's ~50-m.y. surface age, sputtering thus should have removed ~1 m of material, with redeposition of some frost just as it is by the process of frost segregation (see section 6). Sputtering at these high energies is explosive, with up to 1000 water molecules removed from the surface for a single incident high-energy heavy ion (see chapter by Johnson et al.).

Europa's tenuous atmosphere is addressed in chapters by McGrath et al. and Johnson et al. It is mostly due to oxygen sputtered from its surface ice, and is in fact a near-vacuum with a surface number density of O_2 gas of ~10^8–10^9 cm^{-3} (*Ip,* 1996; see chapter by McGrath et al.), equivalent to that at ~400 km altitude in Earth's atmosphere. Atomic sodium is also present in Europa's atmosphere and could come either from an exogenic source such as material ejected volcanically from Io, or perhaps from an endogenic europan source.

5.4. Impacts and Gardening

Europa, like all other solar system bodies, is subject to impacts by comets, asteroids, and smaller objects. Gravitational focusing by Jupiter's large mass enhances the impact flux experienced by Europa. The principal objects impacting Europa and the other bodies in the jovian system are thought to be Jupiter-family comets (*Zahnle et al.,* 1998; *Schenk et al.,* 2004), of which a spectacular example was the impact of Comet Shoemaker-Levy 9 with Jupiter in 1994. Shoemaker-Levy 9 was gravitationally disrupted into multiple pieces during an earlier pass by Jupiter, and then impacted Jupiter in a series of well-documented collisions.

The number of craters on planets and satellites can be used to gauge both relative and, with additional information, absolute surface ages (see chapter by Bierhaus et al.). Europa's surface has only about 15 impact craters with diameters greater than 10 km (*Moore et al.,* 2001), and the majority of small craters seem to be secondary craters formed by debris thrown out from these large events (*Bierhaus et al.,* 2001). This lack of craters implies a young surface, because recent (or perhaps current) geological activity is required to erase them. Bodies with current geological activity such as Io and Earth have very few recognizable impact craters — Io, in fact, is so volcanically active that not a single impact crater, of any size, has been found on its surface to date. The Earth has about 150 recognized craters, but many have been geologically modified and would be difficult to recognize from orbit. On the other hand, geologically inactive bodies with old surfaces, such as Earth's Moon or Callisto, are covered with impact craters of all sizes. The most densely cratered surfaces are said to be in "equilibrium," as each new crater destroys old craters beneath and immediately surrounding it.

Relative ages can be deduced by comparing the number density of craters on different parts of a planet or satellite's surface. Europa's crater density is so low, however, that no part of the surface has been reliably inferred to be younger or older than any other. With respect to absolute age, various efforts have been made to establish impactor fluxes throughout the solar system's history in order to establish the actual age of a body's surface based on its crater density. From calculations of the comet and asteroid fluxes at Jupiter, the age of Europa's surface has been estimated to be only ~50 m.y. old (*Zahnle,* 2001; see also chapter by Bierhaus et al.). By comparison, the majority of Earth's surface, the ocean floors, are typically 100–200 m.y. in age. Despite Europa's geologically young surface, no current geological activity has been found to a resolution of ~2 km over 20% of Europa's surface, in a comparison between Voyager and Galileo spacecraft images over a 20-year time period (*Phillips et al.,* 2000).

Europa's surface is subject to a range of impacts extending in size from those responsible for the largest observed impact structures down to micrometeorites. These impacts together result in the formation of a regolith, a layer of broken impact debris at the surface of an airless world that is continually overturned and thereby mixed through a process often referred to as impact gardening. The depth to which surface material has been mixed after a given length of time is called the gardening depth. Impact gardening can potentially preserve surface material by moving it down below the sputtering depth and even the radiation processing depth. If sputtering dominates, compounds produced by radiation processing at the surface are lost before they have a chance to be buried and perhaps eventually transported to a subsurface ocean layer. However, if gardening dominates, then material may be buried faster than most of it can be removed through sputtering.

Models for gardening on Europa are currently being examined and improved. *Cooper et al.* (2001) scaled from Ganymede regolith depth estimates, as well as a mass flux from studies of rings, to estimate a gardening depth of 2.6 m over a surface age of 50 m.y. *Phillips and Chyba* (2001) used a model of impact gardening based on studies of the lunar regolith (*Shoemaker et al.,* 1969; *Melosh,* 1989), coupled with recent studies of the impactor population in the outer solar system (*Zahnle et al.,* 1998, 1999), to find a similar result. Subsequent work (*Phillips and Grossman,* 2008) focused on revising those estimates to take into account new studies of the numbers of very small craters on Europa (*Bierhaus et al.,* 2001). This approach is based on the regolith formation model of *Shoemaker et al.* (1969), in which the surface overturn time is related to the cumulative number density of craters, the slope of the crater distribution, and the surface age. The regolith depth on Europa is defined as the depth to which the surface has been turned over at least once over the age of the surface, and is closely related to the crater diameter D at which sufficient craters exist to produce a broken-up surface layer down to the depth of this crater size. Because the majority of small craters on Europa are likely to be secondary craters (see chapter by Bierhaus et al.), the depth of this threshold crater size likely is bracketed between 0.1 D, which is the expected depth/diameter ratio for secondary craters, and 0.2 D, which is the value for primary craters on Europa (*Bierhaus et al.*, 2005). The regolith depth varies between about 0.5 cm and 1 cm for the two depth/diameter ratio cases, suggesting an average regolith depth of about a centimeter. It thus appears that sputtering erosion may dominate over gardening for all but the very youngest areas of Europa's surface.

6. THERMAL ICE AND NON-ICE SEGREGATION

6.1. Introduction

Spencer (1987b) modeled the thermal segregation of H_2O ice on Ganymede and Callisto and concluded that sublimation is the most significant process for the redistribution of ice at subkilometer scales. Sublimation processes relevant to (noncomet) icy bodies have been modeled by a number of researchers (*Lebofsky,* 1975; *Purves and Pilcher,* 1980; *Squyres,* 1980; *Spencer,* 1987b; *Colwell et al.,* 1990; *Moore et al.,* 1996). *Purves and Pilcher* (1980) and *Squyres* (1980) both concluded that Ganymede and Callisto should accumulate frost at their poles, at the expense of the equatorial regions where exposed ice should entirely sublimate away. High-resolution (<25 m/pixel) Galileo images of Europa's surface reveal the presence of loose, relatively lower albedo material that has collected in topographic lows. It has been proposed that the formation of this loose dark material is, at least in part, generated by the process of thermal ice segregation (e.g., *Moore et al.,* 1999).

6.2. Observations and Modeling

Concepts on redistribution of surface materials by sublimation developed on the basis of Voyager data by *Spencer* (1987a) have been applied to much-higher-resolution Galileo images of Ganymede by *Prockter et al.* (1998) and of both Ganymede and Callisto by *Moore et al.* (1999). These workers noted that pole-facing slopes in Ganymede's Galileo Regio were brighter than the opposite slopes, and they concluded that the brightening agent was a thin frost veneer, as it had no effect on underlying crater morphology. They surmised that the north-/south-facing slope dichotomy in surficial frost distribution was due to sublimation of ices on the sunnier, warmer equator-facing slopes and precipitation of frost on the less sunny (hence colder) pole-facing slopes, consistent with the hypothesis of *Spencer* (1987a). As ices sublimate, the low-albedo component of the surface materials remains behind to form a refractory-rich sublimation lag deposit. The lag may eventually become thick enough to suppress any further sublimation from underneath.

Topographic thermal models for airless bodies have been constructed by several workers (e.g., *Winter and Krupp,* 1971; *Spencer,* 1990; *Colwell et al.,* 1990). These models consider scattering and reradiation of reflected and thermal radiation from various topographic elements, and some also include diurnal subsurface heating variations. These models show that, as would be expected, poleward-facing slopes at high latitudes are cooler than the average surface, and are natural sites for accumulation of frost deposits (*Spencer and Maloney,* 1984). They also show that at low latitudes, temperatures tend to be higher in depressions, because surfaces in depressions receive thermal and reflected solar radiation from their surroundings. Low-latitude depressions, therefore, are expected to be sites of net sublimation and should develop a lag deposit on an initially icy surface somewhat faster than the plains surfaces of the same albedo. The presence of bright, icy interior crater walls at low latitudes, as is commonly seen on Europa and the other icy Galilean satellites, therefore requires additional explanation. A possibility is that lag deposits are sloughed off these steep slopes by gravity, exposing dark dirty-ice bedrock with high thermal inertia, so that the exposed outcrop acts as a cold trap (which becomes even colder once a layer of frost forms), relative to nearby lag-covered, dark regions of lower thermal inertia.

Consider a vertical icy scarp on Europa at midday, overlooking a horizontal plain covered in dark material, and compare its energy balance to that of a hypothesized horizontal bright icy surface at the top of the scarp (*Spencer,* 1987b; *Moore et al.,* 1999). Figure 8 illustrates our model. Assume unit emissivity for all surfaces, Lambertian scattering of sunlight, isotropic emission of thermal radiation, and negligible thermal inertia for the dark material, so that absorbed solar radiation is immediately reradiated in the infrared. This last assumption is justified to first order by remote thermal measurements that show that noontime temperatures on the icy Galilean satellites are close to equilibrium values. Also ignore warming of the dark material by radiation from the scarp, and do not consider times other than midday.

With these assumptions, the upward energy flux from Europa's dark material at midday is equal to the incident solar flux. Unlike the incident radiation, however, the energy is radiated isotropically, and a fraction ($1-A_d$) of the energy, where A_d is the bolometric albedo of the dark material, has been converted from visible and near-infrared radiation to thermal-infrared radiation. As seen from the vertical scarp of ice, the dark material, which radiates into 2π steradians, subtends a solid angle of π steradians, so on the assumption of isotropic radiation from the dark surface, the incident visible and thermal energy flux onto the scarp is half that reflected and radiated from the dark surface, and the total is half the incident solar flux. The vertical surface, at midday, receives no direct insolation.

Because the horizontal top of the ice block receives the full incident solar flux, it would seem that it would be warmer than the vertical face, which receives only half that flux. However, ice is a good reflector of visible radiation, but a good absorber of thermal infrared radiation (e.g.,

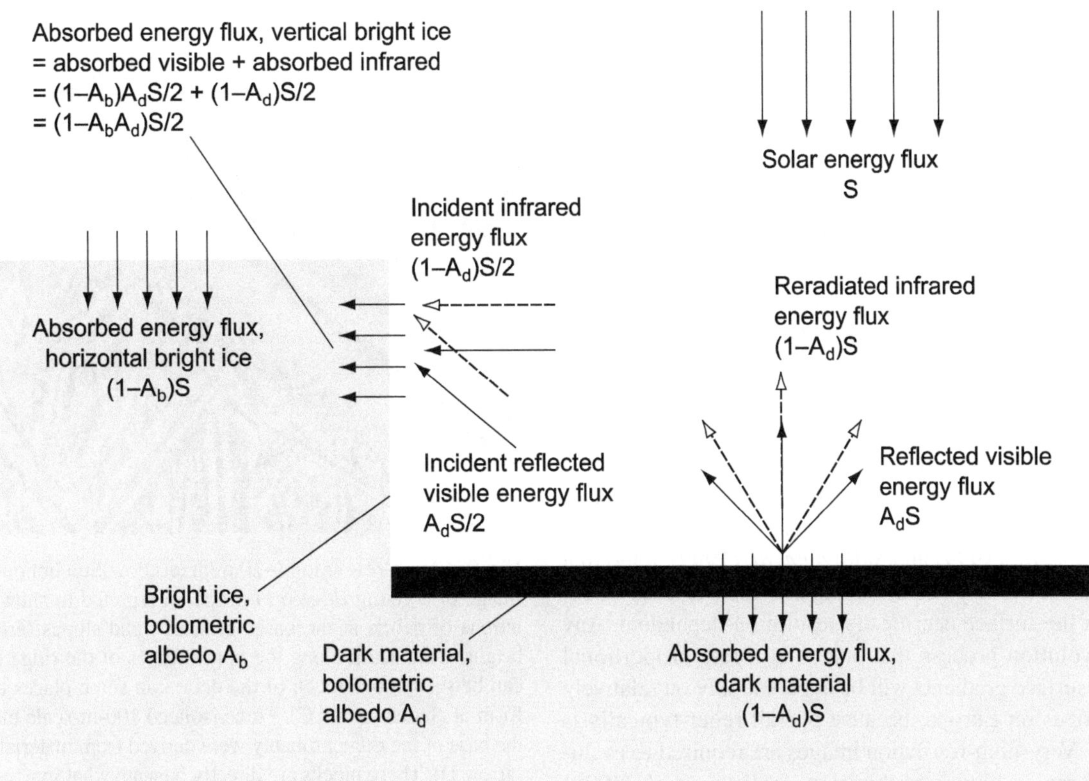

Fig. 8. Energy flux diagram. We consider a vertical icy scarp at midday, overlooking a horizontal plain covered in dark material, and compare its energy balance to that of a hypothesized horizontal bright icy surface at the top of the scarp. We assume unit emissivity for all surfaces, Lambertian scattering of sunlight, isotropic emission of thermal radiation, and negligible thermal inertia for the dark material. See text for discussion. From *Moore et al.* (1999).

Warren, 1984). The assumption of unit emissivity for the ice implies that it absorbs all the incident infrared radiation that it receives from the dark material, while it will only absorb a fraction $(1-A_b)$ of the visible radiation, where A_b is the ice bolometric albedo. From Figure 8, the result is that the energy flux absorbed by the vertical face is $(1-A_b \times A_d) \times S/2$, while the horizontal ice surface absorbs $(1-A_b) \times S$. Adopting $A_d = 0.15$ and $A_b = 0.7$, the vertical face absorbs almost 50% more energy than the horizontal face. If the thermal inertia is similar on the vertical and horizontal ice surfaces, the vertical face will therefore be warmer than the horizontal, and will have a higher sublimation rate at noon. Because of the extreme temperature dependence of sublimation rates, most sublimation occurs near the peak temperature, so it is likely that diurnally averaged sublimation rates are also highest on the vertical ice surface. For a surface with many such ice scarps, there will be a net transfer of ice from the vertical scarps to nearby horizontal surfaces, and the scarps will retreat.

An additional process, not considered in the simple model above, will tend to maintain the steepness of the ice slopes. Some trapping of heat in the concave regions at the base of a slope (as in a crater) is expected, because surfaces here see a large solid angle of surface that can radiate to them, and a relatively small solid angle of deep space. Therefore, the dark material at the base of the scarp will tend to be warmer than the dark material farther away, and in turn will preferentially warm the ice near the base of the scarp. Enhanced sublimation at the scarp base will tend to maintain the steepness of the scarp as it retreats. The steepness of the scarp will prevent accumulation of a lag deposit on it: Dark material released from the ice will accumulate at the base of the slope, and will be left behind as the scarp continues to retreat.

This process provides a natural mechanism for the icy brightening of high-standing topography that is so conspicuous on Europa and the other icy Galilean satellites: Scarps of exposed ice-rich bedrock will tend to retreat, with the liberated H_2O accumulating as frost on nearby horizontal surfaces and the lag accumulating at the bases of the scarps. Ice sublimates from the face of the scarp and accumulates on the top, until oversteepening causes mass wasting, which exposes fresh bedrock (cf. *Howard and Moore,* 2008).

7. EROSION AND MASS WASTING OF SURFACE MATERIALS

Galileo and Voyager images revealed the europan surface to be sparsely cratered and covered with landforms such as ridges and chaos that have relatively youthful ages.

However, other morphological evidence from the highest-resolution images indicates that in many places a fragmental debris layer is present that is thin in comparison with overall surface relief. Although the europan surface is young, a combination of processes that could include micrometeorite impact gardening, sputtering, and sublimation has produced a debris layer of regolith that is substantially thicker than the optical depths sampled by photometry.

Evidence for fragmental regolith on Europa from high-resolution Galileo images was presented in *Moore et al.* (1999) and *Sullivan et al.* (1999). These authors evaluated the highest-resolution images for information on mass wasting and slope evolution on the satellite. Dry mass wasting of steep slopes contributes to slope and landscape evolution by reducing overall surface relief. Dry mass wasting implies the presence of a preexisting surficial debris layer and/or the generation of such a layer.

Moore et al. (1999) and *Sullivan et al.* (1999) noted that the evidence for dry mass wasting and the presence of regolith at the surface is critically resolution-dependent. Any slope evolution process that occurs at a rate proportional to local surface gradients will be apparent only on relatively short slopes on Europa, because overall relief typically is <300 m. Very-high-resolution images are required to evaluate features on slopes of this size. *Sullivan et al.* (1999) found that images better than 35 m pixel^{-1} are required to recognize the morphological evidence for mass-wasted regolith. Their observations and conclusions regarding four of the highest-resolution Galileo imaging observations are summarized below.

Galileo imaging observation E6ESBRTPLN02 (Fig. 9) was obtained at 20 m/pixel on Galileo's sixth orbit, and sampled bright plains that are criss-crossed densely with simple ridges of various sizes (see chapter by Kattenhorn and Hurford). Morphological details along ridge flanks and within ridge troughs indicate gradient reductions have occurred by mass wasting. *Moore et al.* (1999) and *Sullivan et al.* (1999) note that despite the intense tectonism responsible for creating the ridged plains, slopes obviously steeper than probable angles of repose are apparently very rare in the scene. Moreover, darker material distributed subjacent to many of these slopes could be debris shed from higher elevations.

The twelfth orbit of Galileo included a particularly close pass to Europa, when the highest-resolution Europa images were obtained during the mission. The 12ESCHAOS_01 observation imaged at 11 m pixel^{-1} an area of chaotic terrain, where bright plains have broken apart into disjointed plates that have been partly replaced by lower-standing rubble (see chapter by Collins and Nimmo). Two types of steep slopes are well-displayed in this observation (Fig. 10): (1) Stratigraphically younger steep slopes (Type 1) form margins around the surviving remnant blocks of ridged plains, and these slopes have two distinct components: an upper, brighter, steeper component; and a lower, darker talus with isolated bright downslope streaks. (2) Stratigraphically

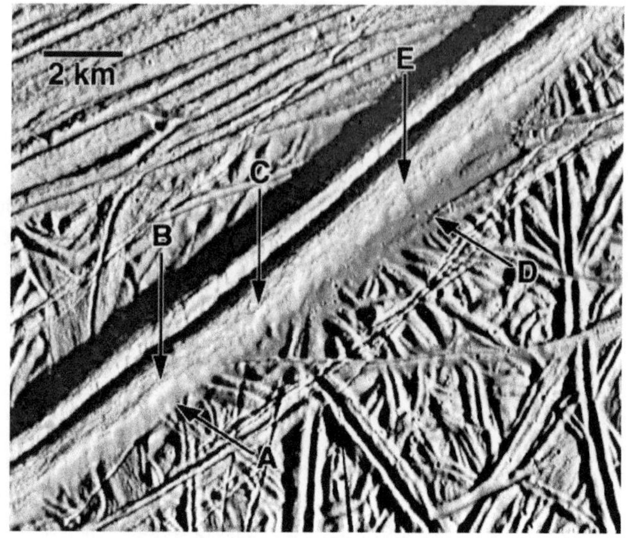

Fig. 9. High-resolution (~20 m/pixel), low-Sun oblique-looking image of a young ridge on Europa, interpreted to show accumulations of debris at the bases of scarps and slopes (arrow A). A bright, steep scarp along the upper slopes of the ridge (arrow B) can be seen above much of the debris, in some places taking the form of chutes (arrow C). Three isolated 100-m-scale blocks near the base of the ridge probably were derived from materials upslope (arrow D). These blocks are directly beneath what may be a down-glided kilometer-sized tabular slab (arrow E). As the large isolated blocks are apparent only here, it is possible that they are associated with the putative movement of this large slab. Illumination is from the lower right, and north is up. Scene center coordinates are 16°N, 173°W (a portion of Galileo image E6E0074 from observation E6ESBRTPLN02). From *Moore et al.* (1999).

older steep slopes (Type 2) are found within the remnant ridged plain blocks, and have only minor gradient differences between an upper, brighter zone, and darker materials immediately subjacent that lack bright slope streaks. *Sullivan et al.* (1999) linked the morphological differences between the two slope types and their relative ages. They concluded that slope evolution is still ongoing for the younger, first type (Type 1) of slope where bright materials are being released as falls down onto the darker talus, creating bright streaks that eventually fade to darker tones with time. In contrast, they inferred that slope evolution has ended for the older, second type (Type 2) of slope, having reached terminal gradients.

More tentative evidence for a similar, active albedo-changing process involving debris generation and mass wasting at the surface was cited by *Sullivan et al.* (1999) in the Galileo 12ESMOTTLE01 and 12ESMOTTLE02 observations that were obtained on Galileo's twelfth orbit at 6 and 12 m pixel^{-1}, respectively. These oblique Galileo images captured an area of intensely ridged plains affected by frost deposition and sublimation. Most surfaces in the area are bright, but darker materials are present in local hollows or along concave-up slope breaks, suggesting that the distribution of frost is topographically and thermally

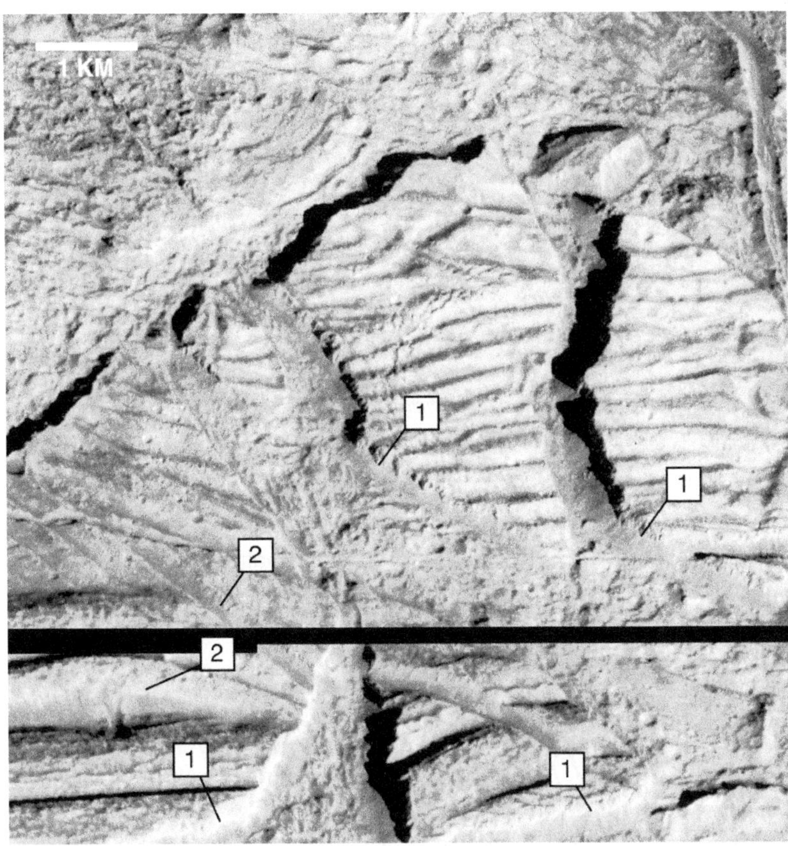

Fig. 10. An image from the 11 m pixel^{-1} Galileo observation 12ESCHAOS_01 showing two different slope types, in an area where bright plains has been disrupted and partly replaced by lower standing rubble. Slopes marked "1" are interpreted by *Sullivan et al.* (1999) to be younger, where the plates have broken apart. These have two-component gradients with upper, brighter, rockface morphology standing more steeply above darker talus below (in some places streaked with bright material). Slopes marked "2" are interpreted as more evolved, with only minor differences in gradient between upper, brighter units and lower, darker units. Type 1 slopes appear to be actively receding currently, generating new regolith. Galileo image 12E0039 from observation 12ESCHAOS_01.

controlled in a manner similar to that proposed initially for Ganymede and Callisto (*Spencer and Maloney,* 1984; *Spencer,* 1987b; *Prockter et al.,* 1998; *Moore et al.,* 1999). *Sullivan et al.* (1999) noted in some places that very narrow, bright albedo markings appear along the centers of trough floors that are otherwise dark. The bright lineaments might represent relatively recent arrivals of brighter, frost-covered debris from upslope, collecting along narrow trough floors, which have yet to fully darken and fade from view.

In summary, evidence for debris generation and mass wasting on steep slopes is not uncommon in the few images available with resolutions better than 35 m/pixel (*Sullivan et al.,* 1999). This process likely continues today, as slope gradients in some settings continue to evolve to lower gradients. Mass wasting and consequent generation of fragmental regolith could be widespread across the europan surface.

8. SUMMARY SYNTHESIS

Putting it all together, our current understanding of Europa's surface at the smallest scales suggests a surface type that is probably uncommon among airless bodies elsewhere in the solar system (Fig. 11). The surface, being relatively young overall, will not have the meters-thick layer of impact-generated regolith typical of ancient bodies. The surficial debris layer is instead probably mostly the product of tectonic fracturing and crushing and attendant downslope mass wasting and accumulation of debris in topographic lows. Indeed, Europa's location within the jovian magnetosphere combined with a youthful surface probably results in high-velocity charged-particle sputtering of ice being dominant over impact gardening. The liberation of water molecules by sputtering and sublimation encourages thermal frost segregation: the reprecipitation of ice on local topographic high points and other local and regional cold traps. Ice reprecipitation may be the dominant cause of the photometric observations that the optical surface is dominated by small transparent ice particles, in most places minimally contaminated by implanted sulfur ions from Io. Although ice dominates the spectrum of Europa, there is a contribution from lower-albedo, relatively refractory material, which appears in high-resolution images to accumulate as detritus in topographic lows. Diurnal temperature measurements are best interpreted to imply that the thermal inertia of the surface is similar to that of other icy Galilean satellites, which is in the range of fine-grained porous material — a result consistent with the photometric observations and sputter modeling.

Groundbased radar, which penetrates several meters into the surface, indicates that the surface layer, to this depth, is porous and highly fractured to a depth of at least a few meters. It is likely that a brittle, fractured realm persists to a depth of a few kilometers (e.g., *Nimmo et al.,* 2003). The brittle-to-ductile transition may occur at a depth of hundreds of meters to several kilometers (*Billings and Kattenhorn,* 2005). Beyond that depth the ice is generally thought to be

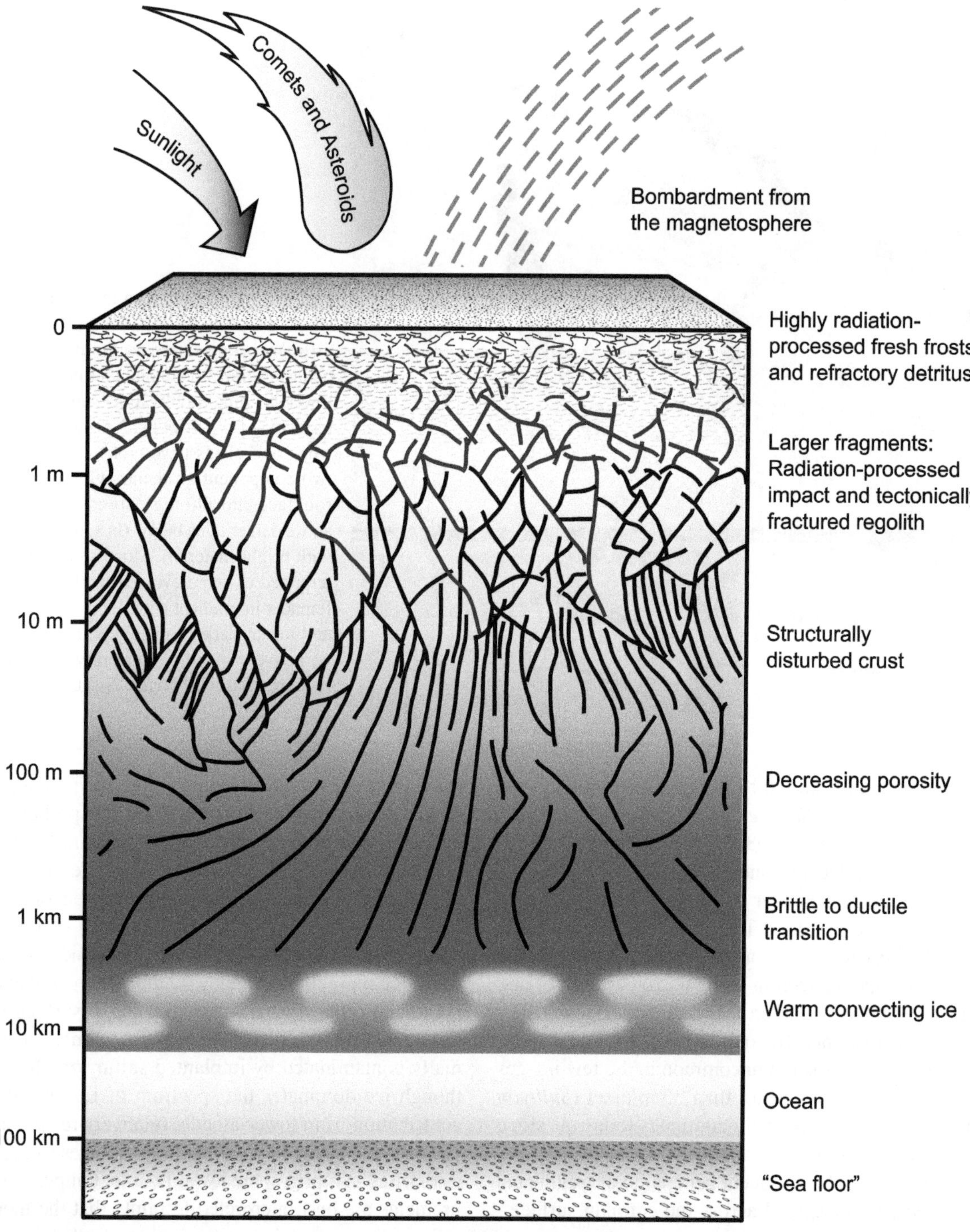

Fig. 11. Schematic visualization of the icy shell of Europa. The vertical scale is exponential. This schematic represents an internally consistent synthesis of photometric and radar observations, combined with high-energy sputtering and impact gardening modeling, as well as models of the deeper (kilometer-scale) crust. Ice redepostion from sputtering and thermal segregation are not explicitly shown here for the sake of diagrammatical clarity. Photometry, even in the infrared, can never penetrate deeper than a few centimeters, and most often only a few micrometers. Micrometer impacts and the charged-particle environment are thought to generally affect only the upper meter or so. A thermal signal ostensibly may, at most, express itself from a number of meters depth. Mass wasting can locally produce many meters of debris, such as at the base of scarps and other topographic lows. Mass wasting conversely may expose fresh ice in places such as along the upper slopes of scarps. Radar can potentially see as deeply as a few kilometers, but its results must be deconvoluted from the "noise" of the intervening radar scatterers, especially those of the surface. The thickness of the brittle layer, the total thickness of the solid ice shell, and the depth of the underlying ocean are somewhat conjectural, but based on information reviewed in other chapters.

ductile and potentially convecting (see chapter by Barr and Showman) down to the ice-ocean interface, which may occur at depths of around 20 km (e.g., *Schenk,* 2002).

Acknowledgments. Thanks to C. Chyba for assistance in developing the gardening section, and to J. Eluskiewicz and P. Helfenstein for helpful reviews. Work carried out by B.B. was performed at the Jet Propulsion Laboratory, California Institute of Technology, under a contract with the National Aeronautics and Space Administration.

REFERENCES

Abramov O. and Spencer J. R. (2008) Numerical modeling of endogenic thermal anomalies on Europa. *Icarus, 195,* 378–385.

Abramov D. and Spencer J. R. (2009) Endogenic heat from Enceladus' south polar fractures: New observations, and models of conductive surface heating. *Icarus, 199,* 189–196.

Auda H. and Emborg C. (1973) Studies on post-irradation degradation in Micrococcus radiodurans, strain RII5. *Rad. Res., 53,* 273–280.

Baron J. E., Tyler G. L., and Simpson R. A. (2003) Three-dimensional numerical modeling of refraction and reflection scattering from icy Galilean satellites. *Icarus, 164,* 404–417.

Bierhaus E. B., Edward B., Chapman C. R., Merline W. J., Brooks S. M., and Asphaug E. (2001) Pwyll secondaries and other small craters on Europa. *Icarus, 153,* 264–276.

Bierhaus E. B., Chapman C. R, and Merline W. J. (2005) Secondary craters on Europa and implications for cratered surfaces. *Nature, 427(20),* 1125–1127.

Billings S. E. and Kattenhorn S. A. (2005) The great thickness debate: Ice shell thickness models for Europa and comparisons with estimates based on flexure at ridges. *Icarus, 177,* 397–412.

Black G. J., Campbell D. B., and Ostro S. J. (2001a) Icy Galilean satellites: 70 cm radar results from Arecibo. *Icarus, 151,* 160–166.

Black G. J., Campbell D. B., and Nicholson P. D. (2001b) Icy Galilean satellites: Modeling radar reflectivities as a coherent backscatter effect. *Icarus, 151,* 167–180.

Blaney D. L., Goguen J. D., Veeder G. J., Johnson T. V., and Matson D. L. (1999) Europa's thermal infrared emission. In *Lunar and Planetary Science XXX,* Abstract #1657. Lunar and Planetary Institute, Houston (CD-ROM).

Brown R. H. and Matson D. (1987) Thermal effects of insolation propagation into the regoliths of airless bodies. *Icarus, 72,* 84.

Buratti B. J. (1985) Application of a radiative transfer model to bright icy satellites. *Icarus, 61,* 208–217.

Buratti B. J. and Golombek M. (1988) Geologic implications of spectrophotometric measurements of Europa. *Icarus, 75,* 437–449.

Buratti B.J. and Veverka J. (1983) Voyager photometry of Europa. *Icarus, 55,* 93–110.

Buratti B. J. and Veverka J. (1985) The photometry of rough planetary surfaces: The importance of multiple scattering. *Icarus, 64,* 320–328.

Buratti B. J., Nelson R. M., and Lane A. L. (1988) Surficial textures of the Galilean satellites. *Nature, 333,* 148–151.

Calvin W. M., Clark R. N., Brown R. H., and Spencer J. R. (1995) Spectra of the icy Galilean satellites from 0.2 to 5 μm: A compilation, new observations, and a recent summary. *J. Geophys. Res., 100,* 19041–19048.

Campbell D. B., Chandler J. F., Pettengill G. H., and Shapiro I. I. (1977) Galilean satellites of Jupiter: 12.6-centimeter radar observations. *Science, 196,* 650–653.

Campbell D. B., Ostro S. J., Pettengill G. H., and Shapiro I. I. (1978) Galilean satellites: 1976 radar results. *Icarus, 34,* 254–267.

Clark B. E., Helfenstein P., Veverka J., Ockert-Bell M., Sullivan R. J., Geissler P. E., Phillips C. B., McEwen A. S., Greeley R., Neukum G., Denk T., and Klaasen K. (1998) Multispectral terrain analysis of Europa from Galileo images. *Icarus, 95,* 95–106.

Clark R. N., Fanale F. P., and Zent A. P. (1983) Frost grain size metamorphism — Implications for remote sensing of planetary surfaces. *Icarus, 56,* 233.

Colwell J. E., Jakosky B. M., Sandor B. J., and Stern S. A. (1990) Evolution of topography on comets. II. Icy craters and trenches. *Icarus, 85,* 205–215.

Cooper J. F., Johnson R. E., Mauk B. H., Garrett H. B., and Gehrels N. (2001) Energetic ion and electron irradiation of the icy Galilean satellites. *Icarus, 149,* 133–159.

Domingue D. L. and Hapke B. (1992) Disk resolved photometric analaysis of europan terrains. *Icarus, 99,* 70–81.

Domingue D. L. and Verbiscer A. (1997) Re-analysis of the solar phase curves of the icy Galilean satellites. *Icarus, 128,* 49–74.

Domingue D. L., Hapke B. W, Lockwood G. W., and Thomson D. T. (1991) Europa's phase curve: Implications for surface structure. *Icarus, 90,* 30–42.

Eluszkiewicz J. (1991) On the microphysical state of the surface of Triton. *J. Geophys. Res., 96,* 19217.

Eshleman V. R. (1986a) Radar glory from buried crater on icy moons. *Science, 234,* 587–590.

Eshleman V. R. (1986b) Mode decoupling during retrorefraction as an explanation for bizarre radar echoes from icy moons. *Nature, 319,* 755–757.

Goldstein R. M. and Green R. R. (1980) Ganymede: Radar surface characteristics. *Science, 207,* 179–180.

Grundy W. M., Buratti B. J., Cheng A. F., Emery J. P., Lunsford A., McKinnon W. B., Moore J. M., Newman S. F., Olkin C. B., Reuter D. C., and 6 colleagues (2007) New Horizons mapping of Europa and Ganymede. *Science, 318,* 234–237.

Gurrola E. M. (1995) Interpretation of radar data from the icy Gali-lean satellites and Triton. Ph.D. thesis, Stanford University.

Gurrola E. M. and Eshleman V. R. (1990) On the angle and wavelength dependencies of the radar backscatter from the icy Galilean moons of Jupiter. *Adv. Space Res., 10,* 195–197.

Hagfors T., Gold T., and Ierkic H. M. (1985) Refraction scattering as origin of the anomalous radar returns of Jupiter's satellites. *Nature, 315,* 637–640.

Hagfors T., Dahlstrom I., Gold T., Hamran S. E., and Hansen R. (1997) Refraction scattering in the anomalous reflections from icy surfaces. *Icarus, 130,* 313–322.

Hansen O. L. (1973) Ten-micron eclipse observations of Io, Europa, and Ganymede. *Icarus, 18,* 237–246.

Hansen G. and 23 colleagues (2005) Ice grain size distribution: Differences between jovian and saturnian icy satellites from Galileo and Cassini measurements. *Bull. Am. Astron. Soc., 37,* 729.

Hapke B. (1981) Bidirectional reflectance spectroscopy. 1. Theory. *J. Geophys. Res., 86,* 3039–3054.

Hapke B. (1984) Bidirectional reflectance spectroscopy. 3. Correction for macroscopic roughness. *Icarus, 59,* 41–59.

Hapke B. (1986) Bidirectional reflectance spectroscopy. 4. The extinction coefficient and the opposition effect. *Icarus, 67,* 264–280.

Hapke B. (1990) Coherent backscatter and the radar characteristics of outer planet satellites. *Icarus, 88,* 407–417.

Hapke B. (2002) Bidirectional reflectance spectroscopy. 5. The coherent backscatter opposition effect and anisotropic scattering. *Icarus, 157,* 523–534.

Hapke B. (2008) Bidirectional reflectance spectroscopy. *Icarus, 195,* 918–926.

Harcke L. J., Simpson R. A., Tyler G. L., and Zebker H. S. (2000) Radar imaging of the icy Galilean satellites during 1999 opposition. In *Lunar and Planetary Science XXXI,* Abstract #1789. Lunar and Planetary Institute, Houston (CD-ROM).

Harcke L. J., Zebker H. A., Jurgens R. F., Slade M. A., Butler B. J., and Harmon J. K. (2001a) Radar observations of the icy Gali-lean satellites during 2000 opposition. In *Lunar and Planetary Science XXXII,* Abstract #1369. Lunar and Planetary Institute, Houston (CD-ROM).

Harcke L. J., Zebker H. A., Tyler G. L., Simpson R. A., Ostro S. J., and Harmon J. K. (2001b) Radar imaging of Europa, Ganymede, and Callisto with the upgraded Arecibo 13 cm radar. *Bull. Am. Astron. Soc., 33,* 918.

Harmon J. K., Arvidson R. E., Guinness E. A., Campbell B. A., and Slade M. A. (1992) Mars mapping with delay-Doppler radar. *J. Geophys. Res., 104,* 14065–14090.

Harmon J. K., Ostro S. J., Chandler J. F., and Hudson R. S. (1994) Radar ranging to Ganymede and Callisto. *Astron. J., 107,* 1175–1181.

Helfenstein P. and Shepard M. K. (1999) Submillimeter-scale topography of the lunar regolith. *Icarus, 141,* 107–131.

Helfenstein P., Currier N., Clark B. E., Veverka J., Bell M., Sullivan R., Klemaszewski J., Greeley R., Pappalardo R. T., Head J. W., Jones T., Klaasen K., Magee K., Geissler P., Greenberg R., McEwen A., Phillips C., Colvin T., Davies M., Denk T., Neukum G., and Belton M. J. S. (1998) Galileo observations of Europa's opposition effect. *Icarus, 135,* 41.

Henyey L. G. and Greenstein J. (1941) Diffuse radiation in the galaxy. *Astrophys. J., 93,* 70–83.

Howard A. D. and Moore J. M. (2008) Sublimation-driven erosion on Callisto: A landform simulation model test. *Geophys. Res. Lett.,* 35, L03203, DOI: 101029/2007GL032618.

Ip W. H. (1996) Europa's oxygen exosphere and its magnetospheric interaction. *Icarus, 120,* 317–325.

Ip W. H., Williams D. J., McEntire R. W., and Mauk B. H. (1998) Ion sputtering and surface erosion at Europa. *Geophys. Res. Lett., 25,* 829.

Irvine W. (1966) The shadowing effect in diffuse reflection. *J. Geophys. Res., 71,* 2931–2937.

Johnson T. V., Soderblom L. A., Mosher J. A., Danielson G. E., Cook A. F., and Kupperman P. (1983) Global multispectral mosaics of the icy Galilean satellites. *J. Geophys. Res., 88,* 5789–5805.

Kivelson M. G., Khurana K. K., Joy S., Russell C. T., Southwood D. J., Walker R. J., and Polanskey C. (1997) Europa's magnetic signature: Report from Galileo's pass on 19 December 1996. *Science, 276,* 1239.

Lebofsky L. A. (1975) Stability of frosts in the solar system. *Icarus, 25,* 205–217.

McEwen A. S. (1986) Exogenic and endogenic albedo and color patterns on Europa. *J. Geophys. Res., 91,* 8077–8097.

Melosh H. J. (1989) *Impact Cratering: A Geologic Process.* Oxford Monographs on Geology and Geophysics, No. 11, Oxford Univ., New York.

Millis R. T. and Thompson D. T. (1975) UBV photometry of the Galilean satellites. *Icarus, 26,* 408–419.

Mishchenko M. I. (1992) Polarization characteristics of the coherent backscatter opposition effect. *Earth Moon Planets, 58,* 127–144.

Moore J. M., Mellon M. T., and Zent A. P. (1996) Mass wasting and ground collapse in terrains of volatile-rich deposits as a solar system process: The pre-Galileo view. *Icarus, 122,* 63–78.

Moore J. M., Asphaug E., Morrison D., Klemaszewski J. E., Sullivan R. J., Chuang F., Greeley R., Bender K. C., Geissler P. E., Chapman C. R., Helfenstein P., Pilcher C. B., Kirk R. L., Giese B., and Spencer J. R. (1999) Mass movement and landform degradation on the icy Galilean satellites: Results from the Galileo nominal mission. *Icarus, 140,* 294–312.

Moore J. M., Asphaug E., Belton M. J. S., Bierhaus E. B., Breneman H. H., Brooks S. M., Chapman C. R., Chuang F. C., Collins G. C., Giese B., Greeley R., Head J. W., Kadel S., Klaasen K. P., Klemaszewski J. E., Magee K. P., Moreau J., Morrison D., Neukum G., Pappalardo R. T., Phillips C. B., Schenk P. M., Senske D. A., Sullivan R. J., Turtle E. P., and Williams K. (2001) Impact features on Europa: Results of the Galileo Europa Mission (GEM). *Icarus, 151,* 93–111.

Morrison D. (1977) Radiometry of satellites and of the rings of Saturn. In *Planetary Satellites* (J. Burns, ed.), pp. 269–301. Univ. of Arizona, Tucson.

Morrison D. and Cruikshank D. (1973) Thermal properties of the Galilean satellites. *Icarus, 18,* 224–236.

Morrison D. and Morrison N. (1977) Photometry of the Galilean satellites. In *Planetary Satellites* (J. Burns, ed.), pp. 363–378. Univ. of Arizona, Tucson.

Morrison D., Morrison N., and Lazarewicz A. (1974) Four-color photometry of the Galilean satellites. *Icarus, 23,* 399–416.

National Research Council Space Studies Board (2000) *Preventing the Forward Contamination of Europa.* National Academy, Washington, DC.

Nelson M. L., McCord T. B., Clark R. N., Johnson T. V., Matson D. L., and Mosher J. A. (1986) Europa: Characterization and interpretation of global spectral surface units. *Icarus, 65,* 129–151.

Nimmo F., Pappalardo R. T., and Giese B. (2003) On the origins of band topography, Europa. *Icarus, 166,* 21–32.

Orton G. S., Spencer J. R., Travis L. D., Martin T. Z., and Tamppari L. K. (1996) Galileo photopolarimeter-radiometer observations of Jupiter and the Galilean satellites. *Science, 274,* 389.

Ostro S. J. (1982) Radar properties of Europa, Ganymede, and Callisto. In *Satellites of Jupiter* (D. Morrison, ed.), pp. 212–236. Univ. of Arizona, Tucson.

Ostro S. J. (1993) Planetary radar astronomy. *Rev. Mod. Phys., 65,* 1235–1279.

Ostro S. J. and Pettengill G. H. (1978) Icy craters on the Galilean satellites? *Icarus, 34,* 268–279.

Ostro S. J. and Shoemaker E. M. (1990) The extraordinary radar echoes from Europa, Ganymede, and Callisto: A geological perspective. *Icarus, 85,* 335–345.

Ostro S. J., Campbell D. B., Pettengill G. H., and Shapiro I. I. (1980) Radar observations of the icy Galilean satellites. *Icarus, 44,* 431–440.

Ostro S. J., Campbell D. B., Simpson R. A., Hudson R. S., Chandler J. F., Rosema K. D., Shapiro I. I., Standish E. M., Winkler R., Yeomans D. K., Velez R., and Goldstein R. M. (1992) Europa, Ganymede, and Callisto: New radar results from Arecibo and Goldstone. *J. Geophys. Res., 97,* 18227–18244.

Peters K. J. (1992) Coherent-backscatter effect — A vector formulation accounting for polarization and absorption effects and small or large scatterers. *Phys. Rev. B, 46,* 801–812.

Phillips C. B. and Chyba C. F. (2001) Impact gardening rates on Europa: Comparison with sputtering. In *Lunar and Planetary Science XXXII,* Abstract #2111. Lunar and Planetary Institute, Houston (CD-ROM).

Phillips C. B. and Grossman L. A. (2008) Impact gardening on Europa. In *The Science of Solar System Ices (ScSSI): A Cross-Disciplinary Workshop,* p. 129. LPI Contribution #1406, Lunar and Planetary Institute, Houston.

Phillips C. B., McEwen A. S., Hoppa G. V., Fagents S. A., Greeley R., Klemaszewski J. E., Pappalardo R. T., Klaasen K. P., and Breneman H. H. (2000) The search for current geologic activity on Europa. *J. Geophys. Res., 105,* 22579–22597.

Pospieszalska M. K. and Johnson R. E. (1989) Magnetospheric ion bombardment profiles of satellites — Europa and Dione. *Icarus, 78,* 1.

Prockter L. M., Head J. W., Pappalardo R. T., Senske D. A., Neukum G., Wagner R., Wolf U., Oberst J., Giese B., Moore J. M., Chapman C. R., Helfenstein P., Greeley R., Breneman H. H., and Belton M. J. S. (1998) Dark terrain on Ganymede: Geological mapping and interpretation of Galileo Regio at high resolution. *Icarus, 135,* 317–344.

Purves N. G. and Pilcher C. B. (1980) Thermal migration of water on the Galilean satellites. *Icarus, 43,* 51–55.

Rignot E. (1995) Backscatter model for the unusual radar properties of the Greenland Ice Sheet. *J. Geophys. Res., 100,* 9389–9400.

Russell E. E., Brown F. G., Chandos R. A., Fincher W. C., Kubel L. F., Lacis A. A., and Travis L. D. (1992) Galileo photopolarimeter/radiometer experiment. *Space Sci. Rev., 60,* 531.

Saari J. M. and Shorthill R. W. (1963) Isotherms of crater regions on the illuminated and eclipsed Moon. *Icarus, 2,* 115.

Schenk P. M. (2002) Thickness constraints on the icy shells of the Galilean satellites from a comparison of crater shapes. *Nature, 417,* 419–421.

Schenk P. M., Chapman C. R., Zahnle K., and Moore J. M. (2004) Age and interiors: The cratering record of the Galilean satellites. *In Jupiter: The Planet, Satellites and Magnetosphere* (F. Bagenal et al., eds.), pp. 427–456. Cambridge Univ., Cambridge.

Shepard M. K. and Campbell B. A. (1998) Shadows on a planetary surface and implications for photometric roughness. *Icarus, 134,* 279–291.

Shepard M.K. and Helfenstein P. (2007) A test of the Hapke photometric model. *J. Geophys. Res., 112,* E03001.

Shkuratov Y. G. and Helfenstein P. (2001) The opposition effect and the quasi-fractal structure of regolith: I. Theory. *Icarus, 152,* 96–116.

Shoemaker E. M., Batson R. M., Holt H. E., Morris E. C., Rennilson J. J., and Whitaker E. A. (1969) Observations of the lunar regolith and the Earth from the television camera on Surveyor 7. *J. Geophys. Res., 74,* 6081–6119.

Simonelli D. P. and Buratti B. J. (2004) Europa's opposition surge in the near-infrared: Interpreting disk-integrated observations by Cassini VIMS. *Icarus, 172,* 149–162.

Spencer J. R. (1987a) The surfaces of Europa, Ganymede, and Callisto: An investigation using Voyager IRIS thermal infrared spectra. Ph.D. thesis, University of Arizona.

Spencer J. R. (1987b) Thermal segregation of water ice on the Galilean satellites. *Icarus, 69,* 297–313.

Spencer J. R. (1990) A rough surface thermophysical model for airless planetary surfaces. *Icarus, 83,* 27–38.

Spencer J. R. and Maloney P. R. (1984) Mobility of water ice on Callisto: Evidence and implications. *Geophys. Res. Lett., 11,* 1223–1226.

Spencer J. R., Tamppari L. K., Martin T. Z. and Travis L. D. (1999) Temperatures on Europa from Galileo photopolarimeter-radiometer: Nighttime thermal anomalies. *Science, 284,* 1514–1516.

Squyres S. W. (1980) Surface temperatures and retention of H_2O frost on Ganymede and Callisto. *Icarus, 44,* 502–510.

Sullivan R., Moore J., and Pappalardo R. (1999) Mass-wasting and slope evolution on Europa. In *Lunar and Planetary Science XXX,* Abstract #1747. Lunar and Planetary Institute, Houston (CD-ROM).

Thompson W. R. and Squyres S. W. (1990) Titan and other icy satellites: Dielectric properties of constituent materials and implications for radar sounding. *Icarus, 86,* 336–354.

Tittemore W. C. and Sinton W. M. (1989) Near-infrared photometry of the Galilean satellites. *Icarus, 77,* 82.

Urquhart M. L. and Jakosky B. M. (1996) Constraints on the solid-state greenhouse effect on the icy Galilean satellites. *J. Geophys. Res., 101,* 21169.

Verbiscer A. and Veverka J. (1990) Scattering properties of natural snow and frost: Comparison with icy satellite photometry. *Icarus, 88,* 418–428.

Veverka J. (1977) Photometry of satellite surfaces. In *Planetary Satellites* (J. Burns, ed.), pp. 171–209. Univ. of Arizona, Tucson.

Warren S. G. (1984) Optical constants of ice from the ultraviolet to the microwave. *Appl. Optics, 23,* 1206–1225.

Wechsler A. E., Glaser P. E., and Fountain J. A. (1972) Thermal properties of granulated materials. In *Thermal Characteristics of the Moon* (J. W. Lucas, ed.), pp. 215–241. MIT, Cambridge.

Wildey R. L., Murray B. C., and Westphal J. A. (1967) Reconnaissance of infrared emission from the lunar nighttime surface. *J. Geophys. Res., 72,* 3743.

Winter D. F. and Krupp J. A. (1971) Directional characteristics of infrared emission from the Moon. *Moon, 2,* 279–292.

Zahnle K. (2001) Cratering rates on Europa. In *Lunar and Planetary Science XXXII,* Abstract #1699. Lunar and Planetary Institute, Houston (CD-ROM).

Zahnle K., Dones L., and Levison H. F. (1998) Cratering rates on the Galilean satellites. *Icarus, 136,* 202–222.

Zahnle K., Levison H., Dones L., and Schenk P. (1999) Cratering rates in the outer solar system. In *Lunar and Planetary Science XXX,* Abstract # 1776. Lunar and Planetary Institute, Houston (CD-ROM).

Part III:

Interior, Icy Shell, and Ocean

Interior of Europa

G. Schubert
University of California, Los Angeles

F. Sohl and H. Hussmann
German Aerospace Center (DLR)

On the basis of Europa's measured density, quadrupole gravitational coefficients, and shape, it is inferred that the satellite is fully differentiated into a metallic core, silicate mantle, and outer water ice/liquid shell. From measurements of the induced magnetic field due to currents in Europa driven by time variations in the jovian-directed component of the planet's magnetic field, it is inferred that there is a global subsurface liquid water ocean within Europa's outer shell. Geological features are consistent with the existence of a liquid water ocean beneath Europa's icy surface. This chapter reviews the evidence for the internal structure model of Europa, describes how the structural model is obtained, explains how the existence of the subsurface liquid water ocean is inferred, and discusses the possible composition of Europa's metallic core, silicate mantle, and subsurface ocean.

1. INTRODUCTION

It is quite remarkable that we can know what is inside a small satellite in the outer solar system. In the case of Europa, this is possible thanks to the Galileo spacecraft, which obtained gravitational, electromagnetic, and shape data on several flybys of the moon. Europa's mass and gravitational quadrupole coefficients are inferred from the Doppler shift of the spacecraft's radio communication signal during flybys of the satellite. Europa's size and shape are obtained from pictures of the moon acquired by the imaging system on the Galileo spacecraft. The mass and size yield the satellite's average density, an important indicator of internal structure and especially composition. Size, mass, and gravitational coefficients yield the moon's moment of inertia (under the assumption that Europa is in hydrostatic equilibrium). The density and moment of inertia allow the construction of simple structural models of the interior, which indicate that Europa has a metallic core surrounded by a silicate mantle and an outer shell of liquid water/ice (Fig. 1). The shape data are consistent with these inferences. The existence of Europa's subsurface liquid ocean is established from the electromagnetic induction signals measured by the Galileo magnetometer and some properties of the ocean (electrical conductivity, depth below the surface, thickness) are constrained by these data. Even with these data our knowledge of Europa's interior is rudimentary. Particularly uncertain are the composition of the core and mantle, the exact size of the core and thickness of the mantle, and the physical state of the core (liquid, solid, partially liquid). In this chapter we describe in more detail how the gravitational, size and shape, and electromagnetic data from the Galileo mission have been used to learn about Europa's interior.

2. GRAVITATIONAL FIELD AND IMPLICATIONS FOR INTERNAL STRUCTURE

During the flyby of a body a spacecraft experiences the tug of the body's gravitational force. The consequent acceleration and deceleration of the spacecraft in its orbit results in a Doppler shift of the spacecraft's radio communication signal with Earth. The Doppler data recorded by stations of the Deep Space Network thus contain a record of the body's gravitational field that can be inverted to reveal the characteristics of the field. If the gravitational field

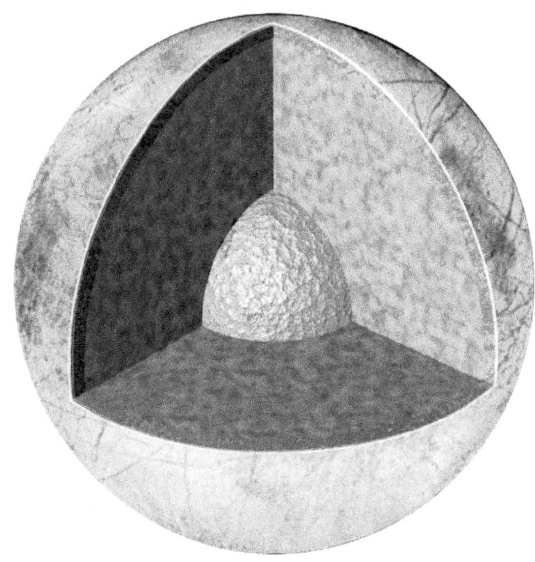

Fig. 1. See Plate 20. Cutaway of Europa's interior.

TABLE 1. Physical parameters of Europa and Jupiter.

Property	Symbol	Value	Reference
	GM	3202.72 ± 0.02 km^3 s^{-2}	*Schubert et al.* (2004)
Gravitational constant	G	6.67259×10^{-11} m^3 kg^{-1} s^{-2}	
Mass	M	$4.79982 \pm 0.00062 \times 10^{22}$ kg	*Anderson et al.* (1998)
Mean radius	R	1562.09 km	*Seidelmann et al.* (2007)
Orbital eccentricity	e	0.0101	*Murray and Dermott* (2000)
Semimajor axis	a_J	670,900 km	*Murray and Dermott* (2000)
Rotation period	T	3.551810 days	*Murray and Dermott* (2000)
Mass of Jupiter	M_J	1.8986×10^{27} kg	*Murray and Dermott* (2000)

external to the body is represented by the spherical harmonic expansion of the gravitational potential V according to

$$V(r, \phi, \lambda) = \frac{GM}{r} \tag{1}$$

$$\left[1 + \sum_{n=2}^{\infty} \sum_{m=0}^{n} \left(\frac{R}{r}\right)^n (C_{nm} \cos m\lambda + S_{nm} \sin m\lambda) P_{nm}(\sin\phi)\right]$$

then this translates into a determination of the gravitational coefficients C_{nm} and S_{nm}. In equation (1), r is the radial distance from the center of mass of the body, ϕ is latitude, λ is longitude, R is the radius of the body, G is the gravitational constant, M is the mass of the body, and P_{nm} are the associated Legendre polynomials of degree n and order m. The number of gravitational coefficients that can be solved for depends on the number of spacecraft flybys and their geometry. A spacecraft in polar orbit around a body is in the best possible configuration for determining the maximum number of gravitational coefficients.

The Doppler data for four flybys of Europa by the Galileo spacecraft have been analyzed to yield GM = (3202.72 ± 0.02) km^3 s^{-2}, $J_2 = -C_{20} = (435.5 \pm 8.2) \times 10^{-6}$, and $C_{22} = (131.5 \pm 2.5) \times 10^{-6}$ (*Anderson et al.*, 1998; *Schubert et al.*, 2004). These values of the gravitational coefficients are based on the *a priori* assumption that Europa is in hydrostatic equilibrium. Such a body, in synchronous rotation about its primary, adopts an ellipsoidal equilibrium shape in response to rotational and tidal forcing with its long axis along a line from its center to the center of the primary, its intermediate axis along its orbital direction, and its short axis parallel to its spin direction. The radii of the triaxial ellipsoid are a, b, c with $a > b > c$. The ellipsoidal distortion of the body is directly responsible for producing the quadrupole gravitational field described by the coefficients J_2 and C_{22}. All four spacecraft flybys of Europa were nearly equatorial, a circumstance that makes it impossible to determine both J_2 and C_{22} independently, without the *a priori* hydrostatic constraint ($J_2 = (10/3)C_{22}$) (see equation (13)). The tidal and rotational forcing of Europa is larger than that of the other Galilean satellites, save Io, a factor in support of the likelihood that Europa's quadrupole gravitational field derives from an ellipsoidally distorted body in hydrostatic equilibrium. Europa's shape, discussed in a later section, is also consistent with its hydrostaticity.

With the assumption that Europa is in hydrostatic equilibrium, it is possible to infer its axial moment of inertia factor (MoI = C/MR^2) (C is the axial moment of inertia) from the gravitational coefficient C_{22}; the MoI of a body is an essential constraint on the radial distribution of mass in its interior. For a hydrostatic satellite in synchronous rotation about its primary, C_{22} is related to the rotational forcing parameter q_r and the fluid potential Love number k_2 by

$$q_r = \frac{C_{22} k_2}{4} \tag{2}$$

where

$$q_r = \frac{\omega^2 R^3}{GM} \tag{3}$$

and ω is the angular frequency of Europa's rotation and mean orbital motion [$\omega = 2\pi/T$ and T is Europa's orbital period of 3.55181 days (Table 1)]. From equations (2), (3), and the measured values of ω, GM, R, and C_{22}, we can find k_2. For Europa, *Schubert et al.* (2004) obtained $k_2 = 1.048 \pm 0.020$. The MoI factor of Europa follows from the value of the fluid Love number and the Radau relation (*Kaula,* 1968)

$$\text{MoI} = \frac{C}{MR^2} = \frac{2}{3}\left\{1 - \frac{2}{5}\frac{(4 - k_2)^{1/2}}{(1 + k_2)^{1/2}}\right\} \tag{4}$$

The value of Europa's MoI factor is 0.346 ± 0.005 (*Anderson et al.,* 1998; *Schubert et al.,* 2004). Since the MoI factor of a body of constant density is 0.4, it is apparent that Europa's mass is significantly concentrated toward its center.

The gravitational data obtained by the Galileo spacecraft flybys of Europa provide two constraints on the internal structure of the moon, its average density and its MoI. With only two numbers to constrain interior models of the satellite, it is sensible to restrict the models to simple two-layer or at most three-layer spherical shell structures. Such simple models are actually good physical representations of the real Europa, which, we will see, is differentiated into a metallic core, rocky mantle, and outer water ice/liquid shell. Because it is differentiated into layers of very different density, and the density within each layer is relatively constant (Europa is small enough that densification with increasing pressure is not a major effect), a three-layer model of Europa is a realistic representation of the satellite. For a three-layer model of Europa, the average density $\bar{\rho}$ and MoI factor impose the following two constraint equations on the models

$$\bar{\rho} = \rho_s + (\rho_c - \rho_m)\left(\frac{r_c}{R}\right)^3 + (\rho_m - \rho_s)\left(\frac{r_m}{R}\right)^3 \quad (5)$$

$$\text{MoI} = \frac{C}{MR^2} = \frac{2}{5}\left\{\frac{\rho_s}{\bar{\rho}} + \left(\frac{\rho_c - \rho_m}{\bar{\rho}}\right)\left(\frac{r_c}{R}\right)^5 + \left(\frac{\rho_m - \rho_s}{\bar{\rho}}\right)\left(\frac{r_m}{R}\right)^5\right\} \quad (6)$$

In equations (5) and (6), ρ_s is the density of the outer (liquid water/ice) shell, ρ_m is the density of the middle shell (silicate mantle), ρ_c is the density of the core, r_c is the radius of the core, and r_m is the outer radius of the middle shell. It is seen in equations (5) and (6) that there are five independent parameters in the three-layer model, three density ratios $\rho_s/\bar{\rho}$, $(\rho_m - \rho_s)/\bar{\rho}$, $(\rho_c - \rho_m)/\bar{\rho}$, and two fractional radii (r_c/R) and (r_m/R). Even the simple three-layer model is underconstrained.

Two- and three-layer models of Europa constrained by equations (5) and (6) have been discussed by *Anderson et al.* (1998), *Sohl et al.* (2002), *Zhang* (2003), *Schubert et al.* (2004), *Sotin and Tobie* (2004), *Kuskov and Kronrod* (2005), and *Hussmann et al.* (2007). Figure 2 describes the set of models obtained by *Anderson et al.* (1998). The figure shows possible models for two assumed values of core density, 5150 kg m^{-3} for an Fe-FeS core and 8000 kg m^{-3} for an Fe core. All models lie on the colored surfaces, and it is seen that values of two additional variables are required to specify a particular model. The two variables could be ice shell density and silicate mantle density or core radius and ice density, for example. The colors of the surfaces give the thickness of the ice shell in the model. Figure 3 provides a two-dimensional representation of Fig. 2 for an ice shell density of 1050 kg m^{-3}. The solid curves (labeled 129, 131.5, and 134; see figure caption) are the loci of possible models satisfying equations (5) and (6). An assumed value of core radius identifies a particular model in these figures (since the icy shell density and core density are assumed in these figures). An alternative representation of possible models is shown in Fig. 4, which has additional information on the relative core mass fraction of a model. Examination of Figs. 2–4 leads to the following conclusions:

1. Europa has a metallic core (*Anderson et al.,* 1998). Models of Europa that lack a metallic core require that the interior is a mixture of rock and metal with a density higher than 3800 kg m^{-3}. Such a density would imply that Europa's interior is enriched in dense metallic phases relative to Io, which has a bulk density of 3529 kg m^{-3}. If the metal is Fe, the enrichment is a factor of 12, and the enrichment is even greater for lower-density metallic phases such as magnetite. Such degrees of enrichment in dense phases are unlikely for a smaller body forming farther out in the protojovian nebula than Io. It is more likely that this mixture would separate into a metallic core and rock mantle, because radiogenic heating in the silicates alone would raise Europa's interior to temperatures high enough for differentiation to occur, and tidal heating is potentially an important additional source of heating in the mantle.

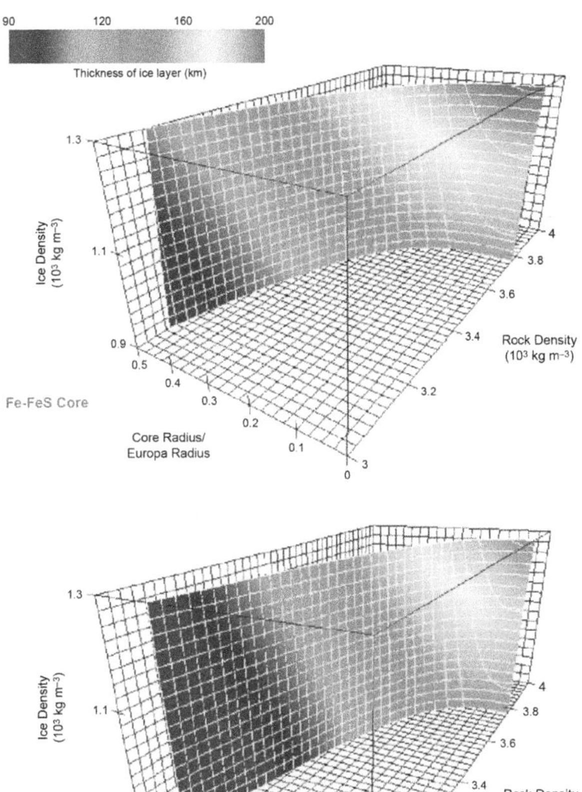

Fig. 2. See Plate 21. Possible three-layer models of Europa consistent with its mean density and axial MoI (MoI = 0.346) (*Anderson et al.,* 1998). Two sets of models are considered, one having Fe cores with density 8000 kg m^{-3} and the other having Fe-FeS cores with density 5150 kg m^{-3}. Any point on one of the surfaces defines an interior structure with properties given by the coordinate axis values and the color of the surface. Ice density refers to the density of the outer spherical shell of the model, which is predominately water in either ice or liquid form with an admixture of some rock. The color of the surface gives the thickness of this outer shell according to the color bar. Rock density refers to the density of the mainly silicate intermediate shell, which may also contain some metal. Possible Europa models are defined by the surfaces whose colors give the thickness of the water ice liquid outer shell. Other model parameters (outer shell or ice density, intermediate shell or rock density, and core radius) are provided by the coordinate axes. Two-layer Europa models are given by the intersection of the model surfaces with the core radius = 0 plane.

2. The minimum thickness of the water ice-liquid outer shell is ~80 km (*Anderson et al.,* 1998). Smaller outer shell thicknesses are possible only for mantle densities smaller than 3000 kg m^{-3}. Such low mantle densities are possible only if the mantle silicates are hydrated. Hydrated silicates break down and release their water at temperatures between

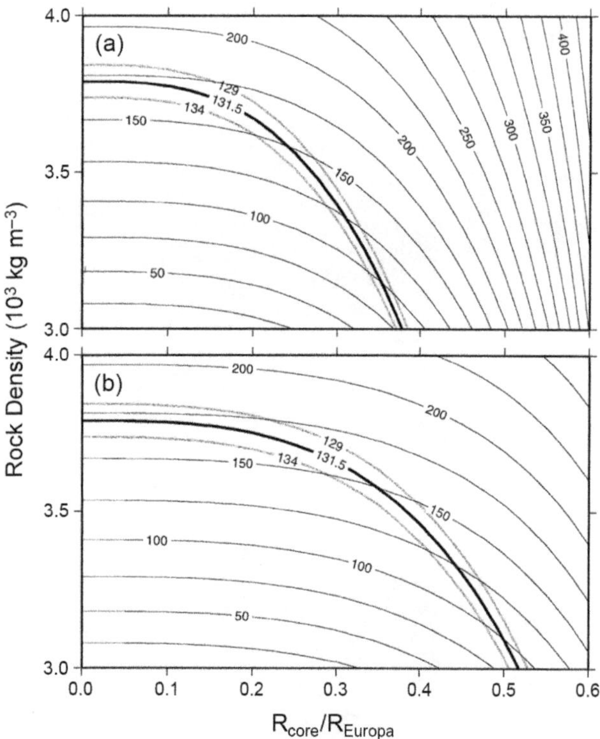

Fig. 3. Details of the intersection of the model surface of Fig. 2 with the horizontal outer shell density = 1050 kg m^{-3} plane (*Anderson et al.,* 1998). Europa three-layer models having an ice density (outer shell density) of 1050 kg m^{-3} are shown for **(a)** an Fe core and **(b)** an Fe-FeS core. The solid curve labeled 131.5($\times 10^{-5}$) defines models constrained by Europa's mean density and the indicated values of C_{22} used in constructing the model surfaces in Fig. 1. The curves designated 129 and 134 ($\times 10^{-6}$) delineate models with the $\pm 1\sigma$ values of C_{22}. The numbers within the curves denote the outer shell thickness (in kilometers).

700° and 800°C at the pressures in Europa's interior, making it unlikely that a thick europan mantle would have an average density lower than 3000 kg m^{-3}. Additionally, it is implausible that Europa would have differentiated a metallic core while retaining a hydrated silicate mantle.

3. The maximum thickness of Europa's outer shell of water ice-liquid is about 170 km (*Anderson et al.,* 1998).

4. The core radius could be as large as about 45% of Europa's radius (if the core composition is that of an Fe-FeS eutectic and if the water ice-liquid shell is about 100 km thick) or only as large as about 13% of Europa's radius (if the core is mainly Fe and if the water ice-liquid shell is 170 km thick) (*Sohl et al.,* 2002).

The gravitational data provide no information on the liquidity or solidity of the outer water ice-liquid shell or the core because of the small density differences involved. However, as we discuss in a following section, magnetic field data from the Galileo spacecraft provide strong evidence that a global liquid water ocean exists beneath the surface ice of Europa. The geology of Europa's surface provides supportive evidence of a subsurface liquid water ocean. The magnetic field data are silent about the physical state of the metallic core. The Galileo magnetometer did not detect an intrinsic magnetic field at Europa (*Schilling et al.,* 2004), which only means that dynamo action is not occurring in the core. Lack of a dynamo could be consistent with a fully liquid core, a completely solid core, or even a partially molten core. Another possibility is that an intrinsic europan magnetic field exists but its intensity is below the detectability limit of the Galileo magnetometer. The physical models of Europa discussed in this chapter have been extended by *Sohl et al.* (2002) and *Kuskov and Kronrod* (2005) to address the composition of the satellite. This exercise requires additional assumptions about the likely composition of Europa to supplement the gravitational constraints. These models are discussed in a later section.

3. FURTHER DISCUSSION OF STRUCTURAL MODELS: COMPOSITIONAL CONSIDERATIONS

In this section we probe the range of possible structural models of Europa's interior and discuss the compositional aspects of the models. All models assume the radius and mass values as given in Table 1, hydrostatic equilibrium, a moment of inertia factor of 0.346 ± 0.005, and constant density layers [as shown by *Sohl et al.* (2002), effects of compression are small in the case of Europa]. The consideration of allowable internal structural models in this section expands on the discussion already presented in section 2.

Two-layer models of Europa consisting of an iron/rock core and an H_2O layer are not ruled out on the basis of the gravitational data alone. We assume three different values (917, 1000, 1050) kg m^{-3} for the density of the H_2O layer representing (1) clean ice-I, (2) liquid water or ice with salts or other impurities, and (3) ice including a silicate dust component or high-pressure ice phases or salt hydrates, respectively. The hypersaline crustal models of *Spaun et al.* (2000) have even denser outer layers. The simultaneous solution of the equations for the moment of inertia factor (equation (6)) and the total mass or mean density (equation (5)) yield unambiguous values for the core radius and core density in the two-layer case for each model. Possible solutions that are consistent with the gravitational data are given in Table 2. A typical model for the two-layer case is shown in Fig. 5.

All solutions yield values of the core density between about 3700 and 3900 kg m^{-3}, significantly higher than the bulk density of Io. The variation in the thickness of the H_2O layer is relatively small, ranging from 144 to 188 km. Correspondingly, the H_2O mass fractions range from 7.7% to 11.2%. Although these models are consistent with the gravitational data, they are unlikely because of the high core density compared to Io's density (see section 2).

A more plausible structure of Europa is the three-layer model consisting of an iron core, a silicate mantle, and an outer H_2O layer (*Anderson et al.,* 1998; *Sohl et al.,* 2002; *Schubert et al.,* 2004). This introduces two additional unknown quantities, and the two equations for the total mass and the moment of inertia do not yield unambiguous solu-

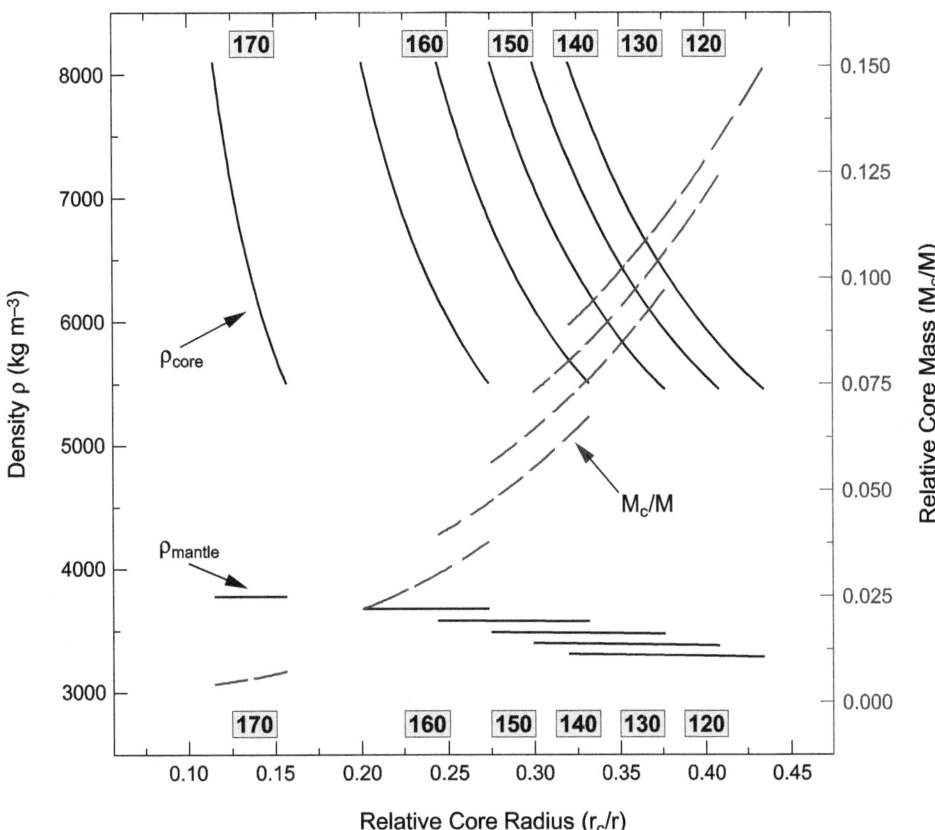

Fig. 4. Mantle and core densities of Europa and core mass fractions M_c/M, where M_c is the mass of the core and M is the mass of the satellite, as a function of relative core radius r_c/r (adapted from *Sohl et al.,* 2002). The models satisfy the mass and moment-of-inertia constraints imposed by the Galileo radio tracking data and are consistent with an olivine silicate mantle mineralogy and an Fe-FeS core composition. From left to right, there are six sets of curves, each consisting of a ρ_{core}-curve (solid and steep), a ρ_{mantle}-curve (solid and flat), and an M_c/M-curve (dashed and steep). Each set of curves is for a particular ice shell thickness value. The ice shell thickness is given as a label and decreases in steps of 10 km from 170 to 120 km.

TABLE 2. Two-layer models of Europa assuming three different density values for the outer H_2O layer.

	Models A H_2O density 917 kg m^{-3}		Models B H_2O density 1000 kg m^{-3}		Models C H_2O density 1050 kg m^{-3}	
Core density (kg m^{-3})	3711	3905	3720	3918	3725	3927
Core radius (km)	1418	1387	1411	1379	1407	1374
Thickness of H_2O layer (km)	144	176	151	183	155	188
H_2O mass fraction (%)	7.7	9.2	8.7	10.4	9.4	11.2
Core mass fraction (%)	92.3	90.8	91.3	89.6	90.6	88.8

The range of the resulting quantities is determined by the uncertainty in the moment of inertia factor. Values on the left within each column correspond to MoI = 0.346 + 0.005 = 0.351 and on the right to MoI = 0.346 – 0.005 = 0.341.

tions. Possible parameter ranges, assuming an H_2O density of 1000 kg m^{-3} and a MoI factor of 0.346, are given in Table 3.

The lower bound on the silicate density of 2500 kg m^{-3} represents hydrated rock. Deep hydrothermal systems on Europa could serpentinize the mantle to considerable depth (*McKinnon and Zolensky,* 2003; *Vance et al.,* 2007). The upper bound of 3760 kg m^{-3} is the maximum possible value for which real solutions are found assuming a core density of 8000 kg m^{-3}. For very low densities the H_2O layer gets extremely thin.

Structural and compositional models of Europa's interior that satisfy the satellite's surface composition, average density, and MoI factor therefore range from hydrated sili-

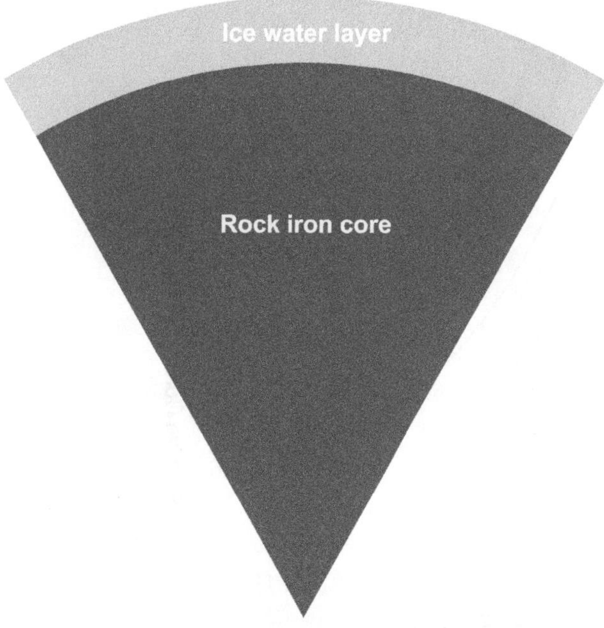

Fig. 5. A typical two-layer model of Europa assuming a MoI factor of 0.346 and a density of the H_2O layer of 1000 kg m^{-3}. The H_2O-layer has a thickness of 167 km. The H_2O mass fraction is 9.6%, corresponding to a density of the rock/iron core of 3816 kg m^{-3}. The model is consistent with the gravitational data assuming hydrostatic equilibrium. However, such a structure is unlikely for Europa's present state mainly because of the high core density required to satisfy the MoI and mass constraints (see text).

cate plus metal interiors beneath relatively thin ice shells to dehydrated interiors underneath thick ice shells. However, although such models are consistent with the gravitational data, there are arguments that rule out some of these solutions. Modest radiogenic and tidal heating would be sufficient for dehydration and subsequent differentiation of an initially uniform rock-metal interior. The process of differentiation and formation of a metallic core would certainly yield sufficient heat to release the water from the silicate mantle. Hydrated rock is therefore inconsistent with full differentiation (*Anderson et al.,* 1998) (see also section 2). Core formation could be accompanied by fracturing caused by volume increase and stress accumulation upon solidification of the growing icy shell (*Squyres and Croft,* 1986).

Sohl et al. (2002) have constructed basic models of the composition of the silicate shells and cores of the Galilean satellites. The silicate shells are assumed to be composed of olivine end members fayalite (Fe-rich) and forsterite (Mg-rich), whereas the main core constituents are iron and iron sulfide. Mass fractions of olivine end members in the mantle and core constituents are calculated from general mass balance constraints, using an isothermal Murnaghan equation of state for the densities corrected for thermal pressure at volume-averaged mantle and core pressures. Bulk iron-to-silicon ratios Fe/Si are then calculated from the amount of iron in the mantle and core constituents and from the silicon abundance in the mantle components, assuming a mean mantle temperature of 75% of the silicate melting temperature (Fig. 6).

The Fe/Si ratio (by mass) has been found to increase with increasing icy shell thickness but is almost independent of core density and chemistry as seen by the small slope of the curves for each individual model. Icy shells approximately 150 km thick give Fe/Si ratios that are subchondritic and similar to those calculated for Io. In that case, however, the mantle composition would differ between the two satellites with Io's mantle, at about 80% forsterite, being less rich in iron than Europa's mantle, at only 60–70% forsterite. Furthermore, it is worth mentioning that the Fe/Si ratio increases with increasing temperature. However, most likely models fall below the Fe/Si ratio value for CI carbonaceous chondrites of 1.7 ± 0.1 (*Kerridge and Matthews,* 1988), which is not obviously consistent with a former conclusion reached by *Lewis* (1982) from chemical evolution considerations that Io's bulk composition should be more similar to that of CV or CM chondrites, which are either

TABLE 3. Three-layer models of Europa assuming three different density values for the silicate-layer.

	Models A Silicate density 2500 kg m^{-3}		Models B Silicate density 3500 kg m^{-3}		Models C Silicate density 3760 kg m^{-3}	
Core density (kg m^{-3})	8000	4700	8000	4700	8000	4700
Core radius (km)	706	1012	437	702	245	409
Mantle radius (km)	1561	1529	1427	1425	1401	1401
Thickness of H_2O layer (km)	0.7	33	135	137	161	162
H_2O mass fraction (%)	0.05	2.1	7.9	8.0	9.3	9.3
Mantle mass fraction (%)	75.4	55.4	86.3	77.8	89.7	87.9
Core mass fraction (%)	24.6	42.6	5.9	14.2	1.0	2.8
(Core + mantle) mass fraction (%)	99.95	97.9	92.1	92.0	90.7	90.7

The H_2O layer has a fixed density of 1000 kg m^{-3} and the MoI factor = 0.346. The range of the resulting quantities is determined by the uncertainty in the core density. Values on the left within each column correspond to an assumed core density of 8000 kg m^{-3}, values on the right to 4700 kg m^{-3}, which is given as a lower bound by *Kuskov and Kronrod* (2005).

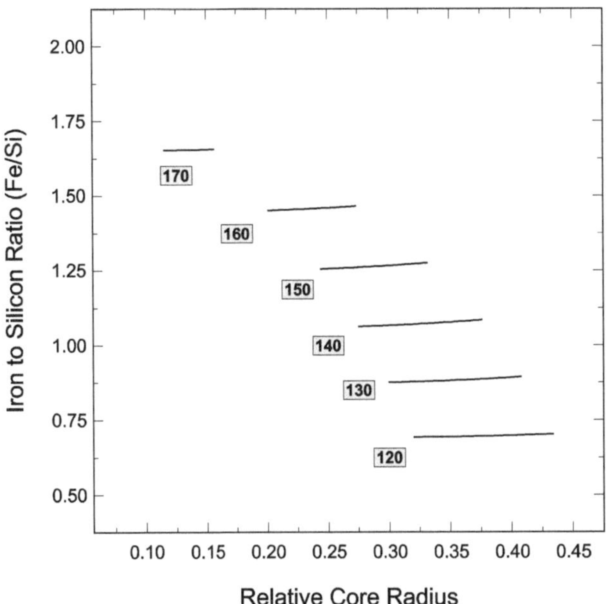

Fig. 6. Bulk iron-to-silicon ratio Fe/Si by weight of Europa's rock-iron core shown in Fig. 7 as a function of the ratio of metallic core radius to surface radius r_c/R (adapted from *Sohl et al.*, 2002). Curve labels in units of kilometers refer to the thickness of the outer ice shell.

slightly depleted or not depleted in iron relative to CI chondrites.

Kuskov and Kronrod (2005) emphasize that the implicit assumption of a three-oxide system (FeO-MgO-SiO_2) may oversimplify the problem and would prevent comparison of satellite bulk chemistry with that of specific meteorite classes. Instead, these authors consider equilibrium phase assemblages in the system Na_2O-TiO_2-CaO-FeO-MgO-Al_2O_3-SiO_2 (NaTiCFMAS), including solid solutions but neglecting oxygen fugacity. The calculation of the model mineralogy is based on the method of free energy minimization using an internally consistent thermodynamic database for minerals and assuming compositions and physical properties of H, L, LL, and CM2 chondrites as analogs for Europa's bulk composition. This gives a density range for the silicate mantle varying from 3320 kg m^{-3} (H chondrite) to 3670 kg m^{-3} (CM2 chondrite). However, the H chondrites can be ruled out because the derived Fe/Si ratios and the amount of metal in the iron core required to fit the gravitational data are not consistent with the composition of H chondrites. Similarly, the CI-chondritic composition can be ruled out because the maximum permissible amount of iron sulfide in the central core (~13 wt.%) required from the gravitational constraints is much lower than the FeS content in CI chondrites (~20%). *Kuskov and Kronrod* (2005) consider the L/LL-chondritic composition as the most likely analog case for Europa's bulk iron-silicate composition. The L/LL-chondritic model gives mean densities and MoI values for Europa's rock-iron core that are closest to the currently accepted values for Io, while carbonaceous chondrite compositional models would yield significantly higher values.

Based on these cosmochemical constraints, a three-layer interior structure, as shown in Fig. 7, can be considered as the most likely one for Europa's present state. For the L/LL-chondrite compositional model of *Kuskov and Kronrod* (2005), core radii are estimated to range between 470 and 640 km (5.3–12.5% of total mass), corresponding to bulk Fe/Si ratios of 0.9–1.3. The thickness of the H_2O layer in this case ranges from about 105 km to 160 km and yields a total H_2O mass fraction of 6–9%. The possible existence of a low-density silicate crust on top of Europa's rock-iron core, which may have formed in the event of pressure-release melting and subsequent melt extraction from the core, could lead to a 20-km-thinner water-ice shell (*Kuskov and Kronrod,* 2005).

With the given density distribution of the model in Fig. 7 it is straightforward to calculate the pressure as a function of the radial distance from the center by solving the hydrostatic equation. The result for the model in Fig. 7 is shown in Fig. 8 (gravitational acceleration) and Fig. 9 (pressure). The pressure at the base of the H_2O shell of 0.18 GPa is not sufficient for high-pressure ice phases, e.g., ice-II or ice-III, to occur. However, high-pressure ice phases may play a role in models with H_2O layer thicknesses near the maximum of about 180 km. As discussed above, such models are not likely if geochemical arguments are taken into account.

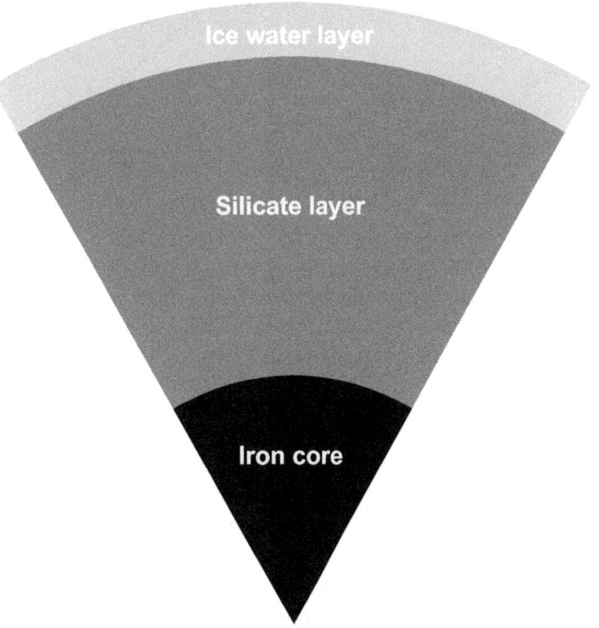

Fig. 7. A typical three-layer model (iron core, silicate mantle, and H_2O shell) of Europa assuming an MoI factor of 0.346 and a density of the H_2O layer of 1000 kg m^{-3}. The H_2O layer has a thickness of 136 km corresponding to an H_2O mass fraction of 8%. The densities of the core and silicate mantle are 5150 kg m^{-3} and 3500 kg m^{-3}, respectively. This implies mass fractions of the Fe-FeS core and the silicate mantle of 11% and 81%, respectively. The model is consistent with the gravitational data assuming hydrostatic equilibrium and with plausible assumptions for Europa's bulk core-silicate composition (see text).

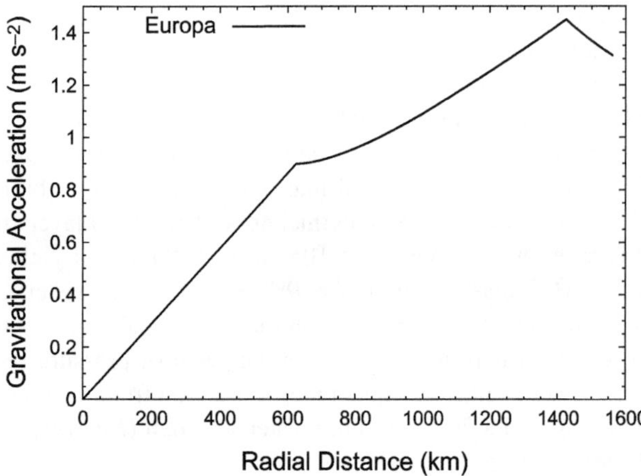

Fig. 8. Gravitational acceleration as a function of radial distance from the center for the model shown in Fig. 7. The surface gravity is 1.31 m s^{-2}. The maximum of 1.45 m s^{-2} is reached at the H_2O silicate boundary.

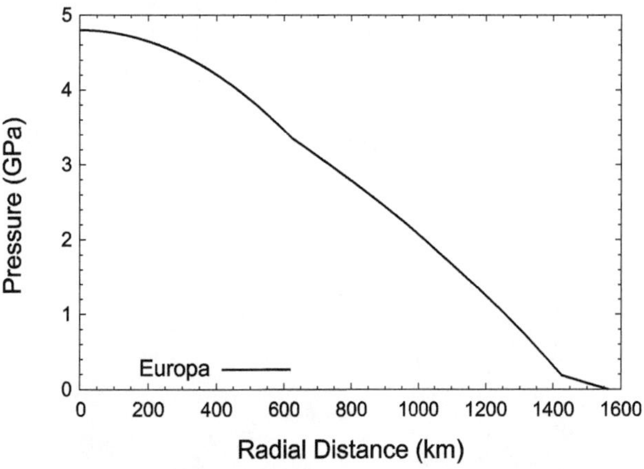

Fig. 9. Hydrostatic pressure P as a function of radial distance from the center for the model shown in Fig. 7. At the base of the H_2O layer P = 0.18 GPa, at the core-silicate boundary P = 3.35 GPa, and at the center P = 4.89 GPa.

An important question is whether part of the H_2O shell can be liquid, forming a global water ocean in Europa. The melting temperature of ice can be calculated as a function of depth (Fig. 10) using parameterizations such as the one of *Chizkov* (1993). From the surface to the base of the ice shell the melting temperature decreases from 273 K to 254 K. This is the minimum decrease in melting temperature. The presence of salts in the ocean would imply a further decrease in melting temperature.

Thermal equilibrium models including tidal heating suggest that there is a sufficient amount of heat available at Europa to keep a substantial part of the ice shell in liquid form (e.g., see chapter by Sotin et al.). This is consistent with the magnetometer data, which provide the strongest evidence for a present-day subsurface ocean on Europa (see section 6).

4. SHAPE AND IMPLICATIONS FOR INTERNAL STRUCTURE

Additionally to the constraints placed by the mean density, radius, and moment of inertia, clues on the interior structure of a satellite are provided by its shape (*Hubbard and Anderson,* 1978; *Dermott,* 1979; *Zharkov et al.,* 1985). In response to the satellite's rotation and the tidal potential due to Jupiter, Europa is deformed into a triaxial ellipsoid with the longest axis pointing toward Jupiter. The ellipsoidal shape is a static distortion depending — aside from the rotation rate and the external tidal potential — on the internal structure of the satellite. Furthermore, accurate measurements of a satellite's shape can provide evidence for (or against) the hydrostatic state obtained by the satellite. In the following we will describe how the internal structure can be constrained from shape measurements and how such an analysis in the case of Europa corresponds to the structure derived from gravitational data (as described in sections 2 and 3).

The rotational forcing parameter q_r defined in equation (3) is the ratio of the centrifugal force at the equator to the gravitational force at the surface; it is a dimensionless parameter characterizing the deformation due to rotation. The equivalent of q_r for the tidal potential q_t is defined by

$$q_t = -3\left(\frac{R}{a_J}\right)^3 \frac{M_J}{M} \tag{7}$$

where a_J is the semimajor axis of Europa's orbit, and M_J is the mass of Jupiter. Equation (7) can be interpreted as the ratio of tidal potential at the subjovian point on the satellite's surface due to Jupiter's gravity $V_t = -GM_JR^2/a_J^3$ and the satellite's gravitational potential GM/R at the surface. The factor of –3 (see below) is due to the fact that the magnitudes of the rotational and tidal deformation differ by a

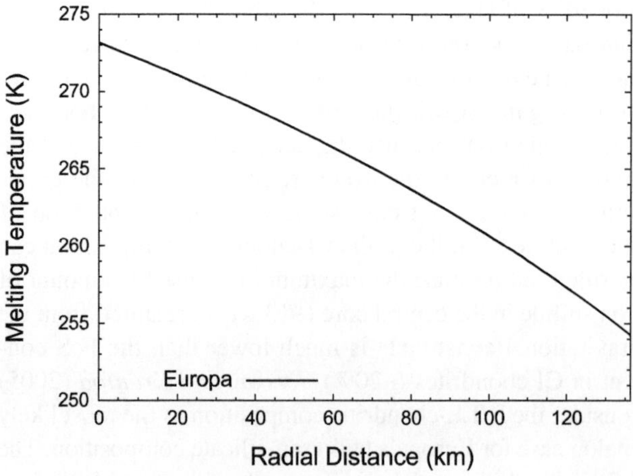

Fig. 10. Melting temperature of pure ice-I as a function of depth within the H_2O layer for the Europa model shown in Fig. 7. The total thickness of the H_2O layer is 136 km.

factor of 3 (e.g., *Dermott,* 1979). The resulting deformations additionally have different axes of symmetry: the polar axis (z axis) in the case of rotation and the subjovian line (x axis) in the case of jovian tides. The opposite sign reflects the fact that the deformation along each axis of symmetry is different. Equipotential surfaces in the case of rotation have smaller radii along the polar axis as compared to the spherical potential without rotation, whereas equipotential surfaces along the subjovian line due to tides have greater radii as compared to the spherical unperturbed case. Applying Kepler's third law, assuming $M_J + M \approx M_J$ and setting $n = \omega$ for a synchronously rotating satellite, we get $a^3 = GM_J/\omega^2$, which yields

$$q_t = -3\frac{R^3\omega^2}{GM} = -3q_r \tag{8}$$

For the parameter values given in Table 1 we obtain $q_r = 4.989 \times 10^{-4}$ and $q_t = -1.4979 \times 10^{-3}$. These quantities are small compared with unity, and therefore a first-order theory can be used to describe Europa's static deformation.

If Europa is in hydrostatic equilibrium, the axes of its triaxial shape can be related to its gravitational coefficients (*Zharkov et al.,* 1985). The tidal and rotational perturbations of the satellite's gravitational potential can be expressed by the quadrupole moment $J_{2_{t,r}}$ and the tesseral moment $C_{22_{t,r}}$, where the indices t and r refer to the tidal and rotational contributions, respectively (see equation (1)). The following expressions are obtained for first-order theory

$$J_{2_t} \approx -\frac{1}{6}q_t k_2 \text{ and } C_{22_t} \approx -\frac{1}{12}q_t k_2 \tag{9}$$

The fluid Love number k_2 is a measure of the contribution to the potential from the displacement of mass inside the satellite in a completely relaxed state, i.e., from the fluid response to the perturbation. For the rotational contribution we obtain

$$J_{2_r} \approx -\frac{1}{9}q_t k_2 \text{ and } C_{22_r} = 0 \tag{10}$$

The latter relation is due to the fact that the rotational perturbation is independent of longitude. In equation (10) we used $q_t = -3q_r$, which is valid for synchronous rotation. Upon addition of the tidal and rotational contributions we obtain the coefficients J_2 and C_{22}

$$J_2 = -\frac{5}{18}q_t k_2 \text{ and } C_{22} = -\frac{1}{12}q_t k_2 \tag{11}$$

For the fluid response, the potential Love number k_n of nth order is related to the corresponding radial displacement Love number h_n by $h_n = 1 + k_n$ (*Zharkov et al.,* 1985) and equation (11) can be rewritten as

$$J_2 = -\frac{5}{18}q_t(h_2 - 1) \text{ and } C_{22} = -\frac{1}{12}q_t(h_2 - 1) \tag{12}$$

Thus, if the satellite is in hydrostatic equilibrium the coefficients J_2 and C_{22} are not independent of each other. From equations (11) or (12), we get

$$J_2 = \frac{10}{3}C_{22} \tag{13}$$

The tidally and rotationally distorted triaxial shape in hydrostatic equilibrium is given by *Zharkov et al.* (1985)

$$\begin{aligned} a &= R\left(1 - \frac{7}{18}q_t h_2\right) = R\left(1 + \frac{7}{6}q_r h_2\right) \\ b &= R\left(1 + \frac{1}{9}q_t h_2\right) = R\left(1 - \frac{1}{3}q_r h_2\right) \\ c &= R\left(1 + \frac{5}{18}q_t h_2\right) = R\left(1 - \frac{5}{6}q_r h_2\right) \end{aligned} \tag{14}$$

where a is the semimajor long axis oriented toward Jupiter (subjovian, equatorial), and b and c are the semimajor axes in the equatorial plane perpendicular to a (along orbit, equatorial) and along the rotational axis (polar), respectively. R is the mean radius corresponding to a sphere with the same volume as the ellipsoid, hence $R = (abc)^{1/3}$. Application of the theory to Europa using $C_{22} = 131.5 \times 10^{-6}$ (*Anderson et al.,* 1998) gives (from equations (11) and (12))

$$k_2 = 1.0535 \text{ and } h_2 = 2.0535 \tag{15}$$

In computing k_2 here, we used the value of R = 1562.09 km from Table 1. The value of k_2 quoted earlier from *Schubert et al.* (2004), $k_2 = 1.048$, used R = 1565 km, an earlier value of R with a large uncertainty. The corresponding axes are given by

$$\begin{aligned} a &= 1.001196R = (1 + 1.196 \times 10^{-3})R = 1563.96 \text{ km} \\ b &= 0.999658R = (1 - 3.418 \times 10^{-4})R = 1561.56 \text{ km} \\ c &= 0.999145R = (1 - 8.544 \times 10^{-4})R = 1560.75 \text{ km} \end{aligned} \tag{16}$$

where again we have used R = 1562.09 km from Table 1. From equation (14) the following relation between the axes is obtained for a satellite in hydrostatic equilibrium (*Dermott,* 1979; *Zharkov et al.,* 1985)

$$\frac{b - c}{a - c} = \frac{1}{4} \tag{17}$$

With the values of a,b,c given in equation (16) we obtain, as expected, (b – c)/(a – c) = 0.25.

Recent estimates of Europa's global shape are collected in Table 4. It is obvious that the values given by *Thomas et al.* (1997), *Seidelmann et al.* (2002), and *Seidelmann et al.* (2007) and the first one of *Nimmo et al.* (2007) deviate significantly from the hydrostatic case. According to the values given by *Seidelmann et al.* (2007), Io is the only Galilean satellite for which the figure is in good agreement with hydrostatic theory, yielding a value of (b – c)/(a – c) = 0.26 for (a, b, c) = (1829.4, 1819.3, 1815.7) km. Since tidal and rotational deformations are much smaller for Ganymede and Callisto, the differences in the axes are too small to yield any constraints on the internal structure of these satellites.

TABLE 4. Estimates of Europa's mean radius R and axes a (subjovian, equatorial), b (along orbit, equatorial), and c (polar).

R (km)	a (km)	b (km)	c (km)	$\frac{b-c}{a-c}=\frac{1}{4}$	Reference
1562.09	1563.96	1561.56	1560.75	0.25	Equation (16)
1562.09	1564.13	1561.23	1560.93	0.09	*Seidelmann et al.* (2002, 2007)
1561.17	1563	1561	1559.5	0.43	*Thomas et al.* (1997)
1560.7	1562.4	1560.2	1559.4	0.26	*Davies et al.* (1998)
1560.7	1562.7	1559.8	1559.5	0.09	*Nimmo et al.* (2007), ellipsoid, nominal
1560.8	1562.6	1560.3	1559.5	0.26	*Nimmo et al.* (2007), ellipsoid, adjusted
1560.8	1562.6	1560.3	1559.5	0.25	*Nimmo et al.* (2007), hydrostatic, adjusted
1562.09	1564.36	1561.44	1560.47	0.25	Equation (14), homogeneous
1562.09	1563.58	1561.67	1561.03	0.25	Equation (18), two-layer, *Dermott* (1979)
1562.09	1563.97	1561.55	1560.75	0.25	Equation (14), three-layer, *Zharkov and Karamurzov* (2006)

Hydrostatic equilibrium corresponding to (b – c)/(a – c) = 0.25 is assumed in lines 1 and 7 to 10. In the first line, the mean radius as given by *Seidelmann et al.* (2007) is assumed, and the axes are calculated based on the gravity measurements (J_2 and C_{22} values under the assumption of hydrostatic equilibrium). Line 2 lists the IAU-recommended values from *Seidelmann et al.* (2007). Lines 3 and 4 are estimates of the radii based on early Galileo and Voyager imaging data. Line 5 is a recent ellipsoidal fit, based on Galileo imaging data without fitting Europa's center of figure to the data. Line 6 is the same ellipsoidal fit including a correction (shift of 0.2 and –0.5 km in x- and y-axis, respectively) of the center of figure. As indicated by (b – c)/(a – c) = 0.26, there are only minor deviations compared to the adjusted hydrostatic fit given in line 7. Lines 8 to 10 are shape models as further described in the text based on the assumption of hydrostatic equilibrium.

For Europa b – c is in a range of only a few hundred meters. This is on the order of observed topography on Europa's surface (e.g., see chapter by Nimmo and Manga), indicating that reliable conclusions for a possible hydrostatic state are difficult to obtain from shape measurements. Deviations from hydrostatic equilibrium generally can arise if the ice shell has sufficient mechanical strength to support internal mass anomalies or if the present shape is a relic of a former dynamical state of the satellite. *Nimmo et al.* (2007) introduced a correction due to the uncertainty of the exact location of the center of the figure in the measurements of Europa's shape. There are six unknown parameters inherent in the data: the three radii and the three coordinates of the center of figure of Europa. Best fits found by *Nimmo et al.* (2007) suggest a shift in the x and y axes on the order of a few hundred meters. Fits were done with and without the assumption of hydrostatic equilibrium and yield the values for the axes listed in Table 4. The adjusted values, including a correction of the order of a few hundred meters in x and y direction, are consistent with the hydrostatic state (compare lines 6 and 7 of Table 4). They conclude that there is no evidence that Europa possesses a nonhydrostatic shape.

The theory of *Zharkov et al.* (1985) allows us to infer Europa's hydrostatic shape independent of the gravitational data if an internal structure is assumed in the calculation of the Love numbers. We can therefore verify whether Europa's shape is consistent with a homogeneous density distribution. In that case the Love numbers are given by $k_2 = 3/2$ and $h_2 = 5/2$. The resulting shape is given in Table 4.

As a second application we consider a two-layer model consisting of a rock core with density ρ_1 and radius R_1 and an icy mantle with density ρ_2 and radius $R_2 = R$. Following the theory of *Dermott* (1979) the axes are given by

$$a = R\left(1 + \frac{7}{6}T_2\right)$$
$$b = R\left(1 - \frac{1}{3}T_2\right) \quad (18)$$
$$c = R\left(1 - \frac{5}{6}T_2\right)$$

where T_2 is given by

$$T_2 = \frac{15}{8}H_h q_r \quad (19)$$

H_h is a constant depending on the densities of the two layers. The subscript h indicates the hydrostatic case for which

$$H_h = \frac{2\bar{\rho}}{5\rho_2}\left(\frac{1 + (3\delta/5\gamma)(R_1/R_2)^2}{\delta + 2\rho_2/5\rho_1 - (9\delta\rho_2/25\gamma\rho_1)(R_1/R_2)^2}\right) \quad (20)$$

γ and δ are defined by

$$\gamma = \frac{2}{5} + \frac{3\rho_2}{5\rho_1} \quad \text{and} \quad \delta = \left(\frac{R_1}{R_2}\right)^3\left(1 - \frac{\rho_2}{\rho_1}\right) \quad (21)$$

and $\bar{\rho}$ is the mean density of the satellite. Assuming density values of $\rho_1 = 3500$ kg m^{-3} for a rock core (including iron) and $\rho_2 = 1000$ kg m^{-3} for an ice/water layer, the core radius $R_1 = R[(\bar{\rho} - \rho_2)/(\rho_1 - \rho_2)]^{1/3} = 0.93\,R = 1451.61$ km. This corresponds to an H_2O layer of 110.5-km thickness.

We used the constraints of total mass and radius given in Table 1, corresponding to a mean density of 3006.2 kg m^{-3}. The values for the semimajor axes are given in Table 4.

As a third case we consider a three-layer model, which is the most likely structure of Europa based on the moment of inertia constraints (see section 2). We assume the following densities: $\rho_1 = 5150$ kg m^{-3} (Fe-FeS core), $\rho_2 = 3300$ kg m^{-3} (silicate mantle), and $\rho_3 = 1000$ kg m^{-3} (H_2O layer). The corresponding radii are $R_1 = 691.52$ km, $R_2 = 1451.61$ km, and $R_3 = R = 1562.09$ km. The thickness of the H_2O layer is the same as in the two-layer case and the core radius is consistent with the total mass. Application of equations (9) and (10) of *Zharkov and Karamurzov* (2006) gives the Love number h_2 for the assumed density profile. We obtain $h_2 = 2.06813$, which corresponds to the values for the axes (equation (14)) given in Table 4.

The differences between the homogeneous and the two-layer model compared to the solution using the C_{22} value of *Anderson et al.* (1998) based on equation (16) (first line in Table 4) are on the order of a few hundred meters only, indicating that the density structure of Europa is difficult to infer from the shape data alone. However, the three-layer model computed using the Love number calculation of *Zharkov and Karamurzov* (2006) is essentially identical to the solution using the C_{22} coefficient derived from the gravitational data. Both solutions do assume hydrostatic equilibrium.

So far we considered only a first-order theory in the small parameters q_t and q_r. Terms of second order are given by *Zharkov and Karamurzov* (2006) and equations (11) and (13) are generalized to

$$C_{22} = \frac{1}{4}k_2 q_r - \frac{16}{21}\left(1 - \frac{5}{16}h_2\right)h_2 q_r^2 \quad (22)$$

$$\frac{J_2}{C_{22}} = \frac{10}{3}\left(1 + \frac{16h_2}{21k_2}(3 - h_2)q_r\right) \quad (23)$$

If we apply this to the three-layer case using the above derived Love number $h_2 = 2.06813$ we obtain $C_{22} = 133.22 \times 10^{-6}$ for the first-order case and $C_{22} = 133.08 \times 10^{-6}$ for the second-order case. Furthermore, we get $J_2/C_{22} = 3.3356$, which differs by only 0.07% from the first-order solution of 10/3. It can be concluded that first-order theory is sufficient in the case of Europa.

Measurement of the shape of a satellite provides complementary data to the gravity analysis. The values derived for Europa are (1) consistent with a hydrostatic state (*Nimmo et al.* 2007) and (2) consistent with the three-layer case, which is considered as the most likely case for Europa based on gravitational data under the assumption of hydrostatic equilibrium (*Anderson et al.*, 1998; *Schubert et al.*, 2004). However, it should be noted that the hydrostatic state has not yet been verified quantitatively from independent measurements in the case of Europa. More accurate topography, e.g., using laser altimetry on a Europa orbiter, will significantly improve the determination of Europa's shape. It will allow verification of whether the satellite is in a hydrostatic state and — if that is the case — it is a means to constrain interior structure models independently from the multipole moments of the potential derived from the gravitational data.

5. GEOLOGICAL IMPLICATIONS FOR INTERIOR STRUCTURE

Since there are neither radar observations nor seismological data available for Europa, the internal structure has to be inferred indirectly. Key to understanding the internal processes and near-surface structure are the geological features observed at the surface. Europa's surface displays a great variety of features that are related to both external and internal processes. Ridges built along tidally generated fractures reflect both external and internal processes. Europa's ice shell has been subject to endogenic resurfacing as indicated by the presence of smooth, bright plains. Resurfacing of the plains could have occurred by outflow of liquid water or warm ice onto the surface or by warming of ice from below, e.g., during a phase of intense tidal heating, and subsequent viscous relaxation of the topography. A thick insulating surface frost layer would raise the surface temperature sufficiently to allow relaxation on short timescales (*Thomas and Schubert,* 1986). Alternatively, resurfacing could have been accomplished tectonically, by successive formation of generation after generation of doublet ridges. The orientation of the planetwide, symmetric fracture pattern is generally believed to be due to nonsynchronous rotation of the icy shell across Europa's tidal bulge. Strike-slip behavior and plate rotation near the antijovian point is taken as the current best geologic evidence that the water ice shell is decoupled from Europa's deep interior due to the existence of a subsurface liquid water ocean or at least a soft ice layer (*Schenk and McKinnon,* 1989). The geology of Europa is described in detail in Part II of this volume. In the following we will therefore focus on the geological features having implications for the interior structure of Europa (for an overview see *Pappalardo et al.,* 1999).

Because of the recent geological activity on Europa the number of large craters is limited. From Galileo data there have been 28 craters >4 km in diameter identified on Europa (*Moore et al.,* 2004; *Greeley et al.,* 2004). Whereas the number of craters is the means to determine the global surface age of Europa, the morphology of large craters can be analyzed to derive physical properties of the near-surface ice layers and the crustal structure (e.g., see chapter by Schenk and Turtle). In general crater morphologies on Europa suggest that there occurred (1) postimpact isostatic adjustment because the craters appear to be anomalously shallow and (2) ductile flow at relatively shallow depths because of concentric rings surrounding the two largest impact features (*Greeley et al.,* 2004). On the basis of stereo imaging, digital terrain models have been obtained from

Galileo SSI data for a few craters (*Giese et al.,* 1999; *Schenk,* 2002). Based on analysis of crater shape, a lower limit of thickness of the icy shell on Europa has been estimated to be 19–25 km (*Schenk,* 2002). A second transition where the ice becomes more ductile as compared to the brittle near-surface ice was identified at a depth of 7–8 km (*Schenk,* 2002). These estimates are roughly consistent with thermal models of Europa's icy shell and a subsurface ocean at a depth of about 20 km or more (e.g., see chapters by Moore and Hussmann and by Nimmo and Manga). The observation of central peaks in some of the largest craters on Europa suggests that the impactors responsible did not penetrate through the icy shell into an underlying ocean, because of the rebound of solid material required to form a central peak. This sets a lower bound on ice thickness of 3–4 km (*Turtle and Pierazzo,* 2001). The limit of penetration depth was later extended to 10–18-km ice thickness based on hydrocode simulations of excavation and crater collapse (*Turtle,* 2004; *Turtle and Ivanov,* 2002).

The above estimates from the morphologies of craters only give lower bounds on the ice thickness. Furthermore, the estimated values are not necessarily valid for the present time, but for the time when the impact occurred. Nevertheless, they do give constraints on the possible ice thickness in Europa's recent past if combined with estimates of the global surface age from the cratering record. This suggests that there has been warm ductile ice present near the surface at the time of impact. Furthermore, the crater morphologies are consistent with the presence of a subsurface ocean if the ice was at least 3–4 km thick at the time of impact. If large craters viscously relaxed and have disappeared from the geological record, it suggests relatively thick ice of about 20 km or more. Besides the craters, which are a consequence of external sources, there is geological evidence for the near-surface structure of Europa's H_2O layer coming from internal dynamical processes in the icy shell. The geological record at the surface includes (1) features that provide evidence for the presence of an ocean without constraining the thickness of the ice and (2) features that are indicative of a certain ice thickness. As discussed in detail by *Pappalardo et al.* (1999) (see also *Greeley et al.,* 2004), the latter is inconclusive because the formation of most of the features (if not all) can be explained with both states of the subsurface layer: liquid water or "warm ice" near the melting point below a cold (average ~100 K) brittle, icy surface. Another possibility to explain the formation of some of the features would be an ice layer with small pockets of liquid water or brines. Furthermore, it is unclear to what extent the visible features are consistent with the present-day state of the icy shell, which may vary in thickness with time (e.g., see also chapters by Moore and Hussmann and by Nimmo and Manga). A conclusive determination of the present-day ice thickness of Europa is therefore not possible on the basis of the geological features alone. Models of the formation of ridges, pits, spots, and domes, chaos terrain, and other surface features on Europa suggest an effective elastic lithosphere thickness of 0.1–0.5 km, an actual brittle-elastic lithosphere thickness of 1–2 km, and a ductile convecting layer from 4 to >20 km and probably ≤40 km thick, plausibly above liquid water (see Table 1 of *Pappalardo et al.,* 1999). This translates into a thickness of the whole icy shell of about 6–40 km. Whereas such an estimate does not constrain the present-day ice thickness very much, it is, however, indicative of a subsurface ocean because the derived ice thickness is significantly less than the minimum thickness of the entire H_2O shell of about 70 km derived from the gravitational data (*Anderson et al.,* 1998).

Europa's most obvious surface features are cracks and lineaments that are visible on a global scale throughout the satellite's surface. Cycloidal ridges (e.g., see chapter by Kattenhorn and Hurford) are unique features among icy satellite surfaces and are a consequence of stresses resulting from tidal forces exerted by Jupiter. On a diurnal timescale (i.e., on the orbital period of 3.55 d), the tidal stresses vary due to the varying distance from Jupiter (radial tides) and due to the varying direction to Jupiter (librational tides; see chapter by Sotin et al.). As a consequence stresses at a given location at Europa's surface change in magnitude and direction continuously during one orbit. The variation of the stresses determines the propagation of fractures and some of the surface features are consistent with preferred directions of cracking dictated by tidal stresses (*Greenberg et al.,* 1998). Diurnal stresses can cause fractures to propagate in arcs, consistent with cycloidal cracks obtained in Voyager and Galileo images (*Hoppa et al.,* 1999). Whereas diurnal tidal stresses can explain many characteristics of the observed cycloidal cracks, better consistency with the observational data is reached if an additional "background" stress pattern due to nonsynchronous rotation of the ice surface is assumed (see chapter by Kattenhorn and Hurford). Nonsynchronous rotation was originally suggested by *Helfenstein and Parmentier* (1985) on the basis of Voyager data. The period of nonsynchronous rotation would probably lie in the range of a few million years and is not detectable from imaging. However, the cycloidal cracks correlating with the combined stress pattern of diurnal tides and nonsynchronous rotation provide strong evidence for a decoupling of the ice shell from the deep interior. The cycloidal cracks therefore suggest the presence of a liquid water layer that would be capable of mechanically decoupling the icy shell from the rocky interior.

The concept of stresses resulting from nonsynchronous rotation was also applied to the fracture history in the bright plains region on Europa (*Kattenhorn,* 2002), leading to consistent results. In contrast to other geological evidence for a subsurface ocean on Europa (e.g., craters) the crack-propagation model due to tidal stresses and/or nonsynchronous rotation requires decoupling by a liquid layer. A solid layer of "warm" low-viscous ice cannot provide the complete decoupling required for the processes to occur. The correlation between the propagation of fractures and the stress pattern (diurnal and on long timescales) therefore strongly suggests the presence of an ocean at the time of formation of these features.

In summary, the gravitational signal obtained from Galileo, which is the most important means to determine the interior structure of Europa, cannot distinguish between liquid water and ice. It can only constrain the thickness of the total H_2O layer. However, the complex surface of Europa suggests the presence of a subsurface ocean. Additionally, there is evidence for liquid water or low-viscosity, "warm" ice relatively close (a few kilometers up to ~40 km) to the surface. This is consistent with the magnetic induction signal, discussed below, which provides the best observational evidence so far for a present-day ocean at Europa.

6. CONSTRAINTS ON THE INTERIOR FROM ELECTROMAGNETIC INDUCTION

The Galileo magnetometer detected magnetic field fluctuations arising from induced electrical currents in Europa (*Khurana et al.,* 1998; *Kivelson et al.,* 1999, 2000; *Zimmer et al.,* 2000; see chapter by Khurana et al.). These induction currents are driven by the time-variable jovian magnetic field experienced by Europa as it orbits Jupiter. The jovian magnetic field is tilted with respect to Jupiter's rotation axis, resulting in a radial magnetic field at Europa that points alternately toward and away from Jupiter during Europa's orbital journey around Jupiter. The temporal variability of this jovian magnetic field component at Jupiter's rotation period induces electric currents inside Europa. The induced electric currents in turn produce magnetic field fluctuations that were detected by the magnetometer on the Galileo spacecraft during several flybys of Europa. The magnitude of the induction-generated magnetic field fluctuations provides information on the electrical conductivity, depth, and thickness of the region in Europa where the induced electrical currents flow. The electrically conducting layer in Europa must lie within about 200 km of the surface and its electrical conductivity must be comparable to or less than that of seawater on Earth if its thickness is at least several kilometers (*Zimmer et al.,* 2000). This is strong evidence for a subsurface liquid salt water ocean in Europa's water ice-liquid outer shell. Ice simply cannot provide the requisite electrical conductivity to explain the magnitude of the induced magnetic fields. Neither is it possible for electrical currents in Europa's core to account for the induction data; core electrical currents are too far below the surface. The constraints on the depth of the internal ocean from the magnetometer data are consistent with the estimate of the thickness of the water ice-liquid outer shell from the gravitational data.

More recent analyses of the magnetometer data place tighter constraints on the depth of the ocean below the surface and on its electrical conductivity (*Schilling et al.,* 2004; *Hand and Chyba,* 2007). *Schilling et al.* (2004) argue that the induced magnetic field is larger in amplitude than earlier estimates (*Zimmer et al.,* 2000). In terms of the parameter A, which measures the ratio of the observed induced dipole strength to that of a perfectly conducting shell, then *Schilling et al.* (2004) obtain $A = 0.97 \pm 0.02$, whereas *Zimmer et al.* (2000) had $0.7 < A < 1$. *Hand and Chyba* (2007) model the induced dipole in a spherical ocean as a function of ocean thickness and electrical conductivity and icy shell thickness. In order to fit the induced field amplitude of *Schilling et al.* (2004), *Hand and Chyba* (2007) require that the icy shell be less than 15 km thick for a NaCl salty ocean, corresponding to a maximum electrical conductivity of the ocean of about 18 S m^{-1}. For an $MgSO_4$ salty ocean the ice thickness must be less than 7 km, corresponding to a maximum electrical conductivity of about 6 S m^{-1}. The salt concentration in Europa's ocean (either NaCl or $MgSO_4$) must be near saturation for it to have the high electrical conductivity required by the amplitude of the measured inductive response (*Hand and Chyba,* 2007).

The existence of a subsurface liquid water ocean on Europa has been advocated on the basis of geological features (*Greeley et al.,* 2004) found on its surface. The thickness of the overlying ice has also been estimated by modeling a number of these features and by modeling the satellite's thermal history. These studies are discussed in detail in other chapters in this book. Although there are other indications of an ocean below the surface of Europa, the electromagnetic induction observations provide the most compelling evidence for its existence. Some interpretations of these data also argue for a salty ocean and a relatively thin overlying icy shell compared to the estimates of several tens of kilometers from some models of surface features and Europa thermal history (see also *Schilling et al.,* 2007, and the chapter by Khurana et al.). Nevertheless, the gravitational and magnetic field data that we have at present cannot provide definitive answers to the many questions about Europa's ocean. How far below the surface is the ocean? How deep is the ocean? Does it reach down to the rocky mantle? How variable is the thickness of the ice covering the ocean? How salty is the ocean? What constituents are dissolved in it? Does the ocean have access to the surface or the rocky interior of Europa? What keeps the ocean from freezing? The definitive answers that we seek to these questions will require further exploration of Europa.

Acknowledgments. G.S. acknowledges support from the NASA Planetary Geology and Geophysics program under grant NNG06GG70G.

REFERENCES

Anderson J. D., Schubert G., Jacobson R. A., Lau E. L., Moore W. B., and Sjogren W. L. (1998) Europa's differentiated internal structure: Inferences from four Galileo encounters. *Science, 281,* 2019–2022.

Chizkov V. E. (1993) Thermodynamic properties and thermal equation of state of high-pressure ice phases. *Prikl. Mekh. i Tekhn. Fizika, 2,* 113–123 (translated).

Davies M. E., Colvin T. R., Oberst J., Zeitler W., Schuster P., Neukum G., McEwen A. S., Phillips C. B., Thomas P. C., Veverka J., Belton M. J. S., and Schubert G. (1998) The control networks of the Galilean satellites and implications for global shape. *Icarus, 135,* 372–376.

Dermott S. F. (1979) Shapes and gravitational moments of satellites and asteroids. *Icarus, 37,* 575–586.

Giese B., Wagner R., Neukum G., Moore J. M., and the Galileo SSI Team (1999) The local topography of Europa's crater Cilix derived from Galileo SSI stereo images (abstract). In *Lunar and Planetary Science XXX,* Abstract #1565. Lunar and Planetary Institute, Houston (CD-ROM).

Greeley R., Chyba C. F., Head J. W., McCord T. B., McKinnon W. B., Pappalardo R. T., and Figueredo P. (2004) Geology of Europa. In *Jupiter: The Planet, Satellites and Magnetosphere* (F. Bagenal et al., eds.), pp. 329–362. Cambridge Univ., Cambridge.

Greenberg R., Geissler P., Hoppa G., Tufts B. R., Durda D. D., Pappalardo R., Head J. W., Greeley R., Sullivan R., and Carr M. H. (1998) Tectonic processes on Europa: Tidal stresses, mechanical response, and visible features. *Icarus, 135,* 64–78.

Hand K. P. and Chyba C. F. (2007) Empirical constraints on the salinity of the europan ocean and implications for a thin ice shell. *Icarus, 189,* 424–438, DOI: 10.1016/j.icarus.2007.02.002.

Helfenstein P. and Parmentier E. M. (1985) Patterns of fracture and tidal stresses due to nonsynchronous rotation: Implications for fracturing on Europa. *Icarus, 61,* 175–184.

Hoppa G. V., Tufts B. R., Greenberg R., and Geissler P. E. (1999) Formation of cycloidal features on Europa. *Science, 285,* 1899–1902.

Hubbard W. B. and Anderson J. D. (1978) Possible flyby measurements of Galilean satellite interior structure. *Icarus, 33,* 336–341.

Hussmann H., Sotin C., and Lunine J. I. (2007) Interiors and evolution of icy satellites. In *Treatise on Geophysics, Vol. 10: Planets and Moons* (G. Schubert, ed.), pp. 509–539. Elsevier, Amsterdam.

Kattenhorn S. A. (2002) Nonsynchronous rotation evidence and fracture history in the Bright Plains Region, Europa. *Icarus, 157,* 490–506, DOI: 10.1006/icar.2002.6825.

Kaula W. M. (1968) *An Introduction to Planetary Physics: The Terrestrial Planets.* Space Science Text Series, Wiley, New York.

Kerridge J. F. and Matthews M. S., eds. (1988) *Meteorites and the Early Solar System.* Univ. of Arizona, Tucson.

Khurana K. K., Kivelson M. G., Stevenson D. J., Schubert G., Russell C. T., Walker R. J., Joy S., and Polanskey C. (1998) Induced magnetic fields as evidence for subsurface oceans in Europa and Callisto. *Nature, 395,* 777–780.

Kivelson M. G., Khurana K. K., Stevenson D. J., Bennett L., Joy S., Russell C. T., Walker R. J., Zimmer C., and Polanskey C. (1999) Europa and Callisto: Induced or intrinsic fields in a periodically varying plasma environment. *J. Geophys. Res., 104,* 4609–4625.

Kivelson M. G., Khurana K. K., Russell C. T., Volwerk M., Walker R. J., and Zimmer C. (2000) Galileo magnetometer measurements: A stronger case for a subsurface ocean at Europa. *Science, 289,* 1340–1343.

Kuskov O. L. and Kronrod V. A. (2005) Internal structure of Europa and Callisto. *Icarus, 177,* 550–569, DOI: 10.1016/j.icarus.2005.04.014.

Lewis J. S. (1982) Geochemistry of sulfur. *Icarus, 50,* 103–114.

McKinnon W. B. and Zolensky M. E. (2003) Sulfate content of Europa's ocean and shell: Evolutionary considerations and some geological and astrobiological implications. *Astrobiology, 3,* 879–897.

Moore J. M., Chapman C. R., Bierhaus E. B., Greeley R., Chuang F. C., Klemaszewski J., Clark R. N., Dalton J. B., Hibbits C. A., Schenk P. M., Spencer J. R., and Wagner R. (2004) Callisto. In *Jupiter: The Planet, Satellites and Magnetosphere* (F. Bagenal et al., eds.), pp. 397–426. Cambridge Univ., Cambridge.

Murray C. D. and Dermott S. F. (2000) *Solar System Dynamics.* Cambridge Univ., Cambridge.

Nimmo F., Thomas P. C., Pappalardo R. T., and Moore W. B. (2007) The global shape of Europa: Constraints on lateral shell thickness variations. *Icarus, 191,* 183–192, DOI: 10.1016/j.icarus.2007.04.021.

Pappalardo R. T., Belton M. J. S., Breneman H. H., Carr M. H., Chapman C. R., Collins G. C., Denk T., Fagents S., Geissler P. E., Giese B., Greeley R., Greenberg R., Head J. W., Helfenstein P., Hoppa G., Kadel S. D., Klaasen K. P., Klemaszewski J. E., Magee K., McEwen A. S., Moore J. M., Moore W. B., Neukum G., Phillips C. B., Prockter L. M., Schubert G., Senske D. A., Sullivan R. J., Tufts B. R., Turtle E. P., Wagner R., and Williams K. K. (1999) Does Europa have a subsurface ocean? Evaluation of the geological evidence. *J. Geophys. Res., 104,* 24015–24055.

Schenk P. M. (2002) Thickness constraints on the icy shells of the galilean satellites from a comparison of crater shapes. *Nature, 417,* 419–421.

Schenk P. M. and McKinnon W. B. (1989) Fault offsets and lateral crustal movement in Europa: Evidence for a mobile ice shell. *Icarus, 79,* 75–100.

Schilling N., Khurana K. K., and Kivelson M. G. (2004) Limits on an intrinsic dipole moment in Europa. *J. Geophys. Res., 109,* E05,006, DOI: 10.1029/2003JE002,166.

Schilling N., Neubauer F. M., and Saur J. (2007) Time-varying interaction of Europa with the jovian magnetosphere: Constraints on the conductivity of Europa's subsurface ocean. *Icarus, 192,* 41–55, DOI: 10.1016/j.icarus.2007.06.024.

Schubert G., Anderson J. D., Spohn T., and McKinnon W. B. (2004) Interior composition, structure and dynamics of the Galilean satellites. In *Jupiter: The Planet, Satellites and Magnetosphere* (F. Bagenal et al., eds.), pp. 281–306. Cambridge Univ., Cambridge.

Seidelmann P. K., Abalakin V. K., Bursa M., Davies M. E., Bergh C. D., Lieske J. H., Oberst J., Simon J. L., Standish E. M., Stooke P., and Thomas P. C. (2002) Report of the IAU/IAG Working Group on Cartographic Coordinates and Rotational Elements of the Planets and Satellites: 2000. *Cel. Mech. Dyn. Astron., 82,* 83–110.

Seidelmann P. K., Archinal B. A., A'Hearn M. E., Conrad A., Consolmagno G. J., Hestroffer D., Hilton J. L., Krasinsky G. A., Neumann G., Oberst J., Stooke P., Tedesco E. E., Tholen D. J., Thomas P. C., and Williams I. P. (2007) Report of the IAU/IAG Working Group on Cartographic Coordinates and Rotational Elements: 2006. *Cel. Mech. Dyn. Astron., 98,* 155–180.

Sohl F., Spohn T., Breuer D., and Nagel K. (2002) Implications from Galileo observations on the interior structure and chemistry of the Galilean satellites. *Icarus, 157,* 104–119, DOI: 10.1006/icar.2002.6282.

Sotin C. and Tobie G. (2004) Internal structure and dynamics of the large icy satellites. *Compt. Rend. Acad. Sci., 5,* 769–780.

Spaun N. A., Parmentier E. M., and Head J. W. (2000) Modeling

Europa's crustal evolution: Implications for an ocean. In *Lunar and Planetary Science XXXI,* Abstract #1039. Lunar and Planetary Institute, Houston (CD-ROM).

Squyres S. W. and Croft S. K. (1986) The tectonics of icy satellites. In *Satellites* (J. A. Burns and M. S. Matthews, eds.), pp. 293–341. Univ. of Arizona, Tucson.

Thomas P. C., Simonelli D., Burns J., Davies M. E., McEwen A. S., and the Galileo SSI Team (1997) Shapes and topography of Galilean satellites from Galileo SSI limb coordinates. In *Lunar and Planetary Science XXVIII,* Abstract #1715. Lunar and Planetary Institute, Houston (CD-ROM).

Thomas P. J. and Schubert G. (1986) Crater relaxation as a probe of Europa's interior. *Proc. Lunar Planet. Sci. Conf. 16th,* in *J. Geophys. Res., 91,* D453–D459.

Turtle E. P. (2004) What Europa's impact craters reveal: Results of numerical simulations (abstract). In *Workshop on Europa's Icy Shell: Past, Present, and Future,* Abstract #7020. LPI Contribution No. 1195, Lunar and Planetary Institute, Houston.

Turtle E. P. and Ivanov B. A. (2002) Numerical simulations of impact crater excavation and collapse on Europa: Implications for ice thickness. In *Lunar and Planetary Science XXXIII,* Abstract #1431. Lunar and Planetary Institute, Houston (CD-ROM).

Turtle E. H. and Pierazzo E. (2001) Thickness of a europan ice shell from impact crater simulations. *Science, 294,* 1326–1328.

Vance S., Harnmeijer J., Kimura J., Hussmann H., Demartin B., and Brown J. M. (2007) Hydrothermal systems in small ocean planets. *Astrobiology, 7,* 987–1005, DOI: 10.1089/ast.2007.0075.

Zhang H. (2003) Internal structure models and dynamical parameters of the Galilean satellites. *Cel. Mech. Dyn. Astron., 87,* 189–195, DOI: 10.1023/A:1026188029,324.

Zharkov V. N. and Karamurzov B. S. (2006) Models, figures, and gravitational moments of Jupiter's satellites Io and Europa. *Astron. Lett., 32,* 495–505.

Zharkov V. N., Leontjev V. V., and Kozenko A. V. (1985) Models, figures, and gravitational moments of the Galilean satellites of Jupiter and icy satellites of Saturn. *Icarus, 61,* 92–100.

Zimmer C., Khurana K. K., and Kivelson M. G. (2000) Subsurface oceans on Europa and Callisto: Constraints from Galileo magnetometer observations. *Icarus, 147,* 329–347.

Thermal Evolution of Europa's Silicate Interior

William B. Moore
University of California, Los Angeles

Hauke Hussmann
DLR Institute of Planetary Research

The thermal evolution of Europa's silicate mantle directly influences the chemistry and dynamics of its ocean and icy shell, and has significant implications for habitability. The current state of understanding of Europa's silicate mantle is reviewed and discussed. Models of the equilibrium thermal structure of the silicate mantle subject to tidal and radiogenic heating are presented. Europa's silicate mantle, like Io's, has two possible equilibrium states, one cold and Moon-like, and one hot (super-solidus) and Io-like. The heat flux out of the mantle in the hot state is ~10^{13} W, less than one-tenth of Io, but an order of magnitude larger than the heat produced in the icy shell. The thermal evolution of Europa's mantle is coupled to its orbital evolution, and models of the coupled system are presented. The dynamics of this system is rich, and includes oscillatory states in which the heat flow varies by orders of magnitude. This behavior would clearly have significant implications for the chemistry of the ocean and the thickness of the overlying icy shell.

1. INTRODUCTION

In order to understand the physics and chemistry of Europa's ocean and icy shell (see chapters by Nimmo and Manga and by Zolotov and Kargel), it is essential to understand the thermal evolution of the silicate portion of Europa's interior. This has been recognized by a number of authors coming at the problem of Europa's internal dynamics from different directions. Those concerned with the composition and chemistry of the ocean (*McCollom,* 1999; *Kargel et al.,* 2000; *Zolotov and Shock,* 2001; *Hand et al.,* 2007) and its possible remnants on the surface (*McCord et al.,* 1999; *Carlson et al.,* 1999) recognize that the level of volcanic activity driven by Europa's internally generated heat is what determines the rates and temperatures of water-rock reactions where the ocean meets the rock. Investigators into the structure of Europa's ice have recognized the importance of the delivery of heat from below (*Ross and Schubert,* 1987; *Ojakangas and Stevenson,* 1989; *Pappalardo et al.,* 1998; *Nimmo and Manga,* 2002; *Moore,* 2006). And the interaction between orbital and thermal evolution has inspired investigation into Europa's thermal state (*Fischer and Spohn,* 1990; *Showman et al.,* 1997; *Hussmann and Spohn,* 2004). In this chapter we will be concerned with the thermal evolution of the silicate mantle of Europa. For a recent review of the structure of Europa's interior, see *Schubert et al.* (2004).

The rocky components that Europa accreted separated from the ice very early in Europa's history and formed a layer occupying the bulk of Europa's volume. It is likely that Europa has formed a metallic core by segregation of molten metal from the rocky component (see chapter by Schubert et al.) and thus the silicates now form a spherical shell between the core and the overlying ocean. In analogy with Earth, this will be referred to as the mantle, although to avoid confusion with other applications of that term it will be more specifically called the silicate mantle. If the rock and metal portion of Europa's interior is constrained to have the same mean density as Io [3527.5 kg m^{-3} from *Anderson et al.* (2001)], then the silicate mantle of density 3300 kg m^{-3} extends from 704 km to 1443 km in radius if an interior model is to match the observed gravity field of Europa (*Anderson et al.,* 1998). The density is not known to within about 10%, thus the depth of the mantle is 700 km with a similar level of uncertainty.

The thermal evolution of the silicate mantle is driven by the heat transport processes that remove internally generated heat. These processes are dynamic and depend strongly on the thermal state of the mantle itself, thus the present state of Europa's interior depends on its past and is not simply predictable on theoretical grounds. Theory may, however, provide considerable illumination until more direct observations are available. In this chapter we discuss the heat generation and heat transport in Europa's silicate mantle and the coupling of Europa's thermal evolution to its orbital evolution through the mechanism of tidal energy dissipation.

2. HEAT GENERATION

Heat generation in Europa's silicate interior has two main sources: decay of radioactive nuclei and tidal dissipation. These sources have distinctly different behaviors over time and will thus influence the thermal history of Europa's interior in different ways.

Radioactive heat production in the silicate part of Europa is, at present, predominantly contributed by three isotopes: ^{40}K, ^{232}Th, and ^{238}U. Earlier in solar system history, ^{235}U

(0.7 G.y. half-life) was important. Shortly after accretion, ^{26}Al and ^{60}Fe may have been important (accretional heating only contributes a few hundred degrees depending on the rate of accumulation). The heat production rates of these short-lived isotopes is so large, and their initial abundances so uncertain (due largely to the unknown time for accretion of Europa), that incorporating them into thermal history models allows nearly any initial state for Europa's silicate interior from extensively molten to quite cold depending on the amount of these isotopes included. Present heat production in chondritic meteorites is ~3–5 × 10^{-12} W/kg and has exponentially declined from a value 4–5 times larger 4 G.y. ago.

The history of radiogenic heating in Europa's interior is not simply a slow decay. The important long-lived isotopes are all incompatible elements in the rock, which means that they are concentrated in the melt phase when the solidus is reached. If the evolution of Europa's interior includes the differentiation of a crust, then this crust, like that of Earth's continents, will be enriched in the heat-producing elements, while the interior will be depleted. This differentiation could have significant implications for the thermal evolution of Europa's silicate interior. Depletion of the interior alters the temperature distribution within Europa's mantle. This is potentially an important effect because of the way tidal dissipation depends on temperature.

Crustal formation is a unidirectional event on most bodies, with two exceptions: Earth, where plate tectonics recycles oceanic crust (although not continental) continuously, and Io, where ongoing burial of crustal materials by new flows results in the remelting of the deepest crust in a shallow recycling loop (*Moore,* 2001). Neither of these processes is an efficient means of remixing the radiogenic heat sources back into the mantle, however, because the formation of continents by remelting the primary basaltic crust effectively sequesters radioactive sources for billions of years on Earth, and Io's melt driven heat transport cycle does not require involvement of the deep mantle. It seems likely, therefore, that if Europa ever experienced significant melting then it has retained a crust enriched in heat-generating elements to the present. For the models presented here, we will consider Europa well-mixed, with no sequestration of radioactive elements in a crust, but future modeling should account for the affect of crust formation and recycling.

The distribution of Europa's heat sources is important because the other major source of heating is tidal dissipation, a viscosity- and therefore temperature-dependent mechanism by which tidal motions deposit heat in the interior. In viscoelastic materials such as silicate rocks, the strain rate $\dot{\varepsilon}$ responds to both the applied stress σ and the applied stress rate $\dot{\sigma}$. For a linear Maxwell-viscoelastic material (a description chosen for its simplicity), the relationship is

$$\frac{\dot{\sigma}}{\mu} + \frac{\sigma}{\eta} = \dot{\varepsilon} \qquad (1)$$

There are two parameters in the Maxwell model, the elastic (shear) modulus μ and the viscosity η. They are often combined into the Maxwell time $\tau = \eta/\mu$, which is the period at which the work done on the material by a periodic forcing is maximized. Thus, when the tidal period is near the Maxwell time, heat production by tidal dissipation peaks. Viscosity is a strong function of temperature in rocks below and above the solidus. Shear modulus is weakly temperature dependent until significant portions (>20%) of the rock begin to melt, at which point the modulus decreases rapidly. At low temperatures, the material is essentially elastic and little deformation occurs. At very high temperatures, the rocks melt, and as a fluid dissipate little or no energy. At intermediate temperatures, while the shear modulus is still large but the viscosity is decreasing, τ approaches the forcing period. For the tidal periods relevant to Europa and the laboratory-derived rheology of rocks, the maximum dissipation occurs at a temperature slightly above the solidus.

3. THERMAL EQUILIBRIUM IN EUROPA'S SILICATE INTERIOR

Due to the nature of thermal equilibrium in tidally heated bodies, we will show that Europa's silicate mantle, like Io's, can be in one of two, very different, equilibrium states (*Fischer and Spohn,* 1990; *Moore,* 2003). One state is determined by the balance between radiogenic heating and convective heat transport, and is the equilibrium believed to be relevant for the Moon and terrestrial planets. The second state occurs when tidal heat generation is balanced by convective heat transport and is characterized by very high (supersolidus) temperatures and heat flows. Io is the only body in the solar system thought to be in the second, high-temperature state. The two stable equilibria are separated by an unstable equilibrium at more moderate temperatures, which drives the evolution of the system toward one of the stable points. The high-temperature equilibrium implies very active volcanism (about one-tenth Io's current activity) and hydrothermal circulation at the ocean-mantle interface, while a Europa in the low-temperature equilibrium is likely to be geologically inactive (similar to the Moon).

Here we review self-consistent models of tidal-convective thermal equilibrium in Europa's mantle, akin to those computed for Io (*Moore,* 2003). In the high-temperature equilibrium, the heat flow at the surface of Europa's mantle (the bottom of Europa's ocean) is about 10^{13} W, while in the low-temperature equilibrium it is on the order of 10^{11} W. Tidal heating in the ice can produce at most a few times 10^{12} W, therefore a measurement of the surface heat flow to within a factor of 10 would determine the thermal state of the mantle, with significant implications for the energetics of a europan biosphere.

3.1. Coupled Tidal-Convective Model

In order to calculate a coupled tidal-convective model, the rheology of the mantle must be accounted for in a self-consistent manner. The processes of convection and tidal deformation operate at very different time-, stress-, and strain-scales, and therefore may be accommodated by different deformation mechanisms. Convective stresses are large (tens of MPa), operate over long time periods (millions of years), and result in very large permanent strains. Silicate rocks under these conditions deform by diffusion creep, dislocation creep, or some combination of the two (*Karato and Wu,* 1993). Tidal stresses are small (about 1 MPa), are periodic on a short timescale (days), and accumulate very little permanent strain. Tidal deformation is therefore most likely accommodated by diffusion creep, although this has not been demonstrated experimentally.

The convecting mantle establishes a temperature distribution in response to internally generated heat that results in the transport of that heat to the surface by advection in the interior and by conduction in the outermost region (often called the conductive lid). The temperature distribution established depends sensitively on the rheology of the mantle (*Solomatov,* 1995; *Solomatov and Moresi,* 2000). For the silicates that make up Europa's mantle, the viscosity η is a strong function of temperature. The Newtonian solid rheology is modeled here using the Arrhenius law

$$\eta = A\exp\left(\frac{E}{T}\right) \tag{2}$$

where A and E are constants that have different values above and below the so-called critical temperature T_{crit}, which corresponds to a few percent melt (*Berckhemer et al.,* 1982). The critical melt fraction employed here is 2%, and the values of A and E are given in Table 1.

This is the same as the rheology used in *Fischer and Spohn* (1990) with the following exceptions (*Moore,* 2003): Above the solidus, the viscosity is multiplied by a term of the form exp $(-B\phi)$, where B is a dimensionless melt fraction coefficient ranging from 10 to 40, and ϕ is the volume melt fraction. Above the disaggregation point (40–60% melt), the Roscoe-Einstein relationship is used to relate the viscosity to the melt fraction: $\eta = \eta_l\ (1.35\phi - 0.35)^{-5/2}$, where $\eta_l = 1 \times 10^{-7} \exp\,(6.2 \times 10^4\ \mathrm{K/T})$ Pa s is the melt viscosity, which applies above the liquidus. The solidus is assumed to be at 1598 K and the liquidus at 1698 K, making the critical temperature (1600 K) correspond to 2% melt. The melt fraction increases linearly from the solidus to the liquidus. Although this is a poor approximation, particularly for wet cases, it makes little difference in the resulting heat flows predicted at equilibrium, because it is the viscosity that sets the equilibrium, not the temperature itself.

The viscosity of silicates is also a strong function of water fugacity, most significantly through altering the melting point and melt production rates (and hence T_{crit}). The wet solidus may be considerably lower, but the wet liquidus is the same, because the solid dehydrates after only a few percent of melt has been generated (*Hirth and Kohlstedt,* 1996). A value of 1498 K will be used for the wet liquidus and the critical temperature is fixed at 2% melt (1502 K). In order to model the maximum effect of water on the viscosity of rocks, the wet viscosity (encompassed in the parameter A) will be set at 50 times less than the dry viscosity (*Hirth and Kohlstedt,* 1996).

TABLE 1. Rheological parameters.

Parameter	$T < T_{crit}$	$T \geq T_{crit}$
A	1.65×10^5 Pa s	0.1536 Pa s
E	4×10^4 K	6.2×10^4 K

The shear modulus is given by

$$\mu = \begin{cases} \mu_0 & T \leq T_c \\ 2.5 \times 10^{-41} \exp(Q/RT)\ \mathrm{Pa} & T_c < T \leq T_d \\ 10^{-7}\ \mathrm{Pa} & T > T_d \end{cases} \tag{3}$$

where μ_0 is a constant (50 GPa), $Q = 1.57 \times 10^6$ J mol^{-1} is an activation energy, T_c is the critical temperature, and T_d is the disaggregation temperature.

3.1.1. Convective heat loss. The heat generated by tidal dissipation within Europa must be carried to the surface and radiated to space if Europa is to achieve thermal equilibrium. In modeling this process, it is typically assumed that convective motions in Europa's mantle are responsible for delivering the heat to the base of the ocean, which quickly delivers it to the surface via the icy shell. Convection in spherical shells has been studied extensively, and parameterizations including the effects of temperature-dependent viscosity and spherical geometry have been derived and fit to the results of numerical calculations (*Reese et al.,* 1999; *Solomatov and Moresi,* 2000).

For fluids with strongly temperature dependent viscosity, a portion of the upper thermal boundary layer stagnates and transports heat only by conduction. An actively convecting boundary layer exists between the stagnant lid above and the well-mixed interior below. The convection in the interior is essentially isoviscous and is driven by a temperature contrast that depends on the temperature dependence of the viscosity η

$$\Delta T_{rh} = \left| \frac{\eta}{d\eta/dT} \right| \tag{4}$$

For a strictly temperature dependent viscosity (pressure dependence should be weak in Europa), $\Delta T_{rh} = T^2/E \sim$ 30–50 K for the rheology given in equation (2).

Because only part of the shell (below the lid) is convecting, we define a Rayleigh number Ra_{rh} for the shell as follows

$$Ra_{rh} = \frac{\alpha \rho g \Delta T_{rh} (r_l - r_b)^3}{\kappa \eta (T_i)} \tag{5}$$

where α is the thermal expansivity (3×10^{-5} K^{-1}), ρ is the density of the mantle (3300 kg m^{-3}), g is the gravitational acceleration (1.45 m s^{-2}), r_l and r_b are the radii of the bottom of the lid and bottom of the shell, respectively, κ is the thermal diffusivity (10^{-6} m^2 s^{-1}), and the viscosity η is evaluated at the mean temperature of the well-mixed interior T_i. The heat flux transported convectively by the shell to the base of the lid is therefore

$$F_{conv} = a_c \frac{k \Delta T_{rh}}{(r_l - r_b)} Ra_{rh}^{\beta} \tag{6}$$

where k is the thermal conductivity (4 W m^{-1} K^{-1}), and a_c =

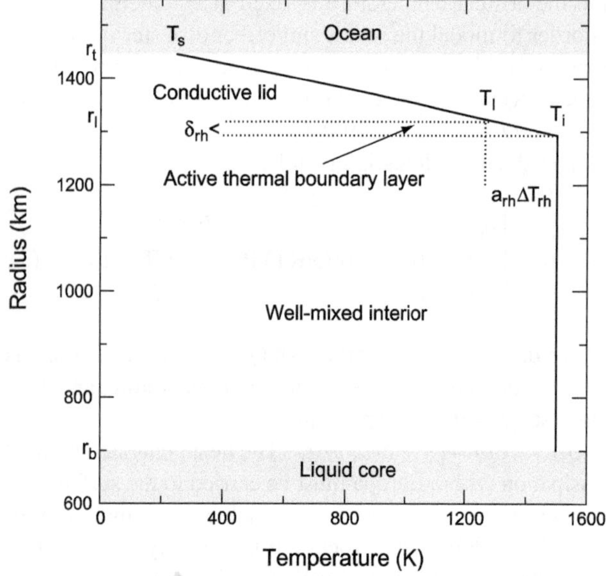

Fig. 1. Example temperature profile implied by the convective parameterization used in this study. Temperatures in the conductive lid are from a solution of the conduction equation using the heat flux from the convecting interior as a boundary condition. The heat flux and thickness of the lid are calculated given the temperature of the well-mixed interior (1500 K in this example).

1.2 (1 + n) and $\beta = n/(n + 2)$ are functions of the strain rate exponent n [*Solomatov and Moresi* (2000)], which is 1 for the Newtonian rheology used here. At equilibrium, this flux must balance the production of heat in the shell

$$F_{tidal} = \frac{\rho H r_l}{3}(1 - r_b^3/r_l^3) \tag{7}$$

where H is the mean volumetric tidal heat production rate. The thickness of the lid is determined from the solution of the conduction equation given the internal heating rate in the lid (assumed here to be equal to the heating rate in the convecting layer, H), the surface temperature T_s, and the flux into the base of the lid F_{conv}. This gives an expression for the temperature at the base of the lid

$$\begin{aligned} T_l &= T_i - a_{rh}\Delta T_{rh} \\ &= T_s + \frac{\rho H}{6k}(r_t^2 - r_l^2) \\ &+ \left(\frac{F_{conv} r_l^2}{k} - \frac{\rho H r_l^3}{3k}\right)\left(\frac{1}{r_l} - \frac{1}{r_t}\right) \end{aligned} \tag{8}$$

where r_t is the radius of the top of the shell (Europa's radius) and T_l has been equated to the temperature of the interior T_i minus the temperature drop $a_{rh}\Delta T_{rh}$ across the active boundary layer (a_{rh} is a constant with a value of 2.4 for the Newtonian rheology used here).

Given T_i, equations (4)–(8) may be solved for the thickness of the lid ($r_t - r_l$) and the heat production rate H, which are in equilibrium with the convective heat transport. Finally, the thickness of the active boundary layer δ_{rh} is given by

$$\delta_{rh} = \frac{k\Delta T_{rh}}{F_{conv}} \tag{9}$$

allowing a complete description of the temperature in the interior of the satellite, as shown in Fig. 1.

3.1.2. Tidal heat production. Given the density, viscosity, and shear modulus as a function of radius within Europa, the tidal heating may be calculated by solving the equations of motion in a layered viscoelastic sphere (*Moore and Schubert,* 2000) subject to the following, time-dependent gravitational potential (*Kaula,* 1964)

$$\begin{aligned} \Phi = r^2\omega^2 e\Big\{ &-\frac{3}{2}P_2^0(\cos\theta)\cos\omega t + \\ &\frac{1}{4}P_2^2(\cos\theta)\,[3\cos\omega t\cos 2\phi + 4\sin\omega t\sin 2\phi]\Big\} \end{aligned} \tag{10}$$

where r is radius from the center of Europa, ω is the orbital angular frequency (2.05×10^{-5} rad s^{-1}), e is the orbital eccentricity (0.0091), θ and ϕ are the colatitude and longitude with zero longitude at the subjovian point, t is time since passage through perijove, and P_2^0 and P_2^2 are associated Legendre polynomials.

The deformation of Europa is solved for given the tidal potential from equation (10) and the r-dependent material parameters. The quasistatic equations of motion and Poisson's equation for the potential are Fourier transformed in time, resulting in

$$u_{i,i} = 0 \tag{11}$$

$$\tau_{ij,j}^{(\partial)} = 0;\ \tau_{ij}^{(\partial)} = \delta_{ij}p^{(\partial)} + \mu(s)(u_{i,j} + u_{j,i}) \tag{12}$$

$$\phi_{,ii}^{(\Delta)} = 0 \tag{13}$$

where each variable is the complex Fourier amplitude, which is a function of the complex frequency s = iω, u is the displacement vector, $\tau^{(\partial)}$ is the isopotential stress tensor, $p^{(\partial)}$ is the isopotential pressure, $g^{(\Delta)}$ is the local gradient of the gravitational potential $\phi^{(\Delta)}$, and indices following a comma imply differentiation with respect to the coordinate corresponding to that index (i.e., $u_{i,j} = \partial u_i/\partial x_j$) and repeated indices imply summation over the range. The isopotential $^{(\partial)}$ and local $^{(\Delta)}$ Lagrangian fields are related to the more conventional material $^{(\delta)}$ fields by

$$f^{(\delta)} = f^{(\partial)} + f_{,k}^0\,(u_k - d_k) \tag{14}$$

$$f^{(\delta)} = f^{(\Delta)} + f_{,k}^0 u_k \tag{15}$$

where f is any field, f^0 is the reference (hydrostatic) value of that field, and d is the displacement of the isopotential surface originally associated with the material element (equivalent to the geoid displacement). The complex rigidity $\mu(s)$ in (equation (12)) is s times the Fourier transform of the Maxwell stress relaxation function

$$\mu(s) = \frac{s\mu}{s + \mu/\eta} \tag{16}$$

where the constant μ on the right side is the shear modulus and η is the viscosity. The ratio μ/η is the inverse Maxwell time. It is the imaginary part of $\mu(s)$ that controls dissipation.

Writing equations (11)–(13) in terms of spherical harmonics and restricting the solutions to spheroidal deformations (tides do not excite toroidal modes in the linear, layered system considered here) results in a sixth-order system that has analytical (power-law) solutions in layers with constant material properties (*Wolf,* 1991). These solutions are propagated through the layers from the center to the surface where boundary conditions are applied to determine the unknown conditions at the center. Liquid layers are represented by zero shear modulus and have no unique solutions for the displacements (*Dahlen,* 1974) (inertial motions have been ignored). Once the deformations are known, the dissipation averaged over an orbital period may be calculated at any point in the body.

3.2. Equilibrium in the Coupled Tidal-Convective Model

Given the temperature profile for a particular T_i defined by the convective parameterization described above, the tidal heating is computed by the method of the preceding section. Although the tidal heating depends on radius in a complex manner (depending on both the viscosity structure and the distribution of tidal strains), the heating rate through the actively convecting part of the mantle is relatively constant because viscosity variations are limited to a factor of 10 at most (as shown below), and a simple treatment is to average the heat production within the mantle to derive a uniform heat production rate that is consistent with the convective parameterization. Although this ignores the possibility that local variations in heat production due to temperature variations could lead to changes in heat transport (*Tobie et al.,* 2003; *Showman and Han,* 2004), there is no quantitative parameterization for convection with viscosity-dependent heating.

Equilibrium states are found by iterating until the computed tidal heating matches the heat transported according to the convective parameterization. Alternatively, the time dependence of the interior temperature can be integrated by equating the net heat retained or lost with the time derivative of the internal energy. This quasistatic approach assumes that the layer can adjust on a timescale rapid compared with the rate of change of the temperature, which can be a problem if the rigid lid is quite thick (see the following section for the results of such models).

The result of computing the tidal heating and heat transport as outlined above is shown in Fig. 2 for both the wet and dry rheologies. The solid lines show the global total heat production as a function of mantle internal temperature. At lower temperatures, tidal heating is negligible, and only the temperature-independent radiogenic heating contributes at a level of about 2×10^{11} W (or 6 mW m^{-2} at the surface). As the temperature increases, the tidal heating increases, beginning at lower temperatures for the weaker, wet rheology but peaking at the same value for both rheologies at a temperature corresponding to 5–10% percent melt. This is the point at which the Maxwell time of the mantle has reached the value (very near the tidal forcing period) that maximizes the tidal dissipation through maximization of the imaginary part of the complex shear modulus (equation (16)). Above this point, the dissipation decreases, first with decreasing viscosity and then decreasing very rapidly above the disaggregation point (40–60% melt) where the shear modulus becomes negligible.

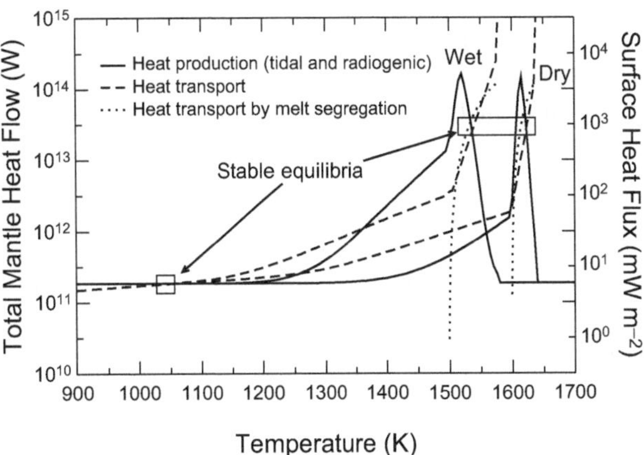

Fig. 2. Total heat production (tidal and radiogenic, solid lines), heat transport (dashed lines), and melt segregation heat transport in the mantle as a function of mantle internal temperature for the wet and dry rheologies as indicated. Stable equilibria are shown within the boxes, as discussed in the text. Heat flux at the surface is given on the right axis.

The heat transport increases with temperature as the viscosity decreases. There is a transition between conductive and convective heat transport at about 1100 K for the wet rheology and at about 1200 K for the dry rheology. Above the solidus the heat transport increases more rapidly as melt along the grain boundaries decreases the viscosity, and above the disaggregation point there is a very rapid decrease as the viscosity decreases suddenly to that of the suspension.

Where the heat transport and heat production are equal, a thermal equilibrium exists, which may be stable or unstable to temperature perturbations. In Fig. 2, three equilibria exist for each rheology. There is a stable, low-temperature equilibrium at a temperature of about 1040 K (indicated by the box) that is insensitive to the rheology. The reason for this is that this equilibrium is achieved by balancing radiogenic heating (tidal heating is miniscule at this low temperature) with conductive heat transport. With no significant tidal heating or convection, this equilibrium is insensitive to variations in the rheology. This equilibrium state is similar to what might be expected of the Moon, which is similar in size and which is heated primarily by radioactive decay at present.

There is also a stable equilibrium at much higher temperatures above the solidus (elongated box). Although the temperatures of this equilibrium are sensitive to the hydration state of the mantle, the viscosity at these temperatures is primarily controlled by the amount of melt present, and occurs at a value of about 20% melt by volume independent of the rheology of the solid. This equilibrium is achieved by balancing the tidal heating (radiogenic heating is negligible in this case) with very active convection carrying 2–3 × 10^{13} W (800 mW m^{-2} at the surface).

At this point it should be noted that above the solidus, melt segregation begins to transport heat across the boundary layer to the surface, if any low-density (basaltic) component remains in the mantle. This heat transport mechanism has been ignored thus far, although it is likely the dominant heat transport mechanism in Io (*Moore,* 2003), because there are no quantitative parameterizations for the heat transport due to melt segregation in a convecting mantle. A very simple estimate of melt segregation heat transport can be derived by assuming the melt segregates from the solid by Darcy flow driven by the buoyancy difference between solid and melt (*Moore,* 2001). The underlying assumption is that convection is able to maintain an effective average melt fraction that erupts in quasisteady state. Of course, melting happens most strongly in upwellings and the nonlinear response of the melt segregation flux makes the effective averaging tenuous. Given these caveats, the following equation for global heat flow F as a function of melt fraction ϕ may be derived

$$F = 4\pi R^2 \frac{b^2(\rho_s - \rho_m)g}{\tau\eta_l} \phi^2 \rho_m [L_f + c_P(T_m - T_s)] \qquad (17)$$

where R is the radius of the top of the mantle, b is a grain size parameter (2 mm), $(\rho_s - \rho_m)$ is the density difference between solid and melt (500 kg m^{-3}), g is the acceleration of gravity (1.4 m s^{-2}), τ is a permeability parameter [200 for the porosity exponent of 3 chosen here (*Wark and Watson,* 1998)], η_l is the viscosity of the melt (100 Pa s), L_f is the latent heat of fusion of the melt (330 kJ kg^{-1}), c_P is the specific heat at constant pressure (1 kJ kg^{-1} K^{-1}), and $(T_m - T_s)$ is the temperature difference between the melt and the surface of the mantle. The values for material properties adopted here are taken from the compilation in Table 4.5 of *Schubert et al.* (2001) except where noted. The melt transport flux for wet and dry melting is shown in Fig. 2 by the dotted lines. Equation (17) does not apply below the solidus or above the disaggregation point so the curves are only shown over that range. The parameters used are sufficiently uncertain that melt segregation may set the equilibrium temperature anywhere from the 20% melt of the convective equilibrium down to several percent melt for extremely efficient melt segregation, with a corresponding range of heat fluxes from 1 to 7 times the convective equilibrium flux (up to several W m^{-2}).

These calculations have all been performed for the present value of the orbital eccentricity and period. As these change over time, the amount of tidal heating will change and the temperature structure of the interior may be altered. The high-temperature equilibrium, for example, requires a certain minimum eccentricity to exist at all and may disappear during periods of low eccentricity. A fully coupled orbital model (e.g., *Fischer and Spohn,* 1990; *Hussmann and Spohn,* 2004) is required to address this behavior, as described in the next section.

4. DYNAMICAL THERMAL ORBITAL COUPLING

The thermal and orbital evolution of Europa cannot be treated independently because of two factors: (1) Tidal heating, albeit less important when compared to Io, is significantly contributing to the satellite's heat budget; and (2) the interaction with the other Galilean moons is coupled by several resonances. Thus, the history of Europa cannot be understood without taking the interaction with Jupiter and Europa's neighboring satellites Io and Ganymede into account. The three inner Galilean satellites are gravitationally coupled and locked in various resonances. The most important ones are a 2:1 mean-motion resonance of Io and Europa, a 2:1 mean-motion resonance of Europa and Ganymede, and the 4:2:1 Laplace resonance of Io, Europa, and Ganymede [for an overview on the resonance couplings, see, e.g., *Greenberg* (1982) and *Peale* (1986)]. Due to these resonances orbital energy and angular momentum are distributed among the satellite system. Because the resonances are stable, the orbital eccentricities of the satellites are forced by the resonances on long timescales (on the order of billions of years, at least as long as the resonances have existed) (*Yoder,* 1979; *Yoder and Peale,* 1981). Therefore, the damping of the orbital eccentricity by tidal dissipation in Europa is avoided by the resonance locking with Io and Ganymede. If Europa were the only satellite of Jupiter its eccentricity would have been damped rapidly and (neglecting effects due to obliquity) tidal heating would have ceased long ago because Europa would then rotate synchronously on a circular orbit. However, the resonances maintain the finite, albeit small, values of eccentricity, thus maintaining tidal flexing in Io and Europa.

The conditions in the Jupiter system are such that the interplay between tides and resonances has a strong impact on the energy available in Europa (see Fig. 3). The small distance of Io from massive Jupiter exerts tidal torques, which transfer angular momentum and orbital energy to Io at a high rate. Part of the energy, the source of which is Jupiter's rotational energy, is dissipated in Io. The orbital energy still gained by Io is distributed further to Europa and to Ganymede via the resonances. Again, part of the orbital energy gained by the other satellites is dissipated in Europa and, to a much lesser extent, in Ganymede.

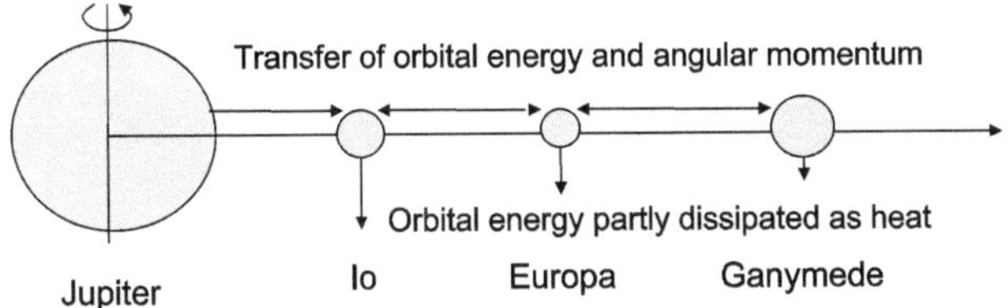

Fig. 3. Angular momentum and rotational energy of Jupiter are transferred to Io via tidal interaction between the satellite and the giant planet. Due to the resonances orbital energy and angular momentum are distributed from Io to Europa and Ganymede. Part of the orbital energy gained by the satellites is dissipated in the moons' interiors because of tidal flexing caused by Jupiter. Dissipation rates depend strongly on the distance to Jupiter and are therefore most important for Io, much smaller but still significant on Europa, and at present negligible at Ganymede (sizes and distances are not to scale).

It is important to note that Io, or more precisely, the Io-Jupiter interaction, is the driving factor, controlling the overall amount of energy available at Europa. Jupiter's rotational energy is the source by which the internal activity of Io and Europa is powered. Because the effect of the moons on Jupiter, which would in principle cause a decrease of the rotation rate of the giant planet, is very small, the rotational energy of Jupiter can be considered as constant. Thus, the amount of energy available for the satellite system will depend principally on Io's orbital characteristics (mainly eccentricity and semimajor axis) and on the proportion of Jupiter's interior being deformed inelastically under the periodic forcing of Io. The efficiency of energy dissipation is usually parameterized by the jovian quality factor Q_J, defined by (*Goldreich,* 1963)

$$Q_J = \frac{2\pi E^*}{\oint (dE/dt)dt} \qquad (18)$$

where E^* is the peak energy stored in the system during one tidal cycle (i.e., one orbital period in the case considered here). The expression in the denominator is the energy dissipated over the complete cycle. The lower the Q_J, the greater the dissipation rate within Jupiter and the stronger the tidal torques involved. The value of the jovian quality factor is highly uncertain. Estimates based on constraints from orbital dynamics range from on the order of 10^4 to 10^6 (*Goldreich and Soter,* 1966; *Gavrilov and Zharkov,* 1977; *Yoder and Peale,* 1981; *Peale,* 2003). Estimates derived from dissipation mechanisms in Jupiter's interior lie in the same range (*Ioannou and Lindzen,* 1993; *Wu,* 2005), but can also yield extreme values of 10^2 or up to 10^{13} if specific processes of dissipation are assumed (*Stevenson,* 1983; *Goldreich and Nicholson,* 1977). Most of the models, however, consider a range between 10^4 and 10^6, which would also be consistent with Io's heatflow of several Wm^{-2} (*Veeder et al.,* 1994, 2004; *Matson et al.,* 2001; *Marchis et al.,* 2005) if an equilibrium between Io's heat flow and the dissipation rate in Jupiter is assumed.

It should be noted that using a constant Q_J value is a simplification. Q_J generally is frequency-dependent (*Wu,* 2005) and it will not necessarily be constant on long timescales (depending on processes in Jupiter's interior). Because dissipation mechanisms in giant planets in response to periodic tidal forcing are not well-constrained, the dependence on frequency is not known, and therefore several approaches have been used in different applications (e.g., Appendix in *Peale,* 2007).

The dissipation rate of a satellite in a synchronous equatorial orbit is given by (*Segatz et al.,* 1988)

$$\dot{E} = -\frac{21}{2}\frac{n^5R^5e^2}{G}\mathrm{Im}(k_2) \qquad (19)$$

where n, e, and R are the satellite's mean motion, eccentricity and radius, respectively. G is the constant of gravitation and $\mathrm{Im}(k_2)$ is the imaginary part of the second-degree potential Love number k_2. The latter quantity describes the inelastic response of the satellite and should be dependent on the generally temperature-dependent rheology of the satellite material (ice or rock), on the internal density distribution, and on the frequency of the forcing. Equation (19) is a general formulation, not introducing a specific rheological model. However, the calculation of $\mathrm{Im}(k_2)$ requires a model for the rheological behavior of the planetary body subject to the periodic tidal forces. As discussed above, the Maxwell model is a simple choice involving few parameters. In the case of Europa, the period of the forcing is the orbital period of 3.55 days. The rigidity is on the order of 50 GPa and 3 GPa for rock and ice, respectively. The period of the forcing would therefore be on the order of the Maxwell time for viscosities of 1.5×10^{16} Pa s and 9.2×10^{14} Pa s for rock and ice, respectively. Such low rock viscosities require the occurrence of partial melt (see section 3.1) in Europa's silicate mantle. It is conceivable, but unknown if such supersolidus states have actually occurred in Europa's evolution.

The viscosity is not only temperature-dependent but will depend on the dominant creep mechanism and on the grain size. Furthermore, it should be kept in mind that the Maxwell rheology is not the only conceivable scenario. In particular, if partial melting occurs, other rheologies may be more appropriate to describe the satellite's response. However, in this case the general formulation of equation (19) is still valid (e.g., *Zschau,* 1977); only the calculation of $\mathrm{Im}(k_2)$ and the material parameters involved would be different.

With a given density profile, it is straightforward to calculate the complex Love numbers if a rheology model is assumed as described above in section 3.3. It is the calculation of the imaginary part of the Love number k_2 using a rheological model like the one of equation (16), which introduces the coupling between the thermal and orbital evolution of Europa. $\mathrm{Im}(k_2)$ depends on both the thermal state because of the temperature dependence of the viscosity and the rigidity and on the orbital state because of the dependence on mean motion. Additionally, the orbital eccentricity and the mean motion enter directly in the calculation of

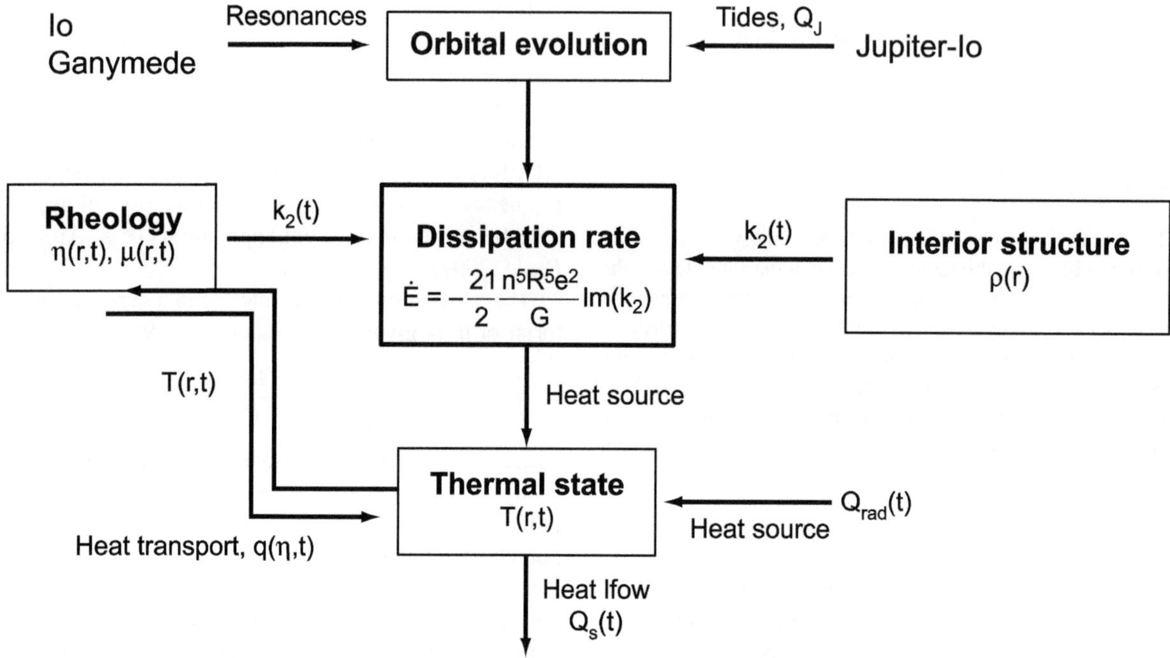

Fig. 4. Europa's dissipation rate provides the link between the satellite's thermal and orbital evolution. The schematic view shows the interplay between several parts of the model by *Hussmann and Spohn* (2004) (see discussion in text).

the dissipation rate. Equation (19) is therefore the link allowing for a coupling of thermal and orbital evolution.

The schematic view in Fig. 4 shows the interplay between several parts of the model by *Hussmann and Spohn* (2004): (1) The orbital evolution of Europa is mainly controlled by the tidal interaction between Io and Jupiter, which determines the overall amount of energy transferred from Jupiter to the satellite system. The crucial parameter here is Jupiter's quality factor Q_J. Furthermore, the 2:1 resonance with Io and the 1:2 resonance with Ganymede forcing the eccentricity of Europa are driving mechanisms for Europa's orbital state. The forced eccentricity e and mean motion n enter in the equation for the dissipation rate. (2) The rheology, here represented by viscosity and rigidity, is a function of the radial distance from the satellite's center r because individual layers have different rheological properties (e.g., silicate mantle and H_2O layer). Furthermore, viscosity and rigidity are functions of the temperature T. Therefore knowledge on the thermal state is required to calculate η and μ, which are required to determine k_2. On the other hand, the viscosity is an important quantity for the heat flow q out of the individual layers (lower arrow between rheology and thermal state). (3) The thermal state will depend on the heat sources, i.e., the radiogenic heating from the silicate mantle Q_{rad} and the dissipation rate, and on the surface heat flow Q_s, which are all functions of time. (4) The interior structure, i.e., the density profile $\rho(r)$, is required to obtain the complex Love number k_2, the imaginary part of which is used to calculate the dissipation rate. Whereas most of the other quantities are functions of the time t, the density is in this model only a function of the radial position r within Europa (e.g., differentiation processes are excluded here.) All four aspects discussed here are connected through the dissipation rate (equation (19)), which depends on density, through $\mathrm{Im}(k_2)$; on rheology, through $\mathrm{Im}(k_2)$; and on orbital quantities, through $\mathrm{Im}(k_2)$ and directly through e and n.

It should be noted that because of the resonances the same coupling as schematically shown in Fig. 4 must be introduced for Io (see also *Ojakangas and Stevenson,* 1986; *Fischer and Spohn,* 1990), which has even stronger implications for the evolution of the satellite system. Both systems are coupled through the resonances. In principle, Ganymede can be included in the same way. However, because Ganymede's dissipation rate yields only a small contribution to the satellite's heat budget, it is sufficient to include this satellite as a point mass in the orbital equations, instead of examining the full thermal-orbital coupling. It should be noted although that the thermal-orbital coupling of Ganymede may have played an important role in the early evolution of the satellites during formation of the Laplace resonance (*Showman and Malhotra,* 1997). Under the assumption that the system remains in the Laplace resonance, the system described schematically in Fig. 4 results in a set of differential equations involving heat balance equations for the thermal part and equations for the conservation of energy and angular momentum for the orbital part (*Fischer and Spohn,* 1990; *Hussmann and Spohn,* 2004). Both parts are coupled through the dissipation rate, i.e., through the calculation of $\mathrm{Im}(k_2)$ of the moons. The integration yields the eccentricities and mean motions of Io, Europa, and Ganymede as a function of time (orbital part) and the tempera-

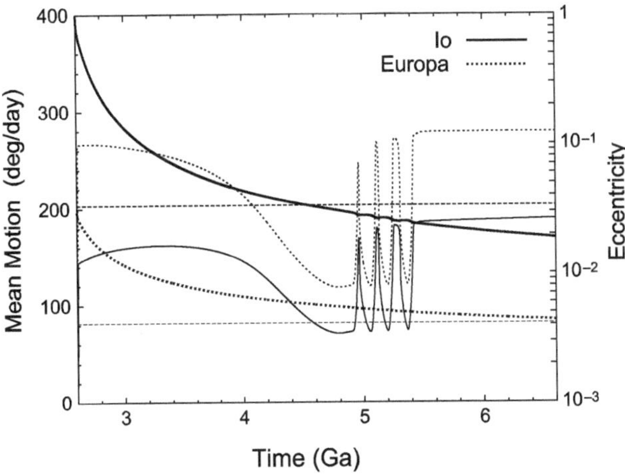

Fig. 5. Mean motion (thick lines) and eccentricities (thin lines) of Io and Europa. Present-day values at 4.6 Ga of Io's mean motion and eccentricity are indicated by horizontal lines. Decreasing mean motion implies that the satellites drift outward while kept in resonance. After *Fischer and Spohn* (1990).

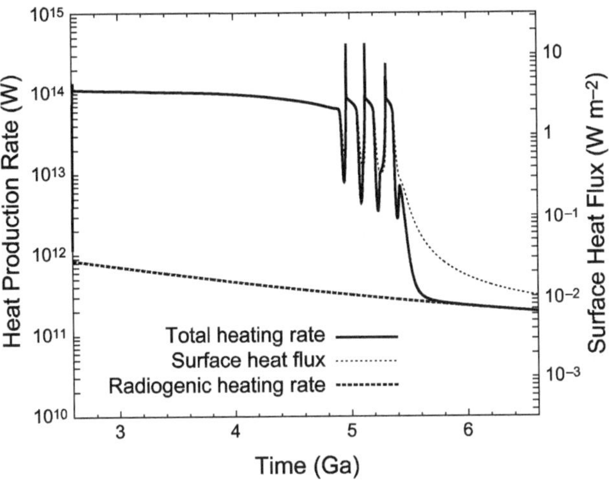

Fig. 6. Total internal heating rate, surface heat flux, and radiogenic heating rate of Io. In the first billion years the total heating rate is dominated by tidal dissipation. After the oscillations, Io gets in a cold elastic state with insignificant tidal heating compared to radiogenic heating. After *Fischer and Spohn* (1990).

tures of the dissipating layers of the satellites as a function of time.

Typical results for Io's evolution are shown in Figs. 5 and 6, which are included here to illustrate the different possible phases of evolution, which are, even for Europa, controlled by Io. The evolution starts here at 2.6 G.y. after formation of the satellite system. Because the stability of the Laplace resonance is assumed in these models, this is assumed to be the time at which the resonance was established.

The value of Q_J determines the amount of energy and the rate at which it is transferred to the satellite system. A small Q_J implies a larger phase lag of Jupiter's tidal bulge in response to Io's forcing and hence greater torques acting on Io. Smaller Q_J values thus imply a faster orbital evolution. There is a bound (around $Q_J = 10^4$) below which the evolution of the satellites would run too fast to be consistent with the satellites' present distances from Jupiter. An upper bound of $Q_J \approx 10^6$ is given because in that case the energy transferred to the satellites would be insufficient to obtain a high-dissipative state of Io. After a small episode of moderate heating Io would cool and would enter the cold equilibrium state, in which tidal heating is negligible compared to radiogenic heating early in its history. This is obviously inconsistent with the vigorous volcanism, high temperatures, and enormous heat flows that are currently observed on Io. An equilibrium heat flow of Io of ≈2 W m^{-2} would imply $Q_J \approx 5 \times 10^4$, which is the approximate value used in the simulations shown here. Again, it should be pointed out that assuming a constant Q_J value throughout the evolution may be an oversimplification (*Peale,* 2007). The simplified scenario can, however, yield qualitative constraints on the evolution of the satellites that could be improved by better knowledge on dissipative processes in Jupiter and in gaseous giant planets in general.

The evolution can be divided into three phases: (1) At the beginning, up to about 5 Ga, Io evolves on a hot branch with the dissipation rate being in equilibrium with the heat flux of about 2 W m^{-2} (Fig. 6). During this phase the dissipation rate is slowly decreasing because of the decreasing trend of the mean motion and eccentricity. In this phase Io's mantle temperature is sufficiently high for partial melting and silicate volcanism to occur. (2) Oscillations set in when the slowly decreasing dissipation rate and hence the mantle temperature drop below a characteristic temperature where partial melting can no longer be present. This will cause a strong decrease in the dissipation rate because the rigid mantle deforms quasielastically if temperatures are low. The runaway process toward a cold state is stopped, however, because a low dissipation rate implies a strong increase in eccentricity. Due to this effect the dissipation rate can increase (see equation (19)), causing partial melting again. The dissipation rate and heat flow are not in equilibrium during this oscillatory phase. After the temperature has been increased to the level in the previous phase, the cycle can start again. The cycle is repeated several times with an overlying general trend of decreasing heating, because of decreasing mean motion (Fig. 5). (3) At a certain point even the effect of increasing eccentricity is insufficient to heat up the mantle again. As a consequence Io ends in a cold rigid state in which heating of the mantle is due to radiogenic heating only. Tidal deformations are quasielastic and friction does not contribute significantly to the heat budget. The heat flow and heating rate are slowly decreasing due to the decreasing radiogenic heat sources in the mantle. It is unknown whether Io is at present in the equilibrium phase evolving along the hot branch or in the oscillation phase. Oscillations involving a feedback mechanism between orbital and thermal energy of Io were first suggested by *Ojakangas and Stevenson* (1986). Both equilibrium models (*Moore,* 2003)

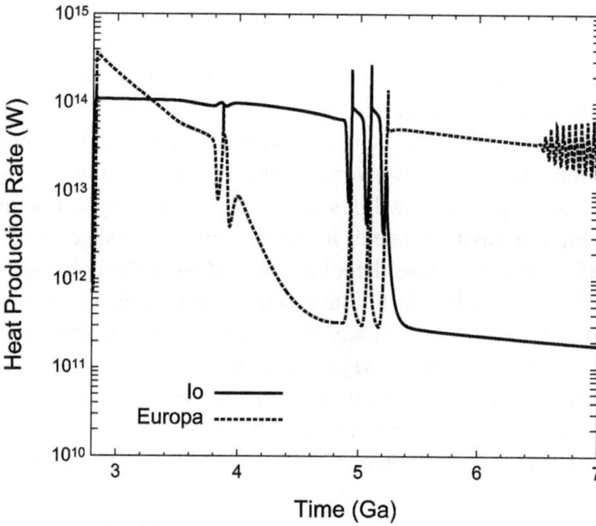

Fig. 7. Heat production of Io and Europa including the thermal-orbital coupling of Europa as shown in Fig. 4. For details on the model, see *Hussmann and Spohn* (2004). The three phases of Io's evolution described previously are also obtained. Additionally, the heat production in Europa's silicate layer is strongly varying with time (see text). From *Hussmann and Spohn* (2004).

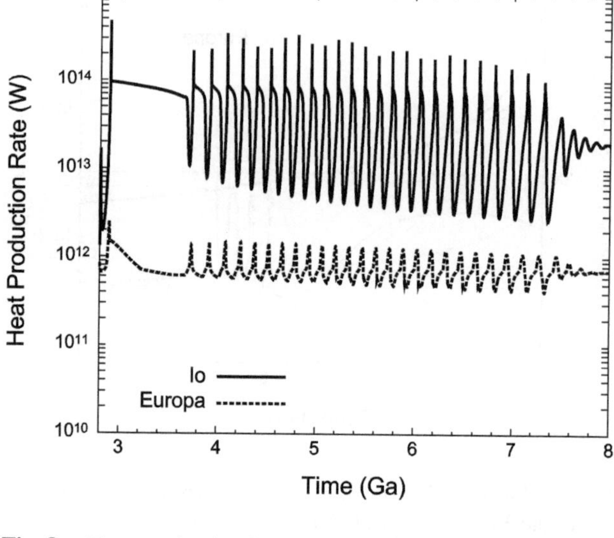

Fig. 8. Heat production in Io and Europa. Here dissipation in Europa is restricted to the ice layer. The silicate shell is assumed to be elastic. In the case shown here the present state at 4.6 Ga is obtained during the oscillatory phase. Dissipation in the icy shell also has consequences on icy shell thickness (see chapter by Nimmo and Manga). From *Hussmann and Spohn* (2004).

and oscillatory models with Io being close to the maximum dissipation rate (*Hussmann and Spohn,* 2004) can explain Io's present-day heat flow. To distinguish between the two states, a measurement of the changes in mean motion and eccentricity would be required. From the presently available data, mainly on eclipses and mutual occultations of the Galilean satellites, no conclusion on the secular variations of orbital elements has been reached (*Lieske,* 1987; *Goldstein and Jacobs,* 1995; *Aksnes and Franklin,* 2001; *Lainey and Tobie,* 2005).

The model introduced so far was further extended, including the full thermal-orbital coupling of Europa as shown in Fig. 4 (*Hussmann and Spohn,* 2004). Typical results show the same three phases, which are now connected with variations in the heat production in Europa. The situation is more complex for Europa because the satellite can dissipate energy in the silicate mantle and/or in the icy shell. In Figs. 7 and 8 models are shown in which the dissipation is restricted to Europa's silicate layer (Fig. 7) or the ice layer (Fig. 8). The evolutionary phases for Io are similar to the previous ones where constant dissipation in Europa was assumed. The coupling does have important consequences for Europa's evolution. In the case of dissipation in the silicate layer, a similar parameterization as for Io is used for Europa's rheological parameters, viscosity and rigidity, and their respective temperature dependence. The main difference from the parameterizations in section 3 is that the disaggregation temperature described in section 3.1 and heat transport by melt (equation (17)) are not included. Above a critical temperature close to the solidus, viscosity and rigidity decrease exponentially as described by *Fischer and Spohn* (1990) and by equations (2) and (3).

As shown in Fig. 7, Europa starts on a hot branch while the planet is cooling in the first billion years, with a short oscillation at about 4 G.y. after formation of the satellite system. The general trend of cooling continues and Europa is at present found in a cold state with low dissipation in the silicate shell. Triggered by Io's oscillations, Europa is heated in steps during the oscillatory phase and can even enter a hot state when Io has reached its final cold state.

The different phases of Europa's evolution can be associated with the equilibrium points shown in Fig. 2. From the initial high temperatures, Europa cools until it reaches the low-temperature equilibrium at about 4.6 Ga, i.e., at the present state. This state corresponds to the low-temperature equilibrium on the left in Fig. 2. Due to the strong variations in Io's heat production rate and the resulting variations in eccentricity of all three satellites locked in resonance, Europa is heated stepwise in the oscillatory phase. Larger eccentricities would shift the heat production curve (solid line in Fig. 2) upward, which would imply a shift of the unstable equilibrium toward the low-temperature equilibrium. Runaway heating initiates if both equilibria coincide and if Europa is perturbed toward high temperature. The stepwise heating of Europa can only be forced by Io shifting Europa's tidal heating rate in Fig. 2 upward because of the resonance. If the unstable equilibrium is reached, runaway heating sets in and the stable high-dissipative state can be reached. In Fig. 7 Europa is in the stable high-temperature equilibrium of Fig. 2 after the oscillation phase.

In the second model where dissipation is restricted to Europa's icy shell, Io and Europa are found to be in the oscillation phase at present. The dissipation rate is besides small oscillations relatively constant at Europa around 10^{12} W. However, such a variation in heat production will cause variations in Europa's ice thickness between about 10 and 40 km (see also the chapter by Nimmo and Manga). Europa's complex surface geology, showing evidence for thick and relatively thin ice, may be a consequence of these variations. The evolutionary paths shown here are consistent with the present orbital state (eccentricity and mean motion) of Io, Europa, and Ganymede and with Io's volcanic activity implying extremely high heat productions in its interior. Some model parameters have to be adjusted (e.g., the jovian quality factor is around a few 10^4 in the models shown here) to obtain consistent results and therefore the solutions are not unique.

5. CONCLUDING REMARKS

Equilibrium in Europa's silicate layer, like in Io, has two stable states. The high-temperature state is supersolidus, implying active volcanism and high-temperature water-rock reactions at the ocean-mantle interface. The heat flux in this state is about 10^{13} W, less than one-tenth of Io, but an order of magnitude larger than the heat produced in the icy shell (*Moore,* 2006). The amount of hydrothermal circulation implied is comparable to that at the mid-ocean ridges on Earth. This may have significant implications for the thickness of the icy shell, since an equilibrium conductive icy shell that transports 10^{13} W is about 3 km thick.

The coupled orbital and thermal evolution of the Galilean satellites has been complex due to the inherent feedback between thermal and orbital state. The mutual coupling through the resonances and the feedback between thermal and orbital state may have played a key role in the evolution of the satellites. Key parameters to constrain the models further would be (1) the determination of Jupiter's quality factor and alternative descriptions of the dissipation in Jupiter; (2) the measurement of the tidal response of Io and Europa (determination of the dynamical Love numbers k_2 and h_2), which in the case of Europa would also be indicative of a subsurface ocean (*Moore and Schubert,* 2000); (3) more precise determination of secular variations in eccentricity and mean motion of Io, Europa, and Ganymede; and (4) the measurement of Europa's heat flow, which could constrain Europa's tidal dissipation rate. Additionally, experiments on the deformation processes of ice and rock on timescales of the tidal periods would be extremely valuable. Such measurements would yield alternatives to the Maxwell model, which would be applicable in differing temperature and creep regimes relevant for the deformation of satellites. Europa is a very special case regarding tidal dissipation. Depending on the internal temperature profile and on the thicknesses of the layers, both the icy shell and the silicate mantle can in principle dissipate substantial amounts of energy. Additionally, dissipative processes in the ocean may contribute to the heat budget. The relative importance of the heating rates in each layer throughout Europa's history — and even at present — and their mutual interactions remain elusive. Future investigations would have to take dissipation in several layers simultaneously into account, which could affect Europa's overall internal energy budget. Presumably, it would also imply an indirect exchange of thermal energy between the silicate mantle, the ocean, and the ice layer not included in the current models. High-dissipation states of Europa's silicate mantle require near-solidus temperatures. A high europan heat flux and large topographic variations at Europa's ocean floor that may be detected by gravity measurements would provide indirect evidence for present or recent activity of Europa's silicate shell.

Acknowledgments. The authors gratefully acknowledge the reviews of S. Hauck and an anonymous reviewer and the editorial efforts of W. McKinnon.

REFERENCES

Aksnes K. and Franklin F. A. (2001) Secular acceleration of Io derived from mutual satellite events. *Astron. J., 122,* 2734–2739.

Anderson J. D., Schubert G., Jacobsen R. A., Lau E. L., and Moore W. B. (1998) Europa's differentiated internal structure: Inferences from four Galileo encounters. *Science, 281,* 2019–2022.

Anderson J. D., Jacobson R. A., Lau E. L., Moore W. B., and Schubert G. (2001) Io's gravity field and interior structure. *J. Geophys. Res., 106,* 32963–32969.

Berckhemer H., Kampfmann W., Aulbach E., and Schmeling H. (1982) Shear modulus and Q of forsterite and dunite near partial melting from forced-oscillation experiments. *Phys. Earth Planet. Inter., 29,* 30–41.

Carlson R. W., Johnson R. E., and Anderson M. S. (1999) Sulfuric acid on Europa and the radiolytic sulfur cycle. *Science, 286,* 97–99.

Dahlen F. A. (1974) On the static deformation of an Earth model with a fluid core. *Geophys. J. Roy. Astron. Soc., 36,* 461–485.

Fischer H. J. and Spohn T. (1990) Thermal-orbital histories of viscoelastic models of Io (J1). *Icarus, 83,* 39–65.

Gavrilov S. and Zharkov V. (1977) Love numbers of the giant planets. *Icarus, 32,* 443–449.

Goldreich P. (1963) On the eccentricity of satellite orbits in the solar system. *Mon. Not. R. Astron. Soc., 126,* 257–268.

Goldreich P. and Nicholson P. D. (1977) Turbulent viscosity and Jupiter's tidal Q. *Icarus, 30,* 301–304.

Goldreich P. and Soter S. (1966) Q in the solar system. *Icarus, 5,* 375–389.

Goldstein S. J. Jr. and Jacobs K. C. (1995) A recalculation of the secular acceleration of Io. *Astron. J., 110,* 3054–3057.

Greenberg R. (1982) Orbital evolution of the Galilean satellites. In *Satellites of Jupiter* (D. Morrison, ed.), pp. 65–92. Univ. of Arizona, Tucson.

Hand K. P., Carlson R. W., and Chyba C. F. (2007) Energy, chemical disequilibrium, and geological constraints on Europa. *Astrobiology, 7,* 1006–1022, DOI: 10.1089/ast.2007.0156.

Hirth G. and Kohlstedt D. L. (1996) Water in the oceanic upper mantle: Implications for rheology, melt extraction and the evolution of the lithosphere. *Earth Planet. Sci. Lett., 144,* 93–108.

Hussmann H. and Spohn T. (2004) Thermal-orbital evolution of Io and Europa. *Icarus, 171,* 391–410.

Ioannou P. J. and Lindzen R. S. (1993) Gravitational tides in the outer planets: 2. Interior calculations and estimation of the tidal dissipation factor. *Astrophys. J., 406,* 266–278.

Karato S. andWu P. (1993) Rheology of the upper mantle: A synthesis. *Science, 260,* 771–778.

Kargel J. S., Kaye J. Z., Head J. W., Marion G. M., Sassen R., Crowley J. K., Ballesteros O. P., Grant S. A., and Hogenboom D. L. (2000) Europa's crust and ocean: Origin, composition, and the prospects for life. *Icarus, 148,* 226–265, DOI: 10.1006/icar.2000.6471.

Kaula W. M. (1964) Tidal dissipation by solid friction and the resulting orbital evolution. *Rev. Geophys., 2,* 661–685.

Lainey V. and Tobie G. (2005) New constraints on Io's and Jupiter's tidal dissipation. *Icarus, 179,* 485–489.

Lieske J. H. (1987) Galilean satellite evolution: Observational evidence for secular changes in mean motions. *Astron. Astrophys., 176,* 146–158.

Marchis F., Le Mignant D., Chaffee F. H., Davies A. G., Kwok S. H., Prangé, R., de Pater I., Amico P., Campbell R., Fusco T., Goodrich R. W., and Conrad A. (2005) Keck AO survey of Io global volcanic activity between 2 and 5 μm. *Icarus, 176,* 96–122, DOI: 10.1016/j.icarus.2004.12.014.

Matson D. L., Johnson T. V., Veeder G. J.., Blaney D. L., and Davies A. G. (2001) Upper bound on Io's heat flow. *J. Geophys. Res., 106,* 33021–33024.

McCollom T. M. (1999) Methanogenesis as a potential source of chemical energy for primary biomass production by autotrophic organisms in hydrothermal systems on Europa. *J. Geophys. Res., 104,* 30729–30742.

McCord T. B., Hansen G. B., Matson D. L., Jonhson T. V., Crowley J. K., Fanale F. P., Carlson R. W., Smythe W. D., Martin P. D., Hibbitts C. A., Granahan J. C., and Ocampo A. (1999) Hydrated salt minerals on Europa's surface from the Galileo near-infrared mapping spectrometer (NIMS) investigation. *J. Geophys. Res., 104,* 11827–11852, DOI: 10.1029/1999JE900005.

Moore W. B. (2001) The thermal state of Io. *Icarus, 154,* 548–550.

Moore W. B. (2003) Tidal heating and convection in Io. *J. Geophys. Res., 108,* 5096, DOI: 10.1029/2002JE001,943.

Moore W. B. (2006) Thermal equilibrium in Europa's ice shell. *Icarus, 180,* 141–146.

Moore W. B. and Schubert G. (2000) The tidal response of Europa. *Icarus, 147,* 317–319.

Nimmo F. and Manga M. (2002) Causes, characteristics and consequences of convective diapirism on Europa. *Geophys. Res. Lett., 29,* Article No. 2109.

Ojakangas G. W. and Stevenson D. J. (1986) Episodic volcanism of tidally heated satellites with application to Io. *Icarus, 66,* 341–358.

Ojakangas G. W. and Stevenson D. J. (1989) Thermal state of an ice shell on Europa. *Icarus, 81,* 220–241.

Pappalardo R. T., Head J. W., Greeley R., Sullivan R. J., Pilcher C., Schubert G., Moore W. B., Carr M. H., Moore J. M., Belton M. J. S., and Goldsby D. L. (1998) Geological evidence for solid-state convection in Europa's ice shell. *Nature, 391,* 365–368.

Peale S. J. (1986) Orbital resonances, unusual configurations and exotic rotation states among planetary satellites. In *Satellites* (J. A. Burns and M. S. Matthews, eds.), pp. 159–223. Univ. of Arizona, Tucson.

Peale S. (2003) Tidally induced volcanism. *Celest. Mech. Dyn. Astron., 87,* 129–155.

Peale S. J. (2007) The origin of the natural satellites. In *Treatise on Geophysics* (G. Schubert, ed.), Vol. 10. Elsevier, Amsterdam.

Reese C. C., Solomatov V. S., Baumgardner J. R., and Yang W. S. (1999) Stagnant lid convection in a spherical shell. *Phys. Earth Planet. Inter., 116,* 1–7.

Ross M. N. and Schubert G. (1987) Tidal heating in an internal ocean model of Europa. *Nature, 325,* 133–134.

Schubert G., Turcotte D. L., and Olson P. (2001) *Mantle Convection in the Earth and Planets.* Cambridge Univ., New York.

Schubert G., Anderson J. D., Spohn T., and McKinnon W. B. (2004) Interior composition, structure and dynamics of the Galilean satellites. In *Jupiter: The Planet, Satellites and Magnetosphere* (F. Bagenal et al., eds.), pp. 281–306. Cambridge Univ., Cambridge.

Segatz M., Spohn T., Ross M. N., and Schubert G. (1988) Tidal dissipation, surface heat flow, and figure of viscoelastic models of Io. *Icarus, 75,* 187–206.

Showman A. P. and Han L. (2004) Numerical simulations of convection in Europa's ice shell: Implications for surface features. *J. Geophys. Res.–Planets, 109,* E01,010, DOI: 10.1029/2003JE002103.

Showman A. P. and Malhotra R. (1997) Tidal evolution into the Laplace resonance and the resurfacing of Ganymede. *Icarus, 127,* 93–111.

Showman A. P., Stevenson D. J., and Malhotra R. (1997) Coupled orbital and thermal evolution of Ganymede. *Icarus, 129,* 367–383.

Solomatov V. S. (1995) Scaling of temperature- and stress-dependent viscosity convection. *Phys. Fluids, 7,* 266–274.

Solomatov V. S. and Moresi L. N. (2000) Scaling of time-dependent stagnant lid convection: Application to small scale convection on Earth and other terrestrial planets. *J. Geophys. Res., 105,* 21795–21817.

Stevenson D. (1983) Anomalous bulk viscosity of two-phase fluids and implications for planetary interiors. *J. Geophys. Res., 88,* 2445–2455.

Tobie G., Choblet G., and Sotin C. (2003) Tidally heated convection: Constraints on Europa's ice shell thickness. *J. Geophys. Res.–Planets, 108,* 5124, DOI: 10.1029/2003JE002099.

Veeder G. J., Matson D. L., Johnson T. V., Blaney D. L., and Gougen J. D. (1994) Io's heat flow from infrared radiometry: 1983–1993. *J. Geophys. Res., 99,* 17095–17162.

Veeder G. J., Matson D. L., Johnson T. V., Davies A. G., and Blaney D. L. (2004) The polar contribution to the heat flow of Io. *Icarus, 169,* 264–270.

Wark D. A. and Watson E. B. (1998) Grain-scale permeabilities of texturally equilibrated, monomineralic rocks. *Earth Planet. Sci. Lett., 164,* 591–605.

Wolf D. (1991) Viscoelastodynamics of a stratified, compressible planet: Incremental field equations and shortand long-time asymptotes. *Geophys. J. Int., 104,* 401–417.

Wu Y. (2005) Origin of tidal dissipation in Jupiter. II. The value of Q. *Astrophys. J., 635,* 688–710, DOI: 10.1086/497355.

Yoder C. F. (1979) How tidal heating in Io drives the Galilean orbital resonance locks. *Nature, 279,* 767–770.

Yoder C. F. and Peale S. J. (1981) The tides of Io. *Icarus, 47,* 1–35.

Zolotov M. Y. and Shock E. L. (2001) Composition and stability of salts on the surface of Europa and their oceanic origin. *J. Geophys. Res., 106,* 32815–32828, DOI: 10.1029/2000JE001413.

Zschau J. (1977) Tidal friction in the solid Earth: Loading tides versus body tides. In *Tidal Friction and the Earth's Rotation* (P. Brosche and J. Sündermann, eds.), p. 62. Springer-Verlag, Berlin.

Geodynamics of Europa's Icy Shell

Francis Nimmo
University of California Santa Cruz

Michael Manga
University of California Berkeley

Processes that operate within Europa's floating icy shell and leave their signature on the surface are largely governed by the thermal and mechanical properties of the shell. We review how geodynamic models for icy shell dynamics can be integrated with observations to constrain the present and past properties of the icy shell. The near-surface exhibits brittle or elastic behavior, while deeper within the shell viscous processes, including tidal heating, lateral flow, and possibly convection, dominate. Given the large amplitude of topography on Europa, the icy shell is probably more than several kilometers thick at present. However, there are both theoretical models and observational evidence suggesting that the icy shell has been thickening over time, explaining the predominance of extensional features and the young (probably ~50 m.y.) surface age. Geophysical measurements on future missions will be able to determine the present thickness of the icy shell, and possibly its thermal structure.

1. INTRODUCTION

Europa's icy shell records a complex tectonic history that reflects an interaction between surface processes, internal structure, and orbital dynamics. The icy shell also separates the hostile surface environment from an inferred subsurface ocean. The nature and history of this shell is thus of great astrobiological, geological, and geophysical importance. In this chapter we will discuss the geodynamics of Europa's icy shell, i.e., its mechanical and chemical properties and evolution, and the processes that are likely to be operating today.

The most direct evidence for an icy shell at present, i.e., a layer of ice separated from a rocky interior by a liquid ocean, is the induced magnetic field produced in a highly electrically conductive (ocean) near-surface region (*Kivelson et al.,* 2000). In addition, models for most of the surface features we see on Europa's surface, including the pits, domes, chaos, bands, and ridges shown in Fig. 1, probably either require the presence of an ocean or involve this water reaching the surface. While the existence of an ocean is generally accepted (see, e.g., chapter by McKinnon et al.), many of the properties and processes in the icy shell remain the subject of considerable uncertainty and debate. Indeed, the Galileo mission, which through its higher imaging resolution identified surface features unseen by the earlier Voyager spacecraft, raised more questions than it answered. Nevertheless, a combination of surface observations and theoretical models allows certain questions to be answered and hypotheses to be tested.

There are two overarching questions on which we focus: (1) What are the origins of the observed surface features? (2) What do the characteristics of surface features tell us about the properties and evolution of the icy shell?

In order to answer these questions, the first two parts of this chapter will consist of theoretical explorations of the likely structure of a floating icy shell, and the different mechanisms available to deform the shell and leave a tectonic record on the surface. Section 4 will then compare these theoretical predictions with observations of Europa's surface to infer the present-day characteristics of the icy shell. Section 5 will carry out a similar exercise focusing on the evolution of the icy shell through time. Section 6 will conclude with a review and suggestions for future work.

Some of the topics we cover in this chapter are discussed in more detail elsewhere in this volume. In particular, Europa's tidal heating, interior structure, thermal evolution and potential icy shell convection are addressed by Sotin et al., Schubert et al., Moore and Hussmann, and Barr and Showman, respectively. Similarly, the global stratigraphy, tectonics, and origin of various landforms are discussed by Doggett et al., Kattenhorn and Hurford, Prockter and Patterson, and Collins and Nimmo. Briefer summaries of the interior and surface geology of Europa may be found in *Schubert et al.* (2004) and *Greeley et al.* (2004), respectively.

Although Europa is primarily composed of silicates and iron (chapter by Schubert et al.), its surface appearance is controlled by the characteristics of the icy shell. Before embarking on a detailed exploration of icy shell geodynamics, it is helpful to consider similarities and differences to the more familiar behavior of the crusts and mantles of silicate bodies.

Ice in terrestrial settings (e.g., glaciers), and even on Mars, is always close to its melting temperature and generally deforms in a ductile fashion. At the ~100-K temperatures relevant to Europa's surface, however, ice is strong and brittle and behaves very much like rock at the surface of the Earth. Thus, common tectonic features on Earth ap-

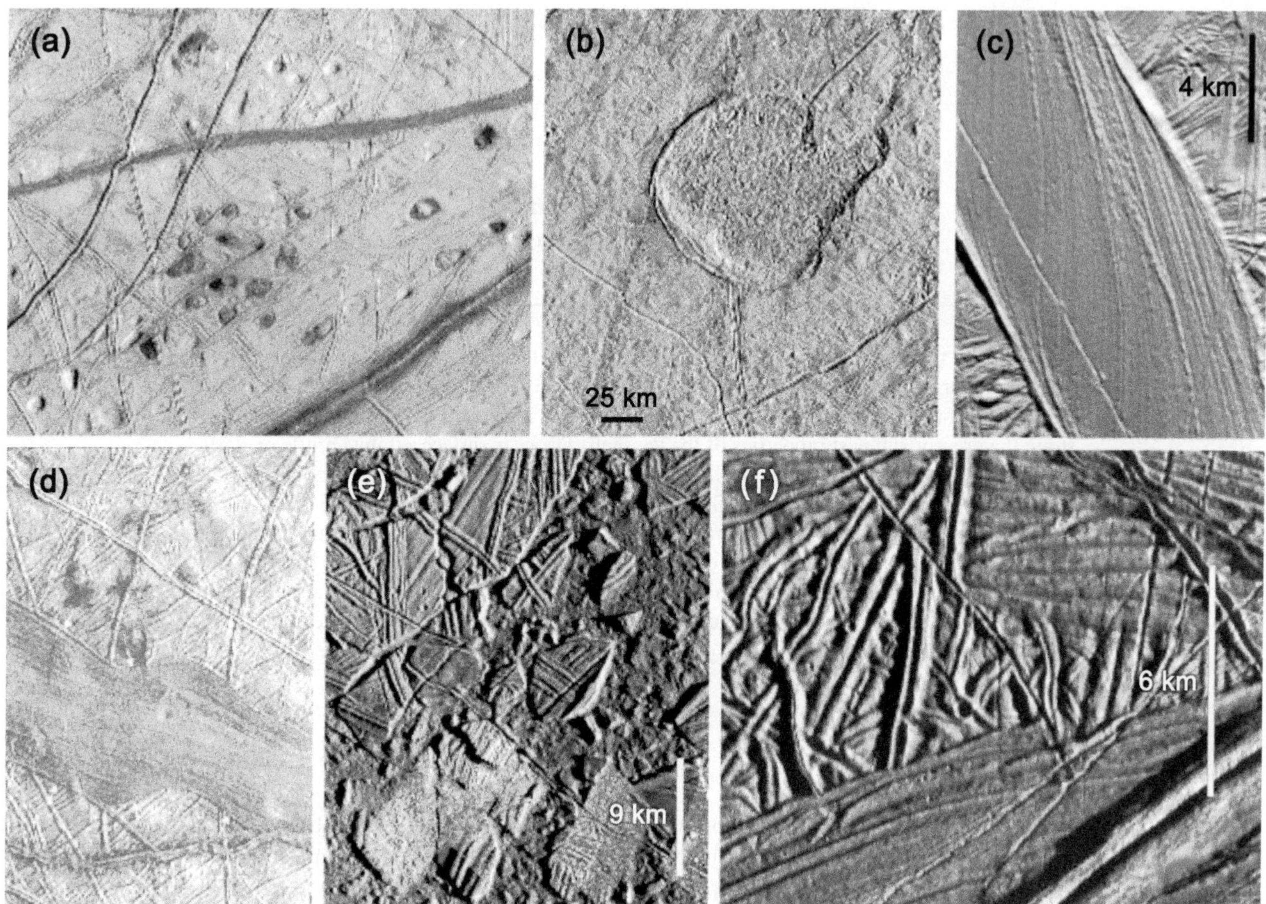

Fig. 1. Examples of surface features whose properties are governed by processes within the icy shell: **(a)** spots, pits and ridges (PIA03878); each pit is about 10 km across; **(b)** the "Mitten" feature (Murias Chaos) (PIA01640); **(c)** gray band that records 60 km of strike-slip motion (PIA01643); **(d)** band showing dilatation (from *Prockter et al.,* 2002); band is about 30 km wide; **(e)** Conamara Chaos showing blocks that have been displaced and rotated within a low-viscosity substrate (PIA01403); **(f)** ridges and fractures (PIA00849).

pear to have europan analogs (see chapter by Kattenhorn and Hurford), including normal faults (*Nimmo and Schenk,* 2006), tail cracks (*Kattenhorn,* 2005), wing cracks (*Schulson,* 2002), anti-cracks (*Kattenhorn and Marshall,* 2006), folds (*Prockter and Pappalardo,* 2000), and strike-slip faults (*Hoppa et al.,* 1999c). An important difference, however, is that ice near its solidus temperature is roughly 10^6 times less viscous than rock near its solidus temperature; thus, ductile flow processes (e.g., convection) within Europa's icy shell happen on much more rapid timescales than similar processes within Earth's crust and mantle even though typical (tidal) stresses on Europa are much smaller than tectonic stresses on Earth (see section 3).

Perhaps the most fundamental difference is that ice, unlike the lithospheric mantle on silicate bodies, is less dense than the underlying material. Here and elsewhere we are referring to ice I — higher-pressure phases of ice are probably not relevant to Europa's shell (see chapter by Schubert et al.), although they are relevant on other large icy satellites, e.g., Ganymede. One immediate consequence of this effect is that melt (i.e., water) generated by heating of ice will generally drain downward and not upward, with important consequences for the likelihood of low-temperature (cryo-) volcanism (section 4.2.4). Another is that the (buoyant) floating icy shell will insulate the underlying ocean, potentially allowing it to persist for geological timescales.

The low density of ice I means that subduction of the icy shell into the underlying ocean is not favored, similar to the resistance of continental material to subduction on Earth. Although some features resembling those at divergent terrestrial plate boundaries are observed on Europa (e.g., *Sullivan et al.,* 1998; *Prockter et al.,* 2002), no evidence has been found for features directly resembling subduction zones.

Although subduction is unlikely to generate significant stresses on Europa, tidal stresses, arising from the varying distance and orientation of Jupiter relative to Europa, are much more important than the Moon's tidal stresses on Earth (e.g., *Greenberg et al.,* 2002). As discussed below, many of the observed surface features are caused directly or indirectly by tidal stresses. Furthermore, friction associated with tidal deformation is more important than decay of radioactive materials in determining the thermal state of Europa's icy shell (e.g., *Cassen et al.,* 1979; *Squyres et al.,*

1983; *Ross and Schubert,* 1987; *Ojakangas and Stevenson,* 1989a). Thus, unlike the terrestrial planets, the orbital evolution of Europa has a direct effect on its thermal evolution, and vice versa (e.g., *Hussmann and Spohn,* 2004).

A final difference between Europa and most silicate bodies is that Europa's surface is much less heavily cratered, and therefore younger, than those of its neighboring satellites (see chapter by Bierhaus et al.). Unfortunately, the absolute age of Europa's surface is disputed. The more generally accepted impact flux model, based primarily on observations of extant Jupiter family comets, results in a mean surface age of only a few tens of millions of years (*Zahnle et al.,* 2003), or roughly 1% of the age of the solar system. An alternative (*Neukum,* 1997), which assumes that the age of Gilgamesh on Ganymede is the same as that of the youngest lunar impact basin (Orientale, 3.8 Ga), and that cratering in the jovian system is dominated by asteroids, results in a surface age of 0.7–2.8 Ga (see *Pappalardo et al.,* 1999). In any event, a significant fraction of Europa's history has no geological record, which makes interpreting its evolution challenging; on the other hand, processes that were only relevant during earliest solar system history (e.g., accretion) cannot have contributed to the sculpting of Europa's surface, and we can thus understand the surface and its evolution in terms of orbital dynamics and internal processes that operate at present.

2. ICY SHELL STRUCTURE

The geodynamic behavior of the icy shell is controlled by its mechanical properties and thermal evolution. As a starting point, we thus begin with a largely theoretical review of the mechanical and thermal properties of ice that govern the deformation and heat transfer within a planet-scale icy shell.

2.1. Rheology

The observable consequences of tectonic stresses depend mainly on the response of the material being stressed, i.e., its rheology. Here we discuss the three main ways in which ice may respond to imposed stresses, and the consequences of these different response mechanisms. At low stresses and strains, ice will deform in an elastic (recoverable) manner. However, at strains greater than roughly 10^{-3}, the ice will undergo irrecoverable deformation. At low temperatures, this deformation will be accomplished by brittle failure, while at higher temperatures the result will be ductile creep.

2.1.1. Brittle deformation. Silicate materials typically undergo brittle failure along preexisting faults when the shear stresses exceed some fraction, typically 0.6, of the normal stresses. This behavior is largely independent of composition and is known as Byerlee's rule. At low sliding velocities and stresses, ice obeys Byerlee's law with a coefficient of friction of 0.55 (*Beeman et al.,* 1988), which is independent of temperature, although at higher sliding velocities the behavior becomes more complex (*Rist,* 1997). At shallow depths where normal stresses are small and temperatures are low, brittle deformation is expected to dominate on icy satellites.

2.1.2. Ductile deformation. At sufficiently high temperatures, ice responds to applied stresses by deforming in a ductile fashion. The response is complicated by the fact that individual ice crystals can deform in several different ways: by diffusion of defects within grain interiors, by sliding of grain boundaries, and by creep of dislocations (*Durham and Stern,* 2001; *Goldsby and Kohlstedt,* 2001). Which mechanism dominates depends on the specific stress and temperature conditions, but each individual mechanism can be described by an equation of the form

$$\dot{\varepsilon} = A\sigma^n d^{-p} \exp\left(-\frac{Q + PV}{RT}\right)$$

Here $\dot{\varepsilon}$ is the resulting strain rate; σ is the differential applied stress; A, n, and p are rheological constants; d is the grain size; Q and V are the activation energy and volume, respectively; R is the gas constant; and P and T are pressure and temperature, respectively. In general, for icy satellites the P,V contribution is small enough to be ignored, and strain rates should increase with increasing temperature and stress and decreasing grain size. When several different mechanisms are operating together, the total strain rate is a superposition (in series and/or parallel) of the individual strain rates (see equation (3) in *Goldsby and Kohlstedt,* 2001).

At low stresses, diffusion creep is expected to dominate and is predicted to result in Newtonian flow (i.e., n = 1) with a grain-size dependence (p = 2). At higher stresses, the dominant creep regimes are basal slip and grain boundary sliding, which result in non-Newtonian behavior (n ~ 2) and grain-size dependence. At even higher stresses, strongly non-Newtonian dislocation creep (n = 4) dominates. Creep rates are enhanced within about 20 K of the melting temperature (*Goldsby and Kohlstedt,* 2001), presumably because of the presence of thin films of water along grain boundaries (e.g., *De La Chapelle et al.,* 1999).

Because stresses and strain rates on icy satellites are expected to be low, the most relevant deformation mechanism is probably diffusion creep (e.g., *Moore,* 2006; *McKinnon,* 2006), but superplastic flow or dislocation creep may dominate in regions undergoing active convection (*Freeman et al.,* 2006). Diffusion creep has the modeling advantage of resulting in Newtonian behavior, but the disadvantage that the viscosity ($\eta = \sigma/\dot{\varepsilon}$) is dependent on the (unknown) grain size, and that the relevant rheological parameters have not yet been measured (*Goldsby and Kohlstedt,* 2001). Ice grain size evolution is poorly understood, because it depends both on the presence of secondary (pinning) phases and because of dynamic recrystallization processes (e.g., *Barr and McKinnon,* 2007). Given the uncertainties, it is often acceptable to assume for modeling purposes that ice has a Newtonian viscosity near its melting temperature in the range 10^{13}–10^{15} Pa s (e.g., *Pappalardo et al.,* 1998).

Although the unknown grain size is the most serious unknown for describing the ductile deformation of ice, other effects can also be important. The presence of even small amounts of fluid significantly enhances creep rates (e.g., *De La Chapelle et al.,* 1999). On the other hand, the presence of rigid impurities (e.g., silicates) and salts serves to increase the viscosity (*Friedson and Stevenson,* 1983; *Durham et al.,* 2005a). Finally, higher-pressure phases of ice, or ices incorporating other chemical species such as methane, tend to have much higher viscosities than pure ice at the same P,T conditions (*Durham et al.,* 1998), although ammonia is an exception to this general rule (*Durham et al.,* 1993).

2.1.3. Elastic deformation. At low stresses and strains, ice will deform elastically and the relationship between stress σ and strain ε depends on the Young's modulus E of the material and is given by Hooke's law: $\sigma = E\varepsilon$.

Although measurement of Young's modulus in small laboratory specimens is straightforward and yields a value of about 9 GPa (*Gammon et al.,* 1983), the effective Young's modulus of large bodies of deformed ice is less obvious (*Nimmo,* 2004c). Observations of ice shelf response to tidal stresses on Earth (*Vaughan,* 1995) give an effective E of ~0.9 GPa, an order of magnitude smaller than the laboratory values. This discrepancy is most likely due to the fact that a large fraction of the ice shelf thickness is responding in a brittle or ductile, rather than an elastic, fashion (e.g., *Schmeltz et al.,* 2002). A similar effect may arise on icy satellites, and will be discussed later.

2.1.4. Viscoelastic behavior. In reality, materials do not behave as entirely elastic or entirely viscous. Rather, they exhibit elastic-like behavior if the timescale over which deformation occurs is short compared to a characteristic timescale of the material, known as the Maxwell time (= η/G, where G is the shear modulus) (*Turcotte and Schubert,* 2002). Conversely, if the deformation timescale is long compared to the Maxwell time, the material behaves in a viscous fashion. Such compound materials are termed viscoelastic.

The viscoelastic model is especially important for icy satellites because it provides a convenient description of tidal heating (e.g., *Ross and Schubert,* 1989; *Tobie et al.,* 2005). Tidal heating is maximized when the Maxwell time equals the forcing (orbital) period, and decreases in a linear fashion with forcing frequency for much larger or much smaller values (e.g., *Tobie et al.,* 2003). Hence, the amount of tidal heating depends critically on the local viscosity of the icy shell, which is in turn determined mainly by the temperature (section 2.1.2). The Maxwell time of ice near its melting point is roughly 10^4–10^5 s, very close to Europa's orbital period of 3.55 d. Thus, tidal dissipation is expected to be important for warm ice on Europa.

2.1.5. Application to icy shells. Temperatures near the surface of icy satellites are sufficiently cold, and overburden pressures sufficiently small, that tectonic stresses are likely to result in brittle deformation. However, at greater depths, temperatures will increase, allowing ductile deformation to dominate. If the principal source of stress is bending, then near the midplane of the bending shell the stresses may be low enough to allow elastic deformation to occur. Thus, in general one would expect a deformed icy shell to consist of the three regions shown in Fig. 2: a brittle near-surface layer, an elastic "core," and a ductile base (e.g., *Watts,* 2001). The interfaces between these zones occur at depths where the stresses due to two competing mechanisms are equal.

The thickness of the near-surface brittle layer in Fig. 2 depends mainly on the temperature gradient, and to a lesser extent on the degree of curvature (bending). Thus, if the brittle layer thickness can be constrained, e.g., by observations of fault spacings (*Jackson and White,* 1989), then the temperature structure at the time of fault formation can be deduced (e.g., *Golombek and Banerdt,* 1986). Similarly, the stress profile shown in Fig. 2 controls the response of the icy shell as a whole to applied bending stresses. The resulting flexural response can be modeled by assuming that the shell is purely elastic with an effective elastic thickness, T_e, which can be derived from topographic observations. This

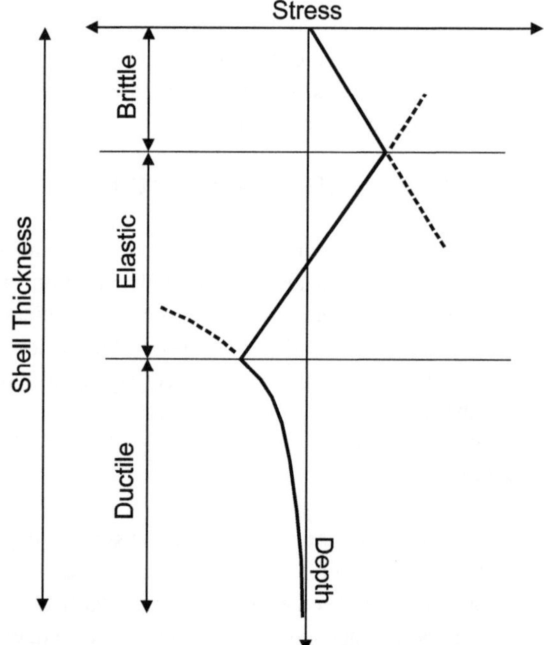

Fig. 2. Schematic stress profile within a generic icy satellite shell. Near the surface the ice is cold and brittle and stress increases in proportion to the overburden pressure. Because the shell is undergoing bending, the elastic bending stresses decrease toward the midpoint of the shell and lead to an elastic "core"; the slope of the elastic stress curve depends on the local curvature of the shell and the Young's modulus (e.g., *Turcotte and Schubert,* 2002). At greater depths, the ice is sufficiently warm that the shell deforms in a ductile fashion. The first moment of the stress profile about the midpoint controls the effective elastic thickness T_e of the icy shell as a whole (e.g., *Watts,* 2001). For typical curvatures seen on Europa, T_e is controlled mainly by the brittle portion (see section 4.1.1). In equilibrium, the positive and negative areas under the composite stress curve must equal each other (e.g., *Luttrell and Sandwell,* 2006).

effective elastic thickness may be related to the real shell structure (Fig. 2) if the local topographic curvature is known (e.g., *Watts,* 2001). Hence, estimates of T_e may be used to determine the thermal structure of icy shells (e.g., *Nimmo and Pappalardo,* 2004). For an icy shell in which heat transfer occurs only by conduction, knowing the thermal structure in turn allows the shell thickness to be deduced (see section 4.1.1). However, if the icy shell is convecting, then only the heat flux, and not the total shell thickness, can be deduced (section 2.3.2).

2.2. Heat Sources

The temperature structure and thickness of the icy shell are determined by the heat being generated, and the rate at which that heat is transported to the surface. There are two principal sources of heat at present: radioactive decay and diurnal tidal dissipation.

Radioactive decay of U, Th, and K is likely confined mainly to silicate minerals and thus occurs within the silicate interior of Europa. Assuming a chondritic composition, the present-day heat production rate results in a surface heat flow of about 7 mW m^{-2} (*Hussman and Spohn,* 2004).

Dissipation due to diurnal tides (see section 3.1) may take place within the silicate interior, within the floating icy shell, or both (dissipation in the ocean is likely small unless the ocean thickness is comparable to the seafloor topography, as on Earth). Dissipation is conventionally modeled as taking place within a viscoelastic body (section 2.1.4) so that the heat generated depends on the viscosity and thus the temperature of the material, as well as the tidal strain rate. Since the temperature in turn depends partly on the heat generation rate, calculation of tidal heating is not straightforward (e.g., *Tobie et al.,* 2005). Furthermore, the presence of a near-surface rigid layer can reduce the overall tidal deformation (*Moore and Schubert,* 2000), and thus the amount of tidal heating in the interior, although this effect is not pronounced for the relatively thin icy shells expected for Europa.

Tidal dissipation within the icy shell depends on, and influences, the shell temperature structure and thickness and is discussed further below. Because of the spatially variable tidal stress field, tidal dissipation within the icy shell also varies spatially, with the maximum stress and heating occurring at the poles (*Ojakangas and Stevenson,* 1989a).

Although tidal heating is normally modeled using a viscoelastic approach, other dissipation mechanisms are possible. For instance, laboratory deformation of silicates shows a power-law frequency dependence of dissipation that is not characteristic of Maxwell viscoelastic materials (e.g., *Gribb and Cooper,* 1998). Perhaps more relevant to Europa is the idea that tidally driven frictional heating on individual faults may also be important in the near-surface (e.g., *Gaidos and Nimmo,* 2000; *Han and Showman,* 2008).

Heat, either tidal or radiogenic, generated within the silicate interior is presumed to be transported efficiently across the ocean. The magnitude of the tidal contribution from the silicate interior is highly uncertain. If this region is cold, tidal dissipation is negligible. If, however, the interior is partially molten (as is thought to be the case for Io), then scaling Io's heat flow to europan orbital conditions and assuming similar internal structures results in a heat generation rate of roughly 4 TW (*Greenberg et al.,* 2002) and suggests a thin (≈4 km), conductive shell. Because of the feedback between dissipation and viscosity, tidal heat production in the silicate interior is likely to be either Io-like or negligible, and not some intermediate value (see chapter by Moore and Hussmann). The uncertainty in the amount of tidal heat generated below the ocean is an important contributor to the uncertainty in the icy shell thickness of Europa; uncertainty in the icy shell heat production is another (see below).

2.3. Temperature Structure

The temperature structure of the icy shell is one of its most important properties, governing as it does the tendency for different deformation mechanisms to operate (section 2.1) and controlling the flow rate and amount of tidal heating. The detailed temperature structure depends on whether the icy shell is convecting (section 4.2.3) or not. If the shell is not convecting, then the temperature profile will be conductive throughout. If the shell is convecting, then the near-surface ice (the so-called stagnant lid) will be sufficiently cold and rigid that heat is transferred only by conduction, unless some kind of plastic yielding occurs (*Showman and Han,* 2005). The convective interior, meanwhile, will be approximately isothermal, with a thin boundary layer at its base if significant heat is being transferred from the subsurface ocean. Note that in both the conductive and convective cases the near-surface ice is cold, rigid, and immobile, but that the thickness of the conductive layer relative to the total icy shell thickness is significantly different in these two cases. It is therefore difficult to use inferences of the rigid (elastic) layer thickness to determine the total thickness of the icy shell.

2.3.1. Conductive portion. For the conductive portion of an icy shell, it is usually an excellent approximation to assume thermal steady state, in which case the temperature structure depends on the thermal conductivity k, internal heat sources, and the boundary conditions. The thermal conductivity of ice depends on the temperature T [a typical approximation is that k = 567/T where T is in K and K has units of W m^{-1} K^{-1} (*Klinger,* 1980)]. Conductivity can be reduced by near-surface porosity (e.g., *Squyres et al.,* 1983) or the presence of species with significantly lower conductivities, such as methane clathrates (e.g., *Nimmo and Giese,* 2005). For a conductive icy shell, the bottom boundary temperature is simply the ocean freezing point, which depends on the composition of the ocean (see section 2.4). If the shell is convecting, the base of the conductive lid occurs where the viscosity is roughly an order of magnitude greater than the viscosity in the convective interior of the shell (*Davaille and Jaupart,* 1994; *Solomatov,* 1995). For typi-

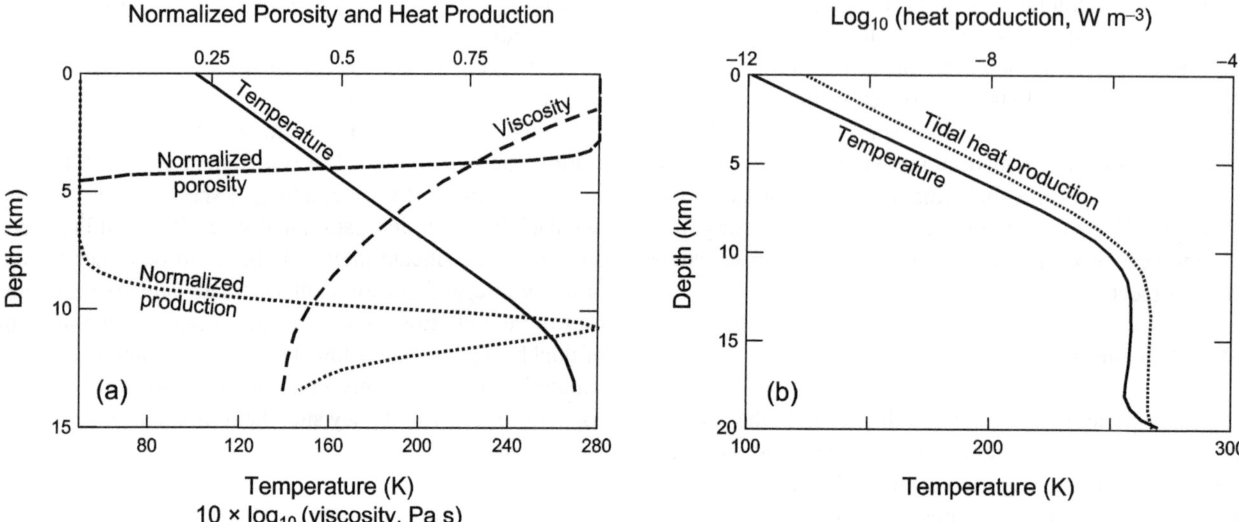

Fig. 3. **(a)** Tidally heated conductive icy shell in thermal equilibrium (steady-state). Temperature, viscosity, and tidal heat production are calculated in a self-consistent manner using the approach of *Nimmo et al.* (2007a). Porosity and heat production are normalized to their maximum values (arbitrary and 2×10^{-5} W m^{-3}, respectively). Reference viscosity is 10^{14} Pa s, mean strain rate is 2.1×10^{-10} s^{-1}, basal heat flux is 7 mW m^{-2}, surface heat flux is 60 mW m^{-2}, thermal conductivity is 4 W m^{-1} K^{-1} and equilibrium shell thickness is 13.2 km. This point is located at 45°N, 45°E; other locations will have different equilibrium shell thicknesses. Porosity evolution is calculated using the method of *Nimmo et al.* (2003a) assuming an evolution timescale of 50 m.y. **(b)** Tidally heated convective shell in thermal equilibrium, redrawn from *Tobie et al.* (2003). Basal viscosity 10^{14} Pa s, ice viscosity ratio 1.2×10^6. Tidal heat production rate was calculated using the temperature profile plotted, but note that the temperature profile was calculated without including the effect of tidal heating.

cal ice activation energies, the corresponding temperature is 230–250 K. In either case the surface temperature varies as a function of latitude, from about 110 K at the equator to ~40 K at the poles (*Ojakangas and Stevenson,* 1989a). The amount of tidal heating depends on the local temperature; thus, the temperature structure and tidal dissipation profile have to be solved for simultaneously.

Figure 3a shows an example equilibrium temperature profile assuming constant thermal conductivity. The inflection toward the base of the layer is because of tidal heating, which only becomes important at the low viscosities characteristic of ice near its melting point. For a purely conductive icy shell, both the bottom heat flux and the temperature are specified, as is the surface temperature. There thus exists a single shell thickness that can satisfy these constraints simultaneously, in this case 13 km. Hence, in the absence of other effects, the thickness of the conductive icy shell will vary spatially by a few tens of percent, because both the surface temperature and the tidal dissipation vary (*Ojakangas and Stevenson,* 1989a; *Nimmo et al.,* 2007a). Variations in the thickness of the icy shell can drive horizontal flows that will reduce thickness variations (e.g., *O'Brien et al.,* 2002; *Nimmo,* 2004a) (section 3.3).

2.3.2. Convective portion. The convective interior of an icy shell is close to its melting temperature, which in general results in enhanced tidal dissipation. Figure 3b shows the temperature structure of an example convecting shell and the corresponding heat production. The top half of the shell is cold, highly viscous, and conductive, while the bottom half of the shell is warm, convecting, and approximately isothermal. As with the purely conductive case, tidal heating is greatest in the low-viscosity ice. Note that the temperature gradients in the conductive portions of Figs. 3a and 3b are almost identical, despite the different total shell thicknesses: This example illustrates the difficulty of inferring total shell thickness from shallow temperature gradients alone.

Just as with the conductive case, there is in general, although not always (*Mitri and Showman,* 2005), only a single shell thickness for which the bottom heat flux and top and bottom temperature conditions can be satisfied in equilibrium (e.g., *Moore,* 2006; *Ruiz et al.,* 2007). Convective shells generally produce more tidal heating than conductive shells because for the former more of the interior is close to the melting point (Fig. 3). Provided the conductive surface layer remains stagnant because of its high viscosity, the temperature differences within the convecting region will be small, $\Delta T \sim 2.2\ RT_i^2/Q$, where T_i is the temperature of the convecting interior (e.g., *Solomatov,* 1995; *Grasset and Parmentier,* 1998). For typical ice rheologies, ΔT is ~ 20 K (e.g., *Nimmo and Manga,* 2002).

An important distinction between conductive and convective icy shells is that for the latter the amount of heat that can be transported across the shell is almost independent of shell thickness (see, e.g., *Moore,* 2006). Thus, if the basal heating exceeds some critical amount, the only stable solution is a conductive icy shell. For both the convective and the conductive cases, the equilibrium shell thickness depends mainly on the basal heating (from the silicate interior) and the icy viscosity, neither of which are well known. Thus, the examples shown in Fig. 3 are only representative and whether or not the icy shell is conductive or convective remains uncertain (see section 4.1.3).

2.4. Porosity and Contaminants

Modeling studies typically assume that the icy shell is homogeneous pure water ice. In reality, the ice is likely to be heterogeneous in at least two important ways. First, the near-surface is brittle and is continuously being fractured by both impacts and tidal stresses. One would therefore expect the near-surface to contain significant fracture porosity: down to depths of a few meters due to impact processes (see chapter by Moore et al.), and perhaps significantly deeper from tidal fracturing. Of course, at greater depths the ice will lose porosity, either through flow at elevated temperatures [typically 160–180 K; e.g., *Nimmo et al.* (2003a); Fig. 3] or through brittle compaction (*Durham et al.,* 2005b). However, portions of the near-surface are likely to be porous, with implications for thermal conductivity, vapor transport, surface topography, radar properties, and strength, among other effects. Figure 3a shows a theoretical porosity profile for a conductive icy shell, demonstrating that the top third of the shell, which is cold and relatively rigid, can retain porosity for millions of years.

Second, icy shells are likely to incorporate contaminants, either during rapid freezing or as surficial additions by meteorites or comets. The interplanetary micrometeorite flux at Europa is 45 g/s (*Cooper et al.,* 2001; *Johnson et al.,* 2004) and the mean flux of material ejected from Io by cometary impacts and delivered to Europa is 6–10 g/s (*Alvarellos et al.,* 2008). Over the 50-Ma age of the icy shell, these fluxes imply a mass fraction of <0.0001% assuming a shell 10 km thick and assuming all the delivered materials remain in the icy shell. Spectroscopic studies of the surface of Europa indicate variable fractions of "non-ice" contaminants (e.g., *McCord et al.,* 1998), possibly salts, some of which are preferentially associated with specific geological features such as bands (e.g., *Geissler et al.,* 1998b) and may therefore have been derived from the satellite interior.

Contaminants can have two important effects. First, they will affect the mechanical properties of the icy shell, such as bulk density or viscosity (section 2.1.2). Second, and potentially more important, they can affect the melting behavior of the ice. In particular, the incorporation of salts can dramatically lower the temperature at which the first melts are generated (e.g., *Kargel,* 1998). Since these first melts are typically salt-rich, a likely way of generating compositional heterogeneity within the icy shell is by partial melting (*Nimmo et al.,* 2003a; *Pappalardo and Barr,* 2004). Equally, the presence of contaminants such as salts or ammonia in the ocean will reduce its freezing temperature. A colder ocean results in lower temperatures and higher viscosities at the base of the icy shell, and thus lessens the likelihood for icy shell convection. These processes have several potentially important applications at Europa (sections 3.4 and 4.3.2).

2.5. Melting and Cryovolcanism

"Cryovolcanism" refers to the generation and emplacement of low-temperature analogs of silicate volcanic products. These products can involve either "cryomagmas" (low-viscosity water, or higher viscosity water-ice slurry) or vapor. A crucial difference between silicate volcanism and the aqueous cryovolcanism predicted for Europa is that in the latter case the melt (water) is more dense than the solid (ice). Thus, the products of melting are expected to percolate downward, and because of the low viscosity of water, the percolation rate is very rapid (*Nimmo and Giese,* 2005). Hence one would generally not expect to detect cryovolcanic eruption products at the surface, nor aqueous "cryomagma chambers" at depth. Perhaps not surprisingly, few features on Europa appear to have been generated by cryovolcanic activity (*Fagents,* 2003) (see section 4.2.4).

Several theoretical mechanisms have been proposed to overcome the density problem. One possibility is that exsolution of dissolved volatiles at shallow depths reduces the volatile-water density to below that of the surrounding ice and drives an eruption (*Crawford and Stevenson,* 1988). Alternatively, the interaction of silicate magma with ice could lead to overpressures sufficient to drive an eruption (*Wilson et al.,* 1997); note, however, that this mechanism requires the silicates to be in contact with the icy shell, which is implausible given the inferred ~100-km ocean thickness. *Showman et al.* (2004) suggested that topographically driven flows can force water or a water-ice mixture upward into topographic lows. High pressures within the subsurface ocean can also be generated as the overlying icy shell progressively freezes and thickens (*Manga and Wang,* 2007; *Kimura et al.,* 2007).

It has been proposed that warm ascending diapirs could cause melting in overlying salt-rich ice (*Head and Pappalardo,* 1999). However, this model does not overcome the difficulty of downward drainage of melt; furthermore, detailed thermal modeling suggests that the ~100-K surface temperature is very effective in preventing near-surface melting (*Nimmo and Giese,* 2005), even taking into account the melting-temperature reduction due to salts.

3. SOURCES OF STRESS

In this section, we examine mechanisms that could potentially account for the deformation observed at the surface of Europa. Doing so is important partly to develop an understanding of the likely stresses and strain rates characteristic of the icy shell, but also because identifying a particular mechanism (e.g., nonsynchronous rotation) provides clues to both the structure and history of the icy shell. Fortunately, the list of potential mechanisms is considerably simplified by the fact that the surface of Europa is geologically young. This means that mechanisms associated with the early accretion and orbital evolution of satellites (despin-

TABLE 1. Sources and magnitudes of stresses from theoretical work and observations of surface features.

Source	Magnitude	Notes/References	Section
Diurnal tides	<~100 kPa	*Greenberg et al.* (2002)	3.1
Librations	Comparable to diurnal?	Bills et al. (this volume); ***Sarid et al.*** (2006)	4.3.2
Nonsynchronous rotation	Several MPa	***McEwen*** (1986); ***Leith and McKinnon*** (1996); ***Geissler et al.*** (1998a); Sotin et al. (this volume)	3.2
Polar wander	Several MPa	*Leith and McKinnon (1996)*; ***Schenk et al.*** (2008)	3.2
Thermal convection	<100 kPa	*McKinnon* (1999); *Nimmo and Manga* (2002), *Tobie et al.* (2003); *Showman and Han* (2004)	3.4, 4.1.3
Compositional convection	<1 MPa	*Pappalardo and Barr* (2004); *Han and Showman* (2005)	3.4, 4.1.3
Thickening of the icy shell	Several MPa (tensile)	*Nimmo* (2004b); *Kimura et al.* (2007)	3.6
Impacts	TPa	Locally; duration ~ tens of seconds	3.7
Cycloid propagation	<40 kPa	***Hoppa et al.*** (1999a)	4.3.1
Normal faults	>6–8 MPa	***Nimmo and Schenk*** (2006)	4.3.3
Band rifting	0.3–2 MPa	***Stempel et al.*** (2005); *Nimmo* (2004d)	4.3.3

References in bold denote observationally constrained values.

TABLE 2. Source and magnitude of strain rates.

Source	Magnitude	Notes/References	Section
Diurnal tides	2×10^{-10} s^{-1}	*Ojakangas and Stevenson* (1989a)	3.1
Opening of bands	10^{-15}–10^{-12} s^{-1}	*Nimmo* (2004d); ***Stempel et al.*** (2005)	4.3.3
Nonsynchronous rotation	<~10^{-14} s^{-1}	***Hoppa et al.*** (1999b)	3.2, 4.3.2
Undeformed craters	<10^{-16} s^{-1}?	Assumes <10% local strain and crater age of 30 m.y.; only applies to postcratering deformation	4.3.5

References in bold denote observationally constrained values.

ning, differentiation, core contraction, giant impacts, etc.) are unlikely to be relevant.

Tables 1 and 2 summarize the results of this section, and also include observational constraints on the stress mechanisms and magnitudes (see section 4.3).

3.1. Diurnal Tidal Stresses

The single most important source of stress arises from the eccentric nature of Europa's 3.55-d orbit (see chapter by Sotin et al.). Because Europa's orbital period equals its rotation period, Jupiter stays almost fixed in the sky as observed from Europa. However, the orbital eccentricity (e = 0.0101) means that Jupiter appears to approach and recede, and also to librate slightly about its mean position in the sky. As a result, the tidal bulge raised by Jupiter (which has a mean amplitude of roughly 1 km) also changes size and position slightly. This time-varying (or diurnal) component of the tide, which scales with the eccentricity, has an amplitude of 30 m if a subsurface ocean is present (e.g., *Moore and Schubert,* 2000) and a period of 3.55 d. It is responsible not only for tidal heating but has also been invoked to explain many of the tectonic features seen at the surface (see below).

Assuming a homogeneous elastic shell, the diurnal tidal stresses and strains are quite low (on the order of 100 kPa and 10^{-5}, respectively) compared with terrestrial near-surface stresses (which are typically tens to hundreds of MPa). However, because of the short orbital periods, the strain rate (2×10^{-10} s^{-1} globally averaged) (*Ojakangas and Stevenson,* 1989a) is much higher than typical terrestrial strain rates. The peak stresses and strains vary with position on the satellite, with the largest values being found at the poles, and minima at the sub- and antijovian points (*Ojakangas and Stevenson,* 1989a).

3.2. Other Tidal Effects

As long as the icy shell surface does not shift position relative to the time-averaged (permanent) tidal bulge, only the relatively small diurnal tidal stresses are associated with Europa's orbit. However, if the icy shell shifts relative to the tidal bulge, either due to a change in rotation rate or a reorientation of the surface relative to the rotation axis (i.e., true polar wander, or TPW), then larger stresses arise (e.g., *Matsuyama and Nimmo,* 2008).

Although Europa is observed to be approximately in synchronous rotation, its eccentric orbit means that the tidal torques have the potential to drive it to rotate very slightly faster than synchronous (*Yoder,* 1979; *Greenberg and Weidenschilling,* 1984). Although these torques can be opposed by torques on mass asymmetries within the icy shell (*Greenberg and Weidenschilling,* 1984), the tendency of ice to flow means that such asymmetries may not persist for geological periods (*Ojakangas and Stevenson,* 1989a). There is thus some theoretical justification for expecting nonsynchronous rotatation (NSR) to occur on Europa, and the absence of leading-trailing crater asymmetries on other outer solar system satellites (e.g., Ganymede and Callisto) (see *Zahnle et al.,* 2001) suggests that either NSR (or TPW) is a common process, or that the current models of impact fluxes are incorrect.

For Europa, the existence of a subsurface ocean means that the icy shell is decoupled from the solid interior. True polar wander therefore requires a change in the moments of inertia of the icy shell. Large impact basins are one way of achieving this (e.g., *Nimmo and Matsuyama,* 2007), but such features are not seen on Europa. A more interesting mechanism is that the lateral thickness variations expected for a conductive icy shell (section 2.3.1) may result in a shell that is rotationally unstable (*Ojakangas and Stevenson,* 1989a,b). In this case, TPW is expected to take place in an approximately continuous fashion and at a rate determined by the thermal diffusion timescale of the icy shell. Observational evidence for NSR and TPW is discussed in section 4.3.2.

Both TPW and NSR involve motion of the permanent tidal bulge with respect to the solid surface, and are therefore described by the same equations (*Melosh,* 1980).The peak tensional stress is given by (e.g., *Leith and McKinnon,* 1996)

$$\sigma = 6f\mu\left(\frac{1+\nu}{5+\nu}\right)\sin\theta$$

where μ is the shear modulus, θ is the angle of reorientation, ν is Poisson's ratio, and f is the effective flattening of the satellite. This flattening depends on the geometry of reorientation (*Matsuyama and Nimmo,* 2008); for instance, flattening is larger along the line joining the pole to the subjovian point than that joining the pole to the center of the leading hemisphere (e.g., *Murray and Dermott,* 1999). Rather than assuming an instantaneous displacement by an angle θ, more recent analyses have compared the reorientation rate with the rate of stress relaxation for a viscoelastic icy shell to determine the stress magnitudes and patterns (*Harada and Kurita,* 2007; *Wahr et al.,* 2008; chapter by Sotin et al.). In either case, the stress patterns generated by NSR or TPW may be calculated and compared with the orientations and styles of geological features.

The stresses generated are large compared with the diurnal tidal stresses, a few MPa compared to roughly 100 kPa, and thus are likely to dominate the tectonic record. However, because the timescales for NSR or TPW are much longer than diurnal timescales, $>10^4$ yr (*Hoppa et al.,* 1999b), the resulting strain rates are small and the tidal dissipation negligible.

3.3. Lateral Shell Thickness Variations

A conductive icy shell may display lateral variations in shell thickness owing to the spatially variable surface temperature and tidal dissipation. Local lateral shell thickness variations may also arise, e.g., due to oceanic plumes of hot water causing local melting of the icy shell (*O'Brien et al.,* 2002; but cf. *Goodman et al.,* 2004). As on Earth, these lateral shell thickness variations result in buoyancy forces that can drive flow from regions where the shell is thick to regions where the shell is thin. The force F per unit length is given by (e.g., *Buck,* 1991)

$$F \approx g\Delta\rho t_c \Delta t_c$$

where g is the acceleration due to gravity; t_c and Δt_c are the shell thickness and shell thickness contrast, respectively; and $\Delta\rho$ is the density contrast between icy and water. The stress, if uniform across the entire icy shell, is given by $g\Delta\rho\Delta t_c$. For a 1-km shell thickness variation, consistent with expectations from variations in tidal heating (section 2.3.1), the resulting stress is ~100 kPa, comparable to the diurnal tidal stresses.

A consequence of these stresses is that the ductile ice near the base of the shell can flow and reduce the shell thickness contrasts. For Newtonian materials the timescale τ for this to happen for a feature of wavelength λ is given by

$$\tau \sim \frac{\eta\lambda^2}{g\delta^3\Delta\rho}$$

where δ is the thickness of the layer in which flow takes place and is proportional to the shell thickness for conductive shells (*Stevenson,* 2000; *Nimmo,* 2004a). The timescale is decreased if the presence of an elastic surface layer is taken into account (*Cathles,* 1975). Because ice has a low viscosity near its melting point, the flow timescale for local shell thickness variations is geologically short unless the shell is only a few kilometers thick. Figure 4 plots the flow timescale as a function of wavelength and shell thickness and shows that a 100-km wavelength feature will disappear over 50 m.y. unless the shell is thinner than 6 km. Global-scale (~4000 km) shell thickness variations, however, can persist for the same period for shells thinner than roughly 60 km. These results are for conductive shells; if the shell is convecting, relaxation will be geologically instantaneous because of the low viscosity of the warm isothermal interior. Thus, lateral shell thickness contrasts can only be maintained in relatively thin, conductive icy shells.

3.4. Density Contrasts and Convection

Lateral variations in density within the icy shell are a possible source of stress. In all cases, the stress σ is given by $\sigma = dg\Delta\rho$ where $\Delta\rho$ is the density contrast and d is the thickness of the anomalous layer.

Thermal and compositional convection are two sources of lateral variations in density. Because the thermal expansivity of ice is relatively low ($\sim 10^{-4}$ K^{-1}), the expected 20-K temperature contrast provided by thermal convection alone (section 2.3.2) only results in a density contrast of 2 kg m^{-3} (0.2%). Assuming a typical convective boundary sublayer thickness of 2 km (*Nimmo and Manga,* 2002) results in maximum stresses of ~5 kPa. If the icy shell is salty,

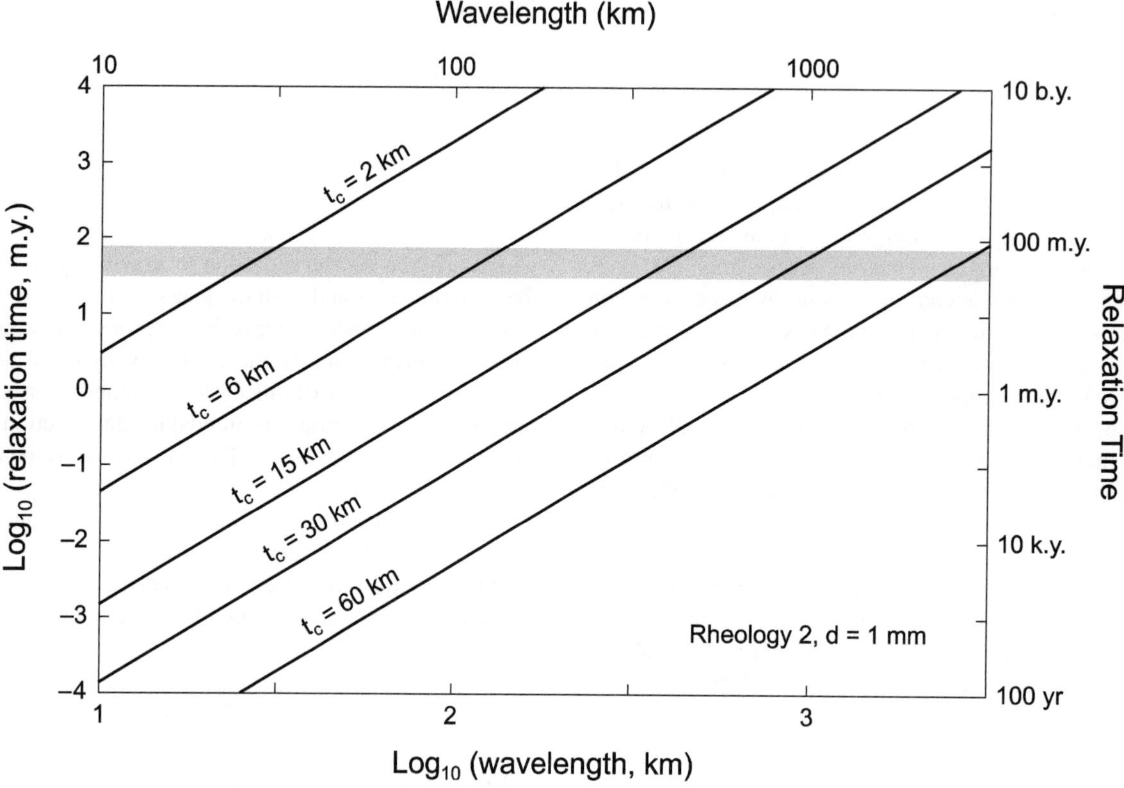

Fig. 4. Relaxation timescale as a function of conductive shell thickness t_c and topographic wavelength. Modified from *Nimmo* (2004a). The shaded band indicates the *Zahnle et al.* (2003) estimated age of Europa's surface; d indicates ice grain size, and the calculations are carried out for a non-Newtonian ice rheology [GBS-assisted basal slip, in the nomenclature of *Goldsby and Kohlstedt* (2001)].

the coupling between the thermal and compositional convection can lead to larger variations in density. The spatial variations of salinity that would drive such thermochemical convection can arise in several ways: segregation of salt through melting within the convecting part of the icy shell (*Pappalardo and Barr,* 2004); injection of salty water into fractures at the base of the icy shell (*Pappalardo and Barr,* 2004); and incorporation of salts into the base of a thickening icy shell (*Spaun and Head,* 2001). Density differences of a few to as much as 10% might be possible (*Han and Showman,* 2005) and would result in correspondingly larger stresses than those produced by thermal convection alone.

3.5. Flexure

If the ice has a finite elastic strength, then the response of the shell to an applied load, whether on the surface or in the subsurface, is distributed over a distance controlled by the elastic thickness T_e of the shell. The stresses applied by the load are partly or completely balanced by elastic bending stresses within the shell, which depend on the shell thickness and Young's modulus (section 2.1.3) and the local topographic curvature. Thus, observations of apparently flexural features can be used to determine the local elastic thickness and stresses present (see section 4.1.1).

3.6. Shell Thickening

Because ice is less dense than water, if the floating icy shell thickens then the surface must move radially outward. Since this outward motion is taking place on a sphere, it is accompanied by global isotropic expansion. This mechanism is an important source of extensional stress for satellites in which the icy shell thickness has increased with time (decreasing thickness leads to corresponding compressional stresses). For thin shells the tangential surface stress σ_t may be derived from *Nimmo* (2004b)

$$\sigma_t \approx \frac{E\Delta t_c \Delta\rho}{(1-\nu)R\rho_w}$$

where Δt_c is the change in shell thickness, ν is Poisson's ratio, R is the satellite radius, and ρ_w is the density of water. For Europa, a 1-km change in shell thickness results in a stress of roughly 600 kPa if E = 6 GPa, larger than diurnal tidal stresses. Shell thickening also generates thermal stresses because of the ice cooling, but these are small compared with the volume-change effect (*Nimmo,* 2004b; *Kimura et al.,* 2007). A further consequence of shell thickening is that the pressure in the underlying ocean will increase, with the pressure being determined from a balance between compression of the ocean water and elastic expansion of the icy shell. *Manga and Wang* (2007) show that while the excess pressure can exceed 10^4 Pa, it will not become large enough to allow liquid water to reach the surface. This issue is discussed further in section 4.3.4 below.

3.7. Impacts

Because of Europa's young apparent surface age, it is relatively lightly cratered and thus impact craters are less important as a source of stress than on most other icy bodies. The peak stress generated by an impact scales as ρv^2, where ρ is the impactor density and v the impact velocity (*Melosh,* 1989). Although the peak stresses are thus in the TPa range, they decay rapidly over lengthscales of a few times the impactor radius and timescales on the order of tens of seconds, and are thus only local stress sources.

The topography generated by a crater also gives rise to buoyancy forces (section 3.3), so craters have a tendency to relax over geological time (e.g., *Dombard and McKinnon,* 2000). Such relaxation typically involves surface deformation; however, one of the most puzzling observations is that europan craters appear very rarely to be tectonized (*Moore et al.,* 2001; chapter by Bierhaus et al.) (section 4.3.5).

4. OBSERVATIONAL CONSTRAINTS ON ICY SHELL CHARACTERISTICS

Voyager and then Galileo provided images that make it possible to identify processes that may operate within the icy shell and to connect models for these processes to the properties and evolution of the icy shell. We will see that answering some of the fundamental questions remains a challenge: How thick is the ice? Does the icy shell convect? Does cryovolcanism occur? In part the challenge arises because we have access only to images of the surface (and topography in some areas). Gravity, seismological, and heat flow data are not available, in contrast to Earth. A second complication arises because many geodynamic processes are controlled by transitions between solid-like and fluid-like behavior, and cannot distinguish between liquid water and viscous ice. A final complication, discussed in section 5, is that the icy shell thickness is likely changing in time.

4.1. Shell Thickness

Perhaps the most important and least-well-known geophysical characteristic of Europa's icy shell is its thickness (*Pappalardo et al.,* 1999). The thickness is important because it controls the degree of exchange between the surface and the ocean, which in turn will affect the ocean's astrobiological potential (*Chyba and Phillips,* 2002) and the probability of detecting any astrobiological activity (*Figueredo et al.,* 2003). The design of future instruments, especially radar (*Chyba et al.,* 1998; *Moore,* 2000), intended to penetrate the ice layer will be significantly affected by the assumed shell thickness. Measuring the shell thickness is also important because it provides clues to Europa's thermal evolution and current state (see section 5). In particular, estimating the shell thickness places strong constraints on the shell's temperature structure and vice versa (section 2.3). Here, we will describe the different methods used to infer Europa's icy shell thickness; an excellent in-depth review may be found in *Billings and Kattenhorn* (2005). As will become obvious, there is currently no consensus on the present-day shell thickness; nonetheless, we will try to reconcile the disparate observations.

4.1.1. Effective elastic thickness. As discussed in section 2.1.5, the amplitude and wavelength of the icy shell's response to an applied load yields the effective elastic thickness T_e of the shell (e.g., *Watts,* 2001). The effective elastic thickness is always less than the total icy shell thickness because parts of the shell deform in a brittle or ductile rather than elastic fashion. Nonetheless, thicker shells tend to generate larger values of T_e and vice versa. Estimating values of T_e on icy satellites is relatively straightforward; converting T_e to total shell thicknesses is considerably less so. Another important point to note is that the elastic thickness obtained is the lowest value since the load was emplaced. Thus, if the shell becomes more rigid with time, the deformation in response to a load applied in the past will only place a lower bound on the present-day elastic thickness. Equally, if the shell has undergone visoelastic relaxation since the load was emplaced (e.g., *Dombard and McKinnon,* 2006), then the estimated value of T_e may differ from that at the time of loading.

Elastic thickness estimates are most often derived using topographic data. On terrestrial planets topography is typically obtained using altimetric radar or lidar. The Galileo spacecraft did not possess such an instrument, and so topography is derived from images using either stereo or photoclinometric (shape-from-shading) techniques. The former technique is less prone to error, except at the longest wavelengths, but is limited because it requires two images of the same region with suitable viewing geometries and resolutions (e.g., *Giese et al.,* 1998). The latter technique has to account for the fact that both slopes and intrinsic albedo contrasts can lead to differences in brightness, and is thus subject to more uncertainties (*Jankowski and Squyres,* 1991; *Kirk et al.,* 2003), but it has an intrinsically higher spatial resolution than the stereo technique.

Various approaches to obtaining T_e values on Europa have been tried, but they all essentially rely on measuring the characteristic wavelength of the shell's response to an applied load (see chapter by Kattenhorn and Hurford). This wavelength is controlled by the flexural parameter, α, which may be directly related to the effective elastic thickness T_e by

$$\alpha = \left(\frac{T_e^3 E}{3g\rho_w(1 - \upsilon^2)} \right)^{1/4}$$

where ρ_w is the density of the underlying fluid. Note that the uncertainty in the correct value of E results in a corresponding uncertainty in T_e: Larger values of E yield smaller

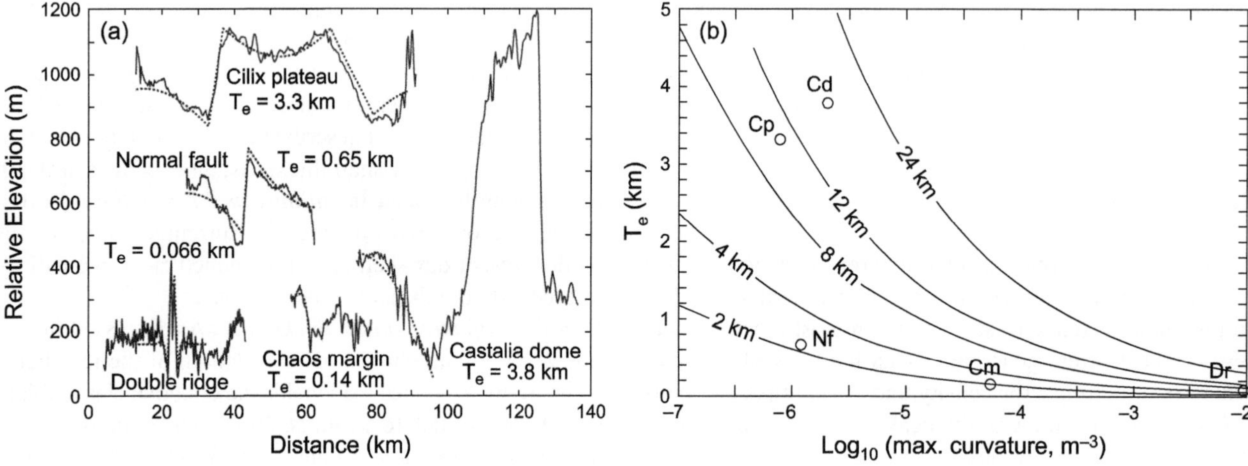

Fig. 5. (**a**) Selected stereo-derived topographic profiles across features on Europa. Solid lines are data; for Cilix, normal fault and chaos margin profiles are averages of multiple individual profiles. Dashed lines are flexural fits, with quoted elastic thickness T_e and an assumed Young's modulus of 6 GPa. References for original profiles are: Cilix (*Nimmo et al.,* 2003b); normal fault (*Nimmo and Schenk,* 2006); Castalia dome (*Prockter and Schenk,* 2005); chaos margin (*Nimmo and Giese,* 2005); double ridge (*Nimmo et al.,* 2003a). (**b**) Effective elastic thickness as a function of maximum curvature and conductive shell thickness (contours). Individual points plotted are of the fitted flexural features shown in (**a**). A strain rate of 10^{-15} s^{-1} was assumed, and the method adopted was that of *Nimmo and Pappalardo* (2004).

T_e values and vice versa. In the figures quoted below, T_e values will be given using E = 6 GPa throughout.

Flexural responses (Fig. 5a) have been identified at four classes of features on Europa: ridges (Figs. 1a,f), chaos regions (Figs. 1b,e), normal faults, and domes/plateaus (Fig. 1a). Studies of ridges yield elastic thicknesses in the range 0.02–0.66 km (*Hurford et al.,* 2005; *Billings and Kattenhorn,* 2005). A small-scale updoming near Conamara Chaos yielded a similar range of T_e (*Williams and Greeley,* 1998), but larger plateaus near Cilix Crater and at Castalia Macula give values of 3.3^{+3}_{-1} km (*Nimmo et al.,* 2003b) and 3.8 km (Fig. 5a), respectively, for E = 6 GPa. Two normal faults gave values of 0.65 km and 0.08 km, respectively, for E = 6 GPa (*Nimmo and Schenk,* 2006). Finally, the margins of the Murias Chaos region (Fig. 1b) yielded a value of 4.6 km, with an uncertainty of ±50% (*Figueredo et al.,* 2002), while the smaller chaos regions imaged in the E15 encounter gave a T_e of 0.06–0.2 km (*Nimmo and Giese,* 2005).

The reasons for this order-of-magnitude variability in T_e estimates are unclear. One possibility is that T_e varies spatially. For instance, ridges may be areas where shear-heating is important (*Gaidos and Nimmo,* 2000; *Han and Showman,* 2008), which could reduce the local rigidity of the ice. On a global scale, variations in the depth to the base of the elastic layer are expected because of spatial variations in tidal heating (section 2.3.1). A second possibility is that T_e varies in time. For instance, if Europa's icy shell is thickening with time, more ancient episodes of deformation would yield a lower value of T_e. Although there are not enough impact craters on the surface to allow variations in surface age to be detected directly, geological observations discussed below (section 5) suggest that a thickening icy shell is possible. A final possibility, which we briefly discuss next, is that the T_e values are affected by the local topographic curvature. In any case, the large variability in T_e estimates results in correspondingly large variations in the estimated shell thickness.

One method of converting T_e estimates to total shell thicknesses is to use the yield-strength envelope approach of *McNutt* (1984). In this technique, the stress-depth profile is calculated for a variety of shell thicknesses (cf. Fig. 2); the second moment of this profile may be related to the effective elastic thickness of the shell.

Figure 5b plots the predicted effective elastic thickness as a function of topographic curvature and conductive icy shell thickness using the yield-strength envelope technique. As expected, thicker shells result in larger values of T_e. However, the topographic curvature also has a significant effect: Higher curvatures result in lower values of T_e for the same shell thickness (because the thickness of the elastic core is reduced — see Fig. 2). This is an important result, because it means that topographic features with different curvatures may produce different T_e estimates even if the underlying shell thickness is the same. Comparison of these theoretical curves with the individual observations from Fig. 5a suggests a conductive shell thickness in the range 2–18 km.

An important feature of the curves shown in Fig. 5b is that they are quite insensitive to the (poorly known) rheological parameters and strain rate. The reason is that the strength envelope is dominated by the brittle portion of the icy shell (*Nimmo and Pappalardo,* 2004). Another feature of these curves is that in the case of convection, the thickness determined is that of the stagnant lid, not the entire shell.

Another approach similar to that outlined above was adopted by *Luttrell and Sandwell* (2006). These authors used the bending moment implied by europan topography to infer the minimum moment associated with the frictional yield-strength envelope (Fig. 2), and thus the minimum thickness of the stagnant part of the icy shell. They obtained a lower bound of 2.5 km, consistent with the range of values derived above. One potential criticism of this work is that the topographic features used were impact craters (*Schenk,* 2002), which may have been modified by processes such as viscous relaxation (e.g., *Dombard and McKinnon,* 2000).

4.1.2. Elevation contrasts and chaos terrain. Looking for flexural signatures is one way of using topographic information to determine shell thicknesses. Another is simply to focus on elevation contrasts. Topography implies vertical stresses. These could be provided by elastic effects (flexure), but the stresses might also arise from dynamic processes (e.g., convection), or static stresses due to density contrasts (section 3.4). Here we will focus on static stresses, while the next section will focus on dynamic effects.

Assuming zero elastic thickness (i.e., isostatic equilibrium), the topography contrast Δh due to a density contrast $\Delta\rho$ within a layer of thickness d is simply $d\Delta\rho/\rho$. Since d cannot exceed the total shell thickness, isostatic elevation contrasts can be used to place bounds on shell thickness. Clearly, it is easier to develop larger topographic contrasts in thick shells rather than thin ones. Furthermore, compositional variations are more likely to result in topographic relief than thermal anomalies because the latter not only generate relatively small density anomalies (section 3.4.1) but are also transient unless a continuous heat source is applied (see *Nimmo et al.,* 2003a).

It is unlikely that $\Delta\rho$ exceeds the ice-water density contrast, roughly 100 kg m^{-3}. Lateral variations in porosity can also generate large density differences, but compaction will limit porosity to the upper quarter to third of the icy shell (e.g., *Nimmo et al.,* 2003a) (see also Fig. 3). A reasonable upper bound for the porosity is 30%, implying maximum depth-averaged density differences of <10%. Since local topographic contrasts on Europa significantly exceed 1 km in some areas (*Schenk and Pappalardo,* 2004) (Fig. 5a), this implies a lower limit on icy shell thickness of 10 km. Put another way, it is hard to envisage a situation in which the shell thickness is as small as the elevation contrasts observed. This logic, although simple, is perhaps the single strongest argument against thin (few kilometers) icy shells. It is, however, predicated on the assumption of zero elastic thickness, which is not necessarily correct (section 4.1.1). Fortunately, in some cases the issue of T_e is irrelevant: The 1-km-high dome at Castalia Macula (Fig. 5a) (*Prockter and Schenk,* 2005) shows a flank trough that implies an elastic thickness of roughly 4 km and a correspondingly larger total shell thickness (about 18 km according to Fig. 5b). Alternatively, an isostatically supported dome 1 km high implies a shell thickness of at least 10 km, using the logic above. Thus, at least at the time features with such large relief formed, the conclusion that the shell was thick appears inescapable.

Global shell thickness contrasts (section 2.3.1) will result in isostatically supported topography and are thus susceptible to the same arguments. However, existing Galileo and Voyager limb profile data reveal no evidence for such long-wavelength topography (*Nimmo et al.,* 2007a). This could be due either to the existence of a thin shell (in which case topographic variations exist but are small), or a thick shell in which lateral flow (section 3.3) has erased initial shell thickness variations.

Isostasy arguments have also been applied to chaos regions (Fig. 1e). These disrupted and topographically variable areas were initially interpreted as blocks floating in liquid, on the basis of their morphology (*Carr et al.,* 1998; *Williams and Greeley,* 1998) and the fact that the blocks have both rotated and translated (*Spaun et al.,* 1998). Block elevation is a few hundred meters; using the same isostatic argument as above yields block thicknesses of 0.2–3 km in Conamara Chaos (*Carr et al.,* 1998; *Williams and Greeley,* 1998). Unfortunately, it is not clear that the melt-through explanation for chaos formation is unique. More recent examinations of chaos topography (*Pappalardo and Barr,* 2004; *Schenk and Pappalardo,* 2004) have invoked diapir-induced deformation. The energy required to generate complete melting of an icy shell would require a concentrated source of heat to be supplied by the underlying ocean, which is dynamically challenging (*Collins et al.,* 2000; *Thomson and Delaney,* 2001; *Goodman et al.,* 2004). Currently neither the melt-through model nor the competing convective diapirism model for chaos formation (*Head and Pappalardo,* 1999) appear to be able to explain all the observations (*Collins et al.,* 2000; *Nimmo and Giese,* 2005). A near-surface low-viscosity substrate is certainly required to allow the chaos blocks to be reoriented and repositioned; however, because of a lack of time information, either ductile ice or liquid water could be the substrate.

4.1.3. Onset of convection. One method of inferring the icy shell thickness relies on identifying circular surface features (domes and pits, Fig. 1a) typically ~10 km in diameter as the products of convection within the icy shell. For an icy shell with a fixed temperature difference between top and bottom, the vigor of convection is governed by the Rayleigh number Ra (see *Solomatov,* 1995; chapter by Barr and Showman), and convection only occurs for Ra > 10^3. Thus, if the other parameters in Ra can be estimated, a lower bound can be placed on the shell thickness for convection to occur (e.g., *Pappalardo et al.,* 1998; *McKinnon,* 1999; *Tobie et al.,* 2003). The main difficulty is in estimating the effective viscosity of the convecting region η, which for ice is both temperature- and stress-dependent, and also depends on the unknown grain size (section 2.1.2). Most models find that a minimum shell thickness of a few tens of kilometers is required to initiate convection (e.g., *McKinnon,* 1999; *Hussmann et al.,* 2002; *Tobie et al.,* 2003; *Moore,* 2006); models incorporating non-Newtonian rheologies tend to require thicker layers (e.g., *Barr et al.,* 2004).

A more fundamental issue is whether the domical features observed are really caused by convection, particularly given that they have amplitudes of several hundred meters (*Schenk and Pappalardo,* 2004). The convective stresses responsible for uplift are controlled by the temperature differences within the convecting interior, which are typically only 20 K (section 2.3.1). The resulting stresses do not exceed 0.1 MPa (e.g., *Tobie et al.,* 2003; *Showman and Han,* 2004), which will generate at most 100-m topography. This value is actually an overestimate, because the rigid stagnant lid significantly reduces the observed surface deformation, typically to ~10 m (*Nimmo and Manga,* 2002), and warm rising material stalls at the base of this lid, kilometers beneath the surface, unless plastic yielding is invoked (*Showman and Han,* 2005). Furthermore, convection simulations show that topographic lows in response to downwellings have larger amplitudes than the topographic highs (*Showman and Han,* 2004), which does not appear to be the case for domes and pits. Thus, simple convective models currently cannot explain this class of surface features on Europa.

One proposed solution is that tidal dissipation may be concentrated in regions of upwelling, leading to locally larger temperature contrasts, higher stresses, and possibly melting (*Sotin et al.,* 2002; *Tobie et al.,* 2003). Another is that thermal convection may be aided by compositional convection (*Pappalardo and Barr,* 2004; *Han and Showman,* 2005), with the compositional contrasts plausibly arising from partial melting. Reconciling surface observations with sufficiently detailed numerical models of convection is likely to be an active area of research in the future. At present, no convincing alternatives to convection as the source of the domes and pits have been proposed. Equally, though, it is clear that the convection does not take the simple form shown in most models, but probably involves thermochemical, plastic yielding, and/or tidal effects.

4.1.4. Fracture penetration. An early argument for the icy shell thickness assumed that extensional features were generated by tidally driven episodic upwelling of the underlying ocean, which required fractures to propagate to the base of the icy shell (*Greenberg et al.,* 1998). Similarly, the arcuate nature of cycloids has been explained by appealing to fractures propagating due to diurnal tidal stresses (*Hoppa et al.,* 1999a).

It is typically assumed that cracks will propagate to a depth where the overburden pressure is similar to the tensile stress relieved by the crack (e.g., *Gaidos and Nimmo,* 2000). For diurnal tidal stresses of 40 kPa, the implied penetration depth is thus ~35 m. Actual tensile cracks will penetrate deeper for two reasons. First, the tensile stress at the crack tip increases as the tensile stress from the fractured region is transferred to the unfractured region. Second, the presence of a traction-free bottom to the icy shell can increase fracture penetration depths compared to the depth they could penetrate in a half-space (*Lee et al.,* 2005; *Qin et al.,* 2007). We caution that all these analyses assume a uniformly stressed icy shell, whereas the lower part of the icy shell, because of its low viscosity, may have much smaller tangential stresses. Because of the low viscosity of ice near the base of the shell, crack initiation (or propagation) there is difficult. However, if such cracks can be initiated, as they propagate upward they will fill with water. The normal stresses on the crack walls are thus reduced, and cracks can much more easily reach the surface (e.g., *Crawford and Stevenson,* 1988). In this case, for tidal stresses of 100 kPa a 2-km-deep fracture could potentially result (*Qin et al.,* 2007). In summary, given the current uncertainties in whether liquid water is required to form surface features, and also in the mechanical behavior of the icy shell, the fracture propagation argument does not place a strong constraint on the shell thickness.

4.1.5. Thermal equilibrium. In theory, the equilibrium thickness of the icy shell on Europa may be determined by balancing the heat transport through the shell (due to conduction or convection) against the heat produced within the shell (by tidal dissipation) and within the silicate interior (*Ojakangas and Stevenson,* 1989a; *Hussmann et al.,* 2002; *Spohn and Schubert,* 2003; *Tobie et al.,* 2003; *Moore,* 2006). Figure 6 shows a typical example of heat production and heat flux as a function of shell thickness. The former increases when convection initiates, because the warm isothermal convecting interior results in a greater volume of ice close to the viscosity at which dissipation is maximized (section 2.1.4). The reduction in heat production for large shell thicknesses is due to the rigidity of the shell reducing the amplitude of the tidal deformation; however, some heat is still generated even when the ocean thickness is zero

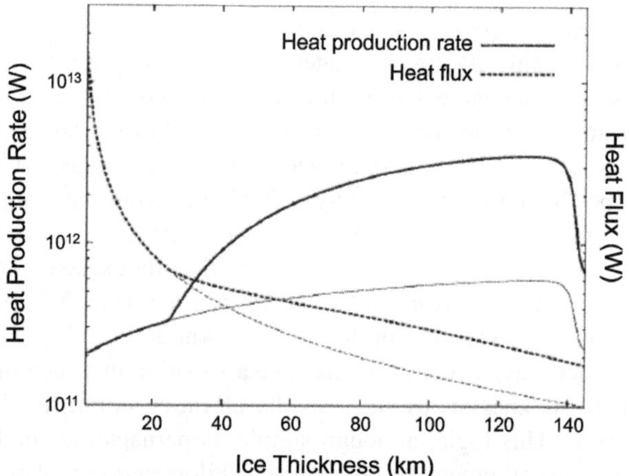

Fig. 6. Heat production and surface heat flux as a function of icy shell thickness, from *Hussmann et al.* (2002). Ice reference viscosity (at 273 K) is 10^{13} Pa s. Convection initiates at 27 km and thermal equilibrium is reached at 35 km shell thickness; initiation of convection may lead to a discontinuous jump in heat flux (*Mitri and Showman,* 2005). Thin lines denote heat flux and production in the (metastable) conductive case. Heat production decreases for thick icy shells because the rigid lid reduces tidal deformation.

(shell thickness 145 km). Heat transport is conductive for shell thicknesses <27 km, and convective thereafter. The convective heat flux decreases with increasing shell thickness because the temperature of the ice at the base of the shell decreases with increasing pressure, leading to an increase in ice viscosity.

Assuming that dissipation within the silicate interior is negligible, equilibrium shell thicknesses are a few tens of kilometers (e.g., *Hussmann et al.,* 2002; *Moore,* 2006) and generally imply convection. However, it is also possible that Europa's mantle resembles that of Io (e.g., *Greenberg and Geissler,* 2002), which is highly dissipative (see section 2.2) and would result in a shell thickness of a few kilometers. Since the state of Europa's silicate mantle is currently unknown (see chapter by Moore and Hussmann), models of the kind shown in Fig. 6 are unlikely to resolve the shell thickness debate.

4.1.6. Impacts. A final method of estimating shell thickness is to use impacts (see chapter by Schenk and Turtle). Large impacts affect deeper regions within the icy shell than smaller impacts; impacts that penetrate the entire shell might be expected to have morphologies distinct from those that do not. A study of crater depth:diameter ratios as a function of crater size yielded two unexpected breaks in slope at diameters D = 8 km and D = 30 km for Europa (*Schenk,* 2002). Although the larger transition diameter was interpreted to imply a shell thickness of 19–25 km, the link between a reduction in depth:diameter ratio and penetration to an underlying ocean has not yet been demonstrated, for instance, by numerical simulations. Numerical simulations that evaluated the preservation of central peaks for europan impact structures placed a lower bound on the shell thickness of 3–4 km (*Turtle and Pierazzo,* 2001), later updated to 5–19 km (*Turtle and Ivanov,* 2002).

4.1.7. Summary. Different methods for inferring the icy shell thickness on Europa generate radically different answers. Part of this variability may be due to temporal or spatial variations in shell properties. Elastic thickness determinations differ partly because of the different topographic curvature of different features, but also suggest a shell thickness range of 2–18 km (Fig. 5b). Shell thicknesses as small as 2 km are supported by geological observations of cycloids (because of the limited fracture depths implied by diurnal tidal stresses) and ridges (from T_e estimates). A shell this thin requires significant tidal heating within Europa's silicate interior. Shell thicknesses of a few tens of kilometers are consistent with observations of impact craters, apparently convective features, and kilometer-scale local elevation contrasts. The latter in particular are a strong argument for a thick shell, which is likely if dissipation in the silicate interior is negligible.

Finally, we note that there are nontectonic constraints on the icy shell thickness. The ocean must be at least a few kilometers thick to explain the magnetometer observations (*Zimmer et al.,* 2000), while *Schilling et al.* (2008) used the same observations to infer an upper bound on the ocean thickness of 100 km. *Hand and Chyba* (2007) obtained an upper bound on the present-day icy shell thickness of 15 km, although this result depends on the extent to which the apparently strong induction response of the ocean can be separated from other effects such as ionospheric noise.

4.2. Icy Shell Structure

Although the thickness of Europa's icy shell is certainly its most important characteristic, various other properties, such as its porosity structure, composition, and lateral variability, may also be constrained by surface observations.

4.2.1. Porosity and mechanical properties. Some bands appear to be elevated with respect to their surroundings, suggesting that they contain lower-density material. *Nimmo et al.* (2003a) investigated several potential sources of reduced density, and concluded that bands contain either substantial porosity or cleaner ice relative to the surroundings. Structural features identified as anti-cracks may have formed through reduction of near-surface porosity (*Kattenhorn and Marshall,* 2006). The existence of substantial near-surface porosity is also consistent with Earth-based radar backscatter data (*Black et al.,* 2001).

A consequence of porosity is that both the tensile strength (*Lee et al.,* 2005) and elastic moduli (*Hessinger et al.,* 1996) of ice are reduced. *Lee et al.* (2005) argue that the reduction in ice strength allows cycloidal fracture propagation due to diurnal tidal stresses; *Nimmo and Schenk* (2006) suggest that normal fault displacement:length ratios are characteristic of materials with shear moduli significantly smaller than that of intact ice. Thus, the near-surface of Europa is likely to show strong gradients in porosity and seismic velocity with depth, which is important for future seismic (*Lee et al.,* 2003; *Kovach and Chyba,* 2001; *Panning et al.,* 2006) and electromagnetic (*Chyba et al.,* 1998; *Eluszkiewicz,* 2004) studies. Additionally, a thick, high (tens of percent) porosity layer, if present, implies low thermal conductivity, which results in elevated temperatures in the subsurface and can in some cases result in melting within a few kilometers of the surface (*Nimmo and Giese,* 2005).

4.2.2. Compositional variations. Since Europa's surface apparently contains non-ice contaminants (*McCord et al.,* 1998), a natural source of density contrasts is compositional variations. As discussed in section 2.4, many of these contaminants lower the ice melting temperature; thus, warming of contaminated ice will initially produce dense, contaminant-rich melts (e.g., *Head and Pappalardo,* 1999; *Nimmo et al.,* 2003a). Drainage of these melts will produce clean, lower-density ice and lead to potential topographic variations. Thus, localized warming may cause uplift in a contaminant-rich icy shell.

There are several settings where such uplift may be occurring. Domical features and chaos regions have both been suggested as areas where melting-induced compositional contrasts may play a role (sections 4.1.2 and 4.1.3), although at least in the case of chaos regions it proves very difficult to generate the required near-surface melting because of the low surface temperatures (*Nimmo and Giese,* 2005).

Another class of features where compositional variations may be important is bands (Figs. 1c,d). As noted above, several bands appear to be elevated with respect to their surroundings (*Prockter et al.,* 2002). Thermal temperature contrasts are likely to decay on million-year timescales, suggesting that compositional variations are a more likely source of the relief (*Nimmo et al.,* 2003a). If bands are the result of complete (deep) fracturing of a contaminant-rich icy shell and subsequent refreezing of the water (*Sullivan et al.,* 1998), the slowly refreezing ice will not incorporate contaminants and may thus be cleaner than the surrounding material. Alternatively, localized shear-heating (*Nimmo and Gaidos,* 2002; *Han and Showman,* 2008) or upwelling of warm ice (*Head et al.,* 1999) could result in partial melting and loss of dense, salt-rich brines, again leading to compositional density contrasts.

4.2.3. Lateral temperature variations. There are several reasons why surface or subsurface temperatures might vary laterally. Tidal heating can be localized, for instance, through focused dissipation in a low-viscosity diapir (*Sotin et al.,* 2002) or due to friction along individual faults (*Gaidos and Nimmo,* 2000). Lateral variations in thermal conductivity or shell thickness would cause lateral temperature contrasts. Extrusion or intrusion of warm material, as may have happened at bands (*Prockter et al.,* 2002) and certain lobate features (*Figueredo et al.,* 2002; *Miyamoto et al.,* 2005), would cause a transient temperature anomaly.

Double ridges (Figs. 1a,f) are features where local temperature anomalies are likely to occur. One possibility is that these ridges form by some combination of linear diapirism (*Head et al.,* 1999) and shear heating (*Nimmo and Gaidos,* 2002). The latter would result in a transient thermal and topographic anomaly, whereas the former would likely involve flow of ice and permanent uplift. Alternatively, ridges may form by complete cracking of the icy shell and tidal pumping of ice masses to the surface (*Greenberg et al.,* 1998). In this case the thermal anomaly (due to warm water rising to near the surface) would be renewed each tidal cycle. If ridges are indeed the source of local thermal anomalies, this effect may be important in interpreting the local elastic thickness, which may therefore be smaller than the background value (see section 4.1.1). Recent support for the idea of warm ridges has come from Enceladus, where the similar-looking "tiger stripes" are measurably warmer than their surroundings (*Spencer et al.,* 2006), perhaps as a result of shear heating (*Nimmo et al.,* 2007b).

Europa's surface temperature is controlled by the solar energy input (about 50 W m^{-2}). To cause an observable temperature anomaly, subsurface heat fluxes would have to exceed roughly 1 W m^{-2} (*Spencer et al.,* 1999). Flows emplaced at the surface will cool conductively, and at time t after emplacement the heat flux is roughly $\rho C_p \Delta T(\kappa/t)^{1/2}$ where C_p is the specific heat capacity, κ is the thermal diffusivity, and ΔT the initial temperature contrast. The thermal anomaly due to a solid-state icy flow will therefore be visible for roughly 10^3 s after emplacement, which makes detection of such an anomaly by a spacecraft very improbable, unless it is being renewed each orbit. More sophisticated models by *Abramov and Spencer* (2008) result in anomaly lifetimes of a few tens or hundreds of years, still small compared to the likely surface age. On the basis of these arguments, the apparent nighttime thermal anomaly detected by *Spencer et al.* (1999) is more likely due to surface thermal inertia variations than an endogenic heat source.

Finally, lateral surface temperature and tidal heating variations might be expected to lead to global shell thickness and topography variations. However, no such variations have been detected in existing limb profile data (*Nimmo et al.,* 2007a) (section 4.1.2).

4.2.4. Cryovolcanism. Despite careful investigation, there is no undisputed evidence for cryovolcanism on Europa. Most candidate features identified from lower-resolution Voyager data were generally revealed as noncryovolcanic when imaged with higher-resolution Galileo instruments (*Fagents,* 2003). Small, smooth low-albedo regions may be the result of minor flooding (and potentially subsequent drainage) by low-viscosity cryovolcanic liquids (*Greeley et al.,* 2000; *Prockter and Schenk,* 2005; *Miyamoto et al.,* 2005). Some linear bands have low-albedo margins that may be the result of explosive cryovolcanism (*Fagents et al.,* 2000) (Fig. 1a), although this interpretation is not unique. Some apparently lobate features, including the ~100-km-wide Murias Chaos region (Fig. 1b), may be the result of extrusive cryovolcanic activity (*Figueredo et al.,* 2002; *Miyamoto et al.,* 2005), although again this interpretation is uncertain.

The lack of unambiguous cryovolcanism on Europa is perhaps not surprising in view of the large density contrast between melt (water) and solid (ice) (section 1). However, the detection of water ice eruption plumes at Enceladus (*Porco et al.,* 2006) serves as a reminder that cryovolcanism can occur, although no such active features were detected at Europa by Galileo (*Phillips et al.,* 2000). Cryovolcanism remains an exciting possibility because it may deliver water from the ocean to the surface.

4.3. Sources and Magnitudes of Stress, Strain, and Strain Rates that Create Tectonic Features

By comparing the morphology of surface features with theoretical models for their formation, we can obtain constraints on the stresses and strain rates that formed them. In some cases, this allows us to test different hypotheses for the formation of surface features, identify the dominant sources of stress, or to constrain the thermal and mechanical properties of the icy shell. Tables 1 and 2 summarize sources and magnitudes of stress and strain rates, respectively.

4.3.1. Diurnal stresses. At a given location of the surface of Europa, the orientation and magnitude of the diurnal tidal stresses changes over time. If the stresses, possibly in combination with other stresses, exceed the tensile strength of ice, a crack can form. *Hoppa et al.* (1999a, 2001) proposed that cycloids, arcuate lineaments, form when these stresses exceed the tensile strength of ice so that a crack forms perpendicular to the smallest horizontal stress. The curved shape is a result of the crack propagating through the time-varying stress field. From a calculated stress field, it is possible to predict the trajectory of synthetic cycloids by specifying the stress at which the cycloid starts to form, the stress at which it stops propagating, and the speed at which it propagates. The theoretical results can only match observations by using very low (apparent) crack propagation speeds, typically a few kilometers per hour (e.g., *Hoppa et al.*, 1999a, 2001; *Greenberg et al.*, 2003), rather than the expected speed of crack propagation of ~1 km/s (e.g., *Lee et al.*, 2005). Cycloids, if formed in this manner, are thus likely the concatenation of many smaller cracks (*Lee et al.*, 2005). Once a crack is formed, the time evolution of the stress field results in the crack experiencing stresses that may allow for strike-slip motion (*Marshall and Kattenhorn*, 2005) or can initiate new cracks (*Kattenhorn and Marshall*, 2006).

The typical starting and stopping stresses for crack propagation in these simulations are ~10^4 Pa, much lower than the tensile strength of intact lab ice, between about 1 and 3 MPa (*Hawkes and Mellor*, 1972; *Schwartz and Weeks*, 1977). One possible explanation for such low stresses causing fracture is that the tensile strength of the europan icy shell is low because it is porous or fractured (*Lee et al.*, 2005) (section 4.2.1). Another more likely possibility, discussed below, is that there are additional sources of extensional stress that bring the icy shell closer to failure.

4.3.2. Other sources of stress needed to explain global fracture patterns. One common conclusion of models that attempt to reproduce the location and shape of global lineaments is that the icy shell must have experienced some reorientation (*Helfenstein and Parmentier*, 1985; *McEwen*, 1986; *Leith and McKinnon*, 1996; *Geissler et al.*, 1998a). Models for cycloid formation by diurnal tides also favor some reorientation (e.g., *Hoppa et al.*, 2001; *Kattenhorn*, 2002; *Sarid et al.*, 2004; *Hurford et al.*, 2007). A recent study by *Schenk et al.* (2008) demonstrated that the location of 2350-km-diameter, 50-km-wide circular troughs on Europa could be explained by ~80° of polar wander, resulting in stresses of up to 3 MPa.

Nonsynchronous rotation has not been observed directly on Europa, but comparison of Voyager with Galileo images suggests that the NSR timescale must exceed 10^4 yr (*Hoppa et al.*, 1999b). The resulting averaged strain rate is thus $<\sim10^{-14}$ s^{-1}. In the absence of dissipation, polar wander can occur rapidly (~10^3 yr) (*Ojakangas and Stevenson*, 1989b), leading to correspondingly higher strain rates (~10^{-13} s^{-1}).

Polar wander caused by variations in icy shell thickness and nonsynchronous rotation generated by small mean tidal torques can both cause reorientation (section 3.2). Both polar wander and nonsynchronous rotation also generate stresses much larger than the diurnal tides. In fact, the stresses are comparable to the tensile strength of intact ice and it may thus not be necessary to appeal to additional processes or properties (e.g., porosity) to weaken ice in order to generate tectonic features.

Thickening of a floating icy shell will add an isotropic extensional stress of up to several MPa to the background stress field (*Nimmo*, 2004b) (section 3.6). This effect will thus have no effect on global stress patterns, but will bring features closer to tensile failure.

The tidal dissipation that acts to synchronize Europa's rotation will also reduce its obliquity, but interactions between the Galilean satellites can force Europa to develop a finite (as of yet unmeasured) obliquity (*Bills*, 2005) and undergo librations (*Van Hoolst et al.*, 2008; chapter by Bills et al.). *Sarid et al.* (2006) show that a small (<1°) finite obliquity can explain why cycloidal cracks cross Europa's equator and can reproduce their observed shapes. Obliquities of this magnitude generate stresses comparable to those caused by diurnal tides.

4.3.3. Stresses and strain rates inferred from extensional features. Extension-related features share many morphological similarities to mid-ocean ridges on Earth (e.g., *Prockter et al.*, 2002) (Figs. 1c,d). Using a mid-ocean-ridge analogy, in which the depth of the brittle-ductile transition is governed by plate cooling, *Stempel et al.* (2005) infer opening rates of 0.3–30 mm/yr and strain rates of 10^{-14}–10^{-12} s^{-1} based on the observed fault spacing. Other rifting models (e.g., *Nimmo*, 2004d; *Manga and Sinton*, 2004) draw roughly similar conclusions but admit the possibility of lower strain rates. The inferred strain rates are much smaller than those from diurnal tides (2×10^{-10} s^{-1}; section 3.1), implying that other sources of strain play a key role.

The magnitude of stress inferred from rifting models also provides constraints on the source of stress. *Stempel et al.* (2005) use their value of the tensile strength, 0.5–2 MPa, to infer that nonsynchronous rotation is the driving mechanism for band opening. *Nimmo and Schenk* (2006) compared the topography produced by normal faults with that of flexure models and infer driving stresses that exceed 6–8 MPa. Extensional stresses this large can only be provided by stresses either from cooling icy shells (section 3.6) or possibly NSR/TPW (section 3.2).

4.3.4. Implications of the paucity of compressive tectonics. A noteworthy aspect of tectonic features on icy satellites such as Europa is the near absence of features uniquely attributed to compressive stresses. One exception on Europa are small-amplitude folds (*Prockter and Pappalardo*, 2000). The strain accommodated by these folds, however, is much smaller than the strain recorded in bands and they are a minor means to compensate for extension documented elsewhere (*Dombard and McKinnon*, 2006). Cross-cutting relationships across some bands have been interpreted to indicate contraction (*Greenberg*, 2004) and convergence may have occurred both along large, predominantly strike-slip faults (*Sarid et al.*, 2002) and some ridges (*Patterson et al.*, 2006). While bands are typically interpreted as being extensional features, *Schulson* (2002) proposed that some wedge-shaped bands are generated by compressive failure.

There are two possible explanations for the small number of compressional features observed on the surface. First, the stress field truly could be truly dominated by extension, which, combined with the lower failure strength of ice under tension than compression, leads to more failure and deformation under extension. Such a situation would occur if, for instance, Europa's shell had been thickening with time (sections 3.6 and 5). However, while the extensional stresses caused by icy shell thickening are large, the global strains are small (few tenths of a percent) (*Nimmo,* 2004b) compared with the estimated global strain across bands of about 5% (*Figueredo and Greeley,* 2004).

Second, compressional tectonic features could be present but difficult to detect. For instance, if each band contained a small compressional component, the global compressional strain accommodated could be significant. The true solution is a probably a combination of these two explanations.

4.3.5. Impact craters. If lower bounds on strain rates associated with bands are global, we would expect significant deformation of at least some impact craters. Assuming a mean age of 10 Ma for craters, and a lower bound on strain rates of 10^{-15} s^{-1}, mean strains should exceed or be close to 1. No impact craters show strains even close to this value (*Moore et al.,* 2001). This implies that surface deformation is highly localized. Moreover, given the ubiquity of tectonic features on Europa and their comparative rarity within craters, the strength of ice below craters may be higher than that typical of the icy shell. Alternatively, most of the tectonism on Europa may have occurred over a short time period during the earliest history preserved on the present icy shell, so that subsequent impact craters remain relatively undisturbed.

5. EVOLUTION OF THE SHELL THROUGH TIME

The manner in which Europa's shell and interior evolved to their present-day states represents a major unsolved problem for at least three reasons. First, the present-day state, especially of the silicate interior, is poorly known. Second, Europa's surface may only record the last 1% of its existence, so there are few constraints on its earlier history. Third, the thermal and orbital evolution of Europa are intimately coupled in a manner that is nontrivial to model (*Hussmann and Spohn,* 2004; see chapter by Moore and Hussmann). Nonetheless, both observational and theoretical approaches are starting to yield important insights into Europa's evolution.

Based on observational approaches, there are now several lines of evidence that Europa's shell is not in steady state. First, although the young surface age could be the result of an equilibrium between the creation of impact features and their destruction (presumably through tectonism), the fact that very few craters are tectonically deformed suggests some kind of rapid resurfacing event followed by relative quiescence (*Figueredo and Greeley,* 2004). A similar scenario has been proposed for Venus but with a much longer time interval between resurfacing events (*Strom et al.,* 1994). Second, as noted above, some of the discrepancies in T_e estimates may be due to temporal variations in shell properties. For example, *Hurford et al.* (2005) find that flexure at younger ridges requires a thicker elastic layer than at older ridges, implying a thickening of the icy shell. Third, regional-scale geological mapping suggests that certain types of features, notably chaos regions, dominate the most recent geologic record preserved on the surface (*Figueredo and Greeley,* 2004; cf. chapter by Bierhaus et al.). While this effect may be due to difficulties in identifying ancient chaos terrains (*Riley et al.,* 2000), it suggests that the geological behavior of Europa has changed with time (*Pappalardo et al.,* 1999). Such geological histories may also help distinguish between models for the formation of surface features. For example, if the shell is thickening, and chaos regions are forming more frequently, then models for chaos formation that appeal to diapirism are more plausible than melt-through models as the latter are favored by thin icy shells and the former favored by thick icy shells.

If Europa's icy shell is currently thickening, there may be important tectonic consequences. As the icy shell thickens, near-surface material moves radially outward, and thus experiences extensional stresses (*Nimmo,* 2004b; *Manga and Wang,* 2007; *Kimura et al.,* 2007). The stresses generated are large enough to dominate other likely sources of stress in the icy shell, and their extensional nature may help to explain the dominance of features formed by extensional on Europa (section 4.3.4).

Coupled thermal-orbital evolution models can be used to investigate whether a thickening icy shell is likely, and the timescales over which any thickening occurs, but such models are complicated. The dissipation and transport of heat within the icy shell are highly uncertain because of uncertainties in the ice rheology. More importantly, dissipation in the silicate interior depends sensitively on the viscosity, and thus temperature, of the interior; however, the interior temperature also depends strongly on the amount of dissipation. Modeling this feedback is also complicated by the fact that the evolution of Europa's orbital parameters depend on the behavior of Io and Ganymede (see chapters by Moore and Hussmann and Sotin et al.). The upshot is that even simplified models show a rich spectrum of behavior (*Ojakangas and Stevenson,* 1986; *Fischer and Spohn,* 1990; *Hussmann and Spohn,* 2004). A particularly interesting result is that Europa's icy shell thickness can show strong oscillations, because of oscillations in Io's behavior (Fig. 7). These arise because of the feedback between dissipation and heat transport in Io's interior, and have a period governed by the ability of Io's mantle to respond to a change in heat production. Although these studies are in their infancy, the importance of these results is that oscillatory behavior is

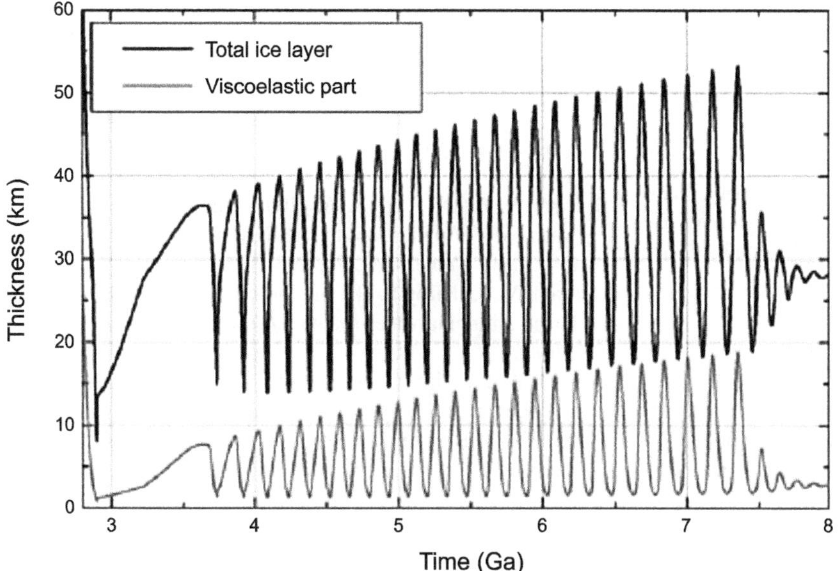

Fig. 7. Variation in Europa shell thickness with time, from *Hussmann and Spohn* (2004). The bold and thin lines refer to the total shell thickness and the viscoelastic part, respectively. The oscillations have a ~ 150 m.y. periodicity and arise primarily because of the coupled thermal-orbital behavior of Io. The model was started with all three bodies in resonance, and 4.5 Ga is the time at the present day.

consistent with the arguments above that Europa's icy shell is not in steady state. As an aside, we note that there are strong arguments that Enceladus, an active saturnian satellite, cannot be in steady state (*Meyer and Wisdom,* 2007). If Europa has undergone oscillatory behavior, this may help to explain why the surface is apparently so young, and also explain why inferences about icy shell thickness and structure based on models for features of different ages are not consistent with each other.

6. SUMMARY, OUTSTANDING PROBLEMS, AND FUTURE WORK

The most important property of Europa's icy shell is its thickness because it governs the rheological structure of the icy shell, which in turn controls its tectonics and formation of surface features. The most compelling argument that the shell is thick at present is the large amplitude of topography (Fig. 5). This does not mean that all the features preserved on the surface of Europa are produced by processes that operated in a thick icy shell, as there is both observational evidence for temporal changes in the style of tectonism and theoretical reasons for the icy shell to change thickness (section 5).

If the shell thickness exceeds 10 km, as suggested by the topography, then it is most likely convecting (section 4.1.3). Convection may be responsible for some surface landforms (e.g., spots and domes), but the topography of such features is hard to explain, at least with current thermal convection models.

Whether the icy shell is presently getting thicker or thinner cannot be determined, but we suggest that given the dominance of extensional features and the fact that thickening icy shells generate large extensional stresses, the icy shell is more likely to be getting thicker. The absence of tectonized craters suggest that this thickening was accompanied by a relatively abrupt change in resurfacing style. The relatively young surface age is consistent with theoretical models of tidally driven oscillatory behavior in which Europa's icy shell thins and then thickens.

Several outstanding questions about the geodynamics and evolution of the icy shell remain, besides determining the present and past thickness of the icy shell: Why is the icy shell so young? Is the icy shell tectonically active today? Does and has liquid water from the ocean reached the surface? Answering these questions may also allow us to determine whether the silicate interior is cold or hot — a feature that in turn will determine whether hydrothermal processes operate within the ocean.

Future missions have the potential of using geophysical techniques to answer some of these questions. Despite the challenging nature of the radiation environment around Europa, radio tracking of an orbiter equipped with a laser or radar altimeter would be able to directly detect diurnal tidal variations and thus conclusively demonstrate the presence or absence of a subsurface ocean, as well as potentially measuring the shell thickness (*Wahr et al.,* 2006). Another method of determining the icy shell thickness is by ice-penetrating radar (*Chyba et al.,* 1998), although ice absorption (*Moore,* 2000) or near-surface porosity (*Eluszkiewicz,* 2004) might make doing so difficult. Global topography and gravity measurements would allow accurate determination of the icy shell elastic thickness, as has been done, e.g., for Mars (*McGovern et al.,* 2002). If the icy shell is currently tectonically active, radar interferometry would be able to detect surface displacements on the order of 10 cm (*Sandwell et al.,* 2004). Although landing a surface package is technically challenging, seismological data, even from a single instrument, would allow determination of Europa's internal structure, including the icy shell thickness (*Kovach and Chyba,* 2001; *Lee et al.,* 2003; *Panning et al.,* 2006). Finally, global multispectral mapping of the kind that Galileo was unable to perform would allow the stratigraphy and history of the icy shell to be resolved with much greater confidence than at present.

Acknowledgments. Funding was provided by NASA's Planetary Geology and Geophysics and Outer Planets Research Programs. We thank H. Hussmann, O. Grasset, and D. Sandwell for careful reviews.

REFERENCES

Abramov O. and Spencer J. R. (2008) Numerical modeling of endogenic thermal anomalies on Europa. *Icarus, 195,* 378–385.

Alvarellos J. L., Zahnle K. J., Dobrovolskis A. R., and Hamill P. (2008) Transfer of mass from Io to Europa and beyond due to cometary impacts. *Icarus, 194,* 636–646.

Barr A. C. and McKinnon W. B. (2007) Convection in ice I shells and mantles with self-consistent grain size. *J. Geophys. Res., 112,* E02012, DOI: 10.1029/2006JE002781.

Barr A. C., Pappalardo R. T., and Zhong S. J. (2004) Convective instability in ice I with non-Newtonian rheology: Application to the icy Galilean satellites. *J. Geophys. Res., 109,* E12008.

Beeman M., Durham W. B., and Kirby S. H. (1988) Friction of ice. *J. Geophys. Res., 93,* 7625–7633.

Billings S. E. and Kattenhorn S. A. (2005) The great thickness debate: Ice shell thickness models for Europa and comparisons with estimates based on flexure at ridges. *Icarus, 177,* 397–412.

Bills B.G. (2005) Free and forced obliquities of the Galilean satellites of Jupiter. *Icarus, 175,* 233–247.

Black G. J., Campbell D. B., and Nicholson P. D. (2001) Icy Galilean satellites: Modeling radar reflectivities as a coherent backscatter effect. *Icarus, 151,* 167–180.

Buck W. R. (1991) Modes of continental lithospheric extension. *J. Geophys. Res., 96,* 20161–20178.

Carr M. H. and 21 colleagues (1998) Evidence for a subsurface ocean on Europa. *Nature, 391,* 363–365.

Cassen P., Reynolds R. T., and Peale S. J. (1979) Is there liquid water on Europa? *Geophys. Res. Lett., 6,* 731–734.

Cathles L. M. (1975) *The Viscosity of the Earth's Mantle.* Princeton Univ., Princeton.

Chyba C. F. and Phillips C. B. (2002) Europa as an abode for life. *Origins Life Evol. Biosphere, 32,* 47–68.

Chyba C. F., Ostro S. J., and Edwards B. C. (1998) Radar detectability of a subsurface ocean on Europa. *Icarus, 134,* 292–302.

Collins G. C., Head J. W., Pappalardo R. T., and Spaun N. A. (2000) Evaluation of models for the formation of chaotic terrain on Europa. *J. Geophys. Res., 105,* 1709–1716.

Cooper J. F., Johnson R. E., Mauk B. H., Garrett H. B., and Gehrels N. (2001) Energetic ion and electron irradiation of the Galilean satellites. *Icarus, 149,* 133–159.

Crawford G. D. and Stevenson D. J. (1988) Gas-driven water volcanism and the resurfacing of Europa. *Icarus, 73,* 66–79.

Davaille A. and Jaupart C. (1994) Onset of thermal convection in fluids with temperature-dependent viscosity — application to the oceanic mantle. *J. Geophys. Res., 99,* 19853–19866.

De La Chapelle S., Satelnau O., Lipenkov V., and Duval P. (1999) Dynamic recrystallization and texture development in ice as revealed by the study of deep ice cores in Antarctica and Greenland. *J. Geophys. Res., 103,* 5091–5105.

Dombard A. J. and McKinnon W. B. (2000) Long-term retention of impact crater topography on Ganymede. *Geophys. Res. Lett., 27,* 3663–3666.

Dombard A. J. and McKinnon W. B. (2006) Folding of Europa's icy lithosphere: An analysis of viscous-plastic buckling and subsequent topographic relaxation. *J. Struct. Geol., 28,* 2259–2269.

Durham W. B. and Stern L. A. (2001) Rheological properties of water ice — Applications to satellites of the outer planets. *Annu. Rev. Earth Planet. Sci., 29,* 295–330.

Durham W. B., Kirby S. H., and Stern L. A. (1993) Flow of ices in the ammonia-water system. *J. Geophys. Res., 98,* 17667–17682.

Durham W. B., Kirby S. H., and Stern L. A. (1998) Rheology of planetary ices. In *Solar System Ices* (B. Schmidt et al., eds.), pp. 63–78. Kluwer, Dordrecht.

Durham W. B., Stern L. A., Kubo T., and Kirby S. H. (2005a) Flow strength of highly hydrated Mg- and Na-sulphate hydrated salts, pure and in mixtures with water ice, with application to Europa. *J. Geophys. Res., 110,* E12010.

Durham W. B., McKinnon W. B., and Stern L. A. (2005b) Cold compaction of water ice. *Geophys. Res. Lett., 32,* L18202.

Eluszkiewicz J. (2004) Dim prospects for radar detection of Europa's ocean. *Icarus, 170,* 234–236.

Fagents S. A. (2003) Considerations for effusive cryovolcanism on Europa: The post-Galileo perspective. *J. Geophys. Res., 108,* DOI: 10.1029/2003JE002128.

Fagents S. A., Greeley R., Sullivan R. J., Pappalardo R. T., and Prockter L. M. (2000) Cryomagmatic mechanisms for the formation of Rhadamanthys Linea, triple band margins, and other low-albedo features on Europa. *Icarus, 144,* 54–88.

Figueredo P. H. and Greeley R. (2004) Resurfacing history of Europa from pole-to-pole geological mapping. *Icarus, 167,* 287–313.

Figueredo P. H., Chuang F. C., Rathbun J., Kirk R. L., and Greeley R. (2002) Geology and origin of Europa's "Mitten" feature (Murias Chaos). *J. Geophys. Res., 107,* 5026.

Figueredo P. H., Greeley R., Neuer S., Irwin, L., and Schulze-Makuch D. (2003) Locating potential biosignatures on Europa from surface geology observations. *Astrobiology, 3,* 851–861.

Fischer H. J. and Spohn T. (1990) Thermal orbital histories of viscoelastic models of Io (J1). *Icarus, 83,* 39–65.

Freeman J., Moresi L., and May D. A. (2006) Thermal convection with a water ice I rheology: Implications for icy satellite evolution. *Icarus, 180,* 251–264.

Friedson A. J. and Stevenson D. J. (1983) Viscosity of rock-ice mixtures and applications to the evolution of icy satellites. *Icarus, 56,* 1–14.

Gaidos E. and Nimmo F. (2000) Tectonics and water of Europa. *Nature, 405,* 637.

Gammon P. H., Kiefte H., Clouter M. J., and Denner W. W. (1983) Elastic constants or artificial and natural ice samples by Brillouin spectroscopy. *J. Glaciol., 29,* 433–459.

Geissler P. E. and 14 colleagues (1998a) Evidence for non-synchronous rotation of Europa. *Nature, 391,* 368–379.

Geissler P. E. and 16 colleagues (1998b) Evolution of lineaments on Europa: Clues from Galileo multispectral imaging observations. *Icarus, 135,* 107–126.

Giese B., Oberst J., Roatsch T., Neukum G., Head J. W., and Pappalardo R. T. (1998) The local topography of Uruk Sulcus and Galileo Regio obtained from stereo data. *Icarus, 135,* 303–316.

Golombek M. P. and Banerdt W. B. (1986) Early thermal profiles and lithospheric strength of Ganymede from extensional tectonic features. *Icarus, 68,* 252–265.

Goldsby D. L. and Kohlstedt D. L. (2001) Superplastic deformation of ice: Experimental observations. *J. Geophys. Res., 106,* 11017–11030.

Goodman J. C., Collins G. C., Marshall J., and Pierrehumbert R. T. (2004) Hydrothermal plume dynamics on Europa: Implications for chaos formation. *J. Geophys. Res., 109,* E03008.

Grasset O. and Parmentier E. M. (1998) Thermal convection in a volumetrically heated, infinite Prandtl number fluid with a strongly temperature-dependent viscosity: Implications for planetary thermal evolution. *J. Geophys. Res., 103,* 18171–18181.

Greeley R., Collins G. C., Spaun N. A., Sullivan R. J., Moore J. M., et al. (2000) Geologic mapping of Europa. *J. Geophys. Res., 105,* 22559–22578.

Greeley R., Chyba C. F., Head J. W., McCord T. B., McKinnon W. B., Pappalardo R. T., and Figueredo P. (2004) Geology of Europa. In *Jupiter: The Planet, Satellites and Magnetospheres* (F. Bagenal et al., eds.), pp. 329–362. Cambridge Univ., Cambridge.

Greenberg R. (2004) The evil twin of Agenor: Tectonic convergence on Europa. *Icarus, 167,* 313–319.

Greenberg R. and Geissler P. (2002) Europa's dynamic icy crust. *Meteoritics & Planet. Sci., 37,* 1685–1710.

Greenberg R. and Weidenschilling S. J. (1984) How fast do Galilean satellites spin? *Icarus, 58,* 186–196.

Greenberg R. and 9 colleagues (1998) Tectonic processes on Europa: Tidal stresses, mechanical response and visible features. *Icarus, 135,* 64–78.

Greenberg R., Geissler P., Hoppa G., and Tufts B. R. (2002) Tidal-tectonic processes and their implications for the character of Europa's icy crust. *Rev. Geophys., 40,* 1004.

Greenberg R., Hoppa G. V., Bart G., and Hurford T. (2003) Tidal stress patterns on Europa's crust. *Cel. Mech. Dyn. Astron., 87,* 171–188.

Gribb T. T. and Cooper R. F. (1998) Low-frequency shear attenuation in polycrystalline olivine: Grain boundary diffusion and the physical significance of the Andrade model for viscoelastic rheology. *J. Geophys. Res., 103,* 27627–27279.

Han L. and Showman A. P. (2005) Thermo-compositional convection in Europa's icy shell with salinity. *Geophys. Res. Lett., 32,* L20201, DOI: 10.1029/2005GL023979.

Han L. J. and Showman A. P. (2008) Implications of shear heating and fracture zones for ridge formation on Europa. *Geophys. Res. Lett., 35,* L03202.

Hand K. P. and Chyba C. F. (2007) Empirical constraints on the salinity of the europan ocean and implications for a thin ice shell. *Icarus, 189,* 424–438.

Harada Y. and Kurita K (2007) Effect of non-synchronous rotation on surface stress upon Europa: Constraints on surface rheology. *Geophys. Res. Lett., 34,* L11204.

Hawkes I. and Mellor M. (1972) Deformation and fracture of ice under uniaxial stress. *J. Glaciol., 11,* 103–131.

Head J. W. and Pappalardo T. T. (1999) Brine mobilization during lithospheric heating on Europa: Implications for formation of chaos terrain, lenticula texture, and color variations. *J. Geophys. Res., 104,* 27143–27155.

Head J. W., Pappalardo R. T., and Sullivan R. (1999) Europa: Morphological characteristics of ridges and triple bands from Galileo data (E4 and E6) and assessment of a linear diapirism model. *J. Geophys. Res., 104,* 18907–18924.

Helfenstein P. and Parmentier E. M. (1985) Patterns of fracture and tidal stresses due to nonsynchronous rotation — implications for fracturing on Europa. *Icarus, 61,* 175–184.

Hoppa G. V., Tufts R. B., Greenberg R., and Geissler P. E. (1999a) Formation of cycloidal features on Europa. *Science, 285,* 1899–1902.

Hoppa G. V., Greenberg R., Geissler P., Tufts B. R., Plassmann J., and Durda D. D. (1999b) Rotation of Europa: Constraints from terminator and limb positions. *Icarus, 137,* 341–347.

Hoppa G., Tufts B. R., Greenberg R., and Geissler P. (1999c) Strike-slip faults on Europa: Global shear patterns driven by tidal stress. *Icarus, 141,* 287–298.

Hoppa G., Tufts B. R., Greenberg R., Hurford T. A., O'Brien D. P., and Geissler P. E. (2001) Europa's rate of rotation derived from the tectonic sequence in the Astypalaea region. *Icarus, 153,* 208–213.

Hurford T. A., Beyer R. A., Schmidt B., Preblich B., Sarid A. R., and Greenberg R. (2005) Flexure of Europa's lithosphere due to ridge-loading. *Icarus, 177,* 80–396.

Hurford T. A., Sarid A. R., and Greenberg R. (2007) Cycloidal cracks on Europa: Improved modeling and non-synchronous rotation implications. *Icarus, 186,* 218–233.

Hessinger J., White B. E., and Pohl R. O. (1996) Elastic properties of amorphous and crystalline ice films. *Planet. Space Sci., 44,* 937–944.

Hussmann H. and Spohn T. (2004) Thermal-orbital evolution of Io and Europa. *Icarus, 171,* 391–410.

Hussmann H., Spohn T., and Wieczerkowski K. (2002) Thermal equilibrium states of Europa's ice shell: Implications for internal ocean thickness and surface heat flow. *Icarus, 156,* 143–151.

Jackson J. A. and White N. J. (1989) Normal faulting in the supper continental crust: Observations from regions of active extension. *J. Struct. Geol., 11,* 15–36.

Jankowski D. G. and Squyres S. W. (1991) Sources of error in planetary photoclinometry. *J. Geophys. Res., 96,* 20907–20922.

Johnson R. E., Carlson R. S., Cooper J. F., Paranicas C., Moore M. H., and Wong M. C. (2004) Radiation effects on the surfaces of the Galilean satellites. In *Jupiter: The Planet, Satellites and Magnetospheres* (F. Bagenal et al., eds.), pp. 485–512. Cambridge Univ., Cambridge.

Kargel J. S. (1998) Physical chemistry of ices in the outer solar system. In *Solar System Ices* (B. Schmidt et al., eds.), pp. 3–32. Kluwer, Dordrecht.

Kattenhorn S. A. (2002) Nonsynchronous rotation evidence and fracture history in the Bright Plains region, Europa. *Icarus, 157,* 490–506.

Kattenhorn S. A. (2005) Strike-slip fault evolution on Europa: Evidence from tailcrack geometries. *Icarus, 172,* 582–602.

Kattenhorn S. A. and Marshall S. T. (2006) Fault-induced perturbed stress fields and associated tensile and compressive deformation at fault tips in the ice shell of Europa: Implications for fault mechanics. *J. Struct. Geol., 28,* 2204–2221.

Kimura J., Yamagishi Y., and Kurita K. (2007) Tectonic history of Europa: Coupling between internal evolution and surface stresses. *Earth Planets Space, 59,* 113–125.

Kirk R. L., Barrett J. M., and Soderblom L. A. (2003) Photoclinometry made simple . . . ? In *Workshop on Advances in Planetary Mapping,* Houston, Texas. Available online at astrogeology.usgs.gov/Projects/ISPRS/MEETINGS/Houston2003/abstracts/Kirk_isprs_mar03.pdf.

Kivelson M. G., Khurana K. K., Russell C. T., Volwerk M., Walker R. J., and Zimmer C. (2000) Galileo magnetometer measurements: A stronger case for a subsurface ocean at Europa. *Science, 289,* 1340–1343.

Klinger J. (1980) Influence of a phase transition of ice on the heat and mass balance of comets. *Science, 209,* 271–272.

Kovach D. L. and Chyba C. F. (2001) Seismic detectability of a subsurface ocean on Europa. *Icarus, 150,* 279–287.

Lee S., Zanolin M., Thode A. M., Pappalardo R. T., and Makris N. C. (2003) Probing Europa's interior with natural sound waves. *Icarus, 165,* 144–167.

Lee S., Pappalardo R. T., and Makris N. C. (2005) Mechanics of tidally driven fractures in Europa's ice shell. *Icarus, 177,* 367–379.

Leith A. C. and McKinnon W. B. (1996) Is there evidence for polar wander on Europa? *Icarus, 120,* 387–398.

Luttrell K. and Sandwell D. (2006) Strength of the lithosphere of the Galilean satellites. *Icarus, 183,* 159–167.

Manga M. and Sinton A. (2004) Formation of bands, ridges and grooves on Europa by cyclic deformation: Insights from analogue wax experiments. *J. Geophys. Res., 109(E9),* E09001, DOI: 10.1029/2004JE002249.

Manga M. and Wang C. Y. (2007) Pressurized oceans and the eruption of liquid water on Europa and Enceladus. *Geophys. Res. Lett., 34,* L07202, DOI: 10.1029/2007GL029297.

Marshall S. T. and Kattenhorn S.A. (2005) A revised model for cycloid growth mechanics on Europa: Evidence from surface morphologies and geometries. *Icarus, 177,* 341–366.

Matsuyama I. and Nimmo F. (2008) Tectonic patterns on reoriented and despun planetary bodies. *Icarus, 195,* 459–473.

McCord T. B. and 12 colleagues (1998) Non-water-ice constituents in the surface material of the icy Galilean satellites from the Galileo near-infrared mapping spectrometer investigation. *J. Geophys. Res., 103,* 8603–8626.

McEwen A. S. (1986) Tidal reorientation and fracturing on Jupiter's moon Europa. *Nature, 321,* 49–51.

McGovern P. J. and 9 colleagues (2002) Localized gravity/topography admittance and correlation spectra on Mars: Implications for regional and global evolution. *J. Geophys. Res., 107,* 5136.

McKinnon W. B. (1999) Convective instability in Europa's floating ice shell. *Geophys. Res. Lett., 26,* 951–954.

McKinnon W. B. (2006) On convection in ice I shells of outer solar system bodies, with detailed application to Callisto. *Icarus, 183,* 435–450.

McNutt M. K. (1984) Lithospheric flexure and thermal anomalies. *J. Geophys. Res., 89,* 1180–1194.

Melosh H. J. (1989) *Impact Cratering: A Geologic Process*. Oxford, New York.

Melosh H. J. (1980) Tectonic patterns on a tidally distorted planet. *Icarus, 43,* 334–337.

Meyer J. and Wisdom J. (2007) Tidal heating in Enceladus. *Icarus, 188,* 535–539.

Mitri G. and Showman A. P. (2005) Convective-conductive transitions and sensitivity of a convecting ice shell to perturbations in heat flux and tidal-heating rate: Implications for Europa. *Icarus, 177,* 447–460.

Miyamoto H., Mitri G., Showman A. P., and Dohm J. M. (2005) Putative ice flows on Europa: Geometric patterns and relation to topography collectively constrain material properties and effusion rates. *Icarus, 177,* 413–424.

Moore J. C. (2000) Models of radar absorption in europan ice. *Icarus, 147,* 292–300.

Moore J. M. and 25 colleagues (2001) Impact features on Europa: Results of the Galileo Europa Mission (GEM). *Icarus, 151,* 93–111.

Moore W. B. (2006) Thermal equilibrium in Europa's ice shell. *Icarus, 180,* 141–146.

Moore W. B. and Schubert G. (2000) The tidal response on Europa. *Icarus, 147,* 317–319.

Murray C. D. and Dermott S. F. (1999) *Solar System Dynamics*. Cambridge Univ., Cambridge.

Neukum G. (1997) Bombardment history of the jovian system. In *The Three Galileos: The Man, the Spacecraft, the Telescope* (C. Barbieri et al., eds.), pp. 201–212. Kluwer, Norwell, Massachusetts.

Nimmo F. (2004a) Non-Newtonian topographic relaxation on Europa. *Icarus, 168,* 205–208.

Nimmo F. (2004b) Stresses generated in cooling viscoelastic ice shells: Application to Europa. *J. Geophys. Res., 109,* E12001.

Nimmo F. (2004c) What is the Young's modulus of ice? In *Workshop on Europa's Icy Shell,* Abstract #7005. Lunar and Planetary Institute, Houston.

Nimmo F. (2004d) Dynamics of rifting and modes of extension on icy satellites. *J. Geophys. Res., 109,* E01003.

Nimmo F. and Gaidos E. (2002) Thermal consequences of strike-slip motion on Europa. *J. Geophys. Res., 107,* 5021, DOI: 10.1029/2000JE001476.

Nimmo F. and Giese B. (2005) Thermal and topographic tests of Europa chaos formation models from Galileo E15 observations. *Icarus, 177,* 327–341.

Nimmo F. and Manga M. (2002) Causes, characteristics and consequences of convective diapirism on Europa. *Geophys. Res. Lett., 29,* DOI: 10.1029/2002GL015754.

Nimmo F. and Matsuyama I. (2007) Reorientation of icy satellites by impact basins. *Geophys. Res. Lett., 34,* L21201.

Nimmo F. and Schenk P. (2006) Normal faulting on Europa: Implications for ice shell properties. *J. Struct. Geol., 28,* 2194–2203.

Nimmo F. and Pappalardo R. T. (2004) Furrow flexure and ancient heat flux on Ganymede. *Geophys. Res. Lett., 31,* L19701.

Nimmo F., Pappalardo R. T., and Giese B. (2003a) On the origins of band topography, Europa. *Icarus, 166,* 21–32.

Nimmo F., Giese B., and Pappalardo R. T. (2003b) Estimates of Europa's ice shell thickness from elastically-supported topography. *Geophys. Res. Lett., 30,* 1233, DOI: 10.1029/2002GL016660.

Nimmo F., Thomas P. C., Pappalardo R. T., and Moore W. B. (2007a) The global shape of Europa: Constraints on lateral shell thickness variations. *Icarus, 191,* 183–192.

Nimmo F., Spencer J. R., Pappalardo R. T., and Mullen M. E. (2007b) Shear heating as the origin of the plumes and heat flux on Enceladus. *Nature, 447,* 289–291.

O'Brien D. P., Geissler P., and Greenberg R. (2002) A melt-through model for chaos formation on Europa. *Icarus, 156,* 152–161.

Ojakangas G. W. and Stevenson D. J. (1986) Episodic volcanism of tidally heated satellites with application to Io. *Icarus, 66,* 341–358.

Ojakangas G. W. and Stevenson D. J. (1989a) Thermal state of an ice shell on Europa. *Icarus, 81,* 220–241.

Ojakangas G. W. and Stevenson D. J. (1989b) Polar wander of an ice shell on Europa. *Icarus, 81,* 242–270.

Panning M., Lekic V., Manga M., Cammarano F., and Romanowicz B. (2006) Long-period seismology on Europa: 2. Predicted seismic response. *J. Geophys. Res., 111,* E12008.

Pappalardo R. T. and Barr A. C.(2004) The origin of domes on Europa: The role of thermally induced compositional diapirism. *Geophys. Res. Lett., 31,* DOI: 10.1029/2003GL019202.

Pappalardo R. T. and 10 colleagues (1998) Geological evidence for solid-state convection in Europa's ice shell. *Nature, 391,* 365–368.

Pappalardo R. T. and 31 colleagues (1999) Does Europa have a subsurface ocean? Evaluation of the geological evidence. *J. Geophys. Res., 104,* 24015–24055.

Patterson G. W., Head J. W., and Pappalardo R. T. (2006) Plate motion on Europa and nonrigid behaviour of the icy lithosphere: The Castalia Macula region. *J. Struct. Geol., 28,* 2237–2258.

Phillips C. B., McEwen A. S., Hoppa G. V., Fagents S. A., Greeley R., Klemaszewski J. E., Pappalardo R. T., Klaasen K. P., and Breneman H. H. (2000) The search for current geologic activity on Europa. *J. Geophys. Res., 105,* 22579–22598.

Porco C. and 24 colleagues (2006) Cassini observes the active south pole of Enceladus. *Science, 311,* 1393–1401.

Prockter L. M. and Pappalardo R. T. (2000) Folds on Europa: Implications for crustal cycling and accommodation of extension. *Science, 289,* 941–943.

Prockter L. and Schenk P. (2005) Origin and evolution of Castalia Macula, an anomalous young depression on Europa. *Icarus, 177,* 305–326.

Prockter L. M., Head J. W., Pappalardo R. T., Sullivan J. R., Clifton A. E., Giese B., Wagner R., and Neukum G. (2002) Morphology of europan bands at high resolution: A mid-ocean ridge-type rift mechanism. *J. Geophys. Res., 107,* DOI: 10.1029/2000JE001458.

Qin R., Buck W. R., and Germanovich L. (2007) Comment on "Mechanics of tidally driven fractures in Europa's ice shell," by S. Lee, R. T. Pappalardo, and N. C. Makris. *Icarus, 189,* 595–597.

Riley J., Hoppa G. V., Greenberg R., Tufts B. R., and Geissler P. (2000) Distribution of chaotic terrain on Europa. *J. Geophys. Res., 105,* 22599–22615.

Rist M. A. (1997) High-stress ice fracture and friction. *J. Phys. Chem., 101,* 6263–6266.

Ross M. N. and Schubert G. (1987) Tidal heating in an internal ocean model of Europa. *Nature, 325,* 133–134.

Ross M. N. and Schubert G. (1989) Viscoelastic models of tidal heating in Enceladus. *Icarus, 78,* 90–101.

Ruiz J., Alvarez-Gomez J. A., Tejero R., and Sanchez N. (2007) Heat flow and thickness of a convective ice shell on Europa for grain size-dependent rheologies. *Icarus, 190,* 145–154.

Sandwell D., Rosen P., Moore W., and Gurrola E. (2004) Radar interferometry for measuring tidal strains across cracks on Europa. *J. Geophys. Res., 109,* E11003.

Sarid A. R., Greenberg R., Hoppa G. V., Hurford T. A., Tufts B. R., and Geissler P. (2002) Polar wander and surface convergence of Europa's ice shell: Evidence from a survey of strike-slip displacement. *Icarus, 158,* 24–41.

Sarid A. R., Greenberg R., Hoppa G. V., Geissler P., and Preblich B. (2004) Crack azimuths on Europa: Time sequence in the southern leading face. *Icarus, 168,* 144–157.

Sarid A. R., Greenberg R., and Hurford T. A. (2006) Crack azimuths on Europa: Sequencing of the northern leading hemisphere. *J. Geophys. Res., 111,* E08004.

Schenk P. M. (2002) Thickness constraints on the icy shells of the Galilean satellites from a comparison of crater shapes. *Nature, 417,* 419–421.

Schenk P. M. and Pappalardo R. T. (2004) Topographic variations in chaos on Europa: Implications for diapiric formation. *Geophys. Res. Lett., 31,* L16703.

Schenk P. M., Matsuyama I., and Nimmo F. (2008) True polar wander on Europa from global-scale small-circle depressions. *Nature, 453,* 368–371.

Schilling N., Neubauer F. M., and Saur J. (2008) Time-varying interaction of Europa with the jovian magnetosphere: Constraints on the conductivity of Europa's subsurface ocean. *Icarus, 192,* 41–55.

Schmeltz M., Rignot E., and Macayeal D. (2002) Tidal flexure along ice-sheet margins: Comparison of InSAR with an elastic-plate model. *Ann. Glaciol., 34,* 202–208.

Schulson E. M. (2002) On the origin of a wedge crack within the icy crust of Europa. *J. Geophys. Res., 107,* 5107, DOI: 10.1029/2001JE001586.

Schubert G., Anderson J. D., Spohn T., and McKinnon W. B. (2004) Interior composition, structure and dynamics of the Galilean satellites. In *Jupiter: The Planet, Satellites and Magnetospheres* (F. Bagenal et al., eds.), pp. 281–306. Cambridge Univ., Cambridge.

Schwartz J. and Weeks W. F. (1977) Engineering properties of sea ice. *J. Glaciol., 19,* 499–531.

Showman A. P. and Han L. J. (2004) Numerical simulations of convection in Europa's ice shell: Implications for surface features. *J. Geophys. Res., 109,* E01010.

Showman A. P. and Han L. J. (2005) Effects of plasticity on convection in an ice shell: Implications for Europa. *Icarus, 177,* 425–437.

Showman A. P., Mosqueira I., and Head J. W. (2004) On the resurfacing of Ganymede by liquid-water volcanism. *Icarus, 172,* 625–640.

Solomatov V. S. (1995) Scaling of temperature- and stress-dependent viscosity convection. *Phys. Fluids, 7,* 266–274.

Sotin C., Head J. W., and Tobie G. (2002) Europa: Tidal heating of upwelling thermal plumes and the origin of lenticulae and chaos melting. *Geophys. Res. Lett., 29,* DOI: 10.1029/2001GL013844.

Spaun N. A. and Head J. W. (2001) A model of Europa's crustal structure: Recent Galileo results and implications for an ocean. *J. Geophys. Res., 106,* 7567–7575.

Spaun N. A., Head J. W., Collins G. C., Prockter L. M., and Pappalardo R. T. (1998) Conamara Chaos region, Europa: Reconstruction of mobile polygonal ice blocks. *Geophys. Res. Lett., 25,* 4277–4280.

Spencer J. R., Tamppari L. K., Martin T. A., and Travis L. D. (1999) Temperatures on Europa from Galileo photopolarimeter-radiometer: Nighttime thermal anomalies. *Science, 284,* 1514–1516.

Spencer J. R., Pearl J. C., Segura M., Flasar F. M., Mamoutkine A., Romani P., Buratti B. J., Hendrix A. R., Spilker L. J., and Lopes R. M. C. (2006) Cassini encounters Enceladus: Background and the discovery of a south polar hot spot. *Science, 311,* 1401–1405.

Spohn T. and Schubert G. (2003) Oceans in the icy Galilean satellites of Jupiter? *Icarus, 161,* 456–467.

Squyres S. W., Reynolds R. T., Cassen P. M., and Peale S. J. (1983) Liquid water and active resurfacing on Europa. *Nature, 301,* 225–226.

Stempel M. M., Barr A. C., and Pappalardo R. T. (2005) Model constraints on the opening rates of bands on Europa. *Icarus, 177,* 297–304.

Stevenson D. J. (2000) Limits on the variation of thickness of Europa's ice shell (abstract). In *Lunar and Planetary Science XXXI,* Abstract #1506. Lunar and Planetary Institute, Houston (CD-ROM).

Strom R. G., Schaber G. G., and Dawson D. D. (1994) The global resurfacing of Venus. *J. Geophys. Res., 99,* 10899–10926.

Sullivan R. and 11 colleagues (1998) Episodic plate separation and fracture infill on the surface of Europa. *Nature, 391,* 371–373.

Thomson R. E. and Delaney J. R. (2001) Evidence for a weakly stratified europan ocean sustained by seafloor heat flux. *J. Geophys. Res., 106,* 12355–12365.

Tobie G., Choblet G., and Sotin C. (2003) Tidally heat convection: Constraints on Europa's ice shell thickness. *J. Geophys. Res., 108,* 5124, DOI: 10.1029/2003JE002099.

Tobie G., Mocquet A., and Sotin C. (2005) Tidal dissipation within large icy satellites: Applications to Europa and Titan. *Icarus, 177,* 534–549.

Turcotte D. L. and Schubert G. (2002) *Geodynamics.* Cambridge Univ., Cambridge. 456 pp.

Turtle E. P. and Ivanov B. A. (2002) Numerical simulations of impact crater excavation and collapse on Europa: Implications for ice thickness (abstract). In *Lunar and Planetary Science XXXIII,* Abstract #1431. Lunar and Planetary Institute, Houston (CD-ROM).

Turtle E. P. and Pierazzo E. (2001) Thickness of a europan ice shell from impact crater simulations. *Science, 294,* 1326–1328.

Van Hoolst T., Rambaux N., Karatekin O., Dehant V., and Rivoldini A. (2008) The librations, shape and icy shell of Europa. *Icarus, 195,* 386–399.

Vaughan D. G. (1995) Tidal flexure at ice shelf margins. *J. Geophys. Res., 100,* 6213–6224.

Wahr J., Zuber M. T., Smith D. E., and Lunine J. I. (2006) Tides on Europa, and the thickness of europa's icy shell. *J. Geophys Res., 111,* E12005, DOI: 10.1029/2006JE002729.

Wahr J., Selvans Z. A., Mullen M. E., Barr A. C., Collins G. C., Selvans M. M., and Pappalardo R. T. (2008) Modelling stresses on satellites due to non-synchronous rotation and orbital eccentricity using gravitational potential theory. *Icarus,* in press.

Watts A. B. (2001) *Isostasy and Flexure of the Lithosphere.* Cambridge Univ., Cambridge.

Williams K. K. and Greeley R. (1998) Estimates of ice thickness in the Conamara Chaos region of Europa. *Geophys. Res. Lett., 25,* 4273–4276.

Wilson L., Head J. W., and Pappalardo R. T. (1997) Eruption of lava flows on Europa: Theory and application to Thrace Macula. *J. Geophys. Res., 102,* 9263–9272.

Yoder C. F. (1979) How tidal heating in Io drives the Galilean orbital resonance locks. *Nature, 279,* 767–770.

Zahnle K., Schenk P., Sobieszczyk S., Dones L., and Levison H. F. (2001) Differential cratering of synchronously rotating satellites by ecliptic comets. *Icarus, 153,* 111–129.

Zahnle K., Schenk P., Levison H., and Dones L. (2003) Cratering rates in the outer solar system. *Icarus, 163,* 263–289.

Zimmer C., Khurana K. K., and Kivelson M. G. (2000) Subsurface oceans on Europa and Callisto: Constraints from Galileo magnetometer observations. *Icarus, 147,* 329–347.

Heat Transfer in Europa's Icy Shell

Amy C. Barr
Southwest Research Institute

Adam P. Showman
University of Arizona

Heat transport across Europa's ice shell controls the thermal evolution of its interior and provides a source of energy to drive resurfacing. Recent improvements in knowledge of ice rheology, the behavior of convection, and the interaction between convection and lithospheric deformation have led to more realistic and complex models of the geodynamics of Europa's icy shell. The possibility of convection complicates efforts to determine the shell thickness because a thin conductive shell can carry the same heat flux as a thick convective shell. Whether convection occurs depends on ice viscosity, which in turn depends on grain size. The grain size may be controlled by internal deformation, or by impurities, depending on shell composition. Creating the observed surface features with steady-state thermal convection is challenging, even with tidal heating, because the near-surface ice is cold and stiff. Convection models that include surface weakening and compositional buoyancy show promise in explaining some chaos terrains, pits, and uplifts, but new spacecraft and laboratory data and geophysical techniques are needed to match theory to observation.

1. INTRODUCTION

Images of the surface of Europa returned by the Galileo spacecraft revealed that, in addition to its global network of linear features, portions of its surface are covered by pits, spots, uplifts, and chaos regions suggestive of convection-driven resurfacing (*Pappalardo et al.,* 1998; *Greeley et al.,* 1998, 2004). The implication that Europa's icy shell could be, or could have been convecting in the past, raised a number of questions about its geodynamical behavior: Can the shell convect at present? How does tidal dissipation affect the convection pattern? Can tidal dissipation and convection drive resurfacing? If the shell convects, can the ocean be thermodynamically stable? What role might compositional heterogeneity play in driving motion in Europa's icy shell?

The idea that solid-state convection may occur within Europa's icy shell dates to the Voyager era when *Consolmagno and Lewis* (1978) suggested that the ice I layers of large icy satellites could convect. *Reynolds and Cassen* (1979) determined that an ice shell on Europa, Ganymede, or Callisto could convect if the shell were thicker than 30 km, and that convection would rapidly freeze any liquid water ocean on a geologically short timescale. In the decades since, advances in our knowledge about planetary convection, coupled with laboratory experiments and field studies of ice deformation, have changed the way we think about convection on Europa.

Recent laboratory and theoretical studies about the fluid-like behavior of solid water ice have clarified the microphysical processes that accommodate deformation in ice (*Goldsby and Kohlstedt,* 2001; *Durham and Stern,* 2001) and the processes that control grain size (*McKinnon,* 1999; *Schmidt and Dalh-Jensen,* 2003; *Barr and McKinnon,* 2007), and thus viscosity and tidal dissipation in Europa's ice shell. Application of numerical terrestrial mantle convection models to Europa has shed light upon the conditions required to trigger convection in Europa's shell (*McKinnon,* 1999; *Ruiz and Tejero,* 2003; *Barr et al.,* 2004; *Barr and Pappalardo,* 2005), the behavior of the icy shell close to the onset of convection (*Mitri and Showman,* 2005, 2008a; *Solomatov and Barr,* 2007), the behavior of tidal dissipation and convection (*Tobie et al.,* 2003, *Mitri and Showman,* 2008b), the convective heat flux (*Freeman et al.,* 2006), the potential for convection-driven resurfacing (*Showman and Han,* 2004, 2005; *Han and Showman,* 2008), and the potential for driving resurfacing with thermochemical convection (*Pappalardo and Barr,* 2004; *Han and Showman,* 2005). Use of scaling relationships between the physical properties of the icy shell, the critical Rayleigh number, and the convective heat flux has helped constrain the conditions under which the ocean can avoid freezing (*Hussmann et al.,* 2002; *Freeman et al.,* 2006; *Moore,* 2006).

Despite these important advances, the modes of heat transfer across Europa's shell and the link between heat transfer and resurfacing remain unclear. Direct numerical simulations of convection-driven resurfacing cannot easily match the observed morphologies of pits, spots, domes, and chaos — simulations find that thermal buoyancy stresses can create uplifts only approximately one-tenth the height of those observed (*Showman and Han,* 2004). Compositional convection in a salty icy shell allows features with the correct heights to be explained (*Han and Showman,* 2005), but determining whether convection can cause chaos formation remains a challenge (*Showman and Han,* 2005). Using the spacing between quasicircular surface features on Europa's surface to constrain the physical properties, heat flux, and thermal structure of the icy shell has proved problematic (e.g., *Nimmo and Manga,* 2002). Compounding the uncertainty, Europa's shell may oscillate between a conductive and convective equilibrium, wreaking havoc on its surface (*Mitri and Showman,* 2005).

Here, we summarize recent breakthroughs in the understanding of solid-state convection, numerical advances in modeling the coupled processes of tidal dissipation and convection, and laboratory experiments clarifying the ductile behavior of ice. We discuss implications of these advances for the modes of heat transport and resurfacing in Europa's icy shell. In section 2, we provide an overview of heat sources and transport mechanisms in a tidally heated ice shell. In section 3, we review recent experiments clarifying how ice deforms in response to an applied stress, which is a long-standing source of uncertainty in europan interior modeling. In section 4, we describe how the icy shell behaves close to the limit of convective stability. In section 5, we discuss how convection and tidal dissipation may contribute to the formation of Europa's rich variety of surface features, including chaos, pits, spots, and uplifts.

2. HEAT GENERATION AND TRANSPORT

Europa's ice shell is heated from beneath by radiogenic (and possibly tidal) heating from its rocky mantle, and from within by tidal dissipation. Radiogenic heating within the rocky mantle of Europa currently supplies roughly $F_r \approx (1/3) \rho_r H_r (R_s - z)(1 - z/R_s)^2 \sim 6$ to 8 mW m^{-2} to its surface heat flux, where $\rho_r = 3000$ kg m^{-3} is the approximate density of Europa's rocky mantle (comparable to its mean density), $H_r = (4.5 \pm 0.5) \times 10^{-12}$ W kg^{-1} is the present chondritic heating rate (*Spohn and Schubert,* 2003), $z \sim 120$ km is the thickness of Europa's H_2O layer, and R_s is Europa's radius.

The surface heat flux due to tidal dissipation in Europa (F_{tidal}) can be estimated as a function of its physical and orbital properties

$$F_{tidal} = \frac{\dot{E}_{tidal}}{4\pi R_s^2} = \frac{21}{8\pi}\left(\frac{k_2}{Q}\right)\frac{R_s^3 G M_J^2 n e^2}{a^6} \tag{1}$$

where $\dot{E}_{tidal}$ is the tidal dissipation rate (*Cassen et al.,* 1979, 1980); R_s is Europa's radius; (k_2/Q) is the ratio between the degree-2 Love number (k_2) and the tidal quality factor (Q), which describe how Europa's interior deforms in response to the jovian tidal potential; M_J is the mass of Jupiter; a is its semimajor axis; e = 0.01 is Europa's orbital eccentricity; $n = 2\pi/P = 2.05 \times 10^{-5}$ s^{-1} is Europa's mean motion; and P is its orbital period. For a nominal value of $k_2 \sim 0.25$ (*Moore and Schubert,* 2000) and Q ~ 100, $F_{tidal} \sim$ 10–100 mW m^{-2}, larger than radiogenic heating (*Tobie et al.,* 2003) (see also the chapter by Sotin et al.).

How does Europa remove its tidal heat? The two possible heat transport mechanisms within the outer ice I shell are conduction and solid-state convection. Figure 1a illustrates the qualitative temperature structures that would accompany each of these states. In both cases, the top layer has a steep, conductive temperature gradient; in the conductive case (Fig. 1a, top), this layer extends to the top of the ocean, whereas in the convective case (Fig. 1a, bottom), a thick, nearly isothermal ice layer lies underneath the conductive lid. Despite greatly different thicknesses, these two solutions can potentially have similar surface heat fluxes. This fundamental ambiguity makes it difficult to infer the ice shell thickness from heat flow measurements. Here, we describe heat transport by conduction, and describe the thermal structure of an icy shell that generates heat with tidal dissipation and removes it by conduction. We then describe the governing equations of solid-state convection in Europa's icy shell and define the fundamental quantities that describe a convecting ice shell, including the Rayleigh number and Nusselt number. More detail about the behavior of a convecting icy shell will be provided in section 4 and section 5.

2.1. Conduction

The simple estimate of how much tidal dissipation occurs in Europa from equation (1) provides us with an estimate of the heat flux carried across Europa's outer ice I shell. If the ice shell is in conductive equilibrium, Fourier's law relates the heat flux to the shell thickness

$$F_{cond} = \frac{k\Delta T}{D} \tag{2}$$

where $\Delta T = T_b - T_s$ is the temperature difference between the temperature at its base (T_b) and its surface (T_s), and k ~ 3.3 W m^{-1} K^{-1} (here assumed constant for simplicity) is a representative thermal conductivity for cold ice. The probable detection of an ocean beneath Europa's ice (*Zimmer et al.,* 2000) suggests that the base of the icy shell should be at the melting temperature of water ice (for pure water ice, T_m = 253–273 K at pressures relevant to Europa's interior). The global average surface temperature on Europa is $T_s \sim$ 100 K (*Ojakangas and Stevenson,* 1989). Setting $F_{cond} = F_{tidal}$ gives an ice shell thickness $D \approx 6$ km (for $F_{tidal} \sim$ 100 mW m^{-2}) and $D \approx 60$ km (for $F_{tidal} \sim 10$ mW m^{-2}).

The thermal conductivity is temperature-dependent; cold ice is a much better conductor of heat than warm ice (see *Petrenko and Whitworth,* 1999, for discussion), so the temperature gradient close to the surface of the icy shell is steeper than at its base. The heat flux across a conductive icy shell with a temperature-dependent thermal conductivity is (cf. *Ojakangas and Stevenson,* 1989)

$$F = \frac{a_c}{D}\ln\left(\frac{T_b}{T_s}\right) \tag{3}$$

where $k = a_c/T$ and $a_c = 621$ W m^{-1} (*Petrenko and Whitworth,* 1999).

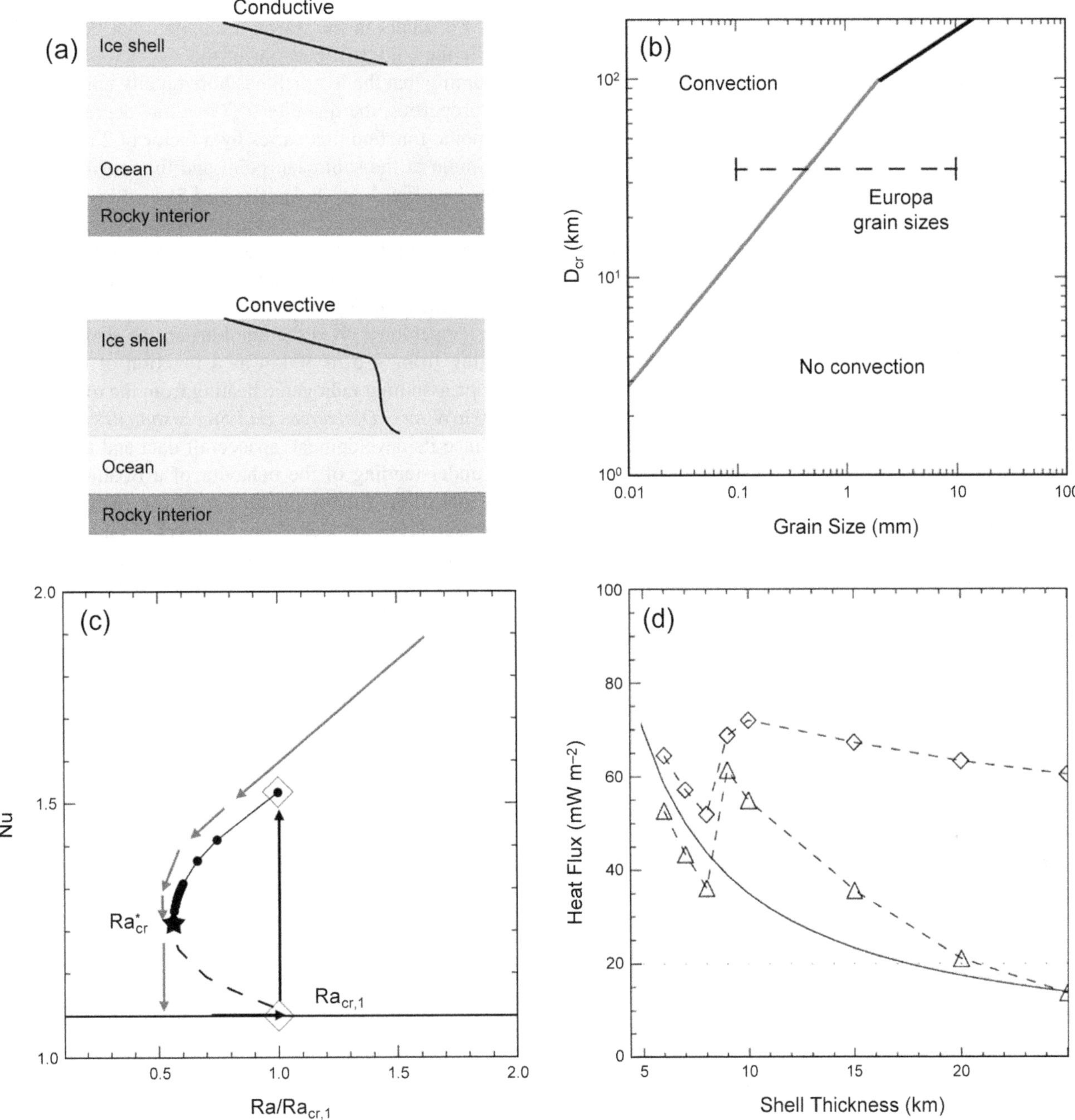

Fig. 1. **(a)** Schematic temperature structures for a conductive (top) and convective (bottom) ice shell on Europa from *Mitri and Showman* (2005). **(b)** Values of ice shell thickness where convection is possible in Europa as a function of grain size. For ice grain sizes d > 2 mm, deformation during the onset of convection is accommodated largely by GSS creep (black), but for smaller grain sizes, deformation is accommodated by volume diffusion (gray). After *Barr and Pappalardo* (2005). **(c)** Behavior of an ice shell in Ra-Nu space during the onset of convection in a basally heated fluid with $\theta = 18$ ($\Delta\eta = 10^8$) (black arrows) and decay of convection (gray arrows). Diamonds illustrate location of simulations of the onset of convection by *Mitri and Showman* (2005), points, solid, dashed lines show locations of simulations of the decay of convection by *Solomatov and Barr* (2007). When convection begins, Nu jumps from 1 to ~1.6–1.7 (see section 4.2.3), depending on the form of temperature perturbation used. When convection stops, Nu can achieve very low values for $Ra < Ra_{cr,1}$, but ultimately stops when $Ra < Ra^*_{cr}$, when Nu ~ 1.1–1.3 for rheological parameters for ice. **(d)** Heat flux as a function of ice shell thickness for equilibrium configurations of Europa's icy shell, illustrating the jump in heat flux at the convective/conductive transition D ~ 9 km. Tidal heating with a tidal-flexing strain amplitude 2×10^{-5} is assumed, and a Newtonian rheology is used with a melting-temperature viscosity of 10^{13} Pa s and a viscosity contrast of 10^6. Triangles and diamonds show heat flux into the bottom and out the top of the ice shell, respectively. Solid curve shows relationship between flux and thickness for a conductive solution with no tidal heating. From *Mitri and Showman* (2005).

2.2. A Tidally Heated Conductive Ice Shell

State-of-the-art models of tidal heating in Europa's icy shell calculate the dissipation occurring in the shell due to its cyclical diurnal tidal flexure by modeling the icy shell as a Maxwell viscoelastic solid. For a general discussion of the Maxwell model, see *Ojakangas and Stevenson* (1989) and *Turcotte and Schubert* (2002). The energy dissipated in a Maxwell solid is maximized when the period of the forcing $T \sim \tau_M = \eta/\mu$, where τ_M is the Maxwell time, η is the viscosity, and μ is the shear modulus of the material. By coincidence (or orbital and geophysical "tuning" in the Jupiter system), the orbital period of Europa (3.5 days) is very close to the Maxwell time for warm ice I ($\mu \sim 3.5 \times 10^9$ Pa and $\eta \sim 10^{15}$ Pa s, which gives $\tau_M \approx 3$ days). A warm, internally heated ice shell on Europa ice shell may be close to a maximally dissipative state.

In a Maxwell viscoelastic solid, the volumetric dissipation rate (q) is proportional to viscosity, $q \sim \eta(T)\dot{\varepsilon}^2$ at high temperatures, and inversely proportional to viscosity, $q \sim \mu^2\dot{\varepsilon}^2/[\omega^2\eta(T)]$, at low temperatures. This implies a strongly temperature-dependent dissipation rate with peak dissipation occurring between ~220 and 270 K depending on the ice grain size (see below). A hot ice shell can therefore be more dissipative than a cold ice shell, and the dissipation should depend on the shell thickness (*Cassen et al.,* 1980). This will affect the existence of equilibria between tidal dissipation and heat transfer.

To date, the most physically realistic model of a tidally heated conductive europan ice shell was proposed by *Ojakangas and Stevenson* (1989), who related the thickness of the ice shell, D, to the heat flux from the deeper interior F_{core}

$$D \approx \frac{\ln(T_b/T_s)}{\left[\left(\frac{2}{a_c}\right)\int_0^{T_b}\frac{q(T)dT}{T} + \left(\frac{F_{core}}{a_c}\right)^2\right]^{1/2}} \tag{4}$$

where $a_c = 621$ W m^{-1} (*Petrenko and Whitworth,* 1999) (see also equation (3)). Physically, this equation describes a conductive equilibrium in an internally heated ice shell with basal heat flux F_{core} and temperature-dependent thermal conductivity. If the ice is modeled as a Maxwell viscoelastic solid, the tidal dissipation rate is (*Ojakangas and Stevenson,* 1989)

$$q = \frac{2\mu\langle\dot{\varepsilon}_{ij}^2\rangle}{\omega}\left[\frac{\omega\tau_M}{1+(\omega\tau_M)^2}\right] \tag{5}$$

where $\langle\dot{\varepsilon}_{ij}^2\rangle$ is the time-average of the square of the second invariant of the strain rate tensor, $\tau_M = \eta(T)/\mu$ is the temperature-dependent Maxwell time, and $\omega = n$. The dissipation rate maximizes when $\omega \sim \tau_M^{-1}$, corresponding to temperatures of ~220–270 K for ice viscosity of ~10^{13}–10^{15} Pa s, is close to plausible viscosities for warm ice (see discussion in section 3). Equation (5) suggests that tidal dissipation occurs in the warmest ice, and that tidal dissipation in the cold, stiff portions of the icy shell is negligible. Assuming that the ice shell has horizontally uniform material properties, the quantity $\langle\dot{\varepsilon}_{ij}^2\rangle$ is a low-degree spherical harmonic function that varies by a factor of 2 between the minimum at the subjovian point and the maximum at the poles (see Fig. 1 of *Ojakangas and Stevenson,* 1989). The surface temperature on Europa also varies significantly, ranging from $T_s \sim 52$ K at the poles to $T_s \sim 110$ K at the equator. Integration of equation (4) with the tidal heat source (equation (5)) and including the spatially varying surface temperature gives the equilibrium ice shell thickness ranging from ~15 to 30 km as a function of location on Europa assuming radiogenic heating from the rocky mantle is 10 mW m^{-2} (*Ojakangas and Stevenson,* 1989).

Since its development, spacecraft data and advances in our understanding of the behavior of a floating ice shell have questioned the applicability of the *Ojakangas and Stevenson* (1989) model. Galileo images of pits, chaos, and uplifts on Europa's surface suggesting a convecting icy shell, and the suggestion that a shell $D \geq 30$ km thick could convect (see section 4), imply that the icy shell could be much thicker and still carry the tidal heat flux. The *Ojakangas and Stevenson* (1989) model, however, is an extremely valuable starting point for study of more complicated tidal/convective systems (e.g., *Tobie et al.,* 2003) because it demonstrates that tidal dissipation in Europa's shell is strongly rheology-dependent, and suggests that variations in tidal dissipation and surface temperature may lead to variations in the activity within the shell and potential for resurfacing.

2.3. Convection

Europa's ice shell may also transport heat by solid-state convection. The density of water ice, like most solids, decreases as a function of increasing temperature. Therefore, an ice shell cooled from its surface and heated from within and beneath is gravitationally unstable: Warm ice rises, cold ice sinks, which transports thermal energy upward to the base of a conductive "lid" at the surface of the icy shell. When this process is self-sustaining over geologically long timescales, it is called solid-state convection.

Thermally driven convection in a highly viscous fluid with no inertial forces is described by the conservation equations for mass, momentum, and energy (cf. *Schubert et al.,* 2001)

$$\nabla \cdot \vec{v} = 0 \tag{6}$$

$$\nabla p - \rho_o\alpha(T - T_o)g\hat{e}_z = \nabla \cdot [\eta(\nabla\vec{v} + \nabla^T\vec{v})] \tag{7}$$

$$\vec{v} \cdot \nabla T + \frac{\partial T}{\partial t} = \kappa\nabla^2 T + \gamma \tag{8}$$

where $\vec{v}$ is the velocity field, T is temperature, t is time, α

is the coefficient of thermal expansion, T_o and ρ_o are reference values of density and temperature, κ is thermal diffusivity (here, assumed to be constant), p is the dynamic pressure (which excludes lithostatic pressure), $\hat{e}_z$ is a unit vector in the vertical (z) direction, g is gravity, η is viscosity, and γ represents heat sources.

In a shell where convection occurs, the heat flux, F_{conv}, is enhanced relative to the conductive heat flux by a factor of Nu > 1

$$F_{conv} = \frac{k\Delta T}{D} Nu \tag{9}$$

where the Nusselt number, Nu, is related to the vigor of convection, expressed by the Rayleigh number

$$Ra = \frac{\rho_o g \alpha \Delta T D^3}{\kappa \eta} \tag{10}$$

As we will describe in more detail in section 4, convection can occur when the Rayleigh number of the icy shell exceeds a critical value, Ra_{cr}. The value of Ra_{cr} depends on how the ice viscosity varies with temperature and stress (*Solomatov,* 1995; *Solomatov and Barr,* 2006, 2007). The relative efficiency of convective heat transport over conduction depends on the vigor of convection, an effect expressed in the relationship between Ra and Nu

$$Nu = cRa^{\beta_n} \tag{11}$$

where the values of c and β_n depend on the variation in ice viscosity as a function of temperature and stress (*Solomatov,* 1995; *Dumoulin et al.,* 1999; *Solomatov and Moresi,* 2000; *Freeman et al.,* 2006) and (Ra/Ra_{cr}), with extremely vigorous convection ($Ra \gg Ra_{cr}$) having different c and β_n than sluggish convection ($Ra \geq Ra_{cr}$) (*Dumoulin et al.,* 1999; *Mitri and Showman,* 2005) (see also section 4.3). Thus, the possibility of convection and the efficiency of convective heat transport depend critically on the viscosity of water ice and its variation as a function of temperature and applied stress.

3. ICE RHEOLOGY AND GRAIN SIZE

A large volume of laboratory experiments and field measurements exist regarding the *rheology* of ice, meaning how solid ice flows in response to an applied stress. *Durham and Stern* (2001) provide an excellent review of recent developments in this area. Here, we focus on advances most relevant to Europa. We discuss the rheology of water ice and the role of impurities in modifying the ice flow law and in controlling ice grain size. We discuss control of ice grain size by secondary phases, dynamic recrystallization, and tidal stresses. It should be noted that despite a number of important advances in the last decade, the deformation mechanisms that accommodate large convective strains in Europa's ice shell and their descriptive parameters are still uncertain. Further laboratory experiments characterizing the behavior of ice at conditions relevant to Europa's ice shell are needed.

3.1. Rheology of Pure Water Ice

Over millions of years, solid water ice, like rock-forming minerals, can behave as a highly viscous fluid. Solid ice is a polycrystalline material composed of individual grains; within each grain, the orientation of the water crystal lattice is constant. Like all polycrystalline solids, deformation in ice is accommodated by the motion of defects in the polycrystal, either within grains, or along grain boundaries. A voluminous literature dating back to the 1900s exists regarding the behavior of water ice, much of it developed by the glaciological community who sought to understand the fluid-like behavior of ice observed in large ice sheets and glaciers.

Ice rheology has traditionally been characterized in two regimes: a high-stress regime (appropriate for glacial flow and the subject of considerable study in field and laboratory settings), and a low-stress regime, wherein the relationship between stress and strain rate has been estimated theoretically. Laboratory and glacial studies typically characterize ice behavior at stresses between 10^{-2} and 10 MPa and strain rates between 10^{-7} s^{-1} to 10^{-4} s^{-1} (*Durham and Stern,* 2001). Typical convective strain rates on Europa are ~10^{-13} s^{-1} (*Tobie et al.,* 2003) and stresses ~10^{-3} MPa (*Tobie et al.,* 2003) (see also section 3.1.1 here), so some extrapolation to lower stresses and strain rates is required to apply laboratory or field results to the study of satellite interiors. The rheology of ice at low stresses may be appropriate for modeling deformation in the warm interiors of convecting ice shells. The laboratory- and field-derived flow laws for ice I at high stresses are most appropriate for modeling, for example, the onset of convection and lithospheric deformation. The boundary between the "high" and "low" stress regimes depends on temperature and grain size (see Fig. 2). Both regimes of behavior may be relevant to calculating tidal dissipation, but we note that the behavior of ice I undergoing cyclical deformation at europan frequencies is not well-characterized: Further laboratory and field characterization is urgently needed to advance our understanding of tidal heating.

3.1.1. Low stress regime: Diffusion creep. The behavior of ice in the low-stress regime is relevant to modeling flow within the warm regions of Europa's icy shell (T > 180 K), in locations where the ice grain size is small (d < 1 mm), and/or where the driving stresses are relatively low (σ < 0.1 MPa). Owing to its low gravity, the thermal buoyancy stresses that drive motion within Europa's icy shell are small. Within the warm, well-mixed interior of a convecting icy shell, thermal buoyancy stresses $\sigma_i \sim \rho g \alpha \Delta T_i \delta_{rh} \sim 10^{-3}$ MPa (equation (25) of *Solomatov and Moresi,* 2000); see also *Tobie et al.,* 2003), where $\Delta T_i \sim 10$ K is the magnitude of temperature fluctuations driving convection and δ_{rh} is the rheological boundary layer thickness (*Solomatov and*

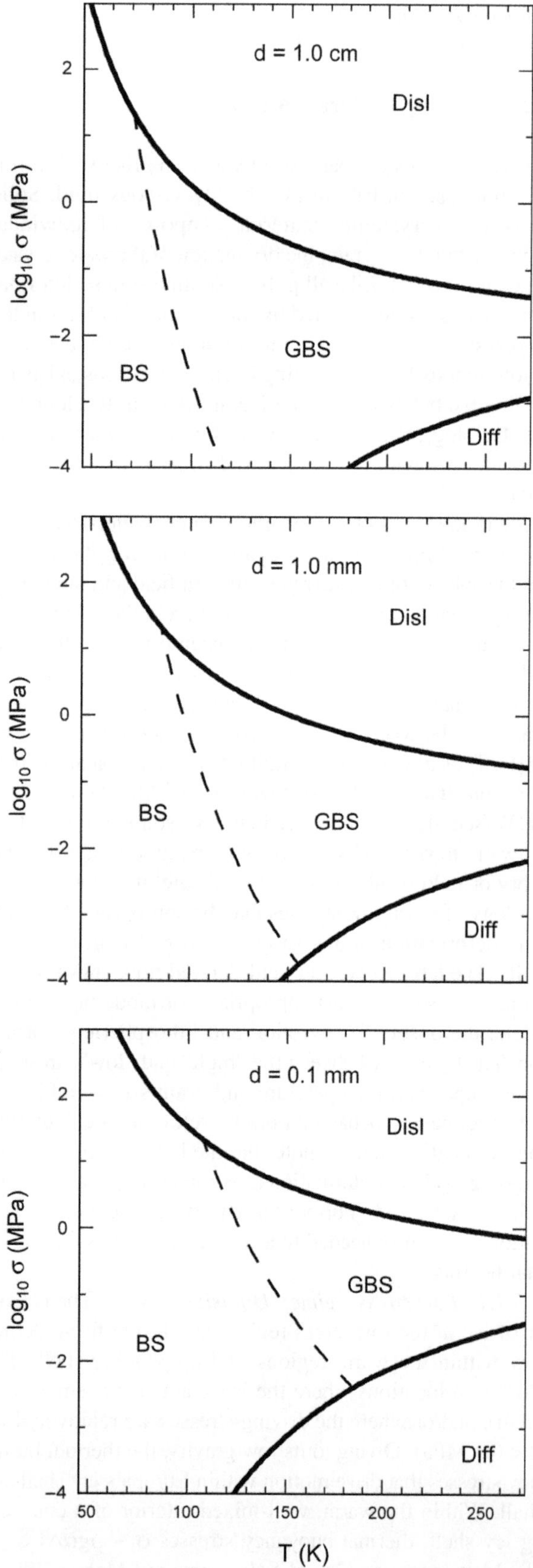

Fig. 2. Deformation maps for ice I using the rheology of *Goldsby and Kohlstedt* (2001), for ice with grain sizes of 1.0 cm, 1.0 mm, and 0.1 mm. Lines on the deformation map represent the transition stress between mechanisms as a function of temperature. From *Barr et al.* (2004).

Moresi, 2000) ~O(1 km). Like most rock-forming minerals, ice is thought to deform by diffusion creep at low stresses, high temperatures, and in materials with small grain size. Diffusion creep occurs by two processes: volume diffusion creep (Nabarro-Herring creep) and grain boundary diffusion creep (Coble creep) (*Goodman et al.,* 1981)

$$\dot{\varepsilon}_{diff} = \frac{42V_m\sigma}{3RTd^2}\left(D_v + \frac{\pi\delta}{d}D_b\right) \tag{12}$$

where σ is differential stress, $R = 8.314\ J\ mol^{-1}\ K^{-1}$ is the gas constant, d is grain size, D_v is the rate of volume diffusion, $\delta = 2b$ is the grain boundary width, b is Burger's vector for ice, V_m is the molar volume, and D_b is the rate of grain boundary diffusion (*Goldsby and Kohlstedt,* 2001). Each diffusion coefficient is strongly temperature-dependent, $D_v = D_{o,v}exp(-Q_v/RT)$, and $D_b = D_{o,b}exp(-Q_b/RT)$. Note also that the strain rate from diffusion creep is grain-size dependent.

To date, diffusion creep in ice has not been directly observed in laboratory experiments, so values of its governing parameters (summarized in Table 1) are calculated based on microphysical models of the diffusion processes. *Goodman et al.* (1981) provide a comprehensive discussion of diffusion processes in ice, section 5.5 of *Goldsby and Kohlstedt* (2001) gives recent updates for governing parameters, and *Goldsby* (2007) provides an update on efforts to observe diffusion creep in the laboratory. At conditions appropriate for a warm convecting ice shell with reasonable grain sizes ~0.1 mm to 1 mm, the deformation rate from diffusion creep is overwhelmingly dominated by the volume diffusion term. At T > 258 K, the rate of Coble creep in ice is expected to increase by a factor of 1000, due to premelting along grain boundaries and triple junctions, which allows for more efficient grain boundary diffusion than a purely solid grain boundary (*Goldsby and Kohlstedt,* 2001). This results in a marked decrease in the viscosity of ice within 10 K of the melting point (see deformation maps of *Durham and Stern,* 2001), an effect that has been largely overlooked in current numerical studies (with the noted exception of *Tobie et al.,* 2003).

An effective viscosity can be calculated from the stress-strain rate relationship (*Durham and Stern,* 2001; *Ranalli,* 1987)

$$\eta = \frac{1}{3^{(n+1)/2}}\frac{\sigma}{\dot{\varepsilon}} \tag{13}$$

where the factor of $3^{(n+1)/2}$, where n is the rheological stress exponent, is included because the stresses that drive deformation have not been resolved into shear and normal components (*Ranalli,* 1987). For diffusion creep, n = 1. This gives an effective viscosity due to volume diffusion

$$\eta_{diff} = \frac{RTd^2}{42V_mD_{o,v}}exp\left(\frac{Q_v^*}{RT}\right) \tag{14}$$

The resulting behavior of ice is said to be "Newtonian," meaning that the effective viscosity is independent of stress,

TABLE 1. Rheological parameters for ice I after *Goldsby and Kohlstedt* (2001).

Parameter	Basal Slip	Grain Boundary Sliding	Dislocation Creep
B (m^p Pa^{-n} s^{-1})	2.2×10^{-7}	6.2×10^{-14}	4.0×10^{-19}
n	2.4	1.8	4.0
p	0	1.4	0
Q^* (kJ mol^{-1})	60	49	60
Parameter	**Name**	**Volume Diffusion**	
V_m (m^3 mol^{-1})	Molar volume	1.97×10^{-5}	
b (m)	Burger's vector	4.52×10^{-10}	
δ (m) = 2b	Grain boundary width	9.04×10^{-10}	
$D_{o,v}$ (m^2 s^{-1})	Volume diffusion constant	9.10×10^{-4}	
Q^*_v (kJ mol^{-1})	Volume diffusion activation energy	59.4	
$D_{o,b}$ (m^2 s^{-1})	GB diffusion constant	7.0×10^{-4}	
Q^*_b (kJ mol^{-1})	GB diffusion activation energy	49	

and that stress and strain rate are linearly related. The volume diffusion flow law and variants of it have been widely applied to study of the interior of Europa's icy shell since the 1980s. It is common to rewrite the flow law as

$$\eta = \eta_o \exp\left[A\left(\frac{T_m}{T} - 1\right)\right] \qquad (15)$$

which is equivalent to equation (14) if $A = Q^*/RT_m \sim 26$ for pure water ice and η_o is equal to $\eta_{diff}(d_o, T_m)$, where d_o is an assumed grain size. The value η_o is commonly assumed to be a free parameter ranging from $\eta_o \sim 10^{13}$ to 10^{15} Pa s, corresponding to grain sizes of 0.1 mm and 1 mm, respectively (using equation (14)). The volume diffusion rheology has been used in all existing calculations of tidal dissipation within Europa's icy shell (*Ojakangas and Stevenson,* 1989; *Tobie et al.,* 2003; *Showman and Han,* 2004; *Mitri and Showman,* 2005, 2008b). We note that, at the stresses associated with Europa's daily tidal flexing, ~0.1 MPa, or during the onset of convection, non-Newtonian deformation mechanisms could be relevant (*Barr et al.,* 2004).

3.1.2. High stress regime: Grain-size-sensitive and dislocation creep. The behavior of ice I at relatively high stresses $\sigma > 0.01$ MPa is well-characterized by laboratory experiments and glacial measurements. Although the stresses in a convecting ice shell on Europa are relatively low, larger stresses ($\sigma \sim 0.01$ MPa) may build up in the icy shell during the onset of convection (see section 4.2.1) or by lithospheric deformation. In ice with a grain size d > 1 mm, flow driven by stresses of this magnitude is accommodated by dislocation creep and grain-size-sensitive creep.

In this regime, the stress-strain rate relationship for water ice is described as

$$\dot{\varepsilon} = \frac{B}{d^p}\sigma^n \exp\left(\frac{-Q^*}{RT}\right) \qquad (16)$$

where B and n are constants determined in laboratory experiments, or from measuring glacial flow, and p is the grain size exponent. A summary of flow laws determined between 1952 and 1979 by *Weertman* (1983) reveals the level of uncertainty in ice rheology during the Voyager era. Values of n ranging from 1.6 to 4 had been measured in different contexts: Creep in polycrystalline ice at low temperature (perhaps most appropriate for Europa) suggested n ~ 3 and activation energies between 60 to 80 kJ mol^{-1} (*Weertman* (1983). Because n > 1, the effective ice viscosity in the high-stress regime depends on stress (i.e., ice is "non-Newtonian," meaning that strain rate depends nonlinearly on stress).

Laboratory experiments reveal that for $\sigma > 1$ MPa, deformation in ice occurs by dislocation creep, characterized by equation (16) with n = 4 and $Q^* = 60$ kJ mol^{-1} (*Goldsby and Kohlstedt,* 2001). Dislocation creep has a high stress exponent n = 4, so the strain rate in cold ice with a large grain size depends strongly on the applied stress (i.e., the ice is highly non-Newtonian) and its strain rate is independent of grain size. Similar to the low-stress regime, strain rates from dislocation creep in ice with T > 258 K are also increased due to premelting at grain boundaries and three- and four-grain junctions in ice (see section 5.4 of *Goldsby and Kohlstedt,* 2001).

Recent laboratory experiments suggest the existence of an intermediate regime for 0.01 MPa < σ < 1 MPa (see Fig. 5 of *Goldsby and Kohlstedt,* 2001), wherein deformation occurs by weakly non-Newtonian deformation mechanism(s) referred to collectively as "grain-size-sensitive" (GSS) creep. GSS creep is characterized by a relatively low-stress exponent n ~ 2, a relatively low grain-size exponent 1 < p < 2, and a modest activation energy $Q^* \sim$ 49–60 kJ mol^{-1} (*Goldsby and Kohlstedt,* 2001; *Durham et al.,* 2001). GSS creep is of particular interest to the glacial community, because most large ice bodies on Earth, which have grain sizes ~1–10 mm and driving stresses ~0.1 MPa, are deforming in the intermediate stress regime.

Although the governing parameters of GSS creep are generally agreed upon, identification of the specific microphysical process that accommodates strain in this regime is an open area of debate. The values of the governing parameters strongly suggest that easy slip (equivalently basal slip), where ice grains deform along the basal planes of their hexagonal crystals, occurs in the intermediate regime (*Goldsby and Kohlstedt,* 2001; *Duval et al.,* 2000). A secondary process such as dislocation creep or grain boundary sliding must operate in tandem with basal slip to accommodate deformation in crystals whose planes are not oriented properly for basal slip to occur (*Goldsby and Kohlstedt,* 2002). The identification of this secondary process has been problematic. Scanning electron microscopy of deformed ice samples allowed *Goldsby and Kohlstedt* (2001) to identify instances of grain switching and occurrence of straight grain boundaries and four-grain junctions, providing evidence that grain boundary sliding accommodates easy slip. This gives rise to a combined flow law in the intermediate regime (*Goldsby and Kohlstedt,* 2001)

$$\dot{\varepsilon}_{gss} = \left(\frac{1}{\dot{\varepsilon}_{gbs}} + \frac{1}{\dot{\varepsilon}_{bs}} \right)^{-1} \quad (17)$$

where gbs stands for grain boundary sliding, and bs stands for basal slip, with the strain rate from grain boundary sliding ($\dot{\varepsilon}_{gbs}$) dominating at conditions relevant to the interior of Europa's ice shell (*Barr et al.,* 2004). Governing parameters for GSS creep effectively controlled by GBS are summarized in Table 1.

However, it has been suggested that grain boundary sliding and basal slip acting together are not able produce the crystal fabric (the coalignment of crystal lattices in adjacent grains) observed in deformed sections of terrestrial ice sheets (*Duval et al.,* 2000; *Duval and Montagnat,* 2002). *Montagnat and Duval* (2000) suggest an alternate hypothesis: that grain boundary migration (essentially, grain growth) and associated recrystallization accommodates basal slip. However, *Goldsby and Kohlstedt* (2002) point out that grain boundary migration does not produce strain and thus is not a deformation mechanism. The identification of the microphysical process accommodating deformation in ice at moderate stresses remains an active area of research.

3.1.3. A combined flow law. *Goldsby and Kohlstedt* (2001) propose that the behavior of ice I across the high-, intermediate-, and low-stress regimes can be described by a single governing equation

$$\dot{\varepsilon}_{total} = \dot{\varepsilon}_{diff} + \dot{\varepsilon}_{gss} + \dot{\varepsilon}_{disl} \quad (18)$$

where, in application to Europa's icy shell, the strain rate due to GSS may be approximated by the governing parameters of grain boundary sliding (*Barr et al.,* 2004). Despite uncertainties in the microphysical mechanisms at work in GSS creep, a combined flow law including both GSS and dislocation creep using governing parameters summarized in Table 1 provides a good match to stress/strain rate/grain size relationships deduced from previous laboratory experiments (*Goldsby and Kohlstedt,* 2001) and glacial measurements (*Peltier et al.,* 2000). The composite flow law can be used to determine the regimes of dominance in stress, temperature, and grain size space for each constituent deformation mechanism. By equating strain rates between pairs of mechanisms, one can construct a deformation map for ice that can be used to predict which rheology is appropriate for a given application (see Fig. 2 and deformation maps of *Durham and Stern,* 2001).

3.2. Effect of Impurities

The presence of substances other than water ice in Europa's icy shell can have an important effect on its rheology. Here we summarize how the presence of various materials, including ammonia, sulfate salts, and dispersed particulates, may affect the rheology of Europa's icy shell.

Ammonia dihydrate, $NH_3 \cdot 2H_2O$, the stable phase in the water-rich, low-pressure region of phase space in the ammonia-water system, melts at T_m = 176 K. Ammonia has been suggested as a possible means to thermodynamically stabilize liquid water oceans beneath convecting ice shells (e.g., *Spohn and Schubert,* 2003) and implicated as a possible component of cryovolcanic magmas on icy satellites. The rheology of ammonia dihydrate has been measured in laboratory experiments by *Durham et al.* (1993). The flow law for ammonia dihydrate, with mole fraction x_{NH_3} = 0.3 (corresponding to a mole fraction of 90% dihydrate), can be expressed in similar form as equation (16), with B = 10^{-15} $Pa^{-5.8}$ s^{-1}, n = 5.8, and Q^* = 102 kJ mol^{-1}. At its melting point and a nominal stress of 0.01 MPa, ammonia dihydrate is 2 orders of magnitude less viscous than water ice, but its large activation energy leads to a rapid increase in viscosity as the temperature is decreased.

Galileo NIMS data suggest that the surface of Europa's icy shell is composed predominantly of water ice and nonice materials that include one or more hydrated materials. Candidates for the latter include hydrated magnesium and/or sodium salts (*McCord et al.,* 1999) or hydrated solidified sulfuric acid (*Carson et al.,* 2005). Geochemical modeling of water-rock chemistry in Europa's ocean also suggest the formation of magnesium and/or sodium salts, supporting the view that the icy shell may be salty throughout (*Zolotov and Shock,* 2001; *McKinnon and Zolensky,* 2003). The sulfate salts were found to have much higher viscosities than pure water ice at comparable temperatures (*Durham et al.,* 2005). For example, the difference in hardness between mirabilite grains and water ice is so high that the dispersed mirabilite particles can act as hard secondary phases, and have a similar effect on the rheology of the bulk material as silicate grains.

Recent laboratory experiments by, e.g., *McCarthy et al.* (2007) have explored dissipation in mixtures of ice and magnesium sulfate. Frozen eutectic mixtures of ice and magnesium sulfate form a lamellar structure with layers of ice and magnesium sulfate sandwiched together. The resulting mixture is stiff but highly dissipative due to the microscale boundaries between layers of magnesium sulfate and water ice. Future laboratory experiments on pure water ice and water ice mixed with other materials may provide alternate models for dissipation than the Maxwell model, and help bridge the gap between theory and data regarding tidal dissipation in Europa.

Europa's icy shell may contain small amounts of silicate dust, but the dust content is limited by its bulk density — the icy shell must maintain a low enough density to remain gravitationally stable atop the $\rho \sim 1000$ kg m^{-3} ocean (or more if the ocean is salty). Small rock particles (with sizes much less than the mean ice grain size) mixed with ice act as a barrier to flow within the ice and disturb the flow pattern — as a result, the effect of particles on ice viscosity depends critically upon the location of the particles within the polycrystalline structure of ice. Laboratory experiments characterizing the viscosity of mixtures of ice with a grain size of ~1 mm and silicate particles with grain sizes ~100 μm show that the increase in viscosity of water ice due to the presence of SiO_2 and SiC/$SiCaCO_3$ grains at low volume fractions relevant to Europa's icy shell, $\phi < 0.1$, is negligible (see Fig. 14 of *Durham et al.,* 1992). However, silicate particles may play a role in inhibiting grain growth in Europa's icy shell, as will be discussed in section 3.3.1.

3.3. Ice Grain Size

Observations of grain size and the processes that control grain size in terrestrial ice sheets experiencing stress and temperature conditions similar to the interior of Europa's icy shell can provide estimates of likely grain sizes. Grain sizes in Europa's shell are commonly assumed to be uniform and between 0.1 mm to 10 mm, by analogy with grain sizes in terrestrial ice sheets (*Budd and Jacka,* 1989) [see also, e.g., *Thorsteinsson et al.* (1997) for a sample ice core grain size profile].

Two recent works cast doubt upon these estimates and have led to a reevaluation of the plausibility of the 0.1–10-mm range for assumed ice grain sizes. *Nimmo and Manga* (2002) estimated the viscosity at the base of the icy shell by using the measured diameters of pits and uplifts (~4–10 km) to infer properties of the underlying convection pattern. They obtained a basal viscosity for the icy shell between 10^{12} and 10^{13} Pa s, and suggested that ice grain sizes at the base of the shell should be between 0.02 and 0.06 mm. Follow-on studies by *Showman and Han* (2004) suggest that near-surface ice grain sizes less than 0.04 mm are required to create small depressions such as those observed on the surface of Europa (see section 5.1). *Schmidt and Dahl-Jensen* (2003) applied a simple model of unimpeded grain growth in the low-stress, high-temperature, and liquid-rich environment at the base of Europa's icy shell and suggested that its grain size may be between 4 cm and 80 m. Because the rate of volume diffusion depends on grain size squared, the range of ice viscosity implied by all the above estimates is 15 orders of magnitude.

However, the estimates of *Schmidt and Dahl-Jensen* (2003) ignore processes known to modify grain sizes within terrestrial ice cores. In addition, the geodynamical studies did not account for near-surface weakening by other processes (e.g., microcracking), which could explain the low effective near-surface viscosity required to create depressions, or the possibility of premelting at the base of Europa's icy shell, which could yield low basal viscosities for a much larger grain size. Because stress and temperature conditions expected within Europa's icy shell are similar to those experienced by many ice bodies on Earth, processes controlling grain sizes within terrestrial ice sheets may control grain size within Europa's ice shell and act to self-regulate and/or limit its ice grain size to values closer to those observed in ice cores. Here, we discuss two possible methods of controlling ice grain size in Europa's icy shell: (1) Zener pinning due to the presence of hard secondary phases (*Kirk and Stevenson,* 1987; *Barr and McKinnon,* 2007), and (2) dynamic recrystallization/tidal flexing (*McKinnon,* 1999; *Barr and McKinnon,* 2007).

3.3.1. Grain size control by secondary phases. Measured grain sizes in many terrestrial ice cores indicate that grain size in impurity-laden ice are invariably smaller than those in clean ice (*Alley et al.,* 1986a,b). In the absence of impurities, grains grow by grain boundary migration driven by the free energy decrease associated with reduction of grain boundary curvature. Non-water-ice materials concentrate on grain boundaries, and can decrease the grain growth rate and in some cases can even halt grain growth altogether (*Poirier,* 1985; *Alley et al.,* 1986a,b). The role that any type of impurity plays in inhibiting or preventing grain growth depends on the location of the impurity within the structure of the ice polycrystal: Impurities concentrated along grain boundaries and at grain junctions can be much more effective at inhibiting grain growth than impurities randomly dispersed in the ice (*Durand et al.,* 2006). Recent advances in SEM imaging of samples from ice cores shows the spatial correlation between silicate microparticles and kinks in grain boundaries, providing compelling evidence that silicate particles can inhibit grain growth (*Weiss et al.,* 2002).

On Europa, one could imagine that silicate or salt particles might act as hard secondary phases (or pinning particles) that could slow or halt grain growth. The effect of pinning particles on grain size was modeled by Zener, who related the drag force exerted by hard secondary phases on grain boundaries to the rate of grain growth dr/dt (*Poirier,* 1985)

$$\frac{dr}{dt} = K_{g,0}\exp\left(\frac{-E_A}{RT}\right)\left(\frac{1}{r} - \frac{P_Z}{\alpha_G\gamma_{gb}}\right) \qquad (19)$$

where $E_A = 46$ kJ mol^{-1} is the activation energy for grain boundary migration, $\gamma_{gb} = 0.065$ J m^{-2} is the grain boundary free energy (*De La Chapelle et al.,* 1998), and $\alpha_G = 0.25$ is a geometric factor. The pinning pressure P_Z exerted on the grain boundary is related to the number of particles on the boundary (fN_x) (*Poirier,* 1985) (where f is the number fraction of particles residing on the grain boundary), $P_Z = \frac{1}{3}\pi\gamma_{gb}r_x rfN_x$, where r_x is the radius of the particles residing on the grain boundary and N_x is the number of particles per unit volume. Grain growth is completely stopped when dr/dt = 0, which gives the Zener limiting grain size (cf. *Durand et al.,* 2006)

$$r_z = \left(\frac{3\alpha_G}{\pi r_x f N_x}\right)^{1/2} \quad (20)$$

where the numerical factor of 3 indicates that the particles reside on grain boundaries, and f ~ 0.25 is estimated by counting the number of particles residing on grain boundaries in SEM images of the GRIP ice core (*Barnes et al.,* 2002; *Weiss et al.,* 2002). If we suppose that the ice shell of Europa is loaded with microscopic silicate particles of density 3000 kg m^{-3} and radii of 10 μm, up to a total volume percentage of 4%, the Zener limiting grain size is

$$r_z = 0.1\ \text{mm}\left(\frac{r_x}{10\ \mu\text{m}}\right)\left(\frac{0.04}{\phi}\right)^{1/2} \quad (21)$$

where ϕ is the volume fraction of silicate in Europa's icy shell and $N_x \sim \phi/(\frac{4}{3}\pi r_x^3)$. The upper limit on ice grain size derived from the Zener pinning model is inversely proportional to impurity content — fewer impurities mean larger grain sizes. We note that *Kirk and Stevenson* (1987) constructed a similar argument to estimate grain sizes in the ice mantle of Ganymede: Their estimate has a similar dependence on radii of the silicate particles but assumes randomly distributed particles and gives $r_z \sim \phi^{-1}$. For Europa, we can put a plausible estimate on the upper limit of silicate content based on the density of the icy shell — if the shell density exceeds 1000 kg m^{-3} (or more for a salty ocean), it will be gravitationally unstable atop the ocean.

Observations of grain sizes between 0.5 and 1 mm in impurity-laden sections of terrestrial ice cores, coupled with our upper limit on ice grain size based on a Zener pinning model, suggest that grain sizes in Europa's icy shell may hover around the 0.1–1-mm range (*McKinnon,* 1999; *Barr and McKinnon,* 2007). At present, terrestrial observations of grain growth and impurity distribution within polycrystalline ice are limited to temperatures between 235 to 273 K. Therefore, knowledge of grain size and processes controlling grain size gained through study of terrestrial cores may be most applicable to the warm and convecting interior ice shell (*Barr and McKinnon,* 2007). Grain sizes closer to the surface of Europa may be controlled by other, non-thermally activated processes such as cyclical tidal deformation (*McKinnon,* 1999).

3.3.2. Dynamic recrystallization. Observations of relatively impurity-free sections of terrestrial ice cores reveal that ice grain sizes are constant as a function of depth (equivalently, time) within sections of the core that have experienced significant strain (*Thorsteinsson et al.,* 1997; *De La Chapelle et al.,* 1998). If ice grains can grow unimpeded, one would expect grain size in the ice sheet to increase as a function of depth and time. This suggests that deformation acts to decrease grain size, thereby competing with natural grain growth and allowing a roughly constant steady-state grain size to be maintained over time. It has been suggested that the accumulated strain due to vertical layer compaction results in grain size reduction due to a process called dynamic recrystallization (*Thorsteinsson et al.,* 1997; *De La Chapelle et al.,* 1998). In dynamic recrystallization, the grain size in a deforming material is controlled by a balance between grain growth and the formation of new grains (nucleation) by a process called subgrain rotation (*Shimizu,* 1998; *DeBresser et al.,* 1998). Subgrain rotation can only occur if deformation is occurring in the material, leading to a threshold strain at which grain sizes in a deforming material achieve their steady-state recrystallized values. For temperature and strain rate conditions appropriate for the GRIP ice core, T ~ 240 K and $\dot{\varepsilon} \sim 10^{-12}\ s^{-1}$, the threshold strain is about 25% (*Thorsteinsson et al.,* 1997). A model of this process has been applied to estimate grain sizes within actively deforming regions of ice shells by *Barr and McKinnon* (2007), who find that in the well-mixed, warm convective interior of an already convecting ice shell, grain sizes will evolve to a steady-state value that depends on the applied stress (*Derby,* 1991; *Shimizu,* 1998; *DeBresser et al.,* 1998; *Barr and McKinnon,* 2007)

$$d_{recrys} = Kb\left(\frac{\sigma}{\mu}\right)^{-m} \quad (22)$$

where K = 1–100 is a grouped material parameter, μ is the ice shear modulus, and m = 1.25. *Barr and McKinnon* (2007) suggest that in the absence of impurities, recrystallized grain sizes in convecting ice shells will be large, d_{recrys} ~ 30–80 mm, which may lead to highly viscous ice and a gradual shut-down of convection as the grains achieve their recrystallized value. The implication is that without impurities to limit grain growth, ice shells may convect sluggishly, and may be limited to a small number of convective overturns before transitioning to a conductive state.

On Europa, however, tidal flexing of the icy shell itself may control the ice grain size. *McKinnon* (1999) hypothesized that if cyclical straining in the presence of convection of Europa's icy shell has the same effect as the continuous strain on terrestrial ice cores (i.e., driving dynamic recrystallization), the grain size in the icy shell would decrease as $d \propto \dot{\varepsilon}^{-1/2}$. The grain size controlled in this manner would have a *maximum* of 1 mm at the warm base of the icy shell. Thus, cyclical tidal flexing may prevent grains in Europa's icy shell from growing to the large values predicted by continuous-deformation dynamic recrystallization models (*Barr and McKinnon,* 2007), exempting Europa's icy shell from being choked while convecting.

4. ICY SHELL CONVECTION

In section 2, we described two possible modes of removing tidal heat generated in Europa's icy shell: conduction and convection. But how do we decide whether convection can happen? Until recently, knowledge of how convection starts and stops in realistic planetary mantles was relatively limited because terrestrial planet mantles are commonly assumed to convect throughout most of their geological history. Although many of the techniques developed for studying terrestrial planet mantle convection apply to Europa, the heat flow history in Europa's icy shell sets it apart from terrestrial planets. Unlike a terrestrial planetary mantle, Europa's icy shell may receive periodic bursts of heat due to tidal dissipation in addition to radiogenic heating from its rocky mantle. As a result, the mode of heat transport across Europa's icy shell may change from conductive to convective many times during its evolution (*Mitri and Showman,* 2005). Here, we summarize several decades' worth of study of the issue of whether convection is possible in Europa's ice shell, how convection may start and stop, and the efficiency of convective heat transport.

4.1. Is Convection Possible?

Early efforts to judge whether convection could happen in Europa's icy shell used two implicit assumptions. First, it was assumed that deformation during the onset of convection would be accommodated by Newtonian diffusion creep, even though field and laboratory measurements of ice viscosity suggested that ice could be non-Newtonian if the grain size is sufficiently large. It was also assumed that the value of the critical Rayleigh number was independent of the type of temperature perturbations available to trigger convection in the icy shell. Here, we discuss the consequences of relaxing these assumptions. Recently developed numerical techniques allow for the study of how convection may be triggered from "realistic" temperature fluctuations in the icy shells; we summarize the results of these studies.

The simplest representation of Europa's icy shell in the language of fluid dynamics is a layer of fluid cooled from above and heated from within and beneath. It is common to assume that a conductive icy shell on Europa is heated mostly at its base by radiogenic and tidal heating because tidal dissipation is likely maximized there (however, this assumption is not necessarily correct; see section 5.2.1). It is also common to model the ice shell as using a two-dimensional Cartesian geometry. Although Europa's icy shell is, truly, a spherical shell, plausible icy shell thicknesses are small compared to the radius of Europa, so for many purposes it is sufficient to think of it as a Cartesian box. At present, most simulations of convection in Europa's icy shell are performed in two dimensions because of limited computing resources.

The question of whether convection can occur in a fluid layer has been studied in a variety of planetary and fluid dynamical contexts. It is a simple geophysical argument: Does the Rayleigh number of the fluid layer exceed a critical value? Using the definition of the Rayleigh number (equation (10)), this can be phrased mathematically as

$$\frac{\rho g \alpha \Delta T D^3}{\kappa \eta} \geq Ra_{cr} \tag{23}$$

where Ra_{cr} is the critical Rayleigh number. The value of Ra_{cr} for convection in any fluid depends on the wavelength of initial temperature perturbation within the fluid layer and the geometry of the layer (see *Turcotte and Schubert,* 2002, for discussion). Because both the thickness of Europa's ice shell and its viscosity are poorly constrained at present, we cannot definitively determine whether convection can occur: The best we can do is determine a critical shell thickness where convection is possible by rearranging equation (20)

$$D_{cr} = \left(\frac{Ra_{cr} \kappa \eta}{\rho g \alpha \Delta T}\right)^{1/3} \tag{24}$$

and constrain values of D_{cr} as a function of physical properties of the icy shell.

4.1.1. Newtonian viscosity. Early works such as that of *Consolmagno and Lewis* (1978) and *Reynolds and Cassen* (1979) approximated ice as a constant-viscosity fluid. In a constant viscosity fluid, $Ra_{cr} \approx 1000$ (*Chandrasekhar,* 1961), so estimates of the mean ice viscosity, thermal and physical parameters, and surface temperature and ice melting temperatures on Europa could be used to determine D_{cr} from equation (24) alone. *Reynolds and Cassen* (1979) determined that convection could occur in a bottom-heated ice I shell on a Europa-like satellite if $D \gtrsim 30$ km.

A source of uncertainty in evaluating D_{cr} for a constant-viscosity ice shell is the appropriate choice of viscosity value. The viscosity of ice is strongly temperature-dependent, so do we evaluate D_{cr} using $\eta(T_s)$, or $\eta(T_m)$, or some well-chosen mean? This is addressed using algebraic ("scaling") relationships between the activation energy in the ice flow law (which controls $\partial\eta/\partial T$) and the critical Rayleigh number for a Newtonian fluid by *Stengel et al.* (1982) (for $n = 1$) and *Solomatov* (1995) (for general n).

The analysis of *Solomatov* (1995) focuses on the behavior of the bottom thermal boundary layer of a basally heated fluid with a strongly temperature-dependent viscosity at the onset of convection. If the viscosity of the fluid depends strongly on temperature [if $\eta(T_s)/\eta(T_m) \gtrsim 10^4$ (*Solomatov,* 1995)], fluid motions in the upper part of the layer are miniscule, and the upper part of the layer forms a lid of cold, high-viscosity fluid (referred to as a "stagnant" lid). In the so-called "stagnant lid regime," convective fluid motions are confined to a warm sublayer at the base of the fluid, where the temperature is approximately constant, and the temperature dependence of ice viscosity can be neglected. With this approximation, the critical Rayleigh number for convection in a fluid with a temperature-dependent viscosity can be estimated by determining when the warm sublayer begins to convect, or determining when the *local* Rayleigh number in the sublayer exceeds 1000. The result is a scaling relationship between the critical Rayleigh number and rheological parameters of the fluid (*Solomatov,* 1995)

$$Ra_{1,cr} = Ra_{cr}(n)\left[\frac{e\theta}{4(n+1)}\right]^{2(n+1)/n} \tag{25}$$

where the subscript 1 indicates that we will compare this critical Rayleigh number to the Rayleigh number at the base of the ice shell (where $T = T_b$), n is the rheological stress exponent, $\theta = \gamma\Delta T$, where the Frank-Kamenetskii parameter $\gamma = -\partial(\ln\eta)/\partial T|_{T_i}) = Q^*/(RT_i^2)$, and $Ra_{cr}(n) \approx Ra_{cr}(1)^{1/n}$ $Ra_{cr}(\infty)^{(n-1)/n}$. In icy satellite convection studies, it is commonly assumed that the warm, well-mixed convective interior of the ice shell has a temperature very close to the

TABLE 2. Critical Rayleigh numbers for stopping and starting convection in a Newtonian ice shell (after *Solomatov and Barr,* 2007).

$\log_{10}(\Delta\eta)$	Equivalent θ	Ra^*_{cr}	$Nu(Ra^*_{cr})$	$Ra_{cr,1}$	$Nu(Ra_{cr,1})$
5	11.5	2.7×10^5	1.30	3.67×10^5	1.56
8	18.4	1.35×10^6	1.22	2.41×10^6	1.53
10	23.0	2.94×10^6	1.19	5.88×10^6	1.49
12	27.6	5.61×10^6	1.17	1.218×10^7	1.47
16	36.8	1.58×10^7	1.14	3.59×10^7	1.43
20	46	3.57×10^7	1.12	9.40×10^7	1.40

ice melting point, so $T_i \approx T_b = T_m$. This gives $\theta \approx (Q^*\Delta T)/(RT_m^2)$ (cf. *McKinnon,* 2006). The value $Ra_{cr}(n)$ represents the critical Rayleigh number for convection in a fluid with a viscosity dependent solely on stress (i.e., $\eta = B\sigma^n$), which is estimated using the value for n = 1, $Ra_{cr}(1) = 1568$, and the limit of $Ra_{cr}(n)$ as $n \rightarrow \infty$, or $Ra_{cr}(\infty) \approx 20$ (see Fig. 5 of *Solomatov,* 1995).

A large volume of work exists regarding the onset of convection in ice I shells of the satellites assuming a Newtonian rheology for ice: In these studies the viscosity of ice depends strongly on temperature, but is independent of stress. For a Newtonian ice rheology, n = 1, and equation (25) reduces to (*Stengel et al.,* 1982; *Solomatov,* 1995)

$$Ra_{1,cr} = 20.9\theta^4 \tag{26}$$

A surface temperature of $T_s = 100$ K and basal temperature of $T_m = 260$ K implies $\theta \approx 18$, which gives $Ra_{cr,1} = 2.2 \times 10^6$ (using equation (26)). Values of $Ra_{cr,1}$ for a range of θ values appropriate for volume diffusion and Europa's icy shell are summarized in Table 2. Evaluating equation (24) using $\rho = 920$ kg m^{-3}, $\kappa = 2.6 \times 10^{-6}$ m^2 s^{-1} as a representative value for warm ice, and evaluating the ice viscosity using coefficients in the volume diffusion flow law (see Table 1), gives an expression for the critical shell thickness for convection (cf. *McKinnon,* 1999)

$$D_{cr,diff} = 31\ \text{km}\left(\frac{d}{0.4\ \text{mm}}\right)^{2/3} \tag{27}$$

where d = 0.3 mm gives a basal ice viscosity of 10^{14} Pa s, midway between the values of 10^{13} to 10^{15} Pa s commonly assumed in Europa studies.

4.1.2. Realistic ice rheology. For ice with small grain sizes d ≲ 1 mm, diffusion creep likely accommodates strain during the onset of convection, and the critical ice shell thickness for convection is given by equation (27). However, the composite flow law for ice suggests that deformation at stresses built up within ice shells during the onset of convection may be accommodated by non-Newtonian GSS creep (*Barr et al.,* 2004). If convection is triggered by temperature fluctuations of $\delta T \sim 5$ K and height $\lambda \sim D$ (see section 4.2.1), the thermal stress due to a plume of this magnitude is approximately $\sigma_{th} \sim \rho g \alpha \delta T \lambda \sim 0.02$ MPa. For grain sizes ≥1 mm, strain due to stresses on the order of σ_{th} may be accommodated by GSS creep (see Fig. 2). When GSS creep accommodates deformation during the onset of convection

$$D_{cr} = \left(\frac{Ra_{a,1}(\kappa d^p)^{(1/n)}\exp\left(\frac{Q^*}{nRTm}\right)}{(3^{(n+1)/2}A)^{1/n}\rho g \alpha \Delta T}\right)^{n/(n+2)} \tag{28}$$

where $Ra_{a,1} = 3.1 \times 10^4$ is the absolute minimum Rayleigh number where convection is possible (for an optimal perturbation) in GBS with $\theta \approx 15$ appropriate for an ice shell with $T_s = 90$ K and $T_m = 260$ K (*Barr and Pappalardo,* 2005). Equation (28), which is for arbitrary n and can be used for non-Newtonian fluids, reduces to equation (24) for n = 1. Evaluating the rheological parameters for GSS using the grain boundary sliding values (see Table 1), and using nominal values for descriptive properties of the ice shell, the critical shell thickness becomes

$$D_{cr,GSS} = 75\ \text{km}\left(\frac{d}{1\ \text{mm}}\right)^{1.4/3.8} \tag{29}$$

The grain size at which GSS becomes the "controlling" rheology at the onset of convection can be determined by setting equations (27) and (29) equal and solving for d. This gives d = 2 mm, indicating that in ice with a grain size ≲2 mm, diffusion creep likely accommodates strain during the onset of convection. For d > 2 mm, GSS creep accommodates strain during the onset of convection. Figure 1b summarizes how the critical ice shell thickness where convection can occur depends on ice grain size in an ice shell with a realistic Newtonian diffusion creep and non-Newtonian GSS rheology (*Barr and Pappalardo,* 2005).

McKinnon (1999) argues that tidal stresses in Europa's ice shell may alter its rheology. In an ice shell deforming by GSS creep, if the tidal stresses are much greater than the thermal buoyancy stresses driving convection, the GSS viscosity law can become effectively "linearized." The effect is similar to the modification of mantle rheology due to interaction between convection and postglacial rebound on Earth (*Schmeling,* 1987). Within Europa's ice shell, the low-stress convective flow field "sees" an effectively Newtonian rheology, but the high-stress tidal field "sees" a non-Newtonian viscosity. *McKinnon* (1999) estimates a basal viscosity of 8×10^{13} Pa s for tidally linearized GBS, which depends on grain size as $\eta \propto d^{1.4/1.8}$. Convection is possible in a 25-km-thick ice shell with such a rheology if the grain size is about 1 mm, which may be the case if grain size is controlled by tides (see section 3.3.2).

Here, we have calculated critical ice shell thicknesses for convection assuming a constant thermal conductivity for the ice shell. The critical ice shell thickness for convection in a variable-conductivity shell is larger than in a constant-conductivity shell. This effect can be estimated by equating the equivalent heat flow F_{conv} across a shell with variable conductivity to F_{conv} with a constant conductivity (*McKinnon,* 1999, 2006; *Tobie et al.,* 2003; *Barr and Pappalardo,* 2005) (see also equations (2) and (3)).

$$\frac{D_{true}}{D_{cr}} = \frac{a_c}{k_c \Delta T} \ln\left(\frac{T_m}{T_s}\right) \qquad (30)$$

where D_{true} is the actual critical shell thickness with variable conductivity taken into account, D_{cr} is the value obtained assuming a constant conductivity of k_c (here, 3.3 W m^{-1} K^{-1}). For T_s = 100 and T_m = 260 K, D_{true}/D_{cr} ~ 1.17.

Finally, we note that although it is beyond the scope of this chapter, it is possible that the microphysical processes that accommodate the first ~10% strain during the onset of convection are entirely different than those that govern well-developed convection (see, e.g., *Birger,* 1998, 2000). Measurements of ice behavior during transient or primary creep (*Glen,* 1955) may be relevant to the question of the onset of convection in addition to flow laws for steady-state creep (*Solomatov and Barr,* 2007).

The results of recent efforts to refine the range of critical ice shell thickness for convection, which predict critical thicknesses from 10 km to a few tens of kilometers, generally agree with the original estimates derived by *Reynolds and Cassen* (1979): D_{cr} ~ 30 km. Although the *value* of critical ice shell thickness may not have changed much in 30 years, the relationship between ice rheology, the critical shell thickness for convection, and ice grain size has been clarified.

4.2. Behavior of the Icy Shell Close to the Critical Rayleigh Number

4.2.1. Starting convection. In the previous section, we described the results of recent studies attempting to narrow the range of conditions where convection is *possible* in Europa's ice shell. In a mathematically idealized scenario, an unperturbed and heated ice shell will sit quiescently unless temperature fluctuations drive flow and trigger convection. Since the earliest work on convective stability, it has been known that the critical Rayleigh number depends on the shape (e.g., *Turcotte and Schubert,* 2002) and amplitude of temperature perturbation within the fluid layer (see, e.g., *Chandrasekhar,* 1961; *Stengel et al.,* 1982, for discussion).

A key open question is whether tidal dissipation can trigger convection in Europa's icy shell. Because the Maxwell time of warm ice near the base of Europa's shell may be close to Europa's orbital period, a purely conductive ice shell may be heated largely at its base. Recent numerical work suggests that maximally effective perturbations for starting convection are concentrated at the base of the fluid layer and have wavelength λ_{cr} ~ $2(n + 3)\theta^{-1}D$ (*Solomatov and Barr,* 2006, 2007). In the absence of temperature fluctuations, zones of weakness, or other means of localizing tidal dissipation, tidal heating in a conductive ice shell is essentially constant over the horizontal scale of convective cells because the r.m.s. strain rate varies by only a factor of ~2 between the equator to pole. One could envision the temperature perturbation due to tidal dissipation as a smoothly varying harmonic function with a wavelength λ_{tidal} ~ R_{Europa}/4 or so (see Fig. 1 of *Ojakangas and Stevenson,* 1989). If tidal heating is the sole cause of the density differences necessary to trigger convection, tidal dissipation must generate temperature perturbations on horizontal length scales λ ~ λ_{cr} to trigger convection (*Barr et al.,* 2004). If $\lambda_{tidal} \gg \lambda_{cr}$, the critical Rayleigh number would increase substantially, perhaps by a factor of 100 or more, because perturbations with such long wavelengths are inefficient at triggering convection. This suggests that tidal dissipation, as envisioned by *Ojakangas and Stevenson* (1989), may not be able to trigger convection in a *purely* conductive icy shell.

Other types of temperature fluctuations, e.g., bursts of heat released close to the surface of the icy shell from large impacts, are essentially useless in triggering convection because they diffuse away too quickly to warm the surrounding ice enough to permit it to flow. Compositional variations present in a realistic icy shell may be able to provide the necessary density contrasts to trigger convection (e.g., *Pappalardo and Barr,* 2004). Understanding how convection begins in Europa's icy shell will require characterization of the types and locations of temperature fluctuations naturally present in the icy shell.

4.2.2. Stopping convection. If the Rayleigh number of Europa's icy shell drops below the value where convection can be maintained, convection will cease. The Ra of the ice shell may change, for example, due to perturbations in the basal heat flux (*Mitri and Showman,* 2005), or due to an increase in ice grain size over time (*Barr and McKinnon,* 2007). Gray arrows in Fig. 1c describe the path in Ra-Nu space taken by an ice shell where convection is stopping. As the Rayleigh number is decreased, Nu decreases until $Ra = Ra^*_{cr}$, the lowest value of Ra where convection is possible. For Newtonian rheologies in the stagnant lid regime ($\theta > 8$), the value of $Ra^*_{cr} \sim \frac{1}{2}Ra_{cr,1}$ (see Table 2). In the vicinity of this point, $(Nu - Nu_{cr}) \propto (Ra - Ra_{cr})^{1/2}$, and at $Ra = Ra^*_{cr}$, $Nu \approx 1.1$ to 1.3, and convective motion is confined to a very thin layer at the base of the ice shell. Values of Ra^*_{cr} and $Nu(Ra^*_{cr})$ for a range of parameters appropriate for a volume diffusion rheology and range of θ appropriate for Europa's icy shell are summarized in Table 2 (see also Table 1 of *Solomatov and Barr,* 2007). When convection stops, very low values of (Nu – 1) can be achieved, and the minimum value scales with θ^{-1} (*Solomatov and Barr,* 2007).

4.2.3. Conductive-convective switching. When convection starts in an icy shell, it results in a reorganization of heat and mass transfer in its interior. The spatially averaged temperature in the shell increases significantly, as does the heat flux across the shell, resulting in thermal stresses that may be sufficient to drive lithospheric deformation (*Mitri and Showman,* 2005). Here, we summarize the results of these recent studies about convective turn-on and turn-off in an ice shell with a Newtonian rheology, and discuss implications for resurfacing of Europa's icy shell.

The black arrows in Fig. 1c describe the path in Ra-Nu space traced out by a Newtonian ice shell during the onset of convection. From a conductive equilibrium (Nu = 1), the ice shell begins convecting when $Ra = Ra_{cr,1}$ (see equation (25)), and the heat flux jumps rapidly to Nu > 1 (*Mitri and Showman,* 2005; *Solomatov and Barr,* 2007). Values of Nu achieved for $Ra = Ra_{cr,1}$ using a volume diffusion rheology for ice and numerical methods described by *Solomatov and Barr,* 2007) are summarized in Table 2.

The value of Nu achieved during the onset of convection depends on the type of temperature perturbation that exists or develops in the conductive ice shell (*Solomatov and Barr,* 2007). As described in section 4.2.1, the most efficient perturbations at starting convection are confined to the base of the ice shell: The optimal perturbation shape (for stagnant lid convection in a two-dimensional Cartesian geometry) is a small convective roll obtained for $Ra = Ra^*_{cr}$ (see section 4.2.2). Using the optimal perturbation shape gives a lower bound on the jump in heat flux when convection begins; for $\theta = 18$, Nu jumps from 1 to 1.56 when convection begins. Using a sinusoidal perturbation that adds $\delta T = 0.175$ to a background conductive equilibrium, *Mitri and Showman* (2005) find that Nu jumps to 1.7 in a tidally heated ice shell. The difference between the two values provides an estimate of the effect of perturbation geometry on the ΔNu associated with the onset of convection: ΔNu changes by ~20% as the wavelength is varied by a factor of ~2. This suggests that careful consideration of the types of realistic temperature perturbations available to trigger convection in icy satellites is required to obtain more accurate estimates of the heat flux jump when convection begins.

A Nu(Ra) plot, as shown in Fig. 1c, is the most straightforward way of summarizing the heat flux across an ice shell during the onset and decay of convection. To understand the geological consequences of the onset of convection, we need to trace variations in the heat flux as a function of ice shell thickness. For Europa, this is most easily done by assuming that the viscosity at the base of the ice shell (i.e., the melting-temperature viscosity) is independent of shell thickness; the Rayleigh number can then be directly translated into a measure of shell thickness. Likewise, with equation (2) we can translate Nu into heat flux. Figure 1d shows an example of such a plot, from *Mitri and Showman* (2005), assuming the melting-temperature viscosity is 10^{13} Pa s. The existence of a heat flux jump implies that, for a range of heat fluxes relevant to Europa (basal heat fluxes between 35–60 mW m^{-2} in this case), two solutions exist for a given heat flux: a thin conductive shell and a thick convective shell (*Mitri and Showman,* 2005). Modest variations in the heat flux can force the shell to switch between these states. This will have important geophysical consequences.

Imagine a thin, conductive ice shell, with its basal temperature held at the local melting temperature of water ice, which is in steady-state with an enormous basal heat flux. Imagine that this basal heat flux gradually declines in time. Such a system would begin in the upper left corner of Fig. 1d and would gradually slide down the conductive branch toward the right as the shell slowly thickened. When the shell reaches 8 km thickness at a basal heat flux of 35 mW m^{-2} (for the parameters in Fig. 1d), a crisis ensues: The critical Rayleigh number is reached and convection initiates, but the convection transports far more heat flux than is available from below. Thus, the shell cannot continue to thicken while remaining in equilibrium. Instead, rapid thickening occurs until the shell reaches a new equilibrium thickness of ~15 km at the heat flux of 35 mW m^{-2}. Any continued reductions in the basal heat flux lead to continued shell thickening on the convective branch. Conversely, suppose the shell lies on the convective branch and experiences a gradually increasing basal heat flux. The shell can gradually thin, maintaining equilibrium, until reaching a thickness of ~9 km at a basal flux of 60 mW m^{-2}. Again, a crisis ensues: The shell cannot continue to thin while maintaining equilibrium with the heat flux available from below. Instead, the shell becomes conductive, and then transports far less heat than is available from below. Melting ensues, which rapidly thins the shell until a new conductive equilibrium is attained at a thickness of ~5 km at the basal flux of 60 mW m^{-2}.

Thus, the heat-flux jump implies that modest variations in the basal heat flux can lead to large, and geologically rapid, changes in the ice shell thickness. These thickness changes occur over a timescale (*Mitri and Showman,* 2005)

$$\tau = \frac{L\rho\Delta D}{\Delta F} \approx 10^7 \text{ yr} \qquad (31)$$

where $L \approx 3 \times 10^5$ J kg^{-1} is the latent heat, $\Delta D \approx 10$ km is the thickness change resulting from the conductive-convective transition, and $\Delta F \approx 20$ mW m^{-2} is the mismatch in fluxes between the heat flux transported by the ice shell and that supplied from below. Importantly for tectonics, this timescale is much shorter than typical orbital and thermal evolution timescales of 10^8–10^9 years.

These rapid changes in shell thickness cause rapid changes in Europa's volume, which can lead to stresses up to ~10 MPa and may induce surface fracture (*Mitri and Showman,* 2005; *Nimmo,* 2004). Thus, conductive-convective switches may have important implications for europan tectonics. The fact that Europa's heat flux could vary episodically (*Hussmann and Spohn,* 2004) introduces the possibility that such conductive-convective switches may have occurred repeatedly in Europa's history.

4.3. Convective Heat Flux

As introduced in section 2.3, the relationship between the convective heat flux and the vigor of convection can be expressed in a relationship of form, Nu = c Ra^{β_n}. For convection in a fluid with a temperature-dependent viscosity, the value of c depends on θ (*Solomatov,* 1995), giving rise to a Ra-Nu relationship of form (cf. *Solomatov and Moresi,* 2000)

$$Nu = a\theta^{-\alpha_n} Ra_i^{\beta_n} \quad (32)$$

where a, α_n, and β_n are constants that depend on n, and Ra_i is the value of Rayleigh number evaluated at T_i (and additionally, a strain rate of κ/D^2 for a non-Newtonian rheology) (*Solomatov and Moresi,* 2000). The grouped $a\theta^{-\alpha_n}$ is analogous to the constant c used in equation (11) and in Voyager-era convection studies. Scalings between Nu and Ra_i can be used for both internally heated and basally heated ice shells provided T_i is properly estimated in the basally heated case (described in detail by *McKinnon,* 2006).

For steady convection in a Newtonian fluid at low Ra (for $Ra/Ra_{cr} \lesssim 10^3$), *Dumoulin et al.* (1999) suggest a = 1.99, α_n = 1, and β_n = 1/5. For vigorous convection, where the velocity and temperature field are time-dependent ($Ra/Ra_{cr} \gtrsim 10^3$), *Solomatov and Moresi* (2000) suggest a = (0.31 + 0.22 n), α_n = 2(n + 1)/(n + 2), and β_n = n/(n + 2), which for Newtonian ice, where n = 1, give a = 0.53, α_n = 4/3, and β_n = 1/3. Numerical simulations of vigorous convection with a multicomponent rheology for water ice including terms from Newtonian diffusion creep and weakly non-Newtonian GBS give a = (0.82 ± 1.69), α_n = 1.07 ± 0.19, and β_n = 0.25 ± 0.02, which are roughly similar to the values for diffusion creep alone (*Freeman et al.,* 2006). The applicability of these relationships to tidally heated ice shells with viscosity-dependent tidal dissipation has not yet been demonstrated explicitly. However, it is likely that the coefficients in the Ra-Nu relationship for tidally heated ice shells will be similar to those derived for uniform internal heating (e.g., *Solomatov and Moresi,* 2000). Because tidal dissipation in the cold stagnant lid at the surface of the ice shell is negligible (*Showman and Han,* 2004), viscosity-dependent tidal heating does not fundamentally alter the value of θ, the behavior of the stagnant lid, or the rheological boundary layer between the stagnant lid and the well-mixed convective interior of the ice shell, which are the key controls on Nu (*Solomatov and Moresi,* 2000; *Barr,* 2008).

5. CONVECTIVE-DRIVEN RESURFACING

Galileo observations showed that Europa's mottled terrain consists predominantly of chaotic terrain, pits, domes, platforms, irregular uplifts, and lobate features (*Carr et al.,* 1998; *Pappalardo et al.,* 1998; *Greeley et al.,* 1998; *Greenberg et al.,* 1999, 2003). Although several formation mechanisms have been suggested, including melt-through of the icy shell (*Greenberg et al.,* 1999; *O'Brien et al.,* 2002; *Melosh et al.,* 2004) and cryovolcanism (*Fagents,* 2003), the most common suggestion is that these features formed from subsurface convection in the icy shell (*Pappalardo et al.,* 1998; *Collins et al.,* 2000; *Head and Pappalardo,* 1999; *Spaun,* 2002; *Spaun et al.,* 2004; *Figueredo et al.,* 2002; *Figueredo and Greeley,* 2004). More generally, tectonic features on Enceladus, Miranda, Triton, and other moons have also been suggested to result, directly or indirectly, from subsurface convection (e.g., *Nimmo and Pappalardo,* 2006; *Pappalardo et al.,* 1997; *Schenk and Jackson,* 1993). Here we review our current understanding of the extent to which subsurface convection can induce surface tectonics on icy satellites.

Two basic routes exist whereby convection may modify a planetary surface. First, the convective stresses and strains below the lithosphere (associated with the convective temperature and motion fields) could directly cause surface fracture and deformation. Candidates for this type of *direct* modification include Europa's chaos, pits, and uplifts; Triton's cantaloupe terrain, and Miranda's coronae. Second, the effects of convection on the internal density structure and long-term evolution could lead, indirectly, to nonconvective stresses that induce surface tectonics. For example, time variation in convective density anomalies could lead to reorientation of the satellite figure relative to the rotation axis; substantial surface stresses would occur as the rotational and tidal bulges shifted across the surface. This may be relevant to Enceladus (*Nimmo and Pappalardo,* 2006). Alternately, convection could lead to changes in the thickness of the icy shell, and hence in satellite volume, leading to global tensional or compressional stresses. This mechanism is potentially relevant to Europa, Ganymede, and other bodies (*Nimmo,* 2004; *Mitri and Showman,* 2005; 2008a) (see also section 4.2).

Here, we focus on the first mechanism and discuss the extent to which convection can produce surface features such as pits, uplifts, and chaos terrains. This problem has been attacked with a variety of approaches, ranging from simplified analytical calculations (*Rathbun et al.,* 1998; *Nimmo and Manga,* 2002) to full numerical simulations of the convection (*Sotin et al.,* 2002; *Tobie et al.,* 2003; *Showman and Han,* 2004, 2005; *Han and Showman,* 2005, 2008). We first address the production of topography (pits and uplifts) and second discuss surface disruption (chaos).

5.1. Pits and Uplifts

A difficulty in explaining large-amplitude topography and surface disruption by convection is the small expected convective stresses on icy satellites (*Showman and Han,* 2004). Thermal-buoyancy stresses associated with convective plumes are $\sigma \sim \rho\alpha g h \delta T$, where ρ, α, and g are density, thermal expansivity, and gravity; δT is the temperature difference between a plume and its surroundings, and h is the vertical height of the plume. The temperature-dependent

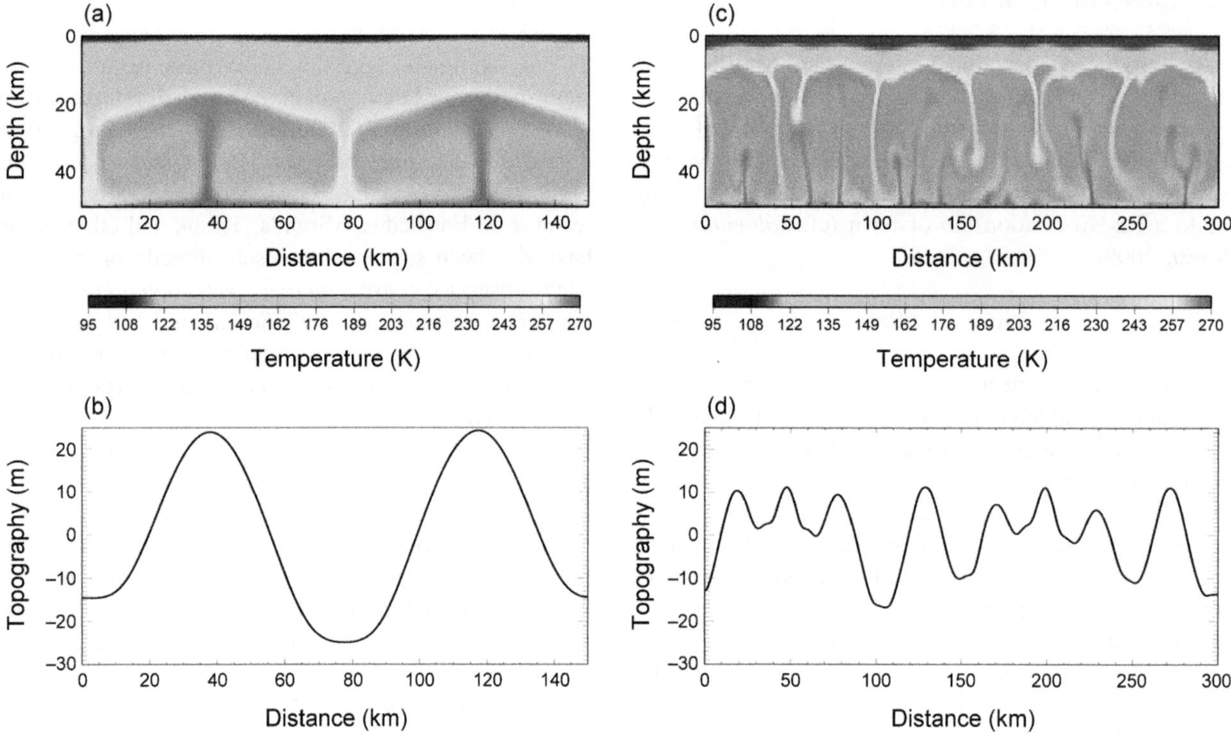

Fig. 3. See Plate 22. **(a),(c)** Temperature and **(b),(d)** dynamic topography from simulations of thermal convection in a 50-km-thick europan ice shell from *Showman and Han* (2004). Rheology is Newtonian with $Q^* = 60$ kJ mol^{-1}, $\eta_o = 10^{13}$ Pa s (right), and $\eta_o = 10^{14}$ Pa s (left), and upper viscosity cutoff of 10^9 η_o. **(a),(b)** A high melting point viscosity leads to sluggish convection beneath a thick stagnant lid, and topography on the order of tens of meters, much lower than observed on Europa. Domain is 150 km wide and 50 km deep. **(c),(d)** Convection in an ice shell melting point viscosity of 10^{13} Pa s leads to vigorous convection characterized by narrow upwellings and a thinner stagnant lid. Domain is 300 km wide by 50 km deep. Although the topography predicted by more vigorous convection has a smaller wavelength, the plume buoyancy is unchanged between the two cases: the dynamic topography is approximately tens of meters.

viscosity leads to the development of a stagnant lid at the surface; the convection then occurs in a nearly isothermal sublayer confined below the stagnant lid. Theoretical studies imply that the total viscosity contrast across such a sublayer is only a factor of ~10 (e.g., *Solomatov and Moresi,* 2000), which for realistic activation energies (Q^*) implies that convective plumes have temperature constrasts of only ~10 K. As emphasized by *Showman and Han* (2004, 2005), this weak thermal buoyancy leads to small stresses of only ~0.01 MPa (*Tobie et al.,* 2003) (see also section 3.1.1). The dynamic (i.e., convectively generated) topography induced by such stresses is only $\sigma/\rho g \sim 10$ m (*Nimmo and Manga,* 2002; *Showman and Han,* 2004), which is far below the ~100–300-m heights of typical europan uplifts. These stresses are also much less than the expected yield stress of ice, suggesting that pure thermal convection cannot easily fracture the surface.

Several authors have performed analytical calculations of the conditions required to explain the properties of pits and uplifts by convection. *Rathburn et al.* (1998) adapted a simple model for the ascent of hot thermal diapirs through a cooler ice shell to study the formation of uplifts in Europa's shell. Based on the fact that diapirs spread laterally as they impinge against the stagnant lid, *Rathburn et al.* (1998) suggested that the initial diapirs must have diameters of several kilometers or less to explain uplifts ~10 km across. They also suggest that, to remain coherent as they ascend, such diapirs must have originated from depths less than a few tens of kilometers. Using boundary-layer theory, *Nimmo and Manga* (2002) carried these arguments further by linking such diapiric behavior to the required convective properties of the ice shell. Hot ascending diapirs presumably originate in the hot convective boundary layer at the bottom of the ice shell, and experimental results show that the initial diapir diameter is ~5 times the thickness of the bottom hot boundary layer (*Manga and Weeraratne,* 1999). Based on this result, *Nimmo and Manga* (2002) infer that the bottom boundary layer thickness needed to explain 10-km-diameter domes is ~1–2 km. This demands a small melting-temperature viscosity of 10^{12}–10^{13} Pa s, implying ice grain sizes of only 0.02–0.06 mm (see section 3.3). *Nimmo and Manga* (2002) also suggest that, for such a diapir to induce surface uplift, the stagnant lid thickness must be <2–4 km, implying heat fluxes of 100–200 mW m^{-2}.

Consistent with the buoyancy arguments described above, their predicted dome heights are only ~5–30 m, far less than the observed heights of typical europan domes.

Although the above analytical studies are valuable, they adopted simplified prescriptions of the dynamics that potentially exclude important effects. To determine whether pit-and-dome-like surface topography can result from the full convective dynamics, *Showman and Han* (2004) performed two-dimensional numerical simulations using a diffusion creep rheology. (We note that use of a non-Newtonian rheology does not modify the fundamental buoyancy argument described above and is unlikely to, by itself, lead to convection-driven resurfacing.) *Showman and Han* (2004) found that, in stagnant-lid convection, ascending and descending plumes have essentially no surface expression — pits and uplifts do not form (see Fig. 3). This results directly from the extremely small temperature contrasts (~10 K) of the ascending and descending plumes. The simulations developed modest surface topography of ~10–30 m, which resulted from long-wavelength lateral variations in the thickness of the stagnant lid rather than from the plumes underlying the lid. *Showman and Han* (2004) found, however, that if the viscosity contrast is $\Delta\eta \sim 10^3$–10^5, then the cold ice at 1–2-km depth deforms enough to participate in the convection ($\Delta\eta$ is small enough that convection occurs in the so-called "sluggish lid" regime), leading to formation of 100–300-m-deep pits over the downwellings. However, none of the simulations produced localized uplifts. The simulated pits range in width from 5 to 100 km depending on the melting-temperature viscosity, thickness of the ice shell, and other properties. Consistent with *Nimmo and Manga* (2002), *Showman and Han* (2004) found that explaining observed pits with diameters of ~5–10 km requires melting-temperature viscosities of ~10^{12} Pa s or less, implying ice grain sizes of ≤0.04 mm. At these viscosities, the maximum heat flux transportable by convection is ~100–150 mW m^{-2}. Explaining pits less than ~4 km in diameter is extremely difficult unless the viscosities are unrealistically small.

Runaway tidal heating in hot convective plumes is sometimes invoked as a mechanism for enhancing the internal temperature contrasts, therefore increasing the amplitude of surface topography. However, *Showman and Han* (2004) pointed out that such runaways, if any (see section 5.2.1), cannot significantly enhance the thermal buoyancy in ascending hot plumes. The mean ice temperature in the convective sublayer is only ~10 K less than the temperature at the bottom of the ice shell, which for Europa is expected to be the ~260-K melting temperature. Even accounting for the pressure-dependence of the melting temperature, this puts a fundamental limit of ~10–20 K on the maximum temperature difference between ascending hot plumes and the background ice through which they rise: Plumes simply cannot be heated to temperatures exceeding the melting temperature. Once a hot plume reaches the melting temperature, any further heating will instead cause partial melting, which would increase the plume's density and therefore *decrease* its thermal buoyancy — lessening the topographic amplitude of any resulting uplifts.

Motivated by the insufficient buoyancy associated with *thermal* density contrasts, several authors have proposed that *compositional* density contrasts are important in generating the large (100–300 m) topography of typical pits and uplifts (*Nimmo et al.,* 2003; *Showman and Han,* 2004; *Pappalardo and Barr,* 2004; *Han and Showman,* 2005). The most plausible scenario for explaining uplifts is one where relatively salt-free (hence low-density) diapirs ascend through a saltier, denser environment. In this case the topography is ~$h\Delta\rho/\rho$, where $\Delta\rho$ is the plume-environment density contrast and h is the height of the plume. For a plume 10 km tall, explaining 300-m-tall uplifts would require a plume/environment density difference of ~30 kg m^{-3}, which could occur if the plume-environment salinity difference were ~5–10% (*Pappalardo and Barr,* 2004), marginally consistent with current estimates of the salinity of Europa's ocean (cf. *McKinnon and Zolensky,* 2003; *Hand and Chyba,* 2007; see also chapter by Zolotov and Kargel). However, as pointed out by *Showman and Han* (2004), it is difficult to understand how strong compositional contrasts can be maintained against mixing if the shell is convecting. Furthermore, any partial melting in the ice would tend to deplete the ice shell of salts (which percolate down into the ocean with the melt), so maintaining such compositional density contrasts over long timescales is difficult (*Showman and Han,* 2004). *Pappalardo and Barr* (2004) proposed that the compositional convection is a transient process that begins with a recent onset of thermal convection and then dies off as the ice shell becomes depleted in salts. If so, then the uplifts would be short-lived and would disappear as the shell became salt-free. However, they also suggested that diking from the base of the ice shell might replenish the shell with salts.

Han and Showman (2005) performed two-dimensional numerical simulations of thermo-compositional convection to test the qualitative scenario of *Pappalardo and Barr* (2004). Because grid-based methods can cause an artificial numerical diffusion of the salinity, *Han and Showman* (2005) treated the salinity using the particle-in-cell method, which allows advection of the salt with essentially no numerical diffusion. Following *Pappalardo and Barr* (2004), they initialized the simulations with a warm salt-poor ice layer underlying a colder, saltier, denser ice layer. In typical simulations, a Rayleigh-Taylor instability developed between the salt-poor and saltier layers, leading to compositionally driven diapirs that generated pits and uplifts with topography of ~300 m or more (see Fig. 4). Because the instability involves the relatively cold near-surface ice, it occurs over a timescale $\eta_0\chi/(g\delta\Delta\rho)$, where η_0 is the melting-temperature viscosity, χ is the assumed viscosity contrast across the ice shell, δ is the thickness of the salty layer, and $\Delta\rho$ is the density difference between the salty and salt-poor layers. This leads to pit-and-uplift formation timescales less than Europa's known surface age of 40–90 m.y. (*Zahnle et*

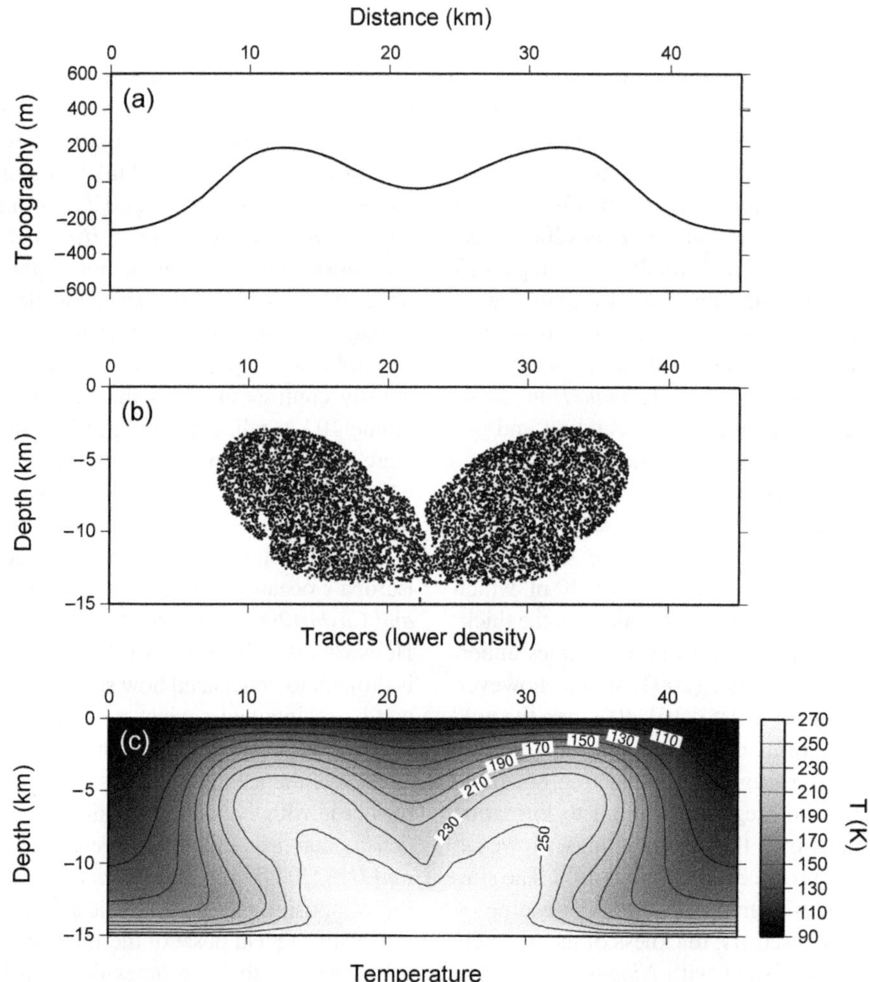

Fig. 4. **(a)** Topography, **(b)** composition, and **(c)** temperature for a numerical simulation of convection with salinity from *Han and Showman* (2005). Domain is 45 km wide and 15 km thick. Black dots in **(b)** are tracers marking the locations of salt-poor, low-density ice, which was initially near the bottom of the ice shell but experiences diapirism. White regions are salty, denser ice. The topography attains 200–300 m with widths of 15–20 km. This suggests that Europa's widest pits and uplifts can form from convection with salinity.

al., 2003; see chapter by Bierhaus et al.) only for viscosity contrasts $<10^7$–10^8. For viscosity contrasts $>10^8$, pits and uplifts cannot form in timescales less than Europa's surface age. The implication is that compositional convection can only produce Europa's pits and uplifts if Europa's surface is weak. The simulated pits and uplifts were 10–30 km wide (see Fig. 4); explaining the many pits and uplifts with diameters <5 km is difficult.

As described above, matching the observed sizes of pits and uplifts remains a challenge. Pits and uplifts range in diameter from ~3 to 50 km (*Greenberg et al.,* 2003; *Spaun,* 2002; *Rathburn et al.,* 1998). Based on an early sampling of Galileo images, *Pappalardo et al.* (1998) suggested that a preferred diameter of ~10 km exists, which *Spaun* (2002) and *Spaun et al.* (2004) revised downward to ~4–8 km after performing an exhaustive survey of images that became available later in the mission. *Greenberg et al.* (2003) also performed an exhaustive survey and suggested that, when a preferred diameter exists at all, it is ~3 km and reflects the limits of image resolution rather than a physical peak. However, these divergent results may reflect differences in analysis methods rather than an actual discrepancy (see *Goodman et al.,* 2004, Appendix A; chapter by Collins and Nimmo). For our purposes, the main point is that, regardless of the preferred diameter, many pits and uplifts are small, with diameters of 3–10 km. Although convection (with salinity) can plausibly produce the largest of these features, explaining the smallest (3–5-km-diameter) features is difficult. It is plausible that multiple origins exist for pits and uplifts, with some of the larger features resulting from convection and the smallest features resulting from some other process.

5.2. Chaos

Based on observations of Europa's chaotic terrain, several authors have proposed that chaos results from convection within the icy shell, possibly aided by partial melting (*Pappalardo et al.,* 1998; *Head and Pappalardo,* 1999; *Collins et al.,* 2000; *Figueredo et al.,* 2002; *Spaun,* 2002; *Schenk and Pappalardo,* 2004; *Spaun et al.,* 2004). Here we review theoretical work that investigates whether convection can lead to surface disruption and whether that disruption has the properties of chaos (see also chapter by Collins and Nimmo).

5.2.1. Runaway heating in convective plumes. Motivated by the suggestion that the disaggregation and tilting of chaos blocks would be aided by partial melting under the lithosphere (*Collins et al.,* 2000), several authors have proposed that runaway tidal heating and localized partial melting occurs within warm, ascending convective plumes, promoting the formation of chaos (*Wang and Stevenson,* 2000; *Sotin et al.,* 2002; *Tobie et al.,* 2003) (see also chapters by Collins and Nimmo and by Sotin et al.). This attractive idea is based on the temperature-dependence of the Maxwell model for tidal heating (equation (5)), which predicts that tidal heating increases strongly with temperature at low temperature (scaling as η^{-1}) and decreases strongly with temperature at high temperature (scaling as η), peaking at intermediate values corresponding to $\omega\tau_M \sim 1$ (at temperatures of 220–270 K for melting-temperature viscosities of 10^{12}–10^{15} Pa s). The low-temperature behavior promotes runaway: Increases in temperature enhance the tidal heating rate, leading to further increases in temperature. For ice grain sizes exceeding ~0.5 mm (i.e., melting-temperature viscosities $\geq 10^{14}$ Pa s), the peak heating occurs at temperatures exceeding the melting temperature, which suggests that under appropriate conditions this runaway can drive temperatures all the way to the melting temperature within a localized, ascending warm plume.

While the idea merits further investigation, several possible roadblocks exist. First, calculations by *Tobie et al.* (2003) and *Nimmo and Giese* (2005) suggest that melting the near-surface ice (at 1–3 km depth) is difficult because of the high power generation that is required. *Tobie et al.*'s (2003) tidally heated convection simulations, for example, produce partial melting only at depths exceeding 10 km, which may be too deep to allow disaggregation of the surface.

A potentially more serious issue is that for ice grain sizes <0.5 mm, the Maxwell model (equation (5)) predicts that the runaway changes sign near the melting temperature. *Tobie et al.* (2003) and *Mitri and Showman* (2005) point out that if the melting-temperature viscosity is 10^{13} Pa s or less (implying ice grain sizes of ≤ 0.1 mm), then for plausible ice-shell temperatures, the greatest tidal heating occurs in *cold* plumes, *not* the warm plumes (see Fig. 3 of *Mitri and Showman,* 2005). In this case, warm plumes would be heated *less* than the background ice, implying a negative feedback that reduces the thermal contrasts of plumes relative to the background. This would effectively preclude runaway heating in warm plumes. Given the small viscosities and grain sizes apparently required to explain pits and uplifts via convection (*Nimmo and Manga,* 2002; *Showman and Han,* 2004), this difficulty is relatively serious. However, the problem might be surmounted in the presence of low-eutectic-temperature contaminants (sulfuric acid or chloride salts), which could allow melting at sufficiently low temperatures for the positive feedback to operate up to the (lowered) melting temperature. Whether large pockets of melt would result [sufficient to mobilize the overlying chaos blocks, as suggested by *Collins et al.* (2000)] would depend on the concentration of the impurities.

Finally, the Maxwell model used in convection studies to date (*Wang and Stevenson,* 2002; *Sotin et al.,* 2002; *Tobie et al.,* 2003), which is essentially that of equation (5), is rigorously appropriate to an ice shell that exhibits no lateral variation in viscosity (i.e., it is a "zero-dimensional" Maxwell model). However, it is unclear whether this is appropriate for the heterogeneous conditions of Europa's icy shell. A warm plume surrounded by colder, stiffer ice may exhibit cyclical tidal stress and strain patterns (hence dissipation as a function of temperature) different from those used to derive equation (5). *Moore* (2001) argued that small-scale structures such as convective plumes would not couple well to the large (hemispheric) scale of the tidal flexing, and that, as a result, minimal lateral variation in the tidal heating rate across convective plumes could occur. If so, the local runaways envisioned by *Wang and Stevenson* (2000), *Sotin et al.* (2002), and *Tobie et al.* (2003) would be ruled out. More detailed analyses are needed to clarify this issue.

Mitri and Showman (2008b) revisited this issue with a two-dimensional model corresponding to a horizontal cross-section through a cylindrical, vertically oriented plume. A Maxwell viscoelastic rheology was adopted; the plume and environment were allowed to take different viscosities and elastic parameters. Given an imposed cyclical tidal-flexing stress and strain field at infinity, *Mitri and Showman* (2008b) solved for the stress, strain, and tidal dissipation within and surrounding the plume. These calculations showed that tidal dissipation does remain strongly temperature-dependent inside a convective plume (even when the background temperature is held constant), broadly supporting the idea that plumes can experience positive (or negative) feedbacks between local temperature and local tidal heating rate. Nevertheless, it would be worthwhile to extend these calculations to three dimensions and explore a broader range of geometries.

In summary, theoretical work to date supports the idea that tidal dissipation depends strongly on temperature in a convective plume, but whether the tidal heating in hot plumes is larger or smaller than the background heating rate depends on the ice grain size. Near the melting temperature, positive feedbacks (runaways) are possible for large grain sizes, but negative feedbacks occur for small grain sizes.

5.2.2. The difficulty of fracturing the surface. Thermal convection causes typical stresses of 10^{-3}–10^{-2} MPa, which is much smaller than the ~1–3-MPa failure strength of unfractured ice. This discrepancy raises difficulties for understanding how solid-state convection can induce surface disruption.

However, several factors could ameliorate this difficulty. First, Europa's surface shows abundant evidence that *tidal* stresses, which reach ~0.1 MPa at Europa's current eccentricity, have fractured the surface (*Greenberg et al.,* 1998; *Hoppa et al.,* 1999). For example, models for the formation of Europa's cycloidal ridges suggest a fracture yield stress of just 0.04 MPa (*Hoppa et al.,* 1999). Field studies on the Ross Ice Shelf, one of Earth's closest analogs to Europa's ice shell, also exhibit failure strengths of order 0.1 MPa (*Kehle,* 1964). While not definitive, these studies are consistent with the idea that Europa's near-surface ice is weak, but it is not clear whether these estimates are appropriate for the failure of Europa's entire lithosphere. If, for example, cycloids are relatively shallow phenomena, the relatively low stress associated with cycloid propagation may be relevant to Europa's near-surface ice only. If Europa's band topography forms in a manner similar to terrestrial mid-ocean ridges (*Prockter et al.,* 2002), the yield strength of Europa's lithosphere at the time and location of band formation is ~0.4 to 2 MPa (*Stempel et al.,* 2005). Another approach may be to consider the effects of microcracking on the rheology of near-surface ice (*Tobie et al.,* 2004). Between a depth of ~15–40 km on Earth, microcracks are expected to play a role in accommodating deformation, and facilitate semibrittle-plastic behavior that is conductive to forming zones of weakness in the crust (*Kohlstedt et al.,* 1995; *Tackley,* 2000a; *Bercovici,* 2003). Further field characterization of relevant terrestrial analogs and studies of flexure and failure on Europa constrained by new spacecraft data are needed to shed light upon this issue.

Second, *Showman and Han* (2005) pointed out that stresses can become greatly enhanced within a thin "stress boundary layer" near the surface, promoting the likelihood of surface fracture. This phenomenon results from the need to balance forces in a lithosphere whose width far exceeds its thickness (*Melosh,* 1977; *Fowler,* 1985, 1993; *McKinnon,* 1998; *Solomatov,* 2004a,b). To illustrate, consider a two-dimensional lithosphere with horizontal dimension x and vertical dimension z. Horizontal force balance in the lithosphere leads to the stress equilibrium condition

$$\frac{\partial \sigma_{xx}}{\partial x} + \frac{\partial \sigma_{xz}}{\partial z} = 0 \qquad (33)$$

where σ_{xx} and σ_{xz} are the horizontal normal and shear stresses, respectively. The shear stress is zero at the surface but due to convection (or other processes) is nonzero in the interior. Suppose the shear stress at the base of the stress boundary layer is σ_b. For a stress boundary layer of thickness h that experiences this shear stress over a horizonal distance L, we have to order of magnitude

$$\sigma_{xx} \sim \frac{L}{h}\sigma_b \qquad (34)$$

The appropriate value for h is the viscosity scale height in the lithosphere (*Solomatov,* 2004a), which is ~1 km for europan conditions. The interior convective stresses should remain coherent over distances L comparable to the interplume spacing, which is similar to the ice-shell thickness. Adopting L ~ 20 km, we thus see that stresses can become enhanced by a factor of ~20 within the stress boundary layer. In agreement with this estimate, numerical simulations by *Showman and Han* (2005) show that, although convective stresses within the ice-shell interior are typically ~10^{-3}–10^{-2} MPa, the normal stresses due to thermal convection can exceed ~0.1 MPa near the surface.

Third, the stresses that occur during compositional convection in a heterogeneous salty ice shell would far exceed those due to thermal convection alone. For the density contrasts needed to explain ~300 m-tall uplifts ($\Delta\rho$ ~ 30 kg m^{-3} over a height range h ~ 10 km), typical convective stresses are ~$\Delta\rho g h$ ~ 0.4 MPa. In the presence of a stress boundary layer, near-surface stresses could be enhanced by an additional order of magnitude or more. These values exceed those needed to fracture ice. Thus, the idea that convection can fracture the surface seems reasonable.

5.2.3. Can convection produce a chaos-like morphology? To test the hypothesis that convection can cause formation of chaos-type terrains, *Showman and Han* (2005) performed two-dimensional numerical simulations of thermal convection in Europa's ice shell including the effects of plasticity, which is a continuum representation for deformation by brittle failure. Plastic deformation occurs when the deviatoric stresses reach a specified yield stress σ_Y; at lower stresses, the rheology corresponds to a Newtonian, temperature-dependent viscosity (cf. *Trompert and Hansen,* 1998; *Moresi and Solomatov,* 1998; *Tackley,* 2000b). Partial melting and salinity were not considered. These simulations showed four regimes of behavior depending on the yield stress, thickness of the ice shell, and other parameters. At large yield stresses (≥0.1 MPa), the stresses never attain necessary values for plastic deformation, and so stagnant-lid convection occurs. At modestly smaller yield stresses (~0.03–0.08 MPa), a thick, cold upper lid remains, but it deforms via plastic deformation (see Fig. 5). *Showman and Han* (2005) dubbed this the "pliable lid" regime. Most of the plastic deformation is confined near the surface, as a result of the stress boundary layer. At even smaller yield stresses (<0.05 MPa), the convection moves away from stagnant-lid regime, exhibiting either episodic foundering and regrowth of the lid (see Fig. 6) or continual recycling of the lid.

What is the connection between these simulations and Europa's chaotic terrain? The formation of chaos requires not only surface fracture but sufficient strains to rotate and translate surviving chaos rafts and disaggregate the inter-

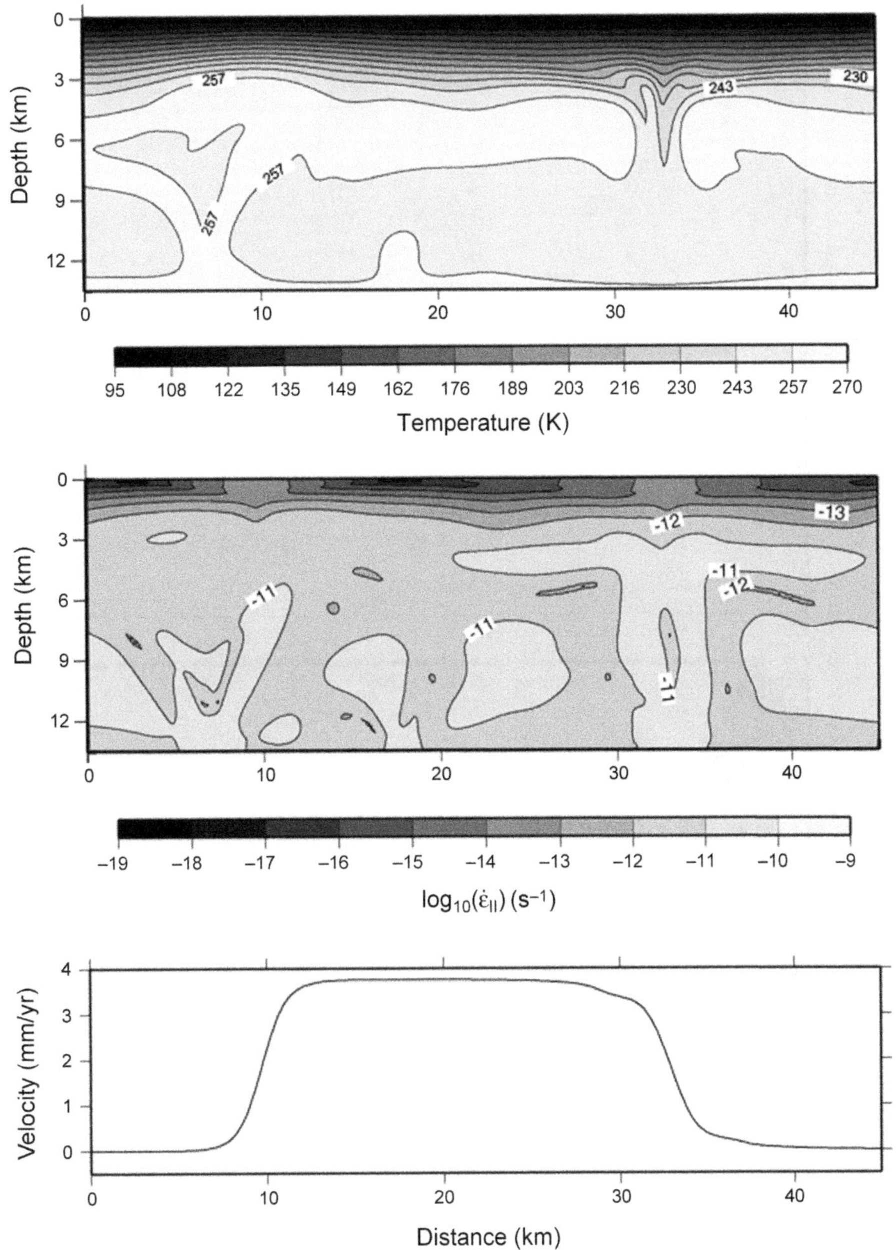

Fig. 5. Simulation of pure thermal convection including plasticity from *Showman and Han* (2005) with a yield stress of 0.03 MPa. Temperature divided by melting temperature (top), second invariant of strain rate (middle), and surface velocity (bottom). Domain is 45 km wide and 15 km deep. Plastic deformation occurs in the upper lid, leading to significant surface deformation. This may be relevant to chaos formation on Europa.

vening matrix. Yet the existence of chaos rafts suggests that, in many cases, surface materials remained near the surface even as they were disrupted. Thus, modes of deformation involving complete foundering of the upper lid (e.g., Fig. 6) appear not to have occurred on Europa. On the other hand, the so-called "pliable lid" regime of *Showman and Han* (2005) (Fig. 5) seems to capture key aspects of the observed behavior. In these simulations, the near-surface strain rates exceed 10^{-14} s^{-1} in localized regions, implying that order-unity strains would occur on timescales of several million years. This is sufficient to disaggregate the surface. The absence of foundering in these simulations suggests that chaos rafts would remain at the surface, as observed. Interestingly, these high-strain regions were localized, occurring in zones only ~5 km wide. This is an encouraging result, although the simulations must be extended to three dimensions to determine whether the disrupted regions would have quasicircular rather than linear (band-like) mor-

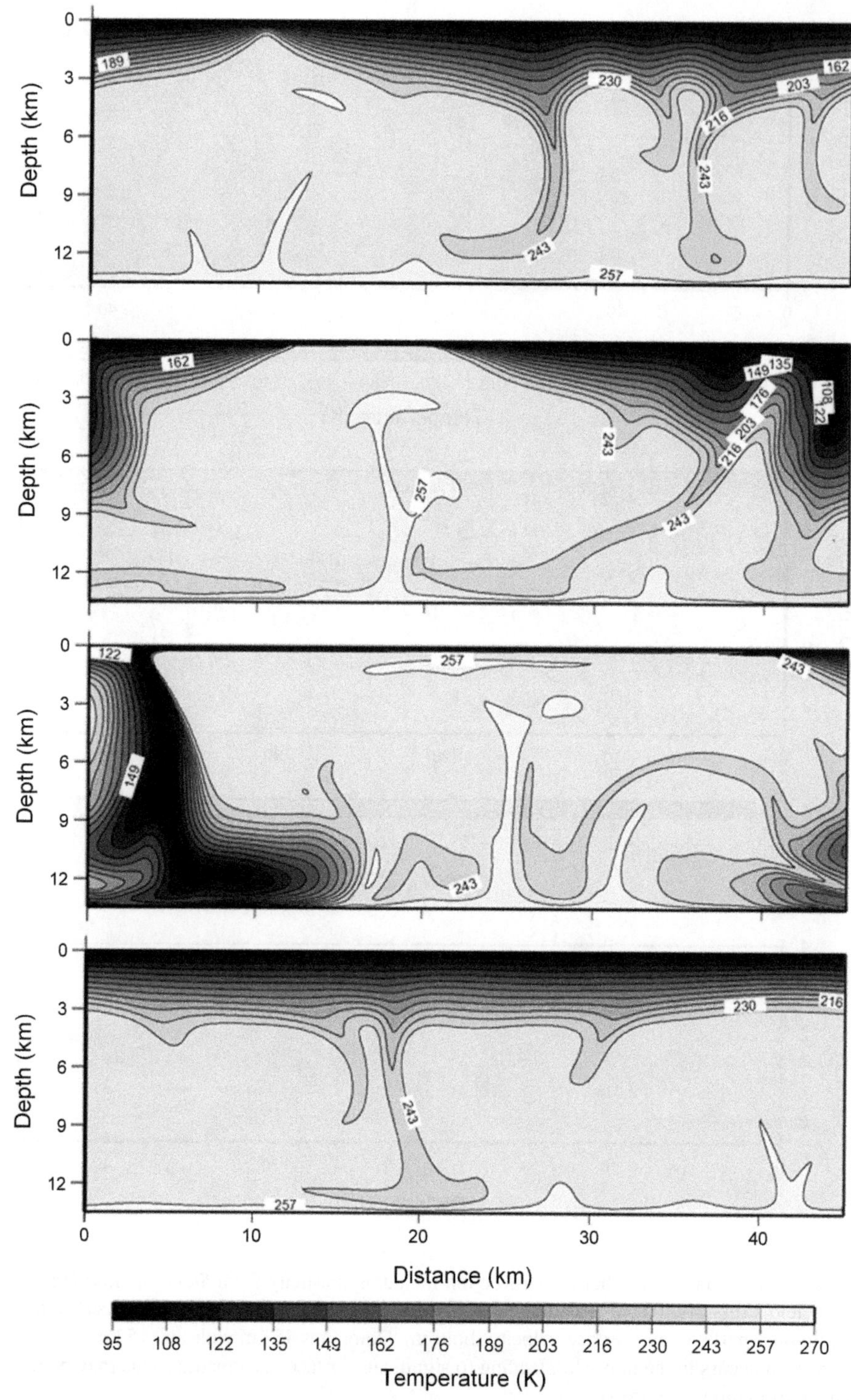

Fig. 6. A time sequence (top to bottom) of a simulation of pure thermal convection including plasticity from *Showman and Han* (2005) with a yield stress of 0.03 MPa. This simulation illustrates necking and overturn of the upper lid, followed by reformation of a cold upper lid by conduction.

phology. However, a difficulty is that the pliable-lid regime occurs over only a narrow range of yield stresses; a yield stress that is slightly too large or small pushes the behavior into stagnant-lid or lithospheric-foundering regimes, respectively. Potentially, partial-melting or porosity in the subsurface could cause a density stratification that would prevent lithospheric foundering (*Collins et al.,* 2000) and allow the observed behavior to occur over a wider range of yield stresses.

Although thermal and/or compositional convection seems to be a viable mechanism for causing at least some chaotic terrain, explaining the specific observed aspects of

Europa's chaos remains a challenge. A cautionary note is provided by attempts to simulate Earth's plate tectonics from first principles, which show that the interaction of convection with brittle deformation can depend sensitively on the adopted formulation for the brittle rheology (*Tackley,* 2000a,b,c; *Bercovici,* 2003). One may expect similar sensitivity in the interaction of convection with brittle deformation on Europa. A full understanding of whether, and how, convection can cause chaotic terrain will require future numerical studies that investigate compositional effects, partial melting, and a wider range of brittle rheologies.

6. DISCUSSION

Despite many advances in the knowledge of ice rheology, the behavior of solid-state convection, and the interaction between convection and lithospheric deformation, several aspects of the behavior of Europa's icy shell remain unexplained. Here, we describe several key gaps in knowledge about Europa, which are required to address the fundamental questions about Europa's icy shell posed in section 1: Can Europa's icy shell convect at present? How does tidal dissipation affect convection? Can convection drive resurfacing? What role does compositional heterogeneity play in driving motion in Europa's shell?

What is the thickness and thermal structure of Europa's icy shell? Further spacecraft data are needed to constrain the true thickness of Europa's icy shell, to characterize the topography inferred to form from convection (at both short and long wavelengths), and to determine the thermal structure of the icy shell (namely, the depth to warm ice). Global geophysical data obtained by an orbiter equipped with, for example, a laser altimeter, radar sounder, a near-infrared mapping spectrometer, and high-resolution imaging system are needed to answer many of the basic questions posed above. Such data would also provide constraints for more sophisticated modeling efforts suggested below.

Can convection cause resurfacing? Models published to date provide encouragement that at least some fraction of Europa's pits, uplifts, and chaos could result from convection in the icy shell, but the models are nevertheless far from explaining the actual observed properties of these features. Compositional convection allows pits and uplifts with the observed topography to be produced, but the simulated features are wider than most of Europa's pits and uplifts unless ice viscosities are extremely small. Convection models including simple parameterizations of brittle failure can produce some chaos-like behaviors, but they also produce behaviors that appear not to occur on Europa. A new generation of three-dimensional models including salinity, partial melting, and more realistic parameterizations of brittle failure can help determine whether pits, uplifts, and chaos can actually result from convection.

How does tidal flexing on Europa affect the microphysical structure and rheological behavior of ice? It is not known how the cyclical tidal flexing of Europa's ice shell affects the ice in its interior. The Maxwell model is perhaps an overly simplistic description of the behavior of Europa's ice. Laboratory experiments are needed to clarify how tidal dissipation occurs on a microphysical scale in ice, and to clarify whether cyclical flexing affects ice microstructure.

How does tidal flexing interact with mechanical, thermal, and compositional heterogeneity in the ice shell? Implicit in our discussion of the effects of tidal flexing on heat transfer has been the assumption that tidal dissipation is heterogeneous, and that tidal heating obeys the Maxwell model (equation (5)). The results of laboratory experiments must be combined with sophisticated geophysical techniques to study the localization of tidal strain and heating in model europan ice shells with thermal, mechanical, and compositional heterogeneity to more accurately model tidal dissipation and its link to resurfacing.

Acknowledgments. A.C.B. acknowledges support from the Southwest Research Institute and NASA OPR grant NNG05GI15G to W. B. McKinnon. A.P.S. acknowledges support from NASA PG&G grant NNX07AR27G. We thank G. Tobie, W. B. McKinnon, and S. Solomatov for helpful comments.

REFERENCES

Alley R. B., Perepezko J. H., and Bentley C. R. (1986a) Grain growth in polar ice: I. Theory. *J. Glaciol., 32,* 415–424.

Alley R. B., Perepezko J. H., and Bentley C. R. (1986b) Grain growth in polar ice: II. Theory. *J. Glaciol., 32,* 425–433.

Barnes P. R. F., Mulvaney R., Robinson K., and Wolff E. W. (2002) Observations of polar ice from the Holocene and the glacial period using the scanning electron microscope. *Ann. Glaciol., 35,* 559–566.

Barr A. C. (2008) Mobile lid convection beneath Enceladus' south polar terrain. *J. Geophys. Res., 113,* E07009, DOI: 10.1029/2008JE003114.

Barr A. C. and McKinnon W. B. (2007) Convection in ice I shells and mantles with self-consistent grain size. *J. Geophys. Res., 112,* E02012, DOI: 10.1029/2006JE002781.

Barr A. C. and Pappalardo R. T. (2005) Onset of convection in the icy Galilean satellites: Influence of rheology. *J. Geophys. Res., 110,* E12005, DOI: 10.1029/2004JE002371.

Barr A. C., Pappalardo R. T., and Zhong S. (2004) Convective instability in ice I with non-Newtonian rheology: Application to the icy Galilean satellites. *J. Geophys. Res., 109,* E12008, DOI: 10.1029/2004JE002296.

Bercovici D. (2003) Frontiers: The generation of plate tectonics from mantle convection. *Earth Planet. Sci. Lett., 205,* 107–121, DOI: 10.1016/S0012-821X(02)01009-9.

Birger B. I. (1998) Rheology of the Earth and a thermoconvective mechanism for sedimentary basin formation. *Geophys. J. Intl., 134,* 1–12.

Birger B. I. (2000) Excitation of thermoconvective waves in the continential lithosphere. *Geophys. J. Intl., 140,* 24–36.

Budd W. F. and Jacka T. H. (1989) A review of ice rheology for ice sheet modelling. *Cold Regions Sci. Tech., 16,* 107–144.

Carlson R. W., Anderson M. S., Mehlman R., and Johnson R. E. (2005) Distribution of hydrate on Europa: Further evidence for sulfuric acid hydrate. *Icarus, 177,* 461–471.

Carr M. H., Belton M. J. S., Chapman C. R., Davies M. E., Geissler P., et al. (1979) Evidence for a subsurface ocean on Europa. *Nature, 391,* 363–365.

Cassen P., Reynolds R. T., and Peale S. J. (1979) Is there liquid water on Europa. *Geophys. Res. Lett., 6,* 731–734.

Cassen P., Peale S. J., and Reynolds R. T. (1980) Tidal dissipation in Europa — A correction. *Geophys. Res. Lett., 7,* 987–988.

Chandrasekhar S. (1961) *Hydrodynamic and Hydromagnetic Stability.* International Series of Monographs on Physics, Oxford, Clarendon.

Collins G. C., Head J. W., Pappalardo R. T. and Spaun N. A. (2000) Evaluation of models for the formation of chaotic terrain on Europa. *J. Geophys. Res., 105,* 1709–1716.

Consolmagno G. J. and Lewis J. S. (1978) The evolution of icy satellite interiors and surfaces. *Icarus, 34,* 280–293.

De Bresser J. H. P., Peach C. J., Reijs J. P. J., and Spiers C. J. (1998) On dynamic recrystallization during solid state flow: Effects of stress and temperature. *Geophys. Res. Lett., 25,* 3457–3460.

De La Chapelle S., Castelnau O., Lipenkov V., and Duval P. (1998) Dynamic recrystallization and texture development in ice as revealed by the study of deep ice cores in Antarctica and Greenland. *J. Geophys. Res., 103,* 5091–5106.

Derby B. (1991) The dependence of grain size on stress during dynamic recrystallization. *Acta Metall. Mat., 39,* 955–962.

Dumoulin C., Doin M.-P., and Fleitout L. (1999) Heat transport in stagnant lid convection with temperature- and pressure-dependent Newtonian or non-Newtonian rheology. *J. Geophys. Res., 104,* 12759–12777.

Durand G. et al. (2006) Effect of impurities on grain growth in cold ice sheets. *J. Geophys. Res., 111,* F01015, DOI: 10.1029/2005JF000320.

Durham W. B. and Stern L. A. (2001) Rheological properties of water ice — Applications to satellites of the outer planets. *Annu. Rev. Earth Planet. Sci., 29,* 295–330.

Durham W. B., Kirby S. H., and Stern L. A. (1992) Effect of dispersed particulates on the rheology of water ice at planetary conditions. *J. Geophys. Res., 97,* 20883–20897.

Durham W. B., Kirby S. H., and Stern L. A. (1993) Flow of ices in the ammonia-water system. *J. Geophys. Res., 98,* 17667–17682.

Durham W. B., Stern L. A., and Kirby S. H. (2001) Rheology of ice I at low stress and elevated confining pressure. *J. Geophys. Res., 106,* 11031–11042.

Durham W. B., Stern L. A., Kubo T., and Kirby S. H. (2005) Flow strength of highly hydrated Mg- and Na-sulfate hydrate salts, pure and in mixtures with water ice, with application to Europa. *J. Geophys. Res., 110,* E12,010, DOI: 10.1029/2005JE002475.

Duval P. and Montagnat M. (2002) Comment on "Superplastic deformation of ice: Experimental observations" by D. L. Goldsby and D. L. Kohlstedt. *J. Geophys. Res., 107,* 2082, DOI: 10.1029/2002JB001842.

Duval P., Arnaud L., Brissaud O., Montagnat M., and de La Chapelle S. (2000) Deformation and recrystallization processes of ice from polar ice sheets. *Ann. Glaciol., 30,* 83–87.

Fagents S. A. (2003) Considerations for effusive cryovolcanism on Europa: The post-Galileo perspective. *J. Geophys. Res., 108,* 5139, DOI: 10.1029/2003JE002128.

Figueredo P. H. and Greeley R. (2004) Resurfacing history of Europa from pole-to-pole geologic mapping. *Icarus, 167,* 287–312, DOI: 10.1016/j.icarus.2003.09.016.

Figueredo P. H., Chuang F. C., Rathbun J., Kirk R. L., and Greeley R. (2002) Geology and origin of Europa's "Mitten" feature (Murias Chaos). *J. Geophys. Res., 107,* 5026, DOI: 10.1029/2001JE001591.

Fowler A. C. (1985) Fast thermoviscous convection. *Stud. Appl. Math., 72,* 189–219.

Fowler A. C. (1993) Boundary layer theory and subduction. *J. Geophys. Res., 98,* 21997–22005.

Freeman J., Moresi L., and May D. A. (2006) Thermal convection with a water ice I rheology: Implications for icy satellite evolution. *Icarus, 180,* 251–264, DOI: 10.1016/j.icarus.2005.07.014.

Glen J. W. (1955) The creep of polycrystalline ice. *Proc. R. Soc. London, Ser. A, 228,* 519–538.

Goldsby D. L. (2007) Diffusion creep of ice: Constraints from laboratory creep experiments. In *Lunar and Planetary Science XXXVIII,* Abstract #2186. Lunar and Planetary Institute, Houston.

Goldsby D. L. and Kohlstedt D. L. (2001) Superplastic deformation of ice: Experimental observations. *J. Geophys. Res., 106,* 11017–11030.

Goldsby D. L. and Kohlstedt D. L. (2002) Reply to comment by P. Duval and M. Montagnat on "Superplastic deformation of ice: Experimental observations." *J. Geophys. Res., 107,* 2313, DOI: 10.1029/2002JB001842.

Goodman D. J., Frost H. J., and Ashby M. F. (1981) The plasticity of polycrystalline ice. *Phil. Mag. A, 43,* 665–695.

Goodman J. C., Collins G. C., Marshall J., and Pierrehumbert R. T. (2004) Hydrothermal plume dynamics on Europa: Implications for chaos formation. *J. Geophys. Res., 109,* 3008, DOI: 10.1029/2003JE002073.

Greeley R., Sullivan R., Klemaszewski J., Head J. W. III, Pappalardo R. T., et al. (1998) Europa: Initial Galileo geological observations. *Icarus, 135,* 4–24.

Greeley R. C., Chyba C., Head J. W., McCord T., McKinnon W. B., and Pappalardo R. T. (2004) Geology of Europa. In *Jupiter: The Planet, Satellites and Magnetosphere* (F. Bagenal et al., eds.), pp. 329–362. Cambridge Univ., New York.

Greenberg R. et al. (1998) Tectonic processes on Europa: Tidal stresses, mechanical response, and visible features. *Icarus, 135,* 64–78, DOI: 10.1006/icar.1998.5986.

Greenberg R., Hoppa G. V., Tufts B. R., Geissler P., Riley J., and Kadel S. (1999) Chaos on Europa. *Icarus, 141,* 263–286, DOI: 10.1006/icar.1999.6187.

Greenberg R., Leake M. A., Hoppa G. V., and Tufts B. R. (2003) Pits and uplifts on Europa. *Icarus, 161,* 102–126, DOI: 10.1016/S0019-1035(02)00013-1.

Han L. and Showman A. P. (2005) Thermo-compositional convection in Europa's icy shell with salinity. *Geophys. Res. Lett., 32,* L20201, DOI: 10.1029/2005GL023979.

Han L. and Showman A. P. (2008) Implications of shear heating and fracture zones for ridge formation on Europa. *Geophys. Res. Lett., 35,* L03202, DOI: 10.1029/2007GL031957.

Hand K. P. and Chyba C. F. (2007) Empirical constraints on the salinity of the europan ocean and implications for a thin ice shell. *Icarus, 189,* 424–438, DOI: 10.1016/j.icarus.2007.02.002.

Head J. W. III and Pappalardo R. T. (1999) Brine mobilization during lithospheric heating on Europa: Implications for formation of chaos terrain, lenticula texture, and color variations. *J. Geophys. Res., 104,* 27143–27155.

Hoppa G. V., Tufts B. R., Greenberg R., and Geissler P. E. (1999) Formation of cycloidal features on Europa. *Science, 285,* 1899–1902.

Hussmann H. and Spohn T. (2004) Thermal-orbital evolution of Io and Europa. *Icarus, 171,* 391–410, DOI: 10.1016/j.icarus.2004.05.020.

Hussmann H., Spohn T., and Wieczerkowski K. (2002) Thermal

equilibrium states of Europa's ice shell: Implications for internal ocean thickness and surface heat flow. *Icarus, 156,* 143–151, DOI: 10.1006/icar.2001.6776.

Kehle R. O. (1964) Deformation of the Ross Ice Shelf, Antarctica. *Geol. Soc. Am. Bull., 75,* 259–286.

Kirk R. L. and Stevenson D. J. (1987) Thermal evolution of a differentiated Ganymede and implications for surface features. *Icarus, 69,* 91–134.

Kohlstedt D. L., Evans B., and Mackwell J. S. (1995) Strength of the lithosphere: Constraints imposed by laboratory experiments. *J. Geophys. Res., 100,* 17587–17602.

Manga M. and Weeraratne D. (1999) Experimental study of non-Boussinesq Rayleigh-Bénard convection at high Rayleigh and Prandtl numbers. *Phys. Fluids, 11,* 2969–2976, DOI: 10.1063/1.870156.

McCarthy C., Goldsby D. L., and Cooper R. F. (2007) Transient and steady-state creep responses of ice-I/magnesium sulfate hydrate eutectic aggregates. In *Lunar and Planetary Science XXXVIII,* Abstract #2429. Lunar and Planetary Institute, Houston.

McCord T. B., Hansen G. B., Matson D. L., Johnson T. V., Crowley J. K., et al. (1999) Hydrated salt minerals on Europa's surface from the Galileo near-infrared mapping spectrometer (NIMS) investigation. *J. Geophys. Res., 104,* 11827–11852.

McKinnon W. (1998) Geodynamics of icy satellites. In *Solar System Ices* (B. Schmitt et al., eds.), pp. 525–550. Astrophysics and Space Science Library, Vol. 227, Kluwer, Dordecht.

McKinnon W. B. (1999) Convective instability in Europa's floating ice shell. *Geophys. Res. Lett., 26,* 951–954.

McKinnon W. B. (2006) On convection in ice I shells of outer solar system bodies, with specific application to Callisto. *Icarus, 183,* 435–450.

McKinnon W. B. and Zolensky M. E. (2003) Sulfate content of Europa's ocean and shell: Evolutionary considerations and some geological and astrobiological implications. *Astrobiology, 3,* 879–897, DOI: 10.1089/153110703322736150.

Melosh H. J. (1997) Shear stress on the base of a lithospheric plate. *Pure Appl. Geophys., 115,* 429–439, DOI: 10.1007/BF01637119.

Melosh H. J., Ekholm A. G., Showman A. P., and Lorenz R. D. (2004) The temperature of Europa's subsurface water ocean. *Icarus, 168,* 498–502, DOI: 10.1016/j.icarus.2003.11.026.

Mitri G. and Showman A. P. (2005) Convective-conductive transitions and sensitivity of a convecting ice shell to perturbations in heat flux and tidal-heating rate: Implications for Europa. *Icarus, 177,* 447–460.

Mitri G. and Showman A. P. (2008a) Thermal convection in ice-I shells of Titan and Enceladus. *Icarus, 193,* 387–396, DOI: 10.1016/j.icarus.2007.07.016.

Mitri G. and Showman A. P. (2008b) A model for the temperature-dependence of tidal dissipation in convective plumes on icy satellites: Implications for Europa and Enceladus. *Icarus, 195,* 758–764, DOI: 10.1016/j.icarus.2008.01.010.

Montagnat M. and Duval P. (2000) Rate controlling processes in the creep of polar ice, influence of grain boundary migration associated with recrystallization. *Earth Planet. Sci. Lett., 183,* 179–186.

Moore W. B. (2001) Coupling Tidal dissipation and convection. abstract number 37.03. *AAS/Division for Planetary Sciences Meeting, 33,* #37.03.

Moore W. B. (2006) Thermal equilibrium in Europa's ice shell. *Icarus, 180,* 141–146, DOI: 10.1016/j.icarus.2005.09.005.

Moore W. B. and Schubert G. (2000) The tidal response of Europa. *Icarus, 147,* 317–319.

Moresi L.-N. and Solomatov V. S. (1998) Mantle convection with a brittle lithosphere: thoughts on the global tectonic styles of Earth and Venus. *Geophys. J. Intl., 133,* 669–682.

Nimmo F. (2004) Stresses generated in cooling viscoelastic ice shells: Application to Europa. *J. Geophys. Res., 109,* E12001, DOI: 10.1029/2004JE002347.

Nimmo F. and Giese B. (2005) Thermal and topographic tests of Europa chaos formation models from Galileo E15 observations. *Icarus, 177,* 327–340, DOI: 10.1016/j.icarus.2004.10.034.

Nimmo F. and Manga M. (2002) Causes, characteristics, and consequences of convective diapirism on Europa. *Geophys. Res. Lett., 29,* 2109, DOI: 10.1029/2002GL015754.

Nimmo F. and Pappalardo R. T. (2006) Diapir-induced reorientation of Saturn's moon Enceladus. *Nature, 441,* 614–616, DOI: 10.1038/nature04821.

Nimmo F., Pappalardo R. T., and Giese B. (2003) On the origins of band topography, Europa. *Icarus, 166,* 21–32, DOI: 10.1016/S0019-1035(03)00236-7.

O'Brien D. P., Geissler P., and Greenberg R. (2002) A melt-through model for chaos formation on Europa. *Icarus, 156,* 152–161.

Ojakangas G. W. and Stevenson D. J. (1989) Thermal state of an ice shell on Europa. *Icarus, 81,* 220–241.

Pappalardo R. T. and Barr A. C. (2004) The origin of domes on Europa: The role of thermally induced compositional diapirism. *Geophys. Res. Lett., 31,* L01701, DOI: 10.1029/2003GL019202.

Pappalardo R. T., Reynolds S. J., and Greeley R. (1997) Extensional tilt blocks on Miranda: Evidence for an upwelling origin of Arden Corona. *J. Geophys. Res., 102,* 13369–13380, DOI: 10.1029/97JE00802.

Pappalardo R. T. et al. (1998) Geological evidence for solid-state convection in Europa's ice shell. *Nature, 391,* 365–368, 1998.

Peltier W. R., Goldsby D. L., Kohlstedt D. L., and Tarasov L. (2000) Ice-age ice-sheet rheology: Constraints from the Last Glacial Maximum form of the Laurentide ice sheet. *Ann. Glaciol., 30,* 163–176.

Petrenko V. F. and Whitworth R. W. (1999) *Physics of Ice.* Oxford Univ., New York.

Poirier J.-P. (1985) *Creep of Crystals.* Cambridge Univ., Cambridge.

Prockter L. M., Head J. W., Pappalardo R. T., Sullivan R. J., Clifton A. E., Giese B., Wagner R., and Neukum G. (2002) Morphology of europan bands at high resolution: A mid-ocean ridge-type rift mechanism. *J. Geophys. Res., 107,* 5028, DOI: 10.1029/2000JE001458.

Ranalli G. (1987) *Rheology of the Earth: Deformation and Flow Processes in Geophysics and Geodynamics.* Allen and Unwin, Boston.

Rathbun J. A., Musser G. S., and Squyres S. W. (1998) Ice diapirs on Europa: Implications for liquid water. *Geophys. Res. Lett., 25,* 4157–4160, DOI: 10.1029/1998GL900135.

Reynolds R. T. and Cassen P. M. (1979) On the internal structure of the major satellites of the outer planets. *Geophys. Res. Lett., 6,* 121–124.

Ruiz J. and Tejero R. (1993) Heat flow, lenticulae spacing, and possibility of convection in the ice shell of Europa. *Icarus, 162,* 362–373.

Schenk P. and Jackson M. P. A. (1993) Diapirism on Triton — A record of crustal layering and instability. *Geology, 21,* 299–302.

Schenk P. M. and Pappalardo R. T. (2004) Topographic variations in chaos on Europa: Implications for diapiric formation. *Geophys. Res. Lett., 31,* L16703, DOI: 10.1029/2004GL019978.

Schmeling H. (1987) On the interaction between small- and large-scale convection and postglacial rebound flow in a power-law mantle. *Earth Planet. Sci. Lett., 84,* 254–262.

Schmidt K. G. and Dahl-Jensen D. (2003) An ice crystal model for Jupiter's moon Europa. *Ann. Glaciol., 37,* 129–133.

Schubert G., Turcotte D. L., and Olson P. (2001) *Mantle Convection in the Earth and Planets.* Cambridge Univ., New York.

Shimizu I. (1998) Stress and temperature dependence of recrystallized grain size: A subgrain misorientation model. *Geophys. Res. Lett., 25,* 4237–4240, DOI: 10.1029/1998GL900136.

Showman A. P. and Han L. (2004) Numerical simulations of convection in Europa's ice shell: Implications for surface features. *J. Geophys. Res.,* E01010, DOI: 10.1029/2003JE002103.

Showman A. P. and Han L. (2005) Effects of plasticity on convection in an ice shell: Implications for Europa. *Icarus, 177,* 425–437, DOI: 10.1016/j.icarus.2005.02.020.

Solomatov V. S. (1995) Scaling of temperature- and stress-dependent viscosity convection. *Phys. Fluids, 7,* 266–274.

Solomatov V. S. (2004a) Initiation of subduction by small-scale convection. *J. Geophys. Res., 109,* B01412, DOI: 10.1029/2003JB002628.

Solomatov V. S. (2004b) Correction to "Initiation of subduction by small-scale convection." *J. Geophys. Res., 109,* B05,408, DOI: 10.1029/2004JB003143.

Solomatov V. S. and Barr A. C. (2006) Onset of convection in fluids with strongly temperature-dependent, power-law viscosity. *Phys. Earth. Planet. Inter., 155,* 140–145.

Solomatov V. S. and Barr A. C. (2007) Onset of convection in fluids with strongly temperature-dependent, power-law viscosity 2. Dependence on the initial perturbation. *Phys. Earth. Planet. Inter., 165,* 1–13.

Solomatov V. S. and Moresi L-N. (2000) Scaling of time-dependent stagnant lid convection: Application to small-scale convection on Earth and other terrestrial planets. *J. Geophys. Res., 105,* 21795–21818.

Sotin C., Head J. W., and Tobie G. (2002) Europa: Tidal heating of upwelling thermal plumes and the origin of lenticulae and chaos melting. *Geophys. Res. Lett., 29,* 1233, DOI: 10.1029/2001GL013844.

Spaun N. A. (2002) Chaos, lenticulae, and lineae on Europa: Implications for geological history, crustal thickness, and the presence of an ocean. Ph.D. thesis, Brown University.

Spaun N. A., Head J. W. III, and Pappalardo R. T. (2004) Europan chaos and lenticulae: A synthesis of size, spacing, and areal density analyses. In *Lunar and Planetary Science XXXV,* Abstract #1409. Lunar and Planetary Institute, Houston.

Spohn T. and Schubert G. (2003) Oceans in the icy Galilean satellites of Jupiter? *Icarus, 161,* 456–467.

Stempel M. M., Barr A. C., and Pappalardo R. T. (2005) Model constraints on the opening rates of bands on Europa. *Icarus, 177,* 297–304.

Stengel K. C., Oliver D. C., and Booker J. R. (1982) Onset of convection in a variable viscosity fluid. *J. Fluid Mech., 120,* 411–431.

Tackley P. J. (2000a) Mantle convection and plate tectonics: Toward an integrated physical and chemical theory. *Science, 288,* 2002–2007.

Tackley P. J. (2000b) Self-consistent generation of tectonic plates in time-dependent, three-dimensional mantle convection simulations: 1. Pseudoplastic yielding. *Geochem. Geophys. Geosyst., 1,* DOI: 10.1029/2000GC000036.

Tackley P. J. (2000c) Self-consistent generation of tectonic plates in time-dependent, three-dimensional mantle convection simulations: 2. Strain weakening and asthenosphere. *Geochem. Geophys., 1,* DOI: 10.1029/2000GC000043.

Thorsteinsson T., Kipfstuhl J., and Miller H. (1997) Textures and fabrics in the GRIP ice core. *J. Geophys. Res., 102,* 26583–26600, DOI: 10.1029/97JC00161.

Tobie G., Choblet G., and Sotin C. (2003) Tidally heated convection: Constraints on Europa's ice shell thickness. *J. Geophys. Res., 108,* 5124, DOI: 10.1029/2003JE002099.

Tobie G., Choblet G., Lunine J., and Sotin C. (2004) Interaction between the convective sublayer and the cold fractured surface of Europa's ice shell. In *Workshop on Europa's Icy Shell: Past, Present, and Future,* Abstract #7033. Lunar and Planetary Institute, Houston.

Trompert R. and Hansen U. (1998) Mantle convection simulations with rheologies that generate plate-like behaviour. *Nature, 395,* 686–689.

Turcotte D. L. and Schubert G. (2002) *Geodynamics: Applications of Continuum Physics to Geological Problems.* Wiley, New York.

Wang H. and Stevenson D. J. (2000) Convection and internal melting of Europa's ice shell. In *Lunar and Planetary Science XXXI,* Abstract #1293. Lunar and Planetary Institute, Houston, Texas.

Weertman J. (1983) Creep deformation of ice. *Annu. Rev. Earth Planet Sci., 11,* 215–240.

Weiss J., Vidot J., Gay M., Arnaud L., Duval P., and Petit J. R. (2002) Dome Concordia ice microstructure: Impurities effect on grain growth. *Ann. Glaciol., 35,* 552–558.

Zahnle K., Schenk P., Levison H., and Dones L. (2003) Cratering rates in the outer solar system. *Icarus, 163,* 263–289, DOI: 10.1016/S0019-1035(03)00048-4.

Zimmer C., Khurana K. K., and Kivelson M. G. (2000) Subsurface oceans on Europa and Callisto: Constraints from Galileo magnetometer observations. *Icarus, 147,* 329–347.

Zolotov M. Y. and Shock E. L. (2001) Composition and stability of salts on the surface of Europa and their oceanic origin. *J. Geophys. Res., 106,* 32815–32828, DOI: 10.1029/2000JE001413.

On the Chemical Composition of Europa's Icy Shell, Ocean, and Underlying Rocks

M. Yu. Zolotov
Arizona State University

J. S. Kargel
University of Arizona

The outer shell of Europa is dominated by water ice, which makes the shell buoyant over a water ocean. The composition of the atmosphere and non-ice surface material may imply a presence of Na sulfates, other hydrated salts, and gases trapped in the icy shell. Europa's ocean likely contains sulfate, Mg, Na, and Cl as major solutes. Seafloor rocks/sediments could contain phyllosilicates, carbonates, Ca sulfates, Fe sulfides, and organic compounds. Oxidation of minerals by water and escape of H_2 likely drove redox evolution of fluids and altered rocks. The ocean could have evolved from a reduced Na-Cl solution toward a Mg sulfate ocean. Radioactive decay and tidal heating affected aqueous, metamorphic, and igneous processes; ocean salinity; and precipitation at the ocean floor and melting in the icy shell. Because chemical information is sparse at present, inferences about the composition of the ice-ocean-rock system must rely on reasonable assumptions and idealized models.

1. INTRODUCTION

Europa's surface is dominated by materials having water absorption features in the near-infrared, including water ice and poorly characterized hydrates (see chapter by Carlson et al.). The strong association of colored hydrated materials with disrupted surface areas (e.g., linea, bands, domes, pits, chaotic terrains, and impact craters) implies their endogenic and secondary origin with respect to ice-dominated areas (*McCord et al.,* 1998b, 1999; *Fanale et al.,* 1999). There are indications of a liquid ocean beneath the ice (*Khurana et al.,* 1998; *Pappalardo et al.,* 1999; *Kivelson et al.,* 2000), a fundamental requirement for which is a buoyant shell, where buoyancy is best achieved by having water ice dominate the shell. Gas hydrates would be the only other cosmochemically likely option to achieve buoyancy, with a less likely alternative being ammonia dihydrate (*Croft et al.,* 1988). Given that most of the observable surface is dominated by water ice absorption features, the general assumption that the major constituent of the shell is water ice is robust.

The significance of Europa's surface and atmospheric materials for interior compositions and processes appears to be considerable but is not clear. Europa's tenuous Na-K-O atmosphere is linked to surface chemistry, including both endogenic and space-weathering-related inputs (e.g., *Hall et al.,* 1995; *Brown and Hill,* 1996; *Brown,* 2001; *Johnson,* 2000), although a relationship to material derived from Io has been difficult to deconvolve from intrinsic Europa material. Atmospheric Na and K, while sputtered from the surface, could have been implanted from Io's volcanos and neutral torii. Surface SO_2 is only abundant on the trailing side of the body and is probably iogenic. Surface oxidants (O_2, H_2O_2) and possible H_2SO_4 hydrate are likely products of radiolysis and photolysis (see chapter by Carlson et al.) and might be present only in the uppermost (~1-m-thick) ice layer, which is gardened by exogenic processes. The distribution of sulfate hydrate(s) correlates with an ultraviolet and near-infrared absorber (probably S allotropes), indicating a complex sulfur cycle at the surface (chapter by Carlson et al.). At least some surface CO_2 (*McCord et al.,* 1998a) could have formed through radiolysis of exogenic C species (*Cooper et al.,* 2001).

Nevertheless, non-ice colored and hydrated materials and CO_2 in disrupted surface areas appear endogenic (*Greeley et al.,* 2004) and could reflect interior processes (*McCord et al.,* 1998a, 1999; *Fanale et al.,* 1999; *Kargel et al.,* 2000; *Zolotov and Shock,* 2001; *Hansen and McCord,* 2008). The latter materials may have been emplaced as aqueous solutions, fractionally crystallized plutonic masses, solid-state diapirs, organic compounds, or other solute/impurity-rich aqueous or icy materials produced in the interior. The abundances of these materials are such that diverse internal processes and material properties of Europa's icy shell and ocean probably are affected (*Kargel,* 1991; *Hogenboom et al.,* 1995; *Kivelson et al.,* 2000; *Zolotov and Shock,* 2001; *Prieto-Ballesteros and Kargel,* 2005; *McCarthy et al.,* 2007). At the surface, these endogenic materials are affected by sputtering by energetic particles and radiolysis, sublimation, frost deposition, and other surface space-weathering processes. For example, H_2SO_4 hydrate could form through sputtering Na from Na sulfate in a H_2O-rich environment. Variable resistance of non-ice materials to sputtering and sublimation would modify these materials with age and could be a cause of chemical fractionation between surface and atmospheric compositions. Note that some materials could be upwardly concentrated and made visible by space-weathering processes, and so may not fully constitute geologically consequential deposits. In some areas, non-ice materials are covered by water ice formed by recondensed water molecules that were sputtered/sublimated from the surface ice.

Although the Galileo gravity data imply a differentiated interior (*Anderson et al.,* 1998; *Sohl et al.,* 2002; chapter by Schubert et al.), they do not provide direct compositional information. Composition is linked to formation and early heating scenarios, but these pathways, the source of material, and timing of accretion remain hypothetical. It is not clear if short-lived radionuclides (e.g., ^{26}Al, ^{60}Fe) were accreted by Europa or were already largely extinct by the time the moon formed. The amount of accreted aqueously processed asteroidal material with phyllosilicates, carbonates, and sulfates [similar to CI/CM carbonaceous chondrites (*Zolensky and McSween,* 1988; *Brearley and Jones,* 1998; *Brearley,* 2006)] remains unknown as well. The precise means by which heat and mass were transferred within Europa and the accounting of chemical phenomena, including oxidation-reduction (redox) processes, the degree of chemical equilibration or segregation of compounds during the early evolution of Europa, and potential impact erosion and its timing relative to differentiation, are all key to the current state and composition of the ocean and icy shell. These aspects are poorly understood and current constraints are insufficient to tell us what actually exists in Europa, much less how it got that way.

The situation is not as dire as may seem, however, so long as we can be satisfied with a set of approximate compositions, because most of the plausible assumptions result in a few generalized chemical evolutionary paths, which together could bracket the actual path taken by Europa. It follows that a number of assumptions, along with physical and chemical models, need to be considered to estimate the composition of Europa's subsurface. Insights from other satellites, parent bodies of chondrites (asteroids), and terrestrial processes may also be invoked.

Throughout the satellite's history, the icy shell, an ocean, and underlying silicate rocks could have been coupled by physical and chemical processes. This plausible assumption leads down particular sets of chemical evolutionary paths dependent mainly on the temperature, degree, and duration of coupling. The current icy shell could have been formed through freezing of oceanic water and likely contains oceanic species. In turn, the oceanic composition was affected by freezing from above, dissolution of minerals in underlying rocks, mineral precipitation, degassing processes, and loss of volatiles to space. Radiolysis, photolysis, vapor cycling, and gardening of the surface materials, and injection of material from Io (both volcanic and impact ejecta) and other jovian satellites, produced surface materials that could have affected the composition of the deeper icy shell and the ocean. Degassing of H_2, N_2, and S and C volatiles into the ocean from the deeper interior, within the ocean, from oceanic water into space through openings of the icy shell, and from the icy shell should have influenced the composition and oxidation state of the whole system. The composition of suboceanic rocks could have been affected by magmatic and metamorphic processes and alteration by circulating oceanic water. Mineral and/or organic precipitation onto the ocean floor could have isolated the suboceanic silicate crust from the ocean, and thus may have influenced the nature of ocean-rock interaction.

Formation of the current atmosphere through sputtering implies that atmospheric chemistry reflects surface composition superposed by radiolytic/sputtering effects. The geologic control of surface material distribution then further suggests that atmospheric and surface compositions are related to the ocean composition and interior processes, including rock-ocean-ice coupling (a "top-down" approach) (*Greeley et al.,* 1999). With this approach, concentrations and relative abundances of atmospheric and surface species (alkalis, salts) can be used to evaluate the composition and oxidation state of the ocean, ocean floor sediments, and underlying rocks. However, this is a controversial point and other possibilities should be considered (e.g., ionian origin of elements and species).

The composition of the icy shell, an ocean, and upper mantle rocks can also be evaluated through modeling of interior processes throughout history (a "bottom-up" method). This approach includes a consideration of dynamics, timing, and sources of matter for the satellite's formation and thermal history. Physical and chemical models of rock-water-ice-gas-organic interactions coupled with thermal evolution could constrain the major periods of compositional evolution of the rocky core and the water (liquid + solid) shell. The reliability of the models can be tested by comparing modeling results (e.g., salt composition) to remote-sensing data. Ideally, "top-down" and "bottom-up" approaches should converge, with due allowance for radiolytic and implantation processes operative as well.

The composition of the interior can also be constrained by geological and geophysical models. As examples, Europa's moment of inertia can be used to estimate the composition and mineralogy of the interior (*Kuskov and Kronrod,* 2001, 2005); complex data on the icy shell thickness and evolution and Galileo magnetometer data can be used to evaluate the salinity of oceanic water; heat transfer and convection models for the icy shell (chapter by Barr and Showman) can be used to constrain the stability and potential distribution of salts, gas hydrates, and brine inclusions. This chapter briefly reviews direct chemical observations and discusses indirect information and models that may constrain the composition of the ice-ocean-rock system and its chemical evolution.

2. COMPOSITION OF THE ICY SHELL

The distribution of non-ice constituents within the icy shell depends on the geological structure of the shell, its thickness, and tectonic processes caused by tidal motions and/or convection. Although the geological structure of the deeper shell remains poorly understood, some speculations can be offered based on surface geological features. If the

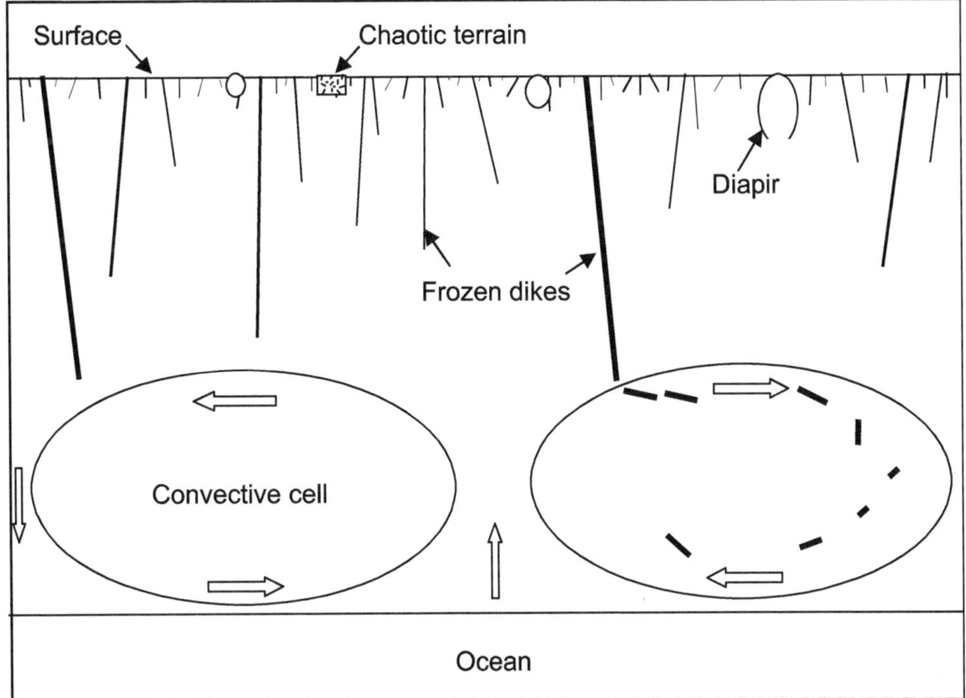

Fig. 1. A hypothetical compositional structure of the current icy shell on Europa. Near-surface layers would be the most abundant in oceanic (salts, CO_2?), exogenic (Io's, chondritic, cometary) and radiolytically formed (O_2, H_2O_2, CO_2, $H_2SO_4 \cdot 4H_2O$) species. Possible formation of the shell by downward freezing implies fewer ice disruptions and lower amount of impurities toward the ocean-ice boundary. If developed, convection in the lower shell may erase rare compositional patterns.

geologically young age of the surface [30–70 m.y. (*Zahnle et al.,* 2003)] reflects the time of significant ice melting/removal, the majority of disruptions could have occurred while the new-grown ice was thin (cf. *Pappalardo et al.,* 1998). Therefore, these disrupted areas with abundant non-ice material could be mostly concentrated in the upper part of the current (thick) shell (Fig. 1). In fact, a thicker shell should have been disrupted less often, and slower freezing would have led to less-efficient capture of oceanic solutes, as discussed below.

2.1. Brine, Salt, and Gas Inclusions

By analogy with terrestrial ices, Europa's icy shell may contain salt, brine (concentrated aqueous solution saturated or nearly saturated with respect to salts), and gas inclusions. Their formation could be attributed to capture of oceanic water and solutions/gases formed in the icy shell. In the upper ice layer, radiolytically formed O_2 and CO_2 could exist in gas inclusions (chapter by Carlson et al.). A typical "oceanic" inclusion could include solids (salts, clathrates) and may contain aqueous (brine) and gas phases. Brine/salt inclusions may also contain abiotic or biogenic organic molecules or even remnants of past or present organisms.

On Earth, sea ice captures solutes from seawater during ice formation (*Thomas and Dieckmann,* 2003; *Horner,* 1985). Cooling of the ice leads to sequential freezing of trapped seawater, precipitation of salts, exsolution of gases, and formation of brine/salt inclusions in channels and pockets. The distribution and composition of brines in sea ice depend on the rate of ice formation, vertical temperature gradient, and age of the ice. With aging, the abundance of brine pockets decreases because of downward migration. Somewhat similar processes are expected to occur during ice formation on Europa, especially when ice forms rapidly near the surface.

Chemical equilibrium models indicate the potential existence of concentrated aqueous solutions in lower parts of Europa's icy shell that cover a salt-bearing ocean (*Zolotov et al.,* 2004; *Marion et al.,* 2005). The uppermost, coldest brine inclusions in the icy shell could represent eutectic brines produced by freezing of oceanic water. The thickness of potential brine-bearing zones depends on the locations of eutectic isotherms, which depend on the composition of trapped water and temperature/pressure distribution. In a nonconvecting shell, brine inclusions are stable only in the lowermost part of the shell. For example, in a 20-km-thick shell, brine inclusions may potentially exist below ~17 km (*Marion et al.,* 2005), as illustrated in Fig. 2. Convection in a thick icy shell (see chapter by Barr and Showman) creates large zones in which brines could exist. Brines can be present in the lower and middle parts of the shell, depending on the location in the convection cell (Fig. 3). In upwelling parts of convective cells, the eutectic represents conditions of complete freezing. In downwellings, it indicates conditions at which brines may form. With or

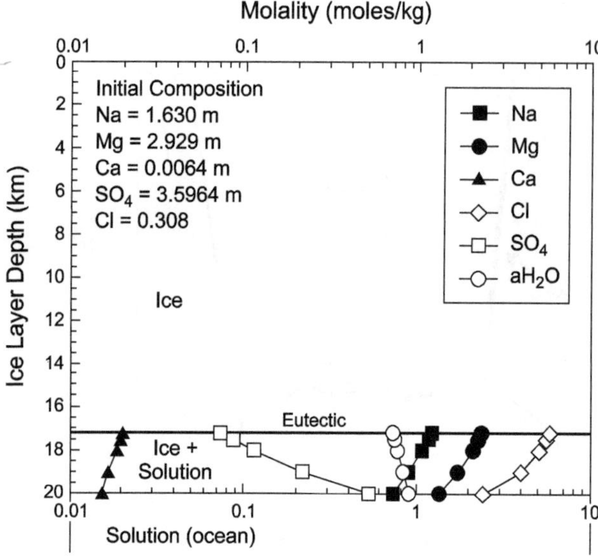

Fig. 2. The ionic composition of a hypothetical brine in a 20-km nonconvecting icy shell. Concentrations are in molalities (m, mol/kg H_2O). aH_2O stands for water activity. For the chosen initial oceanic composition, the eutectic corresponds to temperature ~238 K and pressure 209 bars. Modified from *Marion et al.* (2005).

without convection, decreasing temperature toward the surface leads to concentrated brines. Toward the surface, concentrations of Cl, Na, Ca, and Mg in brines increase (see Fig. 2) and activity of H_2O and volume of brine decrease. Correspondingly, the brine/salt ratio in inclusions decreases toward the eutectic isotherm and the composition of precipitated salts changes. For some oceanic composition models, major precipitated salts are hydrated sulfates of Ca, Na, and Mg, and chlorides of Na and K (e.g., $CaSO_4 \cdot 2H_2O$, $NaCl \cdot 2H_2O$, $Na_2SO_4 \cdot 10H_2O$, $MgSO_4 \cdot 12H_2O$) (*Kargel et al.,* 2000; *Spaun and Head,* 2001; *Zolotov and Shock,* 2001; *Zolotov et al.,* 2004; *Marion and Kargel,* 2008). For the ocean composition models of *Kargel et al.* (2000) and *Zolotov and Shock* (2001), the eutectic temperature is ~238 K.

Although the mere stability of brine inclusions in the lower icy shell does not guarantee their existence, brine pockets should have formed during at least some periods of shell evolution. Formation of the current shell through freezing of oceanic water implies processes that captured oceanic solutes. Some of the colored non-ice material could be frozen brines (*McCord et al.,* 1999, 2002). At any given time, the distribution and amount of brine inclusions were influenced by changing temperature and pressure, downward migration of brines, and ice mass transfer. Downward growth of the icy shell, freezing of oceanic water in fractures (dikes) and chaotic terrains, freezing of cryovolcanic flows, and large impacts could all have been responsible for capture of oceanic water and included impurities into ice. The majority of solutions could have been trapped at early stages of evolution of the current shell that were characterized by fast freezing, often and multiple cracking of thin ice, as well as local melting and freezing in chaotic terrains. Disruptions of the icy shell through cracking followed by upwelling and freezing of oceanic water [i.e., diking (*Crawford and Stevenson,* 1988)] could be a more effective mechanism of water trapping. In contrast, subsequent ice thickening would have led to fewer disruptions (*Figueredo and Greeley,* 2004), slower downward freezing, and reduced efficiency of water trapping into the shell. It is likely that the majority of oceanic salts were trapped in the uppermost part of thickening icy shell, as illustrated in Fig. 1. A progressive decrease in ice temperature with shell thickening led to complete freezing of captured brines, while fewer salts were trapped below. Although ice freezing increased oceanic salinity, freezing 20–30% of a 100-km-thick ocean led to the corresponding moderate increase in salinity and could have not caused more effective capture of oceanic solutes.

Slow solid convection in a thick shell, if it developed (*McKinnon,* 1999; chapter by Barr and Showman), does not favor effective capturing of oceanic water. Even if captured in local high-velocity upwellings of convecting ice, water pockets would move down into the ocean, as observed in terrestrial sea ice. The temperature gradient and density differences between ice and captured water could be the major factors influencing downward migration. Rare diking remains the major mechanism of water capture into a thick shell, where melt-through and impact puncturing is

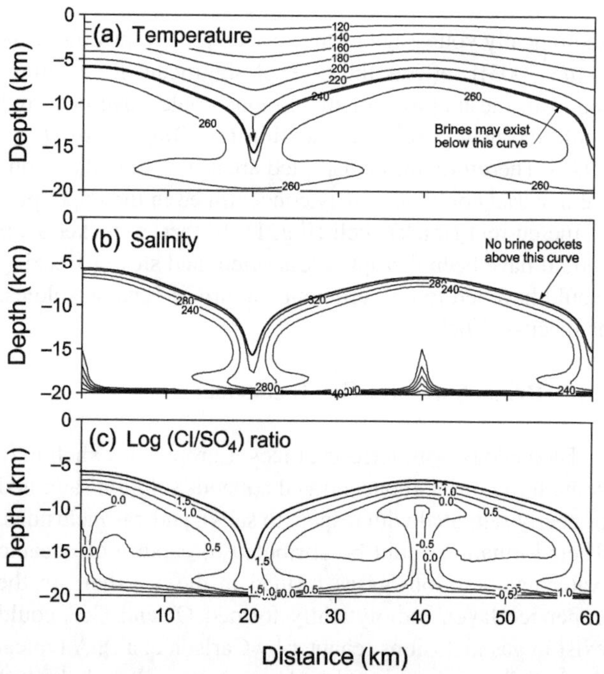

Fig. 3. A hypothetical distribution of brines in a 20-km convective icy shell. **(a)** Distribution of temperature and eutectic conditions (shown by the bold curve); **(b)** salinity of brine pockets, g/kg; **(c)** log Cl/SO_4 mole ratio in brines. Modified from *Zolotov et al.* (2004).

difficult. Convection could disrupt the frozen dikes, leading to redistribution of brine and salt inclusions in lower parts of the shell.

Over time, larger ice crystals might grow (*Schmidt and Dahl-Jensen,* 2003; *Barr and McKinnon,* 2007), causing brine, salt, and gas inclusions to be concentrated at grain boundaries. In convective downwellings, melting should occur at the grain boundaries where impurities are concentrated. Redistribution of inclusions to the boundaries of large crystals may affect rheological properties, limit ice grain size, and make non-Newtonian flow more likely. Once frozen, the physical responses of salt-rich zones to imposed stresses will differ from that of pure ice due to markedly different rheological properties of ice-salt microstructures (*McCarthy et al.,* 2007), as well as the intrinsic strength of salts (*Durham et al.,* 2005b).

Local concentrations of brine/salt inclusions could weaken ice structure and lead to localization of tidal motions and heat production. The size and geometry of a weak zone, migration of brine pockets in temperature gradients, and corresponding changes in ice density, viscosity, and heat release could play roles. Coupled compositional-tidal-convective modeling is needed to best explore the links between tidal forcing and compositional/phase heterogeneity (cf. *Han and Showman,* 2005; chapter by Barr and Showman).

Expelling of dissolved gases from growing ice crystals and lowering water activity in brines with decreasing temperature may cause separation of the gases. However, gas may not separate at higher pressures and/or low concentrations. Most N_2, H_2, CH_4, and light hydrocarbons could have degassed at earlier stages of evolution (section 4), and CO_2 concentration in oceanic water is limited if the ocean is alkaline.

2.2. Clathrate Hydrates?

Gas molecules can form crystalline hydrate species with water, often called clathrates (*Sloan and Koh,* 2007). Although no specific observations indicate solid gas hydrates in Europa's materials, clathrate-forming CO_2, O_2, and SO_2 are observed on the surface (see chapter by Carlson et al.). Formation of O_2-bearing mixed clathrates may account for trapping of radiolytically formed O_2 (*Hand et al.,* 2006). Sulfur dioxide could be gardened in the uppermost surface layer and may also exist in clathrates (*Hand et al.,* 2006). Carbon dioxide in non-ice hydrated surface materials (*Hansen and McCord,* 2008; chapter by Carlson et al.) may manifest dissociated clathrates (although other explanations are also valid). In the higher-pressure parts of the shell, free CO_2 may form clathrates or liquid CO_2 rather than separate in the gas phase. However, without detection of clathrates and estimations of gas abundances in the interior, these suggestions remain speculative.

A broader consideration of other potential clathrate-forming species suggests that clathrates could store a rich variety of organic materials, in addition to other gas species. Multiple cage occupancy of some cages by small molecules, partial cage vacancy, and multicomponent clathrate types, as well as variations in aqueous chemistry, affect the structural type of stable clathrates and their pressure-temperature phase stability fields. In general, lighter guest molecules form less-dense clathrates, and heavier molecules form denser clathrates, but with structure II and structure H clathrates, light, small molecules and large, heavy molecules can form clathrates together. Hence, possibilities for fractionation of guest species can be considerable, with some clathrates potentially stable on the seafloor and others buoyant in the icy shell (*Prieto-Ballesteros et al.,* 2005).

Dynamic conditions of changing pressure, temperature, and fluid phase composition, such as might occur during ascent through the crust of diapirs or sinking or ascendance in the ocean of clathrate crystals, or secular heating and cooling events as tidal heat waxes and wanes, might induce partial dissociation and some fractionation of guest species. Interaction of warm aqueous solutions with clathrate hydrates, or slow thermal dissociation of clathrates, could generate gas pockets or gas-saturated brines. These may erupt in cryoclastic sprays of material, like the gas plumes of Enceladus (*Porco et al.,* 2006), or may produce the needed bubble content to enable liquid water to erupt effusively through the icy shell (*Crawford and Stevenson,* 1988; *Kirk et al.,* 1995; *Croft et al.,* 1995). Not all clathrate species have the high volatility to behave this way, but CH_4 or CO_2 clathrates could.

Any large masses of highly hydrated materials, including clathrate hydrates, could help retain geothermal heat and may localize and sustain geologic activity, such as cryovolcanism and diapirism, and may be involved in formation of chaotic terrain, lenticulae, and other geologic features (*Prieto-Ballesteros and Kargel,* 2005; *Kargel et al.,* 2007). Clathrate hydrates may also affect geologic processes in the shell due to the higher strength of the clathrate than that exhibited by water ice (*Durham et al.,* 2003, 2005a).

If present, clathrate hydrates also may have profound influences on the chemistry of the coupled rock-ocean-ice system. As a storehouse for CO_2, CH_4, SO_2, H_2, and O_2, clathrates potentially could have helped to retain some volatiles that otherwise would likely have been lost from Europa, e.g., H_2 (which forms multiple-occupancy-type clathrates), or may have helped to segregate chemically reactive gases. The hypothetical maximum volumes of guest species are sufficient to convert virtually all of Europa's icy shell into clathrate. The fact that clathrates are not observed spectroscopically could be simply due to the fact that they become unstable when exposed to near-vacuum conditions. Whether clathrates are present in abundance, or not, depends entirely on whether Europa could have eliminated its guest species through physical expulsion into space or by chemical reactions and sequestration in other molecular forms. This uncertainty highlights the very basic and potentially misleading nature of our current understanding of Europa's chemistry.

3. CHEMISTRY OF THE OCEAN-ROCK SYSTEM

3.1. Basic Geochemistry of Ocean-Rock Interactions

This section discusses water-rock reactions in suboceanic rocks based on terrestrial knowledge and their effects on secondary mineralogy and oceanic composition. As on Earth, suboceanic rocks on Europa can be permeable to oceanic water until pressure reaches ~3 kbar (*McKinnon and Zolensky,* 2003). Below a 100-km-thick water layer, that pressure corresponds to a depth of 20–25 km in suboceanic igneous rocks. This thickness could represent an upper limit because the possible presence of fine-grained materials (clay minerals, hydroxides, native sulfur), thick layers of salts, serpentinized rocks, and condensed organic compounds would reduce free pore space and permeability. Water-rock interactions in the permeable layer could be self-limiting due to formation of these pore-closing solids, but they also should have affected the composition of pore fluids. In particular, water-rock reactions should affect salinity, pH, abundances of dissolved gases, and oxidation state of fluids. Solute diffusion in the aqueous phase is important for local fluid mixing in suboceanic rocks. However, tidal pumping in the ocean and hydrothermal circulation in the suboceanic environment could be the main processes mixing solutions.

3.1.1. The nature of suboceanic rocks. Galileo gravity data reveal a differentiated interior with a Fe or Fe-FeS core and a silicate mantle (*Anderson et al.,* 1998; *Sohl et al.,* 2002; chapter by Schubert et al.). The presence of the core implies that temperature of rocks once exceeded that of Fe-FeS eutectic (~990°C) (*Brett and Bell,* 1969). Although this temperature does not cause melting of Fe-Mg silicates, the thermal evolution models of *Consolmagno and Lewis* (1976) indicate a large-scale melting of silicates on Europa through the decay of long-lived radioactive elements. This model shows that the radioactive decay led to a major silicate melting ~2.5 b.y. after accretion in the absence of silicate convection. Accretion of rocks with short-lived radionuclides (*Prialnik and Bar-Nun,* 1990) could have caused silicate melting within only a few million years after formation of Ca-Al-rich inclusions in the solar nebula, as it has also occurred on early-formed asteroids (*McSween et al.,* 2002; *Scott,* 2007; *Qin et al.,* 2008). However, accretion of short-lived radionuclides on Europa is hypothetical and is not consistent with incomplete differentiation of Callisto (*McKinnon,* 2006; chapter by Canup and Ward).

Although upper parts of the mantle could have resisted melting (*Consolmagno and Lewis,* 1976), they might have been affected by silicate volcanism and/or sank into a molten part of the mantle due to gravitational instability. Once core segregation took place, the release of gravitational potential energy added additional heat. Tidal heating might have also caused occasional silicate melting and suboceanic volcanism throughout history (*Hussmann and Spohn,* 2004). If uppermost mantle layers have never been affected by volcanism or melted, suboceanic rocks would be presented by accreted chondritic-type materials (*Lewis,* 1971; *McKinnon and Zolensky,* 2003; *Kuskov and Kronrod,* 2005) that were subjected to aqueous alteration and/or metamorphism (see section 4). Igneous rocks exposed at the oceanic floor would be aqueously altered as well, as discussed below. At deeper levels, heating is expected to have been sufficient to devolatilize rocks.

Partial melting of Europa's rocks with chondritic composition implies formation of mafic and/or ultramafic magmas. Before melting, the accreted material could have been hydrated through low-temperature and hydrothermal reactions, and then at least partially dehydrated (see also section 4). Partial melting of hydrous mantle rocks would have occurred at lower temperatures compared to a dry chondritic substrate. It is possible that elevated water abundances in mantle rocks led to ultramafic magmas/rocks, analogous to terrestrial komatiites (*Abe et al.,* 2000).

Formation of basaltic magmas is the most likely possibility, as we see basalts widely on Earth, Venus, the Moon, Mars, and the eucrite parent body (likely Vesta), and they are the usual predicted composition for melting of peridotites. Other types of suboceanic volcanic rocks could be caused by either unusually low or high degrees of partial melting, by melting of hydrous or CO_2-rich mantle rocks, or by complex liquid lines of descent involving a history of fractional crystallization or crustal anatexis. Compared to Earth, low gravity may favor slow magma ascent, leading to development of large, deep, and long-standing magmatic chambers in which magma differentiated and/or relatively voluminous eruptions, as on the Moon (cf. *Wilson and Head,* 1994). Evolved silicate liquids enriched in silica, alkalis, and other incompatible lithophile elements could present ~10% of Europa's magmas. Larger-scale formation of silica-rich igneous rocks (dacites, rhyolites, granites) is unlikely, unless Earth-like plate tectonics arose.

3.1.2. The nature of oceanic sediments. Oceanic sediments could consist of physically degraded outcrops, altered space debris delivered from the surface, and chemical precipitates from oceanic water and ocean-entering fluids. Space debris could be presented by chondritic and cometary dust (*Pierazzo and Chyba,* 2002) delivered through steady accumulation and larger fragments delivered by impacts. A sizeable fraction could be presented by Io's particles (silicates, sulfur, alkali halides) ejected by impacts and delivered to Europa (*Alvarellos et al.,* 2008), as well as carbonaceous-chondrite-type fragments ejected from outer irregular satellites of Jupiter (*Pollack et al.,* 1978). Low-temperature aqueous alteration of delivered minerals and glasses would lead to complete dissolution (as Io's NaCl particles) (*Postberg et al.,* 2006), hydration (formation of serpentine and clay minerals), and oxidation [formation of Fe(II) phyllosilicates, Fe(III) oxides/hydroxides, pyrrhotite, and/or pyrite]. High-molecular weight organic compounds, typical for chondritic (*Sephton,* 2002) and cometary material, and some inorganic species from space debris (graphite, diamond, spinel, SiC, corundum) may persist at the oceanic floor. Although a significant amount of space materials could have accumulated during late heavy bombardment (e.g., *Gomes et al.,* 2005) ~3.8–3.9 b.y. ago, the thickness of subsequent space sediments may be insignificant. During the last 50 m.y., micrometeorite impacts may have led to the concentration of space materials of ~200 ppm in a ~2.6-m-thick global layer gardened by exogenic processes (see chapter by Carlson et al., but cf. chapter by Moore et al.). If Europa's seafloor is active tectonically or has chemical sedimentation from the ocean or hydrothermal activity, then exogenic debris is apt to be thoroughly ingested into other materials.

Chemical precipitation could be caused by freezing, evaporation, and degassing of oceanic water, changes in temperature, and mixing of ocean-entering solutions with oceanic water, as discussed in following sections. Over wide ranges of icy shell thickness and oceanic redox state, chemical sediments could contain phyllosilicates, Fe sulfides, Fe oxides/oxyhydroxides, Ca-Mg-Fe carbonates, phosphates, native sulfur, condensed organic species, and gas hydrates. In a sulfate-bearing ocean, sparingly soluble Ca sulfates (gypsum) are likely to be present.

3.1.3. Mineral dissolution, secondary mineral precipitation, and pH control. When an igneous rock is exposed to aqueous solution, it is first subjected to dissolution. Each condensed phase in the rock (minerals, glasses, organic compounds) dissolves with a different rate depending on temperature, pressure, solution pH, degree of saturation, and surface area exposed to solution (e.g., *Brantley,* 2004). Silica-poor glasses, which are important constituents of mafic/ultramafic lavas and primordial chondritic materials, dissolve faster than typical igneous minerals (e.g., *Gislason and Oelkers,* 2003). Acidic conditions favor dissolution of major minerals in mafic/ultramafic rocks, while alkaline conditions facilitate dissolution of feldspars (*Brantley,* 2004). Changes in solution pH and composition affect dissolution rates, which in turn influence chemistry of solution. Increase in solution salinity leads to sequential saturation with respect to secondary minerals. Secondary precipitation controls concentrations of specific solutes and pH, and can stabilize salinity and influence dissolution rates of primary phases. In some cases, evolution of the solution may lead to instability of earlier precipitates and cause their dissolution. At each time, the typical mineral assemblage consists of unaltered primary minerals, previously formed secondary phases, newly precipitated solids in equilibrium with solution, and inert species among solids and solutes (e.g., organic compounds). Eventually, alteration of rocks should lead to complete dissolution of the majority of primary phases and the formation of an assemblage of secondary phases in chemical equilibrium with solution. Timing of complete equilibration depends on temperature, reactive surface area of rocks and minerals (e.g., grain size, permeability), and water/rock ratio. If Europa's current ocean is not sealed by an impermeable material, concentrations of many solutes could be controlled by chemical equilibria within rocks/sediments. Solution composition strongly correlates with secondary mineralogy of altered ultramafic rocks, and thermochemical equilibrium models have been used successfully to predict the composition of aqueous solutions (e.g., *Wetzel and Shock,* 2000; *Bruni et al.,* 2002; *Palandri and Reed,* 2004). For a reduced early Europa ocean, equilibrium models demonstrate the formation of phyllosilicates (serpentine, saponite, chlorites), magnetite, chromite, carbonates, and phosphates through low-temperature hydration and oxidation of a material of solar composition (*Zolotov et al.,* 2006a; *Zolotov and Mironenko,* 2007b). Analogous models for basalts reveal the formation of abundant smectites.

Since present-day hydrothermal activity on Europa remains hypothetical, low-temperature (~0°C) alteration of mafic and/or ultramafic rocks would have a major effect on composition of pore solutions and ocean-entering fluids. On Earth, chemical weathering is a major H^+-consuming process, and thus a major mechanism that buffers pH (*Stumm and Morgan,* 1996). For example, dissolution of forsterite can be represented by the H^+-consuming reaction

$$Mg_2SiO_4 + 4H^+ \rightarrow 2Mg^{2+} + H_4SiO_4 \quad (1)$$

Magnesium pyroxenes (e.g., enstatite) are also known to be attacked by aqueous solutions (*Brantley,* 2004). Field observations (e.g., *Gislasson and Eugster,* 1987b; *Stevens and McKinley,* 1995) and experiments (*Gislasson and Eugster,* 1987a) demonstrate that low-temperature anoxic alteration of basalts causes formation of slightly alkaline fluids (pH = ~8–9). Aqueous weathering of ultramafic rocks (serpentization) leads to solutions with pH 9–12 (*Barnes and O'Neil,* 1969; *Barnes et al.,* 1982; *Neal and Stanger,* 1983; *Sader et al.,* 2007).

Alkaline pHs maintained by secondary mafic/ultramafic assemblages favor the conversion of dissolved CO_2 to bicarbonate (HCO_3^-) and carbonate (CO_3^{2-}) ions and precipitation of carbonates, mostly calcite and aragonite (*Bonatti et al.,* 1980; *Kelley et al.,* 2001). If mafic/ultramafic rocks dominate the aqueous chemistry, these inferences imply an alkaline pH of Europa's ocean in which concentrations of at least several elements (Mg, Ca, Fe, Si, Al) are controlled by equilibria with secondary minerals. By analogy with Earth's seawater (e.g., *Holland,* 1978; *Walther,* 2005), long-term oceanic pH could be buffered by acid-base reactions involving carbonates, as well as secondary phyllosilicates. The following reactions of carbonate dissolution and precipitation can affect and possibly control oceanic pH and Ca^{2+} abundance

$$CaCO_3 + H^+ \rightarrow Ca^{2+} + HCO_3^- \quad (2)$$

$$Ca^{2+} + OH^- + HCO_3^- \rightarrow CaCO_3 + H_2O \quad (3)$$

Short-term changes in pH, if they ever occurred, could be buffered by the bicarbonate-carbonate equilibrium ($HCO_3^- \leftrightarrow$

$CO_3^{2-} + H^+$). Concentrations of HCO_3^- and CO_3^{2-} ions could also be restricted by precipitation of carbonates in suboceanic rocks and/or chemical sediments

$$2HCO_3^- + Ca^{2+} \leftrightarrow CaCO_3 + H_2O + CO_2 \quad (4)$$

$$CO_3^{2-} + Ca^{2+} \leftrightarrow CaCO_3 \quad (5)$$

Reaction (4) implies that degassing of CO_2 from oceanic water may cause deposition of a Ca carbonate, as it happens on Earth. In turn, an increase in f_{CO_2} (e.g., with pressure and depth) would decrease pH and make carbonates less stable.

The presence of precipitated solids in sediments would influence and/or control activities and concentrations of corresponding solutes in oceanic water. For example, aCa^{2+} could be controlled by equilibria that include carbonates (reactions (2)–(5)) and/or gypsum

$$Ca^{2+} + SO_4^{2-} + 2H_2O = CaSO_4 \cdot 2H_2O \quad (6)$$

Likewise, precipitation of pyrite may limit activities of bisulfide (HS^-) and Fe(II) solutes

$$Fe^{2+} + 2HS^- = FeS_2 + H_2 \quad (7)$$

Throughout history, temporal changes in tidal heat production and icy shell thickness due to a variable nature of the Laplace resonance (*Hussmann and Spohn,* 2004; *Tobie et al.,* 2005) should have affected oceanic salinity, composition, and temperature (cf. *Melosh et al.,* 2004), as well as mineralogy of suboceanic rocks. Low tidal activity would cause some freezing of the icy shell, increasing salinity of oceanic water, mineral precipitation, and exsolution of dissolved gases. Higher salinities depress freezing temperatures and imply colder oceanic water. In contrast, intensive release of tidal heat would lead to dilution of oceanic water through ice melting and some dehydration of solids. Corresponding changes in volumes of ice, ocean and rocks could have caused tectonic activity and resurfacing (*Nimmo,* 2004). Slow and moderate changes in ice thickness should not have caused variations in oceanic abundances of species that are controlled by solubility and sorption equilibria in altered rocks/sediments. These minerals would include Ca-carbonates and sulfates, pyrite, ferric hydroxides/oxyhydroxides, amorphous silica, clay minerals, and phosphates. Precipitation of these low-solubility solids was likely during the history of the current shell. Concentrations of Cl, Br, I, SO_4^{2-}, Mg, and Na, however, may not be efficiently maintained by solution-mineral reactions. For example, freezing of 50% of oceanic water in unsaturated ocean models (e.g., *Zolotov and Shock,* 2001) would double the salinity of the ocean but would not lead to precipitation of chlorides and sulfates of Mg and Na. Precipitation of abundant salts from near-eutectic oceanic water (below ~–30°C) (*Kargel et al.,* 2000; *Zolotov and Shock,* 2001) can only be possible after freezing of a significant portion of the ocean. A present-day ice thickness of ~20–30 km implies that the liquid portion of the 80–140-km-thick water shell (*Anderson et al.,* 1998) might not be described as a near-eutectic Na-Mg-SO_4-Cl brine, but could be a dilute solution consistent with Galileo magnetometer data (section 3.4).

3.1.4. Hydrothermal processes. Uneven release of heat in upper parts of the silicate mantle could cause hydrothermal circulation of oceanic water in permeable rocks. Localization of heat sources can be caused by tidal motions, magmatism, and exothermic hydration reactions (e.g., serpentinization) (*Grimm and McSween,* 1989; *Kelley et al.,* 2001, 2005) of freshly emplaced rocks. Despite slow fluid movement in low gravity (*Lowell and DuBose,* 2005), Europa's hydrothermal processes should affect the pattern of rock alteration, cause secondary precipitation, and influence the composition of oceanic water. By analogy with terrestrial suboceanic hydrothermal systems (e.g., *Von Damm,* 1990, 1995; *Elderfield and Schultz,* 1996; *Alt,* 1995, 2004; *Staudigel,* 2003), high-temperature fluid upwellings in suboceanic rocks could occur above local heat sources (e.g., magma chambers), driven by buoyancy of warm fluids. The complementary downwellings could occur through diffuse flows of oceanic water over large areas.

Diffuse flows lead to low-temperature alteration of rocks and form solutions at least partially equilibrated with secondary minerals. At low Europa gravity, slow circulation of water favors long-term water-rock interaction. As on Earth, these processes could sequester in solid forms of inorganic C, K, Rb, Cs, and some Na from oceanic water while expelling some Ca and Mg from the rocks (*Staudigel,* 2003). Hydration, anoxic oxidation, and carbonatization are the most likely pathways of low-temperature rock alteration by diffuse flows. By analogy with Earth's oceanic systems, reduction of oceanic water in pore fluids can cause deposition of pyrite and other sulfides, assuming the ocean contains SO_4^{2-}.

Rock alteration by focused hydrothermal flows would cause rapid formation of minerals, depending on temperature, pressure, fluid composition, and water/rock ratio. Elevated temperatures favor stability of chlorite and saponite and these minerals become more abundant in altered mafic and ultramafic rocks, respectively. In addition, high-temperature aqueous conditions are favorable for oxidation of Fe, S, and C species, both thermodynamically and kinetically (section 3.1.5).

Hydrothermal processes favor leaching of K, Li, Rb, and Cs but may cause some trapping of Na in zeolites, albite, and even halite in deep horizons (cf. *Staudigel,* 2003). In turn, hydrothermal precipitation would limit abundances of several oceanic solutes (e.g., S, C, Mg, and Fe species) in oceanic water (*Alt,* 1995, 2004; *Seyfried and Shanks,* 2004). By analogy with terrestrial systems (*German and Von Damm,* 2003), incorporation of Mg into secondary phyllosilicates results in a Mg-depleted composition of hydrothermal fluids vented into the ocean. Leaching of Ca can be compensated by precipitation of carbonates (*Staudigel,* 2003) and sulfates. High-temperature precipitation of anhydrite ($CaSO_4$) from circulating sulfate-bearing oceanic water can affect abundance of oceanic sulfate, especially in a case of large-scale hydrothermal activity. High-temperature hydrothermal reduction of sulfate and formation of sulfides (H_2S, HS^-), then cooling of sulfide-bearing hydrothermal fluids and mixing with cold oceanic water, can cause precipitation of metal sulfides (mostly pyrite), as observed in terrestrial "black smokers" (*Von Damm,* 1990; *Scott,* 1997), thus further drawing down sulfate abundances. Silica and silicate minerals may precipitate from cooling fluids as well. Release of highly alkaline fluids during serpentinization of peridotites causes trapping of oceanic CO_2 to carbonates (*Kelley et al.,* 2001, 2005). In altered oceanic basalts, carbonates form as well.

Mineral dissolution and precipitation affect porosity, permeability, and thermal conductivity of rocks. In turn, these changes could affect temperature, discharge, composition, and oxidation state of hydrothermal fluids and pattern of further alteration. Although these effects may not be significant in basaltic systems (*Alt-Epping and Smith,* 2001), formation of serpentine minerals in ultramafic rocks leads to a significant increase in the volume of solids, which decreases permeability.

3.1.5. Redox processes. Even without strong surface oxidants (O_2, H_2O_2) occasionally delivered to oceanic water (chapter by Hand et al.), many water-rock reactions could involve transfer of electrons (redox reactions). At the ocean-rock interface, oxidation of Fe, S, and C species would dominate over reduction reactions. Water works as a major oxidizing agent in conversion of Fe^0 metal and Fe(II) in silicates to magnetite or ferric oxyhydroxides as exemplified by reactions

$$3Fe^0 + 4H_2O \rightarrow Fe_3O_4 \text{ (magnetite)} + 4H_2 \quad (8)$$

$$1.5Fe_2SiO_4 \text{ (in olivine)} + H_2O \rightarrow Fe_3O_4 + 1.5SiO_2 + H_2 \quad (9)$$

$$Fe^{2+} + 2H_2O \rightarrow FeOOH \text{ (goethite)} + 2H^+ + 0.5H_2 \quad (10)$$

Anoxic oxidation of troilite and pyrrhotite can occur through interactions with H_2S, which is a dominant form of reduced sulfur in acidic solutions, and can be favored by H_2 removal

$$(1-x)FeS \text{ (troilite)} + xH_2S \rightarrow Fe_{1-x}S \text{ (pyrrhotite)} + xH_2 \quad (11)$$

$$Fe_{1-x}S + (1-2x)H_2S \rightarrow (1-x)FeS_2 \text{ (pyrite)} + (1-2x)H_2 \quad (12)$$

In alkaline solutions, more oxidizing conditions are needed for reactions (11) and (12) to proceed. Anoxic oxidation of aqueous species can be exemplified by the formation of sulfate and carbonate ions from sulfide S and organic C, which require elevated temperatures

$$HS^- + 4H_2O \rightarrow SO_4^{2-} + H^+ + 4H_2 \quad (13)$$

$$\text{Organic carbon} + H_2O \rightarrow CO_2 + H_2 \quad (14)$$

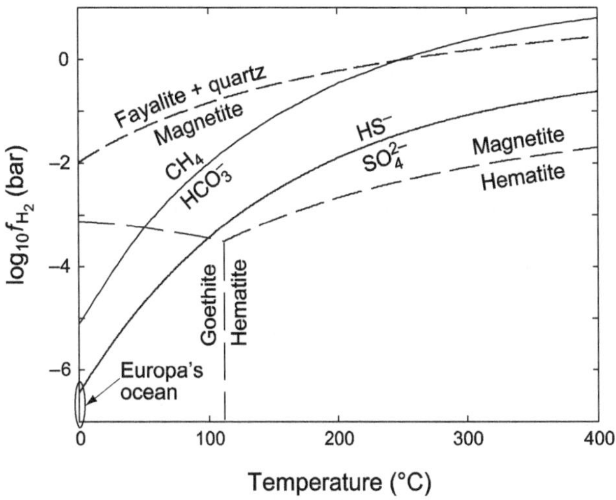

Fig. 4. Stability fields for iron minerals in comparison with stabilities of sulfate and bicarbonate as functions of temperature and H_2 fugacity at pH 10 and 1375 bars. The stability fields of iron sulfides depend on bulk S content and are not shown. The arrow shows hypothetical conditions of Europa's ocean. These conditions correspond to the predominance of SO_4 over dissolved sulfides ($H_2S + HS^-$), and Fe(III) oxyhydroxides (goethite) could be stable at the oceanic floor. Methane, organic compounds, Fe(II) silicates, magnetite, and hematite may not be stable. Higher stabilities of oxidized species at elevated temperatures imply a hydrothermal origin of sulfate in Europa's ocean. Modified from *Zolotov and Shock* (2004).

Figure 4 illustrates that higher temperature favors oxidation of sulfide S and organic C to sulfate and carbonate species, respectively. In addition to thermodynamically plausible oxidation, elevated temperatures (>150°–200°C) make these conversions kinetically possible (e.g., *Ohmoto and Lasaga,* 1982). Note that hydrothermal oxidation of high molecular weight organic compounds would lead to a variety of hydrocarbons and O-bearing organic species that are depleted in H and N compared to precursor molecules/polymers (see section 4.6).

Oxidation reactions (8)–(14) lead to the formation of H_2, which can separate into a gas phase, accumulate below the icy shell, and eventually escape to space via cryovolcanism or through breaches in ice. Convection and diffusion could be responsible for the delivery of H_2-bearing waters to low-pressure horizons in suboceanic rocks and/or ocean, where H_2 gas can exsolve. Removal of H_2 stimulates oxidation of rocks and fluids, as could have occurred at early stages of Europa's evolution (section 4).

Restricted H_2 removal could limit oxidation, however. H_2 removal can be hindered in low-permeability rocks and at pressures above ~1–2 kbar that do not favor gas separation. In addition, low concentrations of dissolved H_2 would prevent gas phase separation, and further oxidation will be limited by the rate of H_2 diffusion out of the rocks. In such a case, redox conditions, which can be expressed by H_2 fugacity (f), could be controlled by mineral assemblages. At pressures more than 1–2 kbar, f_{H_2} can be controlled by equilibrium (reaction (8)), which may characterize the early stages of Europa's evolution (*Zolotov and Mironenko,* 2007b). At more oxidizing conditions, f_{H_2} is likely to be controlled by equilibria between Fe(II) silicates and Fe(III) oxides/oxyhydroxides, as discussed in section 3.2.

Despite net oxidation driven by H_2 escape, some reduction can occur at low-temperature conditions at elevated f_{H_2}, which can be typical in impermeable rocks. Formation of secondary fayalite (Fe_2SiO_4) from magnetite in aqueously processed asteroids (*Krot et al.,* 1998; *Zolotov et al.,* 2006b) or formation of native metals in serpentinites (*Frost,* 1985) can exemplify the reduction. As another example, cooling of H_2-rich hydrothermal fluids creates thermodynamic potential for reduction of Fe, S, and C compounds (e.g., *McCollom,* 1999; *Zolotov and Shock,* 2003a). Figure 4 illustrates that, in cooled fluids at constant of buffered f_{H_2}, Fe(II) silicates, sulfide S, and reduced C (organic) compounds become more stable than ferric Fe, sulfate, and carbonate species. Note that many low-temperature abiotic reduction processes (e.g., back reactions (13) and (14)) may not proceed efficiently because of slow reactions. In fact, redox reactions in C-O-H and S-O-H systems are inhibited at temperature below ~150°–200°C (e.g., *Ohmoto and Lasaga,* 1982; *Seewald et al.,* 2006), and perhaps would not affect redox conditions and speciation in the low-temperature ice-ocean-rock system. Therefore, organic compounds and methane would metastably coexist with sulfate, sulfate with sulfide and native sulfur, and carbonate species with organic compounds.

3.2. Oxidation State of the Ocean-Rock System

Although the oxidation state of the ocean-rock system is not known, some inferences can be drawn from the composition of the non-ice surface material, constraints on the composition of the materials accreted on Europa, and thermochemical evolution models of the interior. The detection of the sulfate group (SO_4^{2-}) on Europa's surface with Galileo near-infrared spectroscopy (*McCord et al.,* 1998b, 1999; *Carlson et al.,* 1999; chapter by Carlson et al.) may indicate a high oxidation state of oceanic water, oceanic sediments, and underlying rocks. The association of the sulfate group (in hydrated sulfate salts and/or sulfuric acid hydrate) with disrupted surface areas indicates an endogenic source for the sulfur, possibly from the underlying ocean (*McCord et al.,* 1998a,b, 1999; *Fanale et al.,* 1999, 2000). Although sulfate observed in the non-ice material can form through hydrolysis of reduced S species at the surface (*Carlson et al.,* 1999, 2002), concentrations of reduced sulfur species (H_2S, HS^-, S^{2-}) in oceanic water should not be high [<~10^{-5} mole (kg $H_2O)^{-1}$] owing to low solubilities of solid sulfides (FeS, $Fe_{1-x}S$, FeS_2) at low temperature (cf. *Zolotov,* 2008). Therefore, traces of sulfide species delivered from a reduced oceanic water may not account for the observed sulfate (*Zolotov and Shock,* 2004). Instead, Mg and Na sulfates are highly soluble in water, and near-infrared spectra of their solid hydrates are consistent with the Galileo spectra of the non-ice material (*McCord et al.,* 1998b, 1999, 2002; *Orlando et al.,* 2005; *Dalton et al.,* 2005; *Dalton,* 2007). Some sulfates ($MgSO_4$) have long lifetimes with respect to surface irradiation (*McCord et al.,* 2001a). It follows that the satellite's subsurface and ocean may contain sulfate, as widely believed (*McCord et al.,* 1998b, 1999, 2002; *Fanale et al.,* 1999, 2000; *Kargel et al.,* 2000; *Zolotov and Shock,* 2001, 2003a, 2004). Although direct observational evidence for an endogenic origin of surface SO_4-bearing compounds is lacking, endogenic formation of sulfate sulfur is a logical consequence of aqueous oxidation of sulfides on water-bearing bodies that experienced internal heating and H_2 escape (see sections 4.2 and 4.5).

The assumed presence of sulfate in the ocean sets a lower limit for the oxidation state of oceanic water, which can be expressed in terms of f_{H_2}. Depending on the pH, sulfate exists stably at f_{H_2} < ~10^{-6} bar at temperature ~0°C and pressure of 1–2 kbar (Fig. 5). These redox conditions correspond to activity (a) of dissolved H_2 < ~10^{-9}. Sulfate becomes the dominant species in oceanic water if f_{H_2} is <~10^{-8} (aH_2 < ~10^{-11}), as illustrated in Fig. 6. These limits for f_{H_2} can be used to evaluate the stability of redox-sensitive species in the ocean and permeable rocks/sediments (cf. *Zolotov and Shock,* 2004). For example, in an alkaline ocean, native sulfur is more likely to stably coexist with sulfate at very high bulk sulfur abundances (Fig. 5b). Pyrite can coexist with oceanic sulfate at specific conditions (Figs. 5c and 6) that also depend on bulk S abundance. In contrast, troilite (FeS) and pyrrhotite ($Fe_{1-x}S$) are not stable in contact with sulfate-rich solution. Primary and secondary Fe(II) silicates (pyroxene, olivine, chlorite, serpentine minerals) and magnetite are not stable and could be oxidized to form ferric hydroxides and oxyhydroxides (e.g., goethite, α-FeOOH) that coexist with sulfate (see Fig. 5c) (*Zolotov and Shock,* 2004). Formation of Fe(III) silicates (e.g., nontronite) can also be possible.

Oxidized inorganic C species (CO_2, HCO_3^-, CO_3^{2-}, carbonates) are stable in sulfate-rich oceanic water and can also exist at more reduced conditions (Figs. 5d and 6). Organic species and CH_4 are not stable in a sulfate-bearing ocean, but can exist metastably. The presence of CO_2 on the surface of Europa (*McCord et al.,* 1998a; *Hansen and McCord,* 2008) may not reflect the oxidation state of oceanic water because of radiolysis of organic compounds (*Cooper et al.,* 2001; chapter by Carlson et al.). In fact, CO_2 observed in the colored non-ice surface material (*Hansen and McCord,* 2008) may represent radiolytically oxidized organic C delivered from interior as well as oceanic CO_2 and/or carbonate species.

By analogy with terrestrial aquifers, the oxidation state of pore fluids near the ocean floor on today's Europa can be controlled by chemical reactions in underlying rocks and/or sediments. Buffering of oceanic redox conditions by minerals is possible, but is uncertain owing to a large oceanic mass and low rates of low-temperature redox reactions. Low permeability would also makes buffering less likely. Independent of buffering, a sulfate-bearing ocean should affect mineralogy of redox-sensitive minerals in permeable rocks.

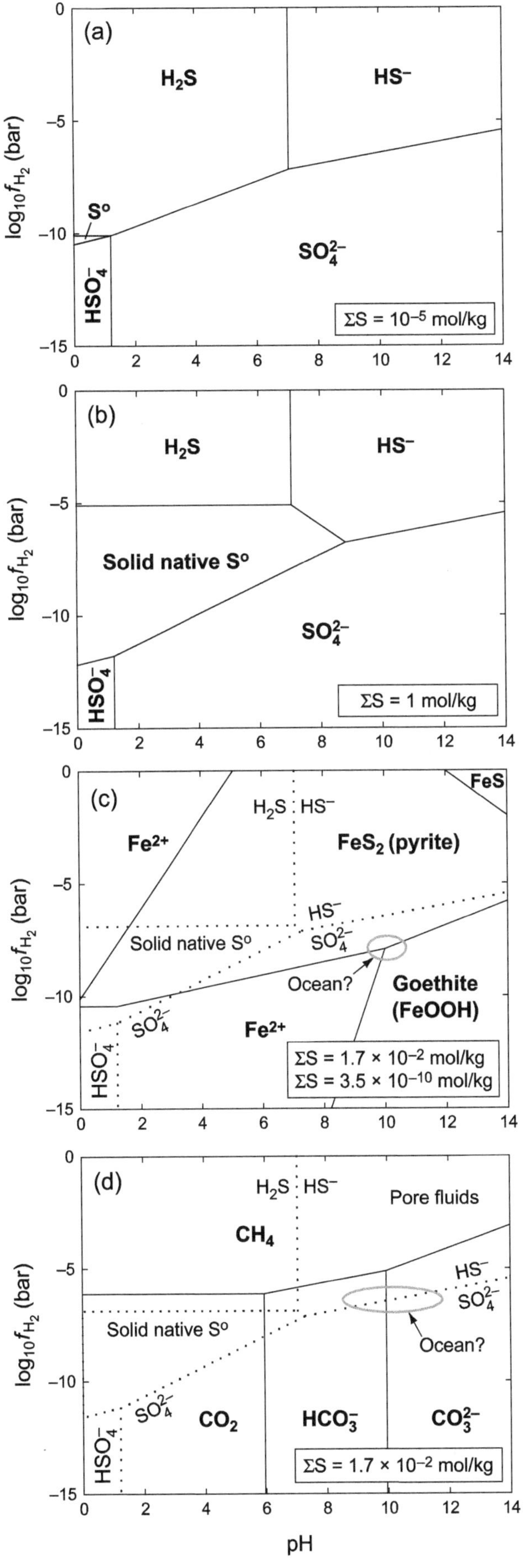

As an example, a pyrite-goethite assemblage can be thermodynamically stable in contact with neutral and alkaline sulfate-bearing oceanic water (Figs. 5c, 6, and 7). Equilibrium thermodynamic models show that pyrite and goethite are the most likely Fe minerals in equilibrium with alkaline sulfate-bearing europan seawater (*Zolotov and Shock,* 2004; *Zolotov,* 2008). If masses of pyrite and goethite are large enough, they may buffer a very low concentration of dissolved H_2 in oceanic water. Further loss of H_2 from an already oxidized Fe^{2+} (in magnetite)-Fe^{3+} (in goethite) system could maintain near-equilibrium f_{H_2} of the following formal equilibria

$$FeOOH + 2HS^- + 2H^+ = FeS_2 + 0.5H_2 + 2H_2O \quad (15)$$

$$FeOOH + 2SO_4^{2-} + 4H^+ + 7.5H_2 = FeS_2 + 10H_2O \quad (16)$$

The conditions of equilibria (reactions (15) and (16)) and some other equilibria with pyrite are close to the conditions at which activities of sulfate and bisulfide (HS^-) ions are equal. Therefore, oceanic water may contain some reduced sulfur (e.g., HS^-, S^{2-}) as well. In a thick (10–20 km) permeable layer, large mass of Fe(II) minerals implies that oceanic f_{H_2} should not be well below the conditions of pyrite-goethite equilibria (Fig. 7). It is possible that oceanic water is not more oxidized than the pyrite-goethite buffer.

Deep in the permeable suboceanic layer, more reduced conditions could be controlled by mineral assemblages that are more reduced than sulfate-bearing oceanic water. These minerals may include ferrous olivine, chlorites, serpentines, magnetite, and pyrrhotite. Considering terrestrial (*Frost and McCammon,* 2008), martian (*McSween,* 2003), and Io's (*Zolotov and Fegley,* 1999, 2000) igneous rocks as examples, the oxidation state of Europa's suboceanic igneous rocks may be between that of the iron-wüstite and quartz-fayalite-magnetite (QFM) buffers. If oceanic sulfates formed through high-temperature hydrothermal processes (*Zolotov and Shock,* 2003a,b, 2004), QFM or a more oxidized mineral buffer could represent the redox state of suboceanic

Fig. 5. Equilibrium speciation of S and Fe as functions of log H_2 fugacity and pH at possible conditions of the oceanic floor on Europa (0°C, 1375 bars). **(a),(b)** Speciation in the S-H-O system. The stability field of native sulfur (S°) is for two different bulk S contents. The stability field of S° becomes broader with increasing S content. **(c)** Speciation in the Fe-S-H-O system superimposed on the speciation in the S-H-O system. Bulk S and Fe contents are 1.7×10^{-2} and 3.5×10^{-10} mole (kg H_2O)$^{-1}$, respectively. The S abundance is chosen to match gypsum saturation and the Fe abundance is used to represent conditions of the FeOOH-FeS_2 equilibrium at pH 10 (*Zolotov and Shock,* 2004). The arrow shows hypothetical conditions in Europa's ocean. **(d)** Speciation in the C-H-O system at 0°C and 1375 bars. The speciation is superimposed on the speciation in the S-H-O system (dotted lines). Carbonate species are stable in a sulfate-bearing ocean. The oval shows conditions that may represent oceanic water. Modified from *Zolotov and Shock* (2004).

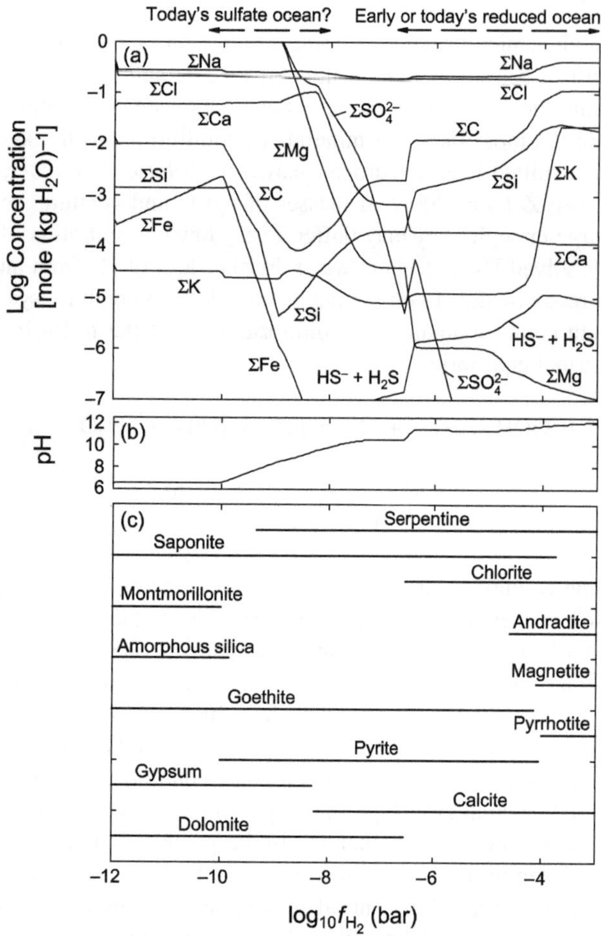

Fig. 6. A modeled oceanic composition and suboceanic mineralogy as functions of H_2 fugacity. The model represents equilibrium composition of the water-rock system at 0°C, 1375 bars, water/rock ratio of 1, and 10% of reacted C in a CI carbonaceous chondrite type rock. In **(c)**, bars with mineral names show f_{H_2} conditions of mineral occurrence in equilibrium assemblages. Axis Y has no units. Modified from *Zolotov* (2008).

rocks. As an example, a system at the QFM buffer (reaction (9)) could control f_{H_2} at about 10^{-2} bar (0°C, 1373 bar) (Fig. 4). A pyrite-pyrrhotite-magnetite (PPM) buffering assemblage can provide similar f_{H_2} values. Sulfide sulfur and organic species could be stable at these moderately reduced conditions and corresponding fluids would contain only traces of the SO_4^{-2} ion (Figs. 4, 5, and 6). After supply of fresh rocks or other disturbing events, low-temperature equilibration could occur slowly. Note that possible redox equilibration among Fe-bearing species in rocks/sediments does not ensure redox equilibria among C and S species, even over geological time.

Terrestrial data demonstrate that serpentinized ultramafic rocks can be more reduced than the QFM and PPM buffers, and may contain Ni-Fe metal alloys and other native metals (*Chamberlain et al.,* 1965; *Frost,* 1985). Formation of H_2 during serpentinization and low permeability of serpentinites could be responsible for the highly reduced conditions (e.g., *Neal and Stanger,* 1983; *Abrajano et al.,* 1990; *Sleep et al.,* 2004). On Europa, high pressure in deep suboceanic rocks may also stabilize native metals in serpentinized chondritic materials and/or igneous ultramafic rocks. If abundant, native metals in a global suboceanic layer may contribute to the observed Galileo magnetometer data. However, native Fe-, Ni-rich metals are not stable in contact with appreciable oceanic sulfate, and may only exist in impermeable rocks isolated from the ocean.

Penetration of surface oxidants (O_2, H_2O_2) deeper into the icy shell is questionable and may not affect the oxidation state of the interior. During a hypothetical large-scale melting of the icy shell only a limited amount of oxidants could be introduced to the ocean. Oxidants could be lost to space from warmed ice and through boiling of oceanic water at near-surface conditions. In oceanic water, surface oxidants would be consumed through abiotic reactions with reduced compounds [sulfides, Fe(II) silicates, organic species] in oceanic water, rocks, and sediments (*Zolotov and Shock,* 2004).

3.3. Alkalis and Halogens

Although alkalis and halogens are not abundant in primary materials of solar composition (*Palme and Jones,* 2003), these elements are likely to have accumulated in Europa's water shell and suboceanic rocks. Igneous processes cause partitioning of lithophile incompatible elements, including alkalis and halogens, into magma and igneous rocks formed in upper parts of silicate mantles. The halogens and alkalis (to some extent) also are partitioned into magmatic

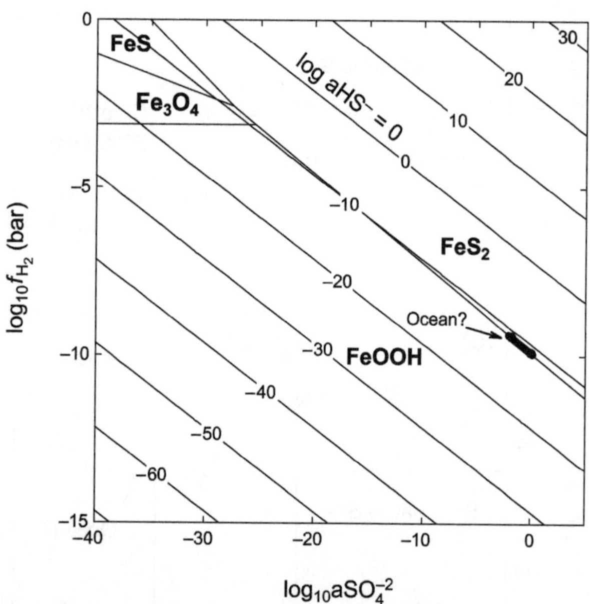

Fig. 7. Stability fields of Fe minerals as functions of H_2 fugacity and activities of SO_4^{2-} and HS^- at 0°C, 1375 bars, and neutral pH of 7.2. Activity of HS^- is shown by counterlines. Hypothetical oceanic conditions are depicted at the line that represents the goethite-pyrite equilibrium (reactions (15) and (16)).

gases and supercritical fluids (e.g., *Luth,* 2003; *Carrol and Webster,* 1994). On Europa, concentrations of alkalis and halogens in unaltered suboceanic igneous rocks could be similar to those in terrestrial and martian counterparts. If they exist, silica-enriched igneous rocks formed through igneous differentiation in magma chambers should be exceptionally enriched in alkalis and halogens.

Efficient aqueous leaching and limited incorporation into secondary solids causes preferential accumulation of Cl, Br, and I in the aqueous phase. Supply of Na via dissolution of silicate glasses and feldspars is only partially compensated by incorporation in secondary minerals. This is consistent with the presence of halite (and sylvite) in some aqueously altered chondrites from asteroids (and possibly comets). Therefore, Na is likely a major cation that balances Cl^- (and Br^- and I^-) to achieve electroneutrality, consistent with geochemical models (*Zolotov and Mironenko,* 2007b; *Zolotov,* 2008). However, K could be notably accumulated in clay minerals in sediments and rocks altered at low temperature. As on Earth, low-temperature sorption and ionic exchange on clays could be the major processes responsible for uptake of K, Rb, and Cs from oceanic water (e.g., *Holland,* 1978; *Staudigel,* 2003; *Walther,* 2005), while hydrothermal fluids favor their release to aqueous phase (*Staudigel,* 2003). These inferences are consistent with endogenic contributions to Europa's atmospheric Na and K (*Brown and Hill,* 1996; *Johnson,* 2000; *Brown,* 2001; *Leblanc et al.,* 2002) and a likely presence of Na sulfate in the non-ice surface material (*McCord et al.,* 1998b, 1999; *Orlando et al.,* 2005; *Dalton,* 2007). In addition to sulfates, Na and K chlorides are expected to be major constituents of the icy shell (*Zolotov and Shock,* 2001). However, Cl species remain to be detected in the atmosphere and on the surface.

Upper limits for alkali and halogen concentrations in oceanic water can be evaluated by assuming total extraction into the aqueous phase. Even at total extraction, a 100-km-thick ocean would not reach saturation with respect to halite and sylvite. Aqueous extraction of 50–90% of the elements (*Zolotov and Shock,* 2001; cf. *Zolotov,* 2007) still results in Na and Cl as major oceanic solutes. Equilibrium and mass balance models for Europa's ocean-rock system lead to oceanic concentrations of Na, K, and Cl in the range of 0.1–0.5, 0.01–0.3, and 0.01–10^{-5} mol/(kg H_2O)$^{-1}$, respectively (*Zolotov et al.,* 2006a; *Zolotov and Mironenko,* 2007b; *Zolotov,* 2008). These evaluations correspond to equilibration of oceanic water with an assumed mass of permeable rocks. However, concentrations could be near the high end of this estimate if the majority of accreted materials interacted with ocean-forming fluids early in Europa's history. The highest Cl values could correspond to (1) complete aqueous extraction from the entire body that may represent dissolution of accreted HCl hydrates (cf. *Zolotov and Mironenko,* 2007a), (2) efficient aqueous leaching of Cl-bearing solids, or (3) complete extraction of Cl into a suboceanic crust produced by thorough melting/degassing of the interior. Overall, leaching of Cl, Br, and I throughout history should have had a larger effect on current oceanic abundances than equilibration between solutes and secondary minerals in permeable rocks. Sodium could have been partially leached into the primordial ocean in amounts equivalent to leached Cl. Because of the solar Na/Cl atomic ratio of ~7 to 11 (*Palme and Jones,* 2003), most Na would have remained in primary/secondary minerals. Sodium could have also remained available also to form Na sulfates, if sulfate-bearing asteroidal material accreted on Europa or formed later through oxidation of sulfides. An uptake of oceanic Na by ultramafic rocks is observed in ophiolites on Earth (*Staudigel,* 2003) and could have occurred in Europa's past.

Elevated concentrations of NaCl in oceanic water and pore solutions lead to higher ionic strength and could decrease activities of Mg^{2+}, Ca^{2+}, and SO_4^{2-} (and affect activities of other solutes), leading to higher solubility of Ca sulfates (e.g., *Newton and Manning,* 2005) and Mg silicates. In other words, extensive leaching of Na (and Cl) into solution could limit trapping of oceanic SO_4^{2-} in solids. Likewise, in parent asteroids of CI carbonaceous chondrites, NaCl-rich solutions could have prevented precipitation of Ca sulfates and caused accumulation of Mg sulfate in solution, which eventually formed epsomite ($MgSO_4 \cdot 7H_2O$).

3.4. Salinity of Oceanic Water

Salinity refers to the mass of dissolved salts per specified mass of aqueous solution. By analogy with terrestrial seawater (*Holland,* 1978; *Millero,* 2003), salinity of Europa's oceanic water could be controlled by a balance of mineral dissolution, secondary precipitation, and ionic exchange with secondary silicates. Compared to Earth, freezing would have a larger effect on oceanic salinity than evaporation (boiling). Salinity could be affected by thickness of the icy shell, the ocean, and permeable rocks. It can also reflect the history, composition, and oxidation state of the ocean. Therefore, salinity estimations depend on various assumptions about oceanic composition and mass balances in the ice-ocean-rock system.

Oceanic salinity could be constrained by the Galileo magnetometer data (see chapter by Khurana et al.), mass balance calculations of water-rock partitioning (*Zolotov and Shock,* 2001), and chemical equilibrium models for water-rock systems (*Kargel et al.,* 2000; *Zolotov et al.,* 2006a; *Zolotov,* 2008). Maximum salinities correspond to eutectic compositions of appropriate highly soluble salts (e.g., $MgSO_4$, Na_2SO_4, $MgCl_2$, $CaCl_2$, NaCl). Freezing computation of assumed oceanic water compositions by *Kargel et al.* (2000) and *Zolotov and Shock* (2001) leads to near eutectic salinities of several hundreds of grams per kilogram. The salinity in the model of *Zolotov and Shock* (2001) implies almost complete freezing and may not reflect contemporary conditions in which the mass of the ocean and icy shell are considered roughly comparable. In contrast, *Kargel et al.* (2000) and *Marion and Kargel* (2008) argued that high salinities can also be reached at low levels of ocean freezing. Lower salinity limits can be evaluated by assuming saturation with respect to sparingly soluble secondary minerals in suboceanic rocks (e.g., phyllosilicates, Ca-carbonates, Ca sulfates).

Although the Galileo magnetometer data indicate the presence of a conducting layer responsible for the observed induced magnetic field (*Khurana et al.,* 1998; *Kivelson et al.,* 2000), evaluations of oceanic salinity are equivocal. Uncertainties in the evaluations are related to the amplitude (A) of the induced magnetic field and possible contribution from non-oceanic conductors. If A ≈ 0.7, the salinity of Europa's global ocean may be less than that of Earth's seawater (35 g/kg) (*Zimmer et al.,* 2000; *Hand and Chyba,* 2007). For a $MgSO_4$-rich ocean and an A value of 0.7, oceanic salinity is estimated to be in the range of 1–16 g/kg (*Hand and Chyba,* 2007). These salinities indicate that oceanic water is not saturated with respect to highly soluble salts such as $MgSO_4$ and Na/Mg/Ca chlorides.

The A value of 0.97 ± 0.02 reported by *Schilling et al.* (2004) implies significantly higher oceanic salinities and/or a presence of other strong conductor(s). *Hand and Chyba* (2007) showed that such a high value for the amplitude may only be explained by a briny ocean saturated with NaCl and/or $MgSO_4$ and covered by a thin (<4–15 km) icy shell. However, the thin icy shell over a ~50–160-km-thick ocean is not consistent with extremely high salinities at the salt saturations, which would correspond to near-complete freezing [except in some of the models of *Kargel et al.* (2000) and *Marion and Kargel* (2008)]. Invoking the presence of highly conductive OH^--rich water expected in some models (Table 1) does not solve the problem. Although conductivity of NaOH solutions is several times higher than that of NaCl and is similar to conductivity of strong acids (*Lobo and Quaresma,* 1989), significant freezing of a NaOH ocean is also needed to account for high conductivity.

A more recent interpretation of the magnetometer data (*Schilling et al.,* 2007) accounts for induction effects in the plasma around Europa and calls into question the A values constraint of *Schilling et al.* (2004). The *Schilling et al.* (2007) result implies an oceanic conductivity >0.5–10 S/m, which is consistent with A > 0.85 and icy shells of ~20 km or less (see Fig. 5 in *Hand and Chyba,* 2007). These conductivity limits are similar to the conductivity of terrestrial seawater (2.75 S/m) and the corresponding $MgSO_4$-NaCl oceanic water would not be saturated with respect to corresponding salts.

Mass-balance evaluations of aqueous extraction from rocks demonstrate that oceanic water may not be well described by a concentrated brine (*Zolotov and Shock,* 2001). Even hypothetical total extraction of Na and Cl from a material of solar composition (~CI chondrites) may not lead to saturation with respect to NaCl. Complete leaching of Na and Cl in a 100-km-thick ocean can only account for salinity of ~70 g/kg H_2O (*Zolotov and Shock,* 2001). If only a portion of Na is leached to achieve electroneutrality with completely extracted Cl, the salinity would be ~5–26 g/kg (cf. *Zolotov,* 2008). Although the presence of Mg sulfate should further contribute to salinity, the amounts of oceanic Mg and SO_4^{2-} could be limited by solubilities of Mg minerals and Ca sulfates (see Fig. 6). As in chondrites and

TABLE 1. Examples of modeled compositions ($mol/kg\ H_2O$) of a primordial europan ocean.

Solute	a	b
H_2	2.5×10^{-1}	3.9×10^{-3}
Na^+	9.0×10^{-2}	3.4×10^{-1}
OH^-	8.6×10^{-2}	3.6×10^{-1}
Cl^-	7.4×10^{-3}	5.2×10^{-2}
K^+	4.7×10^{-3}	1.1×10^{-2}
Ca^{2+}	1.0×10^{-3}	3.5×10^{-2}
NaOH	1.4×10^{-3}	1.7×10^{-2}
$HFeO_2^-$	4.1×10^{-3}	—
$NaHSiO_3$	1.3×10^{-4}	7.2×10^{-2}
NaCl	4.1×10^{-5}	—
CO_3^{2-}	1×10^{-11}	1.7×10^{-4}
$HSiO_3^-$	—	4.5×10^{-3}
KOH	3.6×10^{-5}	2.7×10^{-4}
$CaOH^+$	2.8×10^{-5}	2.5×10^{-3}
HS^-	8.6×10^{-6}	2.7×10^{-4}
$CaCO_3$	2×10^{-12}	3.3×10^{-4}
SiO_2	8.3×10^{-10}	3.7×10^{-8}
$CaHSiO_3^+$	8.5×10^{-8}	2.9×10^{-4}
pH	13.0	13.6
Ionic strength	0.1	0.46
Salinity, g/kg H_2O	4.5	25

Modeling is performed for temperature 0°C, pressure 2 kbar, and water-free composition of a CI chondrite through equilibrium thermodynamic (a) and thermodynamic-kinetic (b) calculations of water-rock interaction (*Zolotov et al.,* 2006a; *Zolotov and Mironenko,* 2007b). In (a), the equilibrium composition represents a closed system with respect to H_2 at water/rock mass ratio of 4. In (b), the oceanic composition is shown at 10^3 yr after ice melting at water/rock ratio of 1. The system is open with respect to H_2. The value f_{H_2} = 127 bar corresponds to H_2 gas separation at the bottom of a 10-km-thick icy shell. Typically, species less abundant than 10^{-5} are not shown.

Earth's oceanic rocks, the majority of Mg is likely to be present in secondary (serpentine, saponite, chlorite, dolomite, sepiolite) and primary (olivine, pyroxene) minerals. Mass-balance calculations demonstrate that a 50–150-km-thick ocean may not be saturated with respect to highly soluble salts such as Na/Ca/Mg chlorides and Mg/Na sulfates.

4. CHEMICAL EVOLUTION OF THE WATER LAYER

4.1. Bulk Composition of Europa

The geological history, geochemistry, and current physical state of Europa depends very much on the type of material originally accreted, including its oxidation state, which currently is a matter of controversy (*Kargel et al.,* 2000; *McKinnon and Zolensky,* 2003; *Zolotov and Shock,* 2001, 2004). According to gas-starved accretion models, the Galilean satellites formed from a mixture of water ice and a rocky component that slowly accumulated from the surrounding solar accretion disk (*Canup and Ward,* 2002; chapter by Canup and Ward). The bulk composition of rocks could be represented by CI carbonaceous chondrites, which roughly corresponds to solar composition, expect C, O, H, and N (*Palme and Jones,* 2003). However, a CM carbonaceous chondrite composition is also a valid possibility, consistent with gravity data (*Kuskov and Kronrod,* 2005). The CI and/or CM carbonaceous chondrite composition of rocks delivered from the Sun's side is supported by the similarity of visible and near-infrared reflectance spectra of these meteorites with spectra of P and D classes of asteroids that are abundant in the outer asteroid belt and Trojan groups (*McKinnon and Zolensky,* 2003). In addition, a similarity of spectra of outer irregular jovian satellites with that of C and D asteroids (*Jewitt et al.,* 2004; *Vilas et al.,* 2006) is consistent with a carbonaceous chondritic bulk composition of rocks accreted to the Galilean satellites. Note, however, that jovian Trojan asteroids, which have similar orbits with Jupiter, could have been captured after the formation of the Galilean satellites (*Morbidelli et al.,* 2005), and the irregular satellites could have been captured later as well (*Nesvorný et al.,* 2007).

Material accreted from the outer part of the solar system (beyond Jupiter) could not contain abundant chondrules (although some chondrule fragments are observed in the Comet Wild 2 dust) and ostensibly was richer in organic matter compared to chondrites. Data obtained for comets (*Bockelée-Morvan et al.,* 2004) are consistent with the volatile-rich bulk composition of that material, generally similar to the solar composition. Accretion of cometary-type ices and mineral particles is also a possibility. Some cometary material should have also accreted during the time of late heavy bombardment observed in the inner solar system (e.g., *Gomes et al.,* 2005).

In a gas-starved jovian disk, accretion of solids as small particles and small planetesimals implies that minerals would not necessarily have been aqueously altered, unlike larger bodies (asteroids). If so, Europa could have accreted from reduced and anhydrous minerals (Fe-Ni metal, FeS, Mg- and Ca-silicates, feldspars, Ca-Al-oxides, amorphous silicates, presolar grains), abundant organic compounds, and predominantly water ice. However, if Europa formed after the asteroids, some contribution of altered asteroidal material remains a possibility. In such a case, hydrated and oxidized phases including phyllosilicates, magnetite, secondary sulfides, phosphates, carbonates, and even sulfates could have contributed to Europa's initial assemblage.

4.2. Insights from Chondrites

Asteroids accreted as a mixture of water ice, reduced and anhydrous rock fragments, and organic material. Solids consisted of solar nebula products (e.g., Fe-Ni metal, Mg- and Ca-silicates, Ca-Al-oxides, FeS) and presolar grains. Decay of short-lived radioactive elements (e.g., ^{26}Al) caused ice melting (*Grimm and McSween,* 1989), and the composition of primary aqueous fluids was affected by dissolution of minerals, glasses, and organic compounds, as well as precipitation of secondary minerals. As alteration progressed, water was consumed in competitive oxidation and hydration reactions. Phyllosilicates (serpentine, saponite) and tochilinite formed through hydration. Oxidation led to formation of ferrous (olivine, serpentine) and ferric (magnetite, cronstedtite, andradite) minerals, secondary sulfides (tochilinite, pyrrhotite, pentlandite), Ni-rich alloys, chromite, and phosphates (*Zolensky and McSween,* 1988; *Brearley and Jones,* 1998; *Brearley,* 2006; *Zolensky et al.,* 2008). Aqueous oxidation of macromolecular organic material led to a variety of species (*Sephton,* 2002) (section 4.6). Hydrogen could have formed through mineral oxidation by water (e.g., reactions (8)–(10), (13)–(14)).

Physical-chemical models for aqueous processes on asteroids (e.g., *Zolensky et al.,* 1989; *Rosenberg et al.,* 2001; *Zolotov et al.,* 2006b) demonstrate that primary solutions had a pH of ~9–12 and were rich in Na, Cl, OH^-, K, and H_2. A lack of fluid convection at low gravity on small (<80 km) asteroids (*Travis and Schubert,* 2005) led to isochemical alteration with respect to minerals and solutes, consistent with observations in aqueously altered chondrites. However, gases (mostly H_2 and some hydrocarbons) would have been able to migrate toward the surface and escape (*Wilson et al.,* 1999). Removal of H_2 and elevated temperatures promoted further oxidation and presumably led to formation of carbonates and sulfates (*Zolotov and Shock,* 2003b). Formation of sulfates was only possible in water-rich, highly porous asteroids, such as parent bodies of CI and CM carbonaceous chondrites. [Note that sulfate veins in CI carbonaceous chondrites may reflect terrestrial remobilization of asteroidal sulfates rather than a terrestrial formation of sulfates (*Gounelle and Zolensky,* 2001).] Filling of pore space during formation of hydrated minerals could have squeezed solutions toward the surface. These solutions may have contained sulfates formed in higher-temperature asteroidal zones. This hypothetic scenario is consistent with a large abundance of sulfates in CI carbonaceous chondrites, in which sulfates are not in equilibrium with more reduced minerals and organic compounds. In turn, low porosities, formation of low-density phyllosilicates (serpentine, saponite), and sealing with ice may have restricted H_2 escape and could have caused chemical reduction in peripheral zones of some bodies. In particular, both low temperatures and elevated f_{H_2} favor the formation of secondary Ni-Fe metal alloys and ferrous silicates (fayalite) though chemical reduction reactions.

Complete consumption of liquid water would have been accompanied by precipitation of alkali halides and thorough separation of H_2 into the gas phase. If heating continued, metamorphic reactions included dehydration of solids and aromatization/graphitization of organic compounds (*Huss et al.,* 2006). Migration of water vapor toward peripheral asteroidal zones would have led to condensation that promoted aqueous processes closer to the surface. Additional supply of water favors further hydration, oxidation, and H_2 escape that would push oxidation further. However, low gravity and a limited water content would have prevented formation of oceans on typical asteroids (Ceres could be an exception). Early formed (<~2 m.y. after formation of Ca-Al-rich inclusions) and some large asteroids (e.g., Vesta, although Vesta was always dry) experienced progressive metamorphism followed by separation of metal cores and even silicate melting (*McSween et al.,* 2002; *Scott,* 2007; *Qin et al.,* 2008).

Europa could have passed through the same aqueous-metamorphic-magmatic sequence of alteration as typical large asteroids. As in asteroids, early aqueous fluids were cold, alkaline, rich in H_2, Cl, and alkalis, and depleted in Mg and S (*Zolotov et al.,* 2006a; *Zolotov and Mironenko,* 2007b). As in asteroids, subsequent high-temperature processes and H_2 escape could have led to oxidation of solids, aqueous solutions, and organic species. However, higher gravity, lack or deficiency of short-lived radionuclides, possibly larger amounts of accreted organic matter, and tidal heating make Europa different. Convective movements of fluids, elevated water/rock ratios in suboceanic environments, and prolonged thermal history over which H_2 could have escaped all support profound net oxidation of at least upper mantle rocks and accumulation of oxidized (carbonate and possibly sulfate) aqueous species in an ocean. In contrast to rapid asteroidal processes (<10–15 m.y.) (*Krot et al.,* 2006), aqueous processes and oxidation of S and C compounds could have occurred over the entire history of Europa.

4.3. Insights from Other Satellites

4.3.1. Io. Io is the extremely volcanically active rocky Galilean satellite powered by tidal heating. Hydrogen-bearing species have not been detected on Io, indicating total dehydration of the interior. However, Io might have had an early aqueous period (cf. *Fanale et al.,* 1974), consistent with Io's high oxidation state (*Zolotov and Fegley,* 1999, 2000), and various S-bearing species and alkali halides accumulated in the crust and emitted by volcanos (*Carlson et al.,* 2007). The Galileo gravity data on Io are consistent with an Fe-rich core, silicate mantle, and low-density crust (*Anderson et al.,* 2001). Io's density (~3528 kg/m^3) and moment of inertia are consistent with a bulk composition similar to L/LL classes of ordinary chondrites (*Kuskov and Kronrod,* 2001). Io's surface consists of volcanic condensates (SO_2, elemental sulfur) with occurrences of fresh silicate lavas. Volcanos emit S, O-S, Cl, and alkali compounds (SO_2, SO, S_n, NaCl, and KCl) (e.g., *McGrath et al.,* 2000; *Lellouch et al.,* 2003). These observations imply accumulation of Cl, Br, I, alkalis, and oxidized sulfur (S^o, SO_2) in crustal materials. An Earth-like oxidation state of Io's volcanic gases may indicate early oxidation of the interior by water followed by H_2 escape (*Lewis,* 1982; *Zolotov and Fegley,* 1999). The lack of detection of C and N species at Io could be accounted for by high-temperature oxidation of organic compounds that led to formation, degassing, and escape of CO_2, CO, CH_4, other hydrocarbons, and N_2 long ago.

Io data and models imply that Europa could have accreted material similar to L/LL-type ordinary chondrites (*Kuskov and Kronrod,* 2005), although this reference is not obviously connected to the compositional inferences in section 4.1. On both early satellites, ultramafic magmas could have formed and erupted. As on Io, Cl, alkalis, and S compounds could have accumulated in magmas, magmatic gases, and aqueous fluids. A longer period of water-rock reactions in Europa's history implies that its mantle could have became even more oxidized than Io's mantle and that halogens (except F) and alkalis were efficiently extracted into the aqueous phase. As on Io, H_2 and N_2 could have formed via oxidation of minerals and organic matter, separated into the gas phase and lost to space early in history. However, high pressures (1–2 kbar) in suboceanic europan environments should have limited magmatic degassing, and degassing of sparsely soluble CO_2 (*Holloway and Blank,* 1994) could result in trapping in an alkaline ocean.

4.3.2. Ganymede and Callisto. Ganymede has a metal core and silicate mantle covered by a water shell that encompasses ~50% of the body's radius (*Sohl et al.,* 2002; *Schubert et al.,* 2004). Galileo magnetometer data suggest a conductive ocean ~150 km below the surface sandwiched between layers of ice (*Kivelson et al.,* 2002). Ganymede may be partially resurfaced by cryo- or liquid-water volcanism (e.g., *Showman et al.,* 2004). Detections of O_2, H_2O_2, and O_3 on Ganymede imply a Europa-like radiolysis and UV photolysis of the icy surface (e.g., *Johnson et al.,* 2004). Hydrated solids, possibly sulfates, are detected in old low-albedo cratered terrains of Ganymede (*McCord et al.,* 2001b). Sulfates, if they are present, could have an oceanic origin (*Kargel et al.,* 2000; *McCord et al.,* 2001b) or could have formed through radiolysis of reduced S species (*Carlson et al.,* 1999). However, an oceanic origin of hydrates on Ganymede is not consistent with the lack of hydrates in younger surface areas. Note that the lack of detection of Na in the atmosphere may not exclude an oceanic source because of geologically inactive surface and a low efficiency of spattering due to some shielding from the intrinsic magnetosphere. The lack of hydrates (sulfates?) in younger areas may indicate resurfacing by low-viscosity ice or by fluids that did not represent the ocean.

Ganymede and Callisto exhibit preferential association of surface CO_2 with low-albedo materials (*Hibbits et al.,* 2000, 2003). This distribution is consistent with CO_2 origin through radiolysis of organic compounds, which could be the case for Europa as well (cf. *Cooper et al.,* 2001). Heavily degraded terrains at Callisto's poles have been interpreted as due to a sublimation-driven process, which implies the presence of polar ices more volatile than H_2O and possibly more volatile than CO_2 (*Moore et al.,* 2004). These unknown ices may be remnants of a primordial crust or may be a younger polar cap; either case suggests outgassing and trapping of volatiles.

4.3.3. Enceladus and Titan. Saturn's icy satellite Enceladus has active degassing processes involving emission of H_2O, CO_2, CH_4, and N_2 and/or CO (*Waite et al.,* 2006). The composition of gases imply early high-temperature oxidation of N-bearing organic compounds, and similar processes could have occurred on Europa (section 4.6). Titan also has a continuing source of CH_4 to supply its atmosphere against its photolytic destruction. Clathrates are discussed as reservoirs of gases in both objects.

4.4. Geochemistry of a Primordial Ocean

Some ice melting in the outer parts of the satellite could have occurred during the late stages of Europa's accretion when the release of impact energy increased (*Stevenson et al.,* 1986; *McKinnon and Zolensky,* 2003), and the amount of accreted ice may have also increased (chapter by Canup and Ward). Subsequent warming of the ice-rock mixture in the moon's interior driven by radioactive decay would have also resulted in more melting of ice, outward migration of water, and accumulation of the aqueous phase in the outer part of the satellite. Although relative contributions of these two ocean-forming processes are not known, the primordial ocean would have a composition consistent with its formation subject to low-temperature (~0°C) aqueous alteration of rocks, often in the presence of partially melted water ice. The composition of aqueous fluids was influenced by dissolution of minerals, glasses, soluble organic species (e.g., methanol), and secondary precipitation. Some species (e.g., HCl, HBr, H_2S, CO_2, NH_3) could have accreted as ices and/or hydrates and released through ice melting. In such a case, very early fluids would have had a low pH.

What happened next is where current geochemical models diverge. The three principle "wild cards" are (1) how extensive was the interaction of aqueous solutions with rocks, (2) at what temperatures did water-rock interaction occur, and (3) were gases retained in ice and/or aqueous solutions or were they efficiently lost to space? In some of the models of *Kargel* (1991), *Spaun and Head* (2001), and *Kargel et al.* (2000), there is not much subsequent water-rock interaction due to isolation of the silicate interior by thick layers of early-precipitating salts. In other models, there is extensive water-rock interaction (*Zolotov et al.,* 2006a; *Zolotov and Mironenko,* 2007b). Below we discuss a scenario in which there was extensive water-rock interaction in a sequence of rising interior temperatures and an increasing drive toward chemical equilibration.

During water-rock separation (early differentiation), the upward migration of solutions caused their sequential interaction with surrounding solids. Highly soluble and easily leachable species (halogens, alkalis) could have been extracted from a large mass of rocky fragments. Concentrations of many other species in aqueous solutions (Fe, Ni, Ca, Mg, Mn, S, and P compounds) would have been mainly controlled by secondary precipitation. Ion exchange (sorption) with secondary phases also affected water-solid partitioning of Mg, K, and Na (to some extent). It is possible that fluid-rock interaction in the uppermost rock layers determined the composition of ocean-forming solutions, except for highly soluble components (Cl, Br, I). Suboceanic rocks would be the most altered because all upwelling fluids would have passed through them. However, rocky material accumulated deeper in the interior was more isolated from aqueous solutions and might have escaped significant alteration. It is likely that water-rock interaction occurred with progressively lower water/rock ratios as depth in the silicate interior increased.

Partial dissolution of primary silicates, some oxidation of Fe^0-metal (reaction (8)), and phosphides and precipitation of secondary minerals (Mg-Fe-phyllosilicates, magnetite, tochilinite, chromite, phosphates) would have been produced reduced (H_2-rich) and alkaline fluids. Kinetic modeling of rock alteration during the formation of an early ocean has demonstrated geologically rapid oxidation of Fe^0-metal (as kamacite) and hydration of silicates (*Zolotov and Mironenko,* 2007b). Models suggest that the composition of an alkaline Na-K-OH-Cl ocean could have been established within ~10^6 yr after ice melting. Equilibrium models for water-rock interaction (*Zolotov et al.,* 2006a) also show that an early ocean would have been essentially a NaCl solution (Table 1). A deficiency of Cl relative to Na in accreted materials (cf. *Palme and Jones,* 2003) would have led to elevated OH^- contents to achieve electroneutrality (Table 1). Ocean-forming solutions and the early ocean would be strongly depleted in Mg, which precipitates in Mg phyllosilicates (serpentine, saponite, sepiolite). Sulfur is not abundant and is mostly in HS^-. Low concentrations of HS^- are accounted for by low solubility of Fe-sulfides (cf. *McKinnon and Zolensky,* 2003). Sulfate ion is not present. Inert behavior of the majority of accreted organic compounds at this stage implies that concentrations of C solutes were limited and carbonates may not have formed (unless CO_2 ice was accreted). Except for the scenarios proposed by *Kargel* (1991) and *Kargel et al.* (2000), models indicate that the primordial ocean would not have been very saline (that is, no much more saline than Earth's ocean), unless subjected to severe freezing/boiling.

Hydrogen produced in oxidation reactions would have exsolved in low-pressure peripheral zones of the body (*Zolotov et al.,* 2006a; *Zolotov and Mironenko,* 2007b) owing to its low solubility in aqueous solutions (Fig. 8). More specifically, it would have accumulated beneath the icy shell, and it eventually escaped owing to impact or tectonic disruptions in the icy shell. H_2 removal was necessary to ensure Fe^0 metal oxidation even at relatively low pressures (~1–2 kbar) in suboceanic rocks. After Fe^0 metal was exhausted, the rate of H_2 production decreased, and the depth of gas separation would have decreased as well. Further H_2 escape and oxidation were limited by the effective permeability of the icy shell and lower concentration of H_2 dissolved in fluids. With limited H_2 escape, redox conditions in altered rocks and pore fluids could have been controlled by mineral equilibria involving Fe-bearing minerals (e.g., reactions (9)–(12)), but many C- and S-bearing species could have not equilibrated at low temperatures (see section 3.1.5).

An alternative scenario of early oceanic evolution assumes accretion of carbonaceous chondritic material that earlier processed on their parent bodies and could have contained leachable salts (e.g., sulfates). Freezing modeling applied to the high-sulfate models of *Kargel et al.* (2000)

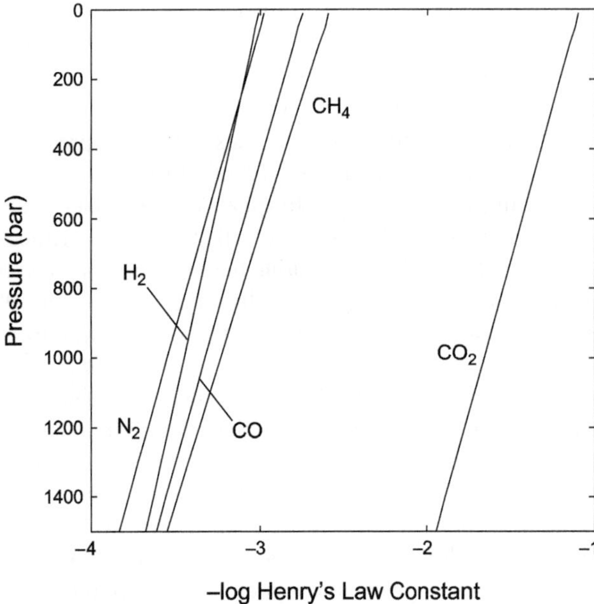

Fig. 8. Henry's law constants for gases that might be present in Europa's ocean. The Henry's law constant is defined as fugacity of gas divided by activity of the gas dissolved in aqueous solution. Larger constants values correspond to lower solubilities. The plot demonstrates higher CO_2 solubility compared to other gases. Low-soluble gases could have separated into the gas phase and escaped. Escape of CO_2 could have been constrained by a low CO_2 content in an alkaline ocean.

and *Marion and Kargel* (2008) starts with an ocean that is saturated in Mg and Na sulfates due to the nature of the primordial leaching/incongruent melting process assumed to have delivered water and solutes to the surface from a carbonaceous chondrite interior. Another model considered by *Kargel* (1991) and *Kargel et al.* (2000) involved subsequent dilution of that initial saturated brine with water derived from dehydration of phyllosilicates, but the initial ocean would still be sulfate-rich, and it would not take much freezing to start massive sulfate precipitation. As a sulfate-rich ocean progressively freezes, the stable precipitates increase their hydration states; eventually chlorides (initially not close to saturation) also become saturated as the ocean nears complete solidification. In these models, tens of kilometers of sulfate salts can be precipitated before any chance of extensive hydrothermal rock/water interaction occurs, which would lead Europa's icy shell and ocean down a totally different path than is familiar on Earth.

4.5. Early High-Temperature Aqueous Processes and Subsequent History

After substantial melting of any ice and stepwise melting of hydrated salts (if they were present) in the suboceanic interior, further decay of radioactive elements led to a gradual temperature increase in the rocky part of the satellite (*Consolmagno and Lewis,* 1976). Before formation of Fe-FeS eutectic melt (~1 b.y. after accretion) (*Greeley et al.,* 2004), the heating led to major stepwise dehydration of earlier formed solids (e.g., serpentine, clay minerals). Upward movement of released steam and fluids affected suboceanic rocks and contributed to oceanic mass and composition (*McKinnon and Zolensky,* 2003). As during ice melting, the composition of ocean-entering fluids was mainly controlled by interactions in suboceanic rock layers. During and after rock dehydration, hydrothermal processes took place as oceanic water circulated through permeable rocks. A release of tidal heat in subjacent rocks enhanced the circulation and magmatic activity (if the Laplace resonance occurred) (cf. *Hussmann and Spohn,* 2004).

High-temperature aqueous processes should have altered the rock mineralogy and oceanic composition. Although some alkalis and halogens could have been added to the ocean, solubility-controlled concentrations of Mg, Fe, Ni, Mn, and P could have not changed dramatically. Delivery of carbonate species (HCO_3^-, CO_3^{2-}) formed via oxidation of organic compounds and some magma degassing could have limited the concentration of Ca solutes through precipitation of carbonates. Mafic suboceanic rocks were subjected to chloritization, leading to mineral assemblages observed at greenschist metamorphic facies in Earth's oceanic crust. Ultramafic rocks were subjected to further serpentinization, which could have decreased permeability. Early high-temperature fluids would have been rich in sulfide species (HS^-, H_2S), and cooling of hydrothermal fluids near the oceanic floor would have led to the precipitation of metal (mostly Fe) sulfides, which could have formed a global layer of massive deposits. In contrast to Earth, where oceanic sulfide deposits are moved from ridges and subducted together with the lithosphere, Europa's hydrothermal sulfides could have accumulated in large amounts.

Hydrothermal processes would have promoted further oxidation of unaltered Fe-Ni metal to magnetite and pentlandite, as suggested for parent bodies of chondrites (*Krot et al.,* 1998, 2000). Separation of Fe-FeS eutectic melt into a core also contributed to oxidation of the mantle (by sequestering Fe^0). Over time, the upward transport (via bubbling, diffusion, convection) and escape of H_2 could have oxidized mantle rocks and hydrothermal fluids. Penetrating impact disturbances of the icy shell should have prevented long-term accumulation of H_2-rich gas below ice. Both high-temperature and low-f_{H_2} conditions increase the carbonate ($HCO_3^- + CO_3^{2-}$) and sulfate concentrations of hydrothermal fluids (cf. *Zolotov and Shock,* 2003a, 2004). Phase separation and escape of H_2 could have driven the consecutive oxidation of Fe sulfides: troilite → (tochilinite?) → pyrrhotite → pyrite. Oxidation of tochilinite and other sulfides can also lead to magnetite formation, as it occurs in Earth's environments. Fe(II) silicates could have been partially oxidized to magnetite (e.g., reaction (9)). The latter process is observed in terrestrial anoxic environments and is accompanied by H_2 production during alteration of basalts (*Stevens and McKinley,* 1995, 2000) and peridotites (e.g., *Neal and Stanger,* 1983; *Abrajano et al.,* 1990; *Sleep et al.,* 2004).

Thorough removal of H_2 from hydrothermal systems could have eventually led to the formation of SO_4-bearing hydrothermal fluids. Accumulation of SO_4^{2-} favors the partial dissolution of Mg phyllosilicates, which would have increased the concentration of Mg^{2+} in solution. The formation of sulfate and release of Mg^{2+} increases the salinity of fluids (see Fig. 6). An increase in SO_4^{2-} content would have also led to elevated Ca^{2+} concentrations, which stabilized after gypsum started to precipitate. This general scenario (e.g., *Zolotov and Shock,* 2003a,b) also applies to asteroids and is consistent with high abundances of Mg sulfates in highly porous CI carbonaceous chondrites (see section 4.2) (but see also *Gounelle and Zolensky,* 2001).

High-temperature transformations of organic compounds would have led to the partial oxidation and release of O- and N-bearing species into buoyant fluids (section 4.6). Involvement of NH_3-bearing solutions in hydrothermal circulation (if NH_3 was accreted in a cometary material) would have led to formation of N_2 (cf. *Matson et al.,* 2007; *Glein et al.,* 2008), which separated into the gas phase (see Fig. 8) and escaped. The accumulation of soluble oxidized C (i.e., carbonate species, organic acids) in the ocean, precipitation of carbonates, and escape of relatively insoluble CH_4 (if formed) would have contributed to the satellite's net oxidation.

If Europa's mantle temperature decreased after a peak in the Laplace resonance (*Hussmann and Spohn,* 2004), this could have led to mineral transformations similar to that observed during retrograde metamorphism of basic and ultrabasic rocks. In suboceanic permeable rocks, rehydration (reserpentinization, rechoritization) would have affected previously dehydrated rocks. In the deeper interior, changes may not be significant owing to a deficiency of fluids.

After magmatic differentiation and corresponding hydrothermal activity, major inorganic species in oceanic water could have equilibrated with hydrated and oxidized suboceanic rocks and sediments. The pattern of subsequent evolution would have been mainly affected by the intensity and location of tidal heating and by magmatic activity in permeable suboceanic rocks, which may also be related to tidal heating. Chemical effects of magmatic and hydrothermal activity at the ocean-mantle interface facilitated hydrothermal circulation and aqueous alteration, as discussed in sections 3.1.4 and 3.1.5. Compositional effects of variations in the icy shell thickness due to the Laplace resonance are considered in section 3.1.3.

The degree of water-rock reactions thoughout history depended on permeability of suboceanic rocks and the nature of material that may separate ocean from the silicate mantle. Thorough serpentinization of fine-grained mineral grains, an accumulation of high molecular weight organic compounds at the oceanic floor, massive precipitation of salts upon considerable freezing, and hypothetical accumulation of dense clathrates would have limited water-rock reactions. In such a case, oceanic pH and redox potential will not be strongly affected by water-rock reactions. The compositional stability of an "isolated" ocean would depend on the nature of isolating material and the amount of rock that remains in the contact with water. On the one hand, an isolated ocean may not be alkaline owing to limited supply of cations and a possibility of endogenic input of CO_2 (and SO_2?), which would decrease alkalinity. On the other hand, low permeability may suppress degassing, and elevated f_{H_2} in suboceanic rocks may account for reduction of Fe, S, and C species, as observed in serpentinites (*Frost,* 1985; *Seyfried et al.,* 2004).

4.6. Fate of Organic Compounds

If Europa accreted from a material similar to carbonaceous chondrites (e.g., *Lewis,* 1971; *McKinnon and Zolensky,* 2003), it should have started with a few weight percent of C as organic compounds. Significantly (10 times) larger amounts of C are expected if Europa accreted material similar to that in the solar photosphere (*Palme and Jones,* 2003), such as cometary-type material (*Bockelée-Morvan et al.,* 2004). By analogy with CI/CM carbonaceous chondrites (*Septhon,* 2002; *Pizzarello et al.,* 2006), the majority of accreted organic compounds was in polymer form, in which small polyaromatic groups are linked by O-, N-, and S-bearing aliphatic units, forming a low-soluble polymer. If accreted, cometary-type materials could have delivered CO_2, CO, CH_4, methanol, ethane, ethene, acetylene, and heavier organic compounds (cf. *Bockelée-Morvan et al.,* 2004). Melting of water ice would have led to dissolution of carbon oxide(s) and methanol in water and accumulation of soluble carbon species in a primordial ocean. Subsequent hydrothermal processes partially broke down organic polymers and other condensed organic species. If accreted, CO would have converted to formic acid, carbonate species, methanol, and some methane, which may not have reached chemical equilibrium with other C compounds (*Seewald et al.,* 2006). Hydrous pyrolysis and oxidation of organic polymers would have partially liberated aromatic and aliphatic molecules and led to the formation of polar O-bearing organic compounds with high solubility in water (carboxylic and amino acids, alcohols), as well as carbonate species. Some of these processes have been suggested for parent bodies of carbonaceous chondrites (e.g., *Cronin and Chang,* 1993; *Sephton,* 2002; *Cody et al.,* 2005; *Alexander et al.,* 2007) and observed in hydrous pyrolysis experiments with chondritic organic matter (*Sephton et al.,* 1999, 2004).

Further increases in internal temperature would have favored thorough oxidation of organic compounds to CO_2, carbonate species, and N_2. By analogy with asteroidal processes (*Cody et al.,* 2005; *Alexander et al.,* 2007), preferential removal and oxidation of aliphatic groups from organic polymers led to higher C/H ratios in the altering polymer, which became also more aromatic. Eventually, some graphitizated C should have been incorporated into a metallic core, forming solid solution with Fe^0. Oxidation of N-bearing organic groups was likely more efficient than oxidation of organic C and would have led to higher C/N ratios in remaining organic matter, as observed in chondrites (*Pearson et al.,* 2006; *Alexander et al.,* 2007). A low-temperature (<100°–200°C) oxidation of N-bearing groups could have caused formation of N-O-bearing species (amino acids, amines, and amides) (cf. *Sephton,* 2002), and higher-temperature oxidation would have converted N species to N_2. However, amino acids, amines, and amides (urea) could have accumulated in oceanic water and could be the most abundant N-bearing species. Note that nitrates and nitrites can exist only in a severely oxidized ocean (e.g., one that does contain much O_2).

Despite net oxidation of primordial organic compounds, hydrothermal alteration of organic compounds at ~100°–300°C could have also caused formation of methane and other aliphatic hydrocarbons via the overall disproportionation reaction

$$\text{Organic matter} + H_2O \rightarrow CO_2 + CH_4 \quad (17)$$

which is suggested to occur in terrestrial sedimentary basins (*Price and DeWitt,* 2001). In addition, mineral-catalyzed Fisher-Tropsch-type synthesis of hydrocarbons (*Anderson,* 1984) could have occurred in H_2-rich localities. The synthesis could have been catalyzed by magnetite (*Berndt et al.,* 1996; *Fu et al.,* 2007), metal sulfides (*Cody et al.,* 2004), Cr oxides (*Foustoukos and Seyfried,* 2004), or secondary Fe-Ni alloys (*Horita and Berndt,* 1999), which can be stable in H_2-rich high-pressure environments (*Frost,* 1985; *Zolotov et al.,* 2006a,b). Presence of a H_2-rich gas phase could have facilitated synthesis of hydrocarbons from carbonate species (*McCollom and Seewald,* 2007). In magmatic processes, some metastable hydrocarbons might also have formed during cooling of CO- and H_2-bearing magmatic gases (*Zolotov and Shock,* 2000). Note that isotopic studies of chondrites do not reveal evidences for Fisher-Tropsch-type synthesis inside asteroids (*Alexander et al.,* 2007). It follows that transformed organic compounds on Europa could be much more abundant than species synthesized from inorganic precursors.

As a result of organic transformations and synthesis, an array of aromatic, aliphatic, and N-, O-, S-bearing organic species and CH_4 could have been delivered into a primordial water ocean in hydrothermal fluids. Highly soluble compounds (acids, alcohols) could have accumulated in oceanic water and could have made multiple passes through hydrothermal systems causing further alteration of organic matter in rocks and solutions. Denser, less-soluble organic compounds (e.g., aromatic compounds with two or more rings) would have exsolved from cooling ocean-entering fluids and precipitated in suboceanic rocks/sediments, limiting inorganic water-rock interactions. In contrast, less-soluble species with lower density (e.g., aliphatic hydrocarbons, benzene, toluene) could have accumulated below the icy shell, forming an oil-paraffin-like layer. Some compounds could have ended up trapped in the ice, and CH_4 (if present) and other small hydrocarbon molecules may have formed clathrates. Throughout history, disruptions and melting of the icy shell should have led to escape of CH_4 and other light organic gases (ethane, propane). At present, the icy shell may not contain CH_4 in any form.

If present, the organic layer below the icy shell should have had a lower viscosity compared to ice, and may have affected the distribution of tidal heat within the shell. In addition, relatively low thermal conductivity of many solid organic compounds could insulate the ocean, leading to higher oceanic temperatures. In a thicker (>~20 km) icy shell, the organic layer below the ice, if such exists, could be involved in solid-state convection. Inside the icy shell, low-viscosity organic (e.g., hydrocarbon) species, as with trapped brine, could localize tidal motions and generation of heat, causing corresponding tectonic activity. The low thermal conductivity of organic compounds, if in sufficient abundance, could have slowed heat transfer toward the surface, increasing vertical temperature gradients. Disruptions of the icy shell may have caused ascent of organic species toward the surface. It follows that organic compounds may appear in ridges and domes. A correlation of CO_2 abundance with non-ice surface materials (*Hansen and McCord,* 2008) may indicate surface oxidation of these compounds.

We note that exogenic organic compounds accumulated at the surface along with cometary/chondritic materials (*Pierrazzo and Chyba,* 2002) are subjected to surface oxidation and could be a less important source compared to the endogenic supply discussed above. Some oxidized organic compounds (e.g., alcohols) can be produced through hydrocarbon irradiation at the surface (chapter by Carlson et al.). Because of efficient surface oxidation, synthesis of organic species during hypervelocity impacts (*Borucki et al.,* 2002) may have no effect on fate of organic matter in the interior.

4.7. Compositional Constraints on Europa's Habitability

Chemical evolution of Europa's interior has affected concentrations of biologically important elements, as well as amounts of nutrients and chemical energy sources for potential metabolism. In the interior, habitable environments should have existed during early low-temperature water-rock interactions, subsequent hydrothermal activity, and in today's aqueous environments (e.g., *McCollom,* 1999; *Kargel et al.,* 2000; *Zolotov and Shock,* 2004; chapter by Hand et al.). Water-rock reactions throughout Earth's history provided necessary elements for life as we know it, and the same may have affected Europa. An early aqueous oxidation of accreted phosphides may have released phosphorus in the terrestrial environment (*Pasek and Lauretta,* 2005), and it is a likely process on early Europa. Transformations of organic matter and some synthesis of organic species undoubtedly led to an array of reduced C and C-N compounds. Formation of amino acids through aqueous transformation of accreted organic matter is suggested for parent bodies of aqueously altered chondrites (*Cronin and Chang,* 1993) and could have occurred on Europa. However, the amount of potential biomass could have been limited by high-temperature oxidation of N-bearing organic compounds followed by N_2 escape. The reduced nature of an early ocean (*McKinnon and Zolensky,* 2003; *Zolotov et al.,* 2006a; *Zolotov and Mironenko,* 2007b) and precipitation of metal sulfides and organic compounds from hydrothermal fluids would have created plausible conditions for prebiotic chemistry and possibly biogenesis at Europa's ocean floor (*Wächtershäuser,* 1990).

Although escape of H_2 favors further oxidation, abiotic redox reequilibration among S and C species should have not occurred at low temperatures. At early stages on Europa, cold, H_2-depleted fluids and oceanic water could have supported metabolism of organisms that oxidize sulfide S, organic C, or reduced N in amine groups, and might have accelerated oxidation of Fe(II) compounds and minerals. If sulfate-bearing fluids were supplied in the ocean later in Europa's history, they would have been out of equilibrium with relatively reduced oceanic solutes and rocks. Low-temperature sulfate is out of chemical equilibrium with any ferrous silicates and magnetite in the rock (*Zolotov and Shock,* 2004). It follows that metabolic energy could have been obtained through reduction of hydrothermally formed sulfate and carbonate species (*McCollom,* 1999; *Zolotov and Shock,* 2003a). Regardless of the origin of sulfates, their coexistence with organic compounds would have provided the potential for biologic sulfate reduction. These and many other low-temperature reactions may support microbial metabolism on Europa at present (*Kargel et al.,* 2000; *Zolotov and Shock,* 2004). If so, Europa's organisms may have adapted to multiple geochemically driven changes of prevailing redox disequilibria (as is believed to have occurred during Earth's history). Given these diverse but limited energy sources, we expect to find, at the most, microbes, and certainly not aquatic animals.

5. CONCLUDING REMARKS

Chemical information on Europa's surface and interior is sparse, and inferences about the composition of its ice-ocean-rock system, as described here, rely on assumptions, analogs, and models. Despite a paucity of compositional data, consideration of the coupled ice-water-rock system throughout Europa's history provides some possible directions to model. Owing to multiple caveats and assumptions, these are more like possible clues, and are not strong or rigid constraints. The compositional estimates and inferences have provided insights into Europa's physical properties (density, permeability, solid-state rheology and tidal dissipative heating, electrical and thermal conductivities) in the ice-ocean-rock system. Compositional evaluations can be used to estimate the speciation and masses of nutrients and chemical energy sources for potential life. Compositional models provide a framework for planning further missions to Europa.

Insights from chondrites, models for formation, interior structure, thermal evolution, and water-rock-organic interactions all suggest (1) hydration and oxidation of primary and igneous rocks in the upper parts of the mantle; (2) preferential accumulation of Cl, Br, I, and Na in a water ocean; (3) an alkaline pH of ocean-entering fluids that were controlled by reactions with minerals, and nonacidic nature of the ocean unless it is isolated from silicate rocks; (4) a chemical nature of oceanic sediments affected by the degree of freezing of the water shell; (5) low-temperature redox disequilibria among solutes, solids, and gases; (6) an accumulation of carbonate and some organic species in the outer layers of the body; and (7) escape of low-solubility volatiles (H_2, N_2, CH_4, noble gases) into space.

Although the oxidation state of the ice-water-rock system is not known, the low solubility of metal sulfides in aqueous solutions implies an endogenic source of sulfate sulfur in the surface non-ice material. The presence of high oxidation states of sulfur in the ocean is consistent with observations of sulfate in some carbonaceous chondrites, of SO_2 on Io, and with unavoidable production and escape of H_2, which favored oxidation of sulfide sulfur at hydrothermal conditions. Although formation of an oxidized sulfate-bearing ocean is the expected pathway of thermal evolution and tidal heating, at least some sulfates could have accreted in aqueously oxidized planetesimals similar to CI/CM carbonaceous chondrites. In the latter case, sulfates could have been leached in a primordial ocean and accumulated at the oceanic floor upon freezing.

Further compositional evaluations could benefit from telescopic and spacecraft spectroscopy, revised and improved models for Europa's accretion, and thermal and tidal evolution. In particular, timing of Europa's formation is a crucial determinant of heat sources and the origin of sulfates. Further models could consider heat production from short-lived radioactive elements and the accumulation of ^{40}K into the ocean and rocks/sediments. Advanced modeling of Galileo, New Horizons, and groundbased spectra could provide new information on the composition of the non-ice materials, including the unknown coloring agent. Further modeling efforts could be focused on coupling thermal evolution and chemical models. In chemical models, coupling chemical kinetics, thermodynamics, mass, and heat transfer are a promising avenue.

Acknowledgments. The authors appreciate comments from W. McKinnon, M. Zolensky, C. Glein, C. Manning, K. Hand, and O. Kuskov. This work benefited from collaborative research with E. Shock, G. Marion, and M. Mironenko. The work of M.Z. is supported by grants from the NASA Outer Planet Research and Cosmochemistry programs.

REFERENCES

Abe Y., Ohtani E., Okuchi T., Righter K., and Drake M. (2000) Water in the early Earth. In *Origin of the Earth and Moon* (R. M. Canup and K. Righter, eds.), pp. 413–434. Univ. of Arizona, Tucson.

Abrajano T. A., Sturchio N. C., Kennedy B. M., Lyon G. L., Muehlenbachs K., and Böhlke K. J. (1990) Geochemistry of reduced gas related to serpentinization of the Zambales Ophiolite, Philippines. *Appl. Geochem., 5,* 625–630.

Alexander C. M. O.'D., Fogel M., Yabuta H., and Cody G. D. (2007) The origin and evolution of chondrites recorded in the elemental and isotopic compositions of their macromolecular organic matter. *Geochim. Cosmochim. Acta, 71,* 4380–4403.

Alt J. C. (1995) Seafloor processes in mid-ocean ridge hydrothermal systems. In *Seafloor Hydrothermal Systems: Physical, Chemical, Biological, and Geological Interactions* (S. E. Humphris et al., eds.), pp. 85–114. Geophysical Monograph Series 91, AGU, Washington, DC.

Alt J. C. (2004) Alteration of the upper oceanic crust: Mineralogy, chemistry, and processes. In *Hydrogeology of the Oceanic Lithosphere* (E. E. Davis and H. Elderfield, eds.), pp. 495–533. Cambridge Univ., New York.

Alt-Epping P. and Smith L. (2001) Computing geochemical mass transfer and water/rock ratios in submarine hydrothermal systems: Implications for estimating the vigour of convection. *Geofluids, 1,* 163–181.

Alvarellos J. L., Zahnle K. J., Dobrovolskis A. R., and Hamill P. (2008) Transfer of mass from Io to Europa and beyond due to cometary impacts. *Icarus, 194,* 636–646.

Anderson J. D., Schubert G., Jacobson R. A., Lau E. L., Moore W. B., and Sjogren W. L. (1998) Europa's differentiated initial structure: Inferences from four Galileo encounters. *Science, 281,* 2019–2022.

Anderson J. D., Jacobson R. A., Lau E. L., Moore W. B., and Schubert G. (2001) Io's gravity field and interior structure. *J. Geophys. Res., 106,* 32963–32970.

Anderson R. B. (1984) *The Fischer–Tropsch Synthesis.* Academic, New York.

Barnes I. and O'Neil J. R. (1969) The relationship between fluids in some fresh alpine-type ultramafics and possible modern serpentinization, Western United States. *Geol. Soc. Am. Bull., 80,* 1947–1960.

Barnes I., Presser T. S., Saines M., Dickson P., and Vangroos A. F. K. (1982) Geochemistry of highly basic calcium hydroxide groundwater in Jordan. *Chem. Geol., 35,* 147–154.

Barr A. C. and McKinnon W. B. (2007) Convection in ice I shells and mantles with self-consistent grain size. *J. Geophys. Res., 112,* E02012.

Berndt M. E., Allen D. E., and Seyfried W. E. Jr. (1996) Reduction of CO_2 during serpentinization of olivine at 300°C and 500 bar. *Geology, 24,* 351–354.

Bockelée-Morvan D., Crovisier J., Mumma M. J., and Weaver H. A. (2004) The composition of cometary volatiles. In *Comets II* (M. C. Festou et al., eds.), pp. 391–423. Univ. of Arizona, Tucson.

Bonatti E., Lawrence J. R., Hamlyn P. R., and Breger D. (1980) Aragonite from deep-sea ultramafic rocks. *Geochim. Cosmochim. Acta, 44,* 1207–1214.

Borucki J. G., Khare B., and Cruikshank D. P. (2002) A new energy source for organic synthesis in Europa's surface ice. *J. Geophys. Res., 107(E11),* CiteID 5114.

Brantley S. L. (2004) Reaction kinetics of primary rock-forming minerals under ambient conditions. In *Treatise on Geochemistry, Vol. 5: Surface and Ground Water, Weathering and Soils* (J. I. Drever, ed.), pp. 73–118. Elsevier-Pergamon, Oxford.

Brearley A. J. (2006) The action of water. In *Meteorites and the Early Solar System II* (D. S. Lauretta and H. Y. McSween Jr., eds.), pp. 587–624. Univ. of Arizona, Tucson.

Brearley A. J. and Jones R. H. (1998) Chondritic meteorites. In *Planetary Materials* (J. J. Papike, ed.), pp. 1–398. Rev. Mineral., Vol. 36, Mineralogical Society of America, Washington, DC.

Brett R. and Bell P. M. (1969) Melting relationships in the Fe-rich portion of the system Fe-FeS at 30 kb pressure. *Earth Planet. Sci. Lett., 6,* 479–482.

Brown M. E. (2001) Potassium in Europa's atmosphere. *Icarus, 151,* 190–195.

Brown M. E. and Hill R. E. (1996) Discovery of an extended sodium atmosphere around Europa. *Nature, 380,* 229.

Bruni J., Canepa M., Chiodini G., Cioni R., Cipolli F., et al. (2002) Irreversible water-rock mass transfer accompanying the generation of the neutral, $Mg\text{-}HCO_3$ and high-pH, Ca-OH spring waters of the Geneva province, Italy. *Appl. Geochem., 17,* 455–474.

Canup R. M. and Ward W. R. (2002) Formation of the Galilean satellites: Conditions of accretion. *Astron. J., 124,* 3404–3423.

Carlson R. W., Johnson R. E., and Anderson M. S. (1999) Sulfuric acid on Europa and the radiolytic sulfur cycle. *Science, 286,* 97–99.

Carlson R. W., Anderson M. S., Johnson R. E., Schulman M. B., and Yavrouian A. H. (2002) Sulfuric acid production on Europa: The radiolysis of sulfur in water ice. *Icarus, 157,* 456–463.

Carlson R. W., Kargel J. S., Doute S., Soderblom L. A., and Dalton J. B. (2007) Io's surface composition. In *Io after Galileo* (M. C. Lopes and J. R. Spencer, eds.), pp. 193–230. Praxis-Springer, Berlin.

Carrol M. R. and Webster J. D. (1994) Solubilities of sulfur, noble gases, nitrogen, chlorine, and fluorine in magmas. *Rev. Mineral., 30,* 231–281.

Chamberlain J. A., McLepod C. R., Traill R. J., and Lachance G. R. (1965) Native metals in the Muskox intrusion. *Can. J. Earth Sci., 2,* 188–215.

Cody G. D., Boctor N. Z., Brandes J. A., Filley T. R., Hazen R. M., and Yoder H. S. (2004) Assaying the catalytic potential of transition metal sulfides for abiotic carbon fixation. *Geochim. Cosmochim. Acta, 68,* 2185–2196.

Cody G. D., Alexander C. M. O., and Tera F. (2005) NMR studies of chemical structural variation of insoluble organic matter from different carbonaceous chondrite groups. *Geochim. Cosmochim. Acta, 69,* 1851–1865.

Consolmagno G. J. and Lewis J. S. (1976) Structural and thermal models of icy Galilean satellites. In *Jupiter: Studies of the Interior, Atmosphere, Magnetosphere and Satellites* (T. Gehrels, ed.), pp. 1035–1051. Univ. of Arizona, Tucson.

Cooper J. F., Johnson R. E., Mauk B. H., Garrett H. B., and Gehrels N. (2001) Energetic ion and electron irradiation of the icy Galilean satellites. *Icarus, 149,* 135–159.

Crawford G. D. and Stevenson D. J. (1988) Gas-driven water volcanism and the resurfacing of Europa. *Icarus, 73,* 66–79.

Croft S. K., Lunine J. I., and Kargel J. S. (1988) Equation of state of ammonia-water liquid — Derivation and planetological applications. *Icarus, 73,* 279–293.

Croft S. K., Kargel J. S., Kirk R. L., Moore J. M., and Strom R. G. (1995) Geology of Triton. In *Neptune and Triton* (D. P. Cruikshank and M. S. Matthews, eds.), pp. 879–948. Univ. of Arizona, Tucson.

Cronin J. R. and Chang S. (1993) Organic matter in meteorites: Molecular and isotopic analyses of the Murchison meteorite. In *The Chemistry of Life's Origins* (J. M. Greenberg et al., eds.), pp. 209–258. Kluwer, Dordrecht.

Dalton J. B. III (2007) Linear mixture modeling of Europa's non-ice material based on cryogenic laboratory spectroscopy. *Geophys. Res. Lett., 34,* L21205.

Dalton J. B., Prieto-Ballesteros O., Kargel J. S., Jamieson C. S., Jolivet J., and Quinn R. (2005) Spectral comparison of heavy hydrated salts with disrupted terrains on Europa. *Icarus, 177,* 472–490.

Durham W. B., Kirby S. H., Stern L. A., and Zhang W. (2003) The strength and rheology of methane clathrate hydrate. *J. Geophys. Res., 108(B4),* 2182.

Durham W., Stern L., Kirby S., and Circone S. (2005a) Rheological comparisons and structural imaging of sI and sII endmember gas hydrates and hydrate/sediment aggregates. *Proc. 5th Intl. Conf. Gas Hydrates,* Trondheim, Norway.

Durham W. B., Stern L. A., Kubo T., and Kirby S. H. (2005b) Flow strength of highly hydrated Mg- and Na-sulfate hydrate salts, pure and in mixtures with water ice, with application to Europa. *J. Geophys. Res., 110,* E12010.

Elderfield H. and Schultz A. (1996) Mid-oceanic ridge hydrothermal fluxes and the chemical composition of the ocean. *Annu. Rev. Earth Planet. Sci., 24,* 191–224.

Fanale F. P., Johnson T. V., and Matson D. L. (1974) Io: A surface evaporite deposit? *Science, 186,* 922–925.

Fanale F. P., et al. (1999) Galileo's multi-instrument spectral view of Europa's surface composition. *Icarus, 139,* 179–188.

Fanale F. P., et al. (2000) Tyre and Pwull: Galileo orbital remote sensing of mineralogy versus morphology at two selected sites on Europa. *J. Geophys. Res., 105,* 22647–22655.

Figueredo P. H. and Greeley R. (2004) Resurfacing history of Europa from pole-to pole geologic mapping. *Icarus, 167,* 287–312.

Foustoukos D. and Seyfried W. E. Jr. (2004) Hydrocarbons in hydrothermal vent fluids: The role of chromium-bearing catalysts. *Science, 304,* 1002–1005.

Frost B. R. (1985) On the stability of sulfides, oxides, and native metals in serpentine. *J. Petrol., 26,* 31–63.

Frost D. J. and McCammon C. A. (2008) The redox state of Earth's mantle. *Annu. Rev. Earth Planet. Sci., 36,* 389–420.

Fu Q., Sherwood Lollar B., Horita J., Lacrampe-Couloume G., and Seyfried W. E. Jr. (2007) Abiotic formation of hydrocarbons under hydrothermal conditions: Constraints from chemical and isotope data. *Geochim. Cosmochim. Acta, 71,* 1982–1998.

German C. R. and Von Damm K. L. (2003) Hydrothermal processes. In *Treatise on Geochemistry, Vol. 6: The Oceans and Marine Geochemistry* (H. Elderfeld, ed.), pp. 181–222. Elsevier-Pergamon, Oxford.

Gislason S. R. and Eugster H. (1987a) Meteoric water-basalt interaction: I. A laboratory study. *Geochim. Cosmochim. Acta, 51,* 2827–2840.

Gislason S. R. and Eugster H. (1987b) Meteoric water-basalt interaction: II. A field study in NE Iceland. *Geochim. Cosmochim. Acta, 51,* 2841–2855.

Gislason S. R. and Oelkers E. H. (2003) The mechanism, rates, and consequences of basaltic glass dissolution: II. An experimental study of the dissolution rates of basaltic glass as a function of pH at temperatures from 6°C to 150°C. *Geochim. Cosmochim. Acta, 67,* 3817–3832.

Glein C. R., Zolotov M. Yu., and Shock E. L. (2008) The oxidation state of hydrothermal systems on early Enceladus. *Icarus, 197,* 157–163.

Gomes R., Levison H. F., Tsiganis K., and Morbidelli A. (2005) Origin of the cataclysmic late heavy bombardment period of the terrestrial planets. *Nature, 435,* 466–469.

Gounelle M. and Zolensky M. E. (2001) A terrestrial origin for sulfate veins in CI1 chondrites. *Meteoritics & Planet. Sci., 36,* 1321–1329.

Greeley R. et al. (1999) *A Science Strategy for the Exploration of Europa.* Committee on Planetary and Lunar Exploration, Space Studies Board, National Research Council, National Academy Press, Washington, DC.

Greeley R., Chyba C., Head J. W., McCord T., McKinnon W. B., Pappalardo R. T., and Figueredo P. (2004) Geology of Europa. In *Jupiter: The Planet, Satellites and Magnetosphere* (F. Bagenal et al., eds.), pp. 329–362. Cambridge Univ., Cambridge.

Grimm R. E. and McSween H. Y. (1989) Water and the thermal evolution of carbonaceous chondrite parent bodies. *Icarus, 82,* 244–280.

Hall D. T., Strobel D. F., Feldman P. D., McGrath M. A., and Weaver H. A. (1995) Discovery of an oxygen atmosphere on Jupiter's moon, Europa. *Nature, 373,* 677–679.

Han L. and Showman A. P. (2005) Thermo-compositional convection in Europa's icy shell with salinity. *Geophys. Res. Lett., 32,* L20201.

Hand K. P. and Chyba C. F. (2007) Empirical constraints on the salinity of the europan ocean and implications for a thin ice shell. *Icarus, 189,* 424–438.

Hand K. P., Chyba C. F., Carlson R. W., and Cooper J. F. (2006) Clathrate hydrates of oxidants in the ice shell of Europa. *Astrobiology, 6,* 463–482.

Hansen G. B. and McCord T. B. (2008) Widespread CO_2 and other non-ice compounds on the anti-jovian and trailing sides of Europa from Galileo/NIMS observations. *Geophys. Res. Lett., 35,* L01202.

Hibbits C. A., McCord T. B., and Hansen G. B. (2000) The distributions of CO_2 and SO_2 on the surface of Callisto. *J. Geophys. Res., 105,* 22541–22557.

Hibbits C. A., Pappalardo R. T., McCord T. B., and Hansen G. B. (2003) Carbon dioxide on Ganymede. *J. Geophys. Res., 108(E5),* 5036.

Hogenboom D. L., Kargel J. S., Ganasan J. P., and Lee L. (1995) Magnesium sulfate-water to 400 MPa using a novel piezometer: Densities, phase equilibria, and planetological applications. *Icarus, 115,* 258–277.

Holland H. D. (1978) *The Chemistry of the Atmosphere and Oceans.* Wiley, New York. 351 pp.

Holloway J. R. and Blank J. G. (1994) Application of experimental results to C–O–H species in natural melts. In *Volatiles in Magma* (M. R. Carroll and J. R. Holloway, eds.), pp. 187–230. Rev. Mineral., Vol. 30, Mineralogical Society of America, Washington, DC.

Horita J. and Berndt M. E. (1999) Abiogenic methane formation and isotopic fractionation under hydrothermal conditions. *Science, 285,* 1055–1057.

Horner R. A., ed. (1985) *Sea Ice Biota.* CRC, Boca Raton. 215 pp.

Huss G. R., Rubin A. E., and Grossman J. N. (2006) Thermal metamorphism in chondrites. In *Meteorites and the Early Solar System II* (D. S. Lauretta and H. Y. McSween Jr., eds.), pp. 567–586. Univ. of Arizona, Tucson.

Hussmann H. and Spohn T. (2004) Thermal-orbital evolution of Io and Europa. *Icarus, 171,* 391–410.

Jewitt D. C., Sheppard S., and Porco C. (2004) Jupiter's outer satellites and Trojans. In *Jupiter: The Planet, Satellites and Magnetosphere* (F. Bagenal et al., eds.), pp. 263–280. Cambridge Univ., Cambridge.

Johnson R. E. (2000) Sodium at Europa. *Icarus, 143,* 429–433.

Johnson R. E., Carlson R. W., Cooper J. F., Paranicas C., Moore M. H., and Wong M. (2004) Radiation effects on the surfaces of the Galilean satellites. In *Jupiter: The Planet, Satellites, and Magnetosphere* (F. Bagenal et al., eds.), pp. 485–512. Cambridge Univ., Cambridge.

Kargel J. S. (1991) Brine volcanism and the interior structures of asteroids and icy satellites. *Icarus, 94,* 368–390.

Kargel J. S., Kaye J. Z., Head J. W., Marion G. M., Sassen R., et al. (2000) Europa's crust and ocean: Origin, composition, and prospects for life. *Icarus, 148,* 226–265.

Kargel J. S., et al. (2007) Martian hydrogeology sustained by thermally insulating gas and salt hydrates. *Geology, 35,* 975–978.

Kelley D. S. et al. (2001) An off-axis hydrothermal vent field near the Mid-Atlantic ridge at 30°N. *Nature, 412,* 145–149.

Kelley D. S. et al. (2005) A serpentine-hosted ecosystem: The Lost City hydrothermal field. *Science, 307,* 1428–1434.

Khurana K. K. et al. (1998) Induced magnetic fields as evidence for subsurface oceans in Europa and Callisto. *Nature, 395,* 777–780.

Kirk R. L., Soderblom L. A., Brown R. H., Kieffer S. W., and Kargel J. S. (1995) Triton's plumes: Discovery, characteristics, and models. In *Neptune and Triton* (D. P. Cruikshank and M. S. Matthews, eds.), pp. 949–990. Univ. of Arizona, Tucson.

Kivelson M. G., Khurana K. K., Russel C. T., Volwerk M., Walker R. J., and Zimmer C. (2000) Galileo magnetometer measurements: A stronger case for a subsurface ocean at Europa. *Science, 289,* 1340–1342.

Kivelson M. G., Khurana K. K., and Volwerk M. (2002) The permanent and inductive magnetic moments of Ganymede. *Icarus, 157,* 507–522.

Krot A. N., Petaev M. I., Scott E. R. D., Choi B-G., Zolensky M. E., and Keil K. (1998) Progressive alteration in CV3 chondrites: More evidence for asteroidal alteration. *Meteoritics & Planet. Sci., 33,* 1065–1085.

Krot A. N., Fegley B. Jr., Lodders K., and Palme H. (2000) Meteoritical and astrophysical constraints on the oxidation states of the solar nebula. In *Protostars and Planets IV* (V. Mannings et al., eds.), pp. 1019–1055. Univ. of Arizona, Tucson.

Krot A. N., Hutcheon I. D., Brearley A. J., Pravdivtseva O. V., Petaev M. I., and Hohenberg C. M. (2006) Timescales and settings for alteration of chondritic meteorites. In *Meteorites and the Early Solar System II* (D. S. Lauretta and H. Y. McSween Jr., eds.), pp. 525–553. Univ. of Arizona, Tucson.

Kuskov O. L. and Kronrod V. A. (2001) Core sizes and internal structure of Earth's and Jupiter's satellites. *Icarus, 151,* 204–227.

Kuskov O. L. and Kronrod V. A. (2005) Internal structure of Europa and Callisto. *Icarus, 177,* 550–569.

Leblanc F., Johnson R. E., and Brown M. E. (2002) Europa's sodium atmosphere: An ocean source? *Icarus, 159,* 132–144.

Lellouch E., Paubert G., Moses J. I., Schneider N. M., and Strobel D. F. (2003) Volcanically emitted sodium chloride as a source for Io's neutral clouds and plasma torus. *Nature, 421,* 45–47.

Lewis J. S. (1971) Satellites of the outer planets: Their physical and chemical nature. *Icarus, 15,* 174–185.

Lewis J. S. (1982) Io: Geochemistry of sulfur. *Icarus, 50,* 103–114.

Lobo V. M. M. and Quaresma J. L. (1989) *Handbook of Electrolyte Solutions, Parts A and B.* Elsevier, Amsterdam. 2354 pp.

Lowell R. P. and DuBose M. (2005) Hydrothermal systems on Europa. *Geophys. Res. Lett., 32,* L05202.

Luth R. W. (2003) Mantle volatiles — Distribution and consequences. In *Treatise on Geochemistry, Vol. 2: The Mantle and Core* (R. W. Carlson, ed.), pp. 319–362. Elsevier-Pergamon, Oxford.

Marion G. M. and Kargel J. S. (2008) *Cold Aqueous Planetary Geochemistry with FREZCHEM: From Modeling to the Search for Life at the Limits.* Springer-Verlag, Berlin. 251 pp.

Marion G. M., Kargel J. S., Catling D. C., and Jakubowski S. D. (2005) Effects of pressure on aqueous chemical equilibria at subzero temperatures with applications to Europa. *Geochim. Cosmochim. Acta, 69,* 259–274.

Matson D. L., Castillo J. C., Lunine J., and Johnson T. V. (2007) Enceladus' plume: Compositional evidence for a hot interior. *Icarus, 187,* 569–573.

McCarthy C., Cooper R. F., Kirby S. H., Rieck K. D., and Stern L. A. (2007) Solidification and microstructures of binary ice-I/hydrate eutectic aggregates. *Am. Mineral., 92,* 1550–1560.

McCollom T. M. (1999) Methanogenesis as a potential source of chemical energy for primary biomass production by autotrophic organisms in hydrothermal systems on Europa. *J. Geophys. Res., 104,* 30729–30742.

McCollom T. and Seewald J. S. (2007) Abiotic synthesis of organic compounds in deep-sea hydrothermal environments. *Chem. Rev., 107,* 382–401.

McCord T. B., Hansen G. B., Clark R. N., Martin P. D., Hibbitts C. A., et al. (1998a) Non-water-ice constituents in the icy Galilean satellites from the Galileo near-infrared mapping spectrometer investigation. *J. Geophys. Res., 103,* 8603–8626.

McCord T. B., Hansen G. B., Fanale F. P., Carlson R. W., Matson D. L., et al. (1998b) Salts on Europa's surface detected by Galileo's near infrared mapping spectrometer. *Science, 280,* 1242–1245.

McCord T. B., Hansen G. B., Matson D. L., Johnson T. V., Crowley J. K., et al. (1999) Hydrated salt minerals on Europa's surface from the Galileo near-infrared mapping spectrometer (NIMS) investigation. *J. Geophys. Res., 104,* 11827–11851.

McCord T. B., Orlando T. M., Teeter G., Hansen G. B., Sieger M. T., et al. (2001a) Thermal and radiation stability of the hydrated salt minerals epsomite, mirabilite, and natron under Europa environmental conditions. *J. Geophys. Res., 106,* 3311–3320.

McCord T. B., Hansen G. B., and Hibbits C. A. (2001b) Hydrated salt minerals on Ganymede's surface: Evidence of an ocean below. *Science, 292,* 1523–1525.

McCord T. B., Teeter G., Hansen G. B., Sieger M. T., and Orlando T. M. (2002) Brines exposed to Europa surface conditions. *J. Geophys. Res., 107,* U61–U66.

McGrath M. A., Belton M. J. S., Spencer J. R., and Sartoretti P. (2000). Spatially resolved spectroscopy of Io's Pele plume and SO atmosphere. *Icarus, 146,* 476–493.

McKinnon W. B. (1999) Convective instability in Europa's floating ice shell. *Geophys. Res. Lett., 26,* 951–954.

McKinnon W. B. (2006) Formation time of the Galilean satellites from Callisto's state of partial differentiation (abstract). In *Lunar and Planetary Science XXXVII,* Abstract #2444. Lunar and Planetary Institute, Houston (CD-ROM).

McKinnon W. B. and Zolensky M. E. (2003) Sulfate content of Europa's ocean and shell: Evolutionary considerations and geological and astrobiological implications. *Astrobiology, 3,* 879–897.

McSween H. Y. (2003) Mars. In *Treatise on Geochemistry, Vol. 1: Meteorites, Comets, and Planets* (A. M. Davis, ed.), pp. 601–621. Elsevier-Pergamon, Oxford.

McSween H. Y. Jr., Ghosh A., Grimm R. E., Wilson L., and Young E. D. (2002) Thermal evolution of asteroids. In *Asteroids III*

(W. F. Bottke Jr. et al., eds.), pp. 559–572. Univ. of Arizona, Tucson.

Melosh H. J., Ekholm A. G., Showman A. P., and Lorenz R. D. (2004) The temperature of Europa's subsurface water ocean. *Icarus, 168,* 498–502.

Millero F. J. (2003) Physicochemical controls on seawater. In *Treatise on Geochemistry, Vol. 6: The Oceans and Marine Geochemistry* (H. Elderfeld, ed.), pp. 1–21. Elsevier-Pergamon, Oxford.

Moore J. M., Chapman C. R., Bierhaus E. B., et al. (2004) Callisto. In *Jupiter: The Planet, Satellites and Magnetosphere* (F. Bagenal et al., eds.), pp. 397–426. Cambridge Univ., Cambridge.

Morbidelli A., Levison H. F., Tsiganis K., and Gomes R. (2005) Chaotic capture of Jupiter's Trojan asteroids in the early solar system. *Nature, 435,* 462–465.

Neal C. and Stanger G. (1983) Hydrogen generation from mantle source rocks in Oman. *Earth Planet. Sci. Lett., 60,* 315–321.

Nesvorný D., Vokrouhlický D., and Morbidelli A. (2007) Capture of irregular satellites during planetary encounters. *Astron. J., 133,* 1962–1976.

Newton R. C. and Manning C. E. (2005) Solubility of anhydrite, $CaSO_4$, in NaCl-H_2O solutions at high pressures and temperatures: Applications to fluid-rock interaction. *J. Petrol., 46,* 701–716.

Nimmo F. (2004) Stresses generated in cooling viscoelastic ice shells: Application to Europa. *J. Geophys. Res., 109,* E12001.

Ohmoto H. and Lasaga A. C. (1982) Kinetics of reactions between aqueous sulfates and sulfides in hydrothermal systems. *Geochim. Cosmochim. Acta, 46,* 1727–1745.

Orlando T. M., McCord T. M., and Grieves R. A. (2005) The chemical nature of Europa surface material and the relation to subsurface ocean. *Icarus, 177,* 528–533.

Palandri J. L. and Reed M. H. (2004) Geochemical models of metasomatism in ultramafic systems: Serpentinization, rodingitization, and sea floor carbonate chimney precipitation. *Geochim. Cosmochim. Acta, 68,* 1115–1133.

Palme H. and Jones A. (2003) Solar system abundances of the elements. In *Treatise on Geochemistry, Vol. 1: Meteorites, Comets, and Planets* (A. M. Davis, ed.), pp. 41–61. Elsevier-Pergamon, Oxford.

Pappalardo R. T., Head J. W., Greeley R., Sullivan R. J., Pilcher C., et al. (1998) Geological evidence for solid-state convection in Europa's ice shell. *Nature, 391,* 365–368.

Pappalardo R. T., Belton M. J. S., Breneman H. H., Carr M. H., Chapman C. R., et al. (1999) Does Europa have a subsurface ocean? Evaluation of the geological evidence. *J. Geophys. Res., 104,* 24015–24055.

Pasek M. A. and Lauretta D. S. (2005) Aqueous corrosion of phosphide minerals from iron meteorites: A highly reactive source of prebiotic phosphorus on the surface of the early Earth. *Astrobiology, 5,* 515–535.

Pearson V. K., Sephton M. A., Franchi I. A., Gibson J. M., and Gilmour I. (2006) Carbon and nitrogen in carbonaceous chondrites. *Meteoritics & Planet. Sci., 41,* 1899–1918.

Pierazzo E. and Chyba C. F. (2002) Cometary delivery of biogenic elements to Europa. *Icarus, 157,* 120–127.

Pizarello S., Cooper G. W., and Flynn G. J. (2006) The nature and distribution of the organic material in carbonaceous chondrites and interplanetary dust particles. In *Meteorites and the Early Solar System II* (D. S. Lauretta and H. Y. McSween Jr., eds.) pp. 625–652. Univ. of Arizona, Tucson.

Pollack J. B., Witteborn F. C., Erickson E. F., Strecker D. W., Baldwin B. J., and Bunch T. E. (1978) Near-infrared spectra of the Galilean satellites: Observations and compositional implications. *Icarus, 36,* 271–303.

Porco C. C., Helfenstein P., Thomas P. C., Ingersoll A. P., Wisdom J., et al. (2006) Cassini observes the active south pole of Enceladus. *Science, 311,* 1393–1401.

Postberg F. K., Kempf S., Srama R., Green S. F., Hillier J. K., McBride N., and Grün E. (2006) Composition of jovian stream particles. *Icarus, 183,* 122–134.

Prialnik D. and Bar-Nun A. (1990) Heating and melting of small icy satellites by the decay of ^{26}Al. *Astrophys. J., 355,* 281–286.

Price L. C. and DeWitt E. (2001) Evidence and characteristics of hydrolytic disproportionation of organic matter during metasomatic processes. *Geochim. Cosmochim. Acta., 65,* 3791–3826.

Prieto-Ballesteros O. and Kargel J. S. (2005) Thermal state and complex geology of a heterogeneous salty crust of Jupiter's satellite, Europa. *Icarus, 173,* 212–221.

Prieto-Ballesteros O., Kargel J. S., Fernandez-Sampedro M., Selsis F., Martinez E. S., and Hogenboom D. L. (2005) Evaluation of the possible presence of clathrate hydrates in Europa's icy shell or seafloor. *Icarus, 177,* 491–505.

Qin L., Dauphas N., Wadhwa M., Masarik J., and Janney P. E. (2008) Rapid accretion and differentiation of iron meteorite parent bodies inferred from ^{182}Hf-^{182}W chronometry and thermal modeling. *Earth Planet Sci. Lett., 273,* 94–104.

Rosenberg N. D., Browning L., and Bourcier W. L. (2001) Modeling aqueous alteration of CM carbonaceous chondrites. *Meteoritics & Planet. Sci., 36,* 239–244.

Sader J. A., Leybourne M. I., McClenaghan M. B., and Hamilton S. M. (2007) Low-temperature serpentinization processes and kimberlite groundwater signatures in the Kirkland Lake and Lake Timiskiming kimberlite fields, Ontario, Canada: Implications for diamond exploration. *Geochemistry: Exploration, Environment, Analysis, 7,* 3–21.

Schilling N., Khurana K., and Kivelson M. G. (2004) Limits on an intrinsic dipole moment in Europa. *J. Geophys. Res., 109,* E05006.

Schilling N., Neubauer F. M., and Saur J. (2007) Time-varying interaction of Europa with the jovian magnetosphere: Constraints on the conductivity of Europa's subsurface ocean. *Icarus, 192,* 41–55.

Schmidt K. G. and Dahl-Jensen D. (2003) An ice crystal model for Jupiter's moon Europa. *Ann. Glaciol., 37,* 129–133.

Schubert G., Anderson J. D., Spohn T., and McKinnon W. B. (2004) Interior composition, structure and dynamics of the Galilean satellites. In *Jupiter: The Planet, Satellites, and Magnetosphere* (F. Bagenal et al., eds.), pp. 281–306. Cambridge Univ., Cambridge.

Scott E. R. D. (2007) Chondrites and the protoplanetary disk. *Annu. Rev. Earth Planet. Sci., 35,* 577–620.

Scott S. D. (1997) Submarine hydrothermal systems and deposits. In *Geochemistry of Hydrothermal Ore Deposits, 3rd edition* (H. L. Barnes, ed.), pp. 797–876. Wiley, New York.

Seewald J., Zolotov M. Yu., and McCollom T. (2006) Experimental investigation of carbon speciation under hydrothermal conditions. *Geochim. Cosmochim. Acta, 70,* 446–460.

Sephton M. A. (2002) Organic compounds in carbonaceous meteorites. *Natl. Prod. Rep., 19,* 292–311.

Sephton M. A., Pillinger C. T., and Gilmour I. (1999) Small-scale hydrous pyrolysis of macromolecular material in meteorites. *Planet. Space Sci., 47,* 181–187.

Sephton M. A. et al. (2004) Hydropyrolysis of insoluble carbonaceous matter in the Murchison meteorite: New insights into its macromolecular structure. *Geochim. Cosmochim. Acta, 68,* 1385–1393.

Seyfried W. E. and Shanks W. C. III (2004) Alteration and mass transport in mid-ocean ridge hydrothermal systems: Controls on the chemical and isotopic evolution of high-temperature crustal fluids. In *Hydrogeology of the Oceanic Lithosphere* (E. E. Davis and H. Elderfield, eds.), pp. 451–494. Cambridge Univ., Cambridge.

Seyfried W. E. Jr., Foustoukos D. I., and Allen D. E. (2004) Ultramafic-hosted hydrothermal systems at mid-ocean ridges: Chemical and physical controls on pH, redox, and carbon reduction reactions. In *Mid-Ocean Ridges: Hydrothermal Interactions Between the Lithosphere and Oceans* (C. R. German et al., eds.), pp. 267–284. Geophysical Monograph Series, Vol. 148, AGU, Washington, DC.

Showman A. P., Mosqueira I., and Head J. W. III (2004) On the resurfacing of Ganymede by liquid-water volcanism. *Icarus, 172,* 625–640.

Sohl F., Spohn T., Breuer D., and Nagel K. (2002) Implications from Galileo observations on the interior structure and chemistry of the Galilean satellites. *Icarus, 157,* 104–119.

Sleep N. H., Meibom A., Fridriksson Th., Coleman R. G., and Bird D. K. (2004) H_2-rich fluids from serpentinization: Geochemical and biotic implications. *Proc. Natl. Acad. Sci., 111,* 12818–12823.

Sloan E. D. Jr. and Koh C. (2007) *Clathrate Hydrates of Natural Gases, 3rd edition.* Chem. Industries Series 119, CRC, Boca Raton.

Spaun N. A. and Head J. W. III (2001) A model of Europa's crustal structure: Recent Galileo results and implications for an ocean. *J. Geophys. Res., 106,* 7567–7576.

Staudigel H. (2003) Hydrothermal alteration processes in the oceanic crust. In *Treatise on Geochemistry, Vol. 3: The Crust* (R. L. Rudick, ed.), pp. 511–535. Elsevier-Pergamon, Oxford.

Stevens T. O. and McKinley J. P. (1995) Lithoautotrophic microbial ecosystems in deep basalt aquifers. *Science, 270,* 450–455.

Stevens T. O. and McKinley J. P. (2000) Abiotic control on H_2 production from basalt-water reactions and implications for aquifer biogeochemistry. *Environ. Sci. Technol., 34,* 826–831.

Stevenson D. J., Harris A. W., and Lunine J. I. (1986) Origins of satellites. In *Satellites* (J. A. Burns and M. S. Matthews, eds.), pp. 39–88. Univ. of Arizona, Tucson.

Stumm W. and Morgan J. J. (1996) *Aquatic Chemistry: Chemical Equilibria and Rates in Natural Waters.* Wiley, New York. 1022 pp.

Thomas D. N. and Dieckmann G. S., eds. (2003) *Sea Ice: An Introduction to Its Physics, Chemistry, Biology, and Geology.* Blackwell Science, Oxford. 416 pp.

Tobie G., Mocquet A., and Sotin C. (2005) Tidal dissipation within large satellites: Applications to Europa and Titan. *Icarus, 177,* 534–549.

Travis B. J. and Schubert G. (2005) Hydrothermal convection in carbonaceous chondrite parent bodies. *Earth Planet. Sci. Lett., 240,* 234–250.

Vilas F., Lederer S. M., Gill S. L., Jarvis K. S., and Thomas-Osip J. E. (2006) Aqueous alteration affecting the irregular outer planets satellites: Evidence from spectral reflectance. *Icarus, 180,* 453–463.

Von Damm K. L. (1990) Seafloor hydrothermal activity: Black smokers and chimneys. *Annu. Rev. Earth Planet. Sci., 18,* 173–204.

Von Damm K. L. (1995) Controls on the chemistry and temporal variability of seafloor hydrothermal fluids. In *Seafloor Hydrothermal Systems: Physical, Chemical, Biological, and Geological Interactions* (S. E. Humphris et al., eds.), pp. 85–114. Geophysical Monograph Series, Vol. 91, AGU, Washington, DC.

Wächtershäuser G. (1990) The case for the chemoautotrophic origin of life in an iron-sulfur world. *Origins Life Evol. Biospheres, 20,* 173–176.

Waite J. H., Combi M. R., Ip W.-H., Cravens T. E., et al. (2006) Cassini ion and neutral mass spectrometer: Enceladus plume composition and structure. *Science, 311,* 1419–1422.

Walther J. V. (2005) *Essentials of Geochemistry.* Jones and Bartlett, Sudbury, Massachusetts. 291 pp.

Wetzel L. R. and Shock E. L. (2000) Distinguishing ultramafic- from basalt-hosted submarine hydrothermal systems by comparing calculated vent fluid compositions. *J. Geophys. Res., 105,* 8319–8340.

Wilson L. and Head J. W. III (1994) Mars: Review and analysis of volcanic eruption theory and relationships to observed landforms. *Rev. Geophys., 32,* 221–263.

Wilson L., Keil K., Browning L. B., Krot A. N., and Bourcier W. (1999) Early aqueous alteration, explosive disruption, and reprocessing of asteroids. *Meteoritics & Planet. Sci., 34,* 541–557.

Zahnle K., Schenk P., Levison H., and Dones L. (2003) Cratering rates in the outer solar system. *Icarus, 163,* 263–289.

Zimmer C., Khurana K. K., and Kivelson M. G. (2000) Subsurface oceans on Europa and Callisto: Constraints from Galileo magnetometer observations. *Icarus, 147,* 329–347.

Zolensky M. E. and McSween H. Y. Jr. (1988) Aqueous alteration. In *Meteorites and the Early Solar System* (J. F. Kerridge and M. S. Matthews, eds.), pp. 114–144. Univ. of Arizona, Tucson.

Zolensky M. E., Bourcier W. L., and Gooding J. L. (1989) Aqueous alteration on the hydrous asteroids — Results of EQ3/6 computer simulations. *Icarus, 78,* 411–425.

Zolensky M. E., Krot A. N., and Benedix G. (2008) Record of low-temperature alteration in asteroids. In *Oxygen in the Solar System* (J. J. Papike and S. Mackwell, eds.), pp. 429–462. Reviews in Mineralogy and Geochemistry, Vol. 68, Mineralogical Society of America, Washington, DC.

Zolotov M. Yu. (2007) An oceanic composition on early and today's Enceladus. *Geophys. Res. Lett., 34,* L23203.

Zolotov M. Yu. (2008) Oceanic composition on Europa: Constraints from mineral solubilities (abstract). In *Lunar and Planetary Science XXXIX,* Abstract #2349. Lunar and Planetary Institute, Houston (CD-ROM).

Zolotov M. Yu. and Fegley B. Jr. (1999) The oxidation state of volcanic gases and interior of Io. *Icarus, 141,* 40–52.

Zolotov M. Yu. and Fegley B. Jr. (2000) Eruption conditions of Pele volcano on Io inferred from chemistry of its volcanic plume. *Geophys. Res. Lett., 27,* 2789–2792.

Zolotov M. Yu. and Mironenko M. V. (2007a) Hydrogen chloride as a source of acid fluids in parent bodies of chondrites (abstract). In *Lunar and Planetary Science XXXVIII,* Abstract #2340. Lunar and Planetary Institute, Houston (CD-ROM).

Zolotov M. Yu. and Mironenko M. V. (2007b) Chemical evolution on an early ocean on Europa: Kinetic-thermodynamic modeling. In *Workshop on Ices, Oceans, and Fire: Satellites of the Outer Solar System,* pp. 157–158. LPI Contribution No. 1357, Lunar and Planetary Institute, Houston.

Zolotov M. Yu. and Shock E. L. (2000) A thermodynamic assessment of the potential synthesis of condensed hydrocarbons during cooling and dilution of volcanic gases. *J. Geophys. Res., 105,* 539–559.

Zolotov M. Yu. and Shock E. L. (2001) Composition and stability of salts on the surface of Europa and their oceanic origin. *J. Geophys. Res., 106,* 32815–32827.

Zolotov M. Yu. and Shock E. L. (2003a) Energy for biologic sulfate reduction in a hydrothermally formed ocean on Europa. *J. Geophys. Res., 108(E4),* 5022.

Zolotov M. Yu. and Shock E. L. (2003b) Aqueous oxidation of parent bodies of carbonaceous chondrites and Galilean satellites driven by hydrogen escape (abstract). In *Lunar and Planetary Science XXXIV,* Abstract #2047. Lunar and Planetary Institute, Houston (CD-ROM).

Zolotov M. Yu. and Shock E. L. (2004) A model for low-temperature biogeochemistry of sulfur, carbon, and iron on Europa. *J. Geophys. Res., 109,* E06003.

Zolotov M. Yu., Shock E. L., Barr A. C., and Pappalardo R. T. (2004) Brine pockets in the icy crust of Europa: Distribution, chemistry, and habitability. In *Workshop on Europa's Icy Shell: Past, Present, and Future,* pp. 100–101. LPI Contribution No. 1195, Lunar and Planetary Institute, Houston.

Zolotov M. Yu., Krieg M. L., Shock E. L., and McKinnon W. B. (2006a) Chemistry of a primordial ocean on Europa (abstract). In *Lunar and Planetary Science XXXVII,* Abstract #1435. Lunar and Planetary Institute, Houston (CD-ROM).

Zolotov M. Yu., Mironenko M. V., and Shock E. L. (2006b) Thermodynamic constraints on fayalite formation on parent bodies of chondrites. *Meteoritics & Planet. Sci., 41,* 1775–1796.

Oceanography of an Ice-Covered Moon

Steve Vance
NASA Jet Propulsion Laboratory/California Institute of Technology

Jason Goodman
Wheaton College

Europa's ocean is the most promising of potentially habitable extraterrestrial environments in the solar system. It is also the interface for transmitting heat and materials between the moon's rocky mantle and its icy shell. Here we review progress in describing the ocean's basic physical parameters and dynamics, with reference to possible ice-shell interactions that might leave clues in Europa's surface geology. Europa's ocean circulation is driven by hydrothermal buoyancy sources at the sea floor. Seafloor microfracture models predict that fluid might penetrate up to an order of magnitude deeper into Europa's crust than on Earth, possibly permitting significant heat and hydrogen production from serpentinization. A review of scaling parameters relevant to ocean circulation shows that Europa's ocean plausibly operates in a similar fluid dynamic regime to Earth's ocean and atmosphere. Rotational confinement allows hydrothermal plumes to reach the ice ceiling with sizes comparable to observed large chaos features on the moon's surface. This free transport of material and heat from seafloor to ice ceiling may be prevented by a stagnant "stratosphere" due to a minimum in water's thermal expansion near the freezing point at low pressures and low salinity. A similar effect occurs if plumes become enriched in salt by uptake at the seafloor; loss of buoyancy at a height proportional to the amount of uptake suggests a tendency toward stratication, possibly leading to diffusive-regime double diffusive convection. Evaluation of constitutive relationships for putative ocean fluids indicates that, while a working theoretical framework exists for calculations of fluid motion, further experimental work is needed at high salinities in the regime of elevated pressure and subzero temperatures for confident treatment of salt precipitation ("snow").

1. INTRODUCTION

In the past few years, discussion has begun to shift from the issue of whether Europa has an ocean to questions regarding its properties. In particular, it is now possible to begin investigating the flows of mass and energy within the ocean, discuss ocean currents both local and global, and to study the impact of heat flows within the ocean on the overlying ice crust.

That is to say, a new scientific field of "planetary oceanography" has been opened. While Earth's oceans possess a diverse array of physical environments, the oceans of ice-covered moons are unlike anything encountered on Earth. Oceanography in extraterrestrial oceans involves regimes of temperature, pressure, and composition not observed in Earth's ocean. Just as meteorology has benefited from comparison of Earth's atmosphere with that of other planets, we now find ourselves in a position to carry out experiments in comparative oceanography. But while Europa's ocean may be as different from Earth's ocean as Jupiter's atmosphere is from Earth's, in all four cases the same basic tools of geophysical fluid dynamics, laboratory experiments, and computer simulation can be usefully applied to gain understanding of these diverse fluids. (The fourth and most important tool of the terrestrial meteorologist or oceanographer, *in situ* observations, is a major challenge in the outer solar system, and seems unlikely to be used in Europa's ocean in the near-term.)

In Europa's ocean, and in other icy satellites that may host deep liquid interior layers, the mode of transport of heat and material from the interior is determined by well-established laws of fluid dynamics. The ocean's composition may have a strong influence on the nature of heat and material transport. We will attempt to describe the necessary physics and chemistry while providing a picture of Europa's ocean as it is described in the current literature. The large scale heights involved suggest an analogy to atmospheric physics. We will use this comparison in the discussion of Europa's ocean that follows.

Given the preliminary nature of constraints on the properties of Europa's ocean, this chapter cannot approach the thoroughness or complexity of a chapter on terrestrial oceanography. Notably absent are treatments of dynamic coupling between the ocean and ice shell or ocean and seafloor. Instead, we cover those topics that are treated in the literature to date, deriving in large part from pioneering work by *Thomson and Delaney* (2001) and *Melosh et al.* (2004). Where appropriate, we provide additional material.

2. CONTEMPORARY QUESTIONS IN EUROPAN OCEANOGRAPHY

This section describes some topics that have been explored for Europa's ocean based on current understanding and by analogy with Earth's ocean. Transport of heat and material recurs as a theme, underscoring the ocean's role as interface between the overlying ice shell and underlying lithosphere.

2.1. Near-Surface Interactions: Formation of Chaos Regions

As discussed in the chapter by Collins and Nimmo, the formation of chaotic terrain remains one of the major outstanding questions in the study of Europa. The melt-through model for chaos formation (*Greenberg et al.,* 1999; *Greenberg and Geissler,* 2002) requires a large amount of heat to be concentrated into a small area at the base of the ice layer. This heat must be communicated from the rocky interior to the ice layer, through the intervening liquid water layer. The behavior of the water layer strongly affects this heat transport, and imposes its own space and timescales on the delivery. Thus, understanding the fluid dynamics of the ocean layer can help us choose between chaos formation models. Are the physical parameters of hydrothermal plumes (dimensions, timescales, heat flux, and velocities) consistent with what is known of the chaos regions?

2.2. Global Ocean Circulation

One key goal of europan oceanographic research is to describe the transport of heat, chemical constituents, and potential biomaterials (organisms and their nutrients) throughout the ocean. Past studies have primarily examined local processes (hydrothermal plumes in particular), while the overall question of the ocean's basic structure and general circulation remains largely unexplored. Is Europa's ocean well-mixed from top to bottom, or divided into "cells" separated by diffusive layers? Does large-scale horizontal circulation occur? What is the effect of seafloor topography (ocean basins), about which we currently have zero information?

2.3. Global Horizontal Circulation

We discuss small-scale vertical convectively driven flows extensively in section 5, but when a terrestrial oceanographer discusses the "ocean circulation," he or she usually is referring to horizontal flows of global scale. Does Europa possess "ocean gyres," "thermohaline circulation," or other global-scale current patterns?

2.4. Seafloor and Subseafloor: The Nature of Europan Hydrothermalism

Separate from their role in the formation of ice raft features on Europa's surface, the generation of hydrothermal plumes has implications for ocean circulation, composition, and habitability. Several authors (*Gaidos et al.,* 1999; *Kargel et al.,* 2000; *Chyba and Phillips,* 2002; *Irwin and Schulze-Makuch,* 2003; *Zolotov and Shock,* 2004; *McCollom,* 2007; *Vance et al.,* 2007) have discussed hydrothermal activity on Europa as a possible energy and nutrient source for life (also see chapters by Kargel and Zolotov and by Hand et al.). Estimates of the extent and depth of putative europan hydrothermal systems are unavoidably based on analogy with Earth's seafloor. To the extent that hydrothermalism away from mid-ocean ridges is not well quantified, the majority of Earth's seafloor also remains unexplored (*Baker and German,* 2004). How does the range of possible fluid-rock interaction depths compare in Europa's seafloor? Progress on these topics is discussed in the next section.

3. HYDROTHERMAL INTERACTIONS IN EUROPA'S SEAFLOOR

As mentioned above, the chemistry and ecology of europan hydrothermal systems have been discussed in some detail by previous authors. In these studies, the depth and areal extent of such systems was assumed to be similar to systems on Earth. Given the sparseness of data pertaining to Europa's interior, this approach is somewhat justified. Further constraints can be placed on the nature of europan hydrothermalism by considering the moon's size and the range of possible thermal inputs, and by assuming a plausible seafloor composition.

Temperature and pressure profiles beneath the seafloor ("europatherms") can be estimated by assuming a composition of pure olivine through which heat is transported by steady-state thermal conduction to hydrothermal depths. These are justified on the basis of the abundance of olivine in primordial meteorites (e.g. *Kargel et al.,* 2000) and estimates of the conductive lid thickness (minimum ~ 300 km) (*Hussmann and Spohn,* 2004; chapter by Moore and Hussmann). The contribution of tidal heating to the seafloor is likely nonzero, and has probably varied throughout Europa's history as a result of its orbital evolution. As these calculations pertain to the depth of hydrothermal circulation, the temperatures discussed here should be considered a minimum.

Temperature increases with depth into a conductively cooling seafloor (the "conductive lid" discussed in the chapter by Moore and Hussmann) as

$$Q = \frac{F}{4\pi(R_E - d)^2} = \lambda T(z) \tag{1}$$

where the linear increase in temperature with depth, ë, is dependent on the thermal conductivity of the seafloor material ($\lambda \approx$ 2–3 W m^{-1} K^{-1} for olivine), heat flux from the interior (F), and depth d from the moon's surface at R_E = 1565 km to the seafloor. Pressure is calculated from the weight of overlying material. For a two-layer model of Europa (*Turcotte and Schubert,* 2001)

$$\begin{aligned} P(z) = \frac{2}{3}\pi G(\rho_{rock}^2(R_m^2 - r_m^2) + \rho_w^2(R_E^2 - R_m^2) + \\ 2\rho_w R_m^3(\rho_{rock} - \rho_w)(1/R_m - 1/R_E)) \end{aligned} \tag{2}$$

in which $R_m = R_E - d$, $r_m = R_m - z$, $\rho_w \approx 10^3$ kg m^{-3}, and $\rho_{rock} \approx 3 \times 10^3$ kg m^{-3} for olivine. Figure 1 shows a europatherm for d = 100 km and λ = 3 W m^{-1} K^{-1}. Depth of fluid-accessible seafloor (z) is discussed in section 3.2.

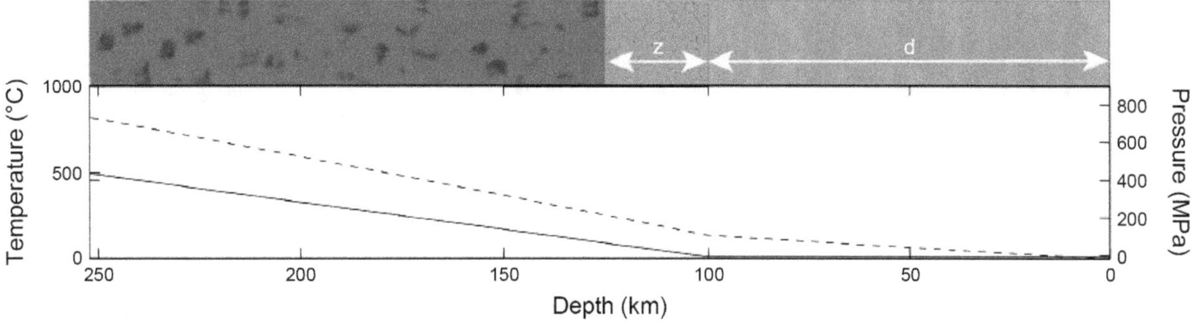

Fig. 1. Depth, z, of fluid-accessible olivine mantle into a europan seafloor of depth d = 100 km, based on the depth of formation of microfractures. In this example, mineral grain size is assumed to be 1 mm. For this ocean depth, and assuming no contributed heat from tidal dissipation in the rocky interior, z = 25 km. From *Vance et al.* (2007).

One likely difference between europan and terrestrial hydrothermal systems is that the formation of water vapor is not a factor constraining the temperature of hydrothermal circulation. Even the shallowest seafloor depth on Europa (~80 km below the surface) (*Anderson et al.,* 1998) is above the critical pressure of 22 MPa. Circulation depth and quartz solubility are other factors governing hydrothermal flow (*Lowell and DuBose,* 2005). The first of these is discussed in section 3.2. The latter enters into the discussion of seafloor permeability in the next section.

In considering the prospects for hydrothermalism in Europa's seafloor, as discussed below, researchers have assumed a basaltic crust similar to Earth's seafloor. Lacking constraints on the nature of the seafloor, they have only speculated on the tectonic nature of the crust. Generally, a homogeneous, i.e., not tectonically altered, seafloor has been assumed. As noted below, this high degree of uncertainty allows for a broad range of permeability and composition depending on the nature of the crust's origin (see chapter by Moore and Hussmann for further discussion of thermal aspects of this problem).

3.1. Number of Hydrothermal Systems

Is Europa's internal heat delivered to the ocean by a large number of very small hydrothermal sources, or a small number of large ones? The buoyancy difference between water leaving the porous rock in a hydrothermal vent and the water entering the rock in the "recharge zones" (Fig. 2) forces fluid through the crust. A large source must pull fluid in from a wider area. Since the available buoyancy force to drive flow is limited, there is an upper limit on the magnitude of the mass flux through a single hydrothermal system.

Using Darcy's Law, *Lowell and DuBose* (2005) estimate the mass flux from high-temperature systems to be

$$\dot{M} = \rho\alpha gkT'A_d/\nu \qquad (3)$$

where A_d is the area of the outflow zone. Other parameters are described in Table 1. *Lowell and DuBose* (2005) assumed values for a high-temperature system on Earth: $A_d \sim 10^3$–10^4 m^2, $\alpha \sim 10^{-3}$ K^{-1}, $\nu \sim 10^{-7}$ m^2 s^{-1}, $T' \approx 400$ K, $k = 10^{-12}$ m^2. For Europa's seafloor, $g \approx 1.7$ m s^{-2}. Permeability (k) in Earth's seafloor may diminish over time due to filling of pore spaces, either by silica that precipitates from ambient fluid (*Stein et al.,* 1995), by other precipitates — native sulfur or SO_2 such as exist on Io (*McKinnon and Zolensky,* 2003) — or by the accumulation of sediment (*Martin and Lowell,* 2000), possibly in the form of salt precipitates (*Kargel et al.,* 2000). Given the uncertainties in terrestrial seafloor permeability and in the structure of Europa's seafloor, a broad range, $k = 10^{-16}$–10^{-9} m^2 s^{-1} (*Fisher and Becker,* 2000), may be consistent with the other hydrothermal system parameters used by *Lowell and DuBose* (2005).

Applying the above range of permeability to *Lowell and Dubose*'s (2005) calculations suggests a range of possible heat output — between 10^2 and 10^8 W (corrected from *Vance et al.,* 2007) — corresponding to a broad range in the estimated number of Earth-like hydrothermal systems in Europa's ocean. Because the gravitational acceleration at Europa's seafloor is about one-seventh the value at Earth's surface, the expected heat flux from a europan hydrothermal system is lower than on Earth. The number of hydrothermal systems needed on Europa to dissipate half the internal heat flux — as is estimated to be the case on Earth — is greater than the equivalent number on Earth (*Lowell and DuBose,* 2005). Whether such systems exist at all is not known. Chemical signatures from hydrothermal fluid-rock interactions may be transported through Europa's ocean (sections 6 and 7), and possibly through the ice shell to the surface (see section 8 and chapter by Collins and Nimmo), where they could be detected by future exploration missions.

3.2. Depth of Water-Rock Interaction

For the estimate of hydrothermal plume temperature, *Lowell and DuBose* (2005) assume a 1-km depth of hydrothermal circulation to a magma lens. This picture probably does not apply to Europa's mantle as a whole. For Europa's present orbital state, a thick thermally conductive upper mantle layer is predicted, even in the case of significant interior tidal dissipation (chapter by Nimmo and Manga; *Hussmann and Spohn,* 2004). Low-temperature hydrother-

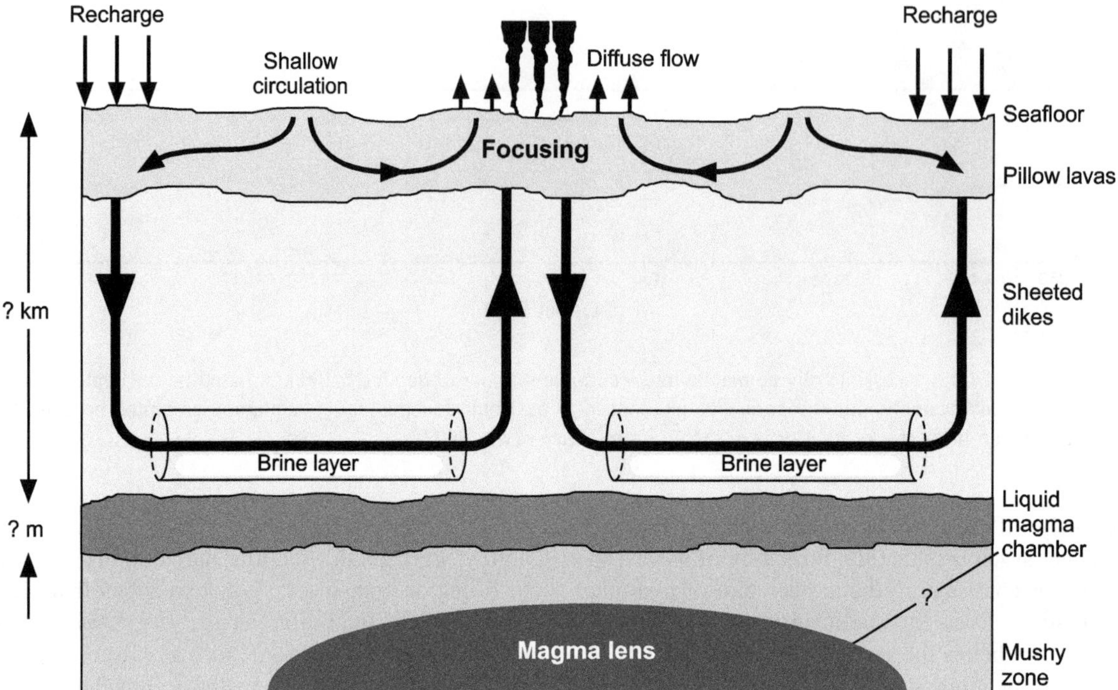

Fig. 2. Schematic model of a terrestrial seafloor hydrothermal system at an ocean ridge. Fluid enters the crust at the seafloor and percolates downward to the top of a magma body. There is flow subhorizontally, extracting heat until it ascends buoyantly toward the seafloor. The spatial scale of these systems on Earth is constrained in part by observed vent spacing and size of the magma lens. In axial systems, appropriate scales are in the range of 1 km for the recharge zone, with the magma chamber thickness on the order of 100 m. At off-axis sites, where fewer systems have been characterized, flow depth and subsurface heating are not as well constrained. In Europa's seafloor, these questions are compounded by uncertainty in seafloor composition, heat flow, and rheology. Modified from *Lowell and DuBose* (2005).

mal activity may be the dominant mode of hydrothermal activity on Europa, including the exothermic hydration of olivine (serpentinization) (see chapter by Zolotov and Kargel).

Rough constraints on the depths and timescales of water-rock interaction based on analysis of microfracture formation (*Vance et al.,* 2007) indicate that in small icy worlds like Europa, the depth of brittle mantle can be much greater than on Earth. In this case, serpentinization may produce significant hydrogen and supplemental heat. The timing of subsequent serpentinization depends on the rate of exposure of fresh rock. If it occurs at all, some amount of serpentinization is expected during initial cooling of Europa's mantle. As the moon's interior cools further, the ultimate depth of fluid penetration may be set by fracturing due to stresses that develop in the rock as a result of thermal expansion anisotropy in mineral grains. In this simple model, cracking depth increases with time as radiogenic heat decreases. Serpentinization may persist to the present era as unreacted rock continues to be exposed.

Cracking depths (z) are calculated from two-layer models, which set P and T in the seafloor as described at the beginning of this section. For Europa, assuming maximum seafloor depth of 170 km (*Anderson et al.,* 1998) and 1-mm grain size typical of seafloor rocks yields cracking depth identical to the Earth's 6-km depth, whereas ten times that depth is predicted for the minimum seafloor depth of 80 km and larger grain size of 10 mm. Figure 1 shows this schematically for an ocean of 100-km depth, for which the cracking depth is 25 km.

Serpentinization of Europa's mantle requires that the mantle has been heated, either during formation or due to subsequent tidal evolution, sufficient to remove mineral-bound water, leaving peridotite as the dominant species. For Europa, such heating is plausibly implied by the moon's differentiated state (*McKinnon and Zolensky,* 2003; *Vance et al.,* 2007). Subsequent tidal heating (*Hussmann and Spohn,* 2004; *Vance et al.,* 2007, chapter by Moore and Hussmann) may cause the cracking front to recede toward the seafloor and, if sufficiently intense, dewater hydrated serpentine minerals. Further work is needed to constrain related permeability, and to predict hydrothermal fluid temperatures, but the comparison of brittle thicknesses for Earth and Europa points to extensive fluid-rock interaction over geological timescales.

4. OCEAN CIRCULATION: REVIEW OF RELEVANT PHYSICS

4.1. Laws of Motion

In considering the laws of fluid motion within an ice-covered ocean, the terrestrial oceanographer is on familiar ground. The ocean is a fluid, so the continuity (conservation of mass) and Navier-Stokes equations apply, which de-

TABLE 1. Variables used throughout this chapter.

Symbol	Name	Value	Dimensions
B	Buoyancy Flux		$m^4\ s^{-3}$
b	Buoyancy		$m\ s^{-2}$
C_P	Specific heat capacity		$J\ kg^{-1}\ K^{-1}$
d	Ocean depth from Europa's surface		m
F	Global heat flux		W
g	Local gravity		$m\ s^{-2}$
H	Fluid ocean depth		m
h	Ice thickness		m
h_{cone}	Plume confinement height		m
k	Rock permeability		$m^2\ s^{-1}$
l, D	Plume diameter		m
$\dot{M}$	Mass flux		$kg\ s^{-1}$
P	Pressure		Pa
Q	Areal heat flux		$W\ m^{-2}$
r_e	Radius of Europa	1565	km
T	Temperature		K
T'	Fluid temperature anomaly		K
U, w	Fluid velocity		$m\ s^{-1}$
v_c	Sound velocity		$m\ s^{-1}$
z	Plume height/Fluid Peroclation Depth into Mantle		m
α	Thermal expansion coefficient		K^{-1}
β	Salinity expansion coefficient		mol^{-1}
λ	Thermal conductivity		$W\ m^{-1}\ K^{-1}$
μ	Volume flux		$m^3\ s^{-1}$
ν	Molecular viscosity		Pa s
Ω	Rotational frequency		Hz
ρ	Density		$kg\ m^{-3}$
θ	Latitude		

scribe Newton's Second Law for a fluid parcel

$$\frac{\partial \rho}{\partial t} + \nabla \cdot (\rho \mathbf{u}) = 0 \tag{4}$$

$$\overbrace{\rho\left(\underbrace{\frac{\partial \mathbf{U}}{\partial t}}_{\text{Unsteady acceleration}} + \underbrace{\mathbf{U} \cdot \nabla \mathbf{U}}_{\text{Convective acceleration}}\right)}^{\text{Inertia}} = \underbrace{-\nabla p}_{\text{Pressure gradient}} + \underbrace{\nu \nabla^2 \mathbf{U}}_{\text{Viscosity}} + \underbrace{\mathbf{f}}_{\text{Other forces}} \tag{5}$$

$$\mathbf{f} = \underbrace{-2\rho\Omega \times \mathbf{v}_{rot}}_{\text{Coriolis}} + \underbrace{\rho\mathbf{g}}_{\text{Gravity}} + \ldots$$

Let us consider the importance of compressibility, viscosity, and planetary rotation on fluid flow by looking at the order of magnitude of the Mach, Reynolds, and Rossby numbers (summarized in Table 2). We will take the ocean depth H — the distance from the ice ceiling to the seafloor — as the length scale of the flows we are interested in. This value is somewhat uncertain as it is subject to limited constraints on the depth of Europa's rocky mantle (*Anderson et al.,* 1998) and clues to ice shell thickness taken from surface geology (see chapter by Nimmo and Manga; and, e.g., *Hoppa et al.,* 1999; *Turtle et al.,* 2001). Following the convention set by previous investigators, we will take H = 100 km.

The speed of sound v_c in water is on the order of 1.5 km s^{-1}; unless the order of magnitude of ocean current speeds U is comparable, the Mach number $M = U/v_c$ is small and we may treat the fluid as incompressible. The molecular viscosity of water is $\nu = 10^{-3}$ Pa s, so the minimum Reynolds number $Re = HU/\nu$ is large: on the order of 10^5 if U = 1 mm s^{-1}, or 10^8 if U = 1 m/s. Thus, fluid flow may be treated as inviscid. Because a high Reynolds number implies highly turbulent flow, the concept of an "eddy viscosity" might be useful for characterizing the transport and dissipation of energy in the smaller-scale flow. The conditions under which eddies behave diffusively, however, are difficult to gauge even in well-measured terrestrial flows,

TABLE 2. Fluid-dynamic quantities used in the present discussion.

Name	Equation
Mach number	$M = U/v_c$
Reynolds number	$Re = HU/\nu$
Coriolis parameter	$f = 2\Omega\sin\theta$
Rossby number	$Ro = U/(fD)$

so this topic must be left for future consideration. Finally, Europa is rapidly rotating, spinning on its axis in synchronous rotation once every 3.55 terrestrial days. The average magnitude of the Coriolis parameter at latitude θ

$$f = 2\Omega\sin\theta \quad (6)$$

is roughly 1.3×10^{-5} s^{-1}, so that the Rossby number for a feature of characteristic size D, Ro = U/(fD), will be very small unless water velocities exceed 10 m s^{-1}, implying Coriolis forces are of major importance.

Thus, flows within Europa's ocean are best considered using the same nonlinear, rapidly rotating, nearly inviscid incompressible Navier-Stokes equations that govern terrestrial oceanic and atmospheric flow. The following ideas, developed to understand terrestrial atmosphere/ocean flows, will be useful.

4.1.1. Convection. Negatively buoyant water sinks; positively buoyant water rises. There is an extensive literature on convection in rapidly rotating environments (e.g., *Fernando et al.,* 1998) that we may draw upon. Throughout this literature, a formal definition of "buoyancy" is used: A fluid parcel of density ρ in an environment of characteristic density ρ_0 has a buoyancy given by

$$b = \frac{\rho_0 - \rho}{\rho_0} g = g' \quad (7)$$

where g is the local gravitational acceleration. b is sometimes called g', the reduced gravity, since g' is generally less than g, and the buoyant fluid parcel's behavior is similar to an object in empty space under the influence of a small gravitational acceleration g'.

4.1.2. Double diffusion. If vertical gradients exist in both temperature and salinity, double-diffusive convection can occur, in which both components contribute to the mechanism of fluid transport owing to the much lower diffusivity of salt relative to heat ($\kappa_c \approx 10^{-2}\kappa_T$). The magnitudes and directions of the gradients determine the style of heat and material transport. If density scales linearly with temperature and salinity (*Turner and Chen,* 1973), two nondimensional numbers characterize the double-diffusive system

$$Rs = R_\rho Ra = \frac{g\beta\Delta S H^3}{\kappa\nu} \quad (8)$$

$$R_\rho \equiv \frac{\beta dS}{dz}\left(\frac{\alpha dT'}{dz}\right)^{-1} \quad (9)$$

Rs is the salinity Rayleigh number, in which β = d ln ρ/dS is the coefficient of saline contraction, analogous to α in Ra. The numbers Ra and Rs are related by the ratio R_ρ.

In the presence of an upward thermal gradient, two regimes of double-diffusive convection are possible depending on the direction of the compositional gradient. If salinity decreases downward, a phenomenon results known as "salt fingering" (*Merryfield,* 2000), in which fluid motion is driven by the negative buoyancy of saline plumes. This phenomenon is observed on the underside of sea ice in Earth's ocean as stalactite-like formations. If salinity also decreases upward, the "diffusive" regime is obtained, in which multiple convective layers are separated vertically by thin diffusive regions. Diffusive regime double diffusive convection in Europa's ocean has been previously discussed in light of available constraints on the properties of Europa's ocean (see section 7.5) (*Vance and Brown,* 2005).

4.1.3. Internal gravity wave. Waves may develop along an interface between fluids of differing density. (To a fluid dynamicist, a "gravity wave" is an ordinary fluid wave, like waves on a beach, for which gravity is the restoring force. There is an unfortunate confusion with the waves of bent space-time predicted by general relativity.) If the buoyancy difference between these fluids is b, the phase speed of these waves is given by

$$c_{grav} = \sqrt{b\frac{H_1H_2}{H_1 + H_2}} \quad (10)$$

where H_1 and H_2 are the depths adjacent layers in the fluid. If $H_1 \sim H_2 = 2H$, $c_{grav} = \sqrt{bH}$. Internal gravity waves are also possible in fluids that are "continuously stratified," with continuously varying buoyancy rather than layers separated by an interface.

4.2. Geostrophic Flow

For a fluid of small Rossby number, the Coriolis force balances the external and body forces acting on the fluid. The most common situation, in which the pressure-gradient force and Coriolis force are in balance, is called "geostrophic flow," and is characteristic of Earth's atmosphere and ocean. In this situation, the fluid flow is perpendicular to both the pressure-gradient force and to the local gravity: Fluid flow is horizontal, following lines of constant pressure (isobars). The speed of this flow is governed by the vertical component of the planetary angular rotation vector: $f \equiv 2\Omega\cos(\theta)$ where θ is the local latitude.

4.2.1. Taylor columns. In an inviscid fluid of uniform density, velocity shear is suppressed along the axis of rotation. Qualitatively, we may say that the fluid behaves as if composed of a large number of vertical columns aligned along the axis of rotation. The fluid moves freely in directions perpendicular to this axis, but is "stiffened" along the axis. These "Taylor columns" resist being stretched or twisted (*Tritton,* 1988).

4.2.2. Thermal wind. A rotating fluid of nonuniform density does support shear. If two fluid layers of differing density are separated by a sloping interface, the pressure gradient caused by that slope will drive geostrophic flow in the layers. The velocity difference ($\vec{u}_1 - \vec{u}_2$) between layers of buoyancy contrast b is given by the "thermal wind equation": $f(\vec{u}_1 - \vec{u}_2) = -b\hat{k} \times \nabla\eta$ where $\hat{k}$ is the local vertical vector and ∇η is the slope of the interface between layers. As with internal gravity waves, one may form a thermal wind equation for continuously stratified fluids.

4.2.3. Rossby radius of deformation. On what time and space scales is the Coriolis effect of rotation important? If an interface between layers of differing densities is perturbed, the perturbation will initially propagate outward along the interface as a gravity wave. But on the timescale of planetary rotation, Coriolis forces will deflect fluid motion, eventually leading to a geostrophic steady state in which the thermal wind equation determines fluid velocity. Thus, the outward propagation of a gravity wave of speed $c_{grav} = \sqrt{bD}$ is impeded after a time f^{-1}. From these we may construct a length scale, the "internal Rossby radius of deformation"

$$r_D = \sqrt{bH}f \tag{11}$$

On spatial scales smaller than the Rossby radius, time-dependent buoyancy-driven forces dominate; on larger scales, Coriolis forces dominate and relatively steady geostrophic flow is more important.

4.2.4. Baroclinic instability. The thermal wind equation above states that horizontal density gradients in a fluid lead to vertically-varying horizontal currents. But these currents push water parcels around, changing the density gradients that created them. Thus, there is a feedback between currents and the density field. When the "thermal wind" is sufficiently strong, this becomes a positive feedback, uncontrollably amplifying small perturbations in density. This is "baroclinic instability." The thermal wind front breaks up into small eddies, whose size and growth rate can be predicted. Linear perturbation analysis (see, e.g., *Pedlosky,* 1987) suggests that these eddies have a radius that scales with the "Rossby Radius of Deformation" (section 4.2.3).

4.3. Constitutive Laws

The above discussion of fluid motion implicitly requires an understanding of how fluid densities change with temperature, pressure, and composition. In addition to accurate constraints on Europa's properties (boundary conditions), we need an accurate way to calculate densities of europan ocean fluids under relevant conditions. For the present calculations, densities are estimated using methods similar to those employed in the chapter by Zolotov and Kargel (see also *Marion et al.,* 2005). Solution volumes are estimated from a semi-empirical model that uses a database of coefficients constructed from measurements of sound velocity and density.

Verifying the accuracy of fluid dynamical computations for a very deep ocean composed of magnesium sulfate or other aqueous species is difficult due to an absence in the literature of equation-of-state data above 100 MPa at nonstandard temperatures. Recent work addressing this problem (*Vance,* 2007) indicates solution densities for $MgSO_4$ (aq) can be calculated based on extrapolation from previous measurements (*Marion et al.,* 2005) to within 1% up to 2 m (~20%) and 200 MPa, although uncertainties remain for temperatures much lower than ~10°C and above ~40°C. This adds validity to a discussion of fluid transport based on density alone. The accuracy of predicting solubility above about 30 MPa remains to be verified at the time of this writing, so our discussion of salt precipitation from buoyant plumes is necessarily somewhat limited.

5. GLOBAL MECHANICAL AND BUOYANCY FORCES

There are three possible driving forces for global scale flow. The first is tidal forcing, discussed below in section 5.2. The second is mechanical surface stresses. On Earth, surface wind stress is a major driver of ocean circulation, but in Europa's ocean this cannot occur. To the extent that Europa's ice shell is stationary, there can be no large-scale surface stress at the ice/water interface (but see chapter by Barr and Showman for a discussion of convective motion within the ice shell). If Europa's ice shell is in nonsynchronous rotation (*Schenk and McKinnon,* 1989; *Kattenhorn,* 2002; chapter by Bills et al.), then a tiny global-scale shear flow must occur as the shell moves with respect to the seafloor, but this flow is likely trivial given the long timescale of rotation.

The third possible driving force is large-scale buoyancy forces caused by density gradients. This is the force that leads to the overturning thermohaline circulation on Earth. Is Europa likely to have a thermohaline circulation? On Earth, strong pole-to-equator temperature gradients result from differential surface heating. But Europa's ocean has a nearly isothermal upper surface boundary condition, since the temperature must be at the freezing point at the ice/water interface. The slight variation in freezing point with pressure might lead to horizontal flow; this is discussed in section 8.1. But apart from this effect, we expect Europa's ocean to be nearly isothermal at the freezing point. Regional variations in basal heat flux cannot lead to strong temperature differences: Since Europa's ocean is nearly inviscid, any surplus heat released by the sea floor will be efficiently transported vertically to the ice/water interface, where it will either transfer into the ice via thermal conduction, or else melt and thin the ice sheet (which leads to more rapid thermal conduction).

Thus, we cannot have strong temperature gradients in Europa's ocean. What about salinity effects? On Earth, salinity is usually an important consideration: Salinity is modified by regional variations in surface evaporation and precipitation. On Europa, we could substitute "melting and freezing" for "evaporation and precipitation," but there is a crucial caveat: For a steady-state haline circulation to be maintained, the salt sources and sinks must be persistent. On Earth, this is possible because the atmosphere transports water from regions of net evaporation to regions of net precipitation. On Europa, this can only occur if ice flows on a global scale from regions of net freezing to regions of net melting. The high viscosity of ice makes water transport more difficult for Europa's ice than for Earth's atmosphere, but little quantitative work has been done to assess the likelihood of global-scale ice flow in Europa's ice shell.

Thus, we are led toward a picture of Europa's ocean with relatively small global-scale temperature and salinity gradients, and thus the buoyancy forces that drive global circulation are probably extremely weak.

5.1. Zonal Jets

There is, however, a fourth way in which global-scale circulation patterns can occur: Rather than being driven by global-scale boundary conditions, they can be forced by small-scale turbulence. Hydrothermal plumes and other small-scale processes will generate geostrophically balanced eddies of small (10–100 km) scale (see section 6.2). Two-dimensional turbulence theory predicts that geostrophically balanced eddies will tend to interact and merge, forming larger and larger eddies, in a process called the "enstrophy cascade" (*Rhines,* 1975). Thus, large-scale circulation can be caused by small-scale turbulent forcing.

However, the eddy merger process cannot continue indefinitely. Remember from section 4.2.1 that in a rapidly rotating fluid, "Taylor columns" of fluid aligned along the axis of rotation tend to resist changes in column length. But on a spherical planet, the length of these columns changes dramatically from pole to equator, as a consequence of the spherical geometry (see Fig. 3). Thus, a fluid parcel cannot easily move from low latitude to high, or vice versa. This limits the north-south scale of turbulence.

However, east-west flow is not restricted; thus, the zonal scale of eddies can continue to grow without bound, while their meridional extent is restricted. Typically, this results in the formation of east-west "jets." This process is crucially important to the formation of zonal jets in gas giant atmospheres (*Kaspi and Flierl,* 2007).

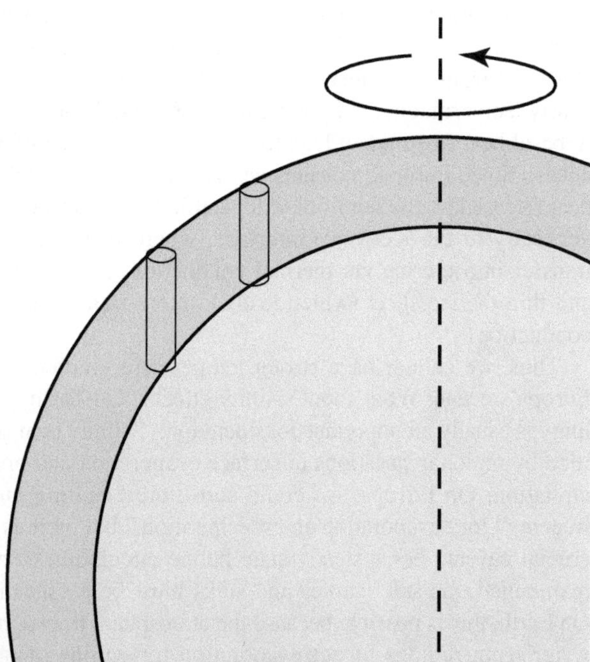

Fig. 3. "Taylor" columns aligned with the axis of planetary rotation do not have the same length, in a fluid of constant depth on a spherical planet. Columns near the equator are longer. Thus, since in a geostrophically-balanced fluid changes to column length are not permitted (See section 4.2.1), large-scale north-south flows are inhibited.

Europa's ocean meets all the necessary criteria for this process, so we might expect to find similar jets there. Any variation in the thickness of the water layer would also cause a change in the length of Taylor columns, thus, these jets might deflect north or south to avoid seafloor topography and/or variations in ice shell thickness.

The maximum north-south scale of turbulent motion is given by the *Rhines* (1975) scale, which can be written as

$$L_\beta = \frac{2\pi}{k_\beta} = 2\pi\sqrt{\frac{Ur_e}{\Omega\cos\theta}}$$

where r_e is the radius of Europa, Ω is its angular rotation rate, θ is the local latitude, and U is the typical fluid velocity. If we assume velocities comparable to predictions by *Goodman et al.* (2004) of hydrothermal plume eddy current speeds (3–9 mm/s), we find a Rhines scale on the order of 150–250 km. Rhines theory predicts that jets will be relatively narrow near the equator, and broader near the poles.

Thus, we are led to the following picture of the large-scale horizontal circulation within Europa's ocean: (1) little to no large-scale buoyancy-driven circulation, due to isothermal upper boundary condition; (2) no surface-stress-driven circulation; and (3) possible circumglobal east-west "jets" with widths of 150–200 km, driven by enstrophy cascade small-scale turbulence.

Flows in Europa's ocean (Fig. 4) might therefore qualitatively resemble the atmosphere of Jupiter, on a vastly smaller scale. However, while we can suggest that jetlike flows might be likely, we cannot as yet place any limits on their velocity, nor identify their position. In fact, since they are driven by random turbulence, the jets may tend to wander in time. Magnetic fields induced by such currents may be detectable using methods for observing currents in Earth's ocean (*Tyler et al.,* 2003), but the signals of such behavior are likely prohibitively small for detection with available technology.

5.2. Tidal Effects

As Europa experiences strong tidal forces due to its orbital motion, we should consider tidal flows, both as sources of direct horizontal circulation and as important sources of turbulent energy that can cause the ocean to be strongly stirred.

Moore and Schubert (2000) argue that provided Europa possesses an ocean, it should have a surface tidal amplitude of approximately 30 m. We can estimate the horizontal currents that would arise from this tide by considering a single tidal bulge covering an area of roughly one-fourth of the moon's surface area, or $A = \pi r_e^2 = 7.6 \times 10^{12}$ m^2. This bulge is lifted by h = 60 m over the course of half a europan day: This corresponds to a gain in volume of roughly 4.6 ×

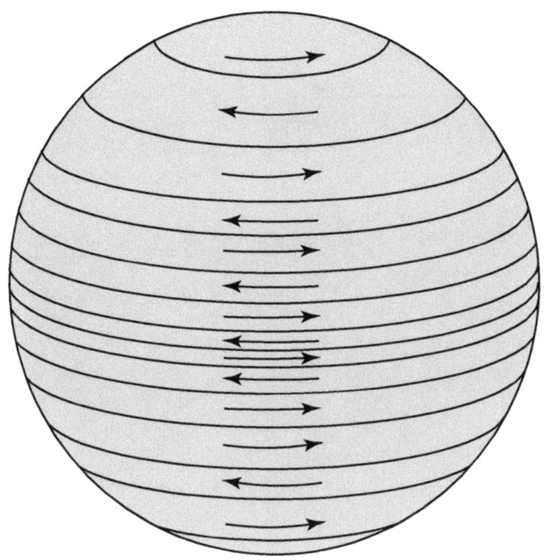

Fig. 4. Speculative schematic of large-scale global circulation in Europa's ocean. In the absence of strong mechanical or global buoyancy forcing, merger of small-scale geostrophic eddies leads to the formation of zonal jets, as in the atmospheres of gas giants. The width of these jets is governed by the Rhines scale (see text); their position and velocity are not constrained. Variations in water layer thickness may lead to meridional deflection of these jets (not shown).

10^{14} m^3. A similar bulge rises on the opposite side of the planet, and the "girdle" in between the bulges loses a compensating amount of volume. This volume change occurs in half a Europa day, or 1.5×10^5 s, implying a volume flux of 3×10^9 m^3 s^{-1}, or 3000 Sverdrups, to use the terrestrial oceanographer's units. The perimeter of the bulge region has a length of $2\pi r_e^2 \cos 30° = 8515$ km: if the ocean is roughly 100 km deep, the tidal volume is flowing through a boundary surface area of 8.5×10^{11} m^2: this implies a typical tidal velocity on the order of 3–4 mm s^{-1}.

A similar calculation performed for Earth, ignoring continents and considering only the M2 tide (r = 6300 km, tidal period 43200 s, tidal amplitude 0.5 m, ocean depth 5 km) predicts typical open-ocean tidal velocities of 1–2 cm s^{-1}, a factor of 5 larger, and very close to typical observed values in the open ocean (*St. Laurent and Garrett,* 2002). Europa's tidal volume flux is similar to Earth's oceans, but since this volume flux is distributed over a very deep ocean, tidal velocities are predicted to be small compared to Earth's.

On Earth, the direct motion of water by tides does little to transport water around the globe. Moving at 1–2 cm s^{-1}, a water mass will only travel a fraction of a kilometer in a single tidal cycle before it reverses course and returns to its starting position. (Europa has slower flow but a longer tidal cycle: this distance-traveled parameter is almost the same for both bodies.) Thus, tidal motions are usually ignored or deliberately filtered out by terrestrial oceanographers interested in global flow patterns. One could argue we should do the same with Europa's tides.

6. OCEAN CONVECTION

In section 6.1, we will consider the fate of hydrothermal fluid once it leaves the sea floor and ascends buoyantly (Fig. 5). But before doing so, we must consider the properties of the fluid into which it ascends.

The ascent of warm fluid from a seafloor source can be halted by either the stratification of the ambient fluid or the presence of a solid boundary. For terrestrial hydrothermal plumes, stratification is the principal impediment: The fluid rises until it is no more buoyant than its surroundings, and then stops. But if a solid boundary impedes the ascent (*Jones and Marshall,* 1993; *Fernando et al.,* 1998), the fluid is forced to spread out against the underside of the "ceiling" rather than at a neutrally buoyant level, and the plume fluid remains positively buoyant. Thus, the presence of a barrier affects both the geometry and the buoyancy of the plume.

Is Europa's liquid layer stratified? Earth's ocean is stratified because it is both heated and cooled at different locations along the upper surface (Fig. 6). Water cooled at the poles slides beneath warm tropical water, forming stable stratification. If the dominant source of buoyancy in Europa's ocean is thermal rather than compositional, the situation is more reminiscent of rising thermal cloud in the atmosphere.

Because of the basal heat input, Europa's ocean should be convectively unstable everywhere, and stable stratification should not occur. As basal heating attempts to place warm water under cold, the warm water rises, mixing turbulently with cold water sinking from above, erasing any vertical temperature gradient. In the inviscid, nonrotating limit, the stratification of a fluid heated from below is zero. Nonzero viscosity or rotation (*Julien et al.,* 1996) can lead to a slightly negative stratification (with dense water overlying light).

The argument above assumes that Europa's ocean is dominated by a buoyancy source (heating) at the base of the fluid, and a buoyancy sink (cooling) at the top. This is true so long as the ocean has a positive coefficient of thermal expansion and a buoyancy dominated by the fluid's temperature rather than its salinity. In section 7 we will consider situations in which one or both of these constraints is broken. For now we will consider the case of an unstratified ocean and discuss the behavior of buoyant hydrothermal plumes in this environment.

6.1. Ocean Convection Based on the Dynamics of Hydrothermal Plumes

The dynamics of convection in Earth's oceans have been considered for two major phenomena: the ascent of buoyant hydrothermal plumes from a seafloor source and the descent of dense surface water, cooled by the atmosphere during wintertime, into the depths (*Goodman et al.,* 2004, and references therein). The dynamics of ascending vs. descending plumes are the same; the key difference between these two phenomena is the size of the buoyancy source.

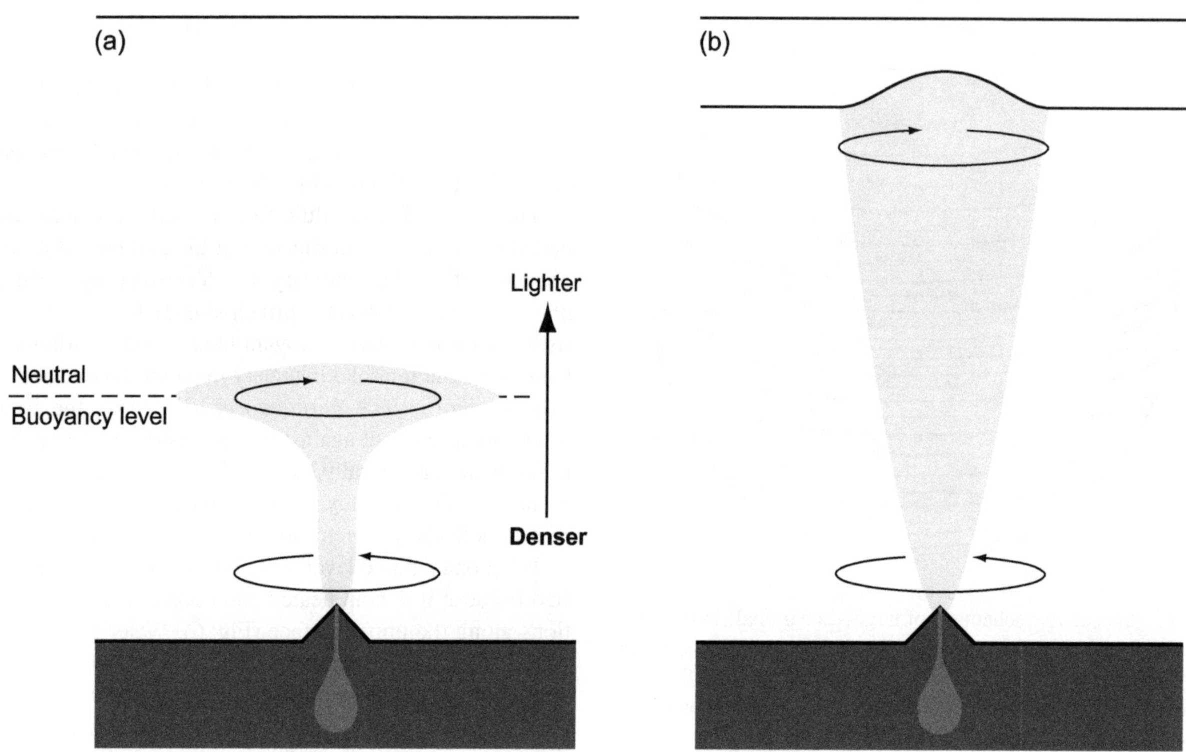

Fig. 5. Qualitative behavior of buoyant plumes from a point source. **(a)** Buoyant plume ascends into a stratified ambient fluid. The plume rises until it reaches a height where it is no longer buoyant relative to its surroundings: at this point, it spreads laterally. Lateral density gradients lead to counter-rotating flow near the plume source and within the "mushroom cap." **(b)** Buoyant plume ascends into an unstratified fluid. The plume is always positively buoyant relative to its surroundings, and ascends until it reaches a solid boundary. Counter-rotating circulation occurs as before.

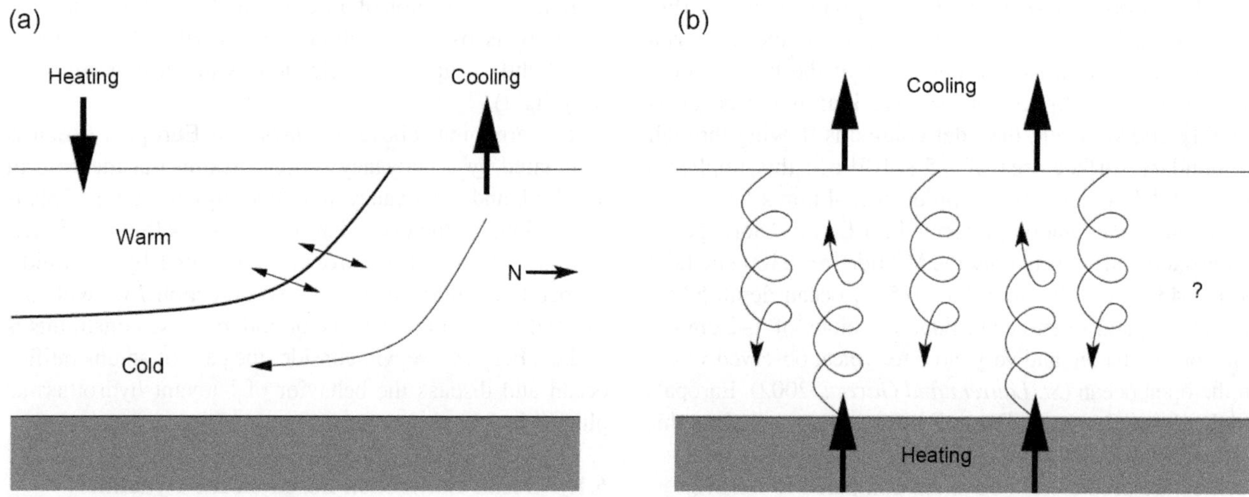

Fig. 6. **(a)** Lateral variation in surface heating/cooling allows cool water to slide beneath warm, causing Earths oceans to become stratified. **(b)** Heating at base, cooling at surface causes instability, turbulent mixing and homogenization of Europa's ocean. Modified from *Goodman et al.* (2004).

In both cases, convective fluid mixes as it rises/falls, forming rotating masses of diluted buoyant fluid whose motion and geometry are controlled by Coriolis forces — those related to the rotation of the planet. Flow becomes more geostrophic (more strongly Coriolis-controlled) at slower velocities (*Gill,* 1982; *Pedlosky,* 1987). The column of plume fluid eventually undergoes baroclinic instability (see section 4.2.4), ejecting swirling blobs of fluid laterally to maintain a steady-state mass balance in the convective zone. The width of these ejected eddies is the Rossby ra-

TABLE 3. Scaling constants for cylindrically confined plumes.

Quantity	Scaling Law	Best-Fit (k)	Best-Fit (k) (*Fernando et al.*, 1998)	Europan Plume Scale
Critical height	$h_c \approx k_h R_o^* H$	4.95	6.0	
Cylindrical plume width	$l_r \approx k_{lr} R_o^* H$	4.8	2.4	
Time to instability	$\tau_{bc} \approx k_\tau (R_o^*)^{-2} f^{-1}$	0.21		
Cone width	$l_{cone} \approx k_{lc} \sqrt{R_o^*} H$	1.79		20–60 km
Drift velocity	$U_{drift} \approx k_{drift} \sqrt{R_o^*} H f$	0.020		3–9 mm s^{-1}

Constants were obtained in tank experiments by *Goodman et al.* (2004). Values from a previous experiment by *Fernando et al.* (1998) are listed for comparison where available. Scales for a hydrothermal plume in Europa's ocean at latitude of Conamara Chaos ($\theta = 10°$) are given for F = 0.1–100 GW.

dius r_D, a scale determined by the ratio of buoyancy forces to the Coriolis effect.

Earth's ocean is stratified: its density increases significantly with depth. In a stratified fluid, a warm hydrothermal plume rises, mixing with its surroundings, until it reaches a neutral buoyancy level, at which its density equals that of the surroundings. At this point, the plume spreads laterally (*Thomson et al.,* 1992; *Speer and Marshall,* 1995), forming a mushroom or anvil shape (Fig. 6).

The plume spreads until it grows wider than the Rossby radius of deformation, r_D, defined as the extent beyond which the baroclinic instability process causes it to break up into smaller eddies (*Helfrich and Battisti,* 1991; *Speer and Marshall,* 1995). These eddies spin away from the plume source. Thus a steady-state plume can be maintained, whose characteristic radius is r_D, which maintains a balance between geothermal heat supply and export via eddy shedding.

A few laboratory experiments and numerical simulations have also been done on convection in an unstratified ambient fluid (*Jones and Marshall,* 1993; *Fernando et al.*, 1998; *Goodman et al.*, 2004). This situation is not generally observed in Earth's oceans, but it may be relevant to europan ocean dynamics.

6.2. Scaling Analysis

The following scaling analysis for cylindrically confined plumes in Europa's ocean, adapted from *Goodman et al.* (2004), demonstrates the application of the material presented so far. Scaling constants are provided in Table 3, along with plume properties based on fluid dynamic measurements (*Fernando et al.*, 1998; *Goodman et al.*, 2004) and constraints on Europa's thermal budget.

At time t = 0, we switch on a point-source of buoyancy, with buoyancy flux

$$B = \frac{g\Delta\rho}{\rho_w}\mu = b\mu \tag{12}$$

where μ is the volume flux of the plume, in units of $m^3 s^{-1}$. B is related to the heat flux from below, F, and b to the temperature anomaly T'

$$B = \frac{g\alpha}{\rho_w C_P}F \tag{13}$$

$$b = g\alpha T' \tag{14}$$

As the plume entrains ambient fluid, its volume flux μ increases while the buoyancy anomaly b declines. However, since B is proportional to the energy flux F, it is the same at every height in the plume.

6.2.1. Initial behavior: Free turbulent convection. In the first few moments after buoyant fluid leaves the source, Coriolis forces caused by planetary rotation are unimportant. Furthermore, the fluid is very far from the upper boundary, and so is unaffected by the finite depth H of the ocean. Thus the buoyancy source B is the only relevant dimensional external parameter (see Fig. 2a). We may form a length scale from B and the time t since the plume began

$$L = (Bt^3)^{-1/4} \tag{15}$$

The plume's current height z above the source, and its width l, are both proportional to this characteristic length scale. Laboratory experiments (*Turner,* 1986) confirm that the plume grows upward and outward in a self-similar fashion, forming a conical plume.

Let us investigate the physical mechanisms that lead to the scaling law (equation (15)). If no other forces act on it, the plume accelerates in response to the buoyancy b; the velocity w and height z of the top surface of the plume are then related to b as

$$\frac{d^2z}{dt^2} = \frac{dw}{dt} = b \tag{16}$$

Now, using equation (12) and noting that the volume flux m equals the cross-sectional area A of the plume head times its average vertical velocity w

$$B = Awb \tag{17}$$

Assuming a conical plume with z ~ L and A ~ L^2, equation (17) becomes (using equation (16))

$$B \sim L^2 \frac{dL}{dt}\frac{d^2L}{dt^2} \tag{18}$$

The scaling law (equation (15)) satisfies this differential equation, with boundary conditions L = 0 at t = 0, w → 0 as t → ∞.

The volume flux μ must be a function of B and z, the only available parameters in the problem. The only dimensionally consistent choice for μ is

$$\mu = k_\mu (Bz^5)^{1/3} \tag{19}$$

where k_μ is an empirically determined constant; $k_\mu \approx 0.15$, according to *List* (1982). This expression may be confirmed by plugging equation (4) into the expression

$$\mu = Aw \sim L^2 \frac{dL}{dt} \tag{20}$$

Equation (15) can be used with equation (19) to find the buoyancy anomaly

$$b = \frac{B}{\mu} \sim (B^2 z^5)^{1/3} \tag{21}$$

This relation for b may be used with equation (14) to find temperatures within the plume.

6.2.2. Influence of rotation: Cylindrical plumes. Once the Europa moon-ocean system has evolved for roughly one rotation period ($t \sim f^{-1} \sim 1$), Coriolis forces become important; both f and B are now important external parameters in the problem. For europan plumes in the energy flux range considered here, one may demonstrate that at $t \sim f^{-1} \sim 1$, the plume's height is still much less than the ocean depth H. At this time, the characteristic length scale for the height and width of the plume (using equation (15)) is $l_{rot}(Bf^{-3})^{1/4}$. Experiments (*Fernando et al.,* 1998) indicate that, as the plume's height and width become larger than l_{rot}, the outward growth of the conical plume ceases. [The plume begins to exhibit "Taylor column" behavior (*Gill,* 1982; *Pedlosky,* 1987). Coriolis forces suppress vertical shear, and the flow changes from fully three-dimensional turbulence to quasi-two-dimensional, rotationally dominated motion. At height h_c, the plume ceases to expand in a cone shape, and begins to ascend as a cylinder of constant width l_r. Assuming zero shear, the rotationally confined plume transmits thermal energy over long distances. As a caution it should be noted that, in Taylor's original experiments, the flow above the moving cylinder continued to separate upstream and flow ceased entirely at a terminal height above the column.] Based on experiments with rotationally confined plumes (*Fernando et al.,* 1998), the relevant scales are

$$h_c \approx 6(Bf^{-3})^{1/4} \pm 20\% \tag{22}$$

$$l_r \approx 2.4(Bf^{-3})^{1/4} \pm 15\% \tag{23}$$

The factor of 2.4 is a correction from the 1.4 given in *Goodman et al.* (2004); the discrepancy comes about due to an error in converting the Coriolis term to frequency for l used by *Fernando et al.* (1998).

These rotationally constrained cylindrical plumes are essentially identical to those found in studies that use a finite-area source of buoyancy (*Jones and Marshall,* 1993; *Maxworthy and Narimousa,* 1994). There, the dilution of plume water by entrainment ceases to change the plume's buoyancy and volume flux above the critical height h_c. Above h_c, $\mu = \mu(z = h_c)$ and $b = b(z = h_c)$

$$\mu_{plume} \approx 0.15(Bh_c^5)^{1/3} = 3.5(B^3 f^{-5})^{1/4} \tag{24}$$

$$\mu_{plume} \approx 6.7(B^3 h^{-5})^{1/3} = 0.30(Bf^5)^{1/4} \tag{25}$$

6.2.3. Natural Rossby number. The cylindrical plume continues to rise until it encounters the upper boundary of the ocean. At this point, the total water depth H enters as a new external parameter, and it becomes possible to define a nondimensional number, the natural Rossby number, from the external parameters B, f, and H

$$h_c/H \sim R_o^* \equiv (Bf^{-3})^{1/4}/H \tag{26}$$

If the natural Rossby number is close to unity ($R_o^* \sim 1$) the plume is controlled by planetary rotation for most of its ascent. If $R_o^* > 1$, the plume reaches the upper boundary before the effects of rotation are felt. Based on the range of possible heat inputs for Europa (as discussed elsewhere in this volume, and by *Goodman et al.,* 2004), and as mentioned in section 4.1, $R_o^* \sim 1$ for hydrothermal plumes on Europa. The scaling laws described above can be recast in terms of R_o^*, H, and f

$$h_c \approx 6R_o^* H \tag{27}$$

$$l_r \approx 2.4R_o^* H \tag{28}$$

$$\mu_{plume} \approx 3.5(R_o^*)^3 H^3 f \tag{29}$$

$$b_{plume} \approx 0.30R_o^* H f^2 \tag{30}$$

6.2.4. Interaction with the upper boundary: Baroclinic cones. When the rising plume encounters the upper surface, it must expand radially outward rather than upward. Experiments by *Fernando et al.* (1998) show that the buoyant fluid spreads laterally over the entire depth, evolving from a cylinder to a straight-sided cone (see Fig. 7c). Coriolis forces create an azimuthal rim current around the boundary of the plume.

The onset of baroclinic instability (Fig. 7d) limits the growth of this cone. Fernando et al. and others find that the plume becomes unstable when its width l_{cone} is of order r_D, the Rossby radius of deformation. At this point, it breaks up into multiple conical eddies.

Different expressions for r_D are appropriate for different fluid density structures. Here, the ambient fluid is unstratified, and the density contrast is a relatively sharp jump between the warm, light water in the plume and the denser ambient fluid. Thus we should use the Rossby radius appropriate for a two-layer fluid

$$r_D = \frac{\sqrt{b_{plume} H}}{f} \tag{31}$$

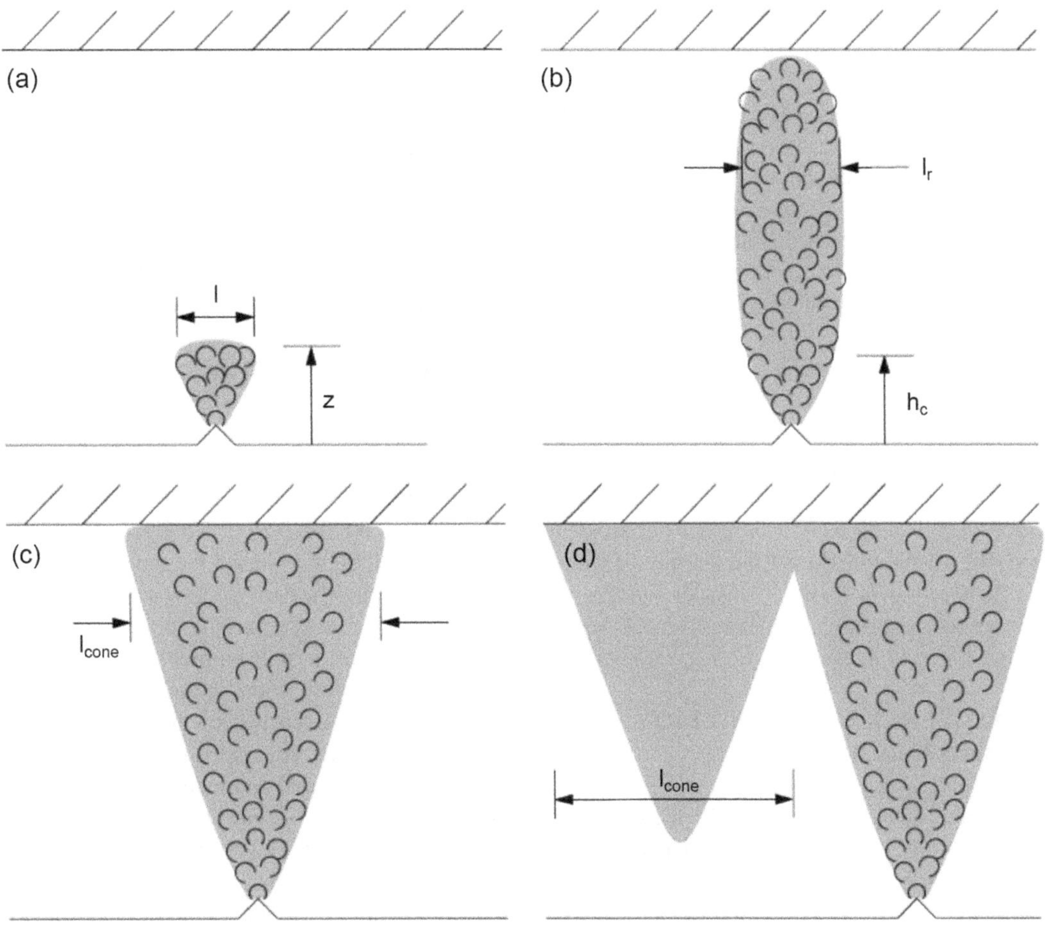

Fig. 7. Stages in the evolution of a buoyant convecting plume. See text for full explanation. **(a)** Free turbulent convection; **(b)** rotationally controlled cylindrical plume; **(c)** baroclinic cone; **(d)** baroclinic instability. From *Goodman et al.* (2004).

where b_{plume} is the buoyancy contrast between the plume and its surroundings (*Pedlosky,* 1987).

While the transition between plume and nonplume fluid is not perfectly sharp, a two-layer treatment is justified, since the density change is substantial, and narrow compared to the ocean depth. Two-layer approximations are quite successful in describing the circulation of Earth's upper ocean, whose density variations are even less sharp than those considered here (*Pedlosky,* 1987).

Recalling that $l_{cone} \sim r_D$ and combining equations (27) and (31)

$$l_{cone} = k_{lc}\sqrt{R_o^*}\,H \tag{32}$$

The time required for the formation of a baroclinic cone is equivalent to the time until baroclinic instability begins. The time to fill a cone is given by the volume of the cone divided by the volume flux into it

$$\tau_{bc} = \frac{V}{\mu} = \frac{\pi}{12}\frac{l_{cone}^2 H}{\mu} = k_\tau (R_o^*)^{-2} f^{-1} \tag{33}$$

The difference in azimuthal velocity between the top and bottom of the cone can be obtained using the thermal wind relation (*Gill,* 1982; *Pedlosky,* 1987). For a two-layer fluid, this relation states

$$U_{top} - U_{bottom} = \frac{b}{f}\frac{d}{dr}h \tag{34}$$

where $\frac{d}{dr}$ is the slope of the interface separating the two layers, measured radially from the center of the plume. In the present case

$$\Delta U \sim \frac{2 b_{plume} H}{f l_{cone}} \approx k_U \sqrt{R_o^*}\,H f \tag{35}$$

This is the difference in the velocities between the two layers. Information is needed on pressure gradients near the surface to compute the actual velocities. However, since fluids likely travel in opposite directions in the two layers (because angular momentum is conserved as the fluid converges at the bottom and diverges near the top), ΔU is the maximum possible velocity in either layer; velocities half this are more likely.

The baroclinic eddies that form during baroclinic instability also have sizes comparable to l_{cone} (*Fernando et al.,* 1998). They are pushed around by currents generated by the convecting plume and by each other, and generally drift away from the source region. The speed U_{drift} at which they move scales with ΔU

$$\Delta U \sim U_{drift} \approx k_{drift}\sqrt{R_o^* H f} \tag{36}$$

Under experimental conditions, the plume maintains its conical shape and diameter l_{cone} after the initial breakup (*Fernando et al.,* 1998). It reaches a steady-state balance, where the accumulation of buoyant fluid in the cone is balanced by the periodic ejection of baroclinic eddies.

6.7. Predicted Scales for Europa's Plumes

All the quantities derived above depend only on the Coriolis parameter f, the water depth H, and the hydrothermal buoyancy flux B. The Coriolis parameter $f = 2\Omega\cos\theta$ is simple to determine. The globally averaged value of $|f|$ is $2\Omega/\pi = 1.3 \times 10^{-5}\ s^{-1}$ for Europa. At the latitude of Conamara Chaos (10°N), $f = 0.71 \times 10^{-5}\ s^{-1}$. Scales for europan ocean plumes based on scaling parameters from *Goodman et al.* (2004) are given in Table 3.

7. INFLUENCE OF SALINITY ON OCEAN CIRCULATION

Europa's ocean is salty, as a result of chemical interaction with the underlying rocky mantle, or earlier water-rock interaction. As on Earth, the presence of salt in Europa's ocean can have important effects on circulation. Lacking strong constraints on the temperatures and oxidation states of the primordial Europa and its ocean (but see *Zolotov and Shock,* 2001; *McKinnon and Zolensky,* 2003, for further discussion), and considering the very few laboratory simulations of chondrite devolitalization, we limit our discussion of saline effects to a single-component ocean of aqueous magnesium sulfate ($MgSO_4$) (see chapter by Zolotov and Kargel for further discussion of ocean composition). We consider the range of possible salinities for this system, from an idealized fresh-water ocean to a eutectic composition of 1.4 ± 0.1 m between zero and –20°C (*Hogenboom et al.,* 1995).

7.1. General Considerations

If Europa's ocean were in a steady-state balance, with uniform heat output everywhere and no net melting or freezing, there would be no saline buoyancy source. But a nonuniform (in space or time) heat output would tend to stratify the ocean, since the salty brine rejected by freezing sinks to the bottom, while the fresh water formed by melting floats at the ice/water interface; this counteracts the tendency of seafloor geothermal heating to remove stratification.

Brine rejection upon freezing represents a negative buoyancy source at the top of the ocean. This is no different from the negative buoyancy that results from cooling as heat is conducted into the ice; it promotes descending turbulently mixing plumes and the removal of stratification. However, melting ice forms a thin layer of fresh water at the ice-water interface. What happens to this layer? Does it lead to a large-scale stratification of the ocean layer?

Since ice effectively contracts as it melts, the buoyant fresh liquid formed by melting a localized patch of ice would be trapped in the melted concavity in the ice. This would prevent lateral outflow of the buoyant melt water, and limit the surface area over which mixing and diffusion can modify the salinity. Thus, only vertical exchanges across the horizontal base of the melt pool need to be considered.

7.2. A Freshwater Ocean

Below a certain threshold of salinity, the anomalous thermal expansion of water leads to a situation in which maximum fluid density is reached at a temperature above the freezing point. An initially buoyant thermal plume, on reaching the height at which the transition occurs, becomes more dense than the cooler ocean water around it and ceases to rise. Pressure suppresses the effect (see Fig. 8) so that it would not occur in an ocean with ceiling pressures higher than about 20 MPa. Thus, in the presence of a moderately thin ice shell, a fresh water europan ocean would have a thermally diffusive "stratosphere" (*Melosh et al.,* 2004), preventing direct thermal and material contact between a convective ocean and the overlying ice.

The departure of the ocean's temperature from its melting point means that the convective portion could be near the temperature of the maximum density rather than that of the melting point. In places where warm ocean water breaks through the stratospheric layer, the difference in temperature between the ocean and ice shell constitutes a reservoir of heat that may contribute to melting or destabilization of the ice shell (see section 8.3 and chapter by Collins and Nimmo).

7.3. Intermediate Ocean Compositions

For moderate ocean salinities — above ~3% for MgSO — or ice overburden pressures exceeding 30 MPa, the ocean behaves as an "ordinary" fluid, with a positive coefficient of thermal expansion. Warm water is buoyant, allowing the type of vertical overturning and top-to-bottom convection discussed in section 6, rather than the "stratosphere" effect of section 7.2.

However, the uptake of a small amount of salt at the seafloor during plume formation can also cause a stagnation effect. In this case, a plume with excess salinity may be initially buoyant, but lose buoyancy as it rises due to changes in ion-ion and ion-water affinities with pressure (Fig. 9) (*Vance and Brown,* 2005).

7.4. A Saturated-Eutectic Ocean

Europa's ocean can acquire saturation salinity if sufficient cooling occurs to extract pure water from the initial solution into the ice shell above. In this case, the degree of

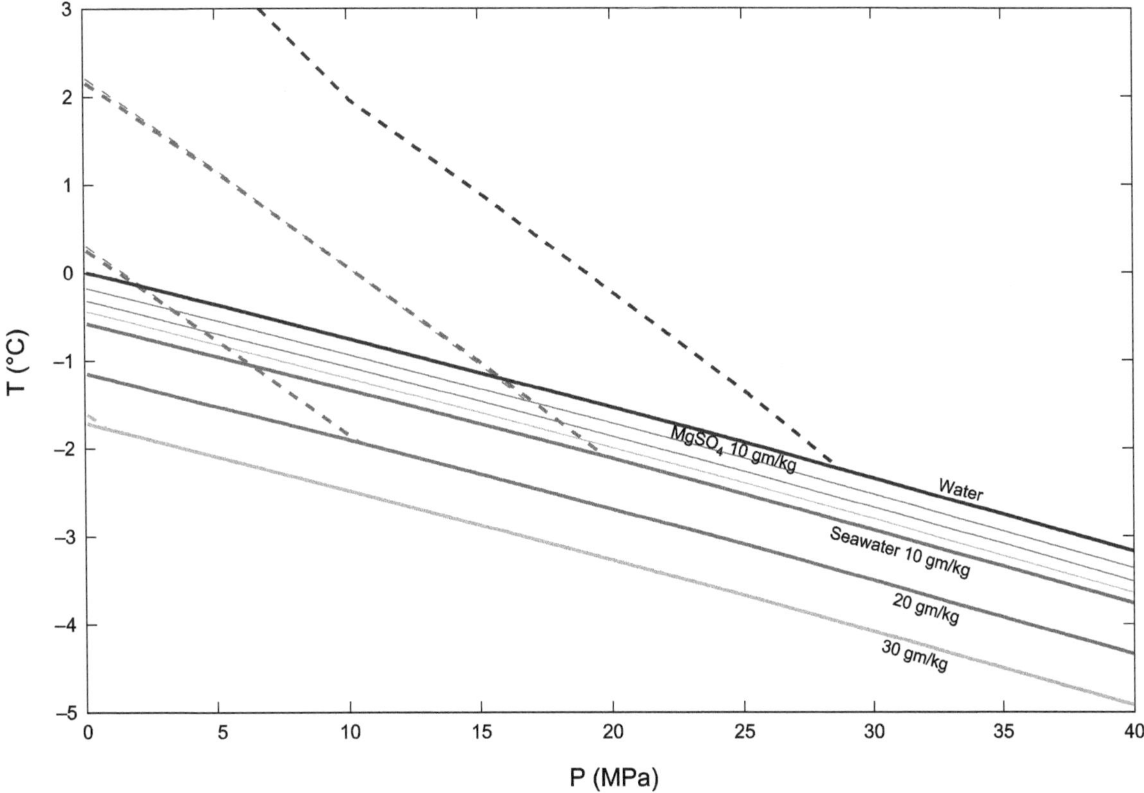

Fig. 8. Minima in the thermal expansion coefficient (dashed lines) for plausible europan ocean compositions, after *Melosh et al.* (2004). For seawater concentrations less than ~0.3 m (~3%) the minimum thermal expansion occurs at temperatures above the melting temperature (solid lines) at modest pressures. Minimum thermal expansion for comparable salinities of $MgSO_4$ is shown for comparison (unlabeled lines indicate 20 and 30 g/kg, downward). The effect on freezing temperature (light dashed line) is nearly identical to that calculated for seawater.

water-rock interaction prior to cooling determines the mineralogical composition of the seafloor. Depending on the initial composition of the proto-ocean, whether it is to the right or the left of the eutectic (minimum solidus temperature) shown in Fig. 10, precipitation upon cooling is salt or ice, respectively. Salt precipitate "snowing" onto the seafloor would affect hydrothermal circulation (as discussed above) and ocean circulation (as discussed below).

For aqueous Mg^{2+} and SO_4^{2+}, and other electrolytes likely to be present in Europa's ocean (Na^+, Ca^{2+}, Cl^-), increasing pressure and temperature with depth depress the eutectic temperature. Thus, if the ocean is well mixed (convecting), the composition can be isohaline but saturated only at the ice-water interface.

A presence of salt precipitate at the seafloor suggests a vertical gradient in salinity as well as temperature, and a tendency toward stratification (see next section).

7.5. Double-Diffusive Convection in Europa's Ocean

Loss of plume buoyancy at a height determined by saline enrichment suggests that Europa's ocean is stratified to some extent. Under certain conditions, linear stratification in both temperature and salinity is consistent with conditions necessary to initiate diffusive regime double-diffusive convection (*Vance and Brown,* 2005).

Diffusive regime double-diffusive convection, as described in section 4.1.2, has been observed on Earth (*Kelley et al.,* 2003) (see below) in regions with aligned upward gradients in temperature and salinity, and in laboratory simulations when $1 \leq R_\rho < 10$ (*Fernando,* 1989). A well-mixed (convecting) lower layer is separated from a well-mixed upper layer, with a step profile in salinity and temperature between them.

For the aqueous $MgSO_4$ system at 271.15 K (*Hogenboom et al.,* 1995), the coefficient of saline contraction is $\beta \sim 0.1\ mol^{-1}$. For a range of difference in potential temperature between the top and bottom of Europa's ocean $T' = 10^{-4}$–10^1 K, the salinity change over the entire ocean to attain $1 \leq R_\rho < 10$ is $\Delta S \sim 10^0$–10^{-6}. Such a ΔS over the entire 100-km ocean depth is 0 to 7 orders of magnitude greater than the 0.5-µm salinity contrast chosen to illustrate the stagnation (Fig. 9).

If double-diffusive convection occurs in an ocean near saturated composition, precipitation from a rising plume would cause salt to precipitate, or "snow," out. Precipitation would enhance the salinity of the ocean below while

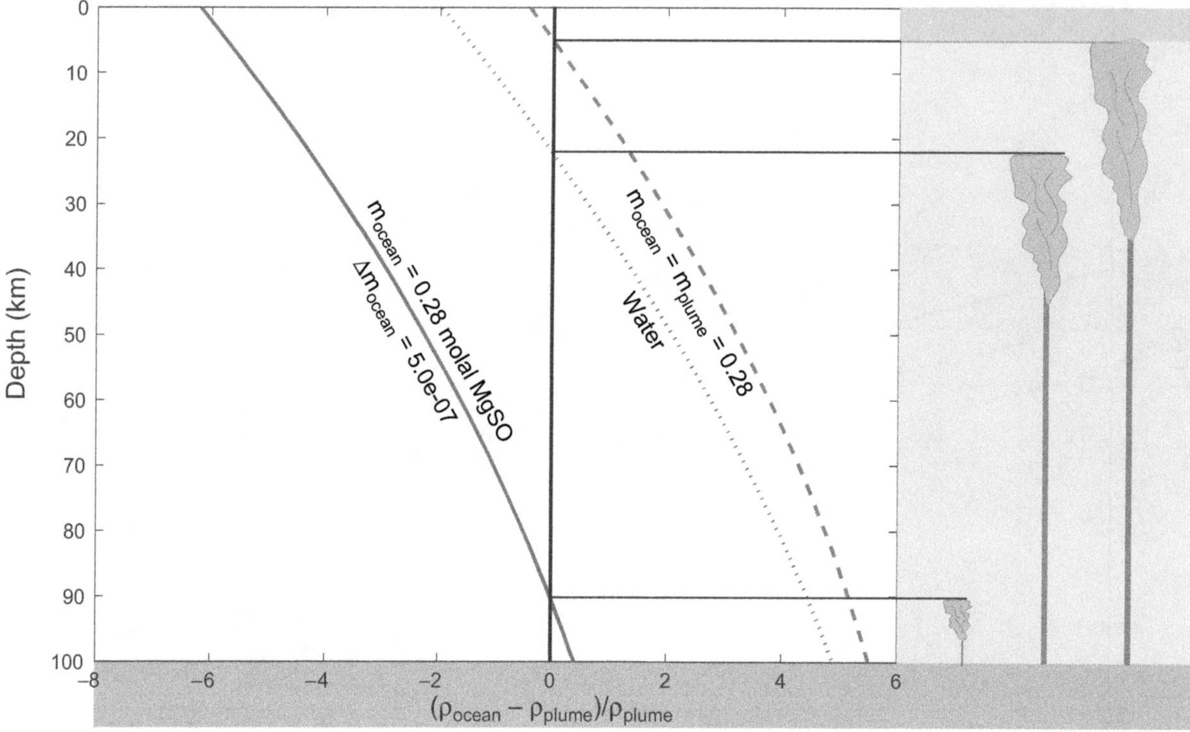

Fig. 9. Density constrasts between a hydrothermal plume in Europa and the ocean surrounding it. Density differences are plotted on a permil basis on the horizontal axis. Depth below Europa's surface, plotted on the y-axis, is calculated from the pressure-and-temperature-dependent weight of overlying fluid. Buoyancy occurs for positive values of B. The image on the right depicts three rotationally confined plumes, with horizontal lines indicating the connection where they intercept the neutral buoyancy line (B = 0). The three plumes are positively buoyant when they form at the seafloor, but cross the vertical line of neutral buoyancy (B = 0) before they reach zero depth. For the pure water ocean and plume (center), the anomalous thermal expansion properties of water create a maximum in density near the freezing temperature. The rightmost plume and ocean contain the minimum addition of magnesium sulfate salt (0.28 m) needed to permit the plume to reach a 5-km-thick ice shell. The leftmost plume depicts the effect on buoyancy from a 0.5 μm addition of salt to the plume such as might be expected from uptake at the seafloor. The stagnation height is inversely proportional to the degree of enrichment in the plume relative to the ocean. From *Vance* (2007).

transferring heat upward rather than keeping it at the level of neutral buoyancy. The resulting disturbance to the relative thermal and saline gradients would increase R_ρ above the range where double-diffusive convection is observed in laboratory experiments.

The existence of stable double-diffusion systems in Earth's waters has been inferred from the step, or staircase, nature in salinity profiles in places such as the Weddel Sea, in the Greenland Sea, below ice island T-3 in the Arctic [all mentioned so far (*Kelley et al.,* 2003)], in ice-covered Lake Vanda in Antarctica (*Hoare,* 1966), and in the Black Sea (*Ozsoy and Unluata,* 1997; *Kelley et al.,* 2003).

In the Black Sea, where temperature and salinity increase with depth and the bottom is heated geothermally, the bottom convecting layer is stable and nearly isothermal, 300–400 m thick, on the order of 1000 years old, and ~0.5 K warmer and 3 wt% more saline than the overlying water (*Ozsoy and Unluata,* 1997). Double diffusive convection is observed directly above (and possibly below) hydrothermal systems in the Red Sea (*Blanc and Anchutz,* 1995; *Anschutz et al.,* 1998). Convective layers in the Red Sea moved up-

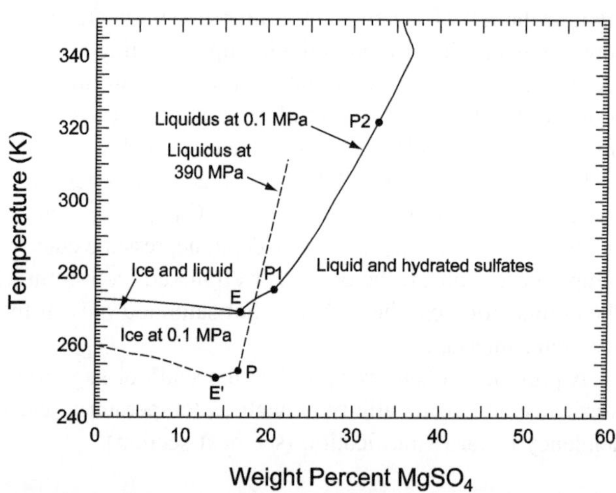

Fig. 10. Phase diagram for aqueous $MgSO_4$. Ice precipitates if the initial composition is to the left of the eutectic (E); hydrated sulfates precipitate if the initial composition is to the right. The dashed line illustrates the effect of elevated pressure on the solidus curve and eutectic (E'). Modified from *Hogenboom et al.* (1995).

ward in the 30 years since they were first observed in the 1960s (*Anschutz et al.,* 1998). No such change has been noted for Lake Vanda. Based on estimates of R_p in areas where salinity and temperature profiles have been measured, *Kelley et al.* (2003) suggest that high-latitude regions on Earth may be susceptible to diffusive regime double-diffusive convection.

Upward progress of convecting layers is not the only possibility. For instance, if cooling from above is very efficient, as would be the case for an ice shell less than a kilometer thick, double-diffusive layers would move downward.

8. OCEAN/ICE-SHELL INTERACTIONS

Thermal evolution of Europa's interior is indicated by the relatively recent occurrence of chaos features (see chapters by Collins and Nimmo and by Nimmo and Manga). Given the above possibilities for a europan ocean that is thermally isolated from the overlying ice by a stratospheric layer, and considering the larger heat capacity of liquid water relative to ice (at 0.1 MPa and 273.15 K, $C_{P,ice}$ = 1960 vs. $C_{P,water}$ = 4220 J K^{-1} kg^{-1}), the ocean represents a thermal mass that is potentially large relative to an overlying ice shell of comparable thickness. Although the ice shell has been completely resurfaced on a timescale approaching 100 m.y. (*Zahnle et al.,* 2007), two factors work against an ocean's ability to melt the ice shell entirely: (1) As the ice thins, it becomes more efficient at dissipating heat through the surface by diffusion. (2) Water has a high enthalpy of fusion (L_{ice} = 334 kJ kg^{-1}), a factor of 80 over the specific heat capacity of the liquid phase. In the context of Europa, this implies that about 80 km of well-mixed ocean at 1° above the melting temperature of the ice would be needed to melt 1 km of ice shell at the freezing temperature. Here, we discuss some possible ocean/ice-shell interactions.

8.1. Pressure-Driven Melting/Freezing at the Ice-Water Interface

In the discussion so far, we have assumed that the base of Europa's ice layer is "flat"; i.e., it is an isobaric surface. However, Europa's surface displays topography of several hundred meters (*Prockter and Schenk,* 2005). If this surface topography is isostatically balanced by variations in shell thickness, the base of Europa's ice shell may vary in depth by several kilometers, with pressure differences of a few megapascals. What effect might this have on ocean circulation?

Let us consider a water parcel, initially in contact with overlying ice, and thus at the local melting point. Suppose the parcel falls adiabatically: Its temperature will change slightly via adiabatic compression. The parcel cools on compression if its pressure and salinity are in the "freshwater" regime described in section 7.2, and warms otherwise. The warming or cooling is relatively modest, ranging from −0.0048 K/MPa for pure water at zero pressure to +0.035 K/MPa for water saturated with NaCl at its eutectic point at 25 km depth (*Fofonoff and Millard,* 1983; *Marion et al.,* 2005).

If the parcel now comes in contact with the ice shell at this new, greater depth, it will no longer be in thermal equilibrium because the melting temperature of water is pressure-dependent, decreasing by about 0.07 K per MPa of pressure (*Fofonoff and Millard,* 1983; *Marion et al.,* 2005), or 0.08 K per km of depth beneath the surface. This is true throughout the range of pressures encountered in Europa's ocean, and the pressure-dependence is only weakly dependent on composition (as illustrated by the solid lines for $MgSO_4$ and seawater in Fig. 8; also note the spacing of the solid and dashed lines in Fig. 10 — they are roughly parallel to the left of the eutectic point, indicating a constant pressure-dependence of melting point). The parcel is warmer than the melting point of the ice it is in contact with. It will thus melt the ice, supplying latent heat of fusion and cooling to equilibrium. What happens to the meltwater? If the surrounding fluid is at all salty, the fresh water will be buoyant, and will flow along the ice/water interface to shallower depth. If the surrounding fluid is fresh, the meltwater will still be buoyant, because it is colder than its surroundings. (Remember that for freshwater near the freezing point at modest pressure, cold water rises.) Either way, as the meltwater ascends to areas of lower pressure, it is colder than the local melting point. Thus, it freezes, accreting to the base of the ice layer; the latent heat of fusion released warms the meltwater parcel toward equilibrium (see Fig. 11).

The overall effects of this process are twofold:

1. "Convective" circulation occurs in the upper ocean completely independent of hydrothermal buoyancy sources

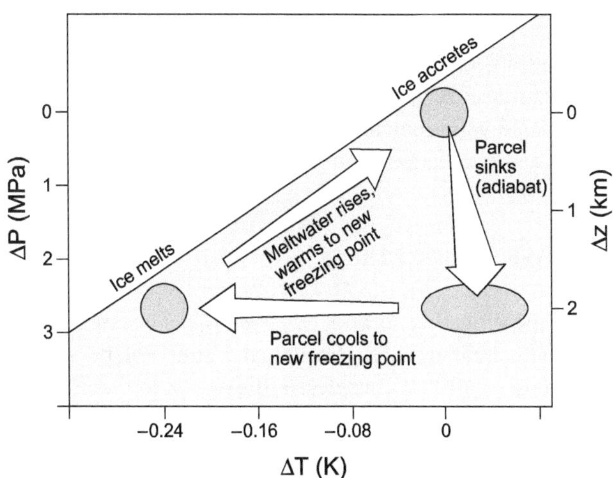

Fig. 11. Pressure-temperature diagram for circulation driven by pressure dependence of the melting point for water ice. ΔP and ΔT represent the excess pressure and temperature, respectively, in a parcel of water just below the ice ceiling. The sloping line represents the variation in melting point as a function of pressure, and can also be interpreted geographically as a sloping ice/water interface. Melting and freezing are caused by variation of the melting point with pressure, leading to an ocean circulation and a tendency to eliminate variations in ice thickness. See text for details.

at the sea floor, and independent of heat flow through the crust. It occurs any time the overlying ice layer has uneven thickness, as a result of the pressure-dependent melting temperature of water.

2. The circulation causes thick regions of the ice shell to melt, and thin regions to thicken by basal accretion. Thus, it tends to even out the thickness of the ice profile, melting away the bumps and filling in the holes.

Terrestrial precedents exist for this sort of flow. Subglacial Lake Vostok in Antarctica is continually accreting ice onto the base of the overlying ice sheet (*Jean-Baptiste et al.,* 2001). However, it is not known whether the pressure dependence of the freezing point is the controlling factor in this case. A clearer example is the large floating ice sheets on the coast of Antartica. Here, seawater above the local melting temperature melts ice on contact with the ice/rock/water junction, eating away at the grounding line of the ice shelf. Meltwater flows up and out away from the grounding line, and freezes as it comes into contact with shallower ice, forming a distinctive layer of bubble-free "blue ice" at the base of the shelf (*Warren et al.,* 1993). (Astrobiologists might be intrigued by the observation of "green ice" as well, which forms when organic materials from seawater are incorporated during this freezing process.)

Pressure-induced melting is not the only process that can normalize between thin and thick ice. Thinner layers conduct heat out of the ocean more rapidly. If basal heating is constant, ice will tend to accrete in the thin spots and melt in the thick spots. Is ice-layer smoothing by pressure-dependence of melting large or small compared to smoothing by thermal conduction? An unknown in answering this question is the velocity and mass flux of the pressure-melting-driven boundary currents in contact with the ice shell, and thus the speed at which water and latent heat of fusion can be transferred from thick spots to thin.

Regardless of their role in reshaping of the ice layer, large-scale variations in ice shell thickness can drive shallow ocean circulation even in the absence of hydrothermal "hot spots."

8.2. Whole-Shell Melting

As mentioned in section 7.5, if Europa's ocean is stably stratified, heat transfer through it might not be steady. *Melosh et al.* (2004) considered this option in their discussion of the freshwater ocean hypothesis (section 7.2). They note that if Europa's mean ocean temperature were raised a few degrees above freezing and its overlying stratosphere destabilized, for example, by intrusion of plumes or by shear (Kelvin-Helmholtz) instabilities, enough heat would be released to (almost) totally melt the overlying ice crust. Based on the balance of basal oceanic heat input against the heat required to melt the ice shell, *Melosh et al.* (2004) found that if overturning and melting are episodic, whole-shell melting events could occur as frequently as every 10 m.y. (assuming a 10-km-thick ice shell and 10 mW m^2 basal heat flux; this timescale is proportionally longer if the heat flux is lower; shorter if the ice shell is thinner). This lower bound is comparable to the estimated surface age of Europa (60–100 m.y.) (*Zahnle et al.,* 2007), leading to the (rather speculative) hypothesis that global ocean overturning might lead to cyclic melting of the whole ice shell on a timescale of tens of millions of years.

8.3. Local Melting

8.3.1. Plume-specific heat transfer. In an unstratified, nonrotating environment, a warm rising plume tends to mix with its surroundings, and its temperature upon reaching the ice/water interface is only a fraction of a millidegree above ambient. In this model, such a tiny temperature difference would have little effect on the overlying ice (*Collins et al.,* 2000).

Accounting for Coriolis effects (as above) (see also *Thomson and Delaney,* 2001; *Goodman et al.,* 2004), the plume turbulently mixes with ambient fluid, but its width is constrained, and it may rise to the ice/water interface. For their choice of source intensity, *Thomson and Delaney* (2001) find plume widths of O(10 km) to O(100 km), in fair agreement with the scales of chaos regions as defined by *Greenberg et al.* (1999). Their calculations suggest that the heat flux per unit area supplied by the plumes is sufficient to melt through the ice layer (assumed to be 25 km thick) in roughly 10^4 years. They note that, given a steady supply of heat, a hydrothermal plume periodically sheds warm baroclinic eddies into its surroundings, and speculate that the satellite lenticulae found near chaos regions may be formed as the warm eddies heat the overlying ice. Finally, they note that ice rafts in Conamara Chaos appear to have drifted in a clockwise direction during chaos formation (*Spaun et al.,* 1998), the expected direction of current flow at the top of a hydrothermal plume at the Chaos' location, and suggest that the blocks were transported by the current. After making an estimate of likely current speeds in the plume, they conclude that the currents could have pushed the blocks into their current orientation if open water existed in the Chaos region for about 22 hours.

8.3.2. Heat flux and melt-through. Are the heat flux produced by a hydrothermal plume sufficient to melt entirely through the ice layer, and if so, how much time is required to do so?

Turbulent mixing of the plume, as discussed here, provides an estimate of the surplus heat flux per unit area applied to the base of the ice layer. The heat flux predicted in section 4.3.3 of *Goodman et al.* (2004) may be used in a simple thermodynamic model of a conducting ice layer to predict the response of the ice layer to these heat flux. For the range of heat flux between 0.1 and 10 W m^2 (F = 0.1–100 GW), a substantial thickness of ice remains unmelted. Conduction and radiation carry away enough heat to bring melting to a halt before melt-through occurs. Equilibrium thickness is roughly inversely proportional to heat flux, ranging from 5.2 km for a 0.1 W m^{-2} flux to 40 m for a 10 W m^{-2} flux (corrected from *Goodman et al.,* 2004).

TABLE 4. Europa's surface temperature (T_s) for assumed basal ice shell heating (F_{base}).

	F_{base} (W m^{-2})	T_s (K)
O'Brien et al. (2002)	32	119
Goodman et al. (2004)	3.3	162

From published ice shell thickness calculations, accounting for Europa's average insolation $\langle F_{solar} \rangle = 8$ W m^{-2}. Neither calculation predicts melting at the surface, even for the very high values of (F_{base}).

To balance thermal radiative emission from the surface to space against heat input via insolation and subsurface heating requires $F_{base} \sim \sigma T^4 - F_{solar}$ with an (open-water) albedo of 0.1; the maximum time-averaged absorbed solar radiation is $F_{solar} = 14$ W m^{-2}. The outgoing thermal radiation for a 273-K blackbody is 315 W m^2. Thus the subsurface heat source must provide $F_{base} = 300$ W m^2 to maintain liquid at the surface. This is far less than the heat flux (0.0110 W m^{-2}) predicted by *Goodman et al.* (2004). The standard cases used by *O'Brien et al.* (2002) (50 and 500 GW over a 200-km-wide patch) are also too small, providing only a maximum of $F_{base} = 3.3$ and 32 W m^{-2} (see Table 4). The balance above does not include latent heat loss by vaporization; this could amount to hundreds of watts per square meter of additional heat loss.

In steady-state balance, then, an ice layer remains. Its thickness can be computed by equating the thermal conduction through the ice layer with the heat source at the base and the heat loss at the surface. *Goodman et al.* (2004) use a one-dimensional vertical diffusive balance, ignoring lateral diffusion, and neglecting latent heat loss from the surface. Ice flow and energy generated by tidal dissipation within the ice shell are also ignored, as *O'Brien et al.* (2002) demonstrate them to be small.

$$k\frac{d}{dz}T = F_{base} = \sigma T^4_{z=0} - F_{solar} \quad (37)$$

$$T = T_f k = b_1/T + b_o \quad (38)$$

where $T_f = 273$ K, $b_1 = 488$ W m^{-1}, and $b_o = 0.468$ W m^{-1} K^{-1} (*Hobbs,* 1974). Solving for surface temperature and ice thickness

$$T_z \equiv T_{z=0} = \left((F_{base} + F_{solar})/\sigma\right)^{1/4} \quad (39)$$

$$h = \frac{b_1 \log(T_f/T_z) + b_o(T_f - T_z)}{F_{base}} \quad (40)$$

Goodman et al. (2004) attribute the zero-thickness result from the model of *O'Brien et al.* (2002) to inadequate spatial resolution. They briefly describe their own model constructed to address this problem, which shows no occurrence of complete melt-through. In their model, minimum shell thickness at the intrusion site is 110 m, with no additional influence found from plausible surface topography of 200 m over an extent of 2 km. If total isostatic adjustment is included, the ice equilibrates to a uniform thin layer with no topography, never thinning to less than 140 m. These two cases (total and zero isostatic adjustment) probably bracket the true behavior of Europa's ice, although tectonic rifting may permit opening in regions where ice becomes sufficiently weak through thinning. Similar results are obtained for a variety of heating intensities and surface elevation profiles, including mesas and deep crevasses. As a general rule, when a model with adequate resolution is used, total melt-through does not occur unless basal heat flux reaches hundreds of watts per square meter.

8.4. A Connection Between Seafloor Plumes and Surface Chaos Features?

The expected equilibrium size of the central plume (45–85 km) is comparable to the size of the Conamara Chaos (75–100 km), and yet much larger than the vast majority of lenticulae or "microchaos" (<15 km; Fig. 12). In the melt-through scenario described by *O'Brien et al.* (2002) (see also *Greenberg et al.,* 1999), the size of the melt-thinned patch always grows rapidly to match the area over which heating is supplied. Since the lenticulae are several times smaller than the pool of warm water produced by a hydrothermal plume, they are probably formed by a different process.

On the other hand, a large hydrothermal plume is the right size to lead to the formation of the entire Conamara Chaos region. Warm eddies will heat the base of the ice out to a distance somewhat larger than l_{cone}, so that the diameter of strong plume heating is comparable to the diameter of the chaos.

One argument against *Goodman et al.*'s (2004) conclusion that lenticulae/microchaoses are too small to be formed by melt-through observes that since heating intensity weakens as one moves away from the axis of the plume, the ice is removed more quickly at the center, and more slowly farther away. Thus the diameter of the melt-through zone increases with time. Perhaps plumes are short-lived, and heat output ceases long before the melt-through diameter exceeds the characteristic width of the plume heating. But such a hypothesis is inconsistent with the observed — albeit incomplete — record of the distribution of microchaos/lenticulae. Figures 7 and 8 of *O'Brien et al.* (2002) show that melt-zone diameter increases very rapidly at first, then levels out as time goes on. Thus only a very narrow range of plume lifetime could produce a small melt-through feature. No melt-through occurs at all below a critical lifetime; 20% longer, and the melt-through diameter is close to the diameter of the heat source. Thus the model of *O'Brien et al.* (2002) implies that small areas of chaotic terrain (<20 km diameter) should be rare. This does not agree with the chaos/lenticulae size distributions measured by *Riley et al.* (2000) or *Spaun and Head* (2001): Features <15 km are the most

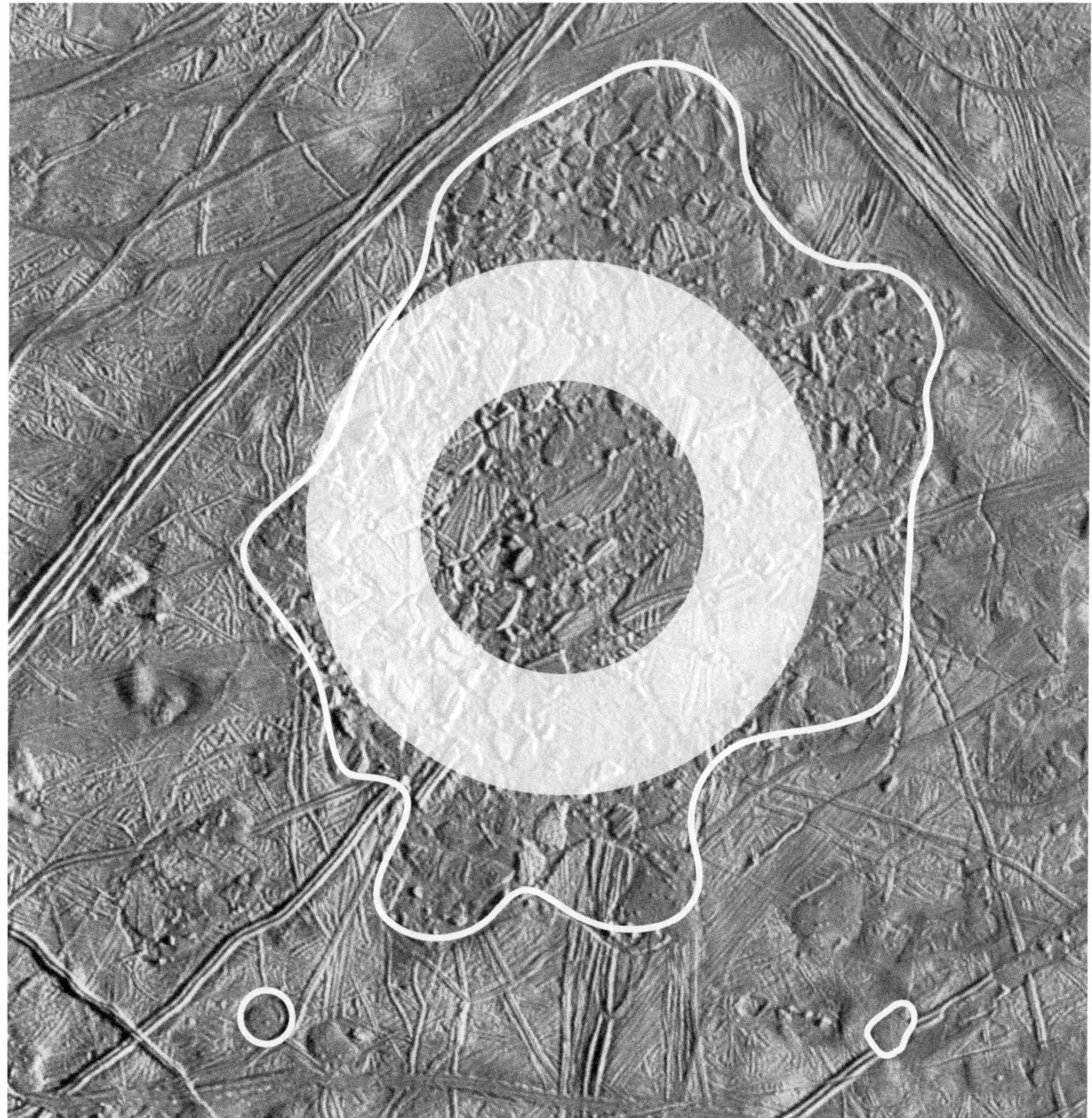

Fig. 12. Size comparison of Conamara Chaos and lenticulae with predicted plume diameter l_{cone}. White outlines show approximate boundaries of Conamara Chaos (large irregular outline at center) and of two representative lenticulae (small round outlines at bottom). Shaded circular zone shows the range of plume diameters (25–50 km) predicted by *Goodman et al.* (2004); the larger overlain region shows the revised range (~45–85 km; section 6.1.1). Base image is from Galileo Orbit E6 imagery. Modified from *Goodman et al.* (2004).

abundant. *Goodman et al.* (2004) demonstrate that the implied size distribution of *O'Brien et al.* (2002) melt-through events is irreconcilably different from the observed distributions. This topic will be of particular interest when high-resolution maps of Europa's entire surface are obtained.

8.4.1. Ice raft drift. *Thomson and Delaney* (2001) reported that a clockwise sense of revolution of the field of rafts in Conamara Chaos (*Spaun et al.,* 1998) is consistent with currents generated by a hydrothermal plume system at the Conamaras location, and argue that 22 hours of ice-free conditions would be sufficient to allow ice rafts to drift into their observed positions as a result of plume-induced currents.

Goodman et al. (2004) caution that uncertainties in the analysis by *Spaun et al.* (1998) make the evidence for circular motion ambiguous. In addition, their much smaller estimate of plume current speeds implies that two weeks to a month are required to move the blocks. Finally, they question whether the area between ice rafts was ever open liquid water exposed to space. They find that even a paper-thin ice layer or a weak ice/water slush between the rafts would be strong enough to prevent them from drifting. In their words, "the predicted stress (~1 Pa) is much less than that exerted by the weight of a cocktail umbrella on the slush in a frozen daiquiri."

8.4.2. Thermal and dynamical stresses. The upward pressure exerted by the plume's buoyancy and its momentum is small, and should be balanced by surface topographic variations of <1 cm (*Thomson and Delaney,* 2001). The plumes discussed here and by *Goodman et al.* (2004) generally have even weaker temperature anomalies, resulting in even smaller values. The direct mechanical effect of the plume's buoyancy on the ice is negligible.

Thermal stresses that would result from warming the ice during a melting event would exceed its brittle strength. However, since the temperature ramps up over a long time (~10^4 yr), stress will develop over a period much longer than the Maxwell time so long as the viscosity of the warmed ice is 10^{20} Pa s or less [assuming rigidity values quoted by *Moore and Schubert* (2000)]. In this case, the thermal expansion can be accommodated by viscous deformation, and no cracking need occur. Effects due to expansion upon thickening of the ice shell may have a role, however, as pointed out by *Manga and Wang* (2007).

8.4.3. Viscous deformation. *Goodman et al.* (2004) argue that melt-through models are unlikely to explain the scales of Europa's small lenticulae, and the motion of ice rafts in Conamara Chaos. Viscous flow of warm, ductile ice beneath the cold, brittle surface is one possible alternative mechanism for chaos formation, which is compatible with hydrothermal plume heating. A small hydrothermal heat source could excite ice diapirism (*Pappalardo et al.,* 1998; *Nimmo and Manga,* 2002; see also chapter by Barr and Showman). A larger heat source could thin the ice sheet through melting; the resulting isostatic adjustment would create a pressure gradient that could push viscous basal ice toward the thin spot [although not with sufficient force to result in explosive cryovolcanism (*Manga and Wang,* 2007)]. This flow might drive ice-raft motion. *O'Brien et al.* (2002) demonstrated that this flow was too slow to counteract melt-through of the ice layer, but perhaps it could transport ice rafts laterally a few kilometers, accounting for the motion observed by *Spaun et al.* (1998). A preliminary calculation suggests that in some cases, ice inflow velocities may exceed 25 cm yr^{-1} at the base of the ice layer.

9. SUMMARY AND FUTURE WORK

Europa's ocean has inaugurated the field of planetary oceanography, providing a unique setting in which to test and extend our understanding of geophysical fluid dynamics.

Examination of feasible properties of Europa's seafloor indicates possibilities for extensive hydrothermalism related to a hydrosphere up to ten times deeper than Earth's, with implications for the production of heat and hydrogen by serpentization, for example, if the microfracture front moves downward into Europa's ultramafic mantle as radiogenic heating diminishes.

The ocean's scale and thermophysical properties suggest a potential for modification of Europa's ice shell by the underlying ocean, although not so much as to easily explain the formation of chaos and other observed surface features. Coriolis (rotational) forces allow hydrothermal plumes to reach the ice ceiling with sizes comparable to observed chaotic features on the moon's surface.

Transport of material and heat from Europa's seafloor to its ice ceiling may be prevented by a stagnant "stratosphere" due to a minimum in water's thermal expansion near the freezing point at low pressures. A similar effect occurs if plumes become enriched in salt by uptake at the seafloor; loss of buoyancy at a height proportional to the amount of uptake suggests a tendency toward stratification, possibly leading to diffusive-regime double diffusive convection.

As noted in the introduction, the study of europan oceanography is necessarily limited, a situation that will hopefully be addressed by future exploration missions. At present, roughly 12% of Europa's surface is mapped by Galileo's cameras at sufficient resolution to identify the distribution of chaos and other features, which remain inadequately characterized. A complete high-resolution map of Europa's surface can reveal more about interior processes related to the ocean. For example, reconstruction of Europa's surface that takes into account nonsynchronous rotation and polar wander of its ice shell can determine whether surface features formed at low-moderate latitudes, consistent with the influence of rotationally confined hydrothermal plumes.

Pending further exploration of Europa, much remains to be determined in the laboratory, on the desktop, and in the field. New measurements of chemistry of putative europan ocean constituents under high pressure and low temperatures can lead to more accurate interpretations of gravity and magnetic field data, and to improved ability to calculate densities and solubility.

Proper treatment of Europa's "ocean climate" — effects from salt precipitation, ice melting and freezing, and turbulence — can be used to assess heat and material transport more fully than has been performed here. Consideration of coupling between the ocean and ice shell and between the ocean and seafloor, including effects from topography, can yield insight info global flow patterns that might be detected remotely.

Analog hydrothermal systems may exist on Earth, among them the Lost City Field near the mid-Atlantic ridge (*Kelley et al.,* 2005) and the sulfur springs of Borup Fjord Pass at Ellesmere Island (*Gleeson et al.,* 2007). Finally, investigation of layered circulation under Arctic and Antarctic ice may yield further insight into the role of similar processes in Europa. In the long term, such comparisons may yield deeper insights into the complex dynamics of Earth's oceans.

Note added in proof: Recent work by *Tyler* (2008) indicates that the dynamic response of Europa's ocean to tides driven by the moon's forced obliquity may give rise to currents four or more orders of magnitude faster than the buoyant plume velocities assumed by *Thomson and Delaney* (2001) and in subsequent efforts by other workers.

Although Europa's ocean should respond rapidly to gravity waves relative to the tidal frequency, *Tyler* (2008) notes that Rossby-Haurwitz waves excited by Coriolis forces are tangentially nondivergent and therefore not mediated by the radial, or up/down, forcing of gravity waves. *Tyler* (2008) obtains an analytical solution to the Laplace tidal equations for the flow induced by the obliquity tide potential, in which flow velocity is proportional to the obliquity θ_o. For the minimum forced obliquity of $\theta_o = 0.1$ (*Bills,* 2005; also see chapter by Bills et al.), the corresponding flow rate is 8.6 cm s^{-1}. Such currents would likely generate magnetic field signatures detectable by an orbiting magnetometer, as has been demonstrated for currents in Earth's ocean (*Tyler et al.,* 2003).

The kinetic energy associated with the obliquity-induced flow is, at a minimum, on the order of 10^{11} W, comparable to current-era radiogenic heating. Quantitatively assessing the dissipation of this energy as heat requires numerous assumptions that cannot be evaluated here, but *Tyler* (2008) suggests that even his conservative minimum assumption surpasses Europa's radiogenic heating value for $\theta_o > 0.16°$.

Acknowledgments. We wish to acknowledge helpful discussions with S. Som and G. Collins. We thank R. Thomson and R. Tyler for their very thorough and helpful reviews. Work by S.V. was funded by a fellowship from the NASA Postdoctoral Program. Research on the part of S.V. required to prepare this publication was carried out at the Jet Propulsion Laboratory, California Institute of Technology, under a contract with the National Aeronautics and Space Administration.

REFERENCES

Anderson J., Schubert G., Jacobson R., Lau E., Moore W., and Sjogren W. (1998) Europa's differentiated internal structure: inferences from four Galileo encounters. *Science, 281(5385),* 2019–2022.

Anschutz P., Turner J. S., and Blanc G. (1998) The development of layering, fluxes through double-diffusive interfaces, and location of hydrothermal sources of brines in the Atlantis II Deep: Red Sea. *J. Geophys. Res.–Oceans, 103(C12),* 27809–27819.

Baker E. T. and German C. R. (2004) On the global distribution of hydrothermal vent fields. In *Mid-Ocean Ridges: Hydrothermal Interactions Between the Lithosphere and Oceans* (C. R. German et al., eds.), AGU Geophysical Monograph Series, Vol. 148, Washington, DC.

Bills B. G. (2005) Free and forced obliquities of the Galilean satellites of Jupiter. *Icarus, 175(1),* 233–247.

Blanc G. and Anchutz P. (1995) New stratification in the hydrothermal brine system of the Atlantis-II Deep, Red Sea. *Geology, 23(6),* 543–546.

Chyba C. F. and Phillips C. B. (2002) Europa as an abode of life. *Origins Life Evol. Biosph., 32(1),* 47–68.

Collins G., Head J. III, Pappalardo R., and Spaun N. (2000) Evaluation of models for the formation of chaotic terrain on Europa. *J. Geophys. Res., 105 E1,* 1709–1716.

Fernando H. J. S. (1989) Oceanographic implications of laboratory experiments on diffusive interfaces. *J. Phys. Ocean., 19(11),* 1707–1715.

Fernando H. J. S., Chen R., and Ayotte B. A. (1998) Development of a point plume in the presence of background rotation. *Phys. Fluids, 10(9),* 2369–2383.

Fisher A. T. and Becker K. (2000) Channelized fluid flow in oceanic crust reconciles heat-flow and permeability data. *Nature, 403(6765),* 71–74.

Fofonoff P. and Millard R. C. (1983) *Algorithms for Computation of Fundamental Properties of Seawater.* UNESCO Technical Report.

Gaidos E. J., Nealson K. H., and Kirschvink J. L. (1999) Biogeochemistry: Life in ice-covered oceans. *Science, 284(5420),* 1631–1633.

Gill A. (1982) *Atmosphere-Ocean Dynamics.* Academic, San Diego.

Gleeson D., Pappalardo R., Grasby S., Templeton A., and Spears J. (2007) Biogeochemical and spectral characterization of a sulfur-rich glacial ecosystem and potential analog to Europa. In *Workshop on Ices, Oceans, and Fire: Satellites of the Outer Solar System,* pp. 46–47. LPI Contribution No. 1357, Lunar and Planetary Institute, Houston.

Goodman J. C., Collins G. C., Marshall J., and Pierrehumbert R. T. (2004) Hydrothermal plume dynamics on Europa: Implications for chaos formation. *J. Geophys. Res.–Planets, 109(E3),* E03008, DOI: 10.1029/2003JE002073.

Greenberg R. and Geissler P. (2002) Europa's dynamic icy crust. *Meteoritics & Planet. Sci., 37(12),* 1685–1710.

Greenberg R., Hoppa G. V., Tufts B. R., Geissler P., and Riley J. (1999) Chaos on Europa. *Icarus, 141(2),* 263–286.

Helfrich K. and Battisti T. (1991) Experiments on baroclinic vortex shedding from hydrothermal plumes. *J. Geophys. Res., 96,* 12511–12518.

Hoare R. A. (1966) Problems of heat transfer in Lake Vanda: A density stratified Antarctic lake. *Nature, 210(5038),* 787.

Hogenboom D. L., Kargel J. S., Ganasan J. P., and Lee L. (1995) Magnesium sulfate-water to 400 MPa using a novel piezometer: Densities, phase equilibria, and planetological implications. *Icarus, 115(2),* 258–277.

Hoppa G. V., Tufts B. R., Greenberg R., and Geissler P. E. (1999) Formation of cycloidal features on Europa. *Science, 285(5435),* 1899–1902.

Hussmann H. and Spohn T. (2004) Thermal-orbital evolution of Io and Europa. *Icarus, 171(2),* 391–410.

Irwin L. N. and Schulze-Makuch D. (2003) Strategy for modeling putative multilevel ecosystems on Europa. *Astrobiology, 3(4),* 813–821.

Jean-Baptiste P., Petit J.-R., Lipenkov V. Y., Raynaud D., and Barkov N. I. (2001) Constraints on hydrothermal processes and water exchange in Lake Vostok from helium isotopes. *Nature, 411,* 460–462.

Jones H. and Marshall J. (1993) Convection with rotation in a neutral ocean: A study of open-ocean deep convection. *J. Phys. Ocean., 23,* 1009–1039.

Julien K., Legg S., McWilliams J., and Werne J. (1996) Rapidly rotating turbluent Rayleigh-Benard convection. *J. Fluid Mech., 322,* 243–273.

Kargel J. S., Kaye J. Z., Head J. W. III, Marion G. M., Sassen R., Crowley J. K., Ballesteros O. P., Grant S. A., and Hogenboom D. L. (2000) Europa's crust and ocean: Origin, composition, and the prospects for life. *Icarus, 148(1),* 226–265.

Kaspi Y. and Flierl G. (2007) Formation of jets by baroclinic instability on gas planet atmospheres. *J. Atmos. Sci., 64(9),* 3177–3194.

Kattenhorn S. (2002) Nonsynchronous rotation evidence and fracture history in the bright plains region, Europa. *Icarus, 157(2),* 490–506.

Kelley D. E., Fernando H. J. S., Gargett A. E., Tanny J., and Ozsoy E. (2003) The diffusive regime of double-diffusive convection. *Progr. Ocean., 56(3-4),* 461–481.

Kelley D. S., Karson J. A., Fruh-Green G. L., Yoerger D. R., Shank T. M., Butterfield D. A., Hayes J. M., Schrenk M. O., Olson E. J., Proskurowski G., Jakuba M., Bradley A., Larson B., Ludwig K., Glickson D., Buckman K., Bradley A. S., Brazelton W. J., Roe K., Elend M. J., Delacour A., Bernasconi S. M., Lilley M. D., Baross J. A., Summons R. T., and Sylva S. P. (2005) A serpentinite-hosted ecosystem: The Lost City hydrothermal field. *Science, 307(5714),* 1428–1434.

List E. J. (1982) Turbulent jets and plumes. *Annu. Rev. Fluid Mech., 14,* 189–212.

Lowell R. P. and DuBose M. (2005) Hydrothermal systems on Europa. *Geophys. Res. Lett., 32(5),* L05202, DOI: 10.1029/2005GL022375.

Manga M. and Wang C.-Y. (2007) Pressurized oceans and the eruption of liquid water on Europa and Enceladus. *Geophys. Res. Lett., 34(L07202),* DOI: 10.1029/2007GL029297.

Marion G. M., Kargel J. S., Catling D. C., and Jakubowski S. D. (2005) Effects of pressure on aqueous chemical equilibria at subzero temperatures with applications to Europa. *Geochim. Cosmochim. Acta, 69(2),* 259–274.

Martin J. and Lowell R. (2000) Precipitation of quartz during high-temperature, fracture controlled hydrothermal upflow at ocean ridges: Equilibrium versus linear kinetics. *J. Geophys. Res.–Solid Earth, 105(B1),* 869–882.

Maxworthy T. and Narimousa S. (1994) Unsteady, turbulent convection into a homogeneous, rotating fluid, with oceanographic applications. *J. Phys. Ocean., 24,* 865–886.

McCollom T. M. (2007) Geochemical constraints on sources of metabolic energy for chemolithoautotrophy in ultramafic-hosted deep-sea hydrothermal systems. *Astrobiology, 7(6),* 933–950.

McKinnon W. B. and Zolensky M. E. (2003) Sulfate content of Europa's ocean and shell: Evolutionary considerations and some geological and astrobiological implications. *Astrobiology, 3(4),* 879–897.

Melosh H. J., Ekholm A. G., Showman A. P., and Lorenz R. D. (2004) The temperature of Europa's subsurface water ocean. *Icarus, 168(2),* 498–502.

Merryfield W. J. (2000) Origin of thermohaline staircases. *J. Phys. Ocean., 30(5),* 1046–1068.

Moore W. B. and Schubert G. (2000) The tidal response of Europa. *Icarus, 147(1),* 317–319.

Nimmo F. and Manga M. (2002) Causes, characteristics and consequences of convective diapirism on Europa. *Geophys. Res. Lett., 29(23),* 2109.

O'Brien D. P., Geissler P., and Greenberg R. (2002) A melt-through model for chaos formation on Europa. *Icarus, 156,* 152–161.

Ozsoy E. and Unluata U. (1997) Oceanography of the Black Sea: A review of some recent results. *Earth Sci. Rev., 42(4),* 231–272.

Pappalardo R. T., Head J. W., Greeley R., Sullivan R. J., Pilcher C., Schubert G., Moore W. B., Carr M. H., Moore J. M., Belton M. J. S., and Goldsby D. L. (1998) Geological evidence for solid-state convection in Europa's ice shell. *Nature, 391(6665),* 365–368.

Pedlosky J. (1987) *Geophysical Fluid Dynamics.* Springer-Verlag, New York.

Prockter L. and Schenk P. (2005) Origin and evolution of Castalia Macula, an anomalous young depression on Europa. *Icarus, 177(2),* 305–326.

Rhines P. (1975) Waves and turbulence on a beta-plane. *J. Fluid Mech., 69,* 417–443.

Riley J., Hoppa G. V., Greenberg R., Tufts B. R., and Geissler P. (2000) Distribution of chaotic terrain on Europa. *J. Geophys. Res.–Planets, 105(E9),* 22599–22615.

Schenk P. M. and McKinnon W. B. (1989) Fault offsets and lateral crustal movement on Europa: Evidence for a mobile ice shell. *Icarus, 79(1),* 75–100.

Spaun N. A. and Head J. W. (2001) A model of Europa's crustal structure: Recent Galileo results and implications for an ocean. *J. Geophys. Res.–Planets, 106(E4),* 7567–7575.

Spaun N. A., Head J. W., Collins G. C., Prockter L. M., and Pappalardo R. T. (1998) Conamara Chaos region, Europa: Reconstruction of mobile polygonal ice blocks. *Geophys. Res. Lett., 25(23),* 4277–4280.

Speer K. G. and Marshall J. (1995) The growth of convective plumes at seafloor hot springs. *J. Mar. Res., 53,* 1025–1057.

St. Laurent L. and Garrett C. (2002) The role of internal tides in mixing the deep ocean. *J. Phys. Ocean., 32(10),* 2882–2899.

Stein C., Stein S., and Pelayo A. (1995) Heat flow and hydrothermal circulation. In *Seafloor Hydrothermal Systems: Physical, Chemical, Biologic and Geological Interactions* (S. E. Humpris et al., eds.), pp. 425–445. AGU Geophysical Monograph Series, Vol. 91, Washington, DC.

Thomson R. E. and Delaney J. R. (2001) Evidence for a weakly stratified europan ocean sustained by seafloor heat flux. *J. Geophys. Res.–Planets, 106(E6),* 12355–12365.

Thomson R. E., Delaney J. R., McDuff R. E., Janecky D. R., and McClain J. S. (1992) Physical characteristics of the Endeavor Ridge hydrothermal plume during July 1988. *Earth Planet. Sci. Lett., 111,* 141–154.

Tritton D. J. (1988) *Physical Fluid Dynamics.* Oxford Univ., New York.

Turcotte D. L. and Schubert G. (2001) *Geodynamics.* Cambridge Univ., Cambridge.

Turner J. S. (1986) Turbulent entrainment: The development of the entrainment assumption, and its application to geophysical flows. *J. Fluid Mech., 173,* 431–471.

Turner J. S. and Chen C. F. (1973) Layer generation in double-diffusive systems. *Bull. Am. Phys. Soc., 18(11),* 1467–1467.

Turtle E. P., Jaeger W. L., Keszthelyi L. P., McEwen A. S., Milazzo M., Moore J., Phillips C. B., Radebaugh J., Simonelli D., Chuang F., and Schuster P. (2001) Mountains on Io: High-resolution Galileo observations, initial interpretations, and formation models. *J. Geophys. Res.–Planets, 106(E12),* 33175–33199.

Tyler R. H. (2008) Strong ocean tidal flow and heating on moons of the outer planets. *Nature, 456,* DOI: 10.1038/nature07571.

Tyler R. H., Maus S., and Luhr H. (2003) Satellite observations of magnetic fields due to ocean tidal flow. *Science, 299(5604),* 239–241.

Vance S. (2007) High pressure and low temperature equations of state for aqueous magnesium sulfate: Applications to the search for life in extraterrestrial oceans, with particular reference to Europa. Ph.D. thesis, University of Washington.

Vance S. and Brown J. (2005) Layering and double-diffusion style convection in Europa's ocean. *Icarus, 177(2),* 506–514.

Vance S., Harnmeijer J., Kimura J., Hussmann H., deMartin B., and Brown J. M. (2007) Hydrothermal systems in small ocean planets. *Astrobiology, 7(6),* 987–1005.

Warren S., Brandt R., and Boime R. (1993) Blue ice and green ice. *Antarct. J. U.S., 28,* 255–256.

Zahnle K., Alvarellos J. L., Dobrovolskis A., and Hamill P. (2007) Secondary and sesquinary craters on Europa. *Icarus, 194(2),* 660–674. DOI: 10.1016/j.icarus.2007.10.024.

Zolotov M. and Shock E. (2001) Composition and stability of salts on the surface of Europa and their oceanic origin. *J. Geophys. Res., 106(E12),* 32815–32828.

Zolotov M. Y. and Shock E. L. (2004) A model for low-temperature biogeochemistry of sulfur, carbon, and iron on Europa. *J. Geophys. Res.–Planets, 109(E6),* E06003.

Part IV:

External Environment

Observations of Europa's Tenuous Atmosphere

M. A. McGrath
NASA Marshall Space Flight Center

C. J. Hansen and A. R. Hendrix
NASA Jet Propulsion Laboratory/California Institute of Technology

Europa is known to possess a predominantly molecular oxygen atmosphere produced by sputtering of its icy surface. This atmosphere, which is diagnostic of surface composition and processes, has been characterized by the Hubble Space Telescope, Galileo, Cassini, and ground-based observations. The primary means of detecting Europa's atmosphere is via emission from the atomic constituents. The relative strengths of the atomic oxygen emission lines allow inference of a dominant O_2 component. Oxygen, sodium, and potassium are the minor constituents detected to date. An ionosphere has also been detected on several occasions by Galileo radio occultation measurements, the presence of which appears to require a sunlit, plasma bombarded (trailing) hemisphere. Neither the spatial distribution of the oxygen emission associated with the atmosphere, nor the obvious variability of the atmospheric emissions and the ionospheric densities, has been adequately explained to date.

1. INTRODUCTION

Europa is one of a growing cadre of solar system objects that possess tenuous atmospheres. Their discovery — at Mercury, the Moon, Io, Europa, Ganymede, Callisto, Enceladus, Triton, and Pluto — has become common in recent years because of the increasing sophistication of remote sensing and *in situ* observing techniques. They are produced by a wide variety of physical processes, including sublimation, sputtering (by both photons and charged particles), micrometeroid bombardment, geysers, and volcanos. This class of atmosphere is important because of the often unique information these atmospheres can provide about the surrounding environment (which also has implications for understanding the magnetospheres of the parent planets), surface processes, and therefore, potentially, interior and surface compositions of these bodies. Although tenuous, these exospheres also produce measurable ionospheres with peak densities of ~10^3–10^4 cm^{-3}, which were one of the first definite indications of the presence of atmospheres associated with these bodies (e.g., *Kliore et al.,* 1974, 1975).

The idea that the Galilean satellites might possess tenuous atmospheres, and that satellites with atmospheres could possess neutral tori, began to be explored in earnest in the early 1970s with a series of important milestones. Water ice was positively identified on the surfaces of Europa, Ganymede, and Callisto by *Pilcher et al.* (1972). An approximately 1-μbar atmosphere was reported on Ganymede from a stellar occultation measurement (*Carlson et al.,* 1973). *McDonough and Brice* (1973) suggested that, as a consequence of their atmospheres, there might be neutral clouds of orbiting atoms associated with Titan and the Galilean satellites because most material removed from the atmospheres does not attain escape velocity from the planet. Detection of hydrogen (H) (*Judge and Carlson,* 1974), sodium (Na) (*Brown,* 1974), and potassium (K) (*Trafton,* 1975) clouds associated with Io seemed to confirm these early expectations. Pioneer 10 observations of Europa made in 1973 using the long wavelength channel (λ < 1400 Å) of the ultraviolet (UV) photometer were first reported as a null result (*Judge et al.,* 1976), but subsequently reported as a detection of oxygen (O) at Europa using the short wavelength (λ < 800 Å) UV photometer channel (*Wu et al.,* 1978). Around this time the Io plasma environment also began to be characterized (*Frank et al.,* 1976; *Kupo et al.,* 1976). The realization that Europa, like Io, is immersed in and impacted by a high flux of particles that could dissociate and sputter water ice (*Brown et al.,* 1977) made the possibility of a tenuous atmosphere and extended neutral clouds associated with Europa seem much more plausible.

The *Carlson et al.* (1973) detection motivated *Yung and McElroy* (1977) to develop a photochemical model of a sublimation-driven water ice atmosphere. Because the H preferentially escapes, such an atmosphere evolves into a stable molecular oxygen atmosphere by photolysis of H_2O. In their model, nonthermal escape of O atoms balances the production of O_2 to yield a surface pressure of ~1 μbar, consistent with the *Carlson et al.* (1973) result. They concluded that Ganymede should have an appreciable oxygen atmosphere and ionosphere, but that the higher albedo of Europa would inhibit sublimation, suppressing the formation of O_2. Based on laboratory data, *Lanzerotti et al.* (1978) suggested that bombardment of the satellite surfaces by the jovian plasma leads to an erosion rate on Ganymede that could support the H_2O partial pressure used by *Yung and McElroy* (1977), and that the rates would be much larger at Europa. Subsequent laboratory data showed that O_2 is directly produced in and ejected from water ice (*Brown et al.*, 1980), a process referred to as radiolysis. When the Voyager 1 Ultraviolet Spectrometer stellar occultation measurements of Ganymede yielded an upper limit on surface pressure of 10^{-5} μbar (*Broadfoot et al.*, 1979), *Kumar* (1982) pointed out that the *Yung and McElroy* (1977) model possessed an additional stable solution with a much lower surface pressure of ~10^{-6} μbar, compatible with the Voyager 1 result.

Finally, using the laboratory sputtering data of *Brown et al.* (1980), *Johnson et al.* (1982) estimated that O_2 sputtered from water ice on Europa could yield a bound atmosphere with a column density of ~(2–3) × 10^{15} cm^{-2}. Since O_2 does not stick efficiently at Europa temperatures and does not escape efficiently, the atmosphere is dominated by O_2 even though the sputtered flux of H_2O molecules is larger than that of O_2. This early work set the stage for the successful detection of oxygen at Europa, which we describe in detail below.

Generally speaking, Europa's tenuous atmosphere is produced by radiolysis and sputtering of its icy surface, with a minor contribution from sublimation of water ice near the subsolar point. As pointed out by *Johnson* (2002), Europa is an example of a surface bounded atmosphere. Once products are liberated from the surface, the diatomic species (primarily O_2, but to a lesser extent H_2) are expected to become the dominant ones because they are noncondensable, neither sticking nor reacting efficiently with the surface and therefore accumulating in the atmosphere, but becoming thermalized through repeated surface encounters. The other water group species are lost either by direct escape because they are light, or by sticking to or reacting with the surface. A summary of the observations acquired to date pertinent to Europa's atmosphere is provided in Table 1. Figure 1 illustrates schematically the species detected, and the range of altitudes over which the detections have been made, where 1 Europa radius (R_E) is 1569 km. It is important to note that the various observations span quite different regimes of the atmosphere, neutral clouds, and torus. In particular, observations of the main constituent of the atmosphere, O_2, and observations of the minor species, Na and K, have no overlap. A rough boundary between the bound atmosphere and neutral clouds is the radius of the Hill sphere of Europa, which occurs at ~8.7 R_E. Inside this boundary, Europa's gravitational field dominates Jupiter's; outside this boundary the opposite is true. We discuss the various observations roughly in order of increasing distance from Europa's surface, considering first the bound O_2 atmosphere, which has a scale height of roughly ~100 km.

We concentrate in this chapter on providing detailed descriptions of the observations made to date of Europa's tenuous atmosphere. The subsequent chapter by Johnson et al. provides a complementary review of the detailed interpretations and modeling these observations have spawned. It includes an in-depth discussion of the wide range of physical characteristics that can be gleaned about Europa and its environment from these observations. We focus here on providing a clear explanation of the quality and limitations of the existing observations and data, and on providing descriptions of the simple models presented in the observational papers required to derive basic quantities such as column densities from the observations. Where possible we provide further details about and new presentations of previously published data. Finally, we summarize outstanding issues, and provide recommendations for future observations that may be helpful in resolving them.

2. OXYGEN ATMOSPHERE

The major constituent of Europa's tenuous atmosphere is now known to be molecular oxygen, which has been inferred via detection of UV line emission from atomic oxygen. Emission from neutral atmosphere constituents are overwhelmed by reflected sunlight at visible wavelengths, whereas at UV wavelengths most planetary bodies have albedos more than an order of magnitude lower than in the visible. Ten sets of ultraviolet observations of Europa now exist, seven made with the Earth-orbiting Hubble Space Telescope (HST), two made with the Cassini spacecraft, and one made with the New Horizons spacecraft. The specifics of the observations are summarized in Table 1. The HST and New Horizons observations acquired in February 2007 are not yet published but preliminary results have been presented by *Retherford et al.* (2007). The HST observations made in April 2007 and June 2008 are not yet published, and no results have been presented to date.

2.1. Hubble Space Telescope Observations

The first unambiguous detection of Europa's atmosphere was made by *Hall et al.* (1995) in June 1994 using HST's Goddard High Resolution Spectrograph (GHRS) with a 1.74" × 1.74" slit centered approximately on Europa's trailing hemisphere. They discovered emission from the semi-forbidden and optically allowed oxygen multiplets O I ($^5S^o$–^3P)1356 Å and O I ($^3S^o$–^3P)1304 Å (bottom panel of Fig. 2). In these and subsequent observations of the UV oxygen emission multiplets there are several possible contributors to the observed emissions:

(1) solar resonance fluorescence scattering by O atoms in the Earth's atmosphere (HST only)

(2) solar emission line and continuum photons reflected from the surface of Europa

(3) solar resonance fluorescence scattering by O atoms in Europa's atmosphere

(4) electron impact excitation of oxygen, $e^- + O \rightarrow O^* \rightarrow O + \gamma$

(5) electron impact dissociative excitation of O_2, $e^- + O_2 \rightarrow O^* \rightarrow O + \gamma$.

TABLE 1. Summary of Europa atmosphere observations.

Obs#	Date	Facility, Instrument	Range to Europa	Subobservation W Longitude, Orbital Phase	λ_{III}	Start (UT) Duration	Integration Time	Reference
Oxygen								
1	2 Jun 1994	HST GHRS	4.57 AU	283–318	156–53	09:52; 8 h 2 min	7290 s	*Hall et al.* (1995)
2	21 Jul 1996	HST GHRS	4.22 AU	85–108	317–139	08:43; 5.5 h	7398 s	*Hall et al.* (1998)
3	30 Jul 1996	HST GHRS	4.27 AU	264–287	280–99	04:59; 5.5 h	5549 s	*Hall et al.* (1998)
4	5 Oct 1999	HST STIS	4.01 AU	245–273	304–156	08:39; 7 h	9360 s	*McGrath et al.* (2004)
5	6 Jan 2001	Cassini UVIS	11.2×10^6 km	201–220, 290–311	159–277	7:30; 4.7 h	17000 s	*Hansen et al.* (2005)
6	12 Jan 2001	Cassini UVIS	15.8×10^6 km	64–103, 173–223	62–180–67	6:30; 11.4 h	41000 s	*Hansen et al.* (2005)
7	26 Feb–3 Mar 2007	New Horizons Alice	46.25–67.08 R_J	N/A	N/A	19:17; several days	6540 s	*Retherford et al.* (2007)
8	27 Feb 2007	HST ACS F125LP	5.42 AU	347–350 (eclipse)	94	12:03; 42 min	2400 s	*Retherford et al.* (2007)
9	18 Apr 2007	HST ACS PR130L	4.66 AU	343–352	296	04:34; 2.33 h	5200 s	unpublished; HST program #11085, P.I. W. Sparks
10	29 Jun 2008	HST ACS PR130L	4.18 AU	74–104	23–252	04:13; 7 h 9 min	12996 s	unpublished; HST program #11186, P.I. J. Saur
Ionosphere (radio occultations; N = entry, X = exit)								
11	19 Dec 1996	Galileo S band radio	1650 km, E4N 4070 km, E4X	345, ~255 165, ~255	109	7:37 7:49	Not available	*Kliore et al.* (1997, 2001, 2006)
12	20 Feb 1997	Galileo S band radio	1490 km, E6aN 4388 km, E6aX	280, ~192 101, ~192	298	17:56 18:08	Not available	*Kliore et al.* (1997, 2001, 2006)
13	25 Feb 1997	Galileo S band radio	2776996 km, E6bNX 2776180 km, E6b	55, ~335 236, ~335	99	15:18 15:24	Not available	*Kliore et al.* (1997, 2001, 2006)
14	1 Feb 1999	Galileo S band radio	Not available, E19N Not available, E19X	Not available, ~144	240	~2:13	Not available	*Kliore et al.* (2006)
15	3 Jan 2000	Galileo S band radio	2531 km, E26N 1096 km, E26X	122, not available 329, not available	Not available	Not available	Not available	*Kliore et al.* (2001, 2006)
Minor Species, Neutral Clouds, Torus								
16	5 Jun 1995	Mt. Bigelow echelle	4.32 AU	91–113	71–247	4:11; 5.5 h	4.33 h	*Brown and Hill* (1996)
17	9 Sep 1998	Keck HIRES	3.97 AU	243–248	134–175	8:01; 1.27 h	1800 s each at 3 positions	*Brown* (2001)
18	28 Dec 1999	Keck HIRES	4.55 AU	98–114	332–93	4:28; 3.5 h	6 sets of scans	*Leblanc et al.* (2002, 2005)
19	28 Nov 2000	KPNO McMath	4.05 AU	285–288	253–28	3:14; ~1 h	1200 s	*Leblanc et al.* (2005)
20	29 Nov 2000	KPNO McMath	4.05 AU	42–46	59–91	6:53; ~1 h	3600 s	*Leblanc et al.* (2005)
21	30 Nov 2000	KPNO McMath	4.05 AU	125–130	333–7	2:40; 1 h	3600 s	*Leblanc et al.* (2005)
22	10 Jan 2001	Cassini ISS	~200 R_J	253 (in eclipse)	95	10:36; >60 m	15 3.2-s exposures	*Porco et al.* (2003) *Cassidy et al.* (2008)
23	1 Jan 2001	Cassini INCA	~140 R_J	N/A	N/A	22:00; 15 h	15 h	*Mauk et al.* (2003)
24	12–13 Feb 2001	Cassini UVIS	$(4.21–4.32) \times 10^7$ km	N/A	N/A	6:08; 29 h 4 min	28 h	*Hansen et al.* (2005)

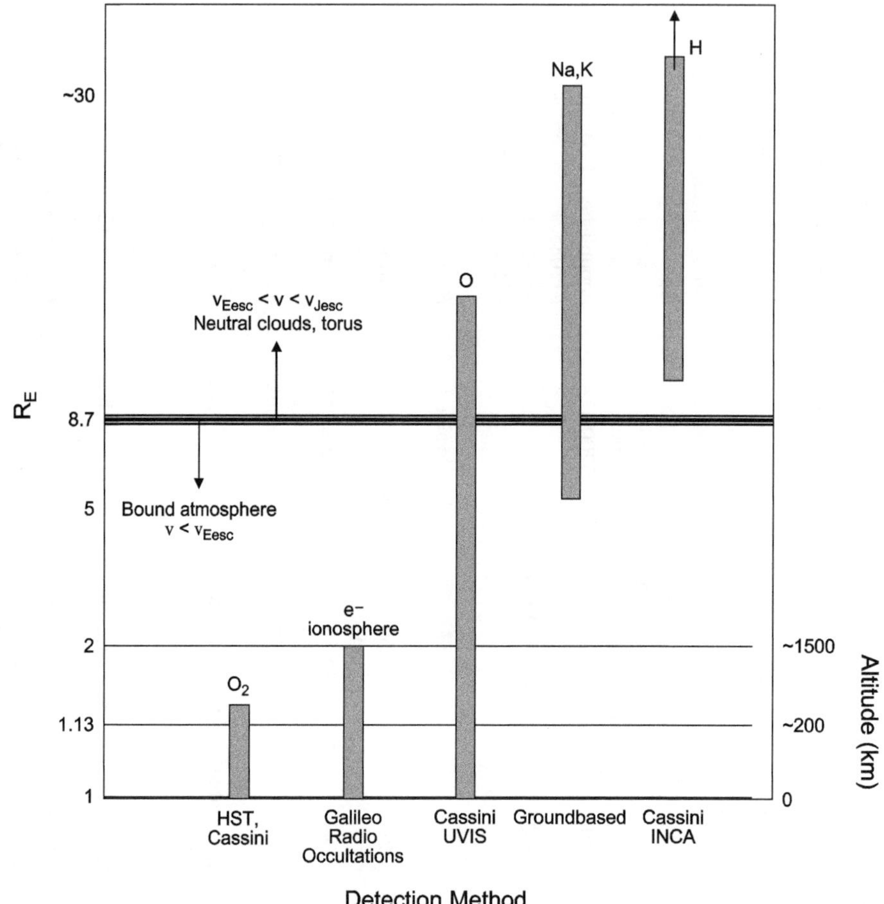

Fig. 1. Schematic diagram showing the constituents detected in Europa's tenuous atmosphere and the regions in which they have been observed. Generally speaking, the bound atmosphere is defined by the region within which Europa's gravity dominates, at $r \leq 8.7\ R_E$ ($1\ R_E = 1569$ km), where the velocity of the constituents is less than Europa's escape velocity (v_{Eesc}) of ~2 km/s. Beyond this region, particles either escape from Jupiter altogether ($v > v_{Jesc}$), or enter orbit around Jupiter as clouds or a torus near Europa's orbital distance.

The Hall et al. observations were performed in Earth shadow, where the contribution from process (1) is minimal; such background emission would fill the slit, and produce flat-topped emission line profiles, which was not observed. Figure 2 shows that in addition to the oxygen emissions, emission from C II 1335 Å is also detected; it is produced by process (2). Although the reflectivity with wavelength is unknown in the UV for Europa, if it is assumed to be constant with wavelength, the C II emission can be used to estimate the contribution to the oxygen emissions from process (2). Modeling the C II emission as reflected sunlight allows a derivation of the geometric albedo of Europa at this wavelength, which was found to be 1.6 ± 0.5% (*Hall et al.,* 1998). The contribution to the spectrum from process (2) is shown in Fig. 2 as a dark solid line underlying the observed spectrum. It is not a significant contributor to the O I 1356 Å emission line, which is a semiforbidden transition, because the Sun does not produce line emission at this wavelength. *Hall et al.* (1995, 1998) showed that process (3) produces a negligibly small contribution to the oxygen emissions at Europa.

The inference that process (5) dominates over process (4), and that Europa's atmosphere is predominantly O_2 and not O, is based on the ratio of the 1356 Å to 1304 Å emission intensities, I(1356)/I(1304). For the June 1994 observations, this ratio is ~1.9:1, after accounting for the contribution from processes (2) and (3). For a Maxwellian distribution of electrons over a broad temperature range this intensity ratio was found to be 2 for process (5) by *Noren et al.* (2001). By contrast, using the O I 1304 Å cross section of *Doering and Yang* (2001), the I(1356)/I(1304) ratio for process (4) has a broad maximum of 0.35 at 4 eV. A molecular oxygen atmosphere with column density 1.5 ± 10^{15} cm^{-2} ($P_o = 2.2 \pm 0.7 \times 10^{-6}$ μbar) was therefore inferred for the trailing hemisphere of Europa by *Hall et al.* (1995), which is consistent with the early estimate of a bound atmosphere by *Johnson et al.* (1982) and the low-pressure solution discussed by *Kumar* (1982). In deriving the O_2 column den-

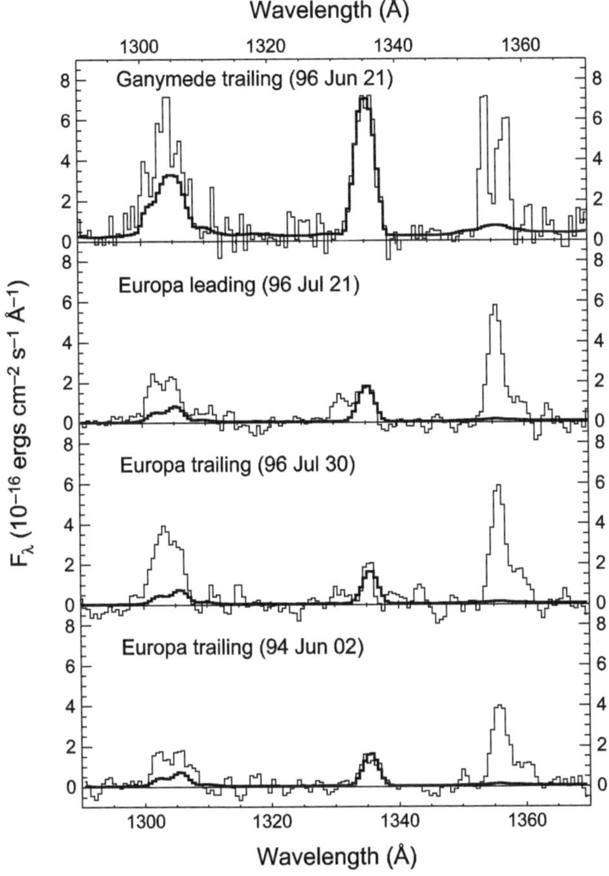

Fig. 2. The *Hall et al.* (1995, 1998) detections of electron excited oxygen emission at 1304 and 1356 Å from Europa, which provided the first direct evidence of an O_2 atmosphere on this satellite. The features at 1335 Å are due to solar C^+ emission reflected from the surface of the satellite. The dark line histogram shows a modeled reflected light spectrum that assumes the albedo is constant with wavelength, normalized by matching the 1335 Å feature. From *Hall et al.* (1998).

sity, *Hall et al.* (1995) assumed that the spatial distribution of Europa's atmosphere is confined to the geometric cross section of the observed hemisphere (i.e., the scale height of the atmosphere is significantly smaller than the radius of Europa); a negligible contribution to the observed flux is emitted from above the tangential limb along the terminator; the Io plasma torus electrons responsible for exciting the observed emissions interact with the atmosphere without energy degradation; and no electrodynamic, sub-Alfvenic interactions such as observed by Voyager at Io (e.g., *Ness et al.,* 1979, *Neubauer,* 1980) were considered.

Two additional sets of GHRS observations were obtained in July 1996: one from the leading hemisphere, and an additional set from the trailing hemisphere (*Hall et al.,* 1998). All three of the GHRS spectra are shown in Fig. 2. With a finite scale height and emission above the limb included, the inferred molecular oxygen column densities are in the range ~(2.4–14) × 10^{14} cm^{-2}. Hall et al. noted that these observations were consistent with no atomic oxygen, and they set 3σ upper limits on the atomic oxygen abundances of (1.6–3.4) × 10^{13} cm^{-2}.

More recent cross-section work by *Kanik et al.* (2003) found that the emission ratio of 1356 to 1304 for process (5) ranges from ~2.5 to 3 for a Maxwellian distribution of T_e = 5 eV to ~2 to 2.5 for T_e > 40 eV. For the four sets of published HST observations (*Hall et al..* 1998; *McGrath et al.,* 2004), the observed ratio varies from 1.3 ± 0.8 to 2.2 ± 1.4. The *Kanik et al.* (2003) cross section ratio therefore "does not support the Europa brightness observation for any energy." As noted by Kanik et al., this may imply that more atomic oxygen is present in Europa's atmosphere than deduced by *Hall et al.* (1998).

In 1999, observations of Europa's trailing hemisphere were obtained using the HST Space Telescope Imaging Spectrograph (STIS) with the 52" × 2" slit (the effective slit length is 25", determined by the size of the detector) and grating G140L, covering the wavelength range 1150–1720 Å (*McGrath et al.,* 2004). Two images were obtained in each of five HST orbits spanning 7 h, with the exception of the second exposure in the second orbit, which failed. Detailed information about the individual exposures is given in Table 1 of *Cassidy et al.* (2007). The data format, illustrated in Fig. 3, shows the type of information available in these observations: monochromatic images of Europa in emission lines of H I Lyman-α 1215.67 Å, O I 1304 Å, C II 1335 Å, and O I 1356 Å, as well as a disk-integrated spectrum of the satellite, which is directly comparable to the previous GHRS and subsequent Cassini observations.

Figure 4 shows the summed, monochromatic images at H I Lyman-α and oxygen wavelengths, as well as a representative visible light image of the corresponding hemisphere of Europa. The H I Lyman-α and O I 1304 Å images have had the geocoronal background (process (1)) subtracted off. The Europa signal in the Lyman-α image consists of solar Lyman-α photons reflected from the surface of the satellite (process (2)). The 1304 Å image consists of both emission from Europa's oxygen atmosphere (processes (4) and (5)), and solar 1304 Å oxygen emission reflected from the surface of Europa (process (2)), which is difficult to subtract from an image due to its unknown spatial distribution. As shown in Fig. 2, and discussed later in section 2.2 with regard to Cassini observations, the contribution of reflected light at 1304 Å does not dominate, but is nonneglibible. By contrast, the reflected component at 1356 Å is negligible because the Sun does not produce O I 1356 Å line emission, so this image includes only emission from the atmosphere of Europa. The O I 1356 image shown in Fig. 4 has a signal to noise ratio (S/N) of only ~3 in the brightest regions, and the O I 1356 Å emission peaks within the disk, not in a ring of emission at the limb of the satellite, as would be expected from plasma interaction with an optically thin atmosphere. It does display the expected limb glow (more easily seen in the middle panel of Fig. 3), but includes a brighter region within the disk on the antijovian hemisphere. Prior to the acquisition of these images, *Saur et al.* (1998) published a detailed model of the plasma in-

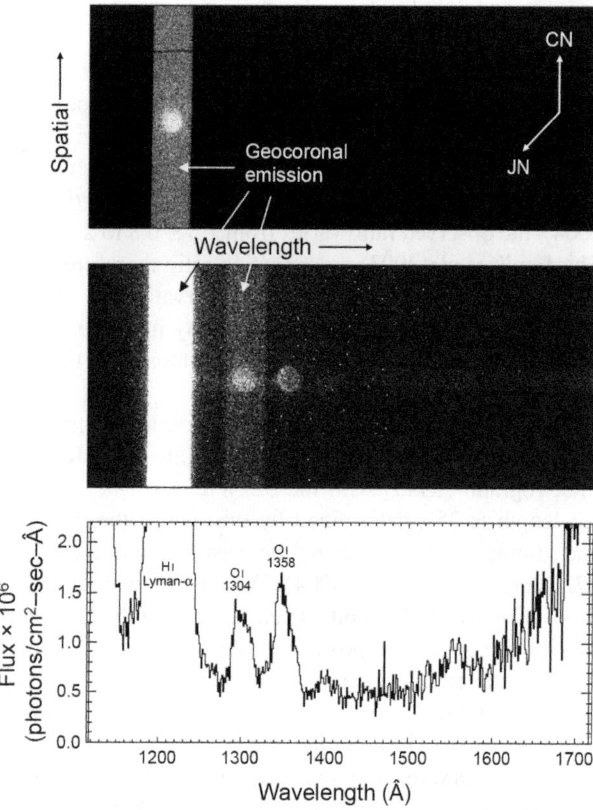

Fig. 3. Observation of Europa obtained in 1999 using the HST STIS, illustrating the data format. The top and middle panels are identical data displayed with a different stretch to emphasize Europa at H I Lyman-α wavelength (top) vs. the oxygen emissions (middle). Emission filling the long slit at H I Lyman-α and O I 1304 is from process (1). The one-dimensional spectrum in the bottom panel is obtained by summing over the spatial rows corresponding to Europa in the middle panel. CN = celestial north; JN = jovian (and Europa) north.

teraction at Europa, including a prediction of the morphology of the O I 1356 Å emission. Their prediction compared with the observation is shown in Fig. 19.10 of *McGrath et al.* (2004); the prediction shows the brightest region off the limb of the subjovian hemisphere (opposite of that observed) and near the equator.

The observed brightness is an integral along the line of sight of the neutral density, the electron density, and an electron temperature dependent excitation rate coefficient. A localized bright region has to be caused by either nonuniform plasma effects (n_e, T_e) exciting the emissions, or nonuniform density of the sputtered neutral gas. It is difficult to understand why the brightest region is on the disk because of the reduced path length there. Since Europa has a weak induced magnetic field (*Kivelson et al.,* 1997) it seems unlikely that it could focus jovian electrons or energetic ions with finite gyro radii to sputter molecules into localized regions. The observation of localized O emission may suggest that the surface is not icy everywhere, but rather that the composition varies considerably with longitude, consistent with visible images and compositional data from Galileo. Comparison of the 1356 Å and visible images shows that the 1356 Å brightness may be correlated with the visibly brighter regions, which are thought to be purer water ice than the visibly darker regions. However, a preliminary analysis of the April 2007 HST observations does not support this hypothesis.

Cassidy et al. (2007) have explored the possibility of nonuniform sticking of O_2 as an explanation for the emission morphology. One caution concerning this interpretation is that the long integration time for the summed images shown in Fig. 4 limits the ability to distinguish between the possibilities of a spatially variable source/loss process or local variability of the plasma exciting the emission. The integration spans 7 h, over more than half of a Jupiter rotation, during which time local plasma conditions, and therefore the emission morphology, might be expected to change significantly due to the undulation of the plasma sheet associated with the tilted jovian magnetic field. In fact, although the region of the surface observed does not change significantly over the 7-h duration of the observations, the emission morphology does appear to vary, as shown in Fig. 5 where the individual images that were summed to produce the Fig. 4 image are shown. Unfortunately, these images have a very low S/N ratio of only ~1 in the brightest regions, so caution is needed to avoid overinterpretation. However, if the variation is real, it would tend to argue against interpretation of the bright region on the anti-Jupiter hemisphere as being associated with the corresponding bright visible region on the surface. Figure 5 shows that the brightest emission is always in the antijovian hemisphere, even though it appears to vary in latitude from north to south between images. As noted by *McGrath et al.* (2000) and *Ballester et al.* (2007), there is not a straightforward correlation between

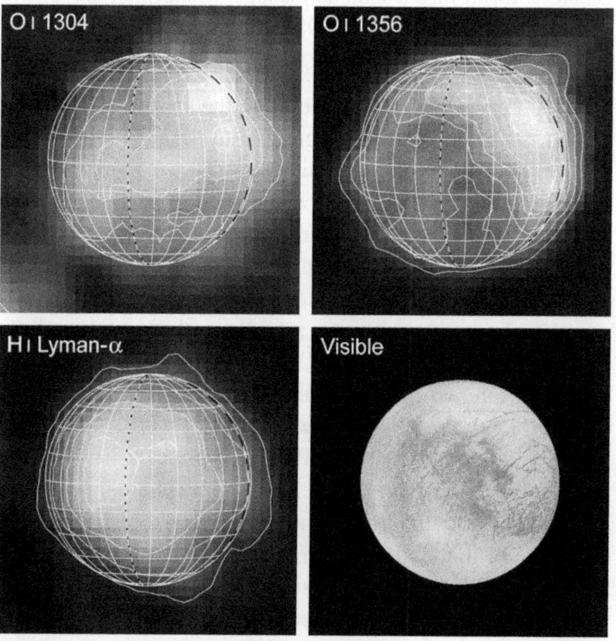

Fig. 4. O I 1304 Å, O I 1356 Å, and H I Lyman-α images from the HST observation shown in Fig. 3, along with a visible light image of the corresponding hemisphere of Europa.

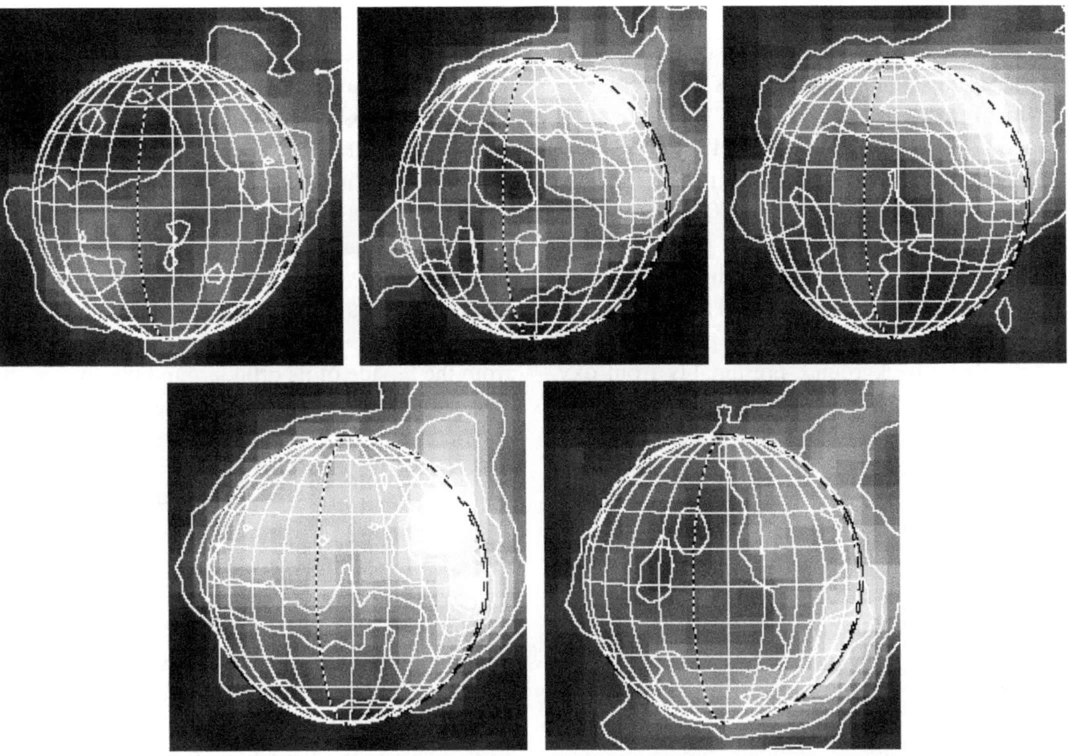

Fig. 5. O I 1356 images (shown with the same stretch) obtained in five consecutive HST orbits on October 5, 1999, over a time period of 7 h. These images were summed to produce the single O I 1356 image shown in Fig. 4. The S/N ratio in these images is only ~1 in the brightest regions. Care is therefore needed to avoid overinterpretation of the spatial and/or temporal variability that is obvious in these images.

the variation in emission location and the orientation of the background jovian magnetic field, as is the case for similar plasma excited emissions on Io.

Simple inspection of the images shown in Fig. 4 reveals an apparent correlation between the Lyman-α bright regions and the dark visible regions. This is likely due to surface albedo variations. The bright visible regions are thought to be composed of purer water ice, which is very dark at UV wavelengths, darker than most non-ice species (*Hendrix and Hansen,* 2008), so it is expected that more Lyman-α is reflected from the surface in the visibly darker surface regions than in the visibly brighter regions. A hydrogen corona such as that detected at Ganymede (*Barth et al.,* 1997; *Feldman et al.,* 2000) has been searched for in the Lyman-α images. It may be marginally present, but not with enough statistical significance to publish (K. D. Retherford, personal communication, 2008).

Four additional sets of UV observations of Europa exist (observations #7, 8, 9, and 10 in Table 1) but have not yet been published, although preliminary results for two of them (#7 and 8 in Table 1) have been presented in an oral talk at the December 2007 American Geophysical Union meeting (*Retherford et al.,* 2007). HST observations were made in February 2007 in conjunction with the New Horizons fly by of Jupiter, during which the New Horizons Alice instrument was also used to observe Europa. The Alice observations produced 17 spectra of Europa in sunlight from ~1250 to 1500 Å, 8 of which provided adequate signal for atmospheric emission measurements. The O I 1304 Å emission is barely detected in the composite spectrum. The brightness ratio of the 1356 Å to 1304 Å emissions is close to 2, as for previous HST observations, implying an O_2 source. There are no obvious trends vs. time in these spectra. Alice observations of Europa in eclipse failed. The accompanying HST observations were done with the Advanced Camera for Surveys (ACS) using the solar blind channel and filter F125LP to acquire 4 10-min exposures while Europa was in eclipse (observation #7 in Table 1), covering the subjovian hemisphere and Europa's magnetospheric wake, as opposed to the trailing (upstream) hemisphere of the satellite captured previously (shown in Figs. 4 and 5). Most of the signal in these images is due to detector dark noise, and because Europa is in eclipse, the location of the Europa disk is uncertain. There is a faint enhancement in emission above the background in these images. Retherford et al. have done limb fits to this faint emission to determine the location of the satellite, which they surmise is located 1.4" from the Fine Guidance Sensor determined pointing, larger than the nominal 1" pointing uncertainty. With this limb fit, they have produced an image that shows a bright region on the disk at northern latitude in the subjovian hemisphere, and an extended bright region extending off the disk that they interpret as wake emission. Wake emission would be surprising because it is expected that Europa's wake would have the lowest electron temperatures and densities of any region near Europa (*Kliore et al.,*

1997; *Saur et al.,* 1998; *Schilling et al.,* 2008). If this interpretation is correct, it tends to further support interpretation of the UV emission morphology as due to plasma, and not surface, effects.

The two additional unpublished sets of HST observations were made using the HST ACS with the prism, PR130L, in April 2007 and June 2008. The April 2007 set (#9 in Table 1) acquired two exposures in sunlight prior to Europa entering eclipse, and two exposures in eclipse. Ultraviolet emission at 1304 Å and 1356 Å is detected above the background in the sunlit images. The two eclipse images also show emission above background, presumably from oxygen at 1356 Å, although its location relative to the Europa disk is uncertain. The detected emission appears to be on the subjovian hemisphere at the limb of the satellite, very similar to the Cassini Na emission images discussed in section 4.1 and shown in Fig. 12. The June 2008 observations (#10 in Table 1) used the same setup as observation set #9 but of the leading hemisphere of Europa in sunlight. These data are still proprietary, with no results yet available.

2.2. Cassini Observations

Observations of the Galilean satellites were acquired by the Cassini spacecraft in December 2000 and January 2001 when it flew by Jupiter on the way to its primary orbital mission at Saturn. Much of the best satellite data were lost when the spacecraft entered safe mode a week before closest approach, however, two UltraViolet Imaging Spectrograph (UVIS) observations of Europa were acquired in the week following closest approach. The objective of these observations was to confirm the *Hall et al.* (1995) detection of a tenuous oxygen atmosphere and to compare these data, acquired at different times and for different orbital geometries, to the HST results.

Europa data pertinent to the atmosphere were acquired with the UVIS far-ultraviolet (FUV) channel (1115–1914 Å) on January 6 and 12, 2001. Details about the observations are given in Table 1. Further details, including graphical representations of the slit orientations for the two observing dates, are given in Table 1 and Fig. 1 of *Hansen et al.* (2005). The UVIS, like the HST STIS, has a two-dimensional detector that collects up to 1024 spectral pixels and 64 spatial pixels. For the Europa observations, the spatial pixels subtended 1.0 mrad, and the spectral pixels 0.25 mrad, while the slit width was 1.5 mrad. Europa subtended 0.28 mrad on January 6, and 0.20 mrad on January 12, meaning it was essentially a point source in both the spatial and spectral dimensions for both observations. Because Europa was boresighted for the Cassini ISS camera, and the FUV channel is offset 0.37 mrad from the ISS boresight, Europa was not centered in the slit for the observations, which results in a slight wavelength shift of the emission lines in the spectral direction, and causes Europa to span spatial rows 31 and 32 in both observations. The total duration of the January 6 observation was 17,000 s, in 17 1000-s integrations, while the January 12 observation collected 41 1000-s images. The observation geometries from Cassini were quite different than the view available to Earth-based telescopes because Europa was observed at greater than 90° phase angle, and portions of both the leading and trailing hemispheres, and sunlit and night sides, of Europa were included on both observing dates. The UVIS slit was oriented perpendicular to Jupiter's equatorial plane for both observations. On January 6 Europa was on the nightside of Jupiter in its orbit, while on January 12 it was on the dayside of Jupiter, near the ansa of its orbit as seen from Cassini, and it was tracked from the farside of its orbit, around the ansa to the nearside over the duration of the observation.

The nonoxygen emissions present in the full spectrum were from the Io plasma torus. The individual lines of the 1304 Å triplet at 1302.2, 1304.9, and 1306.0 Å and the 1356 Å doublet at 1355.6 and 1358.5 Å were all resolved in the UVIS spectrum, consistent with observation of a point source in the spectral direction, as noted above. The 1335 Å feature was used to determine an albedo of ~1% at 94° phase angle (compared with values of 1.3–1.6% from the HST observations), and account for the reflected solar light contribution to Europa's spectrum at 1216, 1304, and 1356 Å, assuming the albedo is flat throughout this region of the spectrum. Subtraction of the reflected light leaves the signal due solely to Europa's atmosphere. Since Europa was well away from the Io torus at the time of the January 12 observation there were no torus emissions to remove from the spectra. Atomic and molecular oxygen abundances derived from Cassini data are tabulated in Table 3. These are derived values, based on the flux observed at the instrument. Assumptions that go into these derived values include the scale height of the atmosphere, and the electron energy and density. *Hall et al.* (1998) derived a molecular oxygen vertical column density of (2.4–14) × 10^{14} cm^{-2}. Cassini molecular oxygen abundances fall within this range. The error bars on the Cassini measurements are ±~15%.

A point source provides the best fit to the shape of the 1356 Å emission feature, consistent with a bound, near-surface O_2 atmosphere with scale height of ~200 km (assuming a temperature of 1000 K). Once the 1356 Å feature was fit, the abundance of molecular oxygen and its contribution to the 1304 Å feature could be calculated. The remaining flux at 1304 Å was then attributed to the resonant scattering of sunlight by atomic oxygen (process (3)), and electron excitation of atomic oxygen (process (4)). Although somewhat subjective, the spectrum was best fit by including ~2% atomic O in the (point source) bound O_2 atmosphere, plus an extended tenuous atomic oxygen component overfilling a pixel. *Hall et al.* (1995, 1998) would not have been able to identify an extended O component, since Europa filled most of the field of view of the GHRS slit, whereas Europa subtends less than one spatial pixel in the UVIS slit, thus more of the surrounding space is measured, enabling the UVIS detection of the extended atomic oxygen component.

The spatial capability of UVIS was used to determine the oxygen profile as a function of distance from Europa. Figure 6a shows the distribution of oxygen in the January 6 dataset and Fig. 6b the January 12 dataset as a function of spatial row. An extended oxygen component is apparent in both sets of observations. The 1356 Å feature is sharply peaked at row 32. The diffuse 1304 Å feature persists across all the illuminated spectral pixels and is detectable in rows 28, 29, and 30. Although the low spatial resolution prevents determination of the size of the diffuse extended atmosphere, and any potential asymmetries such as suggested by *Burger and Johnson* (2004) for Europa's Na cloud, the atomic oxygen cloud appears to be about one pixel in extent, or ~11,000 km (~7 R_E) at Cassini's range. The density of oxygen in Europa's extended atmosphere is estimated to be roughly 1700 (January 6) and 1000 (January 12) atoms cm^{-3}, which was calculated using the simple formula (intensity in rayleighs × 10^6)/(pathlength in cm × probability of emission) where the probability of emission

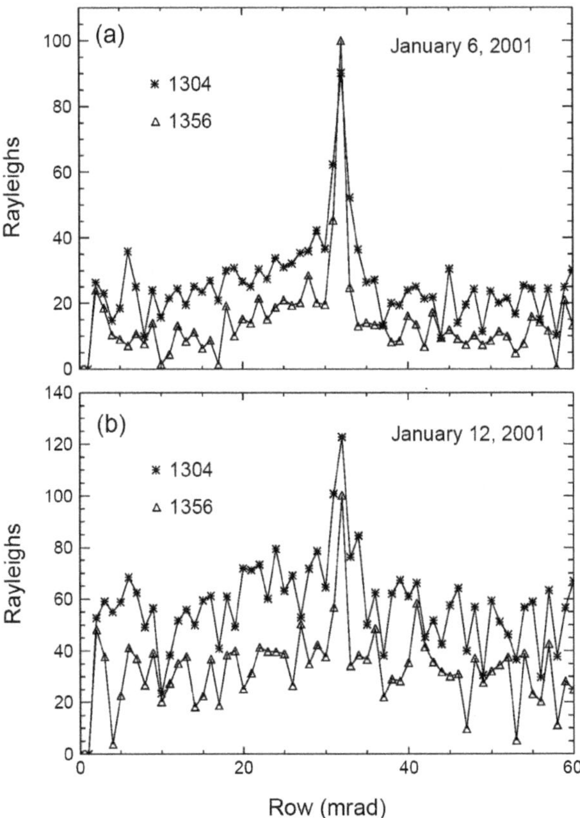

Fig. 6. The distribution of 1304 Å and 1356 Å oxygen emission along the slit in the spatial direction. **(a)** In the January 6, 2001, data the 1356 Å feature is sharply peaked at the position of Europa, while the diffuse 1304 Å feature source persists across all the illuminated spectral pixels; it is detectable in row 28, which corresponds to the opposite side of Europa's orbit, and then gradually drops to the background level. **(b)** The oxygen line emissions for the January 12, 2001, data. Most of these data were collected near the ansa of Europa's orbit as seen from Cassini. Both 1304 Å and 1356 Å are sharply peaked at Europa's position.

(combination for 1302 + 1304 + 1306) is 6.05 × 10^7 and the pathlength is assumed to be 22,000 km. This detection of atomic oxygen as an extended atmosphere adds a new observational constraint to models of the erosion of Europa's surface and the life cycle of its atmosphere in the jovian environment, such as those developed by *Saur et al.* (1998), *Shematovich and Johnson* (2001), and *Shematovich et al.* (2005).

3. IONOSPHERE

In simple terms, an ionosphere is a layer of plasma produced by photo- and/or particle-impact ionization of a neutral atmosphere. In deep space planetary missions, ionospheres are detected by transmitting a one (or two)-way radio signal from the spacecraft to Earth (and back) through the atmosphere of the object of interest along a trajectory that produces an occultation of the spacecraft as seen from Earth. Such an occultation produces both an inbound and an outbound measurement of the target's ionosphere. In these radio occultations, a time series of signal strength and frequency are produced, and via comparison with a time series of predicted frequency computed from a precise spacecraft ephemeris, a set of frequency residuals is derived. The residuals are inverted using standard techniques to obtain the refractivity, which is directly related to electron density, thereby providing an electron density profile vs. altitude above the surface of the object.

The Galileo mission provided four radio occultations of Europa: one in orbit E4, two in orbit E6 (E6a and E6b), and one in orbit E26, as well as a near-occultation in orbit E19. The geometries and results from three of these (E4, E6a, and E6b) are shown in Figs. 1, 2, and 3 of *Kliore et al.* (1997). In addition, the E19 flyby resulted in detections on both the entry and exit paths, and the E26 fly resulted in a nondetection on the entry (designated by "N") leg, and a weak to nondetection on the exit (designated by "X") leg (*Kliore et al.*, 2001, 2006). The results of the 10 inbound and outbound measurements are shown in Fig. 7, and include six strong detections (E4N, E4X, E6aN, E6bN, E19N, E19X), one moderate to weak detection (E6bX), and three weak to nondetections (E6aX, E26N, E26X) of the ionosphere. [Note that the results shown in Fig. 7 and presented in *Kliore et al.* (2006) show lower electron densities and somewhat different profiles for E6b than those published in *Kliore et al.* (1997).] The maximum electron densities of ~10^3–10^4 cm^{-3} occur at or near the surface of Europa (except for E6bN, which peaks at ~100 km altitude) with a plasma scale height of ~200 km below 300 km, and ~400 km above 300 km altitude. Assuming likely candidate constituents such as H_2O, O_2, H, H_2, OH, and O leads to an estimated neutral density on the order of 10^8 cm^{-3}, and a column density on the order of 10^{15} cm^{-2} assuming a neutral scale height of ~100 km.

The E4N detection geometry is looking toward the upstream direction of the plasma flow through the wake,

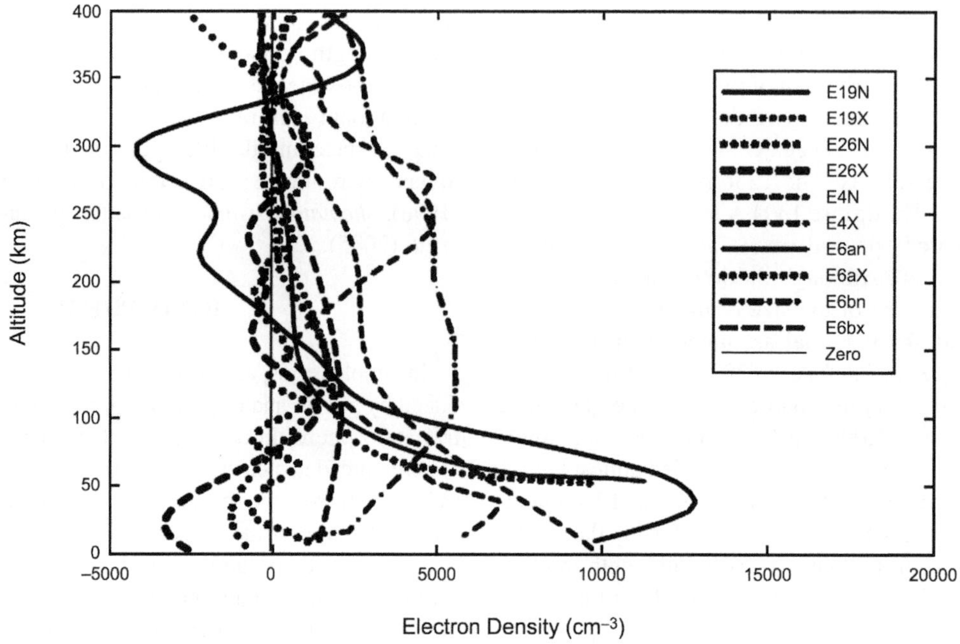

Fig. 7. See Plate 23. Compilation of all the Galileo radio occultation and near-occultation results illustrating the nonuniformity of Europa's ionosphere. Figure courtesy of A. Kliore.

where densities could well be higher. In fact, recent work by *Schilling et al.* (2008) note that the E4 flyby took an oblique trajectory through the Europa's wake. Their models show that the peak density in the wake has a small spatial extent, and produces ion densities in agreement with the plasma results of *Paterson et al.* (1999). By contrast, a broad plasma wake, such as those in the numerical results of *Kabin et al.* (1999) and *Liu et al.* (2000), cannot reproduce the *Paterson et al.* (1999) results. This reconciles the previously noted puzzling lack of higher ionospheric densities in the wake region during the E4 flyby discussed by *McGrath et al.* (2004).

The nondetection in E6aX occurred at night near the middle of the downstream wake region, and the E26 entry occultation nondetection was at high latitude in the wake region. The wake region is shielded from the preferred direction of the plasma flow, which is toward the trailing hemisphere centered at 270 W longitude, and the electron densities there might be expected to be depleted. Also, according to the plasma model of *Saur et al.* (1998), the ionosphere can become detached from the satellite in this region. However, *Kliore et al.* (2001) also conclude, on the basis of the various geometries and detections, that a necessary condition for the detection of Europa's ionosphere may be that the trailing (plasma-impacted) hemisphere is at least partially illuminated, which may indicate that solar photons play a significant role in ionizing the neutral atmosphere, in addition to the magnetospheric plasma electrons.

On the other hand, *Saur et al.* (1998) found that electron impact ionization alone can generate Europa's ionosphere at the electron densities measured by *Kliore et al.* (1997). Using magnetospheric plasma conditions at Europa typical of the Voyager epoch (n_e = 38 cm^{-3} and 2 cm^{-3} at T_e = 20 eV and 250 eV, respectively) the electron impact ionization rate is 1.9×10^{-6} s^{-1}, while a solar maximum photoionization rate is 6×10^{-8} s^{-1}. Given the intrinsic time constants associated with these processes (6 and 190 d, respectively, compared with a 3.6-d orbital period), the diurnally averaged solar photoionization rate would be a factor of ~2 lower, whereas electron impact ionization depends mostly on ambient magnetospheric plasma densities. In order for photoionization to be competitive with electron impact ionization, the magnetospheric electron density would have to be ~1 cm^{-3}. For the Europa radio occultations observed in Galileo orbits E4 and E6, the ambient magnetospheric ion densities were ~25 and 15 cm^{-3}, respectively (*Paterson et al.,* 1999); the corresponding electron densities would be about 50% larger. *Kurth et al.* (2001) suggested a typical torus electron density of 80 cm^{-3} at the orbit of Europa. It is therefore very difficult to understand why the existence of an ionosphere should depend on solar illumination.

In summary, while the existence of an ionosphere at Europa is well established, its origin and characteristics are still very poorly understood.

4. MINOR SPECIES, NEUTRAL CLOUDS, AND TORUS

With the exception of atomic oxygen, discussed above, detection of other minor species associated with Europa has been at distances well above the surface, where there is no overlap with detections of the primary atmospheric species O_2 (see Fig. 1). Minor species in Europa's atmosphere can

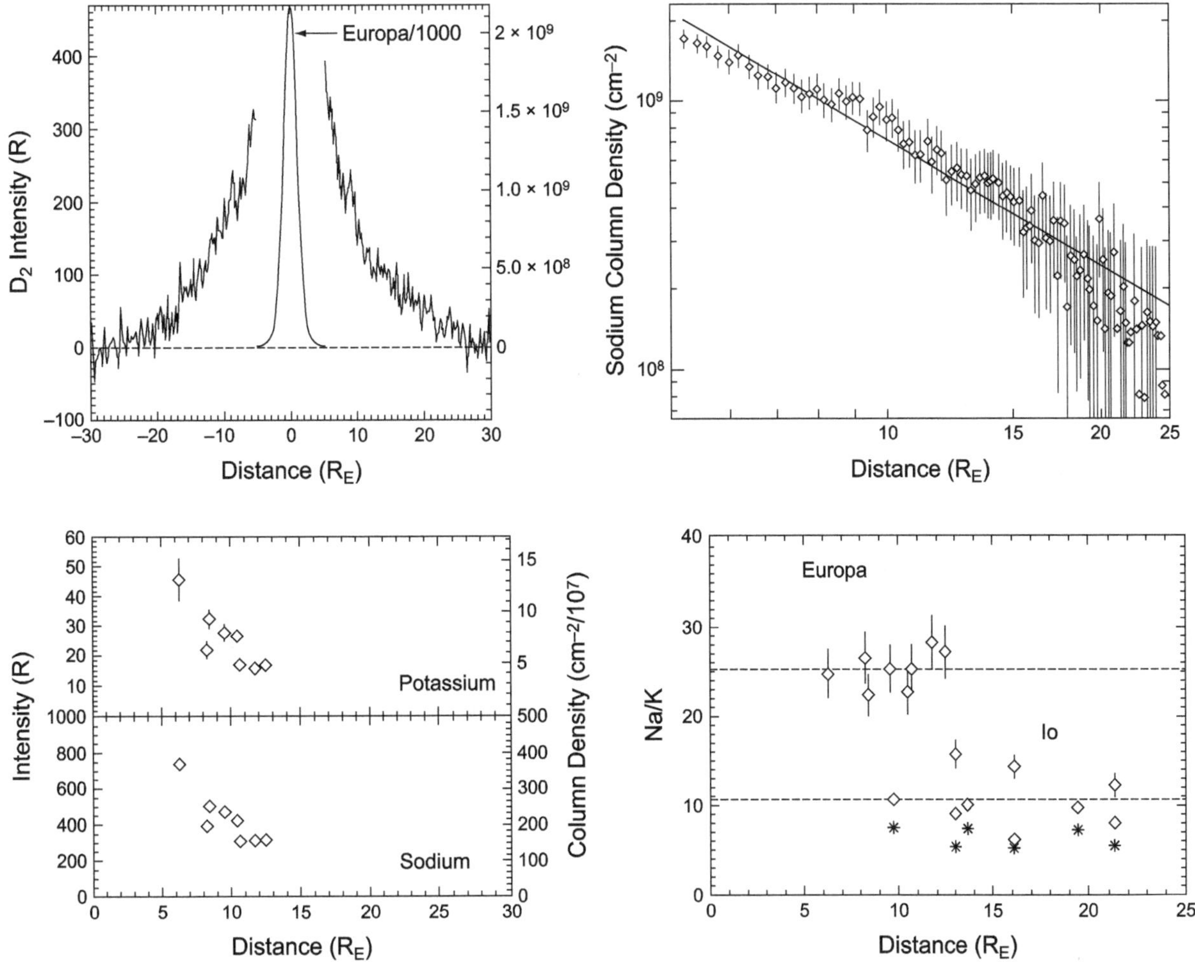

Fig. 8. *Top:* Radial profiles of Na emission from Europa obtained by *Brown and Hill* (1996). Due to a calibration error found after publication, the vertical axes in the top panel need to be multiplied by a factor of 2. *Bottom left:* Radial profiles of Na and K emission from Europa obtained by *Brown* (2001). *Bottom right:* Comparison with identical observations of Io from *Brown* (2001) shows that the Na/K ratio at Europa is about 10 times higher than that from Io. From *Brown and Hill* (1996) and *Brown* (2001).

aid in investigation of surface composition and also serve as a proxy for mapping the distribution of the major atmospheric species far from the surface. The minor species detected to date include Na, K, and some form of H (atomic or molecular). Numerous sets of observations exist of Na, Cassini made several measurements of H, but only a single set of observations has been made of K.

4.1. Sodium and Potassium

The first detection of Na from Europa was made on June 5, 1995 (observation #16 in Table 1) when 10 long-slit spectra covering 5883–5904 Å were obtained at the 1.53-m University of Arizona telescope on Mt. Bigelow with total integration time of 4.33 h (*Brown and Hill,* 1996). The long slit was oriented perpendicular to Europa's orbital plane, and emission intensity along the slit in the north-south direction from Europa was derived (top panel of Fig. 8). No variability was observed among the 10 spectra, so they were all co-added for analysis. Emission from the Na D_1 and D_2 lines at 5895.92 and 5889.95 Å, respectively, was detected with an intensity ratio of 1.70 ± 0.05, consistent with the value of 1.66 expected for optically thin emission from resonant scattering of sunlight by Na in Europa's vicinity. The line of sight column density is therefore directly proportional to the emission intensity; both are plotted vs. radial distance in the north-south direction from Europa in Fig. 8. Note that a factor of 2 calibration error was later found in these published data, so the values along the y axes in the top panel of Fig. 8 need to be multiplied by 2. The emission is seen to be symmetric about the satellite in the north-south direction out to ~10 R_E, and then is slightly brighter in the north than the south from 10–20 R_E. The emission intensity inside ~5 R_E cannot be determined because of the overwhelming brightness of the Europa continuum. The Na emission is detectable out to ~25 R_E, which requires that Na atoms leave the surface with v > 2 km s^{-1}, and that most of the particle velocities are this large or larger to account for the flatness of the profile between 5 and 10 R_E. Since sublimation at the surface temperature of 95 K

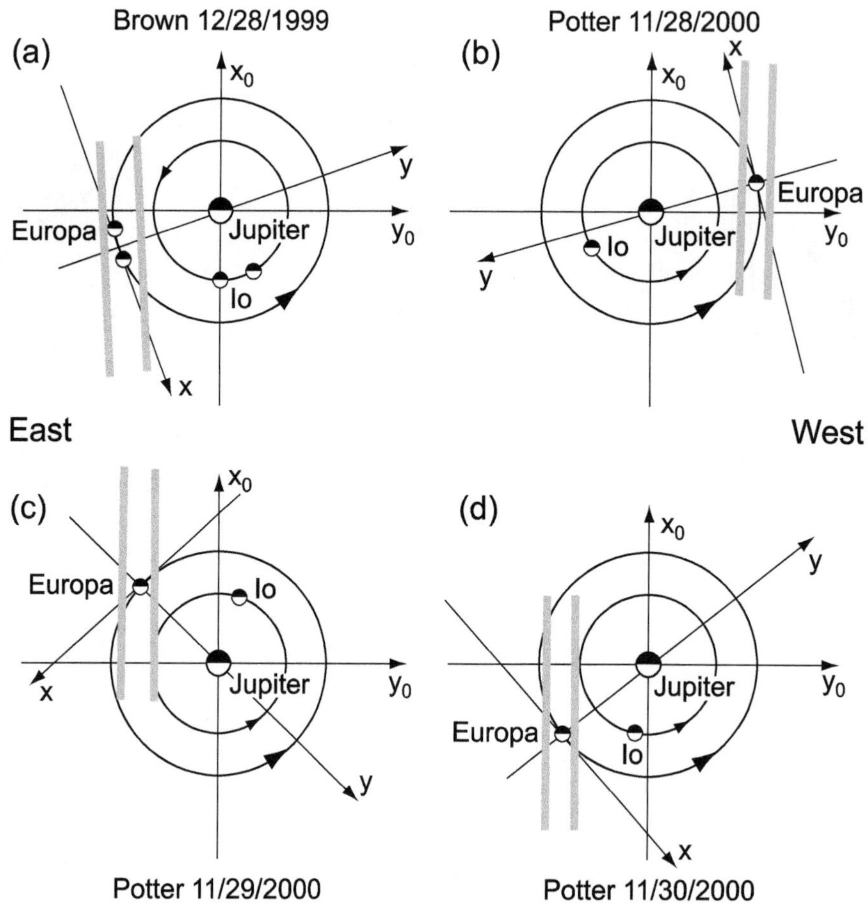

Fig. 9. Geometry of the December 1999 and November 2000 observations of Na emission near Europa reported in *Leblanc et al.* (2002, 2005). From *Leblanc et al.* (2005).

would produce velocities of only 0.3 km s^{-1}, Brown and Hill concluded that the most likely source mechanism is sputtering by energetic magnetospheric particles. The total mass of Na implied by the distribution shown in Fig. 8 is 840 kg, and extrapolation of the profile to the surface implies a surface density of ~100 cm^{-3}, which is 300 times less than the closely bound molecular oxygen atmosphere.

A second set of high-resolution, groundbased spectroscopic measurements was made on September 9, 1998, using the HIRES instrument on Keck (*Brown,* 2001). The Na observations were repeated, but this time simultaneous measurements were also made of the K doublet at 7664.90 and 7698.96 Å. The long slit was oriented perpendicular to Europa's orbital plane, and centered at several different positions east of the satellite. The intensity profile obtained (shown in the bottom panel of Fig. 8) was very similar to that from the *Brown and Hill* (1996) observations. On November 15, 1998, an identical set of observations was also made of Io. These measurements showed that the Europa Na/K ratio of ~25 is both very different from that at Io (Na/K ~ 10), and very different from a meteoritic source or the solar abundance ratios (*Brown,* 2001; *Johnson et al.,* 2002). Iogenic Na implanted into Europa's surface ice was originally suggested as the source of the Na; however, this idea was later revised based on the detection of K, and the measured ratio of Na/K for Europa compared to Io, and an endogenic source is now favored (see chapter by Johnson et al.).

Four subsequent sets of Na observations, reported in *Leblanc et al.* (2002, 2005), were obtained in 1999 by M. Brown and in 2000 by A. Potter (observations #18–21 in Table 1). The orbital geometry of Europa for these four sets of observations is shown in Fig. 9, illustrating the parts of the Na cloud observed on each occasion. The December 28, 1999, observations (reported in *Leblanc et al.,* 2002, 2005) were made with the identical setup as *Brown* (2001) using the Keck HIRES instrument, and comprised the first attempt to obtain detailed maps of the Na emission morphology near Europa. Six sets of scans were made: north-south and east-west sets centered on Europa; east-west sets at 10 R_E north and south of Europa; and east-west sets at 20 R_E north and south of Europa. The consolidated Na intensity information from these scans is shown in Fig. 10. This more complete set of measurements clearly shows that the Na cloud asymmetries are different in the direction perpendicular to the orbital plane (north-south) than in the direction parallel to the orbital plane (east-west). The observed emission morphology is again symmetric, both north-south and east-west,

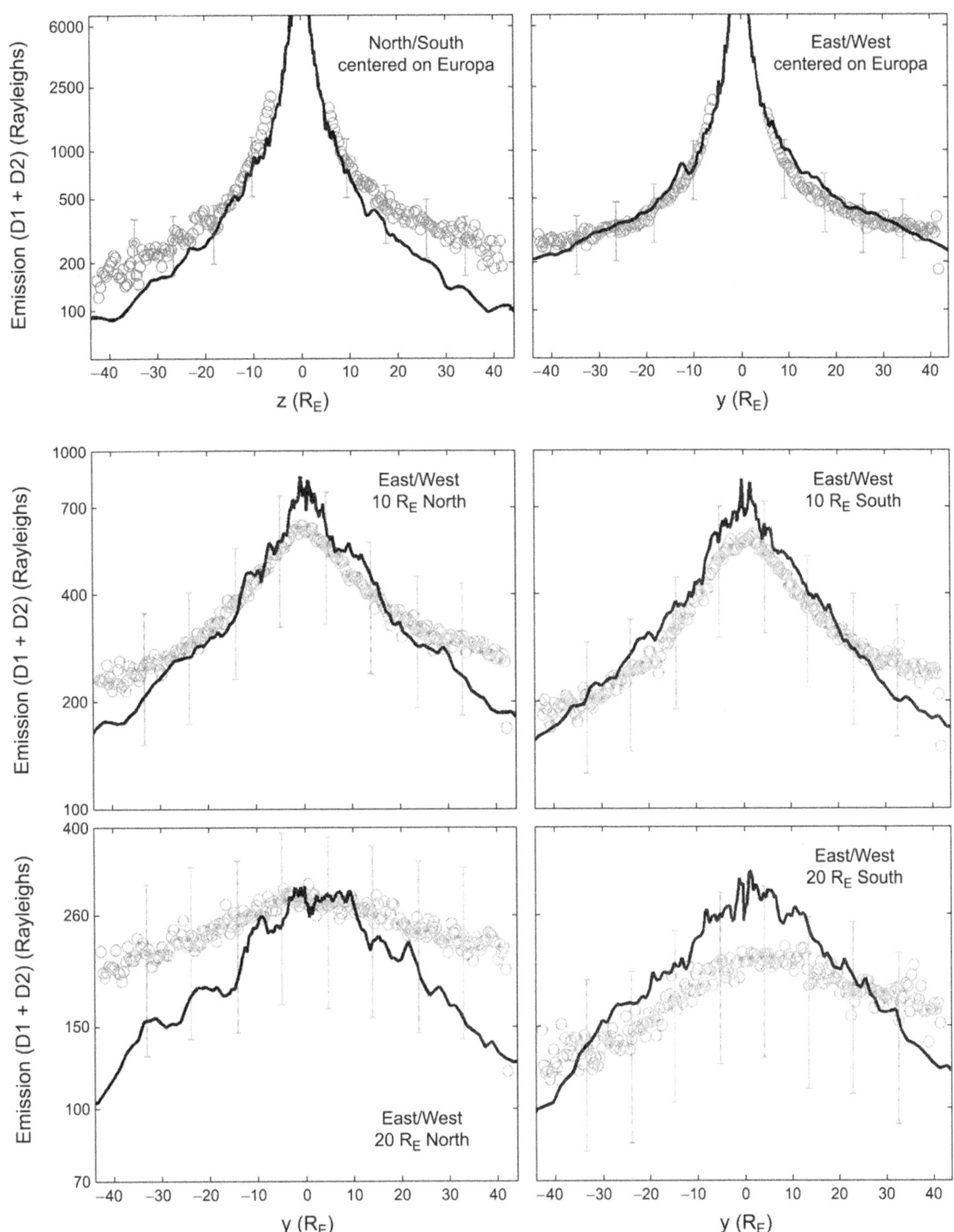

Fig. 10. Na emission profiles observed on December 28, 1999 (open circles), compared with models of *Leblanc et al.* (2002) (solid line). From *Leblanc et al.* (2002).

within ~10 R_E, but then, as with the 1995 observations, somewhat brighter north of Europa than south at d > 10 R_E. However, the overall emission intensity is approximately twice as bright in these 1999 observations than in the 1995 and 1998 observations, and substantial variations in emission intensity at given locations (e.g., 20 R_E north and south of Europa; see Fig. 10) are seen during the course of the observations.

The three sets of Na observations obtained in November 2000 used the McMath-Pierce Solar Telescope at Kitt Peak National Observatory (*Leblanc et al.,* 2005). In these observations, the Na emission was sampled in 1" × 1" regions centered at 2" and 3" from the center of Europa in all directions (north, south, east, and west) on November 28, 2000; and at 2", 3" in the north, south, east, and west directions and 4" north, east, and west on November 29 and

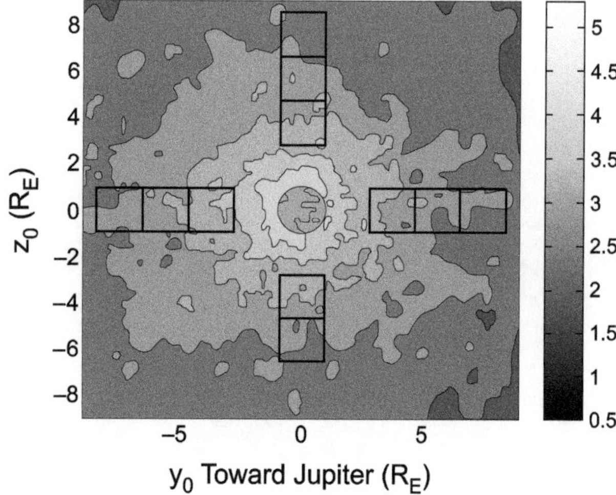

Fig. 11. Square boxes show the locations of the November 29, 2000, observations of Na near Europa superposed on a model of the Na cloud brightness of *Leblanc et al.* (2005). From *Leblanc et al.* (2005).

30, 2000. The mapping locations for the November 29 and 30, 2000, observations are illustrated in Fig. 11. Whereas earlier observations had not been obtained closer than 5 R_E, these 2000 observations obtained measurements as close as ~3.5 R_E. Table 2 of *Leblanc et al.* (2005) provides a detailed compilation of the observed Na intensities at different locations for the four sets of 1999 and 2000 observations. The differences in intensities show that within 2–3" of Europa, there is a leading/trailing asymmetry that *Leblanc et al.* (2005) contend is dominated by the production of atmospheric Na from the trailing hemisphere, and also depends on the solar flux. They also contend that the north-south brightness asymmetries correlate with the centrifugal (magnetic) latitude of Europa, but this is based on only a handful of data points.

A final set of (possible) Na observations add an interesting potential link between the plasma electron excited oxygen emissions and the resonant fluorescent Na emissions just described. A set of Cassini ISS clear filter images obtained on January 10, 2001, while Europa was going into eclipse (*Porco et al.,* 2003) is shown in Fig. 12. In these images, there appears to be a bright region in the northern subjovian hemisphere that persists well after Europa is completely in shadow. Porco et al. noted that Jupiter light reflected from Ganymede is a potential illumination source, but nonuniform bright emission on the Jupiter-facing hemisphere would be inconsistent with such an origin. Serendipitously, the hemisphere captured in these images is almost identical to that of the HST/STIS images shown in Figs. 4 and 5. Although the vantage point and orbital phase are very different for Cassini compared to HST, both sets of observations were viewing the trailing, plasma-bombarded hemisphere. The two sets of images both show enhancements in the northern hemisphere, but the Cassini emission is in the subjovian northern quadrant, while the HST emission is in the antijovian quadrant. Obviously, because the Cassini images were acquired while Europa was in eclipse, resonant fluorescence of sunlight is not a possible source. By analogy with similar images of Io acquired by the Galileo camera (*Geissler et al.,* 2004), *Cassidy et al.* (2008) suggested that the emission in these images may be produced predominantly by electron impact excitation of atomic Na in Europa's atmosphere. Other probable strong emitters in the ISS clear filter wavelength range include O (from O_2 dissociative excitation) and K. Cassini ISS observations were also made with a filter sensitive to O I 6300 Å emission during the same eclipse shown in Fig. 12, but there was no signal above the noise level. The observation of O emission from Europa in eclipse [observation #8 in Table 1, described in section 2.1 above (*Retherford et al.,* 2007)] shows a somewhat similar morphology, which lends credence to the reality of the emission seen in the Cassini eclipse images. These images, along with the variation seen in the O emission morphology shown in Fig. 5, tend to support the interpretation that the various morphologies seen for both the (potentially) electron excited Na and oxygen emissions are caused by variations in the plasma and its interaction with Europa, rather than nonuniform surface properties of the satellite or localized sputtering sources or sinks, as has been put forward by *Cassidy et al.* (2007, 2008).

4.2. Europa Torus

As Cassini approached Jupiter, the Ion and Neutral Camera (INCA), which is a channel of the Magnetospheric Imaging Instrument (MIMI), was used to acquire a 15-hr Energetic Neutral Atom (ENA) image of the Jupiter system beginning on January 1, 2001, when Cassini was at a distance of 140 R_J (Fig. 13). ENAs are produced by charge exchange between neutral atoms and magnetically trapped energetic ions. Bright features were noted in the ENA image on both sides of Jupiter, peaking just outside Europa's orbit at a distance of ~9.5 R_J (*Mauk et al.,* 2003). Mauk et al. interpreted their image as a torus of neutral atoms centered on the orbit of Europa. When magnetically trapped energetic protons in the Jupiter system collide with a neutral atom in such a torus, charge exchange ionizes the neutral atom and neutralizes the original proton, which then leaves the system. The INCA observations cannot distinguish the composition of the original target neturals. The most likely constituents available for charge exchange in the vicinity of Europa are the products of its sputtered water ice surface and oxygen atmosphere: H_2O, H, H_2, O, OH, and O_2. The INCA data are consistent with an emission region with a radial extent and height of 1–5 R_J, symmetric about Europa's orbit, and a total content of ~9 × 10^{33} particles. Assuming a torus radius of 2 R_J gives a density of ~40 particles (atoms or molecules) cm^{-3}. This interpretation is corroborated by the Galileo Energetic Particle Detector (EPD) instrument measurement of a depletion of energetic particle flux in the vicinity of Europa's orbit (*Lagg et al.,* 2003).

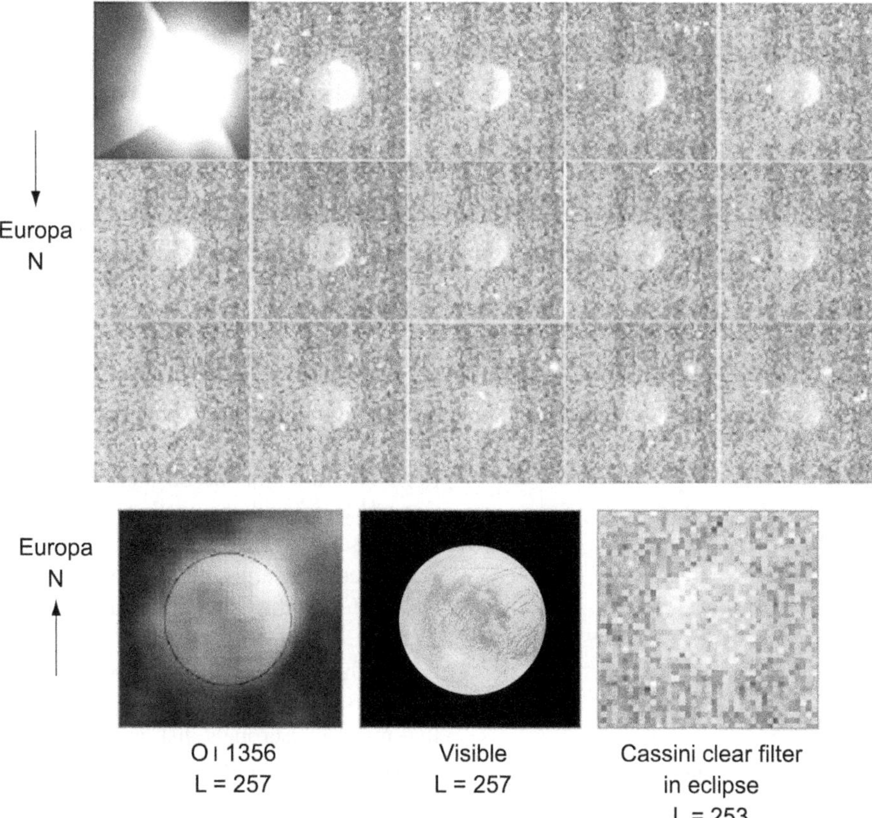

Fig. 12. Top panel shows a series of clear filter Cassini ISS camera images of Europa as it enters Jupiter's shadow, with the time sequence going from left to right, top to bottom. These observations captured the same hemisphere of Europa as the HST observations shown in Figs. 4 and 5. The 11th image from the Cassini sequence (row 3, column 1 of the top panel) is compared to both the HST O I 1356 Å emission image acquired in 1999 (same image as Fig. 4 but without brightness contours), and with a visible image of the same hemisphere of Europa. The HST and Cassini emissions are both brightest in the northern hemisphere of the satellite; however, the brightest HST emission occurs on the antijovian hemisphere, while the brightest Cassini emission occurs on the subjovian hemisphere.

The atomic oxygen in Europa's extended atmosphere is subject to ionization and loss due to photoionization, charge exchange, and electron impact ionization. The rate of ionization from these three processes is ~1.6×10^{-6} per second, thus the lifetime for an oxygen atom in Europa's exosphere is estimated to be 7.2 d. In theory enough oxygen atoms are lost from Europa's atmosphere to account for the total number required by *Mauk et al.* (2003) as neutral species distributed in the magnetosphere. In order to put upper bounds on oxygen in the Europa torus, *Hansen et al.* (2005) analyzed data from a February 2001 outbound Cassini UVIS observation. For this observation the UVIS slit was centered on Jupiter and aligned parallel to Jupiter's equator (see Fig. 6a of *Hansen et al.,* 2005), and the total observation time is ~28 h. During the observation the boresight was slewed slowly from north to south every 30 min, so that the total integration time is ~7 h in any particular direction. Figure 6b of *Hansen et al.* (2005) shows a two-dimensional plot of these data. Atomic oxygen emission from the Io torus is evident, but there is no atomic oxygen emission at or near the position of Europa. If oxygen is a contributor to the torus identified by *Mauk et al.* (2003), its density must be below levels detectable by UVIS. The density that UVIS could have detected is ~8 atoms/cm^3, a factor of 5 less than the value postulated by Mauk et al. for a torus radius of 2 R_J. This suggests that the torus is substantially more extended, possibly out to the 5 R_J considered by Mauk et al., or that it is composed of H and/or H_2. Recent work by *Smyth and Marconi* (2006) argue for H_2 being the most abundant species in the Europa neutral torus (see further details in the chapter by Johnson et al.).

5. VARIABILITY

The multiple observations described above and summarized in Table 1 afford an opportunity to assess, to a limited extent, the degree of variability of Europa's atmosphere and neutral clouds. The major difficulty is our limited ability

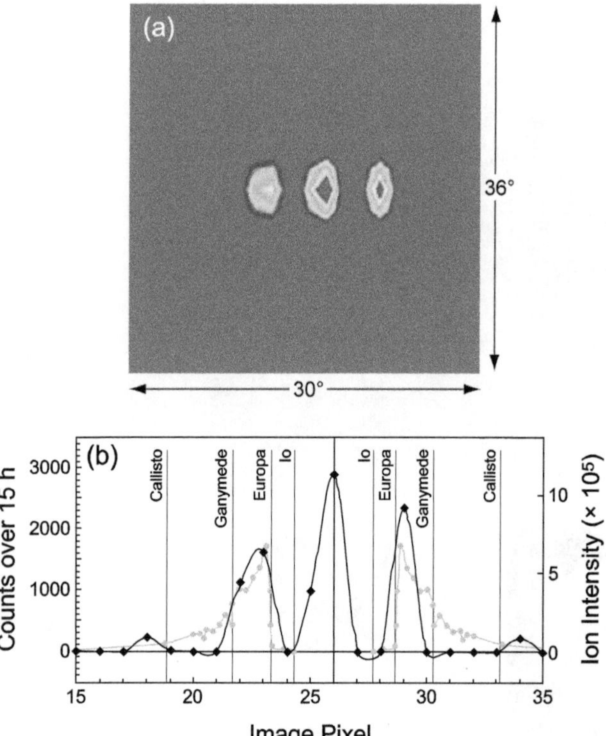

Fig. 13. **(a)** INCA image (in counts) of Jupiter and Europa from *Mauk et al.* (2003) showing an edge-on view of a torus of neutral material (most likely hydrogen) just outside the orbit of Europa. **(b)** The top two-dimensional image collapsed along the vertical axis and plotted as the black line with the intensity of protons, energy-integrated from 50 to 80 keV, plotted in the gray line. The orbits of the Galilean satellites are indicated with vertical lines. From *Mauk et al.* (2003).

to derive meaningful atmospheric neutral abundances because of our scant knowledge about the properties, including variation, of the magnetospheric plasma in the vicinity of Europa. All the observed emissions are an integral along the line of sight, and virtually all inversions of the brightness to derive column densities to date assume a uniform value for electron density and temperature. The only detailed forward modeling to date, by *Saur et al.* (1998), has not been successful in reproducing the oxygen emission morphology. It is important to bear this limitation in mind while considering what solid information can be gleaned from the observed variability.

There are nine sets of observations of the O_2 atmosphere by HST and Cassini (six with quantitative line fluxes), five sets of radio occultations of the ionosphere, and six sets of Na observations taken at different orbital, solar illumination, and magnetic geometries, including multiple observations with roughly the same orientation taken at different times. There are also two observations of the Europa torus, albeit one a nondetection. We consider each in turn.

The disk-averaged spectroscopic line fluxes for O I 1356 Å from the HST and Cassini observations (both the observed fluxes and fluxes normalized to a common distance) are given in Table 2. The observations from Table 1 have been numbered, and are referenced by number in Table 2 and in the discussion that follows. HST has observed Europa's dayside, leading hemisphere twice (observations #2 and #10), its dayside trailing hemisphere three times (observations #1, 3, and 4), and its dayside hemisphere in eclipse twice (observations #8 and 9). Disk average line fluxes are not available for observations #8, 9, and 10. Two of the trailing hemisphere observations (observations #1 and 3) were performed in an identical manner separated by a little over two years; the third trailing hemisphere observation (observation #4) was performed with a different instrument, but the line fluxes derived from it should be directly comparable to the other two. There is no significant variation in the disk-integrated O I 1356 Å line flux between observations #1 and 3, but an ~60% brightening between observations #1 and 3 and observation #4 acquired over three years after observation #3. *Hall et al.* (1995) noted that caution is warranted in interpreting the flux from observation #1 because "during the first HST orbit, the telescope was aligned so that Europa was centered in the aperture; however, during the third orbital exposure, the telescope boresight drifted ~0.8 arcsec from this initial orientation. This drift means that much of Europa's disk may have been excluded from the aperture, possibly artificially depressing detected emission intensities."

Since the oxygen atmosphere is widely accepted to be sputter generated, and the observed emissions are also excited by the plasma, the observed variability seems more likely to be caused by plasma variability than by significant changes in the surface properties of Europa that would affect the sputtering source, unless a currently undetected phenomenon such as surface geysers or surface outgassing are operative. Variability of emission morphology over a 6.5-hr interval also appears to be present in the 1999 HST observations (Fig. 5), although the low S/N ratio hampers the interpretation.

TABLE 2. Variability of O I 1356 Å flux.

Observation #*	Observed† O I 1356 Å	Normalized‡ O I 1356 Å
1	11.1 ± 1.7	0.42
2	12.9 ± 1.3	0.41
3	14.2 ± 1.5	0.45
4	24.0	0.69
Orbit 1	25.2	0.73
Orbit 2	20.6	0.59
Orbit 3	23.1	0.66
Orbit 4	28.9	0.83
Orbit 5	21.0	0.60
5	0.99	0.99
6	0.18	0.36

*Observation # from column 1 of Table 1.

†1, 2, 3, and 4 are in units of 10^{-5} photons/cm²/s; 5 and 6 are in units of photons/cm²/s.

‡In units of photons/cm²/s normalized to 6 Jan 2001 Cassini range of 11.2×10^6 km.

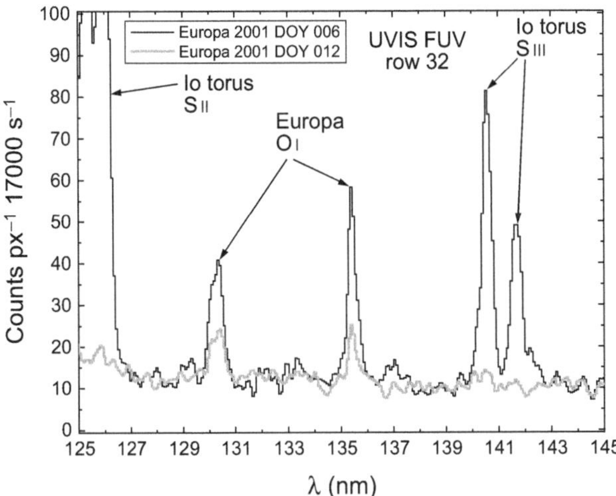

Fig. 14. Comparison of January 6 and January 12, 2001, exposures with the flux from the latter reduced by a factor of ~3. With Europa at the ansa of its orbit as seen from Cassini, Io's torus was not within the UVIS field of view on January 12. From *Hansen et al.* (2005).

The Cassini observations of Europa were at significantly different orbital locations than the HST observations, on the solar (observation #6) and roughly antisolar (observation #5) sides of Jupiter. As with the HST observations, there is a significant variation of the oxygen emission between the two Cassini observations. This is illustrated in Fig. 14, where the difference between the January 6 and January 12 datasets is clearly evident. As can be seen from Table 2, the line fluxes decreased by a factor of 3 in the 6 d between the observations. In order to explore the variability due to viewing geometry, *Hansen et al.* (2005) calculated the emission using the electron environment presented by *Saur et al.* (1998). The Saur et al. model features a steady-state distribution of electron energies around Europa, with plasma flowing in on one side and creating a cavity on the other, producing nonuniform excitation of the oxygen. At the time of the January 12 observation UVIS was looking at the leading side of Europa, at a wake region depleted of electrons. The UVIS observations are qualitatively consistent with the Saur et al. model; however, only a factor of ~1.6 difference in brightness can be attributed to the difference in leading vs. trailing side electron density and electron energy with the Saur model, vs. the 3× difference in flux observed. Using the Saur model for the spatial distribution of electrons and accounting for the viewing geometry, molecular oxygen abundances of 7.4×10^{14} cm^{-2} on January 12 and 12.4×10^{14} cm^{-2} on January 6 were calculated. The spectra show no indication of a variable O I 1356 Å intensity on January 12 over an 11.4-hr time interval (see Fig. 5 of *Hansen et al.,* 2005), which discounts an electron environment variability associated with the Io torus wobble. An increase in oxygen abundance between January 6 and January 12 might be indicative of a transient event, but the significant decrease between the two observations is harder to explain as a transient phenomenon. The more likely explanation is a change in the plasma environment, such as reported by *Frank et al.* (2002).

It is also interesting to compare the Cassini observations with the HST observations. HST observation #1 and Cassini observation #5 captured Europa at almost identical orbital phases (thus the same plasma bombardment orientation) but viewed from different directions. Cassini saw the flank of the plasma flow (30° toward the leading hemisphere), with the half toward the plasma flow being sunlit and the downstream half (with respect to the plasma flow) in darkness, while HST saw the trailing sunlit hemisphere. In this comparison, the less sunlit and less bombarded hemisphere has twice as much flux at 1356 Å. HST observation #2 and Cassini observation #6 saw the same hemisphere of the satellite at very different orbital phases. In this case, HST saw the leading (downstream) sunlit hemisphere, while Cassini saw the leading (downstream) hemisphere with half in sunlight and half in darkness; the HST 1356 Å flux is about 30% larger. It is tempting, as with the ionosphere observations, to associate higher fluxes (densities) with the presence of sunlight; however, there is so little data that this can only be characterized as an indication of a potential trend.

As mentioned briefly in section 2.1, the variation in emission morphology seen in the six consecutive HST 1356 images shown in Fig. 5, for which the sunlight, bombardment, and orbital position change very little over the course of the observations, argues fairly strongly against the emission enhancement in the antijovian hemisphere being associated with surface effects such as locally enhanced sputtering or preferential sticking in visibly dark surface regions, and in favor of variation of the plasma associated with the wobbling of the background jovian magnetic field, as has been seen at Io. However, as pointed out by *McGrath et al.* (2000) and *Ballester et al.* (2007), unlike at Io, there is not a straightforward correlation between the emission morphology and the change in background field orientation.

Finally, regarding the oxygen observations, we should note that as mentioned briefly in section 2.1, the intensity ratio of O I 1356 to O I 1304 — which is indicative of the source process (e.g., process (4) or (5), outlined in section 2.1, or both) — appears to exhibit significant variability. *Hall et al.* (1998) found a range of 1–2, yet argued that this was consistent with no measureable atomic oxygen in all cases. *Hansen et al.* (2005) find a ratio of 1.8–2.1, but argue for the detectable presence of atomic oxygen in the bound atmosphere. Even the higher value of ~2 is only at best marginally consistent with the best available cross-section data and our knowledge of the electron temperature near Europa. It may indeed be the case that variability of the atomic emission line intensities is indicative of variations in both O_2 and O abundances, and variation in the abundance of one relative to the other.

The ionosphere observations show a range of levels of detection, including two nondetections, the E6a exit (observation #12) and the E26 entry (observation #15). It's

TABLE 3. Summary of Europa's atmosphere, neutral clouds, and torus.

Species	Density	Location	Reference
O_2	$(2.4–14) \times 10^{14}$ cm^{-2}	line of sight	1,2,3
	$(3.7–6.2) \times 10^{7}$ cm^{-3}	mean (H = 200 km)	2
	$(5–10) \times 10^{9}$ cm^{-3}	surface	4,5
O	$(1.7–3.1) \times 10^{13}$ cm^{-2}	line of sight	2
	$(0.85–1.5) \times 10^{6}$ cm^{-3}	mean (H = 200 km)	2
	1000–1700 cm^{-3}	<22,000 km	this work
	few × 10^{4} cm^{-3}	surface	4,5
O/O_2	<0.1	H ~ 20–300 km	1
	~0.02	H = 200 km	2
n_e (ionosphere)	peak ~ 10^3–10^4 cm^{-3}	<300 km	6
Na	$(4–0.4) \times 10^{9}$ cm^{-2}	5–25 R_E	7,8
	~100 cm^{-3}	surface	7,8
K	~0.04 × Na	5–13 R_E	8
Na/K	25	5–13 R_E	8
H or H_2 torus	~40 cm^{-3}	r ~ 2 R_J?	9
O torus	<8 cm^{-3}	if r ~ 2 R_J	2

References: [1] *Hall et al.* (1998); [2] *Hansen et al.* (2005); [3] *Saur et al.* (1998); [4] *Shematovich et al.* (2005); [5] *Smyth and Marconi* (2006); [6] *Kliore et al.* (1997, 2006); [7] *Brown and Hill* (1996); [8] *Brown* (2001); [9] *Mauk et al.* (2003).

interesting to note that the E4 ionosphere occultation occurred for a Europa geometry (orbital, sunlight, plasma bombardment) that is nearly identical to that for HST observation #4 (shown in Figs. 4 and 5), except Galileo is looking toward the upstream direction, while HST is looking toward the downstream direction. This is the HST observation that shows the brightest 1356 Å emission region in the antijovian hemisphere, which corresponds to the E4 exit measurement that shows a very weak, low-scale height electron density profile. The geometry (orbital, sunlight, plasma bombardment) for the E6a radio occultation is nearly identical to Cassini observation #6. The E6a entry profile is along the ram direction, which may explain the compressed, low-scale height nature of this profile, whereas the E6a exit profile is along the middle of the wake, and is a nondetection. This Cassini observation corresponds to an oxygen density at the low end of the range observed in the six sets of observations. Finally, the E6b exit profile is over the longitude (256°W) at the center of the HST observation #4 images shown in Fig. 4, where there is no obvious 1356 Å emission.

The variations seen in the Na observations are complex, and have been explored in detail in *Leblanc et al.* (2005). One interesting conclusion they draw is that the total emission intensity from one measurement to another cannot be explained solely by the differences in the geometry of the observations, and it is likely that there was a significant variation in the source rate from November 28, 2000, to November 30, 2000. This is a rich dataset that is continuing to be exploited with ongoing investigations and modeling (e.g., *Cipriani et al.*, 2008). The current thinking concerning the variations seen closest to the satellite is described further in the chapter by Johnson et al. (see their Fig. 2).

6. SUMMARY: OUTSTANDING ISSUES AND FUTURE WORK

We summarize in Table 3 our current knowledge of Europa's tenuous atmosphere, neutral clouds, and torus. Although substantial progress has been made in characterizing them in the past 14 years since the first detection of oxygen emission, the data continues to be very sparse. We have only a rudimentary understanding of the source(s) and sinks for the atmosphere; in particular, it is unclear if localized sources such as geysers are present. A better understanding may come only when additional atmospheric species, such as OH, H_2, H_2O, CO, or CO_2, are detected. For instance, a substantial although transient H_2O sublimation atmosphere will be present in the equatorial regions when the subsolar point is over the icy leading hemisphere but may be absent when the subsolar point is over the trailing hemisphere, which is likely dominated by water of hydration. Such variability will affect the ionosphere. Since the surface materials are decomposed by radiolysis a large number of products other than O_2 and H_2 should be present, like the Na and K components discussed above. Carbon dioxide has been detected in the surface ice (*Smythe et al.*, 1998) and appears to be correlated with the dark terrain on the antijovian and leading hemispheres (*Hansen and McCord,* 2008). Although the CO_2 band depth is comparable to that measured at Callisto, no measurement of CO_2 in vapor form has been made at Europa. *Hansen and McCord* (2008) suggest an internal, subsurface ocean source for the surface CO_2. Searches for CO_2 and its products (CO, C) should be performed at Europa in future. Given that the CO_2 atmosphere on Callisto has not been detected from Earth, detections may be difficult to accomplish by remote sensing. Techniques such as those currently being used on Cassini (including mass spectrometers, stellar occultations, and detailed plasma characterization), as well as high-resolution IR spectroscopy for identification of surface constituents, would be useful tools on future missions to Europa or the jovian system.

Interpretation of the observations is also hampered by the lack of understanding of the local plasma environment, and its variability, at Europa. Further *in situ* characterization is critical because, unlike Io, the plasma near Europa is too tenuous to be observed with current groundbased and Earth-orbit techniques. Until further data are obtained, continued work on detailed interaction models such as those of *Saur et al.* (1998) and *Shilling et al.* (2008) are indispensible in trying to make further progress interpreting the data already in hand.

There is also additional worthwhile work to be done on the existing data. For example, all the GHRS spectra acquired by HST contain temporal information that has yet to be exploited. Upper limits for CO and CO_2 abundances could be estimated from the HST spectra, as has been done for Callisto (*Strobel et al.,* 2002). The HST STIS is scheduled to be repaired during Servicing Mission 4 in 2009. Further observations of the oxygen emissions such as those shown in Figs. 4 and 5 would be very useful. It would be helpful if HST and/or groundbased observations could be done to confirm and/or extend the Cassini eclipse images shown in Fig. 12.

Perhaps the best immediate route for progress is groundbased observing programs aimed at characterizing the Na and K clouds at Europa. Such observations are clearly feasible, and until further observations of the main constituents are obtained, Na and K serve as very useful proxies for understanding the surface/plasma source and loss processes operative at Europa.

Acknowledgments. We would like to sincerely thank A. Campbell for providing Figs. 4 and 5; K. Retherford for sharing results prior to publication; and P. Geissler and A. Kliore for sharing details about their Europa research. The research described in this chapter carried out at the Jet Propulsion Laboratory, California Institute of Technology, was done so under a contract with the National Aeronautics and Space Administration.

REFERENCES

Ballester G. E., Herbert F., Kivelson M., Khurana K. K., and Combi M. R. (2007) Studies of Europa's plasma interactions and atmosphere with HST/STIS far-UV images (abstract). In *Magnetospheres of the Outer Planets,* p. 98. Southwest Research Institute, San Antonio.

Barth C. A., Hord C. W., Stewart A. I. F., Pryor W. R., Simmons K. E., McClintock W. E., Ajello J. M., Naviaux K. L., and Aiello J. J. (1997) Galileo ultraviolet spectrometer observations of atomic hydrogen in the atmosphere at Ganymede. *Geophys. Res. Lett., 24,* 2147–2150.

Broadfoot A. L., Belton M. J., Takacs P. Z., Sandel B. R., Shemansky D. E., Holberg J. B., Ajello J. M., Moos H. W., Atreya S. K., Donahue T. M., Bertaux J. L., Blamont J. E., Strobel D. F., McConnell J. C., Goody R., Dalgarno A., and McElroy M. B. (1979) Extreme ultraviolet observations from Voyager 1 encounter with Jupiter. *Science, 204,* 979–982.

Brown M. E. (2001) Potassium in Europa's atmosphere. *Icarus, 151,* 190–195.

Brown M. E. and Hill R. E. (1996) Discovery of an extended sodium atmosphere around Europa. *Nature, 380,* 229–231.

Brown R. A. (1974) Optical line emission from Io. In *Exploration of the Planetary System* (A. Woszczyk and C. Iwaniszewska, eds.), pp. 527–531. IAU Symposium 65, Reidel, Dordrecht.

Brown W. L., Augustyniak W. M., Lanzerotti L. J., Poate J. M., and Augustyniak W. M. (1977) *Eos Trans. AGU, 58,* 423.

Brown W. L., Augustyniak W. M., Lanzerotti L. J., Johnson R. E., and Evatt R. (1980) Linear and nonlinear processes in the erosion of H_2O ice by fast light ions. *Phys. Rev. Lett., 45,* 1632–1635.

Burger M. H. and Johnson R. E. (2004) Europa's neutral cloud: Morphology and comparisons to Io. *Icarus, 171,* 557–560.

Carlson R. W., Bhattacharyya J. C., Smith B. A., Johnson T. V., Hidayat B., Smith S. A., Taylor G. E., O'Leary B., and Brinkmann R. T. (1973) An atmosphere on Ganymede from its occultation of SAO 186800 on 7 June 1972. *Science, 182,* 53–55.

Cassidy T. A., Johnson R. E., McGrath M. A., Wong M. A., and Cooper J. F. (2007) The spatial morphology of Europa's near-surface O_2 atmosphere. *Icarus, 191,* 755–764.

Cassidy T. A., Johnson R. E., Geissler P. E., and Leblanc F. (2008) Simulation of Na D emission near Europa during eclipse. *J. Geophys. Res., 113,* E02005.

Cipriani F., Leblanc F., and Witasse O. (2008) Sodium recycling at Europa: What do we learn from the sodium cloud variability? *Geophys. Res. Lett., 35,* L19201, DOI: 10.10292008GR035061.

Doering J. P. and Yang J. (2001) Atomic oxygen $^3P–^3S^o$ (λ1304 Å) transition revisited: Cross section near threshold. *J. Geophys. Res., 206,* 203–210.

Feldman P. D., McGrath M. A., Strobel D. F., Moos H. W., Retherford K. D., and Wolven B. C. (2000) HST/STIS ultraviolet imaging of polar aurora on Ganymede. *Astrophys. J., 535,* 1085–1090.

Frank L. A., Ackerson K. L., Wolfe J. H., and Mihalov J. D. (1976) Observations of plasmas in the jovian magnetosphere. *J. Geo-phys. Res., 81,* 457.

Frank L. A., Paterson W. R., and Khurana K. K. (2002) Observations of thermal plasmas in Jupiter's magnetotail. *J. Geophys. Res., 107(A1),* DOI: 10.1029/2001JA000077.

Geissler P., McEwen A., Porco C., Strobel D., Saur J., Ajello J., and West R. (2004) Cassini observations of Io's visible aurorae. *Icarus, 172,* 127–140.

Hall D. T., Strobel D. F., Feldman P. D., McGrath M. A., and Weaver H. A. (1995) Detection of an oxygen atmosphere on Jupiter's moon Europa. *Nature, 373,* 677–679.

Hall D. T., Feldman P. D., McGrath M. A., and Strobel D. F. (1998) The far-ultraviolet oxygen airglow of Europa and Ganymede. *Astrophys. J., 499,* 475–481.

Hansen C. J., Shemansky D., and Hendrix A. (2005) Cassini UVIS Observations of Europa's oxygen atmosphere and torus. *Icarus, 176,* 305–315.

Hansen G. B. and McCord T. B. (2008) Widespread CO_2 and other non-ice compounds on the anti-jovian and trailing sides of Europa from Galileo/NIMS observations. *Geophys. Res. Lett., 35,* L01202.

Hendrix A. R. and Hansen C. J. (2008) Ultraviolet observations of Phoebe from the Cassini UVIS. *Icarus, 193,* 323–333.

Johnson R. E. (2002) Surface boundary layer atmospheres. In *Atmospheres in the Solar System: Comparative Aeronomy* (M. Mendillo et al., eds.), pp. 203–219. AGU Geophysical Monograph 130, Washington, DC.

Johnson R. E., Lanzerotti L. J., and Brown W. L. (1982) Planetary applications of ion induced erosion of condensed gas frosts. *Nucl. Instr. Meth.Phys. Res., A198,* 147–158.

Johnson R. E., Leblanc F., Yakshinskiy B. V., and Madey T. E. (2002) Energy distributions for desorption of sodium and potassium from ice: The Na/K ratio at Europa. *Icarus, 156,* 136–142.

Judge D. L. and Carlson R. W. (1974) Pioneer 10 observations of the ultraviolet glow in the vicinity of Jupiter. *Science, 183,* 317.

Judge D. L., Carlson R. W., Wu F.-M., and Hartmann U. G. (1976) Pioneer 10 and 11 ultraviolet photometer observations of the jovian satellites. In *Jupiter* (T. Gehrels et al., eds.), p. 1068. Univ. of Arizona, Tucson.

Kabin K., Combi M. R., Gombosi T. I., Nagy A. F., DeZeeuw D. L., and Powell K. G. (1999) On Europa's magnetospheric interaction: A MHD simulation of the E4 flyby. *J. Geophys. Res., 104(A9),* 19983–19992.

Kanik I., Noren C., Makarov O. P., Vattipalle P., and Ajello J. M. (2003) Electron impact dissociative excitation of O_2: 2. Absolute emission cross sections of the O I (130.4 nm) and O I (135.6 nm) lines. *J. Geophys. Res., 108,* 5126, DOI: 10.1029/2000JE001423.

Kivelson M. G., Khurana K. K., Coroniti F. V., Joy S., Russell C. T., Walker R. J., Warnecke J., Bennett L., and Polanskey C. (1997) Magnetic field and magnetosphere of Ganymede. *Geo-phys. Res. Lett., 24,* 2155.

Kliore A., Cain D. L., Fjeldbo G., Seidel B. L., and Rasool S. I. (1974) Preliminary results on the atmospheres of Io and Jupiter from the Pioneer 10 S-band occultation experiment. *Science, 183,* 323–324.

Kliore A. J., Fjeldbo G., Seidel B. L., Sweetnam D. N., Sesplaukis T. T., Woiceshyn P. M., and Rasool S. I. (1975) The atmosphere of Io from Pioneer 10 radio occultation measurements. *Icarus, 24,* 407–410.

Kliore A. J., Hinson D. P., Flasar F. M., Nagy A. F., and Cravens T. E. (1997) The ionosphere of Europa from Galileo radio occultations. *Science, 277,* 355–358.

Kliore A. J., Anabtawi A., and Nagy A. F. (2001) The ionospheres of Europa, Ganymede, and Callisto. *Eos Trans. AGU, 82(47),* Fall Meet. Suppl., Abstract #P12B-0506.

Kliore A. J., Nagy A. F., Flasar F. M., Schinder P. J., and Hinson D. P. (2006) Radio science data on the ionospheres of Jupiter, Saturn, and their major satellites. *36th COSPAR Scientific Assembly,* Beijing, China, Abstract #2599.

Kumar S. (1982) Photochemistry of SO_2 in the atmosphere of Io and implications on atmospheric escape. *J. Geophys. Res., 87,* 1677–1684.

Kupo I., Mekler Y., and Eviatar A. (1976) Detection of ionized sulfur in the jovian magnetosphere. *Astrophys. J. Lett., 205,* L51–L53.

Kurth W. S., Gurnett D. A., Persoon A. M., Roux A., Bolton S. J., and Alexander C. J. (2001) The plasma wave environment of Europa. *Planet. Space Sci., 49,* 345–363.

Lagg A, Krupp N., Woch J., and Williams D. J. (2003). In-situ observations of a neutral gas torus at Europa. *Geophys. Res. Lett., 30,* DOI: 10.1029/2003GL017214.1556.

Lanzerotti L. J., Brown W. L., Poate J. M., and Augustyniak W. M. (1978) On the contribution of water products from Galilean satellites to the jovian magnetosphere. *Geophys. Res. Lett., 5,* 155–158.

Leblanc F., Johnson R. E., and Brown M. E. (2002) Europa's sodium atmosphere: An ocean source? *Icarus, 159,* 132.

Leblanc F., Potter A. E., Killen R. M., and Johnson R. E. (2005) Origins of Europa Na cloud and torus. *Icarus, 178,* 367–385.

Liu Y., Nagy A. F., Kabin K., Combi M. R., Dezeeuw D. L., Gombosi T. I., and Powell K. G. (2000) Two-species, 3D, MHD simulation of Europa's interaction with Jupiter's magnetosphere. *Geophys. Res. Lett., 27(12),* 1791–1794.

Mauk B. H., Mitchell D. G., Krimigis S. M., Roelof E. C., and Paranicas C. P. (2003) Energetic neutral atoms from a trans-Europa gas torus at Jupiter. *Nature, 421,* 920–922.

McDonough T. R. and Brice N. M. (1973) A saturnian gas ring and the recycling of Titan's atmosphere. *Icarus, 20,* 136.

McGrath M. A., Feldman P. D., Strobel D. F., Retherford K., Wolven B., and Moos H. W. (2000) HST/STIS ultraviolet imaging of Europa. *Bull. Am. Astron. Soc., 32,* 1056.

McGrath M. A., Lellouch E., Strobel D. F., Feldman P. D., and Johnson R. E. (2004) Satellite atmospheres. In *Jupiter: The Planet, Satellites and Magnetosphere* (F. Bagenal et al., eds.), pp. 457–483. Cambridge Univ., Cambridge.

Ness N. F., Acuna M. H., Lepping R. P., Burlaga L. F., Behannon K. W., and Neubauer F. M. (1979) Magnetic field studies at Jupiter by Voyager 1 — Preliminary results. *Science, 204,* 982–987.

Neubauer F. M. (1980) Nonlinear standing Alfven wave current system at Io — Theory. *J. Geophys. Res., 85,* 1171–1178.

Noren C., Kanik I., Ajello J. M., McCartney P., Makarov O. P., McClintock W. E., and Drake V. A. (2001) Emission cross section of O I (135.6 nm) at 100 eV resulting from electron-impact dissociative excitation of O_2. *Geophys. Res. Lett., 28,* 1379+.

Paterson W. R., Frank L. A., and Ackerson K. L. (1999) Galileo plasma observations at Europa: Ion energy spectra and moments. *J. Geophys. Res., 104,* 22779–22792.

Pilcher C. B., Ridgway S. T., and McCord T. B. (1972) Galilean satellites: Identification of water frost. *Science, 178,* 1087–1089.

Porco C. and 23 colleagues (2003) Cassini imaging of Jupiter's atmosphere, satellites, and rings. *Science, 299,* 1541.

Retherford K. D., Spencer J. R., Gladstone G. R., Stern S. A., Saur J., Strobel D. F., Slater D. C., Steffl A. J., Parker J. W., Versteeg M., Davis M. W., Throop H., and Young L. A. (2007) Icy Galilean satellite UV observations by New Horizons and HST. *Eos Trans. AGU, 88(52),* Fall Meet. Suppl., Abstract #P53C-06.

Saur J., Strobel D. F., and Neubauer F. M. (1998) Interaction of the jovian magnetosphere with Europa: Constraints on the neutral atmosphere. *J. Geophys. Res., 103,* 19947–19962.

Schilling N., Neubauer F. M., and Saur J. (2008) Influence of the internally induced magnetic field on the plasma interaction of Europa. *J. Geophys. Res., 113,* A03203, DOI: 10.1029/2007JA012842.

Shematovich V. I. and Johnson R. E. (2001) Near-surface oxygen atmosphere at Europa. *Adv. Space Res., 27,* 1881–1888.

Shematovich V. I., Johnson R. E., Cooper J. F., and Wong M. C. (2005) Surface-bounded atmosphere of Europa. *Icarus, 173,* 480–498.

Smyth W. H. and Marconi M. L. (2006) Europa's atmosphere, gas tori, and magnetospheric implications. *Icarus, 181,* 510–526.

Smythe W. D., Carlson R. W., Ocampo A., Matson D. L., McCord T. B., and the NIMS Team (1998) Galileo NIMS measurements of the absorption bands at 4.03 and 4.25 microns in distant observations of Europa. *Bull. Am. Astron. Soc., 30,* Abstract #55P.07.

Strobel D. F., Saur J., Feldman P. D., and McGrath M. A. (2002) Hubble Space Telescope imaging spectrograph search for an atmosphere on Callisto: A jovian unipolar inductor. *Astrophys. J. Lett., 581,* L51–L54.

Trafton L. (1975). Detection of a potassium cloud near Io. *Nature, 258,* 690–692.

Wu F.-M., Judge D. L., and Carlson R. W. (1978) Europa: Ultraviolet emissions and the possibility of atomic oxygen and hydrogen clouds. *Astrophys. J., 225,* 325–334.

Yung Y. L. and McElroy M. B. (1977) Stability of an oxygen atmosphere on Ganymede. *Icarus, 30,* 97–103.

Composition and Detection of Europa's Sputter-induced Atmosphere

R. E. Johnson
University of Virginia

M. H. Burger
University of Maryland/NASA Goddard Space Flight Center

T. A. Cassidy
University of Virginia

F. Leblanc
Service d'Aéronomie du Centre National de la Recherce Scientifique

M. Marconi
Prima Basic Research, Inc.

W. H. Smyth
Atmospheric and Environmental Research, Inc.

An intense flux of ions and electrons from the jovian magnetosphere both alters and erodes Europa's surface, producing a tenuous atmosphere. This chapter discusses the physical processes that create and remove the atmosphere, such as ion erosion (sputtering), radiation-induced chemical alteration of the surface (radiolysis), and atmospheric loss processes such as ionization and pickup. Special emphasis is placed on ongoing modeling efforts to connect atmospheric properties to surface composition. Models are at present constrained by a small number of observations, but a future mission to Europa has the potential to detect even trace atmospheric species.

1. INTRODUCTION

Europa has a tenuous atmosphere, which, unlike that of Titan, Io, or Callisto, is only marginally collisional. Because the atmosphere is quasi-collisional, it is often referred to as an exosphere. Alternatively, and more relevant to the present discussion, it is also called a surface boundary-layer atmosphere (*Johnson,* 2002), i.e., as at Mercury, the Moon, and Ganymede, the interaction of the ambient gas with the surface determines its composition, local column density, and morphology. All molecules and atoms in the atmosphere are quickly lost either to space or the surface, and lifetimes are measured in minutes to days, depending on the species. Neutrals that escape Europa's gravity mostly remain gravitationally bound to Jupiter in a toroidal-shaped cloud with peak densities near Europa's orbital path.

Since gas-phase species are readily identified both *in situ* and by remote sensing, Europa's atmosphere is of interest as an extension of its surface. If this atmosphere was only populated by thermal desorption (i.e., evaporation), it would consist only of *trace volatiles* that would be rapidly depleted, plus a small subsolar component, $<\sim 10^{12}$ H_2O/cm^2 depending on the ice fraction in the surface (*Shematovich et al.,* 2005; *Smyth and Marconi,* 2006). However, Europa orbits in a region of the jovian magnetosphere in which the trapped plasma density and temperature are relatively high, and its surface is exposed to the solar EUV flux. This radiation chemically alters and erodes the surface with the ejecta populating the atmosphere, as seen in Fig. 1. These processes are often lumped together using the words sputter-produced atmosphere, by analogy with the industrial process in which gas-phase species are produced from refractory materials by plasma bombardment, but the individual processes are discussed in the following.

Early laboratory sputtering data by Brown, Lanzerotti, and co-workers (e.g., *Brown et al.,* 1982) were used to predict the principal atmospheric component, O_2, and its average column density (*Johnson et al.,* 1982). These predictions were confirmed over a decade later (see chapter by McGrath et al.). Since loss of H_2 accompanies the formation and ejection of O_2 from ice (*Johnson and Quickenden,* 1997) and it escapes more readily than the heavier species, H_2 is the principal species in Europa's neutral torus. There are approximately three times as many molecules and atoms

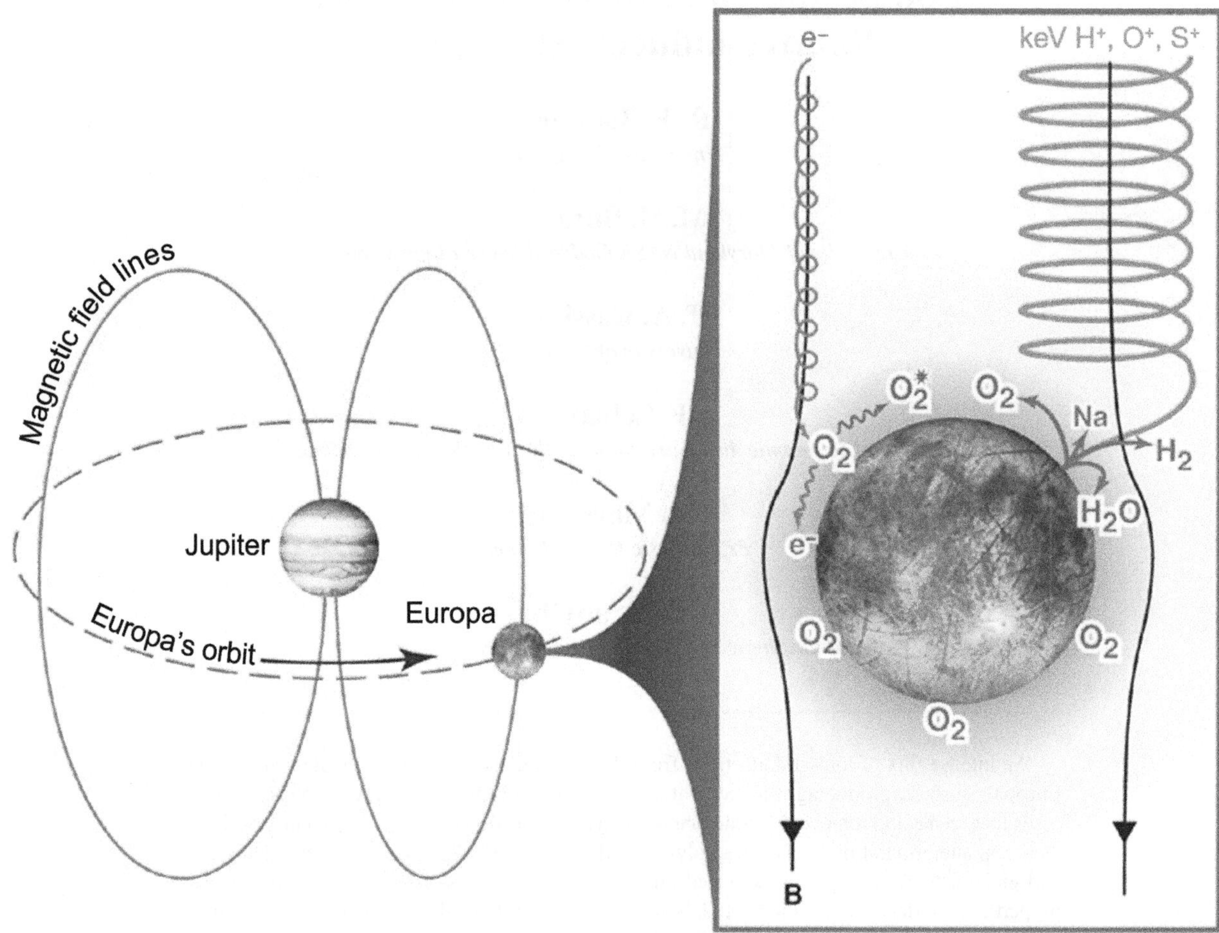

Fig. 1. See Plate 24. Schematic of Europa's interaction with Jupiter's magnetosphere (*Johnson et al.*, 2004). Ions and electrons trapped by Jupiter's magnetic field alter and erode the surface, producing a tenuous atmosphere composed mostly of O_2 with an extended neutral torus of primarily H_2.

in this torus than in Io's neutral torus (*Smyth and Marconi,* 2006). This torus is both a primary source of protons for the jovian magnetosphere and an obstacle to the inward diffusing energetic plasma that significantly affects the properties inside of Europa's orbit (see chapter by Paranicas et al.).

Models for the surface composition, the radiation flux to the surface, and laboratory data have been used in atmospheric simulations in order to interpret the available observations, to refine predictions of Europa's atmospheric structure and composition, and to suggest which sputtered species, other than water products, might be detectable by an orbiting spacecraft. Although the only other neutrals identified to date are sodium and potassium, modeling has shown the importance of the interaction of returning atoms and molecules with Europa's porous regolith (*Leblanc et al.,* 2005; *Cassidy et al.,* 2007), i.e., in a nearly collisionless surface-boundary-layer atmosphere, redistribution and loss to the surface compete with other loss processes such as pick-up (where a newly ionized particle is swept away by the moving magnetospheric plasma), direct escape, and escape of hot (fast) dissociation products. In this chapter we review the modeling of Europa's gravitationally bound atmosphere and neutral cloud. Our principal interest is in the morphology of the atmosphere and the relationship between its composition and Europa's surface composition. The possibility of detection of atmospheric species, either remotely or *in situ* by an orbiting spacecraft, is considered, and their potential relevance to Europa's putative subsurface ocean is discussed.

2. SUMMARY OF PRESENT KNOWLEDGE

2.1. Atmospheric Observations

The first observations of Europa's atmosphere were made by *Hall et al.* (1995) using the Hubble Space Telescope (HST) (see chapter by McGrath et al.). Based on the ratio of atomic oxygen line intensities at 1304 Å and 1356 Å, they concluded that the emissions resulted from electron impact dissociation excitation of O_2 in Europa's atmosphere. Space

Telescope Imaging Spectrometer observations indicated that these emissions were spatially inhomogeneous (*McGrath et al.,* 2004; *Cassidy et al.,* 2007).

The Imaging Science Subsystem on Cassini obtained a series of images of the trailing hemisphere while Europa was in Jupiter's shadow (*Porco et al.,* 2003; see chapter by McGrath et al.). Limb brightening in those images indicated atmospheric emission and the Ultraviolet Imaging Spectrograph (UVIS) detected both hydrogen and oxygen (*Hansen et al.,* 2005). The sharply peaked O I 1356 Å emission line suggested O_2 emission near the surface, while the broader 1304 Å emission line suggested a tenuous extended atomic O component. While no variability was detected in the atmosphere on timescales comparable to the rotation period of Jupiter's magnetic field (10 hours), they did measure intensity differences between observations made six days apart, suggesting temporal variability in the atmosphere or the impacting plasma.

Observations of Europa's ionosphere can constrain both the surface composition and the escape rate, since freshly formed ions are picked-up by the magnetic field and can be swept away (see chapter by Kivelson et al.). The radio signal from the Galileo spacecraft to Earth on the inbound and outbound portions of four occultations showed ionospheric refraction during seven of the eight observation opportunities (*Kliore et al.,* 1997). The ionosphere was found to vary both longitudinally and with Europa's local time (orbital position with respect to the Sun). The maximum electron density of ~10^4 cm^{-3} was observed over the trailing hemisphere. The measured scale height, 240 ± 40 km for alititudes less than 300 km, and 440 ± 60 km for altitudes above 300 km, differed significantly from the 20-km scale height predicted for O_2 (e.g., *Shematovich et al.,* 2005). The nondetection occurred over the downstream hemisphere near the center of the plasma wake. Although not confirmed by modeling, it was suggested that presence of a detectable ionosphere required the trailing (upstream) hemisphere to be partially illuminated. From these observations, Europa has an anisotropic ionosphere with electron densities that reach ~10,000 cm^{-3} near the upstream hemisphere surface.

Using the Galileo PLS instrument, *Paterson et al.* (1999) found maximum ion densities of about 40 cm^{-3} near the closest approach altitudes of 700 km and 600 km on the E4 and E6 flybys. This is several times higher than the average ion densities near Europa's orbit. Pickup ions corresponding to H^+, H_2^+ and mass 16–19 amu (H_2O^+, H_3O^+, OH^+, and O^+ could not be distinguished) were detected downstream of Europa on both flybys. Using the Galileo PWS instrument, strong enhancements in electron density, including a value of 275 cm^{-3} in the downstream wake region during the E11 flyby, were reported (*Kurth et al.,* 2001), as were ion-cyclotron waves associated with pickup ions. Using the Galileo magnetometer, ion-cyclotron waves in Europa's magnetospheric wake were associated with O_2^+, Na^+ and/or Mg^+, Ca^+ and/or K^+, Cl^+, and Cl^- (*Volwerk et al.,* 2001). These ion-cyclotron waves are generated by freshly produced ions in a moving plasma and, since these species were detected near Europa, they are presumably representative of the surface composition.

2.2. Radiation Environment

Properly modeling Europa's atmosphere requires understanding the state of the plasma flowing through it. The undisturbed plasma depends on Europa's position in the Io plasma torus. Because of the 0.13 R_J offset and 9.6° tilt of Jupiter's magnetic dipole, Europa's magnetic coordinate L, defined as the distance from the dipole center that a magnetic field line crosses the magnetic equator, varies from 9.4 R_J to 9.7 R_J as a function of magnetic longitude. In addition, near Europa, plasma is convected back and forth across magnetic field lines as it moves around Jupiter by an east-west electric field, and the magnetic field lines are distorted by an azimuthal current sheet centered on the magnetic equator and stretching to ~50 R_J. Therefore, Europa sees a plasma flux that varies with its local time and magnetic (System III) longitude.

Unlike the cooler, denser plasma near Io, the plasma near Europa is too tenuous to produce emissions bright enough for groundbased observations. Using Voyager data, *Bagenal* (1994) produced a two-dimensional model of the plasma that extends to Europa's orbit. At Europa's mean L shell (9.4 R_J), the electrons were described in the centrifugal equator by a cold component with $n_{e,c}$ = 130 cm^{-3} and $T_{e,c}$ = 18 eV and a hot component with $n_{e,h}$ = 3 cm^{-3} and $T_{e,h}$ = 190 eV. Assuming a longitudinally symmetric torus centered on Jupiter's dipole, the electron density varies between ~70 cm^{-3} and 160 cm^{-3} over Jupiter's rotational period. Nondipolar effects and variable plasma sources affect these numbers: Galileo observations show electron densities varying between 18 cm^{-3} and 250 cm^{-3} with a mean value ~200 cm^{-3} in the centrifugal equator (*Moncuquet et al.,* 2002; *Kivelson,* 2004). On its approach to Jupiter, Cassini UVIS observations of the energy content of the Io plasma torus show a 25% variation over two months and shorter brightening events, lasting about 20 hours, with up to 20% variation (*Steffl et al.,* 2004). Since this outwardly diffusing plasma both supplies and removes Europa's atmosphere, the atmospheric content is likely variable, although there are currently no observations that correlate the plasma variability with atmospheric content.

The ion *number density* is dominated by a cold thermal component with a small population of nonthermal "hot" ions. The thermal plasma is primarily singly and doubly ionized oxygen and sulfur with a significant population of hydrogen ions. Near Europa, O^+ and O^{++} dominate, with S^+, S^{++}, and S^{+++} occurring in roughly equal amounts and the S^{+z}/O^{+z} ratio is ~0.2, but variable in time. Because the mean charge state is ~1.5, the ion density is lower than the electron density (*Bagenal,* 1994; *Steffl et al.,* 2004).

In contrast, the plasma ion *energy density* is dominated by an energetic nonthermal component. Energetic ion spectra from 20 keV to 20 MeV were obtained by the Galileo

TABLE 1. Photoreactions* (from *Huebner et al.,* 1992) and electron impact reactions.

Photoreactions		
Reaction	Rate (s^{-1})	
$O + h\nu \rightarrow O^+ + e$	7.8×10^{-9}	
$O_2 + h\nu \rightarrow O + O$	1.6×10^{-7}	
$O_2 + h\nu \rightarrow O_2^+ + e$	1.7×10^{-8}	
$O_2 + h\nu \rightarrow O + O^+ + e$	4.1×10^{-9}	
$Na + h\nu \rightarrow Na^+ + e$	6.0×10^{-7}	
$H + h\nu \rightarrow H^+ + e$	2.7×10^{-9}	
$H_2 + h\nu \rightarrow H + H$	3.0×10^{-9}	
$H_2 + h\nu \rightarrow H_2^+ + e$	2.0×10^{-9}	
$H_2 + h\nu \rightarrow H + H^+ + e$	3.5×10^{-10}	
$H_2O + h\nu \rightarrow OH + H$	3.8×10^{-7}	
$H_2O + h\nu \rightarrow H_2 + O$	2.2×10^{-8}	
$H_2O + h\nu \rightarrow H + H + O$	2.8×10^{-8}	
$H_2O + h\nu \rightarrow H_2O^+ + e$	1.2×10^{-8}	
$H_2O + h\nu \rightarrow H + OH^+ + e$	2.0×10^{-9}	
$H_2O + h\nu \rightarrow OH + H^+ + e$	4.8×10^{-10}	
Electron Impact Reactions		
Reaction	k(20 eV) ($cm^3\ s^{-1}$)	k(250 eV) ($cm^3\ s^{-1}$)
$O + e \rightarrow O^+ + 2e$ [1]	2.2×10^{-8}	1.1×10^{-7}
$O_2 + e \rightarrow O + O + e$ [2]	1.3×10^{-8}	—
$O_2 + e \rightarrow O_2^+ + 2e$ [3]	2.4×10^{-8}	1.1×10^{-7}
$O_2 + e \rightarrow O + O^+ + 2e$ [3]	8.9×10^{-9}	6.1×10^{-8}
$Na + e \rightarrow Na^+ + 2e$ [4]	1.1×10^{-7}	1.0×10^{-7}
$H + e \rightarrow H^+ + 2e$ [5]	1.3×10^{-8}	2.8×10^{-8}
$H_2 + e \rightarrow H + H + e$ [6]	2.7×10^{-10}	1.3×10^{-11}
$H_2 + e \rightarrow H_2^+ + 2e$ [7]	1.6×10^{-8}	4.4×10^{-8}
$H_2 + e \rightarrow H + H^+ + 2e$ [7]	8.1×10^{-10}	3.4×10^{-9}
$H_2O + e \rightarrow OH + H + e$ [8]	3.8×10^{-8}	—
$H_2O + e \rightarrow H_2 + O + e$ [8]	Unknown†	Unknown†
$H_2O + e \rightarrow H_2O^+ + 2e$ [8]	2.3×10^{-8}	8.0×10^{-8}
$H_2O + e \rightarrow H + OH^+ + 2e$ [8]	5.8×10^{-9}	2.6×10^{-8}
$H_2O + e \rightarrow OH + H^+ + 2e$ [8]	3.3×10^{-9}	2.3×10^{-8}

*Assumes Sun is in "quiet" state, normalized to 5.2 AU.
†According to [8], reaction cross sections have not been measured.
References: [1] *Johnson et al.* (2005); [2] *Cosby* (1993); [3] *Straub et al.* (1996); [4] *Johnston and Burrow* (1995); [5] *Shah et al.* (1987); [6] *De La Haye* (2005); [7] *Lindsay and Mangon* (2003); [8] *Itikawa and Mason* (2005).

Energetic Particle Detector (*Cooper et al.,* 2001; *Paranicas et al.,* 2002; chapter by Paranicas et al.) and at higher energies up to 50 MeV/n by the Galileo Heavy Ion Counter (*Cohen et al.,* 2001; *Paranicas et al.,* 2002). The peak flux occurs near 100 keV and diminishes rapidly at higher energies. The high-energy sulfur and oxygen ion components of this flux are primarily responsible for sputtering the surface and producing the atmosphere.

The surface layer altered by the impacting magnetospheric energetic ions and electrons varies from micrometers on short timescales to meters on timescales of a Gigayear (see chapter by Paranicas et al.). Infrared spectrometry typically probes only to a depth of a few micrometers. The surface is also mixed to meter depths by micrometeoroid impacts. The total impacting plasma energy flux is ~8×10^{10} keV/cm^{2}s, which is comparable to the energy flux due to tidal and radiogenic sources (*Cooper et al.,* 2001). The energetic component of the plasma flux has been observed to vary by up to a factor of 5, with peak flux around a few tens of keV (*Paranicas et al.,* 2002). At such energies, the rate of ejection of molecules due to S and O ion impacts is ~10 H_2O per ion (*Johnson et al.,* 2004). However, the energetic ion flux is significant out to higher energies, where over 1000 H_2O molecules are ejected per ion giving an average resurfacing rate of 0.01 μm to 0.1 μm per year (or 10 to100 m per G.y.) (*Cooper et al.,* 2001; *Paranicas et al.,* 2002). Although this rate is lower than the resurfacing rate induced by meteoroid gardening, sputtering is the principal source of Europa's atmosphere (*Shematovich et al.,* 2005; *Smyth and Marconi,* 2006).

The energetic electrons and the thermal plasma primarily irradiate the trailing hemisphere of Europa, whereas the energetic ion flux is roughly uniform in longitude and latitude (see chapter by Paranicas et al.). Therefore, the total dose rate at 0.1 mm depth may be 100 times smaller on the leading hemisphere and in polar regions than on the trailing hemisphere. This asymmetry, as well as the preferential bombardment of the trailing hemisphere by low-energy sulfur ions from the Io torus (*Pospeiszalska and Johnson,* 1989), may explain the strong asymmetry in the IR signature of the hydrate observed in Europa's surface (see chapter by Carlson et al.).

Not only does the magnetospheric plasma flux produce an atmosphere, it also removes atmosphere. Atmospheric sputtering by ion-neutral collisions, and dissociation and ionization of atmospheric molecules by low-energy (thermal) electrons, are important loss processes (Table 1) (*Saur et al.,* 1998; *Shematovich et al.,* 2005; *Smyth and Marconi,* 2006). This radiation also makes the atmosphere visible by producing the UV emissions from O_2.

2.3. Surface Composition

Europa's surface composition determines the composition of its atmosphere. The surface is predominantly water ice with impact craters, cracks, ridges, possibly melted regions, and trace species determining how its appearance varies with latitude and longitude (see chapter by Carlson et al.). Due to tidal heating, Europa's surface is relatively young, 20–180 m.y. (*Schenk et al.,* 2004), and material from its putative underground ocean may have reached the surface in its recent past. The trace surface species may also be exogenic — implanted as plasma ions from the jovian magnetosphere, as neutrals or grains from Io, or meteoroid and comet impacts.

The most apparent trace species, sulfur, is preferentially seen on its trailing hemisphere as a brownish tinge in the visible, but is also seen in the cracks and crevices as part of a banded structure. Associated with this feature, reflectance spectra in the UV and near IR suggest the presence of SO_2 inclusions in an ice matrix and a hydrated sulfur-containing species: hydrated sulfuric acid possibly containing hydrated sulfate salts (*McCord et al.,* 1998a,b, 1999, 2001, 2002; chapter by Carlson et al.). Spectral features that appear strongest on the leading hemisphere have been associated with CO_2 trapped in an ice matrix (*Smythe et al.,* 1998). Carbonic acid is also seen, but with a more uniform global distribution. These oxygen-rich sulfur and carbon species are produced by the radiolytic processing of implanted or intrinsic materials (*Johnson et al.,* 2004). Radiolytic processing might also be responsible for the oxidized nature of the surface, e.g., the surface is rich in sulfates instead of sulfides. Radiolysis produces O_2 and H_2 from water ice, and the H_2 leaves the surface more quickly. O_2 also leaves the surface, but some of it stays trapped in the ice matrix (*Johnson and Jesser,* 1997; *Teolis et al.,* 2005; *Hand et al.,* 2006). Radiolysis also produces the observed H_2O_2 and O_3 (*Johnson and Jesser,* 1997; *Johnson and Quickenden,* 1997; *Spencer and Calvin,* 2002; *Cooper et al.,* 2003). Therefore, Europa's surface is dominated by oxygen-rich species. Although radiolysis is a surface oxidizing process, Europa may have been oxidized during formation (*McKinnon and Zolensky,* 2003; see chapter by Zolotov and Kargel).

3. SPUTTERING AND VOLATILE PRODUCTION

3.1. Introduction

Following the Pioneer and Voyager spacecraft encounters with Jupiter and Saturn, *Brown et al.* (1982) and others carried out experiments in which water ice and other low temperature, condensed-gas solids were exposed to energetic ions and electrons. In addition to demonstrating that these solids were chemically altered by the incident radiation, they showed that molecules were ejected into the gas phase with surprising efficiencies, producing the atmospheres eventually observed on Europa and Ganymede. Since those early experiments a considerable amount of laboratory data has been accumulated. There are a number of reviews of the data relevant to the production of the thin atmosphere at Europa (*Johnson,* 1990, 1998, 2001; *Baragiola,* 2003; *Fama et al.,* 2008). In order to model this atmosphere, the sputtering efficiencies for ice by ions, electrons and photons are described, followed by a review of the decomposition and production of O_2 and H_2, the energy spectra of the ejected molecules, and estimates for sputtering of trace species.

3.2. Sputtering Yields

When an energetic ion is incident on an ice surface in a laboratory or in space, molecules are ejected (sputtered) due to both momentum transfer collisions between the incident ion and water molecules in the solid and electronic excitations of these molecules. The relative importance depends on the ion velocity and its charge state. Although the database is considerable (*people.virginia.edu/~rej/sputter_surface.html*), experiments have not been carried out on all ions and energies of interest. In order to understand the physics and chemistry occurring in the solid, the sputtering yield, the number of molecules ejected per incident ion, is typically given in terms of the energy that an ion deposits as it passes through a solid, dE/dx, often called the stopping power.

For an energetic ion, dE/dx has two components: one due to the electronic excitation of molecules in the solid, $(dE/dx)_e$, and one due to momentum transfer collisions, written as $(dE/dx)_n$, also called knock-on or elastic nuclear collisions. Since both quantities are proportional to the molecular density, n, one writes: $dE/dx = n (S_n + S_e)$. Estimates of S_n and S_e, called the nuclear and electronic stopping cross sections, can be obtained from the freeware program SRIM (*www.srim.org*) (*Ziegler et al.,* 1985). Using these quantities, the sputtering yields for water ice at low temperatures (<~80 K) are reasonably well fit by

$$Y_s(E, \theta) = [a_1\alpha(m_i)S_n(E) + \eta(v)S_e(E)^2]/\cos\theta^{(1 + x)} \quad (1a)$$

where θ is the angle of incidence measured from the surface normal. The angular term enhances the sputtering yield for ions with shallow angles of incidence (x ~ 0.3–0.7 depending slowly on ion mass and speed). The parameters a_1, α, η(v), and x for water ice *at low incident ion speed,* v, are available in *Fama et al.* (2008) and are used for common ions near Europa in Fig. 2. At high v, the S_e term dominates and the coefficient η decreases with increasing v due to the increase in the radial distribution of the excitations (*Johnson,* 1990).

Using the dependence of S_e on v, a fit to a large amount of sputtering data over a broad range of energies for normal incidence (cosθ = 1) has been given (*Johnson et al.,* 2004, Fig. 20.6). This fit, shown by the solid line in Fig. 2, is obtained from

$$Y = 1/(1/Y_{low} + 1/Y_{high}) \quad (1b)$$

where Y_{low} and Y_{high} are fits in two different regimes of ion energies. Both have the form $Y_i = Z^{2.8}C_1[v/Z^{1/3}]^{C_2}$, where v is the ion speed divided by 2.19×10^8 cm s^{-1} (=1 au, the speed of a ground state electron in the Bohr hydrogen model) and Z is the number of protons in the incident ion. For Y_{low}, $C_1 = 4.2$ and $C_2 = 2.16$; for Y_{high}, $C_1 = 11.22$ and $C_2 = -2.24$. This fit is most useful when the electronic sputtering effects dominate ($S_e \gg S_n$) and the angular dependence is like that in equation (1a).

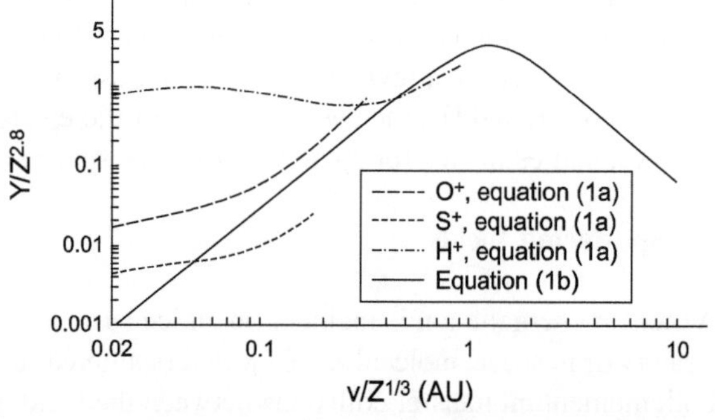

Fig. 2. Sputtering yields of water ice for T < ~100 K as a function of incident ion speed v (divided by 2.19×10^8 cm s^{-1}, or 1 atomic unit of velocity). Solid line: equation (1b), a fit to the data points available at *people.virginia.edu/~rej/sputter_surface.html.* The other lines are low-energy plots of equation (1a) for relevant ions using parameters from *Fama et al.* (2008). Equation (1a) is more accurate than equation (1b) at low ion speeds, where nuclear elastic effects are important.

Brown and coworkers (e.g., *Brown et al.,* 1982) also showed that the yield was dependent on the ice temperature at T > ~100 K. This dependence is due to the chemistry induced by the incident radiation, a process often referred to as radiolysis (see chapter by Carlson et al.). The primary effect for Europa's atmosphere is the decomposition of ice producing H_2 and O_2. An activation energy, E_a ~ 0.03–0.06 eV, determines the temperature dependence of the decomposition rate, which is also dose dependent at low doses (*Johnson et al.,* 2005). After a minimum dose of ~10^{15} ions/cm^2, during which reactants build up, a rough steady state is achieved (*Reimann et al.,* 1984; *Teolis et al.,* 2005). Using the accumulated database the steady state yield in equivalent H_2O molecules ejected can be written

$$Y(T) \approx Y_s[1 + a_2\exp(-E_a/kT)] \quad (2)$$

For low v ions, a_2 ~ 220 and E_a ~ 0.06 eV (e.g., *Fama et al.,* 2008), for energetic heavy ions relevant to Europa, *Reimann et al.* (1984) find E_a ~ 0.05–0.07 eV (see also *Teolis et al.,* 2005). The production of O_2 is equal to the second term in equation (2), and is roughly half that of H_2. For the ion flux at Europa, the H_2O yield is on the order of 10^9/cm^2 s^{-1} and, due to the effect of the regolith, the O_2 yield is of the same order of magnitude (*Shematovich et al.,* 2005).

Writing the yield in this form is somewhat misleading. Sputtering is a surface ejection phenomena, but the H_2 and O_2 are produced throughout the full penetration depth of the incident radiation. Therefore, the temperature-dependent component of the yield is controlled by diffusion and escape of H_2 and O_2 from the irradiated ice (*Reimann et al.,* 1984; *Teolis et al.,* 2005). O_2 formed at depth is also destroyed by continued irradiation. Therefore, in steady state, production rate is often given as the total number of O_2 produced per 100 eV of energy deposited, called a G-value (see chapter by Carlson et al.). Because the sputtering and decomposition are induced by electronic excitations, incident electrons (*Sieger et al.,* 1998) and UV photons (*Westley et al.,* 1995) also produce gas-phase H_2O, H_2, and O_2. Based on the radiation flux to Europa's surface, the photo-production of volatiles is negligible (a factor of about 10^3 less). For incident electrons, the above expression applies with S_n = 0 and α_2 slightly larger because electrons are scattered more efficiently. H, O, and ions are also ejected (*Kimmel and Orlando,* 1995; *Bar-Nun et al.,* 1985), but these yields are typically small.

3.3. Energy Spectra

Modeling of the nearly collisionless atmosphere requires knowing the energies of the ejecta. The few measurements for the steady state yield have been cast in a form typically used for sputtering. The spectra are given in terms of an effective binding energy, U, and becomes a power law at high energies

$$f(E) = c_q[UE^q/(E + U)^{2+q}]F(E) \qquad (3)$$

Here c_q is a normalization constant [such that $\int f(E)\,dE = 1$] and F(E) is a cut-off at large E that depends on the excitation process. In modeling ice sputtering it has been customary to ignore F(E) and use q = 1, c_q = 2, and U = 0.055 eV for the ejection of H_2O, and q = 0, c_q = 1, and U = 0.015 eV for the ejection of O_2 by energetic heavy ions (e.g., *Lanzerotti et al.,* 1983; *Reimann et al.,* 1984; *Shematovich et al.,* 2005). The data used to make these fits is shown in Fig. 3. Although there is prompt ejection of H_2 from the surface layer, most of the H_2 escaping from ice at the relevant temperatures diffuses to the surface and leaves thermally, so that the H_2 is modeled by a thermal flux. In contrast to thermal desorption, equation (3) has a high-energy "tail" proportional to $1/E^2$. Because of this tail, about one-fourth of sputtered H_2O molecules are given enough energy to escape Europa's gravity.

4. ATMOSPHERIC MODELING

4.1. Introduction

Using the data discussed above on the surface composition, radiation flux, and sputtering yields, Europa's atmosphere has been simulated and tested against observations. Species ejected from the surface are tracked analytically or numerically along collisionless or more realistic quasi-collisional trajectories taking into account their initial spatial and energy distributions as well as Europa's gravity, photon and electron impact ionization and dissociation, charge exchange, ion-neutral elastic collisions, and interaction with the surface. Jupiter's gravity and solar radiation pressure also need to be considered when modeling the neutral torus.

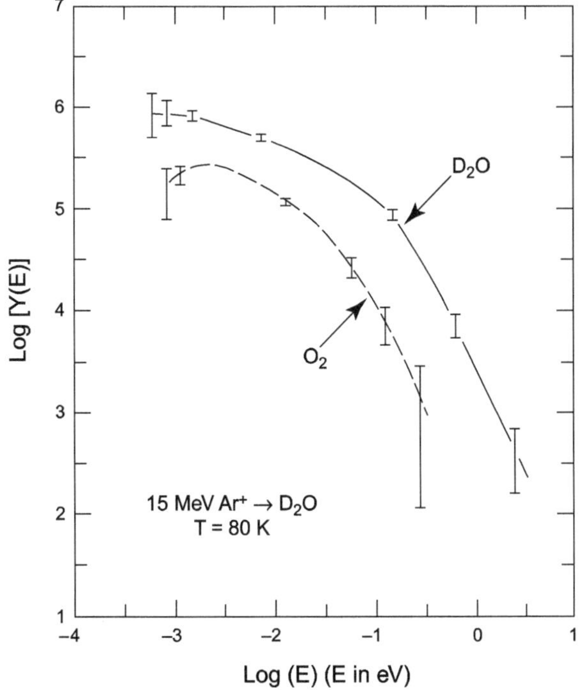

Fig. 3. Laboratory measurements of the energy distribution of sputtered O_2 and D_2O from D_2O ice. From *Lanzerotti et al.* (1983).

Sputtering is the dominant source of H_2O, O_2, and H_2, although sublimation of H_2O is competitive at the subsolar point. Atmospheric loss occurs by gravitational escape, interaction with the ambient plasma and UV photons, or removal through interaction with the surface, e.g., the sticking (freezing) of H_2O on Europa's surface. Due to the high-energy tail of the sputtering distribution in equation (3), ejected molecules can have sufficient energy to escape Europa's Hill sphere, the effective limit of Europa's gravitational influence; outside this distance, Jupiter's gravity dominates. The average radius of the Hill sphere is $r_{Hill} \approx a_E\,(m_E/3m_J)^{1/3} = 8.7\ r_E$, where a_E, r_E, m_E, and m_J are Europa's orbital distance from Jupiter's center, Europa's radius, mass, and Jupiter's mass, respectively. Using the energy spectra above, ~2% of the O_2 and ~24% of the H_2O are ejected with velocity greater than the 1.9 km/s required to reach r_{Hill}. For thermally accommodated O_2 escape is negligible, but for H_2 about 7% escape at the average temperature (~100 K) and about 15% at the subsolar point (~130 K). These percentages were calculated assuming that molecules do not collide in Europa's atmosphere, for in this nearly collisionless atmosphere, molecules with sufficient energy have a good chance of leaving without encountering another molecule.

Atmospheric neutrals are subject to loss by ionization and charge exchange by the local plasma. Photoionization and dissociation also occur, although at a lower rate (Table 1) (*Shematovich et al.,* 2005; *Smyth and Marconi,* 2006). Ionization or charge exchange followed by pick-up leads to the sweeping of ions from the atmosphere into the magnetosphere or into the surface. Incident plasma ions can also collisionally eject neutrals, often called knock-on or atmospheric sputtering. *Saur et al.* (1998) overestimated this loss process by assuming collisions were head-on (*Shematovich et al.,* 2005). Atomic species formed through dissociation or reactions can also escape. For example, atomic oxygen produced by electron impact dissociation of O_2 is estimated to have, on the average, ~0.67 eV (~2.85 km s^{-1}) of excess energy and likewise H formed from dissociation of H_2 has velocities of several tens of km s^{-1} (*Smyth and Marconi,* 2006); if these atoms are produced isotropically, roughly half will escape.

The rates of the plasma-induced processes depends on temperatures and densities of the charged species. For electron impact ionization and excitation the rate is written $\nu_{ei} = n_e k(T_e)$, where k is a rate coefficient, and n_e and T_e are the electron density and temperature. Important rate coefficients are given in Table 1. The charge exchange rate is $\nu_{chx} \approx v_{rel}\,\sigma(v_{rel})\,n_i$, where n_i is the ion density. For interactions with the ambient, undisturbed plasma torus ions, the average rela-

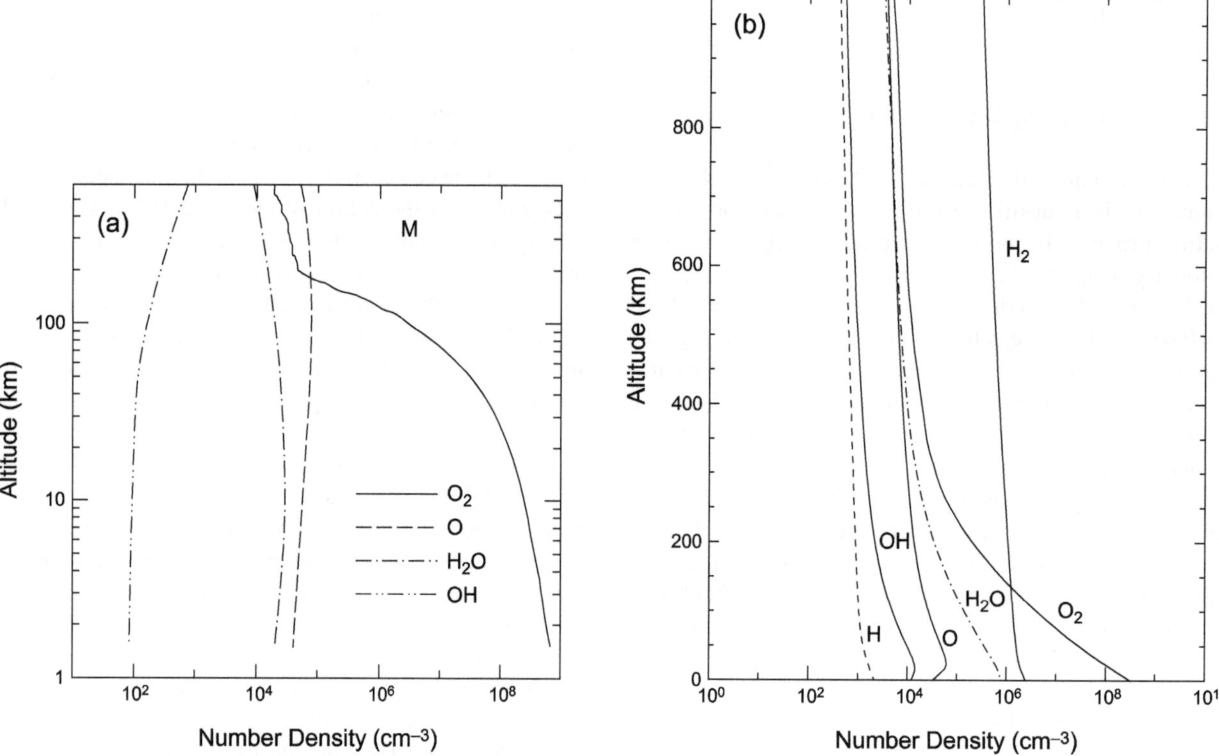

Fig. 4. **(a)** One-dimensional DSMC accounting for photon, electron, and proton impact ionization and dissociation. O and OH are produced by dissociation from O_2 and H_2O; H_2 and H are not included. From *Shematovich et al.* (2005). **(b)** Two-dimensional DSMC globally averaged density. The O_2 rate was set to reproduce the 130.4-nm O brightness of 37 ± 15 Rayleigh observation of *Hall et al.* (1995). Sublimation was taken into account but was unimportant except in the subsolar region. In both simulations, the ejecta energy distributions discussed earlier were used for H_2O and O_2 and thermalization of returning H_2 and O_2 in the regolith is assumed. Sputtering source rates for each model are shown in Table 2. From *Smyth and Marconi* (2006).

tive collision speed is $v_{rel} \approx 10^2$ km/s, the difference between the magnetic field corotation velocity and the Keplerian orbital velocity. In the ionosphere, where the plasma flow velocity is low, v_{chx} is averaged over the ion temperature.

4.2. Analytic Models

For an assumed ballistic atmosphere, analytic expressions are available for calculating the column density, the number of molecules per unit area above Europa's surface (*Johnson,* 1990). Such models were initially used to describe Europa's H_2O and O_2 source and loss rates, including sticking to the surface for H_2O and plasma-induced ionization and dissociation for O_2 (*Johnson et al.,* 1982). The column density of a species, N_i, is

$$N_i \approx (1 - f_i)\Phi_i\tau_i \qquad (4)$$

where Φ_i is the surface source rate given as a flux, f_i is the direct escape fraction, and τ_i is the mean lifetime. For a species like H_2O that sticks to the surface with roughly unit efficiency, τ_i is the ballistic lifetime (~1 hour for bound H_2O). For a species that does not stick, like O_2, $\tau_i^{-1} \approx \nu_i$, the rate for the principal loss process, ionization, which is about six days for O_2. If O_2 is both produced and lost by the ambient plasma, N_{O_2} is *roughly independent* of the plasma density, depending primarily on the local electron temperature, which determines the ionization and dissociation rates (*Johnson et al.,* 1982).

One-dimensional analytic models of the density vs. altitude are also available (e.g., appendix in *Johnson,* 1990). Since O_2 ejected by the incident plasma survives many ballistic hops, it is thermally accommodated to Europa's local surface temperature; i.e., the O_2 desorbs from the surface with a Maxwellian energy distribution. This produces an atmosphere with density that decreases as $\exp[-z/H_{O_2}]$, where z is the height above the surface and $H_{O_2} \approx [kT/m_{O_2}g]$ is called the scale height, with g the gravitational acceleration. For an average temperature of ~100 K, $H_{O_2} \approx 20$ km. Atoms and molecules that do not thermally desorb, such as sodium and H_2O, are sputtered or photodesorbed. The energy distribution for sputtering (equation (3)) can be integrated to give the density vs. altitude. In their ionospheric model, *Saur et al.* (1998) assumed that the O_2 atmosphere was primarily escaping and therefore decayed as the square of the distance from the center of Europa. As can be seen in Fig. 4, that would only apply to a fraction of the atmosphere well above ~400 km.

TABLE 2. Sputtering source rates used in Fig. 4.

	Shematovich et al. (2005)	*Smyth and Marconi* (2006)
H_2O	2×10^9 cm^{-2} s^{-1}	3.3×10^9 cm^{-2} s^{-1}
O_2	2×10^9 cm^{-2} s^{-1}	3.3×10^9 cm^{-2} s^{-1}
H_2	—	6.6×10^9 cm^{-2} s^{-1}
H	—	1.9×10^8 cm^{-2} s^{-1}
O	—	1.9×10^8 cm^{-2} s^{-1}
OH	—	1.9×10^8 cm^{-2} s^{-1}

4.3. Monte Carlo Simulations

Monte Carlo simulations are required in order to study the longitudinal and latitudinal morphology and distribution of dissociated fragments. Because of the complexity of the atmospheric origins and the paucity of collisions, two types of Monte Carlo models have been used: a test particle model and a direct simulation Monte Carlo (DSMC) model. In both models atoms and molecules are tracked subject to interactions with the radiation environment and collisions, and are equivalent to solving the Boltzmann equation. In a DSMC simulation one solves for the motion of each particle given the forces and, at each time step, calculates the effect of all possible collisions within Europa's atmosphere and with the incident magnetospheric plasma. Test particle approaches track individual particles in a collisionless atmosphere *or* in a known background atmosphere. If the atmosphere is marginally collisional, a test particle approach can be used iteratively by using a previously calculated atmosphere (background plus test particles) as a background atmosphere in a new calculation. The principal components in Europa's atmosphere have been calculated using one-dimensional DSMC (*Shematovich et al.,* 2005; *Shematovich,* 2006) or two-dimensional DSMC (*Wong et al.,* 2001; *Smyth and Marconi,* 2006) simulations. The three-dimensional structure of Europa's atmosphere and its time variability with respect to its position within Jupiter's magnetosphere have been described using a test particle simulation for sodium (*Johnson et al.,* 2002; *Leblanc et al.,* 2002; 2005; *Burger and Johnson,* 2004; *Cassidy et al.,* 2008). *Ip* (1996) also applied a test particle approach for O_2 in Europa's atmosphere.

5. SIMULATION RESULTS AND COMPARISON TO DATA

5.1. Introduction

McElroy and Yung (1977) predicted that Ganymede would have a robust O_2 atmosphere formed by water sublimation and gas-phase processes. It was later shown that indeed the icy satellites Europa and Ganymede had atmospheres dominated by O_2, but they are tenuous and produced by radiolysis in the icy surface and subsequent sputtering (*Johnson et al.,* 1982). In other words, even if H_2O is the principal ejecta by sublimation and sputtering, these molecules stick to the surface with unit efficiency at the ambient temperature. This is not the case for O_2 molecules, which although formed and ejected less efficiently, have a significantly longer resident time in the atmosphere. The approximate size of the predicted column of O_2 (~10^{15} cm^{-2}) was eventually confirmed by HST observations (*Hall et al.,* 1995).

5.2. One-Dimensional and Two-Dimensional Simulations: Water, Hydrogen and Oxygen Atmosphere

As shown Fig. 4, one-dimensional and two-dimensional DSMC simulations agree on the globally average structure of Europa's atmosphere. Both simulations constrained the O_2 column density to agree with the column density derived from UV emission (*Hall et al.,* 1998). The differences between Figs. 4a and 4b are due to the choice of sputtering rates (shown in Table 2). *Smyth and Marconi* (2006) chose larger ejection rates for O_2 and H_2O and small additional sources of O, OH, H_2, and H directly ejected from the surface. The larger H_2O (near-surface) density in *Smyth and Marconi* (2006) was due to their use of too small a sublimation energy for H_2O. In both simulations the main origin of O and OH is predominantly due to the dissociation of O_2 and H_2O. At the upper altitudes in Fig. 4b, the H_2 decays as ~$r^{-2.8}$ and the O_2 as ~$r^{-3.5}$, roughly consistent with analytic models (*Johnson et al.,* 1990).

Radio occultation measurements were also used to estimate the atmospheric density by assuming an O_2^+ ionosphere that is lost by dissociative recombination (*Kliore et al.,* 1997). They suggested a surface density ~3×10^8 cm^{-3}, consistent with the results in Fig. 4, but with a very different scale height (~120 km). This would give a column density ~4×10^{15}/cm^2, larger than in the models above, in which case the atmosphere would be collisional. However, ions are primarily removed by magnetic field pickup, either away from Europa or into the surface, not by dissociative recombination.

Observations of the O 135.6-nm line indicative of O_2 dissociation displayed a spatially localized emission peak (see chapter by McGrath et al.). Such a feature can be modeled by a spatially nonuniform electron flux and temperature, a spatially nonuniform O_2 atmosphere, or a combination of these. *Cassidy et al.* (2007) first showed that nonuniform O_2 ejection alone could not produce the required density variation as a result of O_2's long lifetime against loss by ionization or dissociation (*Wong et al.,* 2001). However, a

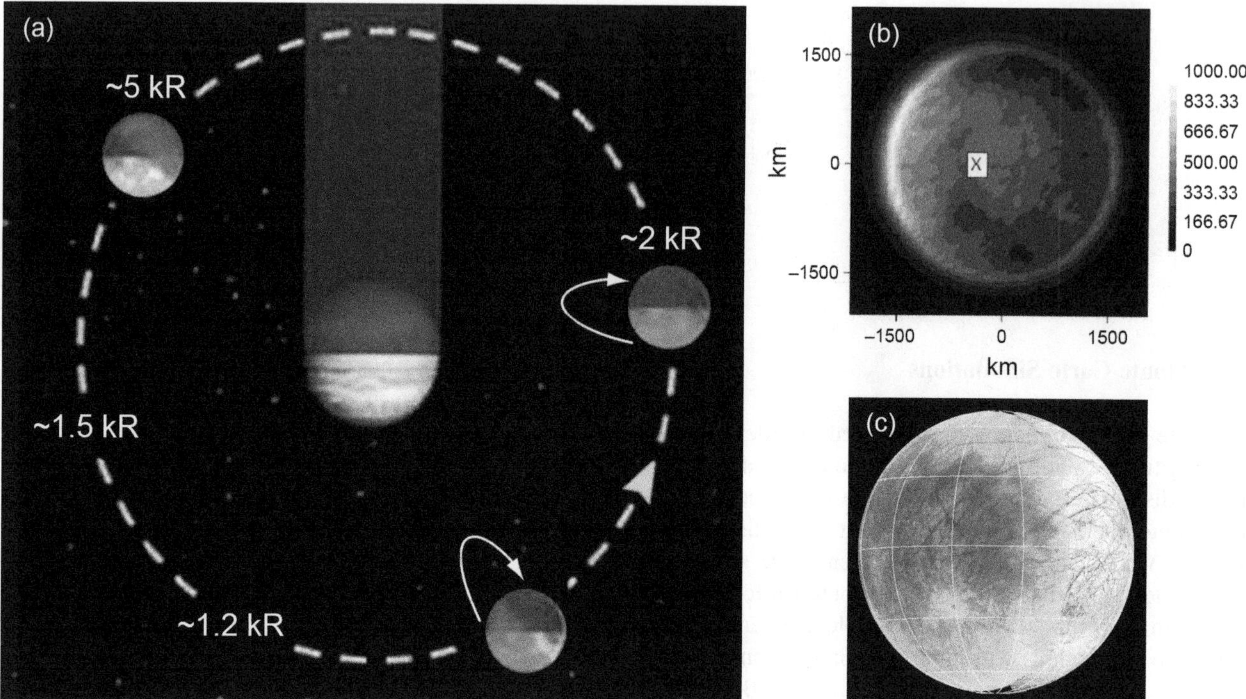

Fig. 5. **(a)** See Plate 25. Average Na emission intensity (in kiloRayleigh) at 4 Europa radii from the surface at different positions around Jupiter (*Leblanc et al.,* 2005). The red part on the surface is the preferentially bombarded trailing hemisphere, whereas the dark part represents the night hemisphere. Also indicated is Jupiter's shadow at Europa's orbit; sizes of Jupiter and Europa are not to scale. White arrows indicate where accumulation of Na atoms on the leading side may occur. **(b)** Simulation of the Cassini observation of Europa in eclipse: brightness in Rayleighs; (x) is apex of trailing hemisphere (*Cassidy et al.,* 2008). **(c)** The surface from the same perspective.

density peak spatially correlated to the observed emission peak could be produced if O_2 is lost via reactions in the regolith in Europa's dark terrain. This terrain has substantial concentrations of sulfur and carbon compounds which may be reactive in a radiation environment. The reactivity was implemented in a test particle model by assigning a probability for a reaction between O_2 and the grains in the dark terrain regolith (*Cassidy et al.,* 2007). This was varied until the spatial distribution of the emission intensity was roughly matched. Because a returning O_2 makes many encounters with grains in the regolith prior to its reentering the atmosphere, the required reaction probability for each encounter with a grain was very small, ranging from 1×10^{-5} to 3×10^{-4}, depending on source and other loss parameters. This small reactivity was sufficient to make a large difference in the atmospheric morphology due to the porous nature of the regolith.

5.3. Trace Species

5.3.1. Sodium atmosphere. Sodium and potassium in Europa's atmosphere have been observed through the resonant scattering of sunlight as discussed. Potassium emission has been only been reported once (*Brown,* 2001), whereas sodium has been observed regularly since its discovery (*Brown and Hill,* 1996; *Brown,* 2001; *Leblanc et al.,* 2002, 2005). Models of the sodium observations have been used to identify the spatial and temporal variability of Europa's extended atmosphere as a function of its orbital position (see Fig. 5a), magnetic longitude, and the angle between the zenith direction and the ram direction of the incident magnetospheric plasma (*Leblanc et al.,* 2002, 2005). A localized emission peak reported by Cassini (*Porco et al.,* 2003) when Europa was in Jupiter's shadow has been associated with sodium preferentially ejected from Europa's dark terrain (*Cassidy et al.,* 2008) (Fig. 5b).

An analytical model (*Johnson,* 2000) was initially used to reproduce the observed vertical distribution of the sodium and to estimate the escape fraction assuming that sodium was carried off with H_2O sputtered from Europa's surface. Comparing source and loss processes, it was suggested that the observed atmospheric sodium is mostly endogenic and therefore could be used as a tracer of Europa's interior composition and, possibly, its putative ocean. *Leblanc et al.* (2002, 2005) developed a three-dimensional test particle simulation of the three-dimensional structure of Europa's sodium atmosphere including its escape component, referred to as the sodium cloud. Sodium was ejected by sputtering (*Johnson et al.,* 2002) and lost by returning to the surface or by ionization. The sputtering rate and its variation with respect to longitude (local time within Jupiter's magnetosphere) and centrifugal latitude (with respect to the Io torus plane) were constrained by observation. Photostimulated and thermal desorption were not included, but their potential effect on the longitudinal variation of the sodium cloud were discussed. Indeed, the content of the extended sodium atmosphere was found to vary by a factor 3 to 4 depending on the orbital position (see Fig. 5a) and again appeared to be primarily endogenic. The significant variation in the sodium density along Europa's orbit was suggested to be due to the variation in the impacting flux of magnetospheric particles, as well as the migration of the sodium atoms from the trailing hemisphere (red portion of Fig. 5a) toward the nightside (dark portion of Fig. 5a), and to their accumulation in a portion of Europa's surface temporarily less irradiated by both photons and energetic particles. This accumulation lasted about a half a Europa orbit until the accumulated sodium was exposed to the solar flux producing the observed emission maximum. This corresponds to the position with an emission of 5 kR in Fig. 5a. *Leblanc et al.* (2002, 2005) also predicted a globally averaged sodium concentration at the surface equal to ~0.01, and suggested that the observed sodium was sputtered from ice and not directly from a mineral or a salt. Recent modeling has improved on this description of the variability and on the Na/K ratios (*Cipriani et al.,* 2008, 2009).

Comparing the simulations to the observations, Europa loses an orbit-averaged flux of ~1.2×10^7 Na cm^{-2} s^{-1}, about an order of magnitude less than at Io (*Leblanc et al.,* 2005). Since Europa receives a flux of ~0.2 to 0.8×10^7 Na cm^2 s^{-1} from the Io torus, the observed sodium atoms are either endogenic or were implanted in epochs when Io was much more active. The endogenic nature of Europa sodium is also suggested by the significant difference of the Na/K ratios at Europa (25 ± 3) and at Io (~10 ± 3) (*Brown,* 2001; chapter by Carlson et al.). Since a sodium atom has 30% higher escape probability than a potassium atom, the Na/K ratio ~25 ± 3 observed far from Europa by *Brown* (2001) implies a Na/K ratio of 20 ± 4 at Europa surface (*Johnson et al.,* 2002). Meteoroid impacts may produce a Na/K ratio of ~13, as measured in meteoroids at Earth, which is still significantly smaller than the required value of 20 ± 4. *Zolotov and Shock* (2001) estimated that the Na/K ratio in the putative internal ocean may be between 14 and 19. Since freezing and dehydration can increase the Na/K ratio, material originating from Europa's ocean that reaches its surface could produce a ratio consistent with that implied by the observations.

5.3.2. Other trace constituents. Both CO_2 and SO_2 have been observed trapped in Europa's surface with apparently different hemispherical distributions, as described earlier. In addition, the surface contains concomitant sulfates and carbonates, molecules with sulfur-sulfur and carbon-carbon bonds, radiation products such as H_2O_2, ion implantation from the Io torus, as well as other endogenic and delivered materials. Because CO_2, CO, SO_2, and SO are more volatile than sodium they should be present in the atmosphere in greater amounts than sodium (*Cassidy et al.,* 2009). However, there is less modeling of the atmospheric content of trace species other than sodium, because of the lack of observations.

Although the sputtering yields of trace species in an ice matrix have not been measured, trace species should be carried away with the large number of water molecules ejected by energetic heavy ions, up to 1000 H_2O ejected per ion. Modeling these trace components requires knowledge of their concentrations in the surface, spatial distribution, and interactions with the surface. The abundance and spatial distribution of a species can be estimated from spectral images (see chapter by Carlson et al.). Assuming ejection is due to the large sputtering yields and the flux of sputtered H_2O is dominated by energetic heavy ion sputtering, then the sputtering rate is nearly independent of latitude and longitude due to large magnetic gyroradii (shown schematically in Fig. 1) (*Cooper et al.,* 2001; chapter by Paranicas et al.). In this approximation, the flux of a trace species will be proportional to its concentration in the surface.

The energy distribution of an ejected trace species is uncertain, but can be approximated by considering that it is ejected along with the H_2O. Assuming that the sodium atoms have the same energy distribution as sputtered H_2O molecules results in a reasonable fit to the sodium data as discussed above. *Wiens et al.* (1998) measured the energy distribution of molecules sputtered from a sodium sulfate solid, one of several candidates for the widespread hydrated substance seen in ice on Europa's surface. They found that the energy distributions for sodium- and sulfur-containing fragments peaked at significantly higher energies than what would be expected for a trace constituent carried off with an H_2O matrix material, *but* with a correspondingly smaller yield. Since any sodium sulfate present on Europa would be hydrated, and in most places is a minor constituent, ejection of trace species with the ice matrix is a reasonable assumption. In regions where the hydrate has a high concentration, ~90% in some places (*Carlson et al.,* 2005), the matrix yield would decrease and the energy of ejecta would increase as the material cohesive energy is larger.

Upon returning to the surface a molecule may "stick" (freeze) or thermally desorb. Modelers have described the probability of sticking as a "sticking coefficient." The closest lab measurement, however, is the residence time on the surface, the average time before thermal desorption. Figure 6 shows the residence time for a variety of species. The temperature range extends from the polar to the subsolar temperatures. The data comes from *Sandford and Allamandola* (1993), except for the O_2 curve from J. Shi (personal communication, 2007). Since there was no data available for SO_2 on H_2O ice, the curve for pure SO_2 was used. It may be appropriate since *Noll et al.* (1995) suggested that SO_2 is segregated from H_2O as inclusions in Europa's surface.

For residence times above the line labeled "lifetime against sputtering," a molecule is more likely to be sputtered before being thermally desorbed. Sodium presumably has a large residence time, but can also be desorbed also by sunlight (*Johnson et al.,* 2002). H_2O_2, sulfur, and other refractory constituents in the ice matrix have large residence times. The residence times for SO_2 and CO_2 cross the sputtering line. For these species, the probability that a molecule will thermally desorb before being sputtered will vary over the europan day. Such a molecule on the nightside is more likely to be sputtered than thermally desorbed, and, when sputtered, will travel much further than a thermally desorbed

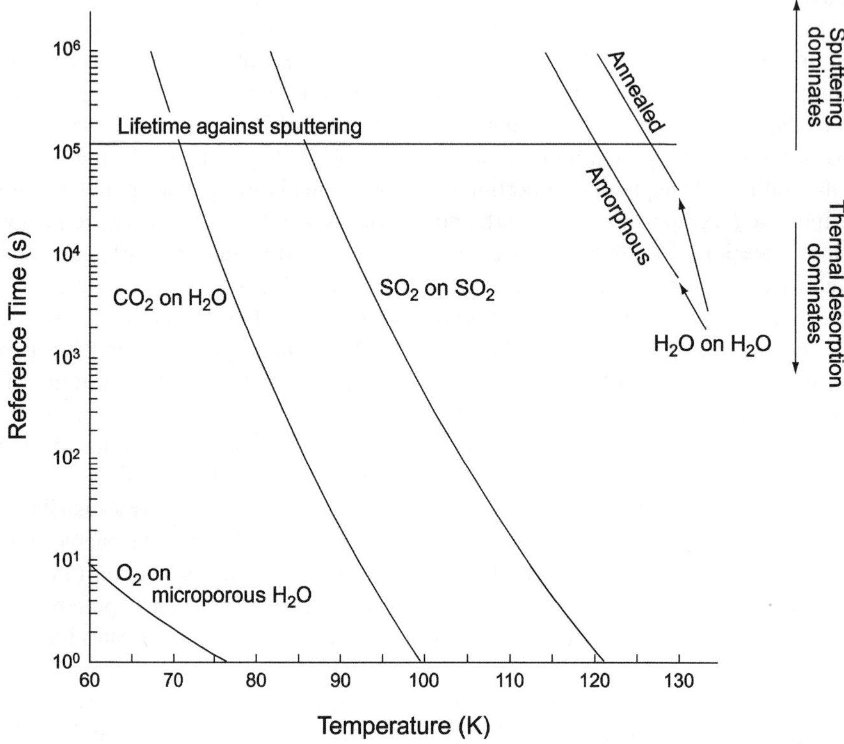

Fig. 6. Residence time vs. temperature. For times above the horizontal line, a molecule is more likely to be sputtered along with H_2O before being thermally desorbed. Data comes from *Sandford and Allamandola* (1993), except for O_2 from J. Shi (personal communication, 2007), and the lifetime against sputtering comes from *Paranicas et al.* (2002).

molecule. Such day-night differences can result in the dynamic redistribution of the atmosphere.

Uncertainties in the residence times are due to lack of data, large error bars, and disagreement between experiments. A potentially larger uncertainty is the porosity of the ice, which has two scales: microporosity or the nanometer scale and regolith porosity on the tens of micrometers scale. Microporous water ice is amorphous and can act like a sponge increasing the residence time by many orders of magnitude (*Raut et al.,* 2007). On the other hand, the time a molecule spends in the regolith due to repeated adsorption and desorption depends sensitively on the thickness of the regolith (*Hodges,* 1980; *Cassidy and Johnson,* 2005; *Cassidy et al.,* 2007). A molecule deep in the regolith can also be protected from sputtering.

6. NEUTRAL CLOUDS AND EXTENDED ATMOSPHERE

Atmospheric molecules can undergo elastic and reactive collisions with the ambient plasma and photochemistry driven by solar radiation. The resulting products will have energy distributions characteristic of their creation and collision processes. Products with sufficient energy can escape the Hill sphere as discussed earlier. Some neutrals, especially those created through charge exchange between plasma ions and atmospheric neutrals, can even escape Jupiter's gravitational field. Most of the escaping particles except H, however, have energies less than Jupiter's escape energy. These constitute the extended atmosphere/neutral cloud as described earlier for sodium. The accumulation of such neutrals, orbiting the planet, is limited by reactions with the magnetospheric plasma and solar radiation. Neutral lifetimes in the magnetosphere are shown in Fig. 7. Neutral lifetimes near Europa are less certain but often assumed to be the same. The morphology and density of the resulting clouds depend upon the interplay of the neutral orbits with the spatially nonuniform magnetospheric plasma and solar radiation. The result is clouds having different morphologies for different species, each containing information about the satellite surface, the ejection process, and the plasma.

Observations of the Europa neutral clouds are limited and include a few observations of the sodium and potassium clouds near Europa, as discussed; a large-scale observation by the Cassini Magnetospheric Imaging Instrument/Ion and Neutral Camera (MIMI/INCA) experiment; and *in situ* plasma measurements. For the sodium and potassium clouds, only that fraction roughly contained in a cylinder located within 40 R_E of Europa has been observed, with a significant fraction contained in the bound component. These observations proved to be highly valuable in defining the source strengths and energy distributions, their variation with time and location on Europa's surface, as well as

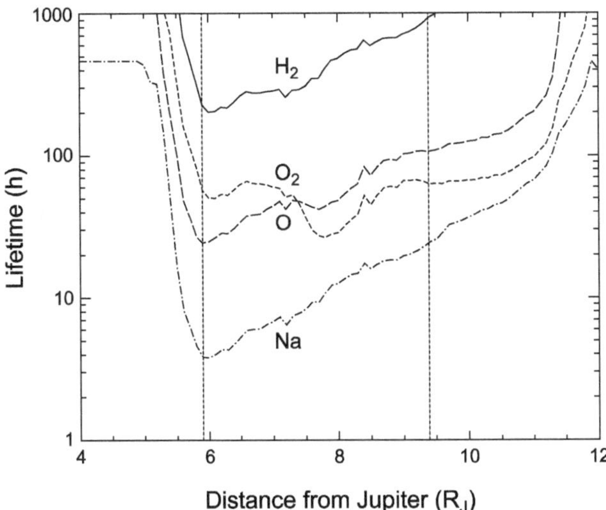

Fig. 7. Mean lifetimes in the equatorial plane of the major neutral cloud species as a function of distance from Jupiter. Updated from *Burger and Johnson* (2004).

the morphology of the local jovian plasma. The Europa sodium cloud is also expected to have a much different morphology from the much-studied Io sodium cloud (*Burger and Johnson,* 2004). As seen in Fig. 8b, the column density was largest in the portion trailing Europa's motion, whereas Io's forward component dominates. This difference is due to the radial morphology of the electron impact ionization lifetime of the sodium, which is determined by the plasma density and temperature. The potassium cloud has been observed and modeled to a far lesser extent (*Brown,* 2001), but is expected to be roughly similar in behavior to sodium after accounting for the larger mass of the potassium atom.

A large-scale, azimuthally averaged observation of the dominant component of Europa's neutral cloud was taken by the ENA imager during the Jupiter Cassini flyby in 2001 (*Mauk et al.,* 2003, 2004). This instrument measures energetic neutrals that are generated by the charge exchange of the energetic component of the plasma with the neutral cloud. Due to the large distance (closest approach ~140 R_J) and one-dimensional nature of the image, little structure was evident, but two distinct enhancements in the 50–80-keV neutral particle flux were noted just outside the orbit of Europa. Determining the neutral composition causing the enhancements is problematic, but by assuming a torus geometry, *Mauk et al.* (2004) estimated a cloud gas content of $(1.2 \pm 0.5) \times 10^{34}$ for a pure H cloud or $(0.6 \pm 0.25) \times 10^{34}$ for a mixture of water-group gases dominated by H.

Several plasma observations near Europa's orbit are also consistent with the presence of a neutral cloud (*Lagg et al.,* 2003; *Mauk et al.,* 2004). The Galileo energetic particle detector observed an abrupt change in energetic plasma properties with decreasing radial distance upon crossing Europa's orbit. In particular, a drop in energetic plasma density, including the almost complete removal of protons in the tens of keV range (*Paranicas et al.,* 2003; *Mitchell et al.,* 2004), an increase in average energy, and a decrease in the 80–220-keV protons with pitch angle 90° at L = 7.8 R_J (*Lagg et al.,* 2003) were interpreted as being caused by charge exchange with Europa's neutral cloud. Assuming a composition of 20% H and 80% O plus O_2, Lagg et al. estimated the cloud had a density of 20 cm^{-3}–50 cm^{-3}, in agreement with the ENA-derived numbers (*Mauk et al.,* 2003, 2004). Starting with Pioneer 10, plasma measurements near Europa's orbit consistently indicated enhanced oxygen composition and higher temperatures, suggesting the presence of an oxygen neutral cloud (e.g., *Bagenal,* 1994, and references therein).

The neutral cloud observations have stimulated several three-dimensional simulations: sodium (*Leblanc et al.,* 2002. 2005), sodium and O_2 (*Burger and Johnson,* 2004), and O and H_2 (*Smyth and Marconi,* 2006). In each case, the species of interest are initialized with a velocity distribution for the escaping component. Test particles are followed along orbits determined by the gravity of Jupiter and Europa, and, for sodium, solar radiation pressure. For the surface-ejected sodium, *Leblanc et al.* (2002) took into account collisions with the background O_2 atmosphere. The escaped neutrals are destroyed by reactions with the plasma, which is more important than the photoionization. The particles are followed until they collide with Europa or Jupiter, depart the computational domain, or are destroyed by an ion or electron. The model for the plasma that is used to react with the neutral cloud differed somewhat in each study.

For sodium atoms a spatially nonisotropic sputter source with a maximum at the apex of the trailing hemisphere was used (*Leblanc et al.,* 2005) as suggested by albedo variations and by *Pospieszalska and Johnson* (1989). Several energy/angle distributions for f(E,Θ) sputtered sodium were used in the model and the results of each were compared to the data. The model that best matched the data used $f(E,\Theta) \sim [2EU/(E + U)^3]\cos(\Theta)$ with U = 0.055 eV, the energy distribution given above for the large sputtering yields for H_2O carrying off an embedded impurity (*Johnson,* 2000). The excitation and ionization of sodium by electrons was based on the radial and latitude dependence of plasma electron density given by *Smyth and Combi* (1988). *Burger and Johnson* (2004) calculated a neutral cloud using a spatially isotropic source and an energy distribution similar to *Leblanc et al.* (2002). Their electron model was based on the *Bagenal et al.* (1994) results at the centrifugal equator and assuming an exponential drop in density with magnetic latitude off the centrifugal equator. The model also accounted for an offset dipole and east-west field. For an initial sodium flux of 3×10^7 cm^{-2} s^{-1}, modeled densities are given in Figs. 8a and 8b. Both clouds are concentrated near Europa and exhibit a dominant trailing cloud. However, the cloud in Fig. 8a is significantly more extended due to the use of a lower electron density and thus longer lifetime.

Smyth and Marconi (2006) simulated the O and H_2 neutral clouds. The H_2 atoms were set in motion using the globally averaged H_2 velocity distribution at 1000 km altitude derived from the previously described two-dimensional DSMC atmospheric model. The initial velocity distribution

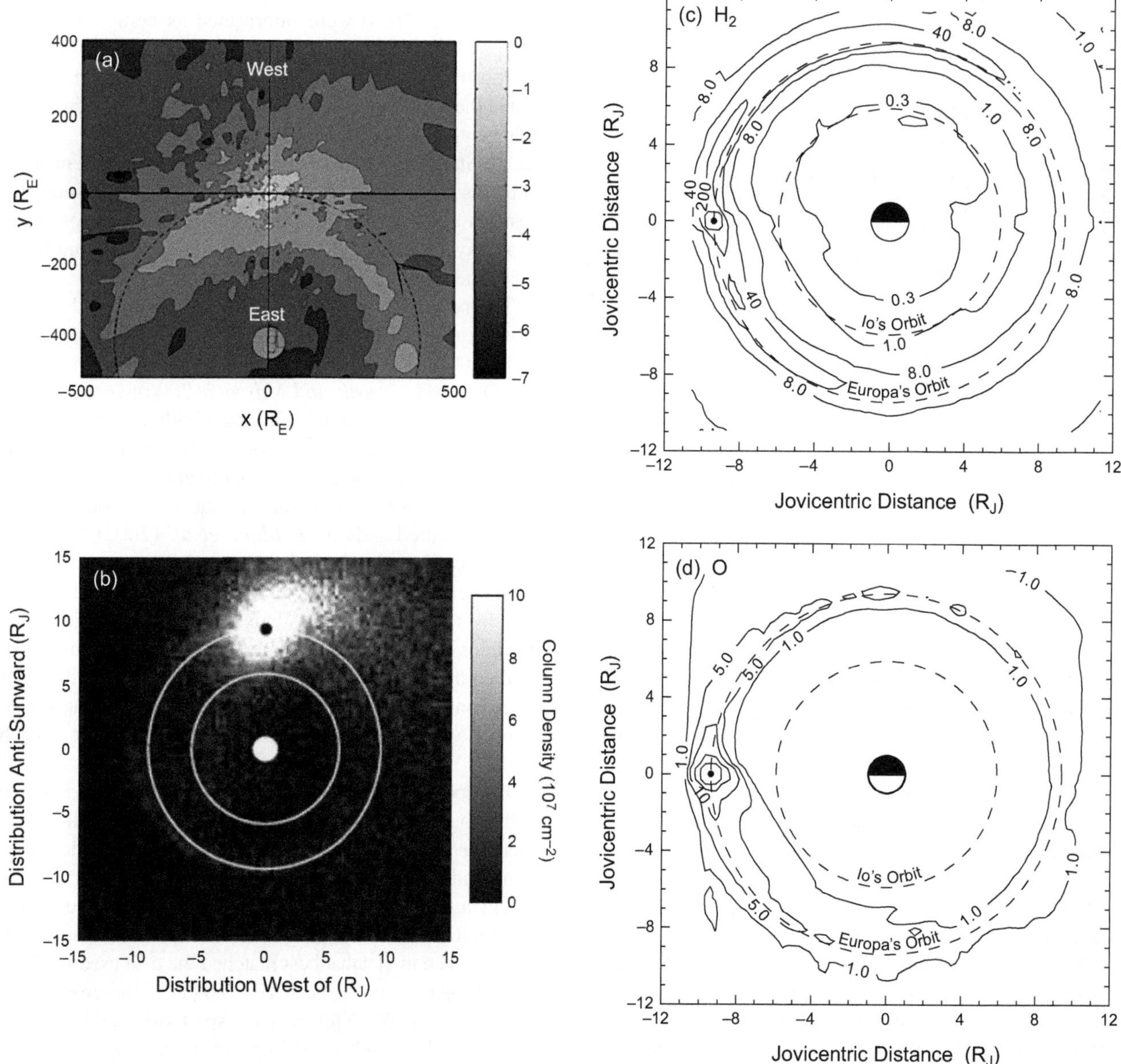

Fig. 8. Neutral cloud (torus) densities or perpendicular column densities viewed from north of the orbit plane with Jupiter in center of each figure. **(a)** Sodium density contours (*Leblanc et al.,* 2002) in a cut centered on Europa's orbit plane at a phase angle of 90°. **(b)** Sodium column density; also roughly the morphology of the O_2 cloud (*Burger and Johnson,* 2004). **(c)** H_2 column density (contour units 1×10^9 cm^{-2}) (*Smyth and Marconi,* 2006). **(d)** Same as **(c)** but for O (contour units 1×10^{10} cm^{-2}).

is a Maxwellian at 120 K with an energetic tail caused by collisions with the fast-moving torus plasma. O atoms were also ejected isotropically, but the energy distribution was, for simplicity, assumed to be monoenergetic with speed 2.5 km s^{-1} (~0.53 eV). Using a velocity distribution for O reduces the escape flux. In the DSMC simulation, the H_2 escape rate was 1.9×10^{27} s^{-1}, while for O, 5.0×10^{26} s^{-1} was used: the sum of the O escape rate and half of the dissociated O_2 (since half of the resulting O atoms likely impact the surface). The plasma model is an update of *Smyth and Combi* (1988) and includes electrons and ions, dipole offset, and east-west field. The plasma density and composition dependence on magnetic latitude is obtained by solving the force-balance equation for the plasma along the field lines. The plasma chemistry includes the electron impact dissociation and ionization of H_2, ionization of O, as well as charge exchange of O with the plasma ions, which dominates atomic oxygen loss within 7 R_J. The results for H_2 and O are shown in Figs. 8c and 8d. The densities are seen to peak near the source, forward and trailing clouds are present, and the neutral cloud is organized about Europa's orbit. The H_2 cloud is broader than the O cloud and both the O and H_2 clouds are more extended than the short-lived sodium cloud (Fig. 7). The neutral clouds for O_2, OH, H_2O, and H were not calculated, but are a smaller fraction of the total since H_2 has, by about an order of magnitude, the high-

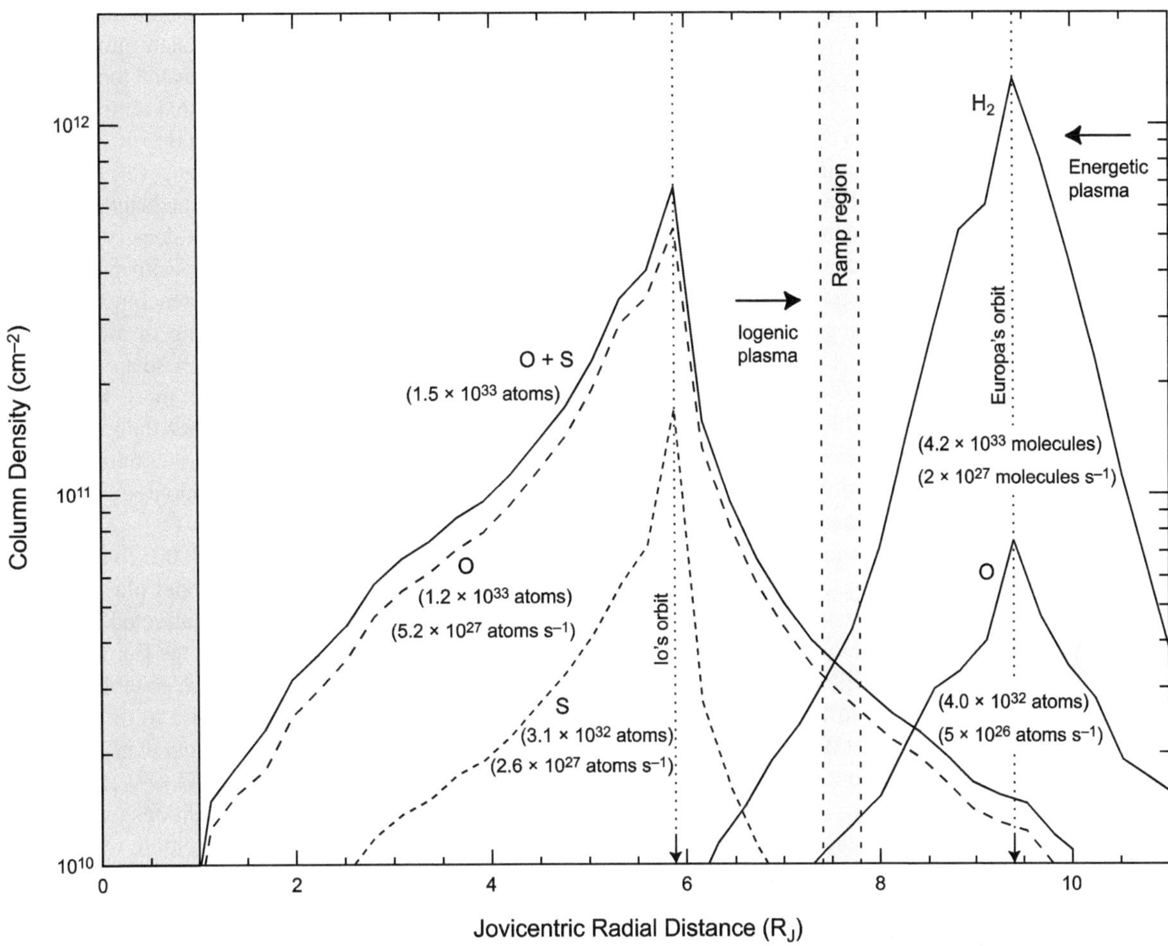

Fig. 9. Comparison of the neutral clouds of Europa and Io. Longitudinally averaged radial column density for Europa's H_2 and O neutral clouds and Io's O, S, and (O + S) clouds (*Smyth and Marconi,* 2006). The total (spatially integrated) cloud population and escape source rate for each neutral cloud as well as the radial location of Jupiter, Io's orbit, and Europa's orbit are indicated.

est escape rate. The neutral cloud identified from the EPD and ENA data is likely dominated by H_2 since H is much more dispersed due to the energy acquired during H_2 dissociation.

Europa belongs to a class of satellites, including Io, Enceladus, and Triton, whose neutral clouds contain more matter than their bound atmospheres (*Smyth and Marconi,* 2006). Estimates of the total content of the O and H_2 clouds, 4.0 × 10^{32} atoms and 4.2 × 10^{33} atoms respectively, are roughly consistent with the estimates discussed above, but exceed by factors of 1000 and 200 their abundances in Europa's bound atmosphere. The ratio for sodium is smaller due to the shorter lifetime. By comparison the ratio at Io is 3 for O and 1 for S (*Wong and Smyth,* 2000; *Smyth and Marconi,* 2003), and the total Europa neutral cloud content, maximum column density, and maximum density exceed that for the total Io neutral cloud (Fig. 9) by a factor of 3. Although Io is has a much denser atmosphere and a larger source rate, the difference is due to the shorter neutral lifetimes near Io (Fig. 7).

The O and H_2 clouds create a source of plasma, primarily H_2^+, with a net mass loading rate of 22–27 kg s^{-1} (*Smyth and Marconi,* 2006). It peaks near Europa's orbit and is about 10% of Io's total contribution, but exceeds the Io source beyond 8 R_J. Although it is not yet clear in what way this source affects the bulk plasma, it has a significant effect on the energetic particles (*Mauk et al.,* 2004). It depletes the energetic plasma, particularly the proton component, alters the pitch angle distribution and characteristic energy, possibly alters the charge state, and is the primary source of the energetic protons.

In much earlier work, *Schreier et al.* (1993) calculated orbitally averaged densities for H_2O, O_2, H_2, OH, O, and H using a box model that accounted for a large number of reactions between ions and neutrals and a neutral sputtering rate. They predicted that Europa's sputtered products would lead to an enhancement in the oxygen ion composition near its orbit, consistent with observations (*Bagenal,* 1994). While they obtained average densities comparable to those given above, they predicted H rather than H_2 to be the densest neutral cloud. This difference with the more recent models described above is due to the use of different sputtering yields and neglect of escape dynamics and neutral transport.

7. IONOSPHERE MODELING

Both solar radiation and torus plasma ionize atoms and molecules in Europa's atmosphere, as discussed, resulting in an ionosphere. Ions sputtered from the surface are a very small source, but may contribute unique ions. The most abundant ionospheric constituent at lower altitudes is likely O_2^+. Thermal electrons from the magnetosphere (temperature ~20 eV) are primarily responsible for the formation of O_2^+ ions from the dominant O_2 atmosphere. Each ionization event is accompanied by the formation of a cold (~0.5 eV) electron (an "ionospheric electron"). This cold ionospheric electron population is one of three electron populations near Europa, in addition to the thermal (~20 eV) magnetospheric electron population and the nonthermal (or "hot") magnetospheric electron population (sometimes approximated as having a temperature ~250 eV). The thermal magnetospheric population is the most important in determining the lifetime of a molecule against ionization (e.g., Table 1). The cold population doesn't contribute to ionization and the nonthermal population is extremely sparse (*Saur et al.,* 1998; *Schilling et al.,* 2007). The cold ionospheric population is, however, the primary population that contributes to dissociative recombination (e.g., $O_2^+ + e \rightarrow O + O$).

At higher altitudes H_2 is more abundant (Fig. 4b) so that H_2^+ may be expected to dominate for some altitude range until the atmosphere is sufficiently tenuous and the magnetospheric ions dominate. Transport of plasma into the surface and out of the atmosphere limits the accumulation of plasma, as does, to a lesser extent, dissociative recombination of the ionospheric ions. The ionosphere of Europa, unlike ionospheres formed in an atmosphere threaded with closed magnetic field lines or a dense atmosphere, such as Earth and Mars, is continuously lost to the plasma torus on a timescale of minutes. As a result, the ionosphere is expected to be rapidly responsive to changes in torus properties. In addition, due to solar radiation and the histories of the flux tubes intercepting Europa, a dynamic and anisotropic ionosphere that feeds europagenic ions into the plasma torus is expected.

Although Europa's ionosphere has been observed by the Galileo spacecraft (section 2.1 above), there have been few efforts modeling the ionosphere. *Saur et al.* (1998) calculated the electron density in Europa's atmosphere. Using a model for the electric and magnetic fields for a magnetized flow interacting with a satellite developed by *Wolf-Gladrow et al.* (1987) they developed a three-dimensional plasma-neutral interaction model for a constant magnetic field. Although more recent models (*Shematovich et al.,* 2005; *Smyth and Maconi,* 2006) predict a bound O_2 atmosphere with a scale height of 20 km near the surface, the neutral atmosphere was assumed to be O_2 with a r^{-2} height dependence. They also chose an O_2 surface density that varied from the trailing to leading hemisphere as prescribed by *Pospieszalska and Johnson* (1989), in contradiction to later estimates of a more uniform production of O_2. Charge exchange and ion-neutral elastic collisions for the ions and electron impact ionization, dissociation, excitation, and recombination for the electrons were taken into account. As mentioned above, they likely overestimated the atmospheric sputtering rate. The electron density and temperature were calculated for a model with a column of 5×10^{14} cm^{-2} and depletion scale height of 145 km, the result of constraining the model to match the oxygen line brightnesses (*Hall et al.,* 1998). The calculated electron density in the equatorial plane, shown in Fig. 10, was anisotropic with larger densities on the inside of the flanks, reaching a maximum of about 9000 cm^{-3}, roughly agreeing in magnitude and anisotropy with *Kliore et al.* (1997). These are mostly ionospheric (cold) electrons. In the wake, there were no electrons, a limitation of the model rather than a prediction. They also suggested that atmospheric conductivity could affect the estimate of the induced-dipole signature attributed to a subsurface ocean.

Kabin et al. (1999) and *Liu et al.* (2000) constructed self-consistent MHD models, i.e., the model plasma produces a magnetic field in addition to being affected by it. Unlike *Saur et al.* (1998), they also included the Europa's induced magnetic dipole, which is thought to be generated by a conductive subsurface ocean as a response to its motion relative to Jupiter's magnetic field (*Khurana et al.,* 1998; chapter by Khurana et al.). *Kabin et al.* (1999) and *Liu et al.* (2000) varied Europa's boundary conditions (superconducting sphere vs. absorbing boundary), dipole moment, mass loading rate, and other parameters to find the best fit to Galileo magnetometer data. They found that the ionosphere/magnetosphere interaction makes a significant contribution to the magnetic field perturbation near Europa, but that the

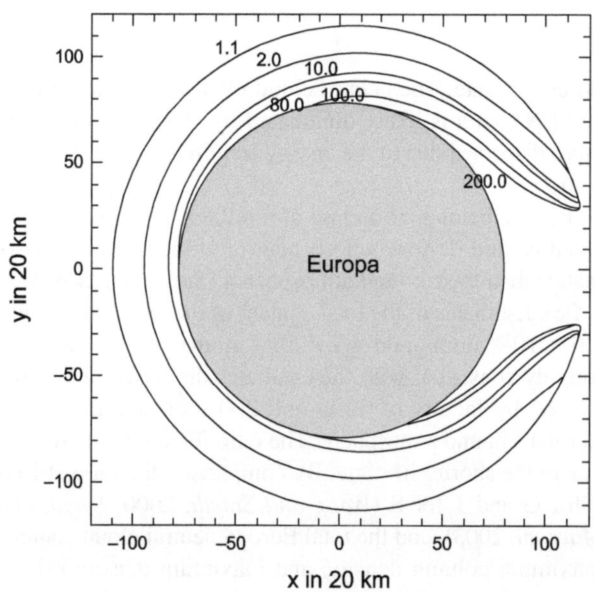

Fig. 10. Model of the density of the equatorial cold electrons density in units of 40 cm^{-3} (*Saur et al.,* 1998). +x is the corotation direction. On the right is Europa's wake, where the electron density is zero in this model.

magnetometer data still requires an induced dipole, such as might be created by a conductive subsurface ocean. Figure 1 shows a schematic of the warped magnetic field near Europa.

Those models used fixed dipoles to approximate Europa's subsurface electromagnetic induction. Most recently, *Schilling et al.* (2007, 2008) modeled the electromagnetic induction self-consistently by including a conductive interior and time-varying magnetic field. In this model, Europa leaves a smaller wake in the magnetospheric plasma than in the models of *Kabin et al.* (1999) and *Liu et al.* (2000). *Schilling et al.* (2008) also predict lower pickup ion densities in the wake than *Kabin et al.* (1999) and *Liu et al.* (2000). This new model wake is in much better agreement with Galileo PLS data (*Paterson et al.,* 1999; chapter by McGrath et al.) than previous models.

8. DETECTION OF EUROPA'S ATMOSPHERE

8.1. Remote Detection

Remote detection is discussed in detail in the chapter by McGrath et al. Even though Earth-based observations of Europa's atmosphere are hampered by the low densities seen in the models and a lack of wavelength coverage, the wealth of likely species suggests that new attempts should be made. Since sodium and potassium are important traces species, continuing observations of the changes in intensity and morphology are important for the modeling of other trace ejecta. The best method for observing the oxygen atmosphere is the FUV O I emission lines at 130.4 nm and 135.6 nm.

8.2. Orbiter: Direct Detection Based on Simulations

The geochemical evolution of the surface and its internal composition are among key objectives of a Europa orbiter. Therefore, *in situ* measurement of Europa's atmosphere, combined with modeling, will give critical compositional information; i.e., from an orbiter placed at 100 km above the surface, trace species should be detectable (*Johnson et al.,* 1998). The 10–100-keV S^+ and O^+ impacting energetic particles act on Europa's surface in a manner similar to the ions beams used in secondary ion mass spectrometry (SIMS), a widely used technique to measure surface composition in the laboratory. SIMS uses a mass spectrometer to measure the ions ejected from the surface. Even though ions are a small fraction of the ejecta they are more readily detected. The yields from Europa's icy surface are large enough to produce a detectable flux of ejected ions, and, as already discussed, the dominant component of ions is produced from neutral atmospheric molecules.

Although the yield from sputtered hydrocarbons is dominated by decomposition products, such as CH_4 and C_2H_x if organics (e.g., gylcerol, POM, L-Leucine) are present in the ice on Europa's surface, whole molecules can be sputtered into the atmosphere (*Johnson and Sundqvist,* 1992). Characteristic ions would also be ejected: e.g., the sputtering of cysteine produces HCN^+, CO_2^+, and even the whole molecules that are protonated (addition of H^+) (*Benninghoven et al.,* 1987). However, organics in the near-surface region are exposed to UV and particle radiation for many years before being sputtered, and thus will be replaced by stable decomposition products such as CO, HCO, HCN, etc. This chemistry is further complicated by the oxidizing surface; e.g., CO produced by decomposition would likely react in the surface to become CO_2 (*Johnson et al.,* 1998). In spite of such complications, near Europa ion detections to date suggest that the local plasma will be rich in species that are characteristic of the surface composition.

The ion densities vs. altitude were roughly estimated for sputtering of species for which laboratory sputtering data were available (*Johnson et al.,* 1998). For simplicity, the neutral column density of impurity sputter products were each scaled to 1% of the O_2 column density if they were volatiles, such as CO_2 and SO_2, or to 1% of the H_2O column density for refractory species such as salts and organics. Combining these neutral densities with ionization lifetimes, CO_2^+, CO^+, NO^+, N_2^+, CH_4^+, NH_3^+, SO_2^+, $NaSO^+$, and HCO^+ could have densities larger than 10 particles/cm^3 at 100 km from Europa's surface. Since a typical ion spectrometer can measure densities smaller than 10^{-3} cm^{-3} if such trace species are present they should be detectable. Similarly, since a typical neutral spectrometer measures densities larger than 1 cm^{-3}, a neutral mass spectrometer will also be useful. A combination of both neutral and ion spectrometers could therefore provide a useful set of surface-related information. Because ejected ions are immediately picked up and can be driven into the surface on the trailing hemisphere, detailed modeling of the near-surface densities is needed. In this way, a low-altitude orbiter could associate local variation of atmospheric composition and density with surface features. At 100 km in altitude, it is expected that a mass spectrometer should provide a spatial resolution at the surface of the same size.

Europa's bulk composition may be also deduced from *in situ* atmospheric measurement. As for the sodium, where the escape rate was used to derive a surface concentration, it should be possible to derive aspects of Europa's interior composition from the full compliment of ejecta. Such measurements would need to energetically characterize the principal escape processes and the implantation rate of exogenic species. These are feasible using both neutral and ion mass and energy spectrometers.

A hydrated substance has been identified on Europa's surface via IR spectroscopy (e.g., *Spencer et al.,* 2006). *Orlando et al.* (2005) suggested that it as a combination of hydrated salts such as $Na_2SO_4^* \times H_2O$ and $MgSO_4^* \times H_2O$, and the radiolysis product $H_2SO_4^* \times H_2O$ (e.g., *Carlson et al.,* 2002). *Wiens et al.* (1997) sputtered pure (nonhydrated) Na_2SO_4 and identified the neutral products via laser ionization. The dominant sputter product was sodium, accompanied by NaO, NaS, Na_2, Na_2O, etc. They also found an extremely low sputtering yield, 0.1 to 0.2 atoms ejected per

3.5 keV Ar^+. The yield for H_2O at comparable heavy ion energy is about 2 orders of magnitude larger. Such low sputtering yields could not account for the sodium source rate needed to model Europa's sodium cloud, but there have been no similar experiments for $Na_2SO_4^* \times H_2O$. The robust source rate seen for sodium suggests that other components of the unidentified hydrated material would likewise be sputtered in amounts comparable to their abundance. Therefore, the hydrated substance could be identified by a mass spectrometer in orbit.

A combined *in situ* measurement of Europa atmosphere by ion and neutral mass and energy spectrometers would also allow a better interpretation of the surface spectral signatures obtained remotely and from the orbiter. Such a spectrometer would need to cover a mass range of up to 400 amu with a mass resolution (m/Δm) between 800 and 1000 if organics are targeted, and would require a typical sensitivity of around 10^{-3} ions/cm^3 and on the order of 100 neutral/cm^3 (*Cassidy et al.,* 2009). In order to distinguish between plasma ions and locally formed ions, an energy range between thermal energies up to few 100 keV with an energy resolution around 10% is required. For the neutral spectrometer to distinguish between bound and escaping molecules, an energy range between thermal energy up to few tens of eV and an energy resolution around 10% would be required. Such spectrometers are presently being designed.

9. CONCLUSIONS

The intense plasma flux onto Europa's water ice surface creates a thin but detectable atmosphere and ionosphere as well as an extensive neutral cloud. Modeling shows these are rich in species derived from Europa's surface. Heavy ions in the jovian plasma erode the surface ejecting up to ~1000 H_2O molecules per ion and any trace species that may be present in the surface. This robust process can eject even refractory components, as suggested by the observed sodium, which might originate in hydrated sulfate salts. Alone, such salts have relatively small sputtering yields, but imbedded in an ice matrix identifiable components are readily ejected by the incident heavy ions. Therefore, most trace species in the surface will be present in the bound and extended atmosphere in some form, so that a sensitive ion and/or neutral mass spectrometer in orbit around Europa would be able to identify many of the surface components.

The incident plasma ions and electrons also break chemical bonds, which then can recombine to form new substances, a process called radiolysis. This produces the O_2 trapped in the surface and that dominates the near-surface atmosphere, as shown by modeling and confirmed by recent Cassini observations. That is, although H_2O is the dominant ejecta it sticks efficiently to surface grains while O_2 does not. Recent modeling shows that O_2 eventually reacts in the regolith or is destroyed (dissociated or ionized and swept away) by the magnetospheric plasma or solar photons. Such interactions are also responsible for the complexity of Europa's ionosphere.

Along with every O_2 created by radiolysis, in steady state two H_2 molecules are also created. These molecules more readily escape Europa's surface material and its gravity to form the most abundant component of Europa's neutral gas cloud, a cloud that, models and observations suggest, contains 10× as many atoms and molecules as Europa's bound atmosphere and possibly more than Io's neutral cloud. This cloud strongly depletes the inward diffusing energetic plasma and is a source of fast H atoms and protons for the magnetosphere. Europa's cloud is, nonetheless, harder to detect by remote sensing than Io's neutral cloud, but a minor component, sodium, has been studied extensively. The modeling results, along with the observations of the less-abundant potassium, have supported the tentative conclusion that Europa's sputtered sodium may be endogenic. Because this result, if proven to be correct, has implications for the presence of an ocean, further alkali observations and concomitant modeling are needed. In addition, *in situ* and remote observations should reveal the presence of many more species as suggested by the modeling described above and by the Galileo measurements of the ion environment. Since those neutrals that are ionized near Europa provide a readily detectable plasma source, we expect a rich ion environment both close to Europa and in the jovian magnetosphere near Europa's orbit.

The modeling described above has shown that the atmospheric structure is complex, but appears to correctly describe the available observations of the oxygen, hydrogen, and sodium content of Europa's gravitationally bound atmosphere and neutral clouds. Based on the physics and chemistry in these models, there is now a critical need for detailed modeling of the desorption of other important trace surface constituents, and for correctly modeling the ionosphere, which is spatially and temporally variable. Such efforts, however, will require a more accurate description of the magnetosphere/satellite interactions, as discussed in the chapter by Kivelson et al., a project that is critically important for the planning of a Europa mission.

REFERENCES

Bagenal F. (1994) Empirical model of the Io plasma torus: Voyager measurements. *J. Geophys. Res., 99,* 11043–11062.

Baragiola R. A. (2003) Water ice on outer solar system surfaces: Basic properties and radiation effects. *Planet. Space Sci., 51,* 953–961.

Bar-Nun A., Herman G., Rappaport M. L., and Mekler Yu. (1985) Ejection of H_2O, O_2, H_2, and H from water ice by 0.5–6 keV H^+ and Ne^+ ion bombardment. *Surface Sci., 150,* 143–156.

Benninghoven A., Rudenauer F. G., and Werner H. W. (1987) *Secondary Ion Mass Spectrometry.* Wiley, New York.

Brown M. E. (2001) Potassium in Europa's atmosphere. *Icarus, 151,* 190–195.

Brown M. E. and Hill R. E. (1996) Discovery of an extended sodium atmosphere around Europa. *Nature, 380,* 229–231.

Brown W. L., Augustyniak W. M., Simmons E., Marcantonio K. J., Lanzerotti L. J., Johnson R. E., Boring J. W., Reimann C. T., Foti G. and Pirronello V. (1982) Erosion and molecular formation in condensed gas films by electronic energy loss of fast ions. *Nucl. Instrum. Meth., 198,* 1.

Burger M. H. and Johnson R. E. (2004) Europa's neutral cloud: Morphology and comparisons to Io. *Icarus, 171,* 557–560.

Carlson R. W., Anderson M. S., Johnson R. E., Schulman M. B., and Yavrouian A. H. (2002) Sulfuric acid production on Europa: The radiolysis of sulphur in water ice. *Icarus, 157,* 456–463.

Carlson R. W., Anderson M. S., Mehlman R., and Johnson R. E. (2005) Distribution of hydrate on Europa: Further evidence for sulfuric acid hydrate. *Icarus, 177,* 461–471.

Cassidy T. A. and Johnson R. E. (2005) Monte Carlo model of sputtering and other ejection processes within a regolith. *Icarus, 176,* 499–507.

Cassidy T. A., Johnson R. E., McGrath M. A., Wong M. C., and Cooper J. F. (2007) The spatial morphology of Europa's near-surface O_2 atmosphere *Icarus, 191,* 755–764.

Cassidy T. A, Johnson R. E., Geissler P. E., and Leblanc F. (2008) Simulation of Na D emission near Europa during eclipse. *J. Geophys. Res.–Planets, 113,* E02005.

Cassidy T. A., Johnson R. E., and Tucker O. J. (2009) Trace constituents of Europa's atmosphere. *Icarus,* in press.

Cipriani F., Leblanc F., Witasse O., and Johnson R. E. (2008) Sodium recycling at Europa: What do we learn from the sodium cloud variability? *Geophys. Res. Lett., 35,* L19201.

Cipriani F., Leblanc F., Witasse O., and Johnson R. E. (2009) Exospheric signatures of alkalis abundances in Europa's regolith. *Geophys. Res. Lett.,* in press.

Cohen C. M. S., Stone E. C., and Selesnick R. S. (2001) Energetic ion observations in the middle jovian magnetosphere. *J. Geophys. Res., 106,* 29871–29882.

Cooper J. F., Johnson R. E., Mauk B. H., Garrett H. B., and Gehrels H. (2001) Energetic ion and electron irradiation of the icy Galilean satellites. *Icarus, 149,* 133–159.

Cooper P. D., Johnson R. E., and Quickenden T. I. (2003) Hydrogen peroxide dimers and the production of O_2 in icy satellite surfaces. *Icarus, 166,* 444–446.

Cosby P. C. (1993) Electron-impact dissociation of oxygen. *J. Chem. Phys., 98,* 9560.

De La Haye V. (2005) Formation and heating efficiencies in Titan's upper atmosphere: Construction of a coupled Io, neutral and thermal structure model to interpret the first INMS Cassini data. Ph.D. thesis, University of Michigan.

Famá M., Shi J., and Baragiola R. A. (2008) Sputtering of ice by low-energy ions. *Surface Sci., 602,* 156.

Hall D. T., Strobel D. F., Feldman P. D., McGrath M. A., and Weaver H. A. (1995) Detection of an oxygen atmosphere on Jupiter's moon Europa. *Nature, 373,* 677–679.

Hall D. T., Feldman P. D., McGrath M. A., and Strobel D. F. (1998) The far-ultraviolet oxygen airglow of Europa and Ganymede. *Astrophys. J., 499,* 475.

Hand K. P., Chyba C. F., Carlson R. W., and Cooper J. F. (2006) Clathrate hydrates of oxidants in the ice shell of Europa. *Astrobiology, 6(3),* 463–482.

Hansen C. J., Shemansky D. E., and Hendrix A. R. (2005) Cassini UVIS observations of Europa's oxygen atmosphere and torus. *Icarus, 176,* 300–315.

Hodges R. R. (1980) Lunar cold traps and their influence on argon-40. *Proc. Lunar Planet. Sci. Conf. 11th,* pp. 2463–2477.

Ip W.-H. (1996) Europa's oxygen exosphere and its magnetospheric interaction. *Icarus, 120,* 317–325.

Itikawa Y. and Mason N. (2005) Cross sections for electron collisions with water molecules. *J. Phys. Chem. Ref. Data, 34,* 1.

Johnson P. V., Kanik I., McConkey J. W., and Tayal S. S. (2005) Collisions of electrons with atomic oxygen: Current status. *Can. J. Phys., 83,* 589–616.

Johnson R. E. (1990) *Energetic Charged Particle Interactions with Atmospheres and Surfaces.* Springer-Verlag, Berlin.

Johnson R. E. (1998) Sputtering and desorption from icy satellite surfaces. In *Solar System Ices* (B. Schmitt and C. deBergh, ed.), pp. 303–334. Kluwer, Dordrecht.

Johnson R. E. (2000) Sodium at Europa. *Icarus, 143,* 429–433.

Johnson R. E. (2001) Surface chemistry in the jovian magnetosphere radiation environment. In *Chemical Dynamics in Extreme Environments, Chapter 8* (R. Dessler, ed.), pp. 390–419. Adv. Ser. Phys. Chem. Vol. 11, World Scientific, Singapore.

Johnson R. E. (2002) Surface boundary layer atmospheres. In *Atmospheres in the Solar System: Comparative Aeronomy* (M. Mendillo et al., eds.), pp. 203–219. AGU Geophysical Monograph 130, Washington, DC.

Johnson R. E. and Jesser W. A. (1997) O_2/O_3 microatmospheres in the surface of Ganymede. *Astrophys. J. Lett., 480,* L79–L82.

Johnson R. E. and Quickenden T. I. (1997) Photolysis and radiolysis of water ice on outer solar system bodies. *J. Geophys. Res., 102,* 10985–10996.

Johnson R. E. and Sundqvist B. U. R. (1992) Electronic sputtering: From atomic physics to continuum mechanics. *Phys. Today, 45(3),* 28–36.

Johnson R. E., Killen R. M., Waite J. H., and Lewis W. S. (1998) Europa's surface composition and sputter-produced ionosphere. *Geophys. Res. Lett., 25,* 3257–3260.

Johnson R. E., Lanzerotti L. J., and Brown W. L. (1982) Planetary applications of ion-induced erosion of condensed-gas frost. *Nucl. Instrum. Meth., 198,* 147–157.

Johnson R. E., Leblanc F., Yakshinskiy B. V., and Madey T. E. (2002) Energy distributions for desorption of sodium and potassium from ice: The Na/K ratio at Europa. *Icarus, 156,* 136–142.

Johnson R. E., Carlson R. W., Cooper J. F., Paranicas C., Moore M. H., and Wong M. C. (2004) Radiation effects on the surface of the Galilean satellites. In *Jupiter: The Planet, Satellites and Magnetosphere* (F. Bagenal et al., eds.), pp. 485–512. Cambridge Univ., Cambridge.

Johnston A. R. and Burrow P. D. (1995) Electron-impact ionization of Na. *Phys. Rev. A, 51,* R1735.

Kabin K., Combi M. R., Gombosi T. I., Nagy A. F., DeZeeuw D. L., and Powell K. G. (1999) On Europa's magnetospheric interaction: A MHD simulation of the E4 flyby. *J. Geophys. Res., 104,* 19983–19992.

Kimmel G. A. and Orlando T. M. (1995) Low-energy (5–120 eV) electron stimulated dissociation of amorphous D_2O ice: D(^{2}S), O(^{3}P), and O(^{1}D) yields and velocity distributions. *Phys. Rev. Lett., 75,* 2606–2609.

Kivelson M. G. (2004) Moon-magnetosphere interactions: A tutorial. *Adv. Space Res., 33,* 2061–2077.

Kliore A. J., Hinson D. P., Flasar F. M., Nagy A. F., and Cravens T. E. (1997) The ionosphere of Europa from Galileo radio occultations. *Science, 277,* 355–358.

Khurana K. K., Kivelson M. G., Stevenson D. J., Schubert G.,

Russell C. T., Walker R. J., and Polanskey C. (1998) Induced magnetic fields as evidence for subsurface oceans in Europa and Callisto. *Nature, 395,* 777.

Kurth W. S., Gurnett D. A., Persoon A. M., Roux A., Bolton S. J., and Alexander C. J. (2001) The plasma wave environment of Europa. *Planet. Space Sci., 49,* 345–363.

Lagg A., Krupp N., Woch J., and Williams D. J. (2003) In-situ observations of a neutral gas torus at Europa. *Geophys. Res. Lett., 30,* DOI: 10.1029/2003GL017214.1556.

Lanzerotti L. J., Maclennan C. G., Brown W. L., Johnson R. E., Barton L. A., Reimann C. T., Garrett J. W., and Boring J. W. (1983) Implications of Voyager data for energetic ion erosion of the icy satellites of Saturn. *J. Geophys., Res. 88,* 8765–8770.

Leblanc F., Johnson R. E., and Brown M. E. (2002) Europa's sodium atmosphere: An ocean source? *Icarus, 159,* 132–144.

Leblanc F., Potter A. E., Killen R. M., and Johnson R. E. (2005) Origins of Europa Na cloud and torus. *Icarus, 178,* 367–385.

Lindsay B. G. and Mangan M. A. (2003) In *Photon and Electron Interactions with Atoms, Molecules and Ions,* pp. 5-1 to 5-77. Londolt-Börnstein Vol. I/17, Subvolume C.

Liu Y., Nagy A. F., Kabin K., Combi M. R., DeZeeuw D. L., Gombosi T. I., and Powell K. G. (2000) Two-species 3D MHD simulation of Europa's interaction with Jupiter's magnetosphere. *Geophys. Res. Lett., 27,* 1791–1794.

Mauk B. H., Mitchell D. G., Paranicas C. P., and Krimigis S. M. (2003) Energetic neutral atoms from a trans-Europa gas torus at Jupiter. *Nature, 421,* 920–922.

Mauk B. H., Mitchell D. G., McEntire R. W., Paranicas C. P., Roelof E. W., Williams D. J., and Krimigis S. M. (2004) Energetic ion characteristics and neutral gas interactions in Jupiter's magnetosphere. *J. Geophys. Res., 109,* A09S12. DOI: 10.1029/2003JA010270.

McCord T. B., Hansen G. B., Fanale F. P., Carlson R. W., Matson D. L., Johnson T. V., Smythe W. D., Crowley J. K., Martin P. D., Ocampo A., Hibbitts C. A., Granahan J. C., and the NIMS Team (1998a) Salts on Europa's surface detected by Galileo's Near Infrared Mapping Spectrometer. *Science, 280,* 1242–1245.

McCord T. B., Hansen G. B., Clark R. N., Martin P. D., Hibbitts C. A., Fanale F. P., Granahan J. C., Segura M., Matson D. L., Johnson T. V., Carlson R. W., Smythe W. D., Danielson G. E., and the NIMS Team (1998b) Non-water-ice constituents in the surface material of the icy Galilean satellites from the Galileo near-infrared mapping spectrometer investigation. *J. Geophys. Res., 103(E4),* 8603–8626.

McCord T. B., Hansen G. B., Matson D. L., Johnson T. V., Crowley J. K., Fanale F. P., Carlson R. W., Smythe W. D., Martin P. D., Hibbitts C. A., Granahan J. C., and Ocampo A. (1999) Hydrated salt minerals on Europa's surface from the Galileo near-infrared mapping spectrometer (NIMS) investigation. *J. Geophys. Res., 104(E5),* 11827–11851.

McCord T. B., Orlando T. M., Teeter G., Hansen G. B., Sieger M. T., Petrik N. G., and Van Keulen L. (2001) Thermal and radiation stability of the hydrated salt minerals epsomite, mirabilite, and natron under Europa environmental conditions. *J. Geophys. Res., 106,* 3311–3319.

McCord T. B., Teeter G., Hansen G. B., Sieger M. T., and Orlando T. M. (2002) Brines exposed to Europa surface conditions. *J. Geophys. Res., 107(E1),* 10.1029/2000JE001453.

McElroy M. B. and Yung Y. L. (1977) Stability of an oxygen atmosphere on Ganymede. *Icarus, 30,* 97–103.

McGrath M. A., Lellouch E., Strobel D. F., Feldman P. D., and Johnson R. E. (2004) Satellite atmospheres. In *Jupiter: Satellites, Atmosphere, Magnetosphere* (F. Bagenal et al., eds.), pp. 457–483. Cambridge Univ., Cambridge.

McKinnon W. B. and Zolensky M. E. (2003) Sulfate content of Europa's ocean and shell: Evolutionary considerations and geological and astrobiological implications. *Astrobiology, 3,* 879–897.

Mitchell D. G., Paranicas C. P., Mauk B. H., Roelof E. C., and Krimigis S. M. (2004) Energetic neutral atoms from Jupiter measured with the Cassini Magnetospheric Imaging Instrument: Time dependence and composition. *J. Geophys. Res., 109,* A09S11. DOI: 10.1029/2003JA010120.

Moncuquet M., Bagenal F., and Meyer-Vernet N. (2002) Latitudinal structure of outer Io plasma torus. *J. Geophys. Res., 107,* A9, 1260. DOI: 10.1029/ 2001JA900124.

Noll K. S., Weaver H. A., and Gonnella A. M. (1995) The albedo spectrum of Europa from 2200 Å to 3300 Å. *J. Geophys. Res., 100,* 19057–19060.

Orlando T. M., McCord T. B., and Grieves G. A. (2005) The chemical nature of Europa surface material and the relation to a subsurface ocean. *Icarus, 177,* 528–533.

Paranicas C., Mauk B. H., Ratliff J. M., Cohen C., and Johnson R. E. (2002) The ion environment near Europa and its role in surface energetics. *Geophys. Res. Lett., 29,* 1074.

Paranicas C., Mauk B. H., McEntire R. W., and Armstrong T. P. (2003) The radiation environment near Io. *Geophys. Res. Lett., 30(18),* 1919, DOI: 10.1029/2003GL017682.

Paterson W. R., Frank L. A., and Ackerson K. L. (1999) Galileo plasma observations at Europa: Ion energy spectra and moments. *J. Geophys. Res., 104,* 22779–22792.

Porco C. C., West R. A., McEwen A., Del Genio A. D., Ingersoll A. P., et al. (2003) Cassini imaging of Jupiter's atmosphere, satellites, and rings. *Science, 299,* 1541–1547.

Pospieszalska M. K. and Johnson R. E. (1989) Magnetospheric ion bombardment profiles of satellites: Europa and Dione. *Icarus, 78,* 1–13.

Raut U, Famá M., Teolis B. D., and Baragiola R. A. (2007) Characterization of porosity in vapor-deposited amorphous solid water from methane adsorption. *J. Chem. Phys., 127,* 204713.

Reimann C. T., Boring J. W., Johnson R. E., Garrett J. W., Farmer K. R., and Brown W. L. (1984) Ion-induced molecular ejection from D_2O ice. *Surface Sci., 147,* 227.

Sandford S. A. and Allamandola L. J. (1993) The condensation and vaporization behavior of ices containing SO_2, H_2S, and CO_2 — Implications for Io. *Icarus, 106,* 478.

Saur J., Strobel D. F., and Neubauer F. M. (1998) Interaction of the jovian magnetosphere with Europa: Constraints on the neutral atmosphere. *J. Geophys. Res., 103,* 19947–19962.

Schenk P. M., Chapman C. R., Zahnle K., and Moore J. M. (2004) Ages and interiors: The cratering record of the Galilean satellites. In *Jupiter: The Planet, Satellites and Magnetosphere* (F. Bagenal et al., eds.), pp. 427–456. Cambridge Univ., New York.

Schilling N, Neubauer F. M., and Saur J. (2007). Time-varying interaction of Europa with the jovian magnetosphere: Constraints on the conductivity of Europa's sub-surface ocean. *Icarus, 192,* 41–55.

Schilling N, Neubauer F. M., and Saur J. (2008) Influence of the internally induced magnetic field on the plasma interaction of Europa. *J. Geophys. Res.–Space Physics, 113,* A03203.

Schreier R., Eviatar A., Vasyliunas V. M., and Richardson J. D. (1993) Modeling the Europa plasma torus. *J. Geophys. Res., 98,* 21231–21243.

Shah M. B., Elliott D. S., and Gilbody H. B. (1987) Pulsed crossed-beam study of the ionisation of atomic hydrogen by electron impact. *J. Phys. B (Atom. Mol. Phys.), 20,* 3501–3514.

Shematovich V. I. (2006) Stochastic models of hot planetary and satellite coronas: Atomic oxygen in Europa's corona. *Solar Sys. Res., 40,* 175–190.

Shematovich V. I., Johnson R. E., Cooper J. F., and Wong M. C. (2005) Surface bound atmosphere of Europa. *Icarus, 173,* 480–498.

Sieger M. T., Simpson W. C., and Orlando T. M. (1998) Production of O_2 on icy satellites by electronic excitation of low-temperature water ice. *Nature, 394,* 554–556.

Smyth W. H. and Combi M. R. (1988) A general model for Io's neutral gas cloud II. Application to the sodium cloud. *Astrophys. J., 328,* 888–918.

Smyth W. H. and Marconi M. L. (2003) Nature of the iogenic plasma source in Jupiter's magnetosphere I. Circumplanetary distribution. *Icarus, 166,* 85–106.

Smyth W. H. and Marconi M. L. (2006) Europa's atmosphere, gas tori, and magnetospheric implications. *Icarus, 181,* 510–526.

Smythe W. D., Carlson R. W., Ocampo A., Matson D., Johnson T. V., McCord T. B., Hansen G. E., Soderblom L. A., and Clark R. N. (1998) Absorption bands in the spectrum of Europa detected by the Galileo NIMS instrument (abstract). In *Lunar and Planetary Science XIX,* Abstract #1532. Lunar and Planetary Institute, Houston (CD-ROM).

Spencer J. R. and Calvin W. M. (2002) Condensed O_2 on Europa and Callisto. *Astron. J., 124,* 3400–3403.

Spencer J. R., Grundy W. M., Dumas C., McCord T. B., Hansen G. B., and Terrile R. J. (2006) The nature of Europa's dark non-ice surface material: Spatially-resolved high spectral resolution spectroscopy from the Keck telescope. *Icarus, 182,* 202–210.

Steffl A. J., Bagenal F., and Stewart A. I. F. (2004) Cassini UVIS observations of the Io plasma torus. II. Radial variations. *Icarus, 172,* 91–103.

Straub H. C., Renault P., Lindsay B. G., Smith K. A., and Stebbings R. F. (1996) Absolute partial cross sections for electron-impact ionization of H_2, N_2, and O_2 from threshold to 1000 eV. *Phys. Rev. A, 54,* 2146–2153.

Teolis B. D., Vidal R. A., Shi J., and Baragiola R. A. (2005) Mechanisms of O_2 sputtering from water ice by keV ions. *Phys. Rev. B, 72,* 245–422.

Volwerk M., Kivelson M. G., and Khurana K. K. (2001) Wave activity in Europa's wake: Implications for ion pickup. *J. Geophys. Res., 106,* 26033–26048.

Westley M. S., Baragiola R. A., Johnson R. E., and Baratta G. A. (1995) Photodesorption from low-temperature water ice in interstellar and circumsolar grains. *Nature, 373,* 405.

Wiens R. C., Burnett D. W., Calaway W. F., Hansen C. S., Lykke K. R., and Pellin M. J. (1997) Sputtering products of sodium sulfate: Implications for Io's surface and for sodium-bearing molecules in the Io torus. *Icarus, 128,* 386–397.

Wiens R. C., Burnett D. S., Calaway W. F., Hansen C. S., Lykke K. R., and Pellin M. J. (1998) Sputtering of sodium sulfate: Implications for Io's surface and sulfur bearing compounds in the Io torus. *Icarus, 128,* 386–397.

Wolf-Gladrow D. A., Neubauer F. M., and Lussem M. (1987) Io's interaction with the plasma torus — A self-consistent model. *J. Geophys. Res., 92,* 9949–9961.

Wong M. C. and Smyth W. H. (2000) Model calculations for Io's atmosphere at eastern and western elongation. *Icarus, 146,* 60–74.

Wong M. C., Carlson R. W., and Johnson R. E. (2001) Model simulations for Europa's atmosphere. *Bull. Am. Astron. Soc., 32,* 1056.

Ziegler J. F., Biersack J. P., and Littmark U. (1985) *The Stopping and Range of Ions in Solids.* Pergamon, New York.

Zolotov M. Y. and Shock E. L. (2001) Composition and stability of salts on the surface of Europa and their oceanic origin. *J. Geophys. Res.–Planets, 106,* 32815–32828.

Europa's Radiation Environment and Its Effects on the Surface

C. Paranicas
Johns Hopkins University Applied Physics Laboratory

J. F. Cooper
NASA Goddard Space Flight Center

H. B. Garrett
NASA Jet Propulsion Laboratory/California Institute of Technology

R. E. Johnson
University of Virginia

S. J. Sturner
NASA Goddard Space Flight Center and
University of Maryland Baltimore County

Europa's orbit in the radiation environment of Jupiter is reviewed as is the influence of the neutral gas torus on the surface weathering of that moon. Data and fits to charged particle intensities in the 1-keV to the tens-of-MeV energy range are provided near Europa. Leading/trailing hemisphere differences are highlighted. Effects of charged particles on the surface of Europa, such as sputtering and chemistry, are reviewed.

1. INTRODUCTION

Charged particles trapped in Jupiter's rotating magnetosphere continuously overtake Europa in its orbit. At sufficiently high energies, these particles are relatively unaffected by the tenuous atmosphere of the satellite and instead bore directly into the ice before losing much energy. For example, energetic electrons and their bremsstrahlung photon products can directly affect the top meter of the icy regolith, which is also processed by meteoritic impact gardening (*Cooper et al.,* 2001; *Chyba and Phillips,* 2001). Charged particle irradiation of ice produces a number of new species as described in the chapter by Carlson et al. Those that are volatile at the ambient temperature, such as O_2, populate the atmosphere, whereas those that are more refractory, such as H_2O_2, can be detected as trace species in the ice. In addition to the chemical weathering of Europa's surface, the bombarding energetic particle flux drives species into the gas phase, a process called "sputtering." This produces a thin atmosphere above Europa's surface as described in the chapters by Johnson et al. and McGrath et al. Extrapolation from laboratory data to the quantification of radiation effects on Europa's surface and atmospheric environment requires modeling of the longitudinal and latitudinal distributions of energy deposited per unit volume vs. depth into the surface. It is also useful to characterize ranges of temporal variation caused by jovian magnetospheric activity and other effects. Therefore a central goal of this chapter is to provide estimates of the average energy vs. depth distributions at representative locations on the surface and to describe variations one might expect. We will also estimate the principal effects produced for different radiation types.

In section 2 of this chapter, we will discuss the jovian radiation environment to provide a context for Europa. In particular, we note the relative levels of radiation among the inner satellites as potentially important for differences in surface weathering. We also point out that Europa's orbit coexists with a cold neutral gas torus. This gas influences all ions up to the few-MeV range because of charge-exchange collisions that create energetic neutral atoms (ENAs) with energies reflective of their parent ion. In section 3, we turn our attention to the environment close to the satellite itself. We provide recent fits to the electron and ion data that describe the intensity of these trapped particles near Europa's orbit. We cover the energy range from about 1 keV to tens of MeV. We will also elaborate on the asymmetric bombardment of Europa by electrons, which has consequences for surface processing.

In section 4 of this chapter, we describe some of the effects of the radiation environment on the surface itself. A good recent summary of the consequences of charged particle weathering of Europa's surface can be found in *Johnson et al.* (2004) and references therein. *Cooper et al.* (2001) provide a table (their Table II) of surface irradiation parameters for the icy satellites. It is not our intention to repeat that material here but to mention highlights and updates since some of the earlier publications. In particular, we will extend some of our previous ideas on the nonuniformity of the surface bombardment. A central reason for improved modeling of space weathering effects on exposed surfaces in space is to determine the chemical compo-

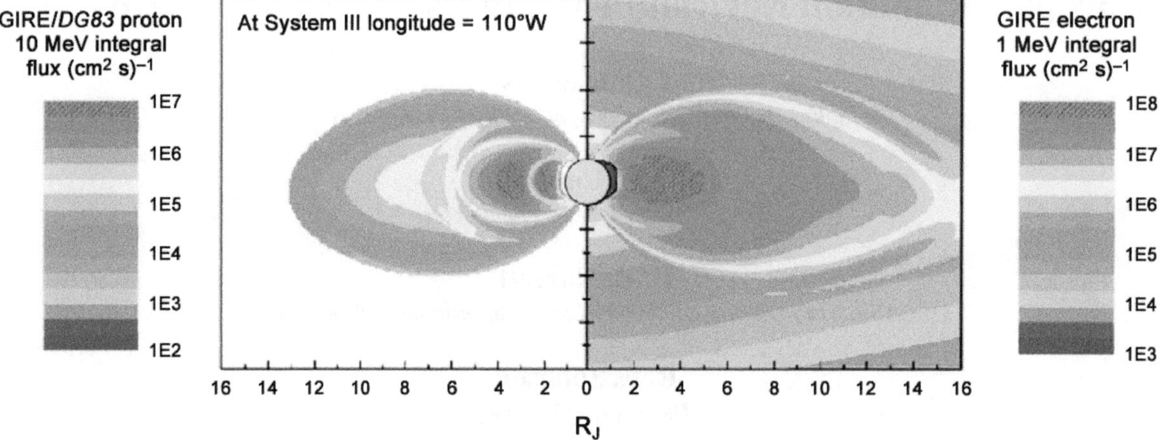

Fig. 1. See Plate 26. Contour plots of the GIRE/*DG83* model E > 10 MeV integral proton (left) and E > 1 MeV electron (right) fluxes for the vovian magnetosphere radiation region. The model provides the flux as a function of position, energy, and pitch angle. The fluxes presented here have been integrated over pitch angle. Note that outside the contour of L = 12, the proton model is set to 0, whereas the electron model fluxes outside L = 16 are only approximate since they are not considered trapped in the model (see *DG83* for details). Figure courtesy of I. Jun.

sition of materials intrinsic to Europa as compared to those originating from the external influence of magnetospheric particle irradiation. For example, the observed concentration (*McCord et al.,* 1998, 1999) of hydrated sulfate materials on the trailing hemisphere, recently confirmed by New Horizons observations (*Grundy et al.,* 2007), is strongly suggestive of connections to the corotating magnetospheric plasma, including iogenic sulfur ions (*Carlson et al.,* 1999, 2002, 2005; *Paranicas et al.,* 2001), but this connection is less clear at smaller scales of surface lineaments (*Geissler et al.,* 1998; *McCord et al.,* 1998) (see also the chapter by Carlson et al.). Galileo Solid State Imaging (SSI) and Near-Infrared Mapping Spectrometer (NIMS) multispectral data indicate compositional differences are present between specific surface features with respect to the surrounding terrain (*Carlson et al.,* 2005). It has also been observed that these features appear to brighten with age (*Geissler et al.,* 1998). This brightening may be the result of long-term exposure to Europa's radiation environment (e.g., *Geissler et al.,* 1998; *Johnson et al.,* 2004). Alternatively, the sulfur may be endogenic (*McCord et al.,* 1998; 1999), for instance, as in salts that have been moved through the ice crust from the subsurface ocean of Europa, but then the hemispheric asymmetry remains to be explained. Comparisons of composition and brightness distributions in highly and minimally irradiated regions is an example of how to potentially settle questions of endogenic vs. exogenic origin.

2. EUROPA IN THE CONTEXT OF JUPITER'S RADIATION ENVIRONMENT

The radiation belt modeling of *Divine and Garrett* (1983; hereafter *DG83*) combined *in situ* data from the Pioneer and Voyager spacecraft with groundbased data to describe the fluxes of energetic charged particles in Jupiter's inner magnetosphere. Their modeling put the peak radiation in >1-MeV electrons close to about L = 3, using a dipolar model of the magnetic field of the planet. In their fits to the various datasets, >10-MeV ion intensities also show a peak close to the planet but fall off rapidly with increasing radial distance. Near Europa's orbit, $R \approx 9.4\ R_J$ (here $R_J = 71{,}492$ km), and keeping in mind the different lower-energy limits, the MeV ions are already substantially reduced from their peak, whereas the MeV electrons are somewhat lower than their peak but still significant. The *DG83* electron fluxes have been compared to Galileo orbiter measurements near Europa, Ganymede, and Callisto (*Cooper et al.,* 2001). The model and Energetic Particles Detector (EPD) fluxes in the 1-MeV range were found to agree well on the decrease in flux with increasing distance from Jupiter. At higher energies the electron data model of *Cooper et al.* (2001) was derived only from *DG83*. The *DG83* modeling for electrons has been superseded by the Galileo Interim Radiation Electron (GIRE) model (*Garrett et al.,* 2002, 2005). Model integral fluxes from GIRE/*DG83* are presented in Fig. 1.

The work of *Jun et al.* (2005) displayed Galileo EPD data, comparing the electron count rates and fluxes for energies ≥1.5, ≥2.0, and ≥11 MeV, for orbits over the whole mission. They found an approximately 2 order-of-magnitude increase in tens of MeV electrons from the orbit of Ganymede ($R \sim 15\ R_J$) to that of Europa. *DG83,* its GIRE update, and *Jun et al.* (2005) also reflect a more or less steady-state structure in the energetic electron belt. This suggests that the population is persistent every time we sample it. Furthermore, from the work of *Jun et al.* (their Fig. 3), it is possible to estimate the variability of this population. Near Europa's orbital distance, the 1σ level of the ≥11-MeV electron flux is about a factor of 2–3 times the mean and the 2σ level is a factor of 10. These data include nearly all the Galileo orbits and are ordered in dipole L shell using the VIP4 model (*Connerney et al.,* 1998). (In a dipole field, the L shell can be calculated from $L = r/\cos^2\lambda$, where r is the distance from the center of the dipole and λ is the magnetic latitude of the point in question.) This spread is probably an upper limit on the variability of that population since, for example, Jun and his colleagues did not separate the data by pitch angle.

Coexisting with the MeV particles, there is dense, cold plasma (see, for example, the chapter by Kivelson et al.) and medium-energy particles in the keV energy range. *Mauk et al.* (2004) have generated fit functions for the tens of keV to tens of MeV ion data obtained by Galileo. *Mauk et al.* (2004) present various moments of the distribution function by radial distance from Jupiter. In Fig. 2, we have used the fits from *Mauk et al.* (2004) to create plots of particle intensity by species and L shell at specific energies. Each panel shows a separate ion in the radial range from Io's orbit to about 20 R_J. The large gaps in coverage are due to the fact that Mauk et al. were only able to compute fits at specific locations in the magnetosphere. Typically we would expect such curves to rise inward toward the planet because as particles are transported inward they are energized and there are typically more charged particles at lower energies. It is notable in Fig. 2 that ions above about 500 keV continue to increase inward across Europa's orbit. On the other hand, lower-energy ions have dramatic changes at or near Europa's orbital distance.

The initial decrease (moving radially inward in L in Fig. 2) in the ion intensities below about 500 keV is likely caused by their loss to the neutral gas torus at Europa's orbit (*Lagg et al.,* 2003; chapter by Johnson et al.). Magnetospheric ions can undergo charge-exchange reactions with neutrals in the gas torus and leave the system as ENAs. Charge exchange cross-sections are large at low energies but begin to fall off rapidly for protons above about 100 keV and for O^+ above several hundred keV (e.g., *Lindsay and Stebbings,* 2005). If charge exchange is the dominant loss process for ions below about 500 keV near Europa's orbit, then these ions are not principally lost to the satellite's surface. It is likely then that the surface is not heavily weathered by these ions, as was believed prior to Galileo. This is not the case for MeV ions or electrons. Their mean intensities continue to rise radially inward to the planet, relatively unaffected by the gas.

One mechanism for populating such a neutral torus is collisions between corotating magnetospheric ions and neutrals in Europa's atmosphere. Neutral modeling by *Smyth and Marconi* (2006) shows high column densities of O and H_2 that peak in the radial dimension at Europa's orbit (see chapter by McGrath et al.). To further support the presence of neutral gas near Europa, *Mauk et al.* (2003) found evidence of ENA emissions from the region of the gas torus, with data from Cassini's Magnetospheric Imaging Instrument (MIMI) during that spacecraft's distant Jupiter flyby. Mauk and his colleagues correlated the ENA signal with the torus and not Europa itself (e.g., there are two emission peaks at the radial distance in question). To summarize our findings then, Europa is heavily weathered by MeV ions and electrons and by keV electrons with much higher intensities than at Ganymede or Callisto. The dominant mechanism of loss for medium energy ions near Europa's orbit is by charge-exchange collisions with neutrals in a gas torus.

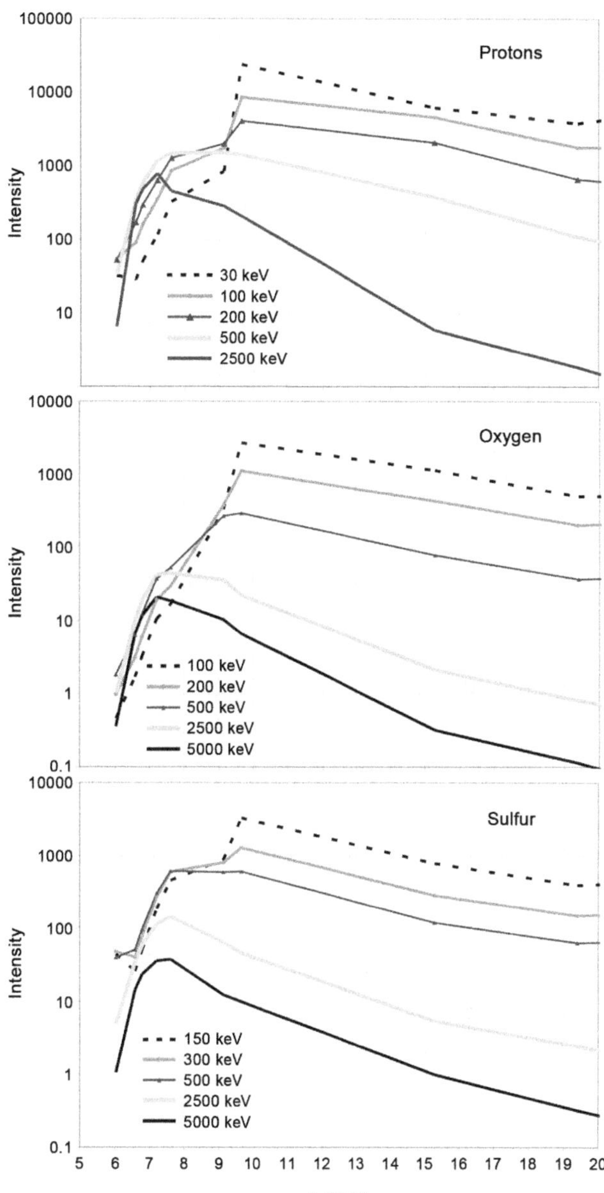

Fig. 2. Ion intensity (ions per cm^2 s sr keV) at constant energy. Values are computed at specific Galileo locations from fits to the full charged particle spectra reported by *Mauk et al.* (2004) and connected by lines for ease of viewing. The fit functions for all points used to create this graph are given in Table 1 of *Mauk et al.* (2004). In a dipole field, the L shell of Europa varies between about 9.3 and 9.7 R_J.

3. EUROPA'S RADIATION ENVIRONMENT

The various data analysis and modeling efforts mentioned above as well as other studies help to place the radiation environment of Europa in the context of the rest of the inner to middle magnetosphere. Europa's radius is ~1561 km and its mean orbital radius is ~9.39 R_J. Its orbit is tilted relative to Jupiter's equator (by about 0.466°) and is slightly eccentric (0.0094) (see *ssd.jpl.nasa.gov*). Because of the tilt of Jupiter's magnetic dipole, Europa makes excursions from the spin plane to approximately 10° north and

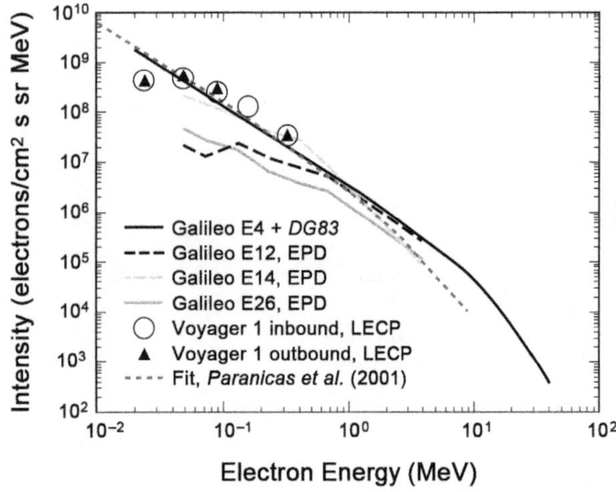

Fig. 3. Energy spectrum of electrons measured near Europa's orbit from various sources including *Cooper et al.* (2001) *Paranicas et al.* (2001) and the JPL model of *Divine and Garrett* (1983) as derived from Pioneer and Voyager data at Jupiter.

south magnetic latitude. The weathering of the moon depends somewhat on its magnetic latitude, since off the equator a fraction of charged particles do not have access to the moon's surface. However, this may be most important for weathering by the cold, heavy plasma. The relative position of Europa to the magnetic equator, which dictates the strength of the induced magnetic field (*Kivelson et al.,* 1999), will be a factor in particle access to the surface. In the following, we focus on the radiation environment at Europa's orbital distance and its variations. The data and fits we present are a compilation from different sources meant to be representative and are not a synthesis of all available jovian data to date.

In Fig. 3, we show a summary of the measured (and fit) electron intensity (electrons per cm^2 s sr MeV) from several sources near Europa or near its orbit. These include the *DG83* model, the Voyager 1 spacecraft Low Energy Charged Particle (LECP) detector data, and Galileo EPD data. Since the *DG83* model is based on Pioneer and Voyager spacecraft data, this set samples the main three generations of spacecraft that crossed Europa's orbit. The plotted fit to the intensity is

$$\text{j(counts per cm}^2\text{ s sr MeV)} = 4.23 \times 10^6\,\text{E(MeV)}^{-1.58}\left(1 + \frac{\text{E(MeV)}}{3.11}\right)^{-1.86} \quad (1)$$

One of the Galileo detectors was severely overdriven in the inner magnetosphere. For several of the data points plotted here, a correction was applied to recover, where possible, the actual rate. The high uncertainty in the rate near the lower energy end of the range probably explains the large variation and it should not be interpreted as variation in the local environment. For this figure, the omnidirectional flux of *DG83* was divided by 4π to obtain intensity per steradian for comparison.

Turning next to energetic ions near Europa, we compare data taken from several different Galileo encounters with Europa. In Fig. 4, we show energy spectra from the dominant ions separately: protons, oxygen, and sulfur. Fits to some of these ion data have been performed using the following function (*Mauk et al.,* 1994)

$$\text{j(counts per cm}^2\text{ s sr keV)} = \text{C} \times \text{E(keV)}\frac{[\text{E}_1 + \text{kT}(1+\gamma_1)]^{-1-\gamma_1}}{1 + \left(\text{E}_1/\text{e}_t\right)^{\gamma_2}} \quad (2)$$

Galileo closest approach distances for these passes are approximately 692 km (E4), 586 km (E6), 201 km (E12), 1439 km (E19), and 351 km (E26) (data from JPL press release, 2003). *Kivelson et al.* (1999) provide details of these Europa passes and *Paranicas et al.* (2000) present energetic

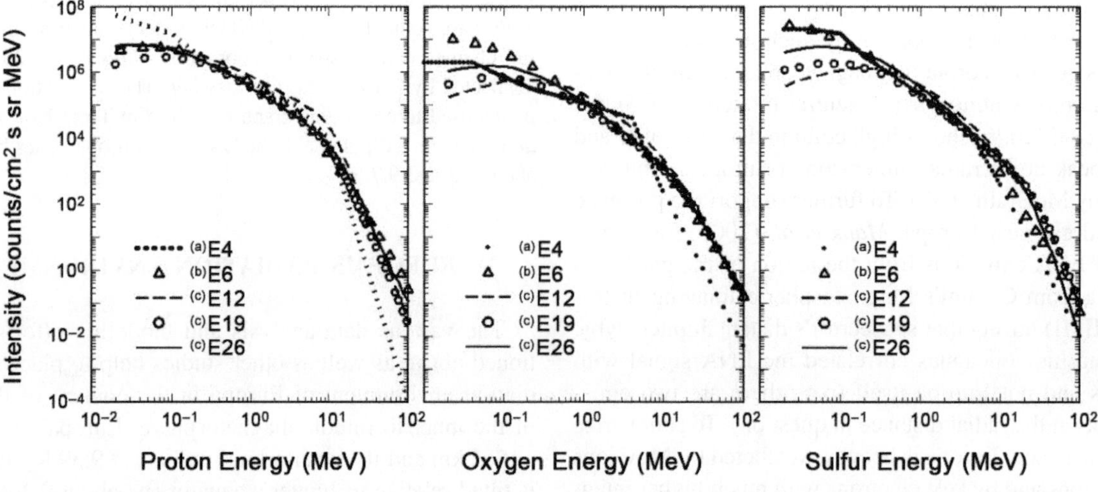

Fig. 4. Energy spectra by ion species computed from Galileo data during various near encounters of the satellite Europa. Sources of data include **(a)** *Cooper et al.* (2001), **(b)** *Mauk et al.* (2004), and **(c)** *Paranicas et al.* (2002).

TABLE 1. Fits to the individual ion energy spectra (by species) detected near Europa from *Mauk et al.* (2004) (to be used with equation (2)).

	H	He	O	S
C	2.79e + 06	8.02e + 04	9.19e + 06	1.42e + 07
kT (keV)	4.5	4.5	8.6	8.6
γ_1	1.213	1.213	1.647	1.599
e_t (keV)	6880	6880	10,838	5700
γ_2	4.177	4.177	2.793	2.616

charged particle data taken during them. For the Galileo E4 encounter, the fit parameters corresponding to equation (2) for various ions are given in Table 1.

On balance, there is a reasonably stable spectral shape with some variable range in absolute fluxes near Europa's orbit. By comparison, the inner magnetosphere of Saturn is injection dominated and some ions are not stably trapped. *Kivelson et al.* (1999) give Europa's position relative to the center of the current sheet during many of the Galileo flybys. They show that during E12, Europa was close to Jupiter's magnetic equator, E19 is off the equator, and E4 and E6 (and E26) are increasingly farther off. In Fig. 4, the ion measurements made near the center of the sheet have some of the highest intensities at high energy, but there is no such trend at lower energy. It is important to keep in mind that several issues are involved in the observed level of variation including the presence of the neutral gas, activity levels in the system (e.g., injections), how the pitch angles are sampled, and instrument aging issues. Furthermore, for the nine Galileo/Europa encounters listed in *Kivelson et al.* (1999), the local time range is confined to 0948 to 1642. A more robust indicator of time variability is provided in the work of *Mauk et al.* (2004). They have calculated an integrated quantity (called the detector current) representing the fluxes of medium energy to few MeV ions over the whole Galileo mission. They find that near Europa there is about a factor of ~2 variation in this integrated quantity in the 4-yr period that is covered (with very sparse sampling). They suggest that a main variation in the detector current is the variations in the neutral gas density.

To complete the description of the energy range of interest to us here, we show model fluxes down to 1 keV in Fig. 5. These are computed near Europa's orbit, but no shielding by Europa that would attenuate the flux is included. Since the range between about 10 keV and 100 keV has been poorly sampled in the past, fluxes are computed as follows. Energetic particle fluxes are derived from the Divine and Garrett model. A cold Maxwellian component is fit at the lower energies. Then a κ distribution is used to join the 1-keV model fit to the more energetic fits. Because there are many more of these particles, particularly electrons, than in the more energetic tail of the distribution function, they are important for processing the surface. The range in water of a 10-keV electron is about 1.6 μm and the range of a 1-MeV electron is 4.2 mm (*Zombeck,* 1982). The range in water of a 10-keV proton is about 0.3 μm and of a 1-MeV proton is about 24 μm (NIST website, *physics.nist.gov/PhysRefData/Star/Text/contents.html*). Therefore the vast majority of ions in the particle distribution functions, i.e., those up to about 10 keV in energy, directly affect only that part of the surface at depths less than about 0.3 μm.

3.1. Injections

To further study the role of variability in Europa's space environment, it is useful to consider transient phenomena. *Mauk et al.* (1999) surveyed over 100 instances of ion and electron injections in Jupiter's magnetosphere. They reported that these injections are most frequently observed between about Europa's orbital distance and 27 R_J and occur at all System III longitudes and local times. Injections of the type Mauk et al. have catalogued, whether they are interchange or another physical process, likely energize ions and electrons as they transport them inward radially, conserving their first adiabatic invariant of motion (see *Walt,* 1994, for a definition and discussion). Injected particles then corotate with the magnetosphere and become dispersed over time. Because injections are localized features, it is expected that they will introduce a level of variability to the background flux, but are not expected to dominate it. In Saturn's inner magnetosphere, by contrast, injections can often be the dominant population because some ambient flux levels are so low. Finally, at Jupiter, injections as described above are certainly a source of the inner, trapped population but it is not well known whether they are the dominant or a secondary source. Other processes such as energization in place by non-adiabatic processes and radial microdiffusion also play a role.

3.2. Bombarding Particles

In this subsection, we turn our attention to how the charged particles actually bombard the moon's surface. Here it is important to review some basics of the particle motion because it is central to what we describe below.

The energy spectrum of all charged particles trapped in Jupiter's magnetosphere is often divided into a plasma range and an energetic charged particle range. This separation is guided, to some extent, by the fact that the cold plasma can be studied with a magnetohydrodynamic (MHD) approximation (e.g., *Kabin et al.,* 1999) in which plasma flows like a fluid onto and around the object. At keV energies and above, the individual particle motions become significant and a single-particle approach is more appropriate. This is partly true because the gyroradius increases with energy and the more energetic particles become sensitive to spatial gradients of the magnetic field and to the curvature of the local field lines. These effects lead to gradient-curvature drifts in planetary longitude as discussed below. Moments of the particle distribution function, such as total mass and total charge, are dominated by the plasma. So, for example, the requirement of quasi-neutrality is typically enforced at plasma energies. Other moments, such as energy flux and

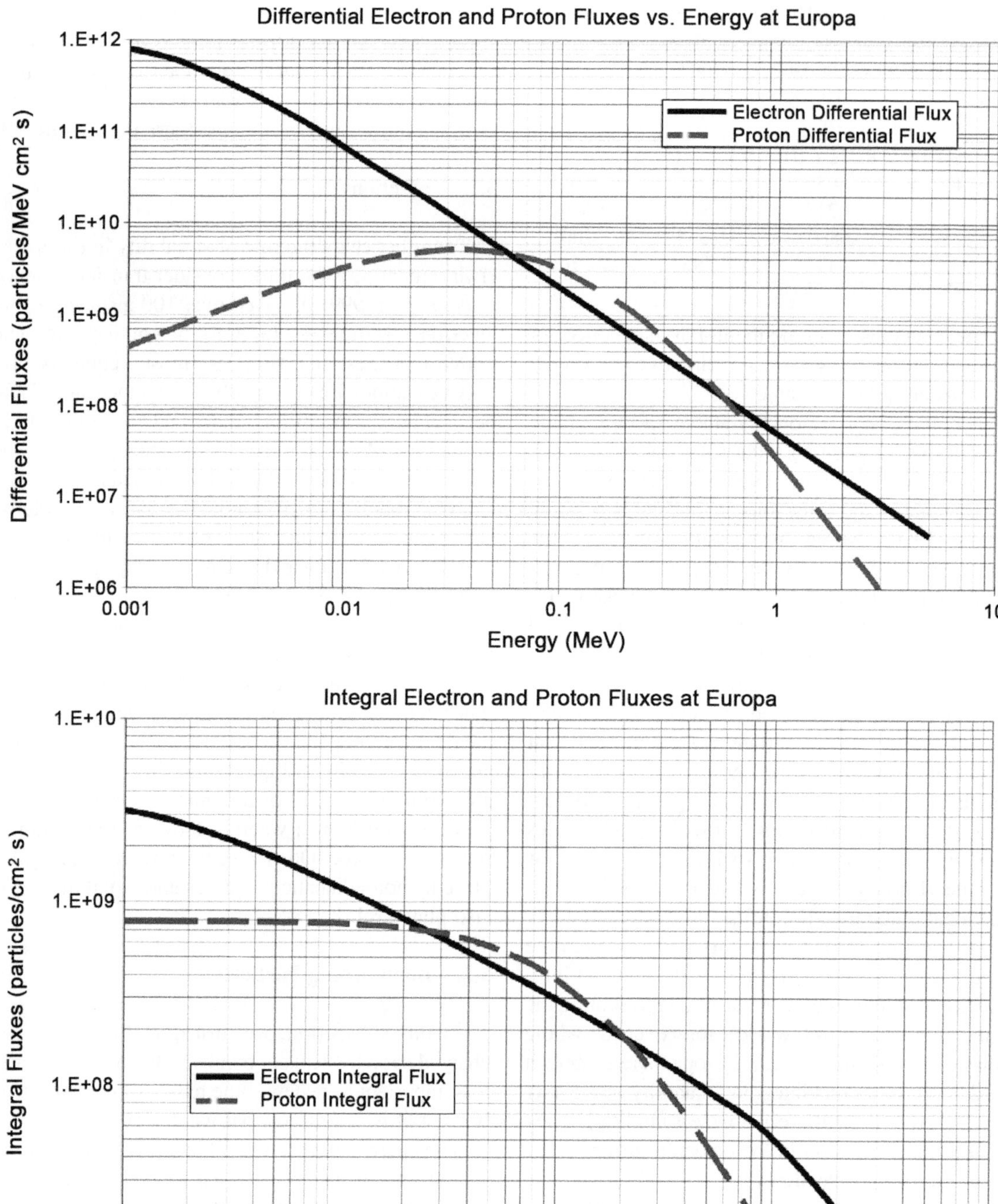

Fig. 5. Electron and proton fluxes computed from the DG radiation and warm plasma models. These correspond to a radial distance of 9.5 R_J.

particle pressure, can have important contributions from both parts of the particle distribution. Since we are interested in energetic charged particles here, we look at these particles from the perspective of individual particle motion.

Charged particles have three principle motions in the inner jovian magnetosphere: the gyration about the magnetic field line, the bounce along the field line between north and south mirror points — where the field becomes strong enough to reflect the particle — and the longitudinal motion. This third motion itself has two contributions: the so-called **E** × **B** drift and the gradient-curvature drift. The first drift dominates for charged particles at plasma energies and corresponds to the plasma flow speed. The **E** × **B** drift is due to the background magnetic field and an outwardly pointing electric field (see *Khurana and Kivelson,* 1993, for a description of the origin of this field). Also, a complete description of charged particle motion in magnetospheres can be found in *Walt* (1994). For a rigidly rotating jovian magnetosphere, the plasma flow speed near Europa's orbit is approximately 118 km/s. Typically detected values are lower; for example, *Paterson et al.* (1999) report a speed that is about 80% of rigid corotation.

For the energies of interest to us here, it is also important to consider the gradient-curvature drift, which is caused by the deviations from a uniform magnetic field. In the corotating frame of the plasma, all ions drift in the same direction as the plasma flow and electrons drift in the opposite direction. This means that in the inertial frame, all ions are traveling slightly faster than the plasma corotation speed and all electrons slower. Above about 25 MeV, the gradient-curvature drift of electrons is comparable to the $\mathbf{E} \times \mathbf{B}$ drift in magnitude but opposite in direction and the electrons consequently have a net azimuthal motion that is retrograde, opposite to the prograde motion of Europa and the plasma flow. All the charged particles, except >25-MeV electrons, would therefore bombard Europa from the trailing hemisphere to the leading hemisphere. Here, by trailing hemisphere we mean the hemisphere that trails Europa in its motion around Jupiter. By convention, the center of the trailing hemisphere is 270°W longitude, where 90°W is the center of the leading hemisphere and 0°W points toward Jupiter.

Some equations that quantify these effects further are provided next. The net azimuthal speed of the charged particle's guiding center with respect to Europa can be expressed as

$$\omega = \Omega_J + \omega_D(L,E,\lambda_m) - \omega_k \qquad (3)$$

Here Ω_J corresponds to Jupiter's rotation rate in rad/s, ω_k is the angular speed of Europa in inertial space, E is the kinetic energy in MeV, λ_m is the particle's mirror latitude, and ω_D is the gradient-curvature drift rate. Following *Thomsen and Van Allen* (1980) this drift rate can be written

$$\omega_D = \pm 6.856 \times 10^{-7} LE \frac{E + 2mc^2}{E + mc^2} \frac{F}{G} \qquad (4)$$

We have modified the leading constant for Jupiter by taking the equatorial field strength as 4.28 G; L is L shell, mc^2 is the ion or electron rest mass in MeV, and ω_D is negative for electrons. We have preserved the *Thomsen and Van Allen* (1980) notation in using a function, "F/G", to express the dependence on the particle's mirror latitude (λ_m)

$$\frac{F}{G} = [1.04675 + 0.45333\sin^2\lambda_m - 0.04675\exp(-6.34568\sin^2\lambda_m)]^{-1} \qquad (5)$$

The net azimuthal speed of the particle's guiding center, ω, is very important for understanding the bombardment of Europa. This speed can be compared with the speed of the particle along the magnetic field line to understand the satellite bombardment. For these purposes, it is useful to define a field line contact time, t_c, as

$$t_c = \frac{2 * \sqrt{R_E^2 - d^2}}{v} \quad d \le R_E$$
$$t_c = 0 \quad d > R_E \qquad (6)$$

Here d is the impact parameter of the guiding center field line to Europa's center of mass and v is the net azimuthal speed of the guiding center field line with respect to Europa in kilometers per second. For charged particles of both species with energies less than about 200 keV, the maximum contact time in a rigidly corotating magnetosphere is about 30 s. This can be compared with the particle's half-bounce time, the time it takes a charged particle to travel from the magnetic equator to its magnetic mirror point and back to the equator. For a 100-keV charged particle with an equatorial pitch angle of 45°, the half-bounce time is about 7 s for an electron and 271 s for a proton (see Table 2). Therefore, for 100-keV protons, the contact time is much shorter than the half-bounce time. This means not all bounce phases have yet come into contact with Europa. Therefore, 100-keV and lower-energy protons are at least in principle capable of bombarding all points on Europa's surface with approximately the same flux.

Pospieszalska and Johnson (1989) presented a detailed analysis of ion bombardment of Europa including gyroradius effects. For a charged particle with an equatorial pitch angle of 45° and a kinetic energy of 100 keV, the ratio of the particle's gyroradius to the moon's radius is 0.001 (electrons) and 0.041 (protons). For electrons, a guiding center approximation is usually sufficient, but for ions it is often important to include the gyroradius explicitly. *Pospieszalska and Johnson* (1989) found that for 1-keV sulfur ions, the bombardment pattern on Europa's surface heavily favored the trailing hemisphere, with some of these ions reaching the leading hemisphere. At higher energies, e.g., for 30-keV and 140-keV sulfur ions, there is some leading-trail-

TABLE 2. Charged particle parameters near Europa; all equatorial pitch angles are 45° and mirror latitudes are 23.13°.

	E (MeV)	$t_b/2$(s)	r_g (km)	t_c (d=0, s)
Electrons	0.1	7.22	1.5	30.1
Electrons	1.0	4.2	6.5	31.51
Protons	0.1	271	62	29.69
Protons	1.0	86	197	27.82
O^+	0.1	1085	250	29.69
O^+	1.0	343	789	27.82
S^+	0.1	1535	353	29.69
S^+	1.0	485	1116	27.82

Formulas are based on the work of *Thomsen and Van Allen* (1980). For the calculation of the contact time, we assume that the magnetosphere is rigidly corotating at Europa's orbit.

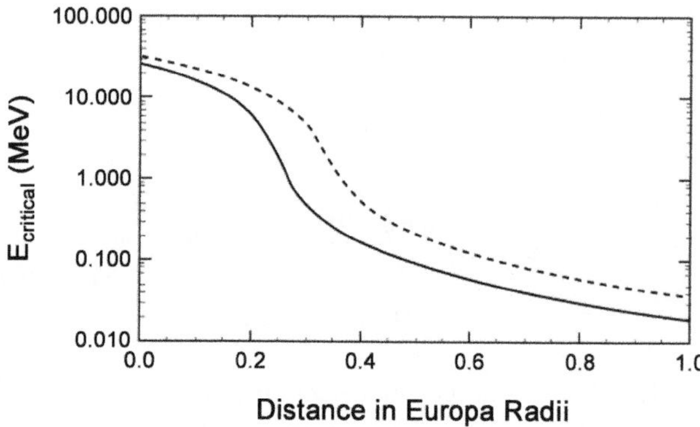

Fig. 6. Trapped electrons below the critical energy can bombard Europa's surface. The horizontal axis is the distance in the projected plane between the first point of contact of the electron's guiding center field line with Europa and the point of interest. The value 1.0 on the x-axis includes the pole. For this plot, electrons with an equatorial pitch angle of ~45° (solid line) and ~10° (dashed line) were used.

ing asymmetry in the bombardment pattern, but it was not a strong effect (see their Fig. 5). Updates to that work can be found in *Johnson et al.* (2004). Like 1-keV sulfur ions, energetic electron bombardment of Europa is expected to be very asymmetric. In the latter case, contact times are very large compared with the particle's half-bounce time. Electrons are therefore expected to bombard the moon close to the point of intersection between the moon and the guiding center field line (*Paranicas et al.,* 2007). For instance, for 5-MeV electrons of the same equatorial pitch angle, the contact time is about 38 s and the half-bounce time is about 4 s. Using the bounce and contact times, we show in Fig. 6 electron energies expected to be present in flux tubes attached to Europa. This critical energy is expressed as a function of distance in the direction of the plasma flow from the first point of contact of the field line with the moon. This distance is in Europa's equatorial plane. So, for example, on the central meridian of the trailing hemisphere, 270°W, at low Europa altitudes, there are very low fluxes of ~200-keV to 25-MeV electrons at north or south latitudes greater than 60°. This is the distance equivalent to 0.5 on the x-axis of Fig. 6 [i.e., $\cos \lambda = 1\text{-distance (in } R_E)$]. Any electron flux present in this energy range would be from diffusive processes because the trapped electrons have already been lost in collisions with the moon's surface. By contrast, low-energy electrons (e.g., $E < 20$ keV) can strike the polar regions.

Using the energy cutoffs in Fig. 6, it is straightforward to compute the energy flux into Europa's surface. At any point on Europa's surface, (λ,φ), the energy flux can be calculated from

$$P(\lambda,\varphi) = \int j(E,\alpha,\xi)\ E\cos\psi(\alpha,\xi)\ \sin\alpha\ d\alpha\ d\xi\ dE \quad (7)$$

where $\psi(\alpha,\xi)$ is the angle at the surface point (λ,φ) between the vector antiparallel to the surface normal and the magnetic field vector at that point; α and ξ are the particle pitch and phase angles. In this general case, the integral at each point on the surface is taken over only those directions corresponding to flux pointing into the surface (see discussion in *Walt,* 1994), here values of ψ between 0° and 90°. In the usual approximation of gyrotropic and isotropic flux, the same integral is written down with a $\cos \alpha$ replacing the $\cos \psi$. The angular integral is then customarily expressed with the value π.

In Fig. 7, we plot P (λ, 270°W) calculated various ways for 80% of rigid plasma corotation. The distribution of energy into the surface depends on the fraction of rigid corotation of the plasma, the pitch angle distribution of the particles and their energy spectrum. In each case displayed in Fig. 7, the falloff of energy flux in latitude is fairly flat near Europa's equator, suggesting that the sensitivity to various approaches is not very important in that region.

In performing these calculations, we have assumed that Europa is an electromagnetically inert body and we do not consider factors such as its jovicentric magnetic latitude or the impact on total magnetic field of the moon's ionospheric currents or other electromagnetic contributions. The electromagnetic fields from Galileo have been well described by a background field and an induced field of the satellite (*Kivelson et al.,* 1999). This induced field varies with Europa's position relative to the jovian plasma sheet, with the greatest induced field at the largest latitudes. The total field, background plus induced, at high moon excursion latitudes has magnetic field lines that do not go straight through the body but are draped around it. (For a description of the full picture, see the chapter by Khurana et al.) This perturbation to the field is not dominant, so that a large deviation from the inert case is not expected. For example, the plasma flow close to the moon measured by the plasma instrument

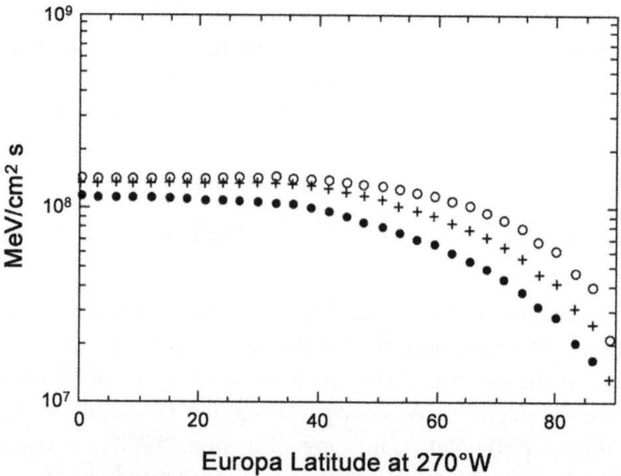

Fig. 7. Electron energy flux into surface as a function Europa latitude at 270°W. Plusses represent the integral using the energy spectrum in equation (1) and assuming an isotropic pitch angle distribution; the open circles use a simple power law, $j = 1.0{*}E^{-2}$, for comparison; and the filled circles use the energy spectrum above assuming the pitch angle distribution goes as $\sin \alpha$.

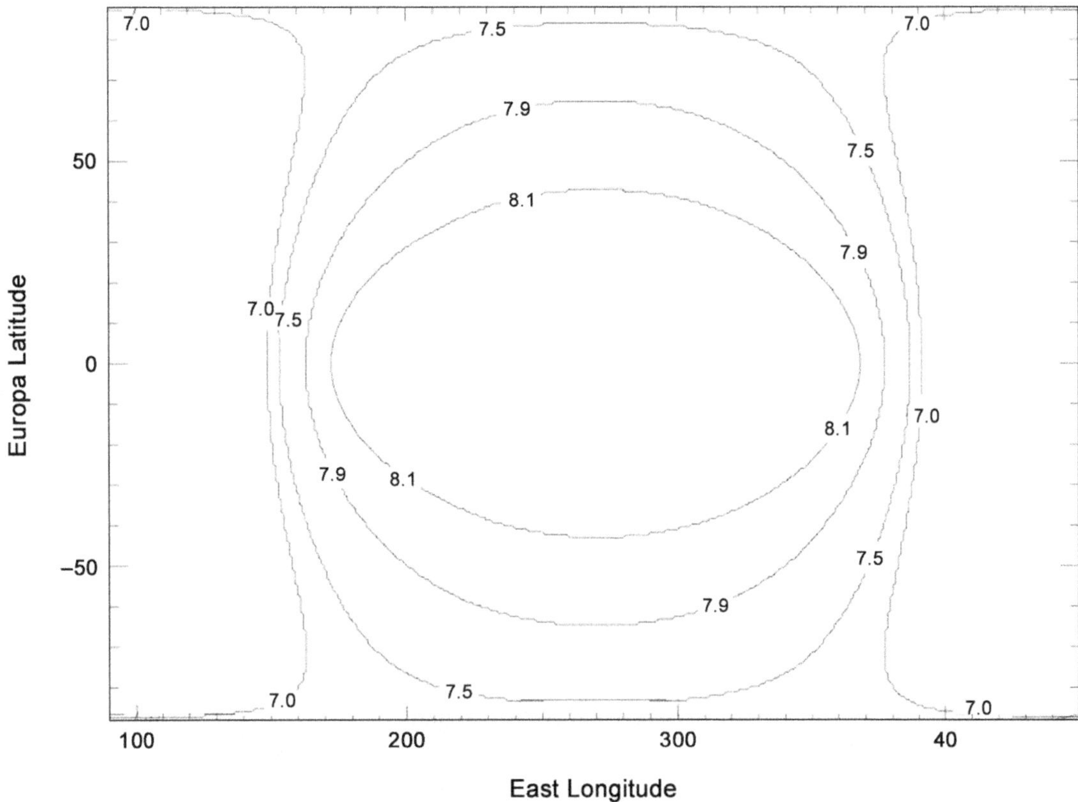

Fig. 8. Power per unit area into Europa surface for 10-keV to ~25-MeV electrons. Contours are labeled in units of log (MeV/cm^2 s). In this longitude system, the center of the trailing hemisphere of Europa is at 270°E.

on Galileo is not very distorted from the nominal corotation direction (*Paterson et al.,* 1999).

For completeness, we also present the value of P (equation (7)) as a function of Europa longitude and latitude. In Fig. 8, we show a contour plot of electron energy per unit area per second based on the energy spectrum given in equation (1). Here we assume the plasma flow is 80% of rigid and that Europa is an inert body whose orbit is in Jupiter's magnetic equatorial plane.

4. EXPECTED SURFACE EFFECTS

As previously reviewed by *Johnson et al.* (2004), the magnetospheric particle population contributes to the surface composition of Europa in three major ways: (1) the low-energy magnetospheric plasma implants plasma ions, most notably iogenic sulfur, and contributes to sputtering; (2) more-energetic ions, the dominant sputtering agents, eject neutrals that contribute to an ambient atmosphere, cause transport of material across Europa's surface, and contribute to the neutral torus; and (3) energetic electrons and light ions, the primary source of ionization energy, drive the surface chemistry. Ignoring possible ionospheric diversion of the flow and effects of the then-unknown induced fields from the ocean, *Pospieszalska and Johnson* (1989) computed a bombardment pattern of 1-keV sulfur ions onto Europa's surface (their Fig. 5). In their calculation, the trailing hemisphere apex (i.e., 270°W, 0°N) received the highest flux and model fluxes fell away toward and onto the leading hemisphere. The implanted sulfur from radiation-induced chemistry and any endogenic sulfur will be in a radiation-induced equilibrium with a sulfate, SO_2, polymeric sulfur. Similarly endogenic or delivered carbon will be in radiation equilibrium with a carbonate, CO_2, and polymerized carbon. In addition, as discussed, the irradiation of the ice matrix will lead to H_2, O_2, and H_2O_2, all detected in either the atmosphere or the surface (see the chapter by Carlson et al. for a discussion of surface species and chemistry). An important goal of future work is to separately determine the role of these effects in producing the darkened terrain on the trailing hemisphere, the production of an atmosphere, and the population of the ambient plasma. In discussing the surface reflectance spectrum, we note that the various spectral signatures (UV, visible, IR) sample different depths. Therefore, implantation primarily affects the very near surface and electrons affect the material to greater depths. However, gardening buries the implanted sulfur, so that a separation by depth may not be straightforward. The similarity of the bombardment patterns of ~1-keV sulfur ions and energetic electrons (*Pospieszalska and Johnson,* 1989; *Paranicas et al.,* 2001) will only lend itself to separation if the sampling depth of the spectral signatures are

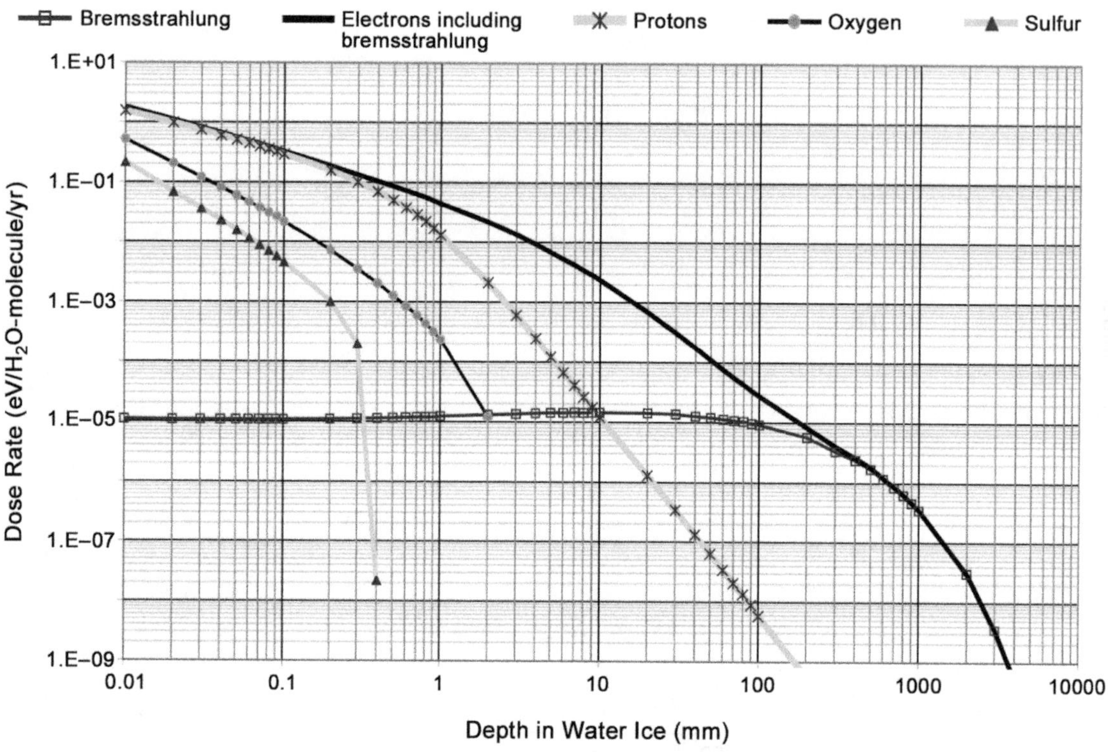

Fig. 9. Predicted dose rate vs. depth at the apex of Europa's trailing hemisphere by charged species using an input spectrum for various species and an ice surface. Heavy ions are stopped almost immediately in the water ice. At large depths, the electron dose rate becomes dominated by the contribution from secondaries (bremsstrahlung). This figure is based on the intensities of medium- to high-energy charged particles. See details in *Paranicas et al.* (2002).

carefully analyzed. Further understanding of the moon/plasma interaction (see the chapters by Kivelson et al. and Khurana et al.) will help us refine our understanding of the sulfur ion bombardment once the effects of the induced field are included. Such comparisons address a major theme of this chapter: the extent to which exogenic processes are understood in producing observable features on Europa.

As noted in the previous section, the penetration depth of charged particles into ice depends on a number of factors, including charged-particle type and energy. A published estimate of the maximum dose vs. depth for charged particles into water ice at Europa's surface is shown in Fig. 9. These curves were created using a Novice radiation transport code and various energy spectra based on spacecraft data as described in *Paranicas et al.* (2002). An important feature of this plot is that the trailing hemisphere dose near Europa's equator is dominated at almost all depths by the electrons. This fact, combined with the asymmetry of the electron dose onto Europa's surface described above, led us to compare the dose pattern with the distribution of hydrated sulfates, potentially sulfuric acid hydrates (see *Paranicas et al.*, 2001, and references therein). The favorable comparison suggested that the surface material in the dark regions, which are primarily on Europa's trailing hemisphere, are radiolytically processed to significant depths by the energetic electrons. This also suggests other leading/trailing asymmetries, such as that of 1-keV sulfur ions, might be a secondary effect.

Impacts of the various radiation dosage components in Fig. 9 must be considered separately for surface effects on the leading and trailing hemispheres. Energetic heavy ions deposit most of their energy very close to the surface. Noting that grain sizes are ~50 μm, sulfur ions are only implanted into the surficial grains and the heavy ions are the primary sputtering agents having the highest rate of energy loss per ion at submicrometer depths. Because of the dynamics of their motion, these heavy ions globally impact the surface in both hemispheres. Therefore, the curves in Fig. 9 for ions can be used as approximate dose-depth curves everywhere on the surface of Europa. The protons and especially energetic electrons lose energy in the ice more slowly and deposit this energy at much larger depths. Energetic electrons deliver the most total energy to the trailing hemisphere and electrons between about 100 keV and 25 MeV have much less impact on the leading hemisphere. Just above 25 MeV, electrons preferentially impact the leading hemisphere and become more uniform over the surface with increasing energy above that.

In Fig. 10, we show a dose-depth presentation with specific energy ranges represented separately. These are based on our implementation (*Sturner et al.*, 2003; *Cooper and Sturner*, 2006, 2007) of the GEANT transport code with

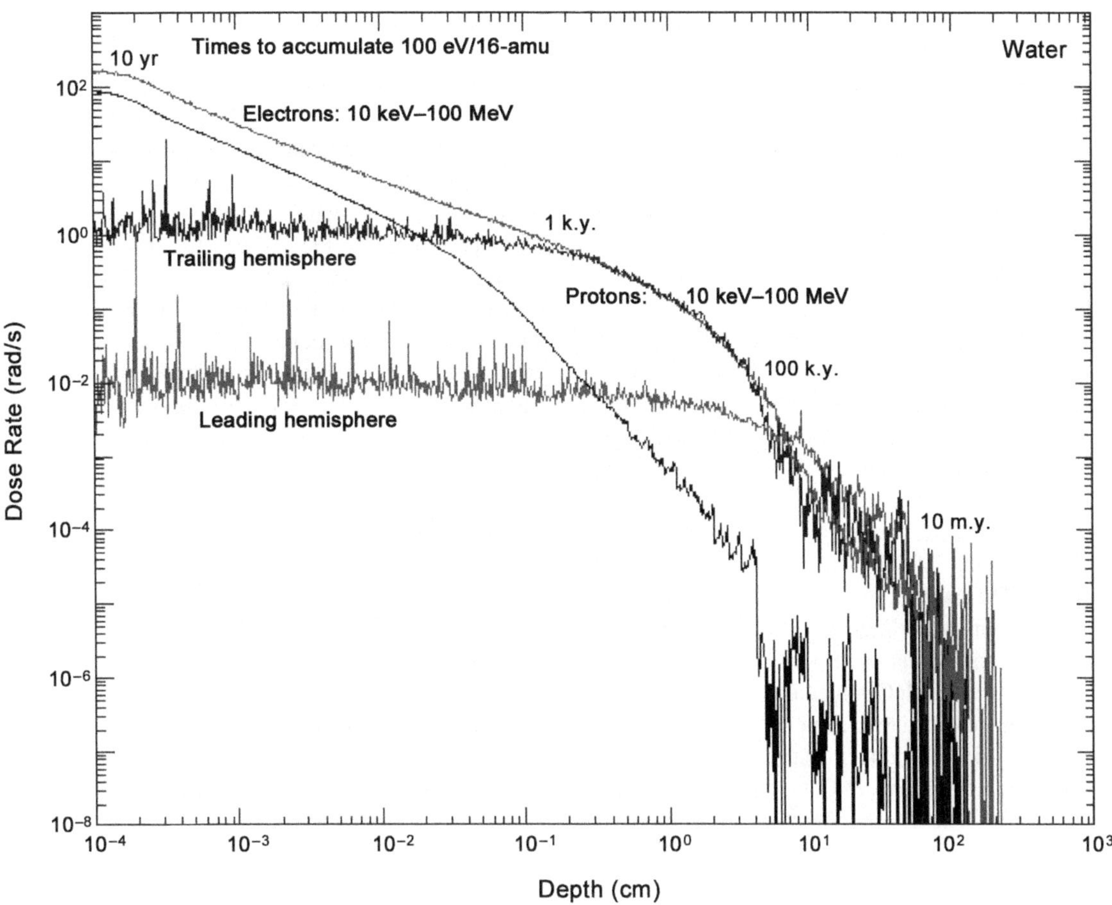

Fig. 10. After *Cooper and Sturner* (2006). Dose rate vs. depth where 1 rad/s is equal to 100 erg/gm/s or about 0.06 eV/H_2O-molecule/yr. The curve labeled "trailing hemisphere" includes the dose rate of 1–20-MeV electrons only, whereas the curve below it labeled leading hemisphere displays the dose rate of 20–40-MeV electrons that drift opposite to corotation. The uppermost of all the curves is the dose rate corresponding to electrons from 10 keV to 100 MeV and the dose rate from protons between 10 keV and 100 MeV follows this curve below it. Spikes and fluctations in the computed curves arise from statistics of limited number of Monte Carlo events used in the simulations and not from physical processes. Times in years are shown to give chemically significant (most bonds are broken at least once) dose of 100-eV/16-amu (60 Gigarads) at selected dose levels.

inclusion of all significant nuclear and electromagnetic (e.g., bremsstrahlung) interactions for primary and later generation particles and γ-ray photons. These researchers further employed the semi-analytical moon interaction model of *Fillius* (1988) to derive energy ranges of electrons primarily impacting the trailing (1–20 MeV) and leading (20–40 MeV) hemispheres of Europa with highest probability. Broadly speaking, the dosage profiles for these two respective energy ranges suggest a drop by 2 orders of magnitude from the trailing to the leading hemisphere for MeV electrons. However, at each surface point and at each depth, care must be taken in assessing which particles dominate the dose. For instance, around the apex of the leading hemisphere, the 10-keV to 100-MeV proton dose dominates over the 20–40-MeV electron dose down to about 3 mm from Fig. 10. Selected times to accumulate a net dose (from all sources) of ~100 eV per water molecule, equivalent to a volume dosage of 60 Gigarads per gram of H_2O, are also indicated. These times are only ~10 yr at the micrometer level but 10^6 yr at tens of centimeters. On multi-billion-year timescales the dosage effects of cosmic rays penetrating from outside the Jupiter magnetosphere for energy deposition to several meter depths in the surface ice, and decay of naturally abundant radioisotopes (e.g., K^{40}) throughout the ice crust to kilometers in depth (*Chyba and Hand,* 2001) become important.

4.1. Radiolysis

As described in the chapter by Carlson et al., the energy deposited by the charged particles and also by the UV photons cause dissociations and chemical reactions in Europa's surface. A principal product is hydrogen peroxide based on the reaction 2 $H_2O \rightarrow H_2O_2 + H_2$. Since H_2 is volatile and mobile in ice at the temperatures relevant to Europa's surface, it diffuses out more readily than other products. The escaping H_2 is light and readily lost to space, as described in the chapter by McGrath et al. Therefore, the ice surface becomes oxygen rich, a principal result of the intense irradiation of the surface. In addition to H_2O_2, O_2 is produced and carbon and sulfur are present in oxidized forms: CO_2 and possibly H_2CO_3, SO_2, and hydrated sulfuric acid respectively (see the chapter by Carlson et al.).

When O_2 is created in the ice, it can diffuse out if it is near the uppermost layer, otherwise it can be trapped at depth (*Johnson and Jesser,* 1997; *Shi et al.,* 2007). Burial by meteoritic impact ejecta, which occurs faster than erosion by ion sputtering (*Cooper et al.,* 2001), can increase steady-state accumulation rates of trapped O_2 and other radiolytic product species. Detection of condensed high-density phase O_2 on the surface (e.g., *Spencer and Calvin,* 2002) indicates that such trapping is occurring in the ice matrix or perhaps, within more stable mixed gas clathrate structures (*Hand et al.,* 2006) in the surface ice of the trailing hemisphere. The O_2 that escapes on production or is caused to diffuse out by the incident plasma is the source of the tenuous oxygen atmosphere over Europa (see chapters by Johnson et al. and McGrath et al.). Under long-term irradiation a rough steady state is achieved with fresh products formed and trapped products destroyed. In this situation, the H_2 and O_2 arising from H_2O are naturally produced in about a 2:1 ratio. The size of the steady-state yield of H_2 and O_2 depends not only on the radiation dose but also on the ambient surface temperature.

Chemical production rates in surface regions with relatively pure water ice can be estimated from the electron energy fluxes and from G values: the number of a new chemical species produced (or destroyed) per 100 eV of deposited energy. As previously reviewed by *Johnson et al.* (2004), these values vary somewhat with radiation type and energy and the presence of trace species in the ice, with a rough average value of G ~ 0.1 for radiolytic yield of H_2O_2 (see chapter by Carlson et al.). Using this G, the average integrated column production rate of H_2O_2 in icy regions of Europa's surface is ~5×10^{10} H_2O_2/cm^2/s (*Cooper et al.,* 2001). The regional rate would be lower on the leading hemisphere but there are significant contributions there from protons and heavier ions as discussed. Higher meteoritic impact rates on the leading hemisphere would give correspondingly higher radiolytic product burial rates and increase the net accumulation. However, in modeling the bound component of the atmosphere, ballistic transport and the fate of the sputtered species on redeposition have to be considered. For instance, *Cassidy et al.* (2007) have modeled atmospheric formation from surface irradiation and find that the darker sulfate regions of the trailing hemisphere may be net sinks, not sources, for O_2. Therefore, the long lifetime of O_2 and the possible dearth of reactive species can result in a comparable O_2 atmosphere on the less-irradiated leading hemisphere.

Oxygen chemistry provides a source of O_2 for the surface and atmospheric environment (*Johnson et al.,* 2003) with possible effects for the crustal chemistry and astrobiological potential of Europa. The surface-trapped O_2 might be conveyed by crustal processes to deeper levels, affecting the rheological properties of the ice crust (*Hand et al.,* 2006), and providing a significant oxidant source for the putative subsurface ocean of Europa (*Cooper et al.,* 2001; *Chyba and Phillips,* 2001; *Chyba and Hand,* 2001). Since oxidants in the irradiated surface environment could destroy organics brought to the surface, future missions to Europa should consider conditions of surface irradiation and oxidant concentration in the choice of surface sites for remote and landed measurements. Sites with relatively low irradiation, e.g., the leading hemisphere and topographically shielded spots in other locations (*Cooper and Sturner,* 2006), and low oxidant (O_2, H_2O_2) concentrations might be preferable in any searches for organic materials. On the other hand, concentrations of CO_2 in association with H_2O_2 might be indicative of ongoing radiolytic oxidation of near-surface organics either emerging from a subsurface repository or earlier deposited by an impact event.

4.2. Sputtering

In comparison to the volume ice effects of radiolysis dominated by electrons and protons, sputtering mainly applies to erosion of upper molecular layers by impact of highly ionizing particles, e.g., the heavy sulfur and oxygen ions originating from the Io torus that have undergone acceleration in the magnetosphere of Jupiter. Sputtering liberates molecules that transiently populate the atmosphere and redistribute surface material or escape Europa's gravity. Lighter molecules, such as H_2, may escape more easily, leaving behind an atmosphere that is richer in heavy molecules such as oxygen (see also the chapter by Johnson et al.). Sputtering primarily carries off water molecules from the icy surface, but also carries off any newly formed and intrinsic species trapped in the ice. For sufficiently high sputtering yields these trapped species could potentially include complex organics of high interest for astrobiology.

As discussed for H_2 above, a fraction of the surface ejecta escapes Europa's gravity; the neutral escape speed is about 2 km/s. The escaping neutrals produce an extended neutral atmosphere that is gravitationally bound to Jupiter (see chapter by Johnson et al.). Components of this nearly toroidal atmosphere (*Mauk et al.,* 2004; *Smyth and Marconi,* 2006) are ionized in a variety of ways. New ions are accelerated by the electromagnetic fields, i.e., they are picked up by the corotating magnetosphere. In this way, Europa's surface is a source of neutrals and ions to the local environment. It has been suggested that Europa is a net source of sputtered Na to the local magnetosphere (*Johnson,* 2000; *Leblanc et al.,* 2002; *Burger and Johnson,* 2004). This finding may extend to other endogenic species. Furthermore, at a much higher end of the ion energy range, *Cohen et al.* (2001) found that MeV carbon ions have a different radial intensity gradient in the magnetosphere than iogenic sulfur. This may originate in part from surface sputtering of Europa, Ganymede, and Callisto, but is complicated by known sources of carbon, such as the solar wind. The study of minor ion species at all energies can provide clues to origin, surface constituents, and energy and mass flow through magnetospheres.

The surface sputtering rate is determined by the energy spectra of ions and the yield, the number of neutrals ejected per incident ion. Furthermore, decomposition and loss of H_2 and O_2 from ice also puts surface material into the gas phase, so the net surface erosion rate has a temperature dependent component that dominates above ~100 K (chapter by Johnson et al.). *Johnson et al.* (2004) give a yield function that is a consolidation of many laboratory studies of the sputtering of water ice by the energetic plasma ions (updated at *www.people.virginia.edu/~rej*). Using their fit, the peak yield is $Y \sim 3.4{*}Z^{2.8}$, where Z is the nuclear charge of the ion. The peak yield occurs at an ion speed that can be expressed as $v^{max} = 2.72 \times 10^6{*}Z^{0.334}$ m/s. This corresponds to a peak yield of ~3.4 for 39-keV protons, Y ~ 1149 for 2.5-MeV O^+, and Y ~ 8000 for 8.1-MeV S^+. Since the heavy ion yields peak in the low MeV, it has been assumed that, due to the falling energy spectrum of the charged particles, the total yield was dominated by the action of keV ions. The most recent estimate of the globally averaged sputtering yield from *in situ* data is ~2.4×10^{27} H_2O molecules/s (*Paranicas et al.,* 2002), ignoring the temperature dependent component and the contributions from ions below 1 keV/amu. Recent reevaluation of the sputtering data for the thermal plasma ions, for which both electronic and collision energy deposition apply, had important effects on the rates in the saturnian magnetosphere (e.g., *Johnson et al.,* 2008) but still indicate that the heavy ions are the dominant sputtering agents at Europa. This gives a globally averaged time of ~6.1×10^4 yr to erode 1 mm of ice (e.g., *Cooper et al.,* 2001). This estimate includes the three major ions of Fig. 4, but the sputtering yield is dominated by energetic sulfur ions. As noted previously from the same referenced work, the meteoritic impact burial time, ~10^3 yr per mm of ice, is shorter, so the Europa surface is buried by ejecta faster than it is eroded by sputtering.

As surface properties are better understood, sputtering rates will need to be refined. For example, microscopic structure and temperature can affect the sputtering yields. The porosity of Europa's regolith will reduce, by approximately a factor of 4, the effective sputtering yield of species such as H_2O that stick to neighboring grains with unit efficiency (*Johnson,* 1990, 1995, 1998; *Cassidy and Johnson,* 2005). Other species with low sticking efficiency, such as H_2 and O_2, more easily escape (see also the chapter by Johnson et al.), but some (O_2, CO_2, SO_2) can be trapped in inclusions (*Johnson and Jesser,* 1997; *Shi et al.,* 2007) or in mixed gas clathrates, wherever these are stable at the surface (*Hand et al.,* 2006).

Another surface property potentially affecting sputtering is surface charging, which can affect the access of low-energy species to the surface or the ejection of ions. Sunlit surfaces charge positively due to photoelectric emission of electrons, whereas other surfaces preferentially bombarded by magnetospheric electrons or protons can respectively accumulate negative or positive charge. Ejection of ions at the same charge as the surface would be enhanced by electrostatic repulsion and diminished for opposite signs. Neutral gas species undergoing electron attachment reactions, e.g., producing O_2^-, would have higher sticking efficiencies in sunlit regions. Maximal intensity electron bombardment of sulfate-rich regions in the trailing hemisphere may increase surface retention of such ions and thereby contribute to the observed non-ice chemistry. Accumulation and potential solid-state mobility of free charges in irradiated surface ices of Europa may also impact the electrical conductivity and electromagnetic properties of these ices with effects on moon-magnetospheric interaction modeling and on deep sounding of the interior with electromagnetic waves for ice radar and oceanic induced magnetic field investigations (*Gudipati et al.,* 2007).

4.3. Radiation Damage

Both penetrating and non-penetrating radiation produces defects in ice. In addition, water vapor is sputtered and redeposited, which can lead to an amorphous layer, depending on the surface temperature. Both processes occur at very low rates, so in principal can be annealed. Reflectance spectra suggest a thin amorphous layer, with crystalline ice forming at ~1 mm (*Hansen and McCord,* 2004). Based on Figs. 9 and 10, such a layer is likely formed by radiation damage produced by the energetic protons and electrons. In addition, icier regions of Europa's surface are relatively bright and light scattering. Light scattering produced by radiation damage in an ice surface has also been correlated with the prior loss of hydrogen that could otherwise form radiation-darkened hydrocarbons (*Strazzulla and Johnson,* 1991; *Strazzulla,* 1998; *Moroz et al.,* 2004). Scattering in the visible is usually associated with multiple defects that aggregate, forming inclusions in the ice. The presence of radiation damage inclusions is manifest not only by the scattering in the visible but also by the presence of molecular oxygen inclusions in Europa's surface produced by the radiation (*Johnson and Quickenden,* 1997; *Johnson and Jesser,* 1997). In these inclusions, the penetrating radiation also produces ozone.

Any organic molecules that reach Europa's surface from endogenic or exogenic sources are also subject to chemical modification and eventual destruction by continuing surface irradiation. In samples containing long chain molecules, chain breaks and cross-linking can occur, changing the character of the organic. Discrete peaks in a molecular mass spectrum can be indicative of biological sources but these peaks can be dispersed by radiation processing. A key issue for future observations is how and where to search for recognizable organic biosignatures that have not been highly degraded by irradiation (see also the chapter by Hand et al.). One approach is to identify regions of relatively low radiation exposure, e.g., on the leading hemisphere and/or in topographically shielded locations.

5. SUMMARY AND CONCLUSIONS

We have shown above that energetic particle fluxes typically increase inward through the middle and inner jovian magnetosphere, with very high intensities at Europa' orbit. It is also found that a part of the ion energy spectrum is heavily depleted because of charge exchange collisions with Europa's neutral gas torus. Many of these energetic ions are lost from the system before they reach Europa or points radially inward from that orbit. We also show that there are specific regions on Europa, particularly on the leading hemisphere, where the bombarding flux of some energetic electrons is relatively small. This is because the ratio of the speed of these particles parallel to the magnetic field line to the speed at which they are carried azimuthally around the magnetosphere is very large.

The radiation environment near Europa varies in time but only to a limited extent. At MeV energies, data reveal less scatter in the stronger magnetic field of the planet, as expected. This suggests that some of the variability due to injections and other types of magnetospheric activity are less likely, than is the case at Saturn, to dominate the fluxes near Europa for many species and energies. In fact, one study of the ≥11-MeV electron channel on Galileo over the entire orbital mission found that the standard deviation from the average value was only a factor of 2–3. In medium energy ions, the total integrated variation in sparse sampling over about 4 yr found a factor of ~2 variation. Future studies of the radiation environment and Europa's place in it would benefit from a more comprehensive measurement of the electron environment. Galileo made many close passes by the moon, but EPD was somewhat limited in its MeV electron coverage and we are learning these are important particles for reactions deep in the ice layer.

Regarding bombardment of Europa, we have argued that almost all charged particles preferentially impact Europa's trailing hemisphere, except >25-MeV electrons, which preferentially impact the leading hemisphere. For each species and energy, the interesting question is how the flux into the surface falls off between the trailing hemisphere apex and the leading one. Previous modeling has shown that 30- and 140-keV sulfur ions can bombard the entire surface by virtue of their gyro, bounce, and drift motions, and there are only small hemispherical differences. On the other hand, 1-keV sulfur ions have strong leading/trailing asymmetries. Energetic electrons between about 100 keV and 25 MeV preferentially bombard the trailing hemisphere. But 20-keV electrons can easily reach the poles at high flux levels. No particle tracing has been done that accounts for Europa's location in the current sheet or accounts for electromagnetic fields that self-consistently include contributions from the ionosphere and/or induced magnetic fields to further substantiate these claims.

In understanding the surface interactions, we have discussed dose vs. depth, radiolysis, sputtering, and radiation damage. We have attempted to link particle species and energies to various processes. For dose rates into the surface, we have emphasized the complexity in understanding the particles that dominate the energy into the layer by surface longitude and latitude and depth. For example, we do not expect many 100-keV electrons to reach Europa's poles, based on a simple picture of electron motion, but electrons of lower energies do reach the poles and contribute to the dose. Once this map of energy into the surface by species at each depth is known, surface properties themselves must be considered. For example, it would be interesting to compare bombardment maps with surface maps of reflectance spectra in a number of different spectral regions having very different sampling depths. We have also described how sputtering depends on ice porosity and temperature and that the sputtered products can escape, stick, etc. We also described how new molecules are formed in ice, how some easily escape and others are trapped. We raised the issue of the net flux of minor species out of a surface as an indicator of surface constituents. Finally, we considered the issue of survivability of species in ice due to radiation damage.

Acknowledgments. We appreciate assistance on this chapter from R. W. Carlson and B. Mauk, and conversations with T. Cassidy, R. B. Decker, D. Haggerty, I. Jun, S. Ohtani, W. Patterson, W. Paterson, and L. Prockter. The portion of this work provided by H.B.G. was carried out at the Jet Propulsion Laboratory, California Institute of Technology, under a contract with the National Aeronautics and Space Administration. J.F.C. acknowledges previous support from the NASA Jovian System Data Analysis Program for Galileo Orbiter data analysis and ongoing support from the NASA Planetary Atmospheres program for Europa surface and atmospheric environment modeling. R.E.J. acknowledges support by NASA Planetary Atmospheres and Planetary Geology Programs. C.P. would like to acknowledge NASA Planetary Atmospheres and Outer Planets Research grant support to JHU/APL.

REFERENCES

Burger M. H. and Johnson R. E. (2004) Europa's cloud: Morphology and comparison to Io. *Icarus, 171,* 557–560.

Carlson R. W., Anderson M. S., Johnson R. E., Schulman M. B., and Yavrouian A. H. (2002) Sulfuric acid production on Europa: The radiolysis of sulfur in water ice. *Icarus, 157,* 456–463.

Carlson R. W., Anderson M. S., Mehlman R., and Johnson R. E. (2005) Distribution of hydrate on Europa: Further evidence for sulfuric acid hydrate. *Icarus, 177,* 461–471.

Carlson R. W., Johnson R. E., and Anderson M. S. (1999) Sulfuric acid on Europa and the radiolytic sulfur cycle. *Science, 286,* 97–99.

Cassidy T. A. and Johnson R. E. (2005) Monte Carlo model of sputtering and other ejection processes within a regolith. *Icarus, 176,* 499–507.

Cassidy T. A., Johnson R. E., McGrath M. A., Wong M. C., and Cooper J. F. (2007) The spatial morphology of Europa's near-surface O_2 atmosphere. *Icarus, 191(2),* 755–764.

Chyba C. F. and Hand K. P. (2001) Life without photosynthesis. *Science, 292,* 2026–2027.

Chyba C. F. and Phillips C. B. (2001) Possible ecosystems and the search for life on Europa. *Proc. Natl. Acad. Sci., 98,* 801–804.

Cohen C. M. S., Stone E. C., and Selesnick R. S. (2001) Energetic ion observations in the middle jovian magnetosphere. *J. Geophys. Res., 106,* 29871–29882.

Connerney J. E., Acuna M. H., Ness N. F., and Satoh T. (1998) New models of Jupiter's magnetic field constrained by the Io flux tube footprint. *J. Geophys. Res., 103,* 11929–11940.

Cooper J. F. and Sturner S. J. (2007) Hemispheric and topographic asymmetry of magnetospheric particle irradiation for icy moon surfaces (abstract). In *Workshop on Ices, Oceans, and Fire: Satellites of the Outer Solar System,* p. 32. LPI Contribution No. 1357, Lunar and Planetary Institute, Houston.

Cooper J. F., Johnson R. E., Mauk B. H., Garrett H. B., and Gehrels N. (2001) Energetic ion and electron irradiation of the icy Galilean satellites. *Icarus, 149,* 133–159.

Cooper J. F., and Sturner S. J. (2006) Europa surface radiation environment for lander assessment (abstract). Paper presented at the Astrobiology Science Conference (AbSciCon) March 26–30, 2006, Washington, DC.

Divine N. and Garrett H. B. (1983) Charged particle distributions in Jupiter's magnetosphere. *J. Geophys. Res., 88,* 6889–6903.

Fillius W. (1988) Toward a comprehensive theory for the sweeping of trapped radiation by inert orbiting matter. *J. Geophys. Res., 93(A12),* 14284–14294.

Garrett H. B., Jun I., Ratliff J. M., Evans R. W., Clough G. A., and McEntire R. W. (2002) Galileo interim radiation model. *Jet Propulsion Laboratory Report D-24811,* Jet Propulsion Laboratory, Pasadena.

Garrett H. B., Levin S. M., Bolton S. J., Evans R. W., and Bhattacharya B. (2005) A revised model of Jupiter's inner electron belts: Updating the Divine radiation model. *Geophys. Res. Lett., 32,* DOI: 10.1029/2004GL021986.

Geissler P. E., Greenberg R., Hoppa G., McEwen A., Tufts R., et al. (1998) Evolution of lineaments on Europa: Clues from Galileo multispectral imaging observations. *Icarus, 135,* 107–126.

Grundy W., Buratti B. J., Cheng A. F., Emery J. P., Lunsford A., et al. (2007) New Horizons mapping and Ganymede. *Science, 318,* 234–237.

Gudipati M. S., Allamandola L. J., Cooper J. F., Sturner S. J., and Johnson R. E. (2007) Consequence of electron mobility in icy grains on solar system objects (abstract). *Eos Trans. AGU, 88(52),* Fall Meeting Supplement, Abstract P53B–1248.

Hand K. P., Chyba C. F., Carlson R. W., and Cooper J. F. (2006) Clathrate hydrates of oxidants in the ice shell of Europa. *Astrobiology, 6(3),* 463–482.

Hansen G. B. and McCord T. B. (2004) Amorphous and crystalline ice on the Galilean satellites: A balance between thermal and radiolytic processes. *J. Geophys. Res., 109,* E01012, DOI: 10.1029/2003JE002149.

Johnson R. E. (1990) *Energetic Charged Particle Interaction with Atmospheres and Surfaces.* Springer-Verlag, New York.

Johnson R. E. (1995) Sputtering of ices in the outer solar system. *Rev. Mod. Phys., 68,* 305–312.

Johnson R. E. (1998) Sputtering and desorption from icy surfaces. In *Solar System Ices* (B. Schmitt et al., eds.), pp. 303–331. Astrophys. Space Sci. Library Vol. 227, Kluwer, Dordrecht.

Johnson R. E. (2000) Sodium at Europa. *Icarus, 143,* 429–433.

Johnson R. E. and Jesser W. A. (1997) O_2/O_3 microatmospheres in the surface of Ganymede. *Astrophys. J. Lett., 480,* L79–L82.

Johnson R. E. and Quickenden T. I. (1997) Photolysis and radiolysis of water ice on outer solar system bodies. *J. Geophys. Res., 102,* 10985–10996.

Johnson R. E., Quickenden T. I., Cooper P. D., McKinley A. J., and Freedman C. G. (2003) The production of oxidants in Europa's surface. *Astrobiology, 3,* 823–850.

Johnson R. E., Carlson R. W., Cooper J. F., Paranicas C., Moore M. H., and Wong M. C. (2004) Radiation effects on the surfaces of the Galilean satellites. In *Jupiter: The Planet, Satellites and Magnetosphere* (F. Bagenal et al., eds.), pp. 485–512. Cambridge Univ., Cambridge.

Johnson R. E., Fama M., Liu M., Baragiola R. A., Sittler E. C. Jr., and Smith H. T. (2008) Sputtering of ice grains and icy satellites in Saturn's inner magnetosphere. *Planet. Space Sci., 56,* DOI: 10.1016/j.pss.2008.04.003.

Jun I., Garrett H. B., Swimm R., Evans R. W., and Clough G. (2005) Statistics of the variations of the high-energy electron population between 7 and 28 jovian radii as measured by the Galileo spacecraft. *Icarus, 178,* 386–394.

Kabin K., Combi M. R., Gombosi T. I., Nagy A. F., DeZeeuw D. L., and Powell K. G. (1999) On Europa's magnetospheric interaction: A MHD simulation of the E4 flyby. *J. Geophys. Res., 104,* 19983–19992.

Khurana K. K. and Kivelson M. G. (1993) Inference of the angular velocity of plasma in the jovian magnetosphere from the sweepback of magnetic field. *J. Geophys. Res., 98,* 67–79.

Kivelson M. G., Khurana K. K., Stevenson D. J., Bennett L., Joy S., Russell C. T., Walker R. J., Zimmer C., and Polanskey C. (1999) Europa and Callisto: Induced or intrinsic fields in a periodically varying plasma environment. *J. Geophys. Res., 104,* 4609–4625.

Lagg A., Krupp N., and Woch J. (2003) In-situ observations of a neutral gas torus at Europa. *Geophys. Res. Lett., 30,* DOI: 10.1029/2003GL017214.

Leblanc F., Johnson R. E., and Brown M. E. (2002) Europa's sodium atmosphere: An ocean source? *Icarus, 159,* 132–144.

Lindsay B. G. and Stebbings R. F. (2005) Charge transfer cross sections for energetic neutral atom data analysis. *J. Geophys. Res., 110,* A12213, DOI: 10.1029/2005JA011298.

Mauk B. H., Williams D. J., McEntire R. W., Khurana K. K., and Roederer J. G. (1999) Storm-like dynamics of Jupiter's inner and middle magnetosphere. *J. Geophys. Res., 104,* 22759–22778.

Mauk B. H., Mitchell D. G., Krimigis S. M., Roelof E. C., and Paranicas C. P. (2003) Energetic neutral atoms from a trans-Europa gas torus at Jupiter. *Nature, 421,* 920–922.

Mauk B. H., Mitchell D. G., McEntire R. W., Paranicas C. P., Roelof E. C., Williams D. J., and Lagg A. (2004) Energetic ion characteristics and neutral gas interactions in Jupiter's magnetosphere. *J. Geophys. Res., 109,* DOI: 10.1029/2003JA010270.

McCord T. B., Hansen G. B., Fanale F. P., Carlson R. W., Matson D. L., et al. (1998) Salts on Europa's surface detected by Galileo's Near Infrared Mapping Spectrometer. *Science, 280,* 1242–1245.

McCord T. B., Hansen G. B., Matson D. L., Johnson T. V., Crowley J. K., et al. (1999) Hydrated salt minerals on Europa's surface from the Galileo near-infrared mapping spectrometer (NIMS) investigation. *J. Geophys. Res., 104,* 11827–11852.

Moroz L. V., Baratta G., Strazzulla G., Starukhina L., Dotto E., Barucci M. A., Arnold G., and Distefano E. (2004) Optical alternation of complex organics induced by ion irradiation: 1. Laboratory experiments suggest unusual space weathering trend. *Icarus, 170,* 214–228.

Paranicas C., McEntire R. W., Cheng A. F., Lagg A., and Williams D. J. (2000) Energetic charged particles near Europa. *J. Geophys. Res., 105,* 16005–16015.

Paranicas C., Carlson R. W., and Johnson R. E. (2001) Electron bombardment of Europa. *Geophys. Res. Lett., 28,* 673–676.

Paranicas C., Ratliff J. M., Mauk B. H., Cohen C., and Johnson R. E. (2002) The ion environment of Europa and its role in surface energetics. *Geophys. Res. Lett., 29,* DOI: 10.1029/2001GL014127.

Paranicas C., Mauk B. H., Khurana K., Jun I., Garrett H., Krupp N., and Roussos E. (2007) Europa's near-surface radiation environment. *Geophys. Res. Lett., 34,* L15103, DOI: 10.1029/2007GL030834.

Paterson W. R., Frank L. A., and Ackerson K. L. (1999) Galileo plasma moments at Europa: Ion energy spectra and moments. *J. Geophys. Res., 104,* 22779–22791.

Pospieszalska M. K. and Johnson R. E. (1989) Magnetospheric ion bombardment profiles of satellites: Europa and Dione. *Icarus, 78,* 1–13.

Shi J., Teolis B. D., and Baragiola R. A. (2007) Irradiation enhanced adsorption and trapping of O_2 on microporous water ice. *AAS/Division for Planetary Sciences Meeting Abstracts, 39,* 38.04.

Smyth W. H. and Marconi M. L. (2006) Europa's atmosphere, gas tori, and magnetospheric implications. *Icarus, 181,* 510–526.

Spencer J. R. and Calvin W. M. (2002) Condensed O_2 on Europa and Callisto. *Astron. J., 124,* 3400–3403.

Strazzulla G. (1998) Chemistry of ice induced by bombardment with energetic charged particles. In *Solar System Ices* (B. Schmitt et al., eds.), pp. 281–301. Astrophys. Space Sci. Library Vol. 227, Kluwer, Dordrecht.

Strazzulla G. and Johnson R. E. (1991) Irradiation effects on comets and cometary debris. In *Comets in the Post Halley Era, Volume 1* (R. L. Newburn Jr. et al., eds.), pp. 243–275. Kluwer, Dordrecht.

Sturner S. J., Shrader C. R., Weidenspointner G., Teegarden B. J., Attié D., et al. (2003) Monte Carlo simulations and generation of the SPI response. *Astron. Astrophys., 411,* L81–L84.

Thomsen M. F. and Van Allen J. A. (1980) Motion of trapped electrons and protons in Saturn's inner magnetosphere. *J. Geophys. Res., 85,* 5831–5834.

Walt M. (1994) *Introduction to Geomagnetically Trapped Radiation.* Cambridge Univ., Cambridge.

Zombeck M. V. (1982) *Handbook of Space Astronomy and Astrophysics.* Cambridge Univ., Cambridge.

Europa's Interaction with the Jovian Magnetosphere

Margaret G. Kivelson and Krishan K. Khurana
University of California, Los Angeles

Martin Volwerk
Österreichische Akademie der Wissenschaften

Europa is embedded in Jupiter's magnetosphere where a rapidly flowing plasma interacts electromagnetically with the moon's surface and its atmosphere. In this chapter, the phenomenology of the interacting system is presented and interpreted using both qualitative and quantitative arguments. Challenges in understanding the plasma environment arise partly because of the diverse scale-lengths that must be considered as well as the nonlinear nature of the interactions. The discussion that follows describes selected aspects of the interacting system. On the scale of gyroradii, we describe the effects of newly ionized particles on fields and flows and their relation to wave generation. On the scale of Europa radii, we discuss the structure of the local interaction. On the scale of the tens of Jupiter radii that separate Europa from Jupiter's ionosphere, we describe the aurora generated near the magnetic footprint of Europa in Jupiter's upper atmosphere. We end by stressing the relevance of plasma measurements to achievement of goals of a future Europa Orbiter mission.

1. INTRODUCTION

The Galilean moons, although small, play a distinctive role in the history of solar system science. Galileo recognized that their motions in periodic orbits around Jupiter were compelling analogs of planetary bodies in a heliocentric system (see chapter by Alexander et al.). The complex orbital interactions of the inner moons were found to account not only for orbital stability (e.g., *Goldreich,* 1965), but also for enhanced tidal heating (*Peale et al.,* 1979), which powers volcanic activity on Io and melting of the ice beneath the surface of Europa. That fluid oceans could be present beneath the icy crusts of the three outer moons was discussed (*Lewis,* 1971) decades before spacecraft observations provided support for (if not full confirmation of; see chapter by Khurana et al.) this speculation for some of the moons, in particular, Europa.

Concurrent with studies of the interior, the particle and fields environments of the moons began to attract attention following the discovery of Io's control of jovian decametric emissions (*Bigg,* 1964). *Goldreich and Lyndon-Bell* (1969) recognized that an electromagnetic link between the moon and Jupiter's ionosphere could explain the observations, a suggestion that implied the presence of plasma along the Io magnetic flux tube. Somewhat later, the existence of an extended nebula around Io's orbit, the Io torus, was established (*Kupo et al.,* 1976; *Mekler and Eviatar,* 1977). Io soon became the focus of *in situ* particle and field measurements by Voyager 1, and the ionian source of heavy ion plasma that shapes the structure of much of the magnetosphere was recognized (*Bagenal and Sullivan,* 1981; *Shemansky and Smith,* 1981; see also review by *Thomas et al.,* 2004). Europa, the smallest of the Galilean moons, was found also to be a plasma source (*Intriligator and Miller,* 1982; *Eviatar and Paranicas,* 2005; *Russell et al.,* 1999), albeit a secondary one. However, Europa's plasma environment received comparatively little attention until it was established by Galileo observations that its geologically young surface lies above what is probably a global ocean (*Khurana et al.,* 1998). This discovery promoted the priority of Europa and its local plasma environment as targets for further planetary exploration. Although only 3 of the 12 flybys of the Galileo prime mission (1995 through 1997) passed close to Europa, the next phase of the mission, designated the Galileo Europa mission, devoted half of its 14 flyby opportunities to Europa. Table 1 summarizes various relevant features of Galileo's flyby trajectories (or "passes") plotted in Fig. 1. The final stage of Galileo's odyssey included a specially designed pass in which magnetometer measurements found a predicted reversal of the orientation of the internal dipole moment, thus confirming the presence of an inductive field at Europa (*Kivelson et al.,* 2000).

This chapter addresses the subject of Europa's interaction with the particles and fields of the jovian magnetosphere. The topic presents a considerable challenge because the moon and its magnetized plasma environment interact nonlinearly. Relevant to the interaction are matters as diverse as the chemical composition of the surface from which particles are sputtered, the properties of the energetic particles responsible for the sputtering, the temporal and spatial characteristics of the magnetospheric plasma near the orbit of Europa, properties of the magnetic field that confines the plasma, and the electromagnetic characteristics of the moon and its ionosphere. Europa's response to

TABLE 1. Characteristics of Galileo's close passes by Europa (boldface emphasizes flybys with full fields and particles data with closest approach at altitudes below 2050 km).

Pass	Date, Time	Radial Distance from Jupiter (R_J)	Local Time (Hours)	Closest Approach (km)	Europa Latitutde	Europa East Longitude	Location of c/a Relative to Europa
E4	**12/19/96 06:52:58**	**9.4–9.5**	**16.6–17.0**	**688.1**	**–1.7**	**322.4**	**Oblique Wake**
E6	02/20/97 17:06:10	—	—	582.3	–17.0	34.7	Recording Lost Except PWS
E11	**11/06/97 20:31:44**	**9.0–9.4**	**10.8–11.9**	**2039.3**	**25.7**	**218.7**	**Oblique Wake**
E12	**12/16/97 12:03:20**	**9.4–9.6**	**14.5–14.8**	**196.0**	**–8.7**	**134.3**	**Upstream**
E14	**03/29/98 13:21:05**	**9.4–9.6**	**14.3–14.7**	**1649.1**	**12.2**	**131.2**	**Upstream**
E15	05/31/98 21:12:56	9.4–9.6	9.9–10.3	2519.5	15.0	225.4	Wake
E16	07/21/98 05:04:43	—	—	1829.5	–25.6	133.6	Recording Lost
E17	09/26/98 03:54:20	9.2–9.6	9.6–10.3	3587.4	–42.4	220.3	Wake
E18	11/22/98 11:44:56	—	—	2276.2	41.7	139.3	Recording Lost
E19	**02/01/99 02:19:50**	**9.2–9.4**	**9.7–10.0**	**1444.4**	**30.5**	**28.2**	**Upstream**
E26	**01/03/00 17:59:43**	**9.2–9.7**	**2.8–3.1**	**348.4**	**–47.1**	**83.4**	**Upstream/Polar**

the currents linking it to the magnetospheric plasma in turn modify the local properties of the plasma and magnetic field (*Kivelson,* 2004; *Kivelson et al.,* 2004).

In the following sections we first summarize the properties of Jupiter's magnetosphere in the vicinity of Europa, and provide a large scale (magnetohydrodynamic) perspective on the interaction between Europa, its ionosphere, and its plasma environment. We then discuss the presence of pickup ions in the local plasma, the special role of energetic particles in the interaction and the modification of the interaction by the inductive magnetic field. The properties of the inductive field itself are discussed in the chapter by Khurana et al. We consider special features of passes upstream and downstream of the moon in the flowing plasma, with particular emphasis on properties of the wake region. Next we describe the link between Europa and the auroral footprint in Jupiter's upper atmosphere. We close with a discussion of expectations for fields and particle measurements as a component of future exploration of the remarkable body that is the subject of this book.

2. OVERVIEW OF FIELD AND PLASMA CONDITIONS NEAR EUROPA'S ORBIT

Europa's orbit, at 9.38 R_J (R_J is the radius of Jupiter, taken as 71,400 km) from the center of Jupiter and effectively in Jupiter's equator, lies at the outer edge of the Io plasma torus within the inner magnetosphere, a region where the magnetospheric magnetic field is quasidipolar. As Jupiter rotates, the 10° tilt of the dipole moment relative to the axis of rotation causes the magnetic equator to sweep up and down over Europa at Jupiter's 11.23-h synodic period with respect to Europa. The changing latitude and varying ambient magnetic field over a jovian rotation period are illustrated in Fig. 2. The time-varying component of the field (whose contribution is principally in the radial direction relative to Jupiter) drives an inductive response within Europa (*Zimmer et al.,* 2000; *Schilling et a.l,* 2008), which produces what can be thought of as a periodically varying internal dipole moment with its axis in Europa's equatorial plane. This changing internal source contributes significantly to the total field near Europa (*Neubauer,* 1999). The properties of the background plasma of the extended Io torus are controlled by a combination of electromagnetic and centrifugal forces; at the orbit of Europa, the plasma density peaks between Jupiter's magnetic equator and its centrifugal equator and decreases markedly above and below; because of the 10° tilt between Jupiter's spin axis and its magnetic dipole axis, plasma properties at Europa are strongly modulated as Jupiter rotates. The most complete survey of plasma density near Europa (Fig. 3, taken from *Kurth et al.,* 2001) makes use of data from Galileo's Plasma Wave System (PWS) (*Gurnett et al.,* 1992). Power at $f_{uh,e}$, the electron upper hybrid frequency, is roughly proportional to the square root of n_e, the electron number density, with a weak dependence on $|\mathbf{B}|$, the magnitude of the magnetic field. The variation of density from pass to pass results largely from the changing location of the Galileo passes relative to Jupiter's magnetic equator, a point to which we will return.

At Europa's orbit, the plasma approximately corotates (meaning that it flows in the azimuthal direction at approximately Jupiter's angular velocity) as a result of electromagnetic coupling between the equatorial plasma and Jupiter's ionosphere. Near the equator at distances beyond ~1.4 R_J, Keplerian speeds are slower than plasma rotational speeds, so the rotating plasma overtakes bodies orbiting Jupiter. Plasma flows onto Europa's trailing hemisphere at a relative speed of roughly 100 km s^{-1} as indicated in Table 2. Slight variations of plasma parameters with local time are observed as Europa orbits Jupiter every 84 h but the dominant temporal variations are at the 11.23-h synodic period during which Jupiter's magnetic equator nods up and down over Europa as the planet rotates. Table 2 (extracted from *Kivelson et al.,* 2004) lists additional key features of the plasma environment near Europa with values given as av-

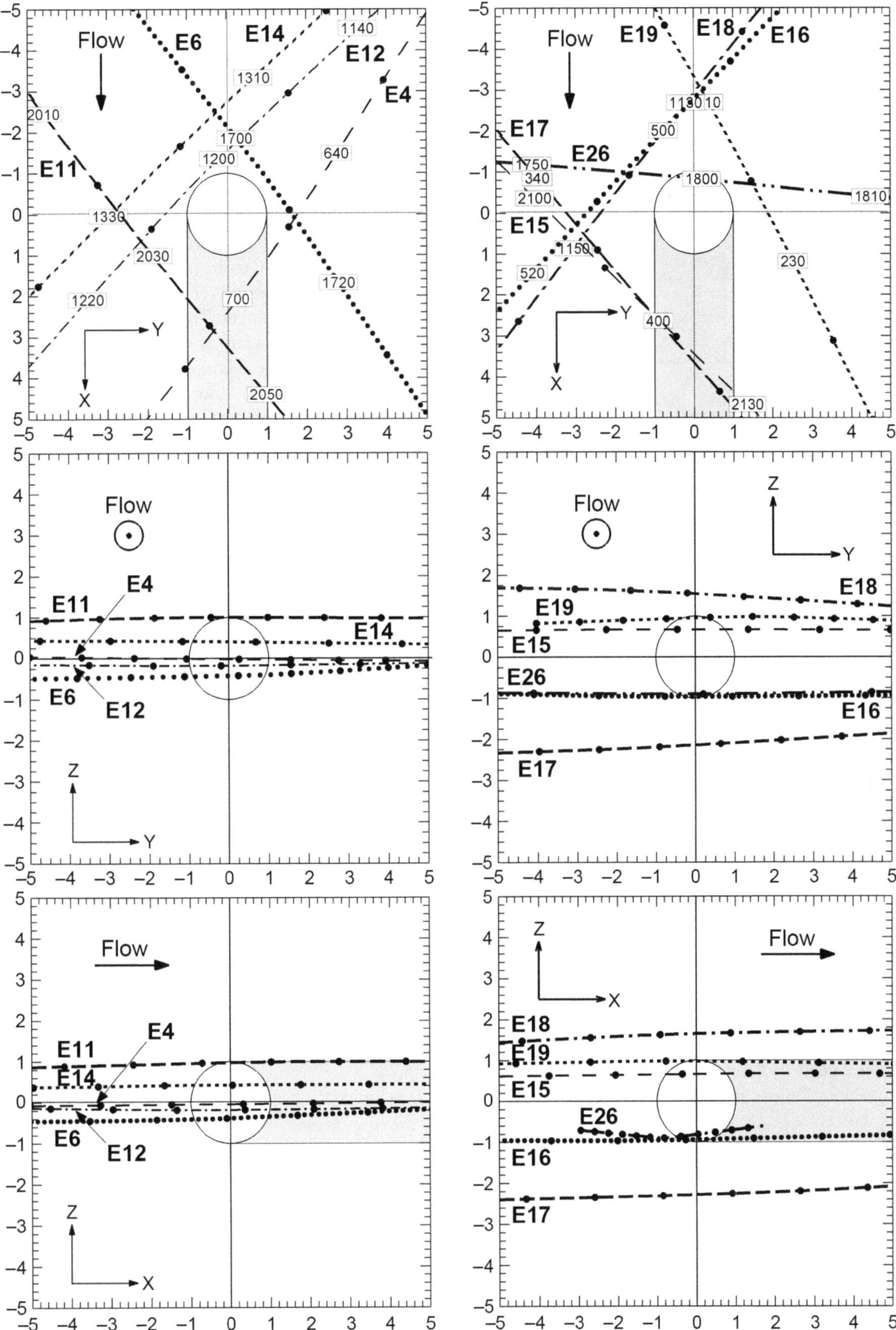

Fig. 1. Galileo flyby trajectories in the EphiO coordinate system with x along the direction of corotation, y radially in to Jupiter, and z parallel to Jupiter's spin axis. To the left, flybys E4–E14. To the right, E15–E26. None of the trajectories crossed over the polar regions.

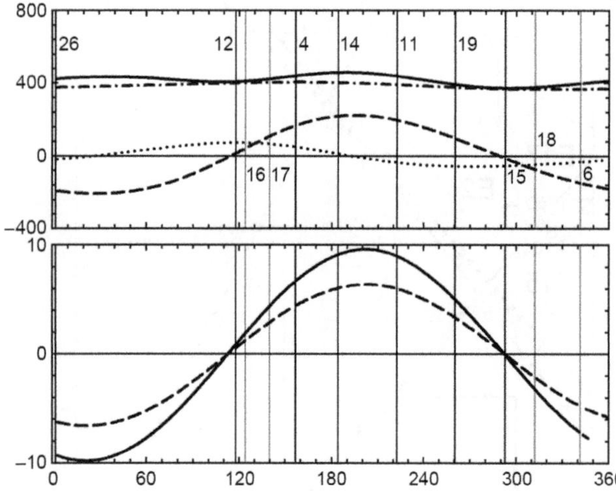

Fig. 2. *Top:* From the *Khurana* (1997) model, the radial (dashed), θ (dash-dot), azimuthal (dotted) components of the magnetic field and its magnitude (solid). *Bottom:* The dipole latitude (solid) and the centrifugal latitude (dashed) of Europa vs. west longitude over one Jupiter rotation period. In this longitude system, Europa's position moves from large to small values.

erages and ranges. Dominant torus ions are low-charge states of iogenic sulfur and oxygen with a mean ion mass of 18.5 m_p (m_p is the mass of a proton) and a mean charge of 1.5 e. Despite the systematic temporal variations of the ambient plasma near Europa at the synodic period, the parameters of the external plasma environment can be considered as stationary on the timescales of tens of minutes required for plasma to flow across Europa and its disturbed surroundings.

Magnetized plasmas interact with a moon in several ways. On a large scale, electrically conducting material within or near the moon diverts the flow and perturbs plasma and field properties. Consequently, measurements made at different locations relative to the moon's equator and from upstream to downstream relative to the flowing magnetospheric plasma are expected to reveal different aspects of the interaction. Furthermore, the conducting region near the moon may vary with solar illumination, implying that aspects of the interaction may change with the spacecraft-moon-Sun angle. Data were acquired at different locations relative to both the flow direction and the spacecraft-moon-Sun angle near closest approach (c/a) on several low-altitude Galileo flybys indicated by the small triangles in Fig. 4 and identified more precisely in Table 1. By convention, latitude is measured relative to an axis parallel to Jupiter's spin axis and Europa longitude is measured from the Jupiter-facing meridian plane; in this paper longitude (phi) is positive in the righthand sense. Closest approach was less than 1000 km on only three passes and less than 2050 km on six passes for which full fields and particle data were acquired. Four of the five relatively close passes approached Europa closely on the side upstream in the flow (or equivalently, trailing relative to orbital motion).

3. MODELS OF THE INTERACTION BETWEEN EUROPA AND THE LOCAL PLASMA ENVIRONMENT

3.1. Magnetohydrodynamic Descriptions

Large-scale features of the interaction between the Europa and its plasma environment are most readily understood by treating the ionized ambient gas as a magnetohydrodynamic (MHD) fluid. MHD is applicable if the length and timescales relevant to the interaction are long compared with the lengths and periods characteristic of single-particle motion in the environment. The radius of Europa is on the order of 1500 km, 2 orders of magnitude larger than the 8–12-km gyroradi of the thermal ions; the period of ion cyclotron motion is on the order of 2 s, short compared with the 30-s time for plasma to flow across the diameter of the moon. Several MHD simulations of the interaction are now available (*Kabin et al.,* 1999; *Liu et al.,* 2000; *Schilling et*

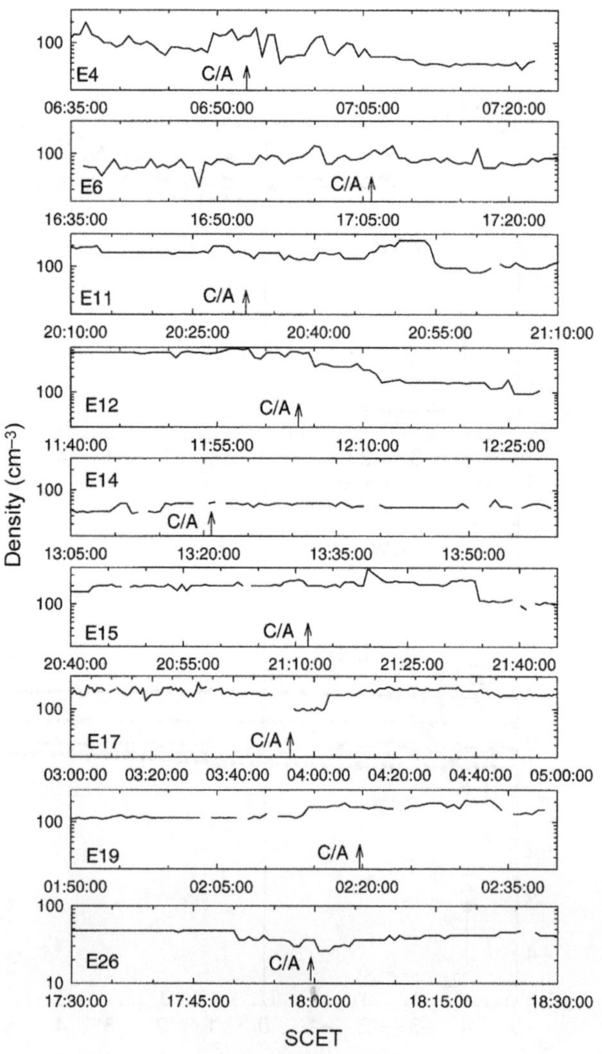

Fig. 3. The electron number density inferred from the upper hybrid resonance line in the plasma wave spectra for all Europa flybys. From *Kurth et al.* (2001).

TABLE 2. Properties of Europa's field and plasma environment.

Symbol (units)	Parameter
B_0(nT), jovian magnetic field, av. min (max)	370 (460)
n_e(elns cm^{-3}), eln. density, eq. av. (range)	200 (18–250)
⟨Z⟩, ion charge, eq. av. (lobe)	1.5 (1.5)
⟨A⟩, ion mass in mp, eq. av. (lobe)	18.5 (17)
n_i(ions cm^{-3}), ion number density, av. (range)	130 (12–170)
ρ_m(amu cm^{-3}), ion mass density, av. (range)	2500 (200–3000)
kT_i(eV), ion temperature, equator (range)	100 (50–400)
kT_e(eV), electron temperature	100
$p_{i,th}$(nPa), pressure thermal plasma, eq. (range)	2.1 (0.10–11)
$p_{i,en}$(nPa), pressure of 20 keV–100 MeV ions	12
p_e(nPa), pressure of "cold" and "hot" electrons	2.4
p(nPa), total pressure, eq. (max)	17 (26)
v_{cr}(km s^{-1}), local corotation velocity	117
v_s(km s^{-1}), satellite orbital velocity	14
v_φ(km s^{-1}) plasma azimuthal vel. (range)	90 (70–100)
u(km s^{-1}), plasma velocity relative to Europa av. (range)	76 (56–86)
v_A(km s^{-1}), Alfvén speed, eq. (range)	160 (145–700)
c_s(km s^{-1}), sound speed, eq. (range)	92 (76–330)
$B_0^2/2\mu_0$(nPa), magnetic pressure, eq. (lobe)	54 (84)
ρu^2(nPa), ram pressure, eq. av. (max)	24 (38)
ρu^2(nPa), lobe ram pressure	2.5

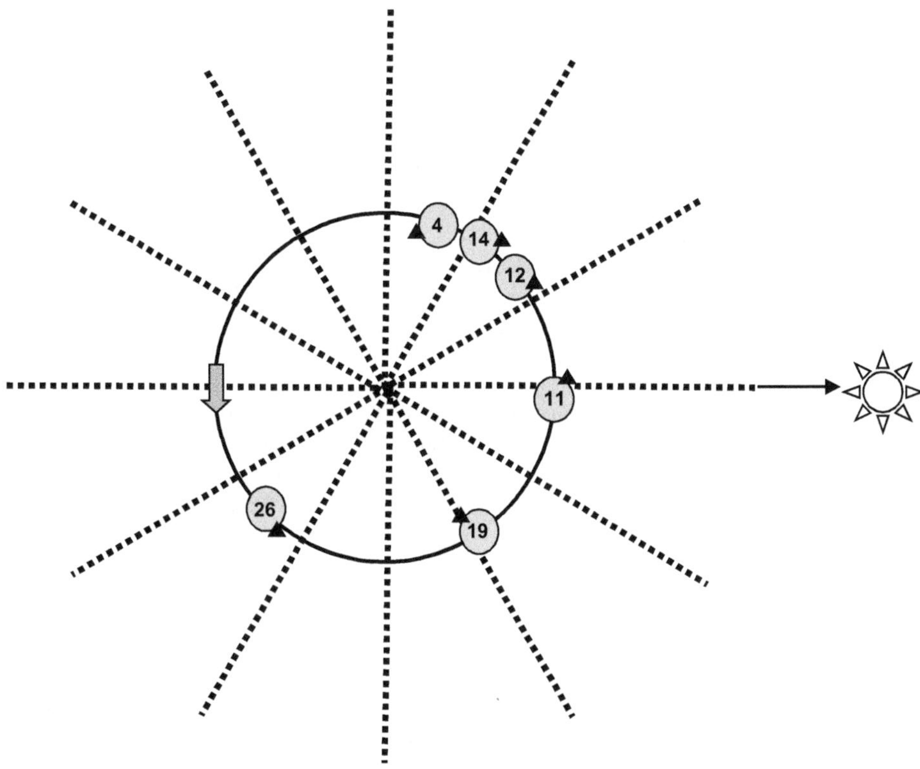

Fig. 4. Location of Europa relative to Jupiter and the Sun for Galileo encounters approaching 2050 km or less above the surface. Numbers identify the Galileo orbit for each flyby. The arrow shows the sense of corotational plasma flow. Triangles identify the location of closest approach.

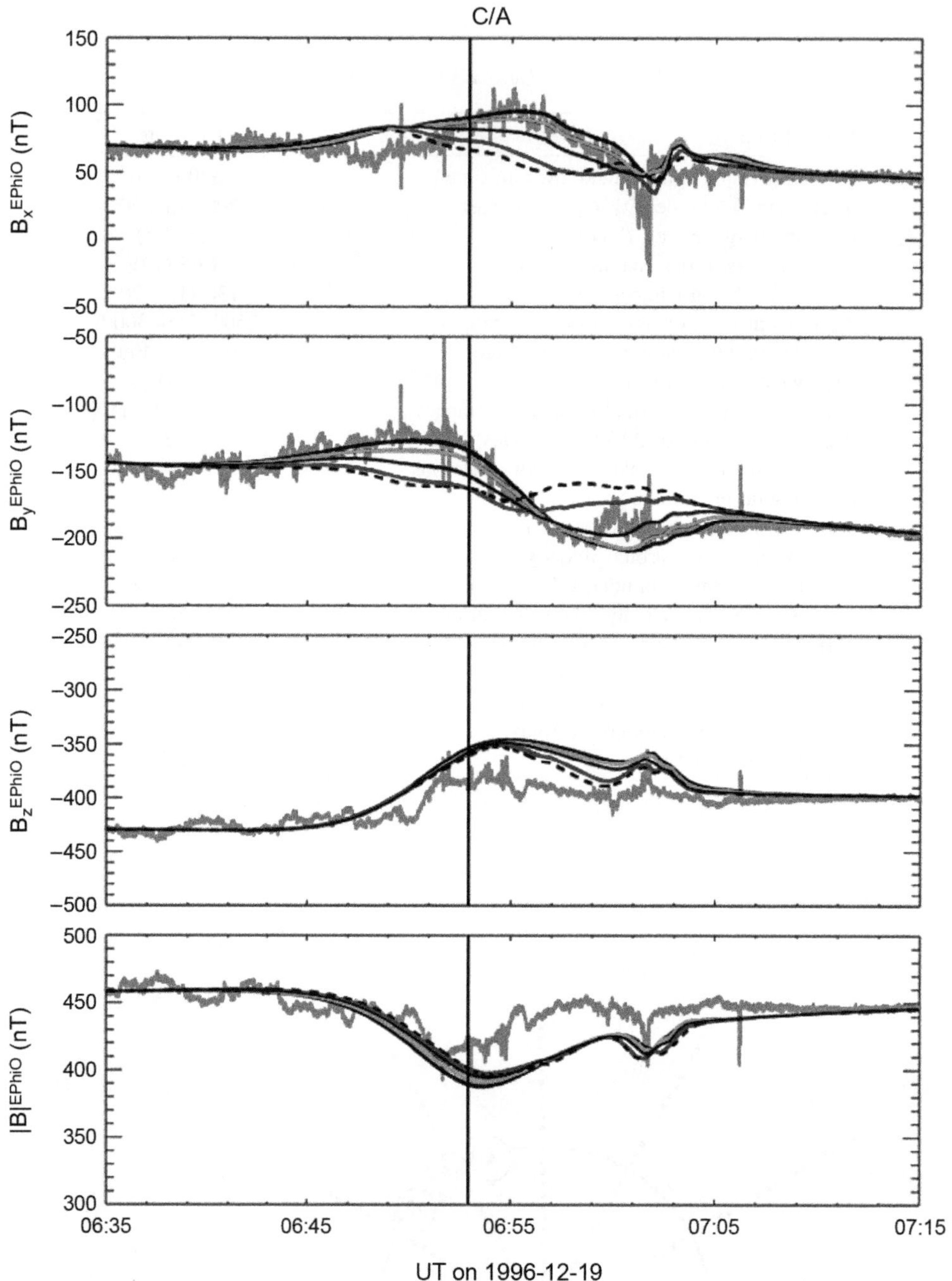

Fig. 5. See Plate 27. Observed and modeled magnetic field for the E4 flyby. Red = measurements of *Kivelson et al.* (1997); dashed black = modeled field with no internally induced field; Blue, green, and black = modeled field including induction in a 100-km-thick ocean lying beneath a crust of 25 km for ocean conductivities of 100, 250, and 500 mS m^{-1}, respectively. From *Schilling et al.* (2007).

al., 2007, 2008). Only *Schilling et al.* (2007, 2008) model the time variation of the system and calculate the induced field by introducing the concept of a "virtual plasma" internal to Europa. It is difficult to assess how this approximate treatment of the induced field modifies the outcome of the calculations, but the signatures along the E4 trajectory capture some key features of the measured magnetic field (*Kivelson et al.,* 1992) as seen in Fig. 5.

Common to the simulation results are aspects of the interaction that can be deduced qualitatively from the linear theory of perturbations of a magnetized fluid and the assumption that there is a region of electrical conductivity at and near Europa. In Europa's frame, the flowing plasma imposes an electric field, $\mathbf{E} = -\mathbf{u} \times \mathbf{B}$, where $\mathbf{u}$ is the flow speed relative to Europa and $\mathbf{B}$ is the magnetic field. Although there may be regions of ionospheric conductivity close to the moon (*Kliore et al.,* 1997), the dominant conductivity arises from ionization of neutrals liberated by sputtering (see chapters by Johnson et al. and McGrath et al.). When a neutral particle is ionized in a flowing plasma, the newly freed ion and electron are accelerated in opposite directions by the flow electric field, creating a transient current aligned with **E**. The conductivity, σ, can be inferred from $\sigma = |j/E|$. Interactions are mediated by the three basic wave modes of the system, two of which are compressional (fast and slow waves) and one (the intermediate or shear Alfvén wave) noncompressional (e.g., *Kivelson,* 1995). From Table 2 it follows that, in Europa's rest frame, the velocity of the magnetospheric plasma is slow compared with the nominal Alfvén speed and the fast magnetosonic speed. This implies that no shock forms upstream of the interaction region, but instead the incident flow is slowed by the action of fast magnetosonic signals. The slowing of the flow builds up a wedge of magnetic and thermal pressure upstream of the obstacle that diverts some of the incident flow. The flow slows first where unperturbed streamlines impact the moon. With the magnetic field frozen into the plasma, the slowing in one portion of the flux tube while remote regions continue in unperturbed motion imposes a kink (or curl) in the background field, which implies that currents are present ($\nabla \times \mathbf{B} = \mu_o \mathbf{j}$ where **j** is the current density). The kink propagates both up and down the field at the Alfvén speed, carried by an Alfvén wave, the only MHD wave mode that carries field-aligned current ($j_{||}$). Thus, to lowest order, the plasma is modified by the presence of a spherical obstacle not merely in the immediately surrounding regions but along a pair of tilted cylinders extending to the north and south. The disturbed regions, bounded by characteristics of the Alfvén wave, are referred to as Alfvén wings. If the plasma flow is perpendicular to the background field, **B**, the angle, θ_A, by which the characteristics are rotated from plus or minus the background field can be expressed in terms of the Alfvén Mach number of the flow, M_A, as $\theta_A = \tan^{-1} M_A$ (*Neubauer,* 1980; *Southwood et al.,* 1980). $M_A = u/v_A$ is defined in terms of u and $v_A = B/(\mu_o\rho)^{1/2}$, the Alfvén speed; here ρ is the mass density. The general structure of the interaction region was described in

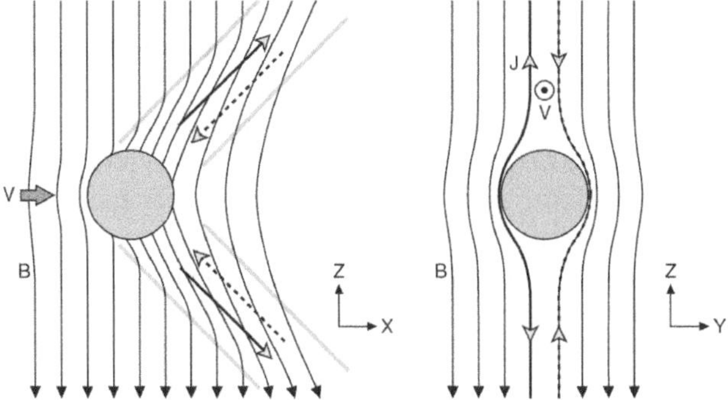

Fig. 6. Structure of the interaction region near a conducting moon. On the left, in the plane containing the field and the unperturbed flow velocity. On the right, a cut through the center of the moon in the plane normal to the unperturbed flow. In this schematic, Jupiter is to the right and only flux tubes lying between the two dark curves with arrows showing the direction of field-aligned current flow actually encounter the moon. Other flux tubes drape around it. Adapted from *Southwood et al.* (1980).

the context of the Io interaction by *Southwood et al.* (1980) and by *Neubauer* (1980) and is illustrated in Fig. 6. In this figure, one sees the field tilted by the interaction to form the Alfvén wings, and one can also infer that a portion of the upstream flow (and the flux tubes that thread this portion of the flow) flows around the moon instead of flowing onto it.

If an internal induced field is present, the symmetry of the Alfvén wings in the direction radial to Jupiter is broken; the Alfvén wings are displaced inward (toward Jupiter) in one hemisphere and outward in the other (*Neubauer,* 1999). Downstream of Europa, further Alfvénic perturbations act to restore the field to its unperturbed orientation. Interaction with the moon reduces the plasma pressure in the downstream wake. The pressure is restored by compressional slow mode perturbations (in which thermal and magnetic pressure are in antiphase).

The overall picture of the flow and field that we have described are clearly evident in Fig. 7, reproduced from the simulation of *Schilling et al.* (2008). In their simulation, the flow is directed toward positive x and the uniform background magnetic field is in the z direction. In Fig. 7a one sees the flow slowing (light color) as it approaches the moon. The diverted flow experiences a Bernoulli effect and speeds up along the flanks (dark color). For $x > 1$, there is a narrow region in the wake of the moon where plasma refilling flux tubes that have interacted with the moon is compressed and flow is very slow. In Fig. 7b the Alfvén wing structure is apparent, with the perturbed flow region bent back along the flow direction as described. The flow is extremely slow not only in the $z = 0$ plane but in other planes at constant z (not shown) where the plasma diverts around a region whose center shifts toward $x > 0$ as z increases. Thus it is not only the moon that perturbs the flow, but also its associated pair of tilted Alfvén wing cylinders. Figure 7c shows the pileup of field in the upstream region of slowed flow. This represents the effect of the fast mode perturbation described. Also evident in Fig. 7c is the change of field orientation imposed by the Alfvén waves whose characteristics bound the Alfvén wings.

The upstream field can be significantly tilted relative to the flow, and this implies that the flow has a component along the background field as well as across the field, which introduces some north-south asymmetry into the solutions. The assumption common to all simulations available is that the background magnetic field is uniform on the scale on the order of a few R_E (1 R_E = the radius of Europa, 1560 km) near Europa. This approximation is good at the 10% level.

3.2. Ion Pickup

The conductivity of the moon's environment is critical in controlling the streamlines of the flow onto and around the surface. Electric currents can flow in Europa's ionosphere, whose properties vary with solar illumination and with the plasma properties of the surroundings. Europa's surface and transient atmosphere are continually bombarded by magnetospheric charged particles that sputter neutrals

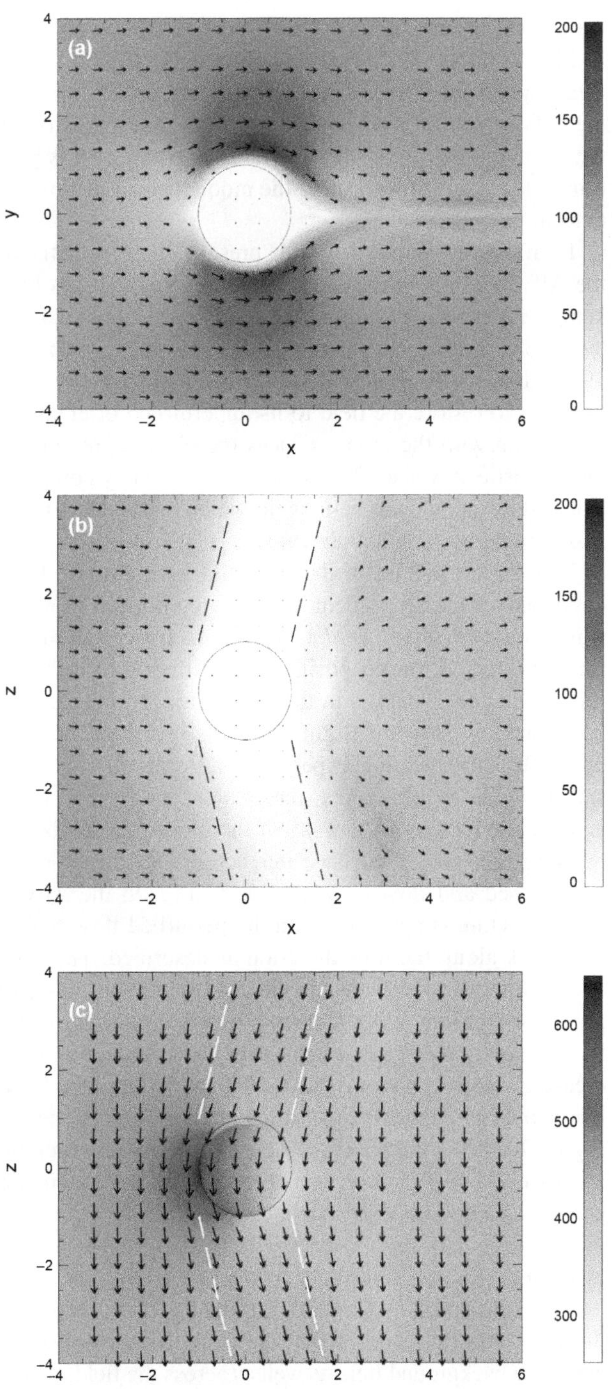

Fig. 7. See Plate 28. The flow speed (**a**) in the x–y plane, (**b**) in the x–z plane, and (c) the magnetic field in the x–z plane. Dashed lines represent boundaries of the northern and southern Alfvén wings. From the MHD simulation of *Schilling et al.* (2008).

from the surface (see chapter by Johnson et al.). Those neutrals are widely distributed near Europa. The neutrals serve as a source of new ions as discussed relative to interactions near Io by *Goertz* (1980). Table 3 of *Luna et al.* (2005) analyzes the processes that produce ions near Europa. Electron impact dominates, with a production rate exceeding that of charge-exchange (photoionization) by a factor ~20 (~60). When first ionized, the ions are, on average, at rest relative to Europa. They must be accelerated to full corotation, a process that creates a drag on the magnetic field lines in addition to the drag of the conducting body and its ionosphere. As mentioned previously, the drag is greatest at the location on the field line where the flux tubes pass closest to Europa. The differential slowing of different points along a field line bends the field as shown Fig. 7b. The bend-back displays itself through a rotation of the field from the z into the x direction. The amount of bend-back is related to the amount of ionization occurring near the moon, which, in turn, varies with the moon's distance from the center of the jovian plasma torus. The kinked field acts to reaccelerate the plasma through the magnetic curvature force.

Instead of describing the curvature of flux tubes, one can describe the phenomenology in terms of the currents generated. Newly liberated charged particles are accelerated by the electric field of the flowing plasma in which they are embedded. The new ions, referred to as "pickup ions," create a pickup current density, j_{pu}, given by

$$j_{pu} = e\dot{n}_i\rho_L = m_i\dot{n}_i u/B \tag{1}$$

Here e is the magnitude of the electron charge, $\dot{n}_i$ is the number density of new ions introduced per second, $\rho_L = um_i/eB$, is the ion Larmor radius of a typical ion of mass m_i in a magnetic field, B, and u is the flow speed. The requirement that current density be divergenceless ($\nabla \cdot \mathbf{j} = 0$) requires the pickup current to close through field-aligned currents (flowing toward Europa's orbit on the side closer to Jupiter and away from it on the other side). The pairs of field-aligned currents produce magnetic perturbations that bend the field below the moon toward the flow direction and above the moon toward the opposite direction ($-\mathbf{u}$), thus producing the previously described Alfvén wing structure. In this description, it is the Lorentz force, $\mathbf{j}_{pu} \times \mathbf{B}$, rather than a curvature force that accelerates the slowed plasma. In a later section, we show examples of passes on which the magnetic perturbations are consistent with generation by pickup currents for reasonable estimates of the quantities that appear in equation (1).

It is now clear that coupling between the local region and the more distant parts of the flux tube is central to imposing the form of the field and the flow near Europa. In turn, the flow patterns in the vicinity of the moon are controlled by the combined effects of the conductivity in the immediate neighborhood of the moon and properties of the background plasma. *Neubauer* (1998) showed that the conductivity of the ionosphere and of the pickup ions can be lumped together by defining a generalized Pedersen conductance, for which we use the symbol Σ_P and, for the purpose of estimates, assume to be constant over a radial distance $(1 + \delta)$ R_E around Europa, where an increment δ accounts for the region of strong pickup.

In order to understand surface sputtering, one needs to consider the flux tubes that actually encounter the moon and give energetic particles access to the surface. Not all the flux tubes in the unperturbed plasma on streamlines directed toward the $2(1 + \delta)$ R_E width of the conducting region near Europa actually intercept that cross section because streamlines diverge as shown in Fig. 7a. The fraction $(1 - f)$ of the upstream fluid that flows into the region of width $2(1 + \delta)$ R_E depends on the Alfvén conductance of the unperturbed plasma, $\Sigma_A = (\mu_o v_A)^{-1/2}$ (*Neubauer,* 1980; *Southwood et al.,* 1980) and on Σ_P according to the relation

$$f = \Sigma_P/(\Sigma_P + 2\Sigma_A) \quad (2)$$

where f is the fraction of the incident flow that avoids the obstacle. In the limit $\Sigma_P \gg \Sigma_A$, none of the upstream flow reaches the surface, whereas all of the flow reaches the surface if the local conductance vanishes. In Europa's case, both Σ_P and Σ_A are in the range of a few to tens of Siemens (1 S = 1 amp volt^{-1}), whereas $\Sigma_A \approx 6S$ using the nominal v_A from Table 2, so some of the flow is diverted but some reaches the surface. The values of both conductances change with Europa's position in the torus because both increase as the plasma density increases. Streamlines obtained from simulations can, in principle, provide insight into the way in which the response varies as local conditions change, but it is important to remember that the specifics of the solutions are extremely sensitive to assumed internal boundary conditions. Conclusions extracted from simulations are instructive but should be viewed with abundant skepticism.

3.3. Beyond Magnetohydrodynamics

Close to the moon, corrections for multifluid phenomena are relevant in understanding some features of the interaction and various models have examined the interaction using approaches that deal with the complexity of the plasma and its interaction with Europa's atmosphere while accepting a non-self-consistent treatment of the perturbations of the magnetic field (e.g., *Saur et al.,* 1998). Streamlines from this solution appear in Fig. 8. As in the MHD treatment, the conductivity of the region surrounding Europa impedes or diverts the plasma flow, allowing only about 20% of the upstream plasma to reach the surface of Europa. The diversion of the flow was clearly observed on Galileo passes upstream and on the flanks of the interaction region (*Paterson et al.,* 1999).

Other aspects of the interaction are illustrated in Fig. 8. The streamlines twist inward toward Jupiter as they move across the polar cap, a consequence of the Hall conductance of the ionosphere. Flux tubes that encounter Europa are slowed in their flow (to about 25% of the unperturbed flow speed), but continue to drift on average in the direction of the corotation flow. Let us assume that the flow speed is reduced to 10 km s^{-1}, implying that it takes on the order of 5 min for a flux tube to flow across the polar cap. Within the flux tubes that pass through Europa, the plasma characteristics are markedly modified. In the simplest picture,

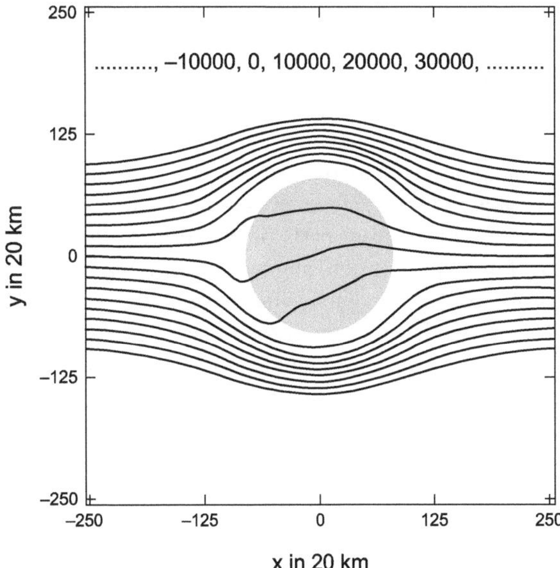

Fig. 8. Lines of equal electric potential or streamlines of electron flow. The spacing between the lines is proportional to the flow speed. About 20% of the upstream plasma encounters Europa in this model, and the flow speed drops to about 25% of the incident flow speed. From *Saur et al.* (1998).

one may imagine that any particle that reaches Europa's surface is lost. In this case, connection with Europa immediately removes all the particles arriving from the opposite hemisphere. To the north of Europa, particles that are moving downward are absorbed, whereas those moving upward continue unaffected, are reflected at a mirror point, and move downward, after which they are absorbed. Thus after half a bounce period, most of the particles initially on a flux tube linked to Europa are gone. The bounce period is energy dependent but scales as $(W/m)^{1/2}$, where W is the particle thermal energy and m is its mass. Imagine that the bulk of the plasma mirrors within, let us say, 2.5 R_J of the equator. Then the relevant timescale for emptying the flux tube is

$$T_b \approx 5R_J/(W/m)^{1/2} = 30s[m(m_e)/W(keV)]^{1/2}$$

This implies that the flux tube is depleted of keV electrons well before it reaches the downstream side of the polar cap. Protons of similar energy persist in the flux tube 40 times longer and heavy ions continue to reach the surface even longer. Thus, it is easy to accept the conclusion of *Paranicas et al.* (2002) that ion bombardment is relatively uniform across the surface of Europa but electron sputtering is localized to Europa's trailing hemisphere (upstream in the flow). (More precise analysis takes into account the fact that drift paths of energetic particles are modified by gradients in the locally perturbed magnetic configuration, as well as finite gyroradius effects.) The bombardment of Europa by the energetic particles lost from the plasma has consequences for the structure of surface ice (*Paranicas et al.,* 2000, 2001, 2002; see also chapter by Paranicas et al.).

In analysis of particle access to Europa's surface, one additional matter should be considered. In addition to the flux tube content and the flow patterns, the geometry of the intersection between the flux tube and the surface also affect the flux per unit area on different parts of Europa's surface. None of the Galileo passes crossed the polar regions, so one must rely on inference and theory to describe particle access to different parts of the surface. One thing is clear: For a uniform field aligned with Europa's spin axis, flux tubes of unit magnetic flux have constant cross-section areas, but the areas on Europa's surface intercepted by the flux tubes vary with latitude (λ) as $1 + \cos\lambda$. This implies that flux tubes carrying constant electron flux deliver fewer electrons per unit area to the near-equatorial surface than to the polar regions. This effect contributes to the variation of sputtering across the surface of Europa. In particular, sputtering by electrons is thereby somewhat reduced in the upstream, low-latitude regions where the flux tubes are still full of energetic electrons, although the compression of the field in the upstream region counters the geometric effect. The regions of most intense sputtering should not vary significantly as the magnetospheric field rocks back and forth over a synodic jovian rotation period because the variations of the external field are largely nullified by the internally induced field.

4. UPSTREAM

Table 1 lists several upstream Galileo passes (E12, E14, E19) with their trajectories illustrated in Fig. 1. The E12 pass is of special interest because it encountered Europa when it was located near the (N–S) center of the torus. On this pass, the plasma density was substantially larger than on other passes (see Fig. 3, noting that the scale shifts for different passes and that n_e on E12 exceeds 900 cm^{-3} shortly before closest approach, whereas it remains below 200 cm^{-3} on the other two passes). Figure 9 shows the magnetic field signature for these upstream passes: E12 (closest approach 196 km), E14 (closest approach 1649 km), and E19 (closest approach 1444 km). The magnetic field perturbations within ~3 R_E of Europa differ markedly on these passes.

On passes E14 and E19 in regions beyond ~4 R_E from Europa, the magnetic field is well described by the *Khurana* (1997) model. [Other models such as those of *Khurana and Schwarzl* (2005) and *Alexeev and Belenkaya* (2005) do not noticably modify field values in the inner magnetosphere.] The electron density, on the order of 100 cm^{-3} or less and close to constant from Fig. 3, is also nominal. Near closest approach, on E19 the density rises to ~200 cm^{-3}, suggesting that local effects roughly double the density. At the same time the increase of field magnitude to a maximum of 481 nT (or 32 nT above the model background field) is consistent with some slowing of the flow. Similar changes of field magnitude upstream of Europa, consistent with increased density, are observed on E14 although no associated density increase is identified by the PWS measurements (Fig. 3).

Pass E12 is more complicated. Downstream of Europa, the field is again well represented by the *Khurana* (1997) model, and the electron density decreases to a nominal 100 cm^{-3}, which appears to be typical of measurements remote from closest approach on several of the passes (*Kurth et al.,* 2001; *Paterson et al.,* 1999). However, in the upstream portion of the pass, the electron density is exceptionally high (~900 cm^{-3}). The very high density can be attributed in part to Europa's location between the centrifugal and magnetic equators, where the background density is highest. The magnetic field is also unusually large; somewhat upstream of Europa the field magnitude reaches 825 nT, a level almost double the nominal background. As local pickup would slow the flow and correspondingly increase the magnitude of the magnetic field, it seems probable that local ionization contributes significantly to the anomalously high electron density and that ionization becomes increasingly significant as the trajectory moves closer to Europa (see Fig. 3).

The combination of high density and large B can arise if ion pickup rates are high. Let us assume that the mass and average charge of the ions added locally is the same as that of the background (Table 2), implying that the ion number density is 0.7 n_e. Here we distinguish between pickup, which adds ions to the plasma, and charge exchange, which does not. Like pickup, which adds new ions that must be accelerated, charge exchange slows the flow because it replaces a moving ion with an ion at rest and this replacement ion must be accelerated to the bulk flow speed. It does not change the ion density because one ion is lost for each ion added. All of the processes increase the field magnitude. Each ion added to the flow acquires a thermal speed and a gyrocenter speed equal to that of the background plasma. At Europa's orbit, the nominal thermal speed exceeds the flow speed, so pickup cools the plasma. As the flow slows, the field magnitude increases. Energy conservation requires the flow kinetic energy of the bulk plasma to decrease by $m_i u^2$ for each new ion added, so the bulk flow speed decreases. (Here we ignore contributions to plasma acceleration imposed by currents connecting the equatorial plasma to Jupiter over times relevant to the interaction.)

Upstream of Europa on E12, the field magnitude is modulated by nearly periodic (~3 min) structures in which the field magnitude decreases and then increases abruptly (*Russell,* 2005). Although it is not possible to establish whether these variations are spatial or temporal structures, the forms have the appearance of periodic pressure pulses propagating upstream with a steep forward edge followed by a relaxation. Pressure pulses would slow the flow and account for the increases of field magnitude. Localized slowing bends the field and the curvature exerts a force like that of a bow string under tension. The curvature force reaccelerates the plasma and reduces the field magnitude. The 3-min recurrence of the pulses has no evident relation to natural periods of the interaction and remains a puzzle.

Although some of the periodic field increases that were measured upstream of Europa on E12 are pulse-like, the

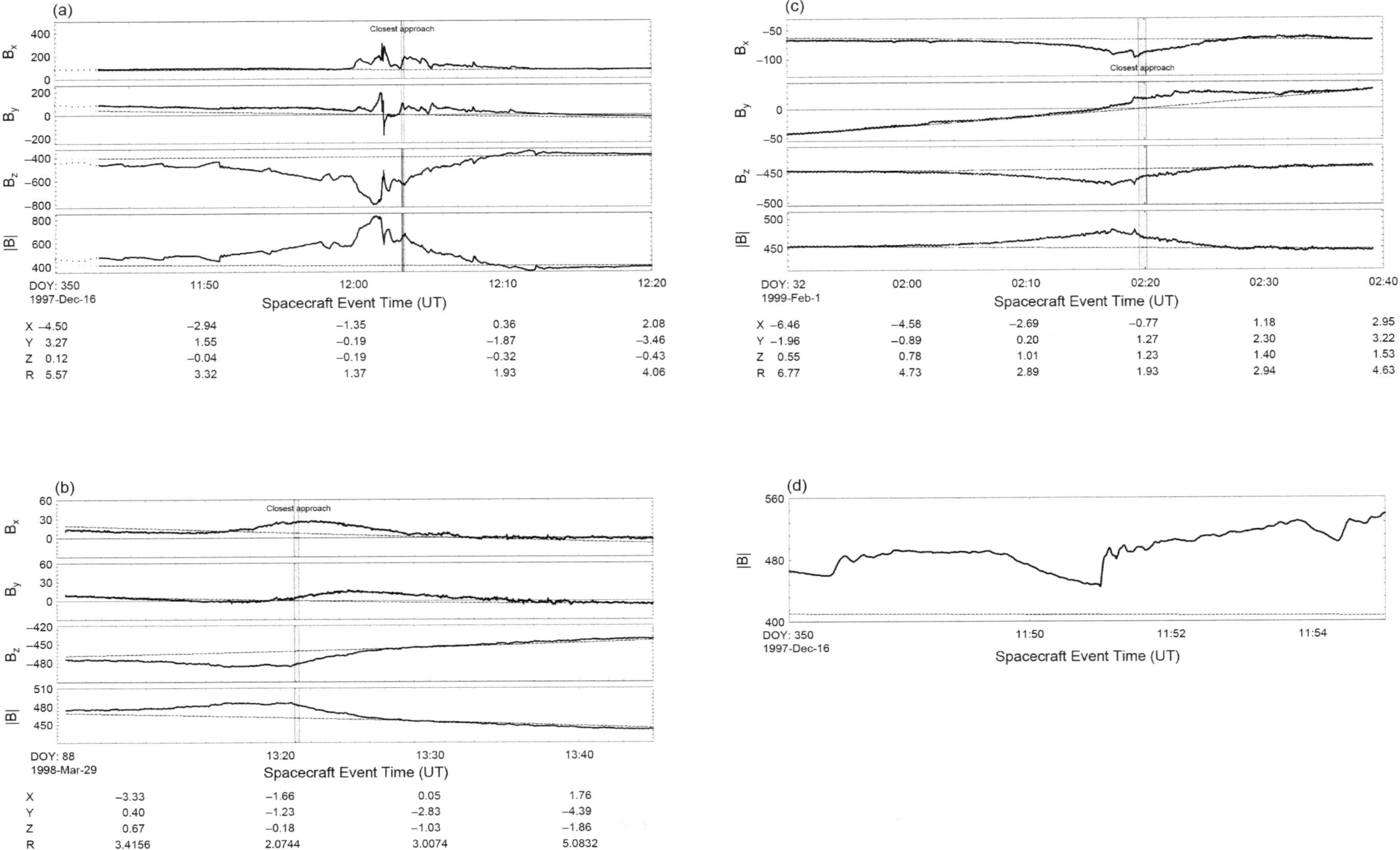

Fig. 9. Magnetic field (EphiB coordinates) for Galileo's **(a)** E12, **(b)** E14, and **(c)** E19 upstream flybys of Europa, very near the center of the plasma sheet. Solid lines = data, dashed lines = model background (*Khurana*, 1997). A shock-like structure in the field magnitude at 11:51 UT on Dec. 16, 1997 is shown expanded in **(d)**.

sudden increase of field magnitude at 11:51 UT is sufficiently abrupt that it may be a shock. The jump occurs in ~4 s, consistent with a thickness of a few ion gyroradii. Assuming that the background plasma has cooled through pickup and that the thermal velocity is small, one can set the fast magnetosonic speed to the Alfvén speed (v_A), which, with B = 480 nT, n_i = 600 cm^{-3}, and m_i = 20 m_p (m_p is the mass of a proton), is 96 km s^{-1}. The field magnitude at 11:51 UT exceeds background by only ~20%, implying that the flow has not yet slowed substantially (there are no published flow estimates for this pass). Table 1 provides a range of 56–86 km s^{-1} for u, but the ranges of Table 1 are not limits as evident from the fact that the E12 electron density of 900 cm^{-3} falls well outside the listed range of 18–250 cm^{-3}. It is then possible to suppose that the exceptionally high densities of this pass reduced the Alfvén speed below the flow speed in portions of the region upstream of Europa. When this happened, the pressure pulses steepened to form weak shocks behind which the flow slowed, causing the field to build up further before reacceleration of the flow decreased it once again.

The exceptionally high plasma density within a few R_E of Europa on E12 calls for a local source of pickup ions. Recognizing that pickup slows the flow and increases the field magnitude, one can confirm that this interpretation is self-consistent. We estimate j_{pu} by assuming that the rate of addition of new ions near Europa is 10% of that near Io, i.e., ~100 kg s^{-1} in a volume on the order of (3 $R_E)^3$. If the pickup current flows across a surface of extent ~2 R_E along the upstream flow, then from equation (1) and Ampere's law, the expected change of B, ΔB, is ~600 nT, roughly in the range observed. Although it seems probable that the exceptionally large pickup rate is caused by the unusually dense plasma encountered on this pass, it is also possible that illumination of the atmosphere on Europa's upstream side also contributes (see chapter by McGrath et al.).

5. THE WAKE REGION

A wake develops downstream of an obstacle in a flowing fluid, whether hydrodynamic or magnetohydrodynamic. One of the earliest scientific sketches of such a region in a fluid flow was made by Leonardo da Vinci (see Fig. 10). Europa's wake lies on its leading side, ahead of the moon in its orbital motion, as described above. In this section we discuss wake structure and describe some features of the wake region, including flux tubes with anomalous plasma content, pickup ions, and nonuniform distribution of energetic ions.

Five Galileo passes crossed the Europa wake but full particles and fields data are available for only four of them. The E4, E11, E15, and E17 flybys encountered Europa in differing locations in the torus (see Fig. 2). Passes E4 and E11 occurred when Europa was at relatively high magnetic latitude, moving toward the (N–S) center of the torus for E4 and exiting it for E11. For E15 (and E17) Europa was in (or relatively near) the center of the plasma torus. This means that plasma responses sensitive to ambient plasma conditions may differ from pass to pass.

Before examining the wake data, we introduce the coordinate systems used, acknowledging that only those immersed deeply in the study of magnetic fields become greatly enamored of coordinate systems. One can separate out some of the effects of the tilt of the background field by an appropriate choice of coordinates. Two systems useful for the analysis of the field observed near Europa (*Kivel-*

Fig. 10. A sketch from Leonardo da Vinci showing the water flow around and in the wake of an obstacle.

son et al., 1992) are both Europa-centered, with **x** along the background flow. In the EphiO coordinate system, **z** is aligned with Europa's spin axis, $\hat{\mathbf{y}} = \hat{\mathbf{z}} \times \hat{\mathbf{x}}$ is positive toward Jupiter, and the background field has nonvanishing x and y components. In the EphiB coordinate system, $\hat{\mathbf{y}} = (-\mathbf{B}/B) \times \hat{\mathbf{x}}$ (again positive toward Jupiter), $\hat{\mathbf{z}} = \hat{\mathbf{x}} \times \hat{\mathbf{y}}$, and the field lies in the x–z plane. When this latter system is used to analyze data near Europa it is referenced to the field orientation at closest approach. Thus, the y component of the field vanishes at closest approach and remains relatively small in the near vicinity of Europa. However, a finite x component along the flow is not removed and can be on the order of 20% of the field magnitude.

The magnetic field data for the wake crossings are shown in Fig. 11 in EphiB coordinates. In the panels, the geometric wake limits are shown. The geometric wake extends over the region $-1 \leq y \leq 1$ R_E. In the EphiB system, one might anticipate symmetry about y = 0, so departures from such symmetry are of interest and the data are not symmetric about the wake center. Part of the asymmetry must result from the changes in downstream distance as Galileo crossed the wake (see Fig. 1). Other contributions to the asymmetry of the wake are discussed below.

Characteristic of the wake region is the "bend-back" of the magnetic field associated with the Alfvén wing, described in section 3. Much of the bendback arises through interaction with pickup currents whose magnitudes decrease as Europa moves away from the center of the torus. Because of the north-south symmetry of the Alfvén wings, the bend-back is most evident on wake passes at $|z_{EphiB}| > 1$ R_E as, for example, in Fig. 11d. Different degrees of bend-back in the data shown in Fig. 11 are plausibly accounted for by a combination of the different values of $|z_{EphiB}|$ and the different plasma density and associated pickup densities associated with the location of Europa relative to the dense center of the plasma torus. The E4 and E11 passes occurred well away from the center of the torus. Bend-back is difficult to assess for E4, which entered the wake region at small $|z_{EphiB}|$. The signature is further obscured by the significant inductive field signature, but it appears to be small except over a very narrow region in the center of the wake. On E11, the wake crossing was relatively distant but the negative excursion of the B_x component near the central portion of the wake is consistent with weak bend-back. Bend-back produces a negative perturbation in B_x for $z_{EphiB} > 0$ and a positive perturbation in B_x for $z_{EphiB} < 0$. Thus, the perturbations of B_x in the geometric wake on passes E15 and E17, which encountered Europa relatively near the center of the torus at $z_{EphiB} > 0$ and < 0, respectively, appear to arise from bend-back.

5.1. Wake Asymmetry

The interaction of Europa with the jovian magnetosphere produces naturally an upstream-downstream asymmetry as evident from the MHD analysis, but to lowest order one might expect symmetry across the flow direction (i.e., in y_{EphiB}). However, the passes plotted in Fig. 11 show distinct asymmetry of the magnetic signatures across the wake. What are the processes that introduce asymmetry across the flow direction? One must distinguish between intrinsic asymmetry and asymmetry because the spacecraft trajectories are not parallel to the y_{EphiB} axis. This means that any analysis should be based on actual trajectories applied to models of the underlying asymmetry.

Neubauer (1999) showed that an inductive field such as that identified at Europa (*Kivelson et al.,* 1999; *Zimmer et al.,* 2000) displaces and shrinks the Alfvén wing and modifies the wake symmetry as illustrated in Fig. 12. Recognizing that an inductive response acts to exclude the time-varying part of the magnetospheric magnetic field from the interior of Europa, the fractional reduction in scale of the Alfvén wing and/or of the wake in the y direction (S) can be estimated as

$$S \approx \frac{B_z}{|B|} \tag{3}$$

where B_z is measured in the EphiO coordinate system. Indeed, this estimate, as well as the predicted displacements of Alfvén-wing-related structures, has been verified in Galileo observations (*Volwerk et al.,* 2007) for a number of passes including E17 (see Fig. 11).

Additional asymmetries in the radial direction can be introduced through ionospheric effects. In an analysis of the Io interaction, *Saur et al.* (1999) showed that the Hall conductance of the ionosphere twists the electric field from radially outward toward the direction of corotation through an angle

$$\theta_{twist} = \frac{\Sigma_H}{\Sigma_P + 2\Sigma_A} \tag{4}$$

in terms of Σ_H, the Hall conductance of Io's ionosphere, and the Pedersen and Alfvén conductances previously defined. Correspondingly, the flow velocity $(\mathbf{E} \times \mathbf{B})/B^2$ twists from azimuthal toward Jupiter through the same angle.

Europa was located at the center of the plasma torus for the E15 pass. At this location, the induced field should have been very small so asymmetry of the wake signature must have resulted from other mechanisms such as the twisted electric field that we have discussed above (*Saur et al.,* 1999) or asymmetries of energetic particle fluxes (discussed below). In order to estimate θ_{twist}, values of the Pederson, Hall, and Alfvén conductances are needed. *Saur et al.* (1998, Figs. 5 and 6) find that the Pederson conductivity dominates the Hall conductivity, with typical values $\Sigma_H \approx 1$ s and $\Sigma_P \approx 10$ S. Σ_A can be obtained from measurements. A characteristic value of v_A is ~160 km s^{-1} (Table 2), which implies an Alfvén conductance of approximately 5 S. Thus, $\theta_{twist} \approx 0.05$ rad (= 3°) rotated from radially outward into the plasma flow direction. According to the predictions of equation (4), the ions should be only slightly deflected by this twist angle in the direction away from Jupiter. Such a small deviation in thermal and pickup plasma density cannot be inferred

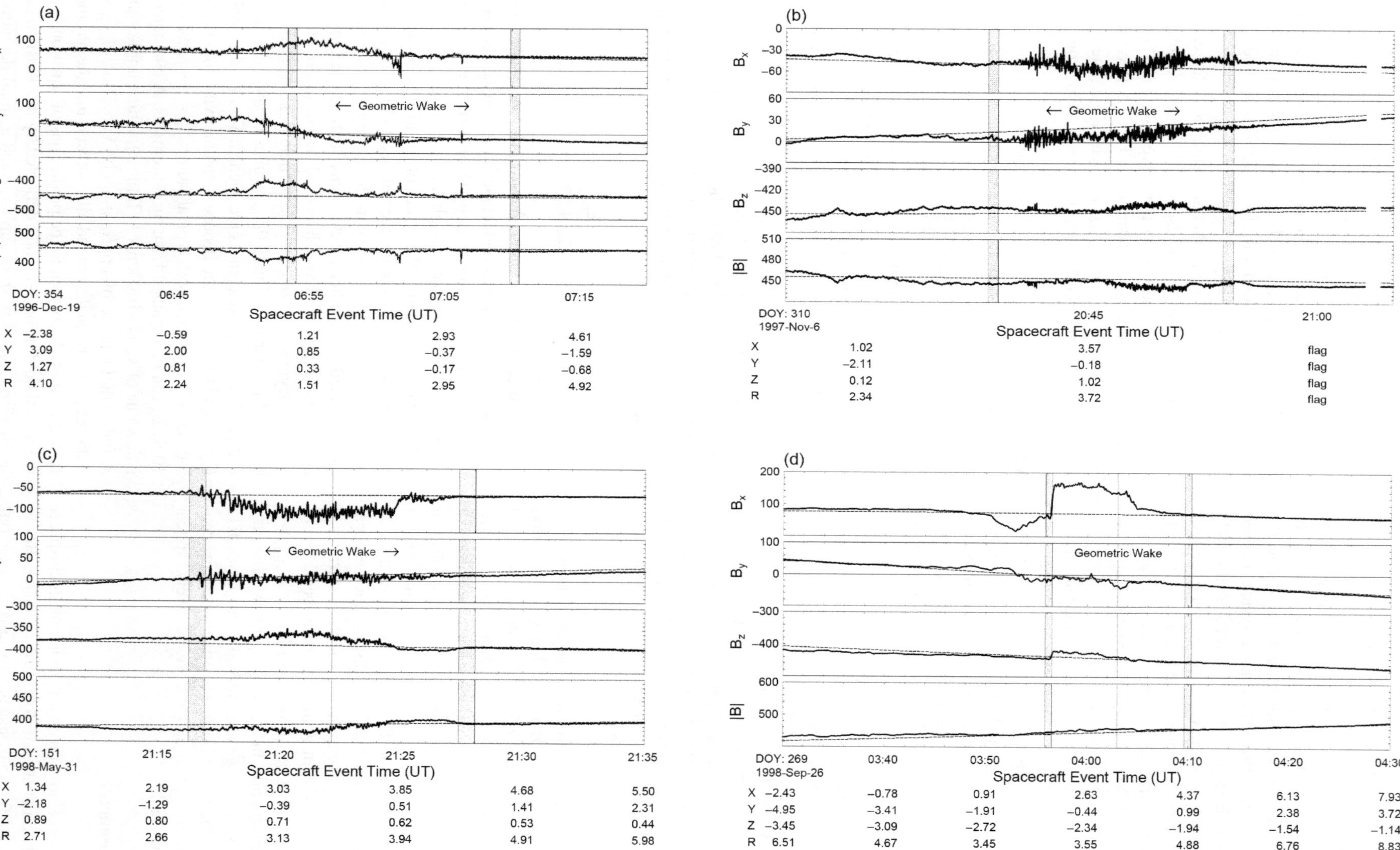

Fig. 11. The magnetic field data for Galileo's wake crossings E4, E11, E15, and E17 in EPhiB coordinates. Data (0.33-s samples except E17 with 24-s samples) are plotted as solid lines. The background magnetic is plotted with a dashed line. The wake ($-1 \leq y \leq 1$) lies between the gray markers and a solid line shows $y = 0$.

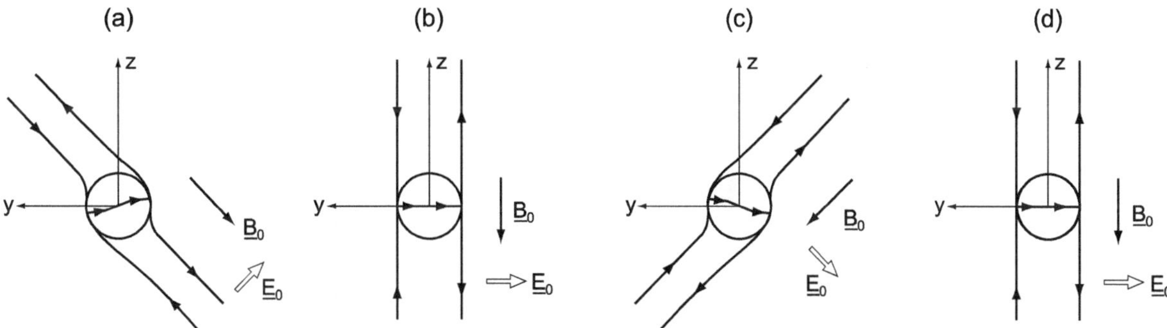

Fig. 12. Schematics of Alfvén wings and associated current systems showing how induction effects introduce asymmetries by shifting the northern and southern wings in opposite directions and simultaneously reducing their cross section areas. The successive images represent forms associated with a rotation period from north through the equator to south and return. (Here the field rotations approximate those at Callisto. The rotation angles would be smaller at Europa.) From Fig. 7 of *Neubauer* (1999).

from the data, but the estimated conductances could be incorrect, implying that larger twists may be imposed. However, although flow paths across the polar cap are modified by the Hall conductance, Fig. 8 shows that near the z = 0 plane, the wake is little skewed.

Some of the wake asymmetry may arise because the fate of heavy ions newly picked up in the flowing plasma differs on the two sides of Europa. The corotation electric field in Europa's frame accelerates newly ionized positive ions outward from Jupiter. Consequently, some of the ions picked up on the subjovian side immediately impact Europa's surface and are lost to the plasma, thus reducing the total mass loading on that side. On the antijovian side, outward acceleration does not lead to impact.

The ion loss through impact on the surface can be modeled using a uniform, vertical magnetic field directed in the negative z direction as appropriate for passes near the center of the plasma torus. We assume that pickup of ions occurs in a small region around Europa, with the rate of pickup decreasing with radial distance from the moon from n_0 at 1.01 R_E to 0.06 n_0 in 10 concentric rings with widths of 0.01 R_E. The magnetic field strength is 400 nT, and the pickup ions have mass 32 amu. The corotation electric field accelerates the particles away from Jupiter, and the fresh ions gyrate around the magnetic field with gyroradii (85 km) determined by the pickup velocity, which is assumed to be 100 km s^{-1}. The total pickup density in this two-dimensional model is calculated in 0.25 × 0.25 R_E bins across the wake. The pickup density, plotted in Fig. 13, develops a clear asymmetry in the wake because of losses on the subjovian side. Near y = 0.6 R_E there is a strong gradient in density in y. Pass E15 near the center of the plasma torus (where induction does not introduce asymmetry) and relatively close to Europa's equator (0.56 $R_E < z_{EphiB} <$ 0.78 R_E across the geometric wake) shows systematic negative B_x, consistent with bend-back that ends abruptly at $y_{EphiB} \sim$ 0.5 R_E. It is possible that the change relates to the orbit, but if pickup ions dominate the plasma density in the wake region, the sharp rotation may be related to the density gradient we propose from the model of Fig. 13. The model also suggests that one should look for a subjovian/antijovian asymmetry of the chemistry of the surface imposed by the reimplanted ions (see chapter by Carlson et al.).

The issue of wake asymmetry has been discussed in connection with simulations of the Europa interaction. *Kabin et al.* (1999, their Fig. 10) modeled the interaction for the conditions of the E4 flyby. *Liu et al.* (2000) also modeled the E4 flyby. This pass was not ideal for tests of asymmetries because it occurred relatively far above the center of the torus and thus contains a strong inductive signature. *Kabin et al.* (1999) had to assume that the incident velocity deviated by 20° from azimuthal in order to obtain reasonable agreement with the observations; *Paterson et al.* (1999) reported that flows deviated from corotation during the E4 encounter with Europa. Prior to closest approach, the flow deviated inward toward Jupiter (as inferred by *Kabin et al.*, 1999) but was unsteady in direction. Thus, the expected orientation of the wake is somewhat uncertain in the simulations and it seems possible that features other than the flow direction may cause a rotation of the wake.

N. Schilling (personal communication, 2008) observes a gradient (in y) of the wake density at y = 0.5 R_E, in simulations of the E12 flyby for which Europa was located very near the center of the plasma torus. Because Schilling uses an MHD code that does not include finite gyroradius effects and the inductive field is small at the epoch of this flyby, it seems possible that the wake asymmetry in his simulation is a response to the Hall conductance (*Saur et al.*, 1999). However, Fig. 8 shows that although flow paths across the polar cap are modified by the Hall conductance, near the z = 0 plane the wake is little skewed so the source of the density asymmetry in the simulation is uncertain.

5.2. Clues to Pickup Ion Composition from Waves in the Wake

Interestingly, on passes E11 and E15 the magnetic field fluctuates considerably through most of the geometrical wake, whereas on pass E4, high-frequency fluctuations appear only in a limited region around y = 0. Data from the E17 pass were acquired at a time resolution (24 s) that is too low to resolve the high-frequency perturbations. Nu-

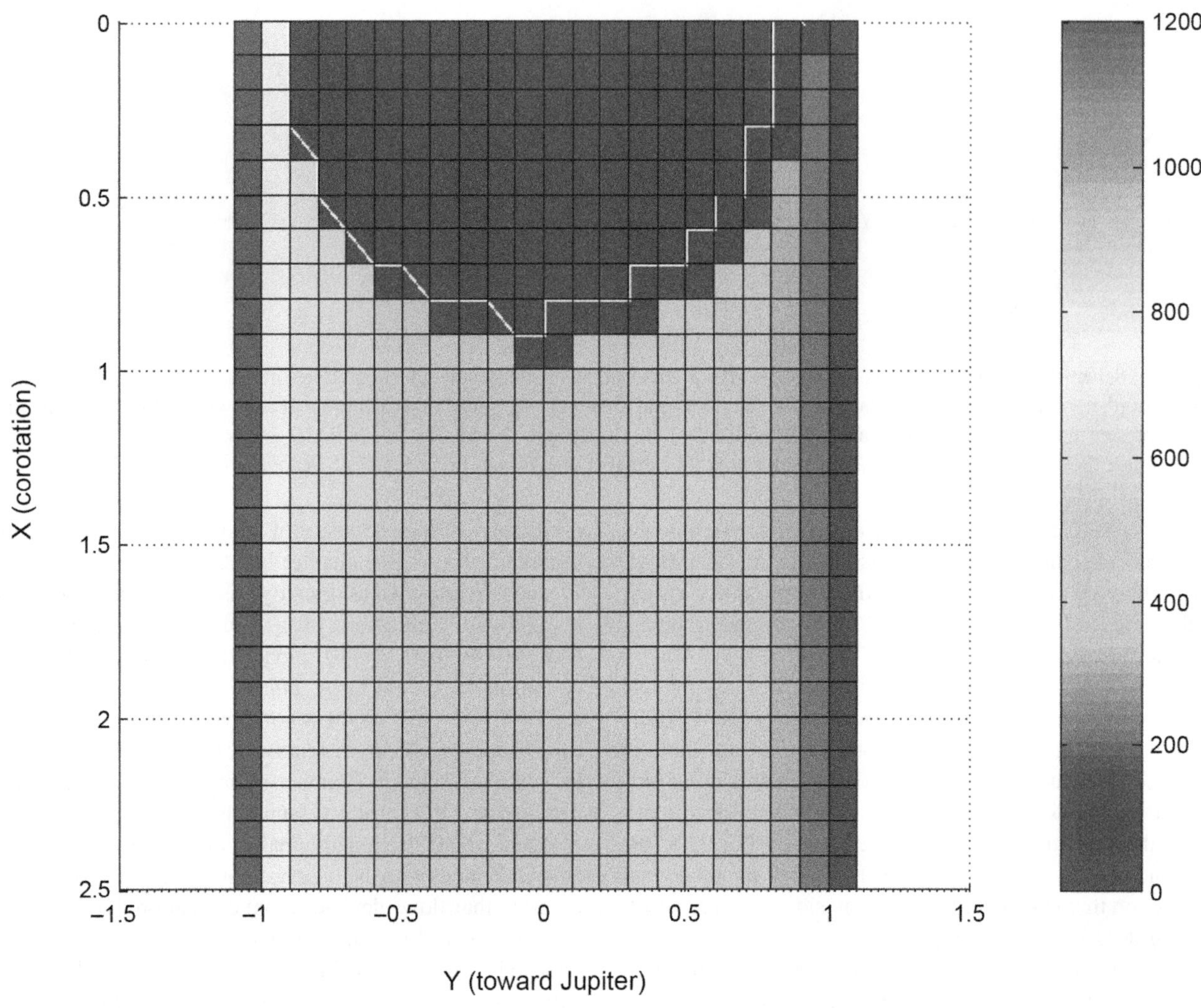

Fig. 13. See Plate 29. A simple model of the ion pickup around Europa, showing an abrupt decrease in density near y = 0.6. Density units are arbitrary.

merical modeling of the interaction of Europa and the jovian magnetosphere by *Schilling et al.* (2007, 2008), whose simulation assume E4 conditions, has shown that the enhanced density downstream of the moon is concentrated in a small region of the wake. It is probable that the newly picked-up ions are concentrated in this dense plasma region and that it corresponds to the interval near the center of the wake where the field magnitude dips and high-frequency fluctuations are found.

Newly picked-up ions form a ring distribution, i.e., particle velocities are distributed in a torus around the field direction in velocity space. The Galileo plasma analyzer (PLS) observed such distributions both in heavy ions (probably oxygen) and lower-mass ions (probably protons) on passes E4 and E6 (*Paterson et al.,* 1999). Similar velocity space distributions were identified near Io (*Frank and Paterson,* 2000, 2001). Such anisotropic distributions may be inherently unstable to the generation of ion cyclotron waves. The magnetic perturbations produced by such waves have been used to characterize the mass per unit charge of pickup ions near Europa (*Volwerk et al.,* 2001) and near Io (*Huddleston et al.,* 1997, 1998).

Ion cyclotron waves grow off the free energy in anisotropic distributions of positive ions and are typically lefthand polarized at frequencies below the ion gyrofrequency. Thus, wave analysis provides a tool for identifying the ions generating the waves. The magnetic field data are transformed to a mean field aligned (MFA) coordinate system, where the mean field is determined by a low-pass filter (for periods longer than 5 min). From the transverse components (B_v and B_ρ) the lefthanded and righthanded polarized components ($B_R = B_v + iB_\rho$, $B_L = B_v - iB_\rho$) are obtained and power spectra can be produced separately for the two polarizations. In Fig. 14 the dynamic spectra are shown for the lefthand and righthand polarized components of the wave power in the frequency band f < 0.5 Hz. The gyrofrequencies of heavy ions and molecular ions in the background field of ~400 nT fall in this range (0.375 for O^+ or S^{++}, 0.1875 for S^+, and 0.0938 for SO_2^+).

The ion source near Europa is an extended cloud of neutrals sputtered from the surface or the atmosphere. A number of elements have been identified on and around Europa. The Galileo Near-Infrared Mapping Spectrometer (NIMS) suggests that Mg is present on the surface (*McCord*

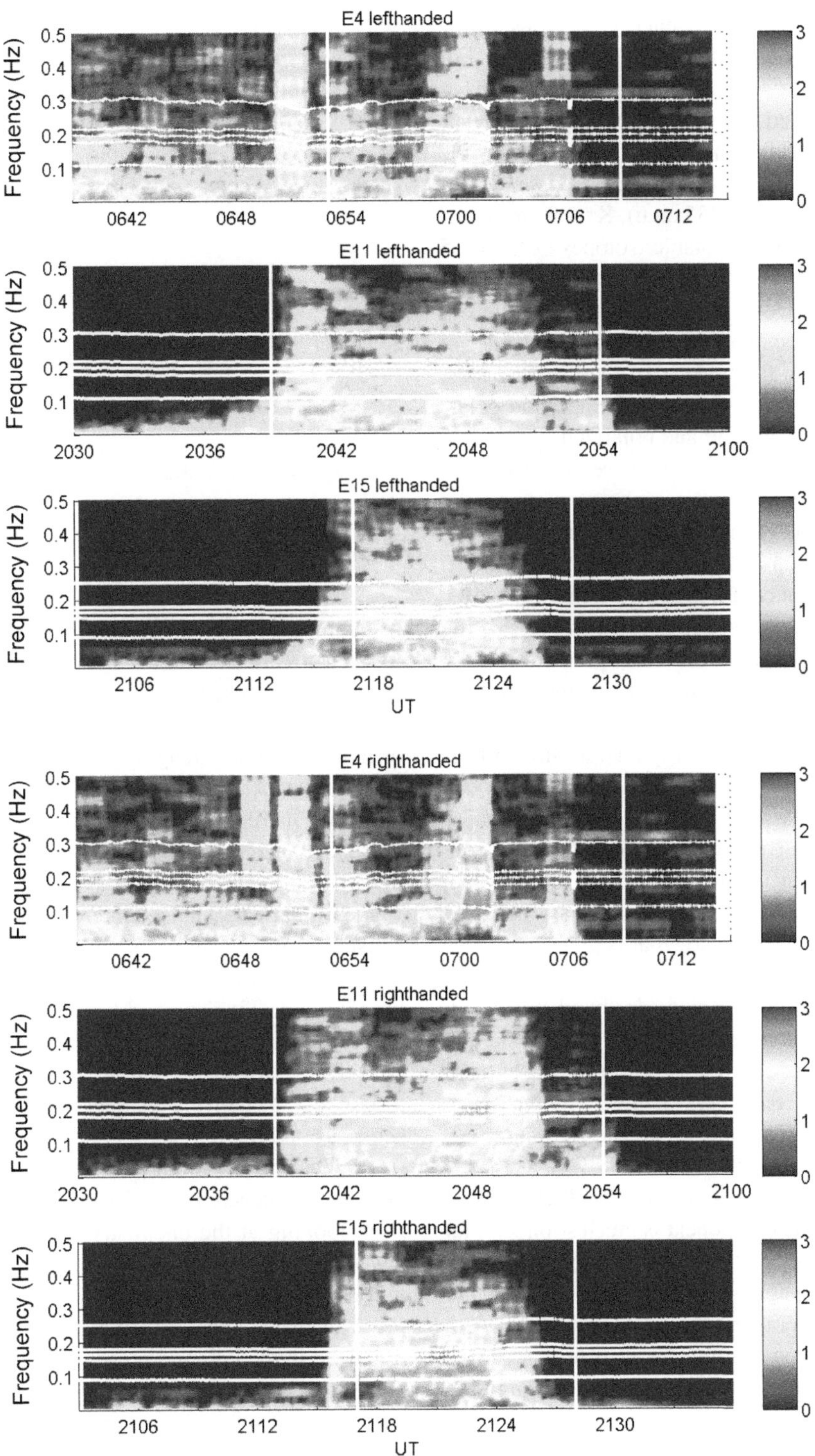

Fig. 14. See Plate 30. Dynamic spectra of the lefthand (top three panels) and righthand (lower three panels) polarized components of the magnetic field for E4, E11, and E15. The white solid traces show the cyclotron frequencies for Na^+ (A = 23), O_2^+ (A = 32), K^+ (A = 40), Cl^+ (A = 35), and SO_2^+ (A = 64). Vertical lines delimit the geometric wake.

et al., 1998; see chapter by Carlson et al.). *Brown and Hill* (1996) observed a neutral Na cloud and *Brown* (2001) reported that K was present in Europa's atmosphere. *Kargel* (1991) and *Kargel et al.* (2000) suggested that Cl would be present on the surface and *Küppers and Schneider* (2000) found spectroscopic evidence supporting this proposal. Ions of these elements are possible constituents of the local plasma. Further details on the properties of Europa's atmosphere are provided by the chapter by McGrath et al.

The power in the spectra of Fig. 14 is intermittently large at frequencies consistent with generation by ions of the heavy elements suggested above. We assume that the background plasma contains only low-energy particles of the same mass per unit charge as the newly ionized ions. For this condition, waves grow fastest at frequencies just below the ion cyclotron frequency of the pickup ions (*Huddleston et al.,* 1997). White lines have been added to the dynamic spectra to identify the cyclotron frequencies of Na^+(23 amu), O_2^+ (32 amu), Cl^+ (35 amu), K^+ (39 amu), and SO_2^+ (64 amu). There exist two stable isotopes of Cl at 35 and 37 amu; the trace in Fig. 14 relates to the 35-amu isotope. Although some wave power extends through the full range of the spectrum, the most intense emission bands are found near the cyclotron frequencies of the heaviest ions. Spectral power is significantly larger in the lefthand polarized component than in the righthand polarized component over much of the interval plotted. The wave power is most intense in the spectrograms in and near the geometric wake for E11 and E15 but for E4 the wave power is less clearly limited to the wake region and may have multiple sources.

The wave power is not continuous across the wake (*Volwerk et al.,* 2001). In the passes illustrated in Fig. 14, the power is most intense in regions with depressed |**B**| (see Fig. 11) where the plasma pressure, and possibly the density, are higher than in the surroundings. Volwerk et al. used spectral analysis to establish the propagation direction of the O_2^+ cyclotron waves and found that during the intervals of enhanced power the waves appeared to originate near the surface of the moon. Tracing the wave paths back from successive locations along Galileo's orbit (see their Fig. 7), they found that only the regions of intense emissions could be mapped back along ray paths to what is plausibly a significant pickup region.

Ion cyclotron waves are typically lefthand polarized. Figure 14 enables us to compare the power in lefthand and righthand polarized waves. Although power in the lefthand polarized waves is larger in most of the frequency-time range shown, there are intervals where the righthand polarized wave is dominant. This happens in E11 at the Cl cyclotron frequency between 20:65 and 20:70 UT, where the power in the righthand component is much stronger than that in the lefthand component. Righthand polarization can be caused by pickup of negatively charged chlorine Cl^- but it is also possible to generate righthand polarized waves in a multicomponent plasma passing through the crossover frequency. We examine these two possibilities starting with the latter.

5.2.1. The crossover frequency. In a multicomponent plasma one defines the crossover frequency ω_X (see, e.g., *Melrose,* 1986, chapter 12.3; *Petkaki and Dougherty,* 2001) from the relation

$$\sum_{\alpha} \frac{N_\alpha}{1-(\omega_X/\Omega_\alpha)^2} = 1 \tag{5}$$

where N_α is the fractional ion density of species α and Ω_α is the gyrofrequency given by

$$N_\alpha = \frac{n_\alpha q_\alpha}{en_e}, \quad \Omega_\alpha = \frac{q_\alpha B}{m_\alpha} \tag{6}$$

with $n_\alpha(n_e)$ the density of α-ions (electrons), $q_\alpha(e)$ the charge on an α-ion (electron), m_α the mass of an α-ion, and B the magnitude of the magnetic field. The introduction of a trace element in a plasma creates a crossover frequency. In a two-component plasma in which the fractional density of the trace species (subscript o) is N, one finds

$$\frac{N}{1-z} + \frac{1-N}{1-\kappa z} = 1, \text{ where } z = \left(\frac{\omega_X}{\Omega_o}\right)^2, \quad \text{and } \kappa = \left(\frac{\Omega_o}{\Omega_1}\right)^2 \tag{7}$$

Solving for z, one finds

$$z = 1 + N\left(\frac{1-\kappa}{\kappa}\right) \tag{8}$$

In the multicomponent plasma, if N is small the crossover frequencies are close to the gyrofrequencies of the trace elements, in this case Cl, Na, and K. Waves traveling through a medium in which there are gradients in composition and/or magnetic field strength can cross ω_X and reverse polarization. *Petkaki and Dougherty* (2001) observed this phenomenon in Jupiter's middle magnetosphere.

5.2.2. Negatively charged chlorine. Chlorine has a particularly strong electron affinity: -3.5×10^5 J mol^{-1}, which suggests that Cl^- can be stable for a finite time in Europa's exosphere and wake region. For example, in the Earth's D layer at 70 km altitude, Cl is the most abundant negative ion (*Turco,* 1977; *Kopp and Fritzenwaller,* 1997). The chemistry in the Earth's D layer is driven by HCl, which could conceivably also be present in the europan atmosphere and ionosphere. Negatively charged ions gyrate in the righthand sense around the magnetic field. Chlorine has been identified in the torus by *Russell and Kivelson* (2001) in the magnetometer data near Io. *Fegley and Zolotov* (2000) suggested from looking at the chemistry of Io's volcanic gases that negatively charged Cl is a component of the torus. The pickup of both Cl^+ and Cl^- would account for the polarization reversals that are observed, with the most of the wave power polarized in the lefthand sense in regions dominated by Cl^+ but with a reversal of polarization where Cl^- dominates.

Either mechanism described above can account for polarization reversals near the gyrofrequency of Cl. Figure 15 shows details of the polarization reversals. On the E11 pass, the polarization becomes relatively linear through the polarization reversals (Fig. 15, left column), but on E15 (Fig. 15, right column) the waves remain circularly polarized. These features are consistent with transitions across the crossover frequency. However, polarization reversals are absent for emissions near the gyrofrequencies of other trace elements such as Na and K. Noting that Cl differs from the other trace ions in its strong affinity for negatively charged ions and that reversed polarization is observed only near the gyrofrequency of Cl ions, we think it most likely that

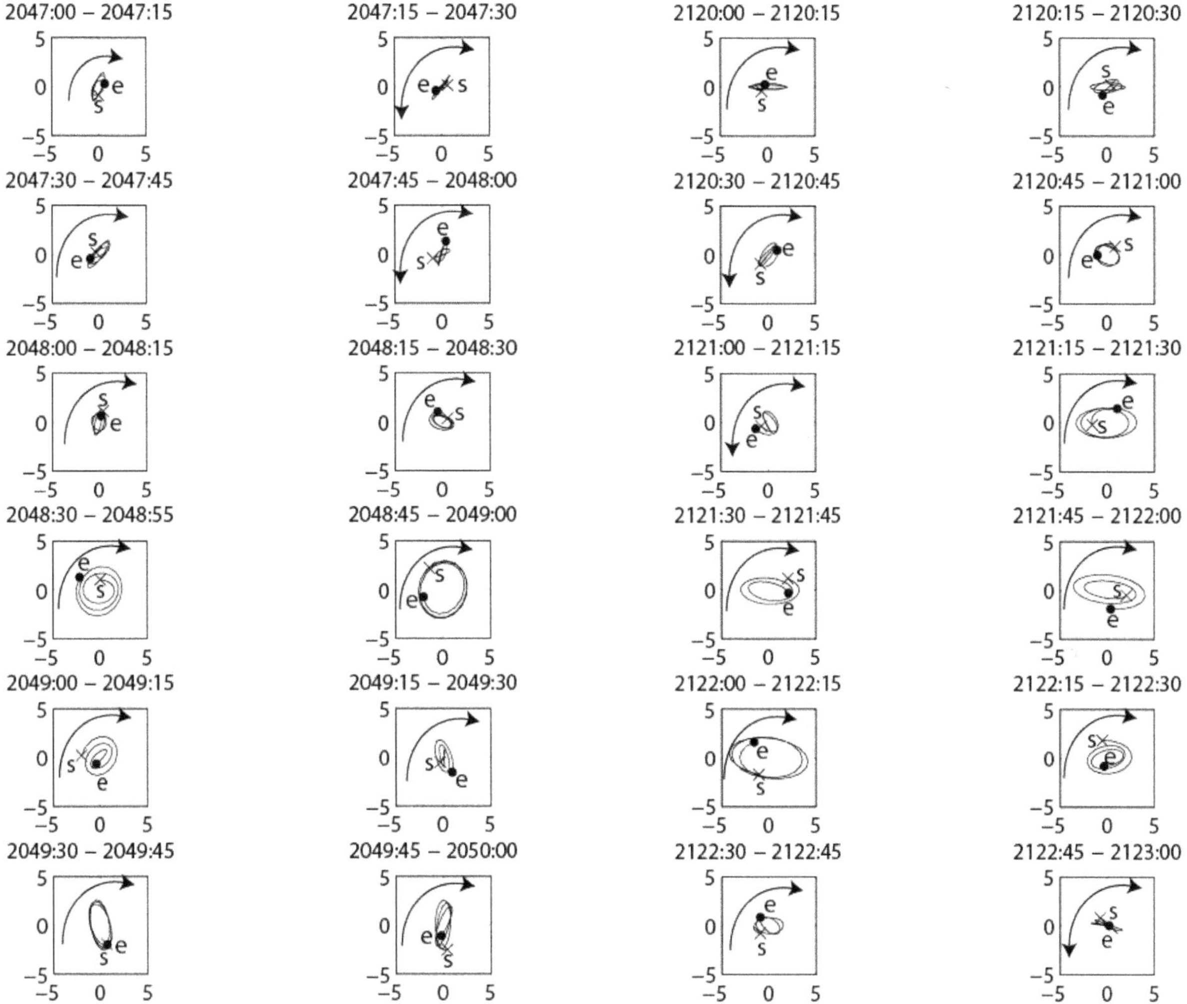

Fig. 15. Hodograms of the magnetic field for E11 (left) and E15 (right) for two short intervals where there are several reversals of the polarization direction of the waves. The data are band-passed filtered around the gyrofrequency of singly charged chlorine.

the polarization reversal indicates that there is a significant trace population of negative Cl ions.

5.3. Energetic Particles in the Wake

The losses of energetic particles in flux tubes flowing over Europa's polar cap were discussed in sections 3.3 and 5.1, where it was noted that the bounce times for electrons and very energetic ions are sufficiently short that one can expect that flux tubes will be partially emptied as they flow across the moon. The signature of the loss should be evident in the energetic particle detector (EPD) data (*Williams et al.,* 1992). In Fig. 16, the data for the following EPD channels are shown: F2 electron channel at 304–527 keV; F3 electron channel at 527–844 keV; A5 total ion channel (H, He, O, S, etc.) for protons at 515–825 keV (higher for other species); and O2 a "pure" oxygen channel (with some S contamination) at 26–51 keV per nucleon. The data shown in Fig. 16 apply to different ranges of pitch angles, α: per-pendicular with $85° \leq \alpha \leq 95°$, and (anti)parallel with $(160° \leq \alpha \leq 180°)$ $0° \leq \alpha \leq 20°$.

The EPD flux drops when Galileo was near and within the geometrical wake (bounded by the vertical lines in Fig. 16). However, losses differ from one channel to another. It would be expected that the most energetic particles would be absorbed most effectively by the moon. Electrons should disappear first; a weaker signature would be expected for ions. For losses depending on the gyroradius of the particles; the signature in the ion flux might be larger than in the electron flux.

The analysis of wake signatures is complicated by changing values of z_{EphiB} and downtail distance as the spacecraft crosses the wake. Decreases of energetic particle fluxes tend to appear abruptly when the wake entry is relatively close to Europa, but gradients become less steep with increasing distance down the wake. Specifically, the plots reveal some of the following behavior.

E4: The flux of perpendicular electrons decreases abruptly near closest approach, before Galileo enters the wake. The decrease occurs in a region where the total electron density (lowest panel in Fig. 16) increases. After the spacecraft enters the wake, it

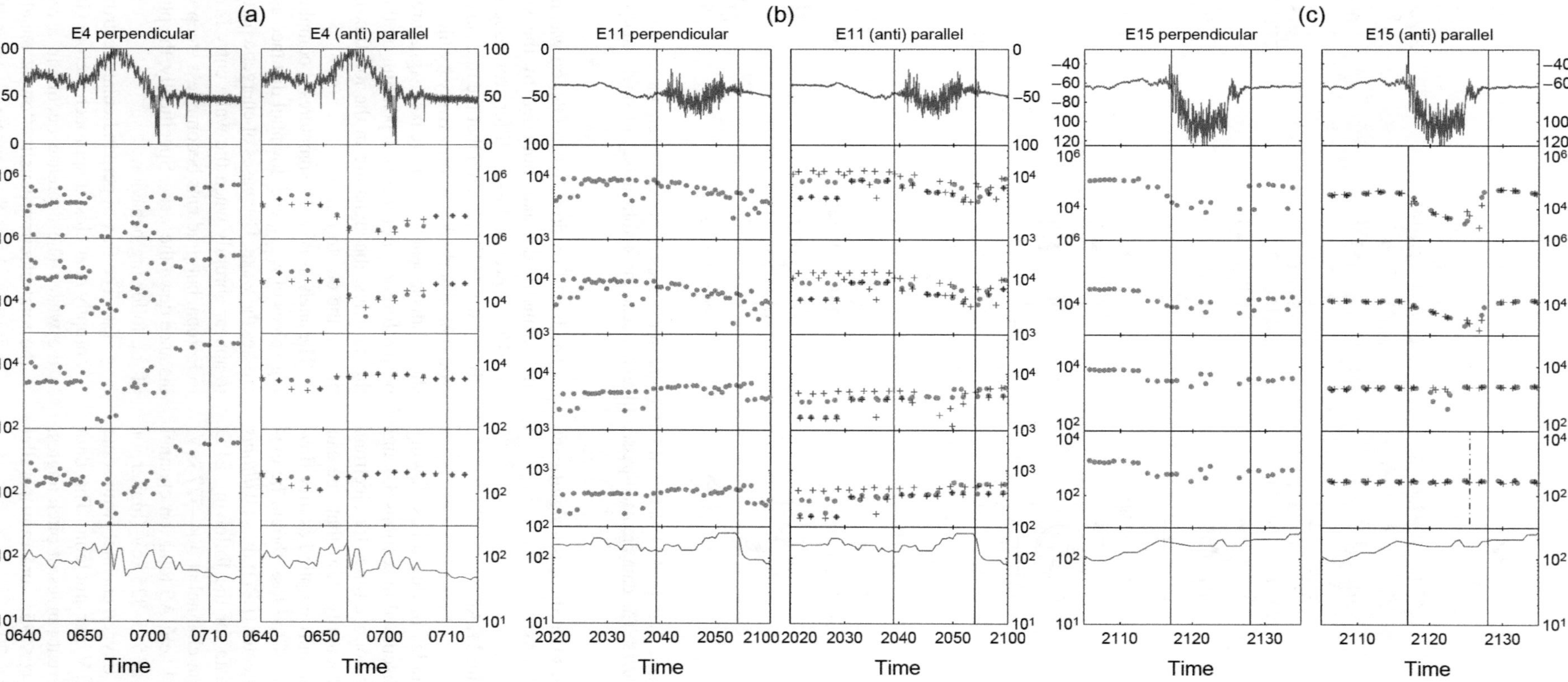

Fig. 16. See Plate 31. B_x, EPD data, and PWS (inferred) electron density for E4, E11, and E15. The energetic particle signatures are plotted separately for perpendicular (85° ≤ α ≤ 95°, left panels) and (anti)parallel (0° ≤ α ≤ 20°, 160° ≤ α ≤ 180°, right panels) pitch angles for the F2, F3, A5, and O2 channels of the EPD. In the (anti)parallel panels the (anti)parallel fluxes are indicated with (red dots) blue plusses. The bottom panel shows the electron density inferred from the PWS instrument. Vertical lines delimit the geometric wake as defined in the text.

moves significantly down the wake (toward large x, see auxiliary data in Fig. 11); and the electron flux slowly increases to preencounter levels. The (anti)parallel electron count rates change less abruptly, but the flux of these electrons also recovers slowly. Perpendicular ion losses are evident on the Jupiter-facing side of Europa prior to Galileo's entry into the wake. In the same region, losses are minimal in the (anti)parallel fluxes, consistent with the loss being related to finite ion gyroradius effects discussed in section 5.1.

E11: The flux signatures are relatively featureless, consistent with the relatively modest variations of the magnetic field in the wake crossing. The flyby took place further downstream of Europa and at a higher z_{EphiB} than the E4 and E15 flybys. In the perpendicular electron data there is a slow drop of counts, starting at the geometrical wake, and continuing over the whole wake, with a minimum outside the geometrical wake.

E15: The perpendicular electron counts decrease slowly but recover quickly on the Jupiter-facing side of the wake as do the (anti)parallel electrons. The ion count rates do not change markedly. Only the antiparallel counts in channel A5 show a slight drop in counts near the center of the wake.

It may be significant that the rapid decrease of flux in several channels on the E4 pass and the rapid increase of flux outbound on the E15 pass, at different downstream distances, both occur on the side of the wake closest to Jupiter ($y > 0$). The effect has not been interpreted in the published literature. However, one can consider processes known to act on energetic particles, asking which might increase fluxes on the Jupiter-facing side of the wake region. Pitch-angle scattering (*Paranicas et al.,* 2000) does not lead to inward displacement of flux tubes. Gradient-curvature drift of energetic particles in the perturbed magnetic geometry surrounding Europa could produce asymmetry favoring inward displacement of the energetic electron wake [see for example, the *Thorne et al.* (1999) analysis of drift paths of highly relativistic electrons for a magnetized Io], but the signature at Europa appears at relatively low energies for which magnetic drift perturbations are small. The interchange process, observed in the Io plasma torus (*Bolton et al.,* 1997; *Kivelson et al.,* 1997; *Thorne et al.,* 1997) leads to inward displacement of flux tubes differing from those in the surroundings by having energy density dominated by energetic particles and number density correspondingly reduced. Although further analysis is called for, it is possible that interchange also can account for the inward displacement of the energetic electron wake.

Another feature of the data in Fig. 16 calls for interpretation. Fluxes in the A5 channel on the E15 pass differ for upward- and downward-moving ions. The antiparallel ion flux (red dots) dips in the center of the wake whereas the parallel ion flux (blue plusses) does not change. With $B_z < 0$, the antiparallel particles are moving toward positive z. With Galileo at $z_{EphiO} \approx 1\ R_E$, the dip in antiparallel ion flux can be understood as the effect of passage over an absorbing body. Above the interaction region, ions coming from positive z would not show an absorption effect, whereas ions coming from negative z would have been absorbed as the flux tubes crossed the moon and would remain depleted in the plasma for some distance downstream of the moon.

5.4. Anomalous Flux Tubes

In the previous section, we alluded to ion pickup in the vicinity of Io. Pickup ions can significantly increase the local mass density and provide the free energy that drives flux tube interchange (*Kivelson et al.,* 1997; *Thorne et al,* 1997; *Bolton et al.,* 1997). The interchange process occurs spontaneously in a rapidly rotating magnetosphere where flux tubes with large integral mass density (flux tube content) move outward, changing places with flux tubes that have a lower flux tube content. Signatures similar to those identified at Io have been found downstream of Europa at the antijovian boundary of the wake (*Volwerk et al.,* 2001).

Figure 17 shows a short interval of magnetic field data acquired close to the exit through the antijovian side of the geometrical wake ($y = -1\ R_E$) on the E4 pass. At about 07:06 UT there is a strong, short-duration (5 s) decrease of the magnetic field strength with simultaneous signatures in B_x and B_y. The perturbation is not a localized rotation of the magnetic field. The sharply bounded field decrease is primarily a change of field magnitude. Assuming perpendicular pressure balance, the localized decrease of magnetic pressure can be interpreted as evidence that a narrow flux tube contains plasma of higher thermal pressure than that in surrounding flux tubes. This is different from the situation at Io where the short duration anomalies were abrupt field magnitude increases that were interpreted as an inward-moving depleted flux tubes. Here the signature can be interpreted as that of an outward-moving, newly loaded flux tube with thermal pressure higher than that of the surroundings. The decrease of the field magnitude at 07:06 UT is close to that measured in the center of the E4 wake in Fig. 11. Thus, it is plausible that the source of the dense flux tube is the high-density region at the wake center where *Paterson et al.* (1999) report a peak density of 70 cm^{-3}, the largest observed during the flyby. The density increase was followed by an increase of temperature (10^6 K in the background plasma). It should be noted that the magnetic field depressions last less than the 1-min resolution of the plasma density and temperature measurements. Pressure balance requires that $B\Delta B = \mu_0 \Delta p$. With the ambient field $B \cong 420$ nT and the field perturbation $\Delta B \cong 30$ nT, the change in magnetic pressure is 10 nPa. Taking values just prior to the density spike in the wake center ($n = 20\ cm^{-3}$, and $T = 2 \times 10^6 K$) from *Paterson et al.* (1999), we calculate $p = nkT = 0.6$ nPa. This implies that there must be a significant contribution of suprathermal particles that provide the additional pressure in the center of the wake that excludes a portion of the ambient field.

Let us suppose that part of the pressure within the wake arises from a localized increase in the pickup rate with re-

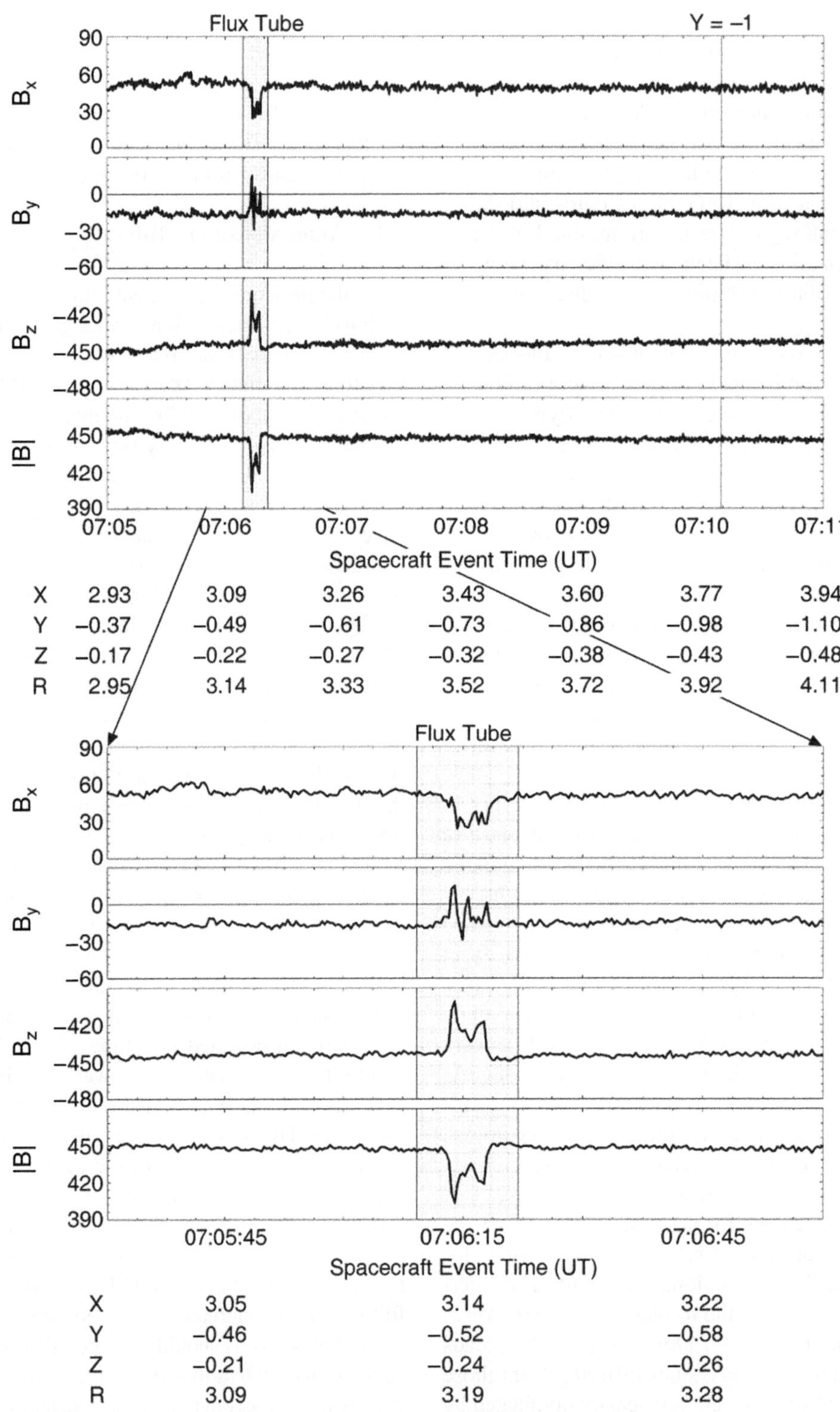

Fig. 17. E4 magnetometer data December 19, 1996: *Top:* From 0705 to 0711 UT. The shaded areas show possible flux tube interchange. *Bottom:* One minute of data (0705:50 to 0706:50) expanded.

spect to the background rate in the bulk plasma. [Even in the background plasma, pickup ions contribute to the reported temperature (*Paterson et al.,* 1999).] Assume that ions are picked up in a background flow at some fraction of local corotation speed, say u(km s^{-1}) = 100 γ, with $\gamma \leq 1$. For $\gamma \geq 1/3$, the thermal energy of the pickup ions would exceed the background temperature. The change of perpendicular pressure from the addition of pickup ions would be $\Delta p_\perp = \Delta n(m_i u^2/2) = \Delta n(m_i u_{co\text{-}rot}{}^2/2\gamma^2) = B\Delta B/\mu_o = 10$ nPa. Solving for the density enhancement, one finds $\Delta n(\mathrm{cm}^{-3}) = 1180/m_i(\mathrm{AMU})\gamma^2$ where m_i(AMU) is the mass of the pickup ion in AMU. For nominal mass 20 ions, $\Delta n \approx 60/$

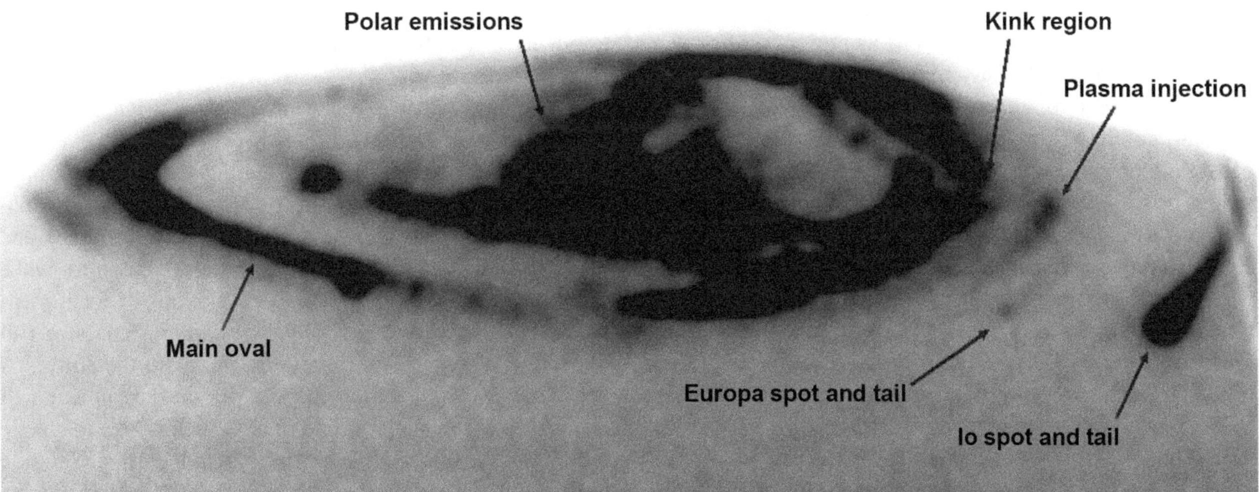

Fig. 18. Three images of the northern polar aurora. Hubble Space Telescope images have been summed and enhanced in contrast in order to display the faint emissions from the tail leading the brighter emission at the footprint of Europa. From *Grodent et al.* (2006).

γ^2. Thus, the change of field magnitude is consistent with the high ion density in the wake if relatively heavy ions are picked up in the wake in a nearly corotational flow. Paterson et al. observed flow only slightly below corotation across the entire wake region.

From this analysis, one is led to suppose that the field depression observed at 07:06 UT, similar in magnitude to that observed in the center of the wake, is produced by a flux tube of small cross section that has been transported outward from the plasma wake. We do not understand why interchange at Io appears to require small depleted flux tubes moving inward in a background of filled flux tubes assumed to be moving slowly outward, whereas the signature at Europa appears closer to the reverse.

6. EUROPA'S SIGNATURES IN THE JOVIAN AURORA

Elsewhere in this chapter, we have discussed how local interactions generate magnetic perturbations, including Alfvénic disturbances that carry field-aligned currents. Such disturbances generated near Europa and other Galilean moons propagate along magnetic field lines from the equator to Jupiter's ionosphere where auroral emissions are observed near the magnetic footprints of Io, Europa, and Ganymede (*Clarke et al.,* 1998, 2002). The auroral emissions associated with Io are most intense near its magnetic footprint and the bright spot tails off into a lengthy tail of reduced intensity along the magnetic mapping of Io's orbit (ahead of Io in the sense of planetary rotation). More recently it has been discovered (*Grodent et al.,* 2006) that Europa's footprint (a localized bright spot near the location in Jupiter's ionosphere magnetically conjugate to Europa) is accompanied by a shorter tail of faint auroral emissions along the magnetic locus of Europa's orbit (Fig. 18). The scale of Europa's localized ionospheric footprint maps to an interaction region on the order of 15 Europa diameters near the equator, requiring that currents between equator and ionosphere flow not only in the immediate vicinity of Europa but also develop throughout the pickup cloud that surrounds it. The auroral tail extends over a distance that maps to more than 70 Europa diameters, or more than 3 R_J.

Alfvén wing theory (*Neubauer,* 1980; *Southwood et al.,* 1980; *Goertz,* 1980) provides the basis for understanding how the flow perturbations arising in near-equatorial regions through the interaction with a conducting moon (or a pickup cloud) drive currents that couple to Jupiter's ionosphere. Field-aligned currents arise not only in the vicinity of the moon, but can also develop in the moon's wake where the flow is slowed below corotation. It is the latter effect that accounts for the emissions seen to extend along the mapped footprint of Europa's orbit. Estimates of the length of the auroral tail for the case of Io are given by *Southwood and Dunlop* (1984), and parallel arguments can be applied to Europa's tail. The field-aligned currents that couple the equator with the ionosphere are carried principally by electrons, which are far more mobile than ions. Details of the physics controlling the currents that link the Galilean moons and the aurora have been considered mainly in the context of the Io interaction (e.g., *Delamere et al.,* 2003), but the considerations that apply to Io are relevant as well to Europa. One critical issue is the distribution of plasma along the flux tubes linking the two regions. Although in the terrestrial magnetosphere charged particle motion from the equator into ever stronger magnetic fields is affected principally by the magnetic mirror force, in the jovian system inertial forces must also be taken into account. In regions near the equator centrifugal forces acting on the rotating plasma impede plasma motion away from the magnetic equator. Similarly, near Jupiter's ionosphere gravitational forces act to concentrate plasma at low altitudes. The plasma density is likely to become very small in regions of a flux tube somewhere between the equator and the ionosphere. Nonetheless, mechanisms must exist to carry the electric currents that couple the two regions, even through regions where the plasma density is low. It is then plausible to expect a field-aligned electric field to develop in order to accelerate the small number of electrons present in the intermediate low-plasma-density region and thereby close the current loop. Electrons accelerated by the field-aligned electric field not only carry current, but also excite the neutrals of the upper atmosphere, which in turn emit the auroral radiation. Using spectroscopic analysis of the auroral emissions, *Dols et al.* (2000) established that the Io footprint signature requires precipitation of electrons with tens of keV energy. The Europa footprint also emits in the extreme ultraviolet (*Clarke et al.,* 2002) supporting the view that there must be field-aligned potential drops on the order of tens of kV on field lines linking Europa to Jupiter's ionosphere.

Recent analysis of auroral signatures at Io's footprint has identified two different excitation mechanisms that must be acting to produce the brightest spots observed in the aurora (*Bonfond et al.,* 2008). Some emissions are found at locations interpreted as being slightly ahead (in the sense of rotation) of the mapped location of Io's footprint. The displacement from the actual magnetic footprint occurs because the signal propagates from equator to ionosphere at a finite speed within a flowing plasma. However, Bonfond et al. also observed emissions at locations closer to the mapped footprint of Io that can link to regions coupled to Io only through signals propagating along the field at speeds well in excess of typical wave speeds. Electrons of greater than or equal to tens of keV energy travel at speeds that are factors of hundreds faster than the Alfvén speed and can carry signals between Jupiter's northern and southern ionosphere in tens of seconds. Thus, it seems that beams of energetic electrons, possibly similar to those observed on Galileo's first pass by Io (*Williams et al.,* 1999; *Frank et al.,* 1999), generate the additional puzzling bright spots. There is good reason to assume that the mechanism that creates the beams at Io is at work near Europa as well, forming electron beams that have not yet been observed.

7. FOR THE FUTURE

One cannot write about Europa in 2008 without thinking about possible future missions to this fascinating solar system body. Studies of Europa missions have been pursued vigorously on both sides of the Atlantic Ocean (*Solar System Exploration Survey,* 2003; *Bignami et al.,* 2005), and, assuming that technological issues are resolved and costs are controlled, it seems likely that the next decade will witness the fruition of the planning, the launch of a Europa Orbiter. The principal objectives of the mission will link to the concept "follow the water," the goal being to find further evidence that there is an ocean under the icy surface and to establish its depth of burial and other properties. In this effort, it will be critical to characterize the surrounding plasma. Signatures of inductive fields that provide the fingerprints of an ocean (see chapter by Khurana et al.) will have to be extracted from a fluctuating background in which magnetic signatures arise from many local sources external to Europa. The goal of understanding Europa's internal structure inevitably demands that we learn a great deal about its interaction with the magnetospheric plasma of Jupiter.

Fields and particles instruments will provide the data needed to improve our knowledge of the inductive field while also adding to our understanding of surface composition by providing fingerprints of pickup ions, seeking evidence of electron beams and clarifying their sources and identifying the aspects of the interaction that control wake structure and wave dissipation. Correlations of local measurements with Hubble-type images should enable us to understand the currents linking Europa with the aurora. A magnetometer supplemented by a plasma instrument will characterize flow patterns and their variation with magnetic latitude. An orbiter regularly probing high europan latitude will provide the data that we lack today on how plasma convects across the polar cap, how quickly flux tubes lose particles of different energies after they first encounter the body, and just where those particles are absorbed on the surface. This information should be of great value in establishing sputtering rates to characterize Europa as a plasma source; it should enable us to investigate the role of energetic particle bombardment in causing surface discoloration (*Khurana et al.,* 2007; see chapter by Carlson et al.), thus enabling us to untangle intrinsic and extrinsic modification. The data from repeated orbits will be acquired at a considerable range of jovian magnetic latitudes, providing a range of high and low external plasma density environments that will enable us to see how the response varies with such local parameters as Alfvén Mach number and sound speed. We will find out how ion pickup rates vary with jovian magnetic latitude. A sensitive, high-time-resolution magnetometer will be able to identify ion cyclotron waves associated with pickup ions of different mass per unit charge, giving evidence of surface composition and the presence and distribution of minor constituents on the surface. Prior to entry into a capture orbit, Europa's wake could be probed at different downstream distances in order to understand the processes that reaccelerate the plasma slowed by its interaction with Europa.

We all remember Arthur C. Clarke's words: "All these worlds are yours, except Europa. Attempt no landing there..." (*Clarke,* 1968), but he never asked us not to go into orbit around it. Let us hope that a return to Europa comes soon and gives us answers to our questions.

Acknowledgments. The authors would like to thank C. Paranicas for making the EPD data available, W. Kurth for making the PWS electron density available, and N. Schilling for useful discussions. Partial support for this work was provided by NASA under grant NNG06GG67G.

REFERENCES

Alexeev I. I. and Belenkaya E. S. (2005) Modeling of the Jovian magnetosphere. *Ann. Geophys., 23,* 809–826, DOI: 2005AnGeo..23..809A.

Bagenal F. and Sullivan J. E. (1981) Direct plasma measurements in the Io torus and inner magnetosphere of Jupiter. *J. Geophys. Res., 86,* 8447–8466.

Biggs E. K. (1964) Influence of the satellite Io on Jupiter's decametric emission. *Nature, 203,* 1008–1010.

Bignami G., Cargill P., Schutz B., and Turon C. (2005) *Cosmic Vision: Space Science for Europe 2015–2025,* ESA Publications Division, Noordwijk, The Netherlands.

Bolton S. J., Thorne R. M., Gurnett D. A., Kurth W. S., and Williams D. J. (1997) Enhanced whistler-mode emissions: Signatures of interchange motion in the Io torus. *Geophys. Res. Lett., 24,* 2123–2126.

Bonfond B., Grodent D., Gérard J.-C., Radioti A., Saur J., and Jacobsen S. (2008) UV Io footprint leading spot: A key feature for understanding the UV Io footprint multiplicity? *Geophys. Res. Lett., 35,* L05107, DOI: 10.1029/2007GL032418.

Brown M. E. (2001) Potassium in Europa's atmosphere. *Icarus, 151,* 190–195, DOI: 10.1006/icar.2001.6612.

Brown M. E. and Hill R. E. (1996) Discovery of an extended sodium atmosphere around Europa. *Nature, 380,* 229–231.

Clarke A. C. (1968) *2001: A Space Odyssey.* New American Library, New York, 221 pp.

Clarke J. T., Ballester G. E., Trauger J., Ajello J., Pryor W., Tobiska K., Connerney J. E. P., Gladstone G. R., Waite J. H. H., Jaffel L. B., and Gerard J.-C. (1998) Hubble Space Telescope imaging of Jupiter's UV aurora during the Galileo orbiter mission. *J. Geophys. Res., 103,* 20217–20236.

Clarke J. T., Ajello J., Ballester G., Jaffel L. B., Connerney J., Gérard J.-C., Gladstone G. R., Grodent D., Pryor W., Trauger J., and Waite J. H. (2002) Ultraviolet auroral emissions from the magnetic footprints of Io, Ganymede, and Europa on Jupiter. *Nature, 415,* 997–1000.

Delamere P. A., Bagenal F., Ergun R., and Su Y-Ju. (2003) Momentum transfer between the Io plasma wake and Jupiter's ionosphere. *J. Geophys. Res., 108,* 1241, DOI: 10.1029/2002JA009530.

Dols V., Gerard J.-C., Clarke J. T., Gustin J., and Grodent D. (2000) Diagnostics of the Jovian aurora deduced from ultraviolet spectroscopy: Model and GHRS observations. *Icarus, 147,* 251–266.

Eviatar A. and Paranicas C. (2005) The plasma plumes of Europa and Callisto. *Icarus, 178,* 360–366.

Fegley B. and Zolotov M. Yu. (2000) Chemistry of sodium, potassium, and chlorine in volcanic gases on Io. *Icarus, 148,* 193–210.

Frank L. A. and Paterson W. R. (1999) Intense electron beams observed at Io with the Galileo spacecraft. *J. Geophys. Res., 104,* 28657–28669.

Frank L. A. and Paterson W. R. (2000) Return to Io by the Galileo spacecraft: Plasma observations. *J. Geophys. Res., 105,* 25363–25378.

Frank L. A. and Paterson W. R. (2001) Passage through Io's ionospheric plasmas by the Galileo spacecraft. *J. Geophys. Res., 106,* 26209–26224.

Goertz C. K. (1980) Io's interaction with the plasma torus. *J. Geophys. Res., 85,* 2949–2956.

Goldreich P. (1965) An explanation of the frequent occurrence of commensurable motions in the solar system. *Mon. Not. R. Astron. Soc., 130,* 159–181.

Goldreich P. and Lynden-Bell D. (1969) Io: A jovian unipolar inductor. *Astrophys. J., 156,* 59–78.

Grodent D., Gérard J.-C., Gustin J., Mauk B. H., Connerney J. E. P., and Clarke J. T. (2006) Europa's FUV auroral tail on Jupiter. *Geophys. Res. Lett., 33,* L06201, DOI: 10.1029/2005GL025487.

Gurnett D. A., Kurth W. S., Shaw R. R., Roux A., Gendrin R., Kennel C. F., Scarf F. L., and Shawhan S. D. (1992) The Galileo plasma wave system. *Space Science Rev., 60,* 341–355.

Huddleston D. E., Strangeway R. J., Warnecke J., Russell C. T., Kivelson M. G., and Bagenal F. (1997) Ion cyclotron waves in the in the Io torus during the Galileo encounter: Warm plasma dispersion analysis. *Geophys. Res. Lett., 24,* 2143–2146.

Huddleston D. E., Strangeway R. J., Warnecke J., Russell C. T., and Kivelson M. G. (1998) Ion cyclotron waves in the Io torus: Wave dispersion, free energy analysis, and SO_2^+ source rate estimates. *J. Geophys. Res., 103,* 19887–19899.

Intriligator D. S. and Miller W. D. (1982) First evidence for a Europa plasma torus. *J. Geophys. Res., 87,* 8081–8090.

Kabin K., Combi M. R., Gombosi T. I., Nagy A. F., DeZeeuw D. L., and Powell K. G. (1999) On Europa's magnetospheric interaction: A MHD simulation of the E4 flyby. *J. Geophys. Res., 104,* 19983–19992.

Kargel J. S. (1991) Brine volcanism and the interior structures of asteroids and icy satellites. *Icarus, 94,* 368–390.

Kargel J. S., Kaye J. Z., Head J. W. III, Marion G. M., Sassen R., Crowley J. K., Ballesteros O. P., Grand S. A., and Hogenboom D. L. (2000) Europa's crust and ocean: Origin, composition, and the prospects of life. *Icarus, 148,* 226–265.

Khurana K. K. (1997) Euler potential models of Jupiter's magnetospheric field. *J. Geophys. Res., 102,* 11295–11306.

Khurana K. and Schwarzl H. (2005) Global structure of Jupiter's magnetospheric current sheet. *J. Geophys. Res., 110,* A07227, DOI: 2005JGRA..11007227K.

Khurana K. K., Kivelson M. G., Stevenson D. J., Schubert G., Russell C. T., Walker R. J., Joy S., and Polanskey C. (1998) Induced magnetic fields as evidence for subsurface oceans in Europa and Callisto. *Nature, 395,* 777–80.

Khurana K. K., Pappalardo R. T., Murphy N., and Denk T. (2007) The origin of Ganymede's polar caps. *Icarus, 191,* 193–202.

Kivelson M. (1995) Pulsations and magnetohydrodynamic waves. In *Introduction to Space Physics* (M. G. Kivelson and C. T. Russell, eds.), pp. 330–355. Cambridge Univ., New York.

Kivelson M. G. (2004) Moon-magnetosphere interactions: A tutorial. *Adv. Space Res., 33,* 2061–2077.

Kivelson M. G., Khurana K. K, Means J. D., Russell C. T., and Snare R. C. (1992) The Galileo magnetic field investigation. *Space Sci. Rev., 60,* 357–383.

Kivelson M. G., Khurana K. K., Russell C. T., and Walker R. J. (1997) Intermittent short-duration magnetic field anomalies in the Io torus: Evidence for plasma interchange? *Geophys. Res. Lett., 24,* 2127–2130.

Kivelson M. G., Khurana K. K., Stevenson D. L., Bennett L., Joy S., Russell C. T., Walker R. J., Zimmer C., and Polansky C. (1999) Europa and Callisto: Induced or intrinsic fields in a periodically varying plasma environment. *J. Geophys. Res., 104,* 4609–4625.

Kivelson M. G., Khurana K. K, Russell C. T., Volwerk M., Walker R. J., and Zimmer C. (2000) Galileo magnetometer measurements strengthen the case for a subsurface ocean at Europa. *Science, 289,* 1340–1343.

Kivelson M. G., Bagenal F., Kurth W. S., Neubauer F. M., Paranicas C., and Saur J. (2004) Magnetospheric interactions with satellites. In *Jupiter: The Planet, Satellites and Magnetosphere* (F. Bagenal et al., eds.), pp. 513–536. Cambridge Univ.,

New York.

Kliore A. J., Hinson D. P., Flasar F. M., Nagy A. F., and Cravens T. E. (1997) The ionosphere of Europa from Galileo radio occultations. *Science, 277,* 355–358.

Kopp E. and Fritzenwaller J. (1997) Chlorine and bromine ions in the D-region. *Adv. Space Res., 20,* 2111–2115.

Kupo I., Mekler Y., and Eviatar A. (1976) Detection of ionized sulphur in the jovian magnetosphere. *Astrophys. J. Lett., 205,* L51–L54.

Küppers M and Schneider N. M. (2000) Discovery of chlorine in the Io torus. *Geophys. Res. Lett., 27,* 513–516.

Kurth W. S., Gurnett D. A., Persoon A. M., Roux A., Bolton S. J., and Alexander C. J. (2001) The plasma wave environment of Europa. *Planet. Space Sci., 49,* 345–363.

Lewis J. S. (1971) Satellites of the outer planets: Their physical and chemical nature. *Icarus, 15,* 174–185.

Liu Y., Nagy A. F., Kabin K., Combi M. R., DeZeeuw D. L., Gombosi T. I., and Powell K. G. (2000) Two-species 3D, MHD simulations of Europa's interaction with Jupiter's magnetosphere. *Geophys. Res. Lett., 27,* 1791–1794.

Luna H., McGrath C., Shah M. B., Johnson R. E., Liu M., Latimer C. J., and Montenegro E. C. (2005) Dissociative charge exchange and ionization of O_2 by fast H^+ and O^+ ions: Energetic ion interactions in Europa's oxygen atmosphere and neutral torus. *Astrophys. J., 628,* 1086–1096.

McCord T. B., Hansen G. B., Fanale F. P., Carlson R. W., Matson D. L., et al. (1998) Salts on the surface detected by Galileo's near infrared mapping spectrometer. *Science, 280,* 1242–1245.

Mekler Y. and Eviatar A. (1977) Jovian sulfur nebula. *J. Geophys. Res., 82,* 2809–2814.

Melrose D. B. (1986) *Instabilities in Space and Laboratory Plasmas.* Cambridge Univ., Cambridge.

Neubauer F. M. (1980) Nonlinear standing Alfvén wave current system at Io: Theory. *J. Geophys. Res., 85,* 1171–1178.

Neubauer F. M. (1998) The sub-Alfvénic interaction of the Galilean satellites with the jovian magnetosphere. *J. Geophys. Res., 103,* 19834–19866.

Neubauer F. M. (1999) Alfvén wings and electromagnetic induction in the interiors: Europa and Callisto. *J. Geophys. Res., 104,* 28671–28684.

Paranicas C., McEntire R. W., Cheng A. F., Lagg A., and Williams D. J. (2000) Energetic charged particles near Europa. *J. Geophys. Res., 105,* 16005–16015.

Paranicas C., Carlson R. W., and Johnson R. E. (2001) Electron bombardment of Europa. *Geophys. Res. Lett., 28,* 673–676.

Paranicas C., Mauk B. H., Ratliff J. M., Cohen C., and Johnson R. E. (2002) The ion environment near Europa and its role in surface energetics. *Geophys. Res. Lett., 29,* DOI: 10.1029/2001GL014127.

Paterson W. R., Frank L. A., and Ackerson K. L. (1999) Galileo plasma observations at Europa: Ion energy spectra and moments. *J. Geophys. Res., 104,* 22779–22791.

Peale S. J., Cassen P., and Reynolds R. T. (1979) Melting of Io by tidal dissipation. *Science, 203,* 892–894.

Petkaki P. and Dougherty M. K. (2001) Waves close to the crossover frequency in the jovian middle magnetosphere. *Geophys. Res. Lett., 28,* 211–214.

Russell C. T. (1999) Interaction of the Galilean moons with their plasma environments. *Planet. Space Sci., 53,* 473–485, DOI: 10.1016/j.pss.2004.05.003.

Russell C. T. and Kivelson M. G. (2001) Evidence for sulfur dioxide, sulfur monoxide, and hydrogen sulfide in the Io exosphere. *J. Geophys. Res., 106,* 33267–33272.

Russell C. T., Huddleston D. E., Khurana K. K., and Kivelson M. G. (1999) The fluctuating magnetic field in the middle jovian magnetosphere: Initial Galileo observations. *Planet. Space Sci., 47,* 133–142.

Saur J., Strobel D. F., and Neubauer F. M. (1998) Interaction of the jovian magnetosphere with Europa: Constraints on the neutral atmosphere. *J. Geophys. Res., 103,* 19947–19962.

Saur J., Neubauer F. M., Strobel D. F., and Summers M. E. (1999) Three-dimensional plasma simulation of Io's interaction with the Io plasma torus: Asymmetric plasma flow. *J. Geophys. Res., 104,* 25105–25126.

Shemansky D. E. and Smith G. R. (1981) The Voyager 1 EUV spectrum of the Io plasma torus. *J. Geophys. Res., 86,* 9179–9192.

Schilling N., Neubauer F. M., and Saur J. (2007) Time-varying interaction of Europa with the jovian magnetosphere: Constraints on the conductivity of Europa's subsurface ocean. *Icarus, 192,* 41–55, DOI: 10.1016/j.icarus.2007.06.024.

Schilling N., Neubauer F. M., and Saur J. (2008) Influence of the internally induced magnetic field on the plasma interaction of Europa. *J. Geophys. Res., 113,* A03203, DOI: 10.1029/2007JA012842.

Solar System Exploration Survey, Space Studies Board (2003) *New Frontiers in the Solar System: An Integrated Exploration Strategy.* National Academy of Sciences, Washington, DC.

Southwood D. J. and Dunlop M. W. (1984) Mass pickkup in sub-Alfvénic plasma flow: A case study for Io. *Planet. Space Sci., 32,* 1079–1086, DOI: 10.1016/0032-0633(84)90133-8.

Southwood D. J., Kivelson M. G., Walker R. J., and Slavin J. A. (1980) Io and its plasma environment. *J. Geophys. Res., 85,* 5959–5968.

Thomas N., Bagenal F., Hill T. W., and Wilson J. K. (2004) The Io neutral clouds and plasma torus. In *Jupiter: The Planet, Satellites and Magnetosphere* (F. Bagenal et al., eds.), pp. 561–591. Cambridge Univ., New York.

Thorne R. M., Williams D. J., McEntire R. W., Armstrong T. P., Stone S., Bolton S., Gurnett D. A., and Kivelson M. G. (1997) Galileo evidence for rapid interchange transport in the Io torus. *Geophys. Res. Lett., 24,* 2131–2134.

Thorne R. T., Williams D. J., Zhang L. D., and Stone S. (1999) Energetic electron butterfly distributions near Io. *J. Geophys. Res., 104,* 14755–14766.

Turco R. P. (1997) On the formation and destruction of chlorine negative ions in the D region. *J. Geophys. Res., 82,* 3585–3592.

Volwerk M., Kivelson M. G., and Khurana K. K. (2001) Wave activity in Europa's wake: Implications for ion pickup. *J. Geophys. Res., 106,* 26033–26048.

Volwerk M., Khurana K., and Kivelson M. (2007) Europa's Alfvén wing shrinkage and displacement influenced by an induced magnetic field. *Ann. Geophys., 25,* 905–914.

Williams D. J., McEntire R. W., Jaskulek S., and Wilken B. (1992) The Galileo energetic particles detector. *Space Science Rev., 60,* 385–412.

Williams D. J., Thorne R. M., and Mauk B. (1999) Energetic electron beams and trapped electrons at Io. *J. Geophys. Res., 104,* 14739–14753.

Zimmer C., Khurana K. K., and Kivelson M. G. (2000) Subsurface oceans on Europa and Callisto: Constraints from Galileo magnetometer observations. *Icarus, 147,* 329–347.

Electromagnetic Induction from Europa's Ocean and the Deep Interior

Krishan K. Khurana and Margaret G. Kivelson
University of California, Los Angeles

Kevin P. Hand
NASA Jet Propulsion Laboratory/California Institute of Technology

Christopher T. Russell
University of California, Los Angeles

An overview of the current status of research on the electromagnetic induction sounding of Europa's ocean and deep interior is provided. After briefly reviewing the history of electromagnetic induction methods used for sounding the interiors of Earth and the Moon, we provide a basic theoretical foundation of electromagnetic wave theory for spherical bodies. Next, evidence of electromagnetic induction field in the magnetic field data from the Galileo spacecraft is presented. Results from several modeling studies and the uncertainties in the fitted parameters are presented. Sources of systematic and random noise in the observations and their effect on the induction signature are highlighted next. The implications of the derived ocean conductivities for the composition of the europan ocean are discussed. Finally, we examine future prospects for multiple-frequency sounding of Europa's interior from orbiting spacecraft and observatories on the surface of Europa.

1. INTRODUCTION AND SCOPE OF THE CHAPTER

Gravity measurements from Galileo Doppler data (*Anderson et al.,* 1998) show that the moment of inertia of Europa (=0.346 × $M_E R_E^2$, where M_E and R_E are the mass and radius of Europa) is substantially less than that expected of a uniform sphere (0.4 × $M_E R_E^2$), implying that Europa's interior is denser than its outer layers. Detailed modeling by Anderson et al. revealed that the most plausible models of Europa's interior have an H_2O layer thickness of 80–170 km overlying a rocky mantle and a metallic core. The physical state of the H_2O layer is uncertain but speculations about a liquid ocean have been made since the realization that tidal stressing of the interior is a major heat source (*Cassen et al.,* 1979, 1980) and the age of Europa's surface is only a few tens of million years (*Shoemaker and Wolfe,* 1982; *Zahnle et al.,* 1998). Models put forward to explain chaotic terrains formed by "mobile icebergs" (*Carr et al.,* 1998), extrusion of new material along lineaments (*Greeley et al.,* 1998) and lenticulae formed by buoyant diapirs (*Pappalardo et al.,* 1998) require liquid water or a low-viscosity layer at the base of these structures. After examining most of the geological evidence available to them, *Pappalardo et al.* (1999) concluded that while a global ocean remains attractive in explaining the geological observations, its current existence remains inconclusive.

Perhaps the strongest empirical evidence for the existence of a subsurface ocean in Europa at the present time comes from the electromagnetic induction measurements provided by the magnetometer (*Khurana et al.,* 1998; *Kivelson et al.,* 1999). The magnetic signal measured near Europa consists of several components. These are a nonvarying uniform field from Jupiter and its magnetosphere in Europa's frame (~500 nT), a cyclical component of the field of Jupiter and its magnetosphere (~250 nT at the synodic spin period of Jupiter and ~20 nT at the orbital period of Europa), Europa's induction response to the cyclical field because of its internal conductivity (~250 nT near Europa's surface), and the field from the moon/plasma interaction currents (typically 20–30 nT when Europa is outside Jupiter's current sheet and up to 200 nT when Europa is located in Jupiter's current sheet). Recent, more-advanced analyses of the induction signatures (*Hand and Chyba,* 2007; *Schilling et al.,* 2007) strongly confirm the existence of a global ocean but also point out that the physical properties of the ocean (location, thickness, conductivity) remain uncertain. In this chapter, our intention is to critically review the current state of our knowledge of Europa's induction signature and assess the effort that has gone into its modeling.

2. HISTORY OF ELECTROMAGNETIC SOUNDING OF EARTH AND THE MOON

2.1. History of Geophysical Electromagnetic Induction

Two seminal works by Arthur Schuster from the late nineteenth century laid the groundwork for electromagnetic sounding of the planetary interiors. Using Gauss' "general theory of geomagnetism," *Schuster* (1886) demonstrated that the daily magnetic variations observed in data from surface observatories on Earth could be separated into external and internal parts. In a subsequent paper (*Schuster,* 1889) using data from four widely separated magnetic observatories and the theory of electromagnetic induction in a sphere from *Lamb* (1883), *Schuster* (1889) deduced that the internal part of the observed daily magnetic variations arose from eddy currents induced in Earth's interior. The next major advance in sounding the interior of Earth was made by *Chapman* (1919), who used diurnal variation data from 21 ground observatories. Using Schuster's inversion technique he concluded that the conductivity of Earth is not uniform and must increase with depth. That Earth's ocean can also generate appreciable inductive fields was first realized around the same time by *Chapman and Whitehead* (1922), who showed that the ocean conductivity seriously impacts the modeling of interior from diurnal variation data. However, it was only after *Chapman and Price* (1930) used much longer wave periods obtainable from storm time disturbances (D_{st}) that it could be proved convincingly that Earth's conductivity continued to increase with depth and exceed 3.6×10^{-2} S/m beyond a depth of 250 km.

The theory of electromagnetic sounding of heterogeneous bodies was spearheaded by *Lahiri and Price* (1939), who derived expressions for spherically conducting objects whose conductivity decreased with radial distance as $\sigma = k\,r^{-m}$ where k and m are arbitrary constants and r is the radial distance. Using the same data as *Chapman and Price* (1930), *Lahiri and Price* (1939) showed that the conductivity jump in Earth's interior actually occurred at a depth of about 700 km and that the conductivity increase was extremely rapid with depth. Much of the modern terminology and techniques used in modern global sounding were introduced by *Banks* (1969), who showed that the dominant inducing field at periods shorter than 1 yr are generated by fluctuations in the strength of Earth's ring current and have the character of P_1^0 spherical harmonic. The response of Earth has now been characterized at periods ranging from a few hours to 11 yr (see Fig. 1).

Determining electromagnetic induction response of Earth from spacecraft data was first demonstrated by *Langel* (1975), who used data from OGO 2, 4, and 6 (see also *Didwall,* 1984). However, the technique really came of age when very accurate data from the low-Earth satellite MAGSAT became available (*Olsen,* 1999a,b). The spacecraft data have now been used to sound Earth's interior to a depth of 2700 km (*Constable and Constable,* 2004). The global, long-duration, high-precision datasets from Orsted, Orsted-2, and CHAMP spacecraft are now routinely analyzed to infer conductivity distributions in both horizontal and vertical dimensions of Earth. For some latest examples of analyses of CHAMP data we refer the reader to *Martinec and McCreadie* (2004) and *Velimsky et al.* (2006).

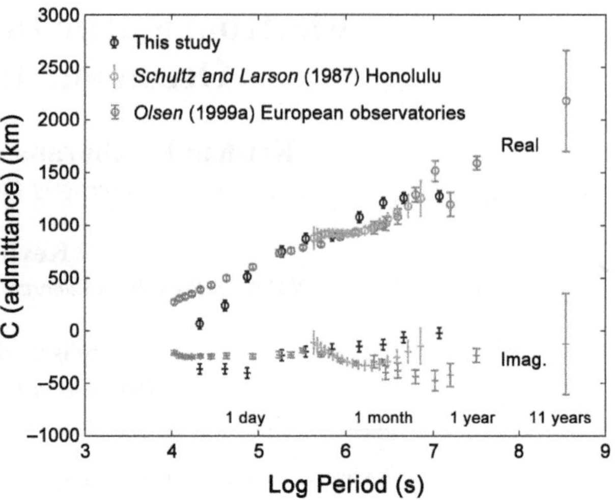

Fig. 1. See Plate 32. The response of Earth computed from data from several European surface observatories (*Olsen,* 1999a), a Hawaiian observatory (*Schultz and Larson,* 1987), and the MAGSAT spacecraft (*Constable and Constable,* 2004). The admittance function C is a rough measure of the depth to which the signal is able to penetrate in a spherical conductor and is given by C = a[l – (l + 1)R]/[l(l +1)(1 + R)], where a is the radius of the Earth, R is the complex response function (ratio of the internal to external signal), and l is the degree of the harmonic. Adapted from *Constable and Constable* (2004).

2.2. Induction Studies of the Moon

The Apollo Moon landings and the accompanying space program in the late 1960s and the early 1970s provided many opportunities for studying the interior of the Moon from electromagnetic induction. Several orbiting spacecraft [Explorer 35 (aposelene = 1.4 RM), Apollo 15 subsatellite (circular orbit at 100 km altitude), and Apollo 16 subsatellite (circular 100 km orbit)] made measurements of the magnetic field around the Moon. Three surface magnetometers (Apollos 12, 15, and 16) were often operated simultaneously with an orbiting spacecraft. An excellent summary of the results from the electromagnetic induction investigations is provided by *Sonett* (1982). *Schubert et al.* (1974) showed that the inductive response of the Moon to the solar wind transients is not detectable at an altitude of 100 km on the dayside because of the confinement of the induced field by the solar wind. Even though several intervals of >1-h duration from surface magnetometers were analyzed by various researchers in the early 1970s, the resulting models of lunar interior have remained nonunique (see *Sonett,* 1982; *Khan et al.,* 2006).

A technique introduced by *Dyal and Parkin* (1973) and used later by *Russell et al.* (1981) on Apollo-era data uses the response of the Moon to a steplike transient. The steplike transient in the primary field arises naturally when the Moon enters the geomagnetotail from the magnetosheath. Because the plasma flows in the magnetotail are submagnetosonic, confinement of the induced field is minimal and the data can be inverted directly for the lunar structure. Using data from 20 orbits of Lunar Prospector magnetic field data when the Moon had just entered Earth's magnetotail, *Hood et al.* (1999) estimated that the size of the lunar core is 340 ± 90 km, consistent with a value of 435 ± 15 km obtained by *Russell et al.* (1981) from the Apollo 15 and 16 data.

TABLE 1. Conductivities of common geophysical materials and their skin depths for a 10-h wave (reproduced from *Khurana et al.,* 2002).

Material	Conductivity (at 0°C) S/m	Conductivity (at 0°C) Reference	Skin depth for a 10-h wave (km)
Water (pure)	10^{-8}	*Holzapfel* (1969)	10^6
Ocean water	2.75	*Montgomery* (1963)	60
Ice	10^{-8}	*Hobbs* (1974)	10^6
Ionosphere (E layer)	2×10^{-4}	*Johnson* (1961)	7×10^3
Granite	10^{-12}–10^{-10}	*Olhoeft* (1989)	10^8–10^7
Basalt	10^{-12}–10^{-9}	*Olhoeft* (1989)	10^8–3×10^6
Magnetite	10^4	*Olhoeft* (1989)	1
Hematite	10^{-2}	*Olhoeft* (1989)	10^3
Graphite	7×10^4	*Olhoeft* (1989)	0.4
Cu	5.9×10^7	*Olhoeft* (1989)	0.01
Fe	1×10^7	*Olhoeft* (1989)	0.03

3. BASIC OVERVIEW OF THE INDUCTION TECHNIQUE

According to Faraday's law of induction, a time-varying magnetic field is accompanied by a curled electric field possibly also changing with time. When a conductor is presented with such a time-varying magnetic field, eddy currents flow on its surface that try to shield the interior of the body from the electric field. The eddy currents generate a secondary induced field, which reduces the primary field inside the conductor. The electromagnetic induction technique relies on the detection and characterization of the secondary field, which contains information on the location, size, shape, and conductivity of the inducing material.

The fundamental equations governing the underlying physics of induction are

Faraday's law

$$\nabla \times \mathbf{E} = -\frac{\partial \mathbf{B}}{\partial t} \tag{1}$$

Ampere's law

$$\nabla \times \mathbf{B} = \mu_0 \mathbf{J} + \mu_0 \varepsilon_0 \frac{\partial \mathbf{E}}{\partial t} \tag{2}$$

Generalized Ohm's law

$$\mathbf{J} = \sigma[\mathbf{E} + \mathbf{V} \times \mathbf{B}] \tag{3}$$

where the vectors **E**, **B**, **J**, and **V** denote electric field, magnetic field, electric current density, and flow velocity, and μ_0 is the magnetic permeability of free space, ε_0 is the permittivity of free space, and σ is the conductivity of the inducing material.

It is straightforward to show that these equations can be combined to yield the electrodynamic equation

$$\nabla^2 \mathbf{B} = \sigma\mu_0 \left[\frac{\partial \mathbf{B}}{\partial t} - \nabla \times (\mathbf{V} \times \mathbf{B}) \right] \tag{4}$$

where we have assumed that there are no spatial variations of conductivity in the primary conductor. In the absence of convection in the conductor, the electrodynamic equation reduces to the well-known diffusion equation

$$\nabla^2 \mathbf{B} = \sigma\mu_0 \left[\frac{\partial \mathbf{B}}{\partial t} \right] \tag{5}$$

3.1. Solutions of the Diffusion Equation in Half-Space Plane and the Concept of Skin Depth

It is instructive to examine the solution of equation (5) for a conducting half-space plane ($z > 0$) in the presence of an oscillating horizontal field ($B = B_0 e^{-iwt}$), which is given by

$$B = B_0 e^{-z/S} e^{-i(wt - z/S)} \tag{6}$$

where $S = (\omega\mu_0\sigma/2)^{-1/2}$ is the skin depth, a distance over which the primary signal decays by an e folding as it travels through the conductor. Equation (6) shows that the skin depth is small when the conductivity of the material is large and/or the frequency of the sounding signal is high. The equation also shows that the primary signal is phase delayed by a radian over a travel distance of one skin depth.

It can be shown that a wave with a period of 10 h (similar to that of Jupiter's spin period) has a skin depth of 30 km in a conductor that possesses a conductivity of 10 S/m (similar to that of a strongly briny solution). If the plane obstacle has a thickness larger than the skin depth of the material, the primary wave cannot significantly penetrate the obstacle and is effectively reflected back, creating the induced field that doubles the amplitude of the primary field outside the conductor. Table 1 lists skin depths for several common geophysical materials for a 10-h wave. Because of the low conductivities of pure water, ice, rocks, and an Earth-like ionosphere, the skin depths of these objects are much larger than the dimension of Europa. Therefore, these materials

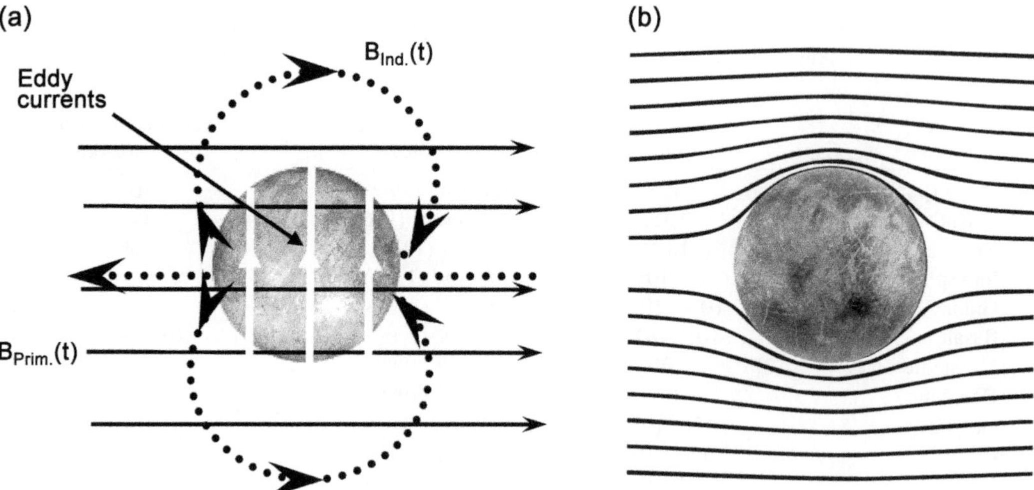

Fig. 2. **(a)** A time-varying primary field (black solid lines) generates eddy currents (white arrows) that flow on the surface of a conductor like Europa's ocean to prevent the primary field from penetrating the interior. The eddy currents generate an induced field (black dotted lines), which tends to reduce the primary field in the interior of the body. **(b)** The primary and induced fields combine such that the lines of force of the varying magnetic field avoid the conducting obstacle.

would be incapable of generating the induction response that was observed (and discussed later in section 4) by the magnetometers onboard the Galileo spacecraft. Highly conducting minerals such as magnetite or graphite and pure metals such as copper and iron have small skin depths but are unlikely to be localized in large amounts in the icy or liquid layers of Europa to produce the observed induction signal. However, a salty subsurface ocean with a conductivity similar to that of Earth's ocean and a thickness of tens of kilometers would be able to produce a significant induction response.

3.2. Diffusion Equation Solutions for Spherical Bodies

For a spherical conductor, an examination of the field at its boundary shows that when the primary field is uniform (i.e., of degree l = 1 in external spherical harmonics), the induced field outside the conductor would also be of degree 1 in the internal spherical harmonics, i.e., it would be dipolar in nature (see *Parkinson,* 1983). The secondary field would have the same frequency as the primary field but can be phase delayed. The primary and secondary fields sum together to form a total field, which avoids the spherical conductor (see Fig. 2). At Europa's location, the primary oscillating field is provided by Jupiter because its dipole axis is tilted by ~10° with respect to its rotational axis. In each jovian rotation, the magnetic equator of Jupiter moves over Europa twice, causing changes in the direction and strength of the field sampled by Europa. In a coordinate system called EϕΩ centered at Europa, in which the x axis points in the direction of plasma flow (Jupiter's azimuthal direction, ϕ), the y axis points toward Jupiter and the z axis points along the rotation axis of Jupiter (Ω), B_z remains relatively constant, whereas both B_x and B_y vary cyclically at the synodic rotation period of Jupiter (11.1 h). The amplitude of the oscillating field is ~200 nT at this frequency. In addition, because of the slight eccentricity (ε = 0.009) of Europa's orbit and local time variations in the jovian magnetospheric field, Europa also experiences variations in the B_z component at the orbital period of Europa (85.2 h). The amplitude of the primary signal at this frequency is estimated to be between 12 and 20 nT, depending on the field and plasma conditions in the magnetosphere.

Following *Zimmer et al.* (2000) and *Khurana et al.* (2002), a three-shell model of Europa (see Fig. 3) can be used to understand Europa's response to the sinusoidal uniform field of Jupiter and its magnetosphere. In this model, the outermost shell of Europa consists of solid ice, has an outer radius r_m equal to that of Europa, and possesses zero conductivity. The middle shell consisting of Europa's ocean is assumed to have an outer radius r_o and conductivity σ. The innermost shell consisting of silicates is again assumed to have negligible conductivity and a radius r_1. As discussed above, because the primary field is uniform (degree 1, external harmonics) and the assumed conductivity distribution has spherical symmetry, the induced field observed outside the conductor ($r > r_o$) would be dipolar (degree 1, internal harmonics) (*Parkinson,* 1983). Thus

$$\mathbf{B}_{sec} = \frac{\mu_0}{4\pi}[3(\mathbf{r} \cdot \mathbf{M})\mathbf{r} - r^2\mathbf{M}]/r^5 \tag{7}$$

The dipole moment of the induced field is opposite in direction to the primary field and is phase delayed

$$\mathbf{M} = -\frac{4\pi}{\mu_0}Ae^{i\phi}\mathbf{B}_{prim}r_m^3/2 \tag{8}$$

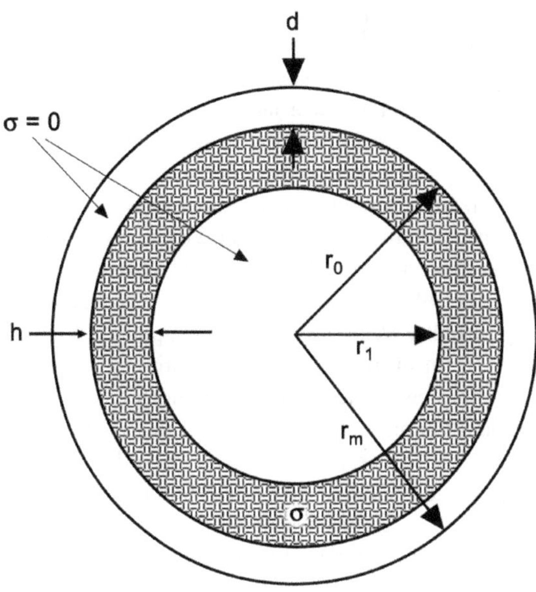

Fig. 3. Three-shell conductivity model of Europa.

Thus

$$\mathbf{B}_{sec} = -Ae^{-i(\omega t - \phi)}B_{prim}[3(\mathbf{r} \cdot \mathbf{e}_0)\mathbf{r} - r^2\mathbf{e}_0]r_m^3/(2r^5) \quad (9)$$

The parameters A (relative amplitude, also known as response function) and ϕ (phase lag) are given by the following complex equations (*Parkinson,* 1983)

$$Ae^{i\phi} = \frac{r_0^3}{r_m^3}\frac{RJ_{5/2}(r_0k) - J_{-5/2}(r_0k)}{RJ_{1/2}(r_0k) - J_{-1/2}(r_0k)} \quad (10)$$

$$R = \frac{r_1kJ_{-5/2}(r_1k)}{3J_{3/2}(r_1k) - r_1kJ_{1/2}(r_1k)} \quad (11)$$

where $k = (1 - i)\sqrt{\mu_0\sigma\omega/2}$ is the (complex) wave vector and J_m is the Bessel function of first kind and order m.

The amplitude response of Europa to the two main frequencies is shown in contour plots of Fig. 4. For plots of phase delay, we refer the reader to *Zimmer et al.* (2000). It was assumed that the wave amplitude of the inducing field at the synodic rotation period of Europa is 250 nT and the wave amplitude at the orbital period is 14 nT. As expected, for higher ocean conductivities and thicker ocean shells, the induction response is higher. For oceans whose height integrated conductance (conductivity × thickness) exceeds 4 × 10^4 S, the amplitude response is close to unity and the phase delay is insignificant (<10°) at the spin frequency of Jupiter. Because of noise in the data from plasma effects, the induction signal is not accurate enough to perform inversion of data using phase delay as information.

It can be seen in Fig. 4 that if the ocean thickness is less than 20 km or the ocean conductivity is less than 0.2 S/m, the response curves of the two frequencies are essentially parallel to each other. In this regime, only the product of the ocean conductivity and its thickness can be determined uniquely. However, when the ocean thickness exceeds 100 km and the conductivity is greater than 0.2 S/m, the response curves of two frequencies intersect each other. For σ > 0.2 S/m and h > 100 km) it is often possible to uniquely determine the ocean thickness and its conductivity if the response factors of 11.1-h and 85.2-h period waves are known. However, even in this parameter range, often only lower limits can be placed on the conductivity and thickness.

4. GALILEO OBSERVATIONS, THEIR INTERPRETATION, AND UNCERTAINTIES

4.1. Field and Plasma Conditions at Europa's Orbit

Europa is located at the outer edge of Io's plasma torus (radial distance 9.4 R_J) where the plasma sheet is thin (half thickness ~2 R_J) and the plasma is mostly derived from Io (*Bagenal and Sullivan,* 1981; *Belcher,* 1983). Because Jupiter's dipole axis is tilted by 9.6° from its rotational axis,

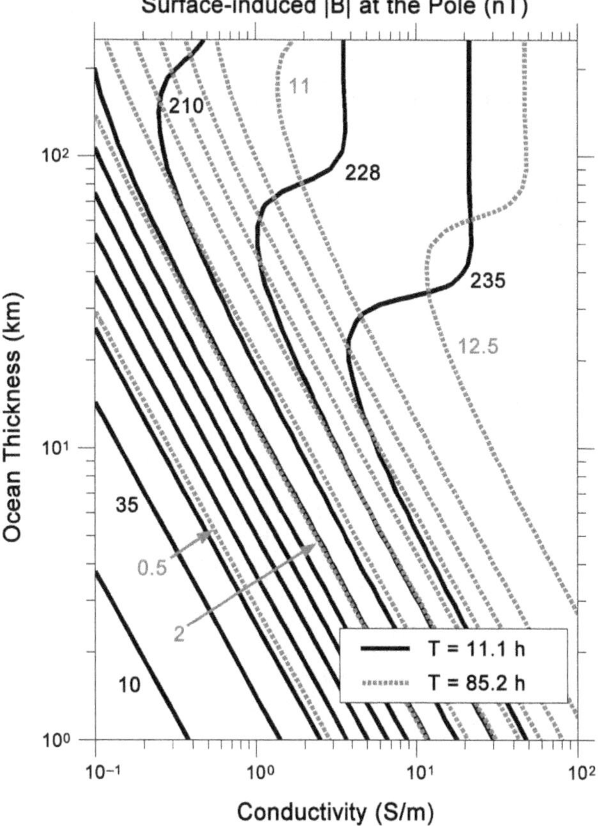

Fig. 4. The dipolar surface induction field created by the interaction of Europa with Jupiter's varying field at the two principal frequencies (T = 11.1 h and T = 85.2 h) for a range of conductivities and ocean shell thicknesses. Adapted from *Khurana et al.* (2002).

the plasma sheet located in the dipole equator near Europa moves up and down relative to Europa at the synodic rotation period of Jupiter (11.1 h). Therefore, the plasma density sampled by Europa changes by almost an order of magnitude at Europa over a jovian rotation with the maximum density observed near the center of the plasma sheet ($n_e \sim 50$ cm^{-3}). The magnetic field strength varies from 400 nT (at the center of the plasma sheet) to ~500 nT (at the edge of the plasma sheet). In the EϕΩ coordinate system introduced above, the dominant component of the background magnetic field is directed in the –z direction and remains fairly constant over the synodic rotation period of Jupiter (11.1 h). However, the x and y components vary with amplitudes of ~60 nT and ~200 nT, respectively, over this period. Figure 5 shows a hodogram of the oscillating field in the x–y plane calculated from the magnetospheric field model of *Khurana* (1997). Also marked on the figure are the instantaneous field conditions for five close flybys of Galileo during which evidence of electromagnetic induction from Europa was observed.

As mentioned earlier, Europa also experiences a nearly harmonic perturbation in the z component of the magnetic field at its orbital period (85.2 h) with an amplitude of 12–20 nT. The main cause of this perturbation field is the day/night asymmetry of Jupiter's magnetospheric field (see *Khurana,* 2001), but the slight eccentricity of Europa's orbit also contributes.

4.2. Interaction of Europa with the Jovian Plasma

As the corotational velocity of the jovian plasma is much larger than the Keplerian velocity of Europa, the plasma continually overtakes Europa in its orbit. Because of its

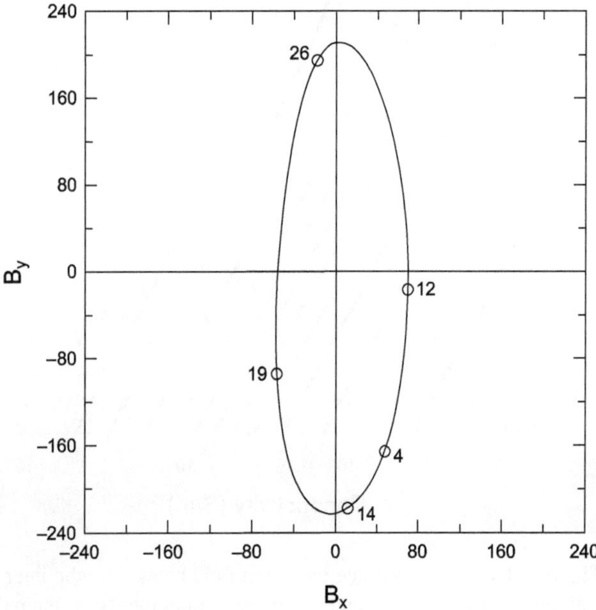

Fig. 5. The time-varying field experienced by Europa in a synodic rotation period of Jupiter.

conducting ionosphere, Europa presents itself as an obstacle to the flowing conducting plasma. The conducting exterior of the satellite extracts some momentum from the flow, slowing it upstream, but a large part of the plasma is diverted around it. Additional slowing of plasma occurs on account of plasma pickup from electron impact ionization, photoionization, and charge exchange between the plasma and the neutral atmosphere (because these processes extract momentum from the background flow). *Goertz* (1980) and *Neubauer* (1998) have shown that plasma pickup processes can be correctly treated by including a plasma pickup conductivity term in the momentum equation.

Currents flow in the conducting regions of Europa (ionosphere and the plasma pickup region) that try to accelerate Europa's ionosphere and the newly picked-up plasma toward corotation. The currents are eventually closed in Jupiter's ionosphere through a pair of Alfvénic disturbances (called the Alfvén wings) that couple the satellite's environment to the northern and southern ionospheres of Jupiter (see Fig. 6 in the chapter by Kivelson et al.). The slowing down of plasma in the upstream region enhances the strength of the frozen-in flux, whereas in the wake region the field strength diminishes because plasma is reaccelerated to corotation. *Neubauer* (1980) showed that the total interaction current flowing in the system is limited by the Alfvén conductance $\Sigma_A = 1/(\mu_0 V_A)$ of plasma to $I = 2\Phi\Sigma_A = 4V_0R_m(\rho/\mu_0)^{1/2}$; here V_A is the Alfvén velocity of the upstream plasma, Φ is the electric potential imposed by the flow (V_0) across the moon, and R_m is the effective size of the moon. *Neubauer* (1999) has shown that the effective size of the obstacle (and therefore the size of the Alfvén wing) is reduced by an electromagnetic induction response from the interior of a moon because induction impedes the closure of field-aligned currents through the ionosphere of the moon by reducing the number of field lines intersecting the moon. *Volwerk et al.* (2007) show that the electromagnetic induction indeed shrinks a cross-section of the Alfvén wing of Europa by as much as 10%.

The strength of the moon/plasma interaction depends strongly on the location of Europa with respect to Jupiter's magnetic equator because both the density of Europa's tenuous sputtered atmosphere and the Alfvén conductance depend strongly on the density of upstream plasma. In order to minimize the effects of plasma interaction currents on the observed magnetic induction signature, flyby times of Europa are favored when it is located outside Jupiter's plasma sheet. In Fig. 5, these times correspond to the situations when the y component of the background field is either strongly positive (Europa located in the southern lobe of Jupiter's magnetosphere) or strongly negative (Europa located in the northern lobe). Conversely, when Europa is located near the center of the plasma sheet ($B_y \approx 0$), the expected dipole moment is weak and the noise from the plasma interaction fields is at its maximum.

The field generated by the moon/plasma interaction currents is a major source of systematic error for the induction signal. The contribution of the moon/plasma interaction currents to the measured field can be assessed from magnetohydrodynamic (MHD) simulations of the interaction. The chapter by Kivelson et al. provides an excellent review of the moon/plasma interaction effects and the available MHD models of this interaction. As discussed by Kivelson et al., the only model that treats the MHD interaction of Europa and the induction from the ocean self-consistently is that of *Schilling et al.* (2007), whose self-consistent model shows that the strength of the interaction field is on the order of 20–30 nT even when Europa is located outside the current sheet. By using prevailing field and plasma conditions for each of the flybys of Galileo in their MHD model, *Schilling et al.* (2007) have greatly improved upon the determination of the Europa induction field, enabling them to place better constraints on the conductivity and the thickness of Europa's ocean. We return to the discussion of results from this model below.

Another source of noise in the observations is perturbations from the MHD waves, which transmit energy between different regions of the magnetospheres and couple them. *Khurana and Kivelson* (1989) have shown that the amplitudes of the compressional and transverse waves peak at the center of Jupiter's plasma sheet. The peak amplitude of MHD waves is not very large (<5 nT) near Europa because of the low-β plasma conditions at the orbital location of Europa. As the waves have high frequencies (wave periods of minutes to tens of minutes), they may not be able to penetrate the icy crust, reducing their usefulness for electromagnetic sounding of the ocean.

Finally, ion cyclotron waves generated by the pickup of plasma in the vicinity of Europa create additional noise in the observed field. *Volwerk et al.* (2001) show that the intensity of these waves can approach 20 nT in the wake region of Europa. Their wave periods are between 3 s and 10 s and some of the most intense waves occur with a period of ~5 s, corresponding to that of O_2^+, a major constituent of Europa's sputtered atmosphere· The effect of ion cyclotron waves can be reduced on the observations by a judicious averaging and/or filtering of the data.

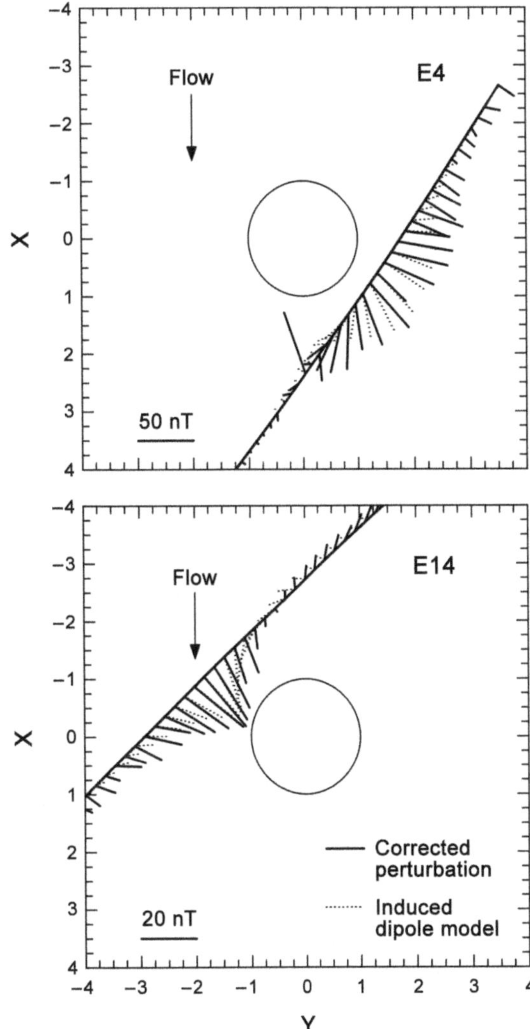

Fig. 6. The perturbation field near Europa measured by Galileo during the E4 and E14 flybys (solid vectors) and that expected from induction from a global perfect conductor (dashed vectors). Adapted from *Khurana et al.* (2002).

4.3. Galileo Observations of Electromagnetic Induction

Galileo encountered Europa 11 times, during which its closest approach altitude was less than 2 R_E. Three of the magnetic field recordings (E6, E16, and E18) were lost because of instrument or spacecraft malfunctions. Out of the remaining eight passes, only five passes (E4, E12, E14, E19, and E26) had Europa altitudes of 2000 km or less, required for an adequate signal-to-noise ratio to decipher the induction field. The magnetic field observations from the E4 and E14 passes formed the basis for the discovery of induction response from Europa (*Khurana et al.*, 1998; *Kivelson et al.*, 1999) and are reproduced in Fig. 6. An examination of these figures shows that the signature is both global and dipolar in character as expected of an induction field from an ocean. It must be mentioned here that *Neubauer et al.* (1998) and K. Kuromoto et al. (1998, unpublished manuscript made available to the magnetometer team) also independently postulated that the source of magnetic perturbations observed near Europa in the Galileo data was electromagnetic induction from an internal source.

The primary field during the E4 and E14 encounters was directed in the negative y direction (see Fig. 5). According to equations (7) and (8), the dipole moment of the secondary field would be directed in the positive y direction for both of these flybys. In order to exclude the possibility that the source of the observed dipole moment was an intrinsic internal dipole tilted toward the y axis, the Galileo magnetometer team designed a flyby (E26) during which the inducing field was directed mainly in the positive y direction (see Fig. 5). The observations from that flyby confirmed that the induced dipole moment had indeed flipped in direction and was directed in the negative direction, confirming that

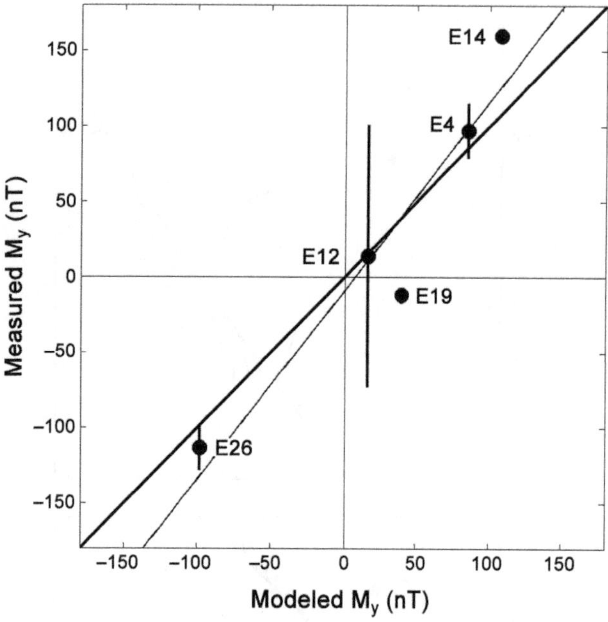

Fig. 7. The y component of the induced dipole moment measured during the five flybys of Galileo (y axis) plotted against the modeled dipole moment. Adapted from *Kivelson et al.* (2000).

the induced field was indeed a response to the changing field of Jupiter (*Kivelson et al.,* 2000). Figure 7 shows the y components of the observed dipole moments from five Galileo flybys plotted against those expected from a perfect spherical conductor with a radius equal to that of Europa. The agreement between the observations and the simple model over a large range of the excitation field provides a compelling evidence of a global conducting layer in Europa. As discussed in section 3, a consideration of various geological materials to explain the internal conductivity naturally leads to the conclusion that a present-day ocean exists in Europa.

4.4. Limits on the Induction Response

Further modeling of Europa data has been performed by *Zimmer et al.* (2000), *Schilling et al.* (2004), and *Schilling et al.* (2007) to place better limits on the induction response from Europa's ocean. The effect of plasma interaction on the field was modeled by *Zimmer et al.* (2000) for the E4 and E14 flybys by using a simple plasma correction model originally suggested by *Khurana et al.* (1998). Their plasma correction model assumes that near the equatorial plane of Europa, the moon/plasma interaction currents produce mainly a compressional signal and no bending of the field is involved. In order to further simplify the data fitting procedure, *Zimmer et al.* (2000) also assumed that the phase delay of the induced signal is zero. Figure 8 shows results from their study for the E14 flyby where models with induction response factor A (ratio of inducing field to induction response) from 0.4 to 1.6 are displayed. It can be seen that models for which $A < 0.7$ or $A > 1.0$, the fits to data are perceptibly poor. Zimmer et al. therefore concluded that the response factor lies between 0.7 and 1.0. The lower limit on the induction factor requires that the conductivity of the ocean must exceed 58 mS/m for an infinitely thick ocean. *Anderson et al.* (1998) place an upper limit of 170 km for the H_2O layer on Europa, which would raise the minimum conductivity of the ocean to 72 mS/m. If the ice and water layers together were only 100 km thick, the minimum conductivity required jumps to 116 mS/m.

Schilling et al. (2004) modeled the Galileo observations from four flybys (E4, E14, E19, and E26) using a Biot-Savart model of the Alfvénic current system and several different models of the permanent and induced internal field. They found that the internal field models that fit the data best and required the least number of fit parameters favored induction from an internal conducting source. In addition, the permanent internal dipole term was found to be quite small (<25 nT at the surface of Europa). The model favored by *Schilling et al.* (2004) yields an induction factor of 0.97 from the internal ocean, suggesting that the ocean water is extremely conducting.

Schilling et al. (2007) have developed a fully self-consistent three-dimensional simulation of temporal interaction of Europa with Jupiter's magnetosphere. Their model simultaneously describes the plasma interaction of Europa's atmosphere with Jupiter's magnetosphere and the electromagnetic induction response of the subsurface ocean to the varying field of Jupiter. The mutual feedbacks — where the plasma interaction currents affect the amplitude of induction, and the induction field affects the plasma interaction by reducing the size of the Alfvén wings — are included. Figure 9 reproduced from their work shows observed and modeled fields for the E14 and E26 flybys. The vastly improved fits to the data allow Schilling et al. to place better-constrained limits on the ocean conductivity and thickness. They find that the conductivity of Europa's ocean would have to be 500 mS/m or larger to explain the observations made by Galileo. In addition, they favor an ocean thickness of <100 km. However, ocean thicknesses greater than 100 km cannot be ruled out because for sufficiently large conductivities, the induction response becomes insensitive to the location of the lower boundary of the ocean. This can be verified by an examination of top right portion of Fig. 4, which shows that the induction response saturates completely for ocean thickness of >100 km (i.e., the amplitude response curves become vertical for large conductivity values). The height-integrated conductivity of the ocean (conductivity × ocean thickness) was found to be $\geq 5 \times 10^4$ S by *Schilling et al.* (2007).

Because of the limited durations of the Galileo flybys of Europa, it has not yet been possible to decompose the observed induction signal into various primary frequencies and various internal and external spherical harmonics. The induction response has been modeled under the assumption that Europa experiences only a single primary frequency (at the synodic rotation period of Jupiter) whose amplitude is computed from a magnetospheric model of Jupiter. As

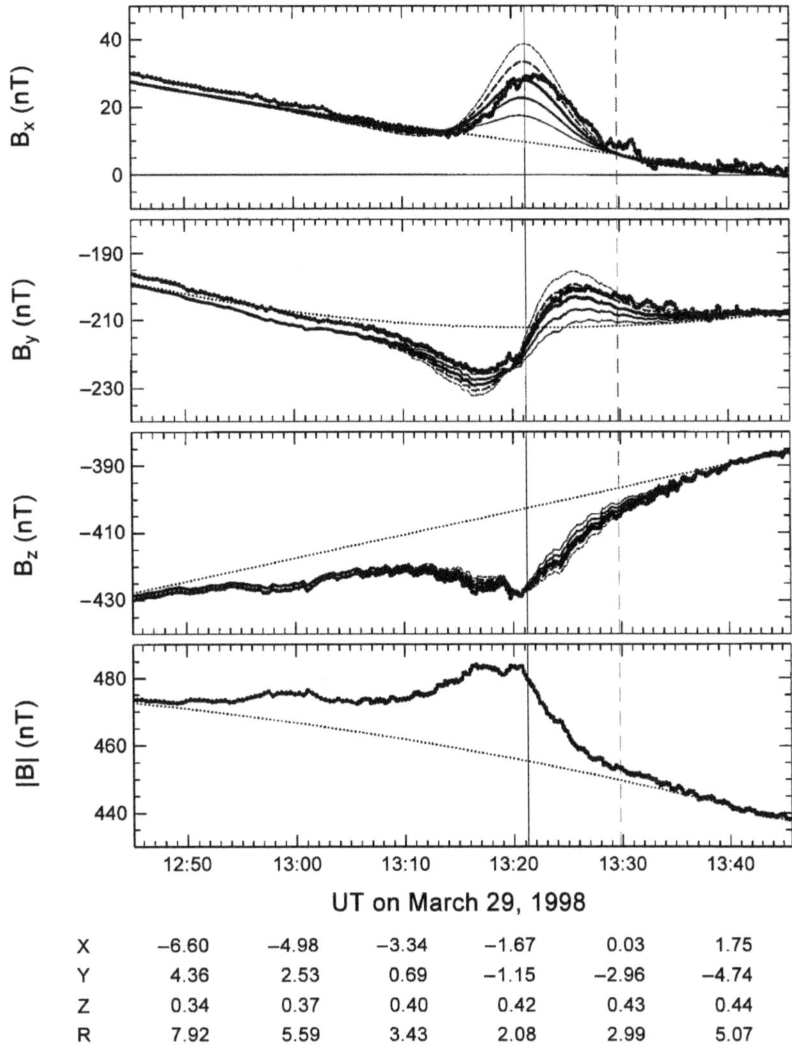

Fig. 8. Data from the E14 flyby (thick dots) and model fits for A = 0 (thin dotted lines), 0.4 (thin solid line), 0.7 (intermediate thickness solid line), 1 (thick solid line), 1.3 (thin dashed line), and 1.6 (thick dashed line). Adapted from *Zimmer et al.* (2000).

discussed above, Europa does experiences a nearly harmonic perturbation in the z component of the magnetic field at its orbital period (85.2 h) with an amplitude of 12–20 nT. However, reliable estimates of the inducing field at this frequency are not yet available. In addition, as discussed in the next section, the induction caused by the ionospheric and surrounding plasma also contributes to the observed induced magnetic field and its effect should be carefully removed from the data.

5. EFFECT OF EUROPA/PLASMA INTERACTION ON THE INDUCTION SIGNATURE

In addition to generating "noise" in the magnetic data through its interaction with the jovian plasma, Europa's interacting ionosphere also contributes to the electromagnetic induction signature through its conductivity. As discussed in the chapter by McGrath et al., the thin ionosphere of Europa has a scale height of ~200 km below an elevation of 300 km with a peak electron density in the range of 10^3–10^4 cm^{-3} near the surface (*Kliore et al.,* 1997, 2001). As discussed by *Zimmer et al.* (2000), in any ionosphere, the Pedersen conductivity is maximized at a location where the electron cyclotron frequency and the electron-neutral collision frequencies are equal and the maximum value is given by $\sigma_P < n_e e/2B$ where n_e is the electron density and e is the electron charge. Using the maximum density of Europa's ionospheric plasma as an input, they find that the Pedersen conductivity is everywhere less than 2.2 mS/m below an altitude of 300 km and less than 0.5 mS/m above 300 km. These are theoretical upper bounds on Pedersen conductivity in Europa's ionosphere; the actual values are expected to be many times lower. For example, a self-consistent three-dimensional neutral and plasma model of Europa's atmosphere by *Saur et al.* (1998) yields height-

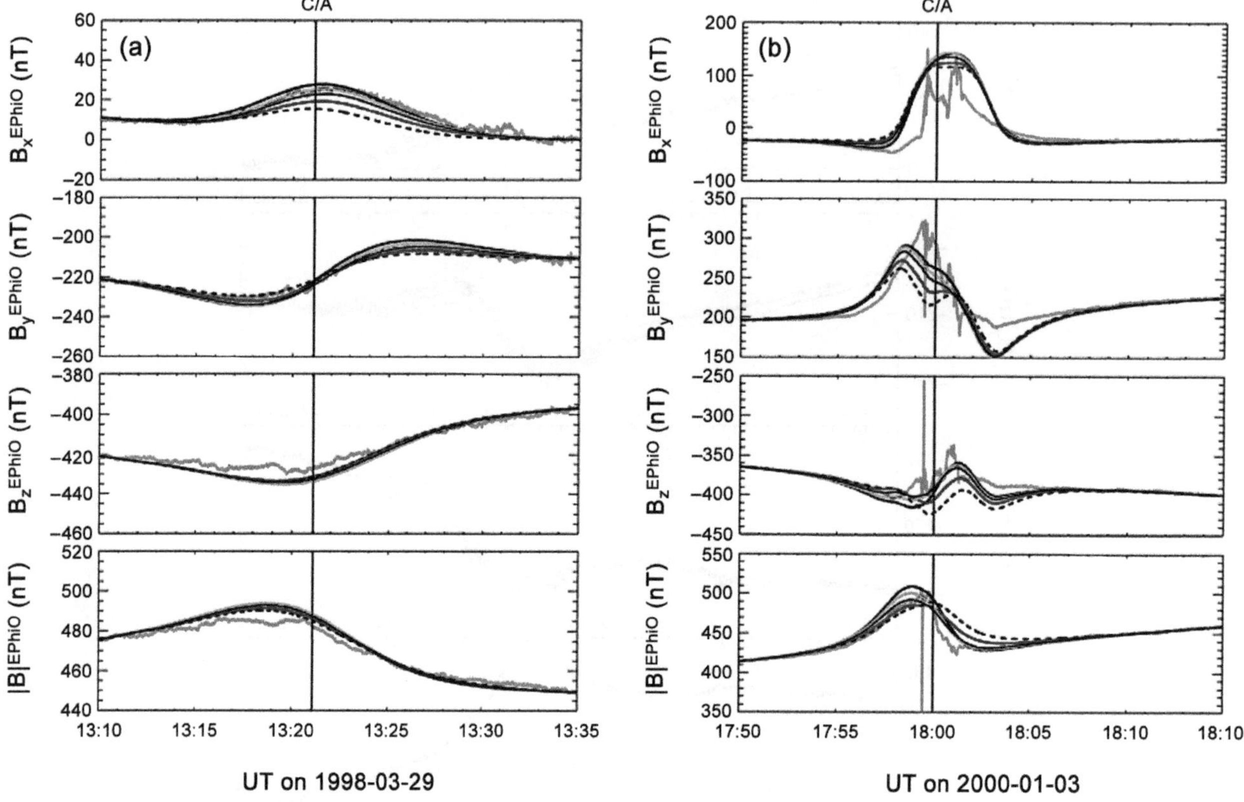

Fig. 9. See Plate 33. Observed and modeled magnetic field for the **(a)** E14 and **(b)** E26 flybys in the EfW coordinate system. Shown in the four panels are the three components and the magnitude of the observed field (red curves), model field with no induction (dashed blue), ocean conductivity 100 mS/m (blue), 250 mS/m (brown), 500 S/m (green), and 5 S/m (black). The ice thickness is 25 km and the ocean thickness is 100 km in these simulations. Adapted from *Schilling et al.* (2007).

integrated ionospheric conductivities in the range of 10–60 S, which are more than an order of magnitude lower than the theoretical upper bounds. This height-integrated conductivity should be compared with that of Europa's ocean (~5 × 10^4 S) derived by *Schilling et al.* (2007). Thus, the contribution of the ionosphere to the observed electromagnetic induction signature is expected to be extremely modest.

Schilling et al. (2007) have computed the time-stationary and time-varying harmonic dipole and quadrupole coefficients of the plasma-induced magnetic fields over Jupiter's synodic period from their three-dimensional simulations. They assumed an ocean thickness of 100 km and an ice crust thickness of 25 km for one of their simulations. The simulation shows that the quadrupole terms dominate over the dipole terms for both time-stationary and time-varying components at least by a factor of 2 (see Fig. 10). The dominant time-stationary term is G_2^1 with an amplitude of –29 nT. For a perfectly symmetric conductor, no plasma-induced dipole component would be expected. However, the Alfvén wings associated with the plasma interaction break the symmetry and helps generate a dipolar response (~12 nT) at zero frequency. The dominant time-varying quadrupolar term is g_2^1, which has an amplitude of 14 nT and a frequency twice that of the synodic rotation period of Jupiter. The power in the induced octupole is much smaller than that of the dipole and quadrupole terms. The strongest time-varying dipolar perturbation occurs when Europa is immersed in the plasma sheet of Jupiter ($\Omega t = 270°$) and is equal to ~12 nT. The amplitudes of the plasma-induced Gauss coefficients are much smaller than those generated in response to the varying magnetic field of Jupiter (~250 nT). However, careful strategies would be required to remove the effects of plasma-generated induction fields from the observations.

6. IMPLICATION OF IMPLIED OCEAN CONDUCTIVITIES FOR THE COMPOSITION OF THE OCEAN

Most of the theoretical and experimental studies of ocean composition suggest that if the salts in Europa's ocean arose from leaching or aqueous alteration of rocks with composition akin to those of carbonaceous chondrites, the dominant salts would be hydrated sulfates of Mg and Na (*Kargel et al.,* 2000; *Zolotov and Shock,* 2001; *Fanale et al.,* 2001). This conclusion is supported by Galileo's Near Infrared Mapping Spectrometer (NIMS) surface spectral analysis of relatively young terrains (*McCord et al.,* 1998, 1999, 2001, 2002), which exhibit water-absorption bands distorted by the presence of hydrated sulfate-bearing minerals. *McKinnon and Zolensky* (2003) caution against a sulfate-rich model, arguing that the original chondritic material may have been much more reducing and sulfidic than assumed in most models. They place an upper limit of $MgSO_4$ concentrations at 100 g $MgSO_4$ kg^{-1} H_2O but suggest scenarios where the

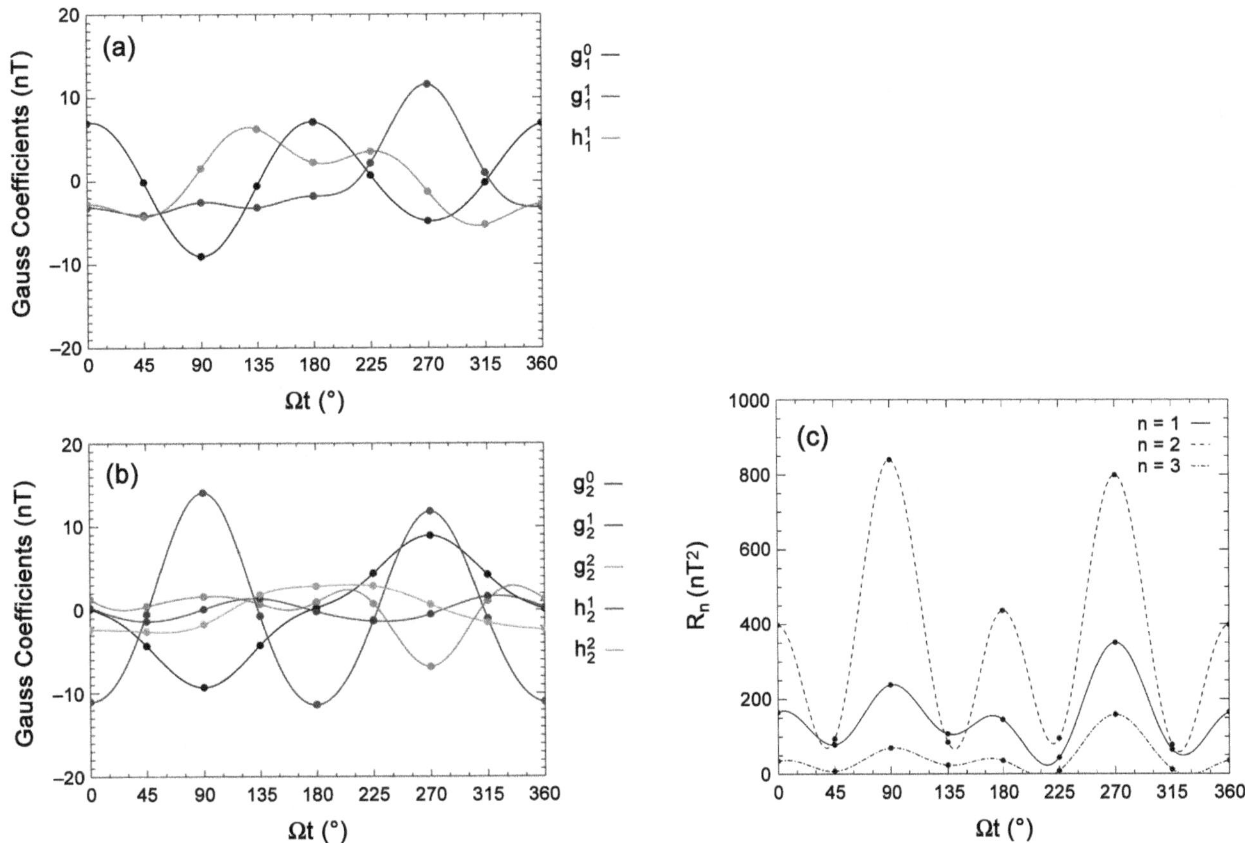

Fig. 10. See Plate 34. The spherical harmonic coefficients for the **(a)** dipolar and **(b)** quadrupolar terms induced by the Europa/plasma interaction over a synodic rotation period of Jupiter. **(c)** Total power contained in the induced field at the first three spherical harmonics. For comparison, the magnetic-field-induced dipole term contains power at the level of 14,000–88,000 nT^2. Adapted from *Schilling et al.* (2007).

$MgSO_4$ concentration may be as little as 0.018 g $MgSO_4$ kg^{-1} H_2O. The models of *Kargel et al.* (2000) on the other hand yield extremely briny solutions with salt concentration as high as 560 g $MgSO_4$ kg^{-1} H_2O, whereas *Zolotov and Shock*'s (2001) "total extraction" model can yield $MgSO_4$ concentration approaching 1000 g $MgSO_4$ kg^{-1} H_2O (see Fig. 11). With such a divergent opinion on the concentration of the most noticeable salt on Europa's surface, it should be clear that there is currently no consensus on the concentrations of salts (or even their composition) in Europa's ocean. Further information on the current status of ocean composition can be found in the chapter by Zolotov and Kargel.

6.1. Relationship Between Salt Concentration and Conductivity

Hand and Chyba (2007) have recently combined the magnetometer-derived ocean conductivities with the interior models of Europa and the laboratory studies of conductivities of salt solutions to place better limits on the salinity of Europa's ocean. In order to obtain an empirical relationship between the salt concentration and the conductivity of the solution usable over a large range of salinity, *Hand and Chyba* (2007) compiled a dataset of experimental values of salt solution conductivities from a wide range of sources and then scaled it to a common temperature of 0°C. Figure 12 reproduced from their work shows the conductivities of water solutions containing both sea salt (solid curve) and $MgSO_4$ (various symbols), which is the most likely salt in Europa's ocean. Also marked on the same figure are the upper and lower limits on the conductivities from the *Zimmer et al.* (2000) work. The lower limit from Zimmer et al. excludes the lowest partial extraction models of *Zolotov and Shock* (2001). The figure also shows that the conductivity of a $MgSO_4$ solution peaks at ~6 S/m at its dissolution saturation limit (282 g kg^{-1} H_2O), whereas for a NaCl solution the peak in conductivity occurs near 18 S/m, corresponding to a saturation limit of (304 g kg^{-1} H_2O).

6.2. Limits on the Salinity of the Ocean from Magnetic Field Observations

In order to further refine the limits on the salinity and thickness of Europa's ocean, *Hand and Chyba* (2007) used multiple-shell models of Europa's interior (ocean, mantle, and core) and its exterior (an ionosphere) to derive the response of Europa to the 11.1-h wave over a range of assumed ocean parameters such as the salinity, thickness, and depth from the surface. Figure 13 shows the relationship between ocean thickness and $MgSO_4$ concentration for an

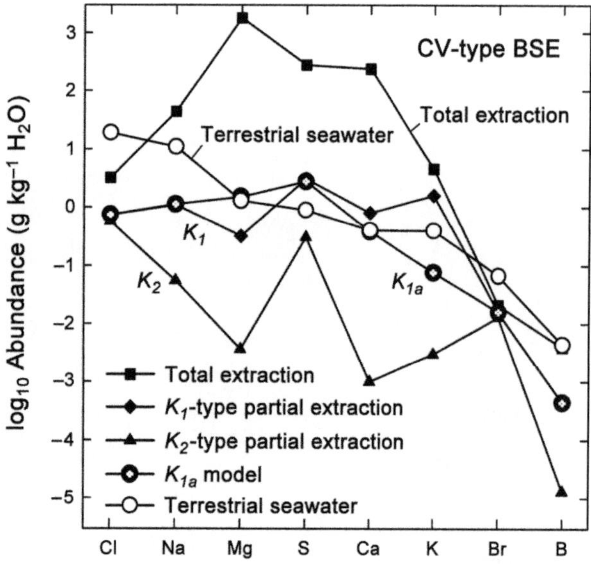

Fig. 11. Abundances of various elements for four different models of partial extraction from a 100-km-thick ocean in the models of *Zolotov and Shock* (2001). Their preferred model (K_{1a}) yields $MGSO_4$ concentration of 7.6 g $MgSO_4$ kg^{-1} H_2O. Adapted from *Zolotov and Shock* (2001).

assumed induction response factor of 0.7 [corresponding to the lower limit of *Zimmer et al.* (2000) estimates]. Calculations were performed for several different values of the icy shell thickness. As expected, there is an inverse relationship between the ocean thickness and the salinity of the ocean. The figure clearly illustrates the nonuniqueness problem arising from single-frequency measurements. The range of allowed parameters includes what would be considered fresh water by terrestrial standards with $MgSO_4$ concentrations below 1 g kg^{-1} H_2O to highly saline solutions containing salt concentrations as high as 16 g kg^{-1} H_2O, corresponding to a 20-km-thick ocean buried under 60 km of ice.

Figure 14 from *Hand and Chyba* (2007) summarizes the response of Europa for a range of ocean conductivities, icy shell thicknesses, and ocean shell thicknesses calculated from the three-layer model introduced above. It can be seen that at the lower end of the response factor ($A \leq 0.75$), the ocean thickness and its conductivity can be traded to get the same response factor. However, as the amplitude of the response factor increases, the thickness of the ocean is not a strong factor in determining the response, but the ocean conductivity is. At the lower end of the response function estimates ($A = 0.75$) a freshwater ocean is clearly possible. However, if the response factor is 0.97 ± 0.02 as suggested by *Schilling et al.* (2004), only very thin ice shells (0–15 km) are allowed and the ocean would have to be hypersaline. *Hand and Chyba* (2007) also found that electromagnetic induction from an exterior ionosphere and/or a conducting inner layer (mantle or core) do not have an appreciable effect on the response from Europa at the 11.1-h wave period. On the other hand, the strong induction signature profoundly affects the strengths of the moon/plasma interaction currents and the magnetic field.

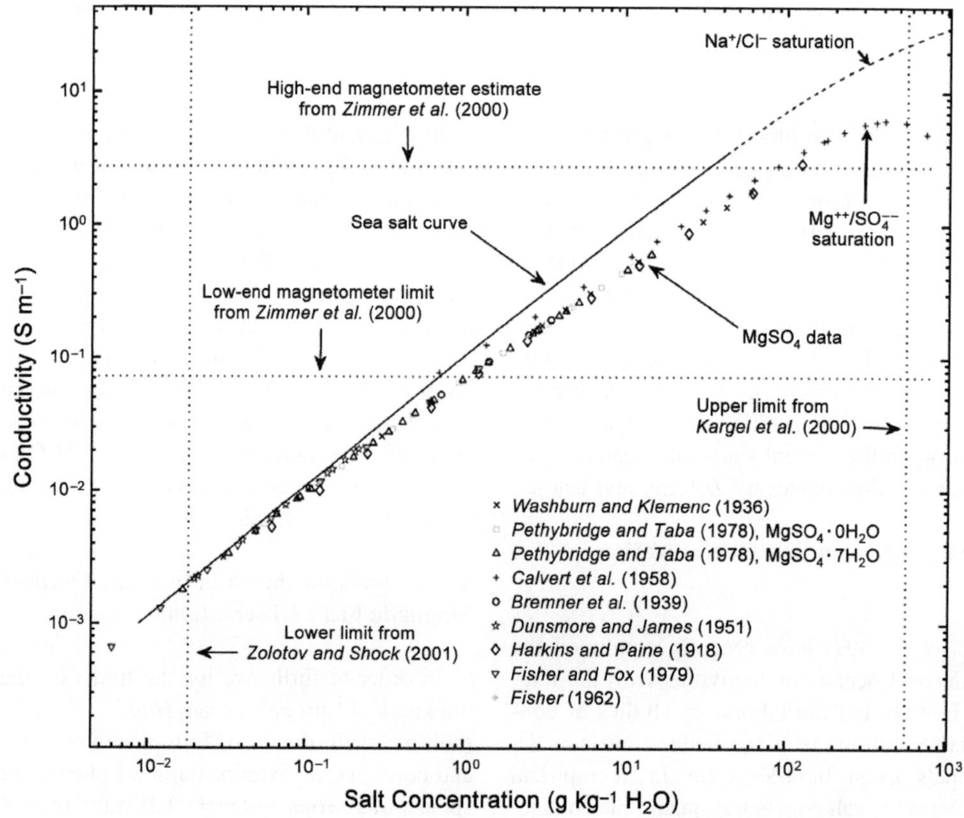

Fig. 12. Conductivities of water solutions containing sea salt (solid curve) and $MgSO_4$ (various symbols). The saturation points for NaCl (304 g kg^{-1} H_2O) and $MgSO_4$ (282 g kg^{-1} H_2O) are marked by short arrows. Adapted from *Hand and Chyba* (2007).

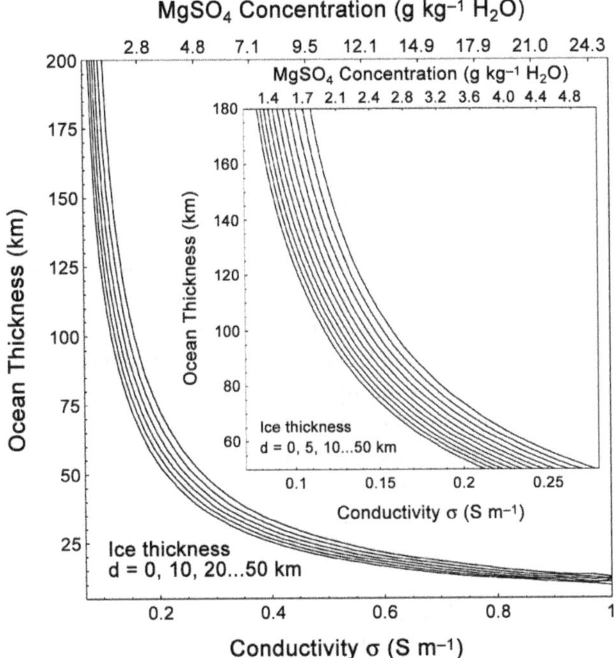

Fig. 13. The relationship between ocean thickness and $MgSO_4$ concentration for a given response factor of 0.7 corresponding to the lower limit of *Zimmer et al.* (2000) estimates. Curves were calculated for six different assumptions of ice shell thicknesses. Adapted from *Hand and Chyba* (2007).

7. SOUNDING THE DEEP INTERIOR OF EUROPA

So far we have focused exclusively on the induction response from the subsurface ocean at two principal frequencies associated with the rotation period of Jupiter and the orbital period of Europa. As the induction response of Europa's ocean at these frequencies is close to unity, most of the signal does not penetrate through the ocean. However, other frequencies must exist in the background magnetospheric field as Europa is located at the outer edge of Io's torus, an active region that responds continuously to changes in the density, magnetic flux, and energy content of the torus. It is believed that Io's torus responds to changes in the volcanic activity of Io over timescales of weeks to months. Thus Europa can be expected to be bathed in many different types of (although nonharmonic) frequencies. As already shown, longer-period waves penetrate a conductor more deeply, so we would like to assess the shielding efficiency of Europa's ocean to longer-period waves.

We have used the three-layer model of Europa's interior illustrated in Fig. 3 to calculate the response of Europa to waves of five different frequencies for a range of ocean thicknesses and an assumed conductivity of 2.75 S/m, similar to that of Earth's ocean (see Fig. 15). We find that for wave periods longer than three weeks, the shielding efficiency of Europa's ocean is less than 50%, suggesting that

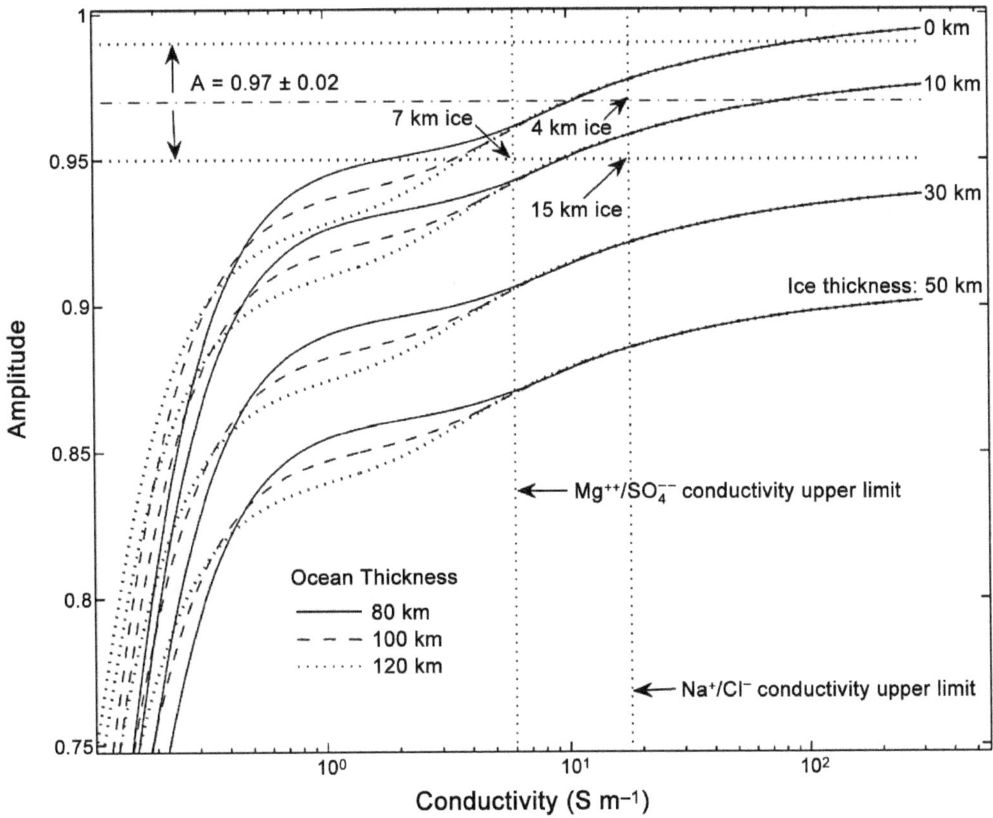

Fig. 14. The induction response factor as a function of conductivity, ocean thickness, and ice shell thickness for the three-layer model. Marked on the figure is the range of response factor deduced by *Schilling et al.* (2004) (horizontal dotted lines). The upper limit imposed on the conductivity of the solution from saturation effects are marked by the two vertical lines. Adapted from *Hand and Chyba* (2007).

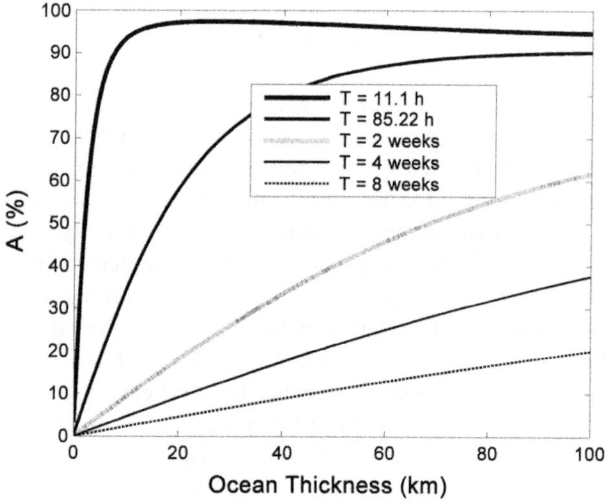

Fig. 15. Shielding efficiency of Europa's ocean to waves of five different periods for a range of ocean thicknesses. The assumed conductivity of the ocean is 2.75 S/m, similar to that of Earth's ocean.

roughly half of the signal is able to penetrate the ocean. Thus, if observations were available over periods of several months, it would become possible to assess Europa's response to deeper layers like the rocky mantle and especially the metallic core. As the conductivity of iron is many orders of magnitude higher than that expected of an ocean (see Table 1), the response factor of the core to wave periods of weeks to months would be close to 100% at its surface. If the core size is 50% of that of Europa (*Anderson et al.,* 1998), one can expect to see a response in the measured induction field at a level of ~6% of the original inducing signal, a small but certainly measurable signal.

8. FUTURE EXPLORATION USING ELECTROMAGNETIC INDUCTION

The exploration of Europa's interior using electromagnetic induction sounding is still in its infancy. The observations and associated modeling have been extremely basic because of the limited nature of observations that have been made so far during brief close flybys of Europa. Future continuous observations from one or more Europa orbiting spacecraft and multiple surface observatories could provide unprecedented capabilities allowing multiple-frequency sounding of Europa's ocean and its deeper interior.

In order to illustrate the power of long-period continuous data, we have computed several synthetic datasets mimicking observations from a Europa orbiter at an altitude of ~200 km. For these simulations we used the three-shell model of Fig. 3 and assumed that the conductivity of the ocean was 2.75 S/m, the inducing signal had a period of 11.1 h, and assigned a thickness to the exterior ice crust of 10 km. In Fig. 16, we plot the expected magnetic field data from three different assumptions about the ocean of Europa: no ocean (thin continuous line), a 3-km-thick ocean (dotted lines), and a deep ocean with a thickness of 100 km (thick continuous lines). First of all, we would like to point out that from continuous time series, it is very easy to distinguish between the primary inducing field (external signal) at the synodic rotation period of Jupiter (large sinusoidal signal in all three components at a period of 11.1 h; thin continuous lines) and the dipolar induction response from the ocean (higher-frequency signal at the orbital frequency of the orbiter, internal signal). Thus, separating the signal into internal and external harmonics is quite straightforward. Next, as expected, we observe that a 100-km-thick ocean does induce a much stronger dipolar response (thick lines) than a 3-km-thick ocean (dotted lines), and data from an orbiter would be easily able to distinguish between the two scenarios. We have also experimented with including other wave periods such as the orbital period of Europa and find that it is possible to distinguish them from the 11.1-h period by Fourier transforming the continuous time series. Finally, we would like to point out that any departure of the ocean from a perfect spherical shape would lead to the generation of higher-degree (quadrupole and octupole) spherical harmonics in the data. Thus an examination of the induction field in terms of its harmonic content would yield information on the shape of the ocean.

Finally, we would like to point out that simultaneous measurements from multiple spacecraft or/and multiple surface sites facilitates the decomposition of the internal and external fields directly in time domain. The decomposed

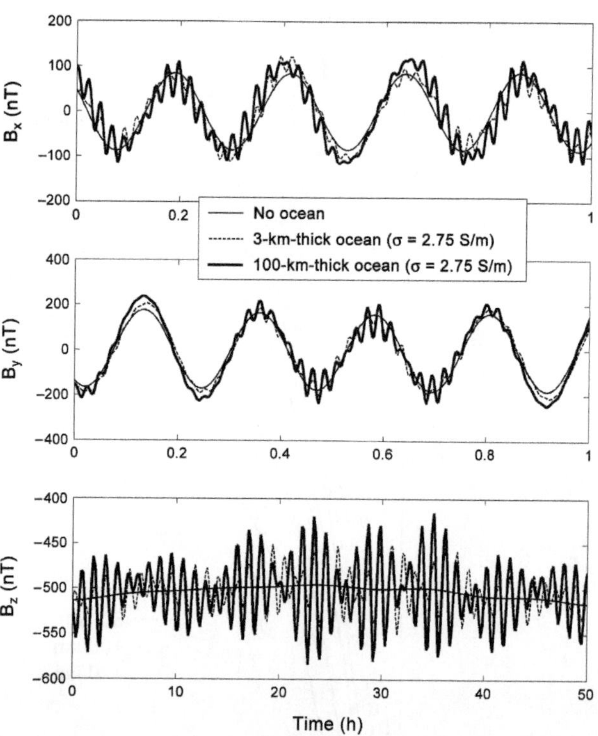

Fig. 16. Simulation of the expected magnetic field from a Europa orbiter orbiting at an altitude of ~200 km for three different assumptions about the ocean of Europa (no ocean, a 3-km-thick ocean, and a deep ocean with a thickness of 100 km). Further information on the properties of the ocean and the model is provided in the text of the chapter.

internal and external field time series can then be Fourier decomposed into the primary field and Europa's response at not only the two prime frequencies but also the weaker nonharmonic frequencies.

Acknowledgments. The authors would like to thank M. Volwerk and N. Schilling for several useful discussions. This work was supported by NASA grant NNX06AB91G. Portions of this work by K.P.H. were performed at the Jet Propulsion Laboratory, California Institute of Technology, under contract to the National Aeronautics and Space Administration.

REFERENCES

Anderson J. D., Schubert G., Jacobson R. A., Lau E. L., Moore W. B., and Sjogren W. L. (1998) Europa's differentiated internal structure: Inferences from four Galileo encounters. *Science, 281,* 2019–2022.

Bagenal F. and Sullivan J. D. (1981) Direct plasma measurements in the Io torus and inner magnetosphere of Jupiter. *J. Geophys. Res., 86,* 8447–8466.

Banks R. J. (1969) Geomagnetic variations and the conductivity of the upper mantle. *Geophys. J. R. Astron. Soc., 17,* 457–487.

Belcher J. W. (1983) The low energy plasma in the jovian magnetosphere. In *Physics of the Jovian Magnetosphere* (A. J. Dessler, ed.), pp. 68–105. Cambridge Univ., Cambridge.

Carr M. H., Belton M. J. S., Chapman C. R., Davies M. E., Geissler P., Greenberg R., et al. (1998) Evidence for a subsurface ocean on Europa. *Nature, 391,* 363–365.

Cassen P. M., Reynolds R. T., and Peale S. J. (1979) Is there liquid water on Europa? *Geophys. Res. Lett., 6,* 731–734.

Cassen P. M., Peale S. J., and Reynolds R. T. (1980) Tidal dissipation in Europa: A correction. *Geophys. Res. Lett., 7,* 987–988.

Chapman S. (1919) Solar and lunar diurnal variations of terrestrial magnetism. *Philos. Trans. Roy. Soc. Lond., A218,* 1–118.

Chapman S. and Price T. A. (1930) The electric and magnetic state of the interiors of the Earth, as inferred from terrestrial magnetic variations. *Philos. Trans. R. Soc. London, A229,* 427–460.

Chapman S. and Whitehead T. T. (1922) Influence of electrically conducting material on phenomena of terrestrial magnetism. *Trans. Camb. Philos. Soc., 22,* 463–482.

Constable S. and Constable C. (2004) Observing geomagnetic induction in magnetic satellite measurements and associated implications for mantle conductivity. *Geochem. Geophys. Geosys.,* Q01006, DOI: 10.1029/2003GC000634.

Didwall E. M. (1984) The electrical conductivity of the upper mantle as estimated from satellite magnetic field data. *J. Geophys. Res., 89,* 537–542.

Dyal P. and Parkin C. W. (1973) Global electromagnetic induction in the moon and planets. *Phys. Earth Planet. Inter., 7,* 251–265.

Fanale F. P., Li Y. H., De Carlo E., Farley C., Sharma S. K., and Horton K. (2001) An experimental estimate of Europa's "ocean" composition independent of Galileo orbital remote sensing. *J. Geophys. Res., 106,* 14595–14600.

Goertz C. K. (1980) Io's interaction with the plasma torus. *J. Geophys. Res., 85,* 2949–2956.

Greeley R., Sullivan R., Klemaszewski J., et al. (1998) Europa: Initial Galileo geological observations. *Icarus, 135,* 4–24.

Hand K. P. and Chyba C. F. (2007) Empirical constraints on the salinity of the europan ocean and implications for a thin ice shell. *Icarus, 189,* 424–438.

Hobbs P. V. (1974) *Ice Physics,* pp. 97–120. Oxford Univ., London.

Holzapfel W. B. (1969) Effect of pressure and temperature on the conductivity and ionic dissociation of water up to 100 kbar and 1000°C. *J. Chemical Phys., 50,* 4424–4428.

Hood L. L., Mitchell D. L., Lin R. P., Acuna M. H., and Binder A. B. (1999) Initial measurements of the lunar induced dipole moment using Lunar Prospector magnetometer data. *Geophys. Res. Lett., 26,* 2327–2330.

Johnson F. S., ed. (1961) *Satellite Environment Handbook.* Stanford Univ., Stanford, California.

Kargel J. S., Kaye J. Z., Head J. W. III, Marion G. M., Sassen R., Crowley J. K., Ballesteros O. P., Grand S. A., and Hogenboom D. L. (2000) Europa's crust and ocean: Origin, composition, and the prospects of life. *Icarus, 148,* 226–265.

Khan A., Connolly J. A. D., Olsen N., and Mosegaard K. (2006) Constraining the composition and thermal state of the Moon from an inversion of electromagnetic lunar day-side transfer functions. *Earth Planet. Sci Lett., 248,* 579–598.

Khurana K. K. (1997) Euler Potential models of Jupiter's magnetospheric field. *J. Geophys. Res., 102,* 11295–11306.

Khurana K. K. (2001)Influence of solar wind on Jupiter's magnetosphere deduced from currents in the equatorial plane. *J. Geophys. Res., 106,* 25999–26016.

Khurana K. K. and Kivelson M. G. (1989) Ultra low frequency MHD waves in Jupiter's middle magnetosphere. *J. Geophys. Res., 94,* 5241.

Khurana K. K., Kivelson M. G., Stevenson D. J., et al. (1998) Induced magnetic fields as evidence for subsurface oceans in Europa and Callisto. *Nature, 395,* 777–780.

Khurana K. K., Kivelson M. G., and Russell C. T. (2002) Searching for liquid water on Europa by using surface observatories. *Astrobiology, 2(1),* 93–103.

Kivelson M. G., Khurana K. K., Stevenson D. J., Bennett. L., Joy S., Russell C. T., Walker R. J., Zimmer C. and Polansky C. (1999) Europa and Callisto: Induced or intrinsic fields in a periodically varying plasma environment. *J. Geophys. Res., 104,* 4609–4625.

Kivelson M. G., Khurana K. K., Russell C. T., Volwerk M., Walker R. J., and Zimmer C. (2000) Galileo magnetometer measurements strengthen the case for a subsurface ocean at Europa. *Science, 289,* 1340.

Kliore A. J., Hinson D. P., Flasar F. M., Nagy A. F., and Cravens T. E. (1997) The ionosphere of Europa from Galileo radio occultations. *Science, 277,* 355–358.

Kliore A. J., Anabtawi A., and Nagy A. F. (2001) The ionospheres of Europa, Ganymede, and Callisto. *Eos Trans. AGU, 82(47),* Fall Meet. Suppl., Abstract B506.

Lahiri B. N. and Price A. T. (1939) Electromagnetic induction in non-uniform conductors, and the determination of the conductivity of the earth from terrestrial magnetic variations. *Philos. Trans. R. Soc. London, A237,* 509–540.

Lamb H. (1883) On electrical motions in a spherical conductor. *Philos. Trans. R. Soc. London, 174,* 519–549.

Langel R. A. (1975) Internal and external storm-time magnetic fields from spacecraft data (abstract). In *IAGA Program Abstracts for XVI General Assembly,* Grenboble, France.

Martinec Z. and McCreadie H. (2004) Electromagnetic induction modelling based on satellite magnetic vector data. *Geophys. J. Inter., 157,* 1045–1060.

McCord T. B., Hansen G. B., Fanale F. P., Carlson R. W., Matson D. L., et al. (1998) Salts on Europa's surface detected by Galileo's near infrared mapping spectrometer. *Science, 280,* 1242–1245.

McCord T. B., Hansen G. B., Matson D. L., Johnson T. V., Crowley J. K., Fanale F. P., Carlson R. W., Smythe W. D., Martin P. D., Hibbitts C. A., Granahan J. C., and Ocampo A. (1999) Hydrated salt minerals on Europa's surface from the Galileo near-infrared mapping spectrometer (NIMS) investigation. *J. Geophys. Res., 104,* 11827–11851.

McCord T. B., Orlando T. M., Teeter G., Hansen G. B., Sieger M. T., Petrik N. G., and Keulen L. V. (2001) Thermal and radiation stability of the hydrated salt minerals epsomite, mirabilite, and natron under Europa environmental conditions. *J. Geophys. Res., 106,* 3311–3319.

McCord T. B., Teeter G., Hansen G. B., Sieger M. T., and Orlando T. M. (2002) Brines exposed to Europa surface conditions. *J. Geophy. Res., 107(E1),* 4.1–4.6, DOI: 10.1029/2000JE001453.

McKinnon W. B. and Zolensky M. E. (2003) Sulfate content of Europa's ocean and shell: Evolutionary considerations and some geological and astrobiological implications. *Astrobiology, 3,* 879–897.

Montgomery R. B. (1963) Oceanographic data. In *American Institute of Physics Handbook* (D. E. Gray, ed.), pp. 125–127. McGraw-Hill, New York.

Neubauer F. M. (1980) Nonlinear standing Alfvén wave current system at Io: Theory. *J. Geophys. Res., 85,* 1171–1178.

Neubauer F. M. (1998) The sub-Alfvénic interaction of the Galilean satellites with the jovian magnetosphere. *J. Geophys. Res., 103,* 19834–19866.

Neubauer F. M. (1999) Alfvén wings and electromagnetic induction in the interiors: Europa and Callisto. *J. Geophys. Res., 104,* 28671–28684.

Olhoeft G. R. (1989) Electrical properties of rocks. In *Physical Properties of Rocks and Minerals* (Y. S. Touloukian et al., eds.), pp. 257–329. Hemisphere Publishing, New York.

Olsen N. (1999a) Long-period (30 days–1 year) electromagnetic sounding and the electrical conductivity of the lower mantle beneath Europe. *Geophys. J. Inter., 138,* 179–187.

Olsen N. (1999b) Induction studies with satellite data. *Surv. Geophys., 20,* 309–340.

Pappalardo R. T., Head J. W., Greeley R., et al. (1998) Geological evidence for solid-state convection in Europa's ice shell. *Nature, 391,* 365–368.

Pappalardo R. T., Belton M. J. S., Breneman H. H., et al. (1999) Does Europa have a subsurface ocean? Evaluation of the geological evidence. *J. Geophys. Res., 104,* 24015–24055.

Parkinson W. D. (1983) *Introduction to Geomagnetism.* Scottish Academic, Edinburgh.

Russell C. T., Coleman P. J. Jr., and Goldstein B. E. (1981) Measurements of the lunar induced magnetic moment in the geomagnetic tail: Evidence for a lunar core? *Proc. Lunar Planet. Sci. Conf. 12th,* pp. 831–836.

Saur J., Strobel D. F., and Neubauer F. M. (1998) Interaction of the jovian magnetosphere with Europa: Constraints on the neutral atmosphere. *J. Geophys. Res., 103,* 19947–19962.

Schilling N., Khurana K. K., and Kivelson M. G. (2004) Limits on an intrinsic dipole moment in Europa. *J. Geophys. Res., 109,* E05006, DOI: 10.1029/2003JE002166.

Schilling N., Neubauer F. M., and Saur J. (2007) Time-varying interaction of Europa with the jovian magnetosphere: Constraints on the conductivity of Europa's subsurface ocean. *Icarus, 192,* 41–55, DOI: 10.1016/j.icarus.2007.06.024.

Schilling N., Neubauer F. M., and Saur J. (2008) Influence of the internally induced magnetic field on the plasma interaction of Europa. *J. Geophys. Res., 113,* A03203, DOI: 10.1029/2007JA012842.

Schultz A. and Larsen J. C. (1987) On the electrical conductivity of the mid-mantle: 1. Calculation of equivalent scalar magnetotelluric response functions. *Geophys. J. R. Astron. Soc., 88,* 733–761.

Schubert G., Lichtenstein B. R., Coleman P. J., and Russell C. T. (1974) *J. Geophys. Res., 79,* 2007–2013.

Schuster A. (1886) On the diurnal period of terrestrial magnetism. *Philos. Mag., 5, Series 21,* S.349–S.359.

Schuster A. (1889) The diurnal variation of terrestrial magnetism. *Philos. Trans. R. Soc., A 180,* 467–518.

Shoemaker E. M. and Wolfe R. F. (1982) Cratering time scales for the Galilean satellites. In *Satellites of Jupiter* (D. Morrison, ed.), pp. 277–339. Univ. of Arizona, Tucson.

Sonett C. P. (1982) Electromagnetic induction in the Moon. *Rev. Geophys. Space Phys., 20,* 411.

Velimsky J., Martinec Z., and Everett M. E. (2006) Electrical conductivity in the Earth's mantle inferred from CHAMP satellite measurements — 1. Data processing and 1-D inversion. *Geophys. J. Inter., 166,* 529–542.

Volwerk M., Kivelson M. G., and Khurana K. K. (2001) Wave activity in Europa's wake: Implications for ion pickup. *J. Geophys. Res., 106,* 26033–26048.

Volwerk M., Khurana K., and Kivelson M. (2007) Europa's Alfvén wings: Shrinkage and displacement influenced by an induced magnetic field. *Ann. Geophys., 25,* 905–914.

Zahnle K., Dones L., and Levison H. F. (1998) Cratering rates on the Galilean satellites. *Icarus, 136,* 202–222.

Zimmer C., Khurana K. K., and Kivelson M. G. (2000) Subsurface oceans on Europa and Callisto: Constraints from Galileo magnetometer observations. *Icarus, 147,* 329–347.

Zolotov M. Y. and Shock E. L. (2001) Composition and stability of salts on the surface of Europa and their oceanic origin. *J. Geophys. Res., 106,* 32815–32827.

Color Section

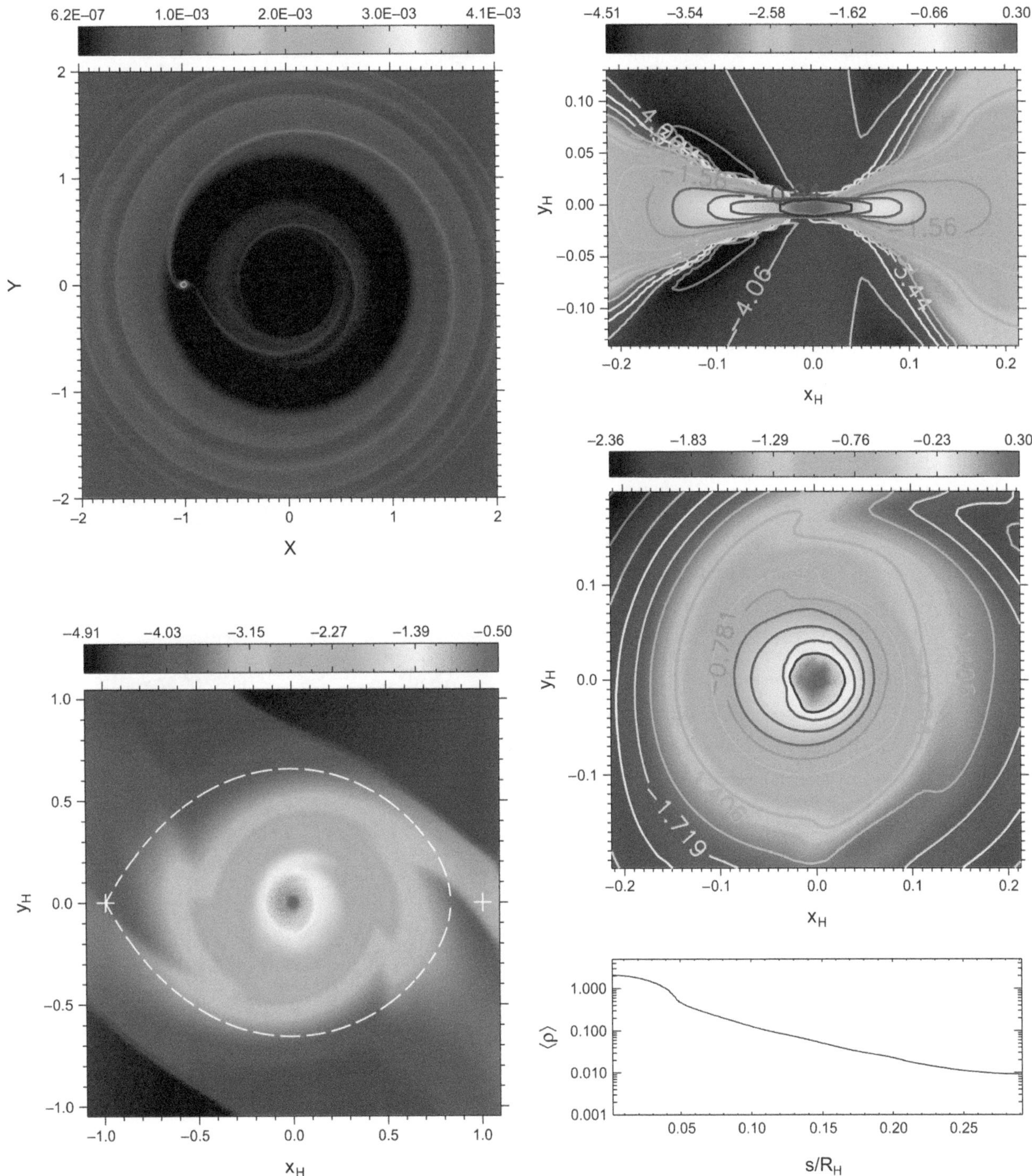

Plate 1. Formation of a circumplanetary disk around a Jupiter-mass planet in three dimensions. The top panel on the left shows the mass density distribution, ρ, in the circumstellar disk's midplane. The bottom panel on the left as well as the top and center panels on the right show density distributions, in logarithmic scale, within the planet's Roche lobe (the teardrop-shaped region marked by the dashed line). Iso-density contours are also indicated in two panels on the right. The logarithm (base 10) of the azimuthally averaged density in the disk's midplane is shown in the bottom panel on the right, where s represents the distance from the planet. The units on the axes are either the planet's orbital radius, r_p (X and Y coordinates), or the Hill radius, R_H (x_H, y_H, and z_H coordinates). The units of ρ are such that 10^{-3} corresponds to 10^{-12} g cm^{-3}.

Accompanies chapter by Estrada et al. (pp. 27–58).

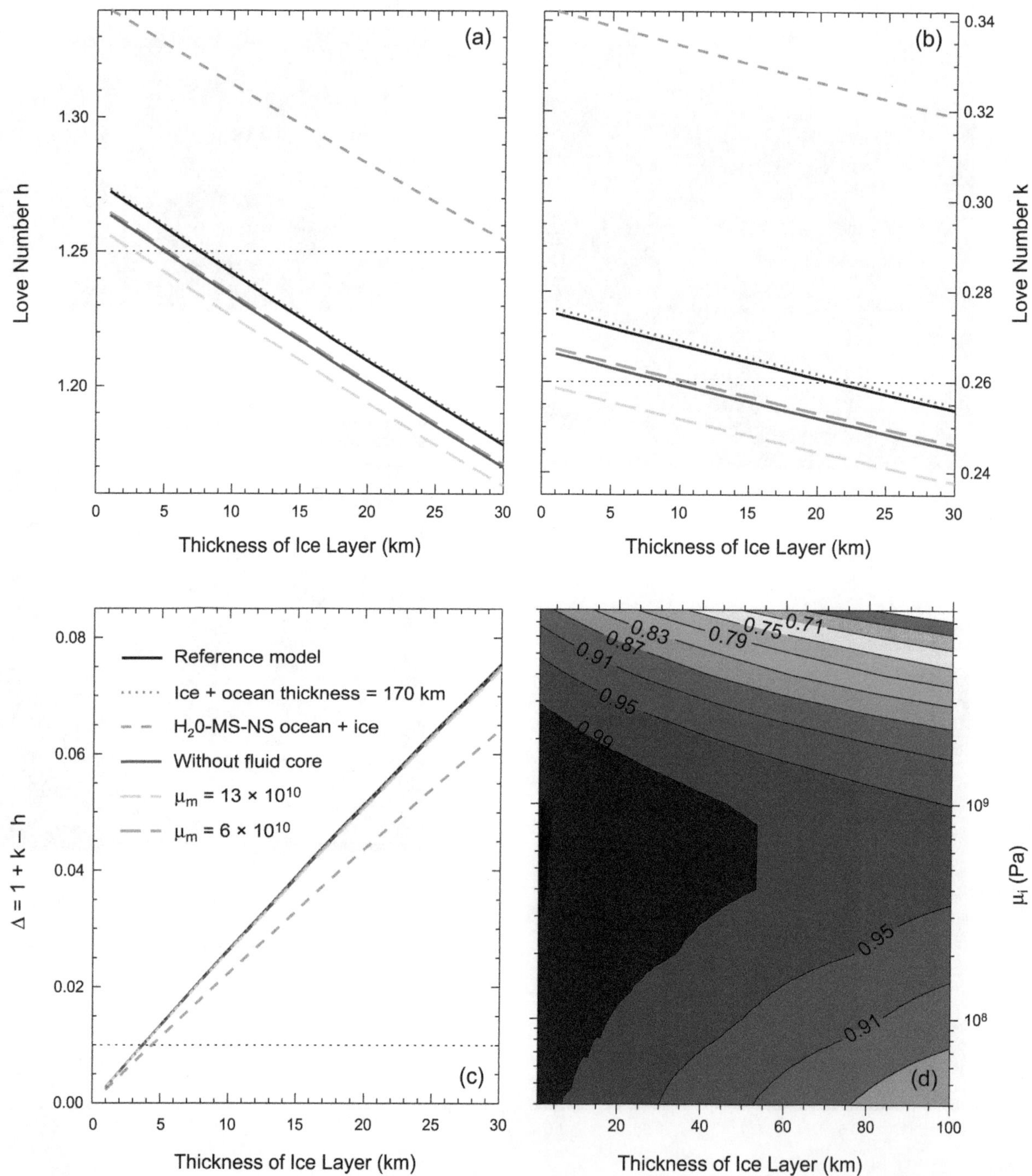

Plate 2. Elastic results for **(a)** Love number h, **(b)** k, and **(c)** $\Delta = 1 + k - h$. All results are plotted vs. the thickness of the icy shell. The different lines correspond to different assumptions about Europa's structural parameters. If not specified in the legend in **(c)**, all parameters are equal to the default values given in Table 2. **(d)** Ratio between our exact numerical results for Δ and the approximation in equation (31), as a function of d and μ_i. The reference parameter values are used for all other parameters. The results show that equation (31) serves as a reasonable approximation to Δ, especially for small values of d.

Accompanies chapter by Sotin et al. (pp. 85–117).

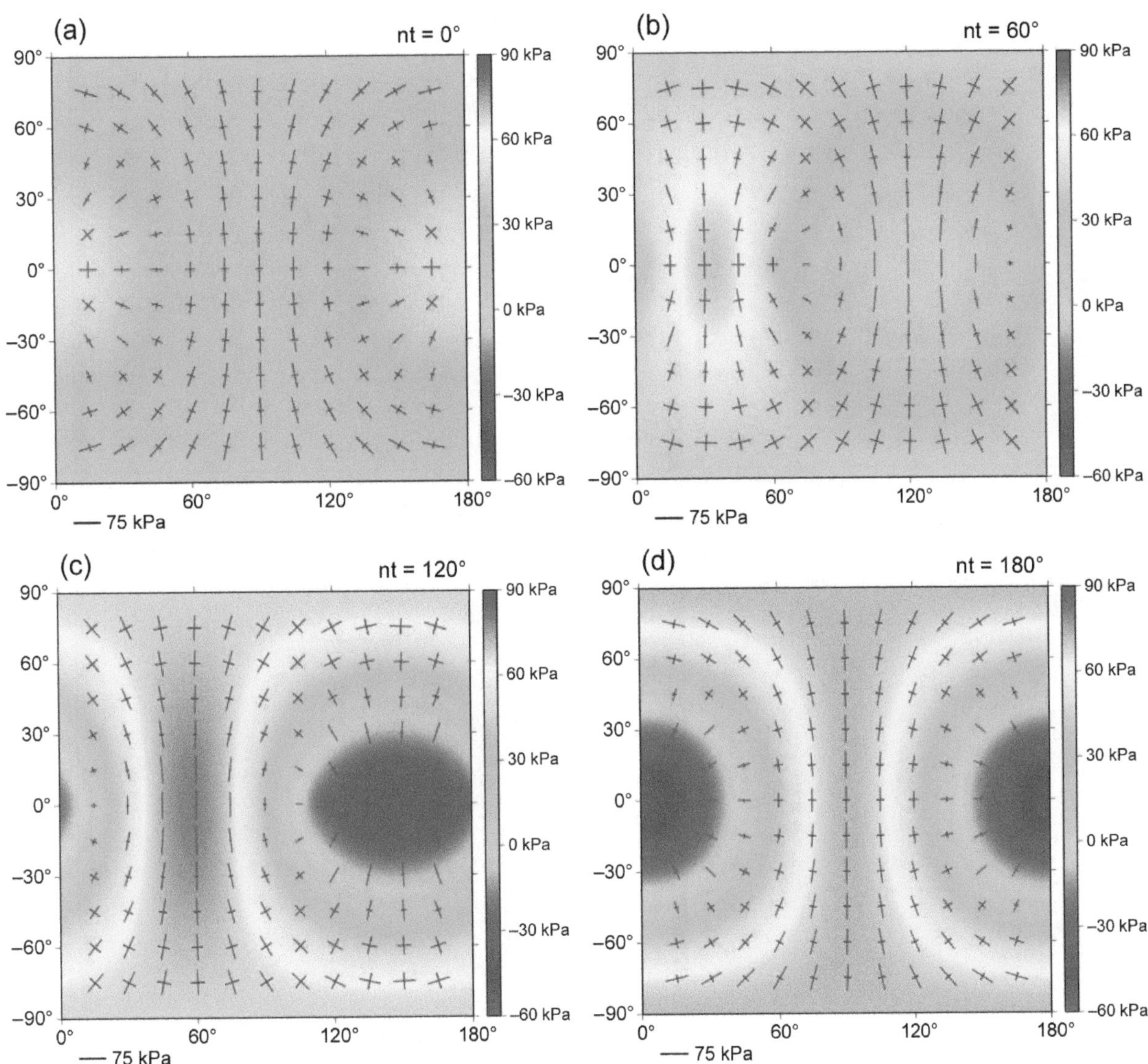

Plate 3. Diurnal stress results from the viscoelastic model, at different values of the angle after perijove (nt in equations (12)–(13)), computed using parameters from Table 2. The subjovian point at perijove is at latitude (y-axis) 0°, longitude (ϕ, x-axis) 0°. East is positive. Results for $180° < \phi < 360°$ are the same as those between 0° and 180°: $\tau(\phi + 180°) = \tau(\phi)$. Stresses in the second half of the orbit ($180° < nt < 360°$) are east-west reflections of those in the first half. Tics show the magnitude and orientation of the principal components of the stresses on the surface of the satellite. Compression (blue) is negative and tension (red) is positive. Background color indicates the magnitude of the most tensile of the two principal components. From *Wahr et al.* (2009); figure provided by Zane Selvans.

Accompanies chapter by Sotin et al. (pp. 85–117).

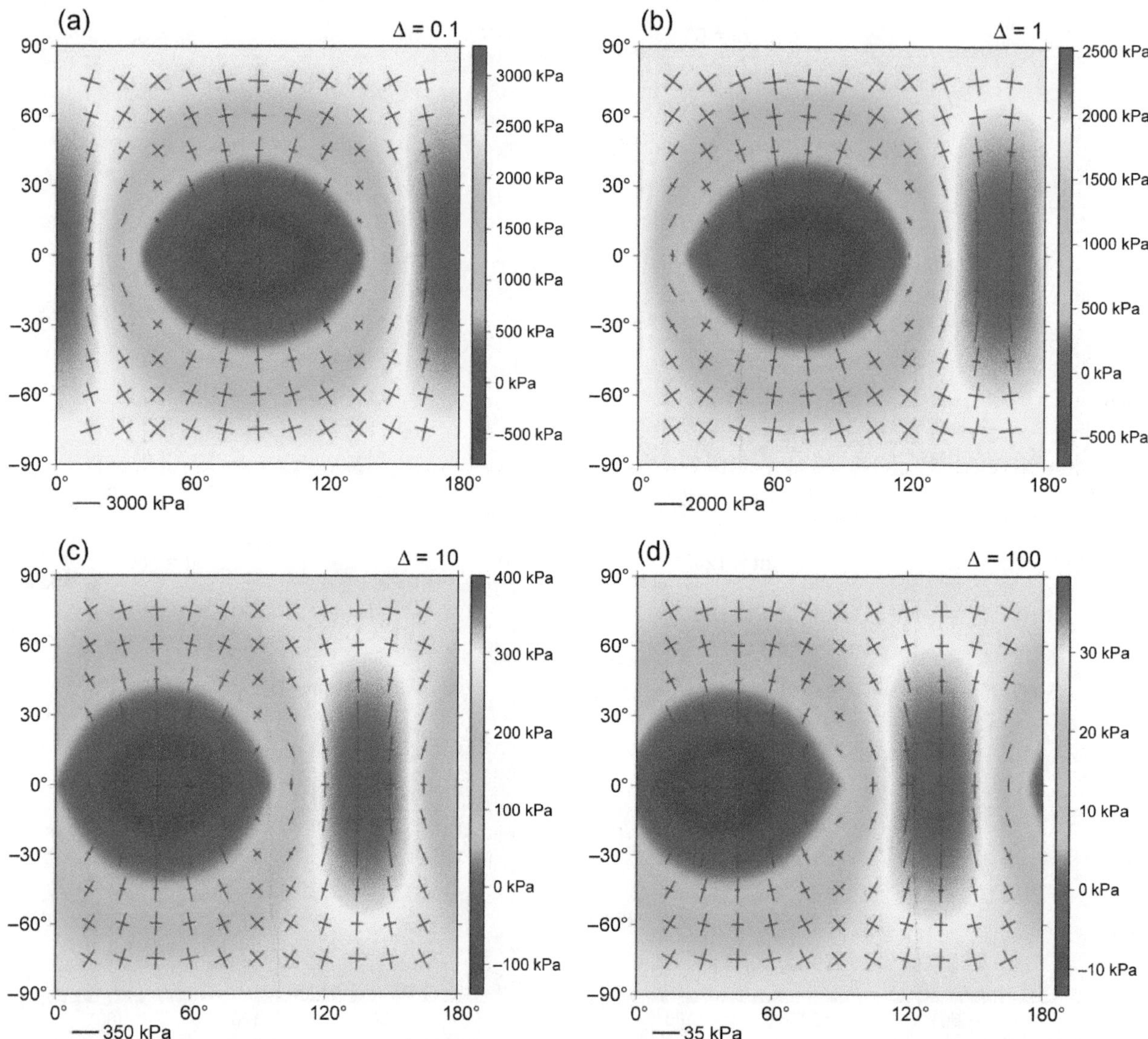

Plate 4. NSR stress results from our viscoelastic model for different values of δ. Plots are similar to those in Fig. 6. The subjovian point is at latitude = longitude = 0°. And $\tau(\phi + 180°) = \tau(\phi)$. Results are computed using the model parameters shown in Table 2, and using a viscosity of 10^{22} Pa s in the outer ice layer. Note that the arrow length and color scales are different in the different panels. From *Wahr et al.* (2009); figure provided by Zane Selvans.

Accompanies chapter by Sotin et al. (pp. 85–117).

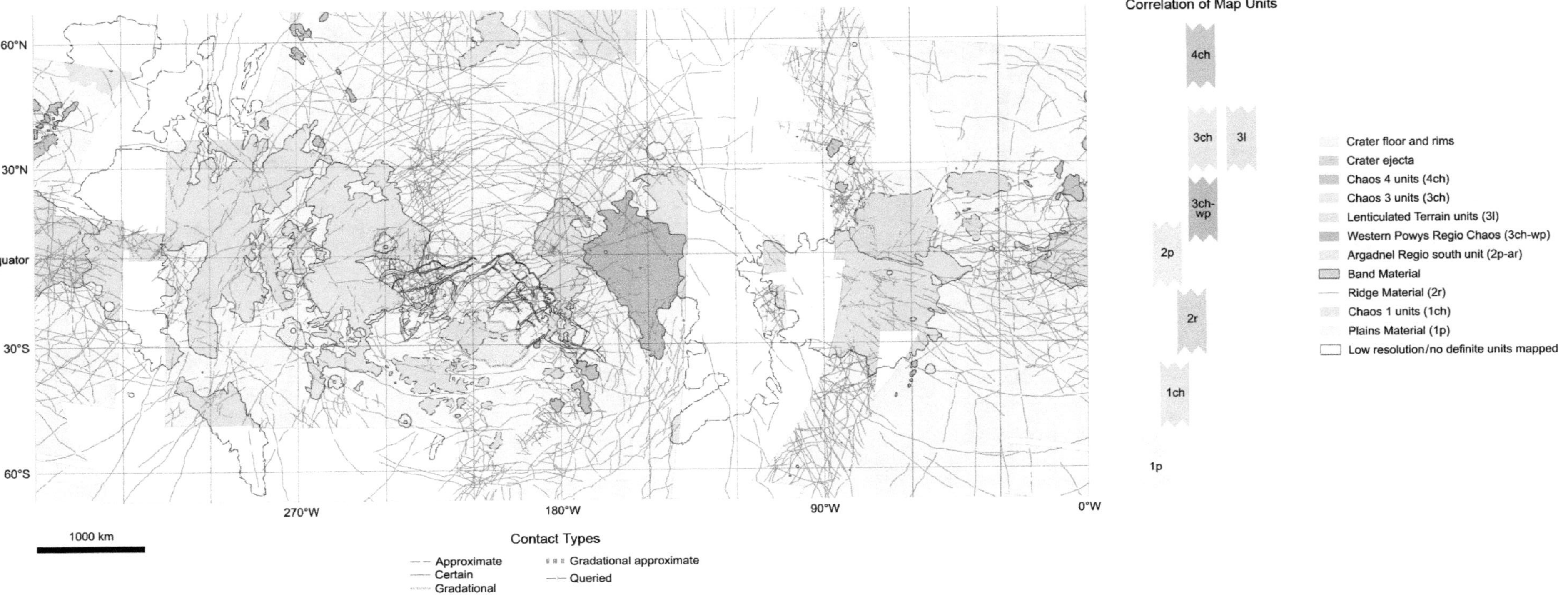

Plate 5. Global geological map of Europa, including stratigraphic column of map units.

Accompanies chapter by Doggett et al. (pp. 137–159).

Plate 6. Cilix (D ~ 19 km) is one of the largest, preserved, and best-imaged complex craters on Europa. High-resolution mosaic shows the basic features including a small and slightly offset central peak complex and one or two terrace scarps. The rimwall is otherwise much narrower than lunar craters of similar size. Note the dark annulus, which may be ejecta or the disruption of surface frosts. Some of this dark material ponds in topographic lows, although this could be due to mass wasting or drainage of mobilized ejecta material. Portions of a pancake ejecta deposit can be seen onlapping ridged plains at upper right and center right. Steep local ridge topography gives this unit an artificial lobate or feathered appearance. Mosaic has been color-coded to show stereo-derived topographic information, showing the classic flat depressed floor and raised rim. Horizontal scale bar is 10 km, vertical topography scale is 1 km.

Accompanies chapter by Schenk and Turtle (pp. 181–198).

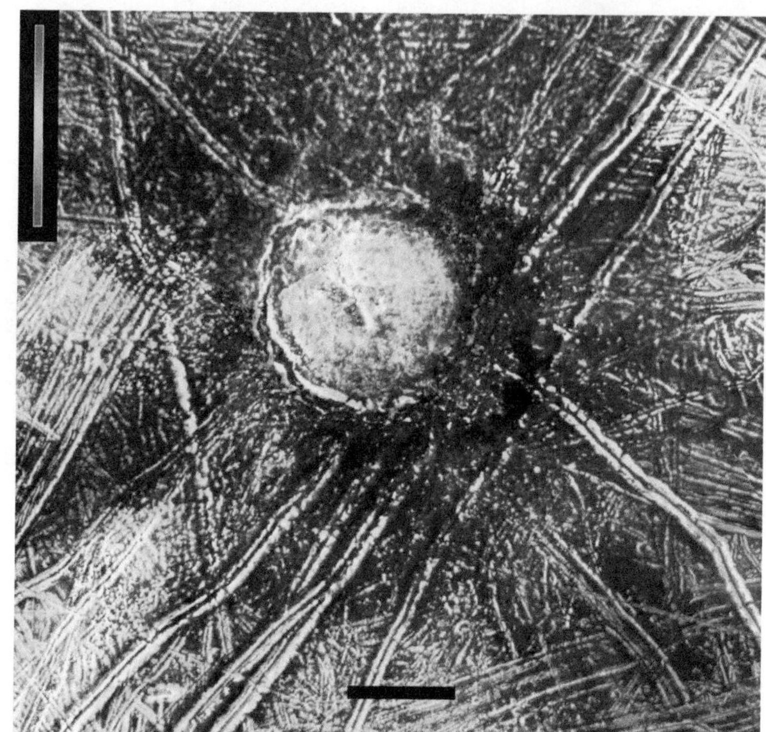

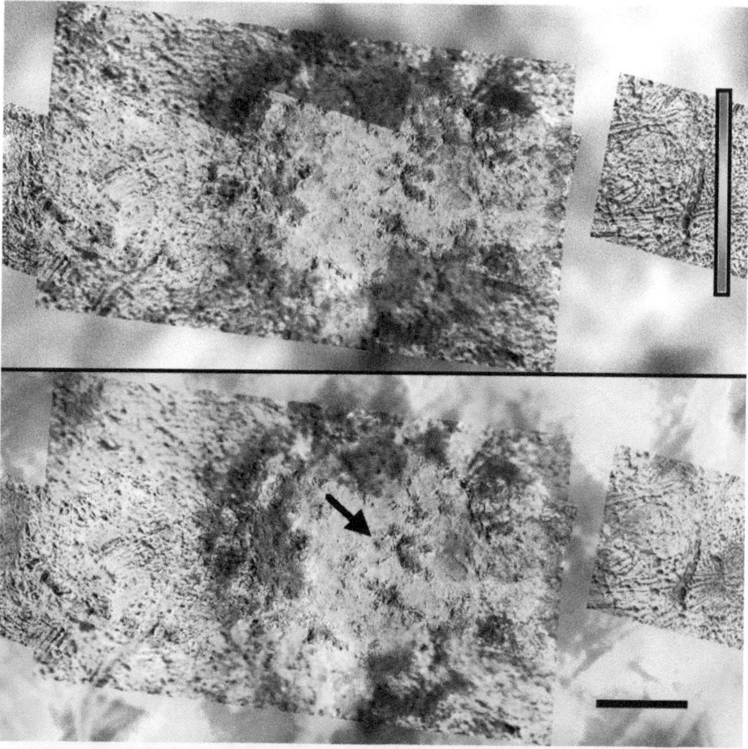

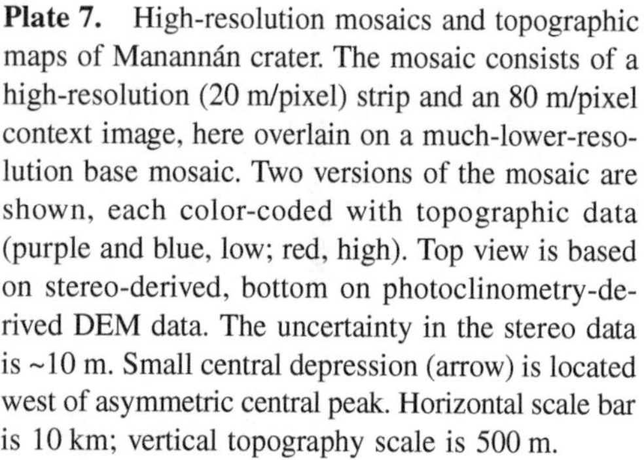

Plate 7. High-resolution mosaics and topographic maps of Manannán crater. The mosaic consists of a high-resolution (20 m/pixel) strip and an 80 m/pixel context image, here overlain on a much-lower-resolution base mosaic. Two versions of the mosaic are shown, each color-coded with topographic data (purple and blue, low; red, high). Top view is based on stereo-derived, bottom on photoclinometry-derived DEM data. The uncertainty in the stereo data is ~10 m. Small central depression (arrow) is located west of asymmetric central peak. Horizontal scale bar is 10 km; vertical topography scale is 500 m.

Accompanies chapter by Schenk and Turtle (pp. 181–198).

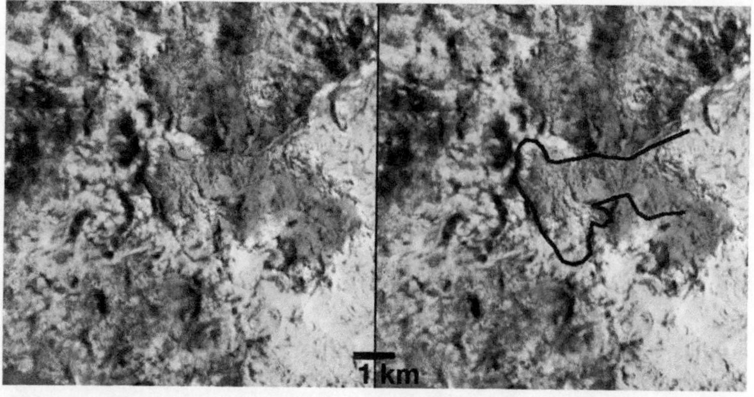

Plate 8. High-resolution view of rim of Manannán showing lobate flow material. Deposit is interpreted as impact-melt-rich material flowing over the rim scarp. View is enlargement of Plate 7 (top). Scene is ~10 km across.

Accompanies chapter by Schenk and Turtle (pp. 181–198).

Plate 9. Three-color mosaic of Manannán obtained by Galileo. The crater floor (deposit inside the dark ring) is covered by bluish and brownish units, some of which appear to correlate with morphologic evidence of impact melt. Color filters are 0.9, 0.6, and 0.4 μm. Image resolution is 80 m.

Accompanies chapter by Schenk and Turtle (pp. 181–198).

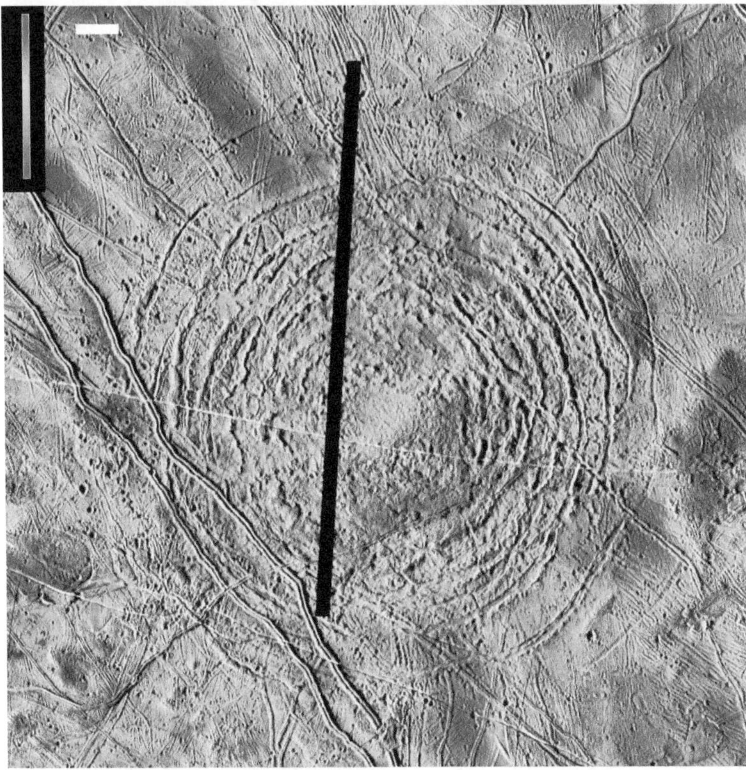

Plate 10. Tyre multiring impact feature, Europa. Color-coding displays topographic data obtained from photoclinometry (purple, low; red, high). Many of the domed and depressed features are real, based on shading information, but the general reliability of longer-wavelength information is unknown. Whether the topographic rise in the southeast portion of Tyre is real, for example, is uncertain, but similar features have been identified in stereo DEMs, where long-wavelength topography is controlled. White horizontal scale bar is 10 km; vertical topography scale is 750 m.

Accompanies chapter by Schenk and Turtle (pp. 181–198).

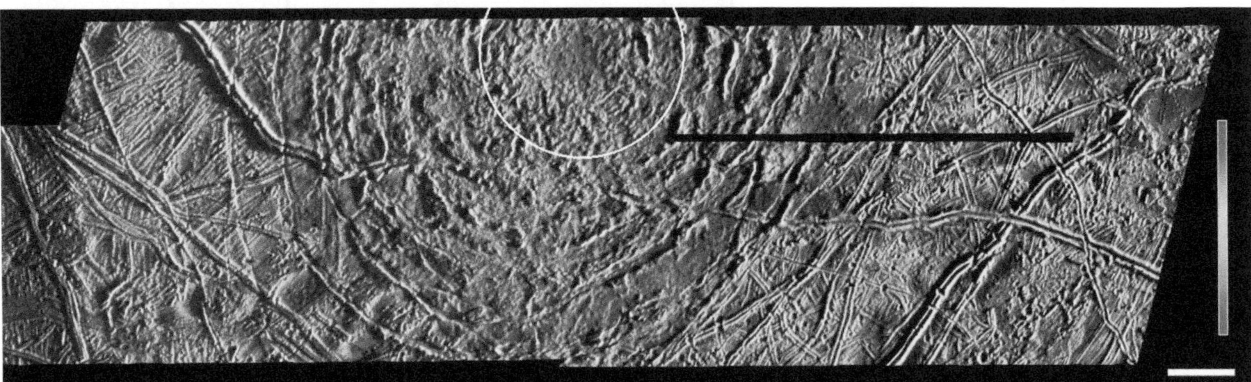

Plate 11. Callanish multiring impact feature, Europa. Position of the final crater diameter estimated from scaling is shown by the white circle. Color-coding displays topographic data obtained from stereo-controlled photoclinometry (in which the stereo component controls long-wavelength topography). Mosaic resolution 120 m per pixel. Horizontal scale bar is 10 km; vertical topography scale bar is 750 m. Vertical DEM precision is ~25 m.

Accompanies chapter by Schenk and Turtle (pp. 181–198).

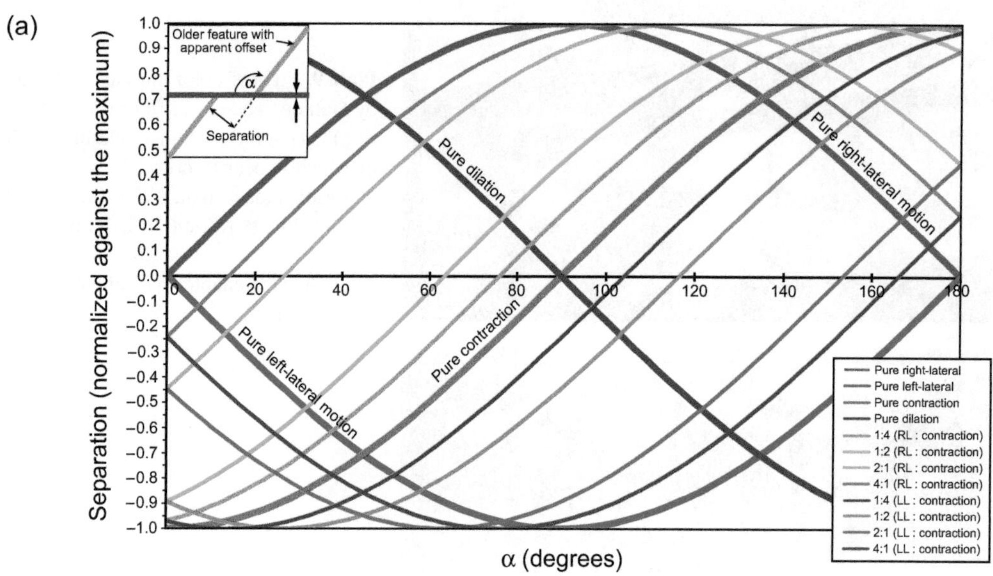

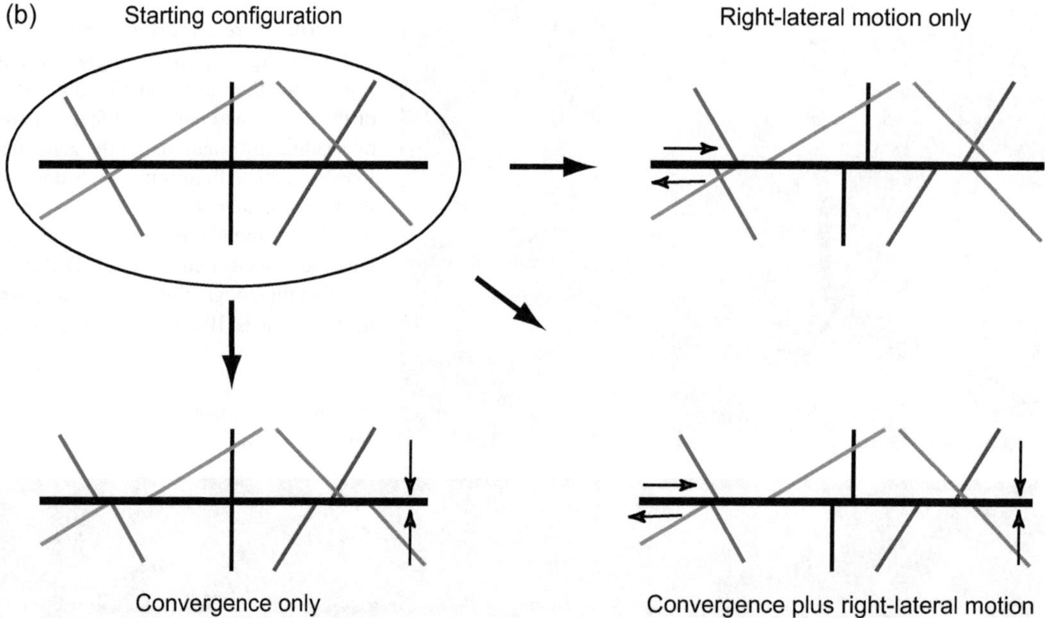

Plate 12. **(a)** Analytical curves can be used to differentiate the relative amounts of strike-slip motion (whether right-lateral or left-lateral) and contraction across a ridge/strike-slip fault based on the amount of separation between two halves of an apparently offset feature (inset, upper left). α is the angle measured from the ridge/strike-slip fault to the offset feature in a clockwise sense. A separate set of curves exists for the case of dilation plus strike-slip motion (*Vetter,* 2005). **(b)** Illustration of various configurations of mismatches of crosscut features in response to pure right-lateral motion, convergence only, or a combination (motion is along the thicker black line, representing a ridge or strike-slip fault). If α is 90°, offsets only occur if there is lateral motion. Pure convergence can produce a mixture of left- and right-lateral offsets, depending on α.

Accompanies chapter by Kattenhorn and Hurford (pp. 199–236).

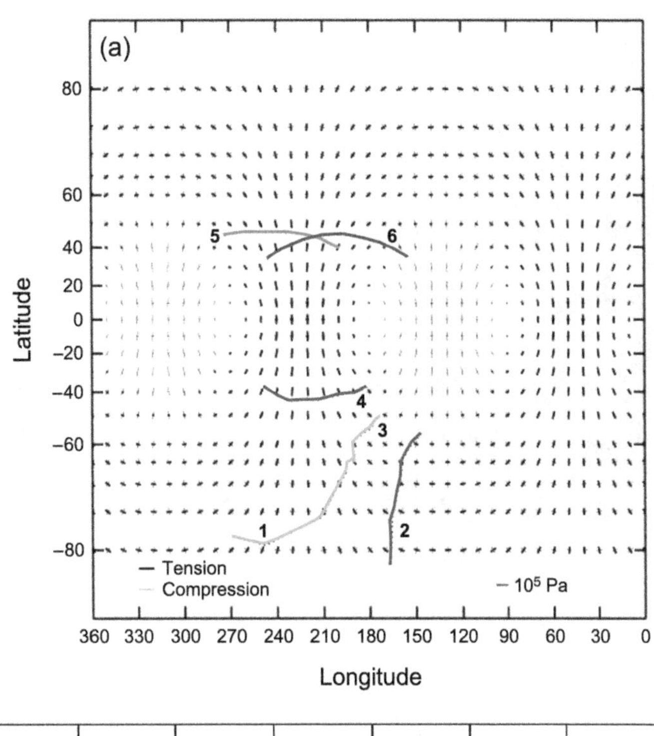

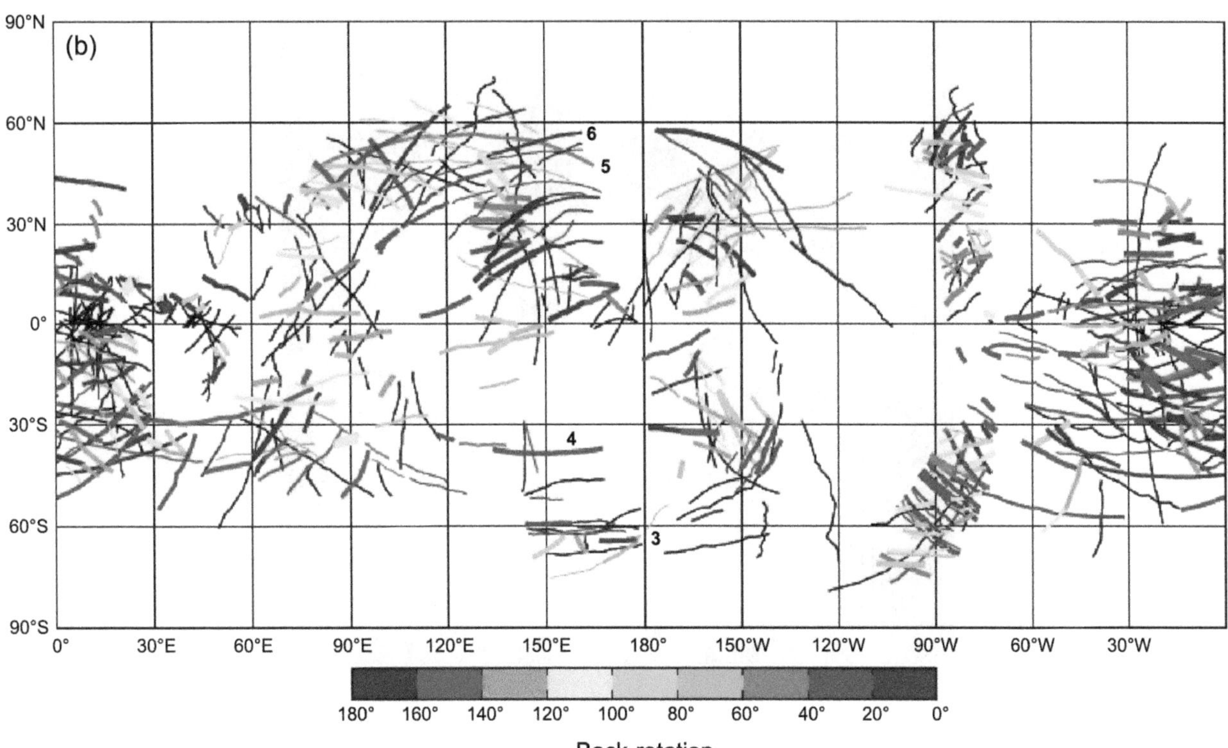

Plate 13. **(a)** The tidal stress-field produced by 1° of nonsynchronous rotation (*Greenberg et al.,* 1998) shows a good fit to the locations and orientations of several lineaments. The lineaments are numbered (1) Astypalaea, (2) Thynia, (3) Libya, (4) Agenor, (5) Udaeus, and (6) Minos Lineae. A better fit is produced if fractures are back-rotated westward in longitude relative to the stress field [by an amount given by their color coding, as defined in **(b)**] such that their orientations are perpendicular to the direction of maximum tension, providing a possible original longitude for the creation of the crack. **(b)** The global pattern of lineaments based on Galileo observations illustrates a range of orientations that cannot all be fitted to the same stress template, such as that shown in **(a)**, because they formed at different times. The lineaments are color coded to indicate how far westward they must be back-rotated in order to form perpendicular to the maximum tension produced by the stress of nonsynchronous rotation [map courtesy of Z. Selvans (Selvans et al., in preparation)]. Some of the numbered lineae in **(a)** are numbered accordingly in **(b)**.

Accompanies chapter by Kattenhorn and Hurford (pp. 199–236).

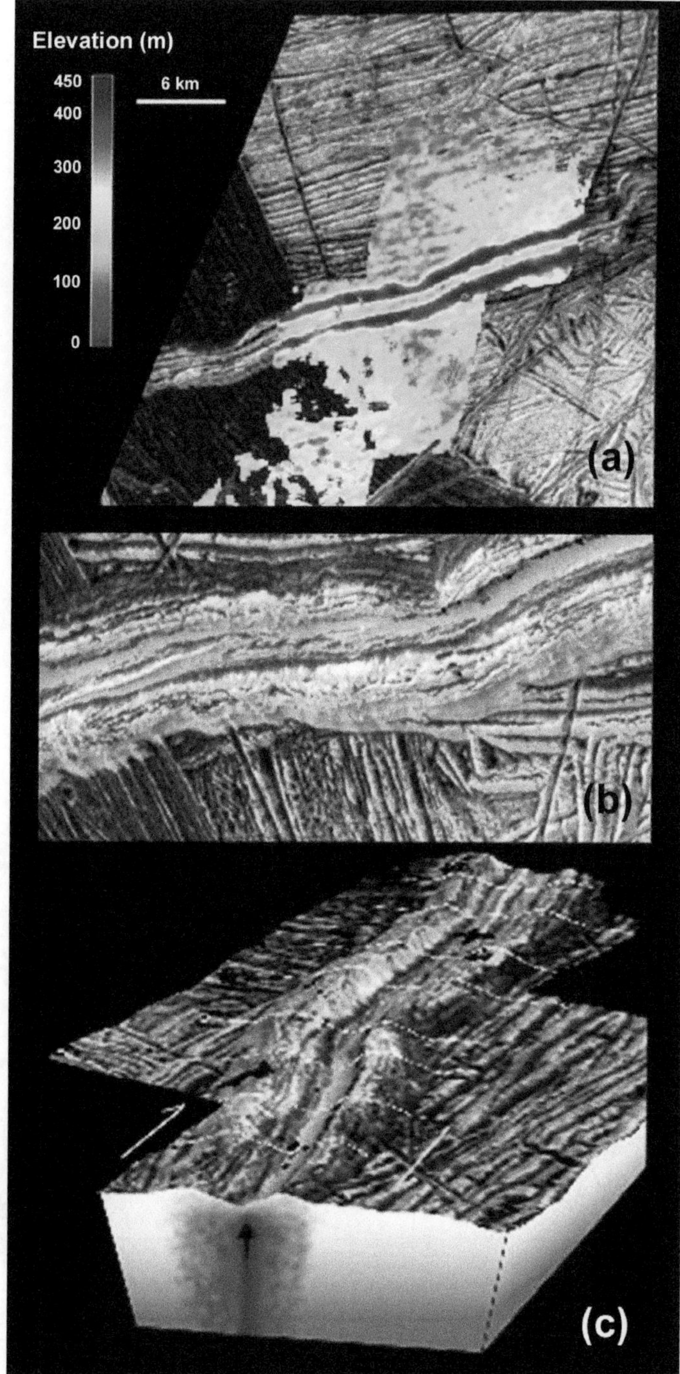

Plate 14. (**a**) Topography derived from stereo imaging across a prominent double ridge observed during the Galileo 12ESWEDGES01 and 02 observations shows that the ridge flanks stand several hundred meters above the surrounding terrain (courtesy of B. Giese, DLR). (**b**) High-resolution image (~12 m/pixel) of the same ridge showing many of the distinctive morphological features indicative of double ridges (image is ~15 km wide). (**c**) A perspective view (courtesy of B. Giese and R. Pappalardo) of the double ridge that illustrates key aspects of the shear heating model for ridge formation (*Gaidos and Nimmo,* 2000; *Nimmo and Gaidos,* 2002). Here, the ridge forms as a result of strike-slip motion along a tensile fracture, causing the ice to warm in a zone surrounding the original trough, with the maximum heating near the brittle-ductile transition depth. The warmed ice is more buoyant than the colder surroundings, leading it to rise up, and/or focus contractional deformation, forming ridge crests that flank a distinct trough.

Accompanies chapter by Prockter and Patterson (pp. 237–258).

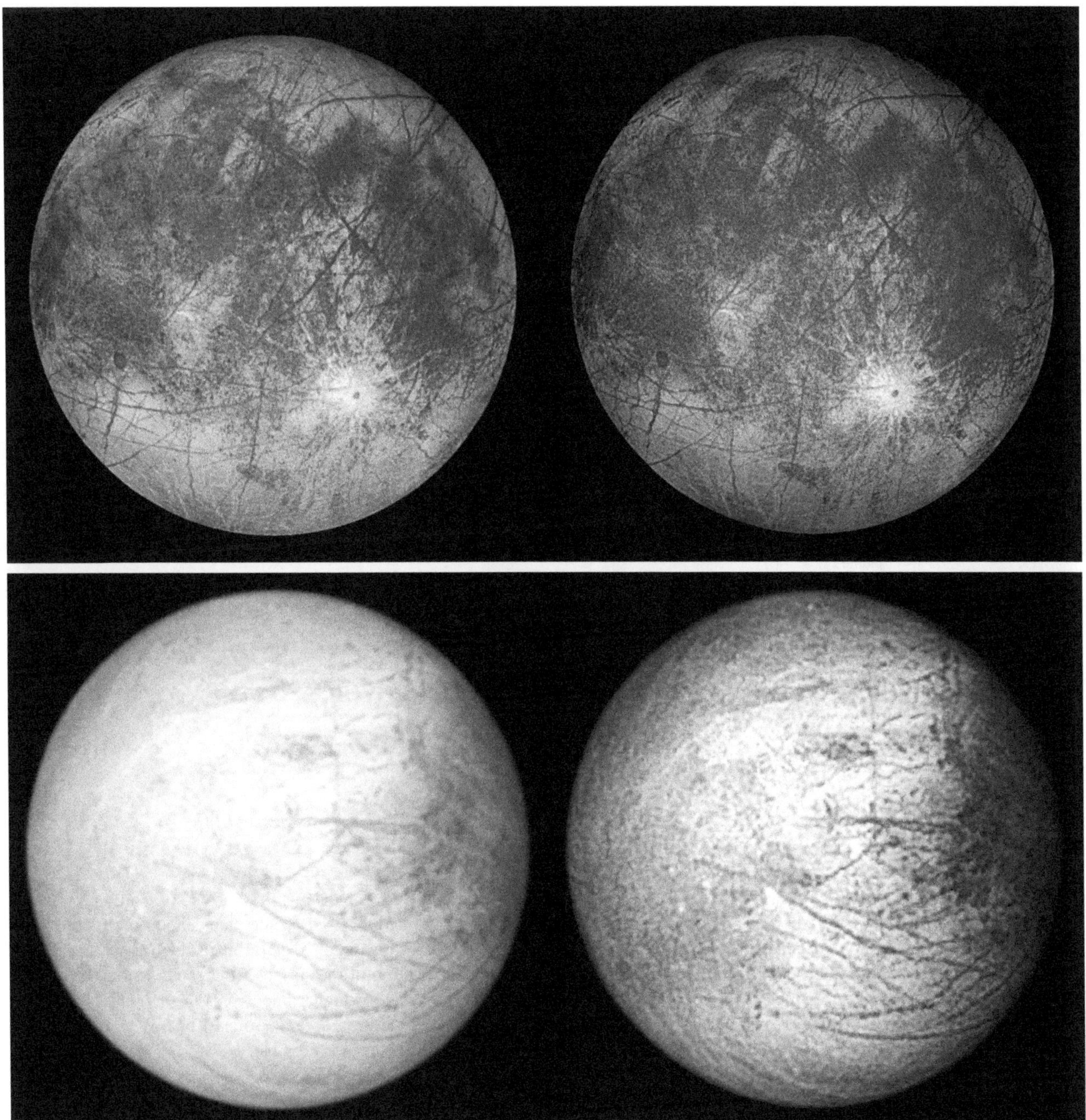

Plate 15. Europa's trailing and leading hemispheres (top and bottom panels, respectively), as imaged by Galileo's Solid State Imaging camera, illustrating the differences between the two sides and the association of the brownish material with geological features. The images at left are natural color, while the images at right are enhanced to show structural detail. The resolutions (and JPL Photojournal number) are 6.9 km/pixel (PIA00502) for the trailing sides and 12.7 km/pixel (PIA01295) for the leading side.

Accompanies chapter by Carlson et al. (pp. 283–327).

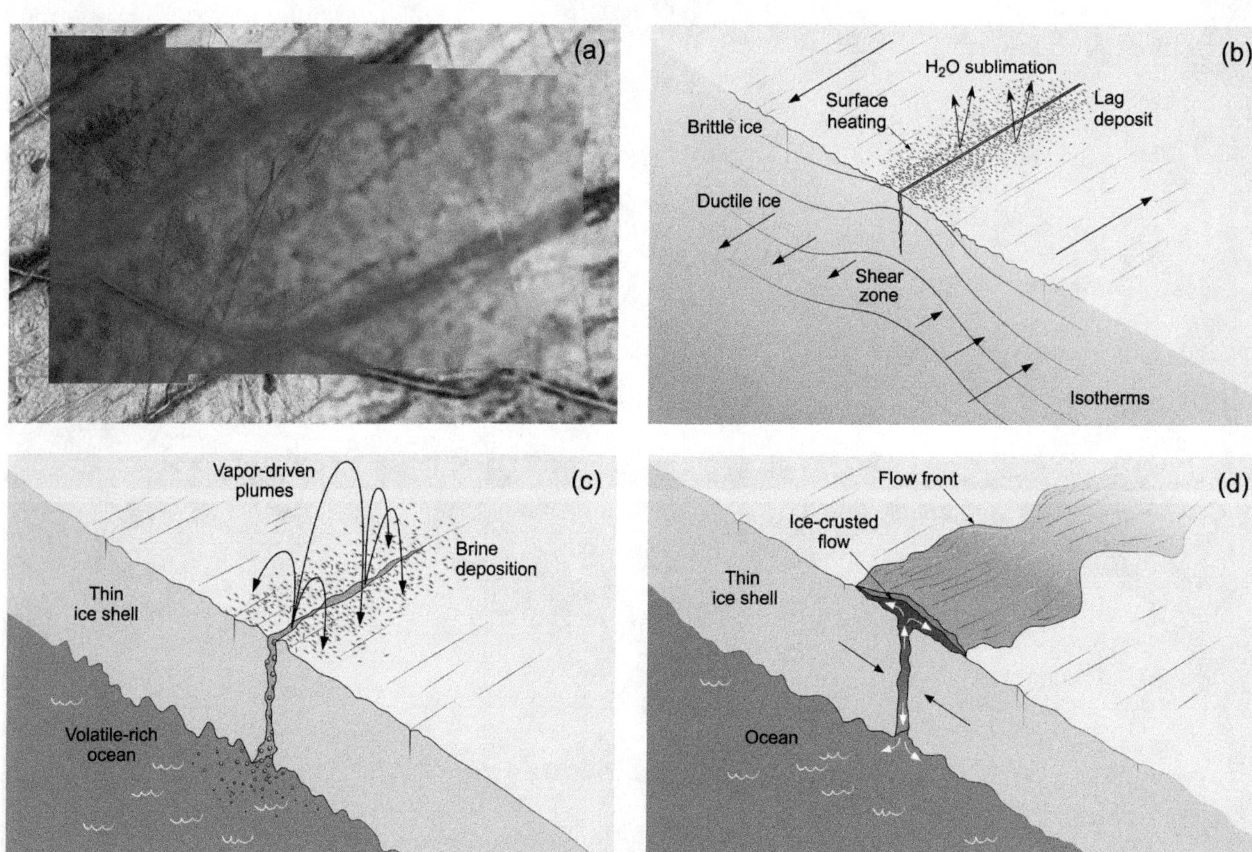

Plate 16. Association of Europa's dark material with lineae and possible geological mechanisms. **(a)** False-color Galileo NIMS observation 11ENCYCLOD01 overlaid upon visible imaging data. Red indicates hydrated materials while blue denotes more pure water ice and frost. Note correlation between red (hydrated) material and lineae. In higher-resolution spectral maps the bright central band of lineae are observed to be water ice (*Carlson et al.,* 2005a). **(b)** Thermal modification of the surface from shear heating of the surface in the region of the linea, sublimating H_2O and leaving enhanced concentrations of higher vapor pressure hydrate and darker, possibly sulfurous material. **(c)** Thin shell volatile-driven explosive cryovolcanism depositing oceanic brine. **(d)** Effusive emplacement of near-surface liquid by tide-induced opening and closing of a crack. Other mechanisms are possible.

Accompanies chapter by Carlson et al. (pp. 283–327).

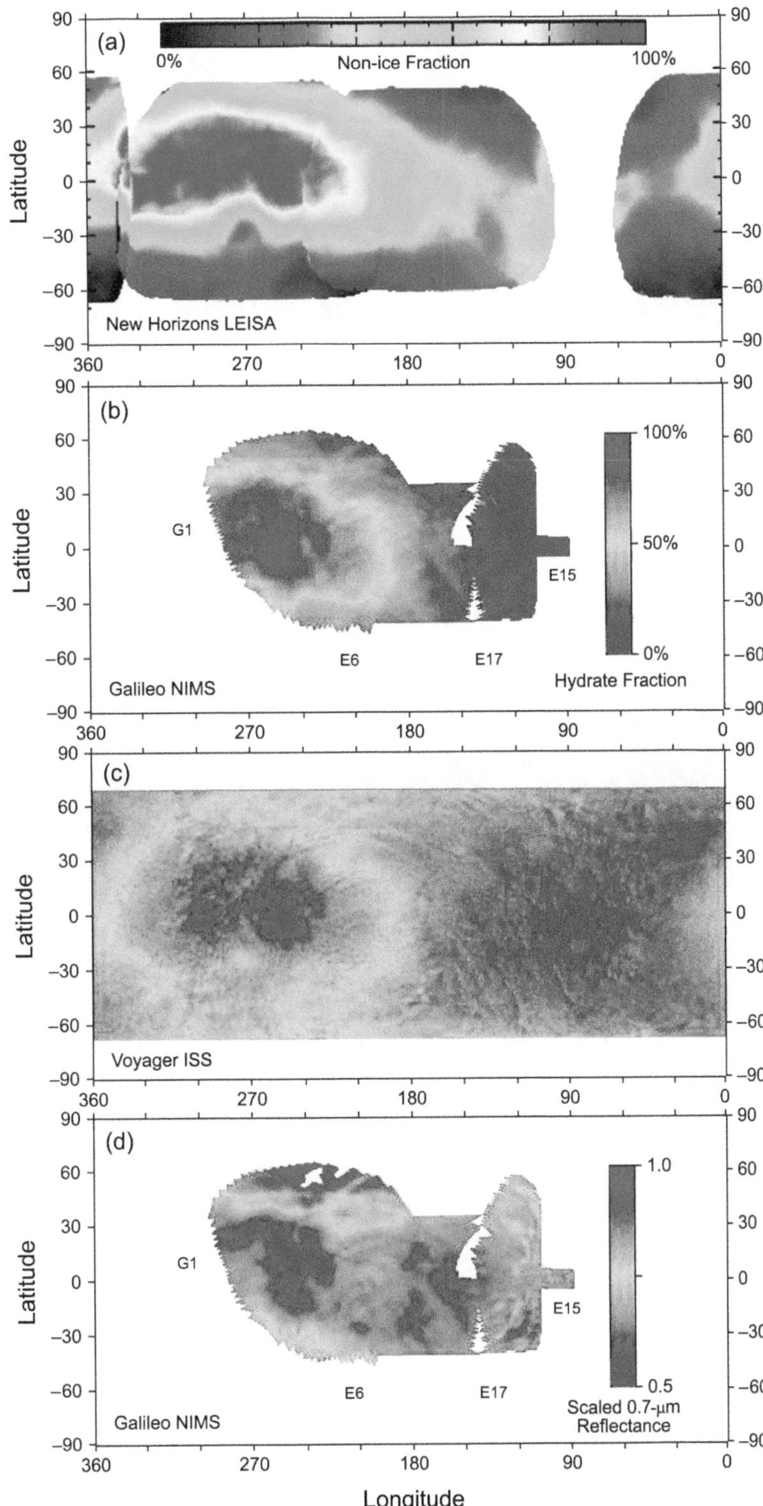

Plate 17. Distribution of hydrate and UV and near-IR albedos. **(a)** Non-ice distribution measured from New Horizons (*Grundy et al.,* 2007). **(b)** Molar distribution of hydrate assuming sulfuric acid hydrate, but representative of other hydrates (*Carlson et al.,* 2005a). **(c)** Voyager UV/V map, from *McEwen* (1986). Blue denotes high UV absorption and corresponding low UV reflectance. (d) Galileo NIMS 0.7 μm/1.2 μm ratio map, scaled (*Carlson et al.,* 2005a). Note the correlation of the non-ice hydrate distribution [**(a)** and **(b)**] with both the UV **(c)** and the near-IR absorber **(d)**, and the trailing-side enhancement of all three.

Accompanies chapter by Carlson et al. (pp. 283–327).

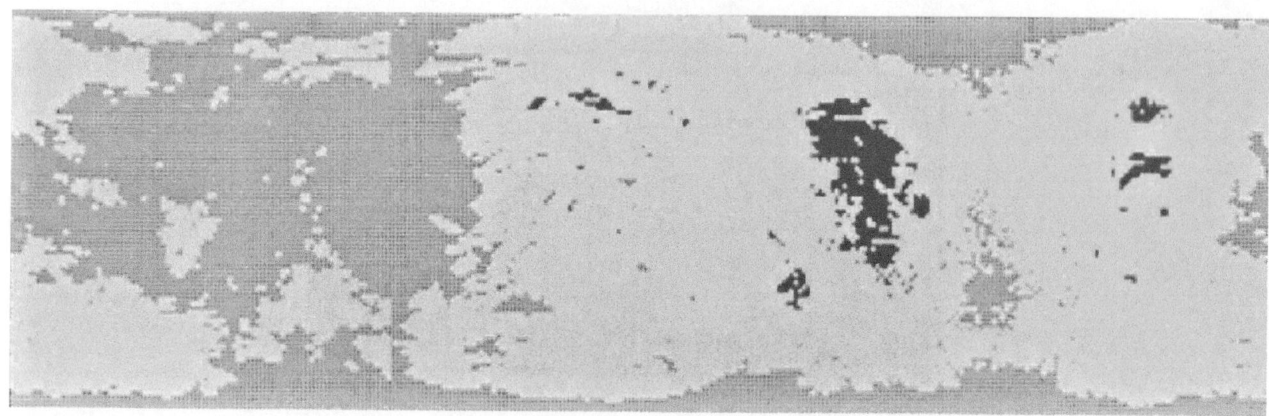

Plate 18. A map of normal reflectances for Europa based on Voyager images (*Johnson et al.,* 1983). The values are green, 36–52%; yellow, 53–70%; red, >71%. The map is a global simple cylindrical projection centered on the antijovian point (0°, 180°W).

Accompanies chapter by Moore et al. (pp. 329–349).

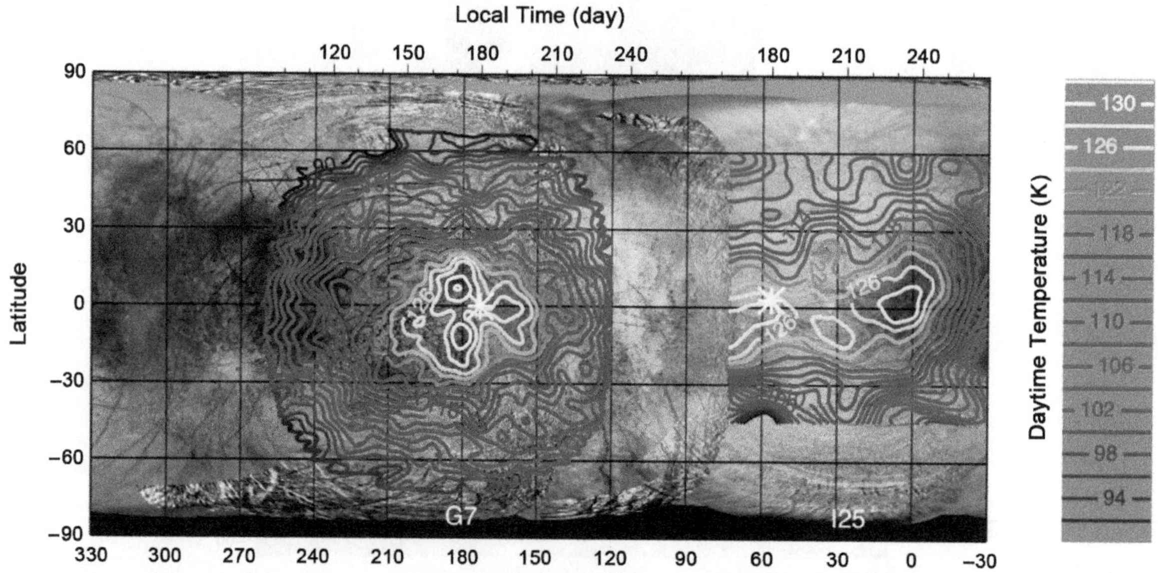

Plate 19. Contours of daytime (top) and nighttime (bottom) brightness temperature distributions on Europa inferred from Galileo PPR data (nighttime from Galileo orbits E17 and I25, and daytime from orbits G7 and I25). Contour interval is 1 K for the nighttime map and 2 K for the daytime map, and the contour color scheme is different for the two maps. Local time (top axis) is presented in degrees of rotation after midnight, and the subsolar points for the daytime data are shown by a white star. Base map is from Galileo and Voyager images.

Accompanies chapter by Moore et al. (pp. 329–349).

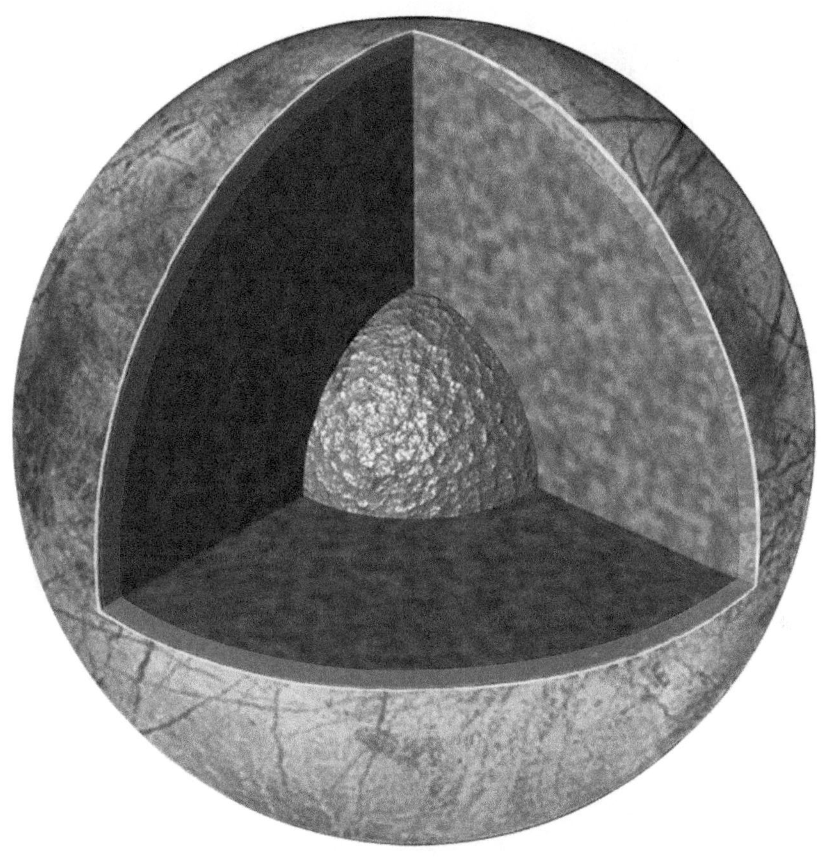

Plate 20. Cutaway of Europa's interior.

Accompanies chapter by Schubert et al. (pp. 353–367).

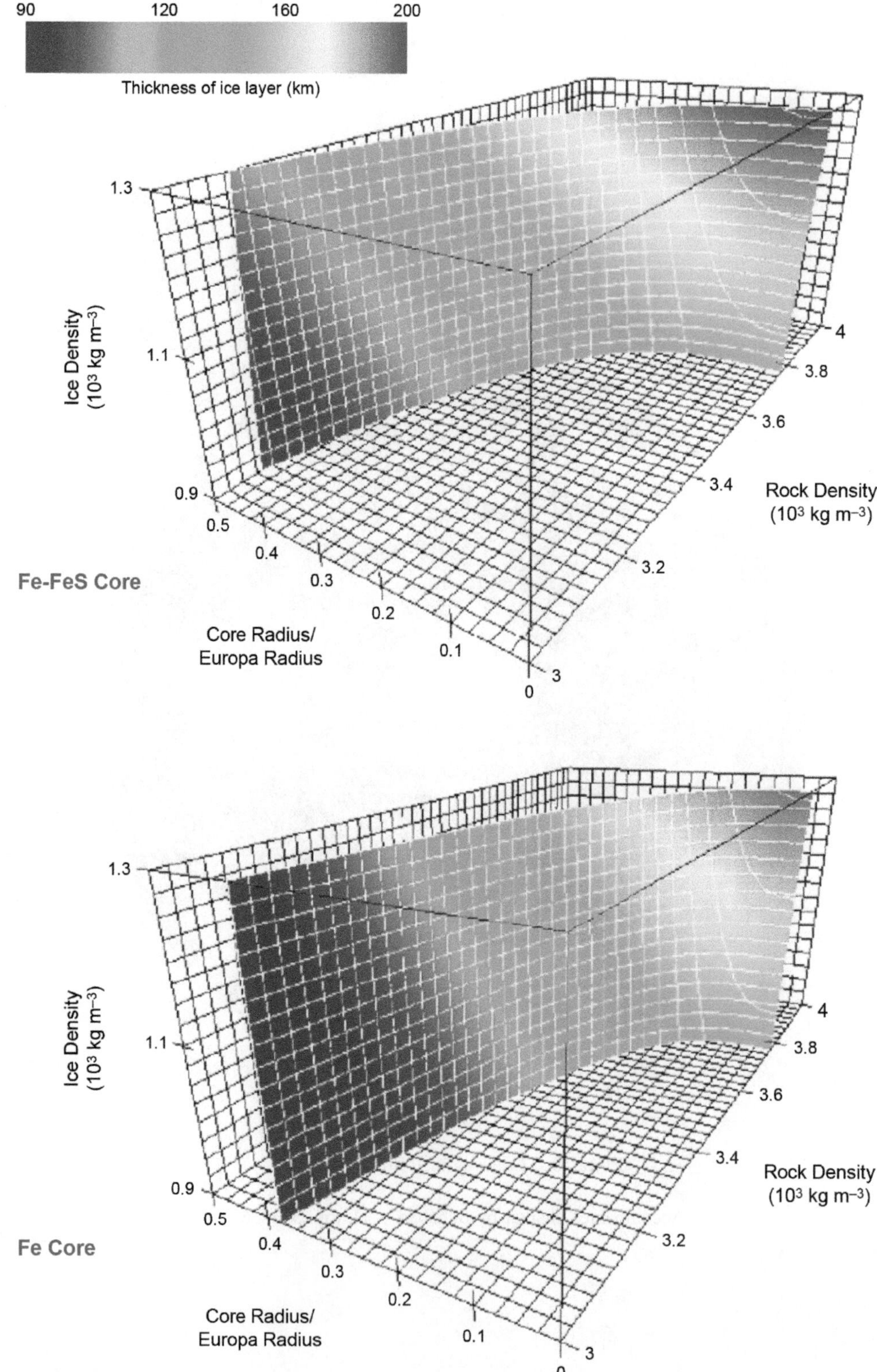

Plate 21. Possible three-layer models of Europa consistent with its mean density and axial MoI (MoI = 0.346) (*Anderson et al.,* 1998). Two sets of models are considered, one having Fe cores with density 8000 kg m^{-3} and the other having Fe-FeS cores with density 5150 kg m^{-3}. Any point on one of the surfaces defines an interior structure with properties given by the coordinate axis values and the color of the surface. Ice density refers to the density of the outer spherical shell of the model, which is predominately water in either ice or liquid form with an admixture of some rock. The color of the surface gives the thickness of this outer shell according to the color bar. Rock density refers to the density of the mainly silicate intermediate shell, which may also contain some metal. Possible Europa models are defined by the surfaces whose colors give the thickness of the water ice liquid outer shell. Other model parameters (outer shell or ice density, intermediate shell or rock density, and core radius) are provided by the coordinate axes. Two-layer Europa models are given by the intersection of the model surfaces with the core radius = 0 plane.

Accompanies chapter by Schubert et al. (pp. 353–367).

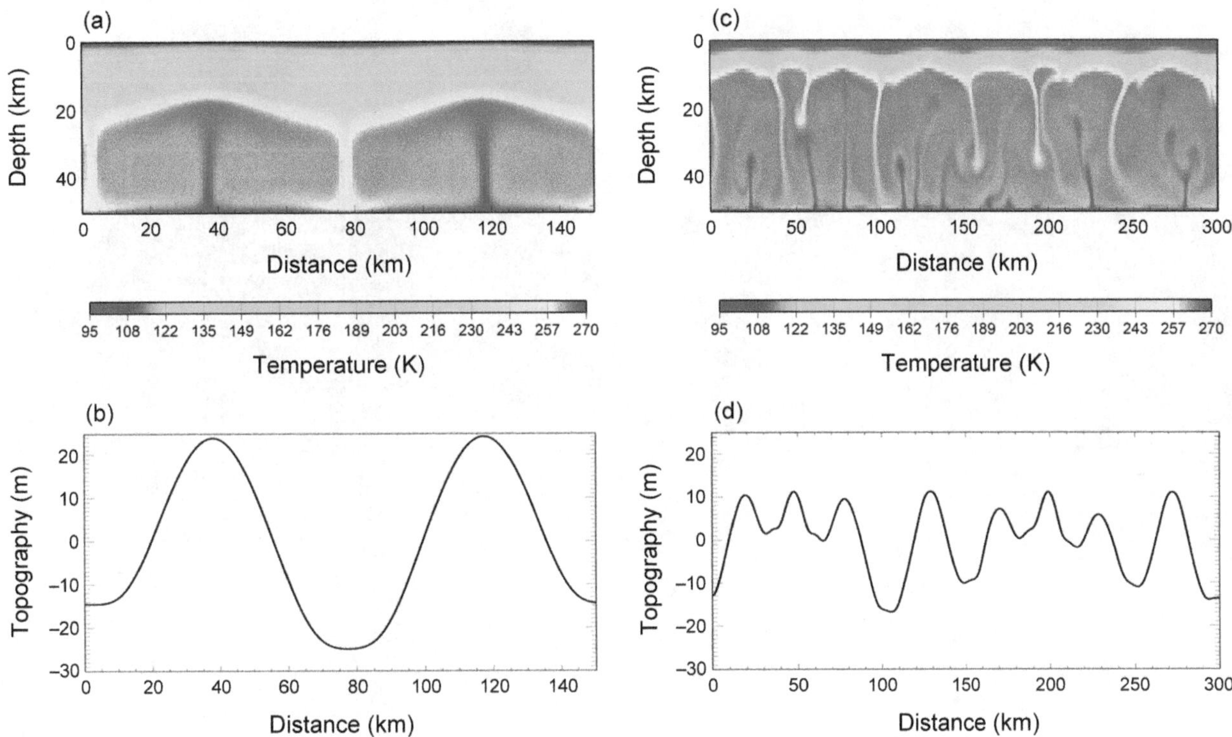

Plate 22. **(a),(c)** Temperature and **(b),(d)** dynamic topography from simulations of thermal convection in a 50-km-thick europan ice shell from *Showman and Han* (2004). Rheology is Newtonian with Q* = 60 kJ mol^{-1}, $\eta_o = 10^{13}$ Pa s (right), and $\eta_o = 10^{14}$ Pa s (left), and upper viscosity cutoff of $10^9\ \eta_o$. **(a),(b)** A high melting point viscosity leads to sluggish convection beneath a thick stagnant lid, and topography on the order of tens of meters, much lower than observed on Europa. Domain is 150 km wide and 50 km deep. **(c),(d)** Convection in an ice shell melting point viscosity of 10^{13} Pa s leads to vigorous convection characterized by narrow upwellings and a thinner stagnant lid. Domain is 300 km wide by 50 km deep. Although the topography predicted by more vigorous convection has a smaller wavelength, the plume buoyancy is unchanged between the two cases: the dynamic topography is approximately tens of meters.

Accompanies chapter by Barr and Showman (pp. 405–430).

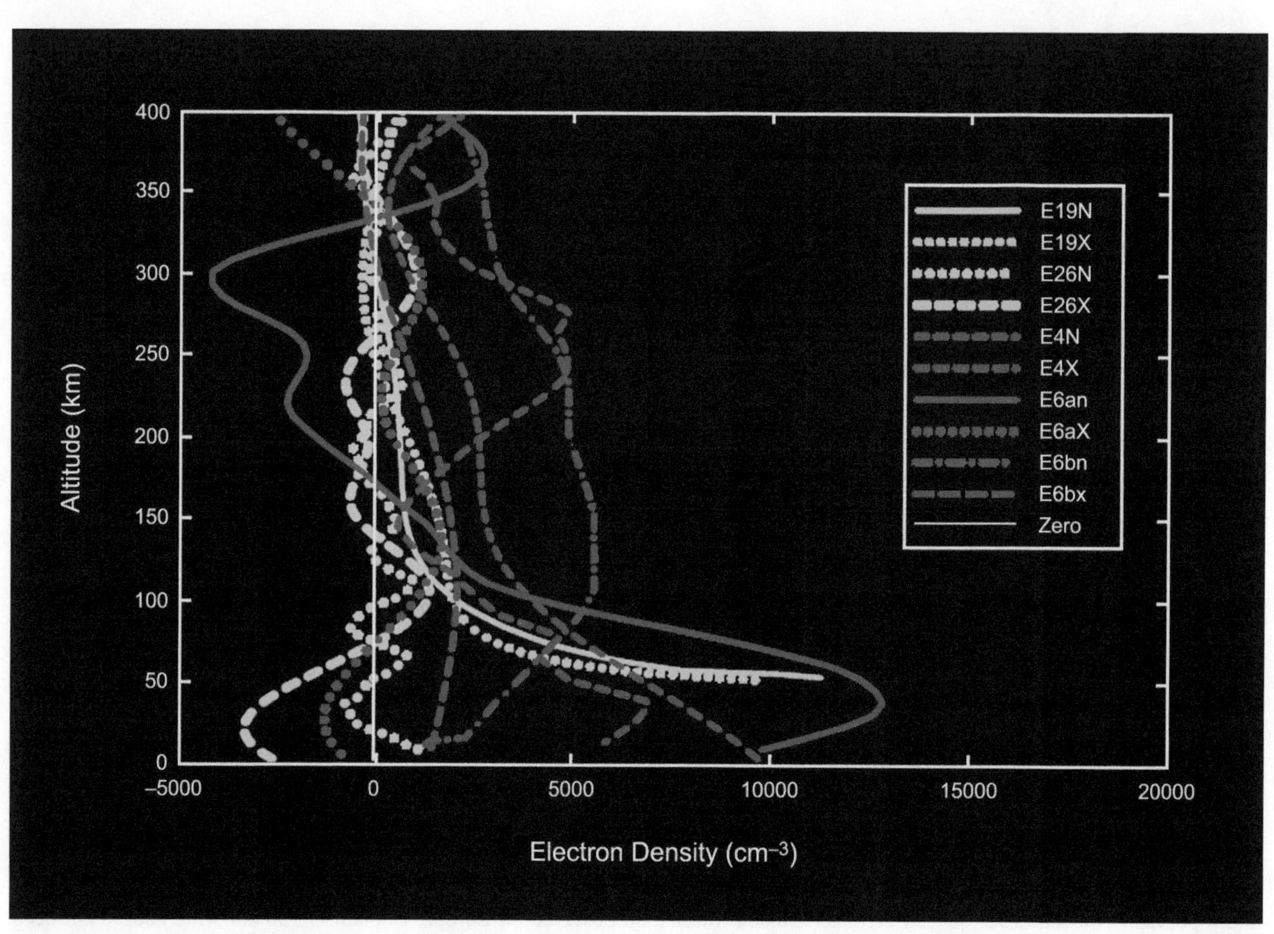

Plate 23. Compilation of all the Galileo radio occultation and near-occultation results illustrating the nonuniformity of Europa's ionosphere. Figure courtesy of A. Kliore.

Accompanies chapter by McGrath et al. (pp. 485–505).

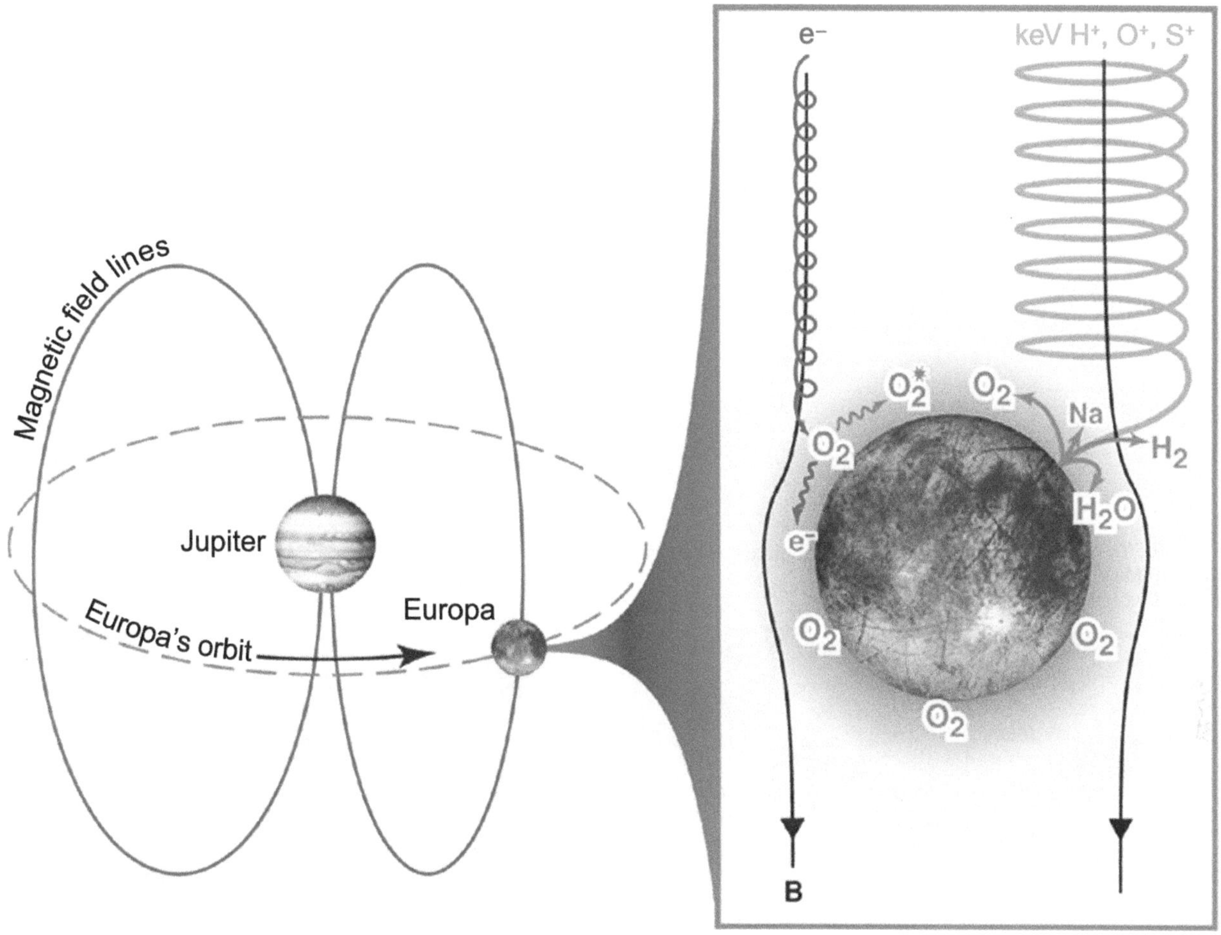

Plate 24. Schematic of Europa's interaction with Jupiter's magnetosphere. Ions and electrons trapped by Jupiter's magnetic field alter and erode the surface, producing a tenuous atmosphere composed mostly of O_2 with an extended neutral torus of primarily H_2.

Accompanies chapter by Johnson et al. (pp. 507–527).

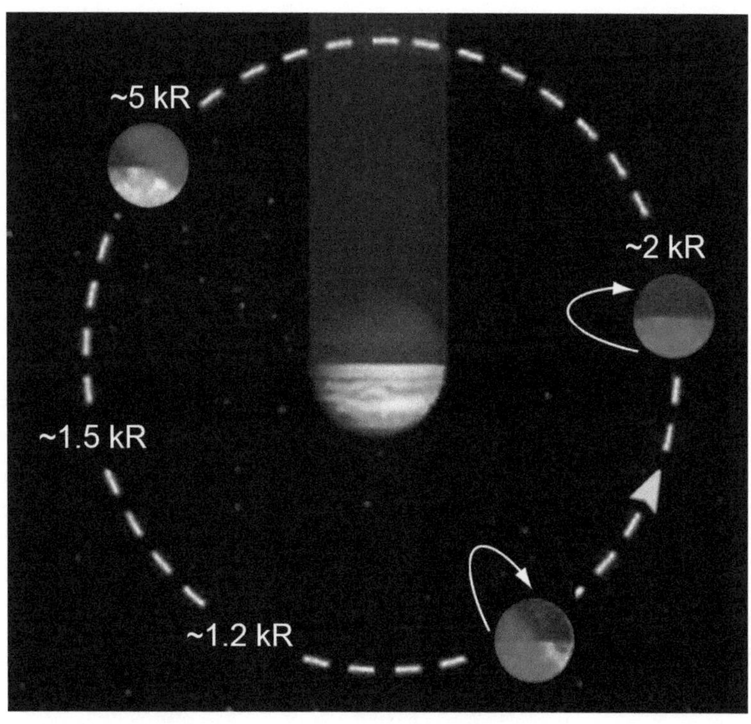

Plate 25. Average Na emission intensity (in kilo-Rayleigh) at 4 Europa radii from the surface at different positions around Jupiter (*Leblanc et al.,* 2005). The red part on the surface is the preferentially bombarded trailing hemisphere, whereas the dark part represents the night hemisphere. Also indicated is Jupiter's shadow at Europa's orbit; sizes of Jupiter and Europa are not to scale. White arrows indicate where accumulation of Na atoms on the leading side may occur.

Accompanies chapter by Johnson et al. (pp. 507–527).

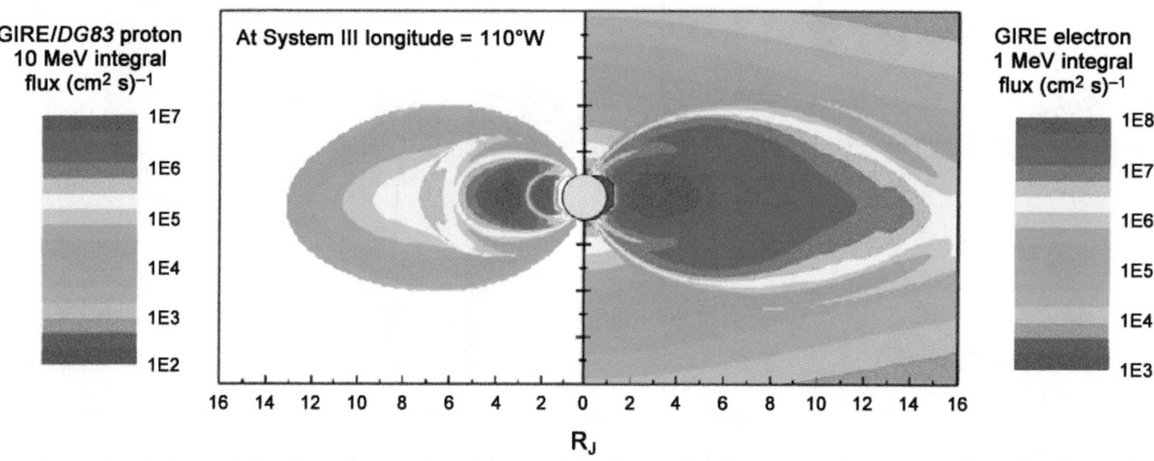

Plate 26. Contour plots of the GIRE/*DG83* model E > 10 MeV integral proton (left) and E > 1 MeV electron (right) fluxes for the vovian magnetosphere radiation region. The model provides the flux as a function of position, energy, and pitch angle. The fluxes presented here have been integrated over pitch angle. Note that outside the contour of L = 12, the proton model is set to 0, whereas the electron model fluxes outside L = 16 are only approximate since they are not considered trapped in the model (see *DG83* for details). Figure courtesy of I. Jun.

Accompanies chapter by Paranicas et al. (pp. 529–544).

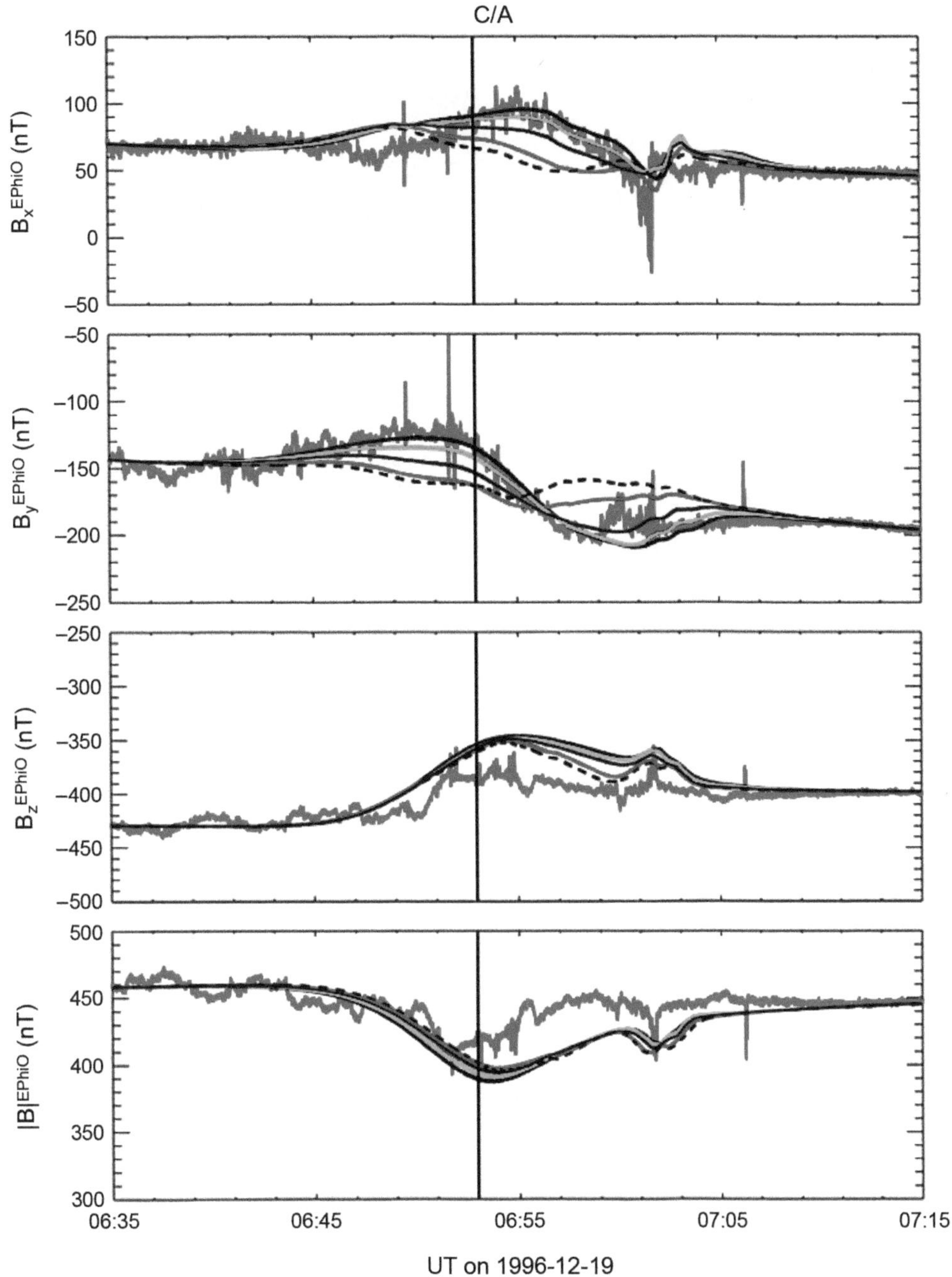

Plate 27. Observed and modeled magnetic field for the E4 flyby. Red = measurements of *Kivelson et al.* (1997); dashed black = modeled field with no internally induced field; Blue, green, and black = modeled field including induction in a 100-km-thick ocean lying beneath a crust of 25 km for ocean conductivities of 100, 250, and 500 mS/m, respectively. From *Schilling et al.* (2007).

Accompanies chapter by Kivelson et al. (pp. 545–570).

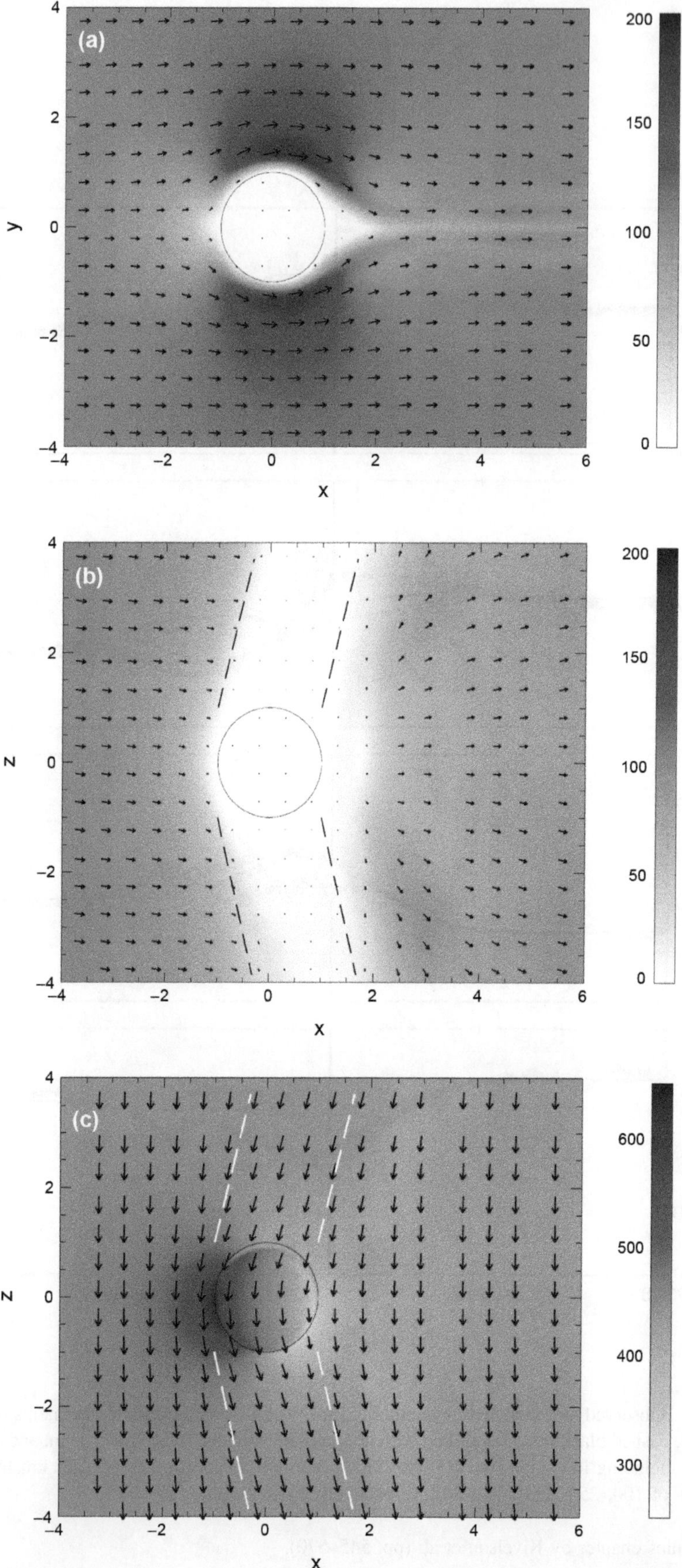

Plate 28. The flow speed **(a)** in the x–y plane, **(b)** in the x–z plane, and **(c)** the magnetic field in the x–z plane. Dashed lines represent boundaries of the northern and southern Alfvén wings. From the MHD simulation of *Schilling et al.* (2008).

Accompanies chapter by Kivelson et al. (pp. 545–570).

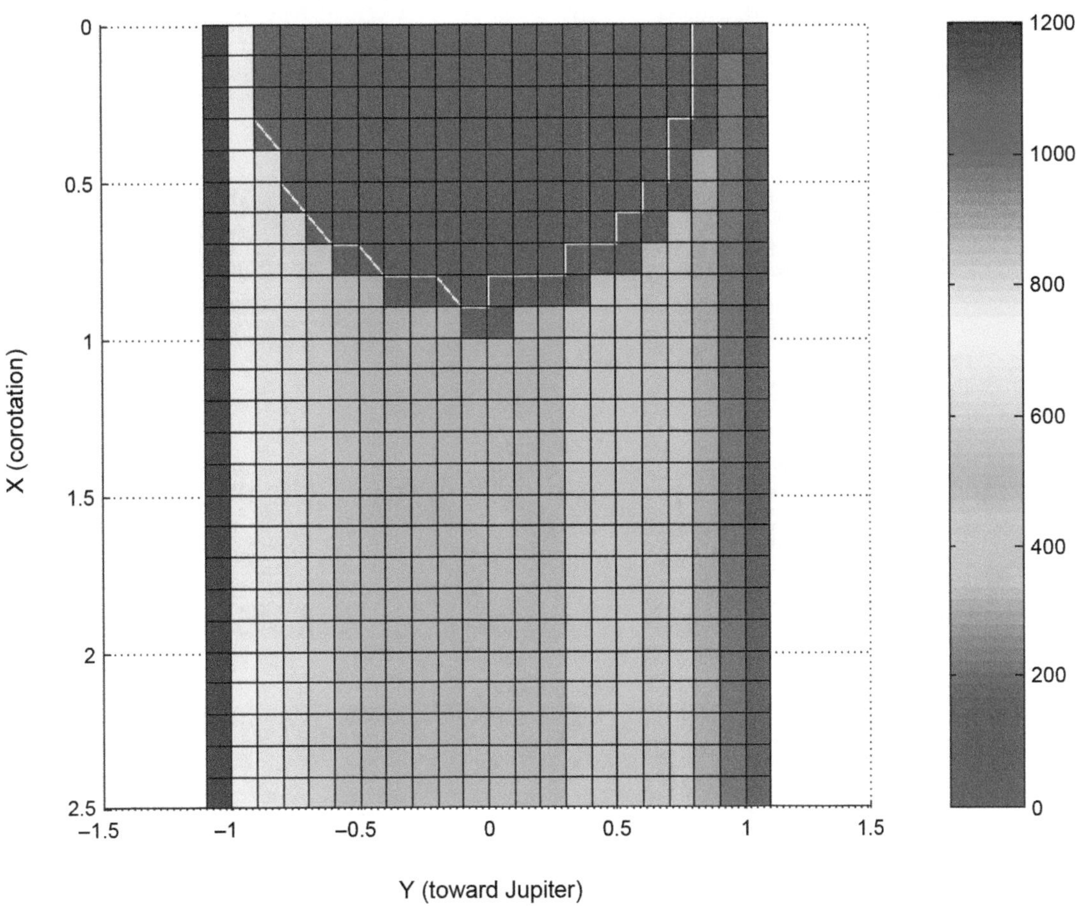

Plate 29. A simple model of the ion pickup around Europa, showing an abrupt decrease in density near y = 0.6. Density units are arbitrary.

Accompanies chapter by Kivelson et al. (pp. 545–570).

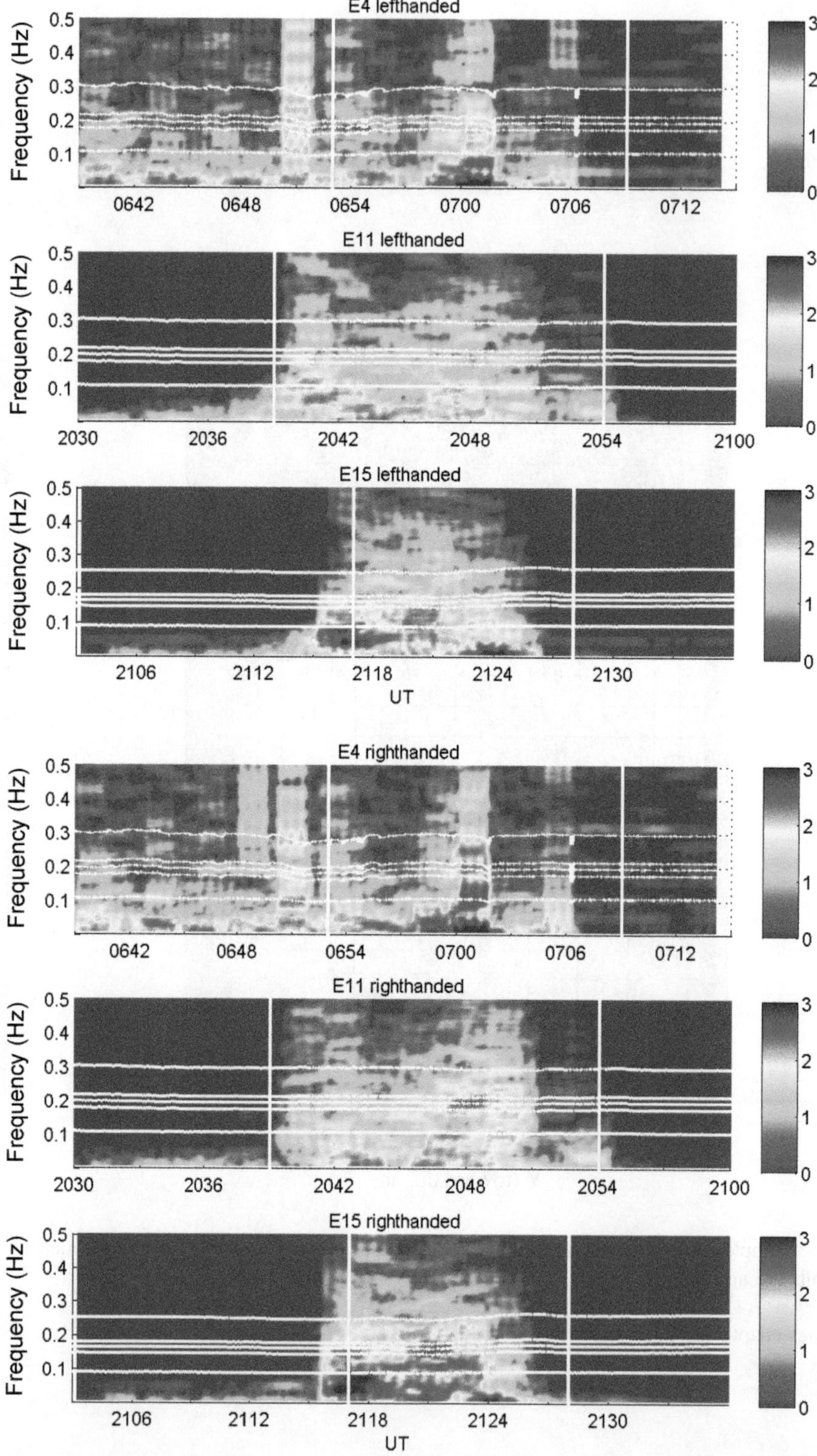

Plate 30. Dynamic spectra of the lefthand (top three panels) and righthand (lower three panels) polarized components of the magnetic field for E4, E11, and E15. The white solid traces show the cyclotron frequencies for Na^+ (A = 23), O_2^+ (A = 32), K^+ (A = 40), Cl^+ (A = 35), and SO_2^+ (A = 64). Vertical lines delimit the geometric wake.

Accompanies chapter by Kivelson et al. (pp. 545–570).

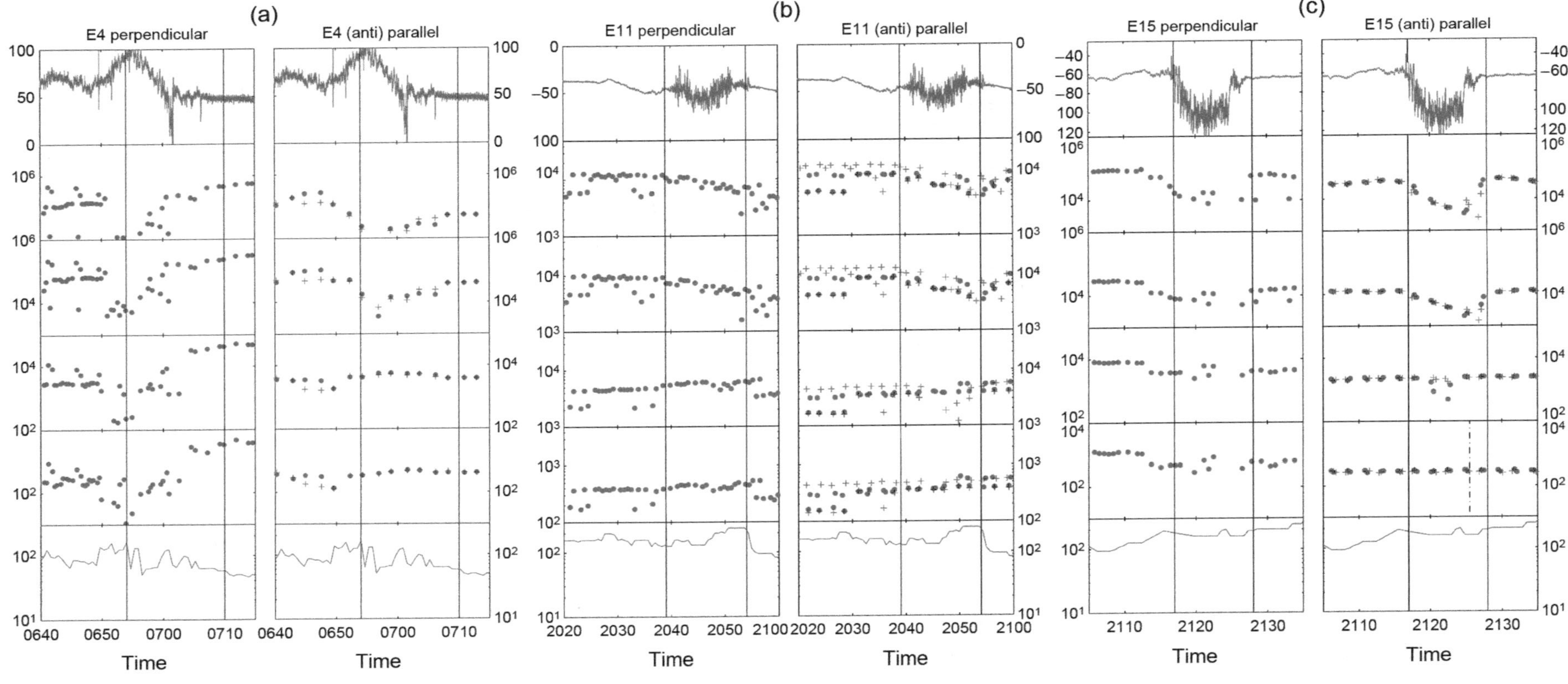

Plate 31. B_x, EPD data, and PWS (inferred) electron density for E4, E11, and E15. The energetic particle signatures are plotted separately for perpendicular ($85° \leq \alpha \leq 95°$, left panels) and (anti)parallel ($0° \leq \alpha \leq 20°$, $160° \leq \alpha \leq 180°$, right panels) pitch angles for the F2, F3, A5, and O2 channels of the EPD. In the (anti)parallel panels the (anti)parallel fluxes are indicated with (red dots) blue plusses. The bottom panel shows the electron density inferred from the PWS instrument. Vertical lines delimit the geometric wake as defined in the text.

Accompanies chapter by Kivelson et al. (pp. 545–570).

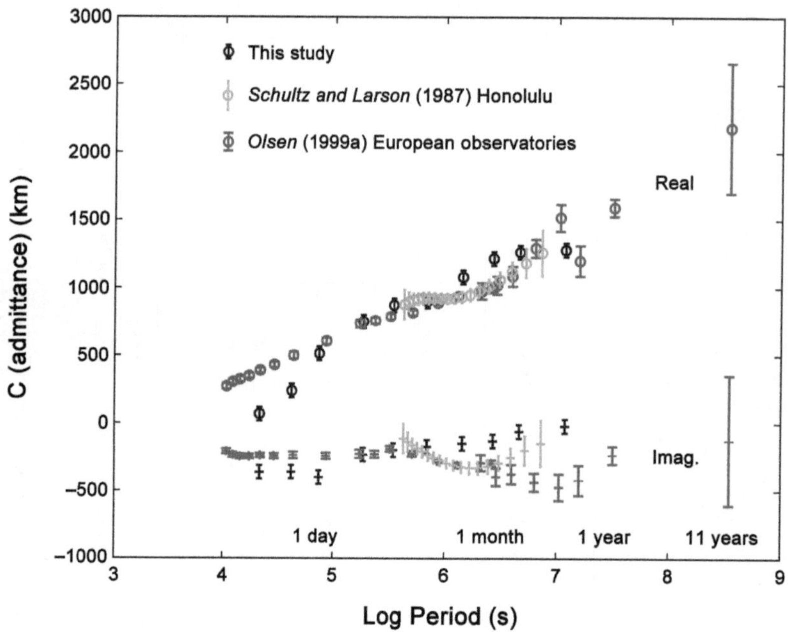

Plate 32. The response of Earth computed from data from several European surface observatories (*Olsen,* 1999a), a Hawaiian observatory (*Schultz and Larson,* 1987), and the MAGSAT spacecraft (*Constable and Constable,* 2004). The admittance function C is a rough measure of the depth to which the signal is able to penetrate in a spherical conductor and is given by $C = a[l - (l + 1)R]/[l(l + 1)(l + R)]$, where a is the radius of the Earth, R is the complex response function (ratio of the internal to external signal), and l is the degree of the harmonic. Adapted from *Constable and Constable* (2004).

Accompanies chapter by Khurana et al. (pp. 571–586).

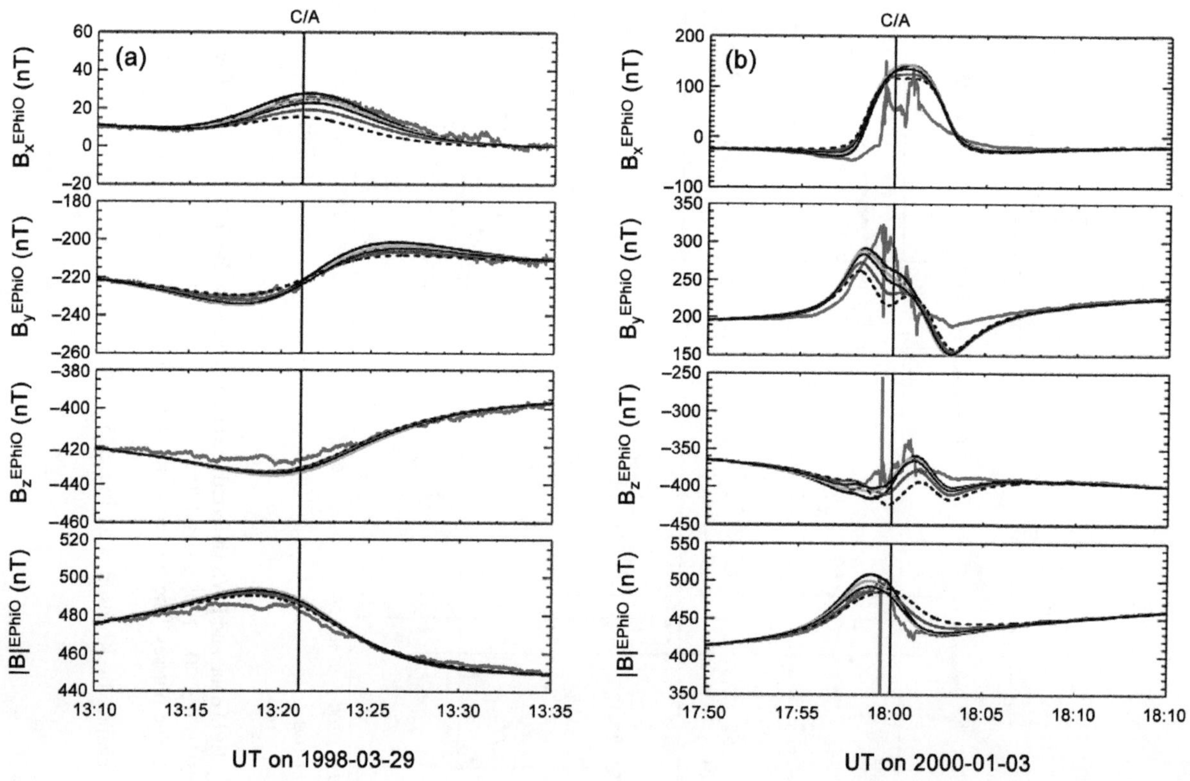

Plate 33. Observed and modeled magnetic field for the **(a)** E14 and **(b)** E26 flybys in the EfW coordinate system. Shown in the four panels are the three components and the magnitude of the observed field (red curves), model field with no induction (dashed blue), ocean conductivity 100 mS/m (blue), 250 mS/m (brown), 500 S/m (green), and 5 S/m (black). The ice thickness is 25 km and the ocean thickness is 100 km in these simulations. Adapted from *Schilling et al.* (2007).

Accompanies chapter by Khurana et al. (pp. 571–586).

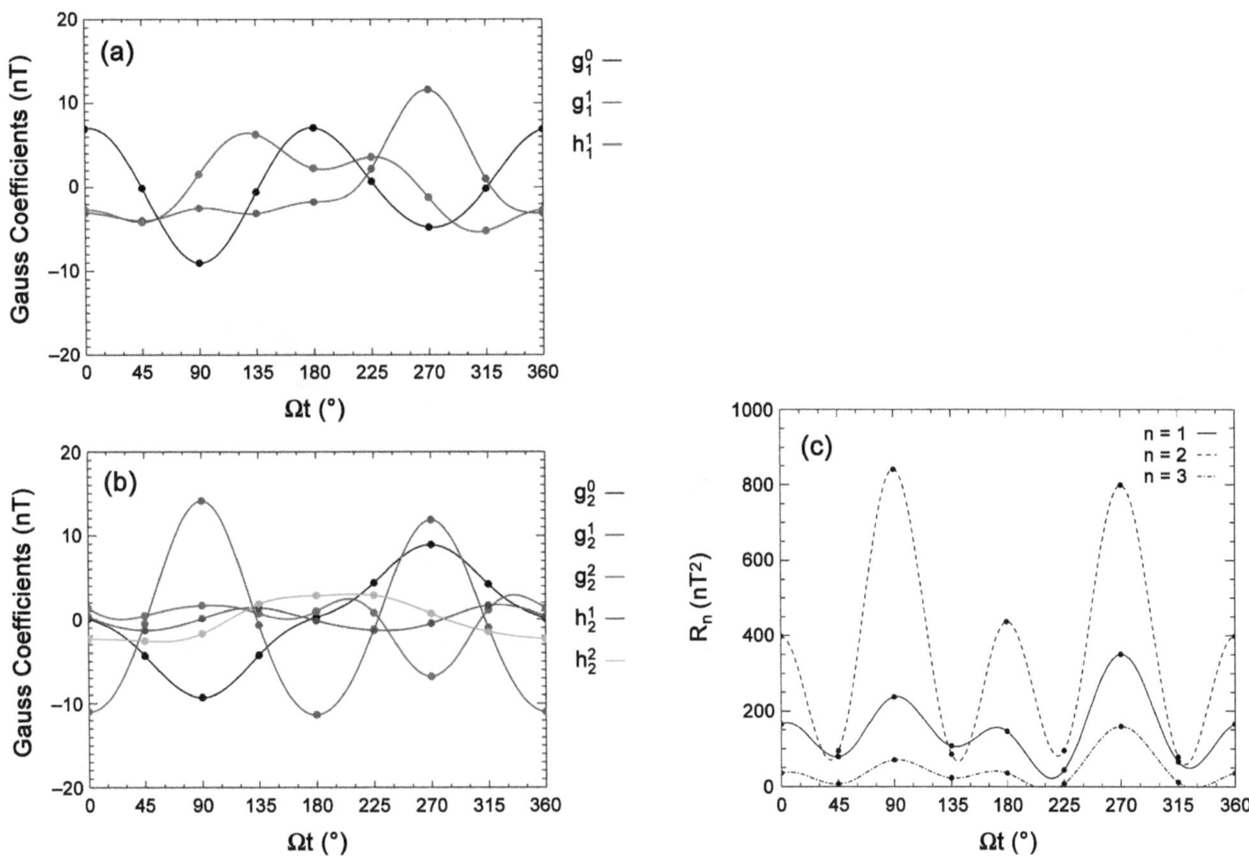

Plate 34. The spherical harmonic coefficients for the **(a)** dipolar and **(b)** quadrupolar terms induced by the Europa/plasma interaction over a synodic rotation period of Jupiter. **(c)** Total power contained in the induced field at the first three spherical harmonics. For comparison, the magnetic-field-induced dipole term contains power at the level of 14,000–88,000 nT^2. Adapted from *Schilling et al.* (2007).

Accompanies chapter by Khurana et al. (pp. 571–586).

Plate 35. Europa experiences a time-varying gravitational potential field in its eccentric orbit (eccentricity = 0.0094), with a 3.551-day (1 eurosol) period. Its tidal amplitude varies proportionally to the gravitational potential, causing Europa to flex as it orbits. This view shows the north pole of Jupiter as Europa orbits counterclockwise with its prime meridian pointed toward Jupiter. Measuring the varying gravity field and tidal amplitude simultaneously allows the interior rigidity structure of Europa to be derived, revealing the properties of its ocean and icy shell (*Moore and Schubert,* 2000).

Accompanies chapter by Greeley et al. (pp. 655–695).

Plate 36. Europa's ice, assuming a thick shell model: Convective diapirs could cause thermal perturbations and partial melting in the overlying rigid ice. Faulting driven by tidal stresses (upper surface) could result in frictional heating. Impact structures might show central refrozen melt pools, surrounded by ejecta.

Accompanies chapter by Greeley et al. (pp. 655–695).

Plate 37. Ganymede's simulated magnetosphere. Field lines are green; perpendicular current is represented by color variation. Note intense currents flow both upstream on the boundary between Jupiter's field and the field lines that close on Ganymede, and downstream in the reconnecting magnetotail region. Figure courtesy of X. Jia.

Accompanies chapter by Greeley et al. (pp. 655–695).

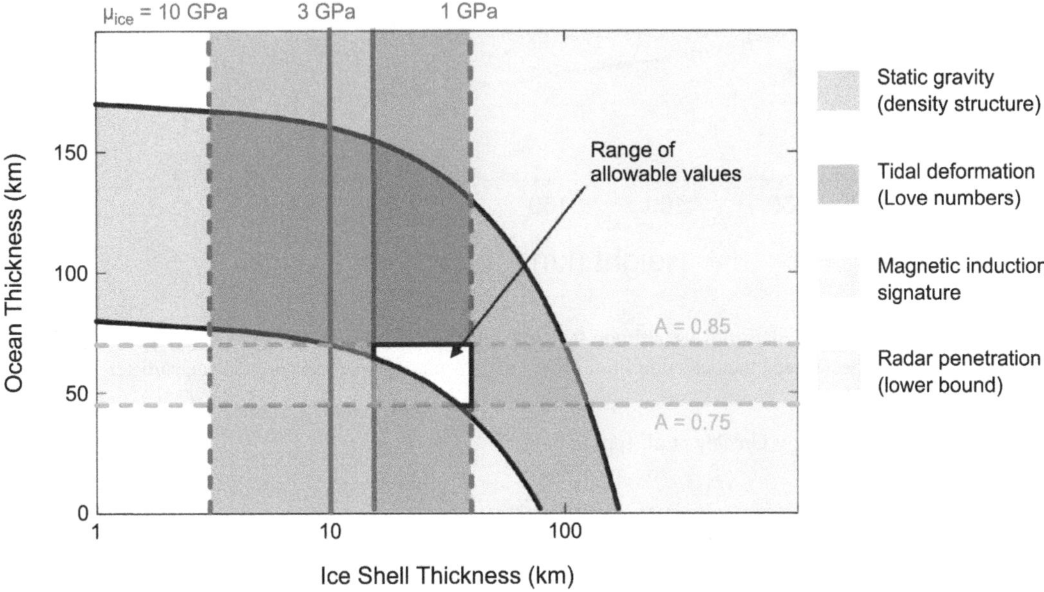

Plate 38. The combination of (hypothetical) JEO measurements can constrain the thickness of the icy shell. Based on the bulk density and moment of inertia (from flybys by JEO and other spacecraft), the thickness of the water + ice layer may be obtained (gray shading) (*Anderson et al.,* 1998a,b); uncertainties arise mainly from lack of knowledge of the rocky interior density (bulk density is already known). Measuring time-variable gravity and topography gives the k_2 and h_2 Love numbers, respectively; hypothetical Love number constraints (red shading) assume observed h_2 and k_2 of 1.202 and 0.245, respectively, and constrain shell thickness as a function of rigidity μ (*Moore and Schubert,* 2000). The hypothetical values assumed here are characteristics of a moderately thick icy shell. In the example shown, the icy shell deformation is sufficiently large that a shell thickness in excess of 40 km is prohibited. Determining both k_2 and h_2 provides additional information. A lower bound on the icy shell thicknesses may be derived from radar data. Here, a tectonic model of icy shell properties is assumed (*Moore,* 2000), resulting in a radar penetration depth (and lower bound on shell thickness) of 15 km (green shading). Multiple frequency (hypothetical) set of observations results in a range of acceptable icy shell thickness (15–40 km) and a range of acceptable ocean thicknesses (45–70 km). A different set of observations would result in different constraints, but the combined constraints are more rigorous than could be achieved by any one technique alone. JEO will be able to provide those constraints to determine the thickness of Europa's icy shell.

Accompanies chapter by Greeley et al. (pp. 655–695).

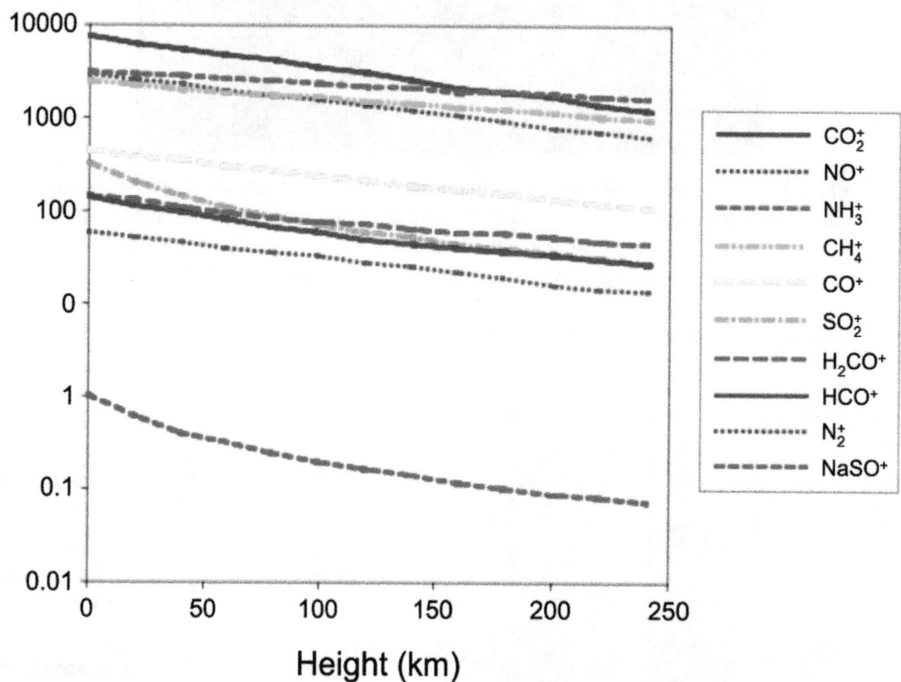

Plate 39. Ionospheric densities vs. altitude for molecules sputtered from the surface (*Johnson et al.*, 1998); all densities exceed the detection limit (10^{-3} cm^{-3}; Y-axis) of a modern mass spectrometer, such as the Cassini INMS.

Accompanies chapter by Greeley et al. (pp. 655–695).

Part V:
Astrobiology and Perspectives

Astrobiology and the Potential for Life on Europa

Kevin P. Hand
NASA Jet Propulsion Laboratory/California Institute of Technology

Christopher F. Chyba
Princeton University

John C. Priscu
Montana State University, Bozeman

Robert W. Carlson
NASA Jet Propulsion Laboratory/California Institute of Technology

Kenneth H. Nealson
University of Southern California

The high likelihood that Europa harbors a contemporary, global, subsurface liquid water ocean makes it a top target in our search for life beyond Earth. Europa's chondritic composition, rocky seafloor, and radiolytically processed surface may also play an important role in the habitability of Europa. Here we review the availability and cycling of biologically essential elements, the availability of energy to power life, and the conditions on Europa that could be conducive to the origin of life. We also address the survivability and detection of possible biosignatures on the surface, and possible search strategies for future orbiting and landed missions.

1. ORIGINS AND HABITABILITY, PAST AND PRESENT

1.1. Introduction

Europa may be the premier place in the solar system to search for both extant life and a second origin of life. The discovery of extant life beyond Earth would provide critical information about the fundamental nature of life and the breadth of possible biological processes. The discovery of a second origin of life would provide a key benchmark to assess evolutionary processes and would lead to a better understanding of life in the broader context of the universe. Were we to discover a second origin, we would transform the universe from a barren expanse in which life is unique to Earth, to a biological universe in which life may arise wherever conditions are suitable. We could begin to separate the contingent from the necessary in our understanding of biological possibility (*Sagan,* 1974). These discoveries lie at the heart of some of humankind's most longstanding and profound questions, and the exploration of Europa has the potential to provide some answers.

Our understanding of life on Earth has led to three key requirements for habitability (Fig. 1). Liquid water, a suite of biologically essential elements, and a source of energy are all prerequisites for life as we know it. In the context of Europa, these three cornerstones may well be satisfied by the attributes listed in the periphery of Fig. 1. This chapter concerns the details of these attributes so that our physical, chemical, and geological understanding of Europa can guide our biological assessment of this world. We do not address in detail the case for liquid water on Europa, as that

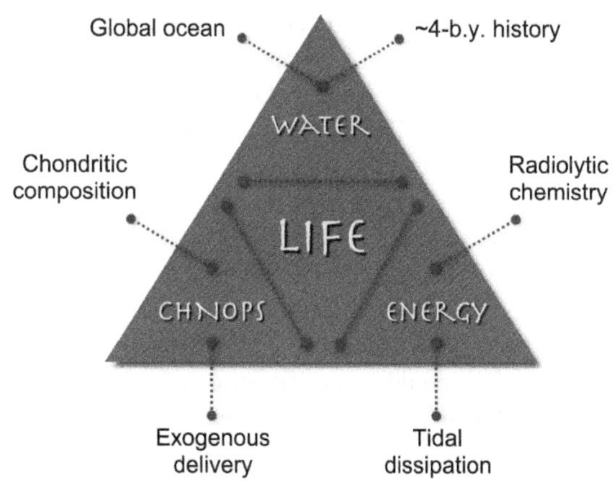

Fig. 1. Representation of the habitability of Europa. At present, our understanding of the conditions necessary for life can be distilled down to three broad requirements: (1) a sustained liquid water environment, (2) a suite of elements critical for building life (e.g., C, H, N, O, P, S, etc.), and (3) a source of energy that can be utilized by life. Here we suggest how these "keystones of habitability" intersect with our current understanding of the conditions on, and within, Europa.

subject is addressed in the chapters by McKinnon et al. and Khurana et al. We also refer the reader to the chapters by Carlson et al. and by Zolotov and Kargel for additional information regarding surface composition and ocean chemistry.

Significant to the issue of extant life on Europa is the fact that the subsurface liquid water reservoir has likely persisted for much of the moon's history. The ocean is predicted to be in contact with a rocky ocean floor, potentially creating a geochemically rich environment suitable for the origin and maintenance of life. Although the ocean lies beneath several to many kilometers of ice (*Pappalardo et al.,* 1998; *Greenberg et al.,* 1999; *Hand and Chyba,* 2007), the global nature of the ocean suggests that much of this world could be inhabited if the chemical conditions are suitable.

The contemporary, global ocean of Europa stands in contrast to several other potentially habitable worlds within our solar system. Mars, for instance, may have harbored a northern hemisphere surface ocean for the first ~2.0 G.y. of the planet's history (*Head et al.,* 1999). On present-day Mars, however, liquid water may be restricted to subsurface aquifers or transient pockets of near-surface meltwater (*Costard et al.,* 2002). The proximity of Mars to Earth could also raise interesting questions of contamination, were life on Mars discovered to be similar to life on Earth (e.g., DNA, RNA, and protein-based). The discovery of such life would certainly be profound, but contamination issues could confuse the discovery of a second origin.

Similarly, the question of contemporary subsurface liquid water on Enceladus remains open (*Spencer et al.,* 2006; *Kieffer et al.,* 2006), but the heat flux and high loss rate from the plumes implies that the observed events are likely transient. Stability of liquid water reservoirs over geological timescales (~10^5 yr or more) is an important consideration for habitability; Enceladus may have contemporary liquid water, but if it is a transient reservoir then the habitability of Enceladus could be compromised.

Titan may contain an ammonia-water mantle, part of which could be liquid (*Lorenz et al.,* 2008). The primary solvent available on Titan's surface and near-subsurface, however, is nonpolar hydrocarbons in liquid methane lakes (*Mitri et al.,* 2007). Such environments could be an interesting place to search for life *unlike* life as we know it (*Schulze-Makuch and Grinspoon,* 2005; *McKay and Smith,* 2005), but it is intellectually more tractable to focus our initial search on environments comparable to those in which we know biology could work. The contemporary ocean of Europa provides just such an environment. Our experience with life here on Earth has taught us that carbon-and-water-based life functions quite well throughout a considerable range of temperature, pressure, and chemical regimes. But given the right set of conditions, the question remains: Will life arise? If we were to discover carbon- and water-based life elsewhere in the solar system, would we see biochemistry comparable to the protein-, RNA-, and DNA-based life we have here on Earth?

Finally, we note that life on Europa could be vastly different in composition and biochemistry from life on Earth. Such "weird life" (*Benner et al.,* 2004) should of course not be excluded from consideration, or from our future search strategies (*NRC,* 2007). However, our working knowledge and scientific understanding is constrained to the carbon- and water-based life found here on Earth. Consequently, in the sections that follow, references to "life" should really be interpreted to mean "life as we know it."

1.2. Habitable Environments

The origin of life on Europa, and the habitability of Europa, are two distinct but related issues. Here we address both questions, starting first with habitability and then moving to the more difficult and less-well-constrained question of life's origin. We focus on conditions relevant to contemporary Europa, but it is important to distinguish Europa in the past from Europa in the present. For instance, *Hand and Chyba* (2007) have argued that the salinity of the contemporary ocean may be habitable by terrestrial standards, but not necessarily conducive to the origin of life. Considering a different aspect of ocean chemistry, however, models of the tidal evolution of Europa indicate that the heat flux may have been much greater in the past (*Hussmann and Spohn,* 2004), perhaps creating a liquid water environment with more hydrothermal activity and thus more chemistry of prebiological potential. We explore these possibilities below.

1.2.1. Inventory and cycling of biologically essential elements. Life requires a suite of biologically essential elements to serve as the basic building blocks and metabolic material for maintenance, growth, and reproduction (*Wackett et al.,* 2004). For life to persist such materials must be available, and their availability must be sustained through cycles within the geological and chemical processes of the ecosystem. On a planetary scale this means that the elemental cycles must be closed-loop systems; if large loss terms exist or the turnover rate is too slow, then life will be constrained by the limiting element.

Carbon-and-water-based life on Earth requires nitrogen, phosphorous, sulfur, and trace amounts of metals. At the orbital distance of Jupiter (5.2 AU), condensation and retention of the volatile elements would have been more efficient than for that of the terrestrial planets (*Lewis,* 1971). As discussed by *McKay* (1991) and *Chyba and Hand* (2006), based on abundance alone, the inner solar system is actually a poor locale for carbon chemistry and hence carbon-based life. High temperatures forced CO_2, NH_3, CO, H_2S, H_2O, and N_2 out beyond the asteroid belt where they could condense onto planetesimals. Although heating in the jovian subnebula also led to dehydration and loss of volatiles, the formation of Europa from chondritic material in the accretion disk of Jupiter likely led to a world rich in many of the biogenic elements (see chapters by Estrada et al. and by Canup and Ward). In Fig. 2 (inspired by *McKay,* 1991) we show the approximate bulk abundance for several biologically essential elements on worlds within our solar system. Values are given relative to solar abundance. To date, carbon, hydrogen, oxygen, and sulfur have all been observed

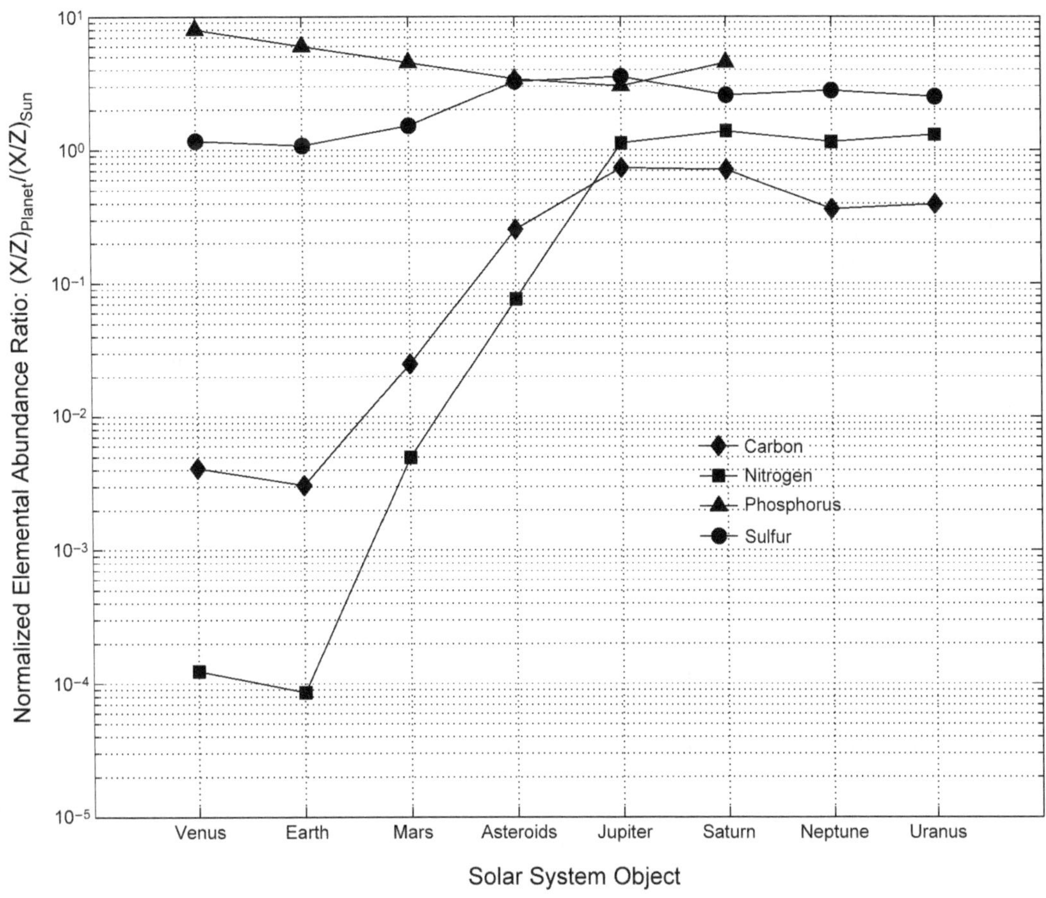

Fig. 2. Ratios of the biologically essential elements C, N, P, S to total heavy elements larger than helium, normalized to solar values (inspired by *McKay,* 1991). With increasing distance from the Sun the relative abundance of C, N, and S increases. This in part reflects the fact that C, N, and S were largely present in volatile species (e.g., CH_4, CO_2, N_2, and H_2S) that condensed into ices in the cold outer regions of the solar system (similar to H_2O). Values presented here were calculated from data and models provided in *Lodders and Fegley* (1998), *Lewis* (1971), *Lodders* (2003), *Taylor et al.* (2004), *Guillot* (2005), *Sotin et al.* (2007) and from references within *McKay* (1991). The values shown here are intended only to represent the broad, order of magnitude, trend in our solar system.

on the surface of Europa (see chapter by Carlson et al.). Here we examine observational and theoretical constraints for the abundance, phase, and cycling of each of these elements.

1.2.1.1. Carbon: The abundance of total carbon (organic + inorganic) incorporated into planets increases by over two orders of magnitude for worlds beyond the asteroid belt (*McKay,* 1991). Despite Earth's carbon-rich biosphere, our planet is depleted in carbon when compared to worlds of the outer solar system. The carbon abundance found in worlds beyond the asteroid belt is still a few orders of magnitude lower than the abundance of carbon found in life on Earth, but it gives us good reason to expect that the abundance of carbon on Europa will not be a limiting factor for autotrophic and heterotrophic biochemistry. Models for the formation of Europa from chondrites (*Crawford and Stevenson,* 1988; *Kargel et al.,* 2000; *Zolotov and Shock,* 2001) put the abundance of carbon at ~0.01–0.7 wt.%, most of which is in CO_2. The observation of 360 ppm CO_2 in the surface ice of Europa leads to 0.028 wt.% carbon on the surface, in good agreement with the models for Europa's bulk composition (*Fanale et al.,* 2001; *Hand et al.,* 2007; see chapter by Carlson et al.).

Other forms of carbon are expected on the surface, but have not yet been observed [e.g., CO and H_2CO_3 (*Hand et al.,* 2007)]. Spectral characterization of Europa by the Galileo spacecraft's Near-Infrared Mapping Spectrometer (NIMS) was limited by radiation noise, but observations of Ganymede and Callisto can serve as a guide to the molecules that might be produced on Europa from meteoritic impact and subsequent radiolysis. Ganymede and Callisto are less geologically active, with correspondingly greater surface ages, and they show more meteoritic debris accumulated on their surfaces. This is in part due to higher escape velocities and lower impact velocities for the large, outer satellites (*Pierazzo and Chyba,* 2002).

Two sets of spectral features of astrobiological interest are observed on Callisto, the first thought to be hydrocarbon C-H stretch features and the second assigned to the nitrile group (C≡N). The latter is also observed on Ganymede. *McCord et al.* (1997, 1998) identified possible CH_2 stretch bands on Callisto at 3.41 ± 0.02 μm and 3.58 ± 0.03 μm (in wavenumbers this corresponds to 2932 ± 20 cm^{-1} and 2793 ± 25 cm^{-1}, respectively). The positions of these bands are close to the 3.44- and 3.53-μm (2907-cm^{-1} and 2833-cm^{-1}) band positions observed on Enceladus (*Brown et al.,* 2006). Similar CH_2 symmetric and asymmetric stretch bands are found in residues produced in the laboratory by ultraviolet and proton-irradiation of ices containing ammonia and alkanes (*Allamandola et al.,* 1998; *Moore and Donn,* 1982). These features have been suggested to arise from aliphatic hydrocarbons dominated by methylene (CH_2) groups (*Allamandola et al.,* 1998). Polymethylene oxides and CH_2 groups associated with C≡N are also possible contributors (*Allamandola et al.,* 1998). These bands are not observed on Europa, although low-noise measurements in this wavelength region were very limited. Using such spectra (see chapter by Carlson et al.) and band strengths from *Moore and Hudson* (1998), we find an upper limit for the number of methylene groups per water molecule of $[CH_2]/[H_2O] < 1.5 \times 10^{-3}$. The number expected for meteoritic infall and burial by gardening, if all the C atoms are associated with methylene, is a factor of 3 less than this upper limit or ~500 ppm. If hydrocarbons are found locally on Europa with concentrations greatly exceeding tenth-of-a-percent levels, they could indicate an oceanic source of organic material, a point discussed in greater detail in section 3.

Cycling of carbon on an ice-covered ocean world, be it inhabited or uninhabited, may not differ considerably from the major attributes of the terrestrial carbon cycle. Most of the carbon on Earth is in geological reservoirs or dissolved in the ocean. Plate tectonics and volcanism are the critical ways to release geologically sequestered carbon back into the atmosphere, where CO_2 is a minor species present at levels comparable to the concentration found in Europa's surface ice. Life on our continents is largely dependent on carbon derived from atmospheric CO_2.

For an ice-covered ocean, however, dissolved inorganic carbon (HCO_3^-, CO_3^{2-}, H_2CO_3), precipitated carbonates, CO_2 clathrates, and organic carbon will be the primary phases of carbon. Tectonics would be required in order to completely recycle carbon, but even on a geologically dead Europa the balance between dissolved carbon and precipitated carbon could lead to a self-sustaining carbon cycle. The critical link for sustaining this cycle, and one key geochemical difference between the oceans of Europa and Earth, is the depth of Europa's ocean and the change in solubility of carbonates as a function of pressure and temperature. On Earth, where the average depth of the ocean is 4 km, the pressure is ~40.1 MPa and carbonates are stable for the bulk ocean pH of 8 and temperature of 4°C. At slightly greater depth (~4.2 km) the increase in pressure raises the solubility of carbon dioxide and precipitated carbonates are subject to dissolution in the unsaturated deep waters. This so-called carbonate compensation depth (CCD) explains why calcite and aragonite are rarely found in the deepest regions of our ocean. Coupled with this, the solubility of CO_2 increases with decreasing temperature. Thus, while carbonates precipitate easily in the warm surface waters of Earth's ocean, the cold deep waters of our ocean enhance dissolution (*Krauskopf and Bird,* 1995).

On Europa, the pressures and temperatures that are expected for the ocean will ensure that much of the inorganic carbon stays in dissolved form. Europa's ice-covered ocean must, on average, hover near the freezing point of water (otherwise the icy shell and ocean could not be stable), and therefore be conducive, at least in terms of temperature and pressure, for retaining inorganic carbon in dissolved forms. Clathrates of CO_2 may be the dominant phase of precipitated carbon (*Crawford and Stevenson,* 1988; *Kargel et al.,* 2000). One interesting consequence of the CO_2 solubility chemistry is that carbonate shells and other biological structures made of carbonate will be unstable at depths of ~30 km or more in Europa's ocean. Were the ice on Europa ~30 km in thickness, then no region of the ocean would be stable for carbonates, save for perhaps localized hot spots and pH extremes on the ocean floor. In other words, carbonate skeletal structures such as those found on Earth are unlikely to be stable throughout much of Europa's ocean. (The consequences for biosignatures on the surface of Europa are discussed later in this chapter.) A carbon cycle on Europa could then be completed by the complementary actions of precipitation of carbonate and organic carbon to depth, followed by dissolution. Ultimately this raises questions about the stability of organic carbon at the ocean floor: Would carbon-based life be stable against dissolution on Europa's seafloor? Evidence on Earth suggests it would. Soft-walled, single-celled foraminifera have been found in sediments at Challenger Deep in the Marianas Trench (*Todo et al.,* 2005), the greatest depth within our ocean (11 km, P = 110.1 MPa, comparable to a depth of roughly 80 km below the surface of Europa).

1.2.1.2. Nitrogen: Nitrogen is also expected to be in abundance at Europa and we have empirical evidence of nitrogen for Europa's outer neighbors. A feature at 4.57 μm in spectra of Ganymede and Callisto may be due to the stretching of a C≡N group, which shows strong, narrow absorption features between 4.4 and 4.9 ìm (*McCord et al.,* 1997, 1998). Most of the simple molecules containing C≡N have features that are inconsistent with the observed features, but polymeric HCN and some tholins possess absorption features in the 4.53- to 4.61-μm region that may be consistent with cyanogens (*Cruikshank et al.,* 1991; *Khare et al.,* 1994). With only one band evident it is impossible to provide unequivocal identifications, and other candidates are plausible, including CO trapped within zeolites and amorphous silica (*Cairon et al.,* 1998) as well as carbon suboxide, C_3O_2 (*Gerakines and Moore,* 2001; *Johnson et al.,* 2004). Spectra at Europa were too noisy to resolve the 4.57-μm feature; thus, while models predict nitrogen to be in abundance, we have at present no direct evidence for nitrogen on Europa.

Modeling estimates for the abundance of ammonia, and hence nitrogen, in primordial Europa are within the range of 5–10 wt.% NH_3 (*Lewis,* 1971; *Grasset and Sotin,* 1996).

Depending on formation mechanisms for the icy shell, ammonia may be excluded from the freezing ice, leading to a surface devoid of ammonia and an ammonia-rich subsurface ocean (*Grasset and Sotin,* 1996; *Spohn and Schubert,* 2003). Even if ammonia did reach the surface (e.g., cryovolcanism), the radiation environment of Europa would radiolyze much of the NH_3 to N_2 and H_2, which are easily lost to space (*Loeffler et al.,* 2006). Could Europa have lost a significant fraction of its nitrogen to space as a result of resurfacing and radiolysis over 4.5 G.y.? Such a loss is unlikely. Even if one assumes complete loss of nitrogen through a 1-m gardened surface layer that is replenished every 10 m.y., the amount of nitrogen lost is only about 0.5 wt.% relative to the starting concentration. Nitrogen should still be in abundance on contemporary Europa.

It has been argued that nitrogen, as opposed to water or carbon, should be used as the primary flag for habitable regions beyond Earth (*Capone et al.,* 2006). In this context, it is interesting to consider the various phases and possible pathways of nitrogen cycling in an ice-covered ocean. Unlike Earth, where atmospheric N_2 serves as the major reservoir for nitrogen, Europa's nitrogen will be in dissolved or solid forms in the ocean, frozen into the ice, or sedimented onto the seafloor. The initial phase of nitrogen during formation was primarily ammonia, but if Europa's waters became oxidized over time then nitrate would have become the dominant phase. Nitrate and the ammonium ion (NH_4^+) are very soluble in water and would be easily incorporated into the icy shell. Neither species was reported in the NIMS spectra, although radiation noise at longer wavelengths may have been a limiting factor for such detection.

The phase of nitrogen has important consequences for biochemistry, as illustrated by terrestrial ecosystems. On Earth, the advent of nitrogen fixation ($N_2 \rightarrow$ organic nitrogen, e.g., C-N-H) may have opened a bottleneck in biological productivity (*Capone et al.,* 2006; *Smil,* 2001). Prior to the evolution of N_2-fixation, life on Earth was limited to the nitrogen found in nitrates in the ocean. This amounts to approximately 0.4% of the available nitrogen (~5.5×10^{21} g), of which ~73% resides in our atmosphere (*Capone et al.,* 2006). Our nitrogen-rich atmosphere resulted largely from the fact that N_2 is chemically inert. Nitrogen accumulated in the atmosphere from volcanic emissions and photolysis, but without biological fixation of N_2, there was no large sink term for biological or geological sequestration. Photolysis and fixation by lightning accounted for ~10^{13} g N fixed per year, roughly two orders of magnitude lower than contemporary biological fixation (*Walker,* 1977). With the advent of biological N_2-fixation, however, nitrogen in biomass and sediments reduced the nitrogen content of the atmosphere by more than 20% over Earth's history (*Walker,* 1977). Maintaining our nitrogen-rich atmosphere, and our global nitrogen cycle, then became partially dependent on the complementary process of biological denitrification ($N_{Organic} \rightarrow NO_3^- \rightarrow N_2$).

On Europa, N_2 is likely to be only a minor component of any abiotic global nitrogen cycle. Some of the primordial ammonia would have gone into ammonium (NH_4^+), nitrite (NO_2^-), and nitrate (NO_3^-), with speciation being largely a function of pH, temperature, and the availability of oxidants. With oxidants available, energetic niches for life exist in the pathways to nitrate, e.g., $NH_3 \rightarrow NO_2^- \rightarrow NO_3^-$. For terrestrial life to incorporate the nitrogen in nitrate into amino acids and macromolecules (e.g., proteins and nucleic acids), nitrate must be reduced to ammonium ions, $NO_3^- \rightarrow NO_2^- \rightarrow NH_4^+ \rightarrow N_{Organic}$. A complete cycle would then be fulfilled by (1) conversion of macromolecular nitrogen back into nitrate by nitrifying organisms, (2) conversion back into ammonia or the ammonium ion by abiotic or biological ammonification, and (3) radiolysis to oxidized species or nitriles on the surface of Europa. Additionally, anaerobic ammonium oxidation (anammox) — an important biological pathway in the nitrogen cycle of Earth's ocean — could serve to oxidize ammonium with nitrite in Europa's ocean, converting the nitrogen back to dissolved molecular nitrogen. Once nitrogen is in organic form it is advantageous for organisms to simply harvest the amino acids and nucleic acids rather than breaking these compounds down any further. This process, however, requires oxygen and would only be possible if radiolytically produced oxygen on the surface of Europa is delivered to the ocean (see section 1.2.2). Were that the case, a biologically active europan ocean could have much of its nitrogen sequestered and cycled in organic forms. Figure 3 shows possible nitrogen cycling and partitioning for three different scenarios in Europa's ocean. Each of these scenarios has important implications for the observation of nitrogen on the surface of Europa and the prospect of distinguishing abiotic organic chemistry from biologically mediated chemistry. These points are discussed in detail in sections 2 and 3.

1.2.1.3. Phosphorous: Phosphorous has not been observed on any of the Galilean satellites but it is seen in trace amounts in the jovian atmosphere (*Taylor et al.,* 2004). The bulk abundance of phosphorous, based on chondritic models for the formation of Europa (*Kargel et al.,* 2000; *Lodders and Fegley,* 1998), is ~0.1 wt.%. *Zolotov and Shock* (2001) estimated that full extraction of phosphorous from Europa's original chondritic material could yield 10–20 g of phosphorous per kilogram of ocean water. However, their more realistic partial extraction models yield concentrations of 0.34 to 4.2×10^{-6} g kg^{-1}, much closer to that found in terrestrial ocean water, 6×10^{-5} g kg^{-1}. The availability of phosphorous is a critical limiting element for life on Earth. The problem is not so much the bulk quantity, but rather the relative insolubility of phosphorous and the geological sequestration of phosphorous in minerals. Much of Earth's phosphorous is believed to reside in the lower mantle or core and only ~0.007 wt.% of the total phosphorous is dissolved in the ocean (*Macia et al.,* 1997). In contrast to nitrogen, phosphorous has no gaseous (atmospheric) reservoir. For the temperature and pressure conditions on Earth, and for those on Europa, phosphorous is found in solid phase as apatite or in relatively insoluble phosphoric acid and related salts. The readily available dissolved ions in water are

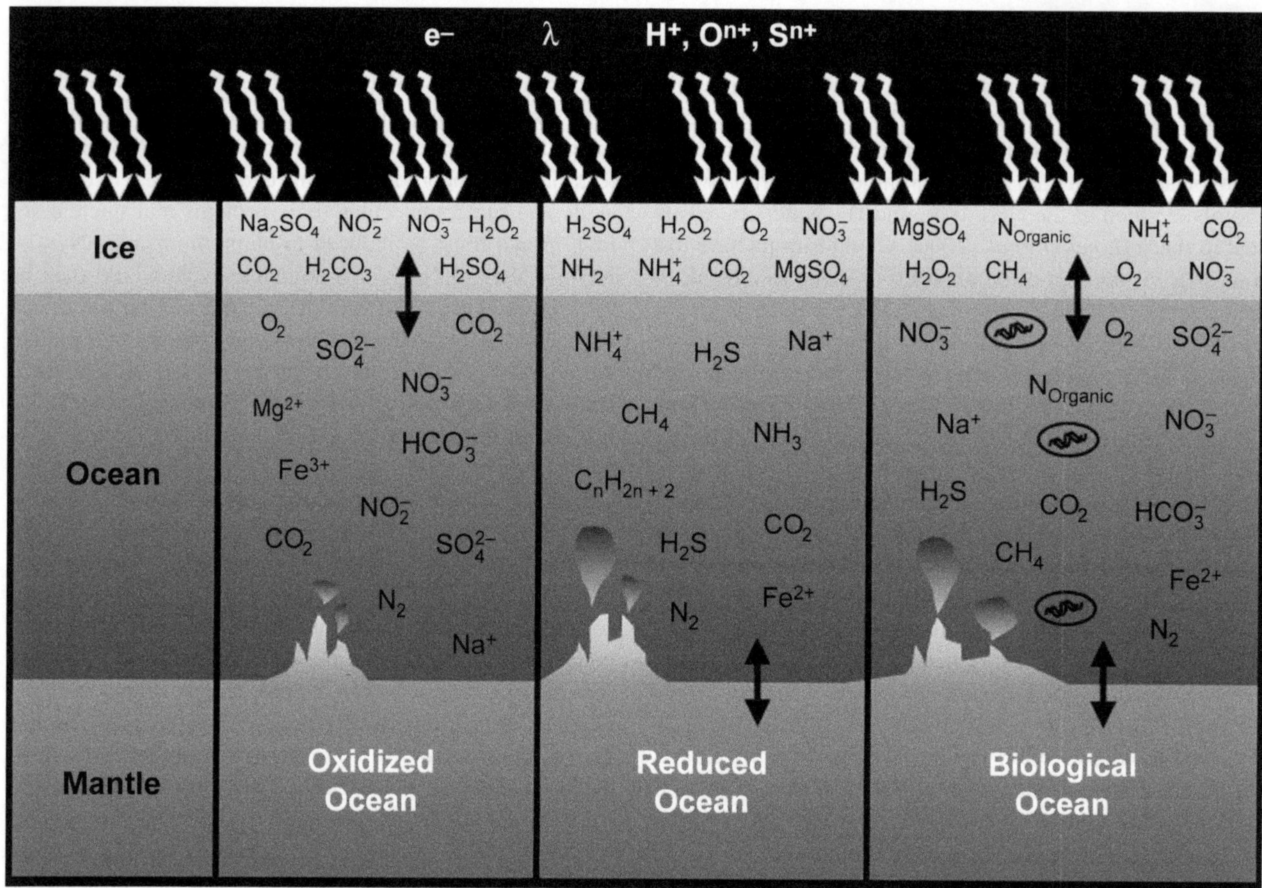

Fig. 3. Consideration of the sequestration of nitrogen on Europa. The sequestration and possible cycling of nitrogen on Europa will be considerably different than that of the Earth, since Europa lacks an N_2-rich atmosphere. Three possibilities are indicated in this diagram: (1) an oxidized ocean, resulting from cycling of the ocean with the radiolytically processed icy shell and dominated by nitrogen in the form of nitrate and nitrite; (2) a reduced high-pH ocean, resulting from cycling with the seafloor and dominated by ammonia and ammonium; and (3) a biological ocean (note the microbe symbols), in which cycling occurs with both the icy shell and seafloor and in which nitrogen is largely in organic form (e.g., amino acids, proteins, nucleic acids). Each scenario holds potentially different implications for nitrogen on the surface of Europa, an important consideration when searching for signs of life.

the phosphate ions PO_4^{3-}, HPO_3^{2-}, and $H_2PO_3^-$, where the former is most stable in basic solutions and the latter is more stable in acidic solutions.

The limiting nature of phosphorous raises the question of why terrestrial biochemistry is so intimately tied to this element. From the phosphodiester bonds of our genetic material to the energy-yielding bonds of ATP, all life on Earth is critically dependent on the availability of phosphorous. Surely there must be some biogeochemical alternative that would not suffer the geological limitations imposed by phosphates. *Westheimer* (1987) explored this question and argued that phosphate and phosphoric acid are well tuned to the needs of genetic biochemistry. Being trivalent, phosphate is able to provide two bonds for building polymers, while the remaining valency yields a negative charge for the molecule. This negative charge serves to keep the molecule within a lipid membrane, and it impedes hydrolytic destruction of the molecule. This latter property is critical for stability; if an information storage molecule is going to be useful for Darwinian selection, it must survive long enough for replication and reproduction. Ester bonds (R–COO–R'), such as those found in ethyl acetate, can survive on average for many months, but information storage requires thousands of such bonds, each of which is important to the integrity of the information stored. For example, if too many bonds are broken, the organism's ability to reproduce could be severely compromised (*Westheimer,* 1987). Here again phosphate is advantageous. Ester bonds with phosphate are typically a little more stable than the alternatives, but phosphate has the added advantage of the negative ionic charge, which protects the bonds from hydrolysis by the surrounding water.

Westheimer (1987) examined citric acid, arsenic acid, and silicic acid as alternatives, but none had the combined advantage of phosphoric acid and the associated phosphate. Sulfate — an anion observed over much of the surface of Europa — is insufficient in both its valency and stability. Sulfate is divalent and capable of making polymers, but the sulfate bonds hydrolyze much faster than comparable phosphate bonds. At the cold temperatures, high pressures, and neutral to alkaline pH [predicted from models for Europa's ocean (*Zolotov and Shock,* 2001)], stability increases and some of these other compounds may become viable alternatives, but phosphate will still be the best. As with our ocean, the ocean on Europa is likely at or near saturation with phosphoric acid and the available phosphorous provides a unique evolutionary advantage for any putative subsurface ecosystem. Finally, we note that for the case of lipids used in biological membranes, sulfolipids have been observed to replace phospholipids in cyanobacteria living in phosphorous deficient regions of the ocean (*Van Mooy et al.,* 2006).

1.2.1.4. Sulfur: Sulfur is unlikely to be a limiting factor for life on Europa. The surface chemistry, and potentially the subsurface chemistry, is dominated by sulfur and water. Formation models involving complete elemental extraction from a bulk silicate Europa calculate an upper limit for endogenous sulfur of 340 g $kg^{-1}_{H_2O}$ (*Zolotov and Shock,* 2001). However, the preferred partial extraction model of that work argues for only ~3 g sulfur $kg^{-1}_{H_2O}$. Coupled with this is the exogenous sulfur delivered from the Io torus (distinct from that of sulfur from micrometeorites, discussed below). Observations put the contemporary flux of iogenic sulfur ions and neutrals to the surface of Europa at 10^7–10^8 and ~10^6 atoms per cm^{-2} s^{-1}, respectively (*Johnson et al.,* 2004). The loss rate is ~10^7 atoms per cm^{-2} s^{-1}. Over ~4 G.y. this amounts to ~6.4 × 10^{10} kg of sulfur, or roughly 20 ng $kg^{-1}_{H_2O}$ added to Europa's water layer. Although perhaps a small net effect, it is important to emphasize the process taking place: volcanism on one world (Io) is having a measurable effect on the chemistry of another world (Europa). On the surface, spectra from the Galileo NIMS indicated regions where hydrated sulfate may constitute upwards of 90% by number of the molecular surface abundance. Considerable debate remains regarding the cation associated with the sulfate (see chapter by Carlson et al.), but radiolysis of the europan surface almost certainly drives a sulfur cycle, creating surface reservoirs of hydrated sulfate, sulfur dioxide, hydrogen sulfide, and various forms of elemental and polymerized sulfur (*Carlson et al.,* 1999a, 2002, 2005).

Biochemically, sulfur is in the amino acids cysteine and methionine, which are components of numerous enzymes and coenzymes that are critical to metabolism. Functionally, the thioester bond (CO–S) mediates electron transfer and group transfer in metabolism (*de Duve,* 2005). Specifically, the thiol within acetyl-coenzyme A plays a key role in the tricarboxylic acid cycle (a.k.a. the Krebs cycle). This cycle is responsible for generating reduced metabolic intermediates and adenosine triphosphate (ATP). ATP is the central molecule that fuels biosynthesis in all life on Earth. In modern metabolic pathways, the protonmotive force is used more than thoiester couplings to drive synthesis of ATP, but the thioester coupling is tied to some of the most primitive metabolic pathways (*de Duve,* 2005). Indeed, the role of sulfur in modern coenzymes and metabolic pathways has motivated some workers to examine whether primordial sulfur geochemistry may have served as the seed for the origin of life on Earth (*Wächtershäuser,* 1988; *Cody,* 2005).

The cycling of sulfur on Europa is potentially of great importance to the habitability of Europa. The oxidation states of sulfur, from H_2S to SO_2, offer a variety of useful biochemical pathways for metabolism. In a cold ocean with an active seafloor, H_2S or SO_4^{2-} will be the dominant form, with the balance being determined by the oxidation state of the bulk ocean. If Europa has an active seafloor but cycling with the icy shell is poor, then H_2S will dominate. Conversely, active cycling with the icy shell but not the seafloor will yield a sulfate-dominated ocean. A detailed discussion of sulfur in Europa's ocean can be found in the chapter by Zolotov and Kargel.

1.2.1.5. Exogenous delivery of biologically essential elements: Delivery of biologically essential elements from comets and micrometeorites over the age of the solar system would have also been an important source for Europa. *Pierazzo and Chyba* (2002) calculated this flux and found that over the past 4.4 G.y. Europa received 9 × 10^{11} to 1 × 10^{13} kg of carbon (the exact phase, e.g., organic or inorganic, depends heavily on the velocity of impact). Even had Europa somehow formed devoid of carbon, the accumulation of cometary carbon over solar system history would have still supplied this baseline reservoir. Similarly, the integrated flux would have led to a total of 2 × 10^{11} to 3 × 10^{12} kg of nitrogen, 2 × 10^{10} to 3 × 10^{11} kg of phosphorous, and 2 × 10^{11} to 2 × 10^{12} kg of sulfur delivered to Europa. These estimates account for much of the material being lost as ejecta upon impact, but they do not account for reaccretion. Therefore, these numbers serve as lower limits for the exogenous delivery of chondritic material. *Johnson et al.* (2004) calculated a micrometeorite flux of ~0.045 kg s^{-1} to Europa, resulting in an integrated flux for the above elements that is roughly an order of magnitude larger than *Pierazzo and Chyba* (2002). The *Johnson et al.* (2004) results do not account for ejecta loss and therefore serve as a good upper estimate.

Table 1 shows a comparison of elemental abundances for several different solar system objects. Also shown in Table 1 are the relative elemental abundances found in life on Earth.

1.2.2. The constraint of chemical energy. Life extracts energy from its environment and, at a minimum, exchanges it for heat energy and entropy. The energy from the environment is used to do the work of cellular maintenance, metabolism, and reproduction. The availability of energy within an environment can be a limiting factor for habitability (*Heijnen and van Dijken,* 1992; *Nealson,* 1997; *Hoehler et al.,* 2001). On Europa, it has been argued that while liquid water and biologically essential elements are available, the chemistry of the ocean may have reached thermodynamic equilibrium, resulting in an ocean devoid of the energy needed for life (*Gaidos et al.,* 1999).

The chemical potential, or molal Gibbs free energy (with units of kJ mol^{-1}), of a solution provides a metric by which we can assess the energetic constraint on habitability. This

TABLE 1. Elemental abundances for solar system objects and life.

C = 1	Life	Earth (bulk)	Earth (crust)	Earth (ocean)	CI chond	CV chond	LL chond	Comets	Jupiter (atm.)	Europa Ocean
H	5.67	10.1	51.5	47,290	7.03	6.34	NA	4.02	952	19.6
N	0.23	0.012	0.125	0.714	0.079	0.013	0.019	0.094	0.384	0.013
O	2.4	5400	1080	23,640	10.1	52.4	96.8	2.02	0.284	9.8
P	0.012	10.4	1.41	0.714	0.012	0.082	0.114	N/A	$2.9x10^{-4}$	0.082
S	0.012	76.1	0.589	12.1	0.588	1.56	2.54	0.071	0.039	1.54
Ref.	*Zubay* (2000)	*Kargel and Lewis* (1993)	*Zubay* (2000)	*Zubay* (2000)	*Lodders and Fegley* (1998)	*Lodders and Fegley* (1998)	*Lodders and Fegley* (1998)	*Lodders and Fegley* (1998) Dust + Ice, Comet P/Halley	*Taylor et al.* (2004)	*Zolotov and Shock* (2001)

Notes: Normalized, by number of atoms, relative to carbon. Numbers for Europa's ocean are based on the *Zolotov and Shock* (2001) 100-km bulk extraction ocean model with Fe-core. Nitrogen and phosphorous for Europa are scaled based on the abundance found in CV chondrites ("chond").

term is the chemical reaction equivalent of gravitational potential energy or electrical potential in circuits (*Krauskopf and Bird,* 1995). The change in Gibbs free energy per mole of a substance can largely be considered a change in energy level for a given phase relative to the initial state. The Gibbs free energy, G, is defined as

$$G \equiv H - TS \tag{1}$$

where H is the enthalpy, T is the temperature in Kelvin, and S is the entropy of the given state. Changes in enthalpy, where H = E + PV, account for changes in heat as a function of pressure P and volume V as well as internal energy. For exothermic reactions $\Delta H < 0$, while for endothermic reactions $\Delta H > 0$. In accord with the second law of thermodynamics, the entropy of the system plus its environment must always increase, until equilibrium is reached, at which point dS = 0 for the system considered. The change in Gibbs free energy is then (assuming constant temperature)

$$\Delta G = G_{Products} - G_{Reactants} = \Delta H - T\Delta S \tag{2}$$

where $\Delta G = 0$ at equilibrium. The value of ΔG (typically in kJ per mole reactant) provides a measure for predicting the tendency of compounds in a system to react. Negative ΔG for a given reaction indicates spontaneous progression of the reaction, while positive ΔG indicates that the reaction requires some input of energy into the system. It is not hard to show that a system with $\Delta G < 0$ and constant T corresponds to $\Delta S > 0$ for the system plus the environment, i.e., is entropically driven. Put another way, a negative ΔG indicates that the reaction can do work on the system (exergonic), while a positive ΔG requires an input of work (endergonic). This capacity for a reaction to do work, or yield free energy, is the difference between the total energy change and the waste heat. In terms of the first law of thermodynamics, this is dW = dQ – dE, where W, Q, and E refer to the work, heat, and energy respectively. It is important to note that while enthalpy is often referred to as "heat content" (indeed, this is consistent with the etymology), ΔH represents both waste heat and some energy to do work (e.g., the familiar PdV = dW term in the differential of dH). Similarly, in a reversible process the $T\Delta S$ term of ΔG represents waste heat, ΔQ, but in practice almost all processes are irreversible, leading to $T\Delta S > \Delta Q$. All known biochemistry is critically dependent on this difference between waste heat and the energy capable of doing work in a given metabolic reaction. Variations of temperature, pressure, and concentrations of relevant chemical species can greatly alter this balance.

A related but distinctly different approach to Gibbs free energy illustrates this relationship. For a given reaction, with reactants X_n, products Y_n, and coefficients a_n and b_n,

$$a_1X_1 + a_2X_2 + \ldots a_nX_n \rightarrow b_1Y_1 + b_2Y_2 + \ldots b_nY_n \tag{3}$$

the equilibrium constant, K, can be found via

$$K = \prod_1^n [Y_m^{b_m}] \Big/ \prod_1^n [X_m^{a_m}] \tag{4}$$

Here brackets denote molal concentrations (mol kg^{-1} solution) of the reactants and products. The Gibbs free energy, can then be quantified by

$$\Delta G = \Delta G^\circ + RT\ln K \tag{5}$$

where R is the gas-law constant (8.314 J mol^{-1} K^{-1}), and ΔG° is the free energy as measured at a reference temperature and pressure (typically 25°C and 0.1 MPa, referred to as standard temperature and pressure, STP).

Two issues arise when addressing the limitations on life as imposed by the need for free energy. First, the availability of reactants, i.e., the left side of equation (3), is a hard constraint on ΔG. Without reactants, there is no reaction. Second, without the right set of reactants and the right set of conditions (e.g., temperature and pressure), the ΔG of the reaction may be insufficient for sustaining life. On Earth, the first limitation is largely solved by plate tectonics and geological cycling of elements important to life. Without this cycling, for instance, much of Earth's carbon would have been permanently sequestered into carbonates (*Kasting and Catling,* 2003). Also significant, especially in the context of icy moons with relatively low gravity, is that Earth does not lose much of its inventory of elements to space. Hydrogen escape is the only significant loss term. On Europa, loss of elements to space may be significant (see chapters by McGrath et al. and Johnson et al.).

The minimum ΔG needed for maintaining life has been a subject of considerable investigation, especially in recent years (*Heijnen and van Dijken,* 1992; *Hoelhler et al.,* 2001). This limit, sometimes referred to as the maintenance energy (ME), typically excludes the energy needed for growth and reproduction and focuses just on the energy needed for maintaining the organism. For the case of methanogenic *Archaea* utilizing the pathway

$$CO_2 + 4H_2 \rightarrow CH_4 + 2H_2O \quad (6)$$

measurements in anoxic sedimentary environments by *Hoehler et al.* (2001) put the limit for metabolizing H_2 at $\Delta G = -10.6 \pm 0.7$ kJ (mol CH_4)$^{-1}$ at 0.1 MPa and 22°C. Similarly they found that sulfate-reducing bacteria, utilizing the pathway

$$SO_4^{2-} + 4H_2 \rightarrow S^{2-} + 4H_2O \quad (7)$$

could be supported by energy yields as low as $\Delta G = -19.1 \pm 1.7$ kJ (mol SO_4^{2-})$^{-1}$ at 0.1 MPa and 22°C. For aerobic heterotrophs, from nematodes to blue whales, the oxidation of hydrocarbons yields values of hundreds of kJ per mole for the net change in Gibbs free energy of the metabolic reaction.

The problem for Europa is that the ocean may be devoid of cycles that could maintain a chemical flux to drive equations (3), leading to what *Gaidos et al.* (1999) term "thermodynamics-driven extinction" and what *Chyba and Hand* (2001) have referred to as the "spectre of entropic death." Even if Europa were hydrothermally active, *Gaidos et al.* (1999) argued, the resulting chemistry would be dominated by reductants and little ΔG would be available from such reactions. Without a source of oxidants to Europa's ocean there will be no chemistry to drive equation (3) and the only free energy for life to harness would be through methanogenesis and sulfate reduction. Both of these metabolic pathways are limited by geological supply at the seafloor. Here we first review the case of chemical energy derived from low-temperature geochemistry in Europa's ocean; we then review the case for a hydrothermally active ocean; and finally we consider the energy available in an ocean that mixes with the icy shell and receives a flux of radiolytically produced oxidants from the surface. Figure 4, based on *Nealson* (1997) but adapted to the specific case of Europa, illustrates the limiting ΔG of oxidant and reductant (redox) pairs. Oxidants and reductants observed on the surface of Europa are shown in outlined boxes.

Zolotov and Shock (2003, 2004) considered the case of low-temperature geochemistry for Europa's ocean and found that biologically useful cycles of sulfur, carbon, and iron could persist provided new rock was periodically made available on the ocean floor. In the model of *Zolotov and Shock* (2003), quenched hydrothermal fluids rich in H_2, CH_4, and/or organics couple with dissolved sulfate to yield

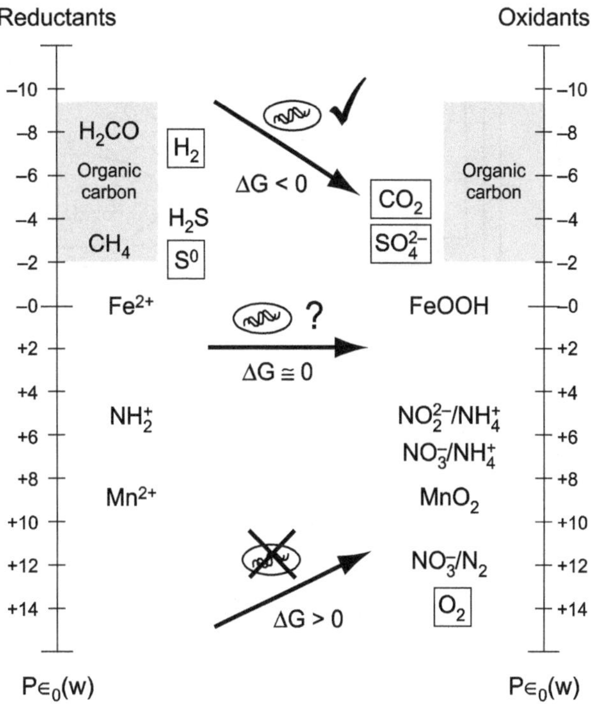

Fig. 4. Gibbs free energy and the energetic limits of life at Europa. Here we show several metabolic pathways that can, and cannot, support life. Life extracts energy from the environment in order to do the work of growth, reproduction, and maintenance. If a given reaction yields a sufficient negative change in Gibbs free energy, then life might be able to utilize that reaction (e.g., the microbe symbol marked with a check sign). The minimum negative change in Gibbs free energy required for life (microbe symbol with a question mark) is a topic of considerable study (see text). Pairing of reductants and oxidants that do not yield a negative change in Gibbs free energy are incapable of supporting life (e.g., the microbe symbol marked "X" for which the change in Gibbs free energy is positive). Reductants and oxidants in outlined boxes have been detected on Europa. Compounds are aligned along a relative redox scale. Pє$_0$ is the negative logarithm of the electron activity. Highly electronegative compounds are plotted at the top of each scale, and the most electropositive compounds are at the bottom. Adapted from *Nealson* (1997).

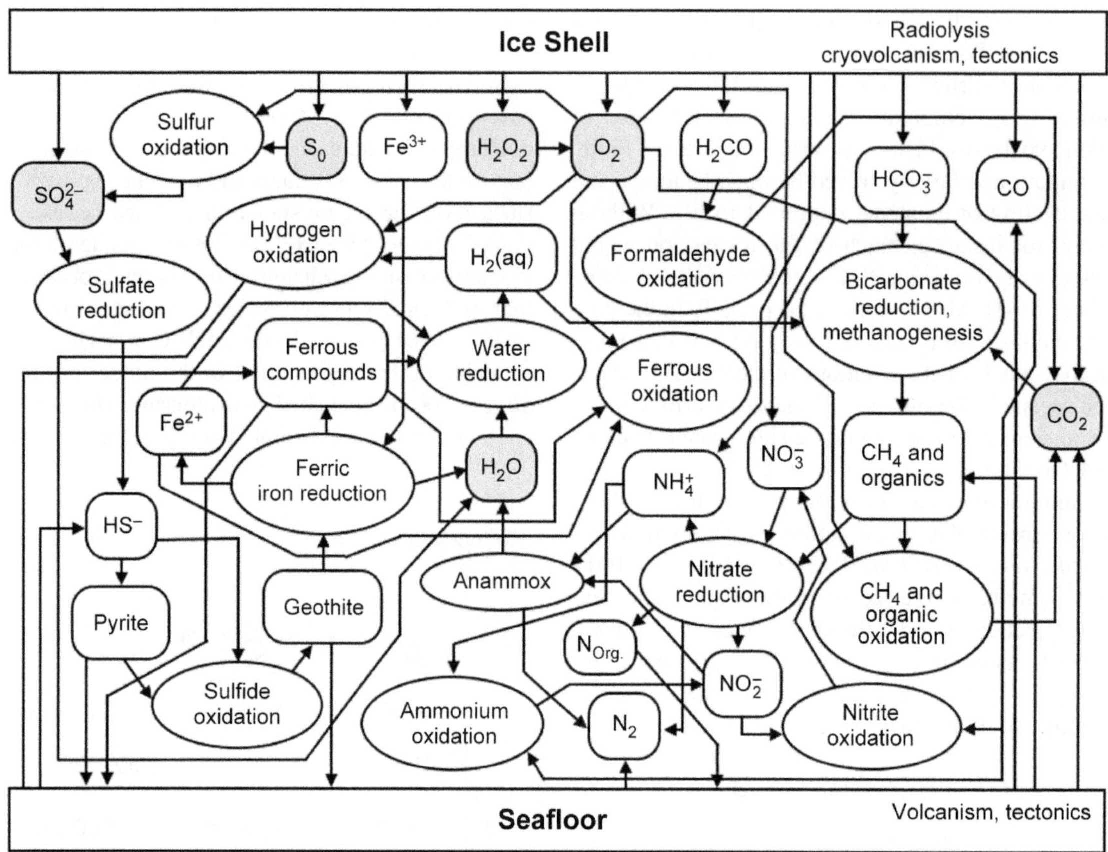

Fig. 5. Possible biogeochemical cycles and energetic niches on Europa. Molecular species are shown in rectangles, and reactions — either abiotic or metabolically mediated — are shown in ovals. Most of the reaction niches indicated are known metabolic pathways for terrestrial microbes. However, direct oxidation of ferrous iron by water, sulfate, or bicarbonate, and oxidation of bisulfide (HS^-) by water, have no known microbial analog. Cycling of surface material and seafloor material would help to maintain oceanic biogeochemical pathways. Compounds in gray have been observed on the europan surface. The diagram is based on Fig. 9 of *Zolotov and Shock* (2004), but here surface oxidants are added.

biologically useful energy. This is comparable to known sulfate-reducing microbial ecosystems in anoxic environments on Earth (*D'Hondt et al.,* 2002; *Amend and Shock,* 2001). Example pathways are

$$SO_4^{2-} + H^+ + 4H_2(aq) \Rightarrow HS^- + 4H_2O(l) \quad (8)$$

$$SO_4^{2-} + 2H^+ + 4H_2(aq) \Rightarrow H_2S(aq) + 4H_2O(l) \quad (9)$$

The equilibrium constant at estimated values for T and P of the europan seafloor (T = 273 K, P = 137.5 MPa) are $\log_{10}K = 13.22–8720T^{-1} + 6.54 \times 10^6T^{-2} – 6.77 \times 10^8T^{-3}$, and $\log_{10}K = 34.36 – 24174T^{-1} + 1.17 \times 10^7T^{-2} – 1.21 \times 10^9T^{-3}$. Under these conditions, the change in Gibbs free energy for the reaction in equation (9) is –263 kJ mol^{-1} (*Zolotov and Shock,* 2003). The low temperature serves to inhibit the abiotic reduction of sulfate, allowing for biological mediation of the reaction. Based solely on estimates for sulfate and hydrogen available in the original hydrothermal fluid after the formation of Europa, *Zolotov and Shock* (2003) calculate that sulfate reduction in a 100-km-deep ocean could yield a total of ~10^{24} J. This corresponds to ~10^{16} kg of total biomass throughout the ocean and over the history of the ocean (*Heijnen and Dijken,* 1992; *McCollom,* 1999). By comparison, terrestrial primary productivity, which is dominated by photosynthesis, yields ~10^{14} kg yr^{-1} (*Field et al.,* 1998; *McCollom,* 1999). Ultimately the above pathways are limited by geological activity at the seafloor supplying H_2, CH_4 and organics to the ocean. A sulfate and carbon flux from the icy shell would serve to maintain a cycle of sulfate reduction on a hydrothermally active Europa.

Considering the combined geochemistry of sulfur, iron, and carbon, in a cold, mildly alkaline europan ocean, *Zolotov and Shock* (2004) examined a host of possible metabolic pathways and biogeochemical cycles that could provide chemical energy for a subsurface biosphere. Figure 5 provides a summary of some of these cycles and possible energetic niches for life. Almost all the niches identified map to metabolic pathways for known terrestrial microbes, but several reactions — including direct oxidation of ferrous iron (dissolved or mineralized) by water, sulfate, or bicarbonate, and oxidation of bisulfide (HS^-) by water — have no known microbial analog. Maintaining the cycles identified by *Zolotov and Shock* (2004) into the present requires periodic delivery of new mantle rock or aqueous fluids from the seafloor.

McCollom (1999) considered a hydrothermally active Europa and used methanogenesis (equation (6)) as a model pathway for assessing biomass production. Both a reduced and oxidized ocean were considered. In the reduced model, CH_4 and H_2S dominate, whereas in the oxidized ocean model SO_4^{2-} and HCO_3^- are in abundance. Carbon dioxide and hydrogen released from fluid-rock interactions were taken to be the sources of oxidants and reductants, and McCollom notes that CO_2 in terrestrial vent fluids is largely derived from leaching of magmatic gases, or from interaction with gases exsolved as the underlying magma chamber solidifies, i.e., it is new carbon, not recycled HCO_3^-. Important to the thermodynamics of both McCollom and *Zolotov and Shock* (2004), abiotic reduction of CO_2 is very slow at the low temperatures expected for Europa's ocean (<300 K). The range of energy available per kilogram of vent fluid in the oxidized ocean and reduced ocean was found to be 250–500 J, capable of yielding ~6–13 mg of dry weight biomass per kg of fluid, based on terrestrial microbial analogs (*McCollom,* 1999; *Heijnen and Dijken,* 1992). Taking Europa's contemporary hydrothermal flux to be 10^{-3} that of Earth's annual 3×10^{13} kg of hydrothermal fluid, *McCollom* (1999) estimates an annual europan biomass productivity of 10^5–10^6 kg yr^{-1}, or about 10^{-5} that of terrestrial hydrothermal biomass productivity. Integrated over 4 G.y., this is comparable to the results of *Zolotov and Shock* (2003) for sulfate reduction.

Cycling of ocean water with the seafloor crust is an important consideration for the supply of reductants on Europa, but the total energy available is greatly limited by the availability of oxidants. Cycling of the ocean with the icy shell may, however, introduce oxidants such as sulfate, carbonic acid, O_2, and H_2O_2 into the ocean, yielding a solution with high chemical potential. The production of oxidants at the surface results from the radiolytic processing of the surface ice by the 7×10^{15} eV cm^{-2} flux of energetic particles, the majority of which are high-energy electrons (*Cooper et al.,* 2001; see chapter by Paranicas et al.).

Using the hydrogen peroxide surface abundance of 0.13% by number relative to water, as constrained by observations from Galileo NIMS (*Carlson et al.,* 1999a), *Chyba* (2000) and *Chyba and Phillips* (2001) examined the possible delivery of hydrogen peroxide to the ocean in the context of habitability. If the surface is gardened (i.e., mixed by meteorites and micrometeorites) to an average depth of 1.3 m over 10^7 yr (*Cooper et al.,* 2001), and the gardened layer contains the above peroxide concentration throughout, then ~4×10^{16} moles of H_2O_2 will be available in the surface reservoir. *Chyba and Phillips* (2001) argue that the resurfacing of Europa could deliver this reservoir to the subsurface ocean, where it would rapidly (~10 yr) decay to O_2 and serve as a useful oxidant for metabolism. Gardening depths may be one to two orders of magnitude lower (see chapter by Moore et al.) and resurfacing rates an order of magnitude higher (*Zahnle et al.,* 2008), leading to delivery of ~10^6–10^7 moles per year of H_2O_2. *Chyba and Phillips* (2001) estimate that radiolysis of CO_2 and CO in ice would yield a flux of ~3×10^5 moles per year of the biologically useful reductant formaldehyde (H_2CO), although that compound was not observed by NIMS. Again, this estimate is decreased by two orders of magnitude if one assumes both shallow gardening and longer resurfacing periods. The calculated abundance of H_2CO is consistent with laboratory measurements of electron radiolysis of CO_2-enriched ice (*Hand et al.,* 2007).

Based on the high-end fluxes, and using the metabolism of the soil bacterium *Hyphomicrobium* as an example,

$$H_2CO + O_2 \rightarrow H_2O + CO_2 \qquad (10)$$

Chyba and Phillips (2001) estimated that delivery of radiolytically produced oxidants (H_2O_2) and reductants (H_2CO) from the surface could alone support an ecosystem of ~10^{23}–10^{24} microbial cells, or ~10^6–10^7 kg of steady-state biomass. If the above reaction were not limited by the availability of H_2CO, then steady-state biomass could reach ~4×10^{11} kg. Considering the gardening and resurfacing uncertainty, steady-state biomass utilizing H_2O_2 as an oxidant ranges from ~10^9 to 10^{11} kg.

Considering only the observed peroxide concentration underestimates the total oxidant abundance on the surface of Europa, and consequently, the possible flux to the ocean. *Cooper et al.* (2001) and *Chyba and Hand* (2001) calculated that radiolytically produced O_2 on the europan surface could lead to combined oxidant fluxes ($H_2O_2 + O_2$) as high as 10^{12} moles per year. Coupled with this is the decay of ^{40}K (half-life = 1.25 G.y.) in both the icy shell and ocean. Scaling potassium concentrations from estimates for salt concentrations given in *Kargel et al.* (2000), *McCord et al.* (1999), and *Fredriksson and Kerridge* (1988), *Chyba and Hand* (2001) find that O_2 produced via ^{40}K decay in a 10-km icy shell could yield 10^7–10^8 mol O_2 yr^{-1}, while that in a 100-km-deep ocean would yield 10^{10} mol O_2 yr^{-1} in the contemporary ocean, and 10 times as much over 4.2 G.y.

Groundbased observations by *Spencer and Calvin* (2002) of absorption at 577.1 nm on Europa indicate solid-phase O_2 trapped in the ice. The 577.1-nm absorption results from interacting pairs of O_2 molecules. Using this quadratic dependence, *Hand et al.* (2006) modeled the radiative transfer of the absorption to calculate O_2 abundances as high as 1.3–4.6% by number relative to water. The lower limit is derived from considering water ice with O_2 as the only non-ice component, while the upper estimate allows for absorption by other compounds in the ice. In either case, the O_2 abundance in the surface ice is more than an order of magnitude greater than the observed hydrogen peroxide (see chapter by Carlson et al.).

Delivery of this much oxygen to the ocean, even at periods of hundred of millions of years, will serve to oxidize a geologically inactive ocean (*Hand et al.,* 2007). If the seafloor is active, then the redox coupling of O_2 and SO_4^{2-}

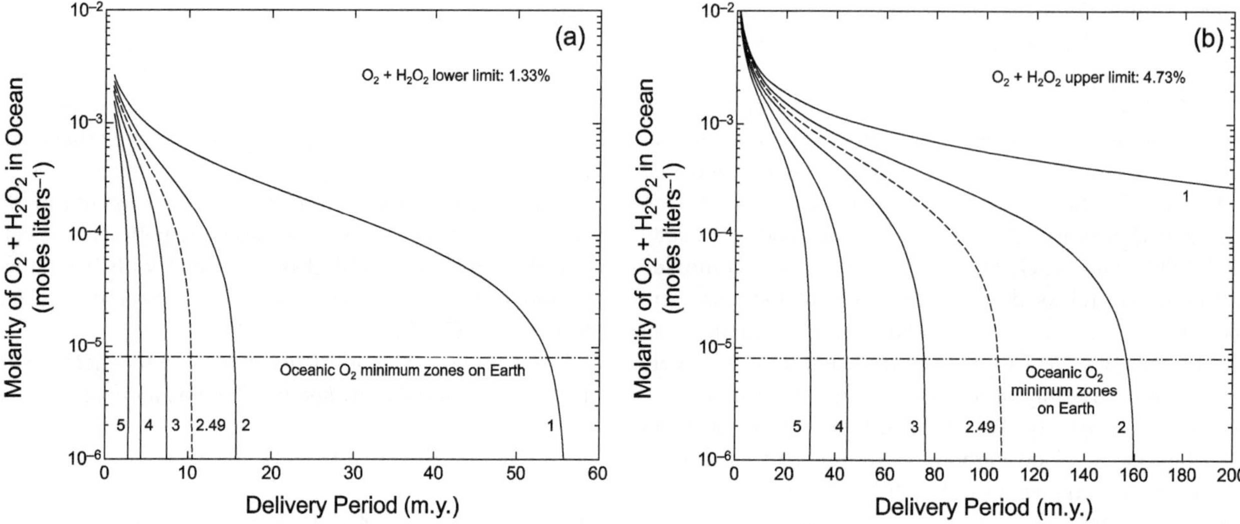

Fig. 6. Molarity of O_2 and H_2O_2 (which rapidly becomes O_2) in Europa's ocean as a function of the time it takes for radiolytically produced surface material to reach the ocean (delivery period). The quantity of O_2 and H_2O_2 delivered accounts for surface gardening (see text) **(a)** for a radiolytically produced surface abundance of 1.33% by number relative to water, and **(b)** for 4.73% by number abundance. In each case, a flux of seafloor reductants is assumed, and the net molarity of O_2 and H_2O_2 in the ocean (assumed 100 km deep) is what is left after reaction with the seafloor reductants. Solid contours, from left to right are for 5, 4, 3, 2, and 1 × 10^9 moles per year of reductants. Dashed line is for 2.49 × 10^9 moles per year, corresponding to the seafloor activity estimate of *McCollom* (1999). The molarity of Earth's oceanic O_2 minimum zones (dash-dotted line) is shown for reference. For a lower O_2 and H_2O_2 concentration **(a)**, a shorter delivery period or a lower reductant flux are required in order for the molarity to exceed that of Earth's O_2 minimum zones. For higher O_2 and H_2O_2 concentration **(b)**, the molarity exceeds that of Earth's minimum zones even for long delivery periods and high reductants fluxes. Sulfate concentrations exceed the O_2 and H_2O_2 by at least an order of magnitude; thus, achieving an oxidized ocean with sulfate is considerably easier than with O_2 and H_2O_2.

from the icy shell and H_2, H_2S, Fe^{2+}, and CH_4 from the seafloor would be a useful energetic resource in the context of habitability. Abiotic oxidation would precipitate magnetite and pyrite, but even for terrestrial banded iron formations biological mediation of these products is believed to be significant (*Canfield,* 1998).

In Fig. 6, adapted from *Hand et al.* (2007), the molarity of O_2 in Europa's ocean is plotted as a function of delivery period, where the delivery period, τ_d, is defined as the average time it takes for material on the surface of Europa to reach the subsurface ocean. In the work of *Chyba* (2000), *Chyba and Phillips* (2001), and *Chyba and Hand* (2001) the delivery period was taken to be roughly equivalent to the observed surface age, t_{ice}, of tens of millions of years (*Zahnle et al.,* 2003, 2008; see chapter by Bierhaus et al.). This was a reasonable assumption, but other alternatives are possible. The molarity of oceanic O_2 was determined by calculating the number of moles delivered to the ocean per year, η_i, for a given delivery period

$$\eta_i = \frac{Ad_g\varepsilon_i}{\tau_d} \tag{11}$$

This flux is a function of A, the surface area of Europa, the gardening depth, d_g, to which surface materials are buried, and ε_i, the concentration of each compound throughout the gardened layer (in units of moles per cubic meter). Here the gardening depth has been calculated based on *Cooper et al.* (2001), but recent work summarized in the chapter by Moore et al. suggests that the gardening depth could be reduced by one to two orders of magnitude. From this flux we have then subtracted a sink term based on the hydrothermal flux of ~10^9 moles per year of reductants as calculated by *McCollom* (1999). By subtracting this sink term we account for a rough estimate of abiotic sinks for oxidants in the ocean. The lines shown in Fig. 6 correspond to sink terms of 1, 2, 3, 4, and 5 × 10^9 mol yr^{-1} reductants. The dashed line for 2.49 × 10^9 mol yr^{-1} is the oxidant consumption based on the reductant flux estimated by *McCollom* (1999). The resulting oceanic molarity is then the net accumulation of O_2 after accounting for hydrothermal sinks. Figure 6a shows the results based on a percent by number abundance of O_2 and H_2O_2 of 1.33%, and Fig. 6b shows the same calculations based on a 4.73% by number abundance. The dotted horizontal line shows the molarity of O_2 in the most oxygen-depleted regions of Earth's ocean. Even in these energetically challenging environments on Earth, multicellular creatures are found, such as the 3.5–12.5-cm-long crustacean *Gnathophausia ingens* (*Childress,* 1968). For the lower limit oxygen concentration in Europa's ice (Fig. 6a), the molarity of O_2 in the ocean exceeds terrestrial O_2 minima levels if the delivery period is <54 m.y. and the sink term is 1 × 10^9 moles per year. If the sink term is doubled, then the delivery period must be <16 m.y. if O_2 molarity is to surpass the minima level. These values for the delivery period are consistent with published values for the surface age (*Zahnle et al.,* 2003). For a sink term of 5 ×

10^9 mol yr^{-1} the delivery period would have to be less than 2.8 m.y. in order to surpass the O_2 minima level. This is clearly much younger than the surface age and thus unlikely.

For the case of $O_2 + H_2O_2$ concentrations of 4.73% by number in the ice, the delivery period can be significantly longer while still achieving an oceanic O_2 molarity suitable for terrestrial macrofauna. With the 5×10^9 mol yr^{-1} sink term subtracted from the oxidant flux, the delivery period need only be <30 m.y. in order to maintain an ocean above terrestrial O_2 minima zone levels. If the reductant sink is decreased, e.g., to 2×10^9 moles per year, then delivery periods can be as long as 160 m.y. and still achieve minima levels. Decreasing the sink term even more allows an O_2-rich ocean to be maintained even with delivery periods of several hundred million years. The delivery periods in the case of a 4.73% by number $O_2 + H_2O_2$ concentrations can greatly exceed the surface age of Europa and still achieve an ocean with high levels of oxygen. We note here that much of our terrestrial ocean is supersaturated with O_2, reaching levels of ~250 μM or roughly 3% supersaturation (*Millero,* 2005). The temperature and pressure environment of Europa's ocean will permit levels of dissolved O_2 in the range of 50 mM (*Lipenkov and Istomin,* 2001). Depending on the delivery period and sink term for oxidants, it is possible that oxygen levels in Europa's ocean could greatly exceed those found in Earth's ocean. While *Gaidos et al.* (1999) argued that habitability may be hindered by limited availability of oxidants, it is possible that the ocean could be highly oxidized with sulfate, oxygen, and hydrogen peroxide if there is even moderate mixing of the ocean with surface material. If this is the case, habitability could actually be inhibited by an overabundance of oxidants. While terrestrial macrofauna have evolved mechanisms to accommodate our oxygen-rich atmosphere, the origin of life on Europa may have been severely compromised if such an oxidant-rich ocean persisted since the formation of Europa's ocean.

1.2.3. Photosynthesis as a possibility. Photosynthesis cannot strictly be ruled out for Europa (*Reynolds et al.,* 1983; *Greenberg et al.,* 2000). Nevertheless, it is considered an unlikely metabolic pathway because the average thickness of the icy shell is generally thought to be at least several hundred meters in thickness (*Carr et al.,* 1998; *Pappalardo et al.,* 1998; *Greenberg et al.,* 2000), and light cannot penetrate this deeply in ice.

The equatorial daytime flux of photosynthetically active radiation (PAR, 375–725 nm) is ~21.87 W m^{-2} (~100 μmol photons m^{-2} s^{-1}) at Europa. For light of wavelength λ and incident intensity I_0, the intensity at depth x is

$$I(x) = AI_0e^{-ax} \tag{12}$$

where A is the local albedo of Europa and a is the absorption coefficient at wavelength λ (*Hobbs,* 1974). With A = 0.67, I_0 = 21.87 W m^{-2}, and a = 8×10^{-4} cm^{-1} for visible light through clear ice with no bubbles (*Hobbs,* 1974), the maximum penetration depth is ~100 m. Adding bubbles and frost to the ice increases a to ~2.8×10^{-4} m^{-1} and reduces the penetration depth to a few meters. Unless regions of the icy shell are thinner than these values, photosynthesis will not be possible.

On Earth, photosynthesis by phytoplankton has been measured beneath up to 7 m of ice at photon flux levels near 0.1% of the incident surface irradiance (~1 μmol photon m^{-2} s^{-1}) in lakes of the McMurdo Dry Valleys, Antarctica (*Priscu,* 1995). Similarly, adequate light penetrates the 3–4-m-thick ice cover of Lake Vanda (McMurdo Dry Valleys) and penetrates the highly transparent waters (extinction coefficient ~0.033 m^{-1}) permitting photosynthesis to depths of ~65 m beneath the ice surface (*Howard-Williams et al.,* 1998). Clearly, viable terrestrial ecosystems demonstrate that photosynthesis can occur at considerable depths beneath ice. In addition to relatively high penetration of PAR, UV radiation can also be transmitted through the ice covers on Antarctic lakes. *Vincent et al.* (1998) found that UV penetration through the ice cover of Lake Vanda can be an impediment to photosynthetic productivity in the upper regions of the liquid water column. The ice covers of Antarctic lakes can themselves harbor an active ecosystem fueled by photosynthesis, providing an oasis for life in what would otherwise seem to be an inhospitable environment (*Priscu et al.,* 1998; *Paerl and Priscu,* 1998; *Priscu et al.,* 2005a).

Reynolds et al. (1983) approached the problem of access to sunlight on Europa by considering crack formation and refreezing. *Buck et al.* (2002) reconsidered this problem and incorporated models for tidal heating of the ice. While formation of the initial surface layer of ice occurs rapidly, the ice growth rate slows as the ice becomes thicker. This allows for periods of days to years during which light may reach the liquid water interface. Using an annual resurfacing area of 5 km^2 based on frost deposition rates from *Squyres et al.* (1983), *Reynolds et al.* (1983) found that 2×10^{15} J yr^{-1} would be available for photosynthesis in the 375–725-nm range. This leads to a total equilibrium biomass (for photosynthesis only) throughout Europa of ~10^4–10^5 kg, or roughly one-billionth of Earth's annual photosynthetic productivity (*Field et al.,* 1998). We note that in the *Reynolds et al.* (1983) model the influence of the crack sidewalls was neglected in the refreezing calculations. This effect would decrease the total power available for photosynthesis by a factor of ~2, depending on the ratio of crack width to depth.

Despite the thickness of the icy shell being an impediment to the emergence of photosynthesis as a metabolic pathway, several aspects of such an ecosystem warrant closer inspection in the context of future robotic exploration. First, the selection effects for life detection on orbiter or lander missions would be biased toward detecting near-surface life utilizing pigments easily detectable with spectroscopy. Photosynthetic organisms satisfy both these conditions. Their need for sunlight would lead them to concentrate in any regions where the ice is thin and thus likely young in age. Second, the molecular pigments used for photosynthesis (e.g., chlorophyll-a, b, c, and various carotenoids and xan-

TABLE 2. The limits for growth and reproduction of life on Earth.

Parameter	Limits	Description
Temperature	–20°C to 121°C	Limits correspond to life in ice veins to hydrothermal systems.
Pressure	0 to 1680 MPa	Spore formation permits survival at low pressures. *Escherichia coli* and *Shewanella oneidensis* can remain metabolically active for ~1 day at the high end limit.
pH	–0.6 to >11	Possible metabolic activity at levels as high as 12.5–13 have been reported.
Salinity	0 to near salt saturation	Upper limit is ~260 g of $MgSO_4$ per kg of water, and to nearly 300 g NaCl per kg of water.
Water activity	>0.6	Survival at activities of zero are possible by forming spores. Unpublished results suggest that metabolically active populations may be possible for activities lower than 0.6.
Ionizing radiation	0 to 6000 Gy	Upper limit corresponds to *D. radiodurans*, which needs to actively repair its DNA. Limits of several tens of Gy (J kg^{-1}) are more typical of bacteria like *E. coli* and *B. subtilis*.

Notes: These limits apply to various types of *Bacteria* and *Archaea*; *Eukaryotes* can only tolerate a subset of the conditions listed. Based on *Nealson* (1997), *Rothschild and Mancinelli* (2001), and *Marion et al.* (2003).

thophylls, in the case of Earth organisms) are evolution's gift to spectroscopists. They are molecules tuned for optimal absorption over a set bandpass. Chlorophyll-a, for example, absorbs sharply at ~400 and 660 nm and is easily detectable from Earth orbit. Chlorophyll-a also has unique fluorescence properties and fluoresces at 683 nm when excited with light near 400 nm. These selection effects for detection of life on Europa imply that while photosynthesis may occupy an unlikely or small niche on Europa, such an ecosystem may satisfy a disproportionately large solution space for our detection techniques. We discuss these issues in greater detail in section 3.

1.2.4. Earth ecosystems as an analog for Europa? Is Europa's ocean an "extreme" environment relative to Earth's ocean? In this section, we begin by considering temperature, pressure, and salinity, finding that Europa's ocean is probably well within the known limits of life. Then we consider some possible extreme terrestrial environmental analogs to Europa.

Europa's ocean is not a particularly extreme environment, in terms of temperature and pressure, at least. The temperature of the water may be suppressed to ~260 K (–13°C) by the dissolution of salts, but it cannot go much lower and still remain liquid (*Kargel et al.,* 2000; *Marion et al.,* 2003; *Melosh et al.,* 2004). In the eutectic limit with 213 g of chlorine per kg H_2O, the freezing point could get considerably lower [238.65 K at 0.1 MPa (*Marion et al.,* 2003)], but this would require ~6×10^{20} kg of chlorine globally, a value hard to reconcile with a chondritic origin for Europa (Cl is present at 700 ppm in CI chondrites) (*Kargel et al.,* 2000; *Lodders and Fegley,* 1998).

The pressure environment within Europa is also not that extreme compared to terrestrial ecosystems. Despite possibly harboring a deeper ocean than that of the Earth, ~100 km on Europa vs. a 4-km average and 11-km maximum depth on the Earth, the pressures are not that great because the gravitational acceleration, g, on Europa is less than one-seventh that of our considerably more massive Earth. For a given depth, d, in kilometers below the surface of Europa, the pressure in MPa is ~1.3 d, whereas on Earth it is ~10 d; therefore, the 110 MPa pressures in the 11-km Marinas Trench on Earth are comparable to the seafloor of a ~100-km icy shell plus ocean on Europa. Many large and complex life forms are found on Earth at the average seafloor depth of 4 km (*Van Dover,* 2000), corresponding to a comparable pressure at a depth of roughly 30 km on Europa. In the sediments of the Marinas Trench and other deep regions of our ocean, soft-walled foraminiferans have been found to flourish (*Gooday et al.,* 2004; *Todo et al.,* 2005). The pressures expected throughout the ocean on Europa are not a challenge for life as we know it.

Aside from a supply of chemical energy, life in the ocean of Europa could be most challenged by the salinity. Table 2 provides a summary of the limits for growth and reproduction of life on Earth based on *Nealson* (1997), *Rothchild and Mancinelli* (2001), and *Marion et al.* (2003). Although still highly uncertain, estimates for Europa ocean salinity range from a few tens of grams of salt per kilogram H_2O to saturation with magnesium sulfate or sodium chloride (*Kargel et al.,* 2000; *Zolotov and Shock,* 2001; *Hand and Chyba,* 2007). Even so, terrestrial halophiles, such as the archaean *Haloferax volcanii* can survive in near-saturated solutions of ~240 g $MgSO_4$ per kg H_2O, when no other ions are present (*Oren,* 1994). Methanogens are capable of surviving in solutions of up to 100–150 g NaCl per kg H_2O (*Oren,* 2001). Water activity is a measure of the availability of water, which, when considering surface water, is simply the water fugacity, and is often invoked as a limitation for life. While brine pockets and cracks in the icy shell are likely to have low water activities, a global ocean saturated with salt will still have an activity of 0.72 or higher and would be suitable for several terrestrial microbes (*Siegel,* 1979; *Nealson,* 1997; *Marion et al.,* 2003).

There are no perfect terrestrial ecosystem analogs for what we might expect to find on Europa. Hydrothermal vent ecosystems are often invoked for comparison. This is certainly a legitimate analogy in that physical conditions (temperature, pressure, lack of light) and chemical conditions (sulfur, iron, methane) could be similar, but the chemistry of the surrounding ambient water remains unknown. Whether deep ocean material mixes with shallower waters on Europa, thereby possibly bringing hydrothermal materials upward toward the ice, is also highly uncertain. In the context of exploration, the bottom of Europa's ocean is arguably the most inaccessible part of Europa's potentially habitable ocean. Spacecraft landers many decades from now may carry submersibles capable of interrogating the seafloor, but for the near-term exploration of Europa our search will be largely restricted to the surface and the ice-shell/ocean interface. The habitable volume of the ocean accessible to these searches will, at best, be the upper few kilometers of water. Considering the habitability of this region of the ocean, the liquid water environments of the recently discovered subglacial lakes in Antarctica provide some intriguing examples of how life on Earth has evolved and adapted to comparable cold and dark environments (*Priscu et al.,* 2005b, 2008).

Direct sampling of the liquid water from Antarctica's Lake Vostok has not yet occurred, but data from the region of the ice core just a few tens of meters above the lake indicate that the lake likely harbors life (*Priscu et al.,* 1999; *Karl et al.,* 1999). This region of the ice, approximately 240 m in thickness, which lies below a depth of 3590 m beneath the ice sheet surface, is accreted ice, so called because it is ice that forms from the lake water and accretes on the bottom of the East Antarctic ice sheet. Cell counts from this ice reveal microbial concentrations of 83–260 cells ml^{-1} (*Priscu et al.,* 1999; *Christner et al.,* 2006), which, based on estimated partitioning coefficients, equates to cell concentrations in the surface waters of Lake Vostok ranging from 150 to 460 cell ml^{-1} using epifluorescence microscopy of DNA stained samples (*Christner et al.,* 2006), and between 10^5 and 10^6 cell ml^{-1} using scanning electron microscopy (*Priscu et al.,* 1999).

Phylogenetic analysis of the 16S rRNA gene sequences of these cells places them among the δ-, γ-, and β-proteobacteria (*Christner et al.,* 2006), leading to the conclusion that the surface waters of Lake Vostok harbor a microbial consortium consisting of chemotrophs and heterotrophs. The Vostok γ-proteobacteria are likely chemolithotrophs harnessing energy from Fe^{2+} oxidation, or reduction of Fe^{3+} and elemental sulfur (S°) with hydrogen. Coupling of S° with Fe^{3+} may also be an energetically useful pathway for these microbes. The Vostok δ-proteobacteria are likely chemotrophic anaerobes yielding energy from reduction of sulfate, S°, Fe^{3+}, and Mn^{4+}. Finally, the Vostok β-proteobacteria appear to be heterotrophs utilizing single carbon compounds such as methanol, formate, and carbon monoxide. The Vostok ecosystem therefore derives much of its energy from redox couples associated with iron and sulfur within the subglacial environment (*Christner et al.,* 2006; *Priscu et al.,* 2006). Dissolved organic carbon, CO_2, N_2, O_2, and other minerals are also advected into the system from the overlying ice sheet as it melts.

Our understanding of the Vostok ecosystem is limited to the accretion ice, and by inference the surface waters. Very little is known about deep-water column processes and the degree to which the deep waters of Vostok mix with the surface waters. Based on comparison with lakes in the McMurdo Dry Valleys, the deep waters of Vostok (below perhaps 400 m) may be a bit saltier than the surface waters and may even be warmer owing to geothermal heating. These waters may also be anaerobic, depending on the gas flux from the melting ice and geothermal sources as well as pos-sible *in situ* respiration. The sediments would be an additional source of minerals and carbon to drive life in the deeper waters. The fact that the accreted ice serves as our only window into the ecosystem of Vostok is perhaps useful for the comparison to Europa. Over the next few decades, our search for life on and within Europa will be limited to the surface ice and, in an optimistic scenario, the water near the ice-water interface. Thus, while sedimentary and hydrothermal ecosystems are interesting in the context of Europa, it is really the near-surface water environments of subglacial lakes that provide the best comparison for what waters we might expect to encounter first with robotic spacecraft. Information on deep convective mixing of this surface layer would also provide information on processes that occur deeper within the water column.

Microbes are also found in veins and brine pockets within the Antarctic ice sheet. Although small — a typical vein might only be 10 μm in diameter — such environments are plenty spacious for a 1-μm-diameter bacterium. Additionally, even in the coldest regions of the ice sheet, thin films of water can exist at mineral or ice surfaces and only approximately three monolayers of water are needed to support metabolism. Given the presence of liquid water and biogenic elements in subglacial lakes and brine pocket veins of the ice sheet, what can be said about the energetic limitations of life in these environments? Based on models by *Jepsen et al.* (2007), most of the available metabolic energy is used for cellular maintenance. This may also be the case for organisms in the upper portion of the ice sheet where temperatures are <–20°C. However, in the deeper portion of the ice sheet, veins are larger and temperatures warm to ~–3°C, which is arguably a clement environment for life. Based on the increase in bacterial density observed in accretion ice, relative to the overlying glacial ice and in concert with the stable and relatively warm liquid water body thought to exist in Vostok, bacteria should be doubling once every 10 to 60 days in these veins.

Metazoan populations (multicellular organisms) may not be present in the Antarctic ice ecosystems. This is consistent with the following ecosystem constraints: (1) not enough

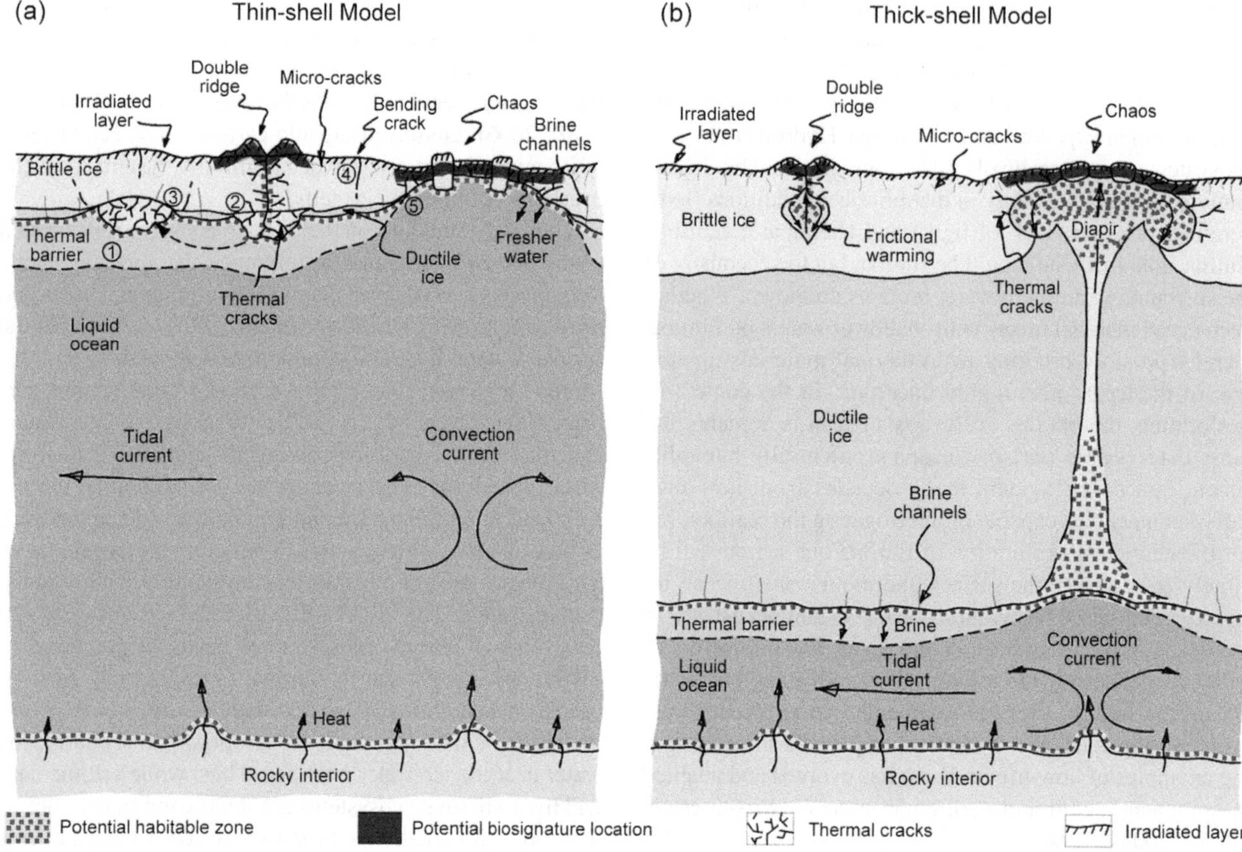

Fig. 7. Schematic cross-sections of Europa's icy shell, ocean, and seafloor showing possible habitats and biosignatures (from *Figueredo et al.,* 2003). **(a)** Thin ice-shell and **(b)** thick ice-shell models are shown for comparison, but in both cases, the regions expected to be most conducive to life are the chemically rich interfaces at the seafloor-ocean boundary and the ice-ocean boundary. Regions within the icy shell may also be habitable. The numbers given in the thin-shell model indicate the following specific features: (1) a sub-ice mixing layer, in which materials are exchanged between the ice and ocean; (2) a crushed-ice "keel," corresponding with isostatic support for surface ridges; (3) migration of subsurface ice material to other regions, causing subsequent upwarping; (4) cracks produced by bending of the icy shell; and (5) cracks resulting from thermal effects at the ice-water interface. In the thick-shell model, convection and the resulting diapirs and brine channels could provide habitable niches. Surface regions where biosignatures may be expressed are also shown.

energy in the primary food base to support higher trophic levels (energy transfer up food webs is not very efficient) and (2) metazoans (eukaryotes for that matter) contain a complex membrane network associated with organelles as well as the plasma membrane. These membranes are made of a mosaic of phospholipid bilayers interspersed with functional proteins (there is often a complex outer membrane matrix as well). Cold temperatures can change the fluidity of the membranes, rendering them metabolically useless (e.g., protonmotive forces like that used to produce ATP can come to a halt at cold temperatures). Finally, ice crystals can form in the membranes, ruining their integrity. Prokaryotic organisms are more capable of surviving freeze/thaw processes and typically dominate icy ecosystems on Earth owing to their paucity of membranes and high metabolic diversity relative to eukaryotic organisms, which can be considered structurally diverse (by virtue of their membrane bound organelles) and metabolically conservative.

The habitability of Europa's ocean will depend on the degree to which, and frequency with which, the ocean water cycles through the mantle rock of the ocean floor and with ice in the icy shell. Cycling through the ocean floor yields a solution rich in reductants while cycling with the icy shell introduces oxidants, both of which are needed to power life. Based on the need for energy, liquid water, and a suite of biologically essential elements, the prime habitats for life on Europa (were life to exist) are likely to be at the seafloor-ocean interface and at the ice-ocean interface. Here chemical energy and useful compounds and elements may combine with liquid water to provide the conditions needed for life. Figure 7, adapted from *Figueredo et al.* (2003), shows a cross section of possible habitats, and their surface expression, on Europa. *Lipps and Rieboldt* (2005) provide a detailed classification and taphonomic analysis of such possible habitats. Ongoing work in the polar regions of the Earth, and in particular in subglacial sulfur-rich springs (*Grasby et al.,* 2003), will help guide our understanding of life in ice and the chemistry and conditions needed to sustain life in such regions.

The availability of chemical energy, as discussed above, addressed the capacity of Europa's ocean to support life, but had little bearing on the issue of life's origin. In that context, a reduced ocean may be highly preferable. We explore our current understanding of life's origin, and the implications for Europa, in the next section.

1.3. Constraints on Origins

The origin of life on Earth remains one of the great, unanswered scientific questions. Lacking a complete understanding of how our own tree of life came to be, we cannot read too much into the chemistry of Europa and possible origins on that world. Nevertheless, the question of life's origin is precisely why Europa is such a prime astrobiological target — it poses a testing ground for a second, independent origin of life in our solar system. Our understanding of terrestrial prebiotic chemistry indicates that life may have also arisen on Europa. There is, as yet, no equation for such a prediction, but informally, the conventional scientific wisdom is that

$$\text{Liquid water} + \text{Biologically essential elements} + \text{Energy} + \text{Catalytic Surfaces} + \text{Time} \Rightarrow \text{Life} \qquad (13)$$

Were we to discover that Europa is not inhabited, but that the conditions on Europa fully satisfied the lefthand side of the above expression — and many of the other conditions thought to be necessary for origin of life (see, e.g., *Chyba and McDonald,* 1995) — we would gain new insight into the unique conditions that lead to the origin of life on Earth. Similarly, were life discovered on Europa, we could then begin a rigorous cross-comparison to investigate the conditions leading to the emergence of life on both worlds.

Laboratory experiments (*Miller,* 1953; *Palm and Calvin,* 1962; *Ponnamperuma et al.,* 1963; see *Wills and Bada,* 2001, for a review) have shown how the lefthand side of equation (13) can readily lead to the production of amino acids and nucleobases, and to some extent, short polymers. Similarly, biologists have sufficiently deconstructed the righthand side of this equation to understand that once the innovation of ribonucleic acids (RNA) occurred, chemistry was well on its way toward mutation, replication, and Darwinian selection — i.e., well on its way toward life itself. While many gaps remain in our understanding of life's origin, the largest gap is arguably that between the formation of monomeric building-blocks [e.g., amino acids and nucleobases as in *Miller and Urey* (1959)] and the emergence of an RNA-world, in which RNA serves the critical dual functions of an information storage molecule and a protein synthesizer (*Joyce,* 1989).

While Europa lacks the tidal pools, warm springs, and the atmospheric chemistry often invoked for terrestrial prebiotic chemistry, several attributes of the europan system make it an attractive host for life's origin. Foremost among these attributes are a rocky seafloor in contact with the ocean water, and a radiolytically processed icy shell that may harbor cracks and pockets of water in which eutectic freezing can concentrate solutes. In this section, we analyze these two attributes in the context of prebiotic chemistry. Specifically, we examine the icy shell and seafloor in the context of (1) the production, stability, and polymerization of amino acids and nucleobases; (2) chemical pathways and protometabolisms for self-organizing biochemical cycles; and (3) formation of membranes and vesicles that could be important for compartmentalization and differentiation of early life. These are three of the most critical aspects of life's origin on Earth and they serve as good starting points for assessing origins elsewhere.

1.3.1. The rock-water interface. An ocean in contact with a rocky seafloor may be an attribute rare among our solar system's four to seven known or probable contemporary liquid water oceans (Earth, Europa, Ganymede, and Callisto, and possibly Titan, Enceladus, and Triton). Ganymede and Callisto, both with oceans of ~10 times the volume of Earth's, likely have seafloors of ice III (*Spohn and Schubert,* 2003). Titan may have ammonia-rich seas trapped between an ice and hydrocarbon-rich crust and an ammonia-hydrate mantle (*Sagan and Dermott,* 1982; *Lorenz et al.,* 2008). Triton too may have a subsurface ocean, but again its liquid water-nitrogen mixture would likely be confined to a region between high-pressure phases of ice or ammonia-hydrate (*Cruickshank et al.,* 1984). Although the south polar plumes of Enceladus are intriguing evidence consistent with a subsurface sea (*Collins and Goodman,* 2007), our knowledge of this world is still too immature to know whether or not a substantial body of liquid is in contact with rock. Europa, the Earth, and potentially an ancient Mars are the only worlds likely to have had vast quantities of liquid water in contact with mantle rocks over geologically-long periods of time. Chemical weathering of these rocks can supply reduced compounds and metals that may be important for the origin of life (*Nisbet and Sleep,* 2001).

On Earth tectonic activity results in convective cooling of magma by flow of ocean water through the seafloor. The deep-sea hydrothermal vents powered by this activity have been repeatedly invoked as a crucible for life's origin on Earth (*Corliss et al.,* 1979, 1981; *Wächtershäuser,* 1988; *Cody,* 2005; *Martin et al.,* 2008). The reasons are manifold. The reductant-rich and warm environments would have been largely sheltered from impact sterilization (*Nisbet and Sleep,* 2001). Additionally, although debated, our top-down approach to understanding the root of Earth's tree of life suggests that hyperthermophiles — i.e., heat-loving microbes — may have been among the earliest life forms on Earth (*Pace,* 1997; *Madigan et al.,* 2003). The chemistry within the vent fluids and chimneys are conducive to early biochemistry, and the chimneys themselves create natural flow-through reactors with mineral-rich pores that could serve to prevent reaction products from diffusing into the ocean (*Russell et al.,* 1988; 1989). We have no knowledge of seafloor activity on Europa, but tidal energy must be dissipated, and the extent to which that energy is dissipated in the mantle will determine seafloor activity. Models for tidal dissipation at the europan seafloor range from 8 mW m^{-2} to more than 100 mW m^{-2} (*Squyres et al.,* 1983; *Thomson and Delaney,* 2001). By comparison, the seafloor flux on Earth is ~80 mW m^{-2} on average and ~200 mW m^{-2} in active regions (*Stein and Stein,* 1994). Although poorly constrained, the existence of hydrothermal systems on Europa seems plausible. Coupled with this, models of the thermal-orbital evolution

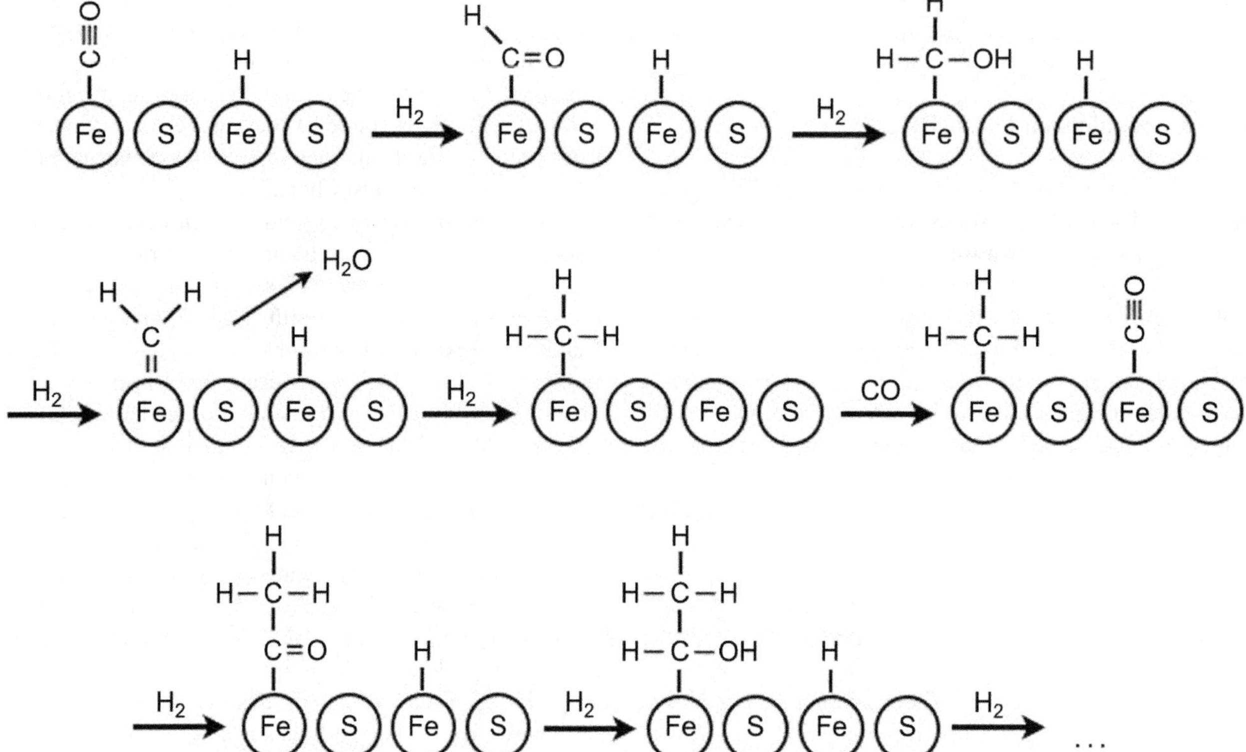

Fig. 8. Fischer-Tropsch-type surface catalysis on the surface of an iron sulfide cluster can yield organics of possible importance to the emergence of life. The positive charge of the mineral surface promotes ionic bonding of electronegative reduced compounds to the surface. Sequential reduction of compounds such as CO_2, CO, and COS leads to organic molecules that could be important for prebiotic chemistry. Adapted from *Cody* (2005).

of Europa make a strong case for a higher heat fluxes billions of years ago (*Hussmann and Spohn,* 2004). Certainly on a young Europa the increased radiogenic flux alone would have contributed to seafloor activity. The contemporary radiogenic flux is between 6 and 8 mW m^{-2}, whereas 4 G.y. ago this would have been roughly an order of magnitude larger. Additional considerations on seafloor activity can be found in the chapter by Vance and Goodman.

But what makes the chemistry of hydrothermal vents amenable to the origin of life? At the most basic level, it is the capacity for the reduced fluids and minerals to synthesize organic compounds from inorganic carbon (*Wächtershäuser,* 1988), although high temperatures at some vent environments may limit the stability of such compounds (*Bada and Lazcano,* 2002). The sulfides H_2S, HS^-, and FeS react with CO and CO_2 to yield hydrocarbons and other organic molecules. Such pathways toward life are referred to as autotrophic origins (*Bada and Lazcano,* 2002), as opposed to heterotrophic pathways that depend on abiotic organic compounds such as carbohydrates, hydrocarbons, and amino acids [e.g., most "prebiotic soup" models (*Miller,* 1953)]. Autotrophic models are attractive for prebiotic chemistry on early Earth, in part because the atmosphere may have been less reducing than initial estimates assumed (*Walker,* 1977; *Holland,* 1984). For Europa, where the ocean may be rich in oxidized forms of carbon, hydrothermal chemistry may also play a critical role in organic synthesis.

The sequential elongation of a hydrocarbon chain by reduction and addition of CO amounts to Fischer-Tropsch-type (FTT) synthesis, which has been well-studied in the laboratory (*Berndt et al.,* 1996; *Heinen and Lauwers,* 1996; *McCollom et al.,* 1999) and in the environment (*Lollar et al.,* 2002). The abiotic methane associated with deep-sea serpentinization sites, e.g., Lost City (*Kelley et al.,* 2001), is derived from a similar pathway. Figure 8, adapted from *Cody* (2005) and described in more detail below, shows a model of FTT surface catalysis on an iron sulfide cluster.

The autotrophic capacity of vent and seafloor chemistry has also been employed to argue for an alternative way of thinking about life's origin. This idea has roots in the work of *Ycas* (1955) and *Corliss et al.* (1979; 1981), but it was the work of *Wächtershäuser* (1988) that was largely responsible for linking the concept, chemistry, and hydrothermal locale in considerable detail. Central to this work is the idea that metabolism would have been the first step toward origins (as opposed to membrane formation or information storage). In its earliest form, life would have consisted of self-sustaining auto-catalytic reaction pathways that served as a protometabolism (*Orgel,* 1998). In the iron-sulfur model championed by *Wächtershäuser* (1988), the metabolic pathway is organic catalysis on a mineral surface in low pH hydrothermal vents, with genetic control and cellular organization arising as a result of a metabolic reaction network evolving toward higher complexity (*Wächter-*

shäuser, 1998). *Goldschmidt* (1952) and *Russell et al.* (1989) contend that hydrogen is the ultimate primordial fuel for life, leading to a preference for low-temperature serpentinization-driven vents as the cauldron for the emergence of life. Both of these autotrophic models may be applicable to the seafloor of Europa, where hot sulfide-rich acidic vents or alkaline low-temperature serpentinization-driven vents are plausible.

Central to the iron-sulfur model is the pyrite-forming reaction

$$FeS + H_2S \rightarrow FeS_2 + 2H^+ + 2e^- \quad (14)$$

Central to the hydrogen model are reactions such as

$$3MgFeSiO_3 + 7H_2O \rightarrow 3SiO_2 + Fe_3O_4 + 3Mg^{2+} + 6OH^- + 4H_2 \quad (15)$$

In the iron-sulfur model for the origin of life, the electrons made available during pyrite formation in equation (14) are used to reduce CO_2, CO, and COS, and in the process organic molecules are created and replicated on the mineral surface in a fashion similar to FTT. Because of the positive charge of the mineral surface, the electronegative reduced compounds become ionically bonded to the surface, allowing for continued synthesis of larger molecules. Pyrite is an attractive host for anionically bound surface reactions because, like many heavy metal sulfides, it has positive surface charges and is capable of forming strong bonds with anionic carboxylate and mercapto groups (*Wächtershäuser,* 1990). Complementing this, fixation of CO_2 results in both carboxylate ($-COO^-$) and mercapto (–SH) groups. As a result, the oxidative pyrite reaction leads directly to products that are capable of ionic bonding. Combining the pyrite forming reaction with CO_2 reduction at standard conditions yields formate

$$FeS + H_2S(aq) + CO_2(aq) \rightarrow FeS_2 + H_2O + HCOOH; \quad \Delta G^\circ = -8.2 \text{ kJ/mol} \quad (16)$$

A host of other reactions have been proposed, some more thermodynamically favorable than others. Below is Wächtershäuser's surface metabolic pathway for the theoretical formation of succinic acid formation, a highly exergonic reaction and an important step toward more complex prebiotic chemistry (*Wächtershäuser,* 1990)

$$4CO_2 + 7FeS + 7H_2S \rightarrow (CH_2COOH)_2 + 7FeS_2 + 4H_2O; \quad \Delta G^\circ = -420 \text{ kJ/mol} \quad (17)$$

Several aspects of the iron-sulfur world chemistry have been demonstrated experimentally, although some of these results are debated (*Schoonen et al.,* 1999). *Heinen and Lauwers* (1996) produced methane thiol (CH_3SH) upon reaction of FeS, H_2S, and water with dissolved CO_2. *Blochl et al.* (1992) used the pyrite reaction to pull nitrate to ammonia, alkynes to alkanes, thioglycolic acid to acetic acid, and phenylpyruvate to cinnamate and phenylproprionate. More recently, *Huber and Wächtershäuser* (1997) have demonstrated conversion of the methane thiol directly to acetic acid by CO insertion using a mixed iron-nickel sulfide, (Ni,Fe)S, or nickel sulfate, $NiSO_4$.

In an effort to further elucidate the role of iron sulfide in the evolution of early metabolic pathways, *Cody et al.* (2000) demonstrated synthesis of pyruvic acid (CH_3–CO–COOH) from formic acid at 250°C and 200 MPa, conditions comparable to what might be expected at hydrothermal sites in the depths of both Earth's ocean and Europa's ocean. Metabolically, pyruvic acid plays a key role in the reductive citric acid cycle and in the production of amino acids and sugars. Synthesis of pyruvic acid via pyrite-pulled reactions has helped build a simplified, but experimentally supported, picture of how life might have originated in an iron-sulfur hydrothermal environment (*Wächtershäuser,* 2000).

In the hydrogen-rich, alkaline model for life's origin, serpentinization reduces CO_2 and yields H_2 (*Russell et al.,* 1989; *Schulte et al.,* 2006; *Bach et al.,* 2006). The hydrogen/carbon dioxide potential, augmented by the natural proton-motive force (pmf) created as a consequence of the acidic (carbonic) ocean water interfacing with the alkaline interiors of hydrothermal compartments through FeS-bearing inorganic membranes, then drives metabolism. The proton gradient across inorganic precipitate membranes aids condensation and polymerization of simple organic molecules. Based on this possibility, *Russell and Hall* (2006) and *Martin and Russell* (2007) have argued that low-temperature (~100°C) alkaline hydrothermal systems on the early Earth would have been conducive to the emergence of an H_2-driven metabolism, production of organic polymers, and the formation of early membranes. In this model acetogenesis and methanogenesis are catalyzed by iron and nickel mineral surfaces (e.g., mackinawite, FeS, and greigite, Fe_5NiS_8). Along with enhanced macromolecular stability in alkaline low-temperature fluids, such systems have the added advantage of favoring phosphate and amine chemistry (*Russell,* 2003). Such models could be applicable to low-temperature serpentinization-driven hydrothermal systems on Europa (*Vance et al.,* 2007). In other words, even if Europa lacked an active seafloor with spreading ridges and hot vents, the exothermic reaction of Europa's ocean water with mantle rock may have provided an environment conducive to the origin of life.

Interestingly, the global ocean of Europa may actually pose problems for the formation of complex molecules important to prebiotic chemistry. The synthesis of peptides and other polymers is impeded in environments rich with liquid water. This is a long-standing problem in origin of life chemistry (*Miller and Bada,* 1988), and is particularly germane to origins in the ocean of Europa. The reaction can be summarized by

$$n \text{ amino acids} \Leftrightarrow \text{peptide} + nH_2O \quad (18)$$

where synthesis of the peptide results in production of water, an unfavorable product in a liquid water solution. With abundant water, the hydrolysis of the peptide — i.e., pushing equation (18) from right to left — is favorable. Consequently, formation of peptides and strands of nucleic acids is impeded in liquid water environments. On the Earth, tidal pools and evaporitic environments provide desiccating conditions that may have allowed equation (18) to proceed from left to right. Comparable environments do not exist on Europa. Instead, mineral surface interactions and hydrothermal vent chemistry may provide a mechanism for circumventing this problem. Formation of peptide bonds linking amino acids under simulated, although alkaline, hydrothermal conditions has been demonstrated (*Huber and Wächtershäuser,* 1998) and may be linked to carbonyl sulfide (COS) within the solution (*Leman et al.,* 2004; *Cody,* 2005). Along with the (Ni,Fe)S results of *Huber and Wächtershäuser* (1998), adsorption of monomers onto mineral surfaces has been shown to promote polymerization (*Ferris et al.,* 1998). Flow-through reactor experiments replicating vent chemistry show that the metal ions and quenching processes in vent fluids can yield oligopeptides (short peptides with fewer than ~10 amino acids) (*Imai et al.,* 1999). Thus, while the vast liquid water ocean of Europa may dilute and inhibit the formation of biopolymers, a hydrothermally active seafloor may provide microenvironments in which such reactions could occur.

Finally, we note that the presence of significant concentrations of salt in Europa's ocean may be an impediment to the origin of life. The induced magnetic field signature observed by Galileo requires salt concentrations of several grams to hundreds of grams (saturation) of salt per kg of water in order to achieve the necessary conductivity (*Hand and Chyba,* 2007). The ionic inorganic solutes influence self-assembly of monocarboxylic acid vesicles, and the nonenzymatic, nontemplated polymerization of activated RNA monomers (*Monnard et al.,* 2002). Sodium chloride or sea salt concentrations as low as 25 mM NaCl (1.46 g per kg of H_2O) reduce oligomerization, and higher concentrations had correspondingly worse effects. If the ratio of the cation to amphiphile molecules falls below ~1, then the excess amphiphile cannot form membranes. This ratio needs to be above ~2 in order for amphiphile to precipitate and form membranes. We note that these experiments have not yet been performed with $MgSO_4$, the "preferred" salt for many Europa models (*Kargel et al.,* 2000; *Zolotov and Shock,* 2001). The *Monnard et al.* (2002) results therefore support the contention that life originated in a freshwater solution, and suggest that the upper range of salinities implied by the Galileo magnetometer results could pose a serious challenge to the origin of life in Europa's bulk contemporary ocean, were abiotic RNA oligomerization or amphiphile membrane formation critical steps toward the origin of life.

If life were to be discovered on Europa, theories for a hydrothermal origin of life would certainly be bolstered, because such environments provide one of the only common geochemical threads between liquid water environments on the Earth and Europa. The other large mineralogical interface on Europa that may provide useful chemistry for origins is the ice-liquid water boundary and melt-water pockets within the icy shell. Similar environments may have existed on the early Earth, particularly given the likelihood of a faint young Sun (*Sagan and Mullen,* 1972; *Newman and Rood,* 1977). Such pathways have not been examined in great detail, but we review this work in the context of Europa in the following section.

1.3.2. The ice-water interface. The origin of life in a cold environment has several advantages over the canonical "warm little pond" of Darwinian fame (*Darwin,* 1871; *Dyson,* 1999; *Nisbet and Sleep,* 2001). Foremost is the ability to preserve large compounds once they are synthesized. Of the amino acids, the aliphatics are the most stable. Even so, temperature variations show a dramatic decrease in stability with increasing temperature. Studies of alanine decomposition via decarboxylation, i.e., conversion of –COOH to CO_2, show a half-life of billions of years for temperatures <25°C but a half-life of only 10 yr at 150°C (*Miller and Orgel,* 1974). Many of the amino acids are known or expected to have a half-life of just a few thousand years at 25°C, but millions to billions of years at temperatures approaching 0°C. Similarly, many purines, pyrimidines, and nucleosides show a decrease in half-life of an order of magnitude to several orders of magnitude as temperatures drop from 25°C to 0°C. Peptides, DNA, and RNA all show a similar trend in stability. At 25°C the phosphodiester bonds in RNA have a hydrolysis half-life of just 30 yr, vs. 900 yr at 0°C (*Miller and Orgel,* 1974). The counterargument to the benefit of enhanced stability of low temperatures is that production rates go up with increasing temperature. This, however, is not necessarily true, especially for the case of reactions with abundant liquid water. Studies of glacial ice (*Price,* 2007) and simulated sea ice (*Trinks et al.,* 2005) have shown that prebiotic chemistry, and even polymerization of substantial nucleotide chains, can be advanced in cold, icy environments.

Ice offers a means through which to concentrate and dehydrate solutes. As liquid water pockets within ice freeze, most nonwater constituents are excluded from the ice matrix, leaving a more concentrated form of the original solution. *Miyakawa et al.* (2002) explored the possibility of ice and eutectic freezing as a means for concentrating HCN and formamide to promote synthesis of prebiotic compounds. Hydrogen cyanide has long been a keystone compound for prebiotic synthesis, dating back to the experiments of *Oró* (1960, 1961) and *Oró and Kamat* (1961), who showed that under Earth-like conditions they could advance reactions such as

$$5\ \mathrm{HCN} \rightarrow \mathrm{Adenine} \qquad (19)$$

Concentration and stability of HCN, however, proves to be a critical limiting factor when moving from the controlled laboratory settings to the natural environment. In dilute aqueous solutions, hydrolysis of HCN leads first to formamide and then to formic acid. Only at high concentrations will HCN lead to the synthesis of nucleobases and amino acids. Formamide is also directly useful for synthesis of prebiotic compounds (*Ochiai et al.,* 1968; *Philipp and Seliger,* 1977), and it has the added benefit of being a good solvent in which to synthesize nucleosides, peptides, and even adenosine-triphosphate (ATP) (*Benner et al.,* 2004). In water these compounds would hydrolyze to ammonia and acids, but in formamide they are stable. *Miyakawa et al.* (2002) found that the low concentrations expected in the ocean of the early Earth would have been insufficient to overcome hydrolysis relative to adenine synthesis, but eutectic freezing in cold-water environments could have provided important niches for enhancing HCN concentrations and promoting polymerization. Miller had the great foresight to prepare, and then freeze (at –20°C and –78°C), solutions of ammonium cyanide (NH_4CN) in 1972. Although at the time the relevance to Europa surely would have been uncertain, the analysis, performed 25 years later (*Levy et al.,* 2000), revealed production of adenine, guanine, and several amino acids, including glycine and alanine.

But achieving HCN polymerization through eutectic concentration still necessitates a nonzero starting concentration for HCN and other prebiotic compounds. The experiments of Oró and colleagues typically involved HCN solutions of >1 M (mole per liter of solution). By comparison, *Sanchez et al.* (1966) froze 0.1-M HCN solutions and did not find adenine, but instead found diamino-maleonitrile (DAMN), an important intermediary in the HCN to adenine pathway. In the NH_4CN experiments of *Levy et al.* (2000), the initial 0.1-M NH_4CN solution yielded adenine at 0.035–0.040% the concentration of NH_4CN, and guanine at about an order of magnitude lower concentration. In other words, even for relatively high initial concentrations, yields are low. By comparison, a warm ocean on the early Earth is estimated to have HCN concentrations of 10^{-10}–10^{-15} M (*Stribling and Miller,* 1987; *Miyakawa et al.,* 2002). This range is based in part on Stribling and Miller's production rate of 100 nmol cm^{-2} yr^{-1} for HCN in spark discharge experiments. *Chyba and Sagan* (1992) calculated HCN production rates an order of magnitude higher. *Miyakawa et al.* (2002) experimentally determined hydrolysis of HCN at different temperatures and found greatly improved stability at low temperatures. Using the Stribling and Miller production rate, those workers argue that a 0°C ocean on Earth could maintain HCN concentrations of 10^{-5}–10^{-6} M, which if concentrated via freezing in an ice pocket, could lead to concentrations high enough to support polymerization. Similar eutectic concentration mechanisms could apply to the icy shell of Europa, but what might we expect for the availability of initial monomers such as HCN?

Pierazzo and Chyba (2002) modeled cometary delivery of amino acids and organics to the surface of Europa, but concluded that most of the inventory is destroyed upon impact or ejected to space. Even so, *Bada et al.* (1994) have pointed out that impacts into frozen oceans on a young Earth could have enhanced prebiotic synthesis as a result of eutectic synthesis and enhanced preservation in ice. Similar arguments can be applied to impact melts on Europa. *Zolotov et al.* (2006) modeled the primordial ocean of Europa and argued that N-bearing organic compounds would have been released into the ocean as a result of differentiation and leaching from an initial composition comparable to CI/CM carbonaceous chondrites. If, as described in section 1, Europa's ocean were to have significantly more ammonia and carbon dissolved in it than Earth's early ocean, then HCN may have been available on primordial Europa. Eutectic freezing in the icy shell could then have provided a pathway toward enhancing concentrations and polymerizing HCN to yield nucleobases and amino acids.

On the early Earth, lightening and UV-photolysis were available to drive prebiotic chemistry in the atmosphere. Europa lacks a thick atmosphere today, but it does have an incessant flux of UV, ions, and electrons bombarding its icy surface. It is a cold, yet highly energetic environment. For pure water ice, the resulting radiolytic chemistry, as detailed in section 1 and in the chapter by Carlson et al., produces the oxidants H_2O_2 and O_2. If other compounds are mixed with the ice, e.g., hydrocarbons and ammonia, then the resulting chemistry can lead to larger compounds of prebiotic interest. Considerable laboratory work has been done on astrophysical (<50 K) and planetary (>50 K) ices (e.g., see *Khare et al.,* 1989; *Bernstein et al.,* 1995; *Hudson and Moore,* 1999; and references therein). Here we highlight select results particularly germane to prebiotic chemistry and to the conditions on Europa. We note, however, that in most laboratory studies, the concentration of nonwater compounds typically exceeds our current understanding of the surface chemistry of Europa.

Khare et al. (1989) used low-energy electrons (mean energy of a few eV) to irradiate methane clathrate at 77 K with a flux of roughly 3×10^{16} eV s^{-1} cm^{-2} for upwards of a week. Subsequent to irradiation, samples were warmed to room temperature and a residue rich in alkanes, aldehydes, and alcohols was found to persist, with perhaps some contribution from alkenes and aromatics. *Hudson and Moore* (1999) found that even with just CO in ice, irradiation with 0.8-MeV protons yielded methanol, formaldehyde, and formic acid. Although their work was conducted at 16 K, a temperature considerably lower than temperatures on Europa, experiments with high-energy electrons and CO_2-rich ice at 100 K show comparable results (*Hand,* 2007). Working with polycyclic aromatic hydrocarbons in 10 K ice, *Bernstein et al.* (1999) produced aromatic alcohols, ketones, and ethers upon irradiation with UV photons at Lyman-α (121.6 nm) and 160-nm wavelengths. Upon warming to 300 K, aliphatic hydrocarbon features were also observed. In summary, despite the simultaneous production of oxidants that might be expected to "burn" any reduced compounds, irradiation of CO_2 or short-chain hydrocarbons in ice under europan conditions appears to consistently lead to complex organics (*Hand,* 2007).

Adding nitrogen to their UV experiments, *Bernstein et al.* (1995) observed that photolysis of methanol and ammonia-rich ices (e.g., $H_2O:CH_3OH:CO:NH_3$ at a ratio of 100:50:10:10) leads to various nitriles (–C≡N) and specifically to vinyl cyanide ($R_2C=CH-C≡N$). When warmed slowly from 12 to 200 K, products were found to include ethanol, formamide, acetamide, and additional nitriles. Those workers found that of the initial carbon and nitrogen in the ice, approximately 20% and 50%, respectively, was retained in the final residue that remained at 300 K. The nitrile band observed by *Bernstein et al.* (1995) and many previous workers at 2165 cm^{-1} (4.62 μm) has long been observed on outer-solar-system bodies (*Cruickshank et al.,* 1991) and may help explain the 4.57-μm feature on Ganymede and Callisto (*McCord et al.,* 1999). Only somewhat recently, however, has the feature observed in the laboratory been confirmed as resulting from the cyanate ion (OCN^-) in ice (*Hudson et al.,* 2001). The proposed formation pathway for OCN^- utilizes CO as a precursor and leaves the ammonium ion in ice

$$\begin{gathered} NH_3 \rightarrow NH_2 \text{ and/or } NH \\ NH_2 + CO \rightarrow H + HNCO \\ NH + CO \rightarrow HNCO \\ HNCO + NH_3 \rightarrow NH_4^+ + OCN^- \end{gathered} \tag{20}$$

While the importance of HCN to prebiotic chemistry is well established, little work on OCN^- has been done in this context. This is an important area for future work since the cyanate ion may be the dominant nitrile form in icy solar system bodies. Upon irradiation of HCN in ice with NH_3, *Gerakines et al.* (2004) produced the cyanate ion along with the ammonium ion, cyanide ion (CN^-), isocyanic acid (HCNO), formamide ($HCONH_2$), and likely HCN-polymers similar to those measured by *Khare et al.* (1994). The connection between OCN^- production in simple, carbon-poor ices (*Hudson et al.,* 2001) and the products produced by irradiation of HCN (*Gerakines et al.,* 2004) must be further examined if we are to advance our understanding of how radiolytic chemistry on icy surfaces may drive prebiotic chemistry.

To examine production of amphiphiles and self-assembly of vesicular structures relevant to the origin of life, *Dworkin et al.* (2001) photolyzed a variety of ice mixtures at 15 K with UV photons. They discovered that with methanol and ammonia as part of their initial mixture (e.g., $H_2O:CH_3OH:CO:NH_3$ at a ratio of 100:50:1:1), a residue containing amphiphilic compounds remained upon warming to 300 K. The amphiphilic fraction was then rehydrated and shown to form membranous vesicles capable of hosting a polar dye. Vesicle sizes ranged from a few to a few tens of micrometers. Similar experiments were conducted by *Dworkin et al.* (2001) with organics extracted from the Murchison meteorite, showing that both in the laboratory and in the solar system environment, formation of molecules needed to achieve compartmentalization in the early stages of biochemistry is possible.

Amino acid production via irradiation of a Titan-analog gas mixture was shown by *Khare et al.* (1986), and *Ponnamperuma et al.* (1963) produced the nucleic acid base adenine with pulsed, 4.5-MeV electrons in a simple mixture of $H_2O + CH_4 + NH_3$ under conditions relevant to the early Earth. The first definitive results of amino acid synthesis in the solid-phase under astrophysical conditions come from *Bernstein et al.* (2002). In that work 15-K ice with methanol, HCN, and NH_3 was irradiated with UV and found to yield glycine and racemic mixtures of alanine and serine. Racemic alanine was also observed after 10-keV electron irradiation of 80-K ice with ammonia and propane ($H_2O:NH_3:C_3H_8$) in a ratio of approximately 5:1:1 (*Hand,* 2007).

Although direct compositional observations of Europa's surface have been very limited, there have been repeated, relatively facile, laboratory production of a host of complex organic compounds of prebiotic interest under conditions ranging from that of the early Earth to interstellar ice grains. This supports the conclusion that the cold but heavily irradiated surface of Europa may play a significant role in providing the subsurface ocean with the building blocks for life, if the ice does mix with the ocean. The largest uncertainty is in the initial composition and concentration of compounds in the ice; if simple compounds (e.g., CH_4 and NH_3) can be concentrated on the surface, then irradiation will likely yield complex organics similar to what is observed in the laboratory. Sputtering effects from heavy ions (S^{n+} and O^{n+}) will likely destroy material in the upper few micrometers, but materials synthesized by the electrons and UV photons that penetrate beneath the sputtering region will be protected. *Miller and Orgel* (1974) long argued that a "cold, concentrated soup would have provided a better environment for the origins of life." Radiolytic products, produced on the surface of Europa but then buried in the icy shell, may have been concentrated in tectonic cracks or fractures where synthesis of peptides and other large molecules could have taken place before mixing into the vast body of liquid water below. Finally, it is interesting to note that formamide as a solution has a density of 1.13 g cm^{-3} and therefore may be buoyant relative to a salty ocean. Many hydrocarbons will also be buoyant in Europa's ocean. Depending on mechanisms for the delivery of compounds from the surface, the ice-water interface on Europa could have been, and may still be, a very interesting and rich environment for prebiotic chemistry.

2. THE SURFACE EXPRESSION OF SUBSURFACE LIFE: POTENTIAL BIOMARKERS ON THE SURFACE OF EUROPA

An inhabited world is of little scientific utility if we have no way of detecting its inhabitants. The upper few meters of ice on Europa are, by any terrestrial metric, certainly uninhabitable. Yet it is this region to which our near-future reconnaissance of Europa will be constrained. The detection of life — were the ocean to be inhabited — therefore depends on the degree to which the surface ice mixes with, and serves as a proxy for, the ocean water. This is essentially an inversion of the resurfacing problem discussed in section 1.2; rather than delivering surface oxidants to the ocean, we are now concerned with delivering potential oceanic life forms to the surface.

The argument can be made that the ocean of Europa never mixes with the surface, and therefore no signs of life will be present. For many of the same reasons discussed in section 1.2, however, we find this hypothesis unlikely. The young surface age (see chapter by Bierhaus et al.) combined with the known sodium escape (*Brown and Hill,* 1996; *Leblanc et al.,* 2002) make a strong case for communication of ocean material with the surface on timescales of less than a few hundred millions years. If life were brought to the surface, its signature on the surface, i.e., its biosignature, would need to survive the surface environment for this period of time in order to permit detection by a spacecraft. Of course, one of the primary goals of a future mission would be to identify young surfaces, thereby reducing the exposure time of any entrained biological material. Here we address terrestrial biosignatures, and their survival through time and under a variety of conditions, in order to better understand possible surface biosignatures on Europa. The term "biosignature" is used here to refer to any sign of life in the environment, whereas the term "biomarker" is used here as a subset of biosignatures and refers to specific compounds that can be connected to life. Often these terms are used interchangeably (*Summons et al.,* 2007), but because we are concerned with detection from orbit, it is useful to differentiate between biomarker compounds that have spectroscopic signatures and those that do not.

Biosignatures on Earth include morphological, molecular, and chemical indicators of past or present life. Morphological biosignatures range from macroscopic fossils to trace fossils and structural indicators of past life, such as stromatalites and microbiolites. With centimeter-scale imagery, such biosignatures might be observable from orbit around Europa. As discussed in section 1.2, large oxygen-utilizing organisms might be possible on Europa if cycling of surface material into the ocean occurs and if heterotrophic food chains exist. Fossil remnants of such organisms on the surface of Europa are perhaps an unlikely, but nevertheless distinct, possibility. The icy shell itself may be modified by biological processes, comparable in concept to microbially mediated reefs structures on Earth. If the ice is chemically rich with radiolytic products, the ice surface at the ice-water interface could be a biologically active region. Overturned blocks of ice in a thin europan shell, one interpretation of chaos features (*Carr et al.,* 1998; *Greenberg et al.,* 1999), would then bring such morphological biosignatures to the surface where they could be detected remotely.

Molecular biomarkers refer to those compounds unique to life and biological processes that, when found in the geological record, can be used to infer the presence of life in the past. DNA is an obvious example, but the survival of DNA is short on geological timescales [~10^3–10^5 yr over the range of surface temperature conditions found on Earth (*Bada and Lazcano,* 2002)]. The efficacy of molecular biomarkers is rooted in the simple fact that in life "an enormous diversity of large molecules are built from a relatively small subset of universal precursors" (*Summons et al.,* 2007). Namely, the nucleobases of DNA and RNA (uracil, thymine, adenine, guanine, and cytosine), 20 amino acids, and two lipid building blocks (acetyl and isopentenyldiphosphate precursors). Alone, such compounds are ambiguous, but in polymer form the structure and complexity helps to build the case for a biogenic origin.

Coupled with the structural subunits is the selectivity and order in which they are polymerized. On the issue of selectivity, this largely refers to chiral, or enantiomeric, preference for subunits (e.g., proteins are constructed from only L-amino-acids). The construction of complex molecules from subunits brings with it the capacity for a large number of permutations for various branches and functional group arrangements of the same molecule. This amounts to a compounded chirality since with each addition new chiral centers are created, making it possible for a larger pool of possible variations on the same molecule. This is referred to as isomerism and a given spatial arrangement for a specific compound is referred to as a diastereomer (*Peters et al.,* 2005; *Summons et al.,* 2007). Biological processes not only preferentially select subunits of a specific chirality, but they also synthesize compounds with a diastereomeric preference.

Lipid biomarkers have arguably been the most robust window into the early history of life on Earth. The acetogenic lipids (e.g., fatty acids) occur in units of methylene ($-CH_2-CH_2-$) and consequently yield gas chromatograph mass spectrometer (GCMS) results patterned in a series of even-numbered carbon compounds (e.g., C_{14}, $C_{16} \ldots C_{28}$, C_{30}, C_{32}). The lipids constructed from isopentenyldiphosphate result in lipids of linked isoprene (C_5H_{10}) units. These are commonly referred to as terpenoids and this family of lipids consists of complex arrangements of linear and ring structures built from the isoprene subunit. It is largely the survival of terpenoid compounds (specifically hopanes and steranes) that have permitted exploration of the history of life in the terrestrial Archaean rock record. These are derivatives of lipids found in the cellular membranes of *Bacteria, Eukaryotes,* and *Archaea.* Through time, heat, and pressure, the starting compounds are reduced to more stable forms that typically have fewer methyl and hydroxyl groups. *Brocks et al.* (2003) identified many of these molecules in 2.5–2.7-G.y. shales from Australia, proving their long-term stability. But the preservation of such molecules typically depends on deposition of the initial biological material into anoxic sediments. The subsequent diagenesis then does little to oxidize the molecule. Conditions on Europa are likely to pose serious challenges for comparable anoxic preservation.

Chemical biosignatures are typically the result of metabolic differentiation or selective production. Common examples include the isotopic fractionation of carbon (biological material on Earth is depleted in ^{13}C relative to the environment), and the use in biological systems of only laevorotatory amino acids and dextrorotatory sugars. Although heavily debated, the relative depletion of ^{13}C ($\delta^{13}C \approx -28‰$) in graphite from 3.8-b.y.-old metasediments of the Isua supracrustal belt in Greenland has been used to argue for signs of life on Earth not long after the late heavy bombardment (*Mojzsis et al.,* 1996; *Rosing,* 1999). The depletion results from kinetic and thermodynamic effects that are largely a function of differences in the mass and quantum characteristics of the element and its isotopes (*Schidlowski et al.,* 1983). As an example, during photosynthesis the Calvin cycle utilizes the carboxylase ribulose–1,5–bisphosphate (RuBP) to fix CO_2. This reaction favors ^{12}C over ^{13}C leading to a ^{13}C depletion of –20‰ to –35‰. Several metabolic pathways and enzymes associated with chemosynthesis (as opposed to photosynthesis) also yield measurable carbon-isotopic fractionations. In particular, methane producing and consuming metabolisms can result in a carbon signature with very strong ^{13}C depletion. Coupling of methanogens feeding methanotrophs, which in turn feed CO_2 to the methanogens, has been invoked to explain an excursion of $\delta^{13}C \approx -60‰$ (*Hayes,* 1994). The chemistry to drive a comparable ecosystem on Europa was explored by *McCollom* (1999). Were such an ecosystem to exist, the carbon isotopic signature of carbon in the surface ice could provide clues to the subsurface life.

Sulfur and sulfur isotopes also play important roles as biosignatures for terrestrial life. Organic sulfur compounds are better preserved, and they are less likely to be degraded by other organisms (*Peters et al.,* 2005; *Summons et al.,* 2007). Additionally, anoxic marine environments are critical for long-term preservation, and such conditions are often rich in sulfide. Sulfur has four stable isotopes (^{32}S, ^{33}S, ^{34}S, and ^{36}S) which makes for a highly complicated isotopic fractionation story, especially in the context of biology. Over a range of temperatures, reduction of sulfate to sulfides by bacteria results in sulfide depleted in ^{34}S by 13–40‰ relative to seawater sulfate (*Canfield et al.,* 2000). These sulfides then form sediments on the ocean floor. Sulfur or sulfide oxidation to sulfate, which could remain in suspension in seawater, leads to enhanced ^{34}S concentrations in the sulfate relative to the initial sulfur; however, progressive oxidation of the sulfide can ultimately lead to ^{34}S depletion (*Philippot et al.,* 2007). Without mineral samples to couple to aqueous sulfate, it is difficult to establish any biogenic fractionation. Given that sulfur dominates much of the observed surface chemistry of Europa, and that it is likely an important constituent in ocean chemistry, sulfur and sulfur isotope biosignatures are important to consider [see *Pilcher* (2003) for an interesting analysis of possible organosulfur biomarkers]. Interpretation of surface samples, however, will be greatly complicated by the radiation environment and the lack of access to sedimentary sulfides.

It is reasonable to expect that isotopic fractionation effects would carry over to alien biochemistry, and thus that isotopic ratios could be a useful biosignature beyond Earth (*Schidlowski,* 1992). However, many terrestrial isotopic biosignatures are heavily mediated by chemical and geological processes and only make sense as biosignatures when taken in the context of the environment. In other words, context is critical for understanding the partitioning of various isotopes. As a tool for astrobiology and life detection, isotopic signatures will be only as useful as our knowledge of the cycles and processes that govern the planetary environment.

On Europa, the morphological, molecular, and chemical biosignatures used here on Earth serve as a valuable but limited guide. Ultimately, the radiation environment of Europa is likely to be the major process controlling preservation of surface biosignatures. Considering irradiation of biomarkers, naturally occurring uranium-enriched shales on Earth offer a unique opportunity for studying the effects of radiation on biomarker survivability. Unlike laboratory experiments that must trade irradiation time for intensity, the so-called "hot shales" (*Dahl et al.,* 1988a) provide the experiment *in situ* on geologic timescales. Several regions are known to have such sediments: the Witwatersrand sequence of South Africa, the Grants Uranium District of New Mexico, Cluff Lake in Saskatchewan, the Arlit formation in Nigeria, and the Oklo formation in Gabon are just a few examples of the best known regions. Here we examine one particularly well-studied region, the Alum Shales of Sweden (*Dahl et al.,* 1988a), in order to gain a better understanding of exactly what is expected from the irradiation of organic matter through time. Uranium concentrations in the Alum Shales have been found to vary between 7 ppm and 250 ppm. For samples with present-day concentrations of 190 ppm, *Dahl et al.* (1988a) calculated that uranium decay has yielded a total dose of 10^7 Gy over the past 500 m.y.

Sundararaman and Dahl (1993) summarize some of the effects of ~10^7 Gy (1 Gy = 1 J kg^{-1}) on organic matter as follows: (1) A decrease in the amount of total extractable organic matter. While uranium decay contributes directly to the destruction of organic compounds, it also dissociates water and thus facilitates oxidation. The radiolytically produced oxidants on Europa's surface may also play a role in the destruction of organic compounds. (2) A decrease in the H/C ratio. Irradiation experiments have been shown to release hydrogen, methane, ethane, and other light hydrocarbons from organic matter while also yielding aldehydes in the remaining material, thus reducing the H/C ratio (*Swallow,* 1963). (3) The concentration of the lipid membrane biomarkers, such as hopanes and steranes, decreases. (4) The ratio of aromatics to saturated hydrocarbons increases. Aromatics are known to be more stable under irradiation than are aliphatics (*Dahl et al.,* 1988b). On Europa, this could complicate the challenge of distinguishing biogenic aromatics from the background flux of exogenously delivered abiotic polyaromatic hydrocarbons. (5) An increase in the $^{13}C/^{12}C$ ratio, i.e., $\delta^{13}C$ trends toward a nonbiological isotopic signature, possibly masking a biogenic origin. The mechanism for this change is likely the liberation of lighter isotopes via irradiation.

The parameters above are just a few of the measured effects of irradiation on terrestrial biosignatures. *Court et al.*

(2006) performed a comparison across several uranium-rich shales and found results consistent with those of *Sundararaman and Dahl* (1993) and *Dahl et al.* (1988a,b). Importantly, for shale samples with ~250 ppm ^{238}U, *Dahl et al.* (1988b) found no measurable hydrocarbons, saturates, or aromatics. Comparing the Alum shale to the surface of Europa is obviously imperfect. The environments of deposition and preservation are completely different. Nevertheless, the results above included a cross-comparison with organics in other terrestrial shales and thus help isolate the effects of irradiation. When comparing terrestrial irradiation in shales via the $^{238}U > ^{206}Pb$ decay series to that of the ion and electron impact environment of Europa, it is important to consider differences that may exist in the radiation particles. For the case of uranium, the radiation is a series of α- and β-decays. Each α-decay corresponds to the ejection of a helium nucleus with approximately 4.7 MeV of energy. The β-decay stages each result in the emission of a ~0.3-MeV electron. The electrons impacting Europa are similar in energy to those of β-decay (keV to MeV), and the H^+ ions are at least comparable in atomic cross-section, if not in total kinetic energy, to the He nuclei from α-decay. The O^{n+}, and S^{n+} ions, however, are much larger — both in mass and in cross-section — than what is seen in the decay of uranium.

For a surface age on Europa of ~10^7 yr, the total radiation dose at 1 cm below the surface would be ~3.6×10^{11} Gy, or ~3×10^3 Gy per month. This is over 100 times the highest dose experienced by the Alum Shales. Consequently, the remnants of any subsurface ocean biota are not likely to be well-preserved for this combination of depth and surface age. For surface ages of just a few thousand to tens of thousands of years, however, the total dose at 1 cm below the surface on Europa is then comparable to, or less than, that of Alum Shale samples that did reveal organics and possible biomarkers. Clearly, identifying young surfaces on Europa is critical from the standpoint of biosignature preservation on the surface.

Trading time for depth, at 1 m below the surface of Europa, the direct radiation dose is reduced to just 1 Gy per month. Using the same resurfacing timescale, biomarkers would be subject to one-tenth the dose experienced by the biomarkers in the Alum Shale with the highest uranium concentrations. Comparable Alum Shale samples are those with roughly 10 ppm uranium. Such samples have saturates and aromatics in the range of ~ 0.5–1.0 mg g^{-1}, well within the limits of detectability for a landed spacecraft with a GCMS. To provide some comparison to the survivability of spores exposed to radiation, we note that *Rivkina et al.* (2005) discovered viable microorganisms in the 3-m.y.-old permafrost of the Kolyma tundra in Siberia, where the estimated radiation dose would have been ~600 Gy. Such organisms could potentially survive for 50 yr on Europa, were they buried 1 m below the surface.

Positive detection of terrestrial biosignatures is largely dependent on geological context. The complex mineralogy of Earth's lithosphere is in part responsible for the challenge faced when attempting to establish biosignatures. On Europa, the radiation environment and ocean chemistry may obscure biosignatures, but the mineralogy is largely limited to ice, clathrate hydrates, and hydrated sulfate phases (see chapters by Carlson et al. and by Zolotov and Kargel; see also *Kargel et al.,* 2000; *Hand et al.,* 2006). The ~100-K surface temperatures would facilitate long-term survival of biomarkers, but the oxidant-rich surface, and gardening and sputtering of that surface by micrometeorites and ions (see chapter by Moore et al.), would serve to degrade biological material. Assessing the remnants of such material with an *in situ* lander would permit the application of a variety of tools and techniques. With solely an orbiting spacecraft, however, remote detection of such degraded material, e.g., through spectroscopy of the surface or mass spectrometry of Europa's sputtered atmosphere, would be our only window to life in the subsurface.

Application of spectroscopic biosignatures to the case of Europa is largely limited to the study of *Dalton et al.* (2003), and more recently to the work of Hand and Carlson (*Hand,* 2007). Noninvasive spectroscopic techniques for characterization and identification of microbes are underutilized and little explored. Much of the existing work stems from industrial applications for food security and sterilization (*Naumann et al.,* 1991; *Kummerle et al.,* 1998). *Dalton et al.* (2003) characterized near-infrared features of DNA, proteins, lipids, and carbohydrates at room temperatures (Fig. 9a). Lipids and proteins display C–H absorption features near 2.3 μm, as well as C–N features at 2.05 and 2.17 μm arising from amide bonds and polypeptide chains. Nucleic acids and carbohydrates have simpler spectra, but all four types of cellular components have strong features associated with water molecules near 1.2, 1.5 and 2.0 μm. To further explore the question of biosignatures on icy worlds, *Dalton et al.* (2003) froze *Sulfolobus shibatae, Deinococcus radiodurans,* and *Escherichia coli* and characterized their spectral signature under europan conditions of high vacuum and low temperature (100 K). In the complex, multiply-scattering and water-dominated cellular environment, the water (much of it bound) dominates the other cellular components and produces spectral features highly similar to those observed by the Galileo spacecraft at Europa. The near-infrared amide features of the microbes were distinguishable in the laboratory spectra, prompting those workers to suggest the possibility of using the 2.05- and 2.17-μm C–N bands as candidate biosignatures (Fig. 9b).

In an effort to establish possible spectroscopic biosignatures that persist after irradiation, Hand and Carlson irradiated spores of the common soil bacterium *Bacillus pumilus,* with 20-keV electrons under Europa-like conditions (*Hand,* 2007). This bacterium was selected because of its relevance to spacecraft sterility and planetary protection. *LaDuc et al.* (2003) studied the Mars Odyssey spacecraft and found *B. pumilus* to be one of the most abundant viable organisms on the spacecraft after the craft was cleaned with H_2O_2, exposed to intense UV, and baked to high temperatures. The fact that *B. pumilus* can tolerate H_2O_2 cleaning and UV irra-

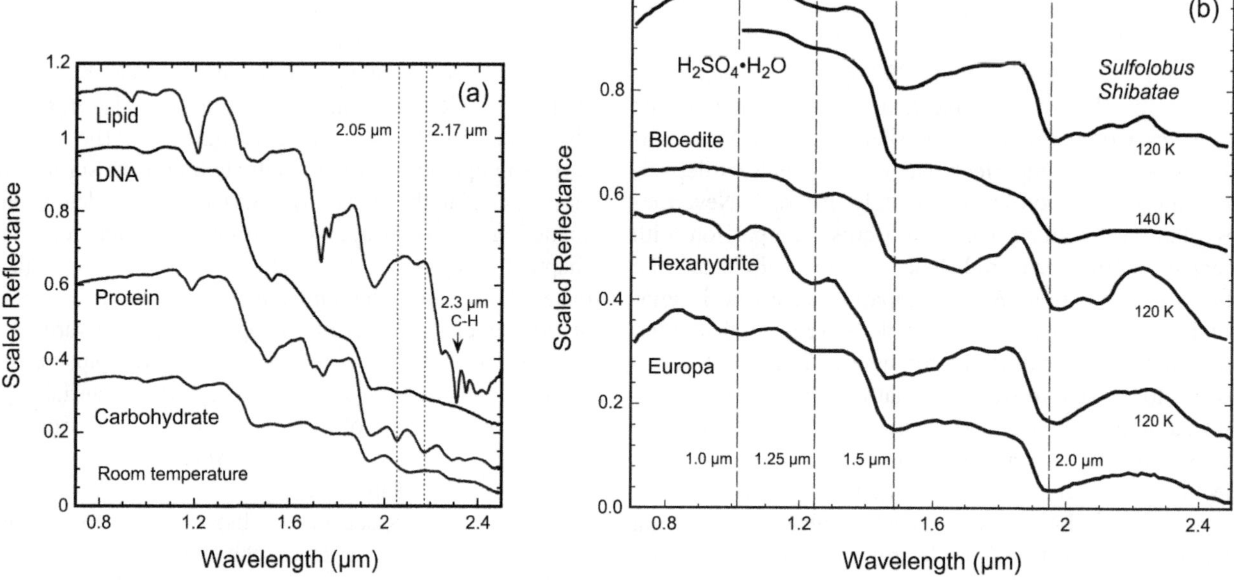

Fig. 9. **(a)** Infrared spectra of infrared-active cellular components, obtained at room temperature by *Dalton et al.* (2003). Vertical dashed lines at 2.05 and 2.17 µm indicate positions of absorption features arising from amide bonds. **(b)** Infrared spectra of Europa's non-ice material and proposed surface constituents, obtained at 120 K by *Dalton et al.* (2003), except for the sulfuric acid spectrum, obtained at 140 K by *Carlson et al.* (1999b). Vertical dashed lines near 1.0, 1.25, 1.5, and 2.0 µm denote positions of asymmetric and distorted spectral features in Europa's dark and disrupted terrains. Features in the spectrum of frozen microbes, *S. shibatae*, show intriguing similarity to the hydration features seen in spectra of Europa's dark surface material.

diation makes it a good candidate for assessing possible chemical or molecular biosignatures under europan surface conditions. Organisms such as *Deinococcus radiodurans* are often touted for their ability to withstand harsh radiation, but in order to do so *D. radiodurans* must actively repair its DNA, something it would not be able to do in the cold, desiccating surface environment of Europa. Radiation tolerant organisms are not particularly relevant to the habitable region of Europa (the ocean and water-filled regions of the icy shell), although spore-forming organisms that can shut down metabolically and survive cold, dry, radiation-intense environments (*Rivkina et al.,* 2005) are interesting in the context of exposure to the surface, followed by reintroduction to the ocean.

Figure 10 shows spectra of *B. pumilus* spores in ice at 100 K before irradiation (lower spectrum) and after irradiation with 20 keV electrons at a current of 1 µA for approximately 24 hr (~10^9 Gy) (middle spectrum). Also shown is a spectrum of abiotic organic chemistry resulting from electron radiolysis (upper spectrum). The asymmetric stretch of $-CH_3$ at 2965 cm^{-1}, the symmetric stretch of $-CH_3$ at 2874 cm^{-1}, and the asymmetric stretch of $-CH_2$ at 2936 cm^{-1} are all indistinguishable from one spectrum to the next. The hydrocarbon chemistry of biology offers little in the way of clues toward a biogenic origin. This is unfortunate as this is the easiest band to observe, and it has been reported for Enceladus at 3.44 µm (2870–2970 cm^{-1}), as well as for numerous other solar system objects (*Hansen et al.,* 2006).

Several other spectral features related to carbon or organic chemistry are also indistinguishable across spectra. The CO_2 (2340 cm^{-1}), CO (2130 cm^{-1}), nitrile band (2164 cm^{-1}), and methane bands (1304 cm^{-1}) are all very similar. The region of the spectrum where life begins to distinguish itself from nonlife is in the region <1700 cm^{-1} (i.e., wavelengths longer than 5.88 µm). In the *B. pumilus* spectrum taken prior to irradiation strong peaks are centered at 1650 cm^{-1} and 1514 cm^{-1}. These peaks correspond to the amide I and amide II bands respectively. The amide bond, and the associated vibrational excitations, specifically refers to the C–N linkage between amino acids and the interaction with the C=O and N–H on either side of that bond. The 1514 cm^{-1} band is sometimes denoted the tyrosine band, as it is a hallmark of the presence of that amino acid in proteins (*Naumann et al.,* 1996; *Maquelin et al.,* 2002). The band sits next to, and in conjunction with, the amide II band at 1520–1550 cm^{-1}, characteristic of the C–N stretch and C–N–H bend in the peptide structures of proteins. The observed decrease in band strength in these experiments indicates destruction and loss of proteins.

Amide bands persist after irradiation and are distinguishable from the abiotic radiolytic products. The critical difference between the abiotic and biogenic is that amino acids in proteins are arranged in ordered sheets and helical structures, whereas amino acids produced radiolytically do not share the peptide structure and associated infrared band shift (*Hand,* 2007). The amide I band of α-helical structures in proteins, resulting from the carbonyl bond in the right-handed spiral of the helical structure, is seen at 1655 cm^{-1}. The NH_3 residue likely consists of carbonyls with a greater diversity of secondary bonds, thereby shifting the peak to-

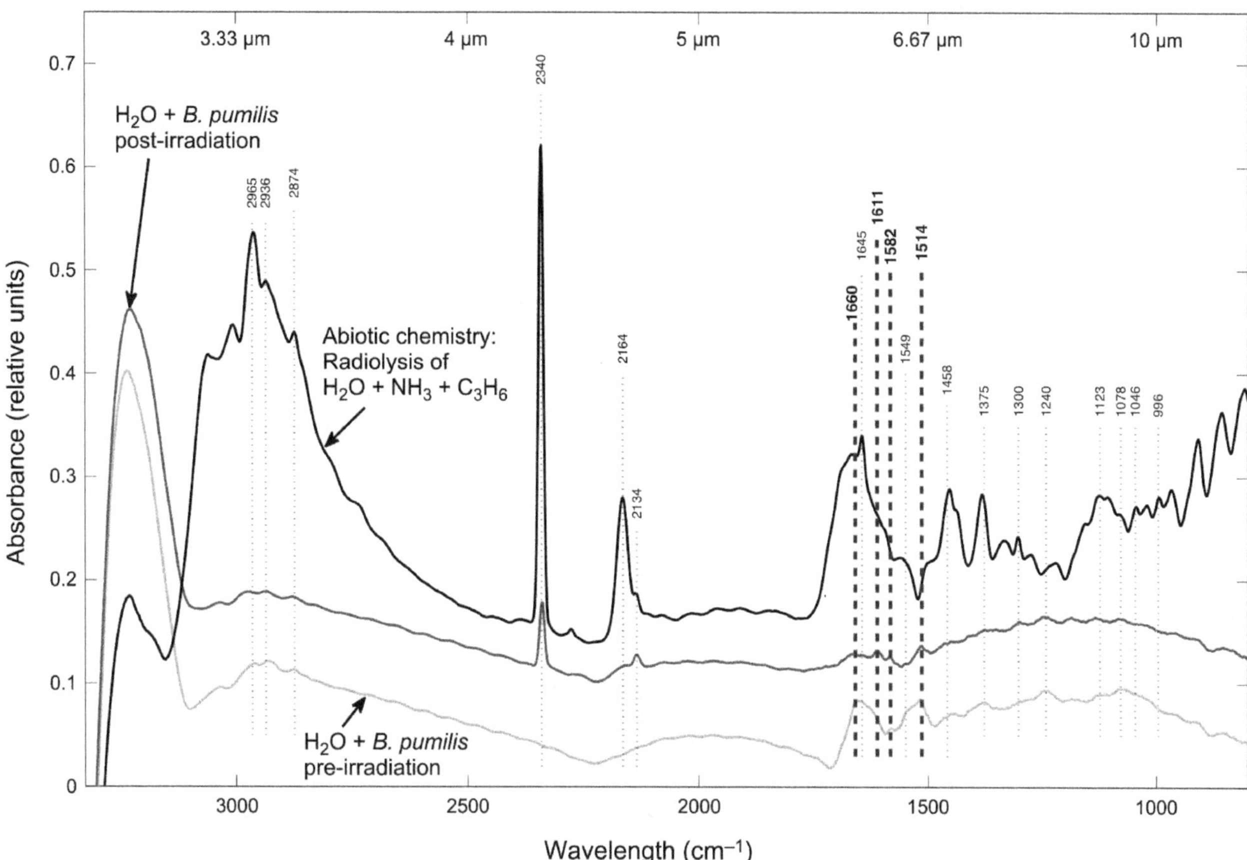

Fig. 10. Possible spectroscopic biosignatures in the amide region of the midinfrared. The bottom spectrum shows spores in 100 K ice before irradiation, the middle spectrum shows similar spores after irradiation with 20-keV electrons, and the top spectrum shows results from irradiation of an abiotic mixture of H_2O, NH_3, and C_3H_6. The spectrum of irradiated microbes is distinguishable from the abiotic radiolysis spectrum primarily due to persistent amide bands (O=C-N-H) at ~6 μm (~1600 cm^{-1}) associated with protein structures (bold dashed lines). From Hand and Carlson (in preparation).

ward the C–C=O carbonyl at 1710 cm^{-1}, as opposed to that of the amide bond (N–C=O) at 1655 cm^{-1}. Also contributing to this difference is the amide I band of the β-pleated sheet structures in proteins. This band, at 1637 cm^{-1} (*Naumann et al.,* 1996), is also due to the carbonyl stretch interacting with the amide N–H, but in this case the strands of linked amino acids form a sheet by hydrogen bonding with neighboring strands. The change in molecular geometry shifts the amide I in β-sheets to the longer wavelength. This band is seen on the shoulder of the α-helix amide I band.

At even longer wavelengths, centered at 1240 cm^{-1}, is a band corresponding to the asymmetric stretch of the P=O bond in a phosphodiester (>PO_2^-) linkage (*Maquelin et al.,* 2002). The phosphodiester bond is responsible for connecting the nucleotides of RNA and DNA. The diesters of ATP occur at slightly shorter wavelengths, ~1260 cm^{-1} (*Liu et al.,* 2005). The second broad band, centered at 1078 cm^{-1}, also corresponds to the phosphodiester bond, but in this case the band results from the symmetric stretch of P=O (*Choo-Smith et al.,* 2001; *Maquelin et al.,* 2002).

The host of biosignatures described above provides some context for the manifestation of preserved biological products through time and under irradiation. The described suite of biosignatures is, however, severely limited by our lack of a comprehensive understanding for how life on Earth is modified and preserved through time. Earth is awash with life, and deconvolving signatures of life in modern environments from those preserved in ancient environments is an ongoing challenge to understanding the history of life on this planet (for extensive reviews, see *Knoll,* 2003; *Schopf,* 1999). In some respects, the surface environment of Europa offers a more homogenous preservational environment than either Earth or Mars. The frozen, sulfate-rich ice covers the surface, and while there is a leading/trailing hemisphere dichotomy for the electron irradiation, the majority of the surface is heavily bombarded by energetic particles (*Cooper et al.,* 2001; see chapter by Paranicas et al.). The traditional rock cycle of igneous, sedimentary, and metamorphic rocks complicates the signature of ancient life on Earth. On Europa, this cycle still applies, but it is likely to be greatly simplified by the fact that one mineral, ice and its polymorphs, dominate all processes. We anticipate that more complex chemical heterogeneity would be revealed with future missions, adding new dimensions to the europan rock cycle, but we also anticipate that such heterogeneity will provide clues to a potentially habitable, or even inhabited, subsurface. We discuss some of these possibilities in the following section.

3. DETECTING LIFE FROM AN ORBITER AND/OR LANDER

The challenge of searching for life on Europa is exacerbated by the fact that we still lack a broadly accepted definition for life (*Cleland and Chyba,* 2002). Without such a definition, establishing a succinct set of search criteria is difficult. Proposed definitions have employed metabolic, biochemical, physiological, genetic, and thermodynamic parameters, but all such definitions face considerable drawbacks (*Sagan,* 1970; *Chyba and McDonald,* 1995; *Chyba and Phillips,* 2002). A popular working definition is that life is "a self-sustained chemical system capable of undergoing Darwinian evolution" (*Joyce,* 1994). While potentially useful on Earth, this definition is difficult to utilize when developing a search strategy. As pointed out by *Chyba and Phillips* (2002), "How long do we wait to determine if a candidate entity is 'capable of undergoing Darwinian evolution'?"

Only once before has a rigorous chemical search for life been conducted *in situ* with robotic spacecraft on another planet. The Viking Landers had three experiments as part of the biology payload (*Klein et al.,* 1972; *Klein,* 1978). The gas exchange experiment involved incubating soil samples with nutrients to prompt growth and gas exchange; the labeled release experiment used ^{14}C-labeled nutrients to monitor production of ^{14}C-gases evolved from the inoculated soil; and finally, the pyrolitic release experiment examined incorporation of ^{14}C into solid biomass within the soil. All three of the experiments implicitly utilized a metabolic definition of life, i.e., it was a search for the effects that living organisms have on the chemistry of their environment. The results of the biology experiments may have been ambiguous had it not been for the gas chromatograph mass spectrometer results showing that no organics were present to the limit of parts-per-billion (*Biemann et al.,* 1977). In this case, a biochemical definition (i.e., organics are needed for carbon-based life) trumped a metabolic definition (*Chyba and Phillips,* 2002).

Based on the lessons learned from the Viking Landers, *Chyba and Phillips* (2001) proposed the following approach for future robotic searches for life: (1) If the payload permits, conduct experiments that assume contrasting definitions for life. (2) Given limited payload, the biochemical definition deserves priority. (3) Establishing the geological and chemical context of the environment is critical. (4) Life-detection experiments should provide valuable information regardless of the biology results. (5) Exploration need not, and often cannot, be hypothesis testing. Planetary missions are often missions of exploration, and therefore the above guidelines must be put in the context of exploration- and discovery-driven science.

On the surface of Europa, our search will largely take the form of a biochemical search. Given the harsh surface conditions, we do not anticipate finding viable life forms on the surface, and thus metabolic or even genetic definitions would prove challenging. Such definitions could be useful with advanced landers capable of excavating to several tens of centimeters in depth, or if freshly exposed subsurface material from the ocean was accessible. For the case of point (3) in the above list, *Figueredo et al.* (2003) have established several geological criteria, specific to the surface of Europa, that can be used to guide the search for biosignatures. A variety of surface units (specifically chaos regions, bands, ridges, craters, and plains) were assessed in the context of the following criteria: (1) evidence for high material mobility; (2) concentration of non-ice components; (3) relative youth; (4) textural roughness (providing a possible shield from the degrading effects of radiation — although obviously complicating engineering considerations for a lander); and (5) evidence for stable or gradually changing environments.

Figueredo et al. (2003) argue that recent, low-albedo bands that fill interplate gaps present very high potential for biosignatures, based on the above criteria (see also Fig. 7). Similarly, smooth, salt-rich plains also rank very high. The evidence for melting and/or material exchange seen in chaos regions, along with their potential for being young in age and salt-rich, merits those surface units a high potential for biosignatures. Large craters and ridges both rank as moderate locales for hosting biosignatures, although we note that impact events could yield abiotic chemistry that complicates detection.

Using the limited high-resolution imagery from Galileo spacecraft, *Figueredo et al.* (2003) present several candidate features that earned the "very high" ranking. In Fig. 11 we show two smooth plain sites and one chaos site. The smooth plains have the added advantage of likely being a more welcoming site for future landed missions.

Ultimately the search for life on Europa is an exercise in geography, with the goal of connecting surface chemistry with the landscape on a local and global scale in order to understand the habitability of the subsurface and the expression of any subsurface life on the surface. A detailed map of surface age, registered with surface chemistry and radiolytic processing, will help determine endogenous vs. exogenous origins and potentially reveal clues about life in the subsurface. Chemical differences between surface features of similar age and type, but spread geographically across the surface of Europa, will serve as one mechanism for assessing radiolytic processing and modification. For example, were we to discover two geologically similar sites both very young in age — one on the leading hemisphere and one on the trailing hemisphere — but find that the trailing hemisphere site was rich with C–H, CO_2, CO, and C≡N, while the leading hemisphere site was rich in O=C–N and N–H, we might infer that the leading hemisphere material experienced less radiolytic processing and thus is more in-

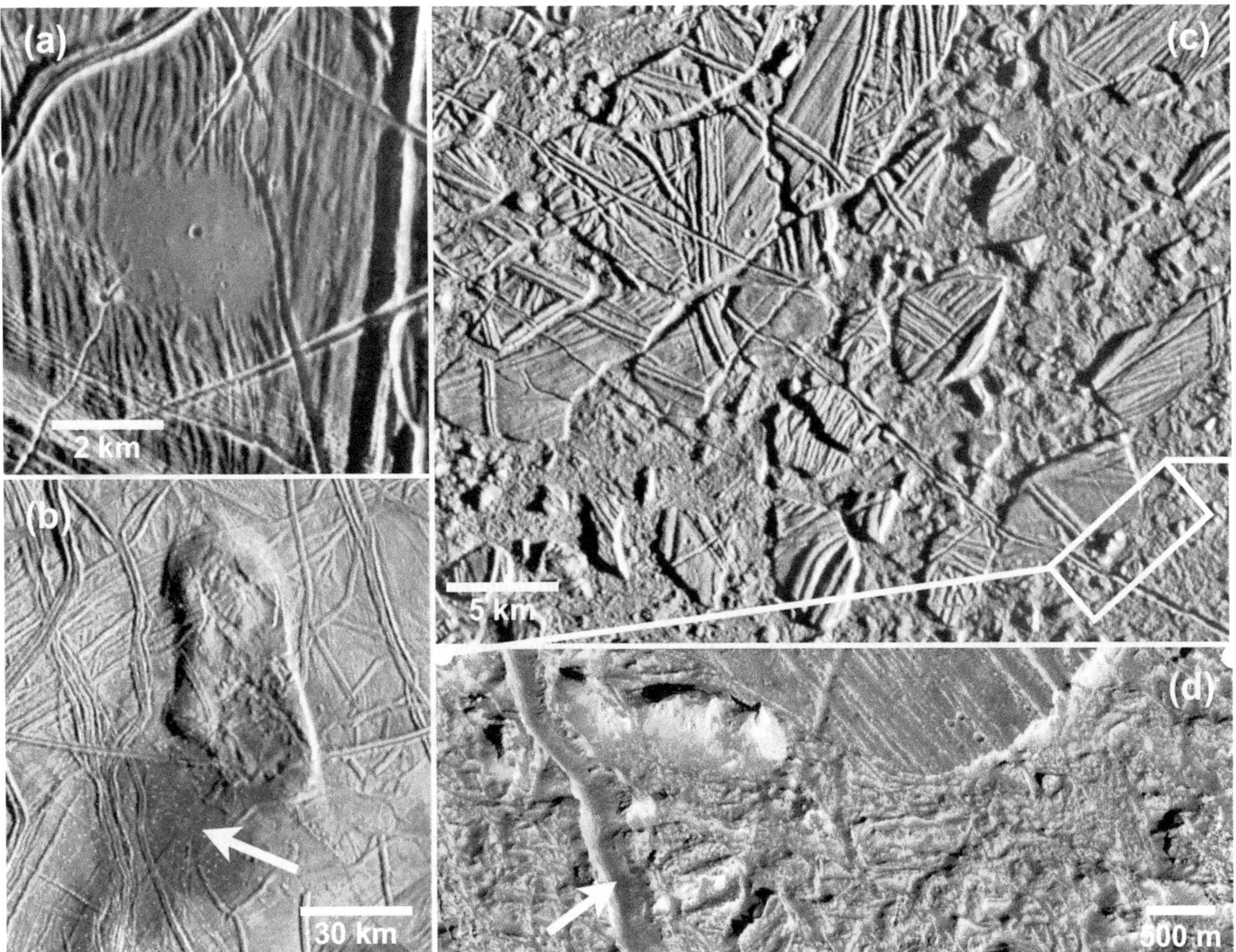

Fig. 11. Surface units on Europa with a "very high" biosignature potential, as determined by the criteria of *Figueredo et al.* (2003). **(a)** A smooth plains "puddle" feature; **(b)** the low-albedo, smooth terrain of Castalia Macula (arrow); **(c)** a portion of Conamara Chaos; and **(d)** a high-resolution version of the boxed inset (arrow indicates possible concentration of non-ice components along a fracture). Smooth plains features [e.g., **(a)** and **(b)**] have the added benefit of possibly offering a relatively benign landing site for future missions.

dicative of the parent endogenous material, possibly of biological origin. Complete radiolytic processing of the surface observable with orbital techniques (the upper ~100 μm of ice) is rapid compared to geological timescales, taking perhaps days to a few years depending on the composition and longitudinal position of the initial material. For a lander on the leading hemisphere, where sputtering erosion rates will also be low (*Paranicas et al.,* 2001; *Cooper et al.,* 2001), surface ages of 10^4–10^5 yr will be gardened to at most 1–10 cm (*Chyba and Phillips,* 2001; see chapter by Moore et al.), thus perhaps eliminating the need to employ drills to reach unprocessed material (melting, scooping, or a soft impact could provide the necessary excavation). As a result, for both orbital and lander measurements, identification of young or active regions is critical to assessing the endogenous chemistry and possible biomarkers on Europa.

Sublimation driven resurfacing may offer one way to access fresh, relatively unprocessed material on the surface of Europa. Regions with surface temperatures of 130–150 K will experience sublimation loss of H_2O at a rate of 1–1000 μm yr^{-1} (*Fagents et al.,* 2000). Non-ice material will be left behind and possibly concentrated as a result of water loss. Shear heating along ridge structures could be one place where thermal upwarping (*Nimmo and Gaidos,* 2002) leads to sublimation and exposure of fresh non-ice compounds. *Carlson et al.* (2005) used Galileo NIMS spectra to argue that such a mechanism, coupled with radiolysis and mass wasting, could explain the strong hydrated sulfate feature seen in the talus and on the flanks of ridges.

Detecting any biosignature on the surface — with a lander or orbiter — could be strongly limited by the concentration of biomass in the surface ice. Table 3 shows cell concentrations for several different environments on Earth. Concentrations in the surface waters of Earth's ocean are in the range of 5×10^3–5×10^5 cells ml^{-1}, or 10^{-6}–10^{-4} g of dry cell mass per kg of water [using 2.3×10^{-13} g $cell^{-1}$ (*Madigan et al.,* 2003)]. The majority of this biomass is photosynthetic, not necessarily appropriate to Europa. The lowest biomass concentrations in the ocean are found in deep-ocean basins devoid of hydrothermal activity. In these

TABLE 3. Bacterial abundance of terrestrial ecosystems, in cells per milliliter.

	Abundance (cells ml^{-1})	References
Ocean, surface	5×10^3–5×10^5	[1,2]
Ocean, deep basins	10^3–10^4	[3]
Hydrothermal vents	10^5–10^9 (in suspension)	[3]
Hot Springs	~10^6	[4]
Microbial mats	10^7–10^9	[6]
Sierra Snowpack	10^3–10^4	[5]
Glacial ice	120	[1]
Vostok accretion ice	83–260	[1]
Vostok water (predicted)	150	[1]

References: [1] *Christner et al.* (2006); [2] *Whitman et al.* (1998); [3] *Van Dover* (2000); [4] *Ellis et al.* (2005); [5] *Painter et al.* (2001); [6] *Sorenson et al.* (2005).

regions, measured values are one to two orders of magnitude lower than surface waters. Cell concentrations in the water around hydrothermal vents can reach 10^9 cells ml^{-1} or roughly 0.2 g kg^{-1} (*Van Dover,* 2000). Of course on Europa, the hydrothermal population would need to somehow translate to a surface concentration, were we to be able to detect it with an orbiter or lander. *Winn et al.* (1986) studied the biological material in plumes above the Juan de Fuca Ridge and found detectable populations of bacteria and ATP for several hundred meters above the plume (Fig. 12). Plume dynamics on Europa could help transport such organisms to the ice-water interface (*Goodman et al.,* 2004; see chapter by Vance and Goodman), but populations would be severely diluted. Subglacial lakes, Vostok accreted ice, and glacial ice are all found to have a few tens to a few hundred cells per milliliter, or ~10^{-8}–10^{-7} g kg^{-1} of water (*Christner et al.,* 2006). Estimates of the total prokaryotic biomass within all Antarctic subglacial lakes, the ice sheet itself, and the suspected subglacial aquifer are 10^{24}, 10^{21}, and 10^{24} cells, which equates to 4.3×10^{-8}, 1.1×10^{-4}, and 1.15 Pg (petagram) of bacterial carbon (1 Pg = 10^{15} g) (*Priscu et al.,* 2008). The majority of this prokaryotic biomass is suspected to be present in the Antarctic subglacial aquifer. This south polar sub-ice biomass rivals many of the surface bacterial carbon pools on Earth (*Priscu and Christner,* 2004; *Priscu et al.,* 2008). Algal blooms in springtime snow can reach several thousands of cells per milliliter (10^{-7}–10^{-6} g kg$_{H_2O}^{-1}$) (*Painter et al.,* 2001). Microbial mats reach cell densities of 10^7–10^9 cells ml^{-1} (*Sorensen et al.,* 2005). In an ideal case for detection on Europa, similar densities would populate niches within and directly beneath the ice layer. Using radiolytic oxidants and carbon availability in the icy shell as limiting factors, *Chyba* (2000) estimated that cell densities could reach ~10^7–10^{10} cells ml^{-1} for ~1 mm of ice melt. Spread globally through a 100-km-thick subsurface ocean, however, this amounts to just 1 cell ml^{-1}.

Does the above range of concentrations permit detection from orbit? As mentioned in section 1.2, photosynthetic pigments such as chlorophyll, carotenoids, and xanthophylls are, through eons of selective pressure, uniquely suited to spectroscopic detection — they are tuned for light absorption and reflection and thus they are particularly sensitive

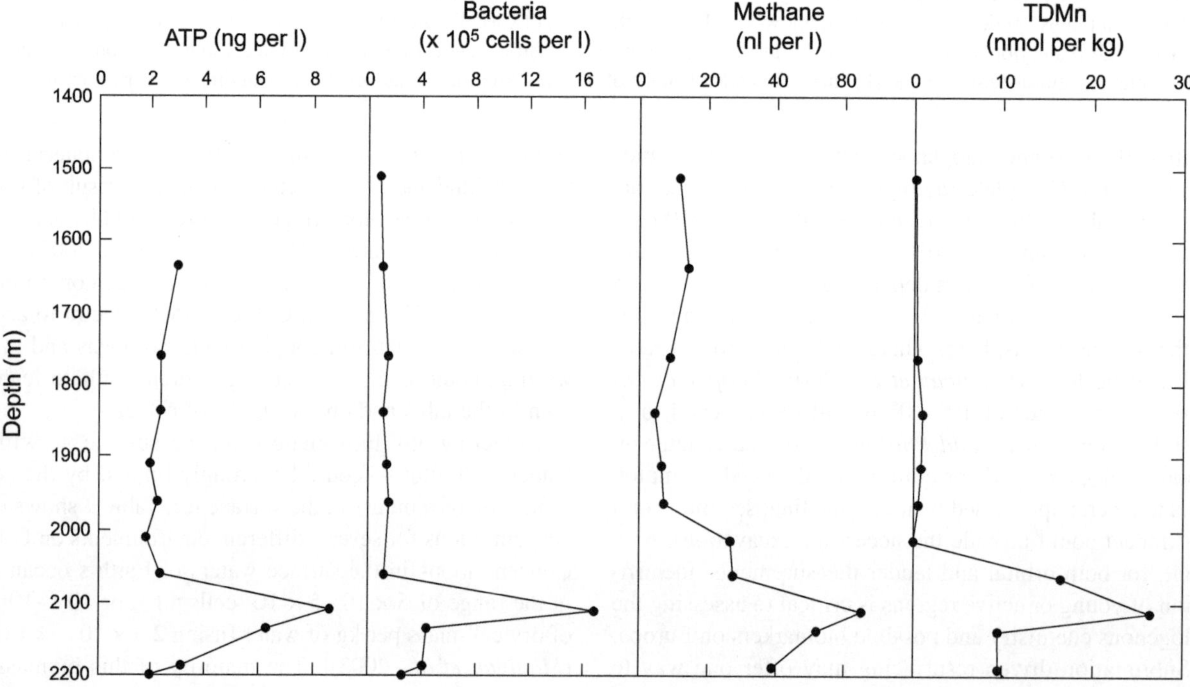

Fig. 12. Distribution of ATP concentration, bacterial cell number, methane, and total dissolved manganese (TDMn) at and above a hydrothermal vent on the Juan de Fuca Ridge, off the coast of the northwestern United States. Reproduced from *Winn and Karl* (1986), courtesy of Nature Publishing Group.

to detection via light spectroscopy. Sea surface phytoplankton populations are easily observable from orbit by monitoring chlorophyll-a absorption features in the 0.4–0.7-μm range (*Montes-Hugo et al.,* 2005). Similarly, using airborne spectroscopy, *Painter et al.* (2001) were able to detect the snow algae *Chlamydomonas nivalis* down to levels of ~2300 cells ml^{-1} by observing carotenoid absorption between 0.4 and 0.64 μm and chlorophyll-a and -b absorption at ~0.68 μm.

But photosynthesis is not a great metabolic model for a world covered in ice. Other colorful biological compounds associated with sulfur-, iron-, or oxygen-metabolizing microbes may serve as better biosignatures. Setting aside such compounds, however, what spectroscopic strategy might be useful based on just the organic chemistry of cell biomass? Consistent with the laboratory results discussed in section 2 (Fig. 10), the first spectroscopic line to look for should be CO_2, as radiolysis of carbon-based life on the surface of Europa will release this product, and it provides a strong absorption line. Nitriles and C–H features are also important, but would be hard to distinguish from abiotic radiolytic chemistry. The amide bands at 2.05, 2.17, 6.11, and 6.6 μm would help build the case for biochemistry. Table 4 shows the various frequencies and wavelengths for detection of biologically relevant molecules and functional groups. Approximately 55% of the dry cell mass of a microbe is attributable to proteins, 9% to lipids, and 20.5% to RNA (DNA is 3.1%) (*Madigan et al.,* 2003; *Loferer-Krössbacher et al.,* 1998). Given the above cell densities and mass fractions, this means that spectroscopic techniques must be sensitive to the ~1 ppb–200 ppm level to detect biochemistry at levels comparable to those found throughout our ocean and in hot springs, microbial mats, and snowpack. For glacial ice and Lake Vostok cell concentrations, sensitivities greater than 1 ppb would be required (*Priscu et al.,* 2006; *Christner et al.,* 2006). On Europa, geological processes such as the sublimation and mass wasting discussed above could help overcome low concentrations in the initial ice.

While infrared spectroscopy from orbit around Europa is hindered by water absorption and is limited to detecting only major species on the surface, naturally occurring ion bombardment (sputtering) constantly ejects molecules from the surface, potentially delivering them to orbital altitude where they could then be measured by an orbiting mass spectrometer or microwave spectrometer (*Johnson et al.,* 1998; *Allen et al.,* 2008). Europa's sputtered atmosphere is produced by ejection of molecules from the surface at velocities of up to ~1 km s^{-1} and these molecules can reach ~300 km in altitude. Both the parent molecule and decomposition products are ejected. For example, in H_2O sputtering, the products are mostly H_2O with a few percent as O_2 and H_2. Large organic molecules can be ejected intact or as decomposition products. If the amino acid glycine were sputtered off the surface, the glycine molecule plus the decomposition products NH_3, CO_2, CH_4, H_2O, and H_2CO, would be in the atmosphere. For a surface concentration of 100 ppm glycine, there could be a total glycine-related column density of 5×10^9 cm^{-2} and a volume density at orbital altitudes of a few hundred per cubic centimeter. This volume density is within the anticipated detection capability of modern, improved flight mass spectrometers (*Waite et al.,* 2006). Highly sensitive rotational-line millimeter-wave spectroscopic measurements of the sputtered atmosphere could make comparable detections (*Allen et al.,* 2008).

Unless the signs of life are readily apparent, e.g., fossils, strong pigment signatures, and other biosynthetic biomarkers, it will be difficult to make a *convincing* case for life on Europa from orbit. A compelling case is possible with the combination of orbital techniques and geography, but ultimately it will take a lander to resolve uncertainties. With a landed gas chromatograph mass spectrometer (GCMS) and microscope a much more certain case could be made for a definitive detection of life. *Chyba and Phillips* (2001) calculated that with onboard fluorescent high performance liquid chromatography roughly 10^2–10^3 kg of ice would need to be melted and filtered to achieve detection at levels of 0.1–1 cell ml^{-1}. For concentrations of 10^2–10^3 cells ml^{-1} the filtering requirement drops to 0.1–1 kg of ice. For ~50–100-g samples of Vostok ice, *Priscu et al.* (2007) were able to detect cell counts down to 10 cell ml^{-1} using the combination of epifluorescence microscopy and DNA stained cells. Because melting, evaporating, and drilling ice all require energy from the spacecraft, these are important engineering considerations (*Chyba and Phillips,* 2002). Additional decontamination protocols would also have to be implemented to ensure sample integrity (e.g., *Christner et al.,* 2005).

Finally, we note that at least in one instance, a spacecraft designed for the study of the Jupiter and the Galilean satellites was capable of discovering life on an inhabited world. The Galileo team famously detected life on Earth during Galileo's Earth flyby (*Sagan et al.,* 1993). Those results were, of course, largely biased toward the photosynthetic and technological signatures of life on Earth and would probably have been ineffective if focused on Earth's polar ice sheets. Nevertheless, the analyses revealed some of the challenges and ambiguities associated with biosignatures known to be derived from life here on Earth. Even on a world ripe with life, detection and confirmation of that life from a spacecraft is not entirely straightforward. Indeed, as we move forward with the challenge of searching for life beyond Earth, and as we work to define convincing signs of life on other worlds, we would do well to keep in mind advice from *Sagan et al.* (1993): "Life is the hypothesis of last resort."

4. PLANETARY PROTECTION

The potential habitability of Europa bestows a great responsibility on our future exploration of the europan surface and subsurface. Planetary protection refers to those policies and practices intended to (1) prevent the contamination of a celestial body by terrestrial microorganisms that could be present on a spacecraft at launch, survive the space

TABLE 4. Biologically important functional groups and molecular assignments for infrared spectroscopy at STP.

Frequency, cm^{-1} (Wavelength, μm)	Functional Groups and Molecular Assignments
25,000 (0.4–0.58)	Carotenoid pigment
16,667 (0.6–0.7)	Chlorophyll-a, -b pigments
10,000 (1.0)	O–H of water of hydration of parent compound ($\bullet H_2O$)
8000 (1.25)	O–H of water of hydration of parent compound ($\bullet H_2O$)
6667 (1.5)	O–H of water of hydration of parent compound ($\bullet H_2O$)
5000 (2.0)	O–H of water of hydration of parent compound ($\bullet H_2O$)
4878 (2.05)	Amide in proteins, N–H vibration with C–N–H bend
4608 (2.17)	Amide in proteins, N–H fundamental with C–N stretch
4348 (2.3)	C–H and methane
~3500 (2.86)	O–H stretch of hydroxyl groups
~3200 (3.1)	N–H stretch (amide A) of proteins
~2955 (3.38)	C–H stretch (a) $-CH_3$ in fatty acids
~2930 (3.4)	C–H stretch (a) $>CH_2$
~2918 (3.43)	C–H stretch (a) of $>CH_2$ in fatty acids
~2898 (3.45)	C–H stretch, CH in methine group
~2870 (3.48)	C–H stretch (s) of $-CH_3$
~2850 (3.51)	C–H stretch (s) of $>CH_2$ in fatty acids
2590–2560 (3.88)	–S–H of thiols
~1740 (5.75)	>C=O stretch of esters
~1715 (5.83)	>C=O stretch of carbonic acid
~1680–1715	>C=O stretch of nucleic acid
~1695 (6.0)	Amide I band components
~1685 (5.93)	Resulting from antiparallel pleated sheets
~1675 (5.97)	Amide I β-turns of proteins
~1655 (6.04)	Amide I of α-helical structures
~1637 (6.11)	Amide I of β-pleated structures
~1550–1520 (6.52)	Amide II
~ 1515 (6.6)	"Tyrosine" specific band
~1468 (6.81)	C–H deformation of $>CH_2$
~1400 (7.14)	C=O stretch (s) of COO^-
~1310–1240 (7.8)	Amide III of proteins
1304 (7.67)	CH_4, methane
~1250–1220 (8.0)	P=O str $>PO_3^-$, phosphodeiesters
1240–180 (8.26)	O–S=O stretch of sulfites
~1200–900 (8–11)	C–O, C–C, str of carbohydrates
~1200–900 (8–11)	C–O–H, C–O–C def. of carbohydrates
~1100–1000 (9.52)	P–O, PO_4^{-3} stretch
~1090 (9.17)	P=O stretch (s) $>PO_2$
~1085 (9.2)	C–O stretch
~1061 (9.4)	C–N and C–C stretch
1140–1080 (9.0)	S–O– stretch of inorganic sulfates
1070–1030 (9.52)	C–S=O of sulfoxides
~1004 (9.96)	Phenylalanine
~852 (11.7)	Tyrosine
~829 (12)	Tyrosine
~785 (12.7)	Cytosine, uracil (ring stretch)

Compiled from *Naumann et al.* (1991), *Naumann et al.* (1996), *Kummerle et al.* (1998), *Smith* (1999), *Choo-Smith et al.* (2001), *Painter et al.* (2001), *Maquelin et al.* (2002), *Dalton et al.* (2003), *Liu et al.* (2005), and *Hand* (2007).

voyage, and grow and multiply on that body ("forward contamination"); and (2) prevent any "back-contamination" of Earth by possible extraterrestrial biota that might be brought to Earth on a sample return mission. The National Research Council (*NRC,* 2006) has critically reviewed the history and current basis of planetary protection policy and practice in the Mars context; this section draws on that discussion, together with an earlier *NRC* (2000) report specific to Europa, and considers implications for preventing the forward contamination of Europa.

The idea of planetary protection dates to the beginning of the space age, with a letter from Nobel laureate Joshua Lederberg in 1957 (*Lederberg,* 1957) and a 1958 resolution by the U.S. National Academy of Sciences (*NAS,* 1958) stating that the Academy "urges that scientists plan lunar and planetary studies with great care and deep concern so that initial operations do not compromise and make impossible forever after critical scientific experiments." In 1964, the international Committee on Space Research (*COSPAR,* 1964) declared that since the forward biological contamination of Mars would make a search for life there far more difficult, "all practical steps should be taken to ensure that Mars be not biologically contaminated until such time as this search can have been satisfactorily carried out." From the beginning, then, planetary protection was concerned with what is sometimes called "protection of the science" rather than "protection of the planet" or protection of a possible alien biosphere *per se.*

The 1967 Outer Space Treaty provides the current legal requirement to conduct planetary protection (*UNOOSA,* 2007). Article IX of the treaty asserts that "States Parties to the Treaty shall pursue studies of outer space, including the moon and other celestial bodies, and conduct exploration of them so as to avoid their harmful contamination . . . and, where necessary, shall adopt appropriate measures for this purpose." The treaty does not define "harmful contamination," and one unofficial legal review of the treaty (*Cypser,* 1993) argues that "harmful contamination" is appropriately interpreted as "harmful to the interests of other states"; therefore, the concern must be with protecting the science that states might wish to pursue. Current NASA policy affirms that the purpose of planetary protection is to protect scientific investigations (e.g., *Rummel and Billings,* 2004), and this has been the approach taken by COSPAR for the past four decades. Notably, however, the National Research Council (*NRC,* 2000) report on *Preventing the Forward Contamination of Europa* explicitly stated that limiting the forward contamination of Europa "is necessary to preserve the scientific integrity of future biological studies and to protect any indigenous life forms." Similarly, the *NRC* (2006) panel on *Preventing the Forward Contamination of Mars* called for an international workshop to discuss whether planetary protection should be extended beyond protecting the science to protecting the destination planet's possible biosphere. As a practical matter, especially in harsh and cold conditions where biological growth and reproduction might be expected to be quite slow, such a commitment might lead to substantially more rigorous planetary protection requirements than would otherwise be the case.

The threat of biological contamination of Europa via spacecraft cannot be easily dismissed. Experiments on NASA's Long Duration Exposure Facility (LDEF) demonstrate that spores of *Bacillus subtilis* survive six years in space at the 1% level, provided that they are shielded from solar ultraviolet light, as any organism within a spacecraft would be (*Horneck et al.,* 1994, 1995). The timescale of the LDEF experiment approximates that required for a spacecraft to reach Jupiter from Earth. Spacecraft assembled within standard NASA class-100,000 clean rooms (meaning a clean room with 100,000 0.5-μm-diameter particles per cubic foot of air) have bacterial spore surface densities of $\sim 10^3$ spores m^{-2} (*Barengoltz,* 2004); total spacecraft surface bioburden is at least two orders of magnitude greater (*NRC,* 2006). Therefore, additional measures must be taken or other factors must come into play to prevent a spacecraft landing or crashing on the europan surface from delivering viable microbes.

Whether such organisms could then reach the ocean in a still-viable state, and grow and multiply within that ocean, is a question whose answer is very difficult to quantify. Such organisms would have had to survive the jovian radiation environment while in orbit around Jupiter and/or Europa, and be buried in Europa's ice quickly enough to survive the surface radiation environment. They would then have to be delivered into Europa's ocean, and be able to survive and reproduce in that environment. Clearly, this adds up to a formidable set of challenges. However, without intervention, a potentially large number of microorganisms of currently poorly known characteristics might be present at the time of launch, so the possibility, while likely remote, cannot currently be excluded.

The Task Group on the Forward Contamination of Europa (*NRC,* 2000) attempted to quantify the probability of forward contamination by considering four classes of microorganisms: common microorganisms, general spore-formers, radiation-resistant spore-formers, and non-spore-forming highly radiation-resistant microbes. Majority opinion in the Task Group argued that NASA should adopt a requirement that the probability P_c of forward contamination of Europa be below 1 in 10,000. They argued that this criterion could be met with minimal groundbased spacecraft bioburden reduction, largely because of postlaunch sterilization effects in Europa's radiation environment.

Greenberg and Tufts (2001) criticized the Task Group's approach, arguing that P_c should be based on a transparent criterion, and recommended that the criterion chosen be that the probability that the space program contaminate Europa be "substantially smaller than the probability that such contamination happens naturally." Certainly P_c should be set low enough so that, integrated over the total number of future missions to Europa (perhaps 10, but maybe 100?), the probability of contamination is small. How small is acceptably small may depend in part on whether one's concern is with "protecting the science" or protecting a possible alien biosphere. The "natural background" criterion of *Greenberg and Tufts* (2001) may prove to set an extremely stringent requirement for Europa, because viable microbial transfer in terrestrial meteorites to Europa is likely very difficult (summarized by *Chyba and Phillips,* 2007). Understanding this problem in detail (*Gladman et al.,* 2006) may be important for providing context for setting planetary protection requirements.

5. CONCLUSIONS

Europa is a prime target for astrobiology. Our current understanding of the geology, geophysics, and chemistry of Europa provide good reason to expect that Europa may be habitable at present, and may have been habitable throughout much of its past. Foremost among the attributes that could make Europa habitable is the likely presence of a global subsurface liquid water ocean with some 2–3 times the volume of all the liquid water on Earth. But while liquid water is a necessary condition for life as we know it, it is not a sufficient condition. Also required is a suite of elements essential to life and a source of energy that can sustain the growth, reproduction, and maintenance of life. In the context of elemental abundance, the formation of Europa in the jovian subnebula likely led to a bulk composition comparable to that of chondritic material, which is rich in carbon, nitrogen, phosphorous, sulfur, and other key elements needed for life.

Energy for life on Europa may be a critical limiting factor. The surface temperature and pressure and radiation environment of Europa make it an inhospitable environment for life, thus photosynthesis on the surface will not be possible. Penetration of sunlight through meters to tens of meters of ice could power photosynthetic life beneath the surface; however, such an ecosystem necessitates very thin ice, a requirement inconsistent with many models. Therefore, harnessing the energy passively received from the Sun is likely a limited niche. Alternatively, chemosynthetic metabolic pathways within the ocean and at the seafloor could present a much larger energetic niche for life in the europan subsurface. Important to maintaining such pathways through time is the active geological cycling of seawater through reducing mantle material on the europan seafloor, and with oxidizing material produced radiolytically on Europa's surface. Reductant delivery via seafloor hydrothermal activity is a well-studied process that helps drive ecosystems here on Earth. Radiolytic production of oxidants in ice, however, has no direct terrestrial analog other than atmospheric photolysis of water. Europa's radiation environment may play a critical role in maintaining the reductant-oxidant (redox) chemical pairing needed to power life.

Life on Europa would likely represent a second, unique origin of life within our solar system. The salinity and oxidation state of the ocean could, however, be an impediment to prebiotic chemistry. Hydrothermal seafloor systems or eutectic melts within the icy shell would provide useful microenvironments for promoting chemistry important to life's origin. Additionally, laboratory work indicates that radiolytic surface chemistry on Europa could result in the synthesis of prebiotic compounds such as amino acids and nucleobases.

A key question, and common theme for understanding the habitability and detectability of life on Europa, persists: Is Europa geologically active? If so, where is the activity and what is the timescale for cycling of material? Tidal dissipation may create an active seafloor and icy shell, driving the chemical cycling needed to maintain life on Europa and needed to bring such life to the surface, where it can be discovered by future missions.

Detection of life on Europa in the coming decades will be limited by the extent to which the surface ice provides a window to the subsurface ocean chemistry. Orbiting spacecraft and surface landers could target young and fresh surfaces where radiolytic processing has not yet destroyed compounds and structures indigenous to the ocean. Chemical bonds indicative of structural complexity and organization, such as those associated with the amide bands and phosphodiester bands of proteins and nucleic acids, could serve as spectroscopic biosignatures. Mass spectrometry of material showing possible biosignatures could be critical for confirming a biochemical detection of life, and microscopic imagery would be essential for confirming morphological and structural biosignatures.

Acknowledgments. K.P.H. acknowledges support from the Jet Propulsion Laboratory, California Institute of Technology, under a contract with the National Aeronautics and Space Administration. J.C.P. was supported by NSF grants OPP-432595, OPP-0237335, OPP-0440943, and OPP-0631494.

REFERENCES

Allamandola L. J., Sandford S. A., and Valero G. J. (1988) Photochemical and thermal evolution of interstellar/precometary ice analogs. *Icarus, 76(2),* 225–252.

Allen M., Beauchamp P., Carlson R., Cooper K., Drouin B., Pearson J., Pickett H., Rodgers D., Siegel P., Skalare A., Gulkis S., and Chattopadhyay S. (2008) Astrobiology from europan orbit. Astrobiology Science Conference 2008, San Jose, California, April 14–17, *Astrobiology 8:2,* 296.

Amend J. P. and Shock E. L. (2001) Energetics of overall metabolic reactions of thermophilic and hyperthermophilic Archaea and bacteria. *FEMS Microbiol. Rev., 25(2),* 175–243.

Bach W., Paulick H., Garrido C. J., Ildefonse B., Meurer W. P., and Humphris S. E. (2006) Unraveling the sequence of serpentinization reactions: Petrography, mineral chemistry, and petrophysics of serpentinites from MAR 15°N (ODP Leg 209, Site 1274). *Geophys. Res. Lett., 33,* L13306, DOI: 10.1029/2006GL025681.

Bada J. L. and Lazcano A. (2002) Origin of life: Some like it hot, but not the first biomolecules. *Science, 296(5575),* 1982–1983.

Bada J. L., Bigham C., and Miller S. L. (1994) Impact melting of frozen oceans on the early Earth: Implications for the origin of life. *Proc. Natl. Acad. Sci., 91(4),* 1248–1250.

Barengoltz J. (2004) Planning for project compliance. In *Planetary Protection: Policies and Practices,* NASA Planetary Protection Office and NASA Astrobiology Institute. NASA, Washington, DC.

Benner S. A., Ricardo A., and Carrigan M. A. (2004) Is there a common chemical model for life in the universe? *Curr. Opin. Chem. Biol., 8(6),* 672–689.

Berndt M. E., Allen D. E., and Seyfried W. E. (1996) Reduction of CO_2 during serpentinization of olivine at 300 degrees C and 500 bar. *Geology, 24(4),* 351–354.

Bernstein M. P., Sandford S. A., Allamandola L. J., Chang S., and Scharberg M. A. (1995) Organic compounds produced by photolysis of realistic interstellar and cometary ice analogs containing methanol. *Astrophys. J., 454,* 327.

Bernstein M. P., Sandford S. A., Allamandola L. J., Gillette J. S., Clemett S. J., and Zare R. N. (1999) UV Irradiation of polycyclic aromatic hydrocarbons in ices: Production of alcohols, quinones, and ethers. *Science, 283,* 1135–1138.

Bernstein M. P., Dworkin J. P., Sandford S. A., Cooper G. W., and Allamandola L. J. (2002) Racemic amino acids from the ultraviolet photolysis of interstellar ice analogues. *Nature, 416,* 401–403.

Biemann K., Óro J., Toulmin P. III, Orgel L. E., Nier A. O., Anderson D. M., Flory D., Diaz A. V., Rushneck D. R., and Simmonds P. G. (1977) The search for organic substances and inorganic volatile compounds in the surface of Mars. *J. Geophys. Res., 82,* 4641–4658.

Blochl E., Keller M., Wächtershäuser G. and Stetter K. O. (1992) Reactions depending on iron sulfide and linking geochemistry with biochemistry. *Proc. Natl. Acad. Sci., 89(17),* 8117–8120.

Brocks J. J., Buick R., Summons R. E. and Logan G. A. (2003) A reconstruction of Archean biological diversity based on molecular fossils from the 2.78 to 2.45 billion-year-old Mount Bruce Supergroup, Hamersley Basin, Western Australia. *Geochim. Cosmochim. Acta, 67(22),* 4321–4335.

Brown M. E. and Hill R. E. (1996) Discovery of an extended sodium atmosphere around Europa. *Nature, 380(6571),* 229–231.

Brown R. H., Clark R. N., Buratti B. J., Cruikshank D. P., Barnes J. W., Mastrapa R. M. E., Bauer J., Newman S., Momary T., Baines K. H. and others (2006) Composition and physical properties of Enceladus' surface. *Science, 311,* 1425–1428.

Buck L., Chyba C. F., Goulet M., Smith A., and Thomas P. (2002) Persistence of thin ice regions in Europa's ice crust. *Geophys. Res. Lett., 29(22),* 12–21.

Cairon O., Chevreau T., and Lavalley J. C. (1998) Bronstead acidity of extraframework debris in steamed Y zeolites from the FTIR study of CO absorption. *J. Chem. Phys., Faraday Trans., 94,* 3039–3047.

Canfield D. E. (1998) A new model for Proterozoic ocean chemistry. *Nature, 396(6710),* 450.

Canfield D. E., Habicht K. S., and Thamdrup B. (2000) The Archean sulfur cycle and the early history of atmospheric oxygen. *Science, 288(5466),* 658–661.

Capone D. G., Popa R., Flood B., and Nealson K. H. (2006) Follow the nitrogen. *Science, 312(5774),* 708–709.

Carlson R. W., Anderson M. S., Johnson R. E., Smythe W. D., Hendrix A. R., Barth C. A., Soderblom L. A., Hansen G. B., McCord T. B., Dalton J. B. and others (1999a) Hydrogen peroxide on the surface of Europa. *Science, 283,* 2062–2064.

Carlson R. W., Johnson R. E., and Anderson M. S. (1999b) Sulfuric acid on Europa and the radiolytic sulfur cycle. *Science, 286,* 97–99.

Carlson R. W., Anderson M. S., Johnson R. E., Schulman M. B., and Yavrouian A. A. (2002) Sulfuric acid production on Europa: The radiolysis of sulfur in water ice. *Icarus, 157,* 456–463.

Carlson R. W., Anderson M. S., Mehlman R., and Johnson R. E. (2005) Distribution of hydrate on Europa: Further evidence for sulfuric acid hydrate. *Icarus, 177,* 461–471.

Carr M. H., Belton M. J., Chapman C. R., Davies M. E., Geissler P., Greenberg R., McEwen A. S., Tufts B. R., Greeley R., Sullivan R. and others (1998) Evidence for a subsurface ocean on Europa. *Nature, 391,* 363–365.

Childress J. J. (1968) Oxygen minimum layer: Vertical distribution and respiration of the Mysid Gnathophausia ingens. *Science, 160,* 1242–1243.

Choo-Smith L.-P. and 16 colleagues (2001) Investigating microbial (micro)colony hetereogeneity by vibrational spectroscopy. *Appl. Environ. Microbiol., 67(4),* 1461–1469.

Christner B. C., Mikucki J. A., Foreman C. M., Denson J. and Priscu J. C. (2005) Glacial ice cores: A model system for developing extraterrestrial decontamination protocols. *Icarus, 174,* 572–584.

Christner B. C., Royston-Bishop G., Foreman C. M., Arnold B. R., Tranter M., Welch K. A., Lyons W. B., Tsapin A. I., Studinger M. and Priscu J. C. (2006) Limnological conditions in Subglacial Lake Vostok. *Antarct. Limnol. Oceanogr., 51(6),* 2485–2501.

Chyba C. F. (2000) Energy for microbial life on Europa. *Nature, 403,* 381–382. See also correction. *Nature, 406,* 368.

Chyba C. F. and Hand K. P. (2001) Life without photosynthesis. *Science, 292,* 2026–2027.

Chyba C. F. and Hand K. P. (2006) Comets and prebiotic organic molecules on early Earth. In *Comets and the Origin and Evolution of Life* (P. J. Thomas et al., eds.), pp. 169–206. Springer, New York.

Chyba C. F. and McDonald G. D. (1995) The origin of life in the solar system: Current issues. *Annu. Rev. Earth Planet. Sci., 23,* 215–249.

Chyba C. F. and Phillips C. B. (2001) Possible ecosystems and the search for life on Europa. *Proc. Natl. Acad. Sci., 98,* 801–804.

Chyba C. F. and Phillips C. B. (2002) Europa as an abode of life. *Origins Life Evol. Biosph., 32,* 47–68.

Chyba C. F. and Phillips C. B. (2007) Europa. In *Planets and Life: The Emerging Science of Astrobiology* (W. T. Sullivan and J. A. Baross, eds.), pp. 388–423. Cambridge Univ., Cambridge.

Chyba C. F. and Sagan C. (1992) Endogenous production, exogenous delivery and impact-shock synthesis: An inventory for the origins of life. *Nature, 355,* 123–130.

Cleland C. E. and Chyba C. F. (2002) Defining "life." *Origins Life Evol. Biosph., 32,* 387–393.

Cody G. D. (2005) Geochemical connections to primitive metabolism. *Elements, 1(3),* 139–142.

Cody G. D., Boctor N. Z., Filley T. R., Hazen R. M., Scott J. H., Sharma A., and Yoder H. S. (2000) Primordial carbonylated iron-sulfur compounds and the synthesis of pyruvate. *Science, 289,* 1337–1340.

Collins G. C. and Goodman J. C. (2007) Enceladus' south polar sea. *Icarus, 189,* 72–82.

Cooper J. F., Johnson R. E., Mauk B. H., Garrett H. B. and Gehrels N. (2001) Energetic ion and electron irradiation of the icy Galilean satellites. *Icarus, 149,* 133–159.

COSPAR (Committee on Space Research) (1964) *COSPAR Resolution No. 26, COSPAR Information Bulletin, No. 20.* COSPAR, Paris.

Corliss J. B., Dymond J., Gordon L. I., Edmond J. M., von Herzen R. P., et al. (1979) Submarine thermal springs on the Galapagos Rift. *Science, 203(4385),* 1073–1083.

Corliss J. B., Baross J. A., and Hoffman S. E. (1981) An hypothesis concerning the relationship between submarine hot springs and the origin of life on Earth. *Proceedings 26th International Geological Congress,* pp. 59–69. Geology of Oceans Symposium, Paris, July 7–17, 1980, Oceanologica Acta Special Publication.

Costard F., Forget F., Mangold N., and Peulvast J. P. (2002) Formation of recent martian debris flows by melting of near-surface ground ice at high obliquity. *Science, 295(5552),* 110–112.

Court R. W., Sephton M. A., Parnell J., and Gilmour I. (2006) The alteration of organic matter in response to ionising irradiation: Chemical trends and implications for extraterrestrial sample analysis. *Geochim. Cosmochim. Acta, 70(4),* 1020–1039.

Crawford G. D. and Stevenson D. J. (1988) Gas-driven water volcanism and the resurfacing of Europa. *Icarus, 73,* 66–79.

Cruikshank D. P., Brown R. H. and Clark R. N. (1984) Nitrogen on Triton. *Icarus, 58(2),* 293–305.

Cruikshank D. P., Allamandola L. J., Hartmann W. K., Tholen D. J., Brown R. H., Matthews C. N. and Bell J. F. (1991) Solid CN bearing material on outer solar system bodies. *Icarus, 94,* 345–353.

Cypser D. A. (1993) International law and policy of extraterrestrial planetary protection. *Jurimetrics J., 33,* 315–339.

Dahl J., Hallberg R., and Kaplan I. R. (1988a) The effect of radioactive decay of uranium on elemental and isotopic ratios of Alum Shale kerogen. *Appl. Geochem., 3,* 583–589.

Dahl J., Hallberg R., and Kaplan I. R. (1988b) Effects of irradiation from uranium decay on extractable organic matter in the Alum Shales of Sweden. *Org. Geochem., 12(6),* 559–571.

Dalton J. B., Mogul R., Kagawa H. K., Chan S. L., and Jamieson C. S. (2003) Near-infrared detection of potential evidence for microscopic organisms on Europa. *Astrobiology, 3(3),* 505–529.

Darwin C. (1871) Some unpublished letters, Ed. Sir Gavin de Beer. *Notes Rec. R. Soc. Lond., 14(1),* 1959.

de Duve C. (2005) *Singularities: Landmarks on the Pathways of Life.* Cambridge Univ., Cambridge.

D'Hondt S., Rutherford S., and Spivack A. J. (2002) Metabolic activity of subsurface life in deep-sea sediments. *Science, 295(5562),* 2067–2070.

Dworkin J. P., Deamer D. W., Sandford S. A., and Allamandola L. J. (2001) Self-assembling amphiphilic molecules: Synthesis in simulated interstellar/precometary ices. *Proc. Natl. Acad. Sci., 98,* 815–819.

Dyson F. J. (1999) *Origins of Life.* Cambridge Univ., New York.

Ellis D., Bizzoco R. L. W., Maezato Y., Baggett J. N., and Kelley S. T. (2005) Microscopic examination of acidic hot springs of Waiotapu, North Island, New Zealand. *N.Z. J. Marine Freshwater Res., 39(5),* 1001–1011.

Fagents S. A., Greeley R., Sullivan R. J., Prockter L. M., and the Galileo SSI Team (2000) Cryomagmatic mechanisms for the formation of Rhadamanthys Linea, triple band margins, and other low-albedo features on Europa. *Icarus, 144,* 54–88.

Fanale F. P., Li Y. H., De Carlo E., Farley C., Sharma S. K., and Horton K. (2001) An experimental estimate of Europa's "ocean" composition independent of Galileo orbital remote sensing. *J. Geophys. Res., 106,* 14595–14600.

Ferris J. P., Hill A. R., Liu R., and Orgel L. E. (1998) Synthesis of long prebiotic oligomers on mineral surfaces. *Nature, 381(6577),* 59–61.

Field C. B., Behrenfeld M. J., Randerson J. T., and Falkowski P. (1998) Primary production of the biosphere: Integrating terrestrial and oceanic components. *Science, 281(5374),* 237–239.

Figueredo P. H., Greeley R., Neuer S., Irwin L., and Schulze-Makuch D. (2003) Locating potential biosignatures on Europa from surface geology observations. *Astrobiology, 3(4),* 851–861.

Fredriksson K. and Kerridge J. F. (1988) Carbonates and sulfates in CI chondrites — Formation by aqueous activity on the parent body. *Meteoritics, 23,* 35–44.

Gaidos E. J., Nealson K. H., and Kirschvink J. L. (1999) Life in ice-covered oceans. *Science, 284,* 1631–1633.

Gerakines P. and Moore M. H. (2001) Carbon suboxide in astrophysical ice analogs. *Icarus, 154,* 372–380.

Gerakines P. A., Moore M. H., and Hudson R. L. (2004) Ultraviolet photolysis and proton irradiation of astrophysical ice analogs containing hydrogen cyanide. *Icarus, 170,* 202–213.

Gladman B., Dones L., Levison H., Burns J., and Gallant J. (2006) Meteoroid transfer to Europa and Titan. In *Lunar and Planetary Science XXXVII,* Abstract #2165, Lunar and Planetary Institute, Houston (CD-ROM).

Goldschmidt V. M. (1952) Geochemical aspects of the origin of complex organic molecules on Earth, as precursors to organic life. *New Biology, 12,* 97–105.

Gooday A. J., Hori S., Todo Y., and Okamoto T., Kitazato H., and Sabbatini A. (2004) Soft-walled, monothalamous benthic foraminiferans in the Pacific, Indian and Atlantic Oceans: Aspects of biodiversity and biogeography. *Deep-Sea Research Part I, 51(1),* 33–53.

Goodman J. C., Collins G. C., Marshall J., and Pierrehumbert R. T. (2004) Hydrothermal plume dynamics on Europa: Implications for chaos formation. *J. Geophys. Res., 109,* 10.1029.

Grasby S. E., Allen C. C., Longazo T. G., Lisle J. T., Griffin D. W., and Beauchamp B. (2003) *Astrobiology, 3(3),* 583–596.

Grasset O. and Sotin C. (1996) The cooling rate of a liquid shell in Titan's interior. *Icarus, 123(1),* 101–112.

Greenberg R. and Tufts B. R. (2001) Standards for prevention of biological contamination of Europa. *Eos Trans. AGU, 82,* 26–28.

Greenberg R., Hoppa G. V., Tufts B. R., Geissler P. E., and Reilly J. (1999) Chaos on Europa. *Icarus, 141,* 263–286.

Greenberg R., Geissler P., Tufts B. R., and Hoppa G. V. (2000) Habitability of Europa's crust: The role of tidal-tectonic processes. *J. Geophys. Res., 105,* 17551–17562.

Guillot T. (2005) The interiors of giant planets — Models and outstanding questions. *Annu. Rev. Earth Planet. Sci., 33(1),* 493–530.

Hand K. P. (2007) On the physics and chemistry of the ice shell and sub-surface ocean of Europa. Ph.D. thesis, Stanford University.

Hand K. P. and Chyba C. F. (2007) Empirical constraints on the salinity of the europan ocean and implications for a thin ice shell. *Icarus, 189(2),* 424–438, DOI: 10.1016/j.icarus.2007.02.002.

Hand K. P., Chyba C. F., Carlson R. W., and Cooper J. F. (2006) Clathrate hydrates of oxidants in the ice shell of Europa. *Astrobiology, 6(3),* 463–482.

Hand K. P., Carlson R. W., and Chyba C. F. (2007) Energy, chemical disequilibrium, and geological constraints on Europa. *Astrobiology, 7(6),* 1–18.

Hansen C. J., Esposito L., Stewart A. I., Colwell J., Hendrix A., Pryor W., Shemansky D., and West R. (2006) Enceladus' water vapor plume. *Science, 311,* 5766.

Hayes J. M. (1994) Global methanotrophy at the Archean-Preoterozoic transition. In *Early Life on Earth* (S. Bengtson, ed.), pp. 220–236. Columbia Univ., New York.

Head J. W. III, Hiesinger H., Ivanov M. A., Kreslavsky M. A., Pratt S. and Thomson B. J. (1999) Possible ancient oceans on Mars: Evidence from Mars Orbiter Laser Altimeter data. *Science, 286(5447),* 2134–2136.

Heijnen J. J. and van Dijken J. P. (1992) In search of a thermodynamic description of biomass yields for the chemotrophic growth of microorganisms. *Biotechnol. Bioengin., 39(8),* 833–858.

Heinen W. and Lauwers A. M. (1996) Organic sulfur compounds resulting from the interaction of iron sulfide, hydrogen sulfide and carbon dioxide in an anaerobic aqueous environment. *Origins Life Evol. Biosph., 26(2),* 131–150.

Hobbs P. V. (1974) *Ice Physics.* Clarendon, Oxford.

Hoehler T. M., Alperin M. J., Albert D. B., and Martens C. S. (2001) Apparent minimum free energy requirements for methanogenic Archaea and sulfate-reducing bacteria in an anoxic marine sediment. *FEMS Microbiol. Ecol., 38(1),* 33–41.

Holland H. D. (1984) *The Chemical Evolution of the Atmosphere and Oceans.* Princeton Univ., Princeton.

Horneck G., Bücker H., and Reitz G. (1994) Long-term survival of bacterial spores in space. *Adv. Space Res., 14,* 41–45.

Horneck G., Eschweiler U., Reitz G., Wehner J., Willimek R., and Strauch K. (1995) Biological responses to space: Results of the experiment "Exobiological Unit" of ERA on Eureca I. *Adv. Space Res., 16,* 105–111.

Howard-Williams C., Schwarz A., Hawes I., and Priscu J. C. (1998) Optical properties of lakes of the McMurdo Dry Valleys. In *Ecosystem Dynamics in a Polar Desert: The McMurdo Dry Valleys, Antarctica* (J. C. Priscu, ed.), pp. 189–204. Antarctic Research Series, Vol. 72, AGU, Washington, DC.

Huber C. and Wächtershäuser G. (1997) Activated acetic acid by carbon fixation on (Fe, Ni) S under primordial conditions. *Science, 276(5310),* 245–247.

Huber C. and Wächtershäuser G. (1998) Peptides by activation of amino acids with CO on (Ni, Fe) S surfaces: Implications for the origin of life. *Science, 281(5377),* 670–673.

Hudson R. L. and Moore M. L. (1999) Laboratory studies of the formation of methanol and other organic molecules by water + carbon monoxide radiolysis: Relevance to comets, icy satellites, and intersellar ices. *Icarus, 140,* 451–461.

Hudson R. L., Moore M. H., and Gerakines P. A. (2001) The formation of cyanate ion (OCN-) in interstellar ice analogs. *Astrophys. J., 550(2),* 1140–1150.

Hussmann H. and Spohn T. (2004) Thermal-orbital evolution of Io and Europa. *Icarus, 171,* 391–410.

Imai E., Honda H., Hatori K., Brack A., and Matsuno K. (1999) Elongation of oligopeptides in a simulated submarine hydrothermal system. *Science, 283(5403),* 831.

Jepsen S. M., Priscu J. C., Grimm R. E., and Bullock M. A. (2007) The potential for lithoautotrophic life on Mars: Application to shallow interfacial water environments. *Astrobiology, 7(2),* 342–354.

Johnson R. E. (1998) Sputtering and desorption from icy satellite surfaces. In *Solar System Ices* (B. Schmitt et al., eds.), pp. 303–334. Kluwer, Dordrecht.

Johnson R., Carlson R., Cooper J., Paranicas C., Moore M., and Wong M. (2004) Radiation effects on the surfaces of the Galilean satellites. In *Jupiter: The Planets, Satellites, and Magnetosphere* (F. Bagenal et al., eds.), pp. 485–512. Cambridge Univ., New York.

Joyce G. F. (1989) RNA evolution and the origins of life. *Nature, 338(6212),* 217–224.

Joyce G. F. (1994) Forward. In *Origins of Life: The Central Concepts* (D. W. Deamer and G. R. Fleischaker, eds.), pp. xi–xii. Jones and Bartlett, Boston.

Karl D. M., Bird D. F., Bjorkman K., Houlihan T., Shackelford R, and Tupas L. (1999) Microorganisms in the accreted ice of Lake Vostok. *Science, 286(5447),* 2144–2147.

Kargel J. S. and Lewis J. S. (1993) The composition and early evolution of Earth. *Icarus, 105(1),* 1–25.

Kargel J. S., Kaye J. Z., Head J. W. I., Marion G. M., Sassen R., Crowley J. K., Ballesteros O. P., Grant S. A., and Hogenboom D. L. (2000) Europa's crust and ocean: Origin, composition, and the prospects for life. *Icarus, 148,* 226–265.

Kasting J. F. and Catling D. (2003) Evolution of a habitable planet. *Annu. Rev. Astron. Astrophys, 41(1),* 429–463.

Kelley D. S., Karson J. A., Blackman D. K., Fruh-Green G. L., Butterfield D. A., et al. (2001) An off-axis hydrothermal vent field near the Mid-Atlantic Ridge at 30°N. *Nature, 412(6843),* 145–149.

Khare B. N., Sagan C., Ogino H., Nagy B., Er C., Schram K. H., and Arakawa E. T. (1986) Amino acids derived from Titan tholins. *Icarus, 68(1),* 176–184.

Khare B. N., Thompson W. R., Murray B. G. J. P. T., Chyba C. F., and Sagan C. (1989) Solid organic residues produced by irradiation of hydrocarbon-containing H_2O and H_2O/NH_3 ices: Infrared spectroscopy and astronomical implications. *Icarus, 79,* 350–361.

Khare B. N., Sagan C., Thompson W. R., Arakawa E. T., Meisse C., and Tuminello P. S. (1994) Optical properties of poly-HCN and their astronomical applications. *Can. J. Chem., 72,* 678–694.

Kieffer S. W., Lu X., Bethke C. M., Spencer J. R., Marshak S., and Navrotsky A. (2006) A clathrate reservoir hypothesis for Enceladus' south polar plume. *Science, 314(5806),* 1764.

Klein H. P. (1978) The Viking biological experiments on Mars. *Icarus, 34(3),* 666–674.

Klein H. P., Lederberg J., and Rich A. (1972) Biological experiments: The Viking Mars lander. *Icarus, 16(1),* 139–146.

Knoll A. H. (2003) *Life on a Young Planet: The First Three Billion Years of Evolution on Earth.* Princeton Univ., Princeton.

Krauskopf K. B. and Bird D. K. (1995) *Introduction to Geochemistry.* McGraw-Hill, New York. 647 pp.

Kummerle M., Scherer S., and Seiler H. (1998) Rapid and reliable identification of food-borne yeasts by Fourier-Transform Infrared Spectroscopy. *Appl. Environ. Microbiol., 64(6),* 2207–2214.

La Duc M. T., Nicholson W., Kern R., and Venkateswaran K. (2003) Microbial characterization of the Mars Odyssey spacecraft and its encapsulation facility. *Environ. Microbiol., 5(10),* 977–985.

Leblanc F., Johnson R. E. and Brown M. E. (2002) Europa's sodium atmosphere: An ocean source? *Icarus, 159(1),* 132–144.

Lederberg J. (1957) Letter to Detlev Bronk, President, National Academy of Sciences, December 24. National Academy of Science, Records Office, Washington, DC.

Leman L., Orgel L., and Ghadiri M. R. (2004) Carbonyl sulfide-mediated prebiotic formation of peptides. *Science, 306(5694),* 283–286.

Levy M., Miller S. L., Brinton K., and Bada J. L. (2000) Prebiotic synthesis of adenine and amino acids under Europa-like conditions. *Icarus, 145(2),* 609–613.

Lewis J. S. (1971) Satellites of the outer planets: Their physical and chemical nature. *Icarus, 15,* 174–185.

Lipenkov V. Y. and Istomin V. A. (2001) On the stability of air clathrate-hydrate crystals in subglacial Lake Vostok, Antarctica. *Mater. Glyatsiol. Issled, 91,* 1–30.

Lipps J. H. and Rieboldt S. (2005) Habitats and taphonomy of Europa. *Icarus, 177(2),* 515–527.

Liu M., Krasteva M., and Barth A. (2005) Interactions of phosphate groups of ATP and aspartyl phosphate with the Sarcoplasmic Reticulum Ca^{2+}-ATP-ase: An FTIR study. *Biophys. J., 89(6),* 4352–4363.

Lodders K. (2003) Solar system abundances and condensation temperatures of the elements. *Astrophys. J., 591(2),* 1220–1247.

Lodders K. and Fegley B. Jr. (1998) *The Planetary Scientist's Companion.* Oxford Univ., New York.

Loeffler M. J., Raut U., and Baragiola R. A. (2006) Enceladus:

A source of nitrogen and an explanation for the water vapor plume observed by Cassini. *Astrophys. J. Lett., 649(2),* L133–L136.

Loferer-Krössbacher M., Klima J., and Psenner R. (1998) Determination of bacterial cell dry mass by transmission electron microscopy and densitometric image analysis. *Appl. Environ. Microbiol., 64(2),* 688.

Lollar B. S., Westgate T. D., Ward J. A., Slater G. F., and Lacrampe-Couloume G. (2002) Abiogenic formation of alkanes in the Earth's crust as a minor source for global hydrocarbon reservoirs. *Nature, 416(6880),* 522–524.

Lorenz R. D., Stiles B. W., Kirk R. L., Allison M. D., del Marmo P. P., Iess L., Lunine J. I., Ostro S. J., and Hensley S. (2008) Titan's rotation reveals an internal ocean and changing zonal winds. *Science, 319,* 1649–1651.

McCollom T. M. (1999) Methanogenesis as a potential source of chemical energy for primary biomass production by autotrophic organisms in hydrothermal systems on Europa. *J. Geophys. Res., 104,* 30729–30742.

McCord T. B., Carlson R. W., Smythe W. D., Hansen G. B., Clark R. N., Hibbitts C. A., Fanale F. P., Granahan J. C., Segura M., Matson D. L., Johnson T. V., and Martin P. D. (1997) Organics and other molecules in the surfaces of Callisto and Ganymede. *Science, 278,* 271–275.

McCord T. B., Hansen G. B., Martin P. D., and Hibbitts C. (1998) Non-water-ice constituents in the surface material of the icy Galilean satellites from the Galileo near infrared mapping spectrometer investigation. *J. Geophys. Res., 103,* 8603–8626.

McCord T. B., Hansen G. B., Matson D. L., Johnson T. V., Crowley J. K., Fanale F. P., Carlson R. W., Smythe W. D., Martin P. D., Hibbitts C. A., Granahan J. C., and Ocampo A. (1999) Hydrated salt minerals on Europa's surface from the Galileo near-infrared mapping spectrometer (NIMS) investigation. *J. Geophys. Res., 104,* 11827–11851.

McKay C. P. (1991) Urey Prize lecture: Planetary evolution and the origin of life. *Icarus, 91(1),* 93–100.

McKay C. P. and Smith H. D. (2005) Possibilities for methanogenic life in liquid methane on the surface of Titan. *Icarus, 178,* 274–276.

Macia E., Hernandez M. V., and Oró J. (1997) Primary sources of phosphorus and phosphates in chemical evolution. *Origins Life Evol. Biosph., 27(5),* 459–480.

Madigan M. T., Martinko J. M., and Parker J. (2003) *Brock Biology of Microorganisms.* Prentice-Hall, Upper Saddle River, New Jersey.

Maquelin K., Kirschner C., Choo-Smith L. P., van den Braak N., Endtz H. P., Naumann D., and Puppels G. J. (2002) Identification of medically relevant microorganisms by vibrational spectroscopy. *J. Microbiol. Meth., 51(3),* 255–271.

Marion G. M., Fritsen C. H. Eicken H., and Payne M. C. (2003) The search for life on Europa: Limiting environmental factors, potential habitats, and Earth analogues. *Astrobiology, 3(4),* 785–811.

Martin W. and Russell M. J. (2007) Review. On the origin of biochemistry at an alkaline hydrothermal vent. *Philos. Trans. R. Soc. London, B362(1486),* 1887–1925.

Martin W., Baross J., Kelley D., and Russell M. J. (2008) Hydrothermal vents and the origin of life. *Nature Rev. Microbiol., 6,* 806–814.

Melosh H. J., Ekholm A. G., Showman A. P., and Lorenz R. D. (2004) The temperature of Europa's subsurface water ocean. *Icarus, 168,* 498–502.

Miller S. L. (1953) Production of amino acids under possible primitive Earth conditions. *Science, 117,* 528–529.

Miller S. L. and Bada J. L. (1988) Submarine hot springs and the origin of life. *Nature, 334(6183),* 609–611.

Miller S. L. and Orgel L. E. (1974) *The Origins of Life on the Earth.* Prentice-Hall, Englewood Cliffs, New Jersey.

Miller S. L. and Urey H. C. (1959) Organic compound synthes on the primitive Earth: Several questions about the origin of life have been answered, but much remains to be studied. *Science, 130(3370),* 245–251.

Millero F.J. (2005) *Chemical Oceanography,* 3rd edition. Academic, New York. 496 pp.

Mitri G., Showman A. P., Lunine J. I., and Lorenz R. D. (2007) Hydrocarbon lakes on Titan. *Icarus, 186(2),* 385–394.

Miyakawa S., James Cleaves H., and Miller S. L. (2002) The cold origin of life: A. Implications based on the hydrolytic stabilities of hydrogen cyanide and formamide. *Origins Life Evol. Biosph., 32(3),* 195–208.

Mojzsis S. J., Arrhenius G., McKeegan K. D., Harrison T. M., Nutman A. P., and Friend C. R. L. (1996) Evidence for life on Earth before 3,800 million years ago. *Nature, 384(6604),* 55–59.

Monnard P.-A., Apel C. L., Kanavarioti A., and Deamer D. W. (2002) Influence of ionic solutes on self-assembly and polymerization processes related to early forms of life: Implications for a prebiotic qaueous medium. *Astrobiology, 2,* 213–219.

Montes-Hugo M. A., Carder K., Foy R., Cannizzaro J., Brown E., and Pegau S. (2005) Estimating phytoplankton biomass in coastal waters of Alaska using airborne remote sensing. *Remote Sens. Environ., 98,* 481–493.

Moore M. H. and Donn B. (1982) The infrared spectrum of a laboratory-synthesized residue: Implications for the 3.4 micron interstellar absorption feature. *Astrophys. J. Lett., 257,* L47–LL50.

Moore M. H. and Hudson R. L. (1998) Infrared study of ion-irradiated water-ice mixtures with hydrocarbons relevant to comets. *Icarus, 135,* 518–527.

NAS (National Academy of Sciences) (1958) Resolution adopted by the Council of the NAS, February 8. Addendum to Minutes of the Meeting of the Council of the National Academy of Sciences, February 8.

Naumann D., Helm D., and Labischinski H. (1991) Microbiological characterizations by FT-IR spectroscopy. *Nature, 351(6321),* 81–82.

Naumann D., Schultz C., and Helm D. (1996) What can infrared spectroscopy tell us about the structure and composition of intact bacterial cells? In *Infrared Spectroscopy of Biomolecules* (H. Mantsch and D. Chapman, eds.), pp. 279–310. Wiley-Liss, New York.

Nealson K. H. (1997) The limits of life on Earth and searching for life on Mars. *J. Geophys. Res., 102,* 23675–23686.

Newman M. J. and Rood R. T. (1977) Implications of solar evolution for the Earth's early atmosphere. *Science, 198(4321),* 1035–1037.

Nimmo F. and Gaidos E. (2002) Strike-slip motion and double ridge formation on Europa. *J. Geophys. Res., 107,* DOI: 10.1029/2000JE001476.

Nisbet E. G. and Sleep N. H. (2001) The habitat and nature of early life. *Nature, 409(6823),* 1083–1091.

NRC (National Research Council) (2000) *Preventing the Forward Contamination of Europa.* National Academy Press, Washington, DC.

NRC (National Research Council) (2006) *Preventing the Forward Contamination of Mars.* National Academy Press, Washington, DC.

NRC (National Research Council) (2007) *The Limits of Organic Life in Planetary Systems.* National Academy Press, Washington, DC.

Ochiai M., Marumoto R., Kobayashi S., Shimazu H., and Morita K. (1968) A facile one-step synthesis of adenine. *Tetrahedron, 24(17),* 5731–5737.

Oren A. (1994) The ecology of the extremely halophilic archea. *FEMS Microbiol. Rev., 13,* 415–440.

Oren A. (2001) The bioenergetic basis for the decrease in metabolic diversity at increasing salt concentrations: Implications for the functioning of salt lake ecosystems. *Hydrobiologia, 466,* 61–72.

Orgel L. E. (1998) The origin of life — A review of facts and speculations. *Trends Biochem. Sci., 23(12),* 491–495.

Óro J. (1960) Synthesis of adenine from ammonium cyanide. *Biochem. Biophys. Res. Commun., 2,* 407–412.

Óro J. (1961) Mechanism of synthesis of adenine from hydrogen cyanide under possible primitive Earth conditions. *Nature, 191(479),* 1193.

Óro J. and Kamat S. S. (1961) Amino-acid synthesis from hydrogen cyanide under possible primitive Earth conditions. *Nature, 190(477),* 442.

Pace N. R. (1997) A molecular view of microbial diversity and the biosphere. *Science, 276(5313),* 734–736.

Paerl H. W. and Priscu J. C. (1998) Microbial phototrophic, heterotrophic and diazotrophic activities associated with aggregates in the permanent ice cover of Lake Bonney, Antarctica. *Microbial Ecol., 36,* 221–230.

Painter T. H., Duval B., Thomas W. H., Mendez M., Heintzelman S., and Dozier J. (2001) Detection and quantification of snow algae using an airborne imaging spectrometer. *Appl. Environ. Microbiol., 67(11),* 5267–5272.

Palm C. and Calvin M. (1962) Primordial organic chemistry. I. Compounds resulting from electron irradiation of $C_{14}H_4$. *J. Am. Chem. Soc., 84(11),* 2115–2121.

Pappalardo R. T., Head J. W., Greeley R., Sulllivan R. J., Pilcher C., Schubert G., Moore W. B., Carr M. H., Moore J. M., and Belton M. J. S. (1998) Geological evidence for solid-state convection in Europa's ice shell. *Nature, 391,* 365–368.

Paranicas C., Carlson R. W., and Johnson R. E. (2001) Electron bombardment of Europa. *Geophys. Res. Lett., 28,* 673–676.

Peters K. E., Walters C. C., and Moldowan J. M. (2005) *The Biomarker Guide.* Cambridge Univ., Cambridge.

Philipp M. and Seliger H. (1977) Spontaneous phosphorylation of nucleosides in formamide — Ammonium phosphate mixtures. *Naturwissensch., 64(5),* 273–273.

Philippot P., Van Zuilen M., Lepot K., Thomazo C., Farquhar J., and Van Kranendonk M. J. (2007) Early Archaean microorganisms preferred elemental sulfur, not sulfate. *Science, 317(5844),* 1534–1537.

Pierazzo E. and Chyba C. F. (2002) Cometary delivery of biogenic elements to Europa. *Icarus, 157(1),* 120–127.

Pilcher C. B. (2003) Biosignatures of early Earths. *Astrobiology, 3(3),* 471–486.

Ponnamperuma C., Lemmon R., Mariner R., and Calvin M. (1963) Formation of adenine by electron irradiation of methane, ammonia, and water. *Proc. Natl. Acad. Sci., 49(5),* 737–740.

Price P. B. (2007) Microbial life in glacial ice and implications for a cold origin of life. *FEMS Microbiol. Ecol., 59(2),* 217–231.

Priscu J. C. (1995) Phytoplankton nutrient deficiency in lakes of the McMurdo Dry Valleys, Antarctica. *Freshwater Biol., 34,* 215–227.

Priscu J. C. and Christner B. C. (2004) Earth's icy biosphere. In *Microbial Biodiversity and Bioprospecting* (A. T. Bull, ed.), pp. 130–145. American Society for Microbiology, Washington, DC.

Priscu J. C., Fritsen C. H., Adams E. E., et al. (1998) Perennial Antarctic lake ice: An oasis for life in a polar desert. *Science, 280,* 2095–2098.

Priscu J. C., Adams E. E., Lyons W. B., Voytek M. A., Mogk D. W., Brown R. L., McKay C. P., Takacs C. D., Welch K. A., Wolf C. F., Kirstein J. D., and Avci R. (1999) Geomicrobiology of sub-glacial ice above Vostok Station. *Science, 286,* 2141–2144.

Priscu J. C., Fritsen C. H., Adams E. E., Paerl H. W., Lisle J. T., Dore J. E., Wolf C. F., and Mikucki J. A. (2005a) Perennial Antarctic lake ice: A refuge for cyanobacteria in an extreme environment. In *Life in Ancient Ice* (J. D. Castello and S. O. Rogers, eds.), pp. 22–49. Princeton Univ., Princeton, New Jersey.

Priscu J. C., Kennicutt M.C. III, Bell R. E., et al. (2005b) Exploring subglacial Antarctic lake environments. *Eos Trans. AGU., 86(193),* 197.

Priscu J. C., Christner B. C., Foreman C. M., and Royston-Bishop G. (2006) Biological material in ice cores. In *Encyclopedia of Quaternary Sciences, Vol. 2* (S. A. Elias, ed.), pp. 1156–66. Elsevier, Oxford, United Kingdom.

Priscu J. C., Tulaczyk. S., Studinger M., Kennicutt M. C., Christner C. F., and Foreman C. M. (2008) Antarctic subglacial water: Origin, evolution and ecology. In *Polar Lakes and Rivers: Limnology of Arctic and Antarctic Aquatic Ecosystems* (J. L. Parry and W. F. Vincent, eds.), pp. 119–136. Oxford, New York.

Reynolds R. T., Squyres S. W., Colburn D. S., and McKay C. P. (1983) On the habitability of Europa, 1993. *Icarus, 56,* 246–254.

Rivkina E., Laurinavichysus K., and Gilichinsky D. A. (2005) Microbial life below the freezing point within permafrost. In *Life in Ancient Ice* (J. D. Castello and S. O. Rogers, eds.), Princeton Univ., Princeton, New Jersey.

Rosing M. T. (1999) ^{13}C-depleted carbon microparticles in >3700-Ma sea-floor sedimentary rocks from West Greenland. *Science, 283(5402),* 674–675.

Rothschild L. and Mancinelli R. L. (2001) Life in extreme environments. *Nature, 409,* 1092–1101.

Rummel J. and Billings L. (2004) Issues in planetary protection: Policy, protocol and implementation. *Space Policy, 20,* 49–54.

Russell M. J. (2003) Geochemistry: The importance of being alkaline. *Science, 302(5645),* 580–581.

Russell M. J. and Hall A. J. (2006) The onset and early evolution of life. In *Evolution of Early Earth's Atmosphere, Hydrosphere, and Biosphere-Constraints from Ore Deposits* (S. E. Kesler and H. Ohmoto, eds.), pp. 1–32. GSA Memoir 198.

Russell M. J., Hall A. J., Cairns-Smith A. G., and Braterman P. S. (1988) Submarine hot springs and the origin of life. *Nature, 336(6195),* 117.

Russell M. J., Hall A. J., and Turner D. (1989) In vitro growth of iron sulphide chimneys: Possible culture chambers for origin-of-life experiments. *Terra Nova, 1,* 238–241.

Sagan C. (1970) Life. In *Encyclopedia Britanica* (1970), Reprinted in *Encyclopedia Britannica 1998, Vol. 22,* 964–981.

Sagan C. (1974) The origin of life in a cosmic context. *Origins of Life, 5(3-4),* 497–505.

Sagan C. and Dermott S. F. (1982) The tide in the seas of Titan. *Nature, 300,* 731–733.

Sagan C. and Mullen G. (1972) Earth and Mars: Evolution of atmospheres and surface temperatures. *Science, 177(4043),* 52–56.

Sagan C., Thompson W. R., Carlson R., Gurnett D., and Hord C. (1993) A search for life on Earth from the Galileo spacecraft. *Nature, 365,* 715.

Sanchez R., Ferris J., and Orgel L. E. (1966) Conditions for purine synthesis: Did prebiotic synthesis occur at low temperatures? *Science, 153(3731),* 72–73.

Schidlowski M. (1992) Stable carbon isotopes: Possible clues to early life on Mars. *Adv. Space Res., 12(4),* 101–110.

Schidlowski M., Hayes J. M., and Kaplan I. R. (1983) Isotopic inferences of ancient biochemistries — Carbon, sulfur, hydrogen, and nitrogen. In *Earth's Earliest Biosphere: Its Origin and Evolution* (J. W. Schopf, ed.), pp. 149–186. Princeton Univ., Princeton, New Jersey.

Schoonen M. A. A., Xu Y., and Bebie J. (1999). Energetics and kinetics of the prebiotic synthesis of simple organic and amino acids with the FeS-H_2/FeS_2 redox couple as reductant. *Origins Life Evol. Biosph., 29,* 5–32.

Schopf J. W. (1999) *Cradle of Life: The Discovery of Earth's Earliest Fossils.* Princeton Univ., Princeton, New Jersey.

Schulte M. D., Blake D. F., Hoehler T. M., and McCollom T. (2006) Serpentinization and its implications for life on the early Earth and Mars. *Astrobiology, 6,* 364–376.

Schulze-Makuch D. and Grinspoon D. H. (2005) Biologically enhanced energy and carbon cycling on Titan? *Astrobiology, 5,* 560–567.

Siegel B. Z. (1979) Life in the calcium chloride environment of Don Juan Pond, Antarctica. *Nature, 280,* 828–829.

Smil V. (2001) *Enriching the Earth.* MIT, Cambridge, Massachusetts.

Smith B. (1999) *Infrared Spectral Interpretation: A Systematic Approach.* CRC, Boca Raton, Florida.

Sorensen K. B., Canfield D. E., Teske A. P., and Oren A. (2005) Community composition of a hypersaline endoevaporitic microbial mat. *Appl. Environ. Microbiol., 71(11),* 7352.

Sotin C., Grasset O., and Mocquet A. (2007) Mass-radius curve for extrasolar Earth-like planets and ocean planets. *Icarus, 191(1),* 337–351.

Spencer J. R. and Calvin W. M. (2002) Condensed O_2 on Europa and Callisto. *Astron. J., 124,* 3400–3403.

Spencer J. R., Pearl J. C., Segura M., Flasar F. M., Mamoutkine A., Romani P., Buratti B. J., Hendrix A. R., Spilker L. J., and Lopes R. M. C. (2006) Cassini encounters Enceladus: Background and the discovery of a south polar hot spot. *Science, 311,* 1401–1405.

Spohn T. and Schubert G. (2003) Oceans in the icy Galilean satellites of Jupiter. *Icarus, 161,* 456–467.

Squyres S. W., Reynolds R. T., Cassen P. M., and Peale S. J. (1983). Liquid water and active resurfacing on Europa. *Nature, 301,* 225–226.

Stein C. and Stein S. (1994) Constraints on hydrothermal heat flux through the oceanic lithosphere from global heat flow. *J. Geophys. Res., 99,* 3081–3095.

Stribling R. and Miller S. L. (1987) Energy yields for hydrogen cyanide and formaldehyde syntheses: The HCN and amino acid concentrations in the primitive ocean. *Origins Life Evol. Biosph., 17(3),* 261–273.

Summons R. E., Albrecht P., McDonald G., and Moldowan J. M. (2007) Molecular biosignatures. *Space Sci. Rev.,* DOI: 10.1007/s11214-007-9256-5.

Sundararamn P. and Dahl J. (1993) Depositional environment, thermal maturity and irradiation effects on porphyrin distribution: Alum Shale, Sweden. *Org. Geochem., 20(3),* 333–337.

Swallow A. J. (1963) *Radiation Chemistry of Organic Compounds.* Pergamon, New York.

Taylor F. W., Atreya S. K., Encrenaz T. H., Hunten D. M., Irwin P. G. J., and Owen T. C. (2004) Radiation effects on the surfaces of the Galilean satellites. In *Jupiter: The Planet, Satellites and Magnetosphere* (F. Bagenal et al., eds.), pp. 485–512. Cambridge Univ., New York.

Thomson R. E. and Delaney J. R. (2001) Evidence for a weakly stratified europan ocean sustained by seafloor heat flux. *J. Geophys. Res., 106(E6),* 12355–12365.

Todo Y., Kitazato H., Hashimoto J., and Gooday A. J. (2005) Simple foraminifera flourish at the ocean's deepest point. *Science, 307(5710),* 689–689.

Trinks H., Schroder W., and Biebricher C. K. (2005) Ice and the origin of life. *Origins Life Evol. Biosph., 35(5),* 429–445.

UNOOSA (United Nations Office for Outer Space Affairs) (2007) *Treaty on Principles Governing the Activities of States in the Exploration and Use of Outer Space, Including the Moon and Other Celestial Bodies.* Available online at www.unoosa.org/oosa/SpaceLaw/outerspt.html.

Vance S., Harnmeijer J., Kimura J., Hussmann H., deMartin B., and Brown J. M. (2007) Hydrothermal systems in small ocean planets. *Astrobiology, 7(6),* 987–1005.

Van Dover C. L. (2000) *The Ecology of Deep-Sea Hydrothermal Vents.* Princeton Univ., Princeton, New Jersey. 424 pp.

Van Mooy B. A. S., Rocap G., Fredricks H. F., Evans C. T., and Devol A. H. (2006) Sulfolipids dramatically decrease phosphorus demand by picocyanobacteria in oligotrophic marine environments. *Proc. Natl. Acad. Sci., 103(23),* 8607–8612.

Vincent W. F., Rae R., Laurion I., Howard-Williams C., and Priscu J. C. (1998) Transparency of Antarctic ice-covered lakes to solar UV radiation. *Limnol. Oceanogr., 43(4),* 618–624.

Wächtershäuser G. (1988) Pyrite formation, the first energy source for life: A hypothesis. *Syst. Appl. Microbiol., 10,* 207–210.

Wächtershäuser G. (1990) Evolution of the first metabolic cycles. *Proc. Natl. Acad. Sci. USA, 87,* 200–204.

Wächtershäuser G. (1998) Origin of life in an iron-sulfur world. In *The Molecular Origins of Life, Assembling Pieces of the Puzzle* (A. Brack, ed.), pp. 206–218. Cambridge Univ., New York.

Wächtershäuser G. (2000) Origin of life: Life as we don't know it. *Science, 289(5483),* 1307.

Wackett L. P., Dodge A. G., and Ellis L. B. M. (2004) Microbial genomics and the periodic table. *Appl. Environ. Microbiol., 70,* 647–655.

Waite J. H., Combi M. R., Ip W. H., Cravens T. E., McNutt R. L., Kasprzak W., Yelle R., Luhmann J., Niemann H., Gell D., et al. (2006) Cassini Ion and Neutral Mass Spectrometer: Enceladus plume composition and structure. *Science, 311(5766),* 1419–1422.

Walker J. C. G. (1977) *Evolution of the Atmosphere.* Macmillan, New York.

Westheimer F. H. (1987) Why nature chose phosphates. *Science, 235(4793),* 1173–1178.

Whitman W. B., Coleman D. C., and Wiebe W. J. (1998) Prokaryotes: The unseen majority. *Proc. Natl. Acad. Sci. USA, 95,* 6578–6583.

Wills C. and Bada J. (2001) *The Spark of Life: Darwin and the Primeval Soup.* Oxford Univ., New York.

Winn C. D., Karl D. M., and Massoth G. J. (1986) Microorganisms in deep-sea hydrothermal plumes. *Nature, 320(6064),* 744–746.

Ycas M. (1955) A note on the origin of life. *Proc. Natl. Acad. Sci. USA, 41(10),* 714–716.

Zahnle K., Schenk P. M., Levison H. F., and Dones L. (2003) Cratering rates in the outer solar system. *Icarus, 163,* 263–289.

Zahnle K., Alvarellos J. L., Dobrovolskis A., and Hamill P. (2008) Secondary and sesquinary craters on Europa. *Icarus, 194,* 660–674.

Zolotov M. Y. and Shock E. L. (2001) Composition and stability of salts on the surface of Europa and their oceanic origin. *J. Geophys. Res., 106,* 32815–32827.

Zolotov M. Y. and Shock E. L. (2003) Energy for biologic sulfate reduction in a hydrothermally formed ocean on Europa. *J. Geophys. Res., 108,* 5022.

Zolotov M. Y. and Shock E. L. (2004) A model for low-temperature biogeochemistry of sulfur, carbon, and iron on Europa. *J. Geophys. Res., 109,* 2003JE002194.

Zolotov M. Y., Krieg M. L., and Shock E. L. (2006) Chemistry of a primordial ocean on Europa. In *Lunar and Planetary Science XXXVII,* Abstract #1435. Lunar and Planetary Institute, Houston (CD-ROM).

Zubay G. (2000) *Origins of Life on the Earth and in the Cosmos.* Academic, New York.

Radar Sounding of Europa's Subsurface Properties and Processes: The View from Earth

Donald D. Blankenship and Duncan A. Young
University of Texas at Austin

William B. Moore
University of California at Los Angeles

John C. Moore
University of Lapland and University of Oulu

A primary objective of future Europa studies will be to characterize the distribution of shallow subsurface water as well as to identify any ice-ocean interface. Other objectives will be to understand the formation of surface and subsurface features associated with interchange processes between any ocean and the surface as well as regional and global heat flow variations. Orbital radar sounding, a now maturing technology, will be an essential tool for this work. We review the hypothesized processes that control the thermal, compositional, and structural properties, and therefore the dielectric character, of the subsurface of Europa's icy shell. We introduce fundamental concepts in radar sounding and then assess analog processes represented by, and sounded in, Earth's ice sheet. We use these Earth analog studies to define the radar imaging approach for Europa's subsurface that will be most useful for testing the hypotheses for the formation of major features.

1. INTRODUCTION

Europa is a hypothesized site of incipient habitability because of its potentially vast subsurface ocean (see chapter by Hand et al.). The presence of this water reservoir has been inferred indirectly from Europa's induced magnetic field (*Kivelson et al.,* 2000; *Hand and Chyba,* 2007; see also chapter by Khurana et al.) and tectonic mapping of its young surface (*Hoppa et al.,* 1999; *Pappalardo et al.,* 1999; see chapter by Kattenhorn and Hurford). Future space-based geodetic measurements of Europa's time varying gravity field would definitively demonstrate the existence of an ocean. However, understanding this ocean's coupling to its overlying crust — key for understanding Europa's astrobiologic potential — will require sounding Europa's third dimension.

Airborne ice-penetrating radar is now a mature tool in terrestrial studies of Earth's ice sheets (*Bingham and Siegert,* 2007), and orbital examples have been successfully deployed at Earth's Moon and Mars. Recent terrestrial examples include the University of Texas' High Capability Airborne Radar Sounder (HiCARS) (*Peters et al.,* 2005), the British Antarctic Survey's PASIN system (*Heliere et al.,* 2007), and the University of Kansas's IPR and CARDS systems (*Gogineni et al.,* 2001). Spaceborne demonstrations include NASA's Apollo 17 ALSE (*Porcello et al.,* 1974), JAXA's LRS system on the Kaguya lunar orbiter (*Ono et al.,* 2008), MARSIS onboard ESA's Mars Express (*Picardi et al.,* 2005), and SHARAD onboard NASA's Mars Reconnaissance Orbiter (*Seu et al.,* 2007). In this chapter we briefly review the target of observations, Europa's ice crust and the ocean that likely lies beneath; summarize the state of the art of radar-sounding systems; survey previous observations made by ice-penetrating radar at Earth; and examine the challenge of operating such a system at Europa.

The key to probing Europa's subsurface lies in the relatively large ratio (the loss tangent, τ) between the real and imaginary components of the electrical permittivity (ε) of water ice (i.e., the ice is a good insulator). Instead of losing energy in coupling with ions and electrically charged defects in the ice lattice structure within the medium, propagating electromagnetic waves can penetrate far into an ice column (*Gudmandsen,* 1971; *Petrenko and Whitworth,* 1999). For frequencies between 1 and 300 MHz, the loss tangent of pure water ice does not vary significantly for a given temperature (*Fujita et al.,* 1993; *MacGregor et al.,* 2007). However, with increasing temperatures and/or contamination, crystal defects and ions increase in number and mobility, and above eutectic temperatures water veins with high impurity content may start to play a role in increasing the conductivity of the ice, absorbing more of the radar wave energy.

Liquid water containing impurities (brine) is an effective conductor of electricity, and hence strongly dissipates electromagnetic energy. In addition, strong contrast in real permittivity at radio frequencies between pure water ice ($\varepsilon \sim 3.15$) and liquid water [$\varepsilon \sim 80$ (*Peters et al.,* 2005; *Campbell,* 2002)] lead to a large dielectric impedance contrast that typically results in a highly reflective interface, with a reflection of half or more the incident power, compared to a thousandth of the incident power reflected by silicate rock. It is this strong contrast that enables exploration for water within Europa's crust.

Our foundation for this chapter will be the hypothesized processes that control the thermal, compositional, and structural (TCS) properties, and therefore the dielectric character, of the subsurface of Europa's icy shell. We will begin by reviewing the processes that may affect TCS properties in Europa's icy shell. This includes thermochemical and mechanical processes that have been linked to observable features at the surface of Europa. Sounding of the TCS properties of the shell may thus be used to test hypotheses for the origin of these features and the evolution of the icy shell. Many of these processes have direct analogs in terrestrial ice sheets. We then introduce basic concepts and methods used in radar sounding of ice, including procedures used to improve range, power, and azimuth resolution. We will then review the spectrum of analog processes and TCS properties represented by Earth's icy bodies including both Arctic and Antarctic ice sheets, ice shelves, and valley glaciers. There will be few complete analogs over the full TCS space, but there are more analog examples than one might imagine for significant portions of this space (e.g., bottom crevasses, marine ice shelf/subglacial lake accretion, surging polythermal glaciers). Our final contribution will be to use the Earth analog studies to describe the radar imaging approach for Europa's subsurface that will be most useful for supporting/refuting the hypotheses for the formation of major surface/subsurface features as well as for "pure" exploration of Europa's icy shell and its interface with the underlying ocean.

2. STRUCTURAL, COMPOSITIONAL, AND THERMAL PROPERTIES OF EUROPA'S ICY SHELL

Europa's icy shell has a geologically young surface that records less than about 100 m.y. of impacts (*Zahnle et al.,* 2003; chapter by Beirhaus et al.), and which shows abundant evidence of modification by endogenic processes. Particularly evident are fractures, primarily strike-slip with some extensional features (bands) (see chapters by Kattenhorn and Hurford, and by Prockter and Patterson). There are a small number of impact craters ranging up to 50 km in diameter (see chapter by Schenk and Turtle). Finally, there are a number of features collectively referred to as lenticulae, but plausibly including chaos regions of all sizes, that have been suggested to be due to thermal modification of the near surface by convective (*Pappalardo et al.,* 1998, *Nimmo and Manga,* 2002) or melt-through (*Greenberg et al.,* 1999) processes (see chapter by Collins and Nimmo). There are a number of candidate processes for producing dielectric horizons within these structures, most of which have good terrestrial analogs. This section will describe what we might expect a radar sounder to be able to reveal about these processes and what sorts of structures we would expect to see. Ultimately, we seek to test hypotheses related to the origin of these structures and their implications for the evolution of Europa's ocean and icy shell.

Four classes of processes are considered: marine, convective, tectonic, and impact.

2.1. Marine Processes

A "marine" europan crust could be formed by processes similar to those for ice that accretes beneath the large ice shelves of Antarctica where frazil ice crystals form directly in the ocean water (*Moore et al.,* 1994). This model is characterized by slow accretion (freezing) or ablation (melting) on the lower side of the icy crust (*Greenberg et al.,* 1999). Impurities present in the ocean tend to be rejected from the ice lattice during the slow freezing process. Temperature gradients for a "marine" ice model are primarily a function of ice thickness and the temperature/depth profile is described by a simple diffusion equation for a conducting ice layer (*Chyba et al.,* 1998). The low-temperature gradients at any ice water interface, combined with impurity rejection from accreted ice, would likely lead to significant dielectric horizons resulting from contrasts in ice crystal fabric and composition. Similarly, the melt-through hypothesis for the formation of lenticulae and chaos on Europa's surface implies that ice will accrete beneath the feature after it forms. This process will result in a sharp boundary between old ice (or rapidly frozen surface ice) and the deeper accreted ice. The amount of accreted ice would be directly related to the time since melt-through occurred and could be compared with the amount expected based on the surface age.

Testing these hypotheses will require measuring the depth of interfaces to a resolution of a few hundred meters, and horizontal resolutions of a fraction of any lid thickness, i.e., a kilometer or so.

2.2. Convective Processes

The thermal structure of the shell (apart from local heat sources) is set by the transport of heat from the interior (see chapter by Barr and Showman). Regardless of the properties of the shell or the overall mechanism of heat transport, the uppermost several kilometers is thermally conductive, cold, and stiff. The thickness of this conductive "lid" is set by the total amount of heat that must be transported and thus a measurement of the thickness of the cold and brittle part of the shell is a powerful constraint on the heat production in the interior. This uppermost part of the shell has experienced continuous input of impurities to the surface that are gradually mixed downward through impact gardening.

The lower, convecting part of the shell (if it exists) is likely to be much cleaner, because regions with impurities should have experienced melting at some point during convective circulation (when the material was brought near the base of the shell) and the melt will segregate downward efficiently, extracting the impurities from the shell (*Pappalardo and Barr,* 2004). Convective instabilities also result

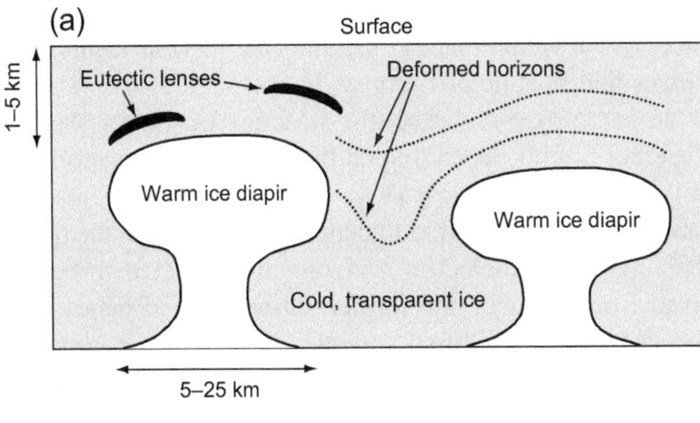

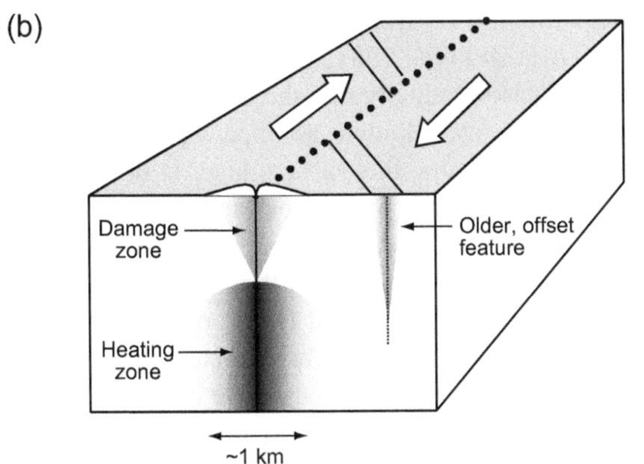

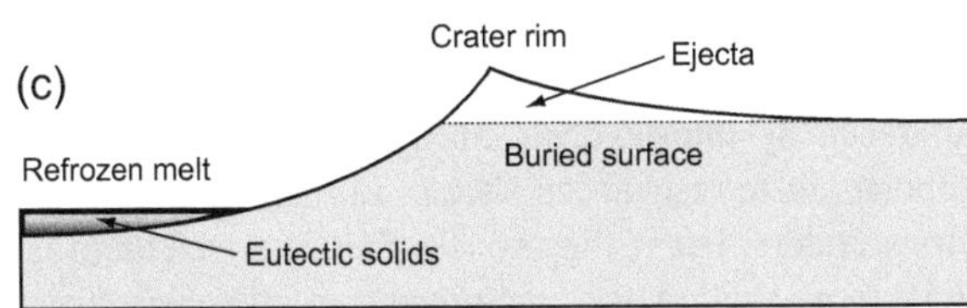

Fig. 1. Illustrations of some of the processes leading to thermal, compositional, and structural horizons in Europa's icy shell. **(a)** Convective diapirs causing thermal perturbations and potentially melting in the overlying rigid ice. **(b)** Tectonic faulting driven by tidal stresses resulting in fault damage and frictional heating. **(c)** An impact crater with a refrozen melt pool at its center and the surrounding ejecta blanket.

in thermal variations in the outer part of the shell (the rigid lid in convective parlance) that may be measured (Fig. 1a). The scale of these instabilites is on the order of hundreds of kilometers, independent of the thickness of the shell, and the temperature variations will be a few tens of Kelvin (*Nimmo and Manga,* 2002). Convective structures would arise through ductile deformation of other horizons (impact or tectonic), or through the effect of these thermal anomalies, which may lead to local regions or highly dissipative (or conversely, highly transparent) ice or even eutectic melting.

A number of features at the surface of Europa (lenticulae and chaos), with scales ranging from 1 km to hundreds of kilometers have been associated with thermal perturbations of the shell (convective or melt-through as described above) (*Pappalardo et al.,* 1998; *Greenberg et al.,* 1999; *Nimmo and Manga,* 2002). When warm, relatively pure ice diapirs from the interior approach the surface, they may be far from the pure-ice melting point, but may be above the eutectic point of many substances, and may therefore create regions of melting within the rigid shell above them as the temperature increases above the flattening diapir (Fig. 1a). The dielectric horizon associated with these melt regions would provide a good measurement of the thickness of the conductive layer.

Given that the cutoff length scale is a key discriminator between models for the formation of lenticulae and chaos (*Pappalardo et al.,* 1998; see chapter by Collins and Nimmo), high horizontal resolution (a few hundred meters) is required to avoid scale-related biases. The ability to sound through regions of rough large-scale terrain will also be essential. Detection of water lenses would require a vertical resolution of at least a few tens of meters.

2.3. Tectonic Processes

Europa's surface is extensively fractured and modified by tectonic processes (see chapters by Kattenhorn and Hurford and by Prockter and Patterson). Many of the offsets are horizontal, but some extensional features (e.g., bands) are seen. Large-scale arcuate troughs form a globally organized pattern (*Schenk et al.,* 2008) that may relate to true polar wander. Limited evidence for contractional tectonism has been inferred. This is a unique tectonic regime in the solar system, and the processes controlling the distribution of strain in Europa's icy shell are uncertain. The ubiquitous double-ridges are commonly associated with strike-slip offset, and the origin of the topography of these features remains debated.

Tectonic processes may result in subsurface structures characterized by rapid freezing of water injected into fracture zones (*Greenberg et al.,* 1998). The large temperature gradients implied by this process could lead to ice with properties similar to terrestrial sea ice (*Weeks et al.,* 1990), although water reaching the surface would be unlike sea ice because of very rapid boiling. Because fractures that extend through the shell require a thin shell, this model would also be characterized by a temperature/depth profile for a simple thermally conducting ice layer.

Other tectonic structures could range from subhorizontal extensional fractures to near-vertical strike-slip features, and will produce dielectric horizons associated primarily with the faulting process itself through formation of pervasively fractured ice, and zones of deformational melt, injection of water, or preferred orientation of crystalline fabric. Some faults may show dielectric contrasts by local alteration including fluid inclusions or simply by juxtaposition of dissimilar regions including offset of preexisting structures through motion on the fault.

While faulting may create dielectric contrasts by relative motion juxtaposing different terrains on either side of the fault, the tidal stress environment of Europa may lead to continuously active faults and significant thermal modification of the ice around the fault (Fig. 1b). Such processes may lead to melting and potentially to communication with the ocean from above. In one model for the triple bands (*Nimmo and Gaidos,* 2002), the double-ridge structure is generated by thermal buoyancy due to frictional heating along the fault. The dielectric structure implied by this model would include a broad region of warm, dissipative ice extending several kilometers on either side of the fault. At the fault surface itself, melt may be generated, which will flow downward. This structure may vary along the fault according to the orientation of the fault relative to the tidal stress field.

There are a number of outstanding questions regarding these tectonic features. A measurement of their depth extent and association with thermal anomalies would strongly constrain models of their origins. In particular, correlation of subsurface structure with surface properties (length, position in the stratigraphic sequence, height and width of the ridges) will test hypotheses for the mechanisms that form the fractures and support the ridges. The observation of melt along the fracture would make these features highly desirable targets for future *in situ* missions.

Extensional structures observed on Europa (bands) have been likened to oceanic spreading centers on Earth (see chapter by Prockter and Patterson). The bookshelf normal faulting known from the Earth case is plausibly responsible for the observed surface morphology of the europan features. These faults are shallow, with depths comparable to their spacing (a few kilometers) and offsets several times less than this (a few hundred meters). Any preexisting layering should be broken up and tilted within each block, but if the analogy with spreading centers is accurate, the material in the band is newly supplied from below and may not have any preexisting horizons. In either case, material must be supplied from below and may have a distinct dielectric signature. The origin of the material in the bands may thus be constrained by sounding the subsurface structure. Bands and ridges typically have length scales of a few kilometers. Horizontal resolutions a factor of 10 higher than this would be required to fully diagnose processes. For extensional structures, the ability to image structures sloping more than a few degrees is also necessary. Again, tens of meters of vertical resolution would be required to image near-surface water associated with these structures.

2.4. Impact Processes

The impact process represents a profound disturbance of the local structure of the shell (see chapter by Schenk and Turtle). Around the impact site, the ice is fractured and heated, and some melt is generated. The surface around the impact is buried to varying degrees with a blanket of ejecta. Finally, the relaxation of the crater creates a zone of tectonism that can include both radial and circumferential faulting. These processes all create subsurface structures that may be sounded electromagnetically. An outstanding mystery on Europa is the process by which old craters are erased from the surface. It may be possible to find the subsurface signature of impacts that are no longer evident at the surface, which would place constraints on the resurfacing processes that operate at Europa.

Impact processes affect the structure of the icy shell to different extents depending on the size of the impactor, and it is possible that Europa's subsurface records events that have penetrated the entire thickness of the shell (at the time). Three types of dielectric horizons are expected to be derived from impact: the former surface buried beneath an ejecta blanket, solidified eutectics in the impact structure itself, and impact-related fractures (e.g., a ring graben).

The surface gradually accumulates contaminants from Jupiter orbit, mostly from Io, but also from micrometeorite impact (*Eviatar et al.,* 1981; *McEwen,* 1986). These micrometeorites also gradually mix the uppermost few meters of the ice in a process known as impact gardening. When a larger impact occurs, this surface layer is buried by many meters of ice derived from deeper and presumably cleaner regions (Fig. 1c). Thus a dielectric discontinuity is created between the ejecta blanket and the old surface. This horizon will be nearly horizontal (some postimpact deformation is expected as the crater relaxes) and should be sharp, because the contaminants accumulate most strongly in the top tens of centimeters. The horizon should approach the surface as the distance from the crater increases.

Multiple, overlapping ejecta blankets may create a complexly layered structure, with several subhorizontal horizons at different depths. This type of structure is similar to that produced by interlayering of snow and ice with volcanic deposits in terrestrial ice sheets. Europa has very few obvious craters, but if the process that removes them is surficial (e.g., shallow tectonic reworking), the deep structures created by previous impacts may be preserved. On the other hand, if the resurfacing process involves the entire shell thickness (melt-through), then such structures will be erased as well. Subsurface sounding will make it possible to test hypotheses for the processes that govern resurfacing on Europa.

In the center of the impact crater, melt will be produced, which will pool at the base of the crater and then freeze out rapidly (Fig. 1c). The melt pool will contain the impurities initially in the ice that was molten during the impact, but as it freezes, the impurities in the pool will concentrate, and various eutectics will be reached depending on the initial composition. We can thus expect a vertically and concentrically layered structure of ice and eutectic compositions with a sharp transition (a few centimeters) between relatively pure ice and the first eutectic solid. Subsequent relaxation of the crater may deform this structure vertically, creating a domed, concentrically zoned structure. A central peak, if formed, would result in an initially ring-shaped melt pool and subsequent rings of eutectic zonation.

Radial fractures near the crater accommodate relatively little motion, but the possibility of melt injection along these fractures (as seen in the Sudbury and Vredefort structures on Earth) may create very strong dielectric contrasts as described above. These structures will be nearly vertical and may extend downward beneath the impact crater itself as well as radially away from the center. The injected dikes may be up to several meters wide near the impact structure and will gradually thin with distance, depending on the size of the impact. The melt will freeze quickly once it has traveled away from the strongly heated region near the crater and may contain enhanced levels of impurities because some water may have been lost during the impact event itself. If sufficient melt is produced, the circumferential fractures may also become melt filled, or melt may be produced by friction along the circumferential fractures during the rapid collapse of the transient crater (analogous to the pseudotachlyte zones associated with the Sudbury structure).

Finally, tectonics is also driven by the impact process. As mentioned above, the large-scale concentric fractures observed on several icy bodies may offset preexisting horizons by hundreds of meters. The scale of motion on these features can be large, up to several kilometers on the largest features (*Moore et al.,* 1998). Although subvertical near the surface, these features are predicted to flatten with depth, and may preferentially sole out into preexisting horizons, a hypothesis that may be tested by sounding.

Vertical resolutions on the scales of a few tens to hundreds of meters will be required to identify ejecta blankets. Detection of at least the edges of steep interfaces would aid in the identification of radial dikes, buried crater walls, and circumferential fractures.

3. FUNDAMENTALS OF RADAR SOUNDING

Radar systems are active remote sensing systems that operate by transmitting an electromagnetic pulse at a given time (t_t) and receiving an echo at a later time (t_r). For a monostatic system, where the transmitting and receiving elements are collocated, the range (r) to the target will be

$$r = c\frac{(t_r - t_t)}{2}$$

where c is the speed of light in the material between the transmitter and the target (Fig. 2). For sounding through ice, the range will be

$$r = c\,\frac{\left((t_{surface} - t_t) + (t_{bed} - t_{surface})/n_{ice}\right)}{2}$$

where $t_{surface}$ is the time of the surface echo, t_{bed} is the time of the basal echo, and n_{ice} is the refractive index of ice. The reflection radar equation for power received for a specular interface, monostatic system, and transmission through free space is

$$P_r = P_t : \left(\frac{\lambda}{4\pi}\right)^2 : \frac{G^2R_{01}}{(2r)^2} \qquad (1)$$

where P_r is the power received, P_t is the power transmitted, λ is the wavelength of the carrier wave, G is the gain (the "focusing") due to the antenna, and R_{01} is the effective reflection coefficient of the surface.

Electromagnetic radiation interacts with surfaces in a number of ways; it can reflect off of a surface flat at wavelength scales, it can scatter off wavelength scale diffractors, and it can be transmitted through the surface. While the scattering radar equation uses an r^4 term [e.g., used for side-looking SAR imaging and observatory-based planetary radar systems (*Campbell,* 2002; *Stiles et al.,* 2006)] with which many may be familiar, the surface return from a nadir-pointed system can be treated as that of a mirrored transmitter, as in the above equation. The implication of this equation is that the inverse square law will reduce the received power, while the gain of the antenna will boost it.

This process is repeated as fast as the travel time will allow as the platform moves along its track. For sounding systems, the resulting waveform records (after processing and coding of intensity) are used as columns in depth profiles known as radargrams (e.g., Figs. 2 and 3). Important parameters in detecting and interpreting these echoes include the system's *signal to noise ratio* (SNR, which represents signal quality and dictates the depth of penetration), its *bandwidth* (which controls the radar's range resolution and hence the vertical resolution of a radargram), its *beam pattern* (which determines the radar's azimuthal resolution and hence the horizontal resolution), and its *center frequency* (which determines its sensitivity to the size of structures and the depth of penetration).

3.1. Signal Quality and Vertical Range Resolution

The first two parameters, SNR and bandwidth, are linked. The SNR is the comparison between the power of the echo of interest and the power from all other sources. Given the large dynamic range of radar receivers, and the geometrically controlled exponential decay rates of radar energy, it is convenient to use a logarithmic scale for expressing power ratios. We use deciBels [i.e., dB; $10\log_{10}\left(\frac{power1}{power2}\right)$] to discuss the ratio of two powers, and similarly use dBm to compare absolute power to a 1 mW reference.

Bandwidth (BW), measured in Hertz, is the range of frequencies over which the system responds. The vertical range resolution for a nadir-pointed system that is implied by a given bandwidth is

$$\delta r = \frac{c}{2BWn_{ice}}$$

Bandwidth is often much lower than the system's center frequency. Higher bandwidth means that the system is capable of resolving smaller time intervals, and hence, smaller range differences. A simple way of generating a received echo with wide inherent bandwidth is to transmit a very short pulse. However, this approach limits the total energy a pulse can carry, and thus reduces the likelihood of detection above electronic noise in the receiver. A long pulse with

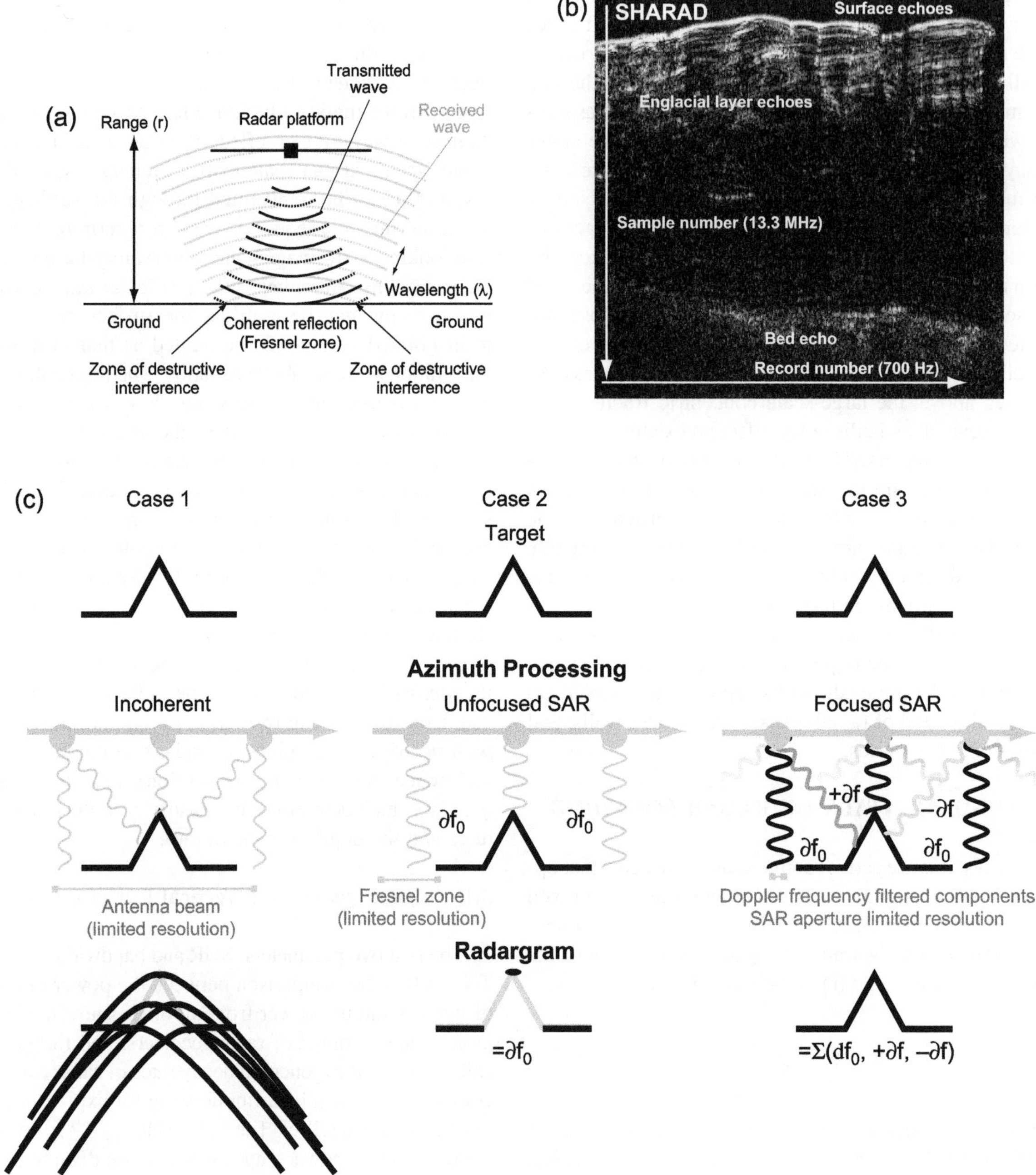

Fig. 2. **(a)** Illustration of the typical geometry of an air- or spaceborne radar altimeter or sounder, illustrating the concept of the Fresnel zone — a region over which adjacent reflecting spots on a flat surface reenforce. **(b)** Example radargram from the SHARAD instrument penetrating Mars' north polar layered deposits, from MRO orbit 1957. Each column of the radargram represents a separate time series sampling the reflected radio pulse. **(c)** Schematic view of the effects of different along-track processing methods on radar sounding. Case 1 shows incoherent processing, where the phase of the signal is ignored and only the total power is considered. Case 2 shows unfocused SAR, which is effectively filtered so that only reflections with zero Doppler shift are passed. Case 3 shows focused SAR, where effectively beams of different Doppler frequency are separated out and recombined at high angular resolution.

narrow bandwidth can contain more total energy and will be more easily detected, but will poorly resolve the target range.

A solution to this conflict has been to use "chirped" pulses, where the frequency is swept across the desired bandwidth with specific phase, over an extended pulse length. The received extended echo is "pulse compressed" through convolution with the original signal, to return the resolution implied by the bandwidth. A side effect of pulse compression are low-amplitude "sidelobes" preceding and following the real echo. In the example of the SHAllow RADar (SHARAD) system on the Mars Reconnaissance Orbiter (*Seu et al.*, 2007) (Fig. 2), pulse compression yields a 27.5-dB power enhancement over a conventional pulse

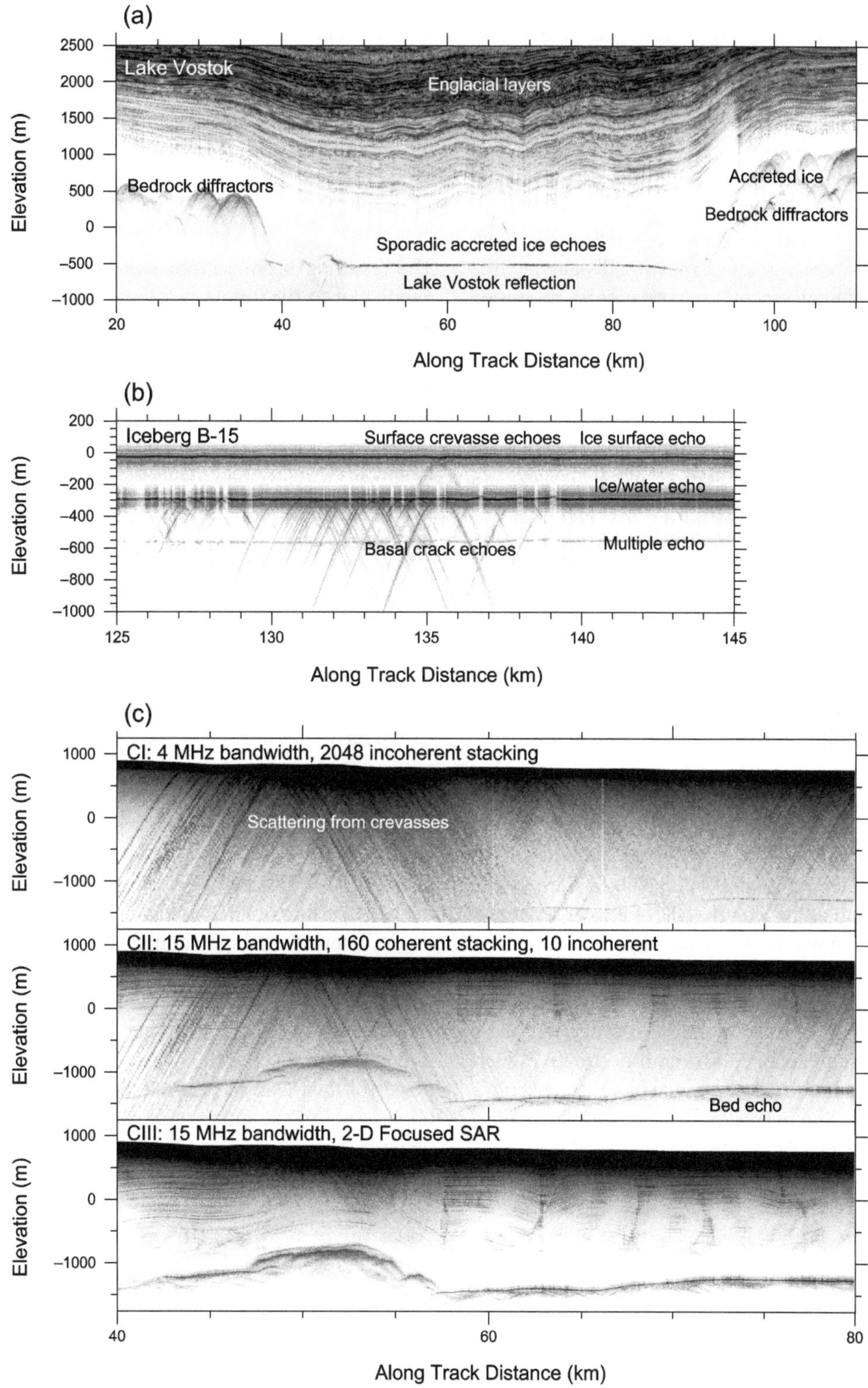

Fig. 3. Three examples of terrestrial radar sounding, all using a 60-MHz radar sounder. **(a)** Example of a radargram from the Lake Vostok airgeophysical survey. The radar is incoherent with 4-MHz bandwidth. The radar data has been converted to absolute magnitudes and logarithmically scaled. **(b)** Example of a radargram from a 15-MHz bandwidth radar across an iceberg (*Peters et al.,* 2007b). These data are incoherently processed. Point corner reflectors (bottom cracks) appear as clearly identifiable hyperbolas as each point is seen by multiple successive pulses. **(c)** Contrasting sounding radars of different bandwidth and processing methods, taken over the onset of Kamb Ice Stream; in CI we show an older, low-bandwidth radar that did not record the phase of waveforms, and thus cannot screen out surface clutter through post-acquisition processing (*Blankenship et al.,* 2001). In CII, we compare a more modern high-bandwidth radar (*Peters et al.,* 2005), with a 3.5-m unfocused aperture length, which reduces some clutter and reveals new bed topography. CIII shows the same data processed using two-dimensional focused SAR and a 2-km aperture length (*Peters et al.,* 2007a). Much more detail in the bed and layers is revealed, while along-track clutter is eliminated.

of the same bandwidth. This enhancement is important, given the –95 dB geometric power loss due to the 330-km-high orbit.

A second method of increasing the signal quality is to increase the pulse repetition frequency (PRF), and add the collected waveforms. The useful PRF is limited by the motion of the radar system with respect to the ground target, and the time delays required to accommodate both the targeted span of ranges and the distance between the radar platform and the target. For a fast moving, distant satellite like the Mars Reconnaissance Orbiter, the pulse repetition frequency is limited to 700 Hz by the pulse travel time (*Seu et al.,* 2007). In the example of the University of Texas' HiCARS radar sounder, mounted on a slow-moving, low-flying aircraft, the PRF used is 6408 Hz (*Peters et al.,* 2005).

Returning signals must be digitized at a rate that correctly samples the returning waveform without aliasing, i.e., at least twice the signal frequency. For VHF frequencies this can be technically challenging. One approach is to use multiple digitizers operating out of phase of one another. As the carrier frequency is often much higher than the bandwidth, a more sophisticated approach is to recognize that much of frequency content of the signal is redundant. Mixing a generated monochromatic waveform back into the signal allows *downconversion* to a manageable frequency range that is only slightly larger than the bandwidth.

3.2. Clutter Reduction and Azimuthal Resolution

The lateral resolution of the system is in part determined by the nature of the target, and the design of the system. For surfaces smooth at the length scale of the carrier wavelength, the resolution will be determined by the *Fresnel zone* (D_F). The Frensel zone is the spot below the radar over which reflected echoes have similar phase, and thus reenforce. As one moves away from the nadir point, the phase of echoes will systematically shift, until destructive interference occurs (see Fig. 2). So far as the radar is concerned, the Frensel zone is illuminated simultaneously. The Fresnel zone for the single layer case can be calculated by

$$D_F = 2\sqrt{\left(r + \frac{\lambda}{4}\right)^2 - r^2} :\sim: 2\sqrt{\frac{r\lambda}{2}}$$

For rougher surfaces, phase coherency will not be preserved. Instead, a weaker echo will be returned from a larger region defined by the transmitted bandwidth of the system. The equation for the single layer *pulse-limited footprint* (D_P) is

$$D_P = 2\sqrt{\left(r + \frac{c}{BW}\right)^2 - r^2}$$

For smooth horizontal interfaces, the earliest echo will represent the nadir point. However, regions of more complexity (and of interest) will suffer from the indiscriminate wide beams of physical radio antennae. Echoes of the same range but different location overlap. This phenomenon is called *clutter* (see Fig. 2a). To reduce clutter from surface features, the radar system needs to narrow its beam size.

For a simple dipole antenna of length L_t, the half-power beam width in radians (*Campbell,* 2002) is given by

$$\theta_{antenna} = \frac{0.88\lambda}{L_t}$$

The resulting beam pattern will have enhanced gain perpendicular to the dipole, while having nulls aligned parallel to the dipole. It can be seen that the beam width can be reduced, and hence the resolution improved, by decreasing the wavelength to antenna length ratio. For wavelengths able to deeply penetrate ice (ranging from 1 to 300 m), this formula implies a large antenna to obtain focusing. Such antennas are usually technically precluded from free-flying systems.

Because of the broad beam pattern obtained from a typical radar antenna without further processing, targets will be detected in multiple records. Point scatterers are smeared out into hyperboloids as the radar passes the target with changing range (Fig. 2c). Energy from such scatterers at the surface can obscure reflections of interest deeper in the ice. To reduce this clutter, the radar beamwidth along-track must be reduced. The motion of the platform allows a means of doing so.

Reducing the angular width of the radar beam, and thus increasing the gain, is done through constructive interference of nadir signals and destructive interference of off-nadir signals. This can either be done physically, by using the real aperture of an antenna, or electronically, by recording waveforms at a high enough sampling rate so as to preserve phase information. Such a radar is known as *coherent.* As the radar platform moves, it can "stack" independent waveforms digitally, a technique known as synthetic aperture radar (SAR). The SAR method, while potentially more powerful than using a real antenna for radar wavelengths, only works in the along-track dimension for a single antenna system.

Unfocused SAR involves coherently adding consecutive waveforms without adjusting their phase (i.e., *Peters et al.,* 2005; *Seu et al.,* 2007). To ensure that only coherently resolved echoes from a contiguous Fresnel zone are being illuminated, the maximum along-track integration distance (the aperture, L) is the same length as the Fresnel zone. The operation is simple enough to be conducted in real time, prior to recording. The result for flat surfaces will be elimination of echoes fore and aft of the nadir spot, and enhancement of the power of the nadir echo that is directly proportional to the number of pulses received over the aperture (see Fig. 2b). The along-track beam width in radians ϕ_x is

$$\phi_x = \frac{\lambda}{2L}$$

A disadvantage to this approach is that echoes from sloping surfaces are canceled out.

With ancillary data regarding platform position and velocity, more sophisticated processing can be performed. Instead of filtering out signals with off-nadir origins, we can use *focused SAR* (*Legarsky et al.,* 2001; *Peters et al.,* 2007a; *Heliere et al.,* 2007) by using the phase history of the returns to establish the target's azimuth in each record (see Fig. 2c). Because of limited bandwidth, the phase of the echo near the nadir will be far more sensitive to basal topography than the apparent range measured from the echo delay. In one-dimensional focused SAR (*Legarsky et al.,* 2001; *Peters et al.,* 2007a; *Heliere et al.,* 2007), a matched filter processor convolves a model for possible phase-azimuth relationships for a given range. However, once the range to a target begins to change significantly, this convolution will fail, limiting the length of the aperture, effectively, to the diameter of the pulse limited footprint. Two-dimensional focused SAR (*Peters et al.,* 2007a) involves a correlation on both phase and range data, allowing much larger apertures to be synthesized, and thus narrower effective beam widths. In theory, the aperture can be as long as the target is in sight of the radar, although in practice it is limited by waveform duration, storage capacity for full phase waveforms and significant processing capacity. Longer apertures also allow steeper specular surfaces to be detected, as echos from shallower angles are retrieved (see CIII in Fig. 3c).

To prevent aliasing, waveforms need to be recorded for SAR processing at more than twice the maximum rate of observed phase change (the Doppler bandwidth) (e.g., *Franceschetti and Lanari,* 1999). The Doppler bandwidth implies the minimum data rate required to achieve the maximum resolution of the system. For an airborne system the Doppler bandwidth is calculated as

$$f_{Doppler} = \frac{2v}{\lambda}$$

where v is the platform velocity.

For a spacecraft in a circular orbit around a slowly rotating spherical body, it would be

$$f_{Doppler} = 2\frac{R}{\lambda\sqrt{\frac{(R+h)^3}{GM}}} \tag{2}$$

where R is the radius of the body, h is the height above the surface, G is the universal gravitational constant of 6.673 × 10^{-11} m^3 kg^1 s^2, and M is the mass of the body. This formula assumes that the antenna beam has little along-track directionality.

3.3. Clutter and the Center Frequency

We have shown that much of the along-track clutter can be removed by using a coherent radar. However, cross-track clutter produced by scattering is difficult to address, especially for orbiting systems. Interpreting sounding radargrams under these conditions requires a good understanding of the scattering processes, as well as being of intrinsic interest (see chapter by Moore et al.). Clutter can be divided into two classes: *random* clutter, which is derived from subresolution, statistically homogenous structures, and *deterministic* clutter, which comes from discrete sources that are resolvable in the radar bandwidth. A second impact of random clutter processes is an attenuation of throughgoing signals, as transmitted energy is scattered away from the receiving antenna.

Random clutter can be separated into three components: quasispecular scattering, diffuse scattering, and volume scattering. In quasispecular scattering, the incoming signal reflects off of facets whose slope variation is small. Therefore there will be coherent summation of adjacent echoes (similar to the case of SAR focusing above), which imparts a strong directionality on the backscattered echo. For wavelength-scale roughness that has a horizontal length scale much smaller than a resolution cell, phase will be uniform in the recorded echoes, and neither constructive or destructive interference will dominate. The backscattered power will be less than the specular return, but the signal will be weakly dependent on across-track angle. *Campbell and Shepard* (2003) point out that for long wavelength systems with high bandwidth, this assumption may break down as the range-defined resolution cells shrink in horizontal size away from nadir.

In the case of volume scattering, point diffractors with dimensions near a wavelength within the sounded medium contribute to the backscatter, again with a low angular dependence. For centimeter-scale wavelengths, this effect can dominate backscatter of icy moons (*Ostro and Shoemaker,* 1990; *Ostro et al.,* 2006; *Wye et al.,* 2007; see chapter by Moore et al.) and even provide apparent reflection coefficients greater than 1, as energy is being returned from a volume instead of a surface.

Resolvable features off nadir introduce deterministic clutter that can complicate the interpretation of subsurface echoes. On the Mars Advanced Radar for Subsurface and Ionospheric Sounding (MARSIS) (*Picardi et al.,* 2005) in orbit around Mars, a receive-only monopole oriented perpendicular to the main dipole was intended to preferentially detect off-nadir echoes, thus allowing cancellation. Low signal-to-noise precluded use of this antenna (*Plaut et al.,* 2007), and indeed it may not have successfully deployed (*Adams and Mobrem,* 2006). However, MARSIS has demonstrated that, if high-resolution digital elevation models (DEMs) of the surface are available, radar models can be constructed that simulate most relevent surface echoes (*Nouvel et al.,* 2004; *Picardi et al.,* 2005). Comparison of such synthetic radargrams with the real radargram allows true subsurface echoes to be isolated and identified. Conversely, echoes can be picked out of a radargram, and mapped back onto a DEM (*Holt et al.,* 2006a).

Most sounder systems cannot distinguish between echoes arriving from the left or right sides of the track. Imaging radar systems [for example, the radars on Magellan (*Saunders et al.,* 1992) and Cassini (*Elachi et al.,* 2005)] resolve this issue by using a high enough frequency so that they can ensure that the entire beam points to one side of nadir. This approach implicitly assumes that only one interface is observable, which makes them inappropriate for airborne or orbital deep sounding. It has been suggested that the interference patterns between the surface and bed signals can be used to unwrap ice thickness in this geometry (*Jezek et al.*, 2006); however, in this case valuable internal structures must be ignored. Side-looking geometries have been demonstrated for groundbased systems, although significant coverage has not been obtained (*Musil and Doake,* 1987; *Gogineni et al.,* 2007).

Ice-penetrating radar systems have traditionally operated in two bands: HF (3–30 MHz; 10–100 m wavelength in free space) and VHF (30–300 MHz; 1–10 m wavelength in free space). The UHF band (300 MHz to 3 GHz; 0.1–1 m) has typically been restricted to shallow ice sounding (e.g., *Spikes et al.,* 2004), although it is now being investigated for Earth orbital ice sounding (*Jezek et al.,* 2006; *Heliere et al.,* 2007), because of limited ionospheric distortion and regulatory availability. As bandwidth is restricted by the carrier frequency, VHF systems have a superior range and azimuth resolution (tens of meters in ice) to HF systems (hundreds of meters). Random clutter is more of an issue for VHF systems, as by Rayleigh's law for structures much smaller than a wavelength, the scattered power intensity increases by the fourth power of the frequency, and must be considered in planetary radar design. Conversely, the difficulty in fielding antennas with high lateral directionality for HF systems means that deterministic clutter will be unavoidable, especially for orbital systems.

A good example of these tradeoffs comes from recent observations of Mars' ice-rich north polar layered deposits (NPLD) (*Phillips et al.,* 2008). The SHARAD system, operating at 20 MHz with 10 MHz of bandwidth, provides an excellent view of the internal layer structure of the upper portion of the NPLD, but cannot penetrate a rough, dust-rich basal unit; the MARSIS system, operating at either 3, 4, or 5 MHz with 1 MHz of bandwidth, poorly resolves internal layering, but retrieves a strong echo beneath the basal unit. A multifrequency approach will be optimal for orbital radar sounding of Europa's icy shell.

4. RADAR OBSERVATIONS OF THERMAL/COMPOSITIONAL HORIZONS AND STRUCTURES IN EARTH'S ICE SHEETS

The fact that Earth's ice sheets are soundable by radar was discovered accidentally during HF ionospheric observations at the United Kingdom's Halley Antarctic station (*Evans,* 1961), because of destructive interference with the floating ice shelf's basal echo, and the failure of UHF aircraft radar altimeters over deep ice (*Waite and Schmidt,* 1962). This property has been subsequently exploited for investigating ice sheet thickness and internal properties (see reviews by *Bingham and Siegert,* 2007; *Dowdeswell and Evans,* 2004). Below we provide an overview of a range of terrestrial targets sounded by radar sounders. The most obvious analog for Europa subsurface targets are subglacial and englacial water bodies and their associated thermal and compositional horizons and structures.

4.1. Flat Subglacial Interfaces

Subglacial lakes in Antarctica were first revealed by their effect on the overlying surface, as the overriding ice flattened through decoupling with the bed, causing landmarks that could be exploited by airborne navigators (*Robin et al.,* 1977). Later airborne surveying revealed that bright continuous basal reflectors with a slope opposing that of the surface (see review by *Siegert,* 2005, and references therein). The most famous example is Lake Vostok (see Fig. 3a).

If the base of the ice is at the pressure melting point, water is present. For a static water-ice interface, the slope at the base of the ice must balance the variation in the the weight of overlying ice, i.e., the lake surface is in hydrostatic equilibrium. This slope criterion can be expressed as

$$\frac{d\left(z_{ice} - h\left(1 - \frac{\rho_{ice}}{\rho_{water}}\right)\right)}{dx} = 0 \quad (3)$$

where z_{ice} is the ice surface elevation with respect to the geoid, h is the ice thickness, ρ_{ice} is the density of ice, ρ_{water} is the density of water, and x is horizontal distance. When this condition is met, there is no local pressure gradient and the water-ice interface will be stress-free. *Carter et al.* (2007), using collocated airborne laser altimetry and radar-sounding profiles, used this condition as a primary criteria for automatically identifying subglacial lake candidates under East Antarctica's ice sheet. Their classification process for these subglacial water bodies is instructive for understanding potential water discrimination on Europa — bounding any stresses being supported by the overlying ice.

4.2. Attenuation Within the Ice

Carter et al. (2007) used a secondary property of the basal interface to classify subglacial lake candidates: the basal echo strength. The ice-water interface should have a reflection coefficient — the ratio of transmitted into the bed to that reflected — of between –3 and 0 dB (*Peters et al.,* 2005).

The measured echo strength will be much less than this, and needs to be corrected using the radar equation (equation (1)), in order to identify the composition associated with a particular dielectric horizon (*Peters et al.,* 2005; *Neal,* 1979). In addition to the factors listed in equation (1), the attenuation because of the finite permittivity of the ice needs to be addressed. *MacGregor et al.* (2007) present a review of ice absorption in terrestrial ice sheets. Ice absorption in terrestrial ice sheets is primarily a function of the conductivity of the ice, and can be expressed as

$$\alpha \approx 0.912\sigma dBkm^{-1}$$

where σ is the conductivity in μSm^{-1} (*Evans,* 1965).

TABLE 1. Depth-integrated attenuation rates for Antarctica.

Location	Two-Way Attenuation	Reference
Ronne Ice Shelf	9 ± 1 dB km^{-1}	*Corr et al.* (1993); measured
Interior Ross Ice Shelf	17.3 dB km^{-1}	*Bentley et al.* (1998); measured
Interior Ross Ice Shelf	18 dB km^{-1}	*Peters et al.* (2005); measured
Dome C, East Antarctica	20.2 dB km^{-1}	S. P. Carter, personal communication, 2008
Siple Dome	20.9 ± 5.7 dB km^{-1}	*MacGregor et al.* (2007); calculated
Siple Dome	25.3 ± 1.1 dB km^{-1}	*MacGregor et al.* (2007); measured
Iceberg B15	22.5 dB km^{-1}	*Peters et al.* (2007b); measured
Taylor Glacier	22–32 dB km^{-1}	*Holt et al.* (2006b); calculated
George V Ice Shelf	26.8 ± 1.5 dB km^{-1}	*Corr et al.* (1993); measured

In turn, conductivity is mainly controlled by the temperature of the ice and impurity content. The conductivity of ice can be expressed as a series of Arrhenius functions (e.g., *Corr et al.,* 1993)

$$\sigma = \sigma_{ice} e^{\frac{E_{ice}}{k}\left(\frac{1}{T}-\frac{1}{T_r}\right)}$$
$$+ : \mu_{acid} : e^{\frac{E_{H^+}}{k}\left(\frac{1}{T}-\frac{1}{T_r}\right)} \times [acid]$$
$$+ : \mu_{ssCl^-} : e^{\frac{E_{Cl^-}}{k}\left(\frac{1}{T}-\frac{1}{T_r}\right)} \times [ssCl^-]$$
$$+ : \mu_{NH_4^-} : e^{\frac{E_{NH_4^-}}{k}\left(\frac{1}{T}-\frac{1}{T_r}\right)} \times [NH_4^-]$$

where σ_{ice} is the conductivity of pure ice; μ_X, E_X, and [X] are the molar conductivity, activation energy, and molar concentration in μMol of ionic component X, respectively; T is the ice temperature in Kelvin, and T_r is a reference temperature of 251 K; and k is Boltzmann's constant. The major impurities important for meteoric ice (ice precipitated from the atmosphere) are acids and sea-salt-derived chlorides, with ammonia being important in the northern hemisphere.

At Siple Dome, Antarctica, where temperature and ionic concentrations are available from an ice core, *MacGregor et al.* (2007) found that the contribution from pure ice dominated loss for ice temperatures above –30°C (243 K); this temperature will be highly dependent on ionic concentrations. Measured and calculated rates of ice attenuation at various locations in the Antarctic Ice Sheet can be found in Table 1. When comparing these values to those calculated for Europa in the next section, one should bear in mind that while the base of the ice is at the pressure melting point, the surface temperature of the ice sheet (210 to 240 K) is dramatically warmer than that of Europa (~100 K).

For Earth's ice sheets, lateral flow of ice, the potential for basal melting, and ice dynamic timescales on the order of major climate variations mean that the internal temperature structure of the ice remains one of the great problems in glaciology. Given the implications of temperature for the radar equation, the search for subsurface water would benefit from a search for the absolute value of the basal reflection coefficient. In fact, it has been common to utilize relative variations in basal echo strength to identify free water (e.g., *Gades et al.,* 2000; *Catania et al.,* 2003) or fix the attenuation rate using a point reference where the basal reflection coefficient is believed to be known (*Shabtaie and Bentley,* 1987; *Bentley et al.,* 1998; *Peters et al.,* 2005).

Carter et al. (2007) used both methods to classify subglacial lakes, combining impurity estimates from distant ice cores and one-dimensional models of ice sheet thermal structure to derive loss values over lake candidates. Lake candidates that were both 2 dB more reflective than their surroundings, and had an absolute reflection coefficient above –10 dB, were classified as definite lakes; if a candidate failed the absolute reflection coefficient critera, it was classified as a "dim" lake. Because these dim lakes were in clusters and in hydrostatic equilibrium it was recognized that the attenuation model overestimated the amount of absorption over these candidates, with implications for thermal history and impurity fluxes in the overlying ice. A similar approach will be necessary for any water search at Europa, although both impurity flux and thermal evolution will need to be hypothesized.

Vertical variations in the dielectric properties of ice lead to radar layering. Englacial radar layers in ice have been long identified in studies of terrestrial ice sheets (see *Dowdeswell and Evans,* 2004) (Fig. 3a). These layers correspond to the chemical, density, or crystal fabric stratigraphy of the ice. On Earth at 1–100 MHz radar-sounding frequencies, chemical variations dominate, with the source of the layers ultimately tied to acidic aerosols from volcanic eruptions, a hypothesis verified through connecting layers to ice cores (e.g., *Jacobel and Welch,* 2006). While radar layers due to particle inclusions are less common on Earth, they have been identified in some locations (*Corr and Vaughan,* 2008).

4.3. Accreted Ice at Ice Shelves and Lakes

While the vast majority of terrestrial ice is meteoric in origin, some originates from liquid reservoirs, with different chemistry. An example is found in West Antarctica's central Ronne Ice Shelf. The results of analog airborne radar sounding from 1969 appeared to show that much of this ice shelf was anomalously thin and had echo strengths much lower than would be expected for a smooth ice water interface (*Robin et al.,* 1983); it was hypothesized that unusual ice dynamics were locally thinning the ice shelf, with the low echo strengths being explained by saline, lossy basal ice, as observed in the Ross Ice Shelf (*Neal,* 1979). A more advanced radar sounder was flown over the ice shelf in 1983–1984 (*Thyssen,* 1998), and found that the basal echo split at the edges of the "thinned" area, with the deeper layer being eventually lost. By applying the isostatic criterion (cf. equation (3))

$$z_{ice} - h\left(1 - \frac{\rho_{ice}}{\rho_{water}}\right) = 0 \qquad (4)$$

it was found that up to 400 m of the ice shelf was composed of a discrete package of lossy ice, a result confirmed by boreholes through the ice (*Engelhardt and Determann,* 1987). This remarkable result was due to the oceanography of a sloping ice shelf. High pressures depress the melting point of water ice; thus, very deep portions of an ice shelf will melt. Being fresher than its surroundings, the melt water will rise up the ice shelf front to regions of lower pressure, where it will promptly recrystallize. As the new crystals are lighter than water, they will settle against the overlying ice shelf, while the rising melt water sucks in more warm deep water against the ice shelf. These phenomena are termed "ice pumps" (*Lewis and Perkin,* 1986).

The result is a slowly compacting layer of ice crystals immersed in very saline, electrically conductive brines termed *marine ice* (see section 2.1) (*Moore et al.,* 1994). Typical attenuation rates for marine ice at the pressure melting point range from 140 to 300 dB km[1] (*Blindow,* 1994). However, the lower interface of an actively accreting ice shelf may also be gradational at radar wavelengths (*Engelhardt and Determann,* 1987), preventing a sharp reflective interface. Timescales available for settling in an ice shelf environment, however, are short; ice shelf spreading rates are on the order of a kilometer per year, so most marine-ice deposits are destroyed within a few centuries, leaving little time for trapped brines to migrate out of the ice. Given ice thickness and surface elevation with respect to the geoid, deviations from equation (4) can be used to map out the distribution of basal accretion (*Fricker et al.,* 2001), a tool that may be useful for Europa.

Accretion is also observed at subglacial lakes. At Lake Vostok (Fig. 3a), enough of a pressure melting point gradient exists to drive an ice pump, and thus accretion at one end of the lake. Fortuitously, we have samples of this ice through the Vostok core (*Souchez et al.,* 2004, 2003). In this case the opposite attenuation properties are seen; conductivity decreases in the accreted section, because of a near complete loss of acids (*De Angelis et al.,* 2004). The interface between accreted and meteoric ice is visible in radargrams (*Oliason and Falola,* 2001; *Bell et al.,* 2002) as a low-amplitude horizon above the main lake reflector. However, lake surface reflection coefficients are not strongly perturbed by the accretion layer (*Carter et al.,* 2007). This lack of impact on the loss profile of the ice column is primarily due to the low salinity of the layer; however, it may be a function of the 25,000 years needed to advect the accreted ice across the lake surface, as residual liquid between crystals drains back into the lake.

4.4. Vertical Structures

Tension fractures generally dominate the grounded ice sheet surface where there are large gradients in ice velocity, whereas tension fractures dominate both the surface and base of the ice where grounded ice sheets (or ice streams) transition to floating ice shelves. The process that controls the distribution of these fractures is the balance between the spatial strain rate gradient and the ability to accommodate these strain rates through annealing (which is a function of temperature). Similarly, pervasive and nearly chaotic shear fractures characterize the lateral boundaries of the ice streams over regions that are many times the ice thickness in width. The ice streaming process that controls the position and width of these zones is dominated by stress concentrations at the boundaries of gravity-driven slab flow.

Vertical structures cannot be directly imaged using a nadir-pointed sounding system. However, scattering points at edges can be detected, and where a vertical plane intersects a horizontal surface, a corner reflector geometry will occur when the platform track is oriented perpendicular to the vertical surface. This property can be exploited to examine both the composition and size of a vertical structure.

The giant Antarctic iceberg B-15 was surveyed in December 2000 and December 2004 (Fig. 3b). The data were acquired using the University of Texas at Austin's high-bandwidth coherent HiCARS system (*Peters et al.,* 2005). The data was analyzed without any SAR processing in order to look at basal diffractors with a range of look angles and identify corner reflectors that were interpreted as the lower edges of basal cracks (*Peters et al.,* 2007b). The classification includes major crevasses filled with seawater and incipient or freezing crevasses that are either small with seawater or larger with marine-ice accretion. The large water-filled crevasses likely exhibit varying amounts of marine-ice accretion with moderate brine inclusions. Crevasse height estimates were obtained under the assumption that all crevasses have interfaces similar to known water-filled crevasses or iceberg edges. These statistics are indicative of the basal dynamics of the seawater/iceberg interface and therefore useful Europa analogs.

4.5. Volume Scattering

Volume scattering has presented a challenge to sounding terrestrial ice. *Peters et al.* (2005) presented a map of derived reflection coefficients for the lower reaches of two West Antarctic ice streams, Whillans Ice Stream and Kamb Ice Stream (formerly "Ice Stream C"). This map (Fig. 4) was derived from airborne data collected in 1987 (*Bentley*

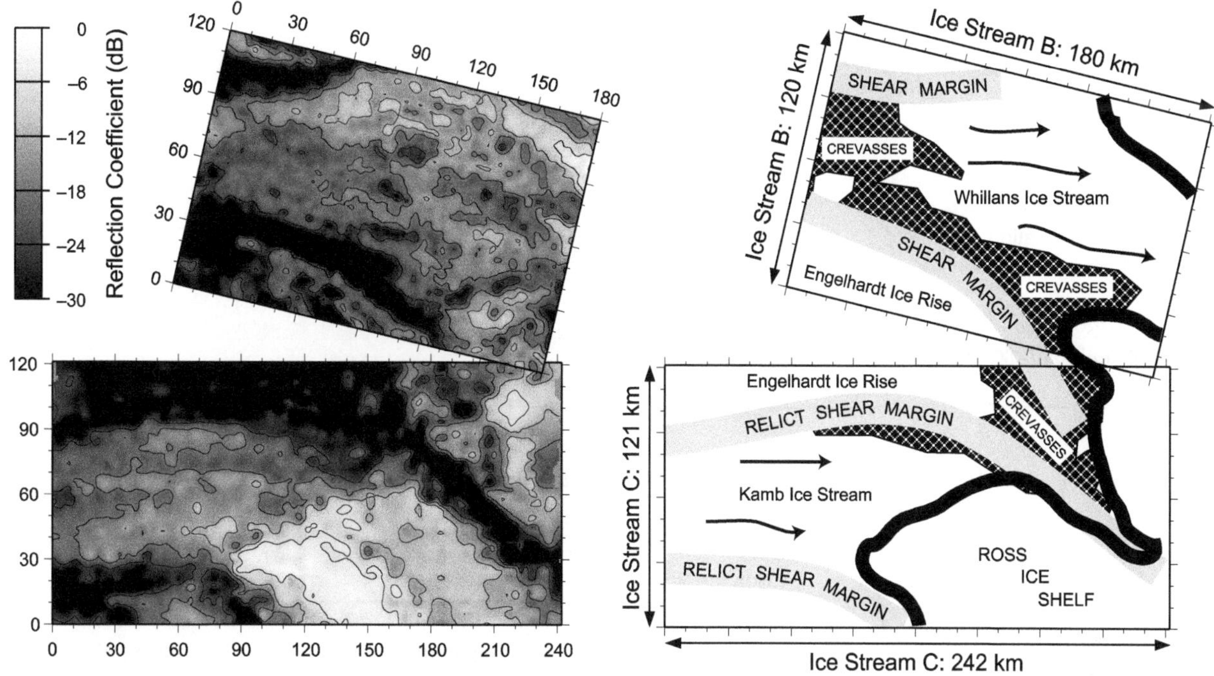

Fig. 4. Map of basal reflections and corresponding sketch map from coherent radar sounding of West Antarctic ice streams (*Peters et al.,* 2005). White regions (high reflection coefficient) correspond with floating ice; narrow dark lanes (low reflection coefficient) correspond to shear margins.

et al., 1988, 1998). The radar sounder was a coherent system with a wavelength of 6 m using an unfocused synthetic aperture of 40 m, resulting in an along-track spot size of ~40 m at the ice surface.

Although seawater is easily identified in these basal reflection maps, in the shear margins of the ice streams the apparent basal reflection coefficients abruptly drop by 20 dB. Pervasive surface crevasses, several meters wide, line the ice streams and effectively attenuate throughgoing signals. The effect of this was demonstrated when the ICESAT laser altimetry experiment, monitoring surface elevation changes, detected an extensive subglacial water system (*Fricker et al.,* 2007) beneath the northern margin of Whillans Ice Stream with a predicted reflection 36 dB higher than observed in these radargrams.

Focused SAR, unavailable for the 1987 radar data, can improve on this performance by allowing more independent "looks" at a single bed target. *Peters et al.* (2007a) demonstrated this algorithm on the upstream portions of Kamb Ice Stream. Coherent waveforms were recorded with a PRF of 6400 Hz that was coherently stacked to 400 Hz (well above the Doppler bandwidth of 28 Hz) and processed with both an unfocused aperture of 70 m and a two-dimensional focused aperture of ~1500 m. The result, shown in Fig. 3c, is an astonishing increase in resolvable structure (and improved resolution) as well as a significant increase in the basal echo strength under active shear margins for the focused SAR case. These techniques form the benchmark for approaches to sounding the subsurface of Europa's icy shell.

Because water pockets exist at the melting point of ice, volume scattering is dominant in temperate terrestrial ice caps and glaciers. For these bodies, pervasive englacial water pockets act as a cloud of diffractors greatly increasing the apparent scattering cross section (*Bamber,* 1988). These pockets also merge to form englacial conduits within the ice (*Bamber,* 1987), and freeze to form ice lenses within the more porous snow column (*Paterson,* 1994; *Pälli et al.,* 2003). Because of these characteristics, temperate (and polythermal glaciers) will be important Earth analogs for sounding hypothesized melt-rich targets on Europa.

5. A RADAR-SOUNDING APPROACH FOR EUROPA

As discussed previously, we consider multiple models to represent ice growth mechanisms that may be present on Europa (*Blankenship et al.,* 1999; *Moore,* 2000). The first of these is a europan crust dominated by marine processes (Fig. 5a) with slow accretion (freezing) or ablation (melting) on the lower side of the icy crust with temperature gradients that are primarily a function of ice thickness (*Squyres et al.,* 1983). A second possible mechanism is the very rapid freezing of ocean water in the linear fracture zones caused by tidal/tectonic processes where large temperature gradients will be present. This process could lead to ice with properties similar to sea ice. This tidal/tectonic model (Fig. 5b) (*Greenberg et al.,* 2002) could represent an oceanic imprint on a primordial europan crust that had been well below the melting point throughout its history, or could

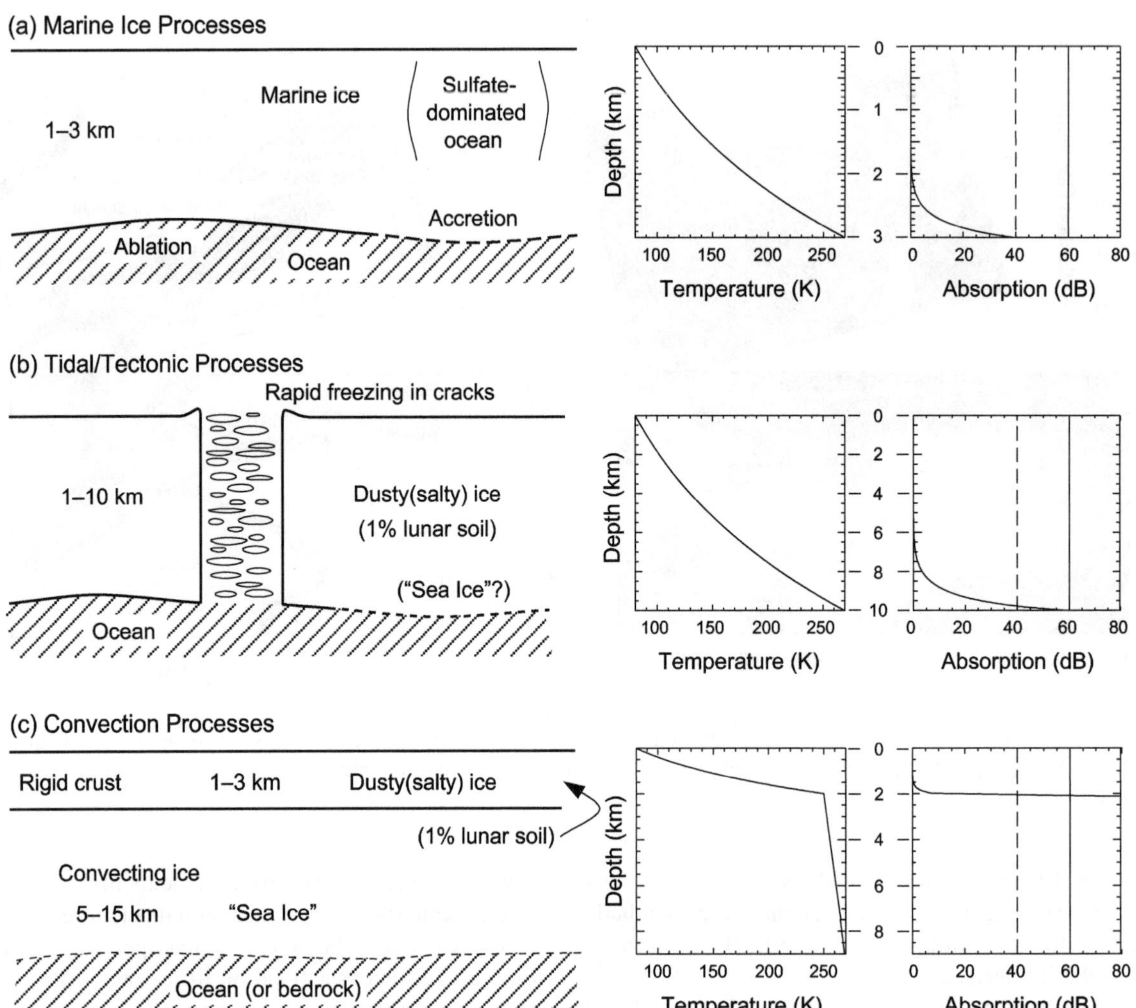

Fig. 5. Schematic diagrams of three ice formation processes that may occur on Europa with expected temperature and radar absorption vs. depth for each assuming a "midlatitude" surface temperature of 80 K (*Blankenship et al.,* 1999). The properties of the component ices and ocean are given in Table 2. **(a)** Model of ice formation similar to that for marine ice on Earth (potentially applicable to Europa's chaotic terrains) with parts of the base subject to melting (ablation) and others to slow freezing (accretion) of frazil ice crystals. **(b)** Ice formation via extrusion into cracks or fissures with rapid freezing (potentially applicable to Europa's ridged terrains). **(c)** Convection scenario with a cold rigid crust underlain by thicker isothermal convecting ice. The dashed and solid lines in the rightmost charts represent approximate system dynamic range for sub- and antijovian Europa sounding with a high-frequency (50 MHz), low-bandwidth (0.85 MHz), and low-power (20 W peak) system.

represent the ice formed entirely by ocean water injected in cracks and then spread over much of the crust by tidally driven tectonic processes. In addition to these two processes, *Pappalardo et al.* (1998) used evidence from Galileo imagery to support the idea of convection in an isothermal layer under a rigid ice crust up to a few kilometers thick that is thermally conducting (Fig. 5c). This convecting ice model could be dominated by ice very similar to that for the tidal/tectonic model although subject to a dramatically different thermal regime.

5.1. Absorption in Europa's Crust

A comprehensive model for electromagnetic absorption in Europa's icy crust was presented by *Moore* (2000). He noted that the species of impurity known to have non-negligible solubility in ice are NH_4^+, H^+, F^-, and Cl^- (*Gross et al.,* 1978) and presented arguments for the dominance of Cl^- in europan ice formation. He then used an activation energy for Cl^- of 0.19 eV (18.4 kJ/mol), measured for marine and meteoric ice on Earth over a wide range of concentrations and at temperatures down to 210 K (*Moore et al.,* 1992), to estimate the absorption of ice derived from a europan ocean with a range of salinities.

Moore (2000) also noted that for impurity concentrations in the ice above a species-dependent solubility limit, the species must be present outside the crystal structure as a liquid or separate salt phase and that radar losses can also come from interfacial polarizations and scattering that occurs whenever the radar waves encounter these mixtures of

TABLE 2. Radar absorption (α) for various ice types and temperatures in Europa's icy shell.

Ice Type	Impurity Content	α(dB m^{-1})	I (dB km^{-1})	II (dB km^{-1})	Notes
Pure ice	Nil	0.0045	1.4–2.4	0.2–0.3	*Glen and Paren* (1975)
Marine-ice (Cl^- Europa)	3.5‰ chlorinity	0.0016	4–7	1.6–2.8	Scaled from Earth; $k_{MI} = 7 \times 10^{-4}$
Marine-ice (SO_4^{-2} Europa)	10‰ chlorinity	0.037	9–16	4–7	*Kargel* (1991) $k_{MI} = 7 \times 10^{-4}$
Dusty/salty ice	1% lunar soil	0.008	5–6	3.6–4.1	*Chyba et al.* (1998) recalculated
Dusty/salty ice	10% lunar soil	0.01	8–9	6–7	*Chyba et al.* (1998) recalculated
Dusty/salty ice	50% lunar soil	0.021	30–33	28–31	*Chyba et al.* (1998) recalculated
Marine-ice (Ronne Ice Shelf)	0.025 chlorinity ice (≈35 ocean)	0.15	36–61	18–31	*Moore et al.* (1994)
Sea ice (Baltic Sea)	≈3 chlorinity ocean	0.85 (at 270 K)	50–85	16–27	*Moore* (2000)

Column I is for a thermally conductive shell with a surface temperature of 50–100 K and its base at the pressure melting point, i.e., over water (≈270 K). Column II is for a thermally conductive shell over "ductile" ice at 250 K. Ice type and impurity content is explained in the text. From *Moore* (2000).

insoluble impurities. Examples of these mixtures were dusty or salty ice (*Chyba et al.,* 1998) and sea ice, where millimeter-scale brine pockets are trapped in the ice by rapid freezing. Because of the large difference in dielectric constant between liquid brine and ice [typical dielectric constants of 86 and 3.2, respectively (*Moore et al.,* 1992)], sea ice is characterized by very large absorption at radar frequencies.

Table 2 reproduces *Moore*'s (2000) estimates of the absorption at radar-sounding frequencies for a range of ices consistent with the europan ice formation processes presented in Fig. 5. End members of the table are pure ice with the conductivity and activation energy given by *Glen and Paren* (1975) and sea ice formed in the relatively low salinity (3‰, parts per thousand) Bay of Bothnia in the Baltic Sea where ice salinities are about 0.5 to ~1‰ and brine pockets are common (*Weeks et al.,* 1990).

Moore's (2000) preferred model for marine-ice formation was based on a straightforward model of the geochemical evolution of Europa (*Kargel,* 1991), which predicts the icy crust to be largely dominated by sulfate salts, mainly $MgSO_4$ and Na_2SO_4, noting that there could also be about 1% by weight Cl, mainly as NaCl and $MgCl_2$. *McKinnon and Zolensky* (2003) have critiqued this view of Europa's geochemistry, using revised meteorite chemistries to suggest a subsaturation concentration of oxidized sulfate in Europa's ocean, and thus a very minor component of Europa's crust; however, this debate has a small impact on our assessment of loss. *Moore*'s (2000) calculation of absorption for the marine-ice models in Table 2 also required an estimate of the distribution coefficient, k, for the Cl^- concentration in ice relative to that in the liquid it is grown in. Table 2 uses a distribution coefficient for marine ice, $k_{MI} = 7 \times 10^{-4}$ (*Moore et al.,* 1994), that is derived from measurements on samples formed under accretion rates observed for the previously described Ronne Ice Shelf in Antarctica.

As an unlikely endmember for marine-ice processes, *Moore et al.* (1994) calculated the absorption losses for the marine ice of the Ronne Ice Shelf using experimental values of conductivity that are well known over the temperature range 200 to 273 K. For marine ice formed by a chloride-dominated europan ocean, *Moore* (2000) used a direct comparison to Earth. Given that the ocean volume to crust ratio for Europa and Earth are about the same, and that the land surface of Earth is about 10 times that of Europa, a simplistic estimate for europan ocean chlorinity is about one-tenth that of Earth's oceans or about 3.5‰. This is consistent with more-detailed geochemical models (*Zolotov and Shock,* 2001) that predict chlorinities ranging from 9.3 to 0.055‰, and with analysis of Galileo magnetometer data that allows for comparatively low salinity for Europa's ocean (*Hand and Chyba,* 2007).

For a sulfate-dominated europan ocean, *Moore* (2000) noted that sulfate ions are not soluble in ice to any significant degree and experiments on ice formed naturally on Earth (and in laboratories) show that SO_4^{2-} seems to play no role in electrical conduction (*Gross et al.,* 1978). He also noted that the permittivity of these solid salts would be similar to that for rock, so the tidal/tectonic processes of Fig. 5b assumed dusty (salty) ice derived from *Chyba et al.* (1998). The 1% soil mixture of *Chyba et al.* (1998) was used as an acceptable representation of the absorption of a sulfate-dominated icy crust on Europa that is being generated and modified by either tidal-tectonic or convection processes.

Given the ice formation processes of Fig. 5, Table 2 also reproduces *Moore*'s (2000) calculated radar absorption averaged over the total ice thickness for a range of possible europan ice. Figure 5 presents the calculated temperature and absorption as a function of depth [T(z) and α(z), respectively] for these ice-formation processes given T_S, the surface temperature, and T_b, the temperature at the base of the thermally conducting layer: The surface temperatures on Europa were assumed to range from 50 to 100 K. For the marine-ice and tidal/tectonic processes of Fig. 5b, T_b was assumed to be close to 270 K for reasonable ice thicknesses.

In the case of convecting processes, where a cold brittle outer crust is underlain by warmer convecting ice that is nearly isothermal at perhaps 235–260 K, *Moore* (2000) chose the the value of T_b for the conducting layer to be the temperature of the isothermal layer. Interestingly, because the radar absorption is governed by the exponential dependence on temperatures, he showed that an average two-way absorption (per kilometer) that is independent of total ice thickness can be found for a given surface temperature. Table 2 reproduces *Moore*'s (2000) averaged two-way radar absorption per kilometer of total ice thickness for two conducting models: one with T_b = 270 K corresponding to marine ice or tidal/tectonic ice in contact with an ocean, and one with T_b = 250 K corresponding to the base of the cold brittle shell of the "convection" model. The range of absorption for each of the conducting models in Table 2 corresponds to a range of surface temperatures (50–100 K) on Europa.

5.2. Noise Environment

Any radar sounding of Europa's icy shell will be constrained by the radio noise environment, which is impacted by two main sources: galactic background noise and emissions from Jupiter's extensive ionosphere. It has been known since the 1950s that Jupiter is a major source of HF radio noise (*Burke and Franklin,* 1955). It was subsequently found that a substantial portion of this decametric emission (DAM) occurs when the electron flux tube connecting Io to Jupiter is dragged over certain longitudes of Jupiter's polar regions (*Bigg,* 1964). The result is an episodic energetic beam emitting at frequencies between 8 MHz and 40 MHz (*Zarka,* 2004). Other, less-well-understood sources, linked to specific sites within Jupiter's magnetic field and tied to Jupiter's rotation, emit narrow episodic beams between 2 and 40 MHz (*Zarka,* 2004). These cyclotron masers abruptly cut out at 40 MHz, and the VHF background is dominated by continuous synchrotron emissions from Jupiter's radiation belts that are ~50–60 dB lower in intensity (Fig. 6).

Observations during Galileo's E12 encounter, which came within 201 km of Europa's antijovian hemisphere, indicate that Europa does screen out RF noise for a portion of the antijovian hemisphere (*Kurth et al.,* 2001). However, given the 100–200-km height of an orbiting spacecraft over Europa, Jupiter's large angular size as seen from Europa, and Europa's synchronous rotation, observations of up to three-quarters of Europa's shell could be exposed to this episodic HF noise, especially the polar regions. The power intensity of episodic DAM events at Europa, scaled from *Zarka et al.* (2004), would be ~–28 dBm at 5 MHz and 1 MHz bandwidth; at 50 MHz and 1 MHz bandwidth, the power intensity is –99 dBm because of continuous synchrotron emissions observed around subjovian Europa (*Zarka,* 2004) and –123 dBm because of the galactic background (*Cane,* 1979), which could be observed from about half of Europa's antijovian hemisphere.

Given the tight beams of the DAM emission, the impact of these events on HF radar observations would vary regularly as Io and Jupiter rotate with respect to Europa. Comparison with Cassini Radar and Plasma Wave Science (RPWS) instrument flyby data (e.g., *Zarka et al.,* 2004) suggest that at 5 MHz, at least 55% of unocculted radar observations would escape interference from DAM emissions. Lastly, the well-shielded portion of Europa's farside includes many important targets for exploration, such as Argadnel Regio and Thrace.

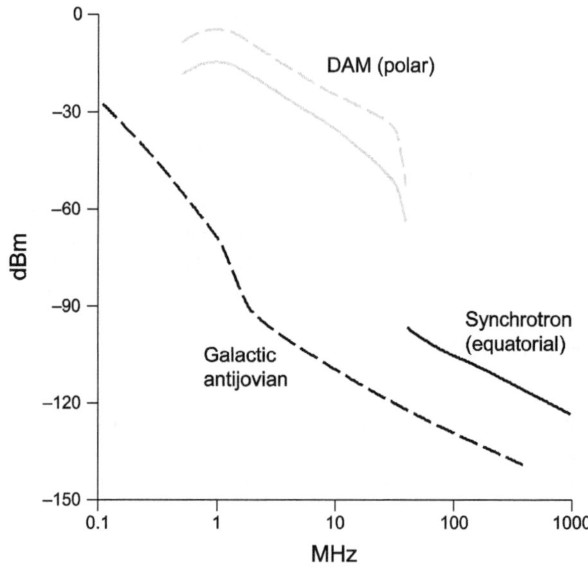

Fig. 6. Radio noise power at Europa, for a 1-MHz bandwidth receiver and an omnidirectional antenna. The data for the subjovian hemisphere (showing episodic polar DAM and continuous equatorial synchrotron noise) is compiled from Cassini and terrestrial observations (*Zarka,* 2004), while the calculation of galactic noise, appropriate for the antijovian region, comes from *Dulk et al.* (2001). The peak (dotted) and typical (solid) intensities of DAM events are shown. The large drop in jovian noise above 40 MHz is very apparent.

5.3. Volume Scattering

At UHF frequencies, acquired by Earth-based radar, the radar cross-section from Europa exceeds unity, suggesting extensive internal scattering (*Ostro and Shoemaker,* 1990); however, this return decreases significantly as the wavelength increases (*Black et al.,* 2001). This suggests that low-frequency returns from Europa may be due to surface scattering mechanisms similar to those operating from Earth's icy surfaces.

Eluszkiewicz (2004) performed a worse-case analysis for volume scatterering, examining the case of a regolith with quarter-wavelength-sized cavities [which, for the 50-MHz frequency favored in some studies (*Chyba et al.,* 1998; *Blankenship et al.,* 1999) for Europa sounding, corresponds to ~1 m]. *Eluszkiewicz* (2004) found that a 5% porosity regolith with such cavities could have a two-way attenuation of 113 dB.

Lee et al. (2005) used analogy to terrestrial ice to constrain a porosity for the regolith, and suggest a typical pore size of the centimeter scale. In this case, HF and VHF radar would not be strongly affected.

5.4. Surface Scattering

The ice at Europa's surface is much colder and stiffer that in Earth's ice sheets, and is in a lower gravity field. The results are that at small scales the surface of Europa is much rougher that that of a typical ice sheet. Combined with the wide beam spot implied by the 50–200-km-high orbit of a Europa sounding mission, surface scattering could be significant. Working to mitigate the effect of this clutter is the low surface reflectivity of water ice (ranging from –11 to –26 dB for a porous regolith), and the fact that the power returned by incoherent clutter drops with a $\frac{1}{r^4}$ spreading loss vs. $\frac{1}{r^2}$ for smooth nadir targets.

In order to predict the nature of scattering we need to know the topography of the surface of Europa. Neither the Voyager or Galileo spacecraft carried explicit altimetry instruments; therefore, the best available data for the large-scale surface characteristics of Jupiter's moons is obtained by using combined stereo and photoclinometric DEMs obtained by using Galileo imagery (e.g., *Schenk and Pappalardo,* 2004; *Giese et al.,* 1998). Because of the failure of Galileo's high-gain antenna, high-resolution stereo coverage is limited. Some early work on analyzing the small-scale topographic spectrum of Europa relevant to radar studies is reported in *Blankenship et al.* (1999), *Schenk* (2005), and *Nimmo and Schenk* (2008).

To sample a variety of terrain types, *Blankenship et al.* (1999) examined DEMs generated for Pwyll crater (300 m to 30 km resolution); the "Wedges" region, Argadnel Regio (100 m); and the Conamara Chaos region (20 m to 3 km). They found that the slope distribution at measurable length scales was consistent with a power law distribution. At Pwyll crater, the slope spectrum was consistent with a self-affine Brownian distribution, implying relatively rough slopes at small scales [RMS slope of ~45° at 1 m; see *Shepard et al.* (2001) for an explanation]; however, at Pwyll crater the minimum horizontal resolution was still much larger than candidate sounding wavelengths. Later work by *Schenk* (2005) and *Nimmo and Schenk* (2008) appear to show that at shorter length scales relevant to radar, the power law is closer to a true fractal for many terrains, implying lower small-scale roughness (RMS slopes of ~11°), which would be "smooth" by many radar criteria. Therefore, only geometric optics scattering, which assumes smooth surfaces at a wavelength scale, will be described further.

Using the slope distributions for the Pwyll region, *Blankenship et al.* (1999) predicted a geometrical optics contribution that decreases from about –10 dB to –20 dB as the incident angle varies from 0° to about 20°. For a 200-km-high orbit, this clutter would overlap subsurface echoes ~20 km below the surface. For incident angles greater than 20°, the geometric optics cross section remains roughly constant at about –20 dB for incident angles smaller than 45°. Modeling the scattering from any europan ice/ocean interface remains problematic because the geometry of any presumed interface is uncertain.

5.5. Sounding Europa's Crust

The radar sounding required to search for any ice-ocean interface on Europa and test first-order predictions of ice-ocean exchange by marine and convective processes must have a very high total sensitivity and multikilometer or better horizontal resolution. For the limited region around the antijovian point, an HF system styled on MARSIS or SHARAD may fulfill this requirement. A 5-MHz system with 1 MHz of bandwidth orbiting 200 km above Europa would have an in-ice vertical resolution of 85 m, with horizontal resolutions at the surface ranging from 20 km (pulse-limited) to ~270 m (one-dimensional focused SAR).

To search for shallower subsurface water, a SHARAD-type system (20 MHz bandwidth and 10 MHz bandwidth) would have theoretical vertical resolutions of 8 m, and horizontal resolutions at the surface ranging from 4.9 km (pulse-limited) to ~153 m (one-dimensional focused SAR) at an orbital height of 100 km. However, the severe but episodic RF noise environment for the other three-quarters of Europa's surface could be an operational challenge for any global mapping program using the HF band.

A 50-MHz system with 10-MHz bandwidth would have horizontal resolutions at the surface down to 62 m (one-dimensional focused SAR) at 100 km height. Such length scales are near those required to test hypotheses concerning shallow europan tectonics and impact processes as well as to characterize shallow subsurface water. While Jupiter's synchrotron radiation remains problematic in the VHF band, a 50-MHz system has a greater probability of global success for targets in the top few kilometers. Ten-megahertz bandwidth waveforms would require greater than 20 MHz sampling, a rate that would likely fill data buffers too quickly to characterize a crust of several tens of kilometers in thickness. Theoretically, two-dimensional focused SAR incorporating 1 μs of range data for every resolution cell could resolve structures as small as a few tens of meters; however, the processing requirements are high, and probably too stringent for continuous onboard use. The Doppler bandwidth from equation (2) for such a system is 900 Hz; current downlink technology preclude returning radar records at this rate. High-rate data recorded over local carefully selected targets and later returned to Earth for deeper processing represents a solution to this problem.

A multifrequency system may be the best approach for fully investigating Europa — comprising a high-bandwidth system for shallow crustal studies, and a low-bandwidth, deep-penetration system, with a low enough information density to record over the full depth of the crust. The results of the deep penetration study will be determined in large part by the absorption profile of Europa's crust.

Figure 5 shows the estimated two-way absorption vs. depth for the three proposed ice formation processes on Europa (*Blankenship et al.,* 1999; *Moore,* 2000), and Fig. 7

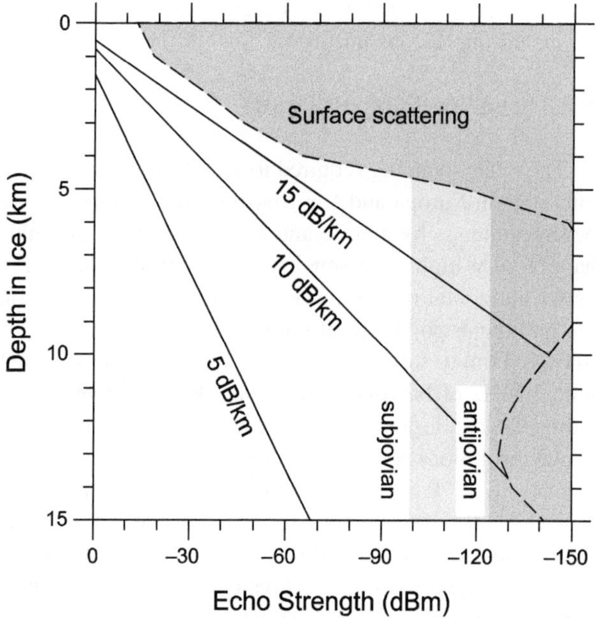

Fig. 7. System performance for a nominal 50-MHz, 0.85-MHz bandwidth Europa radar sounder derived from *Blankenship et al.* (1999) in the context of a Gaussian scattering model (*Peake and Barrick,* 1967) for Europa's surface (dashed lines), and a directional antenna beam pattern. Signal strength is increased by 53 dB through SAR processing and pulse compression. The strength of the return from a specular ice/ocean interface is presented as solid lines with differing ice absorption values (see Table 2). RF noise limits for each side of Europa, derived from Fig. 6, are also presented.

shows the modeled strength of the return from an ice/ocean interface for these processes (*Blankenship et al.,* 1999) for a high-frequency (50 MHz), low-bandwidth (0.85 MHz) system. For both the marine-ice and tidal-tectonic processes (i.e., T_b = 270 K), the ability to characterize any ice/ocean interface or internal layering can be inferred directly from these plots. Recalling that a smooth ice-ocean interface would give negligible reflection losses, Fig. 5 shows that the total absorption is less than a nominal instrumental dynamic range of 60 dB over the full range of total ice thicknesses estimated for both marine and tidal/tectonic ice formation processes on antijovian Europa at 50 MHz. This is also the case for the rougher ice/ocean interface assumed in Fig. 7. In addition, Fig. 7 shows that the range of total ice thickness soundable on subjovian Europa is reduced by about one-third (to a little over 2 km for marine ice and 6 km for tidal/tectonic ice formation processes) compared with antijovian sounding.

For the "convection" processes of Fig. 5, the deeper warm layer is likely to contain brine pockets as the ice temperature will be above the eutectic point of some salts. Because of this, absorption will become very high (similar to the sea ice of the Baltic Sea or even higher; Table 2) and radio waves will be unable to penetrate the convecting ice to any subsurface ocean. If the boundary is reasonably sharp, there will be a radar reflection from it because of the change in dielectric impedance. The magnitude of the reflection will depend on the brine content in the ice and its spatial distribution.

A relevant example on Earth is the reflection coefficient observed at the boundary between cold-dry and temperate-wet glacier ice with an essentially uniform distribution of water pockets, as seen in the Arctic's polythermal glaciers, typically about 20 dB (*Bamber,* 1987). Assuming a similar reflection coefficient for convection processes on Europa, Figs. 5b,c show that it is very likely that a 50-MHz instrument will penetrate a broad range of rigid lid thicknesses and allow characterization of the interface with the convecting layer beneath.

6. DISCUSSION AND CONCLUSIONS

There are strong scientific reasons for studying the subsurface structure of Europa's shell, especially as related to subsurface water and the nature of surface-ice-ocean exchange. The dielectric losses in very cold ice are low, yet highly sensitive to increasing temperature, water, and impurity content; therefore, much can be learned through orbital electromagnetic sounding of the icy shell. This is especially true when subsurface profiling is coupled to observations of both the topography and morphology of surface landforms and placed in the context of both surface composition and subsurface density distribution. Because of Jupiter's strong radio emissions and the unknown size of volume scatterers within Europa's icy shell, the range of sounding frequencies must be carefully matched to the science objectives. The thickness of Europa's icy shell is one of the most important questions left unanswered by Galileo. Determining the icy shell thickness is of fundamental astrobiological significance: It constrains the tidal heat the satellite is generating, whether the silicate interior is Io-like or not, and the means and extent to which the ocean and near-surface are likely to exchange material.

Science objectives for a radar sounder include characterizing the distribution of any shallow subsurface water; detecting the presence of a ice-ocean interface; correlating surface features and subsurface structure to investigate processes governing communication among the surface, icy shell, and ocean; and characterizing regional and global heat flow variations. The subsurface signatures from near-global surveys at high depth resolution combined with surface topography of similar vertical resolution would identify regions of possible ongoing or relatively recent upwelling of liquid water or brines. Orbital subsurface profiling of the top several kilometers of Europa's icy shell is possible, at frequencies slightly above the upper end of Jupiter's radio noise spectrum (i.e., about 40 MHz), to establish the geometry of various thermal, compositional, and structural horizons to a depth resolution of about 10 m (requiring a bandwidth of about 10 MHz). This high-resolution search for shallow water will produce data analogous to that of the SHARAD instrument.

Subsurface signatures from lower resolution but more deeply penetrating near-global surveys might reveal a shallow ice-ocean interface, which could be validated over a region by carefully correlating ice thickness and surface topography. An unequivocally thin icy shell, even within a limited region, would have significant implications for understanding direct exchange between the ocean and the overlying ice. Similarly, the detection of deep subsurface interfaces in these surveys and the presence or absence of shallower interfaces above them could validate hypotheses regarding the convective movement of deep ductile ice into the cold brittle shell implying indirect exchange with any ocean. Additional orbital profiling of the subsurface of Europa to depths approaching 30 km with a vertical resolution of about 100 m is recommended to establish the geometry of any deeper geophysical interfaces, in particular, to search for an ice-ocean interface. Although warm ice is very attenuating, "windows" of cold downwelling material may exist within the icy shell, allowing local penetration to great depths (*McKinnon,* 2005). Moreover, while the presence of meter-scale voids within the icy shell (*Eluszkiewicz,* 2004) would confound sounding efforts at higher frequencies (15 MHz), the presence of such large voids is probably unrealistic.

A deep ocean search would produce data analogous to that of MARSIS. Profiling could establish the geometry of any deeper geophysical interfaces that may correspond to an ice-ocean boundary to a vertical resolution of about 100 m (requiring a bandwidth of about 1 MHz). In particular, frequencies significantly less sensitive to any volume scattering that may be present in the shallow subsurface profiling detailed above (i.e., about 5 MHz) should be used on the antijovian side of Europa, which is substantially shadowed from Jupiter's radio emissions. This low-frequency, low-resolution profiling could complement high-frequency, low-resolution profiling over Europa's subjovian surface (where Jupiter's radio noise is an issue for low-frequency sounding).

Ultimately, observation campaigns targeted to specific features will be required to understand the processes controlling the distribution of any shallow subsurface water and either the direct or indirect exchange of materials between the icy shell and its underlying ocean. The presence of major cracks and faults as well as topographic and compositional anomalies, when correlated with subsurface structures within a particular targeted region, can provide critical information on tidal response and its role in subsurface fluid migration. Important factors include localized heating, the magnitude of tectonic stress, and associated strain release. Similarly, variations of the physical and compositional properties of the near-surface ice may arise because of relative age differences, tectonic deformation, mass wasting, or impact processes.

Because of the complex geometries expected for subsurface structures, full unprocessed subsurface imaging could be obtained along profiles at least 30 km long in any region of targeted study, either to a depth of several kilometers for high-resolution imaging of shallow targets or to a depth of tens of kilometers for lower-resolution imaging of deeper processes, in conjunction with co-located topographic measurements. These targeted subsurface studies would be a pathfinder for future *in situ* astrobiological exploration.

The thermal structure of the shell (apart from local heat sources) is set by the transport of heat from the interior. Regardless of the properties of the shell or the overall mechanism of heat transport, the uppermost several kilometers is thermally conductive, cold, and stiff. The thickness of this conductive "lid" is set by the total amount of heat that must be transported, and thus a measurement of the thickness of the cold and brittle part of the shell will provide a powerful constraint on the heat production in the interior. For a thin icy shell, the ice-ocean interface forms a significant dielectric horizon at the base of the thermally conductive layer.

However, when warm pure-ice diapirs from the interior of a thicker convective shell model approach the surface, they may be far from the pure-ice melting point and above the eutectic point of many substances and may therefore create regions of melting within the rigid shell above them as the temperature increases above the flattening diapir. Any dielectric horizon associated with these melt regions would also provide a good measurement of the thickness of the conductive layer. Global radar profiling of these subsurface thermal horizons to depths of tens of kilometers at a vertical resolution of 100 m will be vital in characterizing both regional and global heat flow variations in Europa's icy shell.

REFERENCES

Adams D. S. and Mobrem M. (2006) MARSIS antenna flight deployment anomaly and resolution. In *47th AIAA/ASME/ASCE/AHS/ASC Structures, Structural Dynamics, and Materials Conference,* No. 2006-1684. American Institute of Aeronautics and Astronautics, Newport, Rhode Island.

Bamber J. L. (1987) Internal reflecting horizons in Spitzbergen glaciers. *Ann. Glaciol., 9,* 5–10.

Bamber J. L. (1988) Enhanced radar scattering from water inclusions in ice. *J. Glaciol., 34(118),* 293–296.

Bell R. E., Studinger M., Tikku A. A., Clarke G. K. C., Gutner M. M., and Meertens C. (2002) Origin and fate of lake vostok water frozen to the base of the east antarctic ice sheet. *Nature, 416(6878),* 307–310.

Bentley C. R., Blankenship D. D., and Moline G. (1988) Electromagnetic studies on the siple coast, 1987–1988. *Antarc. J. U.S., 23(5),* 59.

Bentley C. R., Lord N., and Liu C. (1998) Radar reflections reveal a wet bed beneath stagnant Ice Stream C and a frozen bed beneath ridge BC, West Antarctica. *J. Glaciol., 44(146),* 149–156.

Bigg E. K. (1964) Influence of the satellite Io on Jupiter's decametric emission. *Nature, 203(4949),* 1008–1010, DOI:10.1038/2031008a0.

Bingham R. G. and Siegert M. J. (2007) Radio-echo sounding over polar ice masses. *J. Environ. Engineer. Geophys., 12(1),* 47–62, DOI: 10.2113/JEEG12.1.47.

Black G. J., Campbell D. B., and Ostro S. J. (2001) Icy galilean satellites: 70 cm radar results from Arecibo. *Icarus, 151,* 160–166, DOI: 10.1006/icar.2001.6615.

Blankenship D. D., Edwards B. C., Kim Y, Geissler P. E., Gurnett D. A., Johnson W. T. K., Kofman W. , Moore J. C., Morse D. L., Pappalardo R. T., Picardi G. , Raney R. K., Rodriguez

E. R., Shao X.-M., Weertman J. , Zebker H. A., and van Zyl J. (1999) *Feasibility Study and Design Concept for an Orbiting Ice-Penetrating Radar Sounder to Characterize in Three-Dimensions the Europan Ice Mantle Down to (and Including) any Ice/Ocean Interface.* Tech. Rept. 184, University of Texas Institute for Geophysics, Austin.

Blankenship D. D., Morse D., Finn C. A., Bell R. E., Peters M. E., Kempf S. D., Hodge S. M., Studinger M., Behrendt J. C., and Brozena J. M. (2001) Geological controls on the initiation of rapid basal motion for West Antarctic Ice Streams: A geophysical perspective including new airborne radar sounding and laser altimetry results. In *The West Antarctic Ice Sheet: Behavior and Environment* (R. B. Alley and R. A. Bindschadler, eds.), pp. 105–121. Antarctic Research Series, Vol. 77, American Geophysical Union, Washington, DC.

Blindow N. (1994) The central part of the Filchner-Ronne ice shelf, Antarctica: Internal structures revealed by 40 MHz monopulse RES. *Ann. Glaciol., 20,* 365–371.

Burke B. F. and Franklin K. L. (1955) Observations of a variable radio source associated with the planet Jupiter. *J. Geophys. Res., 60(2),* 213–217.

Campbell B. A. (2002) *Radar Remote Sensing of Planetary Surfaces.* Cambridge Univ., Cambridge.

Campbell B. A. and Shepard M. K. (2003) Coherent and incoherent components in near-nadir radar scattering: Applications to radar sounding of Mars. *J. Geophys. Res., 108(E12),* 5132, DOI: 10.1029/2003JE002164.

Cane H. V. (1979) Spectra of the non-thermal radio radiation from the galactic polar regions. *Mon. Not. R. Astron. Soc., 189,* 465–478.

Carter S. P., Blankenship D. D., Peters M. E., Young D. A., Holt J. W., and Morse D. L. (2007) Radar-based subglacial lake classification in Antarctica. *Geochem. Geophys. Geosys., 8,* Q03016, DOI: 10.1029/2006GC001408.

Catania G. A., Conway H., Gades A. M., Raymond C. F., and Engelhardt H. (2003) Bed reflectivity beneath inactive ice streams in West Antarctica. *Ann. Glaciol., 36,* 287–291.

Chyba C. F., Ostro S. J., and Edwards B. C. (1998) Radar detectability of a subsurface ocean on Europa. *Icarus, 134,* 292–302, DOI: 10.1006/icar.1998.5961.

Corr H. F. J. and Vaughan D. G. (2008) A recent volcanic eruption beneath the West Antarctic ice sheet. *Nature Geosci., 1,* 122–125, DOI: 10.1038/ngeo106.

Corr H., Moore J. C., and Nicholls K. W. (1993) Radar absorption due to impurities in Antarctic ice. *Geophys. Res. Lett., 20,* 1071–1074.

De Angelis M., Petit J. R., Savarino J., Souchez R., and Thiemens M. H. (2004) Contributions of an ancient evaporitic-type reservoir to subglacial Lake Vostok chemistry. *Earth Planet. Sci. Lett., 222(3–4),* 751–765.

Dowdeswell J. A. and Evans S. (2004) Investigations of the form and flow of ice sheets and glaciers using radio-echo sounding. *Rept. Progr. Phys., 67,* 1821–1861, DOI: 10.1088/0034-4885/67/10/R03.

Dulk G. A., Erickson W. C., Manning R., and Bougeret J.-L. (2001) Calibration of low-frequency radio telescopes using the galactic background radiation. *Astron. Astrophys., 365(2),* 294–300, DOI: 10.1051/0004-6361:20000006.

Elachi C., Wall S., Allison M., Anderson Y., Boehmer R., Callahan P., Encrenaz P., Flamini E., Franceschetti G., Gim Y., Hamilton G., Hensley S., Janssen M., Johnson W., Kelleher K., Kirk R., Lopes R., Lorenz R., Lunine J., Muhleman D., Ostro S., Paganelli F., Picardi G., Posa F., Roth L., Seu R., Shaffer S., Soderblom L., Stiles B., Stofan E., Vetrella S., West R., Wood C., Wye L., and Zebker H. (2005) Cassini radar views the surface of Titan. *Science, 308(5724),* 970–974.

Eluszkiewicz J. (2004) Dim prospects for radar detection of Europa's ocean. *Icarus, 170,* 234–236, DOI: 10.1016/j.icarus.2004.02.011.

Engelhardt H. and Determann J. (1987) Borehole evidence for a thick layer of basal ice in the central Ronne Ice Shelf. *Nature, 327(6120),* 318–319.

Evans S. (1961) Polar ionospheric spread echoes and radio frequency properties of ice shelves. *J. Geophys. Res., 66(12),* 4137–4141.

Evans S. (1965) Dielectric properties of ice and snow — A review. *J. Glaciol., 5,* 773–793.

Eviatar A., Siscoe G. L., Johnson T. V., and Matson D. L. (1981) Effects of Io ejecta on Europa. *Icarus, 47,* 75–83, DOI: 10.1016/0019-1035(81)90092-0.

Franceschetti G. and Lanari R. (1999) *Synthetic Aperture Radar Processing.* CRC Press, Boca Raton, Florida.

Fricker H. A., Popov S., Allison I., and Young N. (2001) Distribution of marine ice beneath the Amery ice shelf. *Geophys. Res. Lett., 28(11),* 2241–2244, DOI: 10.1027/2000GL012461.

Fricker H. A., Scambos T., Bindschadler R., and Padman L. (2007) An active subglacial water system in West Antarctica mapped from space. *Science, 315(5818),* 1544–1548, DOI: 10.1126/science.1136897.

Fujita S., Matsuoka T., Morishima S., and Mae S. (1993) The measurement on the dielectric properties of ice at HF, VHF and microwave frequencies. In *Geoscience and Remote Sensing Symposium, 1993. IGARSS '93. Better Understanding of Earth Environment, International,* pp. 1258–1260, DOI: 10.1109/IGARSS.1993.322667.

Gades A. M., Raymond C. F., Conway H., and Jacobel R. W. (2000) Bed properties of Siple Dome and adjacent ice streams, West Antarctica, inferred from radio-echo sounding measurements. *J. Glaciol., 46(152),* 88–94, DOI: 10.3189/172756500781833467.

Giese B., Oberst J., Roatsch T., Neukum G., Head J. W., and Pappalardo R. T. (1998) The local topography of Uruk Sulcus and Galileo Regio obtained from stereo images. *Icarus, 135(1),* 303–316, DOI: 10.1006/icar.1998.5967.

Glen J. W. and Paren J. G. (1975) The electrical properties of snow and ice. *J. Glaciol., 15,* 15–37.

Gogineni S., Tammana D., Braaten D., Leuschen C., Akins T., Legarsky J., Kanagaratnam P., Stiles J., Allen C., and Jezek K. (2001) Coherent radar ice thickness measurements over the Greenland ice sheet. *J. Geophys. Res., 106(D24),* 33761–33772, DOI: 10.1027/2001JD900183.

Gogineni S., Braaten D., Allen C., Paden J., Akins T., Kanagaratnam P., Jezek K., Prescott G., Jayaraman G., Ramasami V., Lewis C., and Dunson D. (2007) Polar Radar for Ice Sheet Measurements (PRISM). *Remote Sensing Environ., 111(2–3),* 204–211.

Greenberg R., Geissler P., Hoppa G., Tufts B. R., Durda D. D., Pappalardo R., Head J. W., Greeley R., Sullivan R., and Carr M. H. (1998) Tectonic processes on Europa: Tidal stresses, mechanical response, and visible features. *Icarus, 135,* 64–78, DOI: 10.1006/icar.1998.5986.

Greenberg R., Hoppa G. V., Tufts B. R., Geissler P., Riley J., and Kadel S. (1999) Chaos on Europa. *Icarus, 141,* 263–286, DOI: 10.1006/icar.1999.6187.

Greenberg R., Geissler P., Hoppa G., and Tufts B. R. (2002) Tidal-tectonic processes and their implications for the character of Europa's icy crust. *Rev. Geophys., 40,* 1–1, DOI: 10.1029/2000RG000096.

Gross G. W., Hayslip I. C., and Hoy R. N. (1978) Electrical conductivity and relaxation in ice crystals with known impurity content. *J. Glaciol., 21(85),* 143–160.

Gudmandsen P. (1971) Electromagnetic probing of ice. In *Electromagnetic Probing in Geophysics* (J. R. Wait, ed.), pp. 321–348. Golem Press, Colorado.

Hand K. P. and Chyba C. F. (2007) Empirical constraints on the salinity of the europan ocean and implications for a thin ice shell. *Icarus, 189(2),* 424–438, DOI: 10.1016/j.icarus.2007.02.002.

Heliere F., Lin C.-C., Corr H., and Vaughan D. (2007) Radio echo sounding of Pine Island Glacier, West Antarctica: Aperture synthesis processing and analysis of feasibility from space. *IEEE Trans. Geosci. Remote Sensing, 45(8),* 2573–2582, DOI: 10.1109/TGRS.2007.897433.

Holt J. W., Peters M. E., Kempf S. D., Morse D. L., and Blankenship D. D. (2006a) Echo source discrimination in single-pass airborne radar sounding data from the Dry Valleys, Antarctica: Implications for orbital sounding of Mars. *J. Geophys. Res., 111(E10),* DOI: 10.1029/2005JE002525.

Holt J. W., Peters M. E., Morse D. L., Blankenship D. D., Lindzey L. E., Kavanaugh J. L., and Cuffey K. M. (2006b) Identifying and characterizing subsurface echoes in airborne radar sounding data from a high clutter environment in Taylor Valley, Antarctica. In *11th International Conference on Ground Penetrating Radar,* June 19–22, Columbus, Ohio.

Hoppa G. V., Tufts B. R., Greenberg R., and Geissler P. E. (1999) Formation of cycloidal features on Europa. *Science, 285,* 1899–1902.

Jacobel R. W. and Welch B. C. (2006) A time marker at 17.5 k.y. BP detected throughout West Antarctica. *Ann. Glaciol., 41(1),* 47–51.

Jezek K. C., Rodrìguez E., Gogineni P., Freeman A., Curlandr J., Wu X., Paden J., and Allen C. (2006) Glaciers and ice sheets mapping orbiter concept. *J. Geophys. Res., 111(E06S20),* DOI: 10.1029/2005JE002572.

Kargel J. S. (1991) Brine volcanism and the interior structures of asteroids and icy satellites. *Icarus, 94(2),* 368–390, DOI: 10.1016/0019-1035(91)90235-L.

Kivelson M. G., Khurana K. K., Russell C. T., Volwerk M., Walker R. J., and Zimmer C. (2000) Galileo magnetometer measurements: A stronger case for a subsurface ocean at Europa. *Science, 289,* 1340–1343.

Kurth W. S., Gurnett D. A., Persoon A. M., Roux A., Bolton S. J., and Alexander C. J. (2001) The plasma wave environment of Europa. *Planet. Space Sci., 49,* 345–363, DOI: 10.1016/S0032-0633(00)00156-2.

Lee S., Pappalardo R. T., and Makris N. C. (2005), Mechanics of tidally driven fractures in Europa's ice shell. *Icarus, 177,* 367–379, DOI: 10.1016/j.icarus.2005.07.003.

Legarsky J. J., Gogineni S. P., and Akins T. L. (2001) Focused synthetic aperture radar processing of ice-sounder data collected over the Greenland ice sheet. *Geosci. Remote Sensing, IEEE Trans., 39(10),* 2109–2117.

Lewis E. L. and Perkin R. G. (1986) Ice pumps and their rates. *J. Geophys. Res., 91(C10),* 11756–11762.

MacGregor J. A., Winebrenner D., Conway H., Matsuoka K., Mayewski P. A., and Clow G. (2007) Modeling englacial radar attenuation at Siple Dome, West Antarctica, using ice chemistry and temperature data. *J. Geophys. Res., 112(F03008),* DOI: 10.1029/2006JF000717.

McEwen A. S. (1986) Exogenic and endogenic albedo and color patterns on Europa. *J. Geophys. Res., 91,* 8077–8097.

McKinnon W. B. (2005) Radar sounding of convecting ice shells in the presence of convection: Application to Europa, Ganymede, and Callisto. In *Workshop on Radar Investigations of Planetary and Terrestrial Environments,* Abstract #6039. Lunar and Planetary Institute, Houston.

McKinnon W. B. and Zolensky M. E. (2003) Sulfate content of Europa's ocean and shell: Evolutionary considerations and some geological and astrobiological implications. *Astrobiology, 3,* 879–897, DOI: 10.1089/153110703322736150.

Moore J. C. (2000) Models of radar absorption in europan ice. *Icarus, 147(1),* 292–300, DOI: 10.1006/icar.2000.6425.

Moore J. C., Paren J. G., and Oerter H. (1992) Sea salt dependent electrical conduction in polar ice. *J. Geophys. Res., 97(B13),* 19803–19812.

Moore J. C., Reid A. P., and Kipfstuhl J. (1994) Microstructure and electrical properties of marine ice and its relationship to meteoric ice and sea ice. *J. Geophys. Res., 99(C3),* 5171–5180, DOI: 10.1029/93JC02832.

Moore J. M., Asphaug E., Sullivan R. J., Klemaszewski J. E., Bender K. C., Greeley R., Geissler P. E., McEwen A. S., Turtle E. P., Phillips C. B., Tufts B. R., Head J. W., Pappalardo R. T., Jones K. B., Chapman C. R., Belton M. J. S., Kirk R. L., and Morrison D. (1998) Large impact features on Europa: Results of the Galileo nominal mission. *Icarus, 135,* 127–145, DOI: 10.1006/icar.1998.5973.

Musil G. J. and Doake C. S. M. (1987) Imaging subglacial topography by a synthetic aperture technique. *Ann. Glaciol., 9,* 170–175.

Neal C. S. (1979) The dynamics of the Ross ice shelf revelad by radio echo sounding. *J. Glaciol., 24(90),* 295–307.

Nimmo F. and Gaidos E. (2002) Strike-slip motion and double ridge formation on Europa. *J. Geophys. Res.–Planets, 107,* 5021, DOI: 10.1029/2000JE001476.

Nimmo F. and Manga M. (2002) Causes, characteristics and consequences of convective diapirism on Europa. *Geophys. Res. Lett., 29(23),* 2109, DOI: 10.1029/2002GL015754.

Nimmo F. and Schenk P. (2008) Stereo and photoclinometric comparisons and topographic roughness of Europa. In *Lunar and Planetary Science XXXIX,* Abstract #1464. Lunar and Planetary Institute, Houston (CD-ROM).

Nouvel J. F., Herique A., Kofman W., and Safaeinili A. (2004) Radar signal simulation: Surface modeling with the Facet Method. *Radio Science, 39,* RS1013, DOI: 10.1029/2003RS002903.

Oliason S. and Falola B. (2001) *Ice Thickness Interpretation Over Lake Vostok, Antarctica.* Applied Research Laboratories Technical Report ARL-LR-DO-01-01, The University of Texas at Austin.

Ono T., Kumamoto A., Yamaguchi Y., Yamaji A., Kobayashi T., Kasahara Y., and Oya H. (2008) Instrumentation and observation target of the Lunar Radar Sounder (LRS) experiment onboard the SELENE spacecraft. *Earth Planets Space, 60(4),* 321–332.

Ostro S. J. and Shoemaker E. M. (1990) The extraordinary radar echoes from Europa, Ganymede, and Callisto — A geological perspective. *Icarus, 85,* 335–345, DOI: 10.1016/0019-1035(90)90121-O.

Ostro S. J., West R. D., Janssen M. A., Lorenz R. D., Zebker H. A., Black G. J., Lunine J. I., Wye L. C., Lopes R. M., Wall

S. D., Elachi C., Roth L., Hensley S., Kelleher K., Hamilton G. A., Gim Y., Anderson Y. Z., Boehmer R. A., and Johnson W. T. K. (2006) Cassini RADAR observations of Enceladus, Tethys, Dione, Rhea, Iapetus, Hyperion, and Phoebe. *Icarus, 183(2),* 479–490, DOI: 10.1016/j.icarus.2006.02.019.

Pälli A., Moore J. C., and Rolstad C. (2003) Firn-ice transition zones of polythermal glaciers. *Ann. Glaciol., 37,* 298–304.

Pappalardo R. T. and Barr A. C. (2004) The origin of domes on Europa: The role of thermally induced compositional diapirism. *Geophys. Res. Lett., 31,* 1701, DOI: 10.1029/2003GL019202.

Pappalardo R. T., Head J. W., Greeley R., Sullivan R. J., Pilcher C., Schubert G., Moore W. B., Carr M. H., Moore J. M., Belton M. J. S., and Goldsby D. L. (1998) Geological evidence for solid-state convection in Europa's ice shell. *Nature, 391,* 365, DOI: 10.1038/34862.

Pappalardo R. T., Belton M. J. S., Breneman H. H., Carr M. H., Chapman C. R., Collins G. C., Denk T., Fagents S., Geissler P. E., Giese B., Greeley R., Greenberg R., Head J. W., Helfenstein P., Hoppa G., Kadel S. D., Klaasen K. P., Klemaszewski J. E., Magee K., McEwen A. S., Moore J. M., Moore W. B., Neukum G., Phillips C. B., Prockter L. M., Schubert G., Senske D. A., Sullivan R. J., Tufts B. R., Turtle E. P., Wagner R., and Williams K. K. (1999) Does Europa have a subsurface ocean? Evaluation of the geological evidence. *J. Geophys. Res., 104,* 24015–24056, DOI: 10.1029/1998JE000628.

Paterson W. S. B. (1994) *The Physics of Glaciers,* 3rd edition, Butterworth Heinmann.

Peake W. H. and Barrick D. E. (1967) *Scattering from Surface with Different Roughness Scales: Analysis and Interpretation.* ElectroScience Laboratory Publication 1388-26, Ohio State University.

Peters M. E., Blankenship D. D., and Morse D. L. (2005) Analysis techniques for coherent airborne radar sounding: Application to West Antarctic ice streams. *J. Geophys. Res., 110(B06303),* DOI: 10.1029/2004JB003222.

Peters M. E., Blankenship D. D., Carter S. P., Young D. A., Kempf S. D., and Holt J. W. (2007a) Along-track focusing of airborne radar sounding data from West Antarctica for improving basal reflection analysis and layer detection. *IEEE Trans. Geosci. Remote Sensing, 45(9),* 2725–2736, DOI: 10.1109/TGRS. 2007.897416.

Peters M. E., Blankenship D. D., Smith D. E., Holt J. W., and Kempf S. D. (2007b) The distribution and classification of bottom crevasses from radar sounding of a large tabular iceberg. *Geosci. Remote Sensing Lett., IEEE, 4(1),* 142–146.

Petrenko V. F. and Whitworth R. W. (1999) *Physics of Ice.* Oxford Univ., New York.

Phillips R. J., Zuber M. T., Smrekar S. E., Mellon M. T., Head J. W., Tanaka K. L., Putzig N. E., Milkovich S. M., Campbell B. A., Plaut J. J., Safaeinili A., Seu R., Biccari D., Carter L. M., Picardi G., Orosei R., Mohit P. S., Heggy E., Zurek R. W., Egan A. F., Giacomoni E., Russo F., Cutigni M., Pettinelli E., Holt J. W., Leuschen C. J., and Marinangeli L. (2008) Mars north polar deposits: Stratigraphy, age, and geodynamical response. *Science, 320(5880),* 1182–1185, DOI: 10.1126/science.1157546.

Picardi G., Plaut J. J., Biccari D., Bombaci O., Calabrese D., Cartacci M., Cicchetti A., Clifford S. M., Edenhofer P., Farrell W. M., Federico C., Frigeri A., Gurnett D. A., Hagfors T., Heggy E., Herique A., Huff R. L., Ivanov A. B., Johnson W. T. K., Jordan R. L., Kirchner D. L., Kofman W., Leuschen C. J., Nielsen E., Orosei R., Pettinelli E., Phillips R. J., Plettemeier D., Safaeinili A., Seu R., Stofan E. R., Vannaroni G., Watters T. R., and Zampolini E. (2005) Radar soundings of the subsurface of Mars. *Science, 310(5756),* 1925–1928.

Plaut J. J., Picardi G., and the MARSIS Team (2007) One Mars year of MARSIS observations: Global reconnaissance of the subsurface and ionosphere of Mars. In *Seventh International Conference on Mars,* Abstract #3341, Lunar and Planetary Institute, Houston (CD-ROM).

Porcello L. J., Jordan R. L., Zelenka J. S., Adams G. F., Phillips R. J., Brown W. E. Jr., Ward S. H., and Jackson P. L. (1974) The Apollo lunar sounder radar system. *Proc. IEEE, 62(6),* 769–783.

Robin G., Drewry D. J., and Meldrum D. T. (1977) International studies of ice sheet and bedrock. *Philos. Trans. R. Soc. B, 279,* 185–196.

Robin G., Doake C. S. M., Kohnen H., Crabtree R. D., Jordan S. R., and Moller D. (1983) Regime of the Filchner-Ronne ice shelves, Antarctica. *Nature, 302(5909),* 582–586.

Saunders R. S., Spear A. J., Allin P. C., Austin R. S., Berman A. L., Chandlee R. C., Clark J. , deCharon A. V., de Jong E. M., Griffith D. G., Gunn J. M., Hensley S., Johnson W. T. K., Kirby C. E., Leung K. S., Lyons D. T., Michaels G. A., Miller J., Morris R. B., Morrison A. D., Piereson R. G., Scott J. F., Shaffer S. J., Slonski J. P., Stofan E. R., Thompson T. W., and Wall S. D. (1992) Magellan mission summary. *J. Geophys. Res., 97(E8),* 13067–13090, DOI: 10.1029/92JE01397.

Schenk P. M. (2005) Landing site characteristics for Europa 1: Topography. In *Lunar and Planetary Science XXXVI,* Abstract #2321. Lunar and Planetary Institute, Houston (CD-ROM).

Schenk P. M. and Pappalardo R. T. (2004) Topographic variations in chaos on Europa: Implications for diapiric formation. *Geophys. Res. Lett., 31,* 16703, DOI: 10.1029/2004GL019978.

Schenk P., Matsuyama I., and Nimmo F. (2008) True polar wander on Europa from global-scale small-circle depressions. *Nature, 453(7193),* 368–371.

Seu R., Phillips R. J., Biccariy D., Orosei R., Masdea A., Picardi G., Safaeinili A., Campbell B. A., Plaut J. J., Marinangeli L. , Smrekar S. E., and Nunes D. C. (2007) SHARAD sounding radar on the Mars Reconnaissance Orbiter. *J. Geophys. Res., 112(E05S05),* DOI: 10.1029/2006JE002745.

Shabtaie S. and Bentley C. R. (1987) West Antarctic ice streams draining into the Ross Ice Shelf: Configuration and mass balance. *J. Geophys. Res., 92(B2),* 1311–1336.

Shepard M. K., Campbell B. A., Bulmer M. H., Farr T. G., Gaddis L. R., and Plaut J. J. (2001) The roughness of natural terrain: A planetary and remote sensing perspective. *J. Geophys. Res., 106(E12),* 32777–32795, DOI: 1029/2001JE001429.

Siegert M. J. (2005) Lakes beneath the ice sheet: The occurrence, analysis, and future exploration of Lake Vostok and other Antarctic subglacial lakes. *Annu. Rev. Earth Planet. Sci., 33(1),* 215–245, DOI: 10.1146/annurev.earth.33.092203.122725.

Souchez R., Jean-Baptiste P., Petit J. R., Lipenkov V. Y., and Jouzel J. (2003) What is the deepest part of the Vostok ice core telling us? *Earth Sci. Rev., 60(1–2),* 131–146, DOI: 10.1016/S0012-8252(02)00090-9.

Souchez R., Petit J. R., Jouzel J., de Angelis M., and Tison J. L. (2004) Reassessing Lake Vostok's behaviour from existing and new ice core data. *Earth Planet. Sci. Lett., 217(1–2),* 163–170.

Spikes V. B., Hamilton G. S., Arcone S. A., Kaspari S., and Mayewski P. A. (2004) Variability in accumulation rates from GPR profiling on the West Antarctic plateau. *Ann. Glaciol., 39(1),* 238–244.

Squyres S. W., Reynolds R. T., and Cassen P. M. (1983) Liquid water and active resurfacing on Europa. *Nature, 301,* 225, DOI: 10.1038/301225a0.

Stiles B., Gim Y., Hamilton G., Hensley S., Johnson W., Shimada J., West R., and Callahan P. (2006) Ground processing of Cassini RADAR imagery of Titan. In *2006 IEEE Conference on Radar,* p. 8, DOI: 10.1109/RADAR.2006.1631767.

Thyssen F. (1998) Special aspects of the central part of Filchner-Ronne Ice Shelf, Antarctic. *Ann. Glaciol., 11,* 173–179.

Waite A. and Schmidt S. (1962) Gross errors in height indication from pulsed radar altimeters operating over thick ice or snow. *Proc. Inst. Radio Engineers, 50(6),* 1515–1520, DOI: 10.1109/JRPROC.1962.288195.

Weeks W. F., Gow A. J., Kosloff P., and Digby-Argus S. (1990) The internal structure, compostion and properties of brackish ice from the Bay of Bothnia. In *Sea Ice Properties and Processes* (S. F. Ackley and W. F. Weeks, eds.), pp. 5–15. CRREL Monograph 90.

Wye L. C., Zebker H. A., Ostro S. J., West R. D., Gim Y., Lorenz R. D., and the Cassini RADAR Team (2007) Electrical properties of Titan's surface from Cassini RADAR scatterometer measurements. *Icarus, 188(2),* 367–385, DOI: 10.1016/j.icarus.2006.12.008.

Zahnle K., Schenk P., Levison H., and Dones L. (2003) Cratering rates in the outer solar system. *Icarus, 163(2),* 263–289.

Zarka P. (2004) Fast radio imaging of Jupiter's magnetosphere at low-frequencies with LOFAR. *Planet. Space Sci., 52(15),* 1455–1467, DOI: 10.1016/j.pss.2004.09.017.

Zarka P., Cecconi B., and Kurth W. S. (2004) Jupiter's low-frequency radio spectrum from Cassini/Radio and Plasma Wave Science (RPWS) absolute flux density measurements. *J. Geophys. Res., 109(A09S15),* DOI: 10.1029/2003JA010260.

Zolotov M. Y. and Shock E. L. (2001) Composition and stability of salts on the surface of Europa and their oceanic origin. *J. Geophys. Res., 106,* 32815–32828, DOI: 10.1029/2000JE001413.

Future Exploration of Europa

Ronald Greeley
Arizona State University

Robert T. Pappalardo
NASA Jet Propulsion Laboratory/California Institute of Technology

Louise M. Prockter
Johns Hopkins University Applied Physics Laboratory

Amanda R. Hendrix and Robert E. Lock
NASA Jet Propulsion Laboratory/California Institute of Technology

---

Reports from NASA, the National Research Council (NRC), the European Space Agency (ESA), and science community groups identify Europa as a priority for outer solar system exploration, especially for astrobiology. From these reports, an international group proposed the Europa Jupiter System Mission, involving a NASA Jupiter Europa Orbiter (JEO, the NASA element), which is the focus of this chapter. Current knowledge of Europa is reviewed, outstanding questions identified, and science objectives formulated. The JEO goal is to "Explore Europa to Investigate its Habitability;" this goal is to be met through objectives to study (in priority-order) (1) Europa's ocean and deep interior structure, (2) the icy shell and its structure, (3) its chemistry and composition, (4) the geology, and (5) the general Jupiter system, including the other major satellites and their atmospheres, the jovian plasma and magnetosphere, the parent planet Jupiter, and the small moons, rings, and dust.

1. INTRODUCTION

Since the first glimpses provided by Voyager, Europa has been recognized as an object worthy of exploration. As reviewed in the chapter by Alexander et al., the Galileo mission confirmed the suspicions that Europa in many ways is unique in the solar system and is a primary target for astrobiology. This chapter draws on results from a study commissioned by NASA and the European Space Agency (ESA) in 2008 for a future Europa Jupiter System Mission (EJSM) in which the Joint Jupiter Science Definition Team (JJSDT) (Table 1) reviewed previous studies for Europa (Table 2), assessed the current state of knowledge, formulated the key questions for the next mission, and identified the measurements that should be made to meet the exploration objectives. A candidate payload was also described, recognizing that the actual payload would be competed with the selection based on the best instruments to answer the key questions.

1.1. The Relevance of Jupiter System Exploration

Jupiter is the archetype for the giant planets of our solar system, and for the numerous planets known to orbit other stars. Jupiter's diverse Galilean satellites — three of which could harbor internal oceans — are the key to understanding the habitability of icy worlds. Thus, the JJSDT has identified "The Emergence of Habitable Worlds Around Gas Giants" as the overarching theme for a combined NASA-ESA mission.

Since the first extrasolar planets were detected in the late 1980s, their discovery has increased tremendously (*Vogt et al.,* 2005) and 10% of all Sun-like stars may have planets. With existing discovery techniques, almost all the known extrasolar planets are giant planets, more akin to Jupiter than to Earth. These bodies are expected to have large icy satellites that formed in their circumplanetary disks, analogous to Jupiter's Galilean satellites. Europa and Ganymede both could be geologically active and harbor internal salt-water oceans. They are straddled by Io and Callisto, key endmembers that tell of the origin and evolution of the Jupiter system. If extrasolar planetary systems are similar, then icy satellites may be the most common habitats in the universe — probably much more abundant than Earthlike habitats, which require very specialized conditions to permit surface oceans.

EJSM would afford the opportunity for detailed scrutiny of the archetype gas giant planet and its four diverse large satellites. EJSM would be invaluable for the insights it could provide into our solar system and into planetary architec-

TABLE 1. Joint Jupiter Science Definition Team.

Team Member	Affiliation	Team Member	Affiliation
Co-Chairs		*United States* (continued)	
Greeley, Ronald	Arizona State University	Showalter, Mark	SETI
Lebreton, Jean-Pierre	ESA/ESTEC	Showman, Adam	Univ. Arizona
		Sogin, Mitch	MBL
Study Scientists		Spencer, John	SWRI
Lebreton, Jean-Pierre	ESA/ESTEC	Waite, Hunter	SWRI
Pappalardo, Robert	Jet Propulsion Laboratory		
		European Union	
United States		Blanc, Michel	École Polytechnique
Anbar, Ariel	Arizona State University	Bruzzonem, Lorenzo	Univ. Trento
Bills, Bruce	NASA-Goddard/UCSD	Dougherty, Michele	Imperial College London
Blaney, Diana	Jet Propulsion Laboratory	Drossart, Pierre	Paris Observatory
Blankenship, Don	Univ. Texas at Austin	Grasset, Olivier	Univ. Nantes
Christensen, Phil	Arizona State University	Hußman, Hauke	DLR, Berlin
Dalton, Brad	Jet Propulsion Laboratory	Krupp, Norbert	Max Planck Inst.
Deming, Jody	Univ. Washington	Mueller-Wodarg, Ingo	Imperial College London
Fletcher, Leigh	Jet Propulsion Laboratory	Prieto-Ballasteros, Olga	INTA
Greenberg, Rick	Univ. Arizona	Prieur, Daniel	Univ Bretagne Occidentale
Hand, Kevin	Jet Propulsion Laboratory	Sohl, Frank	DLR, Berlin
Hendrix, Amanda	Jet Propulsion Laboratory	Tortora, Paolo	Univ. Bologna
Khurana, Krishan	UCLA	Tosi, Federico	IFSI
McCord, Tom	Bear Fight Center	Wurz, Peter	Univ. Bern
McGrath, Melissa	NASA-Marshall		
Moore, Bill	UCLA	*Japan*	
Moore, Jeff	NASA-Ames	Fujimoto, Masaki	ISAS, JAXA
Nimmo, Francis	UCSC	Kasaba, Yasumassa	Tohoku Univ.
Paranicas, Chris	JHU-APL	Sasaki, Sho	NOAJ
Prockter, Louise	JHU-APL	Takahashi, Yukihiro	Tohoku Univ.
Schubert, Jerry	UCLA	Takashima, Takeshi	ISAS, JAXA
Senske, David	Jet Propulsion Laboratory		

TABLE 2. Previous studies of Europa and heritage of science objectives and investigations.

Committee	Report Title	Reference
Europa Orbiter Science Definition Team	Europa Orbiter Mission and Project Description	NASA AO 99-OSS-04 (1999)
Committee on Planetary and Lunar Exploration (COMPLEX)	A Science Strategy for the Exploration of Europa	*SSB* (1999)
NASA Campaign Science Working Group on Prebiotic Chemistry in the Solar System	Europa and Titan: Preliminary Recommendations of the Campaign Science Working Group on Prebiotic Chemistry in the Outer Solar System	*Chyba et al.* (1999)
Solar System Exploration ("Decadal") Survey	New Frontiers in the Solar System: An Integrated Exploration Strategy	*SSB* (2003)
Jupiter Icy Moons Orbiter (JIMO) Science Definition Team	Report of the NASA Science Definition Team for the Jupiter Icy Moons Orbiter (JIMO)	*JIMO SDT* (2004)
Europa Focus Group of the NASA Astrobiology Institute	Europa Science Objectives	*Pappalardo* (2006)
Outer Planets Assessment Group (OPAG)	Scientific Goals and Pathways for Exploration of the Outer Solar System	*OPAG* (2006)
NASA Solar System Exploration Strategic Roadmap Committee	2006 Solar System Exploration Roadmap for NASA's Science Mission Directorate	*NASA* (2006)
Europa Science Definition Team	2007 Europa Explorer Mission Study: Final Report	*Clark et al.* (2007)
Jupiter System Observer Science Definition Team	Jupiter System Observer Mission Study: Final Report	*Kwok et al.* (2007)
The Laplace Team	Laplace: A Mission to Europa and the Jupiter System for ESA's Cosmic Vision Programme	*Blanc et al.* (2007)

ture and habitability throughout the universe. For these reasons, both NASA's Solar System Decadal Survey (*SSB,* 2003) and ESA's Cosmic Vision (*ESA,* 2005) emphasize the exploration of the Jupiter system to investigate the emergence of habitable worlds.

EJSM would include a NASA Jupiter Europa Orbiter (JEO) and an ESA Jupiter Ganymede Orbiter (JGO); a Jupiter Magnetospheric Orbiter (JMO) is also being considered by the Japan Aerospace Exploration Agency (JAXA). While the primary focus of JEO is to orbit Europa, the science return would encompass the entire jovian system with flybys of Io, Ganymede, and Callisto, along with ~2.5 years observing Jupiter's atmosphere, magnetosphere, and rings. Similarly, JGO would investigate Callisto and ultimately orbit Ganymede, and its focused observations of the Jupiter system would complement those of JEO. If it comes to fruition, JAXA's JMO has the potential to focus on particles and fields observations of the jovian magnetosphere. While JEO and JGO are complementary and potentially synergistic, both are designed as "stand-alone" missions as a contingency. The remainder of this chapter focuses on JEO and the potential for future exploration of Europa.

1.2. The Relevance of Europa Exploration

Europa's icy surface is thought to hide a global subsurface ocean with a volume more than twice that of Earth's oceans (see chapter by McKinnon et al.). The moon's surface is young, with an estimated age of about 60 m.y. (*Schenk et al.,* 2004; see chapter by Bierhaus et al.), implying that it is probably geologically active today. The molecular constituents of life have fallen onto Europa throughout solar system history, are potentially created by radiation chemistry at its surface, and may pour from vents at the ocean's floor (*Baross and Hoffmann,* 1985; *Pierazzo and Chyba,* 2002). On Earth, microbial extremophiles take advantage of environmental niches arguably as harsh as those within Europa's subsurface ocean (see chapter by Hand et al.). If the subsurface waters are eventually found to contain life, the discovery would spawn a revolution in our understanding of life in the universe.

It is now recognized that oceans could exist in several icy solar system objects. Titan could have a subsurface ammonia-water ocean (*Lorenz et al.,* 2008; *Tobie et al.,* 2005) sandwiched between ice polymorphs and ice, rather than being in direct contact with the mantle. Enceladus shows jets of water vapor and ice grains streaming from its surface, and emits a measurable heat flux from its south polar region (*Spencer et al.,* 2006), suggesting that pockets of water might exist below the surface (*Porco et al.,* 2006). If Neptune's Triton is a captured Kuiper belt object, it would have experienced tremendous tidal heating during its capture and subsequent orbital evolution (e.g., *Prockter et al.,* 2005). This heating likely produced an internal ocean; the ~100-m.y. crater age of the surface (*Stern and McKinnon,* 2000) suggests that a subsurface ocean might still exist. However, it too might not be in direct contact with the mantle (*Hussman et al.,* 2006). Although some Kuiper belt objects and satellites of Saturn and Uranus could have internal oceans, these are expected to be cold ammonia-rich oceans and energy sources for life are probably lacking.

It is tantalizing to consider whether life might exist in seas of ethane-methane or cold oceans of ammonia-water. Such environments could be fascinating places to search for life *unlike* we know it. However, it is more tractable to focus searches on potential icy habitats comparable to those in which we know biology could work. Experience with Earth shows that carbon-and-water-based life functions well over a wide range of temperature, pressure, and chemical regimes. Thus, Europa is the natural target for the first focused spacecraft investigation of the potential habitability of icy worlds. Its putative thin icy shell, candidate sources of chemical energy for life, and potentially active surface-ocean material exchange make it a top priority for exploration. The JEO would be the first critical step in understanding the potential of icy satellites as abodes for life.

Europa's high astrobiological potential and its complex interrelated processes have been recognized by many groups, including the National Research Council (NRC) and NASA. The NRC's Committee on Planetary and Lunar Exploration (COMPLEX) (*SSB,* 1999) stated that Europa "offers the potential for major new discoveries in planetary geology and geophysics, planetary atmospheres, and, possibly, studies of extraterrestrial life. In light of these possibilities, COMPLEX feels justified in assigning the future exploration of Europa a priority equal to that for the future exploration of Mars." The NRC's *New Frontiers in the Solar System* (referred to as the "Decadal Survey") (*SSB,* 2003) identified a Europa Geophysical Explorer as the top priority "Flagship" mission for the decade 2003–2013, principally because such a mission addresses the fundamental science question: "Where are the habitable zones for life in the solar system, and what are the planetary processes responsible for producing and sustaining habitable worlds?" This recommendation was reaffirmed by the NRC's Committee on Assessing the Solar System Exploration Program (CASSE) (*SSB,* 2007), which recommends "NASA should select a Europa mission concept and secure a new start for the project before 2011."

The NRC recommendations are in turn reflected in the NASA Science Mission Directorate's Solar System Exploration (SSES) Roadmap (*NASA,* 2006), which states "Europa should be the next target for a Flagship mission." NASA's scientific community-based Outer Planets Assessment Group (OPAG) "affirms the findings of the Decadal Survey, COMPLEX, and SSES, that Europa is the top-priority science destination in the outer solar system" (*OPAG,* 2006).

Noting that Europa's neighbors Ganymede and Callisto are also considered to have internal oceans, the NASA Roadmap further states "It is critical to determine how the components of the jovian system operate and interact, leading to potentially habitable environments within icy moons.

By studying the Jupiter system as a whole, we can better understand the type example for habitable planetary systems within and beyond our solar system."

NASA's 2007 Science Plan (*NASA,* 2007) echoes the many previous recommendations, calling Europa "an extremely high-priority target for a future mission." This document acknowledges that several icy satellites are now thought to have subsurface oceans, and states "Although oceans may exist within many of the solar system's large icy satellites, Europa's is extremely compelling for astrobiological exploration. This is because Europa's geology provides evidence for recent communication between the icy surface and ocean, and the ocean might be supplied from above and/or below with the chemical energy necessary to support microbial life." The Science Plan affirms the priority of Europa exploration in addressing fundamental themes of solar system origin, evolution, processes, habitability, and life.

The NASA Astrobiology Roadmap (*Des Marais et al.,* 2003) includes the goal "Explore for past or present habitable environments, prebiotic chemistry, and signs of life elsewhere in our solar system." A subsidiary objective is to "provide scientific guidance for outer solar system missions. Such missions should explore the Galilean moons Europa, Ganymede, and Callisto for habitable environments where liquid water could have supported prebiotic chemical evolution or life." A 2007 letter from the NASA Astrobiology Institute's Executive Council to the previous Europa Explorer SDT reaffirms that a Europa orbiter mission "is in its highest priority mission category for advancing the astrobiological goals of solar system exploration."

The exploration of the Jupiter system and Europa is similarly a high priority of ESA's Cosmic Vision strategic document (*ESA,* 2005). Key questions to be addressed include (1) What are the conditions for planet formation and the emergence of life? This question includes the subtopic "Life and habitability in the solar system," and the goal "Explore *in situ* the surface and subsurface of solid bodies in the solar system most likely to host — or have hosted — life." (2) How does the solar system work? This includes the subtopic "The giant planets and their environments," and the goal "Study Jupiter *in situ*, its atmosphere and internal structure."

If an orbital mission finds that Europa contains a habitable environment today, with active communication between subsurface water and the near surface, then a Europa Astrobiology Lander has been recommended as an important next step in the satellite's exploration (*NASA,* 2006). A Europa orbiter would feed forward to a future landed mission.

All the above recommendations are consistent with the NASA Vision for Space Exploration (*NASA,* 2004), which places high priority on robotic exploration across the solar system, "In particular, to explore Jupiter's moons . . . to search for evidence of life (and) to understand the history of the solar system . . ."

There are many high-priority targets for exploration in our solar system, each offering potential for rich science return. Europa continues to top the priority list for the outer solar system because of its scientific potential, especially related to habitability. The scientific foundation for a mission to Europa has been clearly laid.

2. SCIENCE BACKGROUND

Although scientific studies of Europa predate the space age, the understanding of the satellite has increased greatly in the past dozen years since the Galileo mission. The following summarizes current knowledge for Europa, outlines the broad cross-cutting themes including habitability and planetary processes, and notes the outstanding science issues to be addressed with new data.

2.1. Habitability

Europa's probable subsurface ocean has profound implications in the search for past or present life beyond Earth (see chapter by Hand et al.). Coupled with the discovery of active microbial life in harsh terrestrial environments (*Rothschild and Mancinelli,* 2001), Europa takes on new importance in searching for habitable worlds. Life as we know it depends upon liquid water, a photo- or chemical-energy source, complex organics, and inorganic compounds of N, P, S, and Fe, and various trace elements. Europa appears to meet these requirements and is distinguished by potentially enormous volumes of liquid water and geologic activity that promotes the exchange of surface materials with the subice environment (see chapters by Moore and Hussmann and by Vance and Goodman).

Life on Earth occupies niches supplied by either chemical or solar energy. Europa's ocean has likely persisted from close to the origin of the jovian system to the present (*Cassen et al.,* 1982), although its chemical characteristics likely evolved (*McKinnon and Zolensky,* 2003; see also chapter by Zolotov and Kargel). Inferences from its young surface and models suggest that an ocean and hydrothermal system may lie beneath a sheet of ice a few to tens of kilometers thick (*Greeley et al.,* 2004). Tidal deformation may drive heating and geologic activity within Europa, and there could be brine pockets within the ice, partial melt zones, and clathrates. Hydrothermal systems driven by tidal heating or volcanic activity could serve as a favorable environment for prebiotic chemistry or sustaining microbial chemotrophic organisms.

Cycling of water through and within the icy shell, ocean, and permeable upper rocky mantle could maintain an ocean rich with oxidants and reductants necessary for life. In order to address this aspect of Europa's habitability a better understanding of the mantle and icy shell is needed.

Radiolytic chemistry on the surface is responsible for the production of O_2, H_2O_2, CO_2, SO_2, SO_4, and other yet to be discovered oxidants (see chapter by Johnson et al.). At present, mechanisms and timescales for delivery of these materials to the subsurface are poorly constrained. Similarly, cycling of ocean water through seafloor minerals could replenish the water with biologically useful reductants. If much of the tidal energy dissipation occurs in the mantle (see chapter by Moore and Hussman), then there could be significant cycling between the ocean water and rocky mantle. Conversely, if most of the tidal dissipation occurs in the icy shell, then the ocean water could be depleted in the reductants needed for biochemistry. Chemical cycling of energy on Europa is arguably the greatest uncertainty in our ability to assess Europa's habitability.

Although it is not known if life existed or persists today on Europa with available information, it is possible through new spacecraft data to determine if extant conditions are capable of supporting organisms. Key to this question is the occurrence of liquid water beneath the icy surface and whether the geologic and geophysical properties can support the synthesis of organic compounds and provide the energy and nutrients needed to sustain life.

2.2. Ocean and Interior

Europa's surface suggests recently active processes operating in the icy shell. Jupiter raises gravitational tides on Europa, which contribute to thermal energy in the icy shell and rocky interior (*Ojakangas and Stevenson*, 1989; see chapters by Sotin et al., Schubert et al., and Goodman and Vance), produce near-surface stresses responsible for some surface features, and may drive currents in the ocean. Although little is known about the internal structure, most models include an outer icy shell underlain by liquid water, a silicate mantle, and iron-rich core (*Anderson et al.*, 1998a). Means to constrain these models include measurements of the gravitational and magnetic fields, topographic shape, and rotational state of Europa, each of which includes steady-state and time-dependent components. Additionally, the surface heat flux and local thermal anomalies may yield constraints on internal heat production and activity. Results can be used to characterize the ocean and the overlying icy shell and provide constraints on the deep interior structure and processes.

2.2.1. Gravity. Observations of the gravitational field provide information on the interior mass distribution. For a spherically symmetric body, all points on the surface would have the same gravitational acceleration, while in those regions with more mass, gravity will be greater. Lateral variations in the field strength thus indicate lateral variations in density structure. Within Europa, principal sources of static gravity anomalies can be due to thickness variations of the icy shell, or topography on the ocean floor. If the icy shell is isostatically compensated, it will only yield very small gravity signatures. Gravity anomalies that are not spatially coherent with surface topography are presumed to arise from greater depths.

One of the most diagnostic gravitational features is the amplitude and phase of the time-dependent signal due to tidal deformation (Fig. 1). The forcing from Jupiter is well known, and the response will be much larger if a fluid layer decouples the ice from the interior, permitting unambiguous detection of an ocean and characterization of the ocean and icy shell. With the surface ice decoupled from the rocky

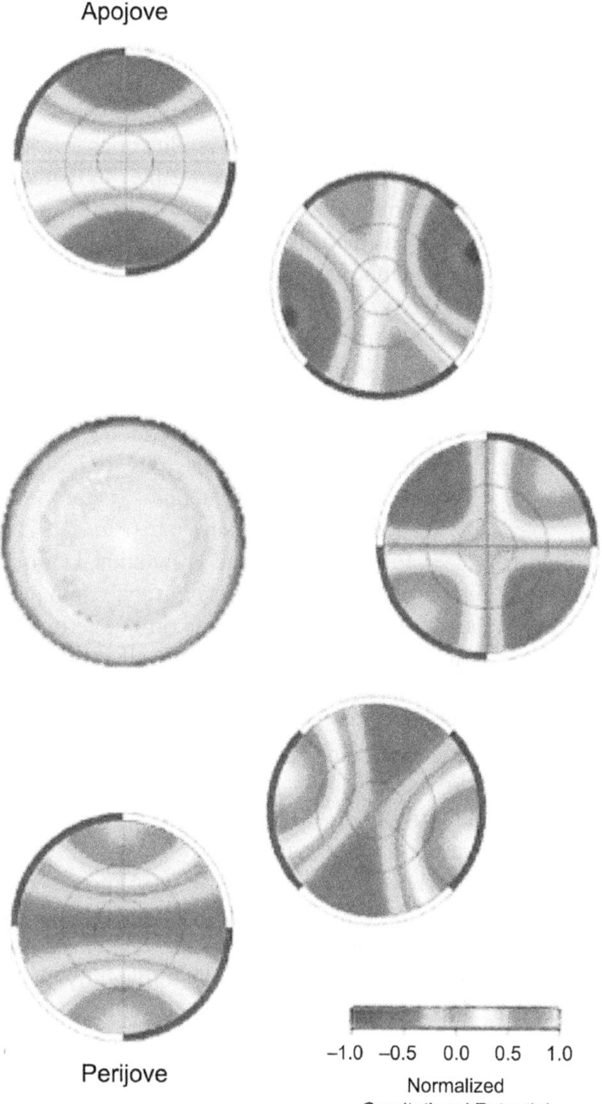

Fig. 1. See Plate 35. Europa experiences a time-varying gravitational potential field in its eccentric orbit (eccentricity = 0.0094), with a 3.551-day (1 eurosol) period. Its tidal amplitude varies proportionally to the gravitational potential, causing Europa to flex as it orbits. This view shows the north pole of Jupiter as Europa orbits counterclockwise with its prime meridian pointed toward Jupiter. Measuring the varying gravity field and tidal amplitude simultaneously allows the interior rigidity structure of Europa to be derived, revealing the properties of its ocean and icy shell (*Moore and Schubert,* 2000).

interior, the amplitude of the semidiurnal tide on Europa is ~30 m, vs. ~1 m in the absence of an ocean (*Moore and Schubert,* 2000).

2.2.2. Topography. At long wavelengths (hemispheric-scale), topography is mainly a response to tides and thickness variations of the icy shell driven by tidal heating (*Ojakangas and Stevenson,* 1989; see chapter by Nimmo and Manga), and is thus diagnostic of internal tidal processes. At intermediate wavelengths (hundreds of kilometers), topographic amplitudes and correlation with gravity are diagnostic of the density and thickness of the icy shell. At the shortest wavelengths (kilometer-scale), small geologic features tend to have topography diagnostic of formational processes.

2.2.3. Rotation. Tidal dissipation probably drives Europa's rotation into equilibrium, with implications for both the direction and rate of rotation (see chapter by Bills et al.). The mean rotation period should nearly match the mean orbital period, so that the sub-Jupiter point will librate in longitude, with an amplitude equal to twice the orbital eccentricity. If the body behaves rigidly, the expected amplitude of this forced libration should be ~100 m (*Comstock and Bills,* 2003), but if the icy shell is mechanically decoupled from the silicate interior, then the libration could be three times larger.

Similar forced librations in latitude are due to the finite obliquity, and are also diagnostic of internal structure. The spin pole is expected to occupy a Cassini state (*Peale,* 1976), similar to that of Earth's Moon. The gravitational torque exerted by Jupiter on Europa will cause Europa's spin pole to precess about the orbit pole, while the orbit pole in turn precesses about Jupiter's spin pole, with all three axes remaining coplanar. The obliquity required for Europa to achieve this state is ~0.1°, but depends upon the moments of inertia, and is thus diagnostic of internal density structure (*Bills,* 2005).

2.2.4. Magnetic field. Magnetic fields interact with conducting matter at scales ranging from atomic to galactic, and are produced when currents flow in response to electric potential differences between interacting conducting fluids or solids. Many planets generate stable magnetic fields in convecting cores or inner shells through dynamos powered by internal heat or gravitational settling of the interior. Europa does not generate its own magnetic field, suggesting that its core has either frozen or is still fluid but not convecting.

Europa, however, responds to the rotating magnetic field of Jupiter through electromagnetic induction (*Khurana et al.,* 1998; see chapter by Khurana et al.). In this process, eddy currents are generated on the surface of a conductor to shield its interior from changing external fields. The eddy currents generate their own magnetic field (the induction field) external to the conductor, as measured by a magnetometer.

The induction technique exploits the fact that the primary alternating magnetic field at Europa is provided by Jupiter because its rotation and magnetic dipole axes are not aligned. It is now believed that the induction signal seen in Galileo data arises within a subsurface ocean. The measured signal remained in phase with the primary field of Jupiter (*Kivelson et al.,* 2000), thus unambiguously proving that the perturbation signal is a response to Jupiter's field.

Modeling the measured induction signal, although indicative of an ocean, suffers from nonuniqueness in the derived parameters because of the limited data from Galileo, forcing certain assumptions. Nevertheless, the analysis of *Zimmer et al.* (2000) reveals that the putative ocean must have a conductivity of at least 0.06 S/m. Recently, *Schilling et al.* (2004) determined the ratio of induction field to primary field at 0.96 ± 0.3, leading *Hand and Chyba* (2007) to infer that the icy shell is <15 km thick and an ocean water conductivity >6 S/m.

To determine the ocean thickness and conductivity, magnetic sounding of the ocean at multiple frequencies is required. The depth to which electromagnetic waves penetrate is inversely proportional to the square root of its frequency. Thus, longer-period waves sound deeper and could provide information on the ocean's thickness, the mantle, and the metallic core.

For Europa, the two dominant frequencies are those of Jupiter's synodic rotation period (~11 h) and Europa's orbital period (~85 h). Observing the induction response at these frequencies could allow determination of both the ocean thickness and the conductivity (Fig. 4 of chapter by Khurana et al.).

Remaining key questions to be addressed regarding Europa's ocean, bulk properties of the icy shell, and deeper interior include the following: (1) Does Europa undoubtedly have a subsurface ocean? (2) What are the salinity and thickness of Europa's ocean? (3) Does Europa exhibit kilometer-scale variations in the thickness of its icy shell? (4) Does Europa have a nonzero obliquity and if so, what controls it? (5) Does Europa possess an Io-like mantle?

2.3. Icy Shell

Understanding the internal structure of the icy shell is essential for assessing the processes that connect the ocean to the surface (see chapters by Nimmo and Manga and by Blankenship et al.). The structure and composition of the surface result from various geologic processes and includes material transport and chemical exchange through the shell. The icy shell may have experienced one or more episodes of thickening and thinning, directly exchanging material with the ocean at its base. Thermal processing could also alter the internal structure of the shell through convection or local melting. Exogenic processes such as cratering influence the surface and deeper structure.

2.3.1. Thermal processing. The thermal structure of the icy shell is governed primarily by heat from the interior (see chapters by Moore and Hussman and by Barr and Showman). Regardless of the properties of the shell or heat transport, the uppermost several kilometers is thermally conductive, cold, and stiff. The thickness of this conductive "lid" is set by the total amount of heat that must be transported, and thus measurement of the thickness of the brittle shell is a constraint on interior heat production. Convective instabilities can result in thermal variations in the shell that may be associated with surface features with scales of 1 km to hundreds of kilometers. When warm, relatively pure ice diapirs from the interior approach the surface, they may be far from the pure-ice melting point, but may be above the eutectic of materials trapped in the lid. This could melt

Fig. 2. See Plate 36. Europa's ice, assuming a thick shell model: Convective diapirs could cause thermal perturbations and partial melting in the overlying rigid ice. Faulting driven by tidal stresses (upper surface) could result in frictional heating. Impact structures might show central refrozen melt pools, surrounded by ejecta.

parts of the shell above the flattening diapir (Fig. 2). The horizon associated with the melt would provide a measure of the conductive layer thickness. Other sources of local heat such as friction on faults may lead to similar melting (*Gaidos and Nimmo,* 2000).

2.3.2. Ice-ocean exchange. Europa's icy shell has likely experienced phases of thickening and thinning, as the orbital evolution alters the internal heating from tides (*Hussmann and Spohn,* 2004). For example, the shell may thicken similar to ice that accretes beneath the ice shelves of Antarctica where ice crystals form directly from the ocean (*Moore et al.,* 1994). This model is characterized by slow accretion (freezing) or ablation (melting) on the lower side of the icy crust (*Greenberg et al.,* 1999). Temperature gradients are primarily a function of ice thickness, and the temperature profile is described by a simple diffusion equation for a conducting ice layer (*Chyba et al.,* 1998). The low temperature gradients at any ice-water interface, combined with impurities, would likely lead to structural horizons resulting from contrasts in ice crystal fabric and composition.

Melt-through of thin ice probably also would lead to ice accretion beneath the melt-features on the surface. This process will result in a sharp boundary between old ice (or rapidly frozen surface ice) and the deeper accreted ice. The amount of accreted ice would be directly related to the time since melt-through and could be compared with the amount expected based on the surface age. Testing various hypotheses of ice-ocean material exchange requires measuring the depth of interfaces to a resolution of a few hundred meters, and horizontal resolutions of approximately kilometers.

2.3.3. Surface and subsurface structure. Europa represents a unique tectonic regime in the solar system, and the processes controlling the distribution of strain in its icy shell are uncertain. Tectonic structures could range from subhorizontal extensional fractures to near-vertical strike-slip features, and will produce structures associated primarily with faulting of fractured ice (see chapter by Kattenhorn and Hurford). This also involves zones of deformational melt, injection of water, or preferred orientation of crystalline fabric. Some faults may show local alteration of preexisting structure including fluid inclusions or juxtaposition of dissimilar regions. There are many outstanding issues regarding tectonic features, including correlation of subsurface structure with surface properties (length, stratigraphic position, height and width of the ridges) to test hypotheses for formation of the fractures and ridges.

Extensional structures observed on Europa (e.g., gray bands) may be particularly important for understanding material exchange processes (see chapter by Prockter and Patterson). If the analogy with terrestrial spreading centers (*Pappalardo and Sullivan,* 1996) is correct, band material is newly supplied from below and may have a distinct structure. Horizontal resolutions of a few hundred meters are needed to discriminate such processes, along with the ability to image structures sloping more than a few degrees. Additionally, tens of meters of vertical resolution are required to image near-surface melt zones.

Impact structures can reflect significant disruption of the shell. At an impact site, the ice is fractured, heated, and parts are melted, and ejecta blankets the surrounding terrain. Rebound of the crater leads to tectonism that can include faulting and other subsurface structures detectable by sounding. An outstanding mystery on Europa is the process by which craters are erased from the surface. It may be possible to find the subsurface signature of impacts that are no longer evident on the surface, which would constrain ideas for the resurfacing processes.

Key questions to be addressed regarding the icy shell by future missions include the following: (1) What is the thickness of the icy shell? (2) What is the structure within the icy shell? (3) Do pockets of liquid water and/or brine exist? (4) Is there evidence for diapiric activity, past or present? (5) Have diapirs or "melt-through" zones provided exchange of material between the ocean and the surface?

2.4. Composition

Surface materials may be ancient, derived from the ocean, altered by radiation, or exogenic in origin. Europa's bulk density and solar system models suggest the presence of both water and silicates. It is likely that differentiation and mixing of water with silicates and carbonaceous materials resulted in chemical alteration and redistribution, with interior transport by melting and/or solid-state convection and diapirism bringing materials to the surface. High-energy particles from Jupiter leave imprints on the surface that provide clues to the exogenic environment, but can also complicate understanding evolution and modification of the surface. Moreover, surface materials can be incorporated into the subsurface and react with the ocean, or can be sputtered from the surface to form Europa's tenuous atmosphere. Thus, characterizing surface composition and chemistry provides fundamental information on the properties and habitability of Europa (see chapters by Carlson et al. and Zolotov and Kargel).

2.4.1. Ice and non-ice composition. Telescopic observations and spacecraft data (e.g., *Kuiper,* 1957; *Moroz,* 1965; *Clark and McCord,* 1980; *Dalton,* 2000; *McCord,* 2000; *Spencer et al.,* 2005) show that Europa's surface is composed primarily of crystalline and amorphous water ice (*Pilcher et al.,* 1972; *Clark and McCord,* 1980; *Hansen and McCord,* 2004). The dark, non-icy materials on the surface help unravel the geological history, and determining their composition is the key to understanding their origin. Spatial distributions and context provide clues to surface processes and the connections to the interior. Understanding this linkage provides constraints on the nature of the interior, potential habitability, and processes and timescales through which interior materials reach the surface. Compositional variations in surface materials may reflect age differences indicative of recent activity, while the discovery of active vents or plumes would demonstrate current connections with the subsurface.

Non-ice components include CO_2, SO_2, H_2O_2, and O_2 based on comparison with laboratory spectra of the relevant compounds (*Lane et al.,* 1981; *Noll et al.,* 1995; *Smythe et al.,* 1998; *Carlson,* 1999, 2001; *Carlson et al.,* 1999a,b; *Spencer and Calvin,* 2002; *Hansen and McCord,* 2008). Spectral observations from the Galileo Near Infrared Mapping Spectrometer (NIMS) of disrupted dark and chaotic terrain indicate water bound in non-ice hydrates. Hydrated materials observed in regions of surface disruption could be magnesium and sodium sulfates that originate from subsurface ocean brines (*McCord et al.,* 1998b, 1999). Alternatively, they may be sulfuric acid hydrates created by radiolysis of sulfur from Io, processing of endogenic SO_2, or from ocean-derived sulfates or other S-bearing species (*Carlson et al.,* 1999b, 2002, 2005). It is also possible that these surfaces have a combination of hydrated sulfate salts and sulfuric acid (*Dalton,* 2000, 2007; *McCord et al.,* 2001, 2002; *Carlson et al.,* 2005; *Orlando et al.,* 2005; *Dalton et al.,* 2005). Thus, an important objective is to resolve the compositions and origins of the hydrated materials.

Earth-based telescopes detected sulfur species thought to be linked to effects of Jupiter's magnetosphere (e.g., *Noll et al.,* 1995). *Brown and Hill* (1996) first reported a cloud of sodium around Europa, and *Brown* (2001) found a cloud of potassium and reported that the Na/K ratio could reflect endogenic sputtering.

A broad suite of additional compounds is predicted for Europa based on observations of other icy satellites, as well as from experiments with irradiated ices, theoretical simulations, and geochemical and cosmochemical arguments. Organic molecular groups, such as CH and CN, occur on the other icy satellites (*McCord et al.,* 1997, 1998a), and their presence or absence on Europa is important to understanding potential habitability. Other compounds that may be detected by high-resolution spectroscopy include H_2S, OCS, O_3, HCHO, H_2CO_3, SO_3, $MgSO_4$, H_2SO_4, H_3O^+, $NaSO_4$, HCOOH, CH_3OH, CH_3COOH, and more complex species (*Moore,* 1984; *Delitsky and Lane,* 1997, 1998; *Hudson and Moore,* 1998; *Moore and Hudson,* 2003; *Brunetto et al.,* 2005; see also chapter by Zolotov and Kargel).

As molecules become more complex their radiation cross-section increases and they are more susceptible to alteration by radiation. Radiolysis and photolysis can alter the original materials and produce highly oxidized species that react with other non-ice materials to form a wide array of compounds. Given the extreme radiation environment of Europa (see chapter by Paranicas et al.), organic molecules or molecular fragments are not expected in older deposits nor in those exposed to greater radiation (*Johnson and Quickenden,* 1997; *Cooper et al.,* 2001). They might, however, survive in younger deposits or in regions of lesser radiation.

Improved spectral observations over broad ranges and high spectral and spatial resolution, together with laboratory studies, are needed to understand Europa's surface chemistry. These data will provide major improvements in the identification of the original and derived compounds, radiation environment, and associated reaction pathways.

2.4.2. Relationship of composition to processes. Galileo's instruments were designed to study surface compositions on regional scales. The association of hydrated and dark materials with certain geologic terrains suggests an endogenic source for the emplaced materials, although these may have been altered by radiolysis. Many surface features with compositionally distinct materials appear to have been formed by tectonic processes, suggesting that the associated materials are derived from the subsurface.

Major open questions include the links between surface composition and the underlying ocean and rocky interior (*Fanale et al.,* 1999; *Kargel et al.,* 2000; *McKinnon and*

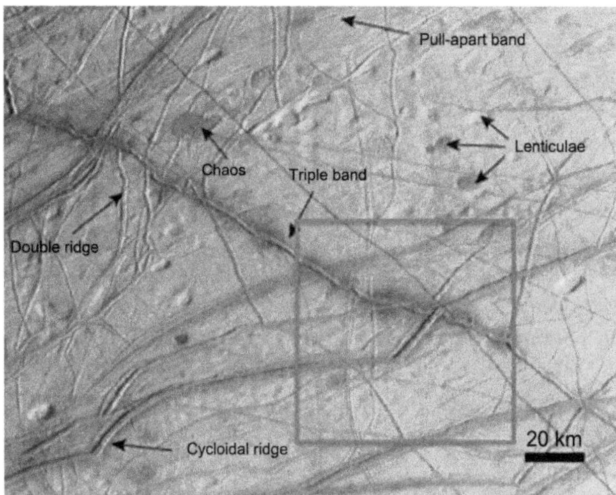

Fig. 3. Europa's diverse surface shows different styles of deformation, which provide clues to its geology, possible connections to tidal processes, and the subsurface ocean. The Galileo NIMS footprint (box) sampled and "mixed" multiple terrain types.

Zolensky, 2003; see chapter by Zolotov and Kargel), and the relative significance of radiolytic processing (*Johnson and Quickenden,* 1997; *Cooper et al.,* 2001; *Carlson et al.,* 2002, 2005). To test these hypotheses, compositional data are required at scales sufficient to resolve geologic features. One of the critical limitations of NIMS data is the low spatial resolution of the high-quality spectra and the limited spatial coverage. The spectra to identify hydrated materials were typically averaged from areas ~75 × 75 km (*McCord et al.,* 1998b; *Carlson et al.,* 1999b) (although a few higher-resolution "postage stamp" datasets were obtained). This typical footprint is shown in Fig. 3, illustrating the problem of "mixing" of terrains. Future observations must resolve non-ice materials at ~100-m scales. In addition, sampling a wide range of latitudes and longitudes is needed to understand global effects such as implantation, temperature dependence, and surface ages. Ultraviolet to IR spectroscopy is needed to identify organic, ice, non-ice, and radiolytically generated materials. Such data, together with images, can provide the spatial correlations necessary to develop models for the origin and history of the surface.

In addition to compositional differences associated with recent geological activity, changes related to exposure age will also provide evidence for sites of recent or current activity. The composition of even the icy parts of Europa is variable in space and time. Polar fine-grained deposits suggest frosts formed from ice sputtered or sublimated from other areas (*Clark et al.,* 1983; *Dalton,* 2000; *Hansen and McCord,* 2004). Equatorial ice regions are more amorphous than crystalline, perhaps due to radiation damage, and mapping ice crystallinity might be used to assess relative age or radiation dose. Venting or transient gaseous activity on Europa would indicate present-day surface activity, and could be detected by UV, IR, or millimeter spectroscopy, similar to those on Enceladus (*Porco et al.,* 2006; *Spencer et al.,* 2006; *Hansen et al.,* 2006; *Waite et al.,* 2006). If a subsurface ocean is present and outgases through fissures, it might result in transient activity, and its composition could provide clues to ocean composition.

Exogenic processes are also important, and much is unknown on the chemistry and sources of implanted materials. Magnetic field measurements of ion-cyclotron waves in the wake of Europa provide evidence of sputtered and recently ionized Cl, O_2, SO_2, and Na ions (*Volwerk et al.,* 2001). Medium-energy ions (tens to hundreds of keV) deposit energy in the upper tens of micrometers; heavier ions, such as those of oxygen and sulfur, have an even shorter depth of penetration, while MeV electrons can penetrate and affect the ice to a depth of more than 1 m (see chapters by Johnson et al. and Paranicas et al.). The energy of these particles breaks bonds to sputter water molecules, molecular oxygen, and impurities within the ice (*Cheng et al.,* 1986), producing the observed atmosphere and contributing to surface erosion.

A major issue is the exogenic vs. endogenic origin of volatiles such as CO_2 and their behavior in time and space. CO_2 was reported on Callisto and Ganymede, with hints of CO_2 (*McCord et al.,* 1998a), SO_2 (*Smythe et al.,* 1998), and H_2O_2 (*Carlson et al.,* 1999b). Recent analyses of NIMS spectra indicate the concentration of CO_2 and other non-ice compounds on the antijovian and trailing sides of Europa (*Hansen and McCord,* 2008), suggesting an endogenic origin. Radiolysis of CO_2 and H_2O ices is expected to produce additional compounds (*Moore,* 1984; *Delitsky and Lane,* 1997, 1998; *Moore and Hudson,* 2003; *Brunetto et al.,* 2005). Determining the presence and source of organic compounds, such as CH and CN groups detected by IR spectroscopy at Callisto and Ganymede (*McCord et al.,* 1997, 1998b) and tentatively identified on Phoebe (*Clark et al.,* 2005), would be important for evaluating the astrobiological potential of Europa, especially if there is demonstrable association with the ocean.

Some surface constituents are directly exogenic. For example, Io's volcanos release SO_2 that is dissociated and ionized, accelerated by Jupiter's magnetic field, and implanted in Europa's ice. Once there it can form new molecules and some of the dark surface components. It is important to separate surface materials formed by implantation from those that are endogenic. For example, the detected Na/K ratio is supportive of an endogenic origin — and perhaps an ocean source — for Na and K (*Brown,* 2001; *Johnson et al.,* 2002; *McCord et al.,* 2002; *Orlando et al.,* 2005).

The relative importance of endogenic vs. exogenic sources of non-ice constituents depends on factors such as the radiation environment. As a result, detailed analysis of spectral observations of disrupted terrain on the leading and trailing hemispheres, which encounter far different radiolytic fluxes, would help to determine radiation effects and unravel the endogenic history.

Some key outstanding questions to be addressed regarding Europa's chemistry and composition include the following: (1) Are endogenic organic materials on the surface?

(2) Is chemical material from depth carried to the surface? (3) Is irradiation the principal cause of alteration of Europa's surface materials? (4) Do materials formed from ion implantation play a major role in surface chemistry?

2.5. Geology

Europa's surface is geologically young, and parts may be active today (see chapter by Bierhaus et al.). This youth is inherently linked to the ocean and the effects of gravitational tides, which trigger processes that include fracturing of the icy shell, resurfacing, and possibly release of materials from the interior. Clues to these and other processes are provided by features such as linear fractures and ridges (see chapters by Kattenhorn and Hurford and by Prockter and Patterson), chaotic terrain (see chapter by Collins and Nimmo), and impact craters (see chapter by Schenk and Turtle).

2.5.1. Linear features. Europa's unusual surface is dominated by tectonic features in the form of linear ridges, bands, and fractures (Fig. 4). Ridges are common and appear to have formed throughout the visible history. They range from 0.1 to >500 km long, are as wide as 2 km, and can be several hundred meters high. Ridges include simple structures, double ridges separated by a trough, and intertwining ridge-complexes. Whether these represent different processes or stages of the same process is unknown. Cycloidal ridges are similar to double ridges, but form chains of linked arcs.

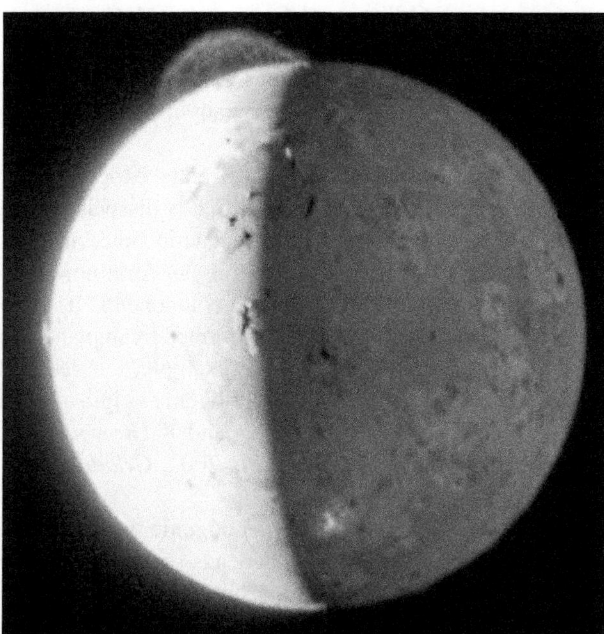

Fig. 4. Volcanic plumes on Io imaged by New Horizons in 2007. The 290-km-high plume from the polar volcano Tvashtar is seen at the top, while the plume from Prometheus is on the left. Beginning with Voyager discoveries, Prometheus has been active during all spacecraft flybys. Long-term observations and flybys with JEO will provide unprecedented detail on Io's active volcanism.

Most models of linear feature formation include fracturing in response to processes within the icy shell (*Greeley et al.,* 2004). Some models suggest that liquid oceanic material or warm mobile subsurface ice squeezes through fractures to form the ridge, while others suggest that ridges form by frictional heating and possibly melting along fracture shear zones. Thus, ridges might represent regions of communication among the surface, icy shell, and ocean, plausibly providing a means for surface oxidants to enter the ocean. Some features, such as cycloidal ridges, appear to be a direct result of Europa's tidal cycle (*Hoppa et al.,* 1999).

Bands reflect fracturing and lithospheric separation, much like seafloor spreading on Earth, and most display bilateral symmetry (e.g. *Sullivan et al.,* 1998). The youngest bands tend to be dark, while older bands are bright, suggesting brightening with time. Geometric reconstruction of bands suggests that a spreading model is appropriate, indicating extension in these areas and possible contact with the ocean (*Tufts et al.,* 2000; *Prockter et al.,* 2002).

Fractures are narrow (hundreds of meters to the ~10-m limit of image resolution) and can exceed 1000 km in length. Some fractures cut across nearly all surface features, indicating that the icy shell is subject to deformation on the most recent timescales. The youngest ridges and fractures could be active today in response to tidal flexing. Subsurface sounding could help identify zones of warm ice coinciding with current or recent activity. Young ridges may be places where the ocean has recently exchanged material with the surface, and would be prime targets as potential habitable niches.

2.5.2. Chaotic terrain. Europa's surface has been disrupted into circular lenticulae and irregularly shaped chaos zones (see chapter by Collins and Nimmo). Lenticulae include pits, spots of dark material, and domes where the surface is upwarped and commonly broken. *Pappalardo et al.* (1998) argued that these features are typically ~10 km across, and possibly formed by upwelling of compositionally or thermally buoyant ice diapirs through the icy shell. In such a case, their size distribution would imply the thickness of the icy shell to be at least 10–20 km at the time of formation (*McKinnon,* 1999). An alternative model suggests that there is no dominant size and that lenticulae are small members of chaos (*Greenberg et al.,* 1999), formed through either direct (melting) or indirect (convection) communication between the ocean and surface (e.g., *Carr et al.,* 1998a).

Chaos is characterized as fractured plates of ice shifted into new positions within a matrix. Much like a jigsaw puzzle, many plates can be fit back together. Some ice blocks appear to have disaggregated and foundered into the surrounding finer-textured matrix, while other chaos areas stand higher than the surrounding terrain. Models of chaos formation suggest whole or partial melting of the icy shell, perhaps enhanced by local pockets of brine (*Head and Pappalardo,* 1999). Chaos and lenticulae commonly have associated dark reddish material thought to be derived from the subsurface, possibly from the ocean. However, these and related models are poorly constrained because the total energy partitioning within Europa is not known, nor are details of the composition of non-ice components. Imaging, subsurface sounding, and topographic mapping are required to understand the formation of chaotic terrain and its implications for habitability.

2.5.3. Impact features. Only 24 impact craters ≥10 km have been identified on Europa (*Schenk et al.,* 2004; see chapter by Schenk and Turtle), reflecting the young surface. This is remarkable in comparison to Earth's Moon, which is only slightly larger but far more heavily cratered. The youngest known europan crater is 24-km-diameter Pwyll, which retains bright rays and likely formed less than 5 m.y. ago (*Zahnle et al.,* 1998; see chapter by Bierhaus et al.). Complete global imaging will allow a more comprehensive determination of Europa's surface age and help identify the very youngest areas.

Crater morphology provides insight into ice thickness at the time of impact. Morphologies vary from bowl-shaped depressions with crisp rims, to shallow depressions with smaller depth-to-diameter ratios. Craters up to 25–30 km in diameter have morphologies consistent with formation in a warm but solid icy shell, while the two largest impacts (Tyre and Callanish) might have punched through brittle ice about 20 km deep into a liquid zone (*Moore et al.,* 2001; *Schenk et al.,* 2004).

2.5.4. Geologic history. Determining the geologic histories of planetary surfaces requires identifying and mapping surface units and structures and placing them into a time sequence. In the absence of absolute ages derived from rock samples, planetary surface ages are assessed from impact crater distributions, with more heavily cratered regions reflecting greater ages. The paucity of impact craters on Europa precludes this technique. Thus, superposition (i.e., younger materials seen "on top" of older materials) and cross-cutting relations are used to assess sequences of formation (*Figueredo and Greeley,* 2004; see chapter by Doggett et al.). Unfortunately, only 10% of Europa has been imaged at sufficient resolution to understand relationships among surface features. For most of Europa, data are both incomplete and disconnected from region to region, making the global surface history difficult to decipher. Where images of sufficient resolution (better than 200 m/pixel) exist, it appears that the style of deformation evolved through time from ridge and band formation to chaotic terrain (*Greeley et al.,* 2004), although there are huge areas of the surface where this sequence is uncertain (e.g., *Riley et al.,* 2000). Europa's surface features generally brighten and become less red through time, so albedo and color can serve as a proxy for age (*Geissler et al.,* 1998).

Quantitative topographic data can provide information on the origin of geologic features and may show trends with age. Profiles across ridges, bands, and disrupted terrains will aid in constraining models of origin. Moreover, flexural signatures are expected to be indicative of local elastic lithosphere thickness at the time of their formation, and may provide evidence of topographic relaxation (e.g., *Nimmo et al.,* 2003; *Billings and Kattenhorn,* 2005).

2.5.5. Landing site characterization. Landers are identified as priority missions if Europa has habitable environments. Landed missions would require high-resolution images (approximately a few meters per pixel or better) for landing site selection. The roughness and overall safety of landing sites can also be characterized through radar data, photometry, thermal inertia, and detailed altimetry. Such data will also illuminate fine-scale regolith and other surface processes (see chapter by Moore et al.). Along with corresponding high-resolution subsurface sounding, these data would help assess likely sites of recent communication with the ocean.

Some outstanding questions for Europa's geology include (1) Do Europa's ridges, bands, chaos, and/or multiringed structures require the near-surface liquid water to form? (2) Where are the youngest regions? (3) Is current geologic activity sufficiently intense that heat flow from the interior is measurable? (4) What is the overall history of the surface?

2.6. Jupiter System

Europa cannot be understood in isolation, but must be considered in the context of the entire jovian system. Europa formed from the jovian nebula and evolved through complex interactions with the other satellites, Jupiter, and Jupiter's magnetosphere (e.g., see chapters by Canup and Ward and Estrada et al.). To understand the development of potential habital environments, knowledge is needed for the origin and evolution of the jovian system, and how the system currently operates. This requires observations of Jupiter and the satellites' magnetosphere and ring system.

2.6.1. Satellite surfaces and interiors. The present environment of Europa depends partly on how it formed and evolved. Europa itself does not record its early surface history, but its neighboring satellites — Io, Ganymede, and Callisto — provide clues to Europa's origin, evolution, and potential habitability, and are interesting on their own.

2.6.1.1. Io. The innermost of the Galilean satellites experiences intense tidally driven volcanism (Fig. 4) and sheds light on Europa's tidal heat engine. Io also provides clues to Europa's silicate interior and could be a major source of contamination on Europa. Io's density suggests a primarily silicate interior (*McEwen et al.,* 2004) while the 4:2:1 Laplace resonance among Io, Europa, and Ganymede as they orbit Jupiter leads to tidal flexing and generation of the heat for global volcanism (*Yoder and Peale,* 1981; see chapter by Sotin et al.).

Galileo data indicate extensive moon-plasma interactions near Io but appear to rule out a magnetic field. Io's moment of inertia suggests that it is differentiated into a metallic core and silicate mantle (*Anderson et al.,* 2001). The inferred Fe-FeS core has a radius slightly less than half of Io and a mass 20% of the satellite. The apparent lack of a magnetic field suggests that the silicate mantle experiences sufficient tidal heating to prevent cooling and a convective dynamo in the core (*Weinbruch and Spohn,* 1995).

Io's mantle may undergo partial melting (*Moore,* 2001) that produces mafic to ultramafic lavas, suggesting an undifferentiated mantle. Silicate volcanism appears to be dominant, although secondary sulfur volcanism may occur locally (*Greeley et al.,* 1984). The heat flux inferred from long-term thermal monitoring exceeds 2 W/m^2, making Io by far the most volcanically active body in the solar system (*Nash et al.,* 1986; *Veeder et al.,* 2004; *McEwen et al.,* 2004; *Lopes and Spencer,* 2007).

Despite the high heat flux, mountains as high as 18 km indicate that the lithosphere is at least 20–30 km thick, rigid, and composed mostly of silicates (e.g., *Carr et al.,* 1998b; *Schenk and Bulmer,* 1998; *Turtle et al.,* 2001; *Jaeger et al.,* 2003). The thick lithosphere can only conduct a small fraction of Io's total heat flux, suggesting magmatic transport of heat through the lithosphere (*O'Reilly and Davies,* 1981; *Carr et al.,* 1998b; *Moore,* 2001).

Silicate lavas, sulfur, and sulfur dioxide on Io interact in complex and intimate ways, with volcanism that includes massive lava eruptions, high-temperature explosions, and overturning lava lakes. Volcanic plumes erupt from central vents and along lava flow fronts where surface volatiles are mobilized. Volcanism and sputtering on Io feed a unique patchy and variable atmosphere, in which S, O, and Na become ionized to form Io's plasma torus, neutral clouds, and aurorae. Sublimation of SO_2 frost is also a source of Io's thin atmosphere but the relative contributions to the atmosphere are not well understood. Electrical currents flow between Io and Jupiter and produces auroral "footprints" in the jovian atmosphere. Near the ionospheric end of the Io flux tube, accelerated electrons interact with the jovian magnetic field and generate decametric radio emissions (*Lopes and Williams,* 2005).

There is an apparent paradox between Io's putative ultramafic volcanism and the widespread intensity of the volcanism on Io. At the current rate, Io would have produced a volume of lava ~40 times the volume of Io over the last 4.5 G.y., resulting in differentiation and consequent eruption of more silicic materials. The resolution of this paradox requires either that Io only recently entered the tidal resonance and became volcanically active, or that whole-scale recycling of Io's lithosphere is sufficient to prevent extreme differentiation (*McEwen et al.,* 2004).

JEO could improve knowledge of Io in several respects. For example, Galileo studies of Io's dynamic processes were hampered by the low data rate and major volcanic events were missed entirely or seen only in disconnected snapshots. JEO would provide a 100-fold increase in data return per Io flyby compared to Galileo, and much more long-term monitoring, which is likely to reveal phenomena not seen previously. Moreover, JEO's superior instruments would allow new investigations, such as high-spatial-resolution spectroscopy of lava flows and *in situ* sampling of its upper atmosphere and plumes.

JEO objectives for Io include (1) understanding Io's heat balance and tidal dissipation, and their relationship to Europa's tidal evolution; (2) monitoring active volcanos and their effect on the surface and atmosphere; (3) determining relationships among volcanism, tectonism, erosion, and deposition; and (4) understanding the silicate and volatile components of Io's crust. Because Io is a dominant source of plasma for the jovian magnetosphere, measurements of trace ion composition in the Io torus and throughout the magnetosphere may reveal details of the internal composition. Knowledge of the composition of material escaping from Io will help distinguish endogenic from Io-derived materials on Europa.

Additional gravity data during flybys would place more stringent constraints on interior structure. New discoveries are likely, such as gravity anomalies similar to those detected by Galileo for Ganymede from a flyby (*Palguta et al.,* 2006). Determination of Io's pole position and changes in the location of the pole would be valuable as constraints on the satellite's shape and thus internal structure. Heat flow determinations would place important constraints on theories of tidal dissipation, internal structure, and thermal and orbital evolution.

2.6.1.2. Ganymede. Ganymede is our largest satellite, exceeding Mercury in diameter, and is the only satellite known to have an intrinsic magnetic field. Its surface is broadly separated into bright and dark terrains (*Shoemaker et al.,* 1982; *McKinnon and Parmentier,* 1986; *Pappalardo et al.,* 2004). Dark terrain covers one-third of the surface and is dominated by impact craters. It is ancient, and appears grossly similar to the surface of Callisto (*Prockter et al.,* 1998). Dark terrain also displays hemisphere-scale concentric furrows, which are probably remnants of vast multiring impact basins.

Bright terrain forms a global network of interconnected lanes, separating dark terrain into polygons, and has a patchwork of smooth surfaces and closely spaced parallel ridges and grooves (Fig. 5). The grooves are extensional tectonic features, and have much in common with terrestrial rift zones (*Parmentier et al.,* 1982; *Pappalardo et al.,* 1998).

Ganymede's surface is dominated by water ice (*McKinnon and Parmentier,* 1986). The polar "caps" appear to follow the magnetospheric boundary between open and closed field lines (*Khurana et al.,* 2007), which provides an opportunity to examine differences in space weathering under different conditions. Dark non-ice materials at lower latitudes could be hydrated brines similar to those inferred for Europa; other minor constituents include CO_2, SO_2, and some sort of tholin material exhibiting CH and CN bonds (*McCord et al.,* 1998b). There is also evidence for trapped O_2 and O_3 in the surface, as well as a thin molecular oxygen atmosphere, and auroral emissions are concentrated near the polar cap boundaries (*McGrath et al.,* 2004), but there are no ionospheric indications from Galileo radio occultation data of an equatorial atmosphere.

Galileo data indicate that Ganymede's moment of inertia is 0.31 MR^2, which is the smallest measured for any solid body in the solar system (*Anderson et al.,* 1996). Three-layer models, constrained by plausible compositions, indicate that Ganymede is differentiated into an outermost ~800-km ice layer and an underlying silicate mantle of density 3000–4000 kg/m^3. A central iron core is allowed, but

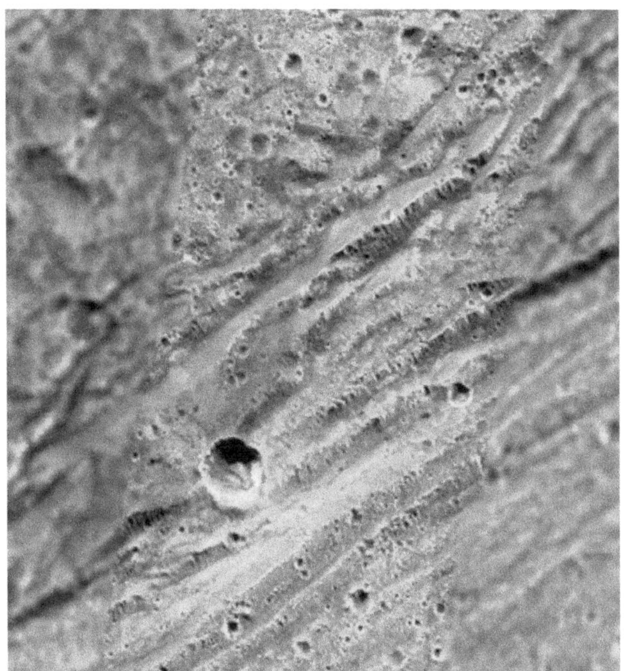

Fig. 5. High-resolution (20 m/pixel) Galileo image of Ganymede's surface; showing an area about 15 × 16 km. Figure courtesy of NASA/JPL/Brown University.

not required, by the gravity data. Ganymede's magnetic field, however, supports the presence of such a metallic core. Galileo gravity data also indicate that Ganymede has internal mass anomalies, possibly related to topography on the ice-rock interface or internal density contrasts (*Anderson et al.,* 2004; *Palguta et al.,* 2006).

Galileo magnetometer data provide tentative evidence for an inductive response at Ganymede, which suggests the presence of a salty internal ocean within 100–200 km of the surface. However, the inference is less robust than for Europa and Callisto, because the data can also be explained by an intrinsic quadrupole magnetic field (superposed on the intrinsic dipole), whose orientation remains fixed in time (*Kivelson et al.,* 2002).

Galileo data show that Ganymede has an intrinsic field strong enough to generate a mini-magnetosphere embedded within the jovian magnetosphere (Fig. 6) (*Kivelson et al.,* 1996). A model with a fixed Ganymede-centered dipole superposed on the ambient jovian field provides a good first-order match to the data and suggests equatorial and polar field strengths of ~719 and 1438 nT, respectively; these values are 6–10 times the 120-nT ambient jovian field at Ganymede's orbit. The most plausible mechanism for generation of the intrinsic field is a dynamo in a liquid-iron core (*Schubert et al.,* 1996).

Multiple flybys of JEO would provide topographic data, subsurface sounding, and high-resolution imaging and spectroscopy for understanding surface formation and evolution. For example, the role of volcanism in modifying the surfaces of icy satellites is poorly understood. Like many other icy satellites, evidence is ambiguous for cryovolcanic processes on Ganymede. Given the physical constraints in eruption of cryovolcanic melt onto the surface (*Showman et al.,* 2004), such deposits would give insight into the interior. Thus, it is critical to learn whether cryovolcanism is widespread or rare on Ganymede, with implications for its role on other icy satellites.

With its mix of old and young terrain, ancient impact basins and fresh craters, and landscapes dominated by tectonism, volcanism, and degradation by space weathering, Ganymede serves as a type example for understanding icy satellites in the outer solar system and would provide insight into how this entire class of worlds evolves differently from terrestrial planets.

2.6.1.3. Callisto. Of the Galilean satellites, Callisto is least affected by tidal heating, thus offering an endmember of icy satellite evolution (*McKinnon and Parmentier,* 1986; *Showman and Malhotra,* 1999; *Moore et al.,* 2004).

Gravity data and the assumption of hydrostatic equilibrium suggest that Callisto's moment of inertia is 0.355 MR^2, suggesting partial differentiation with an ice-rich outer layer <500 km thick, an intermediate ice-rock mixture with a density ~2000 kg/m^3, and a rock/metal core (*Anderson et al.,* 1998b). However, if Callisto's degree-2 gravity structure is not hydrostatically balanced, then Callisto could be more or less differentiated than the moment of inertia suggests. This could have major implications for understanding satellite formation. Pre-Galileo models suggested that Ganymede and Callisto formed from debris in a protojovian disk in ~10^4 yr; however, for Callisto to be relatively undifferen-

Fig. 6. See Plate 37. Ganymede's simulated magnetosphere. Field lines are green; perpendicular current is represented by color variation. Note intense currents flow both upstream on the boundary between Jupiter's field and the field lines that close on Ganymede, and downstream in the reconnecting magnetotail region. Figure courtesy of X. Jia.

tiated, its formation time must have exceeded 10^6 yr (*Canup and Ward,* 2002; *Mosqueira and Estrada,* 2003).

Galileo magnetometer data indicate that Callisto has an inductive magnetic response best explained by a salty ocean within 200 km of the surface (*Khurana et al.,* 1998; *Kivelson et al.,* 1999; *Zimmer et al.,* 2000). Maintaining an ocean today either requires a sufficiently stiff ice rheology to stifle convection and heat loss or existence of "antifreeze" (ammonia or salts) in the ocean (*McKinnon,* 2006). However, reconciling partial differentiation with the existence of the ocean is difficult; some part of the uppermost ice layer must remain at the melting temperature today, while the mixed ice-rock layer must never have attained the melting.

Along with the discovery of Callisto's probable ocean, major discoveries include the absence of cryovolcanic resurfacing and the inference of surface erosion by sublimation. Callisto's landscape at decameter scales is unique among the Galilean satellites, and might be akin to that of cometary nuclei.

Callisto's surface composition is bimodal (water ice and an unidentified non-ice material), with trace constituents in the non-ice material. The color of the non-ice material is similar to C-type asteroids and carbonaceous chondrites. Trace materials detected in the non-ice material include CO_2, C-H, CN, SO_2, and possibly SH (*Carlson et al.,* 1999a; *McCord et al.,* 1998a). Carbon dioxide is detected as an atmosphere and is nonuniformly dispersed over the surface, being concentrated on the trailing hemisphere and more abundant in fresh impact craters (*Hibbitts et al.,* 2002). This asymmetry is similar to that for sulfate hydrates on Europa and is also suggestive of externally induced effects by corotating magnetospheric plasma (*Cooper et al.,* 2001).

The JEO mission would enable many of the key issues for Callisto to be addressed, including determining the distribution of impact craters; studying mass wasting; characterizing the jovian and Callisto magnetospheres to understand the extent and depth of the globally conducting layer; assessing the internal distribution of mass to gain insight into differentiation on large icy satellites; measuring energetic particle fluxes and energies over a long duration and different time periods to understand radiolysis of icy satellites; and mapping surface compositions to understand the relative influences of primordial composition, geological processes, and radiolysis. Results would shed light on similar processes on the other icy satellites.

2.6.2. Satellite atmospheres. Europa's tenuous atmosphere (Figs. 5–12 in chapter by McGrath et al.) is the interface between Jupiter's magnetosphere and the satellite's surface. Composed principally of O_2 with a surface pressure of only ~2×10^{-12} bar (*McGrath et al.,* 2004; see also chapter by McGrath et al.), there is no widely accepted explanation for the nonuniform nature of the atmospheric emissions, and only a single attempt has been made to address this issue (*Cassidy et al.,* 2008). The atmosphere is maintained principally by ion sputtering of the surface, with molecules subsequently dissociated and ionized by electron impact, charge exchange, and solar photons. The abundance and distribution of atmospheric constituents provide clues to surface processes and links to composition. Once released from the surface, some constituents such as Na and K are more readily observed in their gas phase. Their abundance relative to Io provides a discriminator between endogenic and exogenic origin for these species (*Johnson et al.,* 2002). Europa's atmosphere could be in part supplied by active geysers (*Nimmo et al.,* 2007), the discovery of which would provide clues to subsurface processes and interior structure.

Because material from Io is implanted on Europa, it is important to understand Io's atmosphere. Ganymede and Callisto also have tenuous atmospheres, which shed light on the evolutionary paths these satellites have followed. The atmospheric emissions of Ganymede, for example, are reminiscent of classic polar auroral emissions, very different than the case for Europa. Callisto is thought — like Europa and Ganymede — to have a predominantly O_2 atmosphere, but lacks oxygen emissions as seen on Europa, Io, and Ganymede (*Strobel et al.,* 2002). Instead, Callisto has CO_2 emission above the limb, detected by Galileo (*Carlson,* 1999). Although IR limb scans at Europa were not performed, small amounts of CO_2 may be present in its atmosphere, by analogy with Callisto. Callisto's atmosphere may be thicker than either Europa's or Ganymede's (*McGrath et al.,* 2004; *Liang et al.,* 2005), which is reflected by its relatively dense ionosphere (*Kliore et al.,* 2002).

2.6.3. Plasma and magnetospheres. The plasma of Jupiter's rapidly rotating magnetosphere overtakes the satellites in their orbits with flow of charged particles predominantly onto the trailing hemispheres. Energetic ions sputter neutral particles from the surfaces. Many of the liberated particles immediately return to the surface, but some become part of the satellite atmospheres, and some escape to space. A fraction of the neutrals that are no longer bound to a moon can form a circumplanetary neutral torus (*Mauk et al.,* 2003).

Io is the dominant source of particles in Jupiter's magnetosphere (*Thomas et al.,* 2004; *Nozawa et al.,* 2005), but other moons contribute water products and minor species through atmospheric and surface interactions (*Johnson et al.,* 2004); for example, Europa is a source of Na (*Brown,* 2001; *Leblanc et al.,* 2005).

Perturbations of the magnetospheric plasma and electromagnetic fields near the satellites are diagnostic of the satellites themselves. Through such analysis, satellite-induced magnetic fields were detected (e.g., *Kivelson et al.,* 2004), which is the key evidence for subsurface oceans. In turn, magnetospheric particle interactions produce changes in surface chemistry (*Johnson et al.,* 2004).

2.6.4. Jupiter atmosphere. Jupiter contains most of the mass in the jovian system and is the largest object in the solar system after the Sun. Its atmospheric composition reflects the initial nebula conditions, albeit with significant reprocessing, from which the satellites formed (*Ingersoll et al.,* 2004; *West et al.,* 2004; *Taylor et al.,* 2004; *Moses et al.,* 2004; *Yelle and Miller,* 2004). The jovian system provides the best analog for the formation of both our own solar system and the hundreds of exoplanetary systems being discovered around other stars. JEO investigations focused on Jupiter's atmosphere are discussed in section 3.5.4.

2.6.5. Rings, dust, and small moons. A system of small moons and faint rings encircles Jupiter within Io's orbit. Although Saturn's rings are more familiar, faint and dusty rings are more common in the outer solar system. Such rings may represent the evolution of a much denser ring system such as Saturn's. Dusty rings reveal a variety of nongravitational processes that are masked within more massive disks. For example, fine dust grains become electrically charged by solar photons and interactions with Jupiter's plasma. Their orbits are perturbed by solar radiation pressure and Jupiter's magnetic field (e.g., *Burns et al.,* 2004). Thus, a better description of dust dynamics and properties might provide information on Jupiter's plasma and magnetic field within regions that cannot be probed easily by spacecraft.

Jupiter's rings share many of their properties with protoplanetary disks. In both systems, dust and larger bodies co-mingle and interact through various processes. Thus, the ring system provides a dynamic laboratory for understanding the formation of the broader jovian system. JEO investigations of rings, dust, and small moons are described in section 3.5.5.

With regard to the Jupiter system as a whole, some remaining key questions to be addressed include the following: (1) What factors control the different styles of volcanism on Io? (2) Are plasma processes responsible for Ganymede's bright polar caps and if so, how? (3) Has Ganymede experienced cryovolcanism, or does intense tectonism create smooth terrains; what is the distribution and thickness of Callisto's dark component? (4) How does Europa's sputter-produced atmosphere vary? (5) Are Ganymede's and Callisto's atmospheres produced mainly by sputtering or sublimation? (6) How do the sources and dynamics of the fields and plasma in the jovian magnetosphere vary, especially as correlated with Io's activity? (7) How does jovian local atmospheric convection contribute to larger storms?

3. JUPITER EUROPA ORBITER SCIENCE GOAL, OBJECTIVES, AND INVESTIGATIONS

The scientific objectives for JEO were formulated based on previous studies (Table 2) and the science as reviewed in section 2. The goal for JEO is to explore Europa to investigate its habitability, which implies understanding the origin, evolution, and current state of the satellite. This includes addressing the questions outlined above, while also allowing discovery science — unpredicted findings of the type that have often reshaped the very foundations of planetary science. "Habitability" includes confirming the existence of water below Europa's icy shell and determining its characteristics, understanding the possible sources and cycling of chemical and thermal energy, investigating the evolution and composition of the surface and ocean, and evaluating the processes that have affected Europa through time.

Understanding Europa's habitability is intimately tied to investigating the Jupiter system as a whole. Both Ganymede and Callisto may possess subsurface oceans, while Io holds clues to the fundamentals of tidal heating and interactions with the jovian environment. Jupiter can shed light on the initial conditions of the planetary system. Each Galilean satellite can be related to the others, and is intimately tied to Jupiter and the jovian magnetospheric environment. As stated in the 2006 Solar System Exploration Roadmap, "By studying the Jupiter system as a whole, we can better understand the type example for habitable planetary systems within and beyond our Solar System."

Within this context, five primary objectives have been defined for the proposed JEO mission; in priority order these relate to (1) Europa's ocean, (2) Europa's icy shell, (3) Europa's chemistry, (4) Europa's geology, and (5) Jupiter system science.

In the following sections, each objective is described, along with the scientific investigations that are needed to meet the objectives.

3.1. Ocean Objective: Characterize the Ocean and Deeper Interior

The first step in characterizing Europa's ocean is to confirm its existence and extent. If Europa has no ocean and its icy shell is coupled to its rocky mantle, then as it orbits Jupiter the measurable radial tide will vary by only a few meters. On the other hand, if Europa has liquid water beneath a relatively thin icy shell, the tide will vary by ~30 m. Thus, measuring the tides provides a simple and definitive test of the existence of a subice ocean.

In the likely instance that an ocean exists, several geophysical measurements (Fig. 7) will place constraints on its depth, extent, and physical state, as well as provide information on the internal structure of Europa, including the mantle and core. In priority order, investigations would be to (1) determine the amplitude and phase of the gravitational tides, (2) determine the induction response from the ocean over multiple frequencies, (3) characterize surface motion over the tidal cycle, (4) determine the satellite's dynamical rotation state, and (5) investigate the core and rocky mantle.

The gravitational tidal potential from Jupiter varies periodically as Europa orbits (Fig. 8), applying stress that deforms the satellite. The amplitude and phase of the gravitational and topographic tidal responses are determined by the mechanical strength and density of the layered interior. Love numbers are the dimensionless scale factors that parameterize these effects, where k_2 represents effects on the gravitational potential and h_2 represents radial topographic effects. A homogeneous fluid body would have values of $k_2 = 1.5$ and $h_2 = 2.5$. If present, a liquid ocean would dominate the tidal response, while the product of icy shell thickness times icy shell rigidity has a lesser but important effect (Fig. 9).

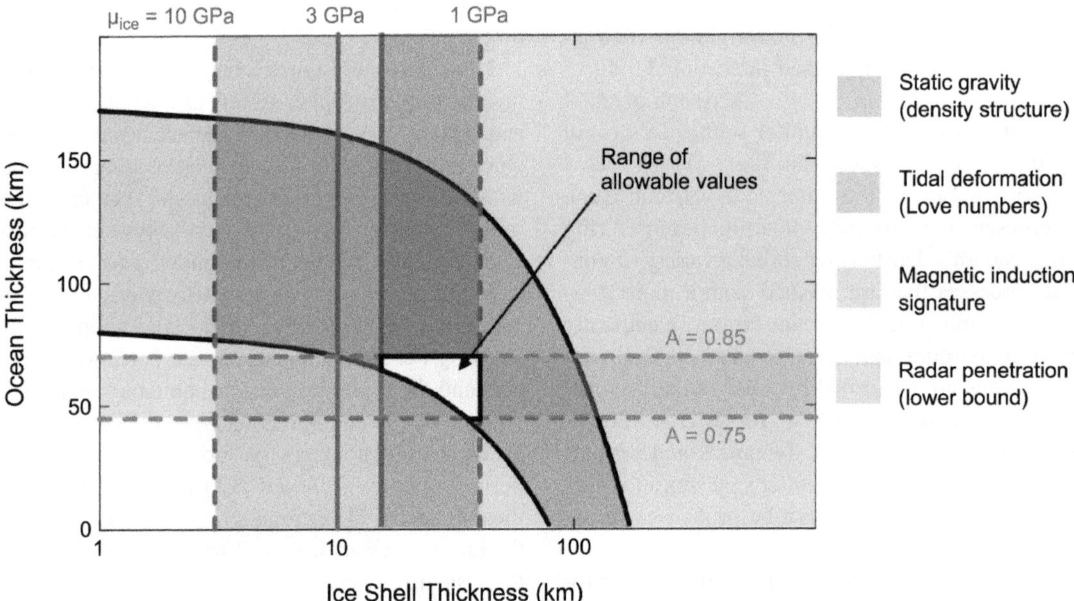

Fig. 7. See Plate 38. The combination of (hypothetical) JEO measurements can constrain the thickness of the icy shell. Based on the bulk density and moment of inertia (from future flybys by JEO and other spacecraft), the thickness of the water + ice layer may be obtained (gray shading) (*Anderson et al.,* 1998a,b); uncertainties arise mainly from lack of knowledge of the rocky interior density (bulk density is already known). Measuring time-variable gravity and topography gives the k_2 and h_2 Love numbers, respectively; hypothetical Love number constraints (red shading) assume observed h_2 and k_2 of 1.202 and 0.245, respectively, and constrain shell thickness as a function of rigidity μ (*Moore and Schubert,* 2000). The hypothetical values assumed here are characteristics of a moderately thick icy shell. In the example shown, the icy shell deformation is sufficiently large that a shell thickness in excess of 40 km is prohibited. Determining both k_2 and h_2 provides additional information. A lower bound on the icy shell thicknesses may be derived from radar data. Here, a tectonic model of icy shell properties is assumed (*Moore,* 2000), resulting in a radar penetration depth (and lower bound on shell thickness) of 15 km (green shading). Multiple frequency (hypothetical) set of observations results in a range of acceptable icy shell thickness (15–40 km) and a range of acceptable ocean thicknesses (45–70 km). A different set of observations would result in different constraints, but the combined constraints are more rigorous than could be achieved by any one technique alone. JEO would be able to provide those constraints to determine the thickness of Europa's icy shell.

Based on plausible internal structures, measurement uncertainties of ±0.0005 for k_2 and ±0.01 for h_2 will permit the actual k_2 and h_2 to be inferred with sufficient accuracy that the combination characterizes the depth of the ocean and constrains the thickness of the icy shell (*Wu et al.,* 2001; *Wahr et al.,* 2006). In turn, icy shell thickness is an important constraint on geologic processes, astrobiology, and heat flux from the silicate interior.

The Love number k_2 is estimated from the time-variable gravitational field of Europa, which is measured by perturbations in the paths of orbiting spacecraft. The component of the velocity change that is in the direction to Earth is measured by a Doppler shift in the radio-frequency communication with the satellite. Because the perturbations are measured only by a single projected component at any given time, a complete resolution of the gravity field requires multiple orbits; moreover, a single profile is difficult to interpret because the same data must be used to determine the spacecraft orbit itself.

At X-band frequencies, velocity measurement accuracies of 0.1 mm/s are typically attained for 60-s averages. At Ka-band the performance is somewhat better and, used together, the two frequencies help reduce interplanetary plasma-induced noise. Figure 8 illustrates the estimated gravitational spectrum for Europa, with separate contributions from an icy shell and a silicate interior, along with simulated error spectra for 30 days of tracking at each of three representative orbital altitudes (cf. *Wu et al.,* 2001), using the X-band-only error estimate. The recovered gravity errors are smaller at lower altitudes because the spacecraft is closer to the anomalies, and thus experiences larger perturbations.

Improving accuracy in the measurements allows better determination of long wavelength features and initial discrimination of some shorter wavelength features. Variations in gravitational signal amplitude and correlation with topography are diagnostic of internal structures. For the model parameters depicted in Fig. 8, the lowest-altitude orbit errors are small enough to resolve part of the transition from the long-wavelength, silicate-dominated part of the spectrum (in which correlation with topography would be poor) into the shorter-wavelength, ice-dominated regime, where topography and gravity should be spatially coherent (*Luttrell and Sandwell,* 2006). This would permit detection of isostatic anomalies in response to topographic variations (such as volcanos) at the silicate-ocean interface. Radio frequency tracking data will provide initial spacecraft orbit

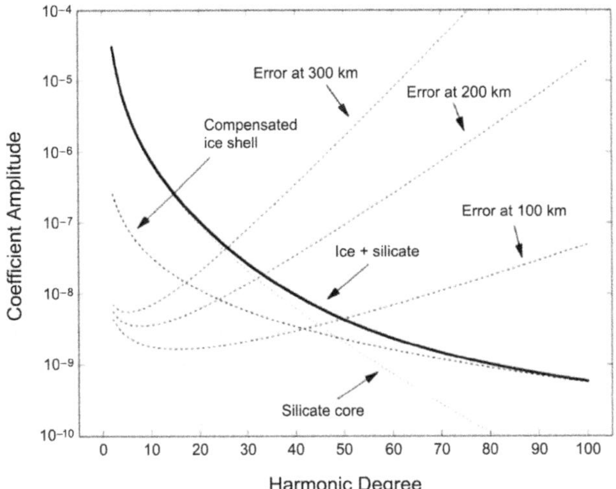

Fig. 8. Models of Europa's gravity spectrum, assuming an icy shell 10 km thick with isostatically compensated topography above an ocean, and a silicate interior with a mean surface 100 km below the ice surface. The variance spectra of the ice topography and silicate gravity are assumed similar to those seen on terrestrial planets (*Bills and Lemoine,* 1995). The signal has contributions from the silicate mantle and icy shell. The error spectra represent 30 days at fixed altitude, and reflect variations in sensitivity with altitude. The error spectra at different orbital altitudes do not have the same shape because the longer wavelength anomalies are attenuated less at higher altitudes. During a few days at these altitudes, the improvement is linear with time; for longer times, repeat sampling leads to improvement proportional to square root of time.

estimates. As the gravity field knowledge improves during the orbital mission, near-real-time orbit position knowledge will also improve. The tracking data will be used, together with spacecraft attitude and altitude information, to estimate simultaneously the static and tidal components of gravity and topography, and the forced rotational variations including libration.

The Love number h_2 is derived by measuring the time-variable topography of Europa, specifically by measuring topography at crossover points (Fig. 10), a technique that has been demonstrated for Earth (*Luthcke et al.,* 2002, 2005) and Mars (*Rowlands et al.,* 1999; *Neumann et al.,* 2001). After ~60 days in orbit about Europa the subspacecraft track will form a reasonably dense grid, comprised of N (~700) great circle segments over the surface of Europa. Each of the N arcs intersects each of the remaining N 1 arcs at two roughly antipodal locations, and at these crossover locations, the static components of gravity and topography should agree. Differences in the measured values at crossover points are equal to a sum of actual change in radius caused by tides and libration, combined with the difference in orbital altitude, along with any errors in range to the center of the body or orbital position (Fig. 10). The errors are dominated by long-wavelength effects and can be represented by four sine and cosine terms in each orbital component (radial, along track, and cross track). The tidal effects in gravity and topography have known spatial and temporal patterns and can each be represented globally by two parameters, an amplitude and phase. The librations are

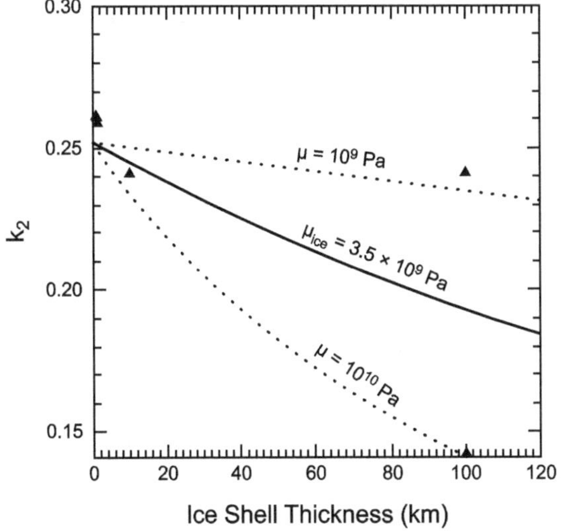

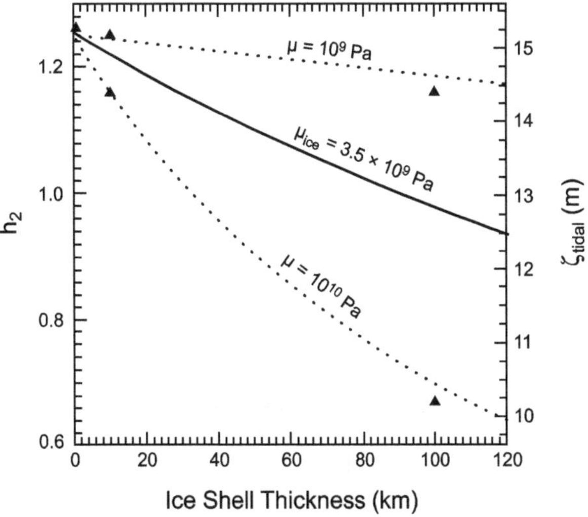

Fig. 9. Sensitivity of Love numbers k_2 (left) and h_2 (right) to thickness and rigidity of the icy shell (assuming a subsurface ocean). For the same curves that depict h_2, the righthand axis shows the amplitude ζtidal (which is half of the total measurable tide) as a function of thickness of the icy shell. For a relatively thin icy shell above an ocean, the tidal amplitude is ζtidal ~ 15 m (total measureable tide ~30 m), while in the absence of an ocean ζtidal ~ 1 m (*Moore and Schubert,* 2000). Solid curves show the h_2 and corresponding ζtidal for an icy shell rigidity of μice = 3.5 × 10^9 Pa, while the dotted lines bound a plausible range for ice rigidity. A rocky core is assumed with a radius 1449 km and rigidity μrock = 10^{11} Pa and the assumed ice + ocean thickness = 120 km. Triangles show the reported values from *Moore and Schubert* (2000), which did not include a core. Figure courtesy of Amy Barr.

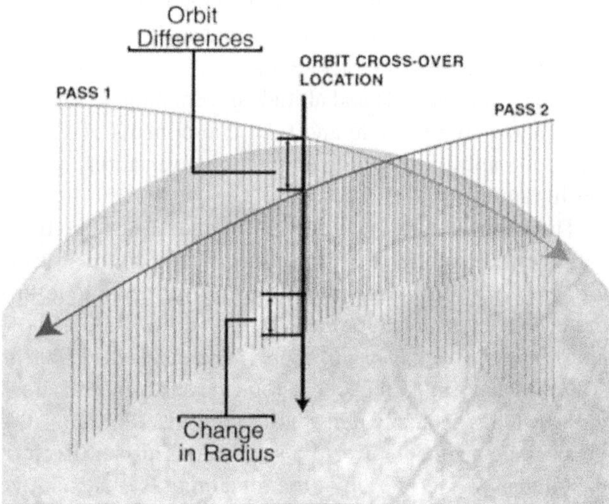

Fig. 10. Illustration of the crossover technique. Actual change in radius of Europa due to tidal and librational motions is determined by measuring altitude from the spacecraft to the surface, and by accounting for the distance of the spacecraft from the center of mass by means of Doppler tracking (*Wahr et al.,* 2006).

effectively periodic rigid rotations with specified axes and periods, and again an amplitude and phase parameter suffices to describe each axis. Thus, there are 12N + 10 parameters to be estimated (12N orbital, 4 tidal, and 6 librational), from 2N*(N 1) crossover points. The accuracy with which the altimetric profiles can be interpolated to the crossover locations depends on range accuracy, surface spot size over which altitude is sampled, and along-track sampling rate. In an ideal case, the surface spots would be small (to minimize topographic variation within spots), and near-contiguous or even overlapping. Those considerations need to be assessed against power and data-rate constraints of an instrument, and the desire to interrogate topography for as much of the surface as possible.

The magnetic induction signal from an ocean within Europa is sensitive to the product of the electrical conductivity and thickness of the ocean (Fig. 4 of chapter by Khurana et al.). Determining the induction response at both the synodic frequency with respect to Jupiter's rotation (T = 11.1 h) and the orbital frequency of Europa (T = 85.2 h) can allow for ocean thickness and conductivity to be determined uniquely. In turn, ocean conductivity constrains its salinity. It is possible that additional longer-period signals, caused by the background fluctuations of the magnetic field (e.g., associated with Io's torus reorganizations), could be used to sound the ocean. This requires the sensitivity of the magnetometry measurements to be 0.1 nT. Magnetometry requires near-continuous observations from Europa orbit for at least 8–10 eurosols, i.e., at least one month. In addition, measurements of plasma density, temperature, and flow are required to quantify the currents generated in Europa's vicinity by the moon-plasma interaction and remove their contribution from the measured magnetic field. This requirement can be met by measuring fluxes of charged particles over a broad energy range (tens of eV to several MeV) over a solid angle of 2π. Because the energies of the sputtering particles are very high ($E \geq 100$ keV) and the energies of the recently picked-up ions is quite low (a few keV), measurements over a broad energy range are desired to quantify the plasma interaction.

Primary data on the deeper interior structure (mantle and core) can be derived from the gravitational and magnetic fields, and the dynamical rotational state, including rate, obliquity, and libration. The amplitude of forced librations in longitude, which are gravitationally forced periodic variations in rotation rate, constrains the combination (B-A)/C for the principal moments of inertia $A < B < C$, as has been done for the Moon (*Newhall and Williams,* 1997).

There may be two librational signals, one from the icy shell, and another from the deeper interior. The shell's signal would be revealed in both gravity and topography data, whereas the deeper signal would appear only in the gravity. Moreover, the tidally damped obliquity, or angular separation between spin and orbit poles, provides a constraint on the polar moment of inertia C (*Ward,* 1975; *Bills and Nimmo,* 2008), which in turn constrains radial density variations. The dynamical rotational state of Europa will be determined using Doppler tracking and altimetry data. Initially assuming both steady rotation and zero obliquity, the crossover analysis described above can be used to adjust the spacecraft orbit estimate and to determine the dynamical rotation and tidal flexing of Europa. Magnetometry data, which measure very low-frequency magnetic variations (periods of several weeks), over durations of several months or longer, will shed light on the magnetic properties of the deep interior, including the core.

3.2. Icy Shell Objective: Characterize the Icy Shell and any Subsurface Water, Including Their Heterogeneity, and the Nature of Surface-Ice-Ocean Exchange

Knowledge of the structure of Europa's icy shell is critical, especially as related to subsurface water and possible surface-ice-ocean exchange. The dielectric losses in very cold ice are low, yet highly sensitive to temperature anomalies, water, and impurities, and much can be learned from electromagnetic sounding of the icy shell. This is especially true when subsurface profiling is coupled to observations of the topography and morphology of landforms and placed in the context of surface composition and subsurface density distributions. In priority order, investigations are to (1) characterize the distribution of any shallow subsurface water; (2) search for an ice-ocean interface; (3) correlate surface features and subsurface structure to investigate processes governing material exchange among the surface, icy shell, and ocean; and (4) characterize regional and global heat flow variations.

High-resolution subsurface profiles from near-global surveys combined with surface topography of similar vertical resolution would enable identification of regions of possible current or recent upwelling of liquid water. Profiling the upper 3 km of the icy shell should be at frequen-

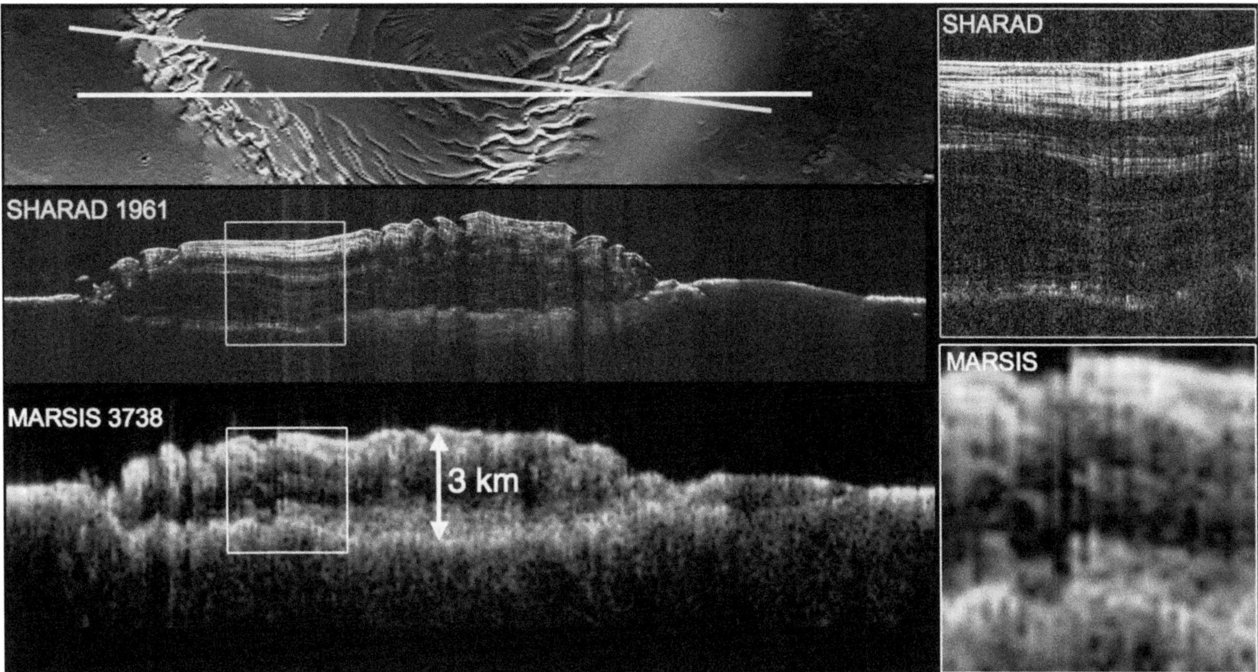

Fig. 11. Orbital subsurface profiling of Mars north polar cap demonstrate the value of complementary perspectives provided by the high-center frequency and high bandwidth of the SHARAD instrument (20 MHz and 10 MHz, respectively), and the low-center frequency and low bandwidth MARSIS (5 MHz and 1 MHz, respectively). In particular, note the clarity of shallow horizons revealed by SHARAD (top right) and the prominence of deep interfaces revealed shown by MARSIS (bottom right). The value of a multifrequency approach on Europa would be enhanced in the presence of strong volume scattering. Figure courtesy of the MARSIS and SHARAD teams.

cies slightly higher than Jupiter's radio noise spectrum (i.e., about 50 MHz) to establish the geometry of various thermal, compositional, and structural horizons to a depth resolution of about 10 m (requiring a bandwidth of about 10 MHz). The search for shallow water will produce data analogous to that of the Shallow Subsurface Radar (SHARAD) instrument on the Mars Reconnaissance Orbiter (Fig. 11). Profiling should be done in conjunction with co-located stereo imaging and altimetry to register topography to a vertical resolution of better than 10 m, permitting surface clutter effects to be removed from the radar data. Such profiling should extend over at least 80% of Europa's surface with spacings no more than twice the hypothesized maximum thicknesses of the icy shell (~50 km).

Subsurface signatures from lower resolution but more deeply penetrating radar might reveal a shallow ice-ocean interface, which could be validated by correlating ice thickness and surface topography. An unequivocally thin icy shell, even within a limited region, would have significant implications for understanding direct exchange between the ocean and the overlying ice. Similarly, the detection of variations in the subsurface interfaces might suggest convective movement of ductile ice into the cold brittle shell, implying indirect exchange with any ocean. Additional profiles of the subsurface to depths of 30 km with a vertical resolution of ~100 m would establish the geometry of any deeper geophysical interfaces such as an ice-ocean interface. Although warm ice is very attenuating (*Chyba et al.,* 1998), "windows" of cold downwelling material may exist within the icy shell, allowing local penetration to great depth (*McKinnon and Gurnis,* 1999). Moreover, while the presence of meter-scale voids within the icy shell would confound sounding efforts at higher frequencies (>15 MHz) (*Eluszkiewicz,* 2004), the presence of such large voids is probably unrealistic (*Lee et al.,* 2005).

Deep ocean searches would produce data analogous to those of the Mars Advanced Radar for Subsurface and Ionosphere Sounding (MARSIS) instrument on the Mars Express spacecraft (Fig. 11). Such profiles could establish the geometry of any deeper geophysical interfaces that may correspond to an ice-ocean boundary to a vertical resolution of ~100 m (requiring a bandwidth of about 1 MHz). Frequencies significantly less sensitive to volume scattering (i.e., about 5 MHz) should be used on the antijovian side of Europa, which is substantially shadowed from Jupiter's radio emissions. This low-frequency low-resolution profiling should be complemented by high-frequency low-resolution profiling over Europa's subjovian surface (where Jupiter's radio noise is an issue for low-frequency sounding). Combined, the deep low-resolution profiling should cover at least 80% of the surface with a minimum profile separation of about 50 km. Profiling should be accompanied with co-located stereo imaging and altimetry of better than 100-m vertical resolution, permitting surface clutter effects to be removed from the radar data.

Targeted radar observations would lead to understanding the processes controlling the distribution of any shallow subsurface water and either the direct or indirect exchange of materials between the icy shell and the ocean. Fractures, topography, and compositional data potentially correlated with subsurface structures can provide information on tidal response and its role in subsurface fluid migration. Similarly, differences in the physical and compositional properties of the near-surface ice may arise due to age differences, tectonic deformation, mass wasting, or impact gardening. Knowledge of surface properties gained from spectroscopy and high-resolution topographic data will be essential for integrated interpretation of subsurface structure, as well as understanding liquid water or ductile ice migration within the icy shell.

Because of the complex geometries expected for subsurface structures, subsurface imaging should be obtained along profiles ~30 km long over targeted areas, either at high resolution for shallow targets or low resolution for deeper features, in conjunction with co-located topographic data. Targeted subsurface sites should be considered a prerequisite for future *in situ* astrobiological exploration. Gravity data can also provide insight into the icy shell.

The thermal structure of the shell is influenced by local heat sources and transport from the interior. Regardless of shell properties or mechanisms of heat transport, the uppermost several kilometers are thermally conductive, cold, and stiff. The thickness of this conductive "lid" is governed by the total heat transported; thus, a measurement of the thickness of the lid will provide insight into heat production in the interior. For a thin icy shell, the ice-ocean interface forms a dielectric horizon at the base of the thermally conductive layer. However, when warm pure-ice diapirs approach the surface, they may be far from the pure-ice melting point and above the eutectic of many substances, and may lead to melt within the shell. Any dielectric horizon associated with the melt regions would also provide a good measurement of the thickness of the conductive layer. Global radar profiles of the subsurface thermal horizons to depths of 30 km at a vertical resolution of 100 m combined with thermal data for the surface enable mapping heat flow variations in the icy shell.

Figure 7 shows how a combination of JEO measurements could be used to constrain the thickness of the icy shell. Based on the bulk density and moment of inertia of the satellite derived from spacecraft flybys, the thickness of the water + ice layer may be obtained (gray shading). Uncertainties arise from lack of knowledge of the rocky interior density. Measuring the time-variable gravity and topography gives the k_2 and h_2 Love numbers, respectively, which in turn constrain the thickness of the icy shell (red shading). The main uncertainty is the effective rigidity of the icy shell, μ_{ice}. In the example shown, deformation of the icy shell is sufficiently large that a thickness in excess of 40 km is prohibited. Determining both k_2 and h_2 provides additional information; the ratio of h_2/k_2 is quite different depending on whether a subsurface ocean exists or not, and provides an additional test of the ocean.

A lower bound on the thickness of the icy shell can be derived using radar. Although the base of the shell is difficult to image because warm ice is radar absorptive, even a nondetection of the ice-water interface provides a lower bound on the thickness (e.g., 15 km in Fig. 7 in green). For relatively thick oceans, magnetometer data can be used to determine both the ocean thickness and water conductivity (*Khurana et al.,* 2002) (blue shading in Fig. 9).

By combining the datasets, a range of acceptable thicknesses of the icy shell and ocean can be derived, yielding more rigorous results than from any one technique. Results include assumptions that, while reasonable, may need additional verification. For example, the moment of inertia is derived by assuming hydrostatic equilibrium; this assumption can be checked independently by measuring the polar as well as the equatorial variation in long-wavelength gravity. Similarly, the calculated values of h_2 and k_2 assume radially symmetric material properties, which may be an oversimplification.

3.3. Chemistry Objective: Determine Global Surface Compositions and Chemistry, Especially as Related to Habitability

Composition enables understanding Europa's potential habitability in the context of geology. Composition also provides clues to the interior and records the evolution of the surface by internal and external processes. Prioritized investigations are to (1) characterize surface organic and inorganic chemistry, including abundances and distributions of materials, with emphasis on indicators of habitability and potential biosignatures; (2) relate compositions to geologic processes, especially communication with the interior; (3) characterize the global radiation environment and the effects of radiation on surface and atmospheric composition, albedo, sputtering, sublimation, and redox chemistry; and (4) characterize the nature of exogenic materials.

Candidate composition measurements and instruments for Europa are listed in Table 3. These investigations require synergistic, coordinated observations of targeted geologic features, along with synoptic near-global remote-sensing data, including multispectral and stereo imaging, radar sounding, and thermal mapping.

The first priority investigation is to identify surface organic and inorganic constituents and map their distribution and association with geologic features. The search for organic materials, including compounds with CH, CO, CC, and CN, is especially relevant to astrobiology, while identifying specific salts may constrain the composition, physical environment, and origin of the ocean (*Kargel et al.,* 2000; *McKinnon and Zolensky,* 2003; see chapter by Zolotov and Kargel). Additional compounds of interest include water ice (crystalline and amorphous phases), products of irradiation, such as H_2O_2, compounds formed by implantation of sulfur and other ions, and other as yet unknown materials.

3.3.1. Surface spectroscopy. The best means to map surface composition at spatial scales relevant to geologic processes is through near-UV to IR imaging spectroscopy.

TABLE 3. Composition measurements for Europa and candidate instruments.

	VIS/NIR Imaging Spectrometer	UV Spectrometer	INMS	IR or Millimeter Spectrometer
Europa science measurements	• Composition of organic and inorganic materials (C1a) • Relationship of surface materials to geologic processes (B3c; C2a, C3c) • Effects of radiation environment (C3a) • Nature of exogenic materials (C4c) • Exposure age (D1f) • Recent activity (D2e)	• Plume composition and regional mapping to surface vents (C1c) • Current activity through spatial and temporal variability of venting (D2b) • Effects of radiation, sputtering (C3b, C3e) • Relationship of surface materials to geologic processes (Imager; B3g; C2c; C3c, D2e) • Exposure age (Imager; D1f) • Nature of exogenic materials (Imager; C4d)	• Composition of organic and inorganic surface materials (C1b) • Effects of radiation, sputtering (C3d) • Nature of exogenic materials (C4b)	• Composition of organic and inorganic surface materials (C1b) • Effects of radiation, sputtering (C3d) • Nature of exogenic materials (C4b) • Recent activity (D2e)
Species of interest				
Identified	H_2O; CO_2, SO_2, H_2O_2, sulfate hydrates, CH compounds, CN compounds, O_2	• H, O (gas emission) • H_2O, SO_2 (solids)	• O_2 (~10^6 cm^{-3}); H_2; Na (~300 cm^{-3}); K; Cl+ (atmosphere) • SO_2 (~1600 cm^{-3}); CO_2 (~700 cm^{-3}); H_2O (~10^5 cm^{-3}) (surface)	• H_2O
Expected	HC, SH, SO, Fe^{2+}, S_8, HCHO, H_2S, $MgSO_4$, H_2SO_4, H_3O^+, $NaSO_4$, Na_2MgSO_4, CH_3OH, CH_3COOH	• OH, C, CO (gas mission) • H_2O (gas absorption)		• OH, C, CO (atmosphere) • H_2O (atmosphere)
Possible	NaHCO, $NaCO_3$, H_2CO_3, $MgCO_3$, $MgCl_2$, NaCl, OCS, HCN, OCN^-, KOH, K_2O, SO_3, CH_2CO	• S, Cl, N (gas emission) • CO_2, SO_2, SO, O_3, hydrocarbons (gas absorption) • Water ice, salts, suflates, acids, tholins (solids)	• H_2O_2 (~200 cm^{-3}); sulfur, sulfate, carbon, carbonate, CN, organics, minerals	• CO_2, SO_2, SO, O_3, hydrocarbons, salts, sulfates, acids, tholins • Sputtered species, e.g., Mg-sulfate ⇒ $MgSO_3$, MgO_2, MgS, MgO, Mg
Detection limits	Surface: 0.1 to 10% abundance, varying with species and environmental conditions	Atmosphere: 1 × 10^{15} cm^{-2} H_2O column	~200 cm^{-3}	Atmosphere: Column abundance 10^{-3} to 30% relative to H_2O vapor for many possible species
Measurement requirements				
Spectral/mass range	0.4–>5 μm (desired) ~1.2–4.8 μm (floor)	EUV (60–110 nm); FUV (110–200 nm); NUV (200–350 nm) (desired) FUV (110–200 (floor)	1 to ≥300 amu	IR: 5–50 μm mm: 110 ± 20 GHz; 560 ± 30 GHz
Spectral/mass resolution	(Grating) 0.4–2.5 μm: 5 nm; 2.5–≥5 μm: 10 nm (desired) 1.2–4.8 μm: 10 nm (floor)	0.5 nm EUV, FUV; 3 nm NUV	• Mass resolution: Dm/m ≥ 500 • Pressure range: 10^{-6}–10^{-17} mbar • Sensitivity: 10^{-5} A/mbar	IR: 1–5 cm^{-1} mm: 100–250 kHz
Spatial/mass resolution	25 m/pixel from 100 km (0.25 mrad) (desired) 100 m/pixel (1 mrad) (floor)	1 mrad/pixel (imager)	100–200 km (comparable to orbital altitude	100–500 m
SNR	≥128 (0.4–2.6 μm), ≥32 (2.6–5 μm)	≥5	N/A	>50
Coverage	Global	Occultation profiles at ≤25 km spacing over >80% of surface	Regional	Regional
Heritage	(Grating) NIMS, VIMS, Hyperion, CRISM, ARTEMIS, M3, Rosetta VIRTIS, Mars Express OMEGA, Dawn VIR	Cassini UVIS; New Horizons ALICE	Cassini INMS; Rosetta ROSINA	CloudSat, EOS-MLS, MIRO, and Herschel-HIFI

Galileo NIMS data of Europa and Cassini VIMS data of the saturnian system demonstrate the wealth of spectral features throughout this spectral range (e.g., *McCord et al.,* 1998a; *Carlson et al.,* 1999a,b; *Clark et al.,* 2005; *Cruikshank et al.,* 2007). Of the materials studied in the laboratory, hydrated sulfates appear to closely reproduce the asymmetric and distorted spectral features observed at Europa. In these compounds, hydration shells around anions and/or cations contain water molecules in various configurations, held in place by hydrogen bonds. Each configuration corresponds to a particular vibrational state, resulting in complex spectral behavior that is diagnostic of composition. These bands become particularly pronounced at temperatures <150 K as the reduced intermolecular coupling causes the individual absorptions that make up these spectral features to become more discrete (*Crowley,* 1991; *Dalton and Clark,* 1998; *Carlson et al.,* 1999b, 2005; *McCord et al.,* 2001, 2002; *Orlando et al.,* 2005; *Dalton et al.,* 2003, 2005; *Dalton,* 2000, 2007). As a result, the spectra of low-temperature materials provide highly diagnostic, narrow features ranging from 10 to 50 nm wide (Fig. 6 in chapter by Carlson et al.).

Cryogenic spectra for the hydrated sulfates and brines display diagnostic absorption features near 1.0, 1.25, 1.5 and 2.0 μm that are endemic to water-bearing compounds (Fig. 7 in chapter by Carlson et al.). These features generally align with those in water ice and the features observed for Europa. Other spectral features arising from the presence of water occurs including features of moderate strength near 1.65, 1.8 and 2.2 μm. An additional absorption common to the hydrates at 1.35 μm arises from the combination of low-frequency lattice modes with the asymmetric O–H stretching mode (*Hunt et al.,* 1971a,b; *Crowley,* 1991; *Dalton and Clark,* 1999). Although weak, this feature is usually present in hydrates and has been used to place upper limits on abundances of hydrates in prior studies (*Dalton and Clark,* 1999; *Dalton,* 2000; *Dalton et al.,* 2003).

Cassini VIMS observations of Phoebe provide additional examples of the wealth of information available in IR spectra. *Clark et al.* (2005) reported 27 individual spectral features indicating H_2O, CO_2, and possibly CN-bearing ices. The 3–5-μm portion of the Phoebe spectrum include absorptions tentatively interpreted as nitrile and hydrocarbon compounds. This spectral range is useful for detecting numerous organic and inorganic species anticipated at Europa.

Obtaining spectra at wavelengths >5 μm would enhance mapping and characterizing organic chemistry and potential biosignatures. In particular, C and N compounds, which are central to the chemistry of known life, have numerous absorption bands associated with C–O, C–C, and C–N bonds in the 5–7-μm region. The strong carbonyl and amide bands at ~5.9 μm might be detected at concentrations of tens of ppm using sufficiently long integration times and large spatial averages. The 5–7-μm region has the added benefit of not being dominated by water features, as is the case for C–H bands at 3.44 μm on the shoulder of the broad 3-μm water band. Although the flux of reflected solar photons decreases toward longer wavelengths, reflected photons out to nearly 8 μm are more abundant than thermal photons for a blackbody surface temperature of ~100 K. For surface temperatures of ~130 K, thermal photons dominate down to ~6 μm. Detection of longer wavelength reflected photons and thermal photons could be achieved using a combination of spectral integration made possible by being in orbit around Europa, spectroscopic techniques (such as Fourier transform spectroscopy, which significantly enhances signal throughout), and reduced spatial resolution "point spectroscopy."

The spectral features of hydrated minerals are not seen in high-spectral-resolution 1.45–1.75-μm Keck telescopic data collected from dark Europa terrain that are >100 km across (*Spencer et al.,* 2006). However, while these regions are dominated by dark materials, ice-rich materials likely occur, and significant spatial mixing and dilution of the spectra may occur. It is also possible that the various hydrated species are mixed in such proportions that their diagnostic features overlap, or the dark terrain is dominated by material without diagnostic features in this range, perhaps due to radiation damage or flash-freezing. In this case there may be smaller regions (perhaps the youngest ones) in which diagnostic features can be found. An excellent example of the importance of spatial resolution is observed for martian dark region spectra, in which telescopic spectra in both the thermal and near-IR (e.g., *Bell,* 1992; *Moersch et al.,* 1997) did not reveal the mineralogic components that were detected once high-spatial-resolution spectra were acquired (e.g., *Christensen et al.,* 2001; *Bibring et al.,* 2006).

Going further, laboratory studies show that at Europa's surface temperature, anticipated materials (in particular hydrates) exhibit fine structure, with the full width at half maximum (FWHM) of spectral features ranging from 7 to 50 nm (*Carlson et al.,* 1999b, 2005; *Dalton,* 2000; *Dalton et al.,* 2003; *Orlando et al.,* 2005). Detection of materials in relatively low abundance, or in mixtures with dark materials, requires signal-to-noise ratios of >128 for shorter wavelengths and 732 for longer wavelengths (Fig. 12). An *ideal* spectral resolution of 2 nm per channel would be sufficient to identify all the features observed in laboratory hydrates (*Dalton,* 2000; *Dalton et al.,* 2003, 2005). This would ensure multiple channels across each known feature of interest. As noted above, however, at Jupiter's distance from the Sun the reflected NIR radiance limits the achievable spectral resolution for high-spatial-resolution mapping. Signal-to-noise performance is further complicated by severe radiation noise at Europa's orbit, which is of course a general concern for any mission to the Jupiter system.

The multiple spectral features and fine (10–50 nm) structure of materials of interest in the 1- to ≥5-μm range in low-temperature spectra are sufficiently unique to allow these materials to be identified even in mixtures of only 5–10 wt.% (*Dalton,* 2007; *Hand,* 2007). The ability to resolve these features fully through high-spectral, high-spatial resolution observations will permit determination of the relative abundances of the astrobiologically relevant molecules on the surface. A spectral sampling of 5 nm through the visible and near-IR wavelengths of 0.4 to ~2.5 μm, and 10 nm from ~2.5 to ≥5 μm would provide the required SNR while maximizing spectral separability (Figs. 12–14) (see

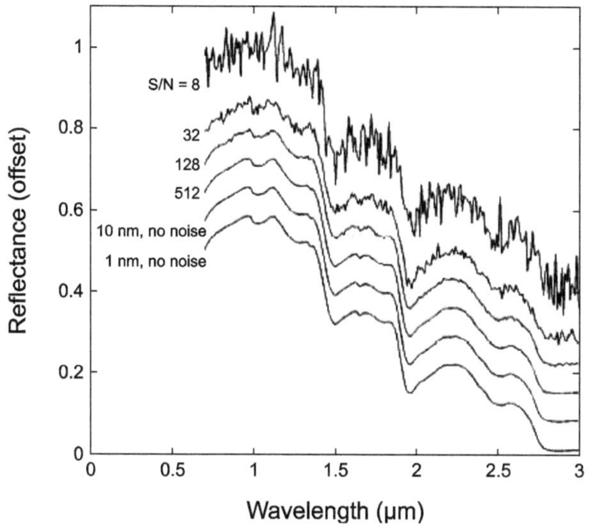

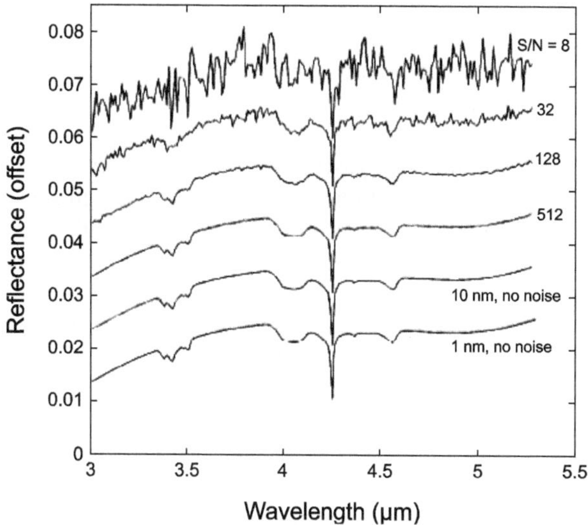

Fig. 12. IR reflectance spectra for a range of signal-to-noise ratios (S/N) show that to detect absorption bands of materials in relatively low abundance, or in mixtures with dark materials, S/N > 128 is desirable in the shorter wavelength range 0.4–2.6 μm, and S/N > 32 is desirable in the longer wavelength range 2.6–5.0 μm. Figure courtesy of Tom McCord.

also Fig. 4 in chapter by Johnson et al.; *Dalton et al.,* 2003; *Dalton,* 2007). Observations should sample across at least 80% of the globe, with targeted imaging observations having better than 100-m/pixel spatial resolution in order to resolve small geologic features, map compositional variations, and search for locations with distinctive compositions. By comparison, the Galileo NIMS observations of Europa had a spectral sampling of 26 nm, a spatial resolution of 2 to >40 km, and an SNR that varied from 5 to 50 in individual spectra. Linear spectral modeling using Galileo NIMS data with cryogenic measurements of hydrate spectra displayed sensitivities to abundances at the 10% level (*Dalton,* 2007). High spectral resolution, coupled with high spatial resolution that can permit sampling of distinct compositional units at 25–100-m scales, will allow identification and quantification of the contributions of hydrated salts, sulfuric acid, sulfur polymers, CO_2, organics, and other compounds anticipated on Europa's surface.

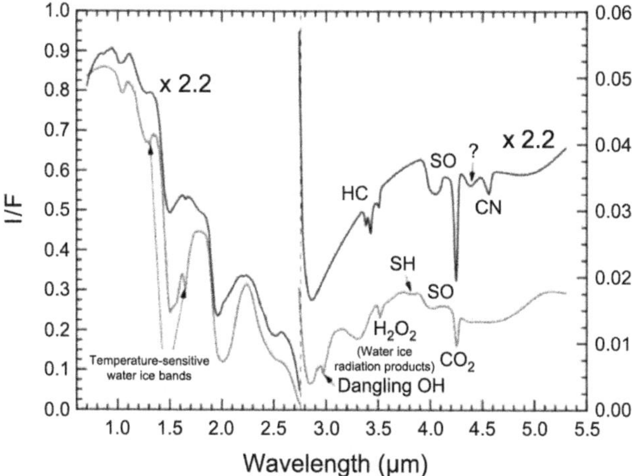

Fig. 13. Notional reflectance spectra for icy (lower) and non-ice (upper) regions on Europa at 6-nm spectral resolution in the 1–5-μm spectral range. Various materials and molecules have been identified or inferred from Galileo. These spectra are composites to illustrate the types and variety of features found or expected, and the detailed spectral structure observed in hydrates is not fully represented. The non-ice spectrum is scaled by 2.2 from the ice spectrum, and the 2.8–5-μm range spectra are scaled by 10 over the shorter wavelength range. Figure courtesy of Tom McCord.

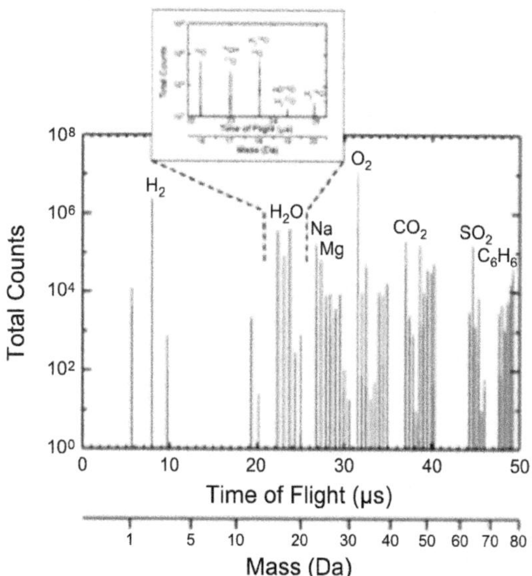

Fig. 14. Simulated mass spectrum of the anticipated Europa Ion Neutral Mass Spectrometer (INMS) results for neutral species at an orbit altitude of 100 km. The simulation is based on a surface composition given by 60% sulfuric acid hydrate: $H_2SO_4 \cdot 8H_2O$; 20% mirabolite: $Na_2SO_4 \cdot 10H_2O$; 10% hexahydrite: $MgSO_4 \cdot 6H_2O$; 5% epsomite: $MgSO_4 \cdot 7H_2O$; and 5% CO_2 combined with the modeled atmospheric composition and 1% heavy organic represented in this case by benzene (but similar for any heavy organic that may be present). This surface composition was used as input for a Monte Carlo model based on *Cassidy et al.* (2008), with results introduced into an instrument model of the Rosina Relectron Time-of-Flight (RTOF) mass spectrometer to produce the simulated spectrum.

3.3.2. Atmospheric spectroscopy. Remote UV through millimeter spectroscopy of the atmosphere would enhance the study of surface composition and the search for current activity at Europa, with ties to the subsurface ocean and habitability.

Venting or transient gaseous activity could occur on Europa. Ultraviolet measurements would provide high sensitivity to very low gas abundances using stellar occultations, as demonstrated in the detection of the Enceladus gas plume (*Spencer et al.,* 2006; *Hansen et al.,* 2006). Ultraviolet imaging of Europa could measure atmospheric density, distribution, and temporal and spatial variations that could be related to surface composition on regional scales. Ultraviolet observations of the icy Galilean moons have detected surface species produced by chemical processing due to charged particle bombardment (*Domingue and Lane,* 1997; *Hendrix et al.,* 1999). Such species include H_2O_2, H_2SO_4 (*Carlson et al.,* 1999a,b), and SO_2 (*Noll et al.,* 1995). Mapping these species would provide insight into surface exposure ages.

Ozone might be present on Ganymede (*Hendrix et al.,* 1999) and possibly Europa. Water ice is present on Europa, but the far-UV H_2O band measured on the icy moons of Saturn (*Hendrix and Hansen,* 2008) was not observed on Europa, possibly due to larger grains or non-ice species that mask UV absorption. Other candidate surface species include salts, sulfates, acids, and tholins. Ultraviolet imaging should have spatial resolution of 1 mrad/pixel to resolve features such as linea and chaos. For comparison, the best resolution obtained by the Galileo UVS was ~50 km/pixel. A UV imager should provide simultaneous EUV (extreme ultraviolet, 60–110 nm), FUV (far ultraviolet, 110–200 nm), and NUV (near ultraviolet, 200–350 nm) with 0.5–3-nm spectral resolution to resolve spectral signatures. A stellar occultation instrument operating only in the FUV could collect data information on the derived atmospheric constituents.

Long-wavelength (IR through millimeter) observations could detect, definitively identify, and determine the abundance of atmospheric species. The rotational-vibrational absorption lines of gases are extremely diagnostic of specific composition, and can provide total column abundances at ppm levels. These observations would provide sensitive detection of plumes at low opacities and indicate the isotopic composition and abundance of major components (e.g., C, O, and N). Millimeter to submillimeter observations have modeled sensitivities of 2%, 3%, 12%, and 36% for NaCl, MgS, NaO, and CH_3CN relative to water vapor for an assumed water column abundance of 5×10^{13} molecules/cm^3 in nadir observations. Improvements by factors of several hundred are possible for limb emission and solar occultation observations respectively. Long-wavelength observations could provide insight into the physical processes that have led to the atmospheres of Europa and the other icy satellites and the processes that underlie the formation of any plumes. These observations could also provide measurements of the temperature of the solid surfaces of the moons. As with the UV measurements, an IR or millimeter instrument operating remotely would be less sensitive to distance from target, and would be effective for assessing compositions for all targets within the jovian system.

TABLE 4. Water vapor components, including isotopes in Europa's atmosphere expected to be measurable by INMS.

Species	Mass	Expected partial pressure (mbar) at 200 km
H_2	2.01	$>10^{-10}$
O_2	31.99	10^{-10}
O	15.9	10^{-12}
H_2O	18.0	5×10^{-13}
$^{18}O^{16}O$	33.9	2×10^{-13}
$^{17}O^{16}O$	32.9	7×10^{-14}

TABLE 5. Calculated densities of sputtered Europa surface materials at 100 km.

Species	Predicted densities at 100 km
Na	60–1600 cm^{-3}
CO_2	170–580 cm^{-3}
SO_2	290–1800 cm^{-3}
H_2O	150–3000 cm^{-3}
L-Leucine	3 × 106F cm^{-3}
(Mass 131)	(F = number-fraction at surface)

Europa's tenuous atmosphere has four observed components: O near the surface (*Hall et al.,* 1995, 1998), Na and K in the region from ~3.5 to 50 R_E (*Brown and Hill,* 1996; *Brown,* 2001; *Leblanc et al.,* 2002, 2005), and H_2 in Europa's co-orbiting gas torus (*Mauk et al.,* 2003). The robust plasma bombardment of the surface might produce many other components (see chapter by Johnson et al.). Because few data are available, models are used to infer the vertical structure and abundances of other species (Fig. 4 in chapter by Johnson et al.).

An Ion Neutral Mars Spectrometer (INMS) would provide data for components in Europa's atmosphere that are derived from the surface by sputtering, outgassing, and sublimation. Most of the postulated atmospheric constituents could be detected at a 100-km orbit based on a recommended detection limit of 200 cm^{-3} (Tables 4 and 5).

An important contribution from an INMS would be measurements of oxygen-isotopic ratios in water vapor. The variations in the $^{17}O/^{16}O$ and $^{18}O/^{16}O$ ratios are the most useful for distinguishing different planetary materials. For example, two gaseous reservoirs, one terrestrial and one ^{16}O rich, are probably required to explain oxygen-isotopic variations in meteorites. The terrestrial fractionation line is due to mass fractionation of the oxygen isotopes in terrestrial materials and the carbonaceous chondrite fractionation line represents mixing between different components. Obtaining similar isotope information (as well as D/H) for Europa would provide important constraints on the origin of water ice in the Galilean satellites.

The sensitivity for ions is much higher than for neutrals, and based on experience at Enceladus and Titan (*Waite et al.,* 2006), ionization will occur by electron impact, photoionization, charge exchange, and electron attachment. Predicted ionization rates for several of these molecules are O_2: 2×10^{-6}/s; H_2O: 3×10^{-6}/s; O: 2×10^{-7}/s; Na: 5×10^{-6}/s; CO_2: 5×10^{-6}/s; SO_2: 10^{-5}/s.

The trace materials (SO_2, CO_2) detected from surface spectroscopy should be readily detected using INMS (see chapter by Johnson et al.). Further characterization of the hydrate and associated dark materials could also be accomplished. For example, Mg should be present in the atmosphere if $MgSO_4$ is present at the surface. Atmospheric emission measurements have confirmed a surface source for Na and K (*Johnson et al.,* 2002; *LeBlanc et al.,* 2002), with some evidence that the Na and K originate in dark regions (*LeBlanc et al.,* 2005; *Cassidy et al.,* 2008). However, these have not yet been detected in surface spectral measurements. Vented material or materials from flows emplaced on the surface are rapidly degraded by the incident radiation, but this process also produces sputtered products that could be detected. Trace organics would also be sputtered from the surface.

Based on a modeled atmosphere and assumed trace salts and organics, predicted atmospheric compositions at 100 km altitude can be generated (Figs. 14–15).

In order to understand Europa's surface composition, it is important to separate the effects of weathering by photons, neutral and charged particles, and micrometeoroids. In particular, radiolytic processes may alter the chemical signature over time, complicating assessment of the original surface composition. This requires detailed spectroscopy with UV-IR, using global and targeted observations. It is also critical to map ion and electron flux on the surface as a function of species and energy. The separation of the primary and alterated surface compositions will be aided by high-spatial-resolution spectra on both leading and trailing hemispheres, in which younger, less-altered materials may be exposed by magmatic, tectonic, or mass-wasting processes.

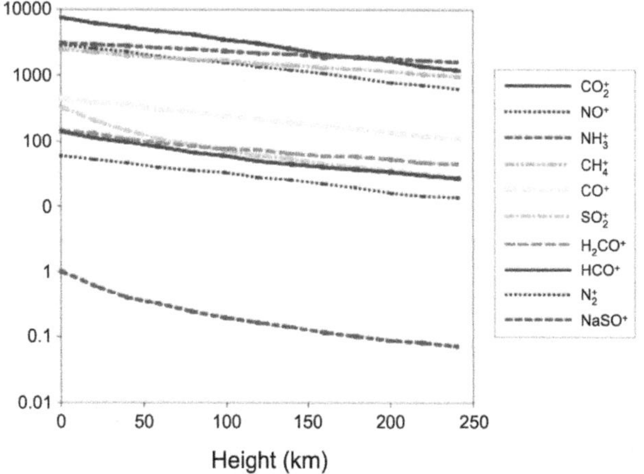

Fig. 15. See Plate 39. Ionospheric densities vs. altitude for molecules sputtered from the surface (*Johnson et al.,* 1998); all densities exceed the detection limit (10^{-3} cm^{-3}; Y-axis) of a modern mass spectrometer, such as the Cassini INMS.

Characterization of the sputter-produced atmosphere with UV or long-wavelength observations of Europa's atmosphere would provide data for the species, abundances, and ion implantation rates. This could be accomplished with FUV stellar occultations and UV imaging of atmospheric emissions; UV imaging through a range of at least 0.10 0.35 μm at equal to or better than 3-nm spectral resolution is needed to map surface species, including radiolytic compounds. Infrared observations from 5 to ~50 μm with a spectral resolution of 1–5 cm^{-1} or mm observations between 100 and 600 GHz with spectral resolution of ~100–250 kHz should be used to investigate the atmospheric components. These observations would require SNR > 100, profiling observations for at least 80% of the surface with spacing no more than 25 km and spatial resolutions of 100–500 m to resolve geologic features.

An INMS would directly measure species sputtered from the surface, which may include organic fragments. It should operate in the mass range from 1 to >300 amu, with a mass resolution (m/Δm) of ≥500, and pressure range of 10^{-6} to 10^{-17} mbar, and should make continuous measurements throughout the mission. For sputtering sources, it is important to measure ions from the plasma energies to about 10 MeV. The number of sputtered molecules is the product of the precipitating ion flux and the neutral yield per ion. For heavy ions, the yield peaks in the few-MeV energy range; thus, it is important to measure ion fluxes into the MeV. Combined with images and geologic maps, such observations allow determination of how surface materials evolve in the radiation environment.

3.3.3. Exogenic materials. Exogenically implanted materials can be characterized by measuring ions. Each ion energy and species has a specific penetration depth in ice; if they reach the surface, cold plasma ions are deposited in the most processed layer. Energetic charged particles can penetrate more deeply into surfaces and therefore will not be removed as readily by processes such as sublimation. A 1-MeV proton has a range of 24 μm depth and a 1-MeV electron has a range of 4.2 mm in water. In addition to sputtering by ions, which adds molecules to the atmosphere of a satellite, electron radiolysis also can create neutral species in the material (e.g., H_2O_2). Such molecules can dissociate in the ice and the lighter byproducts can escape from the surface and enter the atmosphere with a small amount of energy.

To constrain the exogenic materials properly, the flux of the precipitating charged particles over the entire surface should be determined for the energy range between the eV and few MeV, with resolutions ΔE/E ~ 0.1, 15° angular resolution, and basic ion mass discrimination. This is accomplished with a plasma sensor and an energetic charged particle sensor, both with the capacity to measure the upstream flux at the satellite orbits and view (in the precipitating

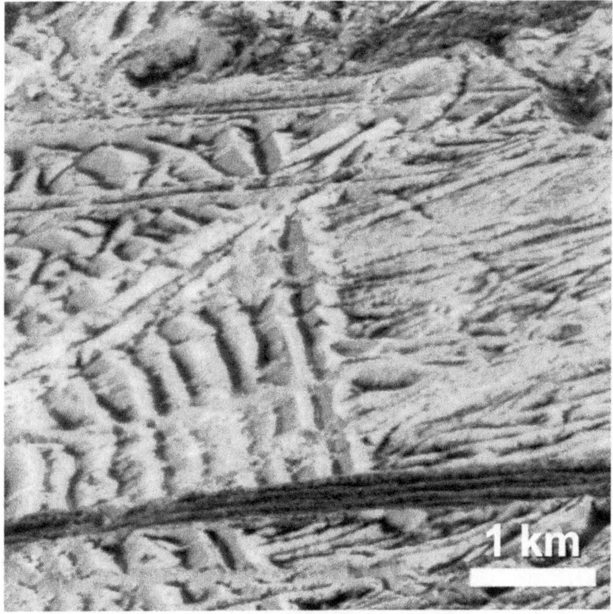

Fig. 16. Galileo image of ridged plains (~6 m/pixel). Some ridges have central troughs containing dark deposits ~100 m wide, flanked by bright, icy walls. Spectral observations with at least 100-m spatial resolution are needed to determine the composition of both the bright and dark materials.

direction) all points in Europa orbit. These measurements should be synthesized with globally distributed IR and UV measurements, along with global multispectral visible images. These data allow materials to be traced from their magnetospheric sources, to the surface, and into the sputter-produced atmosphere. The atmosphere can then be measured directly (e.g., with the INMS) and by remote sensing.

3.3.4. Relationship to geologic processes. The spatial resolution required for compositional mapping is determined by the scale of critical landforms such as bands and ridges (Figs. 3 and 16). Europa displays albedo and morphological heterogeneity at scales of 25–100 m, suggesting that compositional variations also exist at this scale. However, the composition of these features remains unknown because Galileo NIMS observations are averages of light reflected from large areas containing both icy and non-icy terrain units (e.g., *McCord et al.,* 1999; *Fanale et al.,* 1999). Spectra of adjacent regions within an instrument field of view combine to produce an average spectrum, with spectral features from all the materials. However, these composite spectra have potential overlap of spectral features and reduced contrast relative to the spectra of the individual surface units. Spectral mixing and reduced contrast results in an attendant decrease in detectability, and a given target cannot be distinguished from its surroundings. Thus, it is desirable to resolve regions of uniform composition in order to map distinct surface units. While these in turn may be mixtures, spatially resolving dark terrains that have fewer components and are free of the strong and complex absorption features of water ice will greatly facilitate identification of the non-ice materials. For reasonable statistical sampling, it is also desirable to have multiple pixels within a given surface unit. Adjacent measurements can then be compared and averaged to improve the signal and reduce noise.

Galileo images of Europa suggest geologically recent formation for ridges, chaos, and other features. The images also show abundant evidence for much younger materials exposed by mass-wasting (see chapter by Moore et al.). These modification processes have likely affected many surfaces, potentially exposing fresh materials that are less altered than their surroundings. Spectroscopy at a resolution better than 100 m would isolate these surfaces and provide an opportunity to determine the composition of primary materials.

3.4. Geology Objective: Understand the Formation of Surface Features, Sites of Recent or Current Activity, and Candidate Sites for Future Exploration

Europa's landforms are enigmatic and a wide variety of hypotheses have been posed for their formation. The search for geologic activity is especially significant for understanding Europa's potential for habitability. Moreover, characterizing potential landing sites is important for future exploration. Prioritized investigations are to (1) characterize magmatic, tectonic, and impact features; (2) search for areas of recent or current geological activity; (3) investigate global and local heat flow; (4) determine relative surface ages; and (5) characterize the physical properties of the regolith and assess processes of erosion and deposition.

Characterizing the distribution, morphology, and topography of surface features at regional and local scales is critical for understanding surface evolution. Galileo images demonstrate that regional-scale data (~100 m/pixel) are excellent for geologic studies, yet less than 10% of the surface was imaged at better than 250 m/pixel (Fig. 17). Thus, near-global coverage (>80% of the surface) in at least three colors at 100 m/pixel is needed. Galileo images also show the great value of targeted high-resolution (~10 m/pixel) monochromatic images for detailed characterization of selected landforms.

Topographic mapping through stereo images greatly aid morphologic characterization and geologic interpretation. Stereo imaging could be achieved in a single pass with a stereo camera, or through horizontal overlap of adjacent orbits, resulting in ~10-m vertical height accuracy from 100-m/pixel images. Height accuracy further improves by $\approx\sqrt{N}$ by averaging of N overlapping stereo pairs. It is also important to determine topography at high resolution (1 m or better) across representative features. Europa's surface is quite heterogeneous and rough at the decameter scale (Fig. 16), and the same may be true at smaller scales. Very-high-resolution monochromatic images (1 m/pixel; <0.1% of the satellite) could reveal the detailed character of landforms, regolith properties, and erosion-deposition processes. Moreover, images at this scale are critical for characterizing future landing sites.

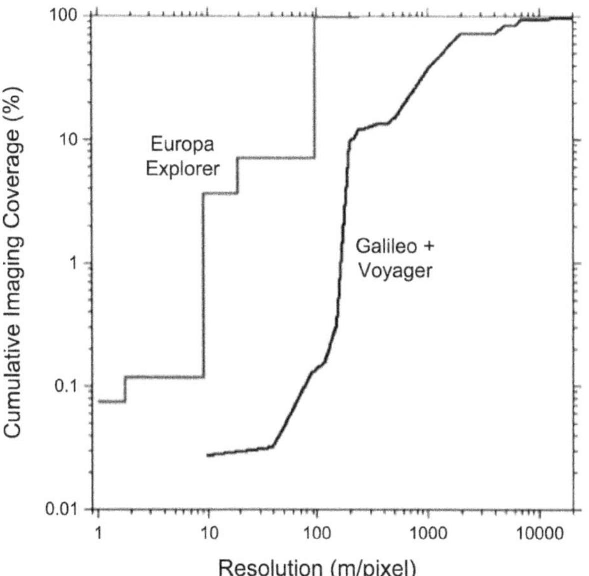

Fig. 17. Cumulative imaging of Europa's surface as a function of resolution, illustrating the 1–2 orders of magnitude improvement of JEO relative to Voyager and Galileo combined. Unlike the opportunistic coverage obtained from earlier flybys, JEO's systematic imaging from orbit will be in discrete resolution steps.

Geologically active sites are the most promising for astrobiology, and will be important to identify. Active processes typically involve high heat flow and may involve plumes detectable by imaging, laser altimetry, or UV occultations. These would also be the most likely locations for near-surface liquid water. Moreover, recently or currently active regions best illustrate the processes involved in the formation of some surface structures, showing pristine morphologies and possible associations with thermal and/or plume activity.

Discoveries of any active regions would be followed by visible and other remote sensing of the inferred source (Fig. 18). In addition, it might be possible to observe surface changes within the timescale of the JEO mission. Imaging at high resolution (10 m/pixel) in stereo, coupled with higher-resolution (~1 m/pixel) images, thermal data, and compositional measurements, would be used to study suspected candidates for recent activity. If age-sensitive chemical or physical indicators are identified (e.g., H_2O frost, ice crystallinity, sulfate hydrates, SO_2, or H_2O_2), then mapping their distribution may reveal past or present active regions.

Constraining the global and local heat flow of Europa is of great importance. High heat fluxes (~1 W/m^2) would be necessary for detection of uniform conductive heat flow (*Spencer et al.,* 1999), but lower levels of endogenic heat flow can be detected if locally concentrated, as on Enceladus (*Spencer et al.,* 2006). A high heat flow could indicate that significant tidal heating and likely volcanic activity is occurring in the mantle. This would have important implications for astrobiology, and potentially for understanding the development of life on Earth, where life might have developed at hydrothermal vents on the seafloor. A search for thermal anomalies could be conducted with thermal-IR detectors. Constraints on global heat flow would also come from subsurface ice temperatures derived from radar and estimates of ice thickness derived from gravity or radar data.

Determining the relative ages of Europa's surface allows the geologic history to be derived. Relative ages are determined from stratigraphy, cross-cutting, embayment, and the density of impact craters. Galileo three-color images taken at low phase angle showed their advantage in stratigraphic studies because features generally brighten and become less red with time (*Geissler et al.,* 1998). However, without global coverage, the relative ages of different regions cannot be correlated because they cannot be linked, and this poses an intractable problem for Europa based on Galileo images. Global color imaging (>80% of the surface) at resolution better than ~100 m/pixel, with near-uniform lighting conditions and phase angle ≤45°, would allow Europa's global stratigraphic sequence to be derived.

Europa's regolith provides information on modification processes occurring on a submeter scale and for understanding means of communication between the radiation-processed, oxidant-rich upper meter of the surface and the sub-

Fig. 18. The plumes of Enceladus illustrate geologic activity, revealed in Cassini high-phase images. Analogous plumes would be ~70 km high when scaled to Europa's gravity (*Nimmo et al.,* 2007), and might contribute to Europa's recently discovered torus (*Mauk et al.,* 2003). A combination of thermal and UV observations would permit a thorough search and characterization of active regions on Europa.

surface. Modification occurs by mass wasting, sputtering, impact gardening, and thermal redistribution of material. Regolith studies are also important in characterizing high-priority sites for future landed missions. Regolith processes can be investigated by targeted imaging at ~1 m/pixel resolution. Magnetometry data are important for understanding sputtering and its effects on regolith evolution; thus, it is valuable to measure ion-cyclotron waves, which can be related to plasma-pickup and erosion processes.

3.5. Jupiter System Objective: Understand Europa in the Context of the Jupiter System

Aside from its intrinsic value, understanding the Jupiter system is critical for placing Europa in context as a member of the jovian system. Jupiter system science includes satellite surfaces and interiors; satellite atmospheres, plasma, and magnetosphere; Jupiter atmosphere; and rings, dust, and small moons. The investigations are prioritized within each of these categories, with the lower priorities having less relevance to the Europa habitability goal of JEO, yet all are pertinent to the overarching theme of EJSM.

3.5.1. Satellite surfaces and interiors. The Galilean satellites probably formed in a similar environment, while the inner three affected each other as their orbits and interiors evolved. Io's silicate and sulfur-rich surface is dominated by processes related to tidal heating, while Europa, Ganymede, and Callisto share similar icy surfaces altered to different extents by external and internal processes. Prioritized investigations are to (1) investigate the nature and magnitude of tidal dissipation and heat loss on the Galilean satellites, particularly Io; (2) study Io's active volcanism for insight into its surface history and its influence on the rest of the Jupiter system; (3) investigate the presence and location of water within, and the extent of differentiation of, Ganymede and Callisto; (4) determine the composition, physical characteristics, distribution, and evolution of surface materials on Ganymede; (5) determine the composition, physical characteristics, distribution, and evolution of surface materials on Callisto; and (6) identify the dynamic processes that cause internal evolution and near-surface tectonics of Ganymede and Callisto.

Tidal heating is most readily studied on Io, where the spatially and temporally variable surface temperature and heat flow can be measured using a thermal mapper. Combined with high-resolution images, topographic data, and visible/NIR observations of volcanic thermal emissions, these measurements would characterize the different styles of volcanic compositions and their controlling factors. For the icy satellites, recent activity is also potentially detectable with a thermal mapper. Otherwise, the strongly temperature-dependent radar absorption of ice allows subsurface radar profiling to determine an approximate temperature gradient and, thus, the heat flux. Predicted long-wavelength variations in stress and heating (due to tides or other processes) could be investigated using either the thermal or the radar approach, and correlated with spatial variations in surface features.

Ganymede and Callisto are both thought to possess oceans sandwiched between two layers of ice. The presence of such oceans might be confirmed by measuring the time-dependent (tidally driven) gravity field or surface topography; these data would also help determine the amount of tidal heating. Magnetometer data will place bounds on the depth, thickness, and salinity of the putative oceans. The deeper structure of these satellites, and in particular the extent to which Callisto is differentiated, may be determined by Doppler-tracking during equatorial and polar flybys. Combining regional images taken at different times will establish the spin pole orientation and potentially detect librations (both of which depend on the internal structure). For Ganymede, characterization of the spatial and temporal variability of its internal magnetic field will constrain dynamo models.

Characterizing surface materials on Ganymede and Callisto will help disentangle the origin of similar materials on Europa. Infrared and UV spectroscopy would identify individual compounds and map their distribution (e.g., hydrated non-ice material and trace constituents such as CO_2). Combined with high-resolution color images, correlations between particular species and geologic features can be tested. Repeat measurements will allow characterization of any surface changes. Visible and IR mapping may be used to determine compositional variability, spatial and temporal distribution, and geological associations of silicates and volatiles at Io.

3.5.2. Satellite atmospheres. Atmospheres above the jovian satellites are extremely tenuous. However, as described in section 2, they are important in the Jupiter system and play critical roles in understanding the satellites. Prioritized investigations are to (1) characterize the composition, variability, and dynamics of Europa's atmosphere and ionosphere; (2) understand the sources and sinks of Io's crustal volatiles and atmosphere; and (3) characterize the sources and sinks of the Ganymede and Callisto atmospheres.

The surfaces of the Galilean satellites continuously exchange material with the atmospheres, so it is important to make direct measurements of the atmospheres and ionospheres. Data of major and minor constituents of the neutral atmospheres will greatly aid geological, compositional, and exospheric studies. In addition to water ice, heavy molecules and molecular fragments can be sputtered and subsequently detected by UV and IR spectroscopy (including limb scans such as those used to detect the CO_2 atmosphere of Callisto) (*Carlson,* 1999), and *in situ* by an INMS (see section 3.2.2). Ultraviolet stellar occultations provide stringent constraints on the extent and structure of the satellites' atmospheres. Measurements of Europa's ionosphere is critical to understanding the higher-order magnetic moments for interpretation of Europa's induced magnetic field. Radio occultations provide a proven technique for sounding the ionospheres of all the Galilean satellites (*Kliore et al.,* 1997, 2001a,b, 2002).

Multiple techniques are required to understand Io's unusual atmosphere, which is produced from a combination

Fig. 19. Artist's rendition of Europa's tenuous atmosphere, supplied chiefly by charged particle bombardment of the surface. If Europa is currently geologically active, plumes may also contribute.

of sources. For example, surface volatiles, especially SO_2, are a major source, as are the volcanic eruptions. The Io plasma torus is a key link in the transfer of material from Io to Europa (Fig. 19). Understanding the sources and sinks of the volatiles on the surface and in the atmosphere that contaminate Europa is a key undertaking of JEO. Eclipse imaging, plume monitoring, far- and mid-UV and IR spectroscopy of the surface and atmosphere with high spatial and temporal resolution, as well as stellar occultations, will provide critical constraints on the flux of materials escaping Io. Monitoring the neutral clouds and plasma torus, together with measurements of dust and neutral and charged particles, will help determine how these materials are dispersed throughout the jovian system and beyond.

3.5.3. Plasma and magnetospheres. Jupiter's magnetosphere is significant throughout the entire system, and is potentially important for Europa, which resides near the outer edge of Io's plasma torus. Studies of the magnetosphere in general are important for understanding the Jupiter system, as well as obtaining data on the related charged particles, auroral signatures, and the stress balance between magnetic and particle pressures. Prioritized investigations are to (1) measure the plasma and neutral ejecta from Europa; (2) characterize the composition of and transport in Io's plasma torus; (3) study the pickup and charge exchange processes in the Jupiter system plasma and neutral tori; (4) study the interactions between Jupiter's magnetosphere and Io, Ganymede, and Callisto; (5) understand the structure, composition, and stress balance of Jupiter's magnetosphere; and (6) determine how plasma and magnetic flux are transported in Jupiter's magnetosphere.

Investigation of Jupiter's magnetosphere requires near-continuous temporal observations, with spatial sampling throughout the magnetosphere, especially in special regions including boundaries and near satellites. For example, Jupiter's magnetopause separates regions of high intensities of electrons from the solar wind. Furthermore, investigations of magnetic reconnection and sporadic plasma flux enhancements in the magnetotail require continuous field and plasma measurements at distances greater than 60 R_J in the nightside sector of Jupiter's magnetosphere. Closer to the planet, the middle magnetosphere contains current systems and electron beams that link the equatorial and polar regions of the magnetosphere and are responsible for auroral emissions. Closer to the planet, the radiation belts must be studied because they are co-located with the orbits of the Galilean satellites and continuously weather their surfaces.

In addition to the volcanic source at Io, neutrals are added to the magnetosphere from surface sputtering and possibly from plumes. The neutrals become ionized quickly at Jupiter and must be accelerated to the co-rotation speed of the local plasma. By measuring the plasma flow field over a wide range of radial distances and latitudes and the current systems (through magnetometer measurements), we will be able to understand the addition of new material to the magnetosphere.

Plasma observations near satellites should be made at various altitudes and orientations with respect to the plasma flow direction and under an assortment of background plasma conditions (determined by the satellite's location relative to Jupiter's current sheet). Satellites interact with the magnetosphere over a large distance (e.g. the so-called "Alfven wing"), and this interaction indicates the conductivity of the body and the electrodynamics of the ionosphere and atmosphere. In addition to magnetometer measurements near the satellites to constrain the fields and current systems, it is also important to detect modifications to the plasma flow field by the satellite.

Because of its critical role in the transfer of material from Io to Europa, the Io plasma torus is an important target for UV spectrometer measurements in the range of ~70–200 nm. For example, during Cassini's flyby, the UVIS obtained perhaps the most illuminating series of synoptic observations of the Io torus ever obtained (*Delamere et al.,* 2004; *Steffl et al.,* 2004, 2006, 2008). Because the torus energy loss is primarily by EUV line radiation from multiple sulfur and oxygen ions, these measurements cannot be made from Earth. JEO could provide a much longer set of synoptic observations during which multiple Io volcanic eruptions are likely to occur, leading to plasma injections into the torus, followed by transport of material to Europa.

For studies of Ganymede's internal field and magnetosphere, JEO's trajectory could complement that of Galileo. Ganymede's surface is weathered differently on open and closed magnetic field lines and it is critical to measure the charged particle populations on those field lines. Elucidating the links between Io's volcanic activity and the dynamics of the jovian magnetosphere requires UV and visible observations of the Io torus, jovian auroral oval, and the satellite auroral footprints at high time and spectral resolution, combined with monitoring of Io activity in the visible and infrared. The jovian ionosphere can be probed at a range of latitudes through radio occultations at multiple frequencies, accomplished using precisely time-referenced Ka- and X-band transmissions.

3.5.4. Jupiter atmosphere. Following the Juno mission many questions will probably remain regarding the dynamics, chemistry, and vertical structure of Jupiter's upper atmosphere. Juno has the potential to increase the understanding of the jet stream depths, the local structure of Jupiter's magnetic field, and the deep abundances of water vapor and ammonia. However, there is insufficient spatial resolution, wavelength coverage, and temporal coverage to map the upper troposphere and stratosphere; such mapping by JEO would constrain local processes and their long-term variations governing the basic structures of the jets, clouds, belts, zones, and vortices. JEO would surpass previous investigations in terms of continuous high-resolution global coverage over many months, with the goal of producing the first comprehensive database with visible imaging and spectroscopic characterization for Jupiter's atmosphere. Prioritized investigations are to (1) characterize the abundance of minor species (especially water and ammonia) in Jupiter's atmosphere to understand the evolution of the jovian system, including Europa; and (2) characterize jovian atmospheric dynamics and structure.

JEO would undertake several studies within these investigations to address some of the puzzles unresolved after the Juno mission. Results will aid the understanding the relations among the upper troposphere, lower troposphere, and internal structure, thereby revealing Jupiter's bulk composition and the evolution of the system as a whole.

3.5.4.1. Jet stream meteorology. High-resolution long-term cloud tracking at VNIR wavelengths at 30 km resolution allows the zonal (east-west) and meridional (north-south) velocities to be obtained, and could allow measurement of the mean-meridional velocity at the cloud level for the first time. Observations at different wavelengths permit studies of the vertical wind shear in the tropospheric jets, which could then be related to the deep structures observed by Juno to determine the vertical coupling between the upper and lower atmospheres. Spectroscopic studies would constrain the thermal and chemical environments in the vicinity of these cloud tracers.

3.5.4.2. Tropospheric hazes and clouds. The characterization of the altitude and global distribution of photochemical hazes and condensation clouds will provide fundamental clues to their origin and the meridional transport in the troposphere. The haze distribution is vital to understanding the details of solar energy deposition in the atmosphere and its role in hemispherical asymmetries. Correlation between temporal variations of cloud properties (size, optical properties, vertical distribution, color, and albedo) with changes of environmental temperatures and composition will be used to determine the causes for major changes in Jupiter's cloud properties. These may provide a fundamental clue for the origin of their various colors. The detection of condensed ammonia, ammonium hydrosulfate, and water ice will provide significant clues to the size and strength of updraft regions in the atmosphere (e.g., *Baines et al.,* 2002).

3.5.4.3. Evolution of discrete cloud features. Weather-layer phenomena such as thermal hotspots, large anticyclonic vortices, turbulent regions, convective plumes, and thunderclouds can be monitored using VNIR imaging over a long temporal baseline. Spatial resolutions of ~30 km provide data to study cloud properties, energetics, and angular momentum of individual storm systems (cf. *Porco et al.,* 2003) and their relation to the global atmospheric circulation. Measurements of thunderstorms on the dayside and lightning on the nightside will constrain the energetics of the atmosphere at depth.

3.5.4.4. Atmospheric waves and the thermosphere. Radio occultations would characterize the detailed vertical temperature structure in the stratosphere and upper troposphere, thus providing a window into stratospheric dynamics. This would allow characterizing the vertical propagation of various atmospheric waves, including local gravity, larger Rossby waves, the quasiperiod stratospheric oscillation, and the altitude dependence of slowly moving thermal waves that are uncorrelated with cloud structures. Radio and stellar occultations can also be used to study the thermosphere and determine the extent to which wave absorption can cause the high thermospheric temperatures.

3.5.4.5. Tropospheric dynamical tracers. To complement the Juno investigation of the lower troposphere, H_2O and NH_3 would be measured in the upper troposphere, in addition to disequilibrium species such as PH_3, GeH_4, and CO as diagnostics of the dynamics associated with jet stream meteorology and discrete cloud features. In particular, high-inclination orbits would map these species and tropospheric aerosols at polar latitudes to determine the relative roles of dynamics and seasonally forced radiation in maintaining Jupiter's cold and hazy polar vortices (cf. *Vincent et al.,* 2000; *Porco et al.,* 2003).

3.5.4.6. Stratospheric temperature and composition. Spectroscopic studies of Jupiter's stratosphere will shed light on the photochemistry and atmospheric motion responsible for the distribution of hydrocarbons and hazes. Thermal monitoring could allow the first detection of tides raised by the Galilean satellites. Detection of material from exogenic sources (HCN, CO, CO_2, H_2O) will permit studies of the changing environment surrounding the satellites and rings.

3.5.5. Rings, dust, and small moons. The "skeleton" that holds the jovian ring system together is its collection of source bodies, which include four inner moons. But the ring also contains abundant embedded meter- to kilometer-sized bodies. Learning the nature and properties of these bodies is critical to understanding the origin, evolution and long-term stability of the system. Prioritized investigations are to (1) characterize the properties of the small moons, ring source bodies, and dust; and (2) identify the processes that define the origin and dynamics of ring dust.

At a certain point, the distinction between a "moon" and a "ring body" should become moot, because the ring contains a continuum of sizes. However, imaging at low phase angles is needed to characterize the radial distribution of the bodies and identify its largest members. A detection threshold for bodies 1 km in diameter was achieved by New Horizons. Future imaging should reduced this threshold to ~100 m (assuming albedos comparable to Adrastea and Metis) to ensure more complete detection and be sensitive to intensities in the range of 10^{-8} for high-quality ring detections. Searches for the source bodies must be conducted primarily at low phase angles, which means that targeting be just off the planet with minimal interference from scattered light. Little is known of the composition of the jovian ring or its ring moons, beyond the fact that they are very dark and red. Visible near-infrared spectrophotometry would provide surface composition data.

The dust component of the system should also be fully characterized. The size distribution of ring dust probably varies with location, and can be derived from light phase curves and spectra. Imaging of all the ring components must be conducted at a full range of phase angles, up to and including within a few degrees of exact forward-scatter. This is where diffraction by the dust particles dominates, yielding the most precise size constraints. Such imaging can only be obtained when passing through Jupiter's shadow. Measurements should be sensitive to rings with optical depths as low as ~10^{-9}, the approximate value for a faint outward extension to Thebe's "gossamer" ring.

The search for jovian dust and moons should not end at the ring boundary. Trojan moons are common in the saturnian system, but a deep, systematic search for small moons orbiting among the Galilean satellites has never been conducted. Any bodies that are found would surely have interesting dynamical histories and place new constraints on how the entire system formed. The detection threshold should be less than 1 km. The camera should also be used for precision astrometry of any bodies that are found. At higher phase angles, the system should be searched for faint dust belts, which might be indicators of unseen small bodies in the system.

In addition to precise measurements of the particle properties, a better understanding of the dust grains' motion is needed. Grains do respond to solar radiation pressure and magnetic forces, which produce the "Lorentz resonances" that distribute much of the dust well out of the ring plane. However, the three-dimensional structure of the system's faint inner "halo" and its outer gossamer rings have never been mapped in detail. This requires imaging at a large variety of viewing and lighting geometries, with sensitivity to $I/F \sim 10^{-8}$ and resolution of finer than 100 km. Such observations may illuminate the dynamics behind some of system's more peculiar features, such as vertical ripples found in some Galileo images (*Ockert-Bell et al.,* 1999).

Because the system has shown clumps and other time-variable phenomena, repeated observations of the system are required. The most rapid phenomena are likely to change in periods of days to months. It is critical to obtain reliable constraints on their timescales to understand the active processes.

4. SCIENCE IMPLEMENTATION

Since the first return of spacecraft data of the Jupiter system from the Pioneer and Voyager flybys, the scientific study of Europa has evolved from addressing first-order questions on its general characteristics, to a wide variety of investigations based mostly on Galileo data (e.g., see chapter by Alexander et al.). In the late 1990s, most studies of Europa were conducted by Galileo science instrument teams that focused on data validation. Through venues such as the Jupiter System Data Analysis Program and the Outer Planets Research Program, datasets were merged and synthesized by the wider scientific community. In the time since the first return of Galileo data, the scientific understanding of Europa has greatly matured, leading to the formulation of sophisticated questions to be addressed through new data.

4.1. Payload Considerations

Developing the proposed JEO planning payload involved identifying the types of instrument that are designed to test specific hypotheses and have the potential for serendipitous discovery and exploration. Solar system exploration is replete with examples of the latter consideration; for example, the initial payload for the Mars Observer/Mars Global Surveyor mission included neither a camera nor a magnetometer, and both were subsequently added. Few would argue that the discoveries from these instruments, including the existence of young water-carved gullies and a remnant magnetic field, revolutionized the understanding of Mars and helped direct future mission. The JEO planning payload (Table 6) enables testing current hypotheses, while providing a broad instrument suite for exploration and response

TABLE 6. Science planning payload instruments for JEO.

Instrument
Wide-Angle Camera (WAC)
Medium-Angle Camera (MAC)
Narrow-Angle Camera (NAC)
Vis-IR Spectrometer (VIRIS)
UV Spectrometer (UVS)
Thermal Instrument (TI)
Laser Altimeter (LA)
Ice-Penetrating Radar (IPR)
Magnetometer (MAG)
Particle and Plasma Instrument (PPI)
Ion-Neutral Mass Spectrometer (INMS)
Telecom System

TABLE 7. Mission constraints imposed by science.

Science Objective	Architecture and Orbit Constraints	Additional Mission Constraints
Ocean	*(See chapters by Alexander et al., Sotin et al., Bills et al., Schubert et al., Vance and Goodman, Khurana et al., and McKinnon et al.)* *Gravity and altimetry*: Orbiter required, low altitude (~100–300 km), orbital inclination of ~40°–85° (or retrograde equivalent) for broad coverage and crossovers. Ground tracks should not exactly repeat (while near-repeat is acceptable), so that different regions are measured. Requires a mission duration of at least several eurosols to sample the time-variability of Europa's tidal cycle. *Magnetometry, particles and plasma, INMS*: Near-continuous measurements near Europa, globally distributed, at altitudes ≤500 km, for a duration of at least 1–3 months.	*Gravity and altimetry*: Knowledge of the spacecraft's orbital position to high accuracy and precision (~meters radially) via two-way Doppler. *Gravity*: Long undisturbed data arcs are required (>12-h periods without spacecraft thrusting; see section 4.3), and momentum wheels to maintain spacecraft stability. *Magnetometry*: Magnetic cleanliness of 0.1 nT at the sensor location, and knowledge of spacecraft orientation to 0.1°. Calibration requires slow spacecraft spins around two orthogonal axes each week to month.
Ice	*(See chapters by Moore and Hussman, Nimmo and Manga, Barr and Showman, and Blankenship et al.)* *Radar sounding*: Low orbit (≤200 km) considering likely instrument power constraints. Near-repeat groundtracks are required to permit targeting of full-resolution observations of previous survey-mode locations. Close spacing of profiles requires a mission duration of months, and near-global coverage implies orbital inclination ≥80°.	*Radar sounding and altimetry*: Datasets need to be co-aligned, and highly desirable to be time-referenced to 10–30 ms accuracy. *Radar sounding*: Raw full-resolution targeted radar data requires ≥900-Mb solid-state recorder. Early flyby of Europa for radar signal processing assessment.
Chemistry	*(See chapters by Estrada et al., Canup and Ward, Carlson et al., Zolotov and Kargel, McGrath et al., Johnson et al., Paranicas et al., and Hand et al.)* *Infrared spectroscopy*: Solar incidence angles of ≤45°, with orbital inclination ≥80° for near-global coverage. Near-circular orbit is desirable. Close spacing of profile-mode data implies a mission duration on the order of months. A near-repeat orbit is desired, to permit targeted observations to overlap previous profiling-mode observations. *INMS:* As low an orbit as feasible is desired, for direct detection of sputtered particles.	*Optical remote sensing*: Boresight co-alignment of all nadir-pointed imaging and profiling instruments is highly desirable.
Geology	*(See chapters by Dogget et al., Bierhaus et al., Schenk and Turtle, Kattenhorn and Hurford, Prockter and Patterson, Collins and Nimmo, and Moore et al.)* *Optical remote sensing*: Near-repeating orbits required to permit regional-scale coverage overlap, follow-up targeting, and stereo; close spacing of profile data implies a mission duration on the order of months; ≥80° orbital inclination to provide near-global coverage. *Imaging*: Solar incidence angles of 45°–60° are best for morphological imaging, while a solar phase angle ≤45° is best for visible color imaging. Near Sun-synchronous and near-circular orbit is highly desired to permit global coverage to be as uniform as practical. Beginning at a higher orbital altitude and reducing to a lower altitude will allow rapid initial areal coverage, followed by improved resolution coverage at low altitude. *Thermal mapping*: Day-night repeat coverage required; afternoon orbit is desirable.	*Optical remote sensing*: Boresight co-alignment of all nadir-pointed imaging and profiling instruments is highly desirable, and also highly desirable to be time-referenced to 10–30 ms accuracy. *Radar sounding and altimetry*: Datasets need to be co-aligned, and highly desirable to be time-referenced to 10–30 ms accuracy. *Magnetometry*: Magnetic cleanliness of 0.1 nT at the sensor location, and knowledge of spacecraft orientation to 0.1°. Calibration requires slow spacecraft spins around two orthogonal axes each week to month. *Ultraviolet spectroscopy*: Atmospheric emissions observations and stellar occultations require a view to the satellite's limb.

to new discoveries. It should be noted, however, that the actual instrument suite would ultimately result from the NASA Announcement of Opportunity selection process, and might or might not include the instruments listed in Table 6.

4.2. Instrument and Mission Requirements

The JEO goal and objectives place certain requirements on the mission design, as summarized on Table 7. Optimizing among these requirements has shaped a mission scenario, which includes ~2.5 years in Jupiter orbit before insertion into orbit around Europa. The spacecraft would then orbit Europa for at least 28 eurosols (1 eurosol = 3.55 Earth days), or ~3 months. The recommended orbit is near-circular, with an inclination of 95° (i.e., equivalent to 85° but retrograde, thus offering slower orbital precession). The optical remote sensing instruments are nadir-pointed and mutually boresighted. The initial planned orbital altitude is 200 km, which is reduced to 100 km after several eurosols to best achieve requirements of gravity, altimetry, magnetometry, and radar data. The planned orbit is not quite Sun-synchronous but precesses slowly, such that the orbit does not exactly repeat the same groundtrack but allows instru-

TABLE 7. (continued).

Science Objective	Architecture and Orbit Constraints	Additional Mission Constraints
Jupiter system	*(See chapter by Kivelson et al.)*	
Satellite surfaces and interiors	*Optical remote sensing*: Up to three flybys of Io with one at low altitude over an active volcanic region; at least five flybys of Ganymede (altitudes of <1000 km with at least four with an altitude <200 km); at least five Callisto flybys including one polar; all with altitudes <1000 km); closest approach distributed globally in latitude and longitude. *Imaging*: Solar incidence angles of 45°–60° are best for morphological imaging, while a solar phase angle ≤45° is best for visible color imaging.	*Optical remote sensing*: Boresight co-alignment of all nadir-pointed imaging and profiling instruments is highly desirable.
Satellite atmospheres	*Radio science:* ≥10 radio occultation observations of the Galilean satellites. *INMS:* At least very close flyby of Io.	*Radio subsystem:* Inclusion of an ultrastable oscillator (USO) is desirable.
Plasma and magnetospheres	*Magnetometry, particles and plasma, INMS:* Near-continuous measurements throughout the tour; dedicated campaign to observe the Io torus; broad distribution of Ganymede-magnetic latitude sampled on both leading and trailing hemispheres; near-continuous measurements near Europa during flybys, globally distributed, at altitudes ≤500 km.	*Magnetometry:* Magnetic cleanliness of 0.1 nT at the sensor location, and knowledge of spacecraft orientation to 0.1°. Calibration requires slow spacecraft spins around two orthogonal axes each week to month. *Particles and plasma, and INMS*: Require observing in the ram direction.
Jupiter atmosphere	*Optical remote sensing:* Coordinated feature-track observations using the entire suite of remote sensing instruments; sufficient time and resources for dedicated campaigns covering at least two full Jupiter rotations; solar, stellar, and radio occultations covering as wide a range of latitudes as possible.	*Optical remote sensing:* Boresight co-alignment of all imaging instruments is highly desirable. *Radio subsystem:* Inclusion of an ultrastable oscillator (USO).
Rings	*Optical remote sensing*: ≥One shadow passage from long range; ≥3° inclination off of the ring plane.	*Optical remote sensing*: Boresight co-alignment of all imaging instruments is highly desirable.

ment fields of view to overlap with previous tracks. Thus, the orbit would be near-repeating after several eurosols, within about 1° of longitude at the equator. The solar incidence angle is nominally 45° (2:30–4:00 p.m. orbit) on average, as the best compromise to the requirements of imaging and spectroscopic optical remote sensing measurements.

4.3. Data Acquisition Strategy

4.3.1. Jovian tour data acquisition. Jupiter system science would be the principal focus of the 2.5-year jovian tour phase of the mission. Monitoring and measuring the plasma environment, magnetosphere, and Jupiter's atmosphere would be accomplished routinely each week. Flybys of the satellites would be highly varied for latitude and lighting, but are opportunistic as the trajectory is optimized for duration, fuel used (Δv), and radiation dose.

The JEO solid-state recorder is likely to provide ~16 Gb of science data storage during the closest approach (1–2 h) for each satellite encounter (Fig. 20). This would enable constant data collection from the magnetometer, plasma instrument, INMS, and laser altimeter along with a radar profile, thermal instrument profile, and several imaging sequences. Real-time data downlink and Doppler data would be collected as well.

4.3.2. Europa orbital mission data acquisition. In Europa's orbit, data-takes are obtained in the global framework campaign, regional processes campaign, and the targeted processes campaign. Several instruments collect data continuously on the day and nightsides of Europa: radio science — gravity (by the telecom subsystem), magnetometer, laser altimeter, particle and plasma instrument, and thermal instrument. The INMS would be operated ~50% of the time.

The UVS would observe stellar occultations from orbit when appropriate stars disappear beneath Europa's limb. A two-orbit repeating scenario is planned for the other remote sensing instruments, which would permit power and data rate equalization. Even-numbered orbits emphasize optical remote sensing by the wide-angle camera, medium-angle camera, and infrared spectrometer, while odd orbits emphasize data collection by the ice-penetrating radar. The radar and infrared spectrometer typically operate in low-data-rate profiling modes, permitting a high degree of areal sampling. These instruments would also operate in higher-data-rate targeted modes, obtaining higher-resolution data of priority features.

Targeted observations are implemented by orbital timing when passing over a feature of interest, coordinated (Fig. 21) among the various optical remote sensing instruments, along with the profiling radar mode, and the continuously operating TI and LA. About 800 such targets of ~200 Mb each could be obtained during the nominal orbital mission, and each remote sensing instrument has additional noncoordinated targeted opportunities during the Europa orbiter phase of the mission.

In the global framework campaign, the orbiter would be at 200 km altitude for 8 eurosols (~28 d), with highest-prior-

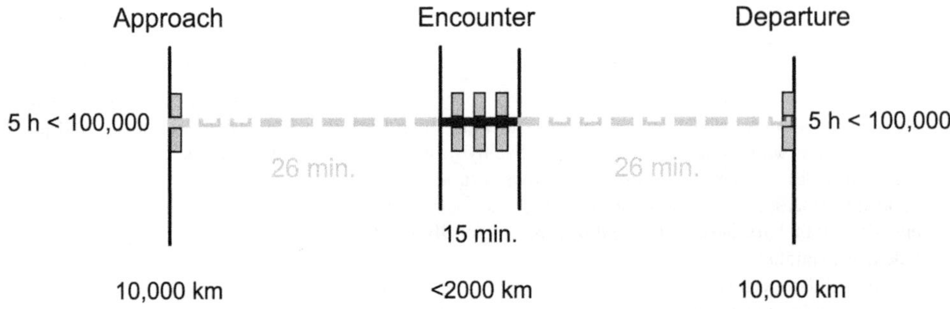

Fig. 20. Example flyby observation timing for a 4-km/s, 100-km closest approach encounter early in the Tour science phase.

ity data acquired during the first 4 eurosols, including gravity, altimetry, and magnetometry to perform a first-order characterization of the ocean and icy shell. In addition, the wide-angle camera would obtain global color data, and the radar would profile the ice for shallow water with additional targeted medium-angle observations. In the next four eurosols, the wide-angle camera would acquire a global stereo map, the radar would perform a deep ocean search, and other remote sensing instruments would continue to acquire profiling and targeted data based on existing Galileo information.

In the regional processes campaign, the orbiter would first lower to 100 km altitude for 12 eurosols (~43 d), thus improving gravity, altimeter, and magnetometer data. Global mapping and shallow water searches are completed initially, followed by stereo mapping by the wide-angle camera and a deep ocean search by the ice-penetrating radar, which (at lower altitude) are greatly improved compared to the first campaign. Optical remote sensing observations continue in profile mode to obtain a denser grid, now at higher spatial resolution. Key discoveries from the first campaign will be targeted for additional observations.

In the targeted processes campaign, coordinated observations (Fig. 21) are targeted with high-data-rate radar data, focusing on specific features at local scales. Profiling observations achieve a grid spacing of <25 km for the optical remote sensing observations, and <50 km for each of the shallow water search and deep ocean search modes of the radar.

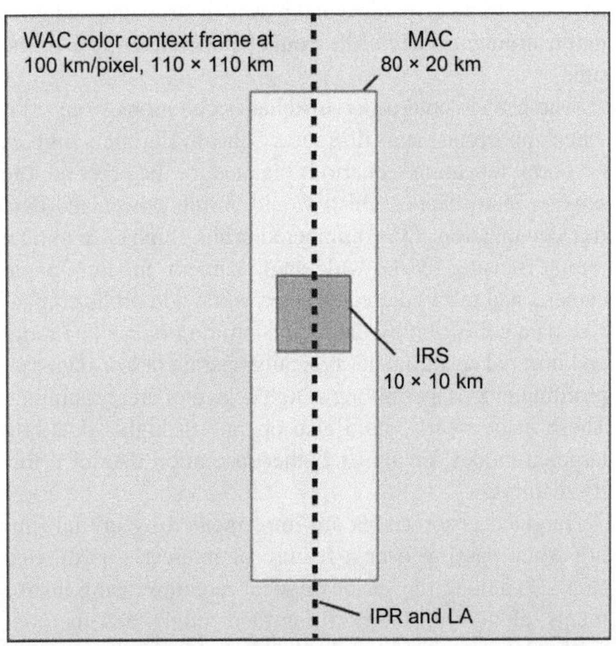

Fig. 21. Coordinated targeted observations, with scales based on an altitude of 100 km; "targeted" observations are set within Wide-Angle Camera (WAC) color context (100 m/pixel, three colors plus broadband monochromatic) with Medium-Angle Camera (MAC) monochromatic imaging (10 m/pixel), infrared spectrometer imaging (25 m/pixel, 400 wavelengths), and a low-data-rate radar profile (IPR, dashed). The laser altimeter (LA) operates continuously, as do the fields and particles instruments.

4.4. Paving the Way for a Future Landed Mission

The inherent values of future lander science include *in situ* measurement of surface chemistry and interrogation of the subsurface with seismometers and magnetometers. Astrobiology science would be advanced through gas chromatography and mass spectroscopy, Raman spectroscopy, and optical microscopy of surface samples. Such investigations also provide "ground truth" for orbital observations of the surface. A lander might also enable investigations beneath the radiolytically processed and gardened layer, and possible concentrate important chemical, molecular, and possibly biological components by filtering meltwater, depending on the lander design (*Chyba and Phillips,* 2001).

To prepare for a future landed mission on Europa, several measurements from JEO would be critical. For astrobiology, identification of young regions rich in interesting surface chemistry (e.g., C–H, C–C, and C–N bonds) are likely to yield important information about potential habitable environments and would be key sites for a future lander.

The technical design of a Europa lander requires an understanding of the meter-scale topography and surface heterogeneity. Information on the depth and porosity of the regolith is required to ensure that the lander is not covered or lost in surface materials. In combination, the JEO planning payload is well-suited for identifying and characterizing potential landing sites.

APPENDIX A: ACRONYMS AND ABBREVIATIONS

CASSE	Committee on Assessing the Solar System Exploration Program
COMPLEX	Committee on Planetary and Lunar Exploration
EJSM	Europa-Jupiter System Mission
ESA	European Space Agency
EUV	Extreme Ultraviolet
FUV	Far Ultraviolet
G.y.	Gigayears (billion years)
INMS	Ion Neutral Mass Spectrometer
IR	Infrared
ISAS	Japan Institute of Space and Astronautical Science
JAXA	Japan Aerospace Exploration Agency
JEO	Jupiter Europa Orbiter
JGO	Jupiter Ganymede Orbiter
JHU/APL	Johns Hopkins University Applied Physics Laboratory
JIMO	Jupiter Icy Moons Orbiter
JJSDT	Joint Jupiter Science Definition Team
JMO	Jupiter Magnetospheric Orbiter
JOI	Jupiter Orbit Insertion
JPL	Jet Propulsion Laboratory
MAC	Medium-Angle Camera
MARSIS	Mars Advanced Radar for Subsurface and Ionospheric Sounding
MBL	Marine Biological Laboratory
NAC	Narrow-Angle Camera
NASA	National Aeronautics and Space Administration
NIMS	Near Infrared Mapping Spectrometer
NIR	Near Infrared
NRC	National Research Council
NUV	Near Ultraviolet
OPAG	Outer Planets Assessment Group
SDT	Science Definition Team
SNR	Signal to Noise Ratio
SSB	Space Studies Board
SSES	Solar System Exploration Survey
SWRI	Southwest Research Institute
TI	Thermal Instrument
UCLA	University of California Los Angeles
UV	Ultraviolet
UVS	Ultraviolet Spectrometer
UVIS	(Cassini) Ultraviolet Imaging Spectrometer
WAC	Wide-Angle Camera

Acknowledgments. This work is drawn primarily from the results of the Science Definition Team for the potential mission to the Jupiter system. The full team contributed to elements in this chapter through reviewing the current understanding of Europa, developing the scientific rationale for the mission, and working with members of the engineering team to derive a mission design that has the potential to address the first-order science questions. We are grateful for these contributions, and for the spirit of cooperation on the part of the technical team. Parts of this work performed by R.T.P. and A.R.H. were carried out at the Jet Propulsion Laboratory, California Institute of Technology, under a contract with the National Aeronautics and Space Administration. On behalf of the full team, we thank C. Niebur for his continued enthusiastic support for the exploration of the outer solar system. We appreciate helpful reviews provided by R. Binzel and W. McKinnon, which substantially improved the chapter.

REFERENCES

Anderson J. D., Lau E. L., Sjogren W. L., Schubert G., and Moore W. B. (1996) Gravitational constraints on the internal structure of Ganymede. *Nature, 384,* 541–543.

Anderson J. D., Schubert G., Jacobson R. A., Lau E. L., Moore W. B., and Sjogren W. L. (1998a) Europa's differential internal structure: Inferences from four Galileo encounters. *Science, 281,* 2019–2022.

Anderson J. D., Schubert G., Jacobson R. A., Lau E. L., Moore W. B., and Sjogren W. L. (1998b) Distribution of rock, metals, and ices on Callisto. *Science, 280,* 1573–1576.

Anderson J. D., Jacobson R. A., Lau E. L., Moore W. B., and Schubert G. (2001) Io's gravity field and interior structure. *J. Geophys. Res., 106(E12),* 32963–32970.

Anderson J. D., Schubert G., Jacobson R. A., Lau E. L., Moore W. B., and Palguta J. L. (2004) Discovery of mass anomalies on Ganymede. *Science, 305,* 989–991.

Baines K. H., Carlson R. W., and Kamp L. W. (2002) Fresh ammonia ice clouds in Jupiter I: Spectroscopic identification, spatial distribution and dynamical implications. *Icarus, 159,* 74–94.

Baross J. A. and Hoffmann S. E. (1985) Submarine hydrothermal vents and associated gradient environments as sites for the origin and evolution of life. *Origins Life, 15(4),* 327–345.

Bell J. F. (1992) Charge-coupled device imaging spectroscopy of Mars 2, results and implications for martian ferric mineralogy. *Icarus, 100(2),* 575–597.

Bibring J.-P., Squyres S. W., and Arvidson R. E. (2006) Merging views on Mars. *Science, 29,* 1899–1901.

Billings S. E. and Kattenhorn S. A. (2005) The great thickness debate: Ice shell thickness models for Europa and comparisons with estimates based on flexure at ridges. *Icarus, 177,* 397–412.

Bills B. G. (2005) Free and forced obliquities of the Galilean satellites of Jupiter. *Icarus, 175,* 233–247.

Bills B. G. and Lemoine F. G. (1995) Gravitational and topographic isotropy of the Earth, Moon, Mars, and Venus. *J. Geophys. Res., 100,* 26275–26295.

Bills B. G. and Nimmo F. (2008) Forced obliquity and moment of inertia of Titan. *Icarus, 196,* 293–297.

Blanc M. et al. (2007) *Laplace: A Mission to Europa and the Jupiter System for ESA's Cosmic Vision Programme.* Available online at jupiter-europa.cesr.fr.

Brown M. E. (2001) Potassium in Europa's atmosphere. *Icarus, 151,* 190–195.

Brown M. E. and Hill R. E. (1996) Discovery of an extended sodium atmosphere around Europa. *Nature, 380,* 229–231.

Brunetto R., Baratta G. A., Domingo M., and Strazzulla G. (2005) Reflectance and transmittance spectra (2.2–2.4 μm) of ion irradiated frozen methanol. *Icarus, 175,* 226–232.

Burns J. A., Simonelli D. P., Showalter M. R., Hamilton D. P., Porco C. C., Throop H., and Esposito L. W. (2004) Jupiter's ring-moon system. In *Jupiter: The Planet, Satellites and Magnetosphere* (F. Bagenal et al., eds.), pp. 241–262. Cambridge Univ., Cambridge.

Canup R. M. and Ward W. R. (2002) Formation of the Galilean satellites: Conditions of accretion. *Astron. J., 124,* 3404–3423.

Carlson R. W. (1999) A tenuous carbon dioxide atmosphere on Jupiter's moon Callisto. *Science, 283,* 820–821.

Carlson R. W. (2001) Spatial distribution of carbon dioxide, hydrogen peroxide, and sulfuric acid on Europa. *Bull. Am. Astron. Soc., 33,* 1125.

Carlson R. W. and 13 colleagues (1999a) Hydrogen peroxide on the surface of Europa. *Science, 283,* 2062–2064.

Carlson R. W., Johnson R. E., and Anderson M. S. (1999b) Sulfuric acid on Europa and the radiolytic sulfur cycle. *Science, 286,* 97–99.

Carlson R. W., Anderson M. S., Johnson R. E., Schulman M. B., and Yavrouian A. H. (2002) Sulfuric acid production on Europa: The radiolysis of sulfur in water ice. *Icarus, 157,* 456–463.

Carlson R. W., Anderson M.S., Mehlman R., and Johnson R. E. (2005) Distribution of hydrated sulfuric acid on Europa. *Icarus, 177,* 461–471.

Carr M. H. and 21 colleagues (1998a) Evidence for a subsurface ocean on Europa. *Nature, 391,* 363–365.

Carr M. H., McEwen A. S., Howard K. A., Chuang F. C., Thomas P., Schuster P., Oberst J., Neukum G., and Schubert G. (1998b) Mountains and calderas on Io: Possible implications for lithosphere structure and magma generation. *Icarus, 135,* 146–165.

Cassen P. M., Peale S. J., and Reynolds R. T. (1982) Structure and thermal evolution of the Galilean satellites. In *Satellites of Jupiter* (D. Morrison, ed.), pp. 93–128. Univ. of Arizona, Tucson.

Cassidy T. A., Johnson R. E., Geissler P. E., and Leblanc F. (2008) Simulation of Na D emission near Europa during eclipse. *J. Geophys. Res., 113(E2),* DOI: 10.1029/2007JE002955.

Cheng K. S., Ho C., and Ruderman M. A. (1986) Energetic radiation from rapidly spinning pulsars. I. Outer magnetosphere gaps. *Astrophys. J., 300,* 500–521.

Christensen P. R. and 25 colleagues (2001) Mars Global Surveyor Thermal Emission Spectrometer experiment: Investigation description and surface science results. *J. Geophys. Res., 106(E10),* 23823–23871.

Chyba C. F. and Phillips C. B. (2001) Possible ecosystems and the search for life on Europa. *Proc. Natl. Acad. Sci., 98,* 801–804.

Chyba C. F., Ostro S. J., and Edwards B. C. (1998) Radar detectability of a subsurface ocean on Europa. *Icarus, 134,* 292–302.

Chyba C. F., McKinnon W. B., Coustenis A., Johnson R. E., Kovach R. L., Khurana K., Lorenz R., McCord T. B., McDonald G. D., Pappalardo R. T., Race M., and Thomson R. (1999) Europa and Titan: Preliminary recommendations of the Campaign Science Working Group on prebiotic chemistry in the outer solar system. In *Lunar and Planetary Science XXX,* Abstract #1537. Lunar and Planetary Institute, Houston (CD-ROM).

Clark K., Greeley R., Pappalardo R., and Jones C. (2007) *2007 Europa Explorer Mission Study: Final Report.* JPL D-38502, Jet Propulsion Laboratory, Pasadena.

Clark R. and McCord T. (1980) The Galilean satellites: New near-infrared spectral reflectance measurements (0.65–25 μm) and a 0.325–5 μm summary. *Icarus, 41,* 323.

Clark R. N., Fanale F. P., and Zent A. P. (1983) Kinetics of ice grain growth: Implications for remote sensing of planetary surfaces. In *Lunar and Planetary Science XIV,* pp. 120–121. Lunar and Planetary Institute, Houston.

Clark R. N. and 25 colleagues (2005) Compositional maps of Saturn's moon Phoebe from imaging spectroscopy. *Nature, 435,* 66–69.

Comstock R. L. and Bills B. G. (2003) A solar system survey of forced librations in longitude. *J. Geophys. Res., 108,* DOI: 10.1029/2003JE002100.

Cooper R. E., Johnson J. F., Mauk B. H., Garrett H. B., and Gehrels N. (2001) Energetic ion and electron irradiation of the icy Galilean satellites. *Icarus, 149,* 133–159.

Crowley J. K. (1991) Visible and near-infrared (0.4–2.5 μm) reflectance spectra of playa evaporite minerals. *J. Geophys. Res., 96,* 16231–16240.

Cruikshank D. P. and 31 colleagues (2007) Surface composition of Hyperion. *Nature, 448,* 54–56.

Dalton J. B. (2000) Constraints on the surface composition of Jupiter's moon Europa based on laboratory and spacecraft data. Ph.D. thesis, Univ. of Colorado, Boulder.

Dalton J. B. (2007) Linear mixture modeling of Europa's non-ice material using cryogenic laboratory spectroscopy. *Geophys. Res. Lett., 34,* 21205.

Dalton J. B. and Clark R. N. (1998) Laboratory spectra of Europa candidate materials at cryogenic temperatures. *Bull. Am. Astron. Soc., 30,* 1081.

Dalton J. B. and Clark R. N. (1999) Observational constraints on Europa's surface composition from Galileo NIMS data. In *Lunar and Planetary Science XXX,* Abstract #2064. Lunar and Planetary Institute, Houston (CD-ROM).

Dalton J. B., Rakesh M., Hiromi K. K., Chan S. L., and Jamieson C. S. (2003) Near-infrared detection of potential evidence for microscopic organisms on Europa. *Astrobiology, 3,* 505–529.

Dalton J. B., Prieto-Ballesteros O., Kargel J. S., Jamieson C. S., Jolivet J., and Quinn R. (2005) Spectral comparison of heavily hydrated salts to disrupted terrains on Europa. *Icarus, 177,* 472–490.

Delamere P.A., Steffl A. J., and Bagenal F. (2004) Modeling temporal variability of plasma conditions in the Io torus during the Cassini era. *J. Geophys. Res., 109,* A10216.

Delitsky M. L. and Lane A. L. (1997) Chemical schemes for surface modification of icy satellites: A road map. *J. Geophys. Res., 102(E7),* 16385–16390.

Delitsky M. L. and Lane A. L. (1998) Ice chemistry on the Galilean satellites. *J. Geophys. Res., 103(E13),* 31391–31403.

Des Marais D. and 19 colleagues (2003) The NASA astrobiology roadmap. *Astrobiology, 2,* 219–235.

Domingue D. L. and Lane A. L. (1997) IUE's view of Callisto: Detection of an SO_2 absorption correlated to possible torus neutral wind alterations. *Geophys. Res. Lett., 24(9),* 1143.

Eluskiewicz J. (2004) Dim prospects for radar detection of Europa's ocean. *Icarus, 170,* 234–236.

ESA (2005) *Cosmic Vision: Space Science for Europe 2015–2025.* ESA Brochure BR-247, ESA Publ. Div., ESTEC, Noordwijk, The Netherlands.

Fanale F. P. and 22 colleagues (1999) Galileo's multi-instrument spectral view of Europa's surface composition. *Icarus, 139,* 179–188.

Figueredo P. H. and Greeley R. (2004) Resurfacing history of Europa from pole-to pole geological mapping. *Icarus, 167,* 287–312.

Gaidos E. and Nimmo F. (2000) Tectonics and water on Europa. *Nature, 405,* 637.

Geissler P. E. and 16 colleagues (1998) Evolution of lineaments on Europa: Clues from Galileo multispectral imaging observations. *Icarus, 135,* 107–126.

Greeley R., Thelig E., and Christensen P. (1984) The Mauna Loa sulfur flow as an analog to secondary sulfur flows on Io. *Icarus, 60,* 189–199.

Greeley R., Chyba C. F., Head J. W. III, McCord T. B., McKinnon W. B., Pappalardo R. T., and Figueredo P. (2004) Geology of Europa. In *Jupiter: The Planet, Satellites and Magnetosphere* (F. Bagenal et al., eds.), pp. 329–362. Cambridge Univ., Cambridge.

Greenberg R., Hoppa G. V., Tufts B. R., Geissler P., Riley J., and Kadel S. (1999) Chaos on Europa. *Icarus, 141,* 263–286.

Hall D. T., Strobel D. F., Feldman P. D., McGrath M. A., and Weaver H. A. (1995) Detection of an oxygen atmosphere on Jupiter's moon Europa. *Nature, 373,* 677–679.

Hall D. T., Feldman P. D., McGrath M. A., and Strobel D. F. (1998) The far-ultraviolet oxygen airglow of Europa and Ganymede. *Astrophys. J., 499,* 475–481.

Hand K. P. (2007) On the physics and chemistry of the ice shell and sub-surface ocean of Europa. Ph.D. thesis, Stanford Univ., Stanford, California.

Hand K. P. and Chyba C. F. (2007) Empirical constraints on the salinity of the europan ocean and implications for a thin ice shell. *Icarus,* DOI: 10.1016/j.icarus.2007.02.002.

Hansen C. J., Esposito L., Stewart A. I. F., Colwell J., Hendrix A., Pryor W., Shemansky D., and West R. (2006) Enceladus' water vapor plume. *Science, 311,* 1422–1425.

Hansen G. and McCord T. B. (2004) Amorphous and crystalline ice on the Galilean satellites: A balance between thermal and radiolytic processes. *J. Geophys. Res., 109(E1),* DOI: 10.1029/2003JE002149.

Hansen G. B. and McCord T. B. (2008) Widespread CO_2 and other non-ice compounds on the anti-jovian and trailing sides of Europa from Galileo/NIMS observations. *Geophys. Res. Lett., 35,* L01202, DOI: 10.1029/2007GL031748.

Head J. W. and Pappalardo R. T. (1999) Brine mobilization during lithospheric heating on Europa: Implications for formation of chaos terrain, lenticula texture, and color variations. *J. Geophys. Res., 104(E11),* 27143–27155.

Hendrix A. R. and Hansen C. J. (2008) Iapetus: New results from Cassini UVIS. In *Lunar and Planetary Science XXXIV,* Abstract #2200. Lunar and Planetary Institute, Houston (CD-ROM).

Hendrix A. R., Barth C. A., and Hord C. W. (1999) Ganymede's ozone-like absorber: Observations by the Galileo ultraviolet spectrometer. *J. Geophys. Res., 104,* 14169–14178.

Hibbitts C. A., Klemaszewski J. E., McCord T. B., Hansen G. B., and Greeley R. (2002) CO_2-rich impact craters on Callisto. *J. Geophys. Res., 107,* E10, 5084.

Hoppa G. V., Tufts B. R., Greenberg R., and Geissler P. E. (1999), Formation of cycloidal features on Europa.*Science, 285,* 1899–1902.

Hudson R. L. and M. H. Moore (1998) Infrared study of ion-irradiated water-ice mixtures with hydrocarbons relevant to comets. *Icarus, 136(2),* 518–527.

Hunt G. R., Salisbury J. W., and Lenhoff C. J. (1971a) Visible and near-infrared spectra of minerals and rocks: III. Oxides and hydroxides. *Mod. Geol., 2,* 195–205.

Hunt G. R., Salisbury J. W., and Lenhoff C. J. (1971b) Visible and near-infrared spectra of minerals and rocks: IV. Sulphides and sulphates. *Mod. Geol., 3,* 1–14.

Hussmann H and Spohn T. (2004) Thermal-orbital evolution of Europa. *Icarus, 171,* 391–410.

Hussmann H., Sohl F., and Spohn T. (2006) Subsurface oceans and deep interiors of medium-sized outer planet satellites and large trans-neptunian objects. *Icarus, 185,* 258–273.

Ingersoll A. P., Dowling T. E., Gierasch P. J., Orton G. S., Read P. L., Sánchez-Lavega A., Showman A. P., Simon-Miller A. A., and Vasavada A. R. (2004) Dynamics of Jupiter's atmosphere. In *Jupiter: The Planet, Satellites and Magnetosphere* (F. Bagenal et al., eds.), pp. 105–128. Cambridge Univ., Cambridge.

Jaeger W. L., Turtle E. P., Keszthelyi L. P., Radebaugh J., McEwen A. S., and Pappalardo R. T. (2003) Orogenic tectonism on Io. *J. Geophys. Res., 108,* E8, 12-1, DOI: 10.1029/2002JE001946.

JIMO SDT (2004) *Report of the NASA Science Definition Team for the Jupiter Icy Moons Orbiter (JIMO).* NASA, Washington, DC. Available online at www.lpi.usra.edu/opag/resources.html.

Johnson R. E. and Quickenden T. I. (1997) Radiolysis and photolysis of low-temperature ice. *J. Geophys. Res., 102,* 10985–10996.

Johnson R. E., Killen R. M., Waite J. H., and Lewis W. S. (1998) Europa's surface composition and sputter-produced ionosphere. *Geophys. Res. Lett., 25(17),* 3257–3260.

Johnson R. E., Leblanc F., Yakshinskiy B. V., and Madey T. E. (2002) Energy distributions for desorption of sodium and potassium from ice: The Na/K ratio at Europa. *Icarus, 156,* 136–142.

Johnson R. E., Carlson R. W., Cooper J. F., Paranicas C., Moore M. H., and Wong M. (2004) Radiation effects on the surfaces of the Galilean satellites. In *Jupiter: The Planet, Satellites and Magnetosphere* (F. Bagenal et al., eds.), pp. 485–512. Cambridge Univ., Cambridge.

Kargel J. S., Kaye J. Z., Head J. W. III, Marion G. M., Sassen R., Crowley J. K., Prieto-Ballesteros O., Grant S. A., and Hogenboom D. L. (2000) Europa's crust and ocean: Origin, composition, and the prospects for life. *Icarus, 148,* 226–265.

Khurana K. K., Kivelson M. G., Stevenson D. J., Schubert G., Russel C. T., Walker R. J., and Polanskey C. (1998) Induced magnetic fields as evidence for subsurface oceans in Europa and Callisto. *Nature, 395,* 777–780.

Khurana K. K., Kivelson M. G., and Russell C. T. (2002) Searching for liquid water in Europa by using surface observations. *Astrobiology, 2,* 93–103.

Khurana K., Pappalardo R. T., Murphy N., and Denk T. (2007) The origin of Ganymede's polar caps. *Icarus, 191,* 193–202.

Kivelson M. G., Khurana K. K., Russell C. T., Walker R. J., Warnecke J., Coroniti F. V., Polanskey C., Southwood D. J., and Schubert C. (1996) Discovery of Ganymede's magnetic field by the Galileo spacecraft. *Nature, 384,* 537.

Kivelson M. G., Khurana K. K., Stevenson D. J., Bennett L., Joy S., Russell C. T., Walker R. J., and Polanskey C. (1999) Europa and Callisto: Induced or intrinsic fields in a periodically varying plasma environment. *J. Geophys. Res., 104,* 4609–4625.

Kivelson M. G., Khurana K. K., Russell C. T., Volwerk M., Walker R. J., and Zimmer C. (2000) Galileo magnetometer measurements: A stronger case for a subsurface ocean at Europa. *Science, 289,* 1340–1343.

Kivelson M. G., Khurana K. K., and Volwerk M. (2002) The permanent and inductive magnetic moments of Ganymede. *Icarus, 157,* 507–522.

Kivelson M. G., Bagenal F., Kurth W. S., Neubauer F. M., Paranicas C., and Saur J. (2004) Magnetospheric interactions with satellites. In *Jupiter: The Planet, Satellites and Magnetosphere* (F. Bagenal et al., eds.), pp. 513–536. Cambridge Univ., Cambridge.

Kliore A. J., Hinson D. P., Flasar F. M., Nagy A. F., and Cravens T. E. (1997) The ionosphere of Europa from Galileo radio occultations. *Science, 277,* 355–358.

Kliore A. J., Anabtawi A., and Nagy A. F. (2001a) The ionospheres of Europa, Ganymede, and Callisto. *Eos Trans. AGU,* B506, Abstract #P12B-0506.

Kliore A. J., Anabtawi A., Nagy A. F., and Galileo Radio Propagation Science Team (2001b) The ionospheres of Ganymede and Callisto from Galileo radio occultations. *Bull. Am. Astron. Soc., 33,* 1084.

Kliore A. J., Anabtawi A., Herrera R. G., Asmar S. W., Nagy A. F., Hinson D. P., and Flasar F. M. (2002) Ionosphere of Callisto from Galileo radio occultation observations. *J. Geophys. Res., 107,* A11, DOI: 10.1029/2002JA009365.

Kuiper G. P. (1957) Infrared observations of planets and satellites. *Astron. J., 62,* 295.

Kwok J., Prockter L., Senske D., and Jones C. (2007) *Jupiter System Observer Mission Study: Final Report.* JPL D-41284, Jet Propulsion Laboratory, Pasadena.

Lane A. L., Nelson R. M., and Matson D. L. (1981) Evidence for sulphur implantation in Europa's UV absorption band. *Nature, 292,* 38–39.

Leblanc F., Potter A. E., Killen R. M., and Johnson R. E. (2002) Origins of Europa Na cloud and torus. *Icarus, 178,* 367–385.

Leblanc F., Potter A. E., Killen R. M., and Johnson R. E. (2005) Origins of Europa Na cloud and torus. *Icarus, 178,* 367–385.

Lee S., Pappalardo B., and Makris N. C. (2005) Mechanics of tidally induced fractures in Europa's ice shell. *Icarus, 177,* 367–379.

Leovy C. B., Friedson A. J., and Orton G. S. (1991) The quasi-quadrennial oscillation of Jupiter's equatorial stratosphere. *Nature, 354,* 380–382.

Liang M.-C., Lane B. F., Pappalardo R. T., Allen M., and Yung Y. L. (2005) Atmosphere of Callisto. *J. Geophys. Res., 110,* E02003, DOI: 10.1029/2004JE002322.

Lopes R. M. and Spencer J. R. (2007) *Io After Galileo: A New View of Jupiter's Volcanic Moon.* Cambridge Univ., Cambridge. 342 pp.

Lopes R. M. C. and D. A. Williams (2005) Io after Galileo. *Rept. Progr. Phys., 68,* 303–340.

Lorenz R. D., Stiles B. W., Kirk R. L., Allison M. D., Persi del Marmo P., Iess L., Lunine J. I., Ostro S. J., and Hensley S. (2008) Titan's rotation reveals an internal ocean and changing zonal winds. *Science, 319,* 1649–1651.

Luthcke S. B., Carabajal C. C., and Rowlands D. D. (2002) Enhanced geolocation of spaceborne laser altimeter surface returns: Parameter calibration from the simultaneious reduction of altimeter range and navigation tracking data. *J. Geodyn., 34,* 447–475.

Luthcke S. B., Rowlands D. D., Williams T. A., and Sirota M. (2005) Reduction of ICES at systematic geolocation errors and the impact on ice sheet elevation change detection. *Geophys. Res. Lett., 32,* L21S05, DOI: 10.1029/2005GL023689.

Luttrell K. and Sandwell D. (2006) Strength of the lithosphere of the Galilean moons. *Icarus, 183,* 159–167.

Mauk B. H., Mitchell D. G., Krimigis S. M., Roelof E. C., and Paranicas C. (2003) Energetic neutral atoms form a trans-Europa gas torus at Jupiter. *Nature, 421,* 920–922.

McCord T. B. (2000) Surface composition reveals icy Galilean satellites' past. *Eos Trans. AGU, 81(19),* 209, DOI: 10.1029/1000EO00141.

McCord T. B. and 11 colleagues (1997) Organics and other molecules in the surfaces of Callisto and Ganymede. *Science, 278,* 271–275.

McCord T. B. and 11 colleagues (1998a) Salts on Europa's surface detected by Galileo's near infrared mapping spectrometer. *Science, 280,* 1242–1245.

McCord T. B. and 12 colleagues (1998b) Non-water-ice constituents in the surface material of the icy Galilean satellites from the Galileo near-infrared mapping spectrometer investigation. *J. Geophys. Res., 103(E4),* 8603–8626, DOI: 10.1029/1098JE00788.

McCord T. B. and 11 colleagues (1999) Hydrated salt minerals on Europa's surface from the Galileo near-infrared mapping spectrometer (NIMS) investigation. *J. Geophys. Res., 104,* 11827–11851.

McCord T. B., Orlando T., Teeter G., Hansen G., Sieger M., Petrik N., and Van Keulen L. (2001) Thermal and radiation stability of the hydrated salt minerals epsomite, mirabilite, and natron under Europa environmental conditions. *J. Geophys. Res., 106(E2),* 3311–3320.

McCord T. B., Teeter G., Hansen G. B., Sieger M. T., and Orlando T. M. (2002) Brines exposed to Europa surface conditions. *J. Geophys. Res., 107,* 5004.

McEwen A. S., Keszthelyi L. P., Lopes R., Schenk P. M., and Spencer J. R. (2004) The lithosphere and surface of Io. In *Jupiter: The Planet, Satellites and Magnetosphere* (F. Bagenal et al., eds.), pp. 307–328. Cambridge Univ., Cambridge.

McGrath M. A., Lellouch E., Strobel D. F., Feldman P. D., and Johnson R. E. (2004) Satellite atmospheres. In *Jupiter: The Planet, Satellites and Magnetosphere* (F. Bagenal et al., eds.), pp. 457–483. Cambridge Univ., Cambridge.

McKinnon W. B. (1999) Convective instability in Europa's floating ice shell. *Geophys. Res. Lett., 26,* 951–954.

McKinnon W. B. (2006) On convection in ice I shells of outer solar system bodies, with detailed application to Callisto. *Icarus, 183,* 435–450.

McKinnon W. B. and Gurnis M. (1999) On initiation of convection in Europa's floating ice shell (snd the existence of the ocean below). In *Lunar and Planetary Science XXX,* Abstract #2058. Lunar and Planetary Institute, Houston (CD-ROM).

McKinnon W. B. and Parmentier E. M. (1986) Ganymede and Callisto. In *Satellites of Jupiter* (D. Morrison, ed.), pp. 718–763. Univ. of Arizona, Tucson.

McKinnon W. B. and Zolensky M. E. (2003) Sulfate content of Europa's ocean and shell: Evolutionary considerations and some geological and astrobiological implications. *Astrobiology, 3(4),* 879–897.

Moersch J., Bell J. III, Carter L., Hayward T., Nicholson P., Squyres S., and Van Cleve J. (1997) What happened to Cerberus? Telescopically observed thermo-physical properties of the martian surface. In *Mars Telescopic Observations Workshop II*, pp. 26–26. LPI Tech. Rpt. 97–03, Lunar and Planetary Institute, Houston.

Moore J. C. (2000) Models of radar absorption in europan ice. *Icarus, 147,* 292–300.

Moore J., Reid A., and Kipfstuhl J. (1994) Microstructure and electrical properties of marine ice and its relationship to meteoric ice and sea ice. *J. Geophys. Res., 99(C3),* 5171–5180.

Moore J. M. and 25 colleagues (2001) Impact features on Europa: Results of the Galileo Europa Mission (GEM). *Icarus, 151,* 93–111.

Moore J. M. and 11 colleagues (2004) Callisto. In *Jupiter: The Planet, Satellites and Magnetosphere* (F. Bagenal et al., eds.), pp. 397–426. Cambridge Univ., Cambridge.

Moore M. (1984) Studies of proton-irradiated SO_2 at low temperatures: Implications for Io. *Icarus, 59,* 114.

Moore M. H. and Hudson R. L. (2003) Infrared study of ion-irradiated N_2-dominated ices relevant to Triton and Pluto: Formation of HCN and HNC. *Icarus, 161(2),* 486–500.

Moore W. B. (2001) The thermal state of Io. *Icarus, 154,* 548–550.

Moore W. B. and Schubert G. (2000) The tidal response of Europa. *Icarus, 147,* 317–319.

Moroz V. I. (1965) Infrared spectrophotometry of the Moon and the Galilean satellites of Jupiter. *Astron. Zh., 42,* 1287.

Moses J. I., Fouchet T., Yelle R. V., Friedson A. J., Orton S. G., Beard B., Drossart P., Gladstone G. R., Kostiuk T., and Livengood T. A. (2004) The stratosphere of Jupiter. In *Jupiter: The Planet, Satellites and Magnetosphere* (F. Bagenal et al., eds.), pp. 129–158. Cambridge Univ., Cambridge.

Mosqueira I. and Estrada P. R. (2003) Formation of the regular satellites of giant planets in an extended gaseous nebula I: Subnebula model and accretion of satellites. *Icarus, 163,* 198–231.

NASA (2004) *The Vision for Space Exploration.* NASA NP-2004-01-334-HQ, Washington, DC.

NASA (2006) *Solar System Exploration Roadmap.* NASA Science Mission Directorate, Washington, DC. Available online at solarsystem.nasa.gov/multimedia/downloads/SSE_RoadMap_2006_Report_FC-A_med.pdf.

NASA (2007) *Science Plan for NASA's Science Mission Directorate 2007–2016.* Available online at science.hq.nasa.gov/strategy.

Nash D. B., Yoder C. F., Carr M. H., Gradie J., and Hunten D. M. (1986) Io. In *Satellites of Jupiter* (D. Morrison, ed.), pp. 629–688. Univ. of Arizona, Tucson.

Neumann G. A., Rowlands D. D., Lemoine F. G., Smith D. E., and Zuber M. T. (2001) Crossover analysis of MOLA altimetric data. *J. Geophys. Res., 106,* 23753–23768.

Newhall X. X. and Williams J. G. (1997) Estimation of the lunar physical librations. *Cel. Mech., 66,* 21–30.

Nimmo F., Giese B., and Pappalardo R. T. (2003) Estimates of Europa's ice shell thickness from elastically supported topography. *Geophys. Res. Lett., 30(5),* 1233, DOI: 10.1029/2003GL016660.

Nimmo F., Spencer J. R, Pappalardo R. T., and Mullen M. E. (2007) A shear-heating mechanism for the generation of vapour plumes and high heat fluxes on Saturn's moon Enceladus. *Nature, 447,* 289–291.

Noll K. S., Weaver H. A., and Gonnella A. M. (1995) The albedo spectrum of Europa from 2200 to 3300. *J. Geophys. Res., 100,* 19057–19060.

Nozawa H., Misawa H., Takahashi S., Morioka A., Okana S., and Sood R. (2005) Relationship between the jovian magnetospheric plasma and Io torus emission. *Geophys. Res. Lett., 32,* L11101, DOI: 10.1029/2005GL022759.

Ockert-Bell M., Burns J. A., Daubar I. J., Thomas P. C., Belton M. J. S., and Klaasen K. P. (1999) The structure of Jupiter's ring system as revealed by the Galileo imaging experiment. *Icarus, 138,* 188–213.

Ojakangas G. W. and Stevenson D. J. (1989) Thermal state of an ice shell on Europa. *Icarus, 81,* 220–241.

OPAG (2006) *Scientific Goals and Pathways for Exploration of the Outer Solar System.* Available online at www.lpi.usra.edu/opag.

O'Reilly T. C. and Davies G. F. (1981) Magma transport of heat on Io — A mechanism allowing a thick lithosphere. *Geophys. Res. Lett., 8,* 313–316.

Orlando T. M., McCord T. B., and Grieves G. A. (2005), The chemical nature of Europa surface material and the relation to a sub-surface ocean. *Icarus, 177,* 528–533, DOI: 10.1016/j.icarus.2005.05.009.

Palguta J., Anderson J. D., Schubert G., and Moore W. B. (2006) Mass anomalies on Ganymede. *Icarus, 180,* 428–441.

Pappalardo R. (2006) *Europa Science Objectives.* NAI Europa Focus Group (Report to OPAG). Available online at www.lpi.usra.edu/opag/may_06_meeting/agenda.html.

Pappalardo R. T. and Sullivan R. J. (1996) Evidence for separation across a gray band on Europa. *Icarus, 123,* 557–567.

Pappalardo R. T., Head J. W., Collins G. C., Kirk R. L., Neukum G., Oberst J., Giese B., Greeley R., Chapman C. R., Helfenstein P., Moore J. M., McEwen A., Tufts B. R., Senske D. A., Breneman H. H., and Klaasen K. K. (1998) Grooved terrain on Ganymede: First results from Galileo high-resolution imaging. *Icarus, 135,* 276–302.

Pappalardo R. T., Collins G. C., Head J. W. III, Helfenstein P., McCord T. B., Moore J. M., Procktor L. M., Schenk P. M., and Spencer J. R. (2004) Geology of Ganymede. In *Jupiter: The Planet, Satellites and Magnetosphere* (F. Bagenal et al., eds.), pp. 363–396. Cambridge Univ., Cambridge.

Parmentier E. M., Squyres S. W., Head J. W., and Allison M. L. (1982) The tectonics of Ganymede. *Nature, 295,* 290–293.

Peale S. J. (1976) Orbital resonances in the solar system. *Annu. Rev. Astron. Astrophys., 14,* 215–246.

Pierazzo E. and Chyba C. F. (2002) Cometary delivery of biogenic elements to Europa. *Icarus, 157,* 1120–1127.

Pilcher C. B., Ridgway S. T., and McCord T. B. (1972) Galilean satellites. *J. Geophys. Res., 103,* 31391–31403.

Porco C. and 23 colleagues (2003) Cassini imaging of Jupiter's atmosphere, satellites and rings. *Science, 299,* 1541–1547.

Porco C. C. and 24 colleagues (2006) Cassini observes the active South Pole of Enceladus. *Science, 311(5766),* 1393–1401.

Prockter L. M. and 13 colleagues (1998) Dark terrain on Ganymede: Geological mapping and interpretation of Galileo Regio at high resolution. *Icarus, 135,* 317–344.

Prockter L. M., Head J. W., Pappalardo R. T., Sullivan R. J., Clifton A. E., Giese B., Wagner R., and Neukum G. (2002) Morphology of europan bands at high resolution: A mid-ocean ridge-type rift mechanism. *J. Geophys. Res., 107(E5),* 5028, DOI: 10.1029/2000JE001458.

Prockter L. M., Pappalardo R. T., and Nimmo F. (2005) A shear heating origin for ridges on Triton. *Geophys. Res. Lett., 32,* L14202, DOI: 10.1029/2005GL022832.

Riley J., Hoppa G. V., Greenberg R., Tufts B. R., and Geissler P. (2000) Distribution of chaotic terrain on Europa. *J. Geophys. Res., 105(E9),* 22599–22616, DOI: 10.1029/1999 JE001164.

Rothschild L. J. and Mancinelli R. L. (2001) Life in extreme environments. *Nature, 409,* 1092–1101.

Rowlands D. D., Pavlis D. E., Lemoine F. G., Neumann G. A., and Luthcke S. B. (1999) The use of laser altimetry in the orbit and attitude determination of Mars Global Surveyor. *Geophys. Res. Lett., 9,* 1191–1194.

Schenk P. M. and Bulmer M. H. (1998) Origin of mountains on Io by thrust faulting and large-scale mass movements. *Science, 279,* 1514.

Schenk P. M., Chapman C. R., Zahnle K., and Moore J. M. (2004) Ages and interiors: The cratering record of the Galilean satellites. In *Jupiter: The Planet, Satellites and Magnetosphere* (F. Bagenal et al., eds.), pp. 427–457. Cambridge Univ., Cambridge.

Schilling N., Khurana K. K., and Kivelson M. G. (2004) Limits on an intrinsic dipole moment in Europa. *J. Geophys. Res., 109,* E05006.

Schubert G., Zhang K., Kivelson M. G., and Anderson J. D. (1996) The magnetic field and internal structure of Ganymede. *Nature, 384,* 544–545.

Shoemaker E. M., Lucchitta B. K., Plescia J. B., Squyres S. W., and Wilhelms D. E. (1982) The geology of Ganymede. In *Satellites of Jupiter* (D. Morrison, ed.), pp. 435–520. Univ. of Arizona, Tucson.

Showman A. P. and Malhotra R. (1999) The Galilean satellites. *Science, 286,* 77–84.

Showman A. P., Mosqueira I., and Head J. W. (2004) On the resurfacing of Ganymede by liquid-water volcanism. *Icarus, 172,* 625–640.

Smythe W. D., Carlson R. W., Ocampo A., Matson D. L., and McCord T. B. (1998) Galileo NIMS measurements of the absorption bands at 4.03 and 4.25 microns in distant observations of Europa. *Bull. Am. Astron. Soc., 30,* Abstract #5507.

Spencer J. R. and Calvin W. M. (2002) Condensed O_2 on Europa and Callisto. *Astron. J., 124(6),* 3400–3403.

Spencer J. R., Tamppari L. K., Martin T. Z., and Travis L. D. (1999) Temperatures on Europa from Galileo PPR: Nighttime thermal anomalies. *Science, 284,* 1514–1516.

Spencer J. R., Lellouch E., Richter M. J., López-Valverde M. A., Jessup K. L., Greathouse T. K., and Flaud J.-M. (2005) Mid-Infrared detection of large longitudinal asymmetries in Io's SO_2 atmosphere. *Icarus, 176,* 283–304.

Spencer J. R. and 9 colleagues (2006) Cassini encounters Enceladus: Background and the discovery of a south polar hot spot. *Science, 311,* 1401–1405.

SSB (1999) *A Science Strategy for the Exploration of Europa.* Space Studies Board, National Research Council, National Academies, Washington, DC. 68 pp.

SSB (2003) *New Frontiers in the Solar System: An Integrated Exploration Strategy.* Space Studies Board, National Research Council, National Academies, Washington, DC. 231 pp.

SSB (Committee on Assessing the Solar System Exploration Program) (2007) *Grading NASA's Solar System Exploration Program: A Midterm Review.* Space Studies Board, National Academies, Washington, DC.

Steffl A. J., Delamere P. A., and Bagenal F. (2004) Cassini UVIS observations of the Io plasma torus I. Initial results. *Icarus, 172,* 78–90.

Steffl A. J., Delamere P. A., and Bagenal F. (2006) Cassini UVIS observations of the Io plasma torus III. Observations of temporal and azimuthal variability. *Icarus, 180,* 124–140.

Steffl A. J., Delamere P. A., and Bagenal F. (2008) Cassini UVIS observations of the Io plasma torus IV. Modeling temporal and azimuthal variability. *Icarus, 194,* 153–165.

Stern S. A. and McKinnon W. B. (2000) Triton's surface age and impactor population revisited in light of Kuiper belt fluxes: Evidence for small Kuiper belt objects and recent geological activity. *Astron. J., 119,* 945–952.

Strobel D. F., Saur J., Feldman P. D., and McGrath M. A. (2002) Hubble Space Telescope Space Telescope Imaging Spectrograph search for an atmosphere on Callisto: A jovian unipolar inductor. *Astrophys. J., 581,* L51–L54.

Sullivan R. and 12 colleagues (1998) Episodic plate separation and fracture infill on the surface of Europa. *Nature, 391,* 371–373.

Taylor F. W., Atreya S. K., Encrenaz T., Hunten D. M., Irwin P. J. G., and Owen T. C. (2004) The composition of the atmosphere of Jupiter. In *Jupiter: The Planet, Satellites and Magnetosphere* (F. Bagenal et al., eds.), pp. 59–78. Cambridge Univ., Cambridge.

Thomas N., Bagenal F., Hill T. W., and Wilson J. K. (2004) The Io neutral clouds and plasma torus. In *Jupiter: The Planet, Satellites and Magnetosphere* (F. Bagenal et al., eds.), pp. 561–591. Cambridge Univ., Cambridge.

Tobie G., Grasset O., Lunine J. I., Mocquet A., and Sotin C. (2005) Titan's internal structure inferred from a coupled thermal-orbital model. *Icarus, 175,* 496–502.

Tufts B. R., Greenberg R., Hoppa G., and Geissler P. (2000) Lithospheric dilation on Europa. *Icarus, 146,* 75–97.

Turtle E. P. and 12 colleagues (2001) Mountains on Io: High-resolution Galileo observations, initial interpretations, and formation models. *J. Geophys Res., 106(E12),* 33175–33200.

Veeder G. J., Matson D. L., Johnson T. V., Davies A. G., and Blaney D. L. (2004) The polar contribution to the heat flow of Io. *Icarus, 169,* 264–270.

Vincent M. and 18 colleagues (2000) Jupiter's polar regions in the ultraviolet as imaged by HST/WFPC2: Auroral-aligned features and zonal motions. *Icarus, 143,* 205–222.

Vogt S. S., Butler R. P., Marcy G. W., Fischer D. A., Henry G. W., Wright J. T., and Johnson J. A. (2005) Five new multi-component planetary systems. *Astrophys. J., 632,* 638.

Volwerk M., Kivelson M., and Khurana K. (2001) Wave activity in Europa's wake: Implications for ion pickup. *J. Geophys. Res., 106(A11),* 26033–26048.

Wahr J. M., Zuber M. T., Smith D. E., and Lunine J. I. (2006) Tides on Europa, and the thickness of Europa's icy shell. *J. Geophys. Res., 111,* E12005, DOI: 10.1029/2006JE002729.

Waite H. Jr. and 14 colleagues (2006) Cassini Ion Neutral Mass Spectrometer: Enceladus plume composition and structure. *Science, 311,* 1419–1422.

Ward W. R. (1975) Past orientation of the lunar spin axis. *Science, 189,* 377–379.

Weinbruch U. and Spohn T. (1995) A self-sustained magnetic field in Io? *Planet Space Sci., 43,* 1045.

West R. A., Baines K. H., Friedson A. J., Banfield D., Ragent B., and Taylor R. W. (2004) Jovian clouds and haze. In *Jupiter: The Planet, Satellites and Magnetosphere* (F. Bagenal et al., eds.), pp. 79–104. Cambridge Univ., Cambridge.

Wu X., Bar-Sever Y. E., Folkner W. M., Williams J. G., and Zumberge J. F. (2001) Probing Europa's hidden ocean from tidal effects on orbital dynamics. *Geophys. Res. Lett., 28,* 2245–2248.

Yelle R. V. and S. Miller (2004) Jupiter's thermosphere and ionosphere. In *Jupiter: The Planet, Satellites and Magnetosphere* (F. Bagenal et al., eds.), pp. 185–218. Cambridge Univ., Cambridge.

Yoder C. F. and Peale S. J. (1981) The tides of Io. *Icarus, 47,* 1–35.

Yoder C. F., Williams J., and Parke M. (1981) Tidal variations of Earth rotation. *J. Geophys. Res., 86,* 881–891.

Zahnle K., Dones L., and Levison H. F. (1998) Cratering rates in the Galilean satellites. *Icarus, 136,* 202–222.

Zimmer C., Khurana K. K., and Kivelson M. G. (2000) Subsurface oceans on Europa and Callisto: Constraints from Galileo magnetometer observations. *Icarus, 147,* 329–347.

Europa: Perspectives on an Ocean World

William B. McKinnon
Washington University in St. Louis

Robert T. Pappalardo
NASA Jet Propulsion Laboratory/California Institute of Technology

Krishan K. Khurana
University of California, Los Angeles

Europa possesses an outer icy shell; this much has been clear since the Voyager flybys. That Europa's shell is also floating is now generally accepted as well, thanks to observations by the Galileo orbiter. Reviews of the geology and geophysics of Europa, from its interior to tectonic style, from its thermal and orbital evolution to the plausibility of bearing an ocean, and much more, can be found in the other chapters of this volume. The existence of a low density outer "H_2O" layer, 80–170 km in thickness, seems well established. Magnetic induction evidence strongly suggests a conducting near-surface layer and/or interior — a saline ocean. Cycloidal ridges, originating as tidally driven cycloidal fractures, apparently formed in a stress regime dominated by diurnal tides, but could not form in a tidally flexing icy shell grounded to the silicate interior — again supporting decoupling by an ocean. These points are not seriously in contention. Beyond this there is less agreement, especially as to the thickness of the icy shell overlying the ocean, its composition and rheology, and whether the icy shell more or less responds passively to tidal strains and heating from Europa's interior, or whether it plays a more active role by means of solid-state convection. We seek to know more: what the shell is made of, what its rheological, mechanical, and structural properties are; how these govern and respond to tectonic forces and impacts; and how the shell and ocean and rocky interior have evolved through geological time. Whether the rocky interior is or has been geologically active is, from an obser-vational standpoint, one of the most poorly constrained aspects of Europa, but it is key to what is perhaps the most fundamental question of all for the satellite: the astrobiological potential of the icy shell and ocean beneath.

This chapter is not a synopsis or summary of this review volume. Rather, it is a guided tour through some of the important issues that have animated Europa science since the Voyager missions. It is also of necessity selective. The individual chapters in this volume are complete and comprehensive, and where appropriate we will indicate to where and why the reader may refer, both for greater detail and in some cases contrasting interpretations.

1. INTRODUCTION

In their original assessment of Europa's geology as revealed by Voyager 2, *Lucchitta and Soderblom* (1982) speculated that if one were to strip away the top few kilometers of Europa's icy surface, the satellite would appear remarkably similar. To the extent that vertical tectonics (ridges and fissures) and horizontal tidal forces dominate Europa, this was a prescient deduction. Voyager images only gave a hint of the lateral mobility that is now understood to be an important part of the europan story (see the chapter by Kattenhorn and Hurford for a excellent summary of europan tectonics and differing tectonic styles, and the chapter by Prockter and Patterson for a focused discussion of "dilational bands"). The disrupted, chaos regions discovered by Galileo were simply unexpected (indeed, the now famous Conamara "mottled terrain," to use the older Voyager terminology, was not even an original target of Galileo's E6 encounter). This lack of expectation was in part due to a general acceptance of (1) conductive models for the icy shell presuming Europa were fully differentiated (*Ojakangas and Stevenson,* 1989a), the shell then acting as a (somewhat) passive recorder of various tectonic and impact events, and (2) extant theoretical models of Europa's rocky interior tidal dissipation that predicted heating no more than a few times the radiogenic value (*Cassen et al.,* 1982; *Squyres et al.,* 1983). In fact, a thin (≤25-km-thick) solid icy shell directly coupled to a hydrated silicate interior (*Ransford et al.,* 1981) arguably remained a viable model until the Galileo era.

Since Galileo, observations of apparent surface flooding by both low-viscosity fluids (e.g., the "pond" at 6°N, 327°W, and the boundaries of Thera and especially Thrace Maculae) and high-viscosity materials (e.g., certain "spots," or lenticulae, and Murias Chaos) have further testified to the dynamic thermal state of Europa's icy shell (most of these features are illustrated in the chapter on chaos terrains by Collins and Nimmo). Estimates of elastic lithosphere thickness developed over the years (e.g., *Pappalardo et al.,* 1999; *Prockter and Pappalardo,* 2000; *Dombard and McKinnon,* 2006; and especially see Table 1 in *Billings and Kattenhorn,* 2005) are quite thin (generally less than a few kilometers), implying locally high heat flows, and possibly, a very thin shell overall (the icy shell and brittle lithosphere are two distinct things, as the chapter by Nimmo and Manga makes clear; these authors also discuss the nuances of lithosphere thickness extimates and the heat flow constraints that are derived from them).

Evidence also accumulated early for the nonsynchronous rotation (NSR) of the icy shell with respect to Europa's tidal deformation axes (*McEwen,* 1986; *Geissler et al.,* 1998; *Figueredo and Greeley,* 2000; see chapter by Kattenhorn and Hurford). The possibility of such nonsynchronous slip was predicted early on by *Greenberg and Weidenschilling* (1984) on general principles, but a specific predictive model, based on tidal heating, requires an ocean in order to work at all (*Ojakangas and Stevenson,* 1989a,b; *Leith and McKinnon,* 1996). Europa's tectonics are notoriously complex, however, and additional stress sources, such as accruing to true polar wander (TPW) of the shell, are now being seriously discussed (*Schenk et al.,* 2008), whereas the role of NSR has been questioned, again on general principles, in the chapter by Bills et al. (which also looks at TPW from a theoretical perspective). The implications of Europa's high heat flow and dynamic geology for cycling of surface and subsurface materials are critical to astrobiological inferences and a continuing focus for research.

In this chapter we discuss aspects relevant to the interaction of the silicate mantle and ocean, and the ocean and icy shell. After discussing evidence for the ocean (section 2) and observational constraints on its composition (section 3), we address Europa's tectonics, focusing on the potential role of solid-state convection (section 4), and then bear down on the perennial question of the thickness of the satellite's icy shell (section 5). We conclude with consideration of tidal energy sources and possible silicate volcanism (section 6), some implications for Europa's astrobiological potential (section 7), and final remarks (section 8).

2. EVIDENCE FOR AN OCEAN

Why is Europa believed to possess a global subsurface ocean? Four close passes by Galileo determined Europa's second-degree gravity term C_{22}, which in turn yielded a normalized moment-of-inertia (MOI) of 0.346 ± 0.005 (1σ) (*Schubert et al.,* 2004). This MOI *assumes* that the tidal and rotational figure is hydrostatic. Nevertheless, even factoring in generous systematic uncertainty, the MOI implies a differentiated Europa, and for cosmochemically plausible rock + metal compositions, a deep ice and/or water layer (*Greeley et al.,* 2004; *Schubert et al.,* 2004; and naturally see the chapter by Schubert et al.). For solar composition rock + metal (specifically, solar Fe/Si and Fe/S) the icy layer is ~135 km thick (*McKinnon and Desai,* 2003).

The induced magnetic field clearly indicates a conducting layer within or close to Europa (as reviewed in the chapter by Khurana et al.). Because the ionosphere of Europa is insufficiently conductive to carry the required currents, the conductive layer must be within the body of Europa. A metallic core is too deep to account for the magnitude of the induced field, which falls off as the radial distance cubed, so the conducting layer must lie within the icy shell or the outermost mantle. Barring an exotic composition for the latter, Europa must possess a conductive ocean beneath the ice or sufficiently hot outer mantle that an ocean, conductive or not, is implied (*Zimmer et al.,* 2000; also reviewed in *Greeley et al.,* 2004) (see section 3).

A key observational argument for Europa's ocean comes from tectonics — specifically, the so-called cycloidal ridges. The evolution of cycloidal ridges from cycloidal cracks, and the diurnal tidal stress cycle needed to generate them, are powerful geological arguments for a floating icy shell [as first pointed out by *Hoppa et al.* (1999a); see also the discussion in *Greeley et al.* (2004)]. The inherent problem of driving motion on fairly deep (>1 km) faults with modest (~0.1 MPa) tidal stresses appears to have been solved as well (see chapters by Nimmo and Manga, by Kattenhorn and Hurford, and by Prockter and Patterson for additional discussion, and illustrations, of this important point; see also section 4.4 below).

The gravity and magnetic data alone rule out the early post-Voyager hypotheses for a thin solid icy shell directly coupled to a hydrated silicate interior (*Ransford et al.,*1981). In this concept, Europa's lineaments were due to stresses arising from convection in the rocky interior and propagated upward into the ice (*Finnerty et al.,* 1981). Indeed, to a degree such thinking represented a barrier to accepting or even considering a "mobilist" Europa [e.g., the discovery of dilational bands in *Schenk and McKinnon* (1989)] and the deeper implications that flow from such a view. It is also true that "extraordinary claims require extraordinary evidence," and well into the 1980s most planetary surfaces (as best as they were characterized) were not known to be either young or active. Our explorations of the outer solar system, Venus, and even Mars have changed this perspective forever.

3. COMPOSITION

Europa's "icy" shell is mostly water ice, but there are other components, especially in areas of recent tectonic activity or impact exhumation (see the chapter by Carlson *et al.,* which is particularly thorough in terms of what is known and what is unknown about Europa's surface composition). Observed near-infrared absorption bands are distorted in a manner characteristic of highly hydrated sulfates. Radiation-processed $MgSO_4 \cdot nH_2O$ and $Na_2SO_4 \cdot nH_2O$ ices are arguably the leading candidates [as in the earlier review of *Greeley et al.* (2004)], but a strong if not likely alternative or additional component is $H_2SO_4 \cdot nH_2O$ ice (*Carlson et al.,* 1999). Ultimately, a complex mixture of frozen sulfate hydrates and other materials may explain Europa's dark, reddish materials (e.g., *Dalton et al.,* 2005). If sulfuric-acid hydrate alone is responsible, then Europa's sulfur could be exogenic (iogenic) in origin (*Johnson et al.,* 2004). Exospheric Na and K are observed as well, and in a ratio that implies they are *not* dominantly iogenic, but the source minerals on Europa (presumably chlorides and/or sulfates) have not been positively identified (see the chapter by Johnson et al., as well as the discussion in Carlson et al. as to whether Europa's Na and K can really be claimed as endogenic). We note that the original potassium detection of *Brown* (2001) has yet to be repeated.

The composition of the icy shell must reflect, to some degree, the composition of the ocean below, albeit after geophysical and radiolytic processing and exogenic deposition (see chapter by Carlson et al.). Theoretical models favor an oxidized, sulfate-bearing ocean (*Kargel et al.,* 2000; *Zolotov and Shock,* 2001; *McKinnon and Zolensky,* 2003) with low (compared with terrestrial) concentrations of alkali salts (*Kargel et al.,* 2000; *Zolotov and Shock,* 2001). Europa's primordial ocean, however, may have initially been reduced and sulfidic and only later evolved to be oxidized and sulfate-bearing (*McKinnon and Zolensky,* 2003). [Note that Earth's ocean has apparently followed a similar path, but for different reasons (e.g., *Frei et al.,* 2009).] This is in contrast to the hypersaline (~saturated) CI-chondrite analog model (*Kargel et al.,* 2000). The conductivity limits implied by the initial analysis of the Galileo magnetometer data (*Zimmer et al.,* 2000) did not provide strong constraints on sulfate concentration. For ocean depths consistent with the gravity data, the minimum conductivity necessary to account for the induced field would be ~0.1 S m^{-1}, which is ~1/25 of that of terrestrial seawater (noted in *Greeley et al.,* 2004; see chapter by Khurana et al.). If the conductivity were due to sulfate alone, ~0.2 wt% would be the minimum implied, based on Fig. 12 in the chapter by Khurana et al.

Zimmer et al. (2000) found the magnetic induction signature of Europa to be consistent with more than 70% of the induced dipole moment expected from a perfectly conducting sphere. *Schilling et al.* (2004) extended this work to two subsequent Galileo flybys (E19 and E26). Although the stated purpose was to place strict limits on any (small) intrinsic magnetic dipole, they also concluded that the inductive response was at least 93% of the theoretical maximum for a conducting sphere. This was quite a change from the previous lower limit of 70%, and effectively ruled out a signal from any conductive source deeper than the ocean. Taken at face value, this limit also implied that the conducting layer (ocean) could lie no deeper than 20 km, and may possesses the conductivity of seawater or higher (*Schilling et al.,* 2004). The latter is difficult to achieve cosmochemically with alkali halides because of aqueous dilution (*Zolotov and Shock,* 2001). The limits of *Schilling et al.* (2004) also formed the basis for the detailed analysis of *Hand and Chyba* (2007), who concluded that the icy shell was no thicker than 15 km and the ocean nearly saturated in salt or sulfate.

As discussed in the chapter by Khurana et al., however, the most complete simulation to date of Europa's magnetospheric interaction is that of *Schilling et al.* (2007), and these results supersede those of *Schilling et al.* (2004). They find that the product of ocean conductivity and thickness must *exceed* 50 S m^{-1} km. We also note that their magnetohydrodynamic (MHD) simulations all assume an icy shell thickness of 25 km. For an ocean thickness of 100 km, the lower limit on ocean conductivity is 0.5 S m^{-1}, and given that the ocean may be up to 50% thicker, the conductivity limit may be correspondingly lower. An upper limit on the oceanic conductivity cannot be set with the available data, whereas the lower limit above is only ~1/5 of that of terrestrial seawater, but with these limits the ocean would at least qualify as brackish. In terms of explaining the conductivity with sulfate alone, ≥1 wt% would be required, and can be met by the preferred partial extraction model (K_{1a}) of *Zolotov and Shock* (2001).

Electromagnetic induction sounding has clearly demonstrated its ability for important discovery at Europa. A future mission, incorporating both flybys and an orbital tour of Europa, offers the possibility of determining both shell thickness and oceanic conductivity to good precision (see the chapter by Khurana et al. for a theoretical assessment and the chapter by Greeley et al. for how this might be implemented on a future mission to Europa). Given suitable measurements over time, even deeper sounding of Europa's rocky mantle and iron-rich core are possible. Both experiments would be significantly aided by simultaneous measurement of plasma conditions at the spacecraft and magnetospheric conditions away from Europa (i.e., dual spacecraft).

As for surface composition, improved spectral analyses, either from existing or future data, will also be critical to progress in indentification, as will direct measurement of sputtered atmospheric species (see chapter by Johnson et al.). For although the thermomechanical properties (diffusivity, viscosity, etc.) of pure water ice are relatively well understood, those of highly hydrated sulfate salts (for example) are not, and are quite different from those of water ice [owing to, among other things, the large unit cells of the sulfate minerals (*Hawthorne et al.,* 2000)]; see also *Durham et al.* (2005) and later experimental work by *McCarthy et al.* (2007) and *Grindrod et al.* (2008). But perhaps most important among all compositional questions for Europa is that of its ocean, for it is the ocean and its interfaces (icy shell and rocky mantle) that provide potential abodes for life.

4. CONVECTION AND TECTONICS

4.1. Rheology

Experimental studies have made such progress in recent years that a fairly complete understanding of the steady-state viscous creep of pure water ice now exists (*Goldsby and Kohlstedt,* 2001; *Durham and Stern,* 2001). We are no longer restricted to simple considerations of constant vs. temperature-dependent viscosity or Newtonian (viscosity independent of stress) vs. non-Newtonian (stress-dependent viscosity) creep. For most temperature and stress regimes of interest to europan geology, nearly Newtonian, grain-size-sensitive, grain-boundary sliding (GBS) describes the dominant flow law. Only for higher stress levels and larger grain sizes (d > 1 cm) is non-Newtonian (and thus power-law) creep law dominant [see deformation maps in *Durham and Stern* (2001) and *Barr and Pappalardo* (2005)]. For very fine grain sizes and very low stresses, creep via diffusion of individual water molecules may become important (*Goldsby and Kohlstedt,* 2001; *McKinnon,* 2006), but such creep has yet to be observed experimentally. Given the importance of grain size (d), a good understanding of what controls grain growth and evolution under planetary conditions is necessary (*Durham and Stern,* 2001; *Barr and McKinnon,* 2007). Largely untapped glaciological understanding should help, notwithstanding some healthy debate within the glaciological community of the appropriateness of GBS as a deformation mechanism [see the point-counterpoint in *Duval and Montagnat* (2002) and *Goldsby and Kohlstedt* (2002)].

Understanding of ice in the brittle regime, where ice fails by faulting, is in contrast less well developed (e.g., *Schulson,* 2001). The situation in the obscure, mixed-mode, "semibrittle" regime [described for rock in *Kohlstedt et al.* (1995) and *White* (2001)] is even bleaker for ice (i.e., nothing is known). Moreover, research on the effects of salt contaminants, gas clathrates, etc., on ice deformation is in its infancy. But overall, a reasonably good working knowledge of ice rheology exists to apply to tectonic and convection problems on Europa (as the chapters by Nimmo and Manga, and by Barr and Showman, respectively attest).

4.2. To Convect or Not to Convect?

Using a GBS rheology and tidal linearization (described below) and the convection theory of Solomatov and coworkers (summarized in the chapter by Barr and Showman), Europa's ice shell has been shown to be unstable to convection for shell thicknesses >20–25 km (for d = 1 mm), with thinner shells unstable for smaller grain sizes (>10 km or so for d = 100 μm) [*McKinnon* (1999), with an updated GBS flow law from *Goldsby and Kohlstedt* (2001)]. Using an older, generic Newtonian rheology and a modified parameterized convection scheme, *Hussmann et al.* (2002) argued that Europa's shell is less likely to convect [i.e., the shell must be thicker in comparison with *McKinnon* (1999) for similar ice viscosities]. As both studies treated the shell as pure ice and bottom heated (i.e., core heat, or tidal heating only in the bottommost, hottest ice), the difference in results (i.e., thickness estimates) stems from the rheologies and convection formalisms employed.

Tidal heating is a game changer for convection overall, as illustrated in the chapter by Sotin et al. There are two different problems: When does convection initiate (for which the shell is effectively bottom heated in most but not all cases of interest; see chapters by Barr and Showman and Sotin et al.); and what is the steady-state condition, if there is one (when the shell is better described as internally heated)? In the latter case, tidal heating is important in the convecting sublayer but may be neglected (with care) in the stagnant lithosphere or "lid" above (*Hussmann et al.,* 2002; *Ruiz and Tejero,* 2003; *Nimmo and Manga,* 2002); this differs from standard treatments of internally heated convection (which generally assume uniform heating, as would be appropriate to radioisotope decay). Using tidally linearized rheologies and a modified parameterized convection scheme for internal heating, *Ruiz and Tejero* (2003) found steady-state solutions with ice shell thicknesses ranging from ~50 km (for GBS creep and d = 0.1 mm) to ~15–20 km [for GBS creep and d = 1 mm as well as for so-called power-law ductile-A creep, for which there is no explicit grain-size dependence (*Durham and Stern,* 2001)]. *Ruiz and Tejero* (2003) favored the high heat flows from the GBS and d = 1 mm case on geological grounds, but it is notable that a steady-state solution for ductile-A-creep dominance was found with a substantially thinner shell than stability (initiation) conditions otherwise indicated (*McKinnon,* 1999). This implied that the evolution of Europa's shell when convection begins could be quite interesting (in this case, the shell would be predicted to thin). Although only one study among many, the implication is more general, as was soon developed in detail by *Mitri and Showman* (2005), as noted below.

Now, tidal linearization refers to the interplay between the diurnally varying tidal stress field and generally much weaker convective buoyancy stresses. The convective stresses are thus a perturbation, one that acts as if the ice were Newtonian, but with a viscosity determined by the tidal cycle. *McKinnon* (1999) argued that it is an important phenomenon on Europa, but it has received little further study, probably due to its complexity [the tidally linearized viscosity is, among other things, anisotropic (*Schmeling,* 1987)]. Somewhat remarkably, it can be shown that Europa's shell is susceptible to convective overturn by GBS creep *without* tidal linearization if the shell thickness exceeds ~40 km (for d = 1 mm) (see Appendix in *McKinnon,* 2006). This may be important during epochs of relatively low orbital eccentricity and weak tidal heating (e.g., *Hussmann and Spohn,* 2004; and as summarized in the chapter by Moore and Hussmann).

In terms of grain size, it has been hypothesized that grain-size evolution in hot, straining ice will lead to approximately equal flow contributions from GBS and power-law creep (*Durham and Stern,* 2001); this is the field-boundary or "De Bresser" hypothesis (*De Bresser et al.,* 1998, 2001). For present-day tidal amplitudes (stresses of ~0.1 MPa), this implies d ~ 1 mm and effective basal viscosities near 10^{14} Pa s (*McKinnon,* 1999), but note this estimate neglects grain-boundary "premelting" (e.g., *Dash et al.,* 1995), which if accounted for raises the estimated grain size to a few millimeters for the same viscosity. These conditions are close to those for maximum tidal heating in the convecting sublayer, assuming a Maxwell or similar model (*McKinnon,* 1999; *Tobie et al.,* 2003; and see the chapters by Barr and Showman and Sotin et al. for a broader physical context). The Maxwell (τ_M) time of ice undergoing GBS creep can be written (again, neglecting premelting) as

$$\tau_M \approx 22\ \text{h}\left(\frac{d}{\text{mm}}\right)^{1.4}\left(\frac{\sigma}{0.1\ \text{MPa}}\right)^{-0.8}\exp 22\left(\frac{260\ \text{K}}{T}-1\right) \quad (1)$$

which can be compared to Europa's orbital (tidal flexing)

period of 85.2 hr (noting that peak heating actually occurs at periods of $2\pi \times \tau_M$). This result is not particularly special to Europa, but is primarily a function of the similarity of the Maxwell time of near-solidus ice and the orbital periods of many outer planet satellites. Moreover, other dissipation models for ice, shown in the chapter by Sotin et al., have dissipation peaks at similar frequencies, so it would seem that Europa and its icy brethren are blessed in terms of this energy source.

Finally, convective evolution models incorporating tidal heating are becoming more sophisticated, at least conceptually (e.g., *Mitri and Showman,* 2005). In particular, the abrupt nature of the conductive to convective transition is now appreciated. Relatively rapid changes in shell thickness ensue, specifically, geologically rapid (~10^7 yr) thickening of the icy shell once heat flows jump to their convective values; convective turn-on may possibly disrupt the cold, conductive, stagnant surface layer (lithosphere) as well (see the chapter by Barr and Showman). Shell thickness variations also imply surface expansion or contraction, depending on the direction of the transition, with convective turn-on and shell thickening possibly being responsible for the generally extensional tectonics seen on Europa [a simple yet important realization (*Nimmo,* 2004b)].

4.3. Pits and Uplifts

Numerous uplifts, breached uplifts, regular and irregular domes, small chaos regions, and genetically related depressions (pits) are seen across Europa, ranging in size from the very large Murias Chaos (*Figueredo et al.,* 2002) to features no more than a few kilometers across (*Carr et al.,* 1998; *Pappalardo et al.,* 1999; *Greenberg et al.,* 2003; *Barr and Pappalardo,* 2005). The structural relationships and geological associations imply an important role for solid-state diapirism (*Pappalardo et al.,* 1998; *Figueredo et al.,* 2002) and for the pits volume loss due to subsurface melting (*Singer et al.,* 2009). Simply put, too many interrelated features exist on Europa's surface for solid-state diapirism to be dismissed as a likely mechanism. Features the scale of Murias (formerly known as "the Mitten") may be due to solid-state upwelling in an icy shell capped by a relatively thick lithosphere (*Figueredo et al.,* 2002). The far more numerous, small-scale features are more likely due to diapiric instability in a bottom thermal boundary layer (the traditional source of plumes in the terrestrial planets), i.e., the hot ice right above the ocean. The smallest uplifts and chaos regions imply the smallest diapirs, which imply a bottom boundary layer thickness of ~1 km or less (*Nimmo and Manga,* 2002). For such a thin layer to be unstable requires a low viscosity, which these authors took to be due to diffusion creep at very fine grain sizes (d ~ 20–60 μm). Alternatively, tidally linearized GBS creep at similar grain sizes will suffice, especially if weakened by grain-boundary melting (*Goldsby and Kohlstedt,* 2001). Are such small grain sizes possible for hot ice, given the thermodynamic drive for grain growth? Perhaps chemical impurities from the ocean, incorporated in the ice as it freezes, impede grain growth, or perhaps the convective strain itself fines grain size. The important issue of grain size evolution is covered in the chapter by Barr and Showman.

At minimum, existence of a lower boundary layer in the shell provides evidence for (and constrains) heat flow from the rocky mantle, as the boundary layer is the conductive interface between the ocean below and the convecting ice above. That is, in steady state if there is no core heat flow, there is no lower boundary layer. The ability of small diapirs to rise a sufficient distance through the convecting sublayer has been questioned by *Greenberg et al.* (2003), referencing the purely Newtonian analysis of *Rathbun et al.* (1998). But it is getting diapirs to pierce the stagnant lid and/or elastic lithosphere that is the real problem. Ascent may be aided by localized tidal heating (*McKinnon,* 1999; *Sotin et al.,* 2002), low viscosity due to partial melt or small grain size, compositional effects (melting and drainage of brine) (*Nimmo and Manga,* 2002; *Pappalardo and Barr,* 2004), or weakening of the lithosphere (by an unspecified mechanism or mechanisms) (*Showman and Han,* 2005). Illustrations of how this might operate can be found in the chapter by Barr and Showman.

4.4. Tectonics and Cycloidal Ridges

Europa's tectonics seems to require a subsurface ocean at some depth below. Specific models for formation of Europa's ubiquitous cracks and ridges vary (see chapters by Kattenhorn and Hurford and by Prockter and Patterson), but the most successful models to date invoke sunstantial diurnal tidal stresses enabled only by a subsurface ocean. Cycloidal features in particular require the action of diurnal tides, which induces stresses that rotate through the orbital cycle (*Hoppa et al.,* 1999a). Diurnal tensile stresses are only ~50–100 kPa if there is an ocean below Europa's icy shell, which is barely sufficient to break the top layer of a prefractured icy shell (*Hoppa et al.,* 1999a; *Pappalardo et al.,* 1999). If there is no ocean present, diurnal stress is orders of magnitude less, with no opportunity to affect geology. Thus, the presence of cycloids argue for an ocean at the time of their formation, which must be very recent based on Europa's youthful surface age. The tidal "walking" model for strike-slip motion (*Hoppa et al.,* 1999b) and the shear heating model for ridge formation (*Nimmo and Gaidos,* 2002) also both argue for a subsurface ocean, in order to deliver the lateral offset necessary to produce significant strike-slip motion and heating along fractures or ridges. Although an ocean is implied by models of diurnal tectonic stressing, the depth to an ocean is not constrained by this tidal stress argument, because tidal Love numbers and thus surface stresses are only weakly sensitive to ice shell thickness (*Moore and Schubert,* 2000; *Wahr et al.,* 2009).

How propagating cracks on Europa can work themselves into ridges is also now somewhat clearer. Stress levels of 0.1 MPa can only drive crevasses ~150 m into Europa's icy surface if the surface is elastically coherent (and not a mega-regolith, as now seems almost certain; see chapter by Moore et al.). But if the elastic lithosphere behaves as a thin plate rather than an elastic half-space, then stresses at the downward propagating crack tip are not relieved in the far field and the crack may continue propagating. The effect is surprisingly strong: 0.1 MPa in tension can split a nonporous lithosphere 600 m thick (*Lee et al.,* 2005). Moreover, with additional stress, e.g., from nonsynchronous rotation, then almost any lithosphere usually considered can crack down to the asthenosphere (ductile ice) or even to the ocean. Once these vertical fractures form, then tidally driven strike-slip motion can generate the heat necessary to raise the ridges (*Nimmo and Gaidos,* 2002). Further theoretical development of this model would be welcome, with attention to such issues as the proper dynamic coefficient of friction for icy fault gouge (cf. *Preblich et al.,* 2007) (see also section 7.4 and the chapter by Sotin et al.).

5. THICKNESS OF EUROPA'S ICY SHELL

The thickness of Europa's icy shell has turned out to be a surprisingly controversial topic. There are two endmember models, typically termed "thin" and "thick." The "thin" ice model (e.g., *Greenberg et al.,* 2002) argues for an icy shell only several kilometers thick (or less), while the "thick" shell model purports an icy shell only 20 km or so thick. Both models envisage an icy shell thinner than that of any other icy satellite, with even the "thick" ice being equivalent to just ~1% of the satellite's radius. The endmember models have, however, very different implications for the geological processes that can operate within the icy shell, therefore for the potential habitability of Europa.

The "thin" ice model was first suggested by *Carr et al.* (1998) and extensively developed by *Greenberg et al.* (1998, 1999, 2000, 2002). The thin ice model presumes ice so penetrable that the subsurface ocean could communicate readily with the surface, through fracturing and melting (if not explosive venting of cold, boiling water). Such thin ice would allow fractures to more readily penetrate through the entire icy shell without the overlying normal stress causing them to terminate or transition to normal faults (*Golombek and Banerdt,* 1986). Fractures that pierce the icy shell could serve as conduits for the direct rise of liquid water to shallow depths, with implications for freezing into ridges (*Greenberg et al.,* 1998) and bands (*Tufts et al.,* 2000). Moreover, thin ice allows Europa's chaotic terrain to form by local complete (or nearly complete) melt-through of the icy shell if a sufficiently concentrated and intense heat source is applied to its base (*Greenberg et al.,* 1999). The icy shell must be less than several kilometers thick in order to be stiff enough to melt completely through without inward flow of ice more rapidly filling the emerging hole (*Stevenson,* 2000; *O'Brien et al.,* 2002; *Nimmo,* 2004a). The model implies a fully conductive thermal profile through the icy shell, as it is not thick enough to convect (see section 4.2 and chapter by Barr and Showman). Thus, the elastic thickness of the lithosphere would be similar to the icy shell's true thickness (*Billings and Kattenhorn,* 2005). The thin ice model is arguably astrobiologically attractive: Radiation-produced chemical oxidants from Europa's surface could enter Europa's ocean relatively easily, a hot mantle would pour additional chemical energy into the base of the ocean, photosynthesis might be permitted at shallow depths in water-filled fractures, and any organisms in the ocean could be sampled at the surface (*Greenberg et al.,* 2000; *Figueredo et al.,* 2003). If this scenario is correct, it might be possible to someday penetrate through thin spots in the icy shell directly into the ocean below.

In the "thick" ice model, the icy shell is deep enough to convect, accounting for Europa's observed pits, spots, and domes, and perhaps also its chaotic terrain (see section 4.3 and the evaluation of models in the chapter by Collins and Nimmo). [In contrast, in the thin ice model, the origin of pits and domes is unexplained, and the taxonomy is instead challenged (*Greenberg et al.,* 2003).] In the thick shell model, a brittle-conductive layer is situated atop a nearly isothermal, ductile ice asthenosphere; therefore, the elastic thickness of the icy shell is significantly less (by at least ~50%) than the total icy shell thickness. In this model, the relatively thin elastic layer can laterally translate above the "warm" ductile asthenosphere, which at a temperature of ~250–260 K, would simulate many of the effects of a fluid layer. In the thick ice model, fractures need penetrate only through the upper brittle layer, ridging occurs within the brittle layer, and bands result from dilation and spreading of the brittle layer atop ductile ice (as described in the chapter by Prockter and Patterson). If compositional impurities permit partial melting within the ice (e.g., *Head and Pappalardo,* 1999), its viscosity would be even less than that of warm ice. In a thick shell model, communication between the ocean and the surface is necessarily indirect, in the form of convecting plumes and local near-surface melting events. The chemical energy that might support subsurface life would be less plentiful, with a cooler rocky mantle, surface oxidants instead drawn down toward the ocean by melting within the icy shell, and oceanic materials dredged slowly upward in icy convective plumes that rise at a rate of a few centimeters per year (*Zolotov et al.,* 2004).

Putting preconceptions aside, independent modeling and observations shed light on Europa's icy shell thickness. Thermal modeling of current-day icy shell thickness that includes radiogenic heating from below and convection and tidal heating within the icy shell (albeit without tidal linearization) predicts an equilibrium icy shell thickness of >16 km (*Moore,* 2006) and nominally a few tens of kilometers thick (*Spohn and Schubert,* 2003; *Hussman et al.,* 2002; *Hussman and Spohn,* 2004). While intense mantle heating is in principal possible, such is probably not a geophysically sustainable state (as argued in the chapters by Moore and Hussmann and by Sotin et al.). These modeling results support the thick ice model today, while predicting changes in the icy shell thickness through time as linked to the coupled thermal-orbital evolution of Io.

Impacts provide an important constraint on icy shell thickness (beautifully depicted in the chapter by Schenk and Turtle and in the color plate section). *Schenk* (2002) found that Europa's largest central peak craters are shallower in depth than similarly sized craters on Ganymede and Callisto, suggesting a transition to warm ductile ice at ~8 km depth on Europa. Moreover, he found that Europa's two multiring impact structures, Tyre and Callanish, have penetrated an icy shell ≥19 km thick to a layer below that behaves as a fluid on the rapid timescale of the impact process. Thus, Europa's impact craters, as sparse as they are, are consistent with a thick ice model.

Europa's icy shell also shows rare examples of surprisingly large topography. *Prockter and Schenk* (2005) show more than a total relief of more than 1.2 km in the Castalia Macula region. Moreover, *Schenk et al.* (2008) identify broad troughs ~1 km deep. This scale of topography is difficult to reconcile with a thin icy shell of only a few kilometers thickness (an argument backed up in the chapter by Nimmo and Manga).

Because Europa's large impact structures and locally large topographic relief are rare, it seems plausible that Europa's icy shell might be locally "thick" but in other places might be "thin." However, modeling of the lateral flow of ice suggests that any regions of thick ice should laterally flow on geologically short timescales, tending to even out short-wavelength ice thickness variations (e.g., *Nimmo,* 2004a). This implies that the thick ice implied by impact craters and regions of high topography are not coincidentally formed on regions of thick ice, but are instead representative of a globally thick icy shell. And it almost goes without saying that a critical experiment to carry on any future mission to Europa to address the fundamental issue of shell thickness, as well as a host of other structural and geophysical questions, would be an ice-penetrating radar sounder. This instrumentation has now been demonstrated at Mars, and is described in the chapter by Blankenship et al.; a strawman radar sounder for Europa is described in the chapter by Greeley et al.

6. THERMAL AND TIDAL EVOLUTION

Prior to the Galileo mission, the idea of oceans within Europa and other large icy satellites experienced several back-and-forth swings of plausibility based on thermal modeling. The first theoretical models of oceans within Europa and other large icy satellites emerged in the early 1970s, when it was realized that radiogenic heat might be sufficient to melt oceans within icy satellites early in solar system history (*Lewis,* 1971; *Consolmagno and Lewis,* 1976) (a journey through this and other scientific history can be found in the chapter by Alexander et al.). Tidal heating was recognized as an additional heat source that might maintain a subsurface ocean (*Cassen et al.,* 1979, 1980). It was soon realized, however, that as satellites cooled and their outer ice shells thickened, a critical thickness would be reached (approximately tens of kilometers) where solid-state convection would initiate in the ice shell, which would (it was believed) freeze an underlying ocean in about 100 m.y. (*Reynolds and Cassen,* 1979; *Cassen et al.,* 1982). Attention turned to whether Europa might have sufficient tidal dissipation to permit an ice shell thin enough that solid-state convection would not occur, possibly allowing an ocean to persist until the present time (*Squyres et al.,* 1983; *Ojakangas and Stevenson,* 1989a).

Early Galileo imaging suggested geological evidence for solid-state convection at Europa, implying that the critical icy shell thickness for convection has been reached (*Pappalardo et al.,* 1998). Did these observations imply that any ocean within Europa had subsequently frozen solid? Improvements in modeling of ice rheology, of parameterized and numerical convection, and of tidal heating now suggest that a convecting icy shell above an underlying ocean are not incompatible (*Hussman et al.,* 2002; *Spohn and Schubert,* 2003) (see the chapter by Moore and Hussmann for conditions that permit a steady-state thickness). In fact, concentrated tidal heating within warm, convecting ice might further heat the icy shell, and could trigger localized runaway of tidal heating and perhaps internal melting of the icy shell (*McKinnon,* 1999; *Wang and Stevenson,* 2000; *Sotin et al.,* 2002).

If Europa's rocky mantle is very warm and extremely dissipative such that tidal heating within it is very large, then the scenario of a very thin (several kilometers) conductive icy shell above an ocean might be a plausible configuration (*Thomson and Delaney,* 2001; *O'Brien et al.,* 2002). While such might have been possible in earlier, high-eccentricity epochs, this situation is, however, not expected for Europa's current-day eccentricity (*Hussmann and Spohn,* 2004; chapter by Moore and Hussmann). Moreover, as the chapter by Sotin et al. makes clear, the ocean on Europa, which is the focus of such keen interest generally, has the perverse effect of decoupling the tidal deformation of the rocky mantle from the icy shell. The former is reduced, and the existence of Io-like interior states is judged to be unlikely, except perhaps as transient episodes (higher-eccentricity epochs).

6.1. Constraints on Heat Flow

Whatever the true thickness of the icy shell, and/or its variations in space and time, it is important to constrain heat flow by whatever observational means available. Many analyses yield estimates of lithosphere thickness, which when properly interpreted give measures of heat flow (subtleties are recounted in the chapter by Nimmo and Manga). In other situations it is less immediately clear what heat flows are implied. In any event, heat flow is a major clue as to what is taking place below Europa's surface. For example, detailed modeling of folds in the near-polar Astypalaea Linea region imply heat flows at the time of fold formation of ~100 mW m^{-2} (*Dombard and McKinnon,* 2006). Similarly large heat flows are implied from widths of circumferential graben that surround (and define) the Tyre and Callanish multiring impact structures (*Ruiz,* 2005; *McKinnon and Schenk,* 2008), tectonic features that are ostensibly the result of prompt crater collapse and icy asthenospheric inflow (see chapter by Schenk and Turtle). Both of these mechanical analyses identify the base of the lithosphere with the brittle-ductile transition, albeit on different timescales. Other measures of heat flow come from the flexing of the elastic lithosphere under various topographic loads (e.g., *Nimmo et al.,* 2003), although in the latter study a somewhat more modest heat flow (~33 mW m^{-2}) was estimated (see section 4.1 in the chapter by Nimmo and Manga). Given the quantity of topographic information now available (e.g., *Hurford et al.,* 2005; *Billings and Kattenhorn,* 2005), we may expect further build-up of a global lithosphere thickness and heat flow dataset.

6.2. Tidal Energy Sources

To account for the thin lithosphere estimates above, and to power chaos formation and band dilation, requires that, at least intermittently, heat flows exceed terrestrial average values (i.e., be >100 mW m^{-2}). As we have described, one possibility is that the lower part of the icy shell is convecting and thus relatively "hot." The Maxwell time of ice near the melting point is close to the tidal forcing period of 3.55 d (section 4.2), so convecting ice may be the most strongly heated. Based on the model of *Ojakangas and Stevenson* (1989a), for the present root mean square (rms) tidal strain rate each kilometer thickness of convecting ice could supply ~15 mW m^{-2} to the surface, subject to the uncertainty derived from the grain-size dependence of the ice viscosity. Assuming a Maxwell dissipation model, *Tobie et al.* (2003) estimated a maximum dissipation rate of about half this amount, as can be seen in the chapter by Sotin et al. In either case, it is clear that high heat flows and a thick shell can be self-consistent provided convection is occurring in the lower portion of the shell.

It has been argued that chaos regions and other surface disruptions may be the result of thermal runaways in the icy shell (for an overview see the chapter by Collins and Nimmo). Essentially, by concentrating heat and buoyancy in the hottest and most buoyant parts of the icy shell, there is the potential of a "china syndrome" in reverse. On the other hand, such heating would tend to diffuse laterally, and tidally driven melting in Europa's shell would result in negative buoyancy (*Tobie et al.,* 2003), so this hypothesis remains unproven. As noted in section 4.3, a particularly important test of any tidally heated model of convection in Europa's shell is for hot, buoyant ice to be able to penetrate the cold, stagnant lithosphere.

The alternative heat source is the tidally heated deeper, rocky interior, despite the difficulties alluded to above. If this deeper interior has differentiated into a rock mantle and a metallic core, the latter is likely to still be liquid, due to the presence of cosmically abundant sulfur [and possibly oxygen; see *Terasaki et al.* (2008)] and possible tidal heating of the enveloping mantle. The presence of such a metallic core has not been independently established, but the analogy with Ganymede is reasonably compelling. Moreover, all published thermal evolution models for Europa's interior indicate temperatures at some time that exceed the iron-sulfur eutectic (lowest melting point) temperature, and thus conditions appropriate for at least partial core formation [the reviews by *Greeley et al.* (2004) and *Schubert et al.* (2004) and the chapter by Schubert et al. are in agreement on this point].

A liquid metallic core allows for greater tidal flexing and heat production in the rock mantle, compared with a uniform, solid rock + metal interior (*Tobie et al.,* 2005), but the enhancement in tidal heating naturally depends on liquid core size, which is in turn dependent on the partitioning of iron between core and mantle, and the quantity of light elements partitioned into the core. Thus, for relatively modest cores, the enhancement in tidal heating is modest at best, whereas for the most sulfur- and oxygen-rich cores, tidal heat flows from the mantle are potentially in excess of the present radiogenic contribution, for Europa's present-day eccentricity [the latter on the order of 10 mW m^{-3} (*Ojakangas and Stevenson,* 1989a)]. Scaling laws for convection (*Solomatov and Moresi,* 2000) predict that for such a total heat flow a peridotitic mantle may undergo partial melting, or at least be close to melting, which further increases dissipation at tidal strains (to a point). But even in this case, it is unlikely that tidal dissipation in the rocky interior can supply all the heat implied by the surface observations. The required effective k_2/Q, or second-degree potential Love number to specific dissipation factor ratio, would have to be Io-like, which would appear inconsistent with the high but hardly Io-like heat flows referenced above.

In contrast, Moore and Hussmann in their chapter argue that the internal nonlinear feedbacks are strong enough that were Europa's rock to reach a partially molten state, the only stable steady state in terms of dissipation *and* heat flow would be Io-like. Such a very hot interior, supplying a tidal power output in excess of 10 TW (>300 mW m^{-2} at Europa's surface), would be consistent with the "thin" ice model in section 5. Whether it is sustainable in an orbital mechanics sense is unclear, but it may have been achieved intermittently in the past (*Hussmann and Spohn,* 2004), and steady-state conditions do not always prevail regardless.

Partial melting of the silicate mantle, at any time in Europa's past, would imply the potential for volcanism at the mantle-ocean interface, and heat flows were almost certainly great enough for volcanism billions of years ago due to radiogenic heating alone. A rather thick volcanic crust may have built up (perhaps episodically) over time. For modestly tidally heated rock mantle, however, the rock lithosphere could be relatively thick, making any plate-tectonic-syle overturn difficult [e.g., even for a relatively low Q, this lithosphere could be ~200 km thick compared to a total rock mantle thickness of ~500–750 km, as seen in Fig. 15.18 in *Greeley et al.* (2004)]. Such a relatively thick lithosphere would not itself be so much of a barrier to volcanism, however, because the overburden pressures at the source depth for melting are similar to 100 km depth in

Earth. And as long as any internal "crust" produced is mafic, there should be sufficient positive buoyancy for new lava to reach the ocean floor.

Given this picture of at least ancient volcanism it is less clear how hydrothermal interaction with deeper, ultramafic rocks (peridotites, serpentinites, etc.) are to be sustained over geological time (but see chapter by Vance and Goodman for a contrasting view). Nevertheless, additional factors add to the plausibility of mafic volcanism on Europa, at least episodically during its geological history, which it is worth remembering extends far beyond the crater retention age of its icy surface (chapter by Bierhaus et al.):

1. Europa's mantle is probably relatively oxidized due to the presence of copious water and hydration reactions during Europa's initial accretion and ice-from-rock differentiation (*McKinnon and Zolensky,* 2003) (see chapter by Zolotov and Kargel for a far greater wealth of detail). As such, its mantle minerals are probably relatively iron-rich compared to Earth's upper mantle (as are the silicates in carbonaceous meteorites), and the solidus (melting curve) is reached at lower temperatures.

2. Convective upwellings are hotter than average, and tidal heating in Europa's mantle will make them hotter and more buoyant still (and for silicates of course, partial melting increases buoyancy *and* dissipation, to a limit that depends on melt fraction; see chapter by Moore and Hussmann).

3. There may be a floor below which the dissipation in Europa's mantle and core will not drop, either because the Maxwell model is inappropriate in the rocky mantle, or because there is dissipation due to turbulence in the fluid boundary layer at the top of the metallic core. The latter was implicit in arguing for a lunar-like Q of 25 for Europa's interior in the floating ice shell model of *Squyres et al.* (1983), and given recent theoretical developments (*Tyler,* 2008), is worth revisiting.

4. Even if Europa is in thermal steady state, its thermal and orbital history are almost certainly time dependent. Tidal heating is an extremely nonlinear process, of course, and Europa, Io, and Ganymede are coupled together in the Laplace resonance. Models indicate that Io is the most unsteady actor in the resonance (*Ojakangas and Stevenson,* 1986), and Europa's eccentricity and tidal heating will be coupled to Io's orbital and thermal excursions (*Hussmann and Spohn,* 2004). Europa's tidal heating input probably varies on 100-m.y. timescales, possibly leading to pulses of volcanic activity. We emphasize again that the latter do not necessarily require steady-state conditions.

With respect to the last factor, *Lainey et al.* (2009) have recently determined from astrometric measurements, after many attempts by others, that Io's mean motion is presently increasing. This means Io's orbit is slowly shrinking, and thus that it is moving away from exact resonance with Europa and Ganymede. This does not mean that the Laplace resonance is unbinding and will break apart, only that the tidal dissipation in Io is so great at present that its semimajor axis and eccentricity are decreasing. Eventually, Io's eccentricity and dissipation will decrease sufficiently so that tidal torques from Jupiter will halt, if not reverse, its secular acceleration. Another implication is that Io's eccentricity, and probably its tidal dissipation, were higher in the immediate geological past. The implication for Europa's present tidal heating are not clear, as the Lainey et al. solution is apparently insensitive to dissipation within Europa (see their supplemental material). Europa's orbit is expanding, however, and Ganymede's orbit even more so (*Lainey et al.,* 2009), so we can say at present that Europa is moving away from exact resonance with both Io and Ganymede. Based on theoretical models (*Hussmann and Spohn,* 2004), it is not implausible that Europa is evolving away as well from a geologically recent state of higher eccentricity and greater tidal dissipation.

Further modeling of Galileo gravity results for Europa combined with cosmochemical constraints should prove useful in determining Europa's internal structure, and which is a primary input into any calculation of tidal heating. In this regard, we point out the later analysis of Galileo's close G2 pass to Ganymede, where regional gravity anomalies were detected (*Palguta et al.,* 2006). The scale of these anomalies is such that if they occurred on Europa they could affect the gravity solution (MOI) at more or less the level of the stated error in *Anderson et al.* (1998). More important, they demonstrate the utility of gravity tracking from orbit — a prime way for a future Europa mission to detect shell thickness and compositional variations, and practically the only way to detect undersea features such as volcanos.

7. ON HABITABILITY

Volcanic heat is the only reasonable mechanism for supplying sufficient energy to create a chaos region in a thin and otherwise conductive icy shell (if the shell is thin, of course). Chemical disequilibria associated with volcanism, due to quenching of chemical reactions and the creation of metastability, is also a logical source of support for simple life forms (if not their origin) (*Shock et al.,* 2000). Plausible europan biota include methane producers (methanogens) and sulfur and sulfate reducers (*McCollom,* 1999; *Zolotov and Shock,* 2004). Some might argue that such "simple" creatures are not as interesting compared with multicellular forms, but we take the view that any life found on Europa would contend for the discovery of the century. Moreover, as significant as the discovery of life anywhere else in the solar system would be, Europa, by virtue of its distance and isolation from Earth, offers something more — the possibility of an independent, or second, origin, and all that that implies (e.g., *Shapiro and Schulze-Makuch,* 2009; see chapter by Hand et al.).

Gaidos et al. (1999) have argued in particular that methanogens could not thrive on Europa even in the presence of submarine hydrothermal systems, because Europa's rock mantle would be *more reducing* than Earth's (due to a lack of subducted oxidants, i.e., water), and higher pressures would shift the CO_2-CH_4 equilibria as a function of oxygen fugacity to higher temperatures. This would ostensibly cause carbon to outgas as CH_4, rather than CO_2, eliminating this metabolic pathway. The pressure effect is small, however, and in fact works in the opposite direction (E. L. Shock, personal communication). More critically, Europa likely accreted from a mixture of rock and ice (as in the chapters by Canup and Ward and by Estrada et al.), and as we argued in section 6, this implies that Europa's interior, and the mantle derived from it, began in or evolved to a relatively oxidized state. Also, the observed oxidation states of volcanic gases on nearby Io are SO_2-S_2 mixtures, not H_2S, which indicate that Io's interior is no more chemically reduced than Earth's upper mantle (*Zolotov and Fegley,* 2000; cf. *McKinnon,* 2007), and one would be hard pressed to argue that Europa's interior is less oxidized that Io's (see chapter by Zolotov and Kargel for more background).

Finally, if volcanic methane were released through cracks in the ice, etc., this methane would cold-trap at Europa's poles and then be radiation darkened (D. J. Stevenson, personal communication), but this is not seen. Rather, it is CO_2 that has been found in Galileo NIMS spectra of Europa's surface, specifically in the dark, potentially endogenic, hydrate-rich terrains (*Hansen and McCord,* 2008). If high-temperature hydrothermal systems exist or existed on Europa, carbon probably primarily vents or vented as CO_2, not CH_4, which would thus fulfill the "Perrier ocean" prediction of *Crawford and Stevenson* (1988). But even if carbon did vent primarily as CH_4 on Europa, the equilibrium fugacity of methane at low temperature is so high that the thermodynamic drive for methanogenesis (from any other carbon source) will always exist. Other oxidants, such as radiolytically produced, surface O_2 and H_2O_2 (*Chyba,* 2000), may not be required for a simple biosphere, although they may obviously prove important — especially if hydrothermal activity is not sustainable over the long term. We conclude that methanogenesis (and sulfate reduction, etc.) remain viable metabolic pathways for europan biota. We also stress that at Europa, because we have already found the water (we think), in terms of looking for life, we should follow the energy (*Hoehler et al.,* 2007) (and see chapter by Hand et al. for a complete discussion).

The preceding implicitly assumes that Europa has acquired carbon in some form. While there is carbon on the surface, it could in principle be derived from meteoritic infall (see chapter by Carlson et al.), and it is unclear if a meteoritic (i.e., dominantly heliocentric) source of carbon, acquired over several billion years, is sufficient to build even a rudimentary biosphere. In older models for satellite formation, in which the jovian subnebula was initially rather dense and hot, carbon was ostensibly converted to CH_4 (*Prinn and Fegley,* 1989), and thus near Jupiter remained in the gas phase and unaccreted. The modern view posits accretion of largely unprocessed heliocentric solid material (see chapter by Canup and Ward). Even if thermally and aqueously altered on parent heliocentric planetesimals, Europa likely accreted a healthy amount of carbonaceous material, which from an astrobiological perspective is fortunate. The key limiting biogenic element may in fact turn out to be nitrogen (see chapter by Zolotov and Kargel), and the full inventory of biogenic elements is discussed in the chapter by Hand et al.

8. CODA

As we have described, the evidence for a current-day ocean at Europa is compelling. Impact craters indicate that Europa's surface is youthful, with a nominal age of just 40–90 m.y. (this latest update to the ongoing effort to establish a chronology for the outer solar system can be found in the chapter by Bierhaus et al.). Thus, if geological evidence suggests an ocean at the time Europa's surface features formed, the ocean is probably still there today. When *Pappalardo et al.* (1999) critically examined the case for a subsurface ocean on Europa, they urged caution with regard to concluding in the affirmative. Since then, however, the two most compelling pieces of evidence for an ocean were published: magnetometer evidence for an induced magnetic field at Europa (*Khurana et al.,* 1998; *Kivelson et al.,* 2000), and modeling of cycloidal features that indicates that an ocean is almost certainly required for their formation (*Hoppa et al.,* 1999a).

Thermal modeling and geological observations point to an icy shell ~20 km thick, with observational evidence for both a change in tectonic style (conditions favoring ridged plains formation giving way to conditions favoring band formation giving way to conditions favoring chaos) and perhaps a secular decrease in geological activity over the age of the surface (see chapters by Prockter and Patterson and by Doggett et al.). While total shell thickness is unlikely to vary significantly over local or regional scales, the brittle lithosphere thickness certainly does. And as enormously successful as Galileo was, it may simply have been the limited data return from that mission, in both type and quantity, that prevented the discovery of ongoing, Enceladus-like activity there [e.g., see the discussion in *Nimmo et al.* (2007) of how plumes would manifest at Europa]. The evolution of thought about Europa, as summarized in the chapters of this book, provides an important lesson to draw today: to not become unduly enamored of any particular hypotheses. This lesson applies to Europa's icy shell (e.g., the "thick" vs. "thin" debate, summarized in section 5) and the compositional nature of the dark, reddish, non-ice component on its surface (well and fairly laid out in the chapter by Carlson et al.), among other europan controversies. One must continue to see the logic and value of alternatives.

We have come a long way from the sentiments of one Francesco Sizzi, who in criticizing Galileo's discovery of the satellites that now bear his name, remarked "These satellites of Jupiter are invisible to the naked eye, and therefore can exercise no influence on the Earth, and therefore would be useless, and therefore do not exist" (*Howe,* 1904). Not only do they exist, but one of them, Europa, bears more than a passing resemblance to a smaller but more water-rich Earth. In time, this ocean world should offer a test of one of science's greatest questions: whether there was a second origin of life.

Acknowledgments. The portion of this work performed by R.T.P. was carried out at the Jet Propulsion Laboratory, California Institute of Technology, under a contract with the National Aeronautics and Space Administration. This research was supported by grants from the NASA Planetary Geology and Geophysics and Outer Planets Research Programs.

REFERENCES

Anderson J. D., Schubert G., Jacobson R. A., Lau E. L., Moore W. B., and Sjogren W. L. (1998) Europa's differentiated internal structure: Inferences from four Galileo encounters. *Science, 281,* 2019–2022.

Barr A. C. and McKinnon W. B. (2007) Convection in ice I shells and mantles with self-consistent grain size. *J. Geophys. Res., 112,* E02012, DOI: 10.1029/2006JE002781.

Barr A. C. and Pappalardo R. T. (2005) Onset of convection in the icy Galilean satellites: Influence of rheology. *J. Geophys. Res., 110,* E12005, DOI: 10.1029/2004JE002371.

Billings S. E. and Kattenhorn S. A. (2005) The great thickness debate: Ice shell thickness models for Europa and comparisons with estimates based on flexure at ridges. *Icarus, 177,* 397–412.

Brown M. E. (2001) Potassium in Europa's atmosphere. *Icarus, 151,* 190–195.

Carlson R. W., Johnson R. E., and Anderson M. S. (1999) Sulfuric acid on Europa and the radiolytic sulfur cycle. *Science, 286,* 97–99.

Carr M. H. and 21 colleagues (1998) Evidence for a subsurface ocean on Europa. *Nature, 391,* 363–365.

Cassen P., Reynolds R. T., and Peale S. J. (1979) Is there liquid water on Europa? *Geophys. Res. Lett., 6,* 731–734.

Cassen P., Peale S. J., and Reynolds R. T. (1980) Tidal dissipation in Europa — A correction. *Geophys. Res. Lett., 7,* 987–988.

Cassen P. M., Peale S. J., and Reynolds R. T. (1982) Structure and thermal evolution of the Galilean satellites. In *Satellites of Jupiter* (D. Morrison, ed.), pp. 93–128. Univ. of Arizona, Tucson.

Chyba C. F. (2000) Energy for microbial life on Europa. *Nature, 403,* 381–382.

Consolmagno G. J. and Lewis J. S. (1976) Structural and thermal models of icy Galilean satellites. In *Jupiter* (T. A. Gehrels, ed.), pp. 1035–1051. Univ. of Arizona, Tucson.

Crawford G. D. and Stevenson D. J. (1988) Gas-driven water volcanism and the resurfacing of Europa. *Icarus, 73,* 66–79.

Dalton J. B., Prieto-Ballesteros O., Kargel J. S., Jamieson C. S., Jolivet J., and Quinn R. (2005) Spectral comparison of heavily hydrated salts to disrupted terrains on Europa. *Icarus, 177,* 472–490.

Dash J. G., Hu H., and Wettlaufer J. S. (1995) The premelting of ice and its environmental consequences. *Rept. Prog. Phys., 58,* 115–169.

De Bresser J. H. P., Peach C. J., Reijs J. P. J., and Spiers C. J. (1998) On dynamic recrystallization during solid state flow: Effects of stress and temperature. *Geophys. Res. Lett., 25,* 3457–3460.

De Bresser J. H. P., Ter Heege J., and Spiers C. (2001) Grain size reduction by dynamic recrystallization: Can it result in major rheological weakening? *Intl. J. Earth Sci., 90,* 28–45.

Dombard A. J. and McKinnon W. B. (2006) Folding of Europa's icy lithosphere: An analysis of viscous-plastic buckling and subsequent topographic relaxation. *J. Struct. Geol., 28,* 2259–2269.

Durham W. B. and Stern L. A. (2001) Rheological properties of water ice — Applications to satellites of the outer planets. *Annu. Rev. Earth Planet. Sci., 29,* 295–330.

Durham W. B., Stern L. A., Kubo T., and Kirby S. H. (2005) Flow strength of highly hydrated Mg- and Na-sulfate hydrate salts, pure and in mixtures with water ice, with application to Europa. *J. Geophys. Res., 110,* E12010, DOI: 10.1029/2005JE002475.

Duval P. and Montagnat M. (2002) Comment on "Superplastic deformation of ice: Experimental observations" by D. L. Goldsby and D. L. Kohlstedt. *J. Geophys. Res., 107,* 2082, DOI: 10.1029/2002JB001842.

Figueredo P. H. and Greeley R. (2000) Geologic mapping of the northern leading hemisphere of Europa from Galileo solid-state imaging data. *J. Geophys. Res., 105,* 22629–22646.

Figueredo P. H., Chuang F. C., Rathbun J., Kirk R. L., and Greeley R. (2002) Geology and origin of Europa's "mitten" feature (Murias Chaos). *J. Geophys. Res., 107,* DOI: 10.1029/2001JE001591.

Figueredo P. H., Greeley R., Neuer S., Irwin L., and Schulze-Makuch D. (2003) Locating potential biosignatures on Europa from surface geology observations. *Astrobiology, 3,* 851–861.

Finnerty A. A., Ransford G. A., Pieri D. C, and Collerson K. D. (1981) Is Europa surface cracking due to thermal evolution. *Nature, 289,* 24–27.

Frei R., Gaucher C., Poulton S. W., and Canfield D. E. (2009) Fluctuations in Precambrian atmospheric oxygenation recorded by chromium isotopes. *Nature, 461,* 250–253.

Gaidos E. J., Nealson K. H., and Kirschvink J. L. (1999) Life in ice-covered oceans. *Science, 284,* 1631–1633.

Geissler P. E. and 14 colleagues (1998) Evidence for non-synchronous rotation of Europa. *Nature, 391,* 368–379.

Goldsby D. L. and Kohlstedt D. L. (2001) Superplastic deformation of ice: Experimental observations. *J. Geophys. Res., 106,* 11017–11030.

Goldsby D. L. and Kohlstedt D. L. (2002) Reply to comment by P. Duval and M. Montagnat on "Superplastic deformation of ice: Experimental observations." *J. Geophys. Res., 107,* 2313, DOI: 10.1029/2002JB001842.

Golombek M. P. and Banerdt W. B. (1986) Early thermal profiles and lithospheric strength of Ganymede from extensional tectonic features. *Icarus, 68,* 252–265.

Greeley R., Chyba C. F., Head J. W., McCord T. B., McKinnon W. B., Pappalardo R. T., and Figueredo P. (2004) Geology of Europa. In *Jupiter: The Planet, Satellites and Magnetosphere* (F. Bagenal et al., eds.), pp. 329–362. Cambridge Univ., Cambridge.

Greenberg R. and Weidenschilling S. J. (1984) How fast do Galilean satellites spin? *Icarus, 58,* 186–196.

Greenberg R., Geissler P., Hoppa G., Tufts B. R., Durda D. D., Pappalardo R., Head J. W., Greeley R., Sullivan R., and Carr M. H. (1998) Tectonic processes on Europa: Tidal stresses, mechanical response, and visible features. *Icarus, 135,* 64–78.

Greenberg R., Hoppa G. V., Tufts B. R., Geissler P., Riley J., and Kadel S. (1999) Chaos on Europa. *Icarus, 141,* 263–286.

Greenberg R., Geissler P., Tufts B. R., and Hoppa G. V. (2000)

Habitability of Europa's crust: The role of tidal-tectonic processes. *J. Geophys. Res., 105,* 17551–17562.

Greenberg R., Geissler P., Hoppa G., and Tufts B. R. (2002) Tidal-tectonic processes and their implications for the character of Europa's icy crust. *Rev. Geophys., 40,* 1.1–1.33.

Greenberg R., Leake M. A., Hoppa G. V., and Tufts B. R. (2003) Pits and uplifts on Europa. *Icarus, 161,* 102–126.

Grindrod P. M., Fortes A. D., Wood I. G., Samonds P. R., Dobson D. P., Middleton C. A., and Vocadlo L. (2008) Synthesis and strength of $MgSO_4 \cdot 11H_2O$ (Meridianiite): Preliminary results from uniaxial and trixial deformation tests. In *Lunar and Planetary Science XXXIX,* Abstract #1199. Lunar and Planetary Institute, Houston (CD-ROM).

Hand K. P. and Chyba C. F. (2007) Empirical constraints on the salinity of the europan ocean and implications for a thin ice shell. *Icarus, 189,* 424–438.

Hansen G. B. and McCord T. B. (2008) Widespread CO_2 and other non-ice compounds on the anti-jovian and trailing sides of Europa from Galileo/NIMS observations. *Geophys. Res. Lett., 35,* L01202, DOI: 10.1029/2007GL031748.

Hawthorne F. C., Krivovichev S. V., and Burns P. C. (2000) The crystal chemistry of the sulfate minerals. *Rev. Min. Geochem., 40,* 1–112.

Head J. W. III and Pappalardo R. T. (1999) Brine mobilization during lithospheric heating on Europa: Implications for formation of chaos terrain, lenticula texture, and color variations. *J. Geophys. Res., 104,* 27143–27155.

Hoehler T. M., Amend J. P., and Shock E. L. (2007) A "follow the energy" approach for astrobiology. *Astrobiology, 7,* 819–823.

Hoppa G. V., Tufts B. R., Greenberg R., and Geissler P. E. (1999a) Formation of cycloidal features on Europa. *Science, 285,* 1899–1902.

Hoppa G., Tufts B. R., Greenberg R., and Geissler P. (1999b) Strike-slip faults on Europa: Global shear patterns driven by tidal stress. *Icarus, 141,* 287–298.

Howe H. (1904) A new life of Galileo. *Popular Astron., 12,* 80–89.

Hurford T. A., Beyer R. A., Schmidt B., Preblich B., Sarid A. R., and Greenberg R. (2005) Flexure of Europa's lithosphere due to ridge-loading. *Icarus, 177,* 80–396.

Hussmann H. and Spohn T. (2004) Thermal-orbital evolution of Io and Europa. *Icarus, 171,* 391–410, DOI: 10.1016/j.icarus.2004.05.020.

Hussmann H., Spohn T., and Wieczerkowski K. (2002) Thermal equilibrium states of Europa's ice shell: Implications for internal ocean thickness and surface heat flow. *Icarus, 156,* 143–151, DOI: 10.1006/icar.2001.6776.

Johnson R. E., Carlson R. W., Cooper J. F., Paranicas C., Moore M. H., and Wong M. C. (2004) Radiation effects on the surface of the Galilean satellites. In *Jupiter: The Planet, Satellites and Magnetosphere* (F. Bagenal et al., eds.), pp. 485–512. Cambridge Univ., Cambridge.

Kargel J. S., Kaye J. Z., Head J. W., Marion G. M., Sassen R., Crowley J. K., Prieto-Ballesteros O., and Hogenboom D. L. (2000) Europa's crust and ocean: Origin, composition, and prospects for life. *Icarus, 148,* 226–265.

Khurana K. K., Kivelson M. G., Stevenson D. J., Schubert G., Russell C. T., Walker R. J., and Polanskey C. (1998) Induced magnetic fields as evidence for subsurface oceans in Europa and Callisto. *Nature, 395,* 777–780.

Kivelson M. G., Khurana K. K, Russell C. T., Volwerk M., Walker R. J., and Zimmer C. (2000) Galileo magnetometer measurements strengthen the case for a subsurface ocean at Europa. *Science, 289,* 1340–1343.

Kohlstedt D. L., Evans B., and Mackwell S. J. (1995) Strength of the lithosphere: Constraints imposed by laboratory experiments. *J. Geophys. Res., 100,* 17587–17602.

Lainey V., Arlot J.-E., Karatekin Ö., and van Hoolst T. (2009) Strong tidal dissipation in Io and Jupiter from astrometric observations. *Nature, 459,* 957–959.

Lee S., Pappalardo R. T., and Makris N. C. (2005) Mechanics of tidally driven fractures in Europa's ice shell. *Icarus, 177,* 367–379.

Leith A. C. and McKinnon W. B. (1996) Is there evidence for polar wander on Europa? *Icarus, 120,* 387–398.

Lewis J. S. (1971) Satellites of the outer planets: Thermal models. *Science, 172,* 1127.

Lucchitta B. K. and Soderblom L. A. (1982) The geology of Europa. In *The Satellites of Jupiter* (D. Morrison, ed.), pp. 521–555. Univ. of Arizona, Tucson.

McCarthy C., Goldsby D. L., and Cooper R. F. (2007) Transient and steady-state creep responses of ice-I/magnesium sulfate hydrate eutectic aggregates. In *Lunar and Planetary Science XXXVIII,* Abstract #2429. Lunar and Planetary Institute, Houston (CD-ROM).

McCollom T. M. (1999) Methanogenesis as a potential source of chemical energy for primary biomass production by autotrophic organisms in hydrothermal systems on Europa. *J. Geophys. Res., 104,* 30729–30742.

McEwen A. S. (1986) Tidal reorientation and the fracturing of Jupiter's moon Europa. *Nature, 321,* 49–51.

McKinnon W. B. (1999) Convective instability in Europa's floating ice shell. *Geophys. Res. Lett., 26,* 951–954.

McKinnon W. B. (2006) On convection in ice I shells of outer solar system bodies, with specific application to Callisto. *Icarus, 183,* 435–450.

McKinnon W. B. (2007) Formation and early evolution of Io. In *Io After Galileo* (R. M. C. Lopes and J. R. Spencer, eds.), pp. 61–88. Springer-Verlag, Berlin.

McKinnon W. B. and Desai S. (2003) Internal structures of the Galilean satellites: What can we really tell? In *Lunar and Planetary Science XXXIV,* Abstract #2104. Lunar and Planetary Institute, Houston (CD-ROM).

McKinnon W. B. and Schenk P. M. (2008) Multiring basins on icy satellites: A post-Galileo view. In *Large Meteorite Impacts and Planetary Evolution IV,* Abstract #3103. LPI Contribution No. 1423, Lunar and Planetary Institute, Houston (CD-ROM).

McKinnon W. B. and Zolensky M. E. (2003) Sulfate content of Europa's ocean and shell: Evolutionary considerations and geological and astrobiological implications. *Astrobiology, 3,* 879–897.

Mitri G. and Showman A. P. (2005) Convective-conductive transitions and sensitivity of a convecting ice shell to perturbations in heat flux and tidal-heating rate: Implications for Europa. *Icarus, 177,* 447–460.

Moore W. B. (2006) Thermal equilibrium in Europa's ice shell. *Icarus, 180,* 141–146.

Moore W. B. and Schubert G. (2000) The tidal response on Europa. *Icarus, 147,* 317–319.

Nimmo F. (2004a) Non-Newtonian topographic relaxation on Europa. *Icarus, 168,* 205–208.

Nimmo F. (2004b) Stresses generated in cooling viscoelastic ice shells: Application to Europa. *J. Geophys. Res., 109,* E12001, DOI: 10.1029/2004JE002347.

Nimmo F. and Gaidos E. (2002) Thermal consequences of strike-slip motion on Europa. *J. Geophys. Res., 107,* 5021, DOI: 10.1029/2000JE001476.

Nimmo F. and Manga M. (2002) Causes, characteristics, and consequences of convective diapirism on Europa. *Geophys. Res. Lett., 29,* 2109, DOI: 10.1029/2002GL015754.

Nimmo F., Giese B., and Pappalardo R. T. (2003) Estimates of Europa's ice shell thickness from elastically-supported topography. *Geophys. Res. Lett., 30,* 1233, DOI: 10.1029/2002GL016660.

Nimmo F., Pappalardo R., and Cuzzi J. (2007) Observational and theoretical constraints on plume activity at Europa. *Eos Trans. AGU, 88(52),* Fall Meet. Suppl., Abstract #P51E-05.

O'Brien D. P., Geissler P., and Greenberg R. (2002) A melt-through model for chaos formation on Europa. *Icarus, 156,* 152–161.

Ojakangas G. W. and Stevenson D. J. (1986) Episodic volcanism of tidally heated satellites with application to Io. *Icarus, 66,* 341–358.

Ojakangas G. W. and Stevenson D. J. (1989a) Thermal state of an ice shell on Europa. *Icarus, 81,* 220–241.

Ojakangas G. W. and Stevenson D. J. (1989b) Polar wander of an ice shell on Europa. *Icarus, 81,* 242–270.

Pappalardo R. T. and Barr A. C. (2004) The origin of domes on Europa: The role of thermally induced compositional diapirism. *Geophys. Res. Lett., 31,* DOI: 10.1029/2003GL019202.

Pappalardo R. T., Head J. W., Greeley R., Sullivan R. J., Pilcher C., Schubert G., Moore W. B., Carr M. H., Moore J. M., Belton M. J. S., and Goldsby D. L. (1998) Geological evidence for solid-state convection in Europa's ice shell. *Nature, 391,* 365–368.

Pappalardo R.T. and 31 colleagues (1999) Does Europa have a subsurface ocean? Evaluation of the geological evidence. *J. Geophys. Res., 104,* E10, 24015–24055.

Palguta J., Anderson J. D., Schubert G., and Moore W. B. (2006) Mass anomalies on Ganymede. *Icarus, 180,* 428–441.

Preblich B., Greenberg R., Riley J., and O'Brien D. (2007) Tidally driven strike-slip displacement on Europa: Viscoelastic modeling. *Planet. Space Sci., 55,* 1225–1245.

Prinn R. G. and Fegley B. Jr. (1989) Solar nebula chemistry: Origins of planetary, satellite and cometary volatiles. In *Origin and Evolution of Planetary and Satellite Atmospheres* (S. K. Atreya et al., eds.), pp. 78–136. Univ. of Arizona, Tucson.

Prockter L. M. and Pappalardo R. T. (2000) Folds on Europa: Implications for crustal cycling and accommodation of extension. *Science, 289,* 941–943.

Prockter L. and Schenk P. (2005) Origin and evolution of Castalia Macula, an anomalous young depression on Europa. *Icarus, 177,* 305–326.

Ransford G. A., Finnerty A. A., and Collerson K. D. (1981) Europa's petrological thermal history. *Nature, 289,* 21–24.

Rathbun J. A., Musser G. S., and Squyres S. W. (1998) Ice diapirs on Europa: Implications for liquid water. *Geophys. Res. Lett., 25,* 4157–4160.

Reynolds R. T. and Cassen P. (1979) On the internal structure of the major satellites of the outer planets. *Geophys. Res. Lett., 6,* 121–124.

Ruiz J. (2005) The heat flow of Europa. *Icarus, 177,* 438–446.

Ruiz J. and Tejero R. (2003) Heat flow, lenticulae spacing, and possibility of convection in the ice shell of Europa. *Icarus, 162,* 362–373.

Schenk P. M. (2002) Thickness constraints on the icy shells of the Galilean satellites from a comparison of crater shapes. *Nature, 417,* 419–421.

Schenk P. and McKinnon W. B. (1989) Fault offsets and lateral crustal movement on Europa: Evidence for a mobile ice shell. *Icarus, 79,* 75–100.

Schenk P., Matsuyama I., and Nimmo F. (2008) True polar wander on Europa from global-scale small-circle depressions. *Nature, 453,* 368–371.

Schilling N., Khurana K. K., and Kivelson M. G. (2004) Limits on an intrinsic dipole moment in Europa. *J. Geophys. Res., 109,* E05006.

Schilling N., Neubauer F. M., and Saur J. (2007) Time-varying interaction of Europa with the jovian magnetosphere: Constraints on the conductivity of Europa's subsurface ocean. *Icarus, 192,* 41–55.

Schmeling H. (1987) On the interaction between small- and large-scale convection and postglacial rebound flow in a power-law mantle. *Earth Planet. Sci. Lett., 84,* 254–262.

Schubert G., Anderson J. D., Spohn T., and McKinnon W. B. (2004) Interior composition, structure and dynamics of the Galilean satellites. In *Jupiter: The Planet, Satellites and Magnetosphere* (F. Bagenal et al., eds.), pp. 281–306. Cambridge Univ., Cambridge.

Schulson E. M. (2001) Brittle failure of ice. *Engrg. Fract. Mech., 68,* 1839–1887.

Shapiro R. and Schulze-Makuch D. (2009) The search for alien life in our solar system: Strategies and priorities. *Astrobiology, 9,* 335–343.

Shock E. L., Amend J. P., and Zolotov M. Yu. (2000) The early Earth vs. the origin of life. In *Origin of the Earth and Moon* (R. M. Canup and K. Righter, eds.), pp. 527–543. Univ. of Arizona, Tucson.

Showman A. P. and Han L. (2005) Effects of plasticity on convection in an ice shell: Implications for Europa. *Icarus, 177,* 425–437.

Singer K. N., McKinnon W. B., and Schenk P. M. (2009) Pits, spots, uplifts, and small chaos regions on Europa: A search for regional variations. In *Lunar and Planetary Science XL,* Abstract #2336. Lunar and Planetary Institute, Houston (CD-ROM).

Solomatov V. S. and Moresi L-N. (2000) Scaling of time-dependent stagnant lid convection: Application to small-scale convection on Earth and other terrestrial planets. *J. Geophys. Res., 105,* 21795–21818.

Sotin C., Head J. W., and Tobie G. (2002) Europa: Tidal heating of upwelling thermal plumes and the origin of lenticulae and chaos melting. *Geophys. Res. Lett., 29,* DOI: 10.1029/2001GL013844.

Spohn T. and Schubert G. (2003) Oceans in the icy Galilean satellites of Jupiter? *Icarus, 161,* 456–467.

Squyres S. W., Reynolds R. T., Cassen P. M., and Peale S. J. (1983) Liquid water and active resurfacing on Europa. *Nature, 301,* 225–226.

Stevenson D. J. (2000) Limits on the variation of thickness of Europa's ice shell. In *Lunar and Planetary Science XXXI,* Abstract #1506. Lunar and Planetary Institute, Houston (CD-ROM).

Terasaki H., Frost D. J., Rubie D. C., and Langenhorst F. (2008) Percolative core formation in planetesimals. *Earth Planet. Sci. Lett., 273,* 132–137.

Thomson R. E. and Delaney J. R. (2001) Evidence for a weakly stratified europan ocean sustained by seafloor heat flux. *J. Geophys. Res., 106,* 12355–12365.

Tobie G., Choblet G., and Sotin C. (2003) Tidally heated convection: Constraints on Europa's ice shell thickness. *J. Geophys. Res., 108,* 5124, DOI: 10.1029/2003JE002099.

Tobie G., Mocquet A., and Sotin C. (2005) Tidal dissipation within large icy satellites: Applications to Europa and Titan. *Icarus, 177,* 534–549.

Tufts B. R., Greenberg R., Hoppa G., and Geissler P. (2000) Lithospheric dilation on Europa. *Icarus, 146,* 75–97.

Tyler R. H. (2008) Strong ocean tidal flow and heating on moons of the outer planets. *Nature, 456,* DOI: 10.1038/nature07571.

Wahr J., Selvans Z. A., Mullen M. E., Barr A. C., Collins G. C., Selvans M. M., and Pappalardo R. T. (2009) Modelling stresses on satellites due to non-synchronous rotation and orbital eccentricity using gravitational potential theory. *Icarus, 200,* 188–206, DOI: 10.1016/j.icarus.2008.11.002.

Wang H. and Stevenson D. J. (2000) Convection and internal melting of Europa's ice shell. In *Lunar and Planetary Science XXXI,* Abstract #1293. Lunar and Planetary Institute, Houston (CD-ROM).

White S. R. (2001) Textural and microstructural evidence for semi-brittle flow in natural fault rocks with varied mica contents. *Intl. J. Earth Sci., 90,* 14–27.

Zimmer C., Khurana K. K., and Kivelson M. G. (2000) Subsurface oceans on Europa and Callisto: Constraints from Galileo magnetometer observations. *Icarus, 147,* 329–347.

Zolotov M. Yu. and Fegley B. Jr. (2000) Eruption conditions of Pele volcano on Io inferred from chemistry of its volcanic plume. *Geophys. Res. Lett., 27,* 2789–2792.

Zolotov M. Yu. and Shock E. L. (2001) Composition and stability of salts on the surface of Europa and their oceanic origin. *J. Geophys. Res., 106,* 32815–32827.

Zolotov M. Yu. and Shock E. L. (2004) A model for low-temperature biogeochemistry of sulfur, carbon, and iron on Europa. *J. Geophys. Res., 109,* E06003.

Zolotov M. Yu., Shock E. L., Barr A. C., and Pappalardo R. T. (2004) Brine pockets in the icy crust of Europa: Distribution, chemistry, and habitability. In *Workshop on Europa's Icy Shell: Past, Present, and Future,* pp. 100–101. LPI Contribution No. 1195, Lunar and Planetary Institute, Houston.

Appendix: Europa Galileo and Voyager Image Mosaic Maps

The image mosaic maps of Europa on the following pages show the four equatorial quadrants and two polar regions from U.S. Geological Survey (USGS) Map I-2757. Map I-2757 is one in a series of maps of the Galilean satellites of Jupiter developed at a nominal scale of 1:15,000,000. The series is based on data from the Galileo Orbiter Solid-State Imaging (SSI) camera and the Voyager 1 and 2 spacecraft.

Projections used for the USGS map of Europa are based on a sphere having a radius of 1562.09 km. The scale is 1:8,388,000 at ±56° latitude for both projections. Longitude increases to the west in accordance with the International Astronomical Union (1971) (*Davies et al.,* 1996). Latitude is planetographic.

The process of creating a geometric control network began with selecting control points on the individual images, making pixel measurements of their locations, using reseau locations to correct for geometric distortions, and converting the measurements to millimeters in the focal plane. These data are combined with the camera focal lengths and navigation solutions as input to a photogrammetric triangulation solution (*Davies et al.,* 1998; *Davies and Katayama,* 1981). The solution used here was computed at the RAND Corporation in June 2000. Solved parameters include the radius (given above) of the best-fitting sphere, the coordinates of the control points, the three orientation angles of the camera at each exposure (right ascension, declination, and twist), and an angle (W_0) that defines the orientation of Europa in space. W_0 — in this solution 36.022° — is the angle along the equator to the east, between the 0° meridian and the equator's intersection with the celestial equator at the standard epoch J2000.0. This solution places the crater Cilix at its defined longitude of 182° west (*Davies et al.,* 1996).

This global map base uses the best image quality and moderate resolution coverage supplied by Galileo SSI and Voyager 1 and 2 (*Batson,* 1987; *Becker et al.,* 1998, 1999, 2001). The digital map was produced using Integrated Software for Imagers and Spectrometers (ISIS) (*Eliason,* 1997; *Gaddis et al.,* 1997; *Torson and Becker,* 1997). The individual images were radiometrically calibrated and photometrically normalized using a Lunar-Lambert function with empirically derived values (*McEwen,* 1991; *Kirk et al.,* 2000). A linear correction based on the statistics of all overlapping areas was then applied to minimize image brightness variations. The image data were selected on the basis of overall image quality, reasonable original input resolution (from 20 km/pixel for gap fill to as much as 40 m/pixel), and availability of moderate emission and incidence angles for topography and albedo. Although consistency was achieved where possible, different filters were included for global image coverage as necessary: clear/blue for Voyager 1 and 2; and clear, near-IR (757 nm), and green (559 nm) for Galileo SSI. Individual images were projected to a Sinusoidal Equal-Area projection at an image resolution of 500 m/pixel. The final constructed Sinusoidal projection mosaic was then reprojected to the Mercator and Polar Stereographic projections included here.

Feature names are approved by the *International Astronomical Union* (1980, 1986, 1999, 2001). Names have been applied for features clearly visible at the scale of this map; for a complete list of nomenclature of Europa, see the Gazetteer of Planetary Nomenclature at (*planetarynames.wr.usgs.gov*). Font color was chosen only for readability. Je 15M CMN: Abbreviation for Jupiter, Europa (satellite): 1:15,000,000 series, controlled mosaic (CM), nomenclature (N) (*Greeley and Batson,* 1990).

Tammy Becker
U.S. Geological Survey

REFERENCES

Batson R. M. (1987) Digital cartography of the planets — New methods, its status, and its future. *Photogrammetric Engrg. Remote Sens., 53(9),* 1211–1218.

Becker T. L., Archinal B., Colvin T. R., Davies M. E., Gitlin A., Kirk R. L., and Weller L. (2001) Final digital global maps of Ganymede, Europa, and Callisto. In *Lunar and Planetary Science XXXII,* Abstract #2009. Lunar and Planetary Institute, Houston (CD-ROM).

Becker T. L, Rosanova T., Cook D., Davies M. E., Colvin T. R., Acton C., Bachman N., Kirk R. L., and Gaddis L. R. (1999) Progress in improvement of geodetic control and production of final image mosaics for Callisto and Ganymede. In *Lunar and Planetary Science XXX,* Abstract #1692. Lunar and Planetary Institute, Houston (CD-ROM).

Becker T. L., Rosanova T., Gaddis L. R., McEwen A. S., Phillips C. B., Davies M. E., and Colvin T. R. (1998) Cartographic processing of the Galileo SSI data — An update on the production of global mosaics of the Galilean satellites. In *Lunar and Planetary Science XXIX,* Abstract #1892. Lunar and Planetary Institute, Houston (CD-ROM).

Davies M. E., Abalakin V. K., Bursa M., Lieske J. H., Morando B., Morrison D., Seidelmann P. K., Sinclair A. T., Yallop B., and Tjuflin Y. S. (1996) Report of the IAU/IAG/COSPAR Working Group on Cartographic Coordinates and Rotational Elements of the Planets and Satellites, 1994. *Cel. Mech. Dyn. Astron., 63,* 127–148.

Davies M. E., Colvin T. R., Oberst J., Zeitler W., Schuster P., Neukum G., McEwen A. S., Phillips C. B., Thomas P. C., Veverka J., Belton M. J. S., and Schubert G. (1998) The control networks of the Galilean satellites and implications for global shape. *Icarus, 135,* 372–376.

Davies M. E. and Katayama F. Y. (1981) Coordinates of features on the Galilean satellites. *J. Geophys. Res., 86(A10),* 8635–8657.

Eliason E. M. (1997) Production of digital image models using the ISIS system. In *Lunar and Planetary Science XXVIII,* p. 331. Lunar and Planetary Institute, Houston.

Gaddis L. R., Anderson J., Becker K., Becker T. L., Cook D., Edwards K., Eliason E. M., Hare T., Kieffer H. H., Lee E. M., Mathews J., Soderblom L. A., Sucharski T., Torson J., McEwen A. S., and Robinson M. (1997) An overview of the Integrated Software for Imaging Spectrometers (ISIS). In *Lunar and Planetary Science XXVIII,* p. 387. Lunar and Planetary Institute, Houston.

Greeley R. and Batson R. M. (1990) *Planetary Mapping,* pp. 274–275. Cambridge Univ., Cambridge.

International Astronomical Union (1971) Commission 16 — Physical study of planets and satellites. In *Proceedings of the 14th General Assembly,* Brighton, 1970, *Trans. IAU, 14B,* 128–137.

International Astronomical Union (1980) Working Group for Planetary System Nomenclature. In *Proceedings of the 17th General Assembly,* Montreal, 1979, *Trans. IAU, 17B,* 300.

International Astronomical Union (1986) Working Group for Planetary System Nomenclature. In *Proceedings of the 19th General Assembly,* New Delhi, 1985, *Trans. IAU, 19B,* 351.

International Astronomical Union (1999) Working Group for Planetary System Nomenclature. In *Proceedings of the 23rd General Assembly,* Kyoto, 1997, *Trans. IAU, 23B,* 234–235.

International Astronomical Union (2001) Working Group for Planetary System Nomenclature. In *Proceedings of the 24th General Assembly,* Manchester, 2000, *Trans. IAU, 24B.*

Kirk R. L., Thompson K. T., Becker T. L., and Lee E. M. (2000) Photometric modeling for planetary cartography. In *Lunar and Planetary Science XXXI,* Abstract #2025. Lunar and Planetary Institute, Houston (CD-ROM).

McEwen A. S. (1991) Photometric functions for photoclinometry and other applications. *Icarus, 92,* 298–311.

Torson J. M. and Becker K. J. (1997) ISIS — A software architecture for processing planetary images. In *Lunar and Planetary Science XXVIII,* p. 1443. Lunar and Planetary Institute, Houston.

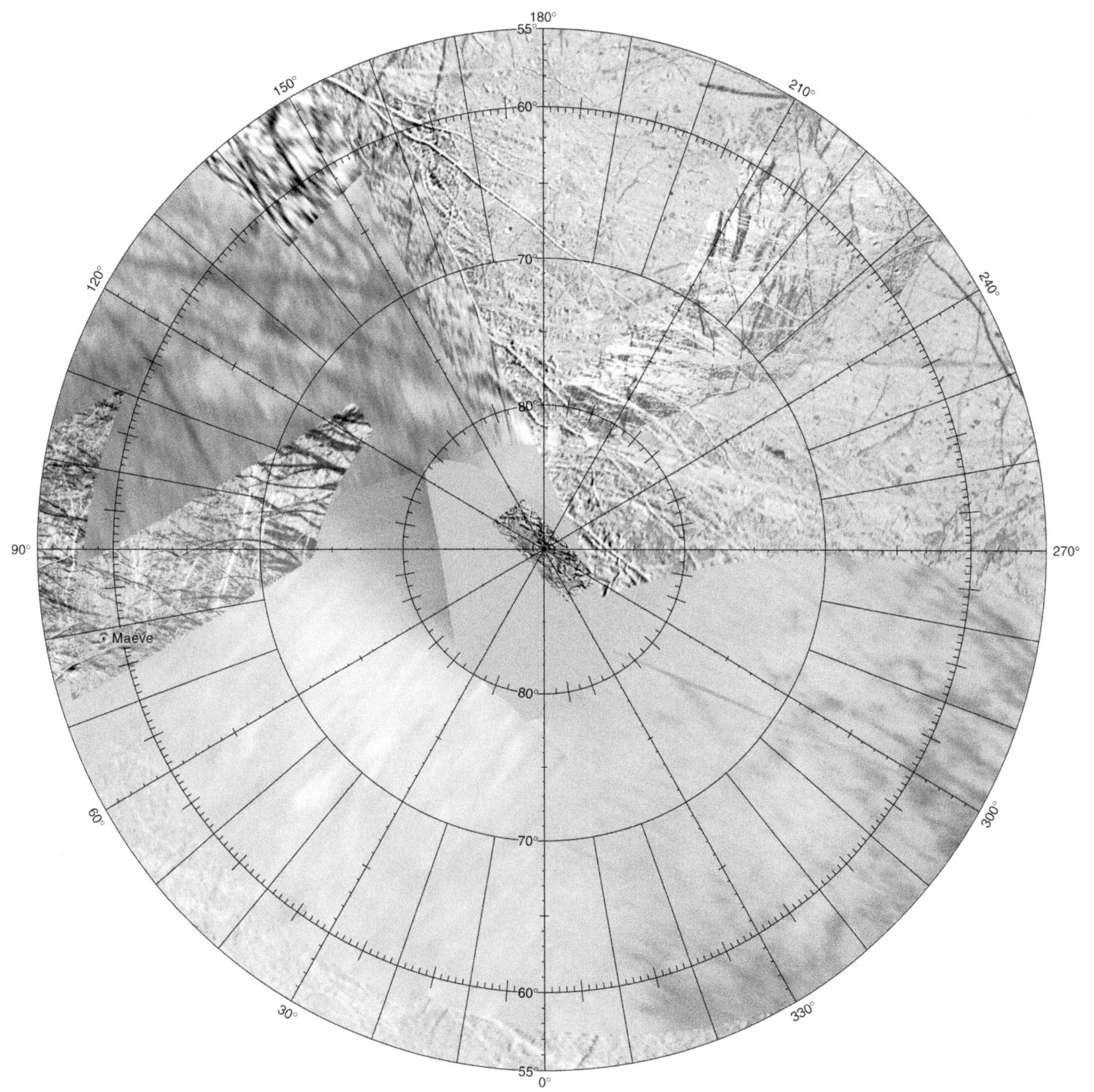

North Polar Region

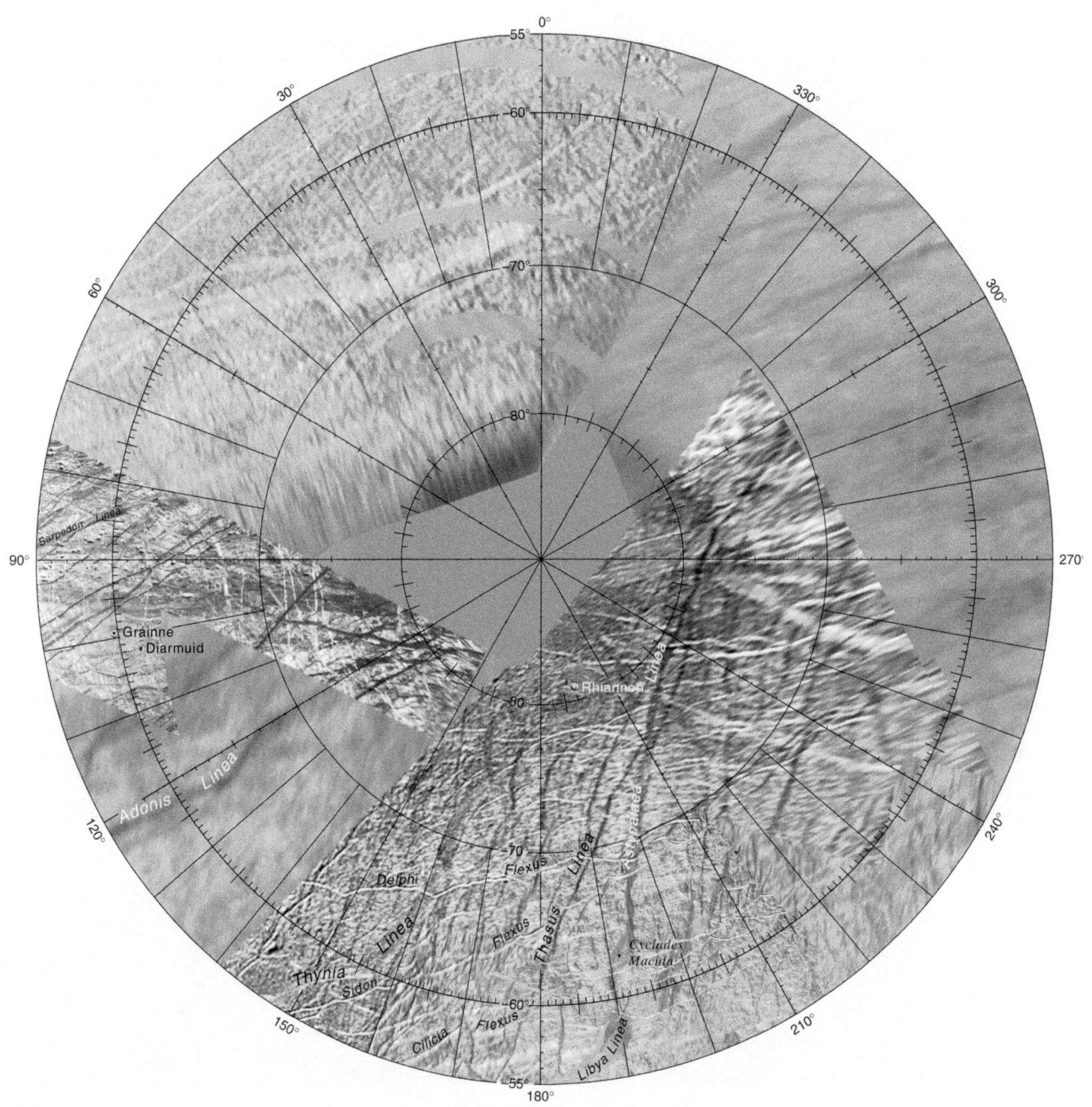

South Polar Region

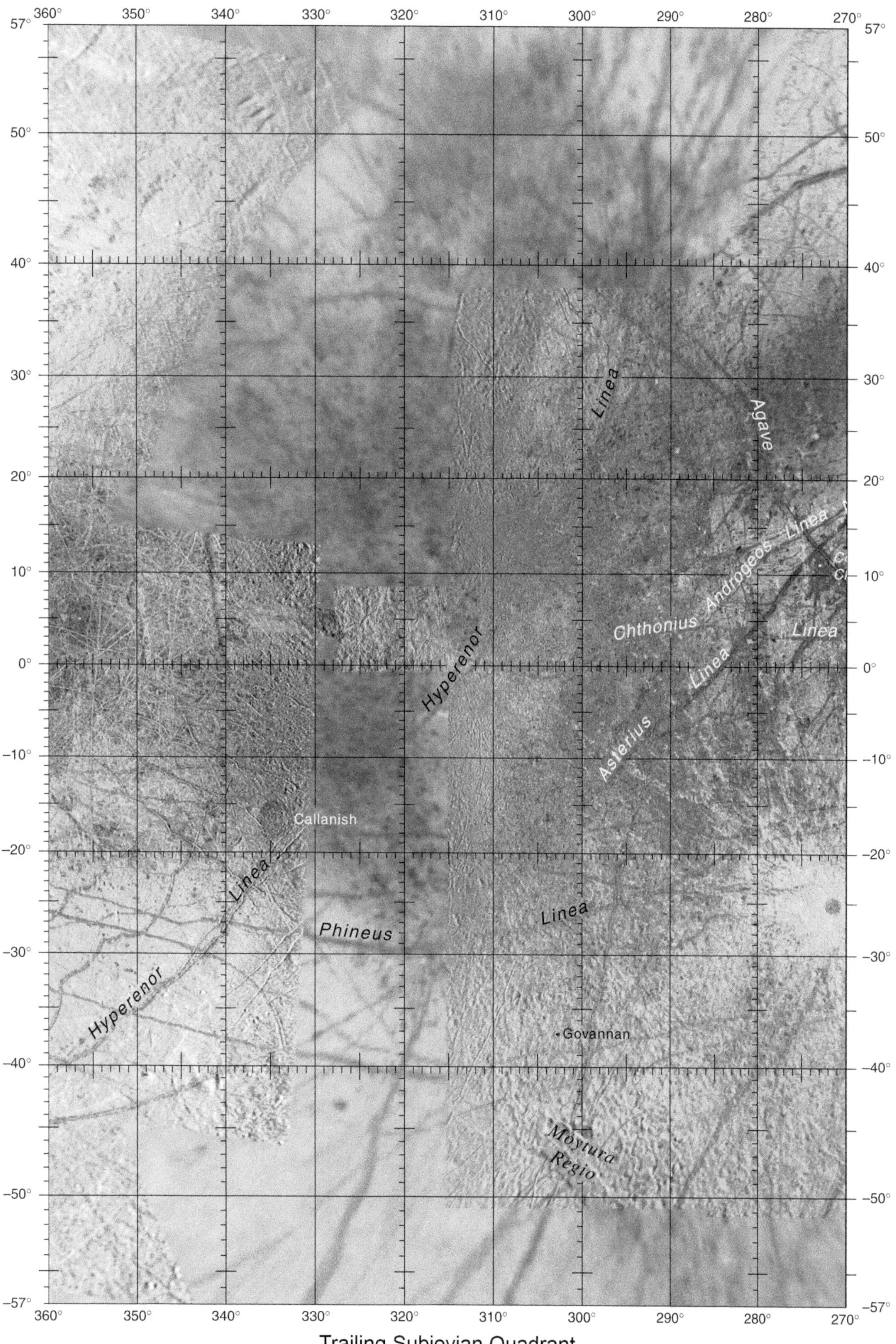

Trailing Subjovian Quadrant

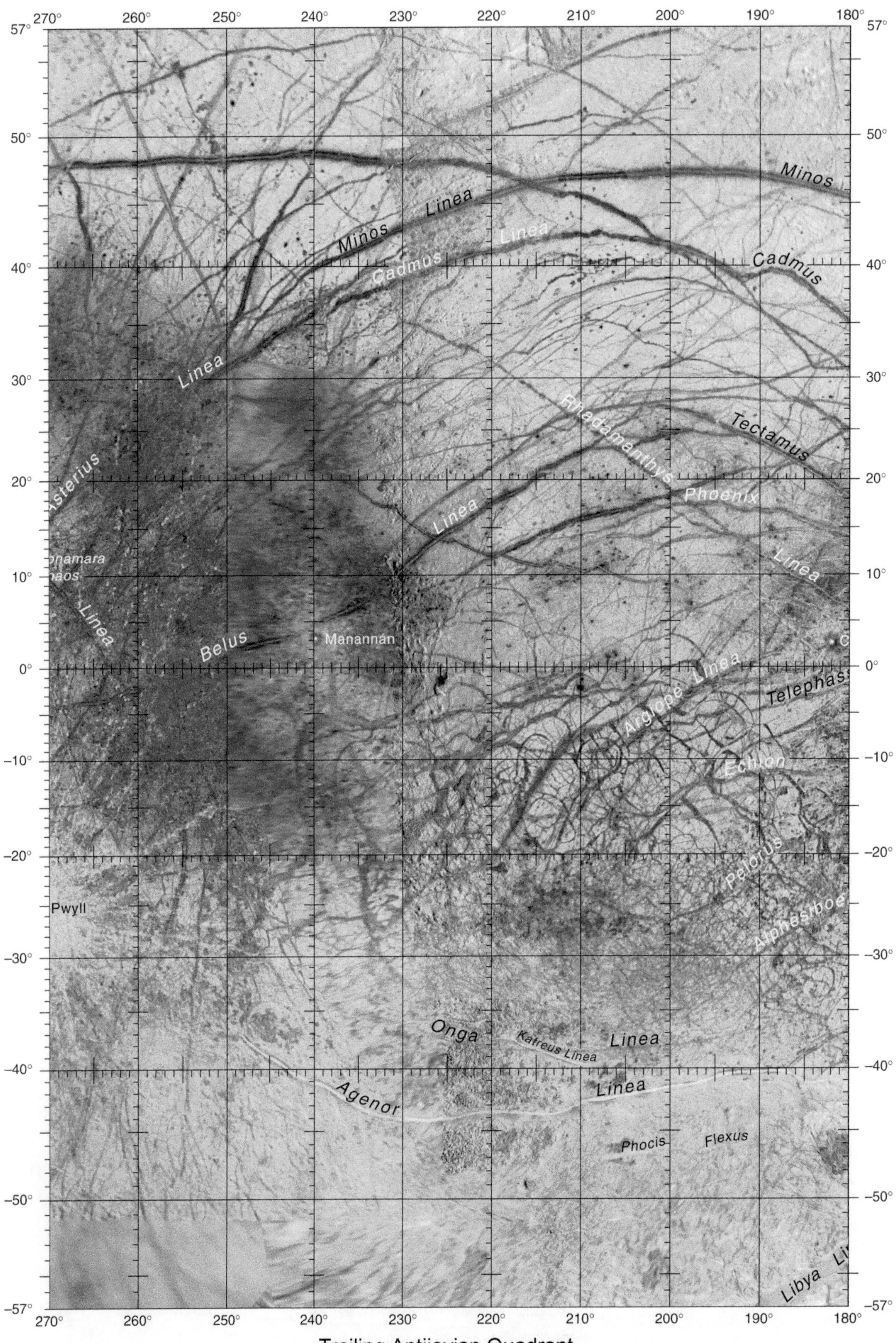

Trailing Antijovian Quadrant

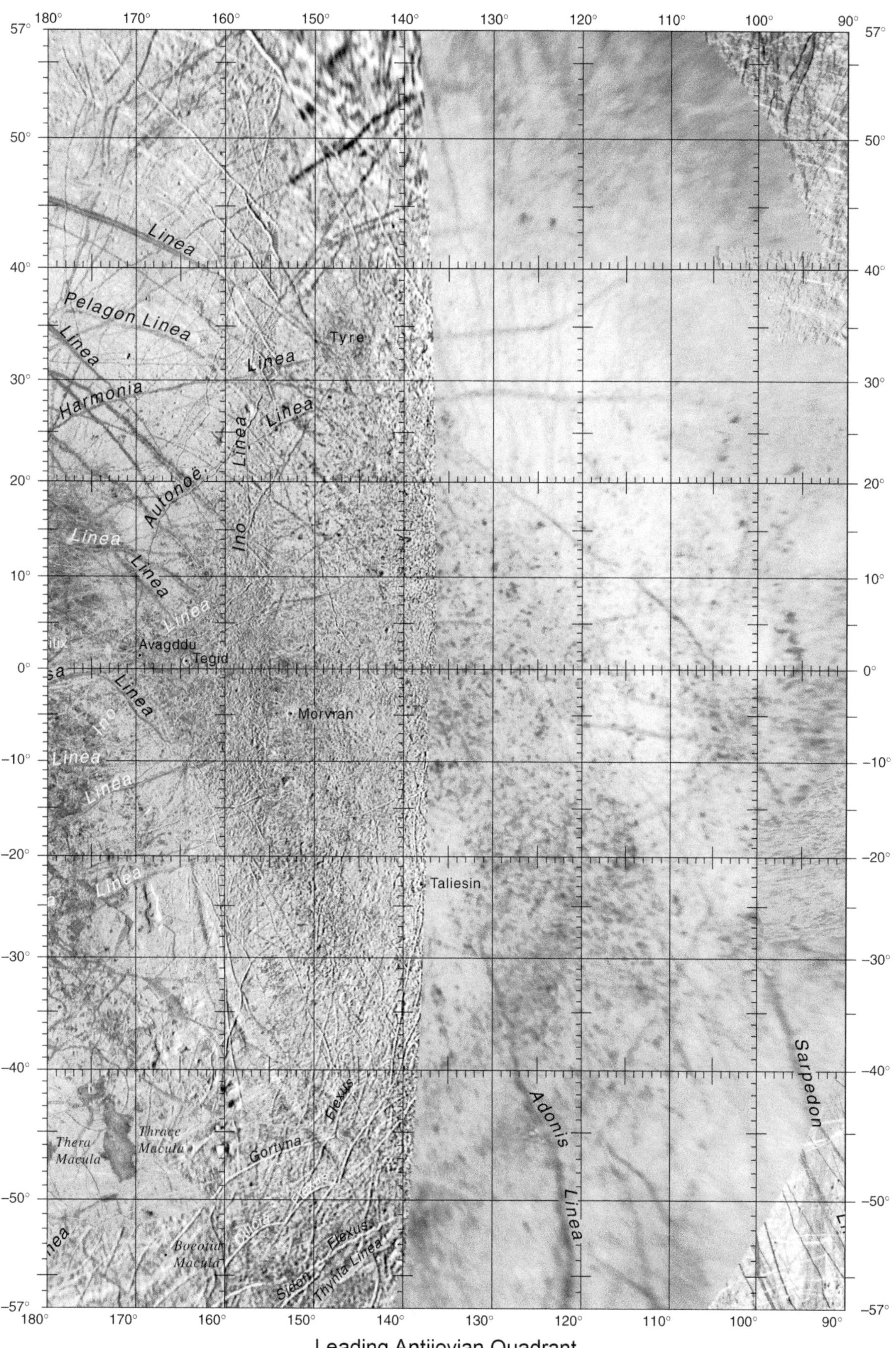

Leading Antijovian Quadrant

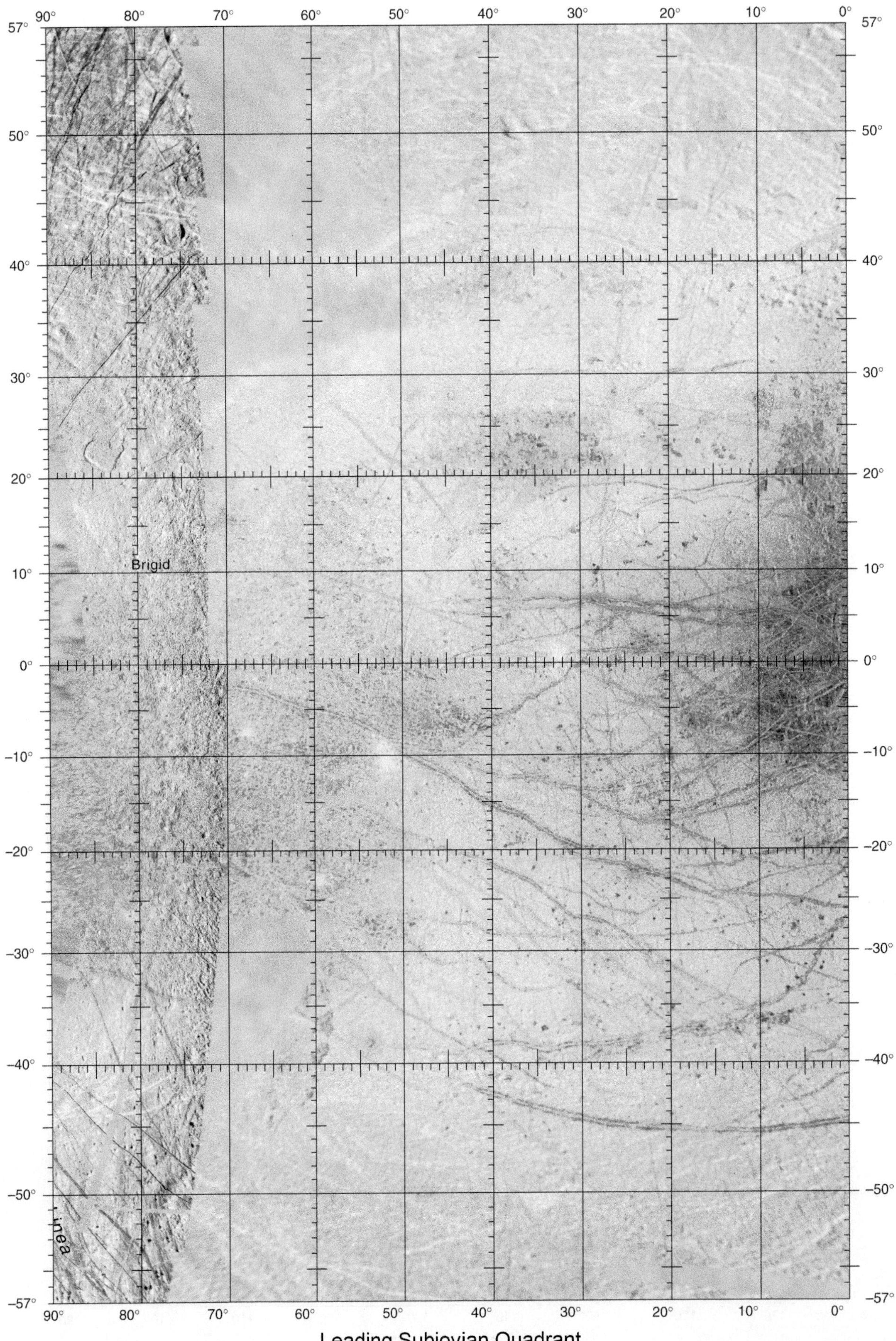
90°
80°
70°
60°
50°
40°
30°
20°
10°
0°
57°
50°
40°
30°
20°
10°
0°
−10°
−20°
−30°
−40°
−50°
−57°
Brigid
Leading Subjovian Quadrant

Index

Page numbers refer to specific pages on which an index term or concept is discussed. "ff" indicates that the term is also discussed on the following pages.